MW01195281

Small Arms Of The World

A BASIC MANUAL OF SMALL ARMS

a basic manual of small arms

SMALL ARMS OF THE WORLD

EDWARD CLINTON EZELL

with research assistance of
Thomas M. Pegg

A completely new and revised version of the classic work by W. H. B. Smith

12th REVISED EDITION

BARNES & NOBLE
BOOKS
NEW YORK

This edition published by Barnes & Noble, Inc.,
by arrangement with Stackpole Books.

1993 Barnes & Noble Books

ISBN 0-88029-601-1

Printed and bound in the United States of America
M 11 10 9 8 7 6 5

In the five years that have passed since the completion of the eleventh edition of *Small Arms of the World,* a number of important changes have taken place. Many of the weapons described as experimental in the eleventh edition are now standard; other previously standard weapons have become obsolete. Thus as we remember the fortieth anniversary of W. H. B. Smith's first edition of *Small Arms of the World* (1943), we have decided that this twelfth edition must be a significantly revamped volume. The coverage given to many obsolete weapons has been reduced; in a few cases some older weapons have been dropped. In this manner we have made room for discussion and illustration of new standard weapons. Some obsolete weapons are still included in the twelfth edition because they are still being used by countries other than the ones that originally employed them. The result is a revised and enlarged volume.

Preface

The twelfth edition of *Small Arms of the World* continues a process started in the eleventh edition. The first chapters (Part I) have been expanded to continue our discussion of small arms development since the end of World War II. In the eleventh edition we introduced these chapters based upon the assumption that the average reader will be more conversant with firearms of the period before 1945 and relying upon the increased availability of books and other printed material on older small arms. Additional historical and technical information about military handguns can be found in our companion volume *Handguns of the World: Military Revolvers and Self-Loaders from 1870–1945.* The forthcoming "Rifles of the World: Military Shoulder Weapons from 1800 to 1945" will in like manner provide additional data on infantry weapons prior to 1945. Part II reflects a continuing attempt to return to a basic manual of standard military small arms, 1939 to 1982. As a result, discussion of developmental weapons has been concentrated in Part I.

We hope this division of material will help the reader understand from where the current small arms came (Part I), and how the standard issue weapons of the world operate and are assembled/disassembled for field maintenance (Part II).

Since each edition of *Small Arms of the World* reflects changing conditions, the current one updates earlier versions. But no one volume of *Small Arms* completely supplants the previous ones. In fact, the serious student of infantry weapons should study all editions. In Part I of the twelfth edition, the reader will find both metric and English measurements given for weights, dimensions, and projectile velocities. Only inch/foot/pound specifications are used in Part II. This partial conversion reflects the status of the metric system in the United States today. It is likely that subsequent editions of *Small Arms of the World* will contain metric measurements only, since that is the trend in the world of small arms. For example, when the US Army let a contract for prototypes of the Squad Automatic Weapon, it called for all drawings and specifications to be made in the metric system. The M249 SAW (FN Minimi) and the M240 armor machine gun (FN MAG) are being manufactured to metric standards. As the world changes, so must this publication.

The twelfth edition of *Small Arms of the World* contains some new categories of weapons (shotguns and shoulder-fired grenade launchers), some new country chapters (Brazil, Singapore, and South Africa), and a new chapter on "Small Arms for Outdoor Sports." Even with these additions, *Small Arms of the World* does not include every military small arm. In the forty years since W. H. B. Smith compiled his first edition of this book there has been a veritable explosion of new small arms. Therefore, there are some weapons we have not covered, have overlooked, or for which we have not been able to obtain data. As in the past, the editor encourages readers to comment on the text and illustrations, but we ask your understanding about some of the limitations with which we must work. Photographs or data on old and new weapons, suggestions for areas to be covered, corrections of errors, and other proposals for improving this book are always welcome. Please address all correspondence to the editor in care of the publisher at P.O. Box 1831, Harrisburg, PA 17105. Insofar as it is possible, correspondence will receive the editor's personal attention, and replies will be made.

EDWARD CLINTON EZELL
Houston, Texas and
Woodbridge, Virginia

Acknowledgments

Walter Harold Black Smith, 1901–1959
Joseph Edward Smith, 1921–1976

Over the years, many individuals, organizations, and business firms have contributed to the preparation of the eleven preceding editions of *Small Arms of the World.* This list is devoted to those who assisted in preparing the twelfth edition.

United States:

Action Arms Ltd., Philadelphia
Mitch Kalter
Waymar Advertising, Inc., New York, NY
Wayne Margulies
ARES, Inc. Port Clinton, OH
Eugene M. Stoner
Ann Paulsen
Beretta U.S.A. Corp., Accontink, MD
Robert Bonaventure
Colt Firearms Division, Colt Industries, Hartford, CT
Robert E. Roy
David A. Behrendt
Deep River Armory, Houston, TX
James B. Hughes, Jr., President
L. Tom Barratt
John Hayes
Eagle Arms Corporation, Mt. Pleasant, SC
Bill Bell, President
Kenneth Dunn
Embassy of Australia, Office of the Counsellor for Defence Supply, Washington, DC
Kevin C. Manie
Embassy of the Federal Republic of Germany, The Attache for Defense Research and Engineering, Washington, DC
Dr. Otto Sailer
Embassy of France, Office of the Military Attache, Washington, DC
Colonel Bernard G. Retat
GAPCO, Wilson, NC
William A. Harvey
Heckler & Koch, Inc., Arlington, VA
Michael W. Iten
Hubert Zink
Interarms, Alexandria, VA
Samuel Cummings
Richard S. Winter
Carl Ring
O. F. Mossberg & Sons, Inc., North Haven, CT
Kenneth J. Fellows
National Defense, American Defense Preparedness Association, Washington, DC
D. Ballou, Editor
Susan S. Coulson
Odin International Ltd., Alexandria, VA
Thomas B. Nelson, President
Carlos G. Davila, Vice President
Charles Palm & Co., Inc., Bloomfield, CT
Bill Clede
RAMO, Inc., Nashville, TN
Pony Maples, President
Kay Horton, Treasurer
SACO Defense Systems, Maremont Corporation, Saco, ME
Berge Tomasian, deceased
Charles M. Roberts
SACO, Inc., Arlington, VA
Jack S. Wood, Jr.
Smith & Wesson, Inc., Springfield, MA
R. G. Jinks
Springfield Armory, Inc., Geneseo, IL
Robert Reese, President
Skillman, Inc.
Charles E. Skillman

Sturm, Ruger Company, Southport, CT
Steve Vogel
James Triggs
The Typing Pond, Deer Park, TX
Frances M. Smith
US Department of Defense, Undersecretary of Defense for Research and Engineering (Research and Advanced Technology)
Ray Thorkildsen
US Air Force, Office of the Chief of History
J. P. Harahan
US Army Foreign Science and Technology Center, Charlottesville, VA
Harold E. Johnson, retired
R. T. Huntington, retired
James Kettrick
US Army Materiel Development and Readiness Command, Alexandria, VA (DARCOM)
US Army Armament Materiel Readiness Command, Rock Island, IL (ARRCOM)
Robert Bouilly
Richard Dorbeck
Dorrell Garrison
Richard Maguire
US Army Armament Research and Development Command, Dover, NJ (ARRADCOM)
Fire Control & Small Caliber Weapons System Laboratory
Chuck Rhoads
Curtis Johnson
Joint Service Small Arms Program
James Ackley
US Navy Surface Weapons Laboratory, Silver Spring, MD
L. H. Dreisonstock
World Public Safety, Inc., Gardnerville, NV
Nan Kraus
Charles Biggs, Pasadena, TX
Jerry Lee Elmore, Houston, TX
Richard S. Smith, Belleville, MI

Argentina:
Ignacio J. Osacar, Rosario
Miguel Enrique Manzo Sal, Cuidad de Cordoba

Austria:
Steyr-Daimler-Puch AG, Firearms Department, Steyr
Wolfgang W. Stadler
Carl Walther

Australia:
Commonwealth Small Arms Factory, Lithgow, NSW
Leader Dynamics, Smithfield, Sydney, NSW
Bryan H. Shaw

Belgium:
Fabrique Nationale, Herstal
Rene Chavee
Claude Gaier
A. Renard
Harry Tintner

Canada:
Diemaco, Inc., Kitchner, Ontario
V. R. Jackson
Jack Krcma, Willowdale, Ontario
Collector Grade Publications, Toronto, Ontario
R. Blake Stevens

Finland:

Finnish Foreign Trade Association, Helsinki
Esko Aaltio
Sotamuseo, Helsinki
Markko Melkku
Markku Palokangas
Valmet Defence Equipment Group, Helsinki
Mikko K. Järvi

France:

Groupment Industriel des Armaments Terrestres (GIAT), Saint-Cloud

West Germany:

Heckler and Koch GmbH, Oberndorf
Walter Lamp
Tilo Möller
Jorge Huber
Wolfhart Fritze
Carl Walther sportwaffenfabrik, Ulm, Donau
Jimbo Terushi, Düsseldorf

Hong Kong:

Conmilit
Alan Sze, Editor

Israel:

Israeli Military Industries, Tel Aviv

Italy:

S.p.A. Giulio Fiocchi, Lecco
Giulio Fiocchi, Executive Director
Luigi Franchi S.p.A., Fornaci, Brescia
Gianni Ricci, Rome

Japan:

Nittoku Metal Industry Co., Inc., Tanashi City, Tokyo
Koichi Seki
Masami Tokoi, Tokyo

Spain:

Astra-Unceta y Cia., S.A., Guernica
Star Bonifacio Echiverria, S.A., Eibar
Instituto Nacional de Industria, Cia. de
Estudios Tecnicos de Materiales
Especiales, S.A., Madrid
The Director
Joaquín Olivé

Switzerland:

Eidgenossische Waffenfabrik, Bern
H. Ditesheim
U. Möhr
A. Wenger
International Defense Review, Geneva
Robert D. M. Furlong, Editor-in-Chief
Schweizerische Industrie Gesellschaft, Neuhasen am Rheinfall

United Kingdom:

Jane's Publishing Company, Limited, London
John Weeks
Christopher Foss
Metropolitan Police of London, Forensic Laboratory
Kevin O'Callaghan
Ministry of Defence, Central Office of Information, London
Parker-Hale Limited, Birmingham
R. C. Hale
Royal Military College of Science, Schrivenham
F. W. A. Hobart, deceased

Royal Small Arms Factory, Enfield, Middlesex
The Director
David Gordon
E. I. Kirkpatrick
The Pattern Room
Herbert J. Woodend
Alan Cooper, deceased
Peter Labbett, Croydon, Surrey

USSR: Mikhail Timofeyevich Kalashnikov, Izhvesk Machine Factory, Urdmurt, Khazak ASSR

My wife, Linda Neuman Ezell, deserves a special word of thanks and acknowledgment of love, because only an author's wife can appreciate the demands writing makes on a marriage.

Finally, Vaclav "Jack" Krcma, Willowdale, Ontario, Canada, whose photographs have contributed to this and earlier editions.

E. C. E.

Contents

Part I Small Arms Developments Since 1945

Part II Description of Small Arms by Nation and a Basic Manual of Current Weapons

Contents

PART 1

Small Arms Developments Since 1945

1 Rifle and Carbine Development

When the Second World War began in 1939, only the Soviet Union and United States—a nonbelligerent—had made firm commitments to the production of self-loading rifles. At the end of the war, the M1 (Garand) Rifle was the only semiautomatic weapon to have successfully withstood the tests of the battlefield. As of September 1945, Springfield Armory and Winchester had delivered 4,024,034 M1 Rifles. Although the Soviets appear to have made greater use of the SVT-38 (Tokarev) Rifle than previously thought, they relied more heavily on the bolt action M1891 rifle and submachine gun than on their semiautomatic rifles. (Soviet weapons factories made 4,450,000 STV-38s, 1,322,085 SVT-40s, and 51,710 SVT-40 sniper versions during World War II.) After World War II, bolt action rifles were relegated to secondary status as most of the world's major armies began to search about for their own solution to the self-loading rifle problem.

Top, **US Rifle, Caliber .30, M1—eight-shot clip;** ***bottom,*** **USSR** ***Samozariadnyia Vintovka Tokareva Obrazets 1938G*** **(SVT-38) 7.62 × 54Rmm—ten-shot magazine. The M1 Rifle was produced from 1936 to 1945 and again during the Korean conflict of the 1950s. The SVT-38 was modified in 1940, and both semiautomatic (SVT-40) and selective fire (AVT-40) versions were produced during World War II.**

In spite of more than thirty years of rapid scientific and technological development in the field of military armaments, the standard weapon of the world's infantrymen is still the rifle. Immediately following the Second World War, many military commentators argued that the foot soldier was obsolete. They expected him to be replaced by tactical nuclear weapons and automated battlefields. A succession of military encounters in Korea, Vietnam and the Middle East, not to mention insurgencies and counterinsurgencies fought around the globe in the same period, have proven the infantrymen to be the match of their colleagues who use more sophisticated arms.

At the end of 1939–1945 war, several important trends were obvious in the small arms of the world. First, the rapid movement of the battlefield, generally typified by the advance of infantry supported by tanks and other armored fighting vehicles, stood in direct contrast to the stalemated battle front of the 1914–1918 conflict. Portability and rapidity of fire were required in all classes of weapons—rifles, submachine guns and machine guns. The German *Wehrmacht* and the Soviet Red Army were notable in their efforts to produce large quantities of pistol caliber submachine guns to provide the desired massed firepower to accompany the offensive thrust of their armored forces, but the striking power of such weapons was limited to close ranges—usually 50–100 meters. Beyond those ranges, all armies during the Second World War relied upon rifles and machine guns firing "full power" cartridges dating from either the last decades of the 19th century or the first decades of this century. World War I had confirmed the desire to have cartridges that were reliably accurate and lethal to 1200 meters and beyond. In both the European and Asian theaters of the 1939–1945 war, the ranges at which effective small arms fire was delivered were usually under 350 meters. Thus, one pressure on weapons designers was the desire of infantry tacticians to acquire small arms that combined the rapid, massed and shocking power of the submachine gun with the lethality of the standard infantry rifle cartridge.

A second fact of life brought home to military planners by World War II was the logistical nightmare created when one's forces had small arms of various types and calibers. The *Wehrmacht* was the worst example. While the German Ordnance Corps struggled mightily to issue only caliber 7.92 × 57mm rifles and machine guns to their front line troops, the variety of weapons chambered for that cartridge was staggering. Each weapon type required a separate spare parts inventory and special training. British and French troops had similar headaches. They started the war with their own national patterns of small arms, but after the evacuation of Dunkirk and the fall of France, American weapons came to play an increasingly important role in the Allied war effort.

Late in the war, the Germans introduced a new class of small arms, the *Sturmgewehr*—or assault rifle, as this class of weapons has been designated in the post-1939–1945 war era. This weapon fired a shortened 7.92mm cartridge (7.92 × 33), with an average velocity of approximately 640 meters per second (m.p.s.) (2100 feet per second [f.p.s.]) versus 840 m.p.s. (2750 f.p.s.) for the standard 7.92 mm round fired from Mauser Kar. 98k. Projectile weight was reduced from 12.8 grams (197.5 grains) to 8.1 grams (125 grains). Although not as lethal as the longer range rifle cartridge, the 7.92 mm *Kurz* was far more effective than the standard 9mm Parabellum pistol cartridge used in submachine guns by the *Wehrmacht*—muzzle velocity, 381 m.p.s. (1250 f.p.s.) for a 7.4-gram (115-grain) projectile. Experimentation with the *Kurz* cartridge led the German army in 1944/45 to plan to replace all existing bolt action and self-loading (semiautomatic) rifles with the MP43/44, *StG 44* series of selective fire assault rifles. However, this plan was thwarted by the crippling effect of the Allied land and air offensives on the German production and supply system. Nevertheless, the German experiment had a profound effect on the post-war thinking of American, Belgian, British, Soviet, Spanish and Swiss small arms designers. Captain H. B. C. Pollard, an English intelligence officer, commented on the virtues of the 7.92 × 33mm cartridge in a 1945 Allied intelligence report: "This is a wholly admirable cartridge which may well be the prototype of the rifle cartridge of the future." He continued his praise, "while as effective for military purposes as the old 8mm Mauser German Army Service Cartridge, [it] is only two-thirds the size and two-thirds the weight. In other words, a man can carry twice the amount of ammunition into action."

As a consequence of the search for rifles with greater firepower and lighter weight during the past thirty years, many new rifles have been proposed, dozens have been tested and a few have become standard weapons, replacing the Second World War generation of small arms. Three basic calibers dominate the scene today—7.62 × 51mm NATO caliber, 7.62 × 39 M43 (Soviet) caliber and 5.56 × 45mm caliber. While each can trace its origins to the assault rifle concept, each of these groups of rifles reflects a particular technical/historic trend in small arms development and as such is described by caliber in this chapter.

POST-WAR RIFLE CARTRIDGES COMPARED WITH THE 7.92MM KURZ AND M1911 .45 CALIBER CARTRIDGES

7.62 × 51mm NATO | **.280/30 (7mm) UK** | **7.62 × 45mm M52 Czech** | **7.62 × 39mm M43 USSR** | **7.92 × 33mm PP 43 Germany** | **.45 (11.43mm) USA**

Two World War II German self-loading rifles. Top, *Gewehr 41 (W)* and bottom, *Gewehr 43.* Both weapons fired the standard 7.92 × 57mm cartridge.

Typical *Sturmgewehr*, the German MP44, which fired the 7.92 × 33 cartridge.

NATO CALIBER RIFLES

The end of the Second World War found the Atlantic Allies armed with an incredible potpourri of small arms. At war's end, the United States, the United Kingdom, France and Canada all began to cast about for new infantry weapons. High on the list was a new rifle. Nationalism, differences in opinion about rifle and ammunition specifications and the desire to standardize weapons among the old allies created a muddled situation that lasted for the better part of a decade, popularly known as the "great rifle controversy."

During the struggle with the Axis powers, the British employed several models of the Short Magazine Lee Enfield (SMLE) Rifle. While this had been the basic pattern for their service rifle since 1902, the desire for a self-loading rifle went back to the pre-1914 period. The successful development of the No. 4 Rifle in the 1930s (an updated SMLE) set back the effort to obtain a self-loading rifle. Some experimentation toward that end was taken during World War II with the Belgian SAFN in 7.92 × 57 (called the 7.92 S.L.E.M.1), but no change was seriously contemplated as long as the war continued.

At the end of the hostilities, the Ministry of Supply created the "Small Arms Ideal Calibre Panel" and assigned it the task of developing a new infantry rifle cartridge. The British sought the lightest rifle and ammunition combination that would be consistent with firing comfort and effectiveness at the reduced maximum range of 600 meters. Commenting on this shortened range, one member of the development team said, "It was recognized that the old .303 over-killed at rifle ranges." They were seeking an "intermediate power" cartridge to be used in an assault-type rifle.

Dr. Richard Beeching, Deputy Chief of the Armaments Design Establishment, and his associates carried out the basic ballistic studies for the "Ideal Calibre Panel," which included an experimental determination of the optimum caliber, muzzle velocity and external and internal ballistics. This work, without equal at the time, led to the conclusion that the ideal caliber was .270. Preliminary talks with American Ordnance representatives led Beeching's staff to increase the caliber to .276, later designated .280, although the diameter of the projectile was not actually altered. In the summer of 1947, Beeching's panel submitted its classified findings in a formal report. The British Army team, dispatched to the United States with this document, encountered for the first time American plans for the development of a lightweight .30 caliber rifle. A difference in calibers was just the first hurdle on the path toward a common rifle cartridge.

At first, the controversy over caliber was difficult to understand, since the desire for an "intermediate power" cartridge was sparked by a common admiration for the 7.92 × 33 *Kurz* round that the Germans employed in their *Sturmgewehr* series. But the Americans displayed an ambivalence in their goals for a new rifle/ammunition system. They wanted the firepower of the *Sturmgewehr,* but they also wanted to keep the long range of their older .30-06 (7.62 × 63) cartridge. As one official statement phrased it:

> The Army is firmly opposed to the adoption of any less effective small caliber cartridge for use in either its present rifle, or in new weapons being developed. . . . Battle experience has proved beyond question the effectiveness of the present rifle ammunition, and there have been no changes in combat tactics which would justify a reduction of rifle caliber and power.

In the spring of 1951, the British Labour Party announced the adoption of the .280 E.M.2 Rifle, and the rifle controversy in the UK divided along party lines. The Labour Party stood for national integrity at the risk of disunity in the NATO alliance (formed in 1949). The Conservatives urged international cooperation based upon the relative power position of the UK in NATO.

Top: Rifle, .280 E.M.2. Bottom: Rifle No. 4 Mark 1. The No. 4 Rifle fired an 11.3-gram (174-grain) projectile at 751 m/s (2465 f.p.s.) from a 592mm (23.3-inch) barrel; while the E.M.2 fired an 8.4-gram (140-grain) projectile at 736 m/s (2415 f.p.s.) from a 622mm (24.5-inch) barrel. Overall length of the E.M.2 was kept short by its bull pup design. The magazine and receiver group are to the rear of the pistol grip, thus permitting the use of a full length barrel in a shorter weapon: E.M.2 914mm (36 inches) vs No. 4 1079mm (42.5 inches).

Stefan K. Janson demonstrating the E.M.2 Rifle he designed at the Royal Small Arms Factory, Enfield.

The latter party asked if Britain were being realistic in expecting the United States with its proportionally larger commitments and greater population to bend to the whims of the British on the caliber of the next American rifle.

By August 1951, the rifle controversy had embroiled the major members of the alliance. The Canadians, in particular, who had been stoically awaiting a solution of the disagreement, began to grow more and more concerned about producing a third cartridge. They were already manufacturing cartridges for both British and American rifles. They did not want to increase the logistical difficulties made evident by the Korean conflict. As a consequence of the Canadian desire for a resolution to the NATO ammunition question, Brooke Claxton, Canada's Defense Minister called for a four-power conference between representatives of Canada, the US, the UK and France. The meeting resolved nothing. A standard cartridge, the American Cal. .30 T65E3 emerged subsequently as the standard round in 1956. A standard rifle was never to be.

FABRIQUE NATIONALE FUSIL AUTOMATIQUE LÉGÈR (FAL)

FN's *Fusil Automatique Légèr* was one of the basic NATO caliber weapons to emerge from the "rifle controversy" of the 1950s. Designers at FN began work on a self-loading rifle before World War II; Dieudonne J. Saive was the principal engineer when the German army invaded Belgium. In 1940, he and several of his associates went to the UK where they continued their work on the rifle at the Royal Small Arms Factory, Enfield. After the war, the rifle was manufactured at FN. Designated the ABL *(Arme Belgique Légèr)* and SAFN *(Saive Automatique, FN)*, it was produced in 7mm, .30-06 and 7.92 calibers.

Building upon his experience with the SAFN, Saive designed a prototype assault rifle that fired the 7.92 × 33mm *Kurz* cartridge. Demonstrated early in 1948, these early FALs were very close to being in concept ideal assault rifles. The short cartridge with its moderate recoil permitted the construction of a compact and relatively light weapon. These initial models were subsequently replaced by prototypes chambered for the British .280 (7mm) cartridge. Two variants of the .280 FAL were developed—a bull pup design and one conventionally stocked. When the US Army rejected the UK cartridge, Saive and Ernest Vervier redesigned the FAL to fire the American experimental 7.62 × 51mm cartridge. During this evolution in design, the rifle gained weight and grew in length. The American, British and Canadian armies dropped the full automatic fire requirement because the weapon was no longer controllable when fired automatically. As ultimately adopted by more than 50 nations, the basic rifle is essentially an advanced semiautomatic rifle with a 20-shot magazine. The heavy barrelled version adopted by several countries as a light squad automatic weapon, replacing older weapons such as the BREN, is neither an assault rifle nor a good light machine gun. Australia's L2A1 heavy barrel FAL, used by several Commonwealth nations, has a "bang, bang, jam" phenomena. Instead of automatic fire, it fires two rounds, and then experiences a failure to feed.

Despite its shortcomings (length, weight and recoil), the FAL has been an exceedingly popular weapon. Once the British discovered that the US Army did not like the E.M.2 Rifle but that there were some American officers who thought the FAL was a good weapon, the British became strong proponents of the FAL. The Belgian weapon was tested extensively by the NATO armies between 1951 and 1956. Two experimental lots of the FAL were manufactured in the United States—Harrington & Richardson (500) and High Standard (13). While the US Army ultimately adopted its own design, the 7.62 × 51mm M14 Rifle, instead of the foreign FAL, the Belgian rifle has seen wide use throughout the world and has been produced in larger quantities than any other NATO caliber rifle since 1945.

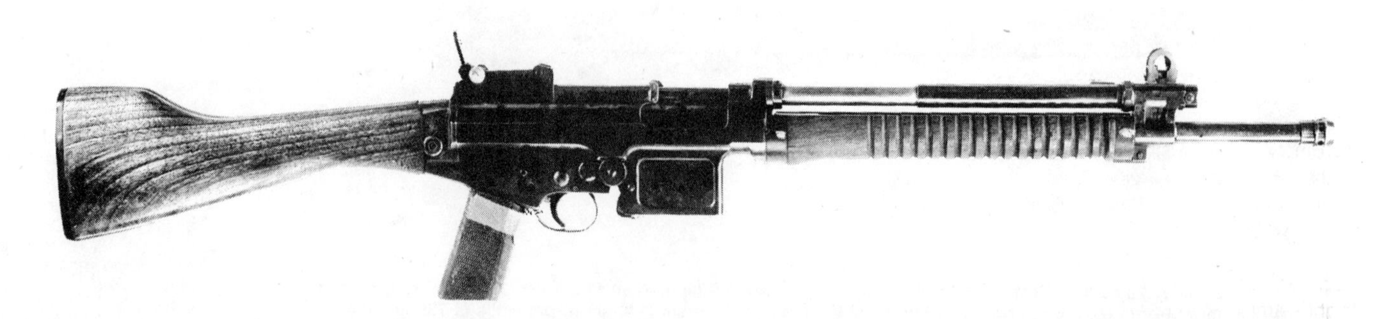

Early FAL prototype chambered for the 7.92 × 33mm *Kurz* cartridge.

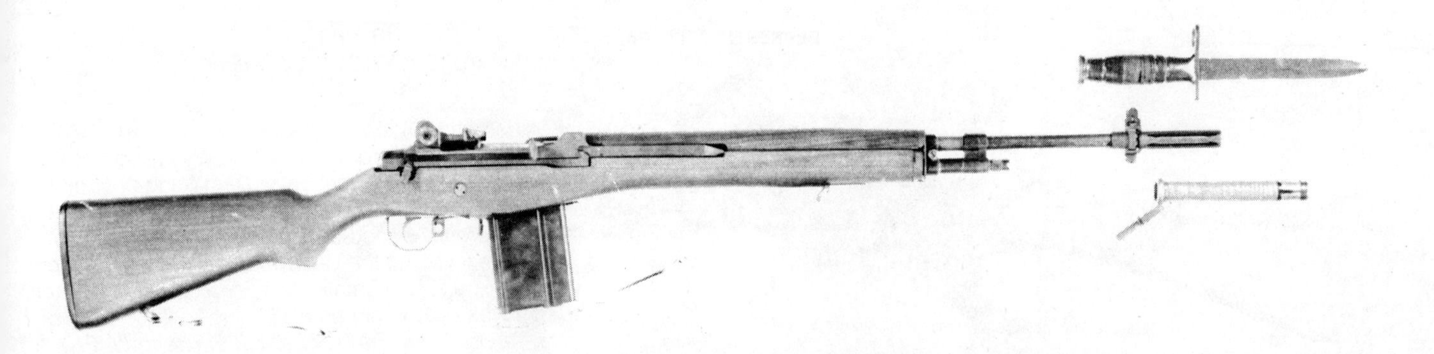

T44

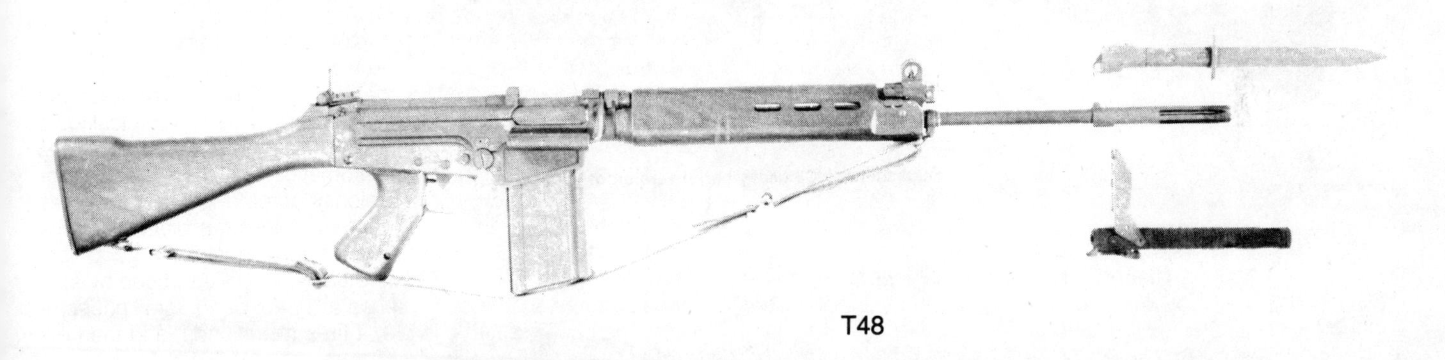

T48

Springfield Armory photograph comparing the American T44 (Prototype M14 Rifle) and the FN FAL (T48) as built for tests in the United States.

SPRINGFIELD ARMORY-ORDNANCE CORPS CHARACTERISTICS RIFLE CAL . . 30, T44 & T48(FN) (LIGHT BARREL)

Characteristics	M1	T44	T48(FN)
Weight - Basic Rifle (w/Empty Mag.-Less Sling)	9.6#	8.7#	9.7#
Weight - Rifle Ready To Fire (Fully Loaded-w/Sling)	10.3# (8 Rds)	10.0# (20 Rds)	11.0# (20 Rds)
Length - w/Flash Hider	43.6″ (w/o F.H.)	44.25″	44.5″
Muzzle Velocity	2800	2800	2800
Weight - Ammunition (20 Rds)	1.18#	1.04#	1.04#
Weight - Bayonet	.88#	.70#	0.63#
Weight - Grenade Launcher	.81#	.31#	0.34#

REPRESENTATIVE FALs

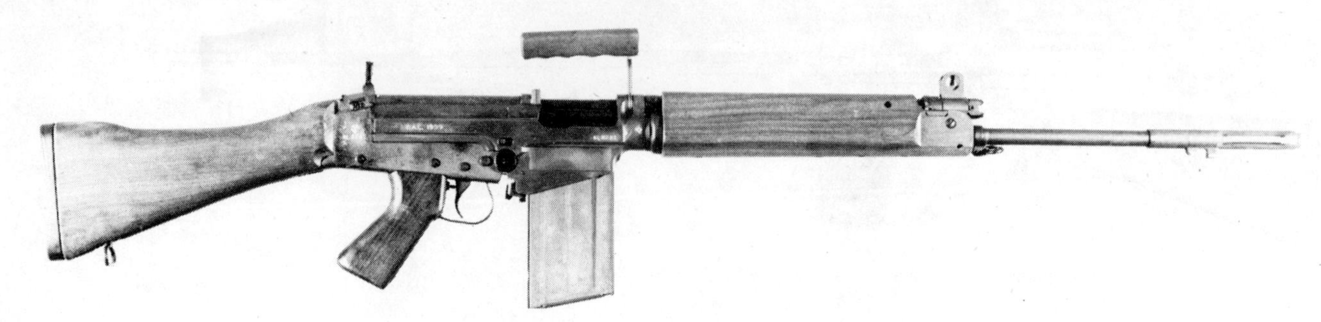

Canadian C1A1 Rifle manufactured by Canadian Arsenals, Limited, 1958.

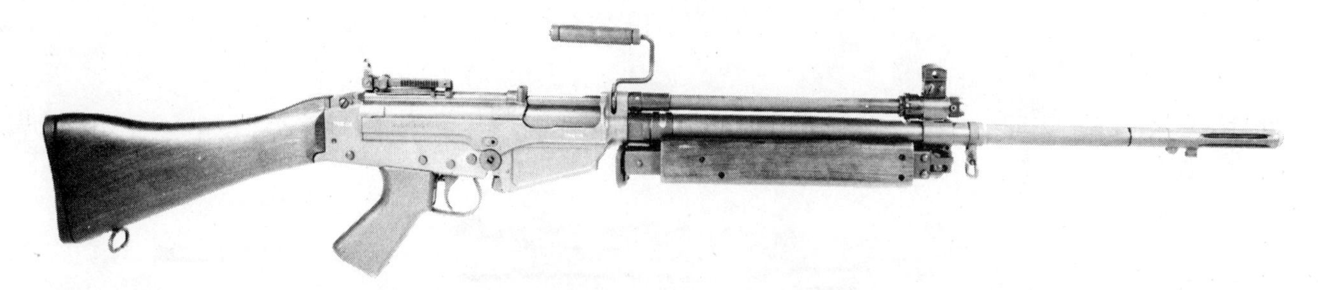

Canadian-made C2 heavy barrel version of the FN FAL.

FN-made FAL for the government of The Netherlands. Note the sheet metal handguard and the combination flash suppressor/ grenade launcher. This rifle also has a unique tunnel-type front sight, and the Dutch were the first to adopt a nonadjustable rear sight.

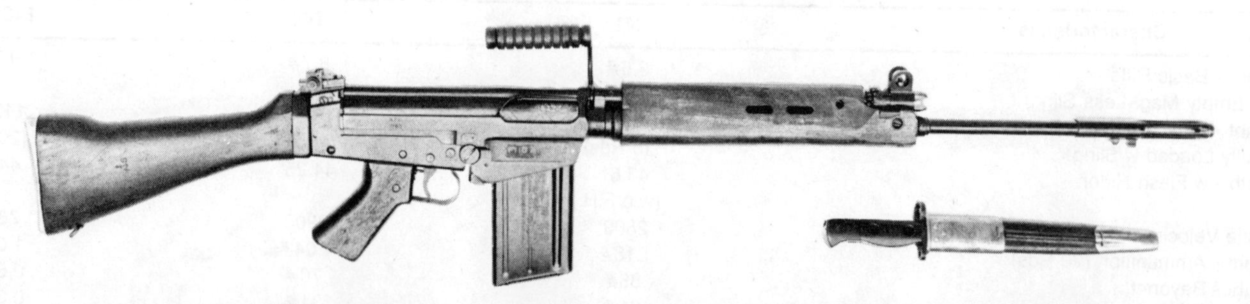

An Indian-made FN FAL marked R.F.I. 1965 (Rifle Factory, Ishapore). This weapon is designated the 1A SL Rifle.

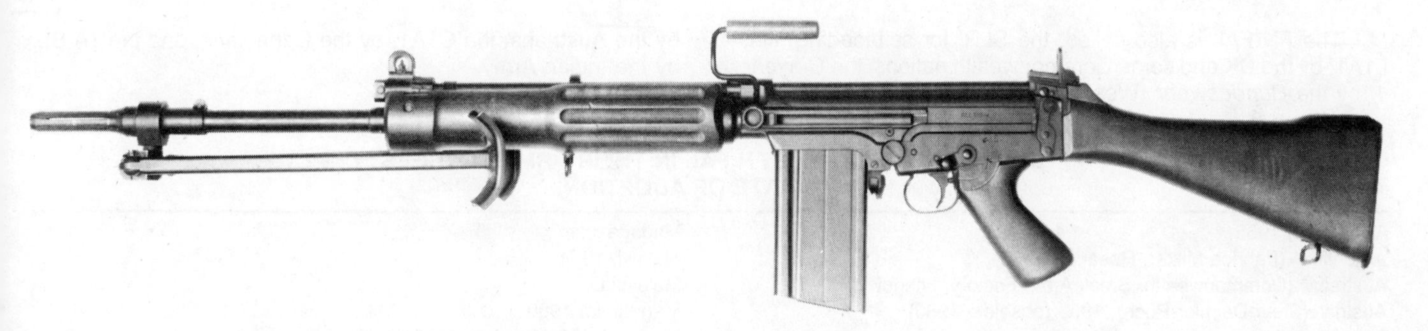

Early FN-made version of the heavy barrel FAL. Note flash suppressor and length of the front handguard.

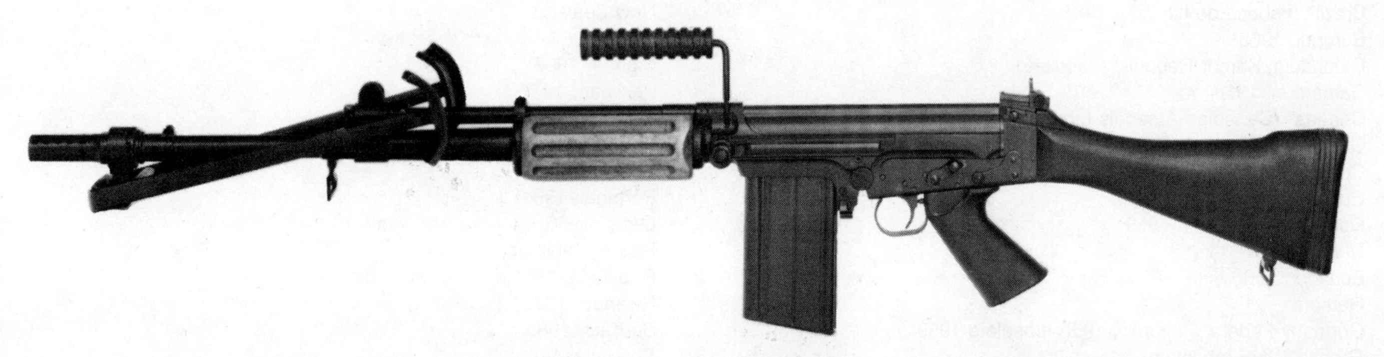

Standard FN heavy barrel FAL, designated the FALO by the FN factory. (FALO is an acronym meaning FAL *lourd*, FAL heavy.)

Israeli FAL. All parts except the receivers were manufactured by Israeli Military Industries. FN manufactured the receivers. Note the strengthened front sight and gas cylinder assembly and the modified handguard. In this model the cocking handle can be used as a forward bolt assist.

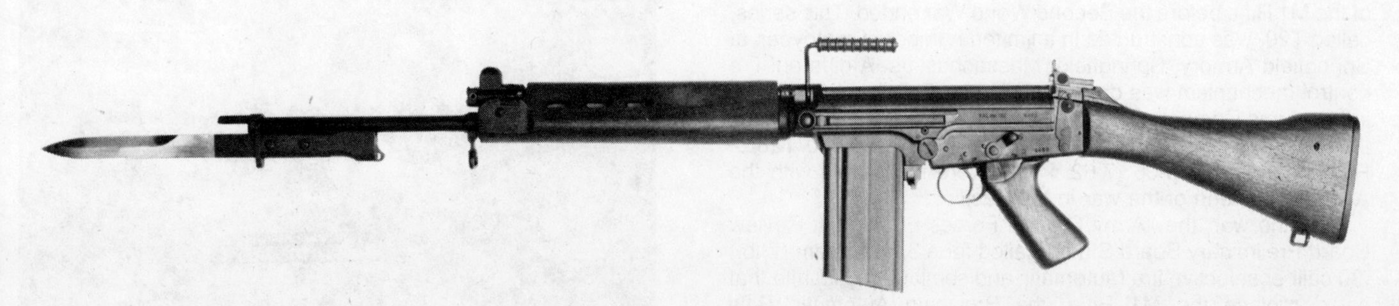

Standard FAL (model 50-00) as manufactured by FN for the government of Peru. This rifle has serial number 4469.

The FN FAL is also called: the SLR, for self-loading rifle; L1A1, by the UK and some Commonwealth nations; the *Gewehr 1,* by the *Bundeswehr* (West Germany); the *Sturmgewehr 58,* by the Austrians; the C1A1, by the Canadians; and the 1A SL, by the Indian Army.

COUNTRIES USING THE FN FAL IN THEIR ARMED FORCES (WITH DATE OF ADOPTION)

Abu Dhabi, 1965
Argentina* (Fabrica Militar, Rosario), 1955
Australia* (Commonwealth, Small Arms Factory, Lithgow)
Austria* (Steyr-Daimler-Puch), 1958 (obsolete 1980)
Bahrein, 1968
Bangladesh
Barbados
Belgium* (Fabrique Nationale Liege), 1954
Bolivia, 1978
Botswana, 1978
Brazil* (Fabrica de Itajubã), 1964
Burundi, 1963
Cambodia, Khmer Republic (obsolete)
Cameroon, 1968
Canada* (Canadian Arsenals Ltd.), 1953
Chile*, 1960
Congo, Republic of, 1956
Cuba, 1959
Dominican Republic, 1959
Dubai, 1969
Ecuador, 1960
Gambia
Germany-Federal Republic, 1956 (obsolete 1959)
Greece, 1965 (obsolete)
Guyana
Haiti, 1968
Honduras, 1969
India* (Ishapore), 1963
Indonesia, 1958
Ireland (Eire), 1961
Israel* (Israeli Military Industries)
Jamaica
Jordan
Kenya, 1966
Kuwait, 1957
Lebanon, 1956
Lesotho, 1971
Liberia, 1963
Libyan Arab Republic, 1955
Luxembourg, 1956
Madagascar
Malawi, 1974
Malaysia
Mauritania, 1980
Mexico, 1968 (obsolete 1980)
Morocco, 1963
Mozambique, 1959
Muscat and Oman, 1960
Nepal
Netherlands, 1961
New Zealand
Niger, 1964
Nigeria, 1967
Pakistan, 1977
Panama, 1961
Paraguay, 1956
Peru, 1958
Portugal, 1961
Qatar, 1956
Ras Al Kahimah
Rhodesia, 1961
Rwanda, 1963
St. Kitts, 1969
St. Lucia, 1963
St. Vincent, 1968
Saudi Arabia, 1960 (obsolete)
Sharjah, 1975
Sierra Leone, 1968
Singapore, Republic of
South Africa, Republic of* (Pretoria), 1960
Sultanate de Raas, 1968
Syria, 1956
Tanzania, 1966
Thailand, 1961
Tunisia, 1967
Ummal Qiwain, 1975
United Kingdom* (BAS and RSAF Enfield), 1954
Upper Volta, 1975
Venezuela, 1954

*Denotes countries that have manufactured their own FALs. See Chapter 9 for the variations in the Fabrique Nationale–produced FALs.

US RIFLE, 7.62mm M14

The American M14 (T44E4 in prototype form) was the major competitor against the FAL in the NATO trials of the 1950s. Designer John Garand had begun work on an automatic version of the M1 Rifle before the Second World War ended. This series, called T20, was constructed in a limited number of protoypes at Springfield Armory, Springfield, Massachusetts. A different fire control mechanism was developed for the Garand by the Remington Arms Company, and that experimental series was called the T22 and T27. Plans for the production of 100,000 T20E2 Rifles in caliber .30-06 (7.62 × 63) were terminated with the August 1945 end of the war in the Pacific.

After the war, the "Army Ground Forces Equipment Review Board Preliminary Board Study" called for a 3.2-kilogram (7-lb.) .30 caliber selective fire (automatic and semiautomatic) rifle that could replace the M1 Rifle, the Browning Automatic Rifle (M1918A2) and all the existing sniper weapons. Subsequently, the US Army altered this requirement so that the new rifle could

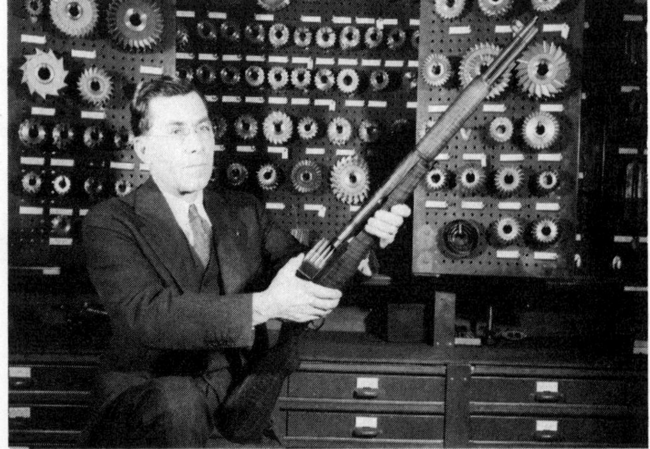
John Garand in his tool room at Springfield Armory holding an M1 Rifle (ca 1943).

Earle M. Harvey, Chief of Small Arms Development at Springfield Armory in the late 1950s and early 1960s, shown here receiving an award from the Commanding Officer of the Armory, Col. O. E. Hurlbut in 1958.

also be used in place of versions of the carbine and all the submachine guns then in use. Three basic weapons came from all this effort: the T25, designed by Earle M. Harvey; the T28, developed by Cyril A. Moore from the unfinished Mauser *Sturmgewehr 45;* and the unconventional T31, designed by John Garand. Only the T25 (later redesignated the T47) got beyond the very limited prototype stage. Building upon design studies conducted during 1942–1944, Earle Harvey had started work on his rifle before the official specifications were laid down in 1946. The first T25s were test fired in 1948. Despite promise, this design fell victim to politics within the US Army. After a series of trials and at least one proposal that it be adopted as the replacement for the M1 Rifle, the T25 was dropped from contention following tests by the infantry at Fort Benning, Georgia, in 1952. Of four candidate rifles, the FN FAL finished first; the new Springfield entry (T44) came in a poor second; and the T25 and British E.M.2 were eliminated from consideration.

Springfield's T44 was essentially a lightened T20E2, or "product improved" M1 Rifle. Lloyd Corbett at the Armory modified the T20E2 by adding a lightweight barrel (T36, November 1949) and adapting the receiver of the weapon so that it would function properly with the shorter T65E3 (7.62 × 51mm) cartridge. As tested in 1951–1952, this weapon (now called the T44) was really a makeshift item that should not have been expected to compete successfully with the more fully developed FAL and T25. When it did perform satisfactorily for a weapon with such a limited development history, the Army decided to press for its improvement. This was only after a Board of Infantry Officers had in August 1952 recommended the limited procurement of the FAL for extensive field trials. During the next five years, the FN FAL and a series of constantly improved T44s were tested, retested and tested again in a wide variety of field conditions—tropical to arctic. After the gunsmoke and political flack finally cleared, the United States Army adopted the T44E4 on 1 May 1957 as the US Rifle, 7.62mm, M14, and the T44E5 (heavy barrel version) as the M15. The latter was never produced in quantity, and on 17 December 1959 was declared obsolete. Subsequently, Captain Durward Gosney of the US Army Infantry

M14A1 Muzzle compensator.

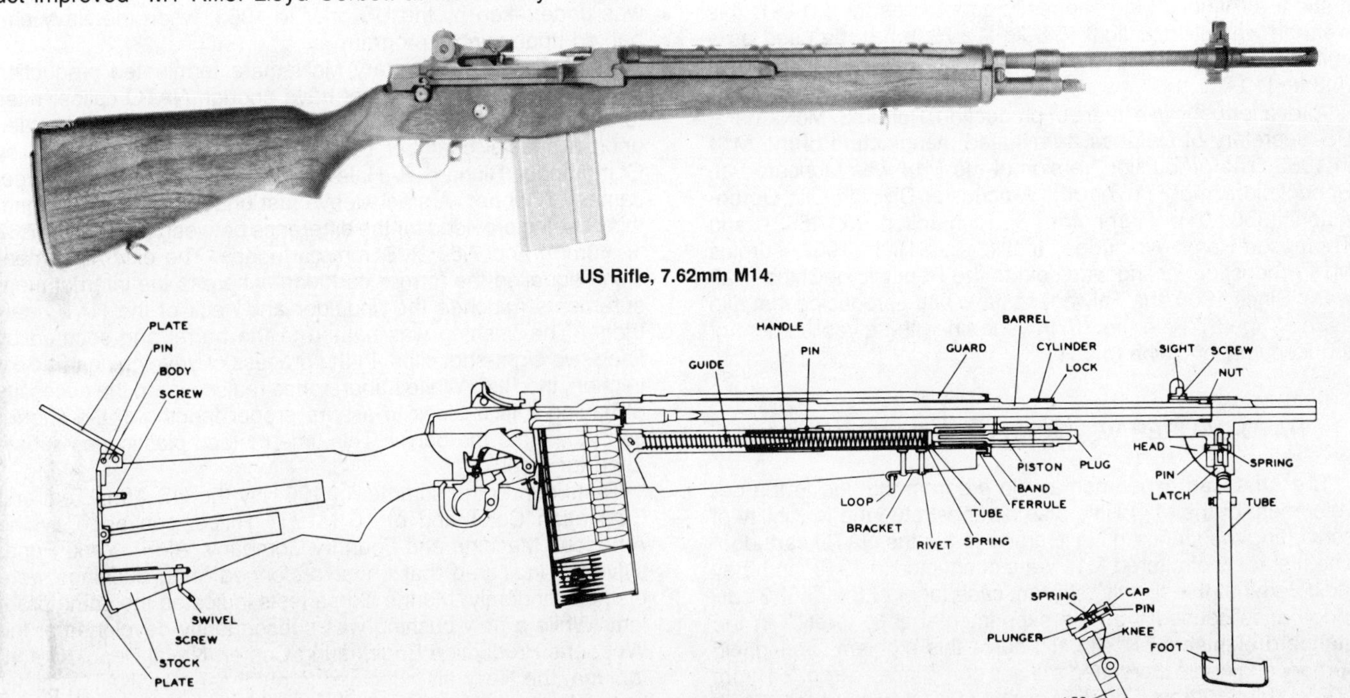

US Rifle, 7.62mm M14.

US Rifle, 7.62mm M15 (obsolete).

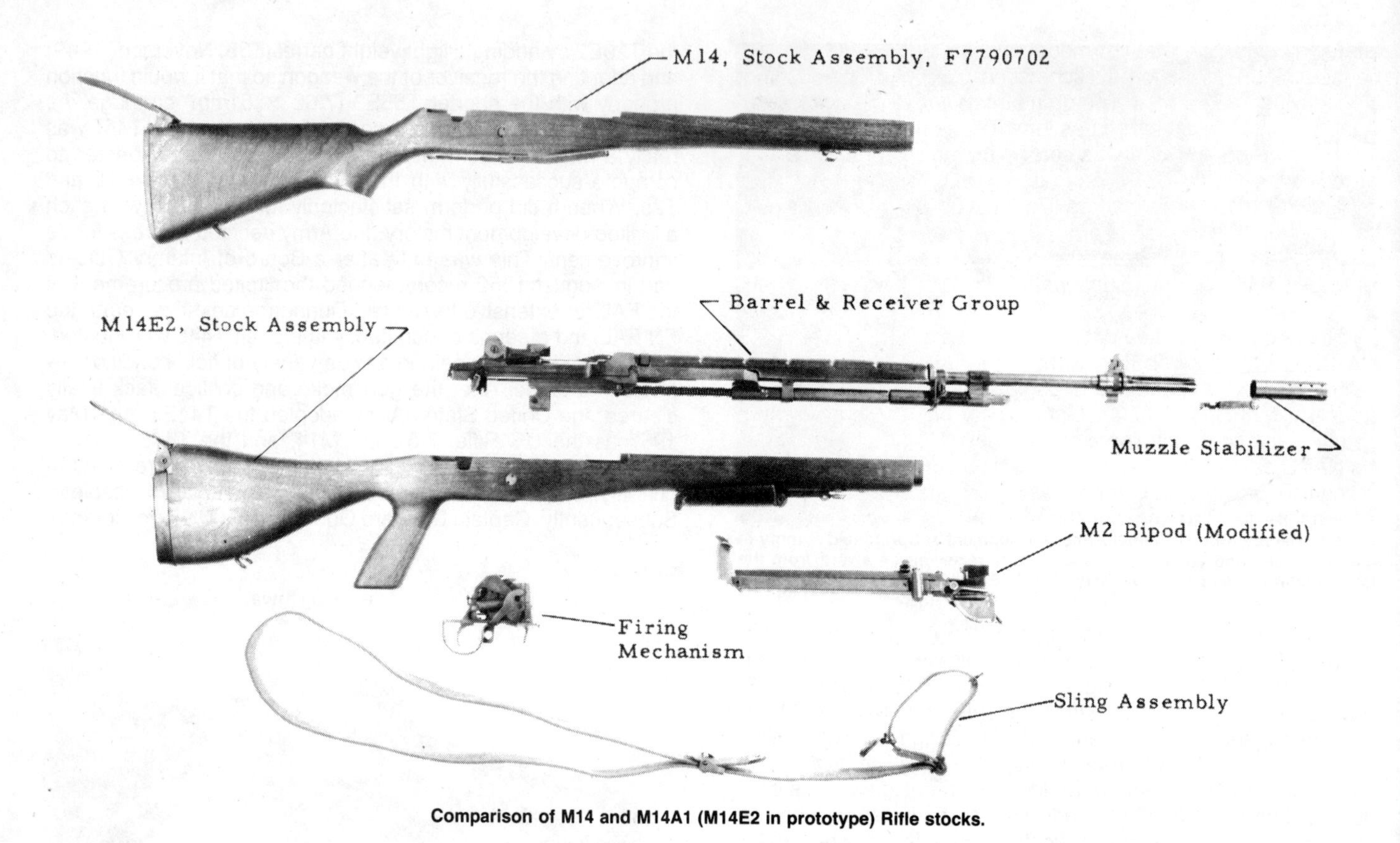

Comparison of M14 and M14A1 (M14E2 in prototype) Rifle stocks.

Board designed a version of the M14 to overcome three specific problems associated with automatic fire—(1) excessive dispersion of bullets, even when fired by an expert rifleman; (2) excessive recoil; and (3) muzzle climb. Gosney's alterations included an "in-line" stock, front and rear pistol grips and a recoil brake that fastened over the standard flash-hider. Combined, these elements made the rifle more manageable during automatic fire. Although subsequently adopted as the M14A1, this weapon was still too light (5.8 kg [12.75 lb.]) to be used as a suitable replacement for the M1918A2 Browning Automatic Rifle (8.8 kg [19.4 lb.]).

After a tortuous five years of production, Robert S. McNamara, US Secretary of Defense, terminated manufacture of the M14 in 1963. The production version of the M14 was fabricated at: Springfield Armory (167,100), Winchester Division Olin Corporation (356,501), Harrington & Richardson (537,582) and Thompson-Ramo-Woolridge, Inc. (319,691). In 1967, surplus M14 production tooling was sold to the Republic of China (Taiwan). Since 1968, the Taiwanese have been producing that rifle as the Type 57 Rifle (i.e., 57th year since the establishment of the Republic of China in 1911).

NATO CALIBER VERSIONS OF THE US M1 RIFLE

The American experimental rifle program that led to the development of the M14 Rifle also demonstrated the feasibility of converting the standard M1 Garand to fire the NATO cartridge. The first tests of altered M1s were conducted in 1948, and they indicated that the slightly different case taper of the T65E3 cartridge case caused the new experimental case to stick in the standard eightshot M1 clip. To solve this problem, Springfield Armory engineers increased the gas port diameter from 2.15mm (.085 in) to 2.22mm (.0875 in). While this increased the power to overcome the sluggishness of the altered rifle, it also increased the strain on the other components. A new barrel was required for the shorter 7.62 × 51mm cartridge, as was a 6.3-gram (.224-oz) steel filler block at the front of the receiver. Experiments with the T35, as this converted M1 was designated, continued through the early 1950s. For various reasons, notably the manufacture of the M14 Rifle, no large alteration of M1 Rifles was undertaken by the US prior to 1963, when the Navy embarked upon such a program.

When Defense Secretary McNamara terminated production of the M14, the Navy did not have enough NATO caliber rifles to equip its shore units. The Bureau of Naval Weapons settled upon a special chamber bushing conceived and patented by Commander Richard F. Haley and a civilian Navy employee, James O'Conner. A steel sleeve just under 25 mm (1 in) long, this bushing provided for the difference between the .30-06 (7.62 × 63mm) and 7.62 × 51mm cartridges. The external dimensions equalled the former cartridge, whereas the internal measurements matched the shoulder and neck of the NATO cartridge. The bushing was seated in the barrel and secured by firing two eight-shot clips. In the process of developing this conversion, the H. P. White Laboratories rediscovered the necessity of having a filler block to assure proper feeding of the shorter cartridges into the barrel. This time molded plastic was substituted for steel.

Tests during the summer of 1964 by the US Army Test and Evaluation Command of 10 M1E14 Rifles, converted by the American Machine and Foundry Company (AMF), York, Pennsylvania, indicated that under prolonged firing bushings were ejected randomly. Marine Corps tests indicated the same problem. While a new bushing was subsequently developed at the Weapons Production Engineering Center, Naval Depot, Crane, Indiana, the Navy also decided to purchase new NATO caliber barrels from Harrington & Richardson for 8,750 M1 Rifles. AMF altered 17,050 with the first model bushing and 5,000 with the new bushing. Harrington & Richardson used the new bushing in modifying 12,250 M1s. The grand total was 30,050 weapons. Congressional critics later questioned the economic wisdom of the Navy program, but it was carried out at a time when there were too few M14s and M16s to go around. All rifles converted in the Navy effort were marked "7.62 NATO" on the left side of the receiver.

While Fabrique Nationale also devised a conversion process for the M1 Rifle, Beretta of Italy went even further and developed an updated NATO caliber M1, the BM59. Early in the 1950s, Beretta began production of the M1 Rifle with technical support from the US. The first rifles went to the Italian Army in 1952. Later, Denmark and Indonesia purchased M1s from the Italian company. More than 100,000 Garands had been produced when Beretta decided to develop a modernized version of that rifle in the NATO caliber. Studies in this direction began in 1958–1959. Domenico Salza and Vittorio Valle were the two key figures responsible for carrying out this project. Many variants were developed—most in prototype form only—with the "BM59 Mark Ital" being adopted by the Italian Army in 1962. That weapon had a Beretta-designed selective fire mechanism—full and semi-automatic—and used detachable 20-shot magazines instead of the Garand eight-shot clip. Further details on variants can be found in Chapter 29. Indonesia and Morocco have produced the BM59 with technical assistance from Beretta. Nigerian plans to produce the rifle appear to have been thwarted by the late civil war.

Total US production of the M1 Rifle between 1936 and 1957 was 6,034,228.

ARMIES USING THE M1 IN .30-06 (7.62 × 63mm)

Austria*	Iran
Bolivia*	Italy (Beretta)
Brazil*	Japan*
Chile*	Korea, Republic of
Republic of China (Taiwan)	Liberia
Costa Rica	Mexico
Cyprus	Norway
Denmark (Beretta)	Pakistan
Ethiopia	Panama
Greece	Philippines
Guatemala	Thailand
Haiti	Turkey
Honduras	United States*
Indonesia*	Vietnam*

* Denotes secondary or obsolete.

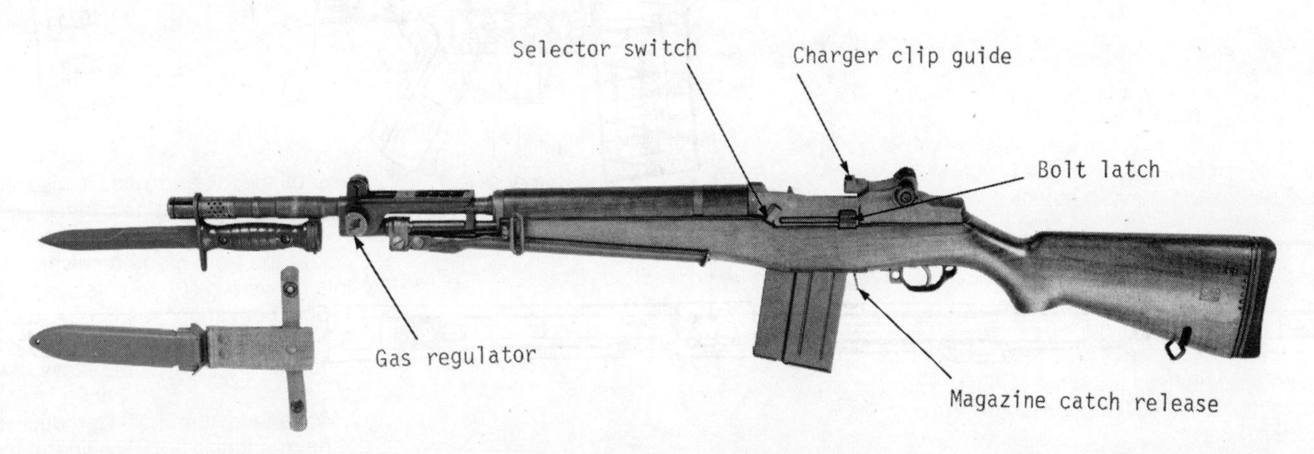

Beretta Mark 1 Ital. *BM59 Versione Normale.*

***BM59 Versione speciale per Paracadutisti* with stock folded and compensator/grenade launcher removed.**

CETME, G-3 AND RELATED ROLLER LOCK RIFLES

A whole family of rifles grew out of work done in 1944–1945 at the *Mauserwerke.* As part of the late Nazi war effort to produce an inexpensive but reliable weapon, engineers at Orberndorf am Neckar developed a delayed blow-back mechanism, using a two-piece bolt with a roller locking system, which provided the necessary delay in opening. This design was not fully developed in 1945, and incomplete weapons were captured by the Allies. The weapon had several designations: StG 45(M), Gerat 06H, MP45(M). (Operational details are presented in Chapter 21 under the G3 Rifle.) As noted above, the American designer C. A. Moore used this concept in his T28 Rifle.

Automatisches Gewehr G 3, Kal. 7,62 mm × 51

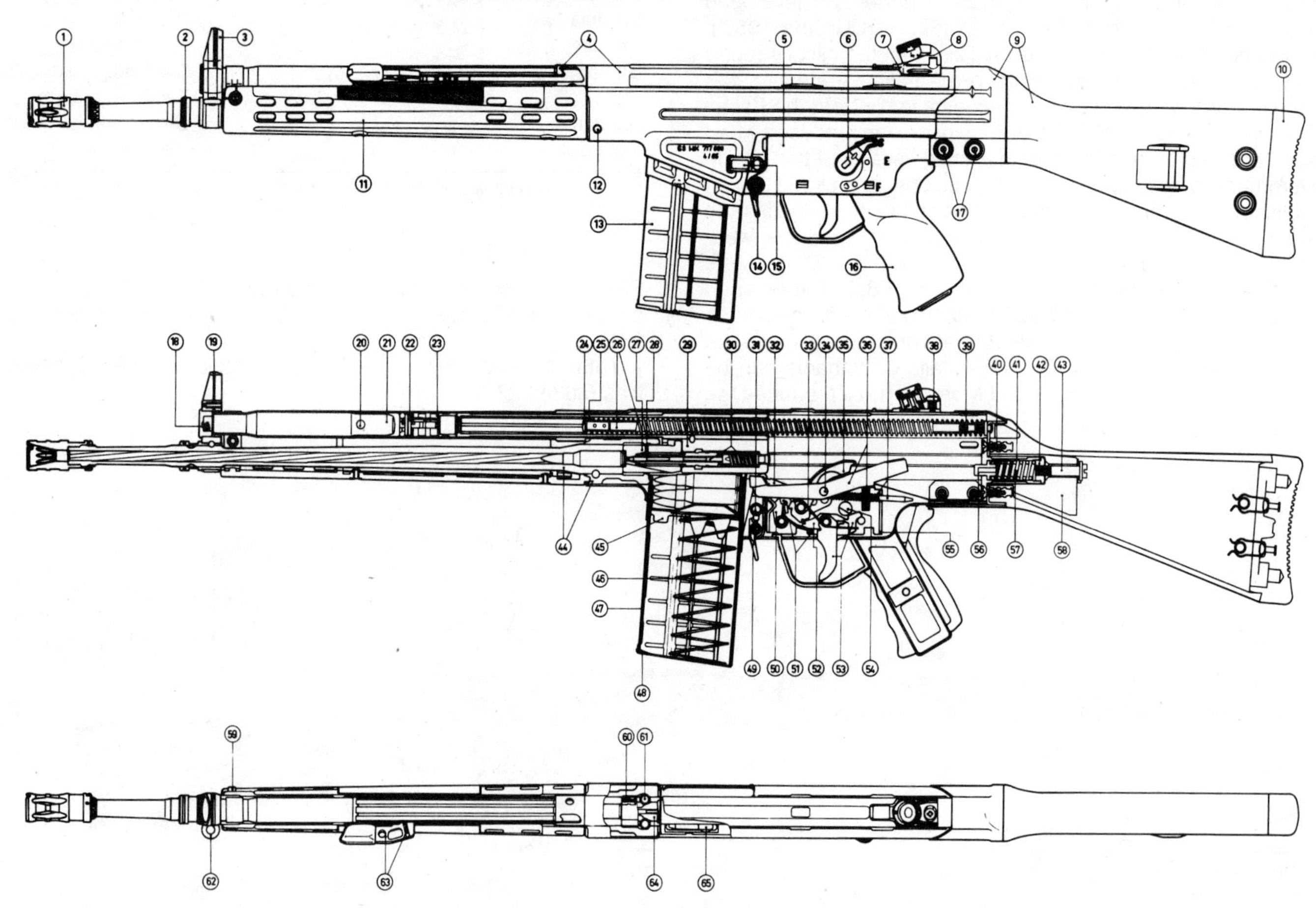

G3 Automatic Rifle, 7.62mm NATO

1. Flash suppressor/ grenade launcher
2. Snap ring
3. Front sight holder
4. Receiver and operating housing
5. Grip assembly
6. Safety
7. Sight base
8. Rotary rear sight
9. Back plate with buttstock
10. Butt plate
11. Handguard
12. Cylindrical pin
13. Magazine
14. Grip assembly locking pin
15. Magazine catch
16. Grip
17. Buttstock locking pins
18. Cap
19. Front sight
20. Stop pin
21. Stop abutment
22. Operating handle spindle
23. Operating handle support
24. Stop pin
25. Recoil spring guide ring
26. Recoil spring tube with recoil spring
27. Bolt head
28. Clamping sleeve and holder for locking roller
29. Bolt body with recoil spring tube
30. Firing pin with firing pin spring
31. Contact piece
32. Release lever
33. Elbow spring for trigger
34. Ejector spindle
35. Hammer
36. Ejector with spring
37. Pressure shank and spring
38. Fixing screw
39. Stop pin for spring guide
40. Countersunk screw
41. Buffer housing
42. Buffer closure
43. Screw for buffer
44. Barrel with barrel extension

45–48. Magazine assembly

49. Magazine release lever
50. Catch
51. Elbow spring with roller
52. Sear
53. Trigger
54. Safety pin
55. Trigger assembly
56. Buffer pin
57. Buffer spring
58. Support for buffer housing
59. Handguard locking pin
60. Extractor with spring
61. Locking roller
62. Eyebolt
63. Operating handle with elbow spring
64. Locking piece
65. Bolt head locking lever

The Mauser engineer Ludwig Vorgrimmler went to France after World War II to work on his roller lock mechanism. While at the French armament center at Mulhouse, he produced two breech mechanisms designed for the American .30 carbine cartridge (7.62 × 33).

In the early 1950s, Vorgrimmler went to Spain where he worked with other German and Spanish engineers at the government *Centro de Estudios Tecnicos de Materiales Especiales* (CETME) in Madrid. Without the immediate pressures of war, the former Mauser engineers were able to give considerable attention to weapons design and ammunition considerations. Included in their efforts was the creation of a 7.9 × 40mm cartridge that had an elongated and very light projectile—6.8 grams (105 grains)—with a muzzle velocity of 800 m.p.s. (2,625 f.p.s.). By 1952, the first prototypes of the CETME Automatic Rifle were ready for testing. After two years of experimentation, the Spanish Government began to look for a company to assist them in establishing a factory for rifle production. In March 1954, Heckler & Koch was invited by the Spanish to attend discussions concerning the advanced development and preparation for manufacture of the CETME Rifle.

Heckler & Koch GMBH, Oberndorf am Neckar, was established in 1949. The founders of the firm had been executive engineers in the *Mauserwerke* at Oberndorf. Most of the key engineers, foremen and skilled machinists had worked for the Mauser factory, which was dismantled at the end of the war. Germany ordered the first 400 rifles in 1956, and they were produced by Heckler & Koch in the NATO caliber, 7.62 × 51mm. In 1958, Spain adopted the improved CETME Rifle, which fired a reduced power version of the NATO cartridge. The following year, the *Bundeswehr* adopted the weapon as the *Automatisches Gewehr G3, Kal. 7.62mm × 51*. The G3 replaced the FAL, which had been used as the G1. A key element in the decision to adopt the G3 was the inability of the German Government to work out a satisfactory licensing agreement with Fabrique Nationale for production of the FAL.

Spanish CETME 7.92mm Assault Rifle, the prototype CETME.

COUNTRIES USING THE G-3 RIFLE

Abu Dhabi	El Salvador	Morocco	Sharjah
Bangladesh	Germany-Federal Republic*	Niger	Spain* (CETME)
Bolivia	Ghana	Nigeria	Sudan
Brazil*	Greece	Norway* (Kongsberg Vapenfabrikk)	Sweden* (FFV)
Brunei	Guyana	Pakistan*	Tanzania
Burma	Haiti	Peru	Thailand
Chad	Indonesia	Philippines	Togo
Chile	Iran*	Portugal*	Turkey
Colombia	Jordan	Qatar	Uganda
Denmark	Kenya	Saudi Arabia	Upper Volta
Dominican Republic	Malawi	Senegal	Zambia
Dubai	Mexico		

* Denotes local manufacturer of the rifle.
The French arms factory MAS, and the Royal Small Arms Factory, Enfield, are manufacturing the G-3 for Heckler & Koch as subcontractors.

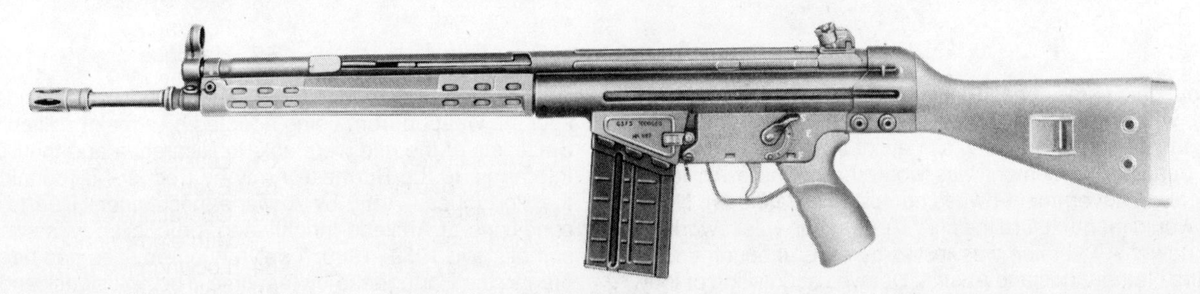

Current manufacture H&K G3A3.

ARMALITE AR-10 and AR-16 RIFLES

The Armalite Division of the Fairchild Engine and Airplane Corporation was established in October 1954, for the express purpose of developing new military firearms using the latest advancements in plastics and non-ferrous metals. While the Armalite firm has gone through several reorganizations, its small development facility has always been located in Costa Mesa, California. Eugene M. Stoner was the key designer with the firm in the early years, while Robert Fremont supervised prototype manufacture and L. James Sullivan oversaw the routine drafting work. Several weapons were undertaken prior to the work on an assault rifle, including:

AR-1—7.62 NATO parasniper rifle, extremely lightweight, using a Mauser-type bolt action mechanism. Only prototypes were built in 1954.

AR-3—7.62 NATO self-loader designed by Stoner into which he incorporated an aluminum receiver, fiberglass stock and a multiple lug locking system of the type later found in the AR-10.

AR-5—.22 Hornet caliber survival rifle developed for the US Air Force and officially designated the MA-1.

AR-7—.22 long rifle self-loader, which comes apart so that the barrel and receiver could be stored in the synthetic stock. Developed in 1959–1960, this rifle is still marketed under the commercial name "Explorer."

AR-9—12 gage (18.5mm) self-loading shotgun with aluminum barrel and receiver, weighing only 2.27 kg (5 lb.). Developed in 1955, it never was produced in quantities.

Work was begun on the AR-10 before Stoner joined the firm. By mid-1956, the Fairchild organization was actively promoting that rifle, despite the fact that it was still in the early stages of development. A third version of the AR-10, with a titanium barrel surrounded with an aluminum jacket, was tested by Springfield Armory in 1956 on the eve of the adoption of the M14 Rifle. When the composite barrel ruptured during the endurance test, there were severe recriminations; the Armalite people thought that their weapon had been mistreated because they were given less than favorable treatment in the report prepared on the test. Subsequently, Stoner, with the assistance of Springfield Armory, designed a new barrel for the AR-10 made of conventional barrel steel.

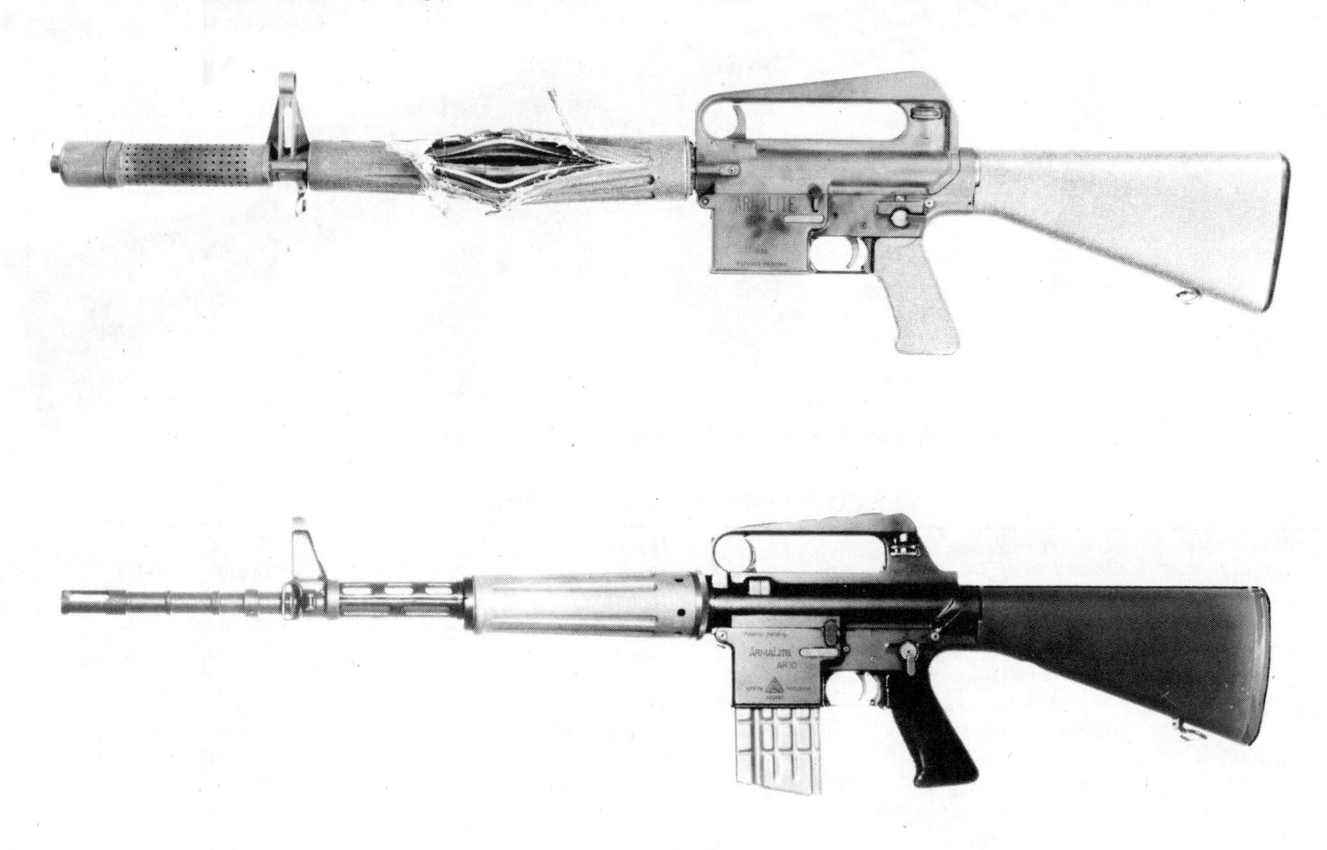

Top: Early Armalite AR-10 with burst barrel. Bottom: AR-10 as manufactured by Artillerie-Inrichtingen.

At about the same time, Richard H. Boutelle, president of Fairchild, was searching for a manufacturing facility that could produce the AR-10. Ultimately, as part of a Fairchild-Fokker of Holland deal, an agreement was worked out where Artillerie-Inrichtingen, a government-owned company in Zaandam, Netherlands, would manufacture the rifle. The weapon was reworked, and the newer A-I version was tested by several countries, including the Netherlands and Austria. Delayed acquisition of tooling with which to produce the AR-10 and political considerations kept the weapon from being adopted by a major military power. Small lots of the rifle were sold to Nicaragua and the Sudan by Interarms, to the Burmese Army by Cooper-Macdonald and to the Portuguese Army by Artillerie-Inrichtingen. Due to chaotic conditions at Artillerie-Inrichtingen, production was suspended sometime in 1959–1960. Twelve hundred rifles had been delivered to the Portuguese by that time. The real significance, however, of the AR-10 is that it led to the AR-15, which after several modifications was adopted by the US Army as the 5.56mm M16/M16A1 Rifle (see section on 5.56 × 45mm rifles).

The AR-16 was another 7.62 NATO caliber rifle designed at Armalite. This weapon appeared following the separation of the company from Fairchild and the departure of the Stoner design team. Whereas the AR-10 utilized aluminum forgings for the upper and lower receivers, the AR-16 was made of sheet metal stampings after the fashion of the German *Sturmgewehr.* Development work went on from 1959 to 1961 but was suspended in favor of a scaled-down version of the AR-16 in 5.56mm. That latter rifle, the AR-18, is discussed later under the 5.56 × 45mm rifles.

A final weapon deserves mention here since it is related in design concept to the other Armalite rifles. The Stoner 62 weapons system, built at the Cadillac Gage Division of Excello Corporation in Warren, Michigan, was another Stoner product. After he left Armalite, Stoner decided that a multipurpose family of weapons could be developed around a common set of basic parts. The Stoner 62 system was an attempt to create a rifle and machine gun family in the NATO caliber. He terminated work on that system when the popularity of the 5.56mm cartridge became apparent. His 5.56 × 45mm Stoner 63 system is also discussed below.

Armalite AR-16 Rifle with folding stock.

SWISS SIG RIFLES 510 AND 542

The Schweizerische Industrie-Gesellschaft at Neuhausen am Rheinfall has been one of the primary sources of sophisticated, precision, self-loading rifles. If anything, their products have tended to be of too high a quality and too expensive to manufacture until recent years. A commercial firm in competition with the government-owned Waffenfabrik, Bern, SIG has been very successful in designing weapons but less so in selling them outside Switzerland. Their sales problems have been due in part to the restrictive export policies of the Swiss Government.

SIG and the Waffenfabrik, Bern, experimented with many designs in the 1940s. The government arsenal designed rifles similar to the German FG-42 to fire intermediate size cartridges in 7.65 and 7.5mm. Shortly after World War II, SIG introduced the SK-46, a gas-operated self-loader designed for such full power cartridges as the 7.5 × 55.5mm Swiss or 7.92 × 57mm German. In this design the operating gases were tapped close to the chamber, and the gas cylinder and piston were positioned to the side of the receiver. The bolt was of the tipping variety with the locking surface located in the upper portion of the receiver. Externally it resembled the straight-pull Schmidt-Rubin Rifle.

SIG subsequently introduced the AK53, a gas-operated selective fire assault rifle. The *Automat Karabin 1953* employed an unusual action in which the barrel moved forward and the fired cartridge case was ejected at the end of the forward cycle.

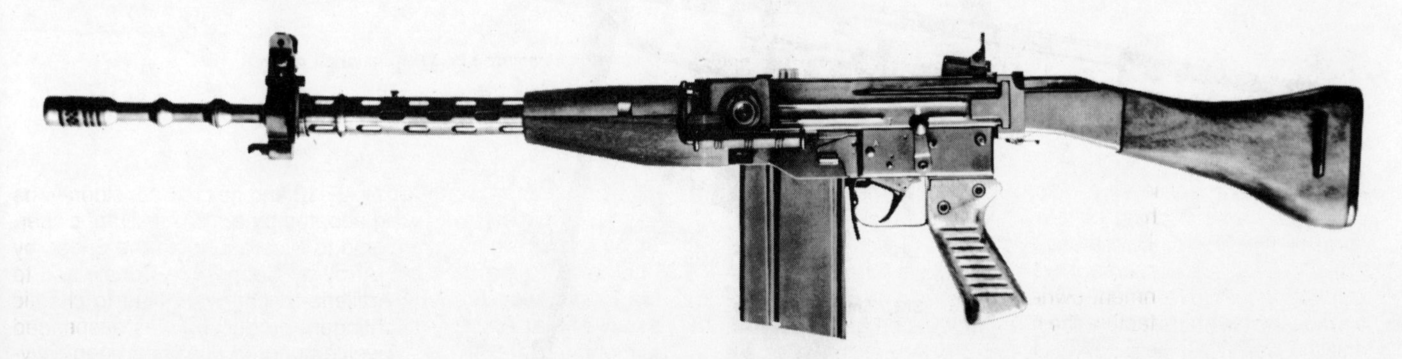

7.62mm NATO SIG Type SG510-4.

SIG Model AK43.

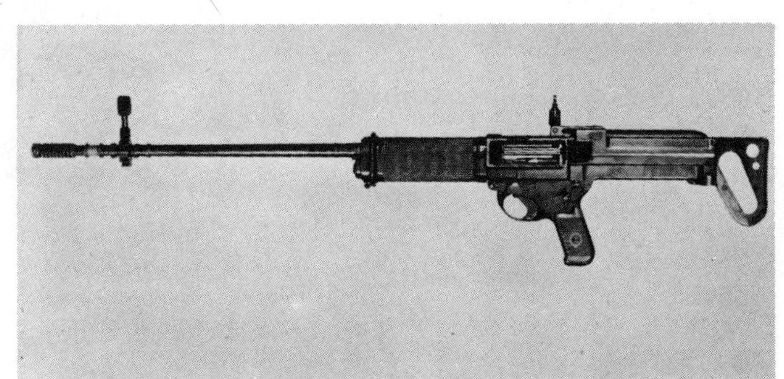

Prototype Waffenfabrik Bern 7.5mm Assault Rifle. Fired standard 7.5 × 55.5mm cartridge.

Prototype Waffenfabrik Bern 7.5mm Short Rifle. Fired experimental short cartridge.

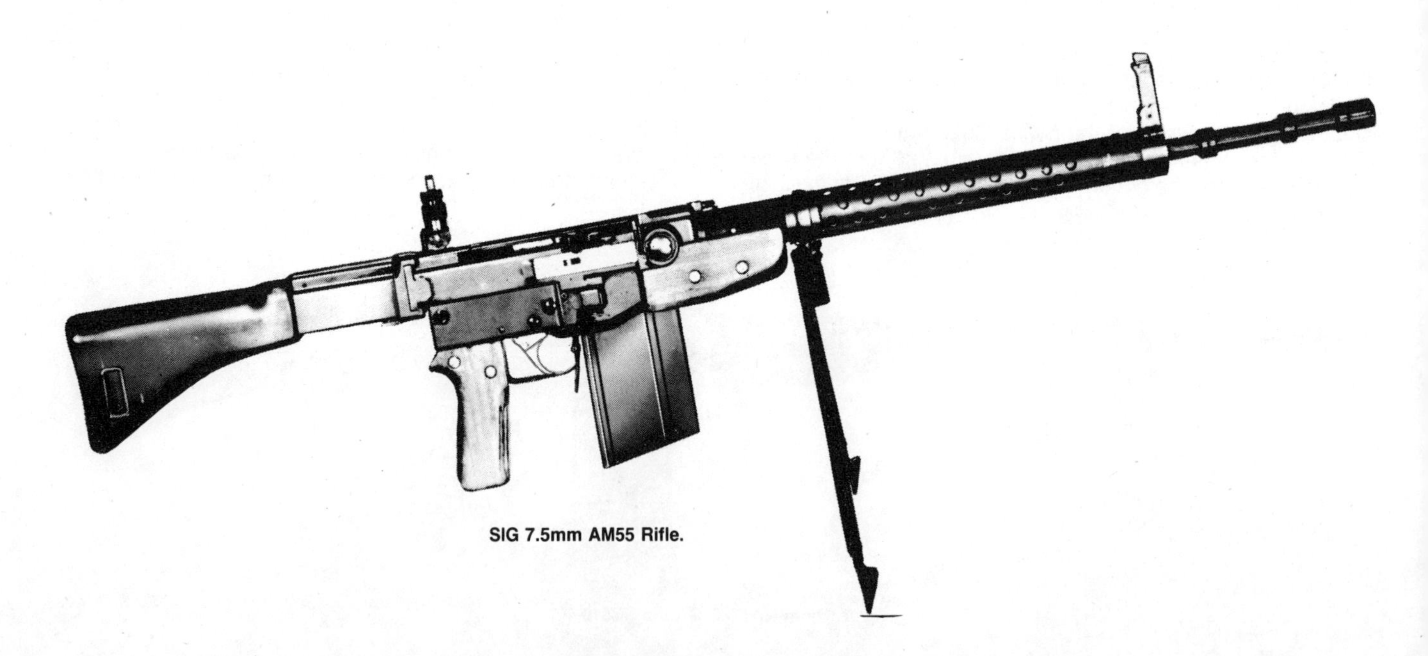

SIG 7.5mm AM55 Rifle.

With this system, a low cyclic rate of about 300 shots per minute was obtained, making the weapon manageable even though it fired the standard 7.5mm Swiss cartridge. This weapon was never put into large scale production.

In the mid-1950s, SIG introduced a new rifle, the AM 55. Subsequently known as the *Sturmgewehr 57* (StG 57) in Swiss Army terminology and SIG 510 in commercial form, it is one additional permutation of the Mauser *Sturmgewehr 45* retarded roller lock breech mechanism. SIG's version was worked out by Rudolf Amsler, the company's technical director. Several 510 series rifles exist:

- 510-0 Swiss Army version in 7.5 × 55.5mm.
- 510-1 Commercial in any standard caliber.
- 510-2 Commercial lightweight.
- 510-3 Commercial in any intermediate caliber; e.g., 7.62 × 39 Soviet.
- 510-4 Commercial in 7.62 × 51 NATO. This weapon has in effect superseded the 510-1.

Although there have been alterations in stock design since 1955, the basic rifle mechanism remains essentially the same. There is no 520 series, but SIG later introduced a 530 series chambered to fire the 5.56 × 45mm cartridge. This weapon incorporated improvements in sheet metal stamping and substituted plastics for wood and metal, but the operating mechanism is still essentially the same as in the 510 Rifles.

Since the early 1970s, SIG has introduced the 540 family of rifles. The 540 and 543 are chambered for the 5.56 × 45mm cartridge; only limited production runs have been made for the 542 Rifle in 7.62 NATO. Outwardly, these rifles bear a strong resemblance to the 530 series, but internally they are quite different. Whereas the 530-1 has a roller locked bolt, the 540 Rifles have a rotating bolt, which is cammed into and out of the locked position by a cam machined into the bolt carrier. The bolt carrier arrangement is quite similar in concept to the Soviet Kalashnikov family of weapons. Unlike the AK, where the rear part of the piston assembly is machined as an integral element of the bolt carrier, on the SIG 540 Rifles the rear end of the piston fits a hole in the bolt carrier, and the operating handle holds it in place by sliding through a slot into the side of the carrier.

Following the current trend in European small arms design, these rifles have a combined flash suppressor/grenade launcher assembly attached to the barrel; they fire three-shot bursts in addition to full and semiautomatic, and they can be supplied with bipod and fixed or folding buttstock.

SIG 530-1 5.56mm Rifle with bayonet.

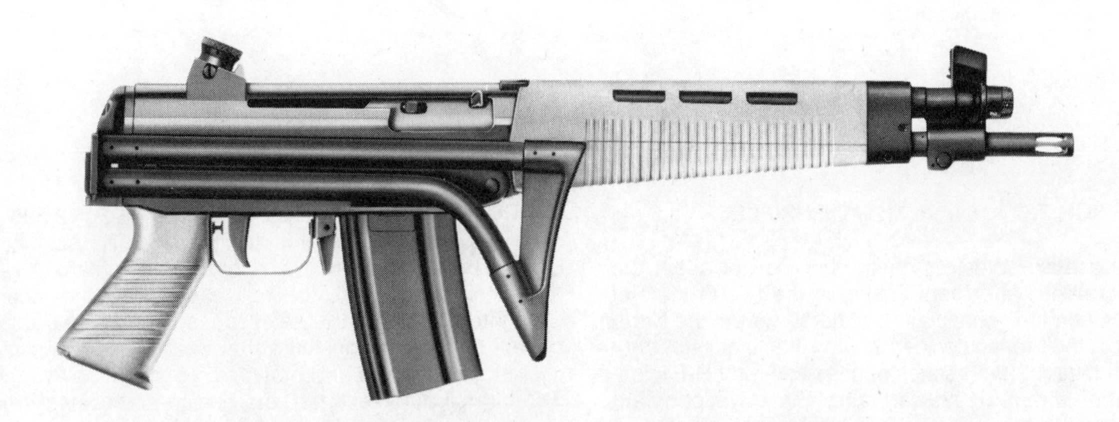

5.56mm SIG, Assault Rifle SG 543 short version with folding butt and 20 rounds magazine.

JAPANESE TYPE 64 RIFLE

While the Japanese Ground Self Defense Forces still use the US M1 Rifle, re-equipment is underway with the Type 64 Rifle, developed by the Howa Machinery Company, Ltd., Nagoya. This weapon, known as the R6E in prototype form, was designed to be used with a reduced charge NATO cartridge. The goal of the designers was to create a standard caliber weapon that could be fired comfortably by their smaller-in-stature troops. A lavender bullet tip coloration indicates the reduced power rifle loading. When full power NATO cartridges are used, the gas regulator can be set to a smaller orifice to prevent overpowering the weapon.

As with the U.S. M14 Rifle, the gas system can be closed off completely for launching grenades from the combination muzzle brake/grenade launcher. The Type 64 has a folding shoulder rest fitted to the top of the butt stock, and all rifles are fitted with bipods.

The 7.62mm Japanese Type 64 rifle as manufactured by Howa. This rifle was designed to fire a reduced charge version of the 7.62 × 51mm NATO cartridge.

FRENCH 7.5 × 54mm M1949/56 RIFLES

France was a major Western nation that did not adopt the 7.62 NATO cartridge. After participating in the NATO trials of the early 1950s and their subsequent withdrawal from the North Atlantic Alliance, the French decided to keep their older standard cartridge. Both their 1949/56 series of rifles and the FR-F1 Sniper Rifle fire the 7.5mm M1929 cartridge. These weapons are described in Chapter 19.

PROTOTYPE NONPRODUCTION NATO CALIBER RIFLES

Several NATO caliber rifles never got beyond the prototype stage. Among these, the 4.81-kg (10.6-lb.) Madsen Light Auto Rifle of Danish design is the most notable. Similar in operating mechanism to the Kalashnikov, it was not produced due to the success of the G3 and to the fact that the Dansk Industri Syndicat (Madsen) went out of the small arms business since it could not compete successfully with other European manufacturers.

The Luigi Franchi LF-59 Rifle is a 4.3-kg (9.5-lb.) selective fire weapon quite similar in design to the FN FAL. It has a tipping bolt like the FAL, and the magazines are interchangeable. The piston is attached to the bolt carrier, and the recoil spring is mounted in a tube that telescopes from the rear of the carrier. Steel stampings constitute the receiver, and plastics are used for the stocks of later prototypes. To make the rifle manageable during full automatic fire, the design embodies a rate reducer, producing a cyclic rate of 610-630 shots per minute. This weapon was intended to be a companion piece to the 9mm LF-57 submachine gun and .30 caliber carbine LF-58.

The Dominican Republic produced in prototype form only the Model 1962 Rifle.

Tactically speaking, none of the 7.62 × 51mm NATO caliber rifles can be called true assault rifles. All of the major weapons in this caliber tend to be too heavy and cumbersome for easy maneuvering in the field. More significant, the recoil produced by the cartridge makes weapons such as the FAL and the M14 all but uncontrollable during bursts of automatic fire. As a consequence, most M14s were issued without the selector switch

Madsen 7.62mm NATO Light Automatic Rifle.

for automatic fire, and many FALs were produced as semiautomatics only. Whereas the next American rifle, the M16, would come closer to the assault rifle ideal, the Soviets, utilizing an intermediate power cartridge, had already introduced a true assault rifle.

7.62 × 39mm M43 (SOVIET) CALIBER RIFLES

Assault rifles as a concept were not a new idea in the Soviet Union. Vladimir Gregoryevich Federov (also spelled Fyodorov) after considerable experimentation introduced his *Avtomaticheskaya Vintovka Federova, 1916g* (Federov 1916 Automatic Rifle) during World War I. He utilized the 6.5 × 50.5SR Japanese cartridge instead of the larger and more powerful rimmed 7.62 × 54R cartridge used in the 1891 Mosin Nagant Rifle, since the Japanese round was better suited for use in a rapid firing rifle. Approximately 3200 M1916 *Avtomats* were fabricated at the Sestroretsk Weapons Factory before the 1917 revolution intervened. These rifles were used toward the end of the 1914–1918 war and during the Russo-Finnish war of 1939–1940.

During the interwar years, the Soviets experimented extensively with self-loading rifles, but for some reason they returned to the 7.62 × 54R cartridge, defeating the lessons learned during World War I. Just as the Americans dropped the .276 cartridge in the 1920s to pacify the twin gods of "standard issue caliber" and "markmanship tradition," the Soviets returned to their old cartridge. Both S. G. Simonov and F. W. Tokarev produced self-loaders to fire the 7.62mm rimmed cartridge. The *Avtomaticheskaya Vintovka Simonova Obrazets 1936g* (AVS36, Automatic Rifle Simonov) recoiled badly, had poor parts durability and demonstrated feeding and extraction difficulties. In 1938, the Red Army adopted a new model of Tokarev's design, the *Samozariadnya Vintovka Tokareva Obrazets 1938g,* SVT38. Tokarev, known for his automatic pistol and his modifications to the Soviet Maxim machine guns, had evolved a rifle bolt mechanism that was quite similar to that which was later used in the FAL. Federov, Simonov and Tokarev weapons were all used during the winter war with Finland. As some of the parts of the SVT38 proved fragile, Tokarev modified his design and produced the SVT40. The most obvious differences between the two models were the use of a one-piece stock and the placement of the cleaning rod beneath the barrel. In the SVT38, a two-piece stock was used, and the cleaning rod fit a groove along the right side of the stock.

SKS45—SAMOZARIDNYA KARABINA SIMONOVA OBRAZETS 1945g

Although the Tokarev rifles were widely used throughout the 1939–1945 war, the Soviets made greater use of the submachine gun than any other country. Infantry massed with large armored units found the firepower of the PPSh41 and PPS43 submachine guns extremely effective. Out of this experience came the requirement for an *Avtomat,* an assault rifle. N. M. Elizarov and B. V. Semin developed an intermediate cartridge, the M43 7.62 × 39mm, for this purpose. Ironically, the first adopted weapon to fire this cartridge was the SKS45 self-loading carbine. Scaled down from the 14.5 × 114mm PTRS self-loading antitank rifle and the 7.62 × 54Rmm SKS41 this 10-shot Simonov weapon represented an anachronism. Despite its large-scale production, the SKS did not fit the Motorized Infantry/Armored Fighting Vehicle tactic that was evolving within the Soviet Union. First field tested in the latter days of World War II, it was adopted as a standard weapon in 1945. The SKS is a secondary weapon in most Warsaw Pact countries today.

COUNTRIES MANUFACTURING THE SKS

USSR
East Germany—*Karabiner-S*
PRC—Type 56 Carbine
North Korea—Type 63 Carbine
Yugoslavia—M59/66 Rifle

THE SOVIET 7.62mm AK ASSAULT RIFLE

A new generation of Soviet small arms emerged from the mind and later the design bureau of Mikhail Timofeyevich Kalashnikov. While on convalescent leave from the Tank Corps due to

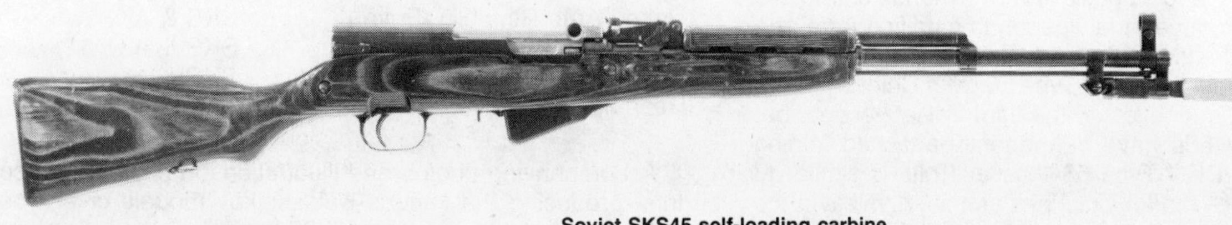

Soviet SKS45 self-loading carbine.

serious wounds received in the battle of Brausk in the fall of 1941, Kalashnikov turned his attentions to small arms design. In 1942, he produced a submachine gun, but it could not compete with such designs as A. I. Sudayev's, which was adopted as the PPS43. Early in 1944, he began work on a turning bolt carbine firing the 7.62 × 39mm cartridge, but little came of that design either. By early 1946, Kalashnikov had completed yet another weapon. Later adopted as the *Avtomat Kalashnikova 1947g* (AK47), this weapon was a true assault rifle.

M. T. Kalashnikov 1919—; designer of the AK47, AKM, RPK, PK and SVD—the latter is a product of the design bureau he heads up at the Izhvesk Machine Factory, Urdmurt, Kazakhstan, USSR.

Even though it was in the 4.3-kg (9.5-lb.) class, unloaded, the AK47 proved to be a highly dependable, highly manageable automatic weapon. This success did not come overnight. A stamped steel receiver model built in the late 1940s and early 1950s did not prove to be durable enough. The machined steel receiver model commonly encountered is actually the third version manufactured by the Soviets. The second model can be distinguished by the angular metal fitting into which the butt stock was mounted. That attachment was in turn pinned to the rear of the receiver. Versions one and two were still in the Soviet inventory in 1952. In addition to the Soviet Union, the People's Republic of China, East Germany, Poland, Bulgaria, Romania, North Korea, Hungary and Yugoslavia have manufactured the AK47. Finland has produced the weapon in modified form, the M60 and M62, while the Israeli Galil 5.56mm Rifle is a derivative design. The initial production PRC Type 56 (with Chinese markings on the selector) is identical to the third Soviet version, but late production Type 56s have permanently attached folding spike bayonets. The PRC Type 56-1 assault rifle is similar to the Soviet folding stock model, but it has prominent rivets in the arms of the stock.

Poland has produced a special grenade launching version, the PMK-DGN-60, to which is attached the 20mm diameter LON-1 grenade launcher. This variant has a gas cutoff valve added to the gas cylinder, a special grenade sight that fastens to the standard rear sight, a recoil absorbing butt pad, a latch added to the recoil spring and a special 10-shot magazine, which will take only the grenade blanks.

Yugoslavia has produced three variations of the AK47—M64 with a longer 508-mm (20-in) barrel and fixed wooden stock; M64A (later redesignated the M70) with a standard 414-mm (16.3-in) barrel and fixed wooden stock; and the M64B (M70A) with standard barrel and folding stock. All models are fitted with a folding grenade launching sight, a muzzle compensator and (unique among AK47s) a bolt hold-open device that catches the bolt in the recoil position after the last cartridge in the magazine has been fired. Finland's M60/M62 series has a special flash suppressor/bayonet mount, front sight mounted on the gas cylinder, aperture, plastic forearm and tubular fixed butt stock. East German AK47s do not have cleaning rods under the barrel or a recess in the butt for cleaning tools. Except for these specific differences and the selector markings, all Eurasian AK47s are similar. Operational and field stripping details are presented in Chapter 43.

Introduction of the modernized Kalashnikov assault rifle (*Modernizirovannyi Automat Kalashnikova,* AKM) in 1959 reflected a shift from the forged and machined receiver to an improved stamped sheet metal construction. The weight of the AKM, 3.15 kg (6.9 lbs.), is about two-thirds that of the AK47. Otherwise, the AK47 and the AKM are mechanically identical, except that the AKM has a cyclic rate reducer in its trigger mechanism, which slightly slows down the automatic firing cycle from the 600 shots per minute of the AK47. Recognition features of the AKM are the sheet metal receiver with a small magazine guide dimple pressed into each side, the grasping rails on the forestock, the ribbed receiver cover, the bayonet lug and the absence of vent holes in the gas tube. AKMs have been produced by the USSR, East Germany, Poland, Hungary, Romania and North Korea. The North Korean weapon does not have a rate reducer. Some late model AKMs are fitted with muzzle compensators. The Kalashnikov series has probably been produced in larger numbers than any other modern small arm; total production has been estimated to be between 30 and 50 million.

COUNTRIES USING THE SKS AND AK FAMILY BUT NOT MANUFACTURING THEM DOMESTICALLY

Country	Weapons
Afghanistan	SKS & AK
Albania	SKS & AK
Chile	AK only
Congo, People's Republic	SKS & AK
Cuba	AK only
Indonesia	SKS & AK
Iraq	SKS & AK
Laos	SKS & AK
Lebanon (para-military forces)	SKS & AK
Mongolia	SKS & AK
Morocco	SKS & AK
Pakistan	AK (PRC) only
Syria	AK only
United Arab Republic (Egypt)	SKS & AK
Vietnam, Socialist Republic of	SKS & AK
Yemen, People's Democratic Republic	SKS & AK

Accompanying photos and illustrations in chapters on countries producing the various Kalashnikov models give a better understanding of model variations.

First model Kalashnikov with stamped steel receiver.

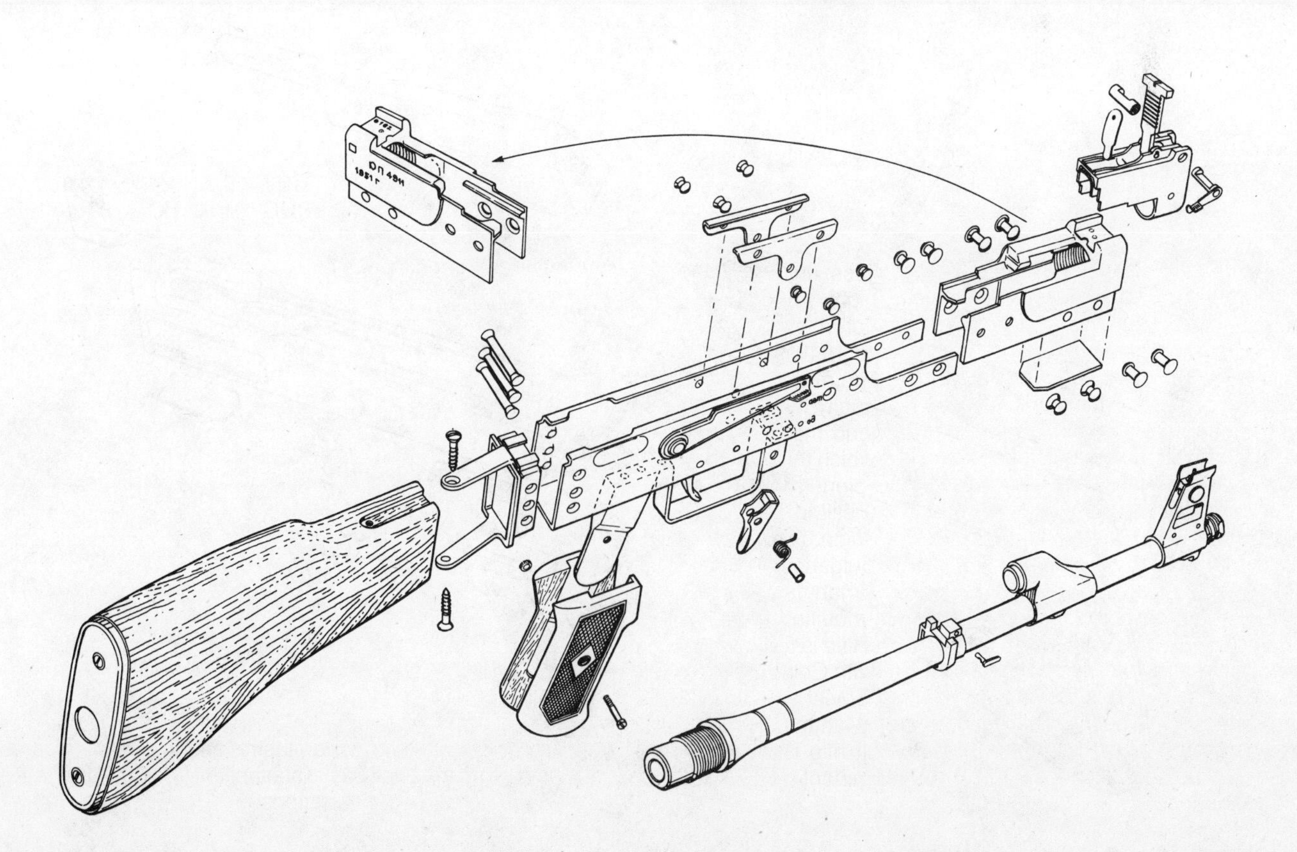

Second model Kalashnikov, which was first to have machined receiver.

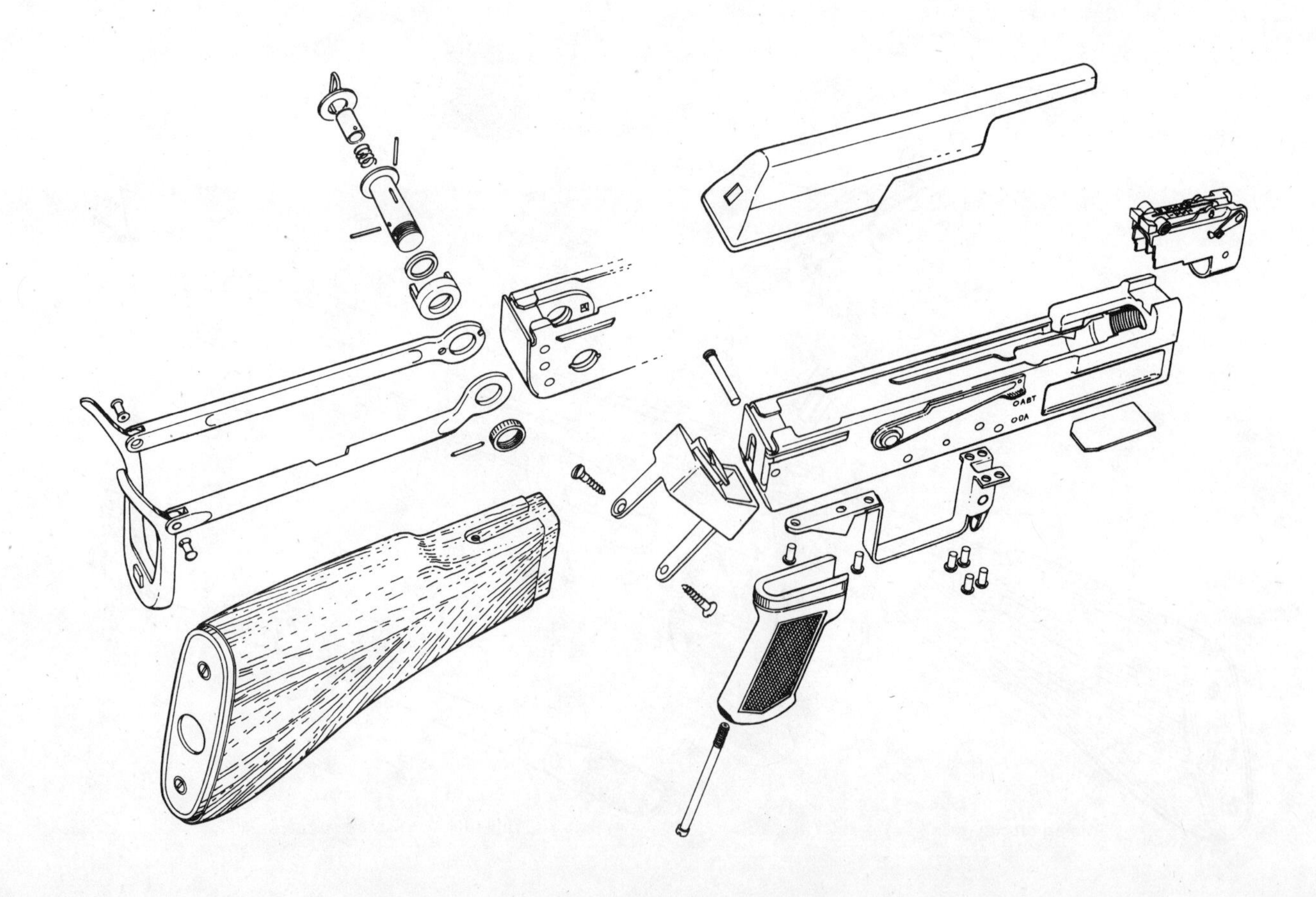

Second (bottom) and third (top) versions of the AK47. Note differences indicated by numbers in the photograph.

Folding stock version of the AKM, called the AKMS by the Soviets. This rifle was manufactured in 1972.

Polish PMK-DGN60 grenade-launching rifle.

Yugoslav 7.62 × 39mm M70AB with an AKM pressed sheet metal receiver.

Finnish copy of the AKM manufactured by Valmet Oy Defense Equipment Group. Note that this M62-76 rifle has a side folding stock.

Soviet AKM.

East German MPiKM assault rifle.

Romanian AKM.

Upper or Full Auto Symbol	Lower or Semi Auto Symbol	Producer	Native Name or Remarks
AB	ОД	Soviet	AK-47, AKM and AKMS
AB	ЕД	Bulgaria	AK-47 and AKM
C	P	Poland	PMK, PMK-DGN, KbK AK
D	E	E. Germany	MPK, MPiKmS - Rifles do not have cleaning rods MPiKM and MPiKMS have cleaning rods
FA	FF	Romania	Has "S" at top for safe position
连	单	Communist China	Early Production
L	D	Communist China	Type 56 and 56-1 Assault Rifle (Late Production)
∞	1	Hungary	
...	.	Finland	RYNNAKOKIVAARI - applies to M60 and M62
련	단	North Korea	Types 58 and 68 Assault Rifle
R	J	Yugoslavia	M64 series - has U at top for safe position
30	1	Czechoslovakia	M58 Assault Rifle

Kalashnikov assault rifle selector markings.

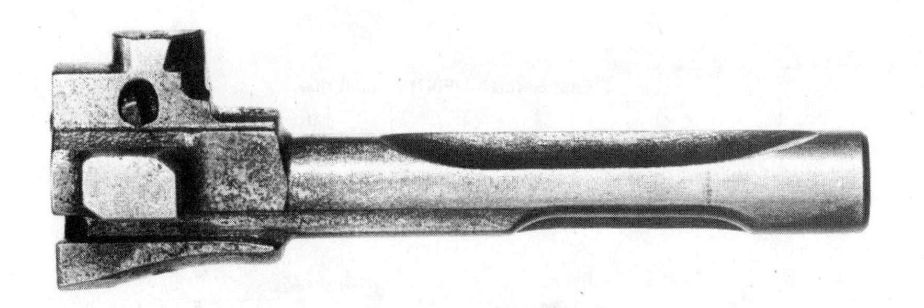

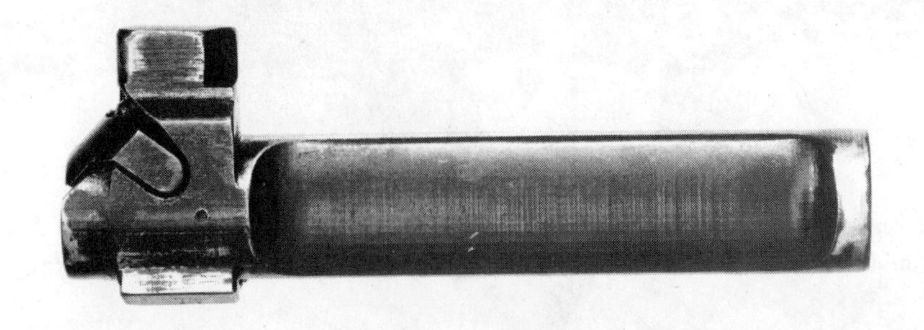

The picture illustrates the AK bolt and the U.S. M1 Carbine bolt, for size comparison.

CZECHOSLOVAK RIFLES

Czechoslovakia has always been a fertile source of small arms. After World War II, the Czech military adopted the Vzor (Model) 52 self-loading carbine chambered for their own 7.62 × 45mm cartridge. Beginning in 1957, many of these carbines were altered to fire the 7.62 × 39mm cartridge, VZ52/57 (M52/57). While obsolete in Czechoslovakia, the M52 and M52/57 carbines have been shipped and sold to third world nations. These weapons are unusual in that they employ a unique concentric gas cylinder, which surrounds the barrel, and the gas operating mechanism is composed of a bearing mounted on the barrel, a gas port in the barrel, a sliding gas cylinder sleeve positioned over the bearing, a connecting semicylinder and an actuator. Upon firing the cartridge, gases force the gas cylinder sleeve, connector and actuator to the rear, an action which in turn forces the bolt and bolt carrier to the rear. In concept, this gas system is similar to the German *MKb42(W) Sturgewehr.* The major complaint with this weapon has been its weight, 4.5 kg (9.8 lbs.), and its somewhat fragile side folding bayonet.

Czech Model 52 Rifle.

Subsequently, the Czechoslovakian Army adopted the VZ58 (M58) assault rifle, which was also a domestic design. This weapon exists in two forms—the M58P with conventional fixed stock and the M58V with folding metal stock. Early production weapons had wooden forearms, pistol grips and, in the case of the M58P, butt stocks. These components are made of a wood fiber/plastic composition material in more recently fabricated weapons. While some of these rifles have been sold abroad, the Czechs are the major user of this design.

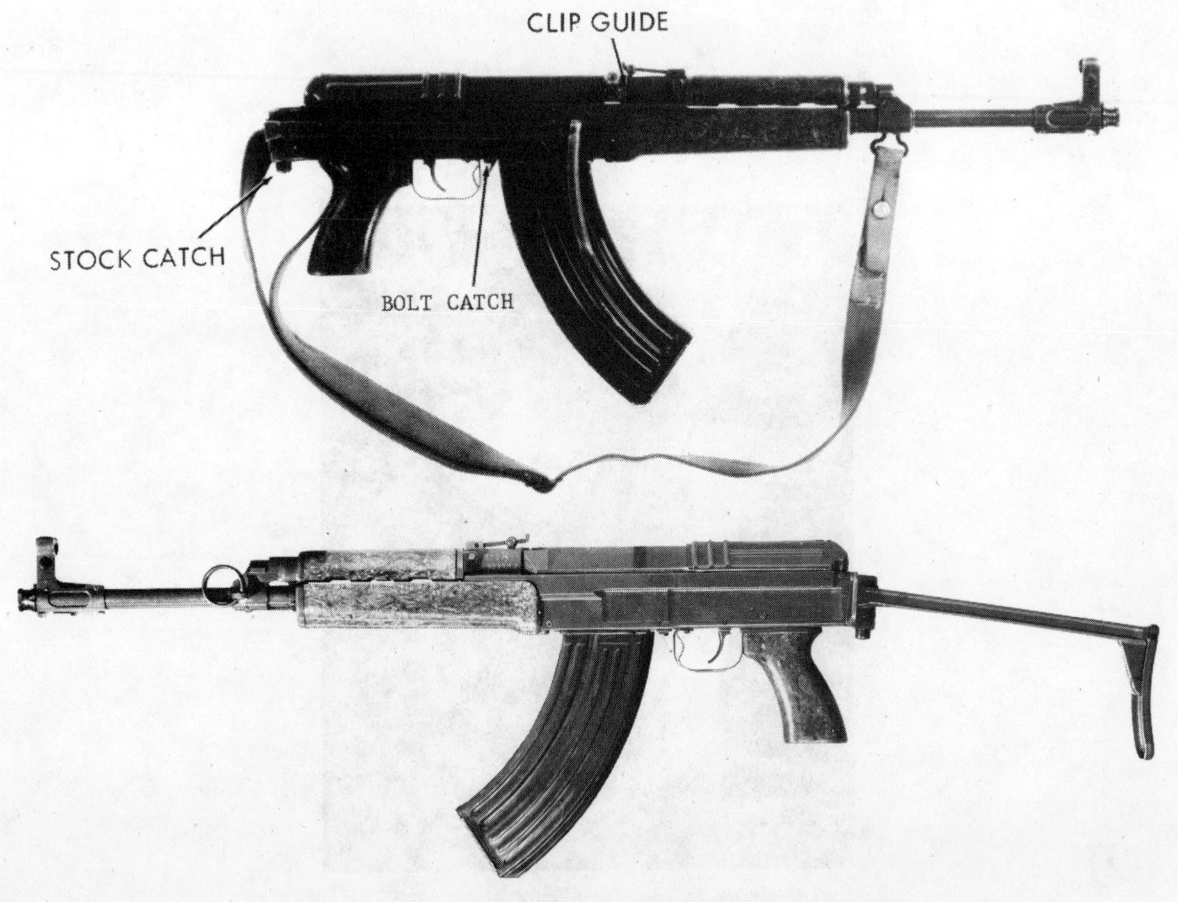

Folding stock version of Model 58.

CHINESE TYPE 68 RIFLE

The Type 68 Rifle, adopted by the People's Republic of China, is yet another basic pattern selective fire weapon adopted in the M43 7.62 × 39mm caliber. This small arm represents a divergence from the assault rifle tactical concept as it has emerged among the Warsaw Pact countries. Evidentally, the People's Liberation Army tacticians decided that its army would not be engaging in the massive armored infantry types of conflicts envisaged for potential European conflicts. Therefore, they adopted a selective fire rifle, the barrel being the same length as the SKS, with a 15-shot magazine, five shots more than the SKS but half the capacity of the AK family. Two versions of the rifle exist. The earlier model has a receiver machined from a steel forging, while the later model has a stamped steel receiver, identifiable by the large rivets at each side of the receiver. There are other minor differences apparent when the weapons are compared.

In the Type 68, the bolt and bolt carrier have evolved from the Kalashnikov design. A major difference lies in the separation of the gas piston and the bolt carrier. The piston rod acts as a tappet against the face of the bolt carrier in a fashion reminiscent of the SKS and SVT 38/40 designs. Unlike the SKS or the AK, the Type 68 has a two-piston gas regulator. A permanently attached spike bayonet similar to the Type 56 carbine and Type 56 assault rifle and an under-the-barrel cleaning rod are standard fixtures. As issued, the Type 68 Rifle has a bolt stop to hold the bolt open after the last shot has been fired. Unless the stop is altered, only the 15-shot magazine will fit; if the stop is ground down, the rifle will also accept the 30-round AK-type magazines.

In 1980, there was word of yet another Chinese 7.62 × 39mm rifle. This weapon, for which a type designation is not known,

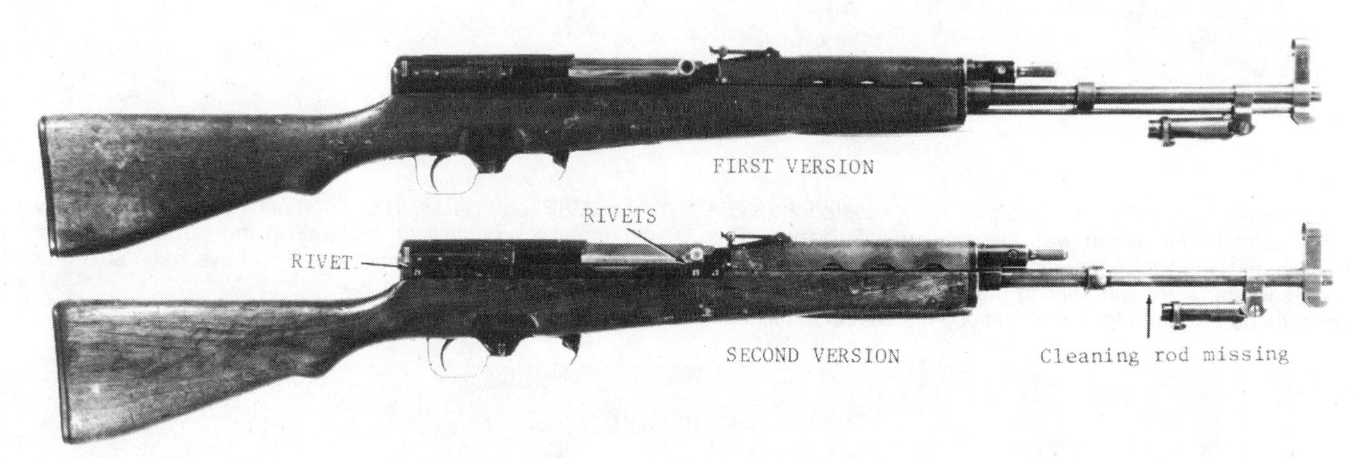

First and second models of the PRC Type 68 selective fire rifles.

Tang Wenlie is credited by Chinese sources with improving the Type 68/73 rifle.

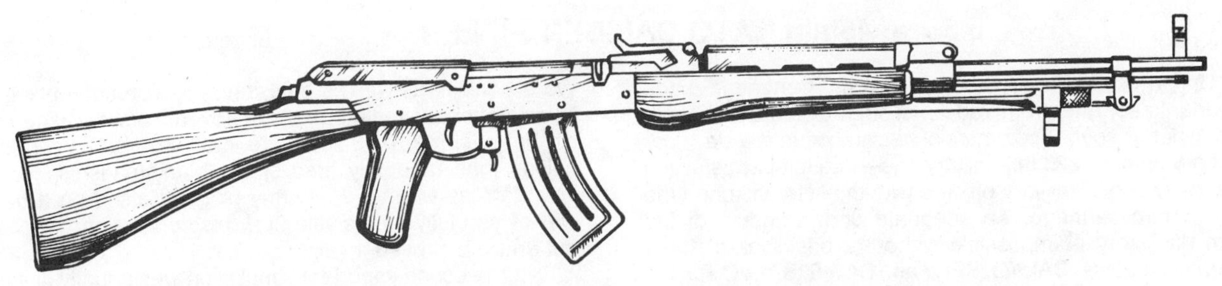

The improved PRC rifle developed by Tang Wenlie, a member of the People's Liberation Army.

was reportedly 0.6 kilograms (1.3 pounds) lighter than the Type 73 Rifle (second model Type 68). Designed by Tang Wenlie, this new rifle is reported to have fewer moving parts than its predecessors and to be more accurate.

MISCELLANEOUS M43 7.62 × 39mm CALIBER RIFLES

In the past two decades several 7.62 × 39 rifles have been produced in limited quantities to appeal to those nations that might have received small arms from the Soviet Bloc. These include the RH-4, made in prototype by Rheinmettal Wehrtechnik of Dusseldorf, West Germany, the HK32 by Heckler & Koch and the SIG 510-3. Only the Egyptians attempted a design for domestic consumption. That rifle, the Rashid, was made in very limited numbers and was a rework of the Ljungman Model 42 Rifle, which has been produced in Egypt (United Arab Republic) as the 7.92 × 57mm Hakim Rifle.

Members of the People's Liberation Army on night maneuvers. The soldier in front carries the Type 56 version of the AK47. Behind him is a PLA-man with a Type 56 light machine gun (RPD)

5.56 × 45mm NATO CALIBER RIFLES

The United States Army Continental Army Command made an important departure from traditional small arms development in 1957 when it sought commercial assistance in the development of a 5.56mm (.223 in.) military rifle—due to dissatisfaction among many senior military officers with the M14 Rifle and the 7.62 × 51mm cartridge. An adequate understanding of the 5.56mm rifle story is impossible without a brief look at three small arms projects—SALVO, SPIW and SAWS. SALVO studies conducted by the Operations Research Office (ORO) at Johns Hopkins University and supported by several contractors gave the impetus for the development of the M16 Rifle. Failure of the radical SPIW (Special Purpose Individual Weapon) concept assured the M16 a permanent place in the US Army's arsenal, and the Small Arms Weapons Study (SAWS), 1966–1967, judged the M16 to be the best small caliber rifle available.

ORO had been created by the US Army in 1948 to analytically study a number of problems associated with ground weapons in the nuclear era. One of ORO's early projects was ALCLAD, a search for better infantry body armor. As that study progressed, ORO and Army specialists discovered just how little was known about how individuals were wounded in combat. ORO looked into several questions regarding the manner in which soldiers were struck by rifle projectiles and shell fragments. Among them were the frequency and distribution of such hits, the types of wounds incurred in combat and the average ranges at which wounds were inflicted. Answers to these questions were obtained by evaluating over three million casualty reports for World Wars I and II, as well as data from the Korean conflict. ORO's investigations revealed that in the overall picture aimed rifle fire did not seem to have any more important role in creating casualties than randomly fired shots. Marksmanship was not as important as volume. For Army officers raised on a traditional diet of carefully aimed rifle fire, this conclusion was heretical, but analysis proved it valid.

ORO's second important conclusion was equally disturbing to the traditionalists. Whereas effective rifle fire had been occasionally delivered at 1200 meters during the 1914–1918 war in the trenches, World War II and Korean war experience indicated that the rifle was seldom effectively employed beyond 300 meters. Even when expert riflemen tried to use their weapons at greater ranges, they discovered that terrain features usually prevented accurate long distance firing. Finally, statistical data indicated that most rifle kills were made at less than 100 meters. These revelations called for some new thinking in rifle design. One fruitful approach appeared to be the development of a light recoil weapon firing a salvo of small caliber projectiles with a controlled dispersion pattern. While Project SALVO was taking form, US ordnance officers were telling their British counterparts that the UK .280 cartridge was too small, but SALVO was considering projectiles as small as 4.2mm (.17 in.).

After following the SALVO work for several years, the Continental Army Command (CONARC) decided to sponsor the development of a .22 caliber military rifle. CONARC commanding officer William G. Wyman asked Winchester and Armalite to

US lightweight rifles, 1957: Top, Winchester .224 Lightweight Military Rifle. Middle, Armalite .223 AR-15. Bottom, Springfield Armory .22 , Rifle.

develop a high velocity 5.56mm rifle. From the outset, the Armalite AR-15 was more popular than the Winchester design. Even Ralph Clarkson, the designer of the Winchester .224 Lightweight Military Rifle (patterned after the M1 Rifle and M1 Carbine), had to admit that the AR-15 had unmistakable "sex appeal." A Springfield Armory contender designed by A. J. Lizza did not get very far either. In fact, ordnance personnel opposed to the small caliber concept forbade the Armory, an Ordnance Corps facility, to participate in CONARC's heretical program.

CONARC specifications for a 5.56mm rifle called for full and semiautomatic fire, a 20-shot magazine, a loaded weight of 2.7 kg (6 lbs.) and penetration of both sides of a standard Army helmet at 500 meters. If possible, the engineers were to keep the trajectory flatter than that of the 7.62mm NATO cartridge. The desire to use the rifle at ranges up to 500 meters indicated a compromise between the SALVO studies and conventional rifle thinking current in the US Army.

M16 RIFLE

Stoner's AR-15 was designed around a slightly enlarged version of the Remington .222 cartridge case. That alteration permitted him to propel a 3.6-gram (55-grain) bullet at 1005 m.p.s. (3300 f.p.s.). The weapon itself was an eclectic design. As other designers before him, Stoner chose the best from earlier designs. For the locking system, he chose a design quite similar in concept to that of the Johnson Semiautomatic Rifle of the 1940s. He also used an "in-line" stock to aide manageability during automatic fire. That stock arrangement permitted him to place the recoil buffer in a tube that ran the length of the stock. A tube type gas system was employed to convey the gas from a port under the front sight, along the top of the barrel and into a space in the bolt carrier assembly.

There were a number of other features reminiscent of earlier weapons, such as the hinged upper and lower receiver mechanism, similar to the FN FAL; the rear sight in the carrying handle, *a la* the British E.M.2; and the ejection port dust cover, which followed the pattern established in the MP44 *Sturmgewehr.* Stoner's achievement in the AR-15 was the combination of all these ideas into an attractive, lethal package that weighed only 3 kg (6.7 lbs.).

In December 1959, Colt Firearms acquired the manufacturing and marketing rights to the AR-15 from Armalite. But selling the rifle proved to be a tough task. Although many people liked the weapon, the Army Ordnance staff was opposed to it. But in 1962, in an end run around the Army, Colt was able to get the Department of Defense's Advanced Research Project Agency (ARPA) to test 1,000 weapons in its Vietnam-oriented Project Agile. ARPA's enthusiastic report led to additional studies by the Department of Defense and the Department of the Army. Despite strong Army opposition, Defense Secretary McNamara ordered 85,000 M16 Rifles for Vietnam and 19,000 for the Air Force. From this beginning, the AR-15 became the M16, and ultimately the M16A1.

Resistance to the adoption of the AR-15/M16 led to serious problems in 1967. Congressional and Department of Defense investigations disclosed that the weapon had been issued without proper operational and maintenance training to some troops and that totally inadequate supplies of cleaning equipment had been provided to the men in the field. Combined with an ammunition/rifle mismatch and the highly corrosive nature of the humid jungle regions of Southeast Asia, this lack of training and cleaning materials led to serious problems. But after training programs were established, cleaning supplies made available and modifications made to the rifle, the M16 performed reliably. Major changes to the rifle included a new buffer mechanism to slow the rate of fire, which was greater with ball type propellants than with the IMR propellants for which the weapon had been designed. A chrome plated chamber—later followed by a chrome plated barrel—solved the rusty chamber problem, which had in turn caused failures to extract. Today (1983) the M16/M16A1 Rifle has become the basic rifle for the United States Army. It was also produced by General Motors and Harrington & Richardson during the Vietnam era; Colt continues to manufacture the rifle for the Army. In addition, with US State Department approval, Singapore, the Philippines and South Korea also produce the weapon under license from Colt.

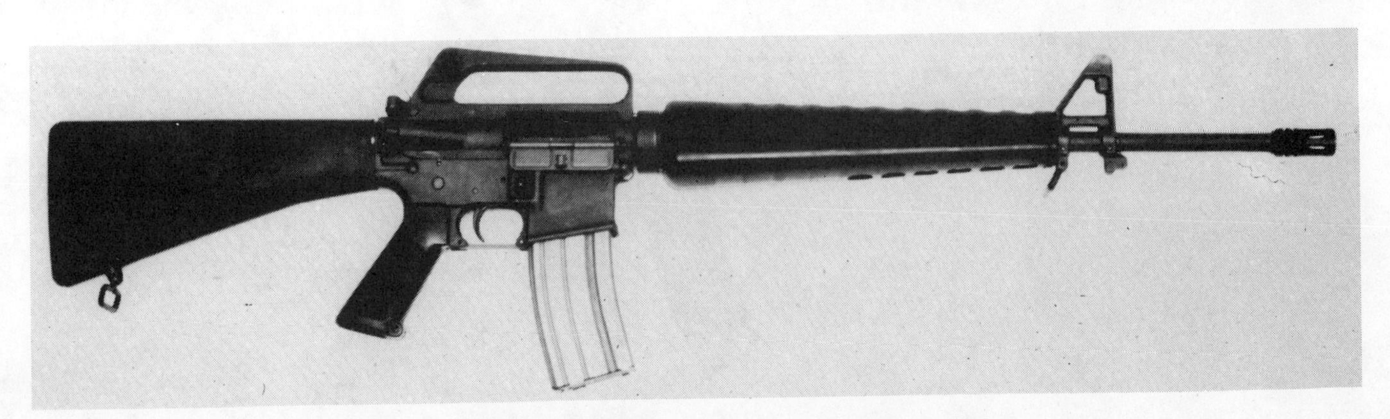

M16A1 with 30-shot magazine and old style buttstock.

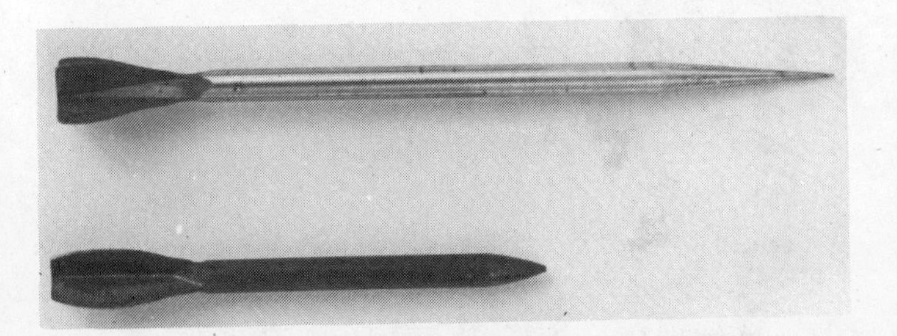

Flechettes: Top, typical rifle version. Bottom, shotgun type.

By the end of 1976, Colt had produced some 3,440,106 M16 Rifles. Harrington & Richardson and General Motors had produced 250,000 each. Total AR15 production through 1976 was 68,211. Of the Colt production, 3,135,227 M16s had gone to the US government and 304,879 had been sold overseas.

SPIW AND SAWS

Before turning to the European 5.56 × 45mm rifles, some mention needs to be made of two other US Army projects—SPIW and SAWS. The former was an attempt to "leapfrog" ahead in small arms development. The latter was a full scale field evaluation of existing small arms.

The Special Purpose Individual Weapon (SPIW) grew out of experimentation by the US Ordnance Corps and Aircraft Armaments Inc. (AAI) with high velocity steel darts called flechettes. Irwin R. Barr pioneered work on flechettes at AAI and proposed a whole range of such projectiles in a February 1951 report. In 1952, AAI produced a 12 gage shotgun shell loaded with flechettes for the Office of Naval Research. A later version was tested by the Army as part of the Project SALVO tests. Large dispersion patterns and limited range led Barr and his AAI colleagues to turn to single flechettes supported by sabots, fired sequentially from a rifle type weapon. After nearly a decade of energetic promotion of the flechette concept by AAI, the US Army decided in March 1962 to develop a Special Purpose Individual Weapon that would combine the flechette projectile with the 40mm grenade cartridge, which emerged from another project called NIBLICK. (See Chapter 45 for more on grenade launcher development.)

In February 1963, contracts for SPIW type weapons were let to AAI, Harrington & Richardson, the Winchester Division of Olin and Springfield Armory with delivery of the prototypes due in February of the next year. This incredibly short development time proved impractical, and although firing models were delivered and tested in 1964, SPIW became a long, drawn out project. After the first rounds of testing, the Harrington & Richardson and Winchester SPIWs were eliminated. Later, the Springfield weapon was dropped following the Armory's closing. Work continued on the AAI model, redesignated the XM19. In 1973, the Army announced that the XM19 could not be made to meet military requirements. A modified and simplified XM19 was tested as the XM70, but for practical purposes the single flechette concept is moribund in the US. No one disputes the lethal nature of these little steel arrows (Witness the multiple flechette artillery projectiles used with devastating effect in Vietnam.), but the SPIW concept died due to complex technical, economic and political problems.

Top to bottom: HK33, HK33K (*kurz*) and HK 53.

In 1964, General Harold K. Johnson, Army Chief of Staff, instituted a comprehensive program to review the major small arms being used and under development. This Small Arms Weapons Systems (SAWS) study was to determine which weapons were most suited for the Army's tactical missions during the years 1967–1980. After more than 18 months of investigation and the preparation of several dozen reports, the SAWS study was completed, and the Chief of Staff's office reviewed the SAWS recommendations, the requirements of the Vietnam war and the state of SPIW development. The Secretary of the Army submitted the Chief of Staff's report to the Secretary of Defense on 17 December 1966. The major suggestions were as follows. First, rifle procurement in the "forseeable" future should be limited to the M16 Rifle. Second, steps should be taken to permit "early replacement" of the M1s and BARs in the Army's inventory. Third, planning over the long term should be based upon the replacement of the M14 with the M16. Fourth, an additional production source for the M16 should be provided in the 1968 budget. And finally, "an active and broadened research and development program should be conducted to bring about further major improvements in the Army's small arms."

In his cover memorandum, the Secretary of the Army made several significant comments about the Chief of Staff's recommendations. He concluded that M16 type weapons were "generally superior for Army combat use." Second, "The current SPIW program is unlikely to result in a satisfactory competitive weapon as early as previously forecast." The Secretary's memo also suggested that some changes might be necessary in the M16, especially the propellant used in the .223 cartridge. These thoughts were passed on to the Secretary of Defense with a request for approval. After considerably more internal debate, Secretary McNamara decided that the Army would use the M16 as its primary rifle but that the M14 would also continue to be considered a standard weapon.

HK33

Heckler and Koch began work on the delayed roller-locked HK33 Rifle in 1963. After a series of modifications, this scaled-down version of the G3 was put into modest production in 1968. Major purchasers of the HK33 include the Brazilian Air Force (15,000), the Malaysian Army (5,000) and the Thai Army. In addition, the Malaysians have assembled 30,000 domestically, and the Thais are manufacturing the rifle at a factory set up for them by Heckler and Koch. There is also a short-barrelled version of this weapon called the HK53, which is designed for use as a submachine gun.

In addition to the M16 and HK33, several other 5.56 weapons have been developed and manufactured in various quantities.

STONER 63 SYSTEM

Gene Stoner, after he left Colt where he had worked as a consultant, joined forces with Cadillac Gage Corporation of suburban Detroit to develop a family of small arms built around a number of common assemblies and parts. Once again, he was assisted by Robert Fremont and James Sullivan. As noted above, Stoner built his 62 system around the 7.62mm NATO cartridge;

XM22E1 Rifle (Stoner 63) Field stripped.

in the 63 system he turned to the 5.56 × 45mm cartridge. Fully matured, the system consisted of six weapons—a rifle, a carbine, two light machine guns (magazine or belt feed), a medium machine gun and a fixed (tank) machine gun. To adequately power the machine guns in the system and to assure reliable belt feed, Stoner used a long stroke piston system instead of the gas tube arrangement utilized in the M16. The rifle (called the XM22 by the US Army) and the carbine (XM23) fired from a closed bolt, with selective fire. The magazines were mounted below the receiver. The machine guns (The belt fed version was tested by the Army as the XM207 and by the Navy as the MK23.) all fired from an open bolt. (See Chapter 2 for more details.) During the mid-1960s, the Stoner 63 system was tested several times by the US military, after which numerous improvements were made based upon their experiments. Only the MK23 was used to any extent in combat, by Navy SEAL teams.

Although Cadillac Gage granted a manufacturing license to Mauser-IWK who later transferred their rights to NWM of the Netherlands, the Stoner system never became popular. In the US, the M16 was too deeply entrenched for the Stoner to make much headway, and in Europe there were several competing designs. The Stoner 63 system was a good concept, but the timing—an important factor—was not auspicious. It was, therefore, never successfully marketed.

ARMALITE AR-18

A scaled-down version of the AR-16, the AR-18, never met military requirements successfully. Throughout its testing history, it had a bad record for parts breakage and feeding difficulties. Since Armalite never developed their own production capabilities, they gave a production license to Howa Machinery Company of Japan in 1967. Military sales to the US military were frustrated by American-Japanese treaty agreements that prohibited Japan from selling military equipment to belligerent nations. The Japanese decided that under this arrangement the rifles could not be sold to the US because this country was engaged in the Vietnam conflict.

In 1974, Armalite and Sterling Ltd. of England concluded a production agreement whereby the AR-18 and the commercial semiautomatic version (AR-180) would be manufactured in the UK. Production began in 1975.

Early Armalite AR-18 Rifle.

STERLING LIGHT AUTOMATIC RIFLE

Before starting production of the AR-18, Sterling's chief designer, Frank Waters, designed a rotating bolt, gas operated rifle to fire the 5.56mm cartridge. Building on the work with this rifle and the AR-18, Sterling engineers evolved a rifle, the SAR 80, which the Singapore government has developed more fully.

BERETTA AR70/.223.

Beretta and SIG began a study of 5.56mm rifles in 1963. After a number of years, the two companies terminated their joint development effort. In 1968, Vittorio Valle at Beretta began work anew, and the Italian firm introduced the AR70/.223 in 1970. At one point in the early 1970s, Beretta negotiated with Colt to produce the M16 in Italy. When these discussions proved fruitless, they began to market the AR70/.223 more aggressively. In this weapon, the designers decided in favor of a conventional gas piston and recoil spring system located above the barrel. After looking at existing locking systems, they adapted the twin lug used in the M1 Carbine and Kalashnikov weapons. Following the pattern of the Soviet AKM rifles, Beretta engineers welded a sleeve into the forward portion of the stamped steel receiver. That sleeve contained locking recesses for the lugs on the bolt.

The only major sale of the Beretta rifle has been to Malaysia (5,000 in 1972; Malaysia also purchased the same number of M16s and HK33s). It would appear that Beretta has abandoned the effort to produce and sell this weapon.

SIG 530 AND 540 RIFLES

As an outgrowth of their joint effort with Beretta in the early 1960s, SIG introduced the 530 series. Unlike their Italian counterparts, the Swiss used a roller lock mechanism evolved from the 510 series. In the 5.56mm rifle, the rollers are actually a locking system instead of a delay system as in the 510 series and the Heckler and Koch weapons. Only limited numbers have been fabricated with few sales.

The latest SIG series, the 540 and 543, represent a departure from the roller lock breech mechanism and a turn to the rotating bolt mechanism. While strikingly similar to the Beretta AR70/.223, the breech mechanism of the 540 series is actually closer to the Kalashnikov family. (See discussion above.)

Beretta AR70/.223.

SIG 530-1.

FN CAL.

Israeli Galil.

The French firm Manurhin, SA is currently manufacturing the SIG 540, 542, and 543 rifles. This arrangement allows SIG to sell their design outside of Switzerland. Swiss export laws regarding military small arms are very stringent. To date at least sixteen countries have purchased rifles in this series.

FN CAL

Fabrique Nationale introduced the *Carbine Automatique Leger* in 1966. While this rifle looks like a smaller version of the FAL, it uses a different bolt than its predecessor. (For more details, see Chapter 9.) Made in very limited quantities, the CAL design embodied a three-shot burst feature in addition to automatic and single fire. FN has terminated active attempts to sell the CAL and are concentrating instead on their newer FNC.

GALIL ASSAULT RIFLE

During the past decade, the FN FAL produced by Israeli Military Industries (IMI) has been the basic rifle in the Israel Defense Force's (IDF's) inventory. The performance of the FAL did not satisfy Israeli military officers during the Six-Day War of 1967, due to malfunctions caused by desert sands and the "bang, bang, jam" malfunction of the heavy barrelled automatic version. Following the war, the IDF tested the M16, the Stoner 63, the HK33 and a native rifle designed by Lt. Col. Uziel Gal, the designer of the UZI submachine gun. All these weapons fired the US 5.56 × 45mm cartridge.

The standard weapon against which the 5.56mm weapons were judged was the AK47. That weapon had performed almost flawlessly in the hands of Israel's adversaries, and it was that performance that the IDF wished to equal or surpass. Tests that ensued were among the most stringent ever conducted for small arms. Indeed, in the effort to simulate the rigors of desert warfare, the battle-hardened Golani Brigade did everything they could to destroy the weapons given them. In nearly every case, they succeeded. IDF authorities decided that the best weapon for their purposes was the AK47.

Israeli Galili (Blashnikov before he changed his name) worked up a 5.56mm version of the Kalashnikov rifle, using a barrel, bolt face parts and magazines from the Stoner system. The resulting weapon showed excellent promise. Meanwhile, IMI

AC-556K version of the Ruger Mini-14.

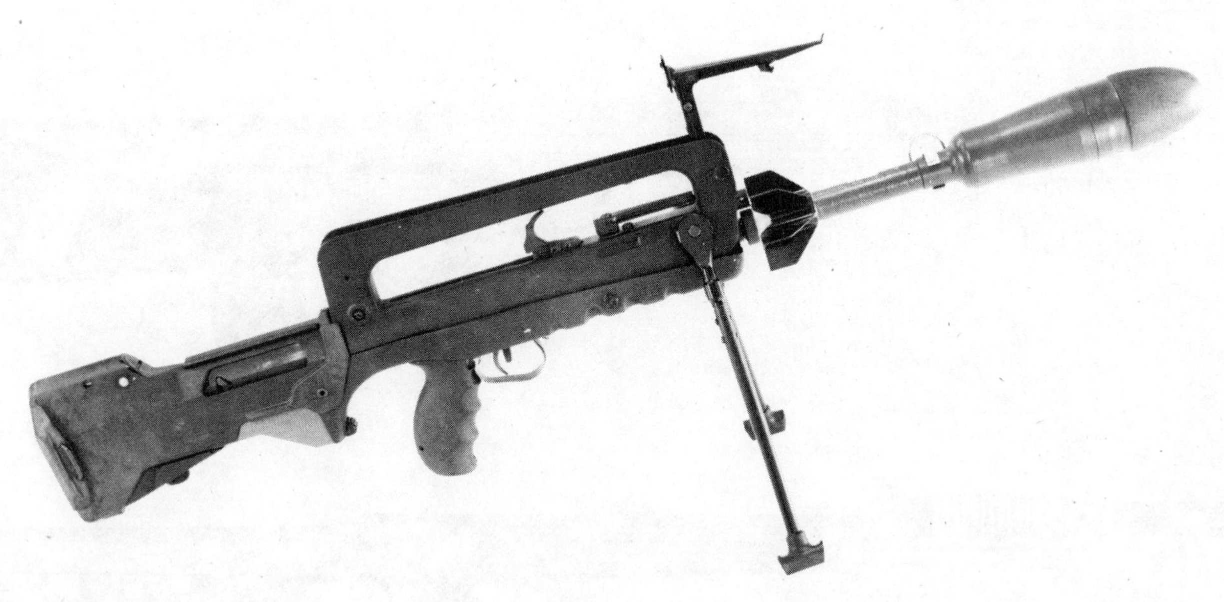

MAS 5.56mm Rifle.

personnel learned from an executive officer of Interarms in the US that Valmet in Finland was producing a copy of the AK47 called the M62. After modifying samples of the M62 purchased from Interarms, IMI, at the direction of Yaacov Lior, head of the small arms branch, purchased unmarked Valmet M62 receivers and mated them to barrel blanks procured from Colt. A modified Stoner magazine was also developed. IMI has produced a hybrid weapon of promise. A folding stock version borrowed the butt assembly from the FAL. The extent of Israeli production of this rifle is unknown, but IMI has offered it for overseas sales. Guatemala has purchased the Galil in unspecified numbers.

The Netherlands submitted the Galil (MN1) for testing in the NATO trials, Sweden developed a version of the Galil as the FFV 890C, and South Africa has adopted the R4, which bears striking resemblance to the Galil.

RUGER MINI-14

Sturm, Ruger and Company introduced the Mini-14 in 1972. As its name suggests, it borrows many characteristics from the NATO caliber M14 Rifle, but there are several differences. It is not simply a scaled-down version of the US military rifle. While Ruger kept the wooden stock, the pattern of which has become a company trademark, the M14 type receiver is fabricated from an investment casting, instead of being machined from a forging. A short barrel, selective fire, folding stock variant—the AC-556K—was introduced in late 1976. Although the resulting product has considerable eye appeal, tests in the Philippines, France and elsewhere indicate that it presently is best suited for use as a police weapon.

MAS 5.56 RIFLE

The French arms factory at St. Etienne—part of the state armament group (*Groupement Industriel des Armaments Terrestres*)—has introduced its own rifle with a bullpup configuration, the MAS, making it one of the shortest weapons in the current crop of 5.56mm assault rifles. The delayed blowback mechanism is also unique among current 5.56mm rifles. (Additional details are given in Chapter 19.) For several years, the French possessed a license to manufacture the HK33. In July 1977 the French General Staff announced the standardization of the MAS rifle and placed an initial order for 236,000 with GIAT. The first weapons were issued to the French armed forces in 1979.

INTERDYNAMIC MKS AND MKR

Made only in a limited number of prototypes, the MKS was the first publicly announced design of Interdynamic AB of Stockholm. The MKS incorporated a rotary bolt mechanism and was gas operated. The designers of the MKS sought to create an extremely short and lightweight weapon in caliber 5.56 × 45mm.

After experimentation with the MKS, the Interdynamic design team introduced a 4.7mm rimfire (but high velocity) caliber rifle and carbine series called the MKR. This weapon was also a

MKS Rifle.

Two versions of the Interdynamic MKR. ***Top,*** **a 5.56 × 45mm model with a 30-shot GAPCO nylon magazine;** ***bottom,*** **a 4.5 × 26mm rimfire caliber model with a 50-shot plastic semicircular magazine.**

bullpup style and embodied a 50-shot crescent-shaped magazine located to the rear of the pistol grip. Development of this series has been limited to date to the creation of test models.

VALMET M76

Valmet Oy of Finland has been more aggressively marketing its family of automatic rifles since the late 1970s. This state-owned engineering and ship building organization has developed a series of model variations around the basic operating mechanism of its M76 assault rifle, which is derived from the Soviet AKM sheet metal receiver avtomat. In addition to the 7.62 × 39mm M76, the Valmet Defense Equipment Group offers the M76 in 5.56 × 45mm (M193) versions. These models (described in more detail in Chapter 18) include tubular fixed and folding stock versions, and plastic and wood stock variants. In 1980, a bullpup model—the Model 255 470—was introduced for armored and airborne personnel. A squad automatic weapon version, the M78 LMG, is available in 5.56 × 45mm, 7.62 × 39mm, and 7.62 × 51mm NATO (see Chapter 2).

Valmet M82 Short (Model 255 470). This 5.56 × 45mm bullpup rifle was designed for use by airborne and armored personnel.

Valmet M76W 7.62 × 39mm assault rifle with wood stock (Model 255 460).

Valmet M76F 5.56 × 45mm assault rifle with side folding stock (Model 254 100).

Valmet M76T 5.56 × 45mm assault rifle with tubular stock (Model 254 060).

Valmet M76P 5.56 × 45mm assault rifle with plastic stock (Model 254 080).

RIFLES—A 1983 STATUS REPORT

The most important development of the past five years has been the switch from 7.62mm caliber rifles to ones firing 5.56mm in NATO and 5.45mm in the Warsaw Pact. In addition to this shift to smaller caliber ammunition, several countries have been added to the list of states designing and manufacturing rifles for their own armed forces, most notably Argentina, Brazil, Singapore, and South Africa.

NATO STANDARDIZES A SECOND CALIBER—5.56 × 45mm

In June 1976, eleven NATO countries* signed a Memorandum of Understanding for the testing, evaluation, and selection of a second NATO standard caliber for small arms ammunition. At the Conference of National Armaments Directors, an agreement was reached whereby only two calibers would be used in the post-1980 family of NATO small caliber weapons. One cartridge was to be the existing 7.62 × 51mm NATO round. A second cartridge was to be chosen from candidates submitted by those nations that had signed the Memorandum of Understanding. The new NATO family of small arms would consist of an individual weapon (rifle), a light support weapon (light machine gun), and a medium support weapon (medium machine gun).

*Belgium, Canada, Denmark, France, the Federal Republic of Germany, Greece, Luxembourg, Netherlands, Norway, the United Kingdom, and the United States.

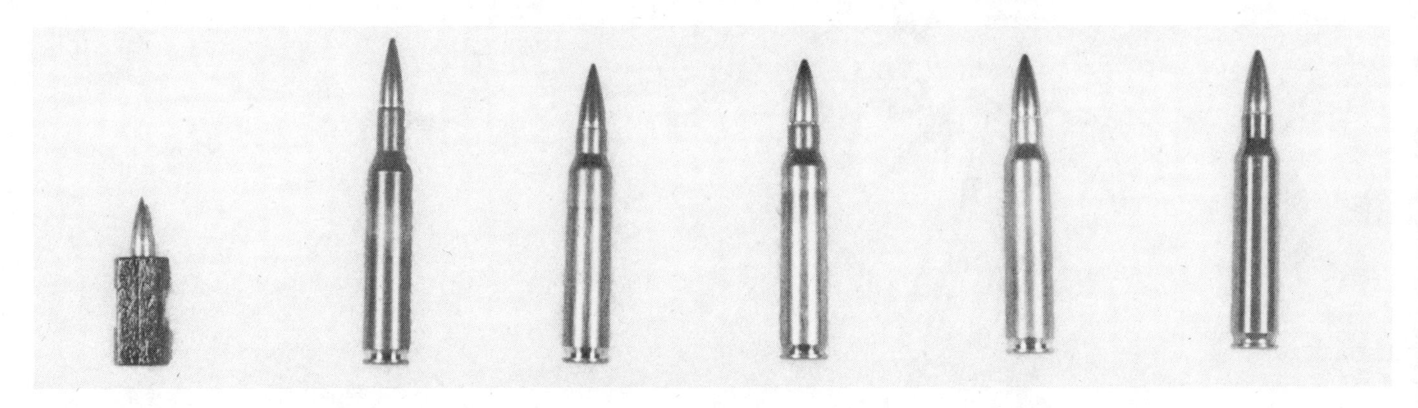

NATO trial cartridges (1977–1979). ***Left to right:*** **4.7 × 21mm German caseless; 4.85 × 49mm UK; 5.56 × 45mm Belgian SS109; 5.56 × 45mm US (brasscase); 5.56 × 45mm French (steel case).**

To define a second cartridge caliber for standardization (and if possible to make recommendations for a standard rifle and light machine gun), a joint NATO test program was created. Test personnel began by evaluating ammunition and weapon candidates submitted by the member countries in early 1977. To supervise the overall management of the small arms trials, a Coordination Panel for the Testing and Evaluation of Small Arms, Ammunition and Weapons was created. This panel's members were part of Subpanel 4, the Subpanel of Experts to Study Infantry Small Arms Weapons Systems for the Post-1980 Period, which is part of the Infantry Weapons Panel (Panel III) of the NATO Army Armaments Group (AC/225). Membership was limited to only the Subpanel 4 representatives whose countries had signed the Memorandum of Understanding. To carry out the actual testing, NATO established a NATO Small Arms Test Control Commission (NSMATCC). Each participating country provided personnel for the test group. The primary representative, called the principal member, had the military rank of colonel or lieutenant colonel or a corresponding civilian equivalent, and staff officers were majors or captains or civilians of equal grades. Test Control Commission personnel were assigned to this project for the period July 1976 to mid-1980 to ensure continuity during the entire operation.

Data generated by the NATO Small Caliber Test Control Commission provided a sound objective basis for making subjective (political) decisions. The tests were divided into two parts. The first consisted of technical tests that provided data relating to the need for further developments of ammunition and weapons and that determined the technical suitability of the ammunition and/or weapon military testing. The technical tests were characterized by the controlled conditions under which they were conducted in the laboratory and on the test field; chances for human error were eliminated where possible. These tests ran from April 1977 to spring 1979, and the data was then statistically analyzed. The second part of the test program included trials conducted under field conditions to determine if the ammunition and/or weapon met the stated requirements and if they were suitable for army use. These tests involved both qualitative observations and the judgment of selected military personnel with suitable field experience. They were conducted by soldiers representative of the troops that would use the materiel in combat. The military tests provided the primary basis for recommending a weapon or ammunition type as suitable for standardization.

The military trials were designed to determine the following:

Hitting performance: number of targets presented, number of targets hit, rounds fired per target, target hits, time to fire from carrying position, and time to first hit.

Training.

Reliability, availability, and maintainability (RAM): rounds between malfunction or stoppage and time to repair.

Human factors: time to negotiate obstacles and troop opinions (questionnaires used).

Safety.

Most of the military testing was conducted at the German Infantry School at Hammelburg, with testing of individual weapons extending from June to November 1978. Tests of light support weapons ran from January to June 1979. The international character of the NATO Small Caliber Tests was unique, and it may set a precedent for the international testing of materiel within the alliance. A comparison of this program with the activities that led to the standardization of the 7.62mm NATO cartridge reveals major differences.

As noted previously, small arms standardization was a goal of the Western Allies even before the creation of NATO in 1949. Talks about the feasibility of standardizing weapons, tactics, and troop training began in the fall of 1946 when Americans, British, and Canadians were trying to overcome the logistical nightmares caused by the multiplicity of weapons and ammunition types in use by the Allies during World War II. Standardization appeared to be a reasonable goal, and by the fall of 1947 some agreements had been reached, particularly in the crucial area of standardizing the thread patterns for nuts and bolts. Weapons, especially small arms, posed a more complicated problem.

At the end of a decade of trials and tests, NATO did have a single standard rifle cartridge—the 7.62 × 51mm NATO cartridge—but a standard rifle was not adopted. The Americans refused to accept the FN *Fusil Automatique Leger* (FAL), which the British had agreed to adopt over the American-designed M14 Rifle. The Bundeswehr adopted the Gewehr 3, an outgrowth of the World War II Mauser Sturmgewehr 45. The Germans had used the FAL as the G1 for the *Bundesgrenzschutz* (border police), but they were unable to obtain a production license from Fabrique Nationale, whose management was hesitant to rearm a former enemy. It was clear to the political and military leaders of the NATO countries that a better approach was necessary to solve the small arms standardization problem.

From the very beginning, there was unhappiness about the NATO small arms cartridge. The British were the least satisfied. They had conducted an extensive series of theoretical studies and practical tests, and their Ideal Calibre Study Panel recommended a 7mm (.276-inch) projectile fired at lower velocities than 7.62mm ammunition. But the United States had sponsored the 7.62mm round, and being numerically the largest partner in the alliance, the United States got its way. However, a decade later (1963), the United States Department of Defense selected the 5.56 × 45mm M16 Rifle for use by American and Allied ground forces in Vietnam. This small caliber rifle was easier to control during rapid bursts of automatic fire, and it was supposedly more lethal. Selection of the M16 began a new period of controversy within United States and foreign military circles. Basic questions about the infantryman and his rifle were raised. Many are still unresolved:

(1) Should the rifleman engage enemy positions with consciously aimed fire against specific individual targets, or should he fire bursts of fire in the general direction of the enemy, thus inflicting casualties with random hits and at the same time keeping the enemy pinned down through the suppressive effects of his fire?

(2) What are acceptable definitions of lethality and incapacitation? Can incapacitation of enemy forces with a rifle be predicted?

(3) At what ranges should rifle projectiles produce lethal and incapacitating effects? How are these effects measured?

These questions likely would have remained a matter of academic concern for NATO had the United States not made the decision in fall 1969 to equip most of the 190,000 American armed forces in Europe with the M16 Rifle. By the fall of 1971, the process of switching from the 7.62mm NATO caliber M14 Rifle to the 5.56mm M16 was complete. There was considerable grumbling in Europe about this change. The British were especially unhappy as they had adopted the 7.62mm cartridge because United States Army ordnance personnel had argued that the smaller British 7mm cartridge was not effective (lethal) enough. The Americans then introduced a nonstandard caliber into the alliance that was still smaller than the British projectile. Older British officers and Ministry of Defense officials were especially outraged when they remembered the acrimonious arguments that had led to their begrudging acceptance of the 7.62mm NATO cartridge.

When the member nations of NATO decided to select a second smaller caliber cartridge, it was apparent to all of the participants that some basic ground rules would have to be established to ensure an efficient and less politicized evaluation process. Despite more than a decade of combat experience with the M16 Rifle and the 5.56mm cartridge in Southeast Asia, basic questions about its suitability—particularly in the European environment—for all military forces remained. While most experts would agree that the recoil produced by the 7.62mm NATO cartridge was too great for use in a rapid firing assault rifle, its long-range effectiveness was a desirable feature. Rifles firing the American 5.56mm cartridge were easy to control, but were they suitably effective at long ranges? NATO had to choose a cartridge that would produce the most lethality at the desired ranges with the least recoil. Of course, before such a selection could be made, a test program would have to be conducted after agreements had been reached about the types of tests that would be run and the performance requirements the candidate cartridges would have to fulfill.

To eliminate personal and national biases, the NATO member states that signed the Memorandum of Understanding agreed that the testing should be under the control of the NATO Army Armaments Group. The tests would be international in nature, and the participants would have to agree on test procedures and performance requirements. Earlier trials had been conducted under the control of national organizations, but these new examinations would be carried out by a NATO body composed of military personnel from the participating nations.

As Colonel Maurice Briot (Belgian Army), the director of the NATO Small Arms Test Control Commission, stated in a June 1979 interview at NATO headquarters, the goal was to approach the tests in a scientific manner so that the results could be verified and reproduced—today, tomorrow, or a decade from now. The conduct of the tests, the collection of the data, and the analysis of the data would be carried out objectively within previously agreed to guidelines. Colonel Briot carefully distinguished between data analysis—the statistical process of compiling the information resulting from months of laboratory tests and field trials—and data evaluation. The analysis would be objective, but the evaluation would be subjective and political in nature. The data base of technical information from which the evaluators worked, however, would be factual, honest, and reproducible.

To ensure the integrity of the trials, the NATO Army Armaments Group agreed to several performance requirements and evaluation procedures documents. The requirements documents outlined desired characteristics for smaller caliber weapons, such as reliability, maintainability, suitability to tactical mission, hit probability, and probability of incapacitation. NATO's requirements were not necessarily the same as those adopted by individual national armed forces. For example, the NATO requirement for the maximum range to be achieved by an individual weapon, a light support weapon, and a medium support weapon is shorter than the United States requirement. The differences are especially noticeable when comparing the desired range requirements for the light and medium machine guns.

NATO's testing procedures were clearly defined in *Evaluation Procedures for Future NATO Weapons Systems* (D14), a document that Colonel Briot indicated was a very important accomplishment in itself. Document 14 was the result of work done by the Group of Experts on a Post-1970 Family of Small Arms, which reported to the Infantry Panel (Panel III). This subpanel was created in 1964 to reach an agreement on common methods of evaluating the various types of small arms and ammunition that might be used in the post-1970 period. It was also given the task of outlining programs for technical and military tests of various weapon and ammunition combinations. Document 14, which outlines the technical and military evaluation procedures, was an evolving document, which has been revised several times since the late 1960s.

In 1970, the Infantry Weapons Panel created Subpanel 4 to consider candidates for a family of small arms for NATO standardization and use in the 1980s. This subpanel subsequently organized its own group of international experts who, under the direction of Mr. Thinat of France, revised the manual (D14) and made it suitable for conducting the 1977–1979 comparative small arms trials. Document 14 was a very wide-ranging test manual, covering the entire spectrum of small arms weapons systems that NATO might seek to evaluate.

(1) Ammunition: both point target and area target ammunition.
(2) Individual weapons: assault rifles with effective ranges of 300–400 meters.
(3) Light support weapons: section/squad automatic weapons firing the same ammunition as the individual weapon, but capable of delivering a high volume of fire to about 800 meters.
(4) Medium support weapons: automatic weapons capable of engaging ground and air targets to a range of about 1,000 meters; should be capable of being mounted on ground vehicles and aircraft.
(5) Heavy support weapons: automatic weapons of heavier caliber than medium support weapons and usable on ground mounts or mounted on ground vehicles or aircraft.
(6) Grenade launching rifles and grenade launchers: part of area target ammunition evaluation.

Test personnel from the several member nations of Subpanel 4 often had difficulty in explaining their country's philosophy of testing materiel and in defining the various tests. For example, a "mud test" might have a different precise meaning for each organization that conducts one. Each test had to be discussed, and acceptable procedures developed. Common agreement as to the translated meanings of technical words and phrases had to be established, too. By creating these test procedures, the members of the evaluation group also created an environment in which they could work together with a minimum of confusion and friction. Although there were lessons waiting to be learned, the evaluations procedures document proved to be a very important milestone for the men conducting the trials.

Candidate Ammunition and Weapons

As each generation of military men has had to learn, ammunition and weapons cannot be considered separately. The two elements must be evaluated together as a system. Three different bore diameters (calibers) were represented in the cartridges submitted to NATO for evaluation: the British 4.85mm, the German 4.7mm, and three 5.56mm projectiles.

The British cartridge had a case 49 millimeters long, compared to the 45-millimeter case of the M16A1 Rifle cartridge. This cartridge case is otherwise based on the American cartridge, which would permit existing 5.56 × 45mm weapons to be converted for use with British ammunition. The British projectile weighs 3.11 grams (48 grains) compared to the 3.56-gram (55-grain) M193 round. Velocity of the British projectile from the 518.5mm (20.4-inch) barrel of the XL64E5 Individual Weapon has been cited as 900 meters per second (2743 f/s) while the M193 projectile velocity at the muzzle is listed as 975 meters per second (2970 f/s) from the 508mm barrel of the M16A1 Rifle.

The German 4.7mm projectile was part of a caseless cartridge and weapon system concept designed jointly by Heckler & Koch of Oberndorf and Dynamit Nobel of Cologne. The Heckler & Koch rifle has been designated the Gewehr 11 and is chambered for the caliber OH 4.7 × 21mm. This length dimension (21mm) refers to the length of the caseless propellant charge. This weapon and ammunition system is discussed in detail later in this chapter.

Although these small caliber cartridges (below 5.56mm) were submitted for the trials, it was not likely that either would be standardized by NATO. The G11 rifle was evaluated only during the technical tests; it was withdrawn afterwards, because the ammunition-weapon combination suffered a cook-off problem. Just as the G11 rifle and OH 4.7 × 21mm ammunition were withdrawn from the trials, the British dropped the development of the 4.85mm cartridge. They converted XL64 weapons series to 5.56 × 45mm. This decision merits an observation. The British once again appear to have pursued an ideal course of development with their 4.85mm cartridge rather than a practical one. Whatever the merits of the 4.85mm, the British were acting in what one must call an unrealistic fashion, considering that the United States Army had 1.3 million M16A1 Rifles in its inventory that it might have to replace if NATO adopted a caliber other than 5.56mm. The replacement cost in 1979 dollars would have been about $360 million. Clearly, interoperability of a new cartridge with existing weapons—either 7.62mm NATO or 5.56 × 45mm—was a tangible piece of practical data that could not be dismissed lightly by the NATO authorities.

Interoperability was one of the major thoughts behind the American development of the XM777 and XM778 projectiles and the Belgian development of the SS109, P112, and L110 projectiles. The United States Army officials started out with the goal of improving the effective range of the 5.56mm cartridge so that it would be suited for use in a Squad Automatic Weapon (light machine gun) as a base-of-fire weapon (see Chapter 2). At the same time, the improved cartridge could be used in the M16A1 Rifle without changes to the weapon or loss of the projectile's basic lethality. The key requirements for the XM777 ball projectile were improved penetration of hard targets at ranges up to 800 meters (penetration of at least one side of the standard United States steel helmet at 800 meters) and the same probability for incapacitation given a hit such as that of the standard M193 ball projectile. For the XM778 tracer projectile, the requirement was a clear daytime trace to 800 meters, a considerable improvement over the standard M196 tracer projectile.

FN Herstal developed its own improved 5.56mm cartridges for use in the FNC Rifle and the Minimi Light Machine Gun. The design team at FN sought an alternative to the American M193 (SS92 in FN terminology) because they saw room for improvement in the long-range effectiveness of the round and because of growing complaints that the M193 projectile can cause unnecessary suffering among those people struck by it. Realizing that there was an increasing possibility that high velocity projectiles of the M193-type might be restrained by international agreements (such as a proposed modernization of the 1949 Geneva agreements), the designers at Fabrique Nationale sought 5.56mm projectiles that would have improved range without the "inhumane" aspects—such as tumbling and breakup—of the M193. The new FN ball bullet, the SS109, has a more sharply tapered form—ogive—and greater weight—4 grams (the M193 weighs 3.56 grams and the XM777 3.53 grams). Like the XM777

The Enfield Weapon System (EWS). ***Top,*** **the 4.85 × 49mm XL64E5 individual weapon as tested in the NATO small caliber trials;** ***middle,*** **the 4.85 × 49mm XL64E5 light support weapon;** ***bottom,*** **after the NATO trials the XL64 series was redesigned and rechambered. This is the 5.56 × 45mm (SS109) XL70E3 individual weapon.**

projectile, the FN SS109 has a combination steel and lead core. In both, the steel insert at the tip of the projectile acts as an armor penetrator, and both the American and Belgian projectiles should be considered semiarmor piercing bullets. To help keep the projectile from tumbling when it hits a human target, the FN engineers changed the rifling twist of the barrel of the FNC rifle. Instead of the bullet making one rotation on its axis in every 12 inches (304.8 millimeters) of travel down the rifle barrel, the SS109 bullet makes one rotation on its axis for every 7 inches (177.8 millimeters). The so-called 1-in-7 twist spins the bullet to a greater extent than the 1-in-12 twist, thus theoretically imparting greater ballistic stability. It should be noted that in small-caliber, high-velocity projectiles improvement in ballistic stability generally is accompanied by a decrease in lethality.*

The SS109 projectile loses some of its effectiveness when fired from the M16A1 Rifle or the FNC with a 1-in-12 twist barrel. If the SS109 projectile was adopted by NATO, the United States Army would have to consider rebarrelling its stock of M16A1 Rifles. As noted, this would be expensive, but it was an issue worthy of consideration.

How do the XM777 and SS109 projectiles compare? This is a difficult question to answer and one of the primary reasons that the NATO small caliber trials were so important. When various organizations compare ammunition or weapons, they often do so to their own advantage. All unilateral data, therefore, had to be considered suspect by NATO, whose analysts examined how different weapons firing a particular cartridge performed under identical environmental circumstances. However, this approach did leave something to be desired in this case. For example, the M16A1 fired M193 control ammunition to establish a performance benchmark. The M16A1 also fired the XM777 round. The M193 and XM777 cartridges could also be fired from the French FAMAS rifle and the Netherlands MN1 (Galil) rifle. But the FNC and the British XL64 cannot fire the XM777 series ammunition, and the M16A1, FAMAS, and the MN1 cannot use the SS109 or the British 4.85 × 49mm cartridge. For control and comparison purposes, it would be useful to have M16A1 Rifles with 1-in-7 twist barrels to fire the SS109 and M16A1 Rifles with 1-in-5 twist barrels to fire the British cartridge.

Accepting the caveat that all nonofficial NATO comparative data is suspect but knowing that the NATO information will not be available until the late 1980s, are there some rough indications of performance that we can examine? Yes, keeping in mind the limitations we have stated. The FN SS109 cartridge appears superior for armor penetration when compared to the other contenders. It acts like a drill on the three NATO penetration targets—NATO plate (a piece of mild steel 3.5 millimeters thick), the Federal Republic of Germany steel helmet, and the United States steel helmet. If penetration of lightly armored targets had been the major criteria, then the SS109 would have been superior, but its reduced lethality also had to be considered. On the other hand, the XM777 was supposedly superior to the SS109 in terms of incapacitation and lethality, but was clearly less effective in defeating armored targets. FN engineers claimed that the design of their steel penetrator keeps production costs equivalent to the M193 (SS92) projectile, while the steel penetrator of the XM777 projectile poses production problems since it could become lodged sideways when inserted into the bullet jacket.

Clearly, the decision about which cartridge should be standardized was affected by a number of complex considerations. For example, what kind of balance need be struck between armor penetration and human incapacitation? Other issues such as the economics of manufacturing the projectile were beyond the scope of the analytical phase of the tests, but they did influence the evaluation phase.

Standardization of a second NATO cartridge, as complicated as it seems, may seem simple when compared to selecting new individual and light support weapons. Three control weapons were selected to provide a basic reference for evaluating contender weapons and ammunition: the Federal Republic of Germany G3 representing standard 7.62 × 51mm NATO class rifles (1-in-12), the United States M16A1 representing 5.56 × 45mm class rifles firing M193-type ammunition (1-in-12), and the Belgian Mitrailleuse a gaz 58 (MAG 58) representing standard 7.62 × 51mm NATO class light machine guns (1-in-12).

Technically, the main point that was not considered by the Test Control Commission was the differences between production and prototype weapons. Whereas the M16A1 Rifle was a battle-proven veteran, the G11 represented an advanced technical concept in prototype form. Only a small number of United Kingdom XL64 weapons had been fabricated; it was thus an advanced prototype. The French FAMAS, the Belgian FNC, and the Dutch MN1 were production models but had been produced in only limited numbers when compared to the M16A1. The FN Minimi machine gun had been under development since the early 1970s, and while it was not in serial production, it was much further along the development path than the British XL64E4 light support weapon. Still, the Minimi was a comparative newcomer when viewed alongside the German MG3E. This weapon is a lightweight version of the World War II German MG42. Colonel Briot noted that the technical tests and military trials were created to provide information about the weapons only as they existed when the evaluation began. This type of analysis would be scientific; weighting the tests to compensate for the different levels of development would have been subjective and speculative.

CANDIDATE WEAPONS

	Individual Weapons (Rifles)	
Belgium:		
5.56 × 45mm	FNC (1-in-7)	SS109 Series
France:		
5.56 × 45mm	FAMAS (1-in-12)	French steel case w/M193-type
Federal Republic of Germany:		
4.7 × 21mm	G11 (1-in-12)	Caseless
Netherlands:		
5.56 × 45mm	MN1 (1-in-12)	M193-type
United Kingdom:		
4.85 × 49mm	Individual weapon XL64E5 (1-in-5)	
United States:		
5.56 × 45mm	M16A1 (1-in-12)	XM777/XM778
	Light Support Weapons (LMG)	
Belgium:		
5.56 × 45mm	Minimi (1-in-7)	SS109 Series
Federal Republic of Germany:		
7.62 × 51mm	MG3E (Shortened and lightened MG3) (1-in-12)	7.62 × 51mm NATO
United Kingdom:		
4.85 × 49mm	Light support weapon, XL64E4 (1-in-5)	

On 28 October 1980, NATO approved the standardization of a second small caliber cartridge for use within the alliance (STANAG 4172):

> To increase the total effectiveness of the family of infantry small arms, several allied governments intend to

*The Swedish government was experimenting with 5.56 × 45mm weapons with 1-in-9 twist barrels for humanitarian reasons.

Top, the French contender in the NATO individual weapon trial—the 5.56 × 45mm FAMAS firing the US M193 cartridge; _middle_, Fabrique Nationale's 5.56 × 45mm FNC barrelled to fire the SS109 projectile (1-in-7 barrel twist); _bottom_, the Israeli Military Industries 5.56 × 45mm Galil, designated the MN1—The Netherlands' entry in the individual weapon trials.

introduce into their forces in the 1980s a new individual weapon. There is general agreement that the present 7.62mm calibre will remain a NATO standard. In order to select a second smaller and lighter weight ammunition for NATO Small Arms to be standardized, an extensive technical and military programme has been organized for the testing of weapons and ammunition presented for evaluation by interested governments. As a result of this exclusively technical and military programme, 5.56mm has been adopted as second standard NATO calibre for small arms and the Belgian SS109 ammunition has been selected as a basis for standardization of ammunition for the second NATO calibre for small arms.

In evaluating the operational effectiveness of the weapons and ammunition offered as candidates for NATO standardization, the commission came to several conclusions:

Small calibre ammunition and weapons tested were smaller and lighter than the 7.62mm control ammunition and weapons. The smaller calibre ammunition is approximately half the weight and size of 7.62mm ammunition; the small calibre individual weapons are an average 0.6kg lighter than the control weapon and the smaller calibre light support weapons were at least 5kg lighter than the control weapon. An infantryman in combat using the new calibre weapons and ammunition can carry more ammunition, especially in the case of the light support weapons, and not increase the load he would carry if he were equipped with 7.62mm ammunition and weapons. Logistic support is eased because the new ammunition is smaller and lighter.

Some NATO nations have differing national concepts of the employment of small arms weapons, particularly of light support weapons. Some countries wish to have the performance of 7.62mm weapons and are prepared to accept the increased weight and size. The tests have shown that based upon the performance of the candidate systems it appears that effective light support weapon systems using the second calibre ammunition can be produced to meet the needs of those nations who wish to employ smaller calibre light support weapons with reduced weight and size.

After studying the final report of the commission, the Coordination Panel for the Testing and Evaluation of Small Arms, Ammunition and Weapons reached the following conclusions based solely upon the technical and military tests conducted during the course of the program. Other factors, such as the economic situations in particular member nations, were not considered. The panel found:

a. **Ammunition**

(1) The reflective test results of candidate ammunitions were not consistent among the various technical and military tests. The contribution of ammunition performance to system effectiveness cannot be evaluated independently since it is more directly influenced by the weapon and fired under field conditions than by the basic quality of the ammunition itself.

(2) Each candidate ammunition has a considerable advantage in weight, size, and cost over the NATO 7.62mm rounds (the weight of each contender ammunition is approximately half that of the NATO 7.62mm). The advantage in weight and size is also true for the contender weapons designed to fire the smaller calibre ammunition. Thus, the second standard NATO calibre ammunition will provide for increased system effectiveness when constrained by weight factors.

(3) There are no significant discriminators in the various candidate rounds for the individual weapon; however, the heavier bullets with a higher spin rate have significantly better terminal ballistics (i.e., penetration, incapacitation) than the lighter bullets at ranges beyond 500 metres.

(4) Each candidate individual weapon ammunition (4.85mm and 5.56mm [type variants of the 5.56mm include SS109, XM777, M193, and French M193]) meets the essential requirements in the operational characteristics . . . for individual weapon system.

(5) Each candidate light support weapon ammunition (4.85mm and 5.56mm [type variants of the 5.56mm include SS109 and XM777]) meets the essential requirements in the operational characteristics . . . for light support weapon systems; however, the SS109 ammunition is significantly better than the other ammunition with regard to the terminal effects at ranges beyond 500m.

(6) Based on the results of the technical and military tests, the SS109 (U.S. designation . . . XM855) ammunition is considered the best candidate as a basis for standardization of ammunition for the second NATO calibre for small arms.

b. **Weapons**

(1) It is not practical to make a proposal for standardization of weapons.

(2) The relative suitability of the individual weapons tested to meet the requirements of the operational characteristics . . . cannot be validly established because the weapons tested varied from prototype to in-service models. Weapons have been further developed since testing was conducted and certain weapons were not tested with the ammunition now proposed. The reliability of the weapon system appears directly related to their maturity. The individual weapons tested generally meet the operational characteristics.

(3) The relative suitability of the light support weapons tested to meet the requirements of the operational characteristics . . . cannot be validly established because the weapons tested varied from prototype to in-service models.

(4) The characteristics and performance demonstrated by the various candidate weapon systems could be used by nations in the selection of weapons to meet specific national requirements.

Considering these findings, the panel recommended the following actions, which were approved in October 1980: (1) that the NATO Army Armaments Group approve the adoption of the 5.56mm caliber as the second standard NATO caliber for small arms; (2) that they approve the SS109 ammunition as the best candidate for standardization for the second caliber; (3) that they direct Panel III to expedite its preparation of STANAG 4172 for an ammunition for both individual and light support weapons; and (4) that they agree that a recommendation for NATO standardization of an individual or light support should not be made.

Based upon their own firing tests, Fabrique Nationale engineers reported that their SS109 projectile will penetrate the NATO armor plate at 640 meters and one side of the American helmet

COMPARATIVE TERMINAL BALLISTICS FOR 7.62 × 51mm AND 5.56 × 45mm NATO BALL CARTRIDGES

Ammunition / Pitch / Target	7.62 × 51-SS77 bullet 12″ (305mm)	5.56 × 45-M193 bullet 12″ (305mm)	5.56 × 45-SS109 bullet (XM855) 7″ (178mm)	5.56 × 45-SS109 bullet (XM855) 12″ (305mm)	5.56 × 45-XM777 bullet 12″ (305mm)
NATO Plate	620m	400m	640m	416m	ca. 410m
FRG Helmet	640m	485m	1150m	Unknown	ca. 600m
US Helmet	800m	515m	1300m	825m	ca. 820m

at ranges of more than 1,300 meters. To produce these results, the SS109 must be fired from a weapon having a 1-in-7-inch (1-in-178mm) twist barrel. Fired from a 1-in-12-inch barrel, the SS109 would not penetrate the NATO plate at ranges greater than 416 meters or the helmet at a distance greater than 825 meters. While this was an improvement over the M193, it did not meet the NATO requirements. And the Americans would have to change the barrel of their M16A1 before the SS109 ammunition could be qualified as interoperable. The competitive XM777 steel-core projectile developed by the United States would penetrate the armor plate at about 410 meters and the helmet at 820 meters. While this was slightly less than the performance of the SS109, some American officials still held out for their round because it could be fired from the M16A1 without requiring any barrel change; it was designed for the 1-in-12-inch twist barrel. The XM777 was also supposed to be less expensive to manufacture according to the estimates of some Americans, although it was not clear how this price comparison was made or what quantities were involved in the estimate. What was clear, however, was the unpopularity of the SS109 among certain groups in the United States and in Europe.

Fabrique Nationale was in an awkward position. Their cartridge won the NATO small arms ammunition competition, but the company was faced with opposition from several directions. Critics of the SS109–including proponents of the 7.62mm NATO cartridge and the new German 4.7mm caseless round—did not dispute the validity of the results within the context of the tests carried out by NATO, but they did question the relevance of those tests. Supporters of the 7.62mm NATO cartridge argued that comparing the 5.56mm SS109 and the 7.62mm SS77 projectiles was not meaningful because the SS109 represented late-1970s technology, while the SS77 reflected the state-of-the-art of the early 1950s. These proponents of the 7.62mm said that a 7.62mm bullet embodying the technology used in the SS109 would have produced a cartridge that would penetrate armor plate and helmets at ranges in considerable excess of 620 and 800 meters. They also pointed out that the 7.62mm projectile has greater total range and greater remaining kinetic energy at those long distances than any 5.56mm projectile. Supporters of the 7.62mm round fell into two camps: those who wanted the 7.62mm for both the rifle and the squad automatic weapon (light machine gun) and those who saw the need for 7.62mm in only the squad automatic weapon. But together they expressed the belief that NATO forces (and United States forces) deployed to such areas as the Middle East would require weapons that could deliver fire at ranges beyond the effective range of any 5.56mm cartridge.

Those experts who favored the German 4.7 × 21mm *ohne Hülse* (caseless) cartridge argued that the 4.7mm projectile would provide satisfactory performance against the type of targets generally engaged by infantry personnel. The kinetic energy produced at the muzzle by the 4.7mm when fired from the Gewehr 11 is about 1,500 joules, as compared to 1,575 joules for the SS109 and 3242 joules for the SS77. Further, German government and industry personnel argued that neither the 4.7 ×21mm OH nor the 5.56 × 45mm is suitable for use in a squad automatic weapon. Only the 7.62mm NATO round—perhaps a product-improved one—would suffice. They thus proposed a 4.7 × 21mm OH rifle and a 7.62 × 51mm squad automatic weapon. A "National Comment" document released in the summer of 1980 by the Federal Republic of Germany relates to the outcome of the NATO Small Arms Test and Evaluation Program: "The FRG welcomes the positive result of the NATO Small Arms Test and Evaluation Programme which has led to recommending the 5.56mm caliber for the new to be standardized round for NATO Small Arms. This result will certainly give those nations having an urgent demand for replacement of their rifles and/or light support weapons the opportunity to take advantage of a lighter and smaller round as compared with the present 7.62mm rounds."

But the Germans had been expecting more performance from the 5.56mm cartridge. The Germans believed that the SS109 did not represent "a very significant quantum jump in technology." They admitted to the considerable advantage in weight and size of the 5.56mm round over the 7.62mm round, but this improvement by itself could not be termed "very significant" operationally and technically. "In order to achieve this an overall new technology will be required such as the caseless ammunition technology as it was presented in this programme (the NATO evaluations) by the FRG," German officials said. They added another important point to their case: "Nations not having to replace their weapons within the very near future will have to consider whether the very high procurement costs for a new generation of infantry weapons in the 5.56mm caliber are to be justified since the conventional technology applied (to the 5.56mm rifle-ammunition system) will be superseded in a relatively short period of time."

Germany would certainly not replace its G3 rifle, since it would be serviceable for at least another ten years, and the German military was assuming that by the second half of the 1980s the caseless technology for small arms ammunition would be ready. Its plans do not call for the procurement of a new rifle system (the G11) until the early 1990s, so the Germans can afford to wait until the G11 and its 4.7 × 21mm ammunition is ready to be standardized.

Although the SS109 cartridge met the technical specifications established by NATO, it obviously did not satisfy the military needs or prejudices of some NATO members. The technical issues resolved by the NATO Small Arms Test Control Commission were only a small part of the overall problem. There remained the several divergent views concerning the tactical utilization of the rifle and the squad automatic weapon. And of paramount concern were the political and economic issues raised by ammunition and weapon standardization. Certain manufacturing concerns stood to profit handsomely if their cartridge or weapon was chosen by NATO. At stake was not just the royalty-free NATO utilization of ammunition and weapon designs, but the licensed production of such materiel in the export markets, as well. In addition, national prestige (lost or gained) and the resultant changes in the international balance of payments would also have to be considered.

Ancillary Issues

U.S. SAW Program. The United States Squad Automatic Weapon (SAW) program entered a new phase in September 1980 with the award of a "maturity phase" contract to Fabrique Nationale (FN) for further development of its 5.56mm XM249 SAW, the Minimi, which has a 1-in-7-inch (1-in-178mm) twist barrel and is designed to fire the SS109 cartridge. After fifteen months of engineering work at FN, the United States Army standardized the Minimi at the M249 (see Chapter 2 for more details).

U.S. M16 PIP. During 1978–1979, the United States Armed Forces and Colt Firearms examined the desirability of developing a product-improved M16A1 Rifle (M16 PIP). Such a project would solve two immediate problems. First the Army and the Marine Corps have an urgent requirement for new rifles to replace their aging M16A1s, and they do not wish to wait years for some future system. And second, a product-improvement program would correct many of the deficiencies in the M16A1 that have already been identified, thus extending the usefulness of the basic rifle. Many of the M16A1 Rifles in the United States inventory have been worn out from use as training weapons, and it is not unusual to find M16A1s that have been fired 40,000 to 50,000 times. Weapons that have seen this kind of wear malfunctioned during United States Marine Corps exercises in Norway in March 1980. The Corps analyzed this and other recent reports of poor performance and concluded that the M16A1 was

***Top,* standard 5.56 × 45mm (M193) M16A1 rifle; *middle,* Colt proposal for product-improved M16A1 (heavier barrel, improved handguards and butt stock); *bottom,* 1982 candidate PIP M16A1 with additional features of improved rear sight and three-shot burst control.**

essentially a good weapon, but that new rifles were needed to replace old and worn out ones. The Marines also concurred with Army conclusions that several M16A1 components could be improved to increase the field life expectancy of the rifle. Most of these improvements had already been suggested by Colt Firearms but had been rejected because of inadequate funds or perhaps because of hostility toward the manufacturer. (This unfortunate attitude toward Colt is a legacy of the manner in which the M16A1 was introduced into United States Army circles during the Vietnam War.)

Product improvements to the M16A1 Rifle include a single-shot fire and 3-shot burst fire, a stronger butt stock, improved front handguards, new tapered slip ring, a new 1-in-7-inch (1-in-178mm) twist barrel with a heavier diameter toward the muzzle end, and a new fully adjustable rear sight. The new butt stock proposed by Army and Colt officials will be made from a new material called super tough nylon. Tests conducted thus far indicate that the improved stock will be ten to twelve times stronger than the currently issued one. Both halves of the new stronger round handguards are the same, eliminating the requirement for stocking left and right handguards. The new barrel contour from the rear of the front sight to the muzzle makes that vulnerable section of the barrel twice as strong as the current M16A1 barrel. The resulting product-improved M16A1 would be much more durable, an especially desirable characteristic for weapons carried by airborne troops. An M16 PIP was tested in the summer of 1981, with special attention being given to the performance of the 5.56mm projectile against targets protected by sand bags and/or flak jackets. These critical target situations were not evaluated during the NATO small caliber trials.

Initially there were many opponents standing in the way of an M16 PIP. For instance, there were individuals within the United States Army Armament Research and Development Command (ARRADCOM) who could not agree on what deficiencies to correct or what changes to make to the rifle. Some Army, Marine, and Colt officials felt that any additional "improvements" beyond those just outlined might delay the fielding of a product-improved weapon and perhaps jeopardize the entire idea, because there are voices within the American defense establishment that call for abandoning the M16A1 altogether—that faction of Army and Marine Corps officers who prefer the 7.62mm NATO caliber, and the group that prefers to wait for some significant improvement in the technology, such as that promised by the new caseless round.

The M16 PIP and the procurement of FN's Minimi forced some serious soul searching within the United States Department of Defense. In 1981, the American defense community had determined the answers to several important questions. An especially important one concerns future technology: Are the Germans really capable of fielding the G11 weapon-ammunition system within the next four to six years, and just how many NATO members will wait for its production or at least defer their decision on a 5.56mm weapon until they see how the new rifle performs? Some American officials feared that the United States would be the odd nation out if it proceeded with the fielding of a product-improved M16 and the Minimi. Others, including an important group of Marine Corps officers, believed the M16 PIP and the SAW were already too late; they already should have been in the hands of troops. The dilemma was clear: Make a decision to adopt the improved rifle and a SAW, and the decision may prove to be the wrong one several years later. Defer the decision and wait to see what the Europeans do, and American troops might be caught without sufficient numbers of conventional ground defense weapons. Both paths were fraught with problems, but a choice had to be made about which alternative would provide American troops with the best weapon systems at the best price.

The Marine Corps was the first to make up its collective mind (although there are minority opinions within it). At a 15 July 1980 briefing to the commandant on the procurement of a squad automatic weapon, Commandant General Robert H. Barrow expressed his frustration about the conflicting and confusing information he was receiving about small caliber weapons. He directed the Marine Corps Development and Education Command to form a special task force to study the requirements for infantry weapons systems. After an intensive four-month study, several important recommendations were made to the commandant, who approved them: (1) Procure the 40mm MK-19, MOD 3, automatic grenade launcher. (2) Reprogram four million dollars in FY1981 funds to purchase equipment that will expedite the production of fuzes for the 40mm M432 HEDP round. (3) Begin fielding the following weapons: the MK-19 (FY1983), the .50 caliber M2 Heavy Barrel Browning machine gun (FY1983), a 5.56mm SAW and ammunition to be determined by analysis of data gathered during the maturity phase contract with FN (FY1984), the M16 PIP (FY1982), and a 9mm Parabellum handgun (FY1983). (4) Cancel testing of 7.62mm NATO caliber SAW candidates. (5) Continue support of the Joint Service Small Arms Program (JSSAP) project to develop improved armor-piercing small caliber projectiles.

Although many may argue with the results of the Corps' analysis, the task force directed by Lt. Col. Richard Maresco approached its study in an interesting manner. The group started with the premise that all the weapons issued by Marine infantry units had to be viewed as part of a whole. Too often in the past, single weapon types—rifles, machine guns, and grenade launchers—had been examined out of the context of the battlefield. Maresco and his associates looked at the seven major types of targets suited for engagement by infantry weapons—personnel, armored personnel carriers (APCs), infantry fighting vehicles (IFVs), general purpose vehicles, tanks, structures, and aircraft (fixed-wing and rotary)—and evaluated each class of weapon available to them as it compared to current weapon threats that the Corps might meet. Deficiencies in current American weapons were also identified, along with near-term weapon options that could help the Marines overcome these deficiencies. As Maresco phrased it, "We need to solve the deficiencies with readily available weapons, and then we need to plan for future developments that will further enhance our infantry weapon inventory." The MK-19, MOD 3, automatic grenade launcher with armor-piercing ammunition would improve the Marine Corps' ability to defeat the BTR (a current Soviet armored personnel carrier) and the BMP (a Soviet infantry fighting vehicle) at ranges greater than 2,000 meters. Adding the .50 caliber Browning Heavy Barrel machine gun at the battalion level would help the Marines counter the 14.5mm machine gun on the BTRs. Firepower in the rifle squad would be increased by the introduction of the 5.56mm squad automatic weapon, and a product-improved M16 would allow the Marines to extend the life of their rifle. The 40mm M203 grenade launcher was identified as one of the most important weapons in the rifle squad for dealing with APCs and IFVs. The primary deficiencies that could not be corrected by the adoption of these weapons were the inability of small arms to successfully engage lightly armored vehicles or to defend against attack helicopters. New weapons will have to be developed to meet these threats. The task force also recommended procuring better one-man portable assault weapons, equipping anti-tank weapons with night acquisition sights, and improving lightweight and medium mortars and mortar ammunition.

Maresco's task force realized the complexity of this study and its political sensitivity from the beginning. Their anlaysis of the SAW issue is illustrative of the analytical process used by the men and their sensitivity to the political problems involved. As Maresco said in his briefing to Commandant General Barrow, "We will discuss this at length due to the volatility of the issue and to highlight the depth, logic and rationale that went into this particular analysis." According to Maresco's team, "The squad automatic weapon (SAW) option introduces a significant new capability by providing suppressive fire against area targets, particularly exposed enemy personnel at ranges of 300 to 800 meters. The SAW has a volume of suppressive fire that surpasses the Soviet 5.45mm AK-74 rifle and is competitive with the next generation Soviet machine gun (RPK-74)." A one-man system, the SAW is lighter than the M60 7.62mm machine gun and can be employed at the squad level, but the Marines' report pointed out that it is heavier than the M16A1 and will demand large quantities of ammunition, which may cause some logistics problems. However, the Marines believed that introducing a SAW would be worth any logistics problems it might cause.

Maresco and his colleagues addressed the question of the best caliber for a squad automatic weapon: "While theoretically a unique caliber could be developed, practical considerations limit the alternatives to 5.56mm and 7.62mm." The task force concluded that a 5.56mm weapon was "roughly comparable" to a 7.62mm weapon in maximum effective range and rate of fire. A 5.56mm weapon-ammunition system would be lighter than a 7.62mm system, but on the basis of relative performance the Marines could find "no clear-cut superiority of one caliber over the other . . . each enjoys some advantages." In the final analysis, a 5.56mm SAW was chosen as the Corps' candidate for several reasons:

(1) 5.56mm permits approximately twice as many rounds as 7.62mm for a given basic load.
(2) 5.56mm selection avoids the disadvantages of unilateral acquisition.
(3) 5.56mm SAW has completed testing and requires only completion of the maturation phase prior to beginning production. The 5.56mm SAW could be fielded as early as 1983; an off-the-shelf 7.62mm SAW, not earlier than 1986.
(4) The lift requirement—combat transport (weight and cube)—for a given number of 5.56mm rounds is less than for the same number of 7.62mm rounds.

(5) 5.56mm SAW affords ammunition and magazine interchangeability with the current Marine infantry rifle.
(6) Although total system life cycle costs have not been developed, 5.56mm SAW ammunition would be less expensive than 7.62mm ammunition.

The XM249 was standardized as the M249 SAW in early 1982.

Marine Corps plans (as of mid-1982) called for the procurement of the 5.56mm product-improved M16A1 (M16A2), the 5.56mm M249 squad automatic weapon, the 40mm MK-19, MOD 3 grenade launcher, and the .50 caliber (12.7mm) M2 Browning heavy barrel machine gun. The accompanying table summarizes proposed procurements:

USMC SMALL ARMS ACQUISITION PLANS

Weapon	FY82	FY83	FY89 (Cumulative Total)
M16A2	30,000	55,000	245,000
M249	2,117	2,907	9,900
MK-19 MOD 3	792	570	2,334
M2 HB MG	400	275	750

G11—WEAPON OF THE FUTURE?

Heckler & Koch engineers Tilo Möller, Günter Kästner, Dieter Ketterer, and Ernst Wössner created the new Gewehr 11, which fires the 4.7 × 21mm *ohne Hulse* (caseless) ammunition developed by Dynamit Nobel AG of Furth-Stadlen. This weapon was tailored to the new round, and it is thus necessary to think in terms of a weapon-ammunition system.

During the NATO small arms ammunition trials, the NATO Small Arms Test Control Commission noted that the Heckler & Koch G11 "was the only weapon system in the tests presenting a new technology." All of the other rifles and ammunition were linear developments from existing concepts. The HK G11 employed caseless ammunition—the metal cartridge case had been eliminated—and the rifle itself was designed around several innovative ideas not commonly encountered in small caliber weapons. Heckler & Koch withdrew the G11 from the NATO tests when they discovered that the weapon tended to fire prematurely as heat built up in the breech mechanism after several rounds had been discharged (cook-off). But this setback did not stop the weapon designers at Heckler & Koch or the ammunition developers at Dynamit Nobel. By mid-1981, they were confident that they had not only solved the cook-off problem but had also significantly improved the G11's mechanism. Their confidence led them to demonstrate the G11 publicly for the first time at the American Defense Preparedness Association's (ADPA) Second International Small Arms Symposium on 14 October 1981. During this event, Heckler & Koch invited the author to visit their factory in Oberndorf/Neckar to test-fire the G11. The following report is based upon an exclusive briefing and shooting session arranged by Heckler & Koch. To date (mid-1982) the author is the only military analyst or journalist to have fired the new weapon.

By simply shooting the G11 at the ADPA Small Arms Symposium, Dieter Rall of the Heckler & Koch factory revealed that the German rifle was a unique weapon. To the casual observer, the G11 did not appear significantly different from the other rifles being demonstrated as they fired single shots, but there was one significant difference. A photographer standing near one of the conventional rifles had his nose bloodied when a fired case flew into his face; from the G11 no cases were being ejected. Its molded-propellant cartridges do not have metal cases: they are caseless. This is the first indication that the G11 is unique.

But the true uniqueness of the G11 becomes clearer when it discharges three-shot bursts. The cyclic rate of the mechanism is 2,000 shots per minute during full-automatic. (For continuous fire, the rate is 600 shots per minute.) Despite this high rate of fire, the noticeable recoil of the G11 is very low and the muzzle movement of the weapon very limited. While one can appreciate these facts by watching someone else fire the G11 and by studying the inventors' comments on the design, only personal experience with the rifle can fully convince the potential user that it represents a major step forward in the design of small caliber rifles. Tilo Möller, director of research and development (*Leiter der Entwicklung*), and his team of designers at Heckler & Koch have re-thought the military rifle. The result is a new concept for shoulder weapons.

Design Assumptions

Möller and his associates began the development of the G11 by analyzing the role of the military rifle on the battlefield of the future. They assumed that future infantry actions would take place in a mobile, fluid environment, with infantrymen engaging their opponents at close range. Once they left their armored personnel carriers, they would have to carry their own supplies of ammunition. Möller and his colleagues, therefore, considered three primary objectives in the design of a new rifle: (1) it must have practical rates of fire that will produce the maximum hit probability; (2) it must be lightweight, so that the infantryman can carry a large amount of ammunition; and (3) its durability must be maximized.

Improved hit probability. Rifles, when fired in the continuous-fire mode, tend to be relatively inaccurate. This fact was clearly demonstrated in American Project SALVO studies conducted in the early 1950s. Since then, arms designers have attempted to reduce the dispersion of the projectiles leaving the muzzle of the rifle by reducing the intrinsic recoil force of the cartridge (e.g., by using smaller caliber projectiles) and by attaching devices to the muzzle of the weapon. The Heckler & Koch design team tried a new approach. In addition to single-shot fire, they incorporated a three-shot SALVO-type burst and full-automatic (continuous-fire) cycles into their design.

In the three-shot burst mode, the G11 fires at a cyclic rate of 2,000 shots per minute. The median dispersion is about 1.2 mil (i.e., 3.6cm at 300m), and the maximum dispersion is 2 mil (6cm

U.S. Patent Dec. 21, 1976 3,997,994

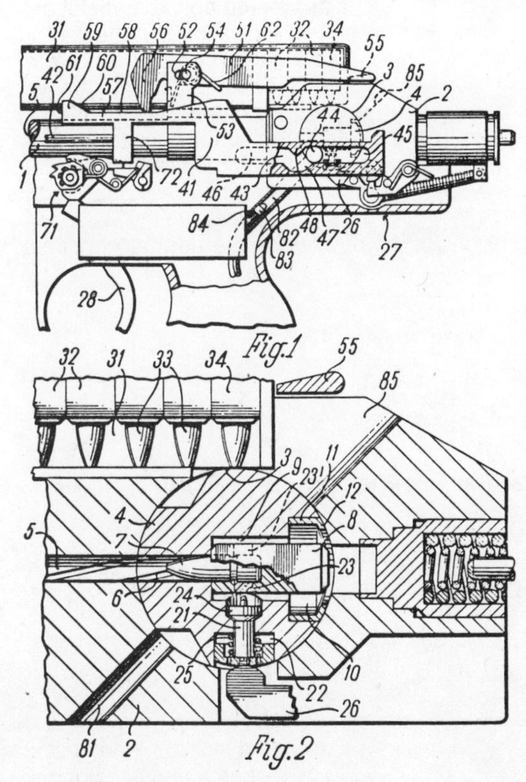

This patent drawing illustrates the early G11. Note the positioning of the hammer and firing pin (numbers 24 and 26) and the forward slanted exhaust port (number 81).

at 300m).* As a Heckler & Koch presentation notes, "A detailed analysis of shooting behavior by the typical infantryman under stressful combat condition shows that the normal single aimed shot misses because of aiming errors by the soldier. The hit probability of the shooter can be significantly increased by ensuring that the weapon has a built-in dispersion which takes into consideration the normal aiming error." Such a built-in dispersion pattern would not be practical in a weapon that fired only single shots or a conventional series of shots. In the G11, the controlled dispersion of the three-shot burst is independent of the shooter. Controlled dispersion is made possible by the G11's extremely high rate of fire and its reduced recoil.

The G11 can also be fired at full automatic, but the rate is only 600 shots per minute. Designers at H&K, knowing that rates above 700 shots per minute tend to waste ammunition and produce limited hits, reasoned that the G11's lower rate of fire, combined with the lower recoil of the cartridge, would provide a dispersion smaller than that generally encountered with conventional rifles. The lower rate of fire would also allow the infantryman to track his target more easily without worrying about inherent weapon accuracy. Rates of fire also affect another basic consideration—consumption of ammunition.

*A mil is the angle whose tangent is 1/1000; i.e., the angle subtended by 1 unit at a distance of 1000 units (e.g., 1 meter at 1000 meters).

Lightweight weapon and cartridge. Over the past century, there has been a continuous reduction in caliber and an increase in velocity of shoulder-weapon projectiles. At the same time, we have seen the introduction of steadily improved small arms mechanisms that allow more ammunition to be fired through them. As a result, battlefield consumption of ammunition has increased dramatically. During the American Revolution (1776–1783), it was estimated that an infantryman had to discharge about seventeen shots from his musket to produce a single casualty among the opposing forces. During the Korean conflict and the Vietnam War (1950–1975), it has been estimated that at least 50,000 shots were discharged to produce one incapacitated enemy soldier. There are a multitude of reasons to explain the increase, but the significant point is that self-loading weapons make it easier to expend ammunition. Taking this into consideration, the Heckler & Koch design team decided it was necessary to keep the weight of the G11 as low as possible, so the user could carry the rifle and a large amount of ammunition comfortably. Since the NATO ammunition trials of 1976, Heckler & Koch engineers have reduced the weight of the G11 from 5.87 kilograms to 4.2 kilograms. Their goal, as they further simplify the weapon and reduce its number of component parts, is a weapon that weighs 3.6 kilograms empty and 4.5 kilograms with one hundred rounds of ammunition. Weight is directly related to weapon durability, another area of concern for Heckler & Koch.

Weapon durability. In combat, small arms take a beating in the hands of infantrymen. Whether the rifle is being bounced on the floor of an armored personnel carrier as it travels cross-country or is dragged along the ground as the soldier advances on his belly, it must be reasonably indestructible. With this in mind, Möller and his colleagues enclosed the G11 in a stamped sheet metal housing which gives the operating parts maximum protection. While this makes the G11 slightly heavier than some other weapons, the weight penalty is justified by the added longevity of the weapon in combat.

Conclusions about the Design

Looking at what was needed in a new weapon for the infantryman, Möller and the others concluded that caseless ammunition would help them reach their design goal. Caseless ammunition would have several advantages: (1) It would contribute less weight per cartridge to the weapon-ammunition system. (2) Short, compact cartridges would permit a shorter operating mechanism. (3) Elimination of extraction and ejection systems would open the way for the manufacture of a nearly sealed mechanism. The new weapon would be short enough and light enough to be an acceptable substitute for both the rifle and the submachine gun.

The concept of a caseless cartridge and gun system is not a new one. Nearly 150 years have passed since Johann Nikolaus von Dreyse introduced his paper-cartridge breech-loading *zundnadelgewehr.* But needle-fire rifles were replaced in the 1860s and 1870s by shoulder weapons that used cased metallic cartridges, because caseless rifles had two major faults. First, the breech seal (obturation) had been difficult to perfect and tended to erode over time, and second, the caseless cartridges, whether made of paper or cloth, could be damaged more easily than brass-cased cartridges. When American ordnance engineers began to experiment with caseless cartridges in the 1950s and 1960s, they added a new problem to the defects traditionally associated with such ammunition. After a few shots were fired from self-loading weapons, the heat that was retained in the chamber section of the barrel would ignite the uninsulated propellant of the caseless cartridges. Obturation, fragility, and cook-offs seemed to be major stumbling blocks that would prevent

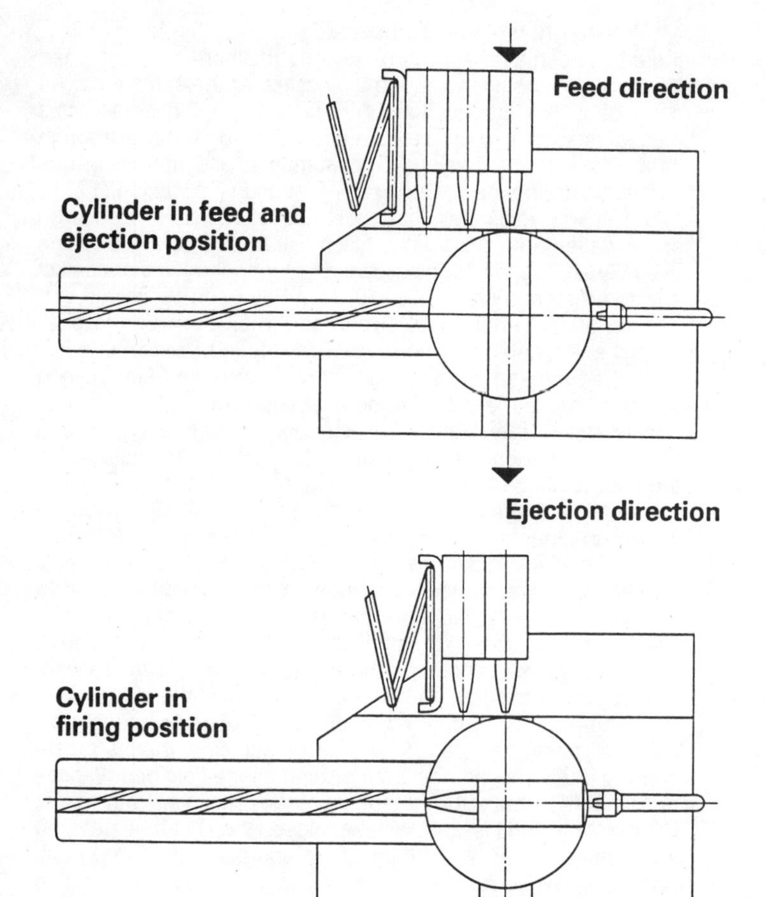

Simplified diagram of the G11 operating mechanism. Note the repositioning of the firing pin and the exhaust port.

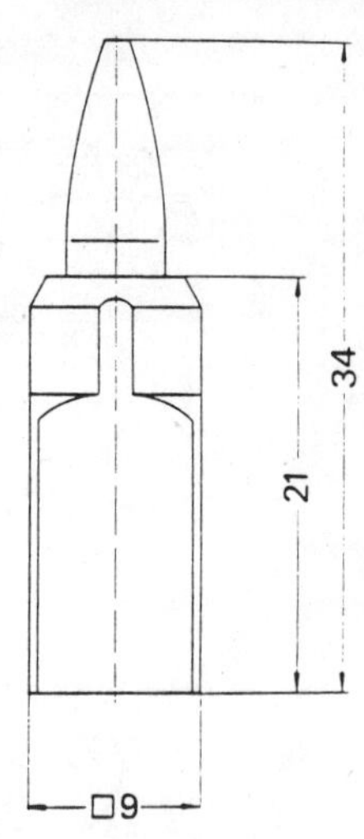

The 1982 version of the G11 cartridge. Note the square cross section.

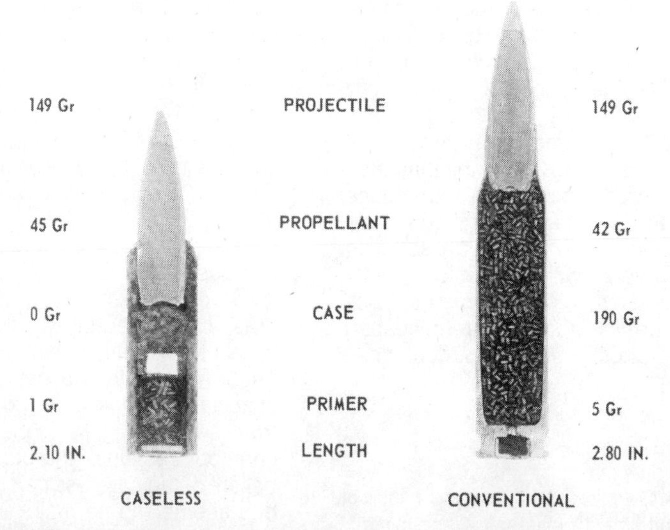

Experimental United States caseless 7.62mm cartridge from the late 1960s. Note the round cross section.

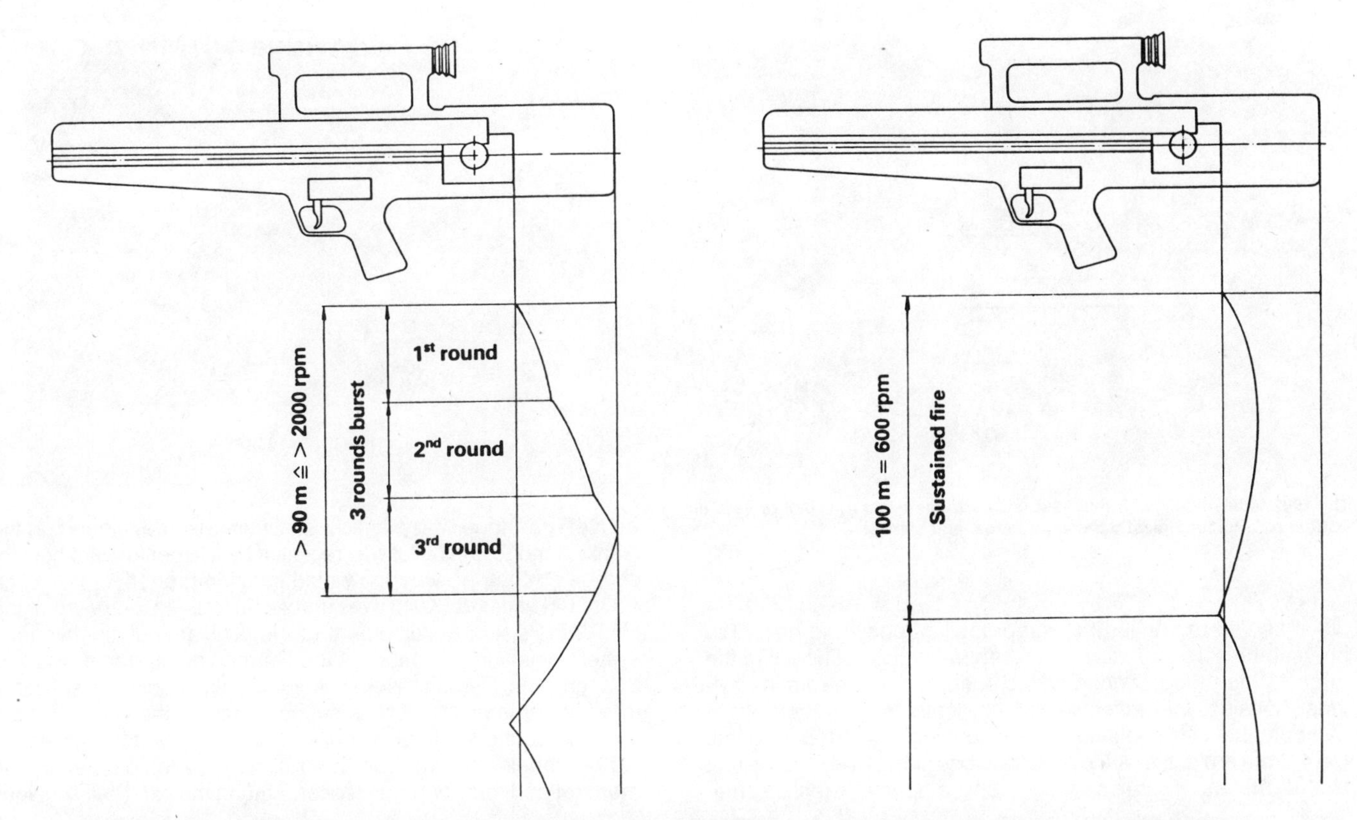

Simplified diagram of the recoiling parts of the G11. Note that the operating mechanism travels once to the rear during the three-shot burst and to the rear with each shot during sustained automatic fire.

the perfection of a caseless ammunition and weapon system in infantry calibers.

These problems challenged the engineers at Heckler & Koch. In 1969, they began to experiment with mechanisms that might be employed to transport, feed, contain, and seal caseless cartridges during the shooting cycle. Their first attempts were essentially the conversions of conventional self-loading mechanisms with linear bolt movements. These first steps sufficiently demonstrated the need for a new type of operating mechanism. From these experiments emerged the cylinder bolt of the G11.

The heart of the G11 is a cylindrical breech piece (*walzenverschluss*) that rotates about an axis at right angles to the bore of the barrel. A cartridge chamber was created by boring a hole across the line of the axis of the breech. When the weapon is loaded, the chamber is aligned with the axis of the barrel. The firing pin in the current models serves a dual function; it detonates the cartridge primer and acts as an obturator that seals off the breech end of the chamber. During the reloading cycle, the breech cylinder rotates ninety degrees clockwise. When it comes to a halt, a cartridge is fed into the chamber, and a cover plate opens on the underside of the rifle allowing any propellant residue or gas to be vented downward. Since it was first introduced in 1973, the cylinder breech has benefited from several refinements. In the earliest versions, the firing pin was mounted at ninety degrees to the axis of the cartridge with the primer located on the downward-facing side of the cartridge.

By the time of the 1977 NATO small arms trials, Heckler & Koch had reached the fifth generation of its G11 prototypes. As noted previously, in the course of those technical trials, premature discharges caused by overheating of the cartridges in the cylinder breech forced the Germans to withdraw the G11 and its ammunition. The Dynamit Nobel cartridges used in 1977 were molded from a nitrocellulose-based propellant (NC), cook-off temperature for which was about 178° C. This temperature was often reached after as few as seven or eight shots.

The 1976 vintage 4.7 × 21mm cartridge had an overall length of 32.5mm (from base of cartridge to tip of projectile), as compared to 57.4mm for the 5.56 × 45mm round. This version of the 4.7 × 21mm OH cartridge had an octagonal cross section; across the widest flat it measured 11.2mm and across the other major flat 8mm. The molded propellant was 20.9mm long; hence, the 21mm designation. The projectile had a diameter of 4.92mm, and the polygonally rifled barrel of the G11 measured 4.74mm across the lands (the high point of the rifling).

The propellant started out as loose powder of a predetermined shape and particle size. After surface treatments and moisture control but without the addition of adhesives or binders, it was compressed under very high pressure into the shape of a cartridge cut in half lengthwise. A prefabricated booster charge was inserted between two halves of the molded propellant before they were sealed together into a single unit by a solvent. The two booster charges served different functions. The one located in front of the primer unit served as an anvil against which the primer could be detonated. A larger charge located behind the bullet acted to disintegrate the propellant charge so that a large surface area was exposed to combustion. The gas pressure curve of this type of cartridge was similar to that of metallic case cartridges with like propellant weights. The entire cartridge was coated with a protective finish (a methatrylade resin) to stabilize it mechanically and environmentally. Its essentially rectangular shape allowed for more efficient packaging in the magazine.

The 3.4-gram (52.5-grain) bullet used in the 4.7 × 21mm OH cartridge was long—23mm—to provide the maximum sectional density. Its ogive offered optimum exterior and target ballistics. Coated inside and out with gilding metal, the projectile had a steel jacket and a lead core. The deep groove at the rear of the

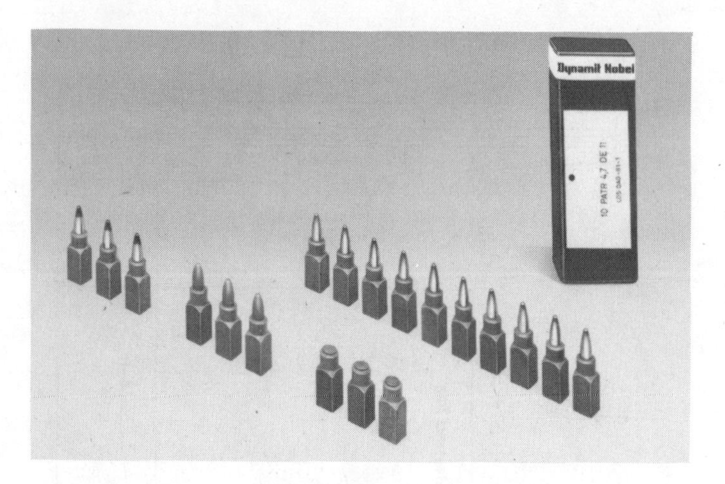

The 1982 family of 4.7 × 21mm caseless cartridges. *Top row,* ball projectiles; *bottom row,* tracers, plastic training rounds, and blanks.

The G11 in 1982.

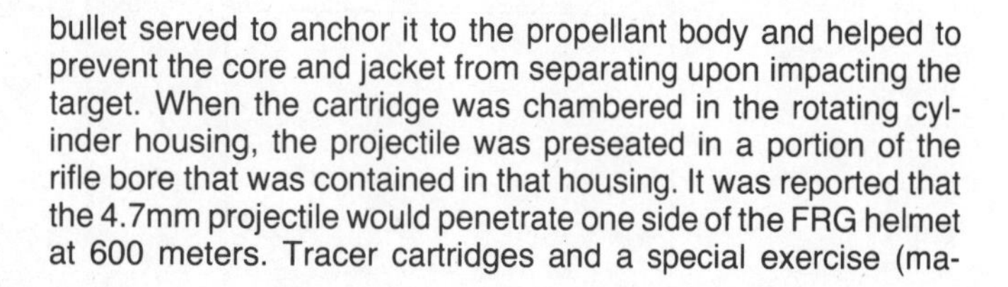

bullet served to anchor it to the propellant body and helped to prevent the core and jacket from separating upon impacting the target. When the cartridge was chambered in the rotating cylinder housing, the projectile was preseated in a portion of the rifle bore that was contained in that housing. It was reported that the 4.7mm projectile would penetrate one side of the FRG helmet at 600 meters. Tracer cartridges and a special exercise (maneuver) cartridge with plastic bullet were also developed. In the blank cartridge, the bullet was replaced by a larger forward booster charge so that the weapon would function properly.

Until Heckler & Koch was forced to pull its entry from the NATO trials, the development of the G11 ammunition-weapon system was a joint venture of the German government, Heckler & Koch, and Dynamit Nobel. After the disappointing showing, the government withdrew its support, leaving the two manufacturers to continue the G11 project with their own funds. Heckler & Koch engineers went on to build more prototypes, each one more refined than its predecessor. Among the externally obvious

G11 Rifle

Left side view of the G11.

changes was the shift from a complicated cocking level assembly beneath the barrel to a simpler one mounted on the left side of the rifle's housing behind the trigger guard. More significant were the internal changes in design that simplified the mechanism and improved the transfer of heat away from the cartridge chamber. Exact details of these changes are carefully guarded by Heckler & Koch because they are in large part the reason for the weapon's improved performance.

After the 1977 trials, the ammunition firm, Dynamit Nobel, set to work on a high-ignition-temperature propellant (HITP) with a cook-off point at least 100° C. higher than that of the earlier NC cartridges. By the time Dynamit Nobel had readied this new ammunition, Heckler & Koch had completed the tenth-generation G11 prototype. The performance of this rifle and the HITP ammunition was markedly improved in terms of both resistance to cook-off and dispersion control, so improved that the G11 again captured the fiscal support of the German government. It is important to note that because of their investment of private resources, Heckler & Koch and Dynamit Nobel have a proprietary control over the exploitation of the G11 technology. By the winter of 1981, Heckler & Koch had reached prototype generation 13.

Generation 13

According to Heckler & Koch, the cylinder breech of the G11 has several advantages over conventional linear bolt mechanisms. These include (1) simplified design with a minimum number of parts; (2) shortened overall weapon length with a longer barrel for the given weapon length (it is not necessary to accommodate the linear travel of a bolt); and (3) the short bolt travel and the simple feed process, which make very high rates of fire possible. There are also other features unique to the G11.

All of the mechanical parts of the weapon, with the exception of the trigger, the firing-mode selector, and the cocking handle, are enclosed in the weapon's sheet metal housing. All of these parts float within that housing. Upon the detonation of the primer and the exit of the projectile from the muzzle, the entire operating mechanism—including the magazine above the barrel—recoils to the rear. At the same time, the cylinder breech rotates 90 degrees to pick up another cartridge. In the three-shot mode, the hammer initiates firing of the second and third cartridges as soon as the reloaded breech is properly aligned. During the three locking and firing cycles, the recoiling parts move to the rear with ever-increasing velocity. The shooter experiences only a single recoil impulse, and that impulse is not as great as that experienced with existing 5.56mm rifles. Still, the recoil of the three-shot burst is greater than that of the G11's single-shot and is distinctive when compared with its continuous-fire mode. The weapon's mechanism returns to the original point of aim relatively easily after each three-shot burst.

When continuous fire is indicated on the ambidextrous selector lever, the hammer remains cocked after the first round has been fired and the weapon has completed a full recoil stroke. Therefore, the rate of fire is not determined by the movement of the breech mechanism; it is controlled by the independent recoil mechanism. The G11 is very easily controlled when fired at the continuous rate of 600 shots per minute.

Möller and his fellow engineers have taken considerable care in the packaging of their G11 operating mechanism, and they have designed the G11 with the infantry rifleman in mind. They sought to minimize the number of protuberances and the number of openings into which dirt, water, and other substances could enter and to eliminate moving parts. And the engineers wanted to produce a weapon that could be aimed instinctively from the shoulder or the hip.

The G11's sheet metal housing surrounds all of the operating parts, including the barrel, but excluding the 50-shot magazine that lies horizontally along the axis of the barrel. With the exception of the barrel's bore opening, the weapon is completely sealed from dirt and moisture; the trigger has a flexible seal, the magazine catch release button is sealed, and the exhaust port on the rifle's underside opens for only a fraction of a second after each shot is fired. The cocking handle, a circular device that rotates 360 degrees to charge the cylinder breech, cannot get in the shooter's way or snag on clothing or other things. This loading device does not rotate during the firing of the weapon. Above the barrel, the horizontal magazine reciprocates with the motion of the barrel, but it, too, is designed to keep foreign matter out of the interior of the rifle. The smooth exterior surface of the G11 offers some other advantages. The sheet metal housing has limited infrared and reflectivity (there is no exposed barrel to provide an IR signature). Additionally, it can be easily decontaminated in the event of exposure to chemical, biological, or nuclear agents. But the most important point for the user is the ease with which the G11 can be carried with one hand, possible because the weapon's center of gravity is slightly behind the pistol grip.

Heckler & Koch's engineers have included a single-power (1 × 1) optical sighting device on the G11 to assist the user with instinctive aiming. Since this sight aid does not magnify, the rifleman can keep both eyes open as he observes the field of fire. The sights currently mounted on the G11 have an illuminated reticle to help acquire targets under poor lighting conditions. To conserve battery energy, this moderately priced sight, which was designed so that it could be mass produced, has a timer in the circuitry to shut off the reticle illumination after three minutes.

4.7 × 21mm OH Ammunition

The G11 is only half the story. Without a reliable and perfected ammunition, the best mechanism would not be suitable as a military weapon. While the molding process for the newest Dynamit Nobel caseless ammunition is essentially the same as that described previously, the exterior and interior designs of the 1981 version of the caseless cartridge were significantly changed from the 1976 ammunition. Instead of having a rectangular cross section, the current ammunition is essentially 9mm square. The improved cartridges were 34mm overall, compared with 32.5mm for the 1976 round. It is impossible to improperly load the cartridge into the magazine because of the square cross section. Whereas the booster charge in the 1976 cartridge was loaded from the bullet end of the cartridge, in the 1981 type it is loaded from the rear. In comparison with the conventional metal cartridge case, the flat surfaces of the caseless cartridges allow it to be loaded into the magazine without the dead spaces. Heckler & Koch specialists also report that the caseless cartridges feed with less friction than that experienced with metal cases. A slightly flattened tip on the projectile aids its movement in the magazine.

Ballistic Performance

The Dynamit Nobel cartridge meets all major NATO criteria for rifle cartridges. Its effective range is in excess of the required 300 meters, and the trajectory elevation at that range (300 meters) is only 0.17 meters (NATO technical specifications call for 0.25 meters). To obtain optimum terminal ballistic performance, the very slender ogive and the long length were retained, which produced a high cross-sectional density (i.e., weight of the projectile divided by the diameter squared).* Because of its high cross-sectional density, the 4.7mm projectile has low sensitivity to crosswinds. For example, the bullet will drift only 0.275 meters at 600 meters in a 1-meter-per-second crosswind. The side drift of the M193 5.56mm ball projectile is about 0.338 meters, and it is even greater for the 7.62mm NATO bullet.

How good is the G11 and its ammunition from a ballistic standpoint? This will certainly be a topic of much debate over the next five years. In basic considerations, such as helmet penetration, the 4.7mm exceeds that of the United States M193 projectile. At the most common combat ranges (under 300 meters), the German cartridge is comparable to the Fabrique Nationale SS109 (XM855), but the 4.7mm cartridge produces a lower kinetic energy than either the M193 or the SS109 (XM855). Since there are still many military people who are not yet convinced that the 5.56mm cartridge is an acceptable substitute for the larger, heavier 7.62mm NATO round, Heckler & Koch personnel may have a difficult time convincing potential users and buyers of the G11 weapon-ammunition system that it has sufficient incapacitating power. Assuring the doubters of the 4.7mm's suitability will surely require a long and arduous process of teaching by the Heckler & Koch–Dynamit Nobel team.

It should also be noted that Heckler & Koch ballisticians, working with their colleagues at Dynamit Nobel, have tailored the terminal ballistics of their 4.7mm projectile so that it satisfies International Red Cross requirements regarding humane wounding effects. Fabrique Nationale took similar steps when they developed the SS109 projectile. The 4.7mm bullet will not distort in human targets, being relatively stable even at short ranges. Some future judges of the G11 system may well question if complying with such guidelines does not run counter to the goal of providing a weapon-ammunition system with maximum incapacitating capabilities. It is precisely the tumbling of the American 5.56mm M193 and the Soviet 5.45mm M74 projectiles that makes them so lethal and devastating. Being marginally stable as they pass through the air, they lose their stability and tumble when they strike the denser medium of human flesh. As they tumble, they tear the flesh, but more significantly they dump their kinetic energy quickly into the target. It is that sudden energy dump that results in serious incapacitating wounds or death. Very stable projectiles pass quickly through the target and do a minimum of traumatic damage from the release of kinetic energy. Humane bullets versus effective incapacitating projectiles is an issue that continues to demand closer scrutiny by military and civilian authorities.

Shooting the G11

At best, these technical details can give the potential user only a partial understanding of the unique character of the G11. When he picks up the new rifle for the first time, however, he will notice at once that it fits the shooting hand very well and that there are some significant differences in the G11. The magazine catch release, the magazine, and the cocking device are all located in unorthodox positions. After a few minutes of handling the weapon, they still may seem strange, but they are also very logically placed. Loading the weapon begins with withdrawal of the 50-shot magazine and loading the feed device with 25-shot charger clips. It is anticipated that the G11 will be issued with one sheet steel magazine. Fifty-shot (two 25-shot chargers back-to-back) loaders will be carried on armored personnel carriers. Men afoot on the battlefield will be issued 10-shot chargers. After two quick strokes, the magazine is loaded and inserted into the rifle. With the left hand, the shooter turns the charger clockwise a full 360 degrees. This loads the chamber and cocks the G11; it is ready to fire.

In the single-shot mode, the G11 behaves much like other small caliber military rifles. The perceived recoil is mild, and the sight is easily brought back on target. It may be disconcerting at first, however, not to have cartridges flying about and getting underfoot. Shooting the G11 in the three-shot burst mode is quite unique. At two thousand shots per minute, the three shots produce what appears to be a lengthy but single report. Unlike a product-improved M16A1 or the FNC, which both have three-shot burst control, the G11 will not give the shooter the sensation of three separate shots being fired.* And, the three rounds are gone before the user has a change to worry about keeping the weapon on target; the recoil is that of a single shot. To the author, the recoil from a three-round burst felt roughly equivalent to the recoil of a single shot fired from the Colt M16A1 or the Fabrique Nationale FNC. Although quite comfortable, the recoil of the three-shot burst was appreciably greater than the recoil of a single shot from the G11 or of the individual shots of the continuous-fire mode. The moderate rate of the continuous-fire cycle (600 shots per minute) allows the shooter to control the rifle from either a sandbag rest or the offhand position.

During the shooting demonstration at Fort Benning and during the author's firing experience in Oberndorf, an improved version of the nitrocellulose-type (NC) caseless cartridge was used. The manufacturers are saving the high-ignition-temperature propellant cartridge for more serious shooting, and they are continuing to guard this HITP ammunition from possible competitors who might wish to analyze its composition. Even with the NC ammunition, there were no cook-offs. At Fort Benning, several magazines were fired from three different rifles. At Oberndorf, one G11 (serial number 64) was fired for more than 200 shots. The shooting varied from a series of single shots to a number of three-shot bursts and several long strings of continuous fire. The Heckler & Koch test shooters reported that there was no reason to worry about cook-offs, and the HITP ammunition would produce even lower probabilities of premature firings. During the Fort Benning and the Oberndorf firing sessions, there were no malfunctions. It is readily apparent that Heckler & Koch has made significant improvements in the performance and safety of the G11 since it was tested by NATO in 1977.

Marketing the G11

Heckler & Koch military sales personnel realize that selling the G11 will be a challenging task. With the recent standardization of the 5.56 × 45mm (SS109-XM855) cartridge as the second NATO infantry caliber, Heckler & Koch and Dynamit Nobel will have to convince the NATO armed forces that the 4.7 × 21mm caseless cartridge merits early consideration as a substitute for the 5.56mm NATO round. However, two steps are being taken that may assist the firms in their effort to sell the G11 and its ammunition: The German armed forces plan to test

*Sectional density (w/d^2) exercises an important influence on the projectile's ability to retain velocity as it travels toward the target. The air pressure retarding the projectile, which builds up in front of the bullet, is greater as the diameter of the projectile is increased. Likewise, the kinetic energy of heavier bullets is greater than it is for lighter bullets traveling the same velocity. The heavier a bullet is in proportion to its diameter, the higher its sectional density and the more efficient it will be in boring through the air. Density and velocity affect the projectile's performance in crosswinds.

*When set on full automatic, the M16A1 and the FNC tend to rise after six or eight rounds have been fired. For this reason, three-shot burst control units were developed for both weapons. When fired in this mode, the rifle can be held on target because the automatic cycle is interrupted before the muzzle begins to climb.

this weapon-ammunition system, and the United States Department of Defense has formalized an agreement that will allow the United States to share the Germans' new small arms technology. The US project is called the Caseless Ammunition Rifle System (CARS).

Development tests of the improved G11 began at the Meppen Proving Ground in the FRG in August 1981. Operational testing is scheduled to start at Hammelburg in early 1983. During the Hammelburg tests, about twenty-five G11s will be subjected to rigorous field trials over twelve to eighteen months. If the rifle and its ammunition are found satisfactory, standardization of the weapon for the *Bundeswehr* would take place in 1984. Limited fielding of the rifle with the *Fernspaher,* a special reconnaissance force, would begin in 1986. Because defense funds for procuring new equipment are limited, general fielding of the G11 would not begin until the late 1980s. Although the German Ministry of Defense is prepared to proceed unilaterally with the standardization and fielding of the new rifle, it hopes for cooperation on this project from its NATO allies.

In November 1981, United States Department of Defense officials began discussing the possibility of participating in the continuing development of Dynamit Nobel's caseless ammunition and Heckler & Koch's "mechanism technology". By the end of November, a draft memorandum of agreement (MOA) between the United States Department of Defense (DOD) and the German Ministry of Defense (MOD) was being circulated in Washington and Bonn. This MOA did not speak directly to the G11 because there was (and is currently) no official American "requirement" for an infantry rifle beyond the M16A1. In the absence of such a requirement, DOD officials can talk only about mechanisms and technology in the abstract. It is understood that the draft MOA involves possible American participation in technical assessments of ammunition and mechanisms (i.e., G11 prototypes), joint U.S.-FRG troop evaluations, the development of a joint U.S.-FRG requirement and specification for a caseless ammunition-weapon system, and finally, a joint project to select an advanced rifle based upon a joint U.S.-FRG requirement and specification.

American and German officials have different motives for pursuing this memorandum. The Americans want to become involved with the Germans at this point so they are not locked out of this revolutionary ammunition-weapon project. The Germans realize that an American decision to adopt a caseless ammunition-weapon would enhance their attempts to sell the system to other members of NATO. But discussions at senior DOD-MOD levels do not always reflect the prevailing opinions held by staffers at lower working levels of these two organizations. In the United States, especially, many people will have to be convinced that the caseless ammunition-weapon concept is viable. As part of an educational campaign, Heckler & Koch demonstrated the G11 weapon and ammunition at a number of American military installations during the spring of 1982. Failures to immediately sell the G11 to the Americans would not kill this development project. On the other hand, if the Americans decide to participate, the whole undertaking will receive an immeasurable boost. But then Heckler & Koch, Dynamit Nobel, and the German Ministry of Defense would have to beware of the Americans' general tendency to join a project and then try to dominate its management.

FABRIQUE NATIONALE FNC

Successor to the FN CAL, the 5.56 × 45mm FNC is major competition for both the M16A1 and the G11. The FNC has a bolt and bolt carrier assembly that is quite similar in concept to that of the Kalashnikov assault rifles. Its upper receiver is stamped sheet metal and its lower receiver is made from an aluminum casting. As with the PIP M16A1 rifle, the FNC has a 3-shot burst setting. It also has a full automatic setting. FNC designers decided to use the M16A1-type magazine to insure interoperability with the M16A1 and the M249 squad automatic weapon (Minimi). Sweden and Indonesia have adopted the FNC as their newest infantry rifle. Additional details are presented in Chapter 9.

HECKLER & KOCH G41

A redesigned version of the 5.56 × 45mm HK33 series, the Gewehr 41 incorporates a longer barrel (480mm [18.9 inches] versus 390mm [15.4 inches]); a M16A1-style handguard; redesigned trigger assembly and pistol grip (single-shot, 3-shot, and full automatic fire modes); dust cover for ejection port; bolt-hold open device, forward bolt assist; carrying handle; and new NATO standard telescopic sight mounting points. The G41 also uses the M16A1 magazine and can take the M16A1 bipod. Additional details are presented in Chapter 21.

THE SINGAPORE ASSAULT RIFLE (SAR 80)

Chartered Industries of Singapore Pte. Ltd., after nearly a decade of producing the American Colt M16A1 Rifle under license, is perfecting a 5.56mm selective-fire rifle and has under development a 5.56mm light machine gun. CIS, as this government-owned manufacturing company is known on the island republic, was created in 1967 to produce military equipment and mint coins. Experience with manufacturing the M16A1 prompted CIS and the government to embark upon a new rifle project to answer long-standing complaints that the Singaporeans had about the American rifle. Many of the raw materials for the weapon—receiver forgings and barrel blanks, as well as smaller finished components—had to be purchased directly from the licensor, Colt Firearms Division of Colt Industries. In addition, the government-to-government agreement that allows CIS to produce the M16A1 requires Singapore to obtain United States State Department approval of sales of the rifle to third parties.

Borrowing a Technology Base

In 1966, the Singapore Armed Forces decided to purchase 5.56mm M16A1 Rifles to replace their 7.62mm NATO-caliber L1A1 Rifles, which were mostly Australian-made. Their first attempt to acquire the American weapon was thwarted when the United States Congress opposed the sale of 20,000 M16A1s to Singapore on the grounds that they were needed more desperately by American and Vietnamese forces in Southeast Asia. After a lengthy negotiation, Colt and the government of Singapore signed a contract permitting the Singaporeans to manufacture up to 150,000 M16A1 Rifles in a facility they were building in the new Jurong Town industrial community on the southwestern coast of the island. Colt trained technicians from Singapore at their factory in Hartford, Connecticut, in 1969 to 1970, and supervisory personnel from Colt reported to Singapore in June 1970 to help set up the production tooling. In March 1972, CIS delivered their first production rifles (50 to 100 pieces).

For the most part, Chartered Industries' rifle factory utilized general-purpose tooling that could be adapted to other uses. By late 1972, the Singaporeans realized that the CIS factory could not be operated economically. Its large number of employees plus the limited quantities of rifles (about eighty thousand) needed by the Singapore Armed Forces combined to keep profits down. Colt was understandably not enthusiastic about allowing CIS to

The Fabrique Nationale FNC field stripped. Note similarity of the bolt and bolt carrier assembly to that of the Kalashnikov weapons.

The Heckler & Koch 5.56 × 45mm HK33 rifle.

The Heckler & Koch G41. Note the longer barrel, redesigned handguard, modified lower receiver, carrying handle, and M16-type magazine.

compete with them for sales throughout Asia, especially when the American company learned that the Ministry of Defense was willing to permit the sale of CIS-made M16A1 Rifles at cost or below to keep the Jurong factory active. When Singapore successfully negotiated the sale of sixty thousand M16A1s to Thailand, they shared the delivery evenly with Colt, but both sides were dissatisfied with the arrangement. Colt officials surmised that Singapore would not be bound by State Department limitations on the sales of Singapore-made M16A1s, giving Singapore a market to which Colt did not have access. And Ministry of Defense personnel knew that Colt could control Singapore's export activities by limiting the availability of raw materials for which they could be charged a premium price.

From their experience with manufacturing the M16A1, CIS reasoned that they had the expertise to produce some other weapon—ideally a weapon that did not require limited-access raw materials and one that could be fabricated with their existing production tooling. Working toward this goal, CIS arranged for engineer Frank Waters of Sterling Armament Ltd. of Dagenham, Essex, in the United Kingdom—holder of production rights for the AR-18, a product of the firm that had created the M16 (AR-15)—to develop a modified version of the Armalite AR-18 5.56mm Rifle. During 1978, CIS was turning out prototypes of the Armalite-Sterling-Waters weapon. These first rifles were field-tested by the School of Infantry Weapons of the Singapore Armed Forces Training Institute (SAFTI), which found that the prototypes demonstrated problems of the type that could be expected with a new rifle.

Despite moderate troop enthusiasm for the Singapore Automatic Rifle (SAR), CIS went on with its program to improve the weapon. Early in 1980, about twenty-eight SAR 80's were gunsmithed together for additional testing. Since that time, CIS personnel have continued to work out the bugs in the SAR 80's design. Plans call for an initial production lot of several thousand rifles.

Design Features

From a design standpoint, there is little new associated with the SAR 80. As with its direct ancestor, the AR-18, the Singapore rifle has a receiver made of stamped sheet metal. (The M16A1 upper and lower receiver components are machined from a special set of aluminum forgings.) A number of other components, such as the gas system regulator-front sight assembly and the flash suppressor, are produced by investment casting. At least 40 percent of the SAR 80's parts—screws, springs, and pins—can be purchased from small machine shops. The SAR 80 can use either of the M16A1 magazines; the 20-shot version is fabricated in Singapore, the 30-shot model is not. SAR 80 ammunition (M193 Ball and M196 Tracer) and the bayonet (the M7) are the same as those used with the Singapore M16A1. Additional details are discussed in Chapter 40.

STEYR AUG

Steyr-Daimler-Puch, in conjunction with the Austrian army, developed the *Armee Universal Gewehr* (AUG) in the mid-1970s as a 5.56 × 45mm replacement for the 7.62 × 51mm NATO caliber Stg. 58 (FAL). Initial serial production began in 1978 with

Sterling Armament Company's AR-18 5.56 × 45mm rifle.

Chartered Industries of Singapore 5.56 × 45mm SAR 80 assault rifle, which evolved from the AR-18.

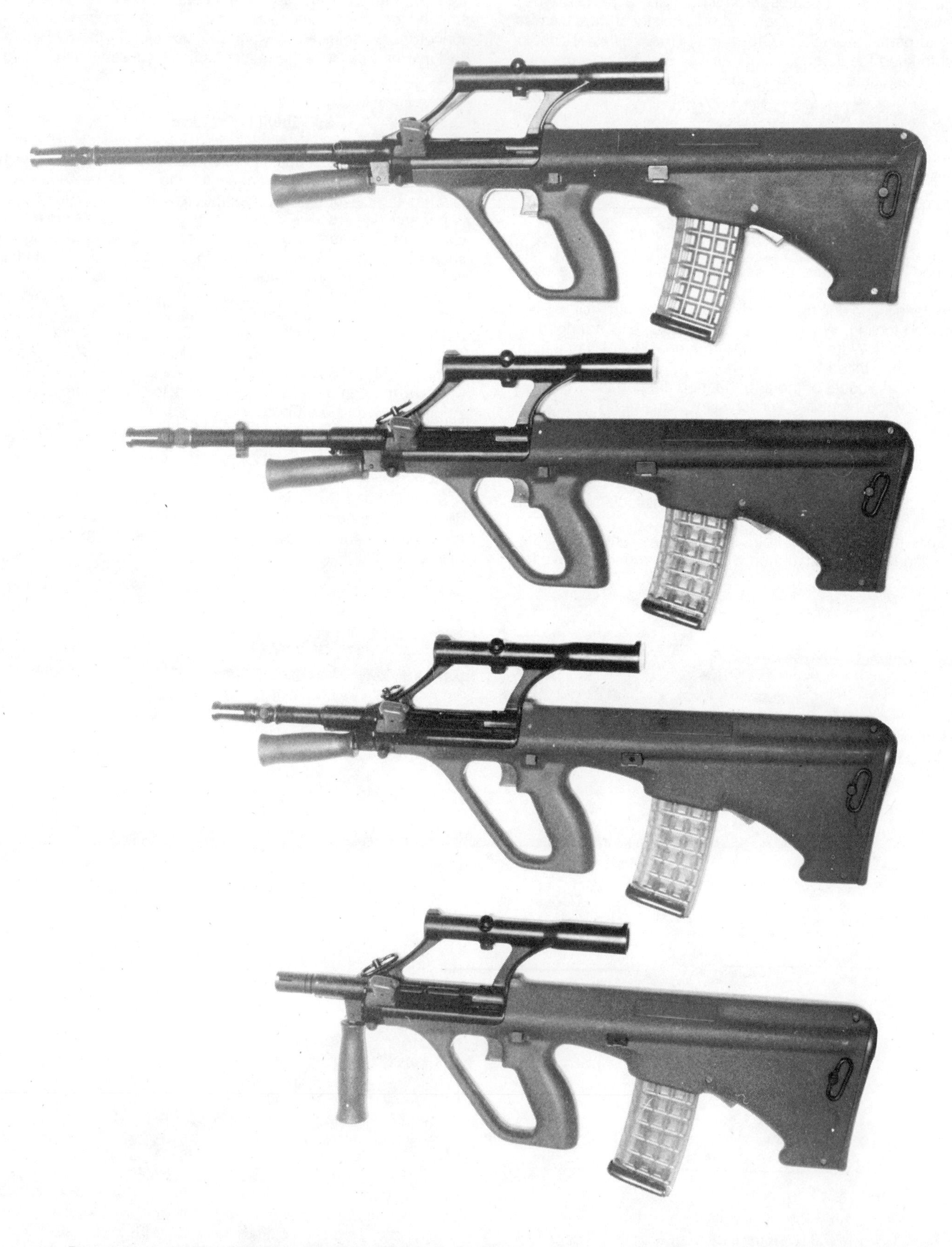

Four versions of the Steyr Armee Universal Gewehr (AUG). ***Top to bottom:*** **light machine gun version with 610mm (24-inch) barrel; rifle variant with 508mm (20-inch) barrel; carbine with 407mm (16-inch) barrel; submachine gun with 305mm (12-inch) barrel.**

the first rifles being consigned to Austrian troops. Steyr has carried out an intensive sales campaign in an effort to sell this rifle to other unaligned nations. There are indications that Steyr has met with some success in the Middle East, especially in the United Arab Republic (Egypt). Attempts to sell to other countries (such as the People's Republic of China) have been opposed by the Soviets who argue that the Austrians must remain neutral under the post–World War II peace treaty.

The Steyr AUG is available in four different barrel lengths: submachine gun, carbine, rifle, and light machine gun. This weapon is also notable for the extensive use of synthetic (plastic) components including a see-through magazine that allows the shooter to tell how many rounds he has remaining. More details are given in Chapter 8.

LEADER DYNAMICS 5.56 × 45MM RIFLE

This 5.56 rifle was developed by Leader Dynamics, a relatively new arms firm established in Smithfield, New South Wales, Australia. Charles George of Sydney was the designer. The Leader Dynamics operating mechanism is quite similar in concept to the Armalite AR-18. There is a bolt and bolt carrier assembly like the AR-18, but the Leader Dynamics rifle bolt has three locking lugs instead of the eight lugs on the AR-18 bolt. Unlike the AR-18, the Leader Dynamics rifle has the operating rod handle mounted near the front sight. A very sturdy weapon, the Leader Dynamics rifle is expected to be a competitor in the forthcoming Australian 5.56mm rifle trials. Other serious contenders are the Colt M16A1 and the Fabrique Nationale FNC. For more details see Chapter 7.

SWISS RIFLES

The Eidgenossische Waffenfabrik, Bern (W+F) has been engaged in small arms design since the late nineteenth century. During that time, the Waffenfabrik has more or less been in competition with the commercial arms factory SIG Neuhausen for production of Swiss military arms. The W+F rifles currently being developed have an operating mechanism conceptually inspired by the Kalashnikov bolt mechanism. There are two calibers being tested, 6.45 × 48mm and 5.60 × 45mm, and two barrel lengths, 850mm (33.46-inch) and 1010mm (39.76-inch). The short barrel 6.45mm weapon is called the MP E21, and MP C41 in 5.60mm. The rifle versions are SG E22 and SG C42 respectively.

COMPARATIVE DATA FOR W+F AMMUNITION

	MP E21	SG E22	MP C41	SG C42
Caliber:	6.45 × 48	6.45 × 48	5.60 × 45	5.60 × 45
Muzzle velocity m/s:	770	990	780	900
(f.p.s.):	(2347)	(3018)	(2377)	(2743)
Energy O J:	1868	2552	1248	1660
(ft. lbs.):	(1378)	(1882)	(920)	(1224)
Weight of projectile:	6.3	6.3	4.1	4.1
(grains):	(97)	(97)	(63)	(63)

In mid-1983, the Swiss Army adopted a much modified SIG 542 rifle, the SG541 in 5.6 × 45mm, as the *Sturmgewehr* 90. Over the next twenty years, 600,000 of these rifles will be manufactured for the Swiss Army.

CETME MODELO "L"

This is a scaled-down version of the 7.62 × 51mm CETME rifle. The model "L" fires the M193 5.56 × 45mm, and it is available in a fixed butt stock version (400mm barrel) and a sliding stock version (320mm barrel). This weapon has been tested by the Spanish armed forces, but to date only small pilot quantities have been manufactured. See Chapter 42 for additional details.

FUSIL DE ASALTO ARGENTINO MODELO 81

Development of the FAA Modelo 81 began in April 1976 at the Fabrica Militar de Armes Portatiles "Domingo Matheu," at Rosario. Development was completed in March 1981 and in the spring of 1982 work began on a pilot lot of fifty rifles that will be used in comparative trials of 5.56 × 45mm rifles. Other leading contenders include the Fabrique Nationale FNC and the Colt M16A1. It is anticipated that the FAA Modelo 81 will be used to supplement the 7.62 × 51mm NATO FAL rifle. If and when this rifle goes into production (not before 1987), it will probably be issued to airborne and commando forces.

The FAA Modelo 81 incorporates a bolt, bolt carrier, and gas piston assembly similar in concept to the Kalashnikov family of rifles. Influence from the design of the Fabrique Nationale FAL, which is manufactured under license in Argentina, can be observed in the style of the upper and lower receivers and in the mechanism of the side-folding stock. The designers of the FAA Modelo 81, CETME Modelo "L", the Waffenfabrik Bern rifles, and the Leader Dynamics rifle all have made extensive use of sheet metal stampings, and synthetic materials have been used for the stocks and handguards. For more details on the FAA Modelo 81, see Chapter 6.

VARIATIONS ON A THEME: NEW VERSIONS OF THE KALASHNIKOV ASSAULT RIFLE

In recent years, American defense officials have been discussing plans for including a capability of qualitative upgrading of materiel after its initial introduction into service. Called preplanned product improvement (P³I), this concept is intended to increase the service life of future systems and to reduce the costs involved in acquiring enhanced capabilities. If adaptability is a significant attribute of weapons systems, then the small arms of the Soviet Union should be examined as an excellent case in point. Nearly forty years have passed since Mikhail Timofeyevich Kalashnikov introduced the first prototypes for his famous assault rifle, and new variants of this basic weapon concept are still being developed within the Eastern Bloc. In this section is a pictorial roundup of some of the more recent additions to the growing family of Kalashnikov-derived weapons.

During the past thirty-five years three different approaches have been taken to the manufacture of the Kalashnikov assault rifles. From 1947 to about 1950, the Soviets tried unsuccessfully to manufacture a rifle, the receiver assembly of which combined sheet metal stampings and machined steel components. These early Kalashnikovs were not satisfactory because the rivets which held the components together were not reliable. Sometime after 1950, the Soviets introduced a Kalashnikov rifle with a receiver assembly that was machined from a steel forging. The second variant was widely copied in the Eastern Bloc and is the rifle that is usually called the AK-47. In 1959, the Soviet Army introduced the *Modernizirovannyi avtomat Kalashnikova* (AKM), which was a new design of a stamped sheet metal receiver. All of the subsequent variants on the Kalashnikov assault rifle have been built around the AKM receiver assembly.

Leader Dynamics Mark V 5.56 × 45mm automatic rifle.

SIG SG541 rifle, the new Swiss Army *Sturmgewehr 90.*

CETME 5.56 × 45mm Modelo "L" assault rifle.

Argentine FAA 81 5.56 × 45mm assault rifle.

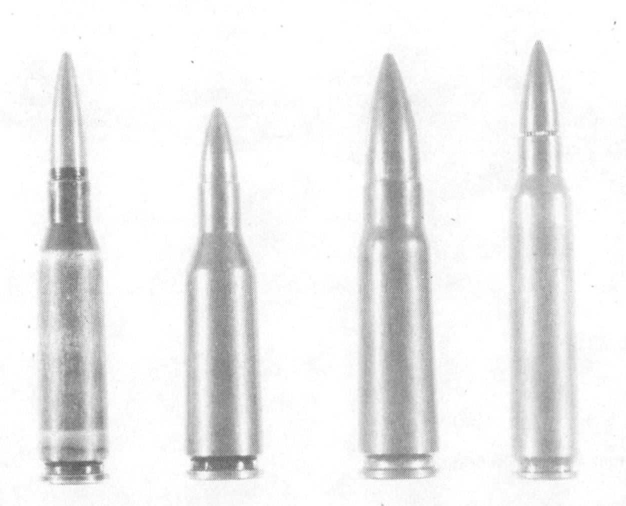

From left to right: A comparative view of the Soviet 5.45 × 39mm M74, the 5.6 × 39mm target and hunting round, the 7.62 × 39mm M43, the US 5.56 × 45mm M193.

The 5.45 × 39mm AKS74 with side-folding stock.

Both the AK74 (*top*) and the RPKS74 fire the 5.45 × 39mm cartridge.

AK-74 Assault Rifle

This new rifle is of singular importance to the European military scene. After twenty-five years of cartridge standardization at the squad level within the Eastern Bloc, the Soviets made the decision in the mid-1970s to replace their 7.62 × 39mm family of weapons—AKM assault rifle and RPK light machine gun—with new versions of those weapons firing a 5.45 × 39mm family of cartridges. When the first reports of the new cartridge and weapons reached western analysts, they assumed that this new materiel would be utilized only by specialized units because of the cost and logistical problems involved in switching calibers and retiring the 7.62 × 39mm caliber weapons. Subsequent observations indicated that the Soviets planned a complete changeover to the smaller caliber. This has caused considerable speculation in the West as to possible tactical implications of the new ammunition.

The muzzle velocity of the Soviet 5.45mm projectile is 900 meters per second, compared to the 947.5 meters per second of the NATO standard SS109 bullet. The muzzle energy of the Soviet bullet is 1,385 joules, versus 1,796 joules for the SS109. As the accompanying chart indicates, the new Soviet 5.45mm cartridge has the lowest energy level of the four cartridges compared. This limited energy level raises some questions about the lethality of this projectile and what the Soviets were attempting to accomplish when they adopted this new caliber ammunition.

The 5.45mm Soviet projectile has a considerably flatter trajectory than the Soviet's 7.62 × 39mm ammunition, which should make it easier to hit targets out to 400 meters. But what will it do to the target when it hits it? The AK-74 rifle has a barrel with tight rifling (1mm to 203mm) twist so that the bullet is spun very rapidly. As a consequence, the bullet is very stable during its entire flight. That stability ends when the bullet enters a medium denser than air—such as the human body—because of its con-

Romanian version of the 7.62 × 39mm AKMS with folding stock.

East German 7.62 × 39mm KmS 72 version of the AKMS.

struction. The basic ball projectile of the 5.45mm cartridge has a gilding metal-plated mild steel jacket. Inside that jacket there is a lead sheath that surrounds a 15mm-long mild steel core. That core is mounted with its base at the base of the jacket. At the nose of the steel core is a 3mm lead plug, which is actually an extension of the lead sheath. In front of the lead plug is an air space about 5mm long. By placing the center of gravity of the projectile toward its rear, the bullet's designer ensured that the bullet would flip when it hit a human body. When the bullet begins to tumble, it rapidly dumps its remaining energy into the target. Lethality is the product of the amount of energy deposited into the target over time. Bullets such as the 7.62mm NATO round tend to be less lethal than the smaller high velocity projectiles because they exit the body carrying with them considerable residual energy. The small caliber bullets deposit most of their energy in the body causing substantially more damage to the target. Individuals who have experimented with the Soviet 5.45mm cartridge by firing it into gelatine blocks have noted that it makes a mess of the target blocks. Some of these people have expressed the hope that the International Red Cross will investigate the humaneness of the 5.45mm Soviet round.

The AK-74 assault rifle itself is essentially the same as the AKM, with the exception of a modified bolt head, which accommodates the smaller diameter case head of the 5.45 × 39mm cartridge, and a new and improved extractor, a new fiberglass reinforced plastic magazine, and very effective muzzle brake. Based upon fluidic principles of gas flow, this muzzle brake not only reduces the blast and noise levels produced when the weapon is fired, but also reduces recoil by counteracting the recoil with a forward and downward movement. As a result, this muzzle device is probably the most effective small caliber recoil reducer ever employed. The AK-74 is currently issued in two forms: one with a standard fixed wooden or plastic stock and one with a folding stock patterned after that used on the RPKS. The latter is designated AKS-74. For additional details see Chapter 46.

RPKS-74 Light Machine Gun

The Soviets have also issued a squad automatic weapon, *Ruchnoi pulemet Kalashnikov* (Kalishnikov light machine gun), chambered to fire the 5.54 × 39mm. Again, with the exception of a modified bolt, new magazine and new muzzle brake, this weapon is the same as the 7.62 × 39mm RPKS. A quick comparison of the muzzle brakes used with the AK-74 and the RPKS-74 disclose their differences. The muzzle device used on the RPKS-74 is patterned after the one used with the United States M16A1 Rifle. Accompanying photographs illustrate the operation of the RPKS-74 folding stock. For additional details see Chapters 2 and 46.

New Romanian Weapons

The Romanian armed forces have introduced three new versions of the AKM assault rifle; a folding stock AKM, a Romanian version of the RPK, and a Romanian-designed sniper rifle built around the operating mechanism of the RPK. These weapons are significant because they demonstrate a domestic small arms design capability in Romania and because at least the sniper rifle appears to have been created for the export trade. All of the new Romanian weapons begin with the basic AKM assault rifle mechanism. The Romanians have manufactured the AKM since the late 1960s, but only recently have they introduced the folding stock version or AKMS. All Romanian AKMs are distinguished by the forward pistol grip, which is fabricated from laminated wood.

Romanian RPK. The Romanian version of the RPF is essentially the same as the standard 7.62 × 39mm Soviet RPK. In the RPKs of both countries a heavier and longer barrel (591mm versus 414mm) has been substituted for the standard AKM barrel. Both RPKs can use a variety of magazines (30-, 40- and 75-shot) and both have bipods. Only the bipod is significantly different. The legs of the Romanian bipod can be adjusted for height, and the method of attaching the bipod is different.

Romanian Sniper Rifle. Technologically the Romanian Sniper Rifle is more interesting than the other two weapons because it fires the M1891 Russian cartridge, which is 15mm longer than the 7.62 × 39mm. Because the bolt of the AKM travels 30mm farther to the rear than is necessary to accommodate the 7.62 × 39mm cartridge, the Romanian designers were able to modify the standard AKM-type mechanism to fire the more powerful 7.62 × 54mm rimmed cartridge. First, they altered the bolt face to accommodate the larger rimmed base of the M1891 rifle cartridge, and they added a new barrel and lengthened the RPK-type gas piston system. The gas system of the Soviet Dragunov sniper rifle is more like that of the World War II Tokarev gas system than the Kalashnikov-type. Second, the Romanians developed their own 10-shot magazine, and they fabricated a skeleton stock from laminated wood. This butt stock, with its molded cheek rest, is probably slightly superior to the one used on the Dragunov. Third, the Romanians have riveted two steel reinforcing plates to the rear of the receiver to help absorb and spread the increased recoil of the more powerful M1981 cartridge. Finally, the Romanian designers have attached a muzzle brake of their own design. The standard AKM wire cutter bayonet can be used with this muzzle attachment. Perhaps the single most interesting feature of this Romanian sniper rifle is the markings on the telescope. The use of "right" and "left" in English on the windage markings seems to indicate that the Romanians plan to export this rifle. For additional details, see Chapters 5 and 39.

Yugoslavian Sniper Rifle

The Yugoslav arms designers have also introduced a sniper rifle based upon the AKM mechanism, but they have chambered it for the 7.9 × 57mm Mauser Rifle cartridge which they also use in their company machine gun. For additional details see Chapters 5 and 50.

German AKMS

The East German Army introduced a folding stock version of their AKM in 1972, but it was identified only recently by Western intelligence analysts. This weapon, the MPiKS 72, has a unique folding stock made of heavy gage wire. Unlike most folding stocks, this one swings to the right and lays along the right side of the weapon instead of the left. For additional details see Chapter 20.

The proliferation of Kalashnikov-type weapons—there is also the PK medium machine gun, which was introduced in 1961 and the new 12.7mm armor machine gun, which is believed to be a Kalashnikov design—indicates the flexibility and adaptability of Kalashnikov's basic rifle design. These variations on a basic theme also indicate that weapons can be given extended life if they can continue to be adapted to tactical requirements. The Soviets seem to have mastered the concept of preplanned product improvement.

COMPARATIVE DATA FOR SELECTED 5.56mm (AND SMALLER CALIBER) RIFLES

Model	Caliber	Barrel mm	Length (inches)	Overall mm	Length (inches)	Weight (Without Magazine) kg	(pounds)	Average Muzzle Velocity m.p.s.	(f.p.s.)
M16A1	5.56 × 45	508	(20)	990	(39)	3.3	(7.3)	991.75	(3250)
AR-18	5.56 × 45	457	(18)	965	(38)	3.1	(6.6)	990	(3250)
Stoner Rifle (XM22)	5.56 × 45	500	(20)	1006.3	(40.25)	3.6	(7.9)	991.75	(3250)
Stoner Carbine (XM23)	5.56 × 45	392.5	(15.7)	896.9	(35.9)	3.5	(7.7)	915	(3000)
HK33A2	5.56 × 45	390	(15.4)	920	(36.25)	3.65	(8.05)	920	(3018)
AR70/.223	5.56 × 45	442.5	(17.4)	925	(36.4)	3.4	(7.5)	969.9	(3182)
SIG 530	5.56 × 45	460	(18.1)	1005	(39.59)	3.55	(7.8)	970	(3183)
SIG 540	5.56 × 45	460	(18.1)	950	(37.4)	3.26	(7.18)	980	(3215)
Ruger Mini 14/20GB	5.56 × 45	470	(18.5)	984	(38.76)	2.89	(6.38)	1058	(3471)
FA-MAS	5.56 × 45	488	(19.23)	757	(29.82)	3.38	(7.45)	960	(3150)
MKS	5.56 × 45	467	(18.4)	868	(34.2)	2.75	(6.06)	9.75	(3200)
MKR	4.5 × 26RF	600	(23.64)	840	(33.1)	3.0	(6.6)	1020	(3347)
U.K. 4.85 I.W. XL64	4.85 × 49	510.3	(20.1)	757.5	(29.85)	3.09	(6.81)	900	(2953)
U.K. 5.56 I.W. XL70	5.56 × 45	518	(20.4)	770	(30.33)	3.12	(6.88)	900	(2953)
Steyr AUG	5.56 × 45	508	(20.02)	790	(31.13)	3.6	(7.93)	ca990	ca(3250)
Galil	5.56 × 45	460	(18.12)	979	(38.57)	3.9	(8.6)	980	(3215)
Taiwan Type 68	5.56 × 45								
FNC	5.56 × 45	450	(17.73)	997	(39.28)	3.8	(8.38)	9.65	(3166)
Valmet M82	5.56 × 45			710	(27.97)	3.3	(7.27)		
HKG11	4.7 × 21	540	(21.27)	750	(29.55)	3.6	(7.93)	930	(3051)
AK74	5.45 × 39	400	(15.76)	930	(36.64)	3.6	(7.93)	900	(2953)
Leader T2-MK5	5.56 × 45	410	(16.15)	910	(35.88)	3.4	(7.5)	975	(3200)
CIS SAR 80	5.56 × 45	459	(18.08)	970	(38.22)	3.7	(8.16)	970	(3183)
FAA 81	5.56 × 45	440	(17.34)	980	(38.6)	3.9	(8.6)	ca970	ca(3183)
HKG41	5.56 × 45	480	(18.9)	997	(39.3)	3.6	(7.94)	ca990	ca(3250)

2 Machine Gun Development

Machine guns can be as confusing a topic as rifles. Armed forces have used machine guns in varying roles since the beginning of this century, and during the past three decades the old distinctions among light, medium and heavy machine guns have blurred considerably. Before the 1939–1945 war, weapons like the UK Vickers and the US Browning, chambered for their respective rifle cartridges, were called heavy machine guns because they were just that. They were used as base-of-fire weapons in essentially fixed positions. The mobile battle fronts of World War II saw these heavy guns become obsolete, as portability became the key consideration in machine gun development. Today, the term light machine gun (LMG) usually refers to a magazine-fed weapon, while medium machine guns (MMG) are generally belt-fed. This distinction can be bewildering when a belt-fed machine gun (e.g., the XM249) is actually lighter than the M1918A2 Browning Automatic Rifle of WWII days. The machine guns discussed in this chapter will be described according to caliber family, dropping the former LMG/MMG distinctions. One new term will be used—Squad Automatic Weapon (SAW), which refers to relatively lightweight magazine or belt-fed automatic weapons used to augment the firepower of the rifle squad. Differentiation between infantry and armored fighting versions of machine guns also will be maintained. To the four basic rifle calibers discussed in Chapter 1, we will add one—the 7.62 × 54R round used in Soviet machine guns.

7.62 × 51mm NATO CALIBER MACHINE GUNS

MG42 AND M60

In addition to the search for a standard rifle, NATO has also looked for a common machine gun. That effort has been just as unsuccessful. As a consequence, there are several major machine guns in the hands of NATO soldiers. The oldest pattern is the UK NATO caliber version of the Bren Gun—the L4 series. Next is the German MG3, an updated version of the WWII MG42 described in Chapter 22. This highly popular weapon is used by Austria, Chile, Denmark, West Germany, Iran, Italy, Norway, Portugal, Spain and Turkey. The MG42/MG3 design has had an impact on American machine gun design as well, its feed mechanism being found in a modified form on the US M60 Machine Gun. After WWII, the US Ordnance Corps undertook a machine gun development program. Out of two designs (the T52 developed by Bridge Tool & Die Co. and the T161 developed by the Inland Division of General Motors), the US Army and its contractors produced an American gun with a heavy Germanic accent. In addition to the MG42 feed system, the M60 had a modified version of the *Fallschirmjäger Gewehr* (FG42) operating mechanism. Whereas the FG42 was equipped to provide either automatic or semiautomatic fire, the M60 is full automatic only. On the other hand, its low cyclic rate allows the gunner to squeeze off single shots without too much difficulty. Standardized in February 1957 as a companion piece to the new NATO caliber rifle (M14), the M60 was produced on a pilot line basis at Springfield Armory and has been subsequently manufactured on a large scale by Maremont Corporation in Saco, Maine (225,000). Australia is also a user of the M60. The Nationalist Chinese Government on Taiwan currently produces the M60 machine gun with US-provided machinery, blueprints etc.

The M60 was the first US machine gun to have a true quick barrel change. A new version, the M60E1, was introduced after a few years to permit even easier barrel removal and to decrease the number of parts. This weapon differs from the M60 as follows: (1) The barrel does not have the bipod or gas cylinder attached to it; they are attached to the gas cylinder tube. (2) The bipod is attached semi-permanently to the rear of the gas cylinder. (3) The gas cylinder has been simplified and has no threads; it has a U-shaped key to retain the gas cylinder extension. (4) The operating rod guide tube has a lug that retains the gas cylinder and bipod on the weapon and eliminates the gas cylinder nut. (5) The modified spool type gas piston has no holes. (6) The modified rear sight has the lateral adjustment increased by 20 mils. (7) The modified die-cast feed cover eliminates parts and allows the cover to be closed whether the bolt is in the forward or cocked position. (8) The modified feed tray eliminates parts. (9) The magazine hanger fitted to the left side of the weapon eliminates parts and can be used with either a modified magazine or a modified bandolier. (10) The new diecast forearm eliminates parts and eases changing the barrel because the absence of a forearm cover allows the carrying handle to be fitted to the barrel. (11) The sling swivels have been relocated to the left side of the forearm and the top rear of the buttstock, improving the ease with which the weapon can be carried. (12) The carrying handle has been increased in diameter.

The infantry versions of the M60 are considerably lighter than the M1919A6 Browning Machine Guns of WWII vintage—10.5 kg (23.1 lbs.) versus 14.74 kg (32.5 lbs.), and only slightly greater than the 20-shot M1918A2 BAR, which weighed 8.8 kg (19.4 lbs.). Other models of the M60 include the M60C and M60D. Both of these weapons are used in aircraft armament roles. The M60C, with the stock removed, is remotely charged and fired. The M60D, which has spade grips and a trigger at the rear of the weapon, has been used as a door gun on US helicopters. And finally, there is the M60E2, which is used as a fixed (coaxial) tank machine gun by the US Marine Corps. This gun has a barrel extension and a gas evacuator tube protruding beyond the gas cylinder. The evacuator system carries the gun smoke forward and out of the armored vehicle. A modified M60E2 used in the Mechanized Infantry Combat Vehicle as its interim armament is designated the XM238.

M73/M219

Two weapons were developed in the 1950s to answer a June 1951 call for a new NATO caliber tank gun—the T153 (M37), an interim modification of the M1919A4, and the T197, which was to become the standard armor machine gun, designed by Richard Colby and Jack Lockhead at Springfield Armory. However, in June 1953, the T197 project was suspended until 1956.

UK L4A1 Bren Gun, 7.62 × 51mm NATO.

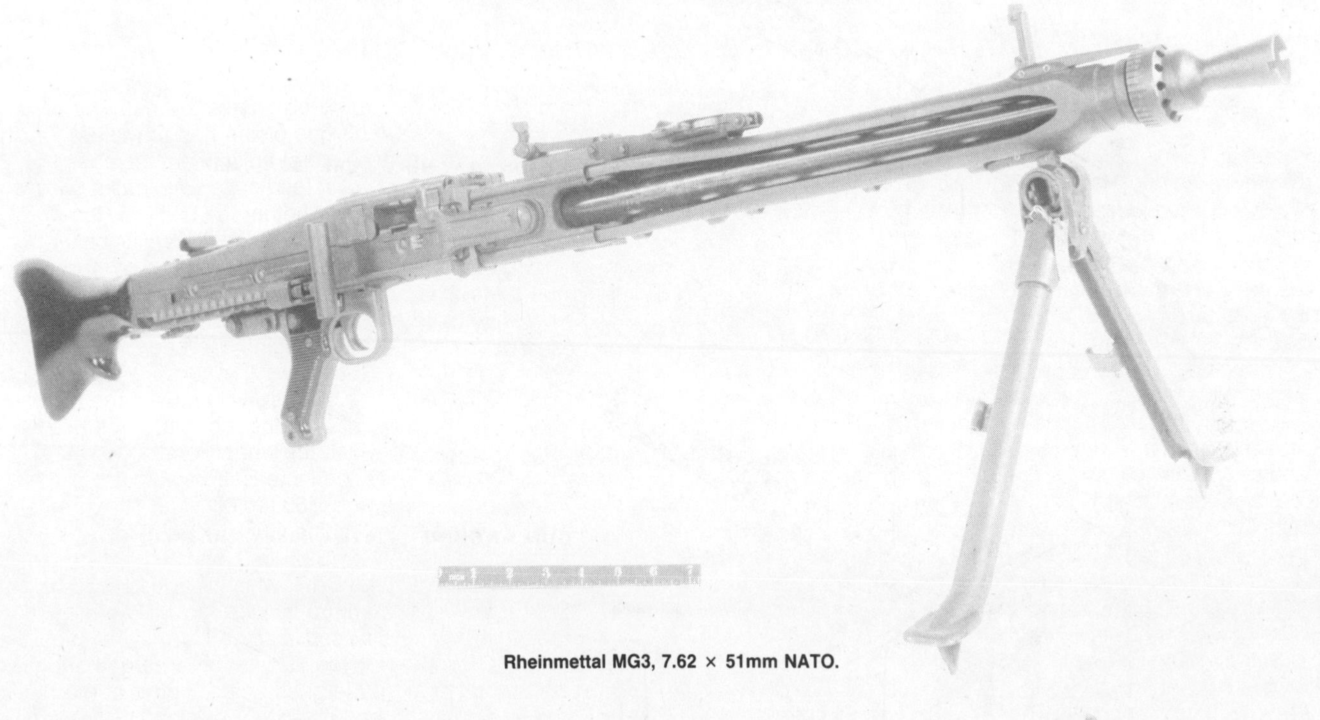

Rheinmettal MG3, 7.62 × 51mm NATO.

US M60 Machine Gun, 7.62 × 51mm NATO.

Several malfunctions, including the tendency of fired cartridges to jam in between the buffer and the lever that actuated the cartridge rammer, had bedeviled the developmental guns. When testing was resumed in June 1957, some functioning problems still showed up. After further research and development work, on 14 May 1959, the T197E2 was standardized as the M73. From 1960 to 1965, Springfield Armory was the sole production source for this weapon.

When problems continued with the M73, Armory engineers began a product improvement program; the M73E1 was the result of their effort. It had fewer parts, a fixed ejector and was easier to clear when it jammed. In December 1970, the M73E1 was standardized and redesignated the M219, since its parts could not be interchanged with the M73. Approximately 13,500 M73/M219s were produced by General Electric and Rock Island Arsenal. The designers sought to provide the following characteristics in the improved weapon: short receiver length, feeding from either side, capability for barrel change from within the armored vehicle, top cover hinged from either side, absence of smoke and fumes, and easy dismounting. Despite continued efforts by GE and Rock Island to improve the performance of their machine guns (Over 40 modifications were made between 1959 and 1974.), the armed forces were never happy with it. When the Israeli Armored Corps had serious problems (the major one being the failure to extract empty shells when the extractor broke off parts of the cartridge rims) with the M73/M219 during the Yom Kippur war of 1973, the US Army began to look for a replacement.

The first post-1945 generation of United States Machine Guns showing relative sizes, feed directions and ejection patterns.

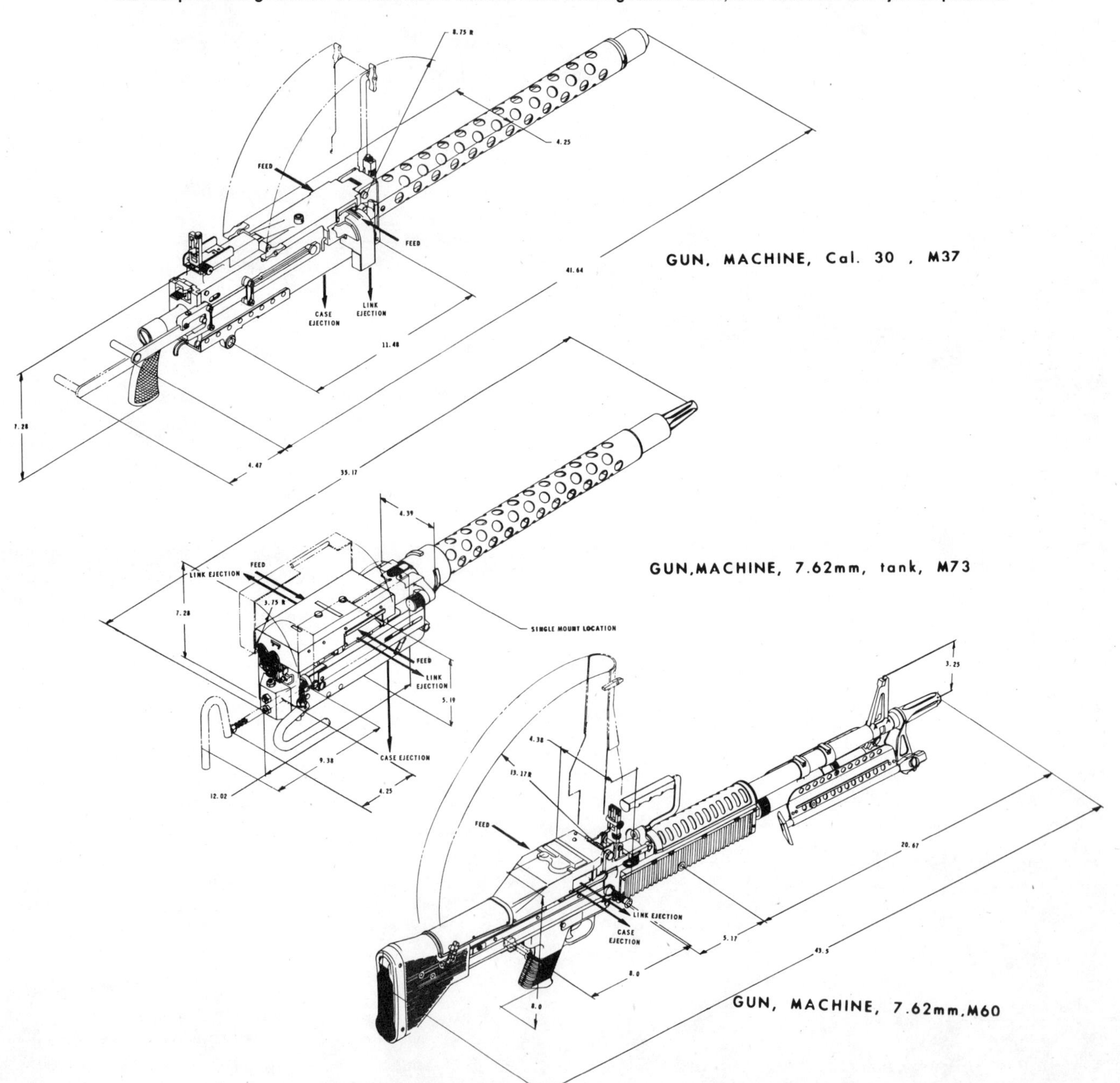

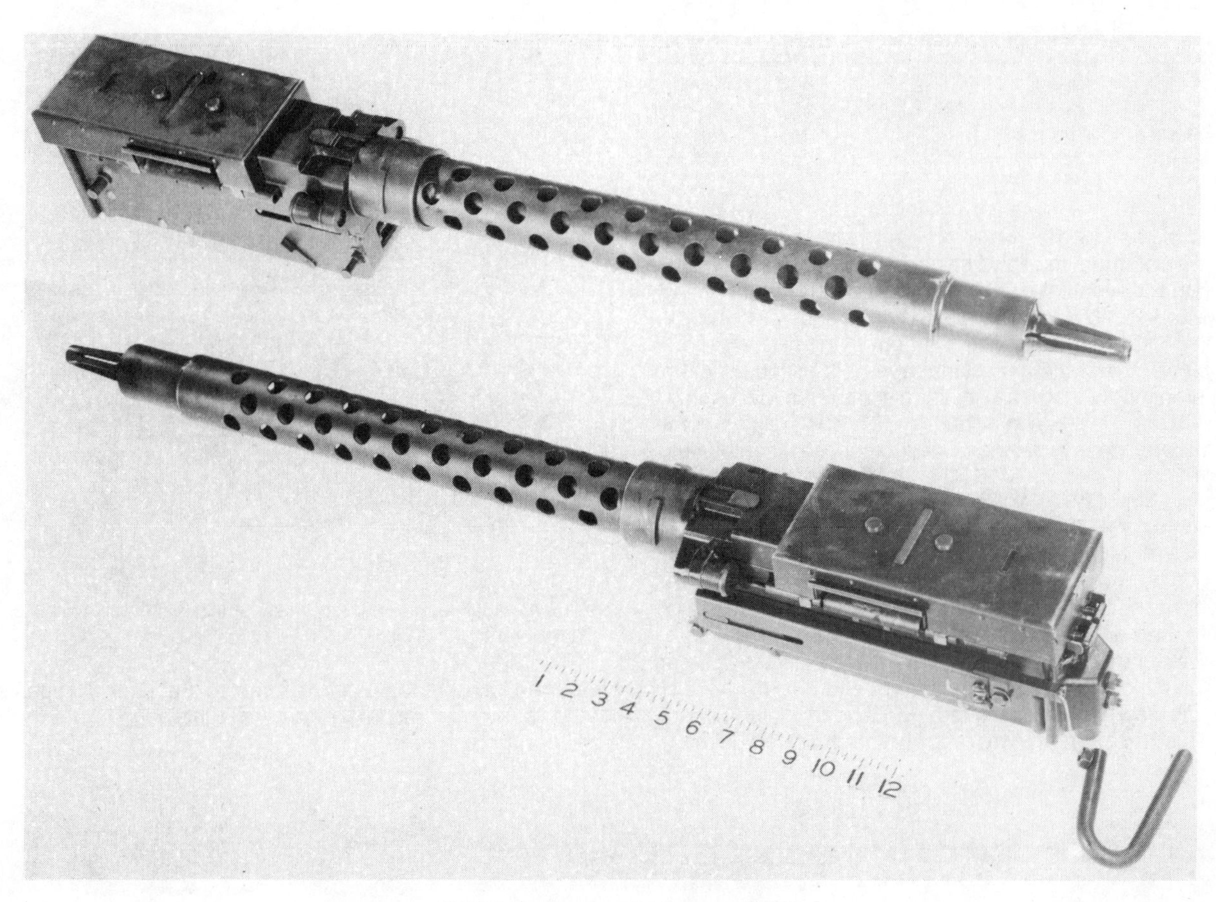

US M73 Machine Gun, 7.62 × 51mm NATO.

MAG58

Besides the M60, M60E2 and M73/M219, six "off-the-shelf" foreign weapons were thoroughly examined to determine their suitability as replacements for the M219—the French *Mle. 1952 AAT NF1,* the Belgian Fabrique Nationale *Mitrailleuse* a gas 58 (MAG58 or GPMG), the UK L8A1 (British version of the MAG58), the Canadian C1 (M1919A4 converted to 7.62mm NATO by Canadian Arsenals Ltd.), the German MG3 and the Soviet Kalashnikov-designed PKM. Beginning in 1974, a series of tests were carried out in two phases (Development Test and Operational Test phase II [DT/OTII] and later DT/OTIII). Operational evaluation of the three US guns was carried out at the Fort Knox Armor Center. Technical laboratory examinations, including test stand firings, were conducted at Aberdeen Proving Grounds for all nine weapons.

Two criteria were given special consideration. These were the so-called Mean Rounds Between Stoppages (MRBS) and Mean Rounds Between Failures (MRBF). The difference between a stoppage and a failure was defined in terms of the time the gun was out of action. Stoppages were less than a minute, while failures jammed the weapon for over a minute or were the result of parts breakage. After a careful test, the machine guns were ranked in order of superiority of performance:

1. M60E2
2. MAG58
3. M219
4. PKM
5. L8A1
6. MG3
7. AAT52
8. M219
9. C1

In March 1975, US Army authorities held a meeting on the coaxial armor machine gun and decided the following. Production and product improvements of the M219 should be terminated; shortcomings in the M60E2 should be corrected; and a new round of tests should be readied. At the end of March, the Army purchased 10 production model coaxial guns from FN. By that time, the Belgian firm and its licensees had manufactured over 700,000 MAGs, which had an operational system similar to US Browning Automatic Rifles. Fewer modifications were necessary to the Belgian weapon to make it fit the coaxial role, and the major alteration involved adding a cocking cable with a return spring. The standard gas regulator was replaced by a special unit that did not have any vents. A different flash suppressor was mounted on the barrels in some versions purchased in Europe. Whereas the barrel extension of the M60E2 protruded through the protective turret mantlet, the MAG barrel was hidden. Gas evacuation was provided by the flash suppressor. The tank version of the MAG could more readily be adapted to ground use by fitting a buttstock and mounting the gun on a lightweight tripod. A final advantage was its ability (with minor parts change-out) to use either the US M13 links or the German DM6 continuous link belts.

The final evaluation phase (DT/OTIII) was divided into three parts—technical testing (parts functioning and operational utilization), human engineering testing (interface between weapon and operator/tank) and RAM-D testing (Reliability, Availability, Maintainability-Durability). The last element, durability and ruggedness, was given the greatest weight in the subsequent evaluations. All of the Maremont M60s displayed a remarkable life span—over 100,000 rounds were fired without any parts problems. FN's guns began to show rivet breakages in the receiver at about 70,000 rounds, but they remained serviceable until well over 90,000 rounds. In the reliability test, the MAG was superior as the following indicates:

Type	No. of rounds fired	MRBS	MRBF
M60E2	50,000	846	1,699
MAG58	50,000	2,962	6,442
Minimum specified		850	2,675
Minimum desired		1,750	5,500

As a result of the US trials at Aberdeen Proving Ground and Fort Carson, Colorado, the MAG has been adopted as the M240 machine gun for eventual use in all American armored fighting vehicles (M60A1, M60A2, M48, M551, MICV and XM1). On 14 January 1977, the US Army and Fabrique Nationale signed a contract for 10,000 M240 machine guns. Subsequently, an American producer will fabricate metric dimensioned M240s from a Technical Data Package provided by FN. The Belgian company will receive a royalty payment on all American made M240s, just as Colt is paid for each M16 Rifle that is made outside their own factory. Early M240s will be installed in the Mechanized Infantry Combat Vehicles (MICVs). Later, the new machine gun will be used in place of the M73/M219, which will be phased out. The US Marine Corps will use the M60E2 in their M60A1 tanks. Over 45 nations currently include the various models of the MAG in their armored forces' inventories.

The process by which the M240 was selected may well set the tone for future NATO standardization efforts. For the first time, the US Army will utilize a foreign designed infantry/tank weapon. This test program was a far cry from the emotionally charged rifle standardization attempt. Fabrique Nationale and other arms manufacturers hope that the adoption of the M240 will set a precedent.

E. Vervier holding a production model of the MAG58 Machine Gun he designed for FN.

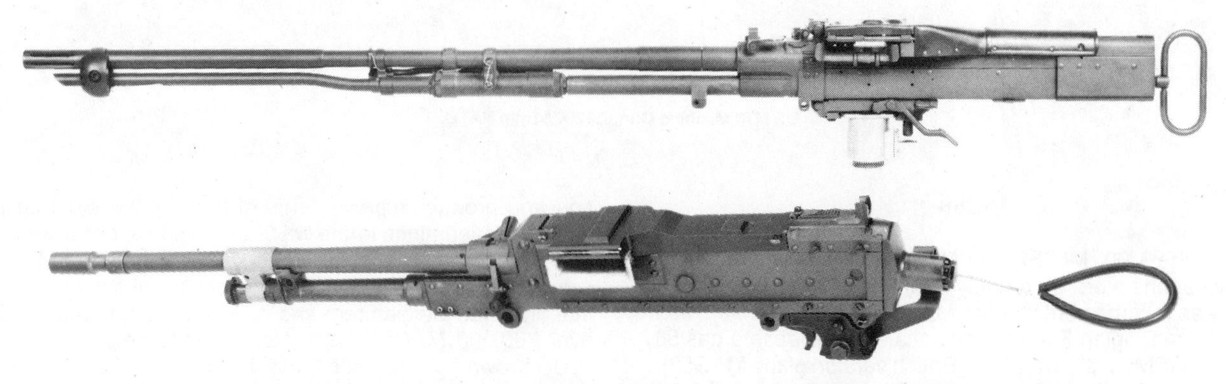

US M60E2 (top) and FN MAG58 [US M240] (bottom) to approximately the same scale.

AAT52

At the end of World War II, the French set out to develop an inexpensive and reliable general purpose machine gun (GPMG). The delayed-blowback *Arme Automatique Transformable Mle. 52* was the product of their search. Fabricated from semi-cylindrical stamped steel shells welded together, the AAT52 is available in standard French 7.5 × 54mm and 7.62mm NATO. The latter is called the *AAT 7.62 NF1.* Since it can be fitted with either a light or heavy barrel, the AAT is a true GPMG. A heavy barrel model (without stock or sights) is used as a coaxial gun in armored fighting vehicles. While this tank version was well received by US field and test personnel, it did not perform as reliably as the MAG.

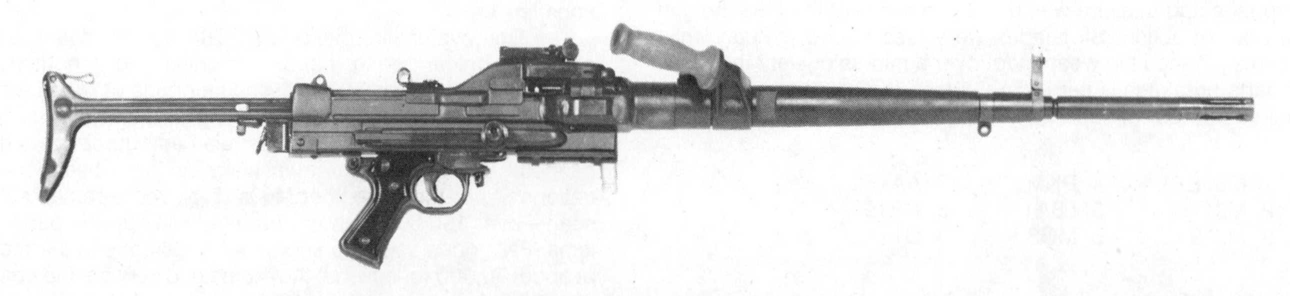

French *AAT Mle. 52*, 7.5 × 54mm.

HK21A1. See Chapter 21.

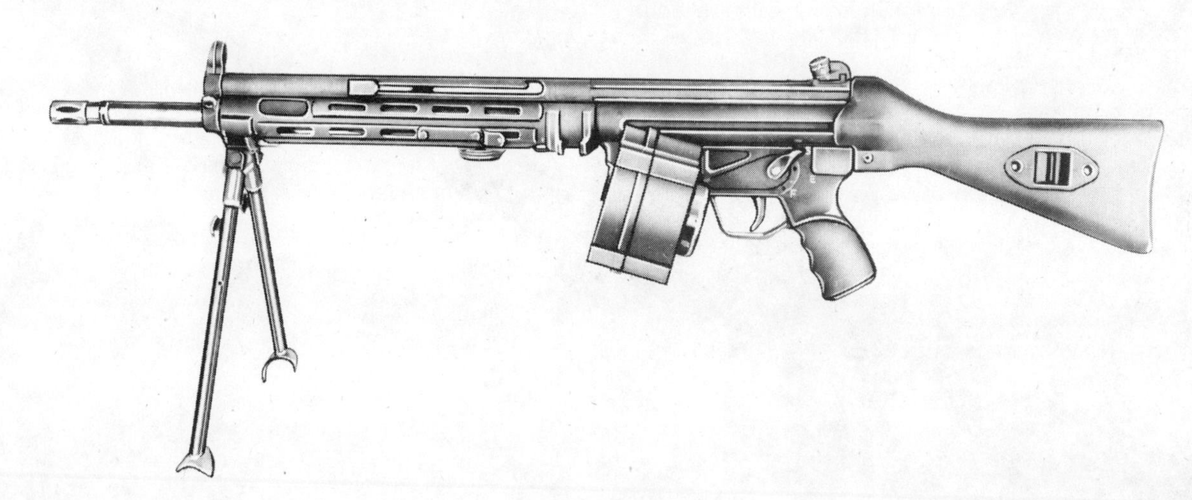

HK11. See Chapter 21.

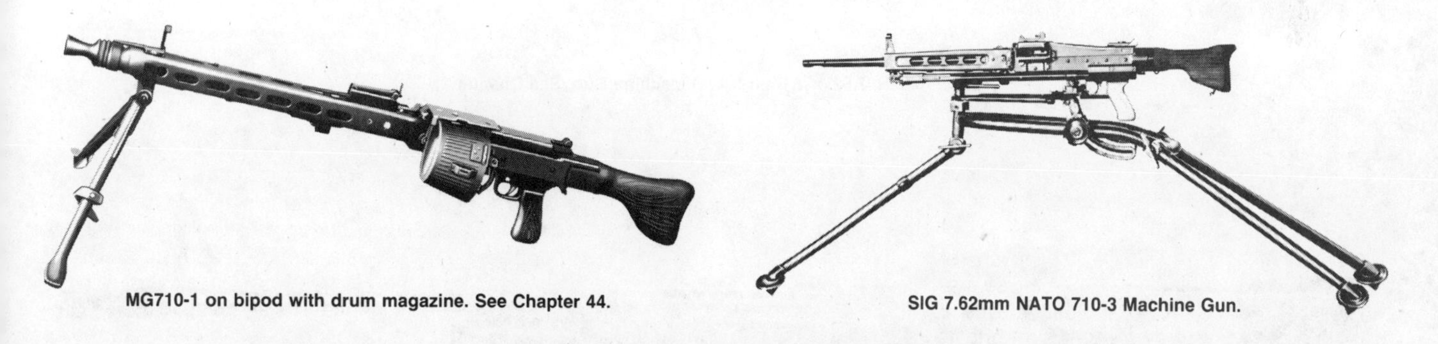

MG710-1 on bipod with drum magazine. See Chapter 44.

SIG 7.62mm NATO 710-3 Machine Gun.

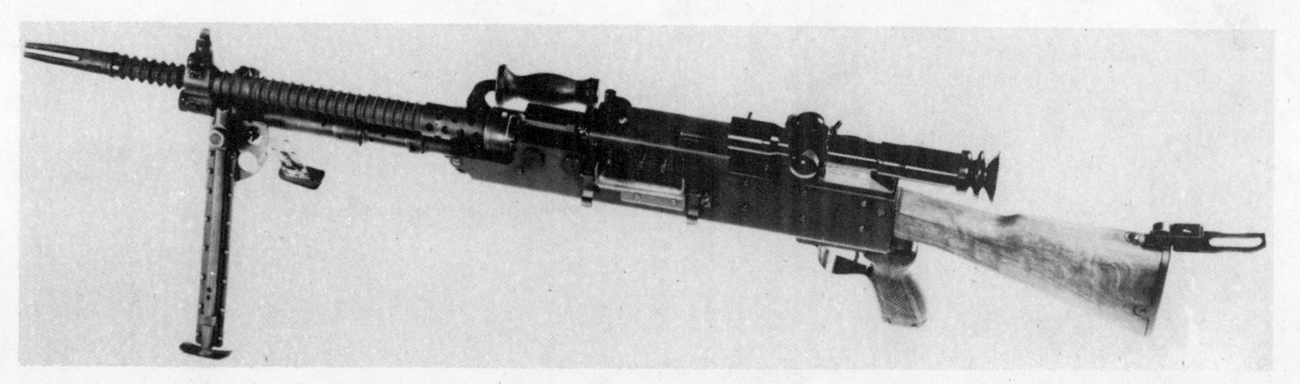

Japanese Type 62. See Chapter 30.

MISCELLANEOUS 7.62 × 51mm MACHINE GUNS

In addition to the above, there are five other NATO caliber machine guns of note. First, there is the Heckler & Koch HK21, which is a belt-fed version of the G-3 Rifle. Presently, the NATO caliber version (Some 7.62 × 39mm and 5.56 × 45mm guns have been made.) is in production, and the weapon is being used by the Portugese armed forces. Second, is the Type 62, adopted by the Japanese Ground Self Defense Forces in 1962. Although somewhat complex in its operating mechanism, the Type 62 has an excellent reputation for reliability and accuracy. A third NATO caliber gun is the Czech M59N, which is a reworked version of their 7.62 × 54Rmm weapon. No large sales of this weapon have been made. SIG has developed the 710-3 machine gun, which fires the NATO cartridge. It is a spinoff from the Mauser MG45 developed toward the end of the Second World War. Although this gun is available for sale, to date there have been no sizeable purchases. Finally, the Maremont Corporation Universal Machine Gun was designed to replace the M60 in the ground role and the M73/M219 in the tank role. The recent adoption of the MAG(M240) probably means that the Maremont UMG will never be sold in large numbers. Indeed, the future of NATO caliber machine guns is unclear as the alliance members move forward with tests of 5.56 × 45mm weapons.

Czech M59N, 7.62 × 51mm NATO Machine Gun. See Chapter 15.

Maremont Corporation's 7.62 × 51mm NATO Prototype Universal Machine Gun.

5.56 × 45mm CALIBER MACHINE GUNS

CMG-2

Colt was the first company to experiment with a 5.56mm machine gun. Their idea was to produce a companion weapon for the M16 Rifle. Early entries into this field were a heavy barrel magazine-fed Colt Automatic Rifle (CAR) and a belt-fed machine gun version of the AR-15. Colt engineers, under the direction of Robert E. Roy, subsequently turned their attention to the development of a true machine gun. The CMG-1 was the fruit of their labor. Using M16 components where possible, this gun was dropped in favor of a completely new design, the Colt Machine Gun-2 (CMG-2) designed and developed by two Colt employees, Henry J. Tatro and George F. Curtis. The US Army tested the first models of the CMG-2 in the fall of 1969. Official reaction to the gun was that it did not have great enough range (even with the improved 4.4-gram [68-grain] projectile) and that it could not yield a high enough rate of sustained fire. This report, dated December 1969, came at a time when the Army was formulating the characteristics for a new Squad Automatic Weapon (See SAW below.), and there was some evidence that the 800-meter range for helmet penetration and tracer visibility later specified were in part the result of prejudice against the 5.56 × 45mm cartridge.

Tatro and Curtis borrowed a number of design concepts from earlier guns for incorporation into the CMG-2. The forward moving pistol grip as cocking mechanism was taken from the Czechs, who have used it in the ZB50, M52 and M59 machine guns. The gas system is essentially the same as the American M60. An M52 type belt feed mechanism was employed and ejection accomplished by a Lewis gun type spring loaded striker, which ejects spent rounds downward out of the gun. The CMG-2 has been ready on the shelf for production since December 1972, but no orders large enough to warrant production have been forthcoming.

Robert E. Roy firing early belt-fed AR-15—note ejected cases which are circled.

CMG-1 mounted on mock-up of a Colt 20mm gun turret. (June 1966)

Later version of belt-fed, heavy barrel AR-15.

STONER 63

Part of the system described in Chapter 45, the Stoner machine guns have also been a system in search for a major market. Despite the fact that the XM207 has been extensively tested by the US Army and was used by the US Navy SEAL teams as the MK23 in Vietnam, the Stoner machine guns have never been well received by the American military. Test reports indicate that the major strike against the weapon has been its unreliable performance under adverse conditions. Navy SEAL teams, who made a virtual ritual of preventive maintenance, were completely sold on the Stoner, but Army officials opposed it because they knew that the average infantry soldier could not be expected to lavish such careful attention on his machine gun. They wanted a soldier-proof weapon. Also, Stoner ran into the same prejudice against a 5.56 × 45mm machine gun in 1968–1972 as had Colt. Until the Army resolved the ammunition question, both the CMG-2 and Stoner were at a disadvantage. Once the ammunition question was settled, the Army had its own candidate weapons.

D. A. Behrendt, Colt technical representative, with CMG-2.

Colt CMG-2, 5.56 × 45mm Machine Gun.

Early Stoner 63, 5.56 × 45mm Machine Gun.

Eugene M. Stoner, designer of the M16 Rifle and the Stoner 63 system.

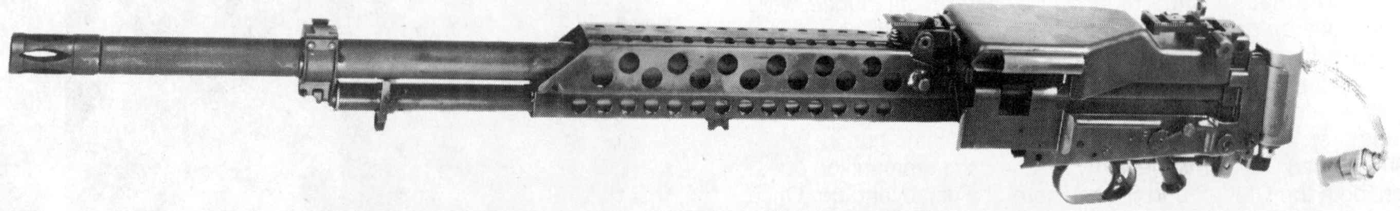

5.56mm Stoner 63 Fixed Machine Gun. Firing solenoid shown as attached to receiver.

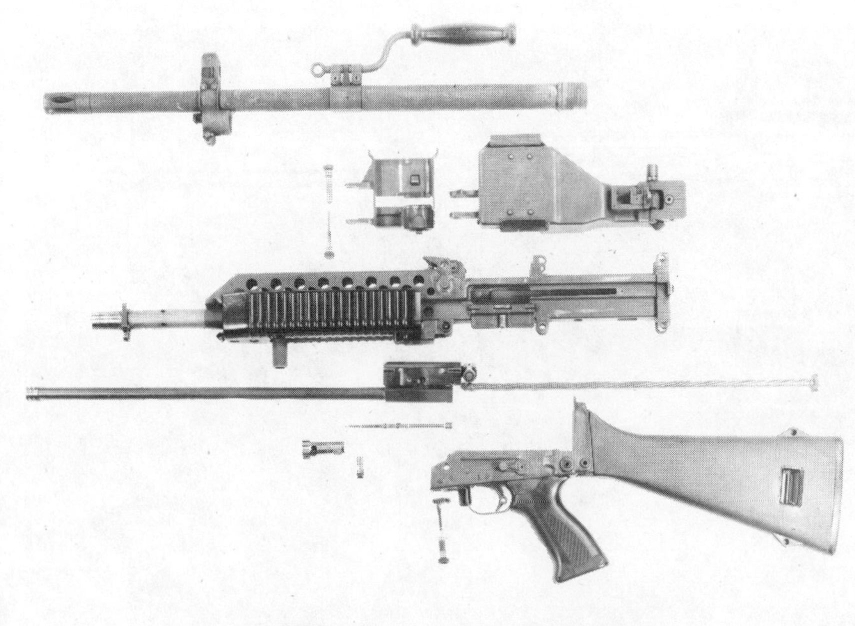

Stoner XM207E1, 5.56 × 45mm Machine Gun field stripped.

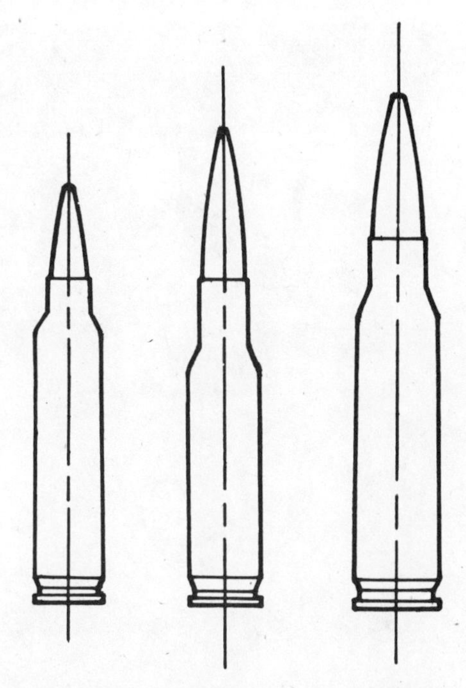

Comparative sizes: 5.56 × 45mm and 7.62 × 51mm NATO with 6.0 × 45mm SAW (Approximately full size).

HECKLER & KOCH 5.56

As yet one more element in their program to fully exploit the roller lock breech mechanism developed in the G3 rifle, Heckler & Koch has developed magazine- and belt-fed (HK13 and HK21) 5.56 × 45mm machine guns. These weapons are essentially the same as the G3 in basic operating principles, and they are described more fully in Chapter 21.

US SQUAD AUTOMATIC WEAPON (SAW) PROGRAM

The United States Army has been studying and evaluating small caliber infantry machine guns for more than a decade. As early as September 1966, the United States Army Small Arms Weapon Study (SAWS) pointed to the need for a more portable base-of-fire weapon than the 10.5-kilogram (25.1-pound) M60 machine gun (which weighs 17.6 kilograms [38.8 pounds] with two hundred rounds of ammunition). Three years later, the Infantry Rifle Unit Study (IRUS) 75, conducted by the Army Combat Development Command, recommended that the dismounted basic infantry element (fire team) be armed with a one-man lightweight machine gun. Such a weapon would supplement the firepower of the 5.56mm M16 rifle and the M60. Although the M60 was commonly employed with the rifle squad, men in the field complained that it was too heavy to fill that role. Since it weighed almost 14 kilograms (30.9 pounds) with one hundred rounds of ammunition, the point was well taken. Having placed increased emphasis on small unit mobility coupled with greater firepower for unit members, United States infantry planners sought a light machine gun, which with two hundred rounds of ammunition would not exceed 10 kilograms (22.0 pounds) but would have a range of about 800 meters.

Once this basic weight specification was established, Army small arms development personnel began considering possible calibers for such a lightweight support weapon. At the upper range, the 7.62mm NATO cartridge was tentatively rejected since it was too heavy in aggregates of two hundred rounds to meet the 10-kilogram (22 pound) weight limit. At the other end of the spectrum, the obvious candidate, the 5.56 × 45mm round, was not an early favorite with United States Army officials. There were both technical and political reasons for its early unpopularity.

Range, for both the 5.56mm ball and tracer projectiles, was the major criticism. Conceived in the mid-1950s to meet battlefield requirements defined by Project SALVO, the 5.56mm round was not intended for use beyond 500 meters. In the late 1960s when the search for a lightweight machine gun began, many military officers and civilian arms experts alike were still not convinced by the statistical evidence that most infantry weapon hits occurred between 150 and 300 meters. Despite a growing body of data that questioned the overall significance of small arms fire in producing combat casualties, small arms personnel raised in a tradition that stressed carefully aimed fire supported by long-range machine gun fire dismissed the M16 rifle and its cartridge for any real wars, notably those that might be fought in Europe. Thus, there began a search for a compromise caliber between 5.56mm and 7.62mm that would be light enough to aggregate numbers to permit a weapon chambered for it to meet the weight limit.*

One of the first steps toward creating a new weapon and a new cartridge was coining a new name. On 6 July 1970 when the "Advanced Development Objective" was approved, the Army dropped the light machine gun label for the new weapon and adopted in its place the designation "squad automatic weapon." This shift was more than cosmetic; the new weapon would be utilized at the squad level as a fire support weapon, and the M60 light machine gun would continue in its role as the company level base-of-fire weapon.

Conceptual Background

At the outset, experts looking for a new lightweight small arm reviewed a large number of current weapons, product-improvement proposals, and conceptual designs that might possibly satisfy the squad automatic weapon requirement for the 1980s. More than a thousand different configurations were considered

*The limited-range problem of the 5.56mm cartridge weighed heavily against two United States contenders—the Stoner 63 and Colt CMG-2 machine guns.

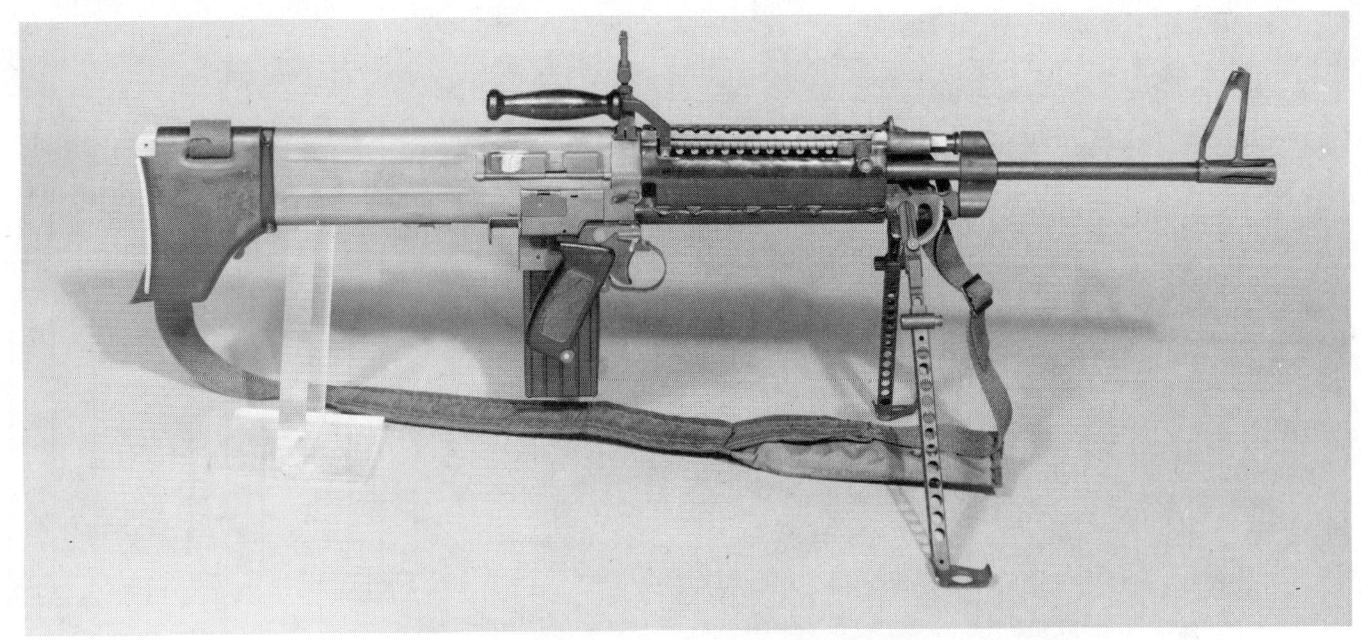

5.56 × 45mm Version of the Rodman Laboratory SAW.

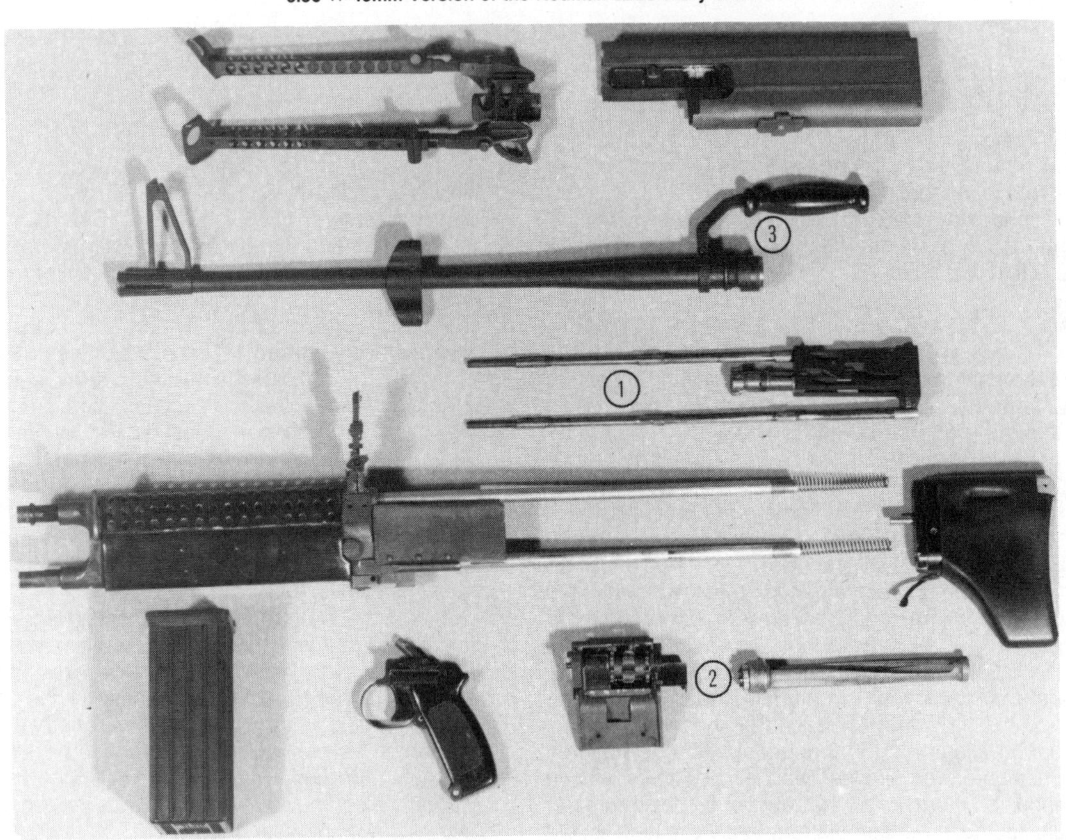

Field stripped view of Rodman Laboratory SAW. Note (1) twin pistons and bolt carrier assembly, (2) long cam for operating the belt feed, and (3) quick change barrel assembly.

in calibers ranging from 5.56mm to 7.62mm. A new 6.0 × 45mm cartridge was the result of a joint user/developer decision; no current or product-improved cartridge could meet the desired characteristics. The new SAW round projected a long, 6.8-gram (105-grain) projectile at 747 meters per second. While this velocity was considerably lower than the 990 meters per second of the M16 cartridge, the projectile weight was nearly double (M16 round equals 5.5 grains [3.0 grams]). A larger diameter projectile had been selected by the United States Army research and development engineers at the Rodman Laboratory, at Rock Island Arsenal, and at Frankford Arsenal so they could produce a reliable tracer projectile. The desired military performance was a fully visible trace in bright daylight to a range of more than 800 meters. In addition, the ball ammunition program sought increased hard target penetration at 800 meters; that is, the equivalent of at least helmet penetration as a measure of defeating hard targets.

After twenty months of preliminary work, during which the performance and dimensional specifications of the cartridge were decided, the Department of the Army issued a "Materiel Need"

Exploded view of XM235 SAW (1976)

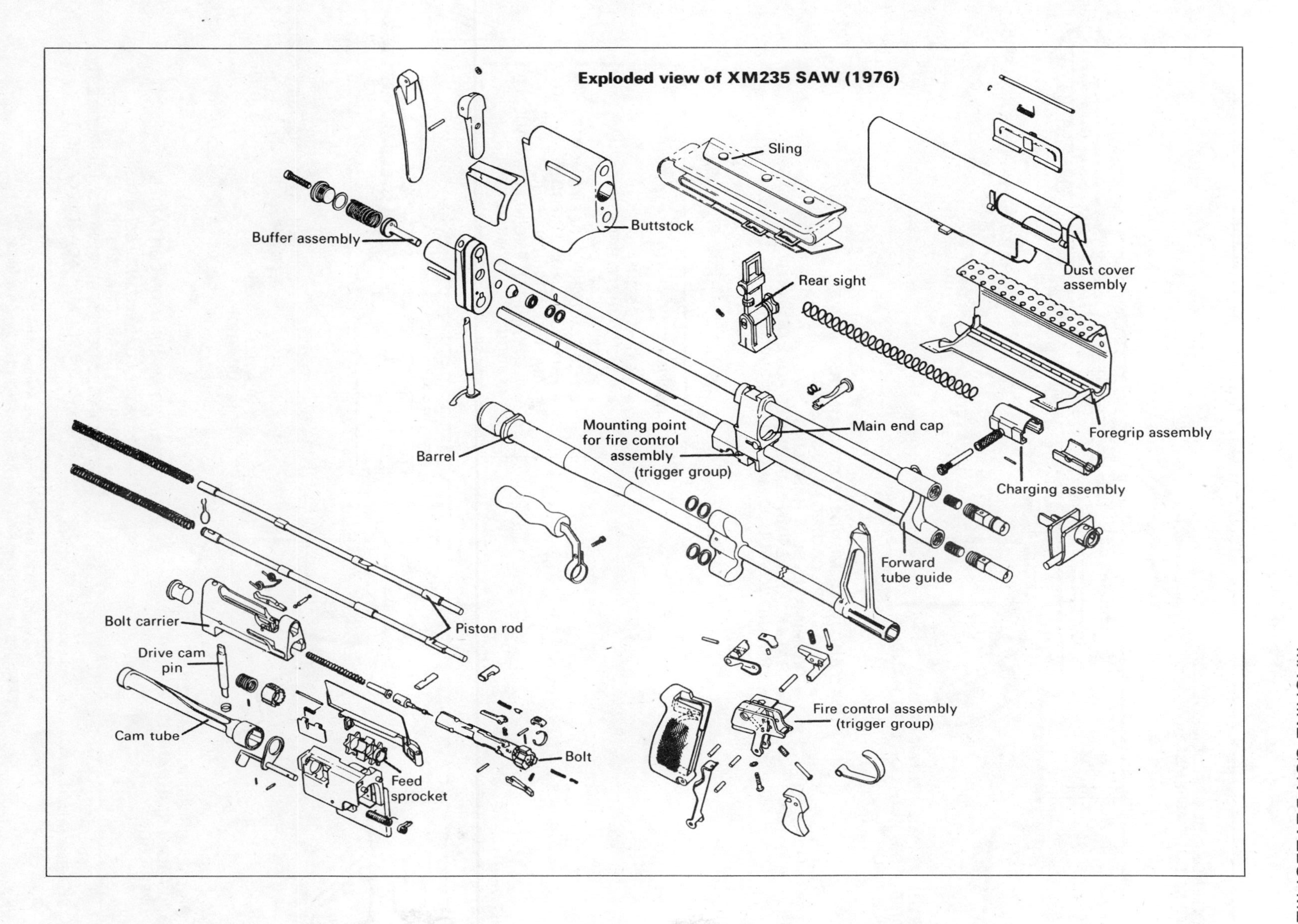

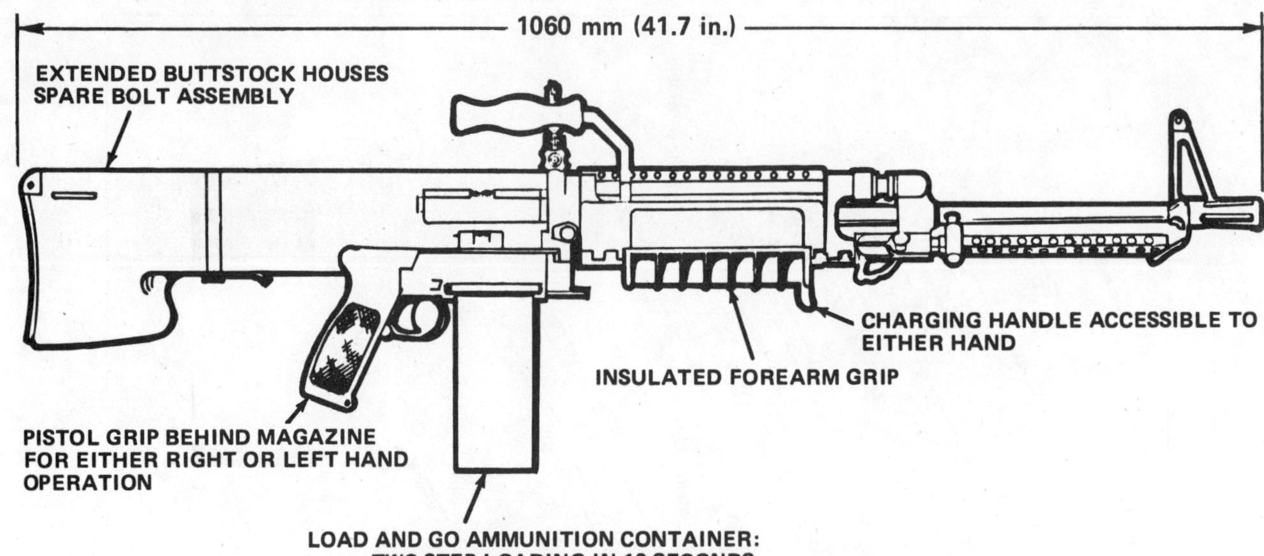

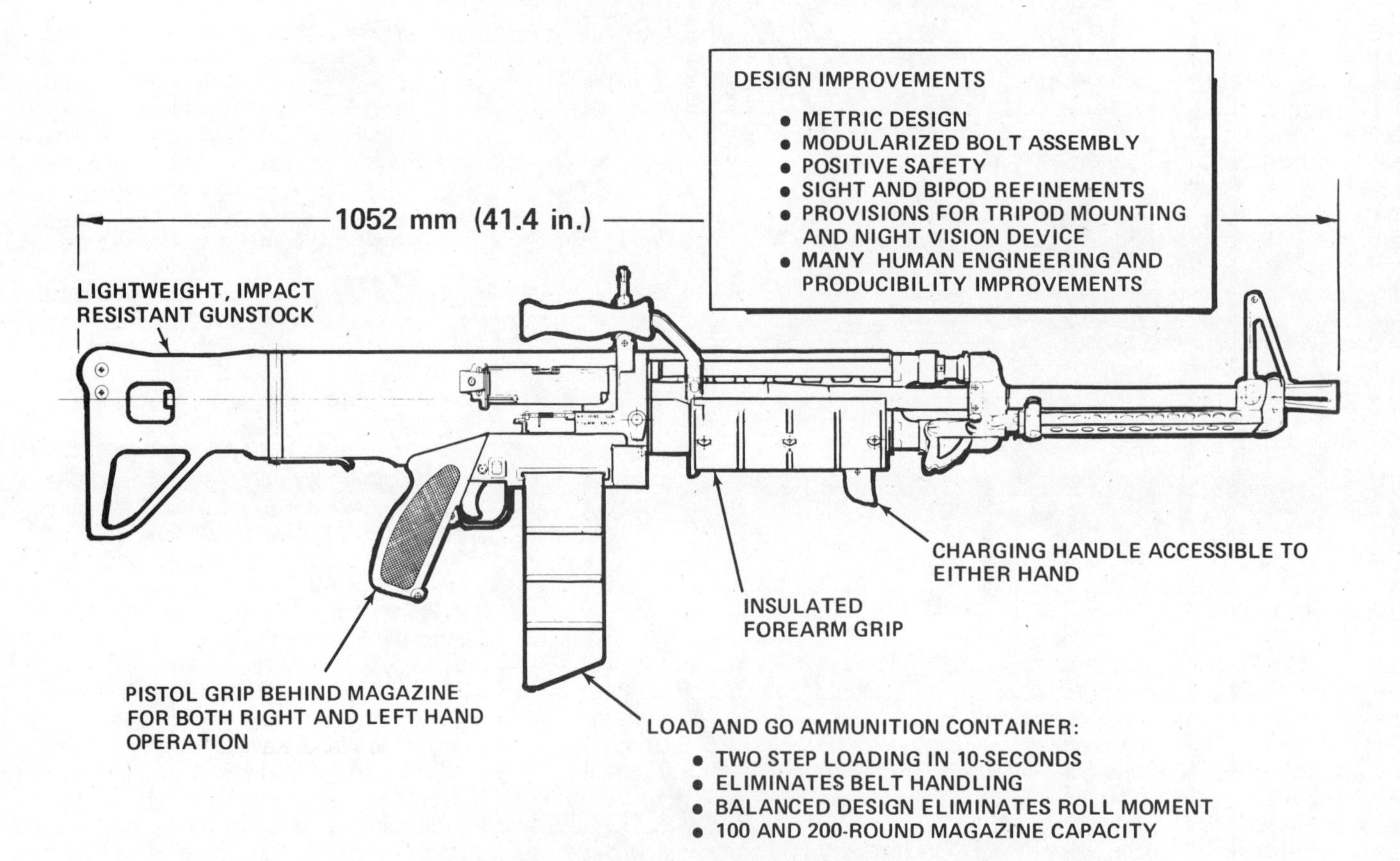

Ford Aerospace XM248 SAW. *Top*, initial Ford rework of the XM235 (Rodman) SAW; *Bottom*, final version of the XM248 as tested in the SAW trials.

The Ford Aerospace XM248 squad automatic weapon.

for a "Squad Automatic Weapon, Light Machine Gun," on 8 March 1972, spelling out in detail the specifications for the weapon. Two contracts were given for the development of prototype weapons—to Maremont Corporation (XM233) and Ford Aerospace and Communications Corporation (XM234). A Rodman Laboratory design team, under the direction of Curtis D. Johnson, developed a third entrant the XM235, for the tests.* After a little more than two years, the Development Test and Operation Test I (engineering and user test) was concluded in December 1974. In addition to the three SAW candidates, the 5.56mm HK23 belt-fed machine gun, the FN Minimi 5.56mm belt-fed machine gun, and the heavy-barrelled M16 were tested. The standard M16 was the yardstick against which the guns were measured. The Heckler & Koch weapon was dropped when it did not pass the safety tests, as was the heavy-barrelled M16, which had a limited magazine capacity—30 shots, versus the 200-round, self-contained belt capacity called for in the Materiel Need.

HK23A-1 Machine Gun

A few words about the HK21 and HK23 weapons are in order to explain not only the problems encountered with the HK23 but also why it was developed in the first place. The HK21 machine gun was introduced in 1961 to complement the 7.62 × 51mm NATO *Gewehr 3* (G3) Heckler & Koch was producing for the *Bundeswehr* and other armies. By early 1978, nearly seven million G3 rifles had been manufactured by Heckler & Koch and its licensees. From the start, the designers at Heckler & Koch sought maximum utilization of G3 concepts and technology in the HK21. They used the basic delayed roller-locked blow-back bolt mechanism and utilized several new features to adapt the weapon to the light machine gun role. For example, to insure versatility, Heckler & Koch provided for conversion of the HK21 from the NATO caliber to the Soviet 7.62 × 39mm or the American 5.56 × 45mm cartridges by substitution of a different caliber quick-change barrel assembly, bolt assembly, and belt-feed insertion unit.

When the United States Army Small Arms Systems Agency (USASASA) decided to test commercial squad automatic weapon designs, weight was a primary consideration. The candidate weapons were supposed to weigh no more than 9.07kg (20 pounds) with two hundred rounds of ammunition. The loaded weight of the HK21A1 with two hundred rounds of 5.56mm ammunition was 11.06kg (24.38 pounds). To meet the weight limit, Heckler & Koch developed an HK21 scaled down to the size required by the 5.56 mm cartridge.

Before Heckler & Koch delivered their HK23A-1 to the United States Army, the Small Arms Systems Agency was abolished (July 1973), and responsiblity for the squad automatic weapon was transferred to the T. J. Rodman Laboratory of the newly organized United States Armament Command Headquarters, Rock Island, Illinois. Heckler & Koch representatives believe that many of the subsequent problems experienced with the HK23A-1 are directly attributable to the series of reorganizations that affected the American small arms community during the next four years.

The tests of the HK23A-1 machine gun illustrated a number of problems—technical and managerial. As a commercial SAW contender (i.e., not United States Government–sponsored as were the XM233, XM234, and XM235), the German gun was not designed to fire the 6.0 × 45mm SAW cartridge. It fired the 5.56 × 45mm cartridge, of which there were several varieties. Besides ammunition trials being conducted within NATO, work was under way within the United States Army to improve the penetration capability of the 3.6-gram (55-grain) M193 Ball and tracer effect of the companion M196 Tracer, the standard 5.56mm cartridges. One experimental variation of the 5.56mm cartridge was the XM287 Ball, with a 4.41-gram (68-grain) projectile, and its companion XM288 Tracer. Growing out of work conducted by Colt with the CMG-2 machine gun and by Cadillac Gage with the Stoner 63 Weapons System, Val Cartier of Canada developed these two projectiles for the United States Army. Parallel work with improved 5.56mm projectiles had been carried out in the early 1970s by the Industrie Werke Karlsruhe (IWK) with the 4.98-gram (77-grain) projectile family and by Fabrique Nationale (FN) with the 4-gram (62-grain) SS109 projectile.

During the Aberdeen Proving Ground trial of the HK23A-1, the weapon was tested by firing only Val Cartier XM287 and XM288 cartridges. Prior to the test, the gun had been tuned at the factory to fire standard M193 and M196 ammunition. Since the brass cartridge cases of this particular lot of Val Cartier loaded cartridges had thinner walls than the standard ammunition, a number of stoppages occurred caused by swollen cartridge case heads and blown out primers. Aberdeen Proving Ground specialists blamed the gun, while Heckler & Koch saw the trouble as ammunition deficiency. A second problem arose when Aberdeen personnel took apart the trigger group mechanism, going beyond the factory-recommended disassembly for

*The co-patent holders included L. D. Antwiler, L. C. McFarland, A. R. Meyer, F. J. Skahill, D. L. White, K. L. Witwer, and R. L. Wulff.

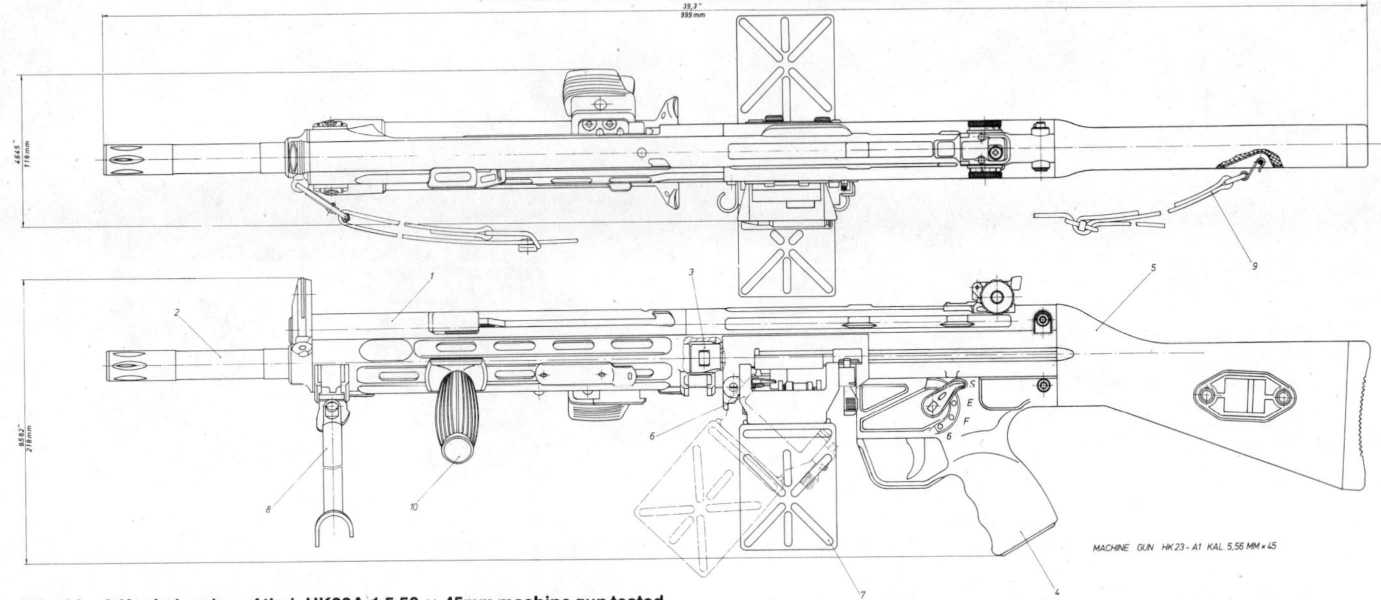

A Heckler & Koch drawing of their HK23A-1 5.56 × 45mm machine gun tested by the United States Army in 1974. Improper handling of this weapon and mismatched ammunition led to functional problems during the trials in 1974.

COMPARISON OF HK21A-1 AND HK23A-1 MACHINE GUNS

		HK21A-1	HK21A-1	HK23A-1
Caliber		7.62×51mm NATO	5.56×45mm	5.56×45mm
Weight of weapon	kg (lbs)	7.69 (16.94)	7.69 (16.94)	5.85 (12.91)
Weight of bipod	kg (lbs)	.59 (1.30)	.59 (1.30)	.50 (1.10)
Weight of ammunition	kg (lbs)	3.69* (8.14)	2.78** (6.13)	2.78** (6.13)
Total weight	kg (lbs)	11.97 (26.39)	11.06 (24.38)	9.13 (20.13)
Overall length	mm (inches)	1028 (40.47)	1028 (40.47)	980 (38.58)
Barrel length	mm (inches)	450 (17.2)	450 (17.2)	450 (17.2)

* Per 100 rounds.
**Per 200 rounds.

maintenance. In the process, a critical component, the hammer catch and release lever, was bent due to improper reassembly. Commenting on this, the Aberdeen Proving Ground report noted, "Although this problem is primarily one of personnel error, it is caused by faulty weapon design, which permits this condition to occur." Needless to say, Heckler & Koch representatives argued that had the basic maintenance instructions been followed, the feeding problems caused by the damaged component would never have developed. Nevertheless, Aberdeen terminated the tests of the gun and declared it unsafe.*

*Aberdeen Proving Ground, "Final Letter Report of Enginner Design Test of the 5.56mm Heckler & Koch Machine Gun, Model 23A1, TECOM Project No. 8-WE-400-SAW-003, Report No. APG-MT-4547," by F. H Miller, 3 Jan 1975, p. 3; and letter Herr Seidel, Managing Director, H&K, to John S. Wood, "HK23A-1/Ammunition," 19 August 1974.

A Heckler & Koch representative reported that a second HK23A-1 demonstrator and test gun had been in the United States since early 1974. "This gun has been fired many times by Heckler & Koch representatives during various demonstrations to Department of Defense officials, as well as having been exhaustively test fired by the United States Navy SEAL Team No. 2 at Little Creek, Virginia. This gun has fired over ten thousand rounds with no problems utilizing all types of standard M193/196 ammunition, as well as various types of 5.56mm foreign ammunition." Furthermore, no safety problems have been reported with the HK23A-1 machine gun.

When the US Army decided to drop the HK23A-1, Heckler & Koch protested. Since this was about the time of the Army Materiel Development and Readiness Command's (DARCOM) reorganization, their complaints were to no avail. Heckler & Koch decided subsequently to terminate development of the special-order 5.56mm HK23A-1 and to concentrate on marketing their proven HK21A-1 machine gun with its caliber conversion feature. At the end of March 1978, Heckler & Koch offered the HK21A-1 to the Small Caliber Weapons Systems Laboratory of the US Army Research and Development Command for testing in the 1979 Development Test–Operational Test of SAW contenders.

Fabrique Nationale Minimi Machine Gun

Fabrique Nationale's 5.56 × 45mm machine gun was another product of the fertile mind of Ernst Vervier. Introduced in 1974, this gun looked like a conventional machine gun, and it had a wood stock, which endeared it to traditional infantrymen weary of fiberglass and other synthetics. In 1974, one strike against the Minimi was its 5.56 × 45mm caliber. Another was its special 4-gram (62-grain) projectile (SS109). Thus, its ammunition was not interchangeable with either the M16 or the three SAW entries. But performance was the key factor, and the Minimi was as reliable as the XM233, XM234, and XM235.

Lieutenant Colonel Bob B. Lukens, SAW project manager at Rock Island Arsenal, prepared an extensive trade-off study in which he evaluated the various weapons. At about the same time, the Department of the Army held a review of SAW, calling for a redirection of the effort. Since the 6.0 mm cartridge would mean the introduction of a third type of ammunition into the United States inventory, the Department of the Army requested

7.62 × 51mm NATO caliber HK21A-1 machine gun.

Quick barrel change feature of the HK21 machine gun series.

the weapons designers to limit their study to 5.56mm and 7.62mm NATO. Army Materiel Development and Readiness Command (DARCOM) personnel were told to consider amending the Materiel Need and think over the international—that is, NATO—implications of any decisions.

Joint studies by DARCOM and the Training and Doctrine Command (TRADOC), which represented the infantry, led to the selection of the 5.56 × 45mm round. This cartridge would permit the development of a machine gun that would weigh not more than 9.6 kilograms (21.2 pounds) when loaded with two hundred rounds of ammunition. In February 1976, TRADOC, commanded by General William E. Depuy, decided to go for improved 5.56mm projectiles fired from the XM235 (Rodman-RIA) and the FN Minimi. General Depuy also wanted a platoon-size comparative test of the two weapons in which they would be measured against the M16 and M60. The new cartridges were the XM777 (ball) and the XM778 (tracer). Whereas the former had better hard

5.56 × 45mm HK23A-1 squad automatic weapon.

target penetration due to design changes, the tracer round almost met the 800-meter requirement.

In June 1976 after a special In-Process Review, DARCOM and TRADOC went to the Department of the Army with two recomendations: (1) Modify the Materiel Need to reflect the change to 5.56mm. (2) Reduce the over-800-meter requirement to up to 800 meters. The Department of the Army approved the revised Materiel Need in October 1976.

DA-APPROVED REVISED MATERIEL NEED HIGHLIGHTS (OCTOBER 1976)

Characteristic	System Requirement
Operator	One-man
Weight	Lightweight (21 lbs. maximum)
Reliability	>M60 machine gun
Sustained fire rate	Machine gun role
Ammunition capacity	100- and 200-round containers
Range (point and area targets)	0 to 800 meters

DARCOM issued a prototype manufacturing request for a proposal to produce eighteen product-improved XM235 SAWs and redesignated the weapon the XM248. Maremont and Ford were the competing bidders, with Ford winning the competition and signing a contract with the Army on 3 February 1977. The XM248 (XM235) included these improvements: (1) two-step loading in ten seconds with a load-and-go system that eliminated belt handling; (2) 100- and 200-round ammunition containers; (3) an insulated forearm; (4) a charging handle accessible to either hand; (5) a spare bolt assembly housed in the butt stock; (6) 100 percent metric design; and (7) substantial human factors analysis.

XM248 (XM235) Design Characteristics

The XM248 (XM235) design team, seeking to create a weapon that would overcome the inherent inaccuracy of machine guns, made several observations: rifles are more accurate because they are easier to aim; single shots or shot bursts usually lead to the projectile having exited the barrel before peak recoil forces are reached; and rifle feed mechanisms are generally simple and do not introduce disruptive forces upon the weapon during each shot. Operating conditions in fully automatic weapons are considerably different. The cumulative effects of burst firing combine to produce a lack of control. The necessity of continuous ammunition feeding during sustained bursts usually involves using recoil force to operate the feed mechanism. Excess recoil force beyond that required to actuate the feed system is absorbed by a buffer, which normally has "high restitution."

As described in the patent covering the XM248 (XM235), "Restitution refers to the proportion of incoming velocity from an impacting mass which is returned to that mass. Thus, if a steel bolt carrier is impacted against a solid steel buffer plate, most of this movement will be returned to the bolt carrier in the opposite direction. A buffer system which impacts solid steel surfaces against similar surfaces has very high restitution, and consequently high recoil . . . through the gun butt stock to the user. Fully automatic weapons of modern design . . . have a fire rate on the order of 650 to 1,000 rounds per minute and apply recoil peak forces of about 500 [to 1,200] pounds [2,225 to 5,338 newtons] or more to the gunner through their butt stocks."

The recoil force naturally disrupts the holding force applied to the weapon by the user, thus affecting his ability to aim or otherwise control the weapon. A succession of projectiles exiting the barrel, the design of the stock and pistol grip, location of the gas system, ammunition belt weight, and pulling force all can contribute to the nature, direction, and duration of the recoil forces. While the dispersion of projectiles was not totally undesirable from a machine gun when it was used as a support weapon, the designers of the XM248 (XM235) sought to reduce it, producing a weapon that could—from an ammunition point of view—yield a more economical means of producing accurate multiple hit probabilities.

Their approach to increasing hit probability while conserving ammunition was to create an automatic weapon mechanism that had a long recoil and soft cycle of operation, which resulted in a rate of fire less than 500 shots per minute, enhancing controllability and weapon-component life expectancy. Subassemblies, or modules as the design team preferred to call them, such as a dual tube receiver, an integral sprocket feed mechanism, and dual gas cylinders and rods, could be removed separately from the weapon as individual modules and replaced with other corresponding subassemblies. Low parts breakage rates and quick-change modules would permit XM248 (XM235) machine guns to be easily maintained in service. Field stripping of the entire weapon required ten to fifteen seconds. The XM248 (XM235) consists of ten modules or subassemblies—receiver, charging mechanism, sight, feed mechanism, operation group (bolt and bolt carrier), butt stock, barrel, fire control mechanism, dust cover, and foregrip. A quick description of these assemblies will provide a better understanding of this unique weapon.

Receiver. The receiver consists of two tubes supported by a main end cap and a forward tube guide. This is the main structure around which all the other parts are gathered. Unlike many other weapons, the exterior parts—dust cover and foregrip assemblies—are not structural; the weapon can function without them. The main end cap serves as the bearing support for the breech end of the barrel (and contains a quick release lever) and provides the mounting point for the trigger group (fire control mechanism) and the rear sight. In addition to supporting the forward end of the barrel, the forward tube guide contains the two gas system assemblies to actuate the weapon. These twin tubes straddle the gun axis, passing through the longitudinal center of gravity. Resulting operating forces are more symmetrical, making the weapon more manageable.

Feed Mechanism. A charging assembly that engages one of the operating rods is mounted on the upper receiver tube. A sprocket feed mechanism fits over and rotates about the lower receiver tube. This assembly contains the sprocket, cartridge guide spring, and anti-rotation and sprocket release mechanisms. It is surrounded by the feed tray housing, which butts against the rear face of the main end cap. The feed mechanism is driven by a cam tube assembly with the drive ratchet and cycloidal cam slot, which is rotated by a cam drive pin that extends downward from the bolt carrier. Due to the long cam groove, the XM248 (XM235) has a long, smooth, continuous feeding cycle instead of the abrupt cycle familiar to most conventional machine guns. In addition this feed mechanism has about one-third the number of parts found in other weapons, such as the M60 or the MG3.

Operating Group. The operating group contains dual interchangeable operating rods attached to the bolt carrier by the cam drive pin, which also powers the feed mechanism. Contained in the bolt are the striker cocking cam and the bolt locking/unlocking cam. The three-lug bolt has a striker (firing pin) mechanism and a conventional rammer, extractor, and ejector. All parts of the operating group reciprocate along the upper receiver tube. The striker mechanism is cocked during recoil and released during the last 2.5 mm (1/10 of an inch) of the counter-recoil cycle to detonate the primer.

Butt Stock. A hydropneumatic buffer is housed in the butt stock and together with the weapon's low restitution results in a greatly reduced recoil. Peak recoil force is about 200 pounds (890 newtons), compared to the 500 to 1,200 pounds (2,225 to 5,338 newtons) of other current machine guns.

Barrel. A quick-change assembly, the barrel includes a flash suppressor, gas housing (with gas ports above and below the barrel), barrel extension containing locking recesses for the bolt, and the barrel proper, which has a chrome-plated bore and chamber.

Fire Control Mechanism. Mounted at the longitudinal center of gravity but offset to the right to permit placement of the ammunition container under the weapon's center of gravity, the fire control mechanism (sear and trigger assemblies) is completely contained in a detachable pistol grip with a double safety, which locks both the trigger and sear system. The trigger guard folds out of the way for use with gloves or arctic mittens.

Dust Cover. This assembly has no function in the alignment of the operating group components as in conventional receiver type mechanisms. As its name indicates, its purpose is to keep dust and other foreign matter away from the moving parts of the weapon.

Foregrip. Constructed from sheet metal, the foregrip is of clamshell construction, which permits it to snap over the receiver tubes for assembly to the weapon. A hinged spring-loaded door provides instant access for quick barrel change. As with the dust cover, this assembly is not a structural element in the weapon.

All of these features combine to make it possible for the XM248 (XM235) to be manufactured at moderate cost. The dual tube receiver concept alone permits a major reduction in machining costs. Simplicity is another key aspect of this design. With 40 percent fewer parts than standard machine guns now in use, spare parts inventories would be smaller, thus adding to the savings expected from component long life. Besides the XM248 (XM235), the Minimi was also being considered for the squad automatic weapon role.

Design Characteristics of the FN Minimi

Fabrique Nationale's Minimi had not changed since its public introduction in 1974. Basically a conventional gas-operated machine gun, the Minimi design embodied a bolt carrier-piston assembly with a two-lug rotating bolt. A cam pin on the top of the bolt carrier actuated the conventional-type belt-feed mechanism. An integral alternative feedway permited the use of the standard M16 rifle magazine should belted ammunition be unavailable. Curtis Johnson and his team had developed a feedway adapter that offered the possibility of using M16 magazines with the XM248 (XM235), as well. Whereas that accessory must be attached to the gun, the alternative feedway on the Minimi requires no parts change.

Other important modifications to the Minimi included a new receiver fabricated from one stamped steel plate. Affixed to the receiver by rivets or welds were the guide rails for the bolt and bolt carrier, the block into which the barrel locks, the gas cylinder support, and the magazine well; the ejection port had a dust cover. A tubular butt stock replaced the wooden stock employed on the model tested by the United States Army in 1974. The 1977 Minimi sported a new bipod design, a new quick release mechanism for the barrel, and a new rear sight. These changes improved an already excellent contender and brought its total weight into line with United States requirements.

In February 1976, General William DePuy, then the commanding general of the Training and Doctrine Command, decided in favor of a side-by-side test of an Advanced Development SAW (XM248 [formerly XM235]) and the FN Minimi chambered for the improved XM777/XM778 5.56 × 45mm ammunition. The reduction of SAW fiscal years 1978 and 1979 funds to zero in December 1976 jeopardized the project's future. In February 1977, the remaining SAW monies were allocated toward the purchase of eighteen XM248 (then the XM235 improved) SAWs from Ford Aerospace and Communications Corporation. Delivery of those weapons was scheduled for June 1978, after which it appeared that the SAW project was to have been terminated without further testing of either the XM248 or an improved Minimi.

In June 1977, the Office of the Director of Defense Research and Engineering (DDR&E) requested a review of the SAW program. On 29 June, the new SAW project officer, Major Robert D. Whittington III, briefed DDR&E personnel and requested $1,945,000 for fiscal year 1978 and $875,000 for 1979 to permit completion of the advanced development phase and to allow a shoot off between the Minimi and the XM248. The Deputy Chief of Staff of the Army for Research, Development and Acquisition (DCSRADA) later notified DARCOM that action had been taken to approve 1978 and 1979 funds for the SAW program.

On 16 August 1977, Major Whittington reviewed plans for the engineering development phase of the squad automatic weapon for DCSRADA. While the engineering development stage was not approved, the following guidelines were issued:

(1) Continue SAW advanced development phase through FY79 (30 September 1979).
(2) After the NATO Ammunition Trial decision (July 1980) on a second (light support) caliber, initiate an eighteen-month design maturity phase with an objective of fielding a SAW at the beginning of FY82 (after 1 October 1982).
(3) Include an M16 Heavy Barrel version of the SAW competition (at the request of the Office of the Deputy Chief of Staff for Operations and Plans).

The additional money for SAW was forthcoming. On 18 November 1977, DARCOM and DCSRADA agreed to provide the funds requested for FY 1978. In mid-December, the United States Marine Corps provided two hundred thousand dollars which was used for the development of the heavy barrel M16.

The T9 5.56 × 45mm FN Minimi tested by the United States Army in 1974.

Fabrique Nationale, 5.56 × 45mm Minimi Machine Gun.

The 5.56 × 45mm FN Minimi as tested in the late 1970s. Designated the XM249 by the United States Army.

XM106 Heavy Barrel M16 SAW

Fabrication of the heavy barrel M16 SAW, designated XM106 in January 1978, was assigned by ARRADCOM to the Ballistic Research Laboratory (BRL), Aberdeen Proving Ground. The XM106 had these features:

(1) Quick-change heavy barrel;
(2) Magazine capacity of not less than eighty rounds. (BRL fabricated 83-shot drum magazines and experimented with three standard 30-shot magazines clipped together as the "Tri-Mag");*
(3) Operation from an open bolt;
(4) Improved bipod with a new mounting point not on the barrel;
(5) A rear sight with settings up to 800 meters—a modified M60 machine gun sight being an acceptable expedient.

A first prototype of the XM106 SAW was completed in May 1978. The quick-change barrel assembly consists of a sleeve 152mm (6 inches) long, which is permanently attached to the upper receiver of the weapon. The barrel slides into this sleeve and is turned 180 degrees to the right to lock it in place. A standard M2 bipod (for the M14 rifle) attaches to this sleeve, which is also fitted with a pistol grip. The new leaf-type rear sight is graduated to 300, 500, 800, and 1,000 meters.

The United States Army completed negotiations with Fabrique Nationale for eighteen XM249 Minimi SAWs during July 1978, with the formal contract being signed on 7 August. On 25 August, after a summer-long debate within Army circles, the SAW project office was told by DARCOM headquarters to include the HK21A-1 in the tests scheduled for 1979. The HK21A-1 was given the experimental nomenclature XM262. Because there was still some opposition to adopting a 5.56 × 45mm machine gun, Heckler & Koch representatives began stressing the convertibility of their weapon. Assuming a basically competitive price—the HK21A-1 cost substantially less than the US M60 or the FN MAG58 machine guns—they pointed out these economic facts:

(1) HK21A-1 5.56mm SAW, with 7.62mm conversion kit to adapt the HK21A-1 as a company level weapon, at 1½ times basic cost;
(2) HK21A-1 5.56mm SAW, plus company level HK21A-1 in 7.62mm, plus .22-caliber Long Rifle conversion kit for training, at about twice the basic cost of the gun;
(3) All of the above, plus 7.62 × 39mm conversion kit for firing enemy ammunition at 2½ times the basic cost of the gun.

With the addition of the XM106 and the HK21A-1 to the competition, the evaluation process became more complex. Although the Heckler & Koch weapon had a proven record of good performance in the field, the fact that the HK21A-1 weighed about a kilogram more than the other SAW contenders, and that it fired from a closed bolt rather than an open bolt, were technical strikes against it. Heckler & Koch engineers argued that the heavy barrel with its heavy barrel extension provided a more than adequate heat sink to prevent cartridge cook-off from an overheated barrel.

The heavy barrel M16 SAW had some more serious shortcomings. At first glance this weapon appeared to be an attractive concept because of the possibility of using existing M16 components and production equipment. But the XM106 design required extensive alteration of the bolt assembly and the barrel and gas assemblies. The drum and multiple box-type magazines were not liked by the infantry officers. Whereas some individuals were extremely critical of the plan to introduce a belt-fed weapon into the infantry squad, those in decision-making positions preferred that approach. (It should be noted that the belt-fed guns, the XM248,* the XM249, and the XM262 [HK21A-1], could be converted to magazine-feed if desired. Heckler & Koch, if required, could adapt their weapon to use the M16 magazine.) Modification of the M16 Rifle for the squad automatic role was the least acceptable alternative in a contest where there were already three strong contenders.

*An earlier Marine Corps candidate for an "interim SAW" had been produced by WAK, Inc. of Medway, Ohio. Modified by Maxwell Atchisson, it used a heavy barrel, stronger buffer, and special muzzle compensator. In addition, the rifle fired from an open bolt to aid cooling and prevent cook-offs in overheated barrels. The 30-shot magazines were held side-by-side with a special metal clip, the Tri-Mag. In March 1977, the Marine Corps decided not to pursue the WAK heavy-barrelled M16 project.

* The XM248 conversion kit existed on paper only.

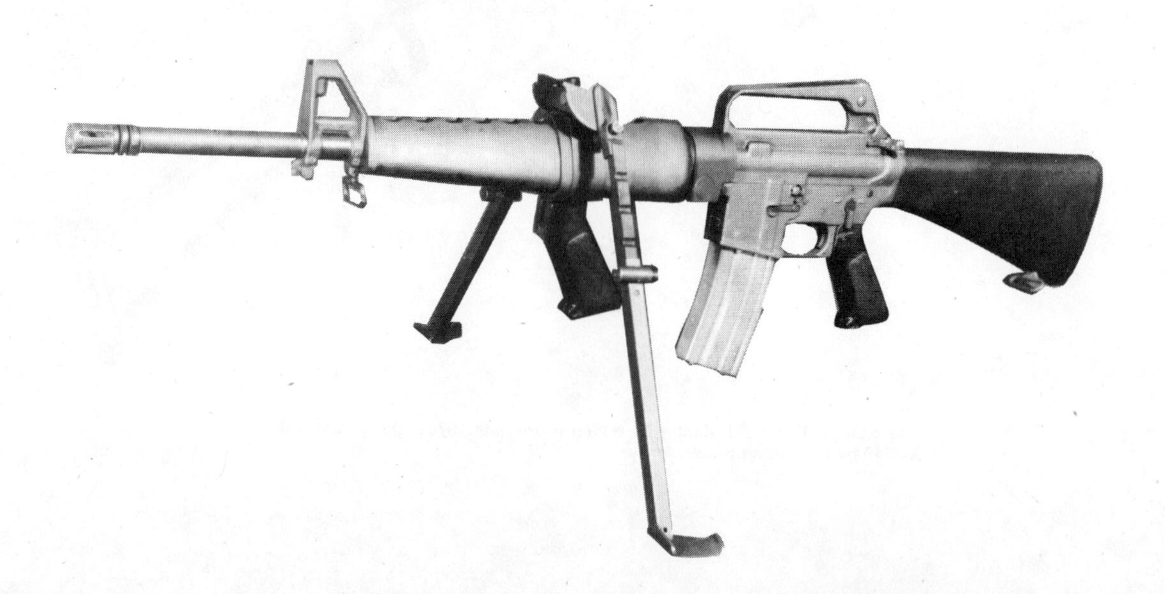

The 5.56 × 45mm XM106 SAW—a quick change, heavy-barrel version of the M16A1 rifle.

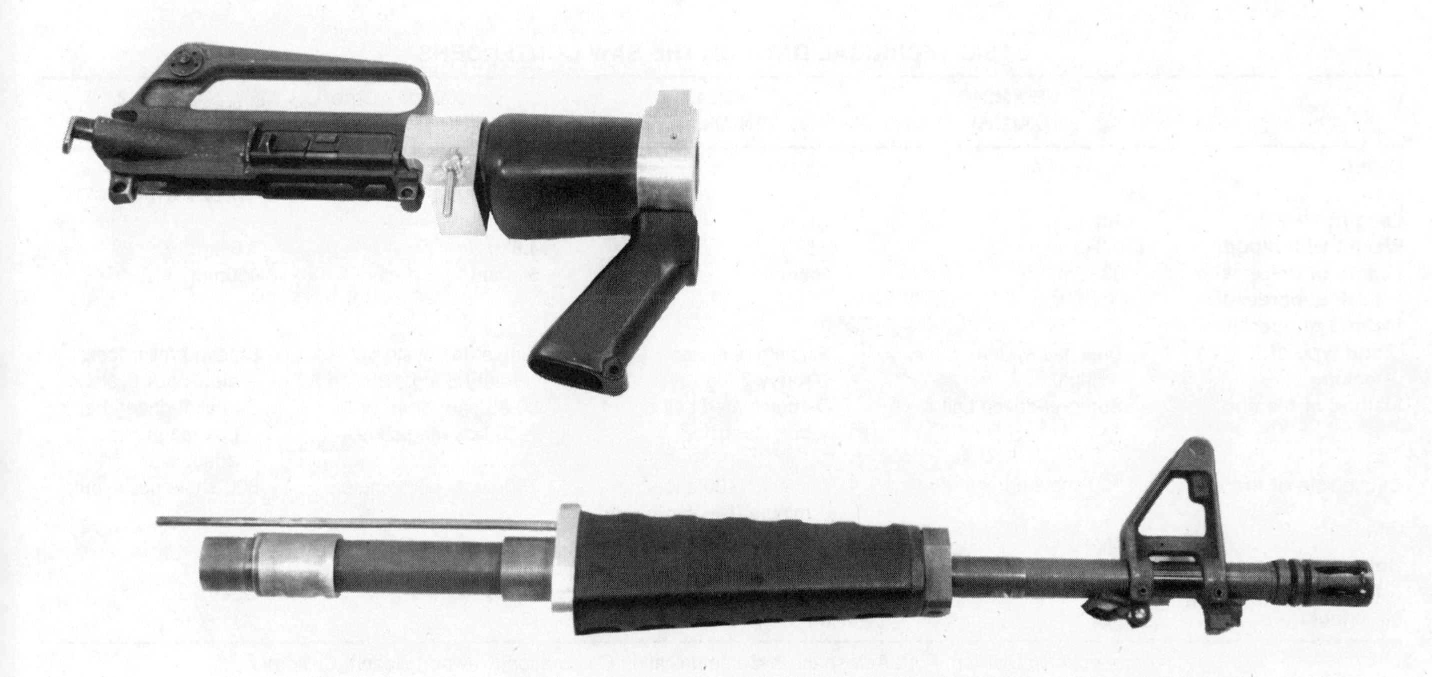

Disassembled view of the upper receiver assembly of the XM106 SAW. Note the extension of the gas tube.

Technical and Field Testing of the XM248, XM249, and XM262

From late 1978 to the summer of 1980, the XM248, XM249, and XM262 went through a series of technical tests (development tests by the United States Army Testing and Evaluation Command at Aberdeen Proving Ground, Maryland) and field tests (operational tests by the United States Army Infantry Center, Fort Benning, Georgia). On 28 May 1980, a DARCOM In-Process review (IPR) was held to decide the best candidate SAW. DARCOM favored the XM249 Minimi on technical per-

The US 5.56 × 45mm M249 SAW, which will be manufactured by Fabrique Nationale.

BASIC TECHNICAL DATA ON THE SAW CONTENDERS

	US XM248 (XM235)	XM249 (FN Minimi)	XM106 (HB M16)	XM262 (HK21A-1)
Caliber:	5.56 × 45mm	5.56 × 45mm	5.56 × 45mm	5.56 × 45mm, 7.62 × 51mm NATO
Length:	1m	1m	1m	1m
Weight with bipod:	5.3kg	6.5kg	4.8kg	7.69kg
Length of barrel with flash suppressor:	609mm	465mm	546mm	450mm
Method of operation and type of locking:	Dual gas system; rotary 3-lug	Single gas system; rotary 2-lug	Single gas system; rotary 8-lug	Delayed roller-lock; blow-back system
Method of feeding:	Sprocket-driven belt	Cam-actuated belt or box magazine	80-83-shot drum or tri-30 box magazine	Sprocket-driven belt or box magazine adapter
Cyclic rate of fire:	500 shots per minute	700 and 1100 shots per minute (two cyclic rate settings)	750 shots per minute	800 shots per minute
Bolt position before firing:	Open	Open	Open	Open
Manufacturers:	(1)	(2)	(3)	(4)

(1) Aeroneutronic Division, Ford Aerospace & Communication Corporation, Newport Beach, California
(2) Fabrique Nationale, Herstal
(3) Ballistic Research Laboratory, Aberdeen Proving Ground, Maryland
(4) Heckler & Koch, Oberndorf

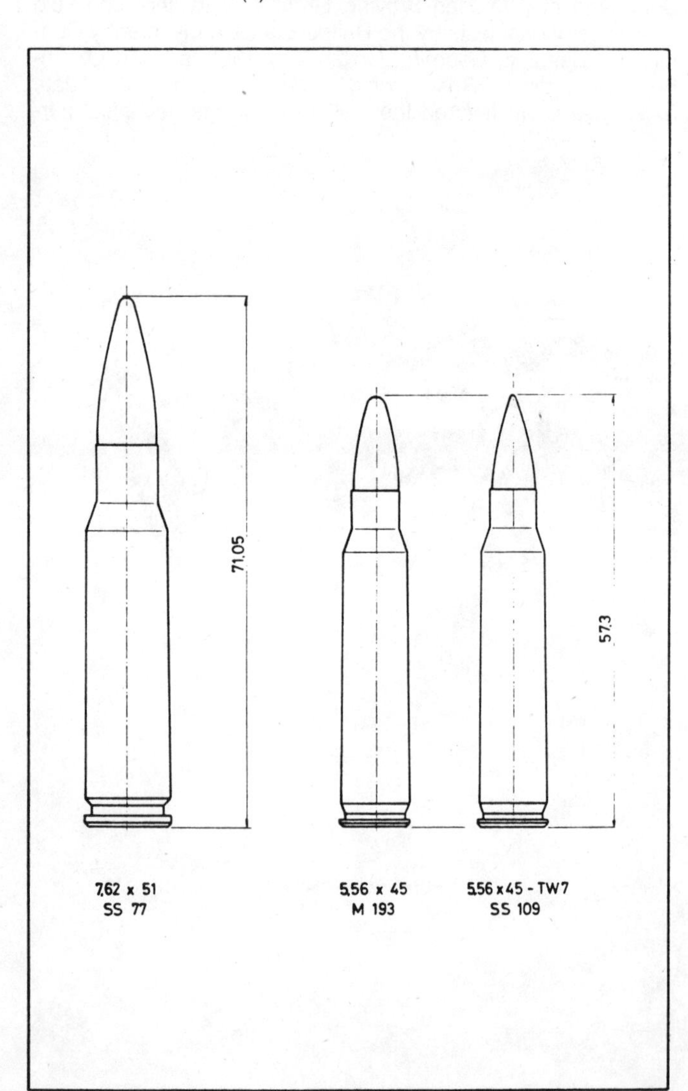

formance and production cost grounds. That decision was not well received by either Ford Aerospace or Heckler & Koch. While the XM248 Ford Aerospace SAW did not live up to the promise shown by the XM235, the Heckler & Koch XM262 had been a close technical match to the XM249 Minimi.

After a summer of intense political maneuvering on the part of all of the contractors, the United States Army decided in September 1980 to pursue the further development of the Fabrique Nationale XM249. FN modified their weapon (XM249E1) and then submitted it for further testing. This phase of testing was successful and was followed by standardization of the Minimi as the M249 Squad Automatic in January 1982. The M249 is chambered for the FN SS109 (M855) 5.56 × 45mm cartridge. As noted in Chapter 1, the United States Marine Corps plans to purchase 8,711 M249s over the next four fiscal years.

During fiscal years 1982 and 1983, the US Army purchased 8, 179 M249 SAWs from Fabrique Nationale in Belgium. Starting in FY 1984, the Army will acquire 40,000 M249s from a contractor in the United States. Contractor selection will be the result of competitive bids.

7.62 × 39mm M43 (SOVIET) CALIBER MACHINE GUNS

RPD AND RPK

The Soviets too have sought companion weapons for their Kalashnikov series of assault rifles. The 7.62 × 39mm *Ruchnoi** *Pulemet Degtyareva* (RPD) evolved from earlier designs by Vasily Alexseyevich Degtyarev (See Chapter 46). Although work was begun on the RPD in 1943, production did not start until after the end of the 1939–1945 conflict. During its service life, several variants of this belt-fed gun were manufactured. The main differences among the five major variants are:

(1) First model: Cup type gas piston; no dust cover; straight reciprocating handle; right-hand windage knob. Most first version guns now have a cylinder sleeve fitted to the gas spigot, so that their gas mechanism resembles the later versions, and have a sliding dust cover fitted over the operating handle slot.

(2) Second model: Plunger type gas piston; no dust covers; straight reciprocating operating handle; left-hand windage knob. Some second version guns have had a sliding dust cover similar to the ones fitted to the first version guns; others have had a bracket riveted to the side of the receiver to accept a nonreciprocating operating handle. This latter type may have a handle that folds upward like the later model RPDs or one that folds forward.

(3) Third model (also PRC Type 56): As for second version, but has dust covers on feed mechanism and has folding nonreciprocating operating handle.

(4) Fourth model (RPDM): As for third version, but with longer gas cylinder, additional roller on piston slide and buffer in butt.

(5) Fifth model (PRC Type 56-1): As for fourth version, but with folding magazine bracket/dust cover and cleaning rod (sectional) carried in butt.

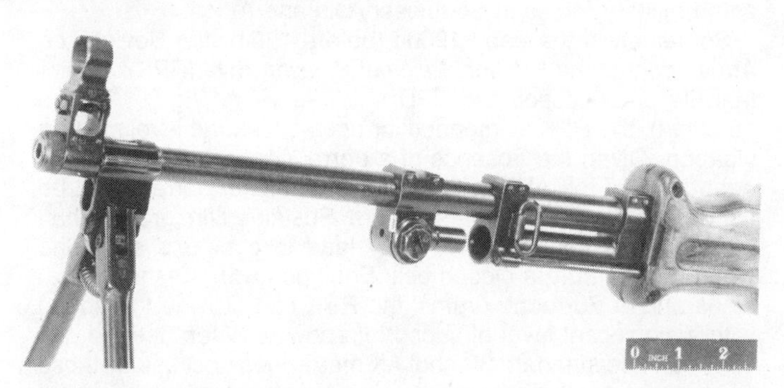

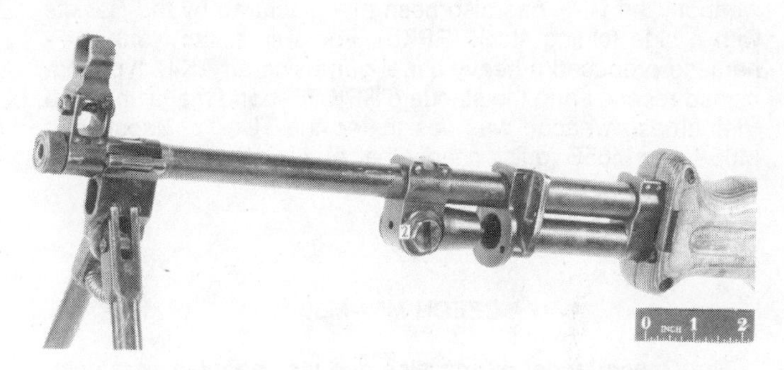

Top, first model RPD. Bottom, second and subsequent models.

These changes have no effect on the gun's operation and very little effect upon its functioning. The RPD has also been manufactured in the People's Republic of China as the Type 56 and Type 56-1 light machine guns and in North Korea as the Type 62 light machine gun. These latter types can be identified by the Chinese or Korean markings on their feed covers.

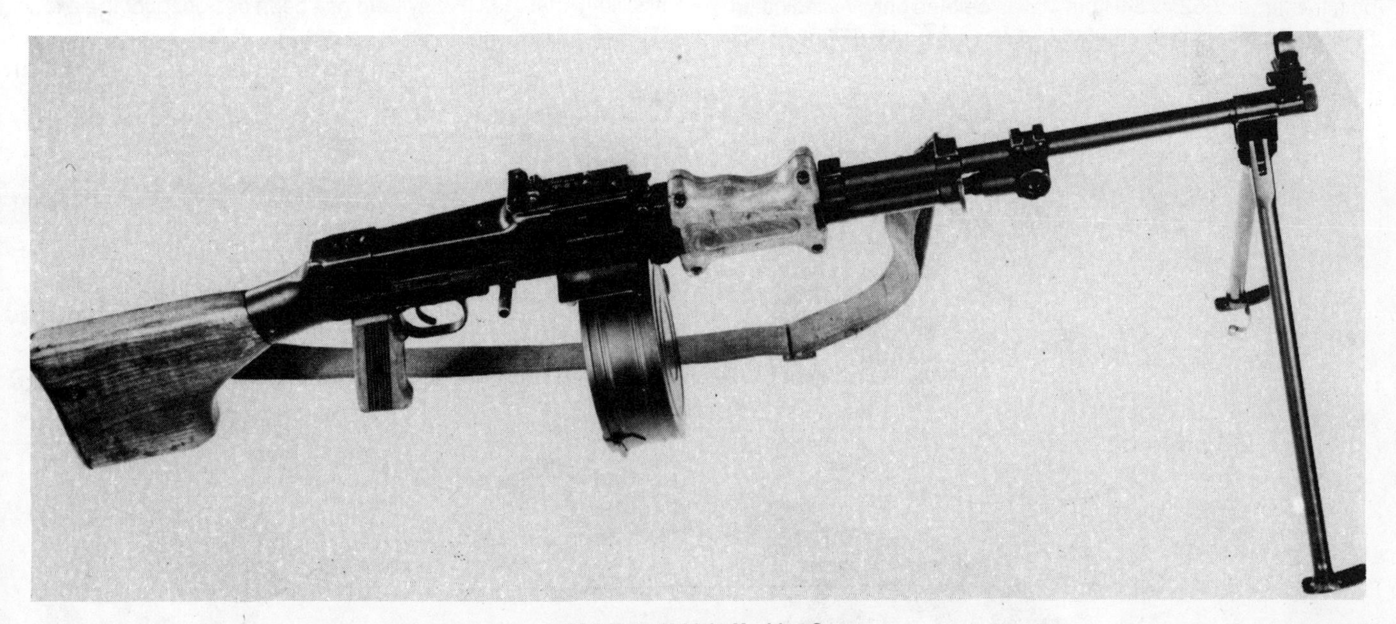

PRC Type 56 Light Machine Gun
(Copy of Soviet *Ruchnol Pulemet Degtyareva*) 7.62 × 39mm.

***Ruchnoi* or the prefix "R" is used by the Soviets to denote shoulder-fired or light machine gun. *Stankovy* or the prefix "S" denotes mounted guns; i.e., on a tripod or vehicle mount.**

Now obsolete in the Warsaw Pact, the RPD is still used by some military forces in Southeast Asia and Africa.

Sometime in the early 1960s (before 1964), the Soviet Red Army adopted the *Ruchnoi Pulemet Kalashnikova* (RPK). Lighter than its predecessor, the RPD (4.99 kg vs 7.1 kg [11 lbs. vs 15.6 lbs]), the RPK is intended for use as a squad level support weapon. Given the absence of a barrel change capability, this adaptation of the AKM with longer and heavier barrel, must be used for relatively short bursts of fire. Sustained fire greater than 80 shots per minute would likely lead to cook-offs since the weapon fires from a closed bolt. Equipped with 40-shot magazines and/or 75-round drums, the RPK can provide the squad with a significant level of support firepower. When needed, the RPK can use standard 30-shot AK magazines because all these feed devices are interchangeable. An attractive and popular weapon, the RPK has also been manufactured by the Soviets with a side folding stock (RPKS). For a time, the North Vietnamese produced a heavy barrel gun using an AK47 type machined receiver and the standard RPK 75-round magazine. The Vietnamese weapon was designated the TUL-1. Yugoslavia's M65A and M65B (quick change barrel) are also designed to be used in the same manner as the RPK.

CZECH M52-M59

The Czech Model 52 machine gun was adopted as a companion to the Model 52 Rifle. Originally chambered for the 7.62 × 45mm cartridge, it was later altered to fire the Soviet M43 round and redesignated the Model 52/57. While an extremely sophisticated weapon using a large number of stampings, it was probably too sophisticated for modern armed forces. It has been replaced by the M59 universal GPMG. The M59 borrowed several design features from its predecessor. It has also been simplified to make it less complicated and easier to manufacture. Whereas the M52 can be fed by either box magazine or linked belt, the larger 7.62 × 54Rmm M59 is belt-fed only. As indicated above, the M59N is chambered for the NATO cartridge.

Seventy-five-shot RPK magazine.

CZECH URZ

Introduced in 1970 by Ceskoslovenská Zbrojovka, Narodní Podnik, the URZ system (universal small arms) was the East European answer to the Stoner system. It consists of an automatic rifle, light machine gun, medium machine gun and tank machine gun. It would seem that the first two weapons use the 7.62 × 39mm cartridge, while the latter two fire the 7.62 × 45Rmm round. Beyond that, limited data is available, but apparently the system is based upon the M59 design. Furthermore, it is likely that the URZ system has been designed for the export market.

Soviet *Ruchnoi Pulemet Kalashnikova* (RPK) with 75-shot magazine.

Soviet RPK with 40-shot magazine.

Vietnamese TUL-1, 7.62 × 39mm Machine Gun.

Yugoslav M65A and M65B, 7.62 × 39mm Machine Guns. The latter has a quick change barrel. These weapons are issued with the standard 30-shot magazines.

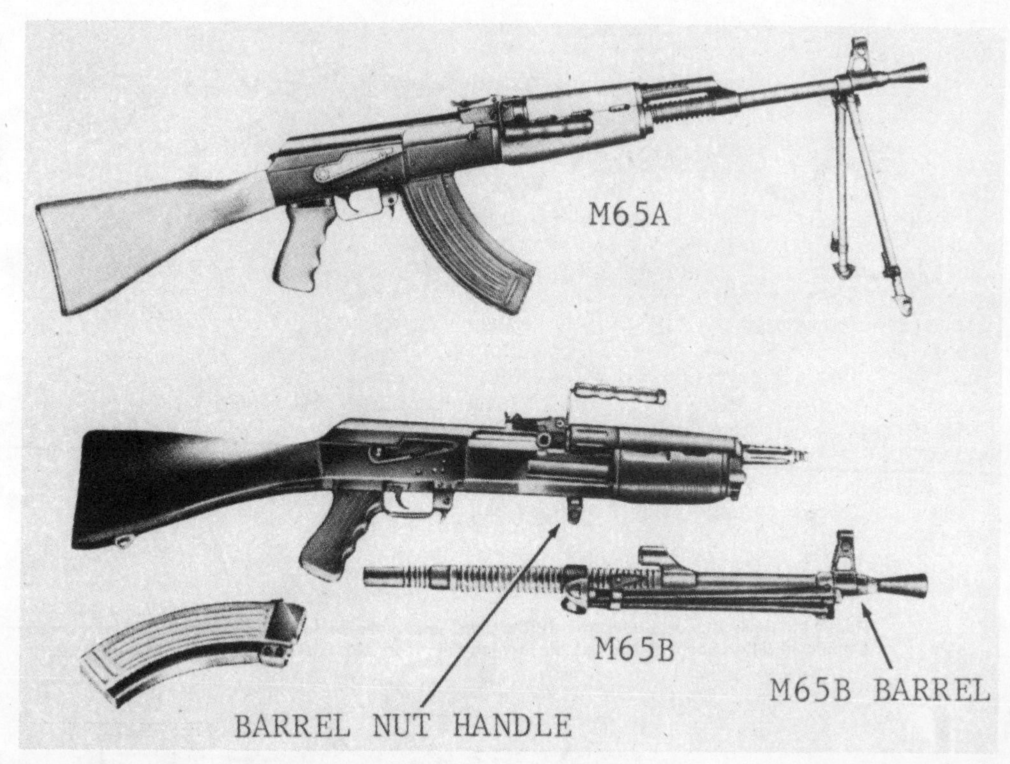

Czech M52 Machine Gun. Originally produced in 7.62 × 45mm, these guns were subsequently altered to fire the Soviet 7.62 × 39mm cartridge. They were redesignated Model 52/57. They can use either a box magazine (as shown) or a nondisintegrating belt.

Standard 7.62 × 39mm RPK squad automatic weapon.

Romanian RPK. This version of the RPK is substantially the same as the Soviet RPK, and it fires the 7.62 × 39mm cartridge. The significant difference is the adjustable legs on the bipod. More details are given in Chapter 39.

RPK 74. As noted in Chapter 1, the Soviet Union has introduced a 5.45 × 39mm version of the RPK. One of the interesting features of this weapon is the side-folding stock. This feature and other details are discussed in Chapter 46.

7.62 × 54mm R (SOVIET) CALIBER MACHINE GUNS

DEGTYAREV MG

Soviet tacticians continue to favor their Model 1891 rimmed cartridge for their base-of-fire weapons. The Soviets have upgraded the infantry machine guns several times since the Degtyarev DP was first manufactured in 1933 (See Chapter 43 for full details.) A. I. Shilin modernized the Degtyarev in 1946 by placing the recoil spring in a straight line with the recoiling parts of the operating mechanism instead of under the barrel and by adding a pistol grip. The weapon was redesignated DPM ("M" for modified). The one major shortcoming of the Degtyarev weapon was its horiziontal 49-shot magazine. More reliable when loaded with 47 cartridges, that magazine severely limited the amount of fire that could be delivered. Toward the end of World War II, Shilin, P. P. Polyakov and A. A. Dubinin devised a belt-fed mechanism for the gun. As standardized in 1946, the RP46 could use either a 250-shot continuous metallic belt or, with a change of the top cover, a flat drum during an assault. Called the "Company" machine gun, it was designed for utilization as a base-of-fire weapon; that is, as a medium rather than light machine gun. Although the Red Army adopted this weapon, it has been more widely used by the PRC (Type 58) and North Korea (Type 64). The RPD described above was the final development in the Degtyarev series.

GORYUNOV MG

A heavy 7.62 × 54mm R Degtyarev machine gun, the M1939 never worked satisfactorily. It was replaced by the SG43 designed by Pytor Maximovich Goryunov. Primarily used as a fixed gun, the Goryunov was often found mounted on tanks, armored personnel carriers and earthwork emplacements. Whereas the SG43 was only .6 kg heavier than the RP46, its barrel weighed 4.8 kg (vs. 3.2 kg for the Degtyarev barrel). This extra weight in the barrel pemitted the weapon to provide more sustained bursts of fire without overheating. All SG43 models (Six different versions are described in Chapter 43.) use the same 250-shot belt as the RP46. SG43s will still be found in use by Warsaw Pact armed forces and in the inventories of nations receiving aid from the Soviet Union and her allies, but the PK series is rapidly displacing the Goryunovs.

Soviet RP46, 7.62 × 54mm R Company Machine Gun.

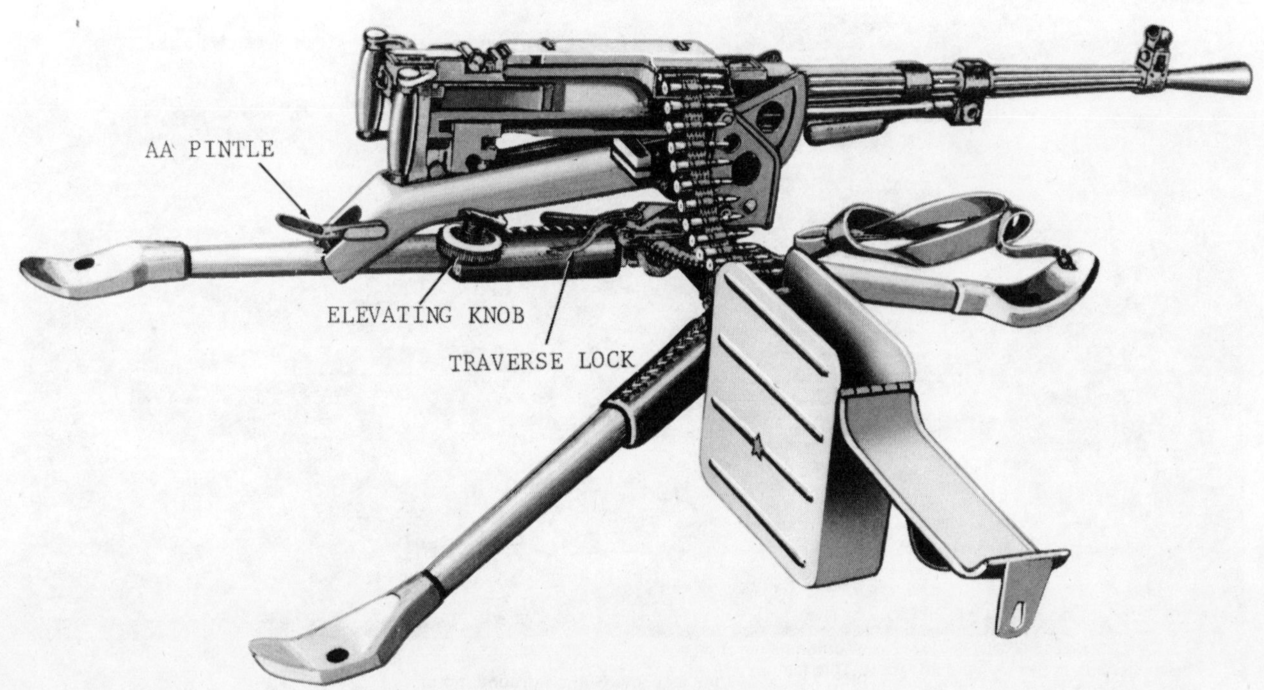

SGM on Sidorenko-Malinovski tripod.

KALASHNIKOV PK SERIES

M. T. Kalashnikov's design bureau has produced yet another excellent weapon in the *Pulemet Kalashnikova.* As the Soviet Union's entry in the GPMG category, the PK borrows a number of components and concepts from earlier weapons. The bolt and bolt carrier are from the AK series and for purposes of a machine gun inverted, with the gas piston mounted below the barrel instead of on top as in Kalashnikov assault rifles. For feeding the standard belt, the cartridge gripper of the Goryunov was incorporated into the PK. The Goryunov barrel change system was also borrowed. In the feed mechanism, the idea of using the piston to power the feed pawls comes from the Czech M52. Finally, the trigger mechanism is an adaptation of the one used on the RPD. This weapon has an excellent record for reliability. During the US search for an M219 replacement, a rebarrelled PK (changed from 7.62 × 54Rmm to 7.62 × 51 NATO) came in fourth. That performance has impressed many American and European technical people.

When the PK first came to the notice of Western intelligence specialists in 1964, it was assumed that the PK would just replace the RP45 Company machine gun. But Kalashnikov's design bureau has shown considerable ingenuity and flexibility. Six variants currently exist.

(1) PK: The basic gun with a heavy, fluted barrel; stamped and machined feed mechanism components; a plain butt plate without shoulder rest; weight = 8.9 kg (19.8 lbs.).

(2) PKS: The basic gun mounted on a ground tripod; can be converted for use against aircraft; weight = 7.4 kg (16.5 lbs.).

(3) PKT: Coaxial version designed for armored applications. Sights, stock, bipod and trigger mechanism have been removed. A longer, heavier and smooth barrel was fitted, and a solenoid was employed to the back of the receiver to permit remote triggering. An emergency manual trigger and safety were also incorporated in the design.

(4) PKM: Modernized or product-improved version; feed cover made entirely from stamped parts. A folding shoulder rest was added to the stock. Weight reduced by about .6 kg (1.3 lbs.).

(5) PKMS: Tripod-mounted PKM.

(6) PKB: Tripod mounted PKM in which the stock and pistol grip have been removed and in their place a set of spade grips substituted and a center thumb operated trigger installed. Quite similar to the SGMB (See Chapter 43.). The PKB is used on armored personnel carriers such as the BRDM, BTR50 and BTR60.

Introduction of the PK machine gun family gives Kalashnikov-designed shoulder weapons a virtual monopoly in the USSR. Even the Dragunov Sniper Rifle (SVD) is a variant of the AK design. Mikhail Timofeyevich Kalashnikov must, therefore, be added to the roll of the world's preeminent weapons designers. Only 62 years old in 1982, Kalashnikov will likely be an important source of small arms concepts for many years to come.

PRC TYPE 67

The PRC Type 67 machine gun is a genuinely eclectic design, which will likely replace both the Type 53 (Soviet SG43) and Type 58 (Soviet RP46) machine guns. This new Chinese weapon borrows design concepts from several earlier weapons—the DP trigger, the Czech ZB30 bolt and piston, a modified Maxim type feed, the SG43 quick change barrel, and RPD type gas regulator,

PKS, 7.62 × 54Rmm Machine Gun on ground mount.

and a Czech M59 push-out type belt. The Type 67 still employs the 7.62 × 54Rmm cartridge. Although far from revolutionary in design, the new Chinese gun will likely provide reliable service since it is based upon well tested design elements.

Kalashnikov PKM Machine Gun.

12.7mm (.50 CALIBER) MACHINE GUNS

There are three basic 12.7mm (.50) caliber guns in use today—the US M2, the US M85 and the USSR DShK-38 series.

M2 and M85

The American .50 caliber M2 machine gun was developed by John M. Browning at the end of World War I. After a series of early water cooled, aircraft and tank models were tested in the 1920, an improved version was adopted in 1933 as the M2 Browning water cooled machine gun. Subsequent models (.50 M2 aircraft gun and M2 heavy barrel gun) using the same receiver were adopted by the Army. During the Second World War, nearly two million M2 guns of all variations were produced.

John Moses Browning, the most successful arms inventor in history.

The last M2s were manufactured in 1946, and despite their incredible record of reliability many of the remaining ones are showing signs of age. Toward the end of 1976, the US Army took steps to begin making the M2 once again. An initial contract for several thousand M2 machine guns, incorporating all the changes made over the past 44 years, was awarded in 1977. The success of the M2, plus the fact that it has not been supplanted for some specific purposes, is a tribute to Browning's design talents.

Although the M2 is an excellent weapon, in the mid-1950s it was decided it was too large for contemporary applications in armored vehicles. After World War II, there was considerable experimentation with .50 caliber tank guns. The T175 series was the most successful design, largely due to its reasonably short receiver. AAI produced several versions of this weapon, and in June 1959 the T175E2 was standardized as the M85 caliber .50 Tank Machine Gun. This gun has been produced by General Electric, Springfield Armory and Rock Island Arsenal. The M85 will continue to be the standard American coaxial .50 caliber machine gun, and the M2 will be used in other roles; for example, as a flexibly mounted gun on armored fighting vehicles. Just over 10,000 M85s have been produced.

Unfortunately the M85 machine gun never lived up to its promise. As a consequence, the designers of the United States Army's new main battle tank, the M1 Abrams, specified the M2 Heavy Barrel Browning machine gun instead of the M85 as the coaxial gun for the M1 tank. Also, in recent years there has been an increased demand for vehicle- and ground-mounted .50 caliber (12.7mm) machine guns to counter the Soviet 14.5mm automatic guns employed on armored personnel carriers (APCs). This need was made forcibly clear during the Yom Kippur War of 1973, when the Israeli Defence Forces encountered APCs armed with 12.7mm and 14.5mm automatic weapons. When the United States supplied M2 HB machine guns on an emergency basis to the Israeli Defence Forces, the American stockpile of M2s was severely depleted. Between 1951 and 1978, the United States Army either gave away or sold 100,000 M2 Browning machine guns; between 1973 and 1978 15,300 were shipped overseas. Coupled with the destruction of several hundred thousand M2s during the post-1945 years, the United States Army

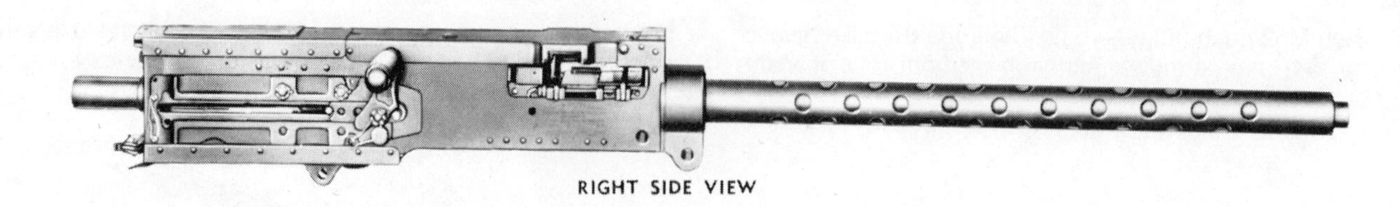

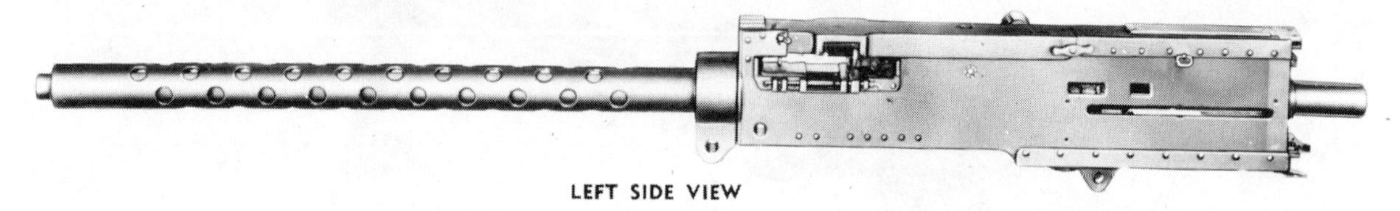

US Aircraft Version of the .50 Caliber Browning Machine Gun.

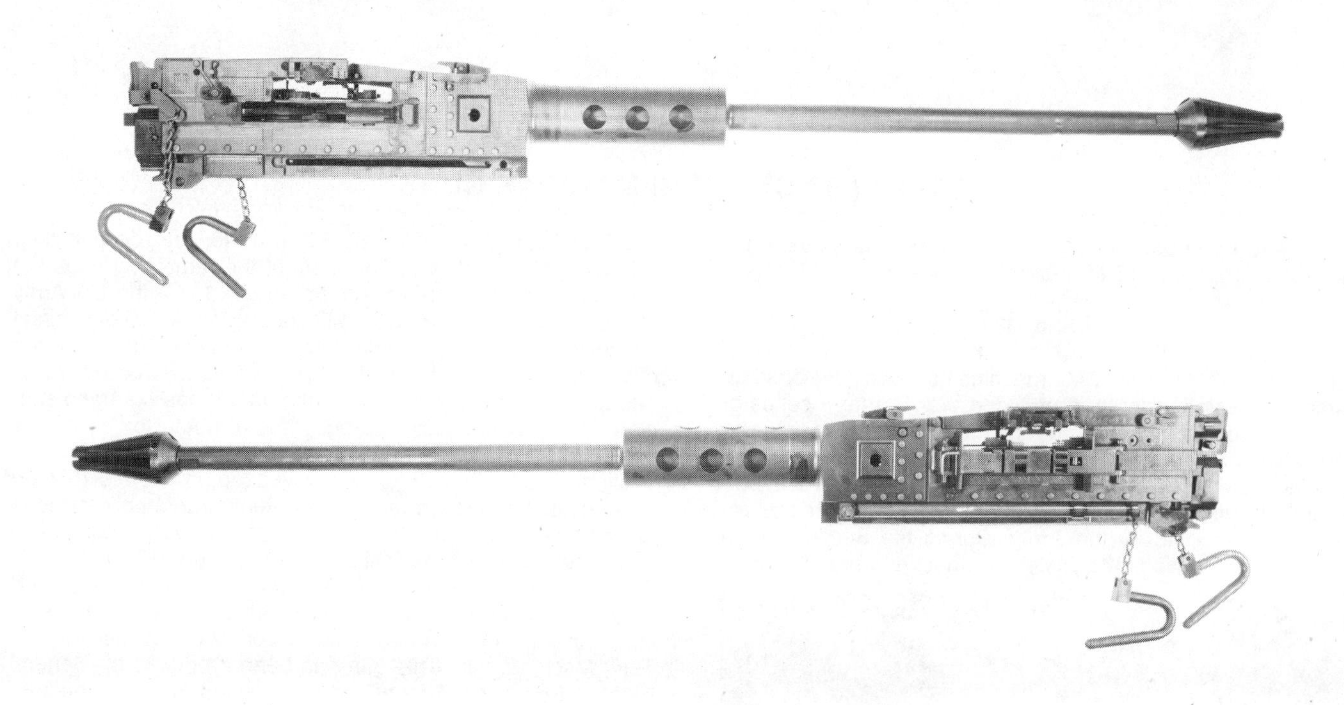

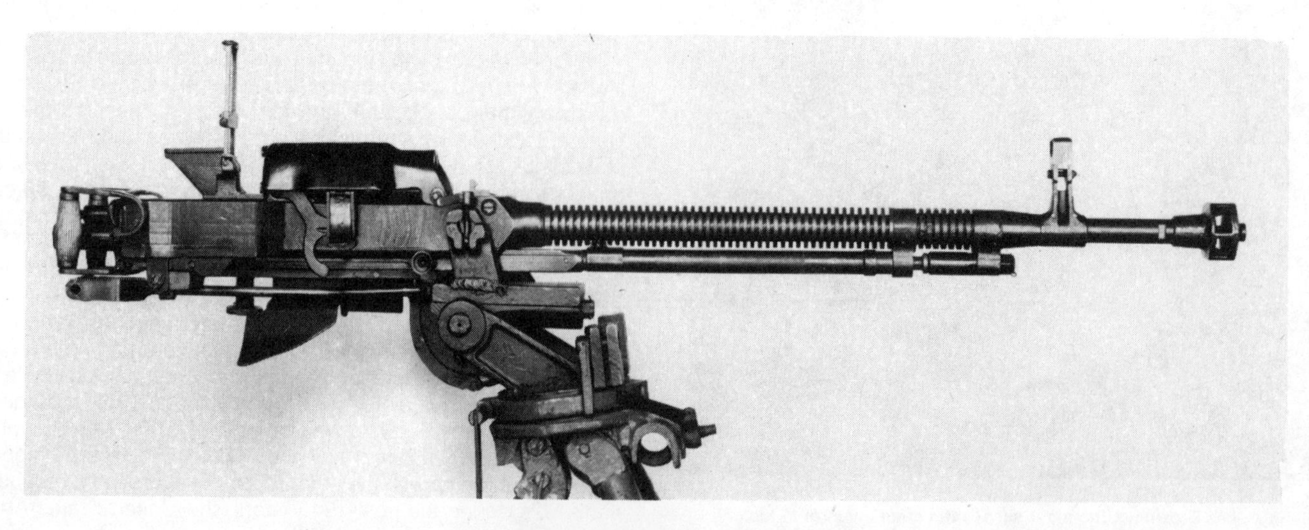

Soviet 12.7mm DShK Machine Gun.

was running low on such weapons by the late 1970s. In 1979, the United States Army needed about 69,000s M2, but only 45,217 were available. This shortage led the United States Government to contract with Maremont Corporation's Saco Defense Systems Division in 1977. The first Maremont M2s were delivered in the fall of 1979 for testing. Since that time Saco has continued to deliver these guns to the United States Army. A significant portion of these have been distributed to friendly nations through the Foreign Military Sales (FMS) program. As noted in Chapter 1, the United States Marine Corps plans to procure 750 M2 HB machine guns between 1982 and 1985. Although that seems like a small number, the price tag will be about $6.5 million.

In response to the growing demand for M2 Browning machine guns from friendly nations not receiving weapons as part of the FMS program, Fabrique Nationale and a new firm (established in 1977), Ramo Incorporated, of Nashville, Tennessee, began rebuilding these guns. Since that time Ramo has shifted to making completely new weapons. When Saco began to feel the pinch of competition from Ramo, it stopped selling barrels to the Nashville firm. In response, Ramo went into the barrel-making business, and it is believed to be the only non-government-supported manufacturer making .50 caliber barrels with stellite liners. Since 1977, Ramo has made more than four thousand .50 caliber M2 heavy barrel machine guns. The Ramo guns are made to the military specifications that existed prior to the Saco contract. Saco guns are made to a revised military specification that was part of the firm's contract. Both firms manufacture quality weapons, but their parts may not interchange.

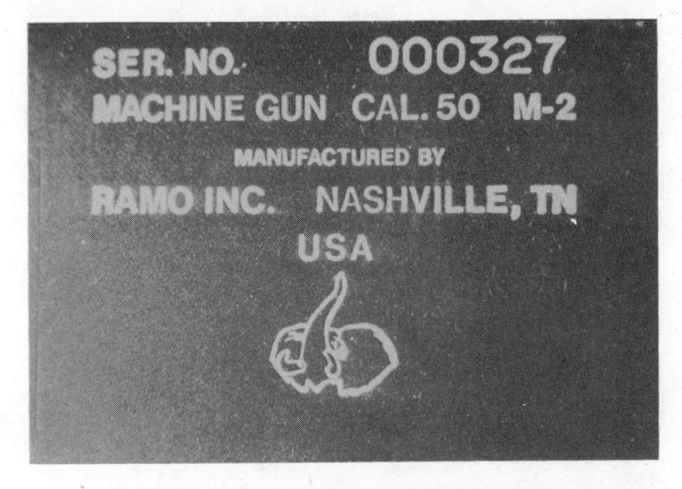

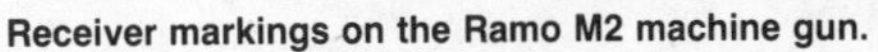
Receiver markings on the Ramo M2 machine gun.

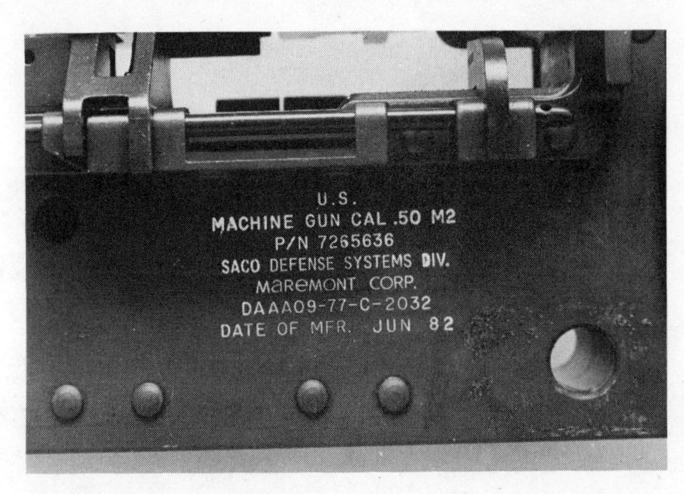

Receiver markings on the Saco M2 machine gun. This weapon is made to a revised military specification.

The .50 caliber (12.7 × 99mm) M2 HB Browning machine gun as manufactured by Ramo, Inc., of Nashville, Tennessee.

DShK38 AND M38/46

The Soviet DShK38 was the product of V. A. Degtyarev (basic gun design) and Georgii Semyonovich Shpagin (feed mechanism). Shpagin is better known for his work on the PPSh41 submachine gun. Shpagin's feed system for the 12.7mm was a rotary-type, in which successive cartridges were removed from the links, fed through a feed plate and then pushed into the chamber by the forward traveling bolt. Since this process was somewhat complicated, there were frequent stoppages with the DShK38 in combat. After the 1939–1945 war, the Soviets introduced the simplified M38/46, which had a scaled-up shuttle type mechanism of the kind found on the RP46. Parts do not interchange between the DShK38 and the M38/46. These guns are still widely used, and a PRC copy of the M38/46 (Type 54 HMG) has also been manufactured in large quantities. It should be noted that American .50 caliber guns fire a 12.7 × 99mm cartridge, while the Soviet guns use a 12.7 × 108mm round.

Sergeant Frederick Lohnes in the foreground is seen firing the 12.7 × 108mm Soviet DShK heavy machine gun during a September 1953 firing demonstration in South Korea. Sergeant First Class Elmer Parsons is firing a US .50 caliber (12.7 × 99mm) M2 HB Browning machine gun.

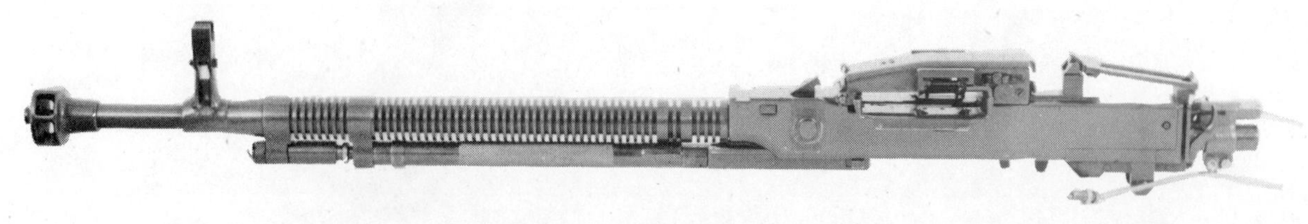

The solenoid-fired version of the 12.7 × 108mm DShK heavy machine gun.

MACHINE GUNS—A 1983 STATUS REPORT

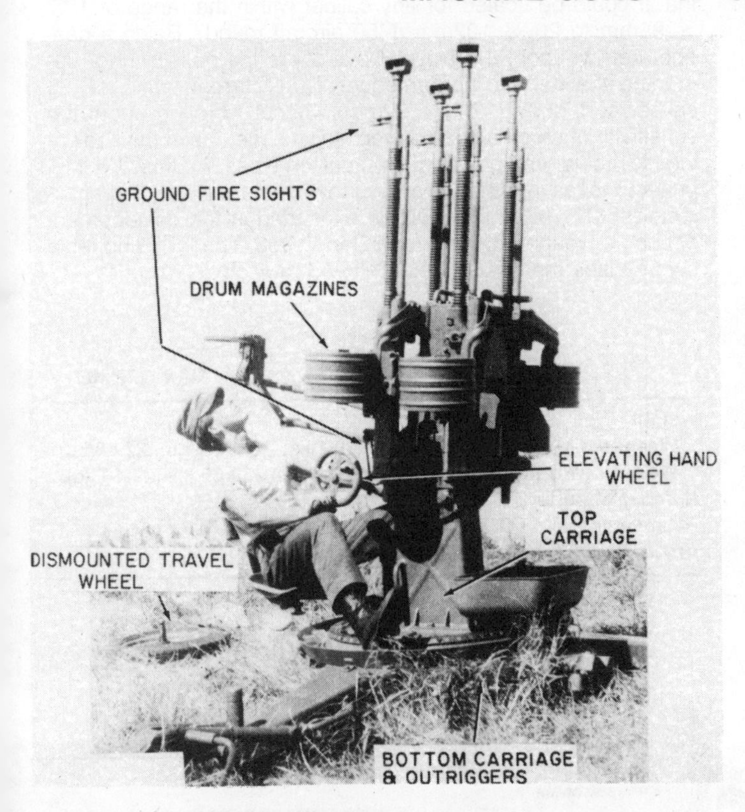

Czechoslovak Quad-mounted M54 12.7mm Machine Guns. These guns are a copy of the Soviet Model 1938/46.

Two important trends in machine guns have appeared during the last five years. First, the caliber of the squad automatic weapons in both NATO and the Warsaw Pact armies has been reduced (from 7.62mm to 5.56mm and 5.45mm). This change has been largely the result of wanting to reduce the weight load of the infantryman while allowing him to carry a greater number of ammunition per pound or kilo. At the platoon level, the forces of NATO and the Warsaw Pact will continue to carry 7.62mm caliber machine guns so a base of fire can be provided with heavier projectiles at longer ranges than can be obtained with the small caliber assault rifle cartridges. Second, both the Warsaw Pact and the NATO ground troops will be supported by heavy machine guns, especially ones mounted on the APCs that accompany the infantry to the forward edge of the battle area. Photographs from the Soviet Union show a new flexibly mounted 12.7mm machine gun on the commander's cupola of the Soviets' more recent vintage main battle tanks. This weapon appears to be a larger scale version of the 7.62 × 54mm R PK machine gun. In the United States, the Joint Service Small Arms Program (JSSAP) is searching for a new .50 caliber (12.7 × 99mm) general purpose heavy machine gun (GPHMG).

The current JSSAP general-purpose heavy machine gun project being carried out by the Fire Control and Small Caliber Weapons Systems Laboratory at the United States Army Armament Research and Development Command in Dover, New Jersey, has been unofficially designated the Dover Devil. This project is exploratory and is a further evolution of the tubular exoskeleton-type receiver mechanism experimented with in the XM235/XM248 SAW. The Dover Devil has a two-piston gas system, dual-feed (right and left sides) mechanism that allows the mixing of am-

US Army Quad-mounted .50 Caliber M2 Machine Guns installed on flat-bed of a truck.

The new Soviet 12.7 × 108mm NSV, Heavy Machine Gun.

munition, and the basic design was created with the thought that it could be scaled for any caliber within the range of 12.7 × 99mm to 20 × 102mm (M50 aircraft round). Being a gas-operated weapon, the Dover Devil offers the possibility of upgrading the ballistic performance of the current family of .50 caliber cartridges (12.7 × 99mm) without worrying about the sensitivity of recoil-actuated mechanisms (i.e., as in the Browning M2) to varying power of ammunition types. As this GPHMG project moves into the advanced development stage, it is anticipated that $14.5 million will be expended in the development of such a weapon between 1982 and 1988. The following table summarizes the goals of the GPHMG project:

	M2 HB	New GPHMG
Weight		
Weapon:	84 lbs. (38.1 kg)	50 lbs. (22.68 kg)
Weapon with tripod:	128 lbs. (58.1 kg)	65 lbs.(29.48 kg)
Number of parts	364	160
Relative cost:	1	.5
Ammo feed capability:	Single	Dual

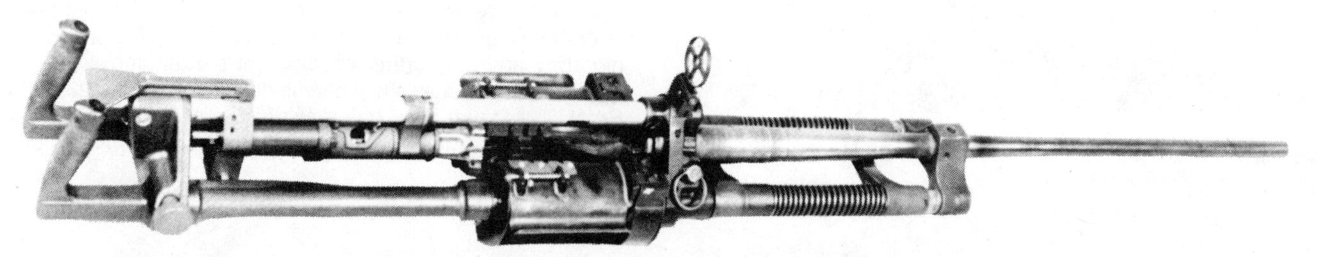

The Dover Devil serial number DD-001 circa 1978–1979. This particular version does not have the sheet metal receiver cover so the open bolt, trigger, and guide tube assembly can be seen.

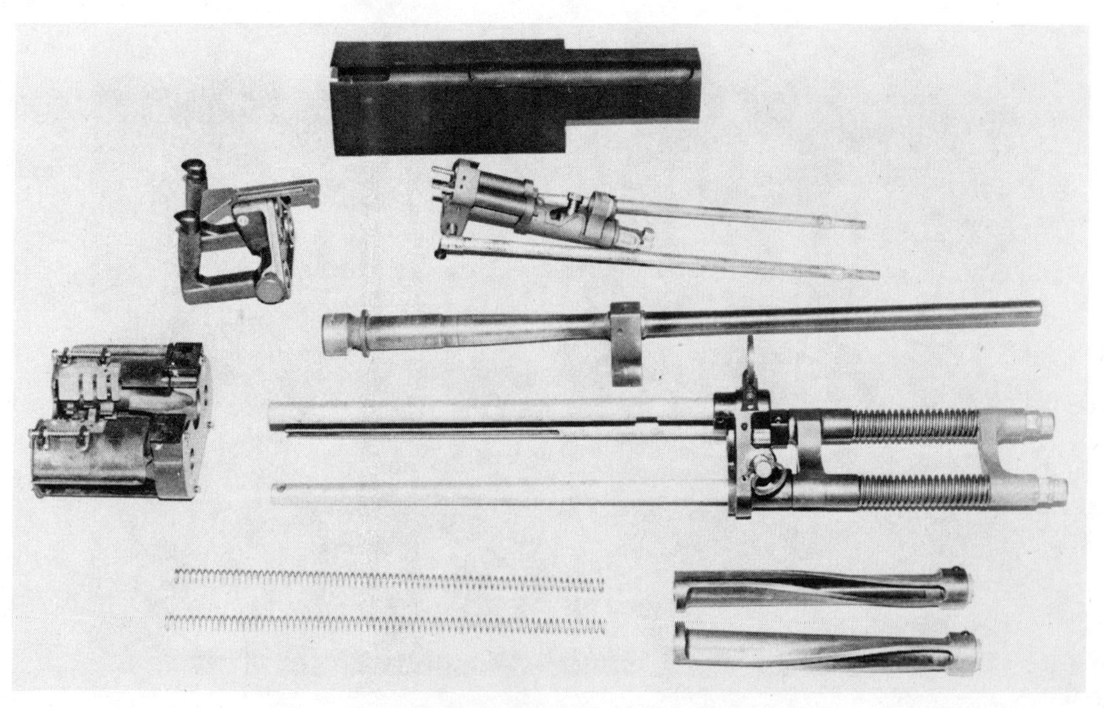

Disassembled view of the Dover Devil GPHMG.

A 1982 version of the 12.7 × 99mm Dover Devil GPHMG.

The 7.62 × 51mm NATO Saco M60 Lightweight. Saco has offered this weapon for foreign sales, and it is currently being tested in several countries. As tested by the US Marine Corps, it was designated the M60E3.

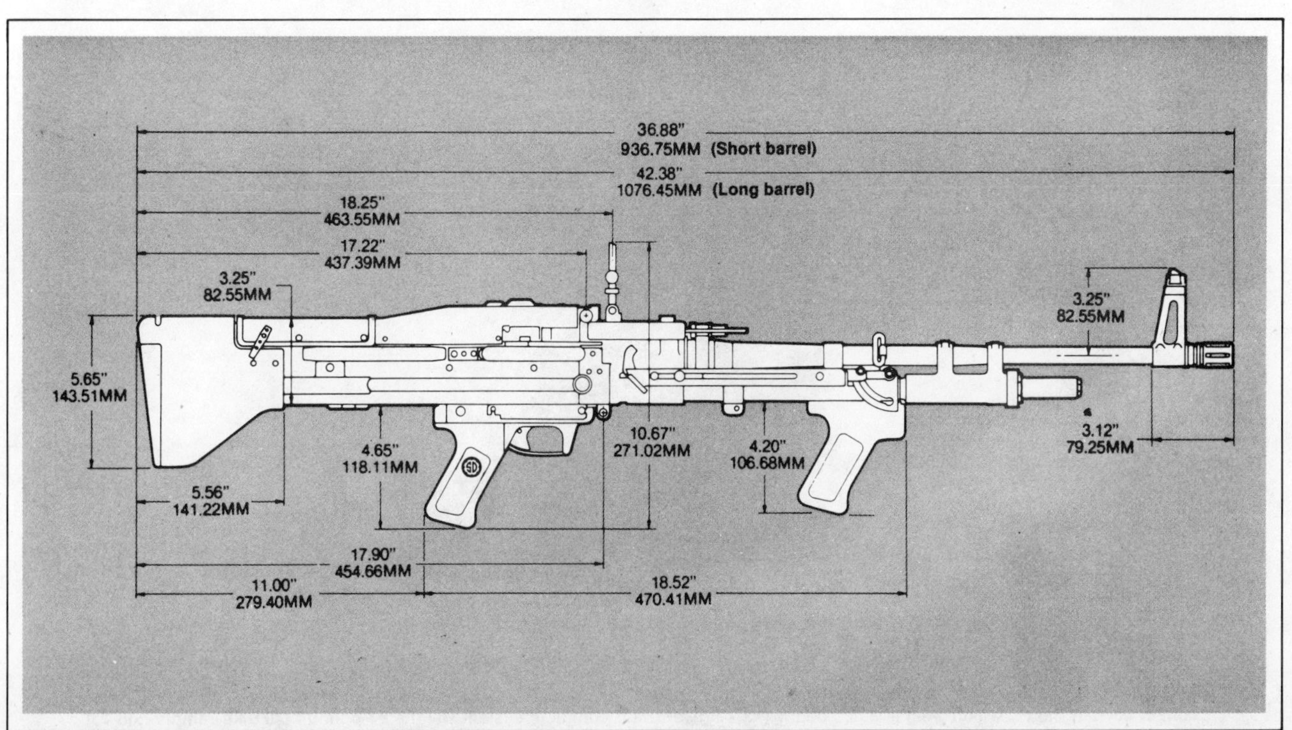

A section view of the Saco Lightweight M60 machine gun.

There are some other discernible trends in machine gun design. Many manufacturers are seeking to reduce the weight of their 7.62mm machine guns. Fabrique Nationale has brought out a reduced weight version of the MAG, which it has offered for sale in markets where troops need to patrol for long periods in jungle or other inhospitable terrain. Saco engineer George Curtis has created a lightweight model of the M60 machine gun that incorporates many of the design improvements that Saco has introduced into its product-improved M60. The Saco M60 PIP activities are intended to yield a minimum component life of twenty-five thousand rounds. Any major improvements will be incorporated into existing United States Army M60s; therefore, the changes must not affect the interchangeability of parts. The major changes include a new barrel lock washer, modified bolt to reduce the likelihood of chipping metal from the top of the bolt, improved extractor and ejector, a new firing pin and firing pin spring and spring guide, and new recoil spring and spring guide. Redesign of the operating rod and feed assembly will permit the shooter to close the feed cover with the bolt forward and still allow him to cock the weapon. Previously, the bolt could not be retracted if the bolt was forward and the feed cover was closed. The M60 PIP and M60 Lightweight will continue to be fired from the open bolt. The lightweight version will weigh 18.5 pounds (8.33 kg) or 25 percent less than the standard M60.

EXTERNALLY POWERED MACHINE GUNS

One final trend in machine guns worthy of note is the appearance of externally powered machine guns (EPMGs). In rifle caliber, the first of these were the multi-barrel miniguns developed by General Electric. These 7.62 × 51mm and 5.56 × 45mm weapons were scaled down versions of the battle-proven 20 × 102mm M61 Vulcan, which was in turn an evolution of the famous Gatling gun. General Electric's externally powered machine guns were designed primarily for use on aircraft. This was almost a necessity because of the large amounts of ammunition they required. Some ground vehicle and infantry applications of the 5.56 × 45mm XM134 Miniguns were experimented with toward the end of the Vietnam conflict, but these projects were terminated in the early 1970s when research and development funds began to dry up.

More recently, the ordnance division of Hughes Helicopters, Inc., of Culver City, California, has developed a 7.62 × 51mm NATO variant of their 25mm and 30mm chain guns. The 7.62mm EX-34 Chain Gun can be used as a coaxial gun in an armored vehicle or can be mounted on an airplane or helicopter. By comparison with the 4000- to 6000-round-per-minute rates achievable with the General Electric miniguns, the 500 shots per minute of the Hughes gun may seem modest, but it is adequate for most roles and permits better management of ammunition supplies. The British Royal Small Arms Factory, Enfield will manufacture the chain gun as the coaxial gun on their next generation main battle tank. Eugene M. Stoner's ARES, Inc., of Port Clinton, Ohio, has also developed a 7.62 × 51mm externally powered machine gun. General Electric also has developed a single-barrel .50 caliber (12.7 × 99mm) motor-driven gun. However, both of these projects have languished because of the lack of an active market. The ARES gun was re-engineered by Saco Defense Systems, but to date it has not had any eager customers.

SUMMARY

With all of the new weapons that have been suggested or adopted over the last decade, machine gun design is far from static. The major limiting factor is the marketplace. With the unit cost of a new 7.62mm machine gun being in the $3,000 to $5,000 price range, many armed forces are content to keep their older serviceable machine guns. As Chapter 10 on Brazil and Chapter 40 on Singapore indicate, several nations are developing new machine guns for their armed forces. Some of these will likely find their way into the international market during the next decade. When they do, they will encounter stiff competition from other weapons offered by more established manufacturers. Price and export regulations will influence the fate of the newcomers.

The Hughes Helicopter 7.62 × 51mm NATO EX-34 Chain Gun. This weapon has an integral 0.3-horsepower (0.22 kw) 26.5-volt direct current motor, which weighs 30.2 pounds (13.7 kg), and the complete system weighs 90 pounds (40.8 kg) empty and 218 pounds (98.9 kg) with 2,000 rounds of ammunition.

3 Submachine Gun Development

The introduction of assault rifles and the increased portability of infantry machine guns have diminished the tactical role of submachine guns. Made by the millions during the Second World War, weapons such as the US M1 Thompson, UK Sten, Soviet PPSh41 and PPS42/43 and German MP38/40 are now considered obsolete. In many instances, thousands upon thousands of these submachine guns lie idle in warehouses. Even though new Sten guns might be available at a modest price from surplus stocks, potential customers seek the M16 or AKM. Not only are they more modern, powerful weapons, they also represent a status symbol—they are the current issue weapons of the two "super powers."

Despite the fact that the submachine gun is a weapon-type in search of a mission, this class of small arms continues to be very popular with designers, students of firearms, police and security forces. Two major subdivisions of submachine guns exist today—those firing full power pistol cartridges (7.62 × 25mm, 9 × 19mm Parabellum, .45 ACP [11.43 × 23mm]) and those firing the less powerful pocket pistol cartridges (.32 ACP [7.65 × 17mm], .380 ACP [9 × 17mm], 9 × 18mm Makarov). In a break with tradition but in an effort to limit confusion, we have designated the former "submachine guns" and the latter "machine pistols" in the discussion that follows. Formerly, the British called submachine guns "machine carbines," but today they also use the American term.

A third category of weapons is sometimes labeled submachine gun but is in reality a short barrelled rifle firing rifle cartridges; e.g., the Colt XM177 series and the Heckler & Koch 53, which are chambered for the 5.56 × 45mm cartridge, or the Hungarian AMD version of the 7.62 × 39mm AKM. In performance, these weapons fall between the submachine gun and the assault rifle. Their most singular feature is their extremely loud report, the result of the shortened barrel. Despite its popularity, the Colt XM177E2 is no longer a standard issue weapon in the United States Army.

By the end of the 1939–1945 war, the designers of submachine guns had evolved several basic types. By far the most popular has been the simple blow back operated weapon. In these submachine guns, the heavy reciprocating bolt once released by pressing the trigger moves forward stripping the cartridge from the magazine and detonating the cartridge primer as the bolt comes in contact with it when the round is seated in the chamber. Recoil energies and gas pressure from the fired cartridge provide the necessary force to drive the heavy bolt to the rear. The firing cycle repeats itself until the trigger is released and the sear stops the bolt from moving forward. Two good examples of the blow back submachine gun are the American M3A1 and the British Sten. First introduced in December 1942 as the M3, the updated and improved M3A1 is still in service with United States armored forces. (See Chapter 48.) The Sten

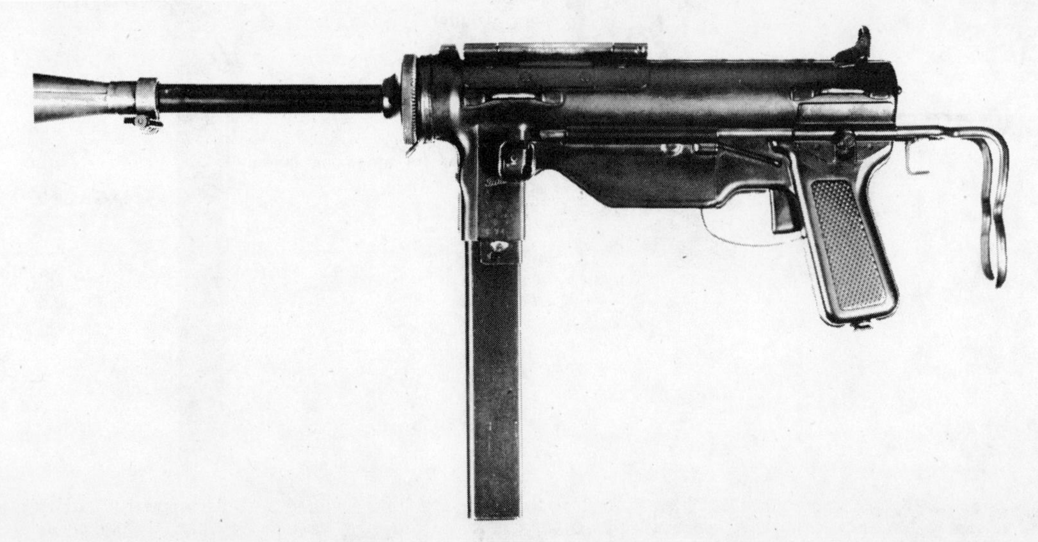

The caliber .45 M3A1 Submachine Gun with flash hider is one of the few World War II era weapons still standard with US forces. It is issued as a personal defense weapon for armored fighting vehicle crews.

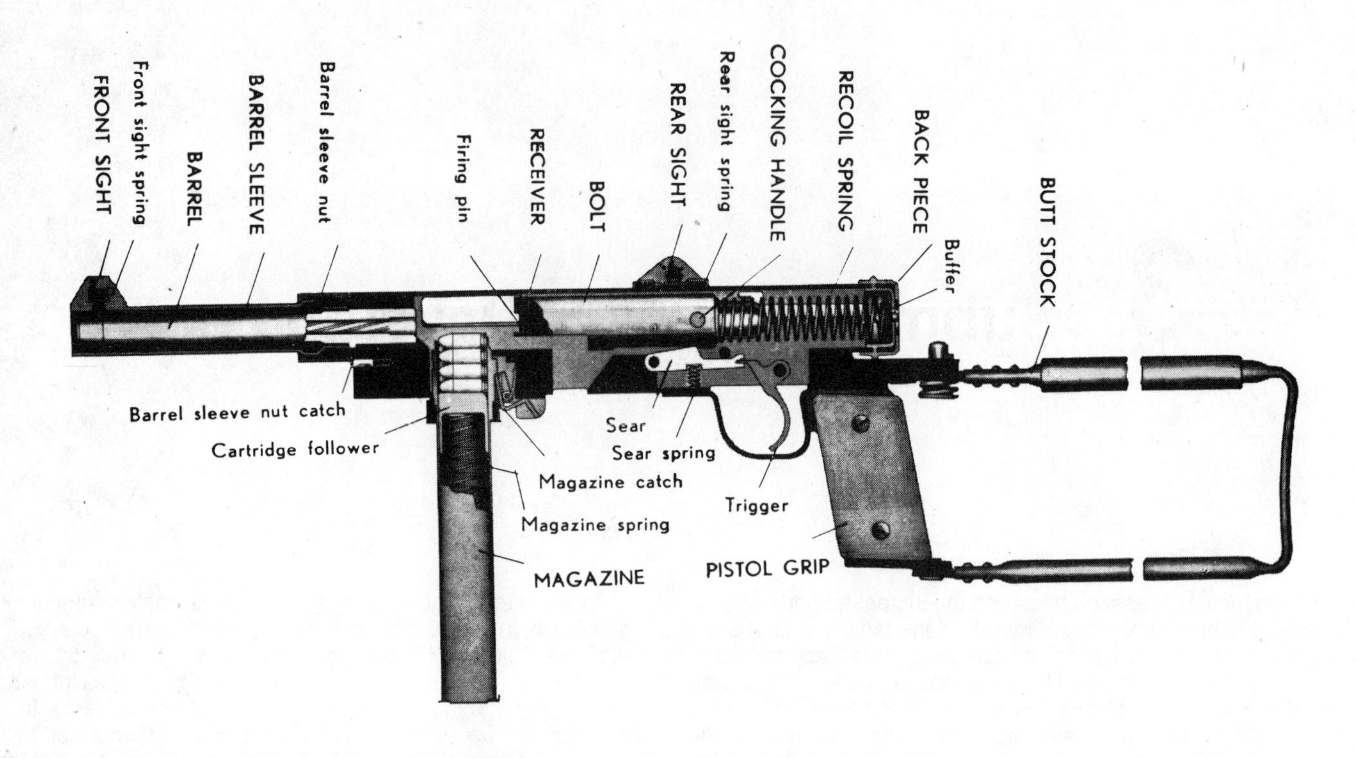

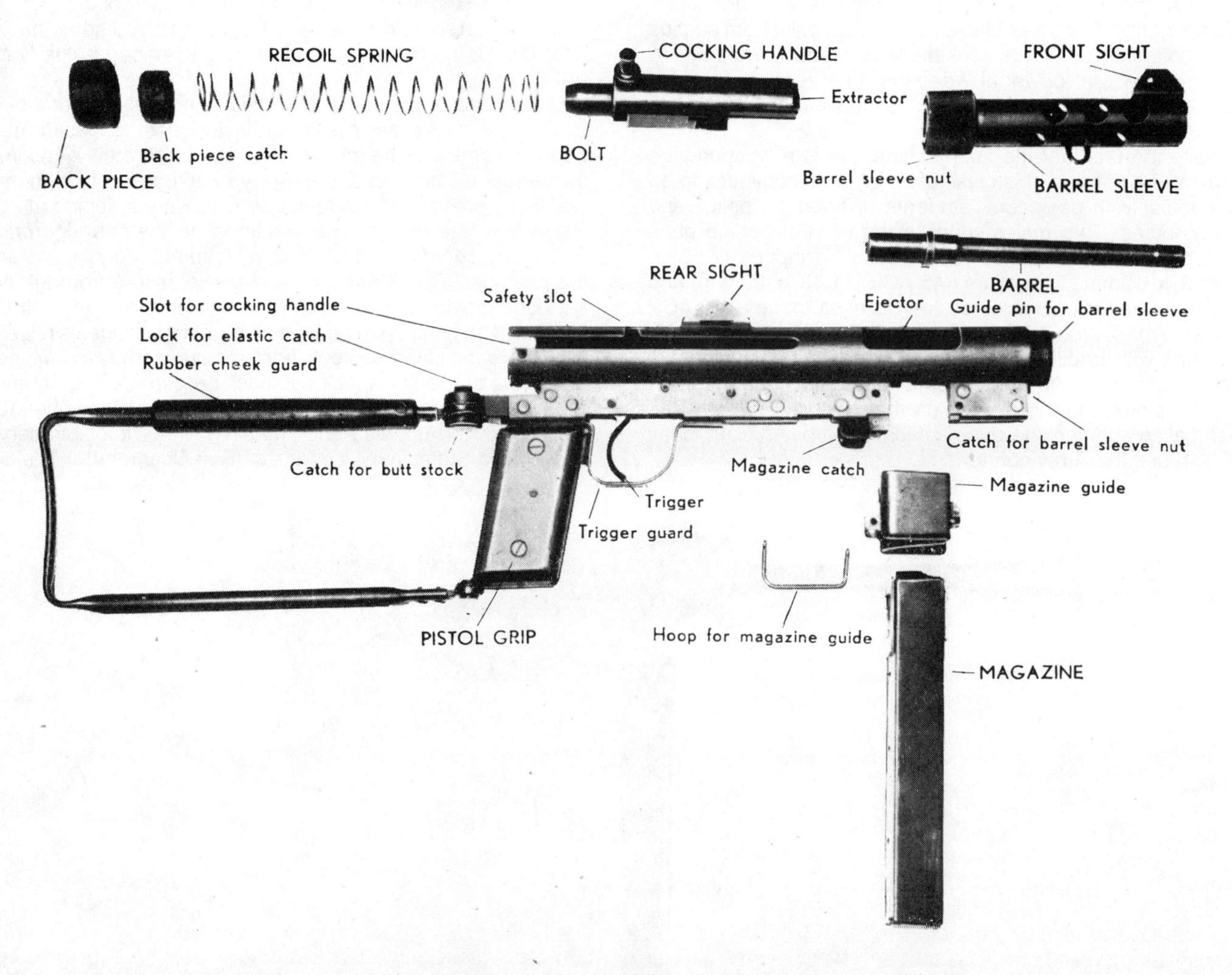

Typical construction for a blow-back submachine gun. This weapon, the Swedish M45 Carl Gustaf, has been one of the more popular post–World War II submachine guns.

Czech Model 24 7.62 × 25mm Submachine Gun.

gun, production of which began in 1941, was one of the simplest and least expensive shoulder weapons ever manufactured. (See Chapter 11.)

More complex are those submachine guns that fire from a closed or locked bolt. A post-war period example of this type of weapon is the Heckler & Koch MP5(HK54). Extending its family of delayed roller locked weapons, Heckler & Koch has marketed for several years this 9mm Parabellum submachine gun, which is based upon a number of standard components used in H&K rifles and machine guns. (See Chapter 21.)

Some representative models of current issue submachine guns and machine pistols are illustrated below.

Israeli 9mm Parabellum Uzi Submachine Gun with folding stock.

Shortly after the Second World War, Československá Zbrojovka, Narodní Podnik, at Uherský Brod, introduced a new submachine gun (two different models in 9mm Parabellum—M23 and M25—and two in 7.62 × 25mm—M24 and M26) that had a significant impact on the subsequent design of such weapons. Whereas earlier blowback submachine guns had relatively short barrels in an effort to keep the overall length of the weapons to a manageable size (The M3A1 had a 203mm [8-in.] barrel.), this new Czech design had a significantly longer barrel—284mm (11.18 in.). Overall length of the Model 25 and 26 folding stock versions was only 445mm (17.5 in.) versus 579mm (22.8 in.) for the M3A1, when the stocks for both weapons were retracted. The designers relied upon a simple design trick to achieve this result. They extended a large part of the barrel into the receiver and machined the bolt in such a fashion that it had a recess for the barrel. When the bolt was forward, it telescoped the rear 159mm (6.25 in.) of the barrel. In addition, the designers included an ejection port in the bolt, as well as in the receiver. When the bolt is in the closed or cocked position, the ejection port in the receiver is closed; thus, in effect, the weapon has nearly full time protection from the entry of dust and other foreign particles. A distinctive feature of this gun is the location of the magazine in the pistol grip. (See Chapter 15.)

While this weapon was only produced for a limited time (About 100,000 were manufactured.), the telescoping bolt concept influenced several subsequent designs—notably the Israeli Uzi and the Steyr-Daimler-Puch Model 69 9mm Parabellum Submachine Gun.

Students of modern small arms generally agree that the Uzi is the best submachine gun ever devised. That consensus, unusual in the world of small arms, is ample testimony to the creative talents of Uziel Gal, the designer of this weapon. First manufactured in 1951 by Israeli Military Industries for the Israeli Defense Forces and subsequently produced under license by Fabrique Nationale, the Uzi has been purchased by several countries, including West Germany, Haiti, Iran, the Netherlands, Panama, Peru and Venezuela. Many police forces use the Uzi for special applications, including the United States Secret Service, whose agents carry the Uzi during their Presidential protection assignments. As with the majority of blowback submachine guns, the Uzi fires from an open bolt. Gal made extensive use of sheet metal stampings and heat-resistant plastics. The barrel, which is held in position by a threaded locking nut, can be quickly changed. The Israeli Defense Forces have even experimented with Colt's Salvo Squeeze Bore ammunition, which contains three projectiles that are reduced from 9mm (.357 in.) to 7.62mm (.30 in.) in a special tapered smooth bore attachment for the Uzi. The Uzi has a 254mm (10-in.) barrel and an overall length, with the stock retracted, of 440mm (17.32 in.); it weighs 3.5kg (7.72 lbs.).

Uziel Gal

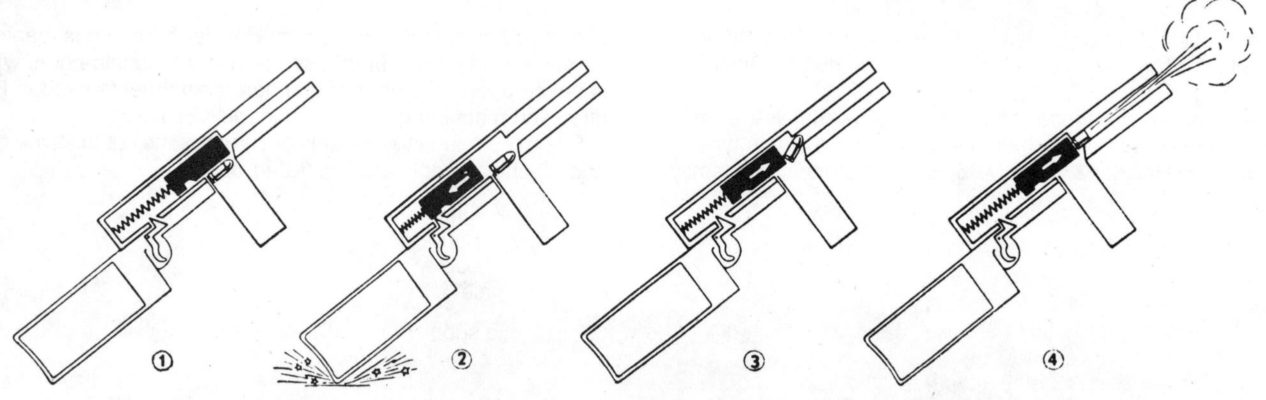

Diagram showing accidental discharge of blowback submachine gun without bolt lock.

Steyr Model 69 9mm Parabellum Submachine Gun.

9mm Ingram Model 10 Submachine Gun.

Current Uzi submachine guns in the Israeli service are designated the No. 2, Mark A. In West Germany, they are called the MP2. As with several of the newer blowback type submachine guns, the Uzi has a grip safety, making it much safer to handle than older weapons, e.g., the Sten, M3A1 and MP40. The Uzi cannot discharge accidentally when dropped because the grip safety locks the sear mechanism, preventing the rearward movement of the bolt and subsequent firing (as illustrated). Special cuts in the bolt also prevent accidental discharge if the shooter's hand slips from the cocking knob before the bolt is pulled fully to the rear in the cocked position. (See Chapter 28.)

The Steyr M69 submachine gun, designed by a team led by Hugo Stowasser, follows in the tradition of the Czech and Israeli weapons. Generally, it is a much simpler design than the Uzi. The 254mm (10-in.) barrel is rifled on the Steyr cold forging equipment. All the processes used to produce this weapon are designed to keep the cost to a minimum. This submachine gun is currently a standard weapon with the Austrian Armed Forces. (See Chapter 8.)

Other submachine gun designs embodying the telescoping bolt concept include the standard issue Argentine FMK(PA3-DM), the Australian Kokoda (prototype), the Mexican Mendoza HM-3 (limited production) and the US Maxwell Atchisson M1957. Before going bankrupt in 1976, Military Armaments Corporation in the US produced the M10 (.45 ACP and 9mm Parabellum) and the M11 (.380 ACP) submachine guns. These weapons, designed by Gordon B. Ingram, were extremely compact and lightweight. Because of the light bolt, these guns had an extremely high rate of fire. The .45 caliber M10 had a cyclic rate of nearly 1200 rounds per minute, whereas the M3A1 has a cyclic rate of about 450 rounds per minute. Both the M10 and M11 were designed to be used with the Sionics noise suppressor. Early in 1977, RPB Industries of Atlanta, Georgia began production of the M10 and M11 submacine guns.

Among the coventional submachine guns currently in service are the following:

Sterling 9mm: Britain (L2A3, Chapter 11), Barbados, Canada, Dubai, Ghana, India, Jamaica, Lebanon, Malawi, Malaysia, Nepal, Oman, Singapore, Tunisia.

Beretta M12 9mm: Italy (Chapter 29), Bahrain, Gabon, Indonesia, Italy, Libya, Nigeria, Venezuela.

Madsen Model 50 9mm and .45 ACP: (Chapter 16), Brazil, Chile, Columbia, Ireland, Paraguay, Thailand.

MAT49 9mm: France (Chapter 19), Lebanon, Morocco, Vietnam (modified to fire 7.62 × 25mm).

M3A1: US (Chapter 48), Nationalist China, Greece, Guatemala, Iran, Turkey.

Other basic submachine gun types are described in the country by country chapters of Part II.

A typical model of the Sterling-type (L2A3) 9mm Parabellum Submachine gun. (See Chapter 11.)

Australian F1 9mm Parabellum Submachine Gun. (See Chapter 7.)

Walther MPL (L = long barrel) 9mm Parabellum introduced in 1963. This weapon has been sold commercially for police use. (See Chapter 21.)

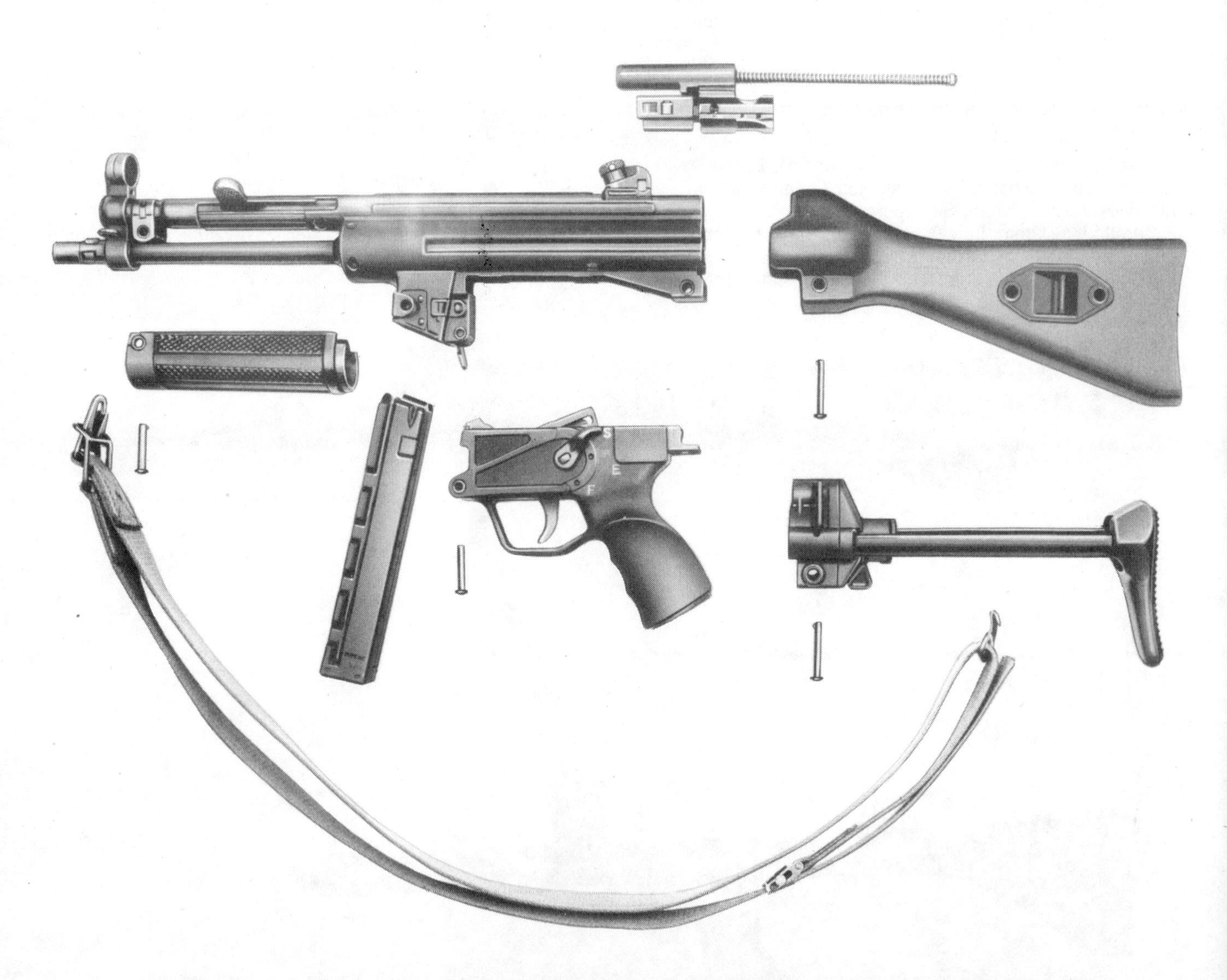

Disassembled view of HK54 adopted by West German border patrol as the MP5. Note the easily interchanged fixed and retractable butt stocks. (See Chapter 21.)

The Argentine 9 × 19mm Parabellum FMK (PA3-DM) submachine gun. See Chapter 6 for more details.

The Beretta 9 × 19mm NATO M12S submachine gun. See Chapter 29 for more details.

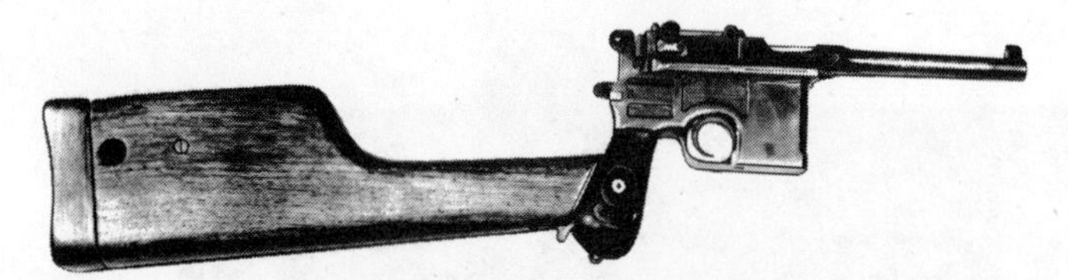

World War 1 Mauser pistol converted to caliber 9mm Parabellum and fitted with detachable shoulder stock and holster. Experimental models were full auto. After the war the design was also issued commercially with a full auto switch. This design is too light for practical true submachine gun use.

Fully automatic pistols were first used during World War I, but in Eastern Europe several new machine pistols have been introduced since World War II. The Czech Model 61 (Skorpion) was one of the earliest models. Designed by Miroslav Rybář at the Zbrojovka Brno Factory, this miniature weapon can be fired on either full automatic or single-shot. The Skorpion, chambered for the 7.65 Browning (.32 ACP) cartridge, has a 10- or 20-shot curved box magazine and fires at a high rate of fire—in excess of 850 rounds per minute. It is available with a silencer attachment. Given the limited power of the cartridge and the weapon's high cyclic rate, this gun is best suited as a personal defense weapon rather than as an offensive weapon. In addition to the Czech Army and Internal Security Forces, the Skorpion has been sold by the Czech export firm Omnipol to a number of African nations. (See Chapter 15.)

Poland, too, has developed a high rate of fire machine pistol. Designed by Professor Wilniewczyc, the Wz63 (Model 63) fires the Soviet 9 × 18mm Makarov cartridge. Unlike the Skorpion, which fires from a closed bolt, the Wz63 fires from an open bold as do conventional submachine guns. The combination of open bolt operation and high rate of fire (probably greater than the specified 600 rounds per minute) causes this weapon to be generally inaccurate. Although best suited as a personal defense weapon, the Polish armed forces issue this weapon to parachute troops as well as to armored fighting vehicle crews. (See Chapter 37.)

For a time, the Soviet Armed Forces used the Stechkin Automatic Pistol. Equipped with a wooden stock/holster, this weapon was similar in concept to the World War I Mauser automatic pistol. The Stechkin fired the 9 × 18mm Makarov cartridge at a rate of approximately 750 rounds per minute. It, too, proved uncontrollable during automatic fire. While the Stechkin is now obsolete in the Soviet Union, it may still be found in the hands of some specialized troops. (See Chapter 46.)

MACHINE PISTOL CHARACTERISTICS

	Model 61	Wz63	Stechkin
Caliber:	7.65 × 17	9 × 18	9 × 18
Overall length:			
Stock extended:	508mm (20 in.)	610mm (24 in.) (approx.)	558mm (22 in.) (approx.)
Stock retracted or removed:	269mm (10.6 in.)	333mm (13.1 in.)	226mm (8.8 in.)
Barrel length:	112mm (4.41 in.)	152mm (6.0 in.)	127mm (5.0 in.)
Weight:	1.55 kg (3.42 lbs.)	1.80 kg (3.97 lbs.)	1.02 kg (2.25 lbs.)
Operation:	Blowback	Blowback	Blowback
Magazine capacity:	10- or 20-shot	15- or 25-shot	20-shot
Muzzle velocity:	305 m.p.s (1000 f.p.s.)	325 m.p.s. (1066 f.p.s.)	340 m.p.s. (1115 f.p.s.)

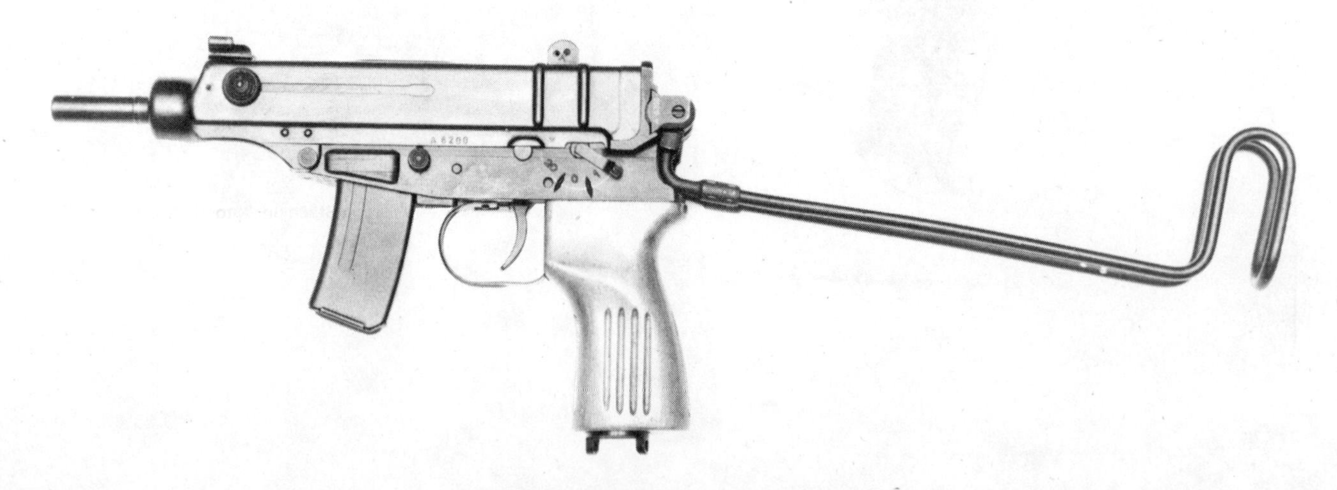

M61 Skorpion 7.65 × 17mm with 10-shot magazine.

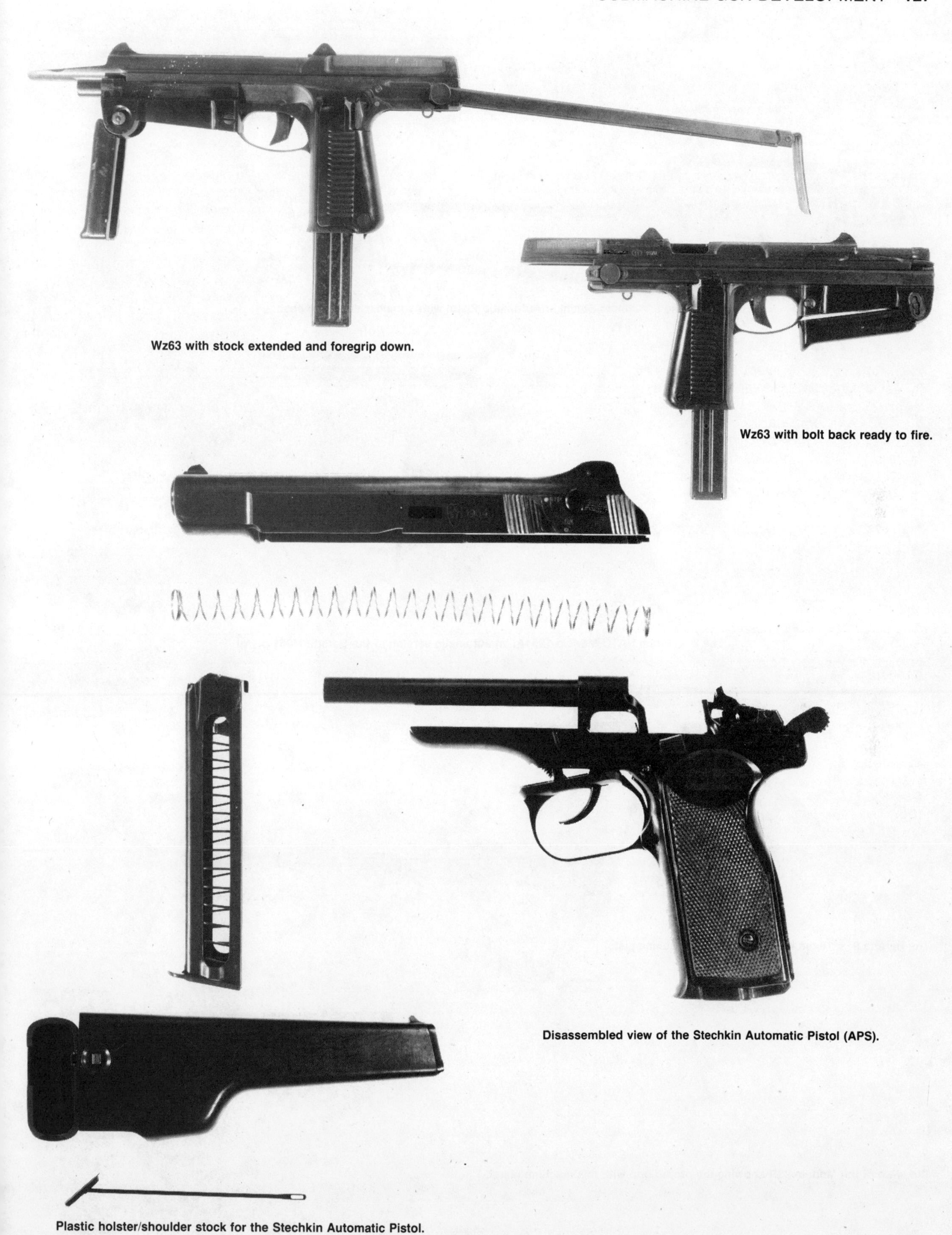

Wz63 with stock extended and foregrip down.

Wz63 with bolt back ready to fire.

Disassembled view of the Stechkin Automatic Pistol (APS).

Plastic holster/shoulder stock for the Stechkin Automatic Pistol.

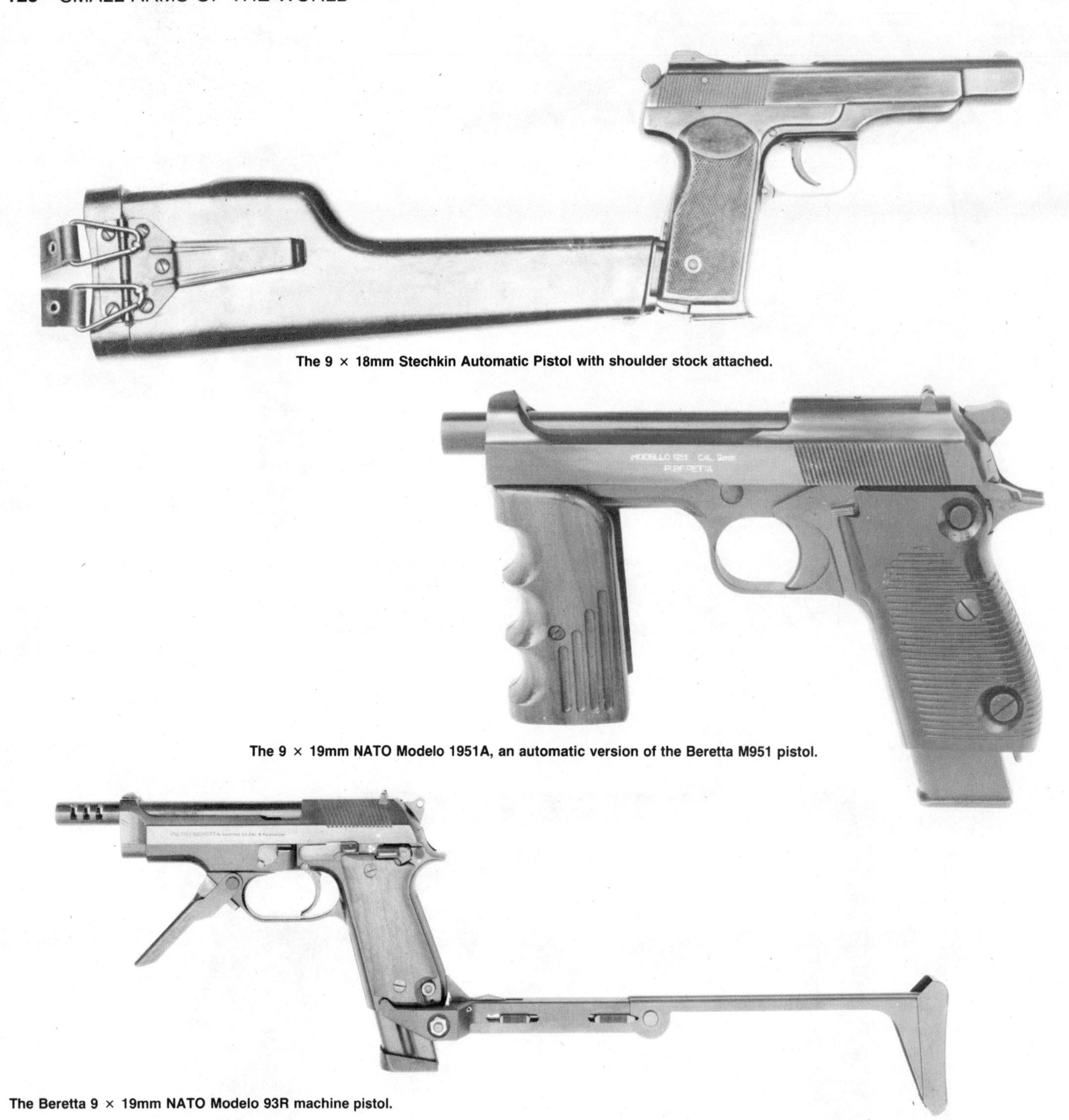

The 9 × 18mm Stechkin Automatic Pistol with shoulder stock attached.

The 9 × 19mm NATO Modelo 1951A, an automatic version of the Beretta M951 pistol.

The Beretta 9 × 19mm NATO Modelo 93R machine pistol.

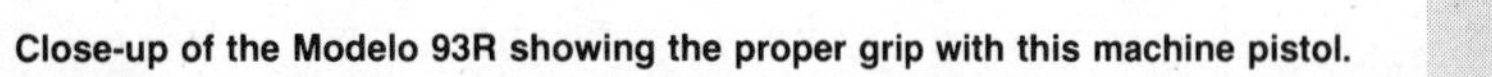

Close-up of the Modelo 93R showing the proper grip with this machine pistol.

Early in the 1970s, Heckler & Koch of West Germany introduced the 9mm Parabellum VP70 automatic pistol. When used without the plastic shoulder stock it will fire only semiautomatically; when the stock is attached, it will fire automatically (2200 rounds per minute) or in three-shot bursts. Even with the shoulder stock, the VP70 is a relatively light weapon (1.6kg[3.5 lbs.] when loaded with 18 rounds) due to the use of plastics for the basic pistol frame. Despite the use of non-traditional synthetics and the high cyclic rate, Heckler & Koch predicts a 30,000-round life span for this weapon.(See Chapter 21.)

SUBMACHINE GUNS—A 1983 STATUS REPORT

Although the traditional submachine gun role has been diminished by the introduction of lightweight assault rifles, these fast-firing weapons are still used widely by paramilitary, security and police forces in situations where a rifle would be too cumbersome or too powerful for safe use. Machine pistols are now quite often found filling roles formerly assigned to handguns. In fact, many military arms designers now speak in terms of personal defense weapons, as a consequence of the blurring of the formerly clear distinctions between machine pistols and handguns. A good example of the confusion that exists is Colt's arm gun developed for the US Air Force. For want of a better nomenclature, Colt has called this weapon the "Lightweight Rifle/Submachine Gun," which is chambered for a caliber .221 IMP cartridge. That round produces a velocity of 732 m.p.s. (2400 f.p.s.), falling between the 390 m.p.s. (1280 f.p.s.) of the Sterling L2A2 and the 1000 m.p.s. (3250 f.p.s.) of the M16 Rifle. As development of small caliber assault rifles, on the one hand, and development of personal defense weapons, on the other hand, continues, the role of submachine guns will likely continue to be restricted. Nevertheless, millions of submachine guns have been manufactured, and they will probably continue to see service. If not as the standard equipment of infantry units, they will be utilized by insurgent and counter-insurgent forces. For example, the need of the United States Coast Guard for a sub-

Lightweight rifle/submachine gun in firing position.

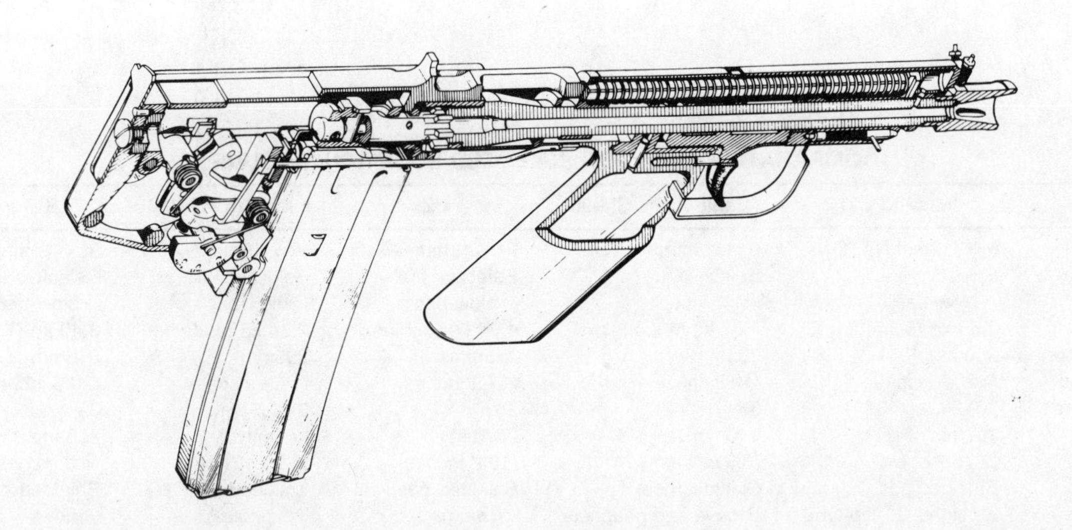

When Colt engineers began their design of an "arm gun," based upon a concept proposed by Dale M. Davis of the Air Force Armament Laboratory at Eglin Air Force Base, they started with the 9mm Parabellum cartridge. After completing one firing prototype, they discovered that they could use a much more powerful cartridge and include an automatic fire feature that would give an airman the equivalent of an assault rifle in a much smaller and more compact unit. Weapons such as this may someday be used for personal defense in place of both submachine guns and handguns.

machine gun has led to experimentation sponsored by the Joint Service Small Arms Program (JSSAP). The requirements for a new submachine gun call for a weapon that can be carried and fired with one hand. This will permit the user to climb ladders on ships or do other activities that require at least one free hand. A successful contender will be used by Army special forces, air crews, armored vehicle crews, special counter-terrorist personnel, and seaborne personnel called upon to board and investigate ships at sea. In addition to being silent and having low flash, a new submachine gun should be lighter and less bulky, and should have improved first shot accuracy, better controllability, single shot and full automatic fire, and a minimum life expectancy of ten thousand shots. The caliber will be 9 × 19mm NATO.

In April 1983, Heckler & Koch received a contract for an early advanced development submachine gun from the United States Navy Naval Weapons Support Center at Crane, Indiana. This weapon, designated the HK54A-1, has an integral turn-on, turn-off silencer, 50-shot drum magazine, single, 3-shot and continuous fire modes, and a locked-closed bolt mechanism for completely silenced shooting. This prototype weapon contains many of the features that one can expect to find on a future submachine gun. Between 1982 and 1988, JSSAP expects to spend about $6.1 million on submachine gun development.

Artist's conception of the Heckler & Koch 9 × 19mm NATO HK54A-1 advanced design submachine gun.

COMPARATIVE DATA FOR SELECTED SUBMACHINE GUNS

	IMI MINI UZI	Eagle Arms SM-90	UZI	Ingram MAC10	S & W 76
Caliber:	9 × 19mm NATO	9 × 19mm NATO	9 × 19mm NATO	9 × 19mm NATO	9 × 19mm NATO
System:	Selective fire blow-back	Selective fire blow-back	Selective fire blow-back	Selective fire blow-back	Selective fire blow-back
Weight:	2.65 kg (5.84 lbs)	1.82 kg (4 lbs.)	4.04 kg (8.9 lbs.)	2.84 kg (6.25 lbs.)	3.40 kg (7.5 lbs)
Length of Weapon w/stock folded:	360mm (14.17 inches)	305mm (12 inches)	455mm (17.9 inches)	295mm (11.6 inches)	516mm (20.3 inches)
Number of rounds:	20, 25, and 32	33	25	34	33
Barrel length:	197mm (7.75 inches)	140mm (5.5 inches)	267mm (10.2 inches)	146mm (5.75 inches)	203mm (8.0 inches)
Front sight:	Shielded post	Shielded post	Shielded post	Shielded post	Shielded post
Rear sight:	Shielded "L" flip-type aperture	Groove and peephole	V (wing)	Peephole	Groove
Muzzle velocity:	350 m/s (1150 f.p.s.)	390 m/s (1280 f.p.s.)	400 m/s (1310 f.p.s.)	390 m/s (1280 f.p.s.)	350 m/s (1150 f.p.s.)
Firing rate:	1200 rounds per minute	750 rounds per minute	650 rounds per minute	1000 rounds per minute	750 rounds per minute

TRENDS IN SUBMACHINE GUN DESIGN

There are some clearly discernible trends in the design of submachine guns. First, there is an increasing demand for submachine guns that are smaller in bulk and lighter in weight. While the standard UZI loaded weighs 8.9 pounds (4.04 kilograms) and is 17.9 inches (455mm) long with the stock folded, the new MINI UZI introduced in 1981 by the Israeli Military Industries (IMI) weighs only 6.31 pounds (3.14 kilograms) and is only 14.7 inches (360mm) long with the stock folded. In a similar vein, the newly introduced Eagle Arms SM-90 submachine gun is only 12.7 inches (333mm) without a stock, and loaded it weighs only 4.5 pounds (2.04 kilograms). Despite the fact that these submachine guns are small and light, they both fire the 9 × 19mm NATO cartridge. Both weapons are easy to handle and control during automatic fire. Also both can be shot with a single hand, which is yet another design trend.

A third trend in submachine gun design is the employment of silencers. While silencers have been around for many years their proliferation is the consequence of the requirement by special police and security forces for weapons that can be employed during counter-terrorist activities. One of the most publicized uses of silenced submachine guns was the successful 1978 Special Air Services (SAS) raid on terrorists at the Iranian Embassy in London. The SAS troopers carried, among other weapons, the Heckler & Koch MP5SD silenced submachine gun. Silenced submachine guns were used by both British and Argentinean forces in the 1982 Falklands/Malvinas War. The accompanying photographs illustrate some of the newer submachine guns.

This photograph illustrates the comparative size of the standard IMI UZI and the IMI MINI-UZI. Both fire the NATO standard 9 × 19mm cartridge.

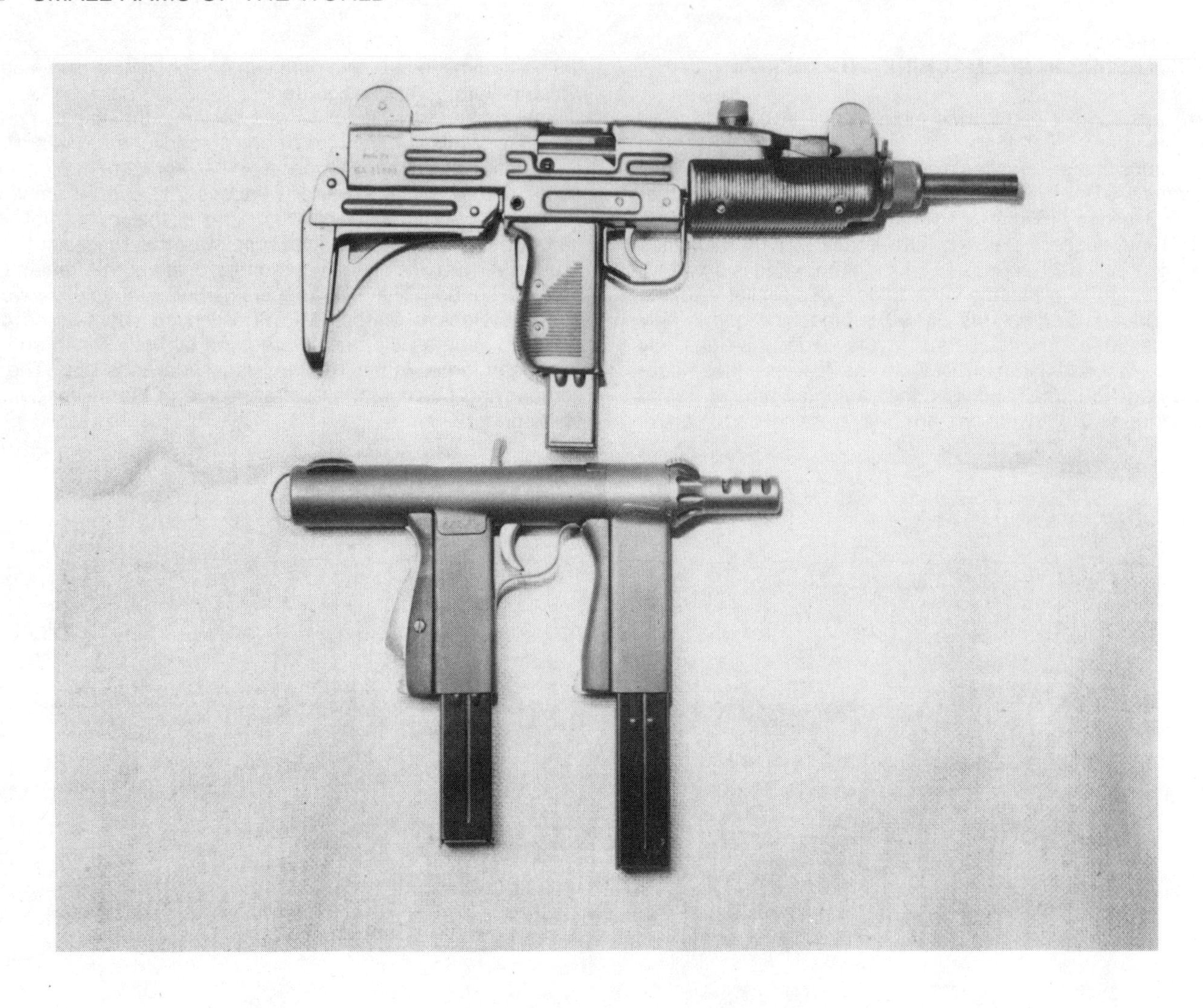

In this photo the relative size of the 9 × 19mm NATO caliber UZI and Eagle Arms SM 90 is clearly indicated.

CASELESS SUBMACHINE GUN AMMUNITION

During the Second World War, the Wehrmacht Waffenamt designers worked on the perfection of a rocket-propelled 9mm projectile that could be fired from a submachine gun or a pistol. This round contained all of the propellant in the interior of the projectile. In concept it was little more than an improvement on the Volcanic self-contained projectile developed by Smith & Wesson in the 1850s. In the 1960s, MBA Associates, a California-based company, perfected its Gyrojet rocket projectile, but as with the World War II German experiments the Gyrojets were short on kinetic energy. In the 1960s and 1970s, Smith & Wesson tried a combustible ammunition for its M76 submachine gun. Most of the M76s sold were chambered for the conventional 9 × 19mm NATO cartridge. As with the Heckler & Koch–Dynamit Nobel G11 weapon system, a new approach to a gun/ammunition combination was needed. In the late 1970s, the Italian firms of Giulio Fiocchi and Benelli Armi joined forces to develop a new weapon/ammunition system.

The 9mm AUPO cartridge developed by Giulio Fiocchi at first appears to be a conventional cartridge. However, it is radically different, and consists of a one-piece hollow extruded brass body. Inside the body there is an annular ring that holds the priming compound. The hollow body is filled with propellant and sealed with a cup-shaped disc of special paper. That cup is crimped in place. This design allows easier manufacture of ammunition and the overall weight of the cartridge is reduced with a slight improvement in ballistic performance. The accompanying illustrations describe the Fiocchi-Benelli ammunition/weapon system.

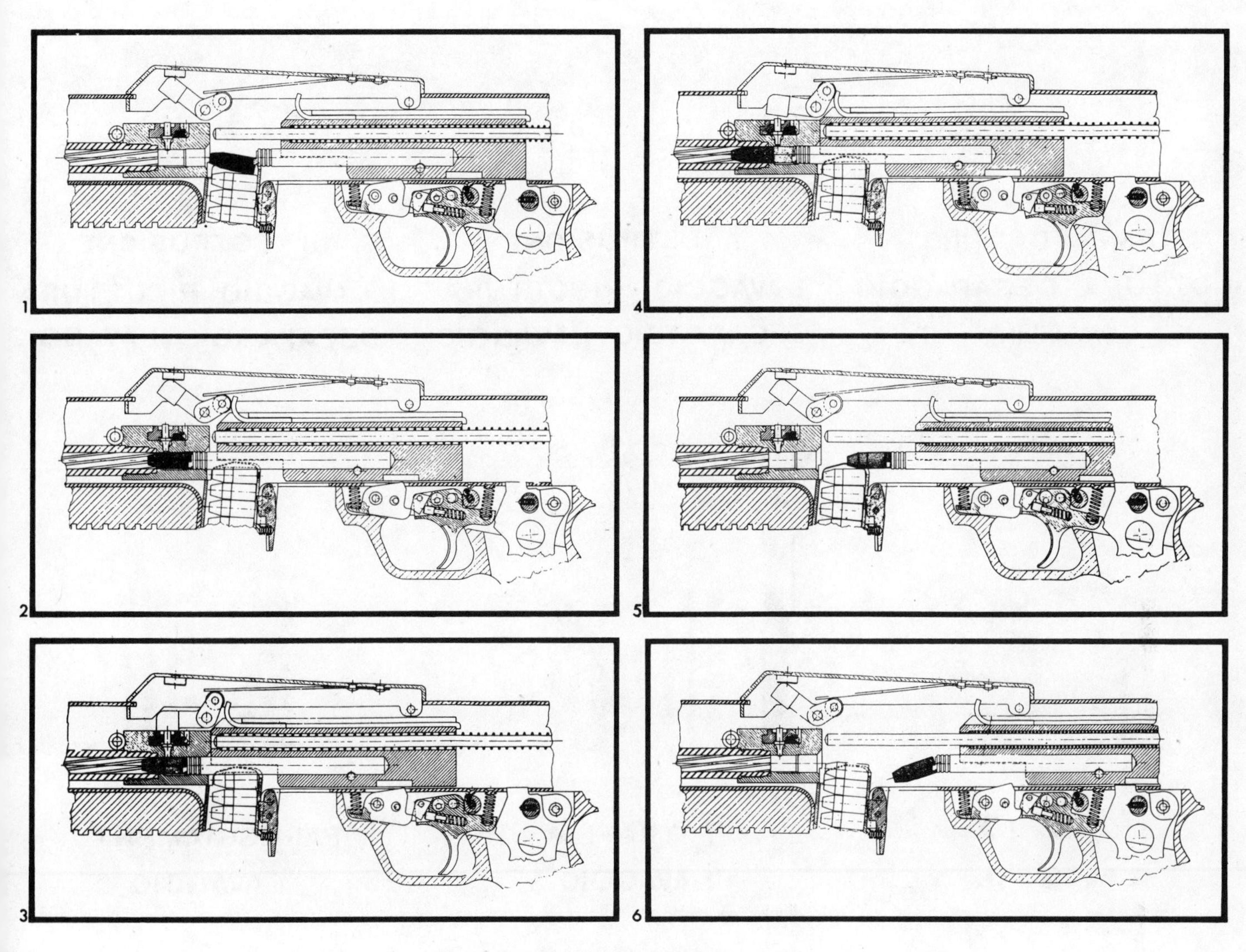

The Fiocchi-Benelli Ammunition/Weapon System.

(1) In the locking phase of the arm the recoiling mass breech-block introduces, by means of the locking pin, the round from the magazine into the barrel; the extractor enters into the rear part of the round.

(2) The breech-block continuing the locking phase of the arm has already fed the round in the barrel and, thanks to the "L"-shaped spring, imparts to the hammer the energy necessary for firing the round; the extractor at the same time has completed the gripping of the round at the inside of the rear locking rim.

(3) The breech-block now completes the locking phase by stopping on the sub-frame face; the round is in the barrel and the hammer is rotated forward on its fulcrum and, thanks to the energy provided by the "L"-shaped spring, hits the firing pin.

(4) After firing the breech-block maintains for an instant the locking position while the pressure of the gas thrusts the round towards the mouth of the barrel releasing it at the same time from the grip of the extractor; the firing pin, free to return to its seat, does not engrave the round and the hammer is returned to its rest position by the return spring.

(5) In the event of a misfire, if a round remains in the barrel with the breech-block locked, it is possible to open the breech-block manually for extracting the projectile still gripped by the extractor.

(6) The breech-block, after completion of the manual opening, carries the round, still gripped by the extractor, to a position at the back of the magazine; the projectile, no longer supported on its point, by the sides of the chamber and by the magazine, is now released by the extractor and drops on the ground through the opposite opening of the arm.

STAMPA DA FILO
RICOTTURA DECAPAGGIO
LAVAGGIO

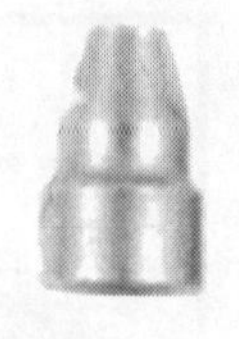

I' ESTRUSIONE
LAVAGGIO RICOTTURA
DECAPAGGIO LAVAGGIO

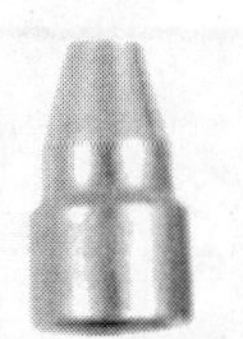

II' ESTRUSIONE
LAVAGGIO RICOTTURA
DECAPAGGIO LAVAGGIO

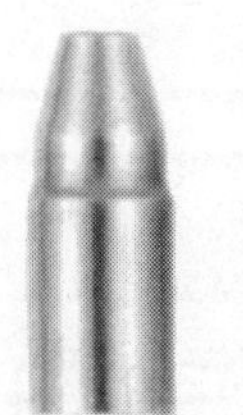
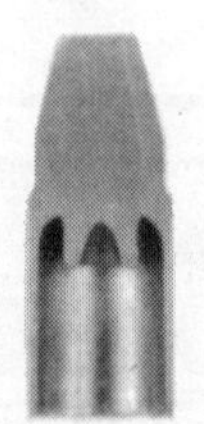

I' TRAFILA
LAVAGGIO

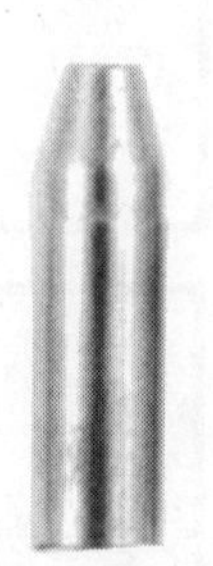

II' TRAFILA
LAVAGGIO

PAREGGIAMENTO
LAVAGGIO

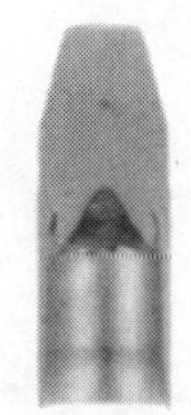

INNESCO

CARICAMENTO
POLVERE CON COPPETTA
E VERNICIATURA

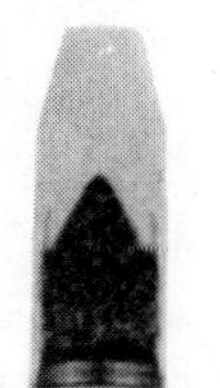

RIVETTATURA

This photograph shows the steps in the manufacture of the Fiocchi 9mm AUPO caseless ammunition. Starting with brass billet, the projectile goes through two extrusions and two draws before the final trim. Then it is primed, the paper cup is inserted and sealed, and the cup is crimped in place. With this ammunition the cartridge case and bullet-forming processes are combined, thus simplifying the fabrication process of the ammunition.

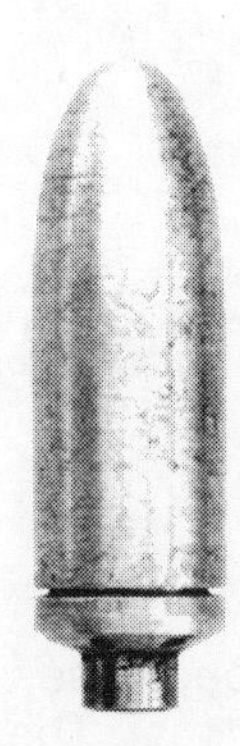
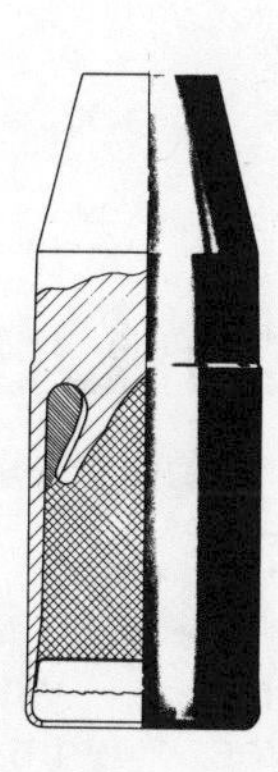
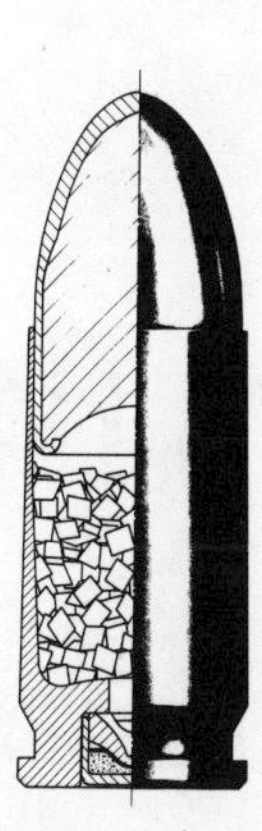

From left to right: **the World War II German 9mm rocket round, the 9mm Fiocchi AUPO cartridge, and the standard 9 × 19mm NATO.**

The Benelli Armi CB-M2 submachine gun which uses the 9mm AUPO cartridge developed by Giulio Fiocchi.

TECHNICAL CHARACTERISTICS OF THE BENELLI/FIOCCHI WEAPON/AMMUNITION SYSTEM

Caliber:	9mm AUPO	
Overall length:	660mm	26 inches
with stock folded:	450mm	17.73 inches
Barrel length:	200mm	7.88 inches
Weight w/o magazine:	3.4 kgs	7.5 pounds
Weight of magazine:	.25 kgs	.55 pounds

COMPARATIVE CARTRIDGE DATA

Caliber:	9mm AUPO	9mm NATO
Projectile weight:	7.05 grams (109 grains)	7.45 grams (115 grains)
Propellant weight:	.40 grams (6.2 grains)	.474 grams (7.3 grains)
Total cartridge weight:	7.55 grams (116.7 grains)	11.74 grams (181.5 grains)
Total length:	24.5mm (.965 inches)	29.15mm (1.15 inches)
Velocity at 10 meters:	390 m/s (1280 f.p.s.)	385 m/s (1263 f.p.s.)
Energy at 10 meters:	536j (396 ft. lbs.)	552j (407 ft. lbs.)

4 Handgun Developments

During the years 1977 to 1982, there was considerable activity in the field of handgun development. In NATO the major developments were in Germany and the United States. The state police departments in Germany adopted a new generation of 9 × 19mm NATO (also called 9 × 19mm Parabellum) caliber handguns. These pistols—P5, P6, and P7—are described in detail in Chapter 21. In the United States an ongoing effort to evaluate and adopt a 9 × 19mm NATO caliber replacement for the M1911A1 .45 Colt pistol remained inconclusive at the end of the summer of 1982. At the same time, many older model handguns were still in service. Two of the major designs—the M1935 Browning High Power (*Grande Puissance*) and the M1911A1—are the product of John M. Browning's fertile imagination. The former, while known by many different names, is essentially the same weapon as introduced by Fabrique Nationale in 1935. More than two dozen nations have adopted this 9mm Parabellum pistol, and Argentina, Belgium, Canada, India, Indonesia and the United Kingdom have produced it. The American M1911A1 .45 caliber pistol is still the standard sidearm in the United States Army after 71 years of continuous service. A dozen or more armies include this Browning design in their inventory, and it has been used by at least twice that many armed forces in the past. The Tula-Tokarev 1933 (TT33) Pistol formerly used by the Soviet Armed Forces is a Browning derivative, and it is still found in the hands of many officers around the world from Afghanistan to Yugoslavia. One other design of World War II vintage that is still standard issue in at least six armies is the post-war version (P1) of the 9mm Walther P38. Presently, this sidearm is manufactured by Carl Walther in West Germany and by Steyr-Daimler-Puch in Austria.

A number of new pistols have been introduced since 1945. Clearly, the trend has been away from revolvers and toward a limited number of calibers for self-loading pistols. By far, the most popular cartridge is the 9 × 19mm Parabellum cartridge, which was introduced in 1902 and adopted by the German Navy in 1904 and by the German Army in 1908. Second, in terms of popular usage, is the US .45 ACP (11.43 × 23mm) cartridge, introduced by John Browning in 1905 and adopted by the US Army in 1911. The Soviet 7.62 × 25mm cartridge (a variant of the 1896 7.63 Mauser pistol round) was previously widely used for both pistols and submachine guns. Since 1945, it has gradually been supplanted by the less powerful 9

Indonesian-made Browning High-Power 9mm Pistol.

Colt-made M1911A1 .45 ACP Pistol.

Current manufacture Walther P1 9mm Parabellum Pistol.

× 18mm Makarov cartridge. The .32 caliber (7.65 × 17mm) Browning cartridge is the other major round used for some smaller pocket pistols. Almost without exception, these are the four basic self-loading cartridges used today. While it would be impossible to cover all the handguns developed since World War II, the text that follows covers some of the major pistols in current service. Additional handguns are described in the country by country chapters of Part II. Older military handguns are described in E. C. Ezell, *Handguns of the World.*

9mm NATO (PARABELLUM) HANDGUNS*

CZECH MODEL 1975 PISTOL

Since the end of World War II, the Czechs have introduced several new self-loading pistols. The best known is the Model 1952 7.62 × 25mm pistol. (See Chapter 14.) In 1975, the Czech export agency Merkuria announced the availability of a new 9mm Parabellum self-loading pistol manufactured by the Ceskoslovenská Zbrojovka factory at Brno. In adding this pistol to their current offerings, CZ borrowed the Browning operating principle. In fact, the weapon looks like a cross between the Browning High-Power and Smith & Wesson's Model 39. Its magazine capacity is 15 rounds, two more than the High-Power. The Model 1975 is unique for a Warsaw Pact country in that it is chambered for 9mm Parabellum instead of 7.62 × 25mm or 9 × 18mm Makarov. Intended for the international market, the Model 1975 has been produced in two versions—commercial and military. (See Chapter 15.)

*The 9 × 19mm Parabellum cartridge has been adopted by NATO as the 9 × 19mm NATO cartridge. The designation 9 × 19mm NATO will be used with handguns tested or adopted by NATO.

Three of the new generation of military handguns. *Top,* the .45 ACP (11.43 × 23mm) SIG-Sauer P220; *middle,* the 9 × 19mm Parabellum Czech vz 75; *bottom,* the 9 × 19mm NATO Beretta Model 92. Variants of the P220 and the Model 92 have been tested in the United States as possible replacements for the M1911A1 pistol.

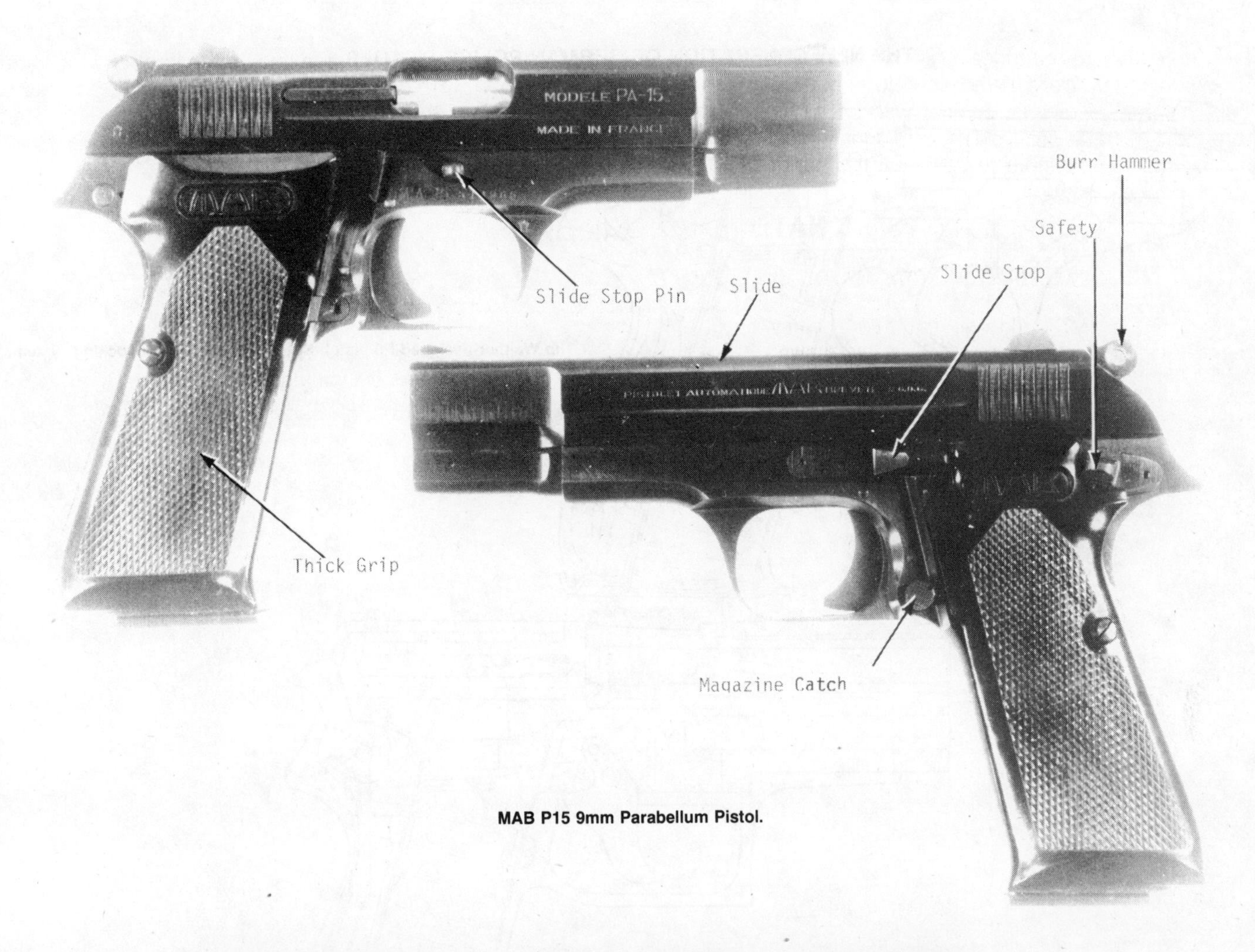

MAB P15 9mm Parabellum Pistol.

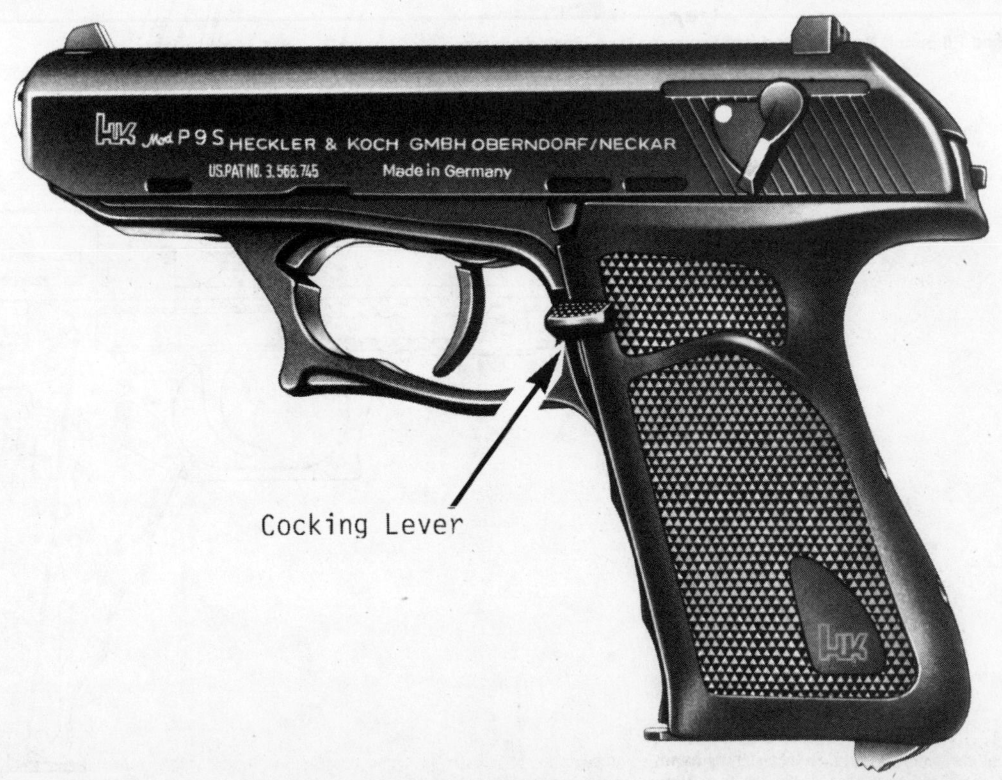

HK P9S 9mm Parabellum Pistol.

THE NEW GENERATION OF GERMAN POLICE PISTOLS

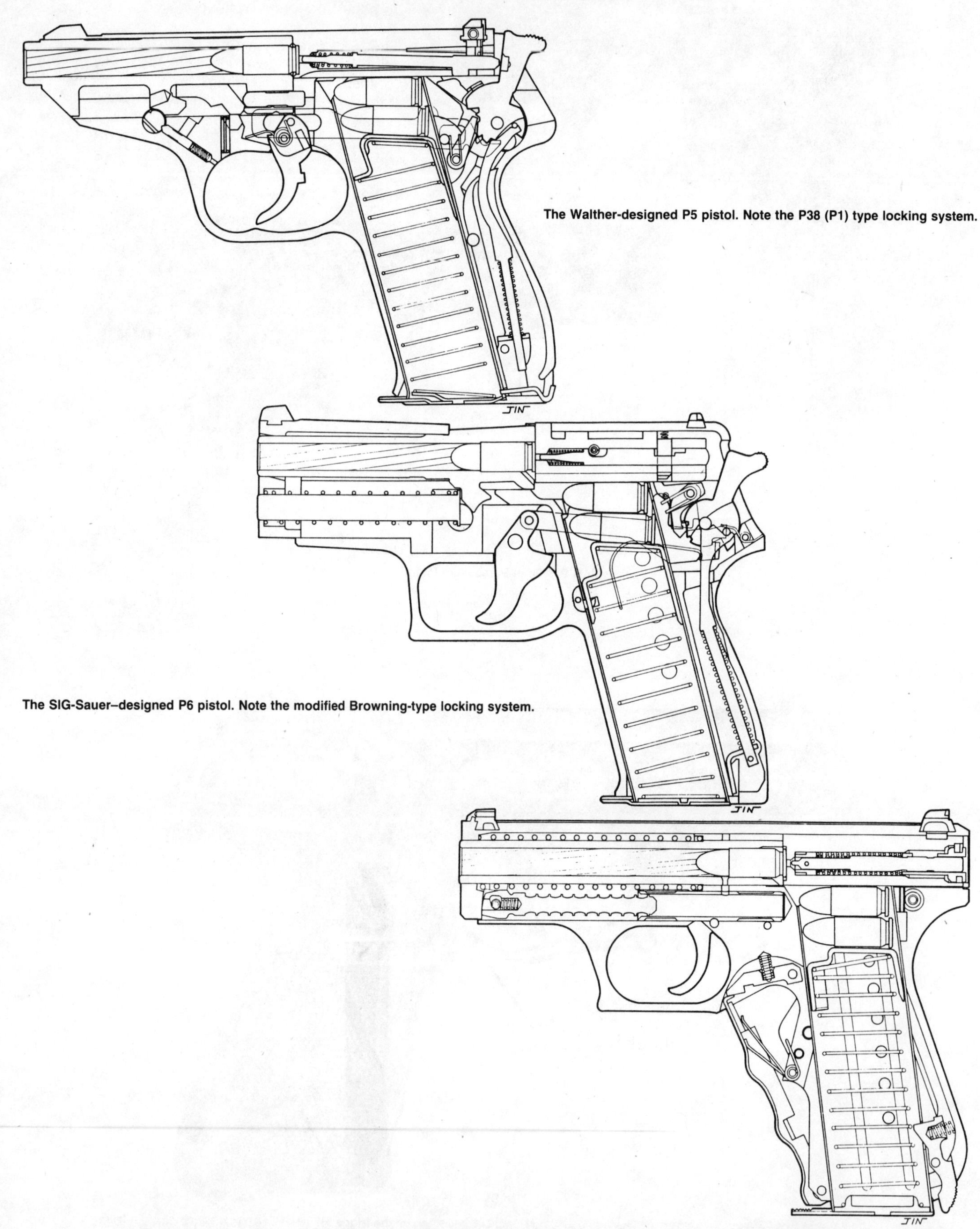

The Walther-designed P5 pistol. Note the P38 (P1) type locking system.

The SIG-Sauer–designed P6 pistol. Note the modified Browning-type locking system.

The Heckler & Koch–designed P7 pistol. Note the striker-type firing pin, squeeze cocker, and gas retarding system. This is essentially a blow-back pistol with a gas pressure delay.

FRENCH MAB P15 PISTOL

The French military arms establishments have also introduced two 9mm Parabellum pistols since 1945. The best known, the MAS 1950, was designed at the *Manufacture d'Armes Automatiques St. Etienne*. This weapon was an updated version of the M1935 Pistol, chambered for the more powerful 9 × 19mm instead of the older and less lethal French 7.65 × 20mm cartridge. Whereas the M1935 and MAS 1950 were basically variations of the Browning design, the MAB P15 is a different design, worked out by the engineers at the *Manufacture d'Armes Automatiques Bayone*. The P15 employs a hesitation lock that retards the rearward thrust of the slide after a cartridge has been fired. A cam lug machined onto the barrel interacts with a cam groove in the slide, slowing its recoil. The cam groove in the slide causes the barrel to rotate, and the interplay between the slide and the barrel delays the rearward motion until the gas pressure from the fired round has been reduced to a safe level. (See Chapter 19.)

The MAB P15 is so designated because of its 15-shot magazine capacity. As with the Browning High-Power and the Czech Model 1975, the large capacity magazine makes the grip quite thick; all of these pistols are a handful. A P-8 (eight-shot) and F-1 target version are also marketed commercially.

GERMAN HECKLER & KOCH PISTOLS

Although the P1 (P38) continues to enjoy extraordinary popularity with German military personnel, the Oberndorf firm of Heckler & Koch has introduced three 9 × 19mm handguns since the mid-1960s—the selective fire VP70 mentioned in Chapter 3, the P9 and P9S series, and the P7 series. The two P9 hand-

The 8-shot P7. Note that the magazine release on this pistol is located to the rear of the magazine floor plate.

The 13-shot P7A13. Note that this handgun has a magazine release lever behind the trigger. It also has an insulating block in front of the trigger to prevent discomfort when a large number of rounds are fired. Heat transfers to this position from the gas used to retard the opening of the slide. This handgun was tested in the United States pistol trials.

guns differ only in their trigger mechanisms; the P9 is single action, and the P9S is double action. Both have internal hammers that can be cocked by an external lever near the trigger guard on the left side of the frame. In further exploitation of the roller lock retarded blow-back breech mechanism (G3 and HK33 rifles and HK54 [MP5] submachine guns), Heckler & Koch has designed a weapon that is unique for two reasons—its locking system and its wide use of stamped metal parts and plastic components. Other features, such as the polygonal rifling, are described in Chapter 21. Magazine capacity of the P9 series is nine shots. The P7 series comes in two variants—the standard P7 with an eight-shot magazine and the P7A13 developed for the United States pistol trials with a thirteen-shot magazine. The P7 series, covered in detail in Chapter 21, is of particular interest because of its squeeze-cocking mechanism that eliminates the mechanical safeties associated with standard double-action hammer-fired pistol mechanisms.

SWISS SIG-SAUER P220 PISTOL

For over three decades, the Schweizerische Industrie-Gesellschaft (SIG) P210 Pistol has been considered the finest machined handgun in the world. Starting with Charles Petter's concepts for an improved Browning-Colt design that first emerged as the French M1935, SIG engineers produced a sidearm made as precisely as a Swiss watch. As the Model 49, the P210 has been the standard Swiss Army pistol for most of the post-war period. Recently, SIG teamed up with the old-line German firm J. P. Sauer & Sohn of Eckenforde to produce two new pistols—the P220 and the P230. These pistols were designed by SIG engineers and are being manufactured by Sauer. The major reason for this across-the-border cooperation is the increasingly stringent Swiss export laws for small arms. While also advertised in US .45 caliber, the P220 is basically a 9mm Parabellum pistol, and the P230 is a modern pocket automatic, produced in 7.65

SIG-SAUER P220 PISTOL SERIES

	P220 (M75)	P225 (P6)	P226 (XM9)
Caliber:	9 × 19mm	9 × 19mm	9 × 19mm
Overall length:	7.79 in. (198mm)	7.09 in. (180mm)	7.72 in. (196mm)
Barrel length:	4.41 in. (112mm)	3.84 in. (97.6mm)	4.41 in. (112mm)
Weight empty:	23.53 oz. (830g)	23.25 oz. (820g)	25.52 oz. (900g)
Magazine capacity:	9	8	15

Sig-Sauer 9 × 19mm P225 Pistol with magazine release at base of magazine well.

Sig-Sauer P226 Pistol with push-button magazine release. This is model tested by United States as part of 9mm handgun trials.

Assembled and disassembled views of the SIG-Sauer P220 9mm Parabellum Pistol.

The 9 × 19mm P225 (P6) pistol.

THE CURRENT GENERATION OF WALTHER HANDGUNS

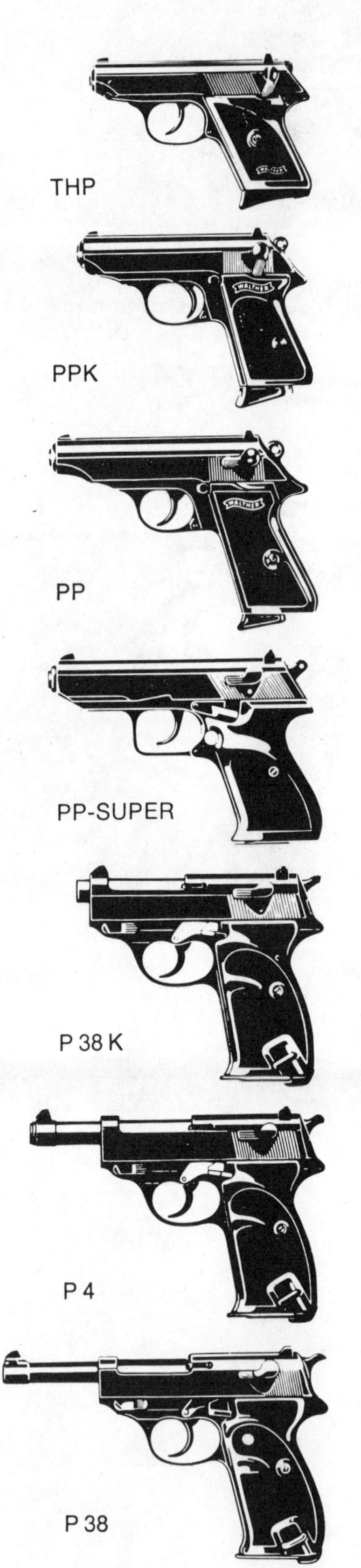

Two views of the 9 × 19mm NATO Walther P5 pistol.

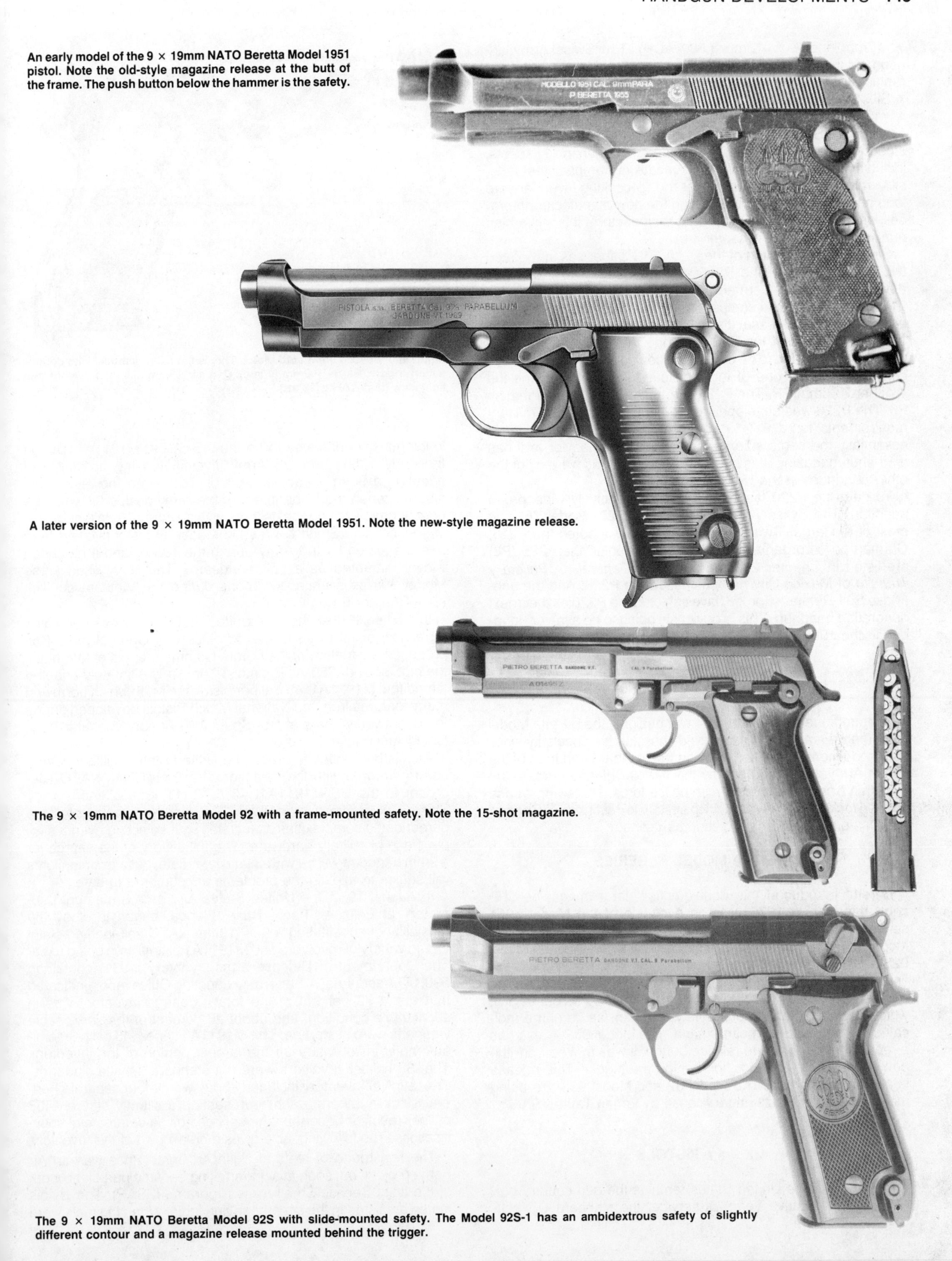

An early model of the 9 × 19mm NATO Beretta Model 1951 pistol. Note the old-style magazine release at the butt of the frame. The push button below the hammer is the safety.

A later version of the 9 × 19mm NATO Beretta Model 1951. Note the new-style magazine release.

The 9 × 19mm NATO Beretta Model 92 with a frame-mounted safety. Note the 15-shot magazine.

The 9 × 19mm NATO Beretta Model 92S with slide-mounted safety. The Model 92S-1 has an ambidextrous safety of slightly different contour and a magazine release mounted behind the trigger.

× 17mm and 9 × 17mm (.380 ACP). The Swiss Army announced the adoption of the P220 as the Model 75. To gradually replace the M49, 10,000 P220s were ordered.

SIG-Sauer's P220 is unique in several respects, but one of its most notable features is the large utilization of stamped steel parts. While SIG was long regarded the last bastion of machined steel components, the P220 is a modern handgun in all aspects, including the techniques used to fabricate its stamped steel slide, slide release lever, trigger, trigger rod, decocking lever, lanyard loop and magazine catch. Although the new manufacturing processes make the pistol cheaper to manufacture, it is still a high quality weapon. (See Chapter 44.)

Since the introduction of the P220, SIG-Sauer has introduced two major variants—the P225 and the P226. The P225 was developed to meet German police specifications (see Chapter 21). It is slightly more compact than the P220 and has a push-button magazine release. It also has an 8-shot magazine instead of the nine shots of the P220. As one might deduce from the size, it was intended to be carried concealed under special police conditions. It was one of three pistols selected to replace the Walther 7.65mm PP. The P225 carries the German designation P6. The P226 was developed to meet the United States military requirements for a 9 × 19mm NATO caliber handgun. It is essentially the same size as the P220, but is heavier and has a 15-shot magazine. It is designated the XM9, as were all of the other handguns in the United States trials.

To date the P220 family has received relatively widespread adoption. The Swiss Army adopted the P220 (Model 75) and most of the larger Swiss police forces have adopted the P225. German police organizations that have selected the P225 (P6) are listed in Chapter 21. The *Direccion General de Policia y Transito* of Mexico City has purchased the P225. And the Japanese Self-Defense Forces have selected the P220 as their next generation handgun. This is obviously going to be a major handgun in the 1980s.

ITALIAN BERETTA MODEL 1951 PISTOL

Although one of the older post-war designs, the Beretta Model 1951 9mm Parabellum has enjoyed considerable popularity since its introduction. In addition to being the standard handgun of the Italian Army, it is also used by the Israeli Defense Forces, the Egyptian Army, and the Nigerian police force. The swing wedge shaped locking block type locking system is described in Chapter 29. This pistol has an eight-shot magazine.

BERETTA MODEL 92 SERIES

Beretta introduced this double-action, large magazine (15-shot) handgun in 1976. It is an evolution of the Model 1951 series and it employs the same P38-style locking system. The Model 92 has a frame-mounted mechanical safety, and the Model 92S has a slide-mounted safety. The latter mechanism locks the firing pin as well as the hammer. The United States Armed Forces have tested the Model 92S-1 (also called Model 92SB), which has an ambidextrous safety that can, as its name indicates, be operated with either hand. The Model 92S-1 also has a repositioned magazine release button; it was moved from the butt of the frame to a position behind the trigger. This release is also ambidextrous. The Model 92 and Model 92S are being manufactured in Brazil under license by Forjas Taurus SA.

US 9mm PISTOLS

In the West, the United States remains the odd country out. After 71 consecutive years of service, the standard US self-

Smith and Wesson 9mm auto pistol. This is the military model with double-action trigger. Barrel 102mm (4 ins.). Overall 188mm (7.4 ins.). Weight with light alloy frame .79 kg (28 oz.).

loading pistol continues to be the .45-caliber M1911A1. During the early 1950s, the US Army investigated the possibility of adopting a 9mm Parabellum pistol. That effort, however, was stopped when the Department of the Army decided it would be uneconomical to change calibers with so many 45-caliber pistols in service. The Smith & Wesson Model 39 pistol evolved from designs tested by the Army during the 1950s, and it has been widely marketed as a police sidearm. The Navy used some Model 39s as silenced weapons during the Vietnam conflict. (See Chapter 5.)

In the late 1960s when the United States Army was searching for a new pistol to be issued to officers of General rank, Colt proposed a variation of their 9mm Commander pistol to replace the old .32 and .380 ACP pistols. The Army ultimately adopted a modified M1911A1 .45-caliber pistol, the M15, which borrowed many features from the National Match Pistols developed during the past twenty years at first Springfield Armory and later Rock Island Arsenal.

As this book goes to press, the United States military continues to struggle with the selection of a 9 × 19mm NATO successor to the US M1911A1 .45 ACP (11.43 × 23mm) pistol. This pistol program, started in 1977, clearly demonstrates the bureaucratic complexities associated with selecting even a simple piece of military hardware. The full history of the search for a 9mm handgun will be written at a later date, but a few highlights will suffice to explain the problems encountered to date.

In August 1977, the United States Air Force Armament Laboratory at Eglin Air Force Base, Florida, began to study the feasibility of substituting a 9 × 19mm NATO self-loading pistol for the variety of revolvers and M1911A1 pistols then being used. The Department of Defense arms inventory (1981) contained 590,177 handguns. A General Accounting Office report indicated that about 73 percent of the total sidearms inventory was in serviceable condition, and about 20 percent of the serviceable weapons were in storage. The M1911A1 series .45 caliber semiautomatic pistol made up the largest portion of the inventory. The .38 caliber revolvers were not standard service sidearms. The revolver inventory included about twenty-four separate Federal stock numbers, each representing a slightly different .38 caliber revolver bought by the services using commercial specifications. The following table gives a breakdown of this inventory.

The first round of tests at Eglin Air Force Base was an Air Force project, but soon this undertaking became part of the work of the Joint Service Small Arms Program (JSSAP). The pistols tested are listed in the accompanying table. When the trials were

COMBINED SERVICE ASSETS OF .45 AND .38 CALIBER SIDEARMS

Sidearm	Serviceable Issued	Serviceable In Storage	Unserviceable In Storage	Totals
.45 Pistol	237,790	63,159	116,499	417,448
.38 2″ Barrel	10,925	13,216	9,209	33,350
.38 3″ Barrel	1,036	—	257	1,293
.38 4″ Barrel	95,307	8,247	34,532	138,086
Total	345,058	84,622	160,497	590,177

completed, and analysis of the test results finished, serious questions were raised about the scientific validity of the results. Poor performance of outdated ammunition during low temperature tests simulating arctic conditions suggested that the results would not be reproducible. Therefore, the exercise was deemed to have been of limited value in selecting a new handgun.

Not unexpectedly, Congress got into the act at this point. At the end of 1978, the House Appropriations Committee Surveys and Investigation Staff argued that prompt action was required to replace the potpourri of handguns and thus eliminate the logistical problems ensuing from this variety of handguns. The following year the committee expressed its dissatisfaction with progress to date. It further stated that should insufficient progress be made in the next year, the committee would consider legislative action.

In November 1979, the Principal Deputy Under Secretary of Defense for Research and Engineering directed the services to undertake "a joint study to determine the minimum number of types of handguns to meet essential service requirements and to determine if the United States should adopt the NATO standard 9 × 19mm handgun cartridge." The Army, as the executive agent for small arms, was designated the lead service responsible for compiling the study.

Guidelines provided by the Secretary directed an analysis of the advantages and disadvantages of a single family of handguns and ammunition, and its impact on such factors as current and future operational requirements, to include the implication of increased numbers of women in the armed forces, domestic production facilities, costs, and NATO's standardization aspects. Only the .38, .45, and 9 × 19mm NATO calibers were to be considered.

The task was passed from the Department of the Army to the Army Materiel Development and Readiness Command in early December 1979, with directions to have the recently formed JSSAP conduct the joint study. JSSAP management committee representatives of the Air Force, Army, Coast Guard, Marine Corps, and Navy coordinated their service inputs.

This study was comprised of two principal parts. The first was the ongoing Air Force evaluation of handguns, which was being conducted at Eglin Air Force Base, Florida, and in which various model 9mm pistols were evaluated against the M1911A1s and Smith & Wesson Model 15 .38 caliber revolvers. The second part of the study considered a review and compilation of all existing historical reports and studies relative to pistols, handguns, and their performance.

For instance, a 1978 survey of Army aircrews was reviewed as was the report of 1978 by the Army's Human Engineering Laboratory on handgun hit probability as part of a test of suitable weapons for women.

In addition, FBI, Secret Service, and the Army's Ballistic Research Laboratory data were also significant considerations. The report of the Illinois State Police, in justification for its adoption of the 9mm pistol, was also studied. Current and future potential production capacity data and cost data were gathered by the JSSAP Support Office, Fire Control and Small Caliber Weapon Systems Laboratory, and the Army Armament Research and Development Command.

There was continuous active participation by the Air Force, Navy, Marine Corps, and Coast Guard. Each contributed materially to the methodology, content, conclusions, and recommendations of the report. Based on input from these services, a draft Joint Service Operational Requirement (JSOR) was prepared. One major factor affecting the study's conclusions was the magnitude of the handgun inventory.

The recommendation of the JSSAP study, dated 5 June 1980, was to have all United States armed services adopt a single family of 9mm caliber semiautomatic handguns. The guns would be in a standard size for general issue, plus a small limited issue in a concealable version of the same type weapon. The NATO standard 9 × 19 parabellum ball cartridge was recommended, along with limited quantities of other rounds, such as subsonic, signal-tracer, and blank.

The study indicated this two gun approach would provide an across-the-board capability for all users plus simplification of training and logistics. Major disadvantages would be the problems of disposing of some 420,000 .45s and 130,000 .38s, many of which are still serviceable, and the need to modify the support system for a completely new handgun.

The selection of the 9mm caliber was an area where emotion clashed with available scientific evidence. The crux of the issue was the need for a bullet with "stopping power." The issue was not whether death is the immediate result of being struck by a pistol bullet. Rather, near instantaneous incapacitation was the desired result, given the fact that a handgun is used generally for last-ditch close-in self-defense. Any handgun is capable of inflicting a fatal wound—the .22 caliber bullet is the largest single bullet killer in the United States—but many handguns do not possess the desired stopping power.

A variety of theories on stopping power were reviewed. These included the Hatcher theory that stopping power is proportional to a bullet's impact momentum times its cross-sectional area, and the 1960s Army-derived theory that incapacitation is a function of the kinetic energy deposited in 6.0 inches (15 centimeters) of gelatin tissue simulate. Also considered were findings done by the Army's Ballistic Research Laboratory in 1973 in response to a request from the National Institute of Law Enforcement and Criminal Justice on terminal effects of police handgun ammunition. Secret Service tests of 1979 and a 1978 FBI report on law enforcement officers killed were also taken into account.

The stopping power issue was resolved largely by the "BRL Computer Man," an elaborate three-dimension computer code of human anatomy and the medical determination of effectiveness of hits on various parts of the body. Ballistic data and bullet cavitation data from experiments on a gelatin target closely resembling human tissue were fed in.

The conclusion of these evaluations was that the most important property of a handgun bullet is its velocity. It was shown, too, that for a given velocity, larger caliber bullets have greater stopping power than smaller ones. However, larger caliber bullets such as the .45 (the current United States standard) need not be as massive to retain stopping power—that is, masses on the order of 158–170 grains (10.2–11 grams) are sufficient.

Perhaps the most significant factors favoring selection of a 9mm handgun were a preference for this type of gun by a segment of military handgun users, a stated preference in a draft Joint Services Operational Requirement, and most importantly, the NATO interoperability aspect. The Coast Guard has been one of the strongest advocates among the services for a changeover as rapidly as possible.

The JSOR has identified some important safety features which are required. These include the capability to completely load, unload, and clear the weapon without actuating the trigger; and lowering the hammer from cocked position to an uncocked position without actuating the trigger. Also, the handgun must be one-handedly and ambidextrously operable, allowing a safe carry with draw and fire by one hand, left or right.

A second round in the selection of a 9mm handgun began in the spring of 1981. Pistol manufacturers were notified about this second attempt to evaluate their products on 22 May 1981; a request for proposals (RFP) for the submission of thirty pistols from each company was issued on 29 June; candidate handguns and relevant documentation were to be delivered by late September. Beretta, Fabrique Nationale, Heckler & Koch, SIG–Saco Defense Systems, and Smith & Wesson submitted pistols. These handguns, officially designated "Pistols, 9mm XM9," meet the following general specifications: maximum load weight, 2.77 pounds (1.26 kg); maximum overall length, 8.7 inches (221mm); minimum barrel length, 4 inches (102mm); ambidextrous safety level if necessary; the magazine falls free of the weapon when the ambidextrous release is pressed; and minimum 10-shot magazine. Beretta's handgun was a variant of its Model 92; Heckler & Koch's submission was a 13-shot version of its 8-shot Pistole 7 (the P7A13); Saco Defense System represented the Swiss firm SIG and its P226; and Smith & Wesson's Model 459 evolved from its commercial Model 39/59 family. Plans formulated for this RFP called for production of a new handgun in FY1983.

Selection of a new pistol ought to have been a relatively easy process once the testing was completed, but the real world of procurement decisionmaking is complicated by many factors, not the least of which are domestic political considerations. Several American legislators, notably Senator Barry Goldwater from Arizona, were reportedly opposed to the issue of "foreign" handguns to American troops. Such mindless chauvinism—made all the more strange by the fact that the winning firm will have to produce the majority of its guns in the United States—will do nothing to reassure NATO allies about America's commitment to RSI—rationalization, standardization, and interoperability. And there were other members of Congress who did not appreciate the fact that handguns are last-chance personal defense weapons for such highly trained personnel as armored vehicle crews, pilots of helicopters and fixed-wing craft, and senior officers; they did not agree that there is a need for a new handgun. As life insurance, handguns represent a small cost, but many lawmakers thought defense money would be better allocated to the purchase of tanks or other more sophisticated materiel.

Additionally, there were legislators who favored conversion of the M1911A1 .45 (11.43mm) pistol to 9mm NATO. Colt Industries submitted an unsolicited proposal for a Product Improvement Program (PIP) for the M1911A1, which would have involved a new slide, barrel, and magazine. Proponents of the PIP M1911A1, including some members of the House of Representatives Subcommittee on Investigations of the Committee on Armed Services, reported that modifying the existing M1911A1 would cost only half as much as procuring a new weapon.* Opponents noted the following shortcomings of a PIP M1911A1: It is single action; lacks an ambidextrous safety or magazine release; has a 7-shot magazine, and loaded in .45 caliber weighs 2.99 pounds (1.36 kg) (a 9mm version would weigh more); the frame and other components used in a PIP M1911A1 would be at least thirty-five years old (none has been produced since 1945); and there are only 300,000 serviceable M1911A1s in the United States inventory (more than 590,000 handguns will be required).

At the October 1981 American Defense Preparedness Association Small Arms Symposium, Lt. Col. Anthony Bizantz, the Army officer in charge of the handgun program, hit the central issue when he noted, "We need arms in the arms rooms of units that may be called into combat." The time for discussion is past. It will cost $125 million to $177 million for the total handgun acquisition program, but it is not clear if the Department of Defense and Congress are willing to pay this price for 9mm handguns.

*Comptroller General of the United States, "Proposed Program for New 9-mm Handguns Should be Re-examined." Report PLRD-82-42 (8 March 1982).

CHARACTERISTICS OF M1969

Caliber: 9mm Parabellum
Barrel length: 104mm (4.1 in.).
Weight: 1.02 kg (36 oz.) (w/magazine & empty).
Trigger pull: 2.04 to 2.95 kg (4½ to 6½ lbs.).
Stocks: American walnut.
Finish: Can be Colt's Standard Blue, Royal Blue or Nickel.
Accuracy: Capable of 100% on silouhette targets at 50 meters.
Design criteria: Same as government specifications utilized by the 1911 Service Pistol.
Magazine capacity: 8 rounds.
Advantages:
—All steel construction; therefore, it can be subjected to all 9mm ammunition regardless of its manufacture.
—Basic components, such as trigger, hammer, sear, disconnector, magazine catch, main spring and housing, etc., are standard Model 1911A1 parts.

Colt General Officers Pistol M1969, 9mm Parabellum.

M15 General Officers .45 caliber Pistol with accessories.

M15 General Officers Pistol, standard issue since 1972.

In late February 1982, the Department abruptly terminated plans to purchase a 9mm NATO caliber handgun. The basic reason given was the failure of any of the candidate handguns to meet a sufficient number of the criteria to be selected. There were seventy-one of these criteria and most pistols failed at least eleven; mud and sand were the biggest problems.

It appears in retrospect that the military wanted too much. Military handguns are generally used as a last-ditch defense. It is the first shot or two that counts. If you do not solve your problem with the first magazine, you are in deep trouble. The requirement that the pistol fire eight hundred shots between major malfunctions was not related to the real world needs of the soldier. No M1911A1 pistol straight off the production lines in World War II would shoot half that number without a malfunction, but those pistols were reputed to be acceptably reliable. Clearly, when the United States military reworks its requirements, some changes will be made. Whatever the outcome of the requirement revision activity, it will be several years before the venerable M1911A1 pistol is supplanted.

UNITED STATES 9mm HANDGUN TRIALS, 1977–1982

	First round	Second round
Beretta Model 92S1	X	X
Colt Stainless Steel Pistol	X	
Fabrique Nationale Browning High PowerM1e1935	X	
Fabrique Nationale Fast Action	X	
Heckler & Koch P9S	X	
Heckler & Koch VP-70	X	
Heckler & Koch P7	X	
Heckler & Koch P7A13		X
SIG-Sauer P226		X
Smith & Wesson 459A	X	X
Star M28	X	

COMPARISON OF M15 GENERAL OFFICERS PISTOL AND M1911A1

	M15 General Officers Pistol	M1911A1
Caliber:	.45	.45
System of operation:	Recoil	Recoil
Length overall:	200mm (7.875 in.)	219mm (8.62 in.)
Barrel length:	108mm (4.25 in.)	127mm (5 in.)
Feed device:	7-shot box magazine	7-shot box magazine
Sights: Front:	NM type blade	Blade
Rear:	NM type square notch	Square notch
Weight:	1.02 kg (2.25 lbs.)	1.1 kg (2.43 lbs.)

Star 9 × 19mm NATO Model 28 of a type similar to the handgun tested by the United States in 1977–1978. See Chapter 42 for more details.

Smith & Wesson 9 × 19mm NATO Model 459 of a type similar to the handgun tested by the United States in 1977–1982. See Chapter 51 for more details.

9 × 18mm MAKAROV HANDGUNS

As part of the overall introduction of new small arms in the post–World War II period, the Soviets added a new sidearm to their inventory, the *Pistolet Makarova*, chambered for the 9 × 18mm cartridge. That cartridge is slightly more powerful than the older .380 ACP (9mm *Kurz*) round that was common in the West before 1945, but it is less lethal than the 9 × 19mm Parabellum. Part of the reduced lethality is due to the lower velocity of the Makarov round, 335 m.p.s., compared to the 350 m.p.s. of the Parabellum. But the projectile of the Soviet cartridge is also somewhat lighter—6.9 grams (106 grains) versus 7.5 grams (115 grains). The Makarov type pistols have been manufactured in the USSR (as the PM), in the German Democratic Republic (as the Pistole M) and in the People's Republic of China (as the Type 59). It has been widely distributed in allied countries, as well. For additional details, see Chapter 46.

Soviet Makarov Pistol PM.

The Polish Armed Forces use another design, the P64, that combines some design features from the Makarov and the Walther PP series of pistols. As with its Soviet counterpart, the P64 is a simple blowback mechanism. It has a six-shot magazine capacity, while the Makarov magazine holds eight rounds. Neither of these pistols are offensive weapons; rather they are small, compact personal defense weapons. (See Chapter 37.)

Perhaps the most unusual development in recent years was the North Korean resurrection of the Browning 1900 pistol as their Type 1964. Chambered for the 7.65 × 17mm (.32 ACP) cartridge, this weapon appears to have been intended for clandestine operations as it has often been found fitted with a silencer. (See Chapter 35.)

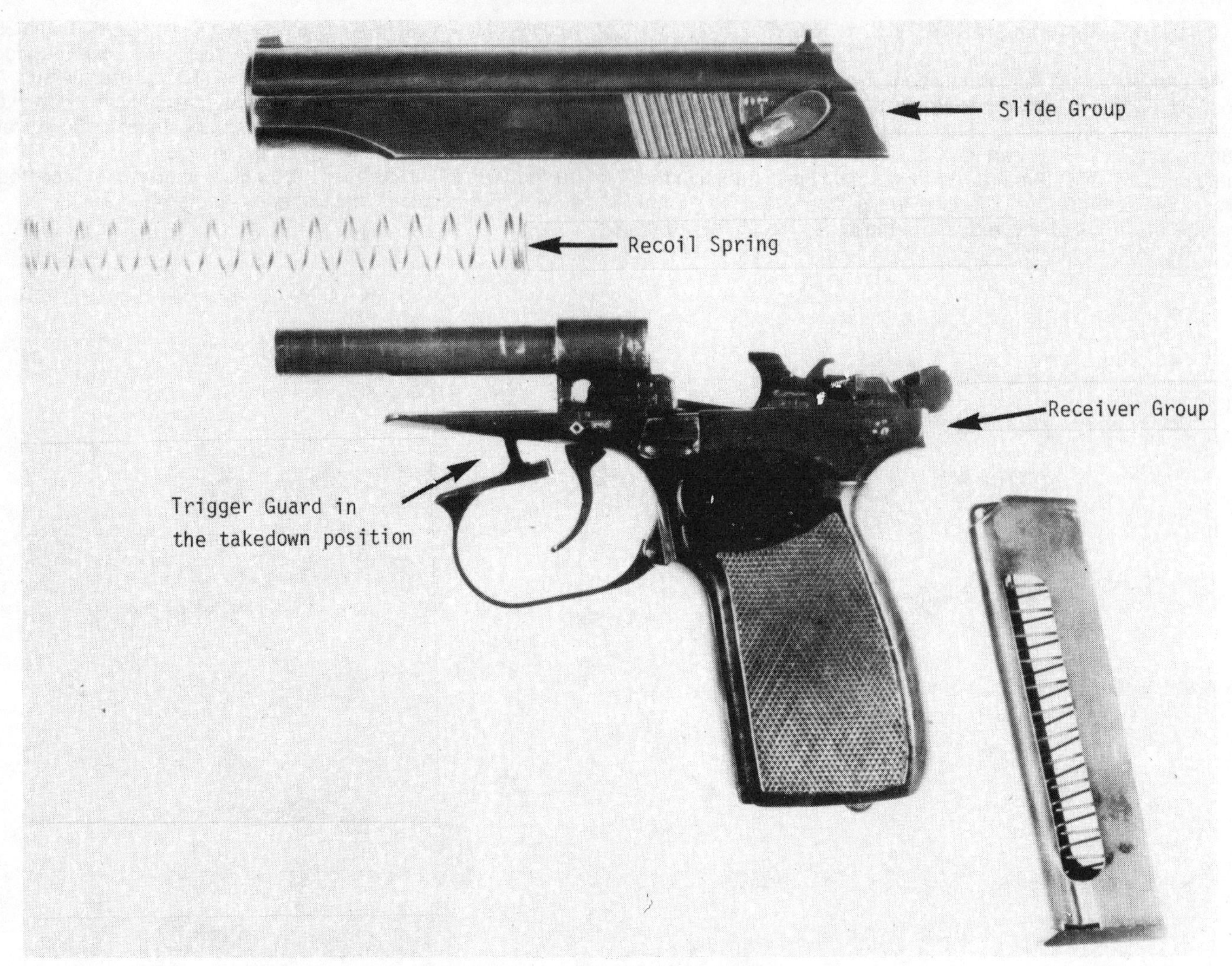

East German *Pistole M* field stripped.

HANDGUNS—A 1983 STATUS REPORT

As does the submachine gun, the handgun appears to be entering a period of transition. With the exception of the United States with its .45-caliber pistols, most armies have already adopted the 9mm diameter bullet. Soviet developments indicate that they no longer consider a high power pistol to be an essential item in their arsenal. Evidentally, they view the 9mm Makarov pistol the last resort in personal defense, the assault rifle being the first line of personal protection.

With the introduction of lighter rifles, most Western armies also are less concerned with handguns. Often a status symbol, for example, the US General Officer Pistol, the handgun will only be used in combat when all else fails. Other new weapons, such as the Firing Port Weapon (Chapter 5), will likely diminish even further the importance of handguns. The small caliber, high velocity assault rifle concept has initiated a period of change that will alter the entire world of small arms in the next several decades.

5 Special Purpose Weapons Development

Changes in the nature of conventional combat since the end of World War II have produced a number of special purpose small arms. This chapter looks at six categories—sniper rifles, firing port weapons, silenced weapons, combat shotguns, shoulder fired grenade launchers, and blank firing attachments.

SNIPER RIFLES

At the end of World War II, most of the world's major armies had special models of their standard service rifle that had been adapted for use by snipers. The basic modification was the addition of a telescopic sight and scope mount, and when possible specially selected components, such as barrels and stocks were chosen to yield a more accurate rifle. The M1C and M1D rifles equipped with telescopic sights replaced the older M1903A4 sniper rifle during the Second World War. In the Soviet Army, both the Model 1891/1930 Mosin bolt action rifle and the Tokarev SVT40 self-loading rifle were modified for sniper use. Since 1945, considerable work has been done in both the East and the West to develop newer sniping rifles. A look at current American and Soviet sniper weapons will give the reader a better idea of post-war developments.

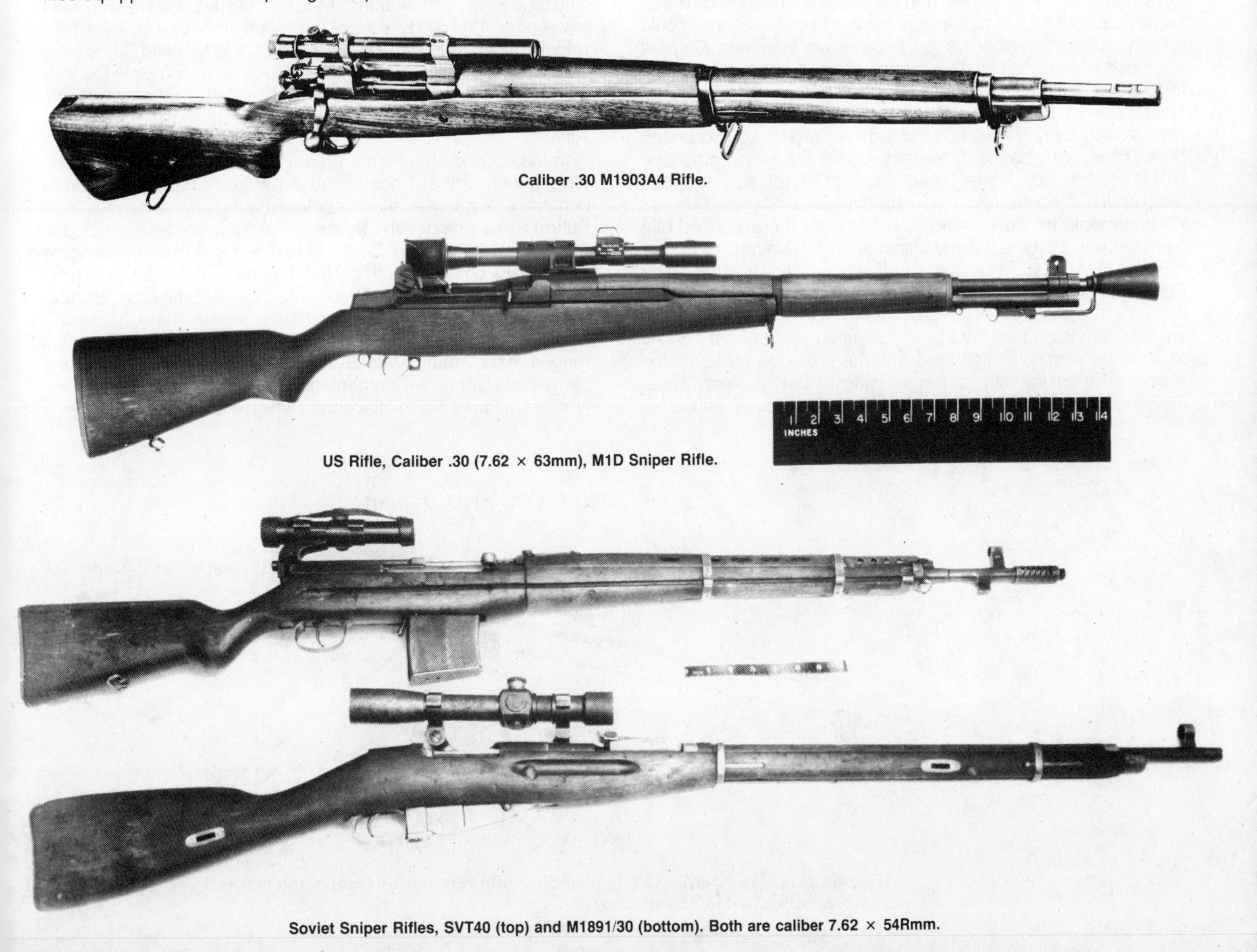

Caliber .30 M1903A4 Rifle.

US Rifle, Caliber .30 (7.62 × 63mm), M1D Sniper Rifle.

Soviet Sniper Rifles, SVT40 (top) and M1891/30 (bottom). Both are caliber 7.62 × 54Rmm.

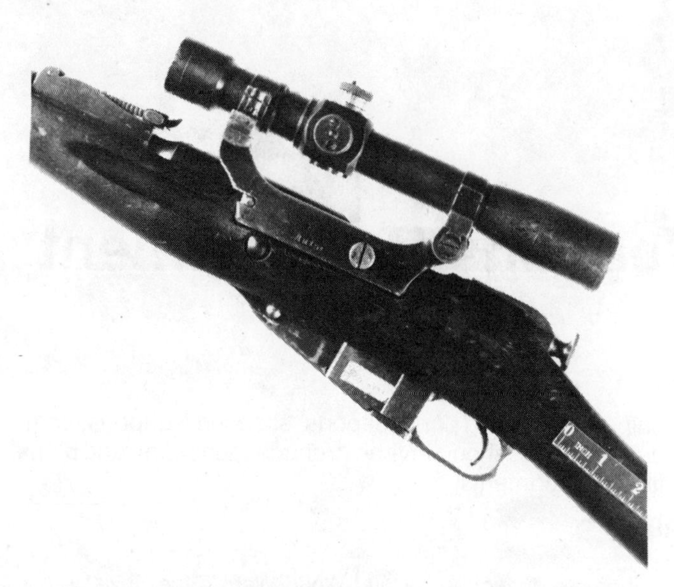

Close-up view of Soviet M1891/30 telescopic scope and mount.

Close-up of Soviet SVT40 telescopic scope and mount.

UNITED STATES SNIPER RIFLES

The long war in Vietnam with its peculiar absence of a stationary or continuous battle front meant that the enemy could literally appear anywhere allied forces were stationed. Snipers equipped with accurized rifles proved to be an excellent addition to the defensive armament of a variety of types of military installations. In September 1968, the Department of the Army directed the Army Materiel Command to provide 1,800 accurized M14 Rifles equipped with the Adjustable Ranging Telescope (ART) designed by James Leatherwood to US forces in Vietnam as ENSURE Requirement 240 (*E*xpediting *N*on *S*tandard *U*rgent *R*equirements for *E*quipment). Standards for the accurized rifle were defined by the US Army Markmanship Training Unit at Fort Benning, Georgia. Telescope specifications were prescribed by Frankford Arsenal based on the Leatherwood design evolved at the US Army Limited War Laboratory. The first 65 weapons shipped to Vietnam came from the Limited War Laboratory. All M14 Rifles were manufactured with the provision for attaching a scope mount directly to the left side of the receiver. Thus, theoretically any M14 in service could be equipped with a telescopic sight in a matter of minutes. But the Army wanted specially selected and very accurate examples fine tuned for sniping. Subsequently, 300 National Match M14s were fitted with M84 telescopes (the M1D scope) and sent to Vietnam on a high priority basis. At Rock Island Arsenal, 1,435 M14NM Rifles were converted to XM21 sniper rifle configuration to complete the total number of 1,800 required by ENSURE 240. At the end of December 1975, the XM21 was standardized as the M21 Sniper Rifle.

In the course of the work on the XM21, the Small Arms Systems Laboratory at Rock Island did an additional evaluation of contemporary sniper rifles to determine the characteristics for a Future Army Sniper Rifle System. The weapons tested included the XM21, the Marine Corps M40 7.62 × 51mm (Remington Model 700 bolt action rifle), the French FRF1 7.62 × 51mm Sniper Rifle (Chapter 18) and the Steyr-Mannlicher Scharfshutzengewehr (SSG) 7.62 × 51mm sniper rifle (Chapter 8). All of these weapons provided excellent accuracy at sniping ranges. The Small Arms Systems Laboratory at Rock Island carried out some experimentation with product improved versions of the M14/M21 Rifle, during which alternative gas systems

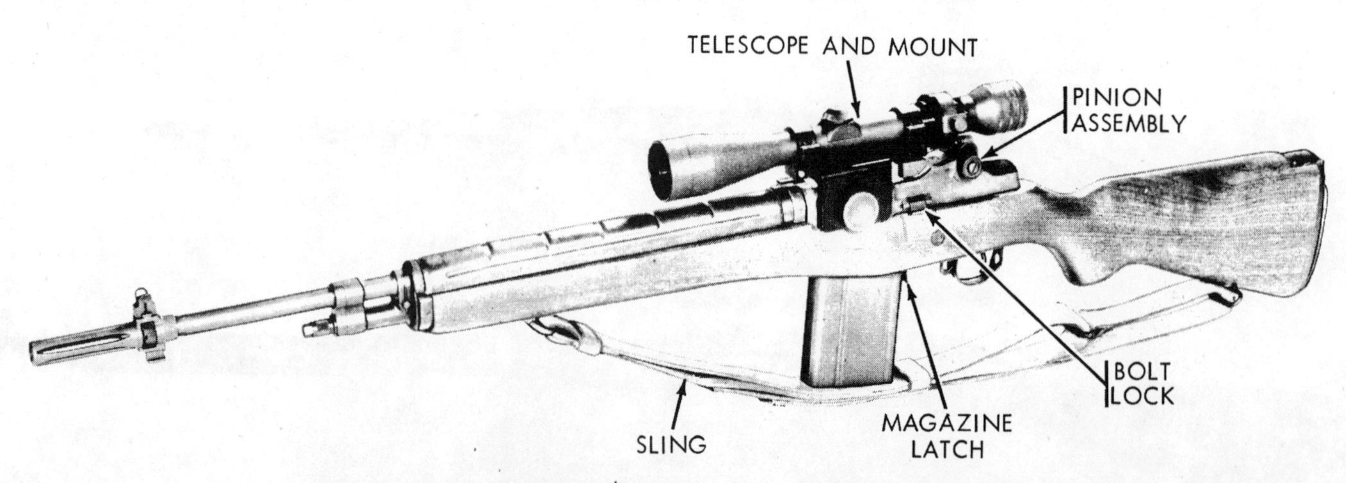

Rifle, Sniper, 7.62-MM, XM21, w/adjustable ranging telescope and mount, sling and controls - left front view.

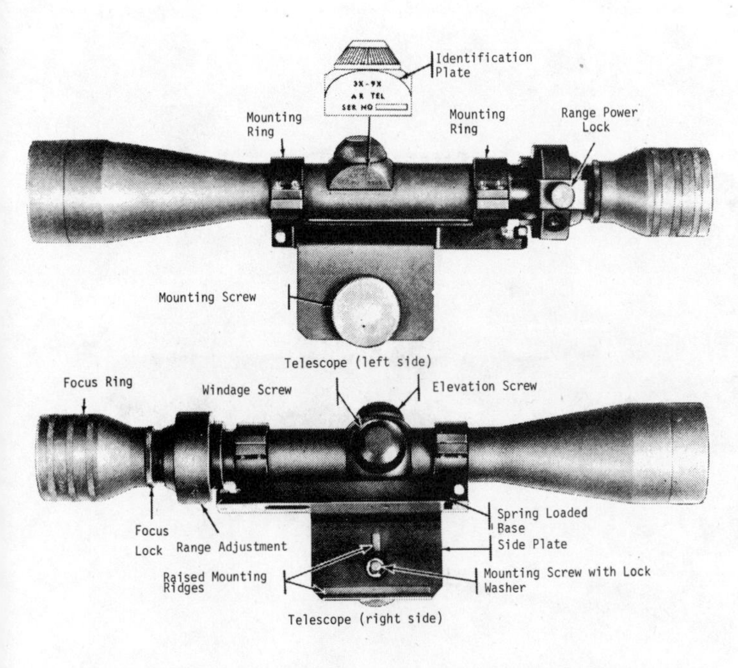

Adjustable Ranging Telescope as used with the M21 Sniper Rifle.

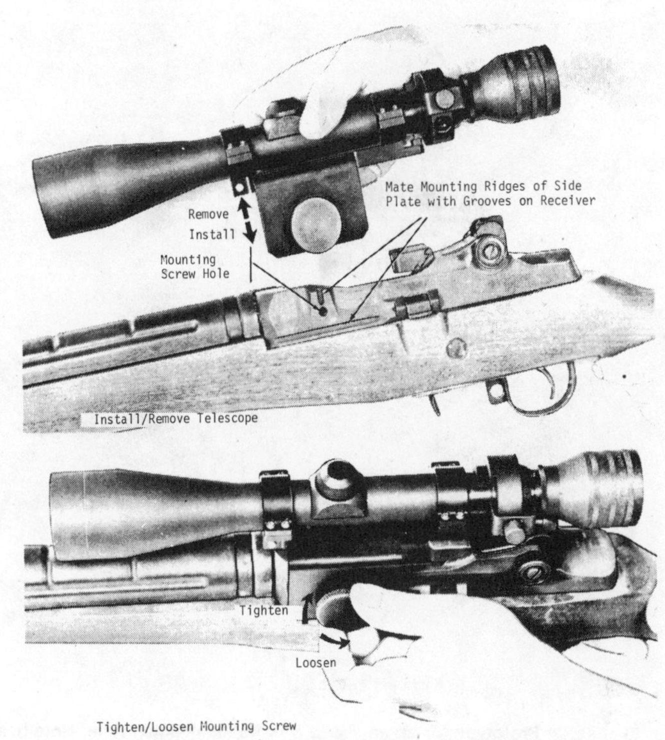

Installation and removal of ART and mount.

and stock configurations were tested. With the winding down of American participation in the war in Indochina, work on newer sniper rifle concepts was terminated. Current US sniper rifles include the M21 and M1C for the Army and the M40 for the Marine Corps.

The telescopic sight used on the M21 deserves some comment. Having variable power magnification—3 to 9—the scope is intended for adjustable ranging between 300 and 900 meters. When set on 3 power and aimed at a target 300 meters distant, the two vertical marks above and below the cross hairs (stadia) mark out a distance of 762mm (30 in.), the approximate distance between the average infantryman's belt buckle and the top of his helmet. By bracketing those features in the stadia on the average human target, the sniper can place the cross hairs at the midpoint of the target's chest. If the range is greater than 300 meters, the magnification ring on the scope is used to increase the size of the image until once again the two vertical stadia rest on the belt and helmet of the target. This adjustable ranging feature removed much of the guesswork from aiming at a target. It thus increased markedly the lethality of the rifle/telescope combination. This ART scope has been ballistically matched with the US M118 NATO match ammunition.

M21 Sniper Rifle with Adjustable Ranging Telescope.

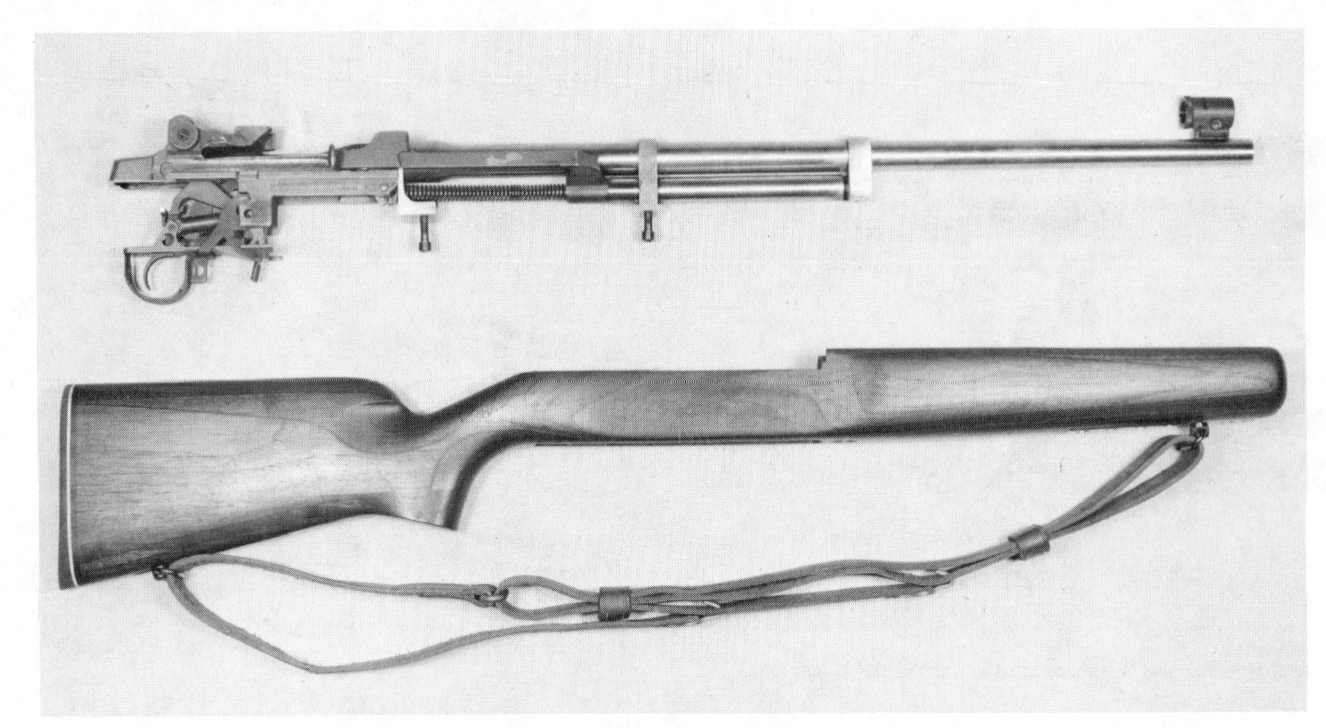

Prototype for an improved M14 National Match Rifle. Note heavy target type barrel, modified gas system, target type stock and two point screw down bedding mounts.

Prototype for an improved M21 Sniper Rifle. Note bullpup type stock, heavy barrel and modified gas system.

Very little officially funded work has been done in the United States with sniper rifles since 1977. In April of that year, Aberdeen Proving Ground tested six sniper rifles, all of which fired the M118 7.62 × 51mm NATO match cartridge. The tested weapons were the AR-10 modified at Rock Island Arsenal, the Remington M40A1 (the standard USMC sniper rifle), a modified M14, the Parker Hale 1200TX (Canadian C3), and the Winchester M70 Match rifle. All of these rifles used the Leatherwood ranging telescopic sight. The US M21 sniper rifle was the control weapon and it was equipped with the standard Adjustable Ranging Telescope. This test indicated that there were several steps that could be taken to improve military sniper rifles. First, a better sighting system, which would improve the ability of the shooter to estimate ranges, could be developed. Probably the most valuable accessory for sniper teams would be a laser range finder that would allow the shooter to determine range accurately. He could then set his scope to a precise number of meters. The other area of improvement would be in the development of more accurate sniping ammunition. The M118 projectile is a good one, but there are better ones available on the commercial market in both the United States and Europe.

EXPERIMENTAL SNIPER RIFLES TESTED BY US ARMY IN 1977

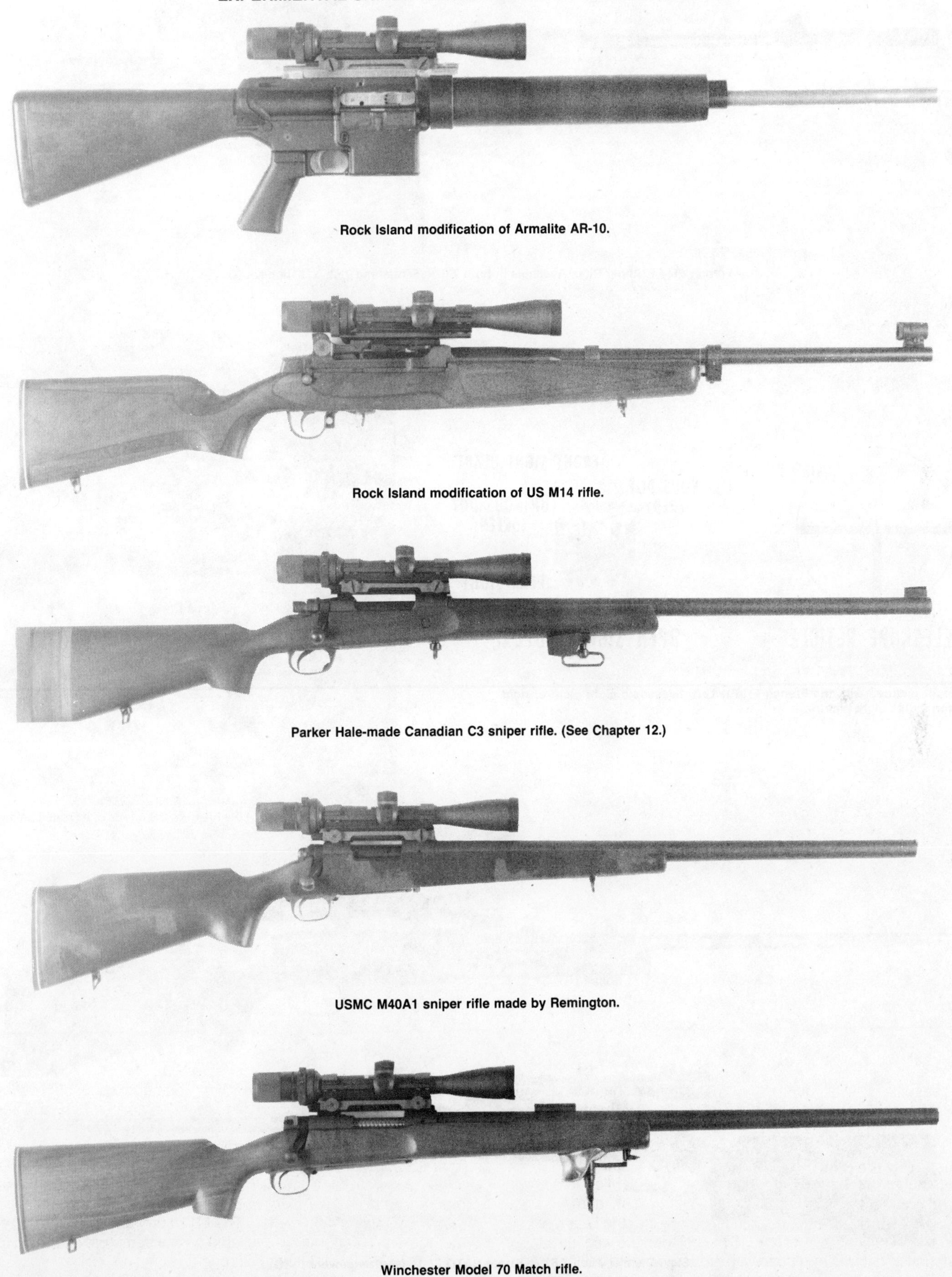

Rock Island modification of Armalite AR-10.

Rock Island modification of US M14 rifle.

Parker Hale-made Canadian C3 sniper rifle. (See Chapter 12.)

USMC M40A1 sniper rifle made by Remington.

Winchester Model 70 Match rifle.

French FR-F1 Sniper Rifle. Available in both 7.5 × 54mm and 7.62 × 51mm NATO.

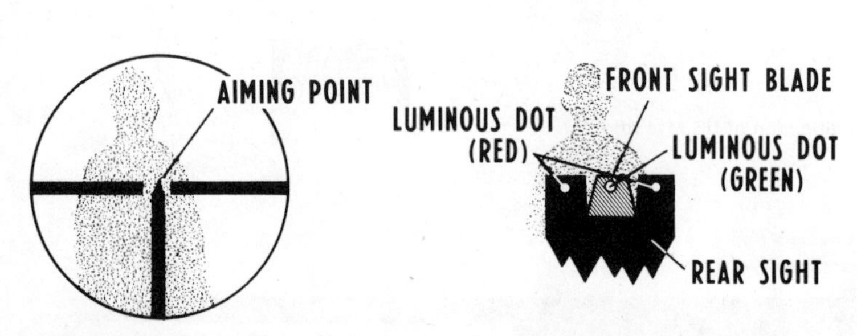

Sight pictures with the French FR-F1. *Left,* telescopic sight picture; *right,* iron sights sight picture.

A French Army photo of the FR-F1 in the hands of a French soldier (É. C. Armées).

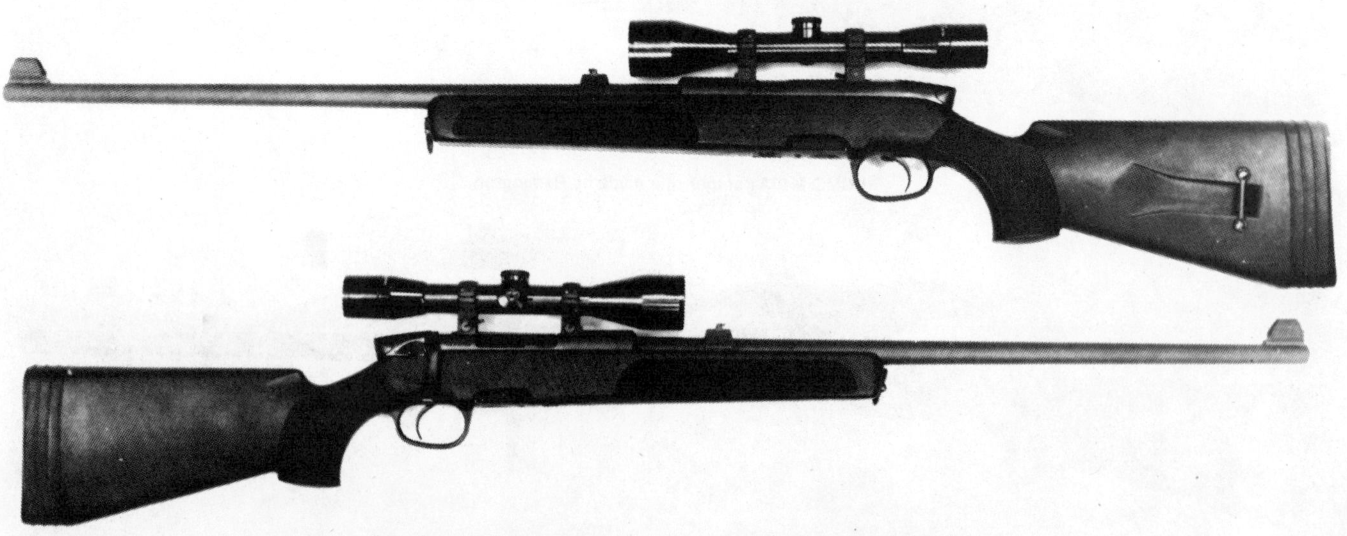

Steyr-Daimier-Puch 7.62 × 51mm NATO *Scharfshutzengewehr* (SSG).

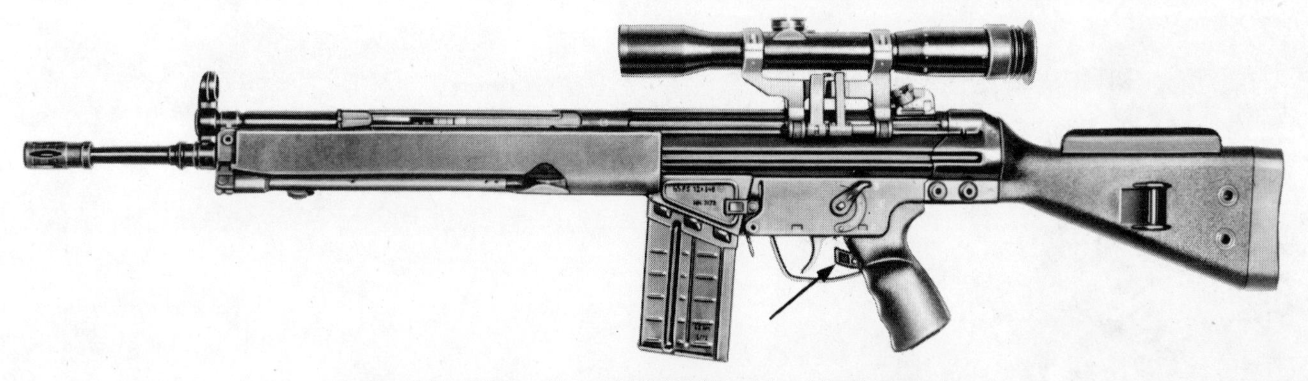

Heckler & Koch 7.62 × 51mm NATO G3SG/1 Sniping Rifle. Note set trigger.

HK33SG/1 5.56 × 45mm Sniper Rifle with set trigger.

HK33A2ZF. Standard 5.56 × 45mm Rifle with telescope but no set trigger.

Heckler & Koch's newest sniper rifle, the PSG-1. (See Chapter 21.)

A member of 40 Commando, Royal Marines, seen in Belfast wearing a flak jacket and carrying a 7.62 × 51mm NATO L42A1 sniper rifle in the spring of 1975.

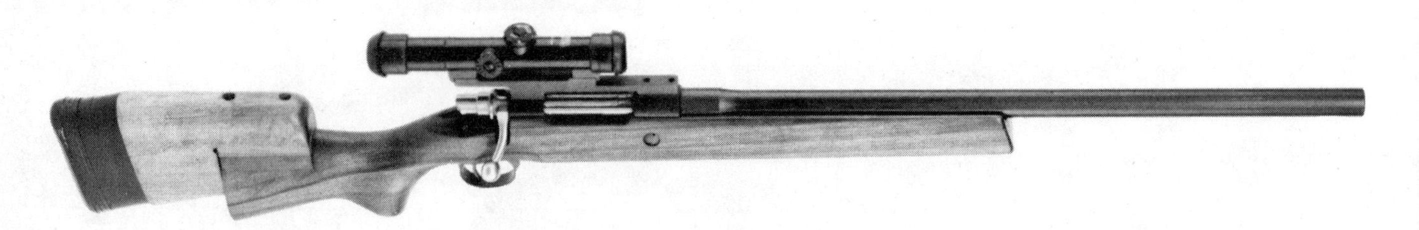

Fabrique Nationale's 7.62 × 51mm NATO sniper rifle. This weapon is built up on a standard Mauser-type receiver.

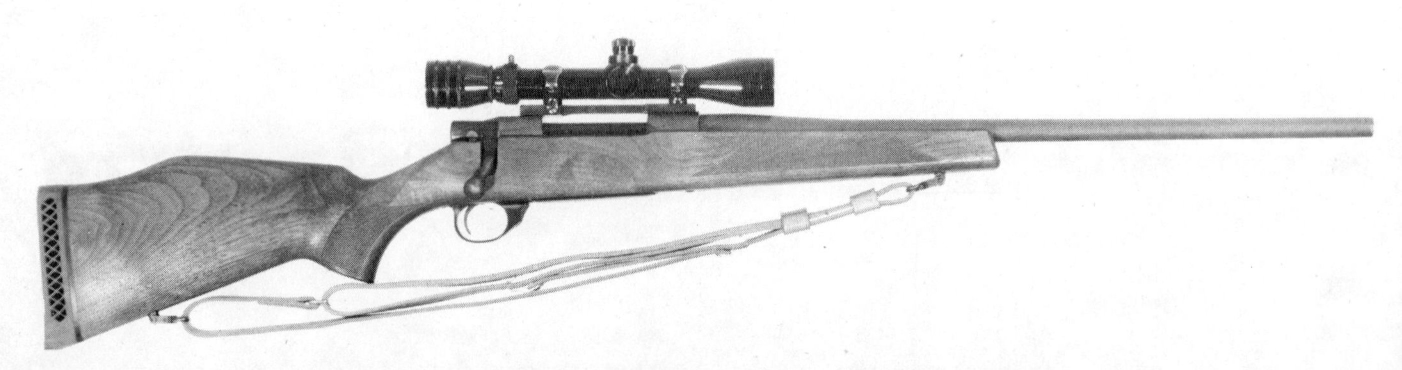

Smith & Wesson's Model 1500 Precision Shooting Rifle in 7.62 × 63mm (.30-06 US).

SOVIET DRAGUNOV SNIPER RIFLE

The Soviet Union adopted a new self-loading sniper rifle in 1963, designed by sporting rifle specialist Yevgeniy Fyedorovich Dragunov. Called the *Samozaridnyia Vintovka Dragunova* (SVD), this 10-shot weapon chambered for the full power 7.62 × 54R cartridge is an evolution of the Kalashnikov operating mechanism. Unlike the AK and PK weapons, the Dragunov does not use a long stroke piston system. That type of gas system tends to be rather heavy for a large caliber rifle, and the motion of that mass during the firing cycle changes the rifle's center of gravity, which in turn reduces the accuracy of the rifle. In the Dragunov, the designer employed a short stroke piston system. The piston, which is separate from the bolt carrier, delivers its impulse to the carrier, which then moves to the rear. The remainder of the operating sequence is quite similar to the Kalashnikov assault rifle. A commercial Soviet rifle, the *Medved* ("Bear"), embodies the basic operating mechanism of the SVD.

While not as sophisticated as the American ART scope, the Soviet PSO-1 telescope, nevertheless, contains some unique features. A range finder scale located in the lower left of the telescope is graduated to the height of a man 1702mm (67 in.). By placing the horizontal line at the foot of the target and following the curved line to the head of the target, one can estimate the range in hundreds of meters. There are alternative aiming marks on the reticle for ranges above 1,000 meters. In addition, the PSO-1 scope has a battery powered element that will illuminate the reticle under poor visibility conditions. And finally, the scope has an infrared detecting filter that can be used at night to detect infrared night vision devices. This filter can also be used with an external infrared spot light to help illuminate targets. (See Chapter 46.)

MISCELLANEOUS SNIPER RIFLES

In addition to the standard Austrian, French, US and Soviet sniper rifles, Fabrique Nationale has produced a new weapon that is essentially a highly accurate Mauser bolt action rifle. FN formerly manufactured a sniper version of the FAL. Heckler & Koch produce a sniper version of the G3, the G3SG/1, which is used by German police forces. This weapon has a special set trigger unit that permits more precise adjustment of the trigger pull. The British use a 7.62 × 51mm NATO caliber version of the older Lee Enfield. This weapon, the L42A1, is essentially a reworked version of the No. 4 Rifle. Other older weapons, such as the Australian Spartco M44 Sniper Rifle, may still be found in use, but most have been out of production for some time.

Soviet *Samozaridnyia Vintovka Dragunova* (SVD) 7.62 × 54Rmm Sniper Rifle.

Close-up view of the Soviet SVD Receiver and telescopic sight.

THREE SNIPER RIFLES BASED UPON THE KALASHNIKOV RIFLE MECHANISM

The Romanian FPK 7.62 × 54mm R sniper rifle. (See Chapter 39.)

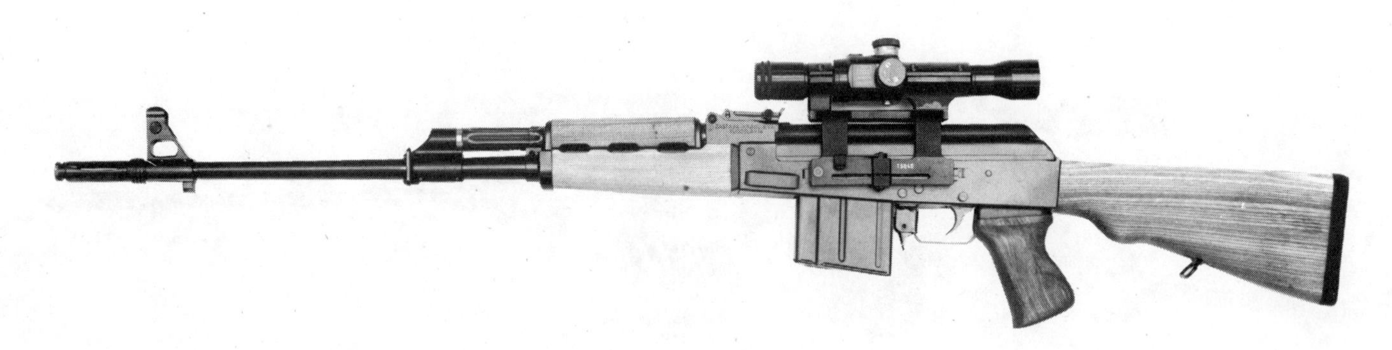

The Yugoslav 7.92 × 57mm sniper rifle made by the Zavodi Crvena Zastava, Kragujevac. (See Chapter 50.)

The Finnish M78 5.56 × 45mm sniper rifle made by Valmet. (See Chapter 18.)

FIRING PORT WEAPONS

Introduction of the Armored Personnel Carrier (APC) as a means of transporting infantry troops has led to the development of modified small arms that can be fired from the inside of the APC. During the 1939–1945 conflict, both Allied and Axis forces discovered the vulnerability of armored fighting vehicles to individuals with hand carried explosive charges. During that war, the German *Wehrmacht* experimented with a special curved barrel attachment (Krummlauf) for the *Sturmgewehr 43*, developed by the Unterluss Laboratory of Rheinmetall-Borsig. Equipped with this barrel and a periscopic sight, the tank crew could fire the *Stg 43* and keep opposing forces off the vehicle's back.

More recently, the US and the USSR have begun using Armored Personnel Carriers. The newer ones include a provision whereby the infantrymen inside can fire out of the vehicle. The Soviets introduced a folding stock version of the modernized Kalashnikov assault rifle—the AKMS—which can be fired through special ports in the sides of vehicles such as the BRDM APC. Cadillac Gage's XM706 Commando Armored Car had firing ports and vision blocks so that the Stoner 63 rifle or carbine could be fired from inside the squad compartment. Passengers could thereby provide perimeter defense for the vehicle. Combat experience demonstrated that conventional weapons produced high levels of carbon monoxide and gun smoke that were both dangerous and discomfiting for the individuals inside the vehicle.

Heckler & Koch's MICV Firing Port Weapon (Modified HK53).

The HK54 9mm Parabellum Submachine Gun mounted in Heckler & Koch's combination vision block/firing port.

UZI Submachine Gun adapted to HK's vision block/firing port.

External view of HK vision block/firing port. Note ball swivel joint.

As a result of these experiences and the requirement for a Firing Port Weapon (FPW) for the XM723 Mechanized Infantry Combat Vehicle (MICV), the Small Arms Systems Laboratory at Rock Island began in 1972 to work on weapons modified so that the gas system would exhaust the operating gases outside the squad compartment. Because of the high cyclic rate requested, the weapons fired from an open bolt. In addition, test weapons were fitted with a barrel sleeve that in turn permitted the weapon to be mounted in the ball and socket mount in the side of the MICV. Two 5.56mm candidates, a modified M16A1 and a modified HK33, and a control weapon, the M3A1 submachine gun, were subjected to a Development Test/Operational Test I. After these tests, a Concept Validation In-Process Review was held in May 1974. The modified M16A1 was selected for advanced development and was designated the XM231.

As a consequence of the DT/I testing, two weapon performance areas (reliability and hit probability) were singled out for further study. The XM231, which then had a 279-mm (11.0-in.) barrel, had a cyclic rate of about 1,050 shots per minute. Experimentation indicated that a reduction of the rate of fire down to 200 rounds per minute improved the probability of hits. However, the overall shock power of six weapons (two on each side and two to the rear), each firing away at over a thousand shots per minute, was sacrificed. Additional study indicated that hit probability could be increased by using all tracer ammunition. In the reliability field, extractors were improved and bolts were modified to reduce incidence of bolt fractures that accompanied the high rate of fire. A DT/OT II was run in late 1976 and early 1977. The XM231s used in the Operational Test/II were equipped with a short handguard, flip-up sights and a M3A1 type collapsible wire buttstock for use as a dismounted weapon. These additions are only expected to permit the weapon to be used as a Personal Defense Weapon when all other means of defense have failed. Some weapons specialists opposed these additions because the XM231 Firing Port Weapon is not really suited for use as a shoulder fired weapon. When the M231 firing port weapon was adopted in December 1979—as part of the standardization of the M2 Infantry Fighting Vehicle (IFV)—the collapsible stock was kept but the front sight was eliminated. This suppressive fire weapon will help keep enemy troops pinned down and may prevent them from posing a close-in threat to the IFV.

Although 65 percent of the parts of the M231 are identical to those used in the M16A1 rifle, substantial changes were made

XM231 Firing Port Weapon with stock collapsed. Note modified buffer.

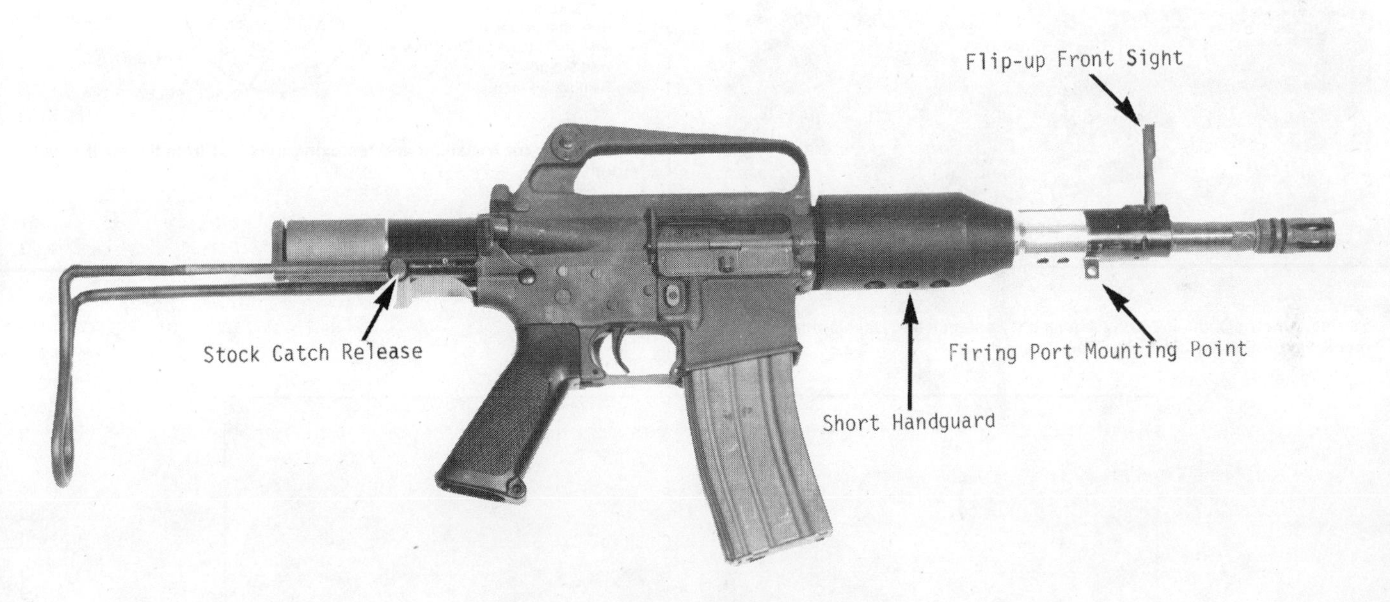

XM231 Firing Port Weapon with stock extended.

COMPARISON OF THE M16A1 AND THE M231

	M16A1 Rifle	M231 Firing Port Weapon
Caliber:	5.56 × 45	5.56 × 45
Length overall:	990.6mm	723.9mm
Length of barrel:	508mm	396.2mm
Loaded weight:	3.35 kg	4.3 kg
Cyclic rate:	700–950 shots/min	1100–1200 shots/min
Modes of fire:	Closed bolt; semi and full automatic	Open bolt; full automatic only

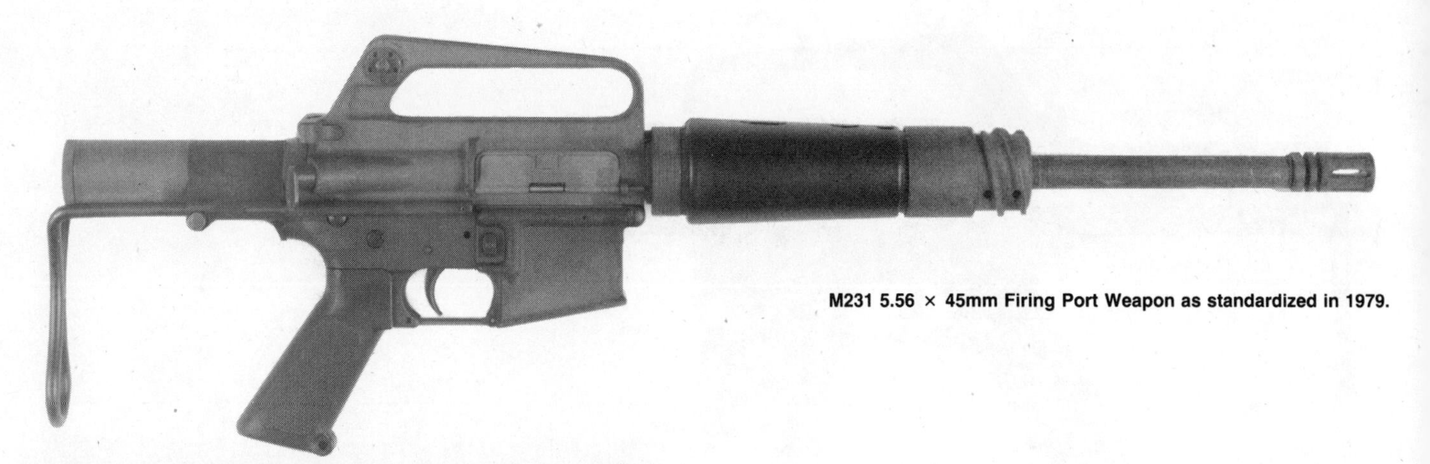

M231 5.56 × 45mm Firing Port Weapon as standardized in 1979.

Sergeant Victor Chan, US Army Armament Research and Development Command, holding the M231 Firing Port Weapon.

INSTALLING

1 With weapon upright, insert barrel thru firing port hole until BARREL COLLAR engages threads of firing port and PIN GROOVE aligns with QUICK RELEASE PIN.

2 2. Rotate weapon one full turn (360°) clockwise until QUICK RELEASE PIN locks into place.

REMOVING

1 Pull out on mount QUICK RELEASE PIN and rotate weapon one full turn (360 degrees) counterclockwise so QUICK RELEASE PIN alines with pin groove.

2 Remove weapon.

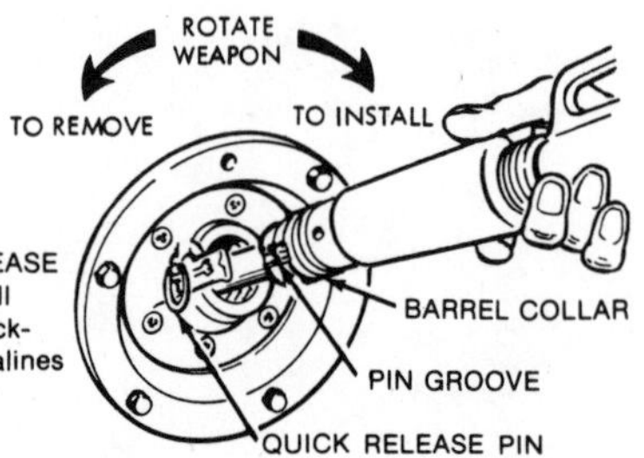

Instructions for installing and removing the M231 from the M2 IFV ball mount.

The M2 Infantry Fighting Vehicle. (Arrows indicate the firing ports.)

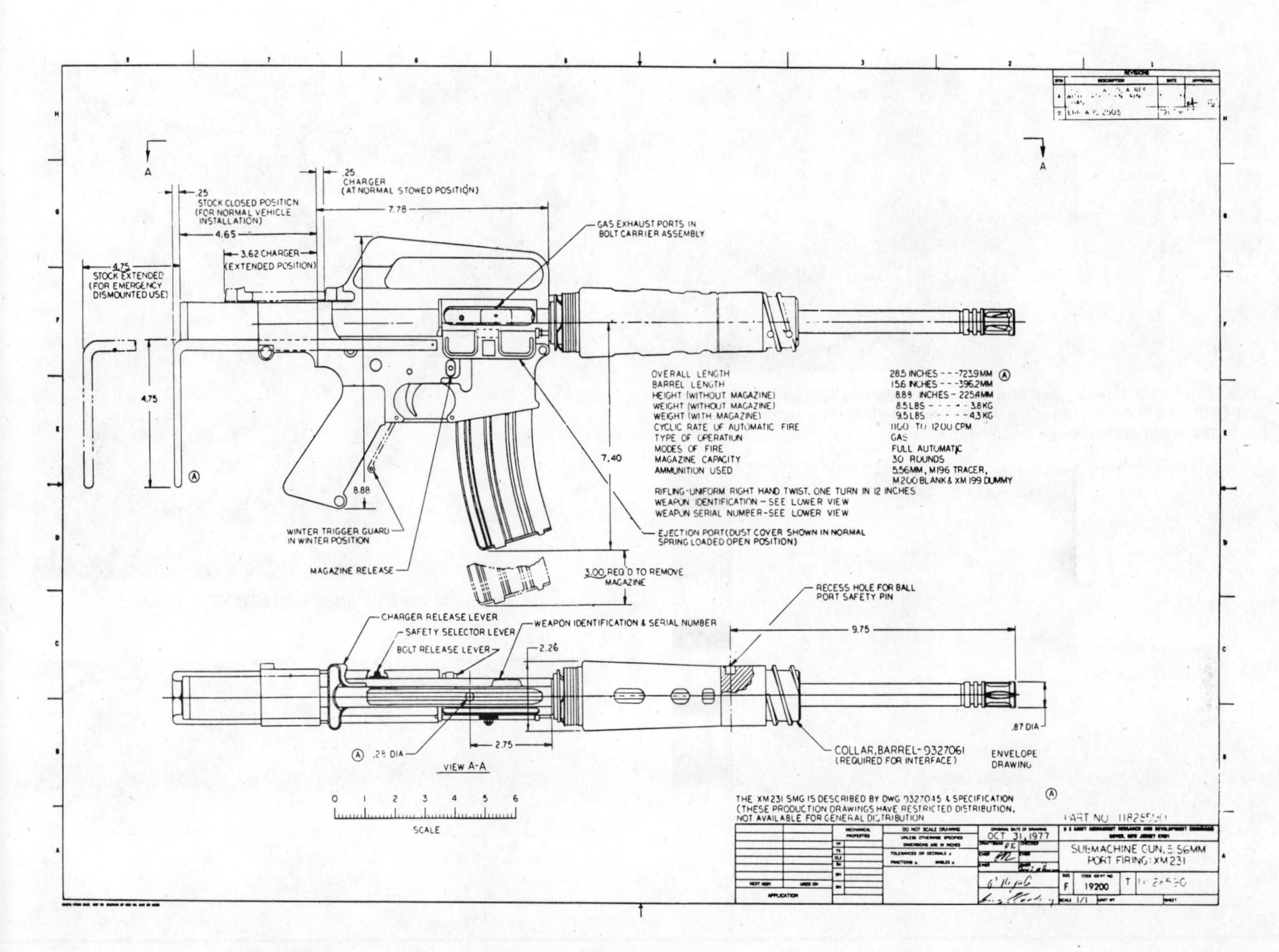

M231 Firing Port Weapon.

to the original design. First, since the M231 fires from the open bolt instead of the closed bolt, the barrel remains cool longer. Some major mechanical alterations were required to permit firing from an open bolt. A floating firing pin, which is propelled against the cartridge primer by a striker, is used in place of the hammer-type mechanism employed in the M16A1. The trigger mechanism has been altered significantly. In addition to the removal of the hammer, the sear has been changed so as to hold the bolt to the rear. When the trigger is pulled, the bolt goes forward, feeding a cartridge into the chamber as it goes, and when the bolt closes, the striker sends the firing pin forward to detonate the primer. Elimination of the hammer and the reduced weight of the modified bolt carrier contribute to the M231's high rate of fire. Second, the interior space available in the M2 IFV dictated the length of the barrel and a shortened receiver extension. Third, a heavier barrel has been used to help fight heat build-up, thus preventing cook-offs and lengthening the life expectancy of the barrel. It is anticipated that the M231 barrel will last 10,000 rounds, despite the exclusive use of tracer ammunition. In addition, the M16A1's forward bolt assist has been eliminated on the M231, and a spent-cartridge collector bag and a gas evacuation system have been added, extras necessary for shooting from inside the IFV.

The M231 can be mounted and dismounted simply and rapidly to the ball swivel mount of the firing port because of a barrel collar with a double-pitch long-lead thread. There are six ports on the M2 IFV; two on each side and two in the rear door. A modified handguard and a collapsible wire buttstock are provided for emergency use of the weapon outside the IFV. The M231 does not have sights. The gunner will aim it by watching through a periscope mounted above the firing port where the tracer projectiles. By contrast, the Soviets use a standard AKM (with folding stock—AKMS) from the firing ports of their BMP fighting vehicle. That weapon is aimed through a vision block of glass that is in alignment with the standard sights of the rifle.

Despite the unorthodox method of bringing the M231 to bear on the target, the weapon is reported to be relatively easy to fire with good effect. It will place rounds in a 300mm circle at 100 meters. Individuals who have fired the M231 say that is uncanny, since its mounting reduces the perceived recoil to zero. After two or three bursts (three to four rounds each), the user can direct fire on the target without difficulty.

The major criticism of the M231 firing port weapon has been the fact that it adds another weapon to the American logistical and training cycle. Whereas the Soviets can use the AKMS with equal effect from the BMP firing port or as an infantry rifle, the American trooper will be equipped with two weapons, the M231 and the M16A1. Through the end of FY 1982, 24,500 M231 firing port weapons have been ordered from Colt Firearms by the US Army.

The Soviet BMP Infantry Fighting Vehicle. In addition to its 76mm cannon, the BMP has firing ports for the AKMS and the PKM. The latter is the 7.62 × 54mm R general-purpose machine gun.

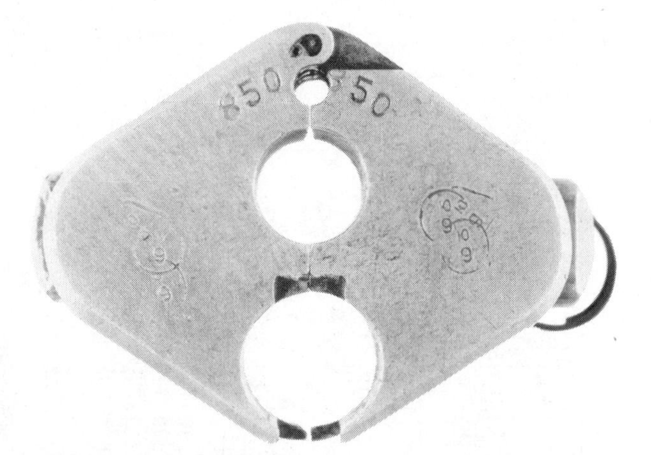

The barrel adapter clamp that holds the AKMS in its firing port.

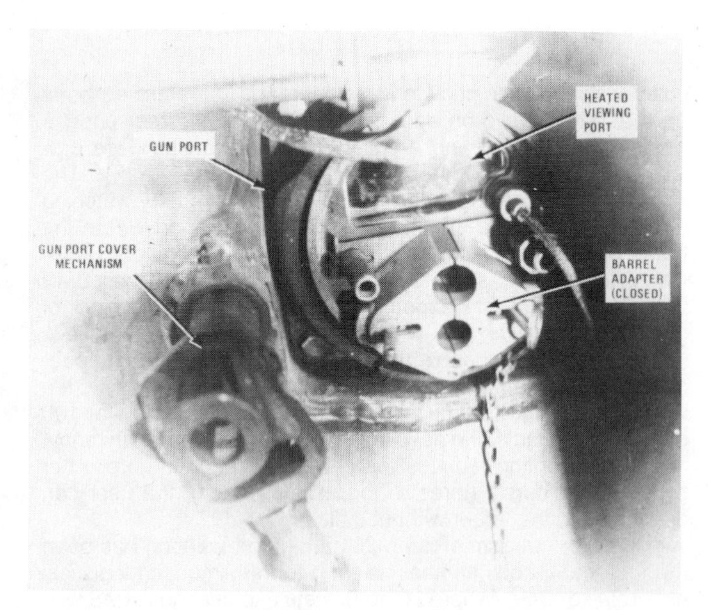

This interior view of the BMP firing port shows its major features.

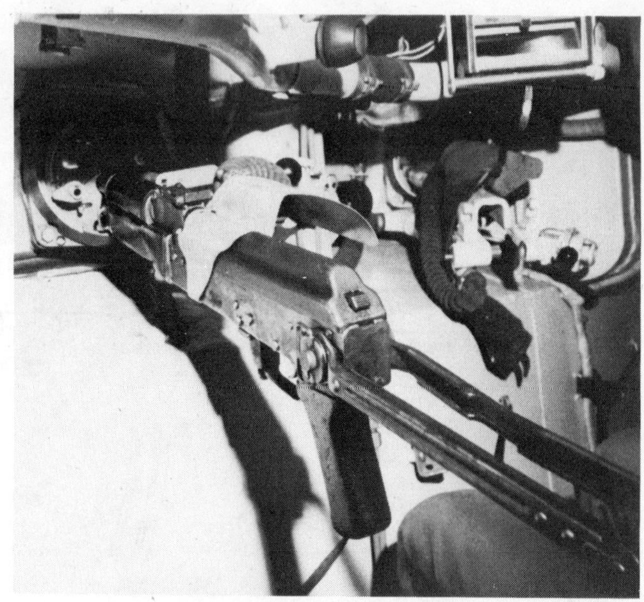

Interior of the BMP showing the AKMS in place.

This interior view of the BMP shows the firing ports, gas mask hook-ups, and the periscopes.

This exterior view of the BMP firing port shows its major features. Note that this is the PKM machine gun port.

SILENCED WEAPONS

While silenced weapons have generally been regarded as the armament of secret agents and Hollywood film script writers, a substantial number of silenced weapons have been developed since 1945 for use by specialized military forces. During the 1939–1945 war, the M3A1 and Sten submachine guns were among the leading weapons fitted with silencers, but they were still noisy due to the movement of the open bolt. Less well known were the Mark I Hand Firing Device and the Sleeve Gun. The former was made in both 7.65mm (.32 ACP) and 9mm, while the latter was produced in 7.65mm only. These weapons, issued to US Office of Strategic Services and British Special Operations Executive personnel, had silenced mechanisms as well as silenced projectiles. Since that time, developments have taken two approaches. First is the true silencer, used with reduced velocity projectiles so that there is no sonic boom as the projectile breaks the sound barrier (335 m.p.s. [1100 f.p.s.]). Second are the noise suppressors, such as the muzzle device used on the Colt XM177E1 and XM177E2 carbines; another suppressor confuses the source of the sound, typified by the Sionics suppressor formerly marketed by Military Armaments Corporation. It should be noted that silenced weapons and silencers are classified as registerable and transfer taxable items under Title II of the US Gun Control Act of 1968.

A review of some of the major developments is all that is possible in this volume.

Two World War II silenced pistols: Top, the Mark I Hand Firing Device (also known as the Welrod). Bottom, the Sleeve Gun. Both weapons fired the .32 Colt (7.65 × 17) Pistol Cartridge.

Czech 7.65mm Model 27 Pistol of wartime manufacture; this pistol has special barrel to be used with silencer.

PEOPLE'S REPUBLIC OF CHINA

The PRC has developed two silenced pistols and a silenced submachine gun since the early 1960s. The Type 64 Silenced Pistol is intended as an assassination weapon. It can be either manually operated or shot as a self-loader. When the complete effect of the silencer is desired, the slide can be locked shut so that it will not recoil open when the pistol is fired. A special 7.65 × 17mm cartridge similar to but not interchangeable with the semi-rimmed .32 ACP round is used with this pistol. Overall length of the Type 64 is 330mm (13 in.), and it weighs 1.27 kg (2. lbs.).

A newer and improved version of the Type 64—the Type 67—has been introduced. The basic action and caliber has been retained, and a new tubular silencer has supplanted the older and more ungainly Type 64 suppressor. (See Chapter 14.)

Disassembled view of the PRC Type 64 Silenced Pistol.

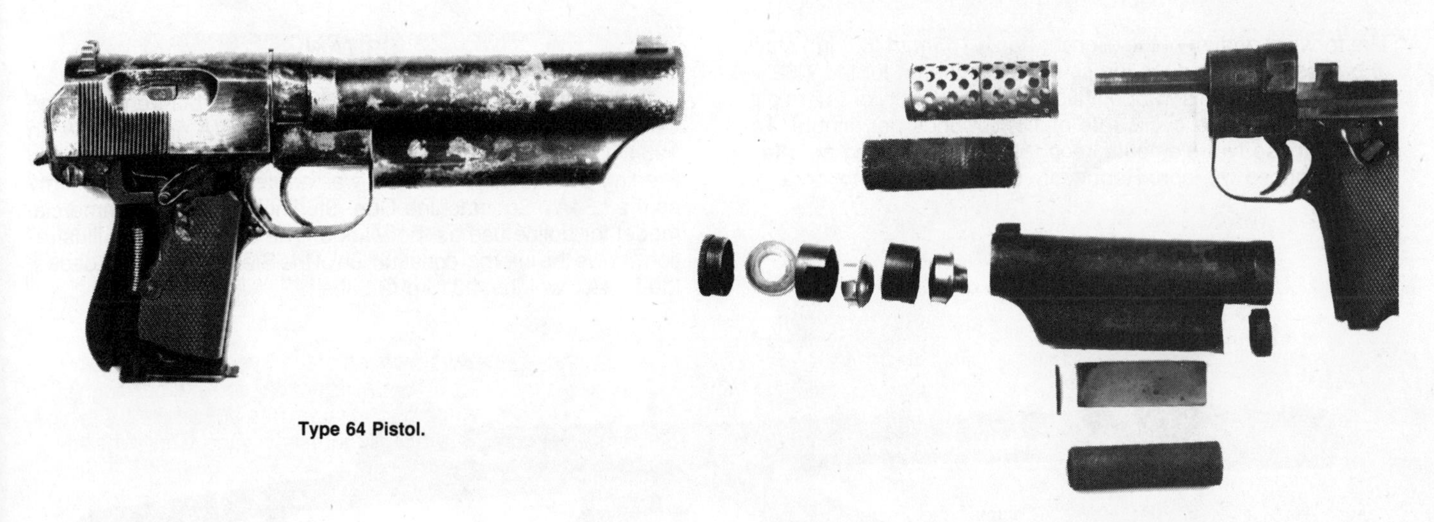

Type 64 Pistol.

Internal view of Type 64 Pistol Silencer.

There is also a Chinese-designed and manufactured silenced submachine gun, the Type 64, that has seen service in recent years. The basic bolt mechanism was borrowed from the Soviet PPS43; the trigger mechanism was patterned after the British Bren gun. The silencer is of the basic Maxim type with a larger tube surrounding a vented barrel. The length of the barrel is 203mm (8 in.), and the length of the tube is 368mm (14.5 in.).

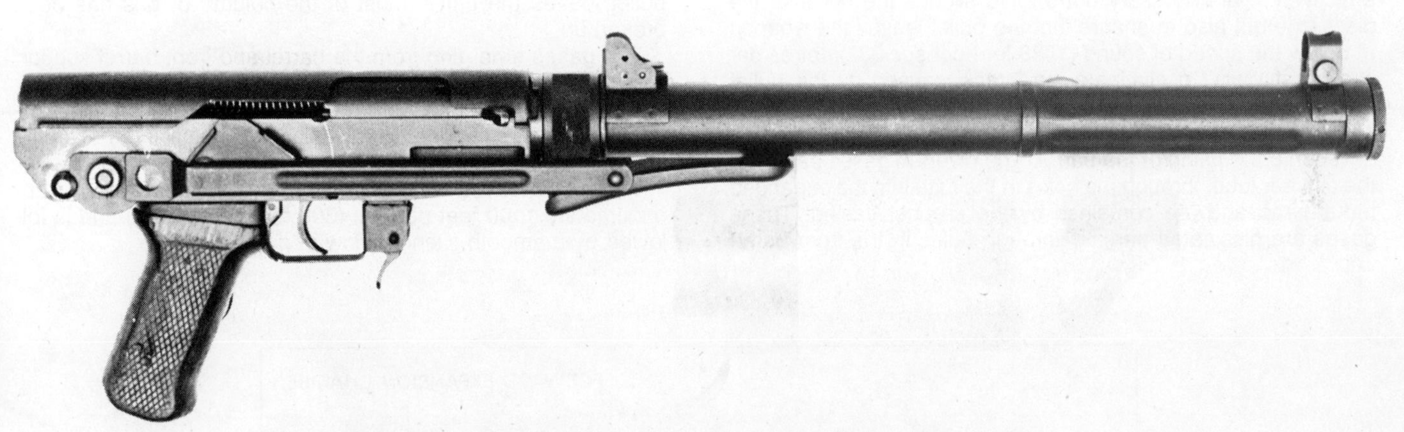

The People's Republic of China Type 64 7.62 × 25mm silenced submachine gun.

Silenced version of the 7.65 × 17mm Czechoslovakian vz 61 Skorpion machine pistol.

The forward portion of the silencer has a 165-mm (6.5-in.) stack of baffles to help reduce the sound. Chambered for the 7.62 × 25mm cartridge, this weapon has a muzzle velocity of 512 m.p.s. (1681 f.p.s.) and a cyclic rate of 1315 rounds per minute. Together, these two elements keep the gun from being an effectively silenced weapon. Reportedly, the Type 64 is no longer in service.

BRITAIN

During World War II, the British issued three different silenced versions of the Sten gun—Mark IIS, Mark IVA and Mark VI. In 1964, George W. Patchett devised a silenced version of the Sterling Submachine Gun that was adopted by the British Army as the L34A1 Submachine Gun. Sterling markets a commercial model for police use as the Mark 5. The accompanying illustration shows the internal construction of the Sterling silencer. Loaded, the L34A1 weighs 4.31 kg (9.5 lbs.).

Sterling Armament Company's Mark 5 Silenced 9mm Parabellum Submachine Gun.

THE OPERATION OF THE SILENCER. To silence a firearm effectively it is necessary not only to silence the noise of the discharge but also to ensure that the bullet leaves the weapon at below the speed of sound (1088 feet per sec/332 metres per sec, in still air), to eliminate the 'crack' caused by the bullet passing through the sound barrier. This is achieved by diverting some of the gases propelling the bullet through 72 small holes drilled into the rifling of the barrel. The diverted gases pass into the diffuser tube, through the holes in the tube into the expanded metal wrap and are contained by the silencer casing. These gases are dissipated partially through holes in the front barrel support and partially by returning to the barrel. By the time the bullet leaves the barrel most of the column of gas has been broken up.

The gases emerging from the barrel and front barrel support are subjected to a swirling motion imparted by the spiral diffuser before passing into the front expansion chamber where the leading gases are deflected backwards by the special internal profile of the front cap.

The bullet leaves the weapon at a subsonic velocity of approximately 1000 feet per sec (308 metres per sec) and is followed by a smooth silenced flow of gas.

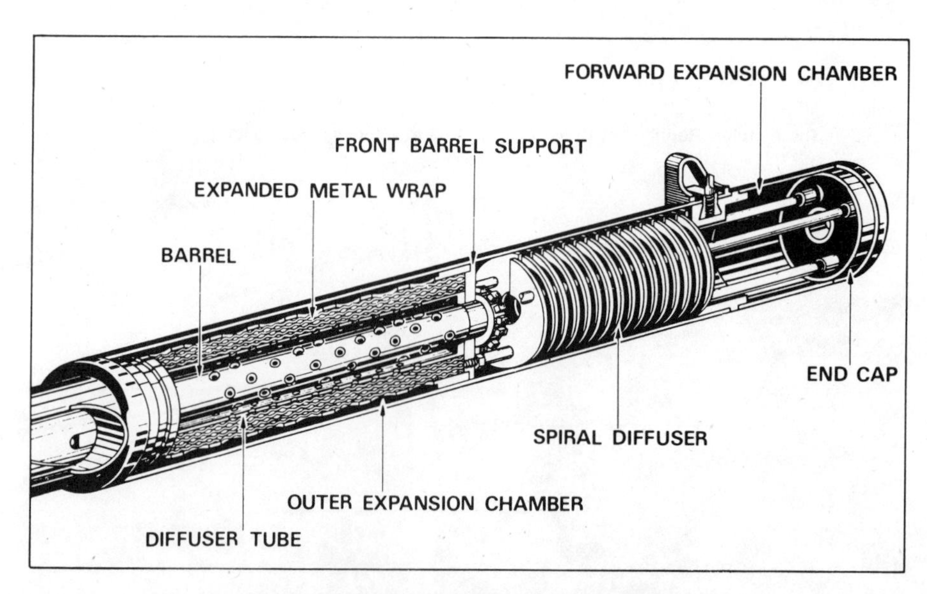

Description of Sterling Submachine Gun Silencer Assembly.

The UK 9 × 19mm NATO L34A1 is silenced submachine gun.

CZECHOSLOVAKIA

The M61 Skorpion has been issued with a silencer attachment that likely is only moderately effective given the high cyclic rate of this weapon.

GERMANY

Both the Walther MPL and Heckler & Koch HK54 submachine guns have been offered with silencers. In the case of the Walther, the silencer is an attachment.

Heckler & Koch's MP5SD2 is a more extensively modified weapon. Like the Chinese Type 64 submachine gun, the MP5SD2 has a vented barrel surrounded by an outer jacket into which the propellant gases can expand. Inside the jacket there is a helix, which slows the gases down. Both the projectile and the gases exit at subsonic velocities. Loaded, this weapon weighs around 3.14 kg (6.9 lbs.), depending on the stock configuration.

Assembled and disassembled views of the HK MP5SD2 Silenced Submachine Gun.

Three versions of the Heckler & Koch silenced submachine gun. *Top,* MP5SD2; *middle,* MP5SD3; *bottom,* MP5SD1.

NORTH KOREA

As noted in Chapter 4, the North Koreans adopted a copy of the Model 1900 Browning Pistol as their 7.62mm Type 64. A version of this weapon has been adapted as a silenced pistol.

North Korean Type 64 Silenced Pistol.

UNITED STATES

During the Vietnam conflict, the US Navy sponsored development of a silenced pistol for use by its SEAL Teams (Sea, Air and Land). Nicknamed the "Hush-Puppy" because of its intended function of killing enemy watch dogs, this modified version of the steel framed Smith & Wesson Model 39 Pistol was put to other clandestine uses as well. Called the Mark 22, Mod. O Pistol by the Navy, the Hush-Puppy had a slide lock to keep the mechanism closed and silent while firing. It fired a special green tipped 9mm Parabellum projectile weighing 10.2 grams (158 grains) that yielded a muzzle velocity of 274 m.p.s. (900 f.p.s.), below the speed of sound. Use of standard supersonic ammunition quickly degrades the effectiveness of the silencer insert. With subsonic ammunition, an insert is good for about 30 rounds; with standard velocity cartridges the insert may have to be replaced after six shots. Official Navy designation for the silencer is Mark 3, Mod. O. Ammunition and replacement silencer parts are supplied as Accessory Kit MK26, Mod. O. Each accessory kit includes 24 9mm Pistol Cartridges MK144, Mod. O. and one silencer tube insert.

All the work on the Model 39 Hush-Puppy was carried out by Smith & Wesson before the end of 1968. Subsequently, Smith & Wesson provided two prototype 13-shot pistols made from stainless steel. These weapons were improved to overcome problems such as extractor breakages, which had been experienced with the Model 39. This modified pistol in a slightly different form was later commercially marketed as the Model 59 Smith & Wesson 9mm Parabellum Pistol.

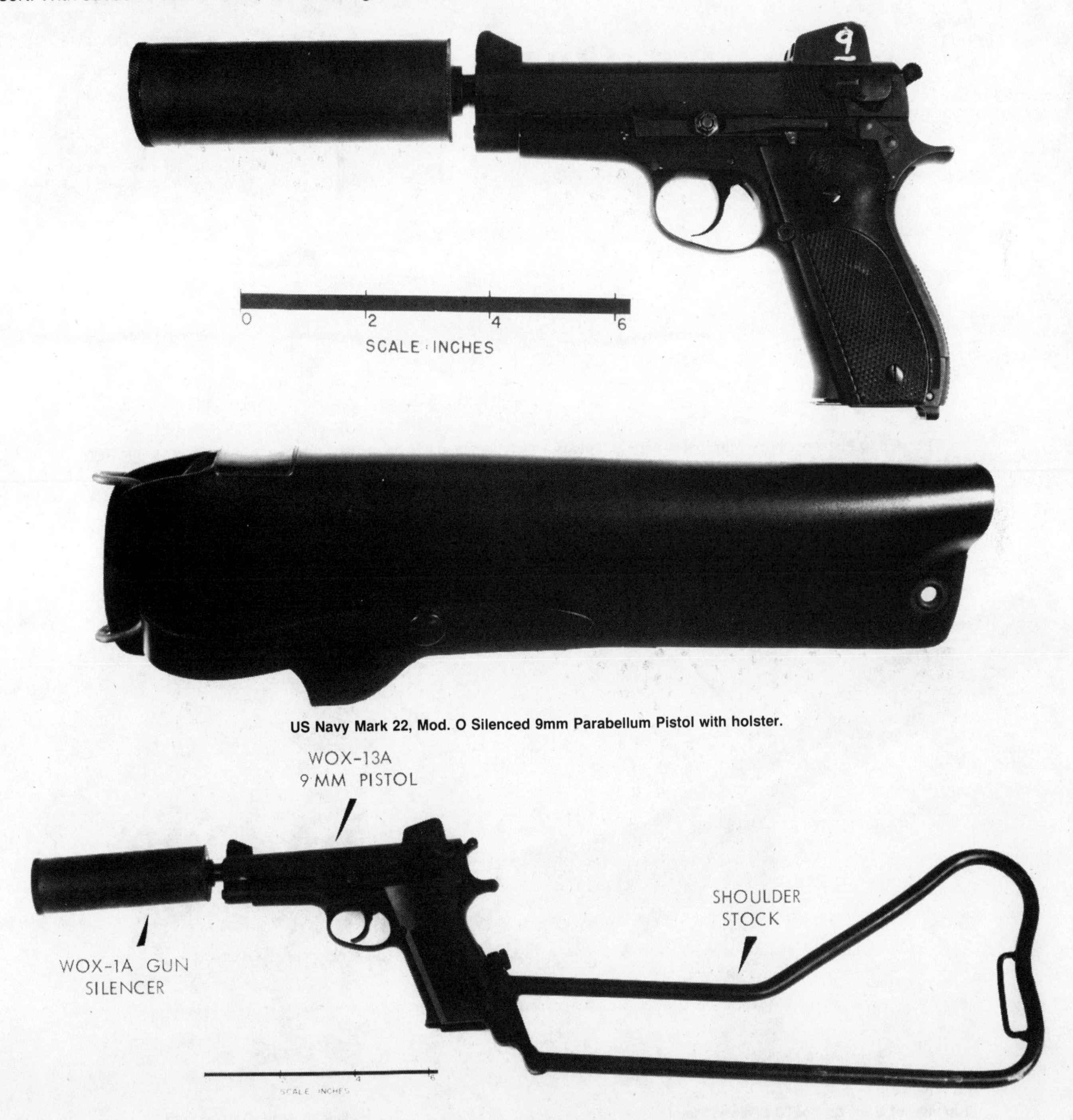

US Navy Mark 22, Mod. O Silenced 9mm Parabellum Pistol with holster.

Mark 22, Mod. O Pistol (Experimental Designation WOX-13A) with shoulder stock.

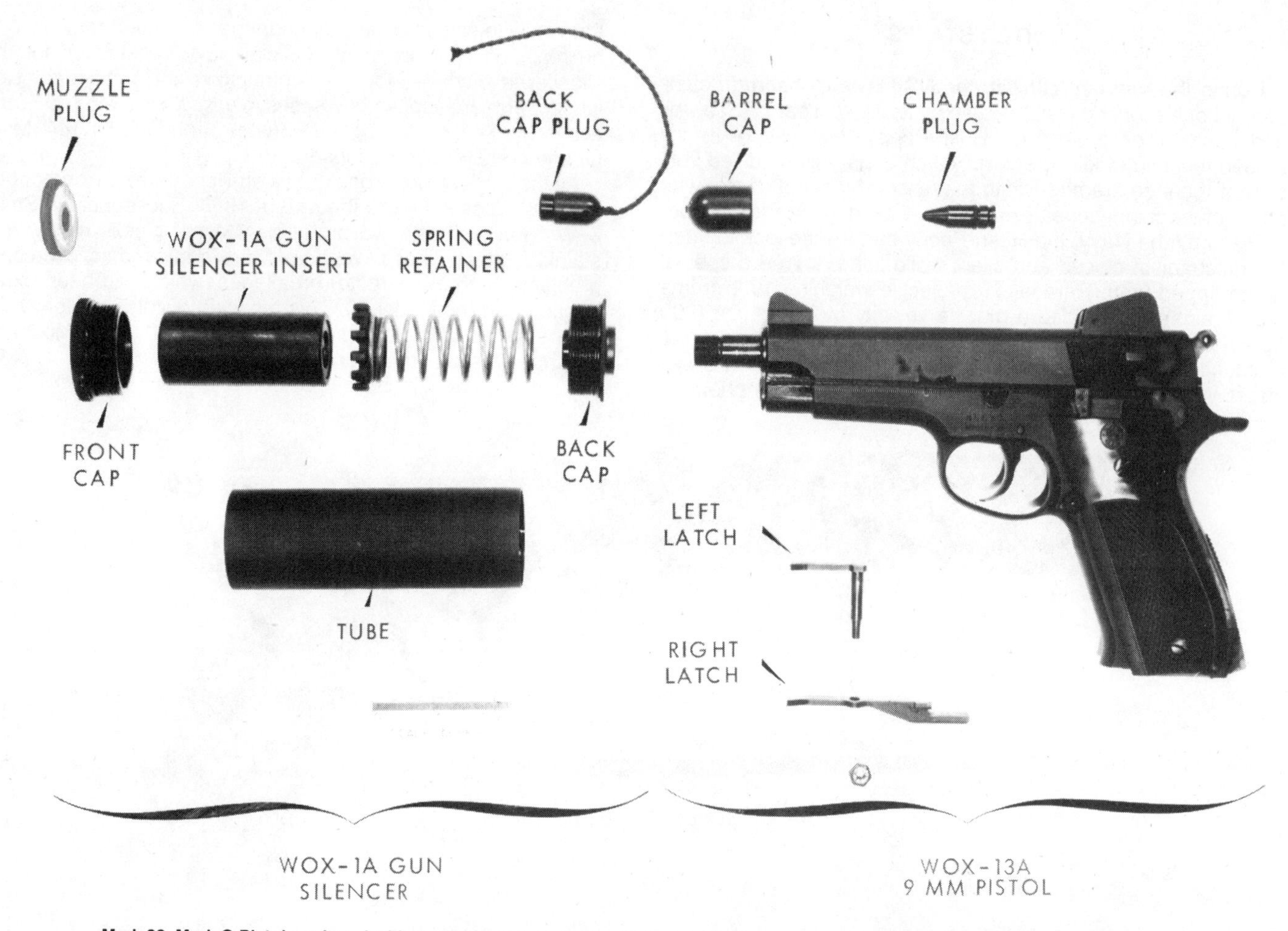

Mark 22, Mod. O Pistol equipped with special caps and plugs, which permitted it to be carried underwater by members of the Navy SEAL Teams.

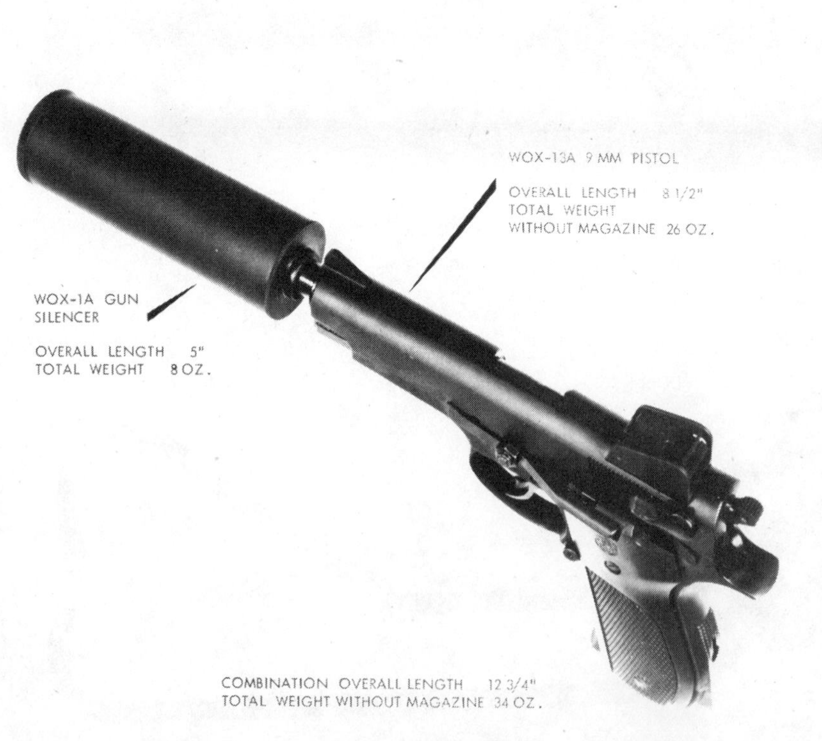

Top view of Mark 22, Mod. O Silenced Pistol.

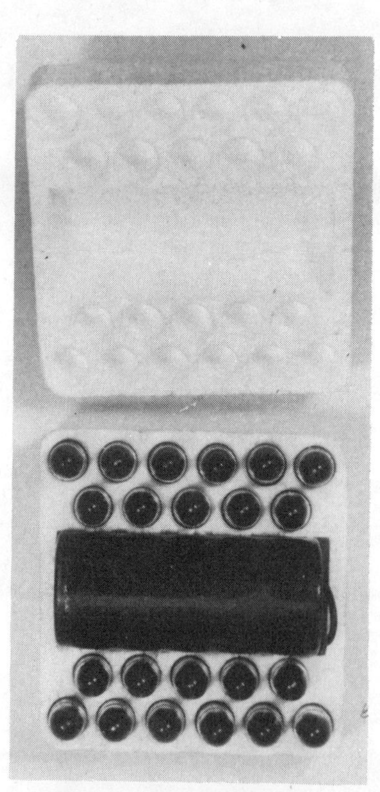

Mark 26, Mod. O Accessory Kit.

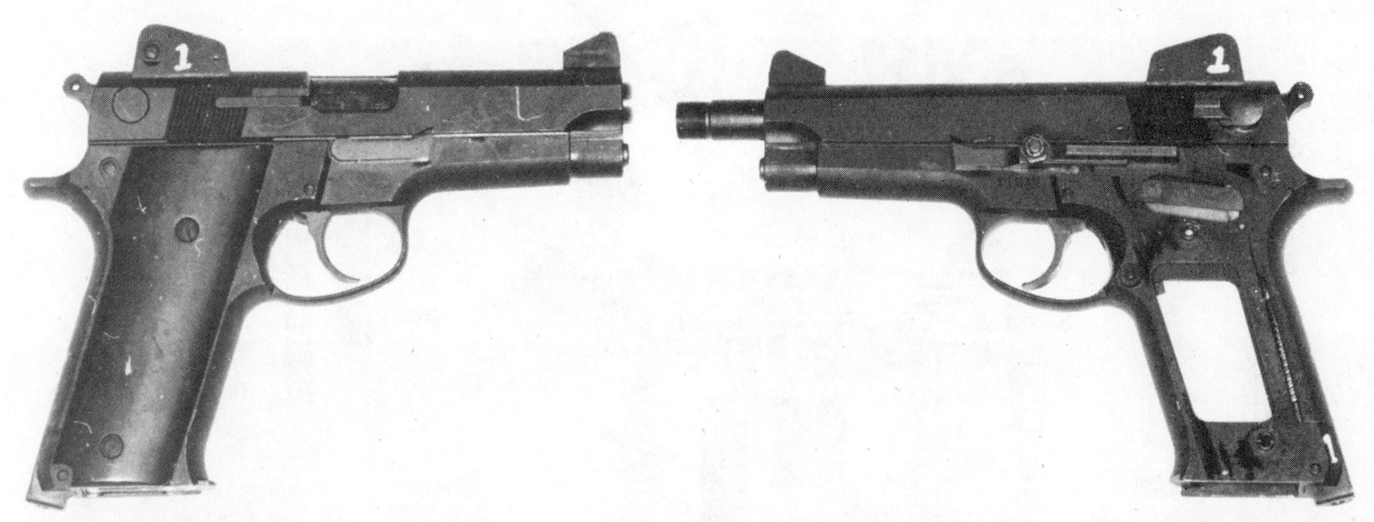

Experimental 15-Shot Smith & Wesson Pistol developed for the US Navy. A later version was marketed as the Model 59. In this prototype, note lock on both sides of the slide.

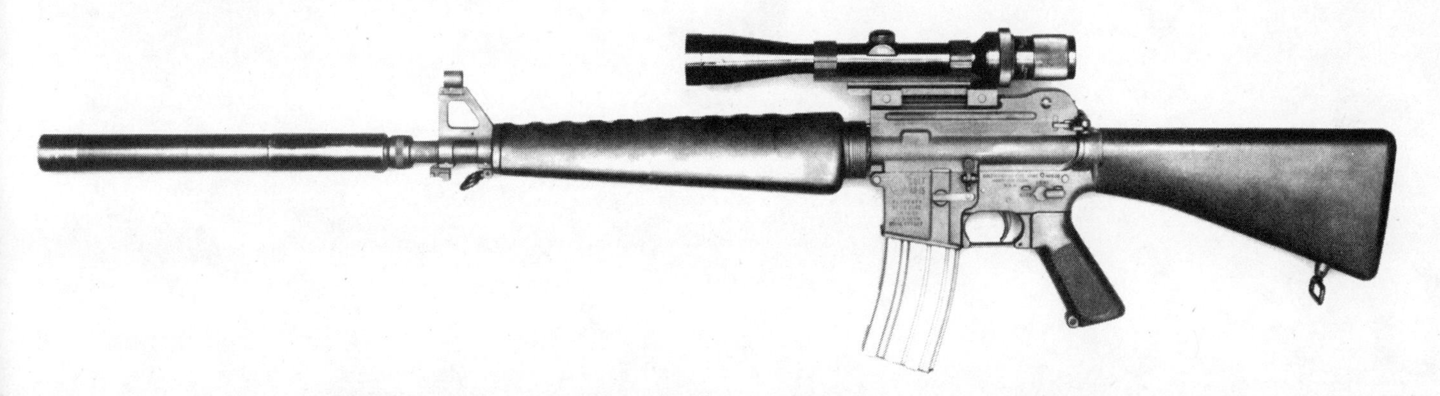

Experimental M16A1 (Note modified upper receiver.) fitted with Sionics Noise Suppressor.

An unofficial, but nevertheless interesting, development is the M10 and M11 Submachine Guns mentioned in Chapter 3. Military Armaments Corporation, while in business, offered both weapons with their own Sionics noise suppressors. The accompanying illustrations show the M11 compared with the Uzi and the interior construction of the Sionics suppressor.

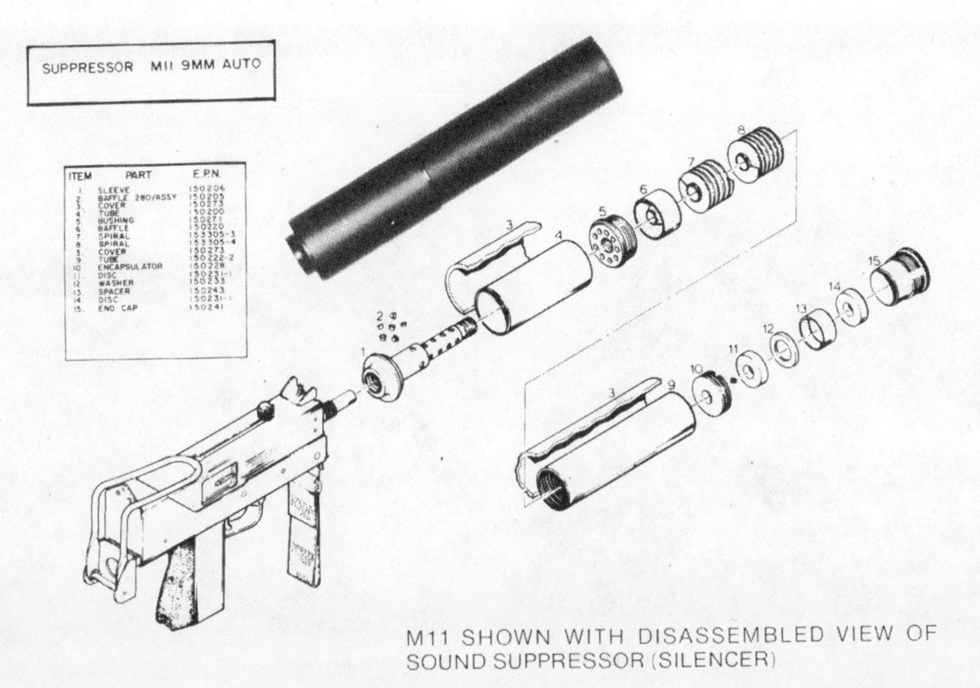

ITEM	PART	E.P.N.
1.	SLEEVE	150206
2.	BAFFLE 280/ASSY	150205
3.	COVER	150273
4.	TUBE	150200
5.	BUSHING	150271
6.	BAFFLE	150220
7.	SPIRAL	153305-3
8.	SPIRAL	153305-4
3.	COVER	150273
9.	TUBE	150222-2
10.	ENCAPSULATOR	150228
11.	DISC	150231-1
12.	WASHER	150233
13.	SPACER	150243
14.	DISC	150231-1
15.	END CAP	150241

M11 SHOWN WITH DISASSEMBLED VIEW OF SOUND SUPPRESSOR (SILENCER)

M11
WITH
Loaded 32 round magazine
COMPARISON SHOWS SIZE AND
POTENTIAL FIRE- POWER OF M11
WHEN COMPARED TO CONVEN-
TIONAL SUBMACHINE GUNS.

TECHNICAL SPECIFICATIONS FOR M11:

	INCHES
Length without stock	8.75 (222 MM)
Length stock extended	18.11 (460 MM)
Barrel length	5.06 (129 MM)
Weight gun with empty magazine	3.50 lbs (1.57 Kg)
Type of fire	Selected semi-automatic or full automatic.
Cyclic rate of fire	Approx. 1200 rounds per minute.
Front sight	Protected post.
Rear sight	Fixed aperture for 50 meters.
Safety arrangements	Two manually operated "safe" positions for locking bolt in open or closed position. The retracting knob will lock the bolt in closed position by rotating the knob 90 degrees. A second safety is located to the side and slightly forward of the trigger guard on the right hand side of the weapon. Pulling the safety lever to the rear locks the bolt in open or closed position.

The weapon employs blowback operation and fires the .380 ACP (9mm Kurz) cartridge.

M21 Sniper Rifle with Sionics Noise Suppressor.

COMBAT SHOTGUNS

The 12-gage (18.5mm) shotgun has been a controversial weapon in the twentieth century. German military authorities condemned its use by American troops in the trenches of the western front during the First World War. During the Second World War and the Korean conflict, the shotgun was carried mainly by troops guarding prisoners of war, but in the Vietnam War and in the Israeli battles in the Middle East this devastating weapon had been employed for close encounters of a hostile kind. Generally, these shotguns were modifications of models available commercially. Among the first of these altered shotguns was the Winchester Model 12 pump which had a barrel jacket added that also allowed the standard rifle bayonet to be used. More recently, the Marine Corps has used the Remington Model 870.

In the late 1970s the US Armed Forces decided that they needed a shotgun that was built not to commercial requirements, but to military specifications. In 1979, O. F. Mossberg & Sons, Incorporated of North Haven, Connecticut received a contract for a Milspec 12-gage shotgun. This weapon was evolved from the Mossberg Model 500 Persuader series of police shotguns. The "Milsgun," as the new weapon is referred to by Mossberg personnel, has a phosphate gray finish on its aluminum receiver and all of its steel parts. It has a tubular 5-shot magazine, 470mm (18.5 inch) barrel, 959mm (37.75 inch) overall length and weighs 2.95 kilograms (6.5 pounds). It is, like all other US military shotguns to date, a pump action weapon.

In an attempt to develop a more sophisticated military shotgun, the Joint Service Small Arms Program managers released a request for proposals for the development of a close assault weapon (CAW). This CAW would be used in military operations such as air base defense, nuclear weapons protection, urban warfare, and other small unit engagements in which a small group of individuals may have to defend a target of very high importance against a much larger assault force. Current shotguns, with their multi-pellet payload, provide a high hit probability at ranges of less than 75 meters. In addition to their short range, current shotguns have the disadvantage of heavy recoil. JSSAP personnel hope that the CAW development project will produce a weapon that will be effective up to 150 meters and which will not kick as badly as the current 12-gage shotguns. Between FY 1982 and FY 1986, JSSAP expects to spend about $8.4 million dollars on the development of a close assault weapon.

At least one commercial firm has begun serious work on such a close assault weapon. Heckler & Koch has completed initial design studies of a 10-shot selective fire 12-gage assault weapon. Weighing 3.96 kilograms (8.8 pounds), the HK weapon will have an effective rate of fire of about 200 shots per minute. HK has teamed up with the Winchester-Western Division of the Olin Corporation. The latter firm is developing an improved 12-gage family of cartridges. While it remains to be seen what the results of the CAW project will be, it is clear that such a weapon will significantly improve the killing power in the hands of the average infantryman.*

*For an in depth study of combat shotguns see: Thomas F. Swearengen, *The World's Fighting Shotguns* (Alexandria, VA: TBN Enterprises, 1979).

A USMC sergeant holds a Remington Model 870 12-gage shotgun developed for the Marines.

Mossberg's 6-shot pump 12-gage shotgun built to US military specifications (Milsgun).

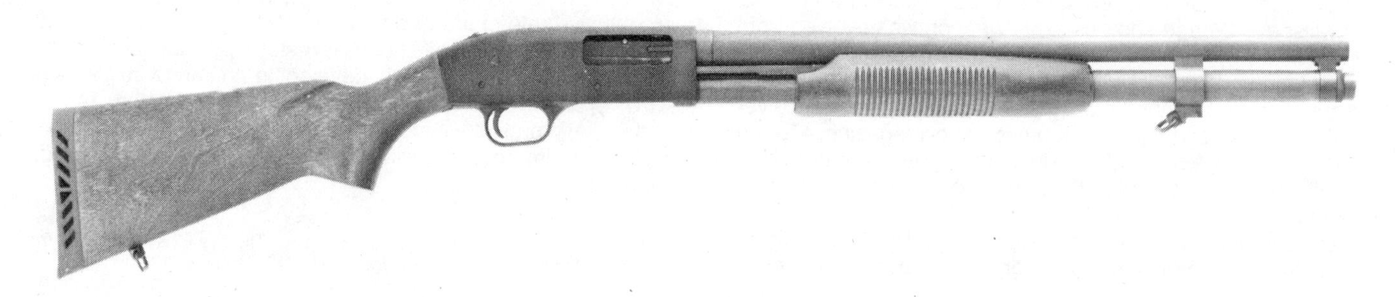

Mossberg's 8-shot Model 500-ATP-8SP pump shotgun with bayonet lug and military finish.

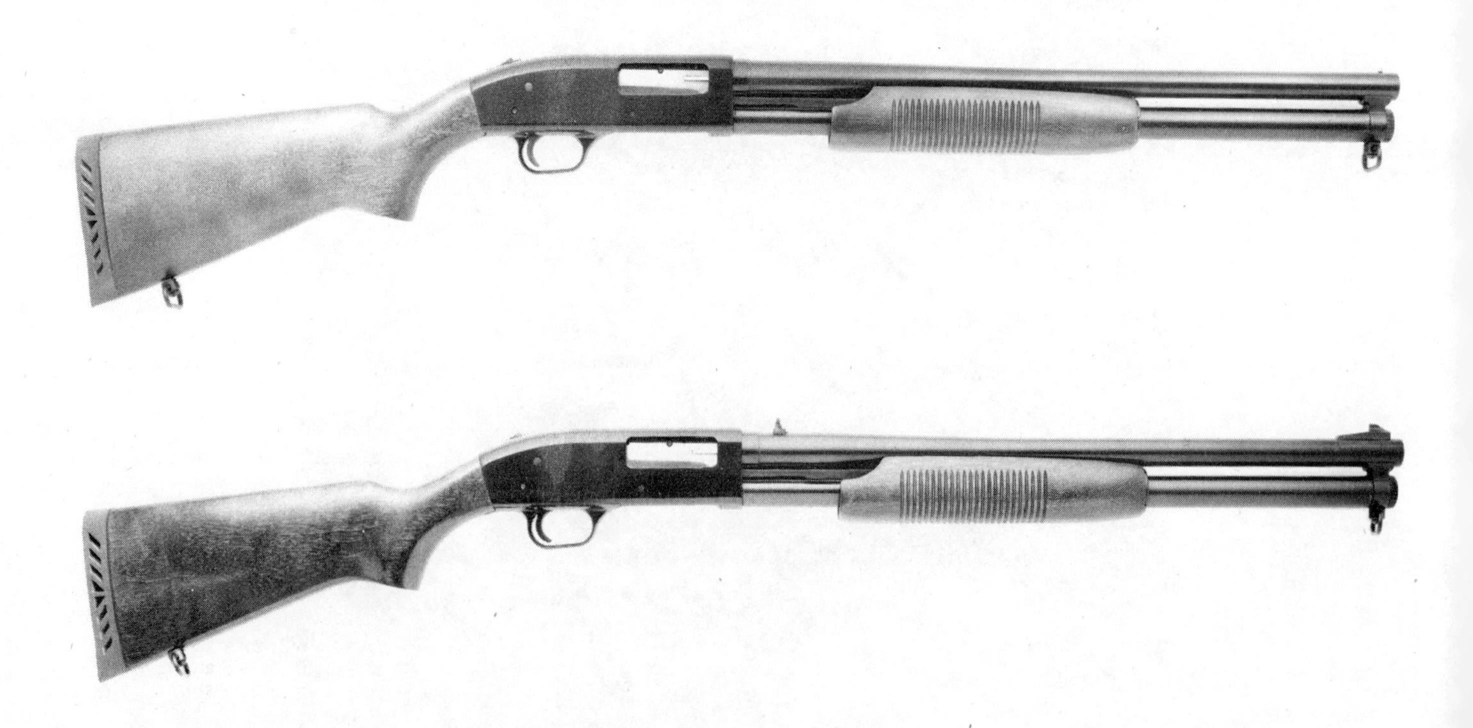

Two versions of the Mossberg Model 500-ATP-8 with blue finish. Both are 8-shot, and the only significant difference is the rifle sights on the bottom gun.

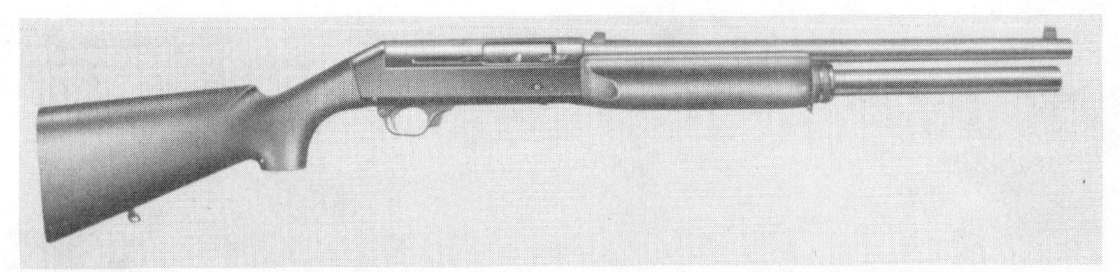

The military and police version of the Benelli 12-gage self-loading shotgun. This is a positive-lock recoil operated weapon. In the United States it is marketed by Heckler & Koch, Inc.

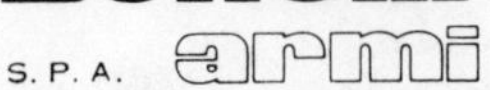

Benelli Part List - 1

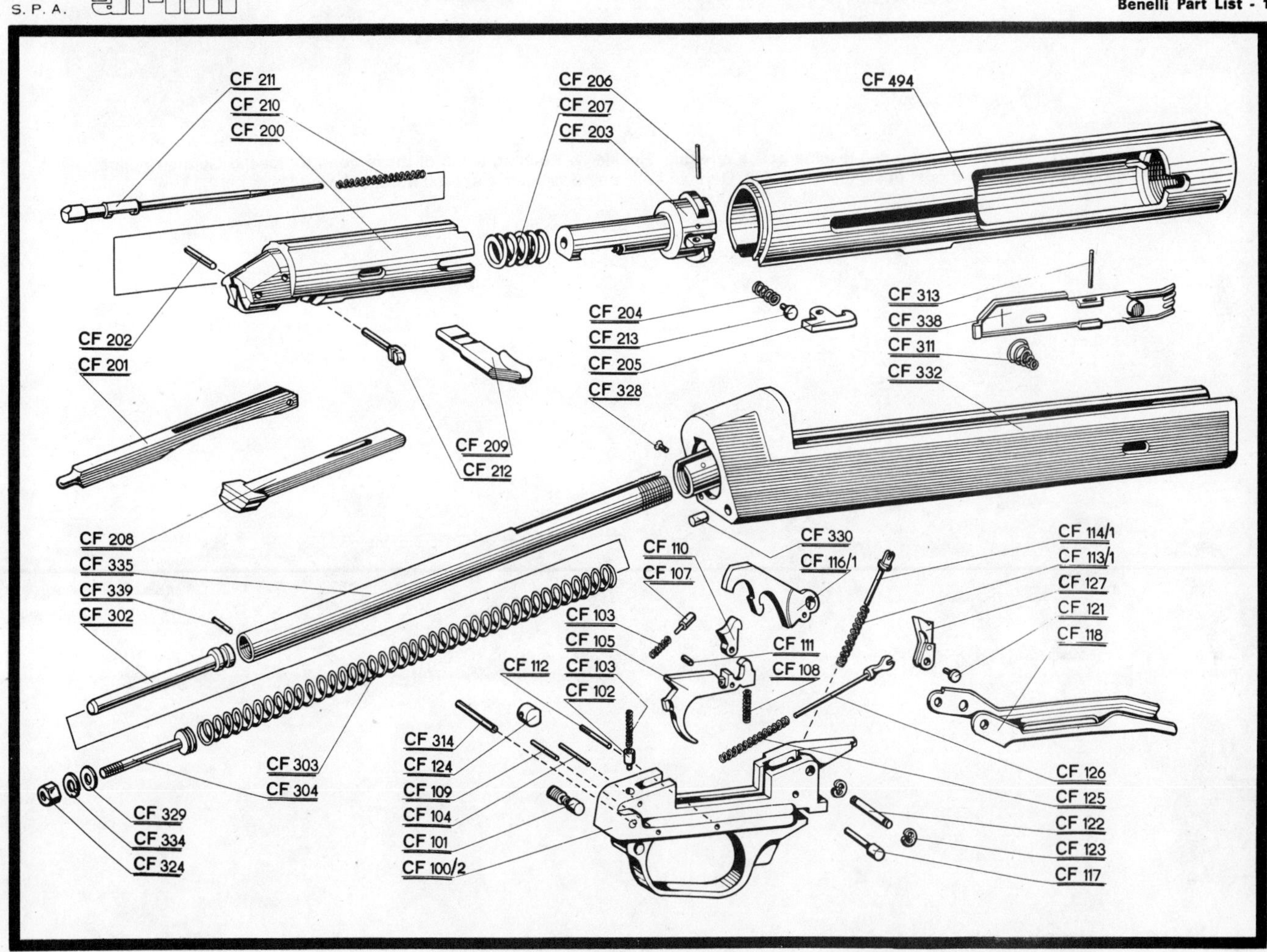

CF—100/1 Trigger housing, small
CF—100/2 Trigger housing
CF—101 Safety
CF—102 Safety plunger
CF—103 Safety plunger spring
CF—104 Safety plunger spring retaining pin
CF—105 Trigger
CF—105/1 Trigger (for use with small guard)
CF—105/2 Trigger, gold
CF—107 Disconnector plunger
CF—108 Trigger spring
CF—109 Trigger stop pin
CF—110 Disconnector
CF—111 Disconnector pin
CF—112 Trigger pin
CF—113/1 Mainspring
CF—114/1 Mainspring plunger
CF—116/1 Hammer
CF—117 Hammer pin
CF—118 Carrier
CF—121 Carrier dog pin
CF—122 Carrier pin
CF—123 Carrier pin washer
CF—124 Carrier spring pawl
CF—125 Carrier spring
CF—126 Carrier spring plunger
CF—127 Carrier dog
CF—200 Bolt body
CF—201 Link
CF—202 Link pin
CF—203/1 Bolt head
CF—204 Extractor spring
CF—205 Extractor
CF—206/1 Extractor pin (rolled)
CF—207 Inertia spring
CF—208 Locking bar
CF—209 Bolt handle
CF—210 Firing pin spring
CF—211/1 Firing pin
CF—212 Firing pin retaining pin
CF—213 Extractor spring plunger
CF—302 Recoil spring plunger
CF—303 Recoil spring
CF—304/1 Recoil spring retainer & tension screw
CF—311 Carrier latch spring
CF—313 Carrier latch pin
CF—314 Trigger guard pin
CF—324/1 Stock retaining bolt
CF—328 Recoil spring tube lock screw
CF—329 Stock screw tension washer
CF—330 Stock orientation pin
CF—332 Lower receiver model SL121
CF—332/1 Lower receiver model SL123
CF—332/2 Lower receiver model extraluxe
CF—334 Stock retaining bolt lock washer
CF—335 Recoil spring tube
CF—338 Carrier latch
CF—339 Recoil spring plunger retaining pin (rolled)
CF—494 Receiver

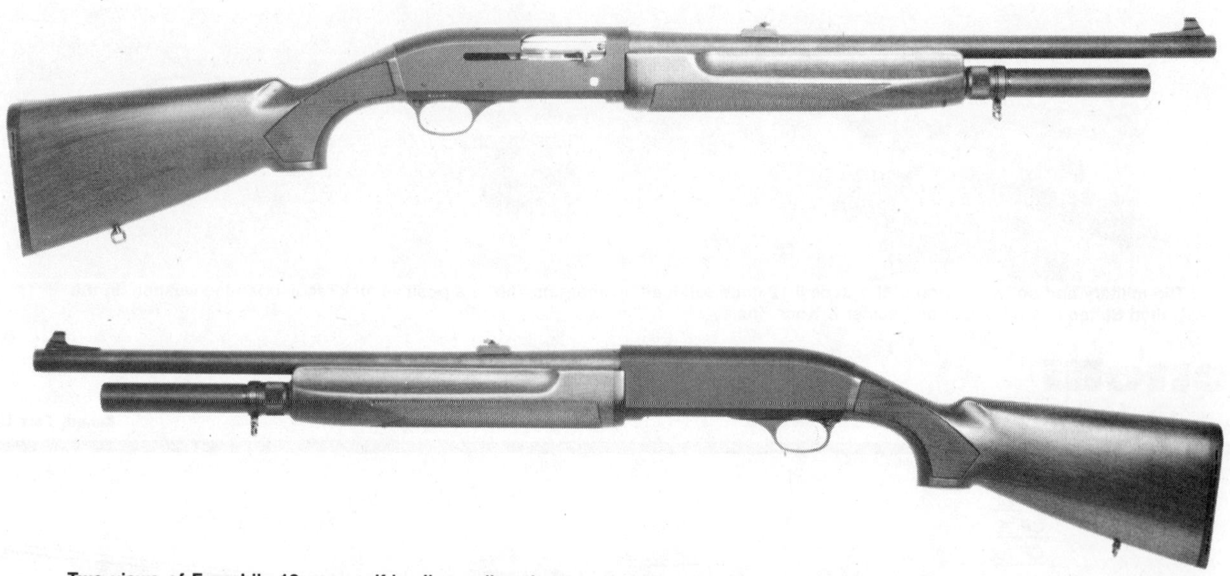

Two views of Franchi's 12-gage self-loading police shotgun. Heckler & Koch sold 500 of these guns to special German police organizations under the model designation HK 502. German police models have a special muzzle device.

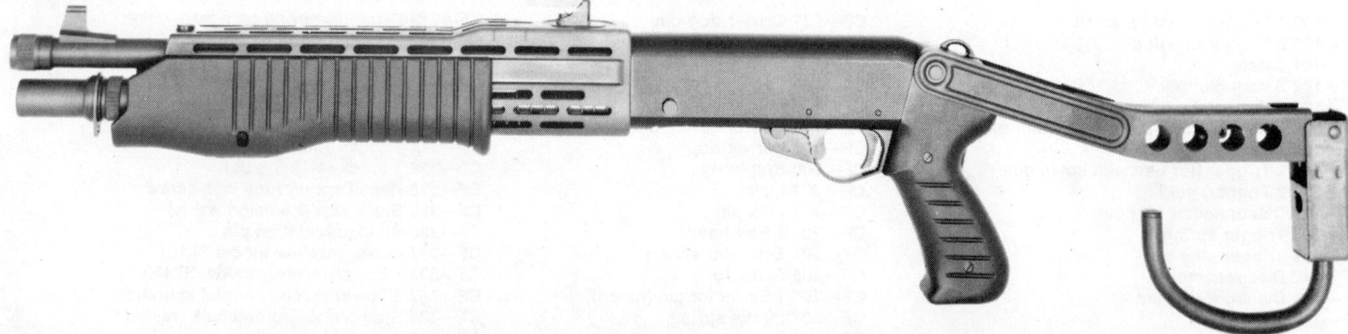

Two views of Franchi's 6-shot 12-gage Special Purpose Automatic Shotgun—SPAS 12. This gun can be fired either as a gas-operated self-loader or as a pump shotgun. The hook on the folding stock is designed to permit the SPAS 12 to be fired as a one-hand gun. The SPAS 12 has an aluminum receiver and weighs 3.65 kilograms (8.05 pounds).

Four variants of Smith & Wesson's 12-gage police pump shotgun. This Model 3000 series is available in a choice of blued or phosphate finishes.

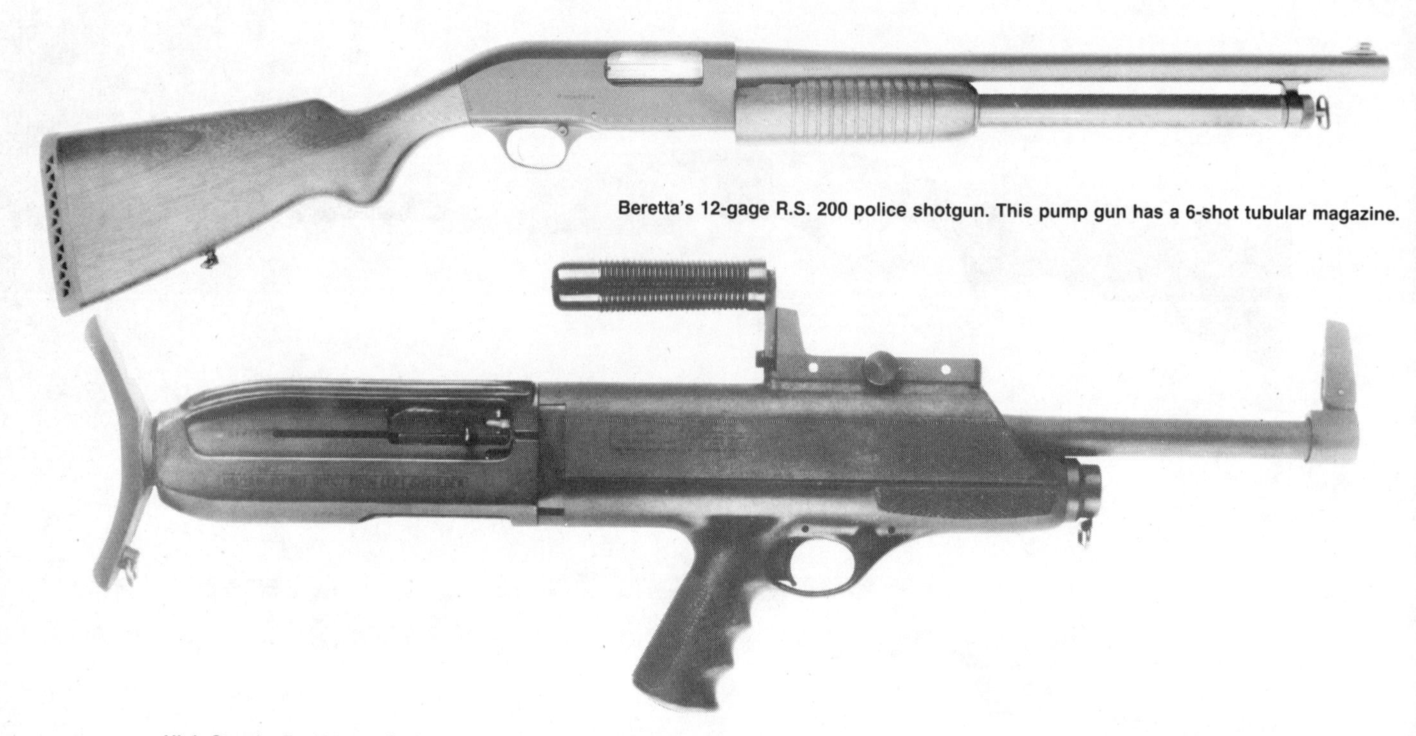

Beretta's 12-gage R.S. 200 police shotgun. This pump gun has a 6-shot tubular magazine.

High Standard's 12-gage Series B, Model Ten Police Shotgun. This gun is self-loading, and it was designed to be handled with one hand.

Test firing the Heckler & Koch close assault weapon. This 12-gage gun will have a 10-shot box magazine and will deliver either single shots or full automatic fire.

GRENADE LAUNCHERS

Rifle-propelled grenades have been common since the First World War. The launcher attachments have generally been of two types—tubes over which the tail tube of the grenade fitted or cup-type launchers into which the grenade body fitted. Rifle-launched grenades continue to be popular in many countries, but in recent years they have been supplanted by the cartridge-type grenade of the US 40 × 46mm SR and the Soviet 30 × 29mm. These grenades are propelled at higher velocities and can attain greater range with greater accuracy. The US M79 was the first of the post-World War II cartridge-type grenade launchers. The M79 itself is a relatively simple weapon, the secret to its success being the high-low pressure system of the cartridge. There is a chamber in the cartridge in which the propellant is ignited and allowed to develop to relatively high pressures. These gases then escape through holes in the high pressure chamber into the low pressure cavity. Initial velocity of the standard M406 high explosive fragmentation 40mm grenade is 75.3 m/s (247 f.p.s.). An experienced M79 grenadier can place a grenade in a house window at 150 meters with regularity.

Where the M79 was a separate shoulder-fired weapon, looking like an oversized single barrel break-open shotgun, it has been superseded in the US service by the M203 grenade launcher. The latter is an attachment for the M16A1, and, while it is a complete weapon, it must be attached to the rifle to be used. Since the late 1960s, several firms—notably Fabrique Nationale, Heckler & Koch and the Argentine Fabrica Militar at Rosario—have developed launchers to use the US family of 40mm grenades. During the Vietnam War, the US military developed a higher velocity for aircraft and vehicular-mounted repeating grenade launchers. This second family of 40mm cartridge-type grenades has a longer cartridge case (40 × 53mm SR) which will not chamber in the shoulder-fired launchers. The effective range of the higher velocity rounds is 2,200 meters; that of the shoulder-fired grenades is 400 meters.

In Vietnam, the US Army experimented with a series of prototype 40mm automatic grenade launchers on vehicles, helicopters, and ground mounts. All of those projects were terminated with the winding down of the war in the early 1970s. During the same era, the US Navy developed a 40mm launcher called the MK 19, which was adopted and further developed by the Israelis. The current version, the MK 19 MOD 3, is being purchased for the US Marine Corps as an area weapon that can be used by the infantryman to counter the threat posed by the Soviet BMP infantry fighting vehicle and its 76mm cannon. The M430 High Explosive Dual Purpose (HEDP) round will inflict significant damage to the BMP at ranges up to two kilometers (1.2 miles). As noted in Chapter 1, each Marine infantry battalion will have 12 MK 19 grenade launchers. The MK 19 weighs 20 kilograms (44 pounds), can fire both single shots and multi-shot bursts, and has a maximum rate of fire of 400 shots per minute.

In the last six or seven years, the Soviets have been fielding an automatic 30 × 29mm grenade launcher, which they call the AGS17 (*Avtomatischeskiy Granatmyot Stankoviy* or tripod-mounted automatic grenade launcher). This weapon has a 30-round belt that fits into a drum attached to the right side of the weapon. The maximum range of the launcher is estimated to be 1,500 meters, with an effective range of about 800 meters. It is believed that the AGS17 has a 300 rounds per minute rate of fire; its effective rate of fire would be about 40 to 60 rounds per minute. Mechanically, the AGS17 is very similar to the US XM174 grenade launcher used in Vietnam. Appearance of the MK19 and the AGS17 indicates that this new class of weapon will be a potent force in any future conflicts in Europe.

SP4 Goes is shown loading an M433 dual-purpose high explosive round into his M79 grenade launcher. This round will penetrate 50mm (2.0 inches) of armor plate.

The 40mm M203 grenade launcher prototype made by AAI. Production units have been manufactured by Colt Firearms.

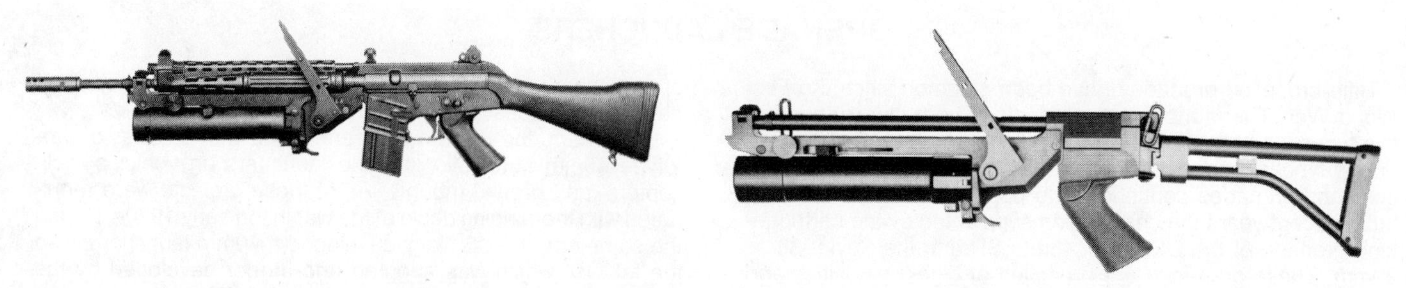

Two views of the FN prototype 40mm grenade launcher: *left,* mounted on the FN CAL; *right,* mounted on a special attachment that allows it to be used as a separate shoulder-fired weapon.

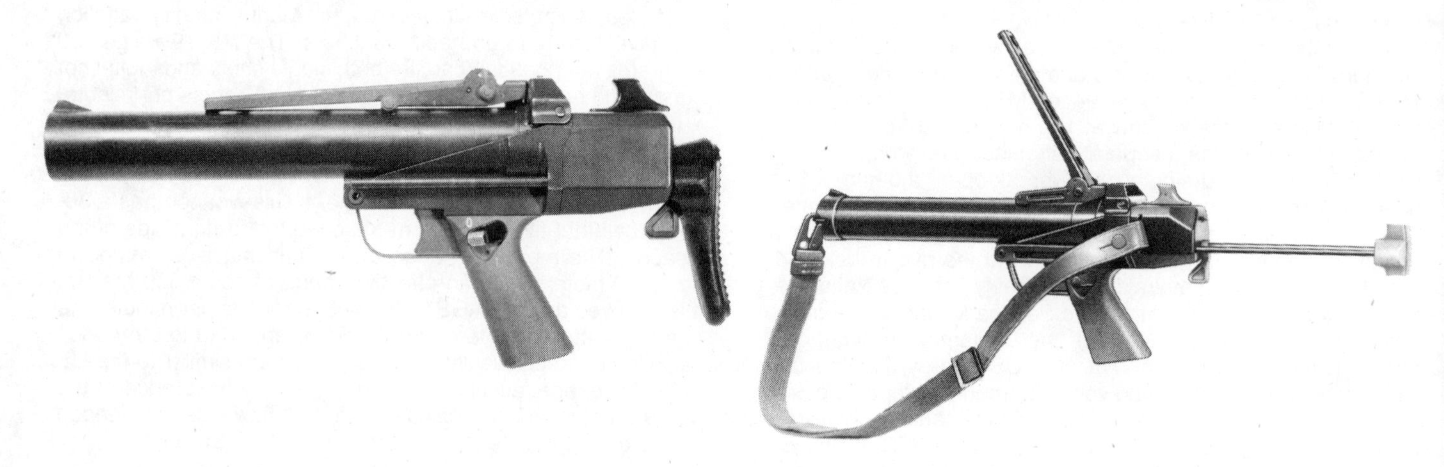

Two views of the Heckler & Koch 40mm HK69 grenade launcher: *left,* the HK69; *right,* the HK69A1.

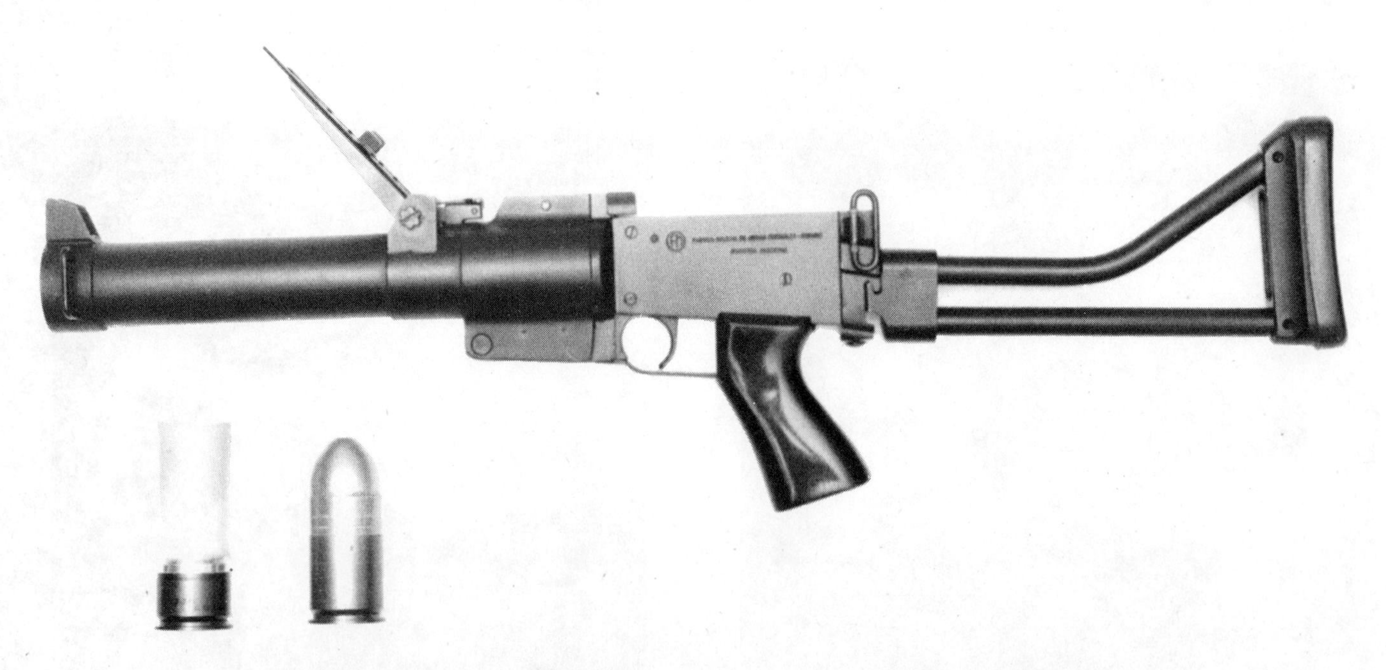

The new Argentine 40mm grenade launcher evolved from their 38.1mm tear gas launcher. (See Chapter 6 for details.)

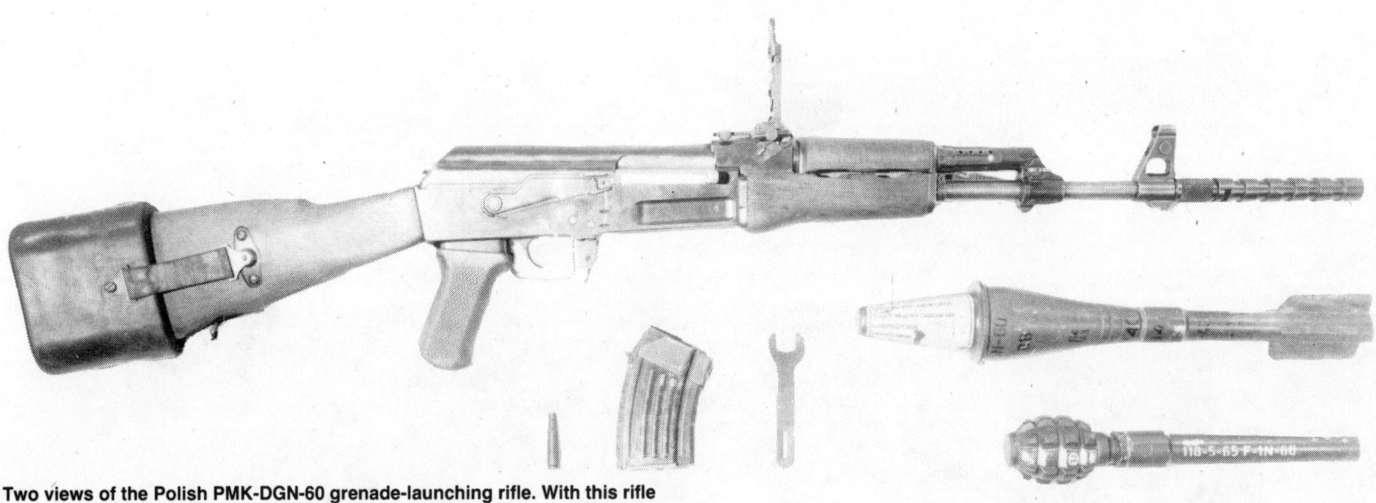

Two views of the Polish PMK-DGN-60 grenade-launching rifle. With this rifle are the F1/N60 fragmentation grenade and the DGN/60 shaped charge antiarmor grenade. The recoil of this weapon is very severe and it should not be fired from the shoulder.

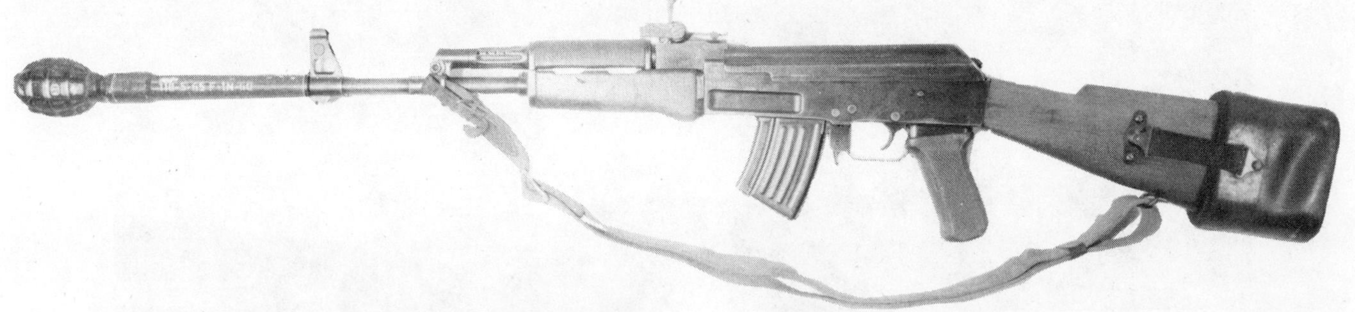

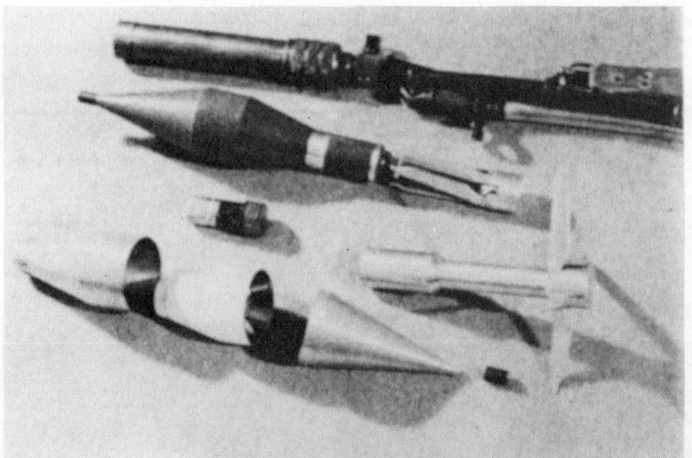

The new rifle grenade of the Chinese People's Liberation Army has folding fins that fit into a cup-type grenade launcher that attaches to the muzzle of the Type 68 and Type 73 rifles. This grenade, standardized in 1979, is reported to be capable of penetrating 150mm of armor.

The 5.56mm French FAMAS with an FN 40mm rifle grenade.

A MK19 MOD O, shown mounted here on an M113 Armored Personnel Carrier of the 4th Mechanized Battalion, 23rd Regiment, 25th Division, returning fire during an ambush in July 1969, in Tay Ninh Province, South Vietnam. US sailors of the River Division 152 are manning the MK19.

The 40mm MK19 MOD3 grenade launcher as it will be employed by the US Marine Corps as a ground support weapon.

The 30mm Soviet AGS17 automatic grenade launcher in the field. This weapon has a three-man crew.

Two views of the Soviet AGS17 grenade cartridge: *left,* external view; *right,* x-ray view showing the fuze, coil of fragmentation wire, and the powder cavity of the cartridge.

BLANK FIRING ATTACHMENTS

In recent years, the nations of NATO and the Warsaw Pact have adopted blank firing attachments for their infantry weapons to improve the realism of their training. All of these attachments work on the principle of restricting the exit of the gas from the barrel thus producing either extra gas or increased recoil impulse to operate the rifle or machine gun. The United States has developed a special interest in blank firing attachments (BFAs), because the new MILES training equipment requires the sound of the blank to discharge the laser transmitter that sends simulated bullets to kill "opponent" forces in training exercises.

In January 1980, the United States Army standardized the M19 and M20 Blank Firing Attachments for the M2 and M85 .50 caliber machine guns. The creation of these accessories was made necessary by the introduction of the MILES (Multiple Integrated Laser Engagement System) training program. MILES provides realistic simulation of combat situations by firing coded laser "bullets" at detectors. When a detector mounted on a soldier or a vehicle picks up the laser bullet it records a kill or a near miss. Hits are coded and graded according to the power of the weapon. All weapons can kill an infantryman, but a rifleman or machine gunner cannot knock out an armored personnel carrier or a tank. Laser systems have been developed for the M16 rifle, M60, M2, and M85 machine guns, the Stinger and TOW antitank weapons, and the 105mm and 152mm tank guns. With each of these weapons, the noise of a blank triggers the firing of the laser. Blank firing attachments were necessary for the small caliber weapons to permit realistic functioning of the weapons. BFAs existed for the M16 and M60, but none had been developed for the .50 caliber machine guns. In addition, the Army had to develop a blank cartridge that would feed reliably into the weapons.

Being recoil operated, the M2 and M85 machine guns required a muzzle device to capture a portion of the muzzle blast produced by the discharge of the blank and direct it against the muzzle of the weapon to enhance its recoil. Although the principle was a relatively simple one, the design of the M2 and M85 machine guns caused some problems. Just before the Second World War, a BFA had been developed for the M3 .50 caliber machine gun. That weapon had a full-length barrel jacket to which a BFA could be easily attached. With both the M2 heavy barrel and M85 coaxial machine guns, the barrel support jacket was relatively short. The length of the barrel jacket on these weapons posed a mounting problem for a blank firing attachment.

In the early 1970s, Robert Schutz at Fort Lewis, Washington devised a BFA for use with tripod-mounted M2 machine guns. In July 1977, Joe Zafian and his colleagues at the Fire Control and Small Caliber Weapons Systems Laboratory, US Army Armament Research and Development Command (FC&SCWSL, ARRADCOM) began work in-house at Picatinny Arsenal on a variant of the Schutz BFA that could be used with flexibly mounted M2 machine guns on the M113 APC and the M60 tank and in the coaxially mounted M85 machine guns on the M60 tank. Their work was coordinated with the Training and Doctrine Command (TRADOC) at Fort Monroe, Virginia, the Project Manager for Training Devices (PM-TRADE) located at the Naval Training Center, Orlando, Florida and Army ammunition specialists.

After a second round of development testing in the final months of 1979, the M19 and M20 blank firing attachments were standardized in January 1980 along with the new blank cartridge, the caliber .50 M1A1. During the test program more than 40,000 rounds of blank ammunition were fired through the BFA without a single failure or stoppage that could be attributed to the attachments themselves. A TRADOC noncommissioned officer, who participated in field trials of the MILES equipment in Germany, commented that the M2 and M85 machine guns seemed to shoot more reliably with the BFAs and blank ammunition than they did with ball ammunition. He also noted that the machine gunners participated in the training exercises with more enthusiasm once they could fire their weapons in a manner that simulated their use in combat realistically.

While identical in concept, the two BFAs are slightly different, both externally and internally. The M19 BFA for the M2 machine gun has a muzzle cap (chamber, in Army nomenclature) that fits over the front end of the barrel. This part is attached to the support assembly by three steel rods. The support assembly is fastened to the barrel support jacket by three clamps that lock into the holes of the barrel jacket. To prevent the inadvertent feeding of ball cartridges, the BFA is issued with a cartridge guide assembly that blocks the projectile of a ball round and which guides the movement of the blank cartridge. The cartridge guide assembly and its quick release storage pin have retaining wires, which are also fastened to the support assembly, to keep those two parts from being lost. During storage, the quick release pin is used to hold the cartridge guide inside the support assembly.

This trooper's rifle is equipped with the MILES laser transmitter and the M17 blank firing attachment. Note the laser detectors on his harness and helmet.

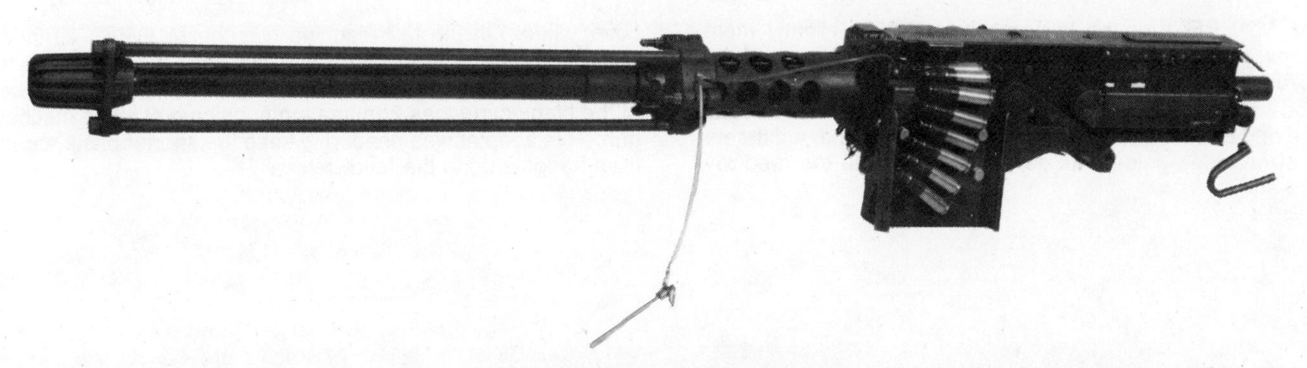

The M2 HB machine gun with the M19 BFA.

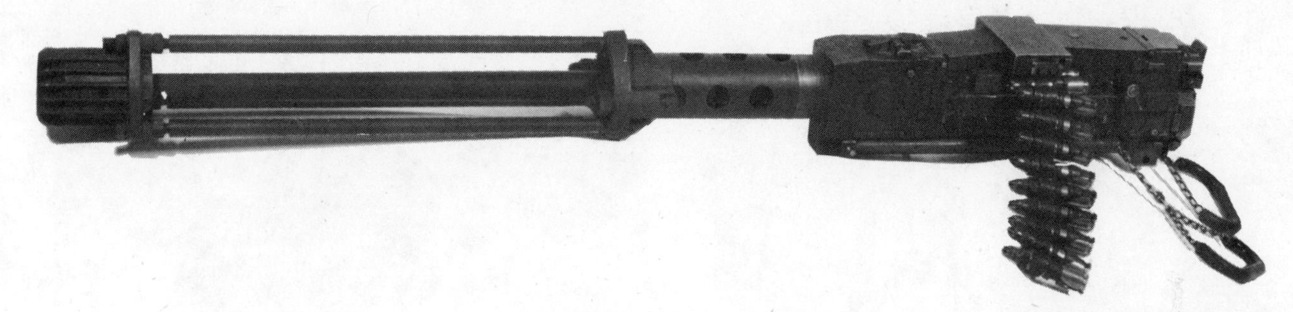

The M85 machine gun with the M20 BFA.

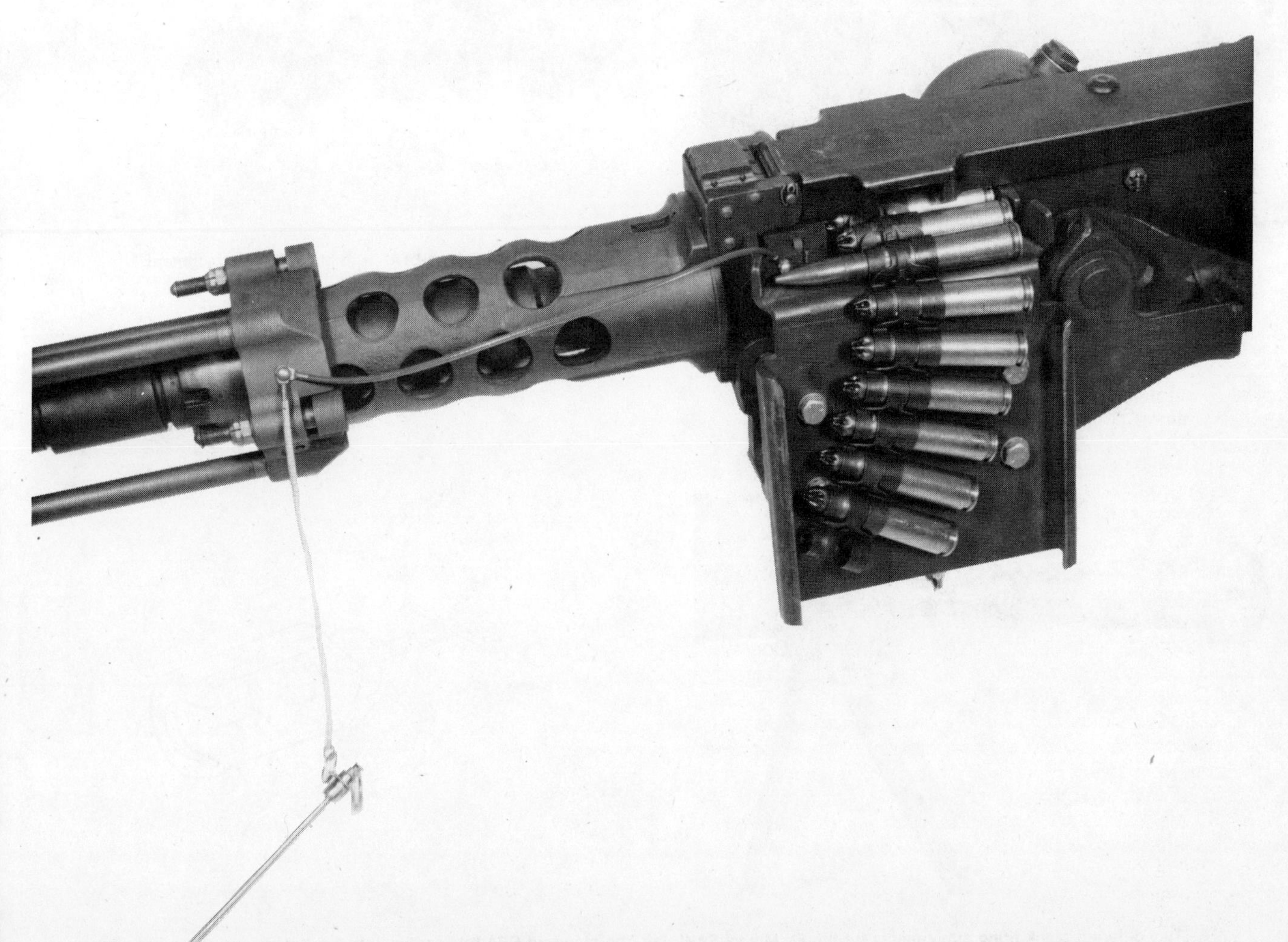

The M2 HB machine gun showing the manner in which the filler block prevents the feeding of ball rounds.

The M20 BFA, for the M85 machine gun, is slightly more complex. First, an adapter was created to fit the muzzle of the M85, because the prong-type flash suppressor had to be removed when the BFA was used. Without the adapter, the threads on the end of the barrel would be damaged. Second, a different type of cartridge guide was developed to fit into the feedway cover plate. On the M2 machine gun the M19 BFA cartridge guide fits into the feed tray. Third, the tank-mounted M85 required the creation of a filler piece to assure the proper feeding of the blank cartridges from the ammunition box to the machine gun. This adapter was needed to keep the shorter blank rounds from hanging up in the feedway.

A West German MG3 with blank firing attachment.

A British L1A1 with blank firing attachment.

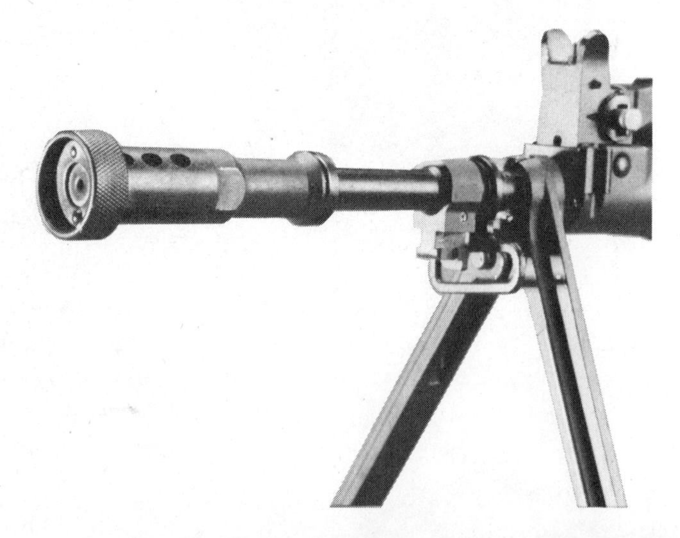

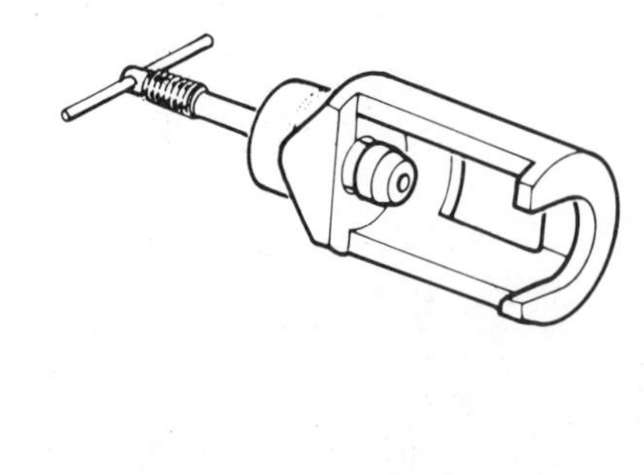

Two different blank firing attachments for the FN Minimi SAW: *left,* the standard BFA that screws into the flash suppressor (FN has used this type BFA on the FAL, FNC, and Minimi); *right,* the FN-developed BFA for use on the US M249 version of the Minimi. (Note that this is very similar to the one used on the M16A1 rifle.)

PART 2

Description of Small Arms by Nation and a Basic Manual of Current Weapons

6 Argentina

Argentina manufactures nearly all of its small caliber weapons. The government small arms factory is the *Fábrica Militar de Armas Portátiles "Domingo Matheu"** (*FMAP "DM"*) located at Rosario in Sante Fe. It is one of eleven military industries operated by the *Dirección General de Fabricaciones Militares* (*DGFM*).

SMALL ARMS IN SERVICE

Handguns: *Pistola Browning*, 9 × 19mm NATO (FN Mle. 1935)
Pistola Sistema Colt, 11.25 × 23mm (Colt M1911A1)

Submachine guns: *Pistola Ametralladora FMK3* (*PA3-DM*), 9 × 19mm NATO and earlier variants
(Some older Halcon models still in use. See *Small Arms of the World,* 11th ed.)

Rifles: *Fusil Automatico Livano* (FAL), 7.62 × 51mm NATO (FN FAL) (Models manufactured include the standard FAL, FAL PARA, PARA III, and FAL IV.)
Steyr SSG Sniper Rifle, 7.62 × 51mm NATO
Beretta BM59, 7.62 × 51mm NATO
Carabina Cabelleria M1909 Mauser, 7.65 × 53.5mm (Obsolescent, limited service)

Machine guns: *Ametralladora MAG,* 7.62 × 51mm NATO (FN MAG 58)
French AAT 52 (ground and tank models) 7.62 × 51mm NATO
Ametralladora, Browning, 12.7 × 99mm (M2 HB purchased from the United States)

A limited number of Star selective fire pistols, IMI UZI submachine guns, and US M1928A1 and M1 submachine guns may still be in service with reserve forces.

*Domingo Matheu (1765–1833) was an Argentine patriot who was a member of the *Primera Junta* established on 25 May 1810. He is best remembered for his work as Director General of *la fabrica de fusiles* in Buenos Aires.

ARGENTINE PISTOLS

***Pistola Browning PD,* 9 × 19mm NATO, serial number 186636 manufactured in Argentina. The left side of the slide is marked Fabrica Militar de Armas Portatiles "D.M." Rosario, D.G.F.M., Licencia F N Browning, Industria Argentina.**

PISTOLA BROWNING PD

The current handgun of the Argentine armed forces is the Pistola Browning PD (FN Browning Modèle 1935 GP). Since 1969, the small arms factory at Rosario has fabricated nearly 183,000 of these handguns. Some of these pistols have been sold commercially in Argentina and abroad. In 1977–78, 5,500 Pistolas Browning were purchased from Fabrique Nationale with FM (Fábrica Militar) markings. (See Chapter 9 for technical data.)

PISTOLA SISTEMA COLT MODELO 1927

Argentina has manufactured copies of the .45 caliber (11.25mm in Argentina) Colt M1911 and M1911A1 pistols, called the M1916 and M1927 respectively. Between 1947 and 1966, the Fábrica Militar at Rosario fabricated 74,866 M1927 pistols. These handguns were still in service with portions of the armed forces as late as the 1982 Falkland/Malvinas conflict. (See Chapter 48 for technical data on the US Colt M1911A1 pistol.)

The Argentine Modelo 1927 Sistema Colt 11.25 × 23mm pistol. This pistol carries two serial numbers, 57504 and 2154–949. The left side of the slide is marked D.G.F.M.—(F.M.A.P.), indicating that it was manufactured at the government small arms factory.

ARGENTINE BALLESTER MOLINA PISTOL

The firm "HAFDASA" of Buenos Aires manufactured the Ballester Molina and Ballester Rigand pistols. This firm is no longer in business, and the pistols are no longer being manufactured. The Ballester Molina caliber .45 pistol is still in wide use as it was manufactured in quantity during World War II.

The Ballester Molina is a slightly modified copy of the Colt .45 M1911A1 pistol. Loading, firing and functioning of the pistol are the same as for the US pistol.

CHARACTERISTICS OF THE BALLESTER MOLINA

Caliber: 45 M1911 automatic pistol cartridge.
System of operation: Recoil, semiautomatic fire only.
Weight, empty: 2.25 lbs.
Length, overall: 8.5 in.
Barrel length: 5 in.
Feed device: 7-round, single column, detachable box magazine.
Sights: Front, blade; rear, notched bar.
Muzzle velocity: 830 f.p.s.

MAJOR DIFFERENCES BETWEEN THE BALLESTER MOLINA AND THE COLT M1911A1

Hammer strut: The hammer strut is much smaller than that of the US M1911A1. It is 0.75 inches in length and 0.158 inches in diameter.

Firing pin stop: The firing pin stop is not recessed on the sides, as it is on the US models.

Safety lock: The safety lock is redesigned and the pin is larger in diameter than the safety lock pin on the US model.

Mainspring housing: The mainspring housing, although arched as in the US Model M1911A1, is an integral part of the receiver.

Trigger: The Argentine pistol has a pivoting trigger. An extension from the trigger, along the right side, cams the disconnector and engages the sear.

Magazine and magazine catch: The magazines are interchangeable. The magazine catch is located in the same place as the catch on the US M1911A1. The assembly of the catch is somewhat different, but it operates in the same manner as the US model.

Slide: There is no slide stop disassembly notch as on the US models.

Field Stripping the Ballester Molina

Disassembly is the same as for the US M1911A1, except that the pin for the hammer and the sear must be driven out. The trigger is held by a trigger pin. Upon removing the trigger and the trigger extension, the disconnector can be removed downward. After the sear has been removed, the main spring can be removed.

7.65mm M1905 MANNLICHER PISTOL

This was formerly the Argentine service pistol, and quantities of them were sold to surplus arms dealers in the US. The pistol was developed at Steyr and introduced commercially in 1901. The weapon is essentially a blowback-operated type and is loaded with a stripper type clip (charger) from the top. The weapon can be unloaded from the top by pulling the slide to the rear and pulling down the catch on the right side of the pistol.

CHARACTERISTICS OF THE 7.65mm M1905 MANNLICHER PISTOL

Caliber: 7.65mm Mannlicher (called 7.63mm Mannlicher in Austria).

Argentine 7.65mm M1905 pistol.

Length overall: 9.62 in.
Barrel length: 6.31 in.
Feed device: 8 round non-detachable box magazine.
Sights: Front, blade; rear, notch.
Muzzle velocity: approx. 1025 f.p.s.
System of operation: Blowback with slight retardation.
Weight: 2 lbs.

ARGENTINE RIFLES

FUSIL AUTOMATICO LIVANO (FAL)

The Fabrique Nationale FAL is the standard rifle of the Argentine armed forces. The accompanying photographs illustrate the currently manufactured Argentine FALs. There are more than 150,000 FALs of all types in the Argentine service; nearly 122,000 have been manufactured at the FMAP "DM" Rosario.

Close-up view of the *Fusil Automatico Livano* (FAL) *Modelo II*, 7.62 × 51mm NATO. This rifle is equivalent to the FN standard Model 50.00. The right side of the receiver is marked with the FM logo and *Fabrica Militar de Armas Portatiles—Rosario, Industria Argentina*. At the rear of the receiver there is the following marking: FSL—Cal. 7.62—002.

Fusil Automatico Livano (FAL) *Modelo II.*

Fusil Automatico Livano (FAL) PARA with 533mm (21-inch) barrel. FN Model 50.61.

Fusil Automatico Livano (FAL) PARA III with 458mm (18-inch) barrel. FN Model 50.63.

Fusil Automatico Pesado (FAP) *Modelo II.*

FN Model FALO 50.41.

OBSOLETE ARGENTINE RIFLES

The M1891 Argentine Mauser is quite similar to the 7.65mm M1890 Turkish Mauser. The M1909 Mauser is a slight modification of the German Gewehr 98 (rifle 98).

The M 1891 Carbine is still used as a police weapon, but the rifles are obsolete. The FN 7.62mm NATO FAL rifle and the 7.62mm NATO heavy-barrel rifles are now standard.

CHARACTERISTICS OF THE M1891 AND M1909 ARGENTINE MAUSERS

	M1891	M1909
Caliber:	**7.65mm rimless.**	**7.65mm rimless.**
System of operation:	Manually operated bolt action.	Manually operated bolt action.
Length, overall:	48.6 in.	49 in.
Barrel length:	29.1 in.	29.1 in.
Feed device:	5-round, single column, box magazine.	5-round, staggered-row box magazine.
Sights: Front	Barleycorn.	Barleycorn.
Rear	Leaf.	Tangent leaf.
Muzzle velocity: (w/spitzer-pointed ball)	2755 f.p.s.	2755 f.p.s.
Weight (empty):	8.58 lbs.	9.2 lbs.

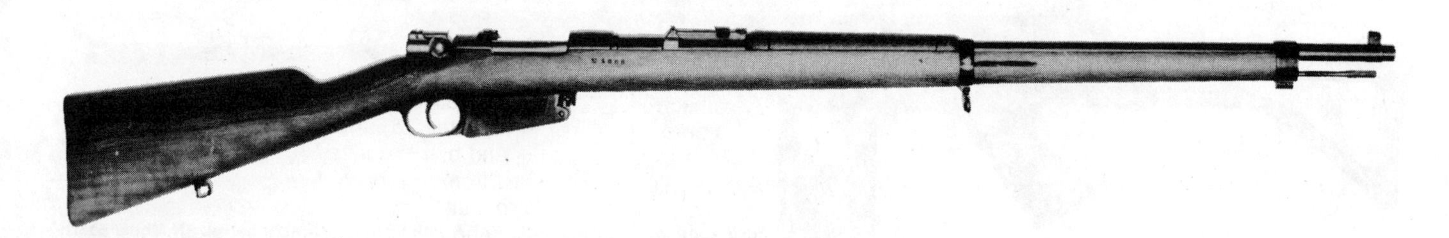

7.65mm Model 1891 Argentine Mauser.

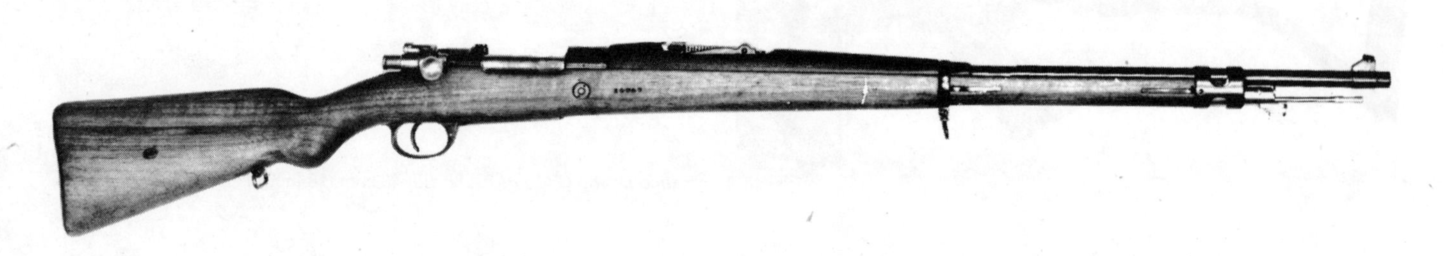

7.65mm Model 1909 Argentine Mauser.

OBSOLETE ARGENTINE CARBINES

Argentina has had a number of 7.65mm Mauser carbines in service. The Model 1891 carbine is stocked to the muzzle, has a 17.6 inch barrel and is 37 inches overall. There are two versions, one with and one without bayonet lug. There are four common versions of the Model 1909 carbine. The cavalry carbine is stocked to the muzzle. Some cavalry carbines have bayonet lugs and others do not. The Engineer carbine has a typical military stock and a bayonet stud, which protrudes horizontally from the upper band like that of the German Mauser 98. A carbine similar to the Engineer carbine, sometimes called the Mountain carbine, has a bayonet stud, which protrudes vertically in front of and below the upper band.

A large number of these weapons have been converted into sporters in Argentina.

ARGENTINE SUBMACHINE GUNS

PISTOLA AMETRALLADORA PAM 1 AND 2

The small arms factory at Rosario began production of the PAM 1 submachine gun in 1955 and produced nearly 30,000 by 1962. This weapon was essentially a copy of the US M3 submachine gun chambered to fire the 9 × 19mm cartridge. In 1961 the small arms designers at the Fábrica Militar modified the PAM 1 to incorporate a safety device that required the shooter to squeeze it with the hand gripping the magazine housing. If the device was not squeezed the gun would not fire. Between 1963 and 1972 nearly 17,000 guns were modified; they were called the "PAM 1 a 2." In 1968, 1,100 PAM 2 submachine guns were fabricated at Rosario. Production of this series was terminated in favor of the FMK 3 and 4 series.

PISTOLA AMETRALLADORA FMK 3 AND 4

Production of the current standard submachine gun, the Pistola Ametralladora FMK 3 (formerly PA3-DM) began at the Fábrica Militar, Rosario in 1971. More than 30,000 had been produced through the end of 1981. The FMK 3 has a retractable stock; the FMK 4 has a fixed plastic buttstock. Similar in concept to the Israeli Military Industries UZI, the FMK submachine guns have a bolt that telescopes over the breech end of the barrel, and the magazine fits into the pistol grip. The round cross-section of the receiver is reminiscent of the US M3 series of submachine guns.

CHARACTERISTICS OF FMK 3

Caliber: 9 × 19 NATO
System of operation: Blowback, selective fire.
Length overall: Stock extended: 27.17 in.
Stock retracted: 20.47 in.
Fixed stock: 27.56 in.
Barrel length: 11.42 in.
Weight w/o magazine: 7.94 lbs.
Feed device: 25- and 40-round, detachable, staggered row box magazines.
Sights: Front, protected post; rear, adjustable for 50 and 100 meters, and for windage.
Muzzle velocity: Approx. 1200 f.p.s.
Cyclic rate: 650 r.p.m.

How the FMK Works

This is a standard blowback submachine gun operating much like the UZI or the US M3. Like the UZI, the FMK has a grip safety that makes it safer to carry when the bolt bas been retracted and the weapon is ready to fire.

Disassembly of the FMK

After removing the magazine and establishing that the submachine gun is not loaded, two large pins are pushed from right to left through the body of the FMK. When these pins have been removed, the top portion of the receiver, containing the barrel and bolt assemblies, can be lifted off the lower receiver that houses the trigger mechanism.

The barrel and bolt are removed by depressing a catch and unscrewing the barrel retaining collar nut. The barrel and bolt come out of the FMK together. When the recoil spring has been removed, the barrel and bolt assembly can be separated. The bolt head is removed from the barrel telescoping bolt tube by removing a large pin that is held in place by a smaller pin. Removal of the bolt head is not necessary for routine maintenance. Assembly is the simple reverse of the disassembly process.

Silenced FMK

As noted in Chapter 5, there is a silenced version of the FMK, but it has been made only in very small quantities through mid-1982.

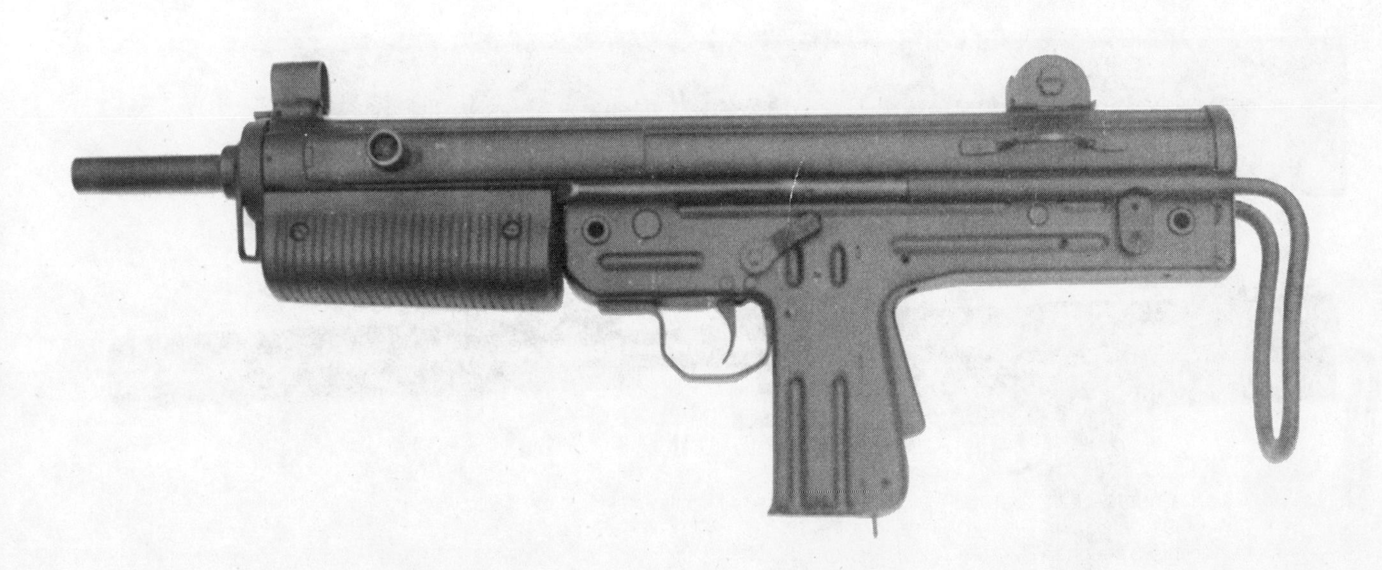

Left side view of the Argentine 9 × 19mm NATO FMK 3 submachine gun.

Right side view of the Pistol Ametralladora FMK 3. This submachine gun is equipped with a plastic handguard.

Special retaining nut required to mount the silencer on the FMK series of submachine guns.

FMK 3 and 4 magazines; 25-shot *above*; 40-shot *below*.

The silenced version of the FMK 3 submachine gun. About fifty of these were fabricated in 1980, the initial year of manufacture. The silencer tube is aluminum, and subsonic ammunition is required for true silenced shooting. Most of these weapons have been issued to Argentine intelligence forces.

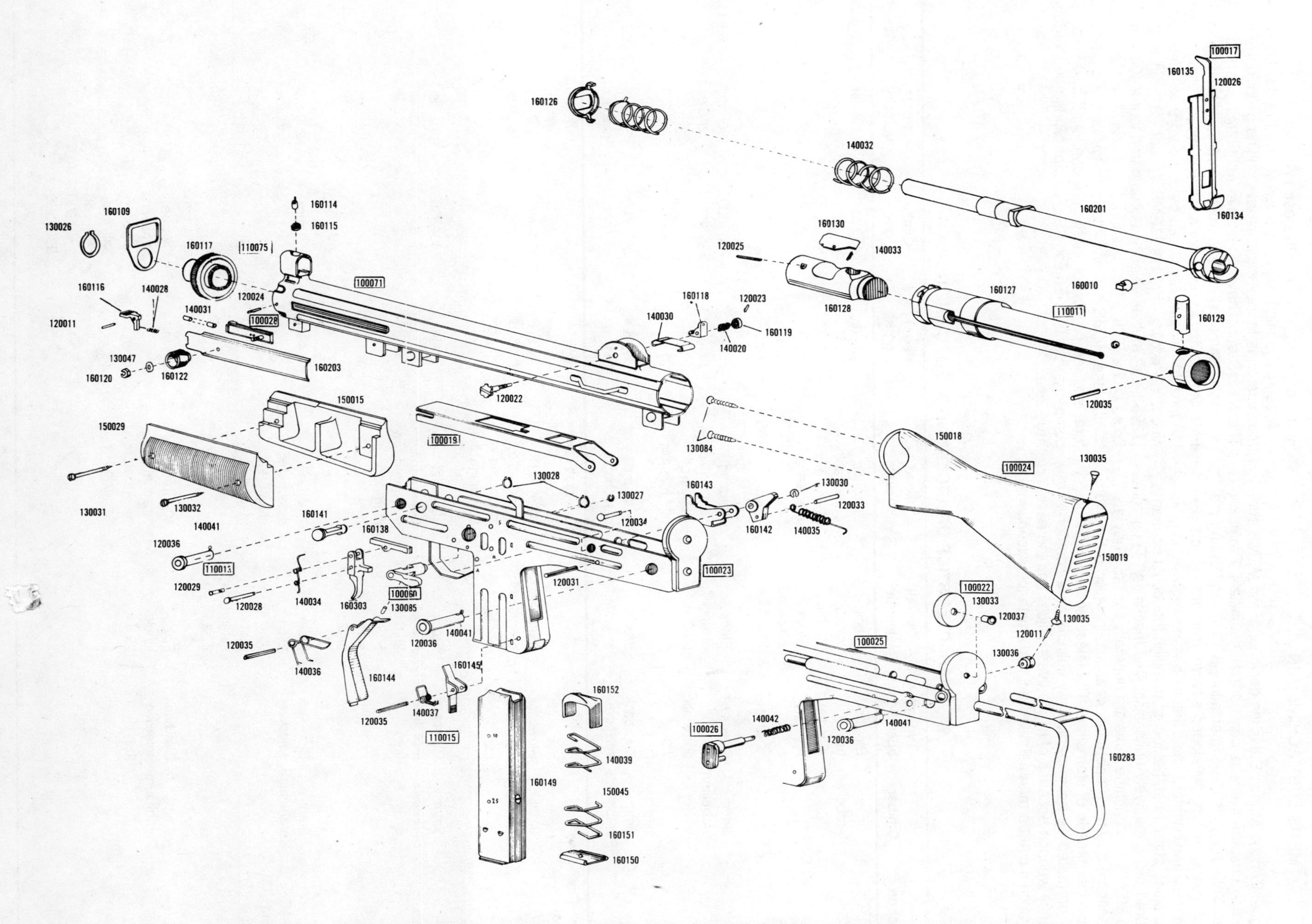

An exploded view of the FMK submachine gun. Note both the sliding stock and the fixed stock.

MEMS SUBMACHINE GUN SERIES

Miguel Enrique Manzo Sal is the designer of the MEMS submachine guns; the name is derived from his initials. Professor Manzo Sal began his experimentation in 1948, and during the last 35 years he has developed several different models. He has had several constant ideas that he has incorporated into his submachine guns. First, he sought a design that could be manufactured using relatively unsophisticated machinery. Thus, all of the MEMS submachine guns are generally similar. They are conventional blowback operation weapons, in which extensive use has been made of stampings. The latest model, the Model 75, is promoted as being particularly suitable as a counter-insurgency weapon because of its relative similicity and the ease with which it can be manufactured. All parts, with the exception of the barrel, are made with loose tolerances.

A second design element present in all of the MEMS submachine guns is the "micro-relieve" rifling used in the barrels. The MEMS guns have 24-groove right-hand twist rifling. This contrasts sharply with the 6-groove right-hand twist rifling of the FMK barrel. Manzo Sal reports that this type of rifling increases the muzzle velocity developed by standard 9 × 19mm ammunition. He reports that his MEMS 75 barrels yield a 425 m/s (1295 f.p.s.) muzzle velocity as compared to the 400 m/s (1200 f.p.s.) of the FMK.

All of the MEMS submachine guns are part of a private venture, and the production rights are controlled by Manzo Sal's company, Armas & Equipos S. R. L. of Ciudad de Cordoba in Argentina.

CHARACTERISTICS OF ARGENTINE SUBMACHINE GUNS

Weapon	Caliber	Type of operation	Overall length	Feed device	Barrel length	Cyclic rate	Muzzle velocity	Weight
PAM 1 PAM 2	9mm Parabellum	Blowback full automatic fire	Stock folded: 21.2 in. Stock extended: 28.6 in.	30-round detachable, in-line box magazine	7.9 in.	450 rpm	1200 fps (approx.)	6.6 lbs.
MEMS Model 52/58	9mm Parabellum	Blowback full automatic fire	Stock extended: 35.2 in.	40-round detachable, staggered box magazine	12 in. (approx.)	750–800 rpm	1200 fps (approx.)	6.8 lbs.
MEMS Model 52/60	9mm Parabellum	Blowback selective fire	Stock extended: 35 in.	40-round detachable, staggered box magazine	12 in. (approx.)	750–800 rpm	1200 fps (approx.)	6.8 lbs.
MEMS Model AR 163	9mm Parabellum	Blowback selective fire	Stock extended: 35 in.	40-round detachable, staggered box magazine	12 in. (approx.)	750–800 rpm	1200 fps (approx.)	6.8 lbs.
MEMS M67	9mm Parabellum	Blowback selective fire	Stock folded: 25.6 in. Stock extended: 34.7 in.	40-round detachable, staggered row magazine	7.9 in.	750–800 rpm	110 fps (approx.)	6.1 lbs.
FMK 3 (PA3-DM) Standard Argentine SMG	9mm Parabellum	Blowback selective	Stock folded: 20.6 in. Stock extended: 27.3 in.	25-round detachable, staggered magazine	11.4 in.	650 rpm	1200 fps (approx.)	7.9 lbs.
MEMS M75/I Infantry Model	9mm Parabellum	Blowback selective fire	Fixed stock: 29.5 in.	40-round detachable, staggered row magazine	9.5 in. w/ compensator	Can be chosen from following: 600–650 750–800 900–950 rpm	1295 f.p.s.	7.3 lbs.

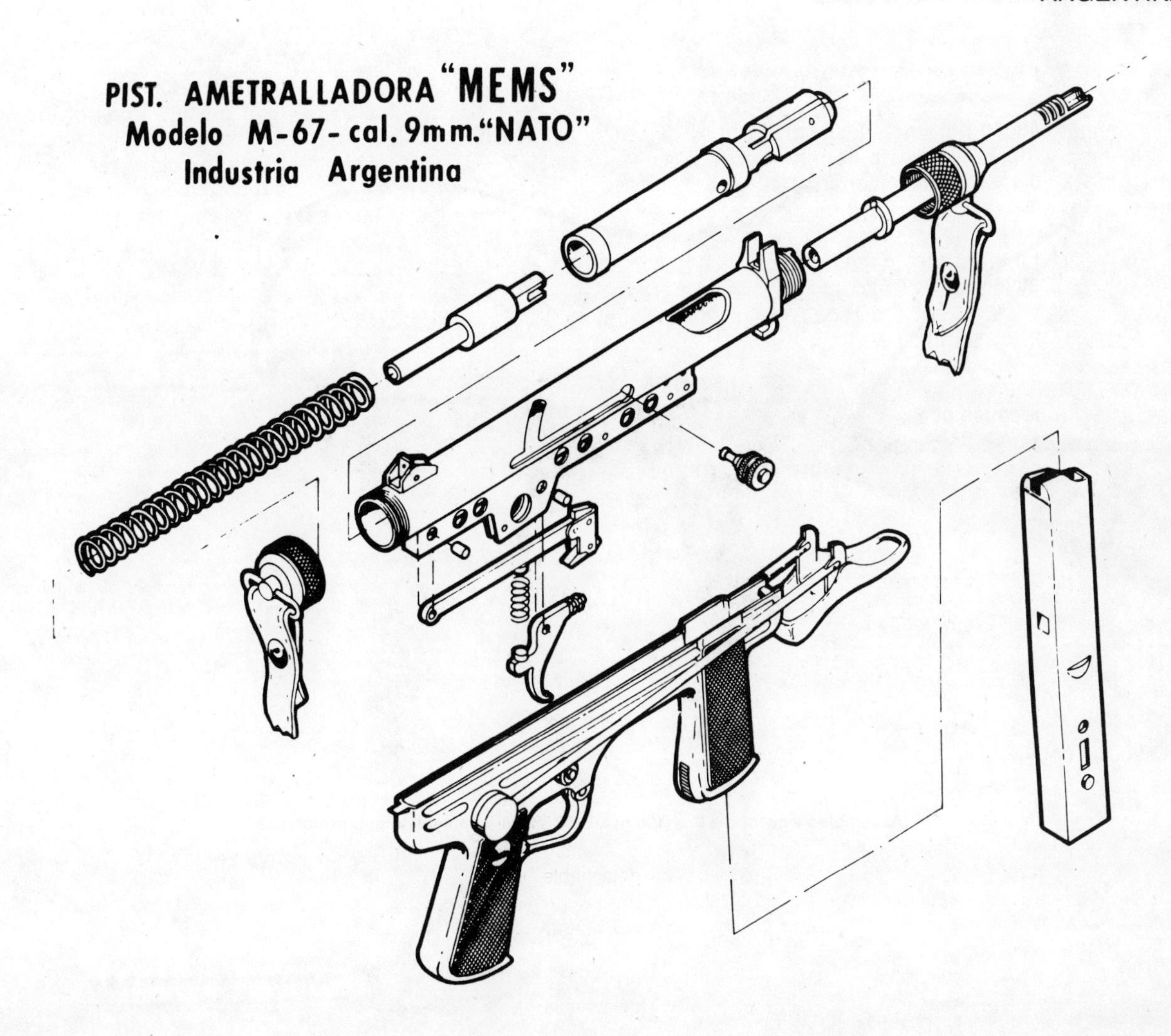

CHARACTERISTICS OF ARGENTINE SUBMACHINE GUNS (continued)

Weapon	Caliber	Type of operation	Overall length	Feed device	Barrel length	Cyclic rate	Muzzle velocity	Weight
MEMS M75/II Airborne Model	9mm Parabellum	Blowback selective fire	Stock folded: 20.5 in. Stock extended: 28.4 in.	40-round detachable, staggered row magazine	9.5 in. w/ compensator	Same as above	1295 f.p.s.	7.5 lbs.
MEMS M75/III Firing port weapon	9mm Parabellum	Blowback selective fire	No stock: 28.4 in.	40-round detachable, staggered row magazine	9.5 in. w/o compensator	900/950 1000/1100 rpm	1325 f.p.s.	6.0 lbs.

Assembled view of the 9 × 19mm MEMS M.75/I submachine gun prototype.

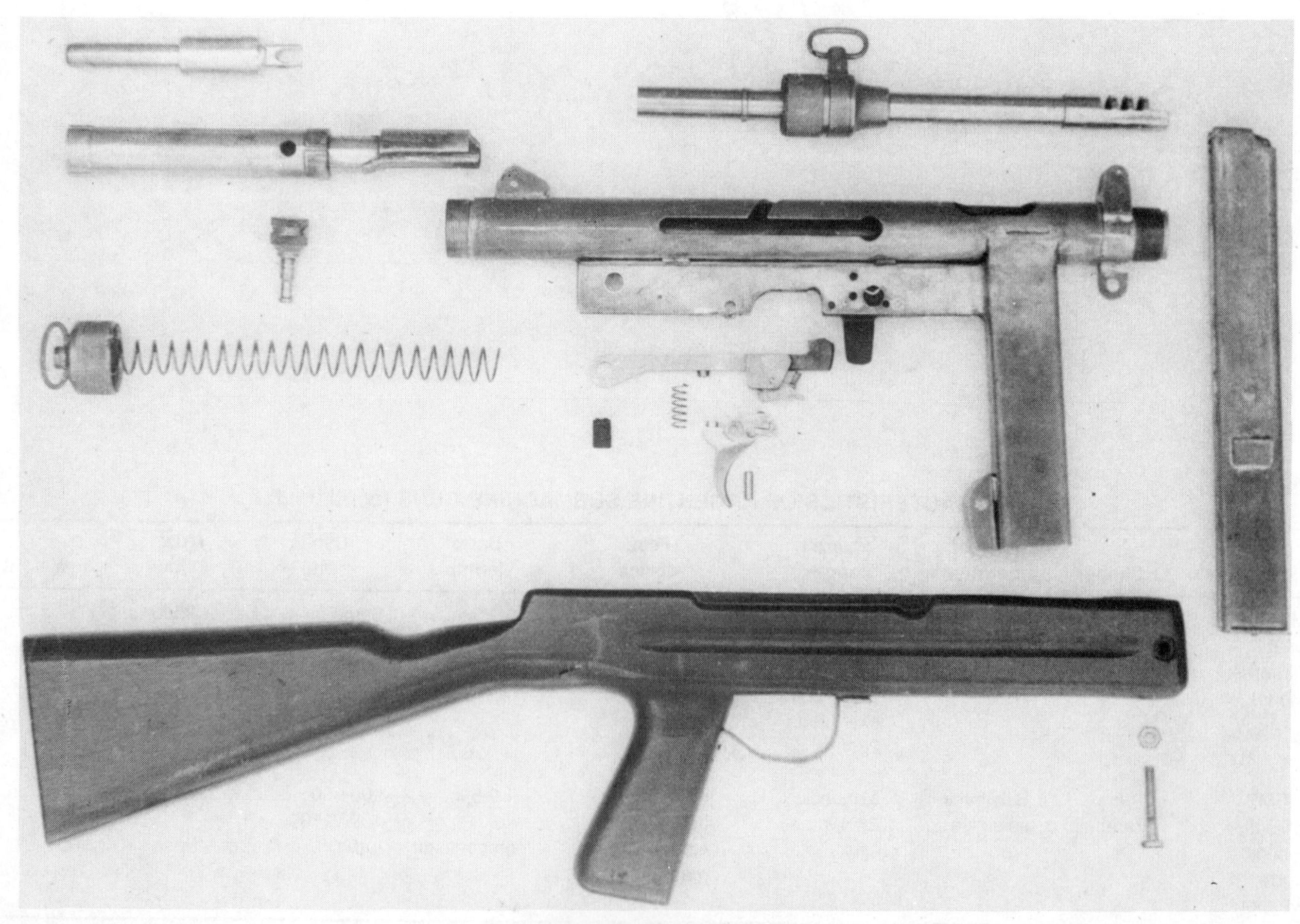

Disassembled view of the MEMS M.75/I.

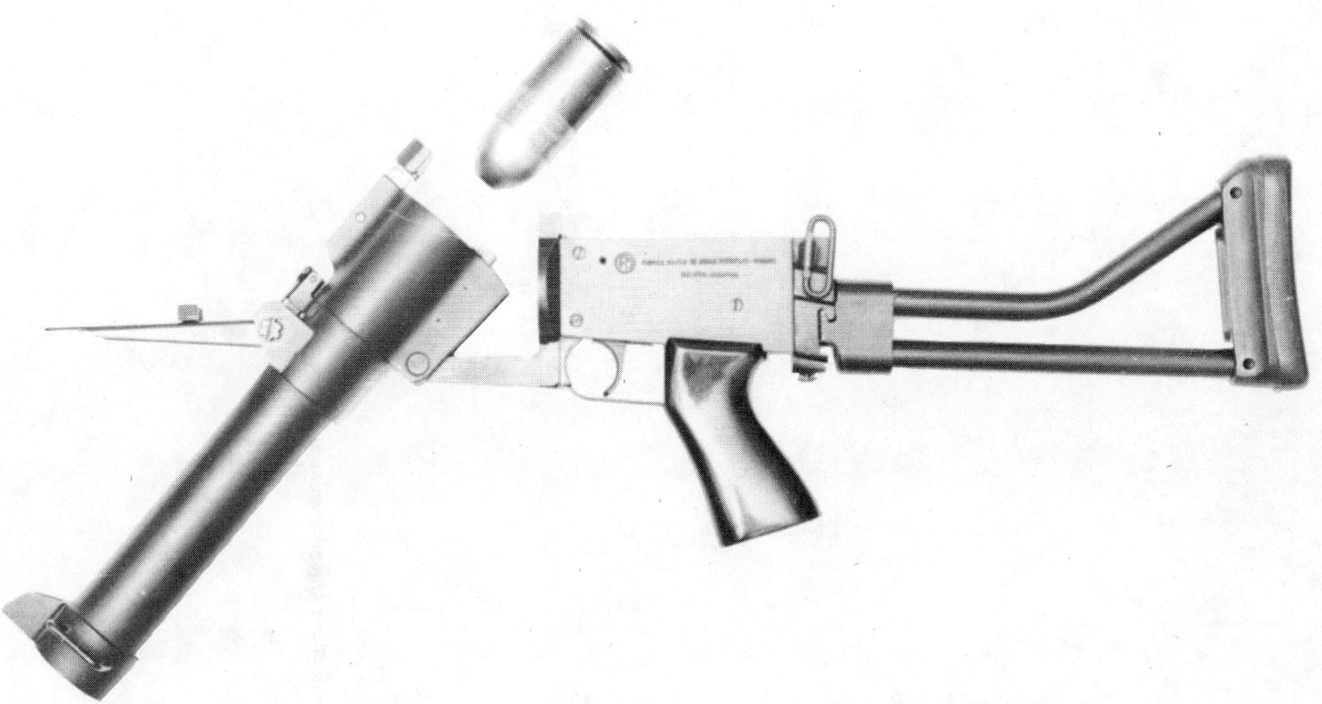

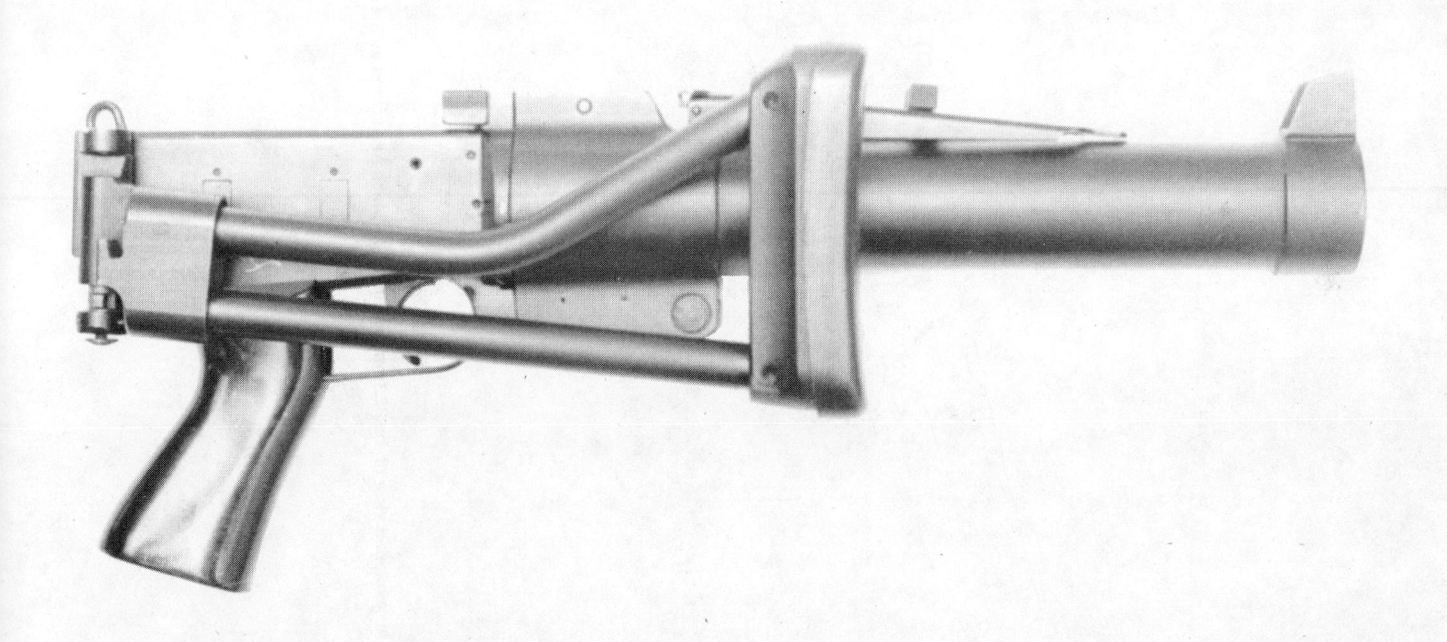

Two views of the prototype 40mm grenade launcher being developed by the Fábrica Militar at Rosario.

ARGENTINE GRENADE LAUNCHERS

In the early 1980s the Fábrica Militar at Rosario undertook the conversion of the *Pistola Lanza-Gases Cal 38.1mm, Modelo Unico* (1.5 inch tear gas launcher) to serve as a 40mm grenade launcher. This modified weapon will fire the standard family of US low velocity grenades, and the Argentines are developing their own grenade series as well. At the end of 1981 a pilot lot of fifty grenade launchers was under fabrication. It is worth noting that the side-folding stock is derived from that of the FN FAL PARA and that the weapon has a double action mechanism that is cocked as the trigger is pulled. Additional details were not available at press time.

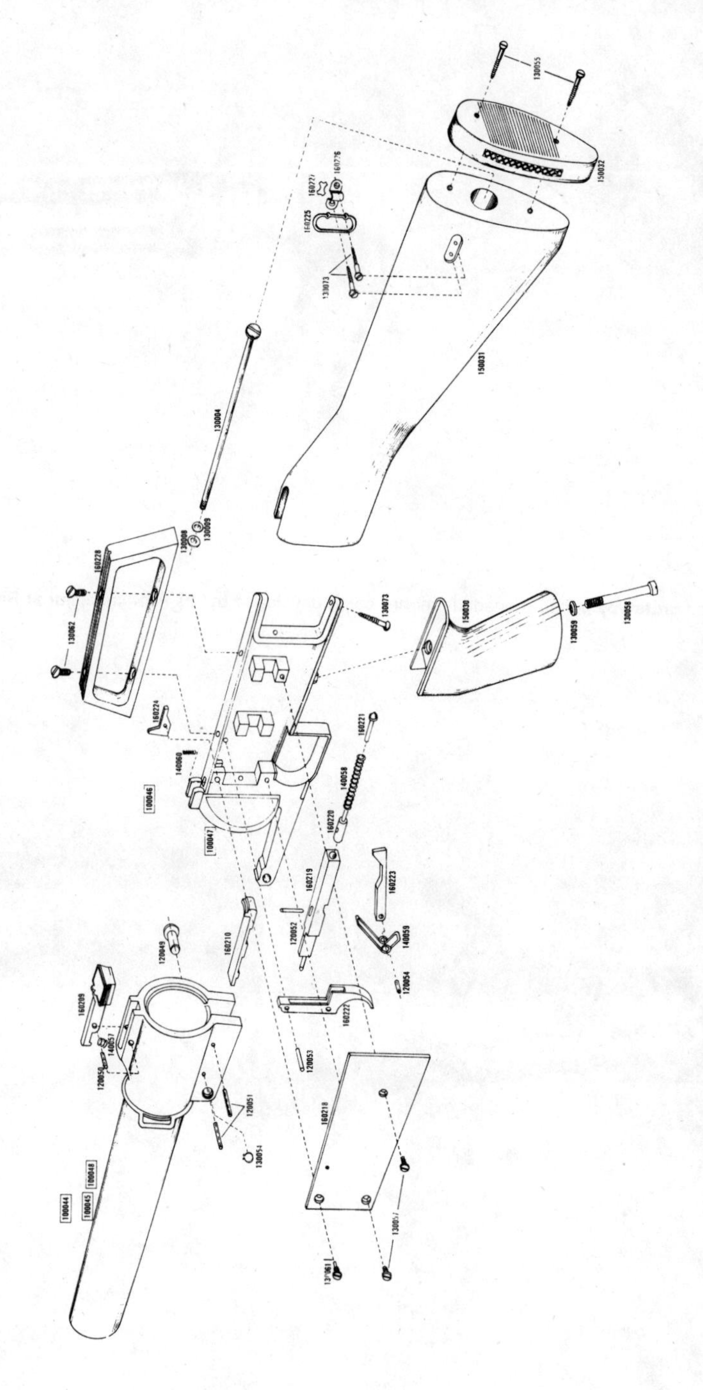

An exploded view of the 38.1mm tear gas gun which has served as the basis for the new 40mm Argentine grenade launcher.

ARGENTINE SMALL ARMS PRODUCTION, 1947–1981

Model	Years Manufactured	Quantity
1. Carabina Mauser Caballeria 1909 (M1909 Mauser Cavalry Carbine) Cal. 7.65 mm	1947–1959	19,072
2. Pistola Sistema Colt (Colt .45 Automatic Pistol) Cal. 11.2 mm	1947–1966	74,866
3. Bayoneta para Mauser 1909 (M1909 Mauser Bayonet)	1951–1952	14,000
4. Ametralladora Alam (Alam Machine Gun) Cal. 7.65 mm	1954–1959	763
5. Pistola Ametralladora PAM 1 (PAM 1 submachine gun) Cal. 9 mm	1955–1962	29,636
6. Modification PAM 1 a PAM 2 (Remanufactured [?] with grip safety)	1963–1972	11,860
7. Pistola Ametralladora PAM 2 (PAM 2 new manufacture) Cal. 9 mm	1968	1,100
8. Modification PAM 1 a PAM 2 (with Navy markings)	1964	5,292
9. Fusil F.A.L. (F.A.L. rifle) Cal. 7.62 mm	1960–1969	63,675
10. Fusil F.A.P. (F.A.P. rifle) Cal. 7.62 mm	1965–1976	4,513
11. Pistola Browning PD (Browning PD Pistol) Cal. 9 mm	1969–1981	178,533
12. Pistola Ametralladora FMK3 (FMK3 submachine gun) Cal. 9 mm	1971–1976	19,953
13. Pistola Browning PD Exportation (Browning PD for export) Cal. 9 mm	1976	3,445
14. Fusil Fal Para (FAL Para Rifle) Cal. 7.62 mm	1976–1978	19,719
15. Pistola Browning "Comando" (Browning Commando Pistol) Cal. 9 mm	1975–1980	415
16. Pistola Browning (Browning imported with F.M. markings) Cal. 9 mm	1977–1978	5,500
17. Fusil F.A.L. Para III (F.A.L. Para Mk III Rifle) Cal. 7.62 mm	1978–1980 (1981 combined Para III & FAL IV)	15,300 (15,000)
18. Fusil FAP II (FAP II Rifle) Cal. 7.62 mm	1979–1981	3,700
19. Pistola Ametralladora FMK3M2 (FMK3M2 Submachine Gun) Cal. 9 mm	1979–1981	20,500
20. Fusil F.A.L. IV (F.A.L. IV Rifle) Cal. 7.62 mm	1980 (1981 see FAL PARA III)	5,940
21. Fusil F.A.L. PARA (F.A.L. imported with F. M. Markings) Cal. 7.62 mm	1980	6,000
22. Ametralladora MAG (MAG machine gun) Cal. 7.62 mm	1980	43

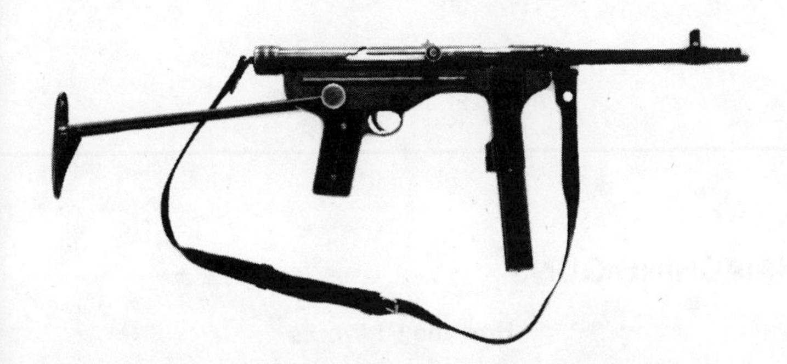

Argentine 9mm Parabellum MEMS Model 52/60 Submachine Gun.

Argentine 9mm Parabellum MEMS Model AR63 Submachine Gun.

Argentine 9mm Parabellum MEMS Model 52/58 Submachine Gun.

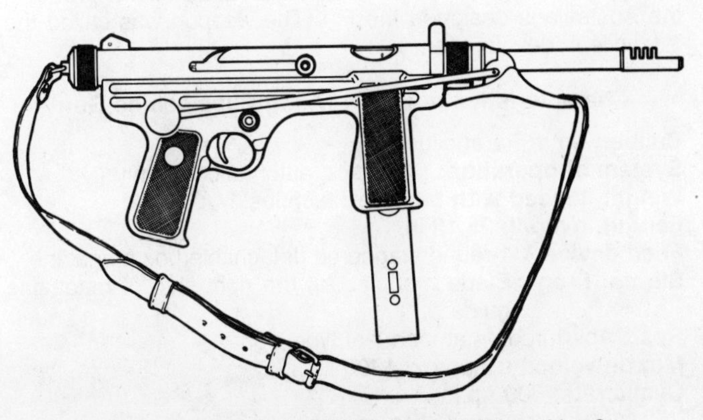

Argentine 9mm Parabellum MEMS Model 67 Submachine Gun.

7 Australia

Australian small arms have been manufactured by the Australian Government Small Arms Factory at Lithgow, New South Wales since 1912. These Australian made weapons have on occasion been supplemented by ones manufactured by domestic companies (as during World War II) and by UK and US factories (as with the L4A4 and M60 machine guns). All of the Browning L9A1 handguns are from foreign sources—John Inglis during World War II and Fabrique Nationale Herstal since then.

SMALL ARMS IN SERVICE

Handguns: 9 × 19mm NATO Browning L9A1 (FN Modele 1935 GP)
Submachine guns: 9 × 19mm NATO F1
Rifles: 7.62 × 51mm NATO L1A1 (FN FAL)
7.62 × 51mm NATO L2A1 (FN FALO)
7.62 × 51mm NATO M82 Sniping Rifle (Parker Hale, Chapter 11)
Machine guns: 7.62 × 51mm NATO L4A4 (Bren, RSAF, Enfield)
7.62 × 51mm NATO M60
12.7 × 99mm Browning M2 HB (US)

AUSTRALIAN SUBMACHINE GUNS

Australia has produced a number of native submachine gun designs. The Austen and the Owen were developed during World War II and were considered excellent designs for their time. Both are obsolete.

9mm SUBMACHINE GUN F1 (AUST)

In a desire to secure a weapon with the reliability of the Owen, but lighter in weight, with lower rate of fire, and easier to produce, the Australians designed the F1. (The weapon was called the X3 while in development.)

Characteristics of the F1 (Aust) Submachine Gun

Caliber: 9mm Parabellum.
System of operation: Blowback, automatic fire only.
Weight, loaded with bayonet: 9.88 lbs.
Length, overall: 28.12 in.
Feed device: 34-round staggered detachable box magazine.
Sights: Front: Blade mounted on the right side of magazine guide.
Rear: Aperture in stamped leaf type sight.
Muzzle velocity: Approx. 1300 f.p.s.
Cyclic rate: 600 r.p.m.

How the F1 Works

The F1 has a separate cocking handle and cover that do not reciprocate with the bolt during firing. The pistol grip and butt plate are the same as those used on the Australian-made FN L1A1 rifle. The top loading magazine of the F1 is the same as that used on the British L2A3 and the Canadian C1 (the Sterling or Patchett submachine gun). The weapon has a bayonet lug on the left side.

Loading and Firing the F1

Pull cocking handle to the rear. Insert a loaded magazine in the guide on top of the receiver. Press trigger, and the weapon will fire. To put the weapon on safe, push safety catch located on the left side of the pistol grip to down position; word "Safe" will be exposed.

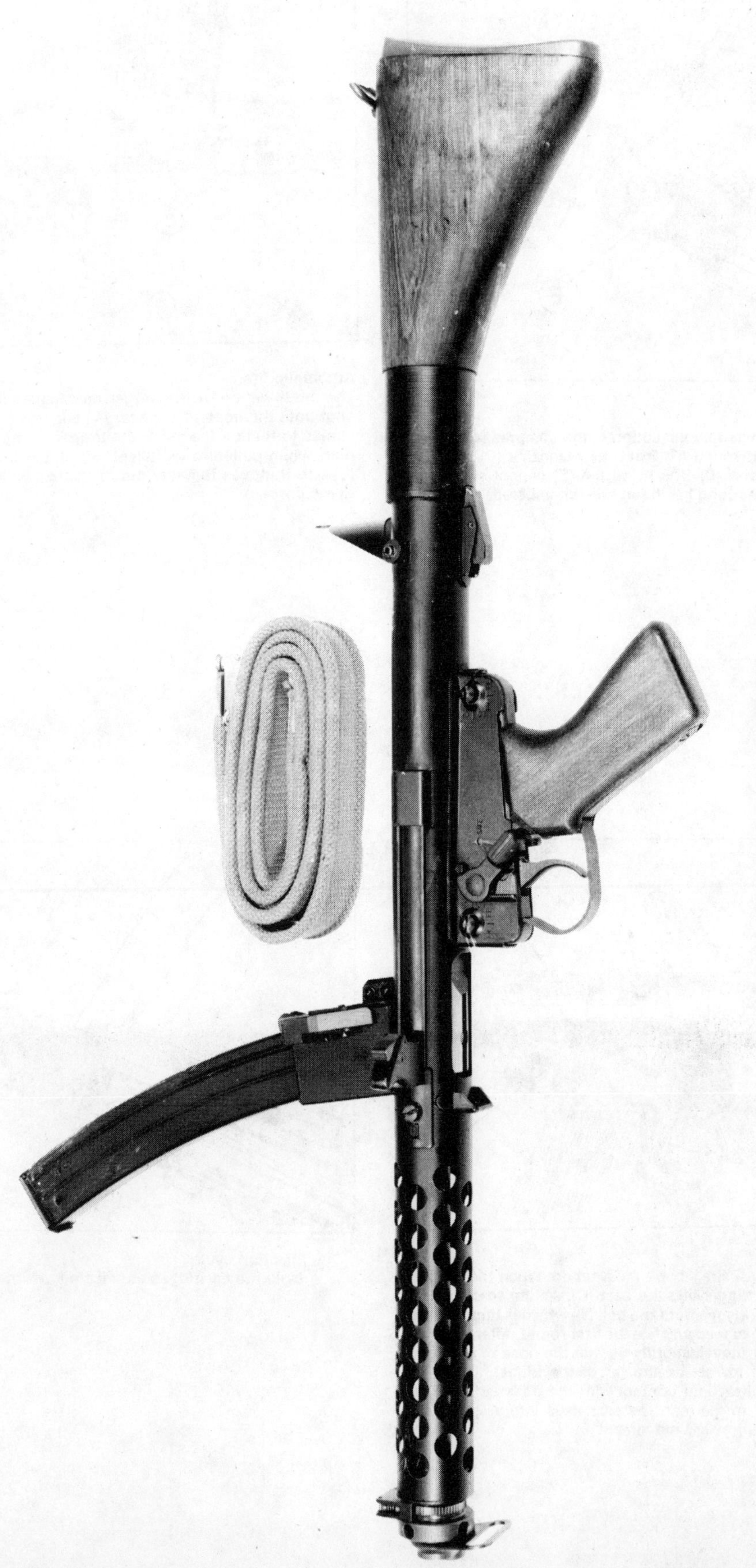

Australian F1 9mm Parabellum Submachine Gun.

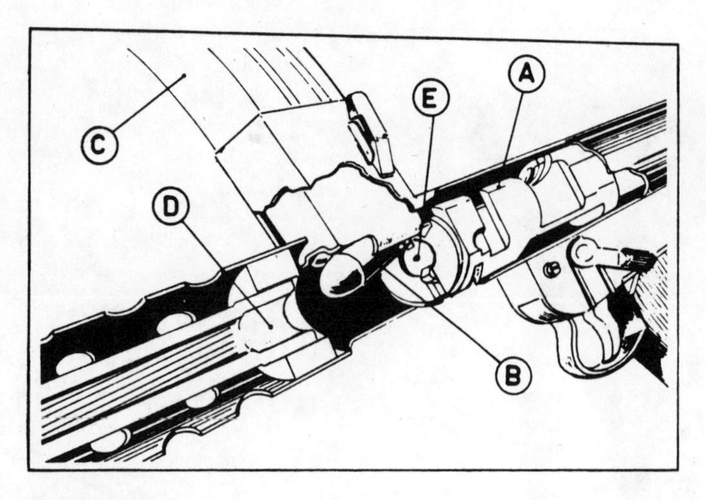

Feeding cycle
When the bolt (A) is moved forward by the action of the compressed return spring, the bolt strips the next round (B) from the magazine (C) and the cartridge is fed into the chamber (D). The firing pin (E) cannot strike the primer of the cartridge until the round has been fully chambered.

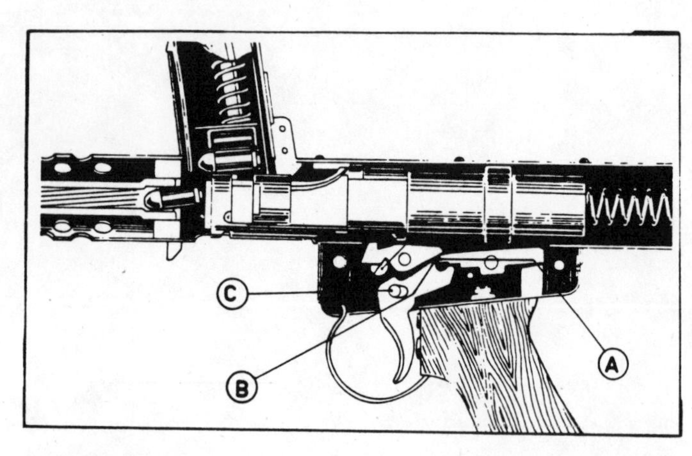

Automatic fire
The initial action of the trigger mechanism is identical to that for a single shot until the nose of the sear (A) slips over the heel (B) of the trigger and comes to rest on the tail of the trigger as the trigger is pulled further to the rear. When pulled to its fullest extent, the trigger rotates about its axis pin (C) and it moves the sear out of contact with the bolt and permits the bolt to reciprocate.

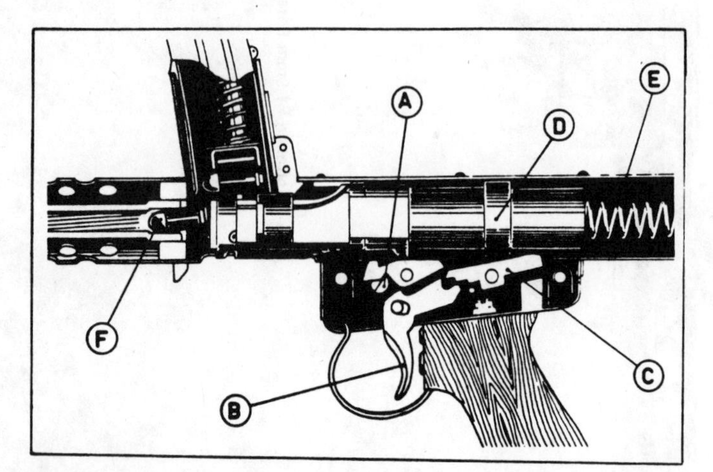

Single shot fire
With the change lever (A) set on "fire", cock the weapon. When the trigger (B) is pressed the heel on the trigger lifts the sear (C). As the sear rotates about its axis pin, the sear disengages from the bolt (D) which is then forced forward by the return spring (E) to feed and fire the first round. After the bolt has moved forward, the trigger moves slightly so that the nose of the sear slips over the heel and comes to rest on the tail of the trigger. The sear spring elevates the tail of the sear into contact with the underside of the bolt. When the bolt moves full to the rear, the sear rises into a space (G) where it restricts the bolt on its forward movement.

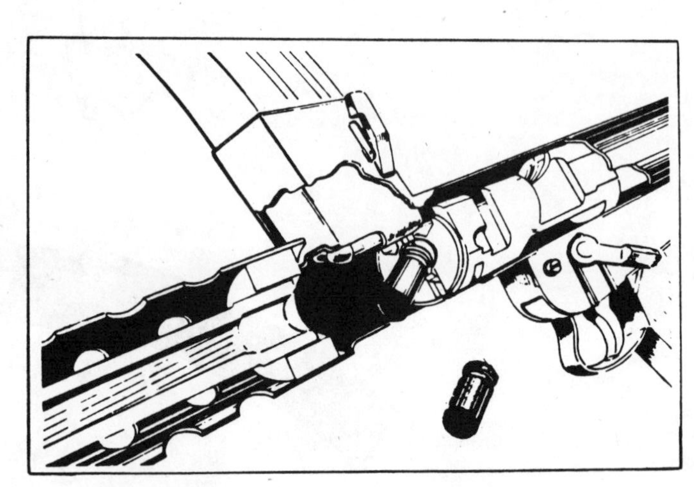

Ejection cycle
Note the cartridges are ejected downward toward the ground.

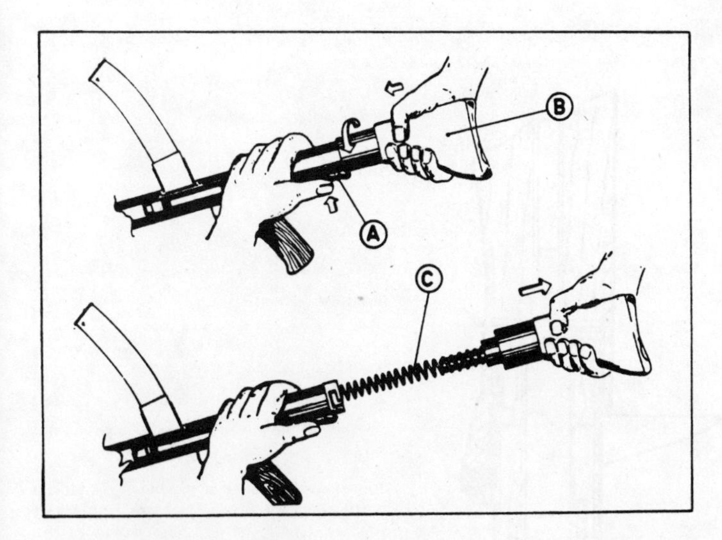

Field stripping and assembly
Set change lever to safe after removing magazine and ensuring that the weapon is unloaded. Depress the butt locking catch (A) with the thumb, rotate the butt (B) in a counterclockwise direction and remove it from the gun. Withdraw the return spring (C). Move the change lever to fire, elevate the muzzle of the gun and restrain the bolt as it drops out of the receiver. If the bolt tends to stick, it may be removed by use of the cocking handle. Reassemble the parts in reverse order. The trigger must be pressed to allow the bolt to go completely forward.

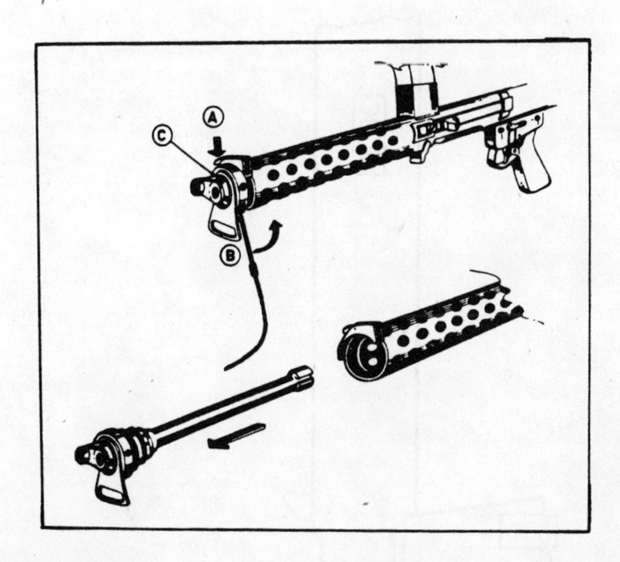

Removal of barrel
Depress the barrel locking catch (A) and apply the weighted end of the pull through cleaning rod as a lever in one of the holes in the barrel nut (C) so it can be unscrewed. The barrel may now be withdrawn from the receiver. Reassembly is the reverse of this procedure. Care must be taken to depress the barrel nut catch until the threaded portion of the nut has engaged the thread of the receiver.

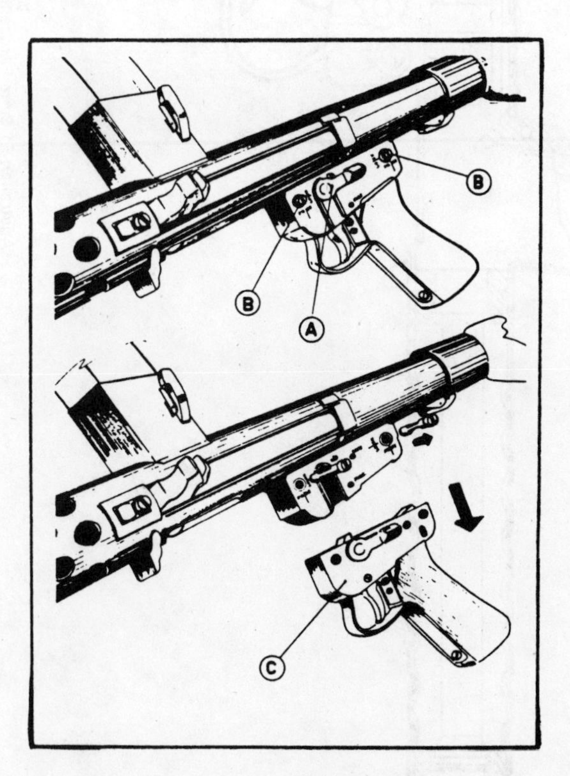

Removal of the trigger assembly
After turning the change lever to safe (A), use the rim of the cartridge to turn the slots in the heads of the retaining pins (B) in line with the words "free" engraved on the left hand side of the gun. Push the pins from the gun (right to left). Remove the trigger mechanism housing (C) by pulling straight downward on the grip. Reassemble in reverse order.

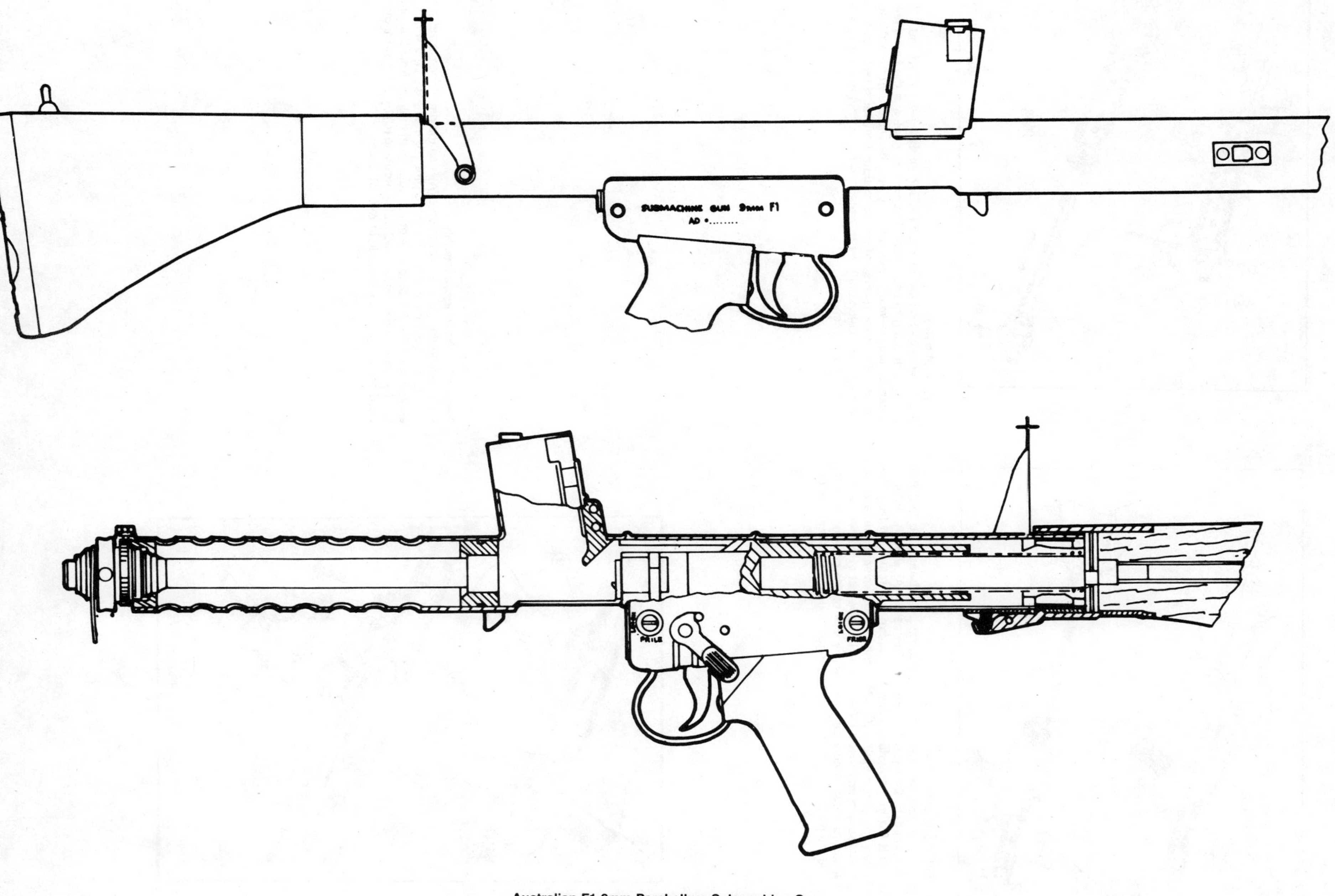

Australian F1 9mm Parabellum Submachine Gun.

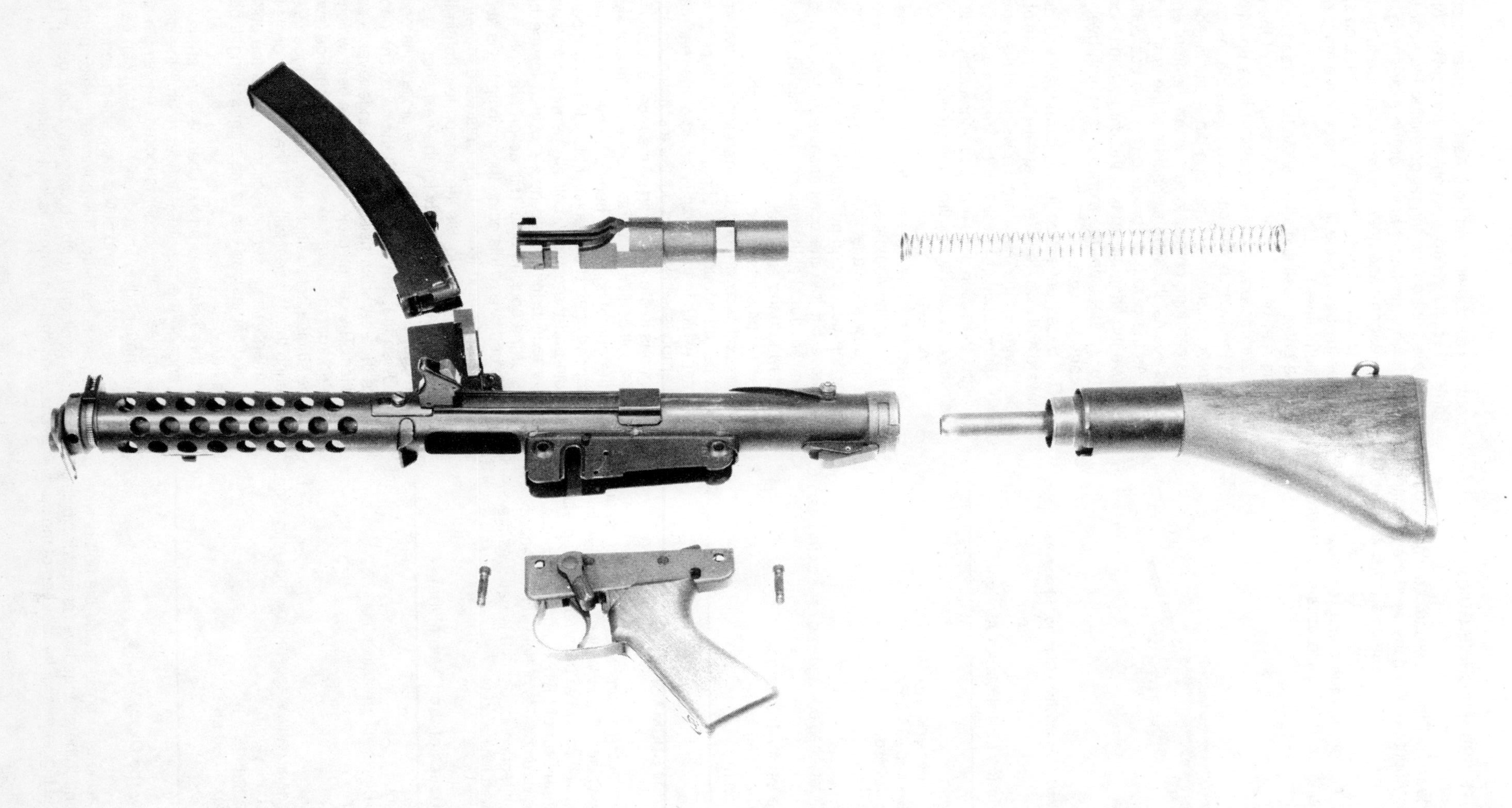

Disassembled view of the Australian 9 × 19mm NATO submachine gun.

THE AUSTEN SUBMACHINE GUNS

These guns were made by Diecasters Ltd. and W. T. Carmichael of Melbourne. About 20,000 guns were made during World War II. Although the Austen resembles the British Sten externally, internally it resembles the German MP 38 and MP 40 (Schmeisser). The Austen has the same telescoping type cover over its recoil spring and firing pin assembly as do the MP 38 and MP 40.

Australian 9mm Austen Mark I Submachine Gun, no longer being made.

Characteristics of the Mark I Austen

Caliber: 9mm Parabellum.
System of operation: Blowback, selective fire.
Length overall: Stock fixed: 33.25 in.
Stock folded: 22 in.
Barrel length: 7.8 in.
Weight: 9.2 lbs.
Feed device: 32-round, detachable, staggered row box magazine.
Sights: Front: Barley corn.
Rear: Aperture set for 100 yards.
Muzzle velocity: Approx. 1280 f.p.s.
Cyclic rate: 500–550 r.p.m.

THE OWEN SUBMACHINE GUN

Over 40,000 Owen submachine guns were made by Lysaghts Newcastle Works, New Castle, South Wales, Australia, during World War II. The Owen is somewhat unusual in having a top-mounted magazine, like the current F1A1, and a quick-change barrel. The ease of barrel removal is of help in maintenance of the weapon, but is not intended for change in battle as are the quick change barrels of machine guns.

Characteristics of the Owen Mark I

Caliber: 9mm Parabellum.
System of operation: Blowback, selective fire.
Length overall: 31.8 in.
Barrel length: 9.8 in.
Weight: 9.37 lbs.
Feed device: 30-round, detachable staggered row, box magazine.
Sights; Front: Off-set barley corn.
Rear: Off-set aperture.
Muzzle velocity: Approx. 1300 f.p.s.
Cyclic rate: 800 r.p.m.

How the Owen Works

When a loaded magazine is in position and the bolt is withdrawn to its fullest extent, if the change lever is set for single-shot fire, pressure on the trigger forces the rear end of the sear out of contact with the bolt. As the operating spring starts to drive the bolt forward, the bent on the front end of the sear slips over the bent on the trigger permitting the sear to return to its former position under action of the sear spring. This leaves it free to engage and hold the bolt when it returns to the rear.

As the bolt is driven forward by the operating spring, the bolt strikes the base of the first cartridge in the magazine and drives it straight ahead.

The bullet nose is guided by the barrel feed into the chamber as the rear of the cartridge clears the lips of the magazine. As the cartridge enters the chamber, it lines up with the bolt, enabling the head of the cartridge to seat in the base of the bolt head recess, in which the firing pin is machined.

The cartridge comes to rest when the front end of the case stops against the square shoulder at the front end of the chamber. As the cartridge comes to rest, the bolt continues forward to drive the firing pin against the primer and discharge the cartridge. At this time, the extractor is sprung over the groove in the cartridge case.

The gases generated in the cartridge case drive the bullet forward and exert rearward pressure to the base of the bolt head. But in view of the much greater weight of the moving parts and the spring tension in relation to the comparatively light bullet weight, the action does not open appreciably until the bullet has left the muzzle. By this time, the breech pressure has dropped to a safe limit.

When the bullet emerges from the muzzle, the gas behind it expands in the compensator. The pressure wave thus created thrusts downward against the solid lower half of the compensator while the gases expanding upward strike against the inclined surfaces cut into the compensator to result in a forward and downward thrust at the muzzle end. This tends to hold the muzzle down during automatic fire.

As the bolt starts to the rear, the empty cartridge case is held in the bolt face gripped by the extractor. When it clears the chamber far enough, the upper face of the cartridge strikes the ejector (which in the Owen is a part of the rear magazine wall). The empty shell is hurled out the ejection opening, which in this weapon is in the lower part of the receiver tube.

The bolt continues to travel to the rear in a straight line pressing the operating spring behind it until the rear face of the bolt is stopped against the receiver plug. In this weapon, the cocking handle and the cocking bolt are permitted to travel still further to the rear in the slotted hole provided for the bolt pin in the head of the cocking bolt. This action prevents a sudden shock on the bolt pin. At this point, the main spring is at practically full compression, and it halts further rearward movement.

The sear spring forces the sear up to catch in the underside of the bolt, holding it ready for the next shot.

Trigger action. There are three projections on the trigger in the Owen that can engage with the sear. The upper projection locks the sear from rising if the change lever is set at the safe position. The central projection forms a bent, which is accurately located from the trigger axis. When the change lever is adjusted to the single-shot position, the rise of the trigger is strictly limited assuring that the sear will hold the bolt back on its first rearward movement. The lower projection engages the underface of the sear and carries one end of the trigger spring. It permits continuous fire.

The three surfaces on the change lever are in a circle and in turn engage the top of the trigger to limit the distance of rise. Thus on "safe," the trigger is locked; on "single shot," the rise of the trigger being limited, only one shot can be fired until the trigger is released and pulled back again; and on automatic fire the trigger is permitted full movement, which permits the bolt to shuttle back and forth so long as cartridges are fed into the chamber and the trigger is kept depressed.

Special Note on the Owen

Two other varieties of this gun were issued. The first is called the "Mark I Wood Butt" type. This is a lightened version of the Mark I in which some of the metal is cut away from the receiver behind the rear grip and which is provided with a wooden butt.

The second type of Owen is the Mark II. In this type, the shape of the receiver to the rear of the rear hand grip is still further modified, resulting in a weight with butt and empty magazine of only eight pounds and three ounces; without the butt it is only eight pounds. The trigger assembly differs from the Mark I.

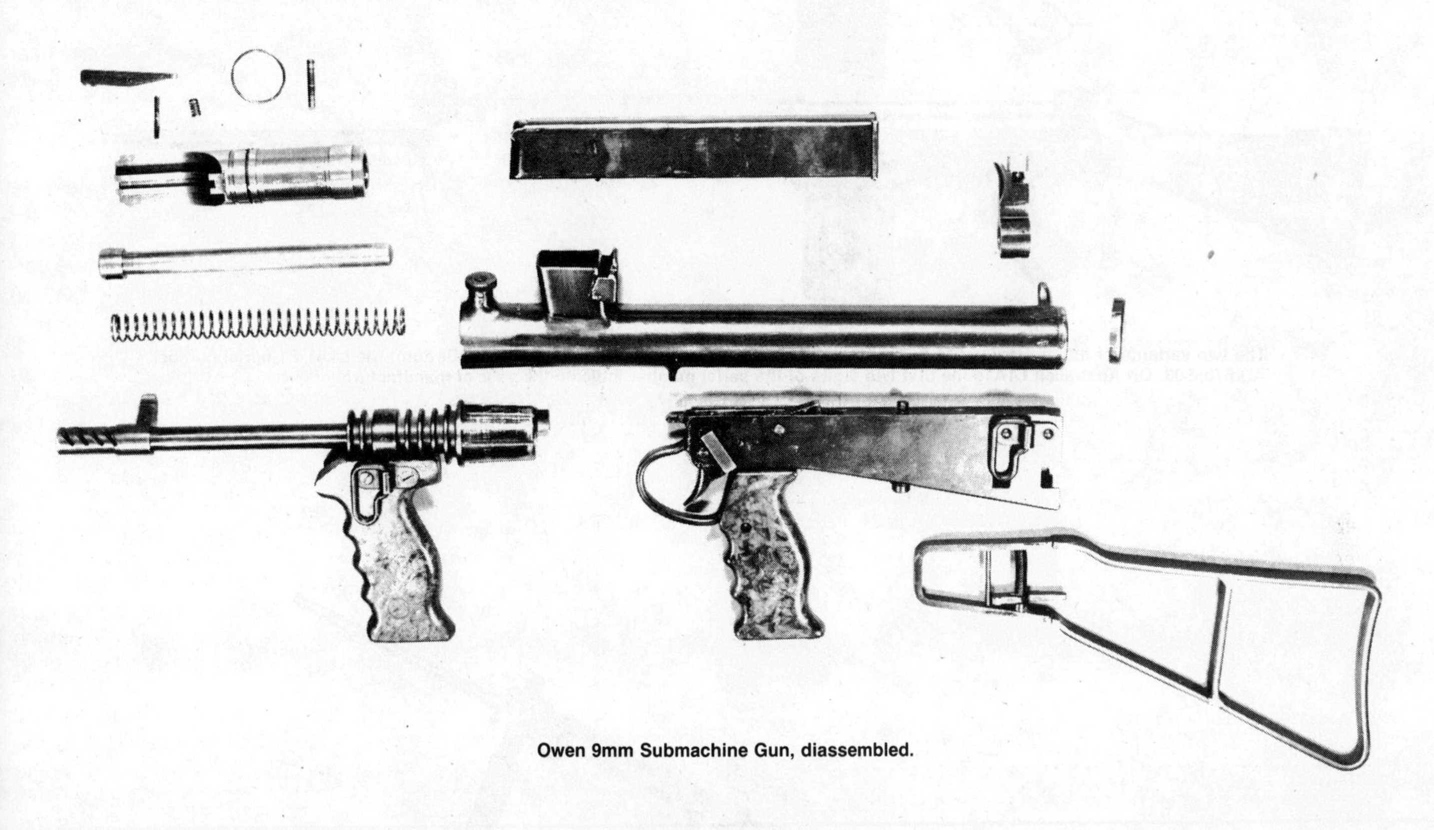

Owen 9mm Submachine Gun, diassembled.

AUSTRALIAN RIFLES

The Australian Government Small Arms Factory at Lithgow has manufactured the FAL since the late 1950s. This weapon is one of the inch-dimension versions of the rifle as standardized by the United Kingdom, Canada and Australia. It is designated the L1A1 Self Loading Rifle. The Australians also have manufactured the heavy barrel L2A1 squad automatic weapon version of the FAL as well. Their most recent addition to the FAL family has been the L1A1 F1, a shortened and lightened variant.

Experience in the field indicated that the standard L1A1 was too heavy and a little awkward for use by soldiers of small stature. The L1A1 F1 was created to do the following:

—reduce the overall length of the weapon without adversely affecting its ballistic performance.
—reduce the length of the muzzle flash suppressor.
—reduce the length of the butt.

The accompanying table compares the three Australian versions of the FAL:

COMPARISON OF AUSTRALIAN MADE FALS.

Model	L1A1	L1A1 F1	L2A1
Caliber (mm):	7.62 × 51 NATO	7.62 × 51 NATO	7.62 × 51 NATO
Overall length (in.):	44.8	42.3	44.8
Barrel length (in.): w/flash suppressor	24.3	21.9	24.3
Weight* (lbs.):	12.0	10.8	15.2
Magazine capacity:	20	20	20 and 30
Type of fire:	Semiautomatic	Semiautomatic	Selective

*Loaded weight.

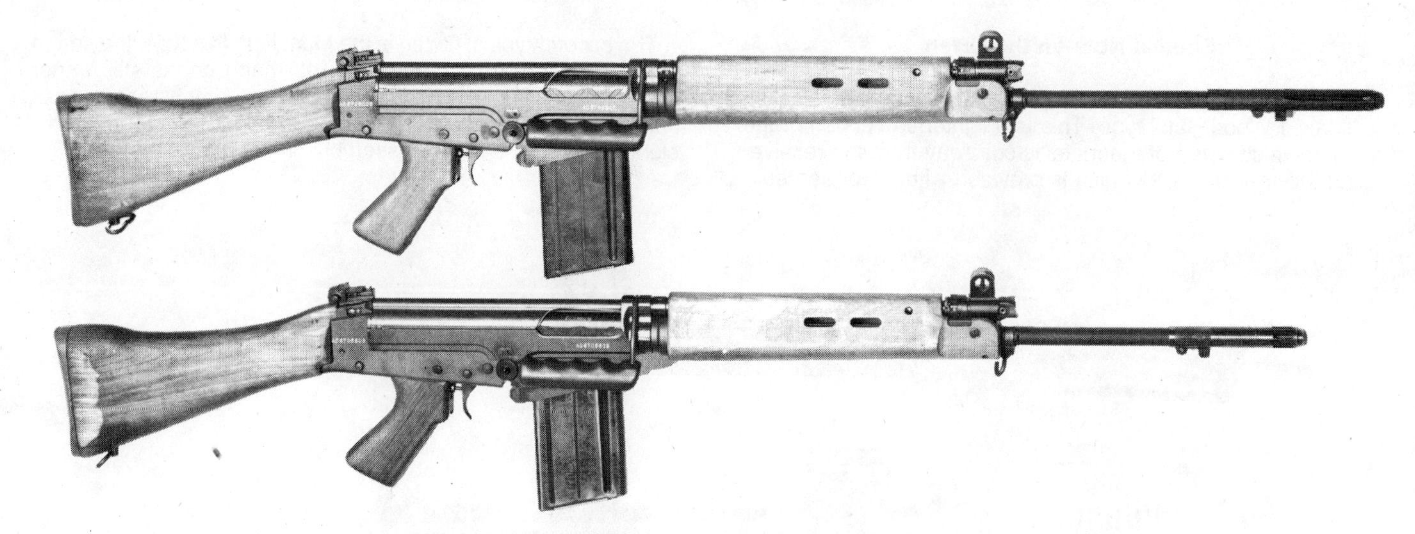

The two variants of the Australian L1A1. Top: the standard L1A1 serial number AD 6705505. Bottom: the L1A1 F1 serial number AD 6705503. On Australian L1A1s the first two digits of the serial number indicate the year of manufacture.

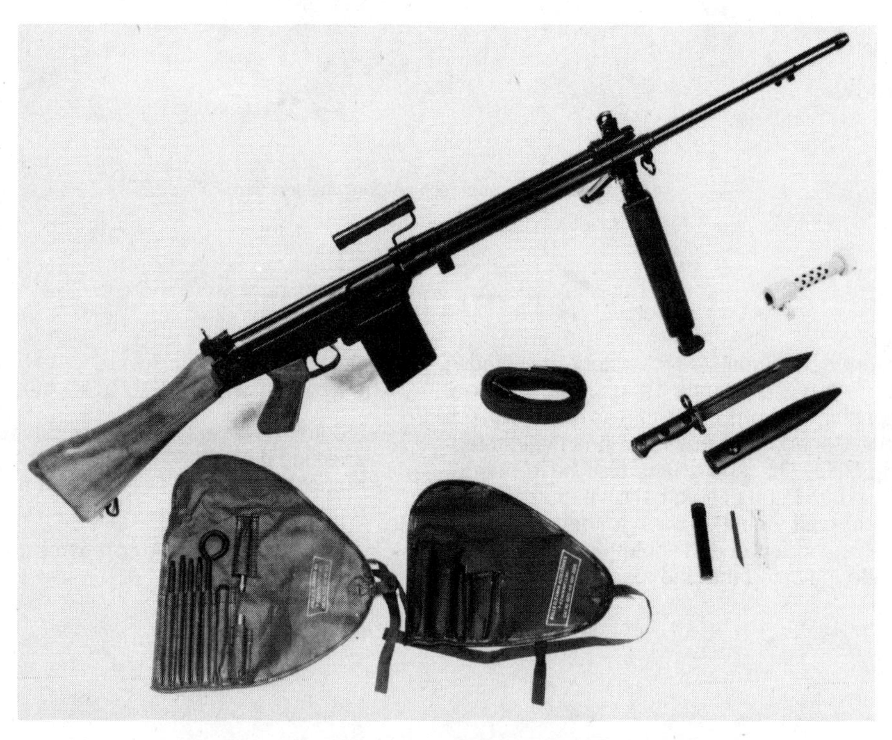

The Australian L2A1 with cleaning kits, sling, bayonet and sheath and blank firing attachment.

THE RIFLE 7.62 M.M. L1 A1—F1

This is a modified version of the standard 7.62 M.M. L1 A1 rifle. The flash eliminator has been used as a sleeve, fitting over the muzzle of the barrel for most of its length and incorporated as a combined flash eliminator/muzzle brake. A short butt is fitted and these two modifications reduce the length of the rifle by 2¾ inches (69.85 mm). It is a comfortable weapon suited to personnel of smaller stature and for operating in close environments.

Design Details

The design details and effect of the modified design are:

(a) A redesigned flash eliminator reduces the length by 2½ inches (63.5 mm) and provides:

(i) Flash elimination (approximately 95 percent efficient)
(ii) Muzzle brake (20 percent reduction in recoil energy).
(iii) A standard fitting for all muzzle attachments; blank firing attachment, grenade launcher, bayonet, etc.

(b) A short butt, ¼ inch (6.35 mm) shorter than the 'normal' butt.

The modified flash eliminator/muzzle brake is as effective as the standard eliminator, producing no more flash than the gas port.

The muzzle brake produced as part of the eliminator design has been developed to produce effective flash elimination combined with an effective reduction in recoil. This creates a minimum concussive effect to the firer. At 45 and 90 degrees to the rifle the pressure/noise level is less than with the standard L1 A1 rifle.

MODIFIED

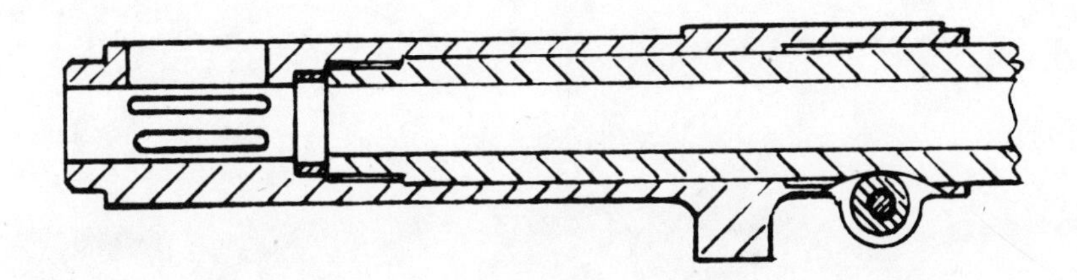

STANDARD

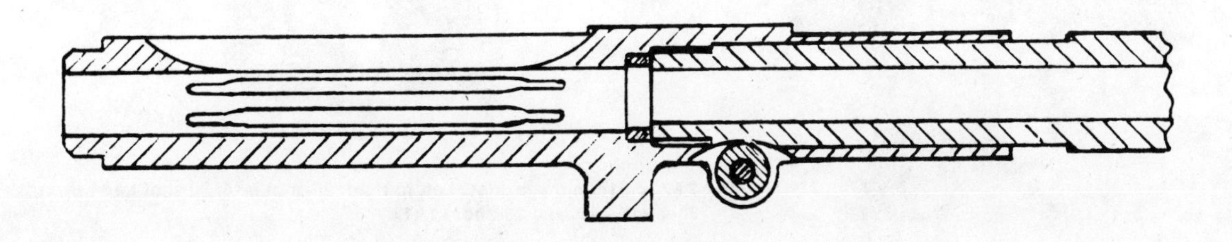

LEADER DYNAMICS 5.56 × 45MM RIFLES

These commercially developed rifles are discussed in Chapter 51.

Magazine Filler, 7.62mm F1.
The Australian Government Small Arms Factory developed a magazine filler to facilitate reloading L1A1, L2A1, L4A4 and M14 magazines. This device also permits cartridges to be stripped from the M13 machine gun links used with the M60 machine gun. The following photographs show this device and its uses. The F1 magazine filler weighs about the same as five 7.62 × 51mm NATO cartridges and is made of nylon coated sheet metal.

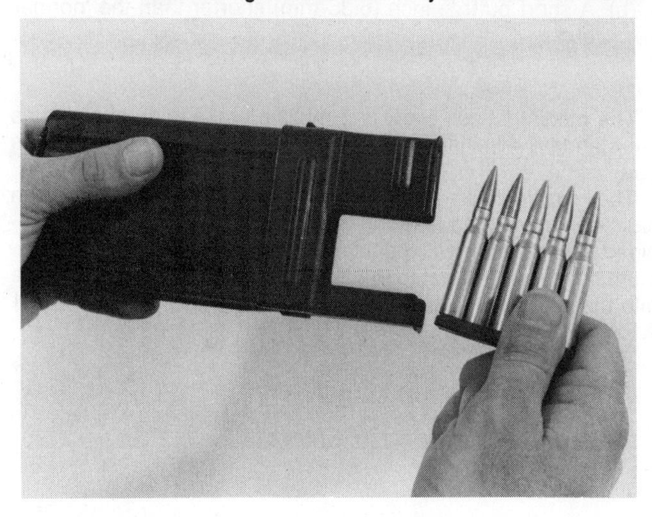

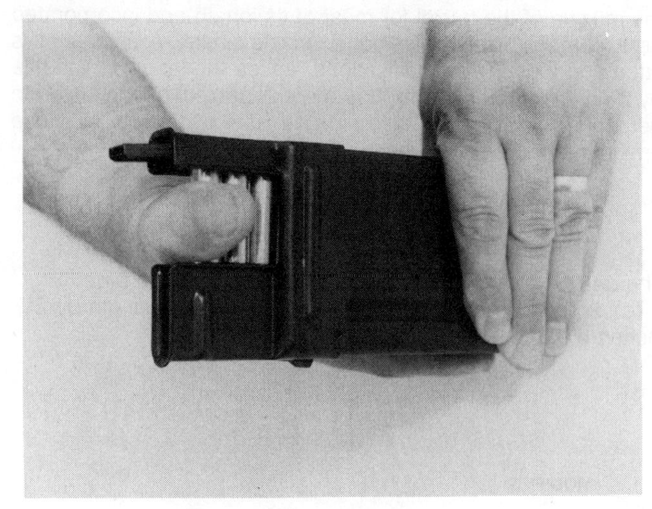

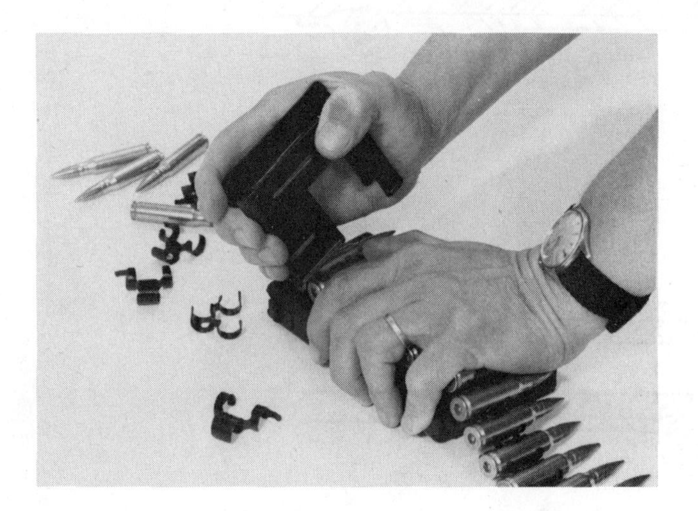

7.62 × 51mm magazines. Left to right: 20-shot M14, 30-shot L4A4 Bren, 30-shot L2A1, and 20-shot L1A1.

The Australian L1A1 and the Australian F1 BFA.

AUSTRALIAN MACHINE GUNS

The Lithgow plant was expanded and tooled to produce the caliber .303 Vickers Machine Gun Mark I between 1925 and 1930. Tooling for the caliber .303 Bren Gun was done in 1938–39. Production of machine guns at Lithgow during World War II amounted to over 12,000 Vickers and 17,000 Bren guns.

Australia adopted the US 7.62mm NATO M60 general purpose machine gun in the late fifties. The .50 M2HB Browning maching gun is also used by the Australian Armed forces.

Vickers .303 Machine Gun Mark I

8 Austria

SMALL ARMS IN SERVICE

Handguns: 9 × 19mm NATO Pistole Glock
9 × 19mm NATO Pistole 38 (Steyr-Daimler-Puch–made P38)
Submachine guns: 9 × 19mm NATO MPi 69 (Steyr-Daimler-Puch)
Rifles: 7.62 × 51mm NATO StG 58 (FN FAL made by S-D-P)
7.62 × 51mm NATO SSG (Sniper rifle made by S-D-P)
5.56 × 45mm StG 77 (Steyr AUG).
Machine guns: 7.62 × 51mm NATO MG42/59, MG3, MG74 (made by Rheinmetall and S-D-P)
12.7 × 99mm M2 HB Browning machine gun (USA)

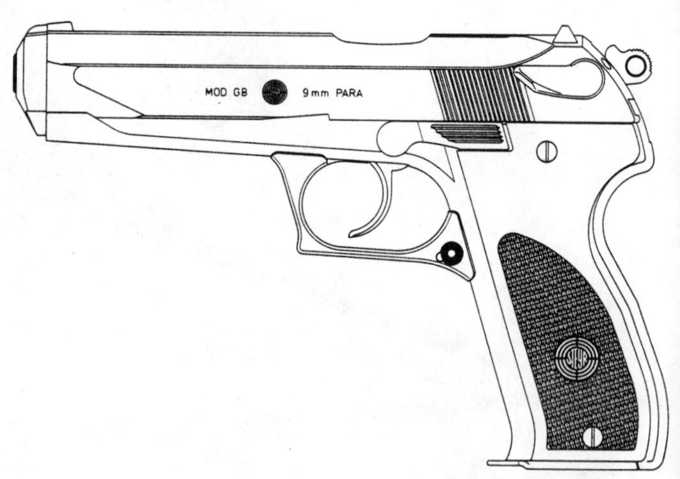

9 × 19mm Steyr Gas Brake (Gazbremse) pistol.

STEYR-DAIMLER-PUCH

In the post–World War II world, Austria has had to maintain essentially a neutral position because of the terms of the peace treaty imposed on that nation in 1945. Steyr-Daimler-Puch maintains a factory for the manufacture of small arms at the traditional weapons making center of Steyr. This firm is the post-war incarnation of the Österreichische Waffenfabrik (Steyr). The factory at Steyr has developed some very interesting weapons in the 1960s and 1970s. More recently, the firm has begun to take a more aggressive stance in the international small arms market. The Steyr 5.56 × 45mm *Armee Universalgewehr (AUG)* is especially worthy of note because of the new technological approach that it represents.

AUSTRIAN PISTOLS

STEYR-DAIMLER-PUCH 9 × 19mm GB PISTOLE

The GB pistol has been under development for a number of years, and it is unusual inasmuch as it uses a small portion of the propellant gases to retard the recoil of the slide until the gas pressure in the barrel has dropped to a safe level. This type of pistol-operating mechanism is categorized as a gas retarded blowback type with a fixed barrel that is screwed into the frame. This pistol also embodies a double-action lock mechanism and a hammer decocking lever that permits the hammer to be lowered safely while there is a loaded cartridge in the barrel.

CHARACTERISTICS OF THE S-D-P GB PISTOLE

Caliber: 9 × 19mm NATO
Overall length: 8.4 in.
Barrel length: 5.4 in.
Weight w/unloaded magazine: 2.2 lbs.
Feed device: 18-shot staggered row detachable box magazine
Muzzle velocity: (Approx.) 1125 f.p.s.
Sights: front, post with luminous dot; rear, rectangular notch with 2 luminous dots

Glock 9 × 19mm Pistol

Early in 1983, the Austrian Army announced the adoption of a new handgun. This weapon was developed by the Vienna firm Glock. Following the decision of the Austrian Army to select a new pistol, the Austrian authorities also decided to limit the competition to Austrian designs. The Glock pistol, which embodies a unique combination of plastics and stamped steel, won out over the Steyr-Daimler-Puch GB Pistole. No additional data was available at the time this edition went to press.

Glock of Vienna is the manufacturer of the new Austrian Army 9 × 19mm handgun.

OBSOLETE AUSTRIAN PISTOLS

AUSTRIAN ROTH STEYR 8mm PISTOL M1907*

This pistol was used by the Austrians in World War I. It was produced by the Österreichische Waffenfabrik (Steyr) and by the Fegyvergyr in Budapest. The design of the pistol is based on patents issued to George Roth, G. Krnka and K. Krnka. Limited quantities of these pistols were apparently used in World War II.

CHARACTERISTICS OF ROTH STEYR 8mm PISTOL M 1907

Caliber: 8mm Roth Steyr, 8mm Steyr M7
Overall length: 9.1 in.
Barrel length: 5.1 in.
Feed device: 10-round in line non-removable magazine.
Muzzle velocity: 1045 f.p.s.
Sights: Blade with notch.
System of operation: Recoil.

Special Feature of Roth Steyr

This weapon was originally designed for use by cavalry. The recoil of the weapon ejects the empty case and strips a new cartridge into the firing chamber as in other automatic pistols. However, it does not cock in the regular fashion. The striker is drawn back and released to fire the cartridge by pulling the trigger, exactly as in hammerless revolvers. This makes the weapon safe to handle but difficult to shoot accurately.

Both pistol and cartridge are generally considered obsolete.

9mm STEYR PISTOL M12

The Steyr Model 12 Pistol was the most widely used of the various pistols used by the Austro-Hungarian forces in World War I. It was also used by Romania and by Germany (in 9mm Parabellum) to a limited extent. There is considerable confusion over the correct nomenclature for this pistol; many call it the Model 1911 or M 11, others call it the Model 1912 or M 12. Both designations are correct—the commercial designation for the weapon is Model 1911; the official Austrian Army nomenclature for the pistol was Selbstlade Pistol M 12. The pistol is also called the Steyr Hahn. During World War II, the Germans rebarreled a number of these weapons for the 9mm Parabellum Cartridge. These weapons can be identified by the "08" stamped on the slide.

Although there were about 250,000 of these pistols made, they are no longer used as service pistols anywhere in the world and have not been made since 1919.

Austrian Roth-Steyr 8mm Pistol M1907.

Austrian 9mm Steyr Pistol M12.

CHARACTERISTICS OF 9mm STEYR PISTOL M12

Caliber: 9mm Steyr.
Overall length: 8.5 in.
Barrel length: 5.2 in.
Weight: 2.12 lbs.
Feed device: 8-round. Located in handle; cartridges must be stripped into it from the top of the pistol.
Capacity: 8 cartridges.
Muzzle Velocity: 1112 f.p.s.
Sights: Blade with notch.
Locked: By cam ribs on barrel, which lock in cam slots on inside of top of slide. As bullet passes down barrel, barrel tends to twist to the right. As barrel and slide move to the rear under recoil, cam rib twists barrel to the left and opens lock permitting slide to continue backward and function the action.
Type of fire: Single-shot only.
Magazine loading arrangement: Clip guide on top of slide permits insertion of loaded clip when action is opened.
Position of slide when last shot is fired: Open.
Safeties: (a) A thumb safety somewhat like that on the Colt .45 Automatic will be found on the left side of the pistol just below the hammer. Turning this up into its notch in the slide makes the pistol safe. (b) An automatic disconnector on the right side of the pistol under the slide prevents this pistol from being fired until the action is wholly closed.

*For additional details see *Handguns of the World.*

Austrian 9mm Steyr Pistol M12—field-stripped.

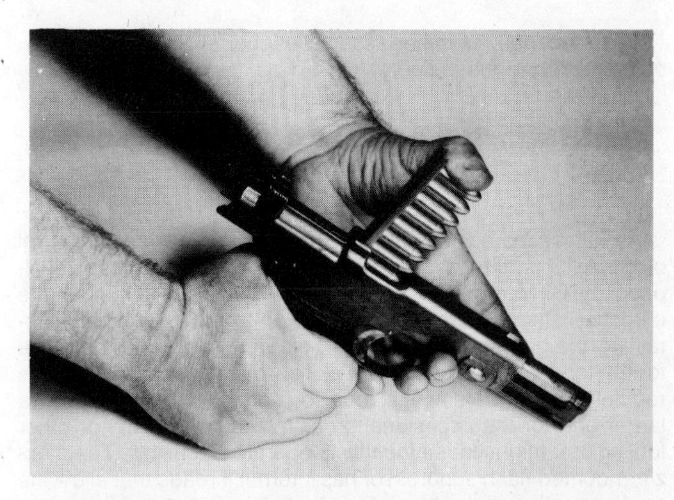
Stripping in a clip of cartridges—loading the Steyr M12 9mm pistol.

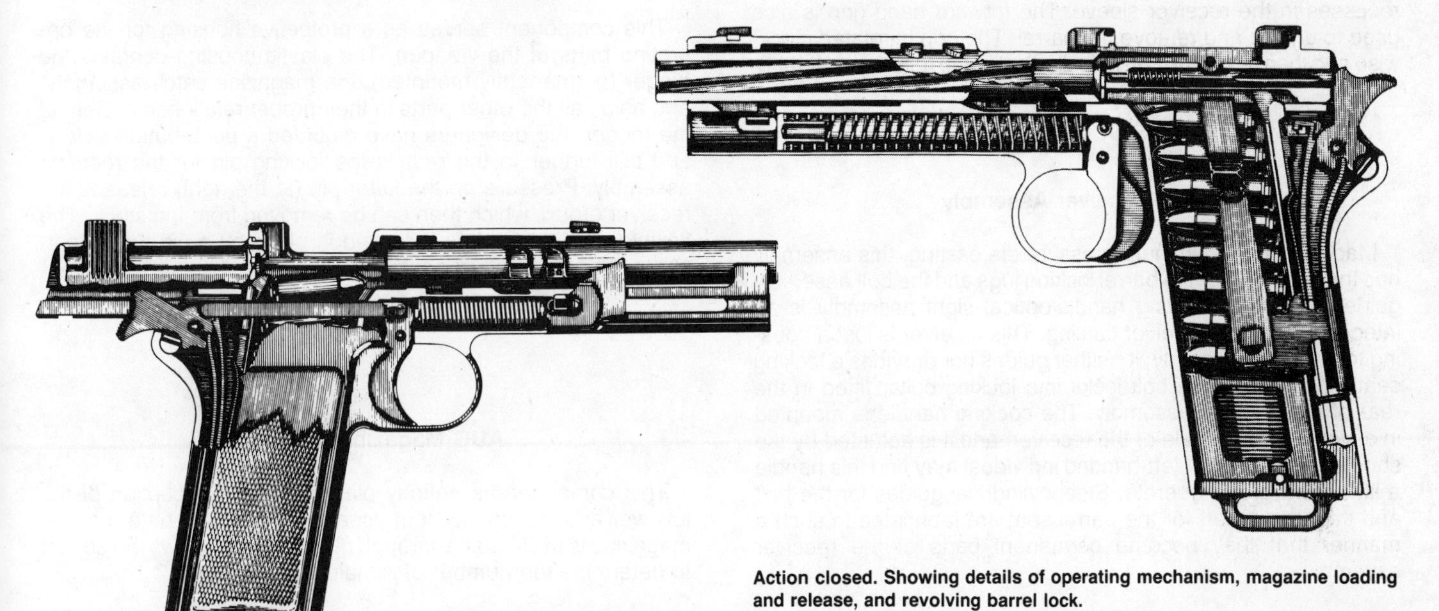
Action closed. Showing details of operating mechanism, magazine loading and release, and revolving barrel lock.

Action open. Showing detail of revolving barrel lock, function of recoil spring, and operation of trigger mechanism.

AUSTRIAN RIFLES

The Austrian armed forces adopted the FAL in 1958 as the *Sturmgewehr 58* (StG 58). An initial 20,000 StG 58s were purchased from Fabrique National Herstal. Subsequent production was carried out at the Steyr-Daimler-Puch arms factory in Steyr under a license granted by FN. All StG 58s used by the Austrians (both FN- and S-D-P–made) had a unique combination flash suppressor–grenade launcher–barbed wire cutter. This muzzle device had three slots that permitted the rifleman to slide the muzzle device over barbed wire and cut it by discharging a bullet. Once the wire was forced into two of the slots, the shooter twisted his weapon, thus forcing the wire to line up across the bore of the rifle. When so aligned the bullet would sever the wire. Major Stoll, the head of the Austrian StG 58 purchasing commission, designed this wire cutter. The StG 58 can also be identified by its sheet metal handguards, though this was not unique to Austrian FALs.

STURMGEWEHR 77/ARMEE UNIVERSAL GEWEHR

Called the AUG for short, this novel weapon represents some new thinking in small arms design. Designers at Steyr created a bullpup mechanism that could use any one of four barrels (12, 16, 20, 24 inch) so that the weapon can be used as a submachine gun, carbine, rifle, and squad automatic weapon. Provision was also made for shooting the AUG from either the right or left shoulder. Left-handed shooters need only change the ejector and move the detachable ejection port cover.

The AUG has six major assemblies: barrel, receiver, trigger, bolt, stock, and magazine.

AUG Barrel Assembly

As with an increasing number of military small arms, the barrel of the AUG is rifled by the cold hammer forging process developed by GFM of Steyr. Both the barrel and the chamber are chrome plated to increase barrel life. An external assembly containing the gas port, gas cylinder, gas regulator, and the hinge for the forward hand grip is shrunk fit around the barrel. The gas regulator has two settings: one for normal conditions, and a larger one for the occasional problems caused by dirt or other fouling that might cause the rifle to operate sluggishly. The muzzle-mounted flash suppressor has internal threads that allow the easy mounting of the blank firing adapter. The barrel locks into the steel sleeve that is part of the receiver assembly. All barrels, no matter their length, have eight lugs that correspond with recesses in the receiver sleeve. The forward hand grip is also used to unlock and remove the barrel. The grip is rotated clockwise and then the barrel can be pulled out of the receiver assembly.

AUG Receiver Assembly

Made from an aluminum pressure die casting, this assembly has the bearings for the barrel locking lugs and the bolt assembly guide rods. The carrying handle/optical sight assembly is an integral part of the receiver casting. This receiver is just a housing for the bolt assembly; it neither guides nor provides a locking seat for the bolt. The bolt locks into locking collar fitted in the rear of the receiver assembly. The cocking handle is mounted in a slot on the left side of the receiver, and it is actuated by the shooter's left hand. A left-handed individual may find this handle a little awkward to operate. Steel cylindrical guides for the bolt and the steel insert for the barrel seat are fabricated in such a manner that they become permanent parts of the receiver assembly.

AUG Trigger Assembly

This group is unique because all of the parts, with the exception of the springs and pins, are made out of plastic; this includes the trigger group housing. This assembly is inserted into the weapon through an opening in the butt of the weapon. Two steel rods transmit the pressure on the trigger to the sear. A plastic hammer strikes the firing pin. The trigger mechanism is two stage; that is, there are no separate settings for full automatic and single shot fire. Individual shots are fired by pulling the trigger a short distance to the first sear stop. Full automatic fire results from pulling the trigger completely to the rear.

AUG Bolt Assembly

This group is rather unusual for a small caliber rifle. It is similar in concept to the bolt system used in the US XM235/XM248 squad automatic weapon. The 7-lug bolt head is carried in a carrier, and the bolt head is rotated by a cam path in the carrier. Because the receiver is an aluminum casting, a steel extension piece is mounted above the breech end of the barrel to operate the cam in the bolt carrier slot. The bolt carrier has two guide rods brazed to it, and these rods travel in the cylindrical tubes pressed into the receiver assembly. These rods contain the return springs for the bolt assembly; these springs are not removed for routine maintenance of the AUG. The left guide rod takes the action of the cocking handle, and these two components can be locked together to permit a forward assist for the bolt if needed. The cocking handle does not reciprocate when the weapon is being fired. The right guide rod acts as the gas piston for the operation of the AUG.

AUG Stock Assembly

This component serves as a protective housing for the operating parts of the weapon. The plastic housing contains the trigger (permanently mounted), the magazine catch assembly, and holds all the other parts in their proper relationship. Behind the trigger, the designers have mounted a push-button safety, and still further to the rear is the locking pin for the receiver assembly. Pressure on the latter pin (to the right) releases the receiver group, which then can be removed from the stock. The flexible plastic buttplate is locked in place by a pin that is part of the rear sling swivel assembly.

AUG Magazine Assembly

This component is entirely plastic with the exception of the follower and the spring. It is especially worthy of note that the magazine is of clear see-through plastic, which allows the soldier to determine the number of remaining cartridges.

AUG Field Stripping

Start by determining that the rifle is unloaded, then place the safety to the "safe" position. After removing the magazine, pull the cocking handle to the rear, push the barrel locking button (release), and rotate the barrel clockwise by grasping the forward grip and turning it. Once the barrel has been removed, the receiver release button can be depressed. This action allows the receiver assembly to be removed from the stock. After the receiver assembly is free of the stock, the bolt, bolt carrier, and associated parts can be removed from the receiver.

The trigger assembly can be removed from the stock by taking off the buttplate. This step is accomplished by pushing with the thumb on the dimple in the buttplate, and then pulling out on the sling swivel. After the buttplate has been lifted off, the trigger assembly can be withdrawn from the stock. The AUG is reassembled in reverse order of these steps.

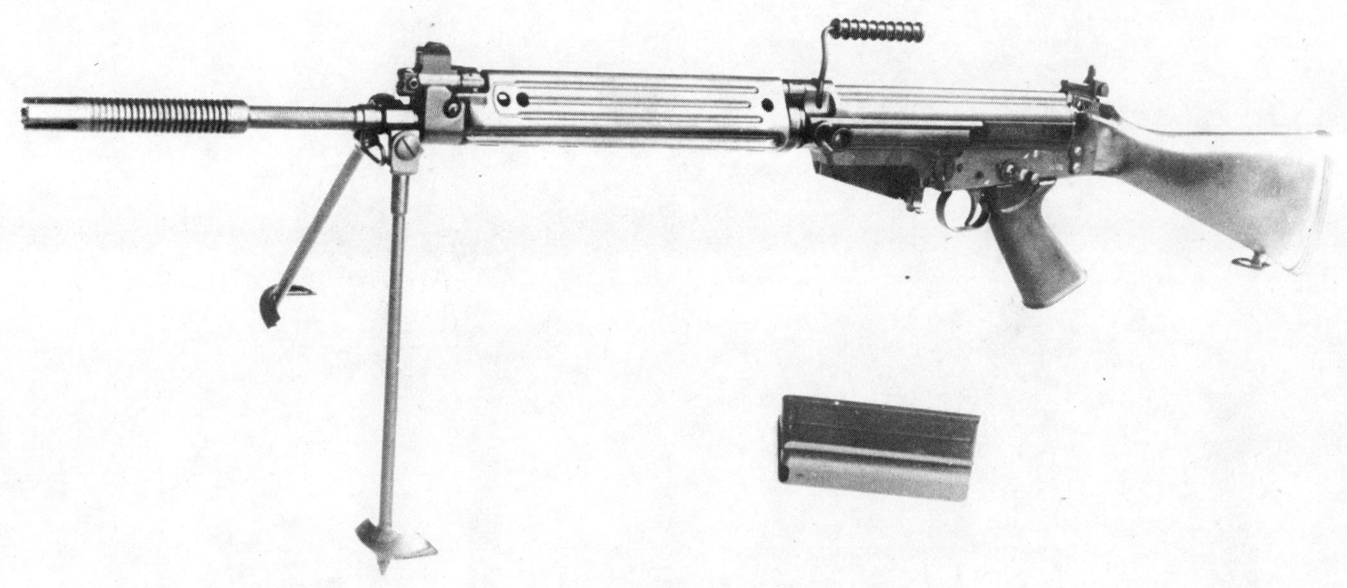

The Austrian StG 58.

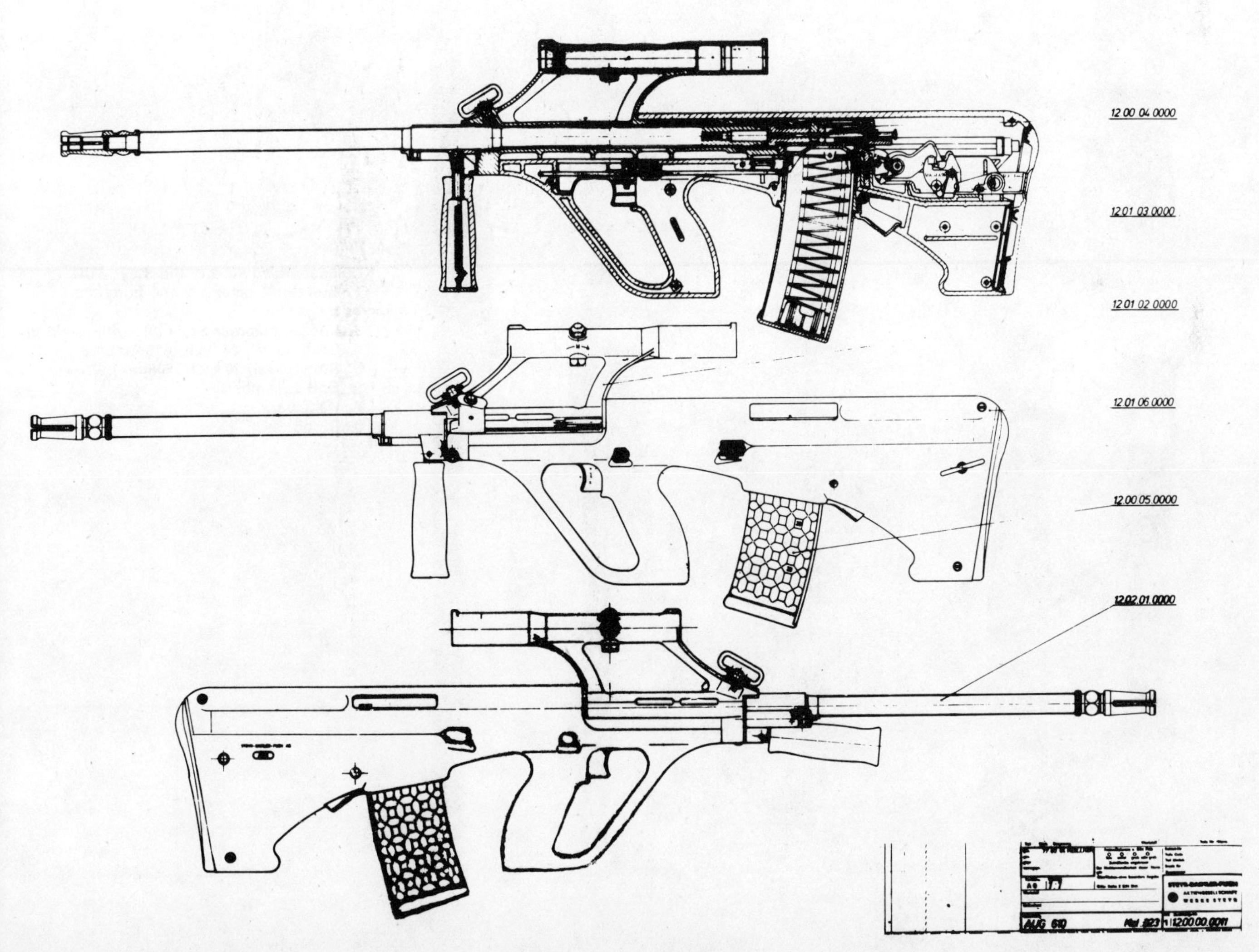

Section view of the Steyr AUG.

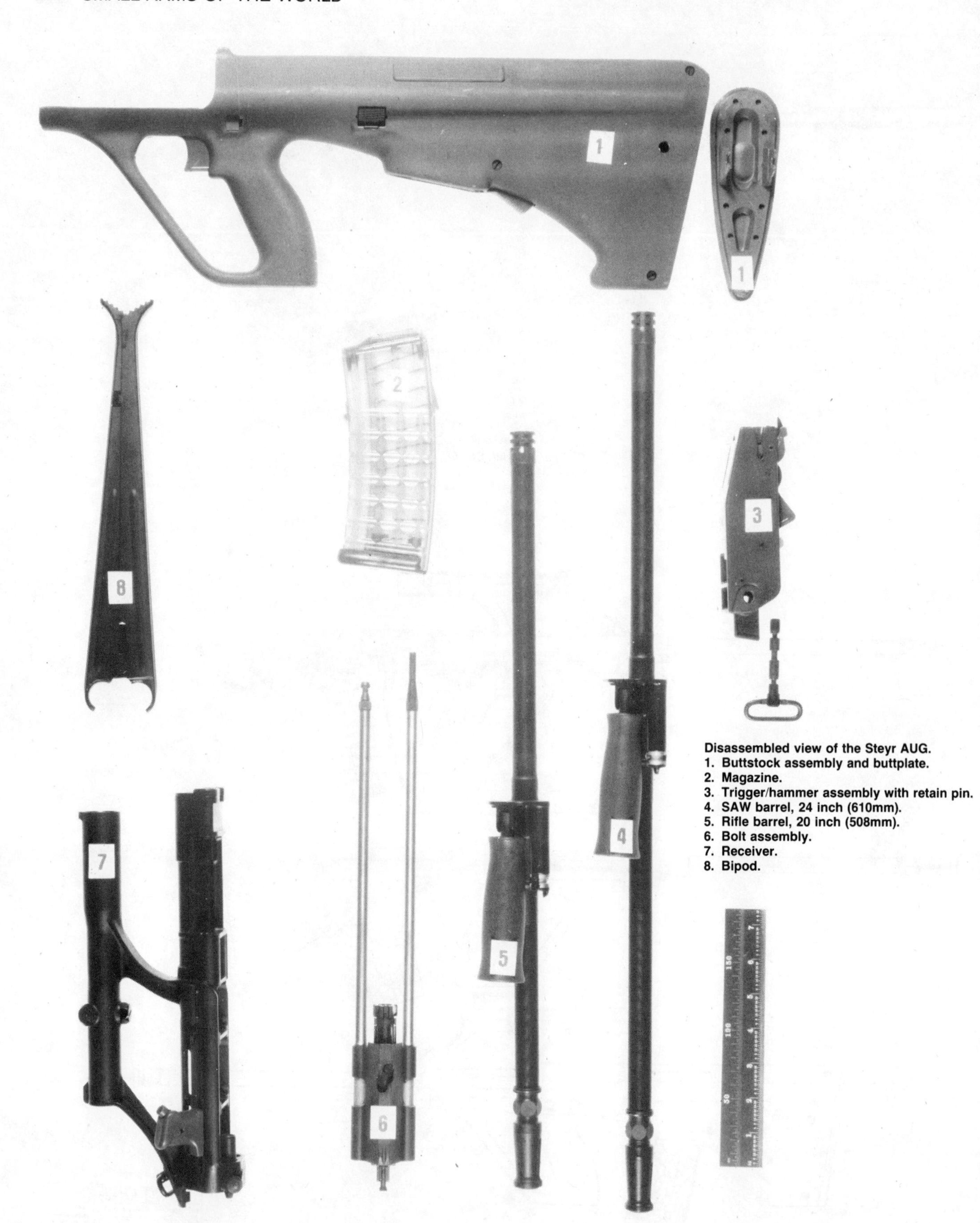

Disassembled view of the Steyr AUG.
1. Buttstock assembly and buttplate.
2. Magazine.
3. Trigger/hammer assembly with retain pin.
4. SAW barrel, 24 inch (610mm).
5. Rifle barrel, 20 inch (508mm).
6. Bolt assembly.
7. Receiver.
8. Bipod.

Right and left side views of the Steyr AUG.

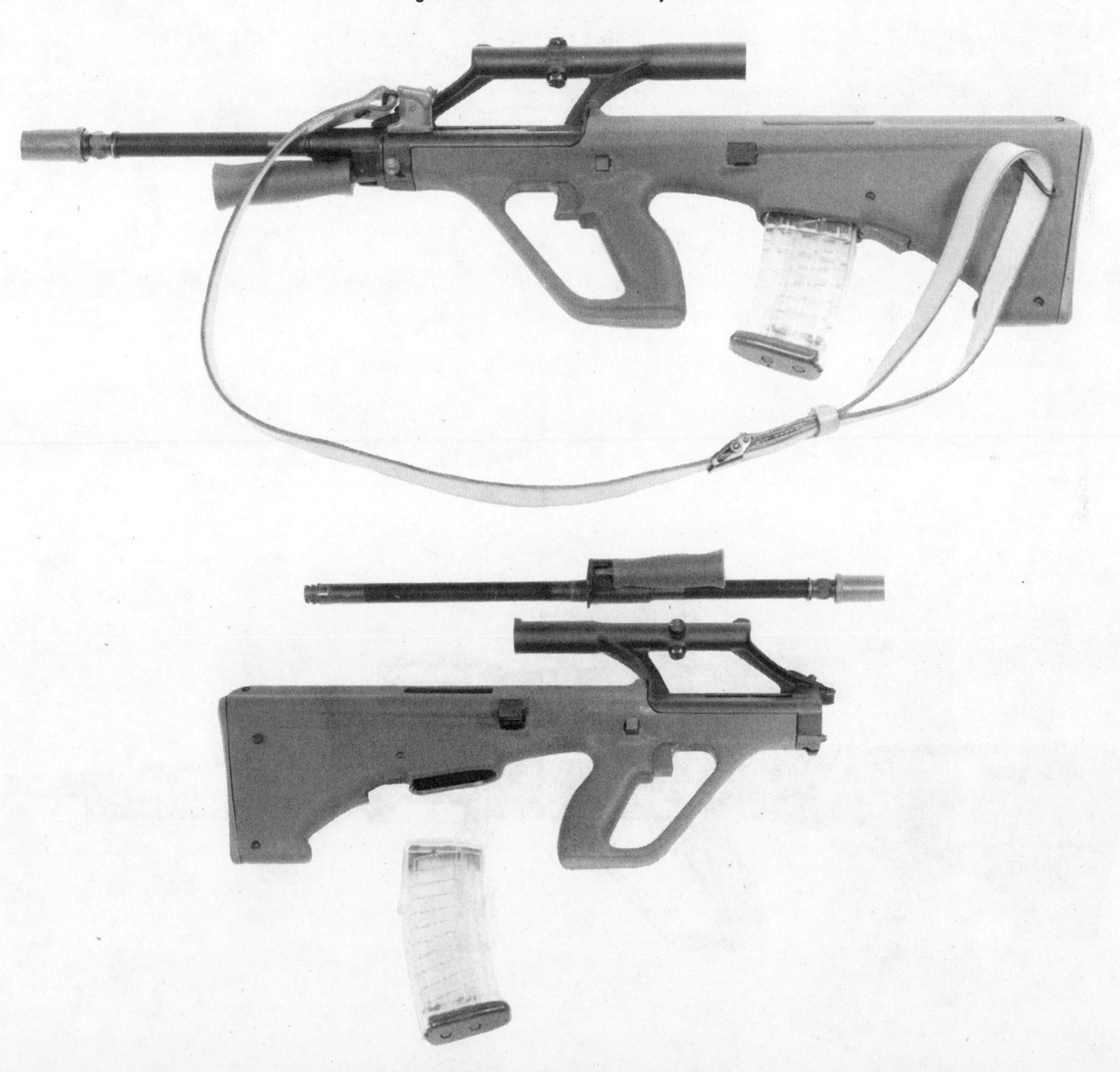

AUG with barrel and magazine removed.

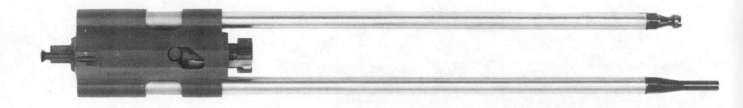

Assembled and disassembled views of the bolt assembly.

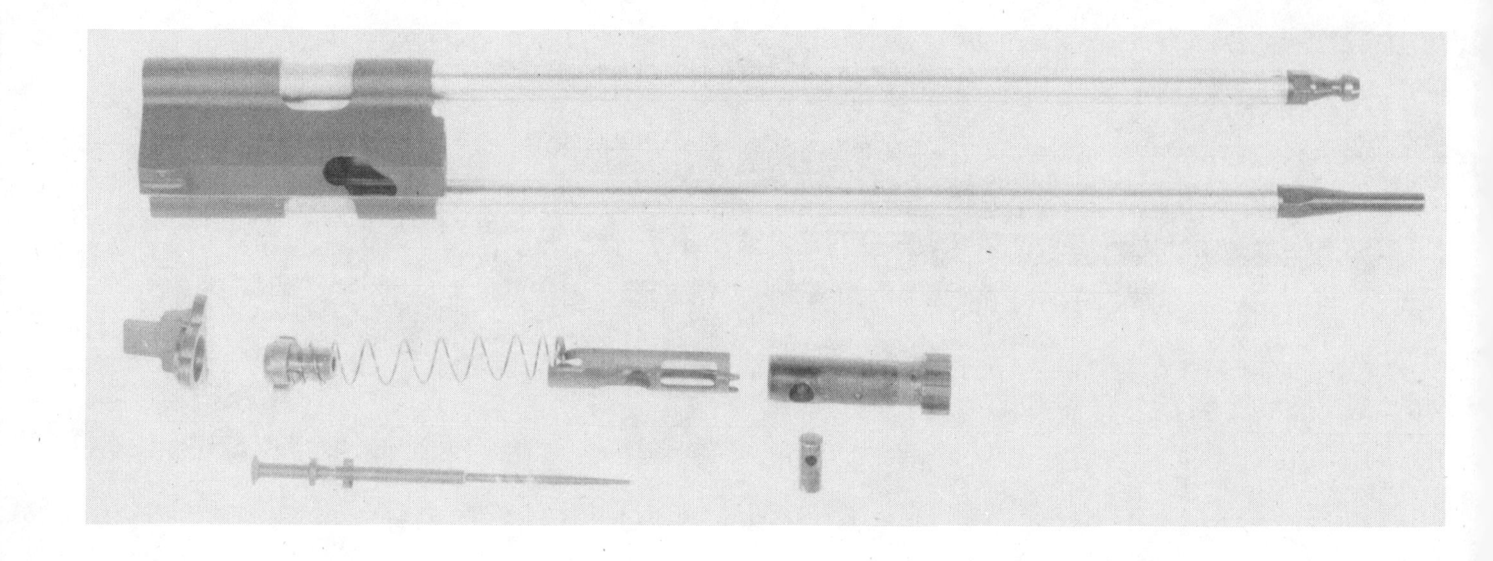

Steyr 5.56 × 45mm HBAR-T, a new light support version of the AUG with 621mm (24-inch) barrel and removable telescopic sight.

Austrian soldier firing the 5.56mm AUG.

STEYR 7.62mm NATO MODEL SSG RIFLE

Steyr Daimler Puch has developed a rifle based on their post war bolt action system. This rifle has been adopted by the Austrian Army as a sniper rifle. It has a heavy, cold hammered barrel. The bolt has six rear-mounted locking lugs. Steyr claims ten round 2.75-inch groups with this rifle at 300 meters.

May also be fitted with Walther micrometer rear and hooded front sight with changeable inserts.

CHARACTERISTICS OF THE STEYR MODEL SSG RIFLE

Caliber: 7.62mm NATO
System of operation: Bolt action
Weight (w/o telescope): 10.25 lbs.
Length overall: 45 in.
Barrel length: 25.6 in.
Feed mechanism: 5-round, removable rotary; 10-round box
Sights: Front: (w/telescope) hooded post
Rear: (w/telescope) notch
Muzzle Velocity: 2800 f.p.s.

OBSOLETE AUSTRIAN MANNLICHER SERVICE RIFLES

None of the Austrian Mannlicher service rifles are used in significant quantity in active military service anywhere today. They were used extensively in World Wars I and II, however, and still exist in large numbers in the hands of collectors. A short description of each model follows.

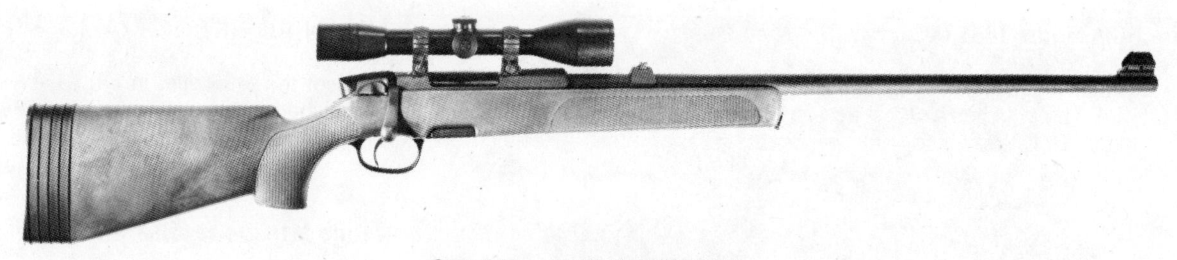

Steyr 7.62mm NATO Model SSG Rifle.

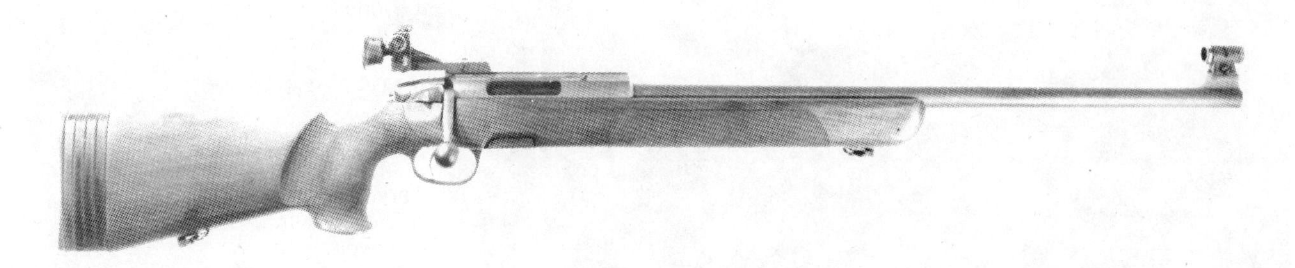

Steyr 7.62mm NATO Model SSG Match rifle.

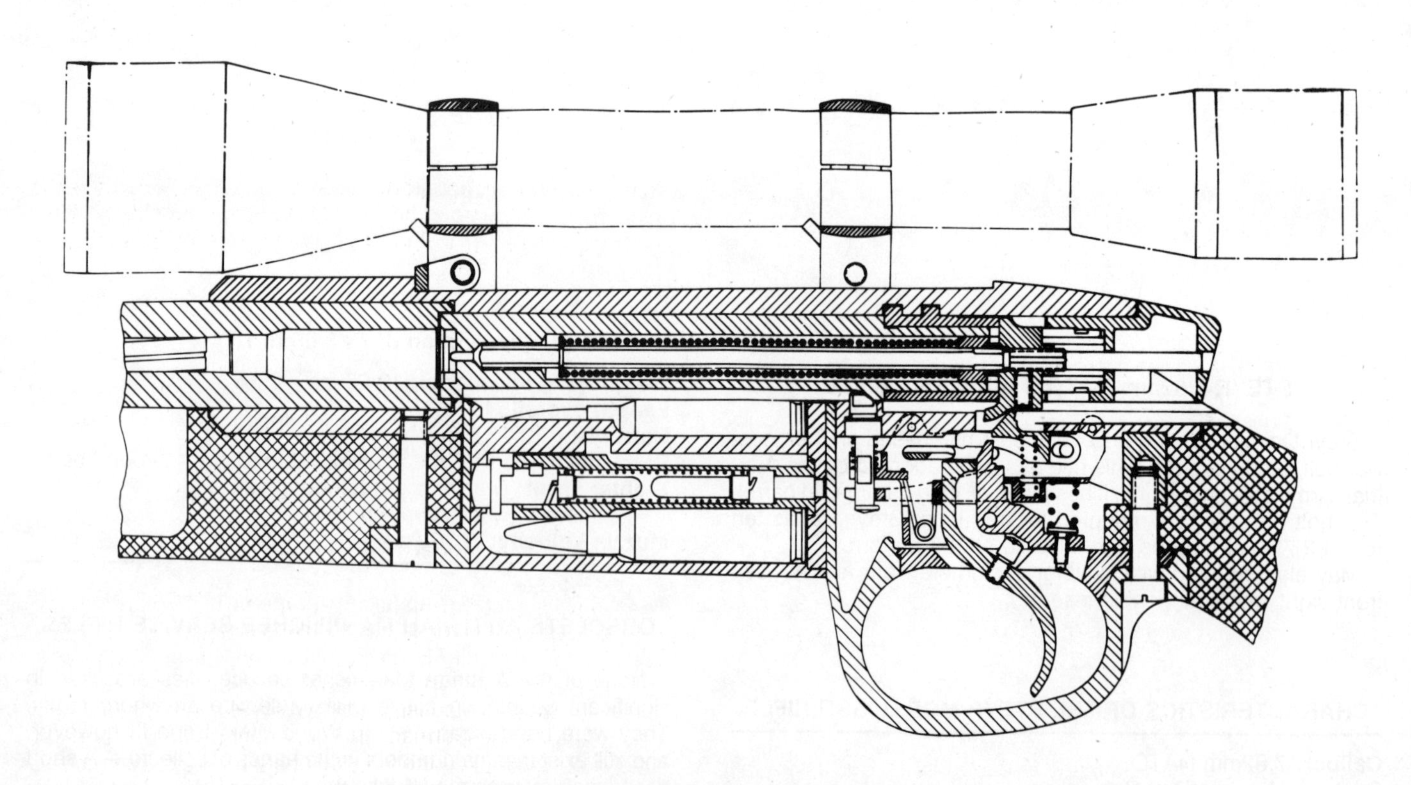

Section view of the Steyr *Scharfschützengewehr*.

11mm Rifle Model 1885 (*Repetier Gewehr M85*)

This was the first magazine rifle to be used by the Austro-Hungarian Empire. The M85 has a straight-pull non-rotating bolt, which is locked by a block pivoted from the underside of the rear section of the bolt. This block abuts against a shoulder in the receiver when in the locked position. The Model 85 introduced the Mannlicher magazine system in which the clip is inserted into the magazine of the rifle and functions as a part of the magazine—as does the clip of the US M1 rifle. The Model 85 was chambered for the 11mm M77 cartridge and was not made in significant quantity. Its design led directly to the next of the Mannlichers.

11mm Rifle Model 1886 (*Repetier Gewehr M86*)

The M86 is similar in most respects to the M85. It is the first of the Austrian service rifles to introduce the feature of the clip dropping out of the bottom of the magazine when the last round is chambered. The 11mm cartridge was improved with the introduction of this rifle, and as a result it had better ballistics than the M85. The sights of the M86, as the M85 and all other Austrian weapons until after World War I, are graduated in "Paces" (one pace equals 29.53 inches), a term similar to the "Arshin" formerly used as a Russian standard of measurement. Approximately 90,000 M86 rifles were made by Steyr.

8mm Rifle Model 1888 (*Repetier Gewehr M88*)

The M88 is chambered for the black-powder M88 cartridge, and its rear sight is graduated for that cartridge; with these exceptions it is the same as the M86.

8mm Rifle Model 1888–90 (*Repetier Gewehr M88-90*)

In 1890 the Austrian 8 × 50mm cartridge with smokeless powder charge was introduced. The sights of the M88 rifle were modified for the new and more powerful cartridge by the addition of new graduation scales, which were engraved on plates and attached to the sides of the sights. Rifles thus modified are called M88-90.

8mm Carbine Model 1890 (*Repetier Carabiner M90*)

This weapon introduced the straight-pull bolt with rotating bolt to the Austrian Service. Although Mannlicher had introduced a rotating straight-pull bolt in 1884, it was not very successful and was never made in quantity. The bolt is of two-piece design. The bolt handle and bolt body are one piece; mounted within the bolt body is the bolt shaft or bolt cylinder. The locking lugs are mounted on the head of the bolt cylinder, and the bolt cylinder rotates within the bolt body during the locking and unlocking process. This bolt system is used with all the later Austrian straight-pull bolt-action Mannlichers and, since it provides for frontal locking, is considered by many to be a stronger system than that of the Models 84, 86 and 88. The magazine system adopted with the M86 is used in the M90 carbine and the later rifles. The M90 carbine has no handguard, and the sight swivels are mounted on the left side of the stock; it is not fitted for a bayonet, and the cocking piece is round. All later models have a thumb-shaped cocking piece. The M90 carbine is a relatively rare piece these days.

8mm Rifle Model 1895 (*Repetier Gewehr M95*)

This weapon, which was made at Budapest as well as Steyr, was the principal Austro-Hungarian rifle of World War I. The M95 rifle was made in tremendous quantities.

8mm Carbine Model 1895 (*Repetier Carabiner M95*)

The carbine version of the M95 rifle, in addition to its short length, can be distinguished by the following: (1) sling swivels on side of stock only; (2) no provision for bayonet lug; and (3) no stacking hook.

8mm Short Rifle M1895 (*Repetier Stutzen M95*)

The M95 "Stutzen" is frequently confused with the M95 carbine. It apparently was designed for use by special troops, i.e. Engineer, Signal, etc., and not for Cavalry, since it is fitted with a bayonet stud and has sling swivels fitted to the underside as well as the side. This weapon also has a stacking hook, which screws into the upper band. When the rifle is fired with bayonet fixed, a blade on top of the bayonet barrel ring is used as the front sight to compensate for changes in center of impact due to the weight of the bayonet on the barrel.

MODIFIED AUSTRIAN SERVICE RIFLES

The Austro-Hungarian Empire, as a loser in World War I, had to provide large amounts of war material to the Allies. Among those countries that benefited from the Austrian war booty were: Italy, Yugoslavia and Greece. Italy used large quantities of the M88-90 and M95 series weapons in World War II without modification and made large quantities of the 8 × 50mm cartridges for those rifles.

Yugoslav 7.92mm M95

Yugoslavia converted many of the M95 weapons to 7.92mm. These weapons can be distinguished by the addition of the stamped letter "M" after "M95," which is on the top of the receiver. These rifles have a clip permanently fixed in their magazine and therefore can be loaded with the standard Mauser five-round charger.

7.92mm M95/24

These rifles were apparently used by Bulgaria. They have the markings "/24" stamped on the receiver after the "M95." They are generally similar to the Yugoslav M95M in their magazine arrangements.

Conversion of M88-90 to 7.92mm

This rifle is believed to have been used by Greece. The barrel has been shortened and a wooden handguard added (the Austrians frequently used a laced canvas handguard on the M88 and M88-90).

7mm Model 1914 and the Model 29

Austria-Hungary was apparently prepared to drop the straight-pull Mannlicher in favor of the 98 Mauser design in 1914. The 7mm Model 1914 rifle is a rare specimen, having the pointed type pistol grip and bands as found on the earlier Mannlicher. The Model 29 was Mauser made by Steyr for export; a quantity were made for the German Air Force during World War II and are marked G29/40 on the left side of the receiver. The receiver ring is marked 660—which was the numerical code for the Steyr plant.

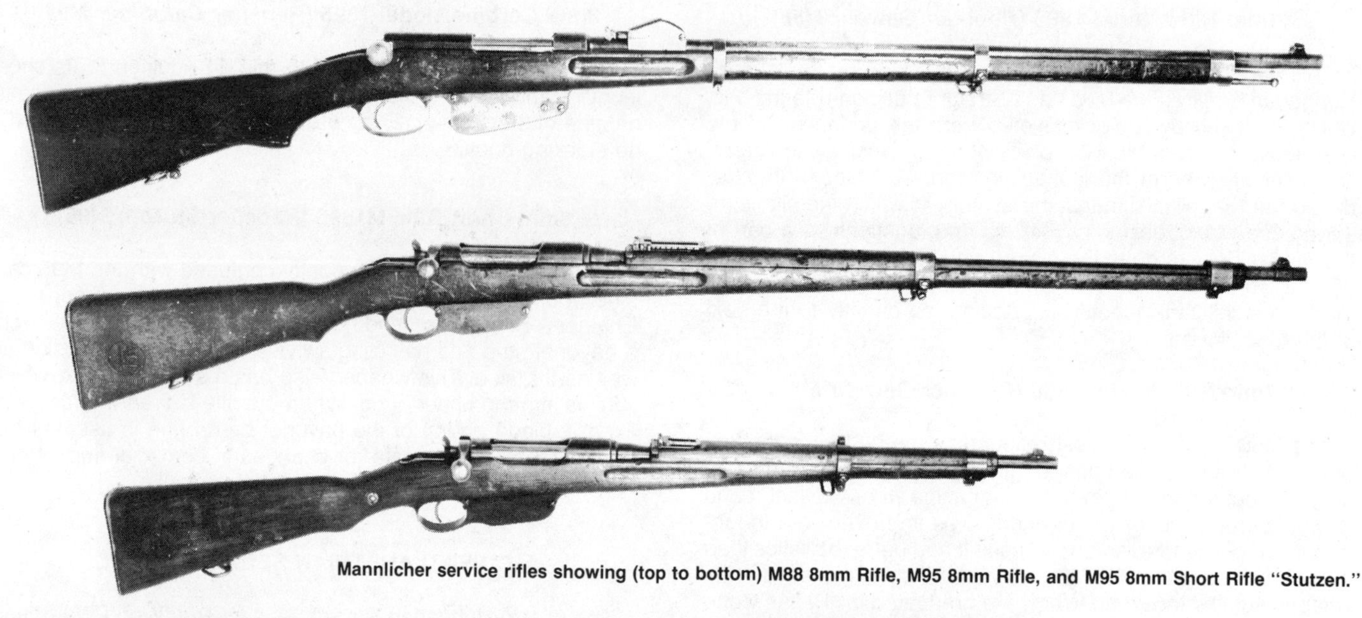

Mannlicher service rifles showing (top to bottom) M88 8mm Rifle, M95 8mm Rifle, and M95 8mm Short Rifle "Stutzen."

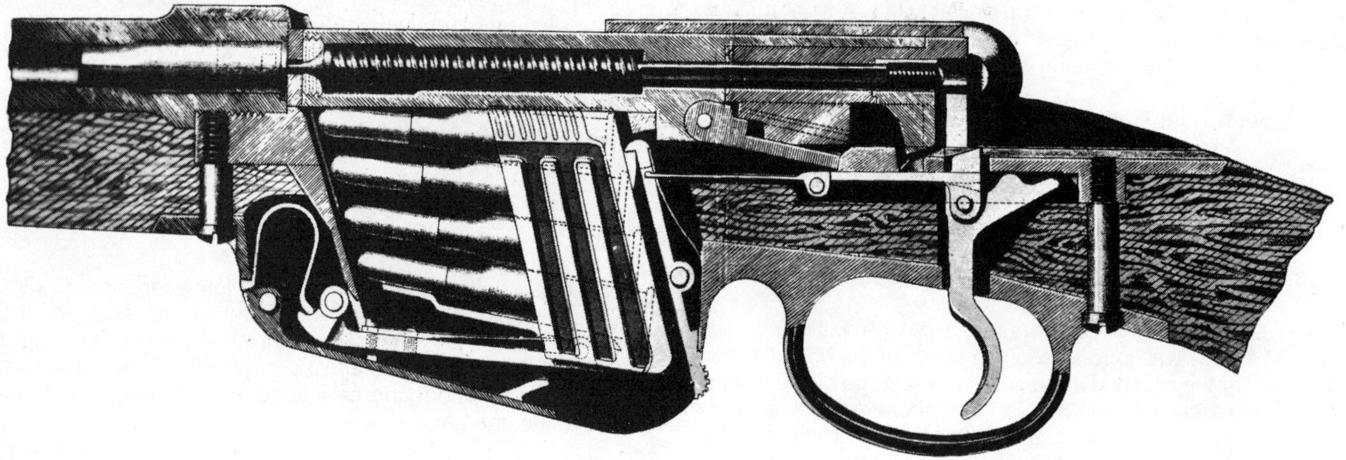

Austrian M1888 and 1888–90. Note engagement of clip catch above clip projection. Locking block is wedged down into locking position in the receiver.

Bolt of the 8mm M90 Carbine.

8mm M95 Stutzen Converted to 8 × 56mm

There are two versions of this weapon extant. The Austrian version appeared in 1930 with the adoption of the Model 30 (8 × 56mm) cartridge. The Model 30 was a large-rimmed cartridge with a pointed bullet—*Spitzgeschoss*—and therefore, a letter "S", twelve millimeters high, was stamped on the receiver to distinguish it from the unconverted weapons. These weapons were used considerably by the German police in World War II, and steel-cased Austrian-made ammunition, bearing the date 1938 plus the German Eagle and Swastika marking, has been found in quantity. In 1931, the 8 × 56mm cartridge was adopted by Hungary, who called it the 31M (Model 31). The Hungarians had large quantities of M95 "Stutzen" on hand and converted many of these to the 31M cartridge; these weapons can be distinguished by the letter "H" stamped on the receiver. It should be noted that M95 rifles rebarreled for the 8 × 56mm cartridge—the M30 or M31—cannot be used with the old conical-nosed 8 × 56mm cartridge. Although both are rimmed, the 8 × 56mm is considerably longer and more powerful—these cartridges are definitely NOT interchangeable. Both the Austrian and Hungarian conversions require special clips.

AMMUNITION FOR AUSTRIAN MANNLICHER SERVICE RIFLES

The 11mm M77 Austrian cartridge is a typical black-powder cartridge. Although it was once made in the United States, it is now basically a collector's item, as is the 8mm M88 black-powder loaded cartridge. The 8mm M90 and its slightly improved version, the M93, may be encountered in quantity on occasion, but it should be born in mind that all military loads are at least 20 years old and all the above rifles chambered for this cartridge require a special clip to be used as magazine loaders. The situation is the same for the 8 × 56mm (M30 or M31) cartridge. This cartridge has not, to the writer's knowledge, been manufactured since before WWII.

The conversions that are chambered for the 7.92mm cartridge present a far simpler problem. They do not require special clips,

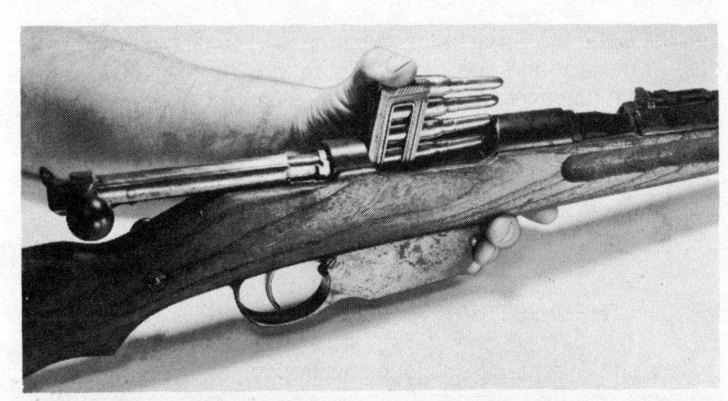

Note that both clip and cartridges are inserted into the magazine when loading the Steyr-Mannlicher Model 95 series 8mm rifles.

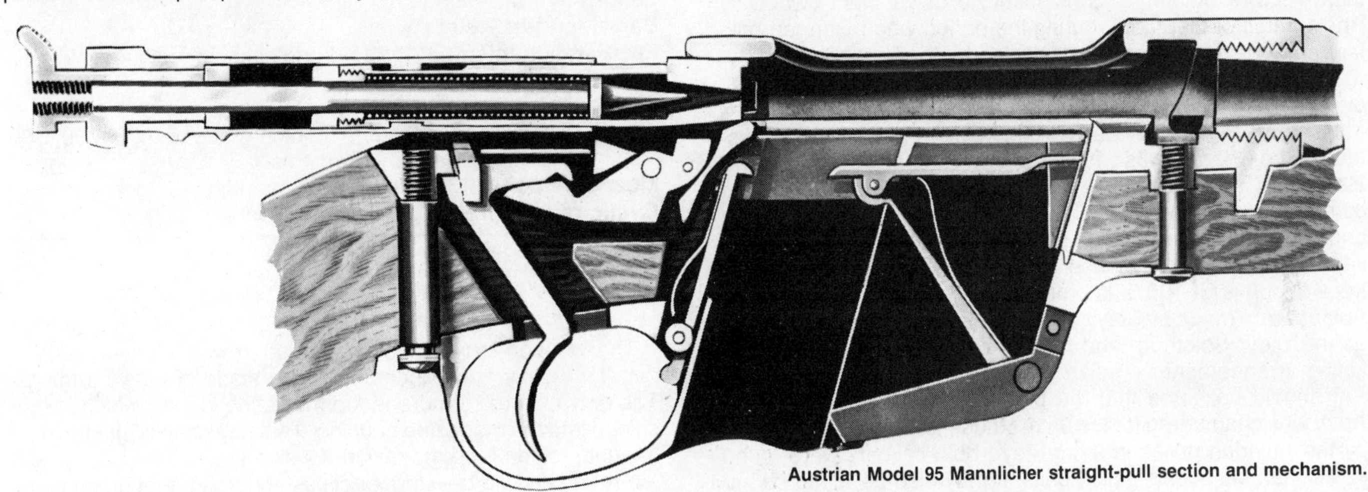

Austrian Model 95 Mannlicher straight-pull section and mechanism.

CHARACTERISTICS OF AUSTRIAN STRAIGHT-PULL MANNLICHER SERVICE RIFLES

Weapon	Method of locking	Overall length	Barrel length	Feed device	Sights	Muzzle velocity	Weight
Rifle: 11 mm M86	Wedge	52 in.	31.75 in.	5-round, single column, fixed box magazine	Front: Barley corn Rear: V notched tangent with long range side sight	1610 f.p.s.	10 lbs.
Rifle: 8mm M88	Wedge	50.38 in.	30.14 in.	5-round, single column, fixed box magazine	Front: Barleycorn Rear: V notched tangent with long range side sight	1750 f.p.s.	9.7 lbs.
Rifle: 8mm M88/90	Wedge	50.38 in	30.14 in.	5-round, single column, fixed box magazine	Front: Barleycorn Rear: V notched tangent with long ranged side sight	2115 f.p.s.	9.7 lbs.
Carbine: 8mm M90	Frontal locking lugs	39.5 in.	19.5 in.	5-round, single column, fixed box magazine	Front: Barleycorn Rear: V notched tangent	1900 f.p.s.	6.9 lbs. (approx.)
Rifle: 8mm M95	Frontal locking lugs	50 in.	30.12 in.	5-round, single column, fixed box magazine	Front: Barleycorn Rear: Leaf	2030 f.p.s.	8.31 lbs.
Short Rifle: 8mm M95	Frontal locking lugs	39.5 in.	19.65 in.	5-round, single column, fixed box magazine	Front: Barleycorn Rear: Leaf	1900 f.p.s.	7.5 lbs.
Carbine: 8mm M95	Frontal locking lugs	39.5 in.	19.65 in.	5-round, single column, fixed box magazine	Front: Barleycorn Rear: Leaf	1900 f.p.s.	7 lbs. (approx.)

and the 7.92mm (8mm Mauser) cartridge is available in quantity both in military and sporting configurations. A word of warning is in order, however; 7.92mm cartridges can be found with pressures up to 55,000 p.s.i. The Model 88-90 converted to 7.92mm was built to take a maximum pressure of about 40,000 p.s.i.; therefore, shooting this weapon with some of the military and commercial cartridges currently available could be EXTREMELY HAZARDOUS! Shooting any of these weapons unless previously checked by a reliable gunsmith can be hazardous, especially if the weapon shows any signs of hard usage.

AUSTRIAN SUBMACHINE GUNS

THE AUSTRIAN MP34 SUBMACHINE GUN

Although Austria did little submachine gun development, they did have one that attained fairly wide usage during World War II. The MP34 is commonly known as the Steyr Solothurn and is a product of German design worked out at Waffenfabrik Solothurn A.G. of Solothurn, Switzerland, a Swiss plant owned by Rheinmetall of Germany, during the period when German military arms development was restricted by the Versailles Treaty.

The MP34 was taken over by the Germans when they took over Austria in 1938 and was called by the Germans MP34 (Ö)—Maschinen Pistole 34 Österreich-, (Österreich meaning Austrian). The weapon was widely used by German police and rear area units. The weapon in various modifications was offered commercially and was also used by Chile, El Salvador, Bolivia and Uruguay. It was used, in extremely limited quantities, by the Japanese in 7.63mm Mauser. The commercial designation for the weapon is SI-100. It is probable that all the MP34s used by Austria were made by Steyr, as from 1930 on the gun was known as the Steyr Solothurn and the two concerns had a joint marketing arrangement.

It should be noted that the MP34 as used by the Austrian Army was chambered for the 9mm Mauser cartridge. It was used by the Austrian police in 9mm Steyr; both calibers were found in German service or police units. It may also be found chambered for the 9mm Parabellum cartridge.

CHARACTERISTICS OF AUSTRIAN MP34 SUBMACHINE GUN

Caliber: 9mm Mauser (Army Model).
System of operation: Blowback.
Weight loaded: 9.87 lbs.
Length overall: 33.5 in.
Barrel length: 7.80 in.
Feed mechanism: 32-round detachable, staggered box magazine.
Sights: Front: Barley corn.
Rear: Tangent with "V" notch graduated from 50–500 meters in 50-meter increments.
Muzzle velocity: 1360 f.p.s. (For 9mm Mauser).
Cyclic rate of fire: 500 rounds per minute.

Unusual Features of MP34

The weapon is typical of the period in which it was made in that it is heavy and expensive, being made of heavy forgings. The only unusual feature is the magazine loader, which is machined into the magazine housing. The magazine is inserted into the underside of the magazine housing and is then loaded with ten-round chargers—stripper clips—through the opening in the top of the magazine housing.

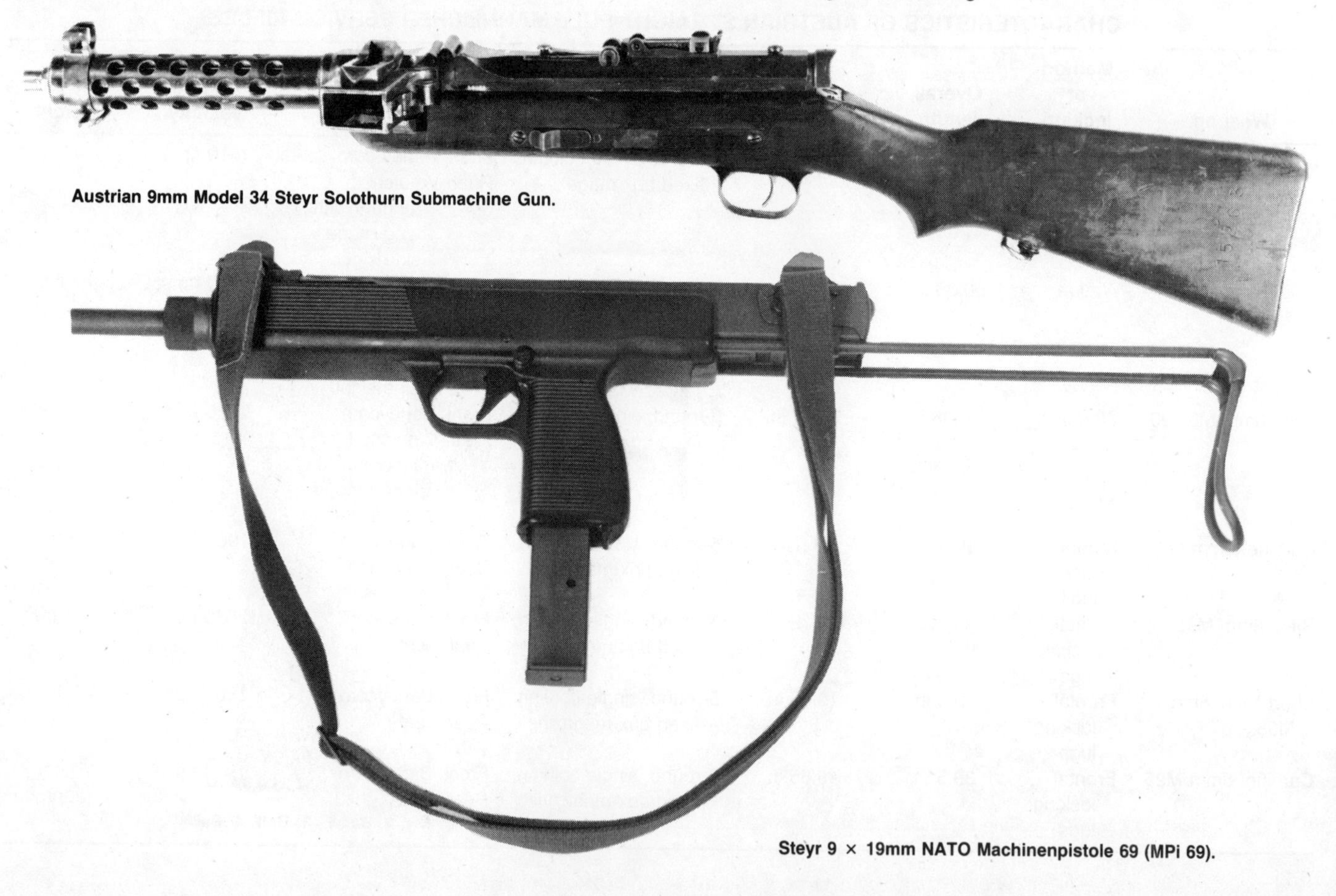

Austrian 9mm Model 34 Steyr Solothurn Submachine Gun.

Steyr 9 × 19mm NATO Machinenpistole 69 (MPi 69).

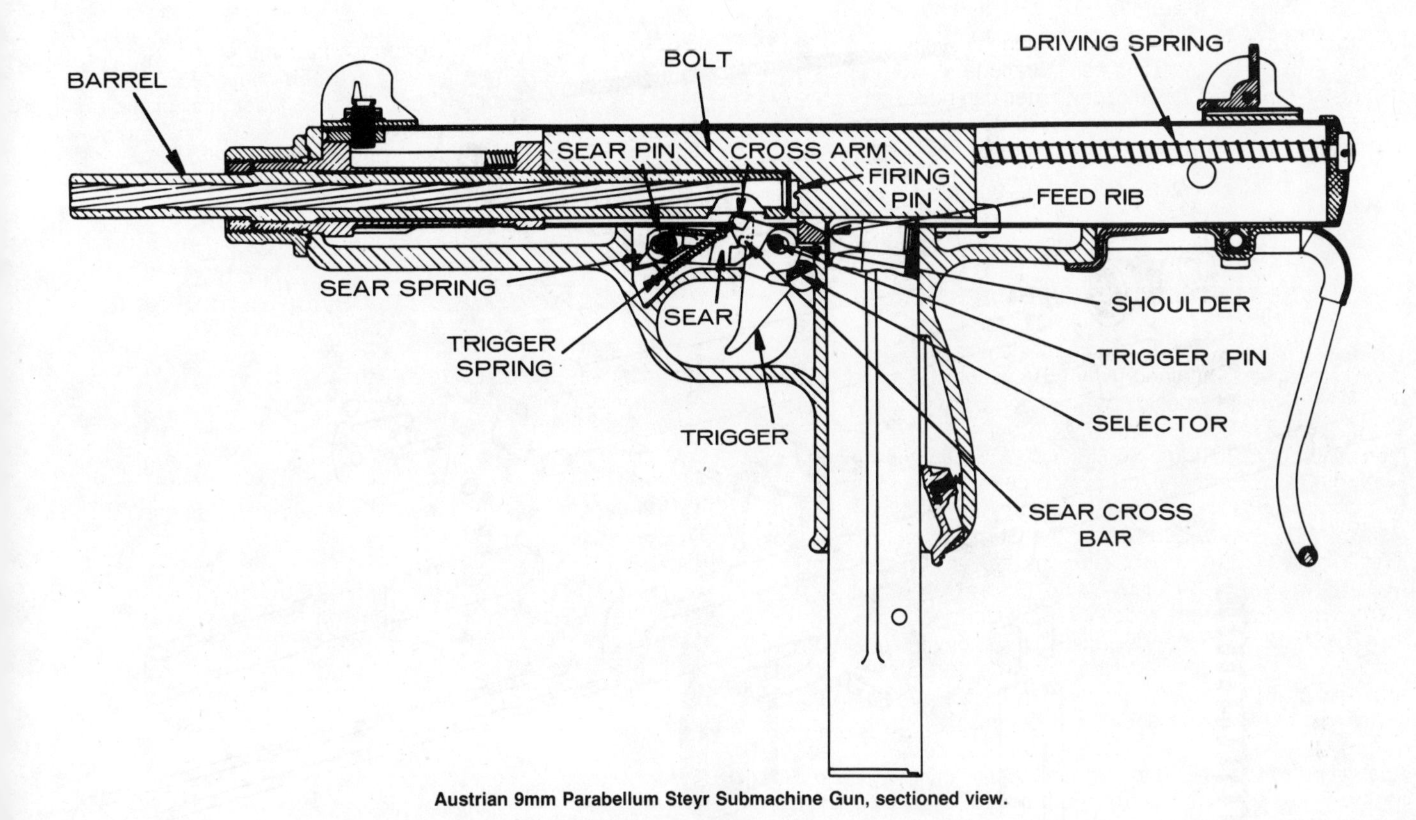

Austrian 9mm Parabellum Steyr Submachine Gun, sectioned view.

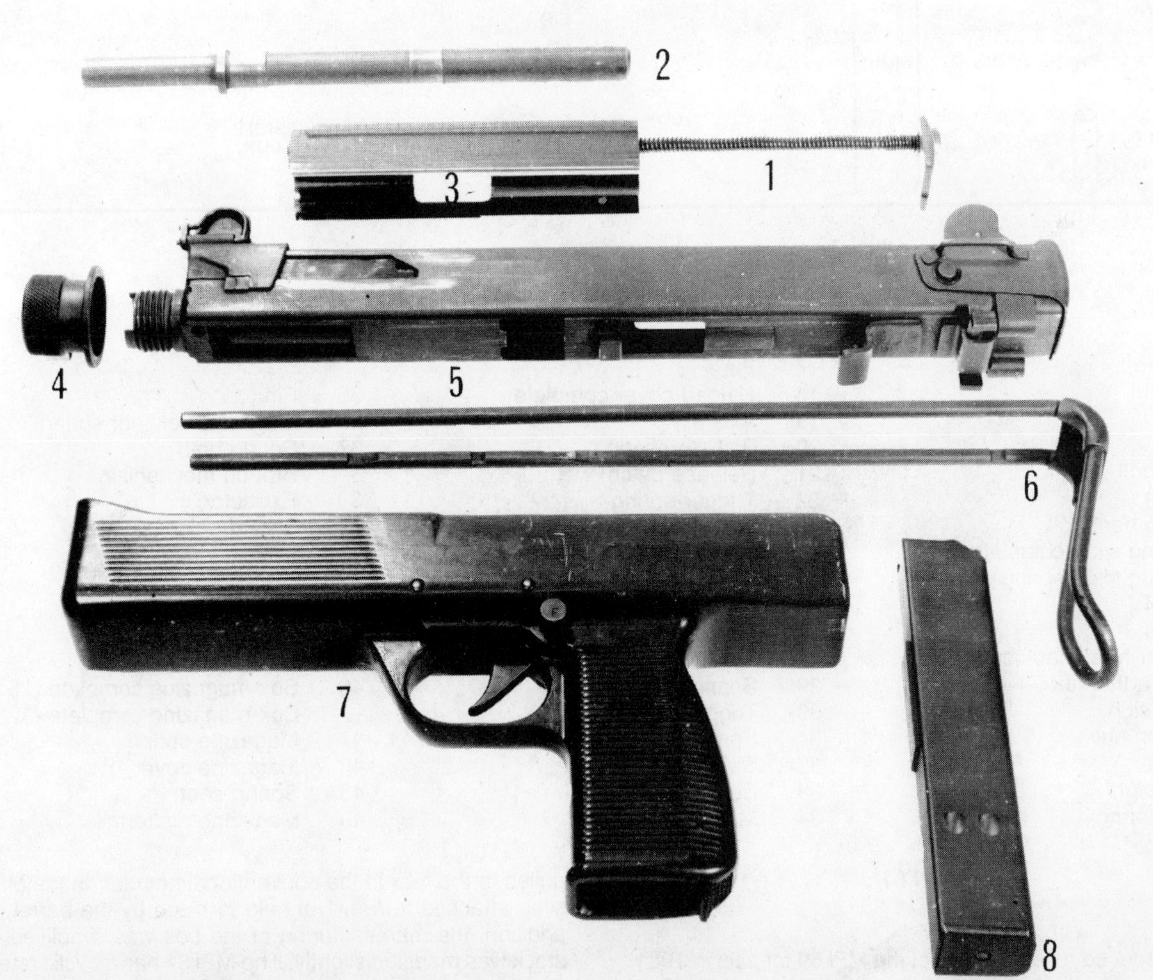

Austrian 9mm Parabellum Steyr Submachine Gun field stripped. 1. Operating spring assembly. 2. Barrel. 3. Bolt. 4. Barrel nut. 5. Receiver. 6. Stock. 7. Trigger housing assembly. 8. Magazine.

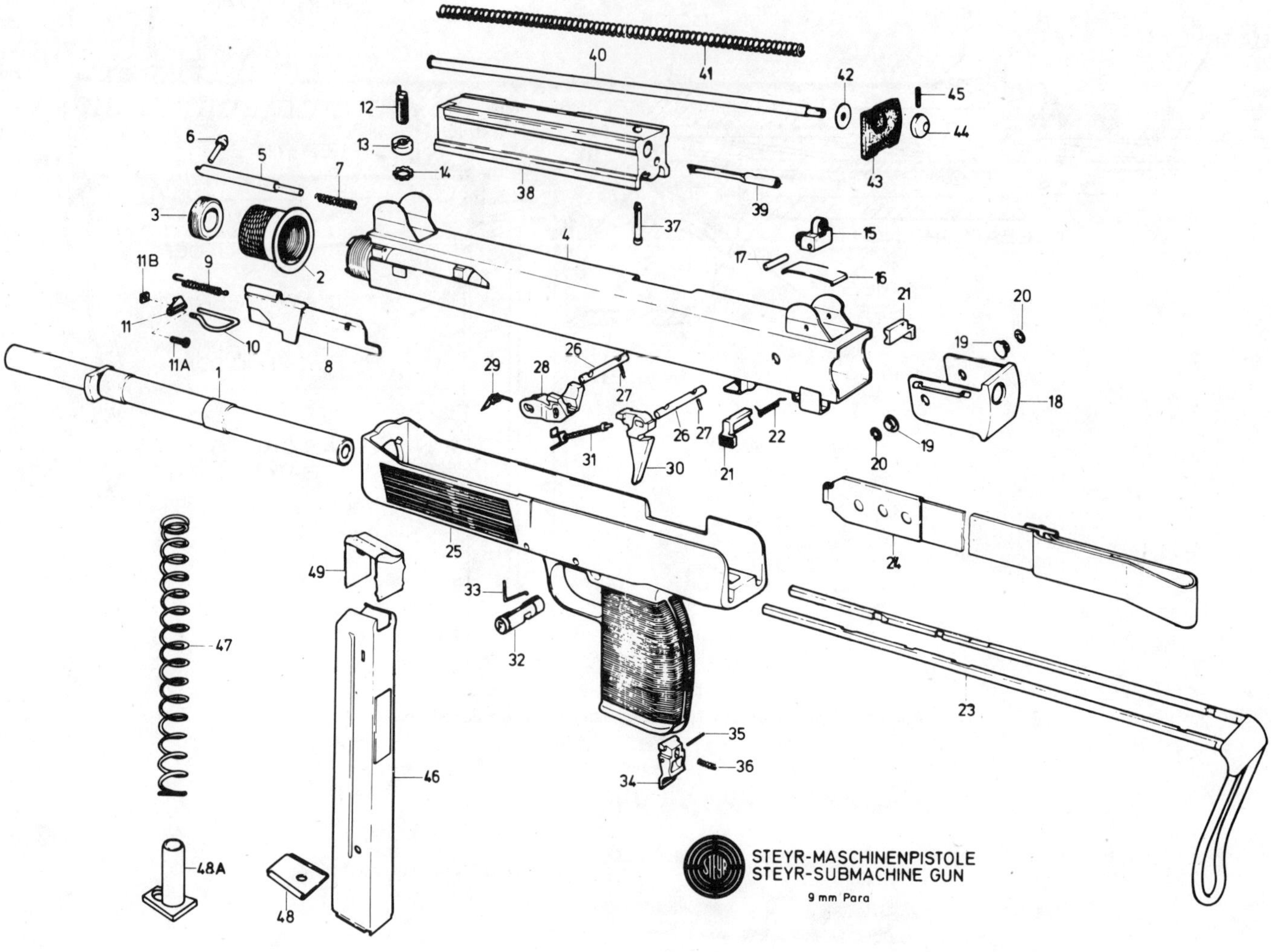

Exploded view of the MPi 69.

1 Barrel
2 Cap nut
3 Nut
4 Receiver complete
5 Bent bolt
6 Button
7 Spring bent bolt
8 Cocking slide complete
9 Cocking slide spring
10 Swivel
11 Detent
11A Fillister head cap screw
11B Square thin nut
12 Front sight
13 Binding nut
14 Washer
15 Rear sight
16 Leaf spring
17 Pin
18 Hinged cover complete
19 Axis pin
20 Bolt stop spring
21 Release catch
22 Trigger spring
23 Butt stock complete
24 Carrying sling
25 Handgrip
26 Pin trigger
27 Pin
28 Lever, sear
29 Spring, lever sear
30 Trigger
31 Spring guide complete
32 Safety bolt
33 Bent spring
34 Magazine retainer
35 Pin
36 Magazine retainer spring
37 Spring bolt
38 Breech mechanism
39 Extractor
40 Spring guide
41 Recoil spring
42 Disc
43 Buffer plate
44 Pressure button
45 Pin
46 Box magazine complete (25 rounds)
Box magazine complete (32 rounds)
47 Magazine spring
48 Magazine cover
48A Spring seat
49 Magazine platform

Steyr MPi 81

Steyr introduced a new version of the MPi 69 for sale in 1981. This weapon does not employ the sling to cock the bolt. Instead of the sling being pulled to retract the bolt, the bolt handle is pulled to the rear in the conventional manner. In the MPi 81 the sling attaches to a swivel held in place by the barrel collar. In addition, the manufacturing of the bolt was simplified, and the stock was modified slightly. The MPi 81 has a cyclic rate of 700–750 shots per minute, while the MPi 69 has a rate of 650. Both versions are offered for sale by S-D-P.

STEYR SUBMACHINE GUN

Steyr Daimler Puch developed a new submachine gun chambered for the 9mm Parabellum cartridge. This was the first new military weapon to be developed in Austria since the 1930s.

The Steyr submachine gun is a compact weapon, resembling the Israeli UZI in several respects. Like the UZI, it has a long barrel and a short overall length, accomplished by having a bolt that telescopes the barrel for about 2/3rds of the barrel length.

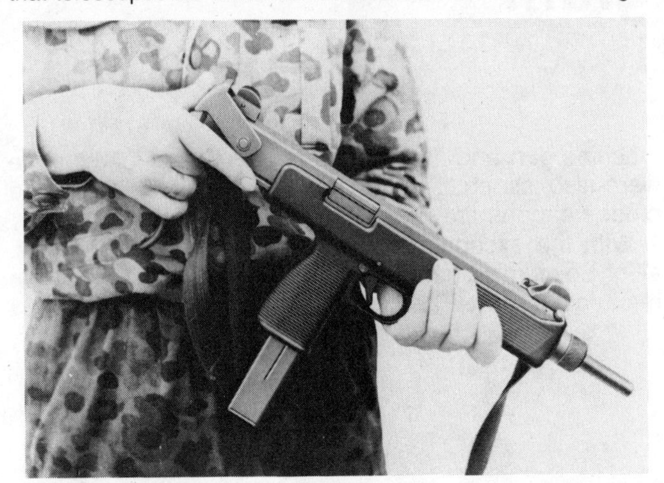

Raising the hinged receiver cover is the first step in the removal of the bolt assembly of the MPi 69.

CHARACTERISTICS OF THE STEYR SUBMACHINE GUN

Caliber: 9mm Parabellum
System of operation: blowback, selective fire
Weight: 6 lbs.
Length overall:
  stock retracted: 18 in.
  stock extended: 25 in.
Barrel length: 10.2 in.
Feed mechanism: 25- or 32-round detachable staggered row box magazine

Austrian 9mm Parabellum Steyr Submachine Gun, inserting magazine. Arrow indicates sling attachment to the operating handle on MPi 69.

Sights—front: post with protecting ears
  rear: "L" w/apertures set for 100 and 200 meters
Muzzle velocity: approx. 1350 f.p.s.
Cyclic rate: 650 r.p.m.

Description of the Mechanism of the Steyr Submachine Gun

The Steyr, as most submachine guns, fires from an open bolt. Pressure on the trigger, which pulls the trigger half of its length of travel, produces semi-automatic fire; pulling the trigger all the way to the rear produces automatic fire. As with the UZI, the ejection port is open only when the bolt moves to the rear with the fired case and on its return to the closed position. The push-button type safety is located on the left side of the trigger housing above and to the rear of the trigger. The magazine catch is located at the bottom rear of the pistol grip. The magazine well is in the pistol grip as on the UZI. There is a cocking safety that prevents the bolt from closing and firing if it slips in cocking. The weapon is easily field stripped. The barrel is moved by depressing the barrel catch lock at the right top front of the receiver and unscrewing the barrel nut. The receiver cover is disengaged by pushing in on a catch at the top rear of the receiver. Before attempting disassembly, the stock must be in the fixed position.

AUSTRIAN MACHINE GUNS

The Austro-Hungarian Empire used the Schwarzlose machine gun in several models. The Schwarzlose, in addition to being used in Austria, was also used in Sweden (6.5mm Model 14), in the Netherlands (08, 08/13 and 08/15), in Czechoslovakia (7.92mm) and in Italy in the form of Austrian war booty (8mm). The Schwarzlose is not in active use in any country, and it is doubtful if this weapon will ever see active service again.

AUSTRIAN 8mm MODEL 30S MACHINE GUN

This weapon was adopted by the Austrians in 1930 and was among the materiel taken from Austria by the Germans after the annexation of Austria. A similar weapon designated as the 31M (Model 31) was adopted by Hungary in 1931. Both guns were marketed by Waffenfabrik Solothurn A.G., a Swiss plant owned by Rheinmetall-Borsig A.G. They were developed from the designs of Louis Stange, a Rheinmetall engineer. Assembling and marketing the gun by Solothurn was a neat dodge by the Germans to avoid Versailles Treaty restrictions. The gun was offered first in 7.92mm as the Solothurn Model 29. In addition to being

Austrian Schwarzlose 8mm Machine Gun M07/12.

adopted by the Austrians and Hungarians as ground guns, the weapon, slightly modified, was adopted in 7.92mm by the Germans as a fixed aircraft gun, the MG15, and as a flexible aircraft gun, the MG17. The current machine gun is the MG 42/59 and its variants MG3 and MG74.

9 Belgium

The Belgian Army was stripped of weapons during World War II and for a period after World War II was equipped mainly with British weapons, i.e., Lee Enfields, Vickers and Bren guns. Re-equipment with FN-made weapons started very rapidly; a caliber .30-06 bolt-action Mauser was among the first of the weapons of native origin to be issued to Belgian troops. The .30-06 FN-made Browning machine gun and the .30-06 FN M1949 self-loading rifle were also issued. The Mausers have been sold, and the other weapons are presumably held in reserve.

With the exception of the Vigneron submachine gun, all of Belgium's post World War II small arms have been manufactured by Fabrique Nationale Herstal.

SMALL ARMS IN SERVICE

Handguns: 9 × 19mm NATO *Pistolet Automatique Browning Modèle 1935 à Grand Puissance*
Submachine guns: 9 × 19mm NATO *Mitraillette UZI*
9 × 19mm NATO *Mitraillete Vigneron*
Rifles: 7.62 × 51mm NATO *Fusil Automatique Légère* (FAL)
Machine guns: 7.62 × 51mm NATO *Mitrailleuse d'Appui General* (MAG)
12.7 × 99mm *Mitrailleuse Browning* (M2 HB)

FABRIQUE NATIONALE HERSTAL

Fabrique Nationale Herstal (FN) Liège is one of the major designers of small arms in the Western World today. The FN Browning G.P. (Grande Puissance—High Power) pistol is used in many countries, as is the FN FAL rifle and the MAG machine gun. FN-made Browning pistols of .32 and .380 caliber, as well as FN-made Browning machine guns, automatic rifles and the FN commercial weapons, are found throughout the Western World.

The FN organization was founded in 1889 by a combine of Liège interests and Ludwig Loewe and Co. of Berlin to manufacture the Model 1889 Mauser rifle for the Belgian government. The company had a stroke of luck early in the present century when John Browning—upset over his financial arrangements with Winchester—decided to take the design of his now renowned long recoil operated shotgun to Liege and to FN. Thus began a close relationship with the outstanding genius of American gun designers, which ended with his death in 1926 in Belgium. Browning brought more than his automatic shotgun to FN. They also produced his automatic pistols—the M1900, M1903, M1906 (.25 automatic), M1907, M1910, M1922 and the M1935 (the High Power). FN produced his commercial semiautomatic rifle and after World War I produced the Browning Automatic Rifle and Browning machine gun, which Browning had developed for his native land during the war. Although John M. Browning is generally known in the US, especially to those who have used his weapons in service, in Europe his name is a household word for automatic pistol. It should be noted that Browning designs similar to those produced and marketed by FN in the Eastern Hemisphere were produced and marketed by American manufacturers in the Western Hemisphere. Colt produced in one form or another successful Browning pistol designs including the still standard US Army caliber .45 Colt automatic pistol. Colt also produced the automatic weapons, Remington and Savage-Stevens produced the shotguns, and of course, Winchester continued manufacture of many of the earlier rifles and is still producing the famed .30-30 Model 94 carbine.

FN, in addition to producing the Browning automatic rifle and Browning machine guns (water cooled, air cooled, aircraft and heavy), began producing Mauser bolt-action rifles based on the Model 98 action in 1924. These weapons were quite successful, and modified forms of these military Mausers were in production as late as 1964 for small Middle Eastern countries that could not afford the more modern semiautomatic and automatic rifles. FN and the Zbrojovka Brno plant of Czechoslovakia had the world's Mauser military rifle market to themselves to a great extent during the period 1924 to 1938; the Belgians, being extremely astute businessmen in addition to manufacturing a fine product, got at least their share of the business. (There is actually no real difference in the quality of the products made by the two concerns during this period; they are among the finest quality military rifles ever built.)

World War II found Belgium again an occupied country. The Germans continued manufacture of weapons suited to their ammunition system, i.e. the 7.92mm Mauser carbine, the 9mm Parabellum High Power pistol (called Pistol 640(b) by the Germans), the 9mm Browning short (.380 ACP) M1922 (called pistol 626(b) by the Germans) and various weapon parts. The weapons made during World War II are not equal in quality to the pre- or post-war products.

Fortunately the FN factory was not completely destroyed by Allied air strikes or by the Germans when they departed in late 1944. The plant was able to set up rapidly for the manufacture of Browning automatic rifle and Browning machine-gun parts for the US Army. At the conclusion of hostilities, FN got a contract to rebuild a large number of US weapons in Europe. The first post-war weapon they marketed was designed prior to the war by D. Saive. It was the FN self-loading rifle, known variously as the ABL, SAFN or Model 1949. This rifle was followed by the Type D Browning Auto rifle, the FAL rifle, and the MAG machine gun.

BELGIAN PISTOLS*

THE 9mm FN BROWNING HIGH POWER PISTOL

The FN Browning High Power is one of the most extensively used military pistols in the world today. It is also widely distributed as a commercial weapon. This pistol was the last developed by John Browning and first appeared in prototype form in 1926. It was introduced to the market in 1935 in two forms—an "Ordinary Model" with fixed sights; and an "Adjustable Rear Sight Model," which had a tangent type rear sight graduated to 500 meters and a slotted grip for attachment of a wooden shoulder stock. This shoulder stock is attached to a leather holster in contrast to the all wooden shoulder-stock holsters made for the Canadian-made FN Browning pistols No. 1 mark 1 and No. 1* during World War II. The High Power, which is called the G.P. (*Grande Puissance*) in Belgium, uses the Browning Colt parallel ruler system of locking but is considerably simplified in many ways in comparison with the US caliber .45 Model 1911.

The High Power has been or is being used as a service pistol by Belgium, Lithuania, Denmark, the Netherlands, Nationalist China, Canada, the United Kingdom, Romania and other countries. It was manufactured by the John Inglis Company of Toronto, Ontario, during World War II. (For further information see under Canada.) Large quantities were manufactured for the Germans during the occupation of Belgium; the High Power was used as a first-line weapon by the Germans because of its caliber, 9mm Parabellum. During the post-war period, the pistol has been manufactured in the Indonesian arsenal at Bandung on the island of Java. The pistol in commercial form is distributed in the US by the Browning Arms Co. As this book went to press, FN Herstal announced the availability of a double action version of the High Power.

*For information on obsolete Belgian pistols, see *Handguns of the World.*

CHARACTERISTICS OF THE FN BROWNING HIGH POWER PISTOL

Caliber: 9mm Parabellum.
System of operation: recoil, semiautomatic.
Magazine: Box type, double line staggered. Capacity 13 cartridges.
Muzzle velocity: 1040 to 1500 feet per second depending on type and manufacture of ammunition.
Barrel length: 4.75 in.
Overall length: 8 in.
Weight: 1.9 lbs.

FN Browning 9 × 19mm Modèle 1935 High Power pistol.

The following are some of the nomenclatures that have been or are being used for the High Power in various countries:

Belgium: Pistolet Automatique Browning Modèle à Grande Puissance (GP).
Canada, UK: Pistol, Browning, FN, 9mm HP No. 1, Marks 1 and 1.* and No. 2, Marks 1 and 1.*
Denmark: 9mm Pistol M/46.
Germany: *Pistole 640(b)*.
Indonesia: Pindad PiA 9-mm
Netherlands: *Pistool*, 9mm Browning, FN, GP.

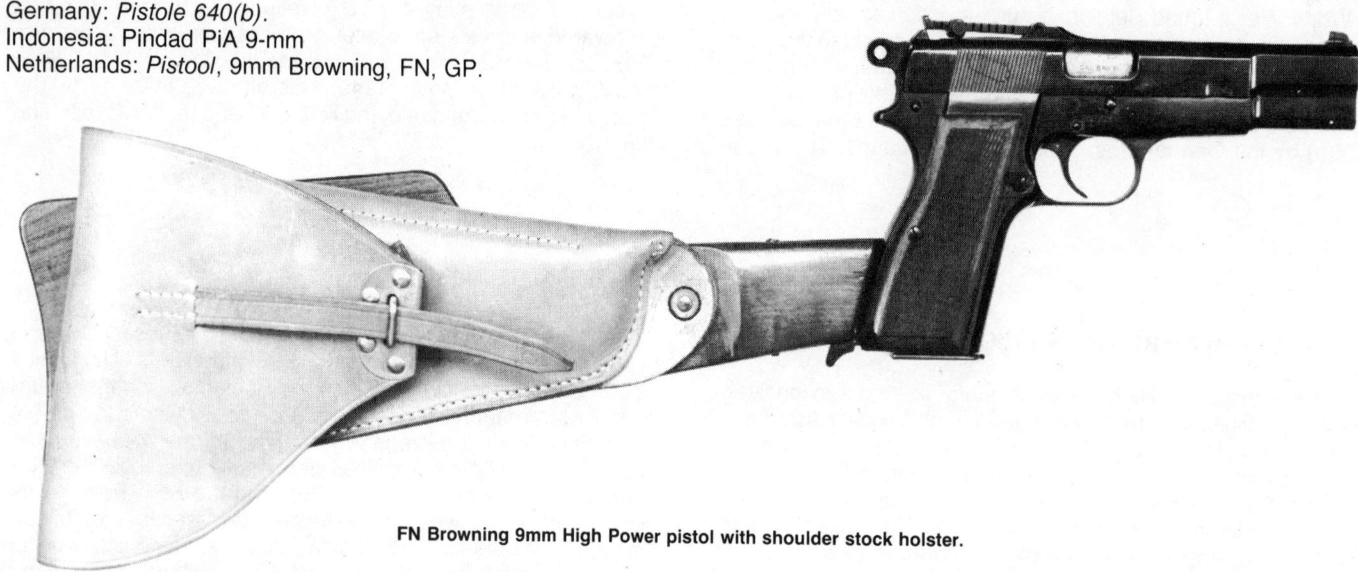

FN Browning 9mm High Power pistol with shoulder stock holster.

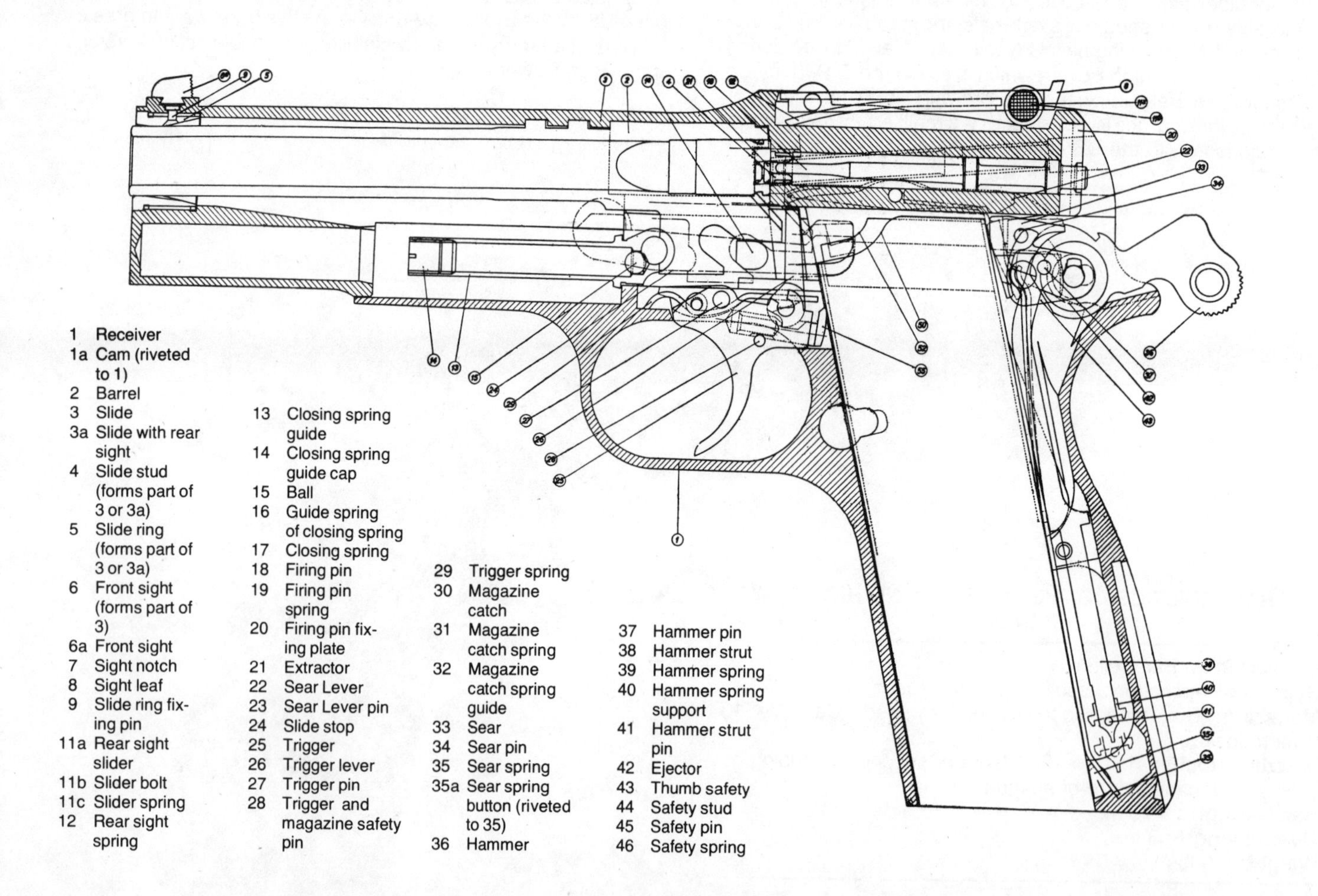

Section view and parts listing—FN Browning High Power pistol.

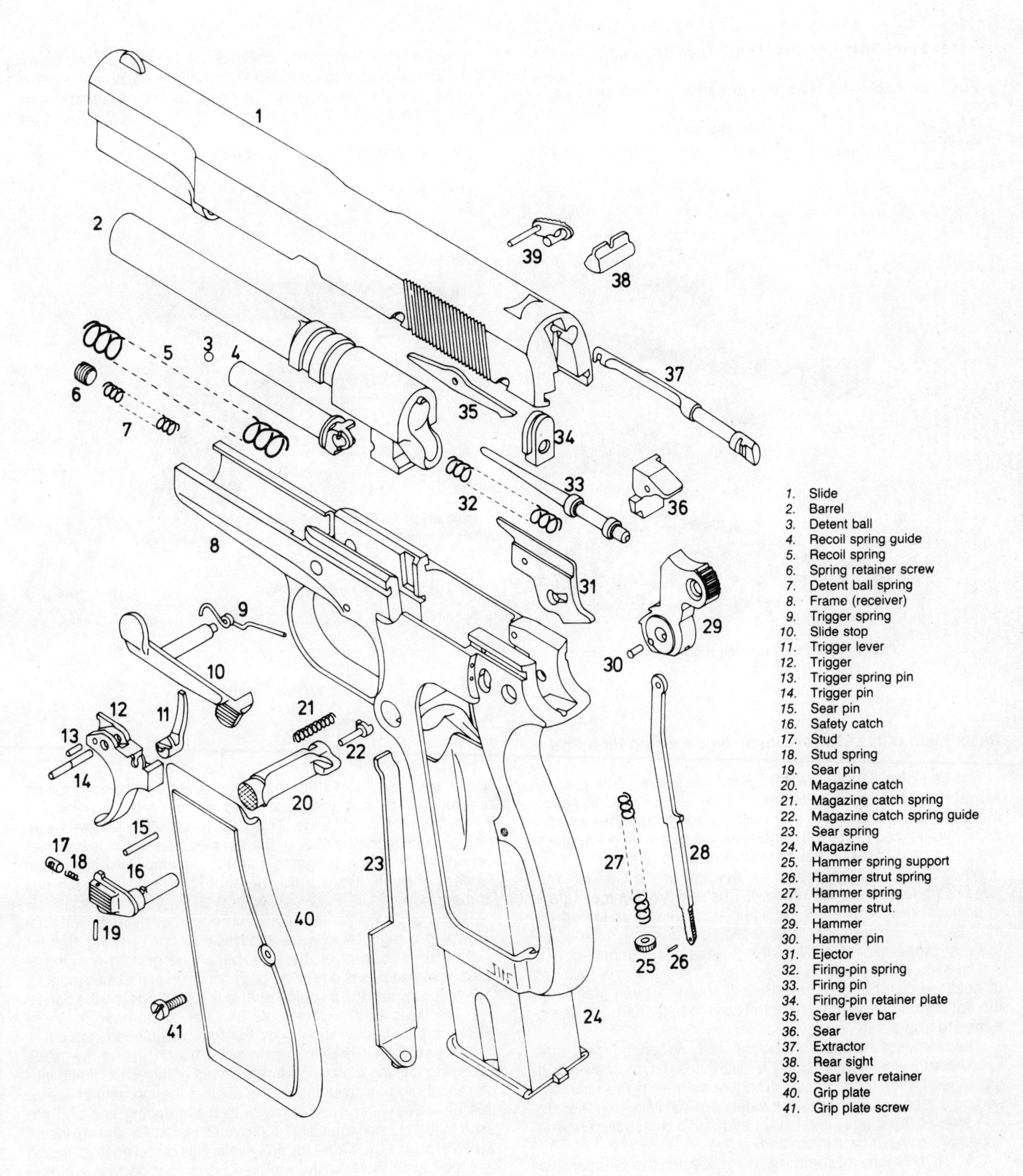

1
2
3
4
5
6
7
8
9
10
11
12
13
14
15
16
17
18
19
20
21
22
23
24
25
26
27
28
29
30
31
32
33
34
35
36
37
38
39
40
41
1. Slide
2. Barrel
3. Detent ball
4. Recoil spring guide
5. Recoil spring
6. Spring retainer screw
7. Detent ball spring
8. Frame (receiver)
9. Trigger spring
10. Slide stop
11. Trigger lever
12. Trigger
13. Trigger spring pin
14. Trigger pin
15. Sear pin
16. Safety catch
17. Stud
18. Stud spring
19. Sear pin
20. Magazine catch
21. Magazine catch spring
22. Magazine catch spring guide
23. Sear spring
24. Magazine
25. Hammer spring support
26. Hammer strut spring
27. Hammer spring
28. Hammer strut.
29. Hammer
30. Hammer pin
31. Ejector
32. Firing-pin spring
33. Firing pin
34. Firing-pin retainer plate
35. Sear lever bar
36. Sear
37. Extractor
38. Rear sight
39. Sear lever retainer
40. Grip plate
41. Grip plate screw

Exploded view of the 9mm Parabellum FN Browning-Saive Modèle 1935 GP pistol. (*Jimbo*)

Field Stripping the High Power Browning

Pull slide back and push thumb safety up into the second notch.

Press magazine catch and withdraw magazine.

Push pin from right side of receiver and lift out the pin and slide stop unit.

Holding firmly to the slide, depress the safety catch and permit the slide assembly to go forward and off the receiver runners.

Holding slide assembly upside down, pull recoil spring toward the muzzle and lift it out of engagement. Remove it and the spring.

Remove barrel from rear of slide.

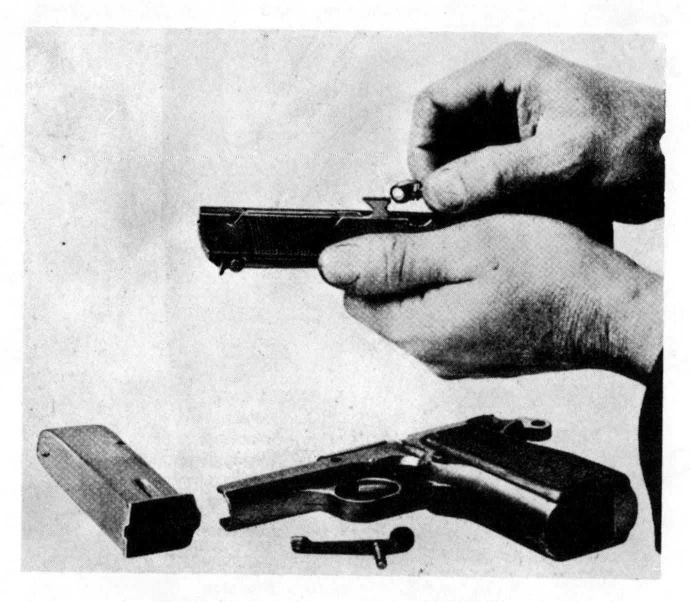

Field stripping the Browning High Power.

Description of the Mechanism of the Browning High Power

The use of a double-row staggered magazine gives greater magazine capacity but at the same time necessarily increases the thickness of the grip. This additional width, together with an arched lower section of the handle section of the receiver, gives the pistol better than usual instinctive pointing qualities.

While in general the design follows that of the Colt .45 Automatic, in detail it is quite different. The positive hammer safety is positioned as in the US service pistol. When it is forced up into its notch in the slide, a projection on its lower side fits at the rear of the sear to prevent the release of the hammer.

The slide stop, which is forced up by the magazine follower to hold the pistol open when the last shot has been fired, and the button magazine release both follow the standard Browning form. No grip safety is provided.

Other features include a magazine safety. When the magazine is withdrawn, the safety spring forces this safety out, swinging the trigger lever forward out of engagement with the sear. Inserting a magazine presses the safety against its spring tension and swings the trigger lever back and under the sear to permit finger pressure to be transmitted.

CAUTION. Many of these pistols made for the Germans in World War II had the magazine safety removed.

In place of the Colt stirrup-type trigger, this weapon is fitted with a comfortable trigger, which, when pressed, forces a trigger lever upward. This rotates the sear lever, which acts upon the sear arm, causing it in turn to swivel and release the hammer.

Unless the slide is fully forward and the barrel securely locked to it, this sear lever remains at the rear, and the trigger lever cannot act upon it. This acts as a positive disconnector to prevent the weapon from being fired. Should the trigger be held back after a shot has been fired, the trigger lever is retained in raised position but is also forced forward by the sear lever on the forward motion of the slide. This prevents the trigger lever from acting upon either the sear or the sear lever, and so the hammer cannot fall until the trigger is permitted to move to its normal forward position. Thus, only one shot is possible on each pull of the trigger.

The slide in particular is an improvement over the original Browning design. Its forward end has only one opening, that for the barrel. The front of the slide below the muzzle opening is solid, doing away with the weak barrel bushing of previous models. The recoil spring and its guide seat in the hollow below the barrel, and the head of the guide sets into the barrel nose (or lock) where it is securely retained by the transverse slide stop pin.

The barrel lock is also improved. While it retains the basic Browning locking idea (that of ribs on top of the barrel fitting into corresponding grooves inside the slide at the moment of firing), this new Browning does away with the swinging link and pin used in the US service pistol. It provides instead a "barrel nose," which is part of the barrel forging itself. This barrel nose is placed directly below the heavily reinforced chamber section and has a guiding slot that is controlled by a cam machined into the receiver. This arrangement gives a much more rigid barrel support and permits simplification of the recoil spring system.

How the Browning High Power Pistol Works

At the moment of discharge, the barrel is locked securely to the slide as the top-locking ribs engage in the locking slots in the slide. As the slide starts back in recoil carrying the barrel with it and the bullet leaves the barrel, the sear lever is disconnected from the trigger lever.

As the barrel pressure drops to safe limits, the lower section of the notch of the barrel nose contacts the cam in the receiver, and the rear end of the barrel is thus drawn down until its ribs are free from the locking slots. At this point, the rearward barrel movement is stopped as the barrel nose brings up sharply against its receiver stop.

The slide continues on backwards, riding over the hammer to cock it. The extractor claw carries the empty cartridge case back until it strikes the ejector and is tossed out of the pistol. Meanwhile the recoil spring is compressed around its guide. Rearward motion of the slide stops when its lower part strikes against the forward end of the receiver.

For the return motion, the recoil spring pushes the slide forward to strip the top cartridge from the magazine into the chamber. The breech end of the slide strikes the barrel; under the action of the cam in the receiver acting against the upper part of the barrel nose notch, the rear of the barrel is brought up into locking position, and its ribs fit firmly into their slots in the slide.

The trigger, sear and hammer mechanisms hook up properly, and the pistol is ready for the next pull of the trigger.

The FN Hi-Power is now being made with a side-mounted pivoting extractor. The spring loaded extractor lies in a cut in the slide behind the ejection port and is visible from the outside, rather than being mounted in a hole cut through from the rear of the slide to the slide breech face.

BELGIAN RIFLES

FUSIL AUTOMATIQUE LÉGÈR (FAL)—FN LIGHT AUTOMATIC RIFLE

CHARACTERISTICS OF FN LIGHT AUTOMATIC RIFLE (FAL)

Caliber: 7.62mm NATO.
Weight w/empty magazine: 9.06 lbs.
Overall length: 40 in.
Barrel length: 21 in.
Method of operation: Gas, selective fire.
Feeding: From detachable box magazine, staggered 20-round capacity.
Sights: Front; Hooded post. Rear aperture graduated in 100 meter steps to 600 meters.
Cyclic rate of fire: 650 to 700 per minute. Most of the weapons in service throughout the world are only used as semiautomatic weapons.

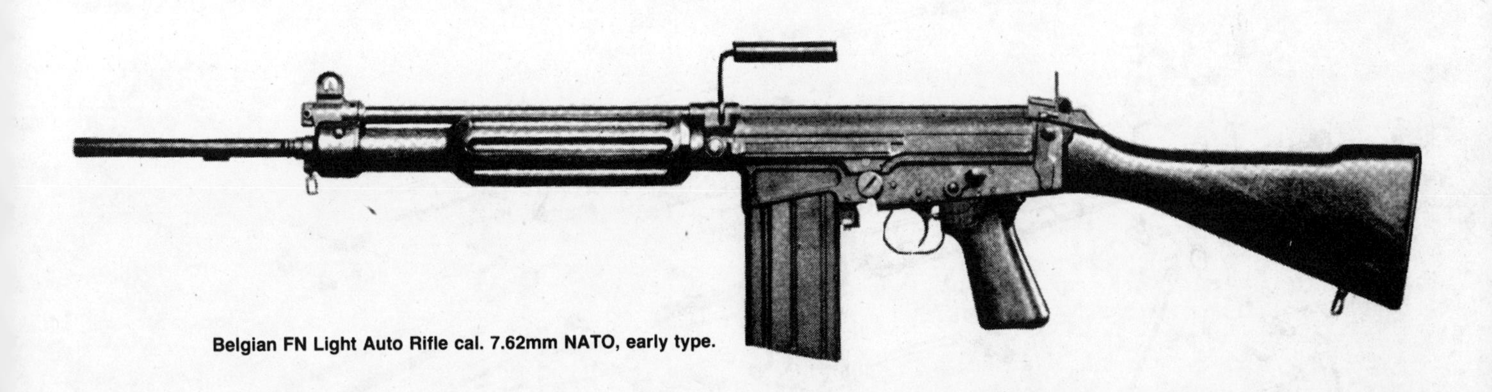

Belgian FN Light Auto Rifle cal. 7.62mm NATO, early type.

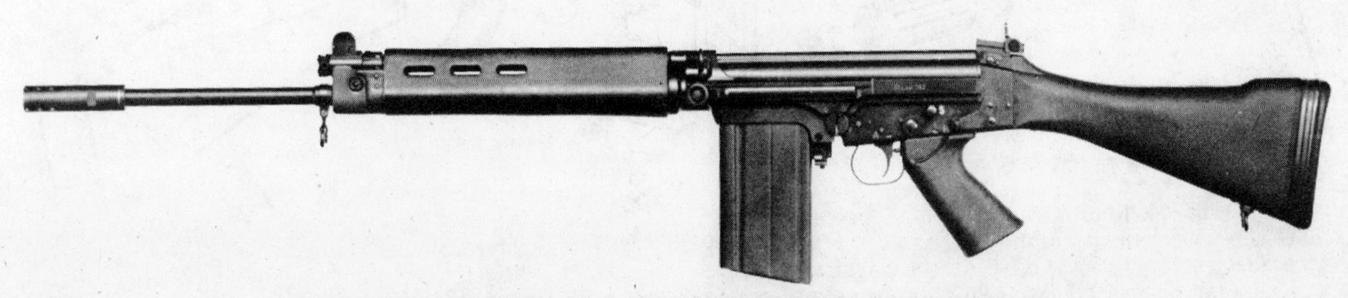

FN 7.62mm NATO FAL Rifle, 1964 pattern with plastic stock, handguard and pistol grip.

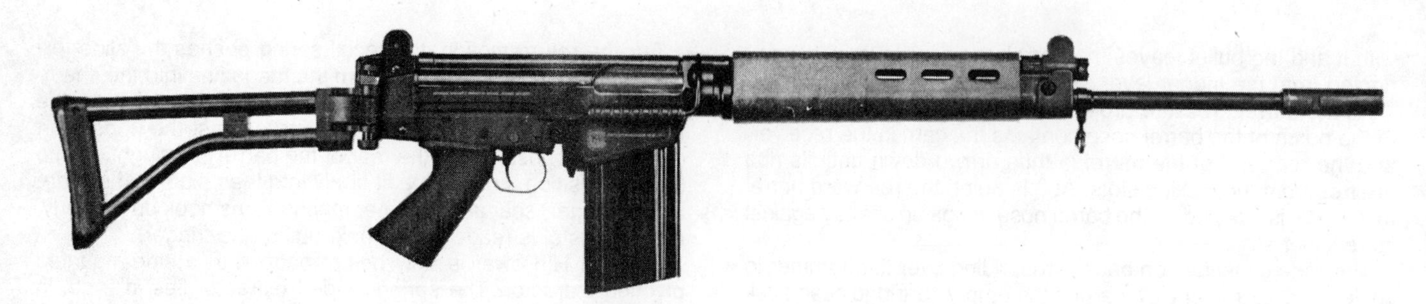

FN 7.62mm NATO FAL Paratroop Rifle, stock extended.

Heavy barrel version of the FN 7.62mm FAL. This rifle is in service in Israel and Peru, among other countries. It is used as squad automatic weapon. A modified version of this weapon is manufactured in Australia and Canada.

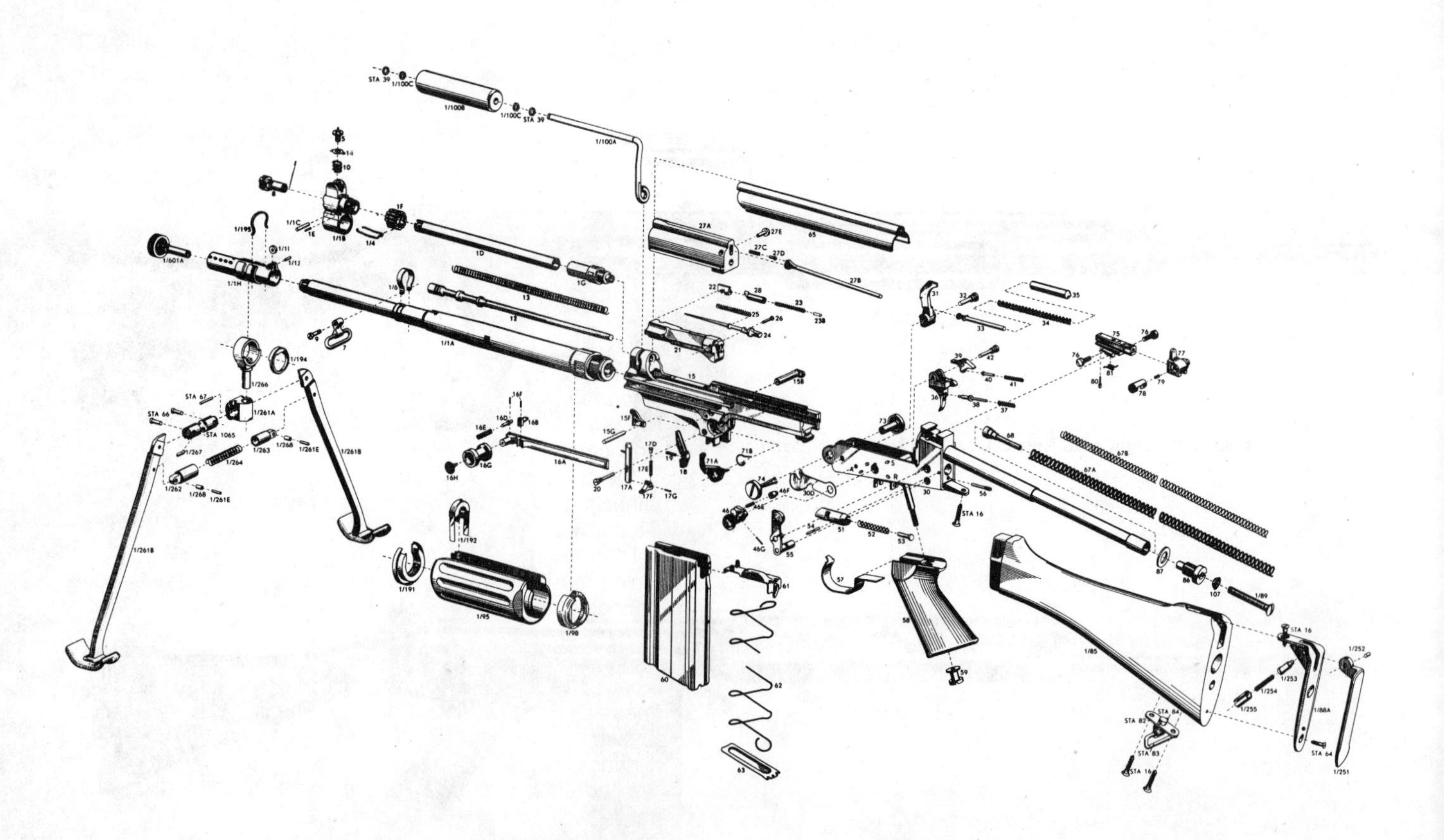

FN Heavy-barrel Automatic Rifle as made for South Africa. Caliber 7.62mm.

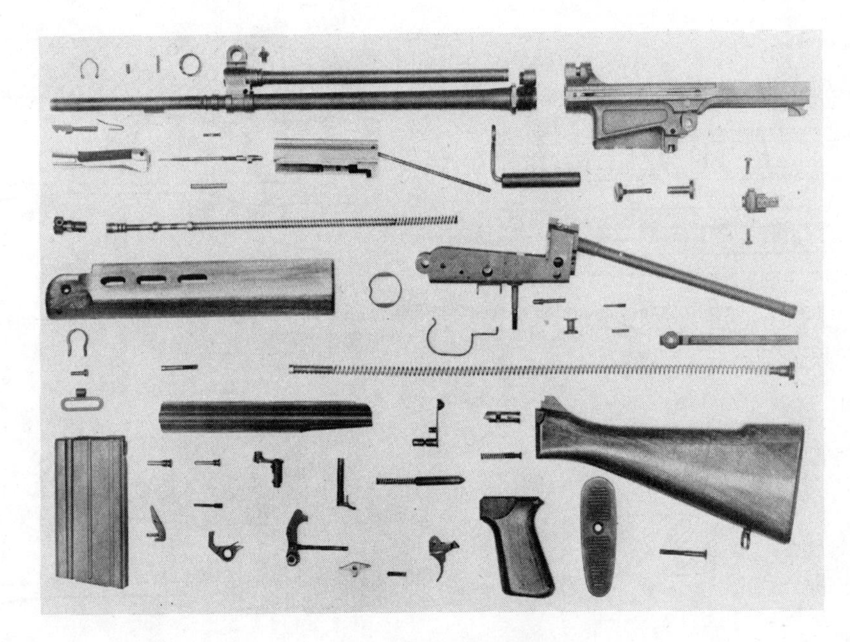

Disassembled view of FAL.

How the FAL Works

Pull cocking handle on left side of receiver to rear as in BAR. This leaves the right hand on the pistol grip ready for firing. The cocking handle does not move during firing, again as in the case of the BAR. This removes danger to the firer's face and does not interfere with aiming.

If there is an empty magazine in the rifle at the end of the cocking handle's stroke, the bolt is held open automatically. A loaded magazine is inserted in the magazine housing and pushed in until the retaining catch secures it.

Pressing the release stud on the left side near the magazine catch releases the bolt, which is driven forward by its compressed return spring.

This return spring is housed in the butt. It acts on the bolt carrier through a rod pivoted at its rear face. As the bolt moves forward, its feed face strips the top round from the magazine in standard fashion and thrusts it up the ramp into the barrel chamber, where the extractor engages the cannelure in the case. The gun is now ready for firing.

Firing the FAL

The change lever on the left side of the receiver may be set for "safe," "single shot" or "full automatic" fire. The positions are widely spaced deliberately so that in the dark one can tell by sense of touch the position of the change lever.

If the change lever is set for single shot fire, pressure on the trigger releases the hammer to strike the rear face of the firing pin and fire the cartridge. Since the weapon fires from a closed breech, there is no such disturbance of aim as occurs in weapons like the submachine gun and the Browning machine rifle, where the mass of the breech mechanism moves forward with the pull on the trigger.

The operation is the standard gas system whose reliability has long been established for this type of weapon. Part of the gases following the bullet down the barrel pass through the port into the forward section of the gas cylinder. A gas regulator, previously adjusted, provides sufficient gas to satisfactorily operate the piston, which is driven back to function the mechanism. Remaining gas passes to the open air through holes in the gas cylinder.

The piston, acting on the tappet principle, is driven to the rear in its tube on top of the barrel. It strikes the bolt carrier and pushes it to the rear.

Ramps machined into the bolt carrier engage a cam on the bolt after the bullet has emerged from the barrel and the pressure has dropped to safe limits. The bolt is cam lifted out of engagement with its locking shoulder in the receiver—this unlocks the action.

The bolt carrier and bolt now travel together to the rear. They ride down the hammer to cock it.

During rearward motion of the bolt, the extractor has carried the empty case out of the chamber with it. The case strikes the ejector and is thrown out the right side of the gun.

After the piston tappet strikes the bolt carrier and imparts the necessary impetus to it, during which travel the spring around the piston has been compressed, the spring operates to return the piston to forward (battery) position. The return spring within the butt is fully compressed during rearward movement of the bolt and carrier in standard fashion. At the end of the recoil stroke, the compressed spring reasserts itself and working through its connecting rod thrusts the bolt and carrier forward. The bolt strips a cartridge from the magazine and chambers it and stops against the face of the barrel. The bolt carrier, which still has continuing movement under thrust from the recoil spring guide and spring, works through the cam and ramps to force the rear locking end of the bolt down its locking recess in the receiver. At the end of the movement of the carrier, it is resting against the receiver above the line of the chamber where it is in line with the rear end of the piston tappet. The extractor, of course, snaps over the cannelure of the cartridge as the bolt chambers the cartridge.

When the change lever is set for automatic firing, the operation is identical with that already described except that the hammer is automatically released to continue fire until pressure on the trigger is released.

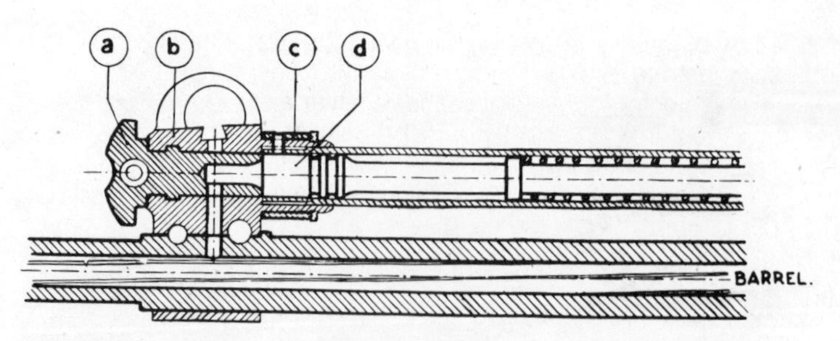

Gas plug (a) is positioned in end of gas cylinder (b). It may be turned to shut off gas when using rifle for grenade launching or as a straight-pull rifle. Gas regulator (c). Head of gas piston (d).

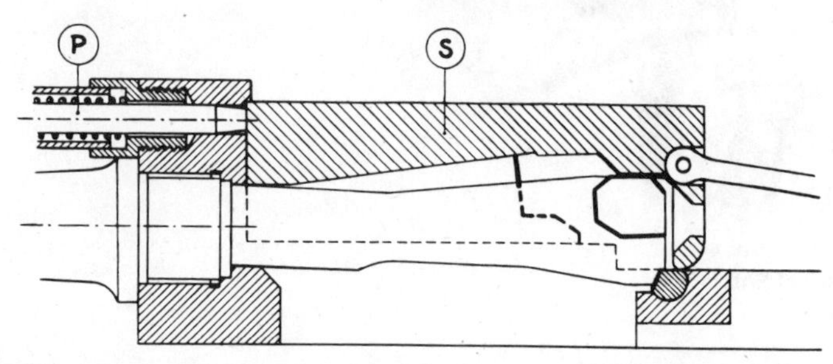

Piston driven back by gas hits movable carrier above locked bolt, driving it back and compressing return spring.

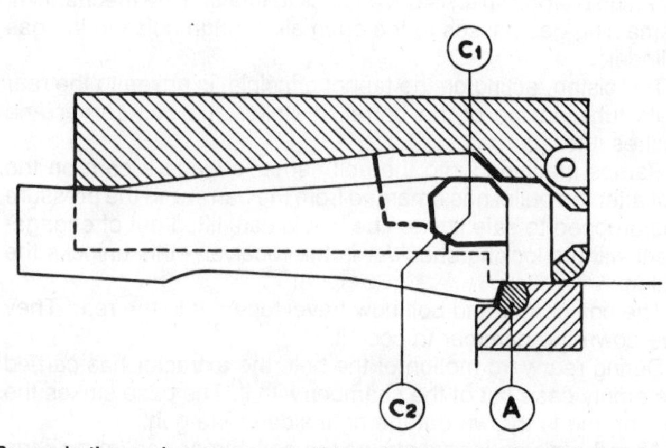

Ramps in the carrier engage cam (C1) on bolt as bullet leaves barrel, thus raising the rear of bolt out of its locking seat (A) in the receiver.

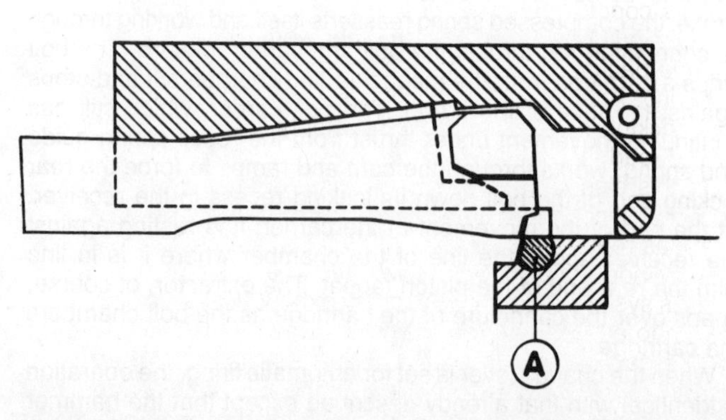

Carrier and bolt travel back together. Piston tappet spring returns piston. Bolt rotates and cocks hammer. Extractor withdraws fired case.

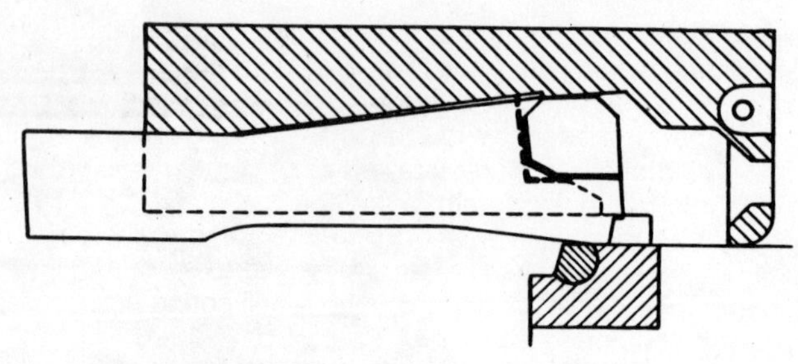

Bolt and carrier reach end of stroke. Case hits ejector. Return spring in butt is fully compressed.

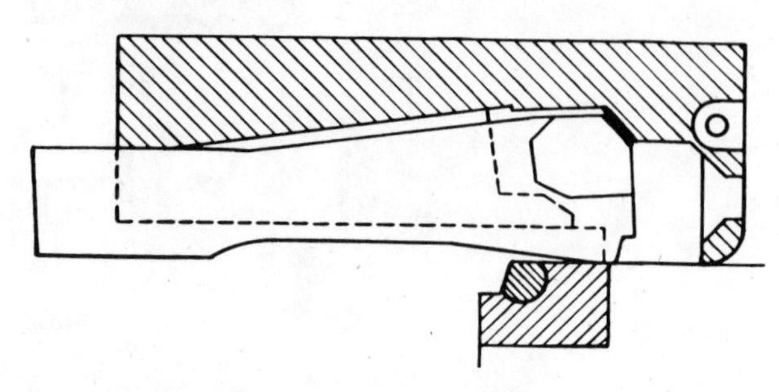

Spring starts carrier and bolt forward. Bolt picks up top cartridge in magazine and starts it toward chamber.

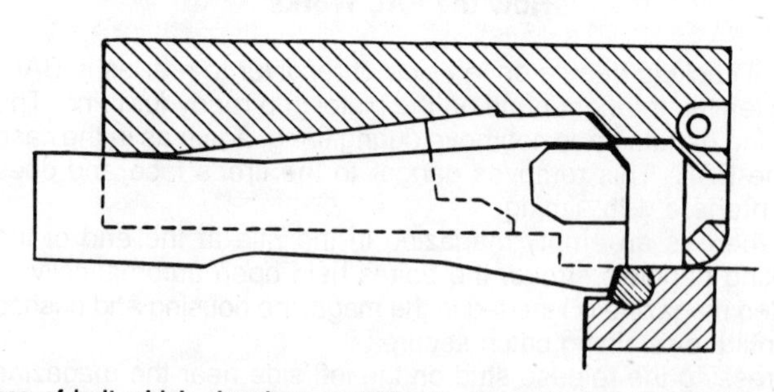

Face of bolt which chambers cartridge is stopped against face of barrel. Carrier is still driven forward, until it cams bolt down into its locking seat in the receiver.

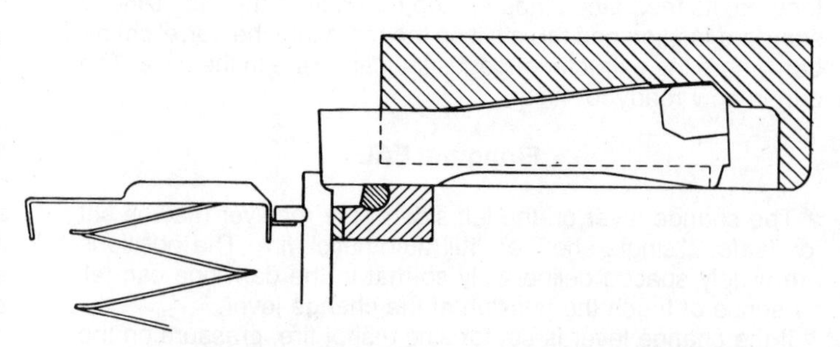

When last shot has been fired, the magazine follower forces a holding plunger up in front of the bolt to hold the action open.

FAL Hold-Open Device

The pin to hold the bolt in rear position when the magazine is empty is mounted in the receiver to the right of the ejector.

It is in the form of a heavy plunger, which is elevated by a nib on the rest of the magazine follower through a small pin working in a slot.

In rest position, this plunger is held down by its spring. However, when the magazine follower rises as the last cartridge is fired, upward pressure of the heavy magazine spring overcomes the lighter spring attached to the pin, and the pin is thrust upwards. The plunger is thereby raised to its complete height and projects into the forward path of the bolt.

When a full magazine is inserted in the gun, drawing the bolt slightly to the rear allows the plunger spring to react and lower the plunger, and the bolt driven by the recoil spring now moves forward to chamber a cartridge.

FAL Trigger Operation

Turning the change lever on the left side of the receiver to the safe position turns a stud into position at the extreme rear of the trigger to lock it mechanically. In this position, trigger movement is impossible.

When the change lever is set for single shot fire, the stud is turned into position to clear the rear of the trigger and to permit a pull on the trigger to draw the tip of the sear attached to the trigger down out of engagement with the hammer notch to release the hammer for firing. There are actually two sears—the rear or firing sear that connects hammer and trigger during single-shot operation and the front or safety sear. This safety sear performs two functions. It blocks the hammer to prevent it from falling except when the bolt and the bolt carrier are fully home. It is struck by the bolt carrier at the final movement of closing travel when it releases the hammer so the firing sear may be operated on single-shot operation to release the trigger, or so the hammer may fire in full automatic fashion if the firing sear is out of engagement due to the automatic fire control being set on the change lever.

The entire operation is extremely simple mechanically and a

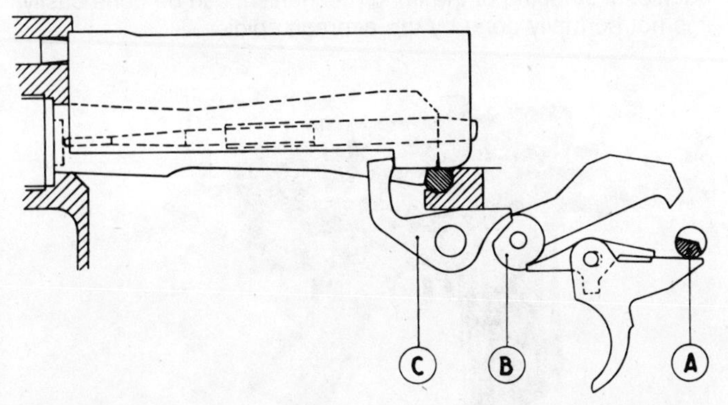

Gun with safety, applied. Change-lever stud (A) is locking trigger to prevent movement.

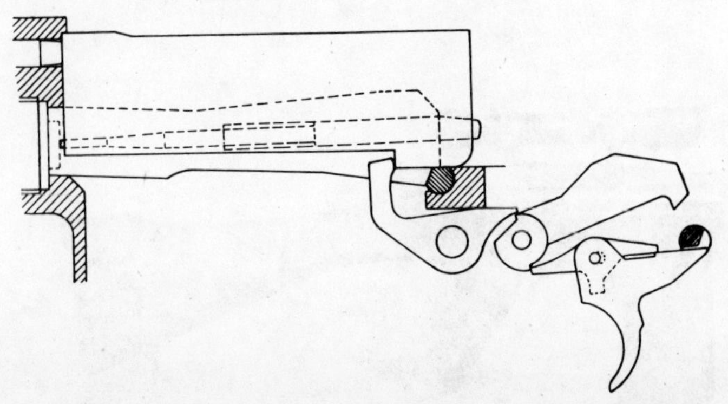

Change-lever stud set for single-shot fire. Pressure on trigger will free sear from hammer contact, but as the rifle action opens, the safety sear will hold the hammer until finger pressure is released.

study of the drawings accompanying the explanation will clearly demonstrate all the mechanical features involved.

The ejector is mounted in the receiver and serves as a guide in its slot in the lower side of the bolt. At the end of the recoil stroke, the cartridge case strikes the ejector as the bolt passes slightly past the nose of the ejector to hurl the empty case out of the gun.

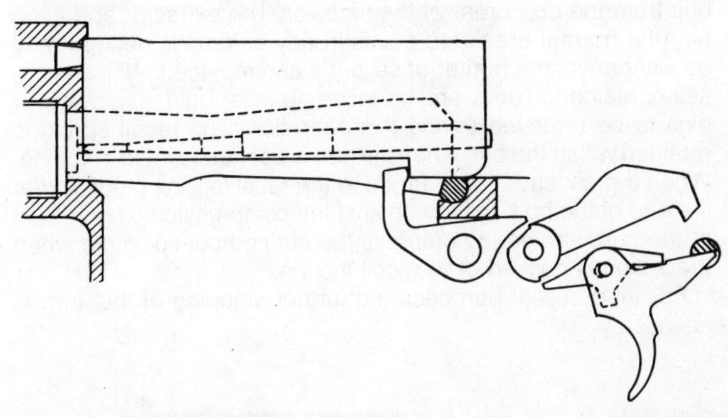

Lever is set for full-auto fire. Sear is out of engagement except when the trigger is released. Safety sear controls firing to prevent premature discharge.

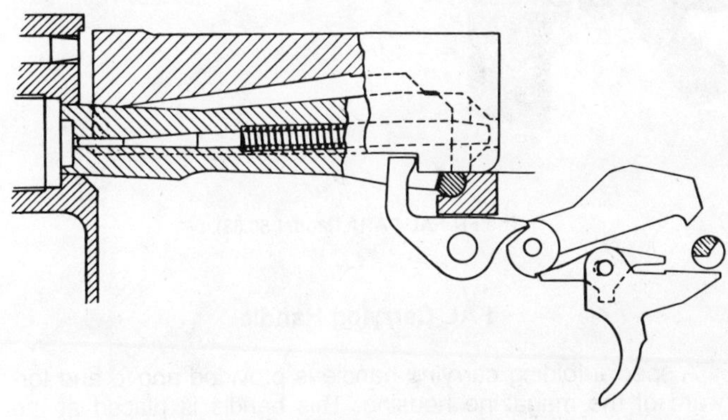

The safety sear mechanism is released by blow of the carrier as it goes home after bolt is locked, thus freeing the hammer to fire.

FAL Gas Regulator

A gas regulator and gas plug control the quantity of gas permitted to reach the piston. The gas plug A is fixed into the end of the gas cylinder B. This plug has two positions. One permits full access of the gas from the barrel directly to the gas cylinder. The other when turned through 180 degrees blocks off all entry of gas. In this condition, the arm will not function automatically. In this condition, however, it can be operated manually as a straight-pull bolt-action rifle. In other words, pulling the cocking handle back and releasing it will load and cock the gun ready for firing. After firing, another full pull to the rear and release of the cocking handle will eject the empty case and load a new round ready for another pull.

In this closed position, the arm may be used as a grenade launcher, since automatic action is not desired for such a purpose.

A gas regulator C consists of a shroud around the end of the gas cylinder. Unscrewing allows gas to escape.

This system of gas regulation by exhaust keeps fouling of the piston to an absolute minimum and allows regulation of power.

The gas cylinder is placed above the barrel. The center of gravity of the weapon is in line with the axis of the barrel. As a result, the recoil does not tend to pull the weapon upwards as much as many semiautomatic rifles.

FAL Field Stripping

Stripping and assembling this weapon for normal maintenance and repairs is done without the aid of tools.

With the magazine removed, the release catch at the rear of the top of the receiver is pressed, and the body of the gun is hinged exactly as in the case of the familiar shotgun. The entire bolt and gas operating assembly may now be withdrawn as a unit from the open rear of the receiver. The extractor and firing pin (the normal breakage points in any automatic design) may be withdrawn in a matter of seconds and replaced. All parts are self-contained. There are no loose springs, guides, screws or pins to be removed during this operation. The recoil spring is retained within the butt when the gun is opened for field stripping. When the assembly is returned to the receiver, the guide rod at the end of the bolt, resting against the compression plug in front of the butt, serves to compress the spring housed within when the action is cocked or in recoil motion.

For all practical purposes, no further stripping of this arm is necessary.

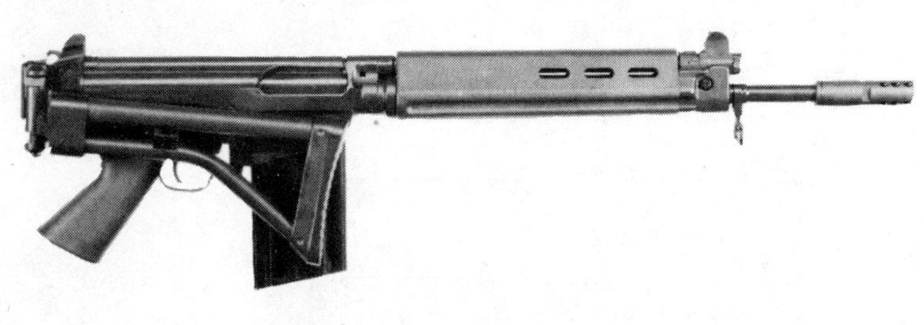

The FN FAL PARA (Model 50.63).

FAL Carrying Handle

A special folding carrying handle is provided above and forward of the magazine housing. This handle is placed at the center of gravity to make the arm easy to carry in rapid advances. It may also be used as a carrying handle for marching and general field order. It can be quickly turned down out of the line of sight and out of the way.

The grip of pistol type design behind the triggerguard aids greatly in stabilizing fire, and since all operations of loading and cocking and change lever adjusting are done on the left side of the receiver full control of the arm may be maintained at all times. In addition, of course, this pistol grip gives the weapon the advantage of ease of operation for firing at waist level under appropriate conditions. The forestock is designed for comfort and secure control as a normal left hand hold for shoulder firing or waist firing.

As special accessories, a muzzle brake and flash hider are available. A grenade launcher is also part of the accessory equipment as is a special bayonet and a detachable folding bipod mount.

FAL Further Dismounting

Remove the magazine. Draw the cocking lever to the rear and check that the chamber is empty. Move the cocking handle back to free the bolt and allow it to go forward to locked closed position.

Push down the butt locking lever on the left rear side of the receiver as far as it will go. Simultaneously push the butt itself downwards. The butt will pivot together with the lower receiver section. This allows the arm to open as in the familiar case of the hinged-frame shotgun.

Pull back the spring rod attached to the bolt; this will pull the bolt and carrier assembly to the rear out of the gun.

Slide the receiver cover to the rear off the receiver.

Lift the front end of the bolt while pressing it forward into the carrier and continue lifting the front end to raise the rear gently out of carrier contact against the pressure of the firing pin spring.

Push out the cross retaining pin while holding onto the rear of the firing pin, and the pressure of its spring will force it out of its housing.

Insert the nose of a bullet under the extractor and pry outwards and upwards to withdraw the extractor.

The gas plug can be removed if desired and a rod passed through the gas cylinder to clean fouling.

The gas plug can be turned with the nose of a bullet for removal. This permits removal of the piston and spring from the gas cylinder. While there are very few remaining parts, and additional stripping of the firing mechanism can be done easily, it is not normally done by the average soldier.

FN 7.62mm FAL Rifle with low-mounted scope.

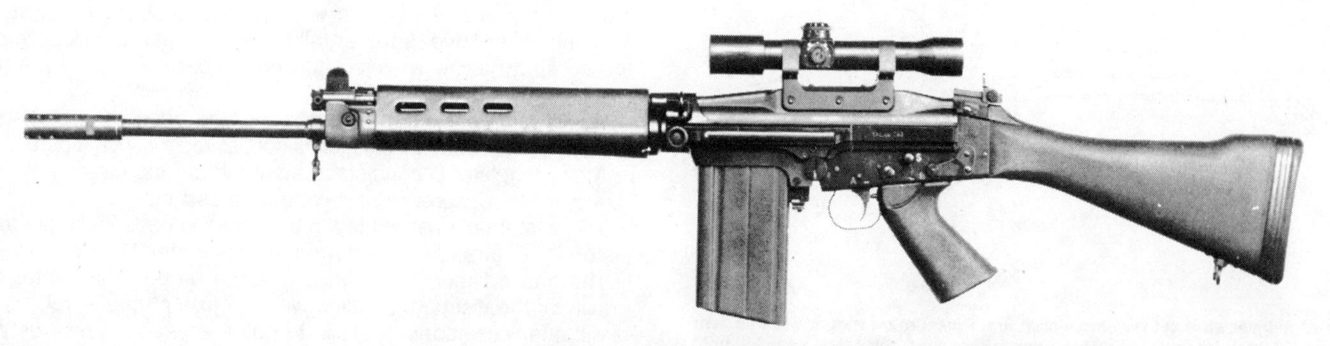

FN 7.62mm NATO FAL Rifle with high-mounted scope.

VARIATIONS AMONG SOME FN LIGHT AUTOMATIC RIFLES PRODUCED AT FN

	Barrel 1. Smooth muzzle 2. Threaded muzzle 3. Threaded muzzle for combined grenade launcher/flash suppressor	**Flash suppressor** 1. Without 2. With 3. Combined with grenade launcher	**Bayonet** 1. Flash suppressor type or normal 2. Tubular type 3. Without	**Extractor in** 1. One piece 2. Two pieces	**Butt stock** 1. Without front socket 2. With front socket	**Butt plate** 1. Without butt trap 2. With butt trap	**Handguard** 1. Wood 2. Metal 3. Molded material	**Bipod** 1. Without 2. With	**Loading** 1. Without charger (Stripper clip) 2. With charger
Austria	3	3	3	2	2	2	2	2	1
Belgium	1	1	1	2	2	1	3	1	2
Cambodia	1	1	1	2	2	1	3	1	1
Chile	2	2	1	2	2	1	3	1	1
Ecuador	2	2	1	2	2	1	3	1	1
Indonesia	1	1	1	2	1	2	1	1	1
Ireland	2	3	2	2	2	2	3	1	1
Israel	1	1	1	1	1	1	1	1	1
Kuwait	1	1	1	2	1	2	1	1	1
Libya	1	1	1	2	2	1	3	1	1
Luxembourg	1	1	1	1	1	2	1	1	2
Netherlands	3	3	2	2	2	1	2	2	1
Paraguay	1	1	1	2	1	2	1	1	1
Peru	1	1	1	2	2	2	1	1	1
Portugal	1	1	1	2	2	1	3	1	1
Qatar	1	1	1	1 & 2	1	2	1	1	1
Santo Domingo	1	1	1	1	1	1	1	1	1
South Africa	1	1	2	2	2	1	3	1	1
Syria	1	1	1	1	1	2	1	1	1
West Germany	1	2	3	2	1	2	2	2	1
Venezuela	2	2	3	1	1	1	1	1	1

There are other variations; as for example, the British-produced L1A1 and the Canadian-produced C1A1.

Identification of FALs can be puzzling. Several modifications can aid in determining the origin of particular FALs. The West German G-1, the Austrian StG58 and the Dutch FAL all have a lightweight, folding metal bipod as part of their metal forestock; the British L1A1 and the Indian Ishapore rifles, capable of semiautomatic fire only, have zigzag dirt clearance cuts in the bolt carriers, folding operating handles and enlarged magazine catches and selectors. In addition, the FAL is found with or without flash suppressors, with different types of bolt covers and with forearms of different styles. It is often difficult to identify the original purchaser of a FAL unless the rifle is stamped with an identifying seal or crest.

More detailed information on the FAL can be found in the following series by R. Blake Stevens:

Stevens, R. Blake. *North American FALs: NATO's Search for a Standard Rifle* (Toronto: Collector Grade Publications, 1979).

Stevens, R. Blake. *UK and Commonwealth FALs* (Toronto: Collector Grade Publications, 1980).

Stevens, R. Blake and Jean E. Van Rutten. *The Metric FAL: The Free World's Right Arm* (Toronto: Collector Grade Publications, 1981).

CARBINE AUTOMATIQUE LÉGÈR (CAL)—LIGHT AUTOMATIC CARBINE, 5.56mm

Engineers at Fabrique Nationale began working on 5.56 × 45mm rifle designs as early as 1963. At first, they considered scaling down the FAL to take 5.56mm ammunition, and FN model makers did build some FALs chambered for the 5.56 × 45mm cartridge. After further experimentation, FN introduced the CAL—the *Carbine Automatique Légère* or light automatic carbine—in 1966. Although the CAL resembled the FAL externally, the FAL has a tilting bolt and bolt carrier system, and the CAL has a cam-operated rotating bolt and bolt carrier system. The CAL bolt-head had an interrupted screw thread-type locking mechanism; there were two buttress-type lugs on the top and bottom of the bolt-head.

Tests of the CAL, especially those conducted in France between 1971 and 1974, indicated that this rifle was too difficult to disassemble and maintain, too costly to manufacture, and its life expectancy under combat conditions was too short. To correct these shortcomings, the FN engineers abandoned the CAL and created a completely new rifle, the FNC.

FN 5.56mm *Carbine Automatique Légèr (CAL).*

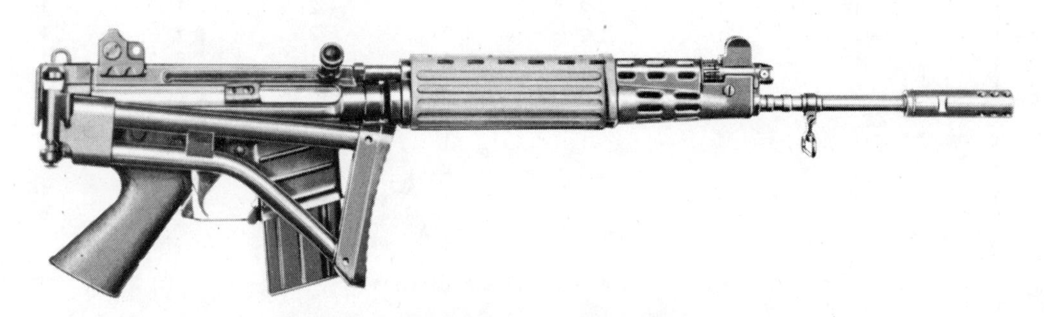

Folding stock version of the CAL.

Release of the magazine can be accomplished without removing the hand from the pistol grip.

CAL fitted with telescopic sight. Note also the three position selector: Safe, single shot, three shot burst and automatic fire.

Field Stripping the FN CAL

Detailed view of folding stock. Push release (A) upwards and push stock assembly (B) downwards. Fold the stock along the left side of the rifle.

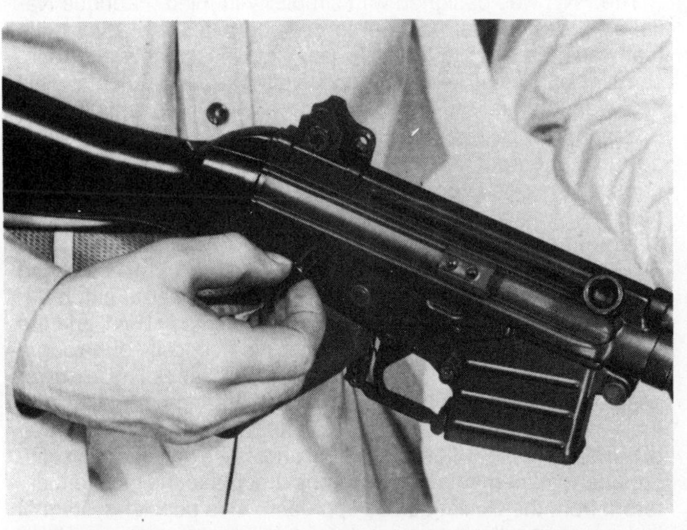

From left to right, press out the take down spring, and then swing open the two halves of the receiver. Pull the operating handle to the rear, remove it and elevate the muzzle, thus allowing the bolt to slide out the rear of the receiver.

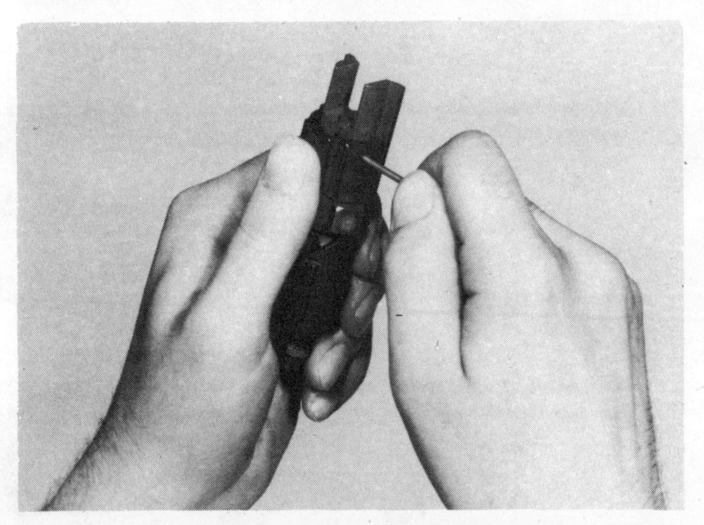

Take hold of the bolt carrier and push the bolt (A) fully to the rear. In this position, the lock pin (B) is in its low and rear position and the retaining pin can be seen in hole at (C). Using the drift pin from the assembly kit push out the firing pin retention pin.

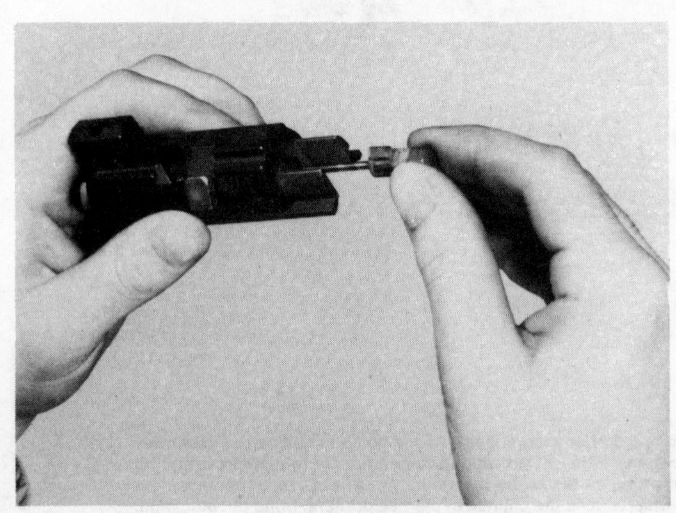

Withdraw firing pin.

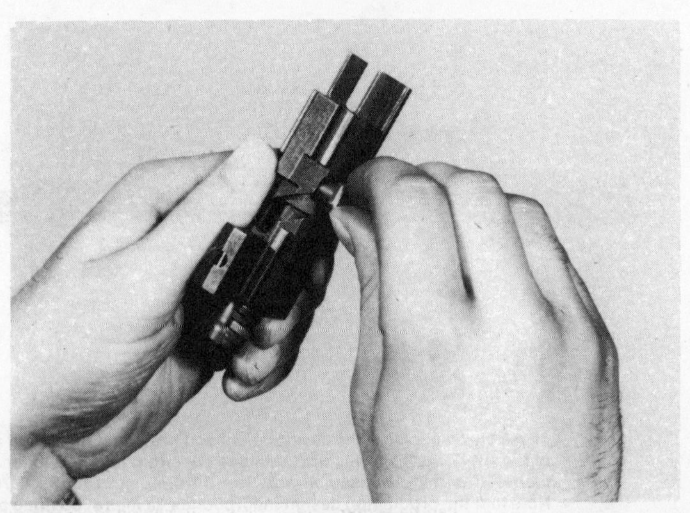

Remove lock pin.

FNC ASSAULT RIFLE

The FNC was designed with simplicity in mind. Fabrique Nationale's design team sought to develop a 5.56 × 45mm rifle that would be easy to maintain, easy to disassemble and reassemble, and most important, relatively easy to fabricate. This new weapon has upper and lower receiver components like the FAL and the CAL, but the FNC upper receiver is of a simplified design made from sheet metal stampings. The upper and lower parts hinge open after the fashion of the FAL. Upon opening the receiver, it is immediately apparent that the bolt assembly is different from that employed in either the FAL or the CAL. The FNC breech is locked by a rotary bolt-head, with two opposing lugs, that locks into the barrel extension. The barrel extension of the FNC (and in other weapons such as the M16A1 and the Kalashnikov assault rifles) eliminates the necessity of having a receiver assembly to take the locking stresses. The receiver assembly becomes a housing for the operating parts. Conceptually, this bolt and bolt carrier assembly is quite similar to that of the Kalashnikov rifles, but the FN engineers have added some important refinements and additions that make their system different from the Soviet design approach. This popular system of locking allows the minimum weight for the operating parts, a short bolt stroke, and a minimum of stress during the firing cycle.

FNC Operating Mechanism

The FNC operating mechanism (reciprocating parts) consists of two main parts: the bolt and the bolt carrier.

- The bolt contains the extractor.
- On the bolt-head there are:
  —the 2 locking lugs, which also guide the bolt in the receiver mounted guide rails.
  —the feed lug, which acts on the top round in the magazine and pushes that cartridge into the chamber.
  —a small lug on the top of the bolt, which initiates the rotary movement of the bolt-head and which ensures primary extraction.
- On the bolt-body there is a stud, which connects the bolt to the bolt carrier and which permits the bolt to be cammed into the locked and unlocked positions.
- The bolt carrier combines the piston and carrier assemblies. It consists of three welded parts—the piston, the piston rod (hollow to contain the return spring assembly), and the bolt carrier (into which the firing pin is mounted and which contains the bolt cam track). The cocking handle is fitted into this assembly. On both sides of the bolt carrier there are two guiding lugs that run in the same receiver guide rails as the bolt-head.
- The return spring is housed in the piston rod. When disas-

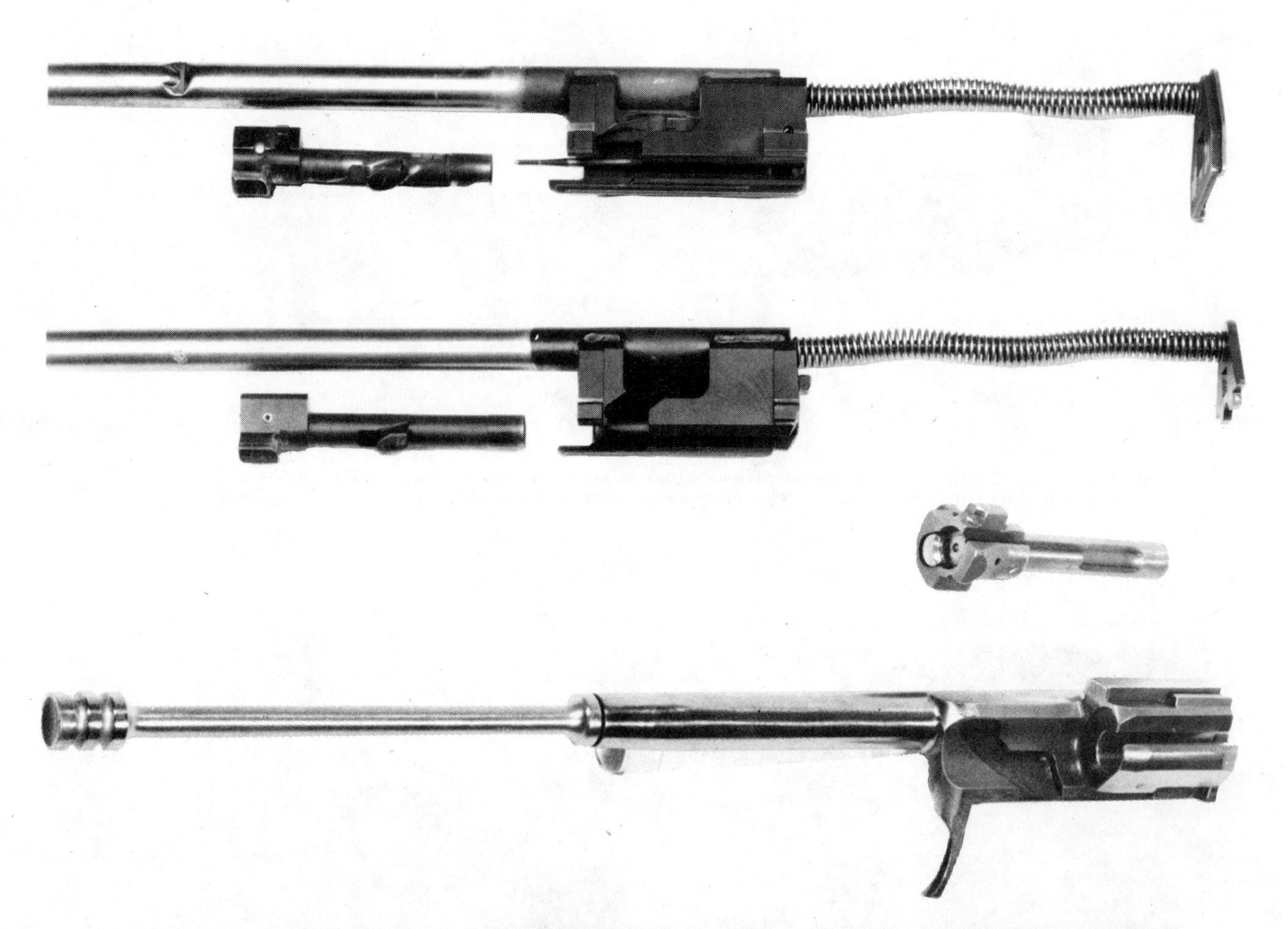

These views compare the development and production FNC bolt and bolt carrier assemblies with the bolt and bolt carrier assembly of the AK47. At the top is the bolt assembly from the production version of the FNC in which the firing pin is retracted by the bolt carrier. The middle assembly was taken from one of the FNC 76 prototypes tested by the NATO small arms trials commission. Note that the production design incorporates an improved firing pin that reduces the chance of pierced primers, firing pin blockage, and gas leakage into the bolt/bolt carrier assembly. The bottom assembly is taken from a Soviet AK47.

Top, **the FNC 76 as tested by NATO;** ***below,*** **the FNC 80 as adopted by Indonesia and Sweden. The production model has a strengthened buttstock to allow grenade launching with the integral flash suppressor/grenade launcher, It also has a simplified handguard assembly, strengthened cocking handle track cover, integral telescopic sight mounting blocks, magazine stop, curved 30-shot magazine, and other improvements incorporated as a consequence of the NATO trials.**

sembling the FNC, the return spring assembly, the piston rod/bolt carrier assembly, and the bolt are removed together from the upper receiver. After removal they are separated. The return spring is retained on its guide rod by a washer and a retaining pin. This pin fits into a recess in the gas piston tube to lock the return spring assembly in place.

FNC Buttstock

The buttstock is fabricated from light alloy tubing to which the butt plate assembly is molded. The upper tube has a plastic coating to insulate it in both tropical and arctic climates. The buttstock folds to the right side of the weapon and is the same pattern as that used on the FAL and the CAL.

FNC Magazine

The FNC magazine is fabricated from steel (including the follower) and it has a 30-shot capacity. This feed device can be used with both the Minimi squad automatic weapon and the M16A1 rifle. The FNC magazine will not actuate the M16A1 bolt hold-open device when the last shot has been fired. The 20-shot and 30-shot M16A1 magazines can be used in the FNC and the Minimi.

FNC Production Technology

Fabrique Nationale designers have created a weapon that can be manufactured using the most up-to-date and automated manufacturing processes (i.e., investment castings, computer numerical control machines, robot welding, and hammer-forged barrels). By employing these processes they have managed to keep the cost of the FNC in line with other competitive 5.56 × 45mm rifles. Thirty parts receive automated heat treatment, including the trigger, the bolt carrier, the bolt, the gas piston, the hammer, and the ejector. The barrel bore, the barrel chamber, the gas port block, and the piston are all hard chrome plated using the FN automated process. Computer numerical control machine tools produce the trigger group frame, the buttstock block support, and the exterior finishing of the barrel after it has been hammer forged. There are a total of 121 components in the FNC. The fabrication of these parts requires 421 machine operations and 98 manual operations.

How to Field Strip the FNC

The accompanying photographs illustrate the steps involved in the disassembly and assembly of the FNC.

FNC CHRONOLOGY

1976–1981	Swedish Arms tests. FNC selected over FFV 890C (modified Galil), SIG 540, Colt M16A1, and HK 33.
1976–1978	NATO tests of FNC and other small arms.
1978–1979	FN carries out redesign of FNC 76 based upon experience in NATO trials.
September 1980	FN begins series production of FNC 80.
March 1981	Swedish Army selects FNC 80.
1981	Indonesian armed forces adopt FNC 80.
Fall 1981–	Australian and Canadian armed forces testing FNC as one of several contenders in national small arms trials.

Field-stripped view of the production of the FNC 80.

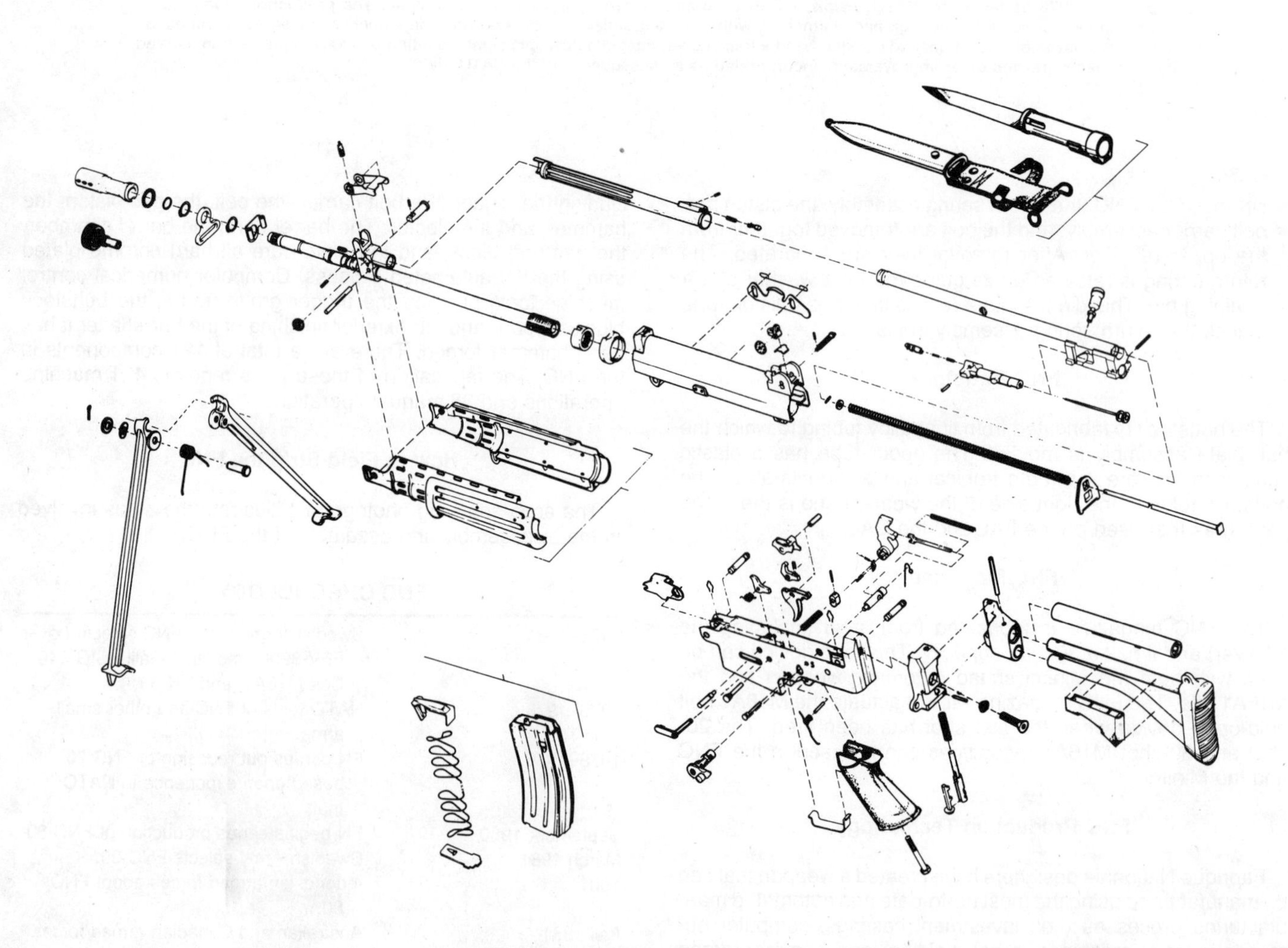

Exploded view of the FNC 80.

The Short Barrel FNC with its 14.29 inch barrel.

CHARACTERISTICS OF THE CAL

Caliber: 5.56 × 45mm (1 in 12 in. twist for M193)
System of operation: gas, selective fire
Length overall: 38.6 in.
Barrel length: 18.4 in.
Feed device: 20-round, detachable, staggered row box magazine
Sights: Front: protected post
Rear: "L" with apertures
Muzzle velocity: 3182 f.p.s.
Weight: 7.3 lbs. loaded w/light alloy magazine
Cyclic rate of fire: approx. 750–800 r.p.m.

CHARACTERISTICS OF THE FNC 80 STANDARD

Caliber: 5.56 × 45mm (1-in-12-in. twist for M193; 1-in-7-in. twist for SS109 (XM855); and 1-in-9-in. for Swedish humane cartridge).
System of operation: gas, selective fire, including 3-shot burst
Length overall: 39.25 in. with stock fixed
30.16 in. with stock folded
Barrel length: 17.68 in.
Feed device: 30-round, detachable, staggered row box magazine (FNC or M16A1).
Sights: front, protected post; rear, "L" with 250m and 400m apertures
Muzzle velocity: 2890 f.p.s.
Weight: 9.61 lbs. loaded w/steel magazine
Cyclic rate of fire: 625—700 r.p.m.

FNC Model Numbers

Standard FNC 80 with 1-in-7-inch twist barrel	2000
Standard FNC 80 with 1-in-12-inch twist barrel	0000
Short barrel FNC 80 with 1-in-7-inch twist barrel	7000
Short barrel FNC 80 with 1-in-12-inch twist barrel	6000
Law enforcement FNC 80 with 1-in-7-inch twist barrel; semiautomatic only	7030
Law enforcement FNC 80 with 1-in-12-inch twist barrel; semiautomatic only	6040

The FNC 80 with a 4X German Hensoldt telescopic sight mounted to the integral scope mounts.

Field Stripping the FNC

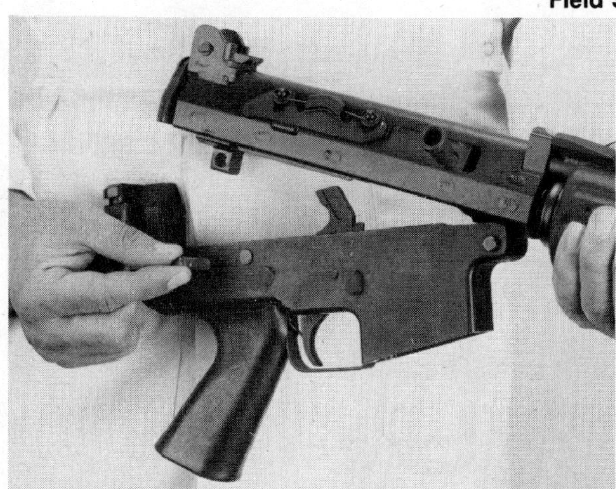

To open the receiver assembly, push the rear pin from the left to the right until it reaches its stop point. Then hinge the receiver assembly open as shown in the photo.

After opening the receiver assembly, pull the cocking handle to the rear until it reaches the channel cover. Then raise the channel cover with the thumb, pull the cocking handle further to the rear and when it stops, pull the cocking handle free.

After the cocking handle has been removed, the bolt and bolt carrier assembly can be withdrawn through the rear of the upper receiver.

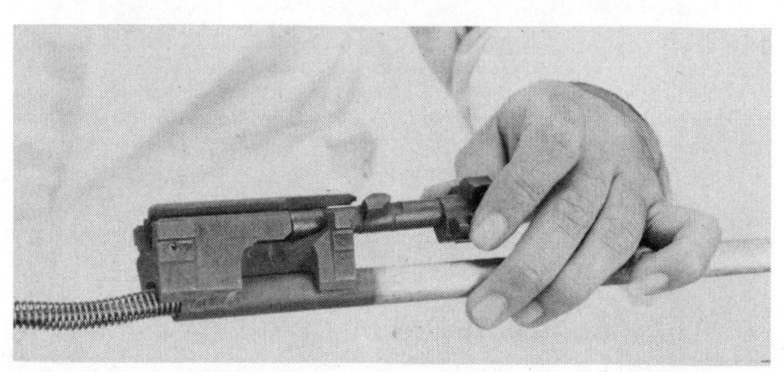

To disassemble the bolt and bolt carrier assembly, press in on the rear plate of the return spring assembly and rotate that plate one quarter turn to the left or right. Then remove the spring assembly by pulling it to the rear. Once the spring assembly has been removed, the bolt can be freed by rotating it and pulling it away from the bolt carrier. *Be careful!* When the bolt has been removed from the slide, the firing pin spring is loose. Take care not to lose it. Except in very special cases, the extractor should not be removed from the bolt. When it is disassembled, the work should be done by an armorer who knows how to do it.

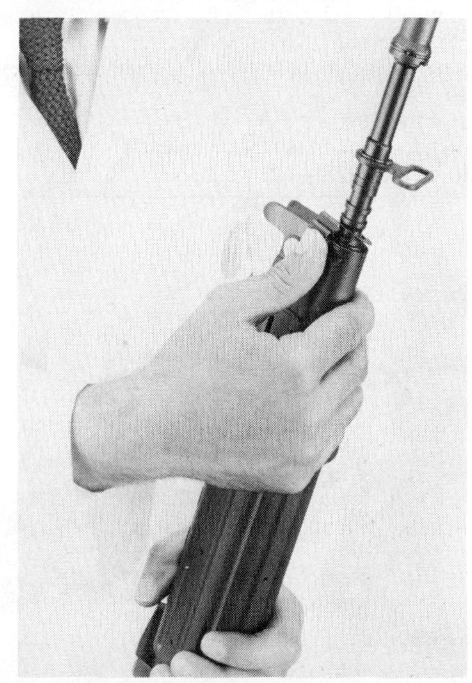

To remove the handguards, force the front retaining clip out of its notch in the handguard with the thumb. This has to be done for both the right and left sides of the handguard.

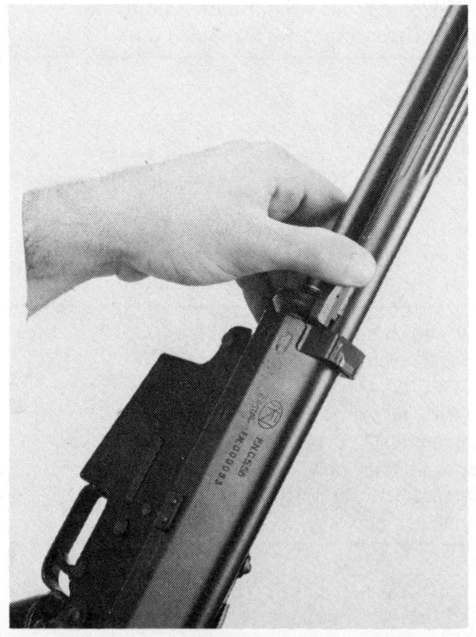

Gas cylinder removal begins with setting the gas regulator to the left. Continuing pressure will cause the gas regulator to continue beyond the "normal" setting point. Continue this rotation until the thumb piece is perpendicular with the front block of the receiver.

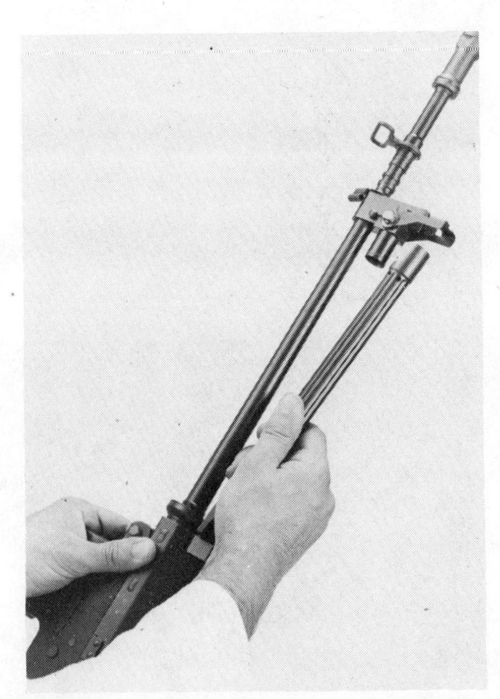

The gas cylinder can then be moved to the rear and lifted up and away from its seat of the gas block portion of the front sight assembly.

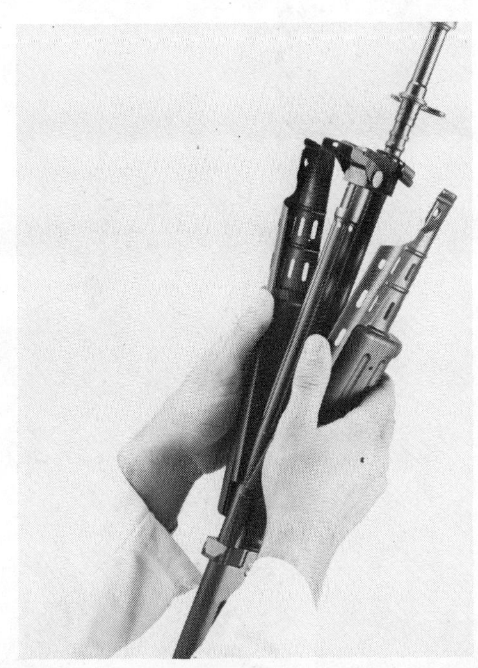

When reassembling the FNC, make certain the grenade launching sight (alidade) is in the vertical position. Replace the gas cylinder in the reverse of the disassembly process. Set the gas regulator to the "normal" setting. Then replace the two halves of the handguard assembly.

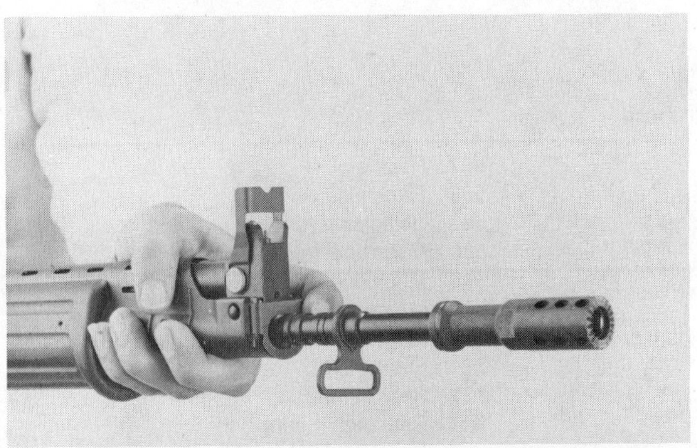

Engage the retaining clip, and the handguards will be locked in place.

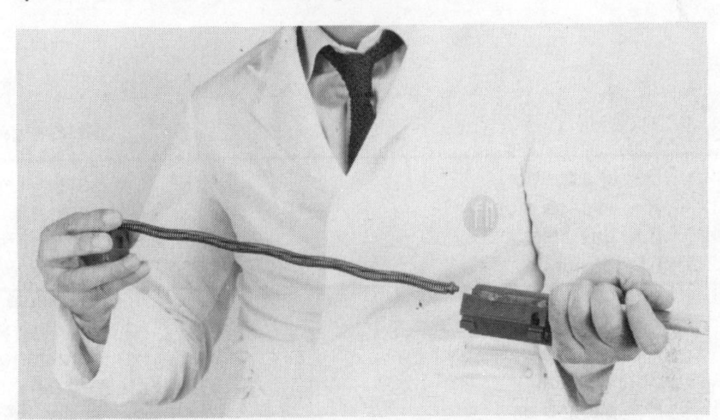

After replacing the firing spring pin on the firing pin, slide the bolt over the firing pin and rotate the bolt into its proper position in the bolt carrier. Insert the return spring into the piston tube of the bolt carrier assembly. Push the return spring about three-quarters of an inch (20mm) and then turn the rear plate one quarter turn to either the left or right to lock the return spring assembly in place.

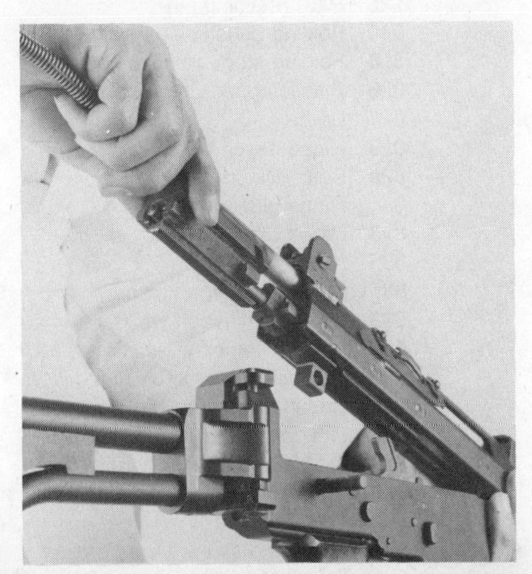

Once the bolt and bolt carrier assembly have been reassembled, they can be inserted into the receiver and slid forward. When inserting this assembly, take care to properly position the locking lugs of the bolt and those of the bolt carrier into the receiver guide rails.

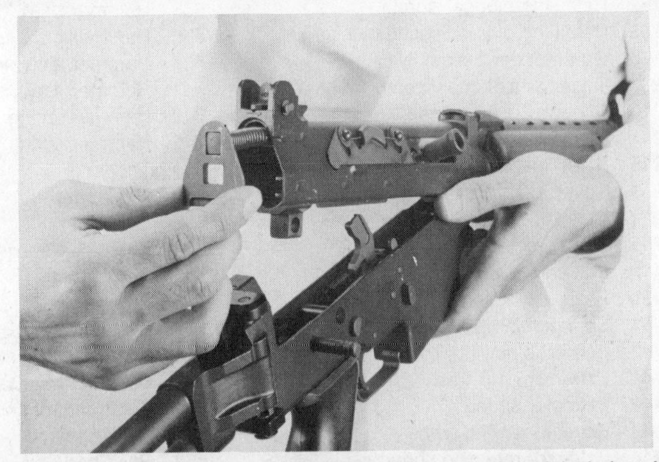

After reassembly of the cocking handle, the forward motion of the bolt and bolt carrier assembly can be completed. Then seat the rear plate of the return spring and close the receiver halves. The final step in assembly of the FNC is the right to left pushing of the assembly pin. The rifle can then be loaded and fired.

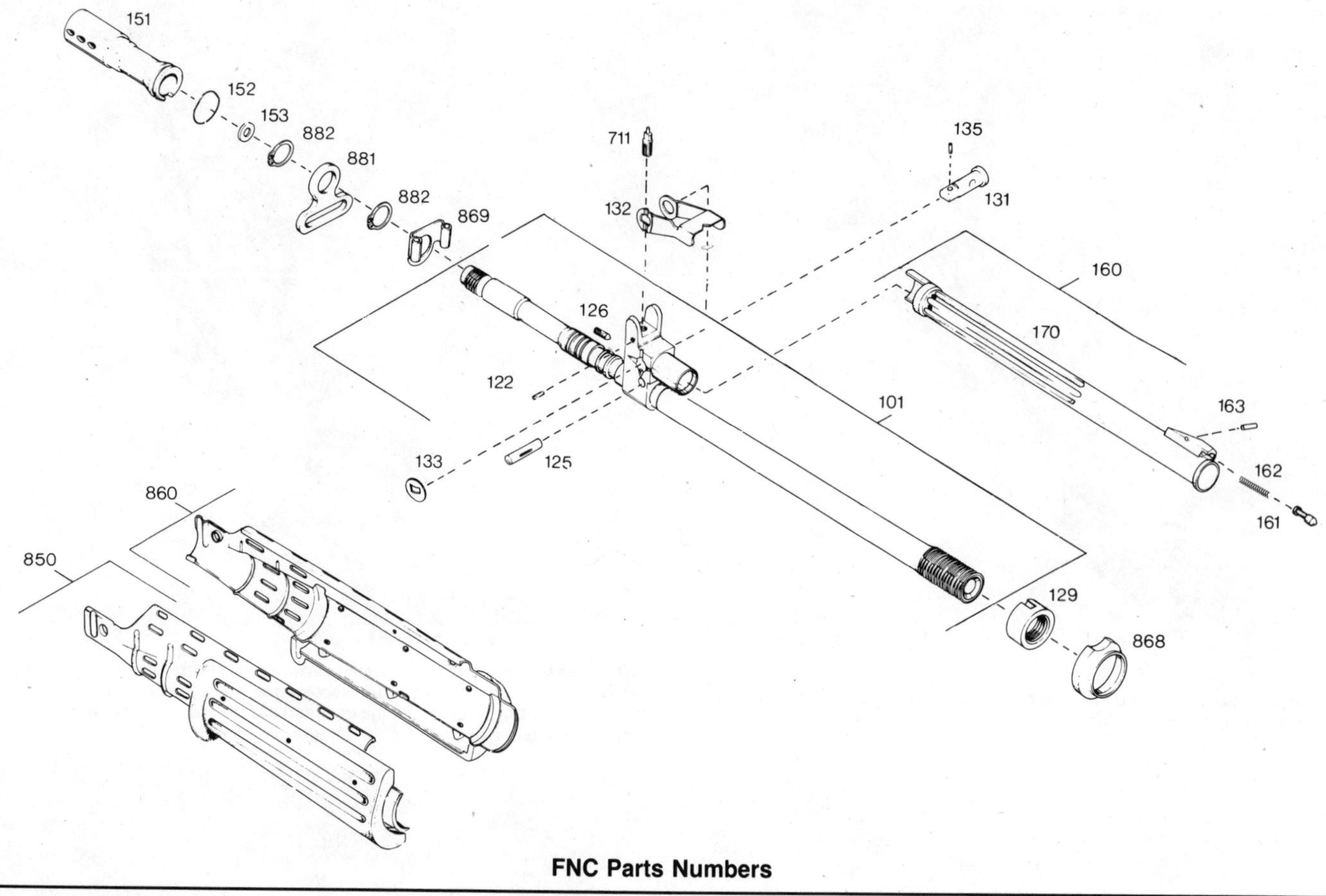

FNC Parts Numbers

101	Barrel assembly
122	Pin, alidade stop
125	Pin, gas block
126	Center punch
129	Breech block adjusting nut
131	Axis, alidade (grenade sight)
132	Alidade (grenade sight)
133	Spring alidade
135	Pin, alidade axis
151	Flash suppressor
152	Retaining spring, grenade
153	Adjusting washer, flash suppressor
160	Gas cylinder assembly complete
161	Indicating plunger
162	Retaining plunger spring
163	Retaining pin, indicating plunger
170	Gas cylinder assembly
201	Receiver assembly complete
210	Receiver components
241	Ejector
245	Rivet, ejector
251	Cover, cocking channel
254	Washer, cocking channel cover
255	Spring, cocking channel cover
410	Bolt carrier assembly
430	Firing pin assembly
433	Spring, firing pin
440	Bolt assembly
441	Extractor
442	Extractor spring
443	Extractor spring guide
444	Extractor retaining spring
445	Pin, extractor, inner
446	Bolt-head
460	Return spring assembly
461	Stop plate
465	Washer, rest
466	Retaining pin, rest washer
471	Return spring guide
481	Cocking handle
511	Trigger group housing
512	Trigger guard retaining pin
514	Retaining pin circlip
516	Index spring
521	Automatic sear
522	Automatic sear axis pin
525	Automatic sear spring
531	Hammer
532	Hammer pawl pin
533	Hammer pawl
536	Hammer pawl spring
538	Hammer axis pin
541	Hammer stay
542	Hammer spring
543	Hammer spring rest
551	Trigger
552	Front sear
553	Rear sear
554	Bush, sears axis pin
555	Sears spring
556	Trigger axis
557	Trigger spring
558	Trigger spring axis
560	Burst control assembly
561	Burst control housing
562	Rack assembly
563	Rack body
564	Rack stud
565	Rack spring
566	Secondary rack
567	Rack stop
568	Rack axis pin
580	Selector lever
590	Magazine catch assembly
595	Magazine catch spring
596	Magazine catch button
597	Magazine catch pin
711	Front sight
721	Rear sight
722	Rear sight spring
723	Rear sight axis pin
725	Rear sight spring pawl
726	Pawl retaining pin
810	Folding buttstock assembly
820	Folding stock components
826	Plastic upper tube sleeve
827	Brace
828	Hinge assembly block
829	Buttstock pin
831	Hinge plunger
833	Hinge retaining pin
835	Buttstock bolt
837	Buttstock support block
838	Buttstock support washer
839	Buttstock support screw
850	Handguard assembly, left
860	Handguard assembly, right
868	Rear barrel collar
869	Handguard spring
871	Pistol grip
872	Trigger guard
876	Pistol grip screw
881	Front sling swivel
882	Sling swivel circlip

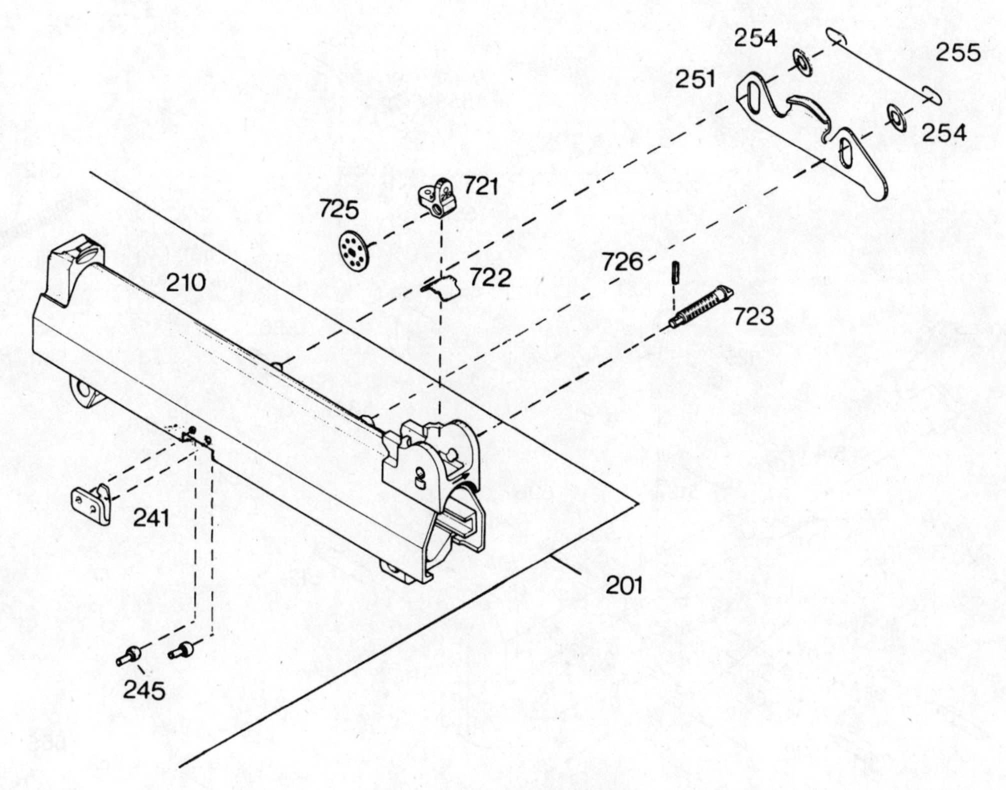

FNC upper receiver assembly.

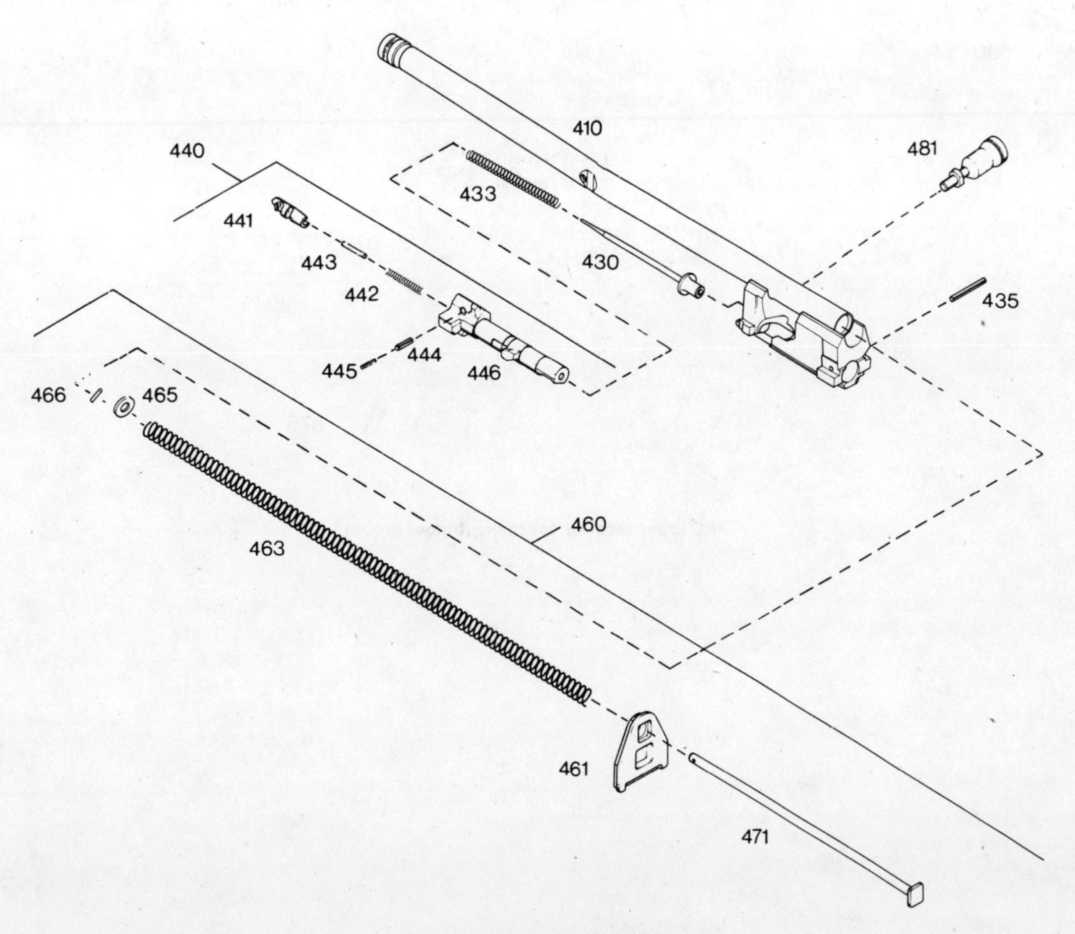

FNC bolt and bolt carrier assembly.

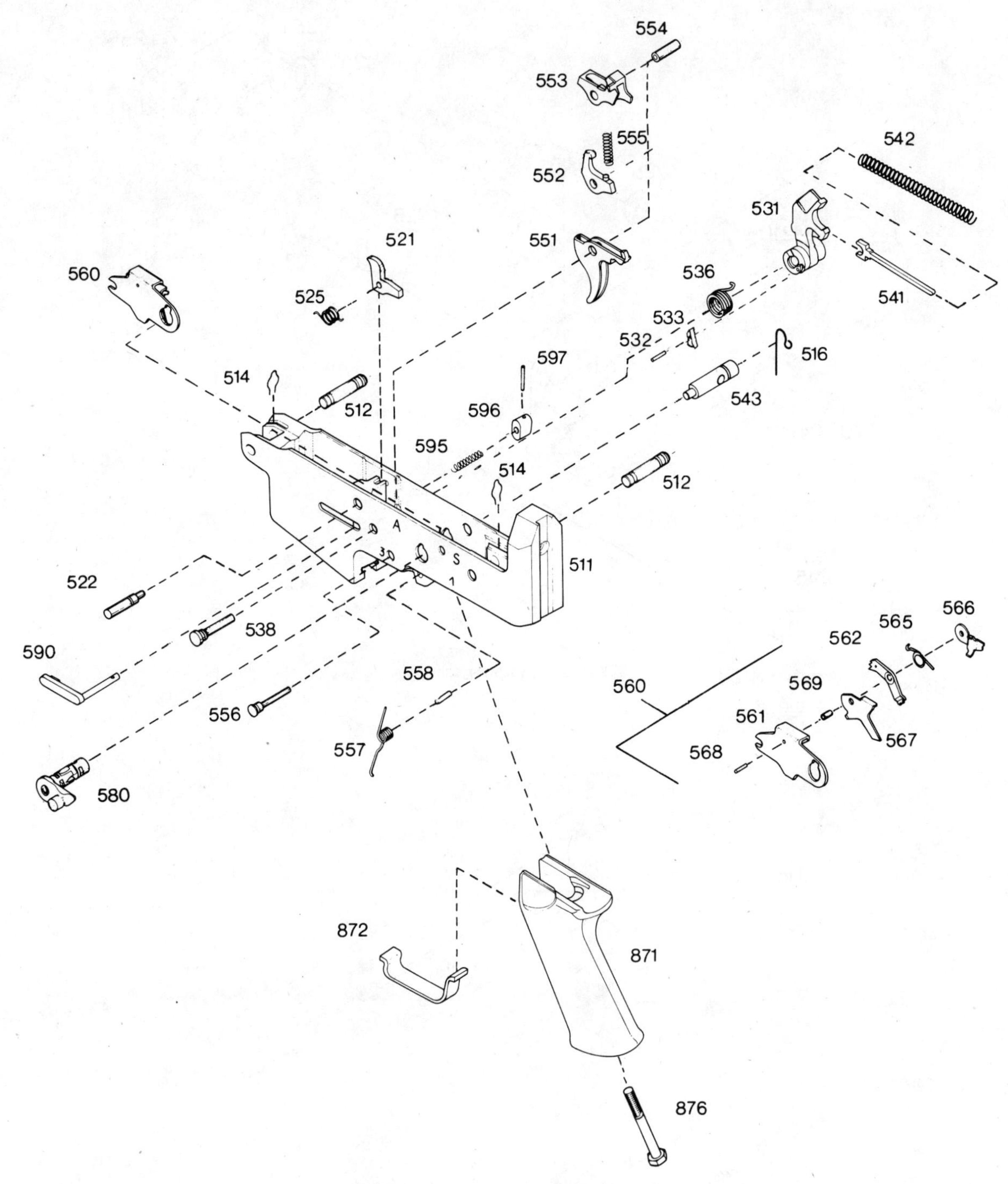

FNC lower receiver assembly (trigger group).

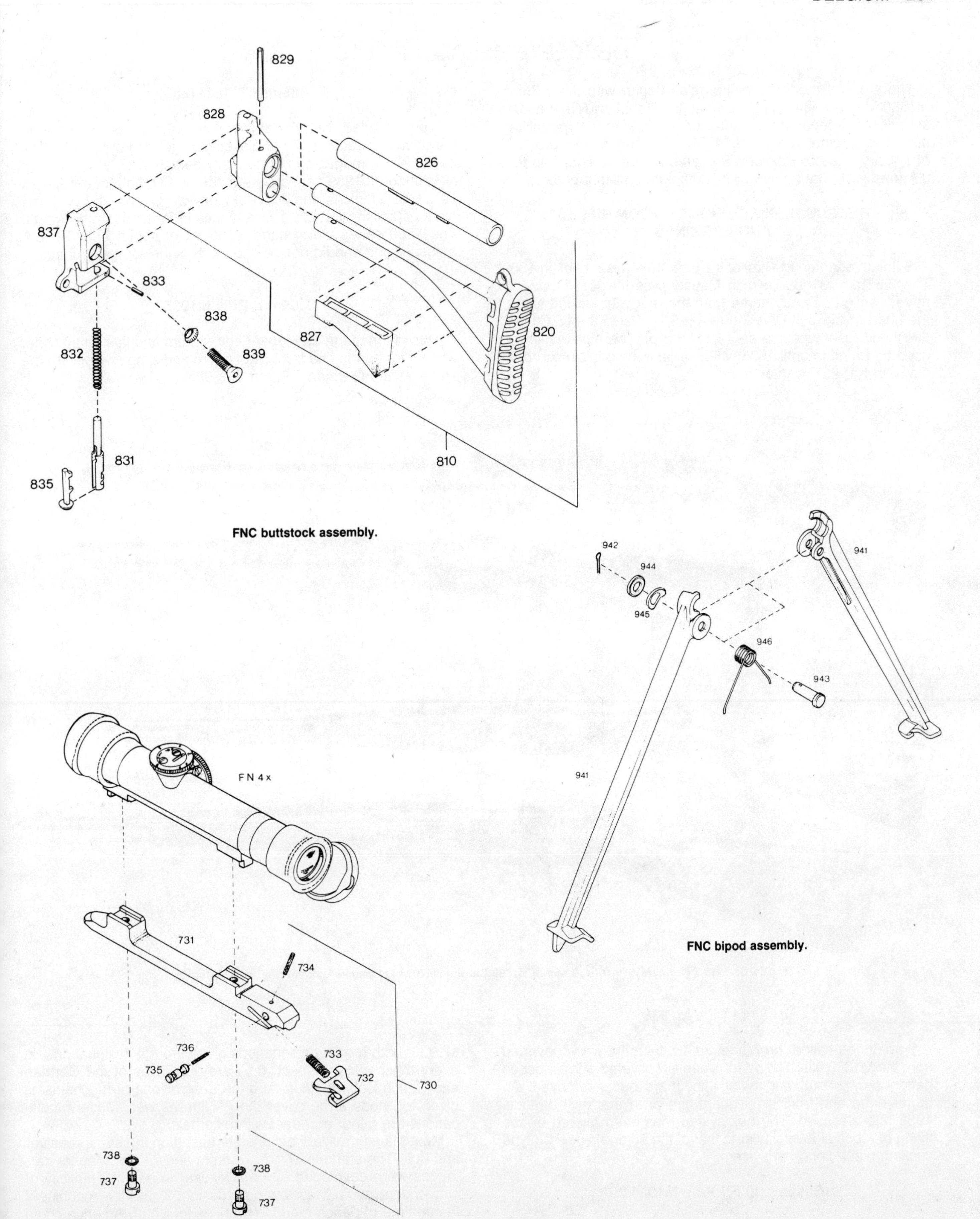

FNC buttstock assembly.

FNC bipod assembly.

FNC telescopic sight mount assembly.

OBSOLETE BELGIAN RIFLES

The story of modern Belgian rifles began with the 7.65mm M1889 Mauser, which is considered the first of the Modern Mauser rifles. FN began the manufacture of Mauser 98 type military rifles and carbines around 1924. FN is the only current producer of military rifles in Belgium; the government arsenal Fabrique d'Armes de L'Etat no longer manufactures military rifles.

BELGIAN SERVICE BOLT-ACTION RIFLES AND CARBINES

Belgium adopted the first of the modern Mauser rifles in 1889. The 7.65mm M1889 Belgian Mauser was the first Mauser to have a solid bolt body bored from the rear with locking lugs at the head of the bolt. The Model 1889 was also the first rifle to use the Mauser type charger (stripper clip). This type of rifle was used by Belgium until 1935 when a rifle generally similar to the German Kar 98K was adopted.

7.65mm Rifle M1889

Long barreled, with metal jacket covering barrel to serve as handguard; straight bolt handle; single line magazine protrudes below stock, magazine not normally detached.

The Model 1889 series were made by FN, Fabrique d'Armes de L'Etat, Hopkins and Allen of Norwich, Conn., and an arms plant established by Belgian refugees in Birmingham, England. The 1889 series were made for loads of about 39,000 P.S.I. pressure and should not be used with higher-powered loads.

7.65mm Rifle M1935

This rifle has the 98 Mauser bolt system and flush magazine. Except for bands, the front sight guard and sling swivels are the same as the German 7.92mm Kar 98K.

Top to Bottom: 7.65mm M1889 Rifle, 7.65mm M1889 Carbine, M1924 FN Mauser-system rifle, M1924 Carbine.

7.65mm Rifle M1936

Sometimes called the Model 89/36, this rifle was converted from the M1889 rifle. The barrel is partially covered with a wooden handguard, the tubular metal handguard being removed. An upper band and front sight guard similar to that of the Model 1935 rifle are used. The bolt system has been altered by the fitting of a bolt sleeve similar to that of the 98 and a new 98 type cocking piece firing pin system.

Caliber .30 FN Rifle M1924/30

Used in limited quantity by the Belgian Army after World War II, it is standard Model 98 type. This rifle is also known as the M1930; both model designations given are FN designations. In every mechanical respect, they are duplicates of the German and Czech Service rifles, and the descriptions of mechanisms given for those arms cover the FN line as well. Model modifications are minor and deal with externals.

Most Mauser military actions are based on the same receiver and bolt. The standard 7mm, 7.65mm and 7.9mm Mauser cartridges were developed by Paul Mauser to permit simplicity of conversion, as well as low cost design. Thus, the machinery that in time of peace made 7mm rifles for South America could be readily converted to 7.9mm German Service when needed.

CHARACTERISTICS OF M1924 RIFLES IN CALIBER 7mm, 7.65mm OR 7.9mm

Length of rifle without bayonet: 43.3 in.
Length of rifle with bayonet: 58.2 in.
Length of barrel: 23.2 in.
Weight without bayonet: 8.5 lbs.
Number of cartridges in magazine: 5.

The Model 1924 and 1924/30 rifles and carbines were made for Argentina, Belgium, Bolivia, Brazil, Chile, China, Columbia, Costa Rica, Ecuador, Iran, Liberia, Lithuania, Luxembourg, Mexico, Paraguay, Peru, Turkey, Uruguay, Venezuela, Yemen and Yugoslavia.

CHARACTERISTICS OF M1924 CARBINES IN CALIBER 7mm, 7.65mm OR 7.9mm

Length of carbine: 37.4 in.
Length of barrel: 17.3 in.
Weight of carbine: 7.3 lbs.
Number of cartridges in magazine: 5.
Lowest rear sight graduation in meters: 200.
Highest rear sight graduation in meters: 1400.

Caliber .30 FN Mauser rifle as made after W.W.II.

CHARACTERISTICS OF PRE-WORLD WAR II BELGIAN SERVICE BOLT-ACTION RIFLES AND CARBINES

	Rifle M1889	Carbine M1889*	Rifle M1935	Rifle M1936
Caliber:	7.65mm Mauser.	7.65mm Mauser.	7.65mm Mauser.	7.65mm Mauser.
Overall length:	50.13 in.	41.16 in.	43.6 in.	Approx. 43 in.
Barrel length:	30.69 in.	21.65 in.	23.5 in.	23.7 in.
Feed device:	5-round in line box magazine.	5-round in line box magazine.	5-round staggered box magazine.	5-round in line box magazine.
Sights: Front:	Barley corn.	Barley corn.	Barley corn.	Barley corn.
Sights: Rear:	V notch, leaf sight.	V notch, leaf sight.	V notch, tangent.	V notch, tangent.
Muzzle velocity: (at date of adoption)	2034 f.p.s.	1900 f.p.s. (approx).	2755 f.p.s.	2378 f.p.s.
Weight:	8.88 lbs.	7.75 lbs.	9.lbs.	8.7 lbs. (approx).

*The other three carbines vary in detail.

FUSIL AUTOMATIQUE MODELE 49 (FN SELF-LOADING RIFLE-SAFN)

Prior to World War II, Dieudonne Saive developed a gas operated rifle that was intended to replace the bolt-action Mausers of the Belgian Army and also to be offered to the armies of the world as a replacement for their bolt-action rifles. It should be remembered that only the US and USSR had adopted semiautomatic rifles as the standard shoulder weapon at that time. The occupation of Belgium by the Germans in 1940 halted work on the self-loading rifle, and it did not appear on the world market until the end of World War II. The rifle was offered in caliber .30, 7mm, 7.65mm and 7.92mm and was adopted by Belgium in caliber .30 model 1949, Egypt in 7.92mm, the former Netherlands East Indies in caliber .30, Brazil in caliber .30, Venezuela in 7mm (model 49), Luxembourg in caliber .30, Argentina in 7.65mm, the former Belgian Congo in caliber .30, and Columbia in caliber .30.

The tilting bolt of this rifle locks on a bar set in the bottom of the receiver, cammed into and out of the locked position by cam slots on the bolt carrier in engagement with lugs on the rear of the bolt. This bolt system is essentially the same as that of the Soviet Tokarev rifles and the FN FAL rifle. Specimens of this rifle were made that had selective fire capability.

This rifle is frequently referred to as the ABL or SAFN. ABL *(Arme Belgique)* is the marking found on the rifles made for the Belgian government; SAFN stands for semiautomatic FN.

CHARACTERISTICS OF FN SELF-LOADING RIFLE (SAFN)

Caliber: .30 M2, 7mm, 7.65mm, and 7.92mm Mauser.
System of operation: Gas, semiautomatic.
Feed device: Projecting steel box. Capacity 10 cartridges. Loaded single shot or from 5-shot clip.
Barrel length: 23.2 in.
Overall length: 43.7 in.
Weight: 9.48 lbs.
Sights: Front, shielded post; rear, tangent.

Operation: Tappet driving back through hole in receiver above line of barrel strikes bolt carrier and starts it to the rear until pressure has dropped. At unlocking point, the housing is machined to cam the bolt up a ramp at its rear end, thus allowing the carrier and bolt to travel to the rear with the bolt carrying the cartridge case in its face, held by the extractor, until it strikes the ejector and is tossed out of the weapon. Recoil springs compressed during this motion start the housing and bolt forward at the end of the recoil stroke. Upon closing, as the cartridge is chambered and the bolt face is against the breech face of the

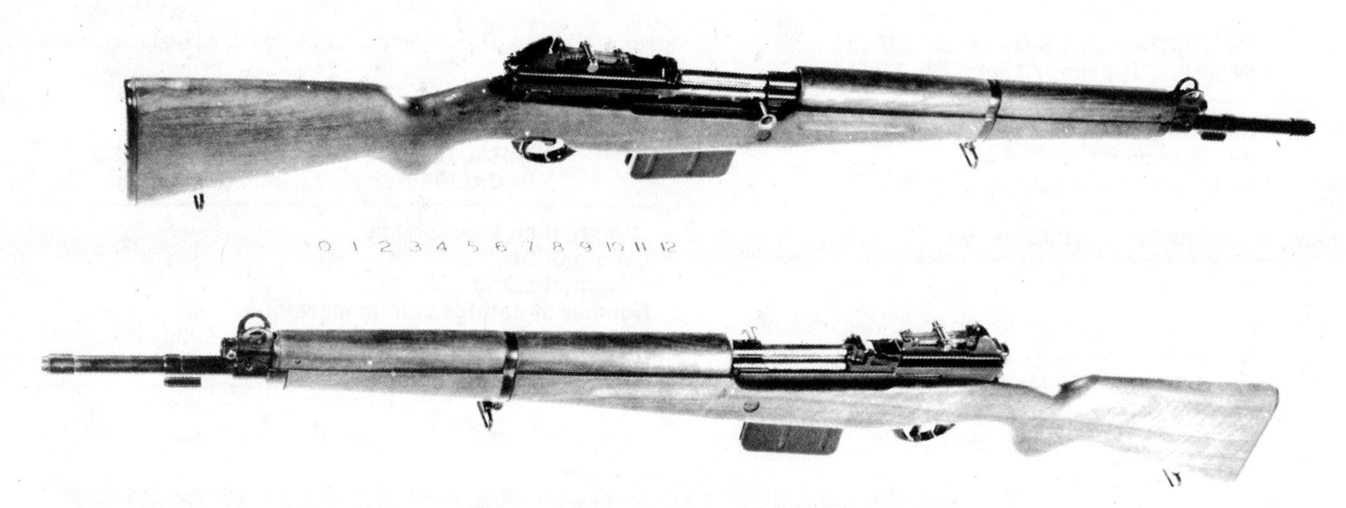

FN Semiautomatic Rifle M 1949 chambered for the US .30-06 (7.62 × 63mm) Cartridge.

barrel, the housing still has continuing forward movement, which enables it, through its machined surfaces, to depress the rear of the bolt into its locking recess with the rear locking surface at the top of the ramp.

How the FN Self-Loader Works

The hammer mechanism in the design is an adaptation of the familiar John Browning hammer hook system as used originally in his automatic shotgun. Variations of this design are encountered in most sporting arms of successful types today, and a minor variant of it is used in the United States Rifle M1 and the Carbine M1.

The hammer hook system is so designed that during recoil the rear upper hook on the hammer is engaged automatically to prevent full automatic fire. When the trigger is deliberately released, the upper hook (or sear) releases; the hammer spring reacts, but the forward holding sear, which is at a lower level than the automatic holding one, grips a lower cut or hook in the hammer. One innovation in the design (which has, however, appeared on German rifles, notably of Walther commercial design) is that the hammer spring guide is designed to protrude slightly below the line of the trigger guard when the hammer is in cocked position. It is remembered that the hammer is concealed in this type of weapon. The FN design permits the holder of the weapon to tell by sight or touch if the hammer is cocked as evidenced by the protruding nose of the mainspring guide.

The safety system also includes a variation of the one utilized in the Garand in that the bolt housing or carrier is so designed that it interferes with the hammer striking the firing pin until both the bolt is locked and the housing itself is in its forward position. Thus, if the housing is not completely closed while the hammer may fall, it can only strike the rear of the housing and cannot fire the cartridge in the chamber. In such an instance, it is necessary to pull back the cocking handle to recock the arm before it can be fired. The firing pin is automatically retracted by a return spring after being driven forward to fire.

The face of the bolt is slotted to allow it to travel back over the ejector, which also helps to serve as a travel guide. As the proper motion nears the end of the recoil stroke, the ejector is far enough out from the face of the bolt so that it can pivot the cartridge out of the gun.

The manual safety is on the right side of the trigger guard and is in the form of a turning lever with a half round block, which not only locks the trigger to prevent any movement but also drops its bar down far enough to interfere with the trigger finger being inserted into the trigger guard as a warning that the safety is applied.

A study of the detail drawings will disclose all the salient features and show the resemblance to the Russian designs in particular. While this is a beautifully constructed rifle, the very nature and quality of the workmanship in it make it a relatively costly one to produce.

How to Load and Fire the FN Self-Loading Rifle

Apply safety by pushing down, and pull bolt operating handle completely to the rear. Bolt will remain to the rear, and two 5-round chargers can be loaded into the magazine in a manner similar to that used when loading a Mauser or Springfield rifle. Pull bolt slightly to the rear and release; the bolt will run forward and chamber a cartridge. The magazine may also be loaded by inserting the cartridges one by one in the magazine. Disengage safety by pushing it back up away from the rear of the trigger. Pressure on the trigger will fire the rifle, and for each individual pull of the trigger a round will be fired until the magazine is exhausted.

How to Field Strip the FN Self-Loading Rifle

At the rear of the receiver is a locking key that seats on the rear end of the operating spring guide, which protrudes through the rear end of the receiver. Making sure that the bolt is forward, turn locking key 180° upward, push receiver cover forward against the pressure of the operating spring lifting rear end of cover to release it from the guide track in the receiver. Pull cover to the rear and remove cover and operating spring assembly. Pull bolt operating handle to the rear until the bolt carrier guides are in line with clearance cut in receiver track. Lift front end of bolt carrier/bolt assembly and remove from receiver. Remove bolt from the receiver. The piston and piston spring are removed by depressing the gas cylinder plug, located at the front of the gas cylinder tube under the front sight, and rotating it 90°. It can then be removed. Tilting the rifle forward will cause the gas piston and spring to slide forward and out of the gas cylinder tube. The magazine can be removed by pushing up the magazine catch with the point of a bullet or some similarly shaped object. The amount of gas let into the rifle can be regulated. Remove gas cylinder plug; then remove front end cap screw and front end cap. Remove front hand guard by swinging its front end upwards. The gas adjusting sleeve can be turned by inserting a bullet or pointed instrument in the holes in its body. To increase gas pressure, screw the sleeve forward; to decrease the gas pressure, screw the sleeve to the rear.

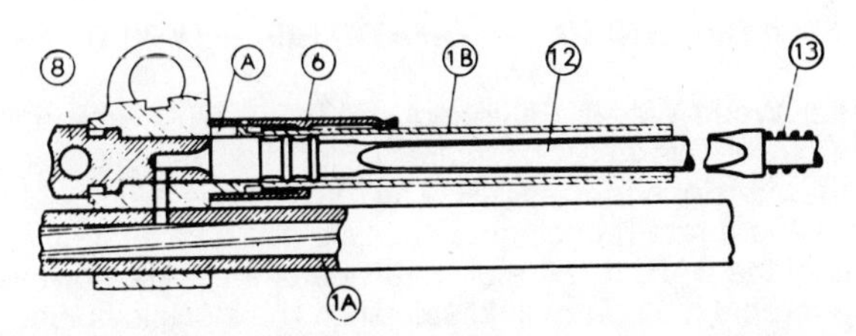

(1) Gas action. May be sealed off by turning nut. Standard gas port top of barrel bleeds gas into cylinder to drive piston back on tappet principle as for German Kar. 43.

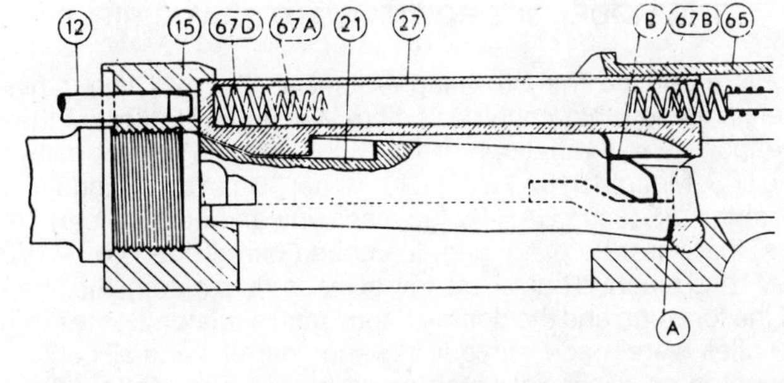

(2) Unlocking action. Bolt locked. Tappet hits bolt carrier starting it back. Spring returns piston to battery. Slide has initial free movement before camming action starts.

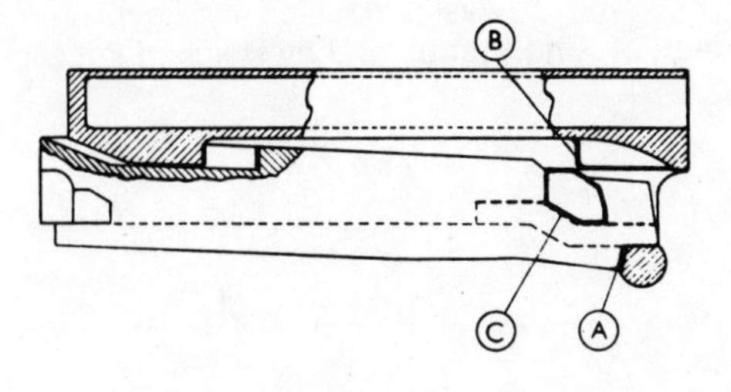

(3) Cams on carrier raise bolt out of receiver engagement as with Russian Tokarev rifle.

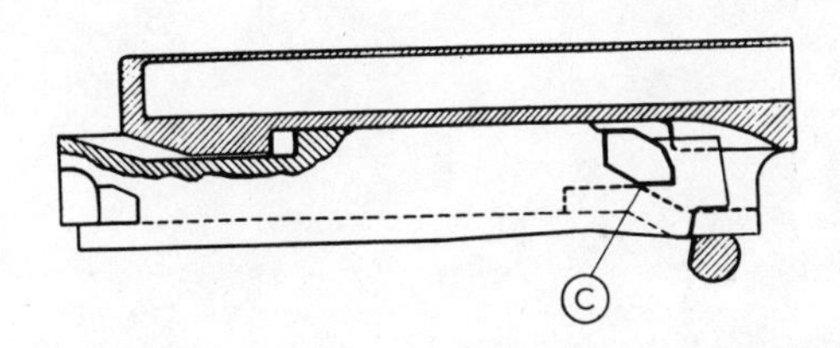

(4) Bolt and carrier go back. Hammer is ridden down to cock. Recoil spring compressed in standard fashion.

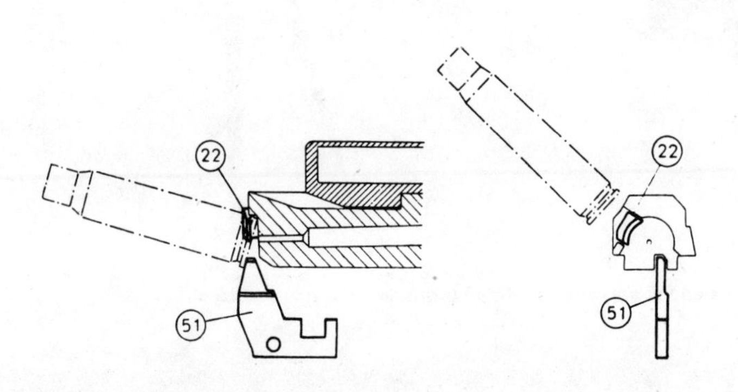

(5) Near end of recoil stroke, the cartridge case held by the extractor hits the ejector and is tossed out of the rifle.

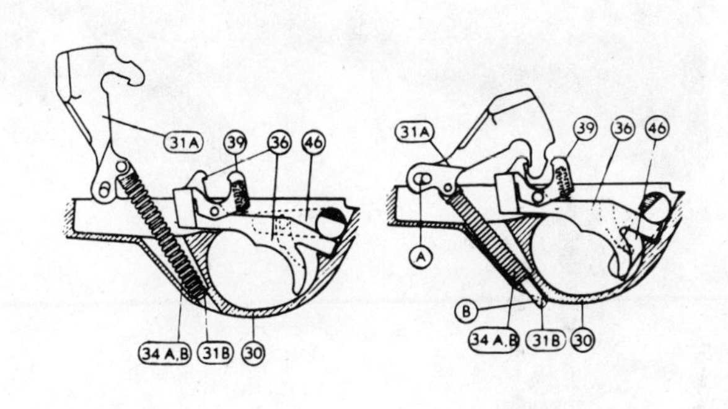

(6) Left-Hammer and mechanism in fired position. On recoil, hammer hook will be engaged Browning-style on rear sear to prevent firing until trigger is released. Right-Hammer held by forward sear ready for firing. Mainspring guide pin projects as signal that hammer is cocked. Walther system.

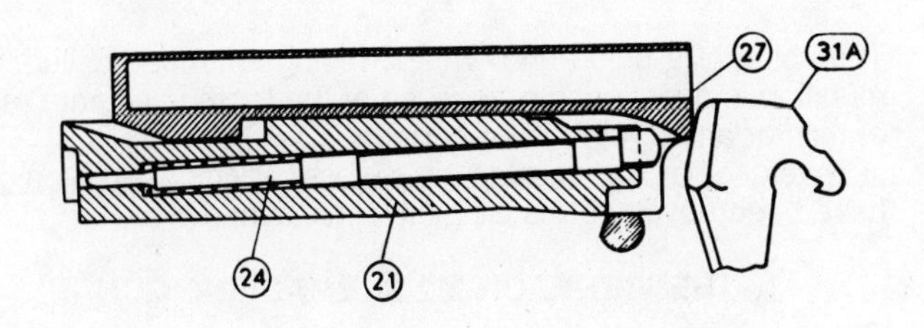

(7) Safety feature. If action is not locked, hammer will hit carrier instead of firing pin.

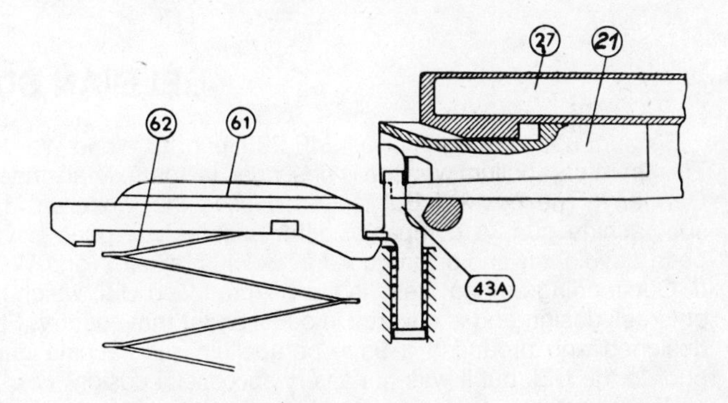

(8) Magazine follower pushed up plunger to hold bolt open when last shot is fired.

Operation of the FN Semiautomatic Rifle.

THE MODEL 30 BROWNING AUTOMATIC RIFLE

FN produced the Browning Automatic rifle for Chile, China, Belgium and other countries before World War II. Most of these weapons were variations of the Model 30 and were similar to the U.S. Model 1918A1 BAR. The Model 30 can be distinguished from the US issue BAR by the magazine and ejection port covers, the separate pistol grip, the ribbed barrel (the U.S. M1922 BAR had a ribbed barrel, but few of these were made), the shape of the fore-end and the dome-shaped gas regulator. Some Model 30 rifles were made with quick change barrels, and all could be mounted on a special tripod made by FN. The Model 30 was made in the follwing calibers: 7mm, 7.65mm and 7.92mm. The Model 30 Browning automatic rifle is an obsolescent weapon and is not likely to be found in the hands of troops today.

BROWNING AUTOMATIC RIFLE TYPE D

After World War II, FN introduced the Type D Browning automatic rifle. This weapon features major improvements not found in other versions of the BAR. The Type D has a quick change barrel and a rate-reducing mechanism as do several other versions of the BAR. It is the only version of the BAR, however, to have a rapid method of field stripping. The stock is hinged, and by removal of the trigger guard assembly pin and butt access pin the piston slide and bolt assembly can be removed. The recoil (operating) spring of the Type D is mounted in the butt rather than in the piston slide assembly, as it is in the US BAR. A clockwork type rate reducer is used with the Type D rather than the buffer type found on the US Browning Automatic Rifle M1918A2. The Type D was purchased by Egypt in 7.92mm during the reign of King Farouk and by Belgium in caliber .30-06.

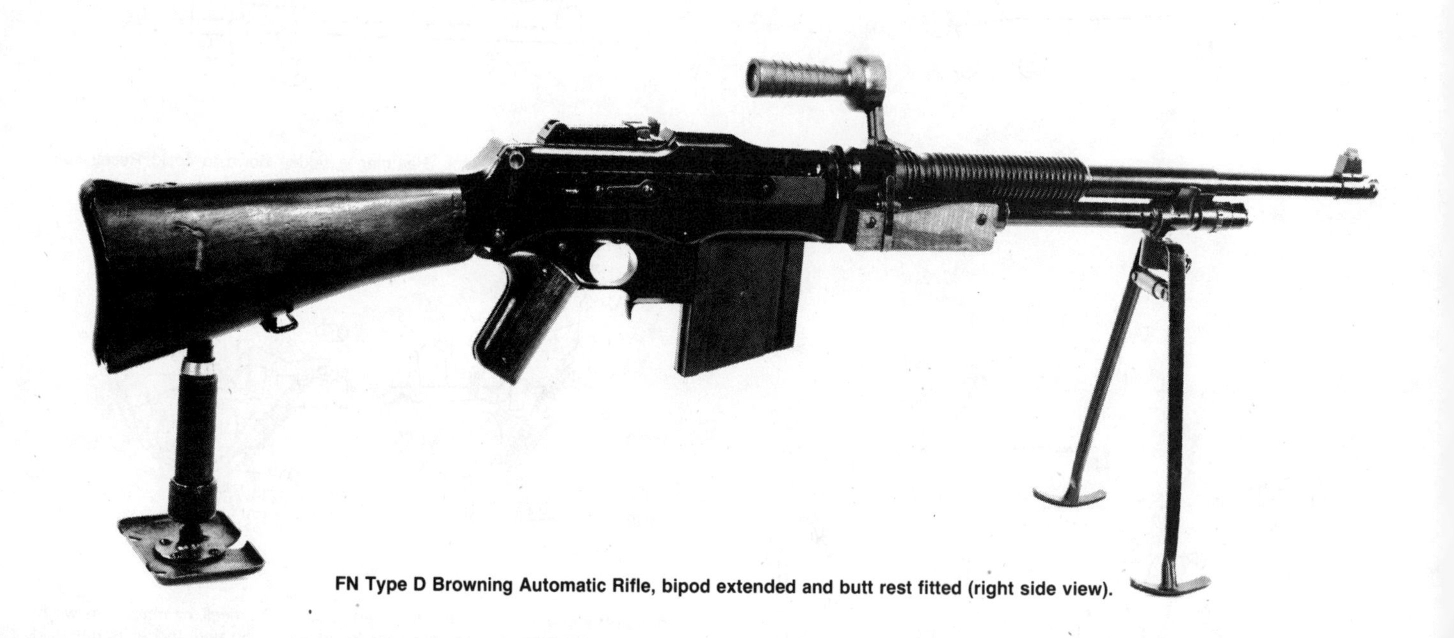

FN Type D Browning Automatic Rifle, bipod extended and butt rest fitted (right side view).

BELGIAN SUBMACHINE GUNS

Belgium used the Schmeisser MP 28 II prior to World War II. This 9mm parabellum weapon is described in some detail under Germany. The postwar Belgian Army used the Vigneron M2 submachine gun. A number of other submachine guns have been developed and/or produced in Belgium since World War II. Outstanding among these is the FN-produced UZI, which is of Israeli design and is covered in detail under that country. FN designed and produced a 9mm parabellum submachine gun prior to the UZI, but it was not a very successful design.

Repousemetal of Belgium developed an interesting weapon in 9mm called the RAN submachine gun. It has an internal cooling system, which uses the bolt as a pneumatic ram to force air through a system of helical grooves around the barrel. It was produced in limited quantities in several versions, including one with a folding bayonet. Another departure from conventional submachine gun design was in the fitting of a bipod to one model. The value of a bipod on a weapon firing 9mm parabellum ammunition is questionable because of the limited accurate range of this, or any other, piston cartridge.

Several other submachine guns—basically modified Sten guns have been developed in Belgium since World War II.

THE VIGNERON M2 SUBMACHINE GUN

The Vigneron is a conventional post World War II submachine gun of stamped construction. Loading and firing of the weapon are basically the same as for the British Sten except that it has a grip safety that must be squeezed to fire the gun and a selector on the left side of the weapon that can be set on semiautomatic, automatic fire or safe. The grip safety prevents accidental discharge if the gun is dropped with a loaded magazine in place and the bolt in the forward position.

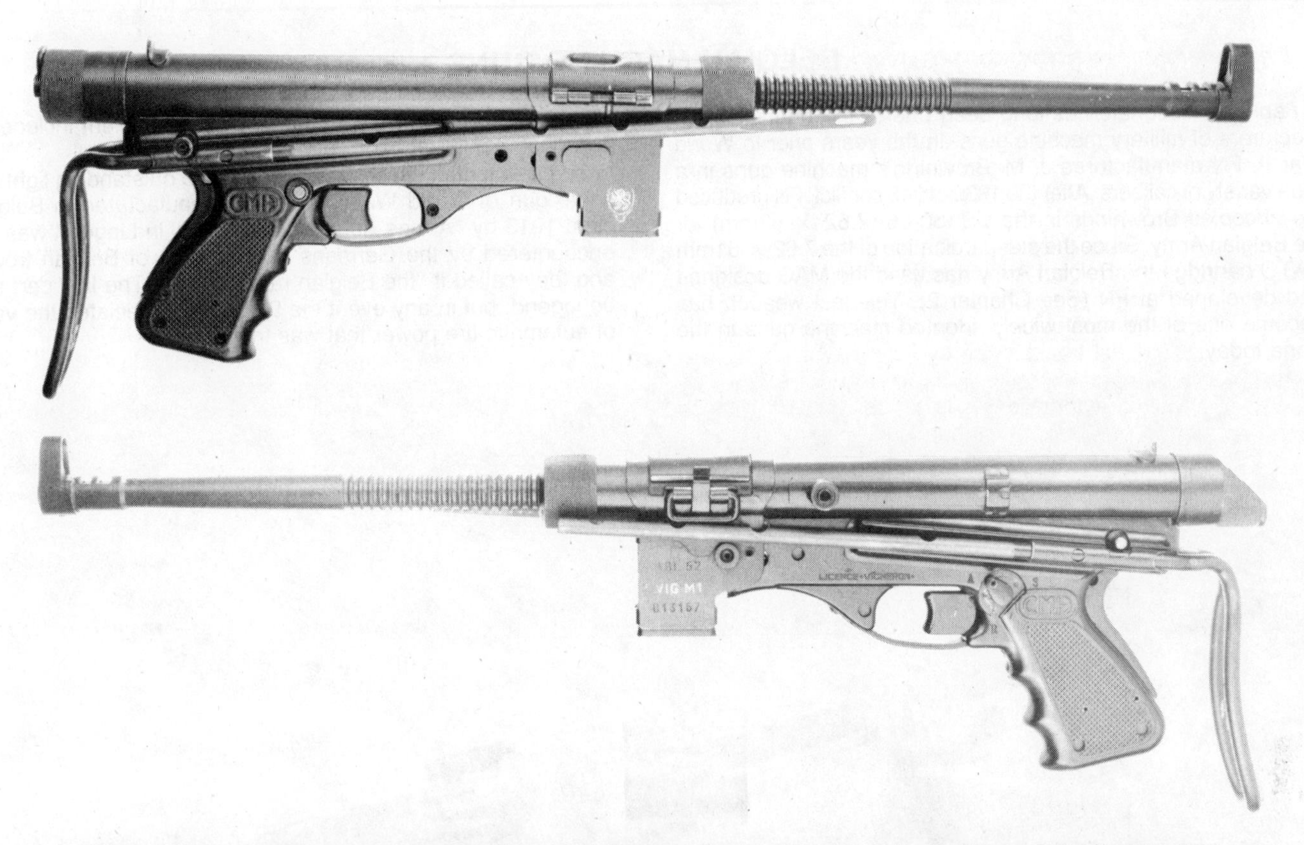

The Belgian 9 × 19mm Vigneron M1 Submachine Gun. This particular specimen is serial number 013167 and carries the date 1952 and the Belgian Army mark A.B.L. (an acronym combining the French *Armée Belge* and the Flemish *Belgisch Leger*, both of which mean "Belgian Army").

CHARACTERISTICS OF THE VIGNERON M2 SUBMACHINE GUN

System of operation: Blowback selective fire.
Weight loaded: 8.74 lbs.
Length overall: 34.9 in. w/stock extended, 24 in. w/stock telescoped.
Barrel length: 12 in.
Feed device: 32-round, detachable, staggered box magazine.
Sights: Front: Blade.
Rear: Nonadjustable aperture.
Muzzle velocity: 1224 f.p.s.
Cyclic rate of fire: 600 r.p.m.

How to Field Strip The Vigneron M2

Remove magazine and check chamber for a cartridge. Unscrew the receiver cap at rear of receiver and remove the bolt. Unscrew barrel nut on front of receiver and remove the barrel. Further disassembly is not recommended. Reassembly is performed by reversing the above steps.

The FN-licensed version of the Israeli Military Industries UZI Submachine Gun with its stock extended.

BELGIAN MACHINE GUNS

Fabrique Nationale has long been one of the premier manufacturers of military machine guns. In the years prior to World War II, FN manufactured J. M. Browning's machine guns in a wide variety of calibers. After the 1939–1945 conflict, FN produced the aircooled Brownings in the US .30-06 (7.62 × 63mm) for the Belgian Army. Since the standardization of the 7.62 × 51mm NATO cartridge the Belgian Army has used the MAG designed and developed at FN (See Chapter 2). This last weapon has become one of the most widely adopted machine guns in the world today.

Non-FN machine guns manufactured in Belgium include the American-designed Lewis gun.

The Lewis gun, which was probably the outstanding light machine gun of World War I, was first manufactured in Belgium circa 1913 by "Armes Automatique Lewis" in Liege. It was first encountered by the Germans in the hands of Belgian troops, and they called it "the Belgian rattle snake." The last part may be legend, but in any event the Belgians appreciated the value of automatic fire power that was truly mobile.

Right and left views of the current production version of the FN 7.62 × 51mm NATO MAG.

MITRAILLEUSE D'APPUI GENERAL (MAG)—GENERAL PURPOSE MACHINE GUN

The MAG is another development of FN, and like most of the products of that concern it demonstrates first-class engineering ability. The gun combines the operating system of the Browning automatic rifle (BAR) with a belt feed mechanism similar to that of the German MG 42. The bolt mechanism of the BAR has been changed in the MAG so that it locks on the bottom of the receiver, rather than on the top as with the BAR. It has a chrome-plated and stellite-lined bore and chamber in its quick change barrel. The MAG, like the German World War II guns and the U.S. M60, is designed to be used on a bipod as a light machine gun and on a tripod as a heavy machine gun. Its rate of fire can be adjusted through the use of its gas regulator from a low cyclic rate of 700 rounds per minute to a high cyclic rate of 1000 rounds per minute.

CHARACTERISTICS OF MAG MACHINE GUN

Caliber: Has been made in 7.62mm NATO and 6.5 Swedish.
System of operation: Gas, automatic only.
Weight, w/butt and bipod: 23.92 lbs.
Weight, w/o butt and bipod: 22.22 lbs.
Weight of FN tripod: 22 lbs. (constructed of aluminum alloy).
Length, overall: 49.21 in. w/flash suppressor.
Barrel length: 21.44 in.
Feed mechanism: Link belt. (Nondisintegrating push-out type links similar to those used on the MG 34 and MG 42 and US M13 disintegrating links may be used.)
Sights: Front: Folding type with blade, or type adjustable for height.
Rear: Combined battle-sight peep and leaf with notch; peep adjustable to 800 meters and leaf adjustable to 1800 meters.
Muzzle velocity: 2800 f.p.s. (approx.) 7.62mm NATO ball cartridge.
Cyclic rate of fire: 700 to 1000 r.p.m.

How to Load and Fire the MAG

Pull the operating handle to the rear; since the MAG fires from an open bolt, the slide and bolt will remain to the rear. Push the button type safety mounted in the pistol grip from the left side, so that the letter "S" is exposed on the right side. Open cover by pressing cover catch at rear of cover. Lay cartridge belt on feedway so that the first cartridge abuts against the cartridge stop. Close cover securely and push safety button from right to left. Squeeze trigger and the weapon will fire.

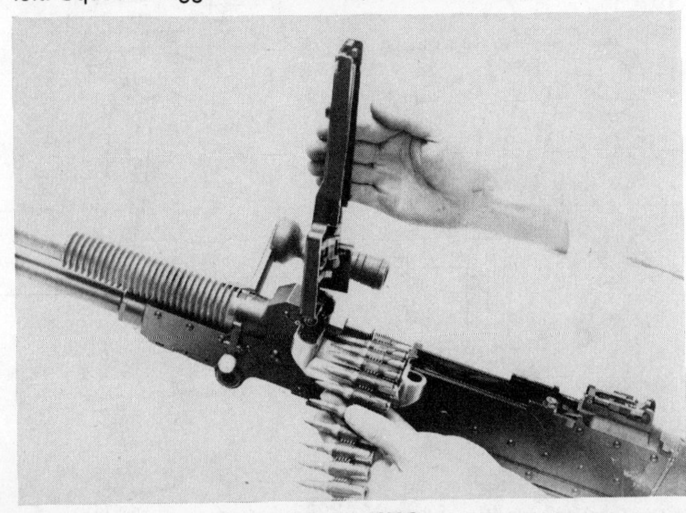

Loading the MAG.

How the MAG Works

Essentially, the operation of the MAG is the same as that of the Browning automatic rifle, with the exception of its belt feeding mechanism and bottom receiver locking. A stud mounted on the top of the bolt operates in a track in the belt feed lever, which moves the belt feed slide back and forth, pulling cartridges into position for ramming by the bolt. The trigger mechanism of this weapon is much simpler than that of the BAR. No rate-reducing mechanism is used with this gun.

Field Stripping the MAG

Open the cover and check to insure that the weapon is not loaded. Push in on stock catch located on front underside of butt and slide butt up and off the receiver. Push in on recoil spring rod and disengage it from the bottom of receiver; remove recoil spring assembly (recoil spring and rod are packaged unit). Pull operating handle to the rear, and the slide and bolt will move to the rear. Grasp slide and bolt assembly by the slide post and withdraw assembly from the receiver. Remove link pin; the link, bolt lock and bolt can be removed from the slide. To remove the pistol grip, pull out the retaining pin from the right side. To remove the cover and the feed tray, pull out the cover pin from the right side. To remove barrel, push barrel lock in (barrel lock is located at left front of receiver), move the barrel handle to the left so that it is in the vertical position and pull barrel straight out. No further disassembly is recommended.

Reverse the above procedure to reassemble the weapon. When reassembling the bolt slide assembly to the receiver, the head of the bolt must be supported so that the forward grooves on the bolt engage the mating ridges on the sides of the receiver.

Special Note on the MAG

The MAG can be used with any rifle cartridge that has the same base dimension as the 7.92mm Mauser; this includes 7.62mm NATO, simply by changing the barrel. The butt and bipod can be removed from the gun for use in transport vehicles or tanks. The weapon can be used on the tripod with the butt removed.

The MAG has proven to be quite a popular machine gun and has been purchased by the following countries: Argentina, Belgium, Cuba, Ecuador, India, Israel, Kuwait, Libya, New Zealand, the Netherlands, Northern Rhodesia. Peru, Quatar, Ruanda, Sierra Leone, Southern Rhodesia, South Africa, Sweden, Tanganyika, Uganda, the United Kingdom (Great Britain), Venezuela and the United States. With the exception of Sweden, which chose 6.5mm all countries adopted the gun in 7.62mm NATO. It is probable that the Swedish guns will also be converted to this caliber in the future. The MAG (machine gun, general purpose L7A1) is being manufactured at the Royal Small Arms Factory at Enfield Lock and at the Argentine small arms factory at Rosario.

The Belgian FN General Purpose Machine Gun. Type MAG.

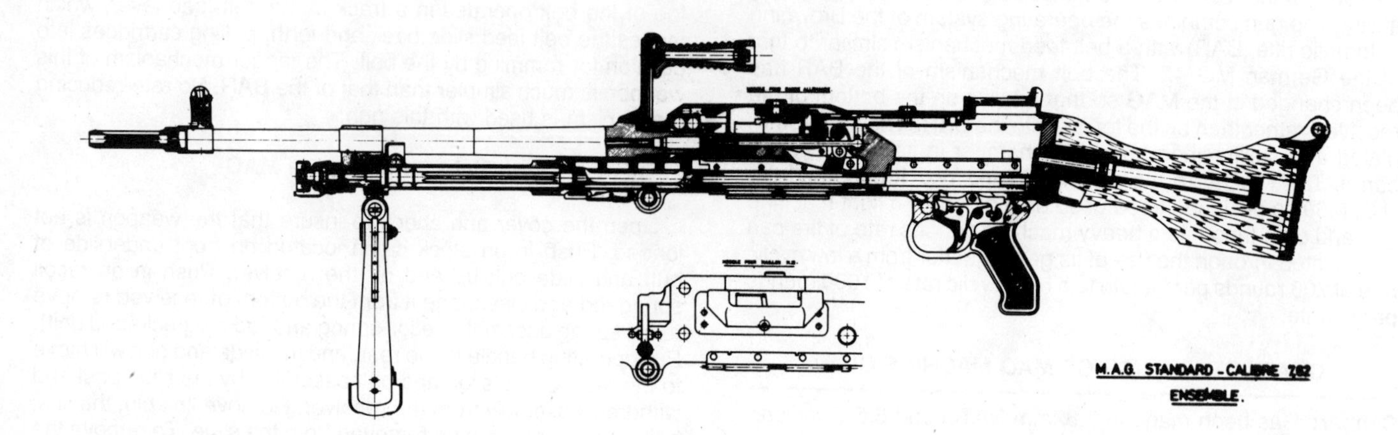

FABRIQUE NATIONALE D'ARMES DE GUERRE - HERSTAL - BELGIQUE.

R
a
b
c
d

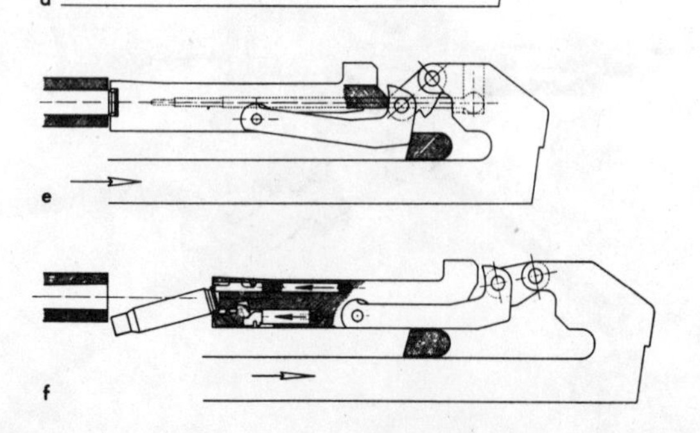

Diagrammatic view of the operation of the MAG bolt mechanism.

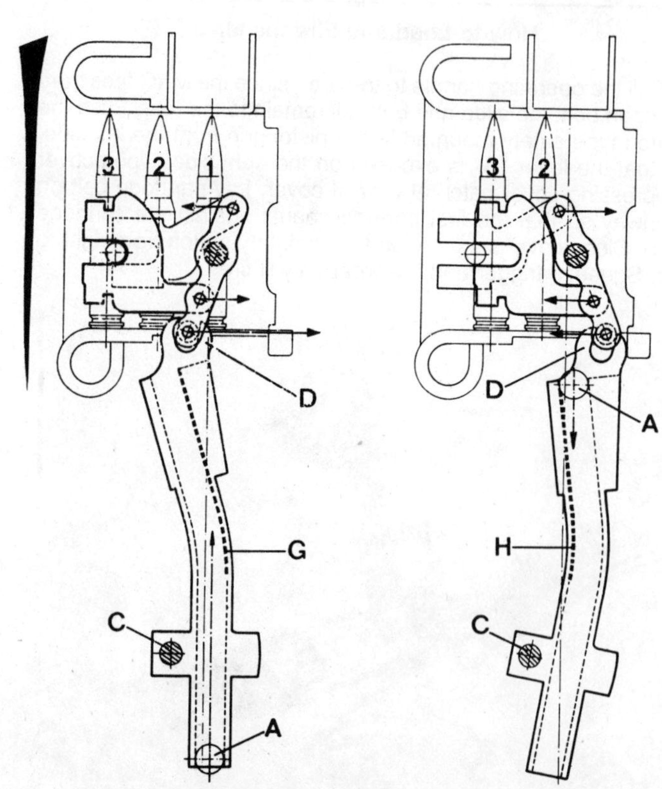

Diagrammatic view of the operation of the MAG feed mechanism.

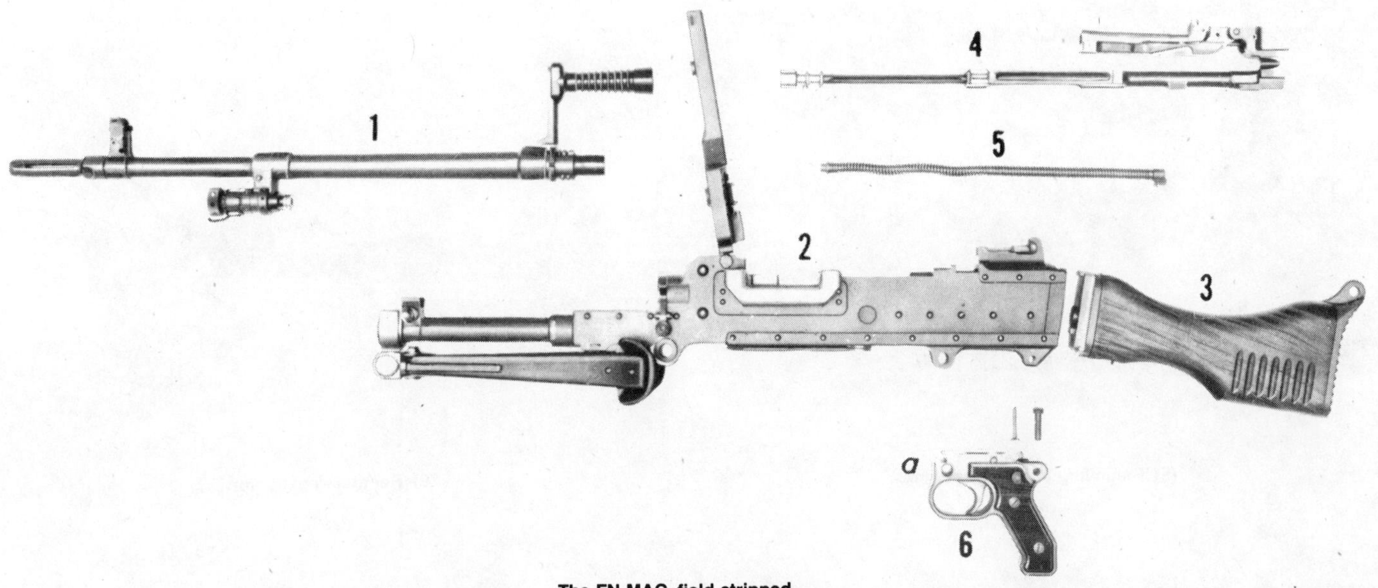

The FN MAG field stripped

1. Barrel assembly.
2. Receiver assembly.
3. Buttstock assembly.
4. Gas piston, bolt carrier, and bolt assembly.
5. Return spring assembly.
6. Trigger assembly.

(1) Removing butt assembly.

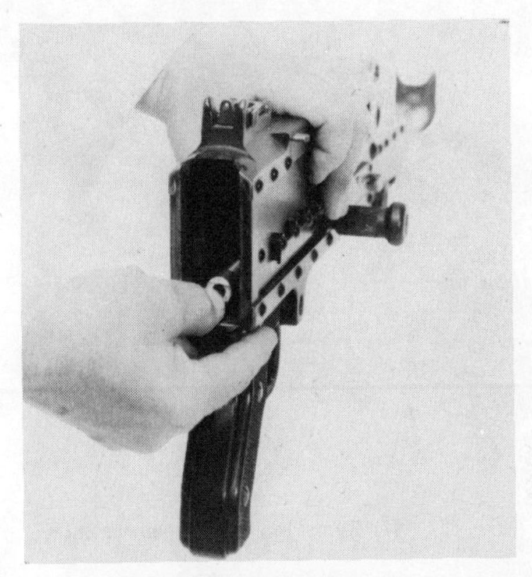

(2) Disengaging recoil spring rod.

(3) Removing recoil spring assembly.

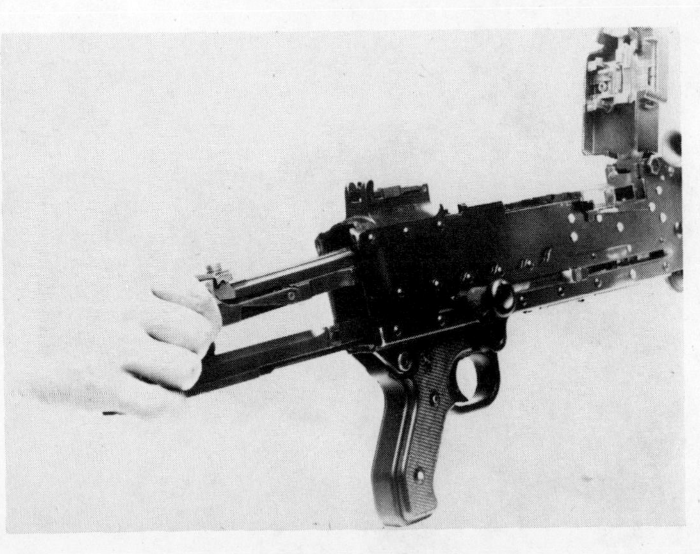

(4) Removing bolt, piston and slide.

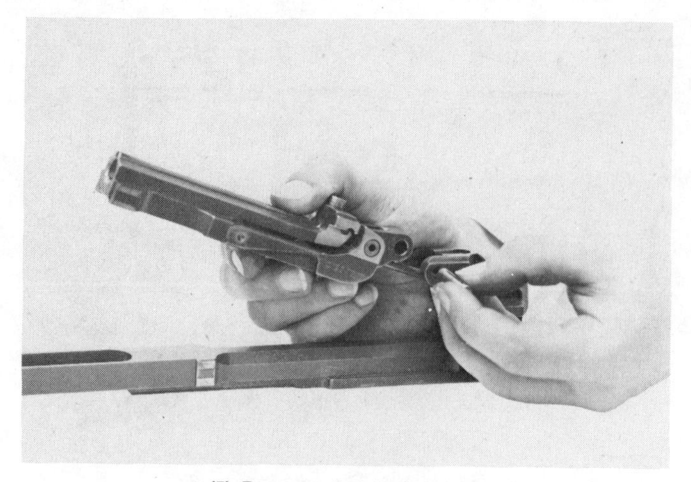

(5) Removing bolt lock and link.

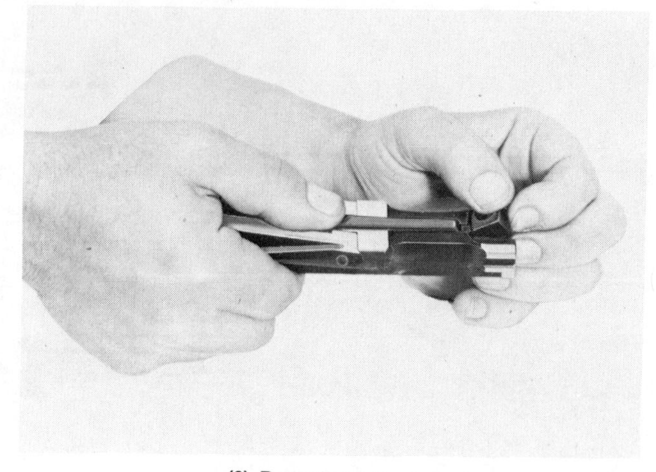

(6) Removing extractor.

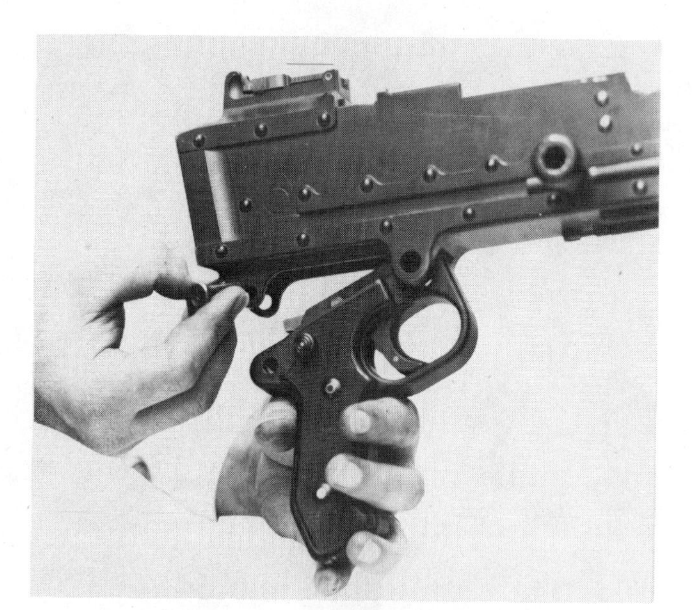

(7) Removing grip trigger assembly.

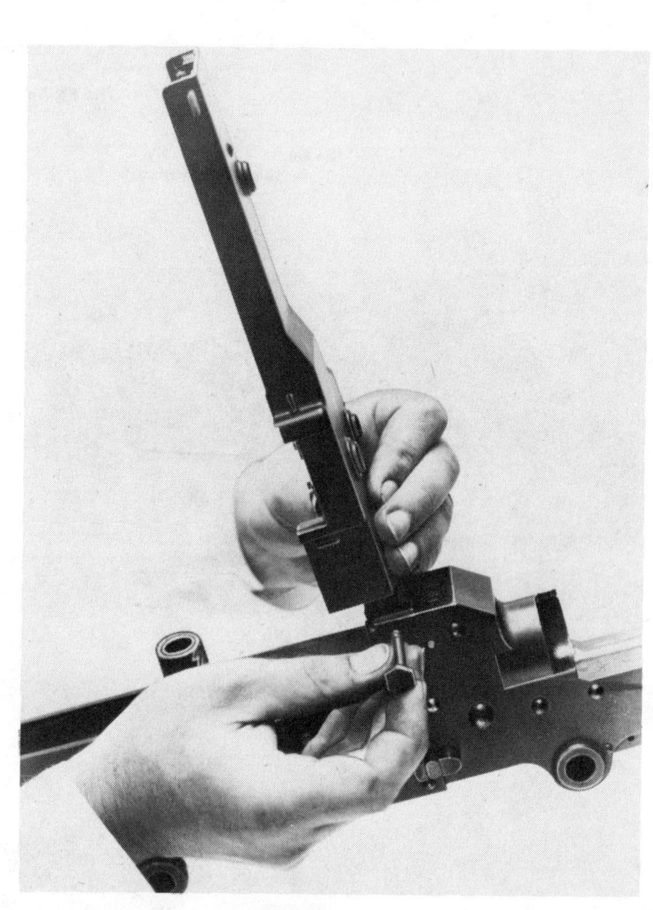

(8) Removing cover.

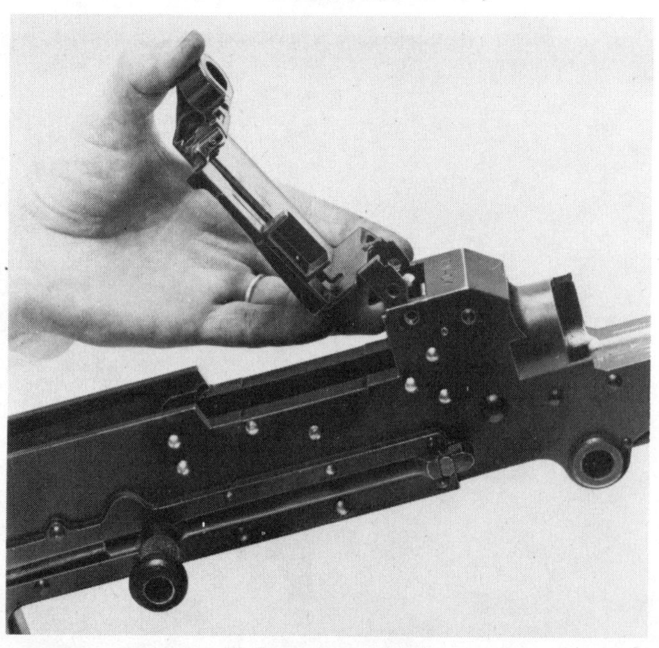

(9) Removing feed plate.

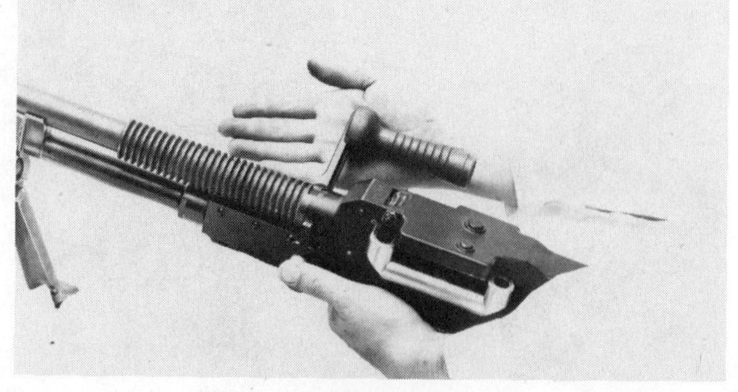

(10) Removing barrel.

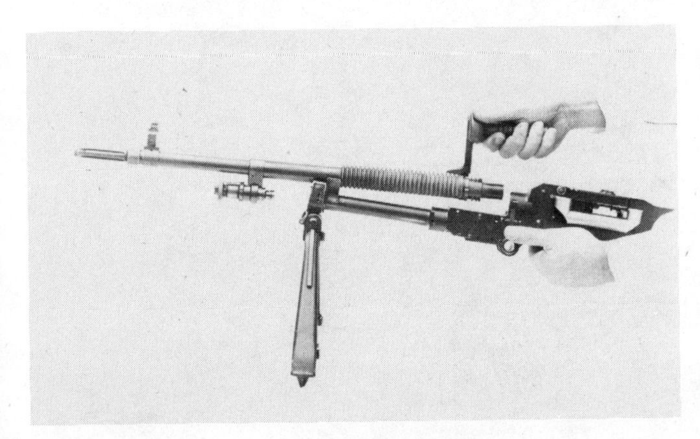

(11) Replacing barrel.

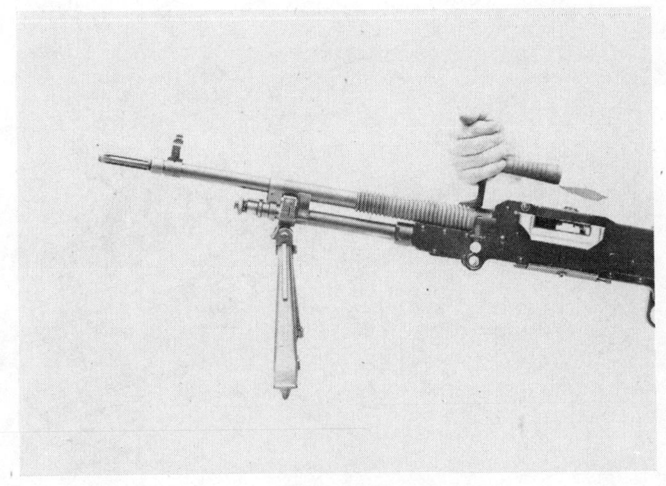

(12) Drawing barrel to rear.

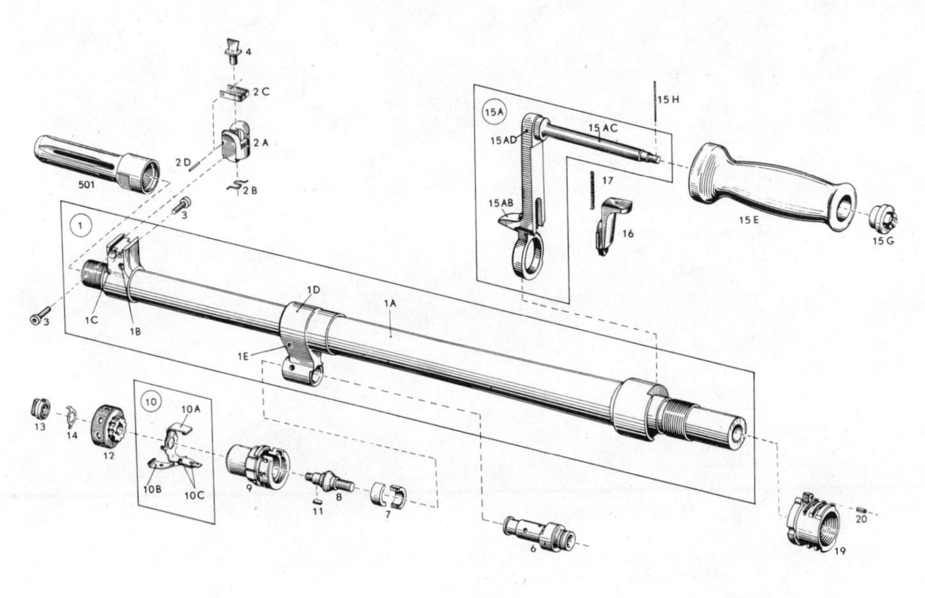

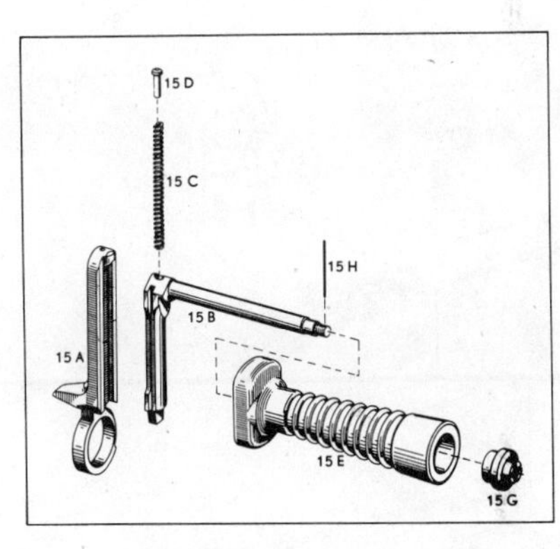

MAG barrel assembly.

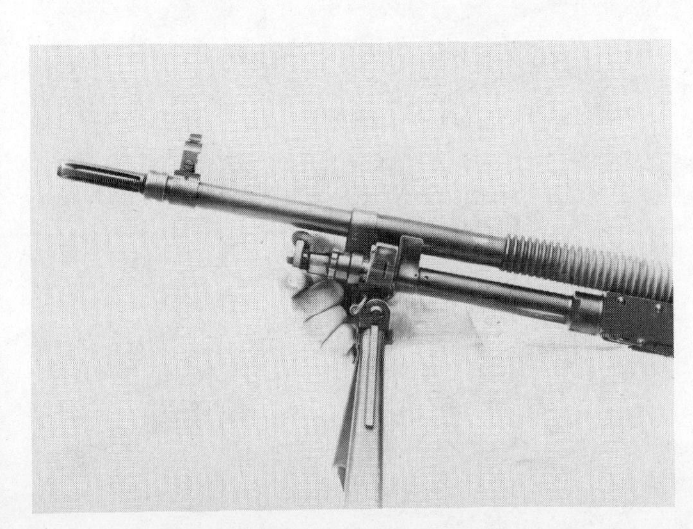

Adjusting gas cylinder aperture.

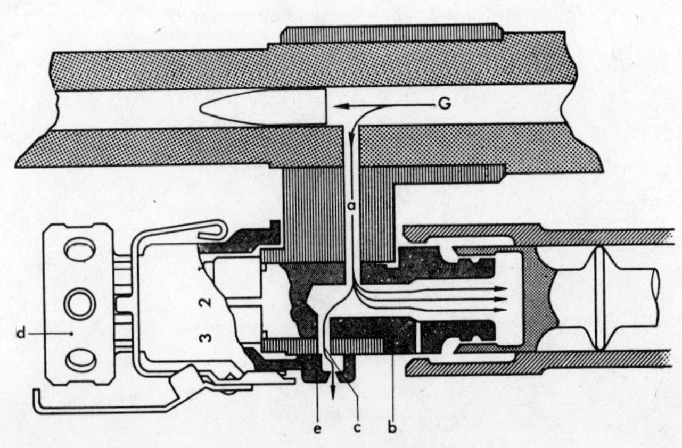

Mag gas regulator assembly.

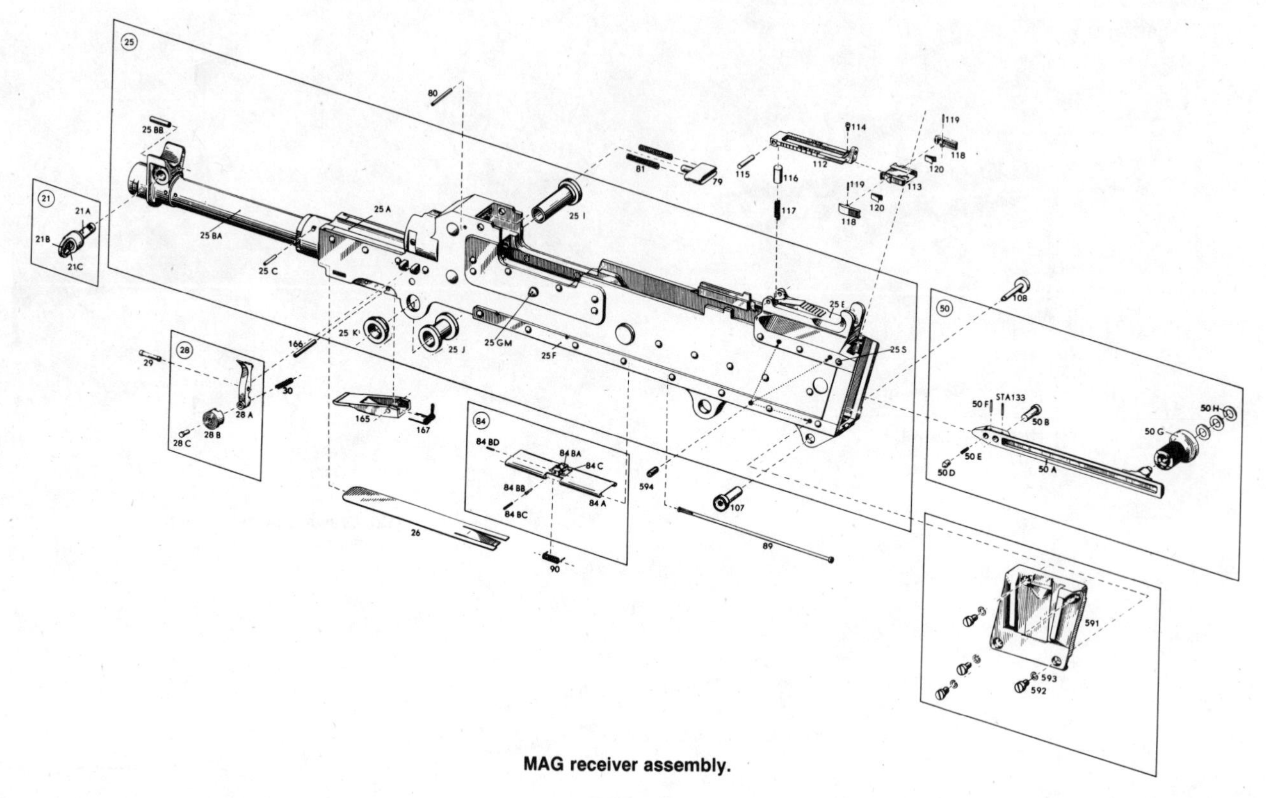

MAG receiver assembly.

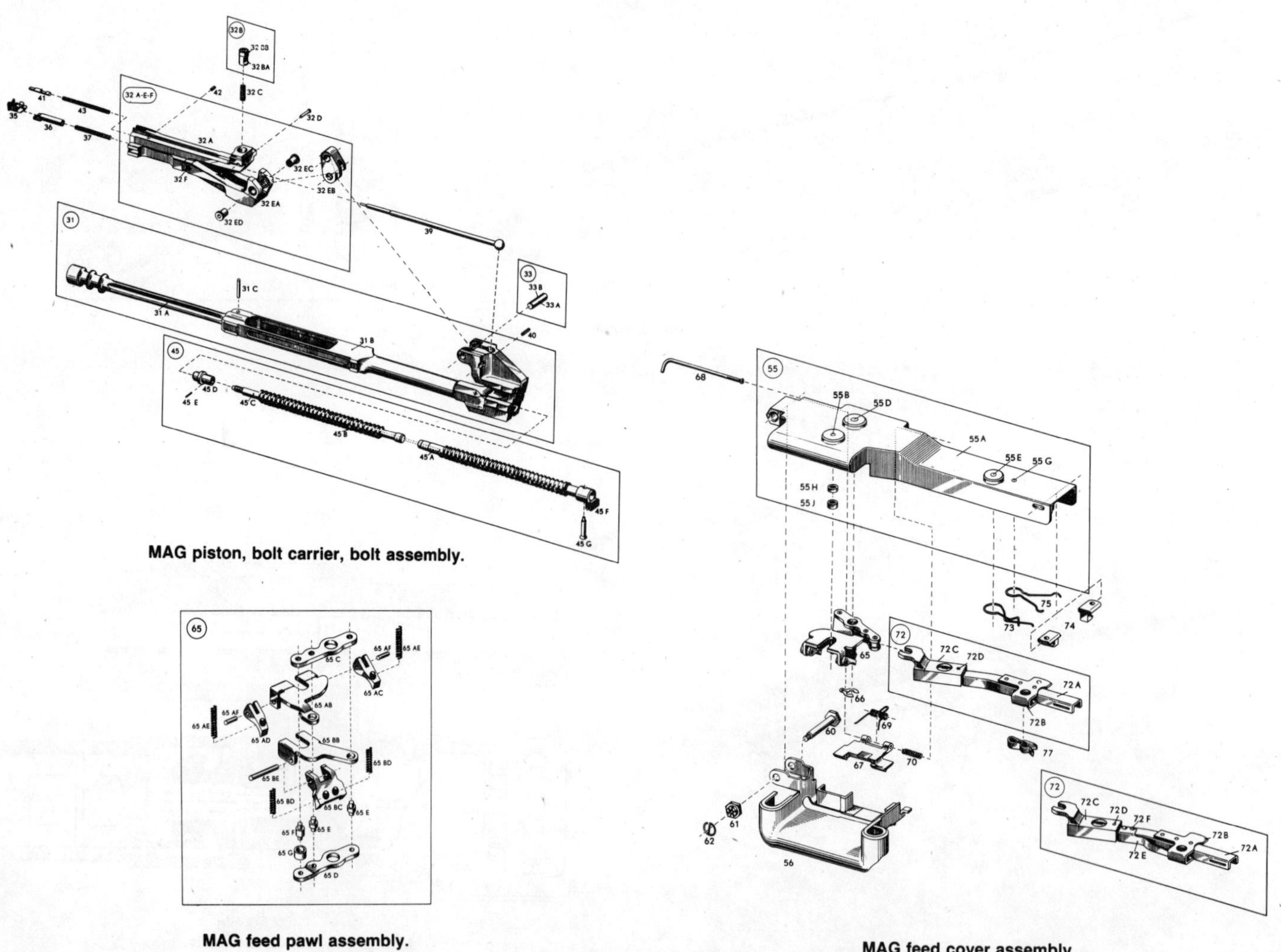

MAG piston, bolt carrier, bolt assembly.

MAG feed pawl assembly.

MAG feed cover assembly.

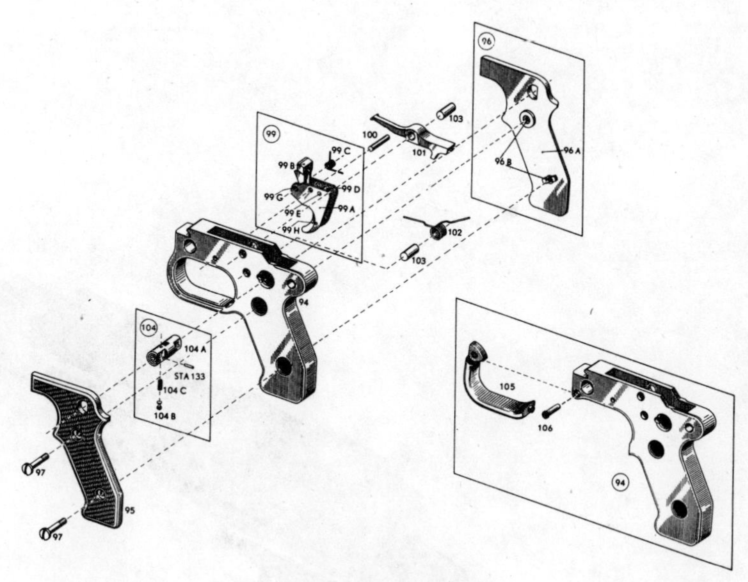

MAG trigger assembly.

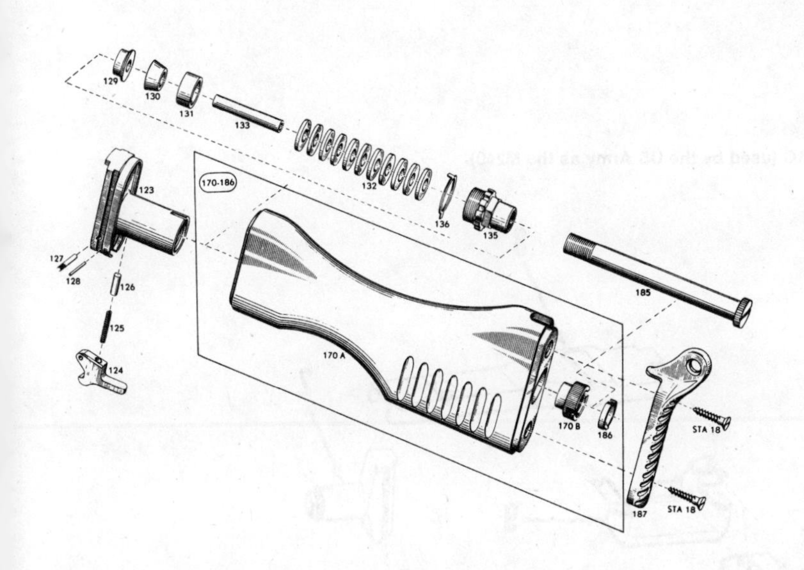

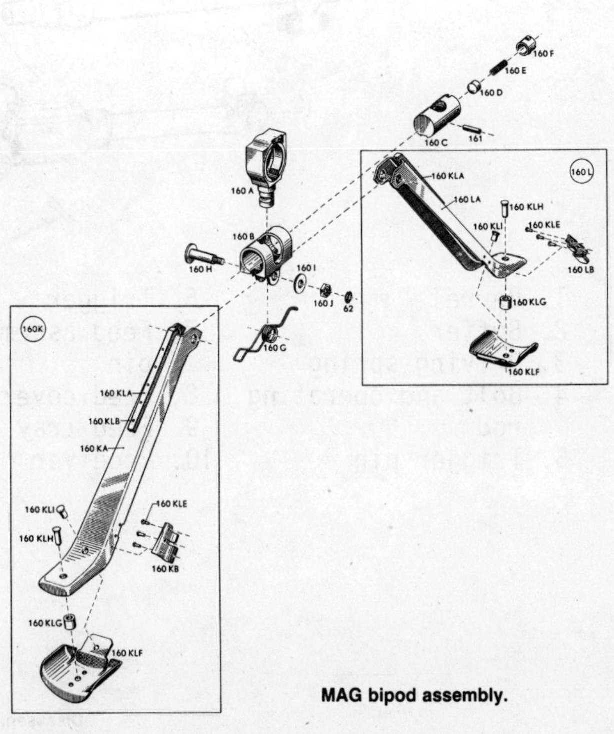

MAG bipod assembly.

Tank version of the MAG (used by the US Army as the M240).

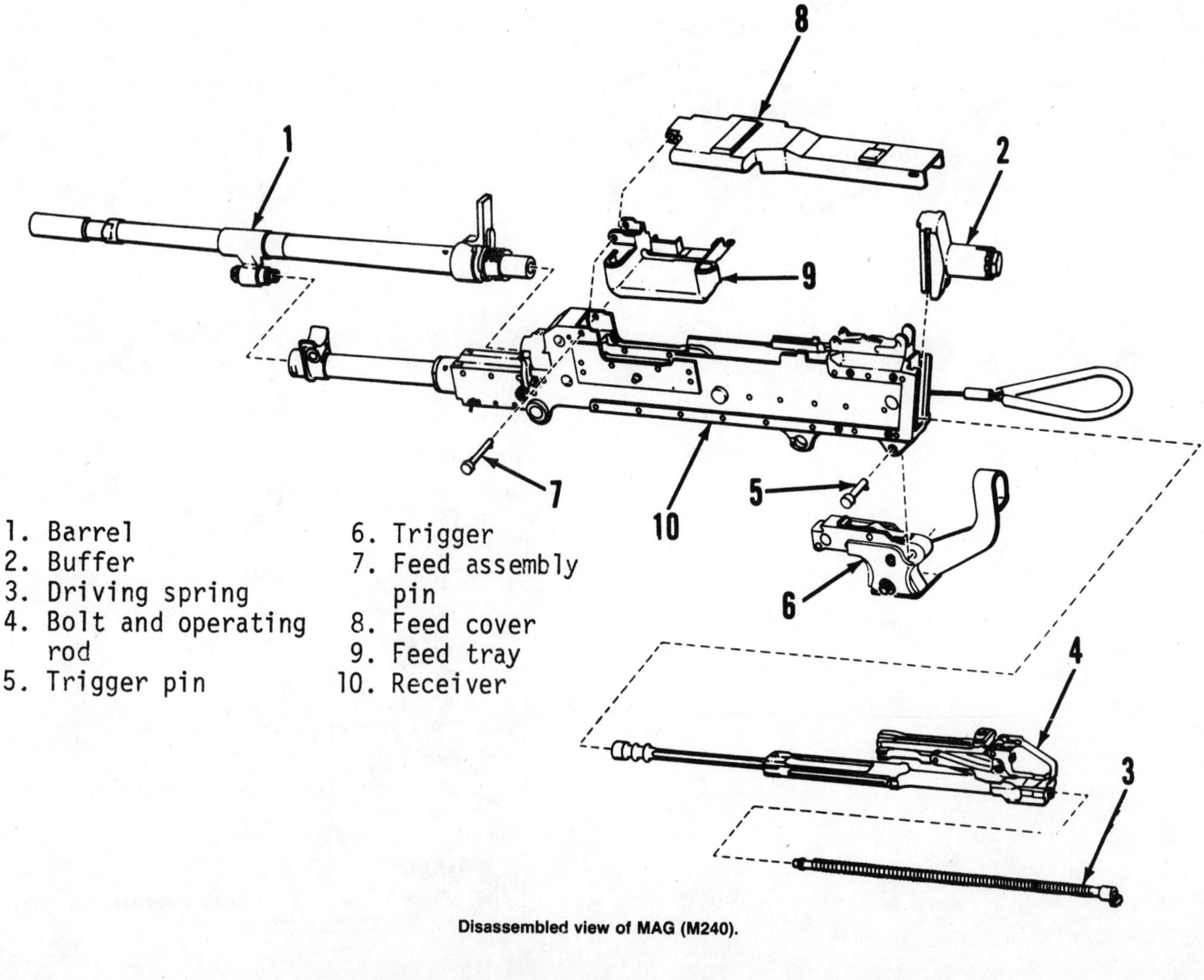

1. Barrel
2. Buffer
3. Driving spring
4. Bolt and operating rod
5. Trigger pin
6. Trigger
7. Feed assembly pin
8. Feed cover
9. Feed tray
10. Receiver

Disassembled view of MAG (M240).

FN MAG Parts Numbers (Partial)

No.	Suffix	Part
1		Barrel gun
1	A	Barrel
1	AM	Sleeve, barrel
1	B	Band, front sight
1	C	Pin straight, headless
1	D	Bush gas hole
1	E	Pin straight, headless
1	I	Pin straight, headless
1	B	Band front sight
1	C	Pin straight headless
2	A	Protector front sight
2	B	Spring, adjusting front sight
2	C	Strap, retaining front sight
2	D	Pin, retaining strap
3		Screw, adjusting front sight
4		Sight, front high or
		Sight, front low
501		Flash-hider
6		Plug gas regulator
7		Collar split
8		Spindle, regulator
9		Sleeve, regulator
10		Indicator
10	A	Body, indicator
10	B	Arm, indicator
10	C	Rivet
11		Key regulator spindle
12		Knob regulator
13		Nut adjusting regulator
14		Washer, adjusting nut
15	A	Bracket, carrying handle
15	B	Stem, handle
15	C	Spring, helical compression
15	D	Plunger, spring stem handle
15	E	Handle
15	G	Nut, retaining handle
15	H	Pin, securing stem nut
15	A	Bracket, carrying handle
15	AB	Bracket, body
15	AC	Stem
15	AD	Pin
15	E	Handle
16		Catch
17		Spring
19		Nut, barrel
20		Pin, spring
21		Swivel assembly, sling
21	A	Stem, swivel eye
21	B	Eye, swivel
21	C	Washer, stem
25		Body
25	A	Block, front
25	BA	Cylinder, gas
25	BB	Pin, retaining, bipod
25	C	Pin, straight headless
25	DA	Plate, right
25	DC	Guide, front right
25	DE	Washer, front guide, right
25	DF	Guide, cocking handle
25	DG	Guide, upper, right
25	DL	Shoulder, locking right
25	E	Bracket, back sight
25	F	Plate bottom
25	GA	Plate, left
25	GC	Guide, front left
25	GG	Guide, upper left
25	GL	Shoulder, locking left
25	GM	Button feed box
25	H	Bracket, catch, sleeve locking barrel
25	I	Bush, mounting front
25	J	Collar, mounting front
25	K	Ring, front mounting bush
25	BA	Cylinder, gas
25	BB	Pin, retaining bipod
25	C	Pin, straight headless
25	GM	Button, feed box
25	I	Bush, mounting front
25	J	Collar, mounting front
26		Cover, dust front
28		Catch, sleeve, barrel locking
28	B	Button, catch
28	C	Rivet, catch button
29		Pin, barrel locking sleeve catch
30		Spring helical, compression
31		Piston assembly
31	A	Piston
31	B	Extension, piston
31	C	Pin, straight headless
32		Block breech assembly
32	A	Block, breech
32	B	Roller, feed, channel
32	C	Spring, helical compression
32	D	Pin, straight headless
32	EA	Lever, locking
32	EB	Link, piston extension
32	EC	Pin, axis, link right
32	ED	Pin, axis, link left
32	F	Pin, straight headless
33		Pin, securing, link
35		Extractor
36		Plunger, extractor
37		Spring, helical compression
39		Pin, firing
40		Pin spring (rollpin)
41		Ejector
42		Pin, spring
43		Spring, helical compression
45		Spring assembly return
45	A & F	Tube outer
45	B	Spring, return
45	C	Stem
45	D	Sleeve, stem
45	E	Pin, straight headless
45	G	Pin, locating guide
50		Handle, cocking
55		Cover, feed mechanism
56		Tray, feed
65		Pawl assembly, feed
66		Clip, retaining feed pawl assembly
67		Pawl, retaining cartridge
68		Pin, cartridge retainer
69		Spring, tension, retainer
70		Spring, helical compression
72		Channel, feed
74		Catch, cover
75		Clip, retaining cover catch
77		Clip, securing feed channel
79		Plunger, positioning cover
80		Pin, plunger, cover positioning

FN MAG Parts Numbers (Partial)—continued

81	Spring, helical, compression	95	Stock, left
84	Cover, ejection opening	96	Stock, right
89	Pin, hinge, ejection, opening cover	99	Trigger assembly
90	Spring, hinge pin, ejection opening cover	100	Sear
94	Frame trigger	104	Catch, fire and safe

MINIMI—SQUAD AUTOMATIC WEAPON

The Minimi was developed by Fabrique Nationale as a companion to their 5.56 × 45mm CAL. By the early 1970s it was apparent that a number of military establishments worldwide would like a 5.56mm companion to their rifles of that caliber. As discussed in Chapter 2, the Minimi was one of the early entrants into the United States Army's Squad Automatic Weapon (SAW) competition.

Minimi mechanism actuation is of a conventional gas piston type. Following the FN preference, there is a gas regulator on the gas cylinder with two settings: one for normal operation and one for operation under adverse conditions. The FAL, CAL, FNC, and MAG also have gas regulators, and that of the Minimi is derived from the one employed on the MAG. The Minimi bolt mechanism is quite similar in concept to that of the FNC, which followed in terms of development chronology. As in the FNC, the Minimi bolt locks into a barrel extension, and the Minimi bolt is rotated by a cam track cut into the bolt carrier.

The normal firing cycle begins with the detonation of the cartridge primer. After the bullet passes the gas port in the barrel, a portion of the gases are tapped to drive the piston and its carrier to the rear. By the time that the cam has begun to rotate the locked bolt, the pressure in the barrel has been reduced to nearly zero. Primary extraction of the fired cartridge case does not begin until the bolt is completely unlocked. This slight delay ensures that there are no residual gases in the barrel and ensures that the cartridge case (which is slightly elastic) is no longer being pressed against the chamber walls by propellant gases. By delaying extraction in this manner, the FN engineers eliminate one of the major problems encountered with small caliber machine guns—failure to extract due to swollen or ruptured cartridge cases.

As in the FNC, the bolt and its carrier have lugs that travel in rails welded inside the receiver. These rails support these two components during their travel, they help ensure smooth functioning, and they direct closing and recoiling movement of the operating mechanism in a completely controlled manner. Reliable functioning was one of the major reasons the Minimi was chosen over the other competitors in the US SAW trials.

One of the unique features of the Minimi is its adaptability to either belted ammunition or loose ammunition in rifle magazines. When using belted cartridges in disintegrating links, the belts can either be used loose or in semitransparent ammunltion boxes that mount to the weapon. These boxes keep the ammunition clean and still allow the gunner to keep track of his remaining cartridges. With either the 100- and 200-round boxes the Minimi can be fired from the hip in the assault role. As with most light machine guns, the Minimi has provision for a quick barrel change, thus allowing the weapon to be fired for extended periods. The Minimi can be fired from its integral bipod, from the US M122 tripod, or from a special firing port mount. In the latter situation, the buttstock is replaced by a special receiver end cap. Further details on the operation and field maintenance of the Minimi are presented in the accompanying photograph captions.

CHARACTERISTICS OF THE MINIMI SAW

Caliber: 5.56 × 45mm (Can be barreled for either the M193 or the SS 109 (XM855) projectiles)
System of operation: Gas, automatic only
Weight, w/butt and bipod: 14.29 lbs.

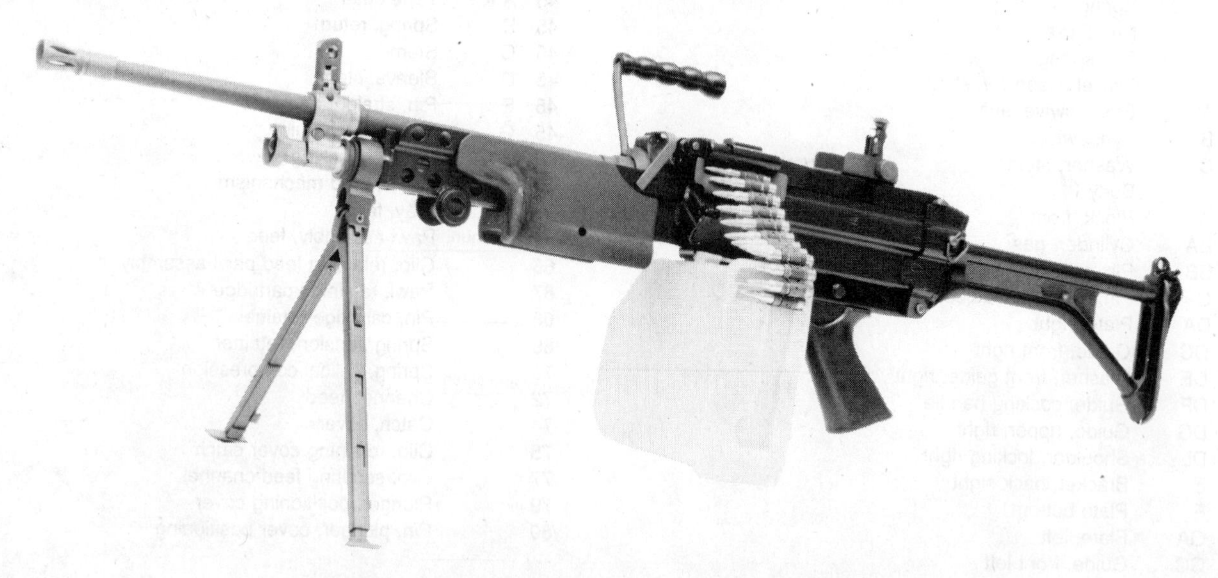

The M249 squad automatic weapon (Minimi) with 200-round feed box.

Overall length: 40.87 in.
Barrel length, w/o barrel extension or flash suppressor: 18.35
Feed mechanism: Link belt (disintegrating XM27) or FNC or M16 A1 magazines
Sights: Front, M249 non-adjustable in the field, Minimi adjustable; Rear, adjustable of different versions; 300–1000 meters
Muzzle velocity: Approx. 2900 f.p.s.
Cyclic rate: 700–1000 r.p.m.

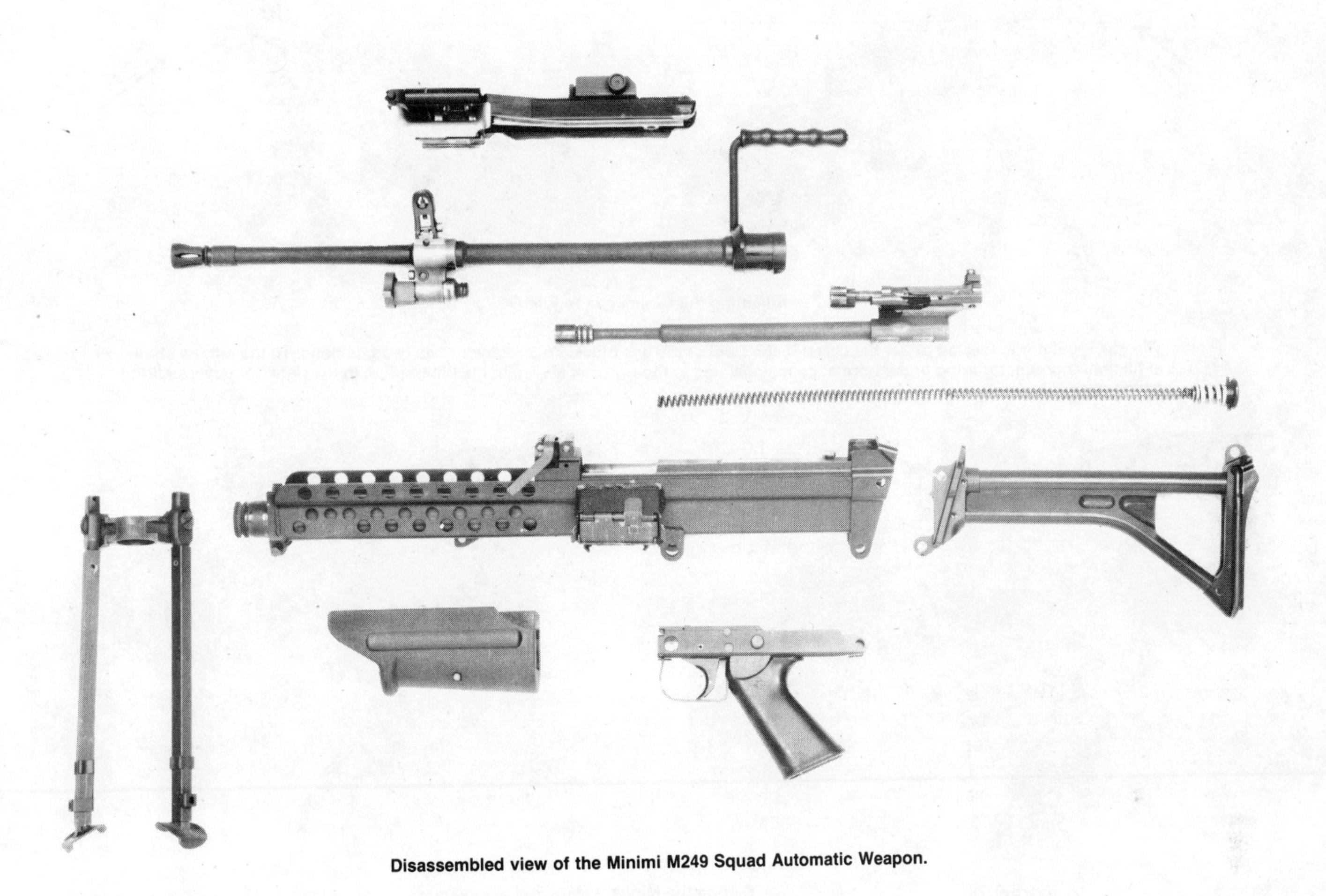

Disassembled view of the Minimi M249 Squad Automatic Weapon.

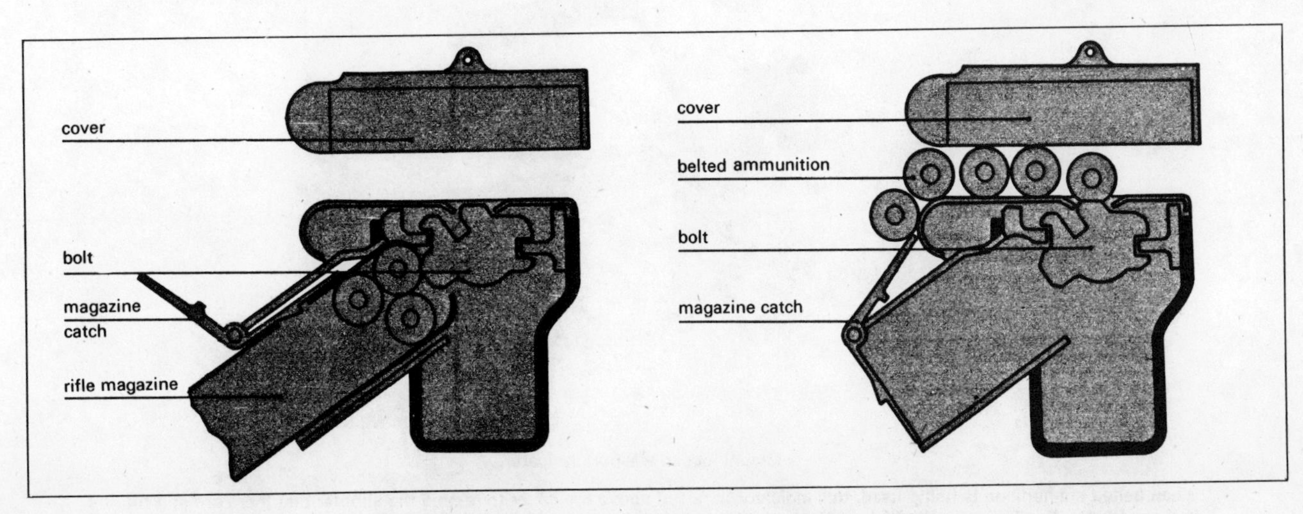

Diagrammatic view of the Minimi belt and magazine feed systems.

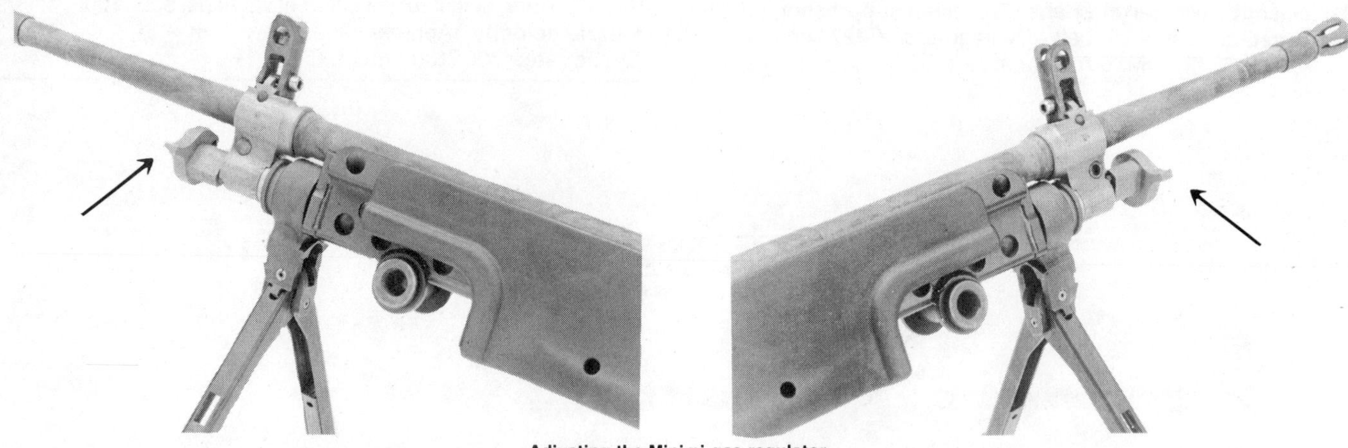

Adjusting the Minimi gas regulator.

The gas regulator is located under the barrel in the front of the gas block. This regulator has two positions. To the left, as shown in this photograph, for firing under normal conditions, and to the right, as shown in the photograph to the right, for adverse firing conditions.

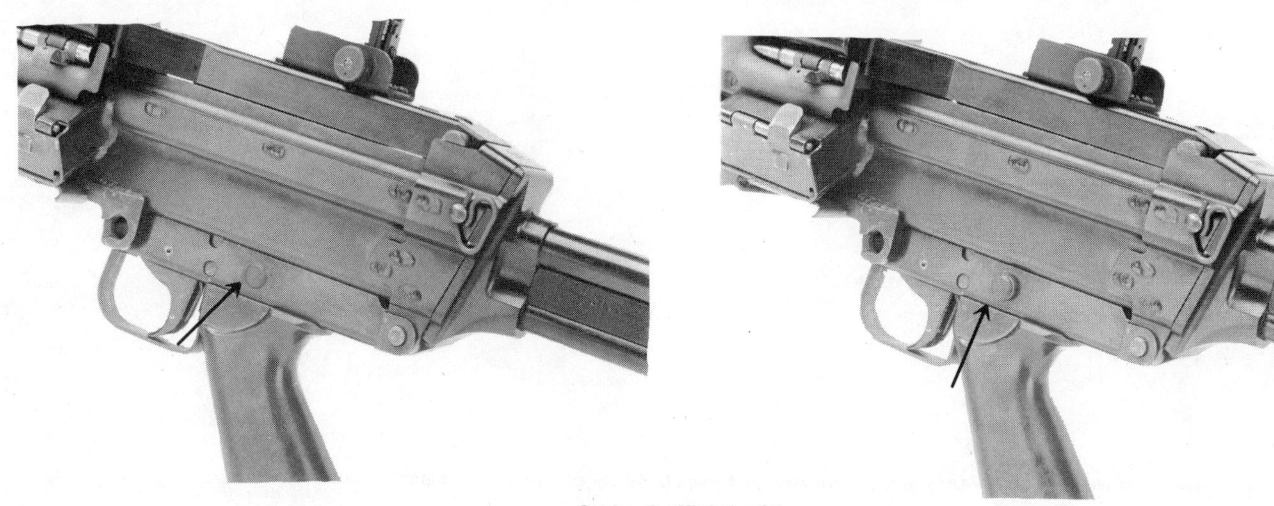

Setting the Minimi safety.

The Minimi safety is the push button type and it is located to the rear of the trigger and above the pistol grip. The safety is on "safe" when it is pushed to the right.

The safety is on "fire" when it is pushed to the left. In the fire position a red ring is visible on the safety. The safety can only be put on "safe" when the weapon is cocked; i.e., the moving parts are to the rear.

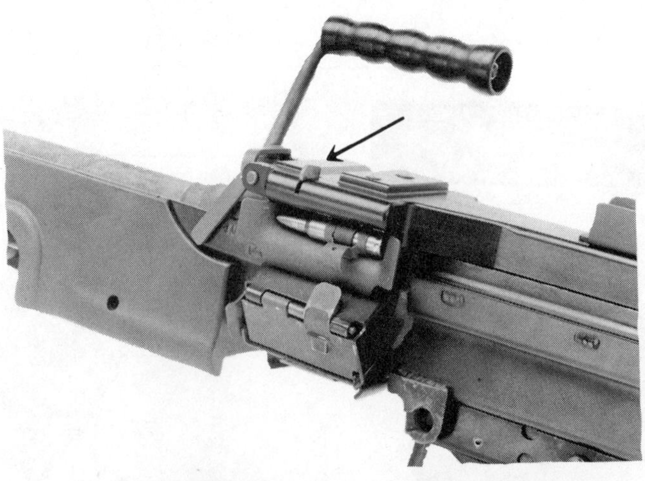

Minimi loaded weapon indicator.

When belted ammunition is being used, this indicator is raised above the cover to remind the shooter that there are still rounds in the feedway. This indicator can be seen in daylight and felt in darkness. *Note:* Three rounds or less cannot be seen easily by the shooter, but this indicator will indicate their presence.

Field Stripping the Minimi

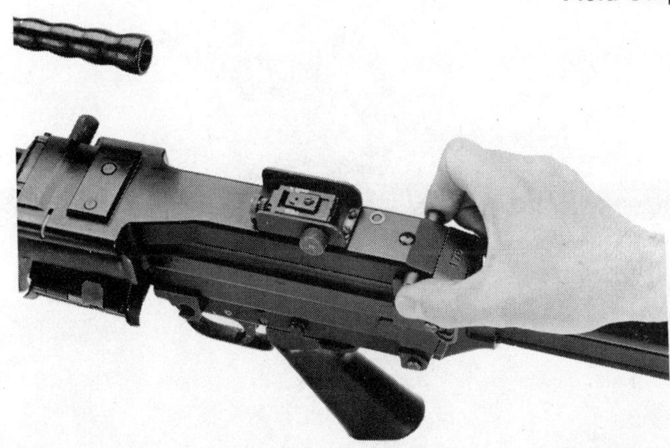

When using belted ammunition, the following steps are to be taken prior to disassembly. Remove the ammunition belt by opening the feed cover. That cover is opened by grasping the two cover latches as shown in the photograph and squeezing and lifting in a coordinated fashion.

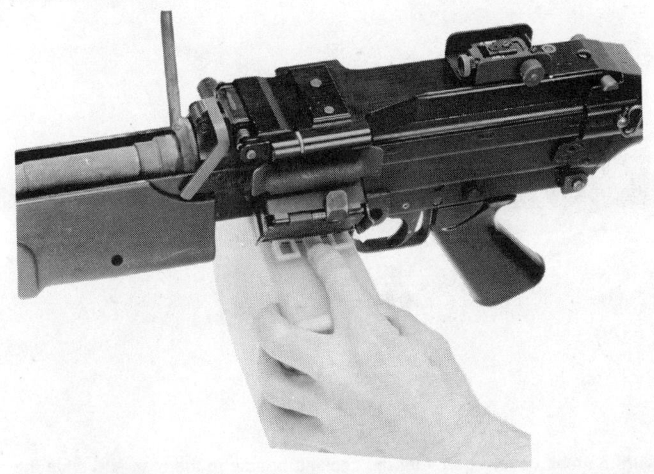

After the weapon has been cleared, the belted ammunition box can be unlatched by pressing with the index finger on the release tab as shown in the photograph. The ammunition box can then be pulled to the left away from the weapon.

When using the box magazine, depress the magazine catch with the index finger as shown in the photograph. Then open the feed cover and check to see if the weapon is unloaded. It should be, but it is always wise to be completely certain.

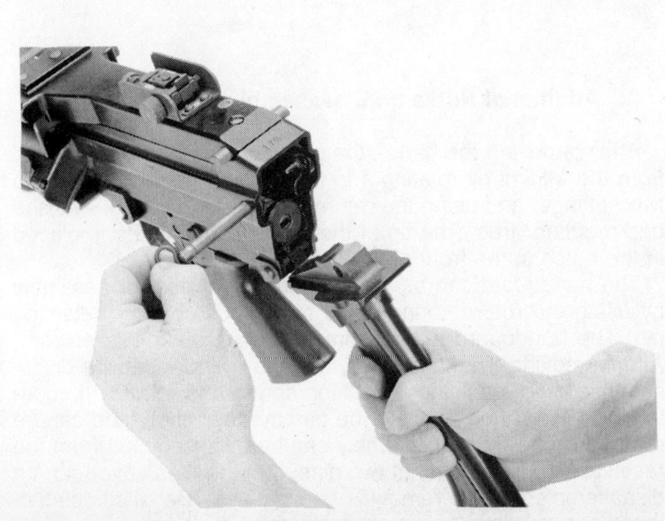

Disassembly should start with the moving components forward. Draw the retaining pin at the rear of the receiver to the left. Hinge the butt assembly downward.

The moving components (bolt and bolt carrier assembly) can be withdrawn by pushing forward and up on the rear end of the return spring assembly. In this way, the return spring is released from its positioning grooves inside the receiver. Withdraw the return spring assembly.

With the feed cover open, pull the cocking handle to the rear, and slide the moving parts out the rear of the receiver.

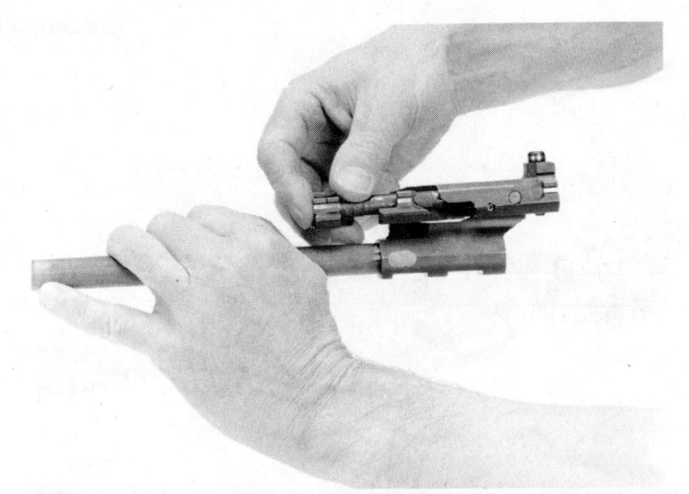

Rotate the bolt to disengage it from the bolt carrier. Note the similarity of this bolt and bolt carrier assembly with the one used in the FNC. *Caution!* When the bolt is removed from the bolt carrier, the firing pin spring is loose. Take care not to misplace it.

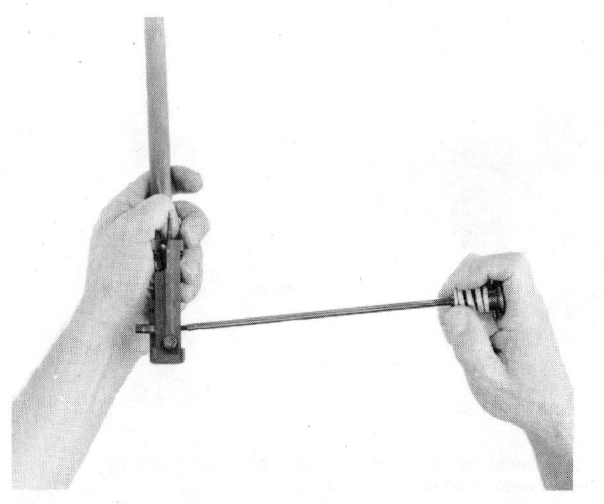

To separate the bolt carrier from the piston rod, press the assembly pin out as shown in the photograph. The return spring guide rod can be used to press out the pin.

To remove the barrel, the moving parts must either be locked to the rear or removed from the weapon. Draw the locking lever to the rear with the index finger of the left hand, and lift the barrel slightly and then slide it forward with the right hand. The operating parts also must be to the rear when replacing the barrel.

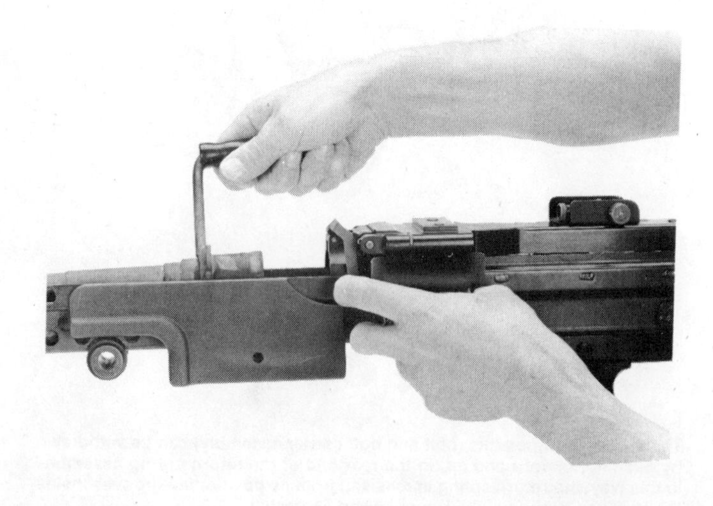

Additional Notes on Disassembly of the Minimi

After removing the barrel, the gas regulator can be removed from the Minimi by rotating it to the midway point between the two settings, and using the return spring guide rod to lever the gas regulator free. The tip of the return spring guide is placed in the notch in the front of the gas block.

The handguard can be removed from the receiver assembly by using the return spring guide rod to push out the retaining pin. The handguard is then removed by rotating it downward. After removing the handguard, the gas cylinder can be disassembled by depressing its locking spring and rotating it counterclockwise. After stripping the gas cylinder, the bipod can be removed. The trigger assembly can be disassembled from the receiver by the removal of two pins, and the feed cover can be disassembled by the removal of a single pin. Detail stripping of these two assemblies is not recommended unless repair is required. Such repairs should be done by specially trained armorers.

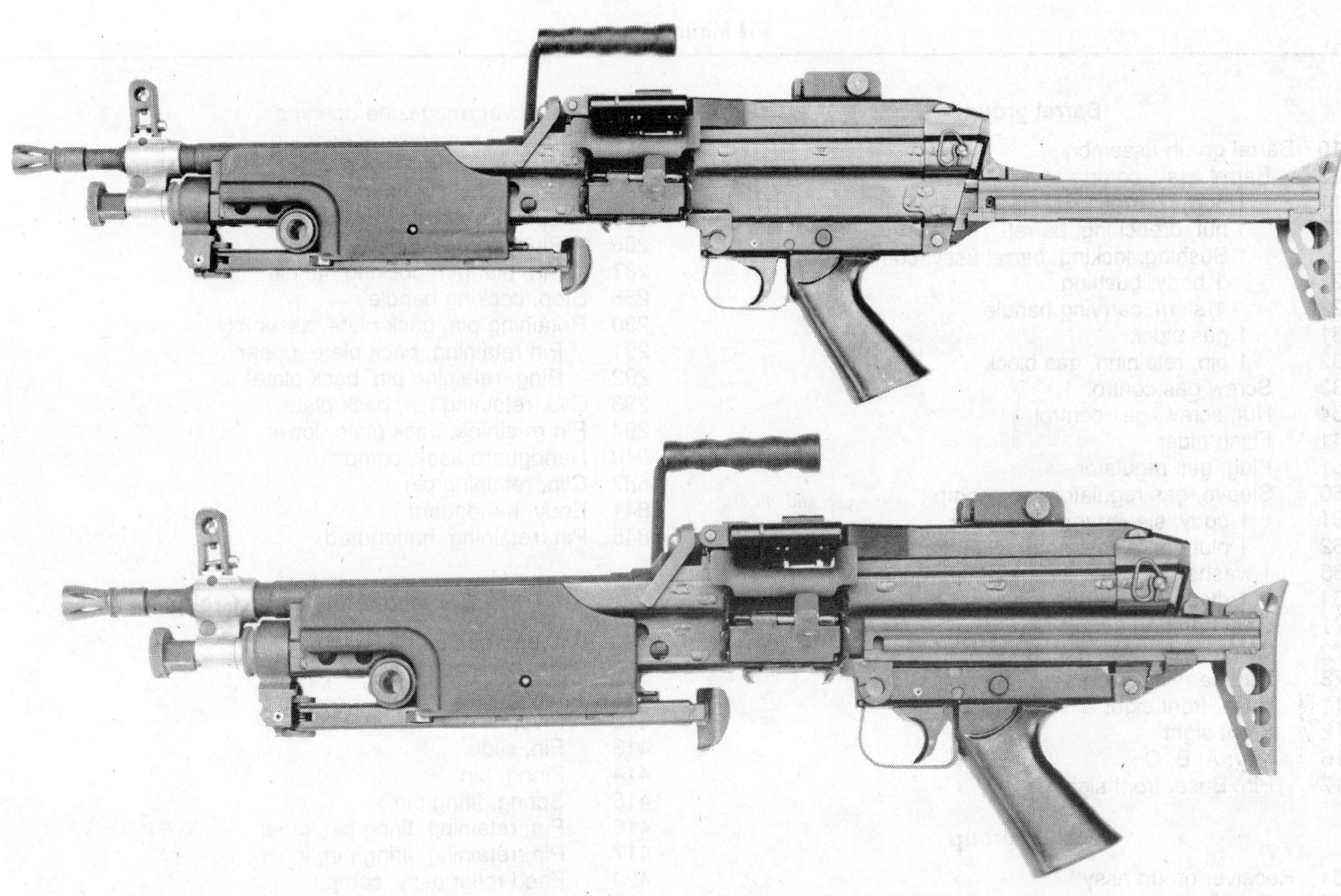

This paratrooper version of the Minimi is offered by FN with a sliding stock and a shortened barrel. With the stock extended it is only 35.5 inches (900mm) and 28.5 inches (725mm) with the stock collapsed. The barrel is 13.8 inches (350mm) without the flash suppressor or locking collar. This model weighs 14.33 pounds (6.56 kgs) unloaded.

This is the vehicle-mounted version of the Minimi. Note the absence of a buttstock and front sights.

Adjusting the Minimi bipod. Note the thumb on the latch of the bipod.

FN Minimi Parts

Barrel group

010 Barrel group assembly
110 Barrel assy. comp.:
111 1 body, barrel
112 1 nut, breeching, barrel
120 1 Bushing, locking, barrel assy. comp.
121 1 body, bushing
122 1 stem, carrying handle
131 1 gas block
132 1 pin, retaining, gas block
133 Screw gas control
134 Nut, screw, gas control
141 Flash hider
151 Plug, gas regulator
160 Sleeve, gas regulator assy comp.:
161 1 body, sleeve, gas regulator
162 1 plunger, sleeve, gas regulator
163 1 washer, keyway, sleeve, gas regulator
164 1 rivet, sleeve, gas regulator
171 Grip, carrying handle
172 Washer, carrying handle
173 Nut, carrying handle
711 Base, front sight
712 Front sight
716 Key: A, B, C
717 Pin, Base, front sight

Receiver group

201 Receiver group assy:
210 Receiver assy. comp.:
211 1 body, receiver
212 1 support, barrel
213 1 support, gas cylinder
214 1 bushing, front
215 1 support, trigger frame
216 1 guide, cocking handle
221 1 guide, slide, left
225 1 guide, slide, right
231 1 block, rear
234 1 magazine sleeve assy. comp.:
235 1 body, magazine sleeve
237 2 bushings, magazine sleeve
239 1 support, feed box
240 Gas cylinder assy. comp.:
241 1 body, gas cylinder
242 1 head, gas cylinder
245 Spring, latch, gas cylinder
247 Axis, front
248 Pin, front axis
249 Support, clip
251 Lever, locking, barrel
253 Spring, locking lever, barrel
254 Pin, locking lever, barrel
256 Ejector
257 Axis, ejector
258 Spring, ejector
260 Cover, ejection opening assy. comp.:
261 1 body, cover ejection opening
262 1 hook, cover, ejection opening
263 2 rivets, cover, ejection opening
265 Axis, cover, ejection opening
266 Spring, cover, ejection opening
271 Cover, magazine opening comp.:
275 Spring, cover, magazine opening
276 Pin, cover, magazine opening
277 Clip, cover, magazine opening
280 Cocking handle assy. comp.:
281 Body, cocking handle
285 Spring, plunger, cocking handle
286 Plunger, cocking handle
287 Pin, plunger, cocking handle
288 Stop, cocking handle
290 Retaining pin, back plate, assembly
291 Pin retaining, back plate, upper
292 Ring, retaining pin, back plate
293 Clip, retaining pin, back plate
294 Pin retaining, back plate, lower
840 Handguard assy. comp.:
687 Clip, retaining pin
841 Body, handguard
845 Pin, retaining, handguard

Bolt and Bolt Carrier Group

401 Bolt and bolt carrier comp.
410 Slide assy. comp.:
411 Slide
412 Pivot, slide
413 Pin, slide
414 Firing, pin
415 Spring, firing pin
416 Pin, retaining, firing pin, outer
417 Pin, retaining, firing pin, inner
420 Feed roller assy. comp.:
421 1 stem, feed roller
422 1 roller, feed
424 1 ring, feed roller
425 Spring, feed roller
426 Pin, feed roller
430 Piston assy. comp.:
431 1 piston
432 1 pin, retaining, piston
440 Operating rod assy. comp.:
441 1 rear part, operating rod
442 1 tube, operating rod
443 1 stem, operating rod
450 Bolt assy. comp.:
451 Bolt
452 Extractor
453 Spring, extractor
454 Guide, extractor spring
455 Axis, extractor
460 Rod, return and buffer assy. comp.:
462 Rod, return spring
463 Guide, buffer spring
464 Spring, buffer
465 Spacer, buffer
466 Pin, spacer, buffer
467 Return spring

Trigger mechanism group

501 Trigger mechanism group
510 Trigger frame assembly
511 Frame, trigger
515 Pin, trigger guard
520 Trigger, assy. comp.:
521 Trigger
522 Lever, tripping

FN Minimi Parts—continued

523	Spring, tripping lever
524	Axis, tripping lever
525	Roller, tripping lever
526	Axis, roller, tripping lever
527	Roller, sear spring
528	Axis, roller, sear spring
535	Axis, trigger
541	Catch, safety
545	Spring, safety catch
561	Spring, sear
562	Axis, sear spring
565	Sear
566	Axis, sear
581	Guard, trigger
582	Grip, pistol
587	Screw, pistol grip

Feed cover and rear sight group

060	Feed cover and rear sight assy. comp.:
620	Cover assembly comp.:
621	1 cover
622	1 pivot, link
623	1 pivot, brackets
626	1 pivot, feed lever
630	Feed pawl assy. comp.:
631	1 link, upper
632	1 link, lower
633	2 spacers, short
634	1 spacer, long
635	1 roller, link
641	1 bracket, upper
642	2 pawls, outer
644	2 axis, outer pawl
645	4 springs, feed pawl
651	1 bracket, lower
652	1 pawl, inner
653	1 axis, inner pawl
655	Retaining clip, feed pawl assy.
661	Indicator, cartridge
662	Spring, cartridge indicator
663	Pawl, retaining, cartridge, front
664	Pawl, retaining, cartridge, rear
665	Spring, pawl, retaining cartridge
666	Cover, link ejection opening
667	Pin, pawls, retaining cartridge
671	Lever, feed
672	Spring, feed lever
673	Clip, retaining, feed lever
674	Cover, cocking lever
675	Axis, cover, cocking lever
676	Spring, cover, cocking lever
677	Latch, cover
678	Spring, cover latch
679	Plug, scope adaptor
680	Feed tray assembly comp.:
681	1 tray, feed
682	1 stop, cartridge
683	2 rivets, cartridge stop
685	Axis, feed tray and cover
686	Spring, hinge, feed cover
687	Clip, retaining, pin
730	Rear sight assy. comp.:
715	Washer, base
731	Base, rear sight
736	Screw, base, rear sight
737	Spring, leaf, rear sight
740	Leaf, rear sight, assy. comp.:
751	Leaf, rear sight
755	Screw, elevation adjustment
756	Knob, elevation adjustment
757	Pin, elevation adjustment
758	Slide, rear sight
759	Half nut, gear-rack, rear sight
761	Peep-hole, rear sight
762	Screw, slide, rearsight
764	Spring, elevation adjustment
765	Screw, windage adjustment
766	Knob, windage adjustment
767	Spring, windage adjustment
768	Plunger, click, windage, rear sight
769	Pin windage knob
771	Scale, windage, rear sight
772	Washer, windage scale
773	Screw, windage scale

Butt group

801	Butt group assy. comp.:
810	Butt assembly comp.:
811	Sleeve, upper tube
812	Piece, distance, butt
815	Back plate
816	Pin, back plate
820	Body, butt assy. comp.:
821	Tube, upper assy.
822	1 body, upper tube
823	1 plug, tube
824	Tube, lower assy.
825	1 body, lower tube
823	1 plug, tube
829	Shoulder-piece

10 Brazil

SMALL ARMS IN SERVICE

Handguns: 9 × 19mm Beretta Model 92	Made by Forjas Taurus SA, Porto Alegre.
9 × 19mm Colt Pst 9 M73	Made by Fábrica de Itajubá, Minas Gerais State.
11.43 × 23mm Colt M1911	Made by Colt Firearms.
Submachine Guns: 11.43 × 23mm INA MB50 & M953	Made by Indústria Nacional de Armas SA, São Paulo. Licensed copy of Madsen Model 1946. See Chapter 16. Limited use by police and reserve forces.
11.43 × 23mm M3	Made in USA. Limited use by Marine Corps.
9 × 19mm M972	Made by Indústria e Comércio Beretta, São Paulo. Licensed copy of Beretta Model 12. Army.
9 × 19mm Walther MPK	Made by Walther. Used by some police forces.
9 × 19mm Mekanika Uru	Made by Mekanika Indústria e Comércio Ltda. Rio de Janeiro. Army, Navy and police forces.
Rifles: 7.62 × 51mm NATO M964 FAL	Made by Fábrica de Itajubá. Licensed copy of FN FAL.
7.62 × 51mm NATO M969A1 (PARA-FAL)	Made by Fábrica de Itajubá. Licensed copy of FN FAL PARA.
7.62 × 51mm NATO M969 FAP	Made by Fábrica de Itajubá. Licensed copy of heavy barrel FAL.
7.62 × 51mm NATO conversion of M1908 Mauser	Made by Fábrica de Itajubá.
7 × 57mm M49	Made by FN, the SAFN 49. See Chapter 9. Reserve forces.
7 × 57mm M1908 Mauser rifle	Made by DWM and Mauser. Obsolete.
5.56 × 45mm HK33	Made by Heckler & Koch for Air Force.
Machine Guns: 7.62 × 51mm NATO M971 MAG	Made by FN. Army and Marine Corps.
7.62 × 51mm NATO Mekanika Uirapuru GPMG	Made by Mekanika Indústria e Comércio Ltda.
12.7 × 99mm M2 HB	Made in USA.

SMALL ARMS IN BRAZIL

In the past decade, the Brazilian armaments industry has matured to the point that Brazil is an exporter of a range of weapons from small arms to armored vehicles, aircraft, and naval vessels. Concurrent with the appearance of this domestic arms manufacturing capability, a number of Brazilian design small arms have appeared. Two of these—the 9 × 19mm Mekanika Uru submachine gun and the 7.62 × 51mm NATO Mekanika Uirapuru General Purpose Machine Gun—will not only be used by Brazilian armed forces, but will likely be offered for international sale. Thus Brazil becomes a nation to watch in the years to come.

BRAZILIAN HANDGUNS

In the years immediately following the Second World War, Brazilian armed forces used US-made Colt 11.43 × 23mm (.45 ACP) M1911A1 pistols, because these had been provided to Brazil as part of their equipment when they fought alongside the United Nations forces in Europe. A domestically produced copy of the M1911A1 was manufactured by the Fábrica de Itajubá. This national arms factory is part of the government arms manufacturing complex called IMBEL. After producing the M1911A1 in 11.43 × 23mm for about eight years, the Fábrica de Itajubá started production of a 9 × 19mm Parabellum model in 1973. This action was taken after the Brazilian armed forces decided to standardize the 9 × 19mm cartridge for all handguns and submachine guns. This 9mm M1911A1 was never produced in significant quantities because the military had decided to adopt the double action, 15-shot Beretta Model 92 pistol. The 9 × 19mm M1911A1 was designated the 9 × 19 mm Pistola Colt 9 M973.

The Beretta Model 92 was initially manufactured in Brazil by a subsidary of the Italian firm—Indústria e Comércio Beretta SA. In mid-1980, this company was acquired by the Brazilian arms maker Forjas Taurus SA. Taurus is best known for its production of revolvers. It continues to manufacture the Beretta Model 92 for the Brazilian armed forces, and since late 1981 has been offering this handgun in the American and European commercial markets as the *PT-92 Pistola Dupla Ação* (Double Action Pistol). A detailed description of the Beretta Model 92 and its variations is presented in Chapter 29.

Because military and police personnel have been allowed to purchase their own side arms—in fact they have been encouraged to make such purchases in many organizations—there is a wide variety of Colt, Smith & Wesson, Taurus, Fabrique Nationale, and other handguns seen in the hands of Brazilian personnel. The military police of Rio de Janeiro still carry a much modified version of the Mauser Schnellfeur. This Policia Militar side arm is called the PASAM (*Pistola Automática e Semiautomática*) and it fires the standard 7.63 × 23mm Mauser cartidge. The PASAM has a 12.6 -inch (320mm) fixed stock and a forward pistol grip to permit it to be more easily controlled during full automatic fire.

BRAZILIAN RIFLES

During the Second World War, the Brazilian Expeditionary Force in Italy was equipped with US M1 rifles. This experience led Brazilian military planners to seek a self-loading rifle for their armed forces in the post-war period. Their first self-loader was the Fabrique Nationale SAFN 49 in 7 × 57mm Mauser (See Chapter 9). In 1964, the Brazilians acquired the production rights for the Fabrique Nationale FAL, and the Fábrica de Itajubá has fabricated it in the standard model, the heavy barrel model—*Fuzil Automatico Pesado* (FAP)—and the folding stock FAL-PARA. It is estimated that more than 200,000 FALs of all types have been manufactured in Brazil through 1982.

Following the adoption of the FN FAL in 7.62 × 51mm NATO, the Fábrica de Itajubá converted an unspecified number of 7 × 57mm M1908 Mauser bolt-action rifles to 7.62 × 51mm for use by military reserve and police forces. In the early 1970s, the Brazilian Air Force decided to replace its US .30 M1 Carbines with the Heckler & Koch HK33 in 5.56 × 45mm.

In 1981, Olympio Vieria de Mello, who also designed the now standard 9 × 19mm Uru submachine gun and the 7.62 × 51mm NATO Uirapuru general purpose machine gun, introduced a 5.56 × 45mm assault rifle. Called the OVM, after the designer's initials, this new rifle has a rotating bolt, is gas operated, and has an aluminum receiver. Prototypes were completed in 1981, and this may become a possible candidate for replacement of the current FALs.

Illustrated are the markings on a Fabrique Nationale–made FAL used in troop trials of this rifle.

Receiver markings on the 964 FAL as manufactured in Brazil by the Fábrica de Armas de Itajubá.

BRAZILIAN SUBMACHINE GUNS

During the years of the Second World War, the Brazilian Army used the M1 and M3 families of submachine guns, which were provided by the US. In the 1950s, the Industria Nacional de Armas SA, of São Paulo, acquired the production rights for the Madsen M50 submachine gun. INA manufactured this weapon for more than ten years in 11.43 × 23mm (.45 ACP). The INA .45 SMG MB50 and Model 953 are described in more detail in Chapter 16.

In the early 1960s, when the Brazilian military authorities decided to switch to 9 × 19mm Parabellum, they selected the Beretta Model 12. This weapon was manufactured by Indústria e Comércio Beretta SA, and continues to be produced by Forjas Taurus. Called the M972 by the military, it is marketed as the MT-12 Metrahadora de mao by Taurus. During the 1970s several experimental submachine guns appeared in Brazil. One of these was the Uru-developed Olympio Vieira de Mello. After five years of experimentation and testing, this weapon was adopted by the Brazilian Army in October 1979. The Uru is a very simple weapon, having only seventeen parts in the gun and four in the magazine. A new manufacturing company was established to fabricate this gun—Mekanika Indústria e Comércio Ltda. in Rio de Janeiro. Conventional in most aspects, the Uru is notable for its simplicity and ease of diassembly and reassembly.

CHARACTERISTICS OF URU SUBMACHINE GUN

Caliber: 9 × 19mm Parabellum/NATO
System of operation: Blowback, selective fire
Length overall: 26.42 in. with fixed stock
Barrel length: 6.89 in.
Weight: 7.28 lbs. unloaded with magazine.
Feed device: 30-round box magazine
Sights: Front, blade; rear, aperture set for 50 meters
Cyclic rate: 750 r.p.m.

BRAZILIAN MACHINE GUNS

Since the mid-1960s, the Fabrique Nationale MAG has been the standard Brazilian general-purpose infantry machine gun. Olympio Vieira de Mello mentioned above is also the designer of the new Brazilian GPMG. He had served on a machine gun task group established by the *Instituto Militar de Engenharia* (Military Engineering Institute) of the Brazilian Army. The IME machine gun project was transferred to private industry in 1972. After four years more work, the Army gave the project over to Olympia Vieira de Mello. After some additional work on the design, the *Instituto de Pesquisa e Desenvolvimento* (Research and Development Institute) ordered two prototypes from the new armsmaker Mekanika Indústria e Comércio Ltda. in December 1977. As with other de Mello weapons, this machine gun, called the Uirapuru, is relatively simple in design and easy to build. This will be the new GPMG of the Brazilian armed forces. Production began in 1982.

CHARACTERISTICS OF THE UIRAPURU GPMG

Caliber: 7.62 × 51mm NATO
System of operation: Gas; automatic fire; dropping lever locking system
Length overall: 51.18 in.
Barrel length: 23.62 in.
Weight: 28.66 lbs. with bipod.
Feed device: Links, standard NATO
Sights: Front, adjustable for windage and elevation; rear, adjustable from 200–600m and 800–1400m
Cyclic rate: 650–700 r.p.m.

11 Britain (United Kingdom)

It is no longer realistic to attempt to collect under Britain all weapons in use in the British Commonwealth, since many of the member nations have taken an independent course. This independence in weaponry should not be too surprising, inasmuch as the United Kingdom does not dictate to the members of the Commonwealth what weapons they will or will not adopt. In the past, these nations have usually adopted the same weapons as the UK, because in most cases (but not all) they were members of defense treaties with the UK and their political, social and economic orientation would put them on the same side as the UK in any major war. Since World War II, however, the defense arrangements have tended to be more general and include nations outside the Commonwealth, notably the United States. It is therefore likely that the future will see many American weapons added to those that are already standard in the British Commonwealth.

SMALL ARMS IN SERVICE

Handguns:	9 × 19mm NATO Pistol Automatic L9A1	FN Mle. 1935 GP
	7.65mm Pistol Automatic L47A1	Walther PP
Submachine guns:	9 × 19mm NATO L2A3	Military version of Sterling Mk 4
	9 × 19mm NATO L34A1	Silenced version of L2A3; similar to Sterling Mk5
Rifles:	7.62 × 51mm NATO L1A1	English version of FN FAL
	7.62 × 51mm NATO L39A1	No. 4 .303 rifle converted to 7.62mm NATO and used for competition shooting
	7.62 × 51mm NATO L42A1	No. 4 .303 rifle converted to 7.62mm NATO and used as sniper rifle
Machine guns:	7.62 × 51mm NATO L7A1	English version of FN MAG
	7.62 × 51mm NATO L7A2	Modified L7A1
	7.62 × 51mm NATO L8A1	Armor version of L7 series
	7.62 × 51mm NATO L8A2	Product-improved version of L8A1
	7.62 × 51mm NATO L20A1 and L20A2	Helicopter version of L7 series
	7.62 × 51mm NATO L36A1	Instructional Skeleton M.G.
	7.62 × 51mm NATO L37A1 and L37A2	Armor version of L8 series that can be used in the infantry role by vehicle personnel
	7.62 × 51mm NATO L41A1	Drill (instructional model) version of L8 series; inoperable
	7.62 × 51mm NATO L44A1	Helicopter gun, Royal Navy
	7.62 × 51mm NATO L45A1	Drill version of L7 series
	7.62 × 51mm NATO L46A1	Drill version of L7 series; inoperable
	7.62 × 51mm NATO L4A4	7.62mm version of Bren
	7.62 × 51mm NATO L4A6	7.62mm version of Bren; obsolete
	7.62 × 51mm NATO L43A1	L7 used as ranging gun for the Scorpion
	7.62 × 63mm L3A3	Browning M1919A4 (fixed); modified rear sear arrangement
	7.62 × 63mm L3A4	Browning M1919A4 (flexible); modified rear sear arrangement
	12.7 × 99mm L1A1	.50 Browning M2 HB; modified and with UK-made barrel
	12.7 × 99mm L2A1	.50 Browning M2 HB Drill; inoperable; also L30A1
	12.7 × 99mm L6A1	.50 Browning M2 HB Ranging; also L11A1 and L21A1; X16E1 is modified M2 for ranging; L6A1 is .50 cal. ranging gun for 105mm tank gun; L21A1 is .50 cal. ranging gun for 120mm tank gun (barrel for L21A1 is 6 inches longer than L6A1)
	12.7 × 99mm X17E1	.50 M85 Ranging; experimental; X17E1 and X17E2 were trial guns only—never an issue model. Both used UK-made barrels fitted to US M85 guns.
	.50 Spotting, M48	.50 ranging gun, L40A1 (US M8C spotting rifle for WOMBAT A/Tank gun

BRITISH SMALL ARMS NOMENCLATURE

The model-designation procedures of the UK are somewhat involved and can be confusing. Prior to World War I, British model designations for small arms were comprised of the word Mark followed by a Roman numeral, e.g., Mark I, Mark II. Two exceptions to this form of designation were the Pattern 13, which was an experimental cal. .276 rifle tested in 1913, and the Pattern 14, which was made in cal. .303 as the production version of the Pattern 13.

Between World Wars I and II, rifles and pistols were given number designations in addition to mark designations. Thus the Rifle Mark III SMLE became Rifle No. 1 Mark III SMLE, and the Pattern 14 became the Rifle No. 3 Mark I. Toward the end of World War II, the British began using Arabic numerals for both the number and mark designations. An additional complication was the star symbol (*), which was frequently found tacked to the end of everything else, e.g., Rifle No. 4 Mark 1*. This star indicates a minor modification from the Mark design.

Since the early fifties, weapons entering the British service have received an L designation, and modifications have been indicated by an A, as in L2A3. The L stands for "Land service." Older weapons that have been considerably modified, such as the Brens rebarreled for the 7.62mm NATO cartridge, also receive an L designation, as do (in most cases) United States weapons adopted as standard or limited standard by the UK.

A revolver is usually called a "pistol" by the British.

CURRENT BRITISH HANDGUNS

The standard handgun of the British armed forces is the FN Browning Mle. 1935 GP, which the British call the Pistol Automatic L9A1. See Chapter 9 for a detailed description of this pistol. Other handguns have been used in limited numbers since the end of World War II. These include the obsolete revolvers described here and the Walther PP. The official designation of the latter is the L47E1.

OBSOLETE BRITISH PISTOLS AND REVOLVERS

The British army was the last major army in the world to use the revolver as a standard service weapon. Lieutenant Winston Churchill of the 21st Lancers used a personally procured Mauser Automatic against the Dervishes at the battle of Omdurman in the Sudan in 1898 and carried the same weapon during his stint as a war correspondent in the Boer War. This put Sir Winston Churchill approximately fifty years ahead of his country's army. After World War II, in which the caliber .38 Enfield revolvers were standard (Webley, Smith and Wesson and Colt revolvers and Canadian-made Browning, Webley, Colt, Star and various other automatic pistols were also used), the UK ran a series of pistol tests. These tests confirmed that a 9mm Parabellum automatic pistol was the best service arm, and the Canadian-made Browning FN H.P. pistol was adopted as standard. Troops were equipped with pistols on hand, which had been used by British paratroop and commando units during the war.

The United Kingdom purchased a quantity of 9mm Parabellum Browning Hi-Power pistols for the RAF from FN. These are the latest type with extractors mounted on the slides and two-piece barrels. They are called Pistol 9mm L9A1 by the British and are now in general use.

WEBLEY REVOLVERS

The British Government used Webley revolvers as standard or limited standard for 60 years. The Mark I was adopted in November 1887, and the last of the standard Webley revolvers, the No. 1 Mark VI, was declared obsolete in 1947. All the standard British issue Webley revolvers were caliber .455, and all were similar in design. The Webley is a top breaking revolver, locked by a heavy stirrup type barrel catch. The first five Marks have "birds head" type grips, and the Mark VI has a square grip. The Mark VI (called No. 1 Mark VI after 1927), which was adopted in May 1915, was made in the greatest quantity, over 300,000 of these revolvers having been made by Webley & Scott at Birmingham during World War I. A quantity of the Mark VI were also made at Enfield Lock after World War I.

After World War I, the British decided that .455 was too heavy for the most effective use and after tests decided upon the use

Note: For additional details on obsolete British pistols and revolvers, see *Handguns of the World*.

of a caliber .38 cartridge based on the .38 Smith & Wesson cartridge case. Webley designed a new pistol using many of the features of their commercial Mark III caliber .38 revolver. The design was taken over by Royal Small Arms Factory and as completed was not compatible with the Webley pistol; parts are not interchangeable. The Webley Mark IV caliber .38 revolver was adopted as limited standard in World War II.

THE WEBLEY .455 "PISTOL" NO. 1 MARK VI

Although this revolver has been obsolete in Great Britain since 1947, it is widely distributed throughout the British Commonwealth and former British territories.

Loading and Firing the Webley Mark VI

Push forward on the curved tail of the pivoted barrel catch, which is on the left side of the revolver just below the hammer. As the catch is pushed, it pivots on its screw drawing the upper latching end back over the barrel strap, freeing the barrel to be tipped down on its hinge. As the barrel is bent down, the extractor will rise on its stem until the revolver is fully opened, at which point the extractor under the influence of its spring will slip back into its place in the cylinder.

Now load the six chambers. With a little practice this may be done two chambers at a time. If the cylinder is to be only partly loaded, remember that the cylinder revolves clockwise; and that the first cartridge must be to the left of the chamber in direct line with the hammer nose when the weapon is closed. Cocking the hammer automatically turns the cylinder the distance of one chamber.

Now turn barrel and loaded cylinder up to the fullest extent. The heavy catch will automatically be sprung over the barrel strap and lock it securely.

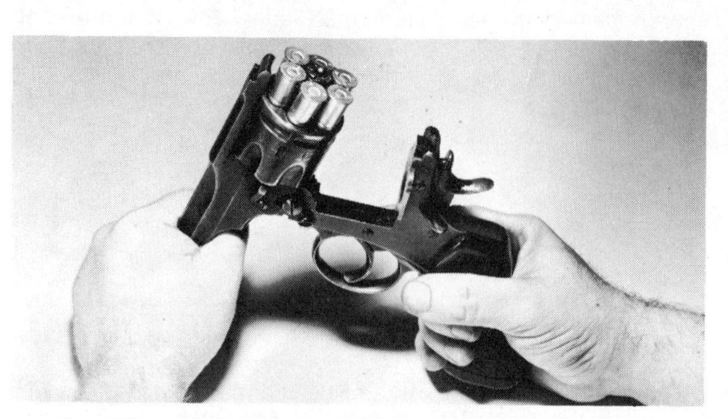

To extract cartridges, break the revolver and push barrel down.

If you have time, and accuracy is desired, always pull back the hammer with the thumb to full cock for each shot. For close quarters or emergency firing, drawing the trigger straight back will raise the hammer to full cock and turn the cylinder and trip the hammer, completing the firing. It is necessary to release the pressure on the trigger after each shot to permit the mechanism to engage for the next shot. Accurate shooting except at close range is difficult when shooting double action.

Webley .455 Pistol No. 1 Mark VI.

Field Stripping the Webley Mark VI

The only stripping necessary and recommended for this revolver is removing the cylinder. The bottom screw at the extreme forward end on the left side of the receiver is the cylinder catch retaining screw. Unscrew this. Now push the bottom of the cylinder catch retainer directly above the screw upwards. This will depress the rear of the catch and permit the cylinder to be lifted out.

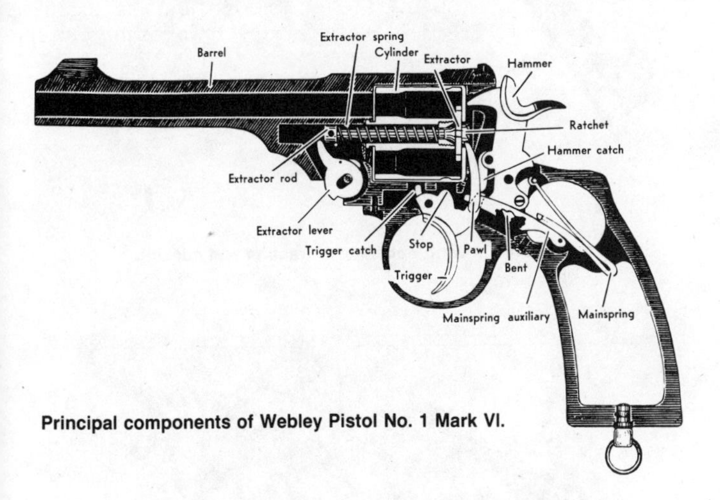

Principal components of Webley Pistol No. 1 Mark VI.

CHARACTERISTICS OF WEBLEY .455 PISTOL NO. 1 MARK VI

Caliber: .455 Webley.
System of operation: Single or double-action, top break revolver.
Weight: 2.37 lbs.
Length: overall: 11.25 in.
Barrel: 6 in.
Feed device: Cylinder with 6 chambers.
Sights: Front: Blade.
Rear: Notch.
Muzzle velocity: 620 f.p.s.
NOTE: Some No. 1 Mark VI revolvers have been rechambered by US gun dealers for the caliber .45 Colt automatic cartridge. These revolvers use the three-round clip used with the Colt and Smith & Wesson Model 1917 revolvers.

ENFIELD REVOLVERS

As noted previously, the British government and Webley & Scott parted company on uniformity of design in 1926 when the No. 2 Mark 1 Enfield pistol was in prototype form. The No. 2 Mark 1 had many of the best features of the .455 Webley Mark VI and in addition had a movable firing pin mounted on the hammer (all the earlier Webley government revolvers had a fixed hammer nose type firing pin) and a removable side plate. These features had appeared in the commercial .38 Webley Mark III.

The revolver, called Pistol .38 No. 2 Mark 1, was produced from 1927 to 1938; it was officially adopted on 2 June 1932. On 22 June 1938, the first modification of this revolver—the No. 2 Mark 1*—was introduced. The modification consisted of the removal of the spur and the single-action cocking notch on the hammer. The No. 2 Mark 1* can therefore only be used double-action. Since this requires lifting of the hammer, firing and rotation of the cylinder by the pulling of the trigger, the trigger pull is very hard. As a result, this revolver is of very limited accurate range. In 1942, another Model was introduced, the pistol No. 2 Mark 1**. This model has no hammer safety stop. As originally issued, the No. 2 Mark 1 had Mark 1 walnut grips of rather square configuration; at a later date Mark 2 black bakelite or walnut grips with thumb recesses were adopted. These are usually seen on the No. 2 Mark 1* and Mark 1** but may occasionally be seen on the No. 2 Mark 1 as well.

Complete revolvers were made by Enfield and by Albion Motors at Glasgow. Singer Sewing Machine of Great Britain made parts, which were assembled into complete revolvers at Enfield. In 1957, the Enfield revolvers were dropped as standard and replaced by the FN Browning Hi-Power automatic. These revolvers are still in extensive use in former British territories and are considered a reserve weapon in the United Kingdom.

The Enfield revolver, as the Mark IV Webley, uses the British .380 (or .38), revolver cartridge. It can also be used with commercial US caliber .38 Smith & Wesson ammunition (not S & W special) but will have a tendency to shoot high with this ammunition, since the issue front sight is set for the heavier British Mark 2 caliber .38 bullet. Higher sight blades can be obtained from arms dealers in the UK, or the blade can be built up by brazing.

CHARACTERISTICS OF ENFIELD REVOLVER

Caliber: .38, .380 in. revolver, .38 S & W, .38 Webley.
System of operation: No. 2 Mark 1 single or double action, No. 2 Mark 1* and No. 2 Mark 1** double action only.
Weight: 1.58 lbs.(1)
Length, overall: 10.25 in.
Barrel length: 5 in.
Feed device: Cylinder with 6 chambers.
Sights: Front: Blade.
Rear: Square notch.
Muzzle velocity: 600.

(1)No. 2 Mark 1* and No. 2 Mark 1** weigh about an ounce less.

Loading and Firing the Enfield

Loading and firing the Enfield Mark 1 are accomplished exactly as in the Webley Mark VI. The Mark 1* is handled in the same manner, except that it can be used only double-action. The barrel latch is the same as on the Mark VI.

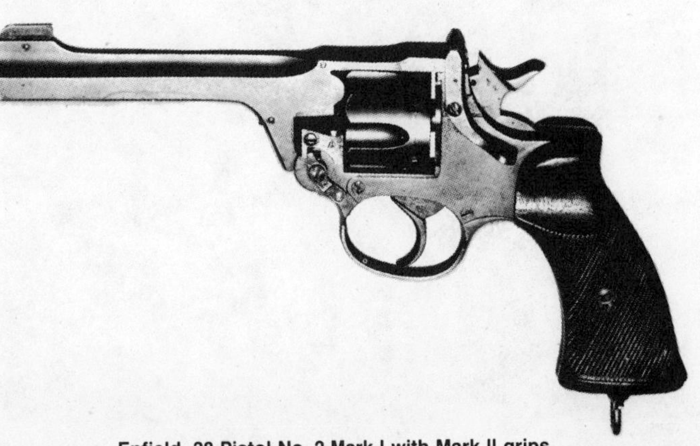

Enfield .38 Pistol No. 2 Mark I with Mark II grips.

Field Stripping the Enfield

Unscrew cam lever fixing screw. Push barrel catch to open pistol and remove the cylinder.

Unscrew the stock and side plate screws on the left side of the weapon; remove the stock and the side plate.

All parts are now exposed and further dismounting is not recommended except by a competent armorer, as springs and parts may be injured unless properly handled.

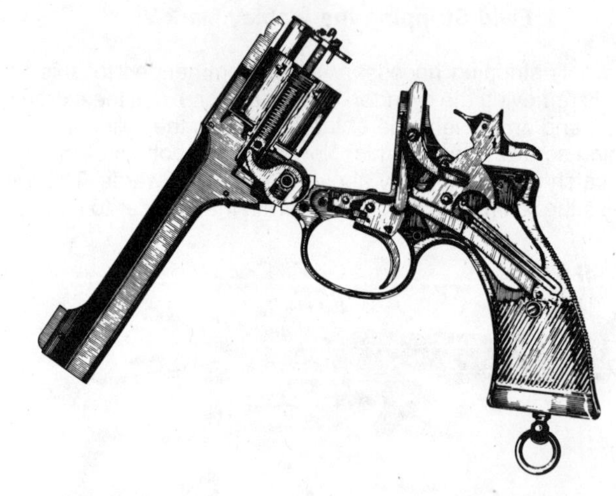

Action open, showing details of extraction and cocking.

SUBSTITUTE STANDARD AND NON-STANDARD BRITISH PISTOLS AND REVOLVERS

In both World War I and World War II, Great Britain found it necessary to obtain automatic pistols and revolvers abroad.

During World War I, Smith & Wesson and Colt made large quantities of caliber .455 revolvers for Britain. Although there were a number of models supplied, the most common models were the Colt New Service and the Smith & Wesson Mark II Hand Ejector, of which 73,650 were supplied to the UK and Canada. Colt caliber .32, .38, .45 and .455 automatic pistols were also purchased by the UK, and many of these weapons have since come back to the US with British proof and broad arrow (signifying government ownership) marks on all major components. Approximately 10,000 .455 M1911 Colt automatics were supplied to the UK during World War I. After the war, these pistols were issued to the R.A.F., and most bear the markings of the organization.

During World War II, the US supplied Great Britain with 20,000 Colt and Smith & Wesson caliber .45 Model 1917 revolvers after Dunkirk. Large quantities of Smith & Wesson .38/200 K200 revolvers and Colt .45 Model M1911A1 automatics were supplied under Lend-Lease. Apparently in 1940, the UK purchased every type of pistol that had any military potential, as they also procured quantities of Smith & Wesson K-38 target revolvers. Ballester Molina caliber .45 automatics were purchased from Argentina, and Star (among other) automatics were purchased from Spain. Since World War II, the United Kingdom has disposed of all of these nonstandard weapons, mainly by sale to surplus arms dealers.

British Smith & Wesson .38 Pistol (.38/200).

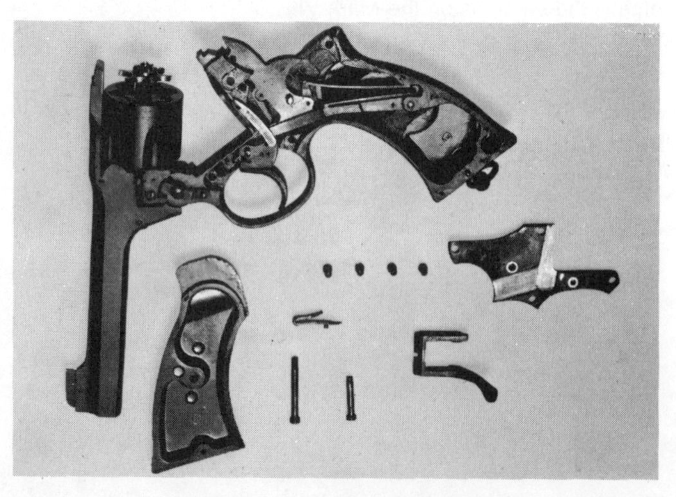

The Enfield No. 2 Mark I* field stripped.

BRITISH SMITH & WESSON .38 PISTOL

CHARACTERISTICS OF BRITISH S & W .38 PISTOL

Caliber: British Service .380 inch. Also .38 S & W, 148- or 200-grain bullet. (Note: British Service Ammunition has metal jacketed bullet).
Cylinder: 6 chambers.
Muzzle velocity: 600 feet per second with British Service Ammunition.
Barrel length: 5 in.
Overall length: 10.2 in.
Weight: 1.81 lbs.
Sights: Front, blade. Rear, square notch.
Other data: Essentially the same in operation and stripping as the United States Smith & Wesson .45 1917 Revolver. This revolver is almost identical to the S & W Military and Police Model.
Note: This revolver, which in British terminology is called a pistol, will not handle the .38 Smith & Wesson Special type cartridges. It will handle the shorter and wider .38 S & W type. These weapons are considered obsolescent in the UK at present.

Hammer back, showing details of lockwork and front and rear cylinder locking mechanism.

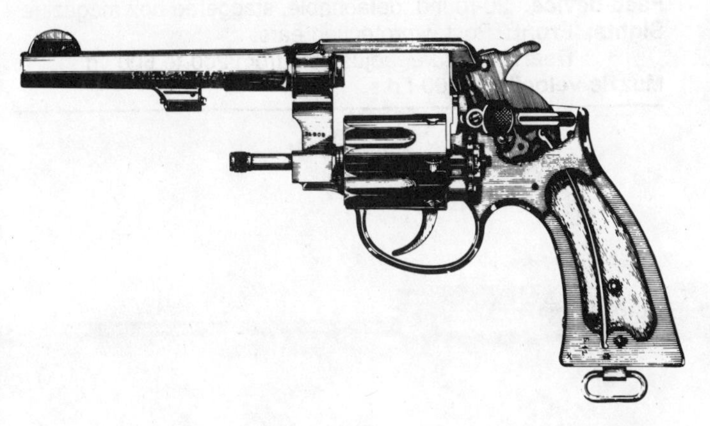

Hammer down, cylinder swung out to show details of extraction. Sideplate cut away to show detail of thumb lock.

THE .455 WEBLEY AUTOMATIC PISTOL

The Webley Automatic pistol was standard issue in the Royal Navy from 1912 until the end of World War II. There are two basic models, the Mark 1 and the Mark 1 No. 2. The Mark 1 No. 2 has a different type rear sight than the Mark 1 and has a different type manual safety. During World War I, some .455 Webley automatics were fitted with shoulder stocks to be used by the Royal Flying Corps.

CHARACTERISTICS OF BRITISH WEBLEY AUTOMATIC PISTOL

Caliber: .455 Webley automatic.
System of operation: Recoil operated
Weight: 2.43 lbs.
Length, overall: 8.5 in.
Barrel length: 5 in.
Feed device: 7-round, in line detachable box magazine.
Sights: Front: Blade.
Rear: Mark 1 fixed notch, Mark 1 No. 2 adjustable.
Muzzle velocity: 710 f.p.s.
Special feature: Magazine is provided with two catch notches in the magazine, one above the other. Push the magazine all the way in and the catch will lock in the lower notch leaving the pistol ready for magazine fire. If the magazine is pushed only part way in so that the catch locks in the upper notch, the pistol can be loaded with single cartridges inserted through the open breech; and action closed by pressing slide release catch. After each shot thus fired, the slide will remain open ready for the next cartridge. Meanwhile the magazine remains loaded in the handle held in reserve. To achieve magazine fire, it is only necessary to push the magazine in until it catches in the second lock notch.

BRITISH COLT .455 AUTOMATIC PISTOL

The standard United States .45 pistol cartridge may be used in this weapon. However, the .455 cartridge will not chamber in the .45 service pistol.

This one way interchangeability of ammunition is occasioned by the fact that the actual bullet diameter of the US .45 Auto cartridge is .4515 inch; while that of the .455 Webley S. L. is actually .455 inch. Thus, while the smaller diameter .45 will chamber in the .455, the reverse is not true. The caliber .455 will be found stamped on the right side of the receiver in this weapon.

Webley .455 Automatic Mark I.

CURRENT SELF-LOADING RIFLES

The L1A1, an English measurement version of the Fabrique Nationale FAL, is still the standard issue rifle of the British armed forces. While similar in nearly all aspects of design with the FN FAL, the parts are dimensioned differently and they are not interchangeable with metric FALs.

Basic satisfaction with the SMLE family of bolt action rifles, production demands imposed by World Wars I and II and the cost of converting to new patterns were all factors that caused the British Ministry of Defence to proceed slowly with the development, testing and adoption of a self-loading rifle. Many models were tested in the 1930s and 1940s, but none were adopted. In the 1950s, the British undertook an extensive ammunition and rifle development effort (see discussion in Chapter I), but due to political and economic considerations they finally settled on the FN FAL (light automatic rifle) and designated it the L1A1. This rifle has been manufactured at both the Royal Small Arms Factory, Enfield Lock (RSAF), and the Birmingham Small Arms Company (B.S.A.).

The British version has been modified in a few components, but basically the weapon is the same as that covered in the chapter on Belgium. Cuts have been added to the bolt carrier to serve as gathering places for dirt and dust that might otherwise enter the action. These cuts are deep enough so that a good deal of foreign matter can accumulate in them without impairing the normal functioning of the weapon. The L1A1 fires the 7.62 × 51mm NATO cartridge.

CHARACTERISTICS OF THE L1A1 (BRITISH VERSION OF THE FN NATO LIGHT RIFLE)

System of operation: Gas, semiautomatic fire only (can be modified to selective fire).
Weight, loaded: 10.48 lbs.
Length, overall: 44.5 in.
Barrel length: 21 in.
Feed device: 20-round, detachable, staggered box magazine.
Sights: Front: Post w/protecting ears.
Rear: Aperture, adjustable from 200 to 600 yd.
Muzzle velocity: 2800 f.p.s.

The currently manufactured L1A1 rifle with SUIT (Sight Unit Infantry Trilux). Note that this weapon has a molded synthetic stock and handguard. Earlier L1A1s had wooden stocks and handguards.

Current L1A1 with Royal Marines pattern arctic sling and muzzle cap.

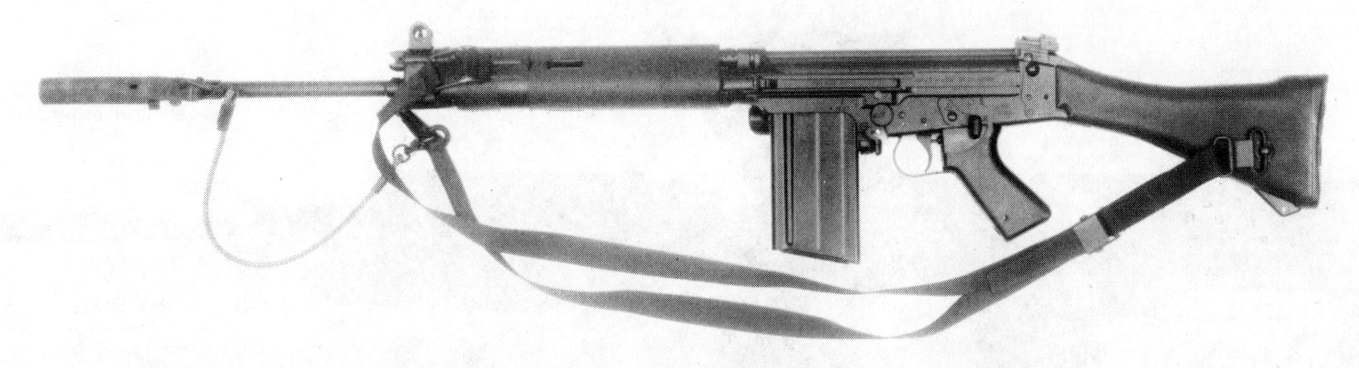

Current manufacture L1A1 with Royal Marines arctic sling and muzzle cap.

Close-up of the receiver of an early British L1A1.

ENFIELD WEAPON SYSTEM

After several years of renewed experimentation with small caliber ammunition, the British selected a 4.85 diameter for their Individual and Light Support Weapons, which were their contenders in the NATO small caliber trials. As noted in Chapter 1, both of these weapons (originally designated XL64 and XL65) were bullpup designs, and 80 percent of their parts interchanged. The Light Support Weapon (LSW) differed from the Individual Weapon (IW) in that it had a longer and heavier barrel, a 30-shot instead of 20-shot magazine, and a bipod. The LSW could not be employed to launch rifle grenades, whereas the IW could. Both weapons were equipped with the experimental issue UK SUSAT (Sight Unit Small Arms Trilux). As tested by NATO the 4.85mm weapons were designated as follows:

IW XL64E5 for right-handed shooters
IW XL68E2 for left-handed shooters
LSW XL65E4 for right-handed shooters
LSW XL69E1 for left-handed shooters

Although the British liked the 4.85 × 49mm cartridge, this ammunition came at the wrong time. Bending to the pressure to adopt a 5.56 × 45mm cartridge, the British modified their system to fire the US M193 cartridge at first. Later it was re-barreled to shoot the SS109 projectile. Redesignated the XL70 series (IW being the XL70E3 and the LSW being the XL70E2), the Enfield Weapon System is likely to become the next British infantry rifle. The Canadians and Australians will also test this weapon in their respective rifle trials.

The Enfield Weapon System (EWS) is a conventional gas-operated weapon locked by a multi-lug rotating bolt engaging corresponding recesses in a barrel collar. This bolt and its bolt-carrier are very similar in design to that of the Armalite AR-18 rifle. The EWS bolt-carrier runs on three guide rods that also carry the return springs. The accompanying illustrations clearly show the differences between the XL64 and XL70 series.

UK 4.85 × 49mm Individual Weapon.

UK 5.56 × 45mm Enfield Weapon System. This is the 1983 version of the Light Support Weapon. Note detachable iron sights, blank firing attachment, and aluminum and nylon M16-type magazines.

CHARACTERISTICS OF THE INDIVIDUAL WEAPON

System of operation: Gas; single-shot and automatic burst.
Weight, loaded: 9.1 lbs.
Length, overall: 30.3 in.
Barrel length: 20.4 in. (including Flash Hider).
Feed device: 20-round magazine; 30-round available.
Sight: Optical sight, ×4.
Muzzle velocity: 2952 f.p.s. (4.85); 2745 f.p.s. (5.56)

CHARACTERISTICS OF THE LIGHT SUPPORT WEAPON

System of operation: Gas; single-shot and automatic burst.
Weight, loaded: 11 lbs., 9.5 oz.
Length, overall: 35.4 in.
Barrel length: 25.4 in. (including Flash Hider).
Feed device: 30-round magazine; 20-round available.
Sight: Optical sight, ×4.
Muzzle velocity: 3051 f.p.s. (4.85); 2880 f.p.s. (5.56)

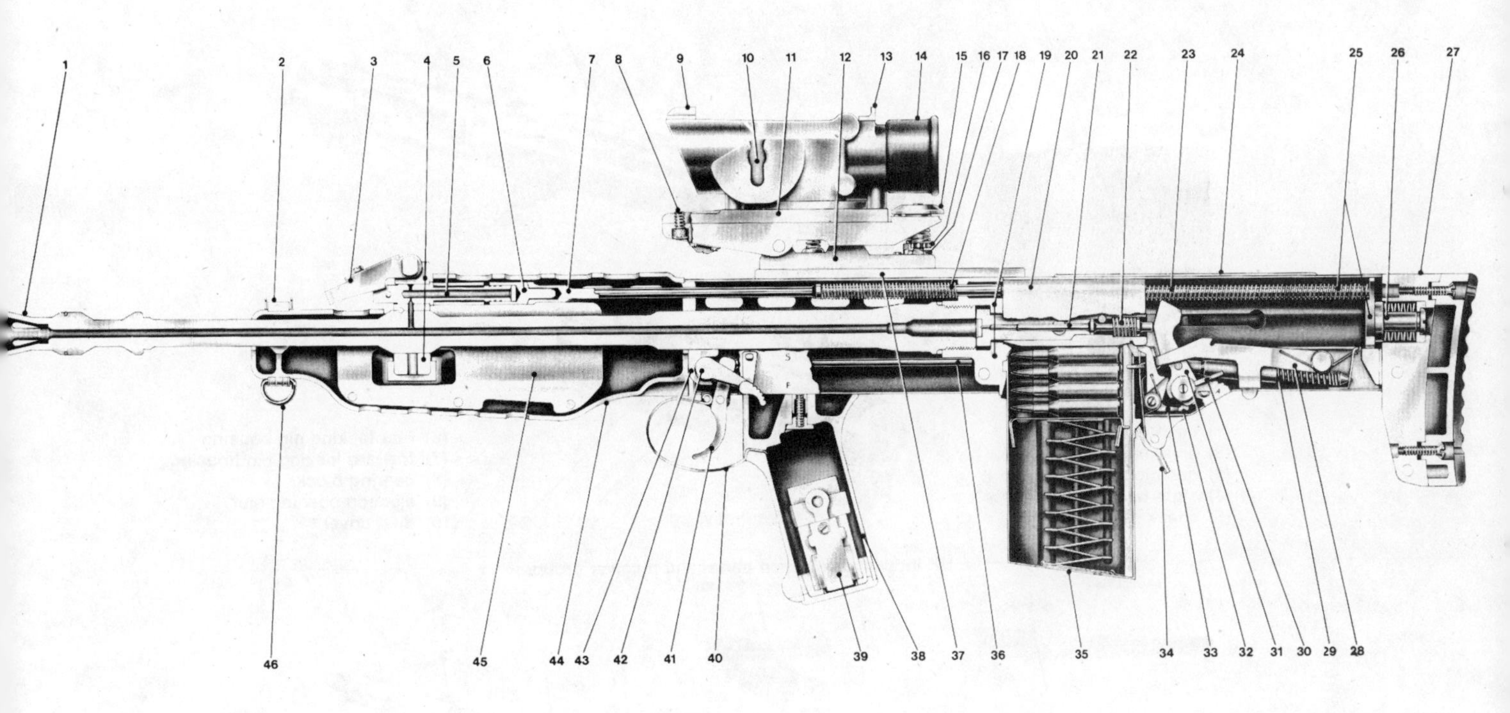

4.85mm Enfield Weapon System (Individual Weapon) XL64 (prototype). (1) flash eliminator (2) toggle catch (3) emergency foresight (4) gas block (5) gas plug (6) gas cylinder (7) piston (8) lateral lock screw (9) battle blade foresight (10) grenade sight housing (11) sight arm (12) sight bracket (13) battle 'Vee' backsight (14) eye guard (15) elevating nut (16) elevation adjusting nut (17) elevation lock nut (18) piston spring (19) barrel extension (20) carrier (21) firing pin (22) firing pin spring (23) return springs (24) cheek rest (25) recoil rod assembly (26) buffer (27) butt assembly (including 15mm extension) (28) holding open device (29) hammer spring (30) change lever (31) hammer (32) sear interceptor (33) safety sear (34) magazine catch (35) magazine (20 rounds) (36) trigger bar (37) sight base (38) pistol grip (39) emergency backsight (40) trigger guard (41) trigger (42) safety catch (43) bearing block (44) handguard (45) heat shield (46) 'D' ring

IMPROVEMENTS TO THE ENFIELD WEAPON SYSTEM

INDIVIDUAL WEAPON AND LIGHT SUPPORT WEAPON

1. **Body/Barrel Group**
   - 1.1 Simplified body shape.
   - 1.2 Lighter sight base dovetail.
   - 1.3 Cold swaged barrel bore and chamber—now 5.56 mm.
   - 1.4 Minor changes to flash hider and fitting.
   - 1.5 New ejection cover—more flush and smooth.
2. **Trigger Mechanism**
   - 2.1 New housing shape for more room and rigidity.
   - 2.2 New fabricate butt plate assembly integral with housing.
   - 2.3 New trigger mechanism—hammer, sears, etc.
   - 2.4 New magazine location to take M16-type magazines.
   - 2.5 New combined bolt-release and hold-open mechanism.
   - 2.6 New safety catch, mag catch, change lever, and extractor.
3. **Gas System and Handguards**
   - 3.1 Minor changes to indexing of gas settings.
   - 3.2 New handguard permanently fitted.
   - 3.3 New hinged H/Gd top cover for access to gas system.
4. **Sighting System**
   - 4.1 New simplified SUSAT sight bracket.
   - 4.2 Minor changes to SUSAT battle sight (emergency).
   - 4.3 New simple iron sights as alternative to SUSAT.
5. **Special to LSW**
   - 5.1 New more effective bipod—better stability and articulation.
   - 5.2 New rear sear mechanism—hammerless.

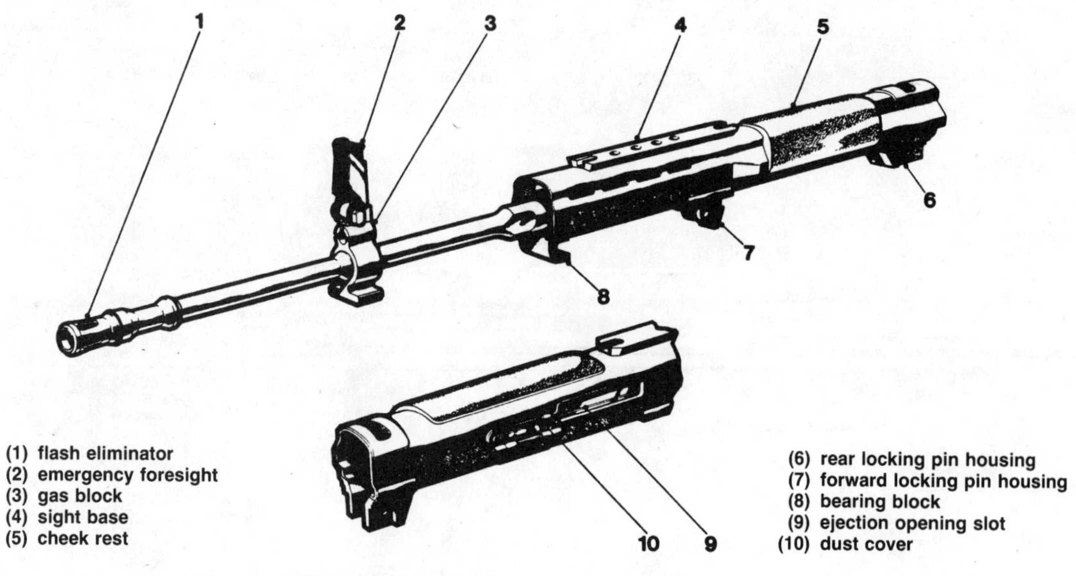

UK Individual Weapon barrel and receiver group.

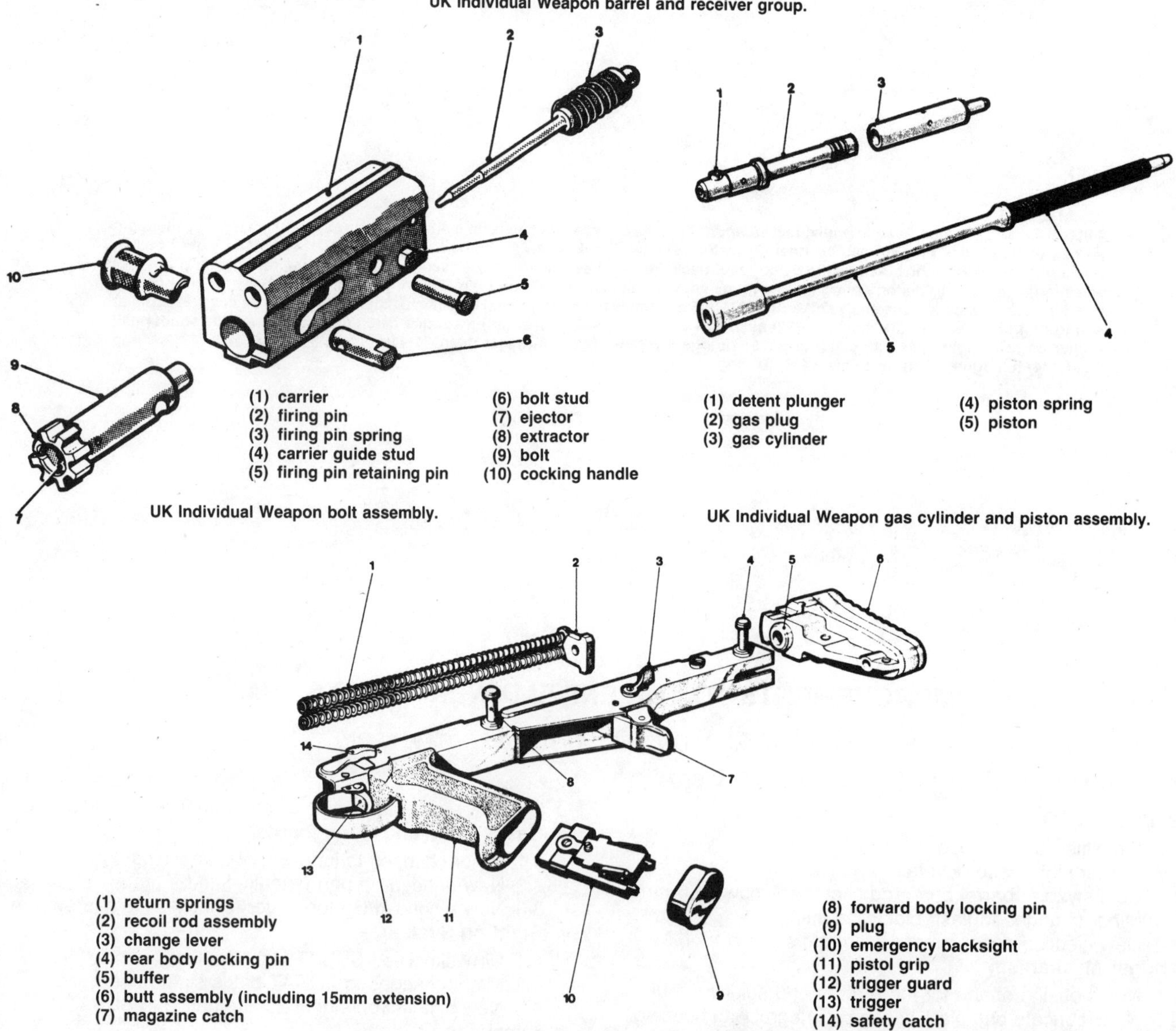

UK Individual Weapon bolt assembly.

UK Individual Weapon gas cylinder and piston assembly.

UK Individual Weapon lower receiver assembly.

Note: These drawings illustrate the 1976 version of the Enfield Weapon System.

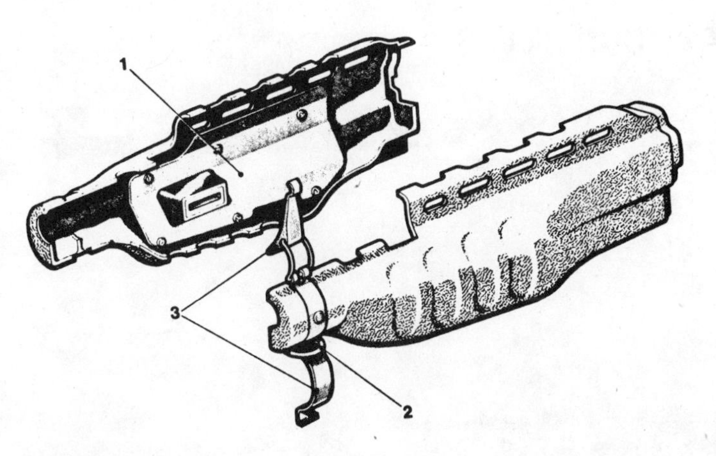

(1) heat shield
(2) 'D' ring
(3) toggle catch

UK Individual Weapon handguard assembly.

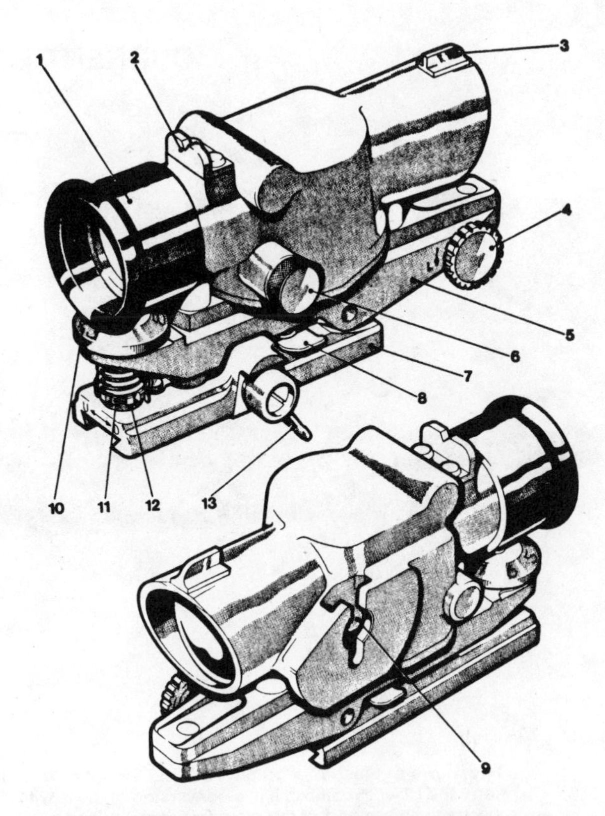

(1) eye guard
(2) battle 'Vee' backsight
(3) battle blade foresight
(4) lateral adjustment knob
(5) sight arm
(6) illuminating knob
(7) sight bracket
(8) detent plunger plate
(9) grenade sight housing
(10) elevation nut
(11) elevation adjusting nut
(12) elevation lock nut
(13) clamping lever

UK Individual Weapon SUSAT (Sight Unit Small Arms Trilux).

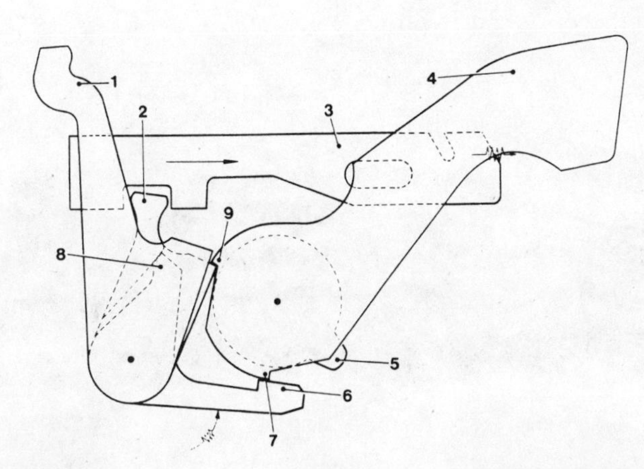

(1) safety sear
(2) sear
(3) trigger bar
(4) hammer
(5) cam lobe
(6) sear interceptor nose
(7) second bent
(8) tail
(9) first bent

The UK Individual Weapon sear mechanism at the moment of trigger release.

CURRENT BOLT ACTION RIFLES

Two views of the 7.62 × 51mm NATO L39A1 rifle. This rifle was developed as a competition rifle for British forces who would carry the L1A1 in combat. It is a conversion of the No. 4 rifle (.303) Marks 1/2 and 2. Those Marks were chosen because the trigger mechanism was part of the receiver assembly and not part of the trigger guard assembly as on the earlier Marks. The L39A1 is assembled at the Royal Small Arms Factory, Enfield. The 27.56-inch (700mm) barrel is of the hammer-forged variety.

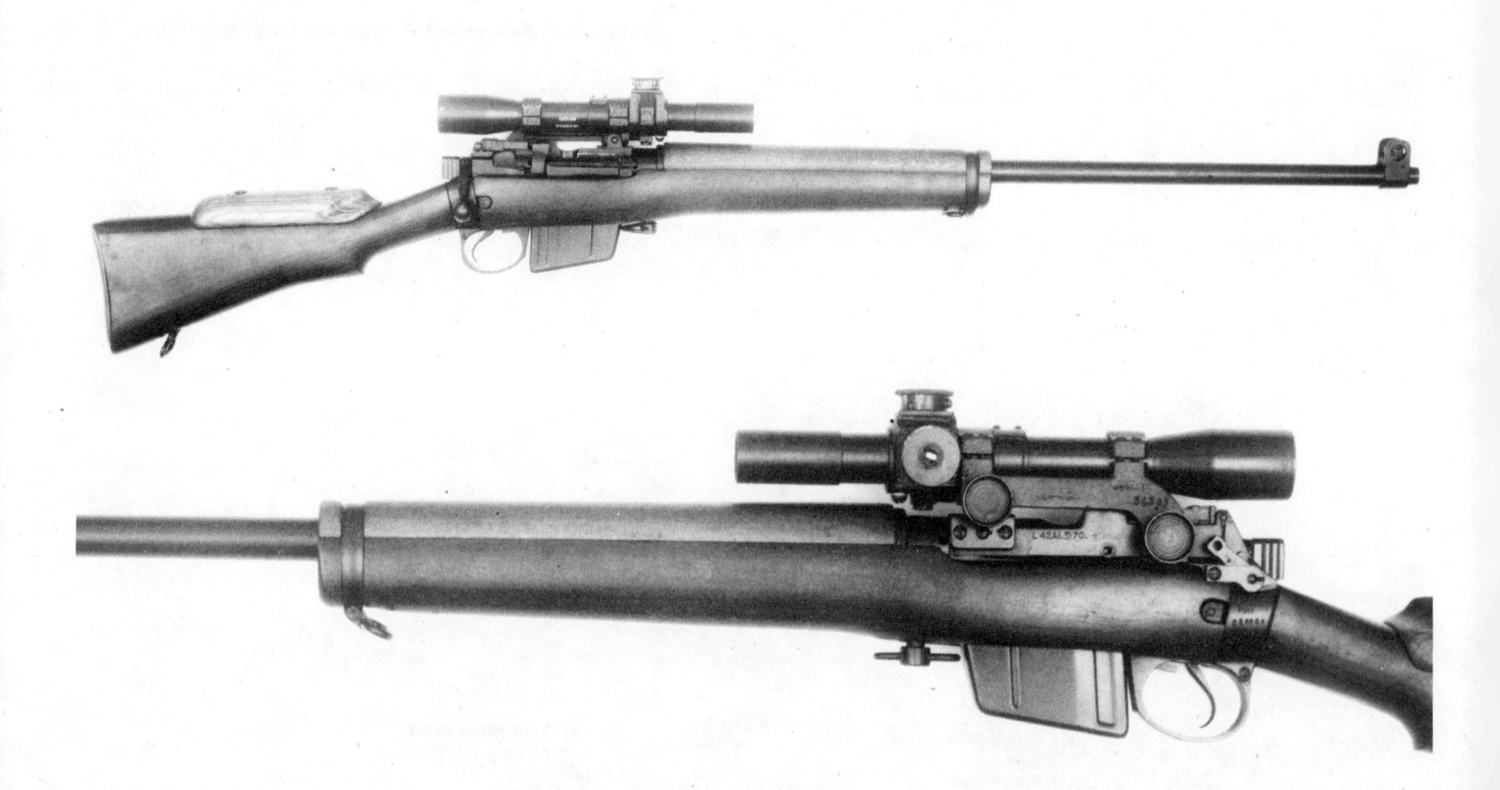

The L42A1 Sniper rifle was created by converting .303 No. 4 rifles originally fitted for sniping—Rifle No. 4 Mark I (T) and Rifle No. 4 Mark I* (T). The conversion process followed is essentially the same as that used in the L39A1 competition rifle.

In addition to the official L39A1 and L42A1 bolt action rifles, there are two other 7.62 × 51mm NATO caliber rifles evolved from the No. 4. The 7.62mm Envoy has been produced for commercial sale to members of the British National Rifle Association by the RSAF, Enfield. A sniper version of the No. 4 rifle has been produced for police use by the RSAF, Enfield. Called the 7.62mm Enforcer Sniper Rifle, this weapon has a 4 to 10 variable Pecar telescopic sight and iron sights.

The well-known firm of Parker-Hale Limited has been manufacturing a family of sniper rifles built on the Mauser 98 rifle action. This rifle has been adopted by the Canadians as the Sniper Rifle C3. The most recent versions of this rifle are designated the Parker-Hale M82 series.

PARKER-HALE M82 RIFLES

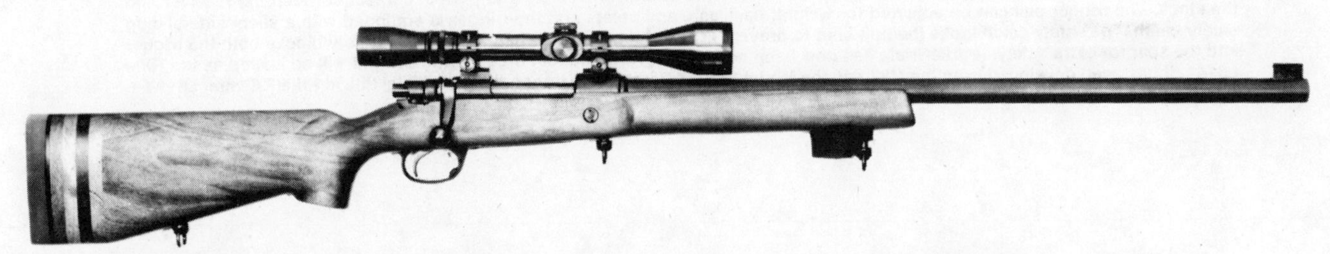

Parker-Hale 7.62 × 51mm NATO M82 equipped as a daylight sniper rifle with a Pecar V2S 4 × 10 power telescopic sight. The butt length can be adjusted by inserting or removing spacers. The barrel of this rifle is 25.98 inches (660mm).

Parker-Hale M82 fitted with the Pilkington PPE Snipe Mk2 passive night vision individual weapon sight. Note the detachable, lightweight folding bipod.

Parker-Hale M82 showing the bipod with legs fully extended. This rifle is fitted with the Smith & Wesson Startron Mk 700 passive night vision sight. The sight mount permits the use of image intensifying sights, daylight scopes, and iron sights as required.

The Parker-Hale M82 fitted with adjustable aperture rear sight PH5E/TX and interchangeable element tunner front sight SM159 for marksmanship competition and target training. The barrels of the Parker-Hale M82 are hammer forged, and they are free floating in the stock. The receiver is bedded in an epoxy resin to ensure complete stability and perfect contact between the receiver and the stock. The trigger pull can be adjusted for weight, backlash, and creep, and the trigger is equipped with a silent side-sliding safety catch. The safety catch locks the bolt shut to prevent accidental opening. When applied, the safety blocks both the trigger and the sear for extra safety. (Parker-Hale has now been adopted for Army Cadet Forces in the UK and will be issued as the Rifle L81A1. It has been developed from the M82 and the Model 1200TX Parker-Hale target rifle. Its official title is Rifle 7.62mm, L81A1—Cadet Target Rifle.

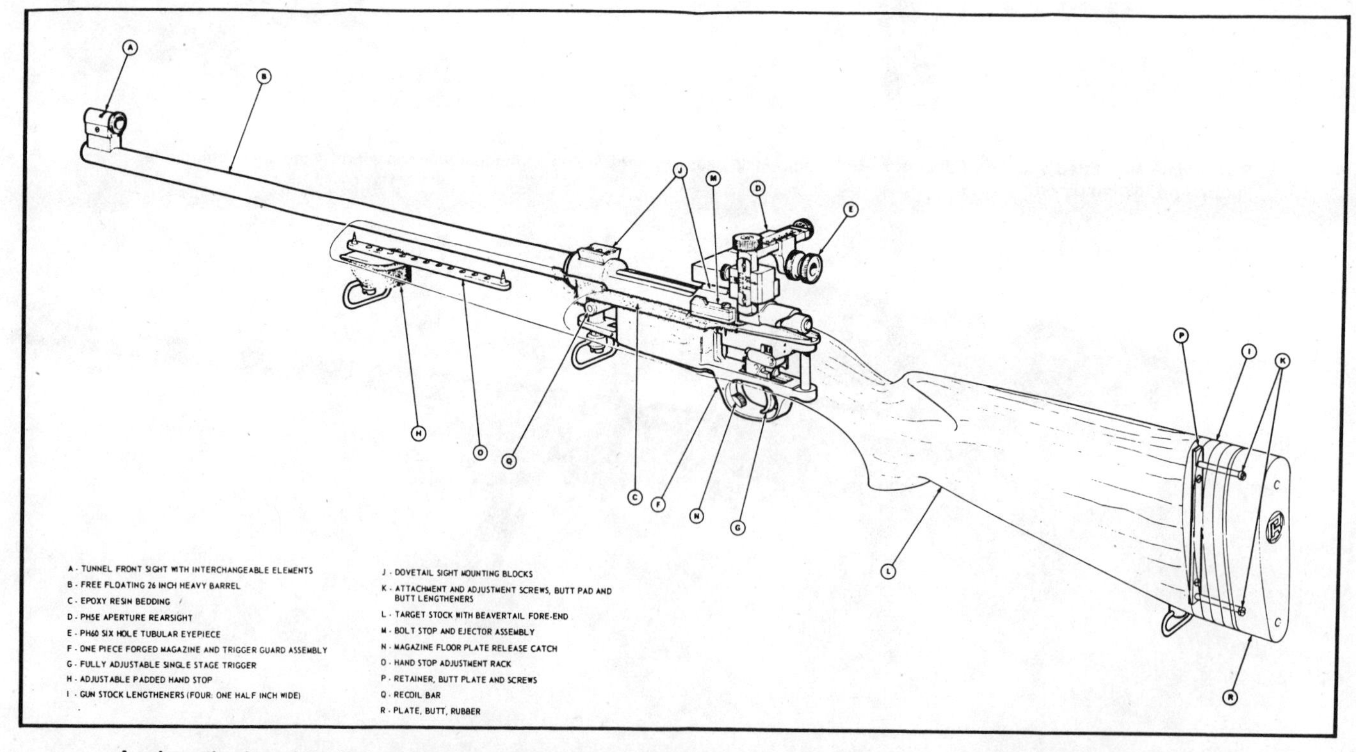

A schematic view of the Parker-Hale sniper rifle as adopted by the Canadian armed forces. Their designation is 7.62 × 51mm NATO Sniper Rifle C3.

OBSOLETE BRITISH RIFLES AND CARBINES

BRITISH BOLT ACTION MILITARY RIFLES FROM 1888 to 1951

The British and Canadian governments have disposed of most of their bolt action rifles since World War II. The caliber .22 No. 7, No. 8 and No. 9 rifles are still used as training rifles. The No. 1, Mark III and Mark III* and the No. 4 Mark I, I* and II are still used throughout the former British territories and in a few other countries. The greater percentage of these rifles have been sold on the US collector market.

The bolt action rifles that follow are not all of the models that have been used by Great Britain by any means; they are the most common. As with pistols, Britain has had to import rifles during both world wars to meet military requirements. During World War I, contracts were let with Remington Arms, Ilion, N.Y.; Remington Arms, Eddystone, Pa.; and Winchester for the production of the .303 Pattern 14 (Rifle No. 3 Mark I*). This rifle was continued in manufacture for the United States in caliber .30-06 as the U.S. Rifle M1917. Although large quantities of Pattern 14 rifles were made, they were apparently used on the battle front in very limited quantities and principally as sniper rifles.

During World War II, the British were desperately short of rifles, especially in 1940–41, due to losses at Dunkirk and temporary loss of industrial plants due to bombing. In order to make up for these losses, Canada gave the UK 70,000 Ross rifles, and the United States supplied 785,000 caliber .30 M1917 Enfield rifles out of United States war reserve stocks, at a price of $7.50 per rifle. The British government contracted with Remington Arms to produce caliber .303 Springfield M1903 rifles; this contract was later taken over by the US government. Stevens Arms of Chicopee, Massachusetts, produced over one million No. 4 Mark I and Mark I* rifles for Great Britain. All contracts let with US manufacturers were supervised by the US government after the introduction of Lend-Lease in 1941. The Ross, M1917 Enfield and Springfield M1903 rifles were used by Home Guard units and have since been disposed of by the British government.

Safety Measures and Inspection Criteria

British rifles are usually well made and, if in good condition, are safe enough. Rifles of earlier marks than the Long Lee Enfield Mark I should not be used with the Mark 7 or other heavily loaded cartridges. The No. 4 and later rifles, if they are in good condition, will safely use any .303 cartridge loaded for rifles. Some United States commercial ammunition is not loaded too heavily in cal. .303 and will not bother any of the Long Lee Enfields or later weapons. Wherever there is any doubt about safety, the weapon should not be fired until checked by a reliable gunsmith.

As an aid to gunsmiths and others who will undoubtedly encounter many British Lee Enfield rifles in the future, some of the inspection criteria for the weapons are listed below.

Headspace. Since the .303 is a rimmed cartridge, headspace is measured from the rear face of the barrel to the face of the bolt. The headspace of the .303 rifle should not exceed .074 inch, although—as a wartime measure—a maximum of 0.80 inch was allowed. Minimum headspace is .064 inch.

Barrel gaging. The bore diameter should be from .301 to .304 inch in a new barrel. To gage a used barrel, plug gages from .303 to .310 inch should be used. The .303 gage should run through the barrel; the .307 gage should not run through the barrel. The .308 gage should not enter the muzzle more than .25 inch, and the .310 gage should not enter the breech more than .25 inch.

Firing pin protrusion. The high for firing pin protrusion for the No. 1's is .055 inch, and the low is .050 inch. The high for the No. 4's and No. 5's is .050 inch, and the low is .040 inch.

Trigger pull. The first pull or slack should be from 3 to 4 pounds. The second pull should be from 5 to 6 pounds. To increase or decrease the trigger pull weight, after the angle of the cocking piece sear notch.

Buttstock lengths. Butts for the No. 1 rifles were made in long and short lengths, and during World War I a special short butt called the Bantam was made. These butts will be marked "L," "S" or "B" on the top of the stock, approximately one inch from the butt plate tang. Butts for the No. 4 and No. 5 rifles come in long, short and normal lengths.

In passing, it is worth noting that the .303 British and .303 Savage are not the same cartridge and are not interchangeable. The .303 British, as currently loaded by one American cartridge manufacturer, has a 215-grain bullet with a muzzle velocity of 2,180 feet per second. This is quite close to the loading of the British Mark 6 cal. .303 cartridge, which had a 215-grain bullet with a velocity of 2,060 feet per second. The Mark 6 cartridge was used with the Long Lee Enfields and the early No. 1s.

The Cal. .303 Lee Metford Rifles and Carbines

Rifle, Magazine, Lee Metford Mark I. Adopted December 1888. Was the first British production Lee. Chambered for the cal. .303 black-powder loaded cartridge. Had an eight-round magazine and a full-length cleaning rod.

Rifle, Magazine, Lee Metford Mark I*. Adopted January 1892. Was a conversion of the Mark I; the sights were changed from "Lewes" and "Welsh" pattern to barleycorn front and V-notch rear sight.

Rifle, Magazine, Lee Metford Mark II. Adopted April 1892. Was the first of the series to be fitted with a 10-round magazine. The bolt was modified, and the outside contour of the barrel was changed. A half-length cleaning rod was fitted to the gun, and the brass marking disk on the buttstock was omitted.

Carbine, Magazine, Lee Metford Mark I. Adopted 1894.

Rifle, Magazine, Lee Metford Mark II*. Adopted 1895. Had a safety catch added to the bolt. Mk I L.M. had safety catch mounted at left side of receiver. Mk I* and Mk II L.M. had no safety.

The Cal. .303 Lee Enfield Rifles and Carbines

Rifle, Magazine, Lee Enfield Mark I. Adopted November 1895. Had the deep Enfield rifling, rather than the shallow Metford rifling used on previous marks. The sights were also modified.

Rifle, Magazine, Lee Enfield Mark I*. Adopted 1899. Had no cleaning rod mounted in the stock.

Carbine, Magazine, Lee Enfield Mark I. Adopted 1896. Same as Lee Metford carbine except for rifling.

Carbine, Magazine, Lee Enfield Mark I*. Same as the Mark I carbine but has no cleaning rod and no sling bar in the left side of the butt.

Carbine, Magazine, Lee Enfield, RIC Model. Adopted in 1905 when 10,000 of this model were made up from Lee Enfield Carbines. The carbine nose cap was removed and the stock cut back and slimmed down to take an upper band with bayonet stud to fit the pattern 88 knife bayonet. These carbines were made up for the Royal Irish Constabulary-RIC, which was disbanded in 1922.

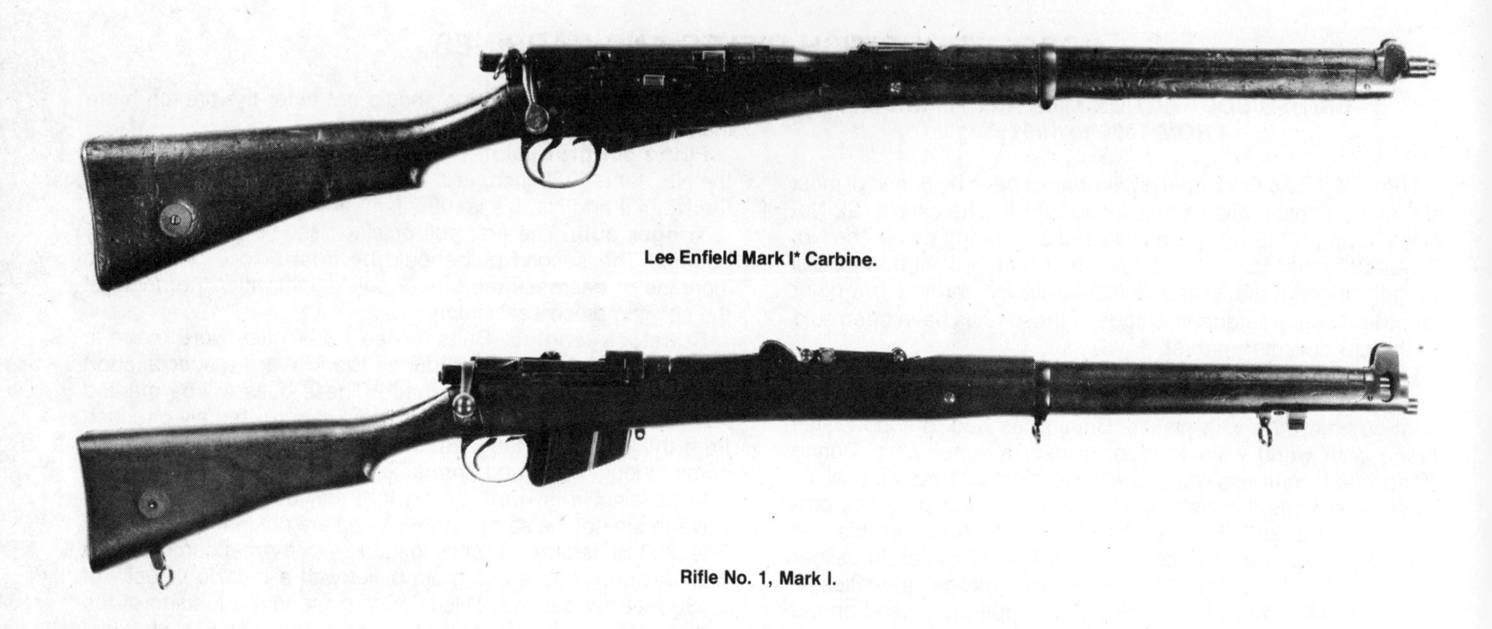

Lee Enfield Mark I* Carbine.

Rifle No. 1, Mark I.

Rifle No. 1, Short Magazine, Lee Enfield Mark I. Adopted December 1902. Was the first of the short rifles (SMLE). Was stocked to the muzzle and charger loaded. The right side charger guide is on the bolt head, and the left charger guide is on the receiver. Has a V-notch rear sight with adjustable windage and a barleycorn front sight. Was the first of what later came to be called the No. 1 series of rifles.

Rifle, No. 1, Short Magazine, Lee Enfield Mark II (COND). Essentially the same as the SMLE No. 1 Mark I, but was converted from earlier Mark II and Mark II* Lee Metford's and Long Lee Enfield's.

Rifle No. 1, Short Magazine, Lee Enfield Mark I*. A minor variant of the SMLE No. 1 Mark I.

Rifle No. 1, Short Magazine, Lee Enfield Mark II*. A minor variant of the No. 1 Mark II SMLE.

Rifle No. 1, Short Magazine, Lee Enfield Mark III. Adopted January 1907. Was the backbone of the British Army in World War I, and was also used extensively in World War II. Is still in use in many of the areas of the British Commonwealth today.

Rifle No. 1, Short Magazine, Lee Enfield Mark IV (COND). Adopted July 1907. Basically the same as the No. 1 Mark III; converted from Long Lee Metford's and Long Lee Enfield's.

Rifle, Charger Loading, Long Lee Metford Mark II. Was converted to charger loading in 1907 for use of the Territorial Army, and converted to rifle charger loading Lee Enfield Mark I* in 1909. Few of these were made.

Rifle, Charger Loading, Long Lee Enfield Mark I. A 1907 conversion of early marks of Long Lee Enfield to charger loading. The Mark I* version is more common. A large number of these weapons were used by British forces in the early days of World War I.

Rifle No. 1, Short Magazine, Lee Enfield Mark III*. Was adopted during World War I and made in very large quantities. Is still in widespread use throughout the world. Does not have the long-range side sights of the Mark III and earlier marks, and does not have a magazine cutoff.

The Royal Ordnance Small Arms Factory at Enfield Lock made over 2 million of this model and the No. 1 Mark III during World War I. During the same period, B.S.A. made 1,601,608 and L.S.A. made several hundred thousand. This rifle was last manufactured in the UK by B.S.A. in 1943. The Australian arsenal at Lithgow and the Indian plant at Ishapore manufactured the Mark III* after the adoption of the No. 4. Lithgow produced 415,800 from 1939 to 1955 when production was switched to the FN rifle.

Rifle No. 1, Short Magazine, Lee Enfield Mark V. Appeared around 1922. The rear sight is mounted on the receiver bridge, and an additional stock band is mounted to the rear of the nose cap.

Rifle No. 1, Short Magazine, Lee Enfield Mark VI. Was developed in the period 1924–1930. Was the forerunner of the No. 4 rifles. Had rear sight on the receiver bridge. Had a lighter nose cap, heavier barrel and smaller bolt head than the earlier marks. Had cut-off, and left receiver wall is cut low as the Mark III.

Note: Abbreviation for "converted" was COND.

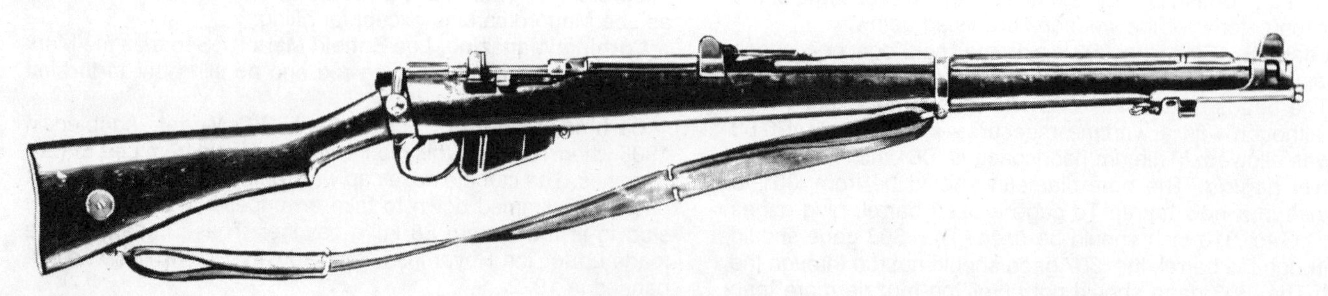

Rifle No. 1, Mark III.

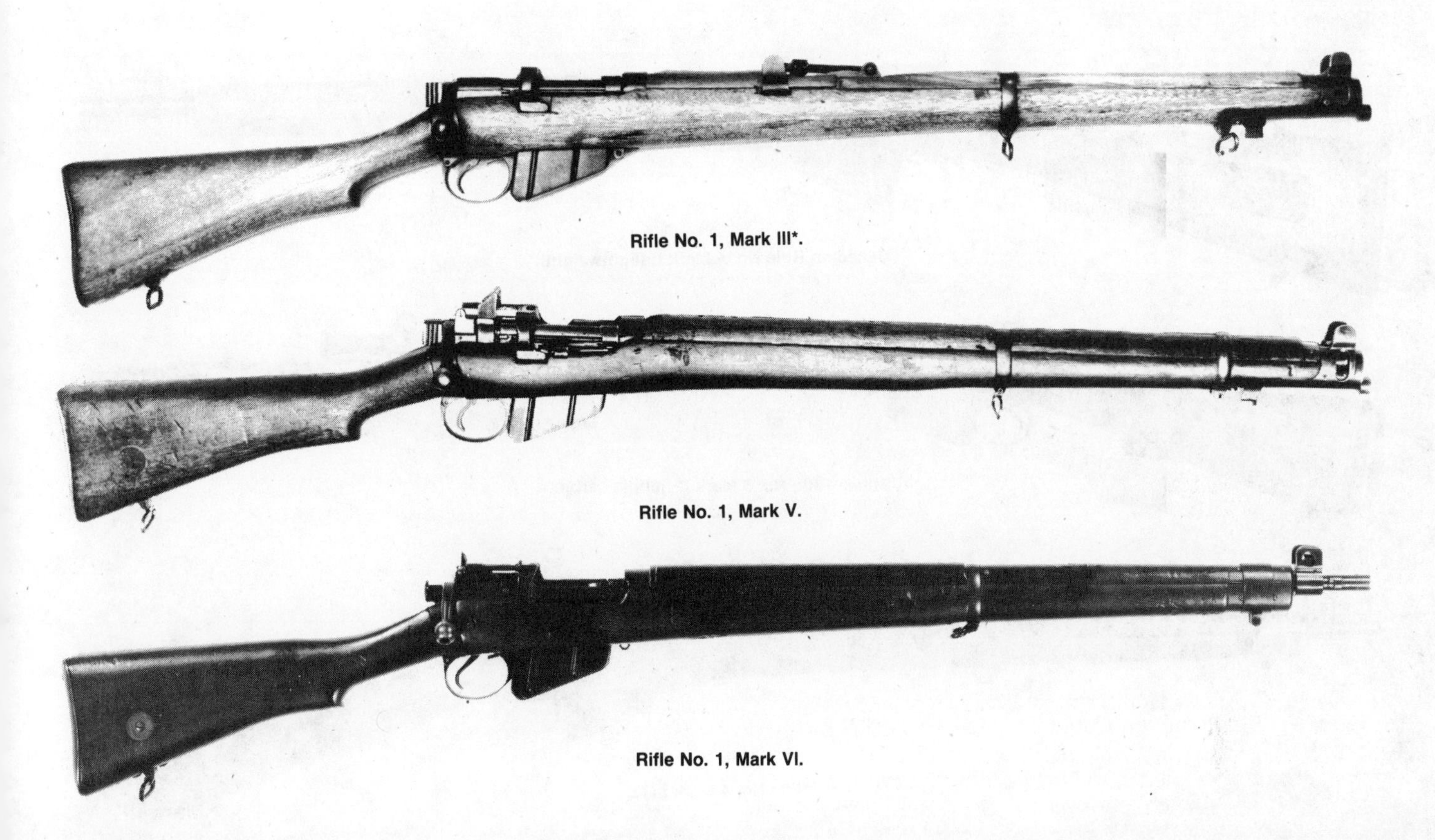

Rifle No. 1, Mark III*.

Rifle No. 1, Mark V.

Rifle No. 1, Mark VI.

Rifle No. 4 Mark 1. Originally appeared in 1931. Was finely made, and was generally similar to the No. 1 Mark VI except that it had a heavier receiver. Was redesigned for mass production around 1939 and became, with the No. 4 Mark 1*, the British "work horse" of World War II. Stamped bands were used, and various manufacturing shortcuts were taken to increase production. Three different marks of rear sights may be found on this weapon, ranging from a finely machined adjustable leaf to a simple L-type. Many of these weapons are still in service in the British Commonwealth and in former British territories.

Rifle No. 4 Mark 1*. Was the North American production version of the No. 4 Mark 1. The principal difference was that the bolthead catch, which was situated behind the receiver bridge on the No. 4 Mark 1 (and earlier Marks), was eliminated on the No. 4 Mark 1*, and a cutout on the bolt head track was used for bolt removal. Over five million No. 4 rifles were made during World War II in the UK, Canada and the United States (Stevens Arms). Australia did not adopt the No. 4 but continued production of the No. 1 Mark III* at Lithgow during World War II.

Canadian Rifle No. 4 Mark I* (light weight). This weapon was produced at the Canadian arsenal at Long Branch in prototype form. It has a one-piece stock, and its trigger is pinned to the receiver. Weight about 6¾ pounds. Barrel length about 23 inches. Overall length about 42½ inches. Receiver wall cut down and stock inletted to reduce weight. Sporting type Hawkins rubber buttplate. Micrometer sights with peep battle sight. Sight adjustable in clicks and 100-yard steps from 100 to 1300 yards. This arm may be used for grenade launching. Has a Mauser type trigger.

Rifle No. 4 Mark 2. Was developed at the end of World War II. Differed from the earlier marks by having its trigger pinned to the receiver rather than to the trigger guard.

Rifle No. 4 Mark 1(T) and No. 4 Mark 1* (T). Are the sniper versions of the No. 4. Are fitted with scope mounts on the left side of the receiver and have a wooden cheek rest screwed to the butt. The No. 32 telescope is used on these weapons. There are also sniper versions of the No. 1 and No. 3 rifles (Pattern 14). The Canadians also used the No. 4 Mark 1*(T) with the Telescope C No. 67 Mark 1.

Rifle No. 4 Mark 1/2 and Rifle No. 4 Mark 1/3: These are conversions of the No. 4 Mark I and No. 4 Mark I* respectively to the pattern of the No. 4 Mark II. These rifles, like the No. 4 Mark II, are still in extensive use and are probably held as reserve weapons by the UK.

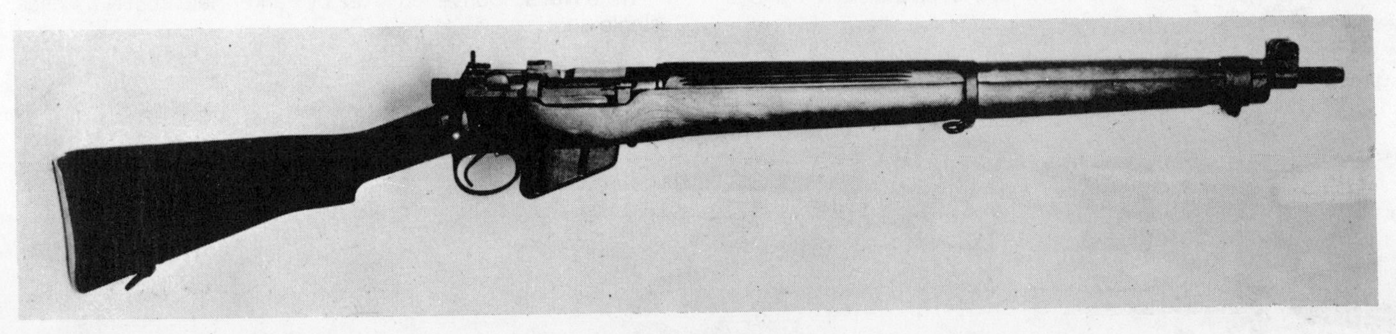

British Rifle No. 4 Mark I*.

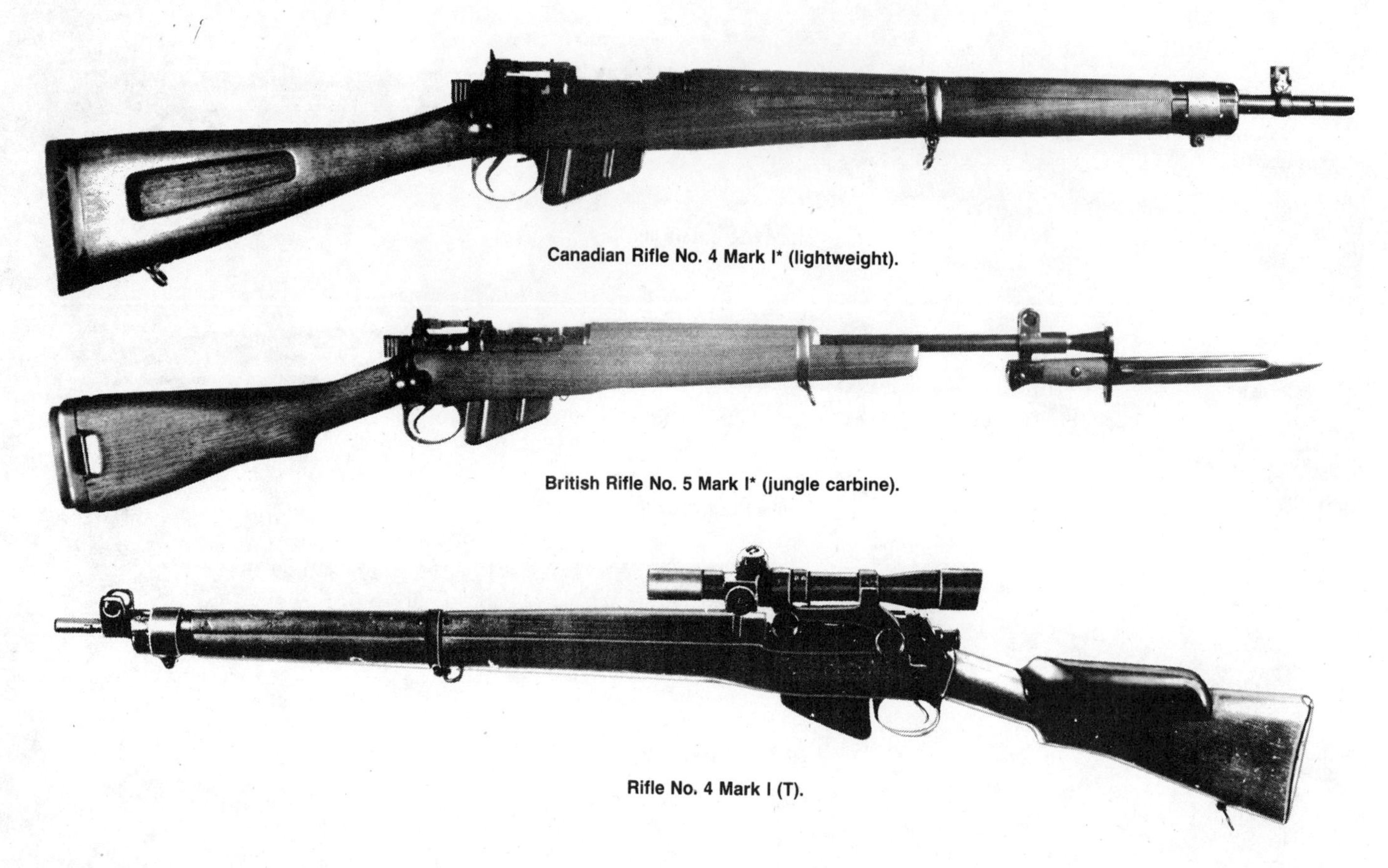

Canadian Rifle No. 4 Mark I* (lightweight).

British Rifle No. 5 Mark I* (jungle carbine).

Rifle No. 4 Mark I (T).

Rifle No. 5 Mark I. Appeared toward the end of World War II. Was a lightweight weapon, and was commonly called the jungle carbine. Has a lightened and shortened barrel, which is fitted with a flash hider. Fore-end has been cut back and rounded, giving weapon the appearance of a sporting rifle. A rubber recoil pad is fitted to the butt.

Rifle No. 6 (Aust). Appeared only as prototype; 18-inch barrel version of No. 1. Developed at Lithgow.

The Cal. .22 Rifles

Cal. .22 R.F. Short Rifle Mark I. Adopted in 1907. This rifle is a conversion of the Lee Metford Mark I* rifle; it is approximately the same length as the SMLE. Sights are an adjustable blade front sight and a tangent type rear sight with adjustment for windage.

Cal. .22 R.F. Long Rifle Mark II. Adopted in 1912. Conversion of Long Lee Enfield to .22 rimfire. The Mark I pattern of .22 R.F. long rifles was converted from Long Lee Metfords.

Cal. .22 R.F. Short Rifle Mark I*. Conversion to .22 rimfire in shortened form of Lee Metford Mark I*. Mark II of this pattern was converted from Lee Metford Mark II.

Cal. .22 R.F. Short Rifle Mark III. Adopted in 1912—this is a conversion of SMLE Marks II and II* to .22 rimfire. A number of different patterns of Lee were converted during World War I to .22 caliber rimfire. Some were fitted with new barrels, and others had their .303 barrels bored out and a caliber .22 liner inserted. Pattern 1914 .22″ was tubed and used the normal .22″ chamber. The .22 R.F. pattern 1918 had "conveyors"—cartridge adaptors or auxillary cartridges, which fed through .303 magazines.

Rifle No. 2 Mark IV. Is a conversion of cal. .303 SMLE's to cal. .22. Some have new .22 barrels, and some were "Parker Rifled," i.e., a .22 liner was placed in a bored-out .303 barrel. A special bolt head was made for these rifles.

Rifle No. 2 Mark IV*. A variant of the No. 2 Mark IV.

Rifle No. 7. Developed at Long Branch; single shot version of No. 4 Mark I*. Called Rifle "C" No. 7, .22 in Mark I. Also has been made by B.S.A. with a 5-shot magazine.

Rifle No. 8 Mark 1. Two variations of this rifle were developed simultaneously. They were called the Infantry Model and the Match Model and differed principally in sights and length of barrel. The Infantry Model has sights similar to the No. 4 rifle and a shorter barrel than the Match Model. The Match Model, in addition to the longer barrel, has match type sights. The Infantry Model was adopted in 1950 as the Rifle No. 8 Mark 1.

Rifle No. 9. Converted to .22 by Parker Hale from No. 4 rifles, single shot.

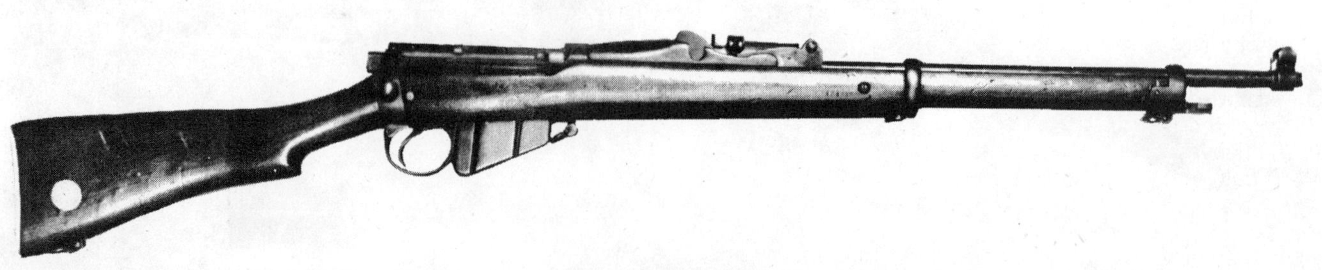

.22 cal. R.F. Short Rifle Mark II.

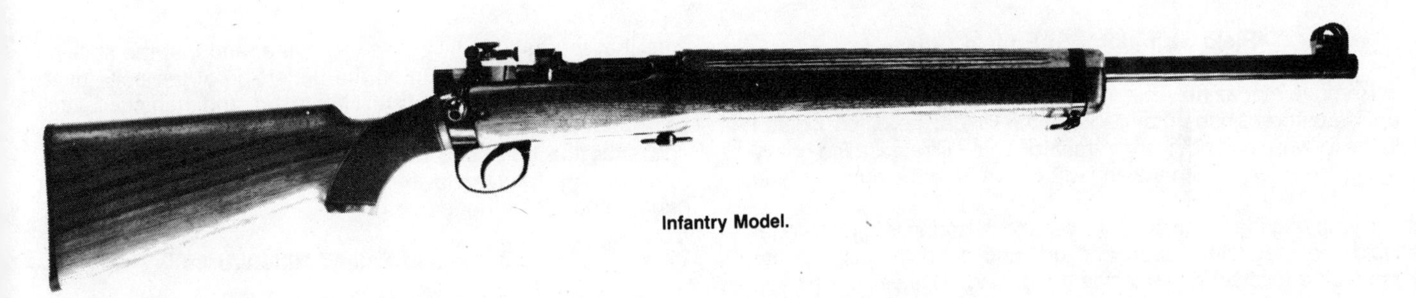

Infantry Model.

Conversion of No. 4 Rifles to 7.62mm NATO

The Royal Small Arms Factory at Enfield Lock has developed a conversion kit for the No. 4 rifles to convert them to use the 7.62mm NATO cartridge. This kit consists of a new barrel, extractor, magazine, charger guide liner, front sight block fixing pin and a barrel breeching washer. This kit can be fitted to an existing No. 4 rifle with normal armorers tools and certain special purpose tools, i.e. a special drift, a taper pin reamer, a breeching gage etc.

Model designations have been assigned to converted rifles as follows: .303 Rifle No. 4 Mark I becomes 7.62mm Rifle L8A4; .303 Rifle No. 4 Mark I* becomes 7.62mm Rifle L8A5; .303 Rifle No. 4 Mark 1/2 becomes 7.62mm Rifle L8A2; .303 Rifle No. 4 Mark 1/3 becomes 7.62mm Rifle L8A3; .303 Rifle No. 4 Mark II becomes 7.62mm Rifle L8A1. A conversion kit also exists for the Rifle No. 5.

B.S.A. has also developed a conversion kit for the No. 4 rifles.

7.62mm Rifle L39A1. is a conversion of the No. 4 MK 1/2 or 2 to 7.62mm NATO using a heavy barrel. A special hand guard is fitted, and the fore-end is cut back in sporting style so that its forward end is just beyond the lower band. A standard No. 4 butt stock is used. This rifle is designed for competitive shooting within the British services. A front sight base is provided, but no front sight or rear sight is provided. Units are supposed to supply their own and a variety of sights are available. The L39A1 uses a 10-round box magazine, but the magazine, which is .303, serves only as a loading platform. The barrel is 27.5 inches long and overall length is 46.5 inches. The trigger pull is lighter than that of the No. 4 L39A1; weighs 9.75 pounds.

7.62mm Rifle L42A1. is the current standard sniper rifle. It has a heavy 7.62mm NATO barrel and the fore-end is cut back the same as the L39A1. Receivers from No. 4 Mark 1 (T) or Mark 1* (T) are used for this rifle. The magazine of the L42A1 is designed for 7.62mm NATO cartridges and has a capacity of 10 rounds. The butt stock has the same type "screw on" wooden check piece as used with the No. 4 Mark 1 (T). The left side of the receiver has a telescope bracket for the telescope No. 32 Mark 3. A leaf type rear sight and a protected blade type front sight are also used.

The Envoy, a 7.62mm match rifle conversion of the No. 4, is also being produced by the Royal Small Arms Factory. This rifle has a swaged, heavy free floating barrel and a full pistol grip type butt stock. It is somewhat similar to the L39A1 and L42A1 but more finely finished and has a match type tunnel front sight and match type aperture rear sight fitted.

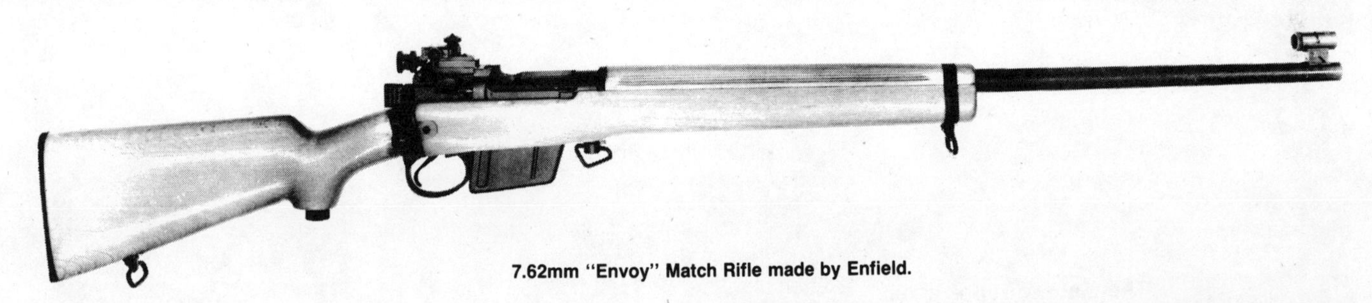

7.62mm "Envoy" Match Rifle made by Enfield.

Loading and Firing Lee Enfield Rifles

Turn bolt handle up as far as it will go and pull it straight back to the limit of travel.

Insert loaded clip in the clip guide in the receiver and strip the cartridges down into the magazine. Remove the empty clip. Insert a second clip, push these cartridges down and remove clip. This will leave the magazine fully charged with 10 cartridges.

Pushing bolt handle fully forward and down loads the firing chamber, cocks, and locks ready for firing with a pull of the trigger.

Unless weapon is to be fired immediately, pull the thumb rocker on the left rear of the receiver to "Safe."

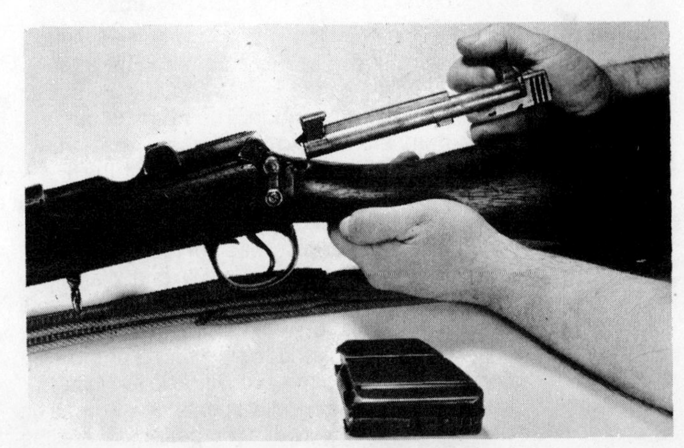

Replacing bolt, No. 1 Lee Enfield Rifle.

Field Stripping Lee Enfield Rifles

Remove magazine. This may be done by pushing in or pulling up, as different rifles may require, the magazine catch located in the forward end of the triggerguard. This will release the heavy sheet steel box, which may be withdrawn from the bottom of the receiver.

Removing the magazine follower and its spring is simply done. Hold the magazine, open end up, and push the rear of the magazine follower down inside the casing. This will permit you to ease the front end of the follower up and out of the casing and remove it and the spring.

In order to remove the bolt, it is first necessary to rock forward the safety catch just above the rear end of the triggerguard, on the left side of the rifle. Then turn the bolt handle up and turn it back as far as it will go. Catch your right forefinger under the head of the bolt. Pull the bolt head up until it is released from its spring catch. Then withdraw it straight to the rear.

Field Stripping for Rifle No. 4 Mark I*. The No. 4 Mark I* rifle has a different method of removal of the bolt from the rifle than do the other Lee Enfield rifles. On the bolt head track—right side of receiver—there is a cut-out; draw bolt back until bolt head is over this cut-out; then lift bolt head straight up and draw bolt out of rifle.

Note on Replacing Bolt. These bolts are not interchangeable, and the number on the bolt should always be checked against the number on the rifle when there has been any possibility of substituting another bolt. Before inserting the bolt, be sure that the head is fully screwed home, and that the cocking piece lines up with the lug on the underside of the bolt. Insert the bolt in the boltway and thrust it forward, and then pull it back as far as it will go until the head touches the resistance shoulders and force the bolt head down over the spring retaining catch. Then push it forward to the forward position. Turn down bolt handle and press trigger.

How the Lee Enfield Rifle Works

Starting with the rifle loaded and cocked, the action is as follows. When the trigger is pressed, it draws down the sear until the sear nose reaches the bottom of the full bent. (This provides the first pull or slack, which is a feature of the best military rifles.) As the trigger pressure continues, the upper part of the sear is drawn still further down until the sear nose clears the bent allowing the cocking piece on the striker to be driven forward by the compressed mainspring. The striker nose, or firing pin, passes through a hole in the face of the bolt head and discharges the cartridge in the firing chamber.

Special Note on the Lee Enfield System. The locking system on this rifle makes it the fastest operating bolt action rifle in the world. The abrupt turning action of the Mauser system will not permit it to attain a speed of operation possible with the Lee Enfield.

This rifle, since it has no locking recesses cut into the receiver, is much easier to clean than the Mauser type and functions well under all battle conditions.

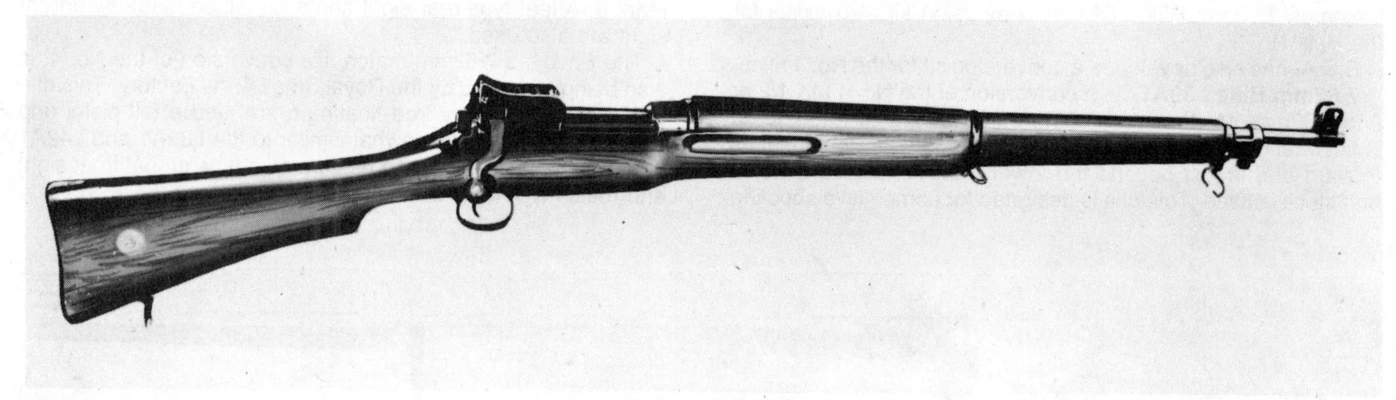

Pattern 14 Rifle.

The Mauser Type Rifles

Pattern 13 (P-13). Tested in 1913. Was a modified Mauser (it cocked on the forward stroke of the bolt), chambered for a large cal. .276 cartridge. The cartridge was remarkably similar to the Canadian cal. .280 Ross cartridge. The rifle was made in comparatively small numbers for field trials.

Pattern 14 (P-14). Was the production model of the P-13. Was made in the United States in cal. .303 for the UK during World War I. The weapon was classed as limited standard in the British Army and except for sniping was not too widely used. Upon the entrance of the United States into World War I, the design was changed to U.S. Cal. .30, and the weapon was produced as the U.S. rifle, cal. 30, M1917 and was commonly known as the Enfield. Between World Wars I and II, the British changed the nomenclature of the P-14 to Rifle No. 3 Mark I.

Pattern 14 Sniper Rifles. The P-14 was extensively used as a sniper rifle in World War I. The two basic patterns were the P-14 (T) and the P-14 (T) A. The former has a Pattern 1918 telescope adjustable for range and windage, and the latter has an Aldis telescope adjustable for range only. In 1926, when all British small arms were given number designations, these weapons were renamed the Rifle No. 3 Mark I* (T) and Rifle No. 3 Mark I* (T) A, respectively.

Springfield .303. This rifle, which apparently has no official nomenclature, is a modification of the U.S. M1903 designed by Remington Arms at the request of Great Britain in 1941. Production of the Springfield rifle by Remington Arms during World War II was initially—until Sept 1941—set up for a British contract. The .303 version was made only as a prototype and development on this rifle stopped in September 1941. The type "C" stock was used with the pistol grip modified to the Enfield type; the bolt face, extractor, follower and magazine were modified to fit the .303 cartridge. The receiver bridge was lengthened. The "L" type rear sight was similar to the Mark 2 sight used on the No. 4s. Note that the bayonet mounting lugs are on the barrel as with the No. 4.

CHARACTERISTICS OF BRITISH BOLT ACTION RIFLES AND CARBINES

	Lee Metford Rifle Mark I*	Lee Metford Rifle Mark II	Lee Metford Carbine Mark I	Lee Enfield Rifle Mark I
Caliber	303	.303	.303	.303
Overall length	49.85 in	49.85 in	40 in	49.5 in
Barrel length	30.19 in	30.19 in	20.75 in	30.19 in
Feed device	8 rd detachable box w/cut-off	10 rd detachable box w/cut-off	6 rd detachable box w/cut-off	10 rd detachable box w/cut-off
Sights:				
Front	Barley corn	Barley corn	Barley corn w/ protecting ears	Barley corn.
Rear	Vertical leaf and ramp	Vertical leaf and ramp	Vertical leaf and ramp	Vertical leaf and ramp
Muzzle velocity (at date of adoption)	2000 FPS	2000 FPS	1940 FPS	2060 FPS
Weight	10.43 lbs.	10.18 lbs	7.43 lbs	9.25 lbs

	Pattern 14 Rifle (Rifle No. 3 MKI*)	Short Lee Enfield Rifle Mark I (Rifle No 1 SMLE MK 1)	Short Lee Enfield Rifle Mark III (Rifle No 1 SMLE MK 3)	Short Lee Enfield Rifle Mark III* (Rifle No 1 SMLE MK 3*)
Caliber	.303	.303	.303	.303
Overall length	46.25 in	44.5 in	44.5 in	44.5 in
Barrel length	26 in	25.19	25.19 in	25.19 in
Feed device	5 rd integral magazine	10 rd detachable box w/cut-off	10 rd detachable box	10 rd detachable box
Sights:				
Front	Blade w/ protecting ears	Barley Corn w/ protecting ears	Blade w/ protecting ears	Blade w/ protecting ears
Rear	Vertical leaf w/ aperture battle sight, long range side sights	Tangent leaf w/notch long range side sights	Tangent leaf w/notch long range side sights	Tangent leaf w/notch
Muzzle velocity (at date of adoption)	Apprx 2500 FPS	2060 FPS	2060 FPS	2440 FPS
Weight	9.62 lbs	8.12 lbs	8.62 lbs	8.62 lbs

	Rifle No 2 Mark 4	Rifle No 4 Mark 1	Rifle No 5 Mark 1	Rifle No 8 Mark 1
Caliber	.22	.303	.303	.22
Overall length	44.5 in	44.5 in	39.5 in	41.05 in
Barrel length	25.2 in	25.2 in	18.7 in	23.3 in
Feed device	Single shot	10 rd detach-able box	10 rd detachable box	Single shot
Sights:				
Front	Blade w/ protecting ears	Blade w/ protecting ears	Blade w/ protecting ears	Blade w/ protecting ears
Rear	Tangent leaf w/notch	Vertical leaf w/ aperture battle sight or L type	Vertical leaf w/ aperture battle sight	Vertical leaf w/ aperture battle sight
Muzzle velocity (at date of adoption)	1050 FPS	2440 FPS	2400 FPS	1050 FPS
Weight	9.19 lbs	8.8 lbs	7.15 lbs	8.87 lbs

Characteristics are listed only for the principal models.
Lengths are with normal butt.

BRITISH SUBMACHINE GUNS

Although B.S.A. had developed a number of modifications of the Thompson Submachine Gun during the 1920s, the British Army did not show much interest in submachine guns until after World War II started. In 1940, large contracts were let for the manufacture of the caliber .45 Thompson Submachine gun M1928A1 by the Auto Ordnance Corporation of Bridgeport, CT.

During World War II, the Lanchester and Sten guns were designed and produced. A number of other submachine guns were produced in prototype form. Among these was the Patchett, developed by the Sterling Engineering Company; in considerably modified form it is the L2A3, the current standard British submachine gun.

THE STERLING (PATCHETT) GUN

This weapon was developed by G.W. Patchett toward the end of World War II at the Sterling Engineering Co., Dagenham, Essex. The weapon was tested by the United Kingdom as the "Patchett" in several different forms and was chosen for extensive field test after the competitive trials held around 1949. It was issued in limited quantities in 1951 and in a modified form was issued as Submachine Gun L2A1 in 1953. The current standard model is the L2A3. The Sterling has been adopted by New Zealand, Canada, India and a number of other countries in addition to the United Kingdom. In addition to the selective fire military version of the Sterling, there is also a semiautomatic version called the Sterling Police Carbine. This weapon was sold quite extensively to planters in Kenya during the Mau Mau uprising and is now available in the United States with 16-inch barrel.

Patchett Machine Carbine

The original gun was considerably different than the weapons in service today. The buttstock was made of heavy flat steel stripping and was of different design than the later weapons. The fire selector/safety is on the front of the trigger housing as opposed to its position on the left top of the pistol grip on the later guns.

9mm Sterling Police Carbine Mark 4.

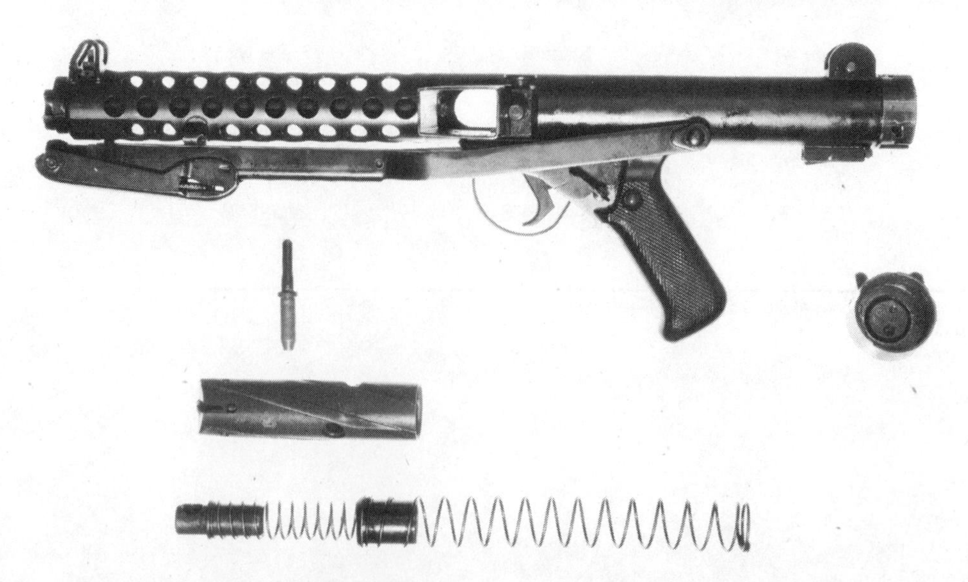

Disassembled view of the 9 × 19mm NATO L2A3 submachine gun. Compare the finish on this military version with that of the commercial Sterling which is illustrated.

L2A1 Submachine Gun

The basic difference between this weapon and the later L2A2 is that the L2A1 had parts—grip screw and cocking (bolt) handle—that could be used for the removal of the barrel screws, and the inner block of the bolt had an extension for removal of the extractor pin.

L2A2 Submachine Gun

Parts were not used as stripping tools; a forward finger guard was added; the rear sight was modified by repositioning of the sight flip-over lever and increasing the size of the 100-yard aperture. The butt was strengthened, a fouling plunger added to the bolt to prevent improper assembly and the chamber modified to feed under adverse conditions.

L2A3 Submachine Gun

This is the standard service gun and differs from the L2A2 as follows: rear sight flip lever has been deleted; the butt has again been redesigned and made as a complete stamping rather than as a fabrication with the butt plate indexing and the position along barrel jacket changed. The chamber was modified to the NATO standard and the trigger guard made removable. The previous model had a special Arctic trigger, which was mounted on the trigger guard. An early model of the Sterling Gun had a folding bayonet; all later models have a bayonet boss and stud for a knife type bayonet.

CHARACTERISTICS OF L2A3

System of operation: Blowback, selective fire.
Weight: **Unloaded, w/o bayonet:** 6 lbs
Loaded, w/bayonet: 8.25 lbs.
Overall length: **Stock extended:** 28 in.
Stock folded: 19 in.
Barrel length: 7.8 in.
Feed device: 34-round, detachable, staggered box magazine.
Sights: **Front:** Blade w/protecting ears.
Rear: Flip-type aperture, graduated for 100 and 200 yd.
Muzzle velocity: 1280 f.p.s. w/British 9mm service ball.
Cyclic rate: 550 r.p.m.

How to Load and Fire the L2A3

Pull the cocking handle to the rear; the bolt will remain to the rear since the weapon fires from an open bolt. Engage the safety by turning the change lever (located on the left side of the pistol grip) to the letter "S." Insert a loaded magazine in the magazine guide, checking to insure that it locks in place. Move change lever to letter "R" for semiautomatic fire or letter "A" for automatic fire. Squeeze the trigger and the weapon will fire.

Field Stripping the L2A3

Elementary Stripping. Before stripping, insure that the weapon is not loaded and remove sling if fitted. Set change lever to "A"; place butt in the folded position and bolt forward.

To Remove Return Spring and Bolt. Press back-cap catch for full depth. Push back-cap forward and rotate counterclockwise until locking lugs disengage from locking recesses. Remove back-cap and draw cocking handle to rear of weapon. Lift cocking handle outward and withdraw return spring assembly from rear of receiver. Remove bolt from rear of receiver. Reassemble in reverse order. The spring-loaded fouling pin will prevent misassembly, since the cocking handle cannot be inserted until this pin is pushed forward by the center pin on the spring assembly. This ensures that the cocking handle must pass through the hole in the center pin.

To Remove Trigger Group. With a small coin or the rim of a cartridge, turn the slot in the head of the trigger group retaining pin until it is in line with the word "FREE" on the right side of the pistol grip. With the nose of a bullet or the blunt end of the cocking handle, push the trigger group retaining pin out and remove. Press the trigger, and pull the trigger group toward rear of weapon, disengaging it from the step in underside of barrel case; then swing front of trigger group out and remove from receiver.

Note: Elementary stripping does not include any further stripping of trigger group.

Assembly

Assemble in the reverse order of stripping.

Applied Safety. When the weapon is cocked, and the change lever is set at the safe position "S," the inner arm of the change lever is positioned directly under the short arm of the tripping lever. When the trigger is pressed, the sear cradle and sear cannot be depressed because the short arm of the tripping lever is held immovable by the inner arm of the change lever.

When the bolt is forward, and the change lever is set at the safe position, the weapon cannot be cocked because the sear is engaged in the safety slot at the rear of the bolt, and the sear cannot be depressed because it is held immovable as described in the previous paragraph.

The Butt Mechanism. To open butt, hold the weapon with the left hand near the rear sight, with the barrel pointing toward the ground. Pull the butt plate outward with the right hand to release the butt catch, and swing the butt to the rear of the weapon. With the thumb of the left hand, press the back-cap catch and snap the butt into engagement with the lugs on the back-cap. Open the butt frame to form a triangle, and the butt catch will engage, to lock.

To close butt, release the butt plate catch and collapse the triangle by pushing the tubular member into the frame. With the thumb of the left hand, press the back-cap catch; at the same time, push the back-cap forward and swing the butt away from the back-cap. Pivot the butt to its folded position, swing the butt plate out to operate the butt catch, to engage in the barrel casing, then fold the butt plate flat to lock in position.

L34A1 Submachine Gun

This weapon is a silenced version of L2A3. The barrel jacket is covered by a silencer casing, which is supported by front and rear supports. The barrel has gas escape holes throughout its length and is threaded at the muzzle. The barrel has a metal wrap and diffuser tube; the extension tube extends beyond the silencer casing and barrel. Beyond the barrel is a spiral diffuser; this is a series of discs and is held in place by tie rods that run from the end cap at the muzzle to the front support. The spiral diffuser has a hole through its center to allow passage of the bullet. L34A1 uses the standard British 9mm Parabellum cartridge.

CHARACTERISTICS OF L34A1

Caliber: 9mm Parabellum
System of operation: blowback, selective fire
Weight: approx. 8 lbs.
Overall length: **Stock extended:** 34 in.
Stock folded: 26 in.
Barrel length: 7.8 in.
Feed device: 34-round, detachable, staggered row, box magazine.
Sights: **Front:** blade w/protecting ears
Rear: "L" type w/apertures
Muzzle velocity: Approx. 1200 f.p.s.
Cyclic rate: 550 r.p.m.

THE FORWARD ACTION

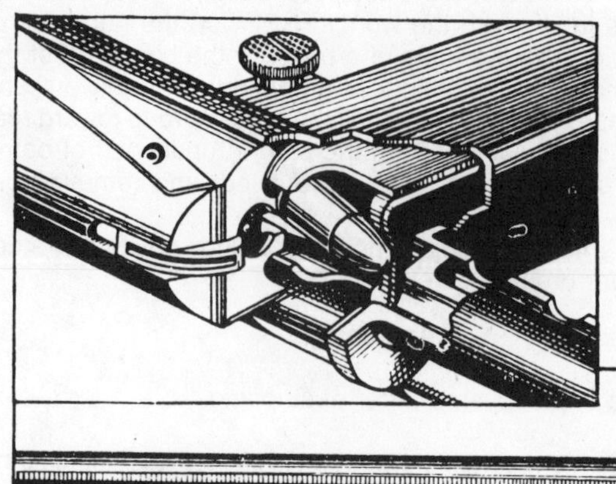

When the bolt reaches the limit of its backward travel it is forced forward by the compression of the return spring. During its forward travel the bolt contacts the top round in the magazine and, guided by the magazine lips, the round is fed into the chamber. The bolt then follows up on the round, feeds it into the chamber, and fires it *just* before the forward movement ceases. During the forward movement of the round from the magazine, the firing pin of the bolt cannot come into line with the percussion cap of the cartridge until the round is actually in the chamber. This provides the mechanical safety for this type of weapon. Upon firing, the backward action again commences.

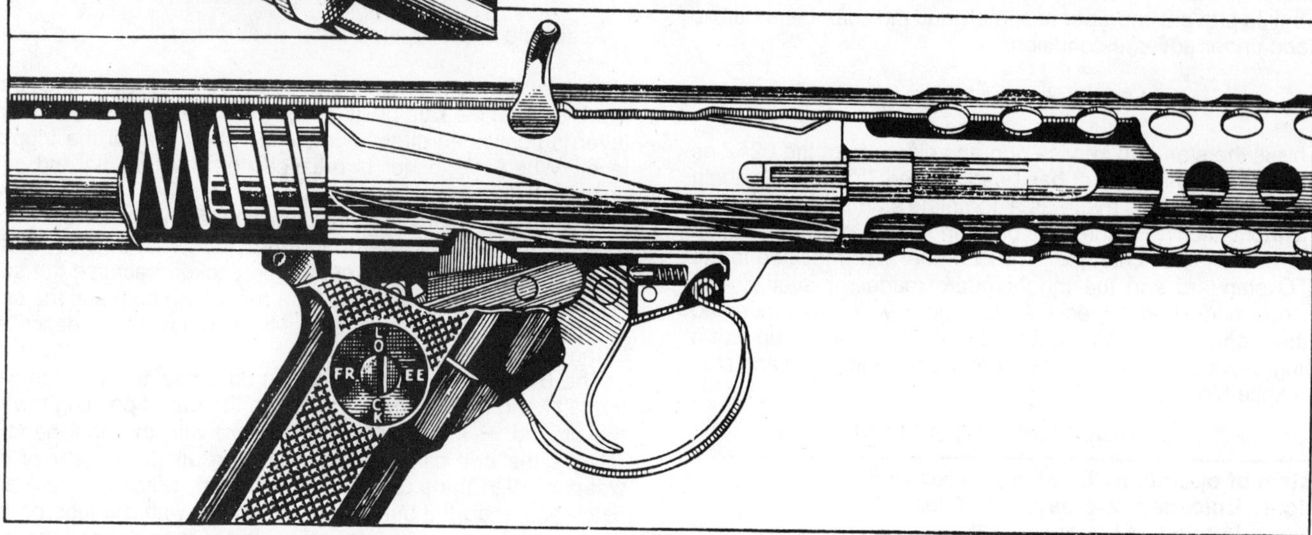

THE BACKWARD ACTION

When the cartridge is fired the propellant gases exert an equal pressure against both the bullet and the cartridge case, the latter being supported by the bolt and the compression of the return spring. The gas pressure accelerates the bullet also the cartridge case and bolt in opposite directions and as the weight of the bullet is considerably less than that of the combined weight of the cartridge case and bolt, the bullet attains a much greater velocity than that of the cartridge case and bolt. When the bullet clears the muzzle all have reached their maximum velocities but the cartridge case has not yet cleared from the chamber, thus preventing the gases escaping from the breech. The cartridge case does not clear the breech until the gases behind the bullet have dispersed into the air, ensuring that pressures are down to safe limits before the breech is unsealed.

The bolt is now being decelerated by the compression of the return spring.

The empty cartridge case, held against the face of the bolt by the extractor, is carried back until it strikes the ejector and is ejected through the opening on the right side of the weapon.

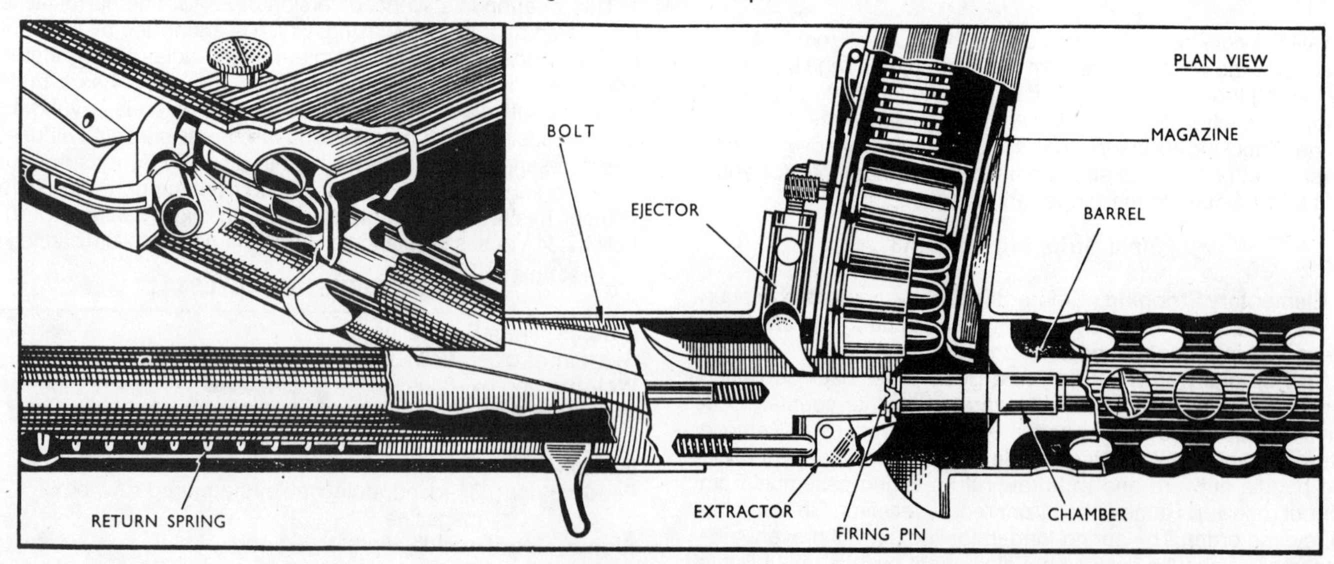

How the L2A3 Works

British 9mm L34A1 Submachine Gun

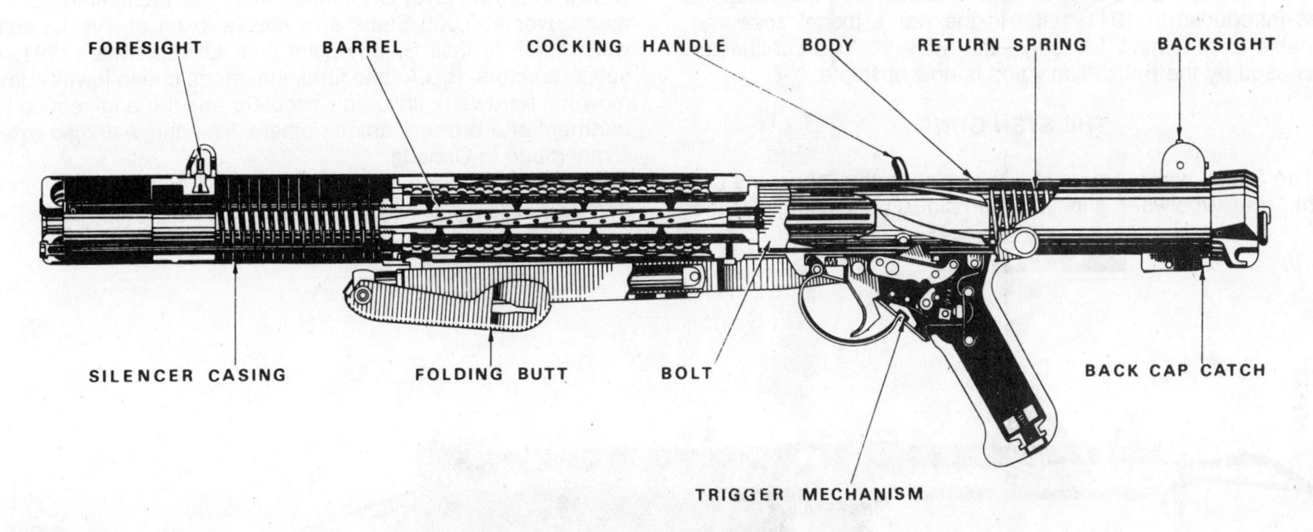

Section view of L34A1 Silenced Submachine Gun.

OBSOLETE SUBMACHINE GUNS

SPECIAL NOTE ON BRITISH SUBMACHINE GUNS

Prior to the mid-fifties the United Kingdom called submachine guns "Machine Carbines." Since that time they have adopted the same terminology as the US, but in Britain it is slightly differently arranged—"Sub-machine gun."

Right side of the Lanchester 9mm Mark I submachine gun, bolt cocked ready for firing. Note recoil spring compressed around end of firing pin unit, which protrudes from rear of bolt. Ejection port is exposed.

THE LANCHESTER MARK I

This submachine gun was designed by G.H. Lanchester; it was manufactured by the Sterling Engineering Company, the same firm that developed the L2A3. The design of the Lanchester is based on that of the German MP 28 II. The selector lever is positioned differently than that of the MP 28 II, and the Lanchester has a bayonet boss and stud for the Mark I (Pattern 1907) bayonet.

The Lanchester is a typical pre-World War II submachine gun in that it is of heavy construction and is relatively expensive and difficult to manufacture. The Mark 1, a selective fire weapon, was introduced in 1941. Later in the war a model appeared capable of automatic fire only—the Mark 1*. The Lanchester was used by the British Navy and is now obsolete.

THE STEN GUNS

The Stens, which are variously known as the "plumbers delight," the "Woolworth gun" and sometimes unflatteringly as the "Stench gun," introduced a new era in submachine gun design and manufacture. The Stens filled the need of the United Kingdom for an easily made, cheap weapon that did not require a large usage of scarce machine tools in their manufacture. Although the early Stens had many shortcomings, they were just as effective in killing people as were more expensive weapons. They have been given the greatest flattery by being copied in Germany, China, Argentina, Belgium and Indonesia.

The Stens were made by the millions by a number of basic manufacturers who in turn were supported by a number of subcontractors. In the United Kingdom, the primary producers were B.S.A. and the Royal Ordnance Factory at Fazakerley. B.S.A. made over 400,000 Stens at a special plant at Tysely; some were made at their Shirley plant prior to September 1941. As subcontractors, B.S.A. had firms that made cheap jewelry, lawn mowers, hardware, children's scooters and the engineering department of a brewery among others. The gun was also extensively made in Canada.

The basic Sten gun was developed at Enfield by R.V. Shepperd and H.J. Turpin, and its name is derived from the first letters

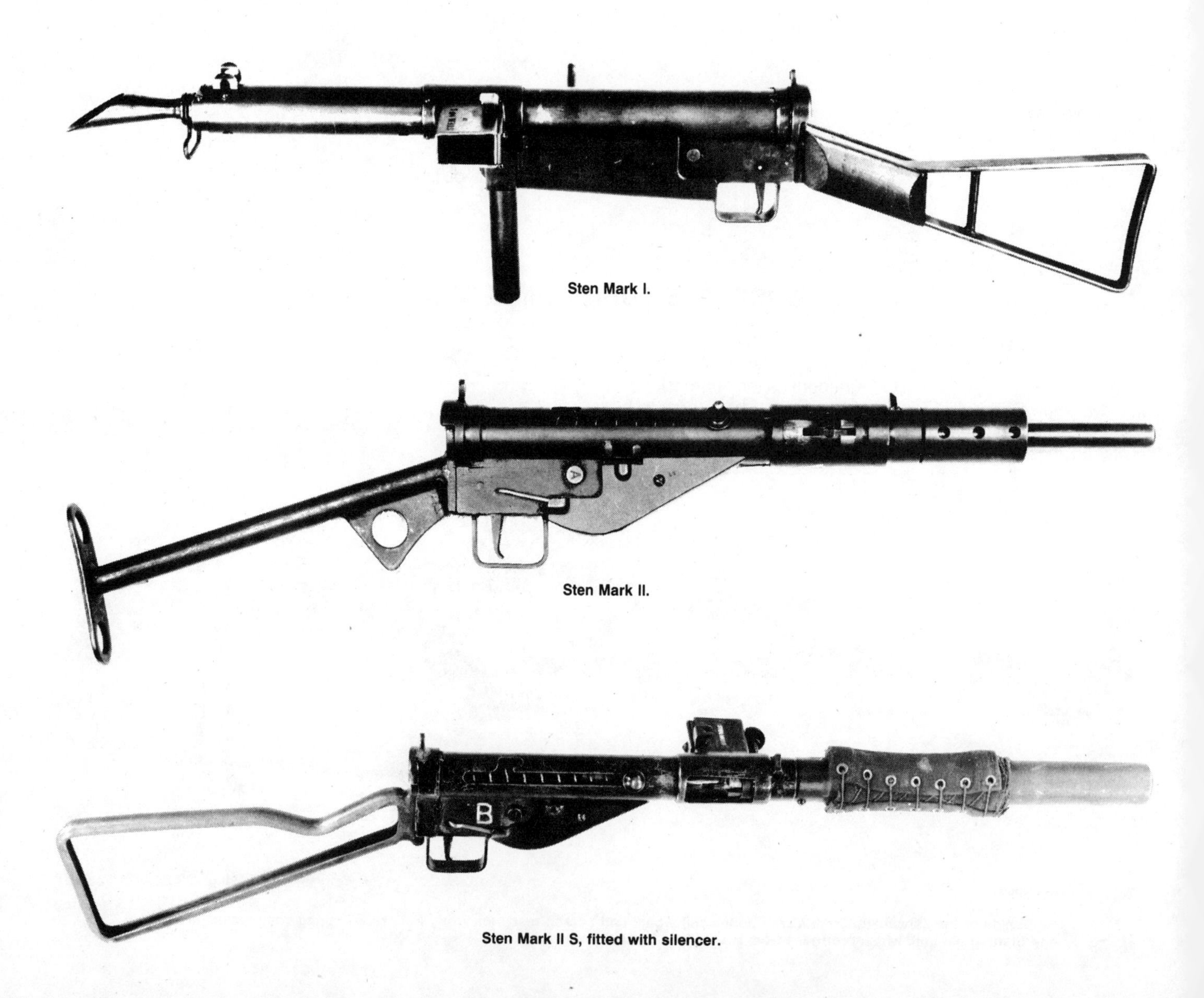

Sten Mark I.

Sten Mark II.

Sten Mark II S, fitted with silencer.

of their last names and the first two letter of Enfield. In addition to being used by the troops of the British Commonwealth, the Sten was dropped in large numbers into occupied Europe during World War II. The later model Stens are still in extensive use throughout the world, but the Stens are no longer used as standard weapons by the United Kingdom.

Sten Mark I

Adopted in 1941, the Mark I has a complete barrel jacket, a flash hider, a wooden fore-end and a vertical fore grip which can be folded up under the barrel jacket. Two basic butt stocks are used with this weapon—the No. 1 Mark I, made of steel with a wooden piece in its forward section. The No. 2 Mark II stock is made of tubular steel and does not have the wooden brace.

Sten Mark I*

A simplification of Mark I without flash hider and wooden fore-end. A stamped steel housing replaces the fore-end. Most of the Mark I* guns do not have a wooden fore grip.

Sten Mark II

The weapon differs from the Mark I only in externals. The barrel and barrel jacket were shortened; the design of the bolt handle was altered, and a simplified buttstock was issued with this gun. The Mark II may be found with a number of different buttstocks as may all of the Sten guns. Butt stocks are interchangeable among the various models. The Mark II Sten magazine housing can be turned on the axis of the receiver so that it acts as a dust cover for the magazine and ejection ports.

Sten Mark II S

This weapon is the Mark II with a shorter barrel, silencer, a lighter bolt and a shorter recoil spring. The weapon should only be used semiautomatic, as automatic fire burns out the silencer very rapidly.

Sten Mark III

The barrel of the Mark III is not detachable as are those of other models. The receiver and barrel jacket are made of one welded steel tube, and the magazine housing is welded to the receiver. The Mark III is probably the most cheaply made of the Sten guns.

Sten Mark IV

This weapon was made in two models—A and B—but very few were manufactured—about 2,000 total. The Mark IV was designed for special units and is a very compact weapon. The Model A has a pistol grip and trigger just to the rear of the magazine port whereas the Model B has the pistol grip and trigger at the rear of the receiver as do the other Stens and has the same type trigger assembly cover as does the Mark II. Both weapons have a flash hider and a very short barrel.

Sten Mark V

This is the last basic design of Sten and was the standard Sten until the adoption of the Sterling (Patchett) in 1953. The Mark V has a number of features not found on most of the earlier Stens. These are: a wooden pistol grip, a wooden stock, a front sight with protective ears (same as that of the Rifle No. 4 Mark I), the barrel has lugs for the No. 7 Mark I and the No. 4 Mark II bayonet. Early specimens had a wooden vertical fore grip.

Sten Mark VI

This weapon is the Mark V fitted with a shortened barrel and a silencer. As with the Mark II S, automatic fire is discouraged.

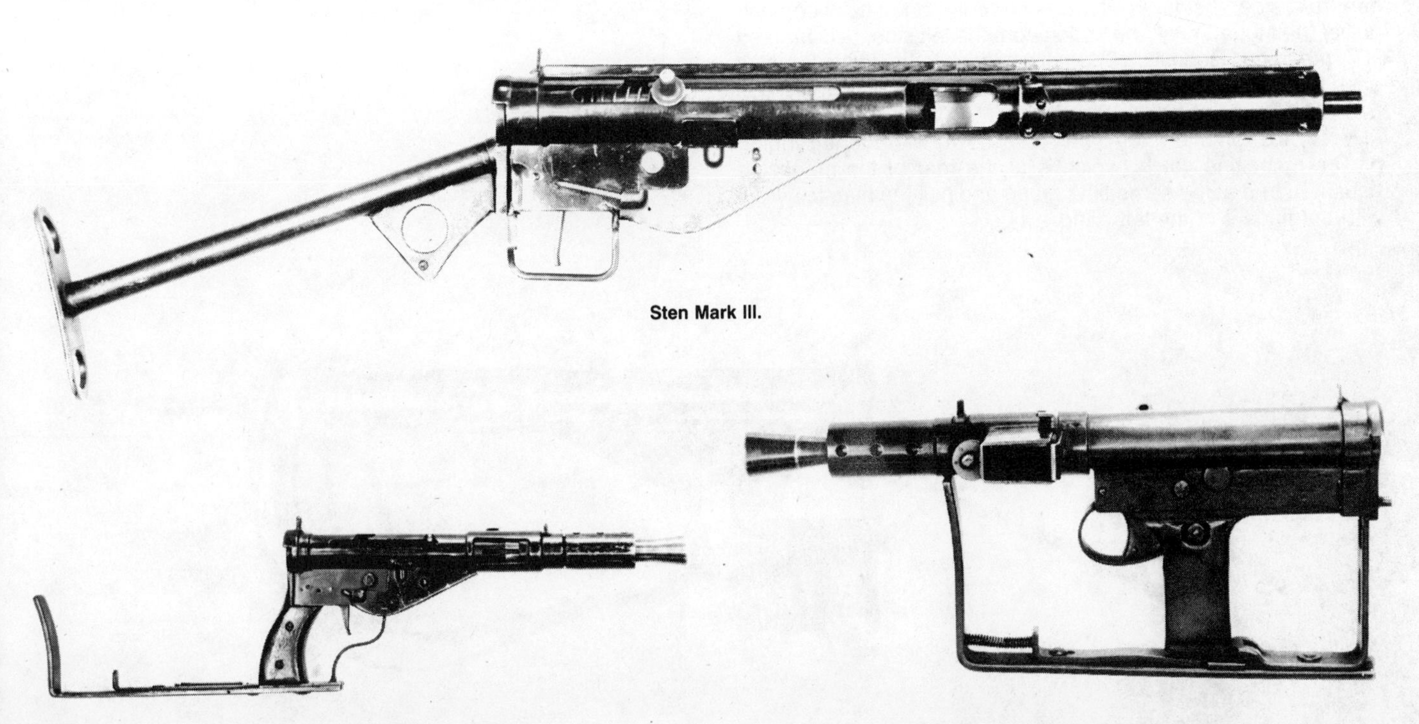

Sten Mark III.

Sten Mark IV, Model A, with stock fixed.

Sten Mark IV, Model B, with stock folded.

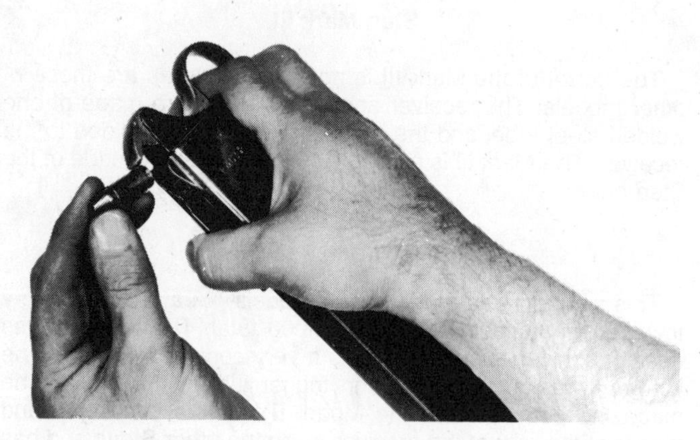

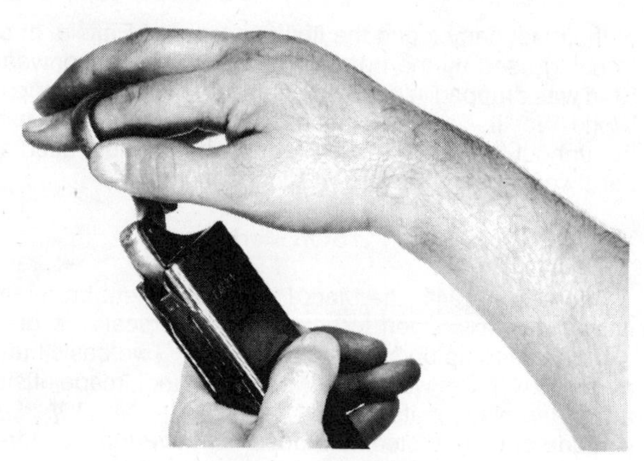

Using special Sten gun magazine loading accessory. Ring down at left to permit cartridge insertion; ring raised up at right to force cartridge down into the magazine.

Loading and Firing Sten Guns

A small special hand loader is provided as part of the equipment of every Sten gun. This is very helpful as compressing cartridges in this magazine is quite difficult due to the cartridge capacity and heavy spring.

The loader is clamped over the mouth of the magazine. The ring is pulled down as illustrated and a cartridge inserted into the mouth of the loader.

The ring is then lifted up to force the cartridge down and back under the magazine lips. It is then brought down to permit insertion of the next cartridge.

Insert loaded magazine, bullets pointing forward, into magazine housing on left side of gun just ahead of forward end of cocking handle slot. Push in until magazine locks with a click.

Pull back cocking handle and turn down into safety slot if the model is Mark I. (If model is Mark II, III, or V, the safety slot is up—so turn cocking handle up into slot.)

When ready to fire, turn cocking handle out of the safety slot.

Directly under the safety slot is a button passing through the gun from side to side. (a) If you wish to fire one shot with each pull of the trigger, push the button from the left side. (It is marked "R", meaning "Repetition."). (b) If you wish to fire full automatic, push the button through on the right side of the gun where it is marked "A," meaning "Automatic."

Note. To remove magazine press down with the left thumb on the magazine catch (which is at the rear of the magazine housing), and at the same time grasp and pull the magazine out with the fingers of the left hand.

How the Sten Gun Works

A loaded magazine being inserted in the magazine housing until it locks, the cocking handle is then pulled back to the cocked position compressing the return spring.

When the trigger is pressed, the heavy breech block is freed and driven forward by the return spring. Feed ribs on the breech block strip the top cartridge from between the lips of the magazine and drive it into the chamber. The extractor, which is attached to the breechblock, snaps into the cannelure in the cartridge case and the firing pin strikes the cartridge primer exploding the powder.

The inertia of the heavy breech block and spring in forward motion keeps the breech closed until the bullet has left the barrel and the breech pressure has dropped to safe limits.

The remaining pressure drives the empty cartridge case and moving parts to the rear. The case strikes against the ejector and is hurled out of the gun. The magazine spring pushes the next cartridge in line for feeding.

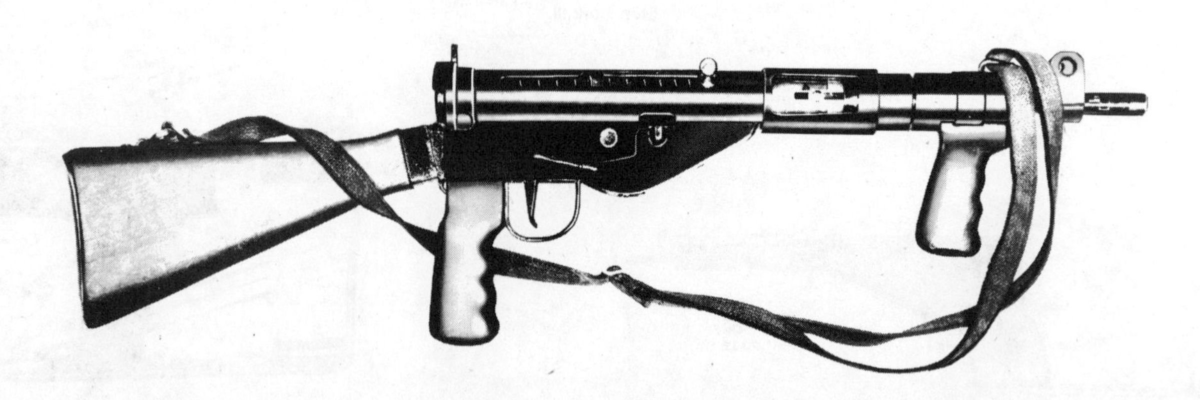

Sten Mark V, early type with fore-grip.

Field Stripping Sten Guns

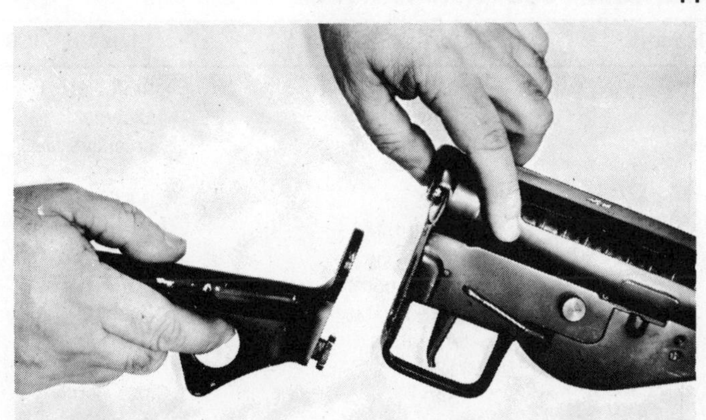

Press in the stud on the return spring housing to clear hole and slide butt down out of its slot.

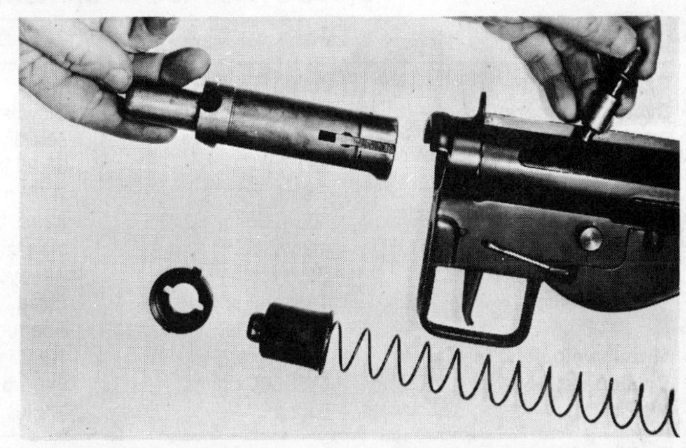

(a) Pull cocking handle back to safety slot. (b) Rotate until it can be pulled out of breechblock. (c) Tip up gun and slide out breechblock.

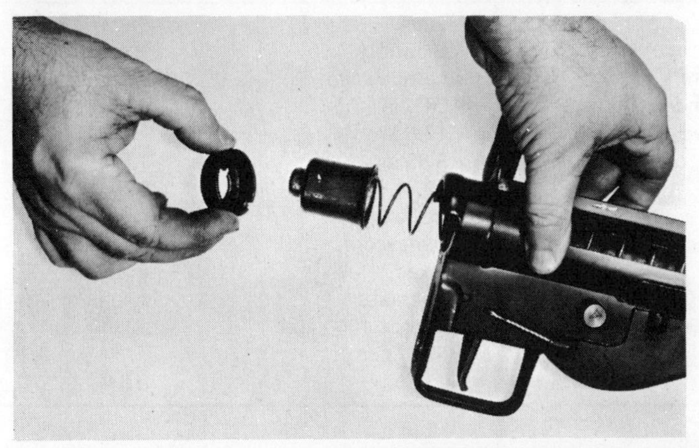

Press in on stud and spring cap and twist to the left to unlock lugs. Ease out spring cap, return spring and return-spring housing and remove.

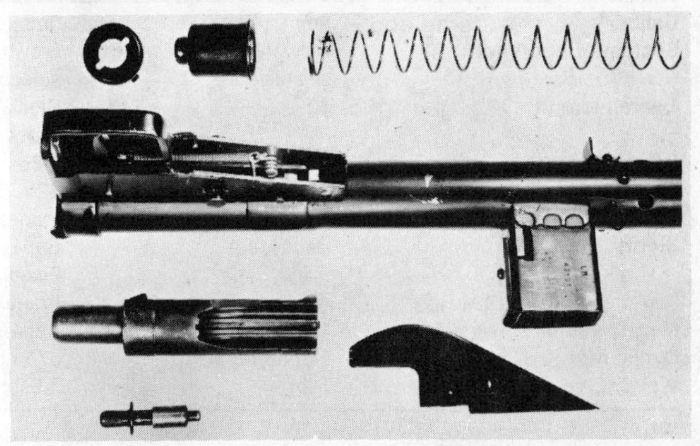

Weapon disassembled showing trigger and feeding mechanism.

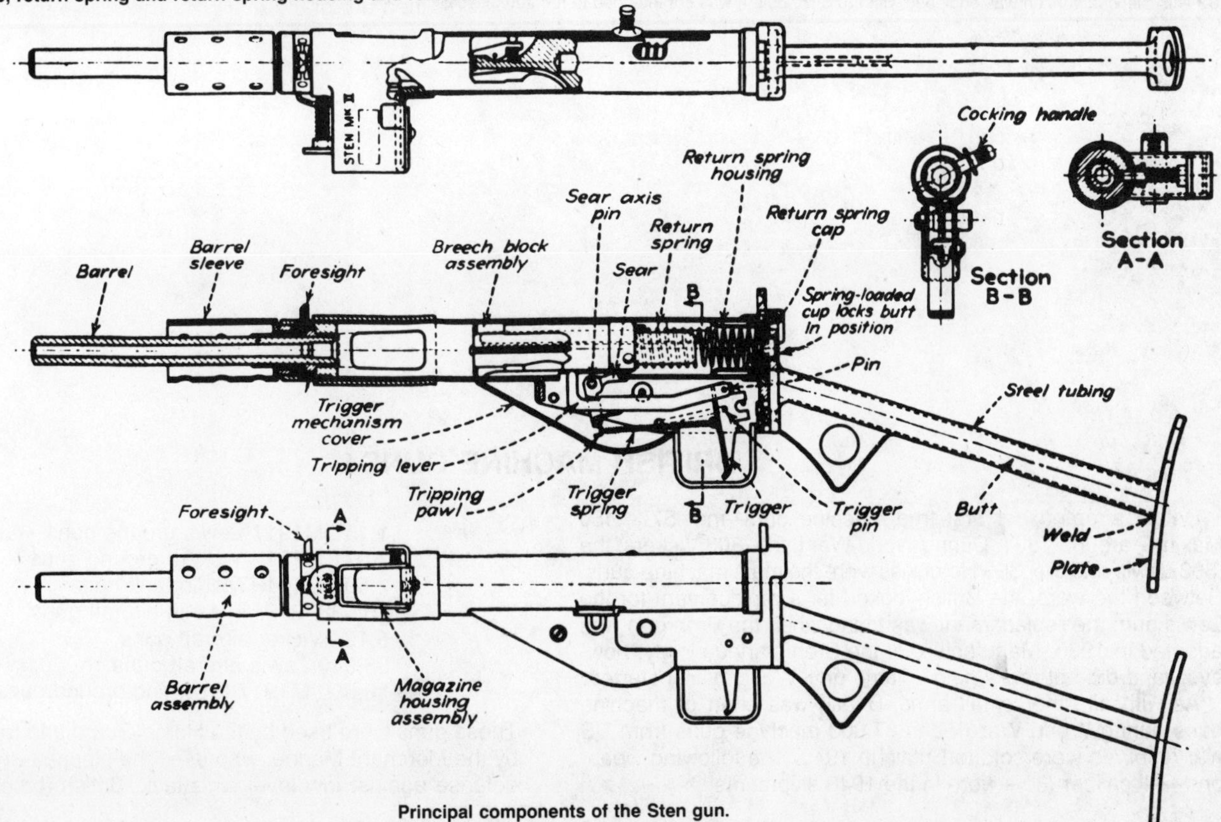

Principal components of the Sten gun.

CHARACTERISTICS OF BRITISH WORLD WAR II SUBMACHINE GUNS

	Lanchester Mark I	Sten Mark I	Sten Mark II	Sten Mark II S
Caliber[1]:	9mm.	9mm.	9mm	9mm.
System of operation:	Blowback, selective fire.	Blowback, selective fire.	Blowback, selective fire.	Blowback, selective fire.
Overall length:	33.5 in.	35.25 in.	30 in.	37 in.
Barrel length:	7.9 in.	7.75 in.	7.75 in.	3.61 in.
Feed device:	50-rd. box magazine.	32-rd. box magazine.	32-rd box magazine.	32-rd box magazine.
Sights: Front:	Barleycorn.	Barleycorn.	Barleycorn.	Barleycorn.
Rear:	Tangent, adj. to 600 yrds.	Fixed. Aperture.[2]	Fixed. Aperture.	Fixed. Aperture.
Muzzle velocity:	1280 f.p.s.	1280 f.p.s.	1280 f.p.s.	1280 f.p.s.
Cyclic rate:	575–600 r.p.m.	540 r.p.m.	540 r.p.m.	[3]
Weight:	9.62 lb.	7.8 lb.	6.62 lb.	7.48 lb.

	Sten Mark III	Sten Mark IV (Model A)	Sten Mark IV (Model B)	Sten Mark V
Caliber[1]:	9mm.	9mm.	9mm.	9mm.
System of operation:	Blowback, selective fire.	Blowback, selective fire.	Blowback, selective fire.	Blowback, selective fire.
Overall length:	30 in.	Stock extended 27.5 in. Stock folded 17.5 in.	24.5 in. 17.5 in.	30 in.
Barrel length:	7.75 in.	3.85 in.	3.85 in.	7.8 in.
Feed device:	32-rd. box magazine	32-rd box magazine.	32-rd. box magazine.	32-rd. box magazine.
Sights: Front:	Barleycorn.	Barleycorn.	Barleycorn.	Barleycorn.
Rear:	Fixed. Aperture.	Fixed. Aperture.	Fixed. Aperture.	Fixed. Aperture.
Muzzle velocity:	1280 f.p.s.	Approx 1200 f.p.s.	Approx 1200 f.p.s.	1280 f.p.s.
Cyclic rate:	540 r.p.m.	575 r.p.m.	575 r.p.m.	575 r.p.m.
Weight:	7 lb.	7.5 lb.	7.5 lb.	8.5 lb.

Notes:
1. All weapons on this chart use the 9mm Parabellum cartridge.
2. The fixed aperture sight on all of the Sten guns is set for 100 yds.
3. This silenced version was equipped with a bronze bolt. It was not intended to fire automatically.

BRITISH MACHINE GUNS

Britain adopted her first true machine gun—the .577–.450 Maxim—around 1891. During World War I, the .303 Vickers, the .303 Lewis and the .303 Hotchkiss were the main machine guns. Between the wars, the British looked for a replacement for the Lewis gun; the replacement was found when the Bren gun was adopted in 1935. Manufacture of the Bren started slowly, however, and did not really gain volume until World War II started.

As with all other small arms, Britain was short of machine guns during World War II, and 87,000 machine guns from US war reserves were sold to Britain in 1940. The following weapons—all caliber .30—were in the 1940 shipments:

1,157 M1917 Lewis ground guns
7,071 M1915 Vickers ground guns
2,602 M1918 Marlin tank guns
15,638 M1917 Marlin aircraft guns
5,124 Vicker aircraft guns
38,040 Lewis aircraft guns
10,000 M1917 Browning ground guns

These guns were used by the Home Guard and to some extent by the Merchant Marine, who used the stripped-down Lewis for defense against low level air attack. British forces in the field used the Mark 1 Vickers, the Bren, the Besa and, to a limited extent, the Vickers gas operated Mark 1, which, although designed as an aircraft gun, was used as a vehicular gun. All the rifle caliber machine guns used in the field were caliber .303 except for the 7.92mm Besa tank guns.

After World War II, Britain looked for a new machine gun to replace the Vickers and, if possible, the Bren. Although the Vickers was a reliable and proven weapon, it was overly heavy and bulky and not as tactically flexible as modern general purpose machine guns. Two of the guns tested by the UK were the B.S.A. general purpose machine gun and the Enfield-developed X11E2. In basic design, both of these guns are quite similar, based on that of the ZB26-Bren family of weapons, excepting the fact that they are belt-fed rather than magazine-fed.

After extensive trials conducted around 1957, the UK decided to adopt the FN 7.62mm NATO MAG machine gun. According to accounts appearing in British papers at the time, the MAG was adopted because it was the best available gun then in production. For reasons of economy the British Government did not want to pay at that time for the industrial engineering and tooling up necessary with the developmental British weapons. The MAG, which was tested as the X15E2, was adopted as the L7A1 machine gun.

THE L7A1 MACHINE GUN

The L7A1 is the British version of the FN MAG 7.62mm NATO machine gun. Enfield has made a few changes in the FN design, particularly in the barrel. In addition, Enfield has developed a tripod for use in the sustained fire role.

Originally a heavy barrel with Stellite liner was to be produced for this gun to use on its role as a tripod mounted sustained fire weapon. Due to manufacturing difficulties, this project was dropped, and L7A1 uses the light barrel in both light and heavy machine gun roles.

CHARACTERISTICS OF L7A2 MACHINE GUN

Caliber: 7.62mm NATO.
System of operation: Gas, automatic only.
Overall length: 49.7in.
Barrel length: W/flash suppressor 24.75 in.
Feed device: Disintegrating link belt.
Sights: Front: Protected blade.
Rear: Peep battle sight of tangent type and leaf.
Weight of gun: 24 lbs. with light barrel.
Weight of tripod L4A1: 29 lbs.
Muzzle velocity: 2800 f.p.s.
Cyclic rate: 700 to 900 r.p.m.

The L7A1 has been modified so that it will have a sear with double nose with slide notched to match. Other versions of this weapon are as follows:

L7A2: has attachment for 50-round belt box on left side of the receiver, double feed pawls and double bent sear with slide machined to match.

L8A1: This is a tank gun; the barrel has a fume extractor—bore evacuator. It has a three-position, non-venting gas regulator, and the trigger is designed for use with a solenoid. It has a folding pistol grip for emergency manual operation, and a feed pawl depressor is fitted. It is used on the "Chieftain" tank.

L8A2: An improved L8A1. The changes include: Stronger and more compact feed tray; cartridge stop modified to prevent link stoppages; improved feed arm made from an investment casting; improved breech-block roller; improved fume extractor; reduced rate of fire.

L20A1: This is a version of the L8 for use in aircraft gun pods. It is capable of left or right feed. It has a hybrid barrel assembly having the L8 gas regulator on the L7 barrel. The front sight and the carrying handle have been removed from the barrel.

L37A2: L37 is a vehicle commander's gun, for "Chieftain" and other AFV's. The L37A2 is a mixture of L7 and L8 components produced to make a gun for armored vehicles other than "Chieftain," which can be removed and used as a normal ground gun. It is basically an L8 with barrel from L7. It may also be found with butt, bipod and trigger group from L7.

L19A1: This weapon has an 8-pound barrel; it has not been issued to troops.

L41A1: Drill (instructional) version of the L8 series. Inoperable.

L43A1: This machine gun was developed to provide ranging information for the gunner of the Scorpion-tracked reconnaissance vehicle. The L43A1 has a special barrel bearing, located between the gas block and the muzzle, which supports the barrel and reduces the mean point of impact (MPI) for projectiles. MPI can vary when a gun has a cold barrel and when its barrel warms up. The barrel bearing and a barrel clamp help to maintain consistent groups.

L46A1: Drill (instructional) version of the L7 series. Inoperable.

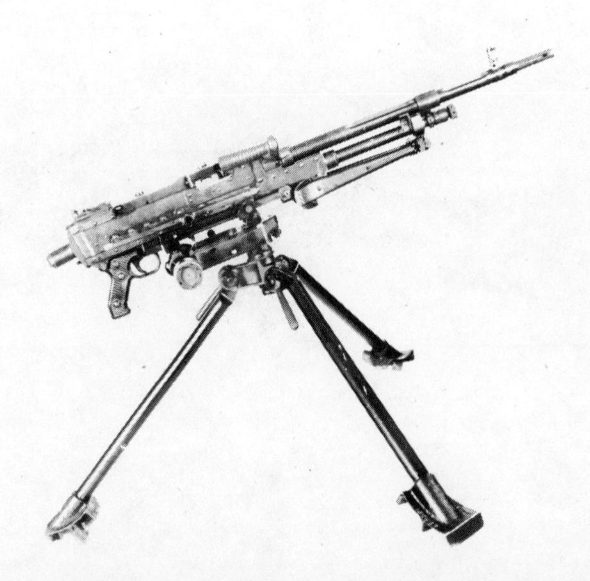

7.62mm Machine Gun L7A1 on L4A1 tripod. The gun shown has a grooved barrel which was developed only in prototype form.

The British 7.62 × 51mm NATO L7A1 general-purpose machine gun. (Note wood stock.) Both L7A1 and L7A2 guns were originally fitted with wood stocks; both models have been updated with plastic stocks.

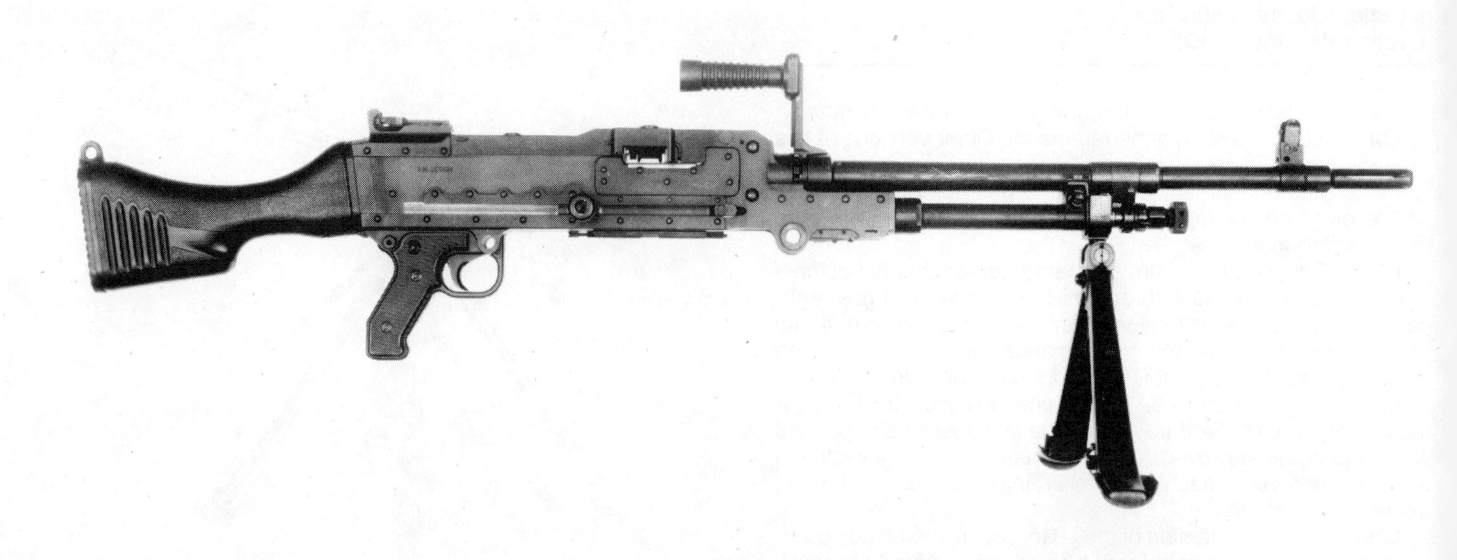

The British 7.62 × 51mm NATO L7A2 general-purpose machine gun. (Note the plastic stock.) The right side of the receiver is marked FN DESIGN.

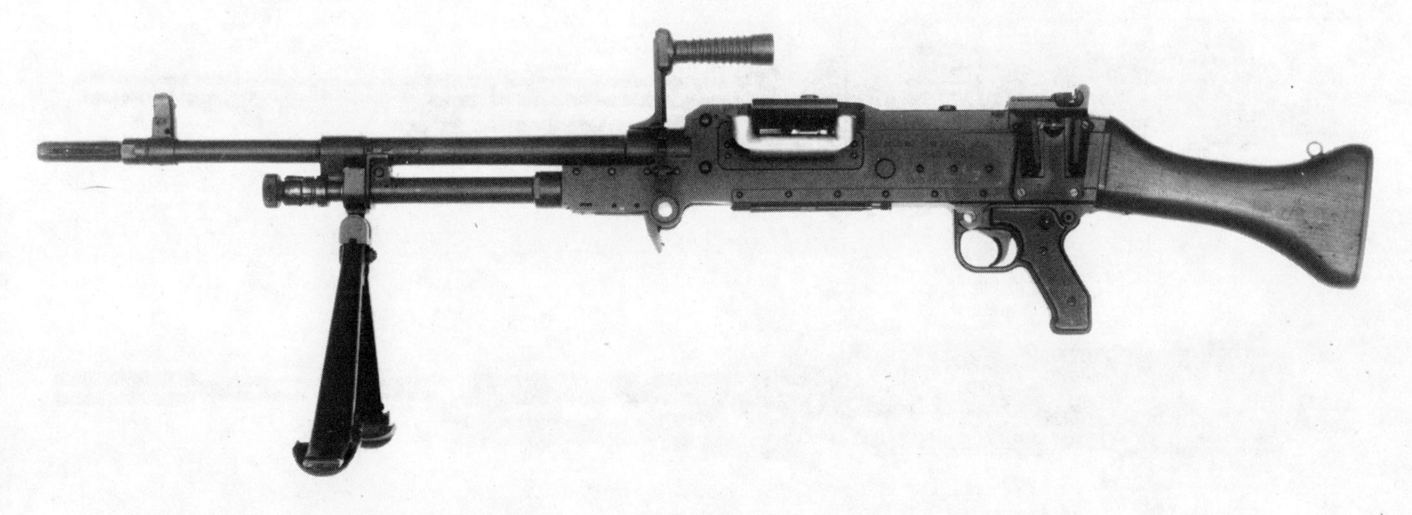

Close-up of the L7A2 GPMG receiver. Note the markings: MACHINE GUN 7.62mm L7A2 BL66 A4106 1005-13-103-2524 (Serial number explanation: BL = Belgian production; 66 = date; A4106 = gun number.)

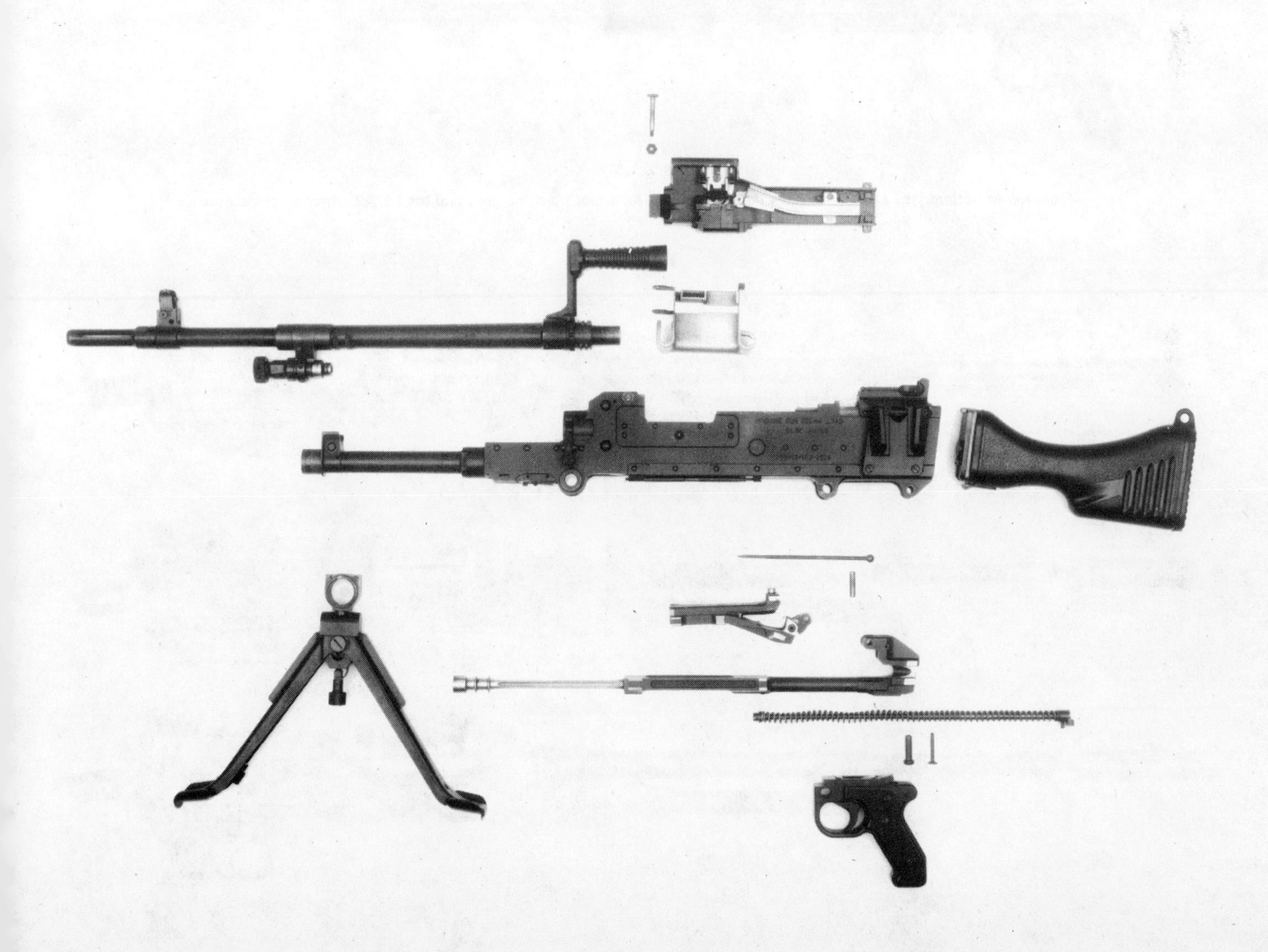

Field-stripped view of the L7A2 GPMG.

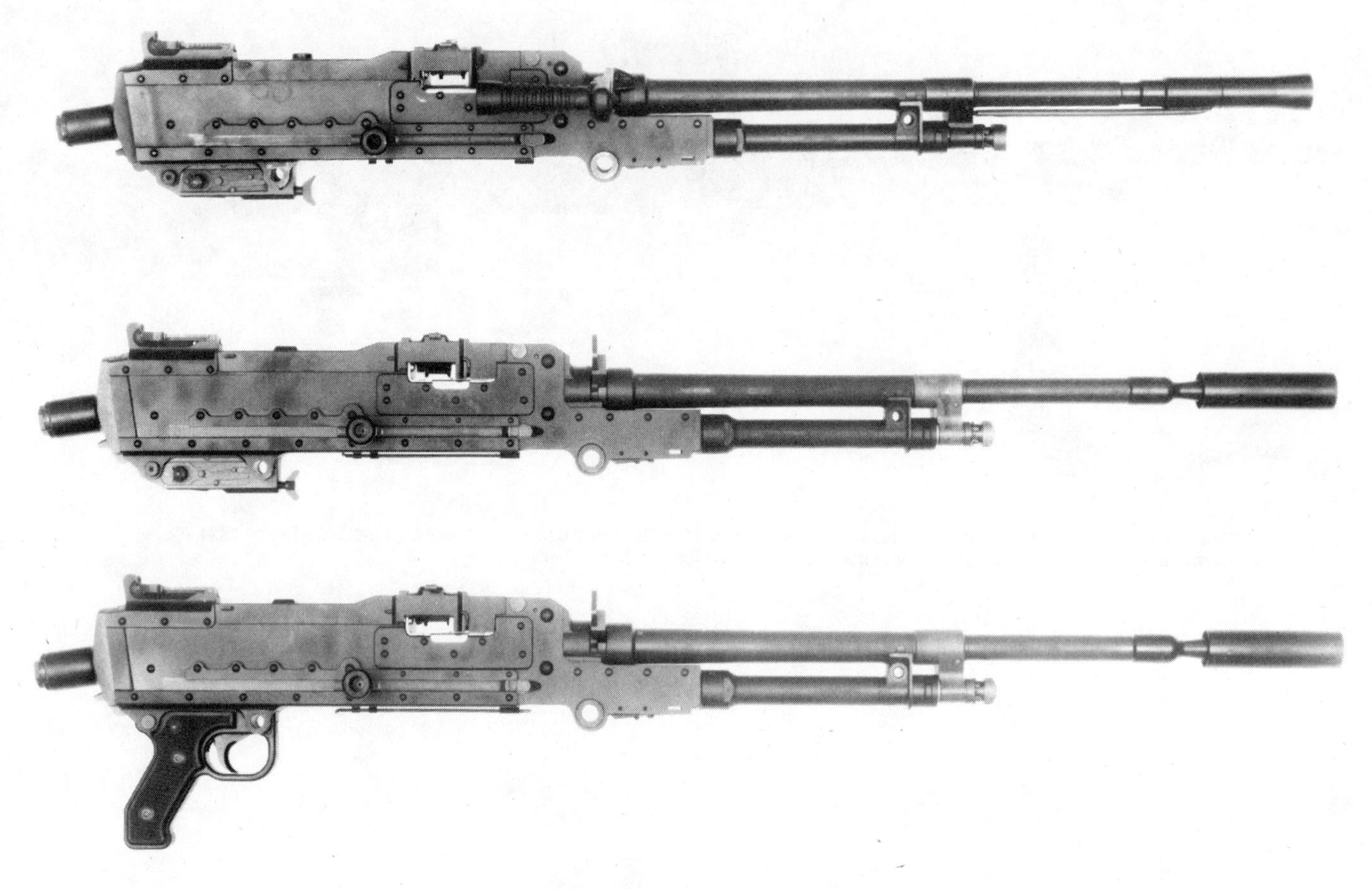

From top to bottom: **the L8A1 armor machine gun, the L8A2 armor machine gun, and the L37A2 armor machine gun.**

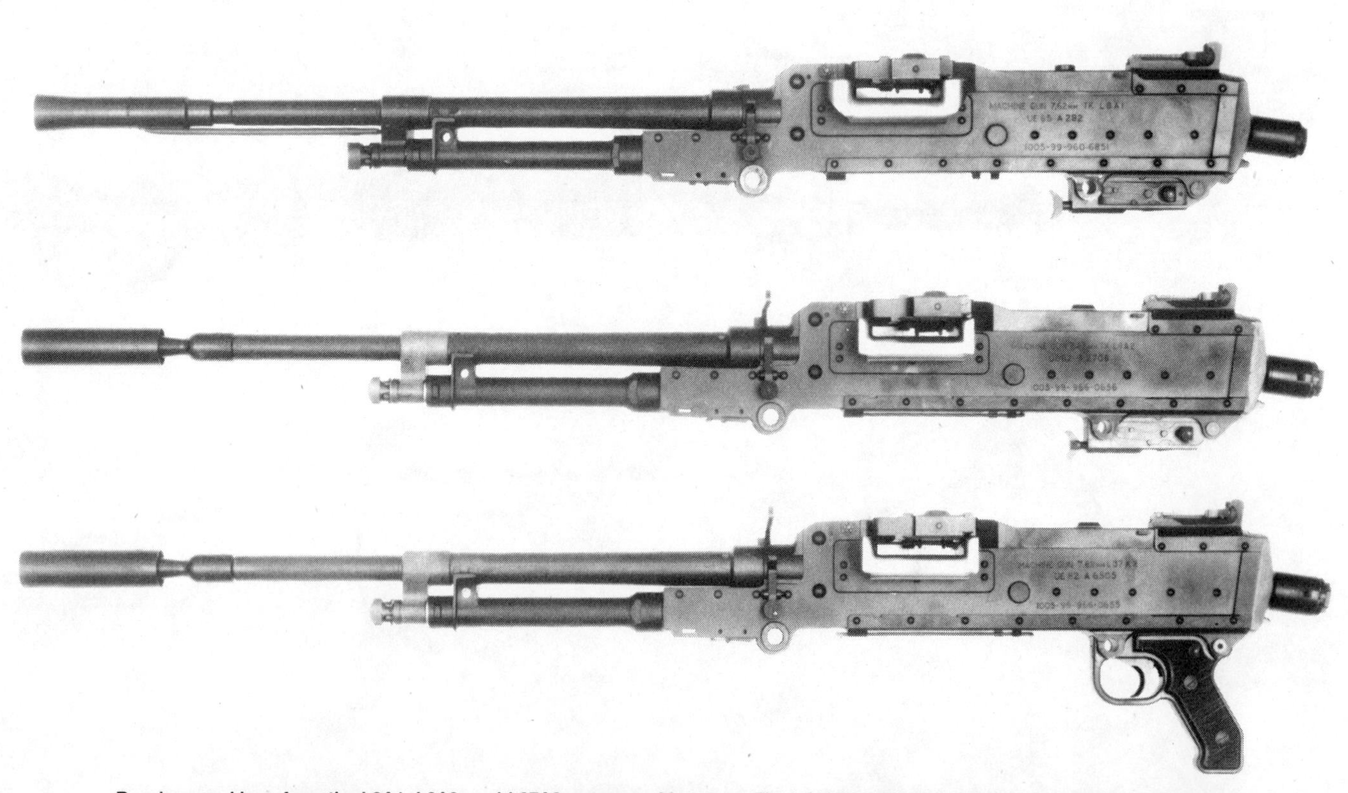

Receiver markings from the L8A1, L8A2, and L37A2 armor machine guns. The L8A1 is marked: MACHINE GUN 7.62mm TK L8A1 UE 65 A282 1005-99-960-6851. The L8A2 is marked: MACHINE GUN 7.62mm TK L8A2 UE 82 A2708 1005-99-966-0656. The L37A2 is marked: MACHINE GUN 7.62mm L37A2 UE 82 A6505 1005-99-966-0655. (Manufacturer's code explanation: UE = Enfield; UB = B.S.A.; US = Sterling; UF = Fazakerley.)

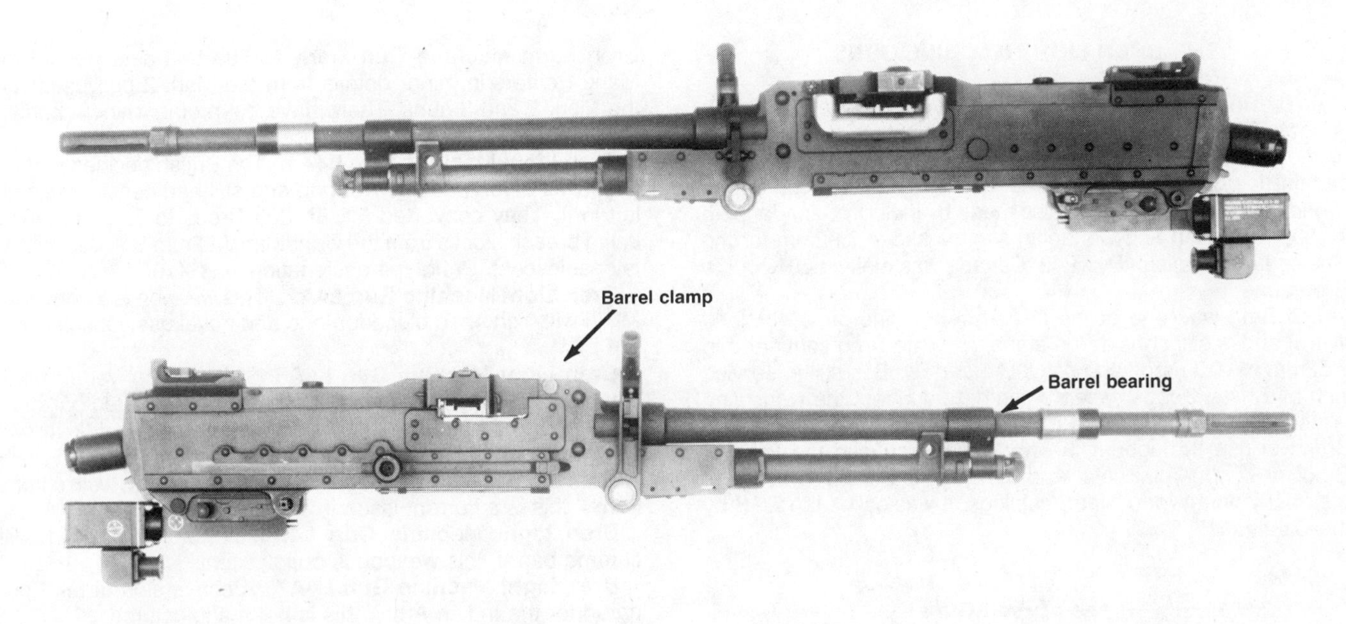

Two views of the L43A1 ranging machine gun. This weapon is marked: GUN RANGING 7.62mm L43A1 UE75 A512 1005-99-964-2619.

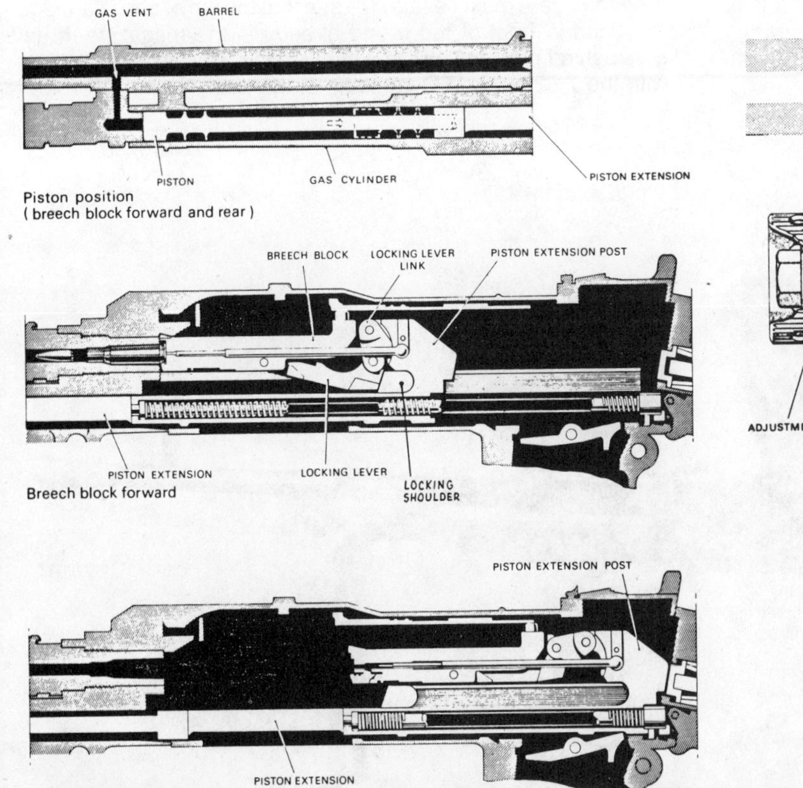

The operating mechanism of the L7-L8 machine gun family. Additional details on the operation, assembly, and maintenance of these machine guns can be found in Chapter 9 under the FN MAG.

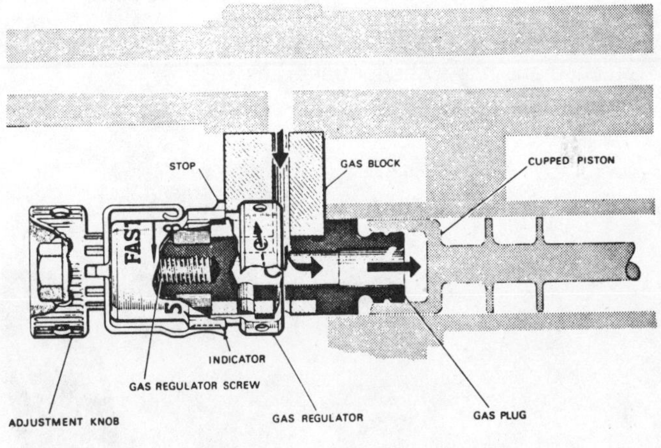

British GPMG gas regulator system.

THE BREN LIGHT MACHINE GUNS

As previously noted, the Bren was developed from the Czech 7.92mm ZB26 by Enfield and ZB in the mid-thirties. Production of the gun started at Enfield in 1937. Most of the production capabilities of Enfield were used to produce Bren guns during World War II. The Bren was also made by Inglis in Canada, both in .303 for British and Canadian service and in 7.92mm for the Chinese Nationalists. By 1943, Canada was making 60% of the Bren guns.

The Bren was one of the best light machine guns of World War II and is still considered a fine gun. The Bren converted to 7.62mm NATO models L4A2, L4A4 and L4A6 is still in service with all British troops except Infantry throughout the world and is used by British Infantry in Asia. There are still many .303 Brens in use throughout the world. ZB produced the Bren at Brno for commercial sale, and the gun was listed in some of their early post-World War II catalogs. It was called the ZGB by the Czechs.

Types of Bren Guns

Bren Light Machine Gun Mark I. The Mark I has a radial type sight, and the butt is shaped differently from the later models. Early versions had a wooden handle, which was hinged under the butt.

Bren Light Machine Gun Mark I (M). This weapon was only manufactured in Canada and differs from the Mark I in the following ways: the bipod legs do not telescope, the gas vent in the barrel has been enlarged, and the stock has been simplified by the removal of the shoulder support (butt strap), simplification of the butt plate and removal of the butt plate buffer spring.

Bren Light Machine Gun Mark 2. This weapon was made in both the UK and Canada. It has the simplified butt and a leaf type rear sight.

Bren Light Machine Gun Mark 3. The Mark 3 has been lightened and has a shorter barrel.

Bren Light Machine Gun Mark 4. The butt assembly of the Mark 4 differs in minor details from the Mark 2 butt used with the Mark 2 and 3 guns. There have been other minor changes as well.

Bren Light Machine Gun L4A1. The British decided that the Bren was a very reliable weapon and still had considerable life left in it. They converted Mk III .303 Brens to 7.62mm NATO using breech blocks from the Canadian 7.92mm Bren. The L4A1 is obsolescent. Prototype designation was X10E1.

Bren Light Machine Gun L4A2. This weapon is a converted Mk III with lightened bipod for land and naval use. Obsolescent. X10E2.

Bren Light Machine Gun L4A3. This is a converted Mk II. Obsolescent with land forces, it has a chrome-plated barrel.

Bren Light Machine Gun L4A4. A converted Mk III with chrome barrel, this is a current issue.

Bren Light Machine Gun L4A5. A converted Mk II with chrome barrel, this is a current issue in the Navy only.

Bren Light Machine Gun L4A6. A converted L4A1 with chrome barrel, this weapon is obsolescent.

Bren Light Machine Gun L4A7. A conversion of Mk I prepared for the Indian Army, this is not manufactured.

Bren Light Machine Gun L4A8. Not produced.

Bren Light Machine Gun L4A9. Modified L4A4 with bracket for A.A. or night sight fitted to left side of receiver.

Loading and Firing the Bren

To load the magazine by hand: The magazine is rested on the thigh or some solid object and cartridges placed in the magazine as for ordinary automatic pistol. They should be inserted with the right hand, and pressed down into place with the thumb of the left hand. Unlike the United States cartridge, the British service cartridge had a rim. In inserting cartridges in magazine, therefore, care must be taken to see that the rim of each cartridge is placed in front of the round already in the magazine. If rim gets behind rim, jams will inevitably result. This is not a problem with the 7.62mm NATO versions of the Bren.

Bren. .303 Light Machine Gun Mark 1.

Prototype of the 7.62 × 51mm NATO L4A1 Bren gun. The original BREN MK III has been cancelled and remarked GUN M/C 7.62mm X10E1. This gun was originally manufactured in 1944.

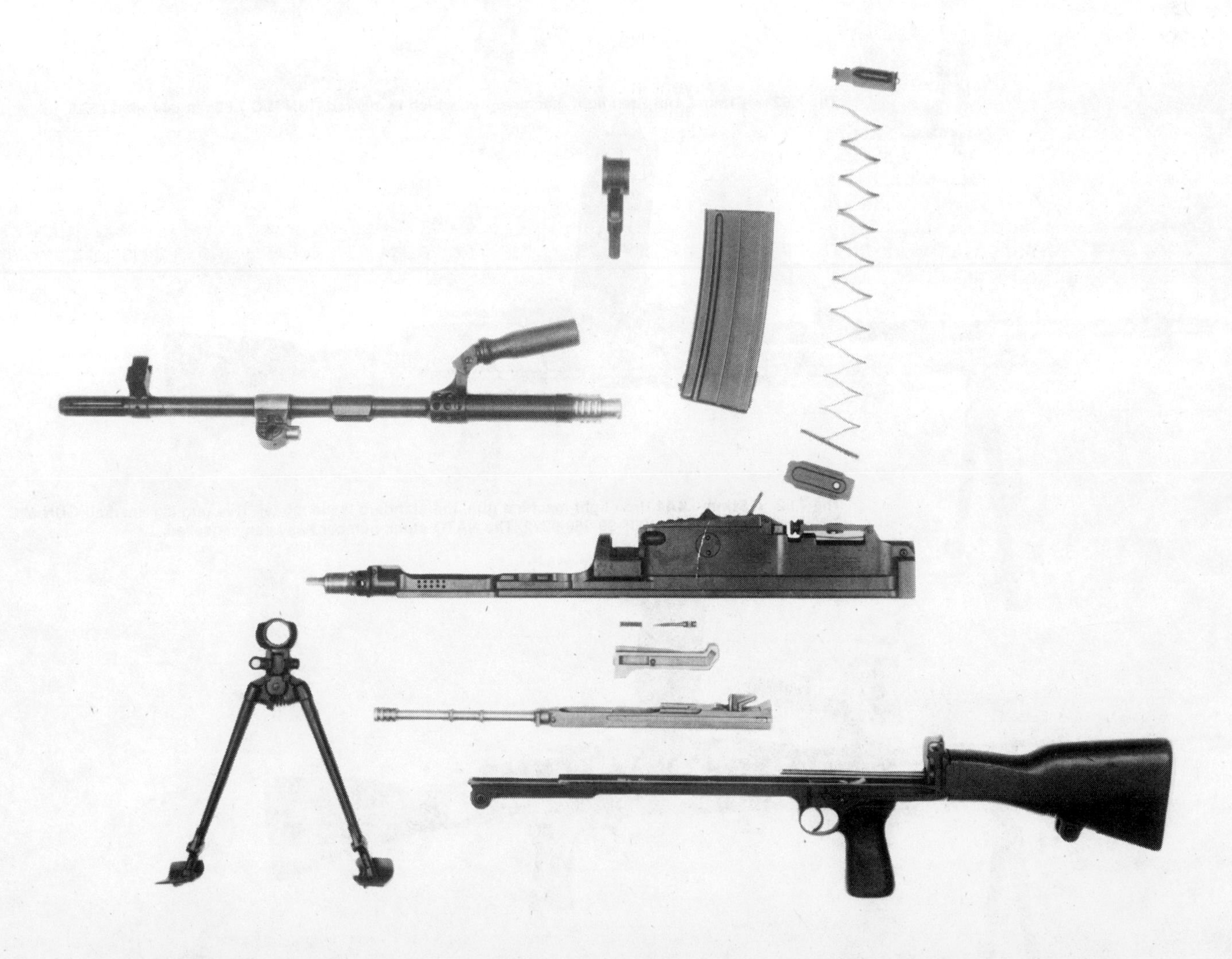

Disassembled view of the 7.62 × 51mm NATO L4A2. The BREN MK III has been cancelled and remarked GUN M/C 7.62mm L4A2 1005-99-960-0222.

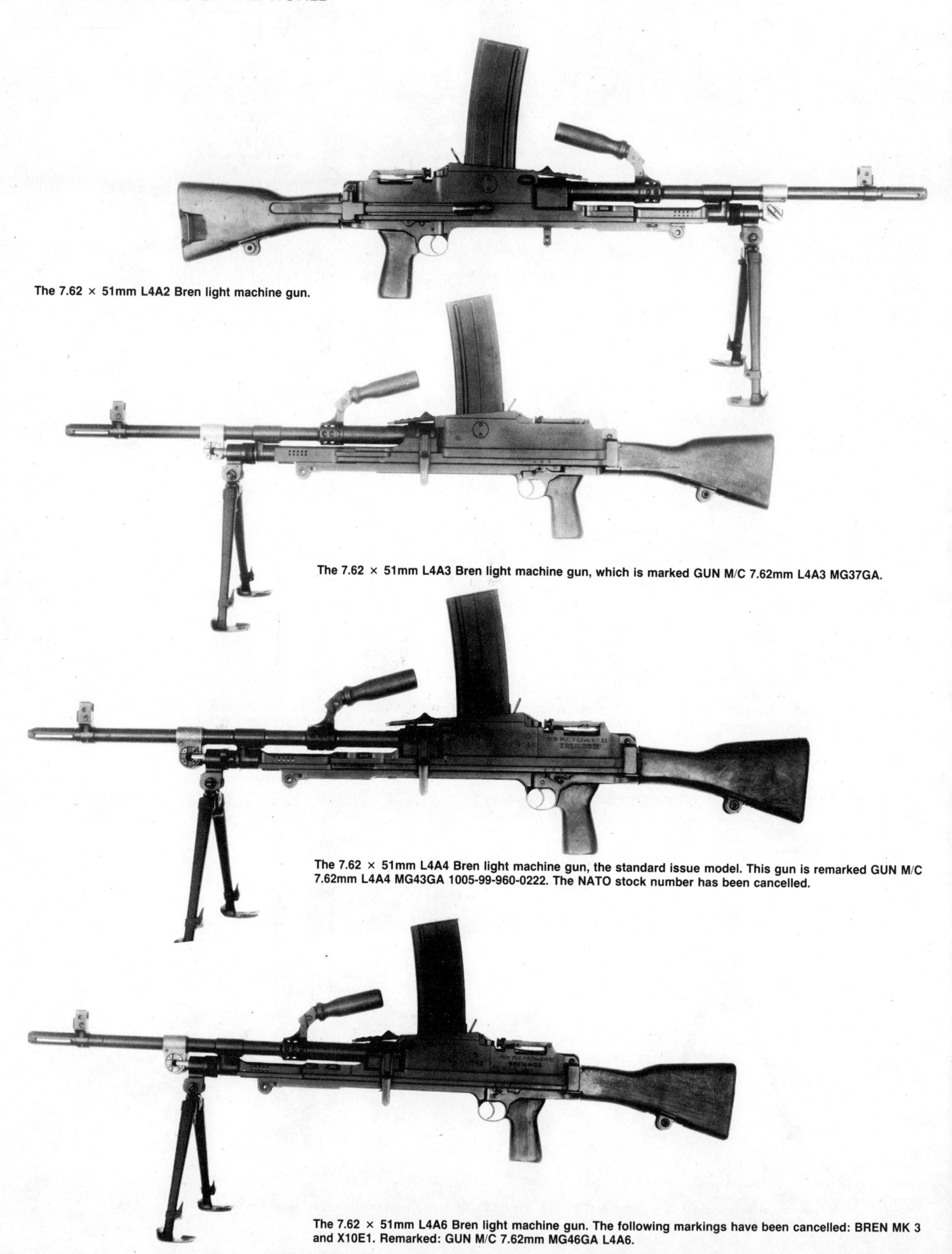

The 7.62 × 51mm L4A2 Bren light machine gun.

The 7.62 × 51mm L4A3 Bren light machine gun, which is marked GUN M/C 7.62mm L4A3 MG37GA.

The 7.62 × 51mm L4A4 Bren light machine gun, the standard issue model. This gun is remarked GUN M/C 7.62mm L4A4 MG43GA 1005-99-960-0222. The NATO stock number has been cancelled.

The 7.62 × 51mm L4A6 Bren light machine gun. The following markings have been cancelled: BREN MK 3 and X10E1. Remarked: GUN M/C 7.62mm MG46GA L4A6.

Magazine Filler. Push the magazine into the mouth of the filler and swing the filling lever as far as it will go to the left. Fill the hopper and push the filling lever over to the right and back to its limit 6 times; this will put 30 rounds into the magazine. If the filler is the small hand type, push magazine in until the magazine catch engages, and then insert a loaded cartridge charger (or clip as it is called in the United States) into the mouth of the filler over the head of the magazine. See that the tip of the operating lever is against the topmost cartridge and push down slowly and firmly with the operating lever.

Note on Magazine. While the magazine capacity is 30, it is better practice to use 27 or 28 cartridges so as not to strain the magazine spring.

The magazine opening on top of the receiver is fitted with a

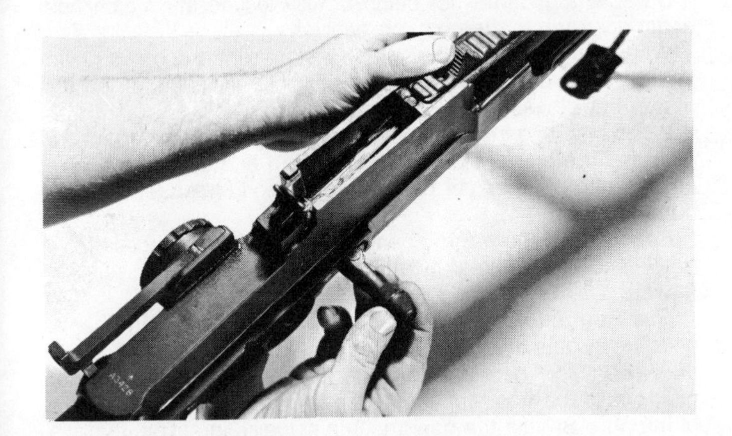

sliding cover; push this opening cover forward as far as it will go.

Holding the magazine mouth downward in the right hand, insert the lip at the front end into the magazine opening and hook it there; then press downward the rear of the magazine until the magazine catch engages on the magazine rim.

Draw the cocking handle back as far as it will go to cock the action and push it forward again. If weapon has a folding cocking handle, fold it over.

Set the change lever on the left side of the receiver at the desired position of "Automatic," "Safe" or "R" for single shot.

Note on Ejection. A cover over the ejection opening will automatically spring open when the trigger is pulled to permit ejection of empty cartridge case.

Caution: Always remember that gun fires from an open bolt. The bolt should never be permitted to go forward while there is a magazine in the gun unless you intend it to fire. The magazine must be removed first, and the action eased forward second in unloading the weapon.

Field Stripping the Bren

Be sure there is no magazine in the gun and all moving parts are forward.

The body locking pin passes through the receiver from right to left directly under the aperture of the rear sight. Push it with the point of a bullet from the left side and withdraw it from the right.

Grasp the back sight drum firmly with the left hand, and with the right pull back the butt group as far as possible. The return spring rod, which is housed in the butt, will now protrude from the butt through the buffer.

With the thumb and forefinger of the left hand, pull the return spring rod to the left out of line with the piston, and with the right hand pull the cocking handle back with a rapid motion. The piston and breechblock will now come out of the receiver and may be removed from the gun.

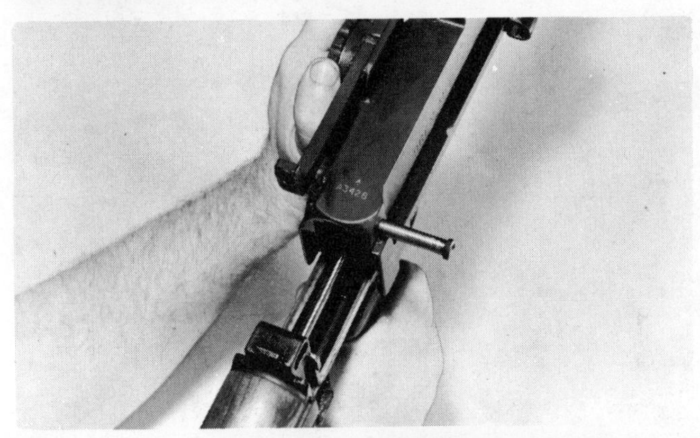

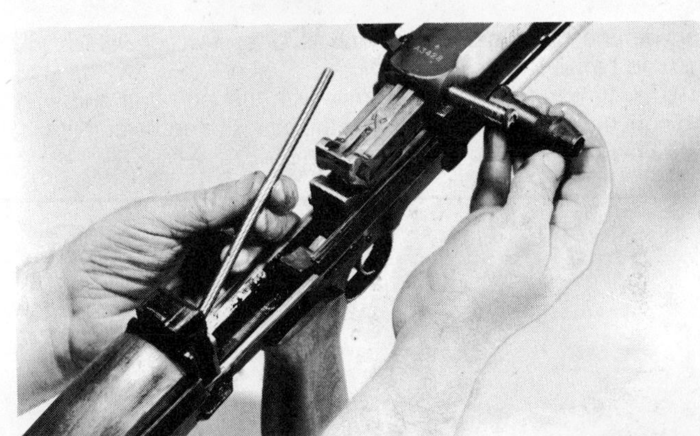

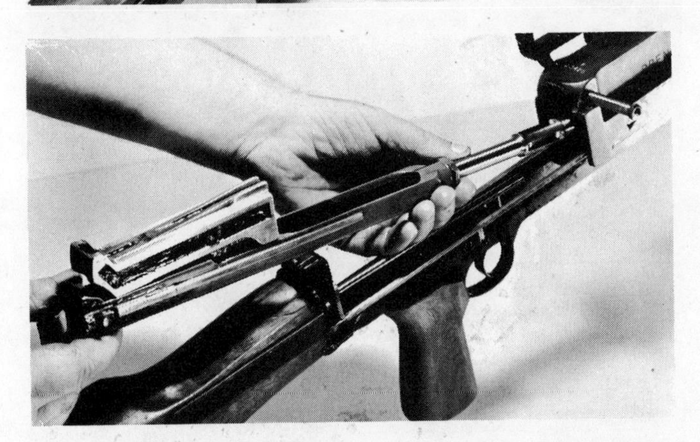

The claws at the front end of the breechblock are in engagement with grooves on the piston, and if the breechblock is slid to the rear it can be lifted out of this engagement and removed.

The barrel nut catch lies on the side of the barrel just ahead of the magazine opening. Force in the spring catch on its un-

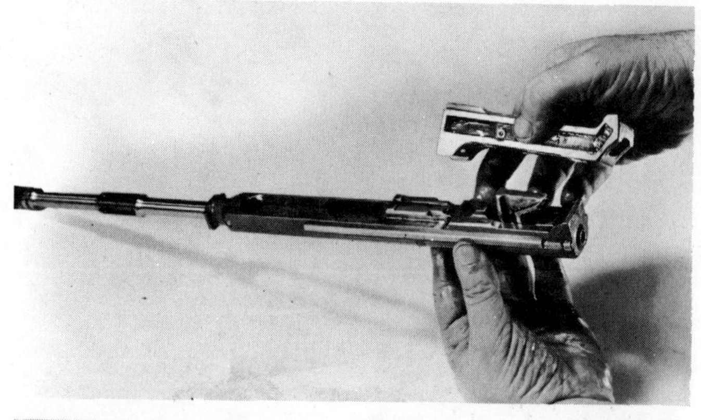

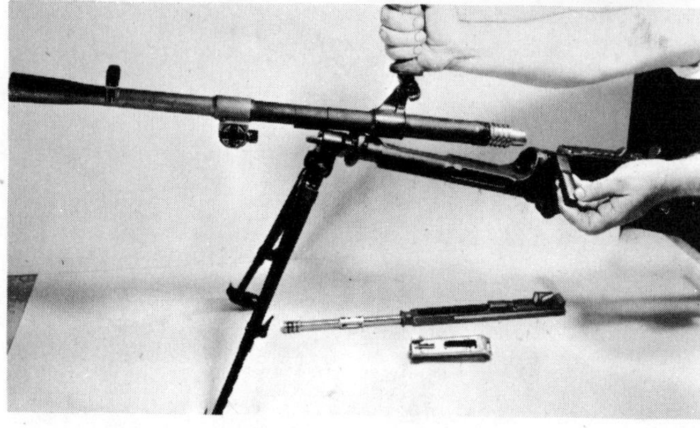

derside and lift the barrel nut catch as far as it will go, which will free the barrel for removal.

Grasp the rear sight drum firmly with the left hand and with the right hand pull directly back on the butt. The entire butt group may now be removed.

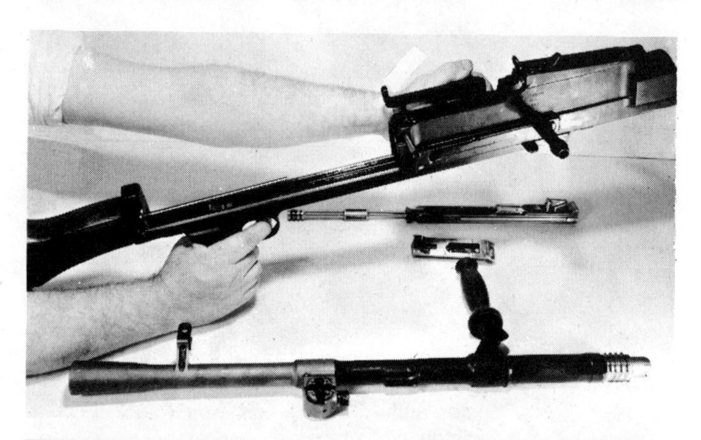

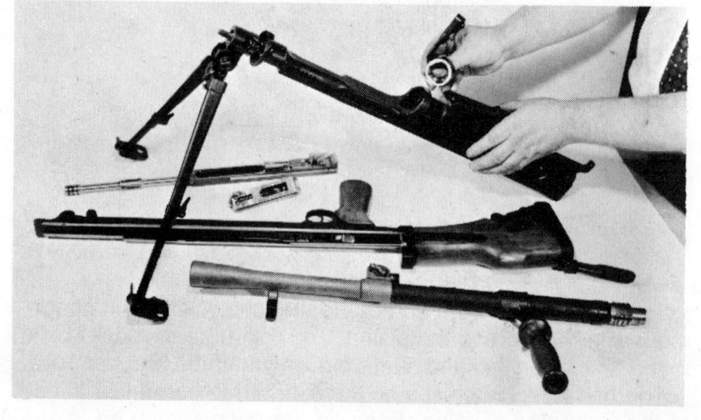

The barrel nut may be removed by lifting the catch as far as it will go and pushing down the small stud in front of the magazine opening cover. The barrel nut is then lifted out vertically.

Now lift the front of the body with the right hand and with the left pull the left leg of the bipod as far forward as possible—slide bipod sleeve off the front end of the gas cylinder.

Notes on Assembling. Reverse the stripping order.

In replacing bipod take care the mount is fully home.

In Mark 1 guns check that the stop on the left of the forward end of the butt group is in front of the barrel nut catch before lowering the catch.

In replacing barrel on Mark 1, make sure the long groove underneath between gas block and carrying handle engages properly with stud on top of receiver.

Be sure the barrel nut catch is fully locked and catch has engaged on rib in the body or receiver.

When replacing breechblock on piston, slide the claws down into the groove as far forward as possible and then let the tail of the breeckblock drop.

When inserting the assembled breechblock and piston, make sure that the breechblock is fully forward and that the two are pushed into the receiver before attempting to push forward the butt group.

Be sure that the return spring rod engages in a recess for it in the end of the piston when the butt group is being pushed forward.

Gas Regulator. The gas regulator is mounted on the barrel near the muzzle. It faces to the left. The correct setting is usually the No. 2 size. There are four different ports. Lifting the retainer pin permits the gas regulator to be turned to increase the size of the port. Should the gun become sluggish in action, the gas regulator is altered to the next larger hole to increase the amount of pressure available.

How the Bren Gun Works

Starting with the gun loaded and cocked the action is as follows. If the change lever is set at "R", pressing the trigger pulls a connecting tripping lever, which in its turn draws down the sear out of engagement with the pin on the piston. This action also compresses the coil sear spring. The compressed return spring, situated in the butt, pushes the rod forward, and this in turn pushes against the seat in the piston driving the piston forward, carrying with it the locking and firing mechanism. Meanwhile the sear spring pushes the sear back into place. The breechblock mounted on the top of the piston is carried forward,

Showing change-lever for Safe, Automatic, or Single-shot fire settings.

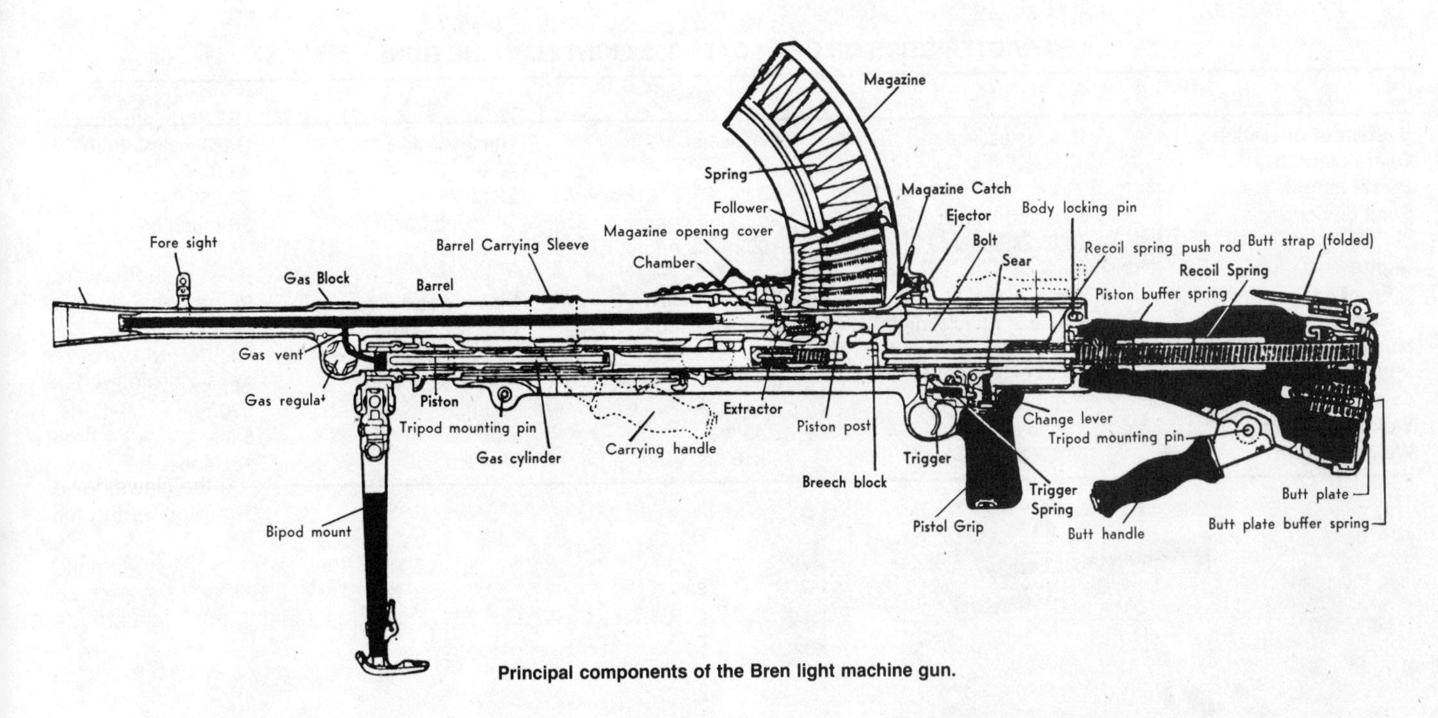

Principal components of the Bren light machine gun.

and the feed piece strikes the base of the first round in the magazine and forces it forward out of lips of the magazine and into the chamber, with the extractor slipping over the rim. The rear end of the breechblock is cammed up into a locking recess in the top of the receiver as the cartridge is properly chambered; in its final move the piston post drives the firing pin against the primer of the cartridge, exploding it.

As the bullet passes over the small gas vent cut in the barrel, a short distance from the muzzle, a small amount of gas under high pressure passes through the vent and through the gas regulator (where the size of the port selected determines the amount of gas to be let in) and escapes into a well where it expands with a hammerlike thrust against the piston. As the piston is driven back in its cylinder, the gas can now escape through holes provided for it.

Meanwhile the sudden thrust on the piston drives it back and forces the return spring rod back into the butt where the return spring is compressed, this action being finally stopped by the piston buffer.

Bren gun on antiaircraft mount.

Details of tripod mount.

The empty cartridge case, gripped by the extractor and carried to the rear in the face of the breechblock, strikes its face against the base of the ejector and is hurled downward through the ejection slot in the piston and out of the weapon. During this rearward action, the upper locking surfaces of the breechblock are forced down into line, so that in its final movement, the piston and breechblock travel together in a straight line.

Note: The buffer spring is in the butt below the line of the return spring.

The Bren Tripod Mark 2 and 2/1

A tripod was issued for use with the Bren light machine gun. Approximately one tripod was issued for every three guns. The tripod is of Czech design and is basically the same as that used with the Czech ZB26 and ZB30 machine guns. A modified form of this tripod is used with the Chinese copy of the US 57mm recoiless rifle.

CHARACTERISTICS OF BREN CAL. .303 LIGHT MACHINE GUNS

	Mark 1	Mark 2	Mark 3	Mark 4
System of operation	Gas, selective fire.	Gas, selective fire.	Gas, selective fire.	Gas, selective fire.
Overall length	45.5 in.	45.6 in.	42.6 in.	42.9 in.
Barrel length	25 in.	25 in.	22.25 in.	22.25 in.
Feed device	30-round box or 100-round drum.	30-round box or 100-round drum.	30-round box.	30-round box.
Sights:				
Front	Blade w/ears.	Blade w/ears.	Blade w/ears.	Blade w/ears.
Rear	Aperture w/radial drum.	Leaf w/aperture.	Leaf w/aperture.	Leaf w/aperture.
Muzzle velocity w/MK 7 ball	2,440 fps.	2,440 fps.	Approx. 2,400 fps.	Approx 2,400 fps.
Cyclic rate	500 rpm.	540 rpm.	480 rpm.	520 rpm.
Weight of barrel	6.28 lbs.	6.46 lbs.	5.09 lbs.	5 lbs.
Weight of gun	22.12 lbs.	23.18 lbs.	19.3 lbs.	19.14 lbs.

Left: **A sentry with the 1st Battalion, The Royal Anglian Regiment, stands watch with his L4A4 Bren during Exercise Atlas Express in northern Norway, March 1976.**

British soldiers on maneuvers in Germany in 1977. From left to right they are armed with the L1A1 rifle, L7A2 GPMG, and the L2A3 submachine gun.

The 5.56 × 45mm XL73E2 Light Support Weapon currently under development to replace the 7.62mm Bren guns.

OBSOLETE BRITISH MACHINE GUNS

THE LEWIS MACHINE GUN

The Lewis gun was the principal light machine gun of the British Army in World War I and was used by the Home Guard and Merchant Marine (for defense against low-level air attack) in World War II. It was also used—in caliber .30—by the US Marine Corps and Navy until World War II. The US gunboat Panay, which was sunk by Japanese bombers in China during 1937, had Lewis guns as part of its antiaircraft armament as did many other US naval vessels.

The Lewis was made in large quantities during World War I, and a few were assembled by Savage Arms early in World War II. B.S.A. made 145,397 guns at Small Heath; Savage Arms Corporation of Utica, N.Y., produced Lewis guns for the UK and Canada and produced 2,500 caliber .30 and 1,050 caliber .303 ground guns for the US. In addition to the ground guns, large numbers of aircraft guns were made as well.

The basic ground gun is the Mark I; during World War I a number of different model ground guns were made, but all were converted to Mark I after the war. The Lewis gun, in one form or another, was used by France, the Netherlands, Norway, Japan, Imperial Russia, Belgium, Portugal, Italy, Honduras and Nicaragua, in addition to the US and the British Commonwealth.

During World War II, many of the caliber .30 aircraft Lewis guns that were sold to the UK by the US were converted to ground use for the Home Guard. An aperture sight fixed for 400 yards was mounted on the rear aircraft gun sight base and either the standard ground gun wooden butt or a steel skeleton stock with wooden cheek rest were substituted for the aircraft type spade grip. They initially had no mounts and were to be laid over walls or fired from the hip; a non-telescoping bipod was later issued. During the same time, a number of British ground guns were modified for antiaircraft use on ships. The radiator casing and radiator were removed, the butt shortened by two inches, a forward hand grip added and a light steel guard fitted over the gas cylinder.

CHARACTERISTICS OF MARK I LEWIS GUN

Caliber: .303 British.
System of operation: Gas, automatic fire only.
Weight: 27 lbs.
Length, overall: 50.5 in.
Barrel length: 26.04 in.
Feed device: 47-round drum magazine, a 97-round drum designed for aircraft use also exists.
Sights: Front: Barley corn.
Rear: Leaf w/aperture.
Muzzle velocity: 2440 f.p.s.

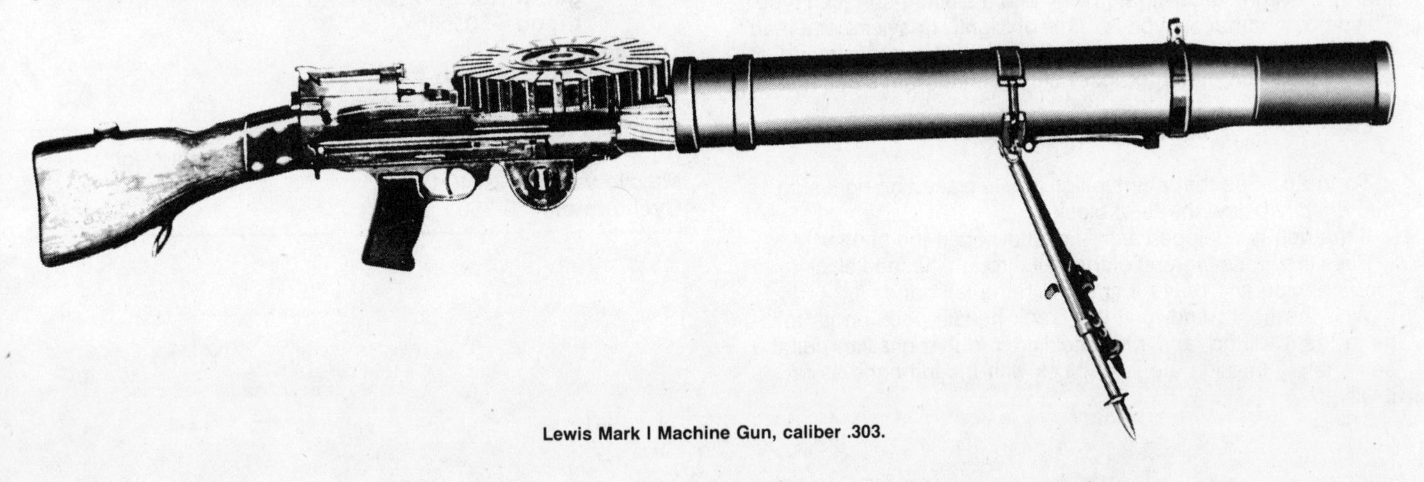

Lewis Mark I Machine Gun, caliber .303.

THE VICKERS MACHINE GUN

The Vickers, originally called the Vickers Maxim, was adopted by Great Britain in 1912. It was their principal heavy rifle-caliber machine gun in both world wars and was a standard weapon until the adoption of the L7A1 general purpose machine gun in the early sixties. The Vickers, which is a modified Maxim gun, has the reputation of being one of the most reliable and rugged machine guns ever built. The weapons used by the UK were made by Vickers at Crayford, Kent, in both ground and aircraft versions. The US had bought some Colt-made Maxims—the Model 1904—in very limited quantities prior to World War I. Colt tooled up to produce the Vickers in World War I; and in caliber .30, it was adopted as the US Machine Gun Model 1915. Due to the emergence of the Browning in 1917 and the limited quantities of Vickers Colt was able to produce (it was a difficult gun to make), US troops received few Vickers ground guns during World War I. Those on hand in 1940 were sold to the UK and were used by the Home Guard. These caliber .30 weapons had a red stripe painted on the receiver, the mouth of the feed block and on the side lever, to distinguish them from the .303 Vickers. The UK made all the Vickers ground guns needed for their forces in both wars.

British .303 Vickers Machine Gun Mark I on tripod. Mark IVB, with ammunition box and steam condensing assembly.

The weapon can be found with two different water jackets (barrel casings): a corrugated type and a smooth-surfaced type. The smooth-surfaced type is made of slightly heavier metal than the corrugated type. The feed block bodies may be made of either steel, gun metal (bronze) or gun metal with steel strips.

Loading the Vickers Gun

To Load. See that ammunition box is placed on right side of gun directly below the feed block.

If the gun is equipped with a shutter, open the shutter.

Pass the brass tag-end of the belt through the feed block from the right side and grasp it firmly with the left hand.

With the right hand, pull the crank handle back on its roller as far as it will go, and while holding it in that position pull the belt sharply through the feed block with the left hand as far as it will go.

CHARACTERISTICS OF VICKERS MACHINE GUN

Caliber: .303 British, Mark 8z ball normally used.
System of operation: Recoil with gas boost from muzzle booster, automatic only.
Weight: **gun w/o water**—33 lbs.
gun w/water—Approx. 40 lbs.
tripod—50 lbs.
Overall length: 43 in.
Barrel length: 28.4 in.
Feed device: 250-round canvas belt.
Sights: **Front:** Hooded blade.
Rear: Leaf with aperture, 400 yard battle sight.
Muzzle velocity 2440 f.p.s.
Cyclic rate: 450-550 r.p.m.

Release the crank handle and let it fly forward under the influence of the spring. This action grips the first cartridge firmly between upper and lower portions of a gib at the top of the extractor. Now pull the crank handle back on its roller once again. Give the belt another sharp tug to the left as far as it will go, and again let the crank handle fly forward under the influence of the spring. This action withdraws the cartridge from the belt, places it in the chamber ready for firing and grips the next cartridge by the gib in the upper part of the extractor.

The gun is now cocked and ready to fire, whenever the safety catch is lifted and the trigger pushed in.

Note on Unloading. Because of the method of feeding, safely unloading this weapon requires special consideration. Without touching the belt, pull the crank handle back onto the roller as far as it will go and release it.

Again pull the crank handle back as far as it will go and permit it to fly forward. The first motion of the crank handle extracts the cartridge from the firing chamber and drops or ejects it through the bottom of the gun. The cartridge in the feed block is withdrawn for positioning by this movement and is fed into the chamber, but as the belt is not moved across no new cartridge is gripped by the top of the extractor. Thus when the crank handle is run back for the second time, the second live cartridge is dropped through the bottom of the gun, leaving the firing mechanism empty.

With the left hand, raise the finger plate of the bottom pawls and simultaneously push down the top pawl by squeezing the pawl grips. Keeping the pawls disengaged, pull the belt out of the block to the right. The pawls hold the belt in position in the feed block, the top pawls being behind the first cartridge and the bottom pawls behind the second. During recoil of the gun, the top pawls feed the cartridge into position while the bottom ones prevent any backward lash of the belt; thus, it is necessary to release the pawls from their position before the belt can be pulled out of the gun.

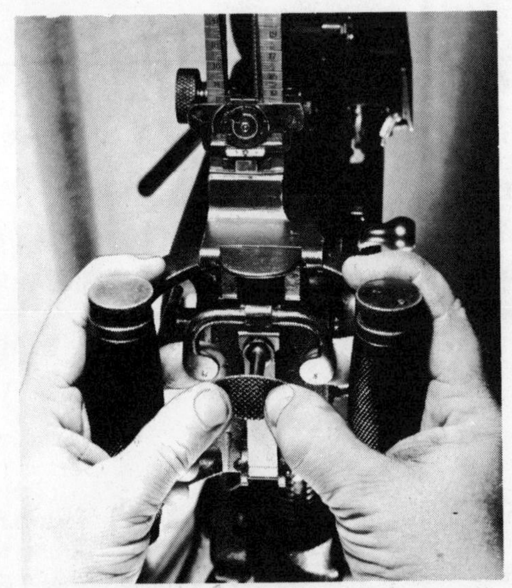

Illustrating correct hand position for firing.

Firing the Vickers Gun

With forefingers wrapped over the top of the traversing handle, raise the safety catch with the second fingers of the hands, wrap the other fingers around the traversing handles and with both thumbs press in on the thumb trigger.

The gun will now fire as long as the trigger is kept pushed in and cartridges are fed into the gun. Releasing either trigger or safety will stop the gun.

Remember that this gun is supported by the tripod and that the hands are intended only for use in firing. Thus, no particular effort is required on the part of the gunner.

How the Vickers Gun Works

Locking Principle. This gun is locked securely at the moment of firing by a toggle joint. This is the principle developed by Hiram Maxim. The Vickers gun is a modification of the Maxim gun. This locking principle was used in Maxim guns throughout the world, notably in Germany and in Russia.

The simplest way to explain this principle is to compare it with the human knee.

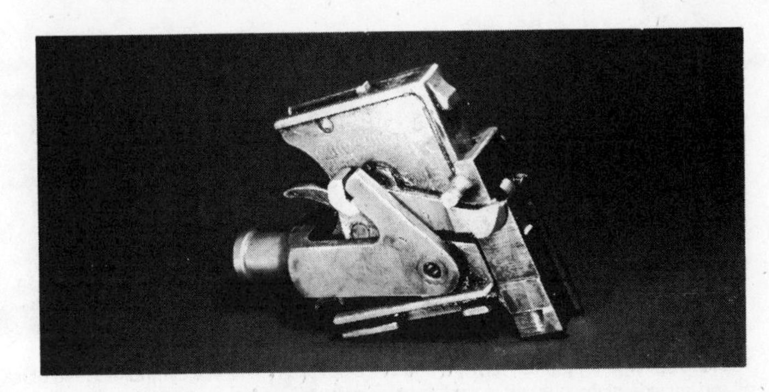

Vickers lock.

When in firing position, the lock on the Vickers gun fits securely against the firing chamber. Now picture the human foot with the heel held firmly in the position of this lock against the head of the cartridge. Pivoted to the lock is the connection rod, a heavy metal bar, thrust straight forward. This connection rod is like the lower part of the leg, but it can buckle at the ankle where it joins the foot. The crank is attached to the connection rod by a hinge pin and extends to the rear. This crank forms a bending knee where it joins the connection rod; it resembles the upper part of the human leg. However, the knee in this mechanical device is actually below the line of the connection rod and crank.

This crank is rigidly supported from below by the inside plates of the weapon, and pressure applied to it by the side levers, during the opening movement of the recoil, merely presses the crank down harder on the plates.

Attached to the crank is a crank handle which travels back with it, and after the gun has recoiled far enough to permit the bullet to leave the barrel, the tail on the lower side of this handle is forced back in contact with a roller, which causes the crank handle to rotate upwards. This raises the axis of the crank pin and permits the knee-like joint to buckle. (Thus, as the human foot is driven backwards, pressure applied to the underside of the knee will buckle the knee but draw the foot straight back.)

The connection rod is locked securely by a twisting motion inside the side lever head, which projects beyond the lower rear end of the lock. As the connection rod buckles, it naturally raises the side lever head with it, and this raises a tumbler, which cocks the lock.

Field Stripping the Vickers Machine Gun

At the rear of the gun above the safety is the rear cover catch. It is held in place by a spring. Push up on the catch and raise the cover up as far as it will go. Now pull back the crank handle against the tension

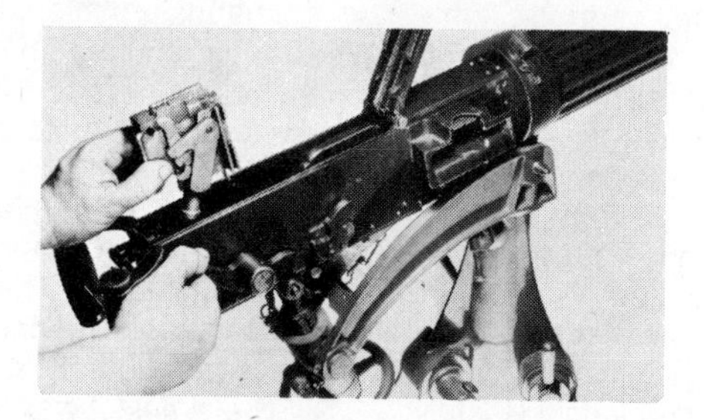

of the fusee spring. Hold it firmly. Reach inside the gun and lift out the lock, which is fastened to the connection rod. Now twist the lock on the connection rod about one-third of a turn to the right, to release it from

the connection rod, which in its turn is connected to the crank. Lift the lock out, ease the crank handle home under the tension of the fusee spring. Then close the cover.

Turn the cover latch (on the forward end of the cover on the left side of the gun) up to the left as far as it will go. This releases the front cover which should now be raised as high as it will go.

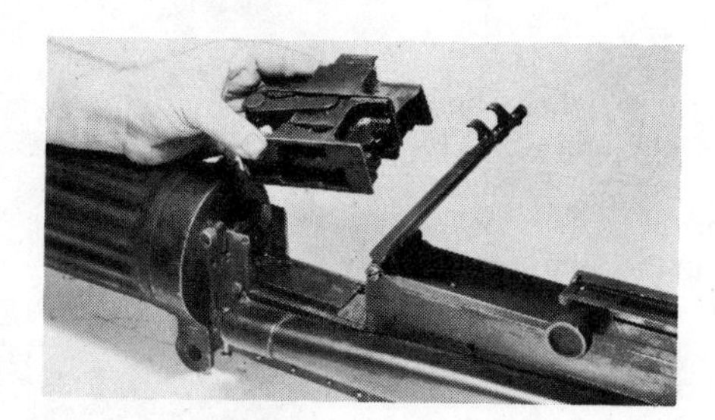

Now lift complete feed block directly up and out of the gun. Go to the forward end of the gun. Pull out the split pin and twist the outer casing through about one-sixth of a turn. It can now be pulled off to the front. The muzzle cup and the front cone may also be unscrewed and removed. (The gland and packing

should be removed only if absolutely necessary.)

Grasping the front end of the spring box with the left hand, push forward on the rear end with the right hand until the hooks which fasten the box at front end and rear can be sprung out of their studs. Disconnect the box from the gun and unhook the fusee spring

from the fusee. The fusee may now be turned until its lugs are free, and then it can be withdrawn with its chain from the left.

Now lift the rear cover and unscrew the large key pin protruding from the left side at the rear of the gun. Pull this pin out to the left, and it will permit the handles and their enclosed mechanism to be swung down to a horizontal position. Slides that travel in the body at the rear may now be pulled straight out. The right slide carries the roller with it. Now pull the crank handle stem directly to the rear, which will withdraw the crank together with the right and left inside plates and the barrel. Disconnect the right and left side plates from the crank and the barrel. This completes field stripping.

THE BESA TANK MACHINE GUNS

The Besa guns were developed by B.S.A. from the Czech ZB53 (Model 37) machine gun and were used by the UK for tank armament. In 1936, B.S.A. signed an agreement with Zbrojovka Brno allowing them to manufacture the 7.92 ZB53. In April 1938, the War Office placed its first order for the gun, and in 1939 production commenced. B.S.A. soon discovered, however, that considerable modification would have to be made to the gun if it was to be capable of mass production; the modified gun was called the Besa. B.S.A. made 59,322 7.92 mm Besa guns during World War II.

The Besa was produced in four different models: Mark 1, Mark 2, Mark 3 and Mark 3*. These weapons differ in minor details, but principally in that the Mark 1 and Mark 2 have two rates of automatic fire, which can be selected by moving the selector lever at the left rear of the receiver ("L" is for low rate and "H" is for high rate), and the Mark 3 and 3* guns have only one rate of fire.

The Besa guns have one unusual feature; although they are gas operated they have a recoiling barrel. The cartridge is chambered and fired before the recoiling barrel is completely in the battery position (fully forward). The recoil of the firing cartridge, therefore, must overcome the inertia of the forward-moving barrel. This action helps in buffing the bolt and reduces the shock on the weapon and the mounting when firing.

The Besa is no longer a standard weapon in Britain.

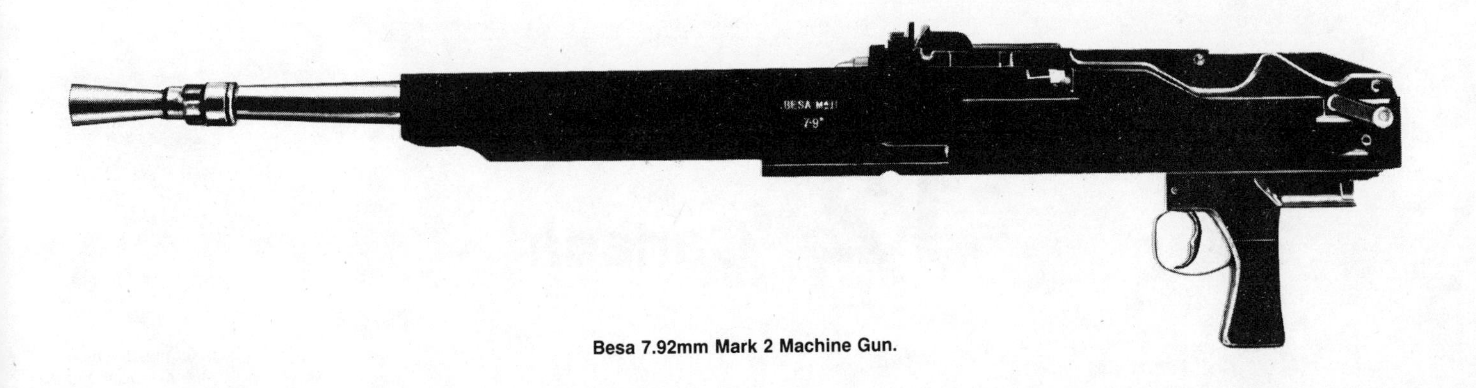

Besa 7.92mm Mark 2 Machine Gun.

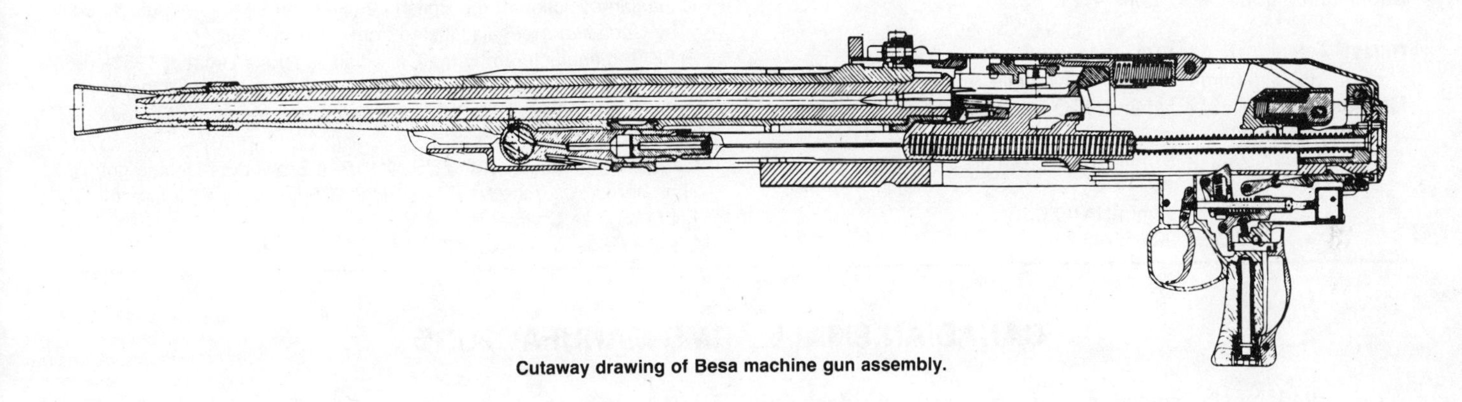

Cutaway drawing of Besa machine gun assembly.

CHARACTERISTICS OF THE BESA TANK MACHINE GUN

Caliber: 7.92mm.
System of operation: Gas, automatic only.
Weight: Mark 1 47 lbs.
Mark 2 48 lbs.
Mark 3 54 lbs.
Mark 3* 53.5 lbs.
Overall length: 43.5 in.
Barrel length: 29 in.
Feed device: 225-round link or metal and (canvas) belt.
Sights: None fitted to gun, telescopic sights used on vehicles.
Muzzle velocity: 2,700 f.p.s.
Cyclic rate: MK 1 MK 2 - 450-750 r.p.m.
MK 3 MK 3* - 450 r.p.m.

12 Canada

SMALL ARMS IN SERVICE

Type	Weapon	Notes
Handguns:	9 × 19mm FN Browning HP	Made by the John Inglis Company during World War II.
Submachine guns:	9 × 19mm C1	Canadian evolution of the British L2 (Sterling) series. Manufactured by Canadian Arsenals Limited from 1958 to 1965.
Rifles:	7.62 × 51mm NATO C1A1	FN FAL manufactured by the Canadian Arsenals Limited, 1957–1968.
	7.62 × 51mm NATO C3 sniping rifle	British-made Parker-Hale Model 1200Tx. Adopted mid-1970s.
Machine guns:	7.62 × 51mm NATO C2A1	Canadian version of the FN heavy barrel FAL manufactured by Canadian Arsenals Limited.
	7.62 × 51mm NATO C1 and C5	Canadian adaptation of the US M1919A4 Browning machine gun chambered to fire the NATO cartridge. C5 is simply an improved C1.
	7.62 × 51mm NATO C6	FN MAG (see Chapter 9).
	12.7 × 99mm M2 HB	US M2 HB (see Chapter 48).

CANADIAN SMALL ARMS MANUFACTURE

Until 1946, most Canadian service arms were manufactured by Small Arms Limited, located at Long Branch, Mississauga, Ontario. On 31 December 1945, Small Arms Limited was taken over by the Canadian government and became the Small Arms Division of Canadian Arsenals Limited. This government-owned Crown Corporation made all of the C1/C1A1 series rifles and C1 submachine guns used by Canadian Forces. In 1976 the Canadian government decided to close the Canadian Arsenals because the factory had not been operating in a cost effective manner. The closing of CAL left the Canadian forces without a reliable source of spare parts and maintenance for in-service small arms or a source of production for future small arms. In March 1976 the Canadian government entered into a ten-year contract with Dimaco, Inc., a private sector company in Kitchener, Ontario, for the manufacture of spare parts, the overhaul of in-service weapons, and the manufacture of the next generation rifle. The latter will be chosen in the fall of 1983 after competitive trials. Production of a new rifle is scheduled for 1985. Dimaco is currently preparing a modern production facility for the production of a new rifle.

CANADIAN PISTOLS

Canada used the British service pistol, the Webley .455 Pistol No. 1, Mark VI to some extent in World War I. However, they extensively used Colt and Smith & Wesson revolvers and Colt automatics. Canada was the first member of the British Commonwealth to adopt and produce a truly modern automatic pistol. The Canadian-made 9mm Parabellum Browning Hi-Power pistol is standard in Canada and in the United Kingdom. This weapon was supplied to the UK during World War II to arm Commandos and paratroop divisions. The pistol was originally put into production for the Chinese Nationalist Army.

THE CANADIAN HI-POWER BROWNING PISTOL

The John Inglis Company of Toronto, Ontario, produced the 9mm Parabellum Browning Hi-Power pistols in several models for the Canadian and Chinese Nationalist Governments during World War II*

*For additional details see *Handguns of the World.*

Pistol, Browning, FN 9mm, HP, No. 1 Mark 1

The butt is machined for a shoulder-stock holster, and a tangent leaf rear sight graduated from 50 to 500 meters is fitted.

Pistol, Browning, FN, 9mm, HP, No. 1 Mark 1*

Also machined for a shoulder-stock holster, but the height of the ejector has been increased, and the tangent-type rear sight (similar to the No. 1, Mark 1) has been machined to accommodate the increased height of the Mark 2 ejector. The extractor also differs and cannot be interchanged with the Mark 1 extractor.

Pistol, Bowning, FN, 9mm, HP, No. 2 Mark 1

Not machined for shoulder stock holster, it has a smaller ejector and a notched, fixed rear sight. Uses Mark 1 extractor and Mark 1 ejector.

Canadian Browning FN 9mm HP No. 2 Mark 1.

Right side of the No. 2 Mark 1.

Canadian lightweight version of the 9mm Browning FN Hi-Power pistol.

Canadian caliber .45 NAACO "Brigadier" pistol.

Pistol, Browning, FN, 9mm, HP, No. 2 Mark 1*

Same as No. 2 Mark 1, except for ejector and extractor. Uses extractor Mark 2 and ejector Mark 2 for which slide clearance is machined.

There are two models—Mark 1 and 2—of hammer and link for these pistols; however, these are interchangeable. The wooden shoulder-stock holster is no longer in common use; these pistols are used now as "one-hand" weapons.

The characteristics of the Canadian Browning FN HP pistol are basically the same as those of this pistol as produced in Belgium and given in the chapter on Belgium.

POST-WAR CANADIAN PISTOLS

After World War II, a number of pistol tests were conducted in Canada, the UK and US. Among the competitors was a lightweight version of the Canadian Hi-Power.

This pistol has lightening cuts on both sides of the slide. The pistol tests turned out to be of little consequence since economics and logistics—the quantity of weapons and ammunition on hand and tooled for—dictated that pistols in hand would be used.

The Brigadier .45 Pistol

NAACO—the North American Arms Corporation of Canada—developed a caliber .45 pistol called the "Brigadier." This pistol is a modified Browning Hi-Power chambered for a new caliber .45 cartridge of considerably more power than the .45 automatic cartridge.

It should be noted that the "Brigadier" has its safety catch on the slide where it blocks the firing pin rather than on the receiver as does the Browning (and the Colt.)

CANADIAN RIFLES

7.62mm AUTOMATIC RIFLES FN C1 and C1A1

Canada was among the first to adopt the FN FAL rifle and was the first to mass produce the weapon. Prior to the adoption of the production model (the C1), Canada, as well as the UK, tested a number of experimental models. The EX 1 model was quite similar to the British-adopted L1A1 in that it could not be fed with chargers and had a rear sight with a fixed-size aperture.

The EX 2 could not be fed with chargers either but had an optical sight similar to that of the British E.M. 2 rifle. Both rifles had barrels without flash suppressors.

Similar rifles were tested by the UK as rifle 7.62mm, FN BR X8E 1, Type A (iron sight type), and rifle 7.62mm, FN BR, X8E2, Type B (optical sight type). Canada made additional modifications and adopted the rifle 7.62mm FN (C1) in June 1955.

The obvious external differences between the C1 and most other versions of the FN FAL family are the rear sight and the charger guide, which allows feeding of the magazine when in the rifle with five-round chargers.

The sight of the C1 is similar to that found on some sporting rifles. A disc containing five differently sized apertures is held in a frame. The edge of the disc is serrated for ease of turning; a flick of the disc turns up the range aperture desired from 200 to 600 yards. The range for which the aperture is set is indicated by numbers from 2 to 6, which are visible in the lower part of the sight. The sight can be folded when not in use.

CHARACTERISTICS OF C1 AND C1A1 RIFLES

Caliber: 7.62mm NATO.
System of operation: Gas, semiautomatic only.
Overall length: 44.75 in.
Barrel length: 21 in.
Weight: 9.4 lbs.
Feed device: 20-round, detachable, staggered row. Box, can be fed with 5-round chargers.
Sights: Front: Protected post.
Rear: Revolving disc with apertures.
Muzzle velocity: 2,750 f.p.s.

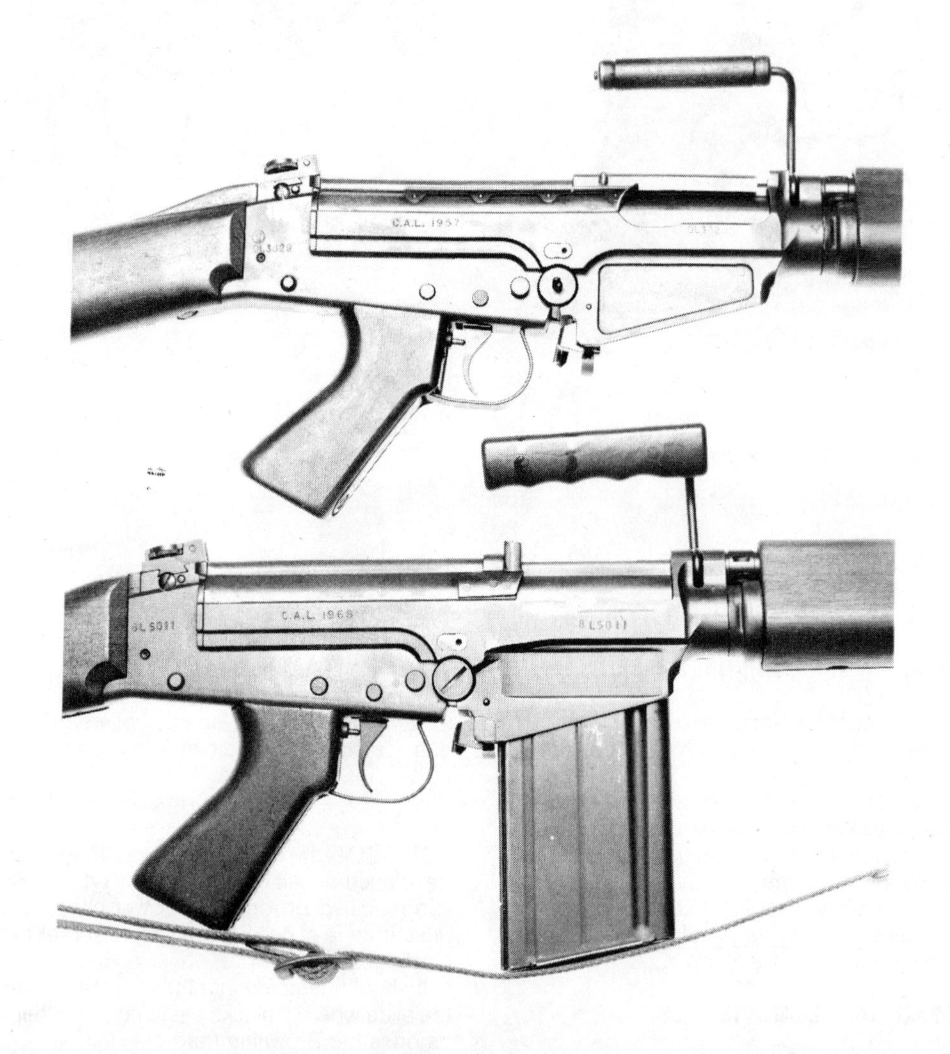

The Canadian FAL. ***Top,*** **the C1 rifle as made by Canadian Arsenals Limited in 1957 (serial number 0L3629);** ***bottom,*** **the C1A1 rifle as made by Canadian Arsenals Limited in 1968 (serial number 8L5011). Note the different carrying handles and the addition of the stripper clip guide on the C1A1. Both rifles have the unusual Canadian rear sight.**

Receiver detail of 7.62mm Rifle FN C1.

Rear sight of 7.62mm FN C1 Rifle.

The current issue C1A1 rifle with Sniper Scope C1 (made by Leitz of Canada). This rifle is equipped with the "XL" long buttstock and an experimental padded sling. (Photo © by Collector Grade Publications.)

7.62mm Rifle FN C1A1.

Modification of C1—the C1A1

The C1 was modified slightly around 1959; the modification is called: rifle 7.62mm FN C1A1. The principal modifications were in the firing pin, which was altered to two-piece configuration, and a new plastic carrying handle, which replaced the wooden type.

Presumably C1 rifles in service have been converted to C1A1 rifles. Both C1 and C1A1 have prong-type flash suppressors fitted to the muzzle of the barrel.

7.62mm AUTOMATIC RIFLE FN C2

CHARACTERISTICS OF 7.62MM AUTOMATIC RIFLE FN C2

Caliber: 7.62mm NATO.
Weight loaded: (30-round magazine); 15.25 lbs.
Overall length: (normal butt): 44.75 in.
Barrel length: 21 in.
Feed device: 20 or 30 round box magazine.
Sights: Front: Protected post.
Rear: Tangent w/aperture graduated from 200–1000 yds.
Muzzle velocity: 2800 f.p.s.
Cyclic rate of fire: 675–750 r.p.m.

The C2 is the heavy-barreled, selective-fire version of the Canadian semiautomatic FN rifle C1. The method of operation and field strip of the C2 are similar to that of the C1. C2 like C1 has a prong-type flash suppressor and can be fitted with a bayonet or grenade launcher. The bipod legs are fitted with wooden strips, allowing the bipod to be used as a fore-end when in the folded position. Differences between the C2 and C1 rifles other than noted above are as follows:

a. The rear sight of C2 is different.
b. C2 has a bipod.
c. C2 has no handguard.
d. C2 change lever has 3 positions: safe, semiautomatic and automatic fire.
e. C2 gas block assembly includes a mounting for the bipod.

This weapon, the Canadian squad automatic weapon, has the advantage of using most of the components of the basic rifle including magainzes.

C2 is, like C1, made at the Canadian government small arms plant at Long Branch, Ontario.

7.62mm AUTOMATIC RIFLE FN C2A1

About 1960, C2A1—a modification of C2—was adopted. C2A1, like C1A1, has a two-piece firing pin and a plastic carrying handle. C2A1 is about one-quarter pound lighter than C2. As with all of the Canadian FN rifles, C2A1 can be fitted with any of three lengths of buttstock; normal, long, and short.

7.62mm Rifle C2 with bipod fixed.

Receiver detail of FN C2 Rifle.

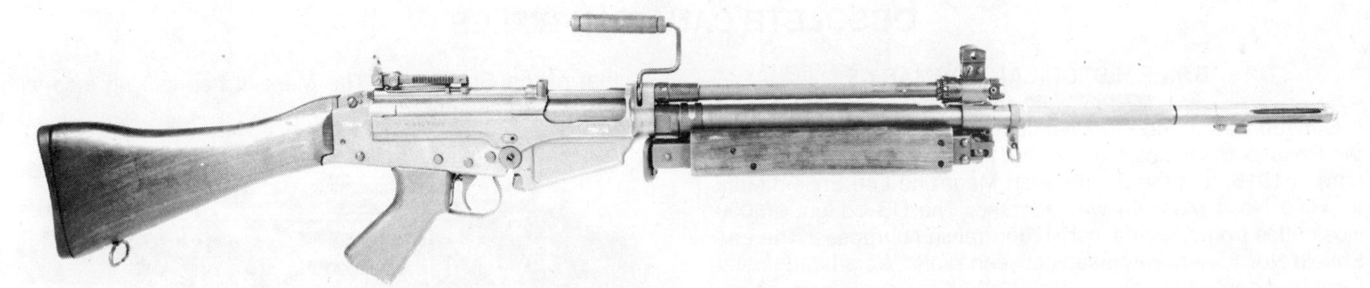

The Canadian C2 with the bipod folded. When folded, the bipod serves as a forward handguard.

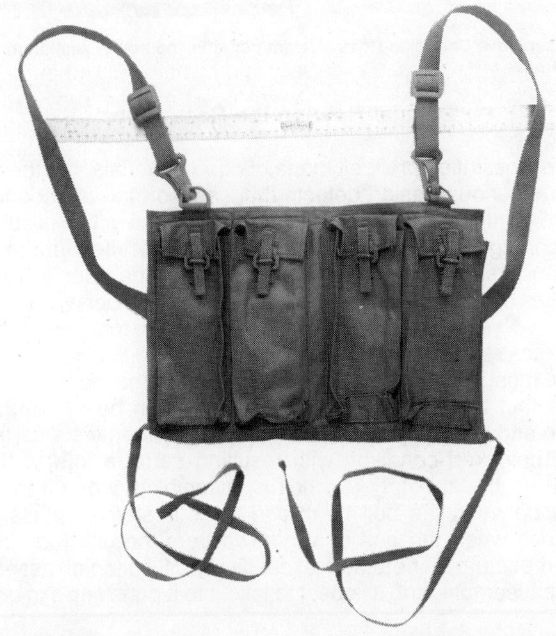

This "gunner's apron" made of waterproofed webbing and plastic fittings was designed to be worn across the chest. It holds four 30-shot magazines for the C2 rifle.

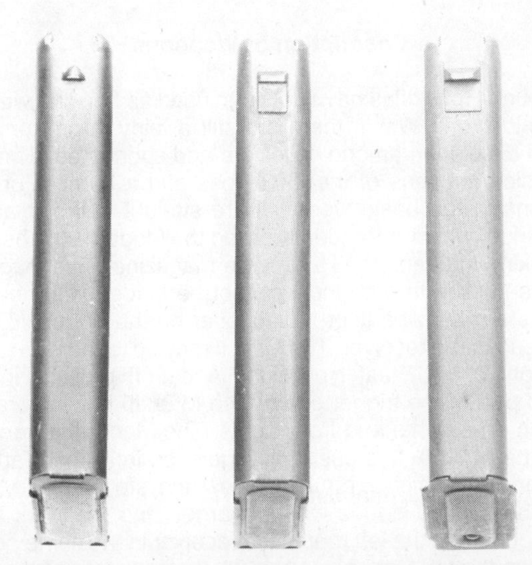

This photograph compares the FN FAL, US T48 (H&R FAL), and C1A1 (CAL) 20-shot magazines. Canadian and British magazines will not fit in FN FALs because of the design of the front lip.

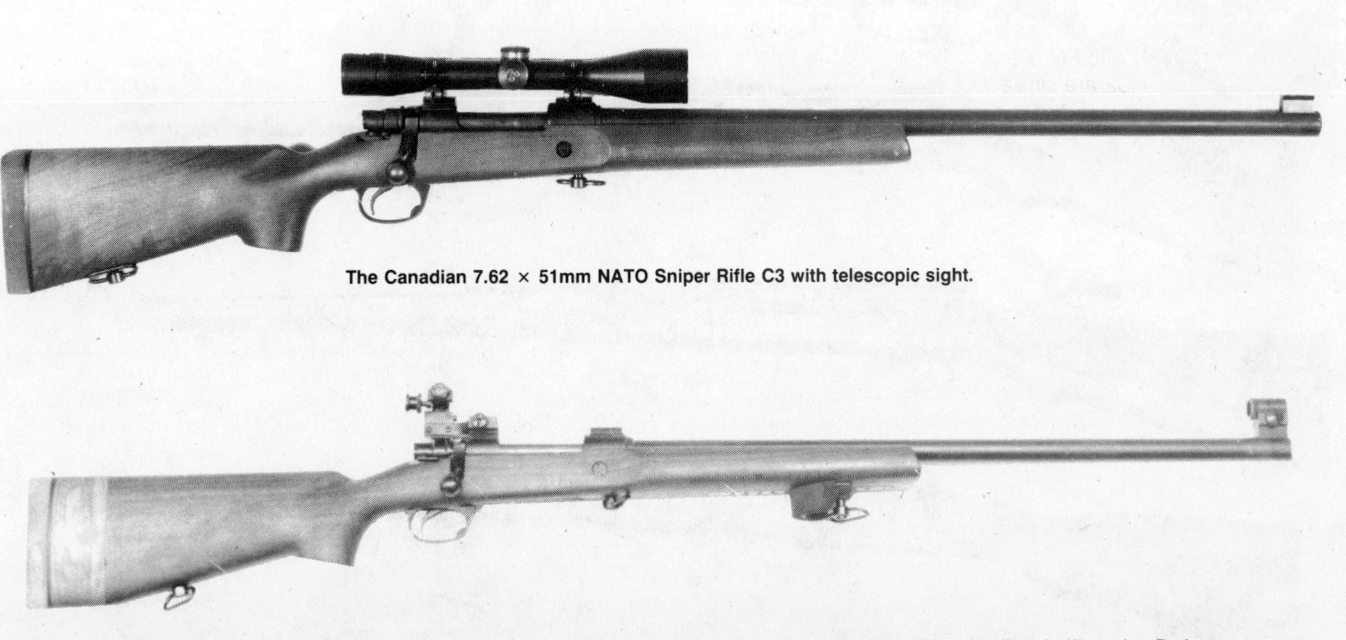

The Canadian 7.62 × 51mm NATO Sniper Rifle C3 with telescopic sight.

The C3 with iron sights. Note the use of spacers to lengthen the stock. These rifles are acquired from the British rifle maker Parker-Hale.

OBSOLETE CANADIAN RIFLES

BRIEF HISTORICAL SUMMARY

Canada had a rifle of native origin for part of World War I—the Ross in its various models. The Ross was dropped as standard in 1916, and the British Short Magazine Lee Enfield Mark III (Rifle No. I Mark III) was adopted. The US bought 20,000 Ross rifles from Canada in 1917 for training purposes. The Lee Enfield No. 1 rifles were used between World Wars I and II, and Canada adopted the No. 4 rifle at about the same time as the UK Long Branch tooled up to produce the No. 4 early in World War II and produced a total of 952,000, of which the greater part by far were No. 4 Mark 1*

ROSS RIFLES

Description of Weapons

Although Ross rifles have not been used as first-line weapons since mid-World War I, there are still a fairly large number of them in circulation among collectors and sportsmen. There are two basic variations of the .303 Ross and a number of minor variations to the basic types; all are straight pull bolt actions. The Mark II, which is frequently called the Model 1905, has solid bolt locking lugs and the Harris type magazine. This magazine, which is flush with the stock, cannot be loaded with chargers. There is a magazine lifter thumb lever on the right side of the rifle ahead of the receiver. The Mark II may be found with tangent type sight or with leaf type sight. A cut-off projects into the forward part of the trigger guard. The Mark III has several variations in rear sights and front sights. The Mark III is frequently called the M1910. It is easily distinguished from the Mark II by its magazine, which protrudes below the stock. The Mark III magazine can be loaded with a charger, and the Mark III cut-off is mounted at the left rear of the receiver in the same position as that on the US Springfield M1903. It also functions the same as that of the Springfield. The Mark III has locking lugs with interrupted screw thread.

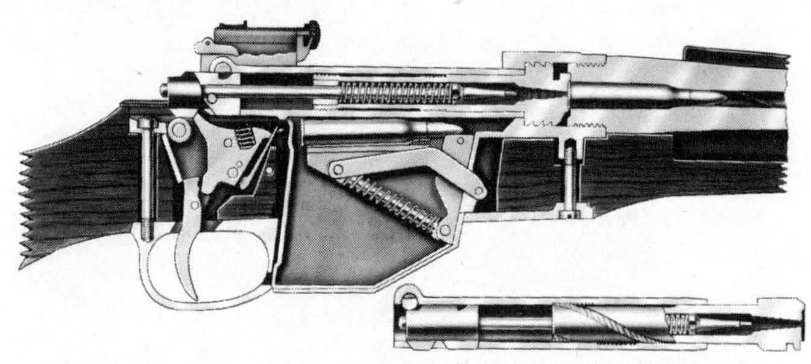

Representative Canadian Ross straight pull rifle and bolt in section to show operation.

Special Note on the Ross Rifles

The Ross rifles are well made of good materials, but they had several serious design defects that caused their abandonment as an infantry weapon by Canada in 1916. The action is suitable for sporting rifles but was found eminently unsuited for the mud of Flanders. In an attempt to make up for the poor extracting qualities of these weapons when dirty, the chamber was relieved slightly. This makes the Ross a rather poor rifle for a hand loader, since cases are badly stretched on firing.

The most serious problem with the Ross for the modern shooter is the fact that on most models the bolt can be reassembled wrong and yet be put in the weapon. It may then fire a cartridge in an unlocked condition with resulting **serious injury, if not death,** to the shooter. The bolt is assembled wrong if the distance between the bolt head and the bolt sleeve is less than one inch when the bolt is withdrawn from the rifle (bolt in unlocked posture). The bolt is exceedingly difficult to disassemble and reassemble, and it is best to take it to a gunsmith if in doubt.

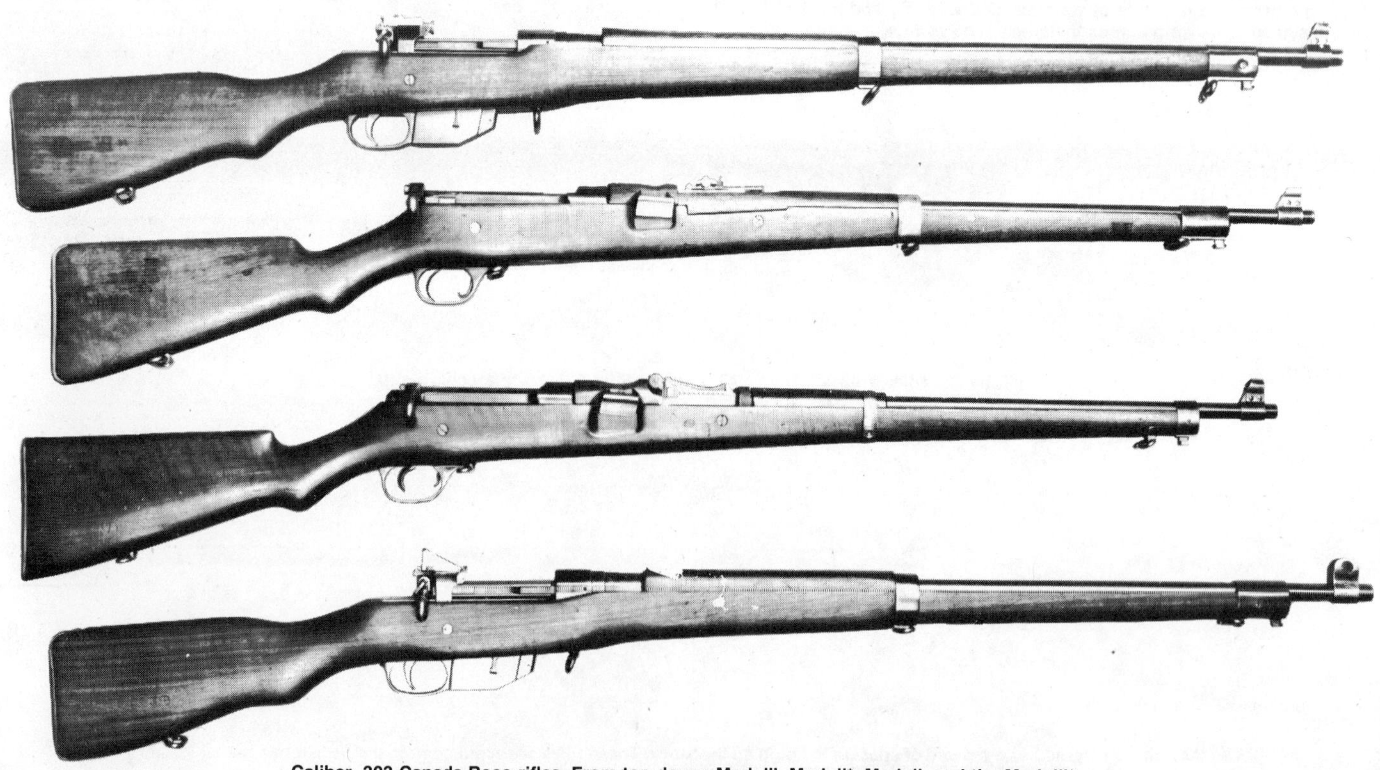

Caliber .303 Canada Ross rifles. From top down: Mark III, Mark II*, Mark II, and the Mark III*.

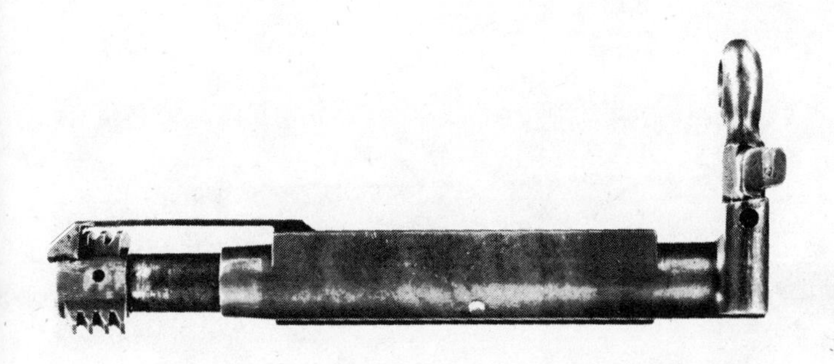

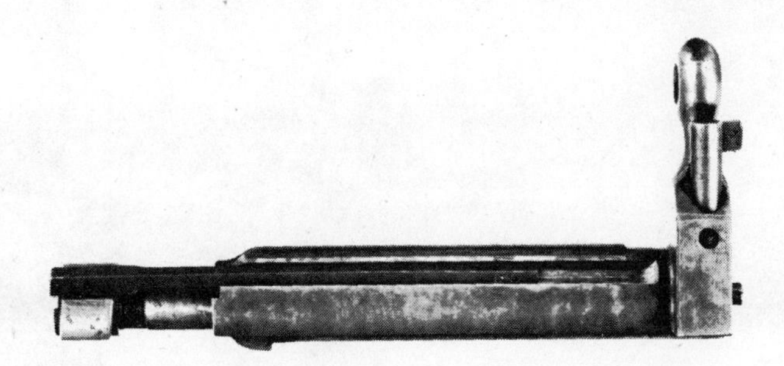

Bolt of the Ross M1910 Mark III Rifle at left and that of the Ross M1905 Mark I Rifle right.

CANADIAN SUBMACHINE GUNS

Canada has adopted the Sterling (Patchett) submachine gun as submachine C4. The Canadian Sterling differs in minor details from the L2A3.

CANADIAN MACHINE GUNS

Canada used the .303 Lewis and Vickers guns in World War II. In World War II, the John Inglis firm in Toronto manufactured large quantities of .303 Mark 1 (M) and Mark 2 Bren guns. In addition, Inglis made substantial quantities of 7.92mm Mark 2 Brens for the Nationalist Chinese. A few .30-06 Brens were made, apparently for experimental work.

The Canadian forces engaged in the Korean conflict were mainly supplied with US weapons in order to simplify the logistics problem. In addition, Canada had considerable stocks of caliber .30 Browning guns, mainly from vehicles left over from World War II. Canada ran tests of several 7.62mm NATO machine guns during the fifties and decided to convert existing stocks of caliber .30 Browning Model 1919A4 machine guns for 7.62mm NATO. The converted gun is the 7.62mm NATO C1 machine gun.

7.62mm NATO C1 Machine Gun

The C1 is supplied in fixed, for vehicular mounting, and flexible, for infantry, versions. In external appearance it differs very little from the US caliber .30 Browning M1919A4. In the ground role, it is used on a light tripod. The principal differences between the 1919A4 and the C1 are the barrel and the feed mechanism. The feed mechanism of the Browning was originally designed for a "pull out" type belt or metallic link. The feed mechanism of the C1 has been modified so that a "push through" type link—the US M13—can be used with a "pull out" type mechanism. The front barrel bearing plug has also been modified to give the action more gas boost in order to provide more power to extract cartridges from the links. A link guide has been installed on the feedway, and a short round stop has been fitted to the receiver in front of the feedway.

In addition to these and several other minor changes, two spring loaded ball bearings have been added to the T slot of the bolt. They prevent rounds from sliding completely through the T slot. The cartridge extractor has also been modified by removal of the ejector finger, spring and pin; the rear sight leaf has been engraved with new graduations in meters to match the ballistics of the 7.62mm NATO round.

The Canadian 7.62 × 51mm NATO C1 machine gun, a modified version of the US 7.62 × 63mm (.30-06) M1919A4.

The Canadian 7.62 × 51mm NATO C5 machine gun on a sled mount. This photograph was taken in February 1976 during the "Atlas Express" exercise.

13 The Republic of China (on Taiwan)

SMALL ARMS IN SERVICE

Type	Weapon	Notes
Handguns:	9 × 19mm Browning HP	John Inglis-made Browning Modèle 1935.
	11.43 × 23mm M1911A1	US-made and supplied M1911A1s.
Submachine guns:	11.43 × 23mm M1921	.45 M1921 Thompson made in China.
	11.43 × 23mm Type 36	.45 copy of the M3A1.
	9 × 19mm Type 37	9mm copy of the M3A1.
Rifles:	7.62 × 63mm M1	.30-06 M1 supplied by US.
	7.62 × 33mm M1	.30 M1 Carbine supplied by US.
	7.62 × 51mm NATO Type 57	7.62 mm NATO M14 made in Taiwan.
	5.56 × 45mm M16A1	5.56mm M16A1 supplied by US.
	5.56 × 45mm Type 68	Domestic design of 5.56mm rifle.
Machine guns:	7.62 × 51mm NATO Type 57	7.62mm NATO M60 supplied by US and manufactured in Taiwan.
	12.7 × 99mm M2 HB	.50 M2 HB supplied by US.

This version of the 5.56 × 45mm Type 68 rifle is marked T85 and has the serial number 82633. The Type 68 weighs 7 pounds (3.2 kg) and is 39 inches overall (990mm) with a 20-inch barrel (508mm).

The Chinese Nationalist Army usually used the term "Type" rather than "Model" for the nomenclature of weapons. The Chinese character for "Type" can be translated as "SHIH" or "SHIKI" and is the same as that used by the Japanese. Since a type designation may be followed by a model designation to indicate a modification (as with the old Japanese system, for example, the Type 34 Model 1), Chinese markings on weapons must be carefully examined to secure the proper nomenclature. On many (but not all) Chinese Nationalist weapons, the Type designation indicates the date of adoption of the weapon in number of years since the Chinese revolution—1911. Thus the Type 36 submachine gun was adopted in 1947; the Type 41 light machine gun was adopted in 1952. The People's Republic of China, however, uses the calendar year designation, i.e., 51 for 1951, etc.

Although for a period of time Nationalist China and the People's Liberation Army used the same types of small arms, this picture has changed considerably since the Korean War. During World War II, the Chinese Nationalists received large quantities of small arms from the United States. After World War II, the Chinese Nationalists started production of some US-type weapons on the mainland of China and also took over Japanese ordnance plants in the Manchurian area. The production of Japanese weapons was continued, as was the production of Japanese small arms ammunition. However, in an effort to standardize on ammunition, some of the Japanese weapons were rebarreled or made for 7.92mm cartridges.

Thus, at the time the Chinese Nationalists were forced to leave the mainland, they had weapons chambered for a multitude of different calibers. Since the Nationalist Army has been on Taiwan (Formosa), it has been able to standardize its weapons and simplify its supply problems. The Nationalist Army has used the US M1903 and M1 Rifles. They have manufactured a modification of the Bren in .30-06, and they are currently producing the US M14 Rifle and M60 Machine Gun on tooling provided by the US.

CHINESE PISTOLS

As with all other small arms, the Chinese had a wide collection of pistols. The Mauser in 7.63mm was a great favorite, and Chinese-made copies of the Mauser in cal. .45 have been encountered. During World War II, the United States supplied the Chinese with cal. .45 M1911A1 automatics, with Colt and Smith & Wesson M1917 revolvers, as well as some cal. .38 revolvers. The Canadians supplied the Chinese with the 9mm Browning Hi-Power pistol made by John Inglis in Toronto.

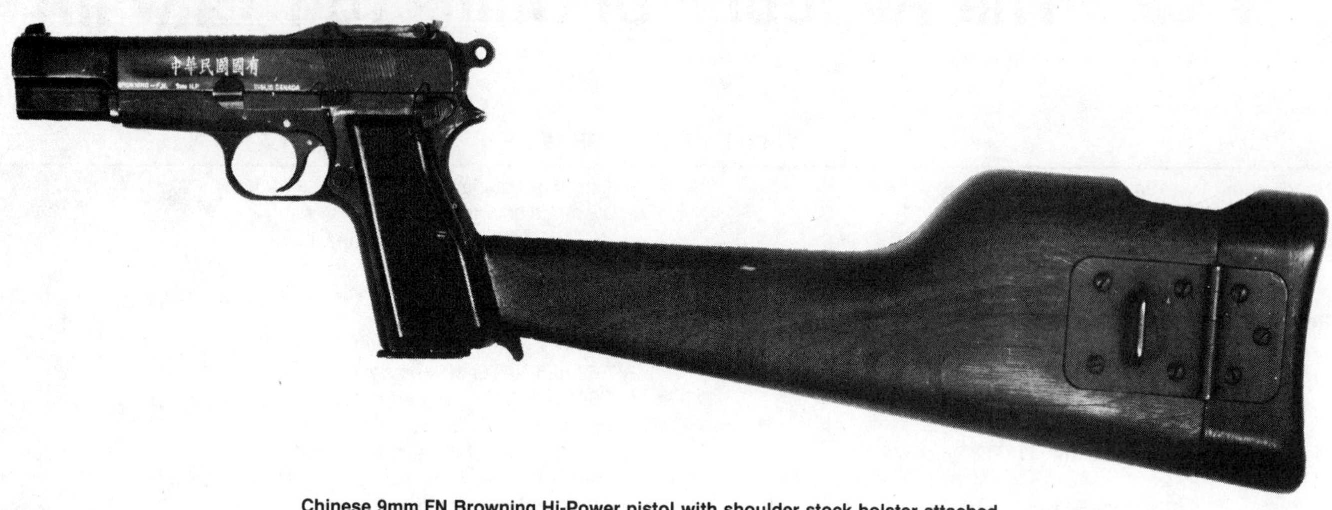

Chinese 9mm FN Browning Hi-Power pistol with shoulder stock holster attached.

CHINESE RIFLES

The pre-World War II Republic of China had many rifles in service. The oldest was probably the Type 88, or Hanyang rifle as it was called from its place of manufacture. This weapon is a copy of the German M1888 rifle, large quantities of which were sold to China by Germany after the Germans adopted the Model 98 rifle. This rifle is chambered for the old 7.92mm × 57mm rimless cartridge and has a .318 bore rather than the .323 bore of the 98 and later 7.92mm weapons. Therefore, this weapon should not be used with 7.92mm × 57mm IS (sometimes called JS) ammunition. The Chinese issued a special conical-nosed ball cartridge for this weapon. It should also be noted that Chinese weapons of pre-World War II manufacture are widely variable in quality of materials and construction. Chinese weapons manufactured since World War II are, so far as can be determined, made of first-class materials and show fine workmanship. Other Chinese rifles of this period are:

7.92mm Belgian FN M1924 and M1930 rifles, and Chinese copies.

7.92mm Czechoslovak Brno M1924 rifles, and Chinese copies.

The Mauser 7.92mm "Standard Model," a Mauser export model copied by the Chinese in 1935, is called the "Generalissimo" or "Chiang Kai Shek" model. This model is now called the Type 79 by the PRC. All these weapons were used by PRC troops in Korea.

The 7.62mm NATO Type 57 Rifle

The Chinese Nationalists manufacture a copy of the US M14 rifle called the Type 57 (1968). This rifle varies only in minor details from the US made rifle. The most noticeable difference is the shape of the stock immediately behind the receiver. The Chinese made rifle has a more pronounced flat surface immediately behind the rear of the receiver than does the US rifle.

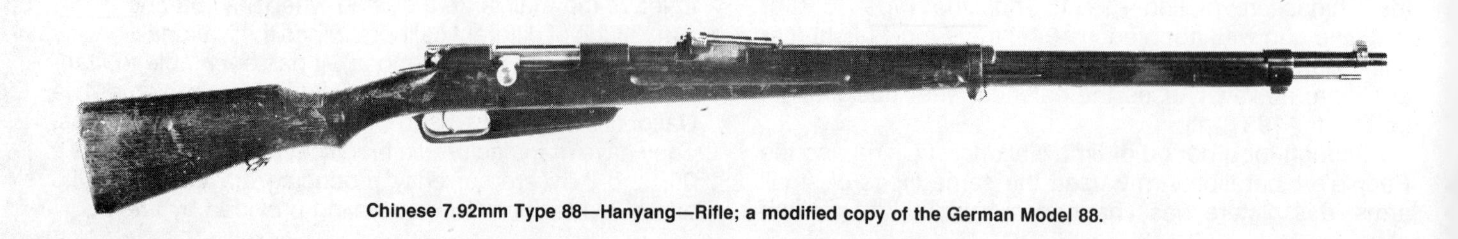

Chinese 7.92mm Type 88—Hanyang—Rifle; a modified copy of the German Model 88.

Below, the receiver markings read left to right: "Rlfle Type 57". Second line, "Made in Republic of China." Third line carries trade mark of 60th Arsenal. Then serial number beginning with 048666.

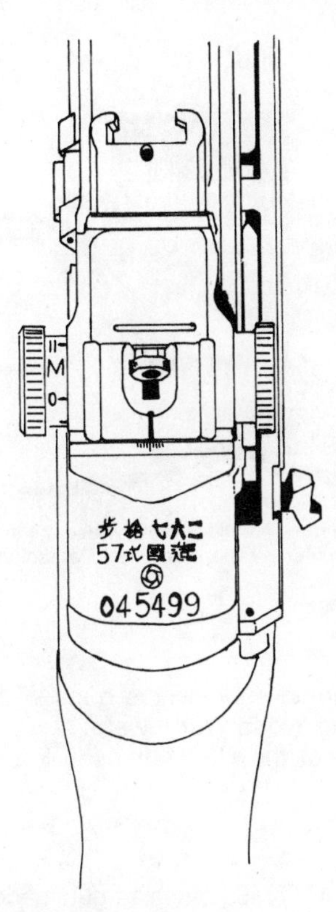

Above, markings read: "Rifle 7.62," first line. "Type 57 Made in China," second line. Trade mark of 60th Arsenal. Serial numbers 000001 to 048665.

The ROC copy of the US M14 rifle, the 7.62mm Type 57 rifle.

The Type 68 Taiwanese rifle is derived from the M16A1 rifle. The major differences visible in this disassembled view include the piston-tappet-type gas system, the modified handguards and buttstock, and alterations to the upper receiver.

CHINESE SUBMACHINE GUNS

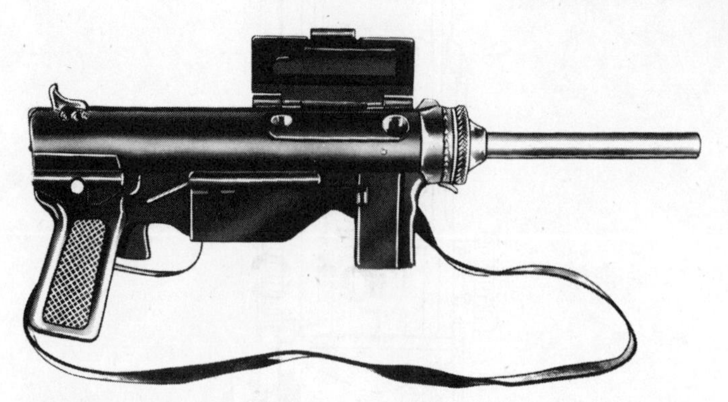

Type 36 submachine gun. A caliber .45 Chinese-made copy of the US M3A1 submachine gun. Similar weapons in 9mm Parabellum were made by the Chinese.

Caliber .45 submachine gun, based on design of Japanese Type 2, believed to be of Chinese manufacture.

The most common submachine guns in the Chinese Army prior to and during World War II were:

All the models of the US Thompson in cal. .45 (frequently called 11mm by the Chinese) plus a Chinese-made copy of the 1928 Thompson.

The US M3 and M3A1, and a Chinese copy of the M3A1 called the Type 36, in cal. .45, and Type 37 in 9mm Parabellum.

British Stens chambered for the 9mm Parabellum cartridge. The Chinese also made some copies of these.

The Thompson, the M3A1 and the Type 36 are still in use in the Chinese Nationalist Army.

CHINESE MACHINE GUNS

The 7.62mm NATO M60 machine gun made in Taiwan is now the standard machine gun of the Nationalist Chinese Army.

The Chinese used a wide variety of machine guns prior to and during World War II. Most of these weapons again appeared in Korea during the early stages of the Chinese Communist commitment. The most common weapons were:

The Chinese-made Type 24 heavy machine gun. This 7.92mm weapon is a modification of the German Maxim Model 08 (MG 08).

Czechoslovak- and Chinese-made 7.92mm ZB 26 and ZB30 light machine guns. These weapons were called Type 26 and Type 30 by the Chinese.

Swiss SIG-made KE 7 light machine guns in 7.92mm caliber.

Danish Madsen-made 7.92mm light machine guns.

Chinese-made 7.92mm Maxim aircooled machine guns.

French-made Hotchkiss M1914 heavy and Model II light machine guns in 7.92mm caliber.

Colt- and FN-made Browning automatic rifles, also chambered for the 7.92mm cartridge. There apparently were not too many of these.

Colt- and FN-made Browning watercooled machine guns in 7.92mm. These were also apparently in short supply.

Chinese-made Type 41 light machine gun. Bren-modified for US caliber .30.

7.92mm Bren guns Mk. 2, made in Canada for the Chinese during World War II.

The list given above is by no means complete. When one considers purchases, Lend-Lease, war booty, etc., the Chinese Army was a gun collector's paradise and an ammunition supply officer's nightmare. All the standard United States machine guns and most of the standard British machine guns were in wide use before the end of World War II.

14 The People's Republic of China (on Chinese mainland)

At the conclusion of World War II, the People's Republic inherited large quantities of Japanese weapons from the Soviets, who had taken them in Manchuria. During the Chinese civil war, they captured materiel from the Chinese Nationalists, including large quantities of US weapons. When the PRC entered the Korean conflict, they had far more Chinese, Japanese and US equipment in the battle area than Soviet materiel. By war's end they were well along the road to standardizing Soviet small arms for their troops.

During the past two decades, they have put a number of those Soviet weapons into production, using their own year of adoption model designations. For example, the PPSh-41 was adopted by the PRC as the Type 50. The M1944 Mosin Nagant carbine was put into production as the Type 50. The DPM light machine gun was called the Type 53. In recent years, the PRC has begun to manufacture a new series of domestically developed small arms. Among these weapons are the 7.62 × 39mm Type 68 rifle, the 7.62 × 54Rmm Type 67 light machine gun and the 7.62 × 25mm Type 64 submachine gun, a silenced weapon. As indicated in the table below, the PRC has developed export models of several of their small caliber weapons.

SERVICE WEAPONS OF THE PEOPLE'S LIBERATION ARMY

SMALL ARMS IN SERVICE

Handguns:	7.62 × 25mm Type 51 (Also called Type 54)	USSR TT 1933	M20 (export designation)
	9 × 18mm Type 59	USSR PM (Makarov)	
	7.65 × 17mm Type 64	Domestic silenced pistol	
	7.65 × 17mm Type 67	Domestic silenced pistol	
Submachine guns:	7.62 × 25mm Type 43	USSR PPS43	
	7.62 × 25mm Type 50	USSR PPSh41	
	7.65 × 25mm Type 64	Domestic silenced submachine gun	
Rifles:	7.62 × 54mm R Type 53 carbine	USSR M1944 carbine	
	7.62 × 39mm Type 56 carbine	USSR SKS45	M21 Export model
	7.62 × 39mm Type 56 assault rifle	USSR AK47	M22 Export model (A woodstock and folding stock version of the AKM is also manufactured in the PRC.)
	7.62 × 39mm Type 68	Domestic design	
	7.62 × 39mm Type 73	Improved Type 68	
	7.62 × 39mm Type?	Further variation of Type 73(?) by Tang Wenlie.	
Machine guns:	7.62 × 54mm R Type 53	USSR DPM	
	7.62 × 39mm Type 56	USSR RPD	M23 Export model
	7.62 × 54mm R Type 57	USSR SGM	
	7.62 × 54mm R Type 58	USSR RP46	
	7.62 × 54mm R Type 67	Domestic design	
	12.7 × 108mm Type 54	USSR DShK 38/46	M17 Export model

Many of the older model Soviet weapons covered later in this text are still in use, and many of these weapons, as well as some of the smaller newer weapons listed above, were shipped to Vietnam.

CHINESE HANDGUNS

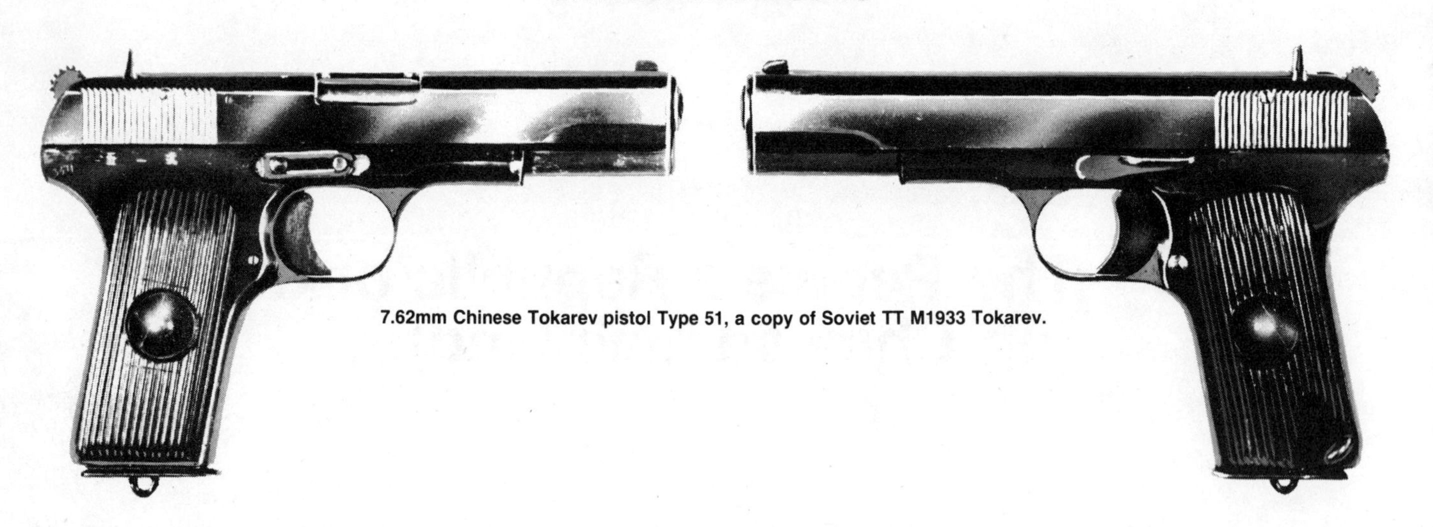

7.62mm Chinese Tokarev pistol Type 51, a copy of Soviet TT M1933 Tokarev.

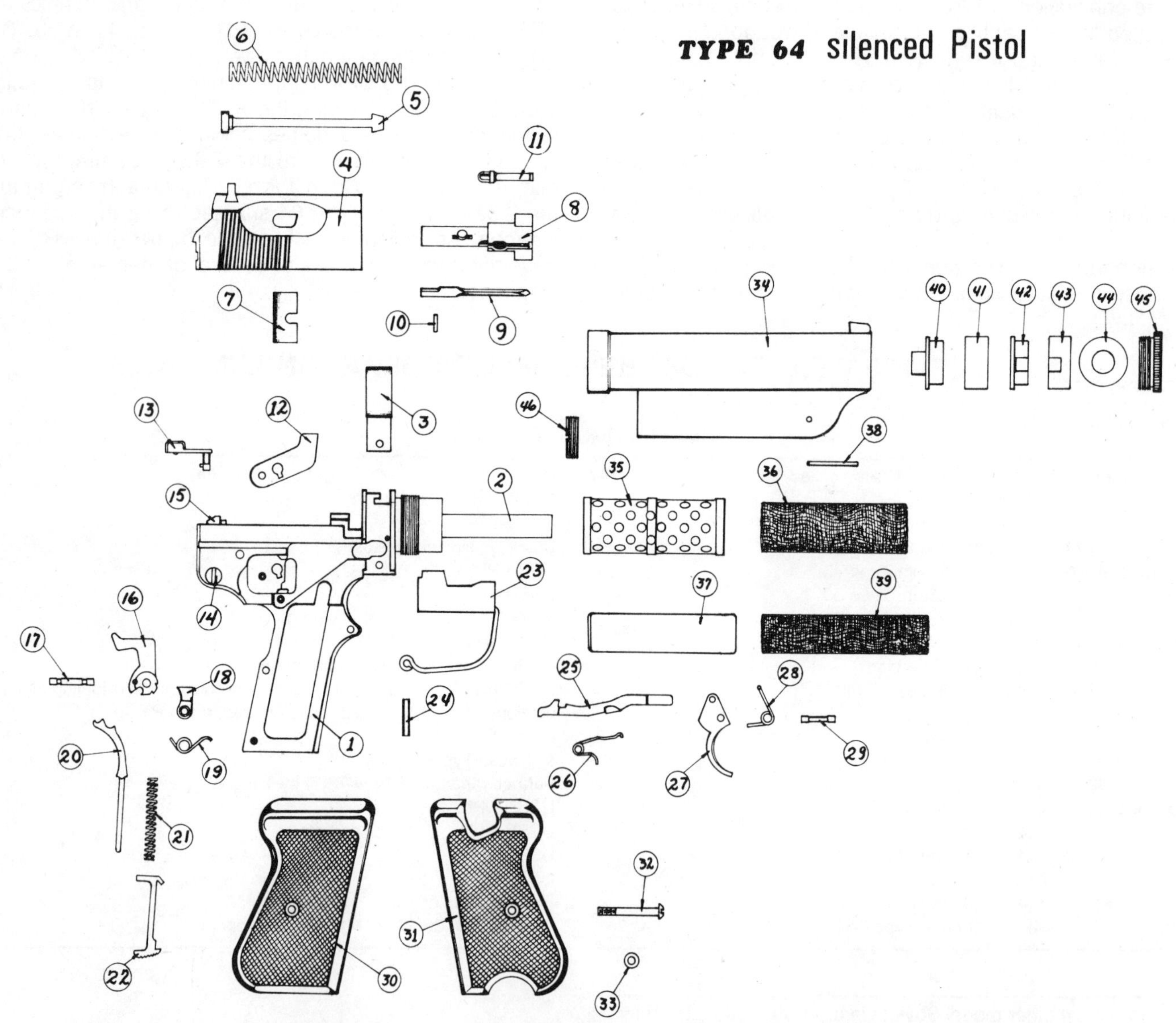

The PRC Type 64 7.65 × 17mm silenced pistol. This is an assassination weapon intended for clandestine operations. When nearly complete silencing is desired, the slide (4) can be locked shut, and the rotating bolt (8) will not unlock until the slide is hand operated. When the selector bar (7) is pushed to the right, the locking lugs of the bolt do not engage, and the pistol operates like a regular blowback pistol.

The Chinese Type 67 7.65 × 17mm silenced pistol. This example bears the serial number 202540 and was made by factory 66 in 1968. This is an evolution from the Type 64 silenced pistol. It is lighter and smaller, but it still has the basic features of the Type 64, including the slide lock feature.

COMPARATIVE CHARACTERISTICS OF PRC SILENCED PISTOLS

	Type 64	Type 67
Caliber (mm):	7.65 × 17	7.65 × 17
System of operation:	Blowback	Blowback
Overall length (ins.):	9.05	8.86
Barrel length (ins.):	4.80	3.31
Weight (lbs):	2.80	2.25
Feed device:	8-shot box magazine	8-shot box magazine

CHINESE RIFLES

CHINESE MODIFICATIONS OF SOVIET DESIGNED WEAPONS

The Chinese have modified several of the Soviet weapons that they manufacture. The standard Soviet 7.62mm AK assault rifle, manufactured in China as the Type 56, has a removable knife type bayonet. While they still use the Type 56, they also make a modification, which has a folding type, cruciform section bayonet attached to the under section of the front sight base. The Chinese are also making the 7.62mm Type 56 Carbine (copy of the Soviet SKS) with a folding bayonet.

The 7.62mm Type 53, which is a copy of the Soviet 7.62mm M1944 Mosin Nagant carbine, has appeared with a rifle grenade launcher. The launcher is of the removable type and has a clamp type lock that engages behind the front sight in a manner similar to the US M7 and M8 rifle grenade launchers. When the rifle grenade launcher is attached, the folding bayonet cannot be fixed.

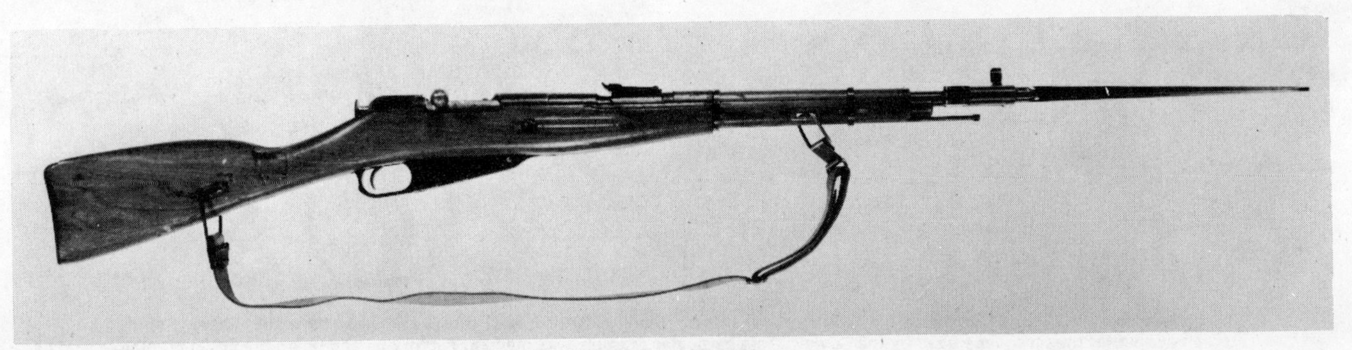

7.62mm Type 53 carbine.

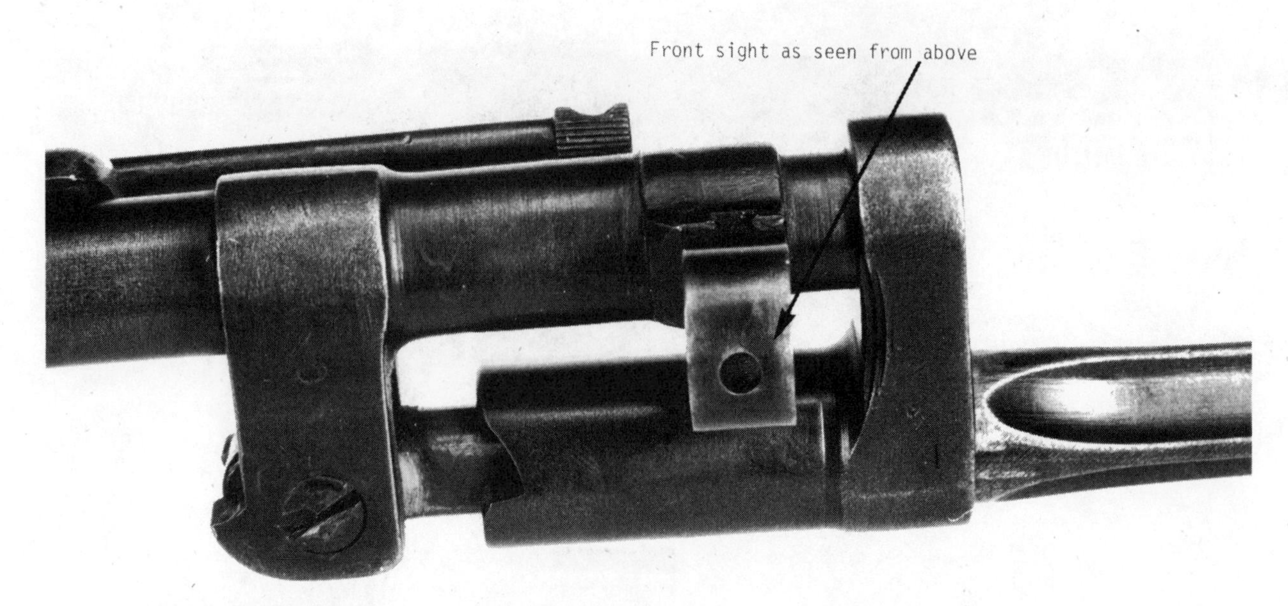

Top view of side mounted folding bayonet on Type 53 carbine copy of Soviet M1944 carbine.

Two views of a PRC Type 56 carbine. This is a copy of the Soviet Siminov SKS45. Note that the Chinese have substituted a spike type bayonet for the blade type used by the Soviets. This particular example was made at Factory 66 and bears the serial number 11343739.

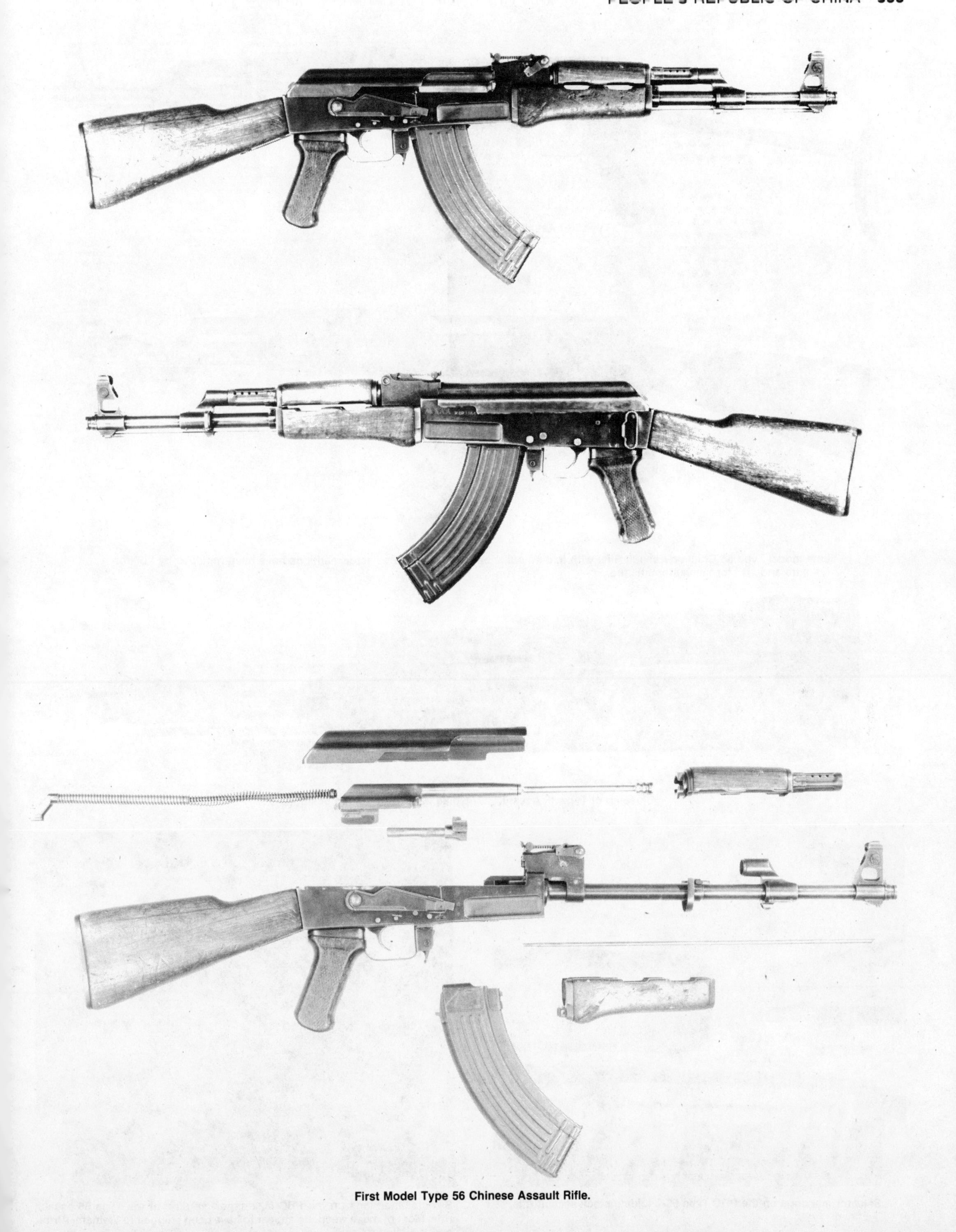

First Model Type 56 Chinese Assault Rifle.

Type 56-1 Assault Rifle.

Late model Type 56 Chinese Assault Rifle with folding spike bayonet. Chinese selector markings have been replaced by "L" for full auto and "D" for semiautomatic fire.

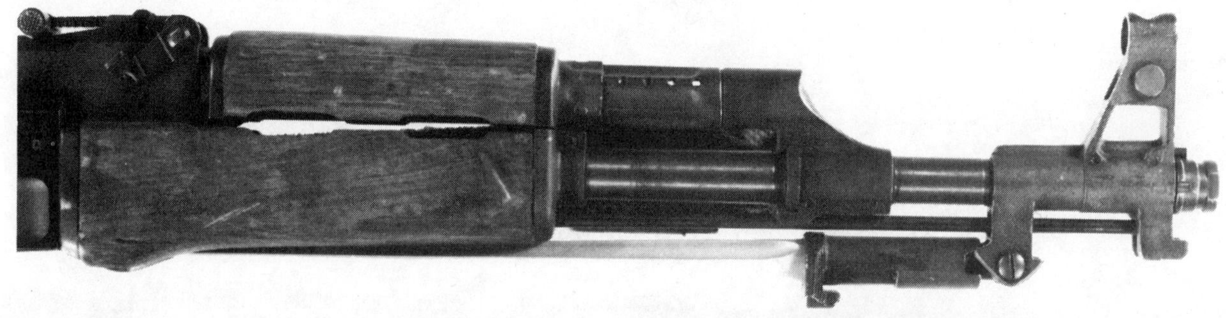

Close-up of Type 56 Assault Rifle spike bayonet in folded position.

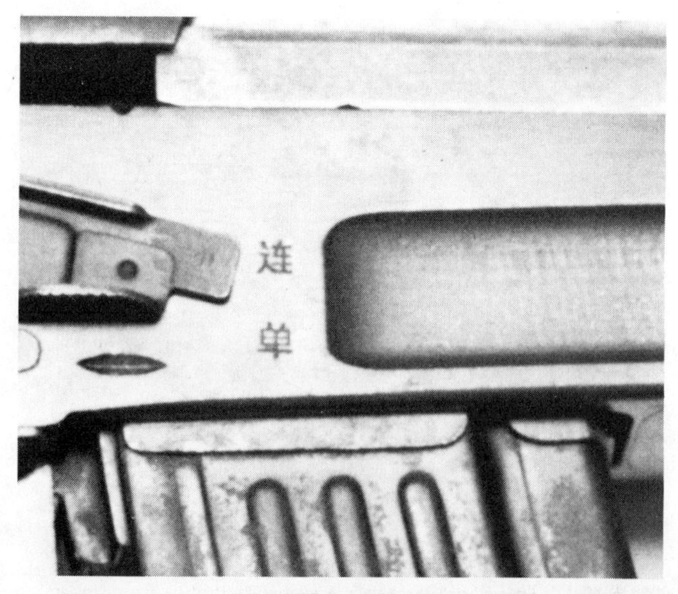

Selector markings on the PRC Type 56-1 folding stock assault rifle.

Selector markings on the PRC M22 export version of the Type 56 assault rifle. Most of these weapons appear to have been shipped to Vietnam, Africa, and the Middle East.

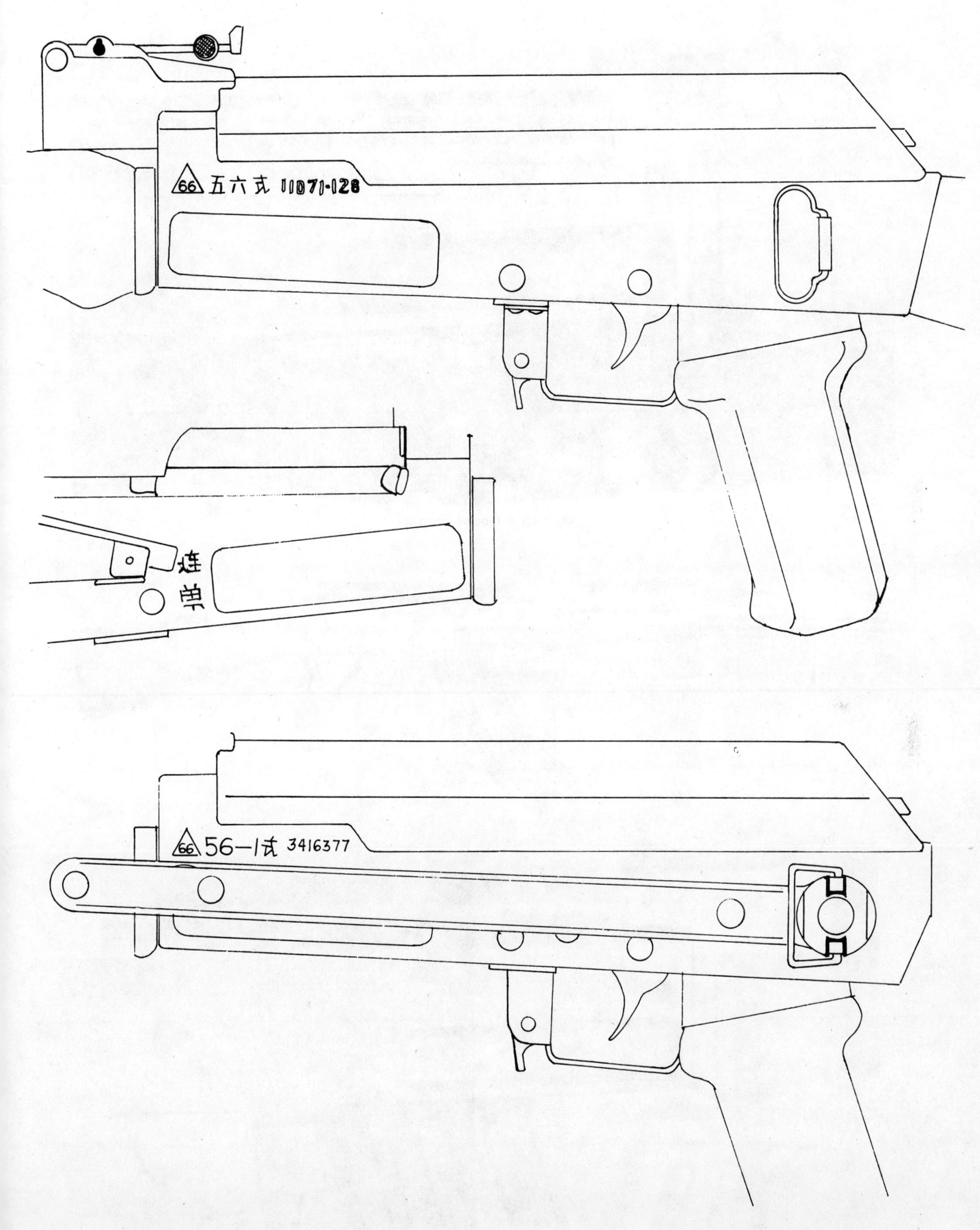

PRC Type 56 Rifle markings. Top, early type 56 with Chinese selector markings. Bottom, Type 56-1, which has L&D selector markings.

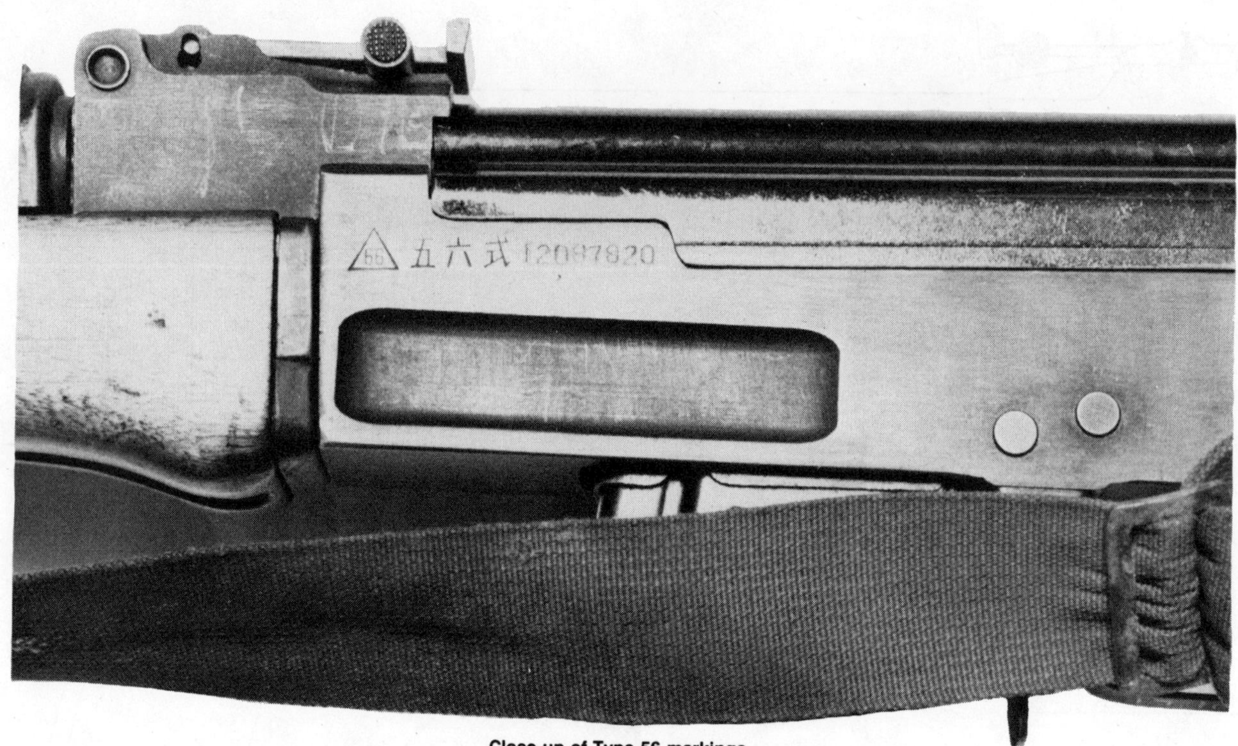

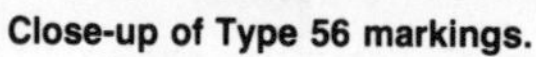
Close-up of Type 56 markings.

Two views of the M22 variant of the Type 56.

CHINESE 7.62mm TYPE 68 RIFLE

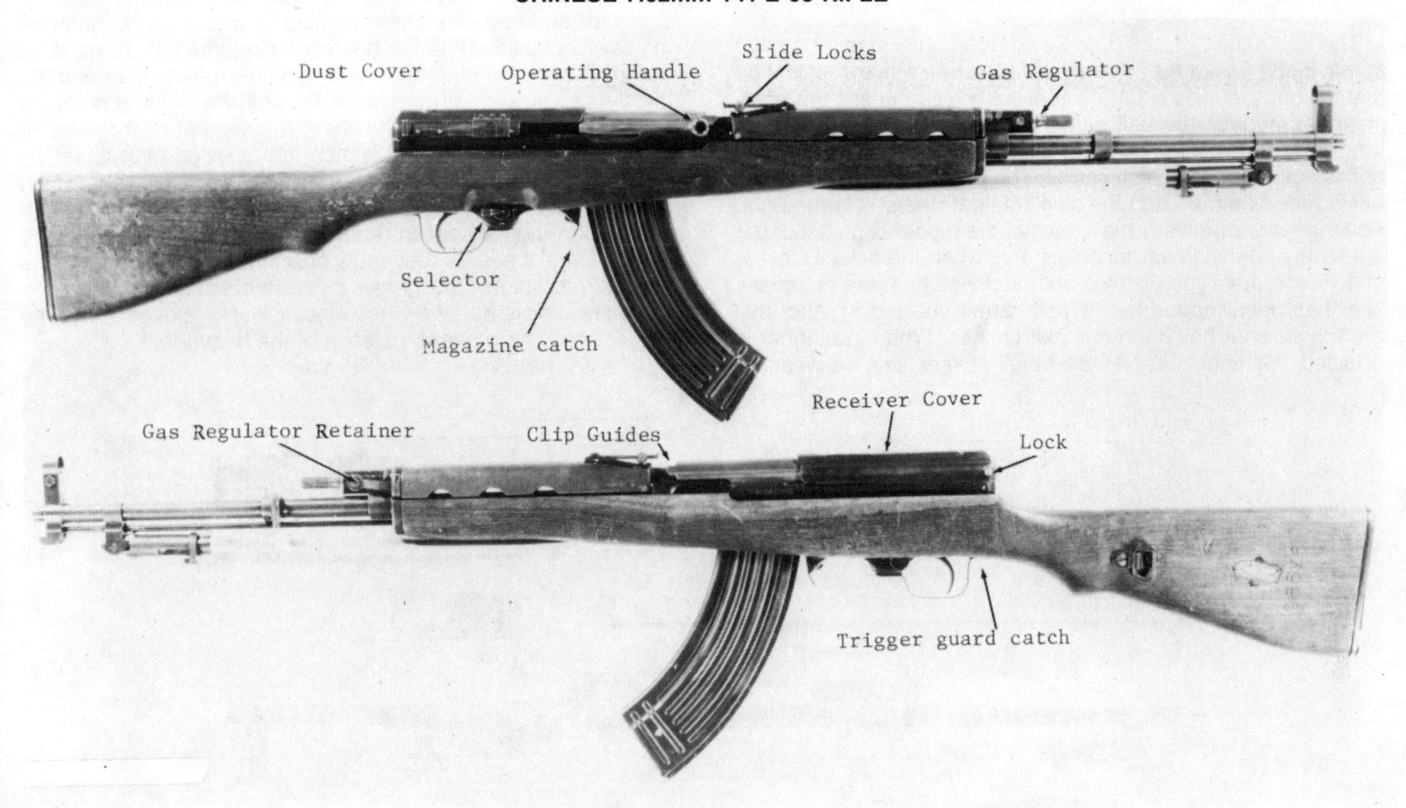

Right and left side views of Type 68 Rifle.

The Type 68 is a native Chinese designed selective fire weapon. Externally it resembles the Soviet SKS carbine modified to use a detachable box magazine. Internally it is quite different; the bolt and bolt carrier mechanism are basically the same as that of the AK, but the piston is a separate piece. The Type 68 is made in two different versions, one having a stamped steel receiver and the other having a machined steel receiver. Unlike the SKS or the AK, the Type 68 has a gas regulator whose retainer is mounted on the left of the gas cylinder beyond the hand guard. The amount of gas fed into the gas cylinder tube can be regulated by pressing the retainer in until it disengages from the hand guard and then rotating the retainer down until it can be pulled out of the gas cylinder. The gas regulator has indicator holes that correspond to the size of the gas ports in the regulator; the indicator hole that is closest to the barrel shows how much gas is being fed to the action.

CHARACTERISTICS OF THE TYPE 68 RIFLE

Caliber: 7.62mm (7.62 × 39)
System of operation: gas, selective fire
Weight: 7.7 lbs.
Length overall: 40.5 in.
Barrel length: 20.5 in.
Feed device: 15-round, staggered row, detachable box magazine
Sights: Front: hooded post
Rear: tangent with notch
Muzzle velocity: 2395 f.p.s.
Cyclic rate: 750 r.p.m.

How to Load and Fire the Type 68

The Type 68 rifle can be loaded either with SKS type chargers (stripper clips) or by removing the magazine and loading the cartridges into it individually. Pulling the operating handle to the rear with a loaded magazine in place will chamber a cartridge. The combination safety/selector lever is on the right side of the rifle in front of the trigger guard. The figure "O" setting is used to put the rifle on safe; the figure 1 setting is used for semi-automatic fire, and the figure 2 setting is used for automatic fire. The bolt remains open after the last round is fired.

How to Field Strip the Type 68

Clear the weapon, press in the lock at the left rear of the receiver and pull the cover rearward off the receiver. If the rifle has a machined receiver, pull the driving spring guide up and out of its seat in the rear of the receiver and remove the driving spring assembly. If the rifle has a stamped receiver, press in lock and push the driving spring and its guide forward until it is clear of its seat. Ease the spring forward; then remove it. Pull the operating handle rearward while keeping downward pressure on the bolt carrier. When the bolt carrier is about one inch from the rear of the receiver, it can be lifted up and out of the receiver. The bolt is removed from the carrier by pushing it to the rear, turning the bolt head and pulling it forward and out of the carrier.

The gas regulator has to be removed in order to remove the gas piston and its spring. It can be removed by turning its retainer down as explained earlier. Lower the muzzle and the gas cylinder, and its spring will slide out. The hand guard can then be slid forward off the gas cylinder. The heat shield, which may be found on the gas cylinder, can be slid to the rear and removed.

The trigger guard catch is at the rear of the trigger guard. Rotate it, and the trigger group is disengaged; the trigger guard can now be pulled free of the rifle. Swing the bayonet downward, and the stock can be disengaged from the barrel and receiver. Reassemble in reverse order.

How the Type 68 Works

The bolt action of the Type 68 operates in a manner similar to that of the Soviet AK assault rifle (Type 56), with the exception that the gas piston is separate from the bolt carrier and therefore impinges against the bolt carrier. The trigger mechanism has three sears. The automatic sear is located in front of the hammer and is actuated by the bolt carrier. The trigger sear is to the right behind the hammer, and the semiautomatic sear is to the right. The trigger sear always moves when the trigger is pressed, but the semiautomatic sear functions only when the selector is set at 1. Since the Type 68 fires from a closed bolt, the automatic sear has been tripped by the bolt carrier on loading, and only the trigger sear holds the hammer cocked. When the trigger is pressed, the trigger sear releases the hammer firing the weapon. As the bolt carrier moves to the rear, it pushes the hammer rearward and releases the automatic sear to hold the hammer in cocked position until the bolt carrier returns fully forward, at which time the full automatic sear releases the hammer and the rifle fires again and will continue to fire until the trigger is released or the magazine emptied. The above happens when the selector lever is set on the figure 2. When the selector lever is set on the figure 1, the selector releases the semiautomatic sear, which is then controlled by the trigger. When the trigger is pressed, it causes the trigger sear to release the hammer and allows the semiautomatic sear to move to a position where it can catch the hammer; therefore, the trigger must be pressed for each shot. If the selector is set at 0, the selector blocks movement of the trigger sear and prevents release of the hammer.

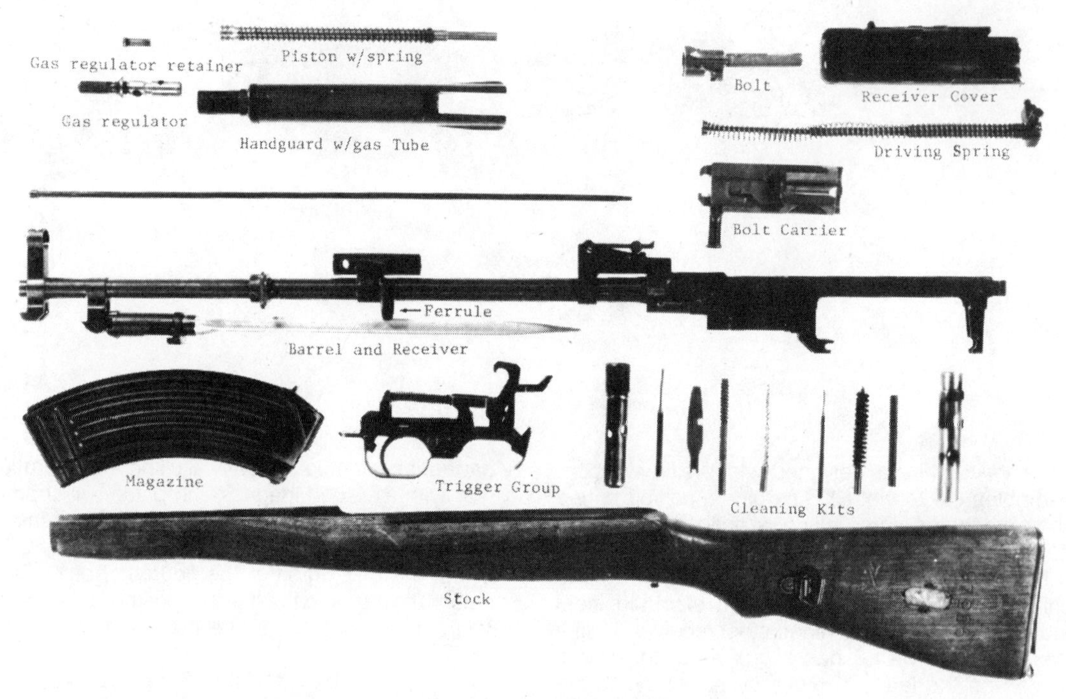

Disassembled view of Type 68 Rifle.

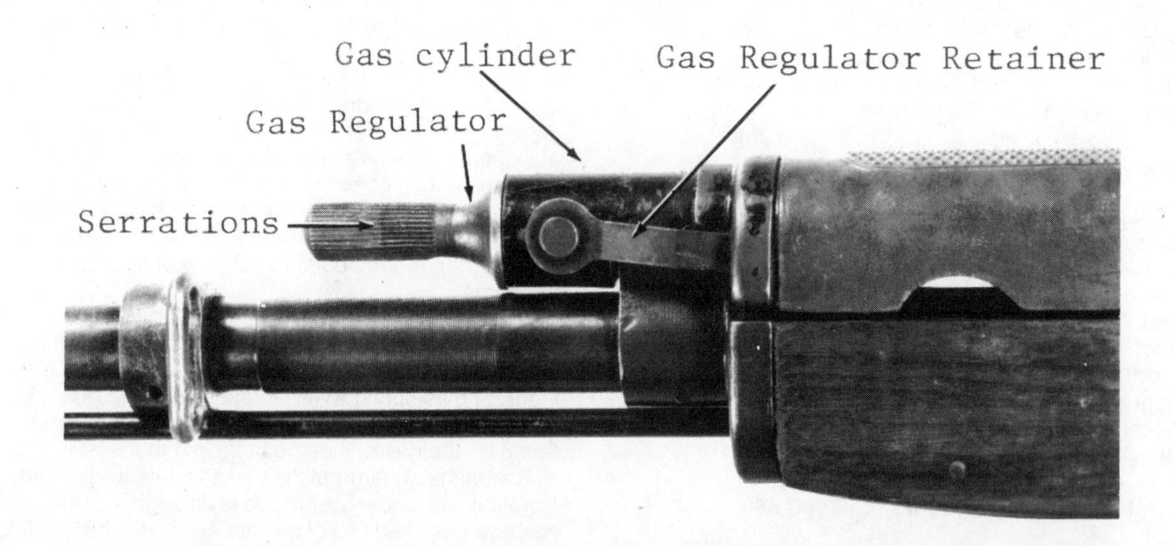

Detailed view of Type 68 Rifle gas regulator.

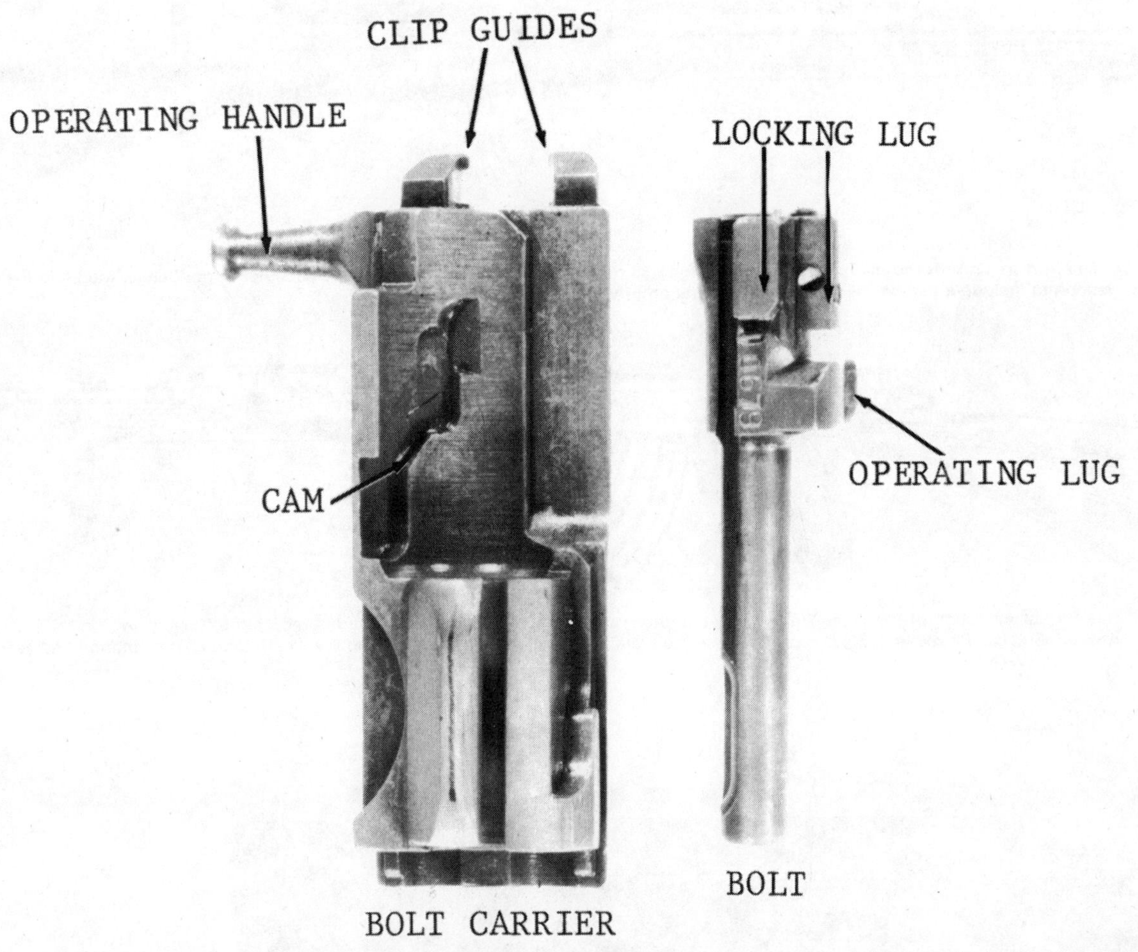

Type 68 Rifle bolt assembly.

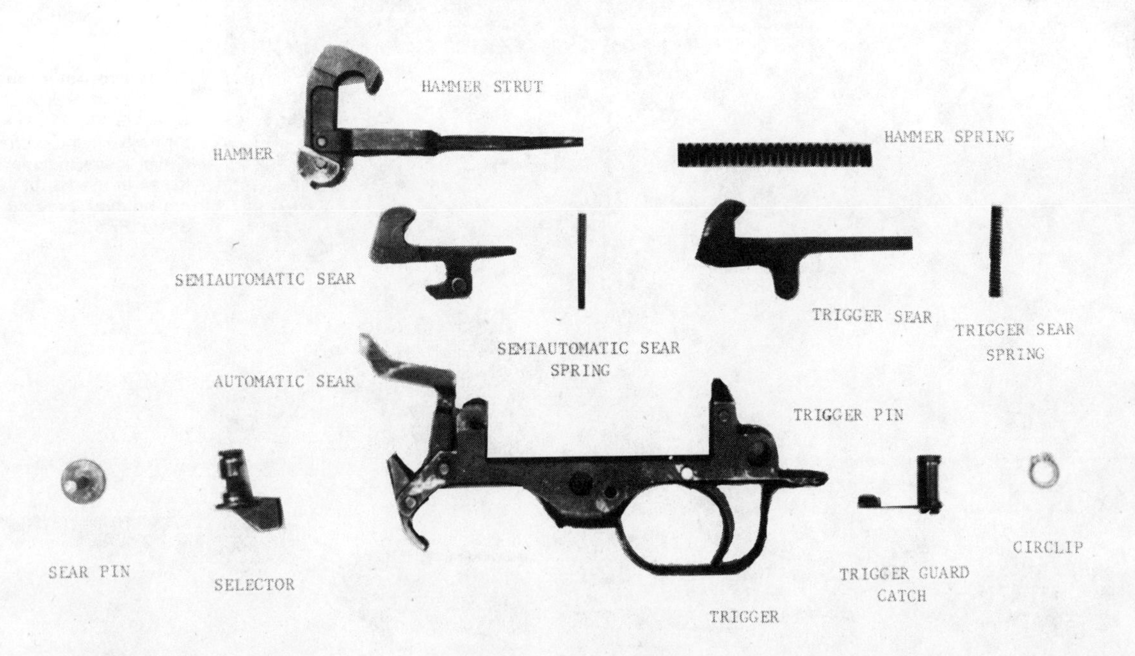

Type 68 Rifle trigger assembly.

Note: The type 68 Rifle has a bolt stop to hold the bolt open after firing the last cartridge in the magazine. If the bolt stop is ground down the standard Type 56 Rifle (AK47) type magazines can be used as well.

In an effort to simplify the manufacturing of the Type 68 rifle, the PRC engineers, under the designer Tang Wenlie, modified the weapon to include a number of stamped steel parts and riveted assemblies.

This further evolution of the Type 68/73 rifle is a lighter version of the earlier models and is reported to have fewer parts. This weapon, first publicized in 1980, is reported to be the work of a team working under Tang Wenlie. Its official designation has yet to be discovered.

These two American soldiers (Sergeant Fred Nakamura and Private David S. Yogi) dressed in PLA uniforms participated in a demonstration of enemy materiel in Seoul, South Korea in the fall of 1953. Both men are holding Type 43 copies of the Soviet PPS 43.

Type 50 7.62mm submachine gun, a copy of the Soviet PPSh M1941.

CHINESE SUBMACHINE GUNS

In addition to their copies of Soviet submachine guns, the Chinese have developed one native submachine gun, the Type 64 silenced machine gun. The bolt mechanism of this weapon is basically the same as the Soviet PPS43, which the Chinese have manufactured as the Type 43. The selective fire mechanism of the trigger system has been adapted from that of the Bren gun. This blowback weapon uses a low velocity variant of the standard 7.62 × 25mm cartridge employed in Eastern Bloc nations for submachine guns. The chamber of this submachine gun has three flutes to ensure satisfactory cartridge extraction.

The sound suppressor used with the Type 64 is an evolution of the basic Maxim type employing a stack of baffles to diffuse the exhausting gases. The barrel of the Type 64 is 7.87 inches, the last 2.36 inches being perforated by four rows of nine holes that follow the spiral of the rifling. The suppressor tube surrounding the barrel extends 6.5 inches beyond the end of the muzzle. With the stock extended, the Type 64 is 33.19 inches overall; with the stock folded, 25 inches. The baffles are of a dished configuration with the concave side toward the rear of the weapon. Each baffle has a 9mm (.35 inch) diameter hole. These baffles are held in place by a locking assembly consisting of two rods and a connecting piece that can be used as an aid to disassembly. The Type 64 is an interesting weapon because it is one of a very limited number that have been created from the start as a silenced weapon.

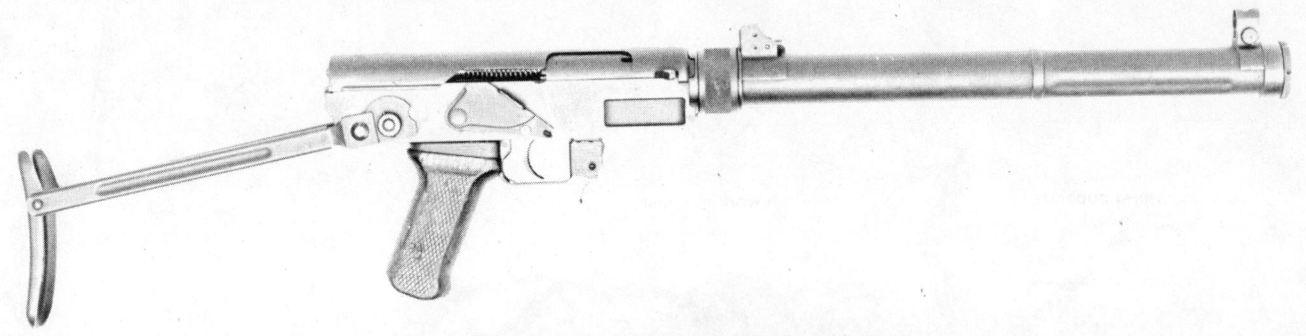

The PRC Type 64 7.62 × 25mm silenced submachine gun.

Close-up of the markings on the Type 64 silenced submachine gun receiver. This weapon was made at Factory 66 and carries the serial number 61090.

Close-up of the selector markings on the Type 64. *To the left,* the Chinese character for single shot fire; *to the right,* the symbol for automatic fire. As set, the gun would fire single shots only.

CHINESE MACHINE GUNS

The Type 56 Light Machine Gun (Copy of the Soviet RPD).

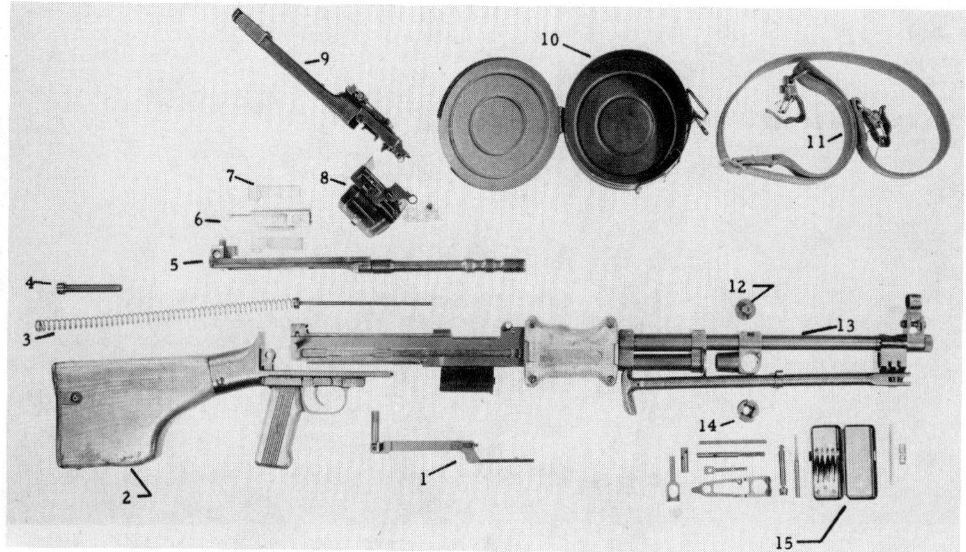

Assembled and disassembled view of the PRC Type 56 light machine gun, which is a copy of the Soviet RPD.

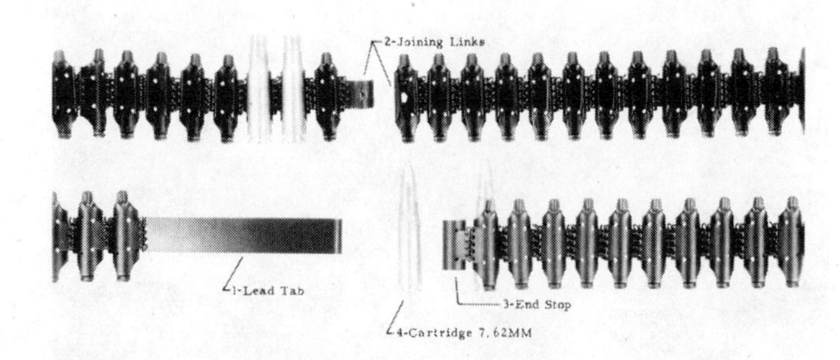

The Type 56 (RPD) link belts are the nondisintegrating type of 50 cartridge capacity. They can be attached at the joining link in order to fill the 100-round drum shown to the right.

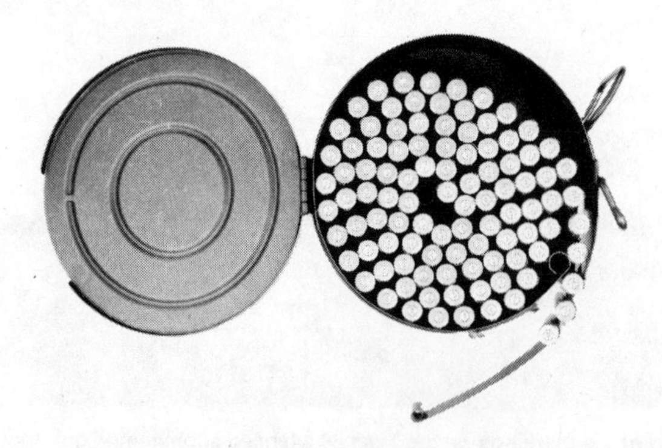

The 100-round Type 56 (RPD) drum magazine.

THE CHINESE TYPE 67 MACHINE GUN

The Type 67 is a rather odd combination of a number of other weapons, some relatively new and some quite old. The bolt mechanism is like that of the Czech ZB30 light machine gun; the trigger mechanism is like that of the Soviet DP; the quick change barrel is like that of the Soviet SG43, and the feed is modeled on the old Maxim gun. The rear sight of the Type 67 is similar to that of the Czech Type 59 machine gun but has a permanently fixed AA rear sight post on its top. A removable speed ring type AA front sight is mounted at the front of the receiver. The Soviet style stock of the Type 67 is made of plastic. The Type 67 also has the screw type headspace adjustment feature of the Soviet SGM machine gun.

The Type 67 is chambered for the 7.62mm × 54mm rimmed cartridge and can be used either on a bipod or a tripod. The tripod can be used for AA fire as well as ground fire.

The 7.62mm Type 67 machine gun will probably replace both the Type 53 light machine gun and the Type 58 company machine gun.

CHARACTERISTICS OF THE TYPE 67 MACHINE GUN

Caliber: 7.62mm (7.62 × 54).
System of operation: gas, automatic fire only.
Weight: 21.8 lbs.
Length overall: 45 in.
Barrel length: 23.5 in.
Feed device: 100 round metallic link belt.
Sights: Front: protected post.
Rear: leaf.
Muzzle velocity: 2,650 f.p.s.
Cycle rate: 600–700 r.p.m.

How to Load and Fire the Type 67

Load the feed belt by placing a cartridge over the link opening, so that the rim of the cartridge is just ahead of the turned down tab on the link. Press the cartridge into the link; the tab must be behind the cartridge base.

Press the cover latch to one side and allow the feed cover to swing open. Unfold the operating handle until it points downward at a 45-degree angle; pull it fully to the rear, then return it fully forward. Place the cartridge belt on the feed tray with the first round in the feedtray slot. Close the cover. The gun is now loaded and ready to fire! If the gun is not to be immediately fired, render it safe by rotating the safety forward. Prior to firing, lift the rear sight to its vertical position and set it for the proper range (calibrated in hundreds of meters). For elevation, up to 1000 meters, one click = 25 meters; beyond 1,000 meters, one click = 20 meters. Turn the right hand knob to change windage; each click changes bullet impact about 25mm (1 inch) for each 100 meters.

To fire, rotate safety to rear, aim and fire. The gun will continue to operate as long as the trigger is pressed and cartridges are in the belt. Best results are obtained firng short bursts of five to eight rounds. The bolt will remain open between bursts and will close on an empty chamber when the last shot has been fired.

To clear or unload the Type 67, reverse the steps for loading. Inspect the chamber to insure that it is clear.

Should the barrel overheat during firing, it may be removed and a cool one substituted. To change barrels, press the cover catch and open the feed cover. Press the barrel lock to the left as far as possible. Use the carrying handle and pull the barrel forward, out of the gun. Insert the spare barrel into the receiver, insuring that the gas cylinder enters the gas cylinder tube. Press the barrel lock in, fully to the right. (Note: There is a screw adjustment on the barrel lock to compensate for wear. This should only be adjusted by an armorer.) Close the cover. Adjust the gas regulator if the gun becomes sluggish during firing. See Soviet RPD regulator description for details.

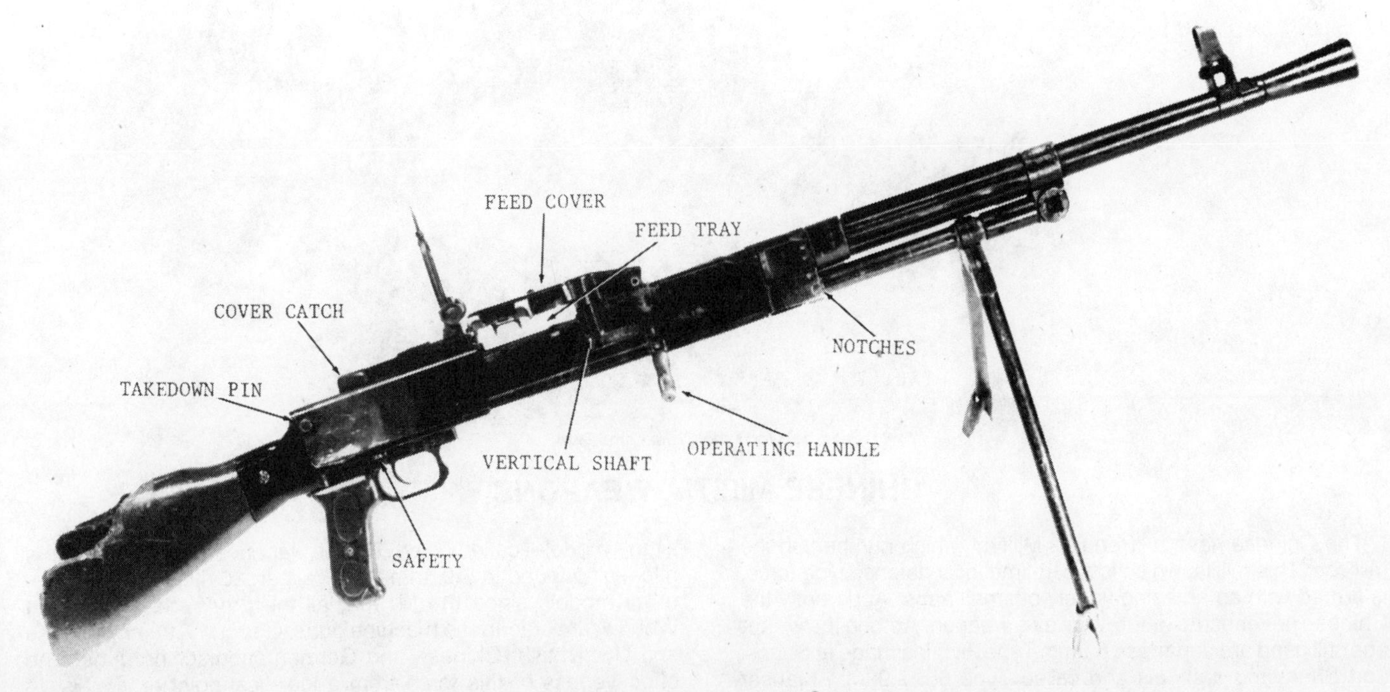

Key features of PRC Type 67 Machine Gun.

How to Field Strip the Type 67

Clear the gun, but do not close the cover or fold the operating handle. Remove the barrel. Press the take-down pin out to the right as far as possible, then pull the stock and trigger group straight to the rear until it comes free. Remove the driving spring and guide. Pull the operating handle rearward; this will move the gas piston, slide and bolt to the rear where they can be removed. Lift the bolt off the slide. Pull the operating handle until it comes clear of the receiver. Unscrew the gas regulator screw and push the regulator to the right until it comes out of its recess in the barrel. (Note: Further disassembly is neither necessary nor desirable.)

To reassemble, first insert the regulator into its recess until the "1" notch lines up with the index pin. Screw the gas regulator screw into the regulator, finger tight. Slide the operating handle onto the lower grooves in the receiver and push the unit as far forward as possible. Place the bolt over the hammer post of the slide and move the bolt forward as far as possible. Insert the bolt and slide, piston leading, into the receiver and press them firmly forward. Insert the driving spring into its tunnel in the slide and then insert the guide into the driving spring. Start the stock and trigger group onto the lower grooves of the receiver until the driving spring guide touches the rear wall. Do not kink the driving spring as you shove the stock and trigger group fully forward. Push the takedown pin fully into the receiver. Insert the barrel, close the cover and fold the operating handle.

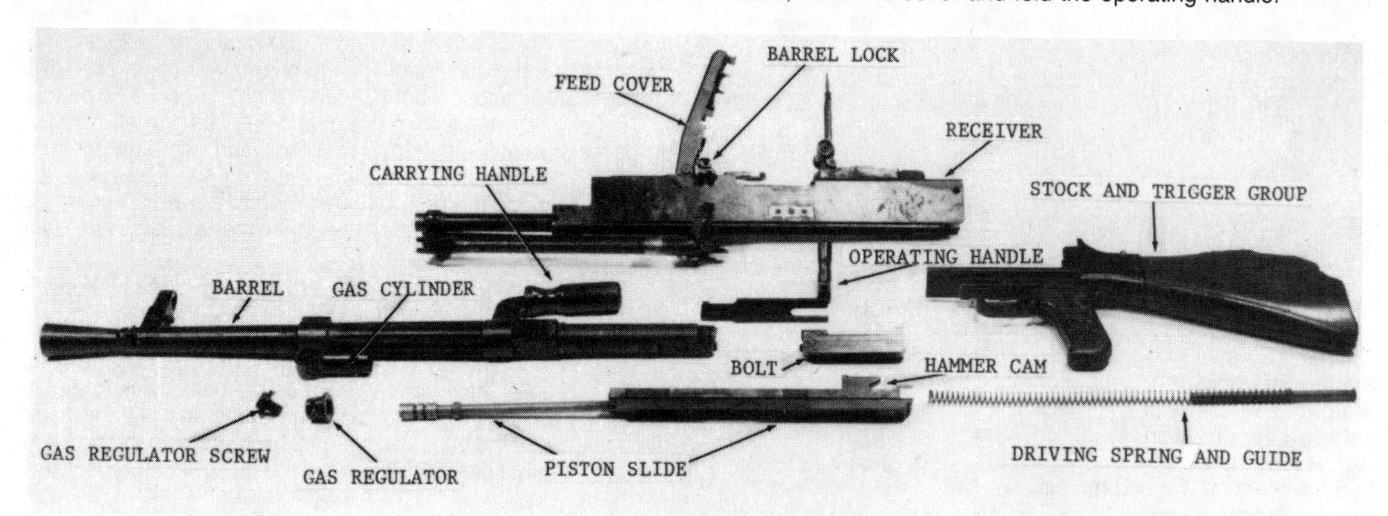

Disassembled view of Type 67 Machine Gun.

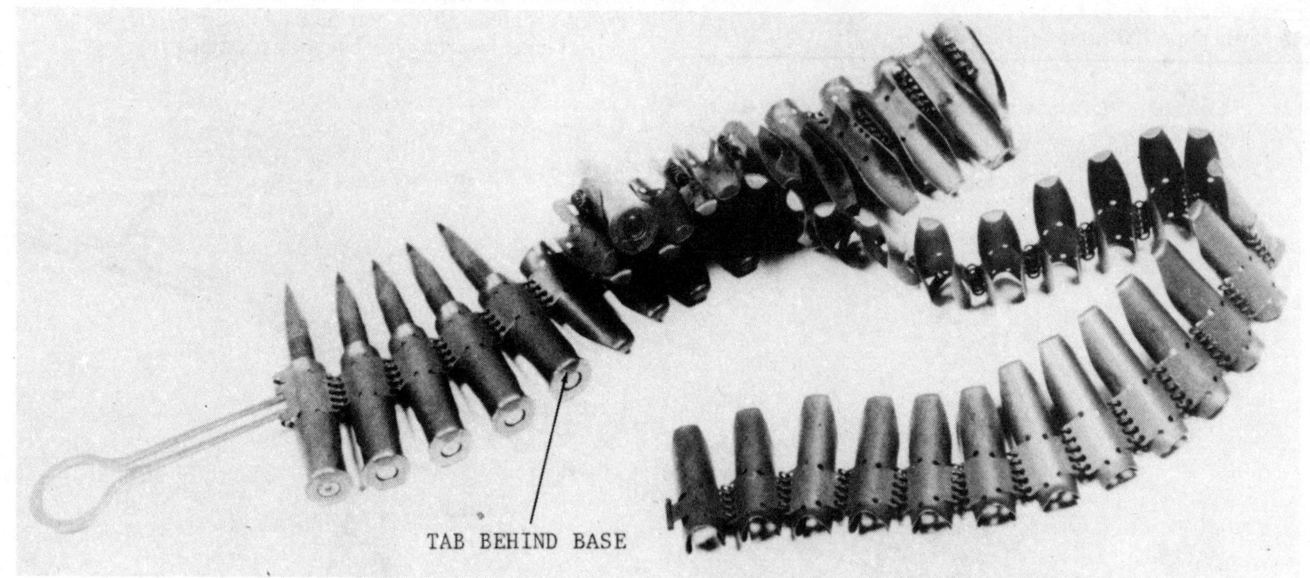

Feed belt for Type 67 Machine Gun.

CHINESE MILITIA WEAPONS

The Chinese have a "People's Militia," which numbers in the millions. This militia, which is a part-time local-defense type force, is armed with an amazing variety of small arms. Apparently the Chinese never scrap or throw away a weapon. Among the weapons still used are Japanese 6.5mm Type 38 rifles made at Shansi and Shenyang arsenals and called Type 65, 7.92mm Mauser "Standard" models—same as Chinese Nationalist "Generalissimo" model—called Type 79, the Japanese Type 99 7.7mm rifle—also found in 7.92mm—US caliber .30 rifles and carbines of all models since the M1903, all the pre-World War II and World War II Japanese machine guns, German 7.92mm MG 34s and US, British, Chinese and German submachine guns. The effectiveness of this force from a logistical point of view is, to put it mildly, dubious.

15 Czechoslovakia

SMALL ARMS IN SERVICE

Handguns: 7.62 × 25mm vz 52 Pistol	Domestic design. No longer used by front line troops.
7.65 × 17mm vz 61 Skorpion	Domestic design. Actually a machine pistol.
Submachine guns: 7.62 × 25mm vz 24	Domestic design. No longer used by front line troops.
7.62 × 25mm vz 26	Domestic design. No longer used by front line troops.
Rifles: 7.62 × 39mm vz 58	Domestic design.
Machine guns: 7.62 × 39mm vz 52	Domestic design. Obsolete.
7.62 × 54mm R vz 59	Domestic design.
12.7 × 108mm DShK 38/46	Soviet heavy machine gun.

Czechoslovakia has been a major source of innovative small arms since the establishment of the country in 1918 following World War I. Despite its satellite relationship with the Soviet Union, the Czechoslovakian armed forces have maintained a hardware independence. Whereas the other members of the Warsaw Pact use variants of the current Soviet small arms, Czech troops continue to carry weapons of domestic design and manufacture. Czechoslovakia is currently using its third generation of post–World War II small arms. Current small arms include the 7.62mm Model 52 pistol, the 7.65mm Model 61 submachine gun, the 7.62mm Model 58 assault rifle, the 7.62mm Model 54 sniper rifle, the 7.62mm Model 52/57 light machine gun, the 7.62mm Model 59 general purpose machine gun and the Czech-made 12.7mm DShK M1938/46 heavy machine gun. All these weapons with the exception of the DShK are of native design. The 9mm Models 23 and 25 and the 7.62mm Models 24 and 26 submachine guns are no longer standard and, like the now also obsolescent 7.62mm Models 52 rifle and machine gun, have been exported in large quantities.

All weapons now in service use standard Soviet type ammunition—the 7.62mm M43 (called M57 in Czechoslovakia) for assault rifle and light machine gun; and the 7.62mm rimmed for the Model 59 machine gun and 12.7mm for the DShK. An exception is the Model 61 submachine gun, which uses the 7.65mm—.32 Colt ACP cartridge. The Czech-designed 7.62mm Model 52 cartridge is no longer standard.

A word on Czech nomenclature: The Czech word for Model is Vzor. This word is usually abbreviated to "vz" in markings on Czech weapons and has frequently been picked up in English language publications as part of the nomenclature, as per example "Light Machine Gun Model vz 26." This of course is the same as saying "Model Model"; in the following text the English term "Model" is often substituted for vz.

POST-1945 CZECH PISTOLS

CZECH 9mm PARABELLUM MODEL 47 PISTOL

The Model 47 was apparently made only in prototype form. It combines the double-action trigger mechanism of the Model 38 with the Browning-Petter locking mechanism.

Czech 9mm Model 47 Pistol.

7.62mm ZKP524 PISTOL

This pistol, chambered for the 7.62mm rimless pistol cartridge, was developed by Francis and Joseph Koucky. It is a recoil operated pistol using the basic Browning type action and has an 8-round magazine. It was produced only as a prototype and was probably designed in competition with the Model 52, which was adopted as the Czech service pistol.

Czech 7.62mm ZKP 524 Automatic Pistol.

CZECH 7.65mm M1950 PISTOL

CHARACTERISTICS, 7.65MM M1950 PISTOL

System of operation: Blowback, semiautomatic fire only
Weight, unloaded, w/magazine 1.5 lbs.
Length, overall: 6.8 in.
Barrel length: 3.8 in.
Feed device: 8 round removable box magazine.
Muzzle velocity: 919 f.p.s.
Sights: Front: Blade.
Rear: Round notch.

Czech 7.65mm M1950 Pistol.

The M1950 is no longer used as a service pistol in Czechoslovakia but is probably still used as a police weapon. It is chambered for the 7.65mm automatic pistol cartridge (cal. .32 ACP). Specimens have been examined and appear to be of good quality manufacture, but there is nothing unusual about the design of the weapon.

CZECH 7.65mm Model 70 Pistol

An updated version of the Model 50, this pistol can be identified by the markings "Vzor 70 CAL 7.65." In addition to police use, this pistol is likely to be carried as a personal defense weapon by senior Czechoslovakian military officers.

How to Load and Fire the M1950/1970 Pistol

Remove magazine from pistol grip by depressing the release button on left side of receiver and pulling magazine out. Fill magazine with eight cartridges. Insert magazine in the well in pistol grip. Draw slide smartly to the rear and release. With the safety off, the trigger may now be squeezed, and the weapon will fire; when the trigger is released, it may be squeezed to fire again. This process may be repeated till the weapon is empty.

The Czech vz 70 (Model 1970) 7.65 × 17mm pocket pistol. Note that there is no backstrap. The plastic grips provide the bearing surface for the heel of the shooting hand.

The safety of the M1950/1970 operates like that of the Walther PP and PPK. The M1950/1970 is a double action pistol; with a cartridge in the chamber and the hammer in the down position, pulling the trigger to the rear will cause the hammer to rise and fall, firing the weapon.

How the M1950/1970 Pistol Works

In its functioning, the M1950/1970 pistol is generally similar to the Walther PP and PPK.

How to Field Strip the M1950/1970 Pistol

The dismounting button is pushed, and the slide is pulled to the rear as far as it will go. The rear end of the slide is then drawn upwards, and the slide is pushed forward and off the weapon. The recoil spring, which encircles the barrel, can be removed. The barrel is fixed to the receiver and should not be removed. No further disassembly is recommended. Reverse the procedure to reassemble the weapon.

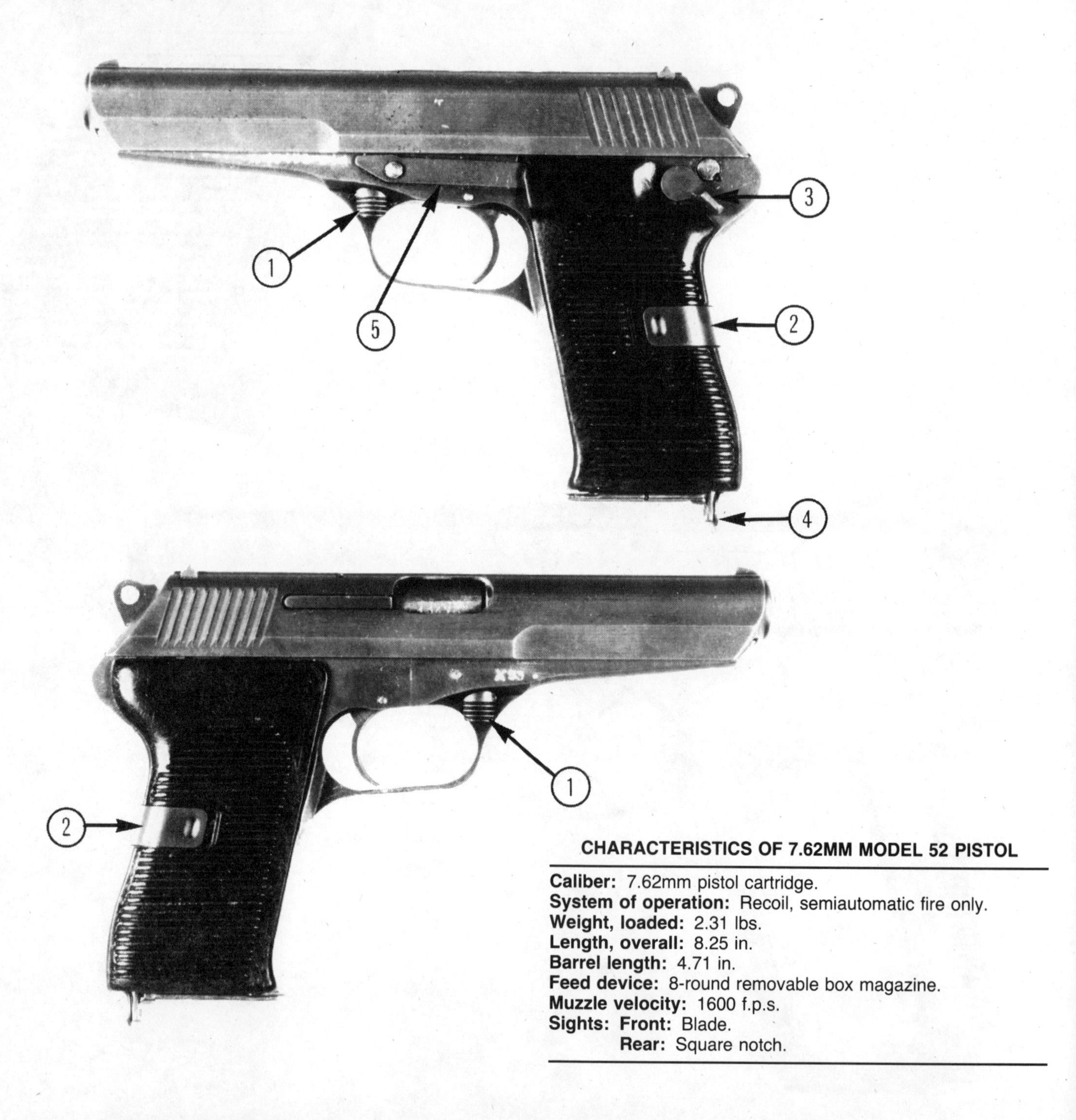

CHARACTERISTICS OF 7.62MM MODEL 52 PISTOL

Caliber: 7.62mm pistol cartridge.
System of operation: Recoil, semiautomatic fire only.
Weight, loaded: 2.31 lbs.
Length, overall: 8.25 in.
Barrel length: 4.71 in.
Feed device: 8-round removable box magazine.
Muzzle velocity: 1600 f.p.s.
Sights: Front: Blade.
Rear: Square notch.

Czech 7.62mm Model 52 Pistol: 1. Takedown catch; 2. Spring retaining clamp holding grips in place; 3. Safety; 4. Magazine catch release; and 5. Slide stop.

CZECH 7.62mm MODEL 52 PISTOL

The Czechoslovaks have used several types of service pistols since the conclusion of World War II. The 7.62mm Model 52 is a current service weapon but obsolescent. It is being supplanted by the vz 61 Skorpion.

The Model 52 pistol is a native Czechoslovak design, which has borrowed its locking system from the German MG42 machine gun. The pistol is chambered for the Czechoslovak-made version of the Soviet 7.62mm Type P pistol cartridge, which the Czechoslovaks call the Model 48. The Soviet and Czechoslovak cartridges are interchangeable with the 7.63mm Mauser but are considerably hotter loadings than are the United States commercial loadings of this cartridge. For this reason, the functioning of Soviet and Czechoslovak weapons with commercially loaded 7.63mm cartridges is, at best, marginal. The Czechoslovak cartridge has a particularly heavy loading, being about 20% heavier than the Soviet. The vz 52 was adopted on 17 May 1952.

The 7.62mm vz 52 service pistol, field stripped.

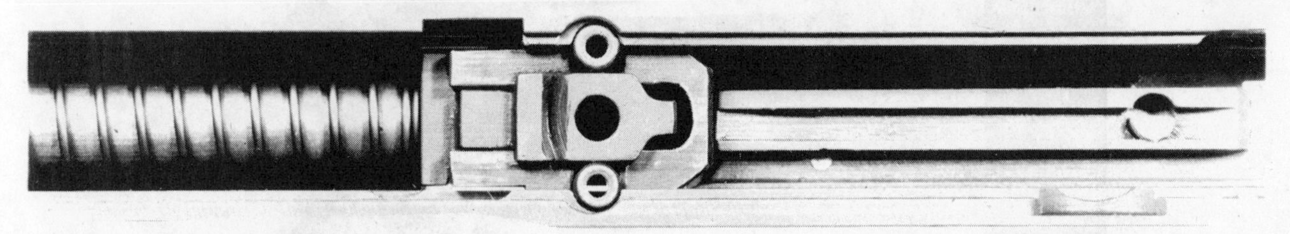

The roller-locking mechanism of the vz 52 in the locked position. Note how the lugs fit into the locking recesses.

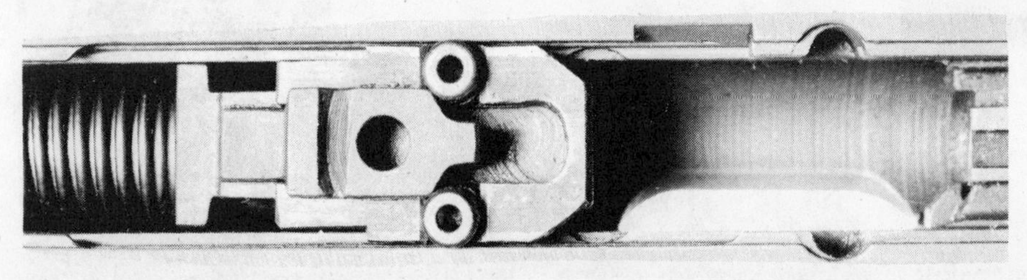

The roller-locking mechanism of the vz 52 in the unlocked position.

CZ 9 × 19mm MODEL 75

This large magazine capacity, double action 9 × 19mm caliber pistol is something of an enigma. To date this handgun appears to be a commercial venture, although it is an excellent design from a military standpoint. The locking system of the Model 75 is essentially an evolution of the Browning locking concept as it appeared in the FN Browning Model 1935. This handgun is reported to have been designed by the Koucký brothers, František and Josef, at the Československá Zbrojovka, Narodní Podnik in Brno.

The Model 75 is somewhat unusual in that the slide rides on rails machined inside the receiver. This system has also been used on the SIG P220 series (SP 47/8) of handguns. In the first version of the Model 75 (1975–1976), the hammer of the pistol had no half cock. This indicates that the gun was intended to be carried with the hammer down on a loaded chamber. Under such a situation the Model 75 could be inadvertently discharged if a sharp blow was given to the hammer, as in the event of the pistol being dropped. More recent production Model 75s have a half cock notch to make the pistol safer to carry. These new Model 75s, introduced in 1980, also have longer side rails—5.5 inches versus 4.5 inches of the earlier models.

Field Stripping the Model 75

Remove the magazine and make certain that the pistol is unloaded. Cock the hammer. Pull the slide to the rear about .25 inch to align the disassembly marks on the left side of the slide and the frame. Depress the slide stop/takedown pin from the right and withdraw it from the left. Pull the slide forward from the frame and remove the recoil spring, guide, and barrel.

How the Model 75 Pistol Works

This handgun operates in the same fashion as the Browning Model 1935 GP described in Chapter 9.

Double Action Mechanism of the Model 75

The double action mechanism of the Model 75 is different from that of the P38, the PPK, or the Smith & Wesson self-loaders. These more familiar double action mechanisms pull the lower portion (toe) of the hammer to retract the hammer through an arc of about 60 degrees to release its connection with the trigger bar hook. At that point the hammer can fly forward under the power of the mainspring. The Model 75 double action mechanism works differently. The Model 75 trigger arrangement employs a wrap-around trigger bar that, instead of pulling the hammer toe, engages an "L"-shaped interrupter pinned to the hammer. This interrupter serves as the connection between the hammer and the trigger bar. Pulling the Model 75 trigger in double action pushes the trigger bar, which in turn pushes the interrupter and hammer to release it at the end of the cycle when the upper rear inclined surface of the trigger bar is forced downward by the shoulder of the sear housing. During the firing cycle, the slide recocks the hammer and forces the trigger bar disconnector (a protuberance on the upper part of the trigger bar) downward to release the connection with the lower part of the sear. This permits the sear to pivot to the rear to re-engage the full cock notch when the hammer is rotated to the rear by the slide.

The first model of the CZ Model 75 9 × 19mm pistol.

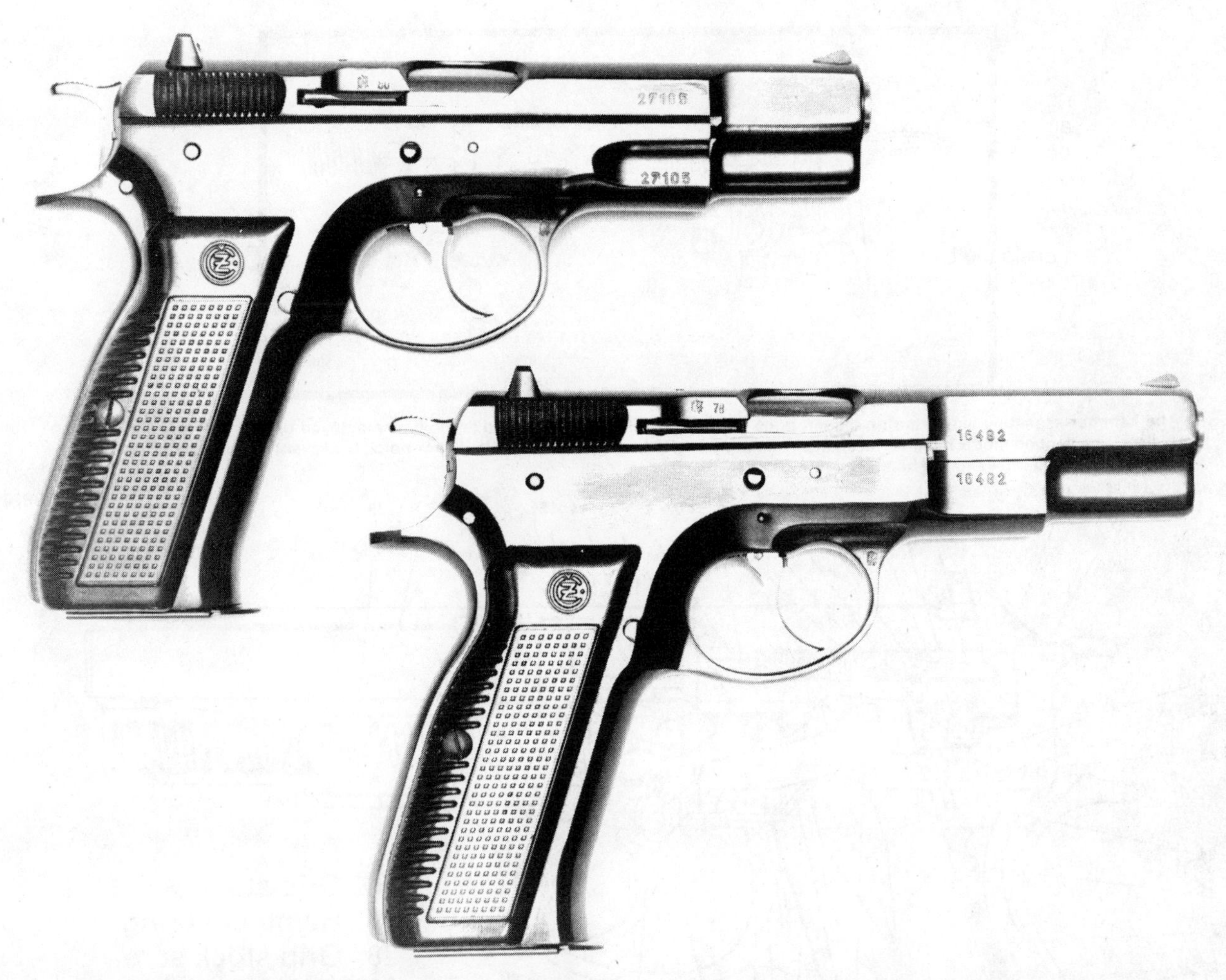

The side rails on the current production CZ Model 75 pistols (*top*) are about 5.5 inches long, which is about one inch longer than those on the early version (*bottom*).

CZ 9mm PARABELLUM MODEL 75 PISTOL

More details are presented in Chapter 4.

CHARACTERISTICS OF CZ 9MM PARABELLUM MODEL 75 PISTOL

Caliber: 9mm Parabellum.
System of operation: Recoil; semiautomatic fire only.
Length, overall: 8 in.
Barrel length: 4.72 in.
Weight, w/empty magazine: 2.2 lbs.
Feed device: 15-round removable box magazine.
Muzzle velocity: 1214 f.p.s.
Sights: Front: Blade.
Rear: Square notch.

Military version of the M75 Pistol.

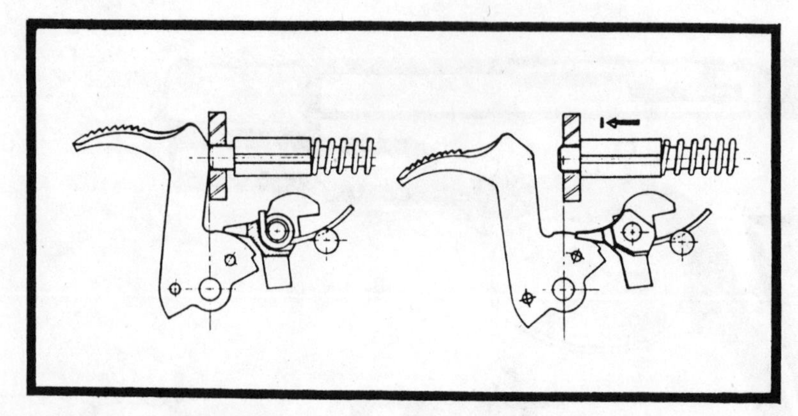

The hammer assembly illustrated on the left does not have a half cock notch and is the type employed on the first model CZ 75 pistols. The second model hammer mechanism illustrated on the right does have a half cock notch to prevent accidental discharges from unexpected blows to the hammer.

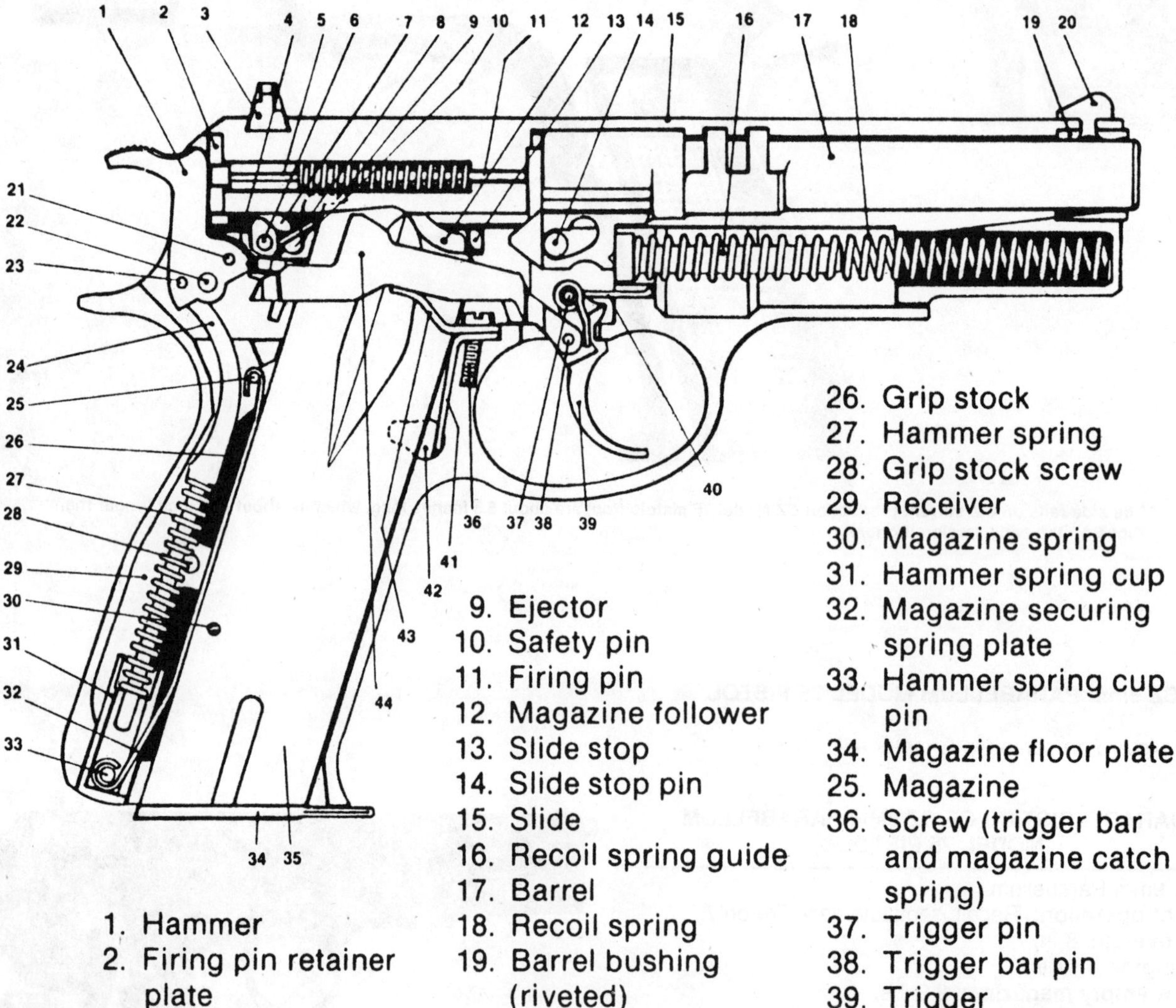

1. Hammer
2. Firing pin retainer plate
3. Rear sight
4. Sear assembly housing
5. Firing pin grooves
6. Sear pin
7. Sear
8. Sear spring
9. Ejector
10. Safety pin
11. Firing pin
12. Magazine follower
13. Slide stop
14. Slide stop pin
15. Slide
16. Recoil spring guide
17. Barrel
18. Recoil spring
19. Barrel bushing (riveted)
20. Front sight
21. Interruptor pin
22. Hammer pin
23. Hammer strut pin
24. Hammer strut
25. Magazine securing spring plate pin.
26. Grip stock
27. Hammer spring
28. Grip stock screw
29. Receiver
30. Magazine spring
31. Hammer spring cup
32. Magazine securing spring plate
33. Hammer spring cup pin
34. Magazine floor plate
25. Magazine
36. Screw (trigger bar and magazine catch spring)
37. Trigger pin
38. Trigger bar pin
39. Trigger
40. Trigger spring
41. Magazine catch spring
42. Magazine catch
43. Trigger bar spring
44. Trigger bar (interruptor)

Section view of the CZ 9 × 19mm Model 75 pistol.

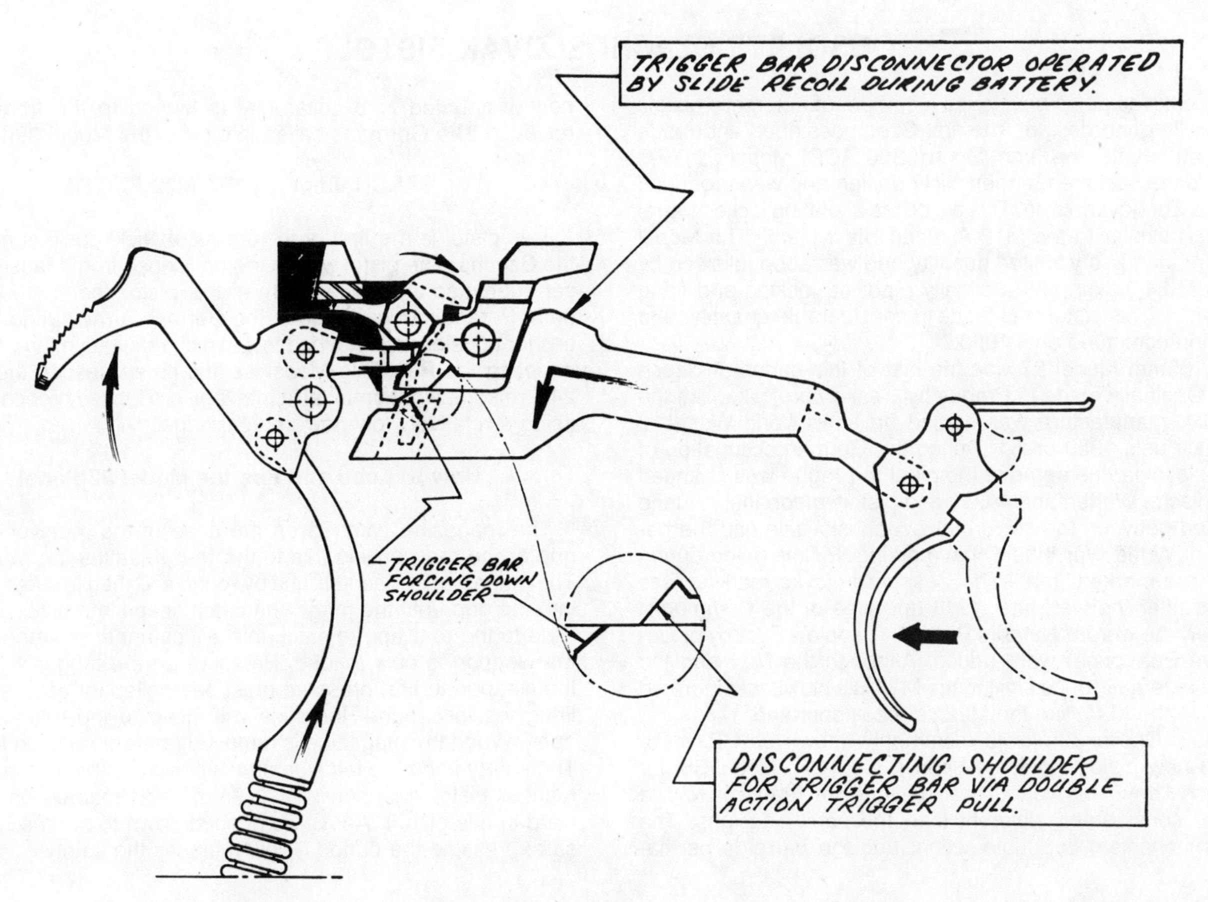

Diagram of the hammer and sear assemblies of the first version of the CZ Model 75 illustrating the double action feature.

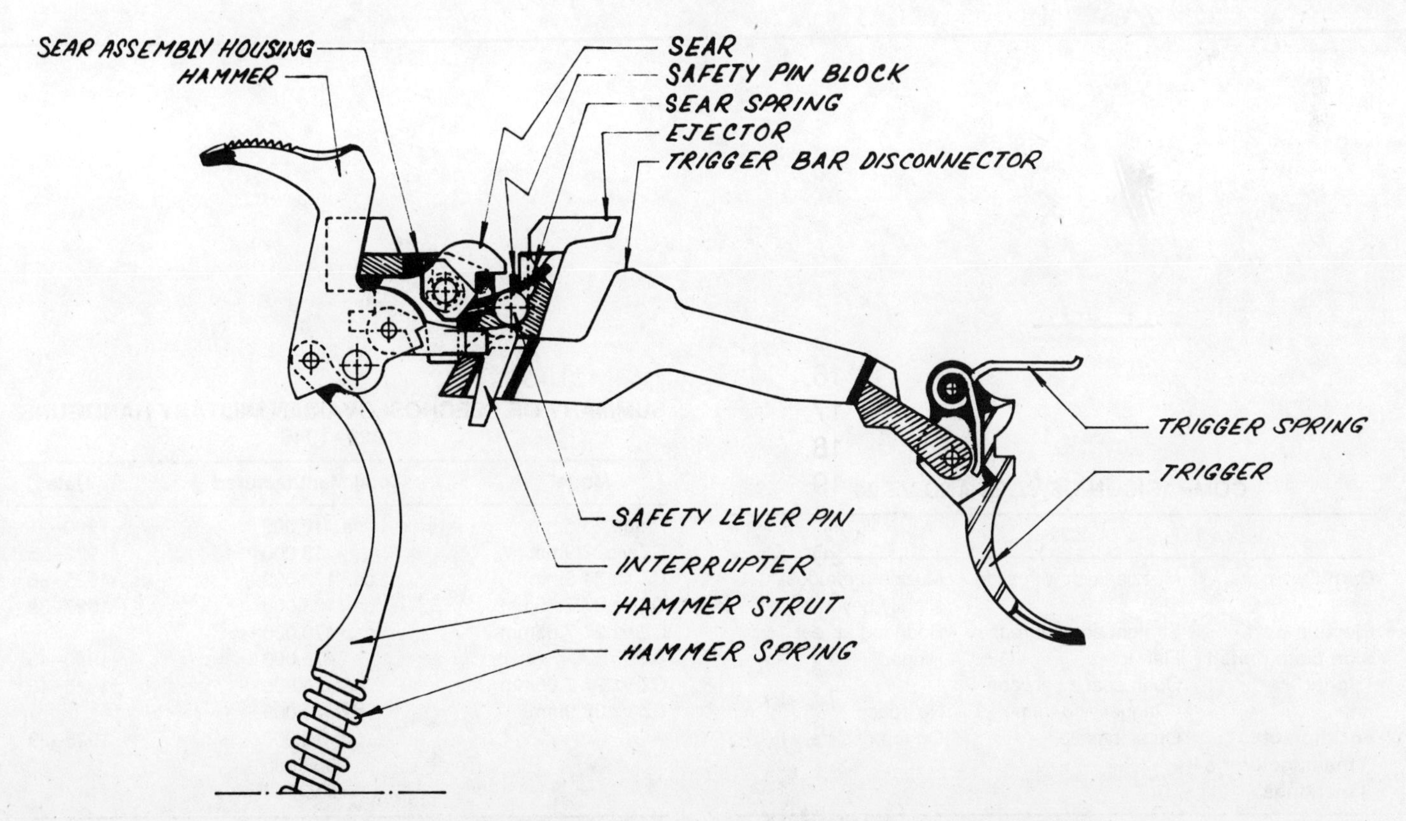

Diagram of the CZ Model 75 hammer and sear assemblies with the components identified.

OBSOLETE CZECHOSLOVAK PISTOLS

Czechoslovakia has developed a number of automatic pistols of generally good design. The first Czech designed and made service automatic, the 9mm Short (.380 ACP) Model 22, was actually based on the German Nickl design and was produced by Česka Zbrojovka, Brno. This pistol has a rotating locked barrel somewhat similar to the M12 Austrian Steyr pistol. The Model 22 was made in very limited quantity and was soon followed by the Model 24, which has a slightly modified locking and firing mechanism. This pistol was made in considerable quantity, and production continued until 1938.*

The 7.65mm Model 27 was the last of this series of Czech pistols. Originally made in Prague by Česka Zbrojovka, like the Model 24, manufacture was carried on after World War II at Strakonice until 1950 or '51. During the German occupation of Czechoslovakia, the name of the plant in Prague was changed to Böhmische Waffenfabrik A.G., and pistols made there during the period between the seizure of Czechoslovakia and the beginning of World War II bear that marking. Pistols made during the war are marked "fnh." The Česka Zbrojovka marking was resumed after the war; after 1948 (the time of the Communist take-over) the words "Narodni Podnik" (People's Factory or Co-operative Enterprise) were added. Although the M24 and the M27 are externally quite similar, the M27 is a blowback operated arm, while the M24, like the M22, is recoil operated.

In 1938, Česka Zbrojovka, commonly known as CZ, introduced a new pistol, the 9mm Short (.380 ACP) Model 38, for the Czech Army. This weapon was designed by Frantisek Myška and was considerably different than the earlier designs. The trigger mechanism is double-action, and the barrel is permanently mounted in a collar that is hinged to the front of the receiver. The Germans called this pistol the Model 39(t).

CZECH 9mm SHORT M22 PISTOL

This pistol is the first of the series of Nickl designs made by the Czechs. The pistol was made on license from Mauser. Mauser submitted a modified form of this pistol chambered for the 9mm Parabellum cartridge to the German Army during the thirties for consideration in the tests, which resulted in the Germans adopting the P38. The Model 22 and its successor, the Model 24, are somewhat unusual in that they are locked breech pistols, using a relatively low-powered cartridge.

How to Load and Fire the Model 22 Pistol

The magazine catch is on the base of the receiver (frame) grip at the rear; pulling this to the rear releases the magazine. The magazine is loaded round by round and then inserted smartly into the grip until the magazine catch snaps into place. Pull the slide to the rear and release; this will chamber a cartridge, and the weapon is now loaded. Pressure on the trigger will cause the weapon to fire; pressure must be applied for each shot. On firing the last round, the slide will move to the rear and stay open. When the magazine is removed, the slide will run forward. The safety is on the left side and is similar to that of the M1910 Mauser pistol (Mauser M1910 9mm short magazines can be used in this pistol). A lever is pushed down to put the pistol on safe; pressing the button below releases the safety.

COMPARISON OF VZ.22 AND VZ.24

	vz.22	vz.24
Barrel	Muzzle flush with slide	Muzzle protrudes slightly beyond slide
Ejection port	Straight edge at rear	Rounded at rear
Side plate on left	Flat	Ridged
Trigger	Open space between trigger and frame	No space
Location of manufacturer's markings	On side plate	On top of slide

*For additional details see *Handguns of the World.*

SUMMARY OF CZECHOSLOVAKIAN MILITARY HANDGUNS, 1920–1946

Model	Total Manufactured	Date
Praga 7.65mm	ca. 10,000	1920–21
CZ vz.22 9mm	18,000	1922–25
CZ vz.24 9mm	ca. 172,000	1925–38
CZ vz.27 7.65mm	ca. 183,000	1927–39
CZ vz.27 7.65mm	ca. 470,000–475,000	1940–45
CZ vz.27 7.65mm	ca. 45,000	1945–46
CZ vz.38 9mm	ca. 10,000–12,000	1938–39
Total	ca. 908,000–915,000	

Field Stripping the Model 22 Pistol

With empty magazine in gun, pull slide to rear. Slide will remain in position; push down on dismounting catch on side of receiver over front of trigger guard. Remove magazine and slide may then be moved forward and lifted off the receiver. The barrel can then be removed by turning barrel bushing 30°; bushing can then be pulled off slide and barrel. Pull barrel out of slide. No further disassembly is recommended. Reassemble by reversing the above procedure.

How the Model 22 Pistol Works

Pressure on the trigger causes the trigger and its springloaded bearing piece to operate the extension sear, which in turn operates the hammer. The hammer strikes the inertia-type springloaded firing pin, causing it to strike the primer of the cartridge, thereby firing the cartridge. The barrel and slide are rigidly locked by two rectangular lugs on either side of the barrel engaging in recesses in the side walls of the slide. A helical-portion camming lug on the barrel underside rides in a corresponding helical groove in a rigidly pinned block on the upper surface of the receiver. Upon functioning of the cartridge, the barrel is rotating clockwise through 22°, causing disengagement of the lugs and permitting further recoil of the slide. Recoil of the slide compresses against the recoil spring mounted on its guide, which passes through the helical camming block below the barrel. During recoil, the extractor, mounted in the slide, has withdrawn the fired case, which is ejected by the ejector, which is mounted over the rear of the magazine well. The slide returns forward under the pressure of the compressed recoil spring and strips a cartridge from the magazine, feeding it into the chamber. Pressure on the trigger again will repeat the above process.

Disconnection—prevention of full automatic fire—is accomplished by the camming down of the trigger-bearing surface so that it cannot contact the trigger extension. The disconnector is mounted in a recess in the slide on the left side.

THE CZECH 9mm SHORT PISTOL MODEL 24

The Model 24 is also a locked-breech pistol and is generally similar to the Model 22.

The pistols vary in the following details:

Locking action: A stopping stud is provided on the under side of the barrel ahead of the helical lug. This stud butts against the helical camming block to act as a stop during recoil. The helical lug is widened and the camming block is machined so that it can be assembled without regard to position. The block of the M1922 has an arrow to indicate assembly position.

Magazine: A magazine safety has been added, which blocks trigger motion and prevents firing the weapon without a magazine in place.

Trigger mechanism: The trigger bearing piece is incorporated into the trigger extension where it serves as a disconnector, and the separate disconnector is no longer used. The trigger extension is loaded with a coil spring rather than a wire form spring, and the trigger mounting and pin differ.

Other parts: The hammer spring is separate from the magazine spring, the former being retained by a screw.

Earlier types have one-piece, unchecked wood grips; later types have one-piece, checked molded-plastic grip with CZ insignia.

Czech 9mm Model 22 Pistol.

THE CZECH 7.65mm M27 PISTOL

The Model 27 was made in the largest quantity of all the pre-World War II Czech automatic pistols. It was extensively used by the Germans, who called it Pistol 27(t), during World War II.

The Model 27 differs from the Models 22 and 24 in caliber and in being blowback operated. The barrel is not rigidly fixed to the receiver as in many blowback operated pistols, such as the Walther PP, but is removable in a fashion similar to the Colt .32 and .380 pocket automatics.

During the course of the war, various simplifications in manufacture were introduced by the Germans. Pistols that bear the marking "fnh" were made with the following modifications: Trigger extension bar, safety, safety release, magazine release and firing pin retaining plate are stamped; magazine catch can be used as lanyard loop, and the retaining screw on side plate is eliminated.

Czech 7.65mm M27 Pistol of post-1948 manufacture.

Side plate on left side is relatively flat, retained by a small screw; 1922 has no screw but has a step projecting from its lower edge to close off the trigger mounting.

The slide rib is matted, and the frame and slide show different machining. A variant of this pistol with a 10-round rather than an 8-round magazine has been reported.

Loading, firing and field stripping of the Model 24 are essentially the same as that of the Model 22.

Manufacture of the pistol was continued by CZ after the war. The modifications made during the war were retained for the most part, and initially the old type Czech markings were used. After 1948, however, the "Narodni Podnik" marking was added.

As can be noted from the photograph, finish is extremely rough, especially for a weapon intended for export as indicated by the marking "made in Czechoslovakia."

The Model 27 is loaded, fired and field stripped in a manner similar to the earlier models with one exception: the barrel is held in place by parallel ribs (similar to Colt pocket pistols) and must be rotated to be removed from the slide.

Czech 9mm Short Model 38 Pistol.

Czech 9mm Short Model 38 Pistol—field-stripped.

THE CZECH 9mm SHORT M38 PISTOL

This pistol was not made in the quantity of the earlier models; it was used by the Germans as Pistole Modell 39(t). There are a number of variations of this pistol and some are fitted with safety catch. Essentially the pistol has the Model 22, 24, 27, trigger mechanism modified to double-action and ordinary blowback operation.

Field Stripping the Model 38 Pistol

Remove magazine, push dismounting catch on left side of receiver above trigger, pull slide to the rear and up. Barrel may be lifted on hinge for cleaning and side plate may be removed for cleaning trigger mechanism. No further disassembly is recommended. To assemble, reverse procedure.

CHARACTERISTICS OF PRE-WORLD WAR II CZECH PISTOLS

	Model 22	Model 24	Model 27	Model 38
Caliber:	9mm Short*	9mm Short*	7.65mm†	9mm Short*
System of operation:	Recoil, semiautomatic only.	Recoil, semiautomatic only.	Blowback, semiautomatic only.	Blowback, semiautomatic only.
Weight:	1.37 lb.	1.5 lb.	1.56 lb.	2 lb.
Length overall:	6 in.	6 in.	6.3 in.	7.8 in.
Barrel length:	3.44 in.	3.56 in.	3.9 in.	4.7 in.
Feed device:	8-round in line. Detachable box magazine.	8-round in line. Detachable box magazine.	8-round in line. Detachable box magazine.	8-round in line. Detachable box magazine.
Muzzle velocity:	984 f.p.s.	984 f.p.s.	919 f.p.s.	1,000 f.p.s.
Sights: Front:	Blade.	Blade.	Blade.	Blade.
Rear:	V notch.	V notch.	V notch.	V notch.

* same as .380 A.C.P. † same as .32 A.C.P.

CURRENT CZECH RIFLES

CZECH 7.62mm MODEL 58 ASSAULT RIFLE

In 1958, a new weapon, the 7.62mm Model 58 assault rifle, appeared. This rifle, a gas-operated selective fire weapon, replaced the Model 52/57 and the 7.62mm Model 24 and 26 submachine guns in the Czech Army.

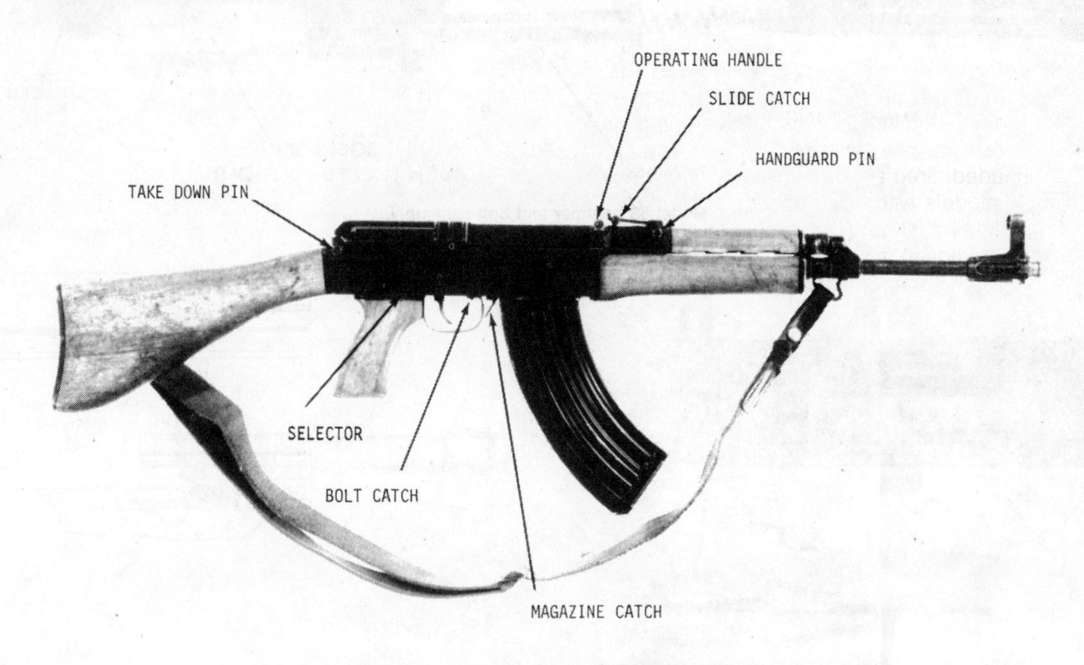

Czech 7.62mm Model 58P Assault Rifle with wooden stock.

Field-stripped view of the Czech 7.62mm Model 58V (vz 58) assault rifle. Note that this is the folding stock model.

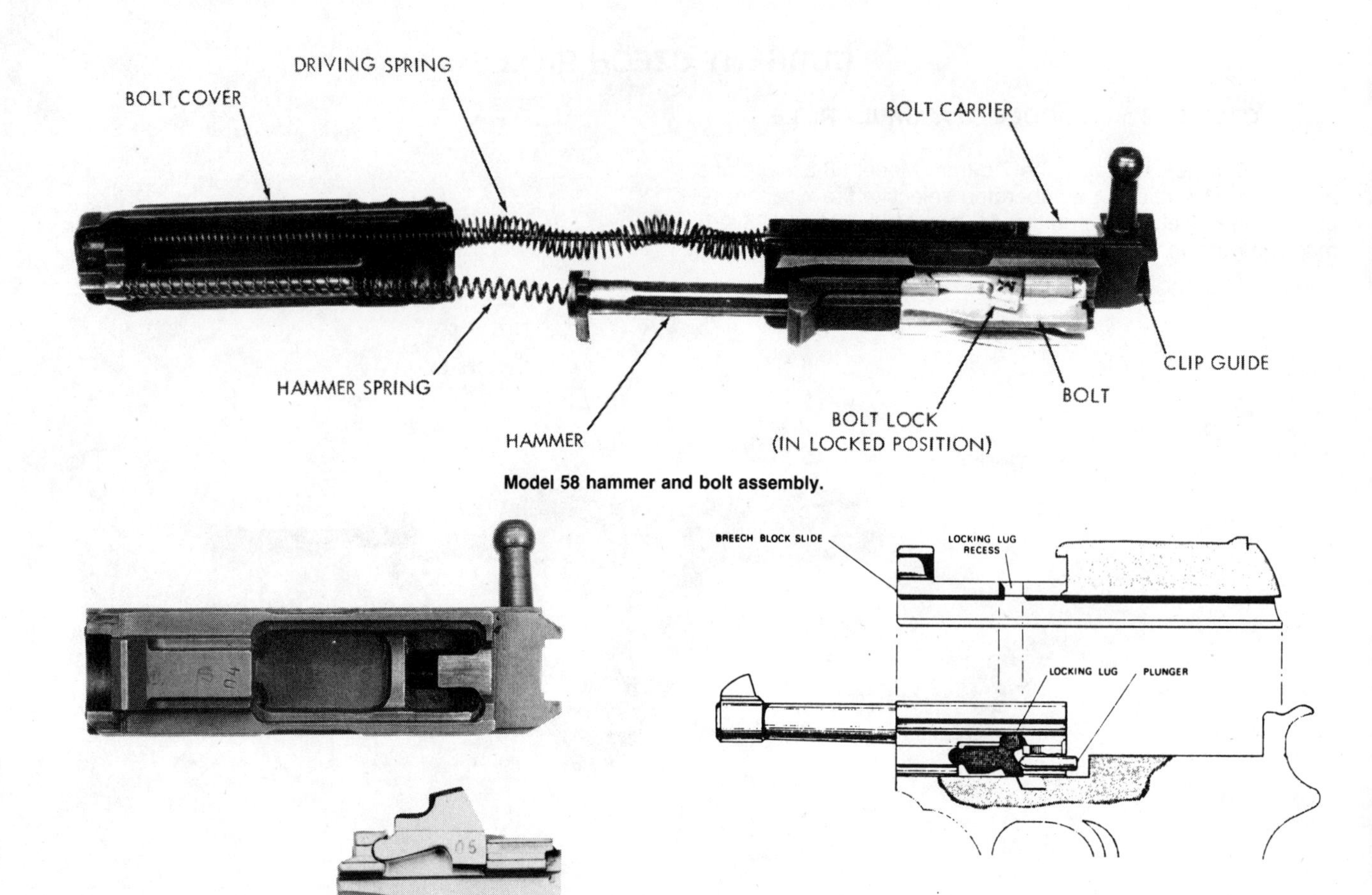

Model 58 hammer and bolt assembly.

The locking system of the Model 58 assault rifle is quite similar in concept to that of the P38. There is a swinging locking block attached to the bolt that locks the bolt and bolt carrier during the early portion of their rearward travel.

The P38 locking system.

CHARACTERISTICS OF MODEL 58 ASSAULT RIFLE

Caliber: 7.62, uses Czech copy of Soviet M1943 rimless cartridge.
System of operation: Gas, selective fire.
Weight, loaded:
  Wooden stock version: 8.75 lbs.
  Metal stock version: 8.75 lbs.
  Weight, empty: 7.31 lbs.
Length, overall:
  Wooden stock version: 33 in.
  Metal stock version, stock folded: 25 in.
  Metal stock version, stock fixed: 33 in.
Barrel length: 15.8 in.
Feed mechanism: 30-round, staggered column, detachable box magazine.
Sights:
  Front: Protected post.
  Rear: Tangent leaf, adjustable in 100-meter increments from 100 to 800 meters.
Cyclic rate: 700–800 r.p.m.
Muzzle velocity: 2300 f.p.s.

The Model 58 assault rifle is the standard shoulder weapon of the Czech Army. It is similar in concept and in external appearance to the Soviet AK47 but is quite different internally. Its outstanding characteristic is its light weight, but this is of dubious value considering its use in automatic fire from the shoulder. Lighter arms rise more rapidly in automatic fire than do heavier arms, even if they do have a straight-line stock configuration as does the Model 58.

The Model 58 may be found with wooden stock and handguards, plastic stock and handguards or with folding steel stock. The version of the Model 58 with fixed stock is called the Model 58P; the version of the Model 58 with folding metal stock is called the Model 58V.

How to Load and Fire the Model 58 Assault Rifle

Push magazine catch in front of trigger guard forward and remove magazine. Fill magazine with cartridges and insert in magazine well. Set safety on safe by turning safety-selector lever to the safe position. Pull operating handle to the rear and release; a cartridge will be chambered. The selector has two fire positions; up for semiautomatic fire and midway for automatic fire; push safety-selector lever to position desired, and the weapon will fire. When last round is fired, the bolt will remain open.

Field Stripping the Model 58 Assault Rifle

Remove magazine and check chamber to insure that weapon is not loaded. Press the trigger to release hammer if cocked. Pull cover assembly retaining pin, mounted at rear of receiver cover, to the right and remove receiver cover. Move bolt and bolt carrier assembly to the rear and lift it up and out of the receiver. Rotate the hammer-striker approximately $\frac{1}{8}$ turn counterclockwise and withdraw to the rear so that the bolt carrier, bolt assembly and locking lugs are separated. Pull handguard retaining pin, located at front of rear sight base to the right and remove handguard by lifting up its rear portion and withdrawing it to the rear. Pull the piston to the rear and pull the piston head up so that piston and piston return spring can be removed. Reassemble by reversing the above procedure.

How the Model 58 Assault Rifle Works

Unlike most weapons of this type, the Model 58 does not have a rotating hammer—it has a linear travel hammer firing pin. Pressure on the trigger causes the sear (there are two sears—semiautomatic and automatic) to release the hammer and fire the cartridge. Gas from the cartridge is tapped off at the gas port into the gas cylinder, and it moves the gas piston to the rear at high speed; the rear end of the spring-loaded piston protrudes through the rear sight base and strikes the top front of the bolt carrier. The bolt carrier moves to the rear, camming the bolt lug out of its locking recess in the receiver and compressing the operating spring. The extractor in the face of the bolt withdraws the empty cartridge case, which is ejected out of the weapon. The operating spring forces the bolt and bolt carrier forward, and the bolt picks up a cartridge from the magazine and chambers it. If the weapon is set on automatic, the automatic sear will remain disengaged—assuming pressure is continued on the trigger—and the weapon will fire again. If the weapon is set on semiautomatic fire, the semiautomatic sear will rise and hold the hammer to the rear until pressure is released on the trigger and the trigger is pressed again.

Model 58V (vz 58) assault rifle bipod.

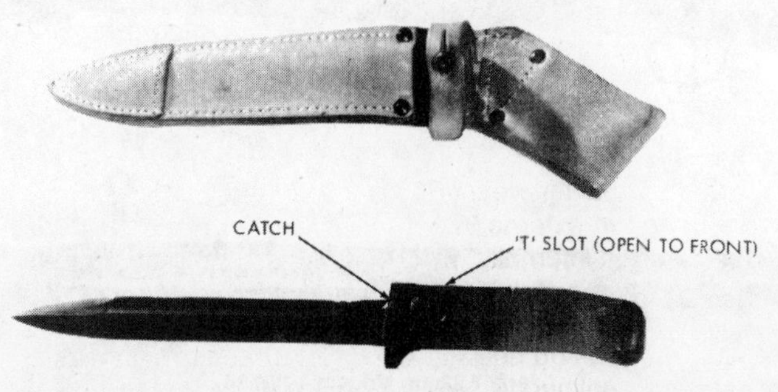

Vz 58 bayonet.

Model 54 Sniper Rifle

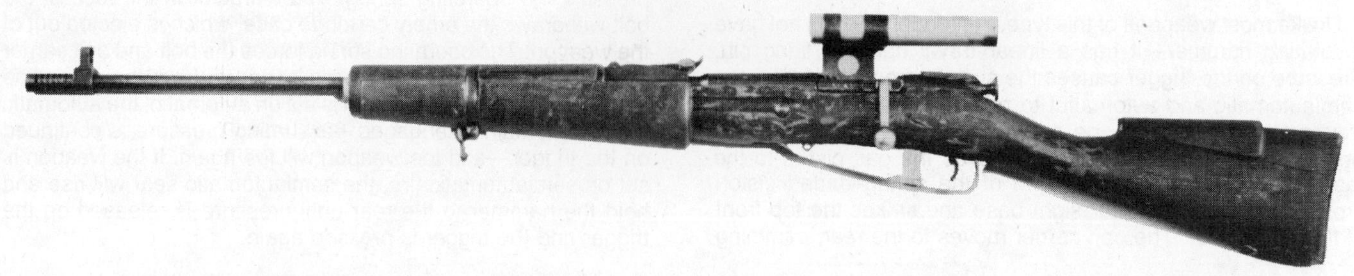

The Czech Model 54 sniper rifle is a more accurate Model 1891 Mosin Nagant 7.62 × 54mm R rifle fitted with a 2.5 power telescopic sight.

OBSOLETE CZECH RIFLES

When the Republic of Czechoslovakia was founded in 1919, it continued to use for a short time the Austro-Hungarian Mannlicher M1895 rifles.

In 1924 Československe Zbrojovka Brno—"ZB"—began the production of Mauser rifles for the Czech Army and for export. The Czechs, in competition with the Belgians of FN, took over a goodly part of the military rifle arms trade of the world.

All the Mausers produced by ZB were based on the M98 action. The models produced were: the Rifle 98/22, Rifle 98/29, Short Rifle 98/29, Carbine 12/33 and the rifle Model 24. The rifle Model 24 was adopted by Czechoslovakia and is covered in detail later in this chapter; the 98/29 weapons are covered in detail under Iran. The 98/22 differs from the 98/29 mainly in the use of front sight guards—-"Sight ears"—on the 98/29. The 98/22 was used by Turkey, among other countries. Romania, Guatemala, Yugoslavia and China used the Model 24 rifle in addition to Czechoslovakia. Model 24(t) was the German Army designation for the Czech Mauser, and eleven divisions of the German Army were equipped with it and other Czech weapons in September 1939 when World War II began.

Czechoslovakia also adopted the 16/33 Carbine, which has a small diameter receiver ring, as the Model 33 rifle. This weapon was kept in production at ZB during World War II with minor modifications as the 33/40 and was issued to German paratroops and mountain troops. During the war, the production of the Model 24 was gradually changed so that the rifles that were being produced at the end of the war were actually Kar 98ks. Those rifles marked 24(t) differ from the standard Model 24 in having a cup type butt plate, a firing pin disassembly disk on the buttstock and a slot through the buttstock for the sling. As finally produced at ZB, the Mauser rifle has a stamped oversize trigger guard, firing pin dismounting hole in the butt plate, slotted buttstock, German-type stamped bands and modified gas escape holes in the bolt. This rifle was produced until circa 1950 and was sold to Israel and Pakistan. It is actually the Kar 98k in its last production version.

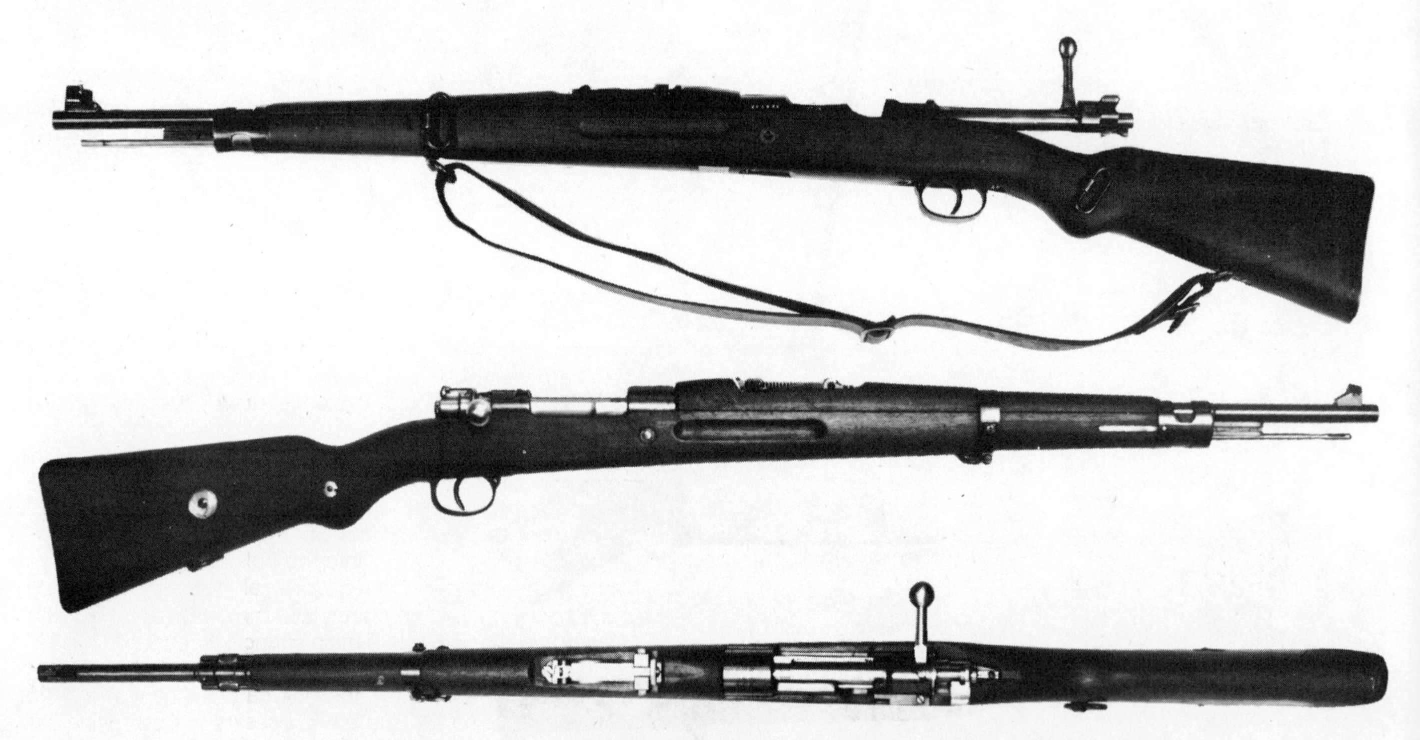

Czech 7.92mm Model 24 (VZ 24).

THE 7.92mm CZECH MODEL 24 RIFLE

The Model 24 Rifle was the standard Czech rifle prior to World War II and as previously stated was widely used by Germany and other countries. It uses the basic Mauser 98 action and differs from the German Kar 98k mainly in fittings, having a full length handguard, sling swivels on both side and underside of the stock and a straight bolt handle.

CHARACTERISTICS OF THE 7.92MM CZECH MODEL 24 RIFLE

Caliber: 7.92mm
System of operation: Bolt action, manually operated.
Weight: 8.98 lbs.
Length overall: 43.3 in.
Barrel length: 23.2 in.
Feed mechanism: 5-round staggered row, non-detachable box magazine.
Sights: Front: Barley corn.
Rear: Tangent graduated from 300–2000 meters by 100-meter steps.
Muzzle velocity: 2700 f.p.s.

Dismounting and Assembling the Bolt, M24

Before removing the bolt from the receiver, turn the safety catch into its central position between its safe and fire position, i.e., with the wing turned upwards. The bolt can be removed from the receiver only when the bolt stop is moved to the side with a slight pressure, whereby the groove in the receiver is (fig. 1) opened to permit bolt passage.

Turn the bolt sleeve to the left, then the latter with the safety catch, the cocking piece and the firing pin may easily be separated from the bolt. While removing these parts, the main spring is still in a state of compression (fig. 2).

For further dismounting, put the striker with its point turned downward against a convenient bearing (for this purpose a hole is provided in the stock sleeve) (fig. 3). Press home the main spring until the cocking piece can be turned 90°, remove the latter and, in releasing the pressure of the spring, take out the firing pin. The safety catch may be dismounted by turning the wing to the right side.

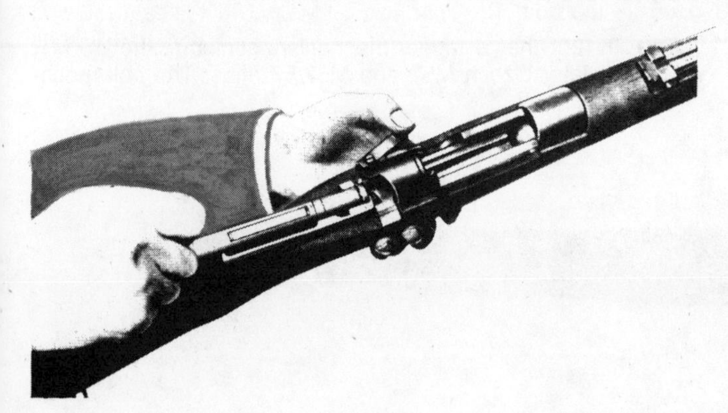

Figure 1.

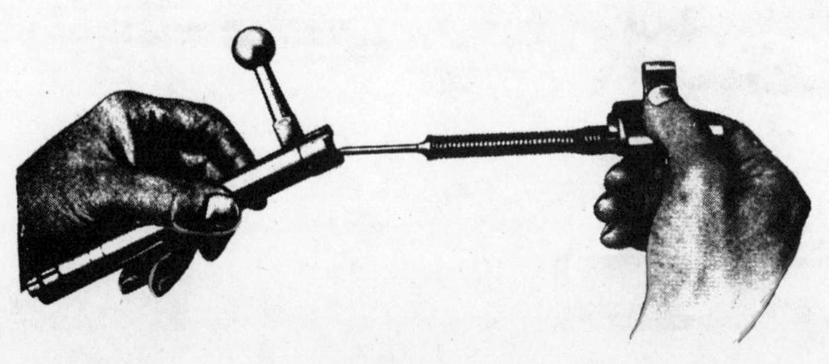

Figure 2.

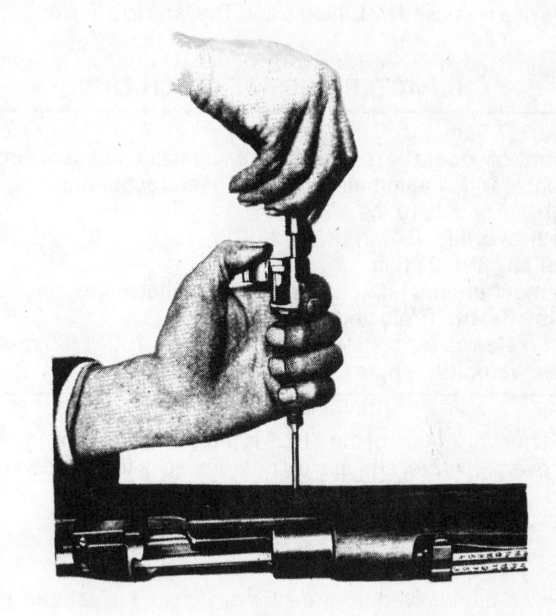

Figure 3.

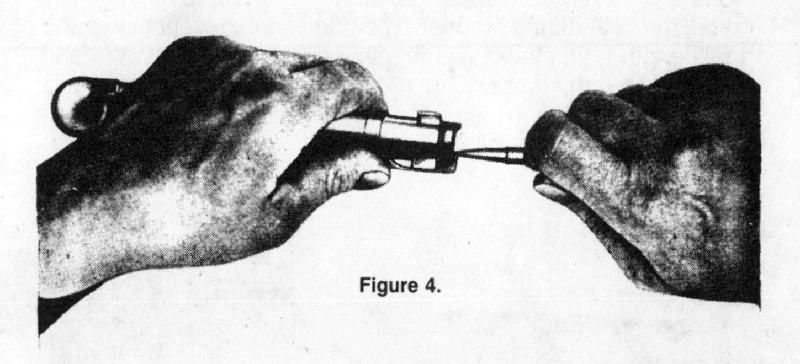

Figure 4.

To remove the bolt sleeve lock, pull it into a position of 90°, where its spindle can be turned in the notch of the bolt lock.

To remove the extractor, pull it to the side with the help of a solid object (for instance with a cartridge) till it disengages the dovetail before the bolt lug (fig. 4). Then turn it to an angle of 90°, and, holding it against a table edge, draw it out of the groove of the extractor ring.

The extractor ring of the bolt should not be dismounted. The extractor and the bolt sleeve lock should seldom be removed.

The assembling is done in the opposite order. Mount the extractor on the extractor ring until it can be turned and engaged in the groove of the bolt. Put into the bolt lock, the bolt sleeve lock spring, then the bolt sleeve lock itself, and turn 90°. Place the main spring on the firing pin, place the safety catch into the bolt sleeve, and compress the main spring, just enough to insert the cocking piece, and turn 90°. The bolt sleeve, thus assembled, is screwed on the bolt. Now the bolt can be pushed into the receiver. When doing so, it is not necessary to push aside the bolt stop, for it disengages itself by the pressure on the bolt.

CZECH SEMIAUTOMATIC RIFLES

The first truly Czech semiautomatic rifle was probably the Netsch, developed at CZ between 1922–24. This 7.92mm rifle was gas operated with a drum magazine. The ZH 29 was the first Czech rifle of this category to become well known and exported in any quantity. The design is credited by the Czechs to Emmanuel Holek of ZB. This rifle, chambered for the .276 Pedersen cartridge, was tested by the United States at Aberdeen Proving Ground in 1929.

This rifle was used by Ethiopia and Thailand to a limited extent.

CHARACTERISTICS OF CZECH ZH29

Caliber: 7.92mm.
System of operation: Gas, semiautomatic fire (specimens capable of full automatic fire have been reported).
Weight: Approx. 10 lbs.
Length overall: 45.5 in.
Barrel length: 21.5 in.
Feed mechanism: 10- or 25-round, detachable box magazine.
Sights: Front: Protected blade.
Rear: Tangent leaf adjustable from 100–1400 meters.
Muzzle velocity: Approx. 2650 f.p.s.

An unusual feature of the ZH 29 is the aluminum cooling jacket fitted over the barrel and gas cylinder ahead of the wooden hand guard.

THE CZECH 7.92mm ZK420 RIFLE

The ZK420 evolved from earlier designs by Josef and František Koucký. This 7.92 × 57mm rifle appeared in 1942. The ZK420 went through a series of prototype versions until the final model—the ZK420S—appeared after World War II. As originally made, the ZK420 had Mauser-type bands and bayonet stud and a full-length type stock. The magazine was not removable and was loaded with chargers. In 1946, the first modification of this design appeared; the 1946 version has a sporter-type stock, a removable magazine, and in general resembles the ZK420S. The outstanding external differences between the 1946 Model and the ZK420S are the configuration of the magazine and the prominent safety catch lever—as that of the US M1 rifle—on the ZK420S. The latter rifle appeared between 1947 and 1950 and was the last of the series.

The ZK420 series of rifles received extensive testing abroad. The 1942 and 1946 Models were tested by Denmark, and the 1946 Model was tested by the United Kingdom, Sweden, Ethiopia, Egypt, Israel and Switzerland. A version of the ZK420S with permanently attached bayonet was made for Israel for test purposes. The 1946 Model was made in 7.92mm, 7mm, .30-06 and 7.5mm Swiss. It is somewhat surprising that the Czech Army never adopted this weapon, since it was extensively developed and certainly advertised throughout the world; the ZK420 appears in a post-1948 ZB catalog. The design in certain features, i.e. bolt and gas system, appear to be at least equal to, if not superior to, that of the rifle finally adopted—the Model 52. The ZK420 series weapons are, however, generally heavier designs and undoubtedly much more expensive to manufacture than is the Model 52.

CHARACTERISTICS OF THE ZK420S RIFLE

Caliber: 7.92mm.
System of operation: Gas, semiautomatic fire only.
Weight: 10.58 lbs.
Length overall: 41.73 in.
Barrel length: 21.65 in.
Feed mechanism: 10-round, double staggered row, detachable box magazine.
Sights: Front: Hooded post.
Rear: Notched tangent with ramp.
Muzzle velocity: Approx. 2700 f.p.s.

The ZK420S has a trigger mechanism similar to the US M1, as do the later Czech M52 and M52/57 rifles. The bolt mech-

Czech 7.92mm Model ZK-420S

Czech ZH 29 Semiautomatic Rifle.

anism of this weapon bears a superficial resemblance to that of the US M1, but it actually more closely resembles the Soviet AK47. The body or shaft of the one-piece bolt is enclosed by the bolt carrier and is cammed by a camway within the bolt carrier. The head of the bolt bears the locking lugs and protrudes forward of the bolt carrier.

CZECH 7.62mm MODEL 52 RIFLE

CHARACTERISTICS OF 7.62 × 45mm MODEL 52

Caliber: 7.62mm, Czechoslovak Model 52.
System of operation: Gas, semiautomatic fire only.
Weight, loaded: 9.8 lbs.
Length, overall: 39.37 in.
Barrel length: 20.66 in.
Feed mechanism: 10-round, double staggered, detachable box magazine (loaded with 5-round chargers).
Sights: Front: Hooded blade (removable hood).
Rear: Notched tangent with curved ramp.
Muzzle velocity: 2440 f.p.s.

The Model 52 is chambered for the Czechoslovak 7.62mm Model 52 cartridge. Its design is a combination of several older designs added to a few native ideas. The gas system is generally similar to that of the German MKb 42W and some of the earlier Walther commercial semiautomatic rifle designs. The trigger mechanism is similar to that of the United States M1 rifle. The bolt appears to be a native design; it is somewhat unusual in that it is a tipping bolt design with frontal locking lugs. For the first time, the Czechoslovaks put a nondetachable bayonet on one of their rifles. The Model 52 makes fairly extensive use of stampings, but, all in all, the design is hardly revolutionary and in some respects is not too remarkable considering the date it appeared—1952.

The Czechs have issued a slightly modified version of the 52 rifle called the Model 52/57, chambered for the Soviet designed 7.62mm M43 cartridge. The Model 52 and Model 52/57 rifle are no longer standard in the Czech Army.

How to Load and Fire the Model 52 Rifle

Pull the operating handle all the way to the rear; since the weapon has a bolt-holding-open device, the bolt and carrier will remain open. Insert the end of a five-round charger in the charger (clip) guide at the front of the receiver cover and force the cartridges down into the magazine with the thumb of the left hand. Repeat this process with another five-round charger; then draw the operating handle slightly to the rear and release. The bolt and carrier will run home, chambering a cartridge. The weapon is now loaded and, with safety off, the trigger will cause it to fire. The trigger must be released after each shot. The safety is similar to that on the United States M1 rifle. When the safety is forward the rifle will fire; when the safety is drawn to the rear, the weapon is on "safe."

Field Stripping the Model 52 Rifle

The magazine is released by pressing forward on the magazine lever. After clearing the weapon to assure that the chamber is empty, disassemble the receiver cover—carefully, as the driving spring is always compressed. Push the receiver cover forward until the side rails are out of the receiver slots. Lift the cover slightly and slide back along the top of the receiver until the driving spring snaps down. Hold the bolt carrier to the right and slide it to the rear until it reaches the disassembly notch. Remove bolt carrier and bolt. To remove bolt from bolt carrier, slide bolt to the rear and press down the front end until it is disengaged from the bolt carrier.

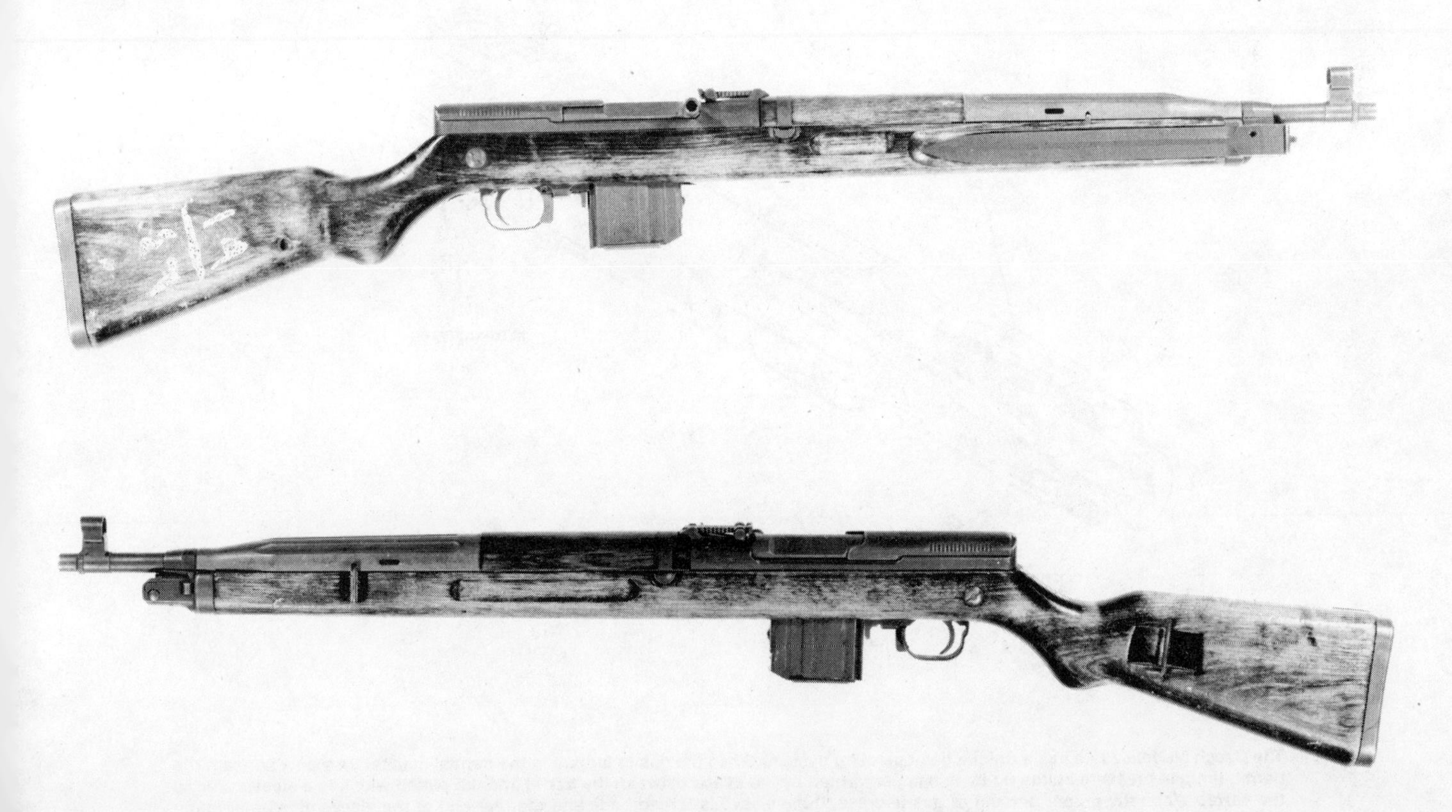

Czechoslovak 7.62 × 45mm Model 52 Semiautomatic Rifle.

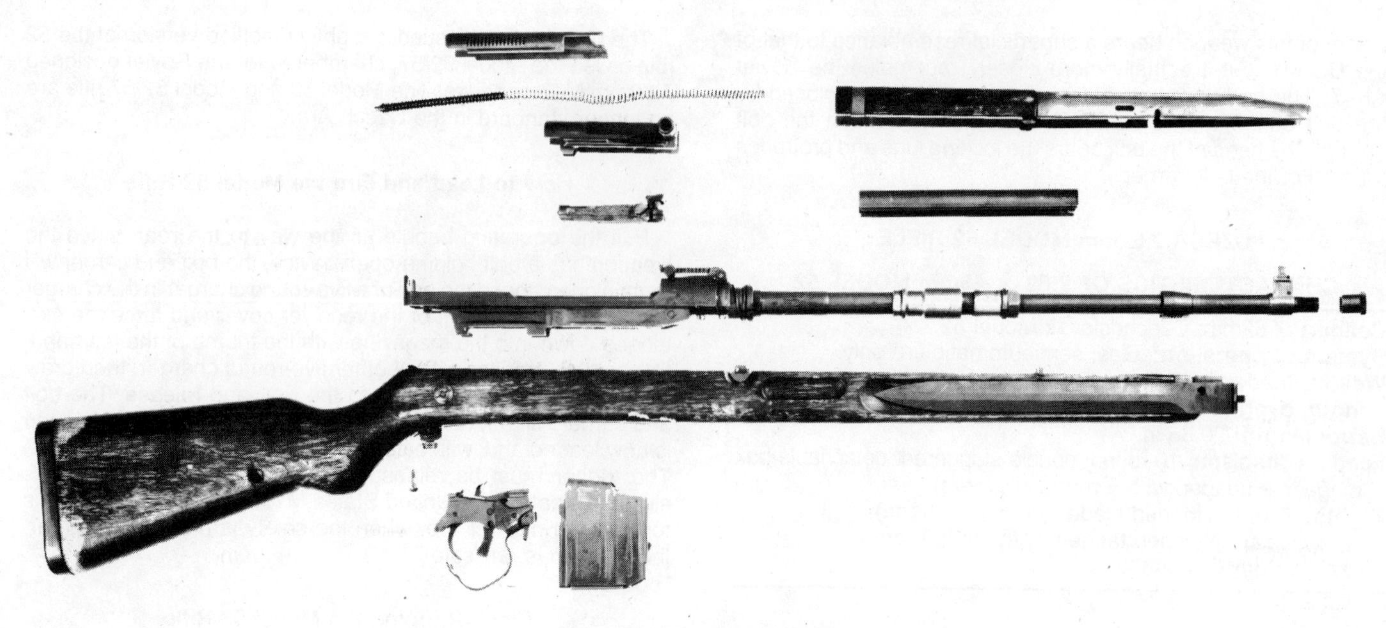

Czech Model 52—field-stripped.

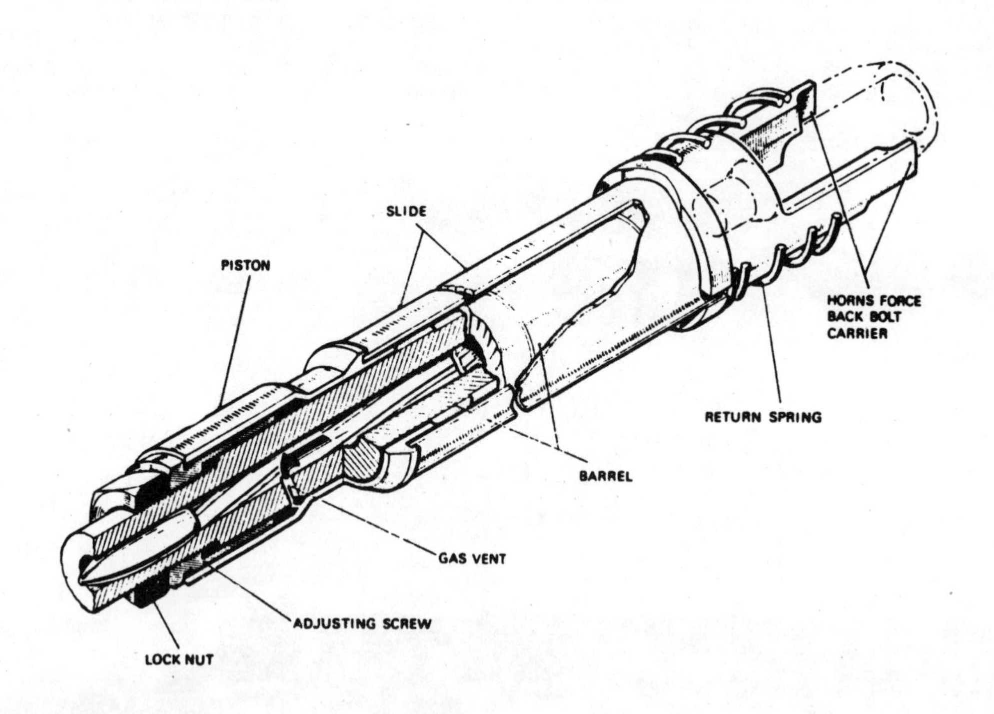

The Czech Model 52 rifle has a unique gas-operating system. While the gas is tapped in the normal manner, in a port beneath the barrel, the gas pressure builds up in a chamber formed by the space between the barrel and the piston which is a sleeve around the barrel. When the proper amount of gas is present, the sleeve is forced back and the rear end of the sleeve acts as a tappet imparting a blow to the bolt carrier. The bolt and bolt carrier travel a short distance before separation. An adjusting screw permits the volume of the chamber to be varied as required by operating conditions.

CZECH SUBMACHINE GUNS

The Czechoslovaks have come out with a large number of submachine gun designs since World War II. Most of these designs appeared only in prototype form and were intended for overseas sales. Among these were the ZK466, CZ247, ZB47 and ZK476. These weapons are not used by the Czechoslovak Army. Before the Communists took over Czechoslovakia, her army used the 9mm Parabellum as the service cartridge for pistols and submachine guns. Two submachine guns chambered for that cartridge were adopted by the Czechoslovak Army: the Model 23 and the Model 25 submachine guns. After the adoption of the Soviet 7.62mm pistol cartridge by the Czechoslovaks, two weapons similar to the Models 23 and 25, but chambered for the 7.62mm pistol cartridge, were produced, the Model 24 and the Model 26. These weapons have been replaced, for most purposes, by the Model 58 assault rifle and in some cases by the 7.65mm Model 61 submachine gun.

THE CZECH MODEL 61 SUBMACHINE GUN (SKORPION)

This weapon might be described as a "machine pistol" since it is designed to be fired from one hand as well as from the shoulder. It could be considered the Czech equivalent of the Soviet Stechkin machine pistol or the M1932 Mauser. It uses a relatively low powered cartridge, the 7.65mm Browning short or the .32 ACP as it is called in the United States, and is relatively easy to control in full automatic fire from the shoulder because of that fact. As a matter of interest, the .32 ACP cartridge as loaded in the United States has less muzzle energy than the .22 Long Rifle cartridge in high velocity loads. With a weapon of the M61 type, however, there is a possibility of obtaining multiple hits on a target. The Model 61 is also manufactured in Yugoslavia where it is called the M61. Czech-made export models of the Skorpion include the vz 64 in 9 × 17mm (.380 ACP), vz 65 in 9 × 18mm (Makarov), and vz 68 in 9 × 19mm Parabellum. The vz 68 is slightly larger than the vz 61.

CHARACTERISTICS OF MODEL 61 SUBMACHINE GUN

Caliber: 7.65mm (.32 ACP).
System of operation: Blowback, selective fire.
Length overall: With stock fixed: 20.55 in.
With stock folded: 10.62 in.
Barrel length: 4.5 in.
Weight: 2.87 lbs.
Sights: Front: Protected post.
Rear: Flip-over notch graduated for 75 and 150 meters.
Feed Device: 10- or 20-round detachable staggered row box magazine.
Muzzle velocity: 1040 f.p.s.
Cyclic rate: 750 r.p.m.

How to Load and Fire the Model 61 Submachine Gun

The magazine can be removed by pressing the button located on the left side of the receiver in front of the trigger guard. Pull bolt to rear; bolt operating knobs are on both sides of the receiver. Set the weapon on safe by pushing the safety-selector lever, located over the pistol grip on the left side of the receiver, down to its central position. Insert a loaded magazine and set safety-selector lever for type of fire desired—forward for automatic fire—to "20" and to the rear for semiautomatic fire to "1." Press trigger and weapon will fire; bolt remains open on the last shot.

Field Stripping the Model 61 Submachine Gun

Remove magazine and pull bolt to the rear to clear gun. Fix shoulder stock. Push out pin—at lower front of the receiver—to the left. Pull receiver foward and hinge upward; remove bolt operating knobs. Remove bolt and operating springs assembly. Further disassembly is not recommended. Reassemble by reversing above procedure.

How the Model 61 Submachine Gun Works

This weapon is a pure blowback, i.e., the cartridge is held in place during firing by the weight of the bolt and the recoil spring. The rate reducer, used to lower the cyclic rate, is a hook at the

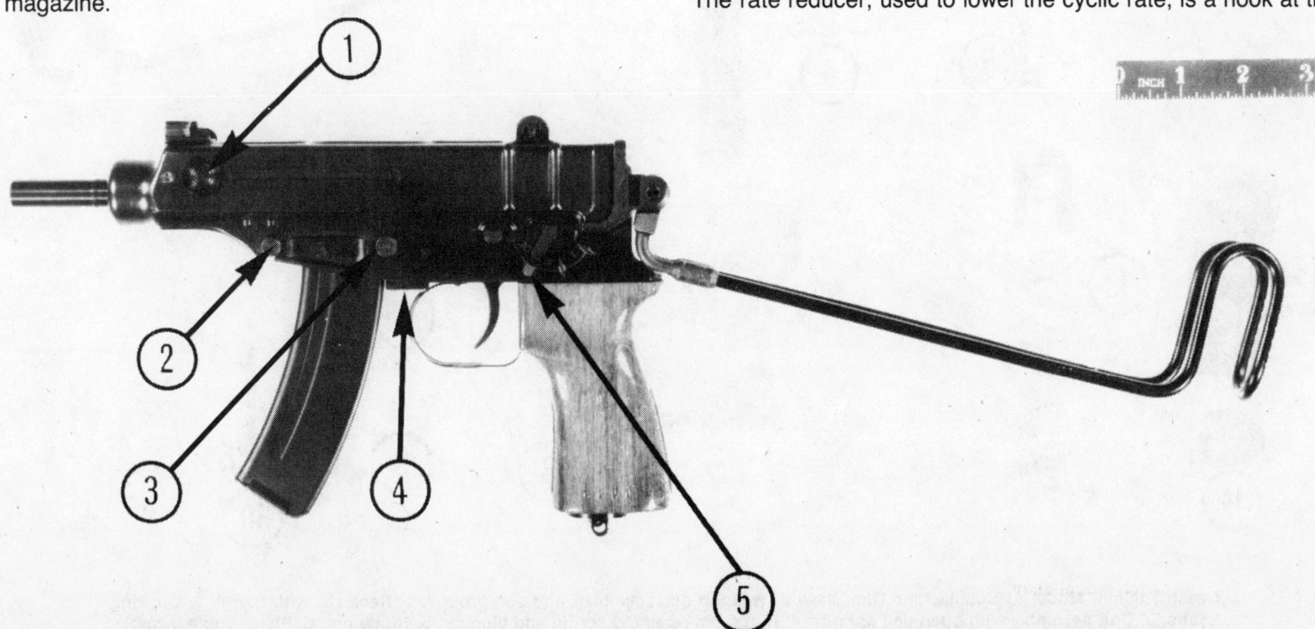

Czech 7.65mm Model 61 Submachine Gun with stock fixed. 1. Cocking knob; 2. Takedown pin; 3. Magazine catch; 4. Bolt stop; and 5. Selector lever.

rear of the receiver, which holds the bolt to the rear momentarily after each shot. This hook is released by pressure from a spring loaded tripper-plunger mounted in the pistol grip. As the bolt comes to the rear, it cams down this tripper-plunger compressing its spring. The tripper-plunger and spring are mounted in a tube and the tripper-plunger travels a considerable distance down the tube before it is pushed up by the spring, causing the hook to tip up and release the bolt. This system, which is quite similar to that used on the Soviet Stechkin machine pistol, has the effect of reducing the cyclic rate of fire. The bolt telescopes the barrel as in the earlier Model 23 series submachine guns but is square in configuration like the Israeli UZI, rather than round like the Model 23.

Special Note on Model 61 Submachine Gun

There is a silencer available for this weapon that should be effective due to the relatively low velocity of the bullet fired. Luminescent night sights are also available.

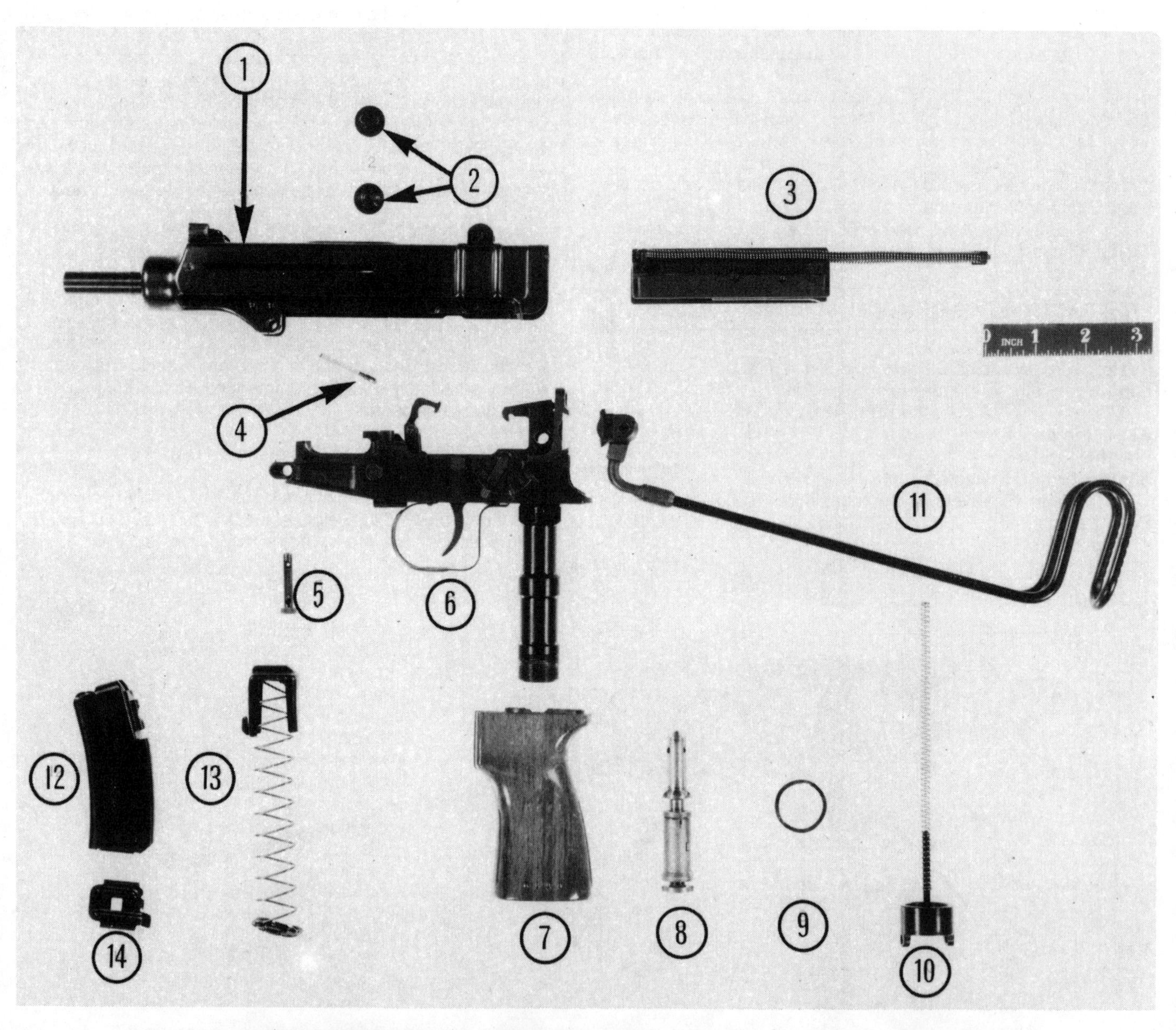

Czech 7.65mm Model 61 Submachine Gun. View of weapon disassembled into sub-groups. 1. Receiver with barrel. 2. Cocking knobs. 3. Bolt assembly with operating springs. 4. Hinge pin retaining spring and plunger. 5. Hinge pin. 6. Pistol grip assembly. 7. Pistol grip. 8. Actuator assembly. 9. Pistol grip cap spring washer. 10. Pistol grip cap and actuator spring. 11. Stock assembly. 12. 10-shot magazine body. 13. Magazine follower and spring. 14. Magazine floorplate.

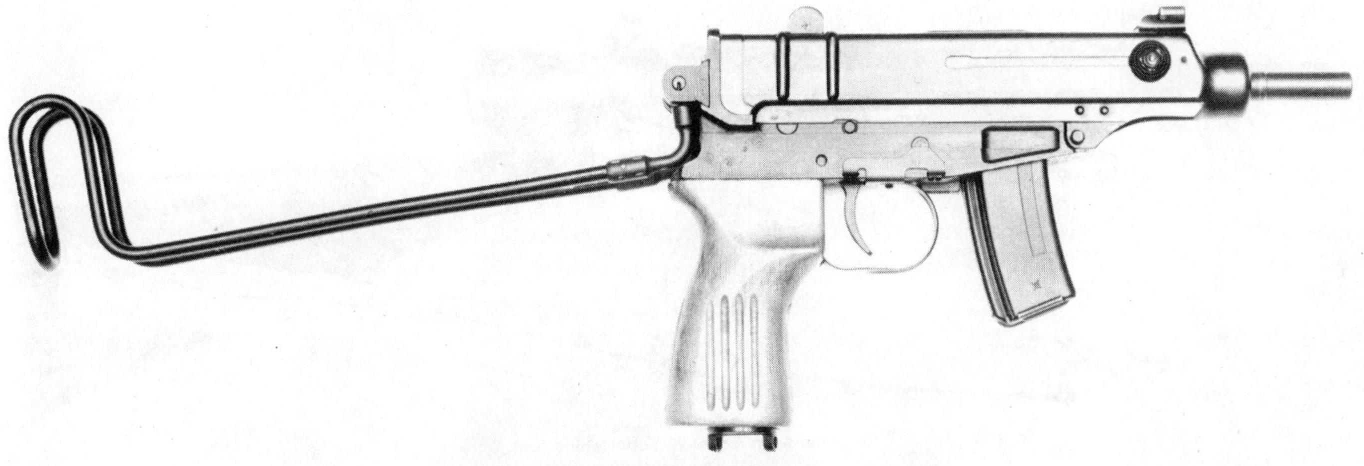

The 7.65mm Model 61 Skorpion with a 10-shot magazine.

The 7.65mm Model 61 Skorpion with silencer.

THE CZECH MODEL 23 AND MODEL 25 SUBMACHINE GUNS

CHARACTERISTICS OF MODELS 23 AND 25

Caliber: 9mm Parabellum.
Weight, M23, loaded: 8 lbs; with 24-round magazine. 8.4 lbs., with 40-round magazine.
M25, loaded: 8.7 lbs. with 24-round magazine. 9.0 lbs., with 40-round magazine.
Length, overall: 27 in. for Model 23 and Model 25 with stock extended; 17.5 in. for Model 25 with stock folded.
Barrel length: 11.2 in.
Feed Mechanism: 24-round or 40-round detachable, staggered, box magazine.
Sights: Front: Hooded barleycorn.
Rear: V-notch on rotary base; adjustable for 100, 200, 300, and 400 meters.
Muzzle velocity: 1470 f.p.s.
Cyclic rate of fire: 600–650 rounds per minute.

The Model 23 and Model 25 are basically the same weapon, but the Model 23 has a wooden stock, and the Model 25 has a folding metal stock. The weapons have several outstanding features and are good examples of modern submachine gun construction. They share most of their unusual features with the Israeli UZI submachine gun. Among these features are:

The magazine well is in the pistol grip. This gives the magazine far better support than is normally obtained with a submachine gun.

The guns have a very short overall length in comparison to their barrel length. This is made possible by a hollow bolt, which telescopes the rear 6 $\frac{1}{8}$ inches of the barrel. Only a 1 $\frac{7}{8}$-inch length of the 8-inch bolt is solid; the remainder is hollowed out to telescope the barrel.

The trigger mechanism on the Czechoslovak weapons is so designed that a short pull on the trigger gives semiautomatic fire, while pulling the trigger all the way to the rear gives full automatic fire.

The ejection port is closed at all times to the entry of dirt, except when the weapon is ejecting.

These weapons also have an unusually simple method of disassembly and assembly.

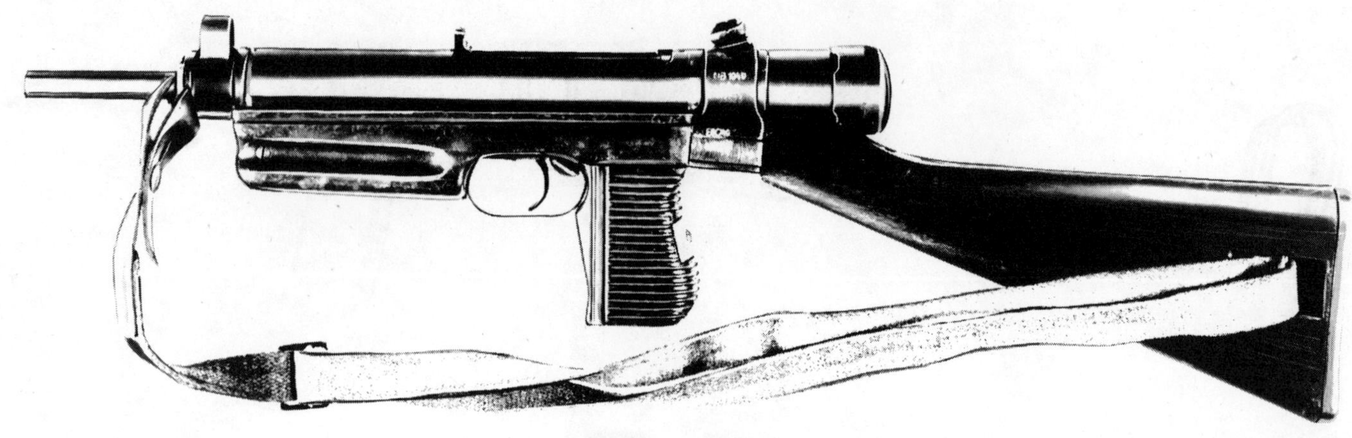

Czech 9mm Model 23 Submachine Gun.

Field Stripping the Model 23 and Model 25 Submachine Guns

Disassembly of Weapon. Remove magazine. Push forward on button in center of receiver barrel jacket cap; at the same time, turn cap $\frac{1}{8}$ turn to right or left and remove to the rear.

Slide bolt assembly rearward by means of the operating handle.

Pull trigger, and slide bolt assembly to the rear, out of receiver/barrel-jacket assembly.

Use bolt assembly as tool to loosen barrel locking nut. Place bolt assembly around barrel, with slots at front of bolt engaging lugs of barrel locking nut. Unscrew locking nut by turning bolt counterclockwise.

Pull barrel forward from receiver/barrel-jacket assembly.

To unfasten stock, remove screw holding the metal neck of stock to the stock support. To remove from receiver/barrel-jacket assembly, pull stock to the rear and left.

Use same procedure for Model 23.

Assembly. Assembly is accomplished by performing the steps for disassembly in reverse order.

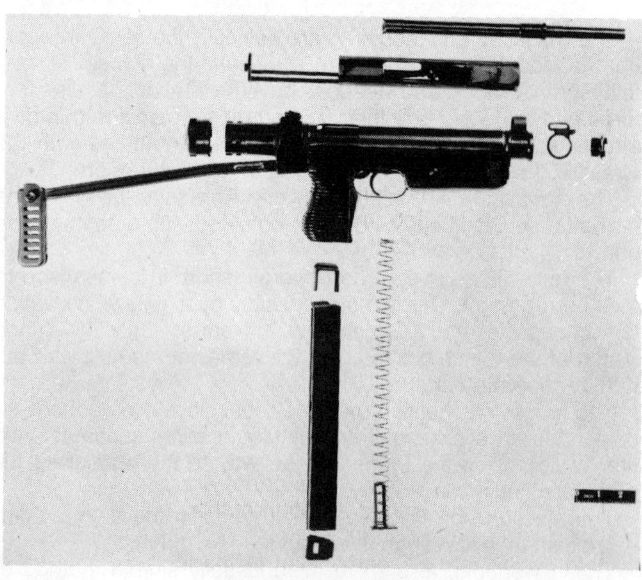

Model 25 field-stripped.

How the Model 23 and Model 25 Submachine Guns Work

Chambering. With the bolt in rearward position, the trigger is pulled. This depresses the sear, releasing the bolt. The operating spring drives the bolt forward. The lower edge of the bolt face strips a round from the magazine and forces it forward into the chamber.

Locking. No locking step takes place, since the weapon is of the straight blowback type.

Firing. Firing occurs when the operating spring drives the bolt forward against the chambered round, and the fixed firing pin strikes the primer.

Extraction. The extractor engages in the extraction groove of the round as it is chambered. When blowback drives the bolt rearward, the extractor pulls the cartridge case out of the chamber.

Ejection. When the cartridge case (held by the recoiling bolt) clears the chamber, it strikes the tip of the stationary ejecting rod and pivots to the right. Simultaneously, the ejection ports of the bolt and receiver move into alignment, allowing the cartridge case to be thrown clear of the weapon.

Cocking. Since the gun fires only from an open bolt, it is cocked by moving the bolt rearward until the sear engages the sear notch in the bolt, thus holding the bolt to the rear against pressure of the operating spring. Cocking may be accomplished either manually or, when firing, by blowback.

Semiautomatic Fire. The sear assembly is pivoted at the front and contains a spring-loaded plunger extending to the rear. The trigger, which is pivoted at the rear, has a forward projection that rests on the sear plunger. As the trigger is pulled, its projection bears against the sear notch in the bolt. The bolt then moves forward to chamber and fire the round. As the trigger is pulled beyond the point at which the bolt is released, the trigger projection rotates out of engagement with the sear plunger, thus allowing the sear to rise under pressure of the sear spring. Upon firing, the bolt is blown back to the rear until the sear engages the sear notch in the bolt.

NOTE. Since no positive disconnector is provided, a slow squeeze on the trigger can result in short bursts.

Full Automatic Fire. In full automatic fire, the trigger is pulled all the way to the rear, beyond the semiautomatic fire position. A projection on the trigger now engages the sear directly, disengaging it from the bolt and allowing full automatic fire until the trigger is released or the magazine emptied.

Czech 9mm Model 25 Submachine Gun.

How to Load and Fire the Model 23 and Model 25 Submachine Guns

Draw the operating handle to the rear. The weapon fires from an open bolt, and therefore will remain to the rear. Insert a loaded magazine in the well of the pistol grip. Squeeze the trigger. If the trigger is pulled about halfway to the rear, the weapon will fire single shots with occasional doubles. If the trigger is pulled all the way to the rear, the weapon will fire automatically until it is empty or until the trigger is released. To remove the magazine, pull the magazine catch (located at the bottom rear of the pistol grip) to the rear, and pull the magazine down and out of the weapon. The safety blocks the rear of the trigger, and also locks the bolt. To put the weapon on SAFE, push the safety lever to the right.

7.62mm Replacement for 9mm Models 23 and 25

The 7.62mm submachine gun, which replaced the 9mm Models 23 and 25 in the Czechoslovak Army, is basically the same as the Models 23 and 25. A few minor changes have been made, but function, loading, firing, assembly and disassembly remain the same. The major change is the chambering of the weapons for the 7.62mm pistol cartridge. The heavy loading of the Czechoslovak-made 7.62mm pistol cartridge gives the weapons a velocity that falls in the class of the United States M1 and M2 carbines.

CHARACTERISTICS OF 7.62MM SUBMACHINE GUN MODEL 24

System of operation: Blowback, selective fire.
Length, overall: 27 in.
Barrel length: 11.2 in.
Feed mechanism: 32-round detachable, staggered-box magazine.
Sights: Front: Hooded blade.
Rear: Square notch on rotary base, adjustable for 100, 200, 300, and 400 meters.
Muzzle velocity: 1800 f.p.s.
Cyclic rate of fire: 600–650 r.p.m.
Weight, loaded: 8.8 lbs.

Note that only the weight and overall length with stock folded will be different for the version with metal stock.

Special Note on Czechoslovak Submachine Guns

The Czechoslovak 9mm Models 23 and 25 submachine guns, as well as the 7.62mm submachine guns, have a magazine filler built into the right side of their plastic fore-ends. The ammunition for these weapons is packed in 8-round chargers (clips). The charger is laid base down in the guide slot of the magazine filler, with the weapon on its left side. The empty magazine is then pushed down the guide, mouth forward (toward the muzzle of the weapon), and the cartridges are stripped into the magazine.

OBSOLETE CZECH SUBMACHINE GUNS

THE CZECH ZK383 SUBMACHINE GUN

The ZK383 was designed and produced in some quantity prior to World War II and was carried in a post-1948 ZB catalog. The weapon was designed by Josef and Frantisek Koucky and was produced in three slightly different designs. The basic ZK383 has a folding bipod attached and was used by Bulgaria and several South American countries. It was continued in production at ZB (then called Waffenwerke Brunn by the Germans) and was apparently used by the Waffen S.S. to some extent during the war. The ZK383P was developed for police use and does not have a bipod or a removable barrel like the other models.

The ZK383H is the third variation of this weapon and differs mainly in having a folding-type magazine housing fitted to the bottom of the weapon rather than to the left side as with the other models.

CHARACTERISTICS OF THE ZK383

Caliber: 9mm Parabellum.
System of operation: Blowback, selective fire.
Length overall: 34.4 in.
Barrel length: 12.8 in.
Weight: 9.6 lbs.
Sights: Front: Protected blade.
Rear: V-notch tangent graduated from 100–800 meters.
Feed mechanism: 30-round detachable, staggered row box magazine,
Muzzle velocity: approx. 1400 f.p.s.
Cyclic rate: 500 r.p.m., normal; 700 r.p.m., accelerated.

The bolt of this weapon has a six-ounce removable block. Removal of this block provides the higher rate of fire.

Czech 9mm ZK383 Submachine Gun with bipod fixed.

CZECH MACHINE GUNS

THE CZECH MODEL 59 MACHINE GUN

The Model 59 is called a "Universal" machine gun by the Czechs. This basically means that it is a general purpose machine gun, similar to the US M60, which is used on a bipod as a light machine gun and on a tripod as a heavy machine gun. The Model 59 comes equipped with heavy and light barrels, which are used in its heavy gun and light gun roles, respectively. When used with the light barrel, this weapon is called the Model 59L.

CHARACTERISTICS OF THE MODEL 59

Caliber: 7.62mm rimmed (7.62mm Russian).
System of operation: Gas, automatic only.
Length overall: W/heavy barrel: 47.9 in.
W/light barrel 43.9 in.
Barrel length:
Heavy: 27.3 in. w/flash hider.
Light: 23.3 in. w/flash hider.
Weight: W/heavy barrel on tripod, 42.4 lbs.
W/light barrel on bipod, 19.1 lbs.
Feed device: 50-round, non-disintegrating metallic link belt.
Muzzle velocity: W/heavy barrel: 2723 f.p.s.
W/light barrel: 2657 f.p.s.
Cyclic rate: 700–800 r.p.m.

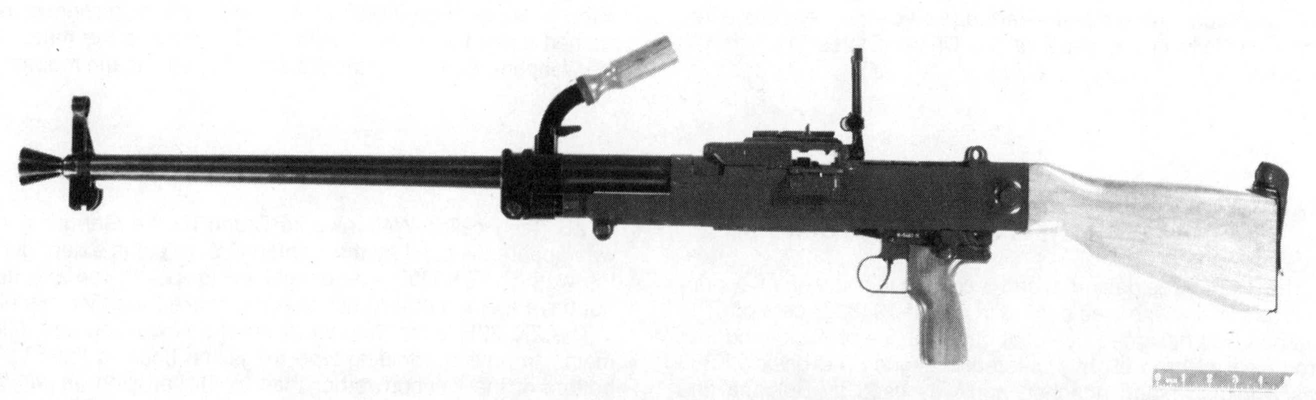

Czech 7.62mm Model 59 Machine Gun without bipod.

Model 59 Machine Gun with telescopic sight.

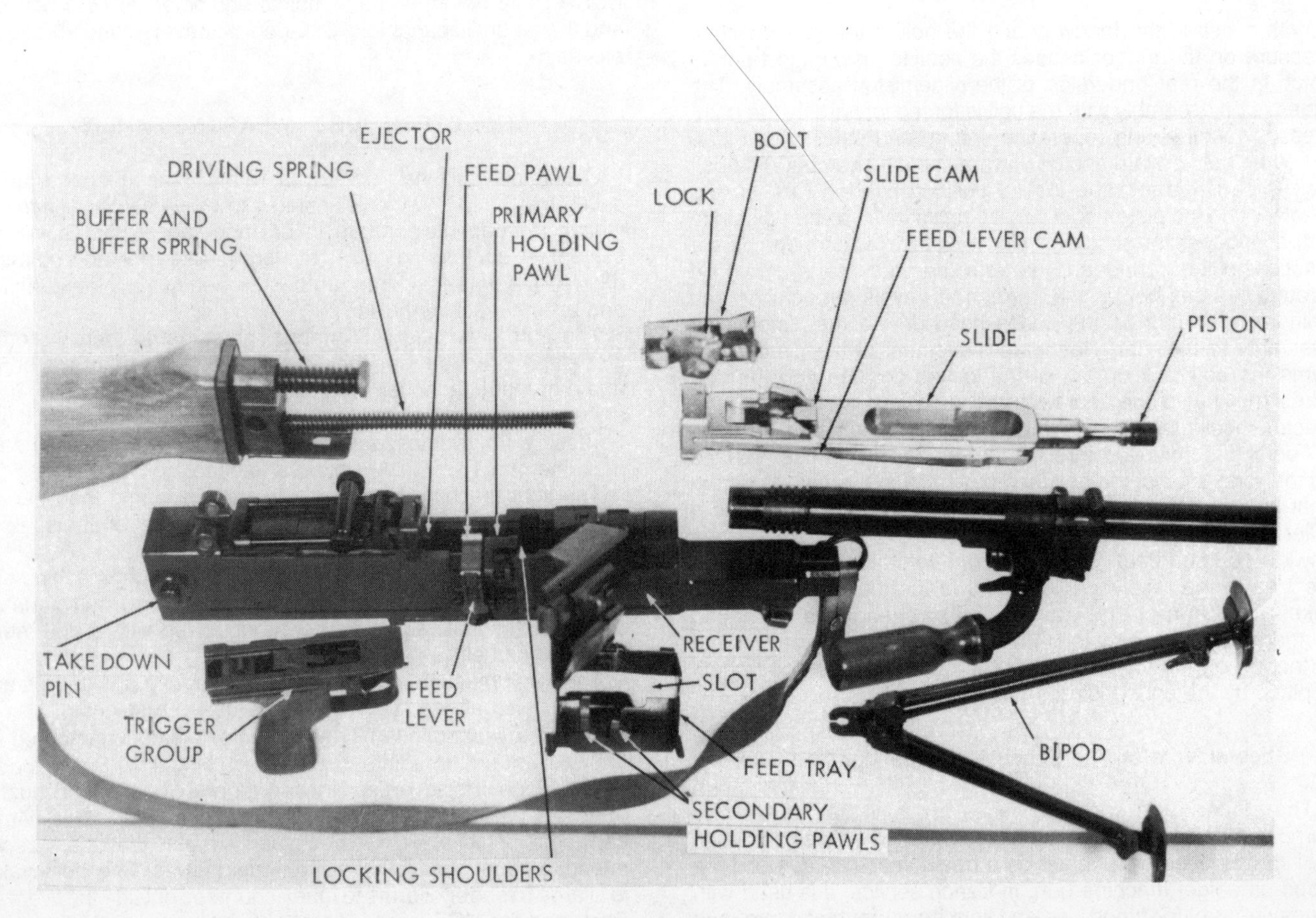

Model 59, field stripped.

How to Load and Fire the Model 59 Machine Gun

The linked cartridges come in 50-round, non-disintegrating link belts; five of these belts are usually joined together for use with the weapon when it is used on the tripod. When used as a light machine gun—on bipod—a box with a 50-round belt may be attached to the side of the receiver. The cover is opened, and the belt—open side of links down—is laid on the feedway from right to left; close cover. If bolt is forward, push lug, which protrudes from the safety mounted on left rear side of the pistol grip/trigger group, down; pull the trigger and push pistol grip as far forward as it will go, release the trigger then draw to the rear. The belt may be inserted whether the bolt is in the forward or rearward position. If the trigger is pulled, the weapon will now fire. To put gun on "safe", push safety lever down. The gas cylinder block has four ports and can be adjusted as required for reliable operation.

Field Stripping the Model 59 Machine Gun

Check to insure weapon is not loaded and let bolt forward. Remove barrel by opening cover and turning one quarter turn to the right and pulling barrel forward out of the receiver. Push pin at middle rear of receiver from left to right. **Bolt must be forward so that tension on recoil spring is released.** Remove butt; recoil spring and buffer spring are mounted in the butt. Withdraw slide/pistol assembly and bolt assembly from the rear of the reciever by pulling pistol grip to the rear. Pistol grip/trigger assembly will also come off the receiver. The feed plate can be lifted up off the receiver. Reassemble in reverse order.

How the Model 59 Machine Gun Works

With a belt in the feedway and the bolt in the rear position, pressure on the trigger causes the sear to disengage from its notch in the rear underside of the slide/piston assembly. The slide/piston assembly with the bolt is forced forward by the compressed recoil spring (operating spring) and strips a cartridge out of the belt forward and downward into the chamber. The bolt lock is cammed into the locked position by cam rails on the receiver and the piston post continuing its travel a short distance further engages the striker causing it to strike the cartridge primer functioning the cartridge. Gas from the cartridge is drawn off through the gas port in the barrel and travels through the gas cylinder impinging on the piston head driving the piston/slide assembly to the rear. Rearward movement of the piston/slide cams the bolt lock up out of the locked position and the bolt starts moving to the rear with the empty cartridge case; the ejector knocks the case downward out of the bottom of the gun.

Feeding is achieved through the operation of a cam surface on the slide against the belt feed pawl, which is mounted on the right side of the receiver. At the bottom of the feed pawl is a roller that contacts the cam surface on the slide causing the pawl to go in and out, i.e., move from right to left, in relation to the side of the receiver. Movement of the feed pawl pulls the linked rounds in one at a time. The pawl engages the linked cartridge as the slide/piston moves to the rear and is then moved to the left on the forward run of the bolt. The cartridge is held in place by the spring loaded belt-feed pawl.

Special Note on the Czech Model 59 Machine Gun

The Model 59 is one of a series of "universal" or general purpose machine guns which have appeared only since World War II. It is used on a tripod as a heavy machine gun or on a bipod as a light machine gun. In Czech service, it is used with a heavy barrel for tripod use and light barrel for bipod use, and a medium weight barrel is advertised as well in trade brochures. The weapon has been made in 7.62mm NATO, in which caliber it is called the Model 59N, to attract sales to non-Communist nations. Barrels, gas piston, gas cylinder block and bolt are chrome plated. There are spring loaded dust covers over the feed, link ejection and case ejection ports. The case ejection port cover is opened and closed by pulling the trigger.

The Model 59 is an interesting gun in many ways. There is no major feature of the weapon that is new in concept. The feed mechanism is basically the same as that of the pre-World War II ZB37 and is also used for belt feed in the Model 52.

The cocking mechanism is also similar to that of the ZB-37 and exactly the same as that of the Model 52. The bolt mechanism is a modification of the basic ZB-26, ZB-37, Model 52; the separate bolt locking lug piece is similar to that used on the Czech Model 58 Assault rifle. The method of barrel removal is the same as that of the Model 52 machine gun. A notable feature of the Model 59 as opposed to the previous Model 52 is the comparatively small number of stampings or fabrications used in its manufacture. The receiver is a milled forging, and most other parts of the weapon except the feed plate and dust covers seem to be forgings as well. The design is basically good, but as pointed out above, hardly novel.

CZECH 12.7mm QUAD DShK M1938/46 HEAVY MACHINE GUN (M54)

This weapon consists of four Czechoslovak-made (but Sovietdesigned) DShK M 1938/46 heavy machine guns on a Czechoslovak-designed, two wheeled antiaircraft mount. (The DShK machine gun is covered in detail in the chapter on the USSR.) The weight of the complete equipment is 1411 pounds, and the mount is capable of 360-degree traverse and 90-degree elevation.

THE CZECH "URZ" WEAPONS FAMILY

Czechoslovakia has come out with a new weapons family called the "URZ"—which translates to Universal Small Arms—which, in concept, appears to be modeled on the Stoner 63 system. A common receiver system is used for an automatic rifle (AP), a light machine gun (LK), a heavy machine gun (TK) and a tank machine gun (T).

The automatic rifle (AP) has a lighter barrel than the other models. A grenade launcher is built onto the end of the barrel and a grenade launcher sight is attached to the rear of the front sight. It uses a fifty-round, light alloy drum type magazine. It can be used with the telescopic sight normally used with the heavy machine gun TK. The AP takes a bayonet.

The light machine gun LK has a heavier barrel than the AP rifle, a folding bayonet, and a chrome lined barrel. It also uses the 50-round drum magazine.

The heavy machine gun TK is actually the light machine gun LK with a belt type feeder mounted on a tripod and normally using a telescopic sight. This weapon is fed with a 250-round metallic link belt.

The Tank machine gun T is intended for use on vehicles and aircraft. It is belt-fed and fired by the use of a solenoid.

The URZ weapons family as advertised is not chambered for the 7.62mm "intermediate" sized Czech Model 57 (Soviet Model 43) cartridge. It is advertised for cartridges of the size and muzzle energy of the 7.62mm NATO or the Soviet rimmed 7.62 × 54mm cartridges. This does not necessarily mean that the Czechs intend to adopt this type of system themselves. Czechoslovakia is a large exporter of arms to many countries outside the Warsaw Pact, and the URZ family is probably intended for this trade.

CHARACTERISTICS OF URZ WEAPONS FAMILY

	Rifle AP	LMG LK	Hvy MG TK	Tank MGT
Caliber:	7.62mm			
System of operation:	gas, selective fire			
Length overall:	39.2 in.	39.2	47.2 in	34.5 in.
Weight: (less magazine)	8.62 lb.	11.5 lb.	24.3 lb. w/tripod mount	12.6 lb.
Feed device:	50 rd drum	50 rd drum	250 rd belt	250 rd belt
Muzzle velocity:	2625 f.p.s.			
Cyclic rate:	800 r.p.m.	800 r.p.m.	800 r.p.m.	1100 r.p.m.

OBSOLETE CZECH MACHINE GUNS

Czechoslovakia has been a prolific producer of machine gun designs since the early twenties. The famed ZB26 (Model 26 light machine gun) is still used extensively throughout the world, as is the ZB30 and ZB53 (Model 37 heavy machine gun). The British Bren was developed from the ZB series of light machine guns and still has an excellent reputation as a light machine gun.

Since World War II, the Czechs have come out with several new machine gun designs. The Model 52, a rather versatile weapon, originally replaced the 7.92mm Model 26 as the standard light machine gun. The Model 52 is chambered for the Czech 7.62mm Model 52 cartridge, but a later modification, the Model 52/57, is chambered for the Czech copy of the Soviet 7.62mm M1943 cartridge. The Czechs abandoned the 7.92mm Model 37 as a heavy machine gun and used the Soviet 7.62mm Goryunov for awhile; this weapon has now been replaced by a new "Universal," i.e., general purpose machine gun, the 7.62mm Model 59. The 15mm Model 38 (ZB60) used prior to the war, and by the Germans during World War II, has been replaced by the Soviet designed 12.7mm DShK M38/46, which is used on vehicles and on a Czech-designed quadruple mount.

CZECH ZB26 AND ZB30 MACHINE GUNS

IN 1924, Václav Holek introduced a belt-fed light machine gun. In the same year, he modified the weapon; it was called the Praga Model 24. This weapon is now known as the ZB26 (although it is frequently still called the Model 24 in Czechoslovakia; the gun is stamped VZ26) and was one of the most popular light machine guns in the world. The Model 26, Model 30 and Model 30J were used in 24 countries throughout the world. Models 26 and 30 have been manufactured in China, the Model 30 in Iran and Romania, and the Model 30J in Yugoslavia. All three models were carried in a post-1948 ZB catalog.

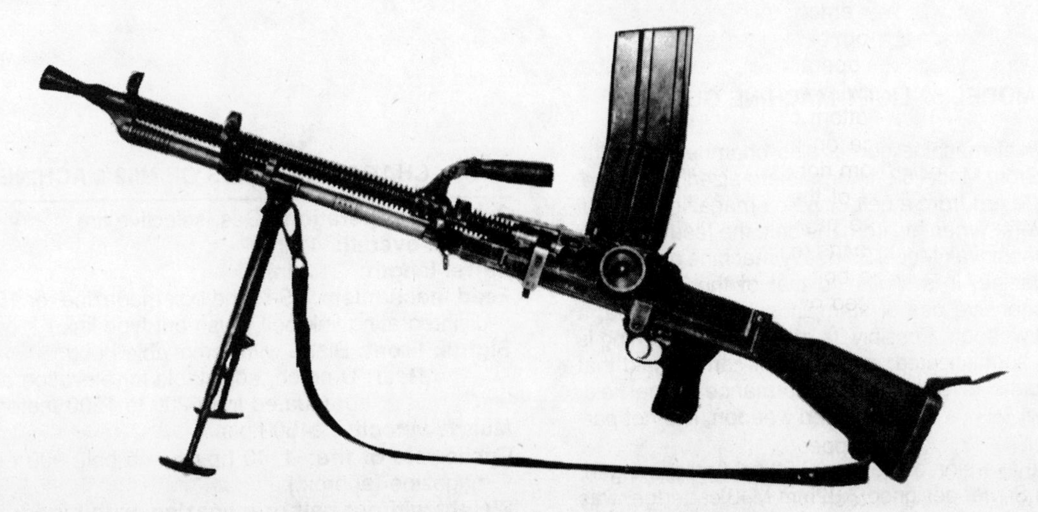

ZB26 (Brno) L.M.G. The magazine is a special oversized one.

CHARACTERISTICS OF ZB26 MACHINE GUN

Caliber: 7.92mm.
System of operation: Gas, selective fire.
Length overall: 45.8 in.
Barrel length: 23.7 in.
Weight loaded: 21.28 lbs.
Sights: Front: Protected blade.
Rear: Radial tangent.
Feed device: 20-round, detachable staggered row box magazine.
Muzzle velocity: 2500 f.p.s.
Cyclic rate: 550 r.p.m.

These weapons can be mounted on a tripod similar to that of the Bren Gun.

Loading and Firing the ZB26

Pull back the bolt handle to cock the bolt and then push the handle forward. Pull back magazine cover and insert loaded magazine mouth end first. Push down until it locks. For semi-automatic fire move the selector on the left side of the trigger group to the rear; for full auto fire, push it forward. When the selector is in the vertical position, it acts as a safety.

Field Stripping the ZB26

Push out receiver locking pin and withdraw the frame group. The slide, bolt and gas piston will now come out the rear of the receiver.

Release the barrel nut catch and lift up to the right as far as it will go. This releases the barrel, which may now be slid off to the front.

Push the butt plate catch and remove the two buffer springs.

Differences Between the ZB26 and ZB30

Outwardly, the two weapons are almost identical. The bolt of the ZB26 does not ride on the piston post as does that of the ZB30 (and the Bren). It is cammed into the locked position by a built-up rear section of the piston/slide assembly but does not "sit" on the post. The ZB30J is similar to the ZB30 but has a knurled section on its barrel to the rear and just in front of the carrying handle.

7.92mm ZB30 L.M.G. of Czech manufacture.

CZECH MODEL 52 LIGHT MACHINE GUN

The Model 52 light machine gun is also chambered for the Czechoslovak 7.62mm Model 52 "intermediate-sized" cartridge. The weapon can be fed from a belt or a box magazine without changing feed covers. When fed from the belt, the feed is similar to that of the Czechoslovak Model 37 heavy machine gun; when fed from the magazine, it is similar to that of the ZB26. The weapon makes extensive use of stampings and is, all told, a very sophisticated weapon. Possibly its greatest shortcoming is that it may be too sophisticated; in general, it can be said that the simplest weapons give the best performance in the field. Although the Model 52 is a well-designed weapon, it is not particularly simple.

A slightly modified version of the model 52 light machine gun, chambered for the Soviet designed 7.62mm M43 cartridge, was adopted in 1957 as the Model 52/57.

CHARACTERISTICS OF M52 MACHINE GUN

System of operation: Gas, selective fire.
Length, overall: 41 in.
Barrel length: 21.3 in.
Feed mechanism: 25-round box magazine, or 100-round non-disintegrating link belt (push-out type link.)
Sights: Front: Blade with removable hood.
Rear: U-notch, adjustable for elevation and windage, graduated from 200 to 1200 meters.
Muzzle velocity: 2450 f.p.s.
Cyclic rate of fire: 1140 r.p.m. with belt, 900 r.p.m. with box magazine (approx.)
Weight without belt or magazine, with bipod: 17.6 lbs.

How to Load and Fire the Model 52 Light Machine Gun

Cock the gun by pressing down on the lug which protrudes from the safety (on top left side of pistol grip) and pull pistol grip to the rear. The bolt will remain to the rear since this weapon fires from an open bolt. If the pistol grip is in the rearward position, press down on the lug and push the pistol grip forward until the slide is engaged by the sear; then pull the pistol grip rearward until the pistol grip catches on the rear lock.

To load the gun with a belt of ammunition, push upward on the feed-way cover. Lay the first cartridge in the belt between the belt holding pawls, and close and lock the cover. The link ejection port cover must be in its raised position or belt movement will be blocked.

To load the gun for magazine fire, press forward on the magazine feed port latch, allowing the cover to spring forward. Invert the magazine and insert it in the magazine opening until the latch engages the magazine. The belt feedway cover and the link ejection port cover should be closed. Pressure on the bottom half moon of the trigger will produce automatic fire; pressure on the top half moon of the trigger will produce semiautomatic fire. To put the weapon on safe, push the safety lever (located on top left side of the pistol grip) down.

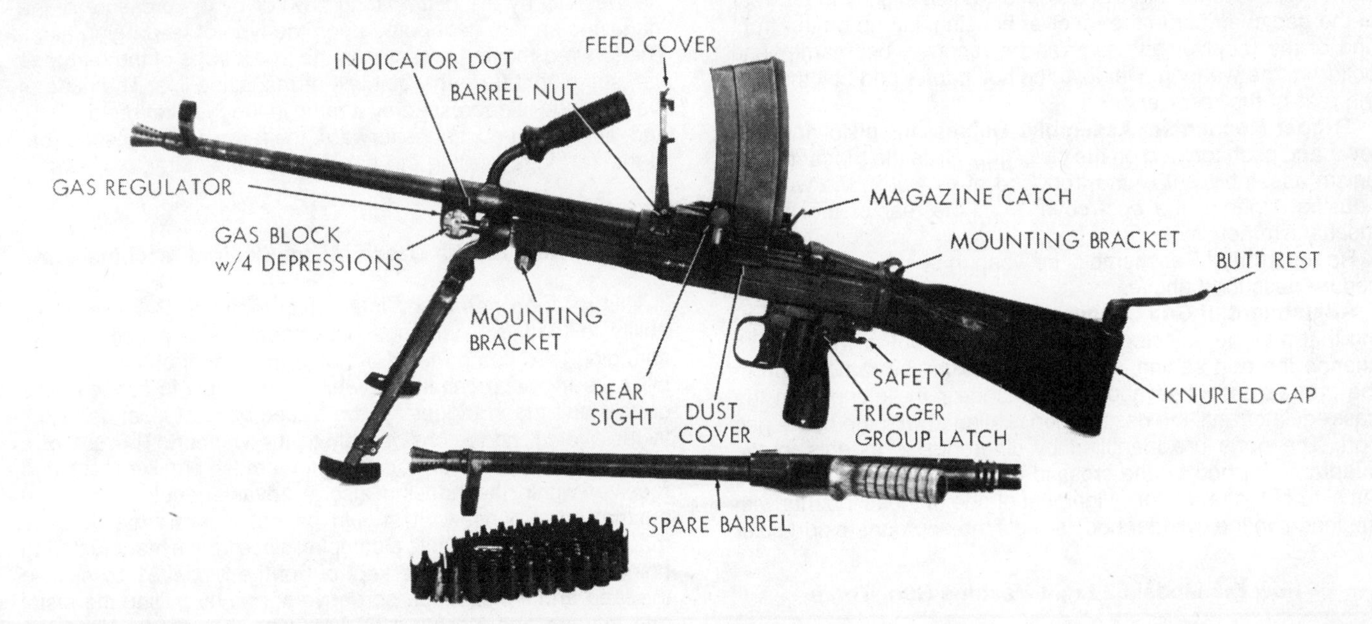

Czech Model 52.

Czech 7.62mm Model 52 Light Machine Gun with magazine feed. Note: 1. Magazine release; 2. Magazine feedway cover in open position; and 3. Gas regulator, lever with position indicator is located on left side of the gun.

Field Stripping the Model 52 Light Machine Gun

Driving Spring and Rod. The driving spring is retained by a cap with internal bayonet-type slots that engage projections of the spring tube within the stock. To remove the spring assembly, press in on the cap, turn it counterclockwise to disengage the slots, and withdraw the spring and rod from the stock.

Barrel and Bipod. Press forward on the magazine feed port latch, allowing the cover to spring open. Using the cover as a handle, turn the barrel lock clockwise and draw the barrel forward out of the receiver. Turn the bipod assembly to disengage its key, and remove the bipod from the weapon housing.

Receiver Assembly, Bolt Carrier and Bolt. The receiver, bolt carrier and bolt are removed simultaneously. Draw the bolt carrier rearward so that the piston is located slightly to the rear of the gas chamber in the receiver housing. Lift up on the front end of the receiver and remove the receiver, bolt carrier and bolt from the weapon. Remove the bolt carrier and bolt through the rear of the receiver.

Trigger Mechanism Assembly. Depress the pistol grip lock lever and push forward on the pistol grip. Slide the trigger mechanism assembly out of the front end of its slot in the weapon housing. Unhook the dust cover from the rear of the trigger mechanism housing.

Reassembly. Reassemble the weapon by reversing the procedure described above.

Adjustment of Gas Cylinder. There are four gas port opening that may be selected to vary the power of the weapon. To change the port setting, the barrel must first be removed from the gun as described above. To change port setting, turn the gas regulator until the desired port is aligned with the barrel gas port. The ports are identified by different-size indents in the quadrants formed by the crossed cylinder locking slots on the left side of the regulator. Alignment of one of these indents with the indent in the cylinder body selects the appropriate port size.

How the Model 52 Light Machine Gun Works

When the gun is fired, gas from the barrel is bled through a port into the gas cylinder and into the gas chamber at the front of the receiver housing. The expanding gas operates on the gas piston to force the bolt carrier rearward in recoil. For a short distance, the bolt carrier travels alone. This short period is followed by an unlocking period, during which the rear of the bolt rotates downward out of engagement with the receiver locking abutments. Upon completion of unlocking, the bolt is carried to the rear by the carrier, the spent cartridge is extracted from the barrel, and the case is ejected forward.

Upon completion of recoil, the residual energy of the bolt and bolt carrier is absorbed by a recoil plate at the rear of the receiver housing, and the bolt carrier moves forward under the force of the compressed driving spring in the stock. If the gun is being fired semiautomatically, the sear will engage the carrier and hold it until the trigger is again pressed. When the trigger is pressed, the carrier will move forward; as the rear of the carrier clears the sear, it will depress the disconnector, releasing the sear so that it can again be in position to engage the carrier.

Feeding takes place on the counterrecoil stroke of the bolt and bolt carrier. The cartridge to be fed is held in the center of the feedway by the belt holding pawl or by the pressure of the magazine spring, depending upon the type of feed being used. The forward-moving bolt strikes the lower edge of the cartridge rim, stripping it from the belt link or magazine lips. The nose of the cartridge is depressed by a ramp in the receiver breech ring, and, as the bolt continues forward, the ramp depresses the rear of the cartridge seating the rim between the extractor claws.

Special Note on the Czech Model 52 Light Machine Gun

This weapon has many interesting features. Because of its ability to feed from either a box magazine or a belt, without changing feed components, it has a great deal of tactical flexibility. Its trigger mechanism, while hardly new in concept (the double half moon trigger can be traced back at least as far as World War I), adds considerably to the weapon. The use of a large stamped receiver housing with a much smaller machined receiver within the housing also is advantageous from an industrial point of view. It should be noted, however, that the machined receiver is quite a complex piece from a manufacturing point of view. Dirt may be kept out of the Model 52 by closing the feed and link ejection port covers, and by pulling the pistol grip forward and locking it by engaging the safety. When the weapon is ready to fire, however, the bottom of the receiver is open to dust and dirt.

The bolt is carried on a post on the slide/piston assembly in a fashion similar to that of ZB26 or the Bren gun, but it does not lock like these guns. The two locking lugs on the rear of the bolt ride in cut-outs in the receiver and are locked in the side of the receiver.

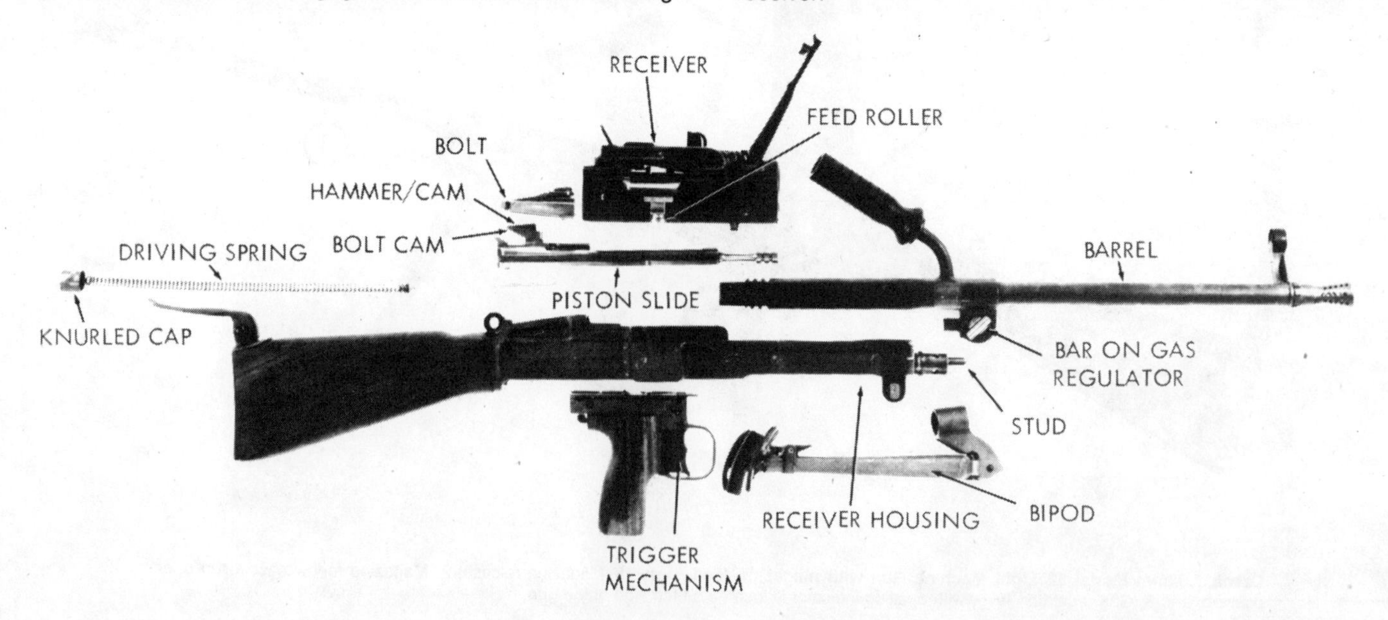

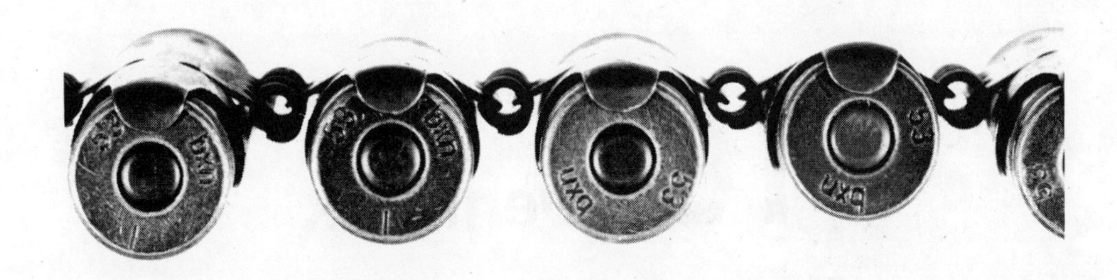

The nondisintegrating link belt used with the Model 52 machine gun. The cartridges shown here are the 7.62 × 45mm Czech.

CZECH MACHINE GUN MODEL 37 (ZB53)

The Model 37 heavy machine gun has been extensively manufactured for export. The military designation for the gun is Model 37 (ZB37); the commercial designation is Model 53 (ZB53).

The Model 37 is an air-cooled, gas-operated weapon with selective slow (500 r.p.m.) and fast (700 r.p.m.) rates of fire. It is fed from the right side by a metal belt of either 100- or 200-round capacity. By use of an attachment to the Model 45 tripod, it can be quickly adapted to antiaircraft fire.

This weapon was the forerunner of the British-made Besa tank machine gun. The main functioning features of the Model 37 are the same as those given for the Besa machine gun, which can be found in the chapter on Britain.

Czechoslovakian Heavy Machine Gun Model 37 (ZB53).

CHARACTERISTICS OF THE MODEL 37

Caliber: 7.92mm.
System of operation: Gas operated, two cyclic rates of fire.
Weight: 41.8 lbs.
Length, overall: 43.5 in.
Barrel length: 26.7 in.
Sights: Front: Blade with guard.
Rear: Folding, leaf, graduated from 300 to 2000 meters in 100-meter increments, fixed 200-meter battle sight.
Cyclic rate: 450–500 r.p.m. or 700 r.p.m.
Feed device: Metallic link belt, 100- or 200-round capacity.
Muzzle velocity: 2600 f.p.s. (approx.).

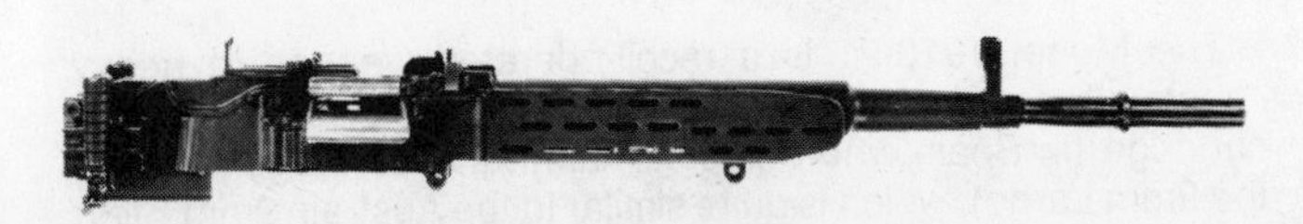

Model 37 with smooth barrel.

16 Denmark

SMALL ARMS IN SERVICE

Handguns:	9 × 19mm P M/40(S)	Swedish Model 40 Lahti (limited) (see Chapter 18).
	9 × 19mm P M/46	FN Modele 1935 GP (see Chapter 9).
	9 × 19mm P M/49	SIG 47/8 (see Chapter 44).
Submachine guns:	9 × 19mm Mp M/49 Hovea	Husqvarna design.
	9 × 19mm Mp M/41	Finnish Suomi (see Chapter 18). Obsolete.
	9 × 19mm Mp M/44	Swedish Model 37-39 Suomi (see Chapter 43). Obsolete.
Rifles:	7.62 × 51mm NATO G M/66	Heckler & Koch G3 (see Chapter 21).
	7.62 × 51mm NATO G M/75	Rheinmetall-made G3s leased from West Germany. Designated M/75 to distinguish them from earlier lot purchased from H&K. M/66s used by Home Guard snipers.
	7.62 × 63mm G M/50	US M1 rifle (see Chapter 48). Obsolete.
	7.62 × 63mm G M/53	US M1917 rifle (see Chapter 48). Obsolete.
Machine guns:	7.62 × 51mm NATO Mg M/42/59	MG42/59 (see Chapter 21).
	7.62 × 63mm Mg M/48	Basically the same as the Madsen M50. Obsolete.
	7.62 × 63mm Mg M/51	SIG Model 50 (see Chapter 43). Obsolete.
	7.62 × 63mm Mg M.52-1	US M1919A4 (see Chapter 48). Obsolete.
	7.62 × 63mm Mg M/52-11	US M1919A5 (see Chapter 48). Obsolete.
	12.7 × 99mm Mg M/50	US M2 HB Browning (see Chapter 48).

DANISH PISTOLS

The Danish Army, which had previously used the Gasser revolver, adopted the Bergmann Bayard automatic pistol Model 1908 in 1911, calling it the Model 1910. This pistol is a slightly modified version of the Bergmann 1903, which was made at the Borgmann plant in Germany. The Model 1908 was made by Anciens Etablissements Pieper of Herstal, Belgium. The weapon was used by Greece and Spain; pistols used by Denmark were made in Belgium or, after 1922, in Denmark.

In 1922, the pistol was slightly changed and the model designation altered to Model 1910/21. All original Model 1910/21 pistols—a total of 2204—were made at the Army Arsenal. Between 1922 and 1935, the 4840 Belgian-made pistols were converted to the 1910/21 pattern.

In 1940, Denmark decided to adopt the 9mm Parabellum FN Hi-Power Browning pistol. A few pistols were delivered to Denmark before it was invaded by the Germans. In 1946, the order was re-instituted, and 1577 pistols were purchased. The Hi-Power is called the Model 46 by the Danes. Among the weapons that the Danish Brigade (troops who had been maintained in Sweden during the war) brought back with them, were Swedish Model 40 Lahti pistols. These pistols were called Model 40S by the Danes. In 1948, Denmark adopted the Swiss S.I.G. 9mm Parabellum Model 47/8, which they call the pistol Model 49. This is the current standard Danish service pistol.*

DANISH 9mm M1910/21 PISTOL

The Model 1910/21 is a recoil-operated weapon of heavy construction. It is chambered for the 9mm Bergmann-Bayard cartridge (in Spain where it is still used, this cartridge is called the 9mm Largo), which is quite similar to the Austrain 9mm Steyr

*For additional details see *Handguns of the World.*

Danish 9mm Model 1910 Pistol.

Danish 9mm Model 1910/21 Pistol converted from Model 1910.

Danish 9mm Model 1910/21 Pistol as made in Denmark; magazine has been removed.

cartridge and the US Super .38 automatic pistol cartridge. It is not made in the United States.

These pistols are quite heavy, weighing about 2.25 pounds, and have a relatively short barrel—about four inches—in comparison to their overall length of about 10 inches. There are six-, eight- and ten-round magazines for these weapons.

One of the strongest and most powerful pistols ever made, the Model 1910/21 incorporates good materials and fine workmanship with a positive lock. However, the design is clumsy and bulky and should be considered obsolete.

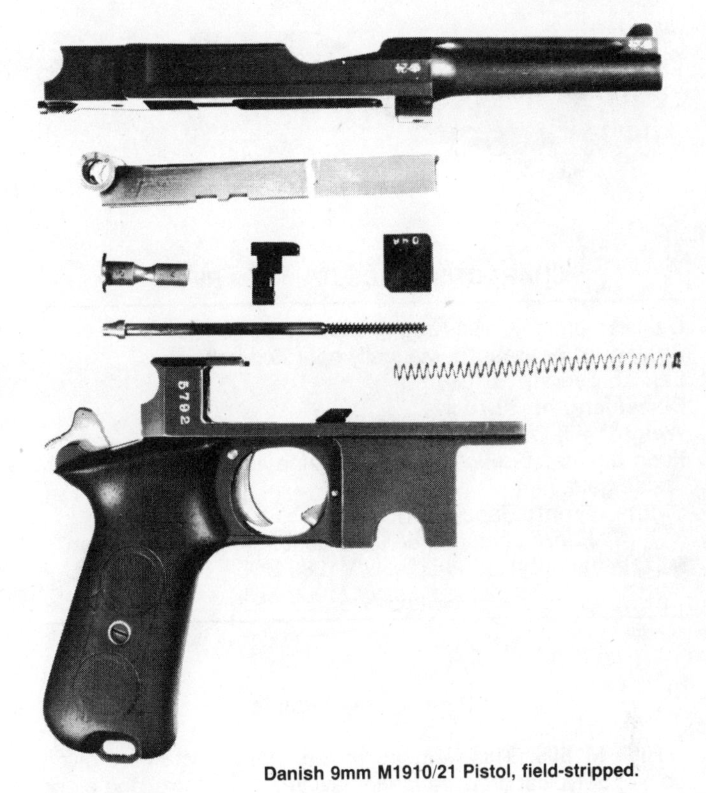

Danish 9mm M1910/21 Pistol, field-stripped.

DANISH RIFLES

Denmark was the first country to adopt the Krag Jorgensen—in 1889—and continued to use this weapon, in a series of models, through World War II. After the war, Denmark was supplied with US, British, and Swedish rifles, and eventually the US caliber .30 M1 was adopted as standard. Recently the German 7.62mm NATO G3 has been adopted, and this rifle will replace all others in the Danish service.

The Danish Home Guard was equipped with Swedish 6.5mm M94, M96 and M38 carbines and rifles until 1953, when these weapons were returned to Sweden. The Home Guard was then issued US caliber .30 M1917 (Enfield) rifles, called Model 53 by the Danes, and US M1 rifles. During World War II, the Danish sporting rifle firm of Schultz and Larsen produced a police carbine (the Model 1942) for the 8mm Danish cartridge. This rifle has four locking lugs on the bolt like the current Schultz and Larsen sporters and a box magazine, which protrudes below the level of the stock.

In 1896, the Danish Navy and Coast Guard adopted the Madsen M1896 recoil operated semiautomatic rifle. Denmark, therefore, was probably the first country to use a semiautomatic service rifle.

DANISH KRAG RIFLES

Denmark was the first country to adopt the rifle developed by Krag and Jorgensen. This weapon in slightly modified form was later adopted by the United States and Norway. The principal points of difference between the Danish and the Norwegian and US Krags are: the loading gate swings out horizontally on the Danish weapon, on the US and Norwegian Krag the loading gate swings down to open and is pushed up to close; the Danish 1889 rifle and several of the carbine models have a metal barrel jacket, the US and Norwegian Krags use wooden handguards. All Danish Krags were chambered for the 8mm rimmed cartridge.

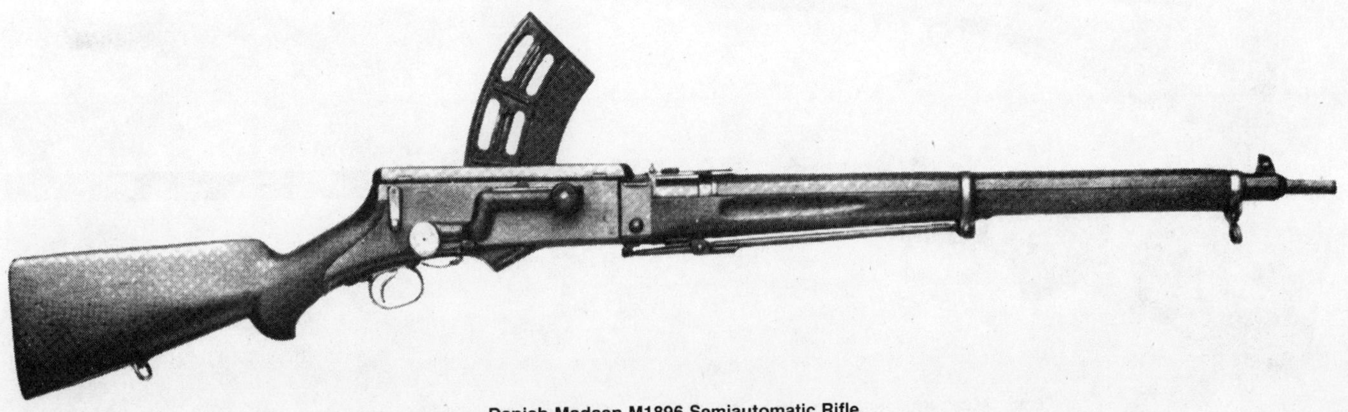

Danish Madsen M1896 Semiautomatic Rifle.

Danish 8mm Rifle Model 1889/10.

CHARACTERISTICS OF M1889 RIFLE

Caliber: 8mm Danish Krag rimmed.
System of operation: Manually operated bolt.
Length overall: 52.28 in.
Barrel length: 32.78 in.
Weight: 9.5 lbs.
Feed device: 5-round in-line magazine, loaded singly through side gate.
Sights: **Front:** Barley corn.
Rear: Leaf, can be used as tangent with leaf down.
Muzzle velocity: 1968 f.p.s. w/M1889 ball.
2460 f.p.s. w/M1908 ball.

Danish Krag Models

Rifle M1889. This weapon, whose characteristics are given above, is typical of its period in having a long barrel and stock without pistol grip. As noted above, a metal handguard encircles the barrel in a fashion similar to that of the Belgian M 1889 rifle. As originally issued, this rifle had no safety catch; a half-cock notch on the cocking piece—firing pin assembly—served this purpose. In 1910, this weapon was modified by the addition of a manual safety, which was placed on the left side of the receiver just behind the closed bolt handle.

8mm Infantry Carbine M1889. Introduced in 1924, this weapon also has a metal barrel jacket and a stud for a bayonet. A tangent-type rear sight is used on this weapon rather than the leaf-type rear sight of the rifle. This carbine has the stamp "F," i.e., *Fodfolk* or Infantry, before the serial number. Overall length is 43.3 inches with a barrel length of 24 inches; the carbine weighs 8.8 pounds. A horizontal-type bolt handle is used.

8mm Artillery Carbine M1889. This carbine was also introduced in 1924. It is generally similar to the Infantry carbine, but it has a turned-down bolt handle, a triangular upper sling swivel and a stud on the left side of the stock. This stud was used to hang the carbine from a leather hanger worn on the gunner's back.

8mm Engineer Carbine M1889. The Engineer carbine was also introduced in 1889. It has a wooden handguard, and the barrel was shortened to 23.6 inches to accommodate the muzzle cap of the Cavalry Rifle M1889. The letter "I" is found before the serial number.

Cavalry Rifle M1889. This weapon was introduced in 1914. The rear sling swivel is mounted on the left side just ahead of the trigger guard. It has a straight-bolt handle and a mounting stud similar to that of the M1889 cavalry carbine on the left side of the stock. The letter "R" is found before the serial number. This rifle is not fitted for a bayonet.

Sniper Rifle M1928. This is an alteration of the rifle M1889 and has a heavier barrel with wooden handguard, a sporting type stock with pistol grip, a turned-down bolt handle, micrometer-type rear sight and a hooded target type front sight. This rifle, which resembles the US caliber .30 Style "T" rifle in general configuration, weighs 11.7 pounds and is 46 inches in length with a 26.3-inch barrel.

8mm M1889 Artillery Carbine, 8mm M1889 Cavalry Rifle and 8mm M1928 Sniper Rifle.

DANSK INDUSTRI SYNDICAT (MADSEN)

After over 60 years of manufacturing weapons of fine quality, Dansk Industri Syndicat—DISA (or Madsen, as it is commonly known)—the manufacturers of the Madsen gun, are out of the small arms business. This company has produced many outstanding weapons and has had much influence on arms design since the early 1900s. DISA fell victim to the East-West arms trade which has seen the USSR and USA dominate small arms sales through competitive prices and subsidized military assistance programs. Of equal impact has been the fact that many nations, which formerly purchased Madsen products, now manufacture their own small arms.

DISA has in recent years continued to design small arms accessories. For example, the firm developed tripods for both the Belgian MAG and the German MG-1. The latter has been adopted by the West German Army.

Madsen Post-War Rifles

Dansk Industri Syndicat developed a number of rifles since World War II. None of these developments were commercially successful.

The Danish M47 Rifle

The Model 47 bolt-action rifle was designed primarily for the smaller races of the world. It weighs about 7.5 pounds and is fitted with a rubber recoil pad. Owing to the availability of large stocks of World War II surplus rifles at very low prices, it was sold in limited quantities only to Columbia.

The Ljungman Rifle

The Ljungman made by Madsen differs from the Swedish and Egyptian varieites of this weapon in one important respect. In the Swedish and Egyptian versions of this rifle, the gas blows through a straight gas cylinder directly against the bolt carrier. In the Madsen-made gun, the gas cylinder is coiled around the barrel; therefore the gas has further to travel before it contacts the bolt carrier. This had the tendency to cool the gas before it hits the bolt carrier and to reduce the thrust of the gas, thereby making the action less abrupt. Possibly because the Ljungman is basically an expensive and rather heavy weapon, this rifle was not a commercial success.

Danish Madsen .30 caliber Model 47 Bolt Action Rifle as supplied to Colombia.

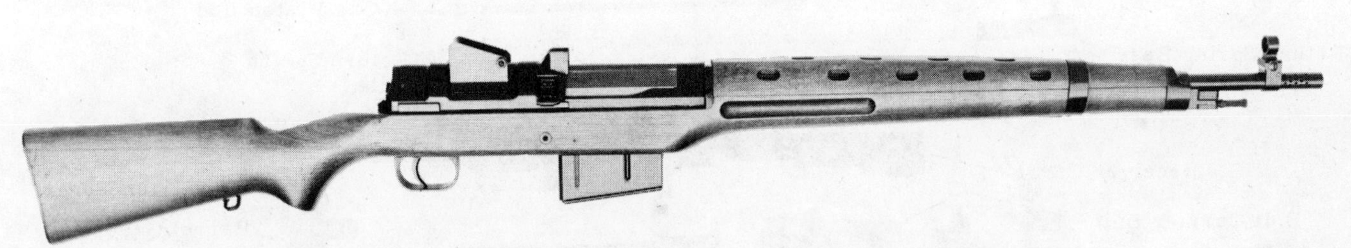

The Danish 7.92 × 57mm version of the Swedish AG 42 Ljungman manufactured by Madsen.

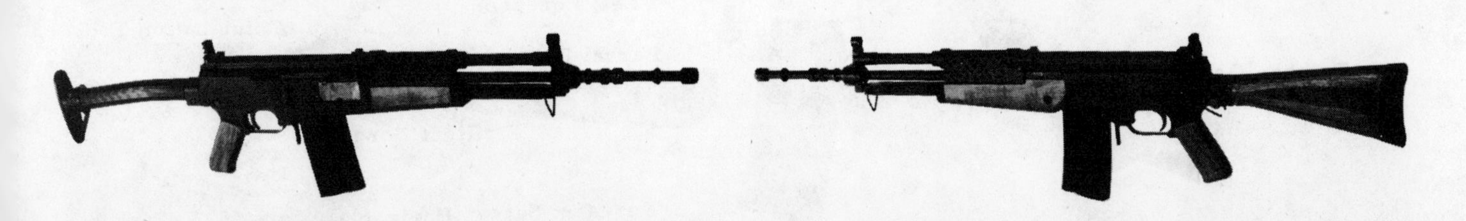

Madsen 7.62mm NATO Light Automatic Rifle with tubular steel stock.

Madsen 7.62 NATO Light Automatic Rifle with wooden stock.

MADSEN LIGHT AUTOMATIC RIFLE

CHARACTERISTICS OF MADSEN LIGHT AUTOMATIC RIFLE

Caliber: 7.62mm NATO
System of operation: Gas, selective fire.
Weight w/loaded magazine: 10.6 lbs.
Length, overall: 42.3 in.
Barrel length: 21.1 in.
Feed device: 20-round detachable, staggered box magazine.
Sights: Front—Hooded, blade.
Rear—Aperture, graduated from 100 to 600 meters.
Muzzle velocity: Approx. 2650 f.p.s.
Cyclic rate: 550–600 rounds per minute.

This weapon was never serially manufactured. Only prototypes were fabricated.

How to Load and Fire the Madsen Light Automatic Rifle

Insert a loaded magazine into the magazine port and push home until locked by the magazine catch. Cock the weapon by pulling the operating handle to the rear as far as it will go, release the handle, and let it run forward. The weapon is now cocked and ready to fire. The safety selector lever can be set on safe, semiautomatic or automatic as desired.

Special Note on the Madsen Automatic Rifle

The Madsen has its return spring mounted above the barrel and circles the piston rod, thus it pulls the bolt forward into battery position rather than pushes it forward as with most weapons. The bolt is similar to that of the Soviet AK 47 in that it rotates to lock and is rotated by means of a lug on the bolt operating in a cam in the bolt carrier. However, the piston rod of the Madsen is not permanently attached to the bolt carrier as with the AK. The piston rod of the Madsen has a ball-shaped end, which fits in a cut-out in the bolt carrier. The trigger mechanism of the Madsen is similar to that of the AK. The Madsen has a grenade launcher built into the end of the barrel. It can also be easily fitted with a bipod and a telescopic sight. There are two versions of the weapon: one with a tubular metal stock and one with a wooden stock.

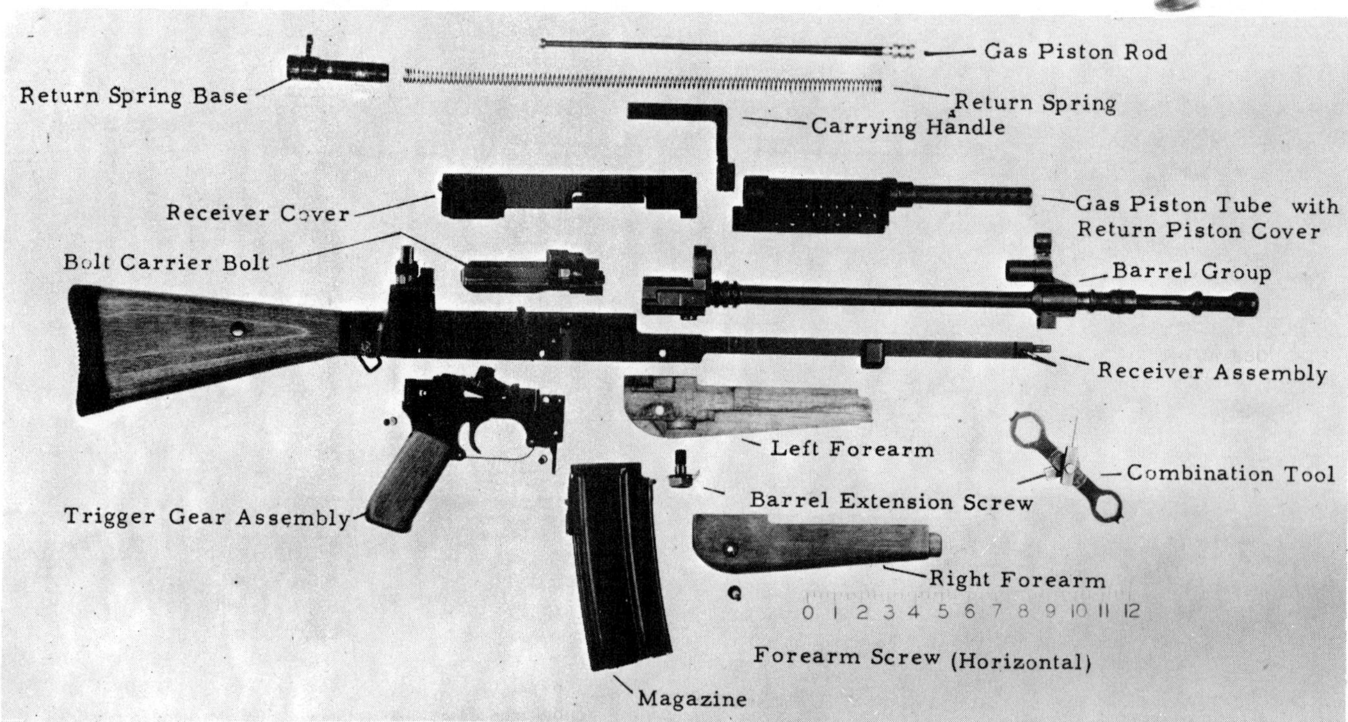

Madsen 7.62mm NATO Light Automatic Rifle, field-stripped.

DANISH SUBMACHINE GUNS

The standard submachine gun of the Danish Army is the 9mm M49 "Hovea." However, several versions—Finnish and Swedish—of the 9mm Suomi are used as well. DISA has developed several submachine guns since World War II. The initial postwar submachine developed by Madsen was the Model 45; although it had several interesting features, it was not a success. The Model 1946 and its successors, the Model 50 and Model 53, have been fairly extensively manufactured and are in use in a number of countries throughout the world.

DANISH MADSEN SUBMACHINE GUN MODEL 1945

While this gun has been rendered obsolete by the improved 1946 and later patterns, it is worthy of some attention because of a few original design factors that may have some future application in other arms.

CHARACTERISTICS OF M1945

Caliber: 9mm Parabellum.
Magazine: Standard box. 50 cartridges mounted below receiver.
Overall length: 31.5 in.
Weight, excluding magazine: 7.1 lbs.
Operation: Blowback. Standard general operation but unusual inertia movements added, selective fire.
Cyclic rate of fire: About 800 r.p.m.
Special feature: Safety pin lock on firing pin.

Despite the fact that this arm uses a rifle type wooden stock and fore-end, the designers achieved the extremely light weight for a submachine gun of only 7.1 pounds.

A cocking cover over the barrel breech in the form of a slide (not unlike that of the typical automatic pistol design) has serrated sides forward of the magazine. Cocking, instead of by customary handle, is by withdrawing this member, which travels with the true breechblock in recoil. The recoil spring is positioned around the barrel below this sliding cover member. Utilization of this spring position, together with the sliding cover, was used to supplement the inertia and mass of a light breechblock, to thereby achieve minimum weight.

In an endeavor to produce a safety factor to overcome accidental discharge during slamming of the breechblock when the gun is dropped or violently put aside, a sear interrupter was introduced in this model. The striker can move forward through the breechblock to fire only when the trigger is depressing the sear. Dropping the weapon will not cause accidental firing.

The spring position around the barrel, of course, subjects the spring to considerable heat under continuous fire, inevitably producing crystallization and spring breakage.

MADSEN M1950 SUBMACHINE GUN

CHARACTERISTICS OF M1950

Caliber: 9mm Parabellum.
System of operation: blowback, automatic fire only.
Magazine: 32-round detachable straight line box.
Weight: 7.6 lbs., excluding magazine.
Overall length: 30.71 in. with folding butt extended.
Barrel length: 7.87 in.
Muzzle velocity: approx. 1200 f.p.s.
Cyclic rate: 500–550 r.p.m.

Construction and Design, M1950

This is one of the most unusual submachine gun designs ever produced. The gun is designed to lend itself to high-speed production at extremely low cost.

Because of its unusual design, we will consider it here in considerable detail. The construction itself is most ingenious. The receiver (or frame) is flat. It is divided vertically in longitudinal section and is hinged at the rear. The pistol grip and magazine guide are a simple stamping.

The barrel is fastened to the receiver by a locking nut which when unscrewed and thrust forward permits the entire left side of the receiver to be folded back exposing the right side in which all the moving parts are housed. The barrel may be lifted out for immediate replacement or for cooling. While this system of design has been applied in Europe to revolvers in the past, this is the first production example of the application of the design principle to the submachine gun. It permits not only simplified manufacture but extreme ease of fitting and assembly.

The stock is a folding metal skeleton design permitting the gun to be used easily either from the shoulder or from the hip.

The stock when folded does not interfere with access to the trigger. The sling swivel positions on the left side of the gun are designed to permit the weapon to be slung across the chest for immediate use at a moment's notice.

A very unusual factor, yet one of extreme simplicity, is the special automatic safety provided. This is a lever positioned to the rear of the magazine housing. When firing, the normal manner of gripping an arm of this type is with the right hand about the grip and the left hand around the magazine or magazine housing. In this position, the firer's left hand in this new design also embraces the safety lever. Should he release this grip at any time, or should he stumble and lose control of it, the lever automatically blocks the path of the breechblock so that it cannot chamber a cartridge from the magazine for firing. Pulling the cocking handle all the way back will put the gun back in readiness for operation.

The receiver is composed of two nearly identical sections of stamped sheet steel. The pistol grip and magazine housing are formed as sections of these individual halves of the receiver. The two frame sections are hinged at the rear. At the front they are secured to the barrel by the barrel bearing nut.

The left side of the receiver serves as an actual cover for the right section in which the moving members are housed. The sling swivels and sights are on the left-hand half. The front sight is positioned at the forward end of the receiver instead of on the barrel, thus eliminating need for accurate barrel positioning, since barrel locking ridges are cylindrical. It is a standard blade design and may be adjusted laterally for windage. The rear sight positioned at the rear of the receiver section is a fixed aperture sight. The gun is sighted in at a practical hundred meters range. While the sighting radius is relatively short, it is still adequate.

A projecting rib at the forward end of the left section of the receiver serves as the ejector. The right side of the receiver is pierced for the ejection port.

Except for the barrel, which is detachable by merely unscrewing its nut, all the other operating parts are housed in the right hand side of the receiver.

Madsen Model 1950 Submachine Gun. One of the finest examples of modern, low cost, fast production stamped designs. Also made in Brazil in caliber .45 A.C.P.

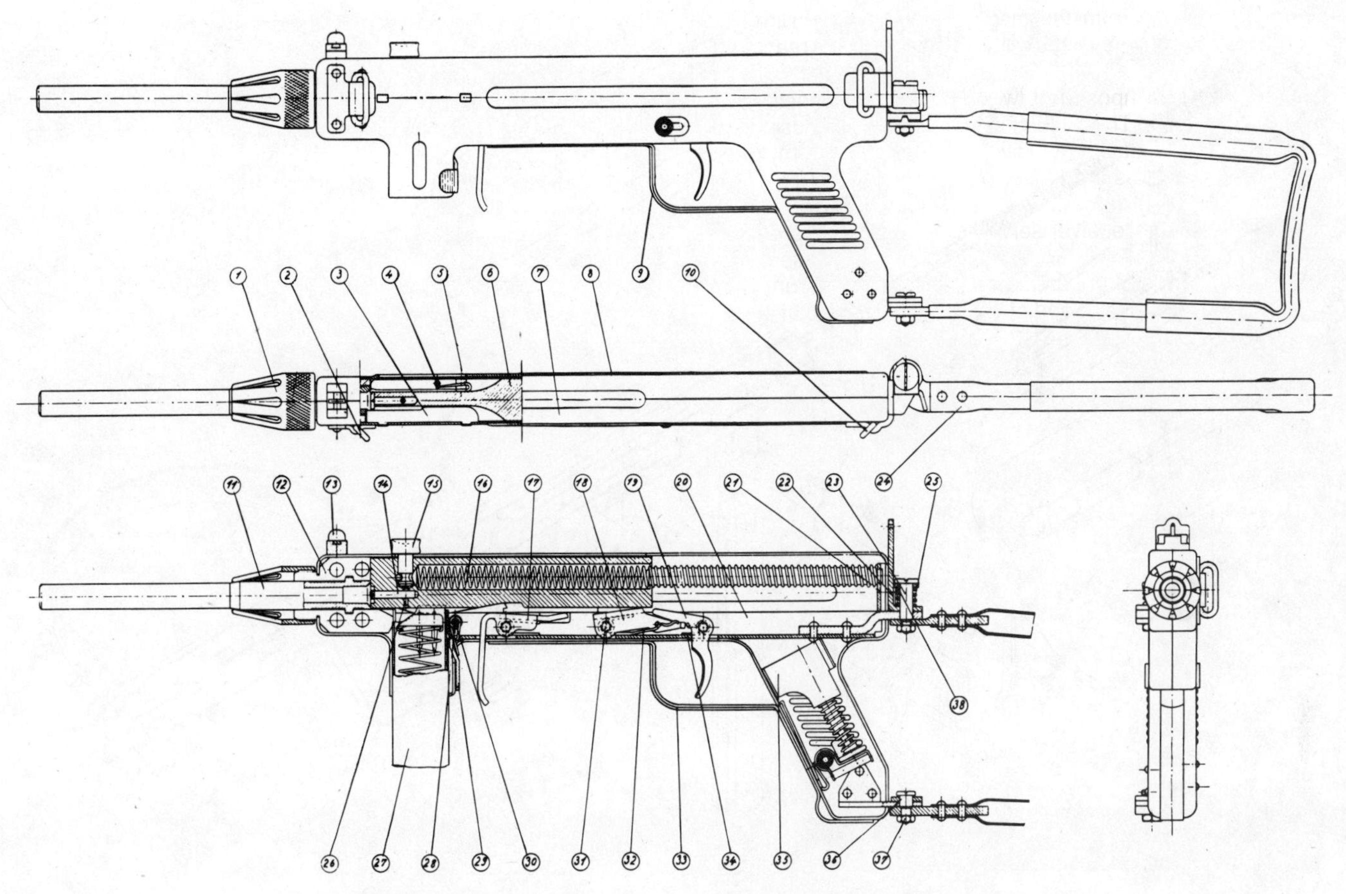

Detail drawings showing construction and parts of the Madsen Model 1950 Submachine Gun.

1. **Barrel-bearing nut**
2. **Sling swivel**
3. **Ejector**
4. **Extractor pin**
5. **Extractor**
6. **Breechblock**
7. **Frame half (left)**
8. **Frame half (right)**
9. **Safety catch**
10. **Sling swivel**
11. **Barrel**
12. **Barrel bearing**
13. **Front sight**
14. **Firing pin**
15. **Cocking handle**
16. **Return spring**
17. **Breech block retainer**
18. **Trigger rod**
19. **Trigger**
20. **Trigger plate**
21. **Return spring guide (complete)**
22. **Rear sight plate**
23. **Shoulder piece bolt (upper)**
24. **Shoulder piece**
25. **Shoulder piece spring**
26. **Firing pin rivet**
27. **Magazine**
28. **Magazine catch pin**
29. **Magazine catch**
30. **Magazine catch spring**
31. **Trigger rod pin**
32. **Trigger rod spring**
33. **Trigger guard**
34. **Trigger spring**
35. **Magazine loading apparatus**
36. **Shoulder piece bolt (lower)**
37. **Shoulder piece bolt nut**
38. **Shoulder piece lock**

The Barrel. No radiating rings or cooling flanges are provided. The barrel is a smooth taper screw machine or lathe-turned unit and is quite light. Because of interchangeability factors, the heavier barrel is not considered essential, though for hard military usage a heavy or flanged barrel could readily be provided.

The breech section of the barrel is housed in the forward breech section of the receiver. It is furnished with an external rib which fits into corresponding grooves in both frame halves to prevent the barrel from moving forward or rearward. A groove cut in the rear of the chamber section of the barrel mates with the ejector rib in the left section of the receiver. This positively prevents any barrel rotation when assembled.

Breechblock. This rectangular unit reciprocates inside the receiver on the flat bottom wall of the right receiver section. The firing pin is an integral part of the breechblock for complete simplicity. The extractor at the front right side section of the breechblock is secured with an elementary vertical pin.

Cocking Handle. This is a separate member, inserted in the top of the breechblock and fastened by a cross pin. It travels in a slot in the top of the frmae between the two halves.

Trigger Mechanism. This is positioned at the bottom of the right receiver section. The unit consists simply of the trigger and arm, the trigger rod, and their respective springs. The rod (or sear member) is thrust upwards by its spring to catch in a notch in the bottom of the breechblock when in cocked position. The trigger arm projects down through the opening in the bottom of the right frame section, where it is protected by the trigger guard.

Safety Catch. This is at the bottom of the frame. It can be locked only when the breechblock is cocked. Pushing back the catch in its short travel slot in the left side of the frame effectively locks the trigger rod in position to prevent any movement of the breechblock.

Magazine. Box design holding 32 cartridges. This is inserted in standard fashion into the magazine housing from below. The magazine retaining catch is in the rear wall of the housing. Pushing the catch backwards releases the magazine. The magazine may be inserted when the breechblock is open or closed. However, if the breechblock is in closed position, more force must be applied to insert the magazine as it must be thrust up enough for the top cartridge (which presses against the underside of the breechblock) to be thrust further down into the magazine itself.

Buttstock. This is a skeleton folding butt of steel tubing partly leather covered for comfort. It is hinged to the rear end of the

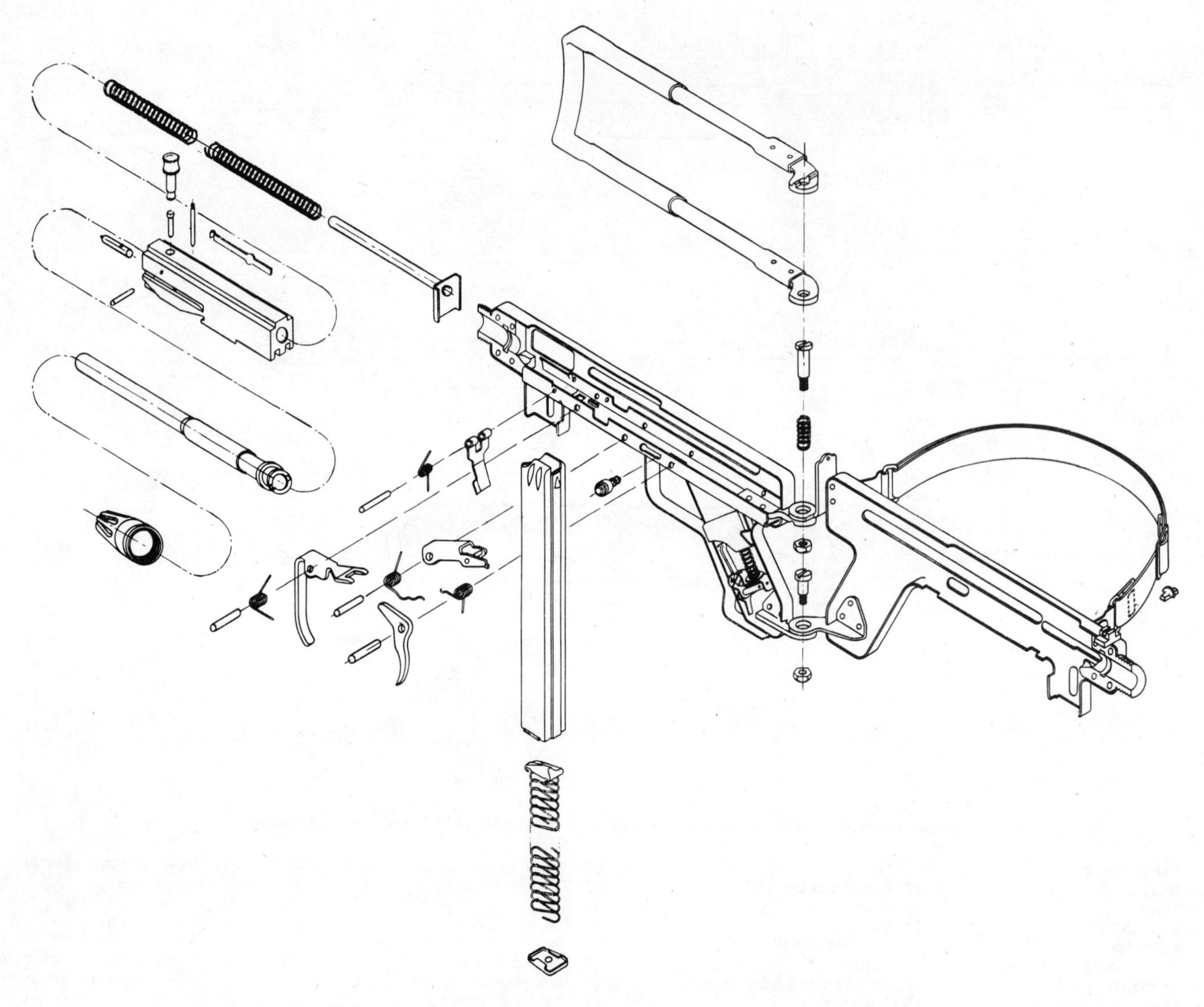

Detail exploded view of the Madsen Model 1950 Submachine Gun as made in Denmark.

right section of the receiver, its lower hinge being behind the pistol grip. In both open and closed positions the stock is held by a notch and a lug in the hinge. A slight jerk will free it to be moved to either position. However, a lock on the upper shoulder piece bolt allows it be securely locked into extended position if desired. When folded forward, it lies along the right side of the frame to make a very compact fold, without interfering with the use of the arm in any way.

Dismounting and Reassembling the M1950

To Dismount. Press the release catch and remove the magazine. Fold the shoulder piece. Ease the breechblock to forward position if it is not already there. Unscrew the barrel bearing nut and slide forward. Place the gun on its right side. Pull the front wing swivel with the right hand while holding the left hand pressing against the barrel. This will raise the left receiver section, freeing the barrel. Withdraw the barrel.

Press the base of the return spring forward and upwards. This will allow it to be withdrawn from the breechblock. Lift the breechblock up and out. No further disassembly is normally required. The magazine bottom may be slid out to remove the spring and follower, although this too is not commonly necessary. Reassembly is merely reversal of this procedure.

Special Note. This gun in Caliber .45 is currently being made in Brazil under Madsen license. Many of the design and manufacturing features lend themselves readily to application to other small arms designs.

Operation of M1950

A magazine is inserted in the housing and thrust in until the magazine catch secures it. The cocking handle on top of the breechblock is drawn to the full rear position to compress the recoil spring. At the end of the stroke, the trigger rod spring thrusts the rod up to catch in its notch in the under side of the breechblock. The safety catch may be pushed to lock the rod into the breechblock if desired. Otherwise the weapon is now ready to fire on pull of trigger.

Firing. Pressing the trigger causes it to pivot, and its forward arm moves the attached trigger rod down out of contact with the breechblock. The compressed recoil spring drives the breechblock ahead. If the left hand is not supporting both the magazine housing and the safety lever behind it, the breechblock will be halted before striking the cartridge in the magazine lips.

Brazilian cal. .45. Similar to the Danish Madsen Model 1946 except for caliber. Presented to Gen. J. Lawton Collins, then Chief of Staff, US Army by the Brazilian Minister of War. Its bolt handle is on the right side rather than on the top.

Danish Madsen Model 1953 9mm Submachine Gun.

Assuming that the supporting hand is pressing in the safety catch towards the magazine housing, the breechblock impelled by the recoil spring is free to move ahead, since the upper section of the safety lever is not in its path. Its feed face strips the top cartridge from between the lips of the magazine and drives it into the chamber. The nose of the bullet is guided by the barrel feed section of the breech into the chamber as the rear section of the cartridge clears the magazine lips. When the cartridge enters the chamber, it lines up with the firing pin, which is fixed in the face of the breechblock. The firing pin strikes the cartridge primer while still moving forward. At the same time the extractor on the right side of the breechblock springs over the cannelure in the cartridge case.

In this quite standard practice for submachine guns, the heavy breechblock and spring members are still moving forward at the actual instant of firing before the chambering is really complete. This serves as an additional inertia factor to offset the recoil of the discharge.

SAFETY

The breech block retainer serves to prevent a round from being fired accidentally owing to an incomplete cocking motion or as a result of a shock to the gun if dropped or laid down hard. — It does not require any attention, however, to release the breech block retainer, because the correct grip for firing the gun, whether kneeling, standing, sitting or prone, is with left hand firmly round the magazine housing, the thumb in the most natural way pressing against the downward lever of the breech block retainer, and with the right hand round the pistol grip.

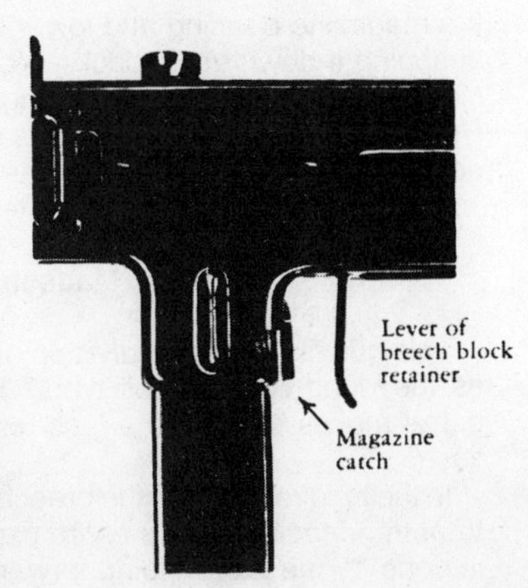

DANISH MADSEN SUBMACHINE GUN M1953

This is a newer model of the submachine gun Model 1946. This weapon was changed in 1950, and again in 1953. This model is called the M53.

The M53 is very similar to the 1950 Model, but it has incorporated several improvements. The most noticeable, at first glance, is the design of the magazine, which is curved instead of being straight as on the 1946 and 1950 models. The curved magazine is considered a better design for feeding purposes.

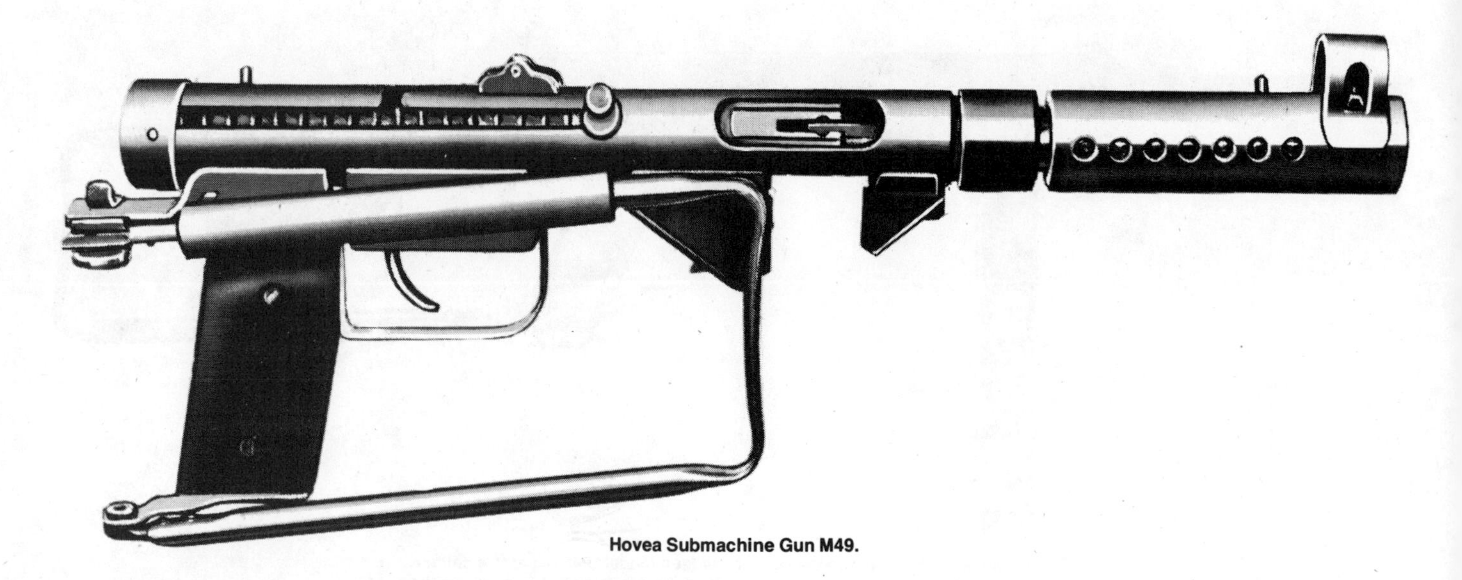

Hovea Submachine Gun M49.

Another new feature in the M53 submachine gun is that the barrel bearing nut now screws onto the barrel, rather than onto the front of the receiver as on the M46 and M50 models. This feature gives the M53 added strength and stability for the barrel. The bolt has also been streamlined to aid in better functioning.

If desired, the weapon can be supplied with a removable barrel jacket to which a special short bayonet can be fixed. All the Madsen models can be made with or without selection-fire features.

The characteristics and field stripping are the same for this weapon as for the 1950 model given earlier in this chapter.

Hovea 9mm Submachine Gun M1949

This weapon, commonly called the Hovea, was developed by Husqvarna in Sweden. It is similar in construction to the Swedish M45 submachine gun. Functioning, disassembly and assembly are similar to that described under the Swedish 9mm M45 submachine gun.

CHARACTERISTICS OF HOVEA 9MM SUBMACHINE GUN M1949

Caliber: 9mm Parabellum
System of operation: Blowback.
Weight: 8.9 lbs. (loaded).
Length, overall: 31.8 in. (w/stock extended)
21.6 in. (w/stock folded)
Barrel length: 8.4 in.
Feed device: 35-round detachable, staggered, box magazine
Sights; Front: Hooded post.
Rear: L-type with setting for 100 and 200 meters
Muzzle velocity: 1263 f.p.s.
Cyclic rate: 600 r.p.m.

DANISH MACHINE GUNS

Denmark has used various models of the Madsen machine gun since 1904 in the standard light gun versions and in aircraft versions as well as heavy (20mm) guns of this design. After World War II, the Danish Army was equipped with British .303 Bren guns. Swedish 6.5mm Model 37 Browning guns and US caliber .30 M1919A4 and A5 and caliber .50 M2 Browning Heavy Barrel machine guns. In 1948, Denmark adopted the last of the true Madsens in caliber .30, and in 1950 they adopted the SIG 50 as the Model 50 in caliber .30. Recently Denmark adopted the German 7.62mm NATO MG1 (MG 42/59) as standard and also adopted the DISA Model F 197 tripod mount for this gun.

THE MADSEN MACHINE GUN

The Madsen machine guns have been among the most popular in the world since their introduction in the early 1900s. They have world-wide distribution and will continue to be encountered in service for many years. There are many variations of this gun in existence, but all operate basically in the same way. The Madsen is an expensive gun to manufacture and requires quality ammunition for reliability of function. These factors limited its use among the major powers during the world wars. In 1926, the Madsen was issued in a water-cooled version; a quantity of these weapons were sold to Chile. The Madsen has been sold to 34 countries in a dozen different calibers.

Loading and Firing the Madsen

Like the Chatellerault and the Bren guns, the Madsen uses a top-loading magazine. This requires the sights to be set off to the side of the gun. The magazine is arc-shaped, for use with rimmed cartridges, which cannot lie flat on top of each other.

Pull the cocking handle back as far as it will go and release it.

Put the forward end of the magazine into the forward end of the magazine opening and lower the rear end down into place, snapping it down until it locks. Now set the selector on the left side of the receiver above the trigger in the fire position. **Note:** Remember that this weapon fires as the bolt goes forward, and no attempt should ever be made to let the action go forward while there is a magazine mounted on top of the gun.

How the Madsen Gun Works

This gun fits into a subdivision of functioning principles known as the "long barrel recoil type." In the short recoil types, the barrel moves backward a less distance than the length of the case.

In the long recoil type, the breech has to move back far enough to permit feeding up the entire cartridge in one operation. This is done by the barrel going forward while the lock is held back

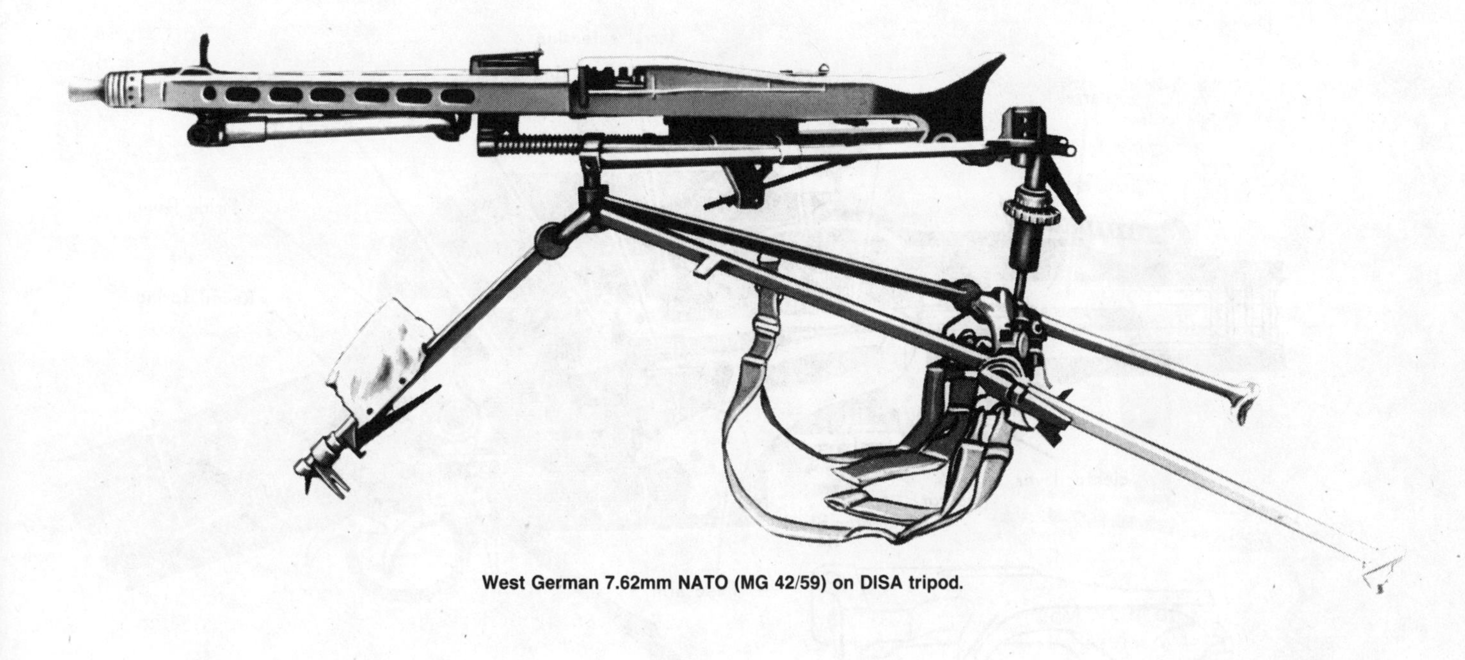

West German 7.62mm NATO (MG 42/59) on DISA tripod.

Madsen 8mm light machine gun.

until the cartridge has partly entered the chamber. In this type of action, the rate of fire is much lower than in the short recoil.

Starting with the gun cocked and in firing position, the action is as follows: As the trigger is pressed, the spring below it is compressed while the trigger nose is pulled down out of the bent of the recoil lever. This permits the recoil spring in the butt to force the lever downward, and, as it is engaged in the rear of the breech mechanism, it thrusts the recoiling parts forward. As the recoiling mechanism nears forward position, the recoil arm is still up somewhat. The hump on the recoil lever now bears on the side of the sear and forces the sear downward. The nose of the sear is thus relieved from the bent of the firing lever and compresses the sear spring. The firing lever is now forced downward by its spring and strikes the tail of the hammer. The front of the hammer drives the firing pin forward to explode the cartridge in the firing chamber. A coiled spring around the firing pin, which is compressed by the hammer movement, pulls the firing pin back into the face of the breech block as the cartridge is fired.

As the recoiling parts are thrust forward by the recoil lever spring, a circular stud in the lower part of the breech block, working in the guide grooves of a switch plate fitted to the non-recoiling portion of the receiver, strikes the rear of the center block in the plate and so guides the breech block downward, leaving the chamber ready for the cartridge to be inserted. As the stud continues forward, it strikes the lower cam surface of the switch plate causing the breech block to rise and close the breech. Now the stud is lined up with the horizontal slot in the switch plate, down which it travels during the final half-inch forward motion, securely locking the breech.

Also during the forward thrust of the recoiling parts, an arm is forced up by a cam on the left side of the receiver, forcing outwards the distributor against the tension of its spring and permitting the first cartridge from the magazine, which was resting on the distributor to drop into the magazine opening.

Meanwhile the front surface of the rear claw of the feed arm engages with the rear surface of the feed arm actuating block, thus rotating the feed arm forward. The arm strikes the head of the cartridge in its seat against the left flange of the breech block, pushing it ahead into the chamber. The rear claw of the feed arm rises up to the rear surface of the feed arm actuating block and travels along its upper surface. The bottom of the rear

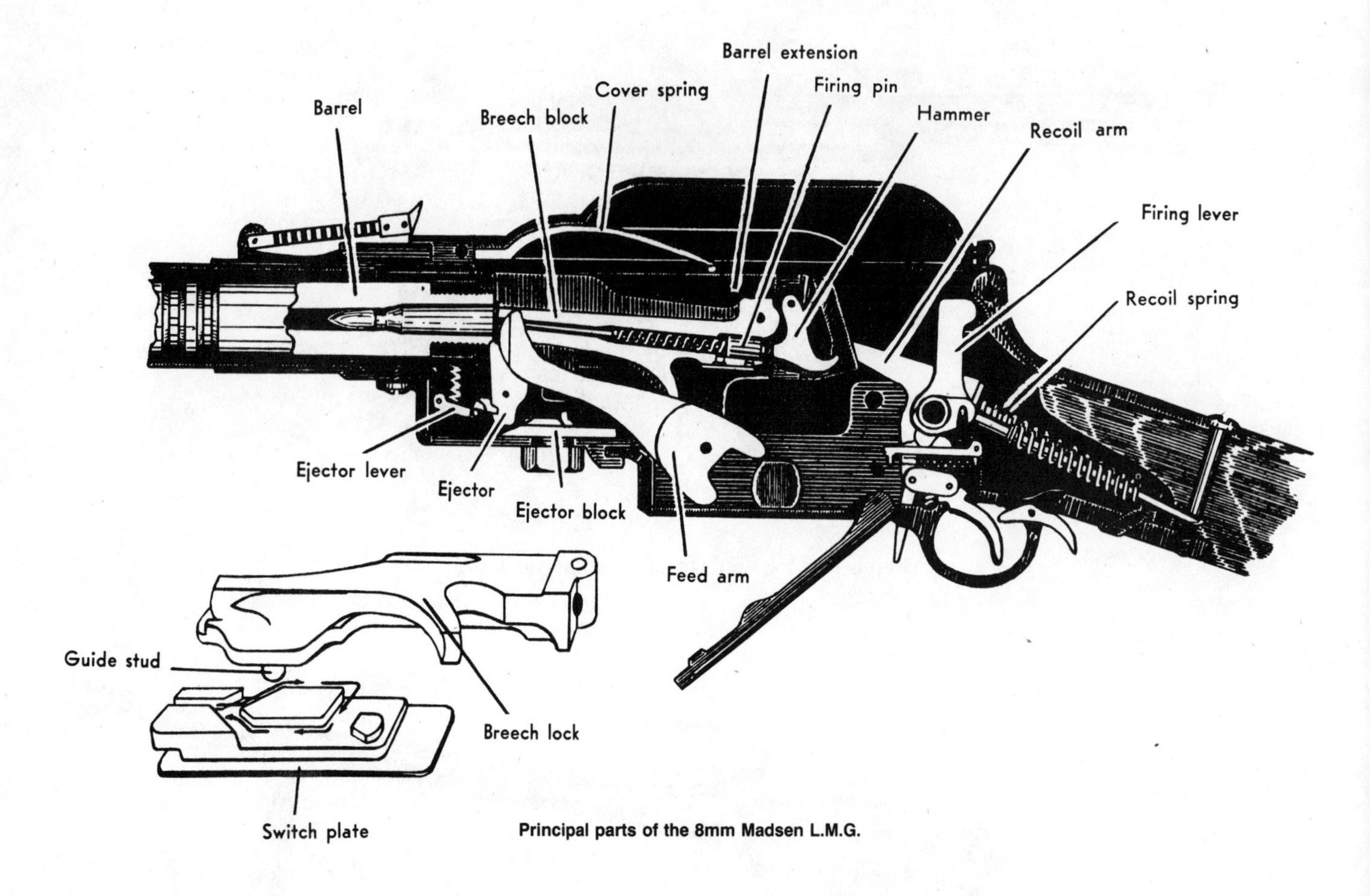

Principal parts of the 8mm Madsen L.M.G.

Madsen M1930/14 Machine Gun.

claw, now being above the feed arm actuating block, moves the cartridge in the chamber to allow the breech block to rise. It also prevents any rebound of the feed arm.

As the forward action starts, the ejector is positioned alongside the rear of the ejector block. As the recoiling parts go forward, a stud on the ejector lever rides down the sloping cam and forces an ejector downward on its spring, thus bringing the lever in contact with the tail of the ejector. As soon as the breech block starts to rise, the tail of the ejector is clear of the ejector block and is raised in position by its lever. As the breech closes, the ejector is able to rise to the vertical under the influence of the ejector lever spring and falls in place just below the chamber with its hook below the rim of the cartridge.

Note that in this type of gun the cartridge is not set into the chamber by the breech block, and until the front of the breech block rises to a complete locking position, the hammer, firing pin and cartridge are not in alignment. Therefore, there can be no accidental discharge during feeding.

Return Movement of the Action. At the breech, the barrel is joined to the breech block casing in which are the hinged breech blocks. These three units recoil together, with the breech remaining closed and locked for about ½". This sudden rearward thrust forces the firing lever up as it is struck by the rear of the breech mechanism and frees it from the hammer, permitting the hammer to pivot back and the firing pin to be withdrawn by the firing pin spring. The guide stud is now passed out of the horizontal groove and travels up the upper cam of the switch plate, which pivots the breech block upwards at its nose to permit ejection. The extra stud travels along the top of the cam, and the cover spring then forces the front of the breech block downward, compelling the stud to drop out of the rear stud of the switch plate.

Note: This gun has no individual extractor. The ejector pulls the empty cartridge out of the chamber and hurls it from the gun.

With the first movement to the rear, the inclined slope in the front of the ejector block raises the ejector, which is held in vertical position by its lever engaged in a recess in the bottom of the ejector. From the influence of this separate movement, a hook on the ejector catches the rim of the empty cartridge case, and the bottom of the ejector lies on top of the front flats of the ejector block.

The stud on the ejector lever runs up the sloping cam on the left of the ejector block to compress the ejector lever spring. This also disengages the ejector lever from the ejector, allowing the ejector tail to be tripped forward by the step on the ejector block, and as the ejector is pivoted about its center, the tripping motion of the tail forces the hook to the rear, pulling the empty cartridge case out and hurling it from the bottom of the gun.

An ejection guide on the breech block guides the empty cartridge case as it is hurled out. The ejector lever stud now rests above the sloping cam holding the ejector levers upwards and free of the ejector, which lies on top of the rear flap of the ejector block beneath the breech block. During the recoil movement, the distributor arm rides down its cam and rotates the distributor inwards and downwards under the influence of the spring. This places the cartridges in the feedway against the left flange of the breech block. Meanwhile, the rear surface of the front feed arm claw engages with the front face of the feed arm actuating block and rotates the feed arm backwards. The bottom of the front claw now rides along the top of the feed arm block, preventing rebound of the feed arm.

Further Note on Recoil System. Some Madsens are fitted with a so-called "recoil increaser," which forms a choke at the muzzle. By reversing the two parts of the increaser, the rear portion forms a collar that forces the gasses escaping at the muzzle to rebound onto the barrel, giving additional thrust to the rearward action. This speeds the gun up greatly.

When the gun is cocked, the claws of the feed arm automatically open the ejector cover. It must be closed by hand on cease fire.

Field Stripping the Madsen

Remove magazine from gun and ease recoiling parts forward.

Lift the locking bolt lever into vertical position and withdraw it to the left.

Push the butt to the right front while gripping the receiver with the left hand; then remove the butt.

Holding forefinger of right hand ahead of feed arm axis bar, with the hand draw back barrel and breech mechanism.

Pull out the barrel. Remove barrel very carefully as it is easy to damage the front end ring. This barrel ring is one of the weak points in the weapon. Handle it carefully.

Remove the breech block bolt.

Pull feed arm to the rear. Lift the front, and lower the end of the breech block; then pivot the front up to vertical position, when the feed arm may be eased forward and lifted out of the block.

Further dismounting need not be attempted.

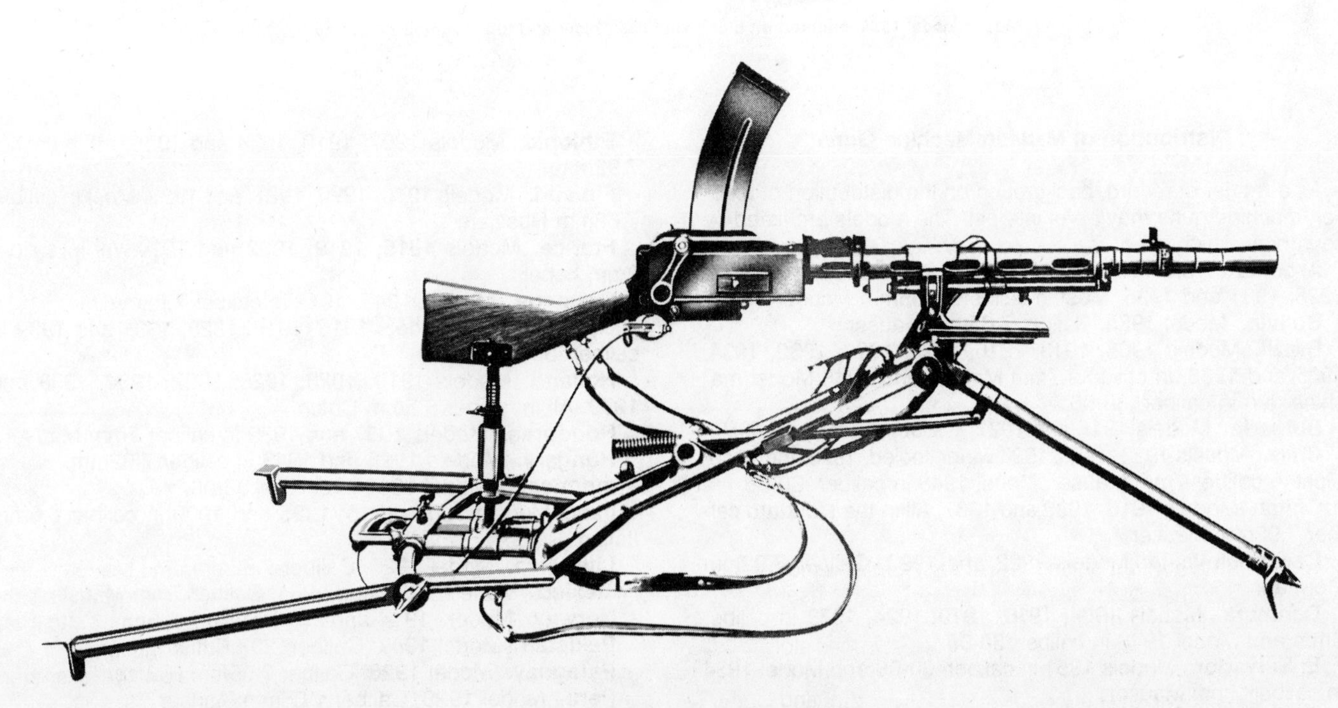

Madsen Model 1950 mounted on tripod.

DANISH MADSEN MACHINE GUN M1950

CHARACTERISTICS OF THE MADSEN MACHINE GUN MODEL 1950

Caliber: Advertised for any rifle cartridge, usually found in .30-06.
System of operation: Recoil, selective fire.
Length overall: 45.9 in.
Barrel length: 18.8 in.
Weight: 22 lbs.
Feed device: 30-round, detachable box magazine.
Sights: Front: Blade.
Rear: Tangent, graduated from 200–1800 meters.
Muzzle velocity: Approx. 2700 f.p.s. w/.30-06.
Cyclic rate: 400 r.p.m.

Special Note on Operating Characteristics of M1950

The operating characteristics of the Model 1950 Madsen machine gun, except for minor variations as required by feed alterations in some instances, are the same as described previously.

This model, while recoil operated on the Madsen principle, is a modification of the earlier production. The change is in the barrel removal. This may be removed without the use of tools or without removing any component parts.

This Madsen may be fired from the bipod with or without the shoulder stock. It may also be fired from a light tripod, which is convertible to an antiaircraft high-angle fire mount. In addition, it may be fired from the shoulder by the rifleman in kneeling, standing, or prone positions.

Madsen Model 1950 mounted on bipod with flash hider and top mounted magazine.

Distribution of Madsen Machine Guns

As a matter of record, background on the distribution of Madsen machine guns may be of interest. The models are listed by country in which used.

Argentina. Madsen machine guns Models 1910, 1925, 1926, 1928, 1931 and 1935. Most in calibers 7.65mm Mauser.

Bolivia. Model 1925. Caliber 7.65mm Mauser.

Brazil. Models 1908, 1913, 1916, 1925, 1928, 1932, 1934, 1935 and 1936, in calibers 7mm Mauser and 1946 Model machine gun in caliber .30-06.

Bulgaria. Models 1915 and 1927. Caliber 8mm (8 × 50R).

Chile. Models 1923, 1925, 1926 water cooled, 1928 and 1940. Most in caliber 7mm Mauser. Model 1946 in caliber .30-06.

China. Models 1916, 1930 and 1937. All in the standard caliber 7.92mm Mauser.

Czechoslovakia. Models 1922 and 1923. Caliber 7.92mm German.

Denmark. Models 1904, 1916, 1919, 1924, 1939 in caliber 8mm and Model 1948 in caliber .30-06.

El Salvador. Models 1951 in caliber .30-06 and Model 1934 in caliber 7mm Mauser.

Esthonia. Models 1925 and 1937 in caliber .303 British.

Ethiopia. Models 1907, 1910, 1934 and 1935. All in caliber 7.92mm.

Finland. Models 1910, 1920, 1921, and 1923. Most in caliber 7.62mm Russian.

France. Models 1915, 1919, 1922 and 1924. All in caliber 8mm Lebel.

Germany. Models 1941, 1942 in caliber 7.92mm.

Great Britain. Models 1915, 1919, 1929, 1931 and 1939 in caliber .303 British.

Holland. Models 1919, 1923, 1926, 1927, 1934, 1938 and 1939. All in caliber 6.5mm Dutch.

Honduras. Models 1937 and 1939 in caliber 7mm Mauser.

Hungary. Models 1925 and 1943 in caliber 7.92mm.

Indonesia. Model 1950 in caliber .30-06.

Italy. Models 1908, 1910, 1925 and 1930 in caliber 6.5mm Italian.

Lithuania. Model 1923. Caliber 7.92mm.

Mexico. Models 1911 and 1934. Caliber 7mm Mauser.

Norway. Models 1914 and 1918. Caliber 6.5mm Mauser.

Pakistan. Model 1947. Caliber .303 British.

Paraguay. Model 1926. Caliber 7.65mm Mauser.

Peru. Model 1929. Caliber 7.65mm Mauser.

Portugal. Models 1930, 1936 and 1952. Caliber 7.7mm. Also

in Portugal, Models 1936, 1940 and 1947 in caliber 7.92mm.
Russia. Models 1904 and 1915 in caliber 7.62mm Mosin-Nagant, Russian.
Spain. Models 1907 and 1922 in caliber 7mm Mauser.
Sweden. Models 1906, 1914 and 1921. Caliber 6.5mm Mauser.
Thailand. Models 1925, 1930, 1934, 1939, 1947 and 1949. Caliber 8mm and Model 1951 in caliber .30-06.
Turkey. Models 1925, 1926, 1935 and 1937. Caliber 7.92mm.
Uruguay. Model 1937. Caliber 7mm Mauser.
Yugoslavia. Various in caliber 7.92mm.

MADSEN/SAETTER RIFLE-CALIBER MACHINE GUN

The Madsen/Saetter machine gun is belt-fed and gas operated. It is usually mounted on a light field tripod but can be fired from a bipod or from the hip. The gun is, in appearance and design, a modern weapon, embodying all the experience gained during the last few years in the development of machine guns. The component parts are easily mass produced by punching, turning and precision casting without detracting from reliability and durability.

The Madsen/Saetter machine gun has gone through three changes. These three models are the Mk I, Mk II and Mk III. The Mk III was not completely developed until 1959. The Mk III is more reliable and shorter than the Mk I or Mk II. Madsen also made a tank machine gun model that has no buttstock and no bipod. The return spring is placed round the gas piston. A special "lightweight" tripod is made for this weapon for use outside the tank.

CHARACTERISTICS OF MADSEN/SAETTER MACHINE GUN

Caliber: Any military rimless cartridge from 6.5mm to 8mm.
System of operation: Gas operated, full automatic fire only.
Weight: 25.6 lbs. w/heavy barrel.
Weight of tripod: 36.2 lbs.
Length, overall: 48 in.
Barrel length: 26 in.
Feed device: By non-disintegrating metallic belts of 50 rounds, which can be joined to any desired length. The weapon can also be supplied with two box magazines that fasten to the receiver directly below the action; one magazine holds 50 rounds, and the other holds 100 rounds, in metal link belts. The feed system used on this weapon is a copy of the German MG42.
Sights: Front: Barley corn.
Rear: Open, with graduations up to 1200 meters.
Muzzle velocity: Standard for ammunition employed.
Cyclic rate: 700 to 1000 r.p.m.

Field Stripping the Madsen/Saetter

Insure that the chamber is empty by opening the feed cover, and pull back the cocking handle.

To remove the buttstock, seize the pistol grip with one hand, and, with the thumb, press the trigger gear housing latch. The buttstock, now released, should be turned 90 degrees with the other hand, and pulled to the rear.

To remove the triggerguard housing, the cocking handle should be pulled fully back, and then pushed forward again. This brings the bolt carrier to the rear. The triggerguard housing can now be turned downward and taken out of engagement.

To remove the gas piston: when the bolt assembly is in its rearmost position, the gas piston head is just outside a clearance at the end of the receiver and can be withdrawn.

To remove the bolt, let it slide to the rear, out of the receiver.

To remove the barrel, turn the barrel handle forward until it is free.

To strip the bolt assembly, press out the bolt carrier pin, pull back the action head, and remove components.

To assemble, follow the above instructions in reverse order.

The Madsen/Saetter light machine gun in caliber .30 is used by and manufactured in Indonesia.

Dansk Industri Syndicat developed a new version of the rifle caliber Madsen/Saetter. The Mark IV is shorter and lighter than the earlier Marks and is used with a lighter tripod. Like the earlier Marks, it can be made for any rimless cartridge from 6.5mm to 7.92mm and, of course, for the US cal. .30 rifle and machine gun cartridge. Any of the Madsen/Saetter machine guns can be easily modified to use disintegrating metallic links.

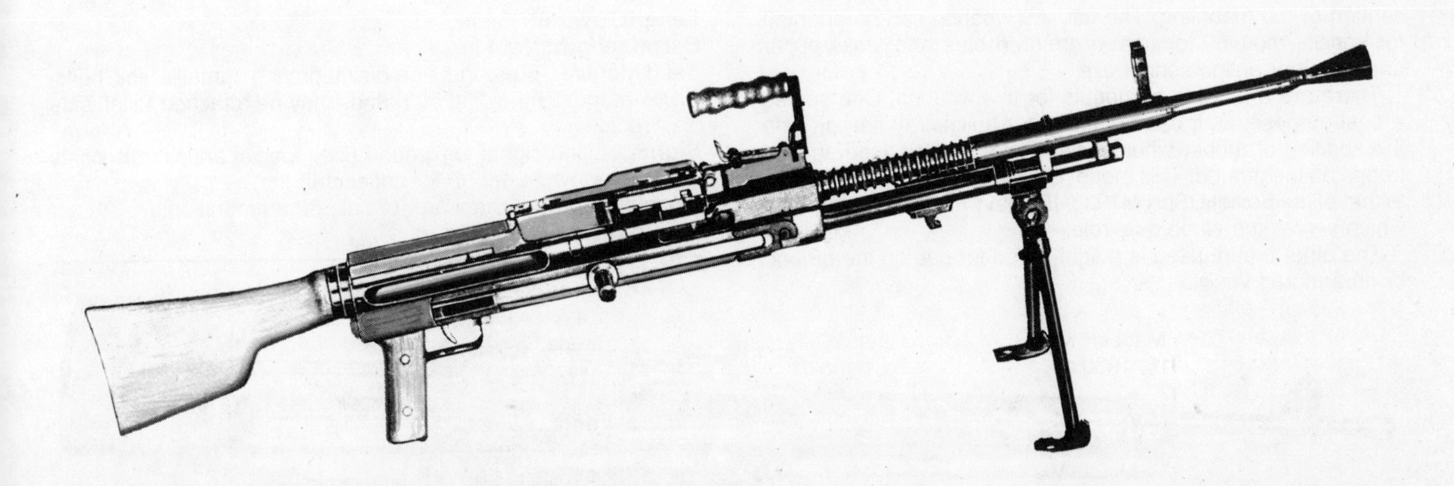

The Danish Madsen/Saetter machine gun (rifle caliber).

MADSEN/SAETTER 7.62mm TANK MACHINE GUN

This is a version of the rifle caliber Madsen/Saetter made specifically for use on tanks and armored vehicles. It can also be used as a ground gun on a tripod. The arrangement of the gas cylinder and piston is interesting. As with the Soviet RPD light machine gun, there is a definite air gap between the gas cylinder and piston when the gun is cocked. This gap serves a useful purpose during functioning—a good deal of the gas bled through the gas port into the gas cylinder is dissipated into the atmosphere after it has served to force the piston to the rear. This has several advantages in that it results in less build-up of carbon in the gas cylinder and piston tube, and, of special importance in a tank machine gun, it cuts down the amount of "operating" gas filters back into the receiver and thereby into the tank.

CHARACTERISTICS OF MADSEN/SAETTER TANK MACHINE GUN

Caliber: 7.62mm NATO (can be made in other calibers).
System of operation: Gas.
Weight: 22.3 lbs.
Length, overall: 38.2 in.
Barrel length w/flash hider: 22.2 in.
Sights: Front: Protected blade.
Rear: Tangent with V-notch.
Feed Device: Non-disintegrating 50-round metallic link belt, which can be joined to other belts. The weapons can be built to use disintegrating links of the M13 (US) type. Belt is normally contained in a box attached to left side of receiver.
Muzzle velocity: 2800 f.p.s.
Cyclic rate: 700–800 r.p.m.

Madsen/Saetter 7.62mm Tank Machine Gun.

DANISH MADSEN/SAETTER CAL. .50 MACHINE GUN

This caliber .50 weapon exists only in prototype form. This machine gun utilizes the same basic system and design features as the Madsen/Saetter rifle-caliber machine gun, with the exception of the mounting. The cal. .50 weapon can be adapted for special mounts, for use in armored cars and tanks or for antiaircraft or antipersonnel use.

There are two types of mounts for this weapon. One serves a dual purpose, as it can be set up for antiaircraft fire, or, with the addition of rubber wheels, can be towed and used against troops on the ground. This mount is generally similar in principle to that of the Soviet DShK M1938/45 heavy machine gun mount, which serves the same dual role.

The other mount used is a light tripod for use on the ground or on armored vehicles.

CHARACTERISTICS OF MADSEN CAL. .50 MACHINE GUN

Caliber: Cal. .50 (12.7mm).
System of operation: Gas operated, full automatic fire only.
Weight: 61.7 lbs.
Length, overall: 64 in.
Barrel length: 39.4 in.
Feed device: 50-round non-disintegrating metallic link belts; box magazine holding 50 rounds may be attached to left side of receiver.
Sights: Open sights for ground use; special antiaircraft sights are used when set up for antiaircraft fire.
Muzzle velocity: Standard for cal. .50 ammunition.
Cyclic rate: 1000 r.p.m.

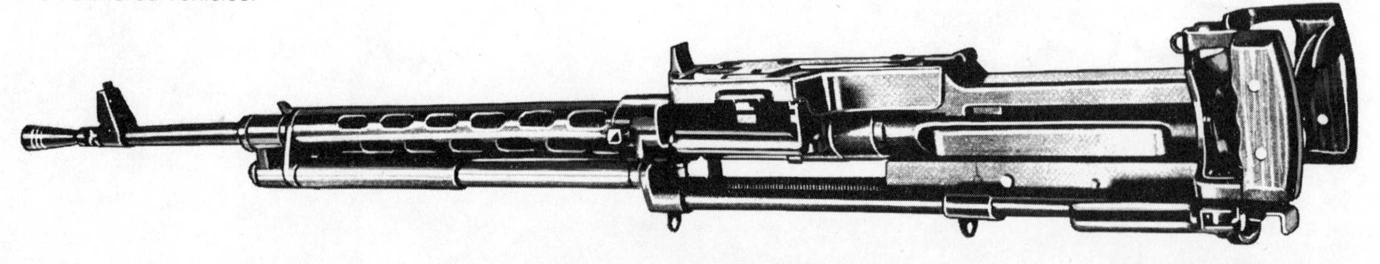

Danish Madsen/Saetter caliber .50 machine gun.

17 Dominican Republic

SMALL ARMS IN SERVICE

Handguns:	11.43 × 23mm M1911A1	US-made M1911A1
	9 × 19mm NATO FN Browning Mle. 1935	Belgian-made
	.38 Smith & Wesson	A variety of models
Submachine guns:	Unknown	
Rifles:	7.62 × 51mm NATO G3	Heckler & Koch-made
	7.62 × 51mm NATO FAL	FN-made
	7.62 × 33m Cristobal Carbine	Armeria San Cristobal
	5.56 × 45mm M16A1	Colt-made
Machine guns:	7.62 × 63mm M1919A4 and M1917A1 Browning Machine Guns	US-made
	12.7 × 99mm M2 HB Browning	US-made

DOMINICAN AUTOMATIC CARBINE CRISTOBAL MODEL 2

The Armeria San Cristobal has manufactured this automatic carbine which is chambered for the US caliber .30 carbine cartridge. This weapon is called the Cristobal Model 2. The Armeria San Cristobal was established with technical assistance from engineers from Pietro Beretta, Italy and technicians from Hungary. Externally the Model 2 bears some striking resemblances to Beretta submachine guns.

Internally, however, there are some significant differences. The Beretta Model 38-series submachine guns are blowback operated; the Model 2 Cristobal carbine is a delayed blowback.

CHARACTERISTICS OF THE CRISTOBAL MODEL 2

Caliber: Cal. .30 (US M1 carbine cartridge.)
System of operation: Delayed blowback, selective fire.
Weight, w/o magazine: 7.75 lbs.
Length, overall: 37.2 in.
Barrel length: 16.1 in.
Feed device: 25- or 30-round, detachable, staggered-row box magazine.
Sights: Front: Hooded blade.
Rear: Notch with elevator.
Muzzle velocity: 1875 f.p.s.
Cyclic rate: 580 r.p.m.

Caliber .30 Cristobal Model 2 Automatic Carbine.

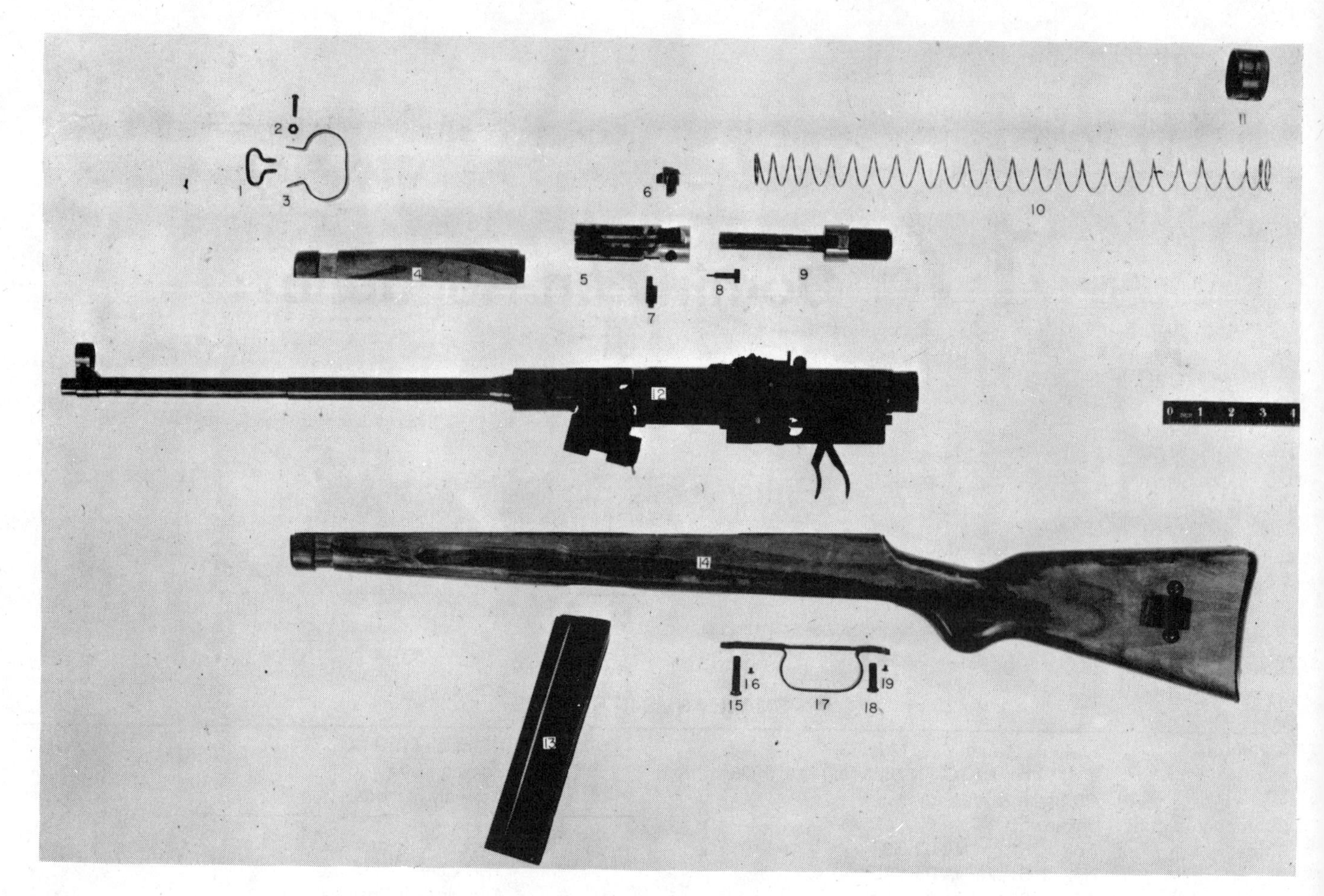

Cristobal carbine, stripped.

How to Load and Fire the Cristobal Model 2

Insert a loaded magazine in the magazine well. Pull the handle on the right side of the receiver to the rear. The bolt will remain to the rear, since this weapon fires from an open bolt. Push the handle forward—it does not reciprocate with the forward and rearward movements of the bolt. If the forward trigger is pulled, the weapon will fire single shots; if the rear trigger is pulled, the weapon will produce automatic fire. To put the weapon on safe, pull the lever mounted on the left side of the receiver to the rear. The safety blocks the sear and the triggers and prevents rearward movement of the bolt. This weapon has no bolt-holding-open device, and therefore the bolt must be pulled to the rear every time a new magazine is loaded.

How to Field Strip the Cristobal Model 2

Remove the magazine by pushing the magazine catch forward. Press in the receiver cap lock and turn the receiver cap, removing it from the rear of the receiver. The recoil spring and the bolt can now be removed from the receiver. The bolt can be disassembled by removing the cross pin at its forward part and slipping off the inertia lock. No further disassembly is recommended. To reassemble the weapon, perform the above steps in reverse order.

How the Cristobal Model 2 Works

The trigger mechanism of this weapon is similar to that of the Beretta Model 38-series of weapons. The bolt consists of two main parts: the bolt body and a heavy part called the striker. These parts are joined by a two-armed inertia lever seated in the rear end of the bolt proper. The upper long arm of the movable inertia lever engages the striker, and the lower short arm of the lever projects down from the bolt. When the bolt is closed, the short arm of the inertia lever stands before a stationary shoulder firmly attached to the bottom of the receiver. When a round is fired, the gases thrust rearward on the cartridge case base, which pushes back against the face of the bolt. Before the bolt can open, however, its inertia must be overcome. When the bolt begins to move rearward, the lower arm of the inertia lever (which could be called an inertia lock) bears against the bottom shoulder of the receiver. The rearward movement of the bolt then causes rearward rotation of the inertia lever, swinging the bottom arm up and out of engagement with the receiver and forcing the upper arm back against the heavy striker. It is claimed that this system of delayed opening offers as great a resistance to the cartridge thrust as that of a bolt three to five times heavier than the one used with this weapon. During the closing of the bolt, another lateral arm of the inertia lever slides on the receiver wall, and retains the inertia lever and the striker in a cocked position long enough for the bolt to run fully home. When the bolt is fully closed, the lateral arm of the inertia lever moves above a slot in the receiver wall, permitting forward rotation of the lever by the striker and the firing of the round. The inertia lever therefore prevents the weapon from firing before the bolt is fully closed.

DOMINICAN CARTRIDGES

The Dominican Republic also manufactures its own cal. .30 rifle and carbine cartridges. These are loaded with Berdan primers, rather than the Boxer type used in the United States and Canada.

DOMINICAN CAL. .30 AUTOMATIC CARBINE M1962

A new model of the Cristobal carbine has been developed. The Model 1962 Automatic carbine may be found either with a fixed wooden stock or a folding metal stock. Its loading, firing, field stripping and functioning are the same as those of the Cristobal Model 2. The main difference in construction, other than the folding steel stock, is the use of a perforated metal barrel jacket on the Model 1962 weapons.

CHARACTERISTICS OF THE M1962 AUTOMATIC CARBINE

Caliber: Cal. .30 (US M1 carbine cartridge).
System of operation: Delayed blowback, selective fire.
Weight, loaded:
W/wooden stock—8.7 lbs.
W/wooden steel stock—8.2 lbs.
Length, overall:
W/Wooden stock—34.1 in.
W/steel stock extended—37 in.
W/steel stock folded—25.6 in.
Barrel length: 12.2 in.
Feed device: 30-round, detachable, staggered-row box magazine.
Sights: Front: Protected blade.
Rear: L type
Muzzle velocity: 1870 f.p.s.

DOMINICAN 7.62mm AUTOMATIC RIFLE MODEL 1962

The Model 1962 automatic rifle combines the gas system of the US M14 rifle with the bolt mechanism of the FN light automatic rifle. This weapon was apparently made only as a prototype.

CHARACTERISTICS OF THE M1962 AUTOMATIC RIFLE

Caliber: 7.62mm NATO.
System of operation: Gas, selective fire
Weight loaded: 10.4 lbs.
Length, overall: 42.5 in.
Barrel length: 21.3 in.
Feed device: 20-round, detachable, staggered row box magazine.
Muzzle velocity: 2700 f.p.s.

INGRAM MARK I SUBMACHINE GUNS

The Armeria San Cristobal has recently undertaken manufacture of the Ingram submachine guns, which they call the Mark I in 9 × 19mm Parabellum and 11.43 × 23mm (.45 ACP). They are known as the Model 10 in the United States. This is a commercial venture, and these submachine guns are not currently used by the Dominican armed forces.

CHARACTERISTICS OF THE SAN CRISTOBAL MARK I INGRAM SUBMACHINE GUNS

Caliber:	11.43 × 23mm (.45 ACP)	9 × 19mm
System of operation:	Blowback	Blowback
Weight w/o magazine:	6.25 lbs.	6.25 lbs.
Length overall		
(stock extended):	19.07 in.	19.07 in.
(stock collapsed):	12.84 in.	12.84 in.
(w/o stock):	10.50 in.	10.50 in.
Barrel length:	5.75 in.	5.75 in.
Feed device:	30-shot	30-shot
Muzzle velocity:	850 f.p.s.	1320 f.p.s.

These submachine guns can be provided with sound suppressors.

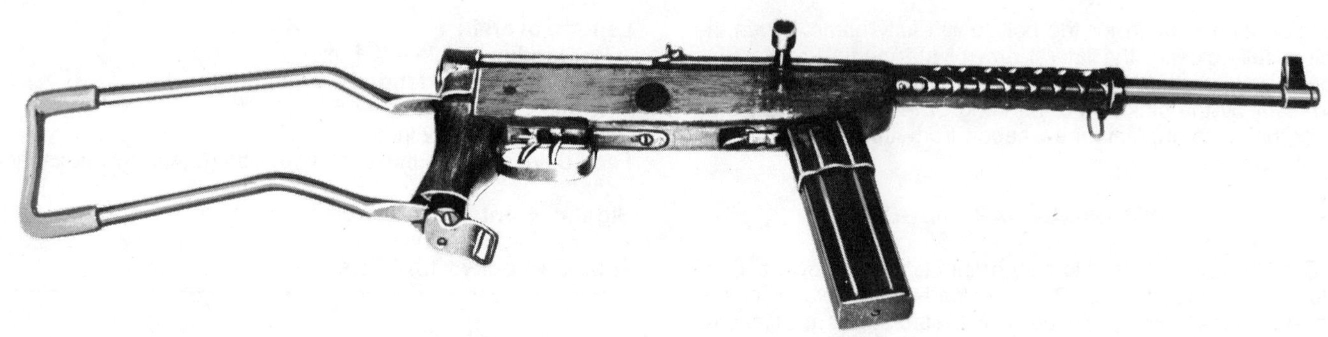

Dominican caliber .30 Automatic Carbine Model 1962 with folding metal stock.

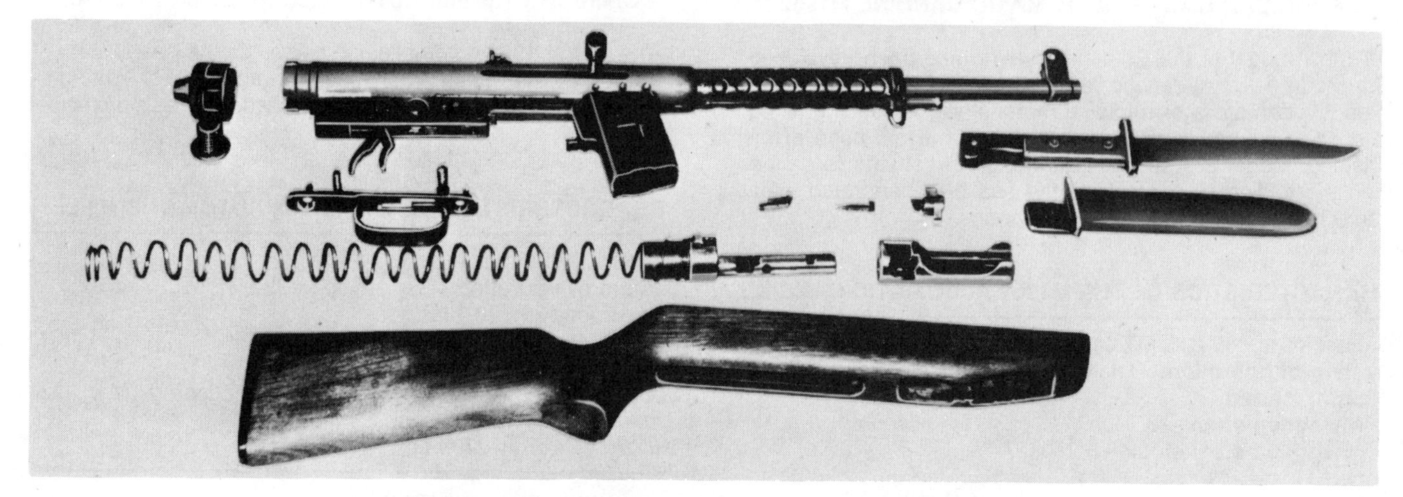

Dominican caliber .30 Automatic Carbine Model 1962, field stripped.

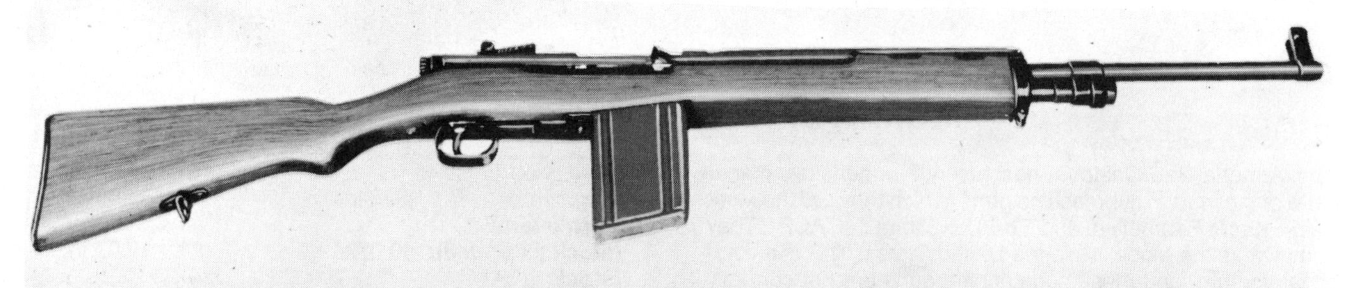

Dominican 7.62mm Automatic Rifle Model 1962.

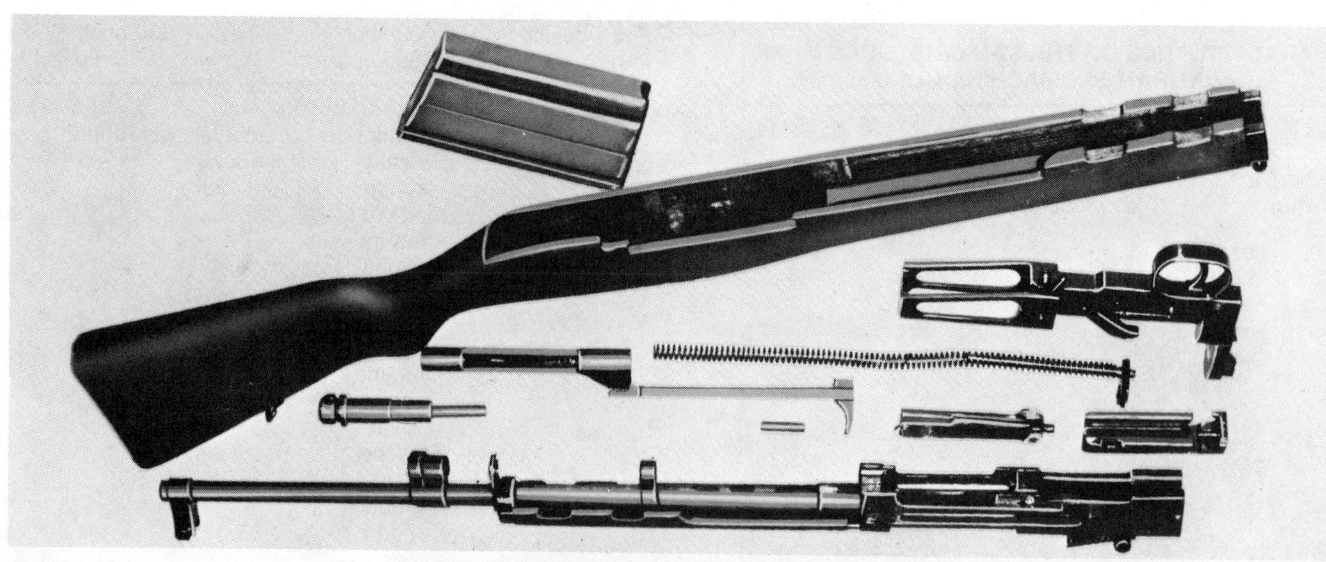

Dominican 7.62mm Automatic Rifle Model 1962, field stripped.

18 Finland

SMALL ARMS IN SERVICE

Handguns:	9 × 19mm Parabellum L35 Lahti	Domestic design made at the Valtion Kivääritehdas (VKT), the State Rifle Factory.
	9 × 19mm Parabellum M23 Parabellum	Reworked DWM PO8s.
	9 × 19mm Parabellum FN Browning Mle. 1935	Purchased in 1940 from FN. Others acquired 1941–43 from Axis powers. Used by Finnish air forces.
Submachine guns:	9 × 19mm Parabellum M44	Finnish adaptation of the Soviet PPS43. Submachine guns are obsolescent in the Finnish armed forces.
Rifles:	7.62 × 54mmR M39	Finnish adaptation of the Mosin Nagant 1891 rifle. Held in reserve for emergencies.
	7.62 × 39mm M62	Valmet-made version of AK47.
	7.62 × 39mm M62-76	Valmet-made version of the AKM.
Machine guns:	7.62 × 39mm M62	Valmet-made domestic design.
	7.62 × 39mm M78	Long barrel squad automatic version of M62/76. Similar to Soviet RPK.
	7.62 × 54mmR M32-33	Finnish-made Maxim.

FINNISH ARMS FACTORIES

Taking advantage of the Russian revolution of November 1917, the Finns declared their independence from the old empire, in which they had had the position of an autonomous Grand Duchy. During the ensuing war for independence, a civil conflict between Finns, a paramilitary organization called the *Suojeluskuntajärjestö* (Defense Corps) was established to promote anti-Bolshevik activities. This organization created its own arms factory, the first in the new nation, which is still known by the name SAKO (*Suojeluskuntain Ase- ja Konepaja Oy*—the Arms and Machine Factory of the Defense Corps). In 1926 the Finnish government created the Valtion Kivääritehdas (VKT—State Rifle Factory) located north of Helsenki at Jyväskylä (pronounced U-vas-kula). A third major factory, Tikkakoski, was also established in this period. At the time of their formation, the Finnish Defense Forces were equipped largely with Russian weapons—something on the order of 80 percent. The Finns decided to standardize on the Model 1891 Mosin Nagant rifle because they had so many carried over from the Imperial period, and because many were available on the international market in Europe. During the years of the Winter War with the Soviet Union, the Finns procured many arms from abroad, but they continued to maintain the 7.62 × 54mmR cartridge (called 7.62 × 53mmR in Finland) as their standard rifle caliber until the adoption of the 7.62 × 39mm M43 Soviet cartridge in the late 1950s.

Following the Second World War, the Allies required the Finns to dismantle their arms industry. Production of military small arms was not resumed until the end of the 1950s. SAKO, VKT, and other organizations switched to consumer goods in the intervening years. VKT became Valmet, a state-owned combine dedicated to the manufacture of a variety of products ranging from ships to streetcars and other items of mass production. At the end of the 1950s work was undertaken to develop a modified version of the Soviet Kalashnikov assault rifle and to develop a new light machine gun. The result was the M62 assault rifle and the M62 light machine gun. By 1975 the Finnish Defense Forces had been completely reequipped with new small arms, and most of the older weapons had been sold off or regulated to storage depots for reserve forces.

FINNISH PISTOLS

THE FINNISH 9mm LAHTI PISTOL MODEL L-35

In 1935, the Lahti pistol was adopted; this pistol was made by VKT at Jyvaskyla, Finland.* (Earlier, in 1923, Finland had adopted the 7.65mm Luger.)

The Finnish Lahti pistol was issued in several variations. The principal difference is in the lock retaining spring; the Lahti uses a yoke-type lock. Early models have a lock retaining spring, but later models do not have this part.

The Finnish Lahti is essentially the same as the Swedish Model 40 Lahti, which is described in detail under Sweden. The principal differences are as follows:

a. The Swedish weapon does not have the lock retaining spring; in this respect it is similar to the Finnish pistols of later manufacture.

b. The Swedish pistol does not have the loaded chamber indicator, which is mounted on the top of the Finnish pistol.

c. The recoil spring is assembled differently on the Swedish weapon. It is assembled on a rod plugged through the grip frame, on a projection of the grip frame that passes into the bolt cavity.

d. The grips of the Finnish pistol are marked "VKT"; those of the Swedish pistol have the trademark of Husqvarna Vapenfabrik of Husqvarna, Sweden.

Finnish 7.65mm Luger.

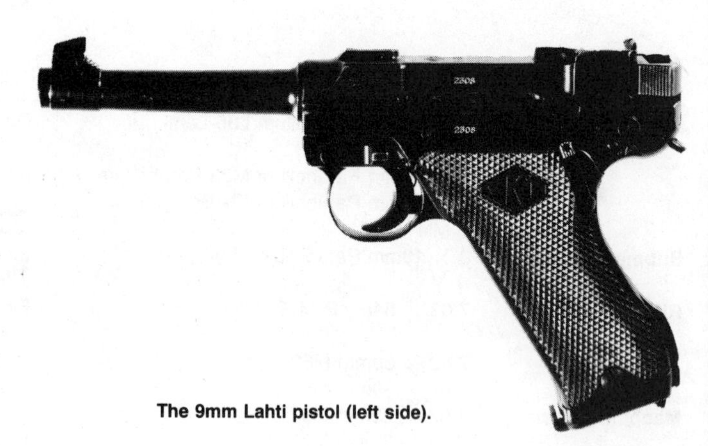

The 9mm Lahti pistol (left side).

The 9mm Lahti pistol (right side).

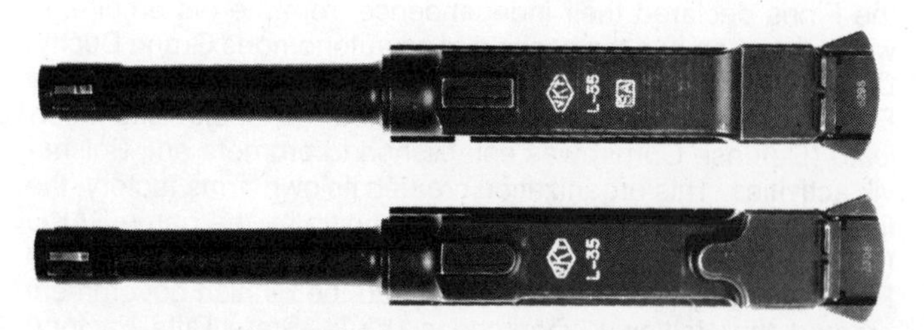

Top of Finnish Lahti of late manufacture (above) with no provision for lock retaining spring; the top of a Finnish Lahti of early manufacture (below) with lock retaining spring. Note larger bulge on top of the latter frame.

*For additional details see *Handguns of the World*.

FINNISH ASSAULT RIFLES

Finnish bicycle troops equipped with Soviet AK47s circa mid-1950s. Note that the fifth man from the right carries a Soviet RPD light machine gun.

The Valmet M60 experimental assault rifle variations tested in 1960–62 by the Finnish Defense Forces. *Top,* this version had a folding bayonet, winter trigger guard that could be opened, and no flash suppressor; *bottom,* this version had a standard knife-type bayonet, no trigger guard, and a flash suppressor.

Two versions of the M62 assault rifle. *Top,* with plain iron sights; *bottom,* with flip-up night sight.

Two versions of the M62-76 assault rifle. *Top,* with standard fixed stock; *bottom,* the M62-76T with folding stock.

The export models of the Valmet assault rifles. *Top*, the 7.62 × 39mm M62S with milled steel receiver; *middle*, the 5.56 × 45mm M71S with sheet metal receiver; *bottom*, the 7.62 × 39mm M71T assault rifle.

M62 ASSAULT RIFLE AND VARIANT MODELS

An experimental series of Finnish versions of the AK47 was developed at Valmet between 1958 and 1960. In 1960, two versions were distributed to the Finnish Defense Forces. They differed in the structure of the trigger guard, the muzzle, and the bayonet and its manner of mounting. After two years of experimentation, a definitive design was adopted as the M62 assault rifle. Large-scale production of M62s for the Finnish Defense Forces was started in 1965, both at Valmet Tourula works in Jyväskylä and SAKO in Riihimäki. During the first four years the M62 was issued with conventional iron sights for daylight shooting—this was the M62PT. Beginning in 1969, these assault rifles were fitted with flip-up nighttime sights which had phosphoric dots. Three years later, 1972, the dots were replaced by two luminescent tritium capsules.

Over the years since its first introduction, the exterior of the M62 rifle has been varied slightly. The buttplate assembly has been modified to make it stronger, and the pistol grip has been reshaped to make it easier to grip. In the mid-1970s the Valmet factory introduced a stamped sheet metal receiver version of the M62. This weapon, which was designated the M62-76, was in effect the Finnish version of the Soviet AKM. In addition to the standard military model, special versions were made for export. These included the M62-76M with a plastic stock, the M62-76P with a wood stock, and the M62-76T with a tubular steel stock that folds to the left side of the receiver.

For the export market, Valmet developed the M71 series. An evolution from th M62-76 series, with a sheet metal receiver, the M71 series had a different front handguard and was available in 5.56 × 45mm as well as the 7.62 × 39mm cartridge. The front sight was mounted at the muzzle and the rear sight was moved to a position above the chamber end of the barrel. Semi-automatic versions of the M62 (the M62S) and the M71 (M71S) were developed for civilian sale in the United States. Those rifles were sold by Interarms.

M76 Assault Rifle Family

Seeking to further exploit the export market (Valmet had made some sales of their rifles in the Middle East, notably to Qatar), the Valmet factory introduced the M76 family in the late 1970s. It has been featured at several military exhibitions, including the Asian Defence Expo 80 held in Kuala Lumpur, Malaysia. The weapons in this series include the following:

Model 255 470	5.56 × 45mm Bullpup model. Now the M82 Short.
Model 254 100	5.56 × 45mm Side folding stock. M76F.
Model 254 060	5.56 × 45mm Tubular stock. M76T.
Model 254 080	5.56 × 45mm Plastic stock. M76P
Model 255 200	5.56 × 45mm Wood stock. M76W.
Model 255 490	7.62 × 39mm. M76 Short.
Model 254 090	7.62 × 39mm. M76F.
Model 253 810	7.62 × 39mm M76T.
Model 254 070	7.62 × 39mm. M76P.
Model 255 460	7.62 × 39mm. M76W.
Model 255 170	5.56 × 45mm light machine gun. M78.
Model 255 160	7.62 × 39mm light machine gun. M78.
Model 255 480	7.62 × 51mm NATO light machine gun. M78.

Disassembly and field maintenance for the Finnish assault rifles is essentially the same as that for the Soviet AK47 and AKM weapons described in Chapter 46. The M76 series is more fully illustrated in Chapter 1.

OBSOLETE FINNISH RIFLES

Former Finnish service rifles were all based on the Russian 1891 Mosin Nagant action. The Finnish models of the Mosin Nagant vary from the Russian mainly in sights, stocks, fittings, etc. In general it can be said that the Finnish-made weapons are of higher quality manufacture than the Russian weapons. Finnish service rifles chambered for the 7.62mm rimmed cartridge are the Model 91 Carbine, Model 27 Rifle, Model 28 Rifle, Model 28-30 Rifle, Model 30 Rifle and the Model 39 Rifle.

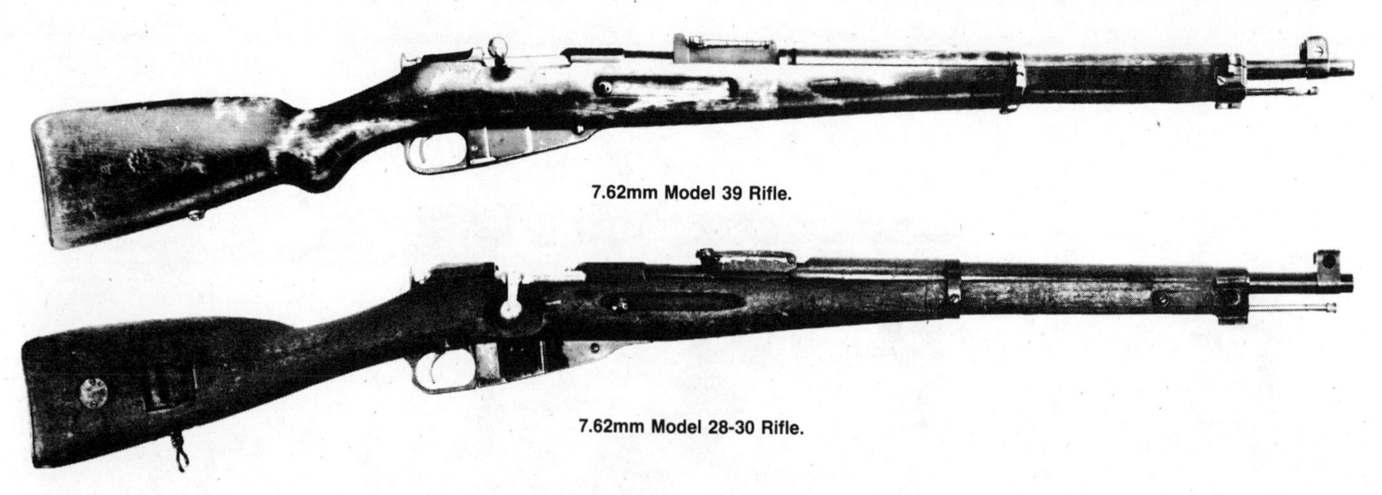

7.62mm Model 39 Rifle.

7.62mm Model 28-30 Rifle.

FINNISH SUBMACHINE GUNS

The submachine gun in Finland is synonymous with the name Lahti. Aimo Johannes Lahti (1896–1970) was one of this century's arms design geniuses. He completed his first submachine gun prototype in 1922, and after some additional experimentation this became the M26. A commercial venture, the M26 was built by Konepistooliosakeyhtiö (Submachine gun, Inc.). An approximate total of 200 M26s were built and the bulk were delivered to the Finnish Defense Forces. The M26 was chambered to fire the 7.65mm Parabellum cartridge.

Lahti continued to perfect the basic design of his submachine gun, and the result of this work was the M31 "Suomi" submachine gun, which was chambered for 9 × 19mm Parabellum. Lahti sold his patent rights to the M31 to the Tikkakoski factory where the gun was produced from 1932 to 1944; about 80,000 M31s were fabricated. The M31 was also used in Sweden where, as made in Sweden, it was known as the Model 37-39 and as imported from Finland, it was called the Model 37-39F. In Switzerland, where the gun was made by Hispano-Suiza and at the Waffenfabrik, Bern, the gun is known as the Model 43/44. In Denmark the Suomi has been made by Madsen, and the Swedish Model 37-39 version of the gun is still in limited use as the Model 44 (37).

Loading and Firing the Suomi

A loaded magazine is inserted below into the magazine housing and pressed up until it locks. The cocking handle protrudes from the rear of the weapon under the milled recoil spring cap. Grip it firmly, pull back to the rear to compress the bolt spring and cock the bolt, and allow to run forward under the influence of its own spring. Pressure on the trigger will now fire the weapon. The bolt will stay back between shots. When a continuous burst of fire is required, maintain a firm stiff pressure.

CHARACTERISTICS OF MODEL 31 SUOMI

Caliber: 9mm Parabellum.
System of operation: Blowback, selective fire.
Length overall: 34 in.
Barrel length: 12.62 in.
Weight: 11.31 lbs. w/empty 50-round box magazine.
Feed device: 70-round drum, 25 or 50-round box magazine.
Sights: Front: Blade.
Rear: Tangent graduated from 100–500 meters.
Cyclic rate: 800–900 r.p.m.
Muzzle velocity: Approx 1300 f.p.s.

Field Stripping the Suomi

At the rear of the receiver is a heavy milled cap. Unscrew this cap and ease out the housing recoil spring guide, and recoil spring.

Drawing back on the cocking handle will now pull the bolt back for removal from the weapon.

Finnish Suomi 9mm submachine gun.

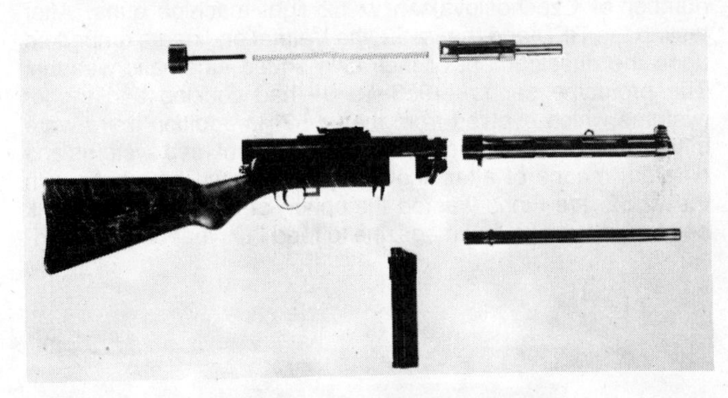

9mm Suomi submachine gun, field stripped.

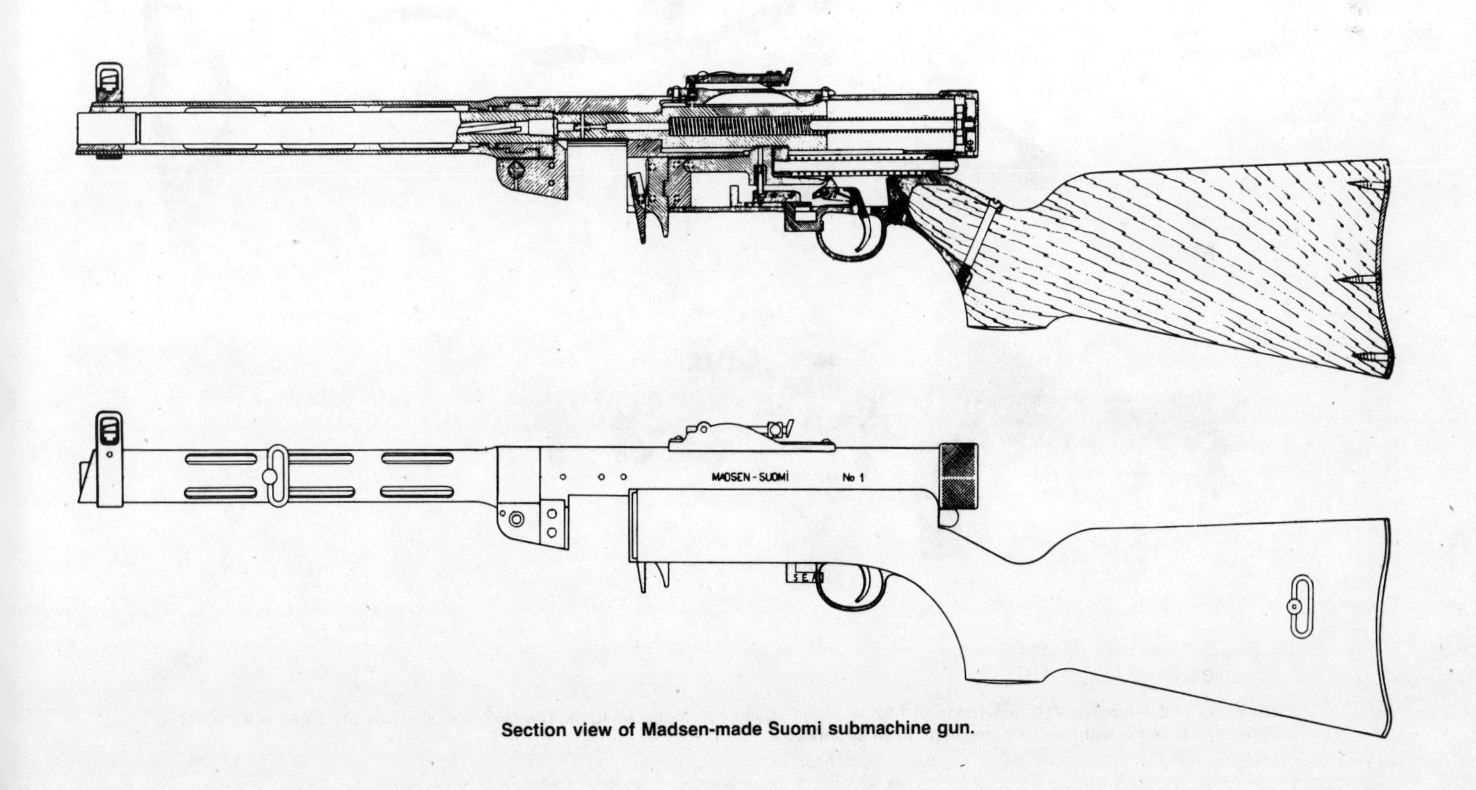

Section view of Madsen-made Suomi submachine gun.

THE M31 AND THE M44

The M31 was Finland's most significant contribution to small arms because it was the first submachine gun to be used in significant numbers in combat. During the Winter War of 1939–1940, the Red Army felt the sting of this gun and discovered the impact of mobile troops on skis equipped with submachine guns and carrying large quantities of ammunition. The Soviet experience in encountering Finnish troops with submachine guns led to their decision to build huge quantities of this class of weapons.

After the introduction of the Soviet Sudayev PPS43 7.62 × 25mm submachine gun, the Finns decided to build a 9 × 19mm version because it was easier and cheaper to manufacture. Tikkakoski built some 10,000 M44s, as this weapon was designated in Finland. A small quantity of unfinished components were assembled in the early 1950s. The last M44s were delivered to the Finnish Defense Forces in 1953. See Chapter 46 for details on the PPS43 (M44) type submachine gun.

FINNISH MACHINE GUNS

FINNISH M62 SQUAD AUTOMATIC WEAPON

After adopting the 7.62 × 39mm cartridge for their assault rifle ammunition, the Finnish Defense Forces decided to develop a squad automatic weapon to fire that same round. Early on they purchased Soviet RPDs, but they did not feel that this weapon would be economically practical for them to manufacture in their own arms factories. The Finns also purchased a small number of Czechoslovakian vz 52 light machine guns. After testing the RPD and the vz 52, the Valmet engineers embarked upon the development of their own squad automatic weapon. The prototype series—1958–1960—had cocking and trigger systems which evolved from the vz 52. In addition there were different models designed to evaluate different feed systems and different means of attaching the quick release barrel. As with the vz 52, the Finns wanted the option of using either the link belt or assault rifle box magazine to feed the weapon. As standardized, the M62 squad automatic weapon uses only a link belt feed. This weapon was manufactured by the Valmet Tourula works from 1965 to 1976. In addition to being used by the Finnish Defense Forces, the M62 has been purchased by Qatar.

CHARACTERISTICS OF M62 SAW

Caliber: 7.62 × 39mm
System of operation: Gas, automatic with tilting bolt locking system.
Length overall: 42.52 in.
Barrel length: 21.18 in.
Weight: 18.74 lbs.
Sights: **Front,** post;
Rear, U notch.
Cyclic rate: 1050 r.p.m.

Two views of the Finnish M60 experimental 7.62 × 39mm squad automatic weapon. The M62 is essentially the same with minor changes to the rear sight and a prong-type flash suppressor.

OBSOLETE FINNISH MACHINE GUNS

The Finns used three 7.62mm heavy machine guns. Apparently, all were Maxim types; their model designations were Model 09, Model 21 and Model 32.

The light machine gun formerly used was the Model 26 Lahti Saloranta. Although this weapon was developed for international sale and was advertised as being suitable for any service caliber, so far as known it has only been made in 7.62mm caliber for the Finnish Army and in 7.92mm for the Chinese prior to World War II.

FINNISH LAHTI SALORANTA LIGHT MACHINE GUN

The Model 26 was one of the first of the post-World War I "true" light machine gun types. It was considered a noteworthy gun at its time of development and, although somewhat lacking in adaptability as compared with the post-World War II guns, is still a basically sound weapon.

CHARACTERISTICS OF THE MODEL 26 LAHTI SALORANTA

Caliber: 7.62mm rimmed.
System of operation: Recoil, selective fire.
Weight, w/loaded 20-rd. magazine: 23 lbs.
Length, overall: 46.5 in.
Feed device: 20-round box, or 75-round drum.
Cyclic rate: 500 r.p.m.
Muzzle velocity: 2625 f.p.s.

How to Load and Fire the Model 26 Lahti Saloranta

Pull the operating handle to the rear, insert a loaded magazine in the magazine port (when using the box magazine) and set the selector on the type of fire desired. Squeeze the trigger and the weapon will fire. The Model 26 fires from an open bolt. To attach the 75-round drum magazine to the gun, remove the magazine support and push the magazine up into position until the holding latch clicks. To remove the barrel, turn the lever 180 degrees. This releases the catch holding the butt to the receiver. Lift the receiver cover, and the barrel, barrel extension, and bolt can be lifted out as a unit. Usually the complete unit is replaced.

Two distinct feed systems are available with this gun. A spring loaded clip which holds 25 cartridges may be used. The alternate is a flat drum magazine mounted below the gun with a capacity of 75 rounds.

Maximum rate of fire is normally set at 500 per minute.

In the Lahti the bolt is automatically held to the rear after the trigger has been released. The effect of this is to keep the action open to prevent a round in the chamber cooking off. This also, of course, permits air circulation through the barrel for cooling purposes.

The barrel is removed by turning the dismounting lever 180 degrees to release the catch, which holds the buttstock to the receiver. The receiver cover is lifted, and the barrel extension together with the barrel and bolt are then pulled out to the rear in a manner not unlike that of the Swiss Furrer.

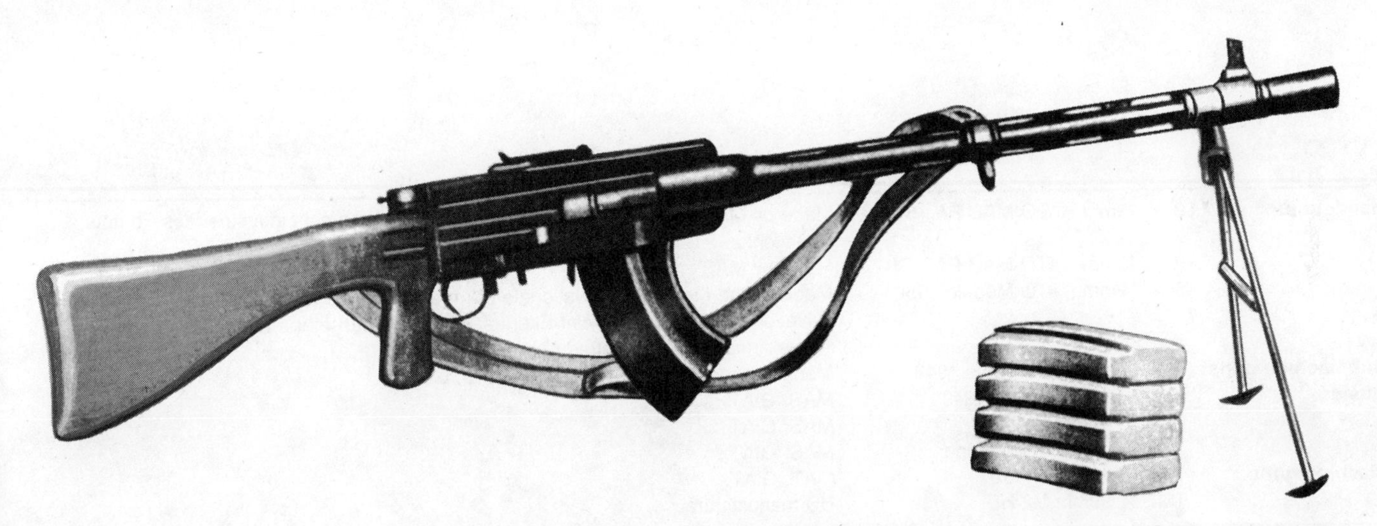

The 7.62mm Model 26 Lahti Saloranta Light Machine Gun.

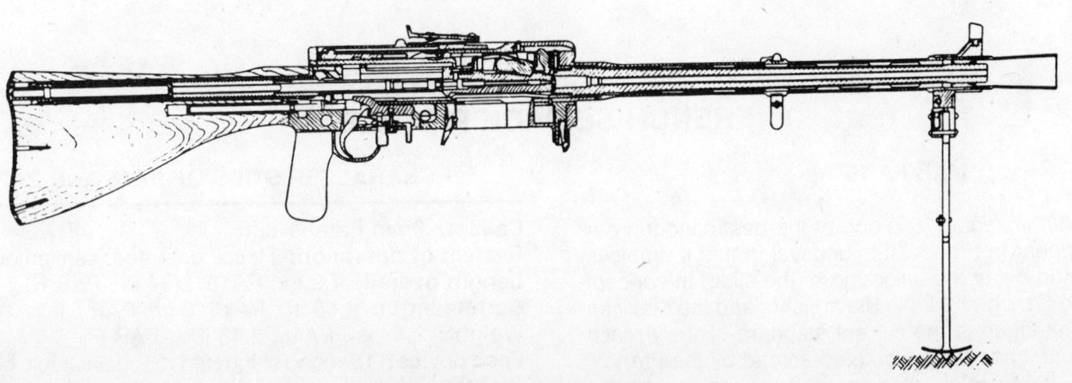

Section view of Lahti light machine gun.

19 France

SMALL ARMS IN SERVICE

Handguns:	9 × 19mm NATO MAB PA 15	Made by Manufacture d'Armes Automatiques, Lotissement industriel des Pontots, Bayonne.
	9 × 19mm NATO MAB PAP F1	Made by MAB.
	9 × 19mm NATO Modèle 1950	Made by the Manufacture Nationale d'Armes de Chatellerault and the Manufacture Nationale d'Armes de Saint Étienne, which are part of the Groupement Industriel des Armaments Terrestres (GIAT).
Submachine guns:	9 × 19mm P-M Modèle 1949	Manufacture Nationale d'Armes de Tulle, GIAT.
Rifles:	7.5 × 54mm Modèle 1949	MAS, GIAT.
	5.56 × 45mm FA MAS	MAS, GIAT.
	7.5 × 54mm FR F1 Sniper	MAS, GIAT.
Machine guns:	7.5 × 54mm AAT 52	MAT, GIAT.
	12.7 × 99mm M2 HB	US manufacture.

FRENCH SERVICE PISTOLS

MAB PA 15

The *Pistolet Automatique 15* is one of the best handgun designs ever to appear in France. It is unusual in that it employs a rotating barrel to delay the unlocking of the slide. In concept it is similar to the Czech vz 24, the Steyr Hahn, and the Mexican Obregon. This handgun is the current standard of the French armed forces, and it has been marketed abroad by the French. A target model—the *Pistolet Automatique Precision F1*—is also manufactured.

CHARACTERISTICS OF PA15 AND PAP F1

Caliber: 9mm Parabellum
System of operation: Recoil operated, semiautomatic
Length overall: 7.9 in., PA15; 9.64 in., PAP F1.
Barrel length: 4.48 in., PA15; 6 in., PAP F1
Weight: 2.4 lbs., PA15; 2.43 lbs., PAP F1
Feed device: 15-round staggered row, detachable box magazine.
Sights: Front, blade; rear, notch adjustable for elevation and windage (PAP F1).

Field stripped view of P15S.

French 9mm MAB P15S Pistol as produced by Manufacture d'Armes Automatiques, Bayonne.

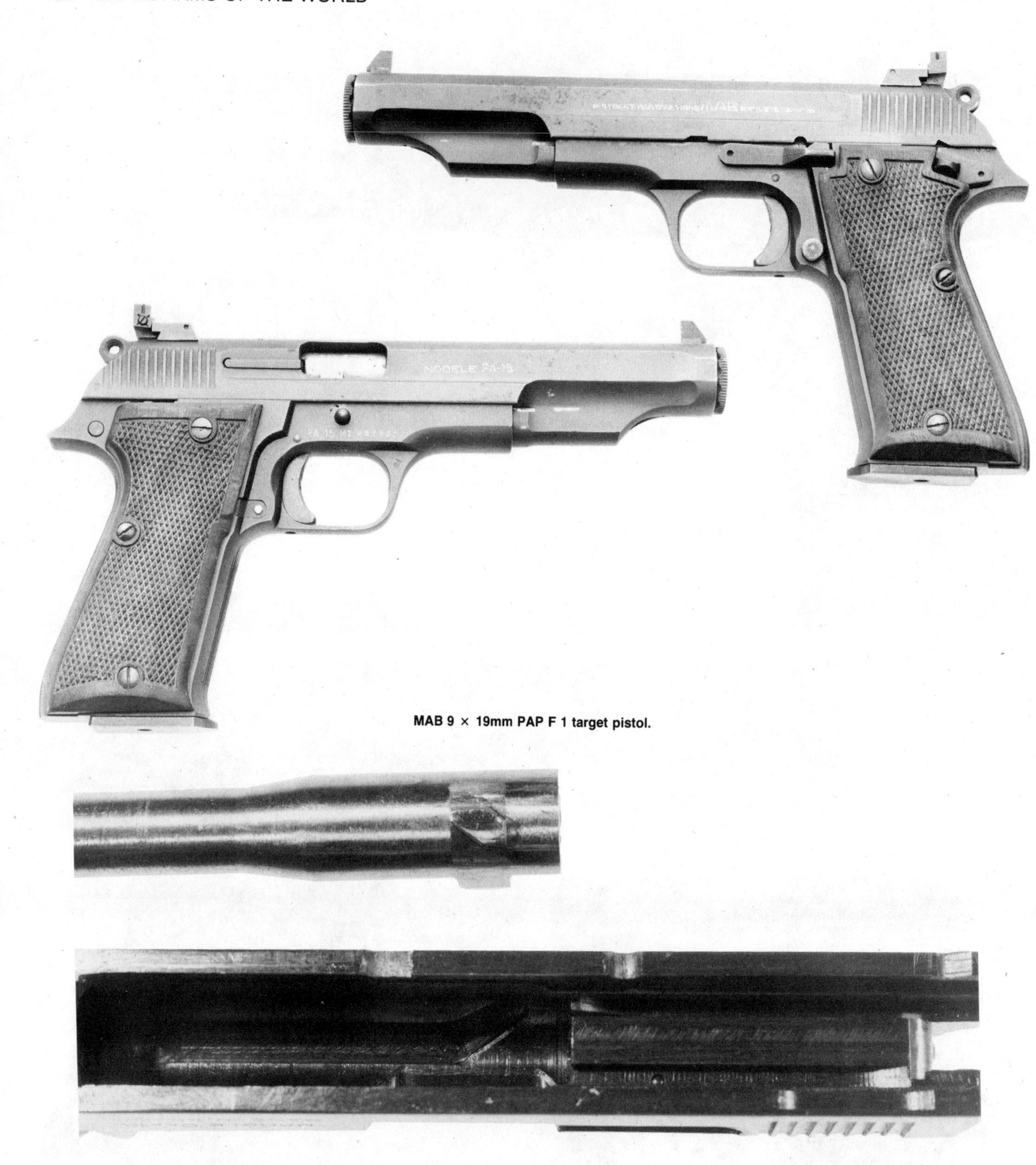

MAB 9 × 19mm PAP F 1 target pistol.

This photo shows the cam lug on the barrel of the PA 15 pistol. This lug rides in the cam path machined into the roof of the slide.

How to Load and Fire the PAP F1

The Model F1 is loaded and fired in the same manner as the US caliber .45 M1911A1 pistol. The pistol has a magazine safety and cannot be fired with the magazine removed.

How to Field Strip the PAP F1

The Model F 1 is field stripped in a manner similar to that of the French 9mm M1950 except that it has a barrel bushing. The barrel bushing is removed by pressing in on the barrel bushing catch, a spring loaded detent mounted on the under side of the muzzle end of the slide, and unscrewing the barrel bushing.

How the PAP F 1 Works

The F 1 has a rotating barrel type locking mechanism. The barrel has lugs mounted on its top and bottom that engage a cutout section in the top of the slide and a cam track cut in a locking piece mounted in the receiver. This piece is held in place in the receiver by the pin portion of the slide stop in a manner reminiscent of the Mexican Obregon pistol; the recoil spring guide rod and recoil spring are mounted on the front portion of the piece. Rearward movement of the slide on firing causes the bottom lug to rotate in the camway of the receiver locking piece, thereby rotating the locking lug on top of the barrel out of its locked position allowing the slide to continue its travel to the rear with the empty cartridge case gripped in the slide mounted extractor. The ejector, mounted on the right side of the receiver, engages the base of the case and forces the case out of the ejection port. At the same time, the slide is forcing the externally mounted hammer to the rear causing the hammer to engage the sear and remain to the rear in cocked position. Rearward movement of the slide has compressed the recoil spring, and the slide reaches its limit of rearward travel; the spring forces the slide forward. The slide, during its forward movement, strips a cartridge from the magazine and feeds it into the chamber. The cam slot in the receiver locking piece begins camming the barrel around so that the locking lug on the barrel is engaged in its recess in the slide when the slide is in its forward position. Pressure on the trigger will fire the pistol again and the whole process is repeated for each separate pull of the trigger until the magazine is empty.

The 9 × 19mm NATO Mle. 1950 pistol.

FRENCH 9 × 19MM PISTOLET MLE. 1950

The Modèle 1950 is a 9 × 19mm caliber version of the Modèle 1935S 7.65mm service handgun. Both of these designs are the result of Charles Petter's efforts to improve the M1911 Colt-Browning–type operating mechanism.*

CHARACTERISTICS OF THE M1950 PISTOL

Caliber: 9mm Parabellum.
System of operation: Recoil, semi-automatic fire only.
Weight: 1.8 lbs.
Length, overall: 7.6 in.
Barrel length: 4.4 in.
Feed device: 9-round, single-column, detachable box magazine.
Sights: Front: Tapered post.
Rear: U-notch; the top of the slide has a matted ramp.
Muzzle velocity: 1156 f.p.s.

*For additional details see *Handguns of the World.*

How to Load and Fire the M1950 Pistol

The M1950 pistol is loaded and fired the same as the US M1911A1 cal. .45 automatic pistol. However, it cannot be fired with the magazine removed. A loaded-chamber indicator is mounted in the slide. The safety catch is mounted on the left rear of the slide.

How to Field Strip the M1950 Pistol

Field stripping is the same as for the US .45, with the following exceptions:

There is no barrel bushing and recoil spring plug; the recoil spring is dismounted after the slide has been removed from the receiver.

The hammer, its spring and lever and the sear assembly are contained in a housing that can be lifted out as one piece when the slide is removed.

How the M1950 Pistol Works

The functioning of the M1950 pistol is essentially the same as that of the US service automatic.

Special Note on the M1950 Pistol

The M1950 pistol incorporates most features of the M1935 French pistols and has a few changes borrowed elsewhere. The safety on the slide was used in the M1935A and M1935S; it blocks the hammer from the firing pin. The internal mounting of the recoil spring is used on practically all the Browning and modified Browning-design pistols in service today except the US cal. .45 and those pistols that are direct copies of the US cal. .45.

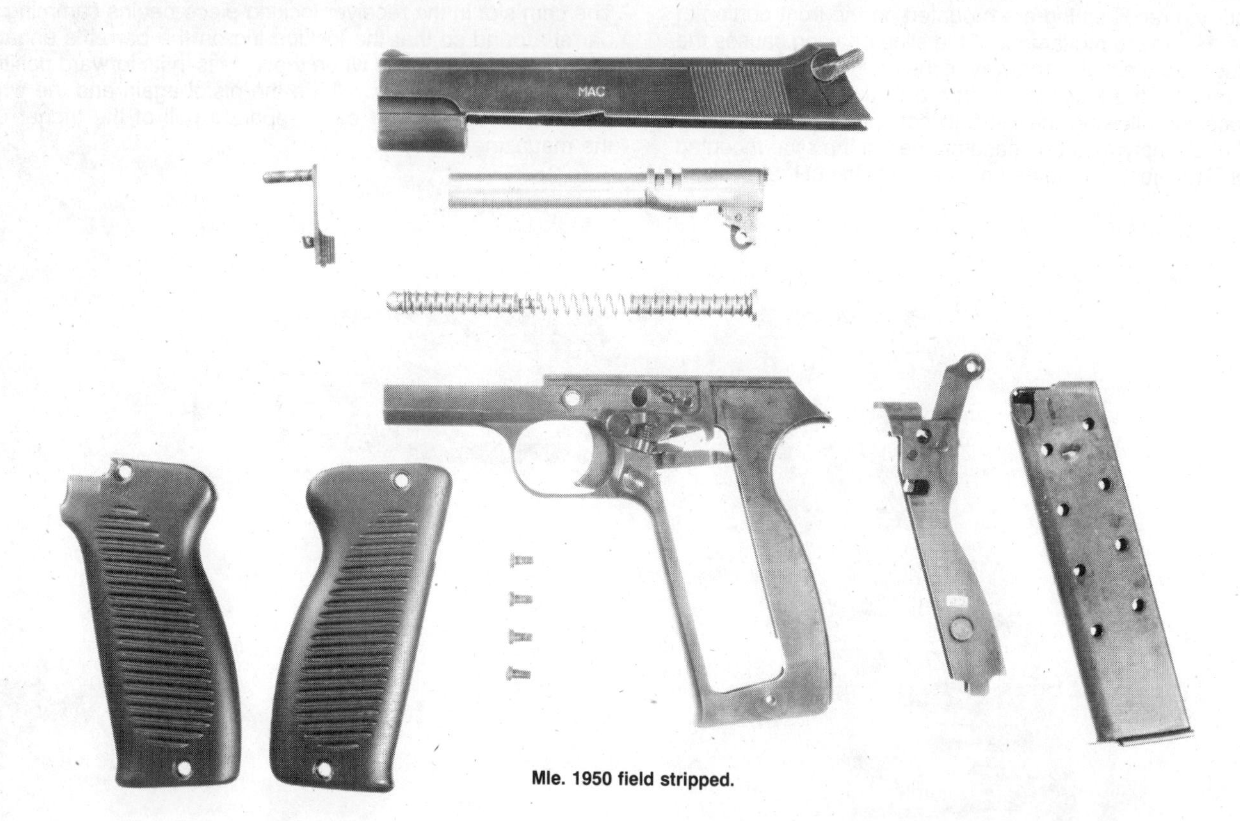

Mle. 1950 field stripped.

THE FRENCH M1935 PISTOLS

The Model 1935A and 1935S pistols are quite similar in design and differ principally in method of locking. The Model 1935A has two lugs on the upper surface of the barrel that lock into mating grooves in the slide in a fashion similar to the US Colt M1911A1. The M1935S, on the other hand, has a step machined on the top of the barrel at a point slightly ahead of the chamber. This step locks into a cut-out section in the slide.

They were made at Chatellerault (MAC), Saint Etienne (MAS), Tulle (MAT), Société Alsacienne de Construction Mecanique (SACM), and Société d'Applications Générales Electriqueset Méchaniques (SAGEM).

CHARACTERISTICS OF THE MODEL 1935 PISTOLS

	Model 1935A	Model 1935S
Caliber:	7.65mm long.	7.65mm long.
System of operation:	Recoil operated, semiautomatic.	
Length overall:	7.6 in.	7.4 in.
Barrel length:	4.3 in.	4.1 in.
Weight:	1.62 lbs.	1.75 lbs.
Feed device:	8-round, detachable, in-line box magazine.	
Sights: Front:	Blade.	Blade.
Rear:	Rounded notch.	Rounded notch.
Muzzle velocity:	1132 f.p.s.	1132 f.p.s.

French 7.65mm Long M1935A Pistol, field stripped.

Loading, firing, and field stripping of these pistols is essentially the same as that of the US caliber .45 M1911A1 or the 9mm M1950. These pistols have packaged hammer, main spring and sear assemblies that can be removed after the slide is dismounted from the receiver.

French 7.65 Long M1935A Pistol

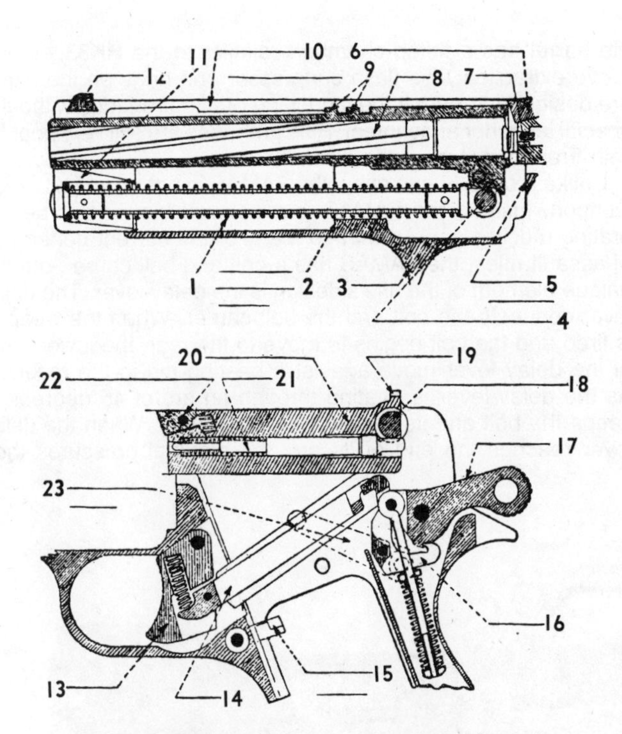

French Pistolet 1935A (Drawing from French Manual).

1. Recoil spring.
2. Receiver stop.
3. Receiver.
4. Barrel pivot pin.
5. Barrel stop.
6. Loading indicator.
7. Face of breechblock.
8. Barrel.
9. Dual barrel locking lugs.
10. Slide-breechblock unit.
11. Front of slide.
12. Front sight.
13. Trigger.
14. Trigger bar.
15. Magazine catch.
16. Sear.
17. Hammer.
18. Manual Safety.
19. Rear sight.
20. Breech section of slide.
21. Firing pin.
22. Loading indicator.
23. Magazine safety.

French 7.65mm Long M1935S Pistol.

FRENCH AUTOMATIC RIFLES

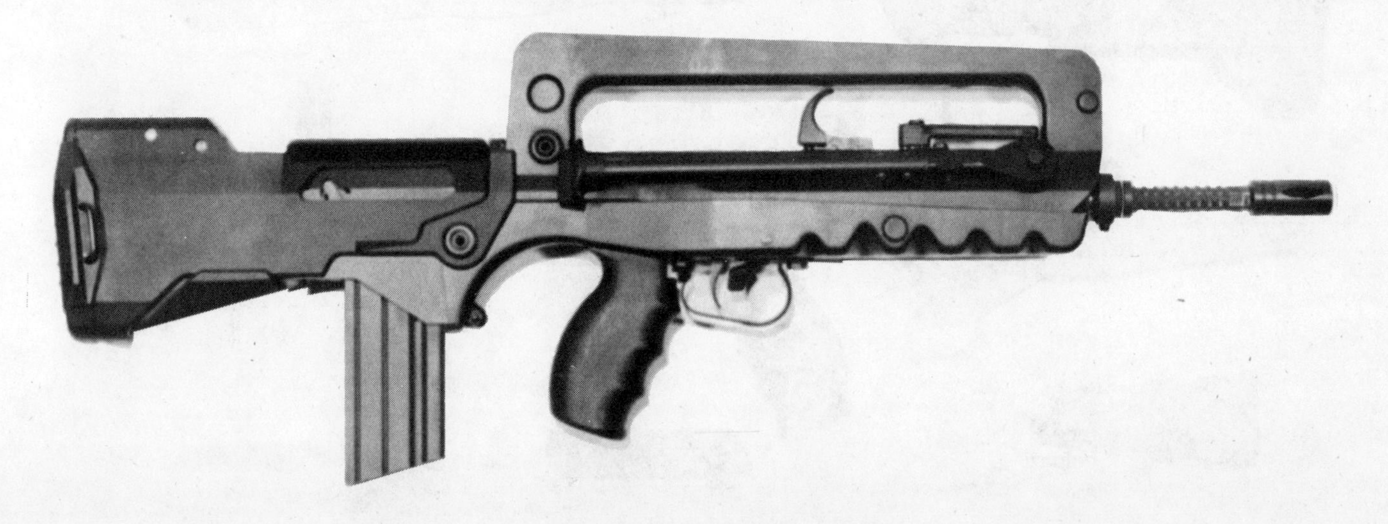

Right side view of the 5.56 × 45mm FAMAS.

Fusil Automatique MAS (FAMAS) 5.56 × 45mm

The French Arsenal at Saint Etienne, now a part of the state industrial combine GIAT, developed the FAMAS after several years of experience with the manufacture of Heckler & Koch's HK33. The designers of the FAMAS were seeking a weapon that had the firepower of the 5.56mm-class assault rifles, but which had the handiness of the submachine gun–type weapons. The result of their work was this bullpup design.

There are six major component groups for the FAMAS—1) the barrel and receiver assembly; 2) the operating parts—bolt, bolt carrier, and delay lever; 3) the trigger assembly; 4) the carrying handle assembly; 5) the buttstock and cheek assembly; 6) the accessory group—bipod, bayonet, and sling. The receiver assembly is made from an aluminum alloy and the nonchrome-plated barrel is pinned into the receiver. It should be noted that the barrel has a fluted chamber—similar to the HK33—to improve extraction. The flash suppressor and rings on the barrel are designed to permit the launching of rifle grenades without a special launcher attachment. Rifle grenades are still very popular with French military authorities.

Unlike most 5.56mm rifles, the FAMAS is not a locked breech weapon. Instead the FAMAS is a delayed blowback-type operating mechanism. As with so many of the current generation of assault rifles, the FAMAS has a bolt and bolt carrier, but the unique element of the rifle's design is the delay lever. The delay lever connects the bolt and the bolt carrier. When the weapon is fired and the bolt begins to move to the rear, the lower arms of the delay lever move against a bearing pin in the receiver. As the delay lever is rotating through an arc of 45 degrees, it keeps the bolt and its carrier from separating. When the delay lever reaches the end of its arc, the residual pressures then

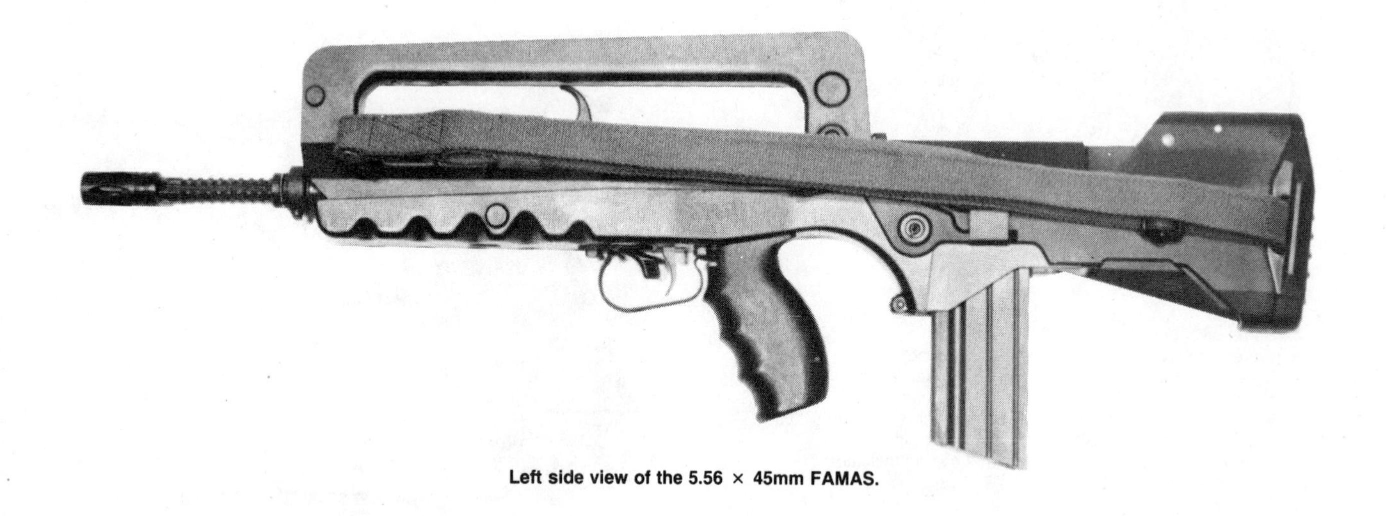

Left side view of the 5.56 × 45mm FAMAS.

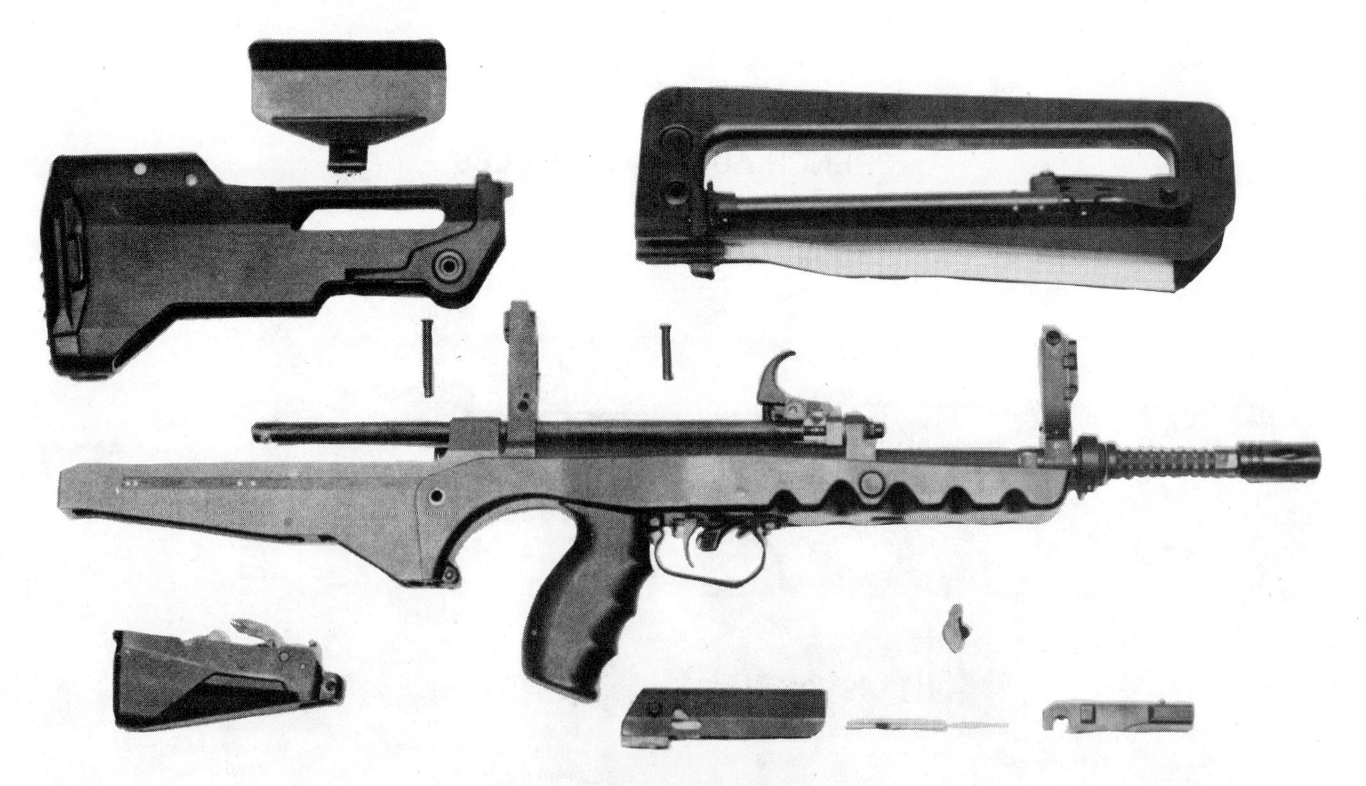

Field-stripped view of the 5.56 × 45mm FAMAS.

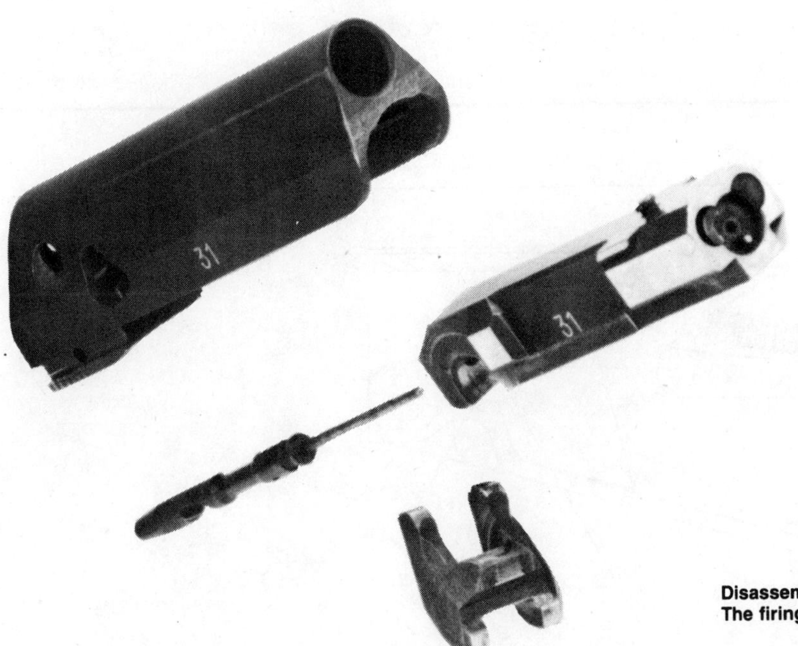

Disassembled view of the FAMAS bolt assembly. At the top is the bolt carrier. The firing pin and bolt are beneath it. At the bottom is the delay lever.

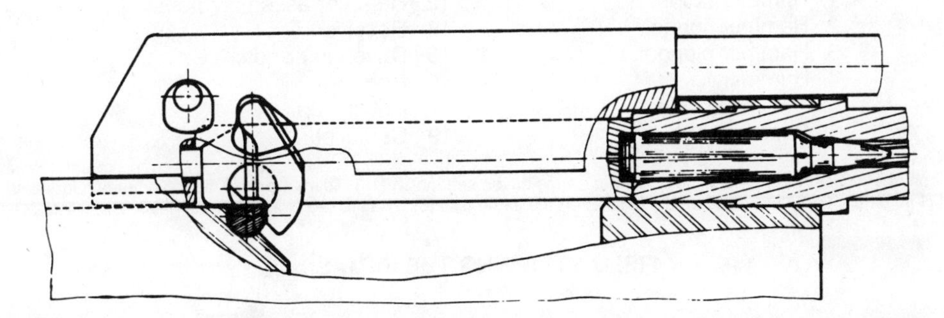

The FAMAS bolt assembly in the ready to fire position. When the bolt is in the forward position, the delay lever is rotated into forward position, too. Note that the lower arms of the delay lever are resting against the bearing pin in the receiver. When the delay lever is in the fire position, it releases the firing pin and the sear. If the bolt is not fully closed, the firing pin and sear are locked, thus preventing out-of-battery firing.

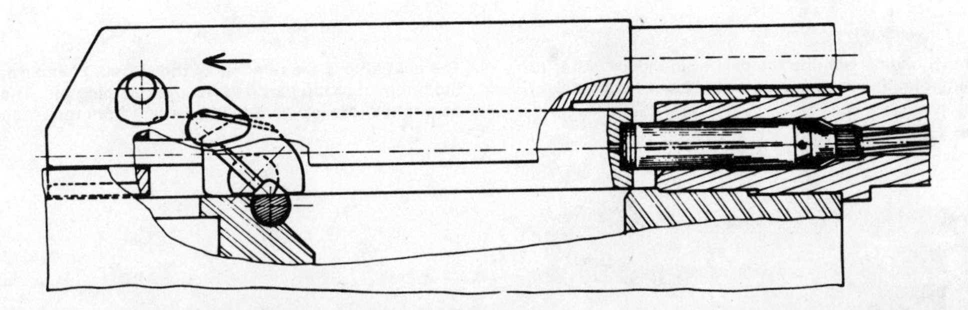

The FAMAS bolt assembly in the partially retracted position. The bolt has begun to move to the rear, and the extraction of the cartridge has begun. Note that the bolt carrier has moved farther to the rear than has the bolt. Note also how the delay lever has rotated away from the bearing pin. At first, this delay lever retards the movement of the bolt. After the lever has arced through 45 degrees, the delay lever behaves as an accelerator.

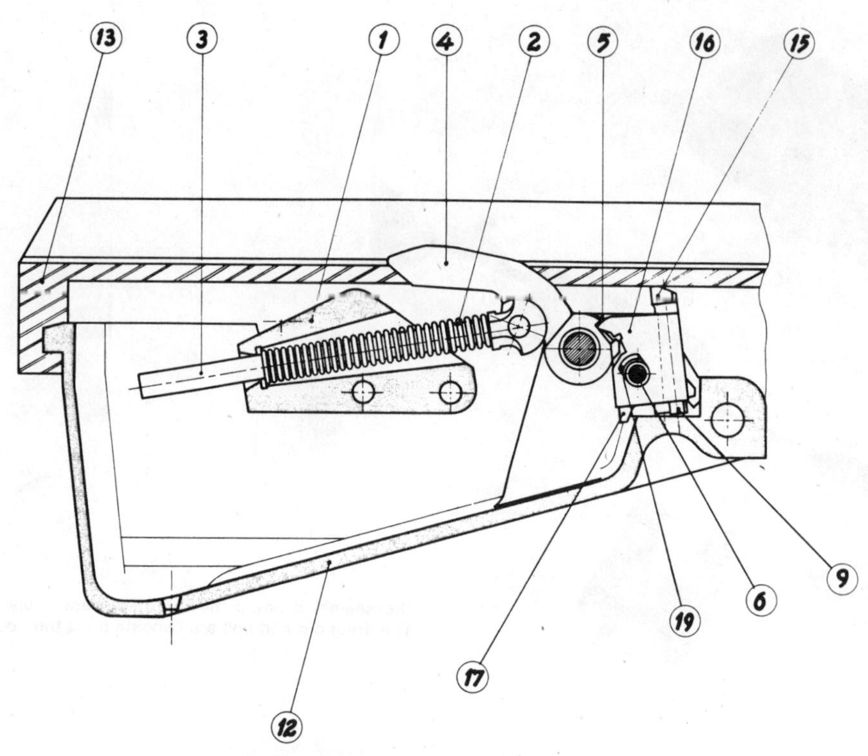

Simplified view of the FAMAS hammer mechanism.

1. Hammer stop.
2. Hammer spring.
3. Hammer plunger.
4. Hammer.
5. Automatic sear.
6. Sear axis.
12. Hammer assembly housing.
13. Receiver.
15. Driven sear notch.
16. Cam.
17. Driven sear.
19. Cam slot.

The linkage from the trigger of the FAMAS pulls forward on the driven sear notch (15). There are three sears; one each for semiautomatic, bursts, and full automatic fire. All three sears are located on the same axis (6). The fire selector brings the desired sear into play.

FIELD STRIPPING THE FAMAS

Disassembly of the FAMAS begins with the removal of the stock. After removing the stock retaining pin, the stock can be pulled free of the receiver.

The next step is the removal of the carrying handle assembly. This requires the removal of the carrying handle retaining pin. The carrying handle must be pushed forward as it is lifted away from the receiver.

The hammer assembly is also retained by a pin. When that pin has been pushed out of the way, the complete hammer mechanism can be removed as a unit. When all of the above assemblies have been removed from the receiver, the bolt, bolt carrier, and related components can be withdrawn from the receiver assembly with ease.

accelerate the bolt to the rear, allowing the positive extraction of the fired cartridge case. This type of mechanism leads to a high cyclic rate of between 900 and 1,000 shots per minute. The nature of the FAMAS design and further information on its operation are presented in the accompanying illustrations.

A French soldier holding a 5.56 × 45mm *Fusil Automatique* MAS.

CHARACTERISTICS OF MAS 5.56MM RIFLE

Caliber: 5.56 mm
System of operation: Delayed blowback.
Weight: 7.5 lbs., w/o magazine.
Length, overall: 29.8 in.
Barrel length: 19.2 in.
Feed device: 25-round box magazine.
Sights: Front: Blade.
Rear: Aperture 0-300 meters.
Cyclic rate: 900–1000 r.p.m., and 3-shot bursts.
Muzzle velocity: 3150 f.p.s.

THE MODEL 1949 AND 1949/56 RIFLES

Little real effort seems to have been expended in France between the wars on semiautomatic rifles. In 1944, a semiautomatic rifle was produced at Saint Etienne—the 7.5mm MAS44. This rifle was further developed into the Model 1949. The current standard rifle, the Model 1949/56, is a modification of the Model 1949.

These gas-operated weapons show considerable designing skill. They are relatively simply in operation and are very easy to strip and clean. They use a tilting bolt system and have no moving parts in the gas system. The forward part of the barrel is used as a grenade launcher on both models, and both have built-in grenade launcher sights. On the M1949, the grenade launcher sights are on top of the barrel, to the rear of the front rifle sight. Both of these weapons have two-piece stocks, as did the M1936 series of weapons. The M1949 does not use a bayonet, but the M1949/56 does.

CHARACTERISTICS OF M1949 AND M1949/59

	M1949	M1949/56
Caliber:	7.5mm French M1929 cartridge, for both models.	
System of operation:	Gas, semiautomatic fire only, for both models	
Weight:	10.4 lbs.	8.6 lb. w/o magazine.
Length, overall:	43.3 in.	43.4 in.
Barrel length:	22.8 in.	20.7 in.
Feed device:	10-round, detachable, staggered-row box magazine, normally loaded with 5-round chargers: for both models.	
Sights: Front:	Blade w/protecting ears, for both models.	
Rear:	Ramp w/aperture, for both models.	
Muzzle velocity:	2705 f.p.s.	2700 f.p.s.

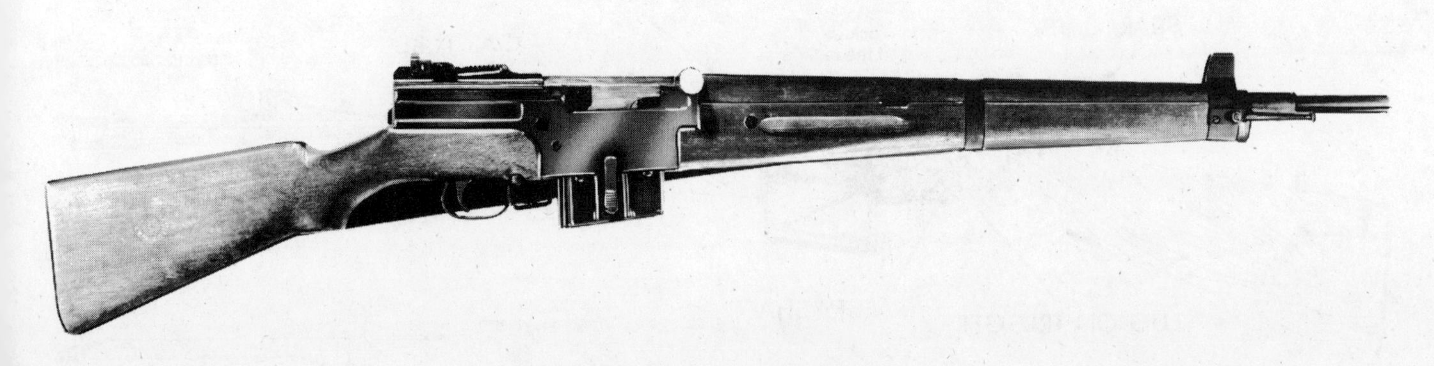

French 7.5mm Rifle M1949 (MAS).

How to Load and Fire the M1949-Series Rifles

The weapon can be loaded by pulling the operating handle to the rear and filling the magazine, using chargers (clips) in a manner similar to that used with the US M1903 rifle, or the magazine may be removed (note that the magazine catch is a part of the magazine, not of the weapon) and then filled by hand. If the bolt is to the rear, it will require a short rearward jerk to release the bolt latch—the bolt remains open after the last round is fired—allowing the bolt to go forward and chamber a cartridge. If the bolt is forward, it will have to be pulled to the rear and released. The safety is located at the right front side of the trigger-guard. Upon firing the last shot, the bolt will remain open.

How to Field Strip the M1949-Series Rifles

Remove magazine by depressing magazine catch on right side of magazine and pulling magazine down and out of the receiver. Release receiver cover by depressing the latch, located at the rear of the receiver; push the cover forward slightly and lift it from its grooves in the receiver. **CAUTION:** The bolt must be in the forward position before attempting to remove the receiver cover. (When the bolt is to the rear, the recoil spring, which is housed in the receiver cover, is compressed.) The bolt and bolt carrier can now be lifted out of the receiver. The trigger group can be removed by removal of one screw. No further disassembly is recommended.

To reassemble the weapon perform the above steps in reverse.

How the M1949-Series Rifles Work

When a cartridge is fired, propellent gas is tapped off through the gas port in the barrel. This gas blows back through the gas tube directly into a hole in the top face of the bolt carrier. The gas tube protrudes a small distance, so that a portion of it actually enters the bolt carrier. The bolt carrier moves to the rear, and after a slight "dwell" time the bolt is cammed up and out of its locked position in the bottom of the receiver and starts to travel back with the bolt carrier. The ejector is a pin type that protrudes from the face of the bolt like that of the US M1 rifle; it ejects the spent case when the case mouth clears the face of the barrel. The trigger mechanism is quite similar to that of the US M1 rifle and operates in a similar fashion. The bolt and bolt carrier compress the recoil spring, which decompresses at the end of their travel and returns the bolt and carrier to the battery position. On the return of the bolt and its carrier, the bolt strips the top cartridge from the magazine and feeds it into the chamber. The bolt carrier cams the lugs of the bolt down, and the rear end of the bolt is brought down into engagement with the locking bar in the bottom of the receiver. The weapon is ready to fire again.

Special Note on the M1949-Series Rifles

The gas system of the M1949-series rifles is similar to that of the Swedish Ljungman Model 42. Its tilting bolt (or propped breech, as it is sometimes called) has been used in a great number of systems, e.g., the Tokarev, the FN, the Ljungman, and so on. These weapons are an illustration of the fact that a good modern weapon can be produced by selectively choosing from proven past designs and modifying them to suit the need.

Both of these weapons have mounting grooves, cut on the left side of the receivers, for the mounting of telescopic sights.

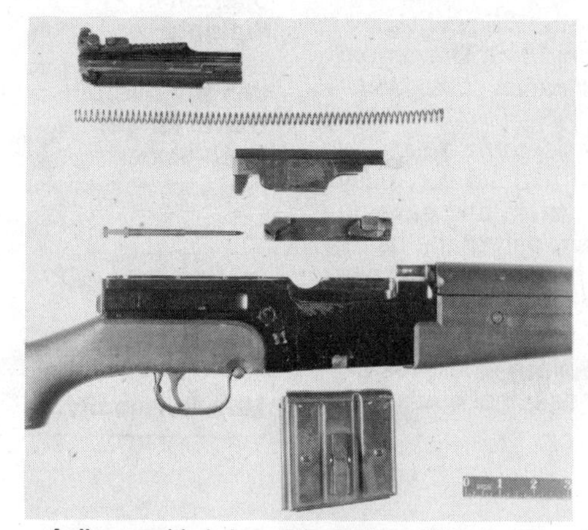

A disassembled view of the 7.65mm MAS 49 rifle.

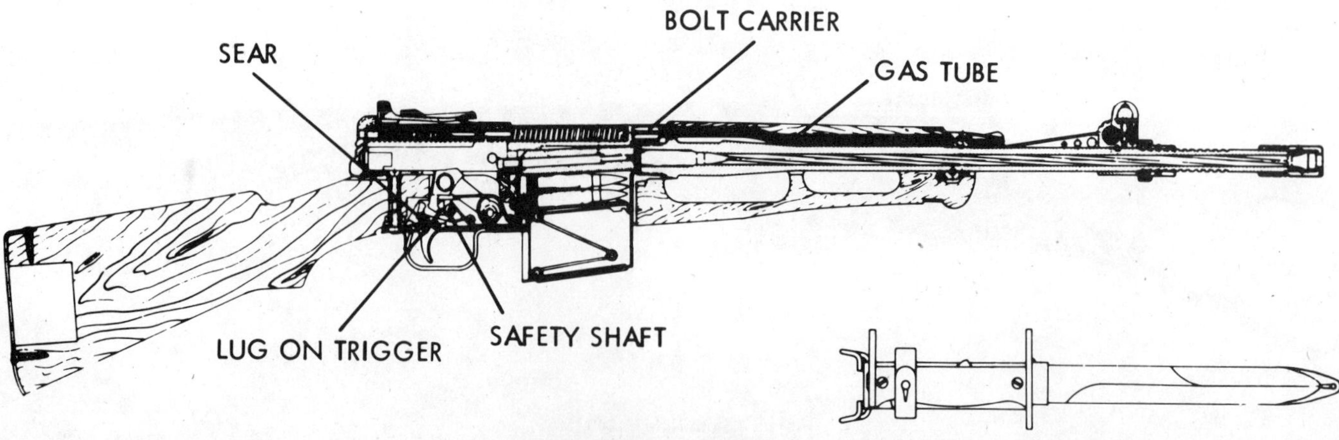

Sectioned view of M49/56.

FRENCH BOLT-ACTION RIFLES AND CARBINES

THE 7.5mm MODEL F 1 RIFLE

The *Fusil a Repetition Modele F 1* (Repeating rifle Model F1) is produced at MAS (St. Étienne) exclusively for the French Army around a modification of the now obsolete MAS 1936 Rifle. It is manufactured in three variations: (1) *Tireur d'Elite* (sniper); (2) *Tir Sportif* (target); and (3) *Grande Chasse* (hunting). The first has a 3.8-power telescopic sight and a non-adjustable 100-meter open sight (with luminous dots for use under poor lighting conditions). The second has metallic type target sights. The hunting version has an APX Model 804 telescopic sight. Only the sniper's rifle is equipped with a bipod. All versions have a muzzle brake/flash suppressor. The length of the butt stock can be adjusted by adding or removing spacers between the end of the stock and the rubber recoil pad.

CHARACTERISTICS OF THE 7.5MM MODEL F1 RIFLE

Caliber: 7.5mm
System of operation: Manually operated bolt action.
Length overall: 44.8 in.
Barrel length: 22.8 in.
Feed device: 10-round, staggered row, detachable box magazine.
Sights: Front: Tunnel type.
Rear: Telescope or micrometer.
Weight: 9.9 lbs.
Muzzle velocity: Approx. 2700 f.p.s.

This rifle has been made in 7.62mm NATO as well as 7.5mm French. The French claim that this rifle will consistently group 10 rounds into a circle smaller than 7.8 inches at 200 meters when used with good ammunition.

Comparison of MAS 1936 and FR-F1 Rifle receivers. Note the new safety and magazine release on the FR-F1 Sniper's Rifle.

Left side view of FR-F1 Sniper's Rifle.

THE FRENCH MAS 1936 RIFLE

The M1936 is a 7.5mm caliber with modified Mauser magazine. Action is shown ready for trigger pull. Note that dual locking lugs are engaged in recesses within the receiver bridge directly above the trigger. Placing the lugs this far back from the chamber permits a short bolt stroke but results in a rifle locking design weaker than the conventional Mauser front-lug type.

This turn bolt rifle has a modified Mauser magazine with quick removable bottom plate. Primary extraction is by action of bolt handle extension working against a cam in the receiver bridge.

The breech locking system of this rifle is somewhat different than the Mauser. Dual lugs at the rear of the bolt body turn into seatings in the receiver bridge as the handle is turned down. No manual safety.

Locking lugs are on rear of bolt cylinder near handle knob; they lock into receiver bridge to rear of magazine. Note cam shape of left face of receiver bridge, which affords leverage for primary extraction cam on modified Mauser system. Rifle is loaded with Mauser-type charger clip. Note cuts in receiver for finger pressing cartridges into magazine. This is necessary because of large bolt diameter.

The 7.5mm M1936 CR39 is the same as the Model 1936 but has a folding aluminum buttstock.

A post-war version of the M1936, the 7.5mm M1936 M51, has a rifle grenade launcher built into the muzzle end of the barrel.

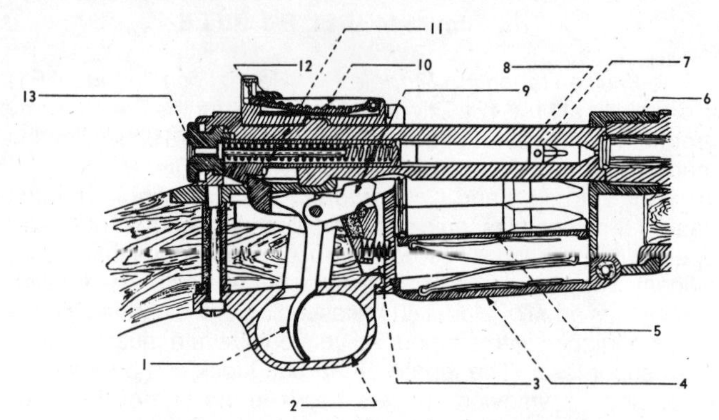

French Fusil 1936 (Drawing from French Manual).

1. Trigger
2. Trigger guard
3. Sear spring
4. Magazine bottom plate
5. Magazine follower
6. Barrel
7. Bolt
8. Firing pin
9. Ejector & Bolt stop unit
10. Receiver bridge
11. Sear
12. Peep sight
13. Cocking piece head.

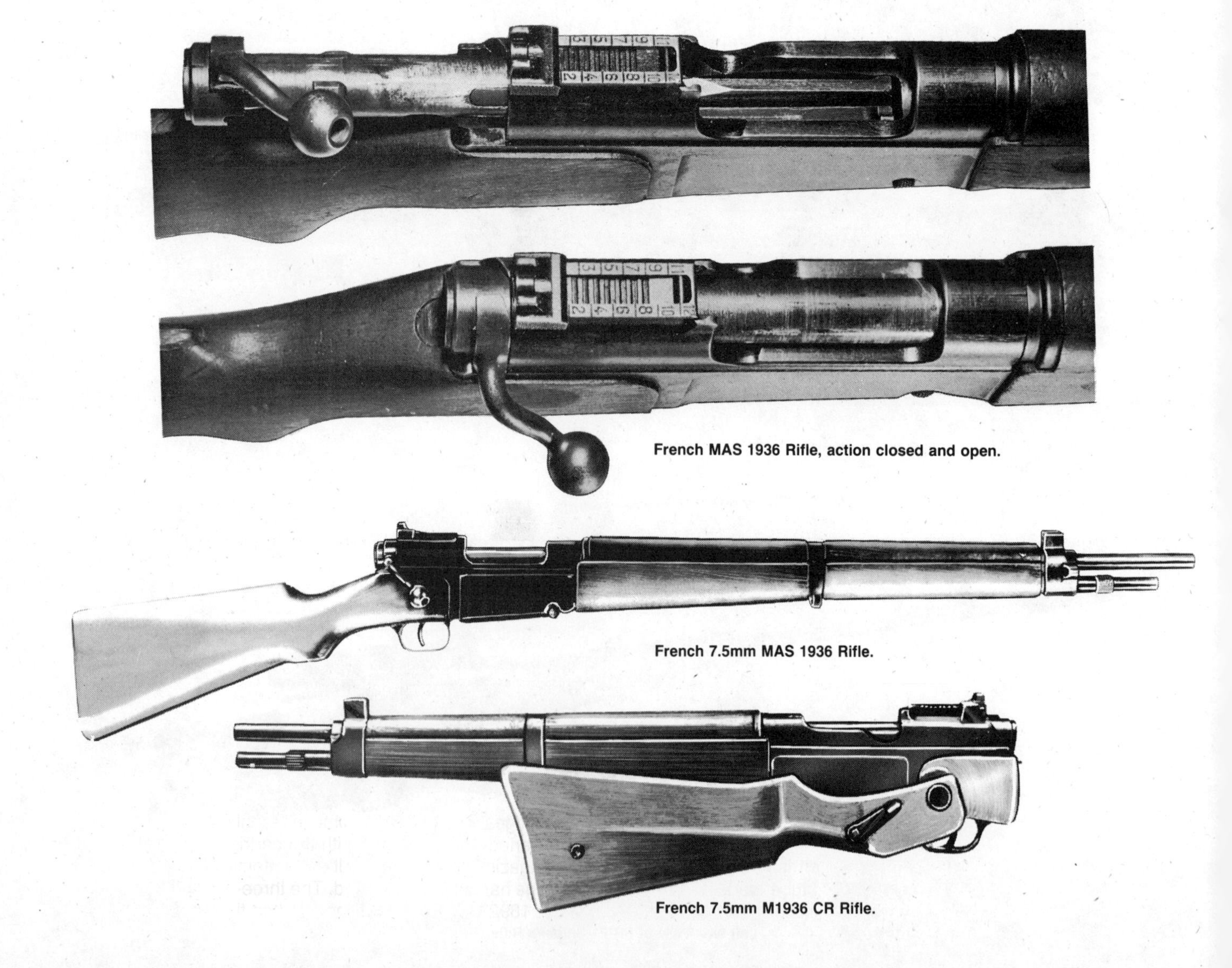

French MAS 1936 Rifle, action closed and open.

French 7.5mm MAS 1936 Rifle.

French 7.5mm M1936 CR Rifle.

OBSOLETE FRENCH RIFLES

BOLT-ACTION CATEGORIES

Basically, the bolt-action French rifles and carbines can be broken down into three categories: (1) the M1886; (2) the M1890; and (3) the M1932 and its production model, the M1936.

The M1886 (Commonly Called the Lebel)

This rifle was introduced in 1886 to use a new 8mm cartridge loaded with the then revolutionary smokeless powder of Paul Vielle. The Lebel bolt is quite similar to that of the 11mm Gras rifle Model 1874, and the feed mechanism is similar to the 11mm Model 1878 Kropatschek rifle used by the French Navy.

The M1886 is the parent of all modern small-bore military rifles. It was slightly modified in 1893 by strengthening the receiver, boring a gas-escape hole in the bolt, changing the rear sight mounting and leaf, and adding a stacking hook to the upper band. It was still in limited service in World War II, while numerous modified rifles were built around many spare parts of the old 86/93.

This is a turn-bolt action rifle of conventional design. The magazine is a tube in the forestock below the barrel, being loaded through the open action. The bolt is a two-piece design with a long detachable bolt head, which carries the dual locking lugs.

To remove the bolt, the action must be opened. A holding screw is then removed from a projecting strap on top of the bolt body. The bolt body can then be pulled out of the receiver, leaving the bolt head in the boltway ahead of the receiver bridge. The head can then be picked out of the boltway. Primary extraction is given as the bolt handle is lifted by the projecting bolt strap working against a cam face in a long overhang on the top of receiver.

As the bolt is withdrawn, it operates an elevator, which raises a cartridge in its trough and lines it up with the bolt. A hook on the bottom of the elevator trough blocks the cartridge in the magazine tube at this point. When the cartridge is chambered, the bolt motion lowers the elevator trough, permitting the magazine spring in the tube to force the next cartridge into the trough ready to be raised on the following rearward bolt stroke.

Modification—1886 M93 R35

This carbine modification was issued in 1935. It is merely a shorter form with different furniture and smaller magazine. Model number is on receiver.

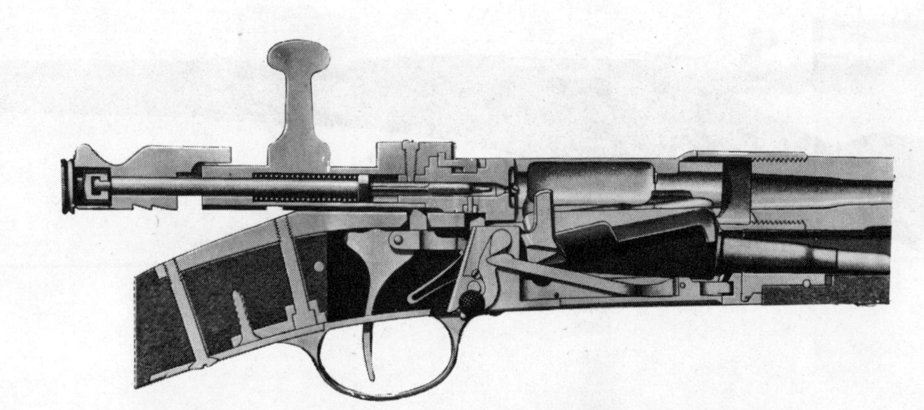

Section of French 1886 (Lebel). Tube repeating rifle showing operating system.

French 8mm Rifle Model 1886 M93.

MANNLICHER BERTHIER CARBINES AND MODIFICATIONS

French 8mm Carbine M1892

This is one of the first Berthier arms. Bolt is two-piece type with removable head as in the earlier Lebel tube loader (1886). It is modified, however, to feed from a Mannlicher-type fixed box magazine.

Cartridges for these arms come in Mannlicher-style clips, which are inserted in the receiver with the cartridges to form a part of the magazine action. Clips fall out bottom of receiver when last cartridge has been chambered. The three-round models use the Model 1892 clip; five round models use the Model 1916 clip.

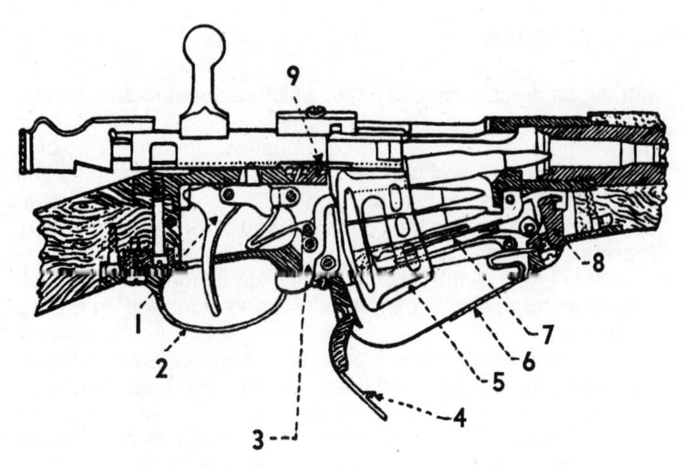

French Fusil 1916 (Drawing from French Manual).

1. Trigger
2. Trigger guard
3. Clip catch
4. Magazine plate
5. Mannlicher-type clip
6. Magazine bottom
7. Follower
8. Follower support
9. Ejector

French Fusil 07/15 M34 (Drawing from French Manual).

1. Trigger guard
2. Trigger
3. Magazine plate lock
4. Bottom plate
5. Follower
6. and 7. Forward magazine section
8. Ejector
9. Sear

Typical 8mm M1892 Carbine as modified in 1927; 3-round capacity.

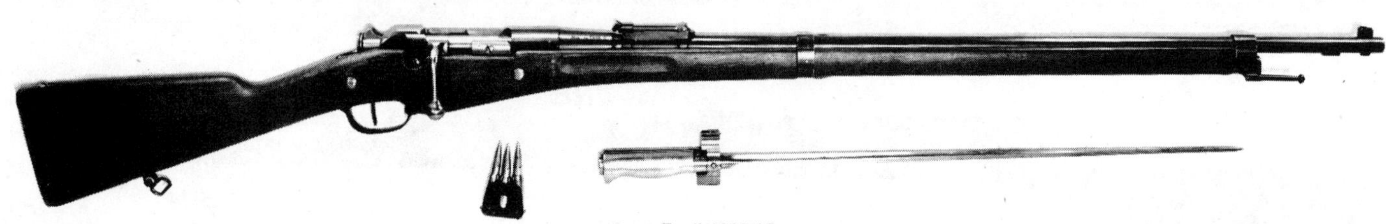

8mm Fusil 1907/15.

All the Mannlicher Berthier carbines (1890, 1892, 1916) were modified in 1927. Modification consisted of removing cleaning rods and adding a stacking swivel.

French 8mm Fusil, M1907/15/34 and M1916

The French drawing (next page) shows details of Fusil (Rifle) 1916. The hinged magazine plate is opened to show how clip is dropped out of magazine when empty. Bolt has been thrust forward far enough to strip cartridge out of clip and guide bullet into chamber. This is a turn bolt rifle. This design uses the Mannlicher-clip loading system.

The 07/15 design was modified in 1934 to handle the 7.5mm rimless cartridge developed in 1929 for light machine guns. The changeover included replacing the old Mannlicher type magazine with a standard 5-shot Mauser type. Cartridges may be loaded into the magazine at any time, staggering from side to side in regular Mauser fashion. The detachable bolt head locks with two heavy lugs into a solid receiver section behind the cartridge head. The Mauser magazine permitting the use of cartridges without clips is an important feature.

FRENCH RIFLES AND CARBINES OF PRE-WORLD WAR II DESIGN

Weapon	System of Operation	Overall Length	Barrel Length	Feed Device	Sights	Muzzle Velocity	Weight
Rifle: 8mm M1886 M93	Manually operated 2-piece bolt	51.3 in.	31.4 in.	8-rd tubular magazine	Front: Notched blade Rear: Leaf	2380 f.p.s.	9.35 lb.
Rifle: 8mm M1886 M93R35	Manually operated 2-piece bolt	37.64 in.	17.7 in.	3-rd tubular magazine	Front Notched blade Rear: Leaf	2080 f.p.s.	7.84 lb.
Carbine: 8mm M1890	Manually operated 2-piece bolt	37.2 in.	17.7 in.	3-rd Mannlicher-type, integral, in-line magazine	Front: Blade Rear: Leaf	2080 f.p.s.	6.83 lb.
Mousquetoon 8mm M1892.	Manually operated 2-piece bolt	37.2 in.	17.7 in.	3-rd Mannlicher-type, integral, in-line magazine	Front: Blade Rear: Leaf	2080 f.p.s.	6.8 lb.
Rifle: 8mm M1902 "Indo-China Model"	Manually operated 2-piece bolt	38.6 in.	24.8 in.	3-rd Mannlicher-type, integral, in-line magazine	Front: Blade Rear: Leaf	2180 f.p.s.	7.9 lb.
Rifle: 8mm M1907 "Colonial Model"	Manually operated 2-piece bolt	52 in.	31.4 in.	3-rd Mannlicher-type, integral, in-line magazine	Front: Blade Rear: Leaf	2380 f.p.s.	8.6 lb.
Rifle: 8mm M1907/15	Manually operated 2-piece bolt	51.42 in.	31.4 in.	3-rd Mannlicher-type, integral, in-line magazine	Front: Notched blade Rear: Leaf	2380 f.p.s.	8.38 lb.
Rifle: 8mm M1916	Manually operated 2-piece bolt	51.42 in.	31.4 in.	5-rd Mannlicher-type, integral, in-line magazine	Front: Notched blade Rear: Leaf	2380 f.p.s.	9.25 lb.
Carbine: 8mm M1916	Manually operated 2-piece bolt	37.2 in.	17.7 in.	5-rd Mannlicher-type, integral, in-line magazine	Front: Blade Rear: Leaf	2080 f.p.s.	7.17 lb.
Rifle: 7.5mm M1907/15 M34	Manually operated 2-piece bolt	43.2 in.	22.8 in.	5-rd integral, staggered-row, box magazine	Front: Blade Rear: Leaf	2700 f.p.s.	7.85 lb.
Rifle: 7.5mm M1932			Prototype of M1936, made in limited quantities				
Rifle: 7.5mm M1936	Manually operated 1-piece bolt	40.13 in.	22.6 in.	5-rd integral, staggered row, box magazine	Front: Barley-corn w/guards Rear: Ramp w/ aperture	2700 f.p.s.	8.29 lb.
Rifle: 7.5mm M1936 CR39	Manually operated 1-piece bolt	Stock extended 34.9 in. Stock folded: 24.3 in.	17.7 in	5-rd integral, staggered-row, box magazine	Front: Barley-corn w/guards Rear: Ramp w/ aperture	2700 f.p.s.	8 lb. (approx.)

FRENCH SUBMACHINE GUNS

France used the German 9mm Parabellum Vollmer Erma submachine gun to a limited extent prior to 1941. French development of a native submachine gun began at MAS (Saint Etienne) during the thirties. In 1935, the first of the MAS weapons developed for the 7.65mm long pistol cartridge appeared. This weapon, called the 7.65mm L Type SE-MAS 1935, was quite similar to the later and more common MAS 1938.

The 7.65mm long MAS 1938 was the standard French submachine gun until 1949. It is still in wide use with French police forces. After World War II, submachine gun development was very active in France. The Hotchkiss firm developed a submachine gun that had some overseas sales and limited service used in Indo-China.

THE FRENCH MODEL 1938 SUBMACHINE GUN

The 7.65mm Long Model 1938 submachine gun, or MAS 38 as it is commonly known, is a relatively simple blowback-operated weapon. It does have a few unusual features, however: the use of a folding trigger for a safety and the angular travel of the bolt.

CHARACTERISTICS OF THE MODEL 1938 SUBMACHINE GUN

Caliber: 7.65mm Long.
System of operation: Blowback, automatic fire only.
Length overall: 24.8 in.
Barrel length: 8.8 in.
Weight: 6.3 lbs.
Feed device: 32-round, detachable staggered row, box magazine.
Sights: Front: Block with notch.
Rear: Two folding leaves, 100 and 200 meter.
Muzzle velocity: Approx. 1200 f.p.s.
Cyclic rate: 700 r.p.m.

French 7.65mm Model 1938 Submachine Gun.

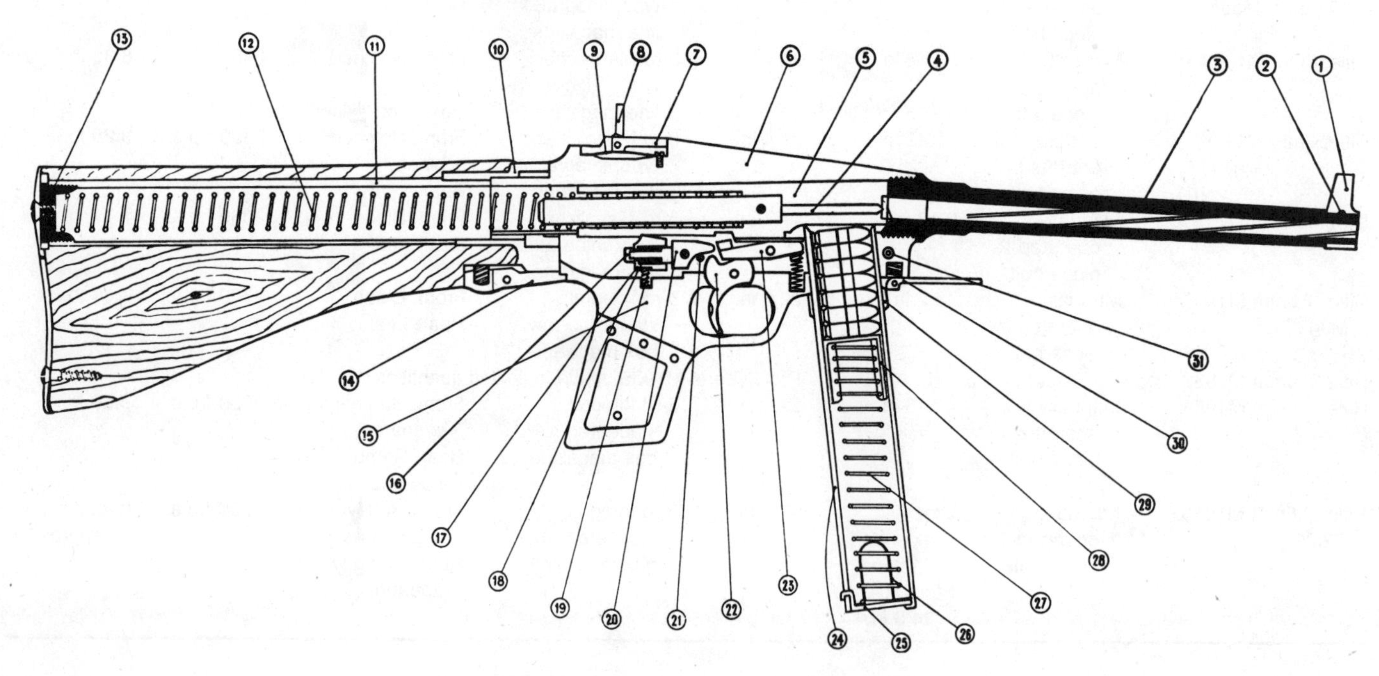

MAS 1938 Submachine Gun (Drawing from French Manual).

1. Front sight.
2. Fixing pin.
3. Barrel.
4. Striker.
5. Breechblock (bolt).
6. Receiver.
7. Rear sight leaf, down position.
8. Rear sight leaf, up.
9. Spring.
10. Recoil spring tube bridge.
11. Recoil spring tube.
12. Recoil (or operating) spring.
13. Tube butt cap.
14. Takedown lever.
15. Grip.
16. Sear nose.
17. Sear.
18. Sear buffer seat.
19. Sear spring.
20. Sear bar.
21. Stop pin.
22. Trigger.
23. Safety lever.
24. Magazine.
25. Floorplate.
26. Spring guide.
27. Spring.
28. Follower.
29. Magazine stop.
30. Opening plate (used to close magazine opening when magazine is withdrawn).
31. Magazine release.

How the MAS Model 38 Operates

A hinged cover seals the mouth of the magazine housing in the bottom of the receiver in this weapon when the magazine is withdrawn. Pushing this forward on its hinge opens the mouth of the housing and permits insertion of a loaded magazine from the bottom. The magazine is pushed in until it locks in standard fashion.

When the bolt handle on the right side of the receiver is pulled to the rear, it draws the bolt back and presses the recoil spring behind it, which extends from the rear of the receiver down into a steel tube inside the shoulder stock. When the bolt has been drawn back far enough, it rides over the sear, which is then forced up by its spring to catch in the bent or notch in the underside of the bolt and held open ready for firing.

Pressure on the trigger will now withdraw the sear from the bolt and permit the compressed recoil spring to drive the bolt up the slightly inclined surface machined into the receiver, stripping a cartridge from the magazine and thrusting it into the chamber.

French 9 × 19mm Mle. 1949 (MAT 49).

Mle. 1949 MAT field stripped.

THE FRENCH MODEL 1949 SUBMACHINE GUN

The M1949 submachine gun, which was built by Tulle (MAT), has a very good reputation with French troops. It has a telescoping steel stock and magazine, which folds up under the gun when not in use; this makes the weapon very handy for armored or airborne troops. The M1949 also has a grip safety and an ejection port cover. The grip safety prevents the gun from firing when accidentally dropped, and the ejection port cover helps to keep dirt out of the internal mechanism of the gun.

CHARACTERISTICS OF THE M1949 SUBMACHINE GUN

Caliber: 9mm Parabellum.
System of operation: Blowback, automatic fire only.
Length, overall:
Stock extended: 28 in.
Stock retracted: 18.3 in.
Barrel length: 9.05 in.
Feed device: 32-round, detachable, staggered-row box magazine.
Sights: Front: Hooded blade.
Rear: L-Type w/apertures for 100 and 200 meters.
Muzzle velocity: 1237 f.p.s.
Cyclic rate: 600 r.p.m.
Weight, loaded: 9.41 lbs.

How to Load and Fire the M1949 (MAT 49) Submachine Gun

Insert a loaded magazine in the well of the magazine housing; if the housing is in the horizontal position, swing it down to the vertical, making sure that the lock located on the underside of the trigger housing engages the magazine housing. Pull the cocking handle, located on the left side of the receiver, to the rear. The bolt will remain to the rear, since this weapon fires from an open bolt. Push the cocking handle to its forward position. Squeeze the grip safety with the rear of the hand, pull the trigger, and the weapon will fire. To remove the magazine, engage the magazine catch located at the bottom rear of the magazine housing. To lift the magazine housing and magazine into the horizontal position under the barrel depress the lock located on the underside of the trigger housing and swing the magazine up till the lug at the forward end of the housing engages the clip on the underside of the barrel jacket. To change the position of—or remove—the stock, depress the catch located on the left side of the trigger housing.

How to Field Strip the M1949 (MAT 49) Submachine Gun

Remove the magazine and clear weapon. Pull tirgger and let bolt go forward. Press the knurled bar located on the rear section of the barrel jacket to the rear. The barrel and receiver assembly can now be pulled upward and forward off the frame. The operating spring and bolt can be removed from the rear of the receiver. Further disassembly is not recommended.

Special Note on the M1949 (MAT 49) Submachine Gun

The M1949 submachine gun is composed mainly of steel stampings. Since its functioning is basically the same as that of any other blowback-operated submachine gun, no extended coverage on functioning is given. The ejection port cover is automatically opened by the forward or rearward movement of the bolt.

FRENCH MACHINE GUNS

FRENCH MODEL 52 MACHINE GUN

The M1952 is called the *Arme Automatique Transformable* (AAT) by the French, since it is designed to be used as both a heavy and light machine gun. The weapon can be used as a light machine gun with a light barrel, a bipod and a butt support. It can also be used as a heavy machine gun on the US M2 tripod with a French adapter. This gun has been made in 7.62mm NATO caliber for test and for export.

How to Load and Fire the M1952 (AAT Mle. 52) Machine Gun

Pull the operating handle to the rear, cocking the weapon. Push the cover catch forward and open the cover; lay the belt on the feedway, so that the first cartridge in the belt is positioned against the cartridge stop. Close the cover and press the trigger; the weapon will fire. An alternate method is to feed the loading tab of the belt into the feed port on the left side of the gun, and pull it through to the right until a cartridge clicks into position. Then pull the charging handle to the rear and fire. The weapon fires from an open bolt.

CHARACTERISTICS OF THE MODEL 52 MACHINE GUN

Caliber: 7.5mm French M1929 rimless.
System of operation: Delayed blowback, full automatic fire only.
Length, overall:
Stock extended: 45.9 in.
Stock retracted: 38.6 in.
Barrel length (Quick-change barrel):
Heavy barrel, w/o flash hider: 23.6 in.
Light barrel, w/o flash hider: 19.3 in.
Feed device: 50-round, metallic, nondisintegrating link belt; when used as a light gun, a box containing one link belt section is hung on the side of the receiver.
Sights: Front: Barley corn
Rear: Tangent leaf w/U-notch.
Muzzle velocity: 2690 f.p.s.
Cyclic rate: 700 r.p.m. (approx.).
Weight, w/light barrel: 21.7 lbs.
w/heavy barrel: 23.28 lbs.

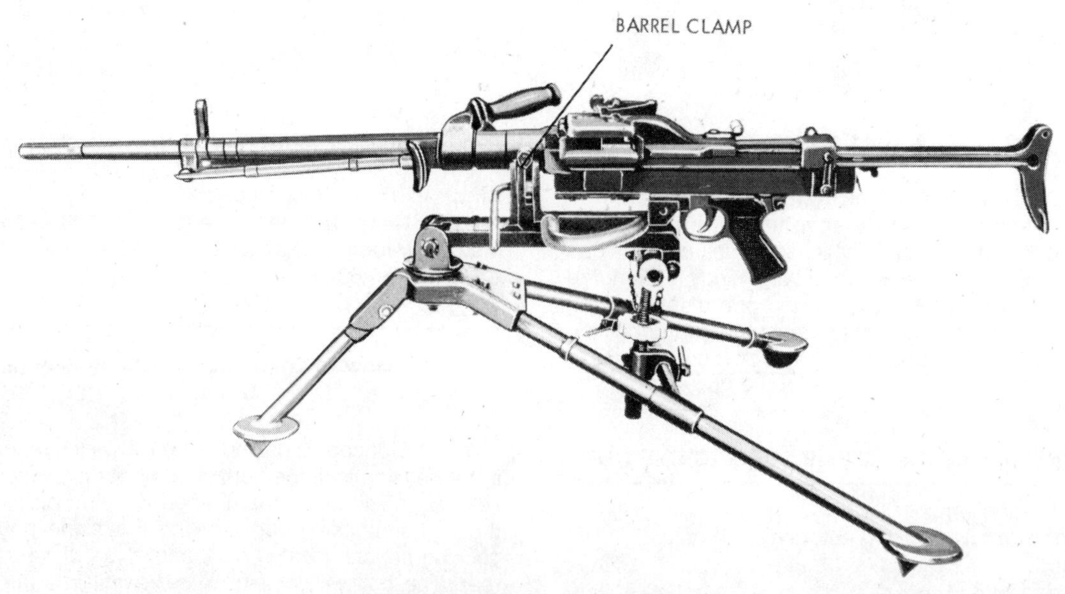

French 7.5mm Model 1952 on US M2 tripod.

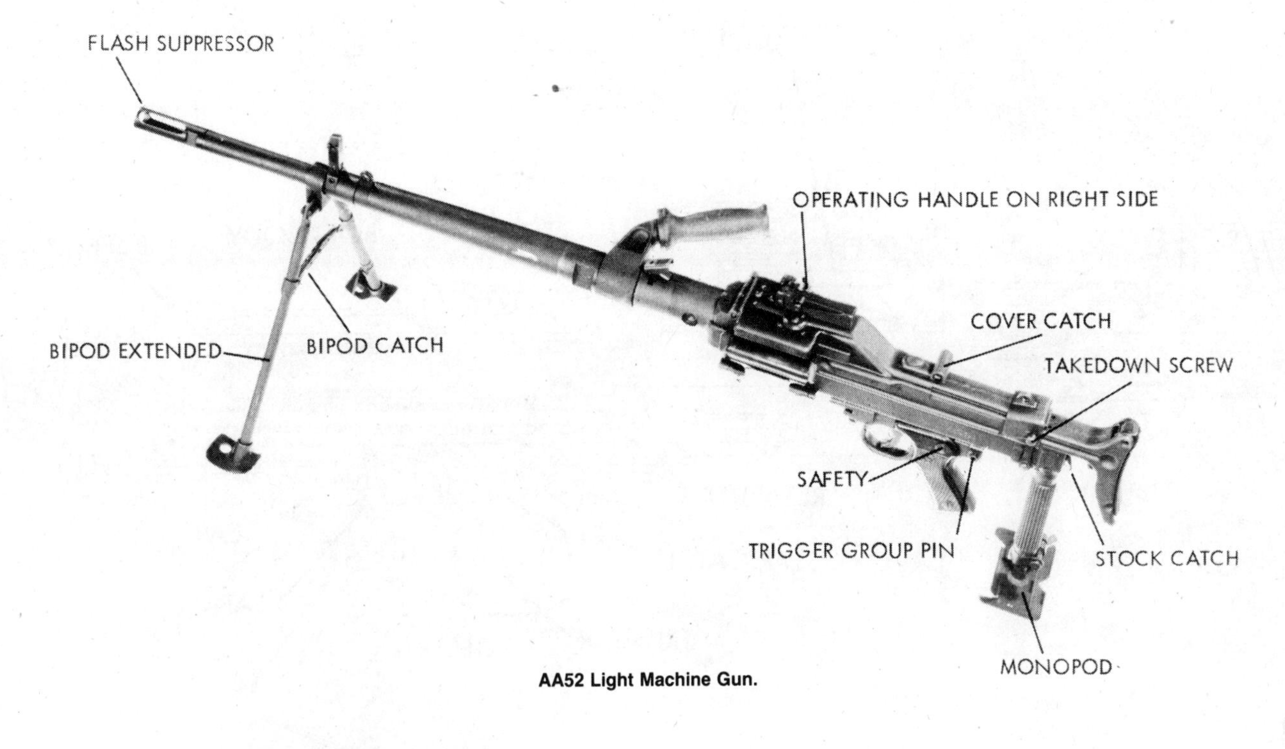

AA52 Light Machine Gun.

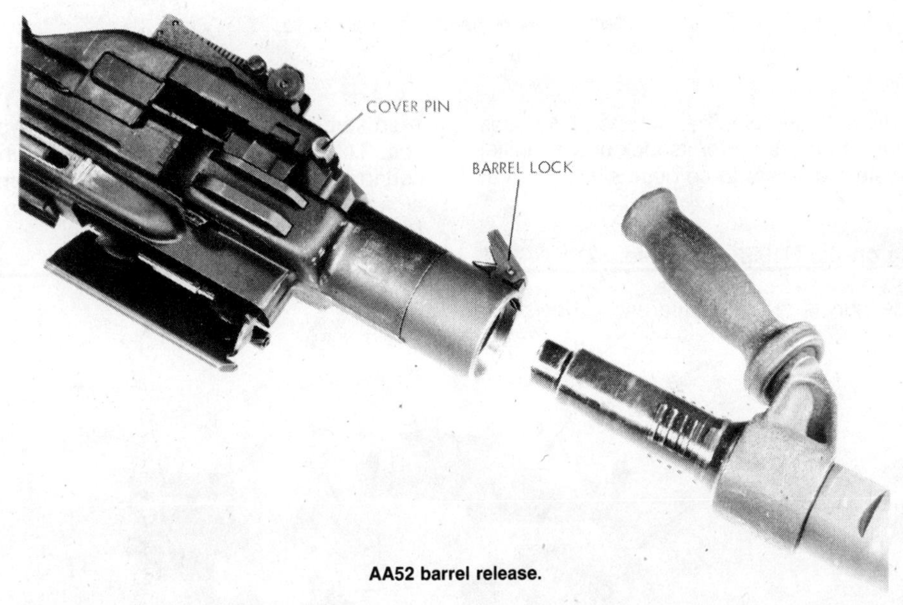

AA52 barrel release.

How to Field Strip the M1952 (AAT Mle. 52) Machine Gun

With the stock in the extended position, press down on the stock bolt lever to remove the two stock bolts; then pull the butt to the rear and remove. Unscrew and remove the assembly pin. Pull back the bolt assembly by swinging it up. Pull the cocking handle to the rear. Remove the recoil spring and the recoil spring guide; then remove the bolt assembly by sliding it out the rear. Remove the trigger guard retaining pin and disengage the trigger guard from the receiver. Pull back the barrel catch with the left hand and, with the right hand, twist the carrying handle 1/6th turn to the right; then pull the barrel forward. Put the cover and the feed plate at a 90-degree angle from the receiver and remove the cover pin. Separate the bolt head by moving the head forward and up. Remove the firing pin and bolt lock from the bolt.

To reassemble the weapon, perform the above steps in reverse order.

How the M1952 (AAT Mle. 52) Works

The delayed-blowback bolt system of the M1952 machine gun is similar in principle to that of the Spanish CETME assault rifle and the Swiss Model 57 assault rifle. The bolt is in two pieces—a head and a body; the head is much smaller than the body. Instead of using two roller bearings for locks, as do the Spanish and Swiss weapons, the M1952 employs a locking lever. The bolt head cannot move to the rear until the pressure is high enough to cause the locking lever to start opening and, in turn,

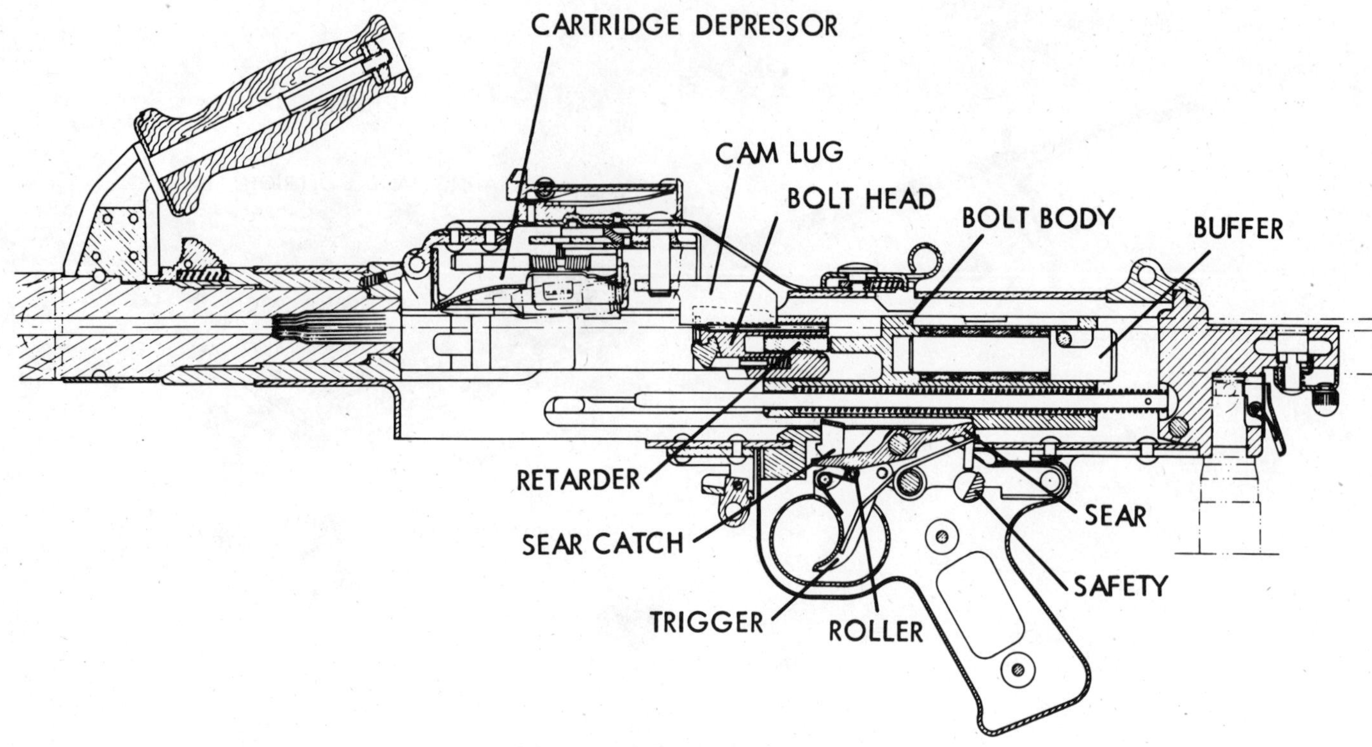

Section view of French 7.5mm Model 52.

force back the bolt body and firing pin. This weapon also uses a partially fluted chamber, to make up for its lack of slow initial extraction. The feed system appears to be quite similar to that of the MG42.

Special Note on the M1952 (AAT Mle. 52)

The M1952 machine gun is quite an interesting design; it represents a great departure from pre-World War II French practice. This machine gun has replaced the M1924 M29 as the squad automatic weapon, since the French refer to the gun as a "fusil Miltrailleur" (automatic rifle) when it is used in the light role on a bipod.

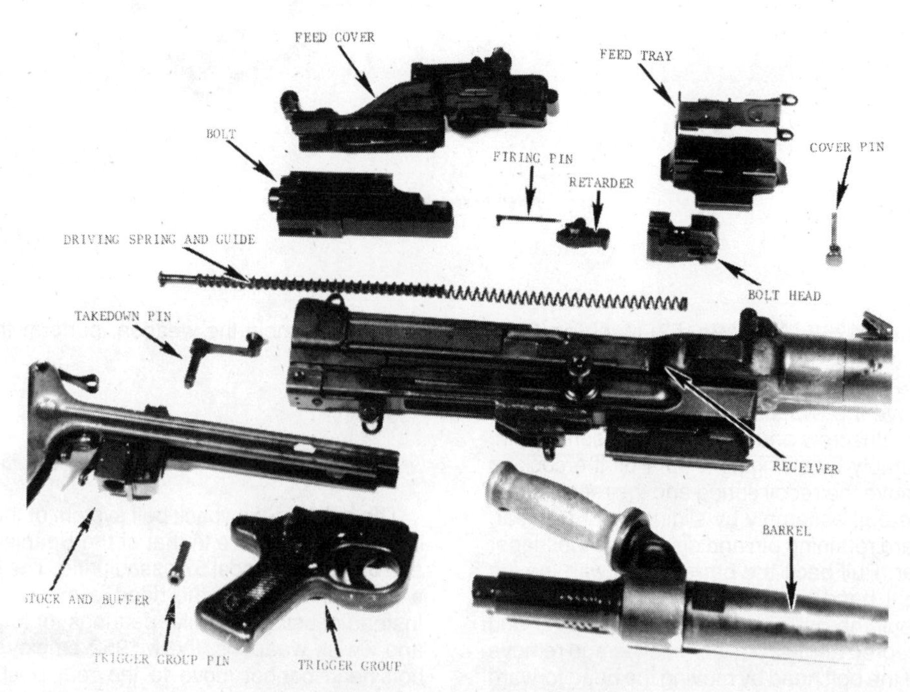

Disassembled view of AAT (Mle. 52).

OBSOLETE FRENCH MACHINE GUNS

THE FRENCH MODEL 1924 M29 LIGHT MACHINE GUN

This weapon is called an automatic rifle by the French and is used as a squad automatic weapon. It has had extensive combat use and is a very popular weapon with French troops. The basic design, except for the top mounted magazine and the double trigger, is quite similar to the US Browning Automatic rifle Model 1918. The original gun, the Model 1924, was issued chambered for the 7.5mm Model 24 cartridge. When the cartridge was shortened, the machine gun was modified to chamber the new cartridge and designated Model 1924 M29. This weapon may be found in any of the former French colonies or mandates.

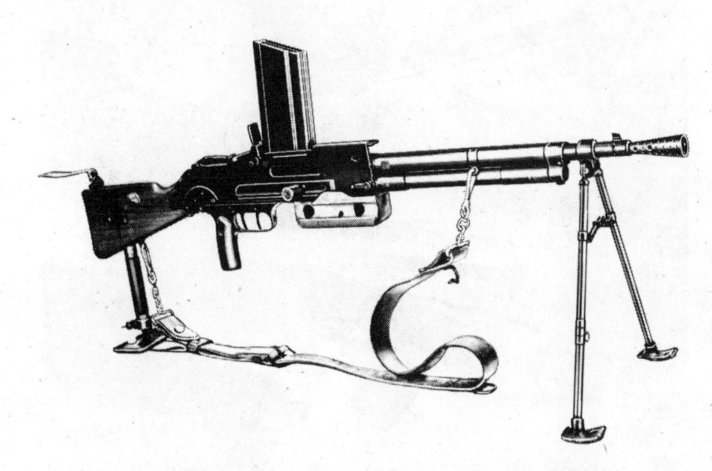

French Chatellerault 7.5mm 1924 M29 Machine Gun.

CHARACTERISTICS OF THE MODEL 1924 M29

Caliber: 7.5mm French.
System of operation: Gas, selective fire.
Length overall: 42.6 in.
Barrel length: 19.7 in.
Weight: 24.51 lbs. w/bipod.
Feed device: 25-round, detachable, staggered row, box magazine.
Sights: Front: Blade.
Rear: Tangent.
Muzzle velocity: 2590 f.p.s.
Cyclic rate: 550 r.p.m.

Field Stripping the M1924/29

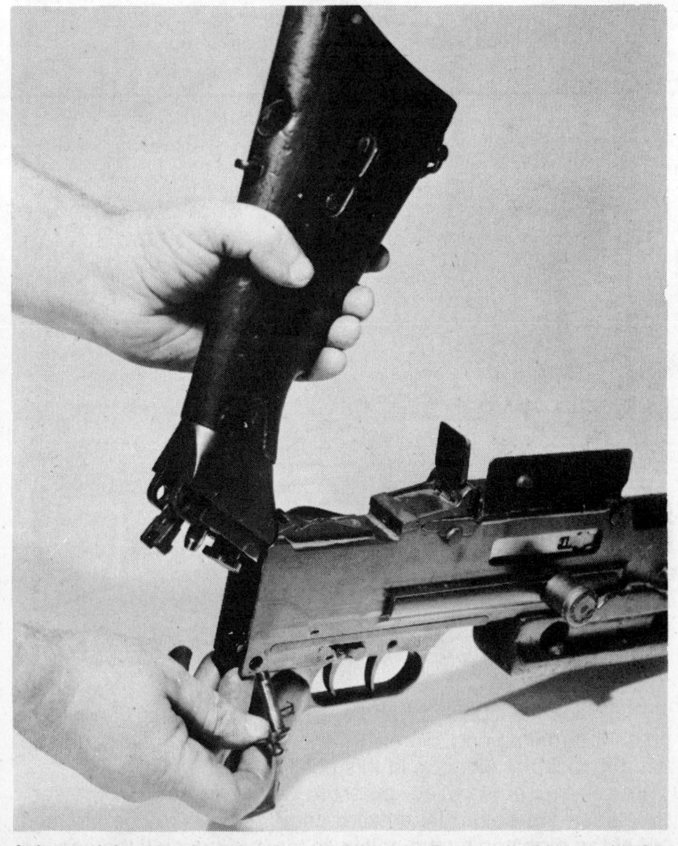

At lower right side of receiver is a retaining pin. Remove it; this permits the rear of the buttstock to be hinged up and back, when it can be lifted out of receiver.

Trigger guard assembly will now swing forward on hinge. Pushing first in, then out, unhook front end and remove.

Pulling head of ejector out of its slot will permit it to be removed from the rear. Remove recoil spring and guide.

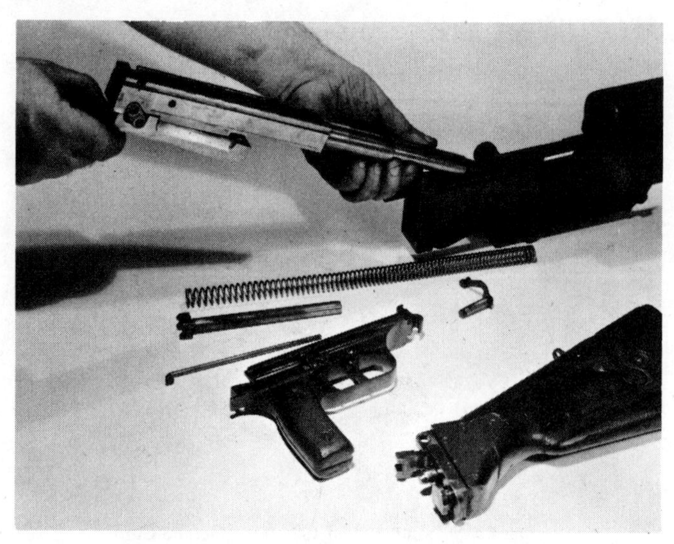

Withdraw bolt and piston with slide out of receiver.

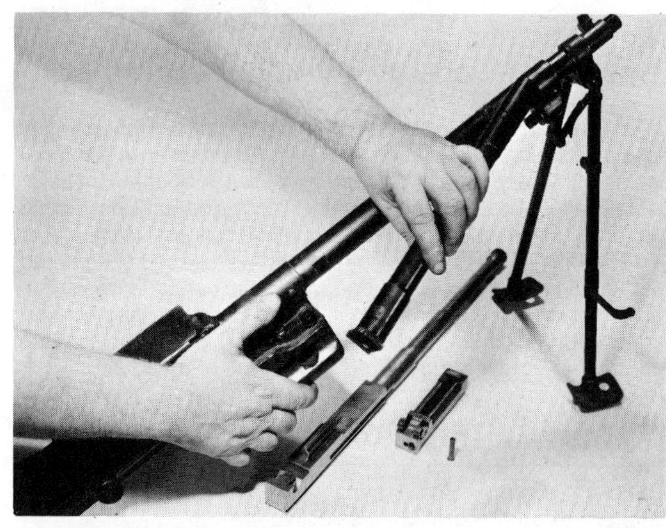

Gas cylinder tube lock is at lower front of receiver. Turn it to "O" mark on receiver, then raise rear end of tube. Tube can now be pulled out of receiver. Barrel lock is on upper front of receiver. Turn it to "O" and unscrew barrel to right.

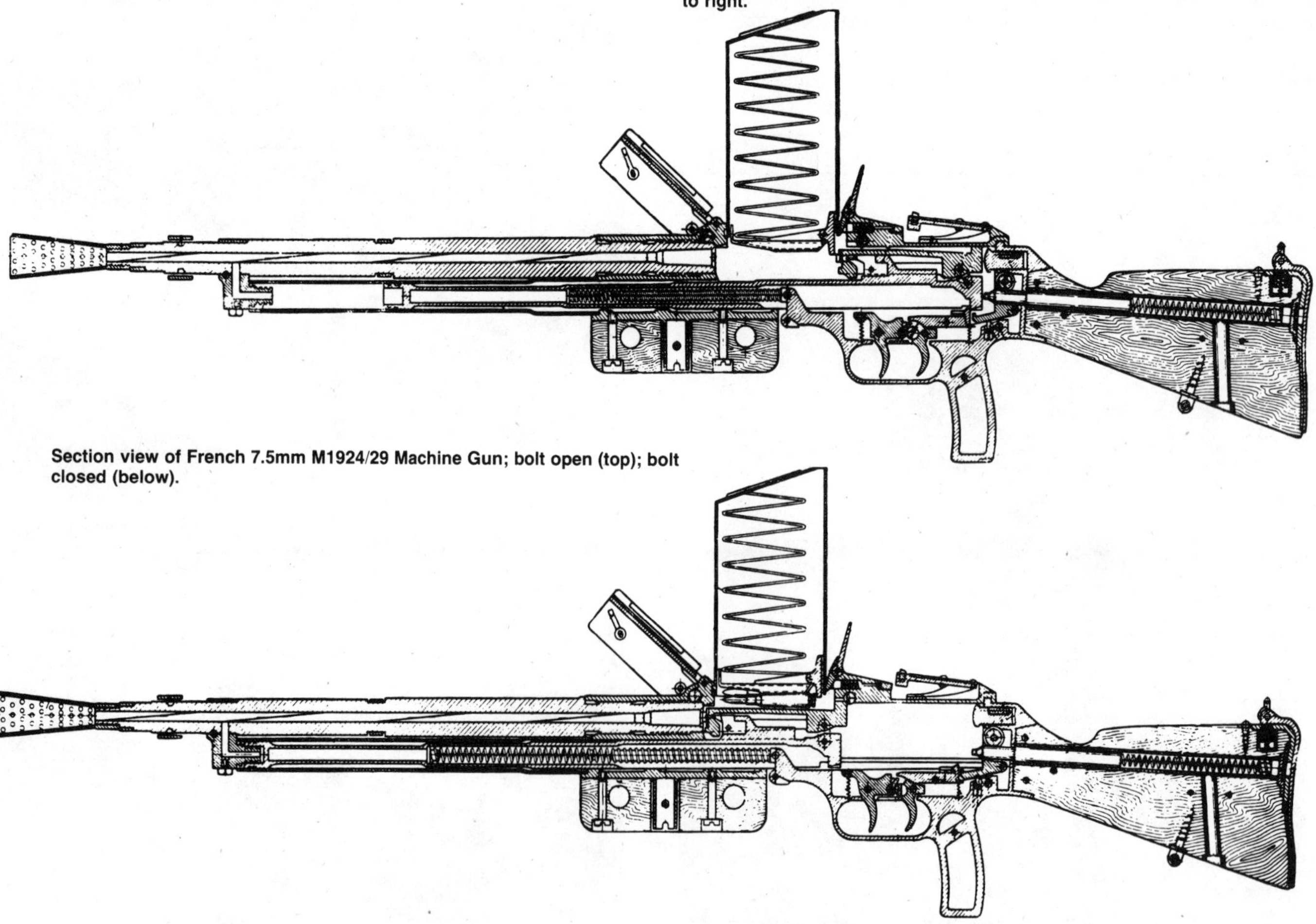

Section view of French 7.5mm M1924/29 Machine Gun; bolt open (top); bolt closed (below).

Loading and Firing the M1924/29

A loaded magazine is inserted vertically in the top of the gun. The magazine opening is fitted with a dust cover, which must be hinged up and forward to expose the magazine opening. The magazine release catch is positioned at the rear opening. Pressing it forward will release the magazine to be lifted out with the hand that is operating the catch.

Pull back the cocking handle to cock the weapon and compress the mainspring. Safety lever on the trigger guard may be used to lock the weapon in this position.

If front trigger is pulled, pressure will be exerted on the sear to release the bolt to fly forward and fire the cartridge stripped out of the magzine by the bolt in its forward travel. If the second trigger is pulled, the sear will not engage as long as the trigger is held back, and the weapon will continue to fire full automatic.

FRENCH MODEL 1931A MACHINE GUN

The French 7.5mm tank and fortress machine gun M1931A was adapted for use on the US cal. 30 M2 tripod because of French post-war shortage of machine guns. This weapon is still in use on French-made armored vehicles. Since it is likely to be encountered by some of the readers of this book, it will be covered in detail even though it is not a post-war weapon.

The M1931A machine gun is the tank and fortress version of the French 7.5mm M1924/29 light machine gun. The weapon has a very heavy barrel, since it is designed for a high rate of sustained fire and does not have a quick change barrel. Two magazines can be used with the gun—a box type or a large drum, which fastens to the side of the gun.

CHARACTERISTICS OF M1931A MACHINE GUN

Caliber: 7.5mm M1929 French rimless.
System of operation: Gas, automatic fire only.
Length, overall: 40.5 in. w/flash hider.
Weight, empty: 27.48 lbs.
Weight of Mounts:
Tripod M2: 16.22 lbs.
Tripod M1945: 32.15 lbs.
Barrel length: 23.5 in.
Feed device: 36-round, detachable, staggered-row box magazine or 150-round drum magazine.
Sights: Front: Blade.
Rear: Tangent leaf w/open notch.
Muzzle Velocity: 2750 f.p.s.
Cyclic Rate: 750 r.p.m.

How to Load and Fire the M1931A Machine Gun

Cock the gun by pulling the operating handle to the rear. If the box magazine is to be used, rotate it into the feed fixture on the right side of the gun, so that the lug on the magazine engages in the recess in the fixture. Then pull the magazine toward the rear of the gun, so that the rear of it is locked in place in the feedway. Pull the trigger, and the weapon will fire. This weapon fires from an open bolt and has no safety. To load the drum-type magazine, fit the magazine on the projection of the receiver, which is just ahead of the feedway. A stud to the rear of the projection seves as an index when assembling the magazine to the gun. Firing is carried on in the same way with the drum as with the box magazine. To remove the box magazine, engage the magazine catch and swing the magazine forward and out. To remove the drum magazine, pull the handle on the outside of the drum, and pull the magazine straight off. When this gun is unloaded, using either magazine, there may be a round in the feedway, unless all the rounds have been fired. To clear the gun, the round must be pushed forward so that it falls free inside the receiver and out the bottom of the gun.

How to Field Strip the M1931A Machine Gun

Check to see that the weapon is not loaded. Depress the springloaded catch in the rear of the backplate. Rotate the lock 90 degrees, and remove the recoil spring and guide. Remove screw at the lower rear of the receiver; a lever attached to the screw assists in its removal. Remove the backplate; the trigger housing and recoiling parts can then be removed. No further disassembly is recommended.

To reassemble the weapon, perform the above steps in reverse order.

How the M1931A Machine Gun Works

The piston and gas cylinder are similar to those on the M1924 M29 light machine gun. However, because the M1931 A ejects through the bottom of the receiver, the piston assembly is cut away to permit cases or complete rounds to pass through. The operating spring is positioned at the rear of the piston assembly, with only a small portion of it inside the piston. Locking is accomplished in the same manner as on the M1924 M29 light machine gun. A spring-loaded lever, positioned in the right side of the bolt, engages the base of the cartridge case for feeding and firing. The ejector operates in a groove in the top of the bolt. The firing pin operates inside the bolt but is held rigidly to a projection on the rear of the piston assembly.

French 7.5mm Machine Gun M1931A on US M2 tripod.

THE FRENCH MODEL 1914 HOTCHKISS MACHINE GUN

The 1914 Hotchkiss was the principal machine gun of the French Army in World War I and was also still in service by the French Army until the fall of France in 1940. The Hotchkiss appeared again, to a very limited extent, during the French campaign in Indo-China (Vietnam). Although a rather heavy and bulky gun, the 1914 Hotchkiss had a good reputation for reliable performance.

CHARACTERISTICS OF THE MODEL 1914 HOTCHKISS

Caliber: 8mm Lebel.
System of operation: Gas, automatic fire only.
Length overall: Approx. 51.6 in.
Barrel length: 31 in.
Weight: Gun, 55.7 lbs; tripod, 60 lbs.
Feed device: 24- and 30-round strip or 250-round belt consisting of articulated strips.
Sights: Front: Blade.
Rear: V-notch
Muzzle velocity: 2325 f.p.s. w/ball 1932N.
Cyclic rate: 450–500 r.p.m.

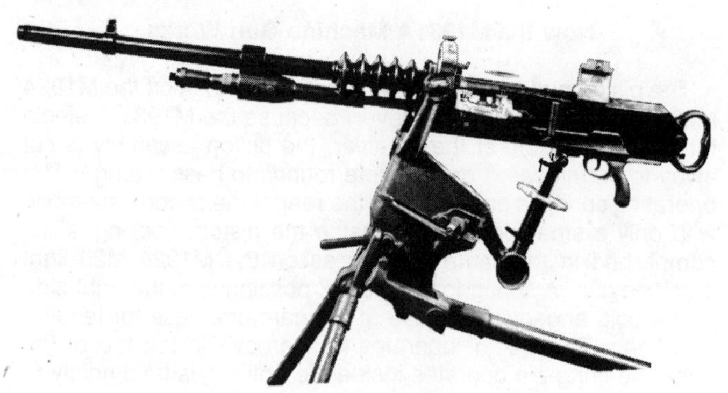

The French 1914 model Hotchkiss, caliber 8mm French.

FRENCH CAL. .30 M1918 LIGHT MACHINE GUN, C.S.R.G.

United States forces used the C.S.R.G., commonly called the Chauchat during World War I. The United States purchased 15,988 of these weapons in 8mm and 19,241 in caliber .30-06. The .30-06 weapons were made by the French specifically for the United States Army and have a straight 16-round box magazine rather than the crescent-shaped magazine required by the rim and taper of the French 8mm cartridge.

The C.S.R.G. was adopted by Belgium in 7.65mm after World War I and by Greece, where it was called the Gladiator, in 8mm Lebel. It was used to some extent in World War II as well. This weapon is called the Chauchard in England.

THE FRENCH MODEL 1915 LIGHT MACHINE GUN C.S.R.G.

Although this weapon is called an automatic rifle by the French, it is usually considered a machine gun in the United States. This weapon was developed by Chauchat, Sutter and Ribeyrolle at Puteaux from the prototype APX 1910 rifle. Some American publications indicate that Chauchat, Ribeyrolle and Sutter were members of the commission that approved the weapon; French publications indicate that they actually worked on the design.

This weapon is long-recoil operated and is of rather poor manufacture, but it was a wartime product and built by a number of subcontractors, who shipped parts to a central plant for assembly. You could consider it a World War I Sten gun insofar as manufacture is concerned. The main problem was that it was not as reliable in performance as the Sten.

CHARACTERISTICS OF THE MODEL 1915 LIGHT MACHINE GUN

Caliber: 8mm Lebel.
System of operation: Long recoil, selective fire.
Length overall: 45.2 in.
Barrel length: 18.5 in.
Weight: 20 lbs. w/bipod.
Feed device: 20-round, detachable, in-line crescent shaped magazine.
Sights: Front: Blade.
Rear: Tangent with notch.
Muzzle velocity: 2300 f.p.s.
Cyclic rate: 240 r.p.m.

Chauchat 8mm Light Machine Gun Model 1915, C.S.R.G. (called Fusil Mitrailleur Chauchat Sutter Ribeyrolle-Gladiator).

20 East Germany: "German Democratic Republic"

The East German Army is currently equipped with most of the Soviet post-war small arms. The 9mm Makarov pistol and 9mm Stechkin machine pistol are used as are the 7.62mm AK47 and AKM assault rifles and the 7.62mm RPD and 7.62mm RPK light machine guns. The 7.62mm RP46 light machine gun and 7.62mm SG43 and SGM heavy machine guns are in service as well as the 12.7mm DShK M1938/46.

The 7.62mm General Purpose PK/PKS machine gun is now in service in the East German Army and replaced the 7.62mm RP 46, SG 43 and SGM machine guns in that Army.

Some of the pre-world War II and World War II Soviet weapons such as the 7.62mm PPSh M1941 submachine gun and the 7.62mm DT and DTM machine guns are still in service with border guards and armored forces respectively (on older armed vehicles).

The workers militia is armed with German World War II weapons such as the 7.92mm Kar 98K and the 7.92mm MP44. They also have older weapons. The East Germans manufacture or have manufactured the 9mm Makarov pistol, the 7.62mm AK 47 and AKM assault rifles, the 7.62mm SKS carbine and the RPK light machine gun. The arms plant in the Suhl area are in production manufacturing sporting arms in addition to military arms.

The East Germans use slightly different nomenclature for the Soviet weapons than do the Soviets. The usual procedure is to use the first letter of the designer's name tacked on to the German abbreviation of the weapon type. Thus the AK 47 assault rifle with wooden stock is the MPi K and with folding stock it is the MPi KmS. The AKM is the MPi KM, the RPK is the LMG K, the RPD is the LMG D, etc.

Some of the Soviet designed weapons manufactured in East Germany differ in minor detail from those made in the USSR. See Chapter 1 for additional details.

East German copy of 9mm Soviet Makarov pistol.

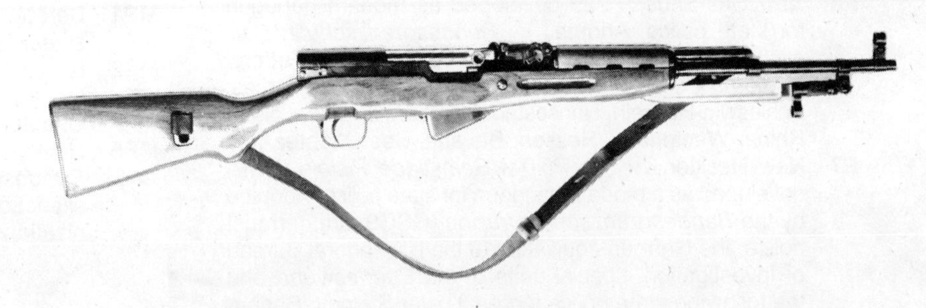

East German 7.62mm copy of Soviet SKS carbine. Note that sling is mounted similar to Kar 98k and that no cleaning rod is carried under barrel.

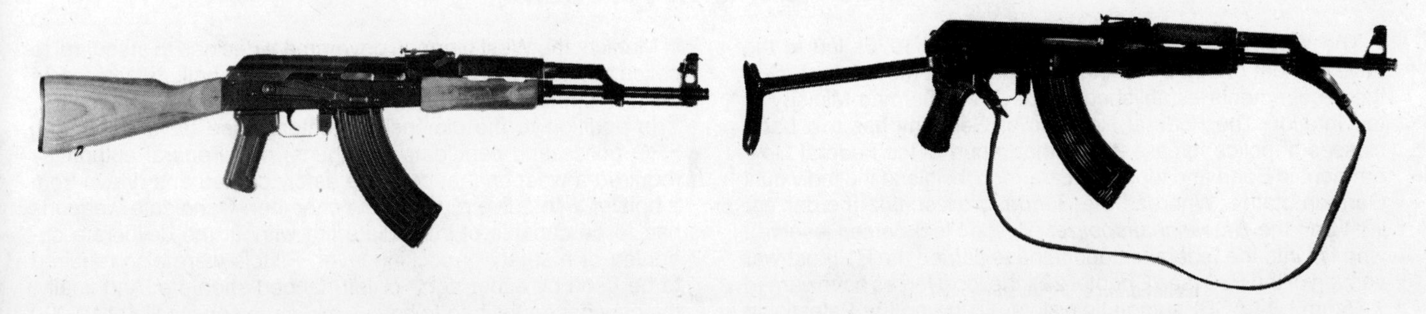

East German 7.62 MPi KM (AKM).

East German 7.62mm MPiK (AK-47) Assault Rifle with folding stock.

21 West Germany: "Federal Republic of Germany"

BUNDESWEHR SMALL ARMS DESIGNATIONS

The *Bundeswehr*, the ground force of the *Bundesrepublik Deutschland* (Federal Republic of Germany) has assigned the following designations to small arms used by its forces, those of the *Bundesgrenzschutz* (border police) and *Land Polizie* (state police forces) since 1945.

PISTOLE (P)

P1 Post-war designation for the P38; see Chapter 22 for details. Standard handgun for *Bundeswehr*, but will be superseded by the P7.
P2 SIG SP47/8 manufactured by SIG at Neuhausen, Switzerland. Obsolete. See Chapter 44.
P3 Astra 600/43. Obsolete. See Chapter 42.
P4 P38K, short-barreled P38.
P5 New Walther handgun derived from P38. Developed as modern handgun for state police. Adopted by Rhineland-Palatinate and Baden-Wurtenberg.
P6 New SIG-Sauer P225 developed as modern handgun for state police. Adopted by *Bundesgrenzschutz*, *Bereitschaftpolizei* (Police Field Force Reserve), Federal Customs Police, and the following state police forces: Schleswig-Holstein, Hansestadt-Hamburg, Bremen, North Rhine–Westphalia, Hessen, Bavaria. See Chapter 44.
P7 New Heckler & Koch *Polizei Selbstlade Pistole* (PSP) developed as a modern handgun for state police. Adopted by the *Bundesgrenzschutzgruppe* (GSG9) anti-terrorist police, the German equivalent of the US Federal Bureau of Investigation, special units of the *Bundeswehr*, and the following state police forces: Lower Saxony, Baden-Wurttenberg, and Bavaria.
P9 Heckler & Koch P9S adopted by the Saarland state police force.
P11 Heckler & Koch HK4 pistol used by police; obsolescent.
P21 Walther PPK used by police; obsolescent.
P31 Walther/Hammerli .22 long rifle training pistol.
P32 Walther OSP .22 long rifle training pistol.

GEWEHR (G—RIFLES)

G1 Fabrique Nationale FAL; obsolete. See Chapter 9.
G2 SIG StG 57, trial quantities only procured. See Chapter 44.
G3 The standard West German rifle fabricated by Heckler & Koch. Also made at one time by Rheinmetall.
G4 Armalite AR-10; trial quantities only procured. See Chapter 1.
G11 Heckler & Koch caseless cartridge rifle. Still under development.
G41 Heckler & Koch commercial rifle; variant of HK33.

MASCHINENPISTOLE (MP)

MP1 Beretta Model 38/49 submachine gun used by *Bundeswehr* in the 1950s and 1960s. See Chapter 29.
MP2 Israeli Military Industries UZI submachine gun. This is the standard submachine gun of the *Bundeswehr*. See Chapter 28.
MP5 Heckler & Koch HK54 submachine gun. Used by the *Bundesgrenzschutz*, GSG9, and state police forces. The silenced version, MP5SD, is also issued to these organizations for special operations.

MASCHINENGEWEHR (MG)

MG42/59 (MG1) Postwar model of the MG42. See Chapter 22 for data on this machine gun and the MG1A1, MG1A2, and MG1A3.
MG3 Current version evolved from the above machine guns.

WEST GERMAN PISTOLS

The West German police trials of the mid-1970s led to the development of three new standard handgun models that meet the requirements established by the West German Ministry of the Interior. The Federal Republic of Germany has two basic classes of police forces—those that report to the Federal Government in Bonn and those that are responsible to the individual German states. Whereas the *Bundesgrenzschutz* (border police) and the *Bereitschaftspolizei* (Police Field Force Reserve), who fall into the federal category, have carried the P1 (post-war version of the P38 (see Chapter 22), the local forces have carried 7.65mm (.32 ACP) automatic pistols. Following the Palestinian terrorist attack on the Israeli athletes at the 1972 Olympic games in Munich, the West German government decided to standardize police handgun ammunition (9 × 19mm Parabellum) and create uniform specifications for law-enforcement handguns.

In addition to the dimensional criteria (see table comparing FRG police and candidate handguns), the Federal authorities required a weapon that could be safely carried and drawn from a holster with a live round in the chamber. Candidate weapons had to be capable of immediate fire without the deliberate operation of a safety or cocking lever. Pistols were also required to be used by either right- or left-handed shooters. And finally, the new handguns had to have a minimum service life of 10,000 shots.

West German 9mm P38 of current manufacture. The P-1 of the West German Army.

The 9 × 19mm Parabellum Walther P5.

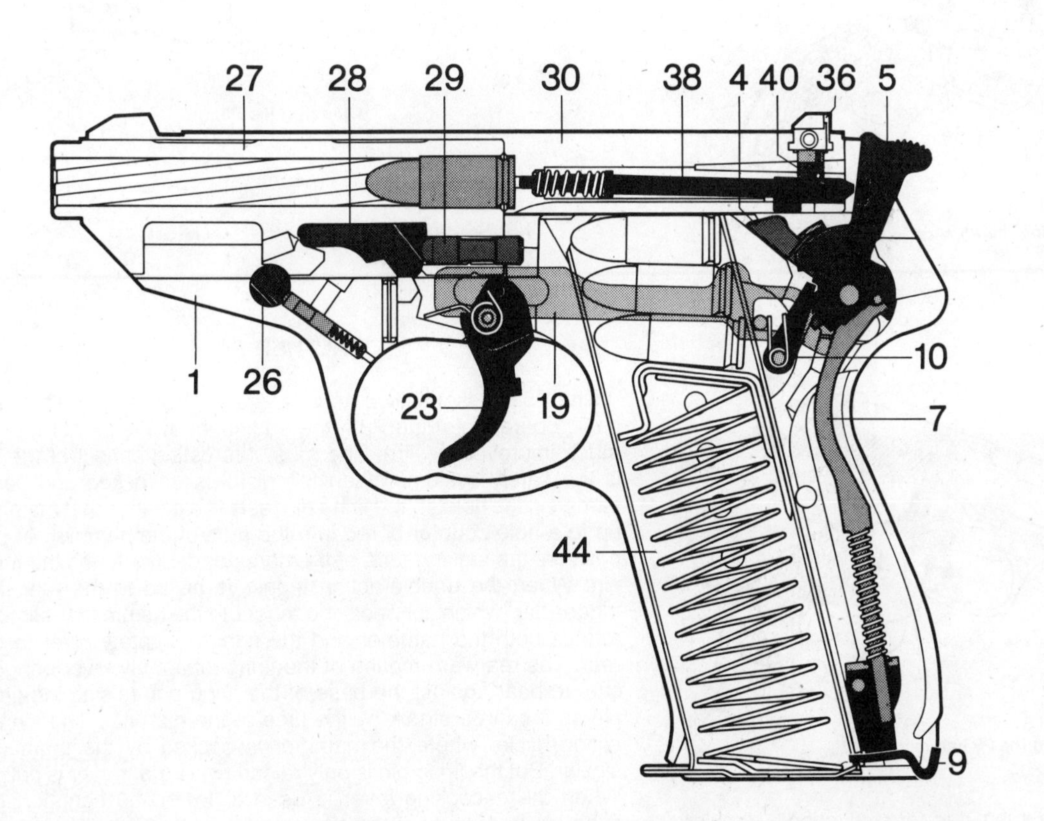

The P5 loaded and ready to be shot.

1. Frame
4. Firing pin lifter
5. Hammer
7. Hammer strut
9. Magazine release
10. Sear
14 and 17. Combination slide release and decocking lever.
19. Trigger bar
23. Trigger
26. Takedown lever
27. Barrel
28. Locking block
29. Unlocking plunger
30. Slide
35. Windage adjustment detent
37. Windage adjustment screw
38. Firing pin
40. Rear sight mount
44. Magazine

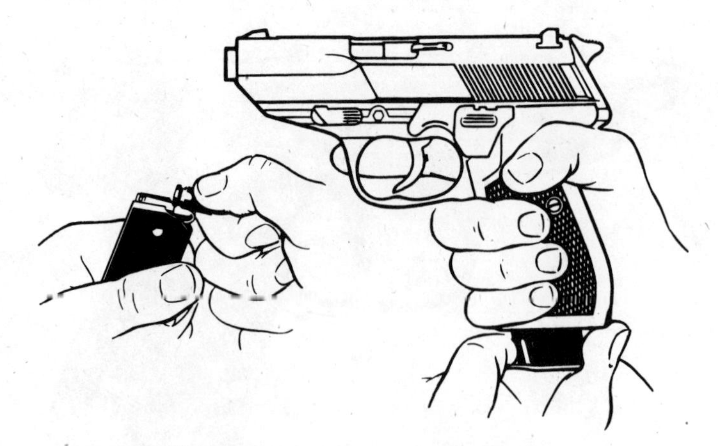

Loading and inserting the P5 magazine.

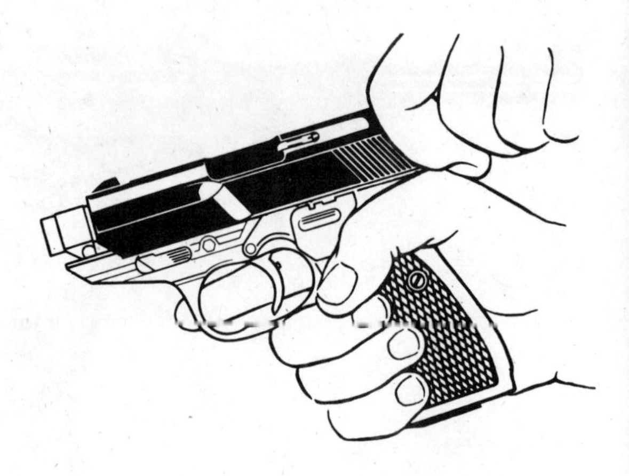

Charging the P5.

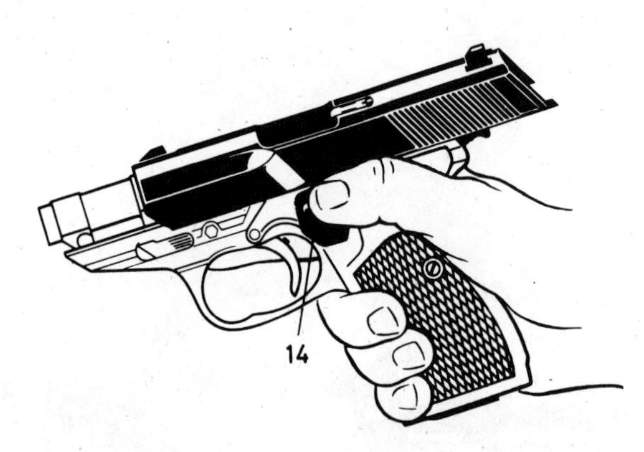

Releasing the P5 slide.

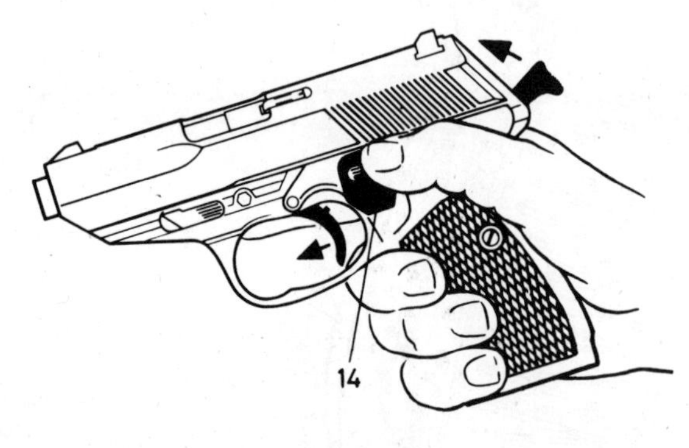

Decocking the P5 hammer.

Originally, four guns were submitted for evaluation—Heckler & Koch *Polizei Selbstlade Pistole* (PSP), Mauser HsP, SIG-Sauer P225, and Walther P5. Early in the trials Mauser withdrew their prototype because the other designs were much more fully developed. All of the remaining pistols were adopted as being suited for police use, and none have a traditional manually operated safety mechanism. Because of the requirement for instant, quick-draw-readiness, the designers at the three arms companies had to develop alternative means of creating a safe, but loaded, handgun.

WALTHER P5

The P5 is essentially a new generation P38. Starting with the basic operation of the P38, the engineers at Walther created a much improved design. The most interesting aspect of the P5 is the safety system. When the pistol is uncocked and being carried in the holster, the firing pin rests in a position that matches up to a hole counter-bored into the face of the hammer. In this position, the striking face of the hammer cannot touch the firing pin. When the double-action trigger is pulled to the rear, the trigger-bar (which connects the trigger to the hammer assembly) rotates both the hammer and the firing-pin safety lever to the rear. The rearward motion of the firing-pin safety lever brings a cam to bear against the base of the firing pin, raising the firing pin so it can be struck by the face of the hammer. The single-action cycle, where the hammer is cocked by the thumb, is similar. But the firing pin is only raised when the trigger is pulled. When the decocking lever is used to lower the hammer, the firing pin is already in the down and safe position. Thus, even with a fast falling hammer, the cartridge in the chamber cannot be fired.

The Walther P5 and the SIG-Sauer P6 (P225) firing-pin safeties are similar in principle, but different in exact details. In function, they serve the same purpose as the transfer bar in modern revolvers. Both mechanical devices eliminate accidental discharge of the handgun by introducing a mechanical linkage that must be consciously engaged to bring the firing pin into proper alignment for firing.

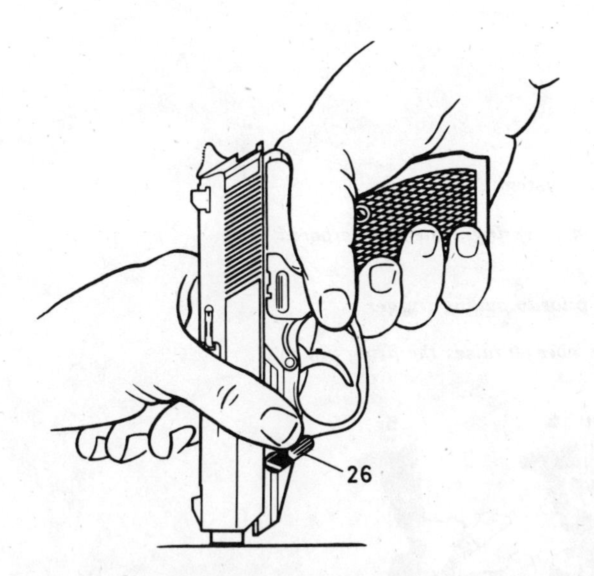

Retract the slide slightly so that the disassembly lever can be rotated from the rear to the front as in the P38.

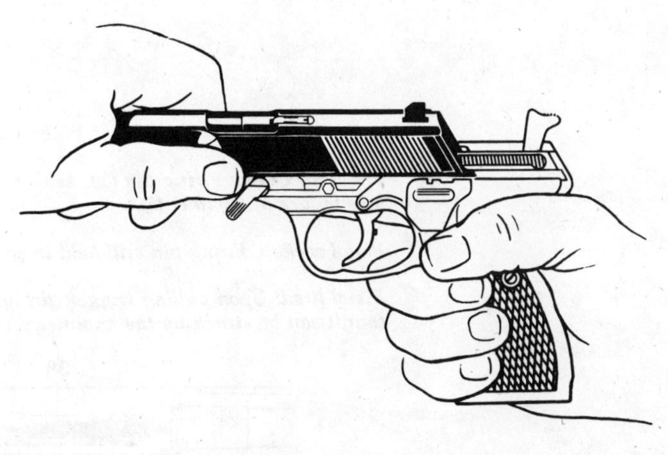

Pull the slide forward off the frame.

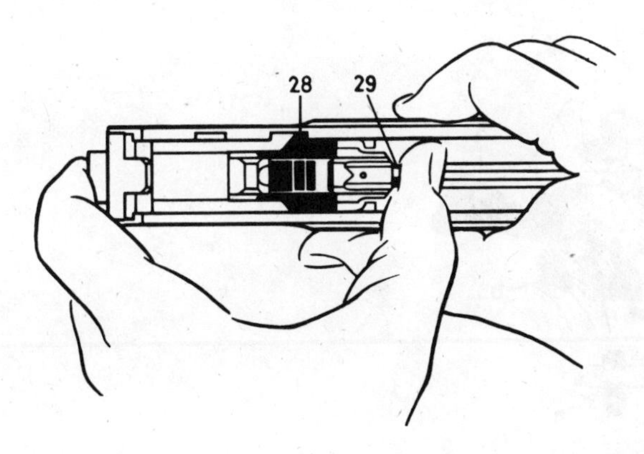

To separate the barrel from the slide depress the unlocking plunger.

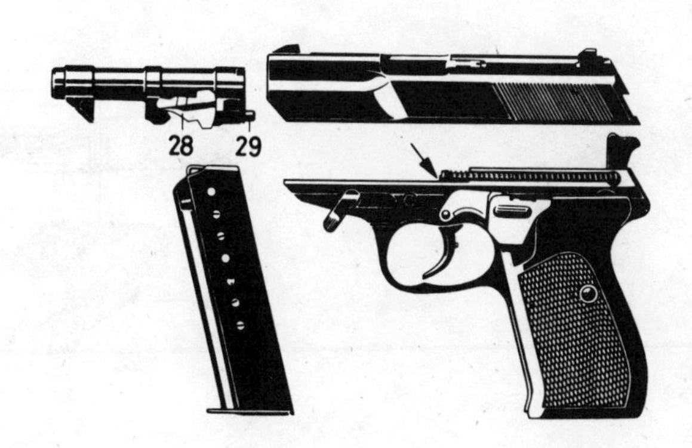

Disassembled view of the P5.

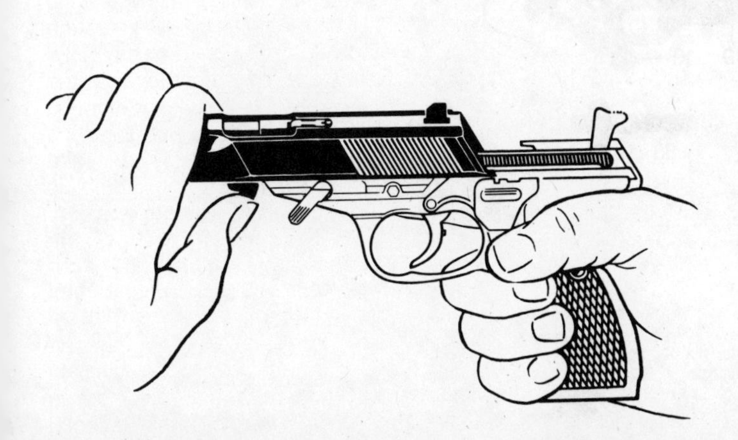

When reassembling the P5, the locking block must be pushed upwards as the slide is moved to the rear.

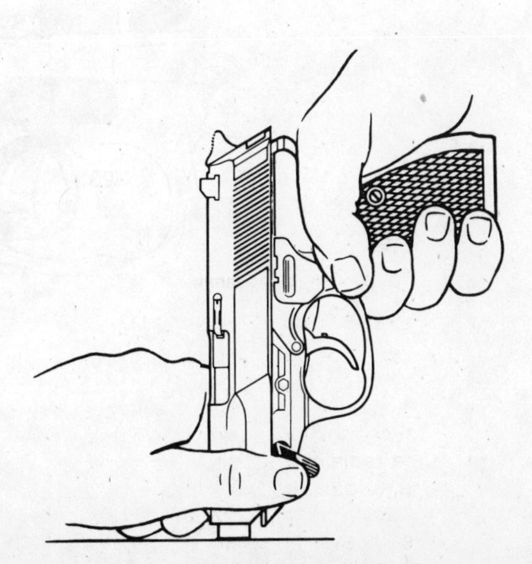

The final step in reassembly is rotating the disassembly lever to the rear.

Walther P5 Firing-pin Safety System

Pistol uncocked: Firing pin (38) held in safe position by spring. Note counterbored hole in face of hammer (5).

Pistol cocked: Firing pin still held in safe position prior to pulling trigger.

Pistol fired: Upon pulling trigger, firing-pin safety lever (4) raises the firing pin so that it can be struck by the hammer.

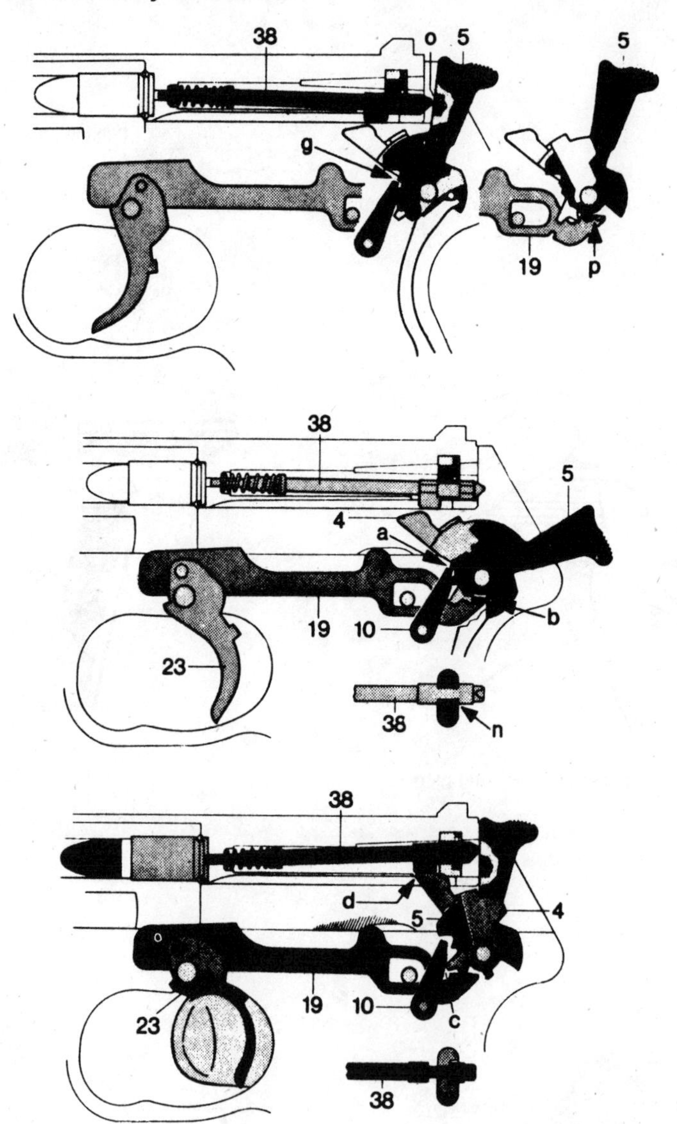

SIG-Sauer P220 and P225 Firing-pin Safety System

Firing-pin safety catch. For maximum safety, the firing pin is locked. Quick readiness is assured since this safety catch is released automatically by pulling on the trigger. The catch is not released until the trigger is consciously pulled.

Decocking lever: This lever permits the hammer to be lowered without danger. The firing pin is blocked during and after decocking.

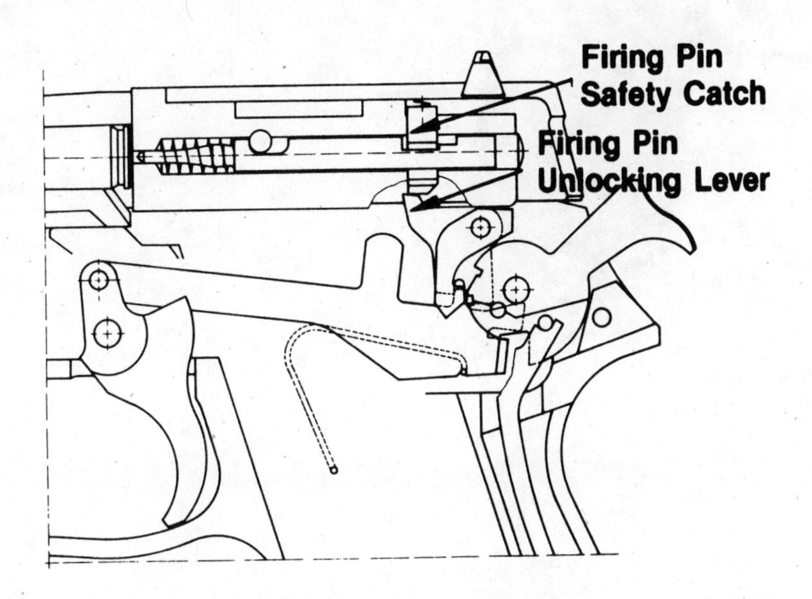

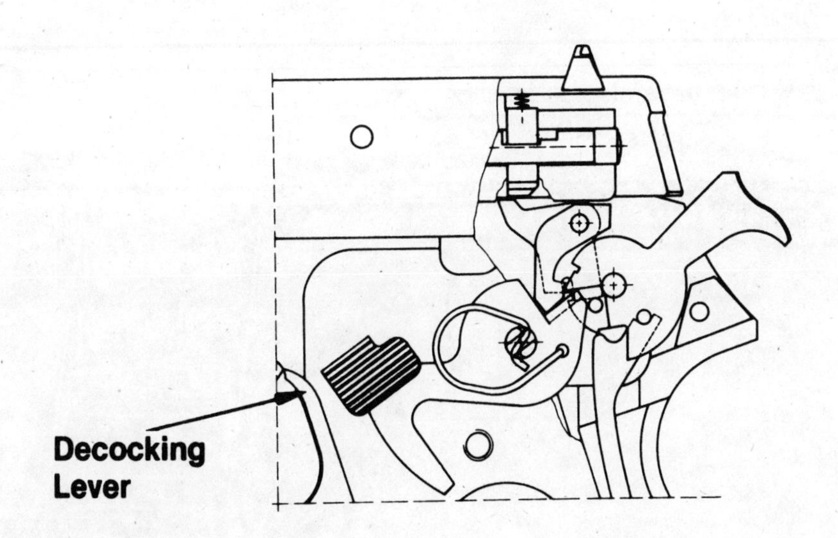

Heckler & Koch P7.

SIG-Sauer P6.

Section view of the P7. Note the gas-retarding piston system and the squeeze cocking device.

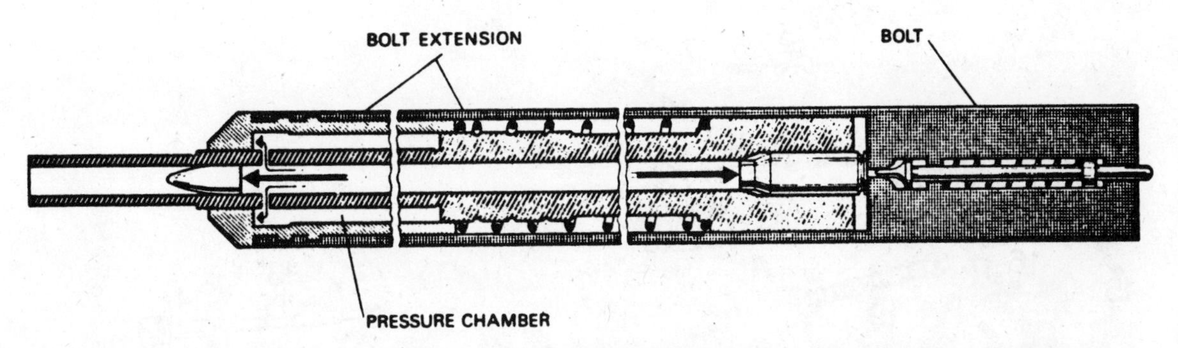

The gas retarding system of the P7 is similar in concept to that employed in the design of the Second World War *Volksturmgewehr*.

SIG-SAUER P6

SIG-Sauer's P6 is basically a smaller version of the P220 that was introduced in 1975. The P6 version is only .3 ounce (.01 kilogram) lighter than the P220. Still it is significantly smaller in size and it has a smaller magazine capacity—8 shots instead of the 9 shots of the P220. The P6 (P220) design is described in Chapter 44. The West German P6 is manufactured by J. P. Sauer & Sohn GmbH of Eckernförde, West Germany.

HECKLER & KOCH P7

Heckler & Koch's P7 has the most radically new design of the new German service pistols. First, it has a gas-retarded blowback operating mechanism. This operating principle was used in the World War II German *Volksturmgewehr* (VG1-5) and in an experimental Swiss self-loading pistol developed during that same era by the *Eidgenossische Waffenfabrik*, Bern. Second, cocking of the P7 is accomplished by squeezing the cocking lever, which runs the full length of the forward portion of the pistol grip.

Although it is much more common to find gas pressure being used to unlock and open the action of a firearm, in the P7 Heckler & Koch engineers have employed a gas-actuated piston to keep the breech shut until the pressure in the chamber has dropped to a safe level. When a loaded cartridge is seated in the P7's chamber, the gas-piston port is directly beneath the mouth of the case. When the cartridge is fired, a portion of the gases generated are diverted into the cylinder below the barrel. The gas pressure acts upon the piston, which is attached to the front of the slide, retarding the rearward motion of the slide until the chamber pressure drops to a safe level. Afterwards, the P7 behaves like a blowback selfloader. Utilizing gas pressure to retard the movement of the slide eliminates the need for a locking mechanism or an excessively heavy slide and powerful springs. Thus a lightweight pistol was possible—34.57 ounces (.98 kilogram) with a steel frame.

Cocking the P7 is a simple process, but it requires some thought. After that it becomes second nature. The P7 is cocked by squeezing the grip to compress the cocking lever located beneath the trigger-guard and hinged at the bottom of the forestrap. Cocking requires about 15 pounds (67 newtons) of pressure, which is not difficult because three fingers of the shooting hand grasp the lever. After the pistol has been cocked, only 1.25 pounds (6 newtons) of force are required to keep it cocked. The P7 striker-type firing pin protrudes from the rear of the slide and thus acts as a cocking indicator. When the P7 is fired, the mechanism is recocked automatically as long as the cocking lever is held in the rear position. When that lever is released, the firing pin is lowered to the uncocked position. The P7 may, of course, be carried safely with a loaded chamber. The German police P7 has the magazine release mounted at the base of the Pistol grip.

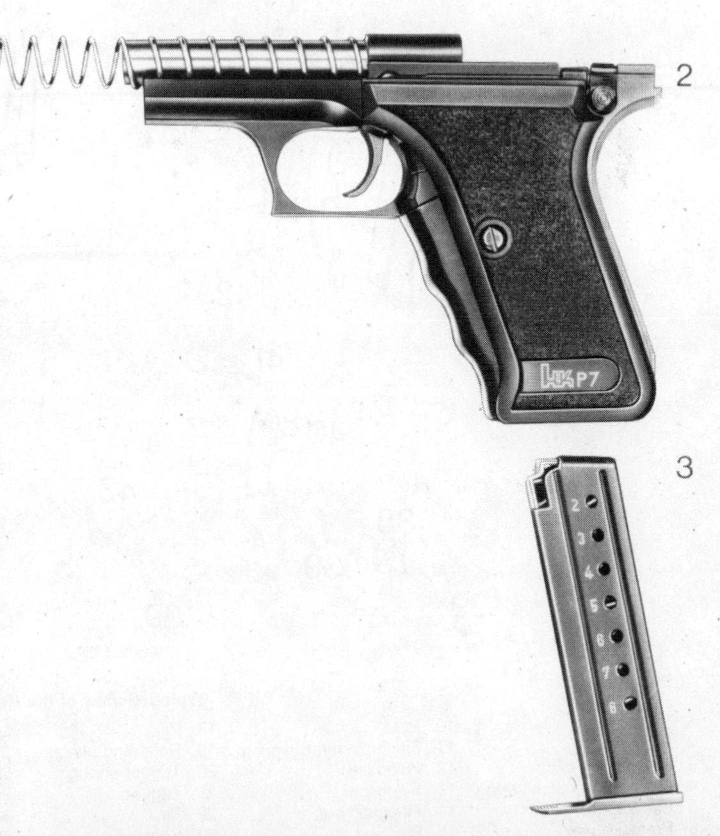

Disassembled view of the P7. The slide is number 1, the frame is number 2, and the magazine is number 3.

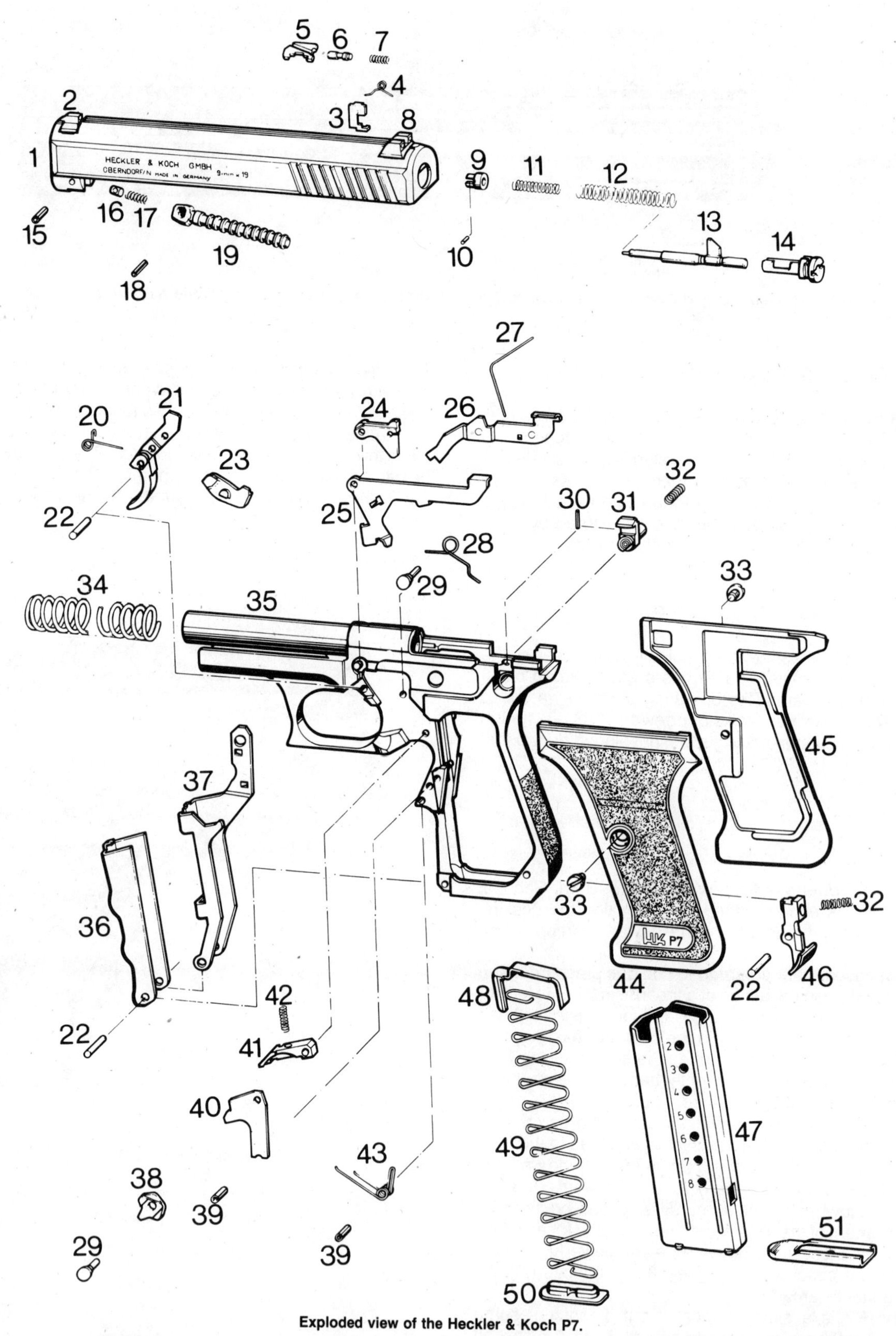

Exploded view of the Heckler & Koch P7.

1. Slide.
2. Front sight.
3. Drop safety catch.
4. Drop safety catch spring.
5. Extractor.
6. Pressure pin.
7. Extractor spring.
8. Rear sight.
9. Spacer ring.
10. Pin.
11. Firing pin return spring.
12. Firing pin.
13. Firing pin.
14. Firing pin nut.
15. Pin.
16. Piston latch.
17. Spring.
18. Pin.
19. Retarding piston.
20. Trigger spring.
21. Trigger.
22. Pin.
23. Sear
24. Ignition block lever.
25. Slide catch lever.
26. Drop safety catch.
27. Spring.
28. Spring.
29. Stop pin.
30. Pin.
31. Takedown catch.
32. Spring.
33. Grip screws.
34. Recoil spring.
35. Frame assembly.
36. Squeeze cocker.
37. Drag lever.
38. Rocker.
39. Roll pins.
40. Cover plate.
41. Pawl.
42. Pawl spring.
43. Spring.
44–45. Grip plates.
46. Magazine catch.
47. Magazine body.
48. Magazine follower.
49. Magazine spring.
50–51. Magazine base assembly.

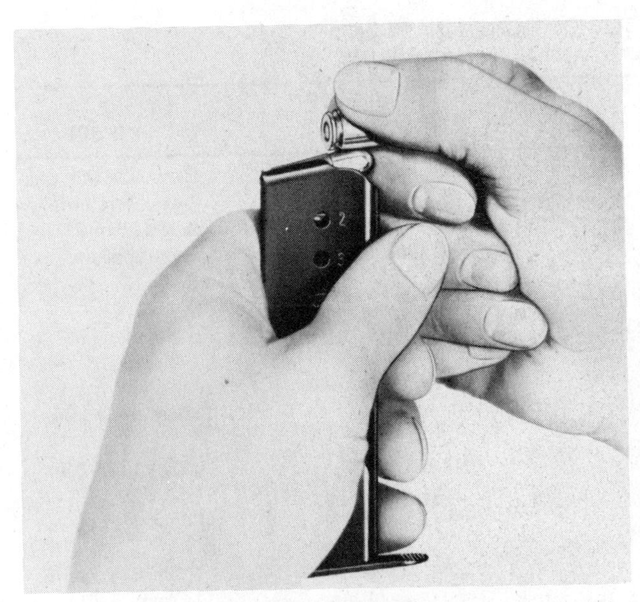

Loading the P7 magazine.

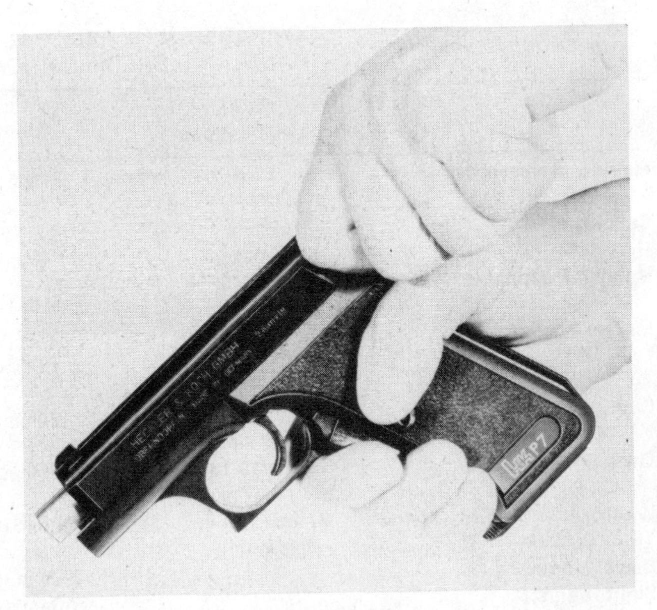

Loading the P7.

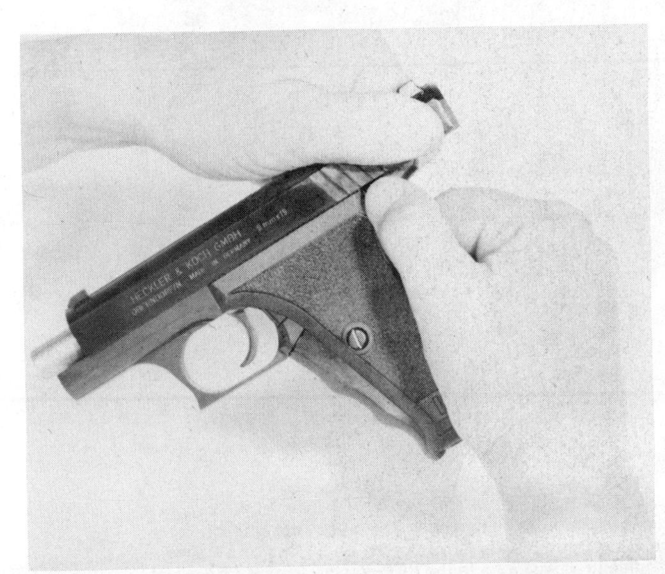

Disassembling the P7. Press the takedown button while pulling the slide to the rear.

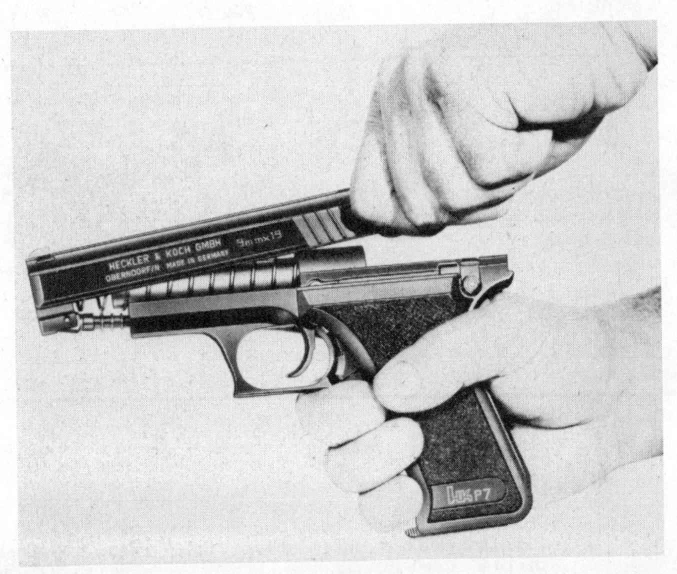

Disassembling the P7. Lift the slide, and then move it forward off the fixed barrel.

Removal and installation of the P7 firing pin. Depress the squeeze cocker until the firing pin is even with the rear of the slide. Press the firing pin nut approximately .02 inch (.5mm) forward with special screwdriver and rotate 90 degrees to the right. Depress squeeze cocker completely and remove the firing pin.

COMPARISON OF FEDERAL REPUBLIC OF GERMANY POLICE HANDGUNS SPECIFICATIONS AND CANDIDATE HANDGUNS

	FRG Specification	P5 (Walther)	P6 (P225)	P7 (PSP)
Method of operation:		Falling block (modified P38)	Browning type	Blowback with gas-piston retardation
Caliber:	9 × 19mm Parabellum	9 × 19mm Parabellum	9 × 19mm Parabellum	9 × 19mm Parabellum
Magazine capacity:	8-shot minimum	8	8	8
Muzzle energy:	500 joules minimum (369 ft-lbs)	ca. 500 joules (369 ft-lbs)	ca. 500 joules (369 ft-lbs)	518 joules (382 ft-lbs)
Overall length:	180mm (7 in) maximum	180mm (7 in)	180mm (7 in)	166m (6.54 in)
Overall height:	130mm (5.13 in) maximum	129mm (5.08 in)	131mm (5.16 in)	125mm (4.92 in)
Width:	34mm (1.34 in) maximum	32mm (1.26 in)	34mm (1.34 in)	28mm (1.1 in)
Barrel length:		90mm (3.54 in)	97.6mm (3.84 in)	105mm (4.13 in)
Sight radius:		134mm (5.28 in)	145mm (5.71 in)	147mm (5.79 in)
Frame material:		Aluminum	Aluminum	Steel
Weight:	1 kg (35.27 oz) maximum	.795 kg (28.04 oz)	.82 kg (28.92 oz)	.98 kg (34.57 oz)

The German Customs Police (*Zollpolizei*) have carried the 7.65 × 17mm HK4 pistol for many years. The official designation is the P11. Illustrated is an HK4/P11 made by the Manufacture Nationale d'Armes de Saint Etienne (MAS) for the Berlin Customs Police—the peace treaty that partitioned Berlin forbade the use of German manufactured weapons in the divided city. The HK4 traces its design lineage to the World War II Mauser HSc.

HECKLER AND KOCH 9mm PISTOL MODEL P9 AND P9S

Heckler and Koch has developed a new self-loading pistol chambered for the 9mm Parabellum cartridge. The P9 uses the same type roller locking system as the G3 rifle series. This pistol makes extensive use of stampings in its manufacture. It has a pin type indicator, which shows that the pistol is cocked. This pin protrudes from the rear of the receiver. The loaded chamber indicator is mounted above and to the rear of the chamber somewhat like that of the Luger. The P9 has a cocking lever mounted on the left side of the receiver. The safety is mounted on the left side of the slide; it operates by blocking the firing pin. The pistol can be cocked or unloaded with the safety in the "on" position.

The only difference between the P9 and P9S is that the former has a single action trigger, and the P9S has a double action trigger. An 11.43 × 23mm (.45 ACP) version is also available.

Heckler and Koch 9mm Parabellum Model P9 Pistol.

CHARACTERISTICS OF THE P9

Caliber: 9mm Parabellum
System of operation: delayed blowback
Weight: 1.93 lbs.
Length overall: 7.5 in.
Barrel length: 4 in.
Feed device: 9-round in-line detachable box magazine
Sights: Front: blade
Rear: square notch
Muzzle velocity: 1180 f.p.s.

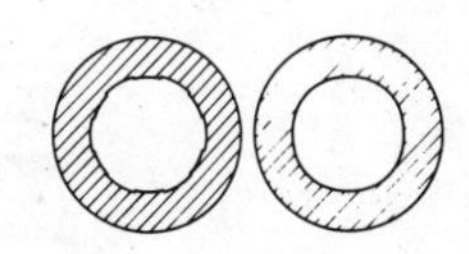

Polygon barrel
As opposed to the grooves and lands of conventional barrels, the bore of the polygon barrel has a highly rounded, rectangular profile. This results in better seating of the bullet in the bore and reduced gas cutting, thereby providing higher penetration power, extremely long service life, and is easy to clean.

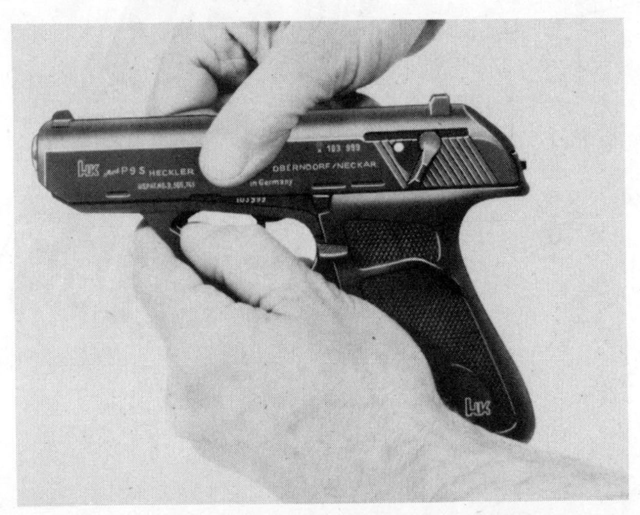

After removing the P9S magazine, depress the barrel latch and then push the slide forward until it stops. Then lift it up and off.

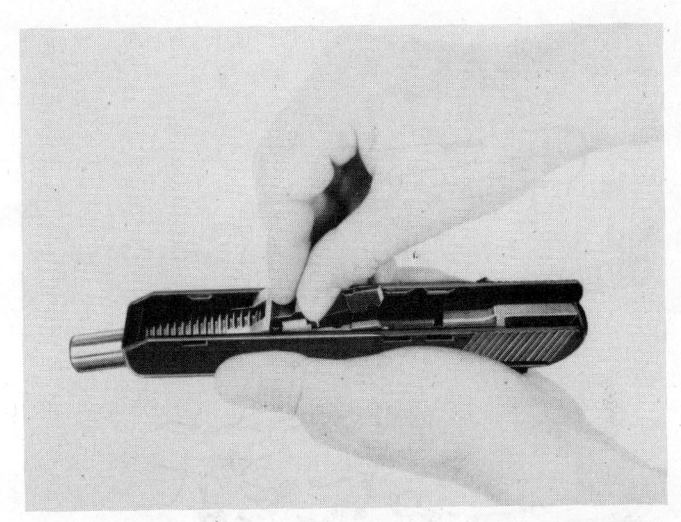

Remove the barrel by pushing it forward until it can be lifted out of the slide.

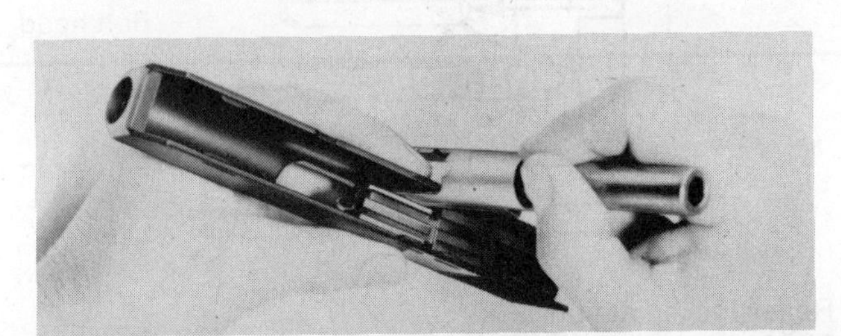

To remove the bolt head, use one of the shanks of the barrel extension to press down on the locking catch between the bolthead and the slide. The bolthead will spring forward.

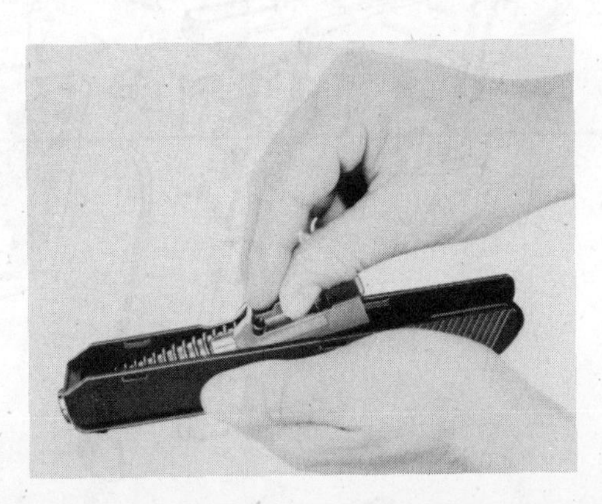

To replace the barrel, push the barrel and recoil spring (large diameter end of spring to the front) through the hole in the slide. Then position the barrel so that the locking rollers engage the barrel extension.

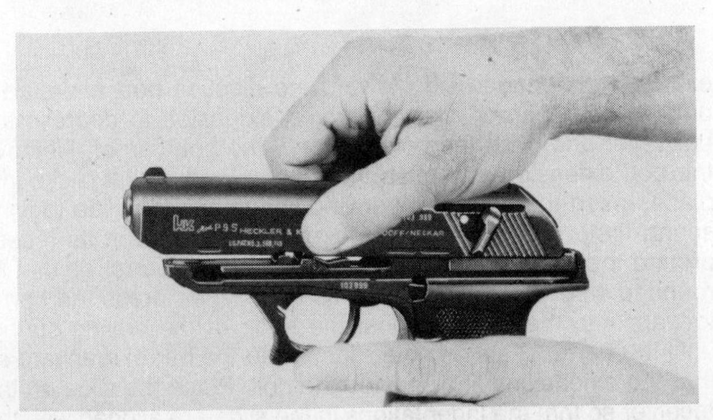

To replace the slide, match the slide so that the guide lugs engage the corresponding notches in the frame. Press down on the slide and draw it to the rear and then release. Decock the pistol and insert the magazine.

Note the indicator that shows that the P9S is loaded. Note also that the safety is set on fire.

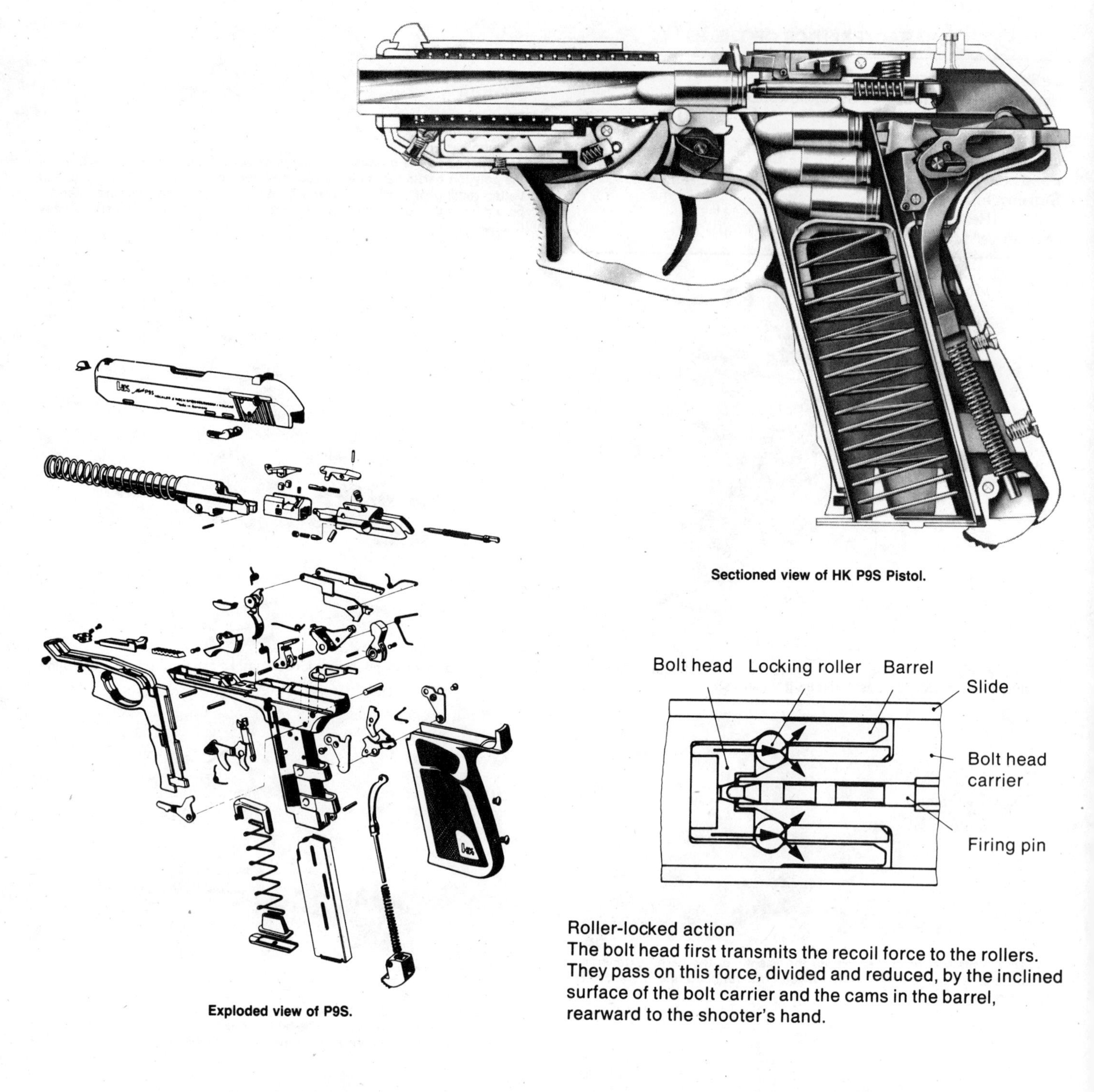

Sectioned view of HK P9S Pistol.

Exploded view of P9S.

Roller-locked action
The bolt head first transmits the recoil force to the rollers. They pass on this force, divided and reduced, by the inclined surface of the bolt carrier and the cams in the barrel, rearward to the shooter's hand.

Disassembly and Assembly

Clear the pistol, but do not insert the magazine. Press the takedown lever into the trigger guard and push the slide forward; then lift the slide up and off the receiver. Press the barrel forward in the slide against the force of the driving spring until the rear of the barrel can be lifted up and out of the slide. Remove the driving spring. Normally no further disassembly is required, but if desired, the bolt can be removed from the slide. To do so, insert one of the barrel extensions into the slide, just forward of the bolt carrier, and press the barrel down to release the bolt lever. Press the bolt forward, off the bolt carrier. No further disassembly should be attempted.

To reassemble the pistol, place the bolt into the slide with its extractor or rounded side toward the ejection port and start it onto the bolt carrier. Use the barrel extension to depress the bolt lever and push the bolt fully onto the bolt carrier. Remove the bolt extension and push the bolt forward until it clicks into place. Insert the barrel with driving spring into the slide (driving spring first), and insure that the spring is seated in its recess around the inner front of the slide. Rotate the barrel so that its rounded side is toward the ejection port; then press the barrel forward into the slide, against the force of the driving spring, until the barrel seats into the slide. Ease the barrel rearward so that the extensions fit alongside the bolt. Place the slide on the receiver so that the indentations in the slide are aligned with the cutaways in the receiver; then pull the slide rearward and release it. The pistol is ready to be loaded like any other automatic pistol.

WEST GERMAN RIFLES

The West Germans were initially equipped with US M1 rifles. They then purchased a quantity of FN 7.62mm NATO FAL rifles, which they called Rifle G1.

Beginning in 1959, the *Bundeswehr* began replacing the G1 with the G3, which had evolved through the Spanish CETME Rifle from the StG 45(M), as described in Chapter 1. Rheinmettal Wehrtechnik of Dusseldorf and Heckler & Koch have produced the G3 for the West German Government. In addition, Heckler & Koch has produced a number of variants of the G3 in the NATO, 7.62 × 39mm and 5.56 × 45mm calibers for the world market.

WEST GERMAN RIFLE 7.62mm G3
CHARACTERISTICS OF THE G3

Caliber: 7.62mm NATO.
System of operation: Delayed blowback, selective fire.
Weight (loaded w/o bipod): 9.9 lbs.
Length, overall: 40.2 in.
Barrel length: 17.7 in.
Feed device: 20-round, detachable, staggered-row box magazine.
Sights: Front: Hooded post.
Rear: Rotary aperture.
Muzzle velocity: 2624 f.p.s.
Cyclic rate: 500–600 r.p.m.

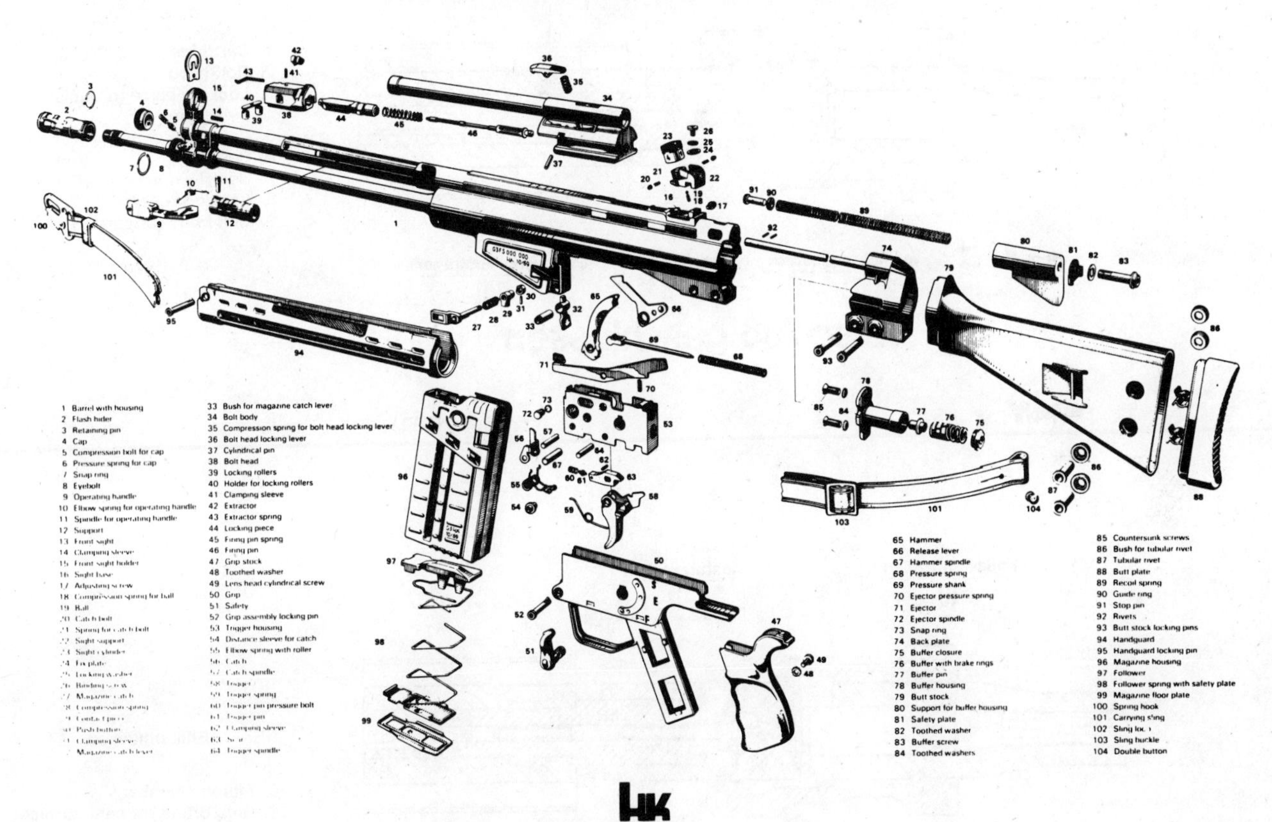

How the G3 Works

The G3 Rifle is a delayed blowback weapon. The rearward thrust of the cartridge case, upon firing, drives the bolt mechanism to the rear, but the rearward movement is delayed by the mechanical arrangement of the two-piece bolt until the chamber pressure has dropped to a safe limit. Rearward movement of the cartridge case is facilitated by flutes cut into the chamber. These flutes allow propellant gases to leak rearward along the cartridge case, providing a film of gas upon which the mouth of the cartridge floats. Cases fired in the G3 (and other weapons such as the Tokarev rifles) are readily identified by the sharply defined gas marks that extend along the neck for about half of the length of the case. When the G3 is ready to fire, a cartridge is chambered; the locking piece in the bolt cams the locking rollers into their recesses, and the hammer is released and strikes the firing pin, which fires the cartridge. The gas pressure drives the cartridge case rearward, and this movement is resisted by the bolt, whose locking rollers are seated in the locking piece attached to the barrel.

Automatisches Gewehr G 3

Kal. 7,62 mm x 51

Vorgang im Verschluß

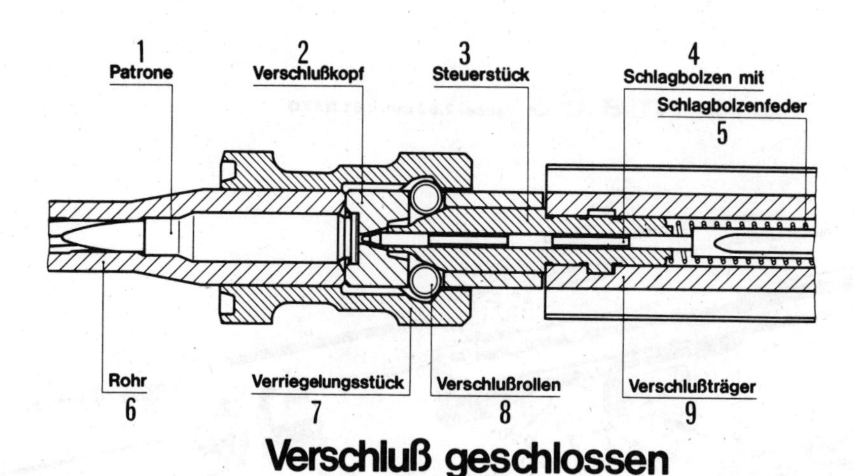

Verschluß geschlossen

Rifle loaded and locked

1. Cartridge
2. Bolt-head
3. Locking piece (or locking cam)
4. Firing pin
5. Firing pin spring
6. Barrel
7. Barrel extension
8. Locking rollers
9. Bolt body (or carrier)

Entlastungsrillen
10

Abstützflächen
11

Rifle unlocked

10. Fluted chamber
11. Supporting (or cam surface)

Verschluß geöffnet

HK

HECKLER & KOCH GMBH OBERNDORF-NECKAR · GERMANY

Operating mechanism of G3 Rifle.

The rearward thrust of the case is sufficient to start the bolt to the rear; this causes the rollers to be forced out of their seats. As they move, the rollers ride on the cam surface of the locking cam (locking piece) and force it rearward; because the cam is locked to the heavy bolt carriers, the bolt carrier is also forced to the rear against the driving spring. The delay, occasioned by the rollers resisting camming the heavy bolt to the rear, allows time for the bullet to leave the muzzle and the pressure to drop to a safe level. Inertia developed and the residual pressure still thrusting the bolt rearward provide sufficient energy to drive the bolt fully to the rear, to compress the driving spring and to cock the hammer.

The extractor pulls the fired case from the chamber and holds it to the bolt face. The rear end of the ejector is struck by the recoiling bolt carrier; this causes the ejector to pivot so that its front end enters the bottom of the bolt head. As the bolt continues to recoil, the cartridge case strikes the ejector, pivots about the extractor and is expelled from the rifle.

The bolt carrier strikes the buffer at the front of the butt stock and stops. The driving spring then drives the carrier forward, and the bolt head forces the top cartridge out of the magazine and into the chamber. The extractor snaps into the cartridge groove, and the forward movement of the bolt head stops when it hits the end of the barrel. The carrier has a lock to keep the bolt head locked forward; this is now tripped by a lug in the receiver. The carrier continues forward, and the locking piece, which travels with the carrier, forces the locking rollers into their recesses.

The rifle's trigger mechanism is similar to that of the FN FAL. The hammer, powered by a coil spring and plunger, is held cocked by a sear that can move back and forth in relation to the trigger. When the hammer is cocked, pressure on the trigger is transmitted to the rear end of the sear through the trigger lever. This causes the front of the sear to move down and release the hammer, which under pressure of its spring, swings forward and strikes the firing pin. Upon firing, the recoiling bolt carrier rocks the hammer back into the cocked position.

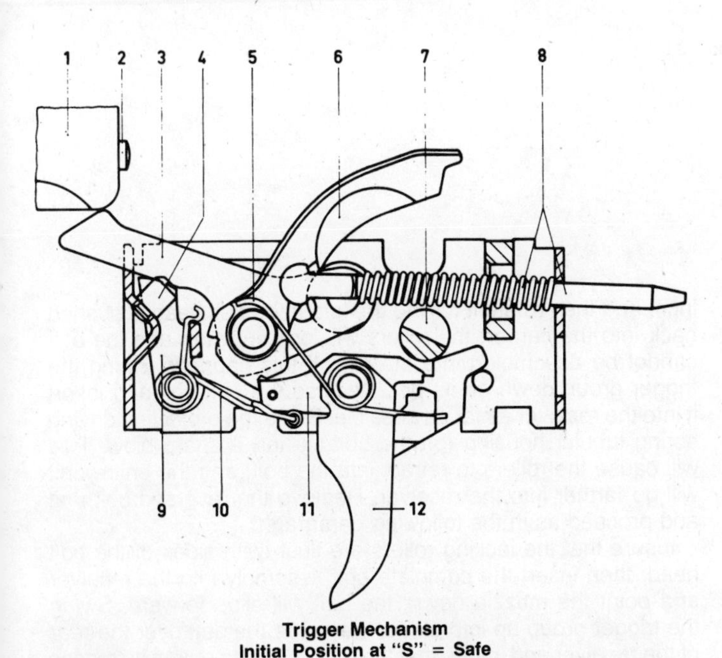

Trigger Mechanism
Initial Position at "S" = Safe

1. **Bolt head carrier**
2. **Firing pin**
3. **Release lever**
4. **Anvil for hammer**
5. **Trigger spring**
6. **Hammer**
7. **Safety pin**
8. **Pressure shank with compression spring for hammer**
9. **Catch**
10. **Elbow spring with roller**
11. **Sear**
12. **Trigger**

When the selector is set for semiautomatic fire—"E"—a cutaway section on the selector shaft limits the upward movement of the trigger lug. When the trigger is pressed and the hammer is released, the sear spring forces the front of the sear forward and upward. As the bolt carrier returns forward, the hammer starts to move forward, and the sear mates with a notch on the hammer. When the strong hammer spring overcomes the weaker sear spring, the sear is forced to the rear against the trigger lever to hold the hammer cocked. By releasing the trigger, the lever is lowered, and the hammer spring, working through the hammer, forces the sear rearward over the lever. Pressure on the trigger will now move the sear and fire another shot.

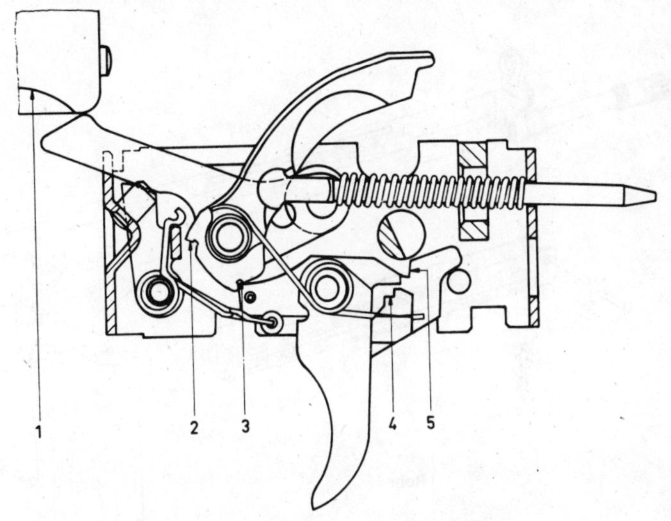

Selective fire lever at "E" = single fire

1. **Cam surface**
2. **Notch for burst**
3. **Notch for single fire**
4. **Notch for the sear**
5. **Recess for single fire**

When the selector is set for automatic fire—"F"—the trigger can rise to its highest point, and the nose of the sear is depressed far enough so that the sear cannot reengage the hammer. The automatic sear holds the hammer cocked, and as the bolt carrier completes its forward travel, it depresses the automatic sear lever; this in turn moves the automatic sear out of engagement with the hammer. The hammer swings forward to fire, and this cycle is repeated until the trigger is released. The sear can then rise, intercept the hammer and interrupt the firing cycle.

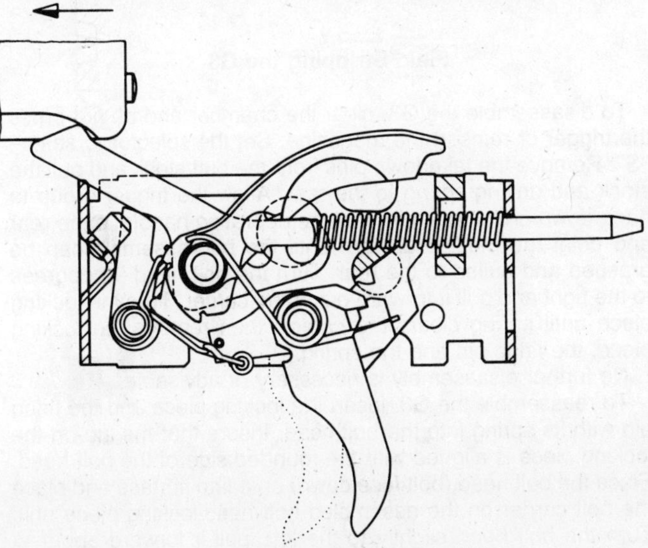

Function in Position "F" = Burst

When set at safe—"S"—the selector places a solid section of its shaft over the trigger lever. This prevents the trigger from moving enough to disengage the sear from the hammer and keeps the rifle from firing.

There are two automatic safeties incorporated into the design of the G3—the automatic sear and the locking piece. If the bolt mechanism is not fully forward, the automatic sear will continue to hold the hammer and prevent firing. The locking piece, unless it is fully forward and holding the locking rollers fully outward, will prevent the firing pin from protruding through the bolt face; thus, the weapon cannot fire unless the bolt is fully locked.

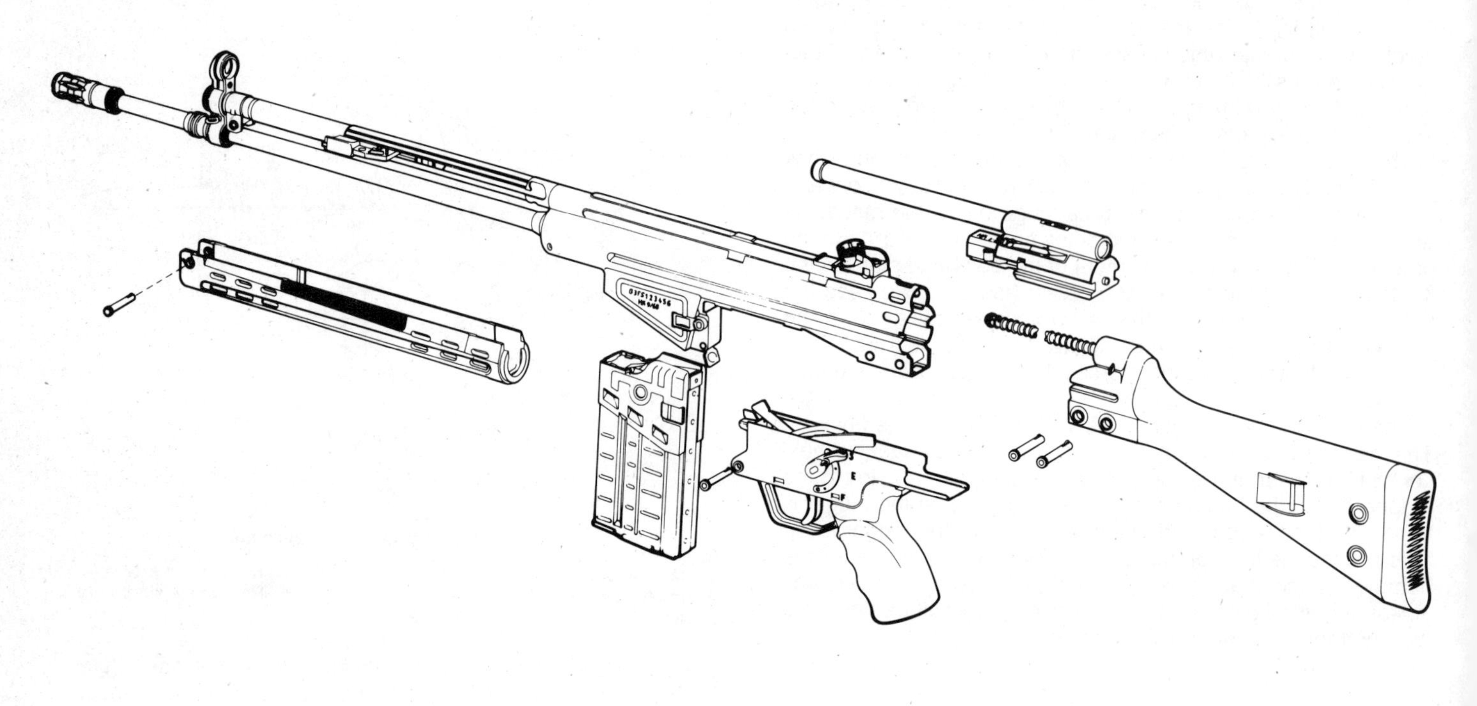

Field stripped G3 Rifle.

Field Stripping the G3

To disassemble the G3, clear the chamber and do not press the trigger or reinsert the magazine. Set the selector to safe—"S". Remove the takedown pins from the butt stock and pull the stock and driving spring to the rear. Allow the trigger group to hang down on its front pin. Pull the operating handle to the rear and point the muzzle upward until the bolt assembly can be grasped and pulled to the rear. Turn the bolt head 90 degrees to the right and pull it forward out of the carrier. Turn the locking piece until its lug clears the carrier; then remove the locking piece, the firing pin and the spring.

No further disassembly is necessary or advisable.

To reassemble the G3, insert the locking piece and the firing pin with its spring into the bolt head. Insure that the lug on the locking piece is aligned with the rounded side of the bolt head. Place the bolt head (bolt face down) on a firm surface and place the bolt carrier on the assembled bolt head/locking piece unit. Turn the bolt head slightly to the left; pull it forward about ¼ inch, and then rotate it fully to the left. If the bolt head is pushed back into the carrier, the rollers will lock outward, and the bolt cannot be assembled into the rifle. If this happens, swing the trigger group down on its pin. Reverse the bolt unit and insert it into the receiver as far as possible. Strike the projecting driving spring tubular housing (on the bolt carrier) a sharp blow. This will cause the rollers to retract into the bolt, and the entire unit will go farther into the receiver. Remove the reversed bolt unit and proceed as in the following paragraph.

Insure that the locking rollers are flush with sides of the bolt head; then insert the complete bolt assembly into the receiver and point the muzzle down; the bolt will slide forward. Swing the trigger group up into place; then slide the butt over the rear of the receiver and insure that the driving spring enters its recess in the bolt carrier. Remove the takedown pins from their storage holes and replace them in the receiver. Insert the magazine, rotate the selector off safe and press the trigger. Replace the magazine.

Barrel with Receiver, Cocking Lever Mechanism and Sights

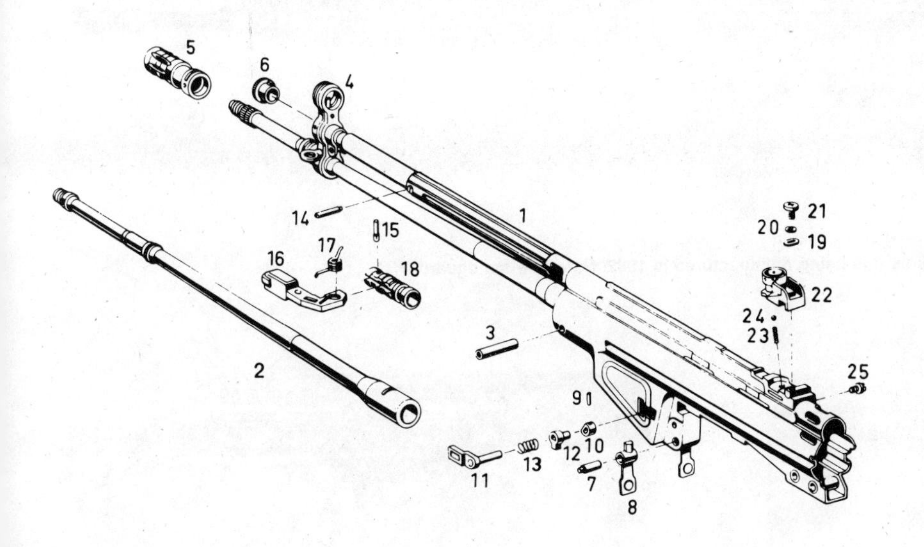

1. Barrel with receiver, cocking lever housing, front sight holder and magazine release lever
2. Barrel
3. Barrel fixing pin
4. Front sight holder
5. Flash suppressor
6. Cap
7. Bush for magazine release lever
8. Magazine release lever
9. Retaining pin
10. Contact button for magazine catch
11. Magazine catch
12. Contact piece for magazine catch
13. Contact spring for magazine catch
14. Pin for stop abutment
15. Axis pin for cocking lever
16. Cocking lever
17. Cocking lever elbow spring
18. Cocking lever support
19. Fix plate for rear sight support
20. Lock washer
21. Binding screw
22. Rotary rear sight
23. Spring for ball catch
24. Ball
25. Adjusting screw

Bolt Assembly

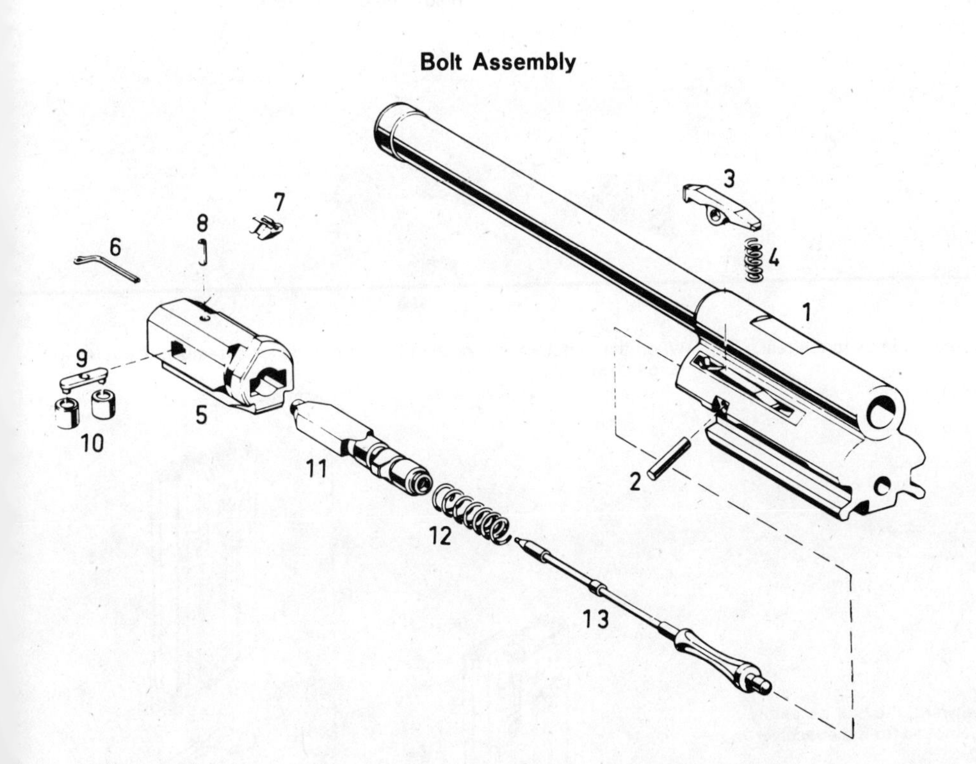

1. Bolt head carrier
2. Cylindrical pin
3. Locking lever
4. Compression spring
5. Bolt head
6. Extractor spring
7. Extractor
8. Cylindrical pin
9. Holder for locking rollers
10. Locking rollers
11. Locking piece
12. Firing pin spring
13. Firing pin

The 7.62mm NATO G3 rifle as originally made for West Germany has a flip over type sight and wooden butt. The G3A1 has a folding type stock and flip over sight. The G3A2 has a rotating type rear sight; the current weapon being manufactured is the G3A3. It has a rotating type rear sight, a modified front sight guard and a prong type flash suppressor. G3A4 is similar, but has a retractable type stock, and G3AZF is the G3A3 with a scope mounted for sniping.

VARIATIONS OF THE G3 RIFLE

Heckler and Koch has produced assault rifle versions of the G3 rifle in caliber .223 (model HK33 and HK33K) 7.62X39 Soviet M1943 cartridge (Model HK32 and 32K).

Semiautomatic versions of the G3 are sold as the HK 91A2 (standard stock) and HK 91A3 (retractable stock); the 5.56mm HK33 is marketed as the HK 93A2 and HK 93A3.

The G3A3 as it is being manufactured in 1982. Note new handguard.

Inserting G3 magazine.

Removing G3 magazine.

After removing the takedown pins, pull the G3 stock sharply to the rear.

When the buttstock clears the tang on the trigger assembly, the latter will hinge downward.

When the trigger assembly has been hinged downward, the bolt assembly will slide to the rear where it can be grasped and removed from the receiver.

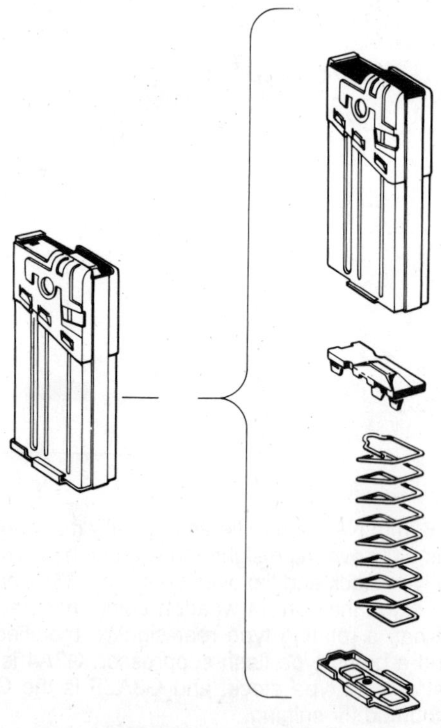

The G3 magazine assembly.

CHARACTERISTICS OF FOUR G3 VARIATIONS

	HK33	HK33K	HK32	HK32K
Caliber:	.233 (5.56 × 45mm)	.233 (5.56 × 45mm)	7.62 × 39mm.	7.62 × 39mm.
Barrel length:	15.35 in.	12.4 in.	15.35 in.	13.6 in.
Length overall:	36.1 in.	Stock retracted, 24.4 in. Stock extended, 32.7 in.	36.1 in.	Stock retracted, 25.6 in. Stock extended, 33.9 in.
Weight:	6.6 lb.	6.6 lb.	6.6 lb.	6.6 lb.
Magazine capacity:	20 rounds	40 rounds.	30 rounds	30 rounds.
Cyclic rate:	600 r.p.m.	600 r.p.m.	600 r.p.m.	600 r.p.m.
Muzzle velocity:	3182 f.p.s.	3018 f.p.s.	2560 f.p.s.	2493 f.p.s.

All the above rifles operate the same as the G3.

West German G3A3 Rifle.

Back Plate with Rigid Butt Stock

1. Butt stock
2. Butt plate
3. Buffer screw
4. Lock washer
5. Safety plate
6. Tubular rivets
7. Bush for tubular rivet
8. Back plate with recoil spring guide tube
9. Recoil spring
10. Guide ring
11. Stop pin
12. Rivets
13. Buffer assembly
14. Locking screw
15. Lock washer
16. Locking pins

Back Plate with Retractable Butt Stock.

1. Butt stock
2. Back plate
3. Recoil spring
4. Guide ring
5. Stop pin
6. Rivets
7. Buffer bolt
8. Buffer spring
9. Buffer screw
10. Spring loaded latch
11. Cover
12. Gripping lever
13. Locking ring with spring
14. Spring ring
15. Locking pins

Heckler and Koch 5.56mm HK 33KA1 Rifle.

HK33 5.56 × 45mm rifle

Heckler & Koch G41 5.56 × 45mm (Updated HK33)

As noted in Chapter 1, the G41 was introduced by Heckler & Koch in 1981 to meet the military requirements and lessons derived from the late 1970s NATO small caliber trials. This weapon has a longer barrel (18.9 in.) than the HK33 (15.35 in.) from which it is derived. New features include a positive bolt closing assist, magazine bolt hold-open device, dust cover on ejection port, use of M16A1-type magazines, single shot, 3-shot and fully automatic fire capability, use of M16A1 bipod, carrying handle, new telescopic sight mounts (STNAG 2324), and provision for use of G3 bayonet.

G41 CHARACTERISTICS

Caliber: 5.56 × 45mm NATO (SS109 with 1 in 7 inch twist)
System of operation: Roller lock, delayed blowback
Overall length: 39.25 in.
Barrel length: 18.9 in.
Weight: w/o magazine 7.94 lbs.
Magazine capacity: 20- or 30-shot M16A1-type magazines.
Cyclic rate: approx. 850 r.p.m.
Muzzle velocity: approx. 3250 f.p.s.

HECKLER & KOCH G11 4.7 × 21mm CASELESS WEAPON

This new rifle is described in Chapter 1. It is still a development-type rifle.

***Präzisionsschutzengewehr 1* (PSG1)**

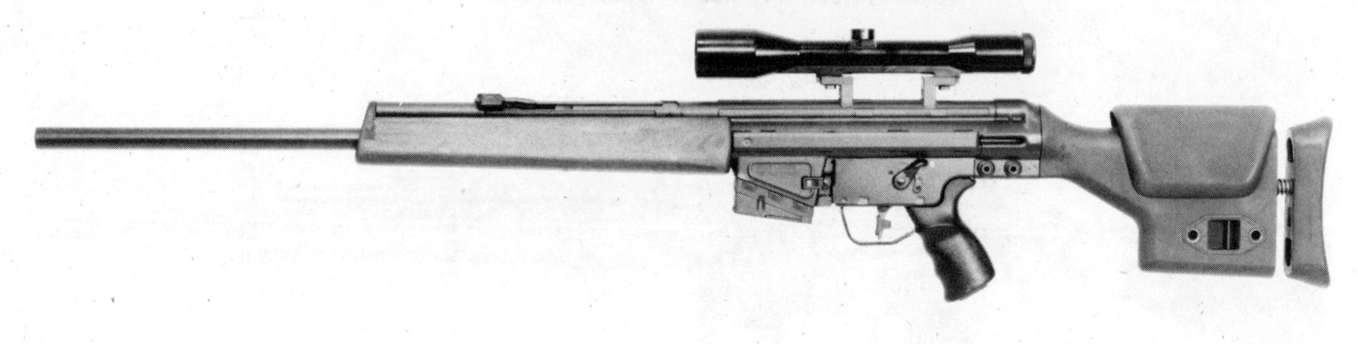

Two versions of the PSG1. *Top,* an early prototype; *bottom,* 1982 version of the PSG1.

PSG 1 grouping, original size

Range . . . . 300 m
Ammunition . . . . Lapua .308 Win. Match
Number of shots 10

The PSG 1 high-precision marksman's rifle achieves the same accuracy as that guaranteed by leading ammunition manufacturers in their acceptance conditions for the rounds concerned.

PRÄZISIONSSCHUTZENGEWEHR PSG1 (PRECISION SHOOTING RIFLE)

As described in Chapter 5, Heckler & Koch has developed this extremely accurate rifle from their G3. This rifle will probably see use by special counter-terrorist units such as the GSG9. Heckler & Koch claims that this weapon will shoot as accurately as the inherent accuracy of the ammunition.

CHARACTERISTICS OF THE PSG1

Caliber: 7.62 × 51mm NATO.
System of Operation: Roller locked, delayed blowback.
Overall length: 47.56 in.
Barrel length: 25.59 in.
Weight w/o magazine or tripod: 15.87 lbs.
Weight of tripod: 2.26 lbs.
Magazine capacity: 5- and 20-shot.
Sighting system: Telescopic sight 6 × 42, with reticle illumination. 6 settings 100 to 600 meters.
Type of fire: Semiautomatic only.

The special Swiss tripod used with the PSG1.

WEST GERMAN SUBMACHINE GUNS

The standard submachine gun of the West German Army is the Israeli-designed UZI, but a number of native West German designs have appeared since the late fifties. Mauser, Walther, Heckler & Koch, Anschütz and Erma have all prepared designs. These weapons have mainly appeared in prototype form and are usually made of stampings and are of short length.

9mm Parabellum Walther MPL submachine gun with stock extended.

DUX 53

The DUX 53 is a weapon designed at the Oviedo Arsenal in Spain by W. Daugs and L. Vorgrimmler, based on the design of the Finnish 9mm Model 44 submachine gun, which in turn was based on the Soviet PPS M1943. A quantity of these weapons was manufactured for the West German Border Police. This weapon and various modifications made by Mauser and Anschütz were tested by the West German Army during the mid-fifties.

THE WALTHER SUBMACHINE GUN

The Walther submachine gun was introduced in 1963. There are two basic versions of this gun: the MPL (long model), and the MPK, (short model). The Walther is made mainly of steel stampings, with a folding steel stock that can be folded to either side. The bolt is guided through the receiver by a guide rod mounted in the receiver above the bolt face.

The safety lever on the Walther gun is fitted on both sides of the receiver.

CHARACTERISTICS OF THE WALTHER SUBMACHINE GUN

	MPL	MPK
Caliber:	9mm Parabellum	
System of operation:	Blowback, full automatic only*	
Length overall: w/stock extended:	29.42 in.	25.96 in.
w/stock folded:	18.1 in.	14.75 in.
Barrel length:	10.25 in.	6.75 in.
Weight:	6.62 lbs.	6.27 lbs.
Feed device:	32-round, staggered row, detachable box magazine.	
Sights: Front:	Protected blade	
Rear:	Flip over, notch for 75m and aperture for 125m.	
Muzzle velocity:	Approx. 1370 f.p.s.	Approx. 1250 f.p.s.
Cyclic rate:	550 r.p.m.	

*Can be made selective fire in special order.

THE HECKLER AND KOCH HK54 (MP5) SUBMACHINE GUN

The HK54 is the submachine gun version of the G3 rifle. It is as the G3 rifle, a delayed blowback operated weapon that fires from a closed bolt. There is a theory that delayed blowback submachine guns have less vibration and rise than blowback operated submachine guns. On the other hand, they are more complex and usually more expensive.

A finer degree of accuracy can be obtained with a gun that fires from a closed bolt, since the only disturbing influence on "hold" is the forward movement of a light hammer and/or firing pin as opposed to the forward movement of a heavy bolt. The "lock time" (the period from trigger/sear release to ignition of primer) is also less on a weapon that fires from a closed bolt than on a weapon that fires from an open bolt; the other side of the coin in this case, however, is the "cook-off" problem. Automatic fire heats up a weapon rather rapidly, and a point is reached when the temperature of the chamber will cause cartridges to function spontaneously. Some designs have solved this problem by firing from an open bolt in automatic fire and from a closed bolt in semiautomatic fire, as is done by the 7.92mm FG42 German World War II paratroop rifle and the US Johnson.

CHARACTERISTICS OF THE HK54 SUBMACHINE GUN

Caliber: 9mm Parabellum.
System of operation: Delayed blowback, selective fire.
Length overall: w/fixed stock, 26 in.
w/retractable stock fixed, 26 in.
w/retractable stock retracted, 19.3 in.
Barrel length: 8.85 in.
Weight 5.5 lbs.
Feed device: 30-round, detachable box magazine.
Sights: Front: Post.
Rear: Flip-type w/"U" notch.
Muzzle velocity: Approx. 1312 f.p.s.
Cyclic rate: 600 r.p.m.

The West German Border Police have adopted the HK54, which with fixed stock is called the MP5 and with retractable metal stock is called the MP5A1. A later version of this weapon with modified front sight, rotating rear sight and modified front section of barrel is called the MP5A2 with fixed stock and MP5A3 with retractable metal stock.

HK MP5A2 (top) and MP5A3 (bottom) as used by the West German Border Guards.

Maschinenpistole 9 mm × 19
·MP5A2·

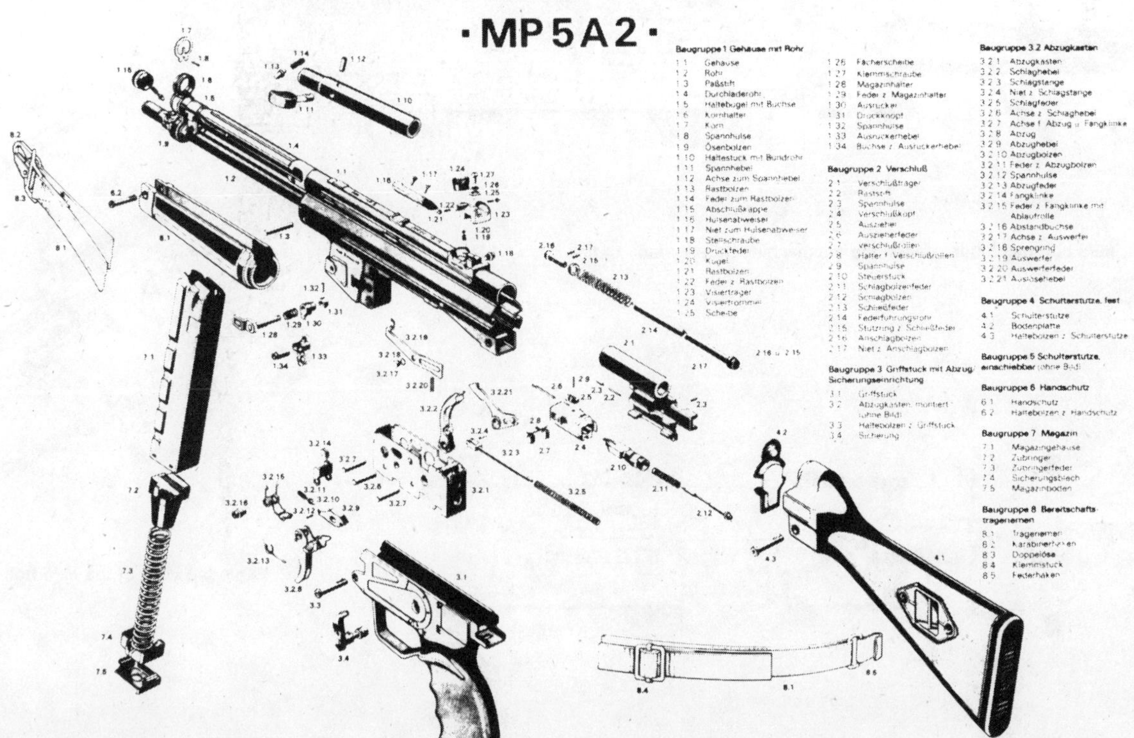

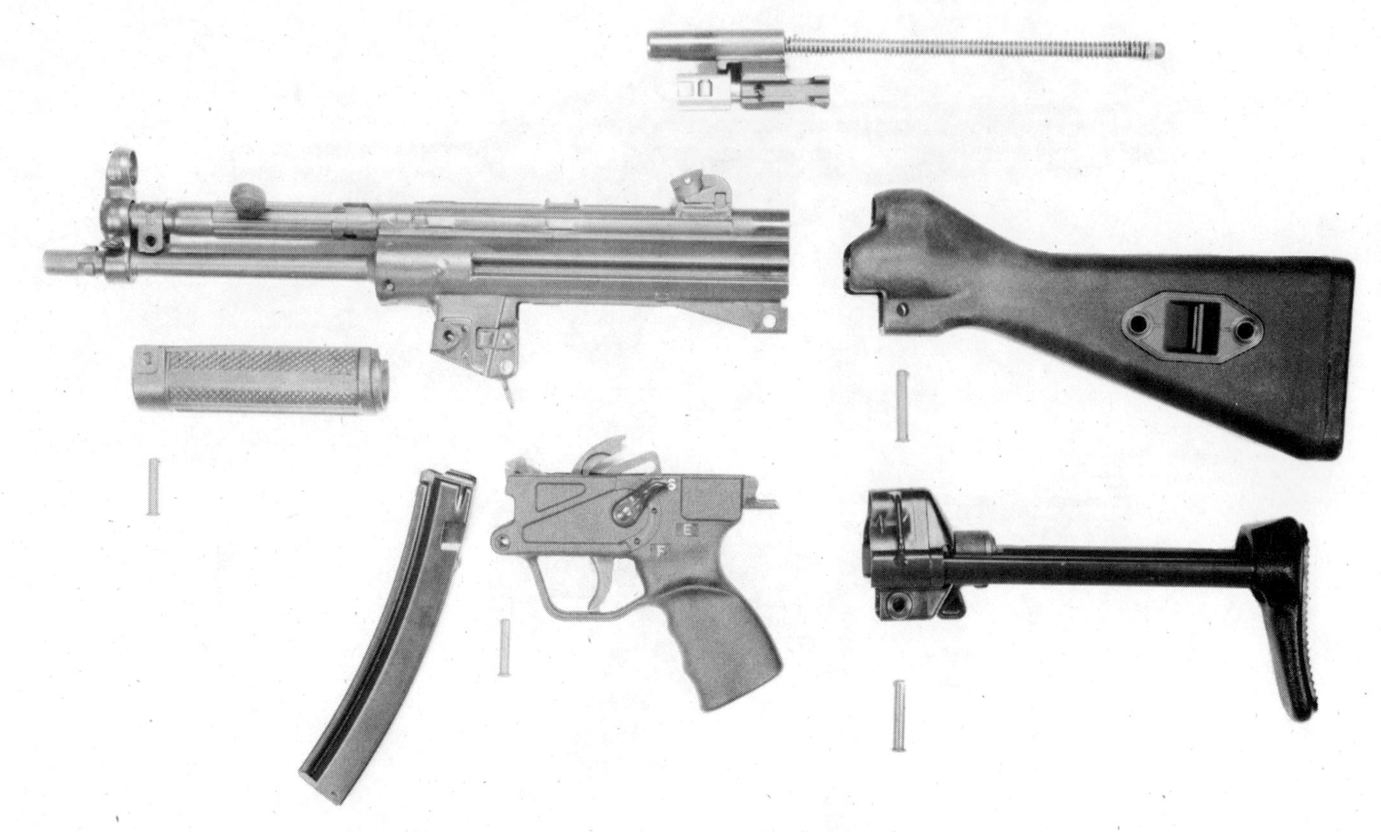

Field-stripped view of the MP5. Note the interchangeable buttstocks.

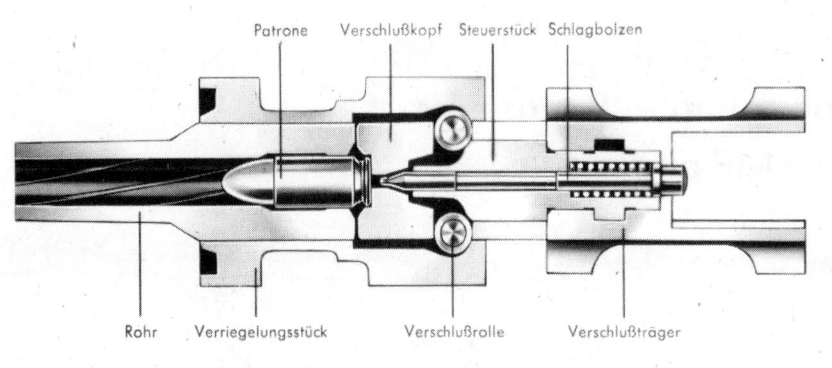

MP5 bolt mechanism when the submachine gun is ready to fire.

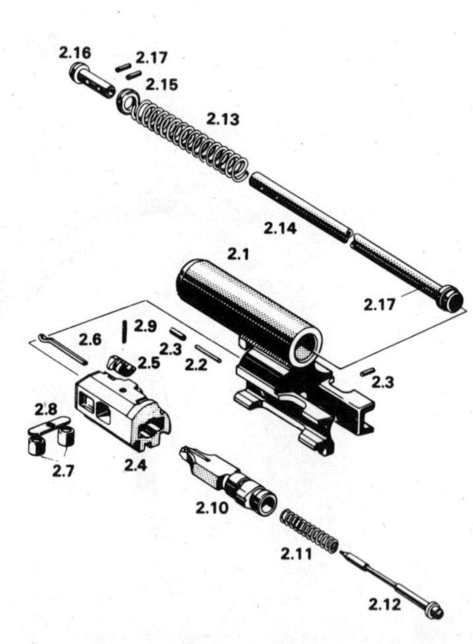

Exploded view of the MP5 bolt assembly.

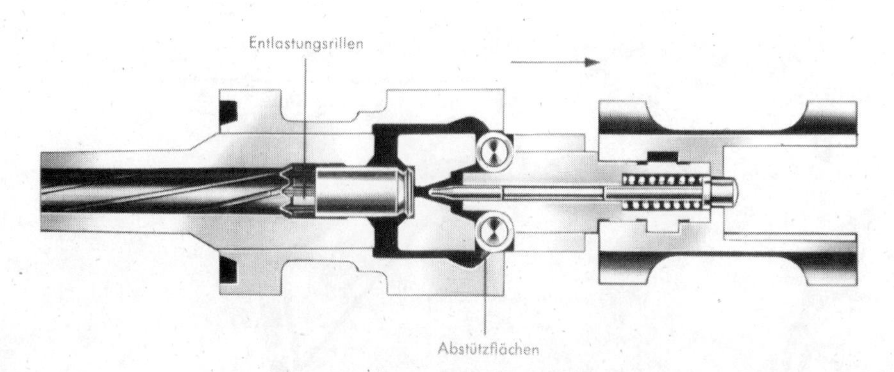

MP5 bolt mechanism immediately after the bolt has begun to recoil.

HECKLER AND KOCH VP70 MACHINE PISTOL

The HK VP70 is a rather unusual weapon in that it is a normal blowback operated pistol when used without its shoulder stock holster and when the shoulder stock holster is fitted it can be used as a full automatic weapon to fire 3 round bursts. The change lever that accomplishes this is mounted on the shoulder stock holster. It operates through a toggle lever, mounted on the top front of the shoulder stock holster, which goes through a slot in the top rear of the pistol receiver. Another somewhat unusual feature is that this pistol fires double action only; for this reason it is not normally fitted with a safety.

CHARACTERISTICS OF THE VP70

Caliber: 9mm Parabellum
System of operation: blowback, selective fire-3-round bursts.
Weight: w/o stock: 2.5 lbs. loaded
w/stock: 3.5 lbs. loaded
Length, overall: w/o stock 8 in.
w/stock 21.5 in.
Feed device: 18-round, detachable, staggered row
Sights: Front: blade
Rear: square notch
Muzzle velocity: 1,180 f.p.s.
Cyclic rate: 2,200 r.p.m.

Loading and Firing the VP70

When the VP70 is used without the shoulder stock holster, it is loaded and fired much like any other self-loading pistol. A loaded magazine is inserted and the slide pulled to the rear and released, chambering a cartridge. Pressure on the trigger will fire the weapon, one round for each separate pull of the trigger. When the VP70 is fired with the shoulder stock holster attached,

Heckler and Koch 9mm VP 70 Pistol.

the selector switch on the left, top front of the shoulder stock determines the mode of fire. If the selector is put on the figure 1, one round is fired for each squeeze of the trigger; if the selector is put on 3, three rounds are fired for each squeeze of the trigger. As pointed out previously, this weapon has no safety catch although Heckler and Koch advise that they can fit it with one if desired. The magazine catch is at the bottom rear of the grip. It is available commercially as VP70Z without the full automatic feature.

Field Stripping of the VP70

Remove magazine and pull slide back to insure that pistol is empty. Release slide and pull the retaining catch, located just to the rear of the trigger guard bow, down and pull the slide to the rear. Lift the slide off upward and to the rear and guide it forward and off the barrel.

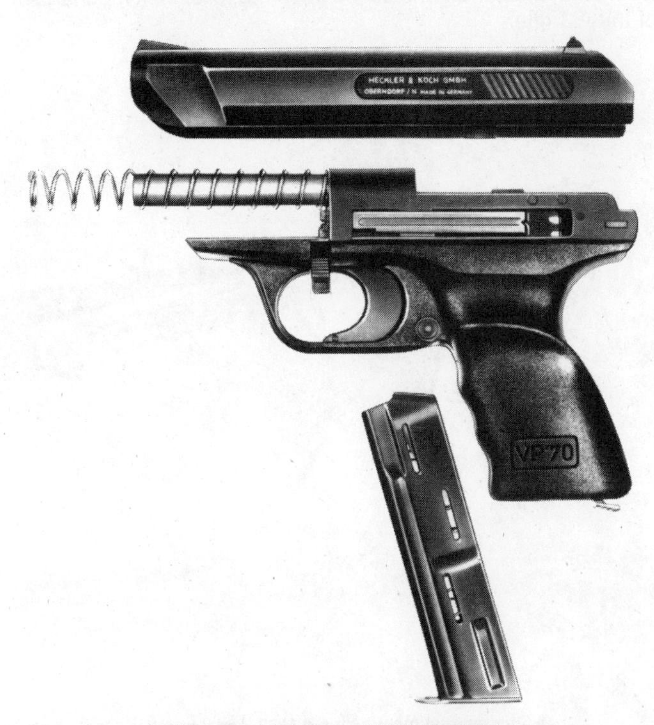

Disassembled view of HK VP70 Pistol.

Sectioned view of VP70. Note arrow indicating selector mechanism which is actuated by the attachment of the holster stock.

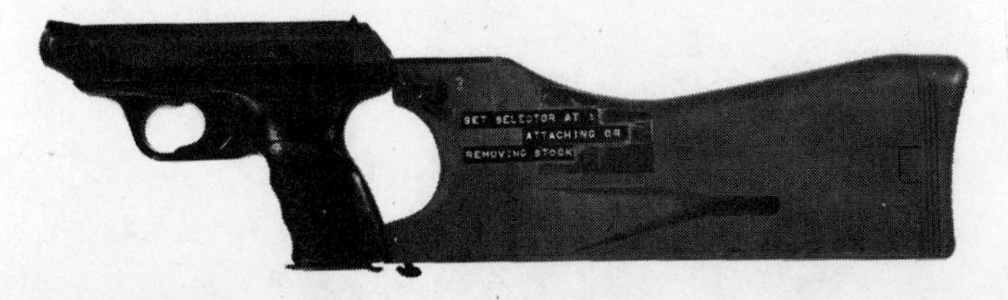

Heckler and Koch 9mm Parabellum Mode V70 machine pistol with shoulder-stock-holster attached.

WEST GERMAN MACHINE GUNS

WEST GERMAN 7.62mm MG42/MG3.

The Bundeswehr uses updated MG42s chambered for the NATO cartridge. These weapons are discussed in Chapter 2 and 22. During the past decade Heckler and Koch has developed a series of machine guns based upon the roller lock mechanism of the G3 rifle.

A Rheinmettal manufactured MG3, serial number 55140, built in March 1967. Note that the antiaircraft sight is folded down on the barrel jacket.

THE HECKLER AND KOCH MODEL 11, 12, 13, 21, 21A1 AND 23 MACHINE GUNS

These weapons are variations on the basic G3 design guns. The Model 11 can be fed with either a saddle drum magazine or a box magazine and is chambered for the 7.62mm NATO cartridge. The Model's 12 and 13 are the same as the Model 11 but are chambered for the Soviet 7.62 × 39mm and the 5.56mm (.223) cartridges, respectively. Models 12 and 13 are also somewhat lighter than the Model 11.

The Model 21 is belt fed but can also, with a special adaptor, use a box magazine. The Model 21 may be chambered for the 7.62mm NATO, 7.62 × 39mm or 5.56mm (.233) cartridges. The Model 21A, which has been adopted by Sweden, differs from the Model 21 in having a hinged rather than a removable feed tray. A reduced weight 5.56mm (.223) version of the Model 21A is the Model 23.

All of the above weapons have quick change barrels. Rear sights are usually of the radial type, but the rotating type aperture, as on the G3A3 rifle, may also be found.

All of these weapons fire from a closed bolt, and all have a quick change barrel. Many of the components of these weapons are interchangeable with the components of the G3 rifle. The concept of having a weapon system consisting of a number of weapons of basically similar construction with different application, i.e., individual weapon, squad automatic weapon and (in some cases) support machine gun, has become quite popular. This type of weapon system design has a definite advantage from a training and logistical point of view but frequently has some technical drawbacks. Briefly, gun design is a series of tradeoffs; what one gains in one area one loses in another, and there is no weapon in any category that is better than all other weapons of its type in all characteristics.

The 7.62 × 51mm NATO HK21A 1 belt-fed machine gun.

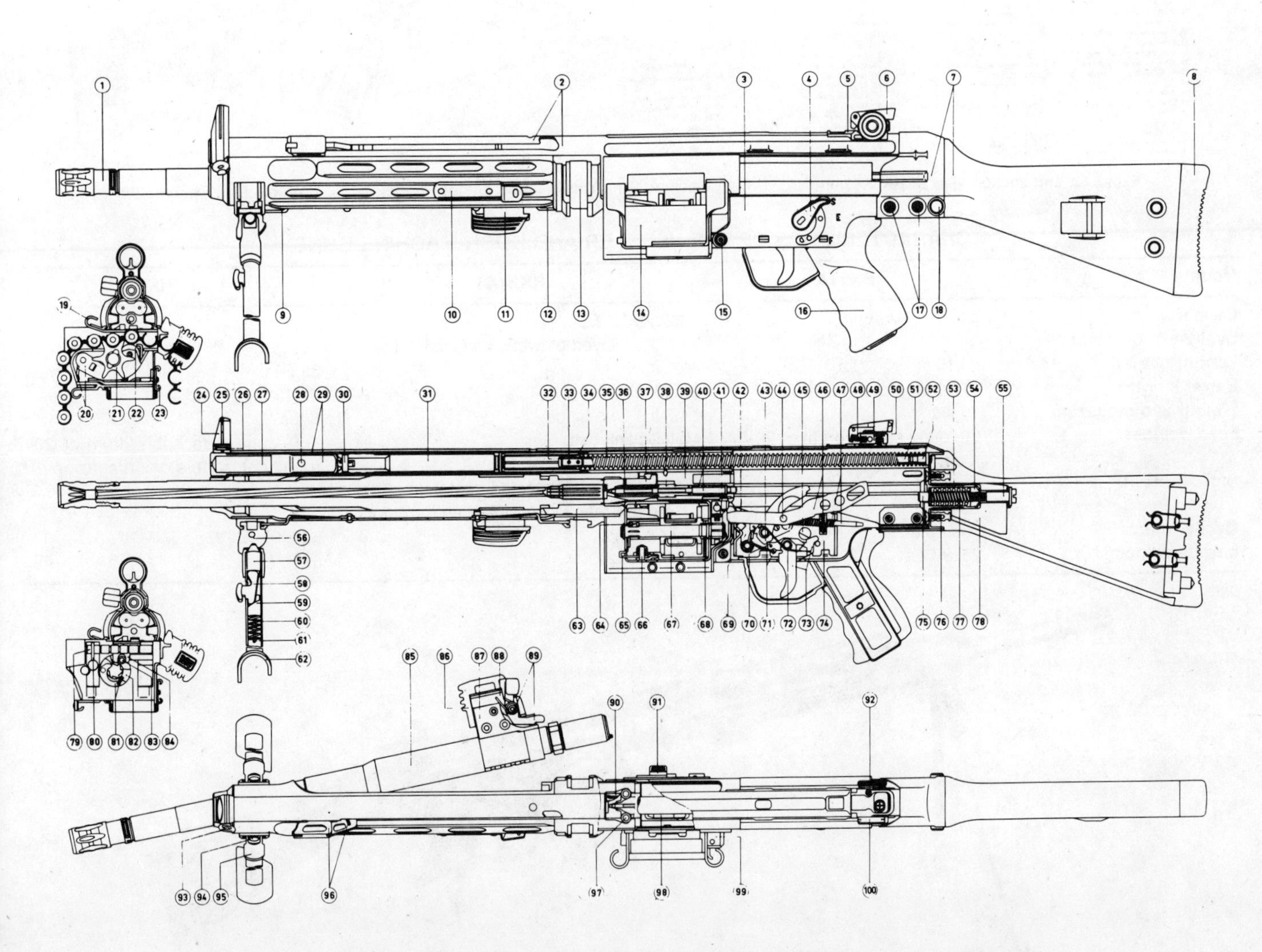
Schematic view of HK21 Machine Gun; note especially that bottom drawing shows the quick barrel change feature.

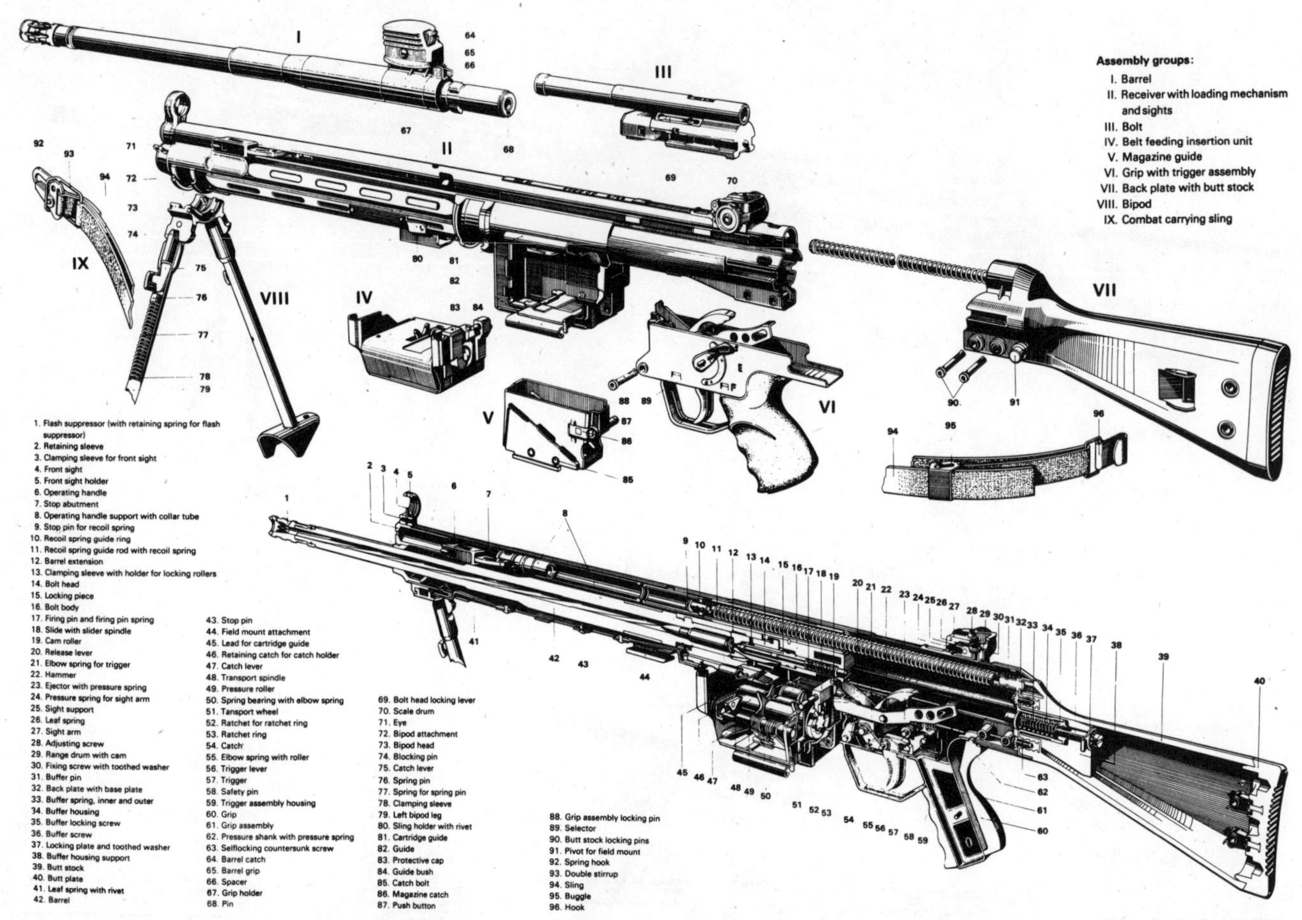

Exploded and section view of HK 7.62mm NATO HK21 Machine Gun. This weapon has been adopted by Portugal (see Chapter 38).

CHARACTERISTICS OF THE HECKLER AND KOCH MACHINE GUNS

Model:	HK11A1	HK21A1	HK13
Caliber:	7.62mm NATO	7.62mm NATO	.223
System of operation:		Delayed blowback, selective fire	
Length overall:	40.15 in.	40.1 in.	38.6 in.
Barrel length:	17.71 in.	22.63 in.	22.13 in.
Weight w/o magazine:	15 lbs.	14.7 lbs. w/o bipod.	8 lbs.
Feed device:	80-round saddle drum or 20-round box	Metallic link belt.	100-round detachable drum or 20-round detachable box magazine
Sights: Front:		Protected post	
Rear:		Rotary rear with V notch and apertures	
Cyclic rate:	850 r.p.m.	750 r.p.m.	600 r.p.m.
Muzzle velocity:	2589 f.p.s.	2625 f.p.s.	3248 f.p.s.

Heckler and Koch .223 HK 13 Machine Gun.

The HK11A 1 7.62 × 51mm NATO 30-shot box magazine-fed machine gun.

22 German World War II Small Arms

Germany manufactured over 13 million small arms during World War II. In addition, they used tremendous quantities of captured arms and arms made in occupied countries. There were never sufficient numbers of standard German small arms on hand at any time during World War II to arm all German forces. Even at the beginning of the war, 11 divisions of the German Army were armed with Czech small arms.

As a result of the continual shortage, the Germans used a great variety of small arms. However, they did standardize to the extent that all front-line units used weapons chambered for standard service cartridges; they also had the highest priority for standard service weapons. Service and police units and other German war organizations used whatever was available.

Considered standard during World War II were the following: the 9mm Walther P38 service pistol (the 9mm Luger 08 was a substitute standard); the 9mm MP38 and MP40 submachine guns; the 7.92mm Mauser Kar98K and 7.92mm G33/40 and semiautomatic Kar43 rifles; and the 7.92mm MG42 machine gun, which became standard at the end of the war, replacing the MG34.

GERMAN PISTOLS

A total of twenty-seven different models of German designed and made pistols (not including caliber .22 pistols) were approved for service use by the German forces between 1914 and 1945. Only a few models were ever purchased in quantity. These pistols were in 6.35mm (.25ACP), 7.65mm (.32 ACP), 9mm short (.380 ACP), 9mm Parabellum, 7.65mm Luger (caliber .30 Luger) and 7.63mm Mauser.*

Toward the end of World War II, Walther and Mauser developed 9mm Parabellum pistols composed mainly of steel stampings, which were to be used to arm the Volksturm. This was a Home Guard type organization composed of those males too old or too young or too infirm for the regular armed forces or those having essential jobs in civilian industry. Both designs were disapproved by the German Army Ordnance Office.

Because of the large number of different types of pistols used by the German Army, only those that were used in large quantities will be covered in detail in this book. The Germans also used many foreign pistols, of which some were captured and some were made in occupied countries. The most common among these were the Polish 9mm Parabellum, Radom M1935, the FN Hi-Power, the Czech 7.65mm Model 27, the Czech 9mm short Model 38, the Hungarian 7.65mm Model 37 and the Belgian 7.65mm and 9mm short FN Browning Model 1922.

The pistol on the left has a special safety feature not embodied in the P-38, though otherwise the pistols are much alike. The P-38 is the mass production version of the Walther HP. Firing pins of the pattern pictured in the left photo are commercial safety types. In these, the pin is retracted to prevent the hammer from reaching it.

THE 9mm P38 PISTOL

The P38 was adopted in 1938. As originally made for the German Army, it bore the Walther marking. During the war, German code letters were used to identify the manufacturers: "ac" for Walther, "cyq" for Spree werke, "byf" and "svw" for Mauser, "dov" for Brunn (Brno) and "ch" for FN, which manufactured approximately 3500 sets of components. More than one million P38s were made during World War II.

*For additional details see *Handguns of the World*.

Field Stripping the P38

(1) Set the safety catch in the "safe" position and pull the slide back over the empty magazine, so that the inside catch on the slide stop will be forced up by the magazine follower and hold the slide open. Then remove the magazine.

(2) At the front end of the frame below the receiver is a lever-type locking pin. Turn this down and around as far as it will go.

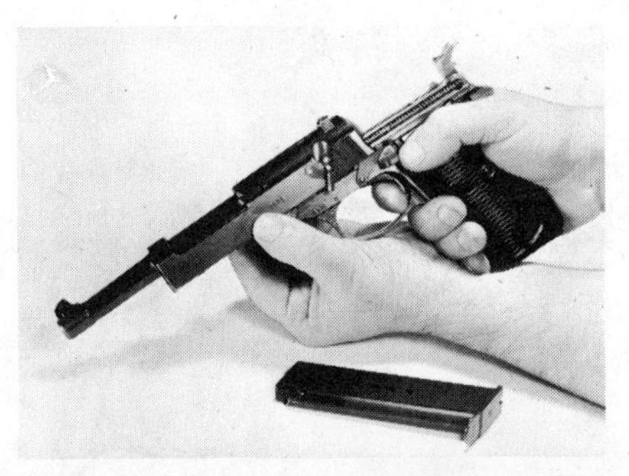

(3) Hold the slide under control with the left hand and with the right thumb push down the slide stop. Now press the trigger and pull the barrel and slide directly forward in their runners on the receiver, sliding them out of the guides.

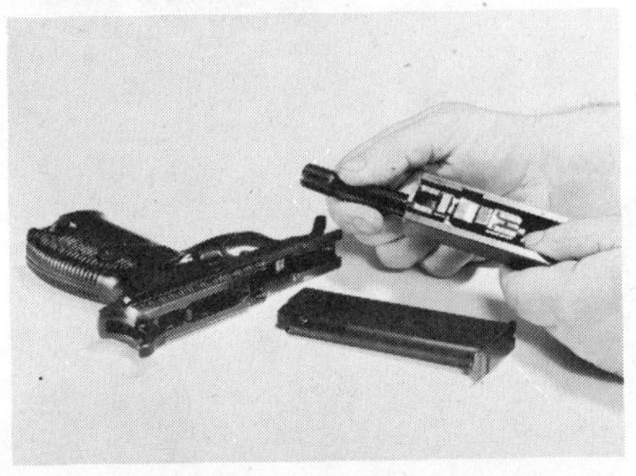

(4) Now turn barrel and slide upside down. A small locking plunger will be seen at the rear of the barrel assembly. Push the plunger and spring out the white metal locking cam block.

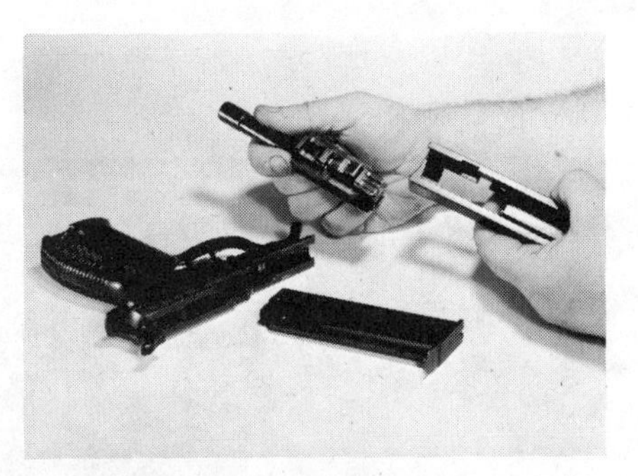

(5) Now slide barrel directly ahead out of slide. Lock will come forward with barrel.

(6) Push forward and up on locking cam block and lift it out of its recess. This completes field stripping.

Protruding pin (arrow) on the P-38 indicates a loaded chamber.

Note on Assembly of P38

Reverse stripping procedure.

When replacing locking block, be sure that its lugs are in line with the wide ribs on both sides of the barrel.

Insert barrel assembly as far as it will go into slide, then push the locking block into its locked position.

Hammer must be uncocked, and the ejector and the safety mechanism levers pushed down to prevent them from catching on the rear end of the slide.

With safety catch at "safe" position, hold the locking block in the locked barrel position, and push the slide and barrel onto the receiver in the guide. Force the slide all the way back against the tension of the springs and raise the stop to catch and retain the slide in the open position.

Turn the locking lever around on its pin as far as it will go to its original position. Press the slide stop and permit the slide to run home. Insert magazine.

Special Note on the P38 Pistol

When the pistol is loaded with a cartridge in the chamber, a floating pin protrudes from the slide above the hammer. A glance, or (in the dark) a touch, will always tell whether the chamber is loaded, making it unnecessary to pull back the slide as in other pistols.

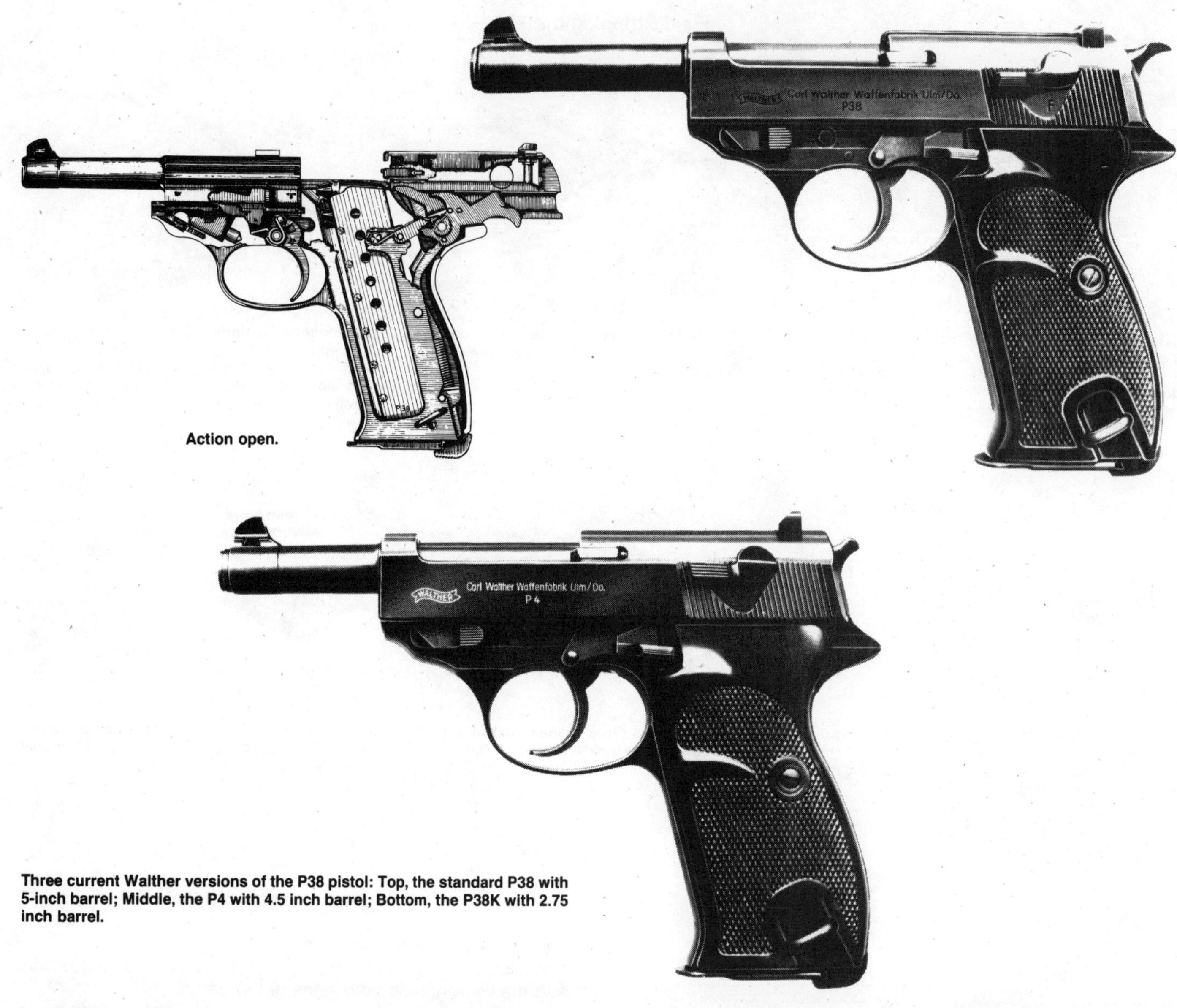

Action open.

Three current Walther versions of the P38 pistol: Top, the standard P38 with 5-inch barrel; Middle, the P4 with 4.5 inch barrel; Bottom, the P38K with 2.75 inch barrel.

The Walther PPKS in .380 ACP. This variant of the PP and PPK family was devised for Interarms to meet current US import regulations.

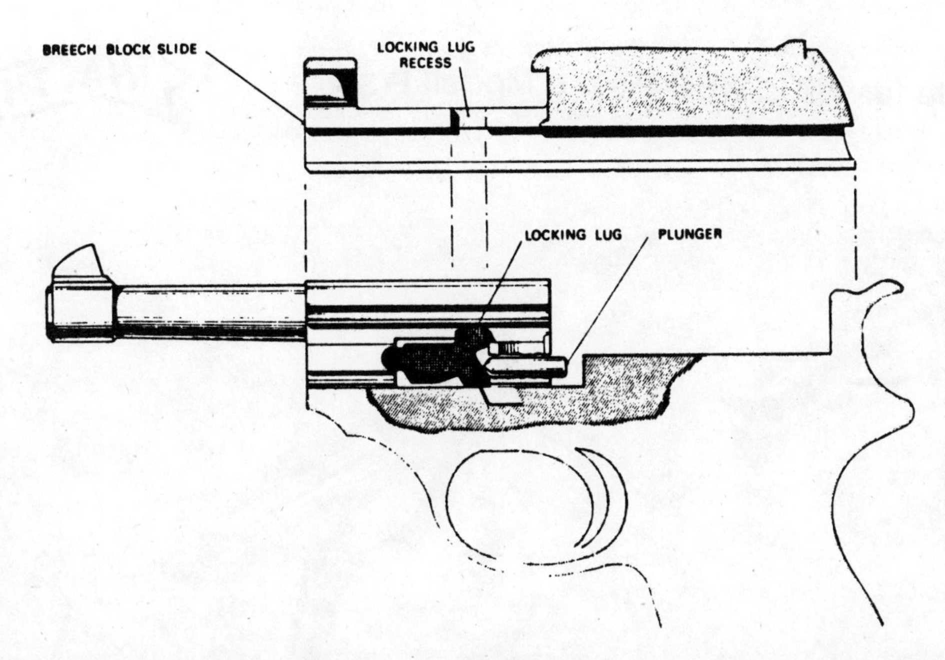

The P38 locking system. There are four main components in the P38 locking system: the barrel, the slide, the frame, and the locking block. The locking block is hinged to the underside of the barrel and is held in the locked position by the frame. When the P38 is fired, the slide and barrel recoil together. After a short travel, the barrel and slide reach a point where the unlocking plunger on the barrel strikes the frame. The forward motion of the plunger strikes the locking block thereby causing it to drop into a recess in the frame. At this point the barrel and slide are no longer locked together. The barrel comes to a stop and the slide continues its path to the rear. On the forward stroke the locking block is forced back to its locked position.

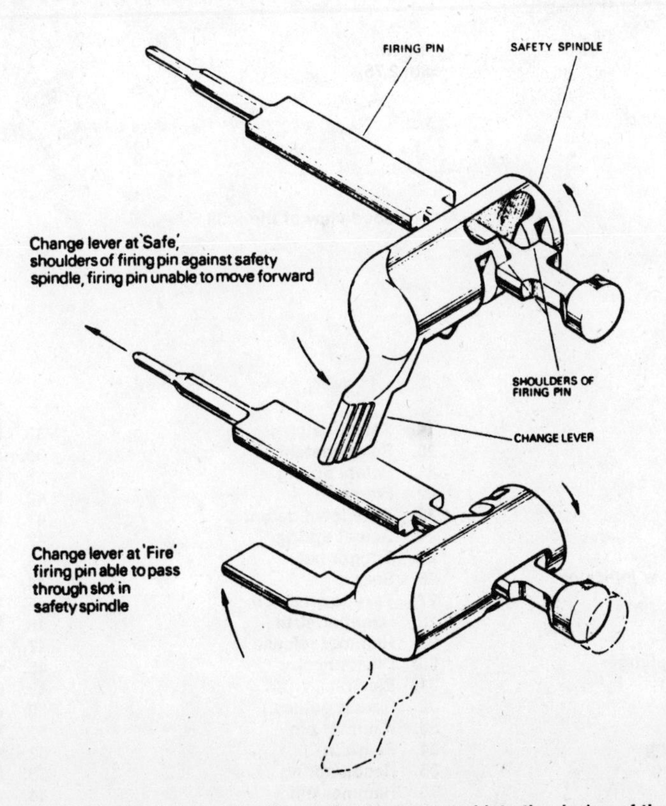

P38 firing pin locking mechanism. Compare this mechanism to that incorporated into the design of the P5 illustrated in Chapter 21.

An exploded view of the P 38.

1. **Barrel**
2. **Front sight**
3. **Locking block**
4. **Unlocking pin**
5. **Block retainer**
6. **Slide**
7. **Slide cover**
8. **Loaded chamber indicator**
9. **Rear sight**
10. **Extractor**
11. **Firing pin**
12. **Disconnector plunger**
13. **Plunger spring**
14. **Plunger**
15. **Retainer pin**
16. **Firing pin spring**
17. **Indicator spring**
18. **Ejector spring**
19. **Safety catch**
20. **Safety detent**
21. **Safety spring**
22. **Frame**
23. **Takedown detent**
24. **Detent spring**
25. **Trigger bar**
26. **Sear**
27. **Takedown catch**
28. **Hammer strut**
29. **Hammer release**
30. **Disconnector**
31. **Ejector**
32. **Spring guides**
33. **Hammer pin**
34. **Sear pin**
35. **Recoil spring**
36. **Hammer pin**
37. **Bar spring**
38. **Sear spring**
39. **Hammer**
40. **Hammer lifter**
41. **Pin**
42. **Spring**
43. **Trigger**
44. **Trigger bushing**
45. **Trigger pin**
46. **Magazine catch**
47. **Hold-open latch**
48. **Grip plate**
49. **Grip plate**
50. **Grip plate screw**
51. **Magazine**
52. **Follower**
53. **Base plate retainer**
54. **Base plate**
55. **Spring**

THE 9mm LUGER (P08)

The Luger is one of the best known pistols in the world, and there are several excellent books in English that deal with this pistol in detail far beyond the scope of this book.* The Luger was first adopted as a service pistol by Switzerland in 1900; this model was chambered for the bottle-necked 7.65mm (called caliber .30 Luger in the US) Luger cartridge. The 9mm Parabellum cartridge version, which introduced the most widely-used pistol and submachine gun cartridge in the world, appeared in 1902 and was adopted by the German Navy in 1904. In 1908, it was adopted by the German Army and remained the standard service pistol until 1938. There were over 400,000 Lugers manufactured for the German Army after the adoption of the P38, and manufacture was continued until 1943. There are at least thirty-five different variations of the Luger in existence, including numerous variations of the basic P08 used by the German Army.

The Luger was manufactured in Germany by Deutsch Waffen und Munitions fabriken (DWM), Simson, Krieghoff, Erfurt Arsenal and Mauser. It has also been manufactured by Vickers in England for the Dutch government and by the Swiss government arsenal at Bern for the Swiss government. Total Lugers manufactured is unknown, but it is probably at least two million and possibly considerably more.

The Luger is a fine-hanging pistol, very pleasant to shoot, and it introduced an exceptionally fine cartridge, the 9mm Parabellum. It is not, however, a standard pistol in any major country today because it is prone to stoppages if mud or sand gets into the action.

The Navy Luger and long-barreled 1908 type (frequently called the Model 1914, Artillery, or Model 1917) were also used in German service. The Navy Luger has a six-inch barrel, adjustable rear sight and is ridged for a wooden shoulder-stock holster. The long-barreled 1908 has an eight-inch barrel and a tangent-type rear sight and was frequently issued with the 32-round "Snail" magazine. The latter type appeared toward the end of World War I and is made of P08 components except for barrel and sights.

*For additional details see *Handguns of the World*.

German 9mm Parabellum Model 08 Pistol (Luger P08).

To Load 32-Shot Magazines

This magazine was issued during World War I. Each magazine comes with a filler. The lever must be wound against the tension of the magazine spring and locked into position by means of a spring loaded catch. The loading tool is slid over the mouth end of the magazine, and by a pumping action the cartridges are forced downward into the magazine. The magazine must be loaded cautiously since the lever is under heavy spring tension and if released it can do serious damage to the fingers.

(Note: this magazine is also used in the early model of the MP 181 Submachine Gun.)

The Long P08 Luger with shoulder stock attachment and 32-shot magazine.

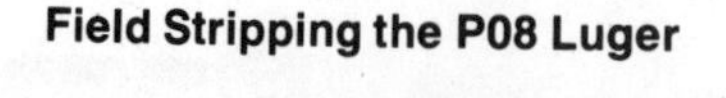

Field Stripping the P08 Luger

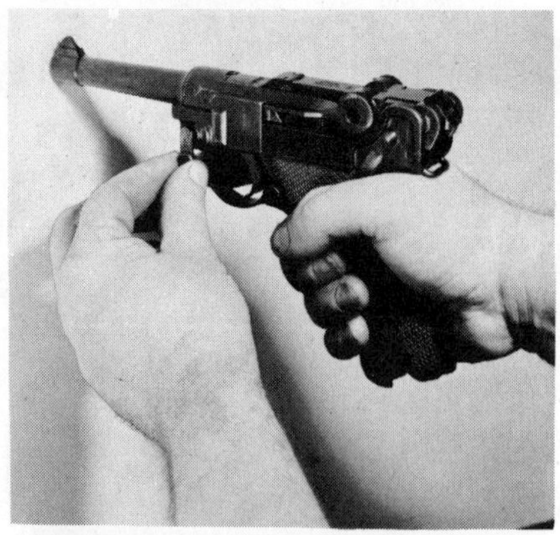

(1) Holding pistol in right hand press muzzle down firmly on a hard surface about ½ inch to release tension on the recoil spring. With the tension removed, the thumb catch on the sideplate may now be turned down to a vertical position.

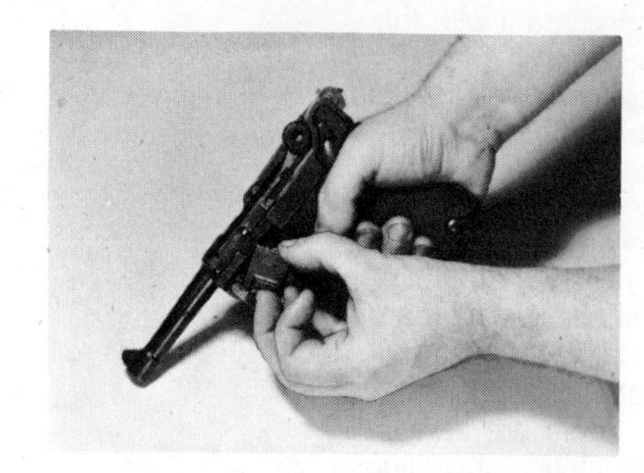

(2) Now lift out the sideplate.

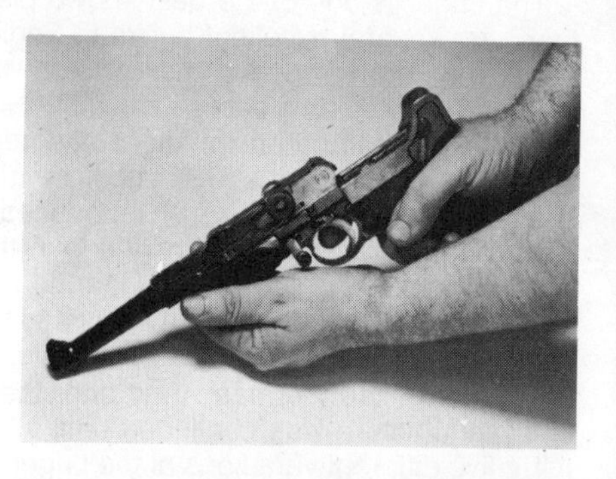

(3) Slide the complete barrel and toggle assembly directly to the front and out of the receiver.

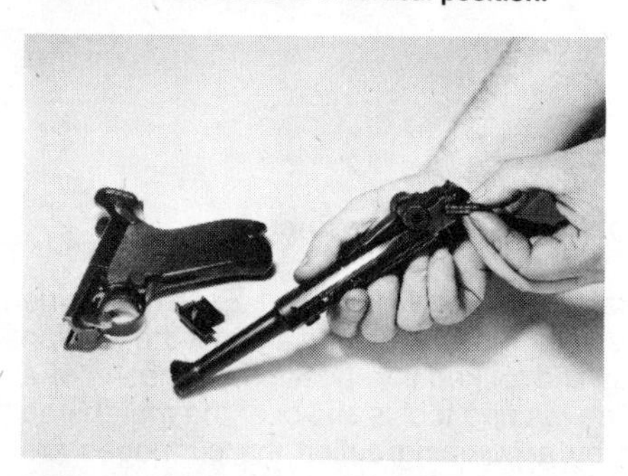

(4) Buckle the toggle slightly to relieve tension and extract retaining pin on the left hand side.

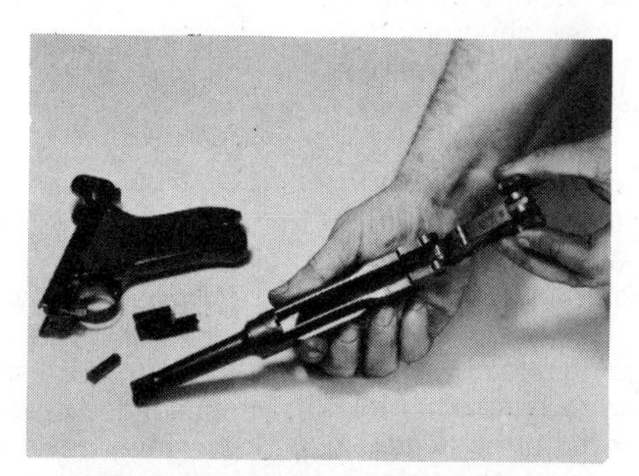

(5) Now pull toggle assembly, breechblock containing firing pin and extractor directly back in their guide and out of the frame. No further dismounting is necessary nor recommended.

Note on Reassembling. Merely reverse stripping procedure. Take care hook suspended from rear of the toggle assembly drops into proper place, which is in front of the inclined ramps. Also note that when replacing the side plate, the tongue on the rear end must be inserted in the recess in the receiver and the projecting section of the trigger bar must fall into the proper slot at the top of the trigger.

Instructions for Loading and Firing the Luger

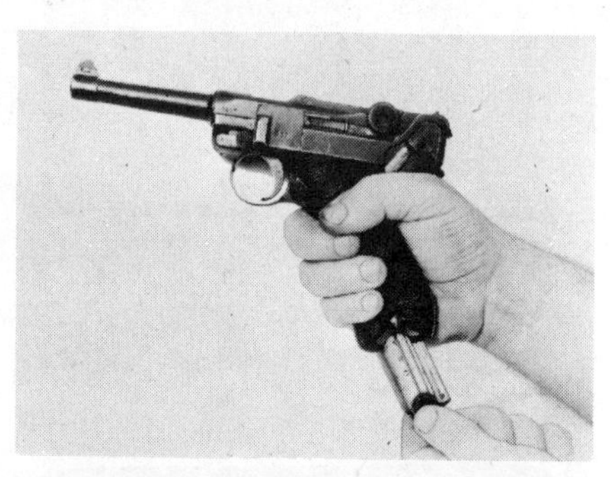

(1) To extract magazine: Press magazine release stud near trigger on left hand side and withdraw magazine from butt of pistol. To load magazine: Hold magazine firmly in left hand. Pull down stud attached to magazine platform. This will compress spring and permit cartridge to be dropped into the magazine.

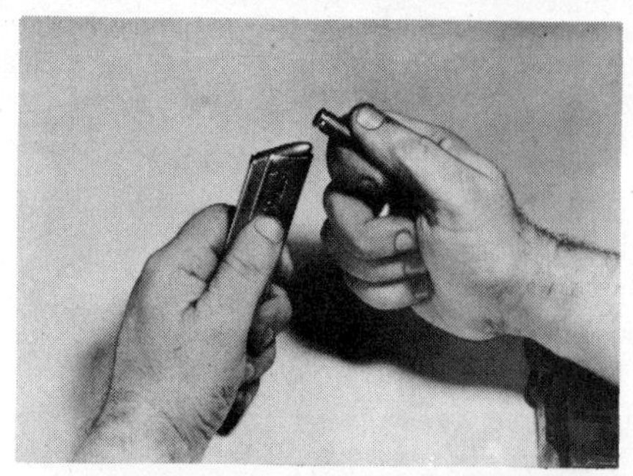

(2) To load chamber: Holding pistol pointed down toward ground with right hand, grip the milled knobs on the toggle and pull up and back as far as the breechblock will go. This compresses the recoil spring in the grip and permits the first cartridge in the magazine to rise in line with the breechblock.

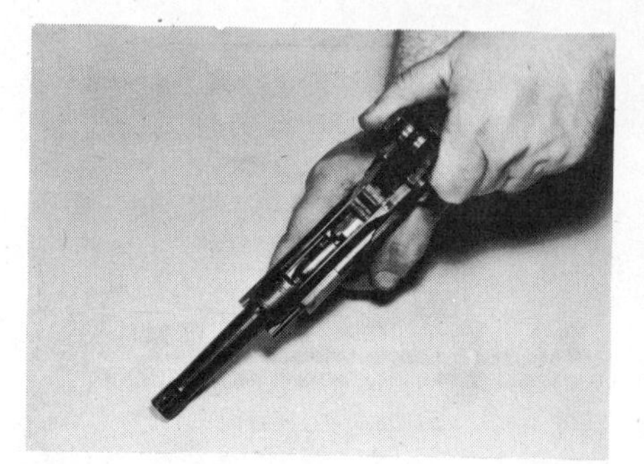

(3) Release grip and spring will force breechblock back into locked position driving a cartridge into the chamber.

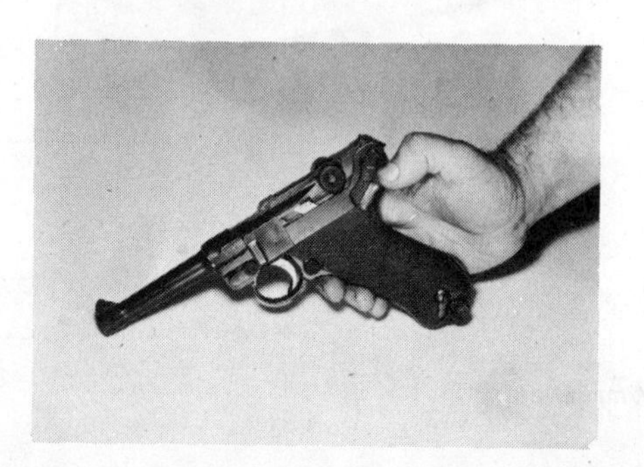

(4) To set thumb safety: Pull thumbpiece back and down. This will expose the German word "Gesichert," "Made safe." At the same time a flat solid steel piece will be seen to rise directly in front of the milled knob on the toggle. This locks the sear so the weapon cannot be fired.

(5) Breechblock stop: When the last cartridge has been fired, the stud of the magazine follower will force the catch up and hold the breech open with toggle joint buckled. Reloading from open breech: (a) Remove empty magazine. (b) Replace with loaded magazine. (c) Pull back on milled surfaces and permit breechblock to drive forward loading chamber.

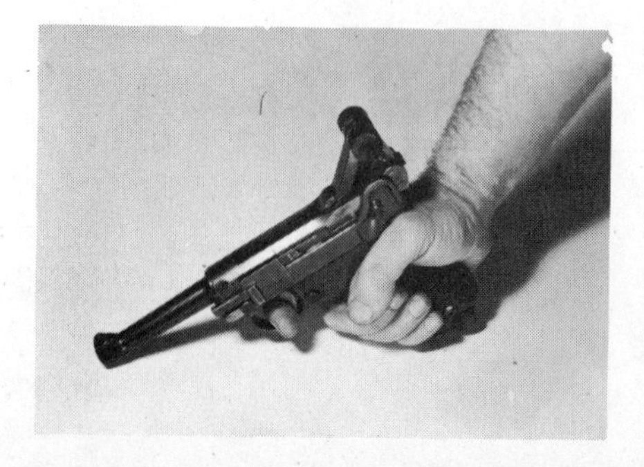

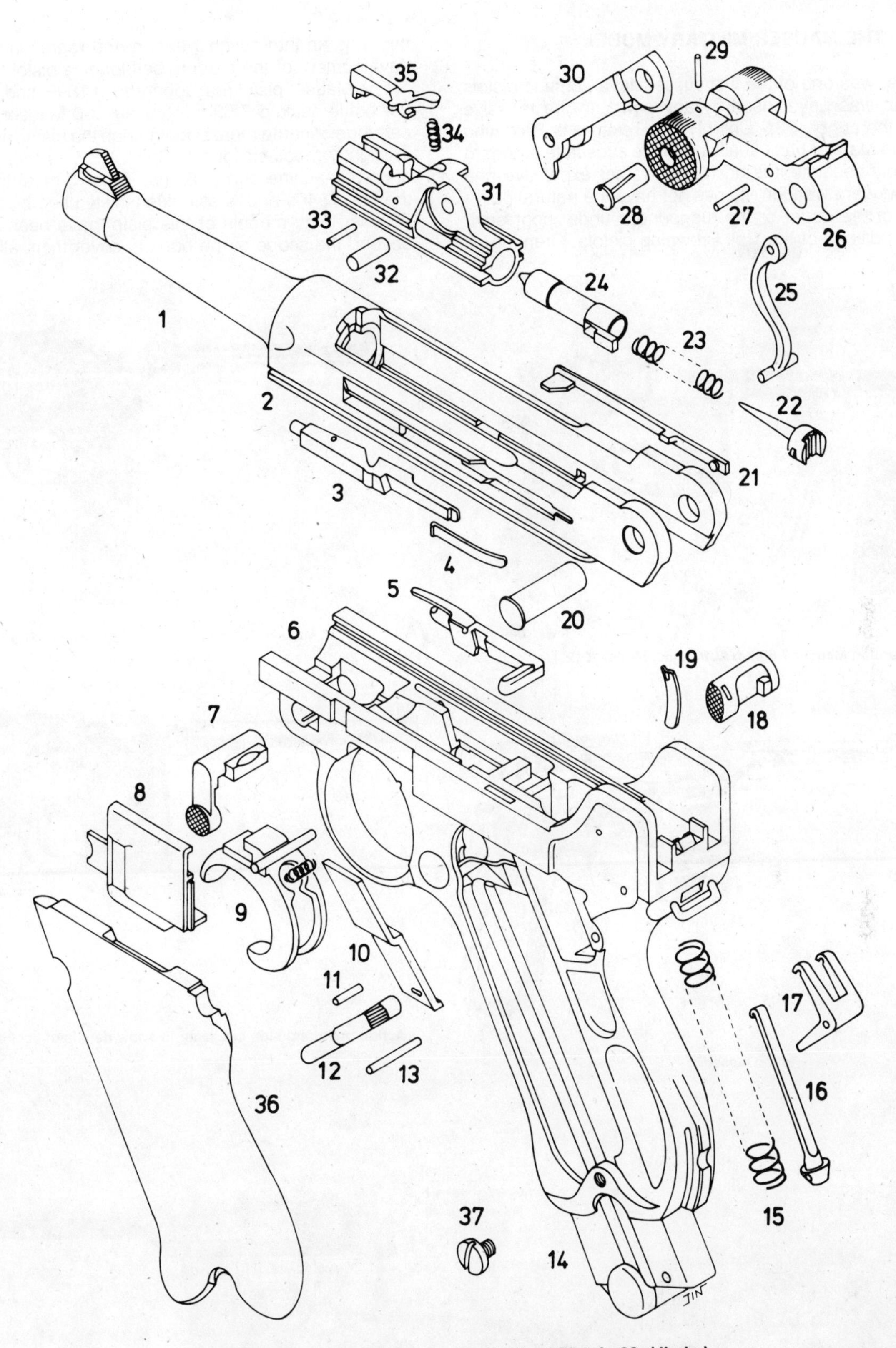

An exploded view of the 9mm Parabellum Pistole 08. (*Jimbo*)

1. Barrel
2. Barrel extension (receiver)
3. Trigger bar
4. Trigger bar spring
5. Hold-open latch
6. Frame
7. Take-down bolt
8. Trigger plate
9. Trigger
10. Safety bar
11. Safety pin
12. Safety lever
13. Recoil lever pin
14. Magazine
15. Mainspring
16. Mainspring guide
17. Recoil lever
18. Magazine catch
19. Magazine catch spring
20. Toggle pin
21. Ejector
22. Firing pin spring guide
23. Firing pin spring
24. Firing pin
25. Coupling link
26. Rear toggle link
27. Coupling link pin
28. Toggle pin
29. Toggle pin retaining pin
30. Forward toggle link
31. Breech block (bolt)
32. Breech block pin
33. Extractor pin
34. Extractor spring
35. Extractor
36. Grip plate
37. Grip plate screw

THE MAUSER MILITARY MODEL

The Mauser was one of the first successful automatic pistols to appear but was never used by any power in the massive quantities of the Luger, P38, Colt M1911, Tokarev or Browning Hi-Power. The Mauser has usually been the substitute standard pistol, probably because it is basically a rather expensive and somewhat awkward weapon. It does not have the natural pointing qualities of the Luger or the ruggedness under poor environmental conditions of the Colt Browning pistols. Granting all this, it is an interesting weapon and represents a stage in the development of the modern self-loading pistol.*

The Mauser pistol first appeared in 1895, and with it appeared the bottle-necked 7.63mm (caliber .30 Mauser) cartridge. The cartridge became more popular than the pistol, as it was adopted in slightly modified form by the USSR as their standard pistol and submachine gun (7.62mm Type P) cartridge from 1930 to the late 1940s and is still extensively used by the Soviet Bloc. At least thirty models of this pistol have been made, and it is beyond the scope of the book to cover them all.

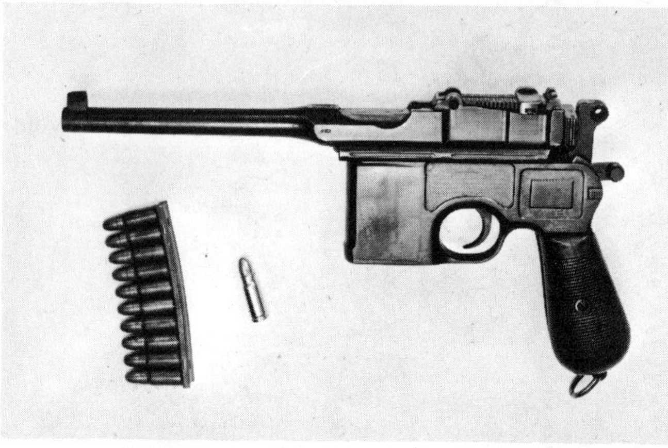

German Mauser 7.63mm Automatic, Model 1912.

Mauser 7.63mm Model 1932 Selective Fire Pistol.

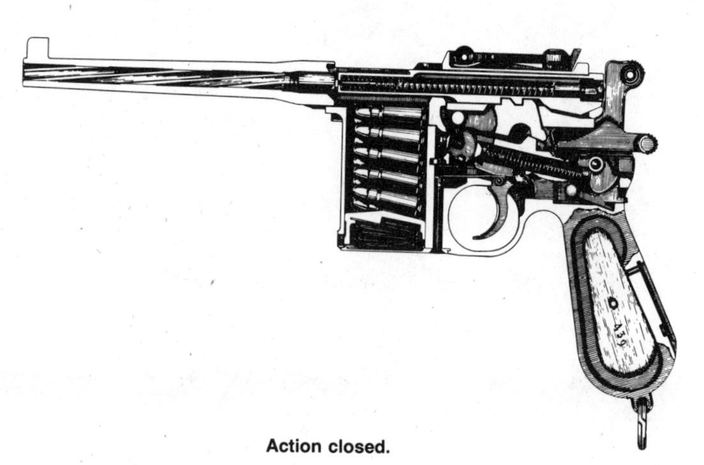

Action closed.

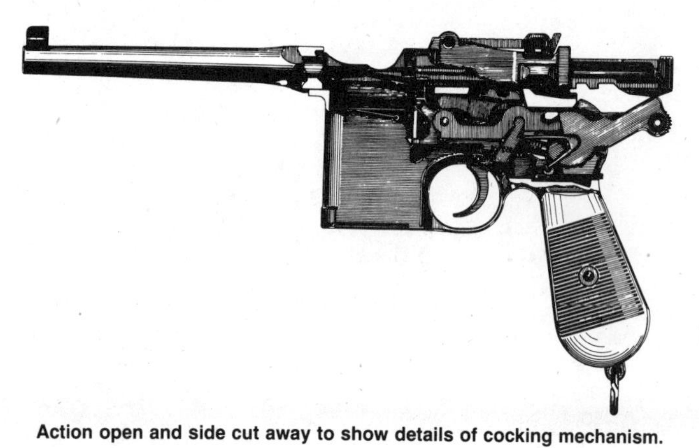

Action open and side cut away to show details of cocking mechanism.

The Mauser 7.63mm automatic pistol with shoulder stock holster.

*For additional details see *Handguns of the World*.

In addition to being made in 7.63mm, the Mauser was made in 9mm Mauser and, for the German Army during World War I, in 9mm Parabellum. The 9mm Mauser cartridge is a rather large cartridge, not interchangeable with any other 9mm cartridge. Copies of the Mauser have been made in Spain and China. Some of the Chinese copies have been in .45 caliber.

The 1912 Mauser was issued in 9mm Parabellum during World War II, equipped with shoulder stock, and was called Model 1916. A large figure "9" is branded on each grip piece of this weapon. The 7.63mm Model 1932, (commercial designation Model 712) was used by the German security units during World War II. This model of the Mauser has a detachable magazine—10- and 20-shot magazines are used—but it can also be loaded with 10-round stripper clips from the top of the receiver. It is a selective-fire weapon. The 7.63mm Mauser cartridge was made for German service use with steel cased cartridges during World War II; this would indicate that there was a considerable quantity of 7.63mm Mauser pistols in service at that time.

MAUSER MODEL 1910 PISTOL

The 1910 Mauser was extensively used as a pistol for service troops during World War I and was used, among others, by SS police units during World War II. The Mauser 1910 was very widely distributed through commercial channels. It is a straight blowback weapon, which has only one really unusual feature. When the last shot is fired, the slide remains open and the insertion of a magazine—after removal of the magazine originally in the weapon—whether loaded or empty causes the slide stop to release the slide and lets it return to the closed position, chambering a cartridge in the process (if a loaded magazine was inserted). The slide will also remain open if it is drawn manually to the rear with an empty magazine in the gun.

In 1934, Mauser modified the 1910. The modification consisted principally of various changes in components to ease manufacture, substitution of stampings for machined parts etc., and the use of streamlined type grips.

Magazine catch is in the bottom of the butt and must be pushed back to release the magazine.

When loaded magazine has been inserted in handle, slide is drawn fully to the rear exactly as in the case of the Colt automatic pistol and then permitted to run forward under the influence of the compressed recoil spring. The indicator pin protrudes from the rear of the breech block when the pistol is cocked, giving warning that the weapon is dangerous.

Pressing down on the milled thumb catch on the left side of the pistol just back of the trigger sets the pistol as "safe." To release this safety, press in the small button directly below the thumb piece.

MAUSER MODEL HSc PISTOL

This pistol was frequently called the Mauser *Pistole neuer Art* (M.n.A.), or Mauser new-type pistol by the Germans. The design was produced in the late thirties and was intended for military, police, and service use. The weapon was widely used by German service and police units during World War II and although the finish varies depending on date of manufacture, it is, generally speaking, a well-designed weapon of good manufacture.

The HSc is a double-action pistol; an enlarged version, the 9mm Parabellum HSv, was Mauser's entry in the German service pistol tests of the late thirties, which resulted in the selection of the Walther P38 as the German service pistol. The basic design of this pistol was revived in the post-war period by Heckler and Koch as the HK4.

Mauser 6.35mm Model 1910 Pistol.

Mauser 7.65mm Model HSc Pistol.

Loading and Firing the Mauser HSc Pistol

The slide is pulled back its full length to cock the hammer and permit the magazine spring to force a cartridge up to the feed way. This movement also cocks the hammer. Releasing the slide permits the recoil spring, which is wrapped around the barrel, to pull the slide forward and load the firing chamber.

When the last shot has been fired, the magazine follower holds the slide open as a notice. When the magazine is removed, and then reinserted, the slide goes forward automatically.

A positive magazine safety is incorporated in this weapon. When the magazine is withdrawn, the trigger cannot be pulled.

A positive disconnector prevents more than one shot being fired for each pull of the trigger.

The action, being a straight blowback, permits the weapon to be a comparatively simple design, as no locking mechanism is necessary to shoot this low-power cartridge.

The exposed hammer may be lowered with the thumb. When it is necessary to fire, a pull on the trigger will function the hammer in the same general way that it does in a double-action revolver.

Field Stripping the HSc Mauser

Set the safety lever at the "Safe" position. Push the small spring supported piece inside the trigger guard directly in front of the trigger. While maintaining this pressure, push the slide forward a short distance.

Move the slide slightly backward. It may then be lifted up and off the receiver.

The barrel may be pushed forward against tension of the recoil spring and lifted up and out of the slide.

Removing the stock screws and lifting off the stocks exposes the firing mechanism.

THE SAUER MODEL 38 PISTOL

The Model 38 is a double-action pistol with internal hammer. It was a very popular weapon with the Germans and has several good features. Like the Walthers, it has a loaded-chamber indicator, which projects from the rear of the slide when the cartridge is in the chamber.

Field Stripping the Model 38 Sauer

When the magazine is removed, the weapon cannot be fired. The thumb safety not only blocks the hammer when it is applied, but pushes it back out of engagement with the sear.

The double-action firing system is one of the best developed. It utilizes a minimum of springs and is entirely enclosed.

One unique feature is an exposed cocking lever that does not move with the action. When the weapon has been loaded and cocked by a movement of the slide, pressing down on this lever will safely lower the hammer so that a cartridge may be carried in the firing chamber in complete safety. Pulling straight through on the trigger will cock and trip the hammer to fire the cartridge. However, pressure on the cocking lever will also recock the hammer should it be desirable to take more deliberate aim.

Sauer 7.65mm Pistol Model 38.

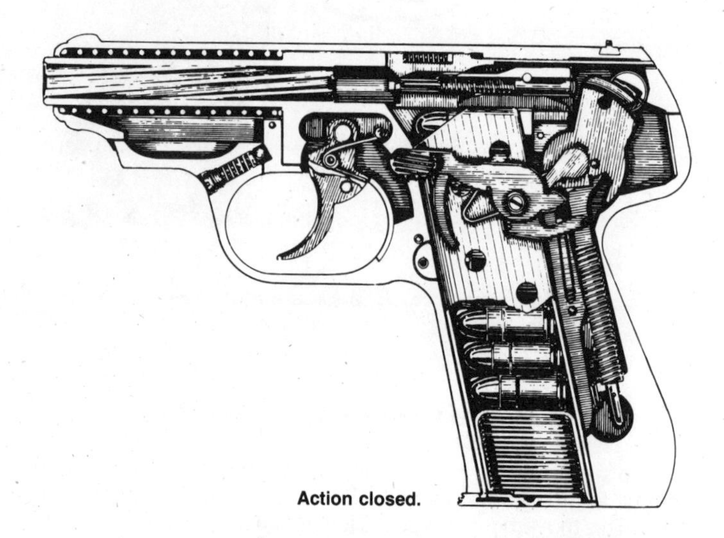

Action closed.

Action open.

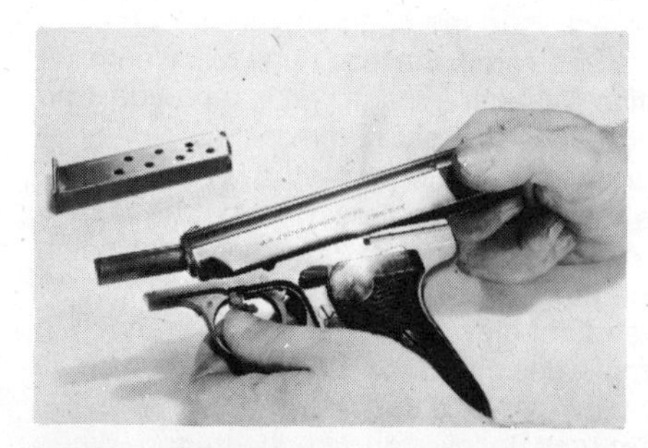

Pull down the locking latch in front and above the trigger. Draw the slide back to its full extent, lifting upward as you draw.

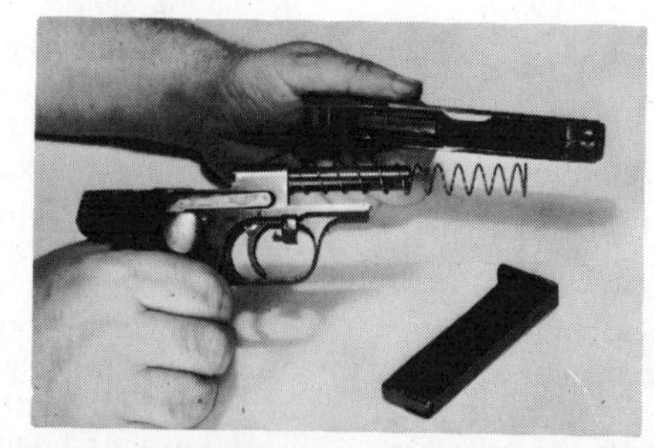

Let the disengaged slide move forward and push it off the barrel.

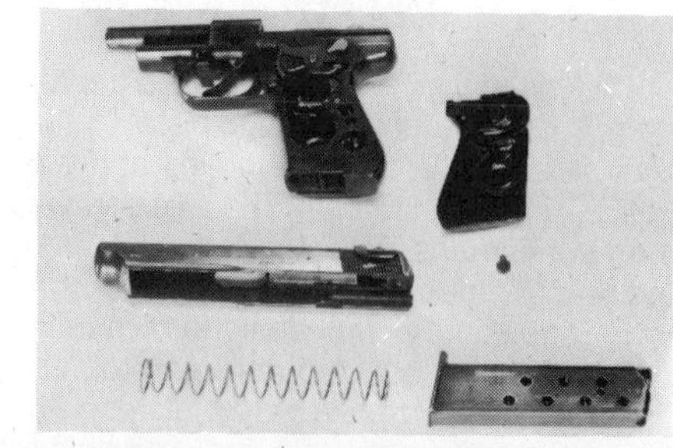

Recoil spring may be drawn off the barrel. Removing the grips exposes all working parts for attention. No further stripping need be done.

THE WALTHER PP AND PPK PISTOLS

These pistols are among the best known in the world and were made in large quantities for German service use and commercial sale. The Model PP (*Polizei Pistole*) was introduced in 1929, and the PPK (*Polizei Pistole Kurz*), a shorter version of the PP, was introduced in 1931. Both were originally chambered for the 7.65mm (.32 ACP) cartridge, but caliber .22, 6.35mm (.25 ACP) and 9mm short (.380 ACP) versions were also produced.

The PP and PPK had considerable influence on pistol design in Germany prior to World War II and throughout the rest of the world since. Copies of the PP have been made in Turkey (Kirikkale), Hungary (M48) and under license in France (Manurhin) since World War II. The PP and PPK may have been the first production pistols made with lightweight alloy receivers, since they were on the market with light receivers during the thirties. The PP and PPK are currently being made by Carl Walther at Ulm a/d, West Germany.

The Model PP was approved for German service use in 9mm short as well as 7.65mm. Some PP pistols made during World War II do not have the pin-type loaded chamber indicator, which normally protrudes from the rear of the slide.

How to Load and Fire the PP and PPK

The magazine is removed by pressing the magazine catch on the left side behind the trigger guard. Fill magazine with cartridges and insert smartly in grip. Push safety catch—on left rear of slide—down; pull slide to the rear and release. Push safety up into the off position, and pistol is ready to fire double action by pulling through on trigger. If a lighter trigger pull is desired, cock hammer and then press trigger. The slide will remain open on the last shot. It can be released on a new loaded magazine by pulling it slightly to the rear and releasing. The hammer will be cocked, and the pistol can be fired by pressure on the trigger or by applying the safety, in which case the hammer will fall and remain in the down position.

How the PP and PPK Work

These are blowback pistols of advanced design with external hammers, double-action triggers and positive manual safeties.

The recoil spring is positioned around the barrel. When the slide is drawn back over the top of a loaded magazine in standard automatic pistol fashion, the rear of the slide runs over and cocks the hammer. The recoil spring is compressed between a shoulder in the front end of the shaped slide (which surrounds the barrel muzzle) and the receiver abutment into which the barrel is secured. Releasing the slide permits the spring to pull it for-

Walther 7.65mm Model PPK Pistol.

ward. The breechblock face of the slide chambers a cartridge from the top of the magazine. The extractor claw in the right side of the breechblock is snapped into cartridge engagement by its spring.

Pulling the trigger will cause an attached trigger bar to draw the sear out of engagement with the hammer to fire the cartridge.

When the chamber is loaded, the front end of a floating pin in the slide is raised. The rear end of the pin projects from the rear of the slide. If this pin can be seen or felt, the chamber is loaded.

Safety Systems. A special hammer block of steel prevents any forward hammer movement until the trigger is deliberately pulled. When the trigger bar pulls the upper section of the rotating sear, a nose on the sear raises the hammer block until it is opposite a cut in the hammer face. Only at this point can the hammer fall far enough to hit the firing pin.

The firing pin is the spring-loaded type, shorter than the length of its stroke. Its spring pulls it back into the breechblock as soon as its forward drive is halted.

The rearward thrust of the gas within the cartridge case drives the slide to the rear to extract and eject and reload in standard blowback fashion.

Disconnector. The opening movement of the slide runs over and forces down a section of the trigger bar. This disconnects the trigger bar from effective sear contact. Until the trigger is released permitting the spring to force it ahead into firing position, the trigger bar attached to the trigger cannot rise into a slide undercut. Only when it can rise in this undercut can the trigger bar tip draw the sear out of engagement to let the hammer fall.

Thus one pull is necessary for each shot fired.

Walther 7.65mm Model PP Pistol.

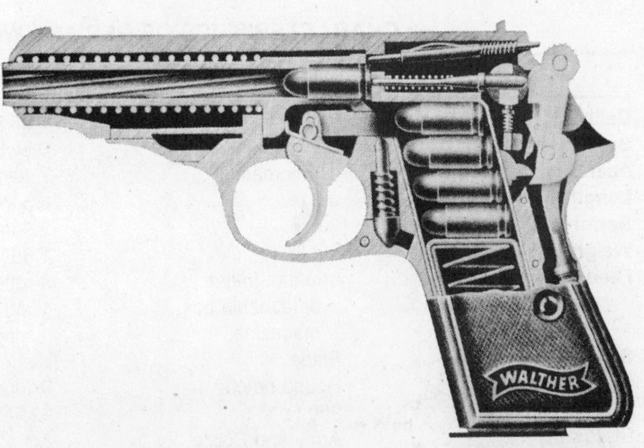

Representative blowback type. The German Walther PP Pistol.

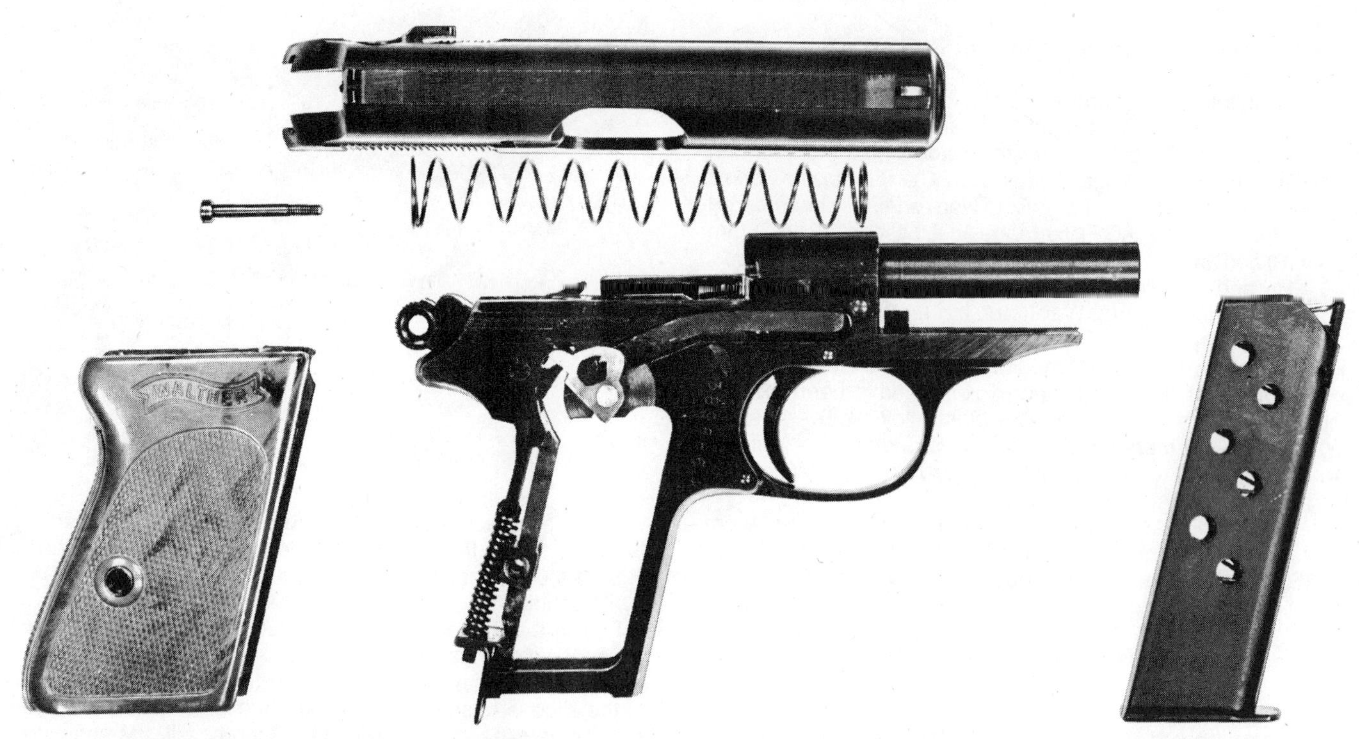

PPK stripped; hammer is cocked. Pull on trigger will draw trigger bar ahead to disengage sear from hammer. Pulling trigger guard down lowers slide locking lug seen under barrel.

CHARACTERISTICS OF GERMAN WORLD WAR II SERVICE PISTOLS

	P38	P08	Mauser Model 1932	Mauser Model 1910
Caliber:	9mm Parabellum.	9mm Parabellum.	7.63mm.	7.65mm.
System of operation:	Recoil, semiautomatic.	Recoil, semiautomatic.	Recoil, selective fire.	Blowback, semiautomatic.
Length overall:	8.6 in.	8.75 in.	11.75 in. w/o stock. 25.5 in. w/stock.	6.2 in.
Barrel length:	4.9 in.	4.06 in.	5.63 in.	3.4 in.
Weight:	2.1 lb.	1.93 lb.	2.93 lb. w/o stock. 3.93 lb. w/stock.	1.3 lb.
Feed device:	8-round, in-line, detachable box magazine.	8-round, in-line, detachable box magazine.	10- or 20-round, staggered row, detachable box magazine.	8-round, in-line, detachable box magazine.
Sights: Front:	Blade.	Blade.	Blade.	Blade.
Rear:	Rounded notch.	V notch.	Tangent leaf.	Round notch.
Muzzle velocity:	1115 f.p.s.	1050 f.p.s.	1575 f.p.s.	950 f.p.s.
Status:	Army Standard.	Army Limited Standard.	Police and S.S. use.	Army and Police use.

*Also made in 9mm short (.380 ACP)

CHARACTERISTICS OF GERMAN WORLD WAR II SERVICE PISTOLS, Continued

	Mauser HSc	Sauer Model 38	Walther PP	Walther PPK
Caliber:	7.65mm.	7.65mm	7.65mm*.	7.65mm*.
System of operation:	Blowback, automatic.	Blowback, semiautomatic.	Blowback, semiautomatic.	Blowback, semiautomatic.
Length overall:	6.5 in.	6.3 in.	6.8 in.	6.1 in.
Barrel length:	3.4 in.	3.5 in.	3.9 in.	3.4 in.
Weight:	1.3 lb.	1.56 lb.	1.5 lb.	1.25 lb.
Feed device:	8-round, in-line, detachable box magazine.	8-round, in-line, detachable box magazine.	8-round, in-line, detachable box magazine.	7-round, in-line, detachable box magazine.
Sights: Front:	Blade.	Blade.	Blade.	Blade.
Rear:	Round notch.	Round notch.	Round notch.	Round notch.
Muzzle velocity:	950 f.p.s.	920 f.p.s.	948 f.p.s.	919 f.p.s.
Status:	Army and Police use.	Army and Police use.	Army and Police use.	Army and Police use.

GERMAN WORLD WAR II RIFLES

The Germans used a great variety of rifles during World War II, but only a few were of German manufacture and considered standard by the German Army. The 7.92mm (called 7.9mm by the Germans) Mauser Kar 98k was the most widely used standard rifle; none of the semiautomatic rifles were ever made in the quantity of the bolt action Kar 98k. The German plan of 1944/45 was to replace the bolt action and semiautomatic rifles with the selective-fire assault rifles of the MP43/44, StG44 series; the conclusion of the war prevented the fulfillment of this plan.

Many of the older German rifles such as the 11mm Mauser Models 1871, 1871/84, Model 98, 98a and the Model 88 were used by *Volksturm* (Home Guard Units), as were large quantities of captured weapons. Some weapons originally of foreign origin were adopted by the Germans, such as the bolt-action Model 33/40, which was basically a slight modification of the Czech Model 33 Carbine, and the Model 98/40, which was an alteration of the Hungarian Model 35 Rifle. Many other foreign rifles were not adopted or standardized in the American sense of the term but were used in quantity, such as the Czech and FN Model 24 Mausers and the numerous varieties of these found in Romania, Yugoslavia, Greece and Bulgaria. It is probable that at least one-third of the rifles brought back to the United States by returning soldiers and referred to as German service rifles are not German at all but in all probability were used by the Germans in one fashion or another. Even the 6.5mm Norwegian Model 1894 Krag was made in limited quantities for the Germans, and some Italian Model 38 Mannlicher Carcano rifles were, around 1943, made in 7.92mm for the Germans. The Polish 7.92mm Model 29 Mauser was also used extensively.

The Germans never seemed, until it was too late, to appreciate the advantage of semiautomatic rifles. They had used two semiautomatic rifles early in World War I—the Mexican-designed, Swiss-made 7mm Mondragon called the Aircraft self-loading carbine Model 1915 and the Mauser 7.92mm aircraft self-loading carbine, both of which were used by the Germans as aircraft guns before their aircraft were fitted with machine guns. A full-stocked version of the Mauser aircraft rifle, the 7.92mm Model 1916, was issued in limited quantities during the war.

Mauser had developed semiautomatic designs as early as 1898. Walther also produced a semiautomatic rifle prior to World War II. Still, the German Army adopted a modified 98 Mauser bolt-action rifle (the Kar 98k) in 1935, one year before the United States adopted the semiautomatic M1 rifle. This dim spot in German weapons technology may have been due to their accent on the machine gun, an area in which they were very advanced indeed.

Desperate measures were taken during the war to make up for the shortage of rifle fire power. A high percentage of each squad was equipped with submachine guns. The 7.92mm Rifle 41(M) and Rifle 41(W) were made in small quantity and were not very successful. The 7.92mm Model 43 Rifle was made in larger quantity but too late to have any real effect on the battlefield.

While the German Ordnance engineers did not exploit the self-loading rifle concept with their characteristic astuteness, they were the first to develop the idea of the comparatively lightweight assault rifle, which would replace the rifle and submachine gun. Development of an "intermediate" sized cartridge was started in Germany prior to World War II, but the conclusion of the war in 1945 prevented the Germans from putting their plan into fruition.

THE 11mm MAUSER MODEL 1871 AND 1871/84 RIFLES

These rifles were not used by any German Army units during World War II but were issued to the Home Guard. The Model 1871 was the first Mauser rifle to be adopted by any country and is a single-shot, black-powder weapon.

The Model 1871/84 is basically the same as the Model 1871 but has a nine-round tubular magazine. The Model 1871, 1871/84 and all their variations use a two-piece bolt. The 11mm Model 71 black powder cartridge, which has a round-nosed bullet, and the 11mm Model 71/84 black-powder cartridge, which has a flat-nosed bullet, were developed for use with these weapons. The Model 1871 rifle was widely used in China for many years and was one of the principal weapons of the Chinese during the Boxer rebellion in 1900. They were also still in wide use by German African colonial troops during World War I.

THE 7.92mm MODEL 1888 RIFLES AND CARBINES

The adoption of the 8mm Lebel, using smokeless powder cartridges, by the French in 1886 caused the Germans to search for a suitable counter weapon. The 7.92mm Model 1888 rifle and carbine were the German answer to the Lebel.

This rifle is frequently called a Mauser and also a Mannlicher; it was actually developed by a German Army Commission and combines the magazine of the Mannlicher with bolt features of the Mauser Model 1871/84. With the introduction of the Model 1888, Germany introduced the 7.92mm cartridge. The 7.92 × 57mm Model 88 cartridge had the same case as found in this cartridge today but had a .318-inch bullet, as opposed to the current .323-inch bullet (the "S" bullet, which was introduced circa 1904-05). Many of these rifles were modified later to use the larger-sized bullet, but as a matter of course the 1888 pattern weapons should not be used with current 7.92mm cartridges. In addition to the bore diameter problem, the chamber pressure of currently available 7.92mm cartridges, especially military rounds, far exceeds that for which the Model 1888 weapons were made.

The Model 1888 uses a five-round Mannlicher clip, which can be loaded with either side down, unlike the Mannlicher 1886 clip, which had to be loaded from one end only. The clip functions as part of the magazine and drops out the bottom of the protruding magazine box when the last round has been chambered.

The Model 1888 rifle and carbine and the Model 1891 rifle do not have wooden handguards. They have a sheet metal barrel jacket, which covers the barrel from the receiver to the muzzle. Theoretically, this barrel jacket provided for better accuracy since it prevented the changes in center of impact frequently caused by the change of bearing on barrels of wooden stocks and handguards, as a result of humidity, etc. In actual practice, the metal barrel jacket suffered from many shortcomings; it was easily dented, water would seep into the joints and rust both the jacket and the outer portion of the barrel, and it was expensive and difficult to replace.

The Model 1888 was made in a rifle version and a carbine version; a carbine with stacking hook was introduced in 1891—for some reason this carbine was called Rifle Model 1891.

Numbers of these weapons were modified by Germany at a later date. There were three basic modifications:

(1). Some were fitted with a plunger and spring to eject the clip out of the top of the magazine after the last round was ejected in a fashion similar to that used by the US M1 rifle. The clip ejection slot in the bottom of the magazine was covered.

(2). Some were modified to use the charger used with the Model 98 rifle by milling a charger guide on the upper front end of the receiver guide and fitting a spring-loaded cartridge retaining rib on the upper side of the magazine.

German 11mm Mauser Model 1871/84 Rifle.

German Gew 88.

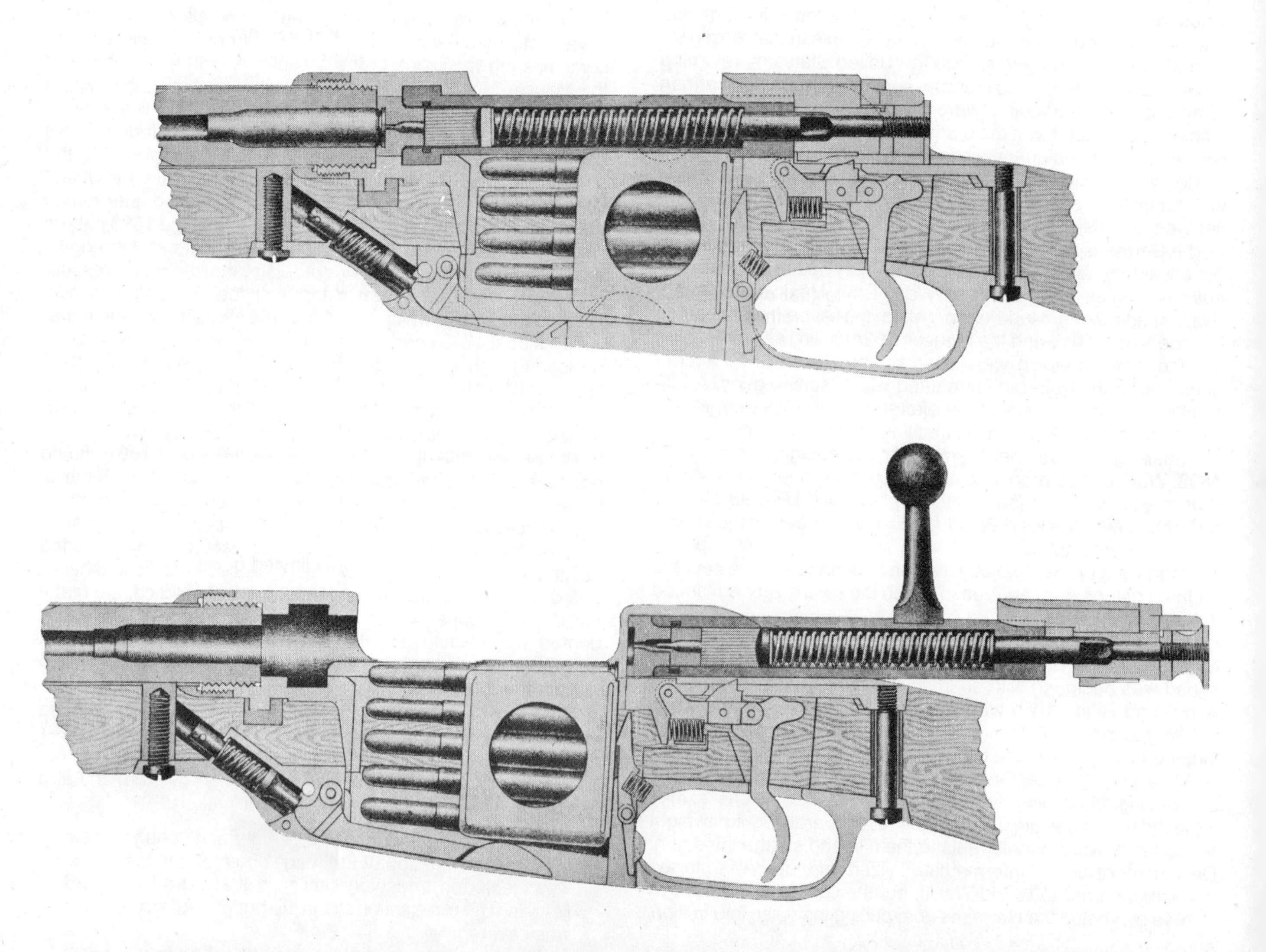

M1888 sectioned to show locking, firing, and magazine systems. Magazine loaded. Bolt-head is detachable. Note relationship of locking lug seats to the face of the breech. Also see magazine follower driving cartridges up between clip walls. This is the Mannlicher clip system.

(3). Many were modified by relieving the chamber neck and forcing cone (lead) to use the "S" (.323 inch) bullet. These weapons are stamped "S" on the receiver.

The various modifications are called 88/05, 88/14 and 88S.

The Model 1888 was made at the German arsenals and by Ludwig Loewe, Haenel, Schilling and Steyr. This rifle was used to some extent by Austria during World War I. China bought many 1888 pattern weapons from Germany and made a modified copy, the Type 88 or "Hanyang" rifle. Yugoslavia and Ethiopia also used the Model 1888 in limited quantities.

THE 7.92mm MODEL 98 AND ITS VARIATIONS

Model 98 Rifle (Gew 98)

The rifle 98 (*Gewehr 98*), introduced in 1898, is the most successful bolt-action design ever produced. In one form or another, the 98 action has been used by most of the countries of the world since 1898. As originally produced in Germany, the 7.92mm Model 98 rifle had the smaller sized (.318) bore of the Model 88 rifle; in 1903 the rifles were altered to use the larger diameter "S" bullet, and bore diameter was set at .323. At the same time, the rear sight was modified to match the ballistics of the "S" bullet.

The rifle 98 was the principal rifle of the German Army in World War I, and a number of variations of the rifle appeared during that war. One of the first was the 98 rifle with turned-down bolt handle used by bicycle troops. Sights were again modified to reduce the battle sight setting from 400 meters to 150 meters; the marking disc on the left side of the buttstock was replaced with a washer type disc used to assist in disassembly of the firing pin. A variant of the 98 (the Model 18) with a sliding breech cover similar in concept to that of the British Lee Metford and the Japanese Type 38 rifle, with detachable 5-, 10- and 25-round box magazines, was developed toward the end of World War I.

The Model 98/17 also had a bolt cover and a square shoulder on the follower to prevent closing the bolt on an empty magazine in the heat of action. The 98/17 had a 100-meter sight setting.

The Model 98 rifle also appeared in a caliber .22 training version, which was made from the standard 98 by fitting a liner in the barrel. Some Model 98 rifles were fitted with tangent-type rear sights.

Model 98 Carbine (Kar 98)

The original Model 98 carbine was apparently never made in quantity since, unlike the Model 98 rifle, it is a rare item these days, and photographs of German troops in World War I rarely show this weapon in evidence. The rear sight is similar to that of the Model 98 rifle, and it has a peculiar stock and band arrangement. The stock runs to the muzzle as with the Model 88 carbine and 91 rifle, but is reduced in diameter at a point about six inches to the rear where a lower band, similar to the upper band of the 98 rifle, complete with bayonet mounting bar is fitted.

Model 98a Carbine (Kar 98a)

Originally called Kar 98, this was the most popular carbine version of the Model 98 in World War I, and it had limited usage in World War II. The Carbine 98a appeared in 1904 and was made in tremendous quantities until approximately 1918. It has been claimed that the appearance of the Mark I Short Magazine Lee Enfield in 1903 and the Springfield of the same year influenced the Germans in the adoption of this weapon. In any event, it is of handy size and was very popular with German troops. It introduced the tangent-type sight in the 98 series and was sighted for the "S" bullet as first issued. The prominent stacking hook, jointed upper band, front sight guard and full length handguards distinguish this weapon from the other German Mauser Service weapons. This weapon served as the model for the Polish Model 98 carbine.

Model 98b Carbine (Kar 98B)

This rifle, although designated a carbine, has the same length as the 98 rifle, but it has a turned-down bolt handle and a tangent sight like that of the Kar 98a, which is graduated for SS (heavy ball) bullet.

Model 98k Carbine (Kar 98k)

This weapon was the standard rifle of the German Army in World War II and has been made in tremendous quantities. It was adopted by Germany in 1935 and has many of the features of the commercial Mauser "Standard Model." The "k" in "98k" stands for *kurz,* meaning "short," and which is somewhat surprising since it is longer than the original Kar 98 and about the same length as the Kar 98a. Kar 98k has a half-length handguard, a tangent-type rear sight, turned down bolt handle and a hole bored through the stock in lieu of rear sling swivel (the Kar 98 and Kar 98a also have this feature).

In addition to the normal wooden furniture, Kar 98k has been issued with laminated and wooden stocks. Kar 98k may still be found in service in various places in the world and rebuilt specimens of this weapon were taken from the Viet Cong in South Vietnam. The Kar 98k was made without bayonet mounting bar and with stamped bands in 1944–45.

Rifle 33/40 (Gew 33/40)

The nomenclature of this weapon is an example of the inconsistency of German nomenclature—this weapon is one of the shortest barreled Mausers used by the Germans, yet it is called a rifle. The 33/40 is the German version (made at Brno in occupied Czechoslovakia) of the Czech Model 33 carbine. It is distinguished by its light weight (the receiver has lightening cuts), short length and the extension of the shoe-type butt plate on the right side of the butt. It was used by German mountain and paratroop divisions, and because of its light weight and short barrel has a very sharp blast and heavy recoil. A folding-stock version of this rifle was made in limited quantity.

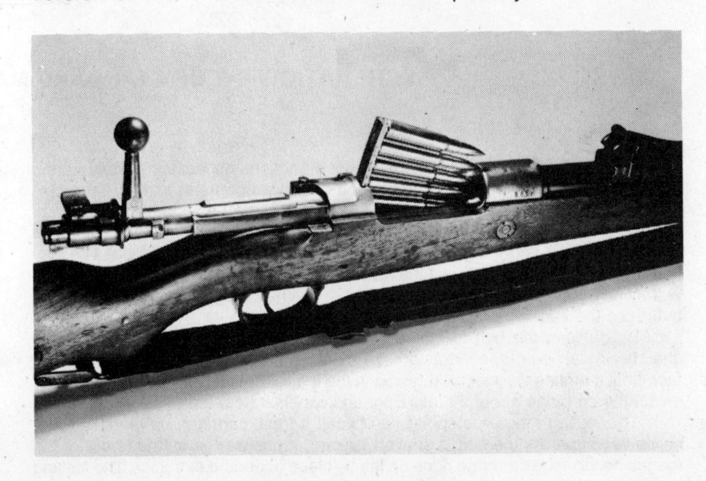

The Kar 98k is clip-loaded.

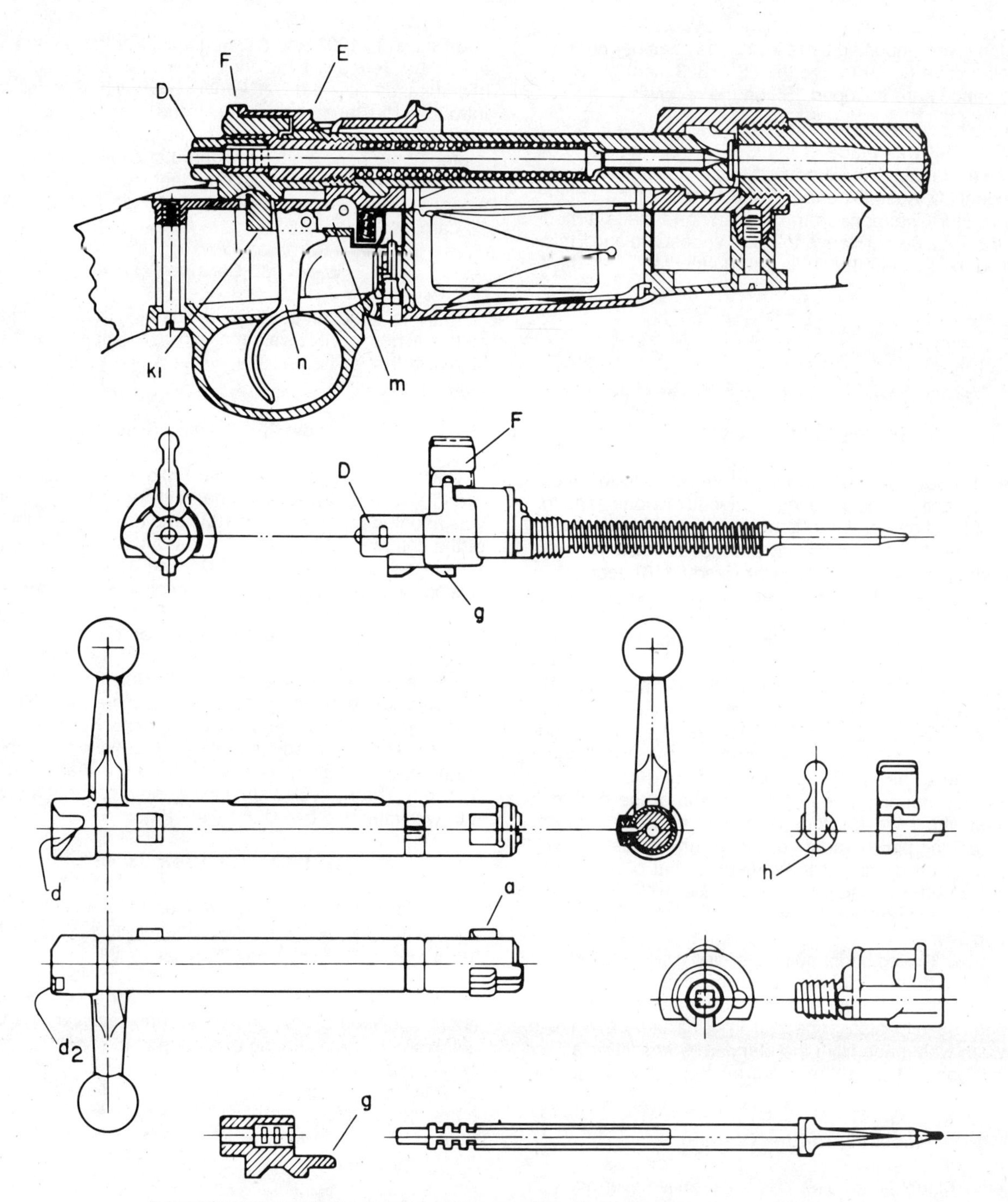

SCHEMATIC VIEW OF STANDARD MAUSER 98 OPERATING MECHANISM.

Functioning

a. The Mauser 98 is manually operated; all actions necessary to remove the fired cartridge case from the chamber and reload with a fresh cartridge are performed by the manipulation of the rifle's mechanism by the shooter.

b. As the bolt handle is turned upward, a cam in the rear of the bolt (d) forces the firing pin nut and firing pin rearward, compressing the firing spring. The root of the handle also cams against the receiver to provide powerful leverage for initial extraction of the fired cartridge. The firing pin unit has a lug on its underside which overrides the sear when the handle is turned fully up.

c. By drawing the bolt to the rear, the empty cartridge is removed from the chamber by action of the extractor. The extractor holds cartridge against the bolt face until it strikes the ejector (housed in the left rear of the receiver). The ejector pivots the cartridge about the extractor and expels it from the rifle.

d. The zigzag magazine spring has forced a fresh cartridge up, under the receiver feed lips. As the bolt is shoved forward, it pushes the cartridge out of the magazine, and the cartridge rides up the bolt face under the extractor. The locking lugs (a) are reseated in their abutments in the receiver by rotating the bolt handle downward.

e. The trigger (n) is pinned to the sear (m) and has two humps on its top where it bears against the bottom of the receiver. When the trigger is pressed, the front hump (closest to the pin) acts as a lever to move the sear down. At this stage, the trigger pull has been very light, but as the second hump at the rear of the trigger contacts the receiver, a definite stop is felt and increased trigger pressure is necessary to completely disengage the sear from the lug on the firing pin nut. When these disengage, the firing pin spring drives the firing pin forward and fires the cartridge.

f. This rifle has two safety features: the manual safety and an automatic safety. The manual safety is operated by swinging it to the left; this interposes a solid portion of the safety (f) in front of the firing pin nut and cams the nut slightly rearward, off the sear. At the same time, a section of the safety shaft rotates into a fore-and-aft cut (d2) in the rear of the bolt; this locks the bolt closed. The automatic safety is the cocking cam (g and d) in the bolt; if the bolt is not completely locked, the cam on the firing pin nut will force the bolt closed by engaging the cam in the bolt. These cams prevent the firing pin from going completely home unless the bolt is rotated to a fully locked position.

7.92mm Model 98 Rifle and Model 98a Carbine.

The Kar 98b.

7.92mm Kar 98k.

The Gew 33/40.

Folding stock version of Gew 33/40.

Rifle 40k (Gew 40k)

This weapon was apparently made in very limited quantity. It is a short weapon with a smaller trigger guard than the Kar 98k. It also has a hole through the bolt handle and does not have a lower band.

Field Stripping the Mauser Carbine 98k

To remove bolt, proceed as for US Rifle Model 1917. With rifle cocked and safety lever vertical, half way between safe and locked position, pull out near end of bolt stop on left side of receiver and draw bolt straight out to the rear.

To dismount bolt, proceed as for Springfield.

To remove magazine mechanism, same as for Springfield.

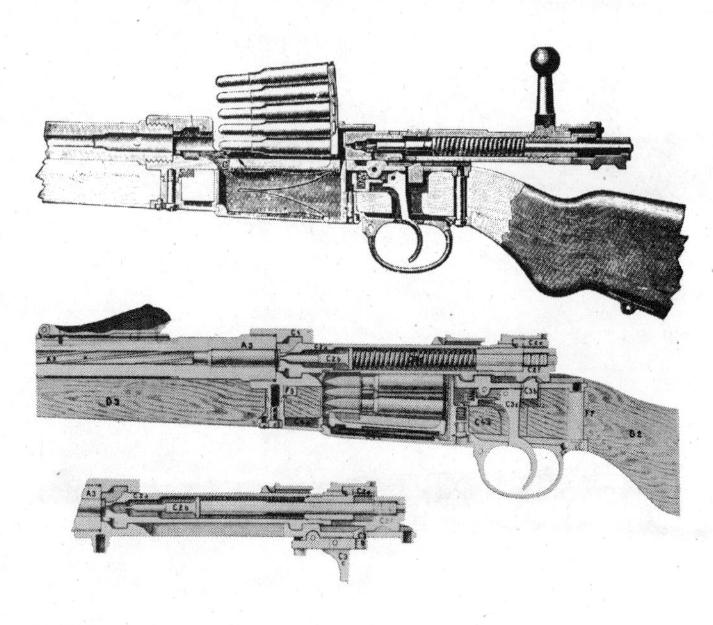

Kar 98k sectioned to show details of locking, firing and magazine systems.

Note position of dual forward locking lugs, which assure maximum support to case head at instant of firing.

The 98/40 Rifle (Gew 98/40)

The 7.92mm 98/40 is not a Mauser. It is based on the design of the 8mm Model 1935 Hungarian Rifle. The 98/40 was made only in Hungary by the Danuvia Arms Works. The 98/40 has a two-piece Mannlicher-type bolt, staggered row box magazine, which is flush with the bottom of the stock, and a two-piece stock similar to that of the Lee Enfield.

The Hungarian Model 43 rifle is quite similar to the 98/40, but the 98/40 can be distinguished from the Model 43 by the bayonet mounting bar under the barrel and the sling mounting slit drilled through the buttstock.

GERMAN VOLKSSTURM RIFLES

Volkssturm Gewehr 1 (VGI)

The 7.92 VG1 was a last ditch weapon made in a number of shops during the closing days of World War II. It has a crudely-made bolt and stock and uses the magazine of the semiautomatic Model 43 rifle.

Firing the VG1 can be a risky affair, since they were made at the low point of German manufacture in World War II.

Volkssturm Karabiner 98 (VK98)

The 7.92mm VK98 uses the Model 98 action combined with miscellaneous barrels from old German and foreign Mausers. The stock is very crude and is of unfinished, unseasoned wood. Most of these weapons are single shot, but some were fitted with the semiautomatic Model 43 rifle magazine.

GERMAN TRAINING RIFLES

The Germans used a number of rifles for training, including service rifles such as the Model 98 rifle converted to caliber .22; it had a conversion unit that could be easily inserted into a service rifle to convert it to .22 caliber; they had similar devices to convert the pistol 08.

They also had two caliber .22 training rifles: the Sport Model 34 and the Small Caliber KKW.

The German Sport Model 34 (DSM34)

This single-shot rifle, which was sold commercially in addition to its military use, was made by most of the standard German rifle makers, i.e., Mauser, Walther, Simson, etc. It has military-type sights and a sling, which is mounted on the left side as was done on the Kar 98k.

The Small Caliber Rifle (KKW)

This rifle is also single shot but is about ½ pound heavier than the DSM34. It has the same type bands as the Kar 98k and has a slightly improved action as compared to the DSM34.

THE 7.92mm MODEL 41(W) SEMIAUTOMATIC RIFLE (GEW41(W))

In 1941, Mauser and Walther both introduced semiautomatic rifles, which were issued in limited quantities to the German Army and used in what was apparently a competitive combat trial. The Walther Model 41(W) was the more successful of the two designs, since it was developed into the Model 43 rifle, which was made in considerable quantity. The 41(W) is quite similar to the Model 43 except for the gas system. The bolt of the Model 41(W) has locking flaps, which are pushed into the locked position by the forward movement of the firing pin and are cammed out of the locked position by the rearward movement of the bolt carrier and the firing pin.

The gas system is a modification of the Bang system. A muzzle cone traps gas, which rebounds against a piston, forcing it to the rear. The piston in turn forces an operating rod to the rear and the operating rod forces the bolt carrier to the rear, thereby unlocking the bolt. The 41(W) has a fixed magazine and is loaded with two five-round chargers. This weapon is a finely-machined weapon and is much better made than the Model 43. There are specimens of this rifle stamped G41. They do not have a bolt-release catch.

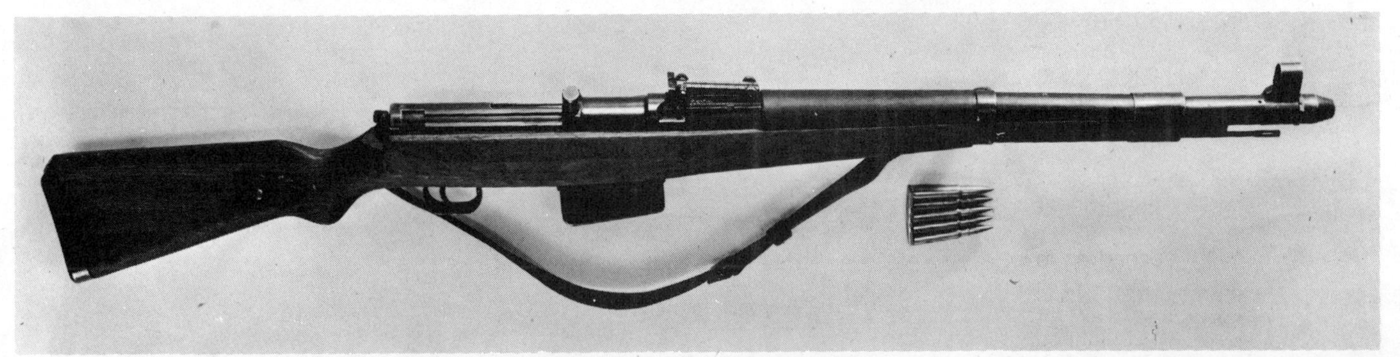

Gew 41(W), right side, bolt closed.

The Gew 41 (above) and the Gew 41(W) (below).

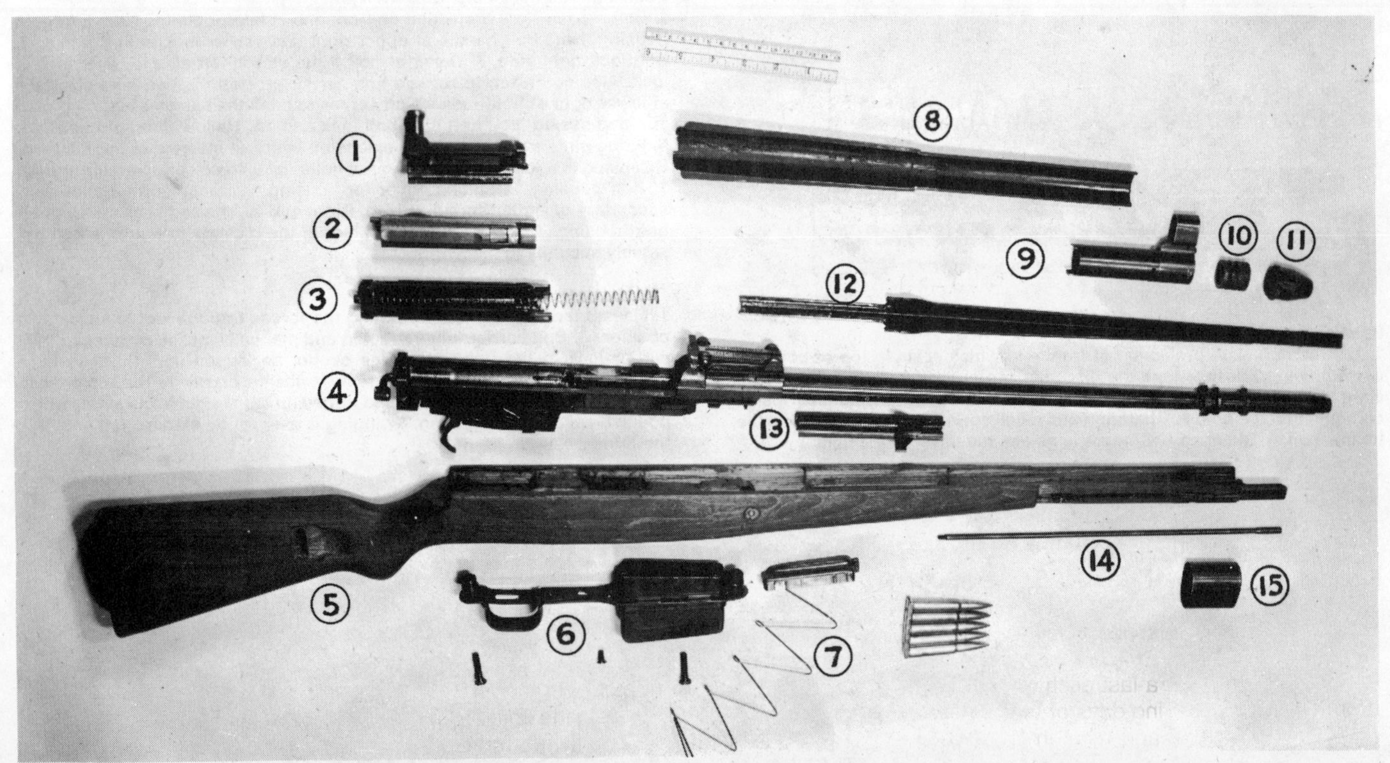

Detailed stripping to show gas and locking systems.
1. Bolt carrier. 2. Bolt. 3. Bolt housing with buffer and recoil springs. 4. Receiver and barrel assembly. 5. Stock. 6. Trigger guard and magazine. 7. Magazine follower and spring. 8. Forearm. 9. Gas cylinder. 10. Gas piston. 11. Gas cone. 12. Operating rod. 13. Operating rod trough and spring. 14. Cleaning rod. 15. Front band.

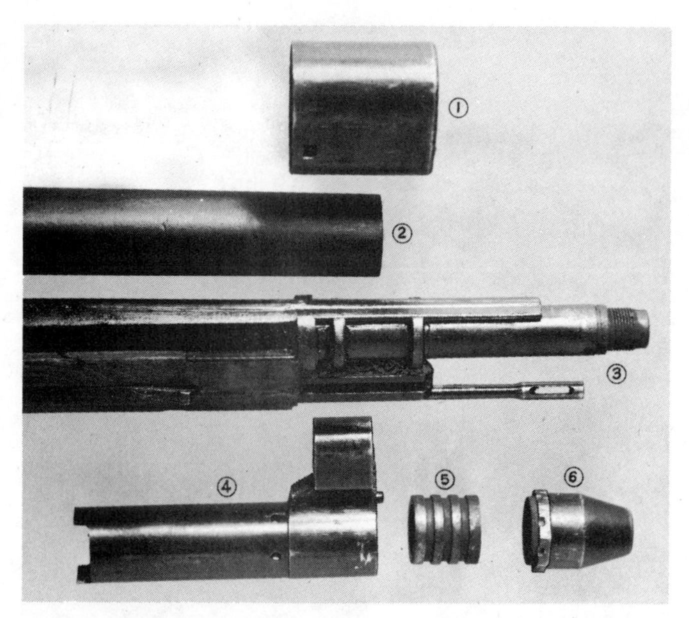

Gas operating assembly, Gew 41(W).
1. Front band lock. 2. Plastic handguard. 3. Barrel threaded for blast cone. Flat operating rod on top of barrel. 4. Operating rod cover which fits over barrel. 5. Piston which mounts around barrel and floats inside operating rod cover. 6. Blast cone which screws on barrel and traps gas to force floating piston back against operating rod.

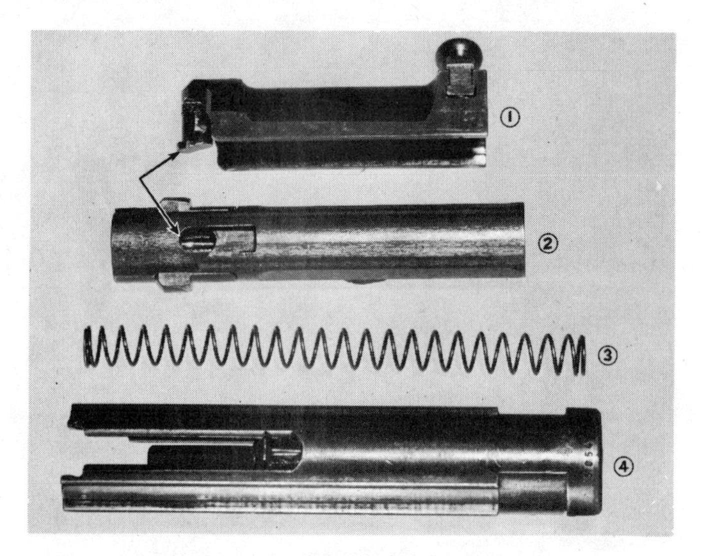

Details of bolt assembly.
1. Bolt carrier. Note projection at front end which seats in top of bolt and when driven back functions the firing-pin carrier to unlock the bolt. 2. Bolt complete with two lugs in locked position. Note firing pin housed inside carrier within the bolt. 3. Operating and recoil spring, which seats inside the hollow bolt. 4. Stamped bolt housing guides the travel of the bolt.

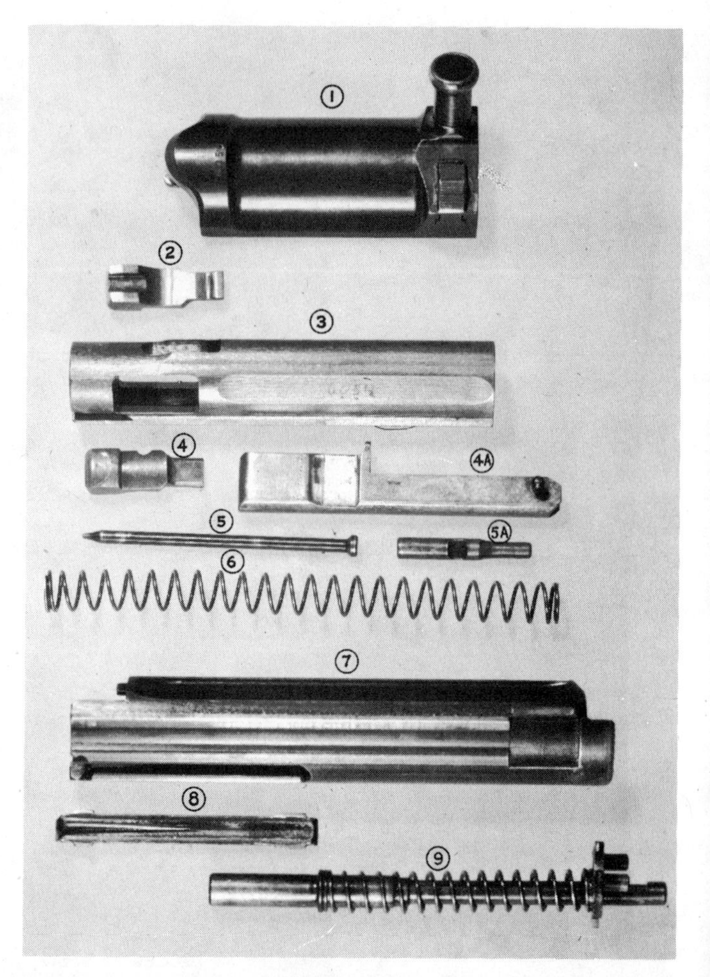

Details of bolt assembly, Gew 41(W).
1. Bolt carrier. Note projection on front end at left for pulling back firing-pin housing. Note bolt handle at upper right and carrier lock in line with it. 2. Bolt lock right side. 3. Detail of bolt. Note slot in left side to receiver bolt lock. Also cut in top to receive bolt carrier projection. Also note cam-face at lower right of bolt body which serves to cock the hammer by riding over and depressing it. 4. Left side bolt lock and 4A. Hollow firing-pin housing. 5. Firing-pin and 5A. Firing-pin extension which is inserted in the housing. Extension is retained by pin at rear of housing. 6. Recoil or operating spring. 7. Bolt housing. 8. Sliding cover for housing. 9. Recoil spring guide with secondary or buffer spring affixed. Right end of this rod projects through bolt housing. The two lugs lock in slot in the receiver to hold the bolt assembly securely in place.

Top receiver details, Gew 41(W).
1. Mauser tangent sight. 2. Operating rod forced through slot to indicate its position. 3. Bolt carrier fully retracted and manually locked by pushing bolt carrier lock to the right. Magazine cannot be loaded until this has been manually locked. 4. Bolt assembly rests inside the carrier. 5. The arm projecting from the rear of the receiver and turned up to the right is the safety which positively blocks sear action. Swinging it over to tbe extreme left sets it in the firing position.

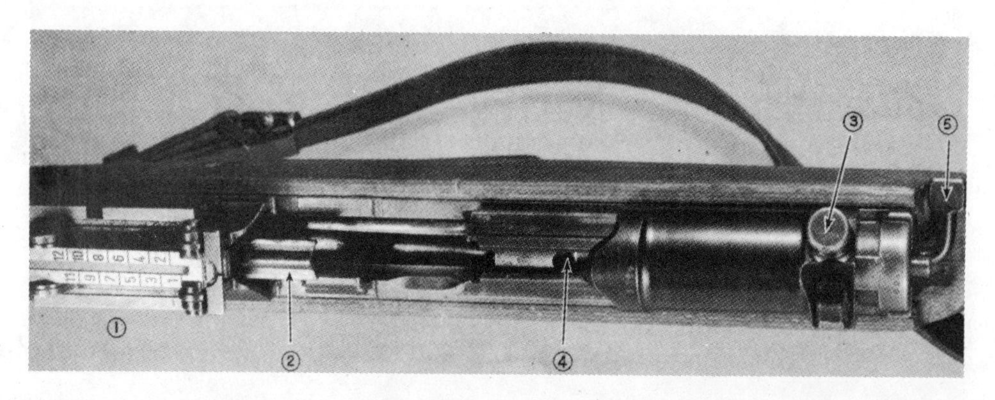

THE 7.92mm MODEL 41(W) (GEW 41(M)) SEMIAUTOMATIC RIFLE

The 41(M), a Mauser development, was not a very successful design, and was abandoned in 1943. Apparently there were not very many of these rifles made, as they are comparatively rare today. The 41(M), like the 41(W), draws its operating gas at the muzzle. The gas rebounds from a muzzle cone and strikes a piston mounted under the barrel. The piston forces back an operating rod, which is connected to the rear section of the two-piece bolt. The rear section of the bolt pulls the front section backward, causing the frontally-mounted bolt locking lugs to be cammed out of their locking recesses in the receiver.

The 41(M) has an operating handle which has the same appearance as a bolt handle on a manually-operated bolt-action rifle and is operated in much the same manner, but it does not reciprocate with the action when the weapon is fired. The magazine of the 41(M) is fixed and loaded with two five-round chargers.

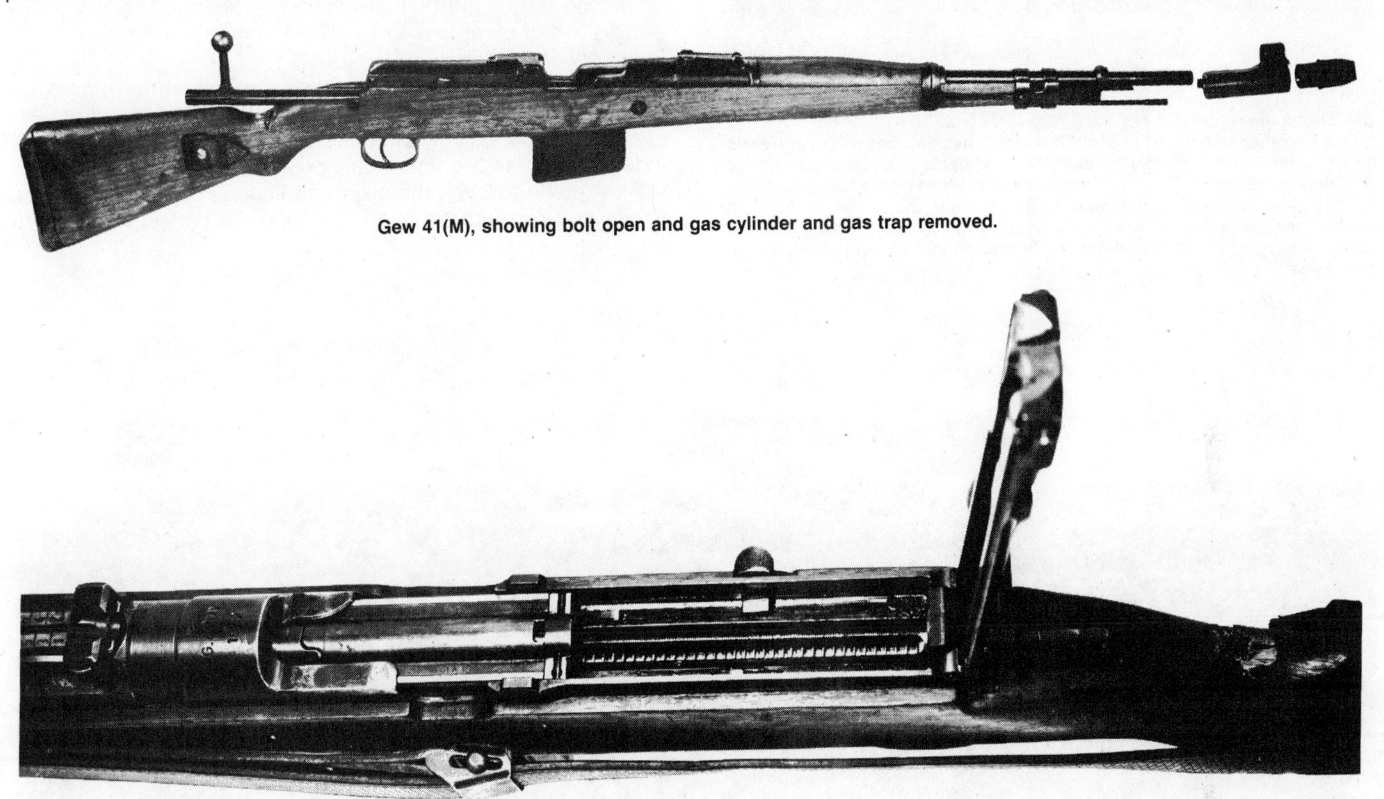

Gew 41(M), showing bolt open and gas cylinder and gas trap removed.

Bolt cover open. Showing detail of bolt and recoil spring. This is the Gew 41(M). It employs a turning bolt-head for locking.

Kar 43.

THE 7.92mm MODEL 43 SEMIAUTOMATIC RIFLE (G43)

In the *Gewehr 43 (G43)*, which is basically the same as the *Karbiner 43* or Kar43, Walther combined the bolt mechanism of the Model 41(W) with a gas system quite similar to that of the Soviet M1940 Tokarev rifle. G43 was made in large quantities, and a number of variations may be found among these rifles. The hand guard may be of wood or plastic, and the bolt carrier latch, which locks the bolt carrier and bolt to the rear, may be on the left or right side of the bolt carrier, or may not exist at all.

The G43 is very roughly made of many stampings, castings and forgings, which are machined only where necessary. All G43 rifles have a scope mounting to be used with the 1½ power ZF41 scope. The G43 has a detachable ten-round magazine, which can be loaded while in the weapon with two five-round chargers.

The G43 was used to a limited extent by the Czech Army for a few years after World War II.

How the *Gewehr 43* Works

This rifle has a gas vent drilled in the barrel about 12½ inches back from the muzzle. It is on top of the barrel and leads into a

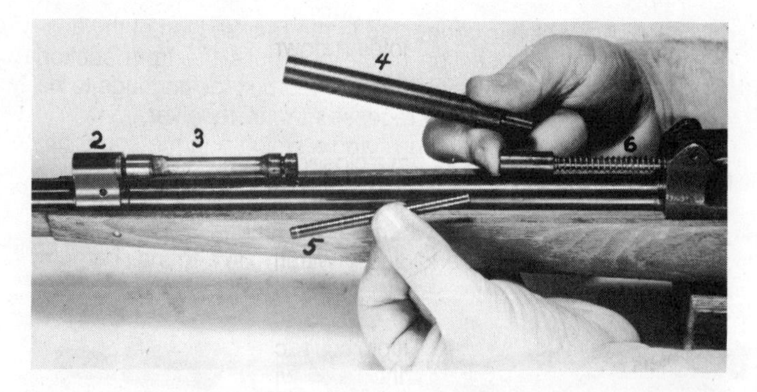

Gas piston operation.
2. Gas chamber above port in barrel into which gas escapes as bullet travels down barrel. 3. Gas cylinder screwed on. 4. Outside gas piston just pulled off the gas cylinder. 5. Connecting tappet piece. When piston is in place over cylinder, the front end of this tappet piece seats in the recess in the front end of the moving piston while its rear seats in the base of the operating rod. 6. Operating rod and spring being held to permit removal of gas piston and tappet.

gas chamber rising above the barrel, which is fitted with a cylinder to receive a very short piston.

As the bullet passes the gas vent in the barrel, a portion of the gas enters the port and expands in the cylinder driving back the short piston violently.

This piston, acting on the tappet principle, strikes the operating rod, which extends backward into the receiver. As the end of this rod passes through a hole drilled in the receiver above the line of the bolt, the thrust imparted to it is transmitted to the bolt carrier on top of the bolt. Meanwhile a spring around the operating rod is compressed to provide energy to return the rod to its forward position.

At the start of its rearward travel, the bolt carrier moves independently, leaving the weapon securely locked until the pressure has dropped. The slide carries the firing pin housing back independently of the bolt at this point also.

After a short travel, the slide and firing pin housing attached to the bolt pick up the bolt. The firing pin housing is so con-

Details of operating rod system, Gew 43.
1. Operating rod being driven to the rear. As it passes through the hole in the receiver above the line of the bore, the spring around it is compressed to store up energy for the forward movement. 1-A. Rear end of operating rod moving back to drive the bolt carrier to the rear. 2. Bolt slide moving back in its tracks. Finger on its underside will carry the firing-pin carrier to cam the bolt locks in so that the bolt may be unlocked and travel back to the rear. 3. Bolt handle used for cocking weapon. 4. Magazine release catch. 5. Ejector. (Note that the receiver and the bolt carrier are steel castings.)

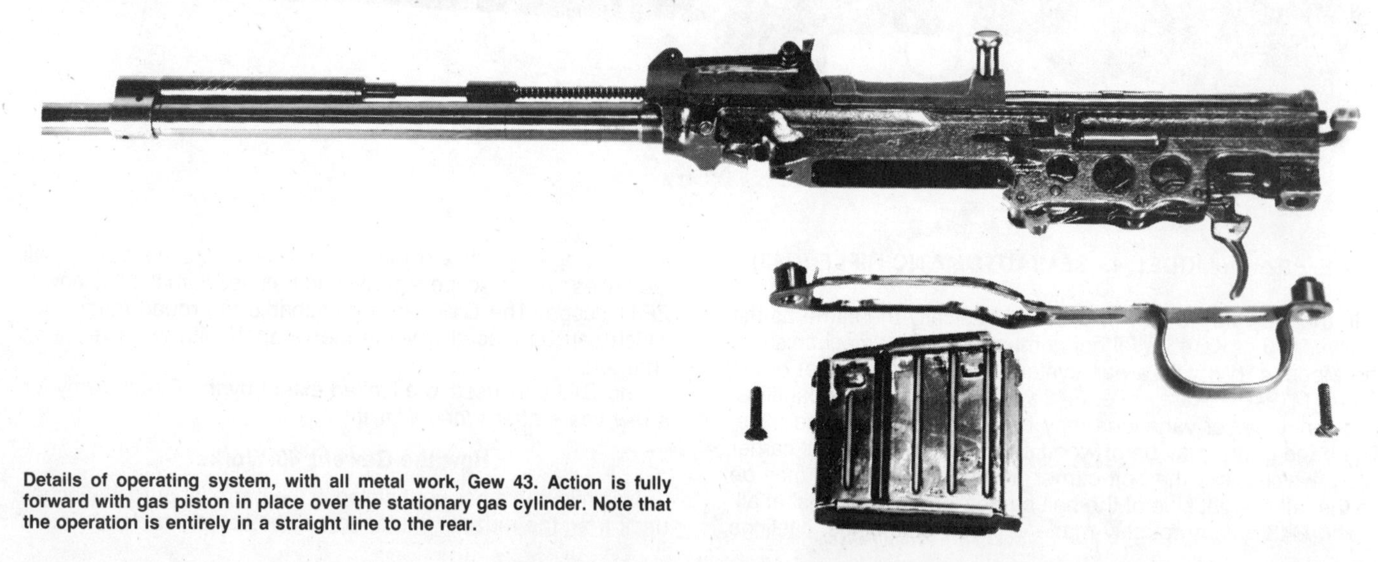

Details of operating system, with all metal work, Gew 43. Action is fully forward with gas piston in place over the stationary gas cylinder. Note that the operation is entirely in a straight line to the rear.

structed that it cams in two locking lugs, drawing them out of their seats in the receiver walls and into the surface of the bolt.

From this point on, the parts travel to the rear together extracting and ejecting the empty shell and compressing the recoil spring around its guide.

Forward Movement of the G43 Action. The recoil spring compressed around its guide now exerts a forward thrust against the bolt assembly. The bolt strips a cartridge from the magazine and chambers it, the extractor set in the face of the bolt snapping into the extracting groove of the cartridge as it is chambered.

As the bolt reaches its fully forward position, it stops against the face of the chamber. The spring still exerts forward pressure and drives the firing pin housing straight ahead independently of the bolt. This is so constructed that it forces the locking lugs, which are loosely set into each side of the bolt, out into the receiver walls in a fashion not unlike that of the Russian Degtyarev light machine gun.

When the bolt is fully home in forward position, the carrier mounted on top of it is forced still farther forward into its niche in the receiver, where it rests against the operating rod hole.

When the trigger is pressed, the sear is rotated away from the hammer (which has been ridden over and cocked by the rearward motion of the bolt). The hammer spring drives the hammer forward to strike the firing pin extension. This in turn strikes the firing pin and drives it against the primer to fire the cartridge in the chamber.

THE 7.92mm PARATROOP RIFLE MODEL 42 (FG42)

One of the most interesting German World War II rifles from a design point of view—the Paratroop Rifle (FG42)—was not adopted by the Army. It was adopted by the Air Force (*Luftwaffe*), who controlled the airborne divisions. The FG42 is a very impressive rifle from many points of view and lives on to some extent in the operating mechanism of the US M60 machine gun.

The *Fallschirmjäger Gewehr (FG42)* was developed by Rheinmettal-Borsig at the request of the German Air Force. It has been reported that only 5,000 of these rifles were made. FG42 had the following good features:

(1) Straight-line stock configuration and muzzle-brake compensator to assist in holding down the weapon in automatic fire from the shoulder.

(2) Reduction of recoil by use of a recoil-spring sliding shoulder-stock system.

(3) The weapon fires from a closed bolt in semiautomatic fire and from an open bolt in automatic fire—this solves the "cook off" problem.

FG42 was designed to replace the rifle, machine gun in the light role, and the submachine gun.

The bolt mechanism of the FG42 was copied from that of the Lewis gun, but a standard multiple coil recoil spring is used rather than the clock work type spring used with the Lewis gun. The FG42-type bolt mechanism is currently used in the US M60 machine gun. The trigger mechanism is cleverly designed and features a swivel mounted sear, which can be moved left or right to engage the semiautomatic or automatic sear notch.

A short gas pistol rod is used with this weapon; the operating handle is connected to the piston rod, which also has a stud to operate the bolt. The stud operates in a camway in the bolt, rotating it into and out of the locked position. FG42 has a spike-type bayonet, which is carried in under the barrel point reversed when not fixed, in a manner similar to that of the French MAS36 rifle. A light stamped bipod, which failed in US tests of the rifle at Aberdeen Proving Ground during World War II, is also used with this weapon.

Some of these rifles are fitted with a stamped steel stock, and others have a wooden stock. All things considered, FG42 was one of the most interesting of the German World War II designs. It did not introduce any revolutionary design principles, but it did combine a number of previously uncombined principles to produce an advanced selective-fire weapon for full-size rifle cartridges. The first US weapon patterned on the FG42 was the 7.92mm T44 light machine gun. This weapon was rather unusual; it is belt-fed—it uses the MG42 type belt-feed mechanism—but the feed cover is mounted on the side of the receiver so that the belt feeds in a vertical position rather than horizontally as is usual. This weapon was developed by the Bridge Tool and Die Manufacturing Corp. under contract with the Ordnance Corps.

It should be noted that several authoritative publications credit the design of FG42 to the Heinrich Krieghoff Plant of Suhl, Saxony, but German publications credit it to Rheinmettal.

German 7.92mm *Fallschirmjaeger Gewehr 42* automatic rifle with steel butt.

7.92mm FG42 with wooden stock.

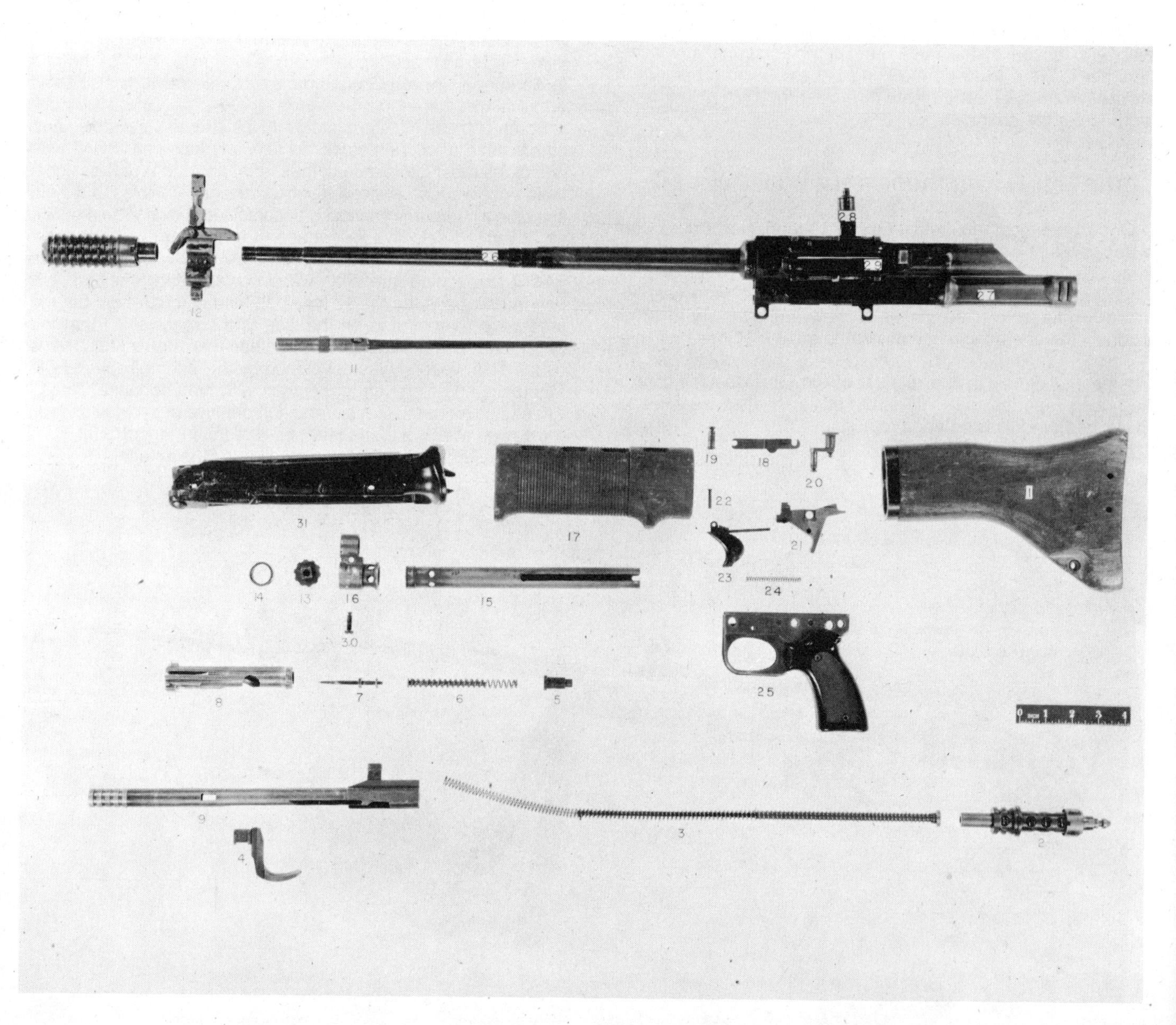

FG42, stripped.

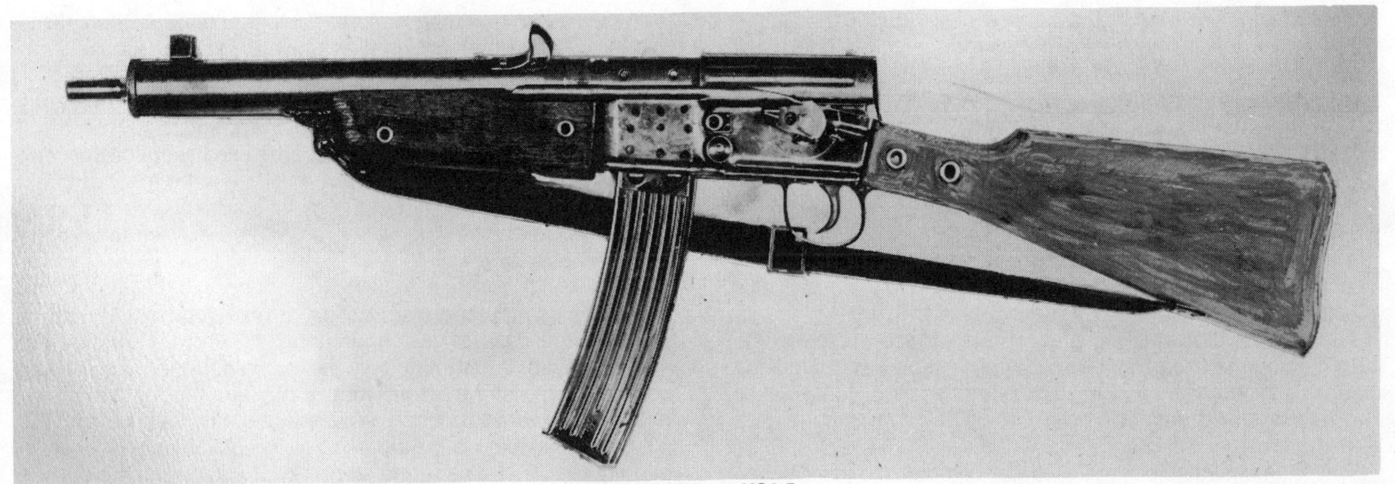

German 7.92mm VG1-5.

THE 7.92mm VG1-5 SEMIAUTOMATIC RIFLE

The *Volkssturm Gewehr 1-5* is a rather unusual weapon in many ways. The first unusual thing about the rifle is that it was never apparently approved by the *Waffenamt* (Ordnance Office) in Berlin and does not bear the usual government acceptance stamps. The VG1-5 was put into production at a time when control was crumbling in Germany, and the Nazis had given the local *Gauleiters* authority to draw up contracts for arming the *Volkssturm* in their own districts with whatever weapons they could beg, borrow or steal.

Main components, VG1-5.
1. Bolt housing and firing assembly. The bolt and slide assembly can move back in this housing when it is in place in the receiver. 2. Bolt and slide assembly. The bolt is riveted into the rear of this hollow sliding member. 3. Safety. 4. Retainer pin. 5. Firing pin and spring. 6. Recoil spring. 7. Gas chamber which locks inside front end of slide when the latter is in place around the barrel. Barrel, stock and receiver assemblies; note slide stop abutment on barrel at forward end of the forearm. Standard MP44 magazine for 7.92mm Short cartridge.

Considering this somewhat dubious ancestry, VG1-5 has some rather good design features. The weapon was designed and produced by the Gustloffwerke at Suhl and was apparently made in limited quantity. Stampings are extensively used in this weapon; the wooden furniture is left rough, and machining is of very simple type.

How the VG1-5 Operates

A standard MP 43-44 magazine loaded with 30 cartridges is inserted in the magazine housing from below and pushed in until it locks.

The bolt handle, which is a heavy steel piece riveted to the left side of the housing, is pulled back. This draws all the moving members to the rear, and the bolt rides over and compresses and cocks the hammer, which is held by a simple sear arrangement.

When the bolt handle is released, the compressed recoil spring pushes the members forward and strips a cartridge from the magazine into the firing chamber, where the extractor grips it as it seats.

Pressure on the trigger is communicated through the sear to release the hammer, which flies forward to strike the firing pin and discharge the cartridge.

As the bullet travels down the barrel, it passes over the 4 gas ports about 2½ inches before reaching the muzzle. Thus it will be seen that gas escapes from the barrel into the space between the removable sleeve and the housing to exert a forward thrust as the bullet continues out of the muzzle. The sleeve is shaped to insure that most of the gas thrust will be exerted toward the concave forward end.

As this is a retarded blowback weapon without a lock, the recoil force of the rearward action of gas against the head of the cartridge case drives the action to the rear. However, the thrust has to overcome not only the inertia and weight of moving parts as in the standard blowbacks but also the forward action of the gases, which have expanded in the moving housing. This check system delays the opening of the action long enough to permit the use of a cartridge so powerful that normally a locking device would be required to permit its use.

Rearward action ejects, recocks and reloads in standard semiautomatic fashion.

THE GERMAN STURMGEWEHR SERIES

The Germans had decided after World War I that their 7.92 × 57mm cartridge was overly powerful for shoulder weapons. Analysis of the average ranges at which rifles were commonly used and the marksmanship capabilities of the average soldier, especially under the stress of battle, led them to the conclusion that a cartridge with considerably less ballistic potential than the 7.92 × 57mm would be adequate and in addition would result in shorter, generally lighter weapons, allow the soldier to carry more cartridges on his person, cause less fatigue from recoil and result in a considerable saving of materials in the manufacture of propellents, cartridge cases and bullets.

The German requirement was solidified in 1934, and prototype cartridges were produced by Gustav Genschow, Rheinisch Wesphalische Sprengstoff (RWS) and Polte. Polte was given the development contract in 1938 and produced the "7.9mm Infanterie Kurz Patrone" by 1941. This cartridge had a case 33mm long with 24.6 grains of propellent as opposed to the 57mm case and 45-50 grains of propellent of the standard full-size 7.92mm rifle cartridge.

To parallel the cartridge development, Haenel was awarded a contract in 1938 for development of a weapon for these cartridges. Hugo Schmeisser of Haenel produced a gas-operated weapon for the 7.92mm Kurz cartridge by 1940, and 50 specimens of the prototype were produced by July 1942. Walther started development of the weapon for the cartridge in 1940, basing their design upon that of an earlier semiautomatic rifle of their conception—the GA115.

MACHINE CARBINE MKb42(H) AND MACHINE CARBINE MKb42(W)

Both the Haenel and Walther designs were produced in limited quantity—approximately 7,800 of each—as machine carbines (Maschinen Karabiner), designated MKb42(H) and MKb42(W), respectively. They were extensively used on the Russian front, and the Haenel design proved superior to that of Walther.

The Walther design MKb42(W) has a somewhat unusual gas system, which was carried on in the design of the Czech Model 52 rifle. Rather than the conventional gas tube, usually found on gas-operated rifles, the gas in this design is confined by a steel jacket around the barrel and drives a piston, which encircles the barrel, and an operating sleeve. There are two gas ports in the barrel, and the bolt is operated by the sleeve. The bolt has frontal locking lugs.

Differences Between MKb42(H) and MP43

The Haenel MKb42(H) is generally similar in internal design to the MP43 series of weapons. The principal differences be-

Walther 7.92mm GA115 semiautomatic rifle.

7.92mm MKb42(W).

tween the MKb42(H) and the MP43 series of weapons are as follows:

(1) The piston of the MKb42(H) is longer than that of the MP43 and is mounted in a separate tube, divided by a visible air space above the barrel. In the MP43, the piston rides in a tunnel immediately above the barrel.

(2) There is a cut-out for the bolt handle on the receiver of the MKb42(H); this cut-out is not present on the MP43.

(3) MKb42(H) has a bayonet lug (its prototype does not). MP43 does not have a bayonet lug.

There are also differences in the stock, fittings, etc.

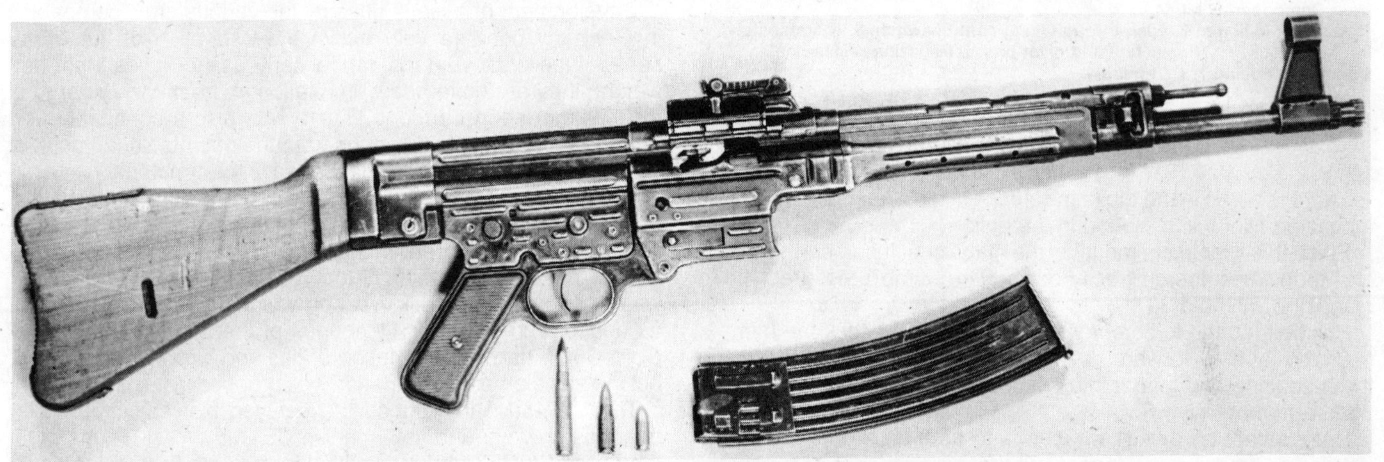

MP43.

THE MP43, MP43/1, MP44, AND StG44 ASSAULT RIFLES

Schmeisser reworked the MKb42(H) in the spring of 1943, and the MP43, 44 series of weapons was born. This was a significant event in current military small arms history, since it introduced for the first time, in large quantities, the concept of the selective fire assault rifle chambered for the "intermediate"-sized cartridge.

The MP43 was adopted by the Waffenamt as a standard weapon and after 1944 was scheduled to replace the rifle, submachine gun and light machine gun in the infantry squad. By February 1944, production of this weapon had risen to about 5,000 per month. Producers of this family of weapons were Haenel, Mauser and Erma; at least seven subcontractors made components. The term "family" as it refers to these weapons means the MP43, MP43/1, MP44 and StG44. All are essentially the same weapon; minor differences are as follows:

(1) MP43/1 is the same as MP43 but has a screw on type grenade launcher rather than the clamp on type used with MP43.

(2) There is no apparent reason for the change in nomenclature from MP43/1 to MP44. Most MP43/1 rifles have the V-type telescope mounting bracket on the right side of the receiver; some MP44s have this bracket, but no MP43s have been found with the bracket.

The change in nomenclature to StG44 was politically inspired, but "Assault Rifle" (StG Sturmgewehr) is more truly descriptive of the role of the weapon than is "Submachine Gun" or "Machine Pistol" (MP-*Maschinen Pistole*).

The StG44(P) and StG44(V) were experimental versions of StG44; they had 90° curved and 40° curved barrels, respectively. Both were rejected by the Waffenamt.

Field Stripping the MP43—MP44 Series

A spring-held pin passes through the receiver and stock from the right. Pull this out from the right side.

The stock may now be withdrawn exposing the recoil spring.

Press the trigger and swing the trigger guard and all its contained units (these are not dismountable) down on its hinge.

Pull back the bolt handle. This will bring back the recoil spring, bolt, bolt carrier and the piston for removal.

With the small steel tool found in the butt trap inserted in the hole of the gas cup protruding from the casing at the front end, the gas cup may be unscrewed and withdrawn. The cylinder may now be cleaned without difficulty.

The bolt may be lifted off the bolt carrier, and the extractor and firing pin removed.

Main Components, MP44.
1. Receiver lock pin. 2. Push-through firing control switch. 3. Thumb safety, in off position. 4. Trigger housing pivot pin. 5. Magazine release catch. 6. Magazine housing 7. Bolt handle.

Loading and Firing the MP43-44

There is a trap in the top of the stock on this weapon. A special magazine filler will be found in the hollow.

Place this filler in the mouth of the magazine. Insert cartridges, and then force them directly down with thumb pressure. The magazine will hold 30.

Pull bolt handle to compress recoil spring and cock the hammer and let it fly forward. It will pick up the top cartridge from the magazine and load it into the chamber. The dust cover will open automatically.

Push safety up unless weapon is to be fired at once. If it is left down, pressing the trigger will fire a single shot or full automatic fire depending on which way the control button has been pushed through the receiver.

How the MP43-MP44 Weapons Work

When the trigger is pressed, the hammer is released to strike the inertia firing pin. This is a wedge-shaped pin that does not have a conventional spring. It is primer retracted.

As the bullet passes the gas port in the barrel, it expands into the gas chamber or cup screwed into the housing around the barrel and on top of it.

The gas impinges on a piston somewhat resembling the old Lewis piston and drives it to the rear. In the start of its rearward travel, the piston (which has attached to it by a fixed pin the bolt carrier) can move without interfering with the secure locking of the weapon. A gas vent in the top of the casing permits the gas to escape as the gas end of the piston clears it.

After a short rearward travel the bolt carrier hook picks up the separate bolt member, mounted below it, and pulls it down and back to perform the unlocking action.

The recoil spring is mounted behind the bolt extending back to the stock. This spring is compressed as the moving members travel to the rear to extract and eject the empty case in normal fashion and to cock the hammer.

This weapon is fitted with a disconnector.

THE StG45(M) AND OTHER PROTOTYPES

Development of assault rifles by the various arms companies in Germany became very active after adoption of the MP43 series. One disadvantage of the early assault rifles was their weight; they are quite heavy in relation to the muzzle energy of the 7.92mm short cartridge. There was also a continual effort at this time to simplify weapons from the manufacturing point of view and from the point of view of saving on materials.

Gustloff Werke, Haenel, Mauser and possibly Erma all developed prototypes, and although none were accepted, the Mauser weapon had a great influence on future weapon developments. The Mauser development, originally called *GerätO6(H)*, was a delayed blowback weapon weighing only 8.18 pounds, as opposed to the over 11 pounds of the StG44. The original *GerätO6(H)* had a combination of gas and blowback operation, but the gas element of operation was dropped in the final design.

The StG45(M) introduced the delayed blowback with roller bearings now used in the Spanish CETME, West German G3, the Swiss StuG57 and in modified form in the French Model 52 machine gun. The construction of this bolt is explained in detail under the G3 Rifle in Chapter 21.

MP44 field-stripped. Note that the gas piston, operating rod, operating handle, spring guide, bolt camming and locking units are actually only two units. A pin secures the rod section to the rear operating section. The bolt is a separate unit.

7.92mm StG45(M) Assault Rifle.

StG45(M), stripped.

CHARACTERISTICS OF GERMAN BOLT-ACTION RIFLES AND CARBINES.

	Rifle 1888	Carbine 1888	Rifle 1891	Rifle 98	Carbine 98
Caliber:	7.92mm.	7.92mm.	7.92mm.	7.92mm.	7.92mm.
Length overall:	48.91 in.	37.4 in.	37.3 in.	49.2 in.	37.4 in.
Barrel length:	29.1 in.	17.6 in.	17.6 in.	29.1 in.	16.9 in.
Feed device:	5-round, in-line, fixed, box magazine.	5-round, in-line, fixed, box magazine.	5-round, in-line, fixed, box magazine.	5-round, staggered row, fixed, box magazine.	5-round, staggered row, fixed, box magazine.
Sights: Front:	Barley corn.	Barley corn.	Barley corn.	Barley corn.	Barley corn.
Rear:	Leaves with V notch.	Leaves with V notch.	Leaves with V notch.	Bridge type tangent or tangent leaf, V notch.	Bridge type tangent w/V notch.
Muzzle velocity: (at date of adoption)	2099 f.p.s.	1935 f.p.s.	1935 f.p.s.	2099 f.p.s.	590 m/s.
Weight:	8.56 lb.	6.88 lb.	6.8 lb.	8.81 lb.	Approx. 7.5 lb.

CHARACTERISTICS OF GERMAN BOLT-ACTION RIFLES AND CARBINES (Cont'd)

	Carbine 98a	Carbine 98b	Carbine 98k	Rifle 40k	Rifle 33/40
Caliber:	7.92mm.	7.92mm.	7.92mm.	7.92mm.	7.92mm.
Length overall:	43.3 in.	49.2 in.	43.6 in.	39.1 in.	39.1 in.
Barrel length:	23.6 in.	29.1 in.	23.6 in.	19.2 in.	19.29 in.
Feed device:	5-round, staggered, fixed, box magazine.	5-round, staggered, fixed, box magazine.	5-round, staggered, fixed, box magazine.	5-round, staggered, fixed, box magazine.	5-round, staggered, fixed, box magazine.
Sights: Front:	Barley corn.	Barley corn.	Barley corn.	Barley corn.	Barley corn.
Rear:	Tangent w/V notch.	Tangent w/V notch.	Tangent w/V notch.	Tangent w/V notch.	Tangent w/V notch.
Muzzle velocity: (at date of adoption)	2853 f.p.s.	2574 f.p.s.	2476 f.p.s.	Approx. 2400 f.p.s.	Approx. 2400 f.p.s.
Weight:	8 lb.	9 lb.	8.6 lb.	8.3 lb.	7.9 lb.

CHARACTERISTICS OF GERMAN BOLT-ACTION RIFLES AND CARBINES (Cont'd)

	Rifle 98/40	VG1	VK98	DSM34	KKW
Caliber:	7.92mm.	7.92mm.	7.92mm.	Cal. .22 L.R.	Cal. .22 L.R.
Length overall:	43.6 in.	43 in.	40.6 in.	43.3 in.	43.7 in.
Barrel length:	23.6 in.	23.2 in.	20.8 in.	25.98 in.	25.98 in.
Feed device:	5-round, staggered row, fixed, box magazine.	10-round, staggered row, detachable, box magazine.	Mostly single shot.	Single shot.	Single shot.
Sights: Front:	Barley corn.	Post.	Barley corn.	Barley corn.	Barley corn.
Rear:	Tangent with V notch.	V notch.	V notch.	Tangent with V notch.	Tangent with V notch.
Muzzle velocity: (at date of adoption)	2476 f.p.s.	2476 f.p.s.	2400 f.p.s. (approx)	1500 f.p.s. (approx)	1500 f.p.s. (approx)
Weight:	8.9 lb.	8.3 lb.	6.9 lb.	7.7 lb.	8.6 lb.

CHARACTERISTICS OF GERMAN WORLD WAR II SEMIAUTOMATIC AND SELECTIVE FIRE RIFLES

	Rifle 41(M)	Rifle 41(W)**	Rifle 43***	FG42	VG1-5
Caliber:	7.92 × 57mm.	7.92 × 57mm.	7.92mm × 57mm.	7.92 × 57mm.	7.92mm Kurz (short).
System of operation:	Gas, semi-automatic only.	Gas, semi-automatic only.	Gas, semi-automatic only.	Gas, selective fire.	Delayed blowback, semiautomatic only.
Overall length:	46.25 in.	44.25 in.	44 in.	37 in.	35 in.
Barrel length:	21.75 in.	21.5 in.	21.62 in.	19.75 in.	14.75 in.
Feed device:	Fixed, 10-round*, staggered row, box magazine.	Fixed, 10-round, staggered row, box magazine.	Detachable, 10-round, staggered row, box magazine.	Detachable, 20-round, staggered row, box magazine.	Detachable, 30-round, staggered row, box magazine.
Sights: Front:	Barley corn.	Barley corn.	Barley corn.	Barley corn on folding base.	Fixed post.
Rear:	Tangent leaf w/U notch.	Tangent leaf w/U notch.	Tangent leaf.	Aperture on folding base.	Non-adjustable U notch.
Muzzle velocity:	Approx. 2550 f.p.s.	Approx. 2550 f.p.s.	Approx. 2550 f.p.s.	Approx. 2500 f.p.s.	2163 f.p.s.
Cyclic rate:	—	—	—	750–800 r.p.m.	—
Weight:	11.25 lb.	11.08 lb.	Approx. 9.5 lb.	9.93 lb.	10.18 lb.

*The magazine can be removed, but is not easily removable for reloading.
**This weapon is also called Rifle 41 (SG41).
***This weapon may be marked G43 or K43.

CHARACTERISTICS OF GERMAN WORLD WAR II SEMIAUTOMATIC AND SELECTIVE FIRE RIFLES (Cont'd)

	MKb42(W)	MKb42(H)	StG44	StG45(M)
Caliber:	7.92mm Kurz.	7.92mm Kurz.	7.92mm Kurz (PP43 m.e.).	7.92mm Kurz (PP43 m.e.).
System of operation:	Gas, selective fire.	Gas, selective fire.	Gas, selective fire.	Delayed blowback, selective fire.
Overall length:	36.75 in.	37 in.	37 in.	35.15 in.
Barrel length:	16.1 in.	14.37 in.	16.5 in.	15.75 in.
Feed device:	Detachable, 30-round, staggered row, box magazine.	Detachable, 30-round, staggered row, box magazine.	Detachable, 30-round, staggered row, box magazine.	Detachable, 30-round, staggered row, box magazine.
Sights: Front:	Hooded barley corn.	Hooded barley corn.	Hooded barley corn.	Hooded barley corn.
Rear:	Tangent w/U notch.	Tangent w/U notch.	Tangent w/U notch.	Tangent w/U notch.
Muzzle velocity:	2132 f.p.s.	Approx. 2100 f.p.s.	2132 f.p.s.	Approx. 2100 f.p.s.
Cyclic rate:	600 r.p.m.	500 r.p.m.	500 r.p.m.	350–450 r.p.m.
Weight:	9.75 lb.	11.06 lb.	11.5 lb.	8.18 lb.

GERMAN SUBMACHINE GUNS

GERMAN MODELS AND MODIFICATIONS OF FOREIGN MODELS

The first German submachine gun was the Bergmann 9mm MP18I which appeared in 1918. This blowback-operated submachine gun was fed with the 32-round "snail"-type magazine developed for the Luger pistol. The MP18I was designed by Hugo Schmeisser, who was by far the best-known of the German submachine gun designers.

The weapon was introduced into battle during the last part of World War I; about 35,000 were made before the war ended. Many of these weapons were used by the German civil police after the war. The MP18I was modified after the war by removing the magazine housing for the "snail"-type magazine and fitting a magazine housing for a box-type magazine. This modification was done by Haenel.

Further modification resulted in the MP28II, which was also produced by Haenel. The 28II has selective fire and a tangent type rear sight. This weapon was extensively used by German police, to include SS Police units, but was never officially adopted by the German Army. This does not mean that army personnel did not use the weapon at one time or another; as with all other small arms, the German Army used almost anything it could get. The MP28II was manufactured in Belgium by Pieper and was adopted by the Belgian Army as the *Mitraillete Model 34*. It was also used by Bolivia, in addition to several other South American countries.

The next German submachine gun produced in quantity was the Bergmann 9mm 34/I. This gun, unlike the earlier Bergmanns, was not designed by Hugo Schmeisser. The prototypes of this gun were made in Denmark, circa 1932; production of the weapon in Germany was at the Walther plant in Zella Mehlis, since Bergmann did not have production facilities. This weapon was not adopted by the German Army but was exported on a limited scale. The Model 34/I can be distinguished by its bolt handle—which resembles, and is operated in a fashion similar to a manually operated bolt-action rifle—and its trigger mechanism. The weapon has two triggers; pressure on the outer trigger produces semiautomatic fire until the inner trigger is engaged, at which time the weapon fires automatically. The Model 34/I was produced in long barrel and short barrel versions.

The 9mm Model 35/I is a modified Model 34/I. This weapon was produced during World War II for the SS by Junker and Ruh. German police manuals of this period refer to this weapon as the Model 35. The Model 35/I was adopted by Ethiopia and Sweden, which called the weapon the "M/39."

The Erfurter Werkzeug and Maschinenfabrik "Erma Werke" of Erfurt produced a submachine gun designed by Heinrich Vollmer; it had fairly wide distribution as the Vollmer Erma. The most common model is called the EMP or MPE. Versions of the Erma were sold to France, Mexico and Yugoslavia. This weapon, as well as many of the earlier German submachine guns, was extensively used in the Spanish Civil War. The Erma submachine gun MPE was used by the German police and Waffen SS Units; it was never adopted by the German Army.

This first submachine gun to be adopted by the German Army after the MP18I was the 9mm MP38. The MP38 and its successor the MP40 were developed by the Erma Werke at the request of the German Army. Although the weapon is commonly called a "Schmeisser," Hugo Schmeisser had little if any connection with its design. The Haenel firm, of which Schmeisser was general manager, was among the producers of the weapon

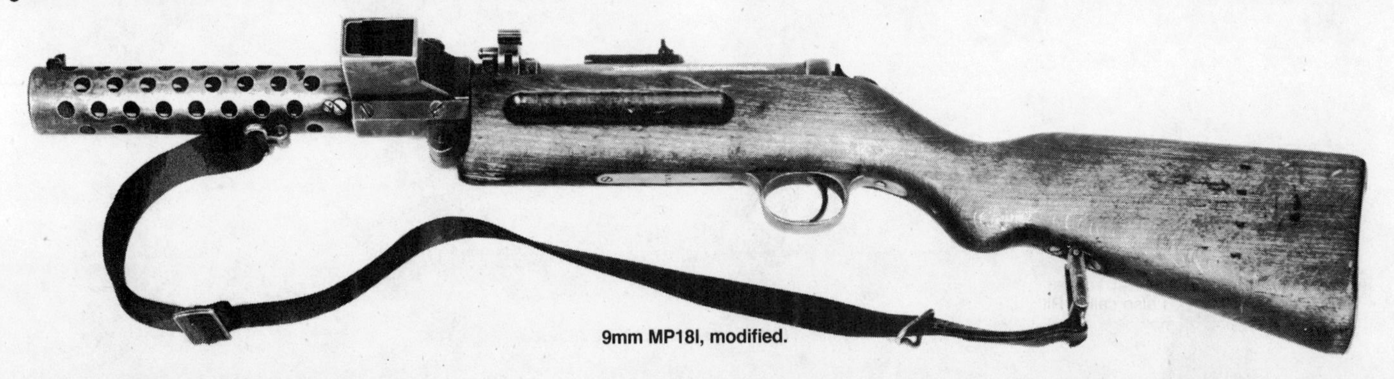

9mm MP18I, modified.

during the war, and Schmeisser developed the MP41, a modification of the MP40. The MP38 and 40 were the standard submachine guns of the German Army during World War II, and over a million of these weapons were made.

During the war several new designs of submachine guns were produced by the Germans in rather limited quantities. In general, their development seemed to be an attempt to produce a cheap, easily-made submachine gun in order to conserve materials and manufacturing facilities.

There are indications, however, that the Mauser-made copy of the British Mark II Sten may have been intended for some clandestine use, since even the British markings were copied. This weapon is called the *Gerät Potsdam.*

The MP3008 is also a copy of the Sten, produced by Mauser and six other firms. This weapon, which appeared in a number of versions, was intended for *Volkssturm* use. Production did not begin until the closing months of the war.

The EMP44 was developed by Erma Werke; it is a relatively simply made weapon consisting mainly of welded steel tubing. It appears to have been made only as a prototype and was probably designed for special use.

The Germans modified a number of Soviet PPSh M1941 submachine guns for their own use by altering them to use the MP38 and 40 magazine and fitting them with 9mm Parabellum barrels. In addition, the Germans used other foreign-made weapons. The Italian 9mm Beretta Model 38/42 was made for the German Army with German markings and acceptance stamps. The Steyr Solothurn submachine gun adopted by Austria was used by the Germans as the MP34 (Ö). Weapons produced prior to 1939 are chambered for the 9mm Mauser cartridge; those produced during 1939 and 1940 were chambered for the 9mm Parabellum cartridge. The Austrian Police used the MP34, to use the Austrian nomenclature, in a version chambered for the 9mm Steyr cartridge; these weapons were used to some extent by the German *Ordnungs polizei*.

The 9mm Model 38, Model 40 and 41 Submachine Gun (MP38, 40 and 41)

The MP38 was the first submachine gun developed for the German Army since the MP18I of World War I. Although the design has been credited to Schmeisser in many publications, it was probably designed by Erma; first production was carried on at that plant. The telescoping, multi-piece recoil spring and firing pin assembly were developed from those used with the Erma submachine gun. The MP38 was made from 1938 to 1940 at the Erma plant.

The plastic receiver housing and aluminum frame and folding steel stock of the MP38 were unique, and the design of this weapon had considerable influence on later submachine guns. The receiver of the MP38 is made of steel tubing.

The MP38 had one serious deficiency—which is shared by most submachine guns—it was not completely safe to handle. The only safety, a cut-out in the receiver into which the bolt handle locked when the gun was cocked, did not allow the gun to be carried safely with the bolt forward and a loaded magazine in the gun. If the gun received a severe jolt, such as falling on its breech end, the bolt could bounce back far enough to pick a round up from the magazine and fire. The MP38 was modified to remedy this defect by the fitting of a two-piece bolt handle and the cutting of a slot above the front of the bolt receiver track to lock the bolt in the forward position. This modification was called the MP38/40.

The MP38 was somewhat expensive to manufacture and the weapon was re-engineered to cut down the use of expensive tooling. The weapon produced as the result of this redesign was called the MP40, which differs from the MP38 in the following: new ejector, magazine release assembly, receiver (ribbing eliminated), grip frame of the MP38 is case aluminum while that of the MP40 is formed, and the stamped middle tube of the recoil spring assembly is drawn and pinched on the MP40. There are

German 9mm Model 38 Machine Pistol.

numerous other minor differences. MP40 was made in much larger quantities than was the MP38 and was manufactured by Steyr, Haenel and Erma with the assistance of a number of subcontractors. Over 1,000,000 MP40s were made from 1940 to 1944.

There were several modifications of the MP40. The most common modification of the MP40 has a stamped, ribbed magazine housing and uses the two-piece bolt handle. This weapon, which apparently was called MP40/I, is far more common than the MP40 itself. A rarer modification is the MP40/II, which is fitted with a magazine housing to accommodate two magazines. The magazines are held in a sliding housing arranged to allow each magazine to feed in turn.

The MP41 was developed at Haenel by Schmeisser. It was made in very limited numbers and it was not used by the German Army or Police; it may have been made for export. MP41 has the receiver and barrel assembly and bolt assembly of the MP40, but the stock and trigger mechanism are modeled on that of the MP28II.

Loading and Firing the MP38

Six spare magazines and a special magazine loader are issued in a web haversack with each one of these guns. The loader is a simple lever device with an attached housing into

Loading the MP38/40

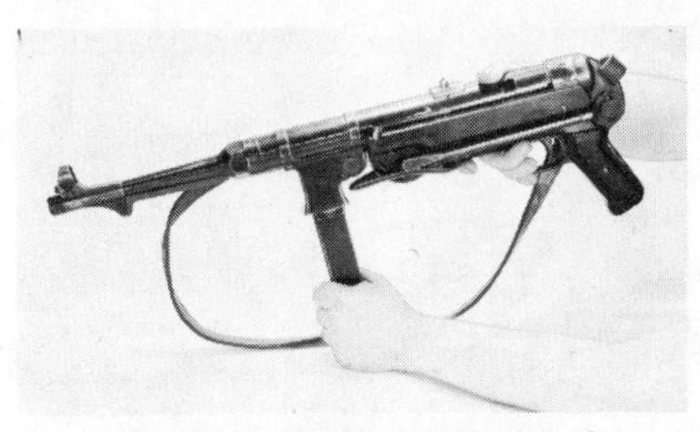

Insert loaded magazine from below into the magazine housing and push up until it locks. Note: a stud on the outside of the magazine will prevent it from going in beyond the proper length.

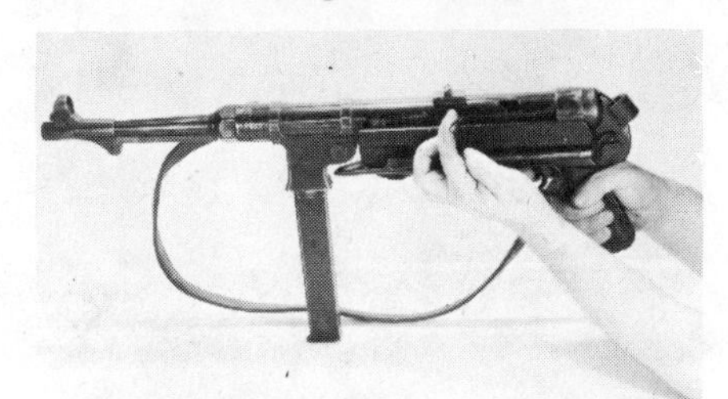

Warning: always remember that this weapon fires when the bolt goes forward! Never, therefore, let the bolt go home while the loaded magazine is in position. Unless you wish to fire the weapon, always remove the magazine before easing the bolt home.

Whenever possible, always use this weapon as a carbine. To do this, press the catch stud as indicated. This will release catch and permit you to unfold the stock and turn the butt piece down into proper place for firing from the shoulder.

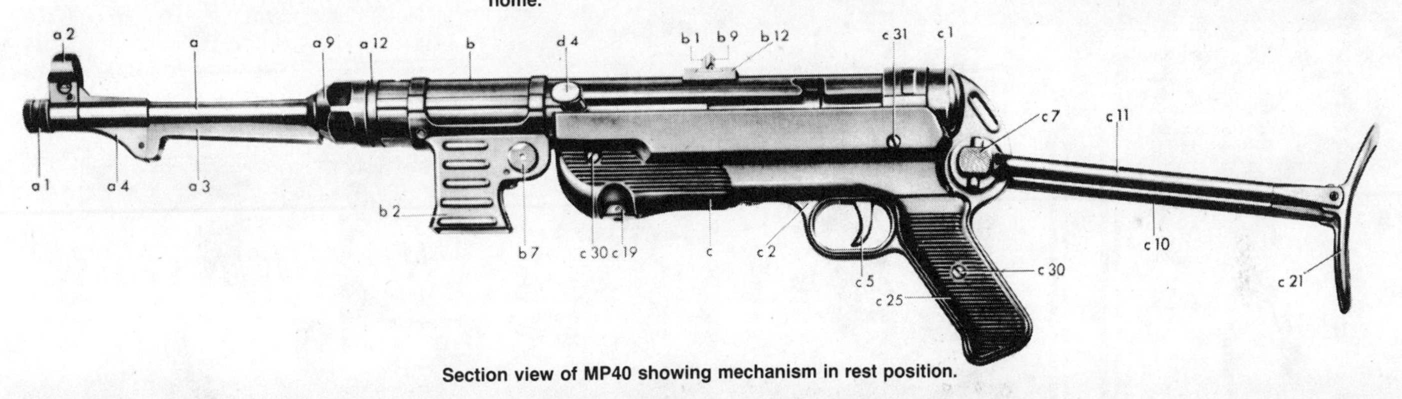

Section view of MP40 showing mechanism in rest position.

CHARACTERISTICS OF GERMAN SERVICE SUBMACHINE GUNS

	MP18I	MP38	MP40	MP40II	MP41
Caliber:	9mm Parabellum.	9mm Parabellum.	9mm Parabellum.	9mm Parabellum.	9mm Parabellum.
System of Operation:	Blowback, full automatic only.	Blowback, full automatic only.	Blowback, full automatic only.	Blowback, full automatic only.	Blowback. Selective fire.
Length:					
Stock extended:	32.1 in.	32.8 in.	32.8 in.	32.8 in.	34 in.
Stock folded:	—	24.8 in.	24.8 in.	24.8 in.	—
Barrel length:	7.88 in.	9.9 in.	9.9 in.	9.9 in.	9.9 in.
Feed device:	32-round, "snail" drum type, detachable, box magazine.	32-round, detachable, staggered row, box magazine.	32-round, detachable, staggered row, box magazine.	64 rounds in 2 detachable, staggered box magazines.	32-round, detachable, staggered row, box magazine.
Sights: Front:	Barley corn.	Hooded barley corn.	Hooded barley corn.	Hooded barley corn.	Hooded barley corn.
Rear:	Notched flip-over leaf.	Notched flip-over leaf.	Notched flip-over leaf.	Notched flip-over leaf.	Notched flip-over leaf.
Weight:	9.2 lbs.	9.5 lbs.	8.87 lbs.	10 lbs.	8.15 lbs.
Cyclic rate:	350–450 r.p.m.	500 r.p.m.	500 r.p.m.	500 r.p.m.	500 r.p.m.
Muzzle velocity:	Approx. 1250 f.p.s.	Approx. 1300 f.p.s.	Approx. 1300 f.p.s.	Approx. 1300 f.p.s.	Approx. 1300 f.p.s.

which the magazine is inserted. Snapping a cartridge into the top of the housing and pushing down firmly on the lever loads the individual cartridge into the magazine. This motion is repeated until the magazine is filled. If no loader is available, cartridges may be inserted by the normal procedure for loading automatic pistol magazines; leverage for inserting the last few cartridges may be exerted by pressure of both thumbs, once the cartridge has been seated.

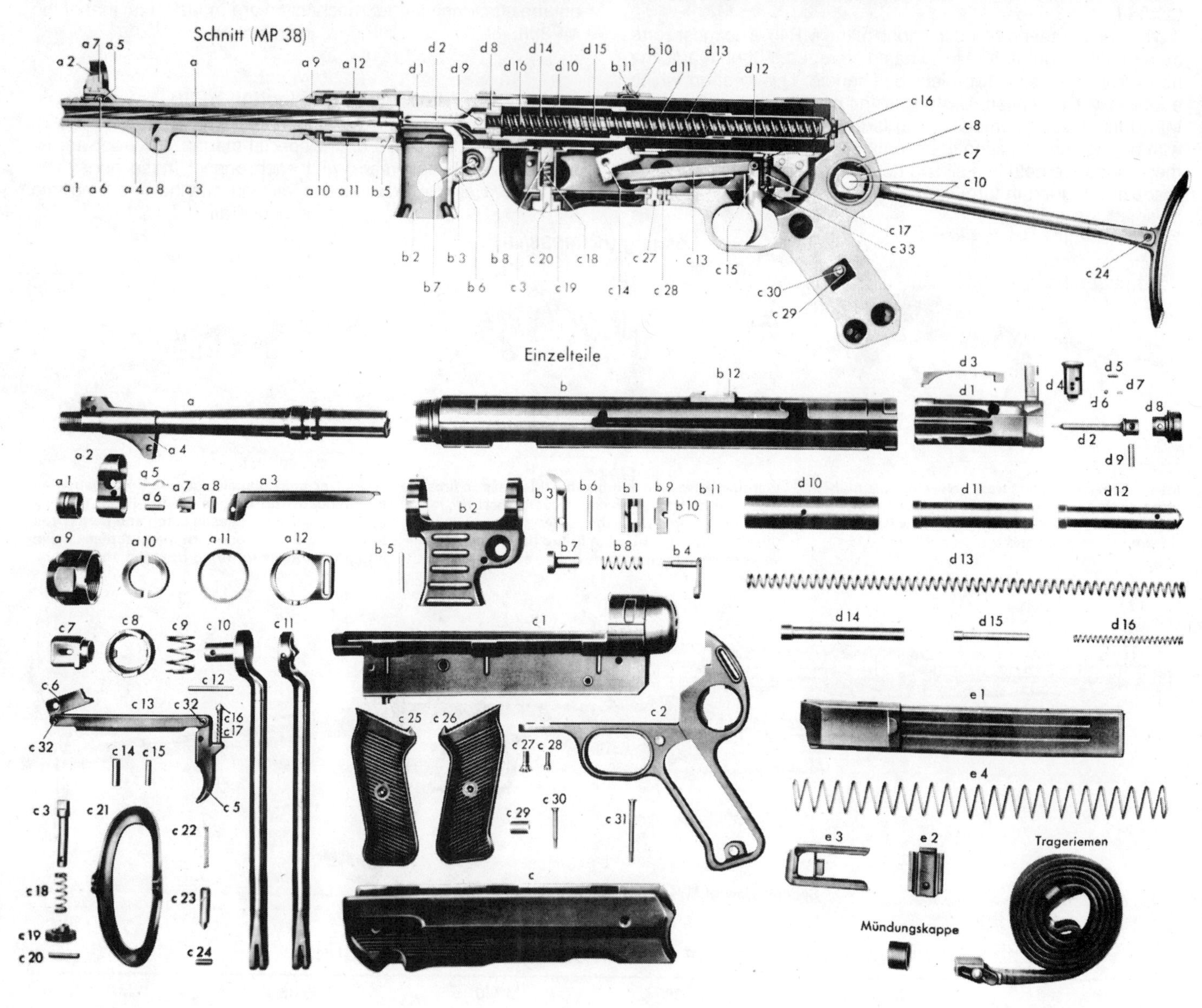

a2 Front cap cover
d10 Recoil spring tube large
a7 Front sight
a5 Front sight retainer
a Barrel
a9 Barrel nut
a12 Collar
d1 Bolt
d2 Firing pin
d9 Firing pin retaining pin
d8 Recoil spring tube end
d14 Recoil spring
d10 Recoil spring tube large
d15 Recoil guide
b10 and b11 Rear sight leaf spring
d11 Recoil spring second tube
d13 Buffer spring
d12 Buffer spring tube
c10 Stock arm
c7 Stock pivot
c8 Stock release
c24 Shoulder piece pivot

a1 Barrel cap
a6 Cover retainer
a4 Barrel jacket
a8 Resting bar pin
a3 Resting and retracting bar
a10 Barrel nut washer
a11 Barrel threads
b2 Magazine guide
b7 Magazine release cap
b3 Magazine release screw
b6 Washer
c3 Receiver lock
c20 Receiver lock screw retainer
c19 Receiver lock screw
c18 Receiver lock spring
c14 Sear
c27 and c28 Frame screws
c13 Sear lever
c15 Trigger axis screw
c17 and c16 Trigger spring
c29 and c30 Grip screws
a Barrel

a1 Barrel cap
a2 Front sight cover
a3 Resting bar
a9 Barrel nut
a12 Collar
b Chamber cover
b2 Magazine guide
b7 Magazine release cap
d4 Bolt handle
b1 and b9 Rear sight leaves
b12 Rear sight base
c31 Lock frame screw
c1 Buffer housing
c7 Stock release button
c10 and c11 Stock arms
c21 Shoulder piece
c30 Fore-end Screw
c19 Dismounting screw
c Fore-end
c2 Trigger guard
c5 Trigger
c25 Pistol grip
c30 Grip screw

How the MP38 Works

The loaded magazine inserted from below is held securely in place by the magazine lock. The firing pin is attached to the forward end of the telescoping housing. It passes through the hole in the center of the bolt, while the abutment behind it lodges into the head of the bolt recess. As the bolt is drawn back by its handle, or forced back by the functioning cartridge, it telescopes the three-piece recoil spring housing (which carries the firing pin) and compresses the recoil spring inside the telescope. The rear of the recoil spring housing rests against the inside of the rounded buffer end of the frame, which is securely locked to the receiver.

When the bolt is in the fully cocked position, the sear locks into the bottom of the bolt and connects with the trigger. Pressing the trigger depresses the sear and permits the bolt to run forward under the influence of the recoil spring acting through the telescopic section to force the bolt forward.

As the feed ribs on the bottom of the bolt strip the top round from the magazine and push it into the firing chamber, the face of the extractor set in the bolt blocks the base of the cartridge. When the cartridge is fully seated, the further forward movement of the bolt pushes the heavy extractor to snap it into the extracting groove. At the same time, the bolt face strikes against the base of the cartridge. The firing pin, a separate unit from bolt, is under the pressure of the recoil spring and functions in the same manner as a fixed firing pin—it protrudes at all times. This pin now strikes the cartridge and discharges it.

During the backward action, the extractor hook withdraws the empty cartridge case, carrying it back until it strikes against the ejector and is ejected. This cycle of operation continues as long as the trigger is held back and there are cartridges left in the magazine.

Field Stripping the MP38/40

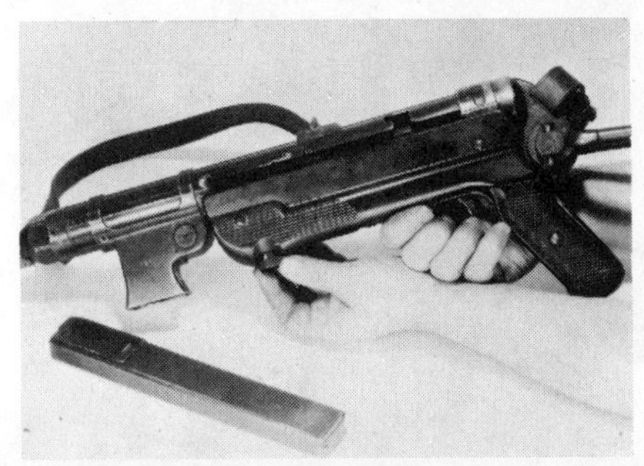

(1) After extracting the magazine, and seeing that the bolt is in its forward position, pull out the receiver lock against the tension of the spring and twist it to keep it locked in the outward position. (This stud is on the bottom of the frame at its forward end.)

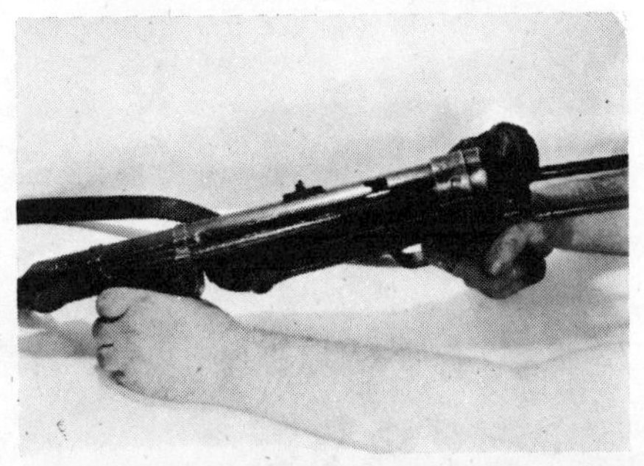

(2) While pressing the trigger with the right forefinger, hold firmly to the magazine housing with the left hand, then twist the pistol grip to the right, about 80°; this will revolve the entire frame assembly and the components.

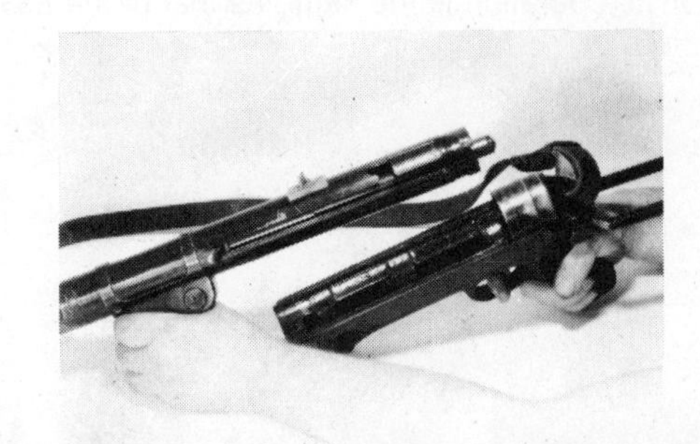

(3) Now draw the frame group back and out of the receiver.

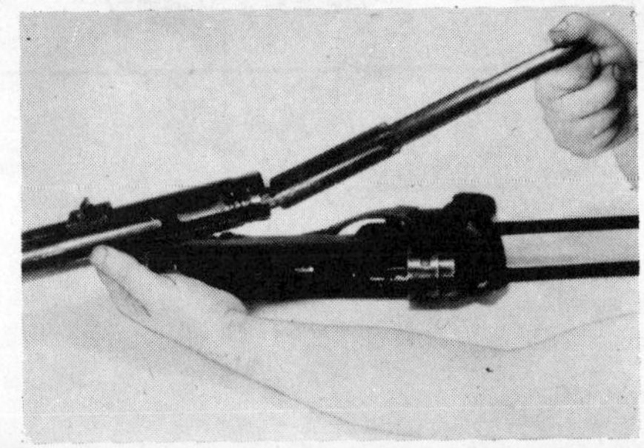

(4) Draw back slightly on cocking handle. This will bring out a telescoping tube inside which is the recoil spring, and at the front of which is the firing pin. Remove this unit.

(5) Now draw straight back on the cocking handle which is a part of the bolt and withdraw the bolt from the receiver. No further stripping is required.

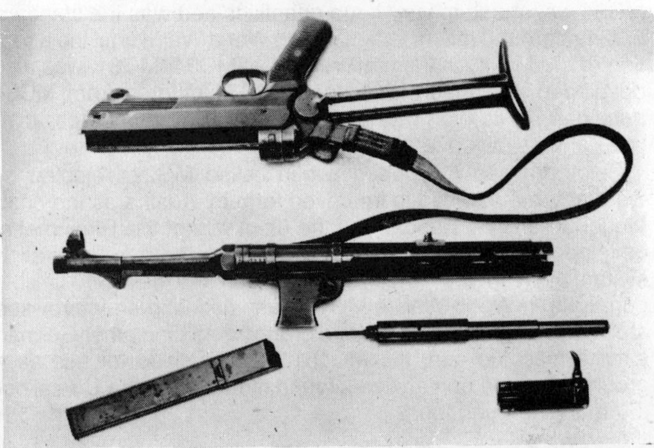

MP40 dismounted, complete field strip.

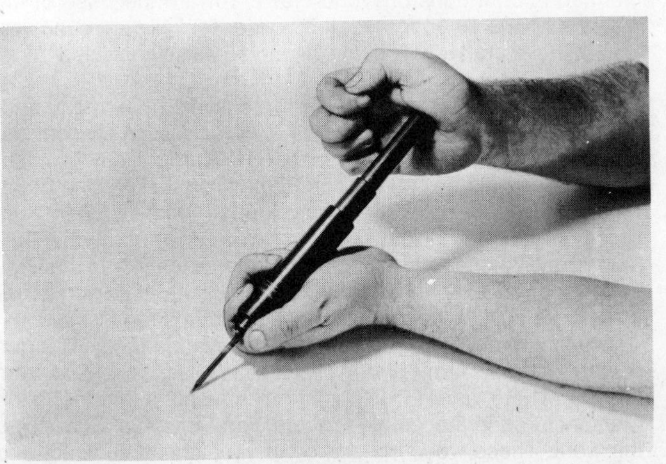

Recoil spring housing, showing firing pin and telescoping of tube.

GERMAN MACHINE GUNS

Germany adopted the Maxim gun about 1899 and in 1908 produced the Maxim gun, which may have the doubtful honor of killing more people than any other military instrument designed by man; it was certainly the most murderous weapon of World War I. The 7.92mm Model 1908 Maxim machine gun (MG08) was the standard German heavy machine gun of World War I and was made in tremendous quantities. A water-cooled weapon, it operated essentially the same as the British Vickers described in detail in the chapter on Britain. It is unlikely that it will be found in service in any country at present.

Although the MG08 was a very effective weapon, it was also quite heavy on the sleigh-type mount used during World War I, and a lightened version, the Model 08/15 (MG08/15), was introduced. The 08/15 is fitted with a shoulder stock, bipod, modified receiver and barrel jacket; its ammunition belt is carried on a reel type drum magazine mounted on the side of the receiver. Its operation is the same as that of the MG08.

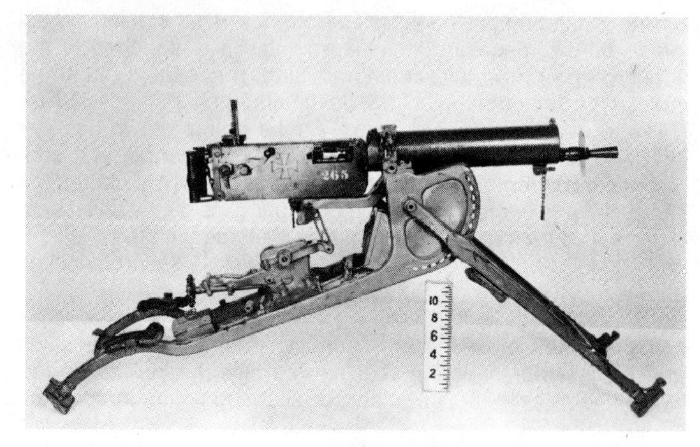

7.92mm MG08.

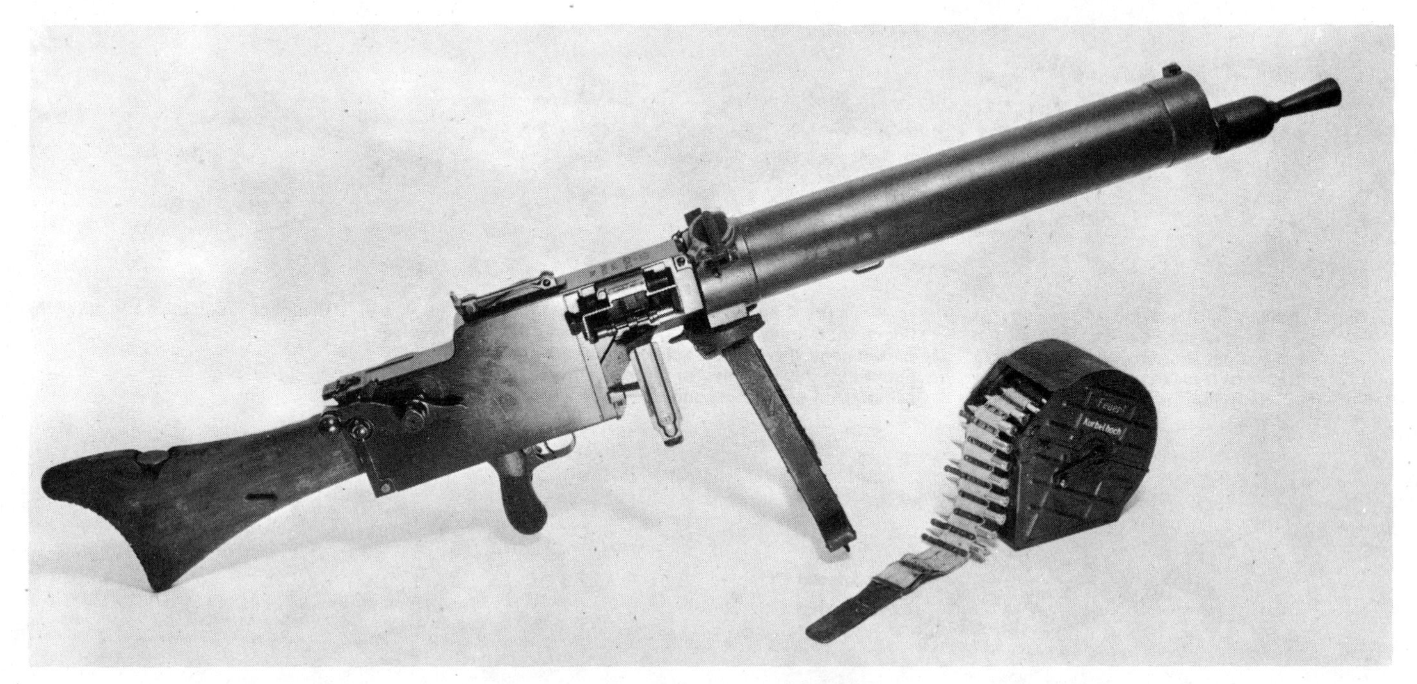
Maxim 7.92mm MG08/15 (light MG).

The MG08/15 was adapted for aircraft use by fitting it with a ventilated type barrel jacket in place of the water jacket. This gun and the Parabellum, a modified Maxim, were the principal German aircraft guns of World War I. Toward the close of the war, the MG08/18 appeared; this was essentially a ground version of the aircraft 80/15, being air cooled with a ventilated barrel jacket.

Machine gun development and production in Germany after World War I was restrained by the Versailles Treaty, but the Germans managed to "keep their hand in" through development done by German-owned firms in foreign countries. None of the ground gun designs produced by foreign firms—Waffenfabrik Solothurn is the principal example—were adopted by the German Army, but they did add to the German capability in that they gave them an experience factor in translating military requirements into design that might otherwise have been lost.

The 7.92mm MG13 was adopted as a standard machine gun by the German Army about 1932. MG13 was made up from Dreyse M1918 water-cooled light machine guns that had been manufactured in the last year of the war, Simson of Suhl doing the work. These weapons existed in very limited numbers and were apparently all sold to Portugal in 1938. The Mauser-developed MG34 was the first true general purpose machine gun made in quantity, i.e., a gun that is used on a bipod as a light machine gun and on a tripod as a heavy machine gun. MG34 was made in very large quantities and was the standard 7.92mm ground machine gun during World War II until the adoption of the MG42. An aircraft version of MG34, MG81, was also developed and made in quantity. MG81 differed from MG34 principally in its high rate of fire—1000–2000 rounds per minute—and its lack of a semiautomatic capability.

Solothurn had developed a gun called MG29, rejected by Germany but adopted in improved form by Austria as the 8mm Model 30 and by Hungary as the 8mm Model 31. Rheinmettal developed two 7.92mm aircraft guns using the basic operating system of the MG30—the MG17 fixed gun and the MG15 flexible gun. Late in World War II, MG15 was fitted with an improvised stock and a bipod and used as a ground gun. One other German aircraft machine gun, the MG151, was modified for use as a ground machine gun and was found mounted on the US caliber .50 machine gun tripod.

The most famous of all German World War II machine guns is the 7.92mm MG42. This weapon, now chambered for the 7.62mm NATO cartridge, is the current standard machine gun of the West German Army. Like the MG34, MG42 is a dual-purpose machine gun and might be considered something of a pace setter in its method of manufacture. The weapon is composed mainly of stampings, and its barrel change system, and feed and locking mechanisms have had considerable influence on post-war machine gun design.

At the close of the war, a delayed blowback machine gun—the MG45 or MG42V—was under development at Mauser. This weapon utilized the same type of bolt mechanism as the prototype StG45(M) assault rifle; the feed mechanism and barrel change was the same as that of the MG42. The SIG MG710-1 is quite similar to the MG45.

As with all other weapons, Germany used foreign machine guns to some extent. Rear area and police units were likely to be found armed with any type of machine gun for which there was sufficient ammunition on hand. The Czech ZB26 and ZB30 were continued in production at Brno after the plant was taken over by the Germans until it was tooled up to produce MG34. The ZB53 (Model 37) was also widely used by the Germans, particularly on vehicles. These Czech guns were considered as limited standard by the Germans, and manuals concerning their usage and maintenance were issued by the German Army.

THE MG34

This weapon was designed by Mauser at the direction of the Waffenamt. It was the first modern general purpose machine gun to be produced in large quantities. The design of MG34 incorporated many of the best features of previously developed weapons and had some outstanding features of its own. Among these were: A good method for changing barrels; a simple method of field stripping, major components being held together with bayonet-type catches; a high impact plastic stock; and a combined recoil booster, flash hider and barrel bearing.

The trigger mechanism of the MG34 is similar to that of the MG13 in that the trigger is pulled at the top for semiautomatic fire and at the bottom for automatic fire. MG34 is frequently confused with the Solothurn MG30. The Solothurn gun is based on patents of Louis Stange, a Rheinmettal engineer. The locking mechanism of the MG30 is a rotating ring, which locks the barrel and bolt together, while in the MG34 the bolt head rotates and locks into a barrel extension, which is permanently attached to the barrel. Stange's design undoubtedly did influence MG34, but it is different.

Various modifications of the basic MG34 were produced during World War II. These modifications and the ways in which they differ from the basic MG34 are as follows:

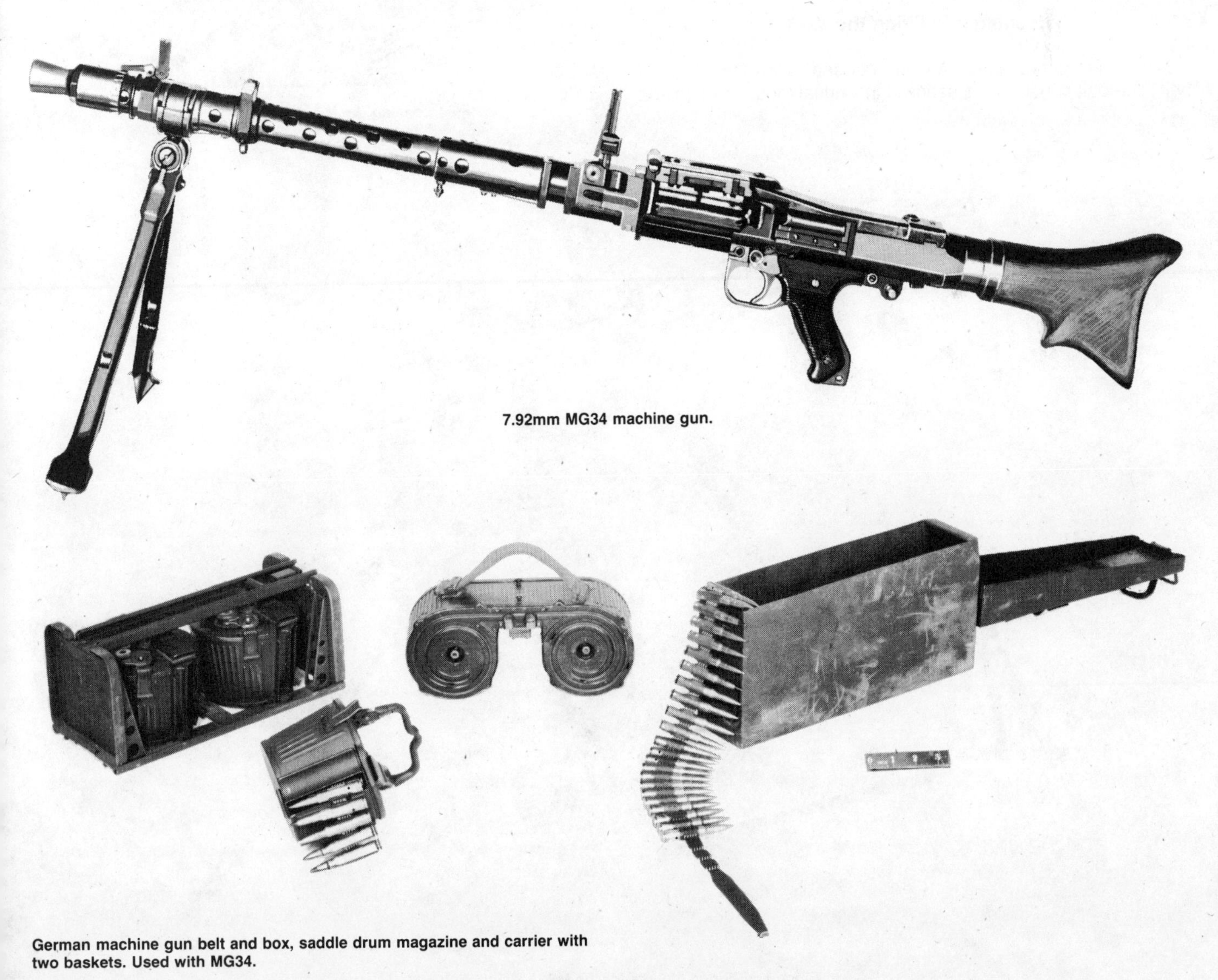

7.92mm MG34 machine gun.

German machine gun belt and box, saddle drum magazine and carrier with two baskets. Used with MG34.

MG34 (modified) has a heavier barrel jacket than MG34, developed for use in armored vehicles.

MG34S and MG34/41: Are several inches shorter than MG34 and have shorter barrels; fire automatic only and have a simple spur-type trigger and simpler trigger mechanism; have a larger buffer; the diameter of the barrel at the muzzle is increased in order to give more surface for the gas trapped in the recoil booster to bear against; the firing pin nut on the rear of the bolt has been eliminated; and minor changes in the feed system.

MG34 has been used since World War II by the Czechs, the Israelis and the French; MG34s have been captured from the Viet Cong in South Vietnam. Of interest is the fact that the United States Army had the MG34 analyzed by the Savage Arms Corporation during World War II. Savage concluded that the weapon would require the use of considerable numbers of machine tools in its manufacture, at a time when machine tools were in short supply; therefore, further investigation was discontinued. MG34 can be cranky on occasion, and the Germans have admitted that it took quite a while to work the bugs out of the weapon when it was first issued. One of the basic problems with the weapon is that it is too finely made, with very close fitting parts. Automatic weapons that have operating parts designed to work with plenty of play operate much better under adverse conditions of dust and mud, since they have plenty of space in which the dirt can lie without causing a malfunction.

Loading and Firing the MG34

A machine is provided but is not needed to load this form of belt. The belt consists of a series of individual metal links, joined

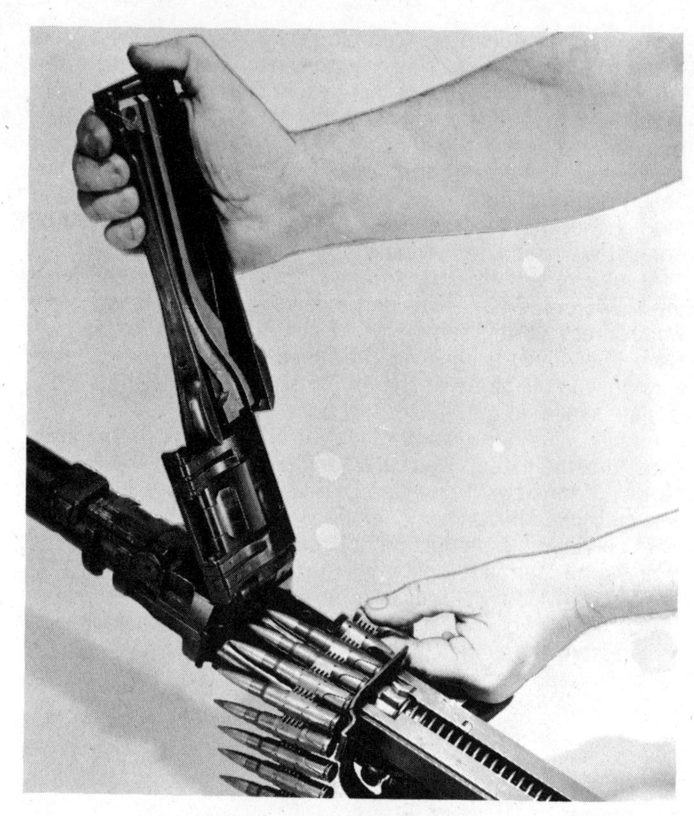

Loading the MG34 Machine Gun.

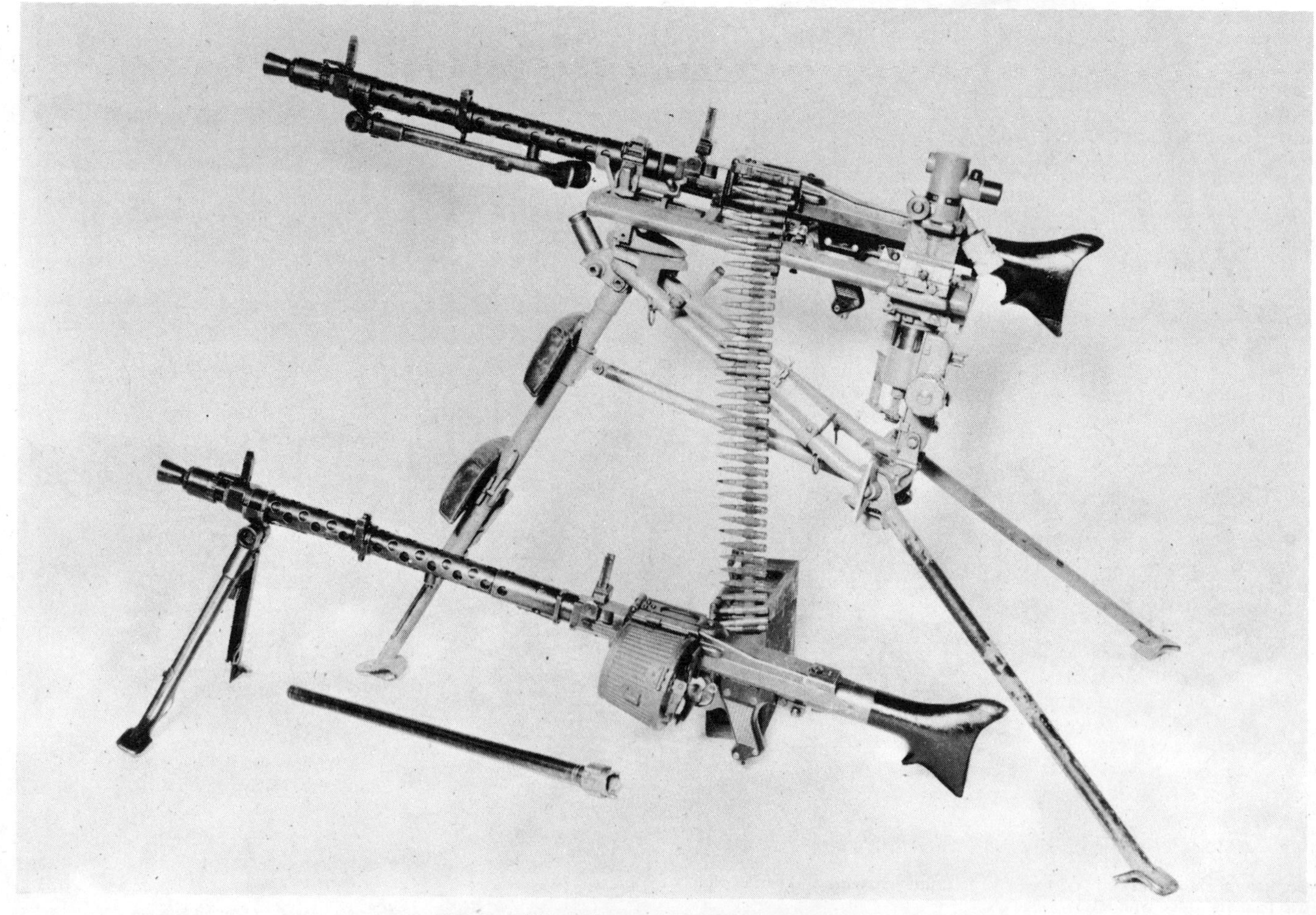

Bipod and Tripod Mounts for the MG34 Machine Gun.

together by small pieces of coiled wire. These links are shaped much like an ordinary pencil clip. Press the cartridge down into the clip so that the spring sides fasten around the cartridge and retain it. A nib at the end of the clip will spring into the cannelure of the cartridge and hold it in the correct position. It will be evident that in this form of belt there can be no malfunction of the type common to web belts, which may expand when wet, and to brass-studded belts, which must pass through a complicated feed mechanism.

In the 50-round drums, the loaded belt is inserted in the drum, being wound around the center piece.

The 75-round, saddle type drums do not use a belt. The drum itself contains the cartridges. The springs force them around into position, one coming alternately from each side.

Tabs are provided on the end of each belt. If several sections are being fastened together, or if no tab is available, then the first two or three cartridges should be removed from the metal belt.

Insert the feeding end of the belt in the feedway on the left side of the receiver, and pull through as far as it will go.

Warning: Unlike the Browning and the Vickers, the belt on this gun lies on top of the cartridges as they pass through the feed block. An alternate way of loading is to push forward the cover catch (which is on top of the receiver at the rear of the gun) and lift the feed cover to vertical position. The belt may then be laid in the feedway; make sure that the first cartridge rests against the stop on the right side of the guide. Close the cover and snap it down in place.

Pull back the cocking handle as far as it will go and the bolt will be caught and held in rearward position by the sear. Now push the cocking handle forward as far as it will go. If this is not done, it will be carried forward as the bolt moves to the front, and this additional weight may cause malfunctioning.

Pressing the upper part of the trigger will now fire a single shot. Pressing the lower part of the trigger will fire the weapon automatically.

NOTE: If the cocking handle will not come back, it indicates that the safety is on. Move the lever to the "Fire" position.

Firing with the 50-Round Drum. Press the catch on the sliding cover of the drum and open the cover so that the tag end of the belt can be pulled out. Insert the tag of the belt in the feedway as for the ordinary belt. The narrow end of the belt is the front end. Engage the hook on the front end with the lug on the rear end of the lower part of the feed plate. Now swing the rear end of the drum around until the spring catch engages with the lug on the rear end of the feedway. Pulling back the cocking handle now leaves the weapon ready for firing.

Firing with the 75-Round Saddle Drum. With this drum the feed cover is removed and a magazine holder is substituted. The feed plate is also removed. Belts are not used in this type of feed. The drum is placed directly over the magazine holder ahead of the trigger guard. Its center piece pushes down the dust cover in the magazine holder. A spring catch at the top center of the connecting piece can be pressed to release the drum and a hand-strap is provided to lift it off the gun.

How the MG34 Works

Starting with the gun loaded and cocked, the action is as follows: Pressing the trigger pulls the sear out of its bent in the bolt and allows it to go forward under the thrust of the compressed recoil spring located in the butt.

A feed piece on the top of the bolt strikes the base of the cartridge in line and pushes it from the belt toward the firing chamber. The feed arm is hollow and is operated by a stud on the top rear end of the bolt, which rides in this hollow groove and causes the feed pawl to push the next cartridge in the direction of the firing chamber.

As the bolt continues forward, two inner rollers on its head strike two cams on a cam sleeve and rotate the head of the bolt from left to right so that threads on the bolt engage threads on the cam sleeve; this effectively locks the bolt to the barrel.

MG34 on AA tripod.

As the cartridge chambers, the extractor in the bolt face slips over the cannelure of the cartridge. Meanwhile the rear of the bolt continues forward, tripping the firing pin lever and allowing the firing pin to go forward through the face of the bolt to strike the primer. The forward movement of the bolt is stopped when a shoulder on its right front side strikes the cocking handle stop, which is in its forward position at the end of its slot. Just before the cartridge is fired, a locking catch on the bolt engages behind the outer roller on the right side of the head of the bolt.

Return Movement of the Action. This gun is fitted at the muzzle with a recoil increaser somewhat resembling that operating on the Vickers gun.

As the bullet leaves the barrel, part of the gas pressure behind it expands in the muzzle attachment and rebounds against the cone to give additional backward thrust to the barrel. This action, together with the rearward thrust of the gas in the firing chamber against the head of the empty cartridge case, which transmits it to the bolt, starts the action to the rear.

Barrel and bolt start back, firmly locked together during the period of high pressure. After a backward travel of about 3/16", the outer rollers on the bolt head again engage with the two cam faces in the forward end of the receiver, thus forcing the bolt head to rotate from right to left, thereby unlocking the bolt from the barrel.

The rearward motion of the barrel is stopped as soon as the

Field Stripping the MG34

(1) Order of stripping: Push the spring catch at the extreme rear of the cover on top of the receiver and lift the cover to a vertical position. Push the cover hinge pin from the right and lift out the cover. The feed block may be lifted off.

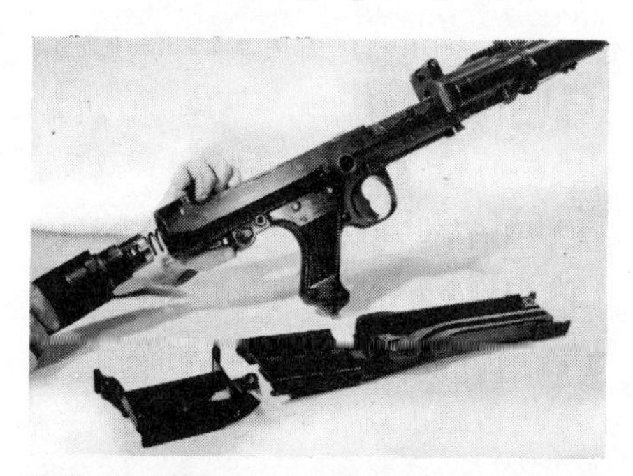

(2) The butt catch is on the underside of the receiver a few inches behind the pistol grip. Press this up with the left thumb. With the right hand, turn the butt a quarter-turn left or right. (Note: the bolt should be in forward position when this stripping motion is being done. Otherwise, the very powerful recoil spring cannot be controlled.) The recoil spring will now force the butt out of the receiver. Now remove the recoil spring.

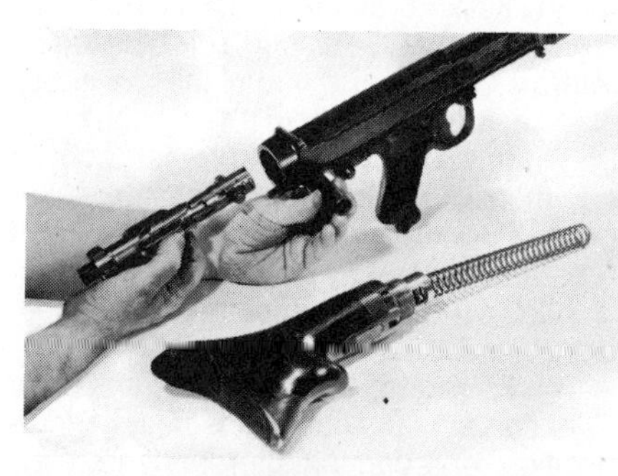

(3) Pull the cocking handle back with a quick motion. (A jerking motion is required here because the action in releasing the bolt twists the barrel extension and the barrel. Watch that the bolt and its carrier do not fly out the back of the receiver.)

(4) Bolt and carrier may now be removed.

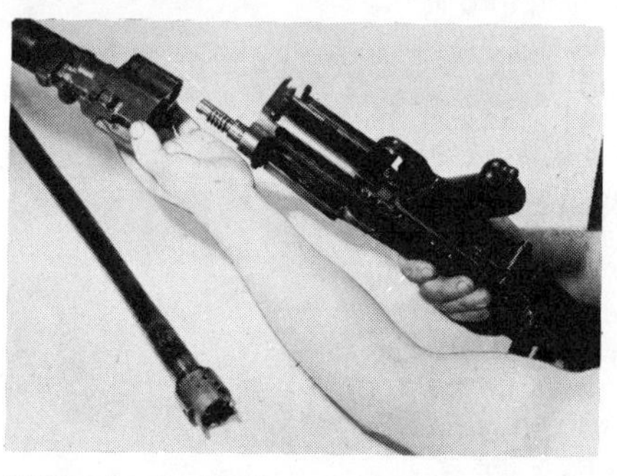

(5) Pressing the locking catch on left of receiver, below and behind rear sight, twist receiver from left to right until it clears the barrel casing. Raise the muzzle and slide the barrel out of the casing. A hinge pin catch will be found on the underside of the barrel casing, near its end and to the right. Press this up and while maintaining pressure twist the receiver, left to right, until it has completed a half-turn. It may now be pulled out to the rear.

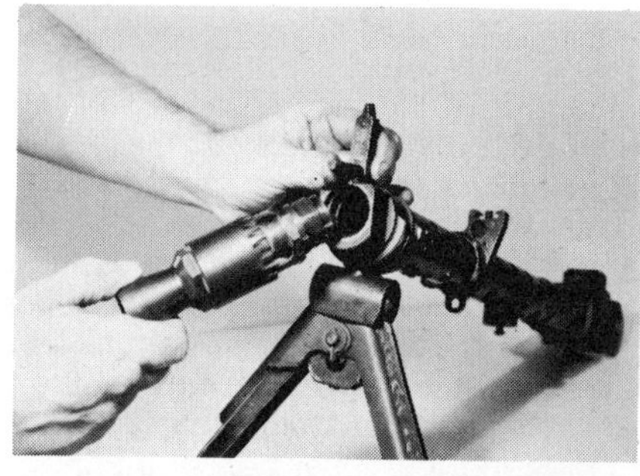

(6) A catch will be found in front of the foresight. Lifting this permits you to unscrew the flash hider over the muzzle. Inside it is a mouthpiece and a recoil cone. Remove them. The trigger assembly is locked to the receiver by two automatic locking pins. Pinching the split ends together permits them to be pulled out. (Removal of this assembly is not recommended without suitable tools.)

unlocking operation is completed, when its cam sleeve strikes against shoulders in the front end of the receiver.

The stud riding straight to the rear on the bolt, its head caught in the groove in the feed arm above it, twists the feed arm, which forces the feed pawl slide to move back, and permits the feed pawl to lock behind the next cartridge in the belt.

The empty case, being drawn from the firing chamber by the extractor in the face of the bolt, is struck by the ejector and hurled out of the gun. The ejector is a pin in the top of the bolt; during the backward movement of the bolt the rear end of this pin strikes against a stop, which forces the front end through its hole in the bolt to hit the base of the empty cartridge case. The ejection is downward. The end of the breech block carrier strikes against the buffer, the compression of the recoil spring is completed, and, if the semi-automatic portion of the trigger is being pulled, the bolt will stop open, engaged from below by the sear forced up by its spring. If the automatic trigger is being pressed, the firing cycle will be completed and continued as long as there are any cartridges left in the belt.

MG42

After nearly three years of combat, the German Army adopted a new machine gun, the MG42. The MG42's design borrowed concepts from several sources. For example, the quick barrel change mechanism (necessary because of the weapon's cyclic rate of fire, 1200 rounds per minute) was an improvement on the Italian Breda machine gun, and the recoil operated locking system was an adaptation of a mechanism devised by Edward Stecke of Warsaw. Reportedly, a mockup of the Stecke gun was captured by the *Wehrmacht* in 1939. Dr. Crunow of Johannus Grossfuss was responsible for incorporating all the various ideas into a single, new weapon, which was made largely from metal stampings and pressings, the first machine gun to be so made. Its design and adoption cut down on machine tool usage during the war years. Since then, the practice of using stampings has become much more common. The MG42's feed mechanism has also been very successful, having been employed in the design of the American M60 machine gun, among others.

During the Second World War, the following factories produced the MG42: Johannus Grossfuss Mettal-und Locierwaffenfabrik, Doblen, Saxony; Mauserwerke, Berlin; Maget, Berlin; Gustlof Co.; Suhl Waffenfabrik, Suhl; and Steyr Daimeler Puch A. G., Vienna. In the post-war period, Rheinmettal, Dusseldorf, has been the basic producer of the updated MG42/59 for the *Bundeswehr*. To date, over 180,000 of these new guns have been manufactured by the German firm alone.

The first post-war MG42/59 machine guns (MG1s), used primarily by West German Border Patrols, were chambered for 7.92 × 57 cartridges. Since then, several modifications have been made by Rheinmettal:

—The MG1A1, converted to 7.62 × 51mm NATO, has a hard chrome plated barrel. The sight has been corrected for the NATO cartridge and the trigger mechanism slightly modified.
—The receiver of the MG1A2 has been modified and case ejection port widened. A heavier bolt has been utilized that reduces the rate of fire to 700–900 shots per minute. The MG1A2 can be used with either the German nondisintegrating link belt or the American M13 disintegrating link belt. The cocking stud on the cocking slide has been shortened and the buffer, recoil booster and flash hider modified. The barrel is similar to the MG1, but the barrel guide sleeve (front barrel bearing) has been modified. This model has been adopted by Italy.
—The MG1A3, with a modified sight, has a barrel similar to the MG1A1's. An additional stud has been added to the bolt housing and other minor modifications made to the bolt, belt-feed lever and feed mechanism. The trigger mechanism has been changed and trigger pull tightened. The stock has been modified in the dimension of the sleeve, which threads on to the buffer.
—Also modified have been the butt plate, mounting screw, bipod, recoil booster and flash hider, which was made in one piece. The barrel and barrel bearing have been chrome plated.
—The current model of this machine gun being used in West Germany is the MG3, which differs in minor details from earlier models. The most noticeable differences are the addition of an anti-aircraft sight in front of the rear sight, the shape of the barrel booster flash hider unit and the fact that the gun can use DMI nondisintegrating belts, DM6 disintegrating belts and US M13 links.
—There are two bolts and two buffers for the MG1 series guns. Use of the light bolt, called V550 (weighing 550 grams) by Rheinmettal, and the Type N buffer produces a cyclic rate of 100–1300 rounds per minute. Use of the heavy bolt, called V950 (weighing 950 grams), and the Type R buffer produces a cyclic rate of 700–900 rounds per minute.

In the post-war period, in addition to the *Bundeswehr*, the Austrian, Chilean, Danish, Iranian, Italian, Norwegian, Portugese, Spanish and Turkish armed forces have been major users of this NATO caliber machine gun. Yugoslavia has built its own version of the MG42 in the older 7.92 × 57mm caliber.

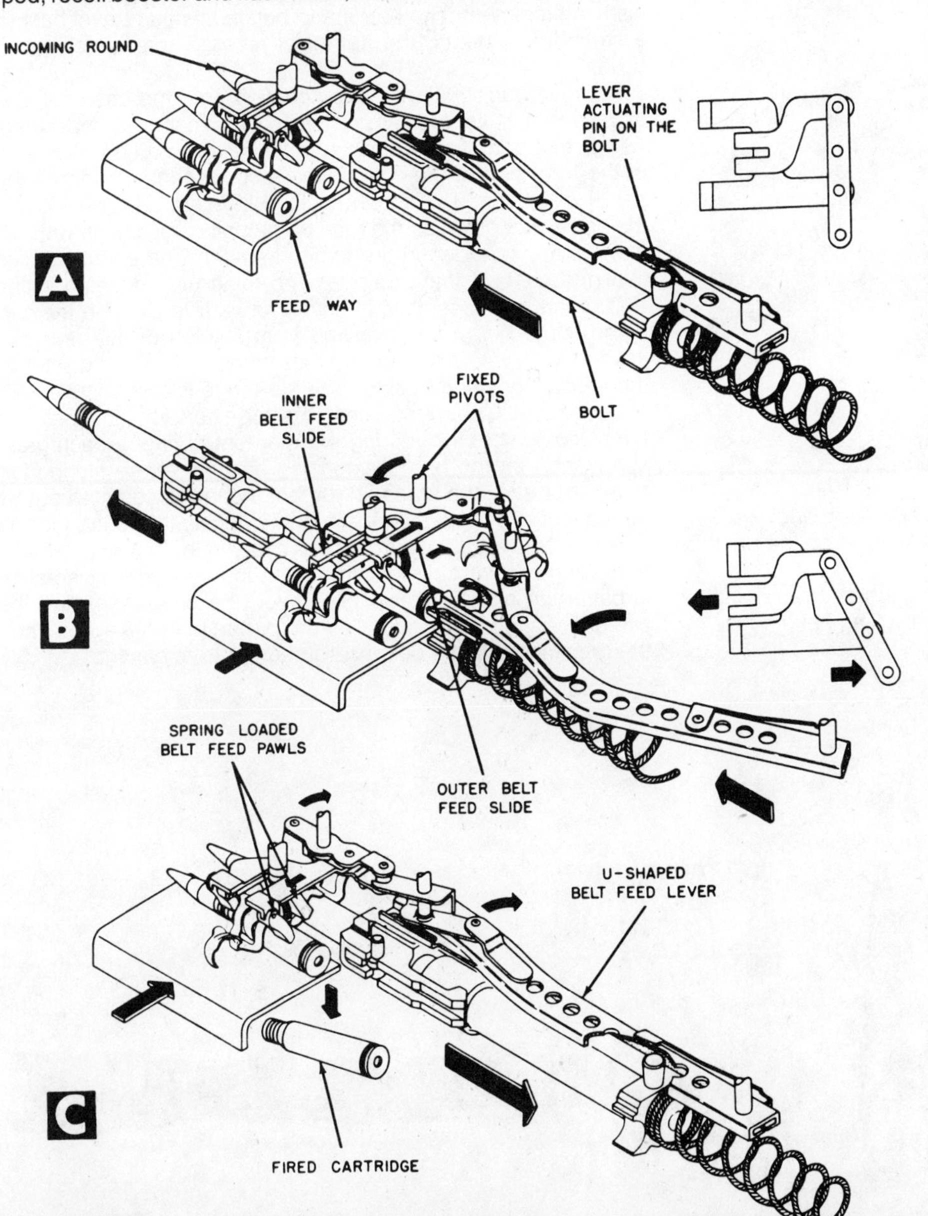

Pin on Bolt Actuates Belt Feed Lever.

Diagrammatic view of MG42 feed mechanism taken from G. M. Chinn, *The Machine Gun*.

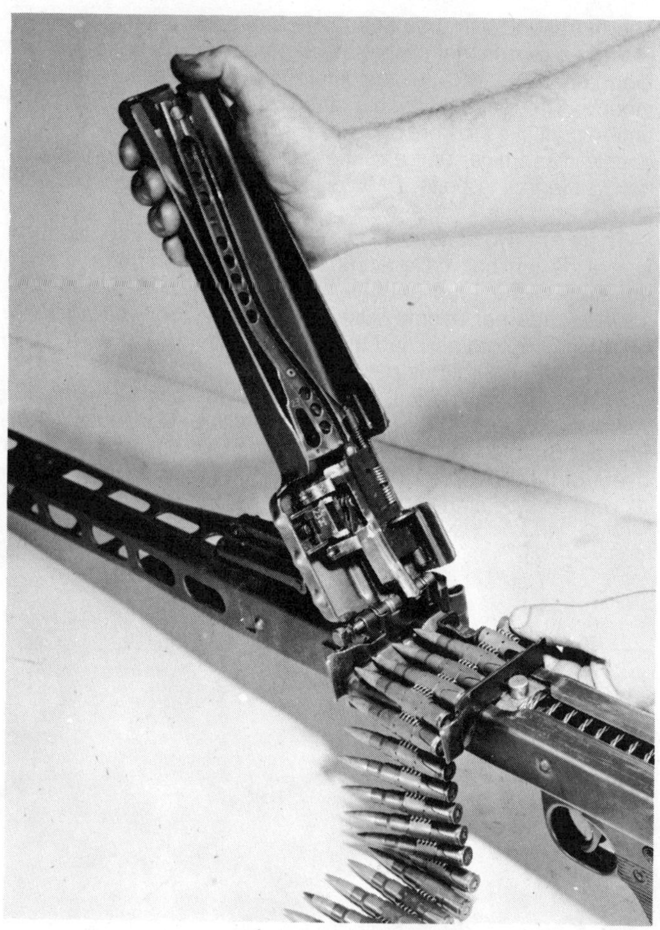

MG42, feed cover open.

Loading and Firing the MG42

To Load: (As for MG34). Feed cover may be open or closed. Be sure that the first cartridge rests against the stop on the right side of the feed guide.

Pull back cocking handle on left side of gun as far as it will go. The bolt will stay open. Then shove the cocking handle fully forward until it clicks.

Unless gun is to be used immediately, set the safety. On this gun the safety is just above the pistol grip. Push the button from the right side and it sets the safety. Push the button from the left as far as it will go and the gun is ready to fire.

NOTES ON UNLOADING. Unloading this gun is a very simple operation. First pull back the cocking handle as far as it will go. Then set the safety. Push forward the cover catch and raise the cover as high as it will go. Lift the belt out of the gun.

It is not good practice to permit the bolt to go home on an empty chamber. Always hold the cocking handle firmly while pressing the trigger and ease the bolt into forward position.

Note that there is a spring cover over the ejection opening in these guns. It flies open when the trigger is pressed. On "Cease Fire" always push it shut. This will keep dirt and dust out of the mechanism.

How the MG42 Works

In general this gun follows the operating detail of the MG34. However, an entirely new design of bolt and locking mechanism is employed. The barrel and bolt in this gun travel back in a straight line during the period of recoil. There is no turning action.

A heavy barrel extension is screwed onto the chamber end of the barrel. In its sides are slots into which cams are machined. As the bolt goes forward, a movable locking stud on each side of the front end of the bolt strikes a corresponding cam in the barrel extension. This forces locking lugs out and into slots in the barrel extension as the face of the bolt comes flush with the base of the cartridge in the firing chamber. The extractor slips over the base of the cartridge. The firing pin, mounted in the rear of the bolt assembly, is driven forward to explode the cartridge. Note that a stud, driving from the top of this rear bolt assembly, travels in a groove in the curved feed arm and shuttles the feed across and back to operate the feed mechanism.

During the recoil movement, barrel extension and bolt are firmly locked together during the moment of high breech pressure. Then as the barrel extension and barrel are stopped in rearward travel, the studs on the bolt head are cammed out by the camming surfaces on the barrel extension, and the locking lugs are thus withdrawn from their seats in the barrel extension permitting rearward direct line motion in the action. This action is patterned after a simple pile-driver, the bolt resembling the pile-driver hammer being pulled up (or out) to its full extent, then the gripping surfaces being cammed out to release it.

German metal belt showing detail of cartridges, locking system and tab.

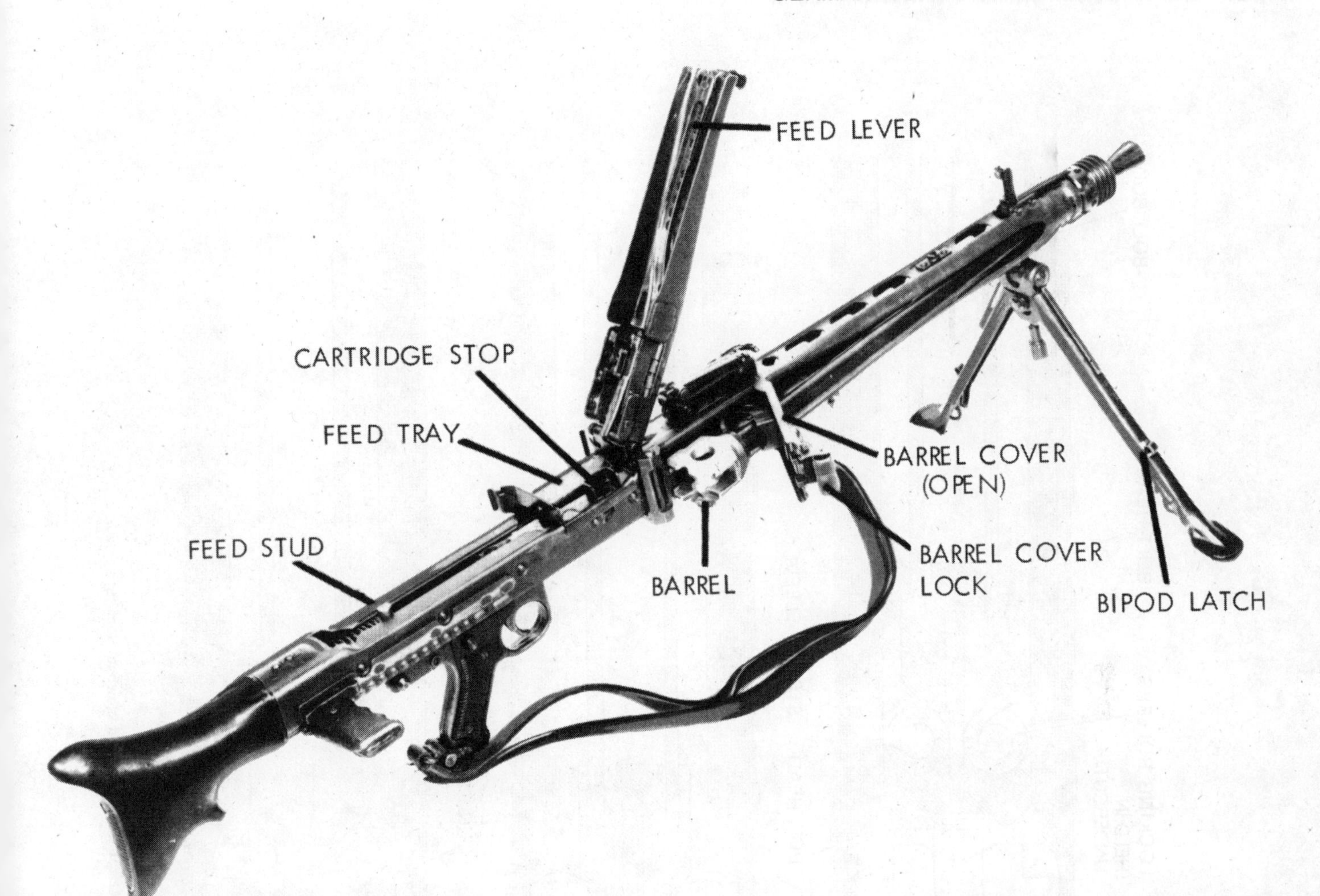

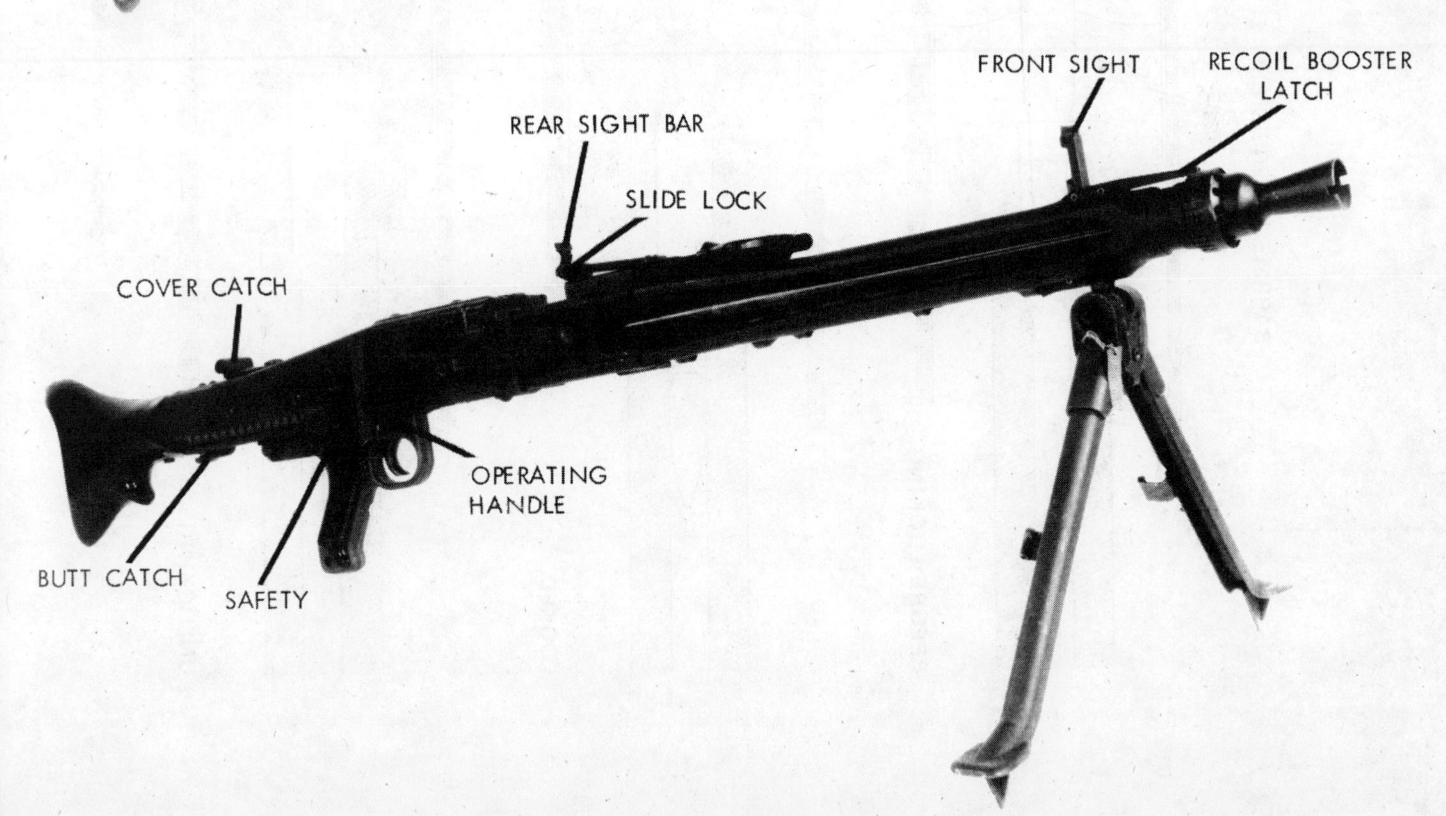

Major features of MG42 (top) and MG42/59 (MG1) (bottom).

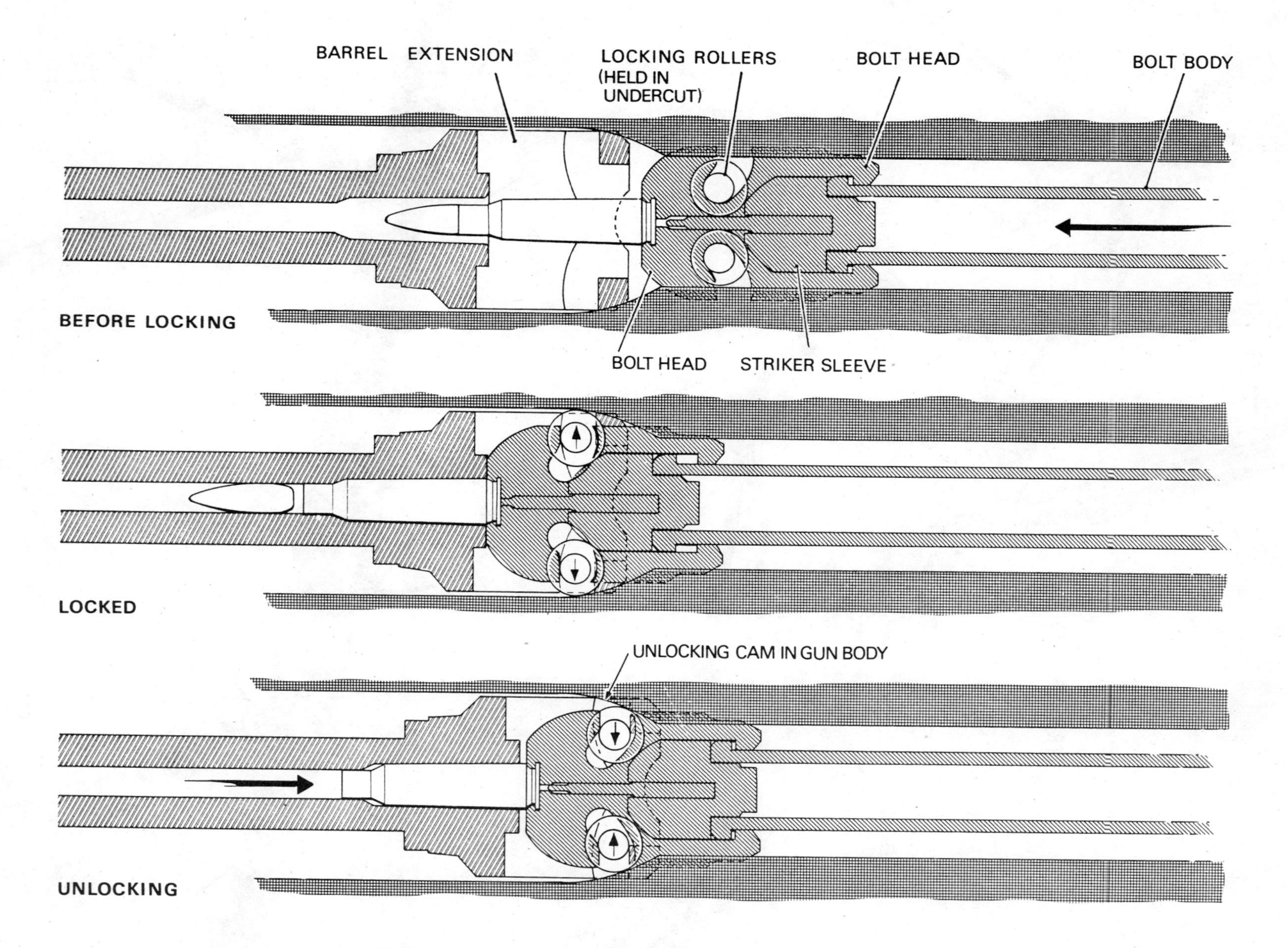

Diagrammatic view of the MG42 roller-lock delayed blowback locking system.

Field Stripping the MG42

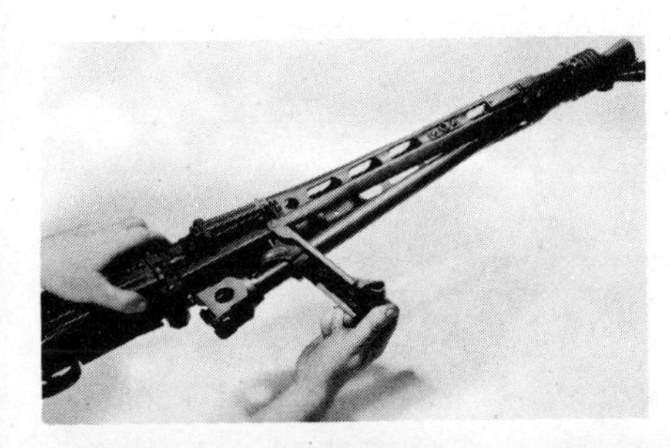

(1) In general field stripping this gun is similar to the MG34. There are, however, some few differences. To remove barrel: It is first necessary to cock the weapon. This is done by pulling back the cocking handle. Then thrust forward and outward on the heavy release catch jutting out from the rear of the barrel extension on the right side of the gun, below the feed block. This draws the rear of the barrel out of its seat and permits it to be drawn from the rear of the gun.

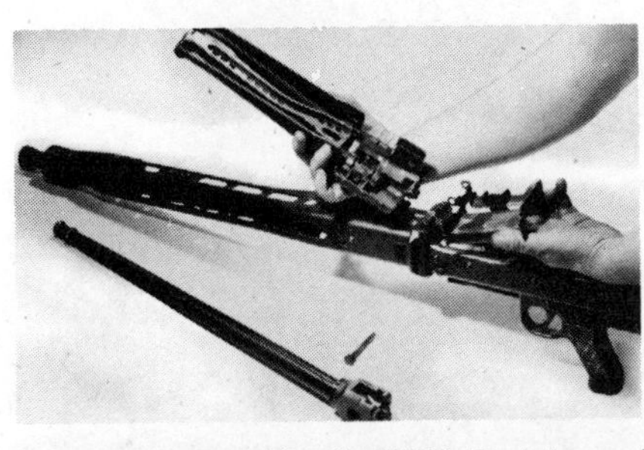

(2) The feed resembles the MG34. Push forward the feed cover catch on top of the gun near the stock and lift the cover. Pull out the feed cover hinge-pin and remove the feed block from the gun. Dismounting this is very simple.

(3) Remove buttstock, same as for MG34. Be sure bolt is in forward position before removing buttstock. Catch is on the underside of the stock. Push it and twist the butt a quarter turn, right or left.

(4) Remove buffer and recoil spring. As in the MG34, the housing catch is on the rear end of the receiver, just back of the pistol grip. Press the catch and control the buffer housing that moves away from the receiver under tension of the powerful spring. Remove the bolt. Press the trigger and strike the cocking handle a sharp rearward blow. This will drive the bolt to the open rear of the receiver where it may be withdrawn. No further stripping is normally necessary.

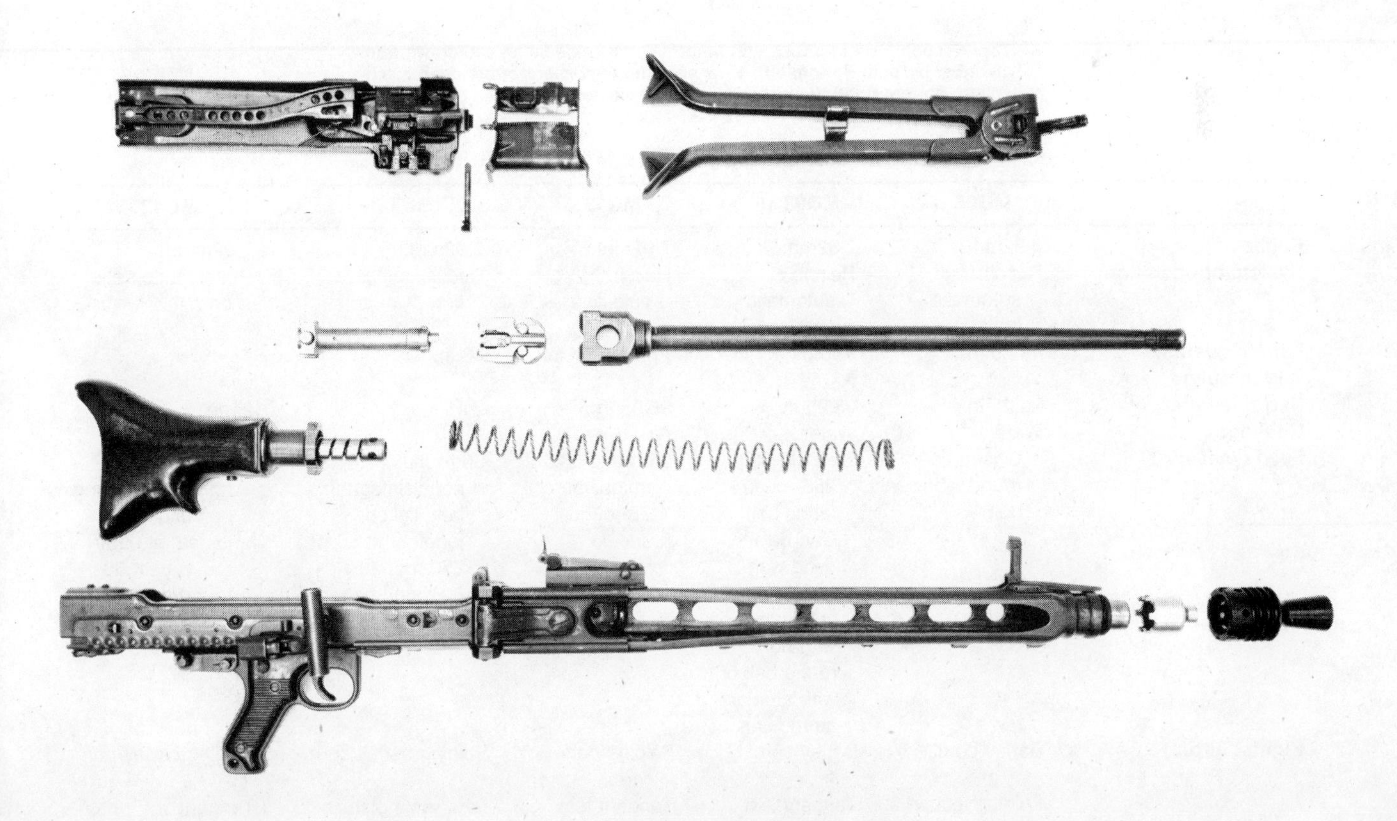

Field-stripped view of the MG42.

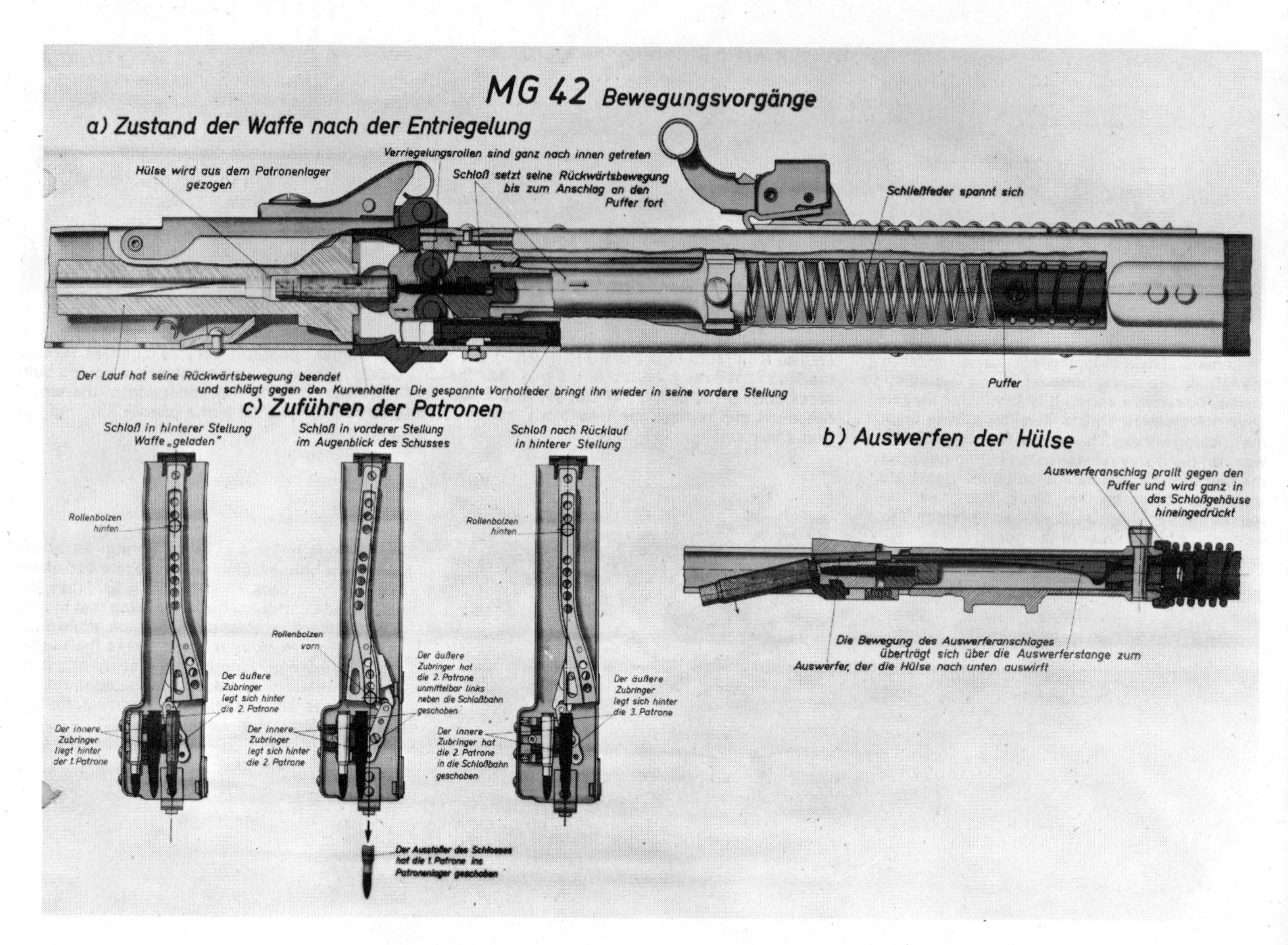

Top: Section view of MG42, gun unlocked, cartridge being withdrawn from chamber by bolt. Bottom left: View showing movement of belt feed mechanism. Bottom right: View shows action of ejector.

STANDARD GERMAN ARMY MACHINE GUNS

	MG08	MG08/15	MG13	MG34*	MG42
Caliber:	7.92mm.	7.92mm.	7.92mm.	7.92mm.	7.92mm.
System of operation:	Recoil, automatic only.	Recoil, automatic only.	Recoil, selective fire.	Recoil, selective fire.	Recoil, automatic only.
Length overall: (gun only)	46.25 in.	Approx. 57 in.	Approx. 57 in.	48 in.	48 in.
Barrel length:	28.25 in.	28.25 in.	28.25 in.	24.6 in.	21 in.
Cooling:	Water.	Water.	Air.	Air.	Air.
Feed device:	100- and 250-round fabric belt.	50-, 100-, & 250-round fabric belt. Can be used with spool type container with 50-round belt or with ordinary ammo can.	25-round box magazine or 75-round saddle drum.	50-round nondisintegrating belt, linked together to form 250-round belt. 50-round belt drum. 75-round saddle drum.	50-round nondisintegrating belt usually joined into 250-round belt, 50-round drum.
Sights: Front:	Barley corn.	Barley corn.	Folding barley corn.	Folding barley corn.	Folding barley corn.
Rear:	Folding leaf w/ V notch.	Tangent leaf w/V notch.	Tangent leaf w/V notch.	Leaf w/V notch.	Tangent w/V notch.

STANDARD GERMAN ARMY MACHINE GUNS—Continued

	MG08	MG08/15	MG13	MG34*	MG42
Muzzle velocity:	2750 f.p.s. (S ball)	2750 f.p.s. (S ball)	2750 f.p.s. (S ball)	Approx. 2500 f.p.s. (sS ball)	2480 f.p.s. (sS ball)
Cyclic rate:	400–500 r.p.m.	400–500 r.p.m.	750 r.p.m.	800–900 r.p.m.	1100–1200 r.p.m.
Weight of gun:	40.5 lb.	31 lb.	26.4 lb. w/ bipod.	26.5 lb. w/bipod.	25.5 lb. w/bipod.
Weight of mount:	83 lb. sled mount. 65.5 lb. tripod mount.	51 lb.	Approx. 25 lb.	42.3 lb.	42.3 lb.

*Data is given for basic MG34; other versions vary as follows:
MG34S and MG34/41—automatic fire only.
Overall length: approx. 46 in.
Barrel length: approx. 22 in.

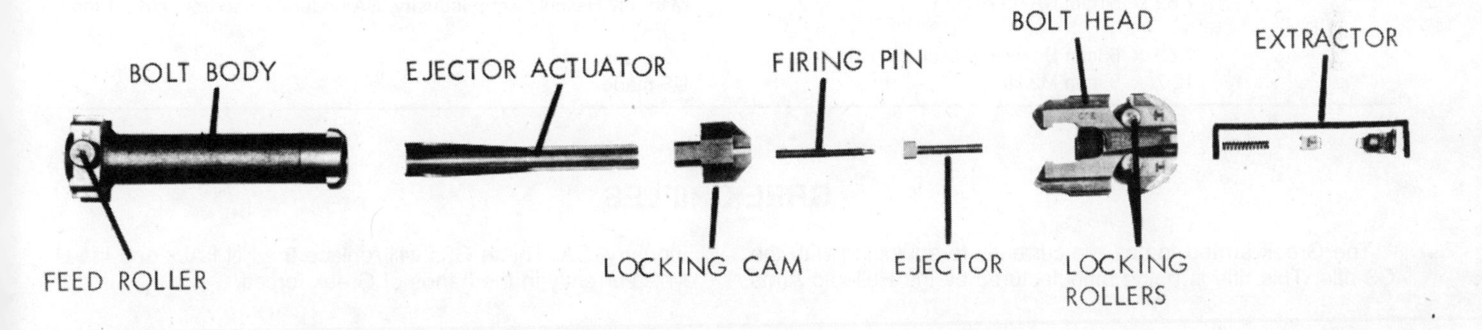

Disassembled MG42 bolt.

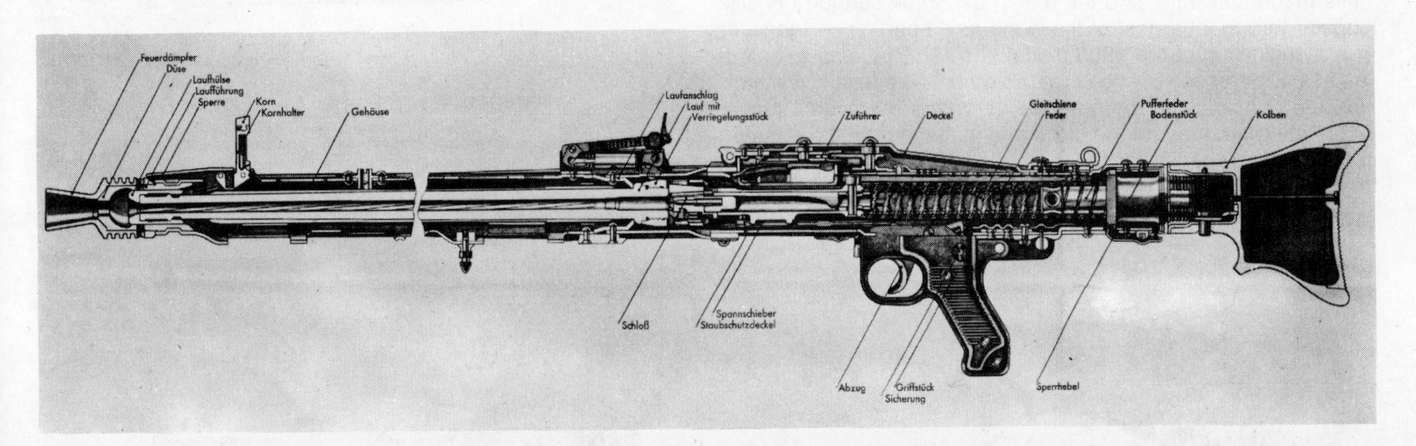

Phantom drawing showing all operating parts in full closed position. View of left side. Nomenclature, left to right, top. 1. Flash hider. 2. Barrel mouth. 3, 4, and 5. Blast cone assembly. 6 and 7. Front sight and spring. 8. Housing. 9. Barrel retainer. 10 and 11. Barrel and barrel locks. 12. Feed. 13. Cover. 14 and 15. Recoil spring and guide. 16. Buffer spring. 17 and 18. Butt assembly and butt. Left to right, bottom: 1. Lock. 2. Bolt. 3. Bolt extension. 4. Trigger. 5. Safety. 6. Grip. 7. Buffer release.

23 Greece

SMALL ARMS IN SERVICE

Handguns:	11.43 × 23mm M1911A1	US-made.
	9 × 19mm NATO FN Browning Mle. 1935	FN-made.
	9 × 19mm NATO MPi69	Steyr-Daimler-Puch-made.
Submachine guns:	11.43 × 23mm M3A1	US-made.
	9 × 19mm NATO SUMAK-9	Made by Hellenic Arms Industry, SA.
Rifles:	7.62 × 51mm NATO G3	Made by Hellenic Arms Industry, SA under license.
	7.62 × 51mm NATO FAL	Made by Fabrique Nationale.
	7.62 × 63mm M1	US-made.
Machine guns:	7.62 × 51mm NATO MAG	Made by Fabrique Nationale.
	7.62 × 51mm NATO EHK11	Made by Hellenic Arms Industry, SA under license from HK for their HK11.
	7.62 × 63mm Browning machine guns	US-made.
	12.7 × 99mm M2 HB	US-made.

GREEK RIFLES

The Greek armed forces are currently re-equipping with the G3 rifle. This rifle is being manufactured by the Hellenic Arms Industry, SA. These G3s will replace the FN FALs and the US M1s currently in the hands of Greek forces.

OBSOLETE GREEK RIFLES

MANNLICHER SCHOENAUER 6.5mm 1903 RIFLE

The Greeks adopted the Mannlicher Schoenauer rifle in 6.5mm in 1903. This rifle combines the Mannlicher two-piece rotating bolt with the rotating spool type magazine developed by Otto Schoenauer. This system is still used in the sporting type Steyr rifles made currently, and the 6.5 × 54 Greek cartridge is still popular throughout most of the world as a sporting cartridge.

A modification of the 1903 model, the 1903/14, was adopted in 1914. Differences between the two models are relatively minor and are principally in the graduations on the rear sight, shape of grasping grooves on the stock and in the length and ease of removal of the handguards.

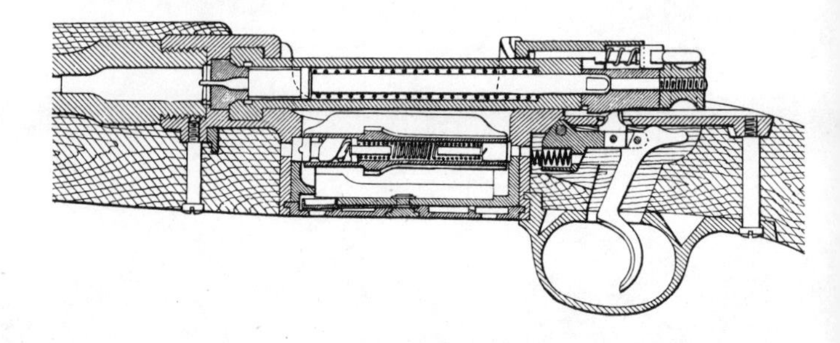

Section of rifle with action closed. Note position of locking lugs in rear of bolt head.

6.5mm Model 1903/14 Carbine.

The G3 as manufactured in Greece by the Hellenic Arms Industry, SA.

OTHER GREEK SHOULDER ARMS

The Greeks had a number of Austrian 8mm M1888/90 and 8mm M1895 rifles converted to 7.92mm. The Model 1895 conversion was called the model 95/24. In 1930, the Greeks adopted the Model 1924 FN Mauser in 7.92mm, which they called the Model 30. This rifle is covered in detail under Belgium.

GREEK SERVICE RIFLES AND CARBINES

	M1903 Rifle	M1903 Carbine	M95/24	M1930
Caliber:	6.5mm.	6.5mm.	7.92mm.	7.92mm.
System of operation:	Turn bolt.	Turn bolt.	Straight-pull bolt.	Turn bolt.
Overall length:	48.3 in.	40.3 in.	39.5 in.	43.3 in.
Barrel length:	28.5 in.	20.5 in.	19.6 in.	23.2 in.
Feed device:	5-round, revolving spool, non-detachable box magazine.	5-round revolving spool, non-detachable box magazine.	5-round, in line* non-detachable box magazine.	5-round, staggered row, non-detachable box magazine.
Sights: Front:	Barley corn.	Barley corn.	Barley corn.	Barley corn.
Rear:	Tangent with 'V'-notch.	Tangent with "V" notch.	Leaf with 'V' notch.	Tangent with 'V' notch.
Weight:	8.31 lbs.	Approx. 8 lbs.	6.8 lbs.	8.5 lbs.
Muzzle velocity:	2225 f.p.s.	Approx. 2125 f.p.s.	Approx. 2410 f.p.s.	2500 f.p.s.

*Magazine is charger fed; charger does not remain in magazine.

GREEK SUBMACHINE GUNS

9 × 19mm NATO SUMAK-9

Little is presently known about this new 9 × 19mm submachine gun. It is a blowback weapon with advanced primer ignition and a bolt that telescopes over the breech end of the barrel.

CHARACTERISTICS OF THE SUMAK-9

Caliber: 9 × 19mm NATO
System of operation: Blowback
Overall length: 25.98 in. with stock extended
17.32 in. with stock collapsed
Barrel length: 10.24 in.
Feed device: 25- or 32-shot box magazines
Weight: 5.51 lbs.
Cyclic rate: 600–650 r.p.m.

24 Hungary

SMALL ARMS IN SERVICE

Handguns:	7.62 × 25mm M48	Hungarian copy of Soviet TT-33 made by Femaru es Szerszamgepgyar NV.
	7.65 × 17mm M48	Hungarian copy of Walther PP made by Femaru es Szerszamgepgyar NV.
Submachine guns:	7.62 × 25mm Spigon	Domestic design blowback submachine gun.
	7.62 × 25mm M48	Hungarian copy of Soviet PPSh41. Obsolescent; reserve forces only.
Rifles:	7.62 × 39mm AK47	Hungarian copy.
	7.62 × 39mm AKM	Hungarian copy.
	7.62 × 39mm AKMD	Shortened version of Hungarian AKM.
	7.62 × 54mm R M48	Hungarian sniper rifle based upon Mosin Nagant 1891/1910. Reserve forces only.
	7.62 × 54mm R SVD	Soviet-made sniper rifle.
Machine guns:	7.62 × 39mm RPK	Hungarian copy.
	7.62 × 54mm R PK	Hungarian copies of PK, PKB, and PKS.
	12.7 × 108mm DShK	Hungarian copy.

HUNGARIAN SERVICE PISTOLS*

THE HUNGARIAN 7.65mm FROMMER STOP PISTOL MODEL 19

Hungary was, until 1918, part of the Austro-Hungarian Empire and in general used the standard weapons of that empire. Because of local development capabilities, and probably local politics, the Hungarians used a different pistol during World War I than the rest of the empire forces. This pistol was the 7.65mm (.32 ACP) Model 19 Frommer Stop pistol. This long recoil-operated pistol was developed in 1912 and used by the Hungarian police and service forces after World War I. A 9mm (.380 ACP) version, which, according to some authorities, was never issued as a military weapon, appeared some time between 1916 and 1919. The designation Model 19 for the 7.65mm Frommer is rather unusual since the weapon appeared in 1912 and was adopted by the Hungarian Army (Honved) prior to World War I, but official Hungarian documents indicate that this is the correct designation. It is basically not a good design, is somewhat delicate for a service weapon and was reportedly not popular with Hungarian troops. The complication of a long recoil system is absolutely wasted on relatively low-powered cartridges, which can be adequately handled by a simple blowback action. A later

Frommer M1939. Caliber .380. Obsolete in Hungary.

9mm Model 29 Pistol.

*For additional details see *Handguns of the World.*

9mm (.380 ACP) version of this pistol is marked "1939M". This was apparently not a service weapon, and the purpose for which it was made is not apparent.

THE HUNGARIAN 7.65mm (AND 9mm) PISTOL MODEL 37

In 1929, the Hungarians adopted a modified Browning blowback design, the 9mm (.380 ACP) Model 29. This weapon has an internal bolt assembly fixed to the slide rather than having the bolt (breech) portion machined in the rear of the slide. This weapon was apparently made in very limited quantity and is overly complicated, as is the Frommer Stop.

A version of the Model 29 with a conventional slide appeared in 1937. The Model 37 is a conventional blowback pistol chambered for both the 7.65mm (.32 ACP) and the 9mm (.380 ACP). The Model 37 was made in large quantities for the Hungarian Army and is probably the most common Hungarian pistol in the US today.

During World War II, Hungary became—for all practical purposes—a satellite of Germany. Hungarian arms plants produced large quantities of weapons for the Germans. Among these weapons was the Model 37 pistol in 7.65mm. The weapon, as made for the Germans, bears German acceptance markings and the three-letter manufacturers code "jhv." The weapons manufactured for the Germans have, in addition to the grip safety fitted to the earlier-made Model 37 pistols, a plate-type safety mounted on the left side of the receiver.

THE HUNGARIAN 7.62mm PISTOL MODEL 48

After World War II, a modified copy of the German 7.65mm Walther PP was introduced in Hungary. This pistol, which has a loaded chamber indicator over the chamber rather than at the rear of the slide as does the Walther, is called the Model 48 by the Hungarians. It is not a military weapon but is used by police and has been sold commercially in 9mm (.380 ACP) as the "Walam." These pistols are three-quarters of an inch longer than the PP.

Loading, firing and field stripping these pistols is the same as for German Walther PP and PPK.

The current Hungarian service pistol, the 7.62mm Model 48, is a copy of the Soviet 7.62mm TT33 and is loaded, fired and field stripped as is that pistol. An aluminum frame version has been introduced and designated the M60. It has been issued to Hungarian military forces.

9mm Model 37 Pistol.

7.65mm Model 48 Pistol.

7.65mm Model 37 Pistol made for the Germans.

9mm Walam 48 Pistol.

CHARACTERISTICS OF HUNGARIAN SERVICE PISTOLS

	Frommer Stop Model 19	7.65mm Model 48	9mm Model 37	7.62mm Model 48
Caliber:	7.65mm (ACP).	7.65mm (.32 ACP).	9mm (.380 ACP)*	7.62mm.
System of operation:	Long recoil, semiautomatic.	Blowback, semiautomatic.	Blowback, semiautomatic.	Recoil, semiautomatic.
Overall length:	6.5 in.	7 in.	6.8 in.	7.68 in.
Barrel length:	3.8 in.	3.94 in.	3.9 in.	4.57 in.
Feed device:	7-round, in line detachable, box magazine.	8-round, in line detachable, box magazine.	7-round, in line detachable, box magazine.	8-round, in line detachable, box magazine.
Sights:				
Front:	Blade.	Blade.	Blade.	Blade.
Rear:	V notch.	U notch.	V notch.	U notch.
Weight:	1.31 lbs.	1.92 lbs.	1.62 lbs.	1.88 lbs.
Muzzle velocity:	920 f.p.s.	920 f.p.s.	984 f.p.s.	1,378 f.p.s.

*Also made in 7.65mm (.32 ACP).

THE TOKAGYPT 58

The Tokagypt 58 was reportedly developed for the Egyptian police. It is a modified copy of the Tokarev TT33 chambered for the 9mm Parabellum cartridge. A safety catch has been added to the basic Tokarev design, which is fitted on the left top side of the receiver.

In addition, a plastic wrap-around type one-piece grip and a finger support on the magazine are found on this pistol. It is not a Hungarian service pistol.

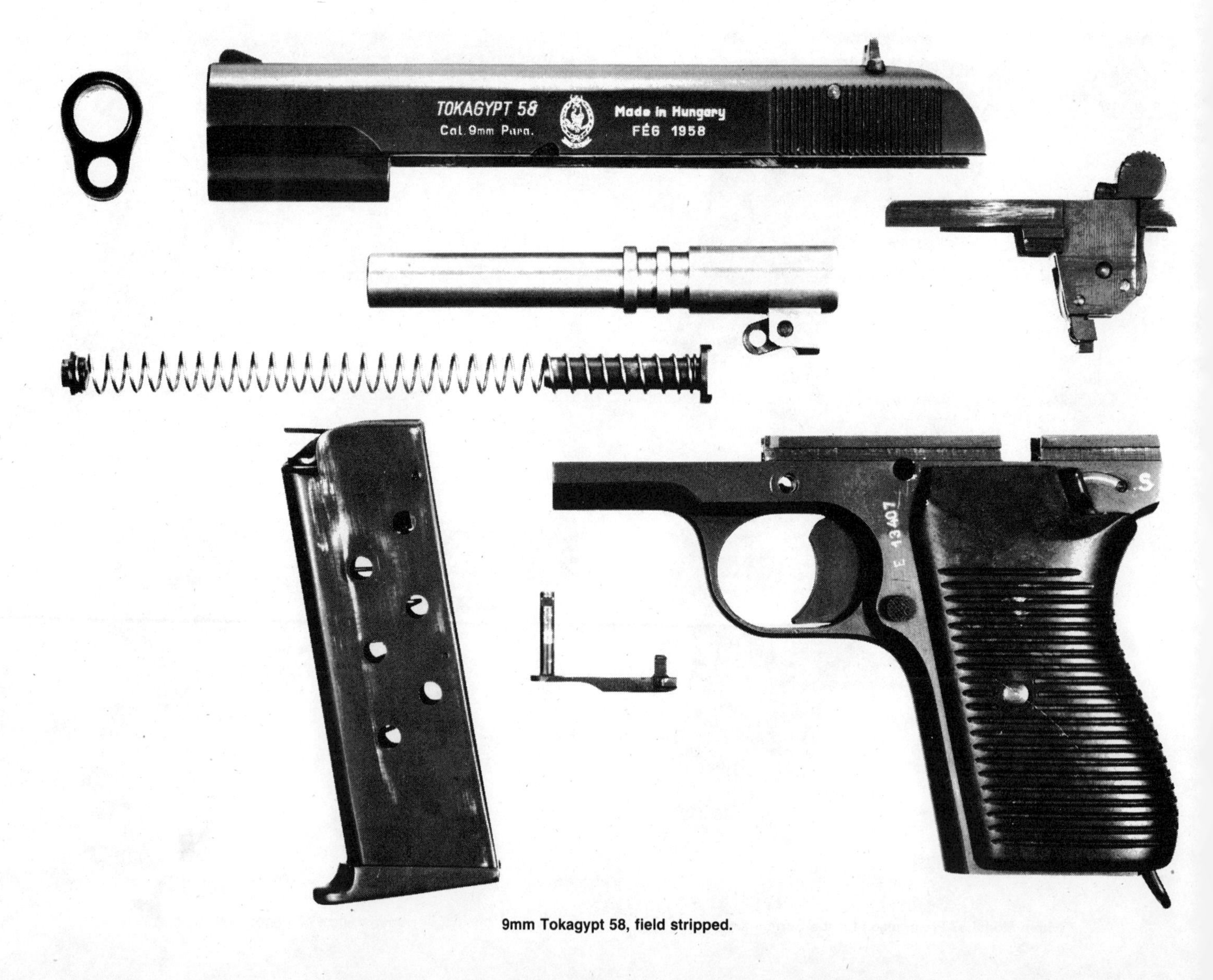

9mm Tokagypt 58, field stripped.

HUNGARIAN RIFLES

HUNGARIAN 8mm STUTZEN RIFLE MODEL 31

The Hungarians were equipped during World War I with various models of the straight pull 8mm Steyr rifle chambered for the 8mm M93 cartridge. After the adoption of the 8 × 56mm rimmed M31 cartridge, a number of the 8mm M95 short rifles (the Stutzen) were modified to take this cartridge. This rifle is called the Model 31 and is identified by a letter "H," 5/16 of an inch high, stamped on the barrel or receiver.

HUNGARIAN 8mm RIFLE MODEL 35

In 1935 a turn-bolt Mannlicher type was adopted, the Model 35, chambered for the 8mm Model 31 cartridge. The Model 35 has a two-piece bolt with the bolt handle positioned ahead of the receiver bridge when the bolt is forward. The magazine is an in-line type, which protrudes below the line of the stock. The Model 35 has a two-piece stock similar in concept to the British Lee Enfield.

HUNGARIAN 7.92mm RIFLE MODEL 43

In 1940, the Model 35 was redesigned for German use. The caliber was changed to 7.92mm, a staggered row Mauser type box magazine flush with the stock was fitted and German type bands and bayonet lug were used on this rifle, which was called the Model 98/40 by the Germans. In 1943, Hungary adopted the 7.92mm cartridge and started issue of a slightly modified 98/40—the Model 43—to the Hungarian Army. This rifle differs from the 98/40 principally in its bands and fittings, i.e., bayonet lug, sling swivels, etc.

HUNGARIAN 7.62mm RIFLE MODEL 48 (COPY OF SOVIET MOSIN NAGANT)

The conclusion of the war found Hungary disarmed. The Hungarian Army was equipped with Soviet 7.62mm Mosin Nagant rifles and carbines. Some copies of these weapons were made in Hungary and called Model 48. Within the past decade, the post-war Soviet rifles and carbines have been adopted, and both the 7.62mm SKS carbine and 7.62mm AK assault rifle have been made in Hungary.

HUNGARIAN 7.62mm MODIFIED SOVIET AKM RIFLE

The Hungarians have made several versions of the 7.62mm AKM. The most common has a plastic stock and pistol grips and a metal hand guard. The forward pistol grip is grasped during fire rather than the hand guard or the magazine. It has no top hand guard. This weapon is loaded, fired and field stripped in a manner similar to the Soviet AKM.

HUNGARIAN AMD ASSAULT RIFLE

The Hungarians have come out with another variation of the AKM. The AMD has a short barrel with a large muzzle brake. It has a single strut steel folding stock. The plastic fore grip and the metal forend are the same as those on the standard Hungarian modification of the AKM. Because of the relatively low energy of the 7.62 × 39mm cartridge, recoil with this weapon is probably not excessive, but the muzzle blast is probably very high. The AMD is 33.5 inches long with stock fixed and 23.5 inches long when the stock is folded. The barrel is 12.6 inches long.

Hungarian 7.92mm Model 43 Rifle.

The 8mm Model 35 Rifle.

Hungarian modification of 7.62mm AKM Assault Rifle.

Hungarian 7.62mm Model AMD Assault Rifle.

HUNGARIAN SERVICE RIFLES

	Model 31	Model 35	Model 43	Model 48	Modified AKM
Caliber:	8 × 56mm.	8 × 56mm.	7.92mm.	7.62 Mosin Nagant.	7.62 M43
System of operation:	Straight pull bolt.	Turn bolt.	Turn bolt.	Turn bolt.	Gas, selective fire.
Overall Length:	39.5 in.	43.7 in.	43 in.	48.5 in.	34.25 in.
Barrel length:	19.65 in.	23.6 in.	23.8 in.	28.7 in.	16.34 in.
Feed device:	5-round single column, fixed box magazine.	5-round single column, fixed box magazine.	5-round staggered row, fixed box magazine.	5-round single column, fixed box magazine.	30-round staggered row, detachable box magazine.
Sights:					
Front:	Barley corn.	Hooded barley corn.	Barley corn.	Hooded post.	Protected post.
Rear:	Leaf with notch.	Tangent with notch.	Tangent with notch.	Tangent with notch.	Tangent with notch.
Weight:	7.5 lbs.	8.9 lbs.	8.6 lbs.	8.7 lbs.	Approx. 10.7 lbs.
Muzzle velocity:	Approx. 2300 f.p.s.	2395 f.p.s.	Approx. 2480 f.p.s. (SS ball)	2800 f.p.s.	2329 f.p.s.
Cyclic rate:					600 r.p.m.

HUNGARIAN SUBMACHINE GUNS

HUNGARIAN 9mm SUBMACHINE GUN MODEL 39

A native-designed submachine gun was produced in Hungary in the late thirties and adopted in 1939. The design of this weapon, which is chambered for the 9mm Mauser cartridge, is credited to Pal D. Kiraly and resembles in many respects the SIG (Swiss) MKMO submachine gun. Kiraly was concerned with the design of the Dominican Cristobal carbine; the bolt design of the Cristobal is similar to that of the Model 39. The folding magazine system of the Model 39 is similar to that of the SIG MKMO.

The standard Model 39 submachine gun has a one-piece stock; a version with a folding wooden butt was produced as

9mm Mauser Model 39 Submachine Gun.

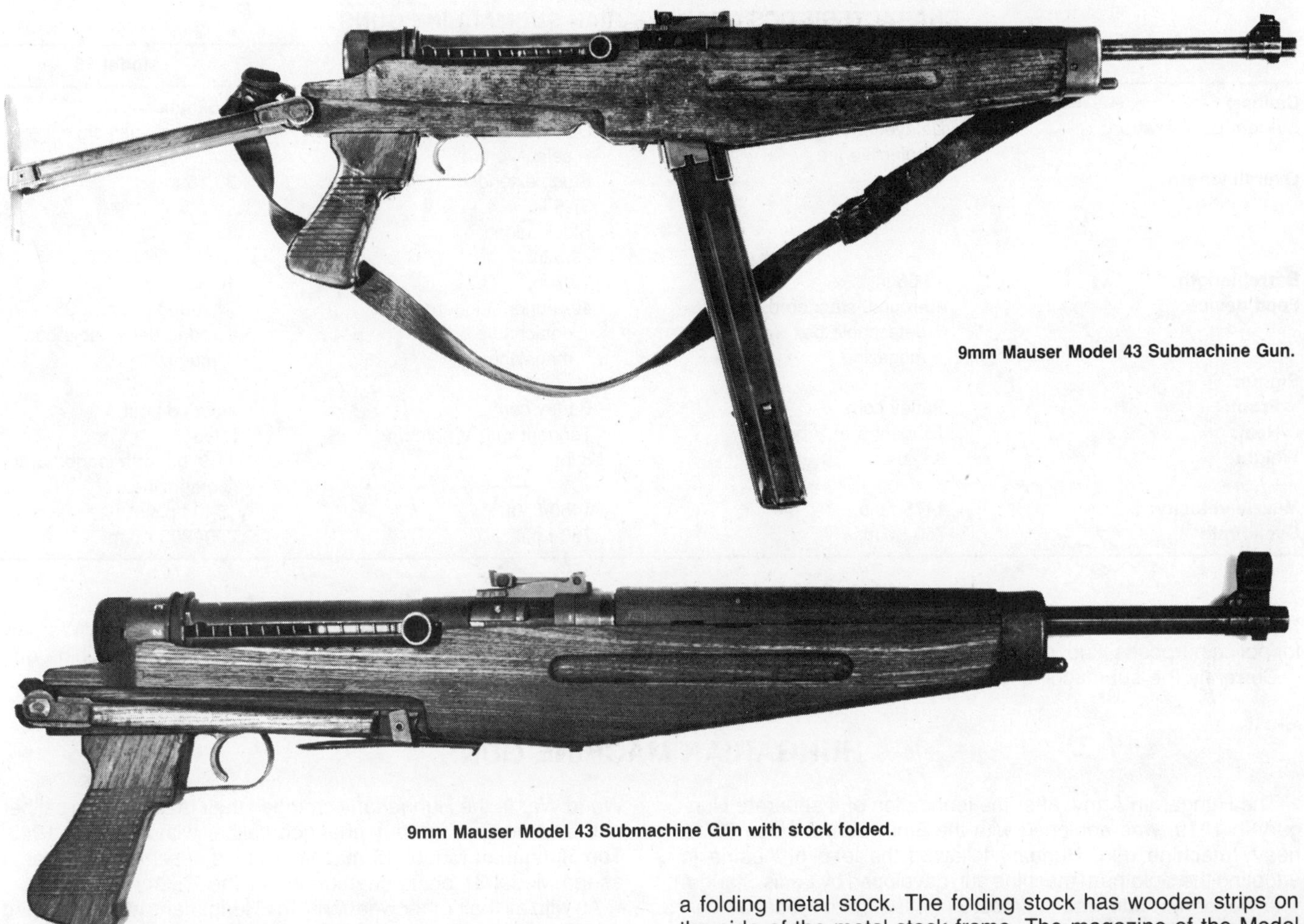

9mm Mauser Model 43 Submachine Gun.

9mm Mauser Model 43 Submachine Gun with stock folded.

the Model 39/A. Both the Model 39 and 39/A were produced in very limited numbers. The fire selector/safety is the circular cap located on the rear of the receiver, operated by rotating the cap to align with one of the three settings: -'E' for semiautomatic fire, 'S' for automatic fire and 'Z' for safe.

HUNGARIAN 9mm SUBMACHINE GUN MODEL 43

The Model 43, which was made in much larger quantity than the Model 39, is essentially the same as the Model 39 but has a folding metal stock. The folding stock has wooden strips on the side of the metal stock frame. The magazine of the Model 43 is canted slightly forward when in the fixed position as opposed to the straight vertical position of the Model 39 magazine, and the barrel of the Model 43 is approximately three inches shorter than the Model 39. The magazines of the Models 39 and 43 are not interchangeable.

HUNGARIAN 7.62mm SUBMACHINE GUN MODEL 48 (COPY OF SOVIET PPSh M1941)

After World War II, the Hungarians were supplied with Soviet weapons and manufactured a copy of the Soviet 7.62mm PPSh

Model 48 7.62mm Submachine Gun.

CHARACTERISTICS OF HUNGARIAN SUBMACHINE GUNS

	Model 39	Model 43	Model 48
Caliber:	9mm Mauser.	9mm Mauser.	7.62mm.
System of operation:	Delayed blowback, selective fire.	Delayed blowback, selective fire.	Blowback, selective fire.
Overall length:	41.25 in.	Stock extended: 37.5 in. Stock folded: 29.5 in.	33.15 in.
Barrel length:	19.65 in.	16.7 in.	10.63 in.
Feed device:	40-round, staggered row, detachable box magazine.	40-round, staggered row, detachable box magazine.	71-round drum or 35-round detachable box magazine.
Sights:			
Front:	Barley corn.	Barley corn.	Hooded post.
Rear:	Tangent with V notch.	Tangent with V notch.	L type.
Weight:	8.2 lbs.	8 lbs.	11.9 lbs. with loaded drum magazine.
Muzzle velocity:	1475 f.p.s.	1450 f.p.s.	1640 f.p.s.
Cyclic rate:	750 r.p.m.	750 r.p.m.	700–900 r.p.m.

M1941 as the Model 48. A submachine gun intended basically for police purposes, known as the Model 54, was also produced.

Currently the submachine gun is used as a reserve weapon or as a Border Patrol weapon in Hungary. As with other Warsaw Pact nations, the assault rifle is replacing the submachine gun in the Army.

HUNGARIAN MACHINE GUNS

The Hungarian Army, after the foundation of a separate Hungary in 1919, was equipped with the 8mm Schwarzlose 07/12 heavy machine gun. Hungary followed the lead of Austria in adopting the Solothurn machine gun developed by Louis Stange. The Hungarian gun was chambered for the 8mm Model 31 cartridge (8 × 56 R) and was called the Model 31. The gun has the same characteristics as the Model 30.

Some Schwarzlose machine guns were modified to use the 8mm Model 31 cartridge; they are called the Model 7/31. The Hungarians also produced an aircraft gun chambered for the Model 31 cartridge, which was called the Model 34/AM. During World War II, the Hungarians changed their machine gun caliber at the same time as they changed caliber with rifles—in 1943. The Hungarian Model 43 machine gun is essentially the same as the Model 31 but is chambered for the 7.92mm cartridge.

As with all their other weapons, the Hungarians were equipped with Soviet machine guns after World War II. Initially they adopted the older Soviet weapons, such as the 7.62mm DP, DPM and DTM machine guns. They currently use the 7.62mm RPD, 7.62mm SGM and 12.7mm DShK 38/46. All these weapons are covered in detail in the chapter on the Soviet Union.

8mm Model 31 Machine Gun.

25 India

SMALL ARMS IN SERVICE

Handguns:	9 × 19mm NATO FN Browning Mle. 1935	FN-made.
Submachine guns:	9 × 19mm NATO Sterling Mk4	Made in India at Cawnpore.
Rifles:	7.62 × 51mm NATO IA SL	FN FAL made at Isapore.
	.303 No. 4	Used as a grenade-launching rifle.
Machine guns:	7.62 × 51mm NATO MAG	FN-made, with some from RSAF, Enfield.
	7.62 × 51mm NATO L4A4	RSAF, Enfield-made.
	12.7 × 99mm M2 HB	US-made.

INDIAN RIFLES

The government small arms factory at Ishapore has been making rifles for a considerable time and produced the Lee Enfield No. 1 Mark III* through World War II. The Ishapore rifles varied in minor details from the rifle as made in the United Kingdom, frequently having no stacking swivels, the stacking swivel mounting lug being left solid. Ishapore made 692,587 Mark III* rifles during World War II. India is now producing the 7.62mm rifle L1A1 (British FN FAL).

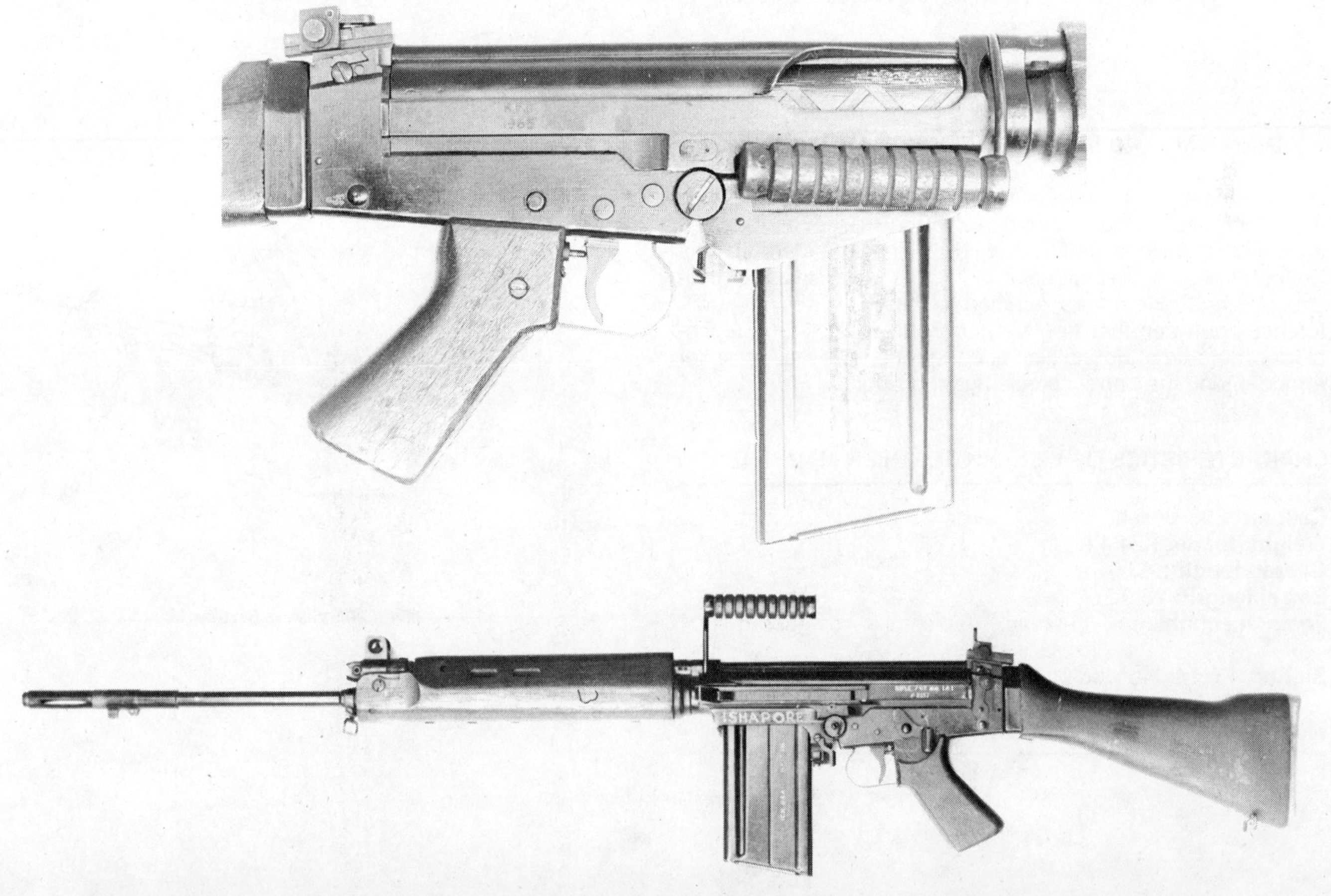

The FAL as manufactured at the Rifle factory at Ishapore. The top view shows one marked R.F.I. 1965 indicating that it was made in that year at Ishapore. Note the distinctive rear sight and the grip retaining screw.

Marking on Short Lee Enfield No. 1 Mark III* made at Ishapore.

INDIAN MACHINE GUNS

India produced 8,357 machine guns during World War II and was preparing to produce Bren Guns at a new factory at Hyderabad when the war ended. The current standard Indian machine gun is the 7.62mm FN MAG general purpose gun. The standard machine gun of the Indian Army during World War II was the Vickers Berthier.

INDIAN CAL. .303 VICKERS BERTHIER MACHINE GUN

During the period between World Wars I and II, the Indian Army, then a semi-autonomous branch of the British Army, adopted the cal. .303 Vickers Berthier Light Machine Gun Mark III. This weapon is still in limited use in India. It is quite similar in construction and external appearance to the Bren. The main differences between the Bren and the Berthier lie in the breechblock, the feed, the holding-open device, the gas cylinder arrangement, the barrel change and the sights.

Caliber .303 Vickers Berthier Mark III.

CHARACTERISTICS OF VICKERS BERTHIER MACHINE GUN

Caliber: .303 British.
Weight, loaded: 24.4 lbs.
Overall length: 45.5 in.
Barrel length: 23.9 in.
Feed mechanism: 30-round, detachable, staggered box magazine.
Sights: Front: Hooded blade.
Rear: Leaf.
Muzzle velocity: 2400 f.p.s. w/Mk 7 ball (approx).

26 Indonesia

SMALL ARMS IN SERVICE

Indonesia has an unusually large variety of service small arms from a diverse number of sources. The US caliber .45 M1911A1, the 9mm FN Hi-Power Browning, the 7.62mm Tokarev and various models of the Walther are some of the pistols used by the Indonesians. Rifles and carbines used include the caliber .30 M1 rifle, caliber .30 M1 carbine, the caliber .30 M1918A2 Browning automatic rifle, the caliber .30 FN semiautomatic rifle (the M1949), the 7.62mm G3 rifle, 7.62mm M59 rifle, the Soviet 7.62mm AK47 assault rifle, the Soviet 7.62 SKS carbine, the British caliber .303 No. 1 and No. 4 rifles and the 7.92mm Kar98k. The submachine guns in service include the 9mm Beretta, 9mm Sten, 9mm M1950 Madsen, the Swedish 9mm M45 Carl Gustaf, the caliber .45 Thompson, the 9mm Owen and the Czech 7.62mm M24 and M26. The standard machine gun is the caliber .30 Madsen Mark II, which is made in Indonesia. The caliber .30 Browning guns, the caliber .303 Bren and Vickers guns, the 7.62 Degtyarev and Goryunov and the 12.7mm DShK M1938/46 are also used.

INDONESIAN HANDGUNS

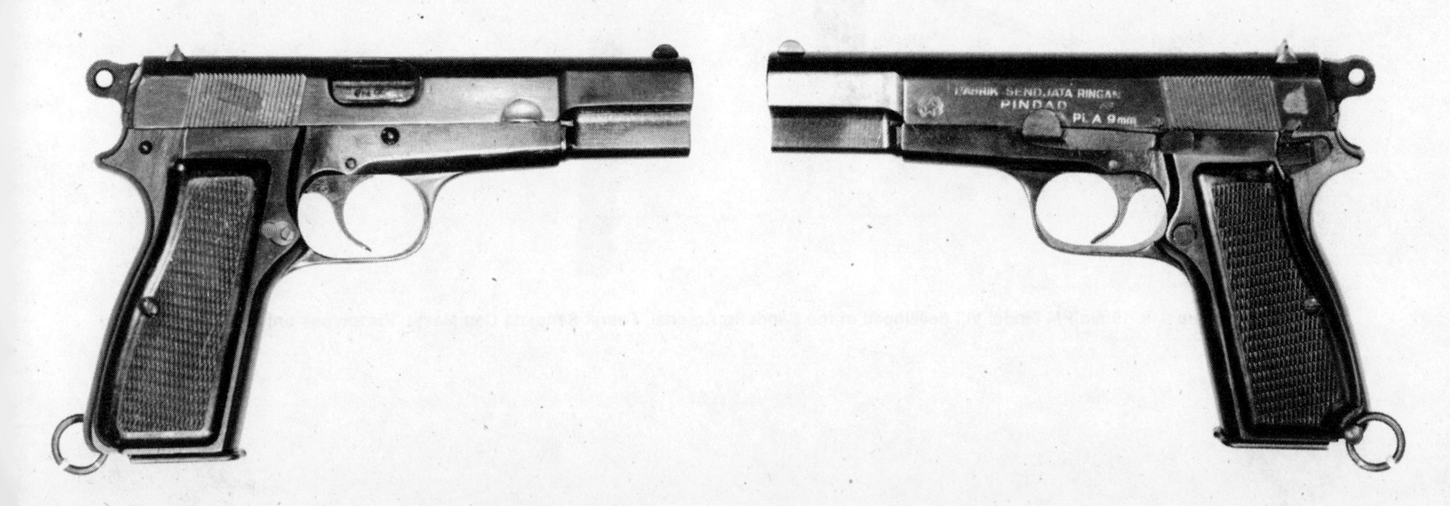

The 9 × 19mm FN Browning Mle. 1935 as manufactured at the Fabrik Sendjata Ringan PINDAD.

INDONESIAN RIFLES

The Indonesians had considerable quantities of Dutch 6.5mm Mannlicher M1895 rifles and carbines, and caliber .303 rifles, at the conclusion of their rebellion against the Dutch. A number of Mannlicher carbines were rebarreled for the .303 cartridge for the Dutch forces during World War II in the then Dutch East Indies. The fall of the Netherlands in 1940 cut off their supply of 6.5mm ammunition. At a later date the Indonesian Army started to standardize on caliber .30 weapons, the principal weapon being the US M1 rifle. The Indonesians purchased some of these from Beretta in Italy and from US sources. Purchases were also made of Soviet 7.62mm AK-47 assault rifles and SKS carbines. The Indonesians have converted some of their M1 rifles to the 7.62mm NATO BM59 (a Beretta design).

INDONESIAN SUBMACHINE GUNS

A number of prototype 9mm Parabellum submachine guns were developed by the Indonesians. None were made in quantity.

Late models of the 9mm Parabellum Model 12 Beretta submachine gun are currently manufactured in Indonesia.

The 9 × 19mm PM Model VIII developed at the Bandung Arsenal, Fabrik Sendjata Dan Mesiu. Prototypes only.

27 Iran (Islamic Republic)

SMALL ARMS IN SERVICE

Handguns:	11.43 × 23mm M1911A1	US-made.
Submachine guns:	9 × 19mm NATO UZI	Made by Israeli Military Industries.
Rifles:	7.62 × 51mm NATO G3	Made in Iran under license from Heckler & Koch.
	7.62 × 63mm M1	Made in USA; obsolete.
	7.62 × 63mm M1919A4	Made in USA; obsolete.
	12.7 × 99mm M2 HB	Made in USA.
Machine guns:	7.62 × 51mm NATO MG1A1	Made by Rheinmetall and made in Iran under license.

OBSOLETE IRANIAN RIFLES

Iran adopted the Czech ZB 7.92mm Model 98/29 short rifle and the Model 98/29 carbine as the Model 1309 and Model 1317 (Moslem calendar years), respectively. The Christian calendar years for these are 1930 and 1938. The rifles were originally made in Czechoslovakia, but the Iranian arsenal produced them as well. Production at the Iranian arsenal probably started during World War II, since Czech supplies were shut off during the war.

The Iranians developed a modification of the Model 1930 short rifle, the Model 1949 (Model 1328 by the Moslem calendar); the Model 1949 differs from the Model 1930 only in bands and sling swivels. These rifles are loaded, fired and field stripped in the same manner as the Czech Model 24 rifle, which is covered in the chapter on Czechoslovakia.

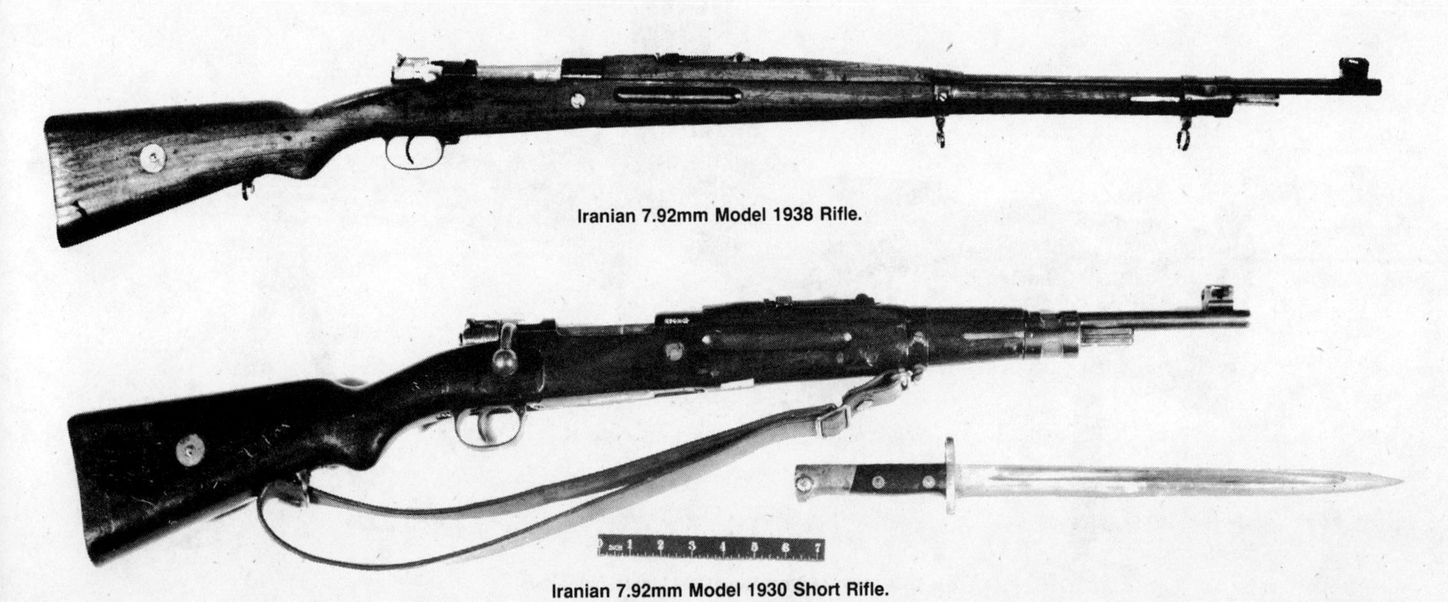

Iranian 7.92mm Model 1938 Rifle.

Iranian 7.92mm Model 1930 Short Rifle.

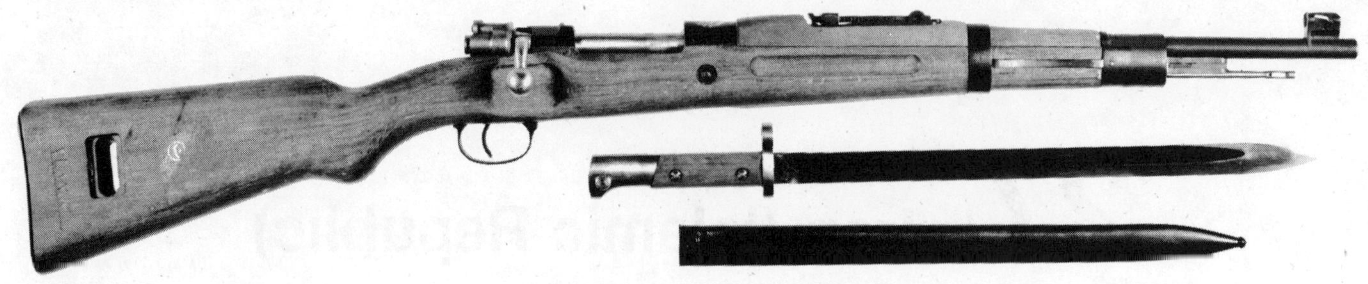

Iranian 7.92mm Model 1949 Short Rifle.

OBSOLETE IRANIAN SUBMACHINE GUNS

Iran has quantities of US caliber .45 M3A1 submachine guns. Iran has also produced copies of the Soviet 7.62mm PPSh M1941 submachine gun. The Iranians began producing the gun during World War II for the Soviets on a contract basis and continued to produce the weapon for their own use. Characteristics of the weapon are the same as those given in the chapter on the Soviet Union.

OBSOLETE IRANIAN MACHINE GUNS

The Iranians have US caliber .30 and .50 Browning machine guns. They also have the Czech ZB30 machine gun, made in Iran. This weapon was originally manufactured in 7.92mm but is currently used by the Iranians in cal. .30. Characteristics, loading, firing, and field stripping of this weapon are basically the same as those given for the ZB30 in the chapter on Czechoslovakia.

Iranian copy of Soviet PPSh M1941 Submachine Gun.

Iranian 7.92mm Model 30 Light Machine Gun.

28 Israel

SMALL ARMS IN SERVICE

Handguns:	9 × 19mm NATO Beretta Model 951	Beretta-made. See Chapter 29.
Submachine guns:	9 × 19mm NATO UZI	Israeli Military Industries.
Rifles:	5.56 × 45mm Galil*	Israeli Military Industries.
	5.56 × 45mm M16A1 rifle and carbine*	Colt Firearms Division, Colt Industries.
	7.62 × 51mm M14	US-made. Issued to Civil Defense guards.
	7.62 × 39mm and AKM*	Soviet and Warsaw Pact–made. Used by front line and police patrols.
	7.62 × 33mm M1 Carbine	US-made. Issued to Civil Defense guards.
	7.62 × 51mm FAL	Fabrique Nationale and assembled by IMI with some parts made there. (Obsolete)
Machine guns:	7.62 × 51mm NATO MAG	Fabrique Nationale.
	7.62 × 51mm NATO FALO (heavy-barrel SAW)	Fabrique Nationale. (Obsolete)
	12.7 × 99mm M2 HB	Made in USA.
	40 × 53mm MK19 MOD 1 Grenade Launcher	Israeli Military Industries.

*Basic infantry rifles.

ISRAELI PISTOLS

The Israelis had a variety of pistols at the time of the foundation of the State. For the most part these were caliber .38 and caliber .455 Enfield and Webley revolvers, but there were odd quantities of 9mm FN Browning Hi-power, Luger and P38 automatics as well. The Israelis developed a modified copy of the Smith & Wesson Military and Police pistol, which is unusual in that it is

Two 9 × 19mm versions of the basic Smith & Wesson revolver. *Top*, a copy made by IMI; *bottom*, a 1964 experimental made by Smith & Wesson. Both required the use of three-shot half-moon clips of the type employed by the US .45 caliber M1917 revolvers.

ISRAELI RIFLES

chambered for the 9mm Parabellum cartridge. This rimless cartridge requires the use of two three-shot clips similar to those used with the US caliber .45 Colt and Smith & Wesson service revolvers. This revolver, which bears Israeli Security Forces markings, is basically a police weapon. The Israelis use the 9mm Parabellum M1951 Beretta pistol as their standard military service pistol. This pistol is covered in detail in the chapter on Italy.

The Israelis were originally equipped mainly with British caliber .303 No. 1 and No. 4 rifles. About 1948, purchases were made of 7.92mm Kar 98k rifles of both World War II German and postwar Czech manufacture. Because of the procurement of 7.92mm machine guns at the same time, the 7.92mm cartridge was adopted as standard. Towards the end of the 50s Israel standardized the 7.62mm NATO cartridge and adopted the FN FAL rifle and the heavy barrel version of this weapon. Israel has converted her 7.92mm Kar 98k Mausers to 7.62mm NATO. These rifles have now been withdrawn from active service.

To the right: **A Jewish settler guarding cattle in Northern Palestine near the Lebanese border in May 1948. He is armed with a British .303 No. 1 Enfield rifle.**

THE ISRAELI 5.56mm GALIL RIFLE

Israel's new rifle, the Galil, was first issued to troops in May 1973. The design lineage of the Galil is described in Chapter I under the 5.56 × 45mm rifles.

The Galil has a folding type steel stock and a bipod that folds between the hand-guards when not in use. It also has a grenade launcher on the end of the barrel. The cyclic rate has been reported as 650 rounds per minute.

The Israelis have announced that the Galil is to be used to replace the FAL and the heavy barrel FAL; it therefore has quite a heavy barrel. A 7.62 × 51mm NATO version of the Galil was introduced in 1981. This latter weapon appears to be intended primarily for export sales.

Two versions of the Galil. *Top,* an Israeli Defence Force model with Hebrew characters for the selector lever; *bottom,* an export model with S, A, R on the selector for "safe," "automatic," and "repeat" or semiautomatic. Note that the bottom Galil has a detachable scope mount.

A comparison of the Galil rifle (18-inch barrel) and the carbine (13-inch barrel).

Typical Galil receiver markings.

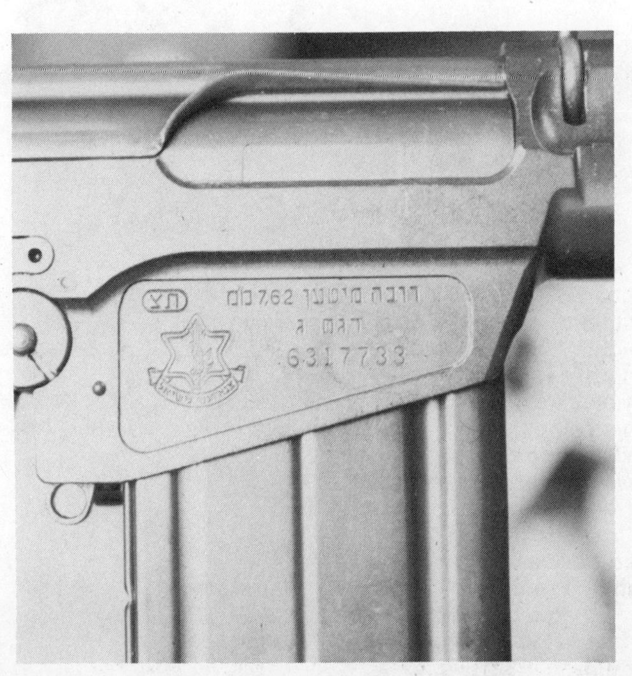

Typical receiver markings on an FN-made FAL.

How to Load and Fire the Galil

Follow the procedures outlined for the Kalashnikov assault rifles in Chapter 46.

Field Stripping the Galil

Follow the procedures outlined for the Kalashnikov assault rifles in Chapter 46.

UNIQUE DESIGN FEATURES OF THE ISRAELI GALIL RIFLE

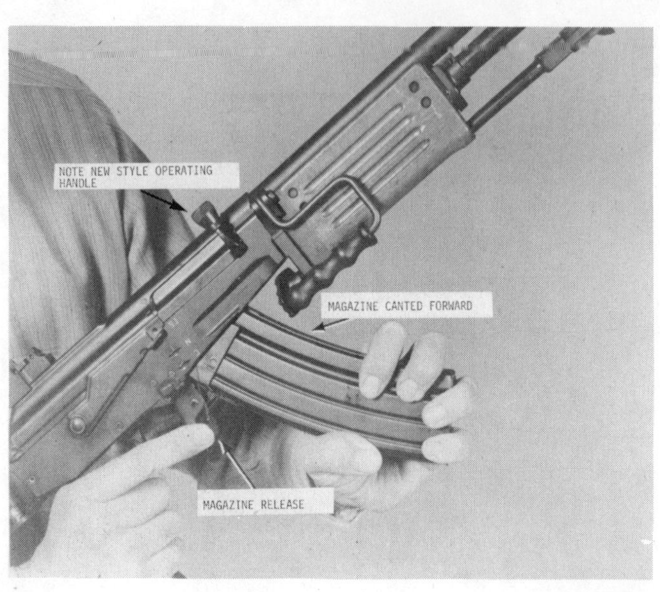

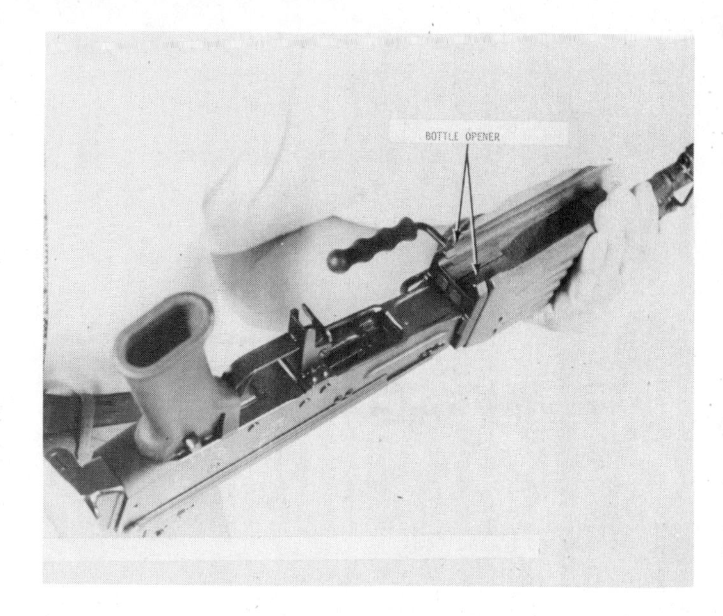

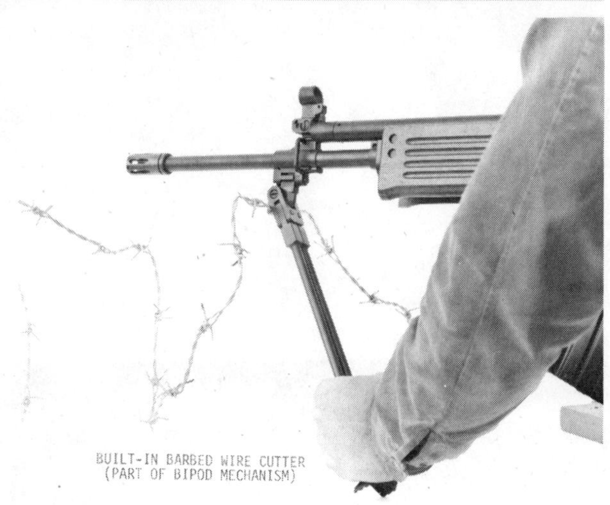

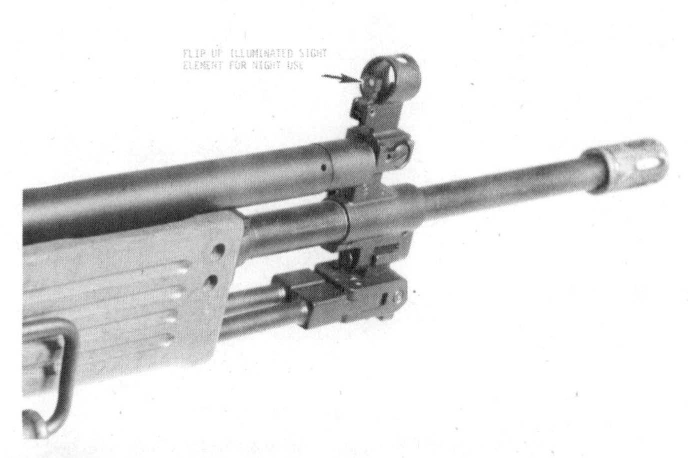

Caliber: 5.56mm.
System of operation: Gas, Kalashnikov type.
Weight: 7.7 to 8.6 lbs.
Length, overall:
w/stock extended, short barrel: 32.3 in.
w/stock extended, long barrel: 38.2 in.
w/stock folded, short barrel: 23.6 in.
w/stock folded, long barrel: 29.1 in.
Barrel length: **short barrel:** 13.0 in.
long barrel: 18.1 in.
Feed devices: 35- and 50-shot magazines.
Sights: **Front:** Post sight set for 100 meters.
Rear: "L" flip type aperture.
Muzzle velocity: **Short barrel:** 3010 f.p.s.
Long barrel: 3215 f.p.s.
Cyclic rate: Approx. 650 r.p.m.

Disassembled view of Galil 5.56mm squad automatic weapon.

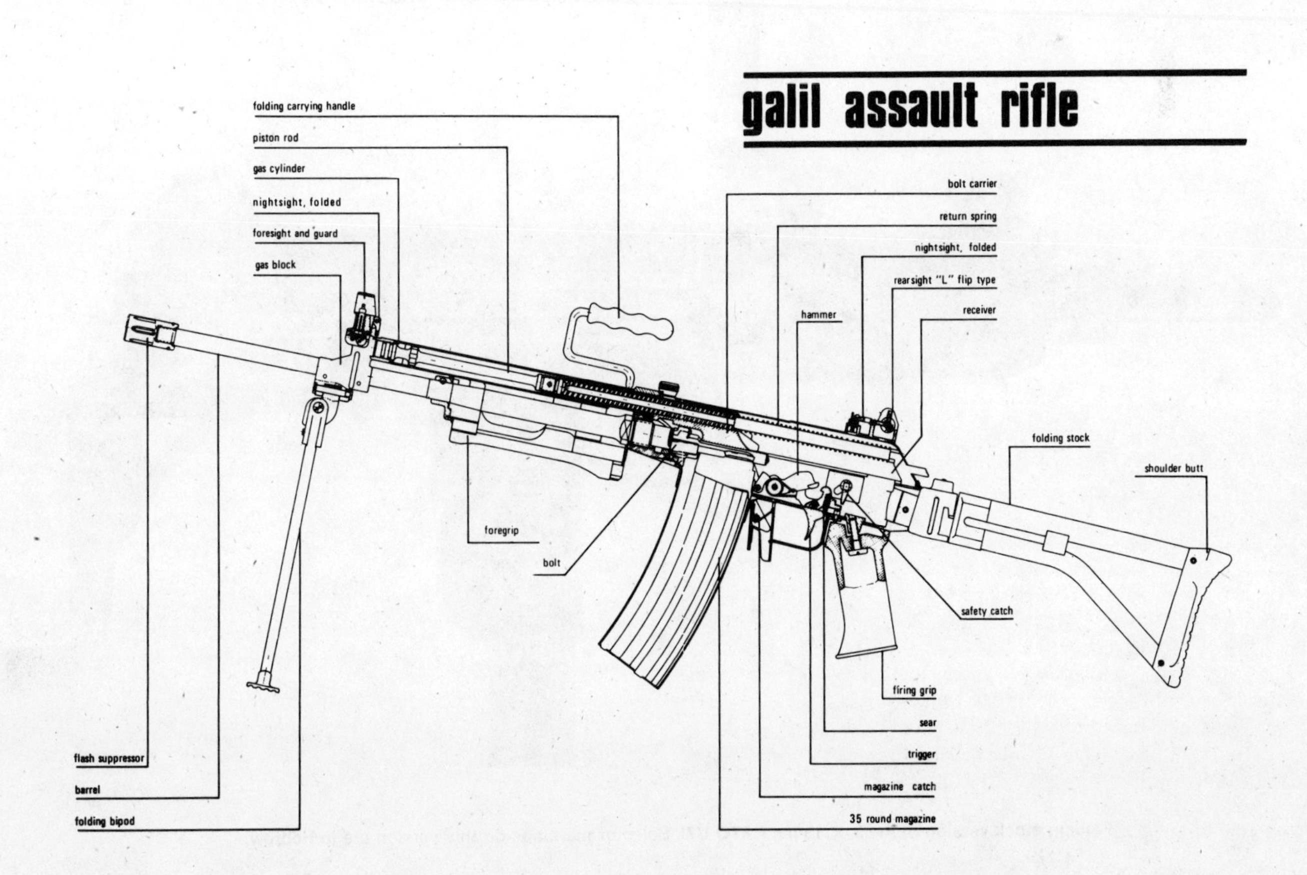

ISRAELI SUBMACHINE GUNS

The Israelis used the 9mm Sten in its various configurations in their early fighting with the British and the Arab States; some German 9mm MP40s were also procured from Central Europe. With the development of a native submachine gun—the UZI—Israel made her greatest impression on the small arms world.

THE ISRAELI 9mm UZI SUBMACHINE GUN

There are several different models of the UZI. The early model guns had a wooden stock and a wooden or fiber fore-end and pistol grip. The model with folding steel stock and plastic fore-end followed soon after the introduction of the gun, circa 1952. About 1960, the UZI was again modified, and a larger bolt-retracting handle was fitted and the fire-selector/safety modified. The model supplied to West Germany has the bolt-handle track in the receiver cover serrated to prevent forward movement of the bolt if accidentally released during retraction. The bolt-retraction handle of the UZI does not reciprocate with the bolt.

The UZI is named for Major Uziel Gal, its developer. It is the standard submachine gun of the Israeli Army and also has been sold to other countries. In basic operating principles, the UZI is quite similar to the Czech ZK 476 and its descendants, the Czech Models 23 and 25 submachine guns. As far back as 1945, the UK had an experimental gun with a bolt that telescoped the barrel—the telescoping bolt is a feature common to all of these weapons and to the Beretta Model 12 as well.

The UZI is a very well-made gun, and its performance leaves little to be desired. FN of Belgium is now marketing this weapon in Europe.

CHARACTERISTICS OF THE UZI SUBMACHINE GUN

Caliber: 9mm Parabellum.
System of operation: Blowback, selective fire.
Weight: 8.9 lbs. w/loaded 25-round magazine and metal butt.
8.8 lbs. w/loaded 25-round magazine and wooden butt.
Length, overall: 25.2 in. (wooden-butt model, and metal-butt weapon with stock extended). The length of the metal-butt model with the stock folded is 17.9 in., and the length of the wooden-butt model with the butt removed is 17,3 in.
Barrel length. 10.2 In.
Feed device: 25-round, 32-round or 40-round, detachable, staggered box magazine.
Sights: Front: Truncated cone w/protecting ears.
Rear: L-type w/setting for 100 and 200 yards.
Muzzle velocity: 1310 f.p.s. w/8-gram 9mm Parabellum bullet.
Cyclic rate: 650 r.p.m.

How to Load and Fire the UZI

Insert a loaded magazine in the magazine well located in the pistol grip. Pull the cocking handle, located on the top of the receiver, to the rear. The UZI fires from an open bolt, and the bolt will remain to the rear. To fire single shots, set the change lever, located on the top left side of the pistol grip, to the letter R (this is the mid-position). To fire automatic fire, set the change lever at the letter A (this is the forward position). To put the weapon on safe, set the change lever at S (this is the rear position). Pressure on the trigger will fire the weapon, if the change lever is set on semiautomatic or automatic fire positions. The grip safety must be squeezed when firing the weapon.

How the UZI Works

With the exception of the trigger mechanism, the functioning of the UZI is quite similar to that of the Czech Model 23 submachine gun. (See Chapter 15.)

Folding stock version of the 9 × 19mm NATO UZI. Selector markings on this version are in Hebrew.

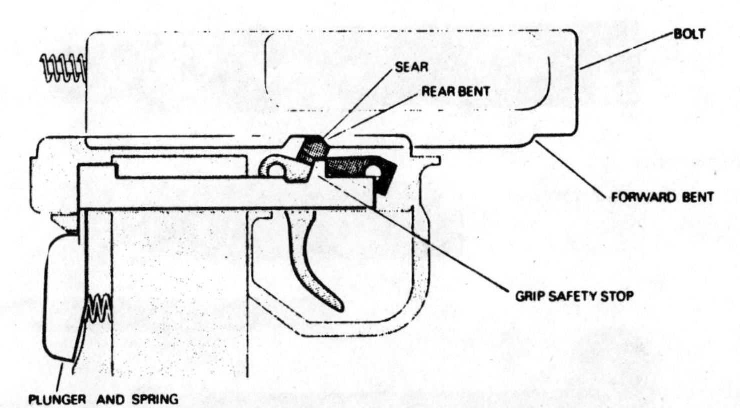

1 Sear engaged in rear bent; gun cannot be cocked until grip safety is operated as in 3

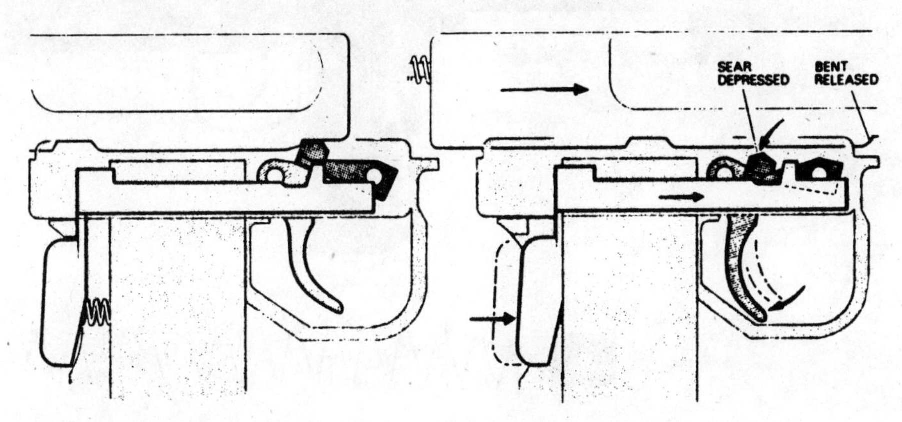

2 Sear engaged in forward bent; gun cocked; grip safety stop operating

3 Grip safety and trigger operated; sear depressed; gun firing

UZI trigger, sear, and grip safety mechanism.

Field Stripping the UZI

To release the cover, press the catch in front of the rear sight housing to the rear. Raise the cover and remove it from the receiver. Raise the bolt from the front; when its forward part is completely disengaged, withdraw it from the receiver together with the recoil spring, giving the whole assembly a short pull forward. To remove the extractor, take out its retaining pin. Press in the lock of the barrel retaining nut; unscrew the nut and withdraw the barrel. Take out the split pin and the sleeve in which it is housed, and remove the trigger group and pistol grip from the assembly. To strip the folding metal butt, unscrew the hexagonal head of the nut retaining the butt (the nut is located in the rear of the receiver). To remove the wooden butt, press in the butt retaining catch, and pull the butt to the rear. No further disassembly is recommended.

To reassemble the weapon, follow the above procedure in reverse order.

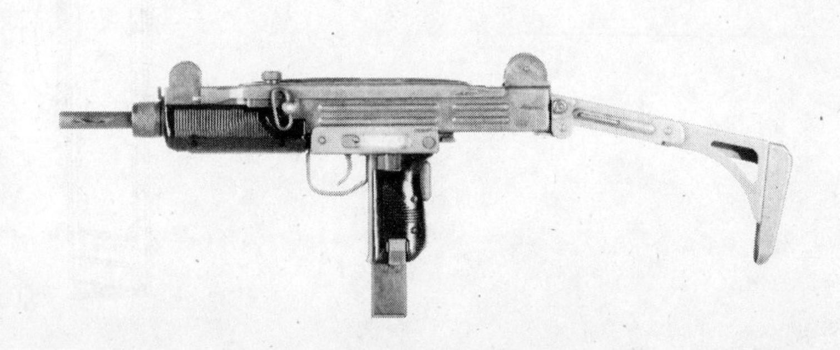

Folding stock UZI with stock extended.

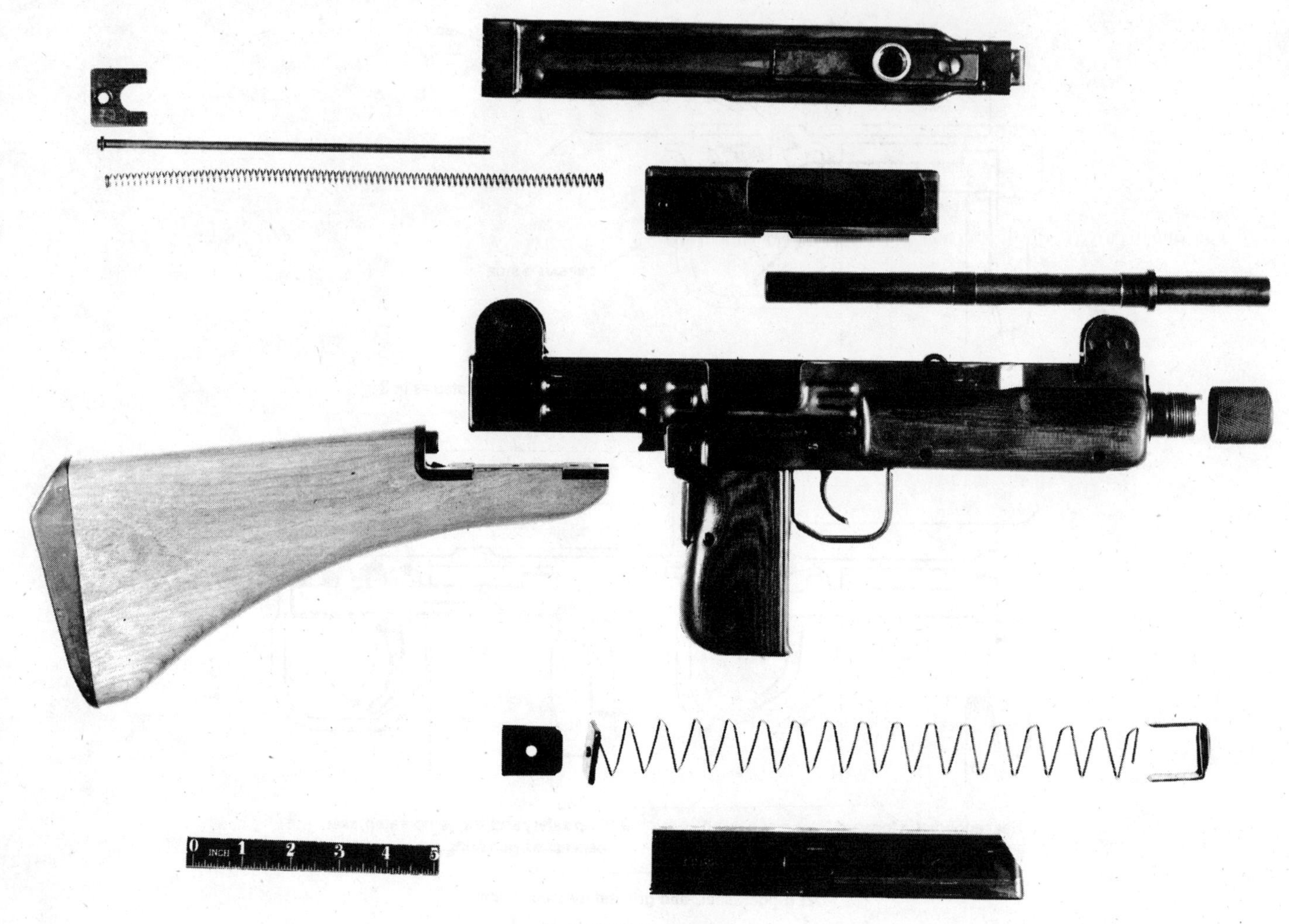

Early UZI field stripped.

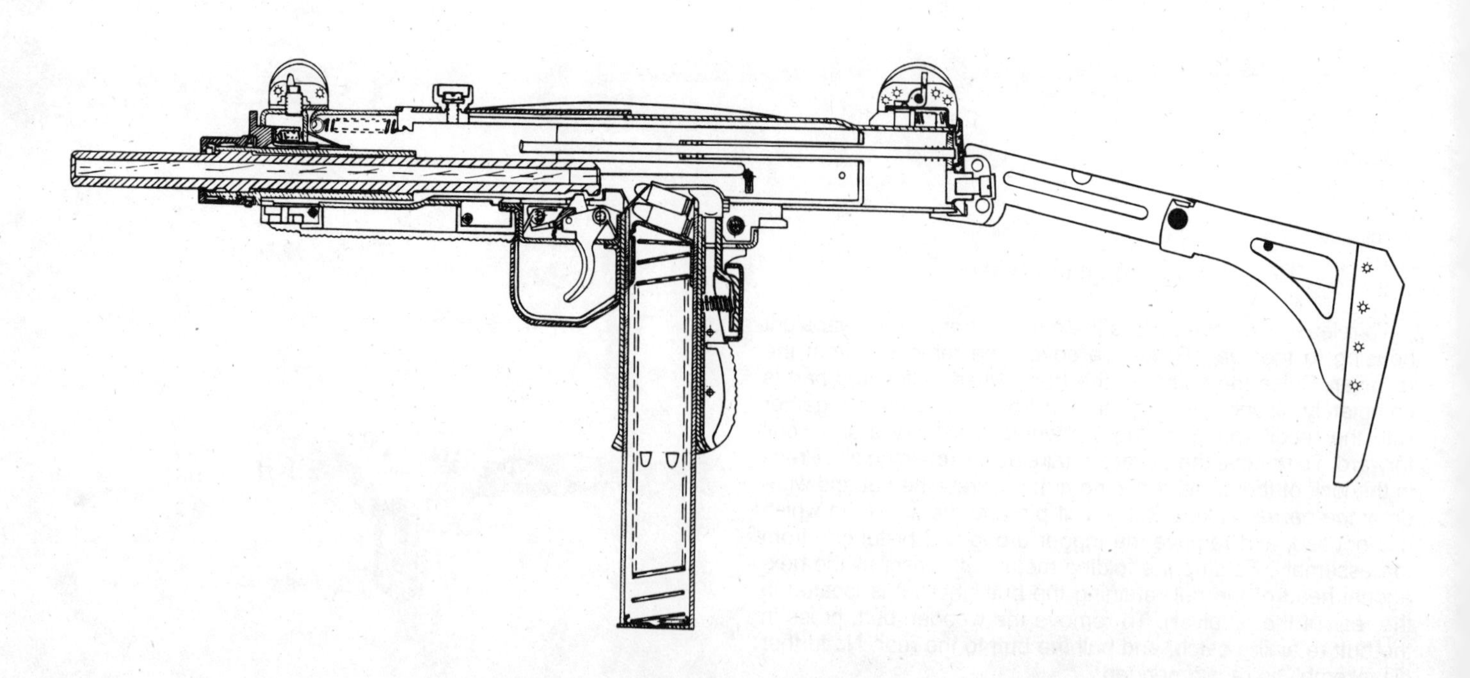

Section view, UZI 9mm submachine gun.

29 Italy

SMALL ARMS IN SERVICE

Handguns:	9 × 19mm NATO M951	Made by Beretta.
	9 × 19mm NATO M92	Made by Beretta.
	9 × 17mm M34	Beretta; limited use only.
Submachine guns:	9 × 19mm NATO M12	Made by Beretta.
	9 × 19mm NATO 57	Made by L. Franchi.
	9 × 19mm NATO MAB 38/49 Models 4 and 5	Made by Beretta.
Rifles:	7.62 × 51mm NATO BM59 series, including the BM59 Mark Ital, BM59 Mark Ital Paratrooper, BM59 Alpini, BM59 Mark II, and BM59 Mark E	Made by Beretta.
	5.56 × 45mm AR70 Mod SC/.223	Made by Beretta. Used by special troops only.
	7.62 × 51mm NATO G3 sniper	Made by Heckler & Koch. Used by Rome police and Carabinieri.
Machine guns:	7.62 × 51mm NATO MG42/59	Made by Beretta Armi Roma, and Whitehead-Motofides, Livorno. Both under license from Rheinmetall.
	7.62 × 51mm NATO M73	Made in USA.
	7.62 × 63mm Browning M1919A4	Made in USA.
	12.7 × 99mm M2 HB	Made in USA.

PIETRO BERETTA S. p. A.

The firm Pietro Beretta, S. p. A., traces its roots back to the 1680s and is today the major Italian manufacturer of military small arms. The firm has been one of the primary small arms suppliers for the Italian military since the unification of the Italian principalities in the 1870s. Beretta specialized for many years in the manufacture of handguns and submachine guns. In the post-1945 era they have also become a leading producer of rifles and machine guns. The handguns, submachine guns, and rifles are fabricated at their factory in Gardone Val Trompia. The MG42/59 is built at their factory in Rome. During the 1970s Beretta operated a factory in Brazil. As noted in Chapter 10, this factory was purchased in 1981 by the Brazilian firm Taurus. Beretta is also manufacturing handguns in the United States at their Accokeek, Maryland facility. They also act as subcontractors for the US subsidiary of Fabrique Nationale, building components for the M240 armor machine gun.

ITALIAN HANDGUNS

THE 9mm BERETTA M1951 PISTOL

The 9 × 19mm NATO Model 1951, also known as Model 951, was developed for the Italian armed forces as its post–World War II sidearm. While this very reliable handgun has an external resemblance to the earlier blowback Berettas—notably the cutaway slide—this weapon embodies a locking, block-locking mechanism derived directly from the Walther P38. Commercially marketed as the Brigadier, this handgun has been adopted by the military of Israel, the United Arab Republic (Egypt), and Nigeria.

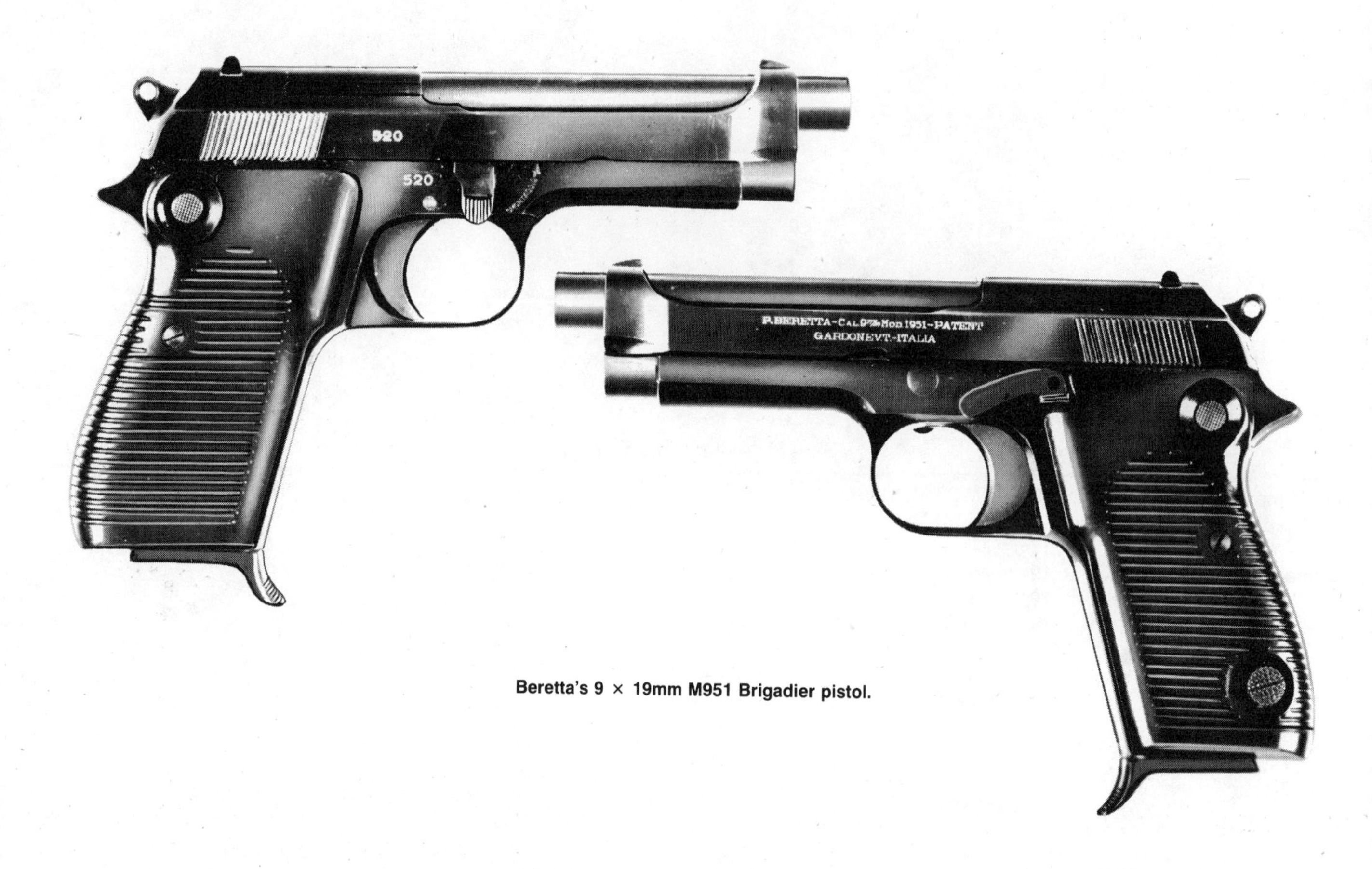

Beretta's 9 × 19mm M951 Brigadier pistol.

How to Load and Fire the M1951 Pistol

The magazine is loaded the same as any single-column pistol magazine. Insert loaded magazine in magazine well located in pistol grip. Pull the slide sharply to the rear, and release; the slide will run home and chamber a cartridge. The weapon may now be fired by squeezing the trigger. The safety is a button type mounted at the top rear of the grip; to put the weapon on "safe," push the button from right to left. The M1951 is not a double action weapon. With a cartridge in the chamber and the hammer down, the weapon must be manually cocked before it will fire.

How the M1951 Pistol Works

Loading, Cocking and Chambering. A magazine containing eight rounds is inserted into the bottom of the grip section of the receiver. The slide is drawn to the rear by hand; it forces the hammer downward and to the rear. The full-cock notch of the hammer engages the nose of the sear, and the hammer remains in its rearward position. When the slide is released, it goes forward under the pressure of the recoil spring. The slide feed rib forces a round from the magazine into the chamber of the barrel.

Locking. As the slide goes forward, the locking wedge is cammed into its locked position in the locking grooves of the slide by the cam surfaces on the receiver. The locking mechanism of this pistol is similar to that of the Walther Pistole 38, covered in the chapter on Germany in World War II.

Firing. When the trigger is pulled, the trigger bar is forced to the rear. The trigger bar pushes the disconnector up into the disconnector notch in the slide. If the weapon is not completely locked, the disconnector will be held down by the slide and will force the trigger bar down and out of contact with the trigger (this arrangement is similar to that of the M1934 Beretta). When the weapon is locked, the trigger bar forces the sear backward against the pressure of the sear spring; the hammer then disengages itself from the sear and moves forward under the pressure of the hammer spring. The hammer strikes the firing pin, forcing it forward against the pressure of the firing pin spring. The firing pin strikes the primer of the cartridge and immediately rebounds under the pressure of the firing pin spring. The M1951 is not a double-action pistol; if a misfire occurs, the hammer must be recocked manually.

Unlocking. The barrel and slide recoil together for a distance of one-half inch; then the unlocking plunger on the rear barrel lug abuts against the receiver and, moving forward, forces the locking wedge out of the locking grooves in the slide and down into the cammed surface on the receiver. Having been disconnected from the barrel, the slide (with the empty cartridge case gripped by the extractor) moves to the rear for an additional two inches, cocking the hammer in the process. The cartridge case is knocked out of the grip of the extractor claw, and out of the pistol, by the nose of the ejector. When the last round is fired, the slide stop is forced up by the magazine follower and holds the slide to the rear.

Field Stripping the M1951 Pistol

Remove magazine by pushing the magazine catch inward.

Draw the slide to the rear and align the dismounting latch with the dismounting notch of the slide.

Push the dismounting latch upward, toward the muzzle end of the pistol.

Pull the slide group forward and off the receiver.

Pull the recoil spring and guide assembly slightly to the rear, and remove it from its position in the slide.

Push the barrel backward and withdraw it from the rear of the slide.

No further disassembly is recommended.

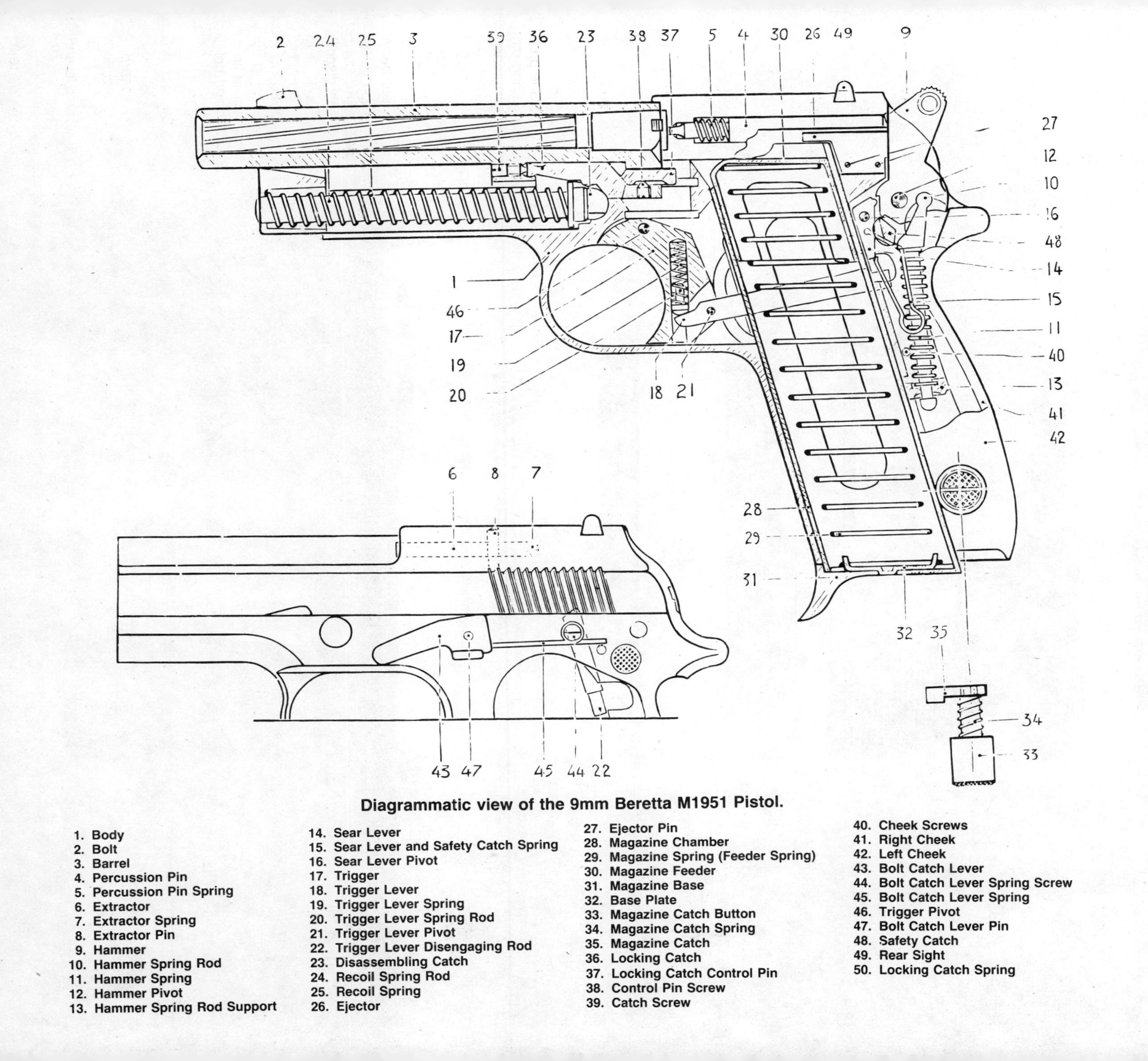

Diagrammatic view of the 9mm Beretta M1951 Pistol.

1. Body
2. Bolt
3. Barrel
4. Percussion Pin
5. Percussion Pin Spring
6. Extractor
7. Extractor Spring
8. Extractor Pin
9. Hammer
10. Hammer Spring Rod
11. Hammer Spring
12. Hammer Pivot
13. Hammer Spring Rod Support
14. Sear Lever
15. Sear Lever and Safety Catch Spring
16. Sear Lever Pivot
17. Trigger
18. Trigger Lever
19. Trigger Lever Spring
20. Trigger Lever Spring Rod
21. Trigger Lever Pivot
22. Trigger Lever Disengaging Rod
23. Disassembling Catch
24. Recoil Spring Rod
25. Recoil Spring
26. Ejector
27. Ejector Pin
28. Magazine Chamber
29. Magazine Spring (Feeder Spring)
30. Magazine Feeder
31. Magazine Base
32. Base Plate
33. Magazine Catch Button
34. Magazine Catch Spring
35. Magazine Catch
36. Locking Catch
37. Locking Catch Control Pin
38. Control Pin Screw
39. Catch Screw
40. Cheek Screws
41. Right Cheek
42. Left Cheek
43. Bolt Catch Lever
44. Bolt Catch Lever Spring Screw
45. Bolt Catch Lever Spring
46. Trigger Pivot
47. Bolt Catch Lever Pin
48. Safety Catch
49. Rear Sight
50. Locking Catch Spring

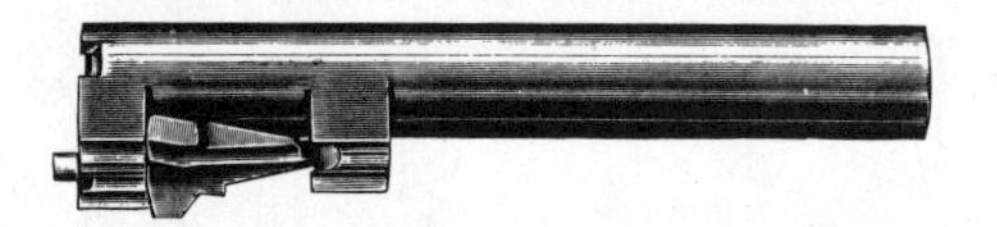

Disassembled view of the 9 × 19mm M951.

Beretta M951A selective fire model.

BERETTA MODEL 92 HANDGUN SERIES

Beretta's latest family of handguns, the Model 92s, evolve from the Model 951. This new series is double action instead of single action, and it has a 15-shot magazine versus the 8-shot magazine of the M951. The basic differences in the various models of the Model 92s lie in the safety mechanisms and the magazine release. The first Model 92s had a frame-mounted safety that locked the hammer and a push-button magazine release mounted at the base of the pistol grip. As with the M951, the Model 92 was introduced with an aluminum frame. Most military users appear to have preferred the steel frame for issue to their armed forces. The following table compares the different versions of the Beretta Model 92.

COMPARISON OF CURRENT ITALIAN HANDGUNS

	M951	M92	M92S	M92S-1**	M92SB**	M92SC	M93R***
Caliber:	9 × 19mm	9 × 19mm	9 × 19mm	9 × 19mm	9 × 19mm	9 × 19mm	9 × 19mm
System of operation:	Short recoil	Short recoil	Short recoil	Short recoil	Short recoil	Short recoil	Short recoil
Length overall (inches):	8.0	8.54	8.54	8.54	8.54	7.76	9.45
Barrel length (inches):	4.51	4.92	4.92	4.92	4.92	4.29	6.14 incl. muzzle brake
Feed device: (detachable box magazine)	8	15	15	15	15	13	15 & 20
Sights: Front:	Blade	Blade	Blade	Blade	Blade	Blade	Blade
Rear:	V-notch	V-notch	V-notch	V-notch	V-notch	V-notch	V-notch
Weight (lbs.):	1.93*	2.09	2.16	2.16	2.16	NA	2.47 w/o stock 3.06 w/stock

* 1.57 lb. with aluminum receiver.
** These are the same basic handgun. The M92S-1 was tested by the US Armed Forces as the XM9.
*** This model has a 3-shot burst feature.

This view of the Model 92 shows the trigger bar that links the trigger and the hammer. Also clearly indicated is the magazine catch.

The 9 × 19mm NATO Beretta Model 92.

The 9 × 19mm NATO Beretta Model 92SB of type similar to that tested by the US Armed Forces.

The 9 × 19mm NATO Beretta Model 92SB-C compact with 13-shot magazine.

The 9 × 19mm NATO Beretta Model 93R. Note the 20-shot magazine and the 3-shot burst indicator on the safety.

Field Stripping the Beretta Model 92

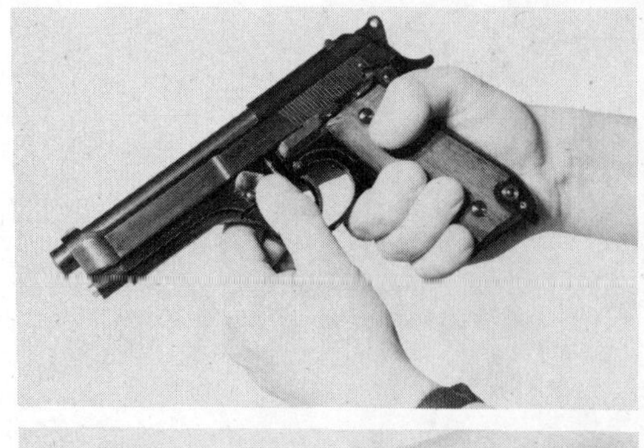

After removing the magazine, grasp the Model 92 in the right hand. With the left forefinger press the disassembly latch release button, and with the left thumb rotate the disassembly latch 90 degrees counterclockwise. It will stop when it is in the proper position.

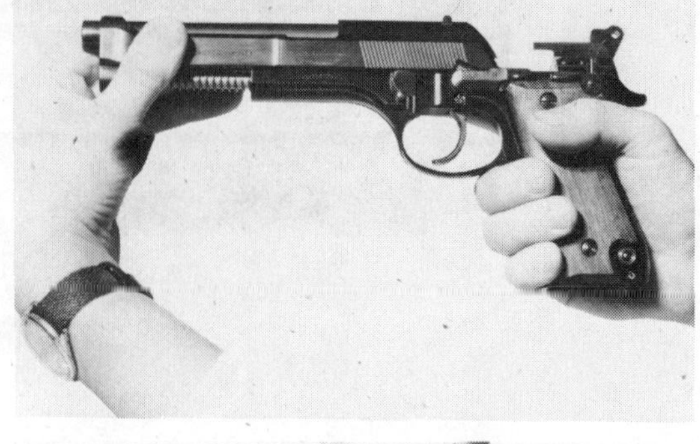

Pull the slide and barrel assembly forward until it is free of the frame.

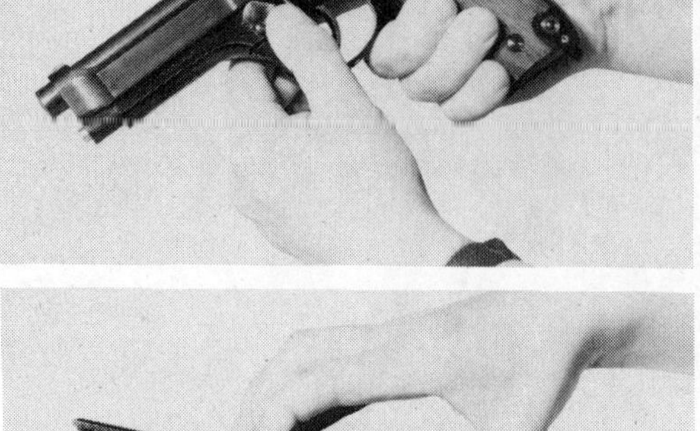

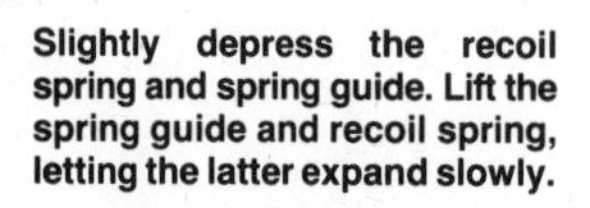

Slightly depress the recoil spring and spring guide. Lift the spring guide and recoil spring, letting the latter expand slowly.

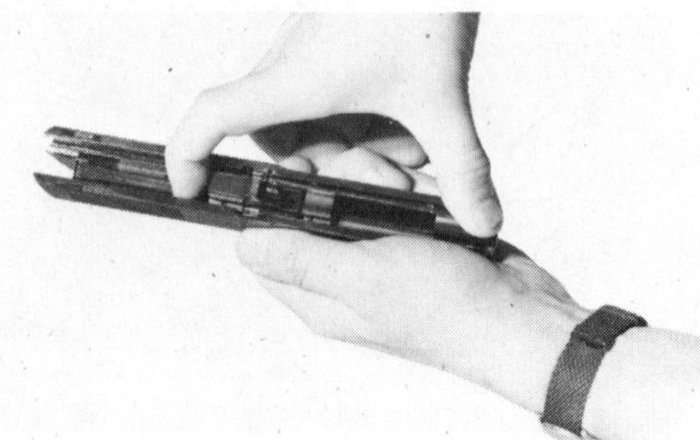

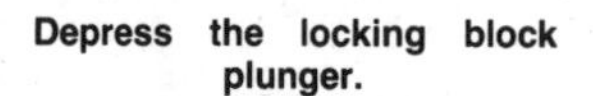

Depress the locking block plunger.

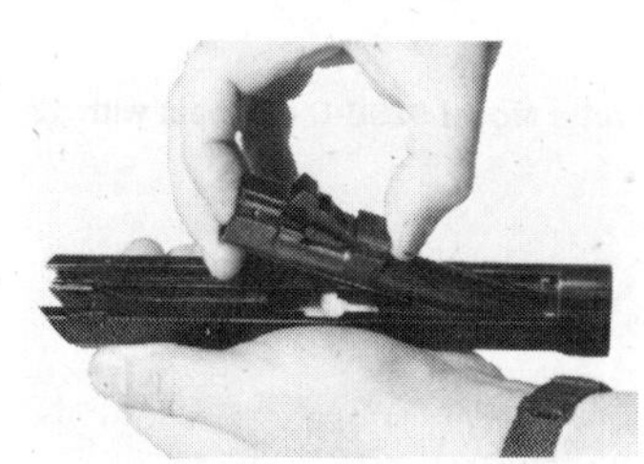

With the plunger depressed, lift the barrel out of the slide. Assemble in reverse order.

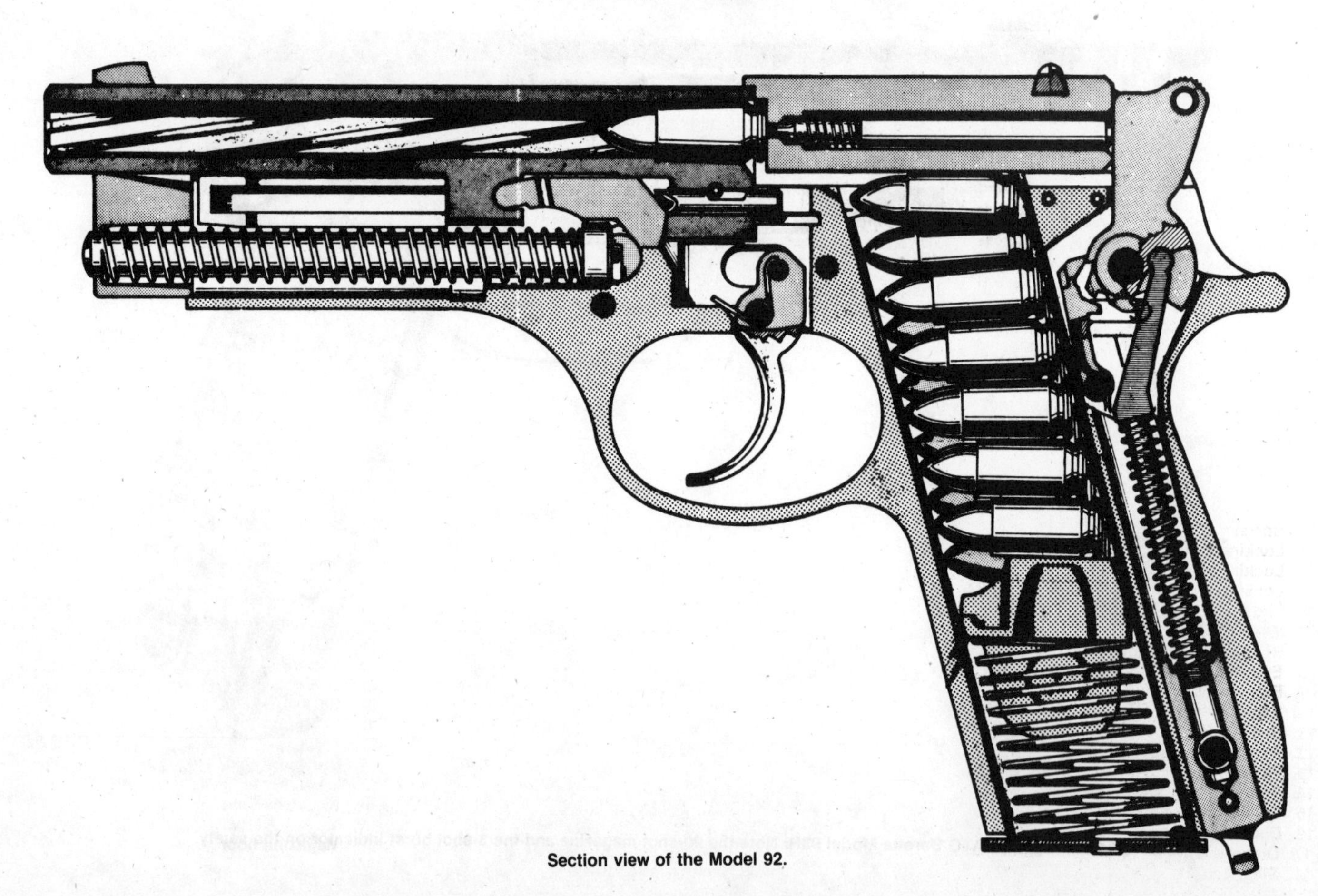

Section view of the Model 92.

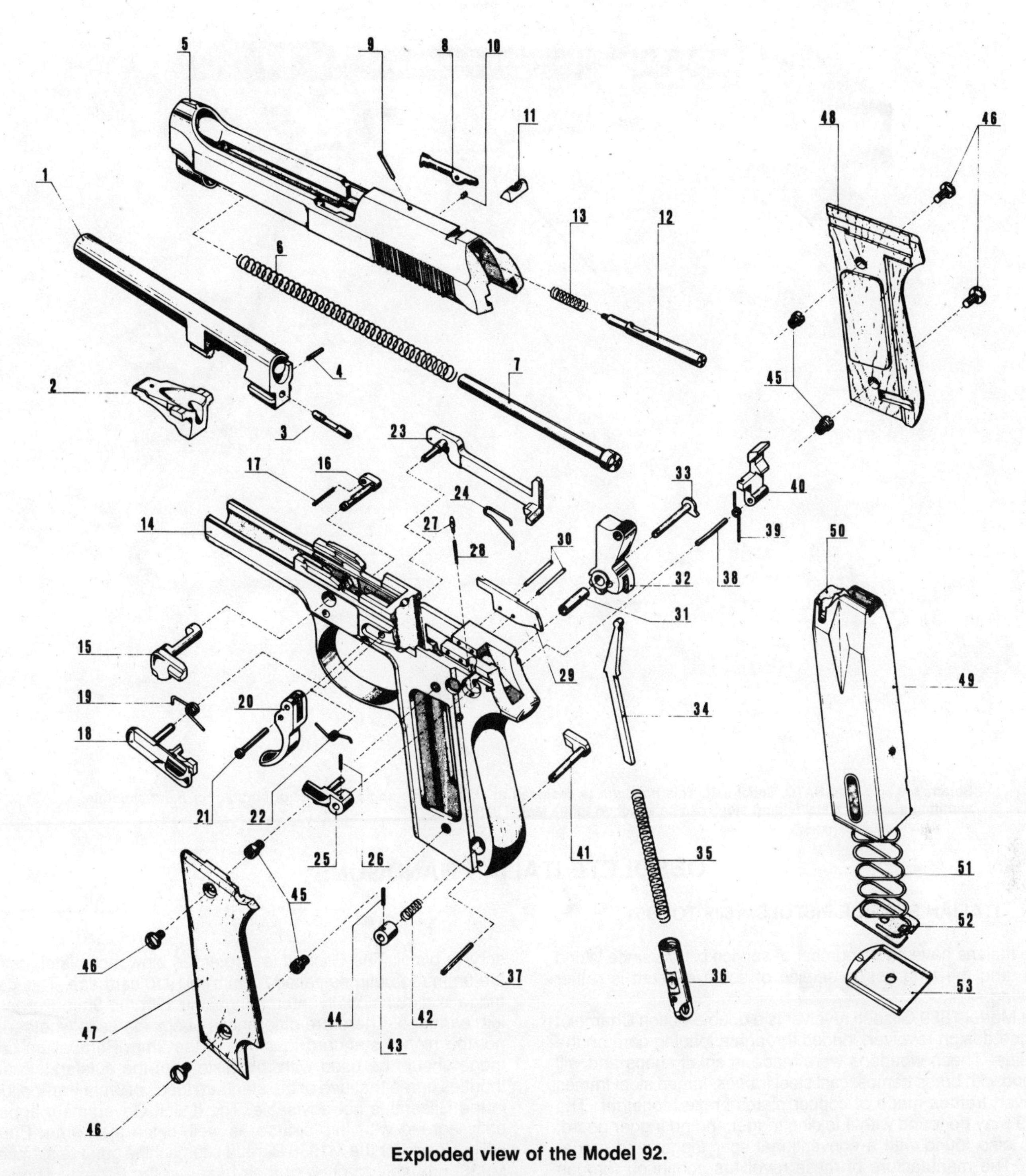

Exploded view of the Model 92.

1. Barrel
2. Locking block
3. Locking block plunger
4. Locking block plunger pin
5. Slide
6. Recoil spring
7. Recoil spring guide
8. Extractor
9. Extractor pin
10. Extractor spring
11. Rear sight
12. Firing pin
13. Firing pin spring
14. Frame
15. Disassembly latch
16. Disassembly latch release button
17. Disassembly latch release button spring
18. Slide catch
19. Slide catch spring
20. Trigger pin
21. Trigger pin
22. Trigger spring
23. Trigger bar
24. Trigger bar spring
25. Safety
26. Safety pin
27. Safety pin spring
28. Safety spring
29. Ejector
30. Ejector pin (2 pieces)
31. Hammer bushing
32. Hammer
33. Safety pin
34. Hammer spring strut
35. Hammer spring
36. Hammer strut guide
37. Hammer strut guide pin
38. Sear pin
39. Sear spring
40. Sear
41. Magazine release button
42. Magazine release button spring
43. Magazine release button bushing
44. Magazine release button pin
45. Grip bushing (4)
46. Grip screw (4)
47. Left grip
48. Right grip
49. Magazine body
50. Magazine follower
51. Magazine follower spring
52. Magazine floor plate lock
53. Magazine floor plate

Beretta's 9 × 19mm NATO Model 93R. This handgun is designed to shoot single shots and 3-shot bursts. For more accurate shooting a special metal folding stock can be attached to the Model 93R.

OBSOLETE ITALIAN HANDGUNS*

ITALIAN SERVICE PISTOLS, 1889 TO 1951

The Italians have had a number of service pistols since World War I, and the long use in service of some of them is rather surprising.

The Model 1889 Glisenti revolver is a double-action Chamelot Delvigne design revolver, loaded through a loading gate on the right side. These weapons were made in small shops and will be found with brass frames, cast steel frames, forged steel frames and even frames made of copper plates brazed together. The M1889 may be found with a folding trigger and no trigger guard, and is also found with a conventional spur trigger and trigger guard. The manufacture of these revolvers continued through the 1920's, and the use of these weapons by the Italians in World War II would be equivalent to the US use of the caliber .45 Model 1873 (Frontier Model) revolver in the same conflict.

The above would suggest that the pistol development in Italy was comatose, but such was not the case. The Brixia automatic pistol appeared in Italy around 1906, and (although it is reported of Swiss origin, developed from patents of Haensler and Roch) it was first manufactured by Metellurgica Bresciana Tempini at Brescia, Italy. This weapon, which is chambered for a low-powered version of the 9mm Parabellum cartridge, was, with slight modification, adopted as the 9mm Model 1910 Glisenti service pistol. The Glisenti is a retarded blowback pistol, using the 9mm (sometimes called 8.9mm) M1910 cartridge. This cartridge is dimensionally interchangeable with the 9mm Parabellum cartridge. There are differences of opinion as to whether or not the higher standard loadings of the 9mm Parabellum cartridge should be used with this pistol, but the outstanding authorities agree that use of the standard loads of 9mm Parabellum in the Glisenti is not advisable. The Italian Government apparently agreed with this course as well, since their 9mm Parabellum loads for the M1938 Beretta submachine gun are stamped M938, probably to prevent their usage in the Glisenti. The Glisenti cartridge has a truncated conical bullet as opposed to the conical nosed bullet used with the 1938 and later Beretta submachine guns.

In 1915, the first Beretta automatic pistol was introduced into Italian service. As of this date, Beretta designed and made automatic pistols are still the standard service pistol in Italy. The most widely distributed Model 1915 is chambered for the 7.65mm cartridge and is a hammerless blowback operated pistol.

A model of the 1915 chambered for the 9mm Glisenti cartridge was also produced; this weapon differs from the 7.65mm only in minor details such as checked wood grips, the positioning of

*For additional details see *Handguns of the World*.

the ejector, the functioning of the magazine as a slide stop and the presence of a small mechanical safety at the rear of the receiver, which can be engaged when the slide is to the rear.

The 7.65mm Model 1915–1919 Beretta is also hammerless but resembles the later Berettas in many respects, such as the larger portion cut out on the top of the slide and the barrel mounting, a grooved block under the chamber, which slides into channels machined in the receiver ahead of the magazine well. This pistol was apparently not issued to the Army, but the specimen shown bears Naval markings.

In 1923, another Beretta was introduced, chambered for the 9mm Glisenti cartridge. The Model 1923 was the first service Beretta to have an external hammer and is, in other respects, the same as the Model 1915–1919. These pistols were used by the Italian Army in limited quantities.

A later Beretta, which closely resembles the well-known Model 1934, is the 7.65mm Model 1931. This pistol has a straighter grip than the Model 34 and issue was limited to the Italian Navy.

The 9mm *corto* (.380 ACP) Model 34 Beretta is probably the best known of the Italian automatic pistols and was one of the best Italian weapons in World War II. It is still used by Italian forces to some extent and is manufactured for commercial sale as the "Cougar." The Model 1934 has also been made in 7.65mm.

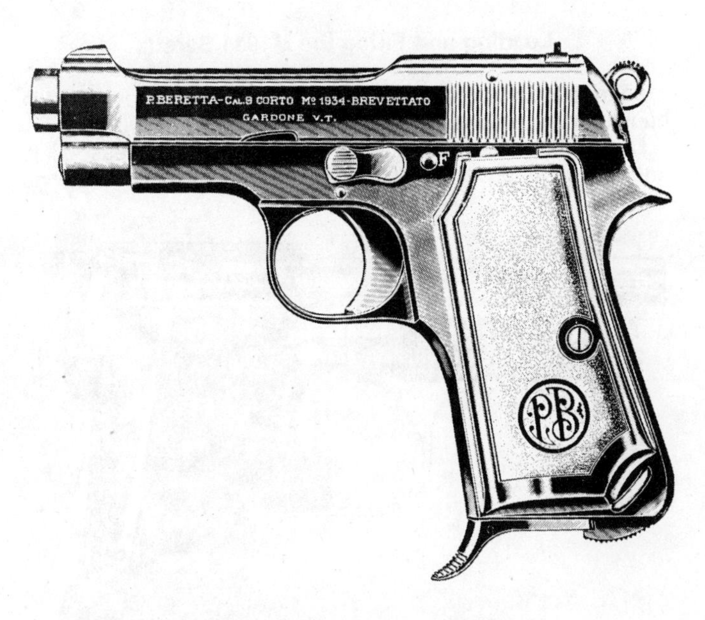

9mm (.380 ACP) Model 1934 Beretta.

THE 9mm BERETTA PISTOL MODEL 1934

The Model 1934 is a very finely-made weapon and was very popular with the Italian Army during World War II, as well as with US troops who managed to acquire them. Specimens marked "RE" (*Regio Esercito*) or "Royal Army" were Italian Army issue; other specimens may be found with the marking "PS," which indicates police or *carabinieri* issue. The Italian service designation for the .380 ACP cartridge is 9mm Model 34.

CHARACTERISTICS OF ITALIAN SERVICE PISTOLS

	Model 1889	Model 1910	Model 1915	Model 1915
Caliber:	10.35mm (10.4mm).	9mm Glisenti (M910).	7.65mm.*	9mm Glisenti (M910).
System of operation:	Double-action revolver.	Delayed blowback, semiautomatic.	Blowback, semiautomatic.	Blowback, semiautomatic.
Length overall:	10.25 in.	8.1 in.	6 in.	6.75 in.
Barrel length:	5.25 in.	3.9 in.	3.4 in.	3.75 in.
Feed device:	6-round, revolving cylinder.	7-round, in-line detachable, box magazine.	7-round, in-line detachable, box magazine.	7-round, in-line detachable, box magazine.
Sights: Front:	Bead.	Barley corn.	Blade.	Blade.
Rear:	V notch.	V notch.	V notch.	V notch.
Weight:	2.2 lb.	1.87 lb.	1.25 lb.	2 lb.
Muzzle velocity:	Approx. 840 f.p.s.	1050 f.p.s.	Approx. 960 f.p.s.	Approx. 1035 f.p.s.

*.32 ACP.

CHARACTERISTICS OF ITALIAN SERVICE PISTOLS (CONT'D)

	Model 1915–19	Model 1923	Model 1934	Model 1951
Caliber:	7.65mm.*	9mm Glisenti (M910).	9mm Corto.**	9mm Parabellum (M1938).
System of operation	Blowback, semiautomatic.	Blowback, semiautomatic.	Blowback, semiautomatic.	Recoil, semiautomatic.
Length overall:	6 in.	6.2 in.	6 in.	8 in.
Barrel length:	3.4 in.	3.88 in.	3.5 in.	4.51 in.
Feed device:	8-round in-line detachable, box magazine.	8-round in-line detachable, box magazine.	7-round in-line detachable, box magazine.	8-round in-line detachable, box magazine.
Sights: Front:	Blade.	Blade.	Blade.	Blade.
Rear:	V notch.	V notch.	V notch.	V notch.
Weight:	1.31 lb.	1.87 lb.	1.25 lb.	1.93 lb.***
Muzzle velocity:	Approx. 960 f.p.s.	Approx. 1035 f.p.s.	970 f.p.s.	1182 f.p.s.

*.32 ACP.
**.380 ACP.
***1.57 lb. with aluminum receiver.

Loading and Firing the M1934 Beretta

Load magazine exactly as for Colt automatic pistol. Insert in handle and push in until it locks.

Draw back slide exactly as for Colt automatic. Release slide and let it drive forward pushing cartridge into firing chamber and closing pistol.

Push safety catch around into locking position unless weapon is to be fired.

(Note: The exposed hammer may be let down gently on the firing pin if care is taken and the operation is done with both hands. It may also be set at half-cock.)

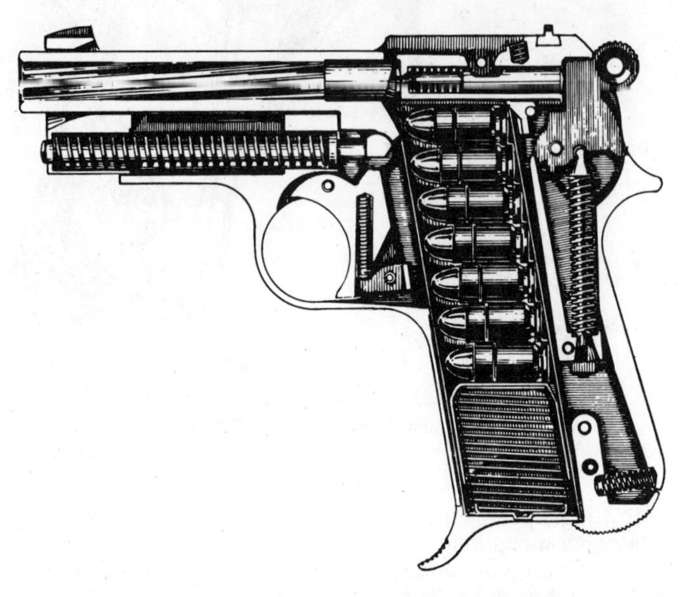

Action closed, showing detail of firing mechanism.

Action open, showing operation of recoil and feeding mechanism.

Field Stripping the M1934 Beretta

(1) With magazine out of weapon, pull back slide as far as it will go; hold it with right hand, and with left thumb push safety catch up into locking notch in underside of slide.

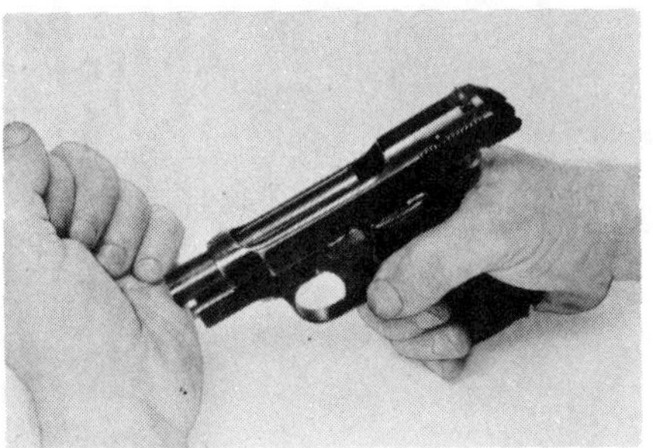

(2) Now push straight back on barrel with palm of hand. This will free barrel from its locking recess.

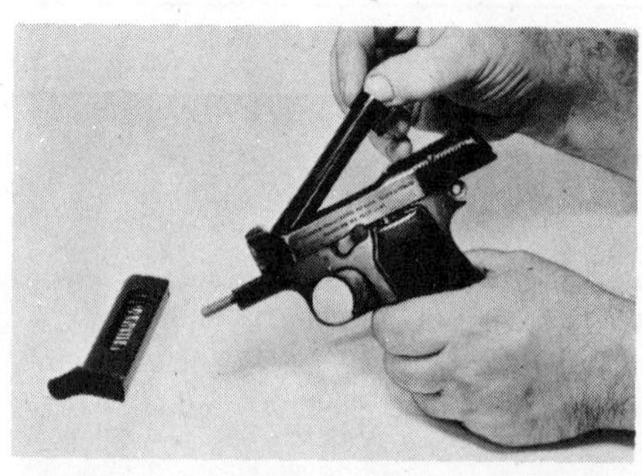

(3) Pull barrel straight back and up, drawing it out of the slide as shown.

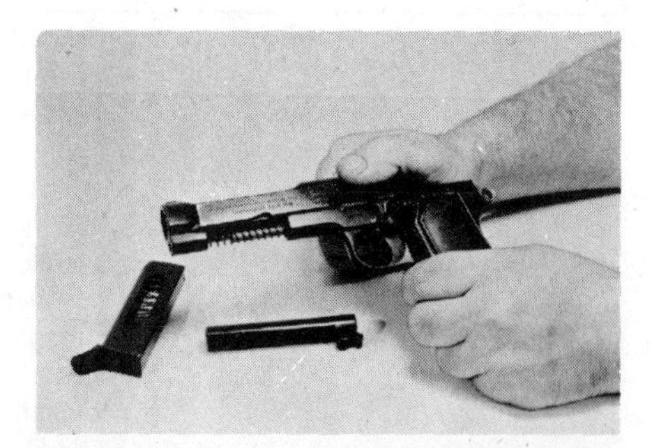

(4) Push slide, recoil spring and recoil spring guide straight forward off the receiver. Spring and guide may now be removed from their seats. Safety locking stud may also be lifted out now. No further dismounting is necessary with this pistol.

ITALIAN RIFLES

THE BERETTA BM59 SERIES OF RIFLES

After a series of prototypes, the basic design of the Beretta conversion emerged as the BM59, which used an M1-type stock and was generally similar externally to the Garand, except for the 20-round magazine, charger guide and the new handguard and gas cylinder arrangement.

BM59D was produced with a straight-in-line stock (with pistol grip) and a rate reducer to facilitate control during full automatic fire.

BM59GL was basically the same as the BM59, except that it had an integral grenade launcher and grenade launcher sight.

BM60CB was offered by Beretta in 1960 with a burst control feature.

Subsequently, Beretta revised the designations for these rifles.

The BM59 Mark "I" and Mark "I"-A

The "I" in these designations stands for "Italian" since these are the weapons adopted by the Italian Government. They are also known as the BM59 Mark Ital and Ital-A rifles. These rifles are equipped with bipods, and grenade launchers and differ from each other mainly in their stocks. The Mark "I"-A is for paratroops and armored troop use.

Loading and Firing the BM59 Mark "I" and Mark "I"-A. Pull operating handle to the rear, locking the weapon. Put safety, located in forward part of trigger guard, on safe by pushing rearwards. Insert a loaded magazine in the magazine well and pull upward and to the rear until the magazine catch engages. If there is already a magazine in the weapon it can be loaded by five-round chargers (stripper clips). Set the selector, located on the far left side of the receiver, on "A" for automatic fire, or "S" for semiautomatic fire. Disengage the safety and press the trigger to fire the weapon. This weapon has a winter trigger, which can be used when heavy gloves are required. It can be swung down into position.

Field Stripping the BM59 Mark "I" and Mark "I"-A

This rifle is field stripped in a manner similar to the US M1 rifle, which is covered in detail in the chapter on the US.

Other Beretta BM59 Models

In addition to the above versions, Beretta has manufactured the following models for the Italian Army. The BM59 Mark Ital TA and BM59 Mark Ital Para were derived from the Mark III and were used respectively by Alpine and Airborne troops. The Mark IV was intended for use as a squad automatic rifle. The BM59 Mark Ital TP is a shortened rifle for paratroopers.

Two views of the Standard BM59 Mark Ital 7.62 × 51mm NATO service rifle.

Beretta 7.62mm NATO BM59 Rifle.

7.62mm NATO BM59 Mark II modified with winter trigger.

7.62mm NATO BM59 Mark I Rifle.

BERETTA .223 MODEL 70/.223 ASSAULT RIFLE

Beretta has developed a selective fire rifle chambered for the 5.56mm cartridge. It resembles the SIG Model 530-1 externally, but the field stripped view indicates that there are significant differences in design. The Model 70/.223 has a combination grenade launcher/flash suppressor built into the barrel. The grenade launcher sight is positioned behind the front sight, pivoting on the front sight base as with those versions of the BM59 that have grenade launchers. There are variations in the weapon; the 70/.223 SC and the 70/.223 LM in addition to the basic 70/.223 AR. They vary in length and weight, the SC having a folding stock and the LM (Light Machine Gun) having a bipod and carrying handle.

Special troops carbine Beretta mod. 70/.223 SC with bayonet.

Field-stripped view of the AR70/.223.

Standard model of the 5.56 × 45mm Beretta AR70/.223.

Two views of the squad automatic weapon version of the 5.56 × 45mm Beretta AR70/.223.

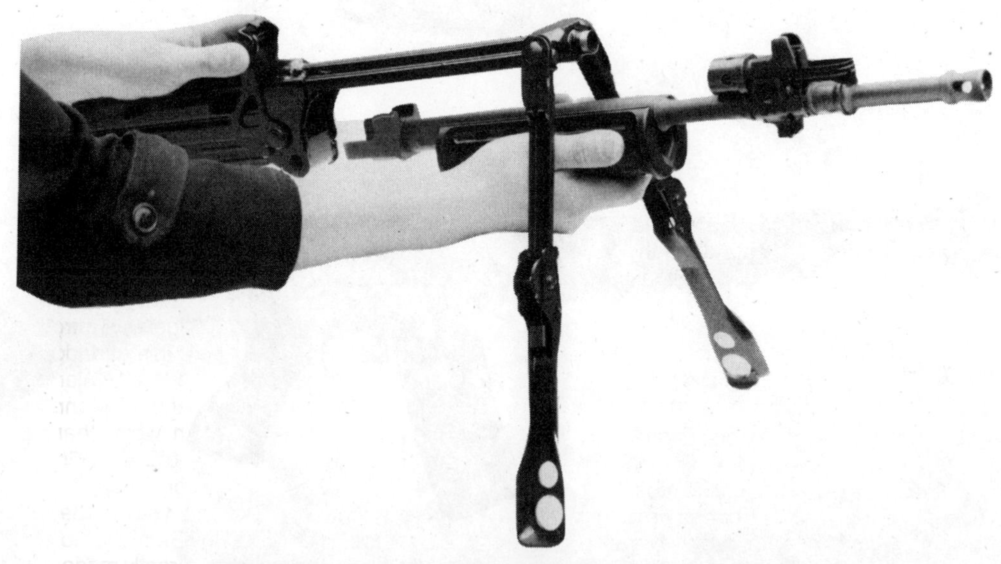

The quick change barrel being removed from the AR70/.223 SAW.

CHARACTERISTICS OF THE BERETTA BM59 RIFLES

	BM59	BM59D	BM59GL	Mark I	Mark II	Mark III	Mark IV	Mark Ital
Caliber:	7.62mm NATO							
System of operation:	Gas, selective fire							
Length overall:	37.2 in.	37.2 in.	43 in.	39.4 in.	48.9 in.*	48.9 in.	42.5 in.	43.1 in.
Barrel length:	17.7 in.	17.7 in.	21 in.	17.7 in.	17.7 in.	17.7 in.	21 in.	19.3 in.
Feed device:	20-round, detachable box magazine							
Sights: Front:	Protected blade							
Rear:	Aperture, adjustable for windage & elevation							
Weight:	8.15 lbs.	9 lbs.	9 lbs.	8.9 lbs.	8.9 lbs.	8.9 lbs.	12 lbs.	10.4 lbs.
Cyclic rate:	750 r.p.m.							
Muzzle velocity:	2620 f.p.s.	2620 f.p.s.	2730 f.p.s.	2620 f.p.s.	2620 f.p.s.	2620 f.p.s.	2730 f.p.s.	2700 f.p.s.

*29.5 in. with stock folded.

CHARACTERISTICS OF THE BERETTA .223 MODEL 70/.223 RIFLES

	70/.223 AR	70/.223 SC	70/.223 LM
Caliber:		.223 (5.56mm)	
System of operation:		gas, selective fire	
Length overall: w/stock fixed:	37.6″	37.8″	37.6″
Length overall: w/stock folded:		28.8″	
Barrel length:		17.8″	
Feed device:		30-round, staggered row, detachable box magazine	
Sights: Front:		protected post	
Rear:		aperture adjustable for 100, 250 and 400 meters	
Weight:	8 lb.	8.5 lb.	8.9 lb.
Cyclic rate:	630 r.p.m.	630 r.p.m.	600 r.p.m.
Muzzle velocity:		3,180 f.p.s.	

OBSOLETE ITALIAN RIFLES

The Italian Army was equipped with a number of different types of rifles prior to and during World War II. Considering everything, the Italian Army was the poorest armed of the major powers in World War II. The 6.5mm M1891 Mannlicher Carcano and its various 6.5mm and 7.35mm variants were the principal weapons, but numerous 8mm M1889 and 1895 Mannlicher rifles, received as war booty from the first war, and even some ancient Vetterli Vitali rifles, rebarreled for 6.5mm rifle, were used.

THE MANNLICHER CARCANO RIFLE

The Mannlicher Carcano, also known occasionally as the Mauser Paravicino, is a modified Mauser. It has a Mauser type one-piece bolt with frontal locking lugs, but, unlike most Mausers, the bolt handle is in front of the receiver bridge when locked. The magazine is a Mannlicher in-line type, fed with a six-round clip, which stays in the magazine until the last round is chambered.

There have been a number of rumors passed around about the safety of the Mannlicher Carcano rifle. When in good condition and used with the proper ammunition, it is as safe as any other military rifle.

The Carcano was developed at the Italian Government arsenal at Turin by M. Carcano. Early models have gain twist rifling, i.e., the twist of the rifling gradually increases toward the muzzle. In 1938, the 7.35mm cartridge was introduced, and a rifle and two carbines chambered for this cartridge were adopted.

Upon Italy's entrance into World War II in 1940, the 6.5mm caliber was re-introduced, and all Mannlicher Carcano's manufactured from that date on were chambered for the 6.5mm cartridge. In addition, many of the 7.35mm weapons already in existence were rebarreled for 6.5mm.

The Mannlicher Carcano was made in limited numbers in 7.92mm, apparently for the Germans, toward the close of World War II. These rifles were apparently made in very limited quantities.

Representative Italian rifles. 1. Model 1891 Rifle. 2. Model 91TS Carbine. 3. Model 91 Carbine. 4. Model 1941 Rifle. 5. Model 91/24 Carbine. 6. Model 38(7.35mm). 7. Model 38TS (7.35mm) Carbine. 8. Model 38 (7.35mm) Carbine. 9. Model 1938 Rifle (caliber 6.5mm). 10. Model Youth Rifle (Moscheto) Ballila, caliber special 6.5mm. Barrel 14.43″. (Weapon does not fire ball cartridges.)

CHARACTERISTICS OF MANNLICHER CARCANO BOLT-ACTION RIFLES AND CARBINES

	Rifle M1891	Carbine M1891	Carbine M1891 TS	Carbine M1891/24
Caliber:	6.5mm.	6.5mm.	6.5mm.	6.5mm.
Overall length:	50.8 in.	36.2 in.	36.2 in.	36.2 in.
Barrel length:	30.7 in.	17.7 in.	17.7 in.	17.7 in.
Feed device:	6-round in line non-detachable box magazine.	6-round in line non-detachable box magazine.	6-round in line non-detachable box magazine.	6-round in line non-detachable box magazine.
Sights: Front:	Barley corn.	Barley corn.	Barley corn.	Barley corn.
Rear:	Tangent w/V notch graduated from 500-2000 M; leaf turned over for battle sight.	Tangent w/V notch graduated from 500-1500 M; leaf turned over for battle sight.	Tangent w/V notch graduated from 500-1500 M; leaf turned over for battle sight.	Tangent w/V notch graduated from 500-1500 M; leaf turned over for battle sight.
Weight:	8.6 lbs.	6.6 lbs.	6.9 lbs.	6.9 lbs.
Muzzle velocity:	2395 f.p.s.	2297 f.p.s.	2297 f.p.s.	2297 f.p.s.
Remarks:	Basic rifle, straight bolt handle, uses knife-type bayonet.	Bayonet permanently attached, bolt handle bent.	Uses knife-type bayonet, bolt handle bent.	Uses knife-type bayonet, bolt handle bent, except for lower band the same as M1891 TS carbine.

	Rifle M1938	Carbine M1938	Carbine M1938TS	Rifle M1941
Caliber:	7.35mm.	7.35mm.	7.35mm.	6.5mm.
Overall length:	40.2 in.	36.2 in.	36.2 in.	46.1 in.
Barrel length:	20.9 in.	17.7 in.	17.7 in.	27.2 in.
Feed device:	6-round in line non-detachable box magazine.	6-round in line non-detachable box magazine.	6-round in line non-detachable box magazine.	6-round in line non-detachable box magazine.
Sights: Front:	Barley corn.	Barley corn.	Barley corn.	Barley corn.
Rear:	Fixed.	Fixed.	Fixed.	Tangent w/V notch graduated from 300-1000 M; leaf turned over for battle sight.
Weight:	7.5 lbs.	6.5 lbs.	6.8 lbs.	8.21 lbs.
Muzzle velocity:	2482 f.p.s.	Approx. 2400 f.p.s.	Approx. 2400 f.p.s.	Approx. 2360 f.p.s.
Remarks:	First of Italian rifles chambered for 7.35mm cartridge, bolt handle bent. Some have knife-type folding bayonet which can be carried folded on rifle or removed.	7.35mm version of M1891 carbine, has permanently attached folding bayonet, bolt handle bent.	7.35mm version of M1891/24 carbine, uses knife-type bayonet, bolt handle bent.	Basically the same as the Rifle M1891 except for length and rear sight.

	Rifle M1938	Carbine M1938	Carbine M1938 TS
Caliber:	6.5mm.	6.5mm.	6.5mm.
Overall length:	40.2 in.	36.2 in.	36.2 in.
Barrel length:	20.9 in.	17.7 in.	17.7 in.
Feed device:	6-round in line non-detachable box magazine.	6-round in line non-detachable box magazine.	6-round in line non-detachable box magazine.
Sights: Front:	Barley corn.	Barley corn.	Barley corn.
Rear:	Fixed.	Fixed.	Fixed.
Weight:	7.6 lbs.	6.6 lbs.	6.9 lbs.
Muzzle velocity:	2320 f.p.s.	2297 f.p.s.	2297 f.p.s.
Remarks:	6.5mm version of 7.35mm Rifle M1938, made after beginning of World War II.	6.5mm version of 7.35mm Carbine M1938, made after beginning of World War II.	6.5mm version of 7.35mm Carbine M1938 TS, made after beginning of World War II.

Notes: (1) All the pre-World War II 6.5mm weapons have right-hand gain twist (progressive) 19.25 to 8.25; 7.35mm weapons have constant right-hand 10-inch twist. (2) All weapons use 6-round Mannlicher type clips. (3) Some of the Model 1938 weapons were made in 7.92 × 57mm Mauser for the Germans during World War II. (4) Caliber is marked on sight base of all 1938 series weapons.

ITALIAN SUBMACHINE GUNS

Italy was the first country to adopt a submachine gun—the 9mm Villar Perosa of 1915. The Villar Perosa as originally produced had no stock and was mounted in dual sets. The weapon was fired with thumb type triggers, similar to those used on machine guns. It was mounted, usually with a shield, on motorcycle side cars, aircraft and on various types of tripods and bipods. At a later date, a stock was added to the gun, and it became a selective-fire individual weapon. The Villar Perosa and all other early Italian submachine guns were chambered for the 9mm Glisenti cartridge—basically a low-powered 9mm Parabellum cartridge.

Beretta produced a modified copy of the Villar Perosa in 1918 and produced a selective-fire weapon called the Beretta Moschetto Automatico, or the M1918 - 1930.

The Model 1918 Beretta was retarded blowback as was the Villar Perosa and was designed by Tullio Marengoni, who designed all the Beretta submachine guns until the late 1950s.

The 9mm Parabellum Model 1938A Beretta was the first of a series of very well-designed finely-made weapons, which were widely distributed in other countries in addition to Italy. The Beretta Model 38A and 38/42 were considered the finest Italian small arms in service in World War II. The first Model Beretta 38 had longitudinal slots in the barrel jacket as opposed to holes in the barrel jackets of later production. Some Model 38As had

9mm O.V.P. Submachine Gun.

Italian Beretta 9mm Submachine Gun Model 38A.

folding bayonets, and early production had bayonet lugs for the mounting of a removable knife type bayonet.

The early Model 38A did not have the push-through type full automatic safety located behind the full automatic trigger and had a dual port compensator rather than the multi-slotted compensator found on later models. The most common variation of the Model 38A was without bayonet or bayonet lug, and with a multi-slotted compensator.

This model was sold to Romania and Argentina and was made in tremendous quantities for the Italian Army. The Model 38A was produced to some extent after World War II, and in 1949 a modification of the weapon, using the cross bolt safety mounted in the stock as used with the Model 38/49 (Model 4), was produced in limited quantity. Although it has generally been considered that the fixed firing pin was introduced with the Model 38/42,

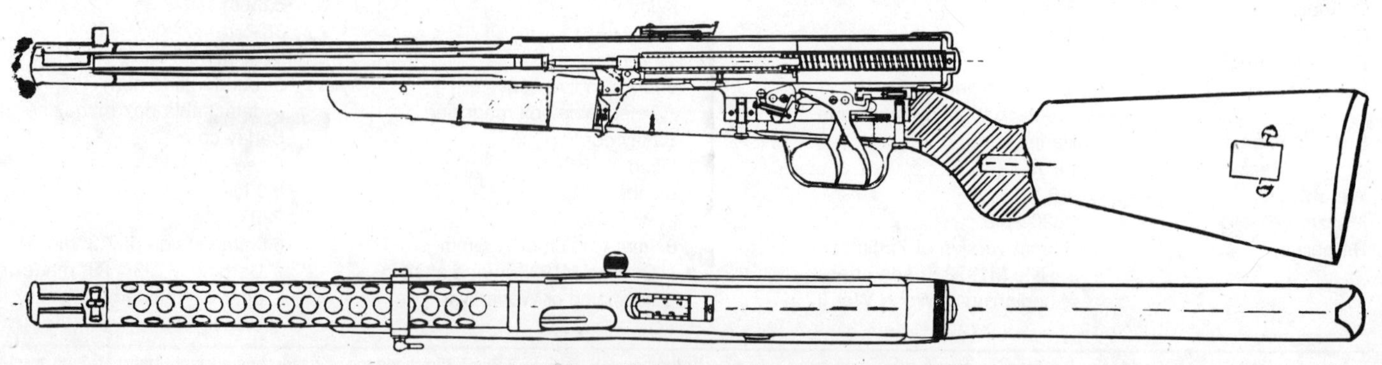

Diagram of the Model 38A from official Italian manual.

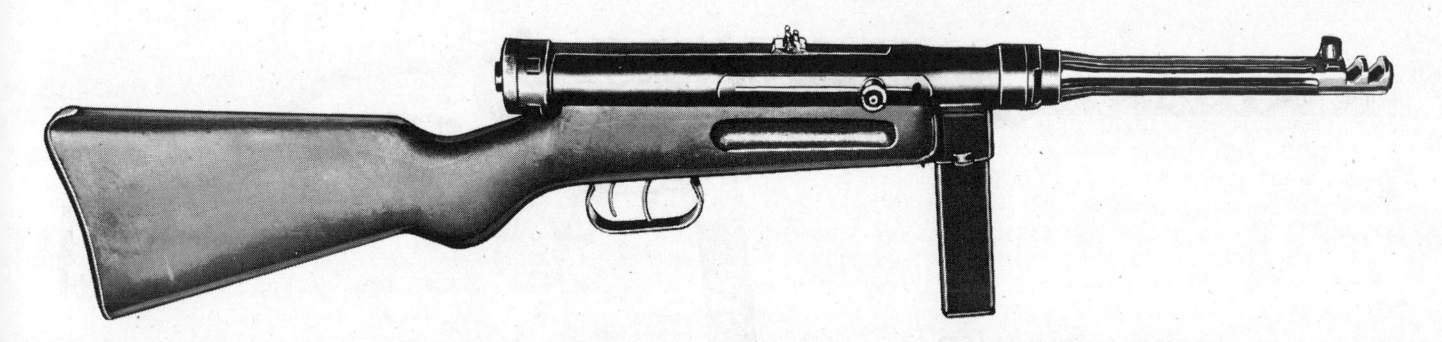

Beretta Model 38/42 Submachine Gun, fluted barrel.

late Model 38As with a fixed firing pin were apparently made. It is quite possible that these weapons were made after the introduction of the Model 38/42, since the Model 38A was made as late as 1950.

The Beretta 9mm Parabellum Model 38/42 is basically a simplified Model 38A; the barrel jacket is not used with the 38/42, and all Model 38/42s have a fixed firing pin. The 38/42 uses a stamped receiver and magazine housing and has a fluted barrel (early production). Models of the 38/42 with smooth barrel are called 38/43 by Beretta. There are three distinctly different models of the 38/42-38/43, and an additional similar weapon, the 38/44, has a shorter bolt but does not have the operating spring guide found on the earlier Berettas. This can be noted by the absence of the recoil spring guide rod head protruding through the cap on the end of the receiver. The external differences can be noted in the pictures. Pakistan, Iraq and Costa Rica purchased Model 38/44 submachine guns.

An unusual Beretta Model 38A submachine gun, which apparently appeared in prototype form, has an aluminum barrel jacket rather than the multi-perforated barrel jacket. Another Beretta, which appeared only in prototype form, was the Model 1. This weapon was developed prior to the Model 38/42 and has a folding stock similar to that of the German MP38; it was apparently an expensive gun to make and was dropped in favor of the Model 38/42. A gun called the 38/44 Special by Beretta closely resembles the Model 1 but does not have the fluted barrel of the Model 1 and does have a cross bolt type safety mounted in the fore-end. This weapon is also called the Model 2.

The Model 38/49, also known as the Model 4, is the current standard submachine gun and is covered in detail separately. The Model 5 Beretta submachine gun is basically the same as the Model 38/49 except that it has a grip safety located in the fore-end.

The Model 5 was introduced in 1957. Another weapon somewhat similar in appearance to the Model 2, and also made only as a prototype, is the Model 4. The Model 4 has a grip safety and a sliding-wire type stock similar to that on the US M3 submachine gun. The weapon illustrated below is similar to the Model 4 but has a wooden fore-end. This weapon is described as a modified 38/44 by Beretta.

Throughout the 1950s, Beretta developed a number of prototype guns, including the Model 6 (1953) and the Model 10 (1957), which were steps in the development of their latest gun, the 9mm Model 12. The Model 12 has a bolt that telescopes the barrel in a fashion similar to the Czech Model 23 and the Israeli UZI. The Model 12 has been adopted by the Italian Government and is covered in detail later in this chapter.

There were other submachine guns developed in Italy in addition to those developed by Beretta, although Beretta has been the most prolific developer. The Fabbrica Nazionale d'Armi of Brescia developed a submachine gun manufactured in limited quantities during the war. This weapon, called the FN A-B Model 1943, is rather unusual for the time in which it was made, since it is machined out of steel forgings. It has a single strut folding steel stock, and the magazine can be folded up under the barrel jacket when not in use. The combination muzzle brake compensator resembles that used on the Soviet PPSh41 submachine gun. The Model 1943 is a retarded blowback, selective-fire weapon. Reportedly 7,000 of these weapons were made for the Italian Army.

The TZ45 is another 9mm Parabellum submachine gun manufactured in limited quantities toward the end of the war. It was designed by Toni and Zorzoli Giandoso in about 1944. The TZ45 is a relatively crudely-made weapon and has a grip safety fitted behind the magazine housing in a fashion similar to the later model Madsen submachine guns. The TZ45 was adopted in a modified form by Burma after the war. It is manufactured in Burma, where it is called the BA52.

The Bernardelli firm produced a modified copy of the Beretta Model 38/42, a few of which were made in 1948 and 1949. This

9mm Beretta Model 38/44 Special (Model 2).

9mm Semiautomatic Model LF-57.

Beretta 9mm Model 38/49 Submachine Gun.

weapon, called the VB, was well made but had no unusual features.

Fabbrica Nazionale d'Armi developed a gun in the 1950s called the X4. The X4 was composed almost entirely of stampings, has a sheet steel retracting slide, which partially surrounds the receiver and is used to retract the bolt. A shorter version of the X4 called the X5, which had a barrel just over 4.5 inches long, was brought out in the mid 1950s.

The firm of Luigi Franchi has also brought out a family of submachine guns called the LF57. The LF57 is another weapon composed mainly of stampings and is of relatively short length. It is 16.52 inches overall with stock folded, using an 8.1 inch barrel. A semiautomatic only version of this weapon with 16-inch barrel was also offered for sale.

THE 9mm BERETTA MODEL 38/49 SUBMACHINE GUN MODEL 4

The Model 38/49 is one of the standard submachine guns of the Italian Army. It is a modification of the Model 38/44; like this weapon, it has no recoil spring guide and a fixed firing pin. The principal difference between the two weapons is the use of a cross-bolt type safety, which is mounted in the stock above the front of the trigger guard.

The 38/49 was sold to Costa Rica, Egypt (with folding bayonet), Yemen, Tunisia, West Germany, Indonesia, Thailand and the Dominican Republic.

How to Load and Fire the Model 38/49

Insert a loaded magazine into the magazine port. Pull bolt retracting handle (located on the right side of the receiver) to the rear and cock the bolt. Push forward retracting handle (it does not reciprocate with the bolt). The safety is engaged by pushing in from the left side. Pressure on the forward trigger will produce semi-automatic fire and pressure on the rear trigger will produce automatic fire.

How to Field Strip the Model 38/49

Remove the magazine and check weapon to insure that a cartridge has not remained in the chamber. Twist the receiver cap one quarter-turn to the left and remove with the operating spring assembly. The bolt may now be pulled to the rear and withdrawn from the rear of the receiver. Further disassembly is not recommended. To assemble, reverse the above procedure.

THE 9mm BERETTA MODEL 12 SUBMACHINE GUN

The Model 12 comes in both folding metal stock and detachable wooden stock models. As previously pointed out, this weapon has a bolt that telescopes the barrel for about three-quarters of the length of the bolt. The weapon therefore appears to have a much shorter barrel than it really has. An excellent feature of the weapon is that the grooves extend the length of the receiver. The grooves serve as dirt catchers and allow operation even with a considerable amount of dirt in the action. The Model 12 has a grip safety.

How to Load and Fire the Model 12

Remove magazine, if an empty magazine is in the weapon, by pushing magazine catch located at the bottom front of the trigger guard forward and (after pulling magazine down and out of the magazine guide) insert loaded magazine and pull bolt retracting handle, located on the left side of the receiver to the rear. The manual safety is a push button located above the grip; to put the weapon on safe, push the button from left to right.

Beretta 9mm Model 12 Submachine Gun.

Beretta Model 12—field stripped.

The fire selector is also of the push-button type, located ahead of the pistol grip. Pushing the button in from the left sets the weapon for semi-automatic; pushing the button from the right produces automatic fire. Push manual safety button from right to left, select type of fire desired, and press trigger.

MODEL 12S

The new Model 12S, illustrated here, has a rotating lever type fire selector, improved rear cap catch, improved sights, new butt plate, and an epoxy resin exterior finish.

Beretta's 9 × 19mm NATO Pistola Mitragliatrice 12 with a plastic stock.

Beretta's 9 × 19mm NATO PM Model 12 with folding stock.

FIELD STRIPPING THE BERETTA MODEL 12

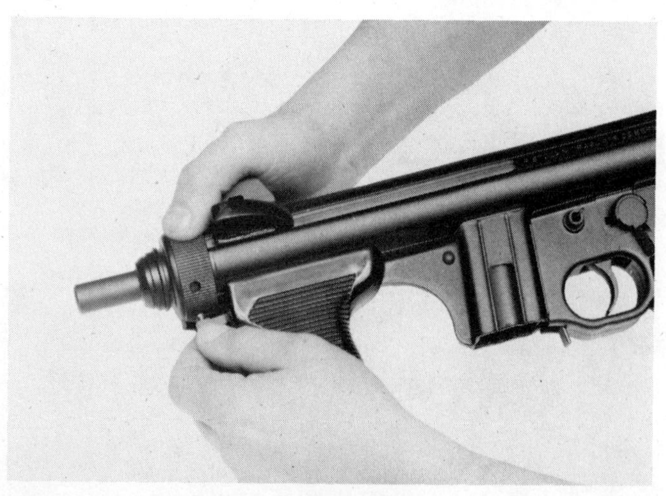

After removing the magazine, pull the barrel-locking nut catch downward and unscrew the nut at the front of the receiver.

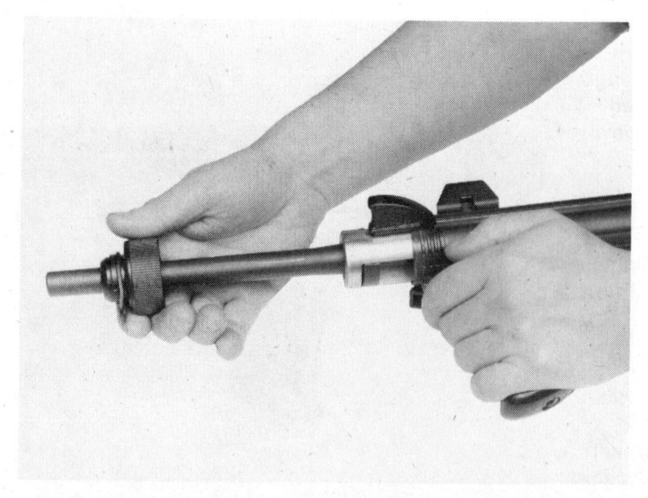

Remove the barrel and bolt assembly from the receiver.

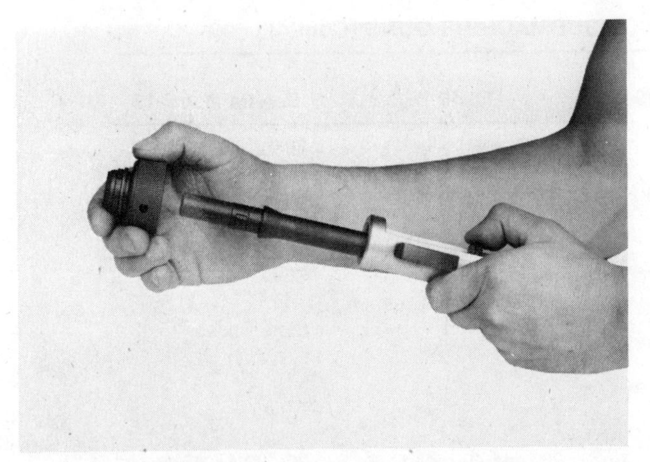

Next remove the barrel-locking nut from the barrel.

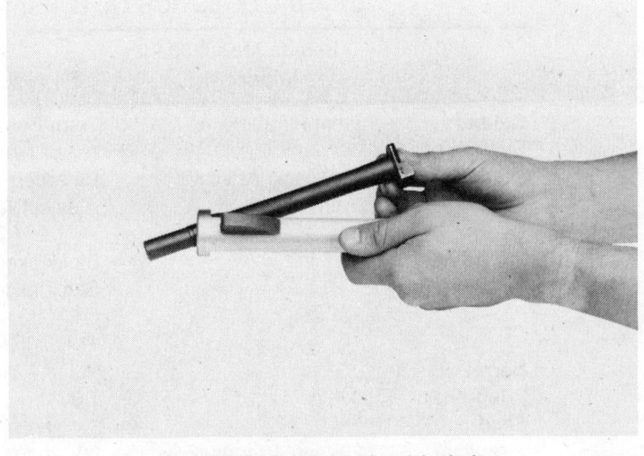

Then lift the barrel free from the bolt.

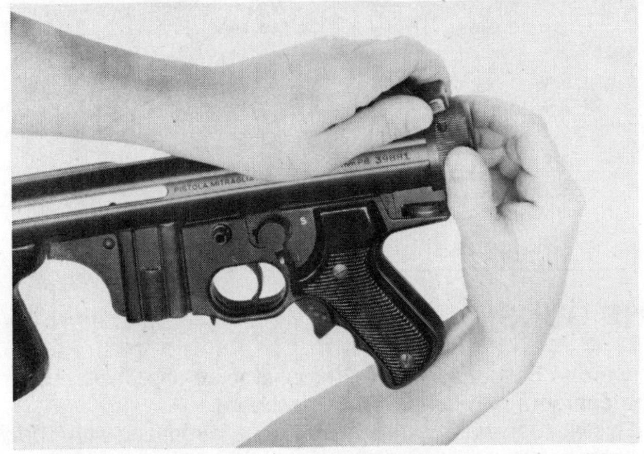

To remove the recoil spring, lift the rear end cap retaining catch.

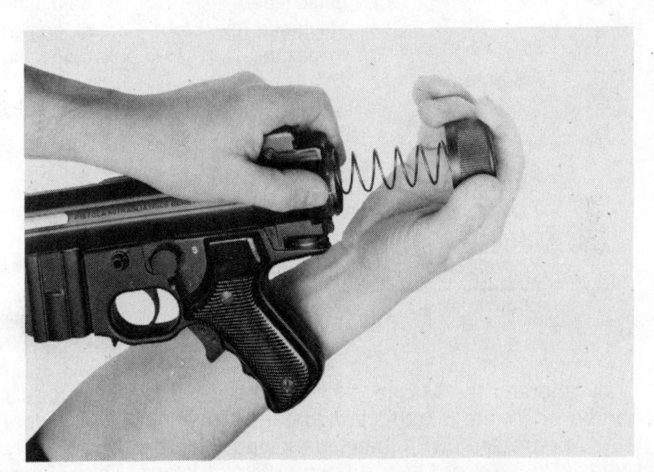

Ease the spring and cap out of the receiver.

CHARACTERISTICS OF ITALIAN SERVICE SUBMACHINE GUNS

	Villar Perosa Model 1915	Villar Perosa (O.V.P.)	Beretta Model 1918	Beretta Model 1938A	Beretta Model 1938/42
Caliber:	9mm Glisenti.	9mm Glisenti.	9mm Glisenti.	9mm Parabellum.	9mm Parabellum.
System of operation:	Retarded blowback, full automatic.	Retarded blowback, selective fire.	Retarded blowback.	Blowback, selective fire.	Blowback, selective fire.
Overall length:	21 in.	35.5 in.	33.5 in.	37.25 in.	31.5 in.
Barrel length:	12.56 in.	11 in.	12.5 in.	12.4 in.	8.4 in.
Feed device:	25-round, staggered row, detachable, box magazine.	25-round, staggered row, detachable, box magazine.	25-round, staggered row, detachable, box magazine.	10, 20, 30, or 40-round, staggered column, detachable, box magazine.	20- or 40-round, detachable, staggered row, box magazine.
Sights:					
Front:	None on some, aperture cut into brace between barrels.	Blade.	Blade.	Blade.	Blade.
Rear:	Y shaped-plate with notch.	U notch.	V notch.	Tangent with V notch.	"L" type flip over with U notch.
Weight:	14.30 lb. for twin gun.	Approx. 8 lb.	7.2 lb.	9.25 lb.	7.2 lb.
Cyclic rate:	1200 r.p.m.	900 r.p.m.	900 r.p.m.	600 r.p.m.	550 r.p.m.
Muzzle velocity:	1312 f.p.s.	1250 f.p.s.	1275 f.p.s.	1378 f.p.s.	1250 f.p.s.

CHARACTERISTICS OF ITALIAN SERVICE SUBMACHINE GUNS (Cont'd)

	Beretta Model 38/49 (Model 4)	FN A-B Model 1943	TZ 45	Beretta Model 12
Caliber:	9mm Parabellum.	9mm Parabellum.	9mm Parabellum.	9mm Parabellum.
System of operation:	Blowback, selective fire.	Retarded blowback, selective fire.	Blowback, selective fire	Blowback, selective fire
Overall length:	31.5 in.	Stock fixed: 31.1 in. Stock folded: 20.7 in.	Stock fixed: 33.5 in. Stock folded 21.5 in.	Stock fixed: 25.39 in. Stock folded: 16.43 in.
Barrel length:	8.4 in.	7.8 in.	9 in.	7.9 in.
Feed device:	20- and 40-round, detachable, staggered row, box magazine.	20- and 40-round, detachable, staggered row, box magazine.	20- and 40-round, detachable, staggered row, box magazine.	20-, 30-, & 40-round, staggered row, detachable, box magazine.
Sights:				
Front:	Blade.	Blade.	Post.	Blade.
Rear:	"L" type flip over with U notch.	V notch.	Aperture.	"L" flip over with U notch.
Weight:	7.2 lb.	7.06 lb.	7.2 lb.	6.6 lb.
Cyclic rate:	550 r.p.m.	400 r.p.m.	550 r.p.m.	550 r.p.m.
Muzzle velocity:	1250 f.p.s.	1225 f.p.s.	1265 f.p.s.	1250 f.p.s.

ITALIAN MACHINE GUNS

Italy adopted the Maxim gun in 1906 and used two models chambered for the 6.5mm cartridge, the Model 1906 and Model 1911. Apparently few of these guns were made, as they rarely appear in photos of Italian troops in World War I. The 6.5mm watercooled Revelli Model 1914 was the first Italian-designed gun to appear on the scene in quantity. This retarded blowback gun was unusual in that it was fed with a magazine composed of ten compartments, each holding five rounds. To provide for the lack of slow initial extraction—a slight bolt turning movement to loosen the case before extraction to the rear begins—the cartridge cases were lubricated by an oil pump built into the receiver. The Revelli was manufactured by Fiat and is frequently known by that name.

During World War I, Italy purchased quantities of the Colt Model 1914, Browning "potato digger" in 6.5mm to supplement their machine gun supply. The French supplied the Italians with large quantities of the M1907 F St. Etienne heavy machine gun chambered for the 8mm Lebel cartridge. At the conclusion of World War I, the Italians received tremendous quantities of 8mm (8 × 50R) Schwarzlose Model 1907/12 from the Austrians as part of their share of war reparations. These weapons were still used by the Italians in World War II.

The Breda 6.5mm Model 1924 was one of the first Italian light machine guns. It is, except in details such as the stock and use of a thumb-operated trigger, basically the same as the 6.5mm Model 1930 Breda. The Model 1930 was the standard Italian light machine gun in World War II. A few thousand of the Model 1924 Breda were made, as well as a few thousand 6.5mm Fiat SAFAT light machine guns—a light version of the Revelli Model 1914. A number of machine guns never made in quantity (such as the Brixia) were introduced during the twenties and thirties.

The 6.5mm Model 1930 Breda was an unusual weapon in many respects. It is a retarded (delayed) blowback-operated gun with recoiling barrel, as is the Model 1914 Revelli. This weapon has a massive bolt with multiple-locking lugs and an oil pump, which lubricates the cartridges prior to chambering. The magazine is permanently attached to the side of the gun and is swung forward to be loaded with a 20-round horse-shoe type charger. The chargers may be brass or cardboard.

This weapon was among the first to be introduced with a quick change barrel. All in all, the Breda Model 30 was hardly a satisfactory weapon. Any weapon that requires lubricated cases will be unreliable in sandy or dusty conditions, because the lubricant will pick up the foreign matter and introduce it into the action. This weapon was sold in 7.92mm to Portugal and Lithuania. It was made in limited quantities in 7.35mm as the Model 38 for Italy.

In 1931, the 13.2mm Breda antiaircraft and tank machine gun appeared. This weapon, unlike earlier Italian machine guns, had nothing in its design that would make it "peculiar" from a machine-gun point of view. The 13.2mm Model 31 was a conventional, gas-operated, magazine-fed gun and was the first step in the development of the best Italian developed machine gun—the 8mm Model 37.

The 8mm Model 37 Breda, although it had some peculiar features, was by far the best of the Italian machine guns used in World War II. A gas-operated gun, it is fed by feed trays from which the cartridge is pushed into the chamber and into which the empty cartridge case is reinserted before the feed tray is ejected. The reason for this tidy arrangement is rather mysterious. The Breda Model 37 was a very reliable weapon, used in many varying climatic conditions with good results. The Model 37 was used after World War II, and there was some discussion about converting the weapon to caliber .30, but this idea was apparently dropped since there is a Model 55 tracer cartridge for the weapon. Although the weapon is gas operated, the cartridges are lubricated by an oil pump since no provision is made for slow initial extraction. The Breda has been replaced in the Italian service by the 7.62mm NATO MG 42/59. The Breda Model 37 was made for Portugal in 7.92mm and is known as the Model 1938.

In 1938, a tank version of the Model 37, chambered for the same Breda 8mm Model 35 cartridge, fed by a top-mounted box magazine, was introduced and adopted.

8mm Breda Model 37 Heavy Machine Gun.

Aircraft versions of the Breda in 7.7mm (.303 British), 7.92mm and 12.7mm were also produced. The 12.7mm gun was also produced as an antiaircraft gun. The Italian government used the 7.7mm and 12.7mm versions of this gun. It might be pertinent to point out that the Italians used seven different machine-gun cartridges during World War II, in addition to four different pistol and submachine gun cartridges.

The Fiat (Revelli) Model 1935 8mm heavy machine gun is essentially a modified Revelli Model 1914; many of the Model 35 Fiats were actually converted Model 1914 weapons. Some of the weapons were of new manufacture. The Model 1935 is air cooled and has a fluted chamber rather than an oil pump (it was also a retarded blowback). It was fed by a non-disintegrating belt rather than by the compartmented magazine of the Model

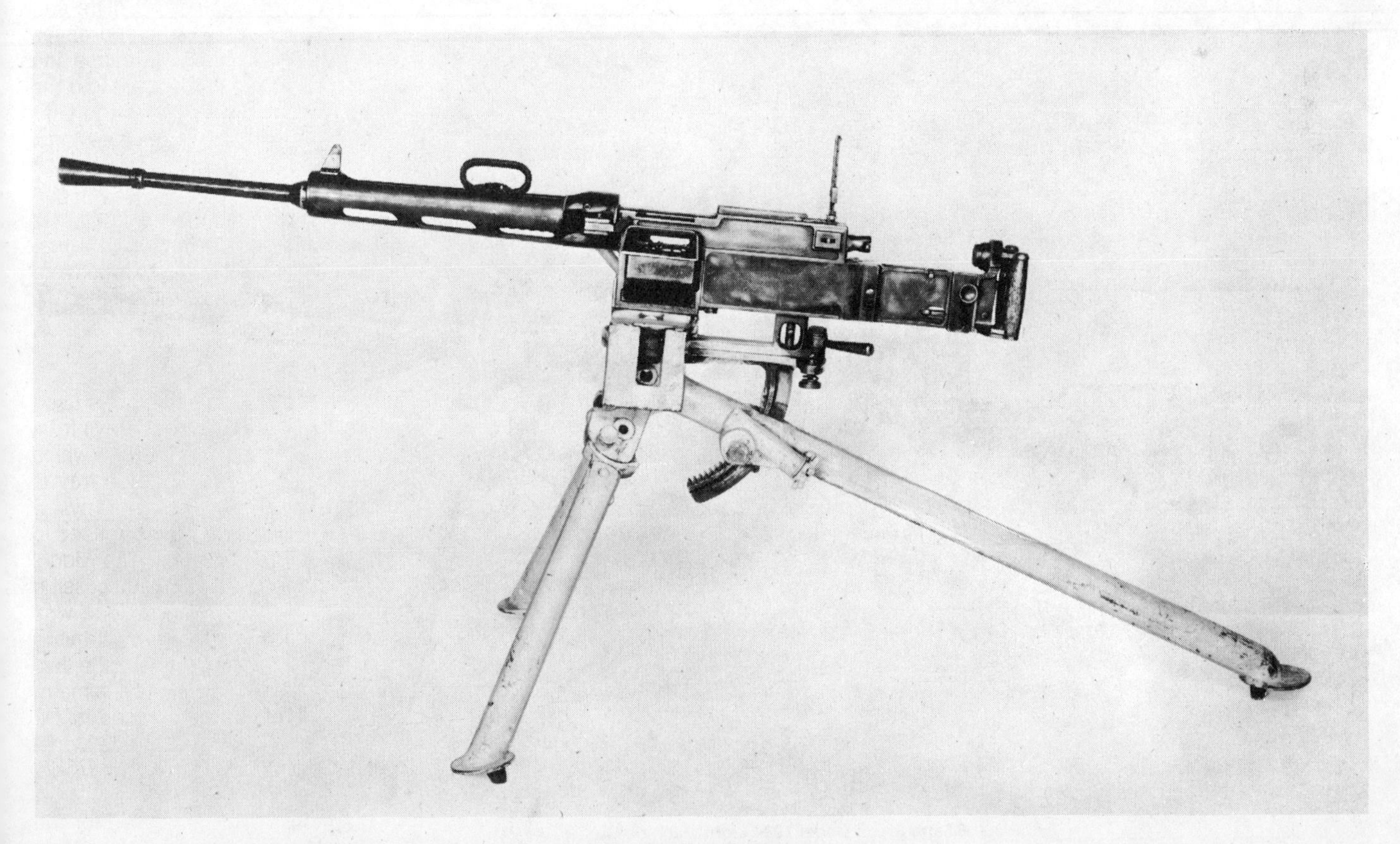

8mm Fiat (Revelli) Model 35.

1914. The Model 35 was not a very successful gun; it was actually worse than the Model 1914. The ammunition still required lubrication for proper function, and the weapon, unlike the Model 1914, fired from a closed bolt with resultant "cook-offs" after periods of sustained fire. This weapon as the 1914 had an operating rod that reciprocated outside to buff against a pad mounted on the front of the back plate. This made clearing a gun with a cartridge in a hot barrel a hazardous proposition and may account for the nickname it received during World War II—the "knucklebuster."

After World War II, Italy was supplied with American and British machine guns. The current standard Italian machine gun is the 7.62mm NATO MG42/59, called the *Mitragliatrice Leggere 42/59* in Italy. This weapon is covered in detail in the chapter on Germany. The MG42/59 is currently being manufactured in Italy.

ITALIAN SERVICE MACHINE GUNS

	Model 1914 Revelli	**Model 1930 Breda***	**8mm Breda Model 37**	**8mm Fiat (Revelli) Model 35**
Caliber:	6.5mm.	6.5mm.	8mm (Italian Model 35 Cartridge)	
System of operation:	Delayed blowback, selective fire.	Delayed blowback, automatic only.	Gas, automatic only.	Delayed blowback, selective fire.
Length overall:	46.5 in.	48.5 in.	50 in.	49.75 in.
Barrel length:	25.75 in.	20.5 in.	Approx. 25 in.	25.75 in.
Feed device:	50-round "mousetrap" type magazine.	20-round, non-detachable magazine, fed by changers.	20-round strip.	300-round non-disintegrating belt.
Sights:				
Front:	Barley corn			
Rear:	Leaf			
Weight:	37.5 lb. (less water) tripod—49.5 lb.	22.75 lb.	42.8 lb. Tripod—41.5 lb.	39.75 lb. Tripod—41.5 lb.
Cyclic rate:	450-500 r.p.m.	450-500 r.p.m.	450 r.p.m.	500 r.p.m.
Muzzle velocity:	2080 f.p.s.	2063 f.p.s.	2600 f.p.s.	2600 f.p.s.

*7.35mm Model 38 has essentially the same characteristics.

6.5mm Breda Model 1924 Light Machine Gun.

30 Japan

SMALL ARMS IN SERVICE

Category	Weapon	Manufacturer
Handguns:	9 × 19mm NATO P220	Early issue will be made by SIG-Sauer.
	11.43 × 23mm M1911A1	US-made.
	.38 Special Model 60 revolver	Shin Chuo Kogo K.K., Tokyo.
Submachine guns:	11.43 × 23mm M3A1	US-made.
Rifles:	7.62 × 51mm NATO Type 64	Howa Machinery, Ltd.
	7.62 × 51mm NATO Type 62 GPMG	Nittoku Metal Industry Co., Ltd.
	7.62 × 51mm NATO Type 74 Armor MG	Nittoku Metal Industry Co., Ltd.
Machine guns:	12.7 × 99mm M2 HB	US-made.

JAPANESE PISTOLS

The firm of Shin Chuo Kogyo K.K. (New Central Industrial Co. Ltd. of Tokyo) has developed a number of pistols since World War II. These are: a modified copy of the US caliber .45 M1911A1, called the Type 57 New Nambu; a smaller caliber .32 automatic, the Type 57B New Nambu; and a caliber .38 special revolver similar to the Smith & Wesson, which is called the New Nambu Type 58.

JAPANESE 9mm TYPE 57 NEW NAMBU AUTOMATIC PISTOL

CHARACTERISTICS OF TYPE 57 PISTOL

Caliber: 9mm Parabellum (also made in .45 cal.).
System of operation: Recoil operated, semiautomatic only.
Weight: 2.12 lbs.
Length, overall: 7.8 in.
Barrel length: 4.6 in.
Feed device: 8-round, single line, detachable box magazine.
Muzzle velocity: 1148 f.p.s.

This weapon, called the "New Nambu," is a modified copy of the US M1911 automatic pistol. It does not have a grip safety and has its magazine catch mounted on the bottom of the grip, but is otherwise much the same as the US pistol.

Loading, firing, field stripping and functioning are the same as the US M1911 pistol with the exception of the features noted above.

CAL. .32 TYPE 57B NEW NAMBU AUTOMATIC PISTOL

CHARACTERISTICS

Caliber: .32 ACP (7.65mm Browning).
System of operation: Blowback, semi-automatic only.
Weight: 1.3 lbs.
Length, overall: 6.3 in.
Feed device: 8-round, detachable in-line box magazine.

The Shin Chuo Kogyo Model 60 .38 Special New Nambu revolver.

9mm Parabellum Model 57A "New Nambu" Pistol.

This weapon is also called a New Nambu. It is basically a Browning type blowback pistol and loading, firing, field stripping and functioning are the same as the Browning M1910 pistol.

A modification of the Model 57, the Model 57A, has been introduced by Shin Chuo Kogyo and differs from the Model 57 in having a button type magazine catch on the left rear of the grip near the bottom, larger stocks, slightly different shaped receiver and slide stop. Although there were indications that the Japanese Self Defense Forces were going to adopt the Model 57, it was produced only as a prototype.

JAPANESE RIFLES

THE 7.62mm NATO TYPE 64 RIFLE

How to Load and Fire the Type 64 Rifle

The magazine can be removed, if in weapon, by pushing in on the magazine catch, located behind the magazine port, and pulling magazine down and out. The safety/selector switch is located over the trigger on the right-hand side of the receiver. Insert a loaded magazine in the magazine port, pushing up until the magazine catch engages. Select type of fire desired and pull operating handle—top front of receiver—to the rear and release it, chambering a cartridge.

CHARACTERISTICS OF THE TYPE 64 RIFLE

Caliber: 7.62mm NATO.
System of operation: Gas, selective fire.
Length overall: 38.97 in.
Barrel length: 17.71 in.
Feed device: 20-round, detachable, box magazine.
Sights: Front: Hooded folding blade.
Rear: Folding aperture adjustable for windage.
Weight: 9.5 lbs. w/o magazine.
Cyclic rate: 450-500 r.p.m.
Muzzle velocity: 2650 f.p.s. with full-charge cartridge.
2347 f.p.s. with reduced-charge cartridge.

Disassembly and Assembly

Clear the rifle but leave the hammer cocked and do not insert the magazine. Set the selector off safe and fold the bipod or sights.

7.62mm Type 64 Rifle.

Pull the takedown pin to the right; then lift up the rear end of the receiver cover until the cover comes off. Pull the driving spring out of the bolt carrier; then pull the carrier to the rear until it can be lifted out of the receiver. Lift the bolt out of the receiver.

No further disassembly is necessary or desirable.

To reassemble, first place the bolt onto the rear of the receiver with the leveled end up and to the rear. Place the bolt carrier over the bolt so that the cam on the bolt carrier mates with the cam cuts in the bolt. Move the carrier slightly back and forth until the carrier seats into the receiver. Push the bolt and carrier fully forward. Insert the driving spring into the bolt carrier.

Place the receiver cover onto the receiver so that the projections on the cover fit into their recesses and the driving spring seats in the cover. Push the takedown pin back into place. Clear the weapon.

Note on Method of Operation

The bolt of the Type 64 is of the tipping type and is held by a bolt carrier, which cams it down into and up out of the locked position in a fashion quite similar to that of the Soviet 7.62mm SKS carbine. The bolt carrier is forced to the rear by a piston rod mounted above the barrel; the pistol is spring loaded and returns to the battery position after firing. The operating spring and guide, which return the bolt to the locked position, are mounted in a tunnel in the top of the bolt carrier.

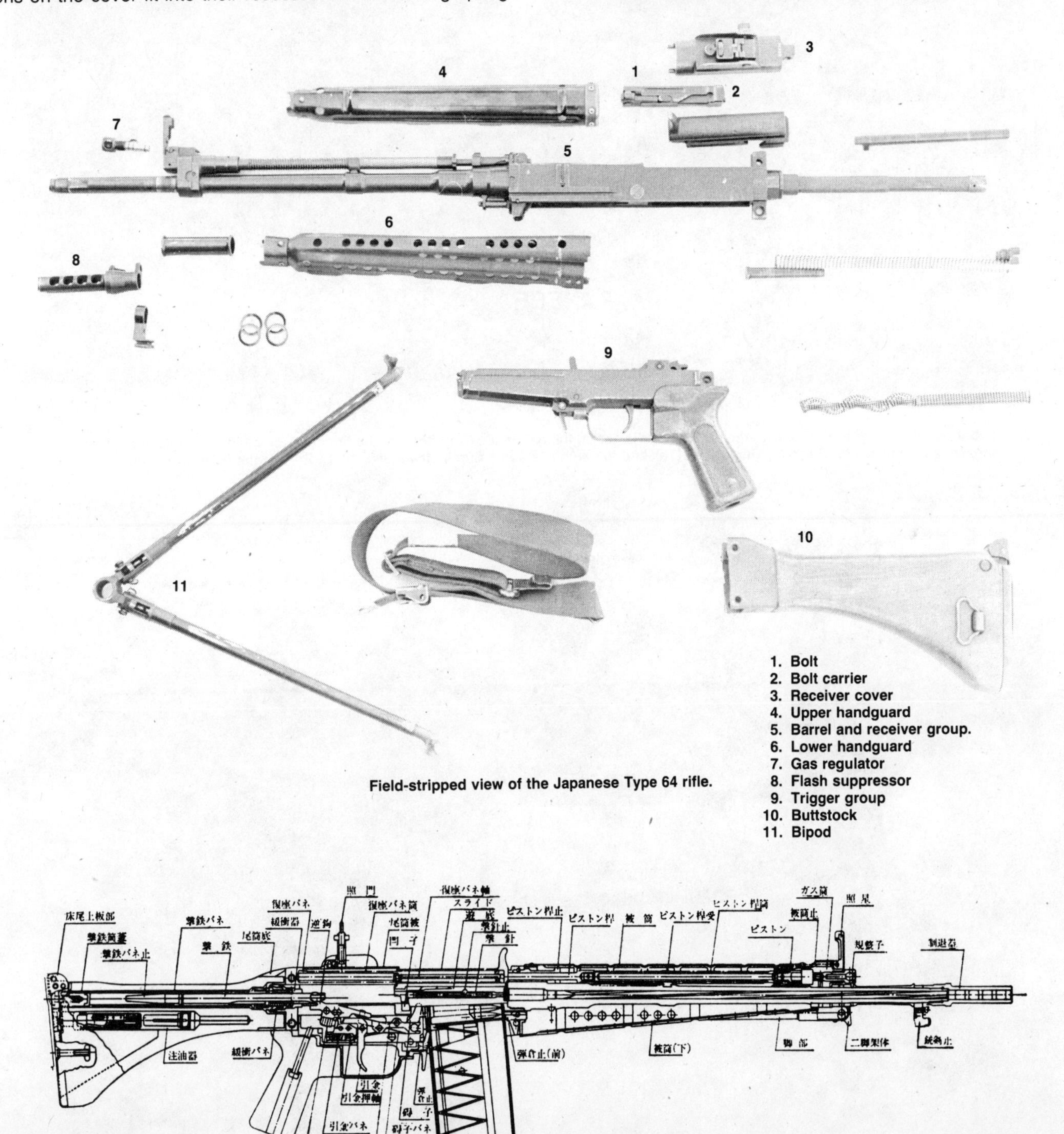

1. Bolt
2. Bolt carrier
3. Receiver cover
4. Upper handguard
5. Barrel and receiver group.
6. Lower handguard
7. Gas regulator
8. Flash suppressor
9. Trigger group
10. Buttstock
11. Bipod

Field-stripped view of the Japanese Type 64 rifle.

Secion view of the 7.62 × 51mm Type 64 rifle.

Two views of the Type 64 rifle. *Top*, the fire selector is set on full automatic (Re); *bottom*, the fire selector is set on single shot (Ta). The Japanese characters for Safe (A), single shot (Ta), and full auto (Re) are a pun in Japanese. "A-Ta-Re" means to hit or strike.

JAPANESE SUBMACHINE GUNS

SHIN CHUO KOGYO SUBMACHINE GUN

This is a conventional 9mm Parabellum blowback operated, selective fire weapon with folding steel stock.

Note that this gun has a grip safety attached to the magazine housing in a manner similar to that of the Madsen and the Italian TOZ. The ejection port has a cover similar to the US M3A1 submachine gun.

CHARACTERISTICS OF THE SHIN CHUO KOGYO SUBMACHINE GUN

Caliber: 9mm Parabellum
System of operation: Blowback
Length overall: Stock folded—30 inches
Stock fixed—19.75 inches
Barrel length: 6 or 7.3 inches
Feed device: 30-round, detachable box magazine
Sights: Front: Hooded post
Rear: "L" type, graduated for 100 and 200 meters
Weight: 8.8 lbs., loaded
Cyclic rate: 600 r.p.m.
Muzzle velocity: 1181 or 1220 f.p.s.

9mm Parabellum Shin Chuo Kogyo Submachine Gun.

JAPANESE MACHINE GUNS

Requirements for a general purpose machine gun to replace the caliber .30 Browning M1919A4 and M1917A1 machine guns were laid down by the JGSDF in 1956. The current gun, the 7.62mm NATO Type 62, is the end result of this requirement. The gun was developed by Nittoku Metal Industry Co. and represents an improvement on the third prototype developed.

THE 7.62mm NATO TYPE 62 MACHINE GUN

The Type 62 is a general purpose gun, i.e., it is used as a light machine gun on a bipod and a heavy machine gun on a tripod. The gun can be used either with full charge 7.62mm NATO cartridge or the reduced charge 7.62mm NATO. It uses the US M13 metallic link. When used on the US M2 tripod, a buffer unit is placed on the mount. There are no really unusual features to this weapon, but it is a basically good design and is of quality manufacture.

The Type 62 has a quick change barrel fitted with a prong-type flash suppressor.

CHARACTERISTICS OF THE 7.62MM NATO TYPE 62 MACHINE GUN

Caliber: 7.62mm NATO.
System of operation: Gas, automatic fire only.
Length overall: 47.3 in.
Barrel length: 23.6 in.
Feed device: Disintegrating metallic link belt.
Sights: Front: Hooded blade.
Rear: Folding leaf with aperture, adjustable for windage.
Weight: 23.6 lbs.
Cyclic rate: 650 r.p.m. with full-charge cartridges.
600 r.p.m. with reduced-charge cartridges.
Muzzle velocity: 2800 f.p.s. with full-charge cartridges.
2530 f.p.s. with reduced-charge cartridges.

7.62mm NATO Type 62 Machine Gun.

Loading and Firing the Type 62 Machine Gun

If the bolt is to the rear, pull trigger and allow bolt to run forward. Insert end link of belt into feedway so that feed pawls grasp the link. Pull bolt retracting handle, located on the right side of the receiver down and to the rear and then return it forward. The safety is located on the trigger guard; when forward, it is on "fire" and when to the rear it is on "safe." Push safety forward and press trigger. The gun will fire. The gun may also be loaded by opening the cover and laying the belt on the feedway.

How to Field Strip the Type 62 Machine Gun

Open cover and pull bolt to rear to insure that there are no cartridges in the weapon. Close cover and allow bolt to run forward. Push stock retaining pin out of the receiver to the left. Remove stock assembly with recoil spring guide, and buffer assembly from rear of receiver. Raise the cover, releasing the barrel-locking plunger, draw bolt handle fully to the rear, and remove. Draw slide and bolt assembly from the rear of the receiver. Align carrying handle dismounting stud with dismounting notch in barrel-locking ring; push barrel-locking plunger rearward and turn carrying handle until dismounting guide lines are aligned, then pull barrel forward out of the receiver. Remove gas piston return spring from receiver. Assemble in reverse order.

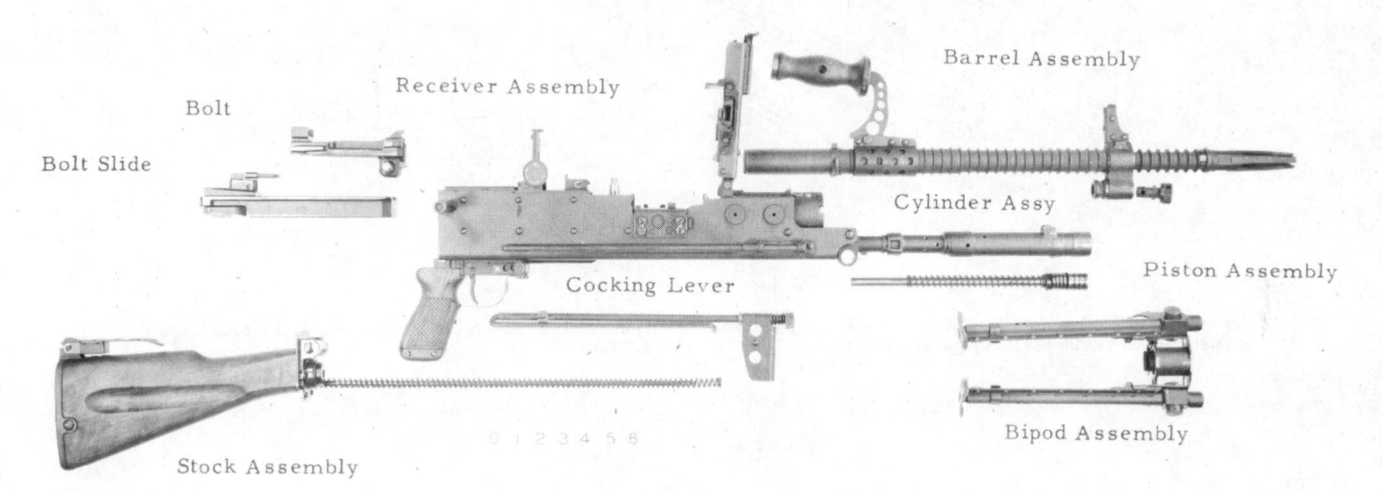

Field-stripped view of Type 62 Machine Gun.

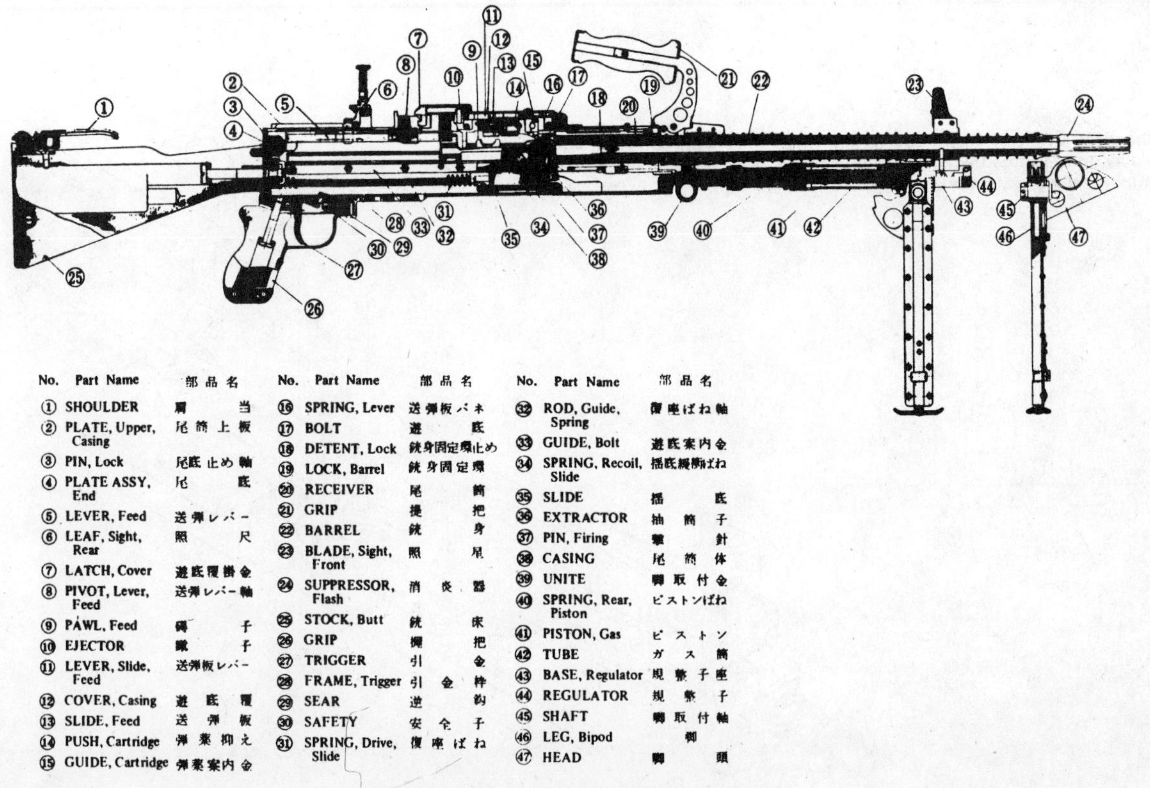

No.	Part Name	部品名
①	SHOULDER	肩当
②	PLATE, Upper, Casing	尾筒上板
③	PIN, Lock	尾底止め軸
④	PLATE ASSY, End	尾底
⑤	LEVER, Feed	送弾レバー
⑥	LEAF, Sight, Rear	照尺
⑦	LATCH, Cover	遊底覆掛金
⑧	PIVOT, Lever, Feed	送弾レバー軸
⑨	PAWL, Feed	[illegible]子
⑩	EJECTOR	蹴子
⑪	LEVER, Slide, Feed	送弾板レバー
⑫	COVER, Casing	遊底覆
⑬	SLIDE, Feed	送弾板
⑭	PUSH, Cartridge	弾薬抑え
⑮	GUIDE, Cartridge	弾薬案内金
⑯	SPRING, Lever	送弾板バネ
⑰	BOLT	遊底
⑱	DETENT, Lock	銃身固定環止め
⑲	LOCK, Barrel	銃身固定環
⑳	RECEIVER	尾筒
㉑	GRIP	提把
㉒	BARREL	銃身
㉓	BLADE, Sight, Front	照星
㉔	SUPPRESSOR, Flash	消炎器
㉕	STOCK, Butt	銃床
㉖	GRIP	握把
㉗	TRIGGER	引金
㉘	FRAME, Trigger	引金枠
㉙	SEAR	逆鈎
㉚	SAFETY	安全子
㉛	SPRING, Drive, Slide	復座ばね
㉜	ROD, Guide, Spring	復座ばね軸
㉝	GUIDE, Bolt	遊底案内金
㉞	SPRING, Recoil, Slide	揺底緩衝ばね
㉟	SLIDE	揺底
㊱	EXTRACTOR	抽筒子
㊲	PIN, Firing	撃針
㊳	CASING	尾筒体
㊴	UNITE	脚取付金
㊵	SPRING, Rear, Piston	ピストンばね
㊶	PISTON, Gas	ピストン
㊷	TUBE	ガス筒
㊸	BASE, Regulator	規整子座
㊹	REGULATOR	規整子
㊺	SHAFT	脚取付軸
㊻	LEG, Bipod	脚
㊼	HEAD	脚頭

Section view of the 7.62mm Type 62 machine gun.

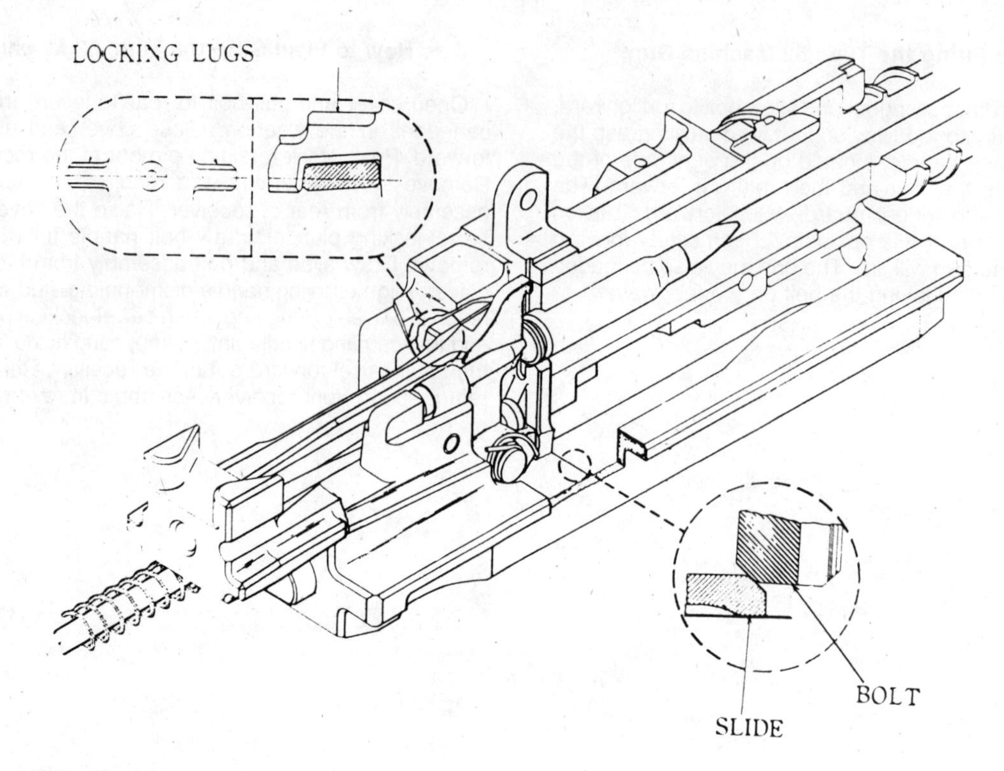

The Type 62 locking system consists of a bolt and bolt carrier assembly in which the bolt is cammed upward as it moves forward against the bolt carrier assembly (piston/slide group).

On the recoil stroke, when the bolt is cammed downward the extractor comes into play and is forced against the extractor groove of the fired case thus insuring positive extraction.

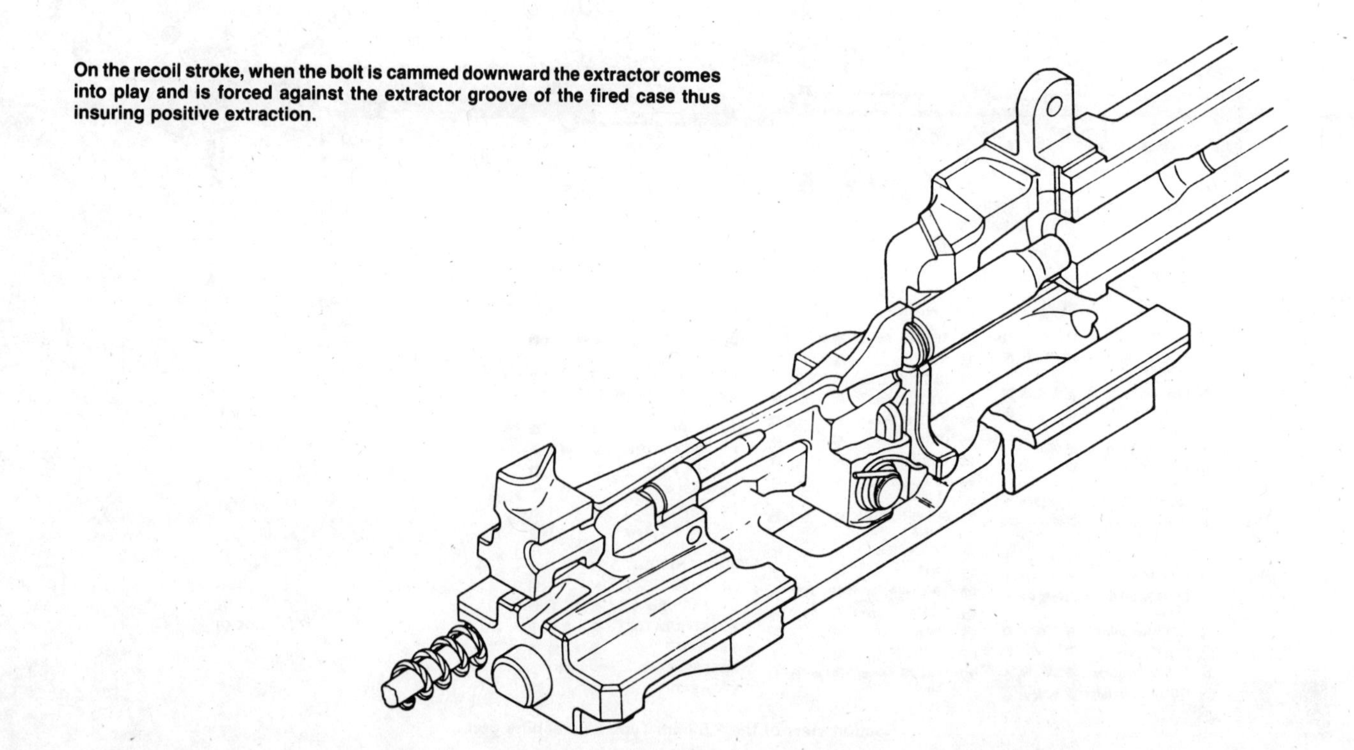

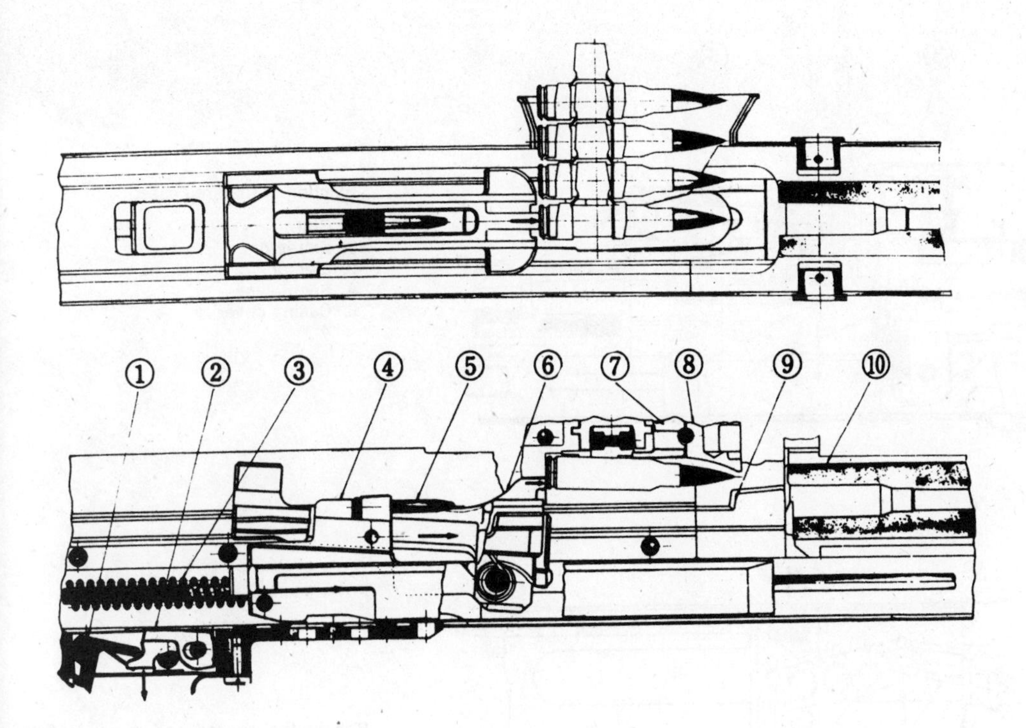

Feeding the cartridge.

1. Trigger
2. Sear
3. Driving spring
4. Slide
5. Firing pin
6. Bolt
7. Cartridge push
8. Cartridge guide
9. Locking room
10. Barrel

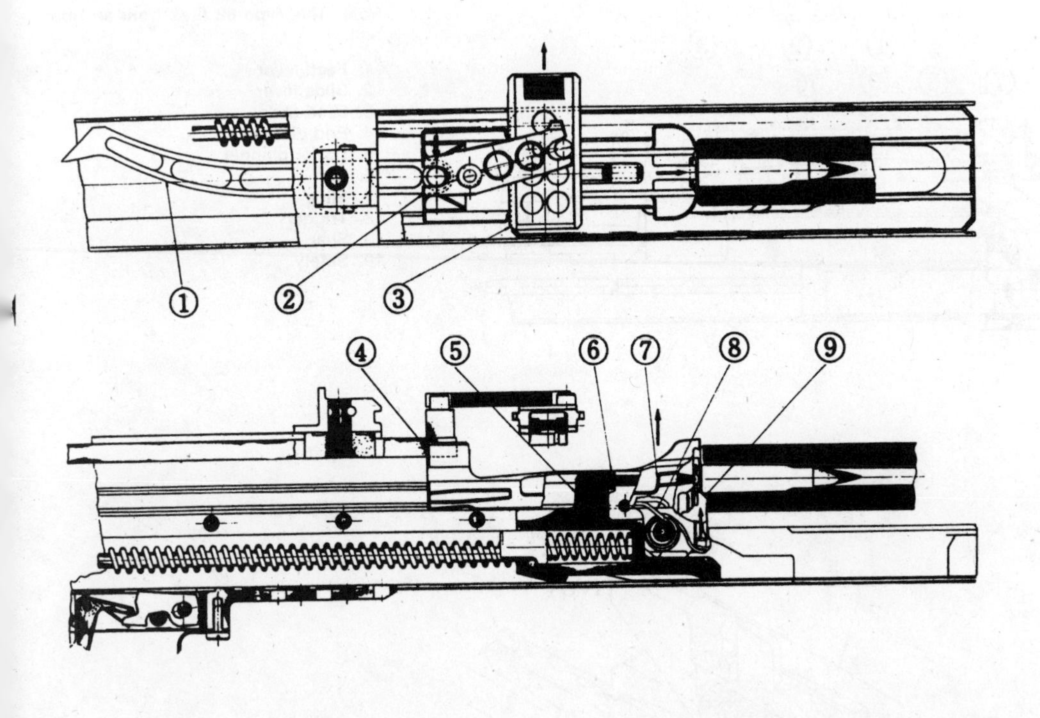

Locking the breech.

1. Feed lever
2. Slide lever
3. Feed slide
4. Bolt
5. Slide
6. Firing pin
7. Locking lug
8. Extractor spring
9. Extractor

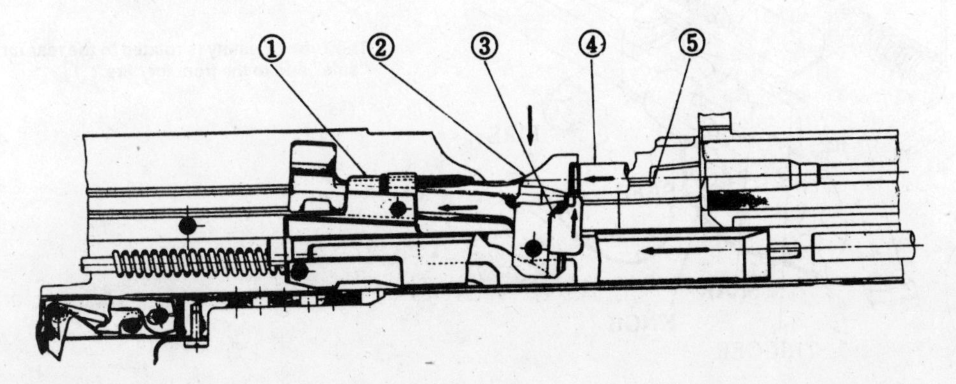

Extracting the cartridge.

1. Slide
2. Bolt
3. Extractor
4. Spent case
5. Locking room

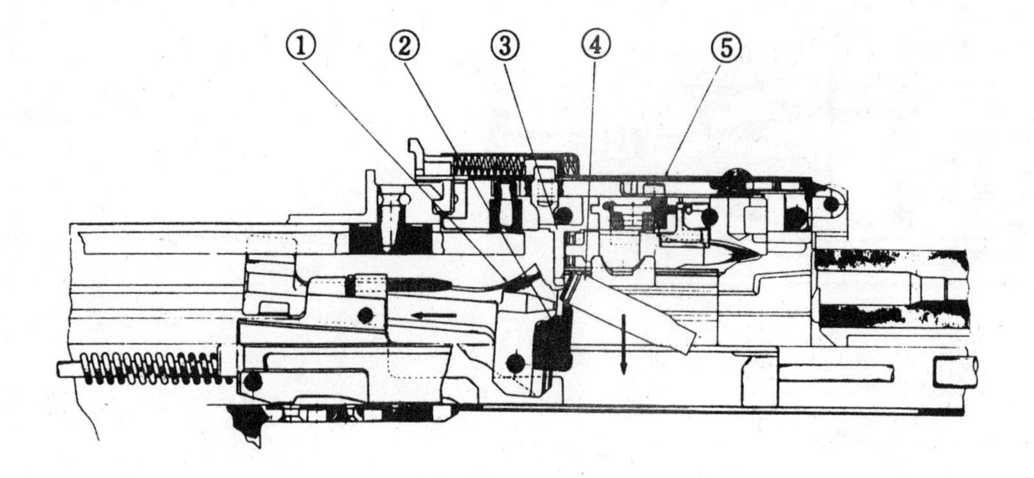

Ejection of the fired case.

1. Bolt
2. Extractor
3. Ejector
4. Spent case
5. Casing cover

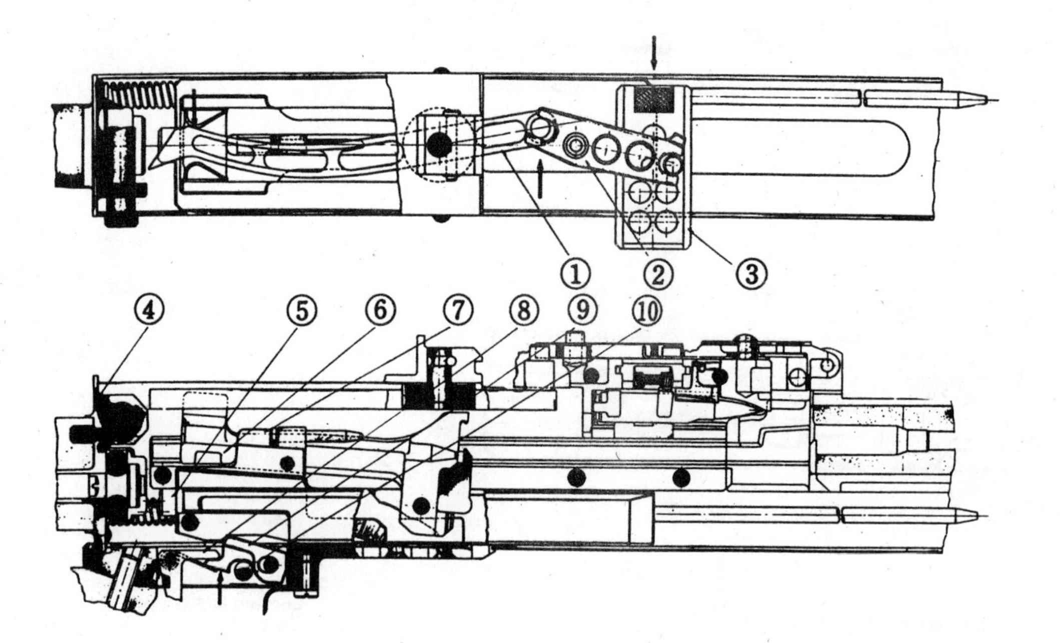

Bolt in the cocked and ready to fire position.
Note: The Type 62 fires from an open bolt.

1. Feed lever
2. Slide lever
3. Feed slide
4. End plate
5. Buffer plunger
6. Bolt
7. Slide
8. Trigger
9. Sear
10. Safety

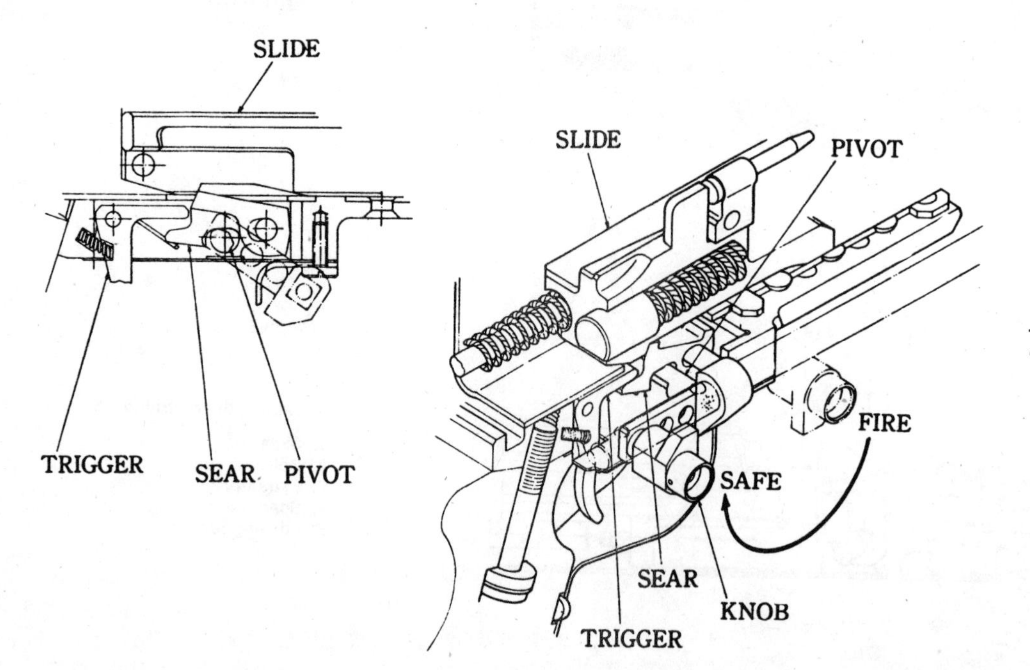

The Type 62 safety is rotated to the rear for "safe" and to the front for "fire."

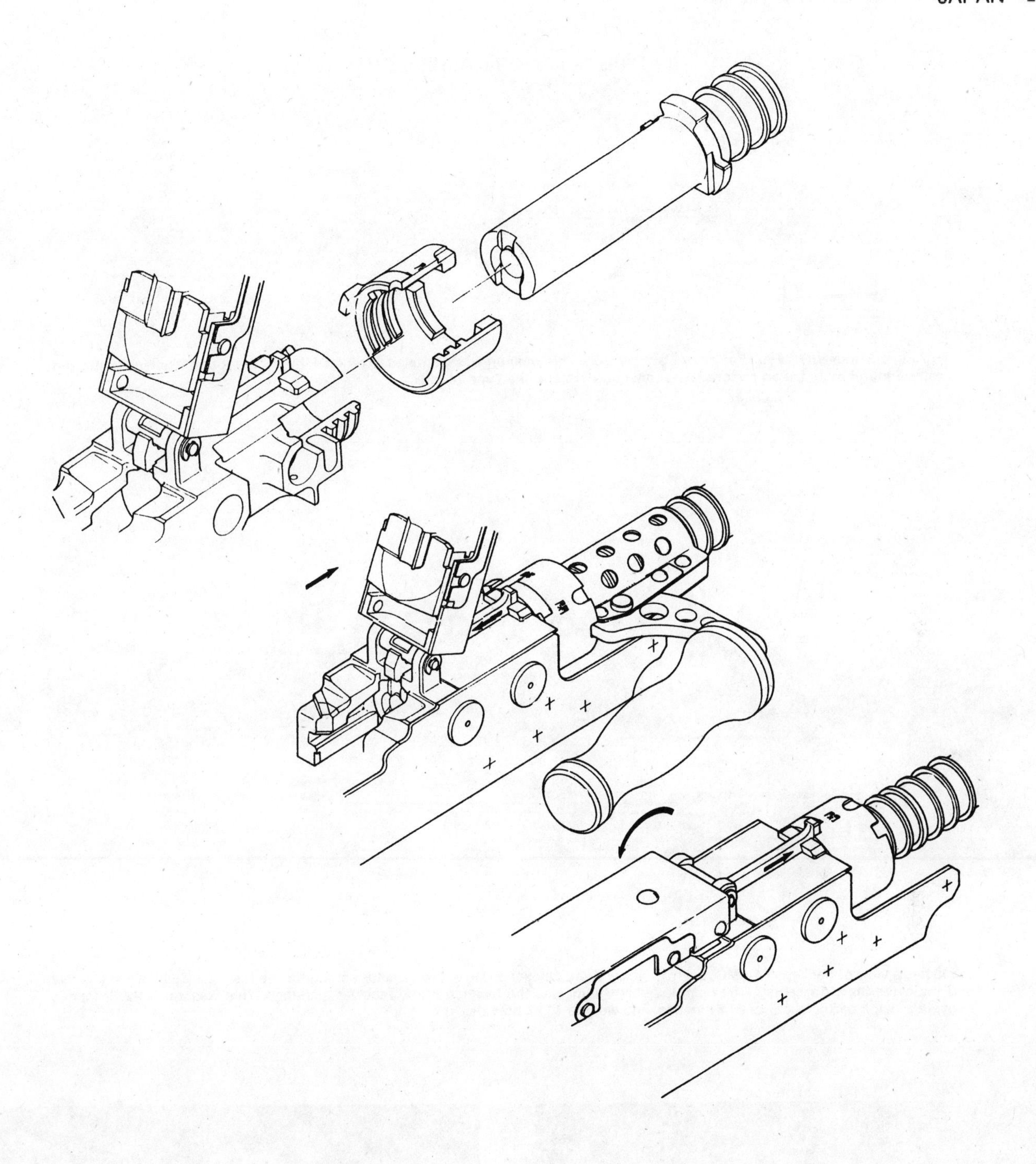

Type 62 barrel removal. The barrel of this machine gun is mounted in a locking collar that can be detached. The barrel and locking collar can be detached from the receiver by pulling to the rear on the collar lock and rotating the carrying handle to the right. Then the barrel assembly can be slid forward and lifted from the weapon.

TYPE 74 ARMOR MACHINE GUN

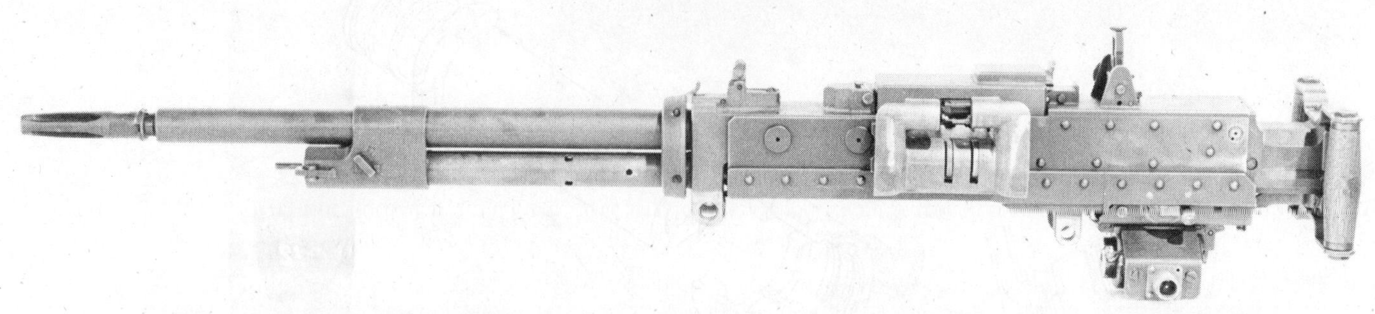

The 7.62 × 51mm NATO Type 74 Armor Machine Gun. This weapon was designed to replace the US Browning machine guns, and it can be found mounted on armored personnel carriers and the Type 74 tank.

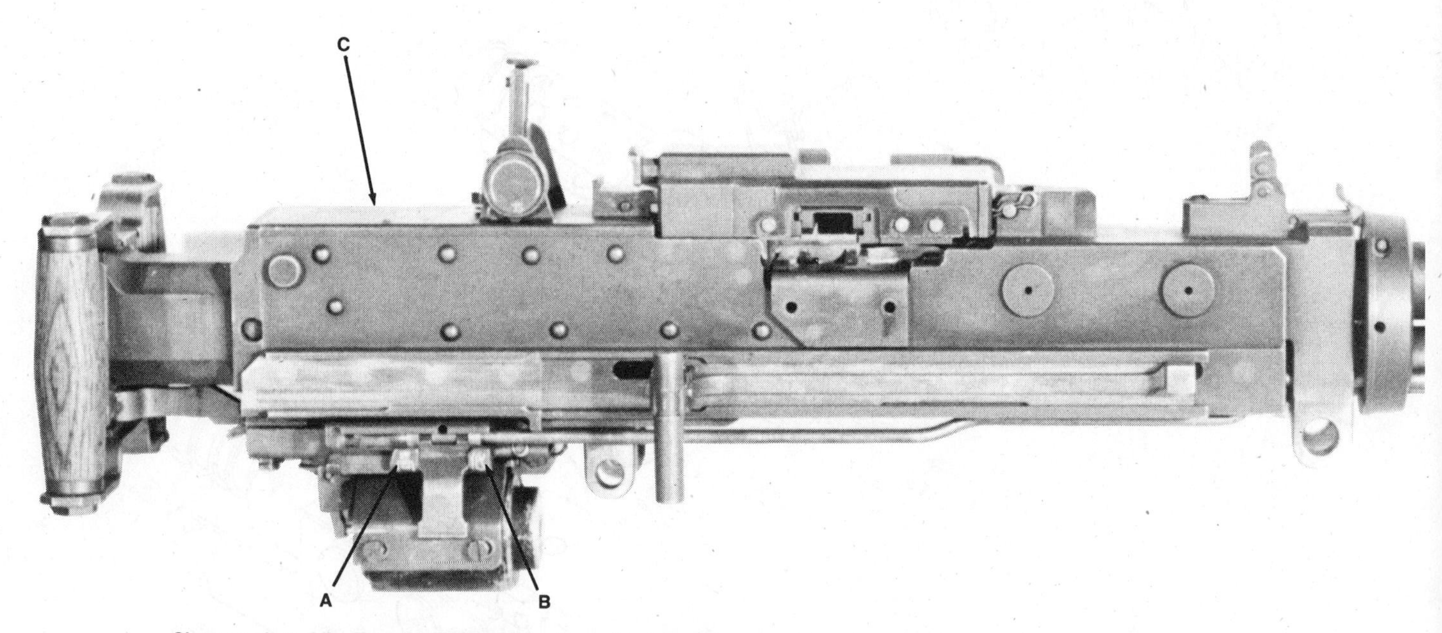

Close-up view of the Type 74 AMG receiver. *Arrow A* indicates the rate of fire selector setting for the fast rate (1,000 r.p.m.). *Arrow B* indicates the slow setting (700 r.p.m.). *Arrow C* indicates the location of the receiver markings. This weapon is 42.72 inches overall, has a barrel length of 24.61 inches, and weighs 44.97 pounds.

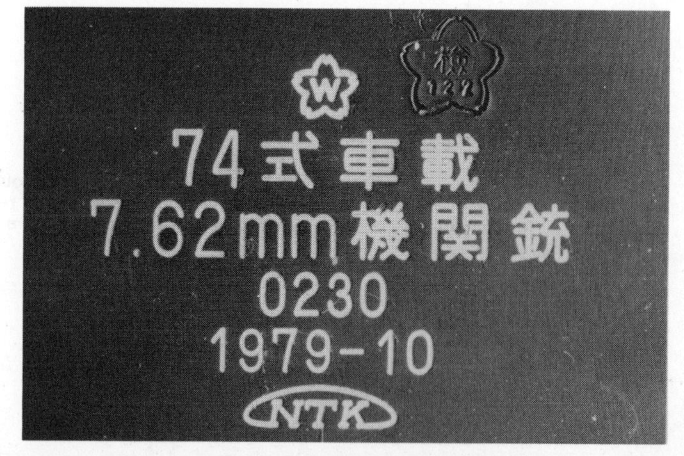

Receiver markings from the Type 62 and Type 74 machine guns. Note that the year and month are indicated on the receiver. Also note that only 3224 Type 62 and 230 Type 74 machine guns had been made through the fall of 1979. The small scale of production contributes to the very high cost of these guns.

TELESCOPIC SIGHTS FOR MACHINE GUNS

The Japanese have traditionally equipped their machine guns with telescopic sights. There are two basic variants for the Type 62 GPMG. Illustrated here is the periscope type. The standard tube type, illustrated earlier in this chapter, is used with the Type 62 when it is bipod fired.

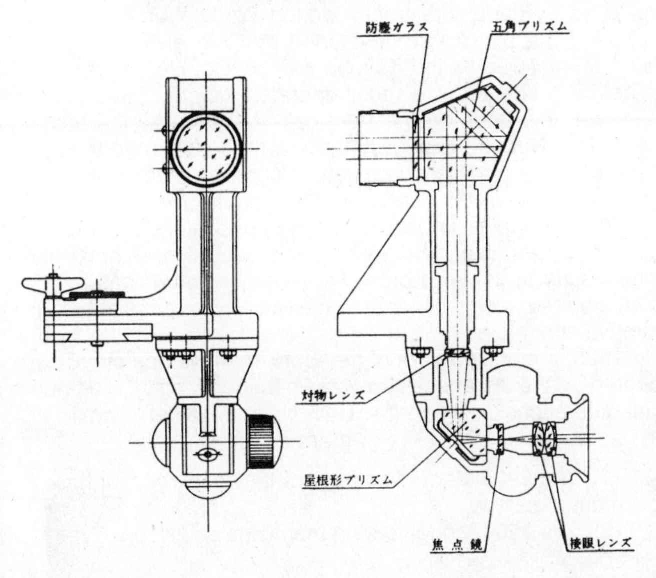

The Japanese periscope-type sight for the tripod mounted Type 62 machine gun.

The Japanese tube-type telescopic sight for the Type 62 bipod mounted machine gun.

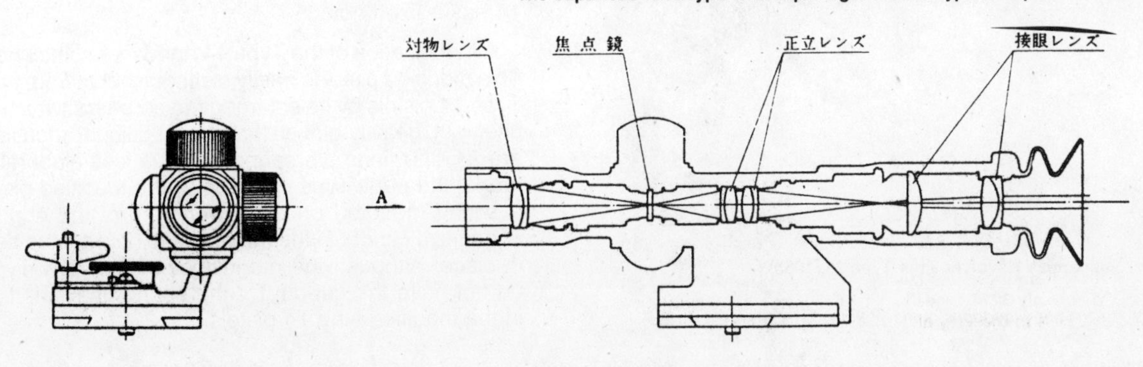

31 Imperial Japanese World War II Weapons

The Imperial Japanese Forces had a heterogenous assortment of weapons during World War II, especially in their collection of automatic weapons. Japan, like Italy, was caught in the midst of a change of rifle and machine gun calibers when the war began. This added considerably to the logistic problems of a country that was using four different rifle caliber cartridges in machine guns. The Japanese had also adopted the German 7.92mm MG15 aircraft gun as the Type 98 and a 12.7mm aircraft gun.

One of the basic reasons for this assortment of so many calibers was the fact that the Army, Navy and Air Force had all proceeded independently, adopting whatever caliber they wanted. No higher authority had forced standardization. In addition, captured material was used; thus 7.92mm FN-made Browning Automatic Rifles captured from the Chinese were used against United States forces in the Philippines. Other captured weapons included: Czech- and Chinese-made 7.92mm ZB26 and ZB30 light machine guns; Dutch 6.5mm Madsen guns; British caliber .303 Bren and Lewis guns; and various Dutch, British and Chinese rifles.

Japanese World War II weapons are no longer standard in the armies of any major power. Their future use is doubtful since, unlike the standard German World War II small arms, they are not chambered for cartridges still manufactured extensively. Limited manufacture of Japanese rifle and machine-gun ammunition was carried on in mainland China as late as the Korean War.

JAPANESE PISTOLS*

The Japanese Army adopted a 9mm revolver, the Type 26, in 1893. This weapon is a break-open type similar to the Smith & Wessons of the period but has lock work similar to the Austrian Rast and Gasser. This revolver is unusual in two respects: it is double-action only (like the later models of the .38 Enfield); and is chambered for a "one-of-a-kind" 9mm rimmed pistol cartridge. Although this revolver was replaced as the standard arm by an automatic in 1925, it was used in quantity during World War II. The only good feature about the weapon is a hinged side plate, which allows for easy exposure of the lock work.

The Nambu 8mm automatic pistol, which appeared in 1904, introduced the bottle-necked 8mm Japanese pistol cartridge. The Nambu was never a standard Japanese service arm, but it was used by Japanese troops during World War II and also exported. The pistol was developed by Colonel Kijiro Nambu and manufactured by the Kayoba Factory Co. Ltd.

Japanese Revolver 9mm Type 26 (1983)

*For additional details see *Handguns of the World*.

The 1904 Nambu is a recoil-operated weapon fitted with a grip safety mounted on the front of the grip below the trigger guard. This weapon is usually found with a slot cut in the rear of the pistol grip to accommodate a shoulder-stock holster. A smaller version of this weapon chambered for a 7mm bottle-necked cartridge is known as the "Baby Nambu." This was apparently a nonstandard weapon although it may have had limited issue in the Air Force. The "Baby Nambu" has a "V"-notch type rear sight, rather than the ramp-type rear sight of the 8mm Nambu.

In 1925, a modified form of the 8mm 1904 Nambu pistol was adopted by the Japanese Army as its standard pistol. The principal differences between the 1925 Nambu, called Type 14 by the Japanese and the 1904 arm are as follows:

(1) The 1904 Model has a grip safety, the Type 14 has a manual safety.

(2) The 1904 Model has no magazine safety, the Type 14 does.

(3) The 1904 Model has a tangent-type sight, the Type 14 has a fixed notch.

A modification of the Type 14 appeared somewhat later. This modified pistol is easily distinguishable from the early Type 14 pistols by its enlarged trigger guard, which allows use with heavily-gloved hands. Additional modifications are: (1) the firing pin spring guide is less elaborate than that of the early Type 14; and (2) the modified pistol has a spring mounted on the lower front of the grip, which engages a cut-out in the magazine to hold it more securely in place. Although the magazines of the early Type 25s do not have the cut-out for the spring, they will function in the modified type 14 pistols.

8mm Nambu Pistol, 1904 Type and 7mm Baby Nambu.

8mm Nambu Type 14 Pistol.

Modified 8mm Type 14 Pistol.

In 1934, a new 8mm pistol was introduced in Japan. This weapon was apparently intended principally for export sale but was used as a service pistol during World War II. This pistol, called the Type 94, is mainly distinguished by having an externally-mounted extension bar sear; it is recoil operated; most specimens show evidence of poor manufacture.

In 1942, work was started on the design of a simplified pistol chambered for the 8mm pistol cartridge. The result of this work was the Type II pistol, approximately 500 being made at Nagoya Arsenal by the end of the war. Although this weapon has been reported as being recoil operated, stripped views indicate that it is blowback operated. In any event, if any are in existence today, they are collectors' items.

Japanese Type 14 (1925) with magazine loaded and action in full forward position.

8mm Type 94 Pistol.

CHARACTERISTICS OF JAPANESE WORLD WAR II PISTOLS

	Type 26	Type 14	Type 94	1904 Nambu*	"Baby Nambu"
Caliber:	9mm.	8mm.	8mm.	8mm.	7mm.
System of operation:	Double-action only, revolver.	Recoil operated semiautomatic.	Recoil operated semiautomatic.	Recoil, semi-automatic.	Recoil, semi-automatic.
Length overall:	9.4 in.	9 in.	7.2 in.	9 in.	6.75 in.
Barrel length:	4.7 in.	4.7 in.	3.8 in.	4.7 in.	3.25 in.
Feed device:	6-round revolving cylinder.	8-round, in-line detachable box magazine.	6-round, in-line detachable box magazine.	8-round, in-line detachable box magazine.	7-round, in-line detachable box magazine.
Sights: Front:	Rounded inverted V.	Barley corn.	Barley corn.	Barley corn.	Barley corn.
Rear:	"V" notch.	Undercut notch.	Square notch.	Tangent w/notch.	"V" notch.
Weight:	2 lb.	2 lb.	1.68 lb.	1.93 lb.	1.43 lb.
Muzzle velocity:	634 f.p.s.	1065 f.p.s.	1000 f.p.s.	1065 f.p.s.	1050 f.p.s.

*Although these pistols were apparently never service pistols, they were used by Japanese Forces.

JAPANESE RIFLES

The first rifle of Japanese design to appear was the Type 13 11mm Murata, a single-shot bolt-action weapon. The Murata was followed by the 8mm Type 20 (1887) rifle and Type 27 Carbine. These were magazine arms chambered for an 8mm cartridge of Japanese design. In 1897, the first Arisaka rifle, the 6.5mm Type 30, appeared and became the standard Japanese rifle in the Russo-Japanese War of 1904-05. A carbine version of this weapon was developed in the same year. The Type 30 introduced to Japan the Mauser action in a modified form and the 6.5mm semi-rimmed cartridge.

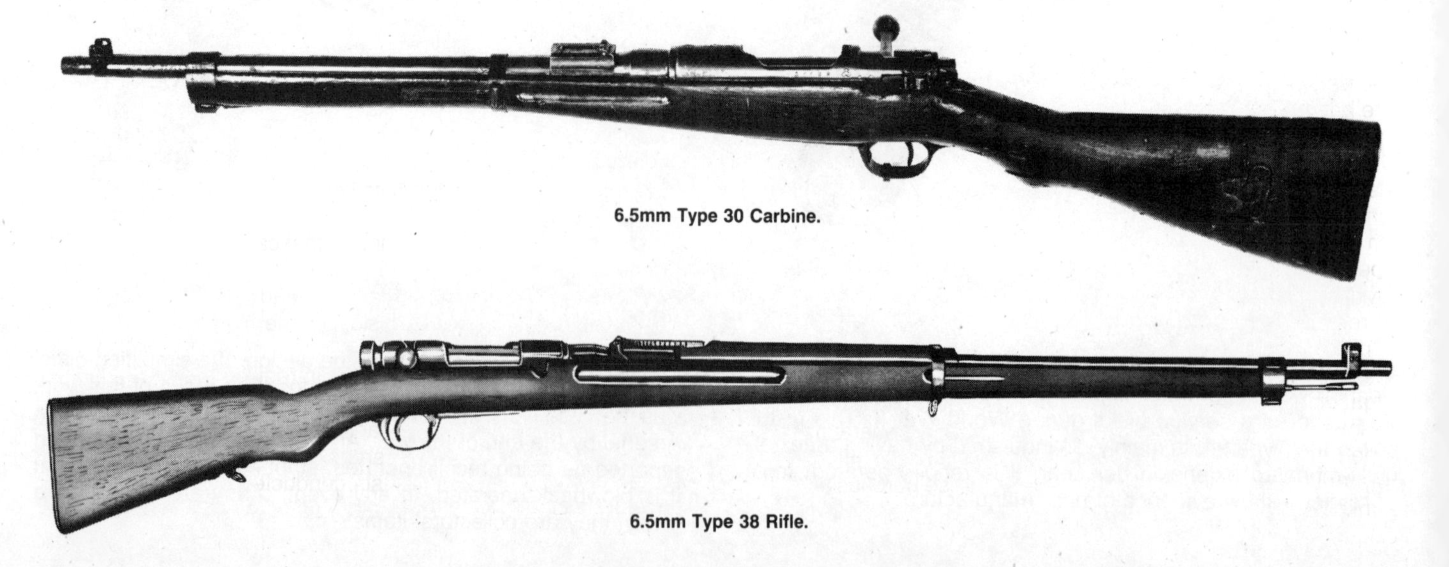

6.5mm Type 30 Carbine.

6.5mm Type 38 Rifle.

6.5mm Type 38 Carbine with hinged stock.

6.5mm Type 44 Carbine.

6.5mm Type 97 Sniper Rifle.

The 6.5mm Type 38 (M1905) was one of the two principal rifles used during World War II. The Type 38 differs from the 98 Mauser action only in the firing pin safety arrangement. The Type 38 has a large knob on the rear of the bolt, which is pushed in and rotated to put the weapon on "safe," rather than the flag-type safety mounted on the bolt sleeve as on the Mauser. The Type 38 also has a sliding bolt cover similar to that of the Lee Metford, but this bolt cover was usually removed from rifles by soldiers in the field because of the amount of noise they made in operation.

The Arisaka cocks on forward movement of the bolt in a fashion similar to that of the Enfield. The Type 38 carbine is a shortened version of the rifle and has a bayonet lug for the standard knife bayonet. A number of type 38 carbines were converted for paratroop use by the fitting of a hinged buttstock.

In 1911, the Type 44 Carbine was introduced; this weapon differs from the Type 38 Carbine in having a permanently attached folding bayonet.

In 1937, a sniping version of the 6.5mm Type 38 rifle—the Type 97—was adopted.

A rifle that was used to some extent by the Japanese Forces but remains very much a mystery is the 6.5mm so-called Type "I" rifle. This rifle, which uses the standard Japanese 6.5mm cartridge, has a Mannlicher Carcano action and a Mauser type magazine. The stock is of typical Japanese two-piece type; i.e., the lower half of the butt and pistol grip is a separate piece pinned and glued to the body of the stock. These rifles were apparently made in Italy, but why the Japanese purchased them is unknown.

Japanese experiences in China showed the need for a cartridge more powerful than the 6.5mm. A 7.7mm semi-rimmed cartridge was already in service with the Type 92 (1932) heavy machine guns. A rimless version of this cartridge was developed for use in rifles. Four trial rifles were submitted, including one each from Nagoya and Kokura arsenals. Several models were patterned after the Type 38 and Type 44 carbines, but tests indicated that the recoil of these weapons was excessive for the short-statured Japanese soldiers. A decision was made to develop a short rifle for cavalry and special troops and a long rifle for infantry. In 1939, the second series of tests was run at Futsu Proving Ground, and the Nagoya designed Rifle "Plan No. 1," which was similar to the Type 38, was adopted. A third series of tests, to dispose of accuracy "bugs" and to check out improved ammunition, was completed, and the Type 99 rifles were adopted in mid-1939. In 1942, a Type 99 rifle with four-power scope was introduced for sniper use.

The Type 99 has the sliding bolt cover of the Type 38 but in addition has a folding wire monopod that can be used to support the rifle when firing from the prone position or from a support. The rear sight has folding antiaircraft lead arms, which can be extended out to 90° at each side of the sight leaf when the weapon is used against aircraft.

In 1943, a substitute Type 99 was introduced made of inferior materials, without bolt cover, sling swivels or chrome-plated bores. The rifles have fixed rear sights. It is inadvisable to fire them, since they can be dangerous. On the subject of material and the strength of actions, tests conducted after World War II showed that the 6.5mm Type 38 action was stronger than the

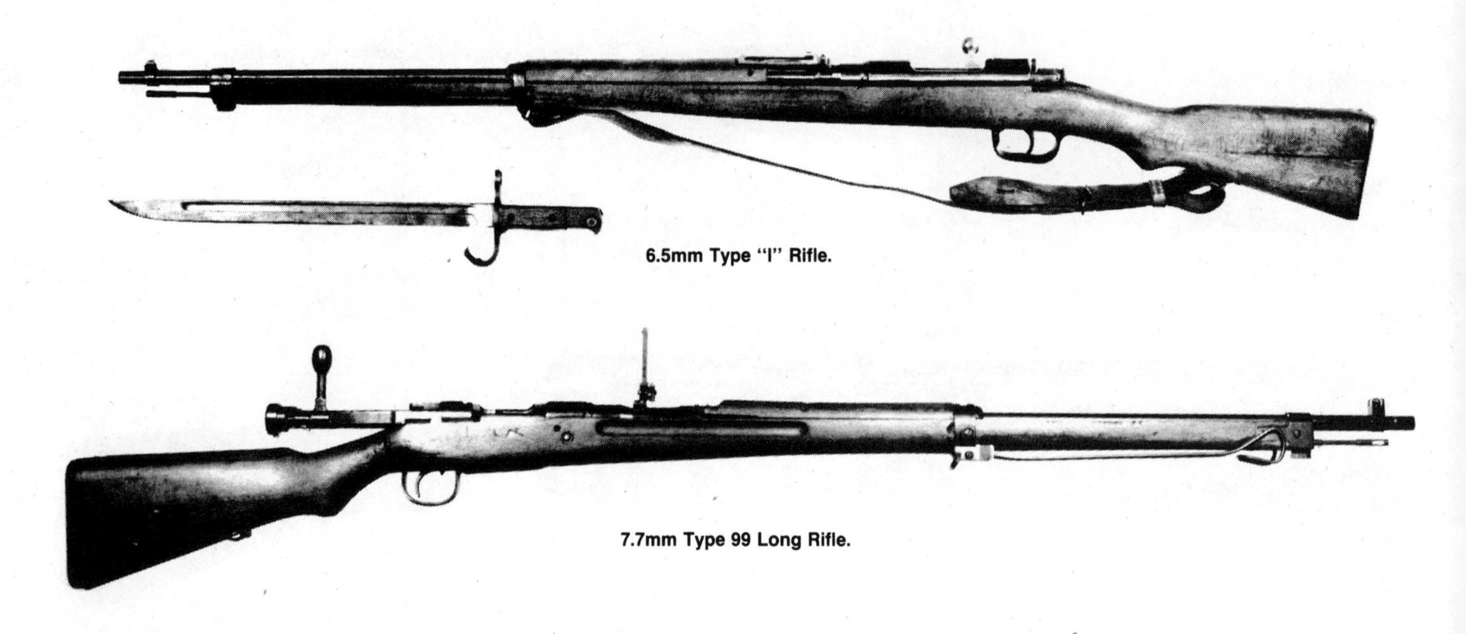

6.5mm Type "I" Rifle.

7.7mm Type 99 Long Rifle.

US Springfield, 1917 Enfield or German Mauser action.

A take-down version of the Type 99 with interrupted screw-type dismounting was introduced in 1942. The barrel of this rifle had a tendency to loosen in service, and further work in this area was done by Japanese Ordnance. The result was the Type 2 takedown rifle with a barrel-locking mechanism similar to that used on some machine guns. A locking key goes through the receiver and engages a slot in the barrel.

Although Japan started experiments with semiautomatic rifles in 1922, no semiautomatic rifle was produced for use in World War II. At the time, the United States was considering adopting the Pedersen rifle in caliber .276; a few copies of this rifle, fitted with a rotary magazine, were made in Japan.

There were many requests from the Japanese Forces in the field during World War II for a semiautomatic rifle to counter the US M1 rifle. The Japanese Navy produced in 1945 a modified copy of the US M1 rifle chambered for the 7.7mm cartridge. This weapon, called the Type 5, used a 10-round box magazine loaded with two 5-round chargers rather than the 8-round en bloc type clip of the M1.

Type 99 rear sight, showing AA lead arms.

Numerous difficulties were encountered with the rifle, principally because of the low quality materials then available. Only about twenty rifles were made.

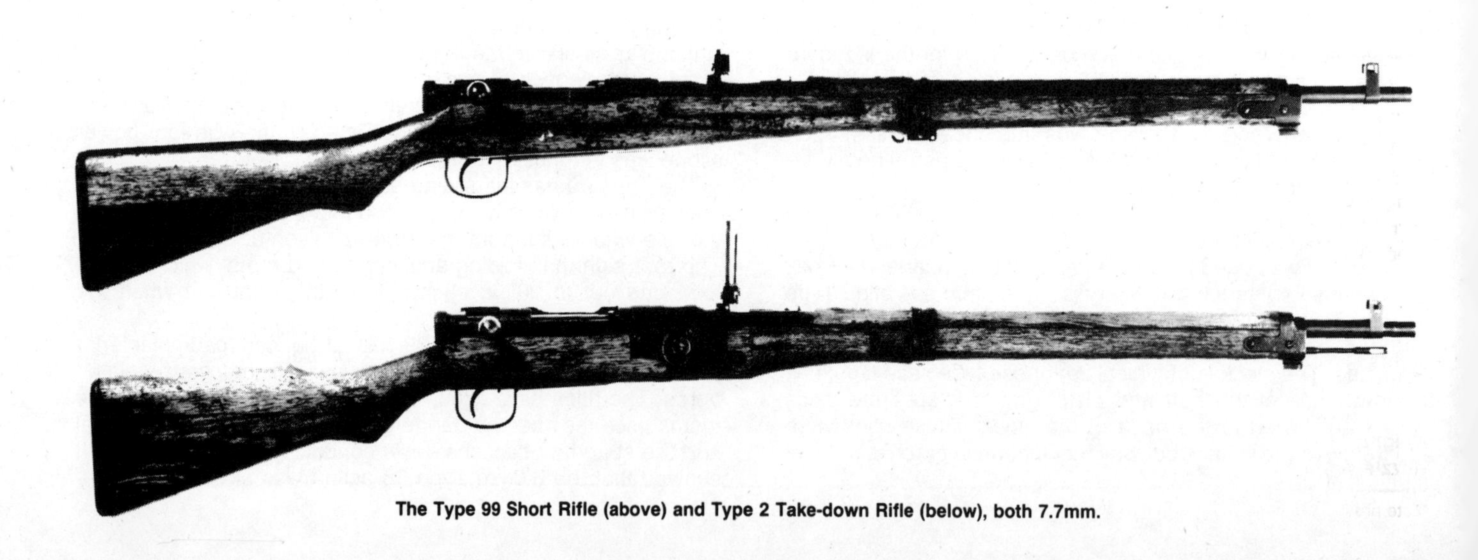

The Type 99 Short Rifle (above) and Type 2 Take-down Rifle (below), both 7.7mm.

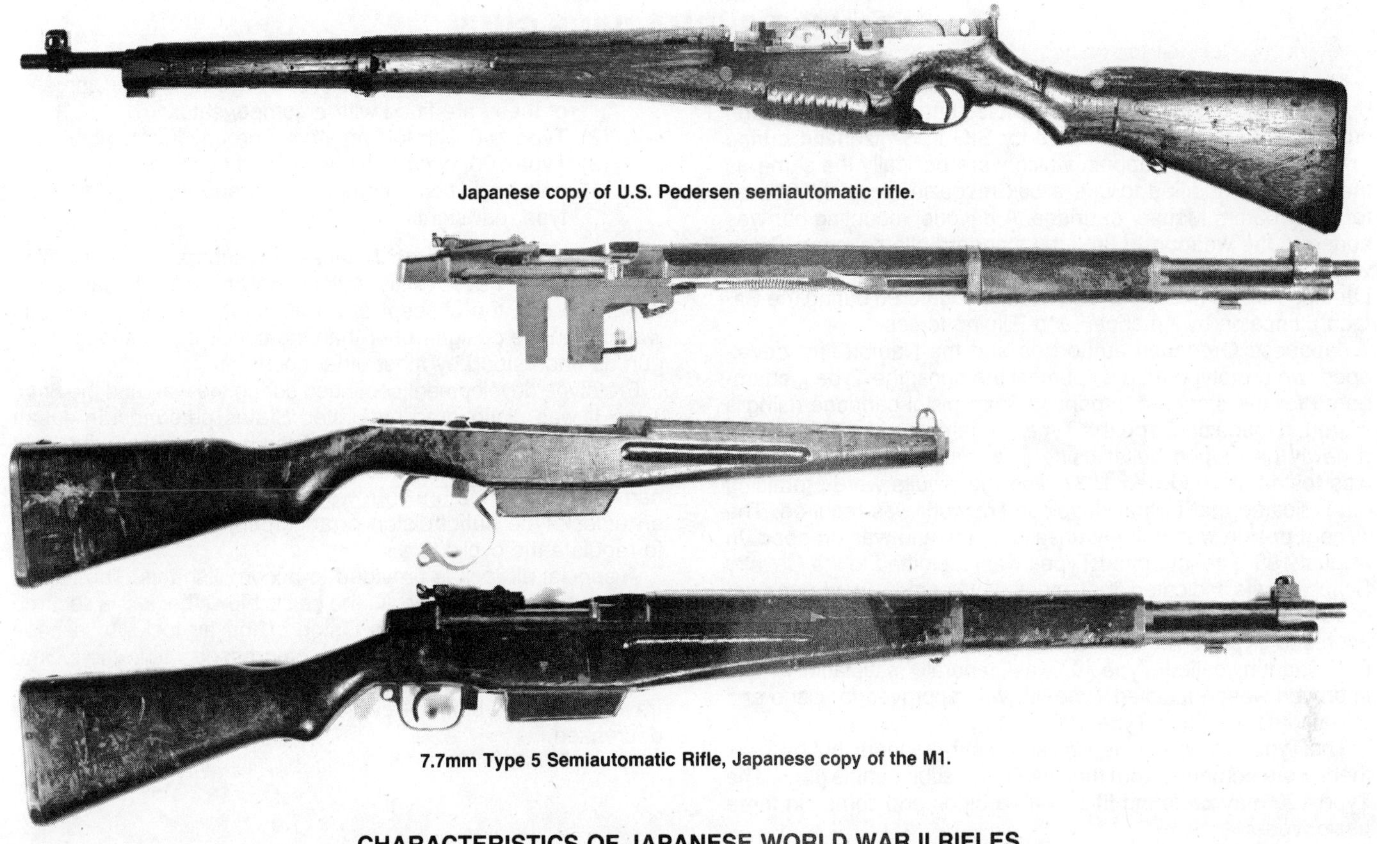

Japanese copy of U.S. Pedersen semiautomatic rifle.

7.7mm Type 5 Semiautomatic Rifle, Japanese copy of the M1.

CHARACTERISTICS OF JAPANESE WORLD WAR II RIFLES

	Type 38 Rifle	Type 38 Carbine	Type 44 Carbine	Type 97 Rifle
Caliber:	6.5mm.	6.5mm.	6.5mm.	6.5mm.
System of operation:	Turn bolt.	Turn bolt.	Turn bolt.	Turn bolt.
Length overall:	50.2 in.	34.2 in.	38.5 in.	50.2 in.
Barrel length:	31.4 in.	19.9 in.	19.2 in.	31.4 in.
Feed device:	5-round, nondetachable, staggered row, box magazine.	5-round, nondetachable, staggered row, box magazine.	5-round, nondetachable, staggered row, box magazine.	5-round, nondetachable, staggered row, box magazine.
Sights: Front:	Barley corn with protecting ears.	Barley corn with protecting ears.	Barley corn with protecting ears.	(Same as Type 38 plus 2.5 × scope.)
Rear:	Leaf.	Leaf.	Leaf.	
Weight:	9.25 lb.	7.3 lb.	8.9 lb.	11.2 lb. w/scope.
Muzzle velocity:	2400 f.p.s.	Approx. 2300 f.p.s.	Approx. 2300 f.p.s.	2400 f.p.s.

	Type 99 Long Rifle	Type 99 Short Rifle	Type 2 Rifle
Caliber:	7.7mm Type 99.	7.7mm Type 99.	7.7mm Type 99.
System of operation:	Turn bolt.	Turn bolt.	Turn bolt.
Length overall:	50 in.	43.9 in.	43.9 in.
Barrel length:	31.4 in.	25.8 in.	25.8 in.
Feed device:	5-round nondetachable, staggered row, box magazine	5-round nondetachable, staggered row, box magazine.	5-round nondetachable staggered row, box magazine.
Sights: Front:	Barley corn with protecting ears.	Barley corn with protecting ears.	Barley corn with protecting ears.
Rear:	Leaf.	Leaf.*	Leaf.
Weight:	9.1 lb.	8.6 lb.	8.9 lb.
Muzzle velocity:	2390 f.p.s.	Approx 2360 f.p.s.	Approx 2360 f.p.s.

*Late production models of this rifle may be found with a fixed notch type rear sight.

JAPANESE SUBMACHINE GUNS

The Japanese did little in the line of submachine gun development until the thirties. They purchased quantities of Bergmann submachine guns manufactured by SIG in Switzerland during the 1920s. These weapons, which were basically the same as the MP 18 I modified to use a box magazine, are chambered for the 7.63mm Mauser cartridge. A bayonet mounting bar was added to the weapon to take the standard rifle bayonet. These weapons were apparently mainly used by the Japanese Special Landing Forces (Marines) and were encountered during the Bataan campaign by American and Filipino forces.

Japanese Ordnance authorities and the Nambu firm developed two prototype model submachine guns: the Type I, chambered for the standard Japanese 8mm pistol cartridge using a 50-round magazine, and the Type II, chambered for 6.5mm with a newly developed bullet using a 30-round magazine. Type I was tested in 1930 and 1937, and the results were promising but indicated that further development work was required. The Type II design was entirely unsatisfactory and was dropped. In August 1937, an improved Type I was submitted to the Cavalry School; tests indicated that more development work was required. In April 1939, the Nambu firm submitted a design for an improved Type I. Testing indicated that with little improvement this weapon, called Type III, was generally satisfactory. The improved weapon, called Type IIIB was approved for issue and adopted as the 8mm Type 100 (1940).

The Type 100 was issued initially to paratroopers, but by 1944 there were demands from the infantry for submachine guns. The Type 100 may be found fitted with a bipod and comes in three basic types:

(1) Type 100 with fixed stock and bayonet lug bar; some of these are fitted with a compensator.
(2) Type 100 with folding stock and bayonet lug bar.
(3) Type 100, circa 1944, with fixed stock, bayonet lug on barrel jacket, compensator, and fixed aperture-type rear sight.

Time caught up with the Japanese on submachine guns. The Type 100 was never really satisfactory, possibly because the Japanese had the concept of an automatic rifle in mind while working on the design rather than the concept of a submachine gun as understood by most other countries.

Prototype development continued during the war, and the 8mm Type II was discovered by United States personnel in Japan after the war was over. The 8mm Type II was a step in the right direction insofar as weight and length were concerned. It also had one unusual feature borrowed from the 1926 Finnish Suomi, an airlock-type buffer/piston arrangement, which can be used to regulate the cyclic rate.

A special air lock is provided to accomplish this. The lock is at the rear of the receiver. As the bolt is blown back, it is secured to an extension arm on the piston of the air lock. An escape valve can be set to allow the air compressed within the lock to escape at different rates. By thus speeding or slowing down the travel rate of the bolt, the cyclic rate of fire can be increased or decreased.

CHARACTERISTICS OF JAPANESE WORLD WAR II SUBMACHINE GUNS

	Bergmann 1920	Type 100	Type 100 (1944 version)
Caliber:	7.63mm.	8mm.	8mm.
System of operation:		Blowback, automatic only	
Length overall:	32 in.	w/stock extended: 34 in. w/stock folded: 22.2 in.	36 in.
Barrel length:	8 in.	9 in.	9.2 in.
Feed device:	50-round, staggered row, detachable box magazine.	30-round, staggered row, detachable box magazine.	30-round, staggered row, detachable box magazine.
Sights: Front:	Barley corn.	Barley corn with protecting ears.	Barley corn.
Rear:	Tangent with notch.	Tangent with notch.	Fixed aperture set for 100 meters.
Weight:	Approx 9.5 lb.	7.3 lb.	8.5 lb.
Muzzle velocity:	Approx 1350 f.p.s.	Approx 1100 f.p.s.	Approx 1100 f.p.s.
Cyclic rate:	600 r.p.m.	450 r.p.m.	800 r.p.m.

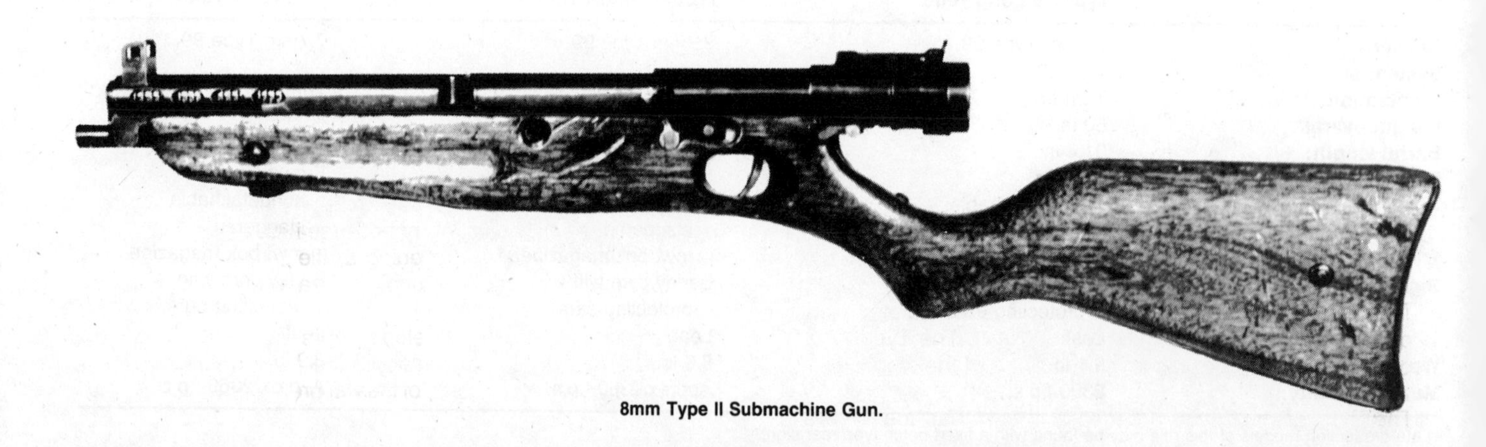

8mm Type II Submachine Gun.

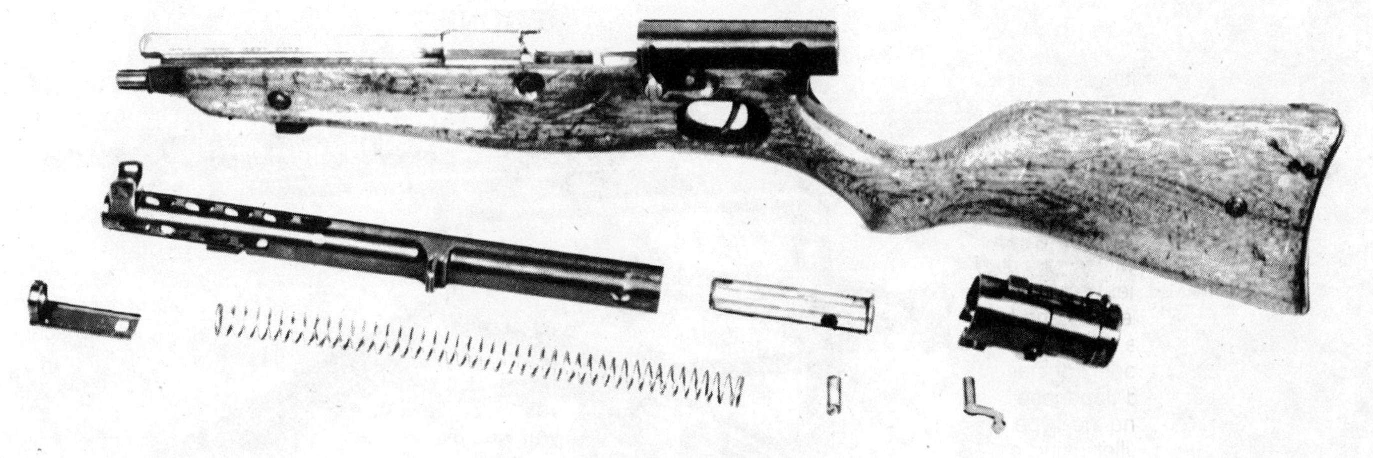

Type II, field stripped.

JAPANESE MACHINE GUNS

The Japanese adopted the Hotchkiss gun at the time of the Russo-Japanese War (1904-05) and also adopted the later 1914 Hotchkiss in a modified form. Both models were chambered for the 6.5mm cartridge, which left something to be desired insofar as long-range performance was concerned, but the initial adoption of the Hotchkiss set the tone for Japanese machine gun development for many years to come.

The Type 3 (1914) is a modified Hotchkiss 1914 developed by General Nambu. Like all later Nambu/Hotchkiss machine guns, it has a gravity-fed oil reservoir, which, through a spring-loaded lubricator, oiled all cartridges before they were fed into the chamber. The Nambus also differ from the Hotchkiss in the method of ejection. They use the ejection system of the Lewis gun.

During World War II, a modification of the Type 3 was encountered; the modified gun was lighter, and the barrel could be removed more rapidly than on the original Type 3. The 6.5mm Type 11 was introduced in 1922 and is designed basically for use as a light machine gun, although it was also used on a tripod.

An unusual feature of the Type 11 is the feed hopper, which is fed with six 5-round chargers (stripper clips). The chargers are the same as those used with the 6.5mm Arisaka rifle.

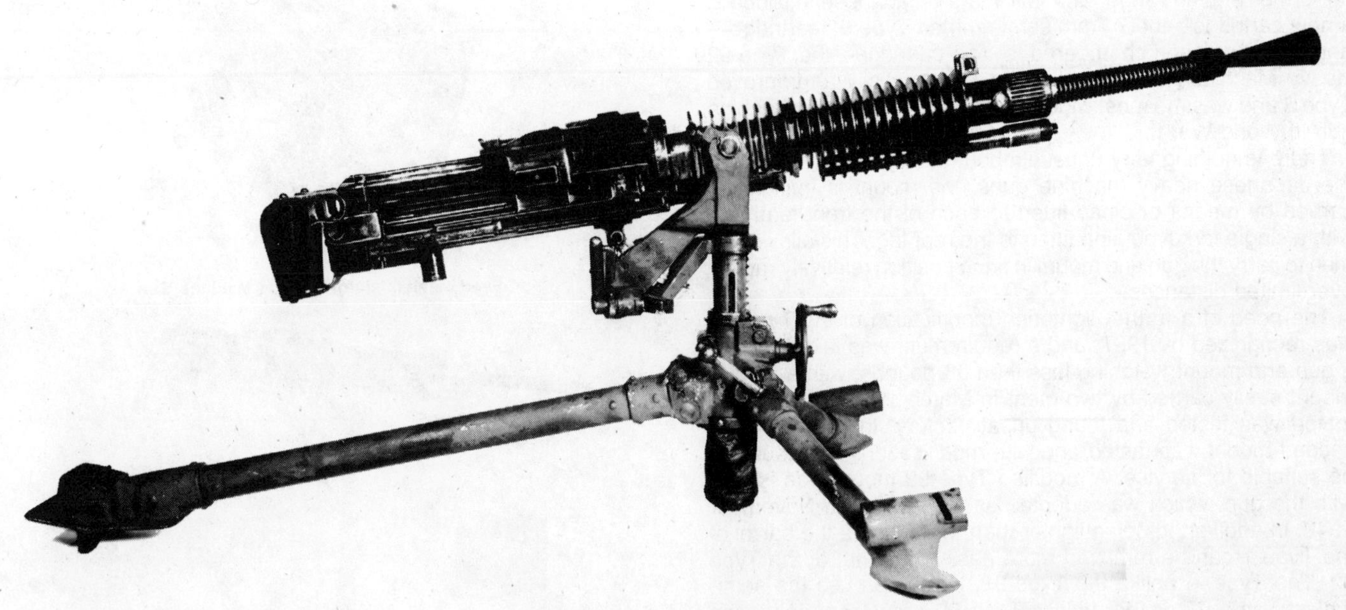

6.5mm Type 3 (1914) Machine Gun.

The Type 11 also appeared in a tank version called the Type 91 (1931). The Type 91 has a telescope mount and a larger feed hopper than the Type 11.

The 7.7mm Type 97 (1937) tank gun was the other standard Japanese tank machine gun of World War II. The Type 97 is a copy of the Czech ZB26 as an examination of the stripped view will show.

The Type 97, although a better tank gun than the Type 91, was not completely satisfactory because it was magazine-fed—hardly an ideal method of feeding a rifle caliber tank machine gun. Research was started during the war to develop a belt-fed gun with a high rate of fire.

A Browning type gun, developed from the Japanese aircraft Browning, was to be introduced as the 7.7mm Tank machine gun Type 4, but the conclusion of the war prevented its introduction.

6.5mm Type 11 on tripod.

Japanese machine gun development was very chaotic since guns were developed by the Army, Navy and for aircraft without any apparent coordination. (The reader will have to forgive the skipping back and forth throughout the text from one type of machine gun to another, but this is apparently what happened in Japanese Ordnance circles at the time. One thing is quite obvious—the Japanese pre-war military had no true apprecia- tion of the logistic difficulties they would encounter in an "all- out" war.)

As pointed out previously, the 6.5mm cartridge was not a good performer at long range, and in 1932 the Japanese introduced a new cartridge—the 7.7mm semi-rimmed Type 92 cartridge— and a new weapon chambered for this cartridge—the Type 92 Heavy Machine gun. This weapon is essentially an improved Type 3 and was the most widely-used Japanese heavy machine gun in World War II.

There is nothing very unusual about the Type 92. As with all the Japanese heavy machine guns, the mount is built to be carried by means of pipes fitted to each of the front legs and with a single fork type pipe fitted to the rear leg. This allows two men to carry the gun and mount in firing position relatively rapidly over limited distances.

The need of a further lightened modification of the Type 92 was recognized by 1937, and a requirement was laid down for a gun and mount weighing less than 88 pounds, with a type of mount easily carried by two men. In March 1940, the first pro- totype was tested and found unsatisfactory. In June 1940, a second model was tested and with modifications was found to be suitable for service. A modified Type 92 mount was issued with the gun, which was adopted as the Type 1 in November 1942. In addition to being lighter than the Type 92 the barrel of the Type 1 can be removed more easily than that of the Type 92. The Type 1, unlike the Type 92, which can use the semi- rimless Type 92 or the rimless Type 99 cartridges, only uses the Type 99 cartridge.

In 1936, the 6.5mm Type 96 machine gun was introduced. This weapon represented a great improvement over the Type 11. The Type 96 does not have a cartridge oiler on the gun, has a quick change barrel and is box magazine-fed. The weapon is frequently found with a 2.5 power telescope mounted on the receiver. The cartridges for the Type 96 are oiled when loaded into the magazine by an oiler built into the magazine loader. Both the Type 11 and Type 96 lacked built in slow initial ex- traction and therefore required oiled cartridges.

Feed mechanism of Type 11 Machine Gun.

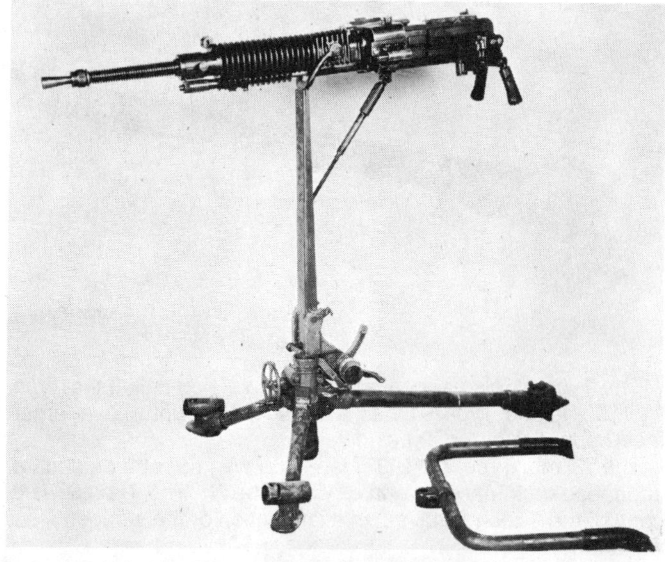

7.7mm Type 92 on an AA mount.

The Type 99 (1939) light machine gun was designed to obviate the necessity for lubricated cartridges. The barrel lock has a head space adjustment nut, and machining of the critical components was held to closer tolerances than with early guns. The Type 99 came into being because of the need for a light machine gun chambered for the 7.7mm rimless Type 99 cartridge. Four different types of prototype guns were built. Type 1 was modeled on the Type 96 light machine gun. Type 2 was a lightened version of the Type 92 with shoulder stock, bipod and quick change barrel. Type 3 was similar to the Type 97 Tank gun, and Type 4 was another light variation of the Type 92 heavy machine gun. The requirement was for a weapon weighing not more than 24.7 pounds with sights graduated to 1500 meters. A flash hider was also required since the muzzle flash of the 7.7mm cartridge was much greater than that of the 6.5mm cartridge. The modified Type 96 prototype was chosen and adopted as the 7.7mm Type 99.

A paratrooper version of the Type 99 machine gun was built at the Nagoya Arsenal in 1943. This weapon was easily broken down into barrel and receiver group, buttstock, barrel and piston and could be rapidly assembled.

The Japanese Navy and Navy Air Corps used the Lewis gun in ground and air versions. These were chambered for 7.7mm rimmed cartridges, the same as the caliber .303 British cartridge. Both guns are called Type 92; note the complication of having two ground machine guns that use different cartridges called by the same model designation. This situation was due to the almost complete lack of coordination between the Army and Navy Ordnance authorities.

The Japanese copied a number of foreign weapons to use as aircraft guns and in one case produced the Browning in a larger caliber—20mm—than it was produced elsewhere. The Japanese also purchased specimens and the right to manufacture a number of German aircraft machine guns, which they made in 7.92mm, including the MG15 called Type 98. The British Vickers aircraft gun was copied as the 7.7mm (.303 British) Type 89. A Japanese copy of the US caliber .50 Browning chambered for the Japanese 12.7mm cartridge was also quite common. This weapon was called the Type 1.

The Japanese also produced various types of blowback operated training machine guns that do not seem to conform to any set pattern. At least five different variations were found in Korea. They are all rather delicate in construction, probably intended for use with a reduced-charge cartridge.

Japanese 7.7mm Heavy Machine Gun, Type I. Standard cartridge strip which is fed into gun from the left side is shown in foreground.

CHARACTERISTICS OF JAPANESE WORLD WAR II ARMY MACHINE GUNS

	Type 3	Type 11	Type 92	Type 96	Type 99
Caliber:	6.5mm.	6.5mm.	7.7mm Type 92 or 99.	6.5mm.	7.7mm Type 99.
System of operation:	Gas, automatic only.	Gas, automatic only.	Gas, automatic only.	Gas, automatic only.	Gas, automatic only.
Length overall:	Approx 47 in.	43.5 in.	45.5 in.	41.5 in.	46.75 in.
Barrel length:	Approx 29 in.	19 in.	Approx 29 in.	21.7 in.	Approx 21.5 in.
Feed device:	30-round strip.	30-round hopper.	30-round strip.	30-round staggered row, detachable, box magazine.	30-round staggered row, detachable, box magazine.
Sights: Front:	Barley corn with protecting ears.	Barley corn with protecting ears.	Barley corn with protecting ears.	Barley corn with protecting ears.	Barley corn with protecting ears.
Rear:	Folding ring A.A. sight and tangent with aperture.	Tangent with "V" notch.	Tangent with aperture or telescope.	Radial wheel tangent with aperture.	Radial wheel tangent with aperture.
Weight:	122 lb. w/tripod.	22.5 lb.	122 lb. w/tripod.	20 lb.	23 lb.
Muzzle velocity:	2434 f.p.s.	Approx 2300 f.p.s.	Approx 2400 f.p.s.	Approx 2400 f.p.s.	Approx 2350 f.p.s.
Cyclic rate:	450-500 r.p.m.	500 r.p.m.	450-500 r.p.m.	550 r.p.m.	850 r.p.m.

All barrel lengths include flash hider length.

Japanese 7.7mm Light Machine Gun Type 99 (1939).

6.5mm Type 96 Light Machine Gun.

CHARACTERISTICS OF JAPANESE WORLD WAR II ARMY MACHINE GUNS (Cont'd)

	Type 1	Type 91	Type 97	Type 93
Caliber:	7.7mm Type 99.	6.5mm.	7.7mm Type 99.	13.2mm.
System of operation:	Gas, automatic only.	Gas, automatic only.	Gas, automatic only.	Gas, automatic only.
Length overall:	42.4 in.	33 in.	46.5 in. w/stock.	95 in. (approx).
Barrel length:	23.2 in.	19.2 in.	28 in.	65 in.
Feed device:	30-round strip.	Hopper.	30-round staggered row, detachable box magazine.	30-round staggered row, detachable, box magazine.
Sights: Front:	Barley corn with protecting ears.	Barley corn with protecting ears.	1.5 × scope.	Barley corn with protecting ears.
Rear:	Tangent with aperture.	Tangent with "V" notch and telescope.		Leaf and speed ring type A.A.
Weight:	70 lb. w/tripod.	22.4 lb.	24.5 lb.	Approx 213 lb. w/ tripod.
Muzzle velocity:	Approx 2400 f.p.s.	Approx 2300 f.p.s.	Approx 2400 f.p.s.	2210 f.p.s.
Cyclic rate:	550 r.p.m.	500 r.p.m.	500 r.p.m.	450-480 r.p.m.

All barrel lengths include flash hider length.

32 Mexico

SMALL ARMS IN SERVICE

Handguns:	11.43 × 23mm M1911A1	US-made.
Submachine guns:	11.43 × 23mm M3A1	US-made.
	9 × 19mm NATO MP5	Heckler & Koch-made. Police issue only.
Rifles:	7.62 × 51mm NATO G3	Mexican-made under license from Heckler & Koch.
	7.62 × 51mm NATO FAL	Mexican assembled from FN provided parts.
	5.56 × 45mm M16A1	Colt-made. Small quantities only.
	7.62 × 33mm M1 Carbine	US-made.
	7.62 × 63mm M1954	Mexican-made.
Machine guns:	7.62 × 51mm NATO MAG	Fabrique Nationale-made.
	7.62 × 63mm M1919A4	US-made.

MEXICAN HANDGUNS*

There has been very limited development of handguns in Mexico. The only significant military-type self-loading pistol was the Obregon, a rotating barrel-locked weapon. This handgun was developed in the 1930s, but the developer was never able to sell his design to the Mexican military authorities. The small number of Obregons, fewer than 800, were generally purchased by military officers, but none are recorded as having been purchased by the Mexican armed forces. The standard sidearm of the Mexican military continues to be the US M1911A1 pistol. Many police organizations use this pistol in the .38 Super caliber.

THE CALIBER .45 OBREGON PISTOL

CHARACTERISTICS OF OBREGON PISTOL

Caliber: .45 Model 1911 (11.43mm).
System of operation: Recoil, semiautomatic.
Length overall: Approx. 8.5 in.
Barrel length: Approx. 5 in.
Feed device: 7-round, in-line, detachable box magazine.
Sights: **Front:** Blade.
Rear: Dove-tailed bar with notch.
Weight: Approx. 2.5 lbs.
Muzzle velocity: 830 f.p.s.

Caliber .45 Obregon pistol.

How to Load and Fire the Obregon Pistol

This pistol is loaded and fired in a manner similar to that of the US Caliber .45 M1911A1, which is covered in detail in the chapter on the United States.

*For additional details see *Handguns of the World*.

MEXICAN MILITARY RIFLES

In June 1979 the Mexican government signed a contract with Heckler & Koch which permitted Mexico to manufacture the G3. The first step was the assembly of components made in Germany, but in late 1980 the Mexicans began manufacturing components for this weapon at the Fabrica Nacional in Mexico City. Much of the equipment formerly used in the assembly of the Fabrique Nationale FAL was converted so it could be used in the production of the G3. The FN FAL was first acquired in 1968, and during the 1970s limited numbers of the M16A1 were purchased by police and military organizations.

The crest of the Mexican Navy on a post-1973 FN-made FAL.

OBSOLETE MEXICAN RIFLES

Mexico adopted a 7mm Mauser in 1895; this rifle is almost identical to the Spanish 7mm Mauser Model 1893. In 1902, another 7mm Mauser was adopted; this rifle has the Model 98 type action. Except for the action, the M1895 and M1902 are almost identical. In 1912, another 7mm Mauser was purchased by Mexico. The Model 1912 is similar to the German Rifle Model 98 but has a tangent type rear sight and a longer handguard than does the Model 98.

7mm Model 1902 Mauser Rifle.

7mm Model 1936 Rifle.

Caliber .30 M1954 Rifle.

During the Mexican revolution, an extended period that began about 1910 and ended about 1920, Mexico secured arms from many sources. At this time, the Mexican Arisaka (the Japanese Type 38 rifle) was purchased from Japan. These are 7mm weapons with the Mexican escutcheon stamped on the receiver. They are relatively rare. In 1936, a Mauser of Mexican design was introduced. The 7mm Model 1936 externally resembles the Springfield Model 1930A1, having a Springfield type cocking piece, bands and stacking swivel, but the action is of the Mauser "short" type. The Model 1936 is of Mexican manufacture and is very well made.

The Model 1954 rifle is patterned on the Springfield Model 1903A3 but also uses the Mauser-type action, i.e., one-piece firing pin. The stock is made of laminated plywood. A carbine version of this weapon, which is chambered for the .30-06 cartridge, has also been reported.

THE MEXICAN MONDRAGON SEMIAUTOMATIC RIFLE

A number of semiautomatic rifles have also been produced in Mexico, the best known being the Mondragon, one of the earliest reasonably successful semiautomatic rifles of military pattern. It was invented by the Mexican General Mondragon in 1908. Manufacture of the weapon, which was a gas operated locked breech rifle, was undertaken in Switzerland by the Schweizerische Industrie Gesellschaft. When World War I broke out, these rifles were shipped to Germany. They did not stand up well in trench service but were among the earliest rifles issued to observers in aircraft as the Model 1915 before the introduction of the machine gun.

CHARACTERISTICS OF OBSOLETE MEXICAN RIFLES

	Model 1895	Model 1902	Model 1912	Model 1936	Model 1954
Caliber:	7mm.	7mm.	7mm.	7mm.	.30-06.
System of operation:	Turn bolt.	Turn bolt.	Turn bolt.	Turn bolt.	Turn bolt.
Length overall:	48.6 in.	49.2 in.	49.1 in.	42.9 in.	Approx 48 in.
Barrel length:	29.1 in.	29.1 in.	29.1 in.	19.29 in.	24 in.
Feed device:	5-round, non-detachable, staggered row box magazine				
Sights: Front:	Barley corn.	Barley corn.	Barley corn.	Hooded barley corn.	Hooded barley corn.
Rear:	Leaf.	Leaf.	Tangent	Tangent w/"V" notch.	Ramp type aperture.
Weight:	8.8 lb.	8.8 lb.	9 lb.	8.3 lb.	Approx 9 lb.
Muzzle velocity:	Approx 2300 f.p.s.	Approx 2300 f.p.s.	Approx 2300 f.p.s.	Approx 2600 f.p.s.	Approx 2800 f.p.s.

MEXICAN SUBMACHINE GUNS

Mexico has used various models of the caliber 45 Thompson submachine gun, and the noted Mexican small arms designer Rafael Mendoza developed a series of simple lightweight submachine guns during the 1950s. These weapons have been made in caliber .45, .38 super and 9mm Parabellum and have limited use in the Mexican service.

MEXICAN MACHINE GUNS

Mexico has used a number of foreign-developed machine guns and developed a number of her own. The Mexicans have used the 7mm Model 1911 and 1934 Madsen guns, the 7mm Model 1896 Hotchkiss gun, the 7mm Browning Model 1919 and 7mm Colt guns.

MEXICAN MENDOZA MACHINE GUNS

In the early 1930s Rafael Mendoza, then working at the Mexican National Arms factory, developed a gas operated light machine gun, which was standardized in the Mexican Army as the Model C-1934.

While the gun itself utilizes principles already established by the Hotchkiss and Lewis guns, it possesses a number of unique developments of its own. The Lewis type action is improved with the incorporation of a double cam slot, which helps to equalize the torque. Locking friction is materially reduced thereby.

While the gas cylinder is modeled somewhat after the Hotchkiss, the piston system of the Hotchkiss is not employed. The gas assembly is in the form of a cup-shaped projection on the barrel. Within it is housed a short piston. The operating stroke is very short, and at its limit the gases are dissipated openly into the air, thereby reversing the operation of the typical gas mechanism of this sort.

Mendoza developed a number of other machine guns, some of which have been produced in limited quantity. Early in World War II a modified version of the Model C-1934, chambered for the 7mm cartridge, was produced. This weapon was designed for use as a light machine gun on a bipod or as a heavy machine gun on a tripod. This weapon was apparently not introduced into Mexican service.

In 1945, Mendoza introduced another light machine gun chambered for the .30-06 cartridge. This weapon is basically an improved Model C-1934 and has the same type quick-change barrel as the Model C-1934, but has a different receiver configuration and a multi-perforated muzzle brake.

The caliber .30 RM-2 is another Mendoza design. It has been especially designed for easy and low cost manufacture.

While it does not have the quick removal barrel feature of the earlier model, it stresses a very simple, rapid and effective takedown. Pressing the release catch at the rear of the receiver allows the gun to be broken on its hinge much in the manner of the Belgian FN automatic rifle. The gas piston and slide with recoil spring and the few operating components may be withdrawn from the rear of the gun. The unusual Mendoza feature of reversible firing pin again appears in this new design but in improved form. In the event of firing pin breakage in the field (a relatively weak point in any machine gun), merely dropping the rear of the gun will allow the mechanism to be drawn back far enough to lift the bolt off the piston slide. The firing pin has a firing point at each end. If one should be broken, merely withdrawing and reversing the pin and reinstalling it in the bolt prepares the gun for firing again.

7mm Light Machine Gun Model C-1934.

Mexican Mendoza Model RM-2 Light Machine Gun, caliber .30-06.

CHARACTERISTICS OF MEXICAN MACHINE GUNS

	Model C1934	Model RM2
Caliber:	7mm.	.30-06.
System of operation:	gas, selective fire.	gas, selective fire.
Length overall:	46. in.	43.3 in.
Barrel length:	Approx 25 in.	24 in.
Feed device:	20-round, detachable box magazine.	
Sights: Front:	Hooded barley corn.	Hooded barley corn.
Rear:	Adjustable aperture.	Adjustable aperture.
Weight:	18.5 lb.	14.1 lb.
Muzzle velocity:	Approx 2700 f.p.s.	Approx 2750 f.p.s.
Cyclic rate:	400-500 r.p.m.	600 r.p.m.

33 Netherlands

SMALL ARMS IN SERVICE

Handguns:	9 × 19mm NATO FN Mle. 1935	Fabrique Nationale-made.
Submachine guns:	9 × 19mm NATO UZI	Israeli Military Industries-made.
Rifles:	7.62 × 51mm NATO FAL	Fabrique Nationale-made.
Machine guns:	7.62 × 51mm NATO MAG	Fabrique Nationale-made.
	12.7 × 99mm M2 HB	USA-made.

NETHERLANDS SERVICE PISTOLS*

The Netherlands Army adopted a 9.4mm service revolver in 1873, which was, with minor modifications, used as late as World War II, especially in the former Dutch East Indies. Prior to World War I, the Netherlands adopted the M1903 Browning 9mm Long pistol and in the early 1920s adopted the 9mm (.380 ACP) Browning M1922 pistol, which they called Pistool M25 No. 2. The Netherlands also used the 9mm Parabellum Luger, some of German pre-war manufacture similar to the P-08, and approximately 9,000 of Vickers (British) production usually called the Model 1920. Whether or not these particular pistols were made by or assembled by Vickers is a controversial question.

The 9 × 19mm NATO FN Hi-Power pistol was adopted by the Netherands after World War II.

NETHERLANDS SERVICE RIFLES

The Netherlands adopted the 6.5mm Mannlicher rimmed cartridge and rifle in 1895. The Netherlands Mannlicher is a turnbolt type, using the standard 5-round Mannlicher type clip, which is loaded in the weapon and drops out when the last round is chambered. The Dutch Mannlicher is basically the same as the Romanian Model 1893 Mannlicher.

Seven carbine versions of this weapon exist, including a model with folding bayonet for gendarmerie use. These carbines vary only in minor details, such as position of sling swivels and length of handguards. The Model 95 No. 1 O.M. carbine has a sporter type stock and uses an old type triangular bayonet. The Model 95 No. 1 N.M., Model 95 No. 3 O.M., Model 95 No. 3 N.M., Model 95 No. 4 O.M. and Model 95 No. 4 N.M. all use knife type bayonets. An unusual feature of the carbines is that most models have a wooden piece covering the left side of the magazine. This piece is glued and doweled to the stock. The No. 3 O.M. and No. 3 N.M. carbines have handguards that extend beyond the stock, almost to the muzzle. Many of these 6.5mm Mannlicher carbines were converted to caliber .303 British by the Indonesians in the early 1950s.

6.5mm Model 95.

6.5mm Model 95, No. 1 N.M. Carbine (above); 6.5mm Model 95, No 3 N.M.Carbine (below).

*For additional details see *Handguns of the World.*

6.5mm Model 95, No. 4 N.M. Carbine.

FN 7.62mm NATO Light Automatic Rifle as made for the Netherlands.

The Dutch submitted the MN-1 (Galil) in 5.56 × 45mm as their candidate in the NATO small arms trials of the mid-1970s.

These weapons are no longer in service in the Netherlands.

In 1917, a version of the Model 95 rifle chambered for the 7.92mm rimmed cartridge was introduced. This weapon was made in very limited quantity.

In 1940, the Netherlands government purchased quantities of the .30-06 Johnson semiautomatic rifle for the East Indies Forces and the Royal Netherlands Navy. These rifles are called the Model 1941. The United States and Great Britain supplied the Netherlands with caliber .30 M1 rifles and caliber .303 Lee Enfield rifles. The Netherlands also purchased a quantity of caliber .30 FN bolt action carbines for police use. About 1960, the 7.62mm NATO FN FAL rifle was adopted. The Netherlands version of the FAL has metal handguards, a folding bipod and protecting ears on the rear sight. The Dutch have also made several modifications in the bolt of the FAL since adopting the weapon.

The government arsenal Artillerie Inrichtingen manufactured the 7.62mm NATO AR-10 rifle in limited quantities on a contract basis. This weapon was never adopted by the Dutch Army.

NETHERLANDS MACHINE GUNS

The Dutch adopted the Schwarzlose machine gun in 7.92 mm rimmed in 1908. This weapon is basically the same as the Austrian 07/12 Schwarzlose and was used by the Dutch Army during World War II. The Dutch were also one of the largest users of the Madsen, having used the Model 1919, 1923, 1926, 1927, 1934, 1938 and 1939—all chambered for the 6.5mm cartridge. The Dutch also adopted the Lewis gun in 6.5mm in 1920. These weapons are not as frequently encountered as are the Dutch Madsens. The Vickers Model 1918 was also purchased in 7.92mm.

After World War II, the Dutch received United States Browning guns and British Bren guns. After extensive tests, the 7.62mm NATO FN MAG machine gun was adopted in the early sixties.

8mm Model 8 Machine Gun.

6.5mm Model 20, Lewis gun.

34 New Zealand

New Zealand forces have always been equipped with British pattern arms, made in the UK, in Australia (Lithgow) or locally. The pistols used have been the caliber .455 Webley Marks V and VI and the Enfield caliber .38 revolvers (No. 2 Marks 1, 1* and 1**). Rifles used in the past have been the No. 1 Mark III and III*. The current rifle is the 7.62mm NATO L1A1 (FN FAL) as made at Lithgow. Submachine guns have been produced in New Zealand. Approximately 10,000 Sten and 2000 Charlton machine guns were made in New Zealand during World War II. The Charlton is a native design.

It should be noted that, in the area of submachine guns and machine guns, New Zealand has conformed more closely to the United Kingdom than to Australia. New Zealand has adopted the 9mm Parabellum L2A3 (Sterling submachine gun)—the standard submachine gun in the United Kingdom—and the FN-developed 7.62mm NATO MAG general purpose machine gun. The MAG, in modified form, is the L7A1 general purpose machine gun in the United Kingdom, and the version of the MAG used in New Zealand is probably the same as the British gun.

THE CHARLTON .303 LIGHT MACHINE GUN AND RIFLE

Philip Charlton, a New Zealander, developed a design for converting the Lee Enfield rifle into a selective fire weapon. He was issued a patent in 1941 and, due to the shortage of automatic weapons, work commenced on the conversion of old long Lee Enfields on hand into Charlton machine guns. A plant was established in Hastings, New Zealand, to carry out this conversion. Approximately 2000 long Lee Enfields had been converted when production ceased. The Charlton weighed 16.5 pounds and had a cyclic rate of fire of about 700 rounds per minute. It is a gas operated weapon fed with a 30-round magazine but could use the standard 10-round Lee Enfield magazine as well.

Essentially the conversion consisted of fitting a cylinder on the right side of the weapon originating at a point approximately midpoint of the barrel and extending beyond the rear of the action. This cylinder acted as a gas cylinder in its forward part and a piston tube for the remainder. The piston rod is attached to a slide which has an inclined cam that operates on the bolt handle stub pulling it to the rear and opening the action. A spring attached to the lower portion of the piston is mounted in a cylinder below the gas cylinder and returns the slide and piston forward. The standard Lee trigger mechanism has been modified by the addition of a disconnector system. A selector is mounted behind the trigger and engages the trigger. Setting it on safe prevents rearward movement of the trigger, setting it on "R" allows the trigger to rotate far enough to the rear to allow single shots to be fired and setting it on "A" lets the trigger come fully to the rear producing full automatic fire.

A hinged cover is placed over the action to protect it from dirt, and a circular fin assembly, to assist in cooling the barrel, is placed over the rear half of the barrel. The semi-pistol grip of the Lee Enfield is cut away, and a full pistol grip is mounted to the rear of the trigger guard. A heavy flash hider with protected front sight is mounted on the front end of the barrel. Australia also converted several thousand Lee Enfields to the Charlton design. The work in Australia was done by the Electrolux Company.

An earlier effort of Charlton's was the conversion of the .303 Rifle No. 1 Mark 3* to a semiautomatic rifle. This design was tested by the United Kingdom but not adopted.

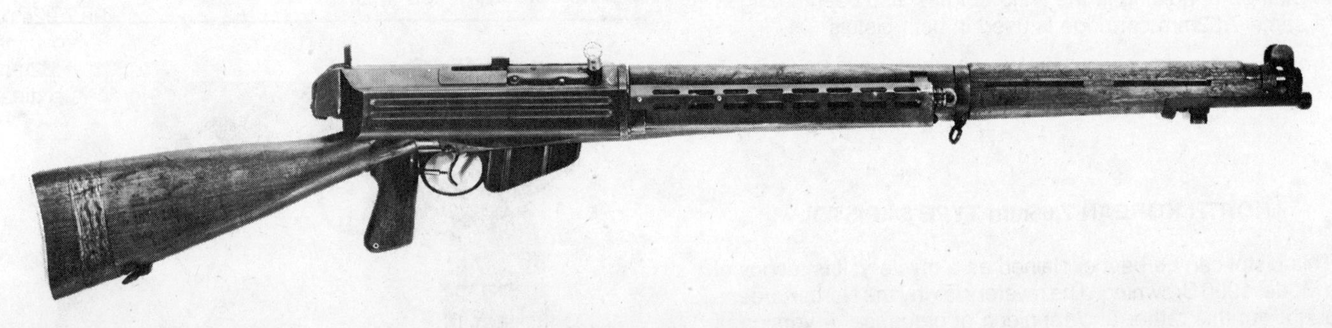

Charlton conversion of .303 Rifle No. 1 SMLE Mark 3*.

35 North Korea

SMALL ARMS IN SERVICE

Handguns:	7.62 × 25mm Type 68	Copy of Soviet TT33.
	7.65 × 17mm Type 64	Copy of FN Browning Mle. 1900.
	7.62 × 17mm Type 64 new model	New blowback pistol of Korean design. Receiver marked 7.62mm, but it is definitely 7.65 × 17mm.
Submachine guns:	7.62 × 25mm Type 49	Copy of Soviet PPSh41.
Rifles:	7.62 × 39mm Type 58	Copy of Soviet AK47.
	7.62 × 39mm Type 63	Copy of Soviet SKS.
	7.62 × 39mm Type 68	Copy of Soviet AKM; both wood and folding metal stocks.
	7.62 × 54mm R Type 1891/30	Soviet sniper rifle.
Machine guns:	7.62 × 54mm R Type 64	Soviet RP46.
	7.62 × 39mm RPD	Soviet RPD.
	7.62 × 54mm R PK, PKB, PKS	Soviet PK series.
	7.62 × 54mm R SGM	Soviet SGM, being supplanted by PK family.
	12.7 × 108mm M38/46	Soviet M38/46.

NORTH KOREAN HANDGUNS

NORTH KOREAN 7.62mm TYPE 68 PISTOL

The Type 68 is a North Korean variation of the Soviet Tokarev TT M1933 pistol. The Type 68 is shorter than the Tokarev and is considerably different in internal construction. The Tokarev, like the Colt M1911, uses a link to pull down the barrel in unlocking. The Type 68 uses the cam type barrel lug as used on the Hi-Power Browning. The magazine catch on the Type 68 has been relocated to the bottom rear of the grip, as opposed to the button type catch on the upper receiver of the Tokarev. The method of holding in the firing pin has also been changed. The same 7.62mm cartridge is used in both pistols.

CHARACTERISTICS OF THE TYPE 68 PISTOL

Caliber: 7.62mm
System of operation: Recoil operated, semiautomatic only
Weight: 1.75 lbs.
Length overall: 7.3 in.
Barrel length: 4.25 in.
Feed device: 8 round, in line, detachable box magazine.
Sights: **Front:** Blade
Rear: Square notch
Muzzle velocity: 1295 f.p.s.

NORTH KOREAN 7.65mm TYPE 64 PISTOL

This pistol can be best explained as a mystery; it is a copy of the Model 1900 Browning. The mystery is why the North Koreans brought out this rather ancient piece of ordnance. A version of this pistol with silencer has a shortened slide and fine thread on the barrel of the attachment of a silencer.

North Korean 7.65mm Type 64 Pistol.

Comparative photo showing differences among North Korean Type 68, PRC Type 51/54 and Soviet Tokarev Pistols.

NORTH KOREAN RIFLES

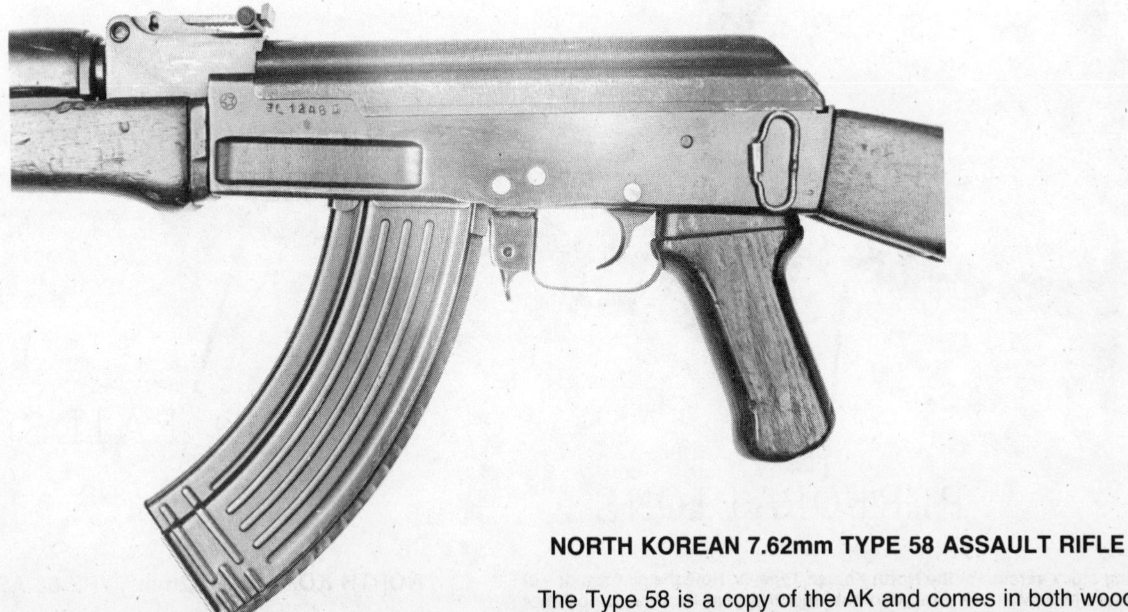

Close-up of the markings on the receiver of the Type 58 copy of the AK47.

NORTH KOREAN 7.62mm TYPE 58 ASSAULT RIFLE

The Type 58 is a copy of the AK and comes in both wooden stocked and folding steel stocked versions. It does not appear to be as finely finished as AKs manufactured in other countries.

Wood-stocked version of the North Korean Type 58.

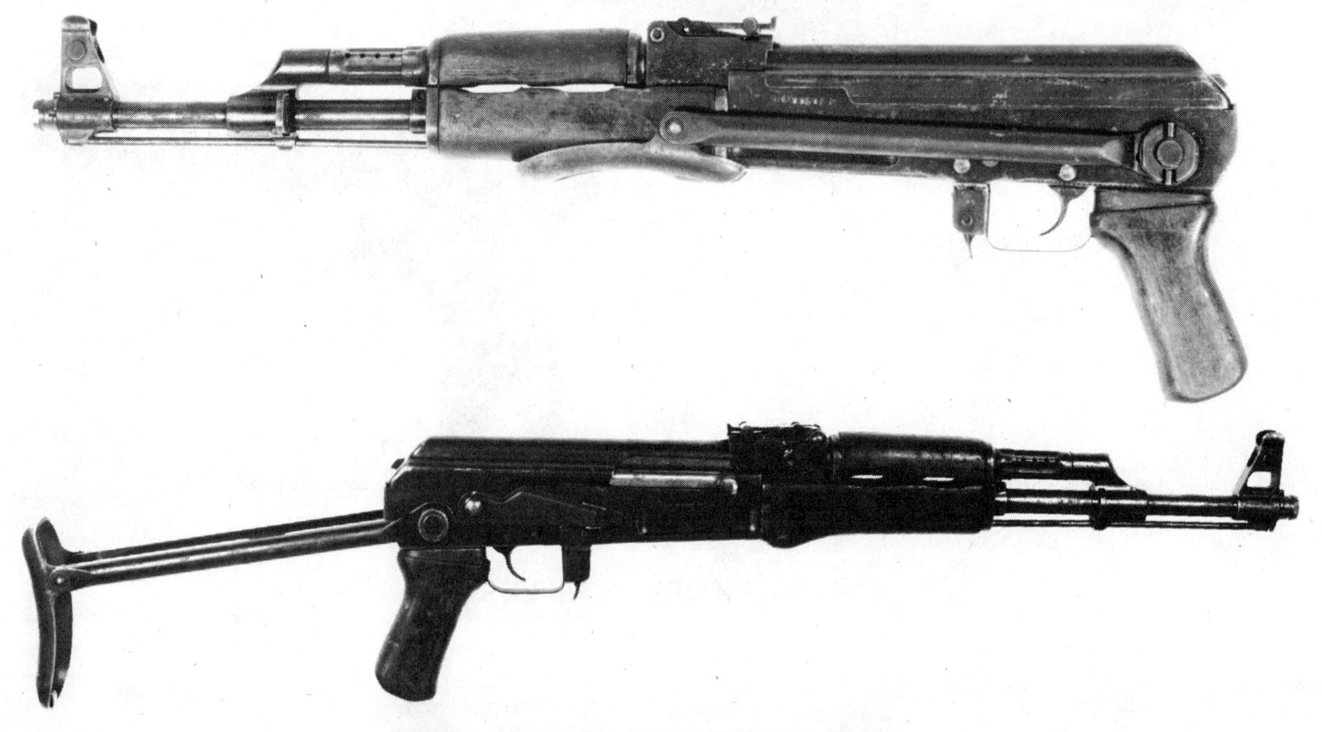

North Korean 7.62mm Type 58 Assault Rifle without magazine.

Folding stock version of the North Korean Type 68. Note the perforations in the stock to lighten the rifle. Also note that the front handguard does not have the gripping rails common to other AKMs.

NORTH KOREAN 7.62mm TYPE 68 ASSAULT RIFLE

This weapon is a copy of the Soviet AKM. It does not have the rate reducer found on the AKM. It uses the standard 7.62 × 39mm "intermediate" sized cartridge.

36 Norway

SMALL ARMS IN SERVICE

Handguns:	9 × 19mm NATO M38	Norwegian-licensed copy of Walther P1.
	11.43 × 23mm M1914 & .45 M1914	Norwegian-licensed copies of Colt M1911.
Submachine guns:	9 × 19mm Parabellum Maskin M40	Copy of German MP40; pre-1945 manufacture.
Rifles:	7.62 × 51mm NATO Gevaer, Automatisk AG3	Norwegian-licensed copy of Heckler & Koch G3.
Machine guns:	7.62 × 51mm NATO Maskin Gevaer MG3	Rheinmetall-made MG3.
	12.7 × 99mm M2 HB	US-made.

NORWEGIAN SERVICE PISTOLS*

Norway adopted a single-action Nagant revolver in 1883, but apparently only 794 of the weapons were procured. In 1893, the 7.5mm Model 1893 Nagant revolver was adopted. This weapon, which is basically the same as the Russian 7.62mm Model 1895 Nagant, was the standard pistol until the adoption of the caliber .45 Model 1914 pistol. The Model 1914 is basically the same as the US caliber .45 M1911 Colt pistol except for the external shape of the slide stop.

The first 300 caliber .45 pistols were procured by Norway from Colt and are unmodified Model 1911s. All of the Model 1914 arms were made at the Norwegian government arsenal at Kongsberg. Since World War II, Norway has obtained quantities of caliber .45 Model 1911A1 automatics from the United States.

In the period after World War II, considerable quantities of 9mm P-08 Luger and P-38 pistols were used by the Norwegian forces.

Norwegian caliber .45 Model 1914 Automatic Pistol.

NORWEGIAN RIFLES AND CARBINES

Norway adopted the Krag Jorgensen rifle, chambered for the 6.5 × 55mm rimless Mauser cartridge, in 1894, and this rifle and its carbine versions were used by the Norwegian Army until the majority of the weapons were lost to the Germans in World War II. Some Model 1894 rifles were made at Steyr, but the majority were made by Kongsberg. A few were made during the German occupation; these bear German *Waffenamt* inspection stamps.

THE NORWEGIAN KRAG

The Norwegian Krag is generally similar to the US caliber .30 Krag but does not have a cutoff. As with the Krag, there is a single frontally-mounted locking lug. It has been stated many times that the Krag action is not suitable for high-pressure cartridges, but the Norwegian Krag was made after World War II (for a limited period of time) chambered for the 7.92 × 57mm cartridge—hardly a low pressured cartridge. This is mainly a matter of metallurgy—Norwegian Krags produced since World War I are obviously made of steel alloys that are better than those used in the United States Krag, which was made from 1892-1902 when metallurgy was not too precise.

Distinctive Details of Norwegian Rifles and Carbines

The various models of rifles and carbines can be distinguished by the following characteristics:

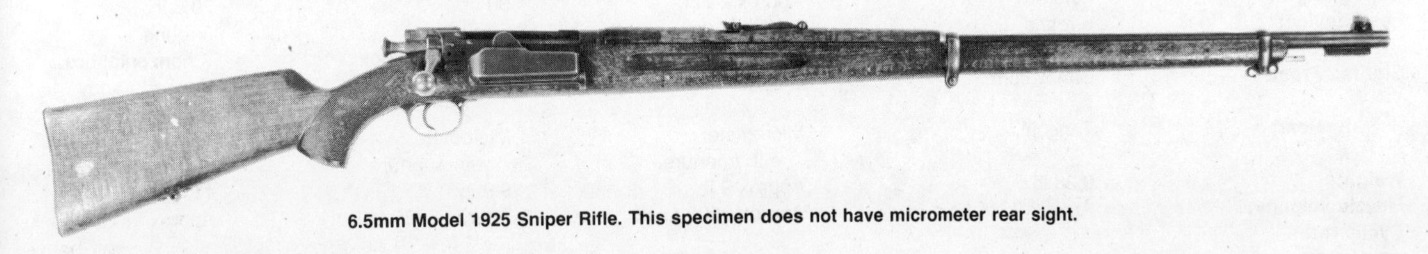

6.5mm Model 1925 Sniper Rifle. This specimen does not have micrometer rear sight.

*For additional details see *Handguns of the World*.

Rifle Model 1894. Stock with pistol grip, half-length cleaning rod mounted in forward part of stock; bayonet lug mounted under the barrel and a half-length handguard is fitted.

Sniper Rifle Model 1923. Full-length stock with full checked pistol grip; full-length handguard, wide upper band/nose cap with bayonet lug, micrometer type rear sight. The rifle is marked M/1894.

Sniper Rifle Model 1925. Basically the same as the Rifle Model 1894, but has checked full pistol grip, micrometer type rear sight, marked M/25.

Sniper Rifle Model 1930. Has sporter type stock with checked full pistol grip, heavy barrel, no bayonet lug, micrometer rear sight, marked M/1894/30.

Carbine Model 1895. Sporter type stock, no bayonet lug, similar in appearance to the United States Krag carbine.

Carbine Model 1897. Same as the Model 1895 carbine, but butt swivel is positioned further toward the rear of the butt.

Carbine Model 1904. Full stock with pistol grip and full length handguard, no bayonet lug.

Carbine Model 1907. Basically the same as the Model 1904, but sling swivels are positioned on rear band and on the butt.

6.5mm Carbine Model 1912. Stocked to the muzzle with full-length handguard, similar to the Sniper Rifle Model 1923. Has bayonet lug mounted on combination upper band nose cap.

After World War II, the Norwegian Army had few Norwegian Krags but many German 7.92 mm Mauser Kar98ks. American and British rifles were also supplied. In 1959, a caliber .30-06 heavy barrel modification of the Mauser with sporter-type stock and micrometer sight was produced for the Norwegian National Rifle Association. A few Krags were produced after the war, but costs were prohibitively high, and there was no intention to equip the armed forces with the Krag, since it can hardly be considered a modern rifle.

In 1964, the Norwegians adopted the West German 7.62mm NATO G3 rifle, and that rifle is now in production at Kongsberg Vapenfabrik. The Norwegians modified the G3 slightly; in early models the bolt operating handle was removed, and the bolt was pulled to the rear by means of a finger hole, as in the US M3A1 submachine gun. A semiautomatic sniper model with telescope has also been adopted. Kongsberg currently manufactures the standard G3 for domestic consumption and sale to West Germany.

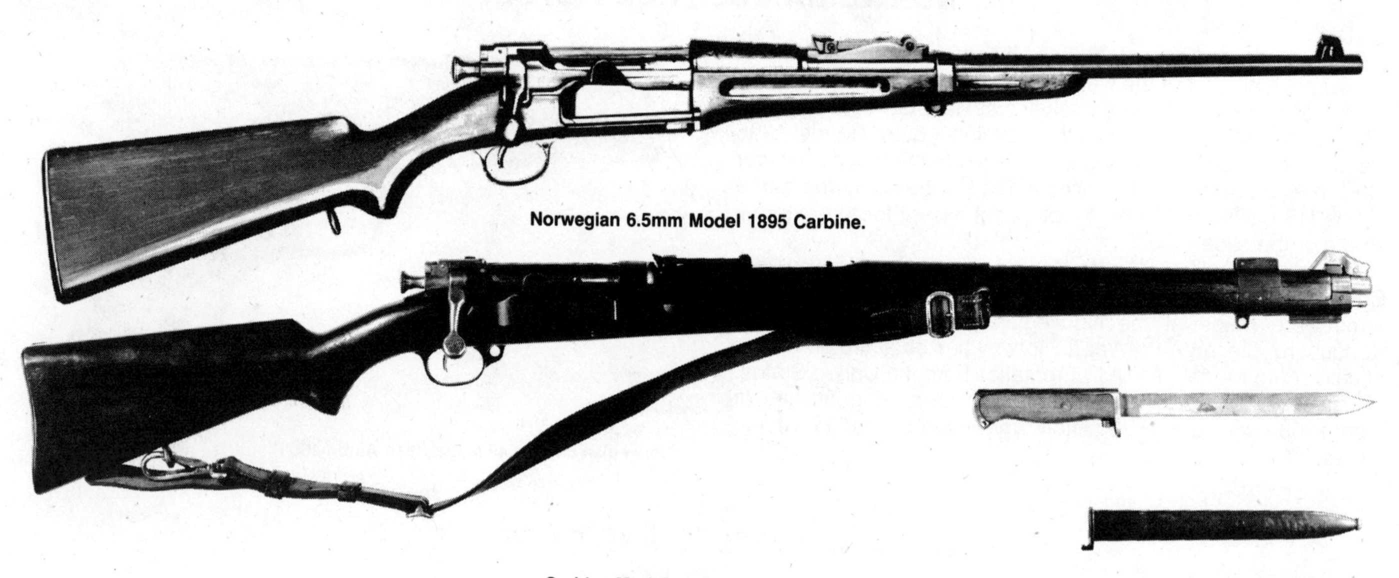

Norwegian 6.5mm Model 1895 Carbine.

Carbine Model 1912, caliber 6.5mm.

CHARACTERISTICS OF NORWEGIAN RIFLES AND CARBINES

	Rifle M1894	Sniper Rifle Model 1923	Sniper Rifle Model 1925	Sniper Rifle Model 1930
Caliber:	6.5 × 55mm.	6.5mm × 55mm.	6.5mm × 55mm.	6.5 × 55mm.
System of operation:	Turn bolt.	Turn bolt.	Turn bolt.	Turn bolt.
Length overall:	49.9 in.	Approx 44 in.	49.7 in.	48 in.
Barrel length:	29.9 in.	Approx 24 in.	30 in.	29.5 in.
Feed device:	5-round horizontal box.	5-round horizontal box.	5-round horizontal box.	5-round horizontal box.
Sights: Front:	Barley corn.	Barley corn.	Barley corn.	Hooded barley corn.
Rear:	Tangent.	Micrometer with aperture.	Micrometer with aperture.	Micrometer with aperture.
Weight:	9.38 lb.	Approx 9 lb.	9.9 lb.	11.46 lb.
Muzzle velocity:	Approx 2625 f.p.s.	Approx 2600 f.p.s.	Approx 2625 f.p.s.	Approx 2625 f.p.s.
Cyclic rate:	—	—	—	—

CHARACTERISTICS OF NORWEGIAN RIFLES AND CARBINES (Cont'd)

	Carbine M1895	Carbine M1904	Carbine M1912	G-3
Caliber:	6.5 × 55mm.	6.5 × 55mm.	6.5 × 55mm.	7.62mm NATO.
System of operation:	Turn bolt.	Turn bolt.	Turn bolt.	Gas, selective fire.
Length overall:	40 in.	40 in.	43.6 in.	40.2 in.
Barrel length:	20.5 in.	20.5 in.	24 in.	17.7 in.
Feed device:	5-round horizontal box.	5-round horizontal box.	5-round horizontal box.	20-round staggered row box magazine.
Sights:				
Front:	Barley corn.	Barley corn.	Barley corn.	Hooded post.
Rear:	Tangent.	Tangent.	Tangent.	Aperture.
Weight:	7.5 lb.	8.4 lb.	8.8 lb.	
Muzzle velocity:	Approx 2575 f.p.s.	Approx 2575 f.p.s.	2600 f.p.s.	2624 f.p.s.
Cyclic rate:	—	—	—	500-600 r.p.m.

Two current models of the 7.62 × 51mm NATO AG3 manufactured at the Kongsberg Vapenfabrik in Norway.

NORWEGIAN MACHINE GUNS

Norway adopted the Hotchkiss gun chambered for the 6.5mm cartridge in 1911 and also used the 6.5mm Model 1914 and 1918 Madsen guns. The Browning water cooled machine gun was also used and called the Model 29. This weapon is similar to the Colt M38B machine gun.

After World War II, Norway obtained American caliber .30 Browning guns and British machine guns and also used the 7.92mm MG34 and MG42 machine guns. The 7.62mm NATO MG42/59 machine gun is now standard in Norway, where it is called the LMG3.

37 Poland

SMALL ARMS IN SERVICE

Handguns:	9 × 18mm Wz63 (machine pistol)	Polish design.
	9 × 18mm P61	Polish design.
	9 × 18mm P65	Polish version of Soviet PM (Pistolet Markarov).
Rifles:	7.62 × 39mm PMK	Polish version of Soviet AK47. Reserve units only.
	7.62 × 39mm PMK-DGN-60	PMK used for grenade launching.
	7.62 × 39mm PMKM	Polish version of AKM.
	7.62 × 54mm R SVD	Polish copy of Dragunov sniping rifle.
Machine guns:	7.62 × 39mm RPD	Soviet-made.
	7.62 × 39mm RPK	Polish copy of Soviet RPK.
	7.62 × 54mmR PK, PKB, PKS, & PKT	Polish copies of Soviet PKs.
	12.7 × 108mm DShK38/46	Soviet-made.

POLISH PISTOLS*

THE POLISH 9mm MODEL 35 ("RADOM") PISTOL

A modification of the Colt Browning locked-breech pistol was designed by the Poles at the Fabryka Broni Radom, a government arms plant. The design is credited to P. Wilniewczyc and J. Skrzypinski. This 9mm Parabellum weapon, called the Vis Model 35, or more popularly the Radom, was the Polish service pistol at the time of WW II, and the Germans continued its manufacture after they occupied Poland. The Model 35 Radom locks and unlocks in a manner quite similar to that of the Browning FN Hi-Power pistol, in that it has a nose forged on the rear underside of the barrel. This serves as a cam to pull the barrel down out of the locked position, rather than using the barrel link as with the US Colt Model 1911 pistol.

As manufactured for the Polish forces, the weapon has the firing pin retracting device mounted on the left rear of the slide and a slide lock, which appears to be a safety, mounted on the left rear side of the receiver. Most of the pistols manufactured during the German occupation do not have the slide lock. The firing pin retracting device is used to lower the hammer on a loaded chamber. This weapon, although it has a grip safety, does not have the conventional Colt-Browning manual-safety catch. The slide lock is used to lock the slide to the rear to assist in field stripping. Pistols manufactured prior to the German occupation are very well made and are marked on the slide with the Polish eagle and Polish markings. The Model 35s manufactured for the Germans are of much rougher construction and bear the German *Waffenamt* proof marks. The Radom Model 35 is no longer in service in any Army.

Manufacture of the 7.62mm Tokarev Model TT M1933 was begun in Poland after World War II; this pistol is being replaced by the 9mm Makarov (PM) pistol and the 9mm Stechkin (APS) machine pistol. Characteristics of these weapons will be found in the chapter on the USSR.

CHARACTERISTICS OF MODEL 1935 PISTOL

Caliber: 9mm Parabellum.
System of operation: Recoil, semiautomatic.
Length overall: 7.8 in.
Barrel length: 4.7 in.
Feed device: 8-round, in line, detachable box magazine.
Sights: Front: Blade.
Rear: "V" notch.
Weight: 2.25 lbs.
Muzzle velocity: Approx. 1150 f.p.s.

*For additional details see *Handguns of the World*.

9mm Model 35 Pistol manufactured for the Polish Army.

9mm Model 35 Pistol manufactured for the Germans without slide lock.

THE POLISH 9mm MODEL 64 PISTOL

This double action, blowback operated pistol is smaller in size than the Soviet Makarov pistol but is chambered for the same 9mm cartridge. In some respects, it seems to resemble the Walther PPK more than it does the Makarov, as for example the trigger bar linkage system and the internally mounted slide stop. The system, however, is actually different than either the Makarov or the Walther.

Polish 9mm Model 64 Pistol

CHARACTERISTICS OF 9MM MODEL 64 PISTOL

Caliber: 9mm Makarov
System of operation: Blowback, semi-automatic only
Length overall: 6.1 in.
Barrel length: 3.3 in.
Feed device: 8-round, in-line, detachable box magazine
Sights: Front: Blade
Rear: Notch
Weight: 1.5 lbs.
Muzzle velocity: 1017 f.p.s.

The Model 64 is disassembled in a manner similar to that of the Walther PP and PPK. The magazine catch is at the bottom rear of the grip, and the safety catch is mounted on the slide and is pushed down to engage. In this position, it blocks the firing pin.

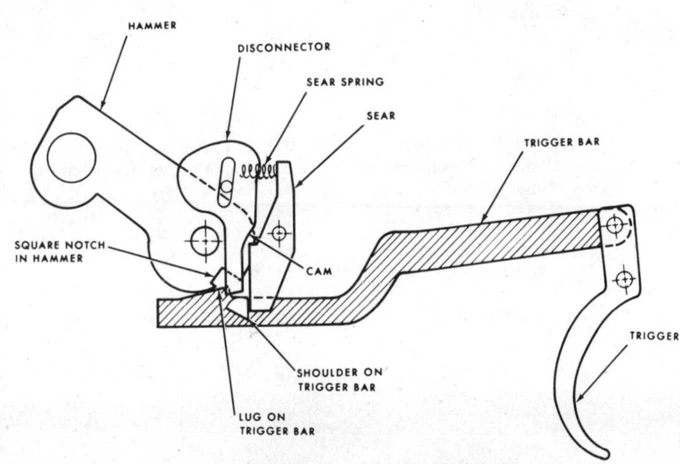

Polish 9mm Model 64 trigger linkage.

Polish 9mm Model 64 field stripped.

POLISH 9mm MODEL 63 MACHINE PISTOL

This weapon is similar to the Soviet 9mm Stechkin and the Czech 7.65mm M61 "Skorpion" machine pistols in concept. It is a selective fire weapon, which can be used as a pistol or as a shoulder weapon. It is seemingly intended for issue to junior officers in combat units and to armored troops.

CHARACTERISTICS OF 9MM MODEL 63 MACHINE PISTOL

Caliber: 9mm Makarov
System of operation: Blowback, automatic only
Length overall: w/folded stock—13.1 in.
Feed device: 15- and 25-round, detachable box magazines
Sights: Front: Blade.
Rear: "L" type.
Weight: 3.96 lbs.
Muzzle velocity: 1065 f.p.s.
Cyclic rate: 600 r.p.m.

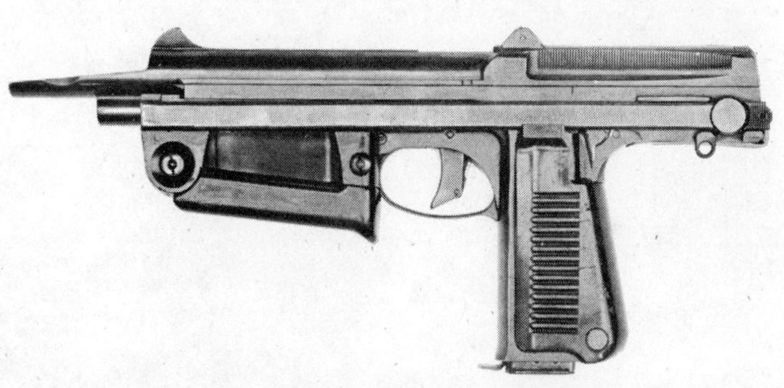

Polish 9mm Model 63 Machine Pistol.

This weapon has a sliding type metal shoulder stock used in conjunction with a vertical fore grip. When the stock is in the retracted position and the vertical fore grip is pushed up, it extends beyond the muzzle of the weapon; the weapon can be carried in a holster and fired with one hand. The butt plate portion of the shoulder stock folds up under the rear of the receiver.

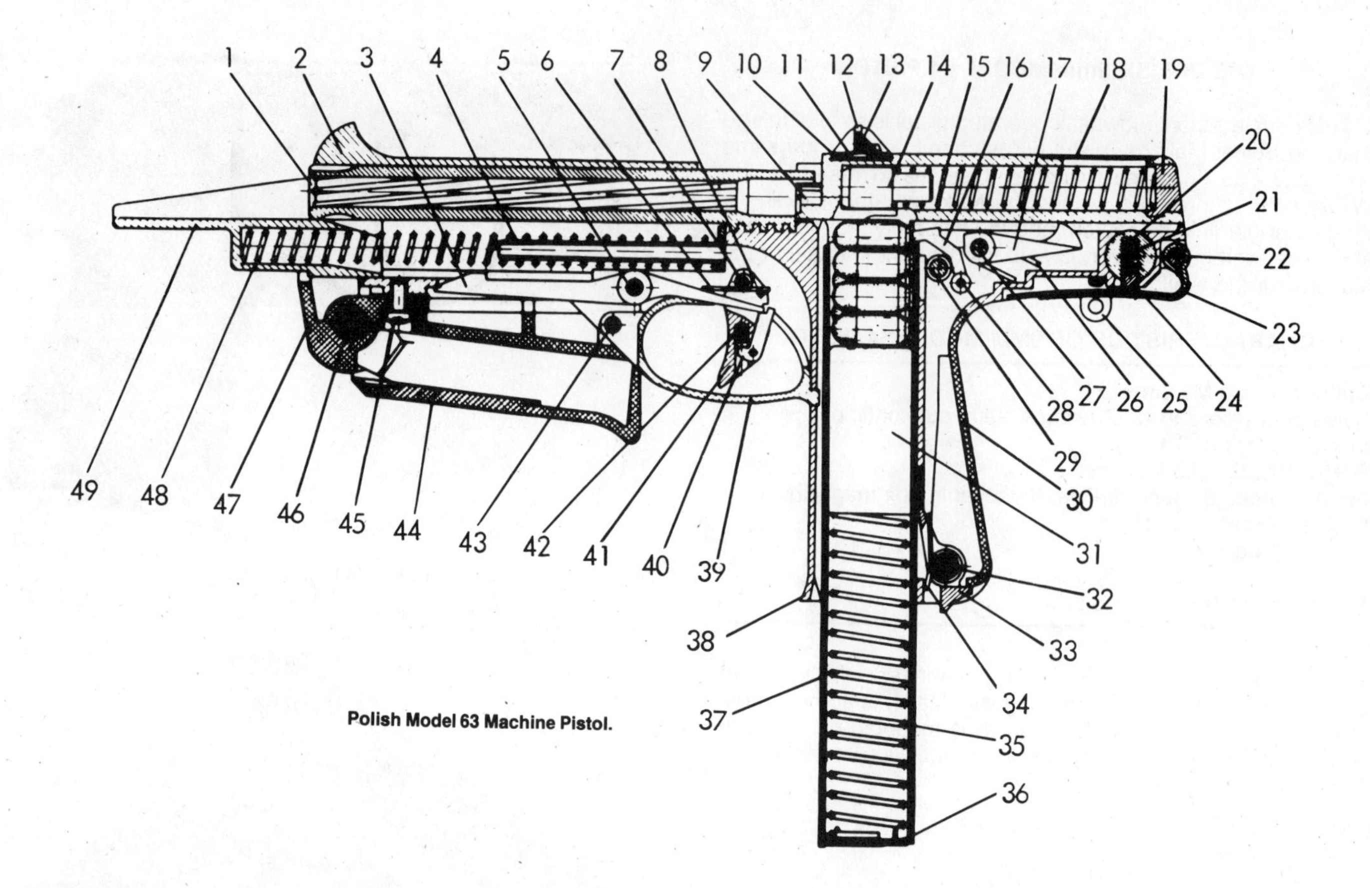

Polish Model 63 Machine Pistol.

1. Barrel
2. Front sight
3. Trigger lever
4. Recoil-spring guide
5. Trigger-lever pin
6. Trigger spring
7.,8. Trigger pins
9. Firing pin
10. Sight-leaf
11. Sight spring
12. Rear sight
13. Rear sight pin
14. Retarder
15. Retard spring
16. Slide stop
17. Retarder lever
18. Bearing latch
19. Bearing
20. Grip mount
21. Butt-latch
22. Butt-plate pin
23. Butt plate
24. Butt latch
25. Lanyard loop
26. Retarder-lever spring
27. Retarder-lever pin
28. Slide-stop axis
29. Safety-lock
30. Stock
31. Magazine follower
32. Magazine-catch pin
33. Magazine catch
34. Magazine-catch spring
35. Magazine spring
36. Magazine cover
37. Magazine body
38. Pistol grip
39. Trigger guard
40. Trigger-catch lever
41. Trigger catch
42. Trigger-catch-lever spring
43. Back screw
44. Grip
45. Front screw
46. Grip catch
47. Frame
48. Recoil spring
49. Compensator

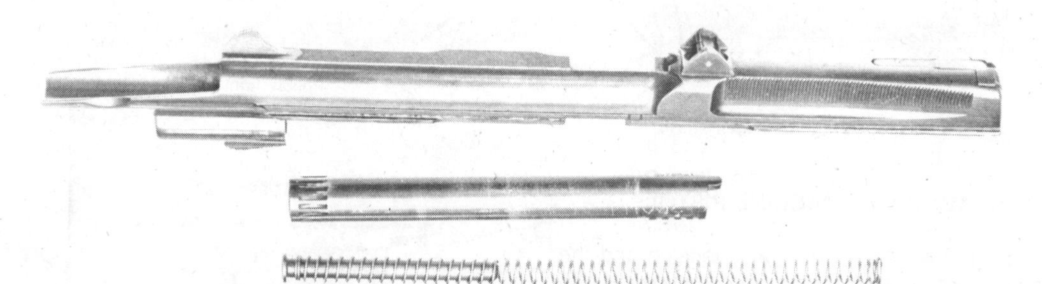

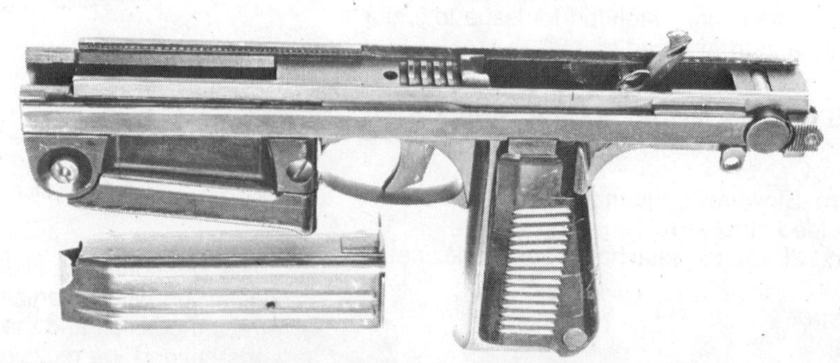

Field-stripped Wz 63 with a view of the Browning-type locking system on the barrel.

The 9 × 18mm Wz 63.

POLISH SUBMACHINE GUNS

The Poles were supplied with the Soviet 7.62mm PPSh M1941 and PPS M1943 submachine guns. They developed a variation of the PPS M1943, which was manufactured in Poland.

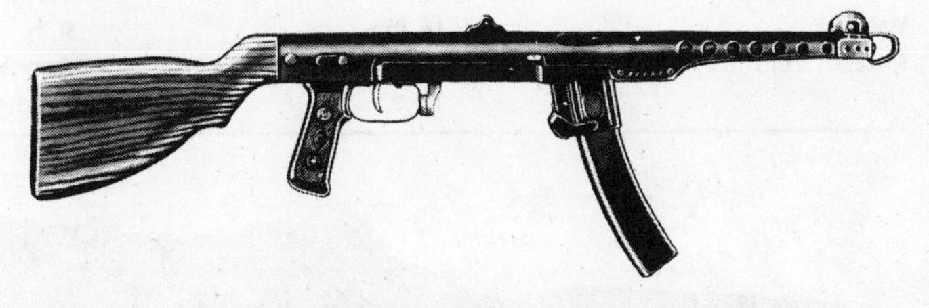

Polish 7.62 M1943/52 Submachine Gun.

THE POLISH 7.62mm M1943/52 SUBMACHINE GUN

The M1943/52 submachine gun is a modification of the Soviet PPS M1943. Its characteristics are basically the same as those of the Soviet weapon, except that it has a wooden stock and is reportedly capable of semiautomatic as well as automatic fire.

Special Note on the Polish M1943/52

Loading, firing, functioning and disassembly of this weapon are the same as for the Soviet 7.62mm submachine gun PPS M1943.

CHARACTERISTICS OF M1943/52 SUBMACHINE GUN

System of operation: Blowback, selective fire.
Length, overall: 32.72 in.
Barrel length: 9.45 in.
Feed device: 35-round, detachable staggered box magazine.
Sights: **Front:** Flat-topped post, adjustable for elevation and windage.
Rear: L-type, open V-notch with setting for 100 and 200 meters.
Muzzle velocity: 1640 f.p.s. w/Soviet 7.62mm Type P pistol cartridge.
Cyclic rate: 600 r.p.m.
Weight: 8 lbs.

POLISH RIFLES AND CARBINES

When Poland gained her freedom after World War I, stocks of Russian 7.62mm M1891 rifles and carbines and 7.92mm Mauser 98 carbines and rifles were available to her.

POLISH-PRODUCED RIFLES

The first rifle produced by the Poles contained features of both Russian and German design. The Polish 7.92mm Model 91/98/25 rifle has the Russian Mosin Nagant action but Mauser-type bands and fittings. The bolt head has been modified to use the 7.92mm cartridge.

The manufacture of Mauser rifles and carbines started at the Warsaw Arsenal soon after World War I. The Polish 7.92mm Karabin 98a is basically the same as the later German rifle 98. Similarly, the Polish Karabinek 98 is basically the same as the German Kar 98a.

The 7.92 mm Karabin 29 is a variant of the Czech Model 24, differing from the Czech weapon mainly in location of the sling swivels and the front sight.

After World War II, the Poles were initially equipped with various models of the Soviet 7.62mm Mosin Nagant rifle and carbine. The Poles now produce the Soviet-designed 7.62mm AKM assault rifle. The characteristics of these weapons are covered in the chapter on the USSR.

7.92mm Model 91/98/25 Rifle.

CHARACTERISTICS OF POLISH RIFLES

	Rifle 91/98/25	Rifle 98a	Carbine 98	Rifle 29
Caliber:	7.92mm			
System of operation:	Turn bolt			
Length overall:	43.3 in.	49.2 in.	43.3 in.	43.4 in.
Barrel length:	23.6 in.	29.1 in.	23.6 in.	23.6 in.
Feed device:	5-round in-line, non-detachable box magazine.	5-round staggered row non-detachable box magazine.	5-round staggered row non-detachable box magazine.	5-round staggered row non-detachable box magazine.
Sights: Front:	Barley corn.	Barley corn.	Barley corn with protecting ears.	Barley corn with protecting ears.
Rear:	Leaf.	Tangent.	Tangent.	Tangent.
Weight:	8.16 lb.	9 lb.	8 lb.	9 lb.
Muzzle velocity:	Approx 2470 f.p.s.	Approx 2570 f.p.s.	Approx 2470 f.p.s.	Approx 2470 f.p.s.

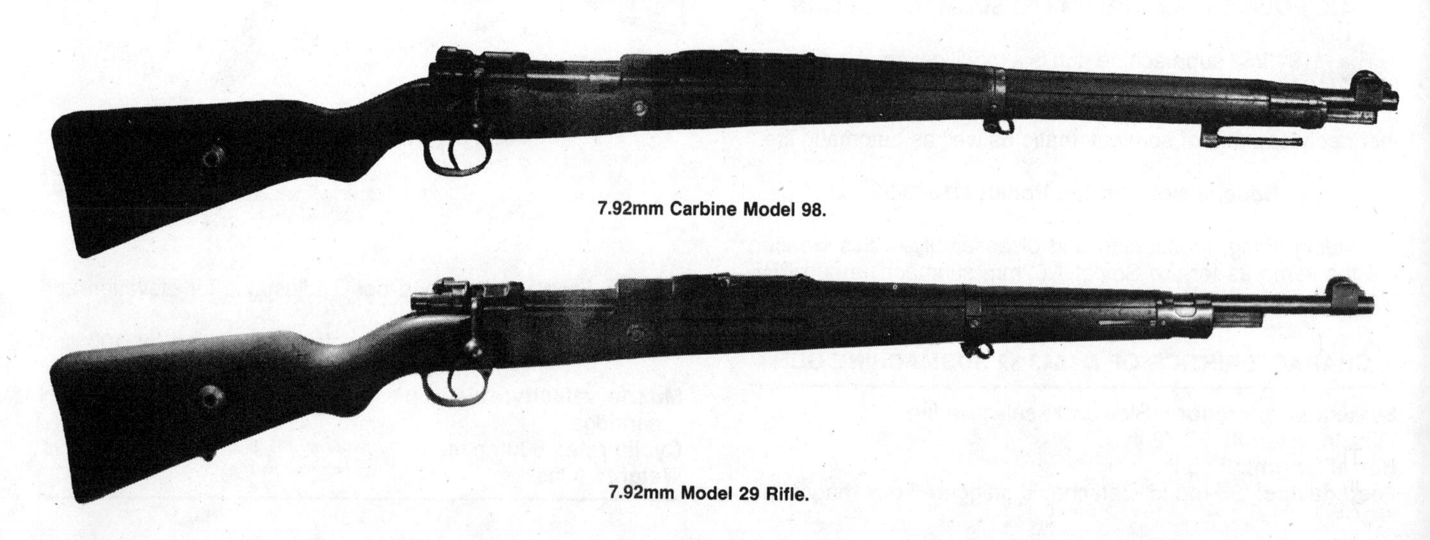

7.92mm Carbine Model 98.

7.92mm Model 29 Rifle.

POLISH 7.62mm AK ASSAULT RIFLE WITH GRENADE LAUNCHER

This weapon is a modification of the AK made in Poland fitted with a rifle grenade launcher. It has been called the PMK-DGN and also the KbKg Model 1960.

The grenade launcher, which is not the same size as those used in the NATO countries, is screwed onto the threaded end of the barrel and locked in place by the spring loaded plunger mounted in front sight base of the AK. A valve is fitted into the right side of the gas cylinder to allow shutting off the gas from the gas port of the barrel when firing grenades. The grenade launcher sight is mounted on the hand-guard retaining pin. A heavy rubber recoil pad is attached to the stock with straps and a special ten-round magazine is used when launching grenades from this rifle. This magazine has a filler block in it that prevents it from being used with bulleted cartridges.

Polish 7.62mm AK Assault Rifle with Grenade launcher (PMK-DGN or KbKg M1960).

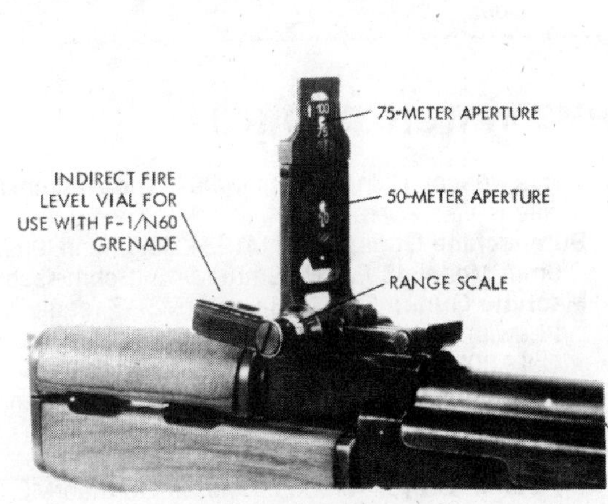

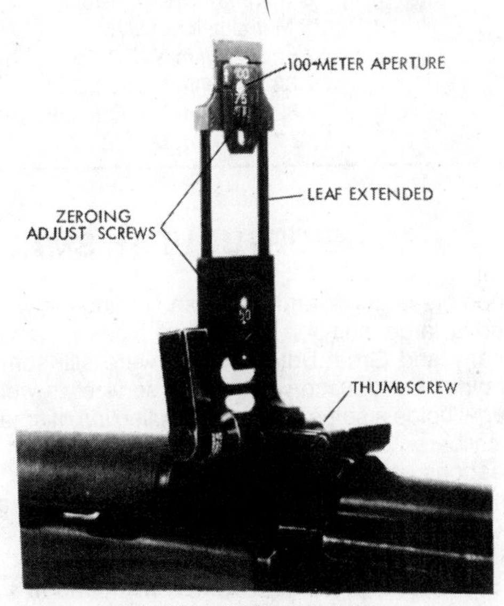

Polish PMK-DGN grenade launcher sights.

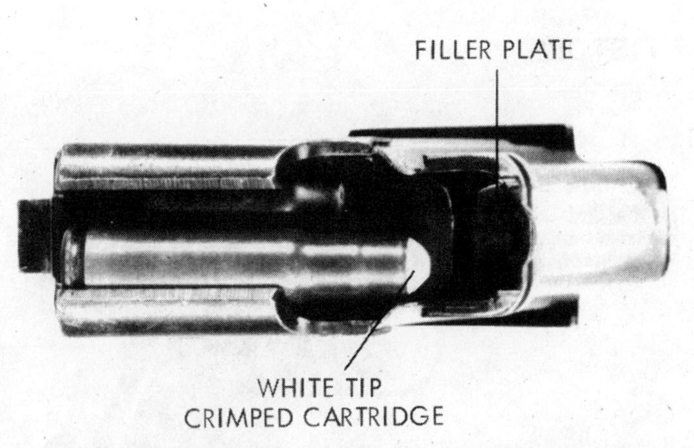

Special 10-shot magazine with filler block to prevent use of ball cartridges.

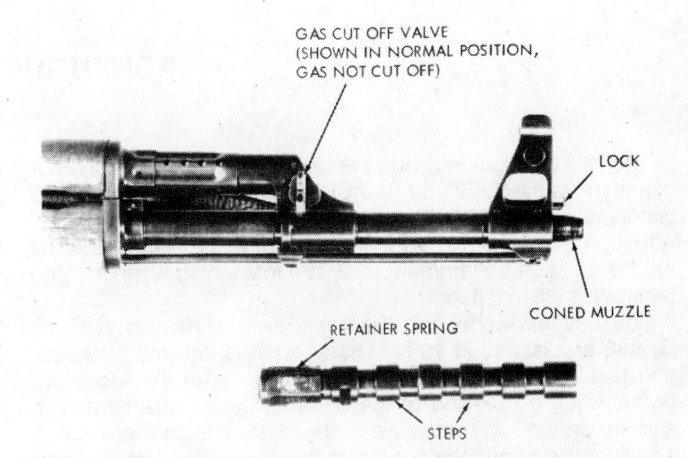

Polish PMK-DGN grenade launcher and gas cut-off valve.

POLISH MACHINE GUNS

The Poles used a 7.92mm water-cooled Browning Machine Gun, which they called the Model 30. They had previously adopted a 7.92mm version of the Browning automatic rifle in 1928. The Polish Browning guns are basically the same as those used by the United States. They are no longer in use in the Polish Army.

38 Portugal

SMALL ARMS IN SERVICE

Handguns:	9 × 19mm NATO P1	P1 made by Walther.
Submachine guns:	9 × 19mm NATO Pistola Metralhadora M/963	Fábrica Militar de Braço de Prata, Lisbon.
	9 × 19mm NATO Pistola Metralhadora M976	Fábrica Militar de Braço de Prata, Lisbon.
Rifles:	7.62 × 51mm NATO G3	FMBP version of G3 made under license from Heckler & Koch.
Machine guns:	7.62 × 51mm NATO M960	Rheinmetall MG42-7.
	7.62 × 51mm NATO M968	FMBP version of HK21 made under license from Heckler & Koch.
	12.7 × 99mm M2 HB	US-made.

PORTUGUESE SMALL ARMS IN WORLD WAR II

During the period preceding World War II and during the war, Portugal procured a large number of different types of small arms from Germany and Great Britain. There were still some quantities of the older Portuguese weapons in service as well. As a result, Portugal holds a somewhat varied collection of small arms in various calibers.

Among the weapons used by Portugal are the following:

Pistols: 7.65mm Savage M/908 and M/915, 9mm M/43 Parabellum(Luger), 9mm M/908 Parabellum (Luger).

Rifles: 7.92mm Mauser M/937 (same as Kar. 98k).
Cal. 303 Lee Enfield M/917 (same as SMLE MkIII and III*).
Portugal has adopted the German G3 (CETME) assault rifle in 7.62mm NATO caliber as standard. The Portuguese call this weapon Rifle 7.62mm M/961. The 7.62mm NATO FN FAL rifle is also used.

Submachine Guns: 9mm M1934 Bergmann, 9mm M/43 (Sten), 9mm Model 48 F.B.P., 9mm M/42 (Schmeisser).

Machine Guns: Cal. .303 Lewis M/917 (same as British Mark I Lewis).
Cal. .303 Vickers M/917 (same as British MK 1 Vickers).
7.92mm Breda M/938 (Breda M1937 in 7.92mm).
7.92 Madsen M/940.
Cal. .303 M/931 (Vickers Berthier).
7.92mm Dreyse M/938 (same as German MG 13).
Cal. .303 M/43 (Bren).
Cal. .303 M/930 (Madsen).

PORTUGUESE PISTOLS*

The only pistol used by the Portuguese not used by any other country was the 7.65mm (.32 ACP) Savage M1907 and M1915 pistols, called the M/908 and M/915 pistols by the Portuguese. Although these pistols have been or are being disposed of by the Portuguese government, they are interesting weapons and bear some description.

The Savage Model 1907 is based on the patents of E. H. Searle, first issued in 1904. The weapon is delayed blowback and has an action similar in some respects to the Model 12 Steyr. A lug on the top of the barrel is butted against a cam surface on the top underside of the slide. The cam surface is initially angled and then parallel to the axis of the bore, causing the barrel to rotate, delaying the action slightly, although the system starts opening from the moment of firing. The M/908 has a rounded and notched cocking lever, which has the appearance of and is mounted in the position of a hammer. The striker of the Savage pistols is spring loaded and is released by the sear. The cocking lever does not strike the firing pin as does the

7.65mm Model 915 Pistol.

*For additional details see *Handguns of the World*.

hammer in the Colt Browning design. The cocking lever of the M/908 should not be let down with a cartridge in the chamber.

PORTUGUESE 7.65mm M/915 PISTOL

The M/915 differs from the M/908 mainly in that it has a spur type cocking lever rather than the rounded type used on the M/908. The M/915 has a greater number of grasping grooves to assist in pulling the slide to the rear than does the M/908. The spur-type cocking lever was introduced in the Savage commercial pistols in 1917. The Model 908 is 6.5 inches overall with a 3.8-inch barrel and weighs 1.2 pounds. It has a 10-round, staggered row, box magazine. Characteristics of the Model 915 are generally similar.

PORTUGUESE RIFLES

PORTUGUESE 6.5mm M1904 MAUSER-VERGUEIRO RIFLE

The only rifle used by Portugal that is unique to that country is the 6.5mm Model 1904 Mauser-Vergueiro. This weapon is somewhat peculiar in that its bolt handle seats ahead of the receiver bridge, like that of the Mannlicher or the Italian Mannlicher Carcano, rather than behind the receiver bridge as with most Mausers. The Model 1904 also has a separate bolt head; the trigger mechanism, bolt stop and ejector are of Mauser 98 type, but this weapon has no bolt sleeve; therefore the safety is attached to the cocking piece.

These rifles were all made by D.W.M. and originally chambered for the 6.5 × 58mm cartridge, which was used only as a military cartridge by Portugal. During the thirties, some of these weapons were converted to 7.92mm. This rifle is not likely to be found in service at present.

Model 1904 6.5mm Rifle.

PORTUGUESE SUBMACHINE GUN

The 9mm Parabellum Model 43 F.B.P. is the only submachine gun of Portuguese design in service with the Portuguese army. All other weapons in service are covered under their country of origin.

THE 9mm MODEL 48 F.B.P.

There are no unusual features in the design of this weapon. It has a massive bolt and a telescoping operating spring assembly as originated by Erma and made famous by the MP38 and MP40. This weapon is made in Portugal.

CHARACTERISTICS OF THE MODEL 48 F.B.P.

Caliber: 9mm Parabellum.
System of operation: Blowback, automatic only.
Weight, empty: 8.2 lbs.
Length overall: Stock extended, 32 in.; stock folded: 25 in.
Barrel length: 9.8 in.
Sights: Front: Blade.
Rear: Fixed aperture graduated for 100 meters.
Feed device: 32-round, detachable staggered row, box magazine.
Muzzle velocity: 2600 f.p.s. (approx).
Cyclic rate: 500 r.p.m.

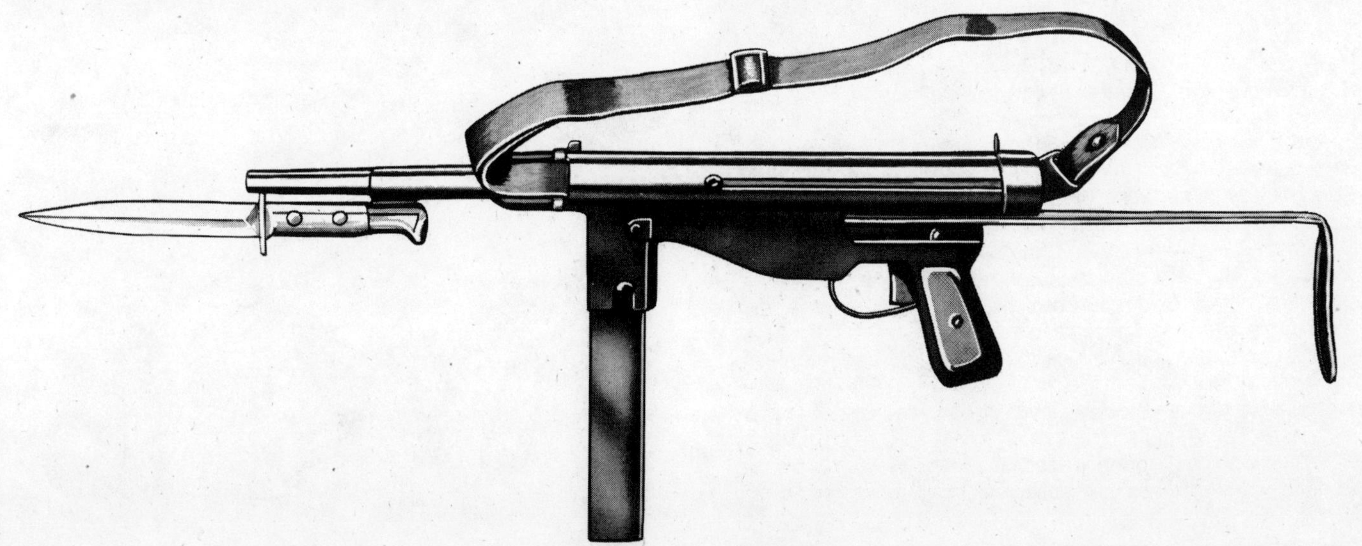

9mm Model 48 F.B.P. Submachine Gun.

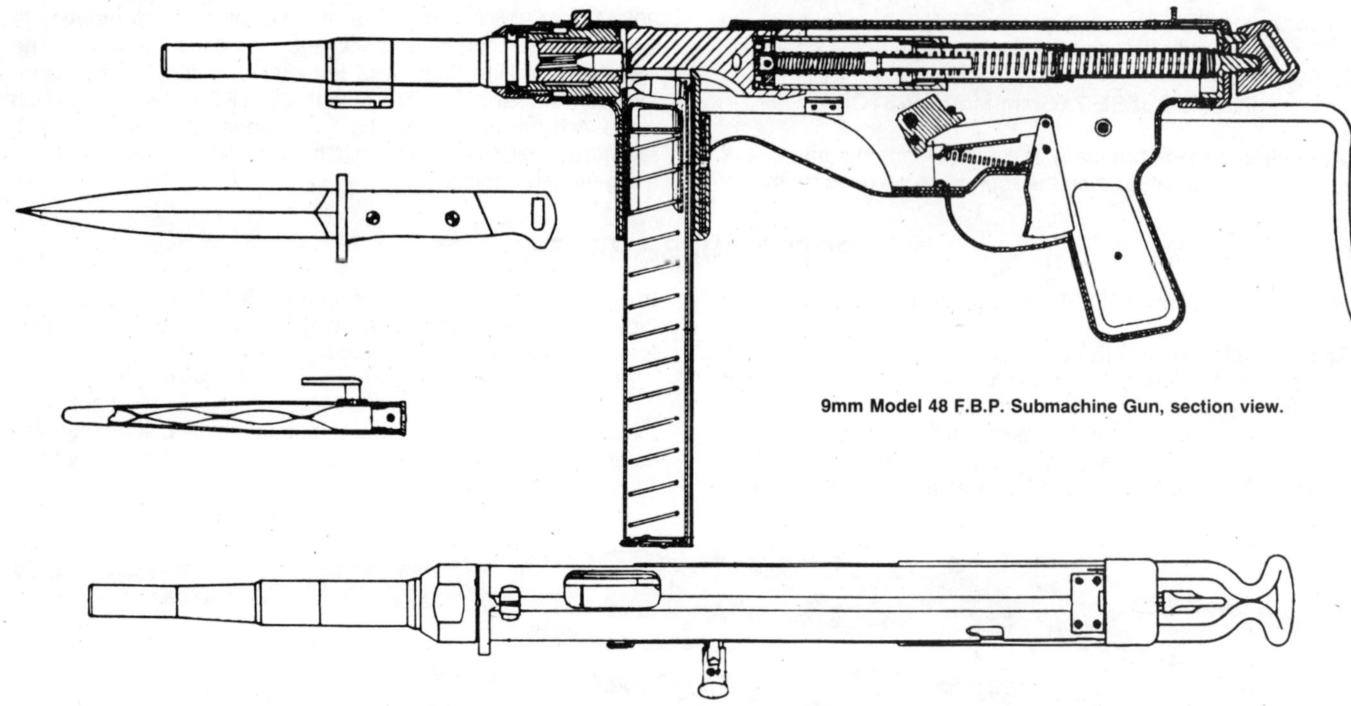

9mm Model 48 F.B.P. Submachine Gun, section view.

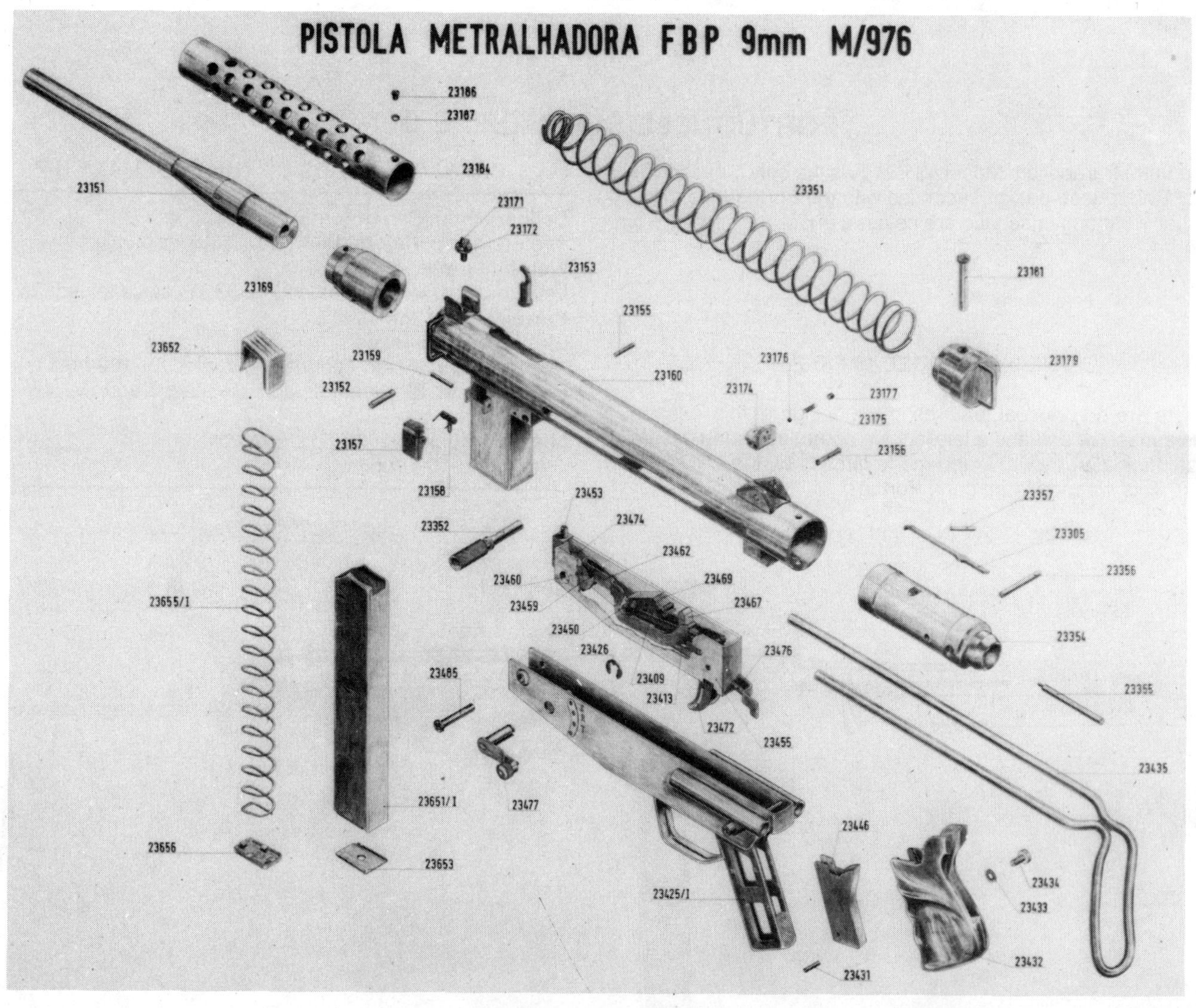

Exploded view of FMBP 9 × 19mm M976 submachine gun.

FMBP 9 × 19MM NATO M976 SUBMACHINE GUN

The Pistola Metalhadora M976 evolved from the M48 through an intermediate gun, the M963. At 25.79 inches with the stock collapsed, the M976 is a relatively long submachine gun. While modern sheet metal stampings are used in this gun, there are still a considerable number of traditionally machined parts. One notable feature of this weapon is the barrel, 9.84 inches, which is rifled using the hammer-forging process. The barrel is made of chrome-molybdenum steel. The standard version of the M976 does not have a barrel jacket, but that attachment can be screwed on without modification to the basic gun. It is expected that a silencer could be attached in the same manner.

Two views of the FMBP 9 × 19mm M976 submachine gun.

PORTUGUESE MACHINE GUNS

There are no machine guns in service in the Portuguese Army of Portuguese origin. The 7.92mm M/938 is not covered in detail under Germany, its country of origin; therefore, some details on this weapon are given here.

THE 7.92mm M/938 MACHINE GUN

This weapon is the German MG13, which was sold by Germany to Portugal in 1938. It was used as the standard German light machine gun until the adoption of the MG34. The weapon is a modified Dreyse M1918.

CHARACTERISTICS OF M/938 MACHINE GUN

Caliber: 7.92mm.
System of operation: Recoil, selective fire.
Weight: 26.4 lbs.
Length, overall: 57 in. (approx).
Barrel length: 28.25 in.
Sights: Front: Folding blade.
Rear: Tangent leaf.
Cyclic rate: 750 r.p.m.
Feed device: 25-round box magazine or 75-round saddle drum.
Muzzle velocity: 2600 f.p.s. (approx).

Special Note on the M/938 Machine Gun

It should be noted that this weapon uses two features also found on the later MG34; the trigger with two half-moon sections, which are used for either automatic or semiautomatic fire, depending on whether the bottom or the top section of the trigger is squeezed; and the saddle drum magazine. (However, the mounting of the saddle drum on the MG34 is different).

Portuguese 7.92mm M/938 machine gun.

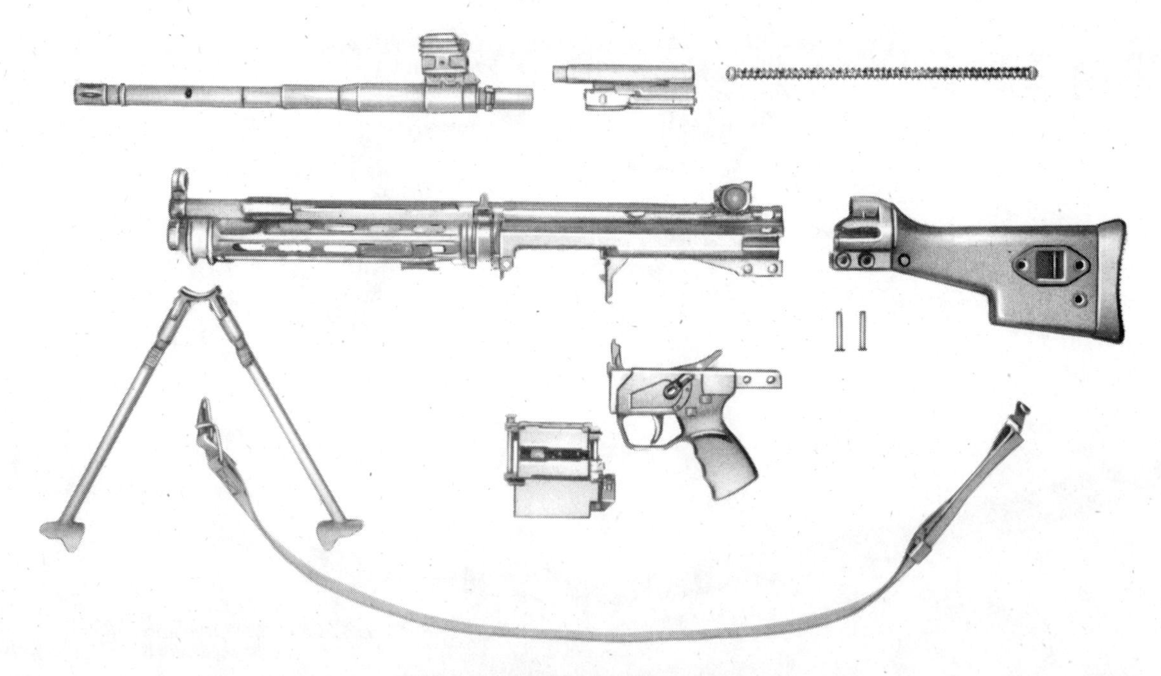

The 7.62 × 51mm M968 (HK21) machine gun.

M968 (HK21) MACHINE GUN FEATURES

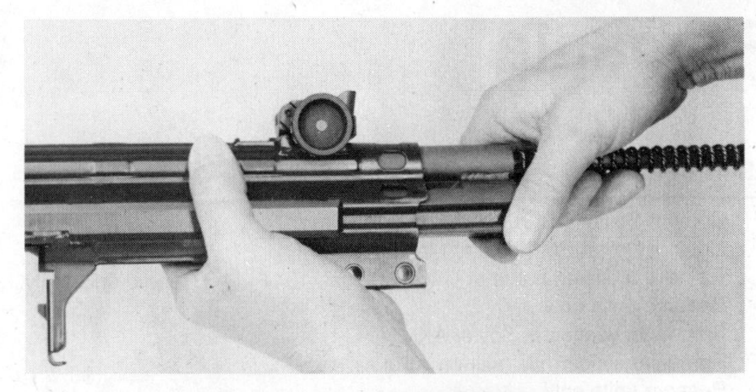

Removing the bolt of the M968 (HK21).

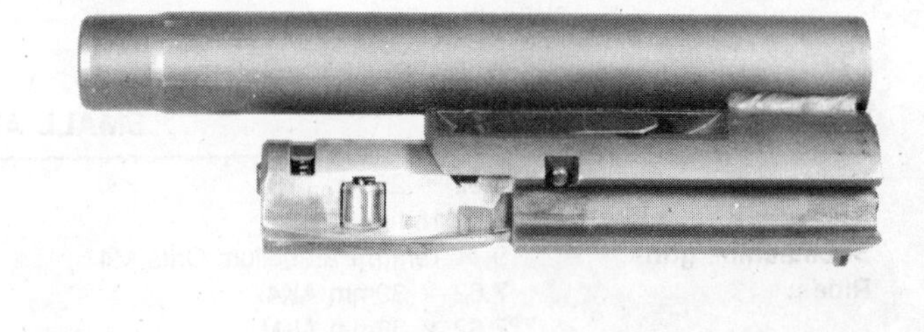

Bolt and bolt carrier assembly.

Disassembled bolt. At the top is the bolt and the roller locking piece. Below is the firing pin and firing pin spring.

Dropping the belt feed unit.

Insertion of the feed belt using the insertion strip.

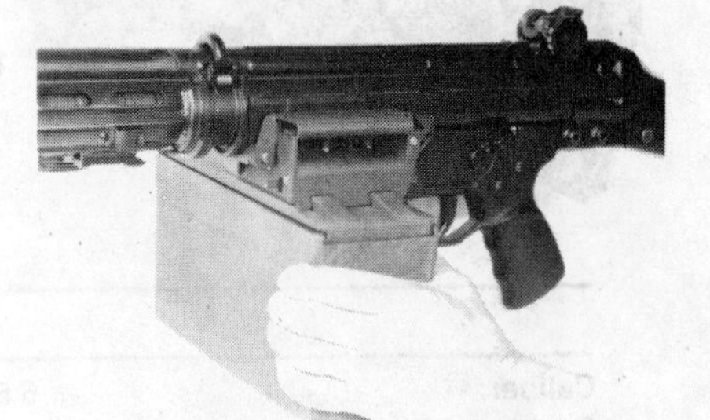

Attaching the ammunition box.

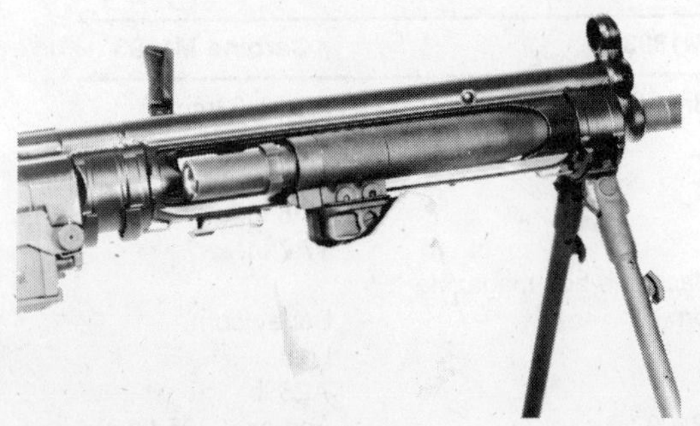

Removing the quick change barrel.

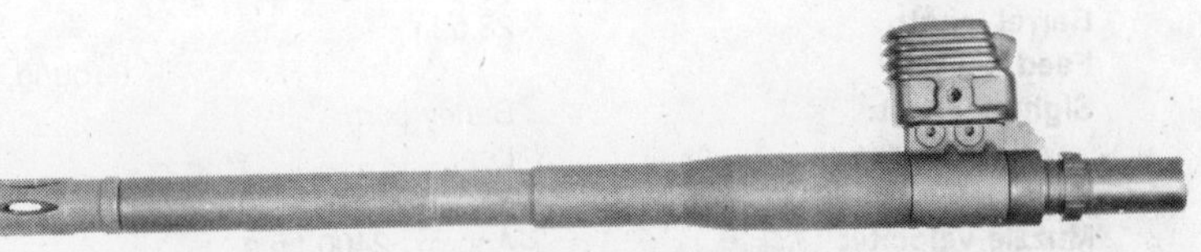

The removable barrel. Note how heavy this barrel is compared to a rifle barrel.

39 Romania

SMALL ARMS IN SERVICE

Handguns:	7.62 × 25mm TT33	Copy of Soviet TT33.
	9 × 18mm PM	Copy of Soviet PM (Pistolet Makarov).
Submachine guns:	9 × 19mm Parabellum Orita M41	Romanian design made at Cugir Arsenal, 1941–44. Reserve units only.
Rifles:	7.62 × 39mm AK47	Reserve units only.
	7.62 × 39mm AKM	Romanian version of Soviet AKM.
	7.62 × 54mm R FPK	Romanian sniper rifle built on AKM receiver.
Machine guns:	7.62 × 39mm RPD	Reserve units only.
	7.62 × 39mm RPK	Romanian version of Soviet RPK.
	7.62 × 54mm R SGM	Reserve units only.
	7.62 × 54mm R PK, PKB, PKS, PKT	Soviet PK family.
	12.7 × 108mm DShK 38/46	Soviet-made.

ROMANIAN RIFLES

In 1892, Romania adopted a 6.5mm Mannlicher Rifle. The Romanian 6.5mm cartridge is the same as that used by the Dutch and is known as the 6.5mm Model 93 or the 6.5 × 53mm R. The Romanian Mannlicher is a turn-bolt type with typical removable bolt head and clip-loading magazine with a clip that functions as part of the magazine, as with the US M1 rifle.

The 1892 and Model 1893 Mannlicher rifles differ only in sight graduations, position of the ejector (on the bolt on the Model 1892 and on the receiver on the Model 1893) and in the presence of a stacking rod on the left side of the upper band of the Model 1893.

The Model 1893 rifle differs from the Dutch Model 1895 only in minor details. A carbine version of the Mannlicher was also used by the Romanians.

After World War I, Romania obtained considerable quantities of French equipment, including 8mm Mannlicher Berthier rifles and carbines. The 7.92mm Czech Model 24 rifle and Austrian 8mm Model 1895 rifles and carbines were also in service. The Model 24 was manufactured in Romania. The Romanian Army was initially equipped with various models of the Soviet 7.62 mm Mosin Nagant rifles. The Soviet 7.62mm SKS carbine and 7.62mm AK assault rifle were the next family of rifles. Characteristics of the rifles other than the Romanian Mannlicher are given under country of origin.

Romania now manufacturers a modified copy of the Soviet 7.62mm AKM. The Romanian AKM has a vertical fore grip that extends downward from the fore-end.

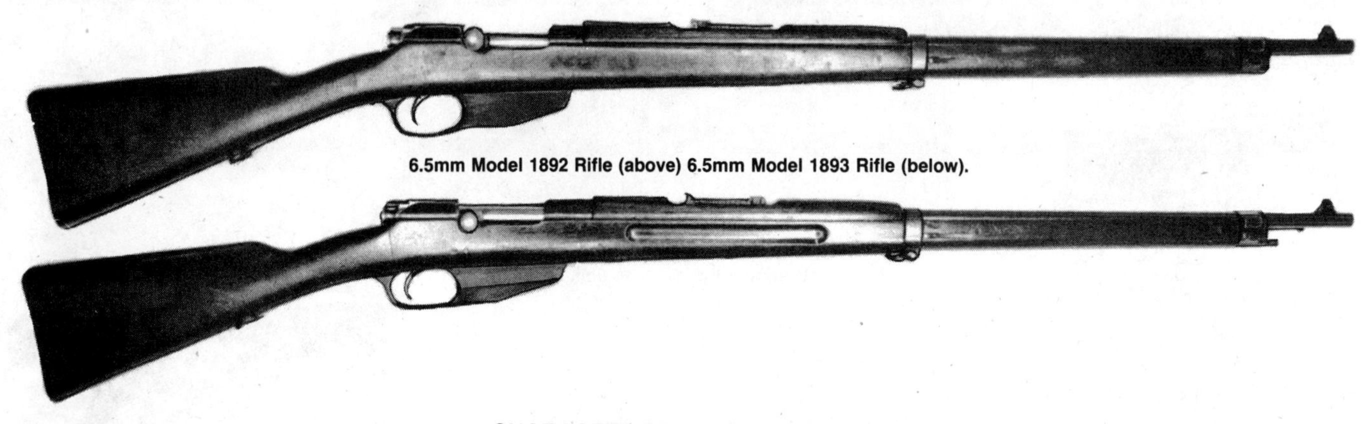

6.5mm Model 1892 Rifle (above) 6.5mm Model 1893 Rifle (below).

CHARACTERISTICS OF ROMANIAN RIFLES

	Rifle M1892	Rifle M1893	Carbine M1893
Caliber:	6.5 × 53mmR.	6.5 × 53mm R.	6.5 × 53mm R.
System of operation:	Turn bolt.	Turn bolt.	Turn bolt.
Length overall:	48.3 in.	48.3 in.	37.5 in.
Barrel length:	28.5 in.	28.6 in.	17.7 in.
Feed device:	5-round, in-line, non-detachable box magazine		
Sights: Front:	Barley corn.	Barley corn.	Barley corn.
Rear:	Leaf.	Leaf.	Leaf.
Weight:	8.9 lb.	9 lb.	7.25 lb.
Muzzle velocity:	Approx. 2400 f.p.s.	Approx. 2400 f.p.s.	Approx. 2325 f.p.s.

Romanian modification of 7.62mm AKM Assault Rifle.

FPK 7.62 × 54mm R SNIPER RIFLE

This new sniper rifle fires the Mosin M1891 cartridge which has a case length that is 15mm longer than the 7.62 × 39mm cartridge. Because the bolt of the AKM travels 30mm (1.18 inch) further to the rear than is necessary to accommodate the 7.62 × 39mm cartridge, the Romanian designers were able to modify the standard AKM-type receiver mechanism to fire the more powerful and longer range 7.62 × 54mm rimmed cartridge. First, they altered the bolt face to take the larger rimmed base of the M1891 cartridge, and they added a new barrel and lengthened RPK-type gas piston system. The gas system of the Soviet SVD (Dragunov) sniping rifle is more like that of the obsolete Tokarev rifle. Second, the Romanians developed their own 10-shot magazine, and they fabricated a skeleton stock from laminated wood (plywood). This buttstock, with its molded cheek rest, is probably slightly better than the one used on the Dragunov. Third, the Romanians have riveted two steel reinforcing plates to the rear of the receiver to help absorb the increased recoil forces of the more powerful M1891 cartridge. Finally, they have attached a muzzle brake of their own design. The standard AKM wire cutter bayonet will attach to this sniper rifle. The rifle illustrated has English language markings on the telescopic sight.

CHARACTERISTICS OF THE FPK

Caliber: 7.62 × 54mm R
Length overall: 45.50 inches
Barrel length: NA
Weight w/ empty magazine and scope: 10.69 pounds
Feed device: 10-round, staggered row, detachable box magazine.

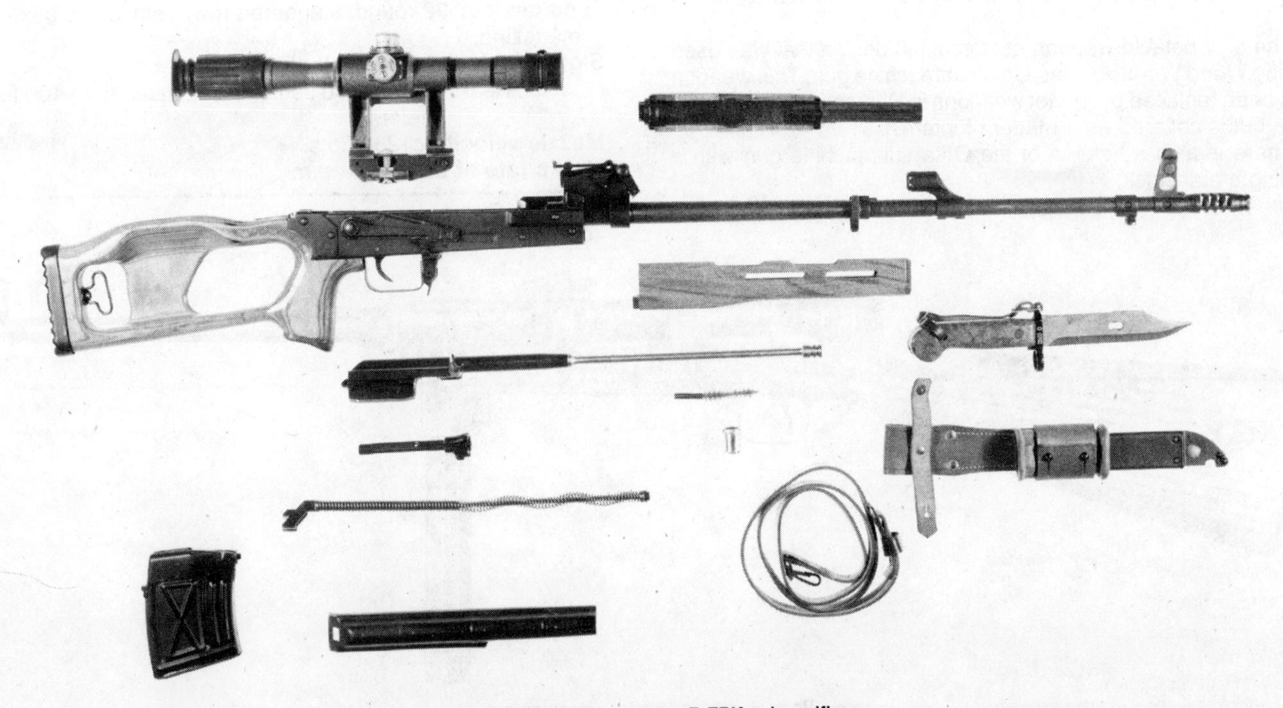

Field-stripped 7.62 × 54mm R FPK sniper rifle.

Romanian FPK sniper rifle.

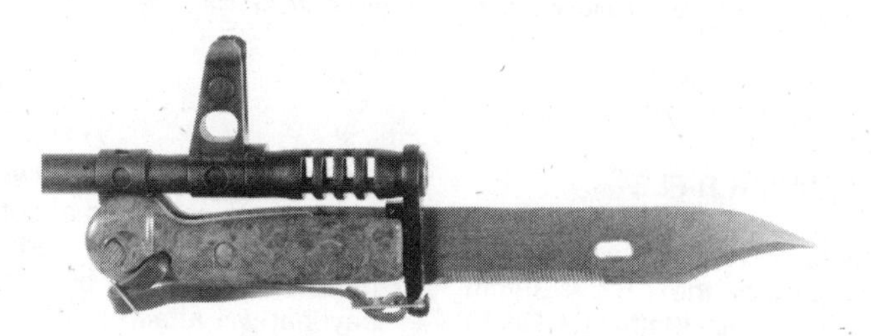

Muzzle brake and AKM bayonet features of FPK.

ROMANIAN SUBMACHINE GUNS

Prior to World War II, the Romanian government purchased 9mm Parabellum Model 1938A Beretta submachine guns and during the war purchased 9mm Parabellum Beretta 38/42 submachine guns.

THE 9mm MODEL 1941 ORITA SUBMACHINE GUN

The only notable weapon of Romanian design that was used during World War II was the Orita submachine gun. This weapon has been replaced by Soviet weapons in Romania at the present time but is covered as a matter of interest.

There is also a version of the Orita submachine gun with a folding metal stock.

CHARACTERISTICS OF MODEL 1941 ORITA SUBMACHINE GUN

Caliber: 9mm Parabellum.
System of operation: Blowback, selective fire.
Weight, loaded: 8.8 lbs.
Length, overall: 35.2 in.
Barrel length: 11.3 in.
Feed device: 32-round, staggered row, detachable box magazine.
Sights: Front: Hooded blade.
Rear: Open V, adjustable, graduated from 100 to 500 meters.
Muzzle velocity: 1280 f.p.s.
Cyclic rate of fire: 400 r.p.m.

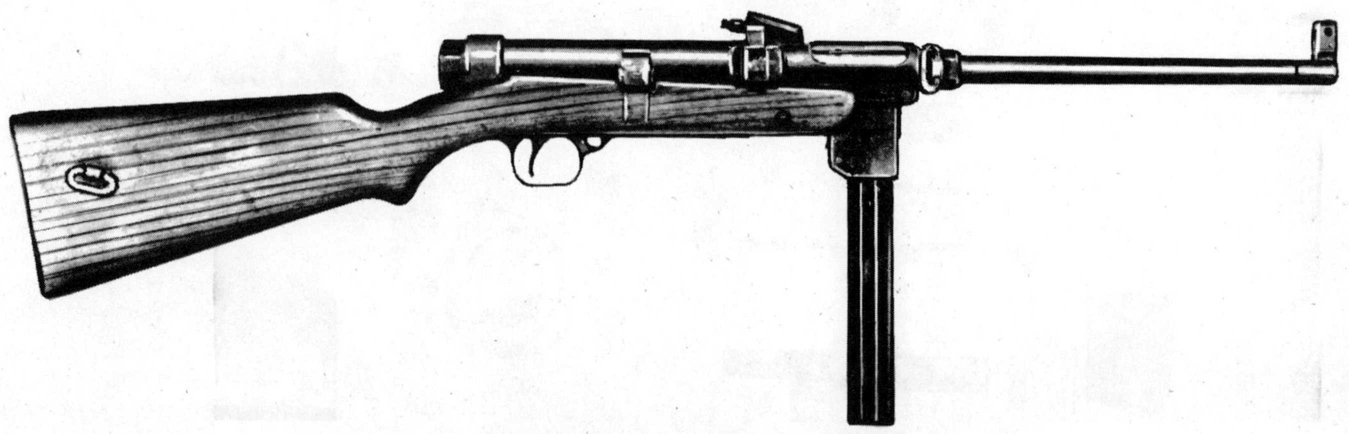

Romanian 9mm Orita submachine gun.

40 Singapore

SMALL ARMS IN SERVICE

Handguns:	9 × 19mm NATO FN Mle. 1935	Fabrique Nationale–made.
Submachine guns:	9 × 19mm NATO Sterling	Sterling-made.
Rifles:	5.56 × 45mm M16A1	Chartered Industries of Singapore, Private Limited under license from Colt.
	5.56 × 45mm SAR 80	Chartered Industries of Singapore.
	7.62 × 51mm NATO L1A1	Commonwealth Small Arms Factory, Lithgow, Australia.
	.303 No. 5 rifle	Reserve units only.
Machine guns:	7.62 × 51mm NATO L7A1	RSAF, Enfield-made.
	5.56 × 45mm Ultimax 100	Chartered Industries of Singapore–developed Squad Automatic Weapon.
	.303 Bren	Reserve units only.

SINGAPORE RIFLES

As indicated in Chapter 1, Chartered Industries of Singapore, Private Limited, a Singapore government chartered and owned enterprise, is responsible, among other things, for the development and manufacture of military small arms for the Singapore Armed Forces. Their first small arms project was the production of the M16A1 which was built under license from Colt Industries. Having had success with that project, CIS has undertaken the development of a new rifle and a squad automatic weapon, both being 5.56 × 45mm. The Singapore Assault Rifle (SAR 80), having as its intermediate ancestor the Sterling Assault Rifle, is a distant relative of the AR-18. The following illustrations will supplement the description of the SAR 80 presented in Chapter 1.

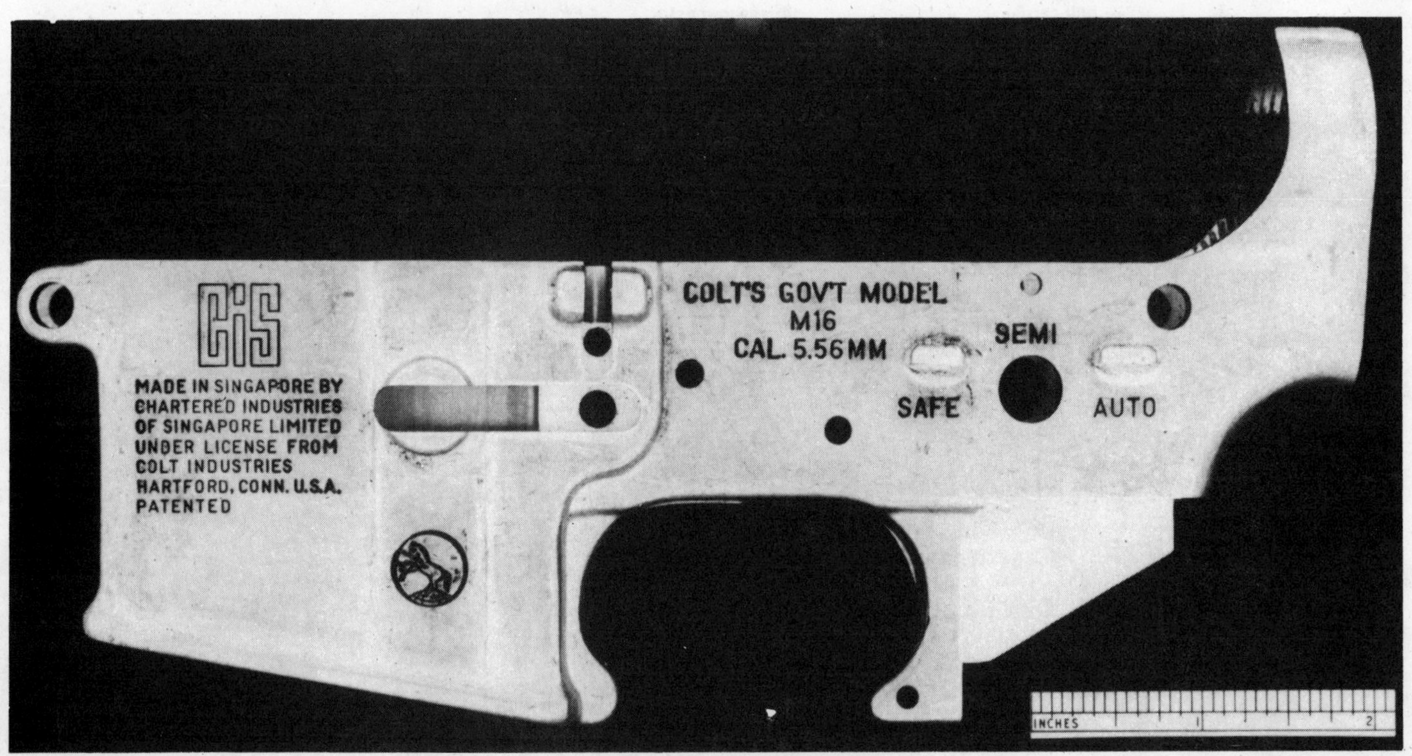

Receiver markings on Chartered Industries of Singapore M16A1.

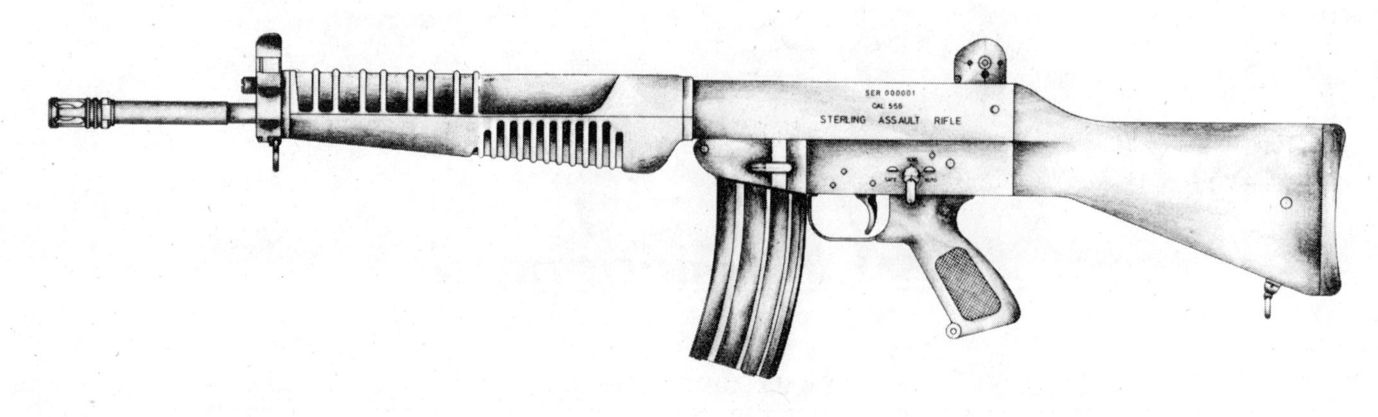

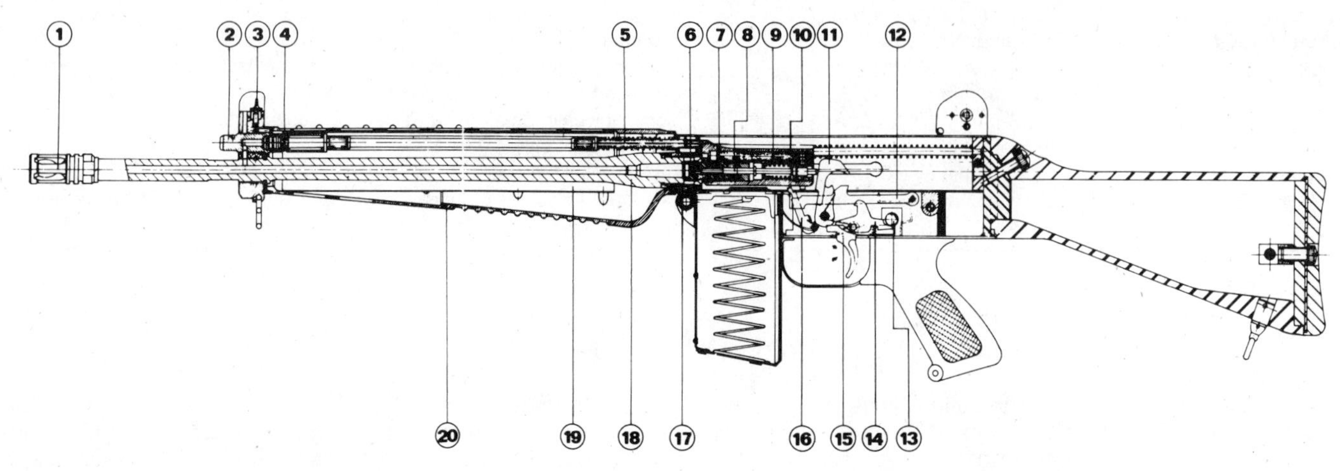

Section view of Sterling Assault Rifle. This drawing is still used by CIS to explain their SAR 80.

1. flash suppressor
2. gas regulator
3. post front sight
4. gas piston
5. gas piston rod
6. bolt
7. bolt carrier
8. firing pin
9. spring firing pin
10. inertial block
11. hammer
12. bolt catch
13. fire-selector control
14. disconnector
15. trigger
16. auto-sear
17. barrel extension
18. barrel
19. handguard liner
20. handguard

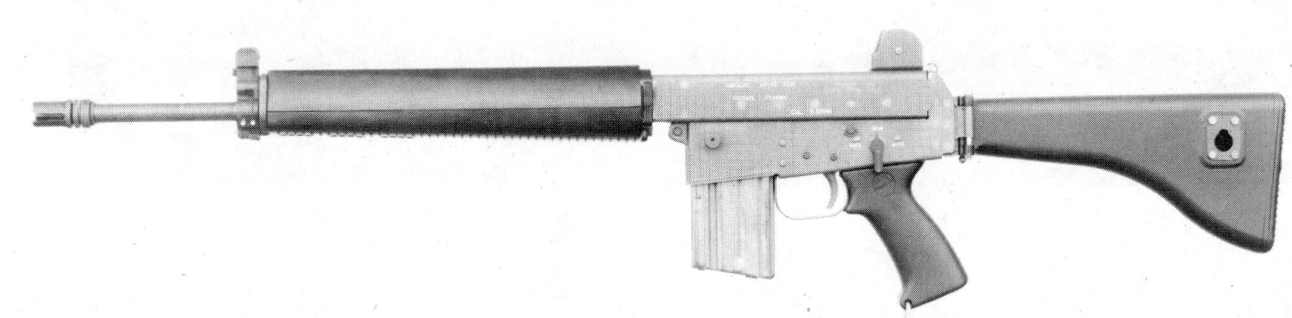

Armalite AR-18, the design grandfather of the SAR 80.

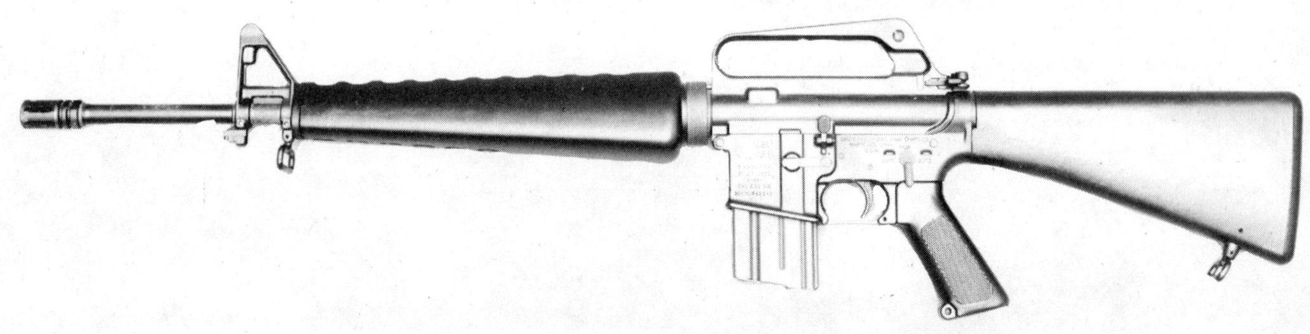

Colt-made M16A1 rifle, which is shown for comparison with the AR-18 and SAR 80.

CIS 5.56 × 45mm SAR 80.

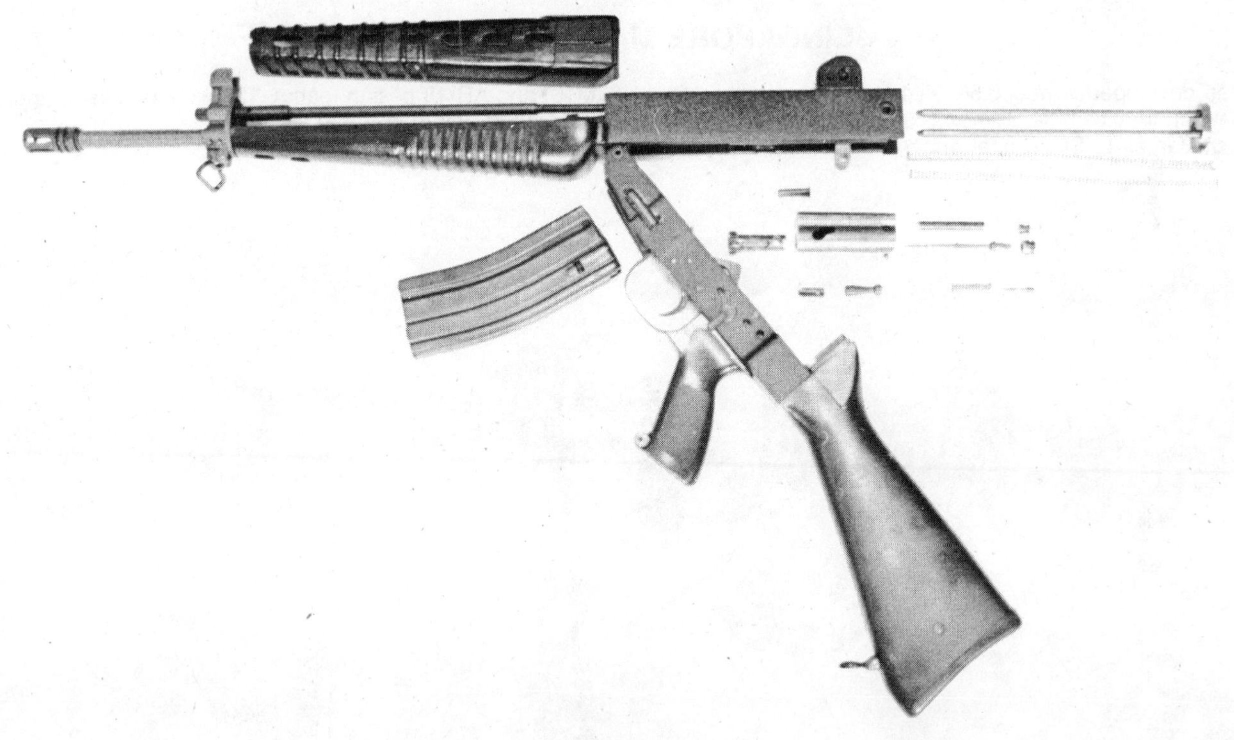

Field-stripped SAR 80.

OPERATION OF THE SAR 80

The gas system utilizes a piston impinging against the bolt carrier which carries the rotating bolt. Rotation is achieved by means of a cam pin projecting out of the cam path in the bolt carrier. The bolt does not rotate as the carrier is projected forward to feed a round into the barrel chamber. Continued movement of the carrier rotates the bolt and locks it to the barrel.

With the rifle armed, the hammer is controlled by a fire-control selector and an additional safety lever, the auto-sear. The auto-sear is so timed as to ensure that the bolt is fully closed before the hammer can be released. In the event of a mechanical failure where the bolt is not fully closed, the hammer would strike the rear of the carrier, thereby dissipating its energy in pushing the carrier forward. No firing would occur.

Upon firing a round, the carrier is forced backwards by the gas piston rod. After a short delay, the cam slot on the carrier engages the cam pin in the bolt. The bolt unlocks, providing both primary extraction and loosening of the empty cartridge case in the chamber. Once the carrier is forced back fully, the empty case is extracted and ejected, and the return spring is compressed. The bolt and carrier return forward, the next round is fed, and if the selector is set on automatic fire and the trigger is still depressed, the weapon is fired again. When the ammunition supply is exhausted, the bolt catch is automatically pushed up by the magazine follower to hold the bolt and carrier open to the rear.

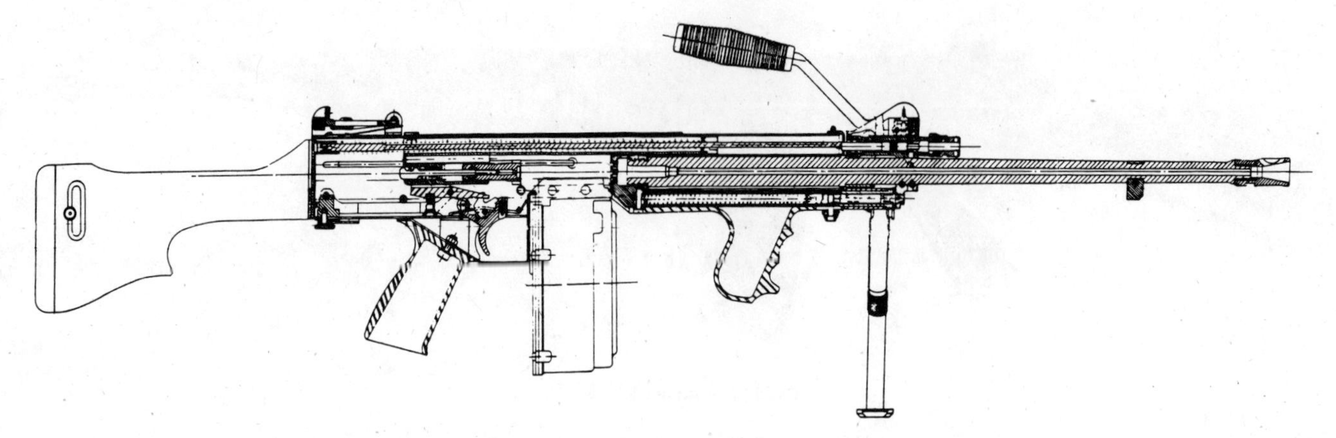

Section view of CIS 5.56 × 45mm Ultima 100 SAW, Mark III with quick change barrel.

SINGAPORE MACHINE GUNS

CIS has developed a new 5.56 × 45mm squad automatic weapon which they call the Ultimax 100. This weapon was developed by James L. Sullivan, a former member of the AR-15 (M16) and AR-18 design teams. The work on this weapon was done in Singapore at the CIS facilities.

CIS 5.56 × 45mm Ultimax 100 squad automatic weapon, Mark III.

COMPARATIVE DATA

	SAR 80 Rifle	M16A1 Rifle	Ultimax 100 SAW
Caliber:	5.56 × 45mm	5.56 × 45mm	5.56 × 45mm
System of operation:	Rotating bolt, gas. Closed bolt.	Rotating bolt, gas. Closed bolt.	Rotating bolt, gas. Open bolt.
Length overall (ins.):	38.2	39.6	40.55
Barrel length (ins.):	18.1	20.3	20.3 (quick change feature on MK III)
Weight (lbs.) empty:	8.1	7.0	10.47
Feed device:	20 or 30 M16 magazine	20 or 30 M16 magazine	100 or 60 drum 20, 30, 60 magazines
Cyclic rate (r.p.m.):	600 to 800	700 to 950	520

41 South Africa

SMALL ARMS IN SERVICE

Handguns:	9 × 19mm NATO FN Mle. 1935	Fabrique Nationale–made.
Submachine guns:	9 × 19mm NATO Sana 77 semiautomatic carbine	Paramilitary forces.
Rifles:	7.62 × 51mm NATO R1	FN FAL–made under license by the state-owned arms factory, Lyttelton Engineering Works of ARMSCOR.
	5.56 × 45mm R4	IMI Galil–made under license by ARMSCOR.
	5.56 × 45mm R5	IMI Galil with 12.99 inch barrel, instead of 18.11 inch barrel of the R4. The R5 is used mainly by armored troops.
Machine guns:	7.62 × 51mm NATO MAG	Fabrique Nationale–made.
	7.62 × 51mm NATO MG4	.30 M1919 LMG converted to 7.62mm NATO.
	7.62 × 51mm NATO Vickers	UK-made.
	12.7 × 99mm M2 HB	US-made.

SOUTH AFRICAN ARMS INDUSTRY

Since the Republic of South Africa has been embargoed for its policy of apartheid, the country has found it necessary to create its own defense industry. The major organization involved in weapons and military electronics production is ARMSCOR, a state-owned establishment. ARMSCOR's small arms factory, Lyttelton Engineering Works, is located in Pretoria. Lyttelton Engineering has manufactured the FN FAL under license as the R1 and the IMI Galil in 5.56 × 45mm as the R4. In 1982 ARMSCOR announced that it was seeking export markets for its small arms, vehicles, and electronic and telecommunications equipment. Little hard data on the ARMSCOR activities is currently available because of strict military control over the release of information.

COMPARATIVE DATA

	R4	R5
Caliber:	5.56 × 45mm	5.56 × 45mm
Length overall w/stock extended (ins.):	38.50	33.46
w/stock folded (ins.):	29.25	24.21
Barrel length (ins.):	18.11	12.99
Weight (lbs.):	11.68	9.48
Feed device:	35- and 50-shot detachable box magazine.	35- and 50-shot detachable box magazine.
Cyclic rate (r.p.m.):	600–750	600–750

SANNA 77 9 × 19mm SEMIAUTOMATIC CARBINE

Dan Pienaar Enterprises (Pty) Ltd. of Johannesburg is marketing a version of the Czech vz 25 submachine gun which they have modified to be a semiautomatic only 9 × 19mm carbine. This gun is designed as a personal defense weapon to be used by farmers and others living in the rural regions of the republic.

42 Spain

SMALL ARMS IN SERVICE

Handguns:	9 × 19mm NATO Super Star	Star Bonifacio Echeverria, SA, Eibar
Submachine guns:	9 × 19mm NATO Z-45	Star Bonifacio Echeverria, SA.
	9 × 19mm NATO Z-62	Star Bonifacio Echeverria, SA.
	9 × 19mm NATO Z-70	Star Bonifacio Echeverria, SA.
Rifles:	7.62 × 51mm NATO CETME Fusil Asalto	Empresa Nacional Santa Barbara de Industrias Militares, SA, Fábrica de Armas de Oviedo.
Machine guns:	7.62 × 51mm NATO MG42/59	EN Santa Barbara.
	7.62 × 51mm NATO MG1A3 and 3S	EN Santa Barbara.
	7.62 × 51mm NATO Browning M1919E1	US-made, Spanish-converted.
	12.7 × 99mm M2 HB	US-made.

SPANISH HANDGUNS*

SPANISH 9mm SUPER STAR PISTOL

The Super Star is the standard Spanish service pistol. Although it outwardly resembles the US caliber .45 M1911A1 pistol, the locking mechanism resembles that of the FN Browning Hi-Power. The barrel is cammed into and out of its locked position in the receiver by the action of the slide stop pin on a camway cut in a lug on the rear underside of the barrel.

The Super Star is field stripped by pushing down and forward on the dismounting lever mounted on the right side of the receiver. The slide, with barrel and recoil spring plug, can then be slid off the receiver from the front. After the slide is removed from the receiver, the recoil spring with guide, recoil spring plug and barrel bushing can be removed from the slide. The Super Star has a pivoting trigger rather than the sliding trigger of the US M1911A1. There is no grip safety on the Super Star. This pistol is of good design and construction and is sold commercially in caliber .380, Super .38 automatic and 9mm Parabellum.

Spanish 9mm Super Star Pistol.

OBSOLETE SPANISH HANDGUNS

The first automatic pistol adopted by Spain was the 9mm Campo Giro Model 1913. This pistol was chambered for the 9mm Bergmann Bayard cartridge, which is called the 9mm Largo in Spain. The Campo Giro was the basic model of the well-known Astra pistol. It is basically a blowback operated pistol with a very heavy recoil spring wrapped around the barrel; it is the product of Unceta and Company, originally of Eibar. The pistol was modified slightly in 1916 and issued as the Model 1913–16.

In 1921, the Astra Model 400 (commercial designation), which had the official Spanish government designation of Model 1921, was adopted. The Astra 400 is a most unusual pistol in that it will chamber and fire the following cartridges: 9mm Largo, 9mm Parabellum, 9mm Steyr, 9mm Browning long and the .38 super automatic cartridge.** The military issue pistol has wooden grips and does not have the full commercial markings. The commercial Astra 400 has hard rubber grips, commercial markings, and many of the small components are nickeled or chromed. The commercial model was sold widely throughout the world.

A later model, the 9mm Parabellum Astra Model 600, never used as a Spanish service weapon, was made in limited quantities for the Germans during World War II; it differs from the Model 400 only in minor details. A post-World War II version of the 9mm Parabellum Model 600—the Condor—has been introduced by Astra. This pistol, although not a standard service pistol, is easily adaptable for such use. Unlike the earlier Astra Models 400 and 600, it has an external hammer.

Unceta and Company produced a copy of the 7.63mm Mauser military model pistol after World War I, several of which were selective-fire weapons. The semiautomatic Astra Model 900 is a copy of the Mauser, with shoulder-stock holster. The Model 902, also known as the Model "F", is a selective-fire weapon, generally similar to the Model 1932 Mauser. These weapons have been used by police and gendarmerie units throughout the world, but they were not used as Spanish service pistols.

*For additional details see *Handguns of the World*.

**The prudent shooter will not shoot the Astra with any cartridge except the 9mm Largo, because there are often chamber variations that make it questionable to fire the other cartridges with complete safety.

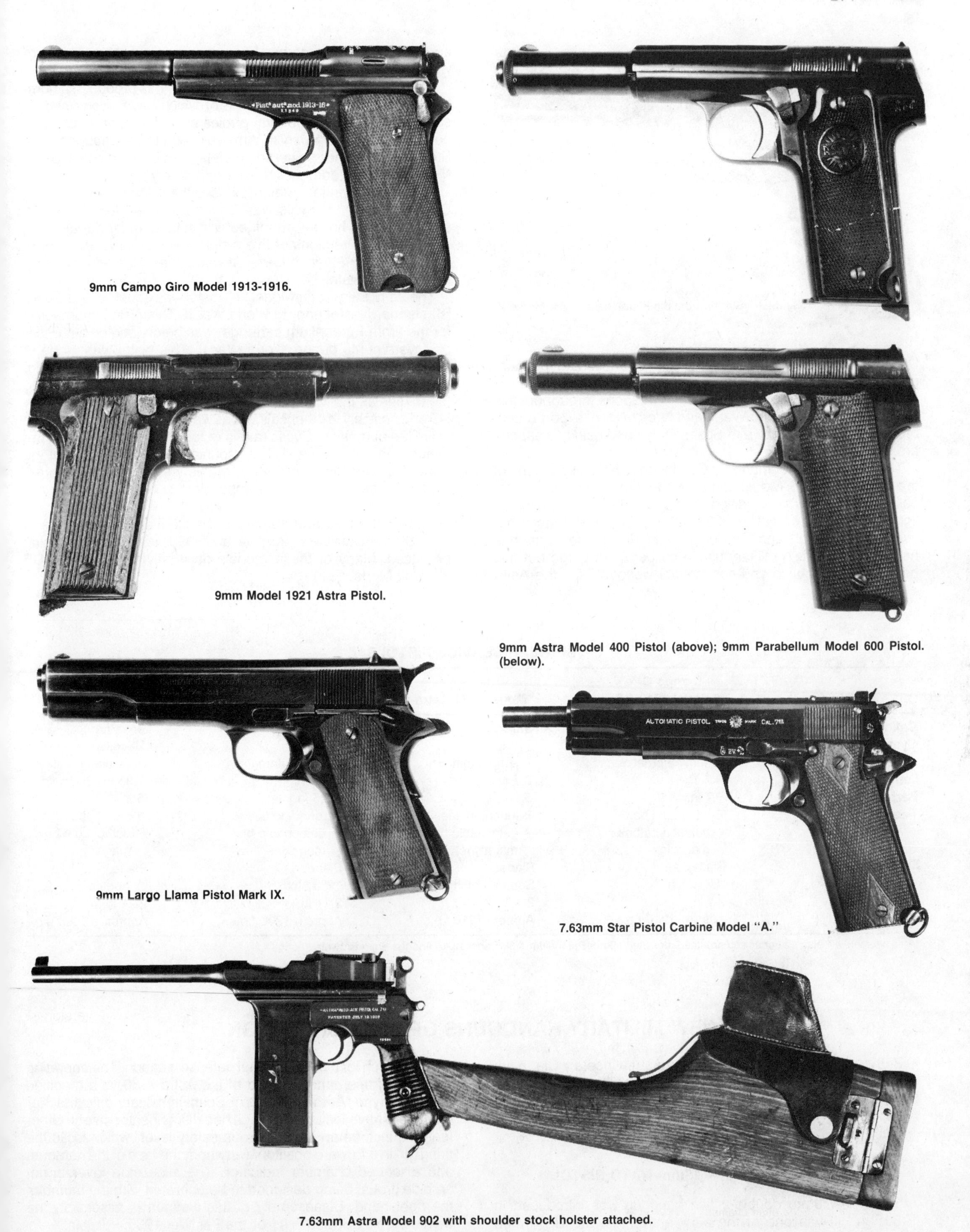

9mm Campo Giro Model 1913-1916.

9mm Model 1921 Astra Pistol.

9mm Astra Model 400 Pistol (above); 9mm Parabellum Model 600 Pistol. (below).

9mm Largo Llama Pistol Mark IX.

7.63mm Star Pistol Carbine Model "A."

7.63mm Astra Model 902 with shoulder stock holster attached.

Spanish "Star" Automatic Pistol, caliber .45, Model M.

Unceta, the manufacturers of Astra pistols; Gabilondo, the manufacturers of Ruby revolvers and Llama pistols; and Star Bonifacio Echeverria S.A. all made quality weapons throughout this period and still do. World War II and large military export orders brought tighter proof laws to Spain. Only good quality weapons are exported currently.

The "Llama" pistol made by Gabilondo is basically a copy of the US caliber .45 Model 1911 Colt automatic in the heavier calibers and blowback edition of the Colt Model 1911 in the lighter calibers, i.e., caliber .22, .32 and .380. Differences between these pistols and the Colt are superficial. The Llama has had limited Spanish military use in caliber 9mm Largo but has been mainly a police and commercial weapon. Like the Astra and the Star, the Llama has been used as a substitute standard wartime pistol by countries other than Spain.

Star Bonifacio Echeverria S.A., of Eibar, has produced a number of military type pistols, some of which have been used as substitute standard in foreign armies and several of which have been used by the Spanish Army as regulation weapons. The Star, in the higher-powered models, is similar to the US .45 Model 1911 but does not have a grip safety.

During the twenties, Star introduced the caliber 7.63mm Pistol Carbine Model "A". This weapon was adapted for use with a shoulder-stock holster and fitted with a tangent type rear sight. There are two versions of this pistol: one with a five-inch barrel and one with a 6.5-inch barrel. It was never adopted as a standard service pistol.

The 9mm Largo Star Model "A" pistol was adopted as a Spanish standard pistol prior to World War II. This pistol, chambered for the 9mm Parabellum cartridge, was purchased by a number of governments during World War II. This pistol was made in very large quantities and is quite common throughout the world.

A selective-fire version of the Model "A" was produced during the thirties as the Model "M". This weapon was made in caliber .45 and possibly in other calibers as well. It was sold to Nicaragua in limited quantities. Cyclic rate is reported to be 800 rounds per minute. The selector switch is mounted on the right side of the slide. The selective-fire feature of this weapon is very impracticable, since the weapon is impossible to control in automatic fire.

As with the Llama, a number of simple blowback versions of the "Star" in calibers .22, .32 and .380 were and are being produced. Many of these models outwardly resemble the US .45 Colt Model 1911.

SPANISH SERVICE PISTOLS

	Campo Grio Model 1913–16	Model 1921 Astra	Super Star	Star Model "A"
Caliber:	9mm Largo.	9mm Largo.*	9mm Largo.	9mm Largo.
System of operation:	Blowback, semiautomatic.	Blowback, semiautomatic.	Recoil, semiautomatic.	Recoil, semiautomatic.
Length overall:	9.7 in.	8.7 in.	8.03 in.	7.95 in.
Barrel length:	6.7 in.	5.9 in.	5.25 in.	5 in.
Feed device:	8-round, in line, detachable box magazine.	8-round, in line, detachable box magazine.	9-round, in line, detachable box magazine.	8-round, in line, detachable box magazine.
Sights: Front:	Blade.	Blade.	Blade.	Blade.
Rear:	"U" notch.	Square notch.	"V" notch.	"V" notch.
Weight:	2.1 lb.	2.1 lb.	2.21 lb.	2.21 lb.
Muzzle velocity:	Approx. 1210 f.p.s.	Approx. 1210 f.p.s.	Approx. 1200 f.p.s.	Approx. 1200 f.p.s.

*This weapon will also chamber and fire: the 9mm Steyr, 9mm Parabellum, .38 Super Auto, and the 9mm Browning Long.

NEW MILITARY HANDGUNS OF SPANISH DESIGN

Both Star Bonifacio Echeverria and Astra-Unceta y Cia. have introduced new double-action 9mm NATO caliber handguns for the 1980s. These are the Star Model 28 and the Astra Model A-80.

STAR MODEL 28 9 × 19mm NATO PISTOL

Star's new 9mm double-action handgun was introduced in 1980, and it was one of the early entrants in the US armed forces search for a new personal defense weapon. The operating principle of this handgun, and of the Astra A-80, is a modified Browning-type. At the request of Spanish military officials, Star engineers have included a magazine disconnector safety. Other features include an ambidextrous safety lever, which locks the firing pin in a forward position away from the face of the hammer, and a loaded chamber indicator. The slide stop lever is not ambidextrous, being designed to be operated with the thumb of the right hand. Disassembly of this weapon is essentially the same as the US M1911A1 or the FN Mle. 1935.

Star 9 × 19mm NATO Model 28.

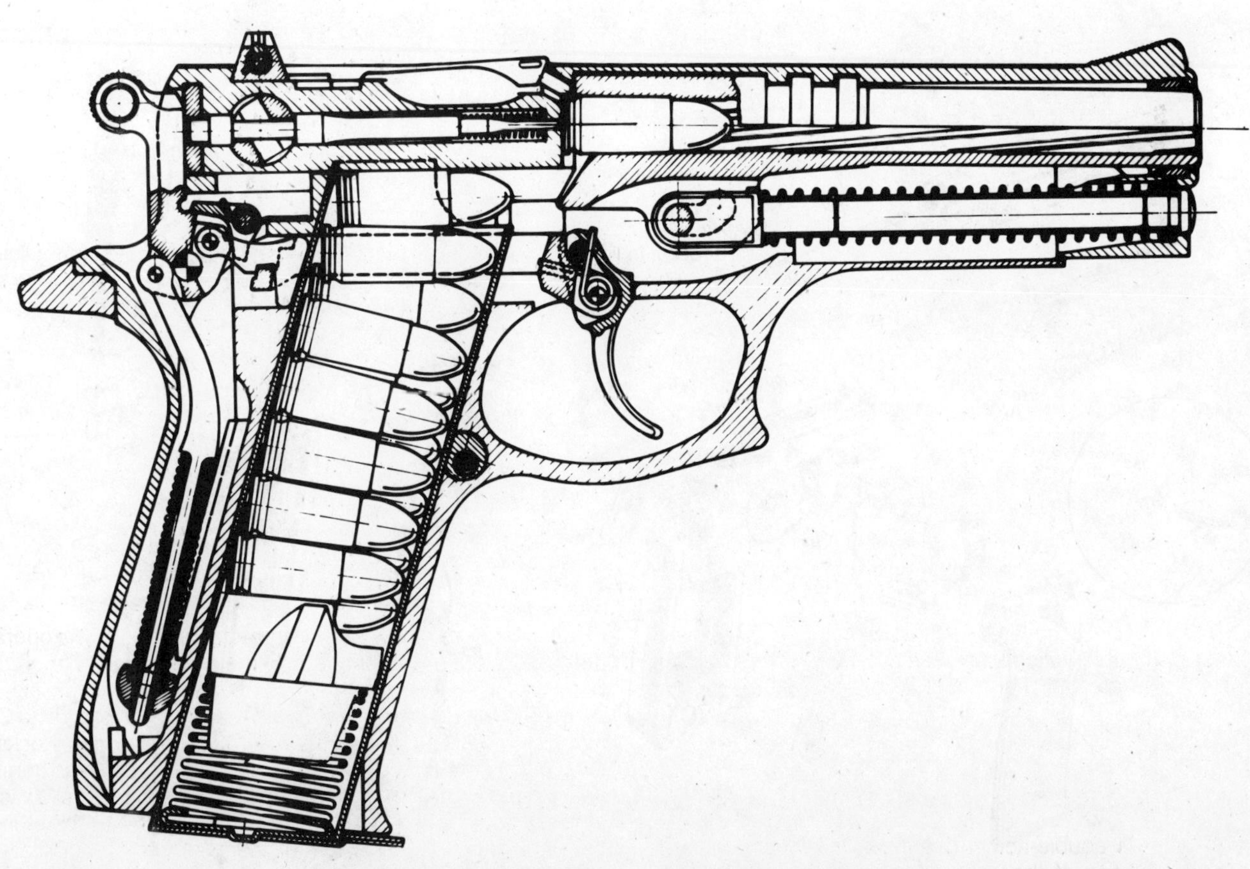

Star Model 28 section view. Note the modified Browning locking system.

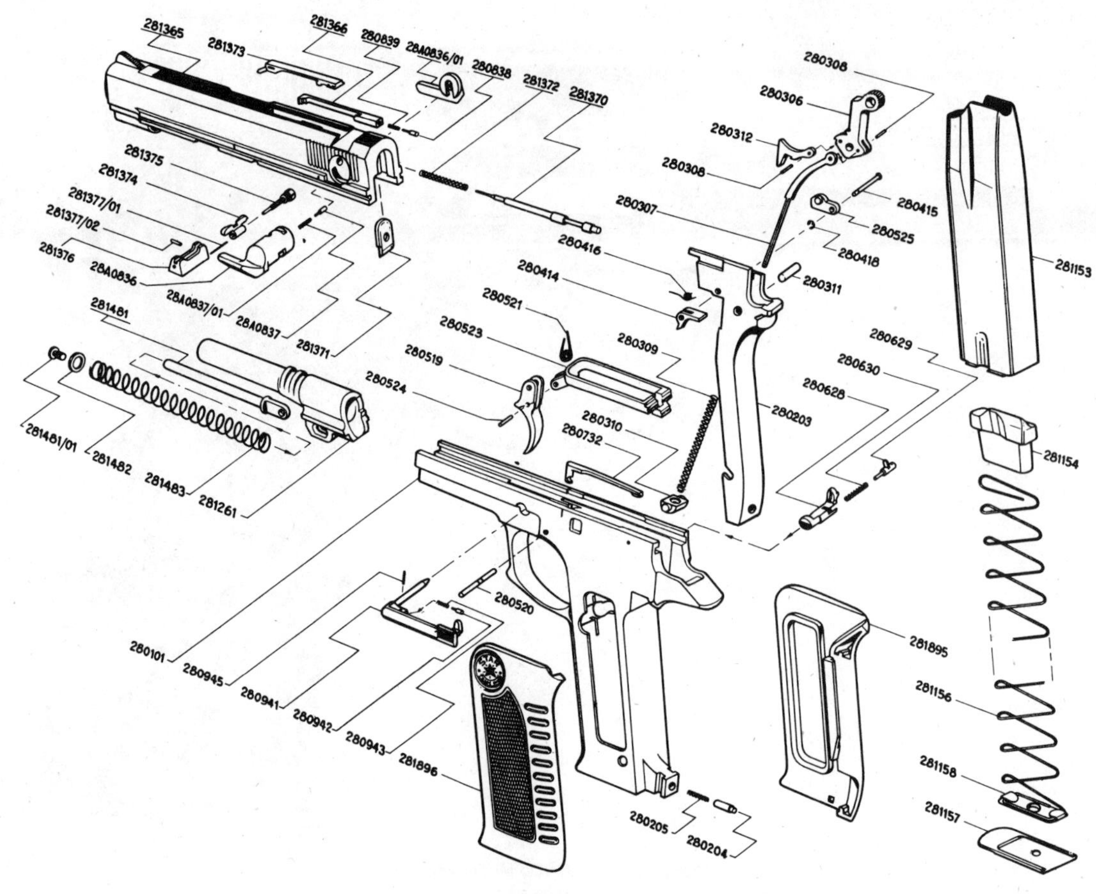

Exploded view of Star Model 28.

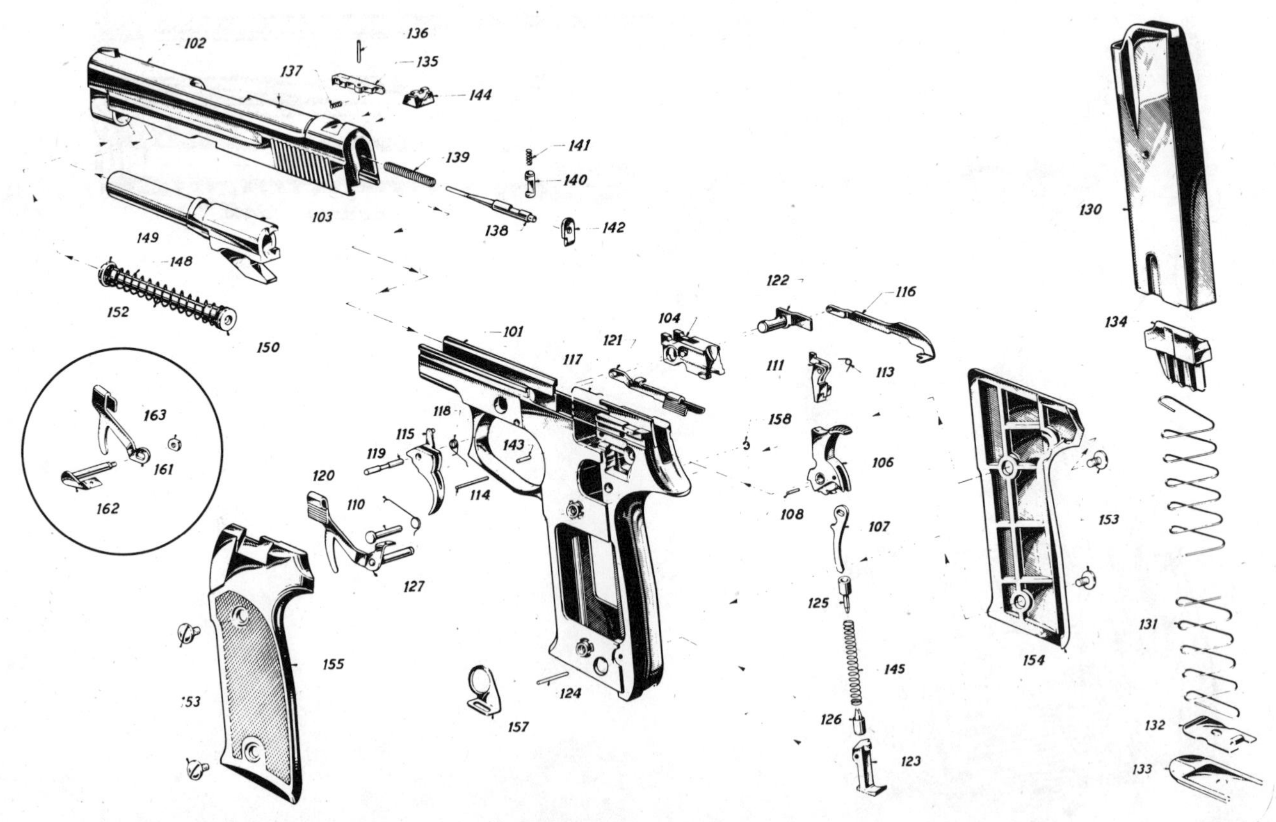

Exploded view of Astra A-80.

Star Model 28 Safety Features

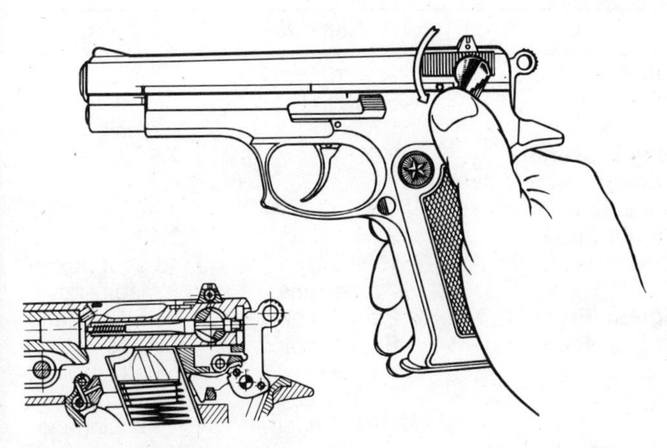

To put the safety ON.

When handling the pistol, always put the ambidextrous safety, located on both left and right rear side of the slide, in ON position, i.e., push it downward until left safety lever covers the red dot on the left rear side of the frame. Through this action, the firing pin is forced in and remains locked inside the slide. The pistol may thus be safely carried and handled, and an accidental discharge of an eventual chambered cartridge prevented, even if the trigger is pulled.

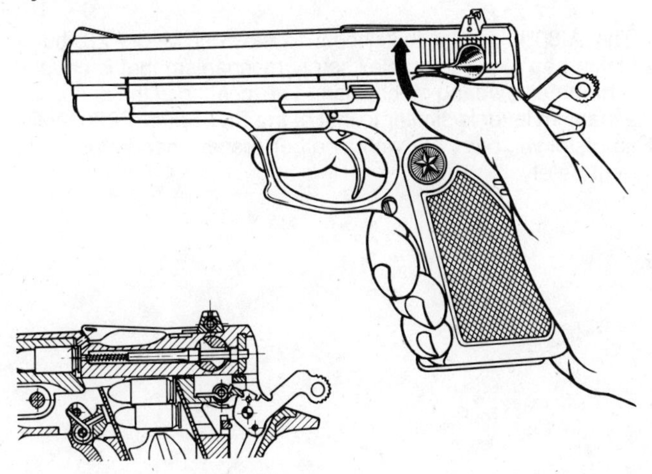

Safety off.

ASTRA A-80 SAFETY FEATURES.

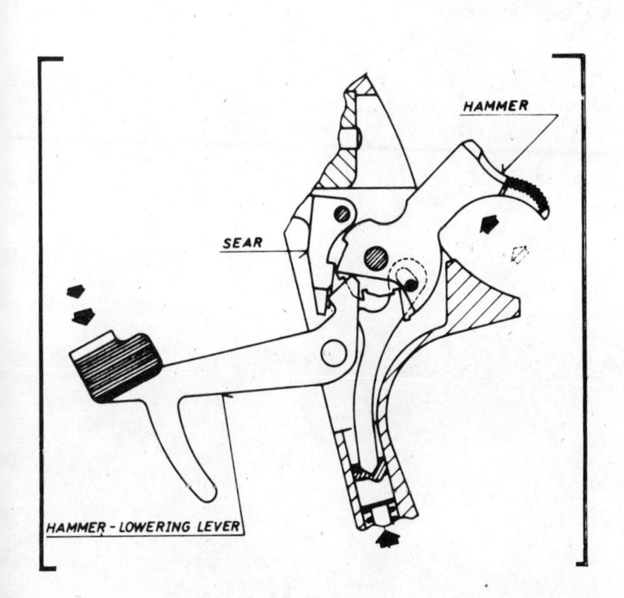

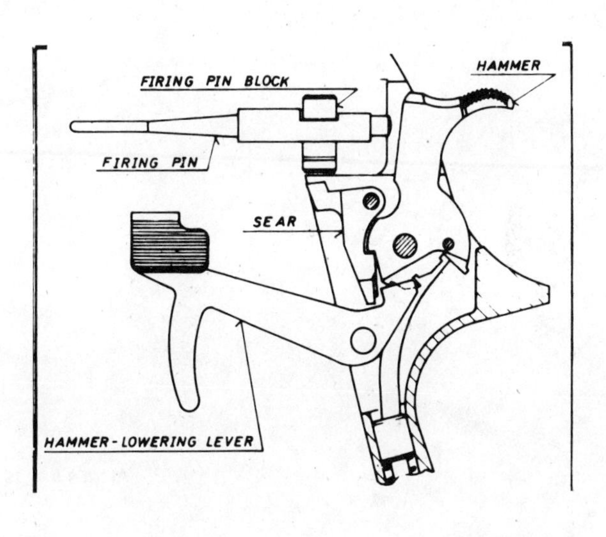

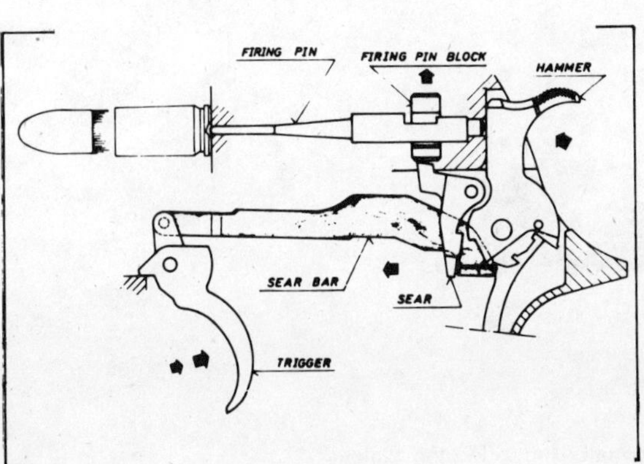

Astra engineers have incorporated a hammer lowering device and a firing pin block in their A-80. When the hammer lowering mechanism is actuated, the sear is activated, and it positions the firing pin block, thus preventing the firing pin from detonating the cartridge. When the hammer is in the lowered position, the firing pin block and the sear lock the firing pin and keep the hammer away from the firing pin. When the trigger is pulled (double-action) or the hammer is cocked manually (single-action), the hammer block is lifted out of the way so the pistol can be fired.

ASTRA A-80 9 × 19mm NATO PISTOL

The A-80 is similar in concept to the Star Model 28, but in addition has a more complex safety mechanism that is coupled to a hammer lowering mechanism. The location of the decocking mechanism lever is similar to that of the SIG-Sauer P220 family. This pistol will compete with all other present handguns for inherent safety.

COMPARISON OF STAR MODEL 28 AND ASTRA A-80

	Model 28	A-80
Caliber:	9 × 19mm	9-19mm
System of operation:	Browning recoil	Browning recoil
Length overall (ins.):	8.07	7.5
Barrel length (ins.):	4.33	3.94
Weight (lbs.):	2.38	2.17
Feed device:	15-shot box magazine	15-shot box magazine
Sights: Front:	Patridge blade	Patridge blade
Rear:	Square notch, screw adjustable for windage	Square notch, drift adjustable for windage

Astra's 9 × 19mm NATO A-80.

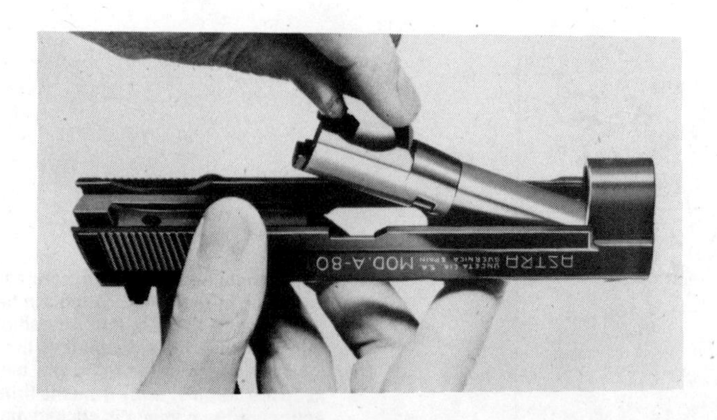

Barrel being removed from the A-80. Note the modified FN Browning Mle. 35 locking system.

THE 7.62mm NATO CETME MODEL 58 ASSAULT RIFLE

CETME, the Centro de Estudios Tecnicos de Materials Especiales, a Spanish government research establishment, has produced a number of interesting weapon designs, but the CETME assault rifle is by far the most successful of their efforts. See Chapter 1 for further background.

How to Load and Fire the CETME Assault Rifle

Insert a loaded magazine in the magazine well. Pull the operating handle located on the left side of the operating rod (above the barrel) to the rear and release it. If the selector lever has been set on the letter "T," the bolt will run home, chambering a cartridge, and the weapon will fire single rounds when the trigger is pulled. If the selector has been set on the letter "R," the bolt will remain to the rear when cocked. When the trigger is pulled, the weapon will deliver automatic fire. The CETME commences semiautomatic fire from a closed bolt and automatic fire from an open bolt. The operating handle does not reciprocate with the bolt. To put the weapon on SAFE, set the selector lever on the letter "S."

How to Field Strip the CETME Assault Rifle

Remove the magazine by pushing on the magazine catch. Remove the two pins at the rear of the pistol grip-trigger assembly and pull off the stock to the rear. Pull the operating handle to the rear and remove the bolt assembly. No further disassembly is recommended. To reassemble the weapon, perform the above steps in reverse order.

How the CETME Assault Rifle Works

The bolt in this weapon is composed of two major parts: the head and the rear section. The head, which contains the locking rollers, is considerably smaller than the rear section, which contains the firing pin. The two bolt sections are joined together so that they can move horizontally with respect to each other. The large rear section of the bolt has a nose end that is pointed to fit in the head of the bolt. When the bolt runs home, this nose end enters the head of the bolt and cams the locking rollers into their locking recesses in the receiver. When the cartridge is fired, the pressure on the base of the cartridge, after a period of time, forces the locking rollers back against the nose of the rear bolt section. The rear bolt section, which is held in position only by the energy of the recoil spring, starts to the rear. The bolt is unlocked and the firing pin drawn to the rear, away from the primer, by the rearward movement of the rear bolt section.

Special Note on the CETME Assault Rifle

The CETME assault rifle is worthy of note in several respects. It is composed mainly of stampings, and CETME claims that the weapon can be made in a total working time of nine hours in series manufacture. There are fewer than twenty parts that require machining. The weapon is fitted with a bipod, which, when folded, serves as a fore-end. When supported on the bipod, the gun can be used as a squad automatic weapon. There is, however, one unfortunate aspect to the delayed-blowback system, and this applies to other weapons as well as to the CETME. The system makes no allowance for slow initial extraction, i.e., for a small turning or a short rearward movement of the bolt to

The CETME Modelo 2 chambered for the special long bullet 7.92 × 57mm cartridge.

Two views of the CETME 7.62 × 51mm Modelo A.

"Sport" model of 7.6mm NATO CETME Assault Rifle; fires semiautomatic only.

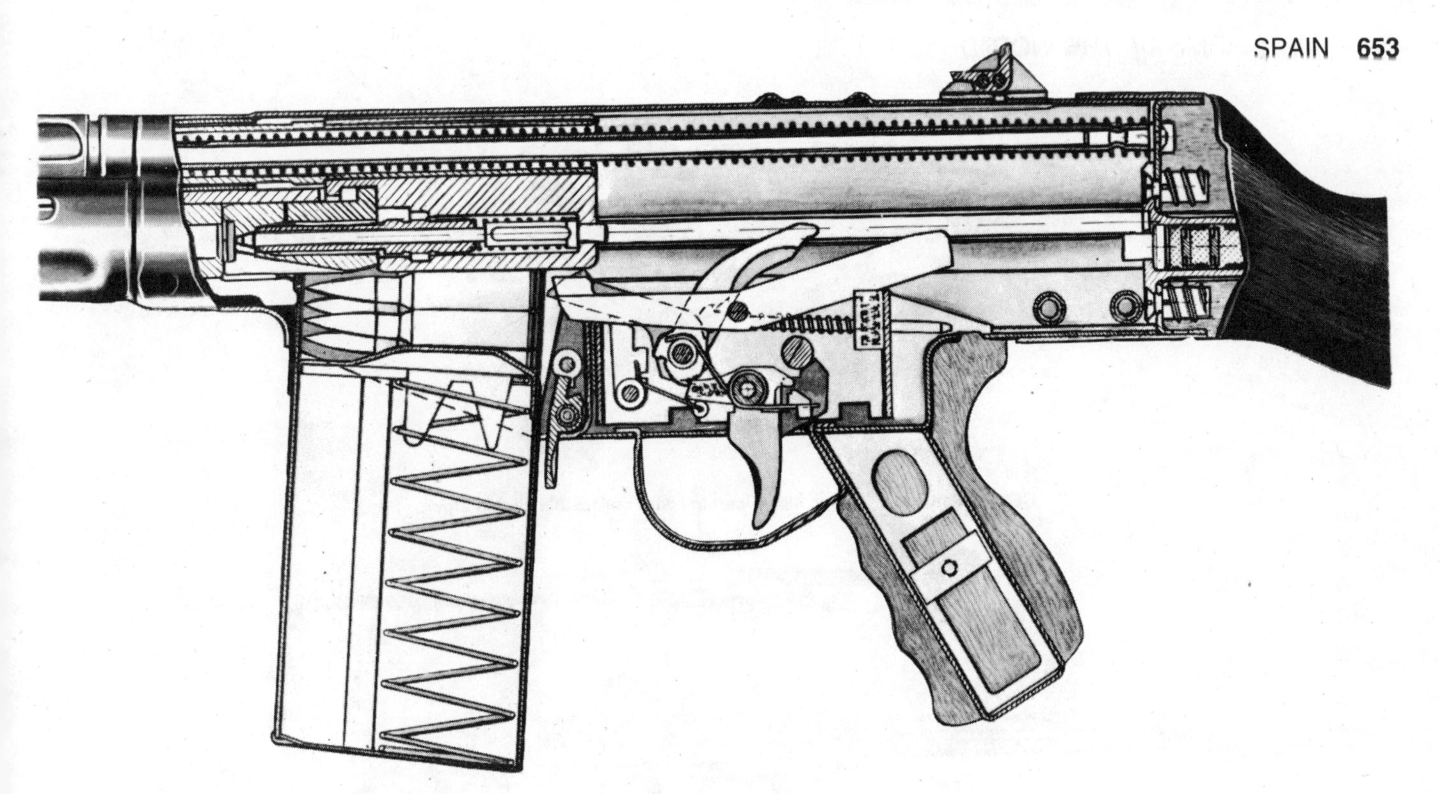

Section view of CETME.

loosen up the spent case before it is withdrawn from the receiver. Delayed-blowback action bolts start to the rear with considerable velocity when the pressure arrives at their opening level. Therefore, delayed-blowback weapons depend on lubricated (oiled) cases, as with the Schwarzlose and Breda Model 30, or on fluted chambers, to make up for their lack of slow initial extraction. Lubricating prevents the case from expanding fully against the walls of the chamber and a fluted chamber allows the forward part of the case to float on gas and therefore not expand completely to the walls of the chamber. Without lubricated cases or fluted chambers (the degree of fluting varies from three-quarters the length of the chamber to only the length of the neck and shoulder), a delayed-blowback may pull the base off the case in extraction. These weapons must therefore be designed around brass cartridges of a certain hardness. If a weapon with a fluted chamber is used with brass considerably softer than that for which it was designed, the brass may be engraved by the flutes, and the bolt may lose enough kinetic energy in extracting the case to prevent it from going into full recoil, with resultant feed jams. It is probable, however, that a delayed blowback designed around the softest brass likely to be encountered in service will do very well with all sorts of brass, hard or soft. Delayed-blowback weapons do not usually give any trouble with steel cases.

The 5.56 × 45mm caliber CETME Modelo L. This rifle is 36.22 inches overall with a 15.75 inch barrel, and it weighs 7.275 pounds.

CETME Modelo LC 5.56 × 45mm carbine with collapsible sliding stock.

Field-stripped CETME MODELO L.

The Arma Automatica CETME Modelo R firing port weapon in its ball mount. This 7.62 × 51mm NATO caliber weapon is 28.16 inches long and weighs 14.11 pounds with its special barrel collar.

***Fusil de Precision CETME*, a 7.62 × 51mm NATO sniping rifle built upon a Spanish Mauser receiver. This weapon is 45.47 inches overall with a 25.20 inch barrel. The rifle weighs 10.36 pounds.**

OBSOLETE SPANISH RIFLES

Spain adopted its first Mauser in 1891, chambered for the 7.65mm Mauser cartridge. A rifle and carbine version of this weapon, closely resembling the Argentine Model 1891, were each produced. A considerable quantity of these weapons, especially the carbines, were captured by United States troops in Cuba during the Spanish-American War. The 1891 has an inline type magazine, which protrudes below the stock. The 1892 Mauser adopted by Spain introduced the 7mm cartridge and the non-rotating extractor attached to the bolt by a collar, which is found on later Mausers and the US Springfield Model 1903.

The most famous and most significant of the Spanish Mausers was the 7mm Model 1893. The Model 1893 introduced the integral staggered-row box magazine used in all future Mausers. The Model 1893 also had a simplified safety lock and an improved bolt stop. A carbine version of this weapon is the 7mm Model 1895, which is stocked to the muzzle as was the fashion with Mauser carbines in those days.

The Model 1893 bolt does not have a third or safety lug or the enlarged shield on the bolt sleeve, as does the Model 98 bolt and bolt sleeve. A number of variations of the Model 1893, in addition to the Model 1895 carbine, were made.

A short rifle version of the Model 1893, and a 1916 short rifle, were also issued. The Model 1916 short rifle was made in very large quantities during the Spanish Civil war. Stocks of 7mm M1893 rifles and 7mm M1916 short rifles on hand were converted to 7.62mm NATO, probably for issue to reserve forces.

During the Spanish Civil war, large quantities of French, Soviet, Italian and German rifles came into Spain. The "Standard Model" Mauser was among the weapons procured by Spain at this time. It was chambered for the 7.92mm cartridge, the standard rifle and machine-gun service cartridge, until 1957.

In 1943, Spain adopted a modified copy of the German 7.92mm Kar 98k, which in the Spanish service is called the Model 1943. This rifle has been replaced by the 7.62mm NATO CETME Model 58 assault rifle, which was adopted in September 1957.

Spanish 7mm Model 1893 Mauser Rifle.

7mm Model 1916 Short Rifle.

7.92mm Model 1943 Rifle.

CHARACTERISTICS OF SPANISH RIFLES

	Model 1893	Model 1893 Short Rifle	Model 1895 Carbine	Model 1916 Short Rifle
Caliber:	7mm.	7mm.	7mm.	7mm.
System of operation:	Turn bolt.	Turn bolt.	Turn bolt.	Turn bolt.
Length overall:	48.6 in.	41.3 in.	37 in.	40.9 in.
Barrel length:	29.1 in.	21.75 in.	17.56 in.	Approx 23.6 in.
Feed device:	5-round, staggered row, non-detachable, box magazine.	5-round, staggered row, non-detachable, box magazine.	5-round, staggered row, non-detachable, box magazine.	5-round, staggered row, non-detachable, box magazine.
Sights: Front:	Barley corn.	Barley corn.	Barley corn w/ears.	Barley corn w/ears.
Rear:	Leaf.	Leaf.	Leaf.	Tangent.
Weight:	8.8 lb.	8.3 lb.	7.5 lb.	8.4 lb.
Muzzle velocity:	2650 f.p.s.	Approx 2650 f.p.s.	Approx 2575 f.p.s.	Approx 2625 f.p.s.
Cyclic rate:	—	—	—	—

CHARACTERISTICS OF SPANISH RIFLES (Cont'd)

	Mauser Standard Rifle	Model 1943	CETME Model 58
Caliber:	7.92mm.	7.92mm.	7.62mm. NATO.
System of operation:	Turn bolt.	Turn bolt.	Delayed blowback selective fire.
Length overall:	43.6 in.	43.7 in.	39.37 in.
Barrel length:	23.62 in.	Approx 23.7 in.	Approx 17 in.
Feed device:	5-round, staggered row, non-detachable, box magazine.	5-round, staggered row, non-detachable, box magazine.	20-round, staggered row, detachable, box magazine.
Sights: Front:	Barley corn.	Barley corn.	Hooded blade.
Rear:	Tangent.	Tangent.	Tangent.
Weight:	8.8 lb.	8.8 lb.	11.32 lb.
Muzzle velocity:	2360 f.p.s.	2360 f.p.s.	2493 f.p.s.
Cyclic rate:	—	—	600 r.p.m.

SPANISH SUBMACHINE GUNS

A considerable number of submachine guns have been developed in Spain, and copies of foreign submachine guns have been made, as well. Prior to World War II, a modified copy of the Bergmann MP28II, chambered for the 9mm Largo (Bergmann Bayard), was made in Spain in limited numbers. This weapon is distinguishable from the Standard MP2811 by its oversized bolt handle, sling swivels and the brass trigger guard and magazine housing.

The Spaniards also manufactured a modified copy of the Vollmer Erma chambered for the 9mm Largo cartridge, which they called the Model 1941/44. The weapon was made at La Coruna until the mid-fifties. The principal difference between the German-produced weapon and the Spanish-produced weapon is that the Spanish weapon has a completely different type of safety from that used on the German weapon.

The Star firm designed a series of submachine guns in the mid-thirties known as the S135, RU35 and TN35. The S135 had an adjustable cyclic rate of fire, 300 or 700 rounds per minute. The RU35 has a cyclic rate of 300 rounds per minute and the TN35 has a cyclic rate of 700 rounds per minute. A limited number of the S135 and RU35 guns were used in the Spanish Civil war, but the weapon was never standard in the Spanish Army.

The Star TN35 was tested by the United States as the "Atlantic" submachine gun in 1942 but was not found acceptable for service. The British and Germans also apparently tested guns of this series but were not interested. All activity on these guns ceased about 1942.

The Labora 1938, sometimes known as the Fontbernat, is another Spanish designed submachine gun that was made in limited quantities during the Spanish Civil war. Other than being made out of expensive machined material during a war and having a push-button type selector, the Labora has no unusual or unique features. In 1953, another submachine gun developed by CETME was introduced. This weapon, known as the A.D.A.S.A. Model 1953, will fire either the 9mm Largo cartridge or the 9mm Parabellum. The weapon resembles the Soviet PPS M1943 and was apparently made in very limited quantity for the Spanish Air Force.

In 1959, the Parinco Model 3R was patented. This is another blowback-operated weapon that has appeared in several forms. The Parinco has a plastic frame and trigger housing and a grip

CHARACTERISTICS OF SPANISH SERVICE SUBMACHINE GUNS

	Model Z-45	Parinco Model 3R	A.D.A.S.A. Model 1953	Star Model Z-62
Caliber:	9mm Largo.	9mm Largo.	9mm Largo or 9mm Parabellum.	9mm Largo, also made in 9mm Parabellum.
System of operation:	Blowback, selective fire.	Blowback, selective fire.	Blowback, selective fire.	Blowback, selective fire.
Length overall:				
Stock extended:	33.1 in.	31.4 in.	31.9 in.	27.6 in.
Stock folded:	23 in.	24.4 in.	24 in.	18.9 in.
Barrel length:	7.9 in.	10 in.	9.85 in.	7.9 in.
Feed device:	30-round, staggered-row detachable box magazine.	32-round, staggered row detachable box magazine.	30-round in-line, detachable box magazine.	20-, 30- or 40-round. detachable box magazine.
Sights: Front:	Blade w/protecting ears.	Blade.	Hooded blade.	Blade.
Rear:	L-type w/notch	Notch.	L-type w/notch.	"L" w/apertures for 100 and 200 meters
Weight:	10 lb. (loaded).	6.8 lb.	7 lb.	6.3 lb w/empty 30-round magazine
Muzzle velocity:	1250 f.p.s.	Approx. 1270 f.p.s.	Approx. 1270 f.p.s.	Approx. 1200-1800 f.p.s.
Cyclic rate:	450 r.p.m.	600 r.p.m.	600 r.p.m.	550 r.p.m.

safety. The trigger group can be rotated on a pin mounted behind the magazine housing and swung down away from the receiver for maintenance and cleaning. The Parinco has been made in 9mm Parabellum and 9mm Largo.

The standard submachine gun of the Spanish Army is the Star Model Z45, which is modeled on the MP40 and differs from that weapon only in the following:

(1) The bolt handle of the Z45 is on the right rather than on the left as on the MP40.

(2) The Z45 has a metal barrel jacket; the MP40 has none.

(3) The barrel and jacket of the Z45 can be removed rapidly by twisting the compensator. The barrel of the MP40 is not easily removable.

(4) The Z45 is a selective-fire weapon; the MP40 has only automatic fire capability.

(5) The Z45 has wooden grips and handguard; the grips and handguard of the MP40 are plastic.

The Z45 has been made with a fixed wooden stock in addition to the more common folding steel stock.

The Z45 has been supplied to Chile, Cuba, Portugal and Saudi Arabia.

Star has developed a later weapon, the Z62, which is a very compact weapon with a tubular steel combined barrel jacket/receiver similar to the British L2A3 Sterling submachine gun.

Characteristics are limited to current Spanish service submachine guns of Spanish origin; other submachine guns used by Spain are covered under country of origin.

THE STAR Z-62 SUBMACHINE GUN

This weapon was adopted by the *Guardia Civil* and the Army of Spain. It had a number of interesting features and was a considerable improvement over the Z-45 in lightness and bulk, safety, maintenance and reliability under rigorous service conditions, i.e., sand, mud, etc.

How to Load and Fire the Z-62

The magazine is loaded in the normal fashion. If the weapon has a magazine in place, it can be removed by depressing the magazine catch, which is on the left side of the magazine housing. Insert loaded magazine in magazine housing as far as it will go. Unlock—pull rearward—the cocking handle located on the left forward side of the barrel jacket. Pull cocking handle completely to the rear and release. The cocking handle will return to the forward position, fold and lock itself to the barrel jacket. The weapon is now cocked and ready to fire; pressure on the lower part of the trigger produces semiautomatic fire, and pressure on the top of the trigger produces automatic fire. To put the weapon on "safe", push the safety button, located in the upper section of the pistol grip, from right to left. This blocks the sear with the bolt in either open or closed (battery) position. If the bolt is forward when the safety is engaged, it cannot be drawn to the rear.

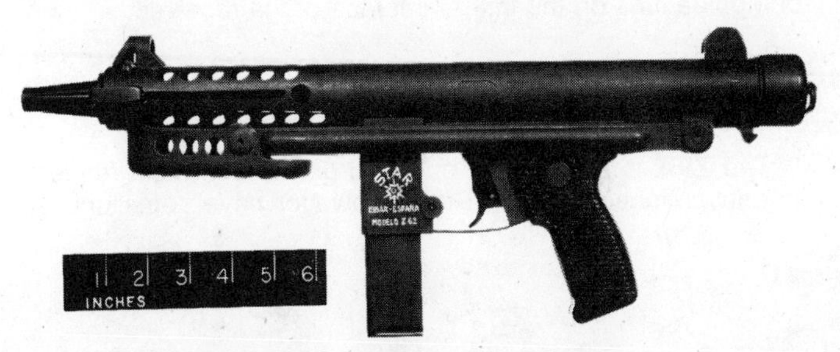

9mm Star Model Z-62 Submachine Gun.

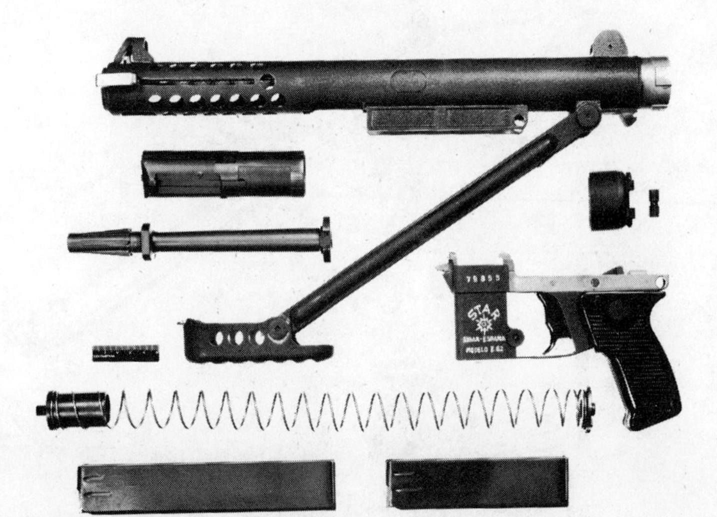

9mm Star Model Z-62 Submarine Gun field stripped.

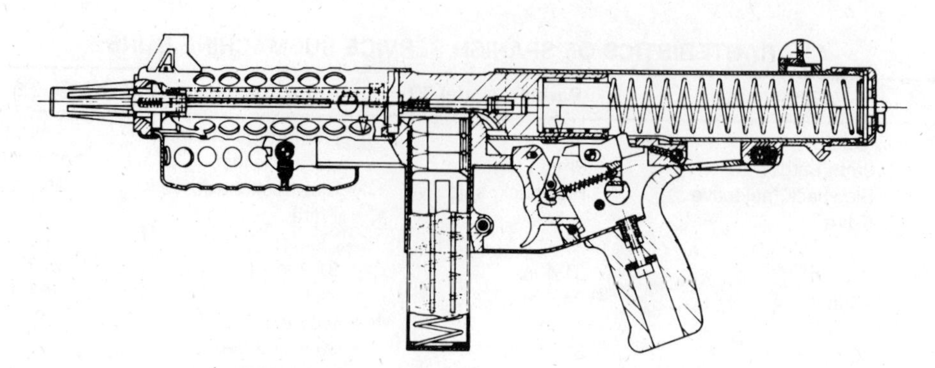

Sectional view of 9mm Star Model Z-62 Submachine Gun.

How To Field Strip the Z-62

Remove the magazine as described above. Push the recoil spring stop, which protrudes through the middle of the receiver cap, in and rotate the receiver cap either way until it disengages from its mounting lugs on the receiver. Carefully remove cap—which is under the tension of the operating spring—and withdraw recoil spring with recoil spring stop and guide. Point the weapon downward and draw cocking handle to the rear; release cocking handle, tilt gun up and the bolt will slide out the rear. The pistol grip/frame can be removed by pushing out the pin mounted at the top rear. The stock must be fixed before the pistol grip/frame can be removed. After the pin, which has a ball type retaining catch, has been removed, the pistol grip/frame can be swung down and disengaged from its mounting in the lower front of the receiver. Grasp barrel at the muzzle and rotate it clockwise a quarter of a turn and tilt the weapon down and slightly to the right. Let the barrel slide back and out of the receiver, ensuring that the "U" shaped notch below the breech is in line with the bolt guide lugs on the inner right face of the receiver.

How the Z-62 Works

The Z-62 is a blowback operated gun and in most respects is conventional enough. It does however have some unusual features which are very interesting. Most submachine guns are potentially dangerous if dropped when the bolt is forward and a loaded magazine is in place. Unless the bolt is locked in place by a grip safety or manual safety, the bolt may go to the rear just far enough to pick up and fire a cartridge. The Z-62 has solved this problem in an unusual way. In the bolt, behind the spring loaded firing pin, are mounted a hammer and a spring loaded inertia bolt safety. The hammer engages the firing pin when a rod-shaped piece, called the tripping lever, engages the hammer. The tripping lever protrudes through the bolt, and when the bolt engages the rear of the barrel it is forced rearward striking the hammer, which pivots and strikes the firing pin driving it forward to function the cartridge. The bolt inertia safety, which is mounted on the same pin as the hammer, is pushed outward into a recess in the receiver/barrel jacket when the bolt is forward. This prevents the bolt from coming to the rear if the weapon is dropped. The tripping lever disengages this lock at the same time as it strikes the hammer, thereby preventing it from interfering with rearward movement of the bolt after firing.

The spring loaded cocking handle and slide assembly do not reciprocate with the bolt, and the automatic folding of the cocking handle makes for ease of handling during firing. Considering everything, the Z-62 was a very well made gun and was easy to maintain in the field.

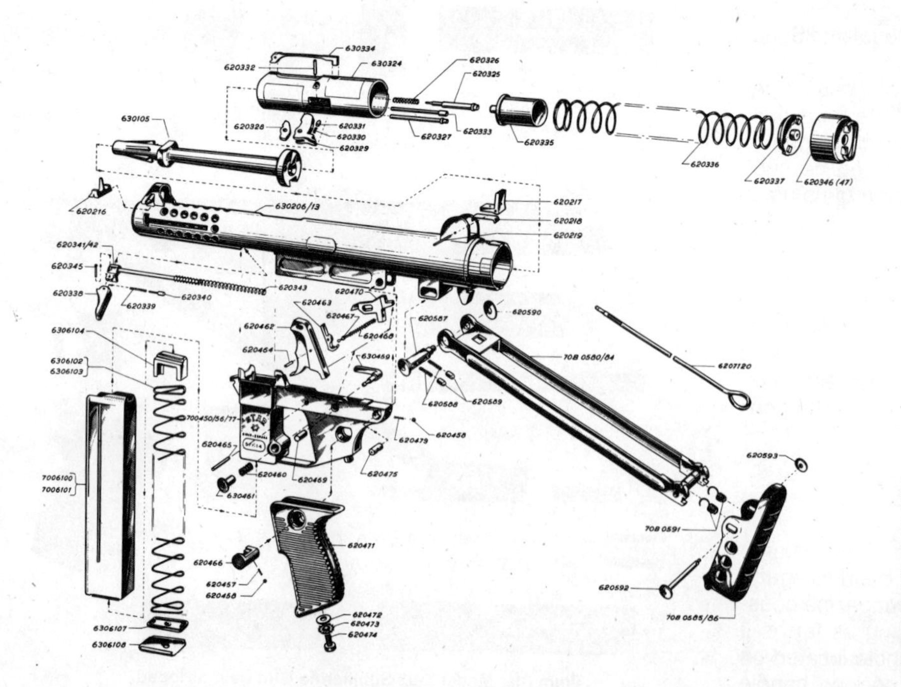

Exploded view of the 9 × 19mm Z-62 submachine gun. With the exception of an improved trigger system, the current Z-70 is the same as the Z-62. In the Z-62, single-shot fire was obtained by pulling the lower portion of the trigger, and full automatic fire was obtained by pulling the upper portion of the trigger. This two-stage trigger mechanism proved troublesome, and in the Z-70 it has been replaced by a more conventional trigger mechanism and a mechanical selector mechanism.

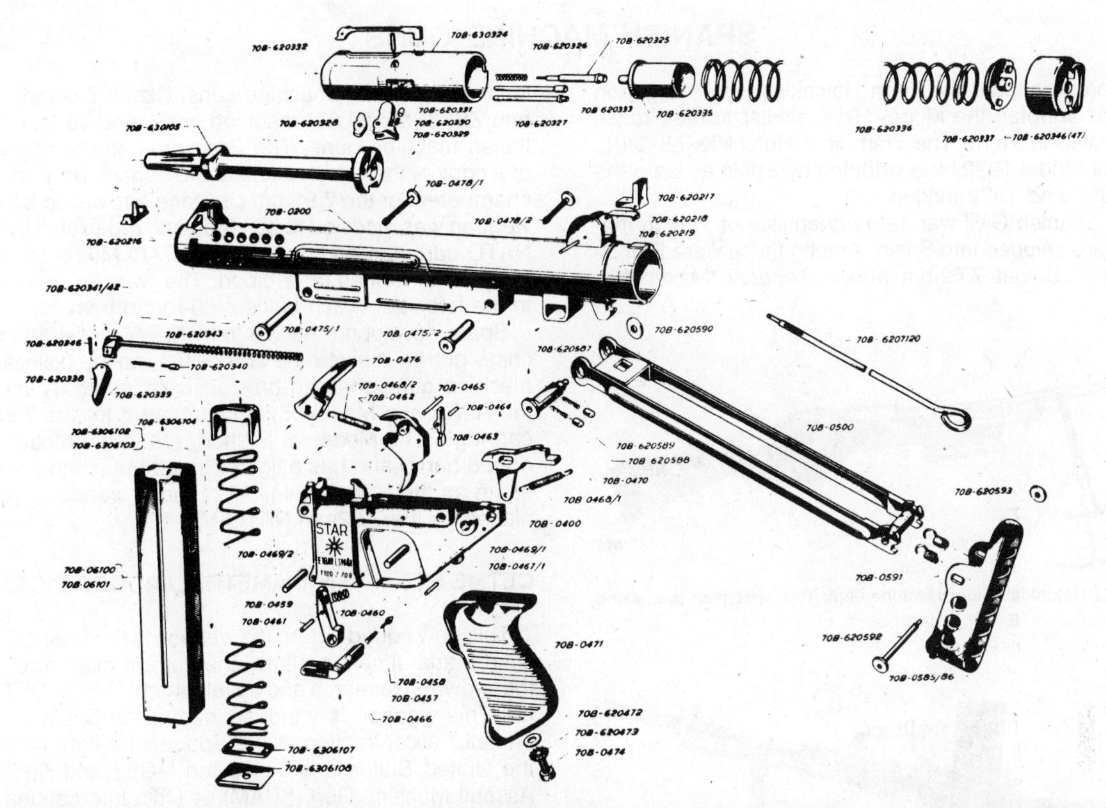

Exploded view of the 9 × 19mm Z-70. Compare the trigger mechanism of this weapon with that of the Z-62 on the preceding page.

SUBFUSIL CETME C-2

CETME developed a submachine gun in the mid-1960s—the Subfusil CETME C-2—in 9 × 19mm and 9 × 23mm (Bergmann-Bayard), but success in selling this Sterling-like design has eluded CETME.

SPANISH MACHINE GUNS

The Spaniards adopted the 7mm Hotchkiss machine gun in 1907 and later adopted the Model 1914, similar to the French Model 1914, also in 7mm. The 7mm light Hotchkiss Model II, also called the Model 1922, was adopted by Spain as were the Madsen in 1907 and 1922 models.

During the Spanish Civil war, large quantities of foreign machine guns were shipped into Spain. Among these were French Hotchkiss guns, Soviet 7.62mm Maxim Tokarev, Maxim Koleshnikov, and DP machine guns, Czech 7.92mm ZB26, ZB30, and ZB53 (Model 37) machine guns, and various German and Italian machine guns. The Spaniards started the manufacture of a copy of the Czech ZB26 at the Fabrica de Armas de Oviedo chambered for the 7.92mm cartridge and called it the FAO. This weapon was modified to belt feed and rebarreled for the 7.62mm NATO cartridge and is called the FAO Model 59. It is used on a tripod as well as on a bipod. The weapon uses a drum fitted to the left side, which holds a 50-round belt.

Spain developed the 7.92mm ALFA Model 1944 heavy machine gun to replace the somewhat motley collection of heavy machine guns she had previously collected. A later version of the ALFA, the Model 55, is chambered for the 7.62mm NATO cartridge. The Model 55 is shorter than the Model 1944, has a ribbed barrel and has a lighter tripod but is otherwise much the same as the Model 1944. All of these weapons have been replaced by the 7.62mm NATO MG 42/59.

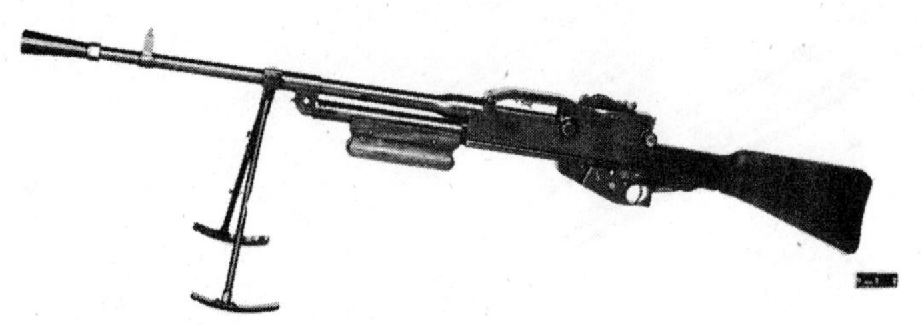

7mm Model 1922 Hotchkiss Light Machine Gun. This specimen is missing a pistol grip.

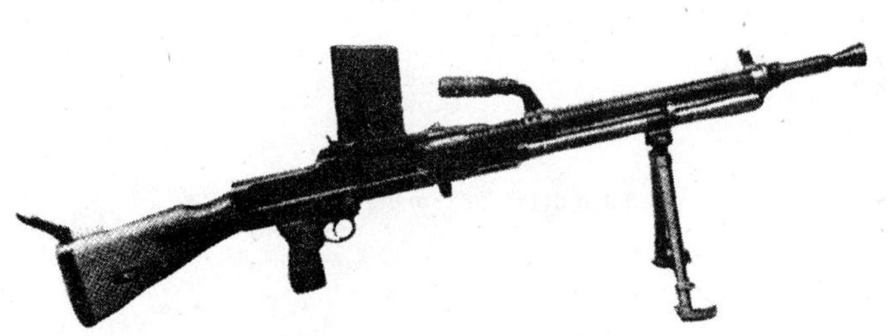

7.92mm FAO Light Machine Gun.

CETME 5.56 × 45mm AMETRALLADORA LIGERA "AMELI"

This new squad automatic weapon is the brainchild of Colonel Jose Maria Jimenez Alfaro, the current director of the government-owned research and development facility, CETME, in Madrid. This weapon is variously named. In Spain it is called the "Ameli," a contraction of the Spanish for light machine gun; in the United States it is called the MG82 and Special Purpose Assault Machine Gun (SPAM) by Odin International. The Ameli is a direct descendant of the 7.92mm MG42 and 7.62mm NATO MG42/59. In addition to exploiting the proven MG42 roller-locked retarded recoil locking system, the MG82 incorporates a muzzle booster to insure the necessary operating forces from the 5.56

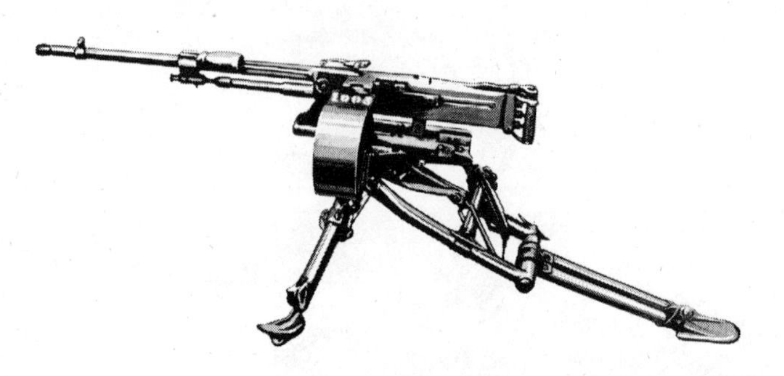

Spanish 7.92mm Alfa Machine Gun M1944.

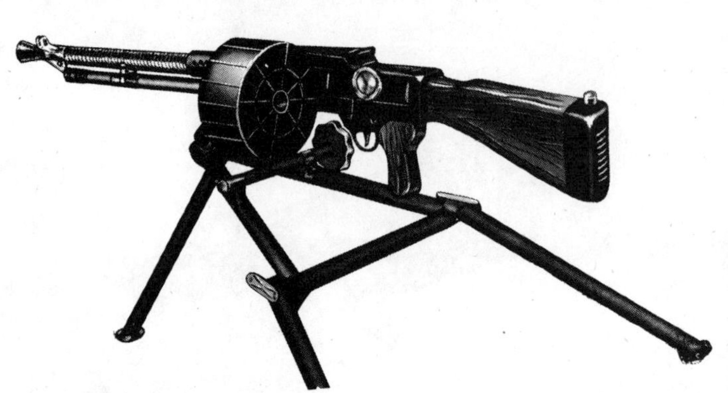

7.62mm NATO FAO Model 59 Light Machine Gun.

CHARACTERISTICS OF OBSOLETE SPANISH SERVICE MACHINE GUNS

	FAO Model 59	ALFA Model 1944	ALFA Model 55
Caliber:	7.62mm NATO.	7.92mm.	7.62mm NATO.
System of operation:	Gas, automatic only.	Gas, selective fire.	Gas, selective fire.
Length overall:	43.3 in.	Approx. 57 in. (gun).	43.4 in. (gun).
Barrel length:	21.8 in.	29.53 in.	Approx. 24 in.
Feed device:	50-round, metallic link belt, loaded in drum.	100-round, metallic link belt, loaded in drum.	100-round, metallic link belt, loaded in drum.
Sights: Front:	Hooded blade.	Blade.	Blade.
Rear:	Radial tangent.	Leaf.	Leaf.
Weight:	Approx. 20 lb.	28.66 lb.	28.6 lb.
Tripod weight:	14.3 lb.	59.5 lb.	59.52 lb.
Muzzle velocity:	Approx. 2800 f.p.s.	2493 f.p.s.	2825 f.p.s.
Cyclic rate:	650 r.p.m.	780 r.p.m.	780 r.p.m.

Two views of the CETME 5.56 × 45mm Ameli.

× 45mm cartridge. The barrel and chamber of the weapon are hard chrome plated, and the chamber has flutes to assist extraction. The metal parts of the Ameli have an olive drab baked-on surface finish, and the olive drab stock is made from a high impact–type nylon material. The Ameli is a very interesting and robust design. It weighs only 14.8 pounds empty, is 36.6 inches long, and has a cyclic rate between 900 and 1,250 shots per minute.

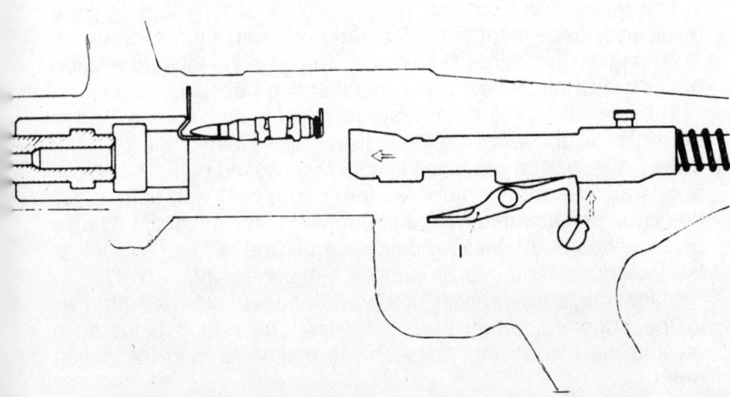

The CETME Ameli cocked and ready to fire from the open bolt.

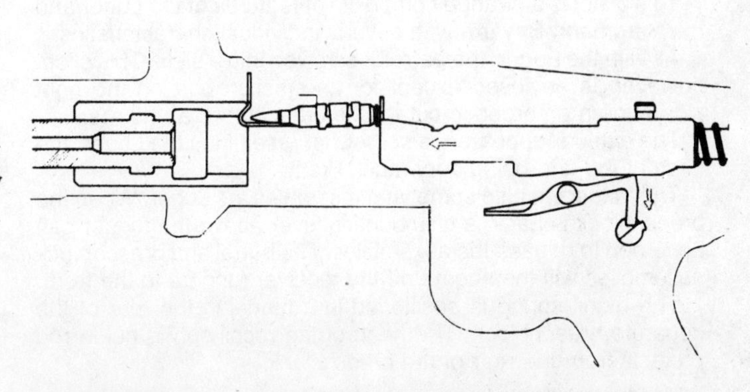

The CETME Ameli at the moment that the bolt strips the cartridge from the M27-type metallic link.

43 Sweden

SMALL ARMS IN SERVICE

Handguns:	9 × 19mm Parabellum Lahti Model 1940	Made by Husqvarna Vapenfabrik.
Submachine guns:	9 × 19mm Parabellum Model 1945 (Carl Gustav)	Made by the FFV Ordnance Division at Eskiltuna.
Rifles:	7.62 × 51mm NATO AK4	G3 made under license by FFV Ordnance Division.
Machine guns:	7.62 × 51mm NATO Kulspruta 58 MAG	Fabrique Nationale MAG made under license by FFV Ordnance Division.
	7.62 × 51mm NATO Kulspruta M42B	Recalibered Browning, similar to US M1919A6.
	7.62 × 51mm NATO Kulspruta M36	Recalibered Browning, similar to US M1917A1 water cooled.
	7.62 × 51mm NATO HK21A1	Source unknown.

Swedish Handguns*

The Swedes adopted a 7.5mm Nagant revolver known as the Model 87. These revolvers remained in limited service until after World War II. Sweden brought the FN 9mm Browning Long M1903 automatic pistol into service as the Model 07. This relatively low-powered, blowback operated pistol was the standard service pistol until the adoption of the 9mm Parabellum Lahti M40 pistol in 1940. A limited number of 9mm Walther P38 pistols were purchased in 1939 and are known in Sweden as the Model 39. The principal Swedish pistol has been the Lahti Model 40.

SWEDISH 9mm LAHTI MODEL 1940 PISTOL

This pistol is basically the same as the Finnish Model 1935 Lahti but differs slightly from the later designed Finnish Lahti in the mounting of the recoil spring. The Model 40 has its grip grooved for a shoulder stock. The Swedish Lahti was manufactured at the Husqvarna Vapenfabrik A.B.

9mm Parabellum Model 40 Pistol.

How the Model 40 Works

This pistol is a strange composite of features of the Luger and the Bergmann-Bayard, with several individual characteristics.

As with the Luger, the barrel is screwed into a slide. This slide, however, is enclosed except for an ejection port on the right side and an emergence cut for the breechblock at the rear.

The exterior appearance somewhat resembles the Luger, the pitch of the grip being very much like it.

The takedown also somewhat resembles the Luger. When the breechblock is back, a dismounting lever above the trigger can be turned to release the assemblies. The barrel and breechblock assemblies will then come off the receiver runners to the front.

The recoil spring is positioned in a tunnel in the rear of the separate breechblock. The head of the recoil spring guide rod projects from the rear of the breechblock.

*For additional details see *Handguns of the World*.

Wings with finger grips are machined at the sides of the rear of the breechblock to permit pulling back for loading.

The breechblock travels in the slide. A surface is machined on its under face to force back and ride over an internal hammer. The firing pin is mounted in a sloped tunnel in the forward section of the breechblock below the recoil spring position.

The breech lock is a separate removable unit. It is housed in the bulge in the extreme rear of the slide. When the pistol is locked, the locking piece is in its slide recesses on right and left and is also in engagement (its inner surfaces) with cuts in the breechblock. During rearward and forward movement, cam faces on the sides of the locking piece are utilized to raise and lower the lock out of and into breechblock engagement.

Magazine is the familiar box type. It has a follower button as in the Luger to make loading easier. The button actuates a pivoted stop when the magazine is empty, to hold the action open.

1. **Barrel**
2. **Accelerator retainer spring**
3. **Accelerator retainer**
4. **Accelerator**
5. **Recoil spring**
6. **Recoil spring guide**
7. **Frame**
8. **Takedown catch spring**
9. **Takedown catch**
10. **Hold-open catch**
11. **Hold-open catch spring**
12. **Trigger spring pin**
13. **Trigger**
14. **Trigger pin**
15. **Trigger spring**
16. **Trigger bar pin**
17. **Disconnector**
18. **Trigger bar**
19. **Trigger bar spring**
20. **Safety lever and catch**
21. **Magazine catch pin**
22. **Magazine catch spring**
23. **Magazine catch**
24. **Magazine**
25. **Hammer spring**
26. **Hammer spring plunger**
27. **Hammer strut**
28. **Hammer**
29. **Sear**
30. **Sear spring**
31. **Sear pin**
32. **Hammer strut pin**
33. **Hammer pin**
34. **Spring guide nut**
35. **Firing pin spring**
36. **Firing pin**
37. **Bolt**
38. **Firing pin retainer pin**
39. **Extractor**
40. **Ejector**
41. **Yoke (locking block)**

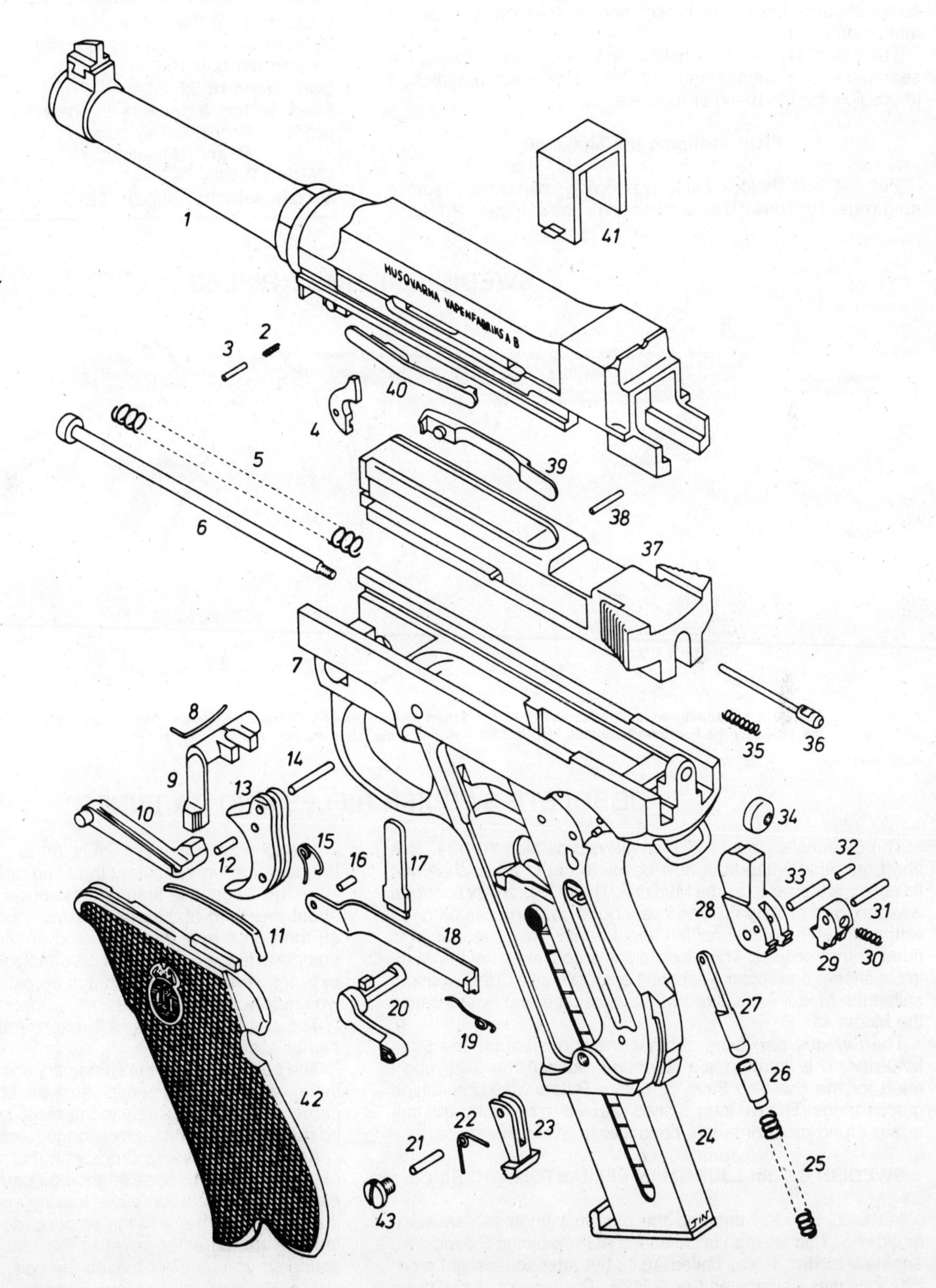

Exploded view of the Swedish m/40 9mm Parabellum Lahti L-35 self-loading pistol.

The trigger is pivoted. It has a bar crossing the grip that transmits the pull on the trigger to release the sear. The internal hammer has the conventional hammer strut, which compresses the coil mainspring below it in the handle. The magazine release catch works off the mainspring.

An unusual device is the accelerator in the forward end of the receiver. It is the Browning MG pivoted type. When barrel travel halts, the barrel flips up the point of the accelerator to deliver an accelerating blow to the breechblock to speed up its rearward movement.

The safety is a positive mechanical block, which forces the sear into locked engagement with the hammer notch, making it impossible for the hammer to rotate.

Field Stripping the Model 40

Pull the breechblock back over empty magazine. Remove magazine. Turn down dismounting lever above trigger. Pull back breechblock to free it from stop. Ease assemblies forward off receiver. Remove breechblock from slide.

Note. Do not unscrew barrel unless necessary.

Removing grips will give access to firing mechanism if necessary.

CHARACTERISTICS OF THE MODEL 40

Caliber: 9mm Parabellum.
System of operation: Recoil operated, semiautomatic.
Length overall: 10.7 in.
Barrel length: 5.5 in.
Feed device: 8-round in-line detachable box magazine.
Sights: **Front:** Barley corn.
Rear: "U" notch.
Weight: 2.4 lbs.
Muzzle velocity: Approx. 1250 f.p.s.

SWEDISH MILITARY RIFLES

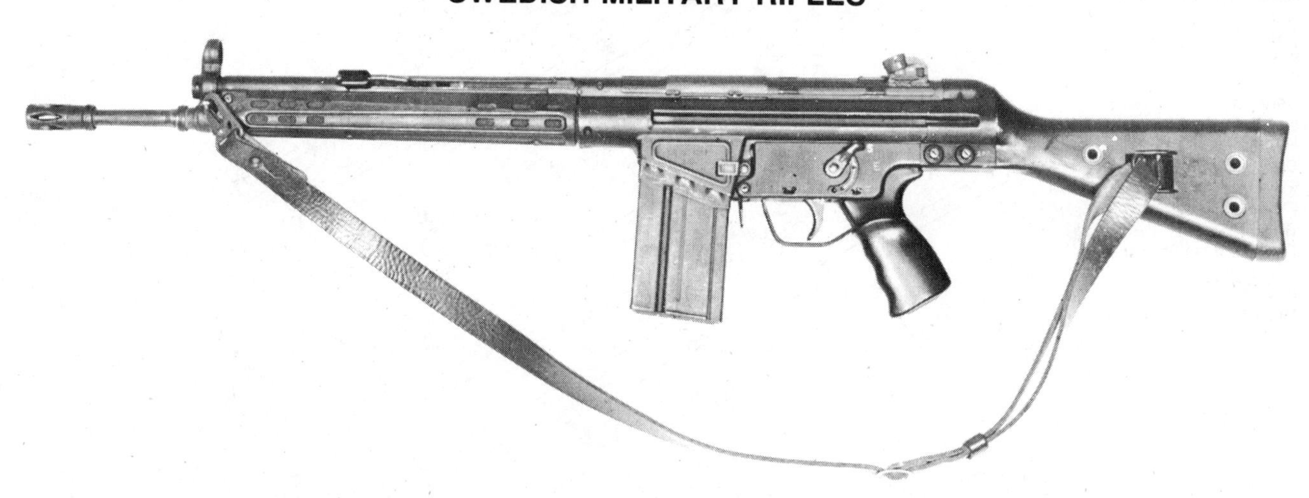

The G3 is manufactured in Sweden as the 7.62 × 51mm AK4 by the FFV Ordnance Division. This weapon will be replaced during the 1980s by the Fabrique Nationale FNC in 5.56 × 45mm. This latter weapon will be made at FFV.

OBSOLETE SWEDISH RIFLES AND CARBINES

The Swedes adopted a 6.5mm Mauser carbine in 1894. This short and light weapon is now obsolete in the Swedish Army. Its action is the same as the M96 rifle. The M96 is a very accurate weapon for a military rifle, and very good shooting can be done with it, if it is in good condition and if a micrometer rear sight is fitted to the receiver. The Model 38 is a conversion of the M96, into a shorter and lighter weapon. A sniper version of this weapon, called the Model 41, is fitted with a telescope that is also called the Model 41.

The Swedes also used another shoulder weapon, the 8mm M40 rifle. This weapon is a German-produced Kar. 98k rebarreled for the Swedish 8mm × 63mm Bofors M1932 machine gun cartridge. The weapon is fitted with a muzzle brake and has a four-round magazine. It is no longer in Swedish service.

SWEDISH 6.5mm LJUNGMAN SEMIAUTOMATIC RIFLE

In 1942, Sweden adopted the 6.5mm Ljungman semiautomatic rifle. This weapon is called the AG42B by the Swedes. As originally issued, it was called AG42 but after some slight modifications was designated the AG42B. The Ljungman has been manufactured in Egypt in caliber 7.92mm.

The Ljungman gas system is unusual in that it seeks to avoid the use of an intermediary thrusting piston for its actuation.

As the weapon is fired and the bullet passes by the gas port about one-third of the distance from the muzzle, gas is tapped off through a hole into the gas cylinder on top of the barrel in standard fashion. The gas is directed back through the gas cylinder where it impinges upon an extension of the bolt carrier whose forward end passes into a mating receiver cut.

The gas thus delivers its thrust directly to the face of the bolt carrier itself.

The bolt carrier has the customary short free travel in its guides in the receiver, during which time the bolt itself, lying below the carrier, is locked securely to the receiver by lugs at its rear end, being depressed into corresponding receiver cuts.

As the gas pressure drops with the passing of the bullet out of the barrel, cam faces on the bolt carrier engage the bolt and elevate its rear section out of locking engagement. At this point, the bolt and carrier with the empty cartridge case gripped in the face of the extractor travel to the rear to compress the recoil springs in the receiver behind the bolt carrier. During the recoil stroke, the magazine follower and spring elevate the next cartridge into line for feeding. The bolt and carrier at end of stroke move ahead under the thrust of the counter recoiling spring. The top cartridge is stripped from the magazine and fed up the ramp into the chamber. The extractor snaps into the cannelure of the cartridge case. At this point, where the bolt face is halted against the breech face of the barrel, the bolt carrier still has forward travel under the impetus of the spring behind it. Its cam faces thrust the rear of the bolt down into locking recesses in the receiver. The bolt carrier continues forward as a free member from this point. The front face of the bolt carrier rides into a cupshaped port in the receiver, which forms the emergence port for the gas at the next discharge.

The rifle is equipped with bayonet. The muzzle brake built into the rifle by slotting across the top of the barrel serves to hold the muzzle down to some degree during firing.

The gas system of the Ljungman, in which the gas blows directly back upon the bolt or the bolt carrier, is also used in the French M1949 and M1949/56 and in the United States (Stoner-developed) AR-10 and AR-15 rifles. The Ljungman never replaced the Mauser in Swedish service and was issued on a basis of several per squad.

In 1965, Sweden adopted the 7.62mm NATO G3 rifle with some modifications. This rifle replaced the AG42B, the Mausers, and the Model 37 Browning automatic rifle. It was made at the government Carl Gustafs Stads (FFV Ordnance Division) plant at Eskilstuna and at the Husqvarna firm. The Swedish rifle, which is called the AK4, has a plastic handguard and stock, a modified magazine, and has an aperture type rear sight. The detailed characteristics of the G3 rifle are given in Chapter 21 on West Germany.

CHARACTERISTICS OF OBSOLETE SWEDISH RIFLES AND CARBINES

	Model 94 Carbine	Model 96 Rifle	Model 38 Rifle
Caliber:	6.5 × 55mm	6.5 × 55mm.	6.5 × 55mm.
System of operation:	Turn bolt.	Turn bolt.	Turn bolt.
Length overall:	37.6 in.	49.6 in.	44.1 in.
Barrel length:	17.7 in.	29.1 in.	23.6 in.
Feed device:	5-round, staggered row, non-detachable, box magazine.	5-round, staggered row, non-detachable, box magazine.	5-round, staggered row, non-detachable, box magazine.
Sights: Front:	Barley corn.	Barley corn.	Barley corn.
Rear:	Leaf.	Leaf.	Leaf.
Weight:	7.6 lb.	9.1 lb.	8.5 lb.
Muzzle velocity:	2313 f.p.s.	2625 f.p.s.	2460 f.p.s.

CHARACTERISTICS OF OBSOLETE SWEDISH RIFLES AND CARBINES (Cont'd)

	Model 40 Rifle	Model 41 Rifle	Model 42B Rifle
Caliber:	8 × 63mm.	6.5 × 55mm.	6.5 × 55mm.
System of operation:	Turn bolt.	Turn bolt.	Gas-operated, semiautomatic.
Length overall:	Approx 49.2 in.	49.6 in.	47.8 in.
Barrel length:	Approx 29.1 in.	29.1 in.	24.5 in.
Feed device:	4-round, staggered row, non-detachable, box magazine.	5-round, staggered row, non-detachable, box magazine.	10-round, staggered row, non-detachable, box magazine.
Sights: Front:	Barley corn.	3X or 4x telescopic.	Hooded post.
Rear:	Tangent.		Tangent.
Weight:	Approx. 9.5 lb.	11.1 lb.	10.4 lb.
Muzzle velocity:	2428 f.p.s.	2625 f.p.s.	2460 f.p.s.

OBSOLETE SWEDISH AUTOMATIC RIFLES

SWEDISH 6.5mm AUTOMATIC RIFLE MODEL 21

Sweden adopted the Browning automatic rifle in 1921, at which time she bought a quantity of these weapons from Colt. The M1921 is now obsolete in Sweden. The weapon is quite similar to the US M1918 BAR except that it has a separate pistol grip and dust covers for the ejection port and the magazine well.

CHARACTERISTICS OF M21

Caliber: 6.5 mm × 55mm Mauser rimless cartridge.
System of operation: Gas, selective fire.
Weight: 19.2 lbs.
Length, overall: 44 in. (approx.).
Barrel length: 26.4 in.
Feed device: 20-round, detachable box magazine.
Sights: Front: Blade.
Rear: Leaf w/notch.
Muzzle velocity: 2460 f.p.s. (approx.).
Cyclic rate: 500 r.p.m.

6.5mm Model 94 Carbine.

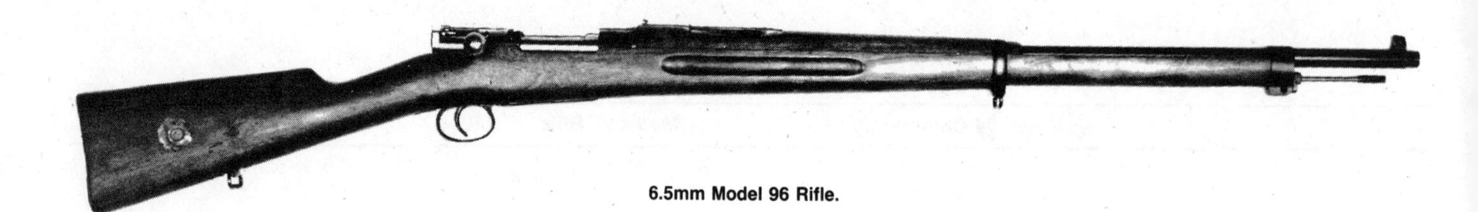

6.5mm Model 96 Rifle.

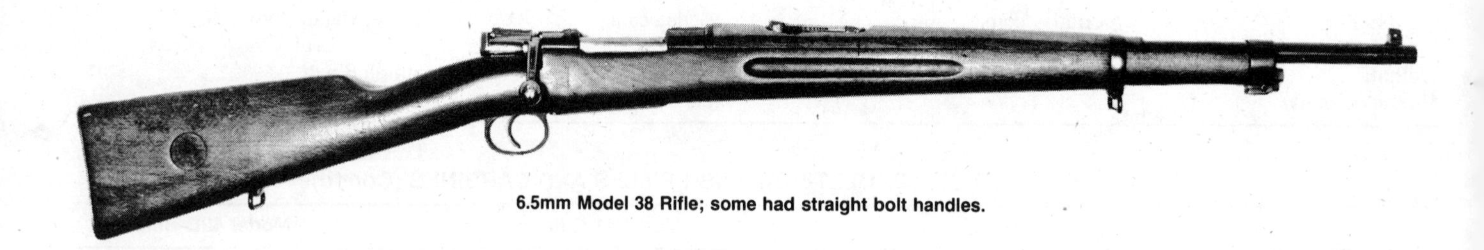

6.5mm Model 38 Rifle; some had straight bolt handles.

8mm Model 40 Rifle. Note muzzle brake.

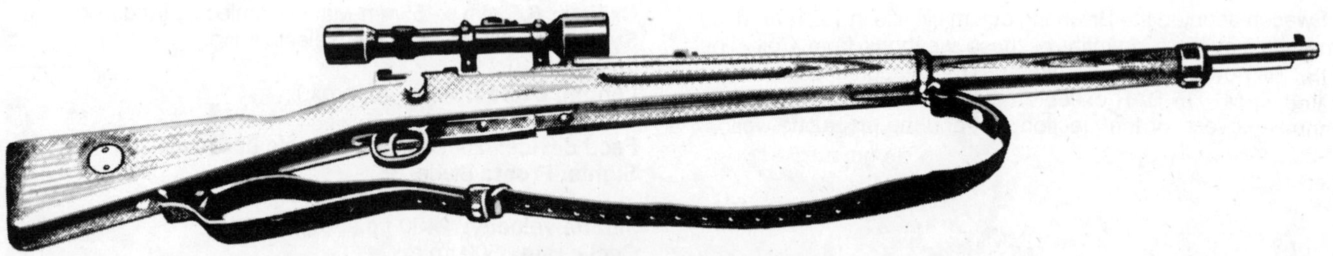

6.5mm Model 41 Rifle. Note scope.

6.5mm Model 37 Automatic Rifle.

SWEDISH 6.5mm AUTOMATIC RIFLE MODEL 37

CHARACTERISTICS OF MODEL 37

Caliber: 6.5mm.
System of operation: Gas, selective fire.
Weight: 20.9 lbs.
Length, overall: 46.1 in.
Barrel length: 24 in. (approx.).
Feed device: 20-round, detachable, staggered-row box magazine.
Sights: Front: Hooded blade.
Rear: Leaf w/aperture graduated to 1200 meters, aperture battle sight.
Muzzle velocity: 2460 f.p.s. (approx.).
Cyclic rate: 480 r.p.m.

This modification of the Browning Automatic rifle was developed by the Swedish government arsenal, Carl Gustaf State Arms Factory. It was replaced in the Swedish Army by the 7.62mm Model 58 machine gun.

How to Load and Fire the Model 37

Cock the weapon by pulling the operating handle completely to the rear—push operating handle back to the forward position. The bolt will remain open, since the weapon fires from an open bolt. The safety selector lever is on the left side of the receiver at the top of the pistol grip. Settings are marked "P"-semiautomatic fire; "A"-full automatic fire; and "S"-safety. Insert loaded magazine in magazine well in underside of receiver. If the trigger is pulled and safety selector is set on "P" or "A", the weapon will fire.

Barrel Change. Beginning with an empty weapon, cock gun; push barrel latch (located at top left front of receiver) to the rear. Lift carrying handle and turn ¼ turn to the right. Pull barrel out.

Field Stripping the Model 37

Remove the trigger guard retaining pin, located at left forward end of trigger guard. Pull pistol grip backward and upward and remove. The recoil spring guide bears against the rear of the slide; gently push it back and disengage it from the slide. Line up the hammer pin with the holes in the side of the receiver and with the hole in the operating lever. Push hammer pin out from left to right. Remove operating lever by pulling straight back and remove hammer from the receiver. Push out bolt guide with screwdriver blade and remove bolt link, bolt lock and bolt. Push barrel latch to rear and remove as described above. Remove gas cylinder retaining pin at forward left side of receiver and slide gas cylinder and bipod straight ahead. Pull slide and gas piston assembly out of receiver.

To reassemble the weapon, reverse the above procedure.

How the Model 37 Works

The Model 37 works basically the same as the US Browning Automatic Rifle M1918 covered in the chapter on the United States. The recoil spring in the Model 37 is located in the butt as with the F.N. Type D Browning Automatic Rifle.

SWEDISH SUBMACHINE GUNS

The Swedish Army has used a number of submachine guns in the past twenty years. The US-made Thompson was used in limited numbers by the Swedes as the 11mm M40, and the Finnish-designed Suomi was used in two versions, the Model 37-39 and the Model 37-39F. The 9mm Bergmann M34 was used as the Model 39.

9mm Model 37-39 Submachine Gun.

9mm Model 37-39F Submachine Gun.

THE SWEDISH 9mm CARL GUSTAF SUBMACHINE GUN MODEL 45

The 9mm Model 45 (commonly known as the Carl Gustaf) is the current standard submachine gun in the Swedish Army. This weapon has also been produced in Egypt. It was designed and is produced in Sweden at the Carl Gustafs Stads Gevärsfaktori in Eskilstuna. All submachine guns in service fire the 9mm Parabellum cartridge.

CHARACTERISTICS OF M45 SUBMACHINE GUN

Caliber: 9mm Parabellum
Magazine: Detachable box, staggered dual line.
Magazine capacity: 36 cartridges.
Overall length: w/stock extended: 31.8 in.
w/stock folded: 21.7 in.
Barrel length: 8 in.
Weight: 9.25 lbs. (loaded).
Cyclic rate of fire: 550 to 600 per minute.
Type of fire: Automatic.
Operation: Elementary blowback. Actuation of bolt mechanism is by projection of spent case.
Sights: Front: Protected post.
Rear: L type.
Muzzle velocity: 1200 f.p.s.

How to Load and Fire the Model 45

Pull bolt handle to the rear; bolt and bolt handle will remain to the rear since the gun fires from an open bolt. The weapon can be put on safe when the gun is cocked by pulling the bolt handle beyond the sear and engaging it in the cutout section above the bolt handle track of the receiver. Insert magazine, take weapon off safe, pull trigger, and weapon will fire. To put the weapon on safe with the bolt forward, push down on bolt handle.

How to Field Strip the Model 45

Remove magazine; bolt should be forward. Depress catch in center of the receiver cap (rear of receiver), turn receiver cap slightly counter clockwise, and cap will come off. Remove recoil spring and bolt. Push in on barrel jacket nut catch with a drift and unscrew barrel jacket nut. Remove barrel jacket and barrel. To reassemble, carry out in reverse the steps outlined above.

Swedish 9mm M45 Submachine Gun with fixed magazine guide, stock fixed.

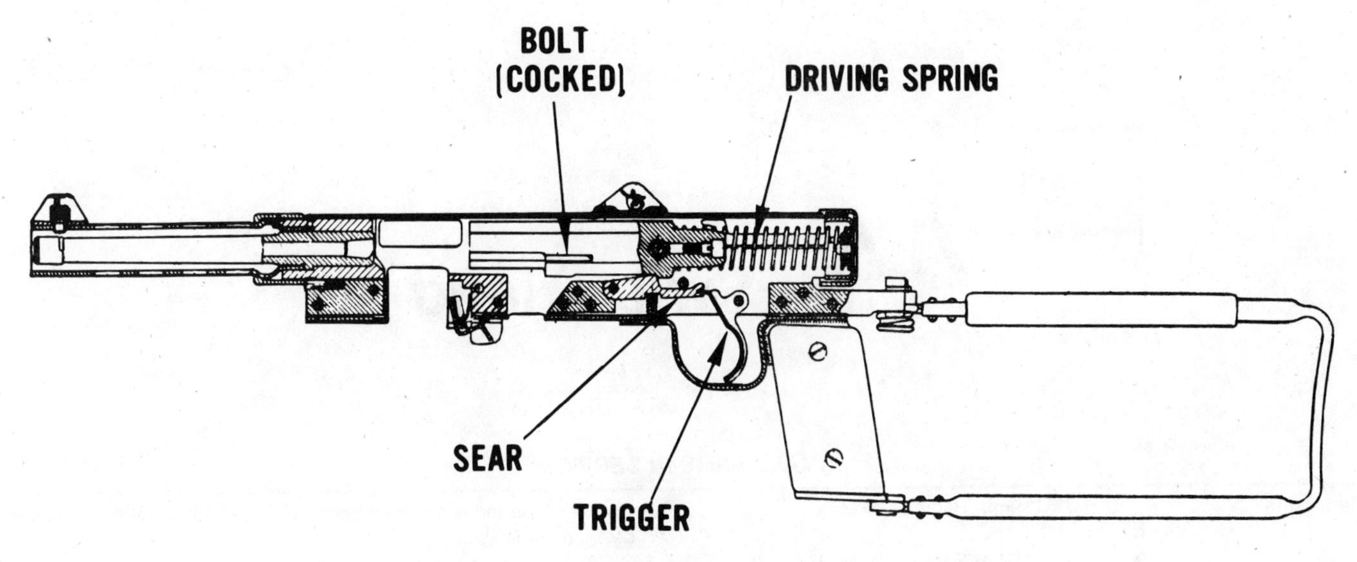

Section view of M45 9mm Submachine Gun.

Special Note on the Model 45

There have been at least three different variations of this weapon since it first appeared in 1945. The variations are in the barrel jacket and magazine guide. Some of the M45s have a removable magazine guide. When the guide is removed, the weapon can be used with the thicker Suomi submachine gun magazine. The weapon is principally made of stampings and has a folding steel stock.

SWEDISH MACHINE GUNS

Since the Second World War, the Swedish armed forces have rationalized their wide variety of machine guns. Sweden was one of the first nations to adopt the Fabrique Nationale MAG and called it the MAG 58. At first these MAGs were chambered to fire the 6.5 × 55mm Mauser cartridge. By the late 1970s, the Swedes had standardized on the 7.62 × 51mm NATO cartridge for their machine guns. As part of this standardization, the Swedes have converted their 6.5mm Browning machine guns to 7.62mm NATO as well. Some obsolete models include the following: The Schwarzlose in 6.5mm, Model 14 and Model 14-29 was used by Sweden; this weapon is out of service at present. A series of rifle-caliber Browning guns were used, chambered for either the 6.5mm cartridge or the 8mm M1932 cartridge (8mm × 63mm). The Model 36 is a water-cooled gun, and the Model 42 is an air-cooled gun. Both of these guns are similar in loading, firing, functioning and field stripping to the US cal. .30 Browning guns, although they have a somewhat higher cyclic rate of fire. The Czech-made ZB26 light machine gun in caliber 6.5mm was also used in Sweden as the Model 39. The German-made Knorr Brense in 6.5mm was also used in limited numbers as the Model 40.

6.5mm MODEL 36 AND MODEL 42B MACHINE GUNS

The Swedish Brownings are basically the same as the United States caliber .30 Brownings. The Model 36 is a water-cooled gun similar to the United States Model 1917A1, and the Model 42B is an air-cooled gun similar to the United States Model 1919A6; it can be used on a bipod with a shoulder stock. As originally made, this weapon was called the Model 42. The Swedish Brownings use spade type grips rather than the pistol type of the United States Brownings. A Model 39 aircraft Browning gun was also used.

The loading, firing, field stripping and functioning of the Swedish Brownings are essentially the same as those of the United States Brownings.

44 Switzerland

SMALL ARMS IN SERVICE

Handguns:	9 × 19mm Parabellum M75	Schweizerische Industrie-Gesellschaft (SIG) an Rheinfall. Commercial designation is P220.
	9 × 19mm Parabellum M49	SIG P210.
	7.65 × 22mm Parabellum 06/29	Obsolete; some reserve usage.
Submachine guns:	9 × 19mm Parabellum M43/44	Obsolete; reserve use possible.
Rifles:	7.5 × 55.4mm Sturmgewehr M57	SIG, with some assembly and component manufacture by the Waffenfabrik, Bern—the State arms factory.
	7.5 × 55.4mm Zf Kar55	Sniper rifle; also called M31/55.
Machine guns:	7.5 × 55.4mm MG51	Waffenfabrik, Bern.
	7.5 × 55.4mm MG25	Waffenfabrik, Bern; obsolete. Some reserve use.
	12.7 × 99mm M2 HB	US-made.

SWISS HANDGUNS*

7.65mm Model 1900 Luger.

*For additional details see *Handguns of the World*.

The Swiss adopted a 7.5mm revolver in 1882; this double-action weapon was modified in 1929 and the model designation changed to Model 1882/29. Although these revolvers are no longer in service, they remained a service weapon of the Swiss Army for an extraordinarily long time.

In 1900, Switzerland adopted the first military production model Parabellum. The Swiss Model 1900 Parabellum is chambered for the 7.65mm Parabellum cartridge (called caliber .30 Luger in the United States) and has a grip safety. In 1906, a modification of this pistol was adopted, the Model 1906, which differs from the Model 1900 in its extractor, firing pin, recoil spring and in various minor details. Another Parabellum was adopted by Switzerland in 1929; this weapon may be called the Model 1906/29 or Model 1929. This weapon has a different side plate and take-down catch than the Model 1906 and differs from that model in other minor points. The 1929 modifications were designed by Waffenfabrik Bern, and the weapons were manufactured at that government arsenal.

M75, 9mm PARABELLUM PISTOL

The latest addition to the Swiss pistol family is the SIG-Sauer P220, which has been adopted as the Model 1975 Pistol by the Swiss Armed Forces. This pistol departs from the standard Browning type action. The blow-back from the detonated cartridge thrusts the barrel and slide to the rear. After these parts have recoiled a distance of about .12 inch, the lock between the barrel and the slide (number 24 in the accompanying illustration) is released; the barrel is swung down and held. In the loading cycle just before reaching the full forward position, the barrel is locked to the slide again. As with many European handguns, this pistol has several safety features. There is a firing pin lock that guarantees absolute safety, even if the loaded weapon is dropped with the hammer cocked. With this feature and the double action mechanism, the designers eliminated the standard manual safety. A de-cocking lever permits hazardless lowering of the hammer into a safety notch without touching the trigger. All these combined features make this pistol an excellent military weapon. (See Chapters 4 and 21 for more details.)

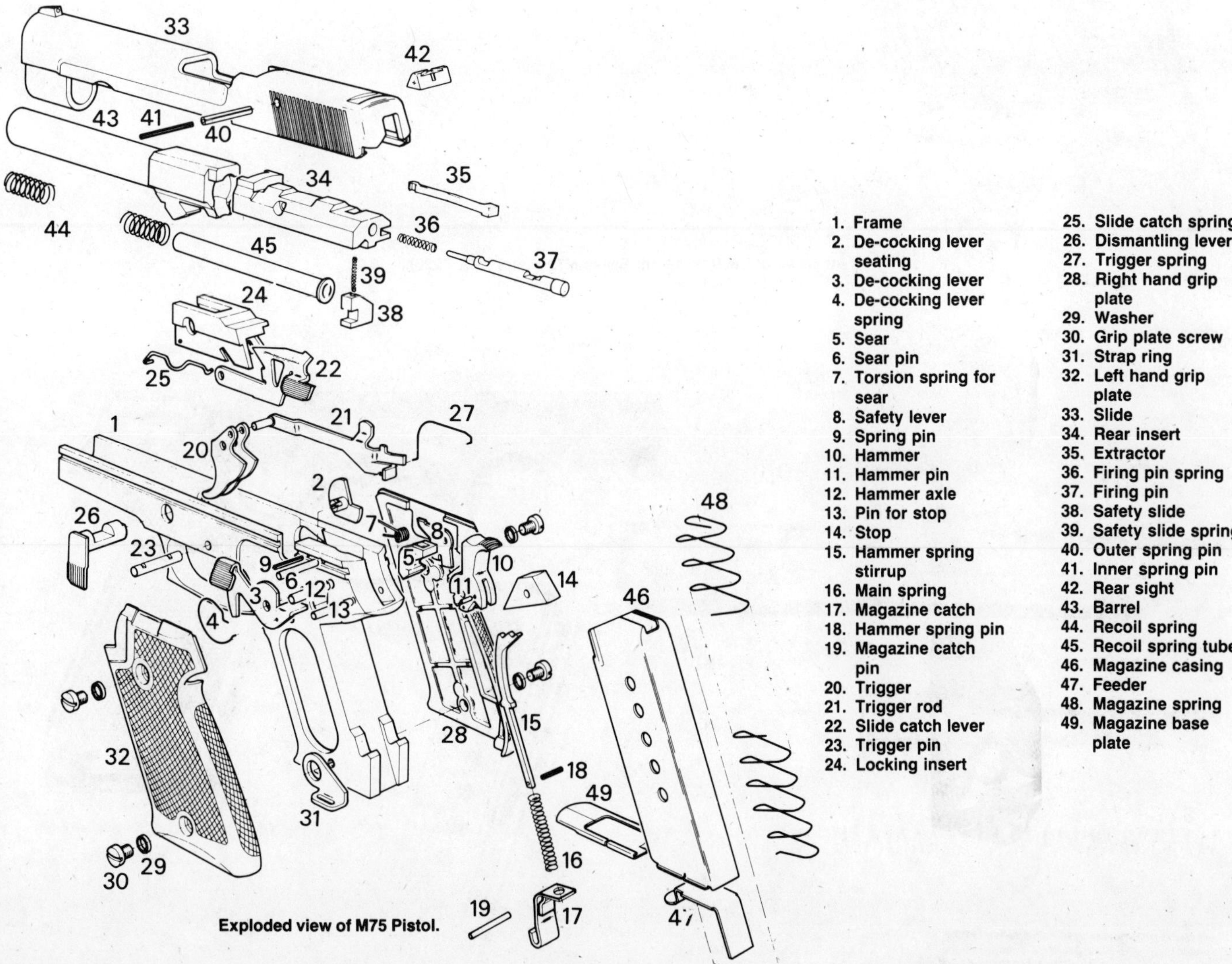

Exploded view of M75 Pistol.

Two views of the 9 × 19mm Swiss M79 Pistol (SIG P220).

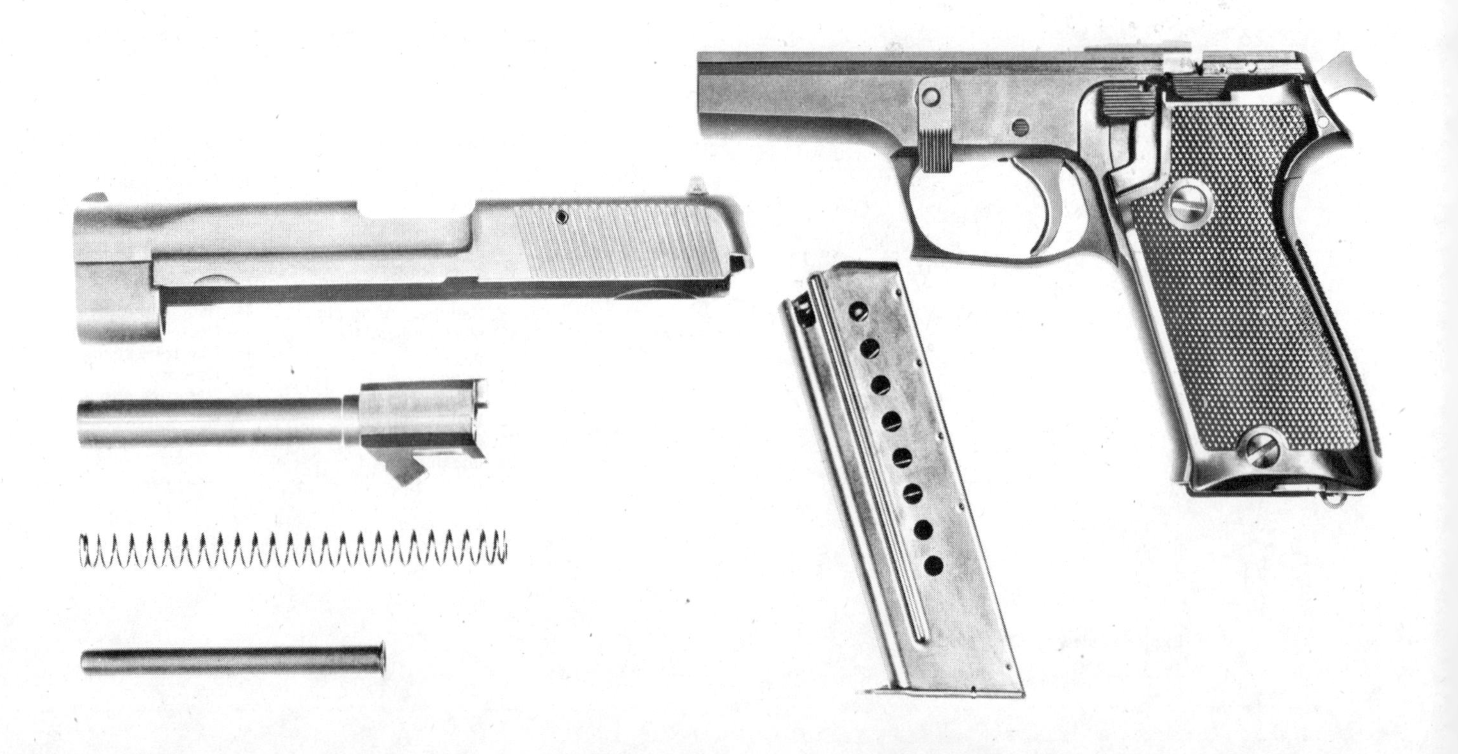

THE SIG P210

The Schweizerische Industrie-Gesellschaft (SIG) Neuhausen am Rheinfall, Switzerland, obtained a license for the Petter System (see Chapter 18) and their P210 pistol was based upon that modified Browning concept. Development of the P210 continued over many years (1938–1946) and included the 9mm SIG Model 44/16 as one of the prototypes in that series. In 1949, the P210

Selfloading Pistol Neuhausen Model 44/16.
Locked breech, small exposed hammer. Thumb safety and bottom magazine release. Magazine capacity 16 cartridges. Calibre 9mm Parabellum (Luger). This was a forerunner of the SP 47/8 (9mm Model 49) and only very few specimens of this weapon were made.

(formerly designated the SP47/8) replaced the Swiss Luger Model 29 as the official service weapon of the Swiss Army. In 1975, the *Pistole 49* was replaced by the *Pistole 75* (or SIG-Sauer P220). Nearly 200,000 *Pistole 49* were made for the Swiss Army, most are still in use, and SIG still offers the P210 commercially. The P210 has been used by the Danish Army and the West German Border Guard (*Bundesgrenzschutz*).

Design and Construction

The P210 (*Pistole 49*) has several features common with other handguns derived from the basic Browning concept. A nose on the underside of the barrel is slotted to permit the slide stop pin to pass through, locking it firmly to the receiver. During recoil, the barrel and slide are firmly locked together for approximately ¼ inch of travel. At this point, the cam slot in the locking nose below the barrel, working against the transverse slide stop pin, serves to cam the barrel down to draw its locking lugs out of their engaging seats in the underside of the breech of the slide.

This weapon is beautifully made in the finest Swiss tradition. Because of the care taken in fitting the barrel, it demonstrates considerable accuracy.

The takedown is quite simple. In general, it follows the standard Colt-Browning procedure except for the firing mechanism.

How to Field Strip the Model 49

Hold pistol in right hand. Push magazine catch at bottom of butt with left thumb and withdraw magazine.

Grip right hand around rear of receiver and slide and draw slide back about ½ inch. With finger of left hand, push slide stop from right side. Withdraw slide stop from left. This can be done when the thumb piece is lined up with its dismounting notch in the slide.

Ease slide, barrel and recoil spring forward.

Grip hammer with thumb and forefinger and work free and up out of receiver. This unit carries the hammer, disconnector, sear and mainspring with its guide compression rod. These may be dismounted if desired, but this is not necessary.

Press recoil spring guide forward from rear and lift out. Spring is captive on its guide rod and will not come loose.

Grip barrel lug and work barrel slightly back and up when it can be lifted out of the slide.

Firing pin may be removed if desired by pushing in on its head and drawing stop down out of slide cut as in Colt Govt. .45.

9mm SIG Pistol P210-2, Swiss Army Model 49, this pistol was formerly called the SP47/8 by SIG.

How the P210 Operates

A loaded magazine is inserted in the butt in standard fashion and pushed in until the catch holds it. The slide is then withdrawn to the full length of its recoil stroke to allow magazine spring to force a cartridge up in line for chambering. Releasing the slide permits the compressed recoil spring to drive the slide forward for chambering and locking.

Applying the thumb safety, which is in a particularly advantageous position for easy operation, locks both the trigger and the trigger bar to prevent firing. This design does not have an automatic grip safety, but it does have, however, the automatic disconnector mechanism which is also standard for the Colt-Browning design. If the slide is not fully forward and the barrel not fully locked into position, the slide is automatically depressing the disconnector, which in turn is pushing the trigger bar down out of possible contact with the trigger. When the slide is fully forward and locked, the spring can force the disconnector up into its cut in the underside of the slide. This allows the trigger, the trigger bar and the sear to function for firing. As the action opens, the slide automatically cams the disconnector down, forcing the breakage in the trigger contact so that on forward motion of the slide the chambered cartridge will not fire. Thus the trigger must be definitely released before contact can again be made for firing the next shot.

This design has several features superior to the Colt Government Model. The forward bushing in the receiver is eliminated, resulting in more machining difficulties but in an inherently simpler design. The recoil spring is under tension in position around its guide rod and during disassembly and reassembly functions as a unit. The unlocking lug on the bottom of the barrel affords much more rigidity during the operation of the pistol. This tends, theoretically at least, to make for more accurate shooting than with the swinging link design. Combining most of the firing mechanism in a single unit (as in the case of the Russian Tokarev pistol) results in a considerable simplification of both design and machining.

To convert the pistol from 9mm to 7.65 Parabellum, it is only necessary to replace the recoil spring and the barrel. The magazine and all other elements will function with either cartridge, since the 9mm was developed directly from the bottle-necked 7.65mm.

A special .22 caliber conversion training unit is also available for use with this pistol for practice shooting. It consists of a lighter weight slide, a special barrel with recoil spring and a magazine. This conversion, of course, does not require a locked breech, hence the barrel is not equipped with locking lugs and does not have any movement. Once installed, it is locked in place. The operation is otherwise identical with that of the heavier caliber.

Considerable attention was given to pitch of grip in the design of this pistol. The center of gravity is at about the trigger point. The locking cam system of barrel operation, as opposed to the swinging link system, prevents movement of the barrel out of the firing axis while the bullet is still in the barrel. This together with the lack of a front bushing makes for a much more solid barrel support.

Field stripped view of an early P210 (SP47/8)pistol. Note the removable hammer assembly.

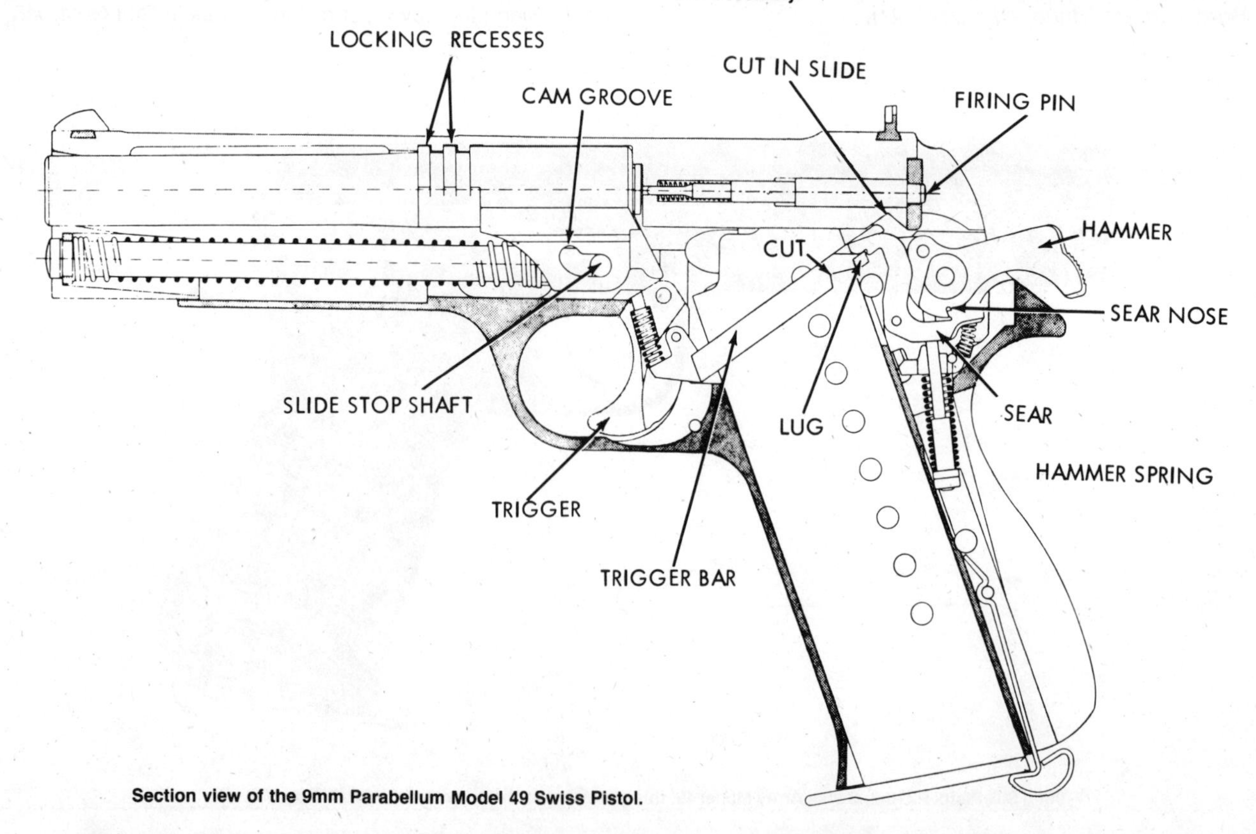

Section view of the 9mm Parabellum Model 49 Swiss Pistol.

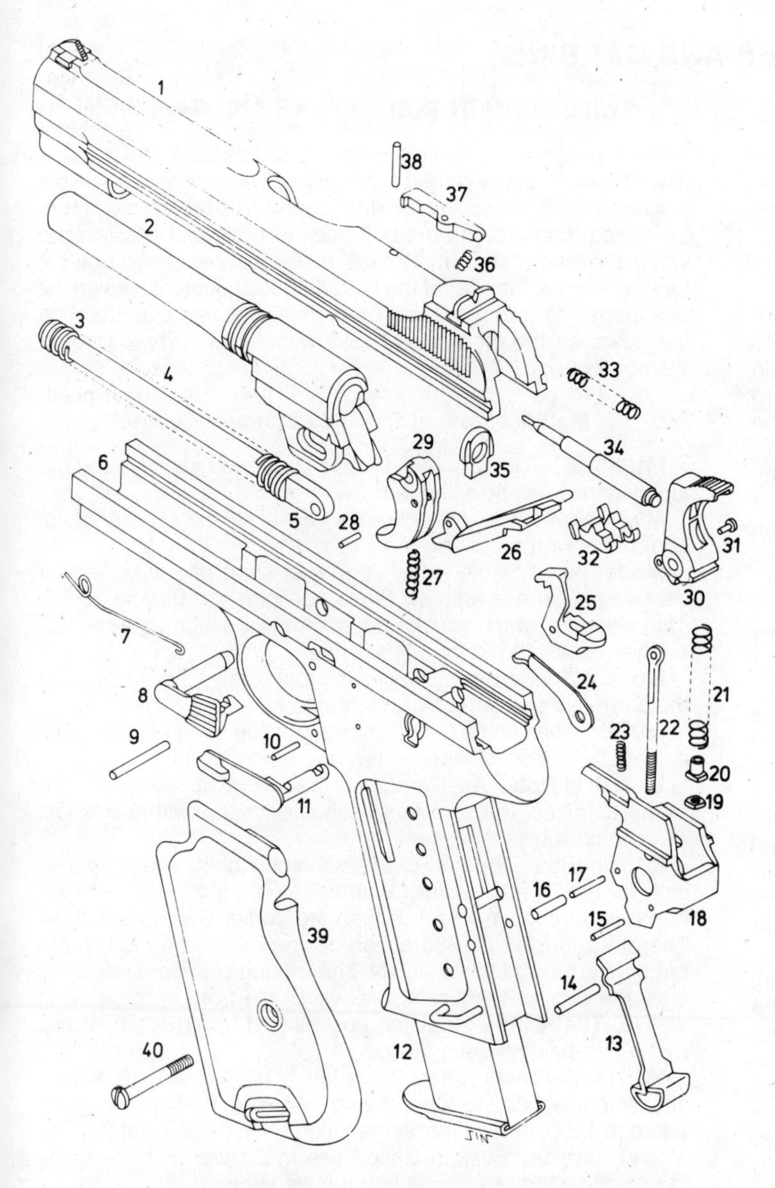

Exploded view of the 9 × 19mm Swiss M49 (SIG P210).

1. Slide
2. Barrel
3. Recoil spring
4. Recoil spring guide
5. Recoil spring guide tip
6. Frame
7. Slide catch spring
8. Slide catch (stop)
9. Trigger pin
10. Pin
11. Safety lever
12. Magazine
13. Magazine catch
14. Magazine catch pin
15–17. Hammer assembly pins
18. Hammer assembly housing
19. Hammer strut nut
20. Hammer strut pressure plate
21. Hammer strut spring
22. Hammer strut
23. Spring
24. Magazine disconnect safety
25. Sear
26. Trigger bar
27. Trigger spring
28. Trigger pin
29. Trigger
30. Hammer
31. Hammer strut pin
32. Sear
33. Firing pin spring
34. Firing pin
35. Firing pin retaining plate
36. Extractor spring
37. Extractor
38. Extractor pin
39. Grip plate (2)
40. Grip plate screw

CHARACTERISTICS OF SWISS SERVICE PISTOLS

	Model 1900	Model 1906	Model 1929	Model 49	Model 75
Caliber:	7.65mm Luger.	7.65mm Luger.	7.65mm Luger.	9mm Parabellum.	9mm Parabellum.
System of operation:	Recoil, semiautomatic.	Recoil, semiautomatic.	Recoil, semiautomatic.	Recoil, semiautomatic.	Recoil, semiautomatic.
Length overall:	9.5 in.	9.5 in.	9.5 in.	8.5 in.	7.8 in.
Barrel length:	4.75 in.	4.75 in.	4.75 in.	4.72 in.	4.41in.
Feed device:	8-round, in-line detachable box magazine.	8-round, in-line detachable box magazine.	8-round, in-line detachable box magazine.	8-round, in-line detachable box magazine.	9-round, in-line detachable box magazine.
Sights: Front:	Blade.	Blade.	Blade.	Blade.	Blade.
Rear:	V-notch.	V-notch.	U-notch.	V-notch.	Square notch.
Weight:	2.25 lbs.	2 lbs. (approx.).	1.98 lbs.	2.14 lbs.	1.94 lbs.
Muzzle velocity:	1200 f.p.s. (approx.).	1200 f.p.s. (approx.).	1200 f.p.s. (approx.).	1150 f.p.s.	1132 f.p.s.

SWISS RIFLES AND CARBINES

SWISS VETTERLI

While Mauser invented the first successful military bolt action, his was not of course the first attempt in the field. The Swiss were actually first in the field to combine the turn bolt and the copper metallic cartridge.

Swiss ordnance men, particularly Frederic Vetterli, had studied closely the American Volcanic and its successful metallic cartridge counterpart, the Henry repeater, as well as the Spencer. They had watched, too, the growing might and arrogance of the Prussian Junkers. They saw Schleswig-Holstein overrun with the aid of the terrible new Dreyse Needle gun; they also saw Austria fall before it, and saw France brace itself for the coming attack, the disaster of 1870–71.

The Swiss set about to devise an arm that might outshoot the German weapon. At this late date, it is difficult to visualize the terror spread by the Dreyse, the first widely used European breechloader. Long before any major European power thought in terms of metallic cartridges, the Swiss actually had a repeater on the drawing boards.

As their whole psychology revolved around defense of their homeland, the rifle-minded Swiss approached the design of their rifle from a far different viewpoint than either the Americans or the other Europeans. Since the American rifle was primarily a hunting or mounted-use arm, US inventors turned in the direction of the lever-action repeater. The typical European military mind outside of Germany thought in terms of better needle guns or conversions to salvage the tremendous stocks of muzzle-loaders. Germany was oversold by the success of the Dreyse, and Mauser had considerable trouble selling his first rifles! Characteristically, once the Prussians adopted the Mauser, they went all out for further development, right up to the day the United States introduced the Garand.

However, a few Swiss and Austrian designers saw the military value of the bolt repeater very early in the game. Since the vertical magazine system of Lee and Mauser and Mannlicher had not yet appeared, and since the successful American systems were tube repeaters, it is understandable that early European magazine design centered around tubes. Basically what the Swiss did was adapt the Henry-Winchester to their needs. They produced an improved rimfire .41 cartridge with copper case. They altered the King loading gate and the Henry cartridge carrier and applied them to turn-bolt actions, which could be better handled prone than could the lever action.

The Federal Assembly of the Swiss Confederation in July and December 1866 officially approved the adoption of the repeating principle. Production of the Vetterli started in 1867, but it took two years to iron out manufacture. In 1871 and 1878, new models were produced at Waffenfabrik Bern. The rifle was used officially until 1889. As issued, it had a 12-shot magazine. The caliber was 10.4mm. With a 313-grain lead bullet, the rifle achieved a muzzle velocity of 1338 feet per second. Maximum range was given at about 3000 meters.

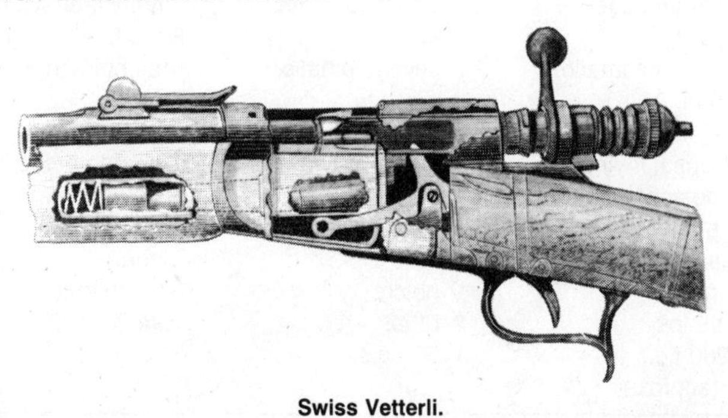

Swiss Vetterli.

SWISS SCHMIDT RUBIN RIFLES AND CARBINES

The next development of interest in Switzerland was in 1883, when Major Rubin designed a straight-pull bolt system. A special revolving cam turned the locking lugs out of receiver engagement and then brought the bolt back in a straight line to eject when the operating handle was pulled to the rear. Lugs for locking were at the rear of the bolt. A special detachable vertical box magazine was provided. The design was improved through the years as the model 1911 and model 1931. The caliber is 7.5mm Swiss.

Modifications of the 1889 Schmidt Rubin

M1889/96. The 1889 action was shortened slightly, and the lug system was strengthened.

M97 Cadet Rifle. A single-shot version of the M1889/96 action made for the training of cadets.

Model 89/00 Short Rifle. A special short rifle designed for the use of Fortress Artillery, Signal, Balloon and Bicycle troops. This weapon, which weighs approximately seven pounds, has a lower velocity than the M1889 rifle.

Model 1905 Carbine. Weighs approximately 7.5 pounds; uses the same action as the M1889/96 rifle.

NOTE: None of the above rifles should be used with the more powerfully loaded Model 1911 7.5 × 55mm cartridges.

M1896/11 Rifle. An alteration of the M1889/96 rifle, and some of the carbines, to make them suitable for use with the Model 11 ammunition.

M1911 Rifle. The first of the Schmidt Rubins made specifically for the 7.5mm Model 11 cartridge. The Model 11 cartridge has a much higher pressure than the earlier Swiss cartridges. Therefore the basic 1889 action design had to be modified to handle the new cartridge safely. The locking lugs were relocated from the rear of the locking sleeve to the front of the locking sleeve. The M1896/11 action is similar. The M1911 rifle also had a bolt-holding-open device.

M1911 Carbine. Differs from the M1911 rifle by having a shorter barrel and stock and being lighter. Its rear sight is graduated to 1500 meters, rather than to the rifle's 2000 meters. The M1911 carbine is still in limited use in Switzerland. It can be identified by the red plastic bolt handle knobs.

M1931 Carbine. Commonly known as the K31. This weapon represents the last basic change in the design of the Schmidt Rubin. The basic difference between this weapon and the earlier Schmidt Rubins is the elimination of the long bolt extension. All earlier Schmidt Rubin bolts had a long bolt extension that actually rammed the cartridge into the chamber, supported its base and extracted it from the chamber. It could be considered that this piece was the bolt body. This piece did not rotate in unlocking; the bolt sleeve rotated, instead. The bolt sleeve, a piece that supported the rear of the extension, had the locking lugs. (In the rifles before 1911, the lugs were at the rear; in the rifles after 1911, at the front.) Thus, an abnormally long receiver was

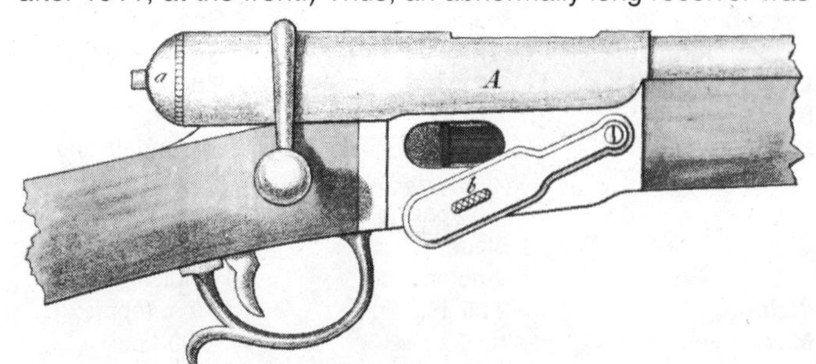

Vetterli Loading Gate, modified from American King type.

Rifle M1911 (above) and Carbine M1911 (below).

Model 31 (K31) Schmidt Rubin. Best of the designs.

needed for these weapons, and the locking lugs were a long way to the rear of the point where they were actually needed. It was not a very satisfactory design from a military point of view, since it meant a great deal of extra length and weight and still did not offer the best type of bolt support. In the K31, the part formerly considered the bolt sleeve became the bolt body itself, and the locking lugs were mounted at its head in the position where they were most needed. This shortening of the bolt assembly has resulted in a much better weapon, which, although it is about the same overall length as the M1911 carbine, has a barrel 2.5 inches longer (made possible by the shorter receiver). The earlier weapons had their magazines placed a considerable distance in front of the trigger guard, because of the position of the long bolt extension. The magazine of the K31, on the other hand, is immediately ahead of the trigger guard. All in all, the K31 is a vastly superior weapon to the earlier Schmidt Rubins.

Model 31/42 Carbine, K31/42. One of the two sniper versions of the K31. It has a 1.8-power scope permanently built into the left side of the receiver. The head of this scope resembles a periscope, and when in use, is lifted up; when not in use, it is folded down against the side of the carbine. The weapon has a tangent type metallic rear sight, graduated from 100 to 1000 meters.

Model 31/43 Carbine, K31/43. Equipped with a 2.8-power scope mounted like that of the K31/42. Its metallic rear sight is graduated from 100 to 700 meters.

Model 31/55 Sniper Rifle. Note bipod position and telescopic sight.

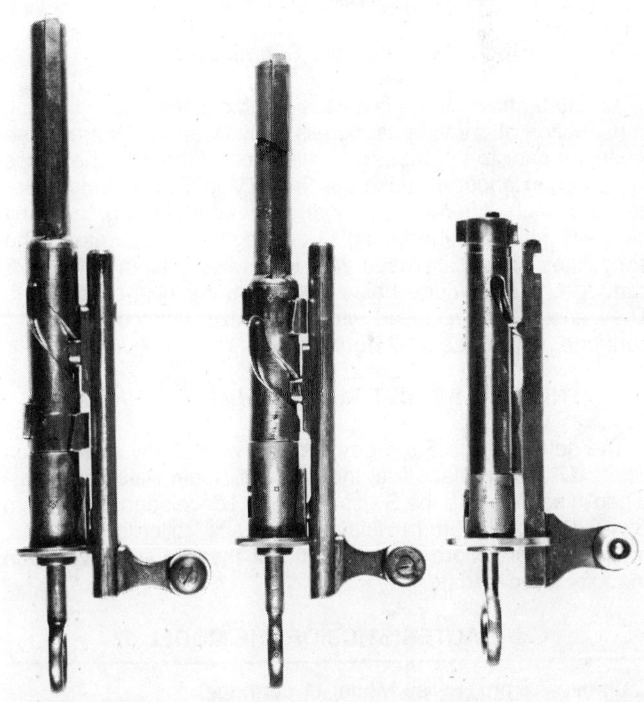

Schmidt Rubin bolts. Left: M1889 with locking lugs at rear of sleeve. Center: M1911 with locking lugs at front of sleeve. Right: M31 with locking lugs at front of bolt.

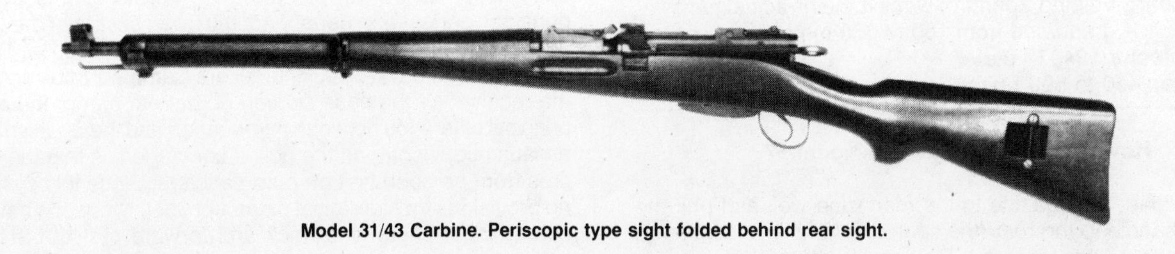

Model 31/43 Carbine. Periscopic type sight folded behind rear sight.

CHARACTERISTICS OF THE SIGNIFICANT SCHMIDT RUBINS

	M1889 Rifle	M1911 Rifle	M1911 Carbine	M1931 Carbine
Caliber:	7.5mm.	7.54mm.	7.54mm.	7.51mm.
Land diameter:	.295 in.	.2968 in.	.2968 in.	.2956 in.
Weight:	9.8 lb.	10.15 lb.	8.6 lb.	8.83 lb.
Length, overall:	51.25 in.	51.6 in.	43.4 in.	43.5 in.
Barrel length:	30.7 in.	30.7 in.	23.3 in.	25.67 in.
Magazine:	12-rd. detachable box.	6-rd. detachable box.	6-rd. detachable box.	6-rd. detachable box.
Chamber pressure:	38,400 p.s.i.	45,500 p.s.i.	45,500 p.s.i.	45,500 p.s.i.
Muzzle velocity:	2033 f.p.s.	2640 f.p.s.	2490 f.p.s.	2560 f.p.s.

Schmidt Rubin Model 31/55 Sniper Rifle.

Special Note on the Schmidt Rubins

As noted above, there is a considerable difference in the bolt construction of the various models of this weapon. None of the weapons prior to the M1896/11 rifle is considered by the Swiss to be strong enough to use the Swiss Model 11 cartridge. Because of slight differences in bore diameter and rifling, the early rifles will not shoot the Model 11 cartridge very accurately. The early rifles should be used with the Swiss 7.5mm M1890/23 cartridge; this cartridge has a 190-grain, conical-nosed bullet. They should not be used with the Spitzer-pointed Model 11 cartridge, which has a 174 grain bullet.

THE SIG ASSAULT RIFLE *STURMGEWEHR 57*

The selective-fire *Sturmgewehr 57,* Swiss Army designation *Stgw. 57,* has replaced all the Schmidt Rubin Rifles and Carbines in service with the Swiss Army. SIG developed this 7.5mm Swiss (7.5 × 53.5mm) delayed-blowback, roller-locked automatic weapon. More than 600,000 of these rifles have been manufactured to date.

CHARACTERISTICS OF THE MODEL 57

Caliber: 7.5mm (Swiss Model 11 cartridge).
System of operation: Delayed blowback, selective fire.
Weight, empty: 12.32 lb.
Length, overall: 43.4 in.
Barrel length: 23 in., w/muzzle brake.
Feed device: 24-round, detachable, staggered-row box magazine.
Sights: Front: Folding blade, w/protecting ears.
Rear: Folding aperture, w/micrometer adjustment graduated from 100 to 650 meters.
Muzzle velocity: 2493 f.p.s.
Cyclic rate: 450 to 500 r.p.m.

How to Load and Fire the Model 57

Insert a loaded magazine in the magazine well and pull the operating handle to the rear (the operating handle is located on the right side of the receiver). Release the operating handle, and the bolt will go home, chambering a cartridge. Set the safety selector on the letter "E" if semiautomatic fire is desired, or on the letter "M" if full automatic fire is desired. Pressure on the trigger will fire the weapon. To set the weapon on safe, set the safety selector on the letter "S."

How to Field Strip the Model 57

Remove the magazine by pushing forward the magazine catch, located at the forward end of the trigger guard. Depress the butt retaining catch, located on the forward underside of the butt; turn the butt approximately one-quarter turn to the right, and remove the butt and recoil spring. Pull the operating handle to the rear, and remove the bolt assembly. No further disassembly is recommended.

To reassemble the weapon, perform the above steps in reverse order.

How the Model 57 Works

When the trigger is pulled, the hammer is released and hits the firing lever mounted on the rear of the bolt. The firing lever strikes the firing pin, which hits the primer of the cartridge. The gas pressure forces the base of the cartridge against the head of the two-piece bolt. When a certain pressure has been reached, the locking rollers begin forcing rearward the nose of the rear section of the bolt. The locking rollers are then free to slip back into their recesses in the bolt head, and the weapon unlocks. The bolt travels to the rear, compressing the recoil spring, extracting and ejecting the spent case, and cocking the hammer. The recoil spring drives the bolt forward again, and the bolt picks up and chambers another round. The nose of the rear section of the bolt cams the locking rollers into their locking recesses in the receiver. If the weapon has been set on semiautomatic fire, the trigger must be pulled again. This weapon fires from a closed bolt on both automatic and semiautomatic fire. If the weapon has been set on automatic fire, it will continue to fire as long as the trigger is held to the rear and there are cartridges remaining in the magazine.

Special Note on the Model 57

The Model 57 has many interesting features. Its delayed blowback action is worthy of note. The bolt is composed of two principal parts—the head and the rear section. The head is approximately one-fourth as heavy as the rear section. The head section contains two locking rollers cammed into recesses in the receiver by the nose section of the rear part of the bolt. The principal difference between this weapon and the Spanish CETME assault rifle, insofar as the bolt is concerned, is that the CETME fires from an open bolt on automatic fire. Since this system has no provisions for slow initial extraction, both weapons have fluted chambers to "float" the neck and forward part of the case on gas.

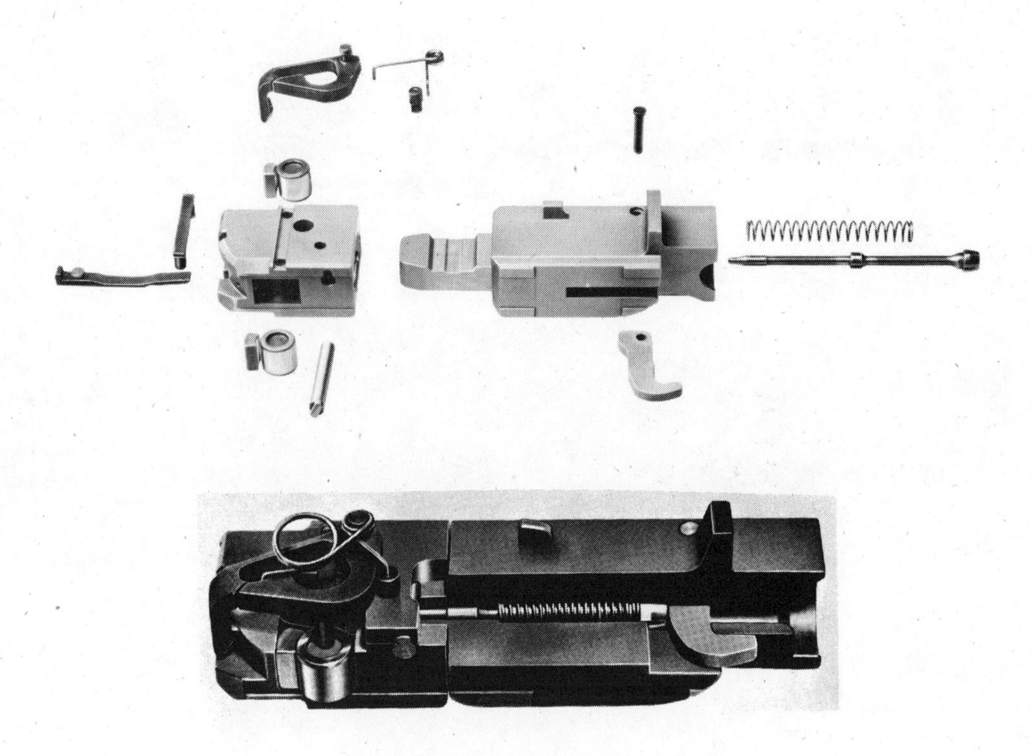

Model 57 bolt.

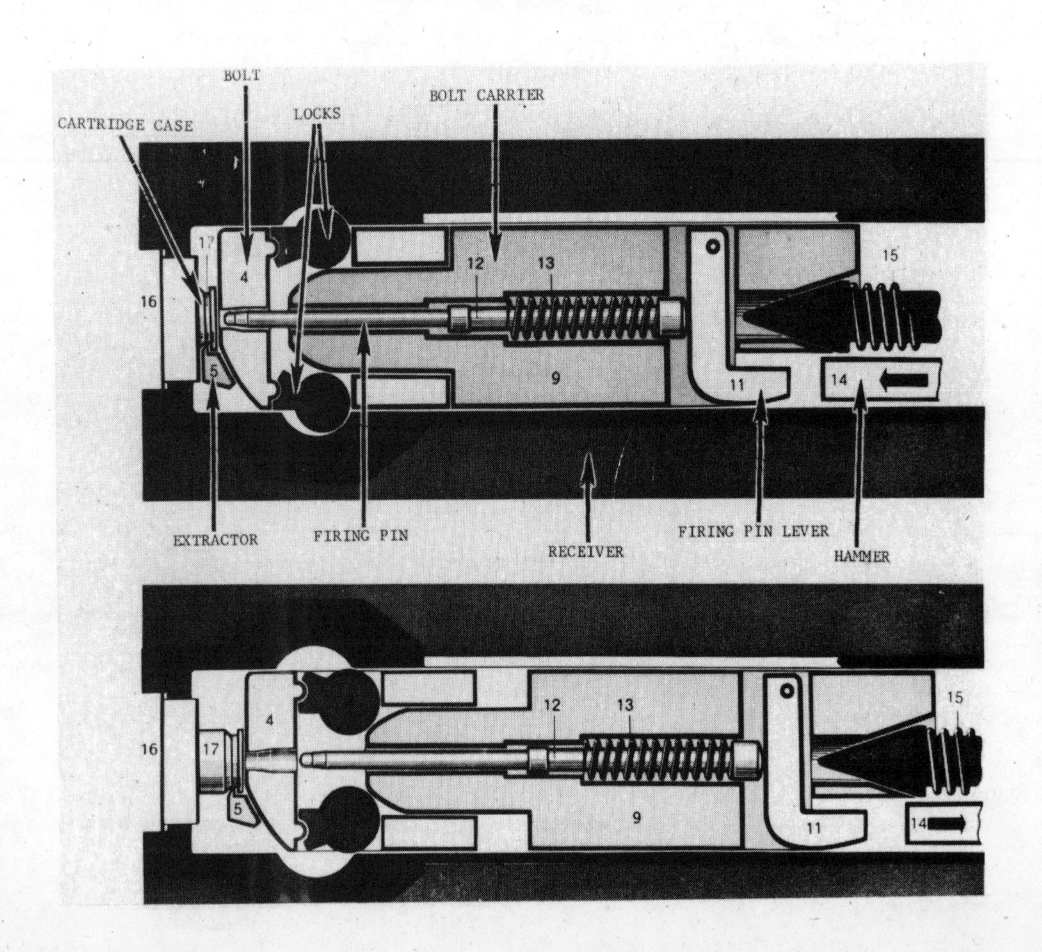

Operation of the Model 57 Roller-locked bolt mechanism. The delay, which results from the time required for the rollers to unseat and force the bolt carrier to the rear, permits the bullet to exit the barrel and the chamber pressures to fall to a safe level.

Swiss 7.5mm Model 57 Assault Rifle.

Model 57 field-stripped.

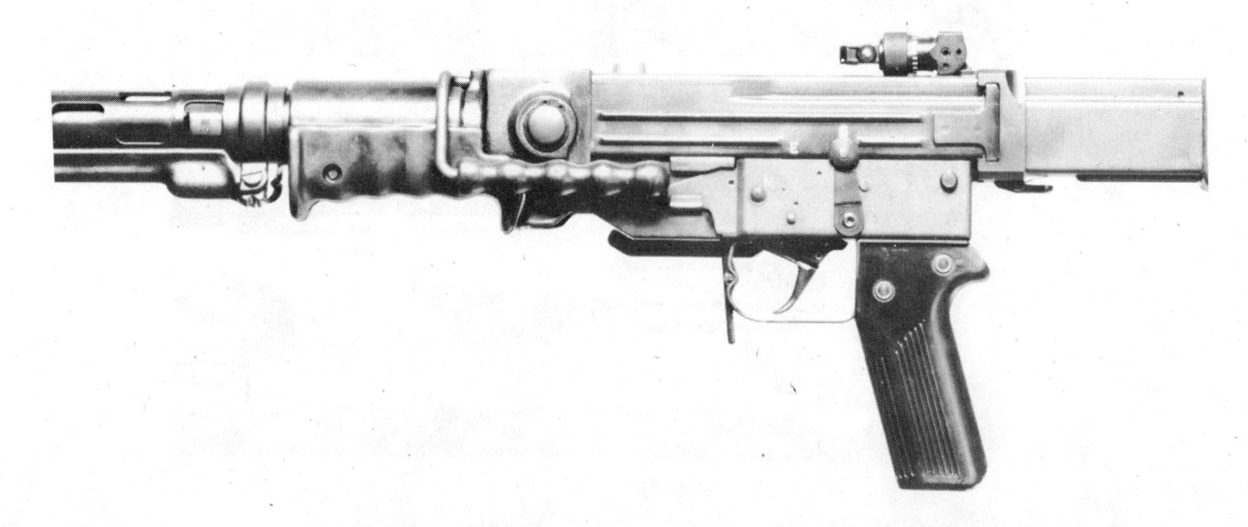

Model 57 Assault Rifle, left side view, current production.

Detailed disassembly of the SG530-1 Rifle.

THE .223 SIG MODEL SG530-1 RIFLE

SIG has developed a new rifle chambered for the .223 (5.56mm) cartridge. This rifle uses a modification of the 7.5mm Model 57 rifle action altered to gas operation. It is made mainly of stampings and fabrications and has a plastic stock, pistol grip and fore-end. In addition to the fixed plastic stock version, there is a folding metal stock version.

CHARACTERISTICS OF THE .223 SIG MODEL SG530-1 RIFLE

Caliber: 5.56mm (.223)
System of operation: Gas operated, selective fire
Length overall: w/fixed stock—39.5 in.
w/folding stock, stock folded—30.8 in.
Barrel length: 18.1 in.
Feed device: 30-round, staggered row, detachable box magazine.
Sights: Front: Protected post
Rear: Aperture adjustable from 100 to 400 meters
Weight: w/fixed stock—8.35 lbs.
w/folding stock—8.44 lbs.
Cyclic rate: 550-650 r.p.m.
Muzzle velocity: 3150 f.p.s. with US M193 ball cartrige.

How to Load and Fire the SG530-1

The magazine catch is immediately to the rear of the magazine; push forward to remove magazine. The magazine is loaded in the normal fashion. Set rifle on safe by turning safety/selector lever, located on the left side of the receiver above the pistol grip, to "S". Insert loaded magazine and pull operating handle, located to the rear of the front sight on the left top side of the gas cylinder tube over the barrel, to the rear and release. The cocking handle will return to its forward position, and the weapon is loaded. If semiautomatic fire is desired, set the safety/selector lever on the number "1"; if automatic fire is desired, set the safety/selector lever on the number "30". Pressure on the trigger will now fire the rifle.

How to Field Strip the SG530-1

Remove magazine. Push receiver catch, located at the top rear of the receiver, in and pivot stock and trigger housing assembly down. The weapon breaks like a shotgun in a fashion similar to the FN FAL and the US M16A1. Withdraw operating spring assembly, bolt and operating rod/bolt carrier. Remove bolt from bolt carrier. To assemble, perform the above steps in reverse.

How the SG530-1 Works

The SG530-1 is gas operated but uses a roller bearing locking system somewhat similar to the StG 57. The bolt carrier is forced to the rear by the gas piston. This movement pulls the striker, which is mounted on the bolt carrier right behind the bolt to the rear and allows the firing pin, which holds the roller bearings in their locked position in the receiver, to move to the rear. Thus unlocking occurs after firing; locking occurs when the bolt carrier, under the pressure of the compressed mainspring, forces the firing pin forward camming the locks into locked position.

The receiver of the rifle is grooved for a scope mount that locks into the front of the rear sight base. A prong type flash suppressor is used, and the fore part of the barrel and the flash suppressor can be used as a rifle grenade launcher.

Swiss SIG 5.56mm Model SG530-1 Rifle.

Swiss SIG 5.56mm SG530-1 Rifle with folding butt stock.

THE SIG MODEL SG540 and 542 RIFLES

SIG has produced two new rifles—the SG540 chambered for the 5.56mm cartridge and the SG542 chambered for the 7.62mm NATO cartridge. Outwardly these rifles are generally similar in appearance to the SG530-1, but internally they are quite different. The SG530-1 has a roller locked bolt, and the SG540 and 542 rifles have rotating bolts cammed into and out of the locked position by a cam slot on the bolt carrier. The bolt system is quite similar in concept to that of the AK.

The rear end of the piston fits into a hole in the bolt carrier, held in place by the operating handle, the end of which fits through a slot cut in the side of the bolt carrier.

The 540 series rifles have flash suppressor/grenade launchers attached to their barrels and can be set to fire three round bursts in addition to full automatic and semiautomatic. They can be supplied with bipod and with fixed or folding butt.

A modified SG540, the SG541 was adopted by the Swiss Army as the Stgw 90 in the fall of 1983.

CHARACTERISTICS OF THE SIG540 SERIES RIFLES

	SG540	SG541	SG542
Caliber:	5.56mm	5.6mm	7.62mm NATO
System of operation:		gas, selective fire	
Weight loaded (20-round magazine):	7.1 lbs.	9.5 lbs.	8 lbs.
Length overall w/stock fixed:	37.5 in.	39.4 in.	38.9 in.
w/stock folded:	28.7 in.	30.6 in.	30.1 in.
Barrel length:	19.3 in.	21.0 in.	19.5 in.
Feed device:	20 or 30-round detachable box magazine	20 or 30-round detachable box magazine	20-round detachable box magazine
Sights: Front:		protected post	
Rear:		rotating drum w/apertures	
Muzzle velocity:	3215 f.p.s.	2952 f.p.s.	2690 f.p.s.
Cyclic rate:	650–800 r.p.m.	700–850 r.p.m.	650–800 r.p.m.

SWISS SUBMACHINE GUNS

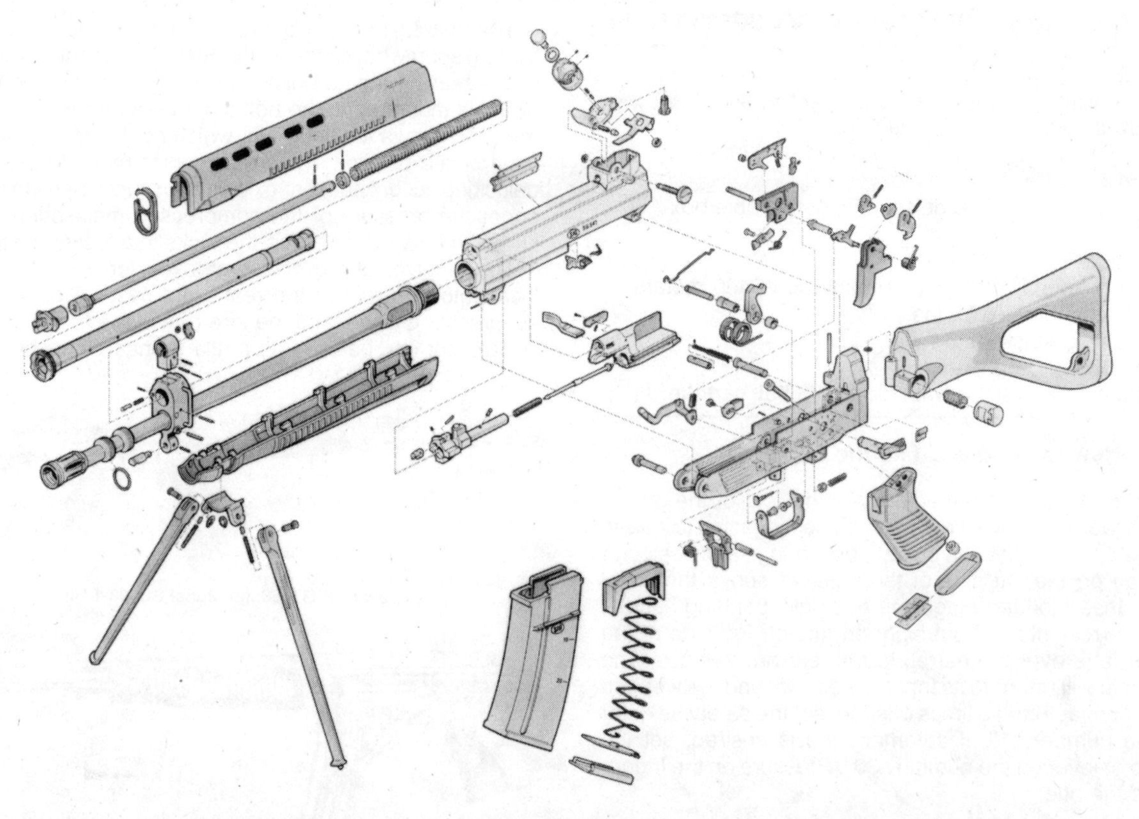

Exploded view of SIG SG541 5.56 × 45mm rifle, which will be procured by the Swiss Army as the *Sturmgewehr 90*. This is an improved version of the SG540.

SIG started production of submachine guns in 1920 with the modified Bergmann MP181. Some of these weapons, chambered for the 7.65mm Luger or 7.63mm Mauser cartridge, were sold to Finland, China, and Japan.

Experience with the Bergmann led the SIG engineers into a development program of their own.

At the beginning of World War II, the Swiss Army requested SIG and the Waffenfabrik Bern to take part in the development of a new submachine gun. Out of this work came the SIG MP 41 blowback weapon chambered for the 9mm Parabellum cartridge and the Furrer Model 41/44 submachine gun developed by the Waffenfabrik Bern. The latter weapon was adopted by the Swiss Army and SIG terminated work on their MP 41.

The Swiss 9mm Parabellum Model 41/44 submachine gun was developed by Waffenfabrik Bern and is a most unusual submachine gun in several respects. The weapon is recoil-operated and has a toggle joint action similar to the Model 25 Furrer light machine gun. The vertical fore grip can be folded up under the barrel. It is an extremely expensive weapon to manufacture and unduly complicated. Unlike the Model 41, the Model 41/44 has a bayonet lug.

The Swiss government also purchased 5,000 Suomi Model 43 submachine guns from Finland. This 9mm Parabellum weapon was the same as that used by the Finnish Army. Hispano Suiza obtained a license to manufacture the weapon in Switzerland. The Suomi as manufactured by Hispano Suiza has a flip-over type rear sight and a bayonet lug; it is called the MP43/44.

9mm Parabellum Model 43/44 Submachine Gun.

Swiss 9mm Parabellum Model 41/44 Submachine Gun developed by the Federal Waffenfabrik at Bern.

Disassembled view of 5.6 × 45mm SG541 (Stgw 90).

SWISS SERVICE SUBMACHINE GUNS

	Model 41/44	Model 43/44
Caliber:	9mm Parabellum.	9mm Parabellum.
System of operation:	Recoil, selective fire.	Blowback, selective fire.
Length overall:	30.5 in.	33.9 in.
Barrel length:	9.8 in.	12.4 in.
Feed device:	40-round, staggered row detachable box magazine.	50-round, in-line, detachable box magazine— has double column divided by central wall.
Sights: Front:	Blade with ears.	Blade.
Rear:	Flip-type w/"U" notch.	Flip-type w/"U" notch.
Weight:	11.4 lb.	10.5 lb.
Muzzle velocity:	1312 f.p.s.	Approx. 1350 f.p.s.
Cyclic rate:	900 r.p.m.	800 r.p.m.

SWISS MACHINE GUNS

Switzerland adopted the Maxim gun in 1894. Later models used were the Model 1900 and Model 1911. The 7.5mm Model 1911 was used as the standard heavy machine gun until 1951.

Colonel Furrer of the Swiss government arms plant at Bern developed the 7.5mm Model 25 light machine gun, also known as the Fusil Furrer. The Model 25 has a toggle-joint action similar to the Luger pistol but which breaks to the side rather than to the top as does the Luger pistol. This weapon was quite impressive in its day but is being replaced by the 7.5mm Model 57 assault rifle. A link belt-fed 13m aircraft gun version of the Model 25 was also produced.

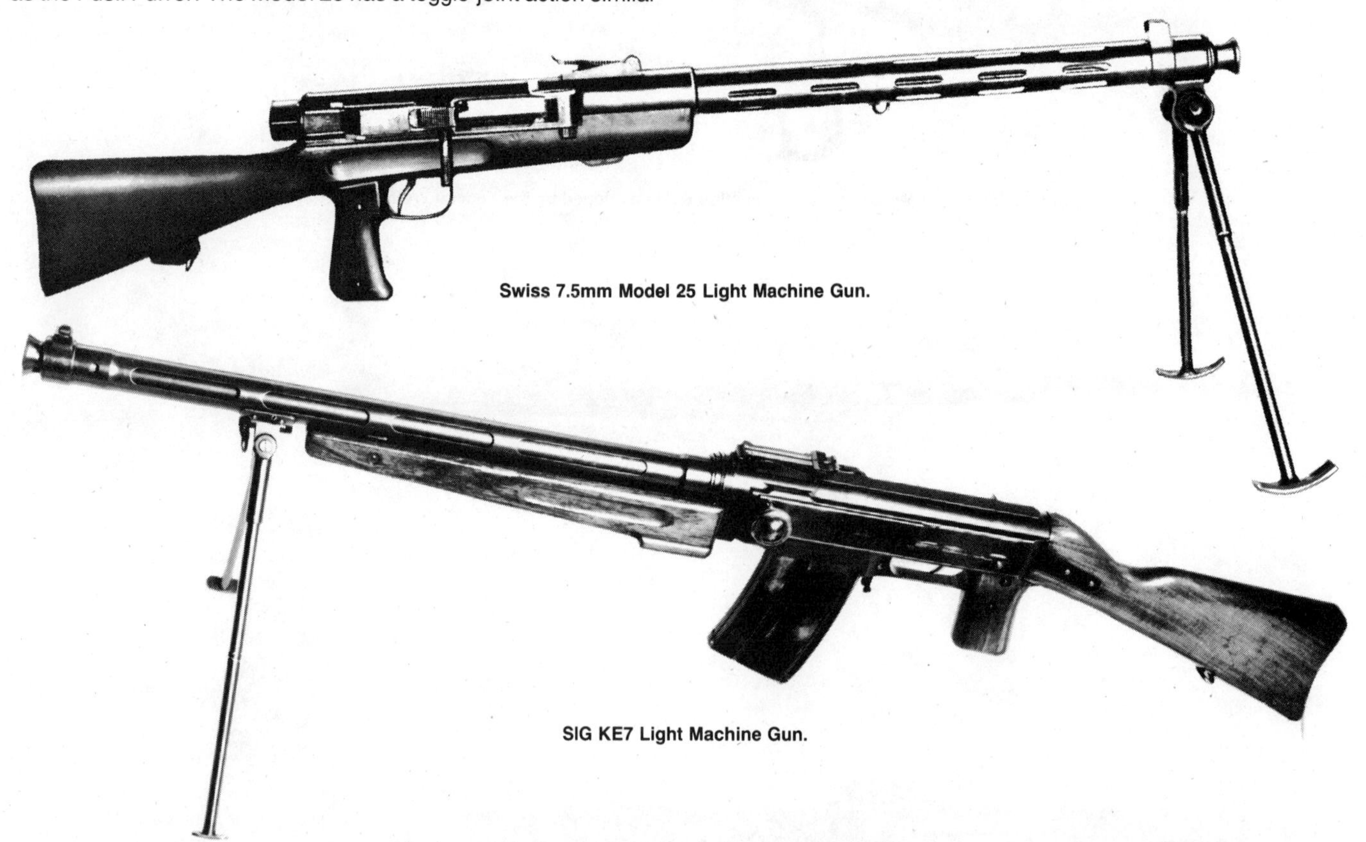

Swiss 7.5mm Model 25 Light Machine Gun.

SIG KE7 Light Machine Gun.

The next machine gun to be developed in Switzerland was a commercial development of SIG—the KE7 light machine gun. This weapon had some success in oversea sales (it was sold to China), but was not adopted by the Swiss government. The KE7 is a recoil-operated gun, which fires from an open bolt. It is selective fire and is relatively light, weighing 17.25 lbs. without magazine.

During the post-war period, the Swiss government arms plant at Bern developed a modified copy of the German MG42 adopted by the Swiss Army as the MG51. The Swiss seem to have defeated much of the purpose of the MG42, however, by making the MG51 mainly out of heavy milled parts rather than stampings. MG51 has locking flaps in its bolt head, which engage locking recesses in the receiver, rather than locking rollers of the MG42. Barrel change and field stripping of the MG51 is generally similar to the MG42.

Tripod Mounted MG M51 with Optical Sight. The Swiss still prefer such weapons for prepared defensive positions.

Hispano Suiza, a well-known manufacturer of aircraft and anti-aircraft cannon, developed a rifle caliber machine gun after World War II. This weapon was not a financial success and was made only as a prototype. It was known as the Type HSS808.

SIG developed a weapon to place in the competition held by the Swiss for a weapon to replace the Model 11 Maxim gun. The SIG Model 50 is a gas-operated weapon designed to be used as a general purpose weapon. It has a quick change barrel and is fed by a non-disintegrating metallic link belt similar to that of the MG42 or by a 50-round drum. This weapon is not used

SIG MG710-2 Machine Gun.

by Switzerland but was adopted by Denmark in caliber .30 as the Model 51 machine gun. It is being replaced in that country by the 7.62mm NATO MG42/59.

After the MG50, SIG developed a series of delayed blowback machine guns using the same locking principles as the Model 57 assault rifle. Originally advertised as the Model 55-1 and 55-2, the first two guns of the series, now called the 710-1 and 710-2, are similar in all respects excepting the barrel, barrel support and the presence of a barrel jacket on the 710-1. The feed mechanism is similar to that of the MG42 and a link belt or a 50-round drum similar to that of MG42 is used.

The SIG MG710-1 is remarkably similar to the MG42V, which was under development in Germany at the close of World War II. The barrel change on this gun is the same as that on MG42.

The SIG 710-2 does not have a barrel jacket, and the barrel is held in the receiver by a bayonet-type joint. The barrel handle is used to change the barrel. The 710-2, as the 710-1, is designed to be used as a general purpose machine gun.

The third weapon in this series is the MG710-3. This weapon has a partial barrel jacket, and the barrel is released for removal by a catch on the top of the barrel jacket. A large barrel-removal handle is fitted to the right side of the barrel and the barrel is pulled to the right and rear to remove. The 710-3 will use a disintegrating or a non-disintegrating link belt. The 710-3 uses more stampings than do the 710-1 and 710-2 and is a bit lighter than those guns. It is advertised for the 7.62mm NATO cartridge only, while the earlier guns were advertised chambered for the 6.5mm, 7.92mm and 7.62mm NATO cartridges.

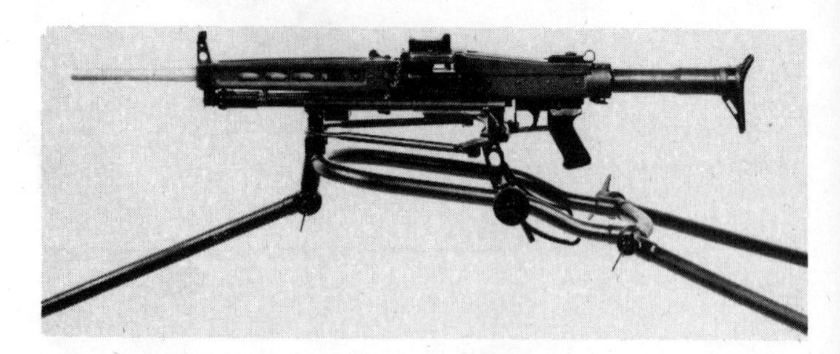

The SIG MG 710-3 Machine Gun in 7.62mm NATO cartridge weighs 20.5 pounds and is notable for its quick barrel change.

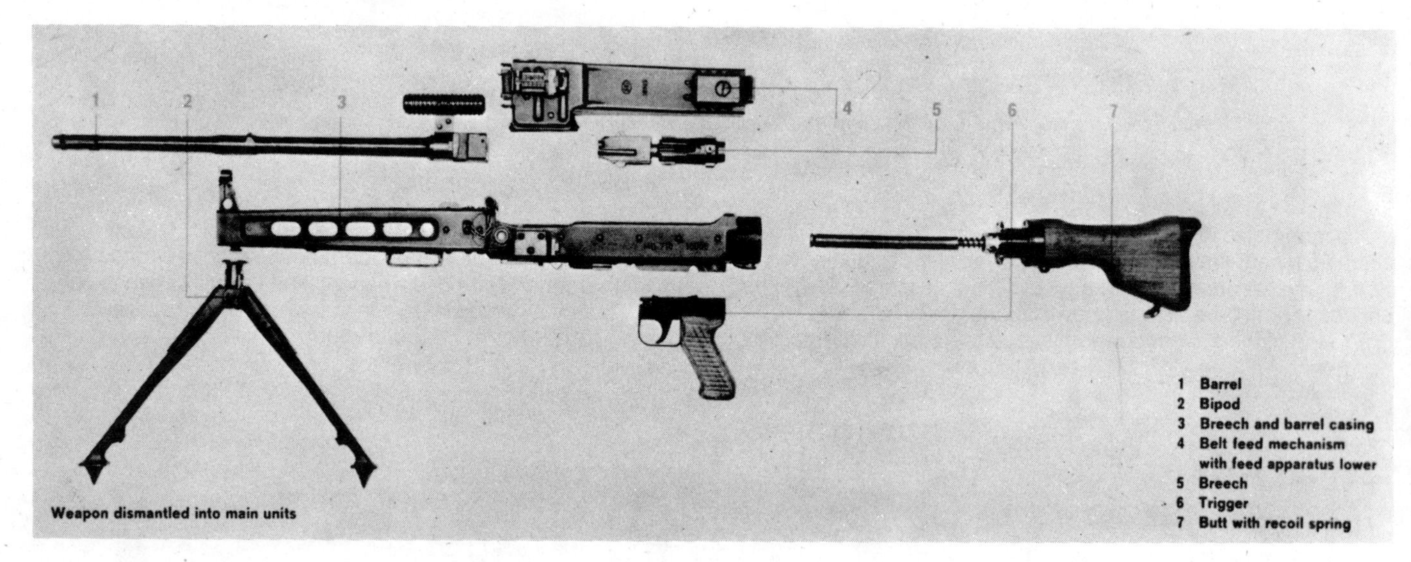

SIG 710-3, field stripped.

CHARACTERISTICS OF SWISS SERVICE MACHINE GUNS

	Model 11 Heavy	Model 25 Light	Model 51
Caliber:	7.5mm.	7.5mm.	7.5mm.
System of operation:	Recoil, automatic only.	Recoil, selective fire.	Recoil, automatic only.
Length overall:	42.4 in.	45.8 in.	50.1 in.
Barrel length:	28.4 in.	23 in.	22.2 in.
Feed device:	250-round web belt.	30-round, staggered row, detachable box magazine.	Non-disintegrating metallic link belt, in 50- and 200-round segments, 50-round drum.
Sights: Front:	Blade.	Blade w/protecting ears.	Folding blade.
Rear:	Leaf.	Tangent w/notch.	Tangent, optical sight used on tripod.
Weight:	40.8 lb. (gun only).	23.69 lb w/bipod.	35.3 lb. w/bipod. 57.3 lb. w/tripod.
Muzzle velocity:	2590 f.p.s.	2460 f.p.s.	2460 f.p.s.
Cyclic rate:	Approx. 500 r.p.m.	450 r.p.m.	1000 r.p.m.

45 Turkey

SMALL ARMS IN SERVICE

Handguns:	7.65 × 17mm MKE	Walther PP made by Makina ve Kimya Endustrisi Kurumu (MKE). Also in 9 × 17mm.
	11.43 × 23mm M1911A1	US-made.
Submachine guns:	11.43 × 23mm M3A1	US-made.
Rifles:	7.62 × 51mm NATO G3 Otomatik Piyade Tüfegi	Made under license from Heckler & Koch by MKE.
Machine guns:	7.62 × 51mm NATO MG3 Makinali Tüfek (Tam Otomatik)	Rheinmetall- and MKE-made.
	7.62 × 63mm (.30-06) M1919A6 Browning	US-made.
	12.7 × 99mm M2 HB	US-made.

TURKISH PISTOLS

A native-made pistol is the 9mm Kirikkale. This weapon, which is made at the Kirkkale arms factory in Turkey, is a copy of the Walther PP. It is identical to the late commercial-type Walther, except for variations in machining and the finger-rest extension on the removable magazine floor plate. For data on the Walther PP, see the chapter on German World War II materiel. The Kirikkale is chambered for the 9mm Browning short (.380 ACP) or the 7.65mm (.32 ACP) cartridge.

MKE Pistol produced by Makina ve Kimya Endustrisi Kurumu located at Kirikkale, Ankara.

TURKISH RIFLES

Turkey responded very rapidly to the advantages of the military small caliber repeating rifle and adopted a 7.65mm Mauser in 1890—the Model 1890. This rifle, which was chambered for the 7.65mm cartridge introduced with the Belgian Mauser in 1889, has an in-line magazine and bolt similar to the Belgian Model 1889 excepting the buttress threads on the bolt sleeve. The Model 1890 does not have the metal barrel jacket of the Model 1889 and introduced the stepped barrel, which was used in Mausers from that time on.

A number of these rifles were obtained by Yugoslavia after World War I. The Yugoslavs rebarreled with a shorter 7.92mm barrel and fitted a longer handguard. This weapon was called the Model 90T by the Yugoslavs.

The Model 1893 7.65mm Mauser is essentially the same as the 7mm Spanish M1893 Mauser, with the addition of a cutoff to the magazine.

The Model 1903 7.65mm rifle is essentially the same as the German Rifle Model 98 except for the rear sight, hand guard, upper band, longer cocking piece and firing pin and modified bolt stop. There is a carbine Model 1905 and Model 1890 stocked to the muzzle.

During World War I, the Turks were supplied by Germany with many 7.92mm German Mauser rifles and carbines. After the war, Turkey purchased quantities of Czech 7.92mm Model 1924 Mausers and rebarreled many of the earlier 7.65mm rifles and carbines for 7.92mm.

Only the characteristics of rifles designed specifically for Turkey are given. Characteristics of other rifles used by Turkey will be found under country of origin.

CHARACTERISTICS OF TURKISH RIFLES

	Model 1890 Rifle	Model 1893 Rifle	Model 1903 Rifle
Caliber:	7.65mm Mauser.	7.65mm Mauser.	7.65mm Mauser.
System of operation:	Turn bolt.	Turn bolt.	Turn bolt.
Length overall:	48.6 in.	48.6 in.	49 in.
Barrel length:	29.1 in.	29.1 in.	29.1 in.
Feed device:	5-round, in line, non-detachable box magazine.	5-round, staggered row, non-detachable box magazine.	5-round, staggered row, non-detachable box magazine.
Sights: Front:	Barley corn.	Barley corn.	Barley corn.
Rear:	Leaf.	Leaf.	Tangent.
Weight:	8.8 lb.	8.8 lb.	9.2 lb.
Muzzle velocity:	2132 f.p.s.	2132 f.p.s	2132 f.p.s.

46 Union of Soviet Socialist Republics (USSR) ("Russia")

SMALL ARMS IN SERVICE

Handguns:	9 × 18mm Makarov (PM)	State arms factories.
	9 × 18mm Stechkin (APS)	State arms factories.
Rifles:	5.45 × 39mm AK74 and AK74S	State arms factories.
	7.62 × 39mm AKM	State arms factories.
	7.62 × 39mm AK47	State arms factories. Obsolete.
	7.62 × 54mm R SVD Sniper Rifle	State arms factories.
	7.62 × 39mm SKS	State arms factories. Obsolete.
Machine guns:	5.45 × 39mm RPK74 and RPKS74	State arms factories.
	7.62 × 39mm RPK and RPKS	State arms factories.
	7.62 × 54mm R PK, PKS, PKT, PKB	State arms factories.
	12.7 × 108mm DShK 38/46	State arms factories.

SOVIET PISTOLS AND REVOLVERS*

The Russians adopted the 7.62mm Nagant revolver in 1895. This revolver was produced in both single-action and double-action versions and is somewhat unusual in that the cylinder moves forward before the hammer falls and the forward end of the chamber aligned for fire telescopes the barrel. The cartridge,

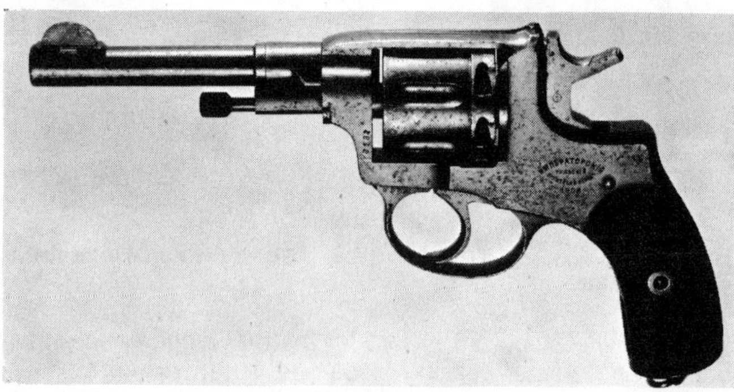

7.62mm Model 1895 Revolver.

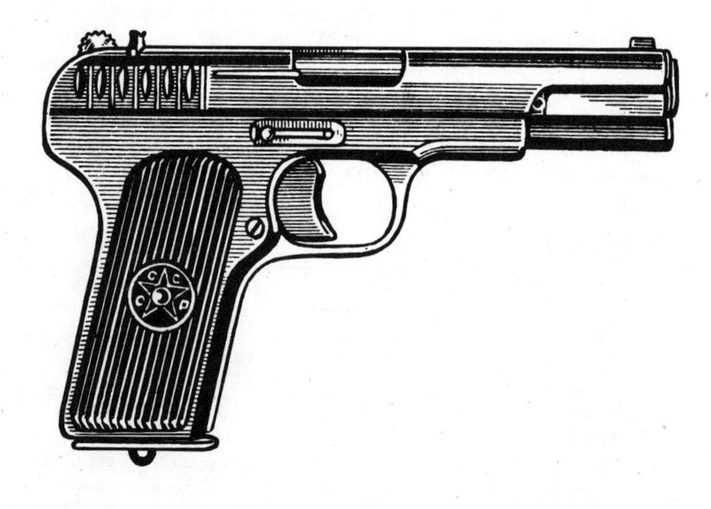

Caliber .22 Model R-3 Pistol.

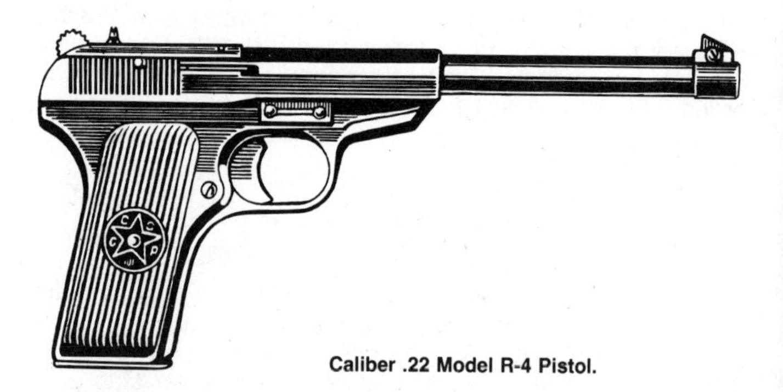

Caliber .22 Model R-4 Pistol.

which outwardly resembles a blank cartridge, has its bullet seated below the cartridge case mouth. The purpose of these design features is to prevent gas leakage at the joint between the cylinder and barrel. It is doubtful if this complicated system is worth the effort.

Although the Model 1895 was manufactured as late as World War II, it is no longer in military service. A somewhat smaller version of this revolver was made for police use, and a caliber .22 version was made for training purposes.

In 1930, the first model of the 7.62mm TT Tokarev automatic pistol was adopted; a slightly modified version was adopted in 1933. These weapons are no longer in Soviet service, but are widespread throughout the world and are, therefore, covered in detail in this chapter. A caliber .22 training version of the Tokarev called the Model R-3 and a caliber .22 target version called the Model R-4 have also been manufactured.

The 7.62mm Tokarev has been replaced in the Soviet service, and in several Soviet satellites as well, by the 9mm Makarov (PM) pistol and the 9mm Stechkin Machine pistol. These weapons are chambered for a new 9mm cartridge, intermediate in size and power to the 9mm Browning Short (.380 ACP) and the 9mm Parabellum cartridge. It is quite similar to the 9mm Ultra prototype cartridge developed for use in a version of the Walther PP to be used by the Luftwaffe. The 9mm Makarov cartridge has a 94-grain bullet.

The Soviets produced a small 6.35mm (.25 ACP) automatic called the TK (Tula Korovin). This is not a military weapon, but may be used by police and para-military organizations.

*For additional details see *Handguns of the World*.

6.35mm TK Pistol.

7.62mm TT M1933 Pistol.

THE 7.62mm TOKAREV TT M1930 AND M1933 PISTOLS

The 7.62mm Tokarev, or Tula Tokarev-TT, was designed by Fedor V. Tokarev. It is a slightly modified Colt-Browning design, notable for its "packaged" type sear and hammer assembly, which is removed as a unit. This type of construction is not unique and has been used in a number of pistols. The TT33 has been manufactured in Hungary as the Model 48 and in China as the Type 51. The Soviet 7.62mm Model 1930 Type P cartridge is almost identical to the 7.63mm Mauser cartridge and is interchangeable with that cartridge.

Description of Tokarev Mechanism

The slide closely resembles the Colt Browning type. However, the barrel bushing, which is inserted in the front end of the slide to support the barrel and retain the recoil spring, is a heavily forged bushing with one large hole for the barrel and a small hole below it in which the steel disc at the end of the recoil spring seats. This is much stronger than the conventional horseshoe shape bushing and plug.

The firing pin unit is retained in the slide by a simple split pin driven through a hole drilled in the slide, the pin passing through

Field Stripping the Tokarev

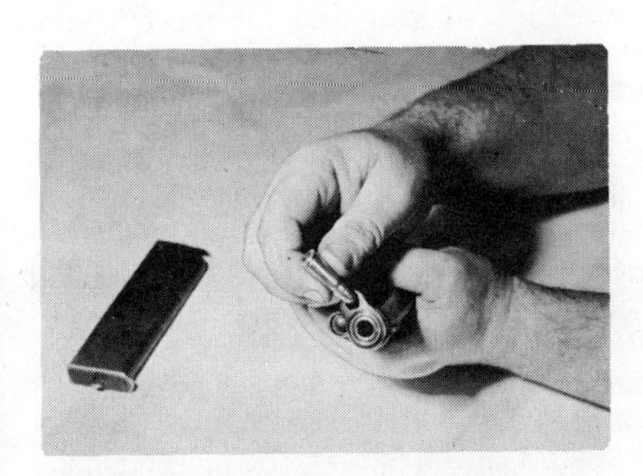

(1) Insert nose of cartridge through hole in bushing below barrel and compress spring until bushing is free to turn.

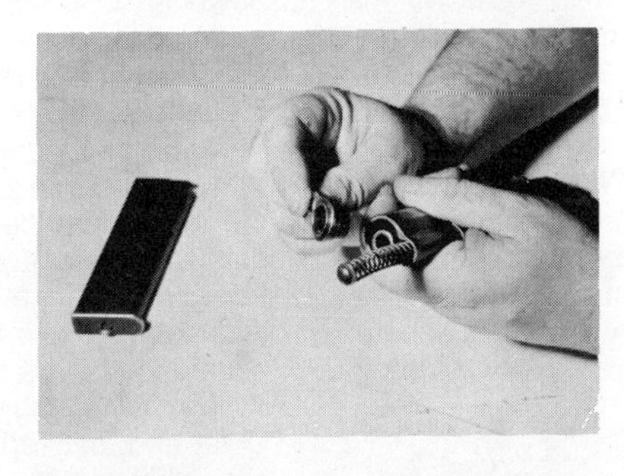

(2) Swing bushing up to the right until its locking lugs disengage. Remove it from the slide. Let the recoil spring protrude.

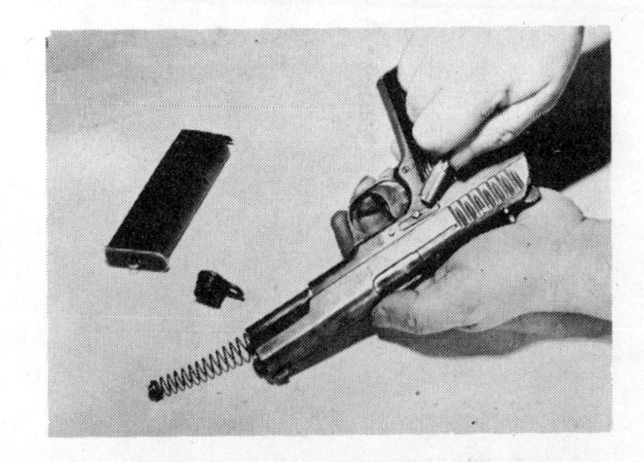

(3) Using a cartridge or the bottom of the magazine, pull back the spring locking clip on the right side of the pistol. This frees the slide-stop pin.

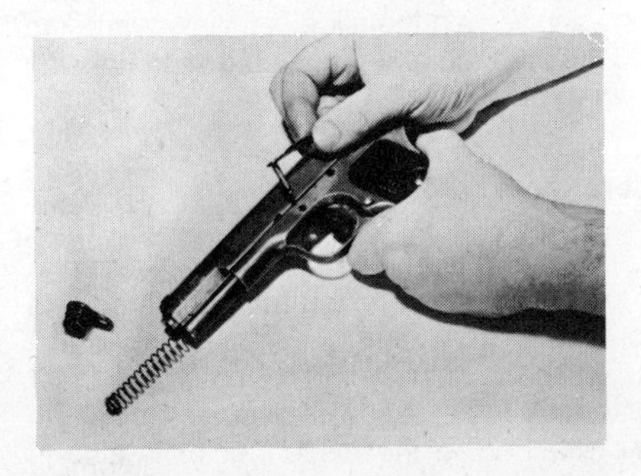

(4) Withdraw the slide-stop barrel-locking pin from the left.

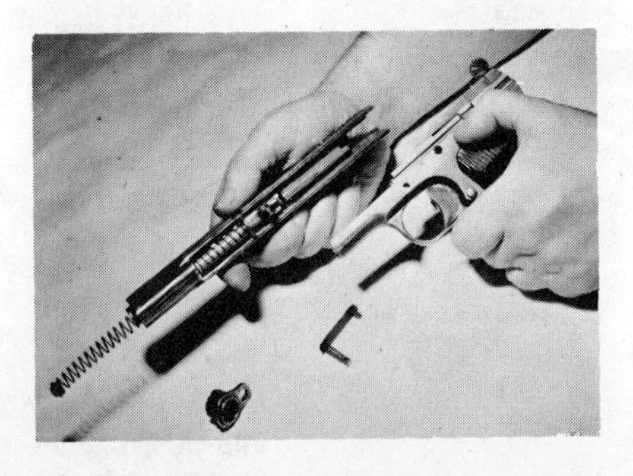

(5) Push the slide and barrel assembly forward out of the guides in the receiver.

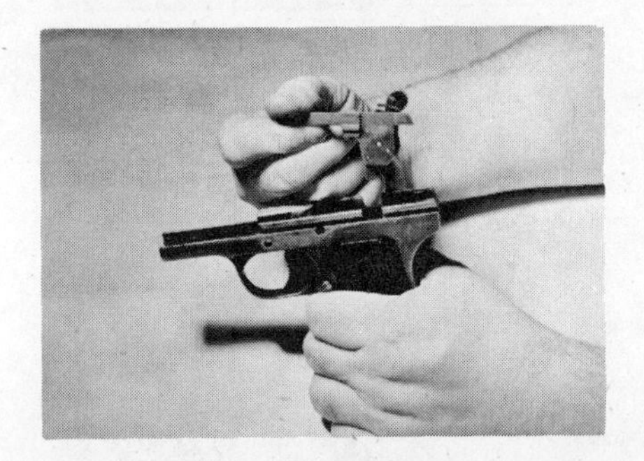

(6) Lift the receiver sub-assembly and hammer mechanism out of the receiver.

(7) Remove barrel, recoil spring and guide from front of slide.

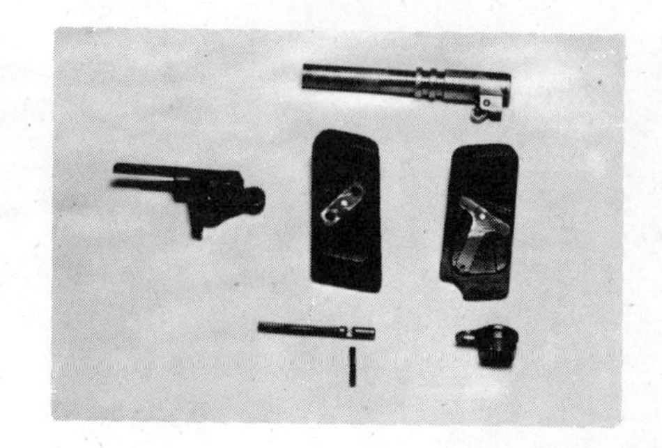

(8) No further dismounting is normally necessary. However stocks may be removed by reaching inside the handle and turning the metal locking buttons. Firing pin may be removed by punching out split pin from right side of slide. Hammer unit may be completely dismounted by pushing out the three retaining pins. The magazine release catch is a split pin which may be driven out from the right side of the receiver. This permits removing trigger and spring to complete entire disassembly.

a slot on top of the firing pin. This pin not only retains the floating firing pin but also determines the distance it can fly forward when struck by the hammer.

The magazine is particularly worthy of note in that it can be easily taken apart for cleaning or repairs. The follower, or platform on which the first cartridge rests, is supported from below by a typical magazine spring. The bottom of this spring, however, rests on a flat platform, which in turn rests on the magazine bottom. The bottom is tongued into grooves in the magazine case, being slid in from the front. It is retained by a small nib on the platform, which protrudes through a hole in the magazine bottom. Pushing this nib in permits the bottom to be pulled forward. The platform, spring and follower can then be removed.

While the barrel locks in the same manner as the Colt Browning, its manufacture has been greatly simplified. Where the Colt has two locking ribs on the top of the barrel only, the TT33 Tokarev locking ribs run entirely around the barrel circumference. The lower sections have no locking function but permit simpler barrel manufacture. They also provide, together with a thickened breech end, a strong support for the high powered cartridge used.

The stocks are of black plastic and require neither screws nor machined-in supports. A pivoted flat spring is riveted on the inner face of each stock. When turned crosswise, each end locks in the frame. Turning the strip disengages it from the frame and permits the stock to be lifted off.

The magazine release catch is a split pin. When punched out, it permits easy removal of the stirrup type trigger and the flat spring in the handle, which acts as the trigger return spring.

The slide stop and barrel locking pin function as in the Colt Browning, but the retaining system has been simplified. Where the pin emerges on the right side of the receiver, a sliding spring clip is provided. When pushed forward, teeth on it engage in notches in the end of the pin. Pulling this back frees the pin to permit removal.

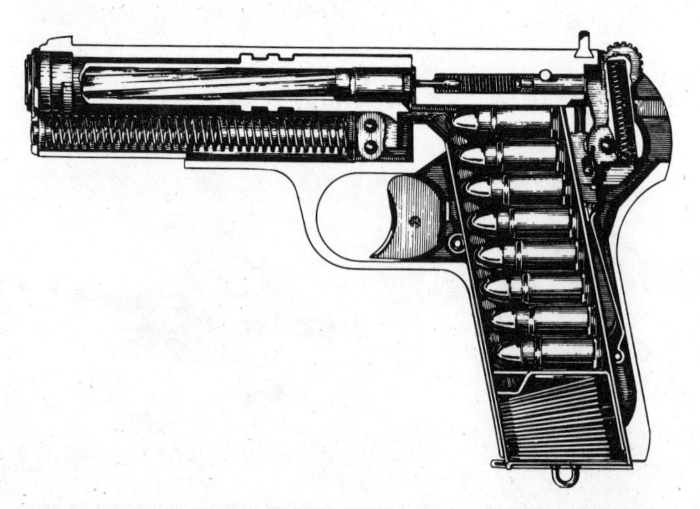

Action closed, showing details of mainspring in hammer and firing mechanism.

Action open, cut away to show details of recoil system and cocking operation.

The startling advance of this pistol over all preceding types, however, is in its hammer mechanism. This mechanism, consisting of a simple hammer, sear, disconnector, sear spring and mainspring, is housed in a sub-assembly, which forms part of the receiver and carries part of the slide grooves.

This sub-assembly block has two arms of unequal length. While the outside surfaces line up with the regular receiver guides, the inside surfaces are specially grooved to act as cartridge guides to assure proper feeding as each cartridge is stripped from between the very flat lips of the magazine. The effect of this feature is to make the feeding much more positive, as magazine mouths are normally weak points in feeding systems.

The mainspring is a coil housed inside the hammer itself. Its lower end rests against a pin driven through the housing in the assembly. As the hammer is rocked back by the slide during recoil, this spring is compressed against the supporting pin to provide the energy source for firing the next shot.

The sear, sear spring and disconnector are mounted together on a third pin in the housing. When the slide is fully forward, the disconnector rises into a slot machined for it in the slide housing below the line of the firing pin.

Operation of the Tokarev

When the trigger is pressed, its stirrup forces the flat spring in the handle back to provide energy for returning the trigger to firing position when pressure is released.

The lower end of the disconnector is riding on top of the trigger stirrup while its upper end is in its seat in the breech block. In this position, the sear attached to the disconnector is also in contact with the rear of the trigger stirrups. The trigger forces back against the sear and presses the upper end of the sear

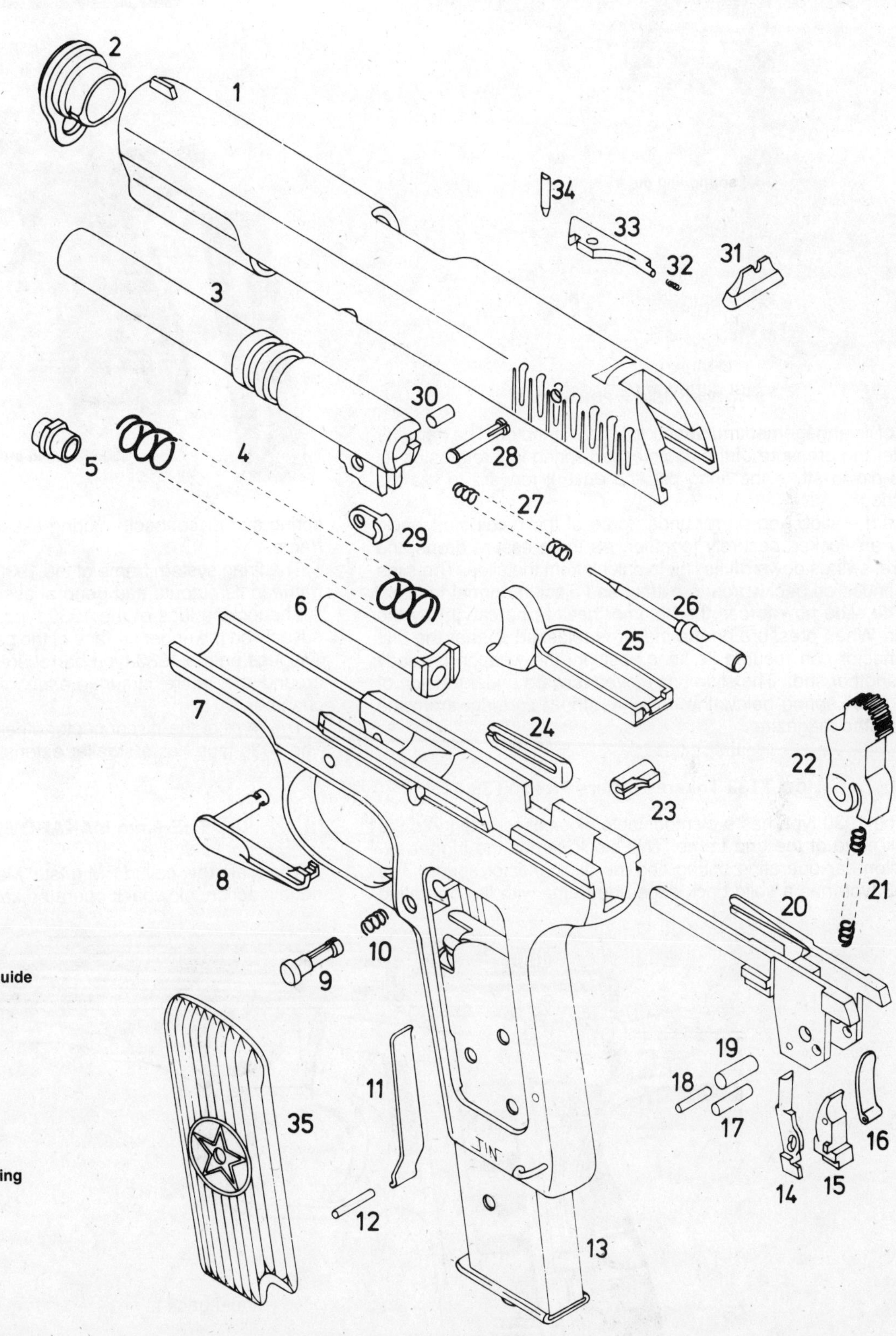

1. Slide
2. Barrel bushing
3. Barrel
4. Recoil spring
5. Recoil spring retainer
6. Recoil spring guide
7. Frame
8. Slide stop
9. Magazine catch spring guide
10. Magazine catch spring
11. Trigger return spring
12. Return spring pin
13. Magazine
14. Disconnector
15. Sear
16. Sear spring
17. Spring retainer pin
18. Sear pin
19. Hammer pin
20. Hammer assembly housing
21. Hammer spring
22. Hammer
23. Magazine catch
24. Slide-stop retainer clip
25. Trigger
26. Firing pin
27. Firing pin spring
28. Firing pin retainer pin
29. Barrel link
30. Barrel link pin
31. Rear sight
32. Extractor spring
33. Extractor
34. Extractor pin
35. Left-hand grip

Exploded view of the Tula-Tokarev 30-33 pistol. (*Jimbo*)

Soviet 9mm Makarov Pistol (PM).

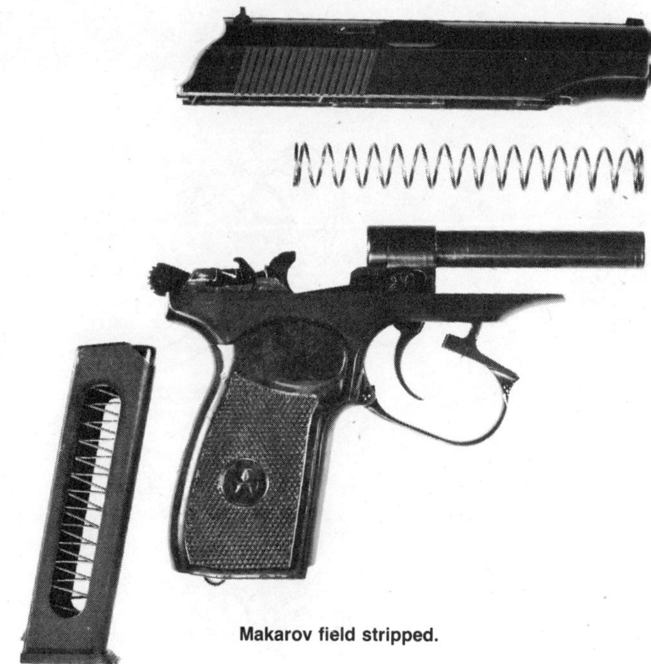

Makarov field stripped.

out of its engagement in the second hammer notch. The hammer under the pressure of the compressed spring in it rotates on its axis pin to strike the firing pin and drive it forward to fire the cartridge.

As the slide and barrel under force of the recoil start back, they are locked securely together. As the pressure drops, the barrel swings down on its link to unlock from the slide. The slide continues on backwards to extract and eject in normal fashion.

The slide now forces the hammer back to be caught by the sear. When pressure on the trigger is released so that the disconnector can resume its firing position, the weapon is ready for another shot. The slide has moved forward under tension of the recoil spring below the barrel to strip a cartridge from the top of the magazine.

How TT33 Tokarev Differs From TT30

The 1930 type has a dismountable block assembled into the back edge of the grip frame. This block carries the trigger extension bar operating spring and the disconnector spring. The 1933 type has a solid back edge grip frame with the operating spring and disconnector spring assembled directly into the grip frame.

The firing system frame of the 1930 type differs from the 1933 frame in its cutouts and general assembly to the grip frame.

The locking lugs of the 1930 type are machined and ground out only on the upper surface of the barrel ahead of the chamber. The lugs on the 1933 type barrel are machined and ground out around the entire circumference of the barrel ahead of the chamber.

The form of the disconnector differs between the two pistols. The 1930 type has a smaller extension bar contacting surface.

THE 9mm MAKAROV (PM) PISTOL

The Makarov, called PM (pistol Makarov) by the Soviets, is a double action, blowback operated self-loading pistol, which out-

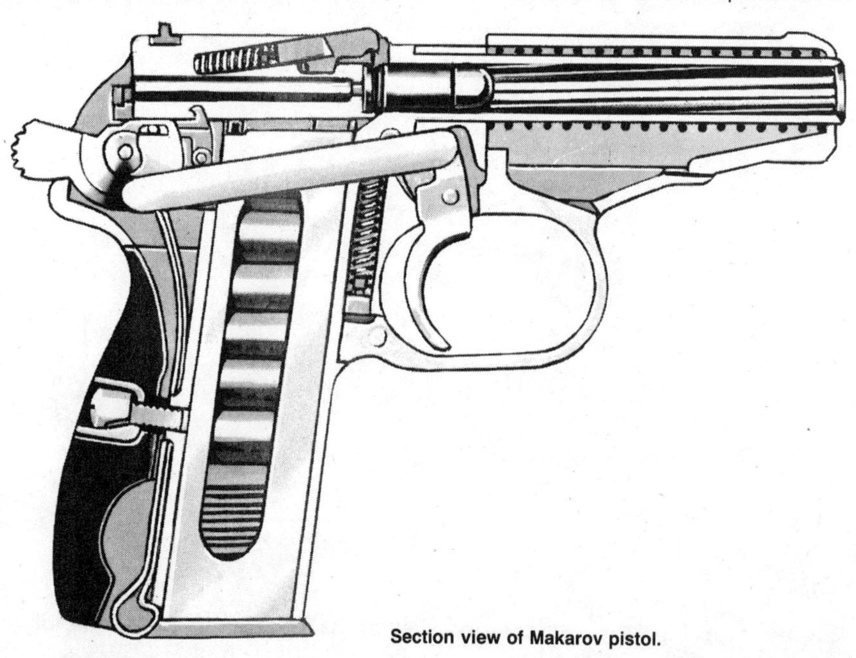

Section view of Makarov pistol.

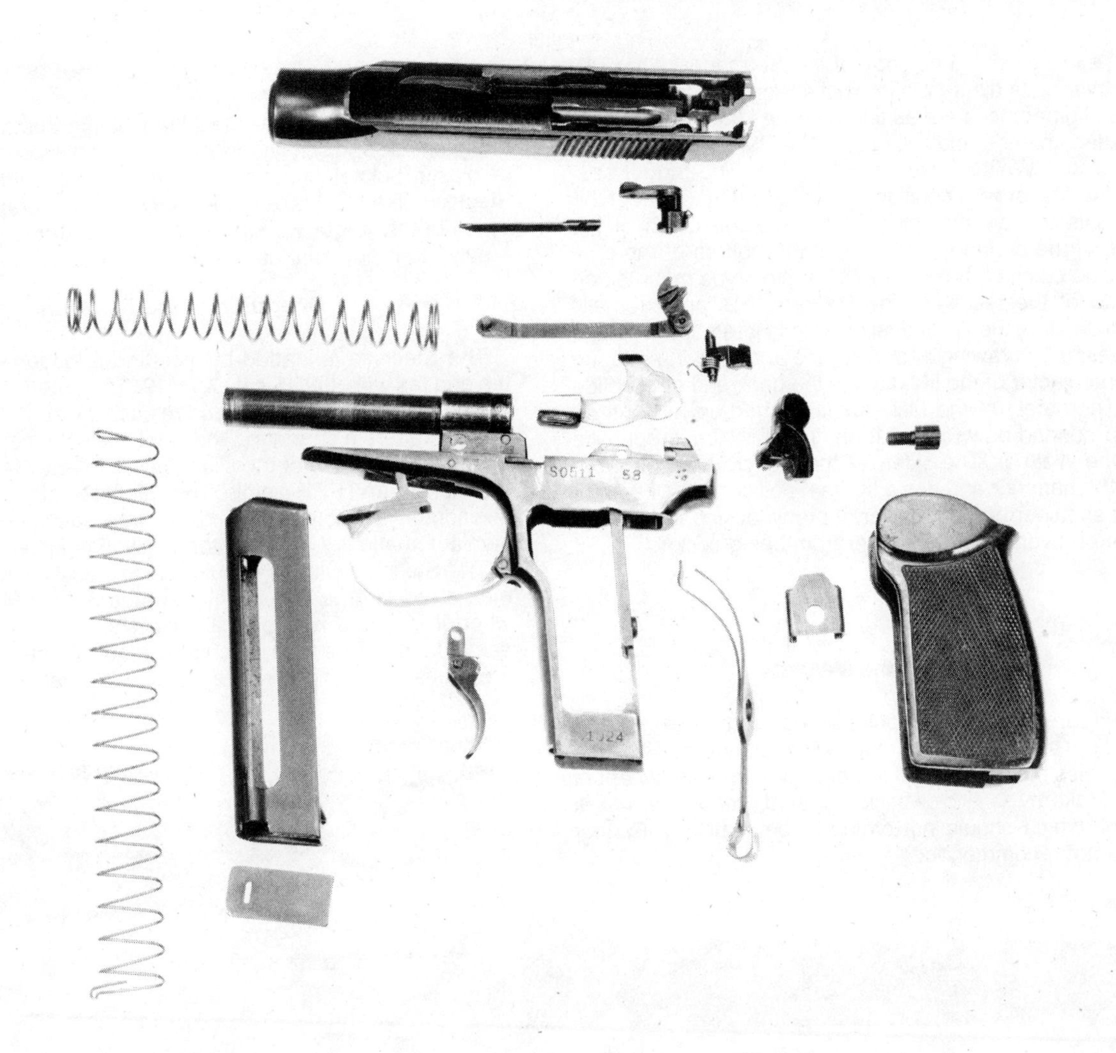

Complete disassembly of the 9 × 19mm Pistolet Makarov.

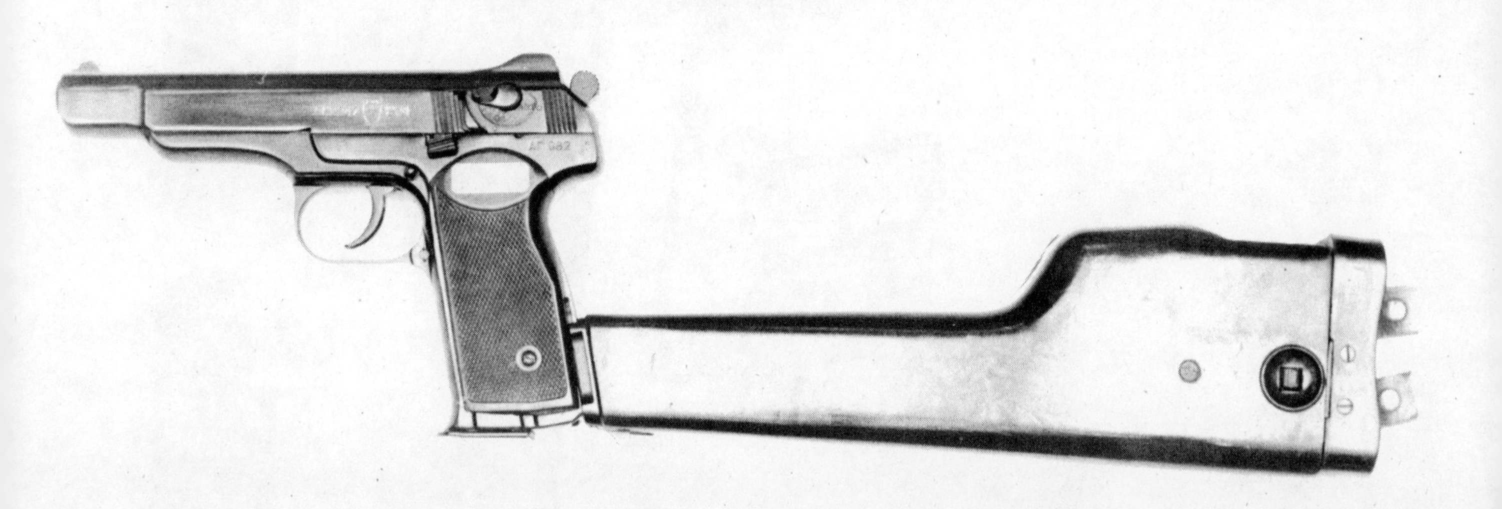

Soviet 9 × 18 Stechkin Machine Pistol (APS) with plastic shoulder stock/holster.

wardly looks like a scaled up copy of the German Walther PP. The Makarov is quite different in internal design from the Walther. The principal differences are as follows. The Makarov does not have a loaded chamber indicating pin. The PM uses a leaf type main spring; the Walther used a coil spring. The trigger-sear linkage of the Makarov is considerably different than that of the Walther, as is the disconnector. The magazine catch of the Makarov is of the spring type and is at the bottom of the grip. The magazine catch of the Walther is a button type mounted on the left side of the receiver. The Makarov has an externally mounted slide stop; the Walther slide stop is internally mounted and is released by drawing back the slide and allowing it to run forward. The ejector of the Makarov is the back end of the slide stop bar. The safety of the Makarov is pushed up to put it on "Safe" and pushed down to put it on "Fire," which is opposite to that of the Walther. The safety of the Makarov places a bar in front of the hammer and has a lug that positions itself in front of the rear shoulder of the receiver, thereby locking the slide in position and preventing the hammer from being cocked.

Stripping of the Makarov

Pull down the trigger guard; pull the slide to the rear and lift its rear end up; then ease forward the slide and the recoil spring (which encircles the barrel), and remove them from the weapon. Since the Makarov is a blowback-operated weapon, it has a fixed barrel, which should not ordinarily be removed. Further stripping is not recommended.

SOVIET 9mm STECHKIN MACHINE PISTOL (APS)

The Stechkin is a true machine pistol in that it is capable of full automatic and semiautomatic fire. It is equipped with a wooden or plastic holster that can be used as a shoulder stock. The Stechkin, like the Makarov, is blowback in operation, but it is a considerably larger weapon than the Makarov and has considerably more potential as a weapon.

Operation of the Stechkin

The Stechkin is loaded by inserting a loaded magazine into the grip (as with the US cal. .45 M1911A1 automatic pistol), then pulling the slide to the rear and releasing it so that it runs home and chambers a cartridge. The safety selector catch is mounted on the left rear side of the slide, as it is on the Makarov and the Walther PP and PPK pistols. The safety selector catch has three positions. The bottom position is safe, the middle position is semiautomatic fire, and the top position is full automatic fire.

The Stechkin with its shoulder-stock holster attached can hit man-sized targets consistently at ranges of 100 to 150 yards, when it is in the hands of a good shot. This, of course, is in semiautomatic fire. The Stechkin's usable range in full automatic fire probably does not exceed twenty-five yards.

Stechkin, with shoulder stock holster.

Soviet 9mm Stechkin Machine Pistol (APS), stripped.

176. АВТОМАТИЧЕСКИЙ ПИСТОЛЕТ

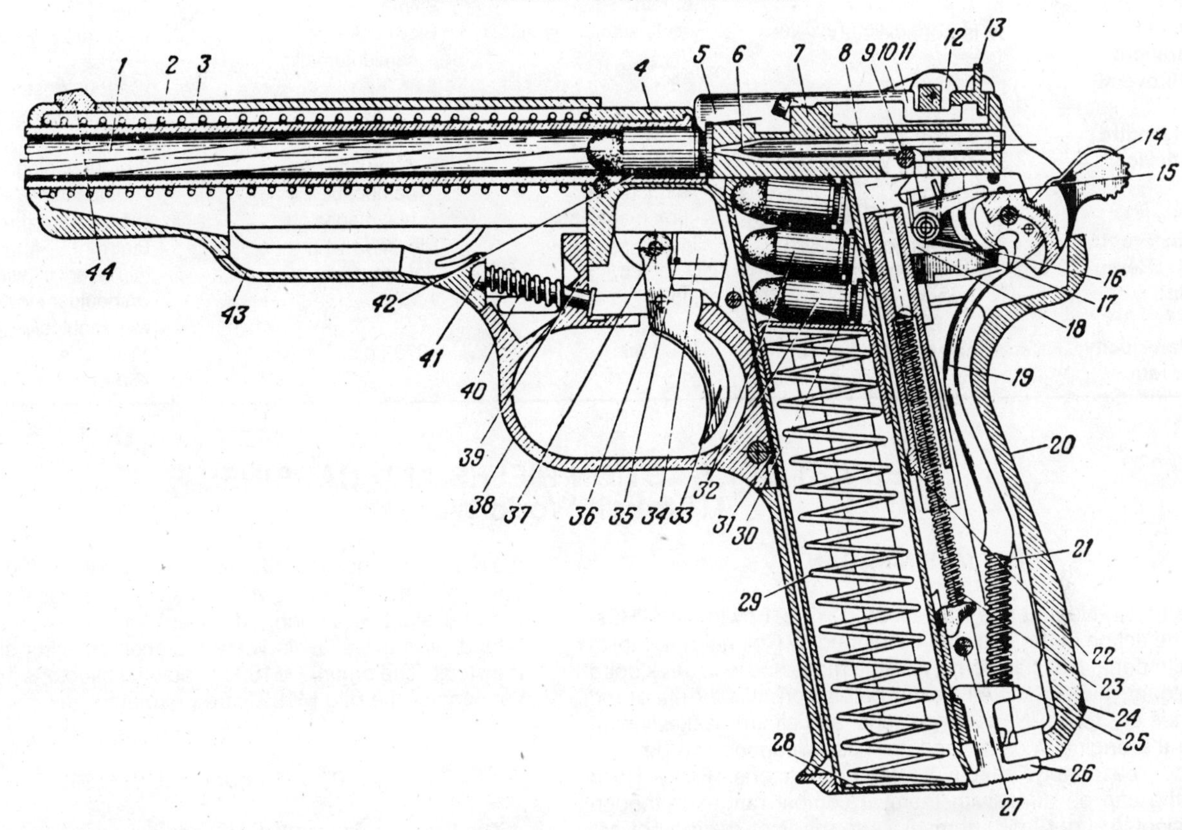

1 ствол
2 возвратная пружина
3 затвор
4 стойка (ствола)
5 выбрасыватель
6 боек
7 пружина выбрасывателя
8 ударник
9 переводчик-предохранитель
10—13 прицел:
11 прицельная планка
12 установочный барабанчик
13 гривка с прорезью;
14 курок
15 шептало
16 пружина шептала
17 разобщитель
18 затворная задержка
19 толкатель
20 рукоятка пистолета
21 боевая пружина
22 замедлитель
23 пружина замедлителя
24 выступ для присоединения кобуры-приклада
25 направляющий стержень боевой пружины
26 защелка магазина
27 отверстие для винта щечек
28 крышка магазина
29 пружина подавателя
30 подаватель
31 корпус магазина
32 пистолетные патроны
33 спусковой крючок
34 спусковая тяга
35 пружина спускового крючка
36 цапфа
37 штифт спускового крючка
38 спусковая скоба
39 стойка (спусковой скобы)
40 пружина стопора
41 стопор
42 патронник с патроном
43 рамка
44 мушка.

Automatic Pistol Stechkin (APS) 9 × 18mm

1. Barrel
2. Recoil spring
3. Slide
4. Barrel bracket
5. Extractor
6. Firing pin tip
7. Extractor spring
8. Firing pin
9. Safety and fire selector

10.-13. Rear sight mechanism
14. Hammer
15. Sear
16. Sear spring
17. Disconnector
18. Slide stop
19. Hammer strut
20. Pistol grip
21. Mainspring
22. Retarder
23. Retarder spring
24. Stock lug
25. Mainspring guide
26. Magazine catch
27. Grip screw hole

28.-31. Magazine assembly
32. Cartridges
33. Trigger
34. Trigger bar
35. Trigger spring
36. Trigger pin axis
37. Trigger pin
38. Trigger guard
39. Trigger guard post

40.-41. Retainer and spring
42. Chamber with cartridge
43. Receiver
44. Front sight

CHARACTERISTICS OF SOVIET PISTOLS AND REVOLVERS

	Model 1895	TT Model 1933	Makarov (PM)	Stechkin (APS)
Caliber:	7.62mm.	7.62mm.	9mm.	9mm.
System of operation:	Double-action revolver.	Recoil, semiautomatic.	Blowback, semiautomatic.	Blowback, selective fire.
Length overall:	9.06 in.	7.68 in.	6.34 in.	w/shoulder stock—21.25 in. w/o shoulder stock—8.85 in.
Barrel length:	4.33 in.	4.57 in.	3.83 in.	5 in.
Feed device:	7-round cylinder.	8-round, in-line detachable box magazine.	8-round, in-line detachable box magazine.	20-round, staggered row, detachable box magazine.
Sights: Front:	Blade.	Blade.	Blade.	Blade.
Rear:	"U" notch.	"U" notch.	Square notch	Flip over "L" with notches.
Weight:	1.65 lb.	1.88 lb.	1.56 lb.	w/shoulder stock—3.92 lb. w/o shoulder stock—1.7 lb.
Muzzle velocity:	892 f.p.s.	1378 f.p.s.	1070 f.p.s.	1100 f.p.s.
Cyclic rate:	—	—	—	750 r.p.m.

SOVIET BOLT-ACTION RIFLES AND CARBINES (Through World War II)

THE MOSIN-NAGANT

The Mosin-Nagant rifle was adopted in 1891 by Imperial Russia. The action of the rifle was developed by Colonel S. I. Mosin of the Imperial Russian Army, and the magazine was developed by Nagant, a Belgian. All Soviet bolt-action military rifles and carbines are Mosin-Nagant weapons, and all are basically similar to the original Mosin-Nagant. These weapons can be considered reasonably effective infantry weapons. Fairly good shooting can be done with them at combat ranges, although their sights do not lend themselves to the finer degrees of accuracy obtained with similar United States weapons. They suffer from an over-complicated bolt but in other respects are relatively simple to service and maintain. The safety, in that it is extremely hard to engage and disengage, represents a shortcoming of the Mosin-Nagant weapons.

The M1891 Mosin-Nagant

The original rifle M1891 was considerably different than later versions of the same model. The original rifle M1891 had no handguard, was fitted with sling swivels instead of the sling slots used on later versions and had a leaf rear sight designed for the old conical-nosed 7.62mm ball cartridge. In 1908, the Spitzer pointed light ball round (which is still used) was introduced, and the rear sight was changed. About this time handguards were added, and the swivels were replaced by sling slots bored in the stock. The original M1891 is now a collector's item. The later versions of the rifle M1891 are obsolete.

The Dragoon Rifle M1891

The Dragoon rifle M1891 was originally developed as a weapon for heavy cavalry. Manufacture of this rifle was dicontinued about 1930, when it was replaced by the rifle M1891/30. The Dragoon rifle M1891 is obsolete. Both the M1891 rifle and the Dragoon rifle have hexagonal receivers.

These rifles are all caliber 7.62mm (.30) for rim cartridges. Mechanically they are all basically the same.

The design differs considerably from the Mauser type on which US bolt-action rifles are all built; the following descriptive matter is set down to provide an understanding of the Russian design.

Bolt. The bolt is an entirely original design. It has never been imitated. It is a 2-piece design in which a removable bolt head carries the locking lugs. The two locking lugs engage horizontally in the receiver, the opposite of most types. The bolt head turns

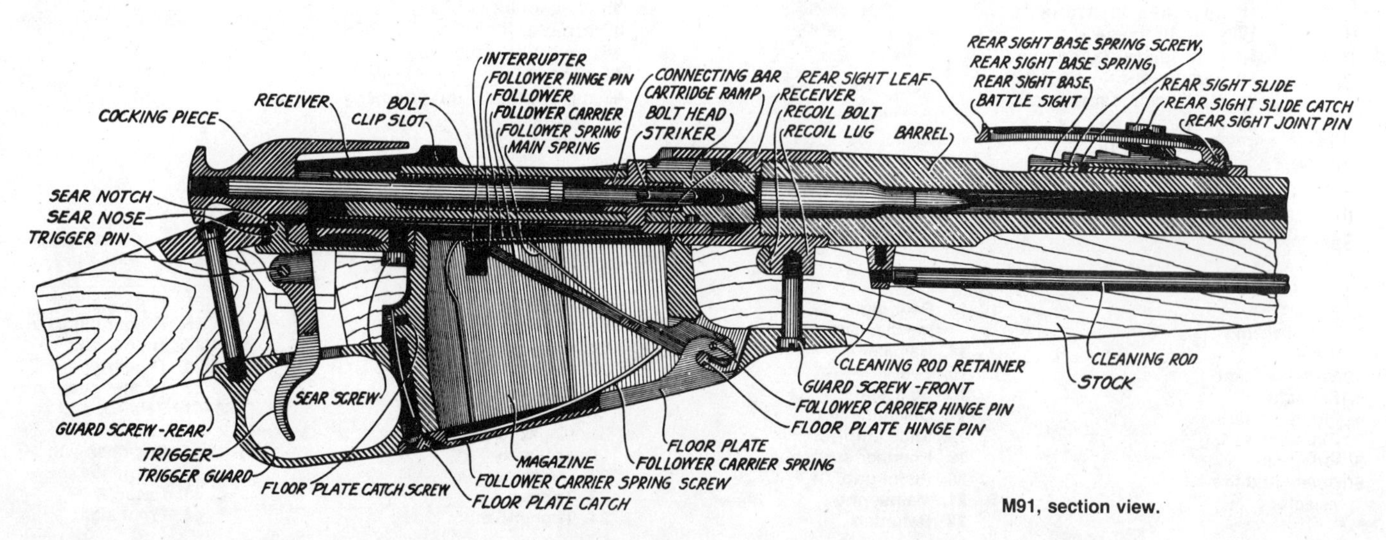

M91, section view.

Export model of the Model 1891 Mosin Nagant made at the Sestroresk Weapons Factory. Note the opened magazine showing the follower and spring.

with the bolt. If this head is accidentally left out when assembling the rifle, the weapon cannot be fired. The extractor is carried by the bolt head.

A special connecting bar joins the bolt head to the bolt itself. This bar does not turn with the bolt; it is positioned in the underside of the bolt assembly and lies in a receiver cut in the boltway. This bar also acts as a guide for the cocking piece and assists in keeping the bolt securely in the receiver. The bar has a projection rising from its forward end. This projection holds a small hollow cylinder. The rear end of this cylinder nests in the hollow front of the bolt, while the forward end of the cylinder nests in the rear of the bolt head.

The handle is part of the bolt forging. A rib extends forward from the hollowed-out bolt. A cut on the underside of this rib takes the projection on the connecting bar. As the bolt handle is lifted, the connecting bar stays fixed in its groove in the receiver below, but its upper projection in the bolt recess holds the bolt assembly securely together.

A small lug on the front end of the connecting bar fits into a groove inside the bolt head, and a small lug on the rear of the bolt head catches in a recess at the front end of the bolt extension rib. Thus as the bolt turns, the bolt head is compelled to turn with it.

The striker and mainspring are inserted from the front end of the hollow bolt. The rear of the striker passes through a small rear hole in the bolt, and a cocking piece is screwed onto it. A collar at the front end of the striker serves as forward mainspring compression point, and the firing pin end, which is forward of this collar, passes through the hollow cylinder on top of the connecting bar. Flat surfaces on the striker at this point and corresponding surfaces within the cylinder prevent the striker from turning.

Cocking Piece. This is a simple and effective design. When it is screwed onto the end of the striker projecting from the bolt, an upper extension of the cocking piece passes along the top of the bolt and through the cut in the top of the receiver bridge.

A small cocking nose is machined on the rear lower surface of the cocking piece.

Safety. When the cocking piece is pulled back slightly, it draws the attached striker back far enough for the flat surfaces on the striker to pass out of the connecting bar cylinder; at this point the striker and cocking piece can be turned to the left. A lug on the underside of the cocking piece extension is thus turned into a locking recess at the rear of the bolt.

Hence by pulling back and turning the cocking piece, the following safety results are obtained: (a) The cocking nose on the cocking piece is turned out of contact with the sear—hence a trigger pull cannot affect it. (b) The striker has been turned so that if it were possible for it to slip it could not hit the cartridge primer, because the flat surfaces at its forward end are turned out of line with the corresponding surfaces inside the connecting bar cylinder through which the striker must pass to hit the primer. (c) The lug on the underside of the cocking piece extension is locked into a cut in the rear of the bolt to prevent any forward movement. (d) The long cocking piece extension is turned to the rear of the cut in the receiver bridge and cannot move forward since the receiver bridge is in its path.

Magazine. The magazine is a projecting single-line box holding five rim cartridges. It forms part of the trigger guard. The magazine floor plate hinges forward and is kept closed by a spring catch. When the catch in the floor plate is drawn back, the magazine follower carrier spring forces the plate down and permits removal of cartridges.

As the accompanying drawing shows, the magazine has not merely a follower on which the first cartridge rests, together with its spring, but also a special "follower carrier" and its spring as part of the cartridge lifting system.

Interrupter-Ejector. This unit is also called "distributor-ejector" and "cartridge valve-ejector." No other rifle needs or uses this device.

The interrupter is a specially shaped plate attached to the left side of the receiver. At its rear is a flat spring held to the receiver by a screw. The thin plate section passes through a cut in the receiver wall above the magazine way. A point on this plate acts as ejector.

The interrupter is operated by action of the bolt. It prevents double feeding. When the top cartridge in the magazine has been chambered, the following cartridge cannot rise because the interrupter is holding it down. Only when the bolt is locked and the extractor engaged around the cartridge case head does a cam-shaped groove in the bolt bear against the interrupter and force it out against the interrupter spring. As the holding projection on the interrupter is forced away from the cartridge, the follower spring can push the top cartridge into line to be ready to feed.

When the bolt handle is raised to open the action, the cam groove in the bolt allows a point on the interrupter to enter the magazine and press against the second cartridge from the top.

An ejecting point on the edge of the interrupter is now in line with a groove in the bolt head. Thus, as the bolt is pulled back, the empty case is struck against this point and ejected from the rifle.

Operation. Lifting bolt handle rotates bolt, bolt head and extractor as a unit. A cam cut at the rear of the bolt forces against the cocking nose on the underside of the cocking piece. The cocking piece and striker assembly are forced back, thereby pulling the firing pin point back away from the primer. The mainspring is partly compressed.

The rib extending forward from the bolt proper passes along a special cam surface on the top of the overhanging receiver ring. (The bolt head is entirely enclosed by the receiver ring.) This action forces the entire bolt assembly back to give primary extraction, the rearward thrust beginning as soon as the bolt has turned far enough for the locking lugs on the bolt head to clear their seats in the receiver ring.

The turning bolt operates the interrupter as described, permitting the interrupter spring to force the interrupter in to hold the next to the top cartridge in the magazine and to line up the ejector point.

The short extractor in the bolt head carries the empty case back and strikes it against the ejector. At this point a lug on top of the trigger hits against metal at the end of a groove in the connecting bar. (The bar has been traveling back as part of the bolt assembly, once the rearward action started.) This acts as a bolt stop.

The upright bolt handle passes back through the cut in the receiver bridge on rearward bolt pull.

As the bolt is pushed forward the bolt head starts the cartridge in its path towards the chamber. At the point where the forward extension rib on the bolt hits the cam face in the receiver ring, the sear nose catches and holds the cocking piece nose. The bolt head locking lugs are at the cam grooves leading to their locking seats in the receiver ring.

As the bolt handle is turned down, the locking lugs are forced along their cam grooves. This draws the bolt forward as the bolt head lugs are locked. Since the sear nose is holding the cocking piece back, this forward pull completes compression of the mainspring. The interrupter works as already described to permit the cartridge to rise below the bolt.

The Rifle M1891/30

The rifle M1891/30 is about the same length as the M1891 Dragoon, but it represents many improvements over the Dragoon. The sights used on the M1891/30 are superior to those of the Dragoon, and, because the metric system of measurement was adopted in Russia during this period, the sights of the M1891/30 are calibrated in meters rather than in arshins. (One arshin equals 0.71 meters or 0.78 yards.) Manufacture of the M1891/30 was initiated in 1930. This model was used in large numbers in the Soviet Army, but was replaced by the carbine M1944 as the standard Soviet infantry shoulder weapon at the end of World War II. It is still in use in some of the satellite countries.

The Sniper Rifle M1891/30

The sniper rifle M1891/30, which is basically the M1891/30 adapted for use with a telescope, is still a standard weapon in some satellite armies. The telescopes employed are somewhat similar to those used on United States hunting rifles.

The Carbine M1910

Although Imperial Russia adopted the Mosin-Nagant rifle in 1891, a true carbine did not appear until 1910. The carbine M1910, with its leaf sight and sling slots, had characteristics of both the original and later versions of the rifle M1891. The carbine M1910 has a hexagonal receiver and does not take a bayonet. This model is comparatively rare.

The Carbine M1938

The carbine M1938 replaced the M1910. It is similar in many respects to the rifle M1891/30. It has a tangent-type rear sight, hooded front sight, and rounded receiver. It does not take a bayonet. This model may be encountered in satellite forces, although it is not manufactured at present.

The Carbine M1944

The carbine M1944, introduced during the latter part of World War II, was the last of the Mosin-Nagants. The permanently fixed bayonet folds down along the right side of the carbine stock when not in use. Except for a slightly longer barrel and the addition of the bayonet, the carbine M1944 is identical to the M1938. It is still in use in various Soviet satellites. See Chapter 13 for illustration of PRC copy of M1944 with folding stock.

CHARACTERISTICS OF SOVIET 7.62MM MOSIN-NAGANT BOLT ACTION RIFLES AND CARBINES

	Rifle M1891	Dragoon Rifle M1891	Rifle M1891/30	Sniper Rifle M1891/30	Carbine M1910	Carbine M1938	Carbine M1944
Weight:							
w/o bayonet & sling:	9.62 lb.	8.75 lb.	8.7 lb.	11.3 lb.	7.5 lb.	7.62 lb.	
w/bayonet & sling:	10.63 lb.	9.7 lb.	9.7 lb.		7.7 lb.		8.9 lb.
Length:							
w/o bayonet:	51.37 in.	48.75 in.	48.5 in.	48.5 in.	40 in.	40 in.	40 in. (folded)
w/bayonet:	68.2 in.	65.5 in.	65.4 in.	65.4 in.			52.25 in. (extended)
Barrel length:	31.6 in.	28.8 in.	28.7 in.	28.7 in.	20 in.	20 in.	20.4 in.
Magazine capacity:	5 rounds.	5 rounds.	5 rounds.	5 rounds.	5 rounds.	5 rounds	5 rounds
Instrumental velocity at 78 ft. w/hvy ball:	2660 f.p.s.	2660 f.p.s.	2660 f.p.s.	2660 f.p.s.	2514 f.p.s.	2514 f.p.s.	2514 f.p.s.
Rate of fire:	8–10 r.p.m	8–10 r.p.m.	8–10 r.p.m.	8–10 r.p.m.	8–10 r.p.m.	8–10 r.p.m.	8–10 r.p.m.
Maximum sighting range:	3200 arshins (2496 yd.)	3200 arshins (2496 yd.)	2000 meters (2200 yd.)	2000 meters* (2200 yd.)	2000 arshins (1560 yd.)	1000 meters (1100 yd.)	1000 meters (1100 yd.)
Front sight:	Unprotected blade.	Unprotected blade.	Hooded post.	Hooded post.	Unprotected blade.	Hooded post.	Hooded post.
Rear sight:	Leaf.	Leaf.	Tangent.	Tangent.	Leaf.	Tangent.	Tangent.
Ammunition**							

*For iron sights when scope is dismounted. Maximum sighting range for the telescopic sight on this weapon is: PE scope—1400m (1540 yd.); PU scope—1300m (1420 yd.).
**Soviet 7.62mm rifle and ground machine gun rimmed ammunition.

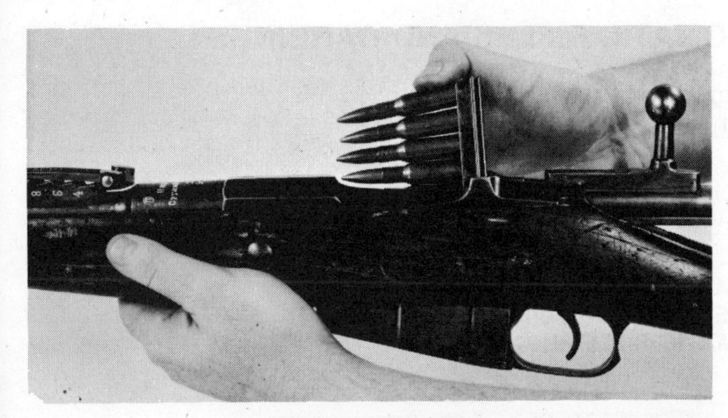

Loading Mosin-Nagant rifle.

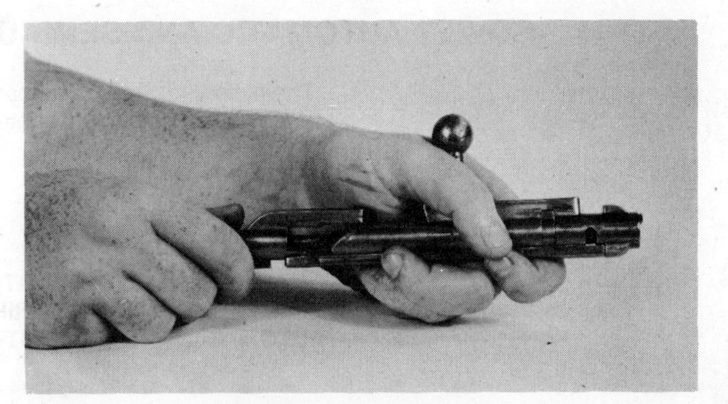

Drawing back cocking piece.

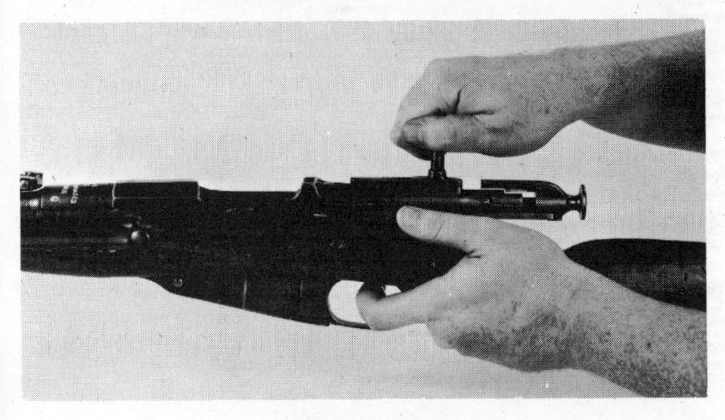

Removing bolt.

How to Load and Fire the Mosin-Nagant Rifles and Carbines

The Rifle M1891. To set the safety, draw back the cocking piece and turn it to the left. This prevents the bolt from opening. To put off safe, pull the cocking piece back, turn it to the right and allow it to move forward.

The rifle M1891 is loaded in the same manner as the United States Springfield or any Mauser rifle. Open the bolt, place a clip of cartridges in the clip guides and press the rounds down into the magazine. Close the bolt; the clip will then fall out of the clip guides onto the ground. The weapon is now ready to fire.

To unload the rifle M1891, open the magazine floor plate and remove the cartridges. The magazine floor plate catch is located on the lower rear part of the magazine, forward of the trigger guard. Press the catch rearward; the follower and floor plate will swing down and forward on a pivot pin, and the cartridges will spill out. Open the bolt and extract the round from the chamber.

The M1891 bayonet is attached by a locking ring; if the M1891/30 bayonet is used, a spring-loaded catch holds the bayonet in place.

The Dragoon Rifle M1891. Operating instructions for Dragoon M1891 are the same as for rifle M1891.

The bayonet of the rifle M1891 or M1891/30 is attached to the Dragoon M1891 in the same manner as described for the rifle M1891.

The Rifle M1891/30. Operating instructions for the rifle M1891/30 are the same as those for the rifle M1891.

The bayonet is attached by a spring-loaded catch.

The Sniper Rifle M1891/30. Operating instructions for this rifle are the same as those for the rifle M1891. The bayonet for the rifle M1891/30 is attached to the sniper rifle by means of a spring-loaded catch.

The Carbine M1910. Operating instructions for the carbine M1910 are the same as those for the rifle M1891; however, bayonets are not provided for this carbine.

The Carbine M1938. Operating instructions for the carbine M1938 are the same as those for the rifle M1891; bayonets are not provided for this carbine.

The Carbine M1944. Operating instructions for the carbine M1944 are the same as those for the rifle M1891; however, this carbine has a nondetachable bayonet, which may be folded or extended by forcing the spring-loaded bayonet tube away from the pivot pin, and then swinging the bayonet to either folded or fixed position.

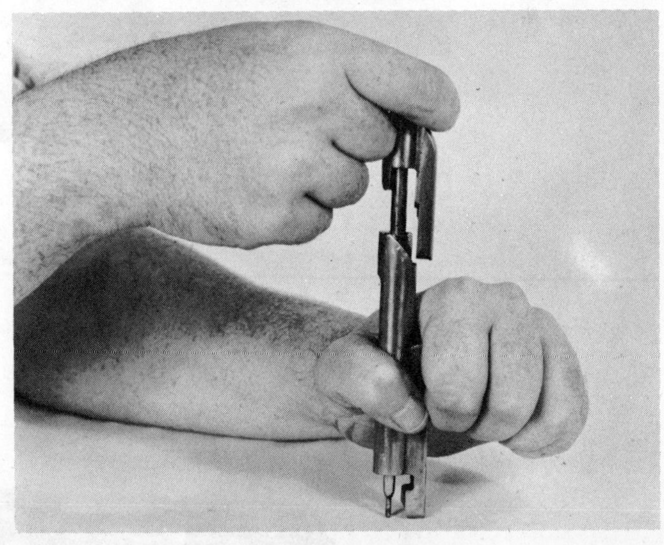

Removal of firing pin.

How to Field Strip the Mosin-Nagant Rifles and Carbines

Open bolt and draw to the rear, pull trigger all the way to the rear and remove bolt from the receiver.

Pull cocking piece to the rear and turn it to the left to relieve the tension of the main spring. Remove bolt head and guide.

Place the firing pin on a solid surface, preferably a block of soft wood, push the bolt body down and unscrew the cocking piece, then remove the firing pin and spring.

The magazine follower can be removed by pushing the magazine floor plate catch, on the bottom of the floor plate just forward of the trigger guard, rearward swinging down the floor plate, follower spring and follower and pressing together, with thumb and forefinger, the follower and floor plate and pulling down to remove them.

Reassembly is in reverse order of disassembly. Care must be exercised when screwing the firing pin into the bolt, that the rear of the firing pin is flush with the cocking piece and that the marks on the rear of the firing pin are aligned with those on the cocking piece in order to assure correct firing pin protrusion.

SOVIET AUTOMATIC AND SEMIAUTOMATIC RIFLES AND CARBINES

The history of Soviet self-loading rifle and automatic rifle development is detailed in Chapter 1.

TOKAREV 7.62mm SEMIAUTOMATIC RIFLE M1938 (SVT), 7.62mm SEMIAUTOMATIC RIFLE M1940 (SVT) and 7.62mm AUTOMATIC AND SEMIAUTOMATIC RIFLE M1940 (AVT)

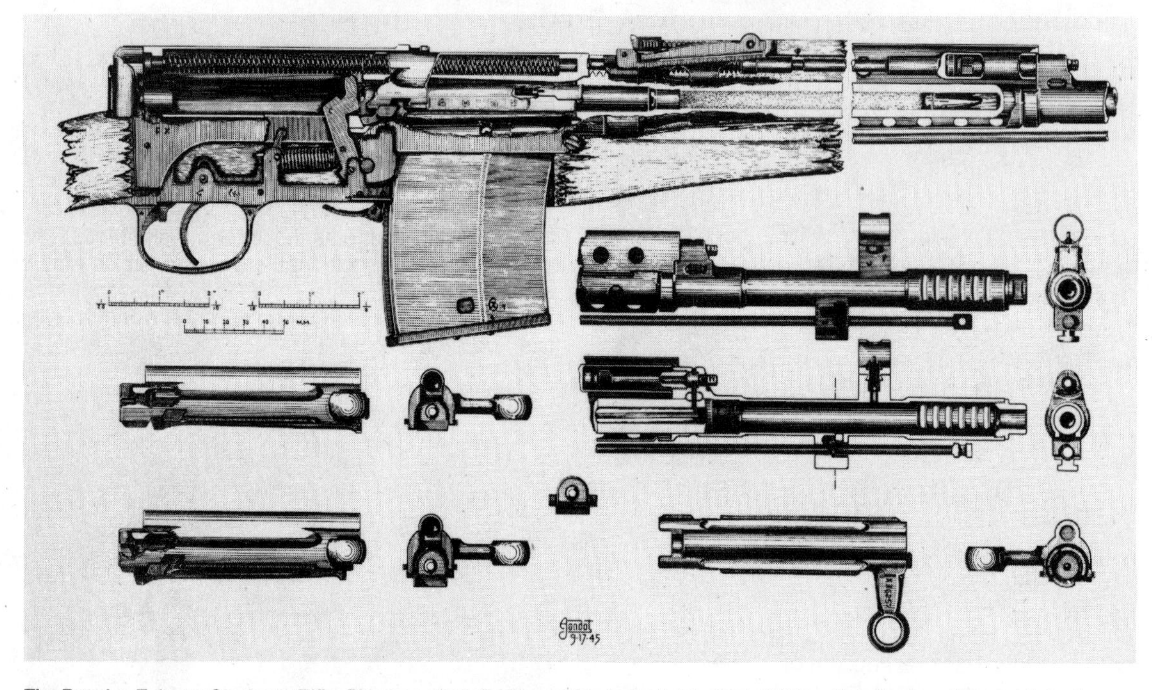

The Russian Tokarev Semiauto Rifle. Phantom view showing action locked at instant of firing. Details show the muzzle brake, gas operation and bolt locking systems.

M1938 7.62mm Tokarev Rifle (SVT), action open.

M1938 Tokarev stock illustrating two-piece construction.

How the M1940 Tokarev Rifle Works

A loaded magazine is inserted in the bottom of the receiver until its catch locks it. The bolt handle is drawn back and released to cock the hammer and strip a cartridge into the firing chamber.

The rear end of the bolt is forced down by the bolt carrier into locking position. The carrier proceeds to ride over projections on the bolt to hold it in the locked position.

When the trigger is pressed, it forces a bar forward to rotate the sear. The sear is fitted vertically in the trigger guard. It now engages a bent in the head of the hammer. The hammer is rotated forward to strike the firing pin and explode the cartridge.

As the bullet passes down the barrel, a portion of the gas escapes through a port drilled in the top of the barrel under the upper section of the wooden fore-end. (The gas regulator has five positions for adjustment.) The expanding gas impinges on the head of the piston rod (which is positioned on top of the barrel) driving it backward for a stroke of about $1\frac{1}{2}$ inches.

The thrust is transmitted to the operating rod, which passes it on to the bolt carrier.

The operating rod is mounted above the barrel, its rear end passing through a hole in the receiver directly above the forward face of the locked bolt and in line with the bolt carrier, which rides on top of the bolt. The operating rod return spring is mounted around the rod itself.

The carrier travels back about $\frac{1}{4}$ of an inch before unlocking. The bolt travels on back carrying with it the empty cartridge case, which is ejected to the right. It also compresses the dual springs mounted in the top of the receiver to store up energy for the return motion. At the same time, it forces the hammer to rotate on its axis until it is caught and held by the sear.

When the last shot has been fired, a catch in the bottom center of the bolt way is operated by a projection of the magazine platform and holds the bolt open.

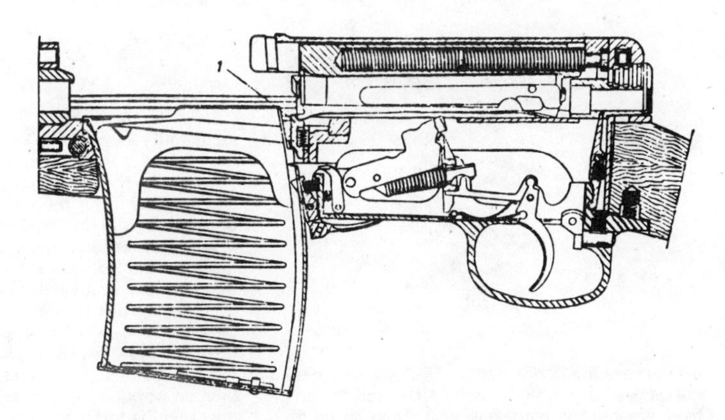

Action open on firing last shot. Recoil spring compressed, hammer cocked, ready for reloading.

Tokarev 7.62mm M1940 Rifle (SVT).

Field Stripping the Tokarev M1940

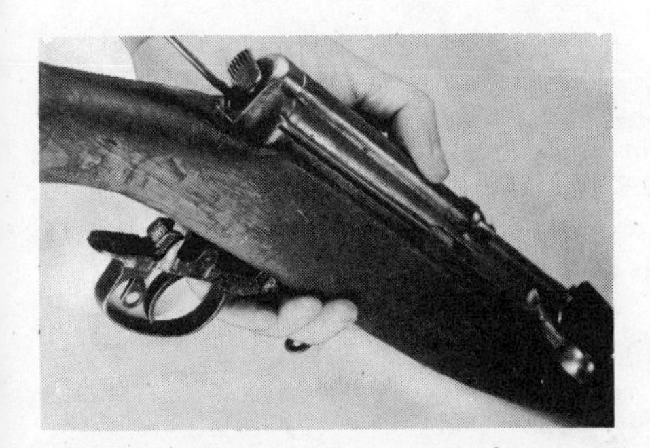

(1) Remove trigger assembly. First turn the thumb latch at the extreme rear of the receiver up to the left as far as it will go. This will expose a cylindrical hole in which the trigger guard locking collar rests. With a screwdriver or a bullet point, push the spring-loaded locking collar in. This will pivot the locking bar in the rear of the receiver and will free the rear end of the trigger guard assembly which may then be removed from the receiver. (Note: when replacing this unit it is necessary that the bolt be back far enough from the chamber so the disconnector is able to rise. Otherwise the front end of the trigger guard unit will not seat in the receiver notches.)

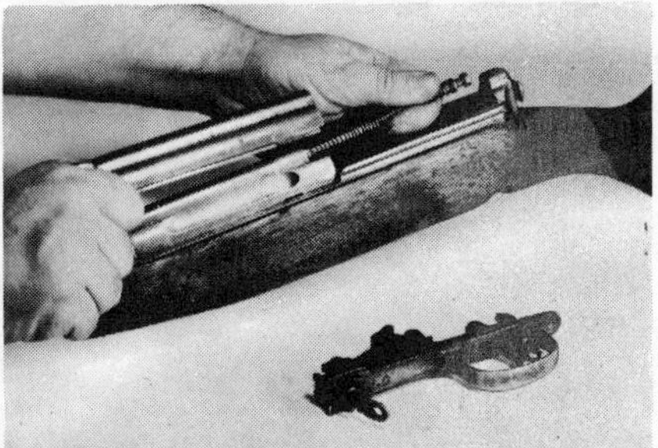

(2) Remove operating spring dust cover. First rest the muzzle of the rifle on the floor, as considerable force must be applied. Then pull the steel cover to the rear of the bolt straight down toward the muzzle. This action will force the buffer and operating springs down inside the bolt, as the buffer guide rod is seated in a groove machined for it at the extreme rear of the cover. When the cover has been pulled down as far as it will go in its guides in the receiver, grasp it firmly with the left hand and squeeze it tightly against the bolt. While retaining this grasp with the left hand, with the thumb of the right hand push the buffer rod down far enough to free its head from the seating in the cover.

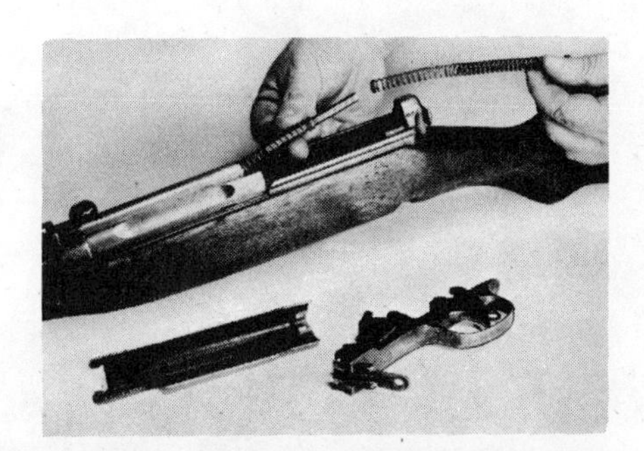

(3) Then push it carefully out away from the cover and ease it and the springs up out of the bolt hole. The cover is free to come off at this point. Remove operating and buffer spring assemblies. Holding the protruding buffer spring and rod close to the bolt hole, ease them up and to your right so the rod will clear the rear of the receiver. When all tension is off the springs, the buffer spring and its guide rod may be lifted out of engagement with the hollow operating spring rod. The operating spring and its rod may now be removed from the bolt hole.

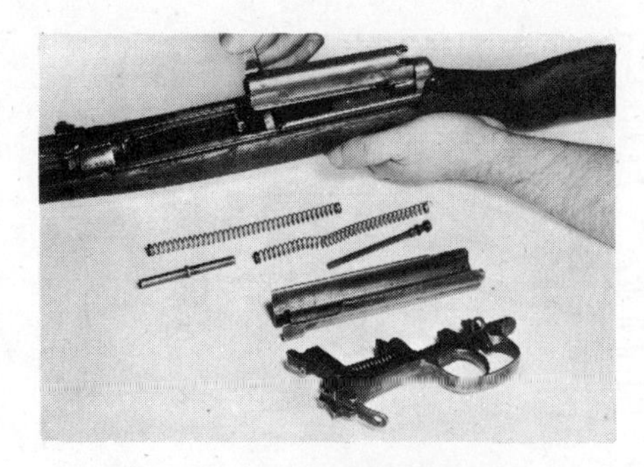

(4) Remove the bolt assembly. Pull the bolt handle back slowly until the bolt carrier and the bolt have cleared the magazine well. Then lift up on the bolt handle and continue to pull back and up. At this point the bolt assembly is free of its receiver guides and may be tilted and lifted up out of the receiver. (Note: When re-assembling this unit, the assembly will drop into its receiver tracks at just the one proper point. Do not exert force, but feel for this point if you cannot locate it immediately.) Remove the bolt from its carrier. Hold the assembly in the palm of the left hand upside down. Place the forefinger over the firing-pin hole and the thumb (of the right hand) over the rear of the firing pin and push the bolt slowly ahead inside the carrier. At the proper point the cam faces on the bolt will mate with the proper surfaces on the bolt carrier and the bolt may then be lifted up out of the carrier. Remove the gas piston. Press in the lock on the front band and push the band forward. This will permit the perforated metal upper cover to be lifted off. The pierced wooden forearm covering the gas piston may then be lifted off. (Note: If it is necessary to remove the stock, unscrew the bolt from the right side of the receiver below the chamber and pull it out. The lower metal fore-end piece may then be removed and the receiver and barrel assemblies lifted out of the stock complete.) Pull the piston back until it clears the face of the gas cylinder over which it is mounted.

(5) Holding the piston at this point, pull the operating rod in which the piston seats back far enough to permit the piston to be lifted off. Then ease the rod forward against the tension of its spring, and remove the rod and its spring from their seating in the receiver below the line of the rear sight. Remove compensator and gas cylinder assembly. Drift the retainer out preferably from the right. The compensator and assembly may then be unscrewed from the barrel and removed.

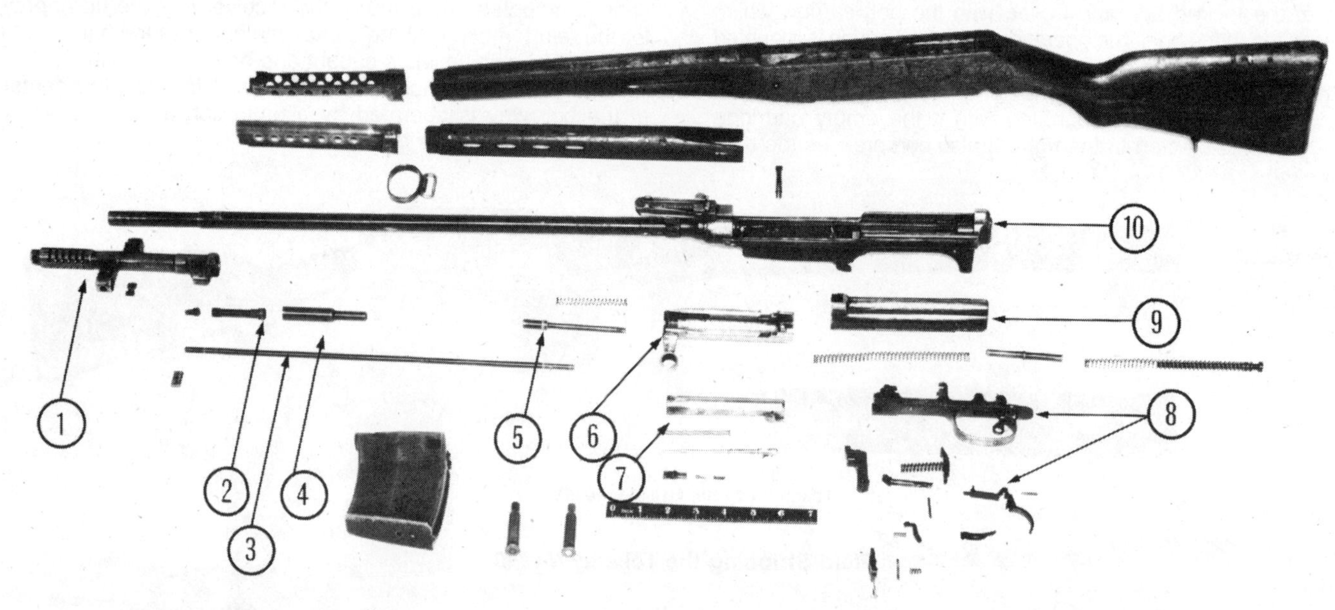

Disassembled M1940 Tokarev Rifle.
1. Muzzle brake, front sight and gas cylinder assembly. 2. Gas cylinder. 3. Operating rod. 4. Gas piston. 5. Operating rod tappet. 6. Bolt carrier. 7. Bolt, disassembled. 8. Trigger group, disassembled. 9. Receiver cover. 10. Barrel and receiver group.

Metal parts removed from stock to show operation of mechanism and dust cover removed. Operating rod is being driven back through hole in the receiver and has pushed bolt carrier back until carrier has unlocked bolt and is carrying it back with it.

CHARACTERISTICS OF SOVIET PRE-WORLD WAR II 7.62mm AUTOMATIC AND SEMIAUTOMATIC RIFLES

	Automatic Rifle M1936	Semiautomatic Rifle M1938	Semiautomatic Rifle M1940	Automatic Rifle M1940	Semiautomatic Sniper Rifle M1940	Semiautomatic Sniper Rifle M1938
Weight:						
w/o bayonet & magazine:	8.93 lb.	8.70 lb.	8.59 lb.	8.35 lb.	9.18 lb.	9.52 lb.
w/bayonet & magazine:	—	10.8 lb.	9.48 lb.	9.24 lb.	—	—
Length:						
w/o bayonet:	48.6 in.	48.1 in.	48.1 in.	48.1 in.	48.1 in.	48.1 in.
w/bayonet:	59.3 in.	60.84 in.	57.1 in.	57.1 in.	57.1 in.	60.84 in.
Barrel length:	24.16 in.	25 in.	24.6 in.	24.6 in.	24.6 in.	25 in.
Magazine capacity:	15 rounds.	10 rounds.	10 rounds.	10 rounds.	10 rounds.	10 rounds.
Instrumental velocity at 78 ft. w/hvy ball:	2519 f.p.s.	2519 f.p.s.	2519 f.p.s.	2519 f.p.s.	2519 f.p.s.	2519 f.p.s.
Rate of fire: (semiautomatic)	30-40 r.p.m.	25 r.p.m.	25 r.p.m.	30-40 r.p.m.	25 r.p.m.	25 r.p.m.
Maximum sighting range:	1500 meters.	1500 meters.	1500 meters.	1500 meters	Iron sights: 600 m. (660 yd.) Telescope: 1300 m. (1430 yd.)	Iron sights: 600 m. (660 yd.) Telescope: 1300 m (1430 yd.)
Front sight:	Open guard blade.	Hooded post.	Hooded post.	Hooded post.	Hooded post.	Hooded post.
Rear sight:	Tangent.	Tangent.	Tangent.	Tangent.	Tangent and telescope.	Tangent and telescope.
Principle of operation:	Gas.	Gas.	Gas.	Gas.	Gas.	Gas.
Ammunition:	*	*	*	*	*	*

*7.62mm USSR rifle and ground machine gun rimmed ammunition.

POST-WORLD WAR II SOVIET SHOULDER WEAPONS

SOVIET 7.62mm SKS CARBINE

The SKS, like the AK automatic rifle and RPD machine gun, is chambered for the new Soviet "intermediate-sized" M1943 cartridge previously described. Its disassembly is not quite as simple as is the AK's but is still relatively easy. The SKS is distinguished by its folding bayonet, which—unlike earlier Soviet folding bayonets—is of blade section and folds under the weapon rather than to the side of the weapon. The SKS is no longer used in first-line Soviet units.

NOTE: The Soviets call the SKS a carbine because of its short barrel length. It is not intended to be nor is it used as a replacement for the pistol (as the US carbine was designed to be). The SKS was intended to be used as the tactical equivalent to the US M1 or M14.

Field Stripping the SKS

Turn off the safety by rotating it downward. Swing the bayonet upward so that it projects at a 90-degree angle from the muzzle

Soviet 7.62mm Simonov Semiautomatic Carbine (SKS).

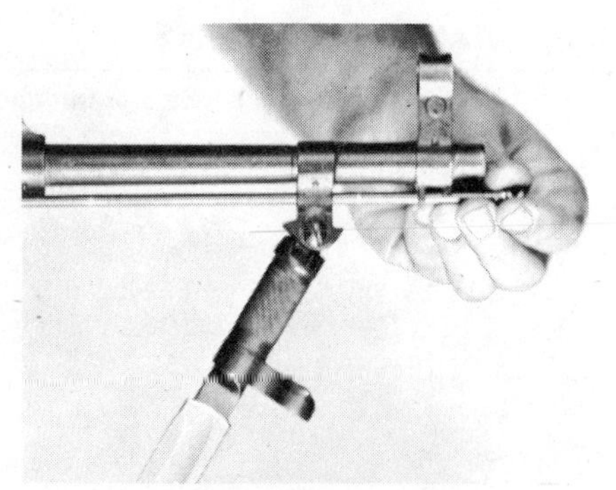
(1) Removing cleaning rod.

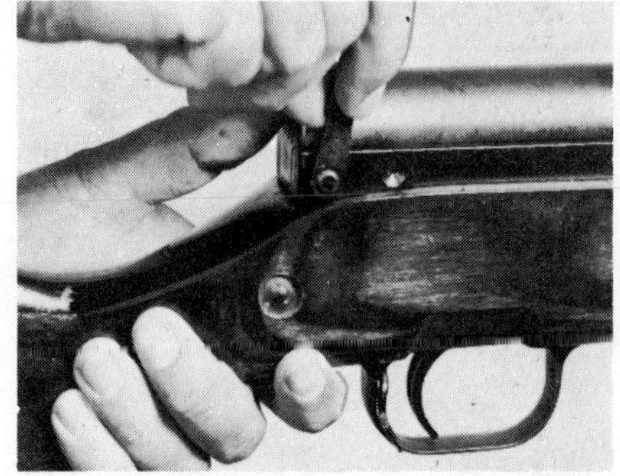
(2) Loosening receiver cover retaining pin.

(3) Removing receiver cover.

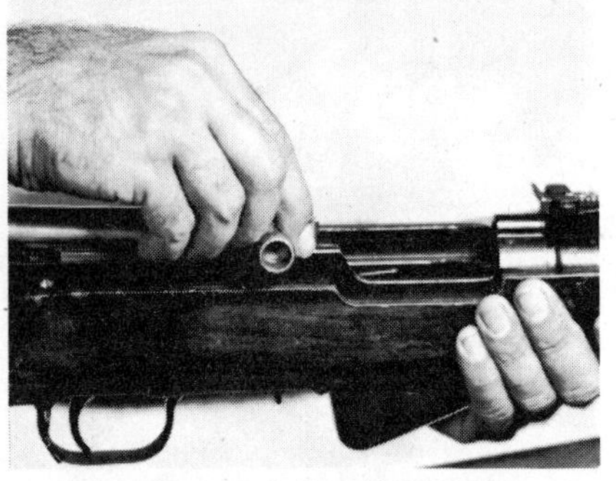
(4) Removing bolt and bolt carrier.

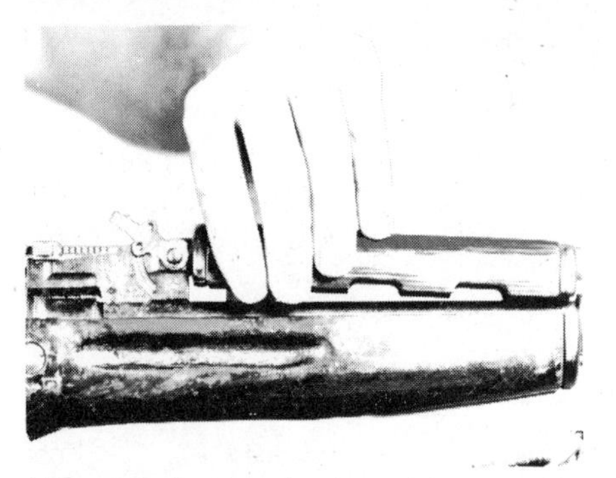
(5) Removing handguard and gas piston assembly.

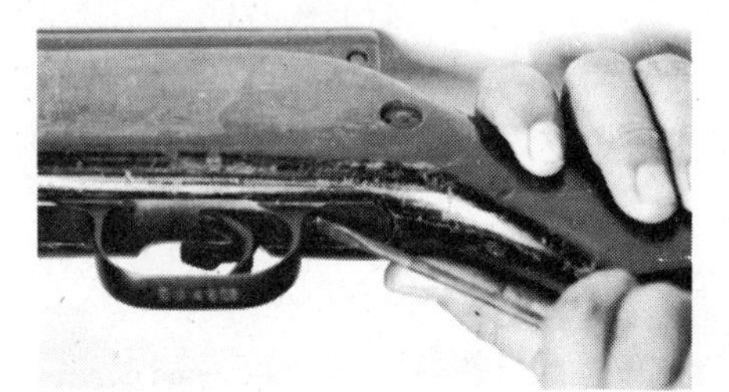
(6) Unlocking trigger guard.

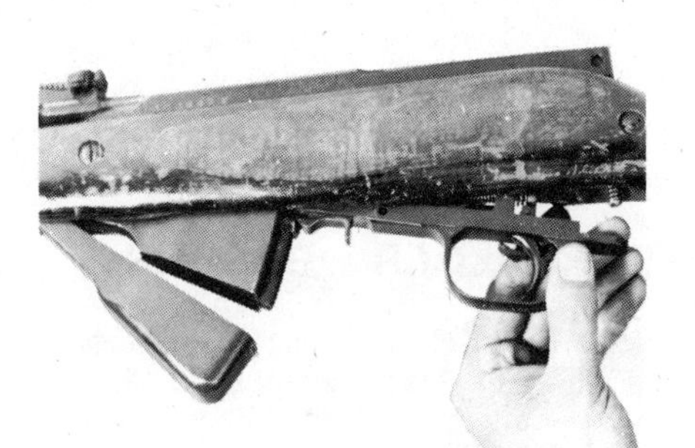
(7) Removing trigger group.

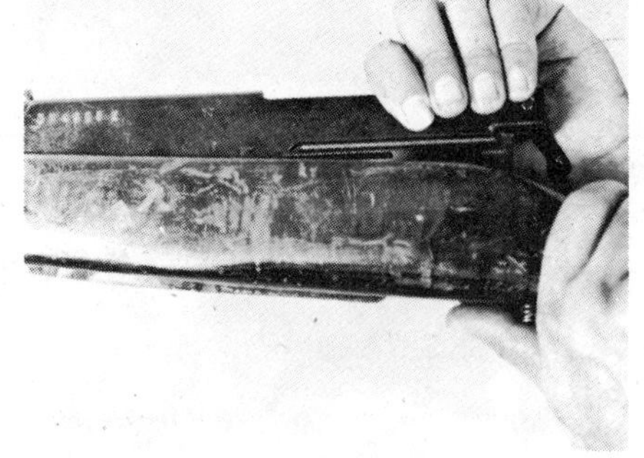
(8) Removing barrel and receiver.

of the rifle. Remove the cleaning rod by disengaging its head from the lugs on the underside of the barrel and pulling the rod forward. Rotate the bayonet downward until it locks in its folded position.

Hold the small of the stock with the left hand and, with the right hand, rotate the receiver cover retaining pin arm (located at the right rear of the receiver) upward; then, pressing on the receiver cover with the thumb of the left hand, move the pin to the right as far as possible, and remove the cover from the receiver. Remove the recoil spring assembly from the bolt carrier. Note that the recoil spring assembly is, like that of the AK, a packaged unit.

To remove the bolt and bolt carrier from the receiver, pull back on the operating handle; the whole assembly will lift out, and the bolt can be removed from the bolt carrier.

Hold the fore-end of the carbine with the left hand and, with the right hand, rotate the gas cylinder tube lock lever upward (the lock is at the right front side of the rear sight base) until the lock lever lug is stopped by the upper wall of the cutout in the rear sight base. Then the gas cylinder tube and handguard assembly can be slightly raised and pulled rearward off the weapon. The gas piston will slide out of the gas cylinder tube.

To assemble the weapon, follow the above steps in reverse order.

Loading and Firing the SKS

Set the weapon on "Safe" by turning the safety lever up as far as it will go. The safety lever is at the rear of the trigger guard. Pull the operating handle to the rear as far as it will go; the bolt and bolt carrier will be held to the rear by the bolt-holding-open device. If the cartridges are assembled in the ten-round charger normally used, insert one end of the charger in the charger guide machined into the top forward end of the bolt carrier. Push down on the cartridges with the thumb until all the cartridges are loaded into the magazine; then remove the empty charger guide. Pull back on the operating handle and release it; the bolt and bolt carrier will now go forward and chamber a cartridge. Disengage the safety by turning it down as far as it will go. The rifle is now ready to fire.

If all ten rounds are fired, the bolt and bolt carrier will stay to the rear, leaving the action open for reloading. If only a few rounds are fired and it is desired to unload the weapon, the following procedure should be observed. To the rear of the mag-

SKS, field stripped.

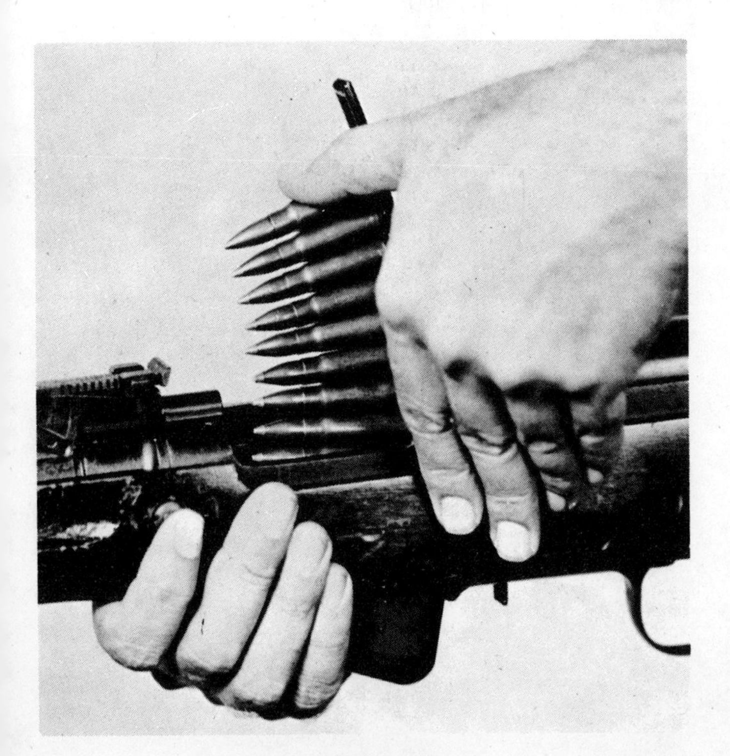

Loading the SKS.

azine, on the underside of the weapon, is a catch; pull this catch rearward. The magazine body will now swing downward, and the rounds in the magazine will spill out. The round in the chamber is removed by pulling the operating handle to the rear.

To Fix the Bayonet

To fix the bayonet, the bayonet handle is pulled to the rear, and the bayonet is rotated upward until the bayonet muzzle ring snaps over the muzzle of the weapon. To fold the bayonet, pull the bayonet handle upward until the bayonet muzzle ring clears the muzzle; then swing the bayonet downward and into its groove in the stock.

How the SKS Works

When the trigger is pressed, the trigger rotates and moves the trigger arm forward. The trigger arm moves the sear forward and disengages it from the hammer. The hammer spring, now being free to expand, rotates the hammer, which strikes the firing pin and drives it through the bolt channel. When the hammer is rotated, the hammer heel lowers the forward end of the disconnector and the forward end of the trigger arm. At this time, the trigger arm is disengaged from the sear, which, under the action of the sear spring, returns to the rear position. The firing pin moves through the bolt channel until the firing pin tip emerges from the bolt face and strikes the primer. After the bullet passes the gas port in the wall of the barrel, the gases enter the gas cylinder, exert pressure on the piston and move the bolt carrier to the rear by means of the piston rod.

The bolt carrier, after traveling a path of eight mm, raises the rear end of the bolt, disengages the bolt locking surface from the receiver lug and brings the bolt to the rear. As the bolt moves to the rear, it extracts the empty cartridge case, which is held in the bolt face by the extractor lug. After traveling a distance of 70 mm, the base of the cartridge case meets the ejector, and the case is ejected from the weapon. In moving to the rear, the bolt carrier, and then the bolt, cocks the hammer. The forward end of the disconnector, under the action of the hammer spring, is raised upward, and the hammer is positioned on the disconnector notch at the same time. The forward end of the trigger arm is located between the base of the trigger guard and the sear. Under the action of the sear spring, the sear, which is in the extreme rear position, is positioned under the hammer cock notch. Under the action of the follower lever spring, the follower positions the next round for the feed rib. The piston, together with the piston rod, pushes the bolt a distance of 20 mm. The bolt thereafter continues by inertia to the extreme rear position, at the same time compressing the recoil spring, and the piston and piston rod return to the forward position under the action of the piston rod spring.

Under the action of the expanding recoil spring, the bolt moves forward, and the bolt feed rib grasps the next round in the magazine and sends it into the chamber. The bolt carrier lowers the rear end of the bolt upon approaching the barrel face, and the bolt locking surface is positioned in front of the receiver locking lug to seal the bore. The bolt, in lowering, depresses the protruding front end of the disconnector, at which time the hammer is disengaged from the disconnector and is then cocked. As it approaches the barrel face, the extractor, which is now sliding along the base of the cartridge case, is forced to the right to grasp the cartridge case rim. The extractor spring is compressed at this time. Under the action of the follower lever spring, the follower raises the next round until it meets the bolt. Before the next round is fired, it is necessary to release the trigger; under the action of the trigger spring, the trigger returns to the forward position, and the front end of the trigger lever is disengaged from the sear and is positioned opposite the sear shoulders.

To fire the next round, it is necessary to press the trigger. The hammer can be released from the cocked position only when the bolt is in the extreme forward position and the bore is completely sealed. If the bolt is in the rear position (on the stop), or if the bore is not completely sealed (bolt in extreme forward position), the forward end of the disconnector is not depressed; the forward end of the trigger lever presses against the guide lugs for the sear in the trigger guard groove, and the hammer cannot be released. If the bolt travels past the locking lug and depresses the forward end of the disconnector and the bolt carrier has not attained the extreme forward position, the weapon still cannot be fired. In this event, the hammer, in moving forward will not strike the firing pin but will strike the vertical surface of the bolt carrier, which cocks the hammer.

THE SOVIET 7.62mm AK ASSAULT RIFLE

The development of this weapon is discussed in Chapter 1.

Loading and Firing the AK

The thirty-round magazine is loaded by hand by pressing the cartridges down into the mouth of the magazine with the thumb. Insert the magazine into the underside of the receiver, forward end first; then draw up the rear end of the magazine until a click is heard or until the magazine catch is felt to engage its slot at the rear of the magazine. Pull the operating handle (which protrudes from the right side of the receiver) smartly to the rear and

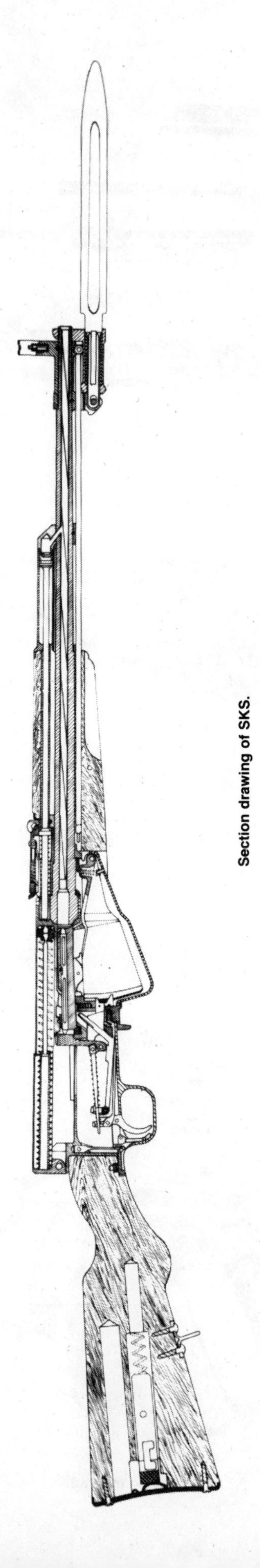

Section drawing of SKS.

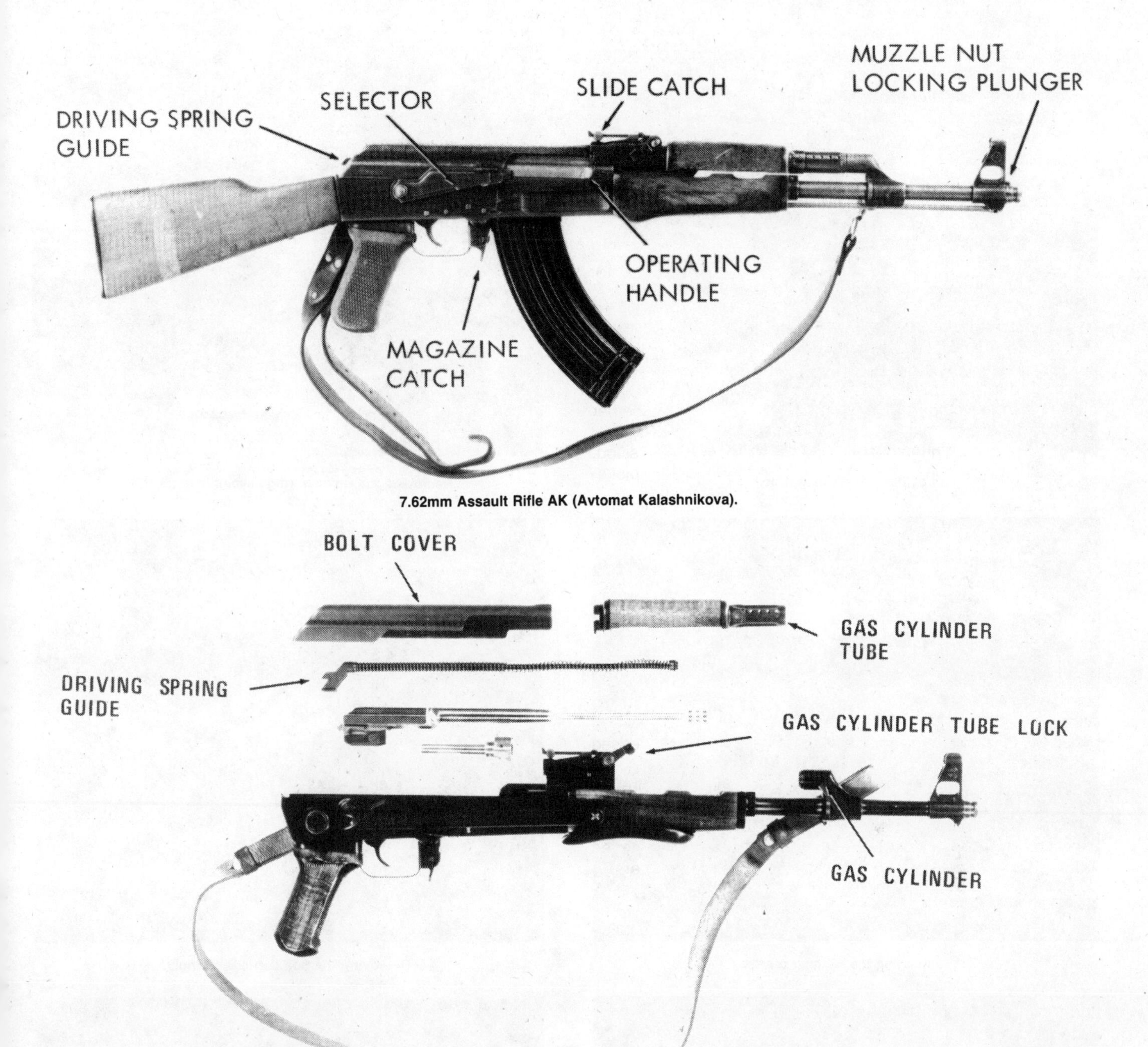

7.62mm Assault Rifle AK (Avtomat Kalashnikova).

AK Rifle with folding stock field stripped.

release it. The bolt and bolt carrier will go forward under the pressure of the recoil spring and will chamber a cartridge. To put the weapon on safe, push the safety selector catch (mounted on the right side of the receiver) as far up as it will go. When on safe, the safety selector blocks rearward movement of the operating handle and covers a gap to the rear of the bolt carrier. For semiautomatic fire, push the end of the safety selector lever all the way down (so that its forward end is opposite the bottom two Cyrillic letters), aim and squeeze (do not pull) the trigger. For full automatic fire, push the safety selector lever to the middle position marked by the two Cyrillic letters "AB" (which stand for "automatic"). Although the AK is quite heavy, it climbs rapidly in automatic fire; it is therefore necessary to get a good grip on the weapon before squeezing the trigger for automatic fire.

Field Stripping the AK

Remove the magazine and pull the operating handle to the rear to clear weapon of any live ammunition. At the rear of the stamped receiver cover, a serrated catch protrudes; this is the rear end of the recoil spring guide rod. Push the catch in with the finger and at the same time pull the receiver cover upward and backward from the receiver. After removing the receiver cover, push the recoil spring guide forward, so that the bottom rear lug section of the guide clears the dovetail slot at the rear of the receiver. Remove recoil spring. Note that the recoil spring and its guide rod are a packaged unit; i.e., they are held together as one piece by a collar fitted to the end of the guide rod. This is similar to the recoil spring used on the cal. .50 Browning

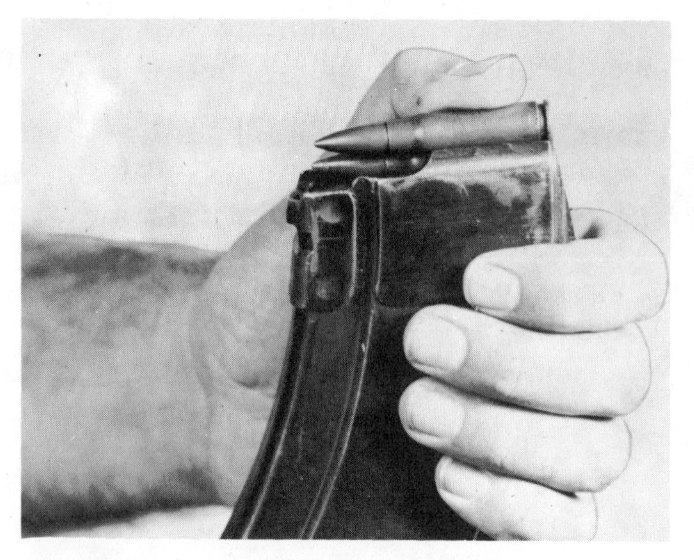

Loading the AK magazine.

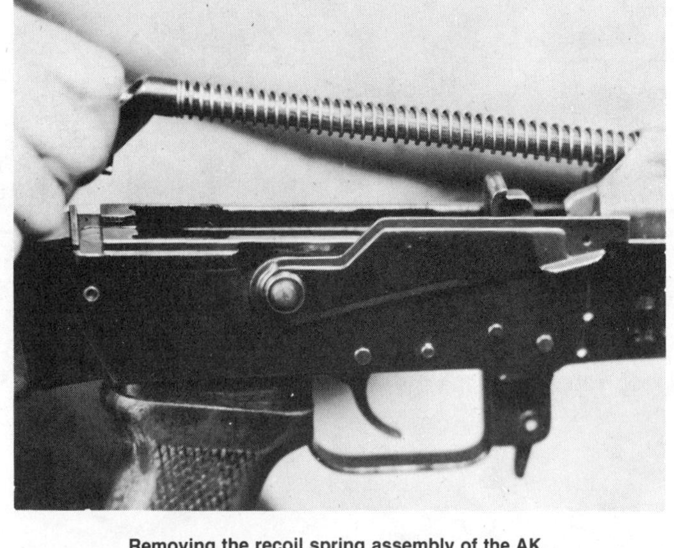

Removing the recoil spring assembly of the AK.

Inserting the AK magazine.

Removing the bolt carrier assembly.

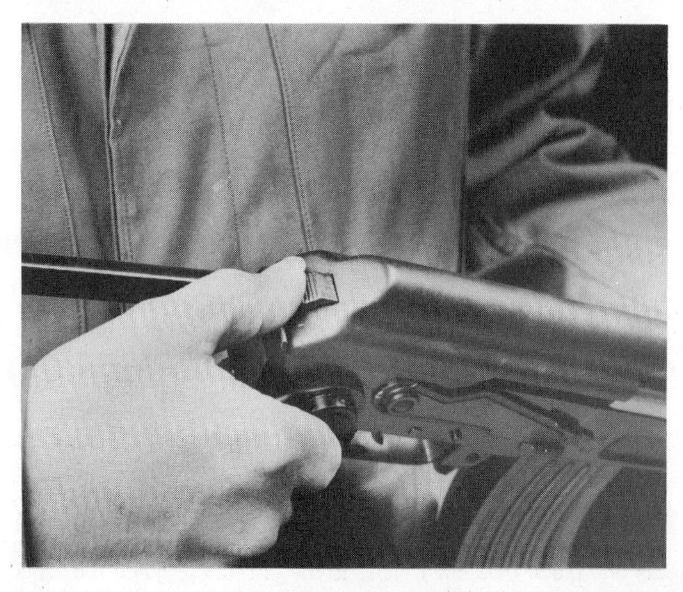

Removing the AK receiver cover.

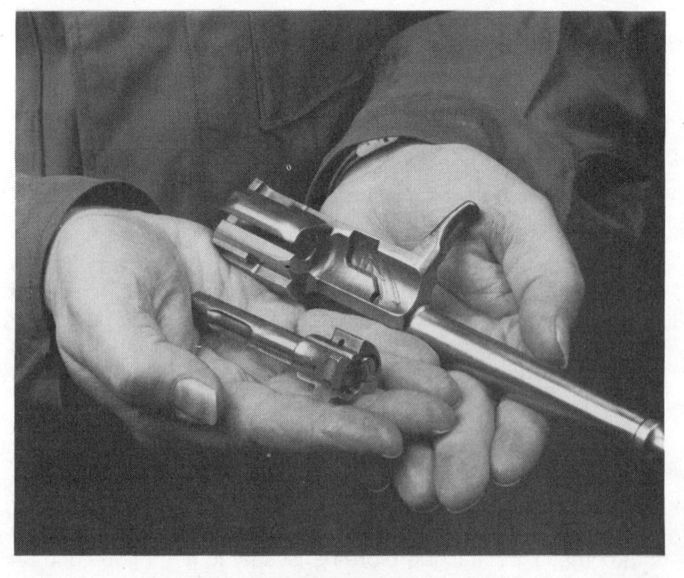

Removing bolt.

machine gun. The bolt carrier, piston and bolt can now be removed as a unit by pulling them back to the cutout section in the receiver track and lifting them upward and rearward. To remove the bolt from the bolt carrier, turn its head so that the guide lug aligns with the cam surface of the bolt carrier, move the bolt as far to rear as possible, rotate the bolt so that the guide lug leaves the camway of the bolt carrier, and pull the bolt forward and out of the bolt carrier. This operation is not as complicated as it seems; the bolt can actually be shaken out of the bolt carrier. Rotate the handguard lock lever (on the forward right side of the rear sight base) upward, and the handguard-piston tube assembly can be lifted out. No further disassembly is recommended.

To assemble the weapon, carry out in reverse the steps outlined above.

How the AK Works

Full Automatic Fire. With the safety selector set on full automatic fire, a cartridge in the chamber (to chamber a cartridge initially, since the AK fires from a closed bolt, it is necessary to pull the operating handle to the rear and release it) and a loaded magazine in the weapon, the following actions occur when the trigger is pulled. The safety selector lug is far enough to the rear to release the rear end of the trigger but stays directly above the rear end of the disconnector. Thus, while the trigger is free to rotate, the safety selector lug prevents the disconnector from rotating. One large multi-stranded spring is used as both hammer spring and trigger spring in the AK.

Semiautomatic Fire. To fire a single round from the rifle, set the weapon for semiautomatic fire by rotating the safety selector as far downward as possible, and then press the trigger.

When the trigger is pressed, the semiautomatic sear and disconnector rotate. The rear end of the sear (which is actually part of the trigger) raises the ends of the hammer-and-trigger spring. As the trigger rotates, the semiautomatic sear releases the hammer cock notch.

The hammer-and-trigger spring rotates the hammer forward, and the hammer strikes the rear end of the firing pin, pushing it forward so that it strikes the primer of the cartridge. The cartridge fires, and gases from the cartridge flow through the gas port in the barrel into the gas cylinder and force the piston and bolt carrier assembly to the rear. As the piston and bolt carrier assembly move to the rear, the recoil spring is compressed. The bevel in the bolt carrier cam acts on the bolt guide lug, rotating the bolt to the left and thereby unlocking the bolt.

After the bolt is unlocked, the bolt carrier and bolt move to the rear together. The bolt carrier rotates the hammer to the rear, compressing the hammer-and-trigger spring. As the hammer rotates, it rotates the disconnector. When the head of the hammer has passed the notch in the disconnector, the disconnector spring forces the disconnector to engage the disconnector notch in the hammer. This holds the hammer at full cock.

The full automatic sear spring rotates the full automatic sear into engagement with the full automatic sear notch in the hammer. The full automatic sear, however, does not hold the hammer in the cocked position, since the disconnector is already performing this function. As the full automatic sear rotates, the upper end of the sear rises to obstruct the passage of the full automatic disconnector.

As the bolt moves to the rear, the extractor pulls the cartridge case from the chamber. When the case strikes the ejector, it is ejected from the receiver.

The top round in the magazine is forced upward by the follower until it is arrested by the magazine flange.

The rearward movement of the bolt carrier and bolt is arrested by the rear wall of the receiver. Forward movement of these parts is caused by the decompression of the recoil spring.

As the bolt carrier moves forward, the top cartridge in the magazine is stripped from the magazine and forced into the chamber.

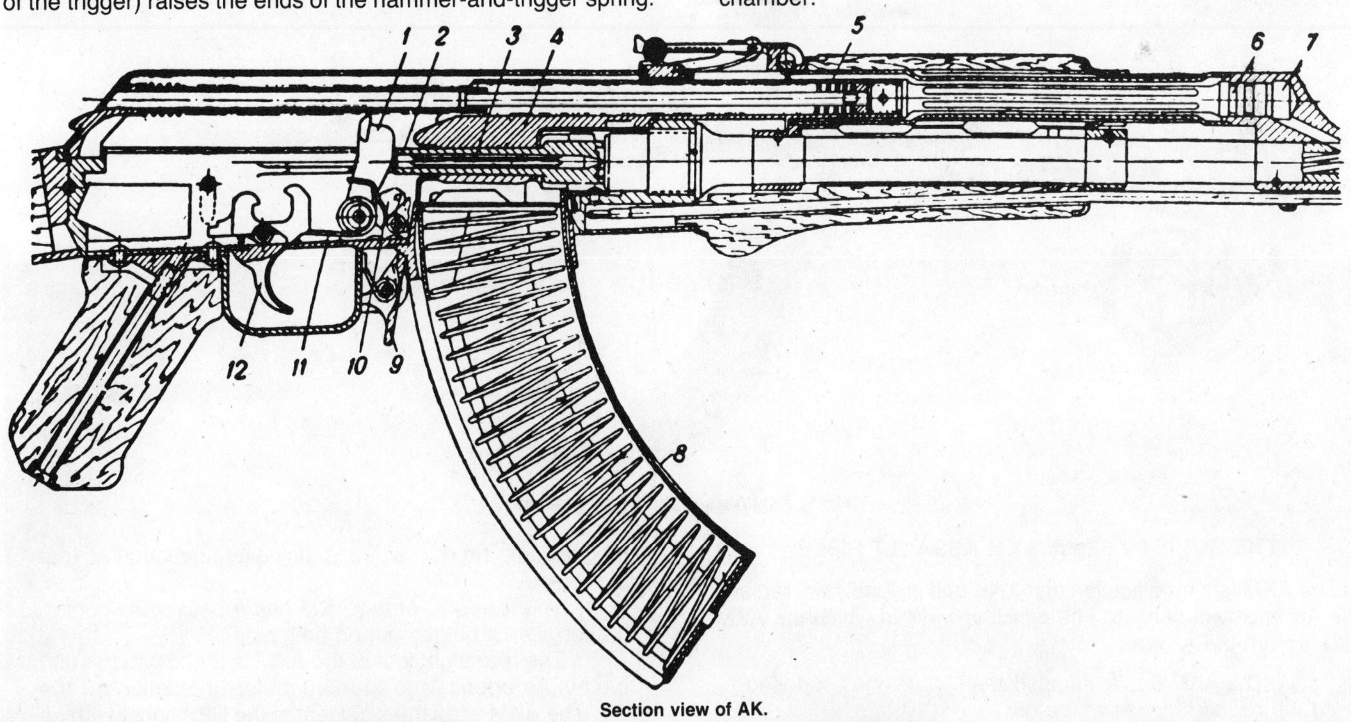

Section view of AK.

1. Hammer
2. Bolt
3. Firing pin
4. Operatin rod
5. Recoil spring
6. Gas piston
7. Gas cylinder
8. Magazine
9. Magazine catch
10. Full automatic sear
11. Trigger and hammer spring
12. Trigger

As the bolt approaches the barrel face, the first stage in the rotation of the bolt to the right takes place. At the same time, the extractor engages the extractor groove of the cartridge case. As the bolt carrier moves to the extreme forward position, it produces the final rotation of the bolt to the right, locking the bolt.

After the bolt is locked but while the bolt carrier is still a short distance from the extreme forward position, the full automatic disconnector (which is integral with the bolt carrier) strikes the upper end of the full automatic sear, and rotates the sear forward. This action moves the full automatic sear away from the hammer, so that the hammer will not be prevented from rotating.

The next round is fired by releasing the trigger, and then pressing it again. When the trigger is released, the hammer-and-trigger spring rotates the disconnector and semiautomatic sear to the rear, disengaging the disconnector from the disconnector notch in the hammer. The trigger-and-hammer spring rotates the hammer until the hammer cocknotch engages the semiautomatic sear. This is accompanied by an audible click.

When the trigger is again pressed, the semiautomatic sear releases the hammer cock notch. The hammer once again strikes the firing pin, and the entire operating cycle of the automatic mechanism is repeated.

Automatic Fire To fire the rifle automatically, set the selector at full automatic fire by rotating the indicator until it is opposite the Cyrillic letters AB on the receiver; press the trigger.

When the trigger is pressed, it rotates on the trigger pin. The disconnector, because it is prevented from rotating by the selector lever, does not engage the disconnector notch in the hammer.

As the trigger rotates, the semiautomatic sear releases the hammer cock notch. The hammer-and-trigger spring rotates the hammer, which strikes the firing pin forcibly. The round is fired. The powder gases act on the gas piston, thrusting the operating rod to the rear, opening the bolt, extracting and ejecting the cartridge case and cocking the hammer.

The full automatic sear engages the full automatic sear notch in the hammer, holding the hammer at full cock.

The top round in the magazine is raised by the follower.

As the bolt carrier and bolt are moved forward by the recoil spring, the round is fed into the chamber, and the bolt is locked.

When the bolt carrier is a short distance from the extreme forward position, the full automatic disconnector strikes the upper end of the full automatic sear and rotates the sear, releasing the hammer. The hammer strikes the firing pin, firing the next round. The entire operating cycle of the automatic mechanism is repeated.

The rifle will continue firing until the last round in the magazine is expended or until the firer releases pressure on the trigger. In the first case, the bolt carrier and bolt will remain in the forward position and the hammer will not be cocked (the AK has no bolt-holding-open device to hold the bolt to the rear after the last round is fired). In the second case, the rifle will be loaded and ready to fire again if the rifleman ceases fire before expending all the rounds in the magazine.

7.62mm AKM Assault Rifle, field stripped.

THE SOVIET 7.62mm AKM ASSAULT RIFLE

The AKM is a modification of the AK and probably will replace the AK in Soviet service. The principal ways in which the AKM differs from the AK are:

(1) The AKM has a stamped steel receiver as opposed to the milled receiver of the AK.

(2) The gas relief holes in the AKM gas cylinder tube are semicircular cutouts at the forward end of the tube, which match similar cutouts in the gas cylinder block. The gas relief holes on the AK are cut into the body of the gas cylinder tube—four on each side.

(3) The AKM has a rate reducer attached to the trigger mechanism; the rate of fire is, however, the same as that of the AK.

(4) The fore-end of the AKM has a beaver-tail configuration, i.e., it bulges out on both sides.

(5) The rear sight leaf of the AKM is graduated to 1000 meters as opposed to the 800-meter graduation on the AK. The AKM uses the sight leaf of the RPK light machine gun.

(6) The AKM stock and fore-end is made of laminated wood; those of the AK are usually made of ordinary beech or birch.

(7) The bolt and bolt carrier of the AKM are parkerized; those of the AK are bright steel.

The AKMS. This folding stock model is designed for the firing ports of the BMP armored personnel carrier.

The 7.62 × 39mm AKMS with stock folded.

The 5.45 × 39mm AK74 which is replacing the 7.62 × 39mm Kalashnikovs.

The AKMS is the folding metal stock version of the AKM. The steel struts of the stock are of stamped steel rather than machined steel as used on the folding stock AK. A short compensator is now frequently seen on the muzzle of the AKM and AKMS. The purpose of the compensator is to hold the barrel down in automatic fire. There is now a light alloy magazine weighing .4 pound available for the AK, AKM series weapons.

THE 7.62mm RPK LIGHT MACHINE GUN

The RPK is basically the same as the AKM assault rifle, with longer barrel and a bipod. It uses a 75-round drum magazine, a 40-round box magazine or the 30-round magazine of the AK and AKM. This weapon is replacing the RPD as the squad automatic weapon (base of fire) of the Soviet Army.

Adoption of the RPK by the Soviets eases their logistical and training problems since the RPK uses for the most part the same parts as the AKM and is operated in the same manner.

RPK does not have a quick change barrel, and as a squad automatic weapon it is not designed for long periods of sustained fire.

THE 7.62mm RPKS LIGHT MACHINE GUN

The RPKS is a folding stock version of the RPK light machine gun. The stock folds to the left side of the receiver, and although the weapon can be fired with box magazines with the stock folded it is doubtful that the 75-round drum magazine can be used in this model. The RPKS is 40.9 inches long with the stock fixed and approximately 32 inches long with stock folded.

The 7.62 × 39mm RPK with 40-shot magazine.

The 5.45 × 39mm RPKS74 with 40-shot magazine and side-folding stock.

The 7.62 × 39mm RPK field stripped.

The 5.45 × 39mm RPKS74 field stripped.

RPKS74 stock partially folded to illustrate the locking mechanism.

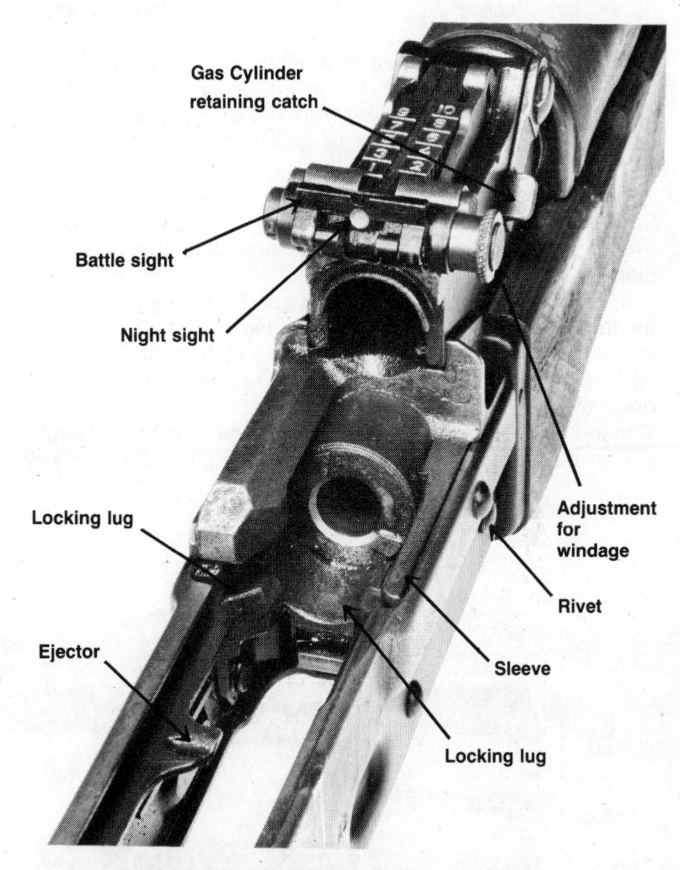

A close-up view of the lower receiver of the RPK.

THE SOVIET 7.62mm SVD SNIPER RIFLE

The SVD or Dragunov (SVD means Self loading Rifle, Dragunov) rifle is the replacement for the M1891/30 sniper rifle. It is chambered for the 7.62mm rimmed cartridge (7.62 × 53 R). The rifle is fitted with a four-power scope Model PSO-1 and has a somewhat unusual stock in that a large section has been cut out of it immediately to the rear of the pistol grip. This lightens the weight of the rifle considerably. The Dragunov uses an action which closely resembles that of the AK. It has a prong type flash suppressor similar to those used on current US small arms.

How to Load and Fire the SVD

The box magazine is loaded in the normal manner; it is removed from the rifle by pushing the magazine catch—located behind the magazine port—forward and pulling down on the magazine. Insert loaded magazine in the magazine port pushing it upward till it securely locks in place. Pull bolt-operating handle which protrudes from the right side of the receiver to the rear and release, thus chambering a cartridge. The weapon is now loaded and will fire one round for each pull of the trigger until the magazine is empty. The safety is similar to that of the AK/AKM assault rifle and is mounted on the right side of the receiver. To put the rifle on safe, push the lever upwards.

How the SVD Works

The bolt operation of the SVD is essentially the same as that of the AK/AKM in semiautomatic fire. The principal difference is that the SVD has a spring-loaded piston rod, which is a separate assembly; it is not attached to the bolt carrier as is that of the AK/AKM. The trigger mechanism is relatively simple consisting of 12 parts including the fabricated trigger housing/trigger guard. It varies from the AK/AKM trigger mechanism in that, among other things, it does not have a full automatic sear or full automatic disconnector, and it has a separate trigger spring in addition to the hammer spring as opposed to the one spring that performs both functions in the AK/AKM.

Disassembly and Assembly

Clear the SVD but do not set the safety to safe or replace the magazine. Release the telescope sight catch and pull the telescope off the rear. Place the telescope in its carrying case. Remove the cheek pad.

Press the receiver cover catch upward and pull the receiver cover and driving spring upward and off the receiver. Pull the operating handle to the rear; then lift the bolt carrier and bolt out of the rifle. Push the bolt into the bolt carrier until it can be rotated so that the operating lug on the bolt is free of its cam recess in the bolt carrier. Pull the bolt out of the carrier.

Rotate the safety until it is vertical; then pull it to the right out of the receiver. Pull the trigger group out of the receiver.

Press the handguard catch in until it is free of the handguard ferrule; then rotate the catch to the right. Push the ferrule forward; then pull the handguard down and out to remove them. It may be necessary to pry tight handguards off the barrel.

Pull the operating rod to the rear against pressure of its spring; then gently move the front of the rod to one side. Pull the piston off the gas block; then ease the operating rod forward and remove it and its spring.

Further disassembly is neither necessary nor desirable.

To reassemble the SVD, first insert the operating rod and its spring in its recess in the rear sight base. Press the rod to one side and slip the piston onto the gas block; then insert the operating rod into the piston.

Slip the rear end of the handguards into their seat, swing the front ends together and push the ferrule to hold the handguards in place. Rotate the ferrule until the handguard catch snaps into place.

Insert the trigger group into its recess in the receiver and when the trigger and receiver are lined up, insert the shaft of the safety into its hole in the receiver. Turn the safety so that its arm is vertical, fully seat the safety into the receiver and then turn it down to its position.

Insert the bolt spindle into the bolt carrier until the operating lug on the bolt can be turned into the cam groove in the bolt carrier. Pull the bolt fully forward in the carrier. Mate the lugs on the bolt carrier with the cutaways in the receiver and insert the bolt carrier and bolt into the receiver; then move them fully forward.

Insert the driving spring into the bolt carrier and while holding the receiver catch up, slide the receiver cover onto the receiver.

Replace the cheek pad and while holding the telescope sight catch, slide the sight onto its seat from the rear. Replace the magazine, press the trigger and move the safety upward.

The main accessory for the SVD is its PSO-1 telescope. This scope has an illuminated reticle powered by a small dry cell. The battery housing is located at the bottom rear of the telescopic sight mount. To change batteries, press in and rotate the battery housing counterclockwise. Remove the old battery and replace with the same type. The reticle lamp can be replaced by unscrewing its housing and removing the bulb. The reticle light is turned on or off by its switch. The lens cap should always be in place except when actually using the telescope for aiming.

Two covers are issued with each rifle; one is for the telescope sight alone; the other covers the sight and breech of the rifle. A belt pouch is provided for carrying the telescope when dismounted from the rifle, four magazines, a cleaning kit and an extra battery and lamp for the telescopic sight.

If the open sights are to be used, set the rear sight by pressing in the locks on the rear sight slide; then move the slide along the rear sight leaf. The front edge of the slide should be aligned with the numeral that corresponds to the range in hundreds of meters. Use the same sight picture as for firing a pistol.

If the PSO-1 telescopic sight is used, rotate the elevation knob until the figure that corresponds to the range in hundreds of meters is aligned with its index. The range can be fairly accurately determined by use of the range finder located in the lower left of the telescopic reticle. This range finder is graduated to the height of a man (5′7″). Look through the telescope and place the horizontal line at the bottom line of the target. Move the telescope until the upper (curved) line just touches the top of the target's head. The number indicates the range in hundreds of meters; if the target falls between numbers, the remaining

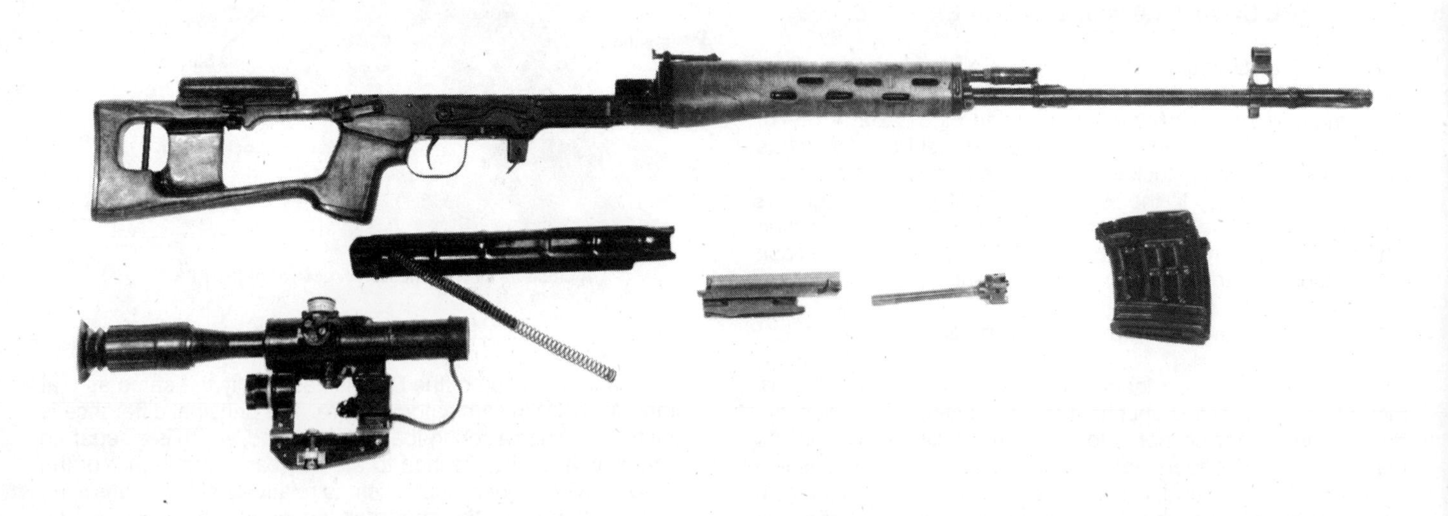

Disassembled view of Soviet Dragunov (SVD) Sniping Rifle.

distance then must be estimated. When the range is determined and set into the elevation knob, use the point of the top chevron on the reticle as an aiming point. The three lower chevrons are used for firing at 1100, 1200 and 1300 meters with the elevation knob set at 10. The horizontal scale extending out from the sides of the top chevron are used for hasty wind corrections; deliberate changes are made by rotating the windage knob.

For firing when the light is dim, illuminate the reticle by turning on the switch in the telescopic sight mount. If active infrared light sources are believed to be in use by the opponents, set the range drum at "4" and switch the infrared detector into place. Scan the area to the front; if any active infrared light sources are in use, they will appear as orange-red blobs in the telescope. Align the point of the reticle on the light and fire. Turn off the reticle light when not in use to conserve the battery and swing the infrared detector out of the way so that it will be activated by light during the day.

A commercial sporting version of the Dragunov—the "Medved" (bear)—is also available.

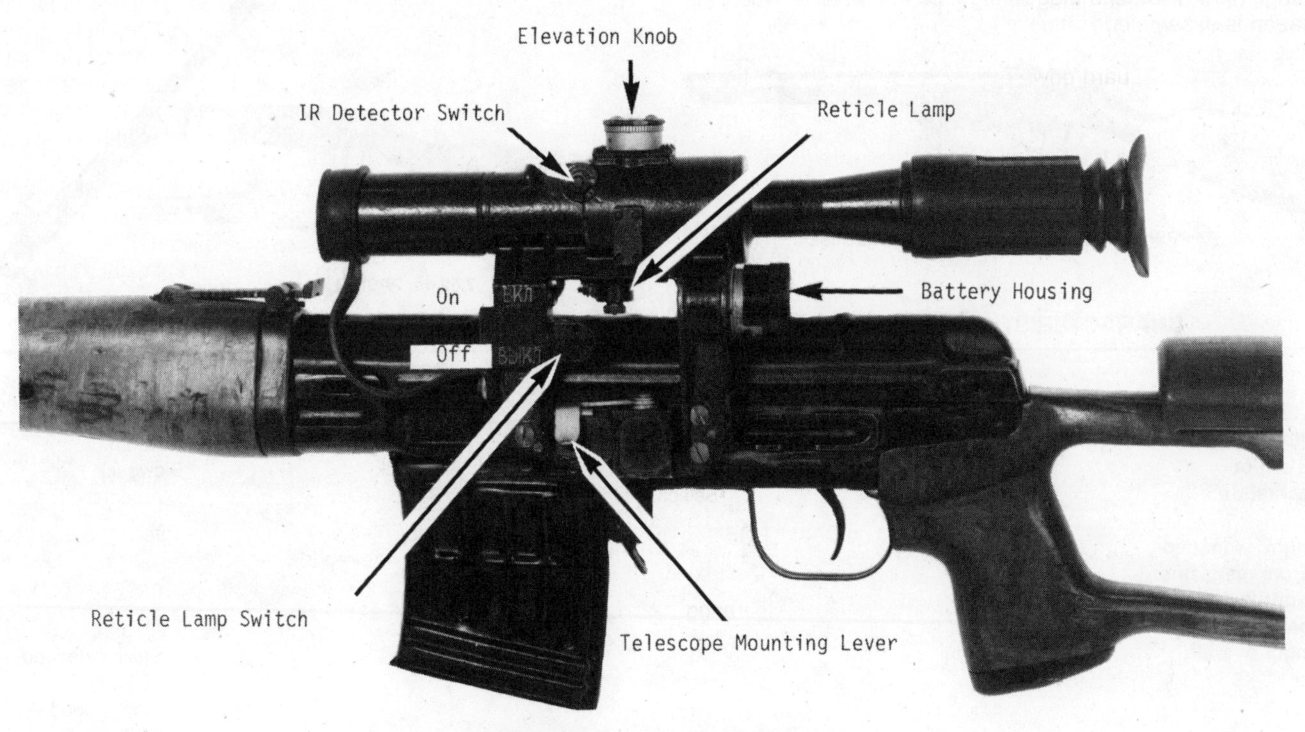

Left side view of PSO-1 telescopic sight.

CHARACTERISTICS OF POST-WORLD WAR II SOVIET RIFLES AND CARBINES

	SKS Carbine	AK Assault Rifle	AKM Assault Rifle	SVD
Caliber:	7.62mm M43 (7.62 × 39mm)	7.62mm M43 (7.62 × 39mm)	7.62mm M43 (7.62 × 39mm)	7.62mm rimmed.
System of operation:	Gas, semiautomatic.	Gas, selective fire.	Gas, selective fire.	Gas, semiautomatic.
Length overall:	40.16 in.	34.25 in.	34.25 in.	48.2 in.
Barrel length:	20.47 in.	16.34 in.	16.34 in.	24 in.
Feed device:	10-round, staggered row non-detachable box magazine.	30-round, staggered row detachable box magazine.	30-round, staggered row detachable box magazine.	10-round, staggered row detachable box magazine.
Sights: Front:	Hooded post.	Post w/protecting ears.	Post w/protecting ears.	Hooded post.
Rear:	Tangent, graduated to 1000 meters.	Tangent, graduated to 800 meters.	Tangent, graduated to 1000 meters.	Tangent w/notch.
Weight:	8.8 lb.	10.58 lb.	8.87 lb.	9.5 lb.
Muzzle velocity:	2410 f.p.s.	2330 f.p.s.	2330 f.p.s.	2720 f.p.s.
Cyclic rate:	—	600 r.p.m.	600 r.p.m.	—

SOVIET SUBMACHINE GUNS

All pre-World War II and World War II Soviet submachine guns were chambered for the Soviet 7.62mm pistol cartridge Type P. Although the two latest types, the PBSh M1941, and the PPS M1943, are still in use in some Warsaw Pact countries, all of these guns are obsolete in the USSR.

THE 7.62mm PPSh SUBMACHINE GUN M1941

Note that the bolt is a very simple machined piece. The recoil spring guide is resting in a hole in the rear of the bolt with the spring around it, and a plastic buffer at its rear end. This spring arrangement also serves to spring lock the receiver when the weapon is assembled.

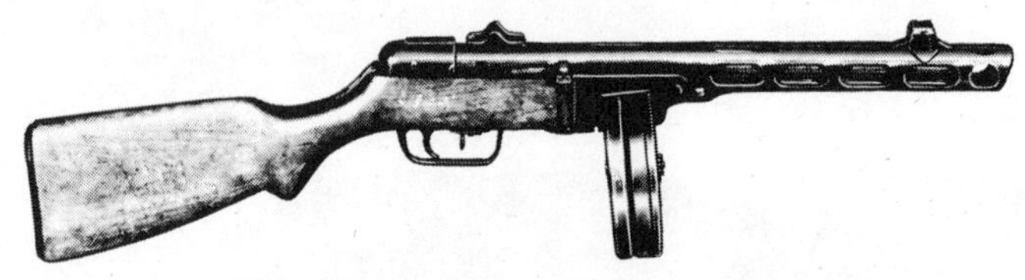

PPSh M1941 with 71-round drum magazine.

Field Stripping The PPSh M1941

Push forward on receiver catch and hinge barrel and barrel casing down.

Draw straight back on bolt handle a short distance, meanwhile exerting an upward pull. The bolt may be lifted out.

Remove recoil spring and buffer.

Barrel is removable from casing. Further dismounting is seldom necessary.

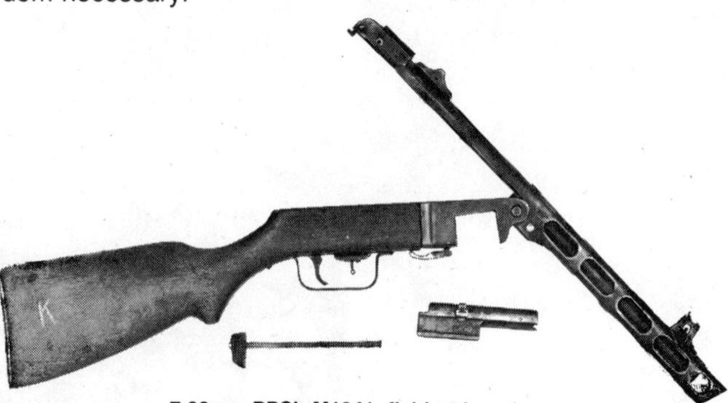

7.62mm PPSh M1941, field stripped.

CHARACTERISTICS OF PRE-WORLD WAR II AND WORLD WAR II SOVIET SUBMACHINE GUNS

	PPD Model 1934/38	PPD Model 1940	PPSh Model 1941	PPS Model 1943
Caliber:	7.62mm.	7.62mm.	7.62mm.	7.62mm.
System of operation:	Blowback. Selective fire.	Blowback. Selective fire.	Blowback. Selective fire.	Blowback. Automatic fire only.
Weight: w/loaded drum magazine:	11.5 lb.	11.90 lb.	11.99 lb.	
Weight: w/loaded box magazine:	8.25 lb.		9.26 lb.	7.98 lb.
Length overall:	30.63 in.	30.63 in.	33.15 in.	Stock extended—32.72 in. Stock folded—24.25 in.
Barrel length:	10.63 in.	10.63 in.	10.63 in.	9.45 in.
Feed device:	71-rd. drum or 25-rd. box.	71-rd drum.	71-rd. drum. or 35-rd. box.	35-rd. box.
Sights: Front:	Blade.	Hooded blade.	Hooded post.	Post w/ears.
Rear:	Tangent leaf.	Tangent leaf.	Tangent leaf or L-type.	L-type.
Cyclic rate:	900 r.p.m.	900-1100 r.p.m.	700-900 r.p.m.	650 r.p.m.
Muzzle velocity:	1640 f.p.s.	1640 f.p.s.	1640 f.p.s.	1608 f.p.s.

NOTE: A 1942 version of the PPS M1943 also exists, but differs only in minor details from the Model 1943.

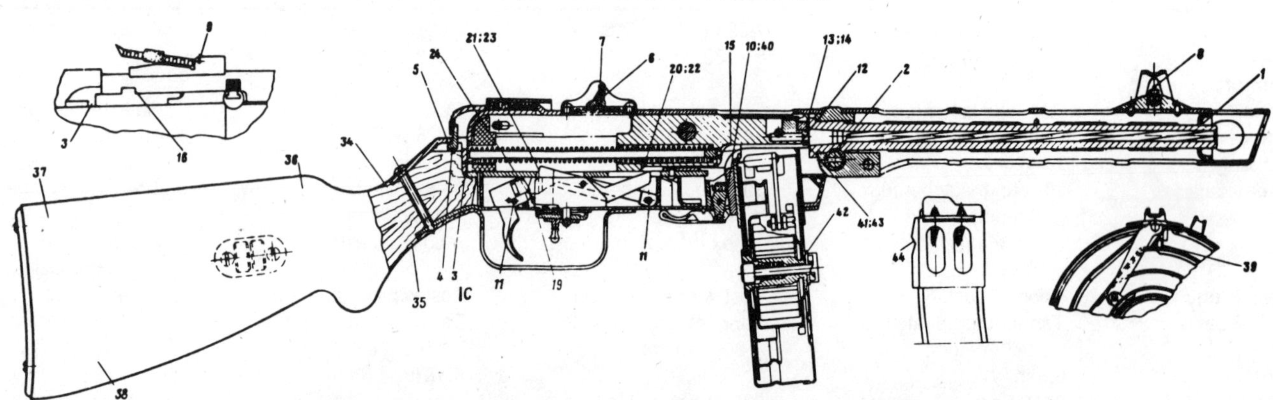

Section view of Soviet PPSh41 with 71-round magazine in place. Upper left drawing illustrates the early type rear sight. Lower right views illustrate differences between the 25-round box magazine and the 71-round drum.

From top down: M34/38, PPD M1940, PPSh M1941 and PPS M1943.

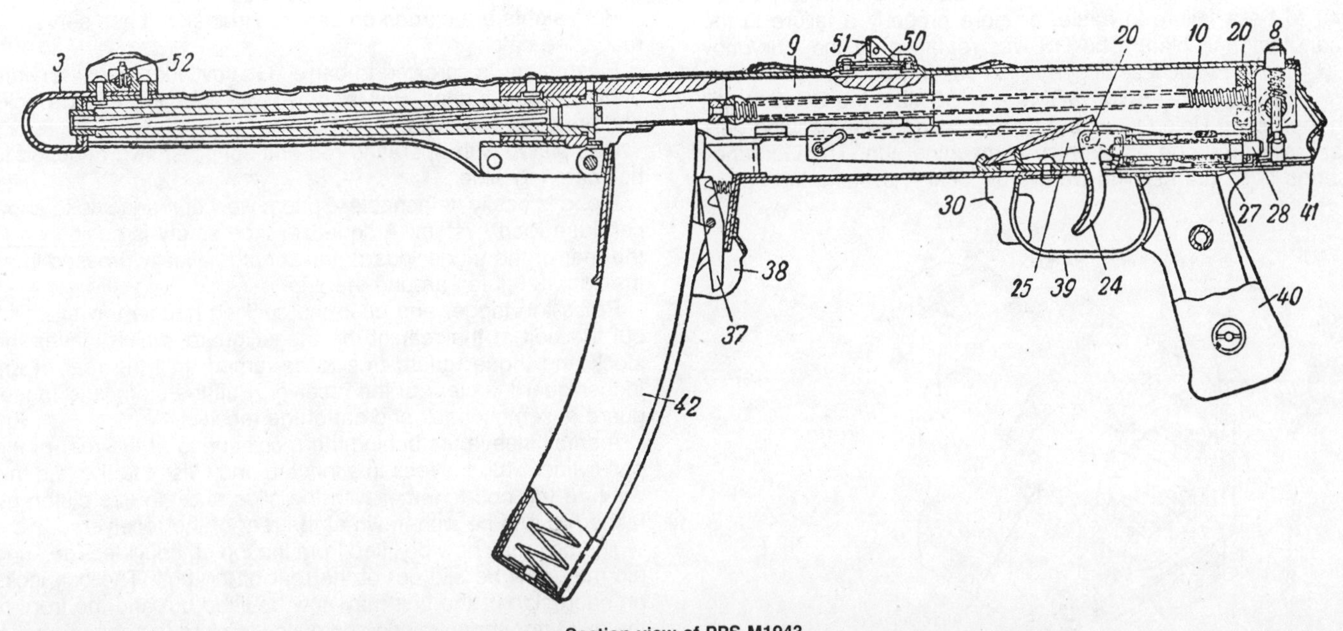

Section view of PPS M1943.
3—Muzzle brake (compensator) 8—Stock release catch 9—Bolt 10—Recoil spring 20—Buffer 24—Trigger 25—Sear 26—Sear pin 27—Trigger spring 28—Trigger spring guide 30—Safety 37—Magazine latch 38—Magazine guide 39—Trigger guard 40—Plastic grip 41—Takedown latch 42—Magazine 50—Rear sight guard 51—"L" type flip sight 52—Front sight post.

SOVIET MACHINE GUNS

Prior to 1900, Imperial Russia adopted the Maxim gun, and Russian troops in the Russo-Japanese War (1904-1905) used Maxims against Japanese Hotchkiss guns. The 7.62mm Model 1905 Maxim was the first machine gun manufactured in Russia. As all the early Maxims, this heavy weapon had numerous brass fittings and a bronze water jacket.

The Imperial government adopted the 7.62mm Model 1902 Madsen as the standard light machine gun. During World War I, the Russians purchased 7.62mm Maxim and Colt Model 1914 (modified Colt Model 1895 "potato digger") machine guns from Colt in the US. They also acquired a considerable number of Lewis guns.

In 1910, the Russians modified the Maxim; the 7.62mm Model 1910 Maxim (SPM) had a ribber water jacket similar to that of the British Vickers. This weapon, used and manufactured through World War II, was modified by the addition of a tractor type water entry port.

After World War I, the new Soviet Army started developing its own weapons. The 7.62mm Maxim Tokarev and Maxim Koleshnikov were among the first efforts. These air-cooled Maxims fitted with bipods were used in large quantities during the Spanish civil war. The first originally developed Soviet machine gun was the 7.62mm DP—Degtyarev Infantry—which appeared in 1926. The DP introduced the modified Kjellman Frijberg locking system to the Soviets, which is still in service in several machine guns. The DP was the first of a series of Degtyarev machine guns adopted by the Soviet Union. A tank version called the DT may still be found on older Soviet armored vehicles in use among Soviet allies. An aircraft version, the 7.62mm DA, was also produced.

The DP, DT and DA have their operating springs coiled on the piston rod, which is seated under the barrel. The heating of the barrel caused distortion of the spring with resultant malfunctions. During World War II, a modification of the DP—the DPM—was put into service. It is basically the same as the DP, with the recoil spring mounted in a tube that projects to the rear of the receiver. A tank version of this weapon, the DTM, is still in service on pre-1949 Soviet armored vehicles.

A heavy machine gun version of the DP, the 1939 DS, turned out to be a failure in battle, or more properly a failure in the manufacturing plant. The DS was replaced by the Goryunov SG43, a very successful gun still in use in Soviet allied armies. In modified form—SGM, SGMT and SGMB—the SG43 is still in the Soviet Union inventory as a battalion level machine gun, tank machine gun and vehicular machine gun. The Goryunov series does not use the Degtyarev locking system, the bolt of the Goryunov being cammed to the side to lock in a side wall of the receiver.

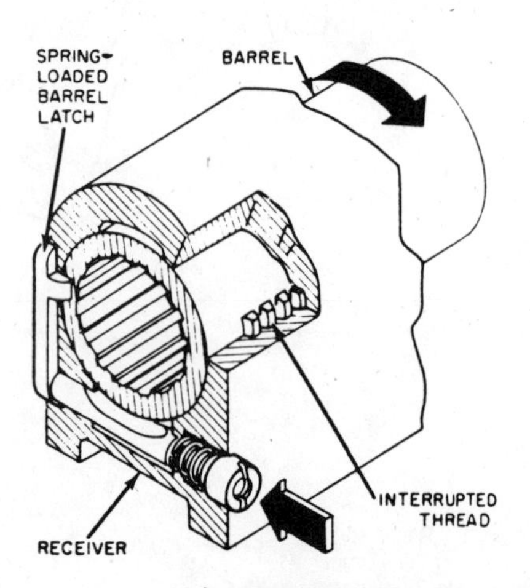

Degtyarev barrel removal.

The 12.7mm DShK Model 1938 was the first Soviet heavy caliber machine gun to be produced in quantity. It was preceded by the 12.7mm DK, which appeared around 1934, apparently not a very successful gun. The DShK uses the Degtyarev locking system.

The 7.62mm RPD machine gun is also basically a Degtyarev weapon, having a bolt system generally similar to the earlier Degtyarevs. It has been replaced in Soviet service by the 7.62mm RPK machine gun, similar to the AK assault rifle in most respects.

For further design details, see Chapter 2.

7.62mm DP AND DPM LIGHT MACHINE GUNS

Degtyarev machine gun variations:

- DP— infantry light machine gun; 49-shot drum.
- DA— aircraft mounted machine gun; DA-2 twin gun mount.
- DT— tank mounted; 60-shot drum.
- DPM— modernized infantry light machine gun; also PRC Type 53.
- DTM— modernized tank gun.
- RP46— belt-fed company machine gun; PRC Type 58; North Korean Type 64.

DP, DPM and DTM guns have sliding dust covers over the feed openings at the top of the receivers. DP, DPM and RP46 all have quick change barrels. The tank models do not. All Soviet Degtyarev guns have adjustable gas systems.

Field Stripping the DP

To remove the barrel, pull the cocking handle to the rear to cock the weapon. The barrel locking stud is on the left side near the front of the receiver. Press this stud in, which will release the barrel, then twist the barrel up one quarter turn to the right. Now slide the barrel straight forward out of the receiver.

Both sights are carried on casing. Rear sight base serves as magazine catch.

Flash hider is screwed to barrel. Gas cylinder is just forward of barrel cooling rings. Bolt is shown with right and left side locks and firing pin removed.

Gas piston with operating rod and spring shown attached to bolt carrying slide.

Bipod is easily detachable. Bottom view of magazine to show cartridge feed system. A squeezer-type safety is positioned in the rear of the trigger guard and is automatically pressed in as the fingers tighten around the grip.

Press the trigger and ease the cocking handle forward. Pull out the bolt at the rear of the trigger guard, which leaves the stock and trigger guard free to be turned until the rear of the trigger guard is clear of the receiver. Pull the stock and trigger guard assembly back and out of the receiver.

A small sleeve fits behind the recoil spring at the rear of the gas cylinder tube. Press this forward and twist it to the left; this will free the bolt together with the slide and the gas piston attached to it to be withdrawn at the rear of the receiver.

The bolt may now be lifted from the top of the slide. The firing pin may now be slid out of the rear of the bolt. The bolt locks on each side of the bolt may now be lifted out and the front of the extractor spring raised and pulled forward to permit removal of it and the extractor.

This completes field stripping. Assembling the gun is equally simple and merely calls for reversing the stripping procedure.

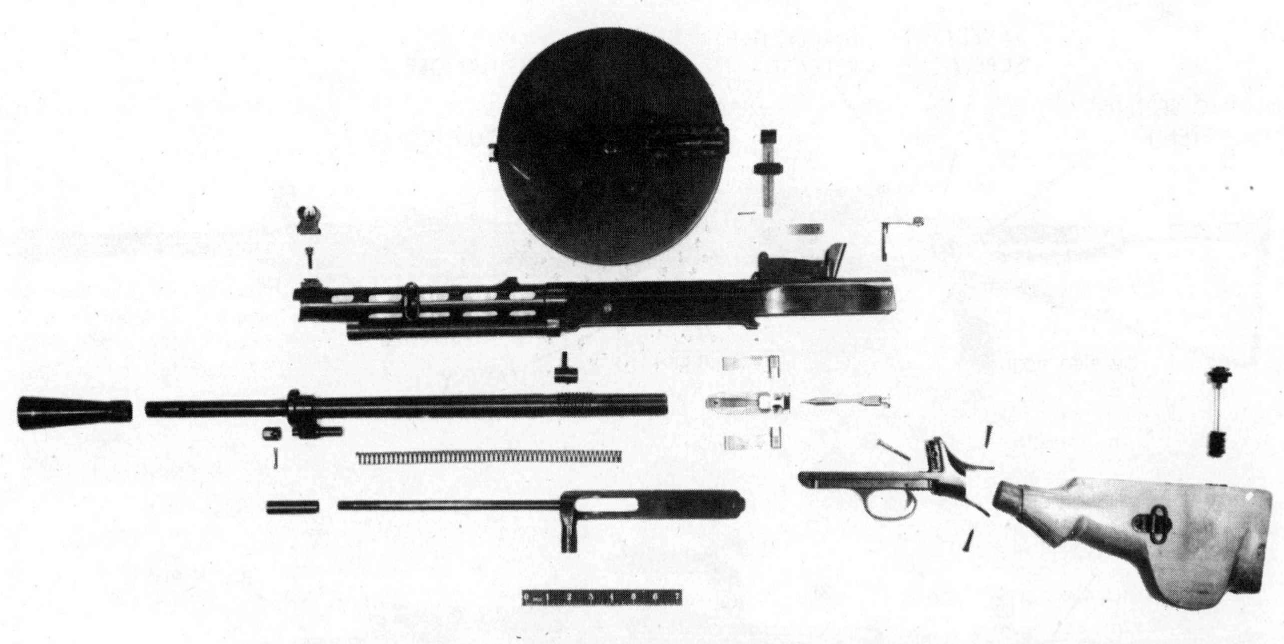

DP Machine Gun field stripped.

Loading and Firing the DP and DPM

The Magazine. This type drum differs radically from the Lewis type. The inner center rotates, while the outer rim is fastened securely to the gun. The cartridges lie in single line around the inside of the pan.

To Prepare for Firing. Mount a loaded magazine on top of the receiver and press firmly down until it is caught by the magazine catch. (The magazine catch is mounted in the front end of the rear sight base.)

Now pull back the cocking handle as far as it will go. It will stay open. Pressing the trigger will now fire the gun. Full automatic fire will ensue as long as the trigger is held down and there are cartridges in the magazine.

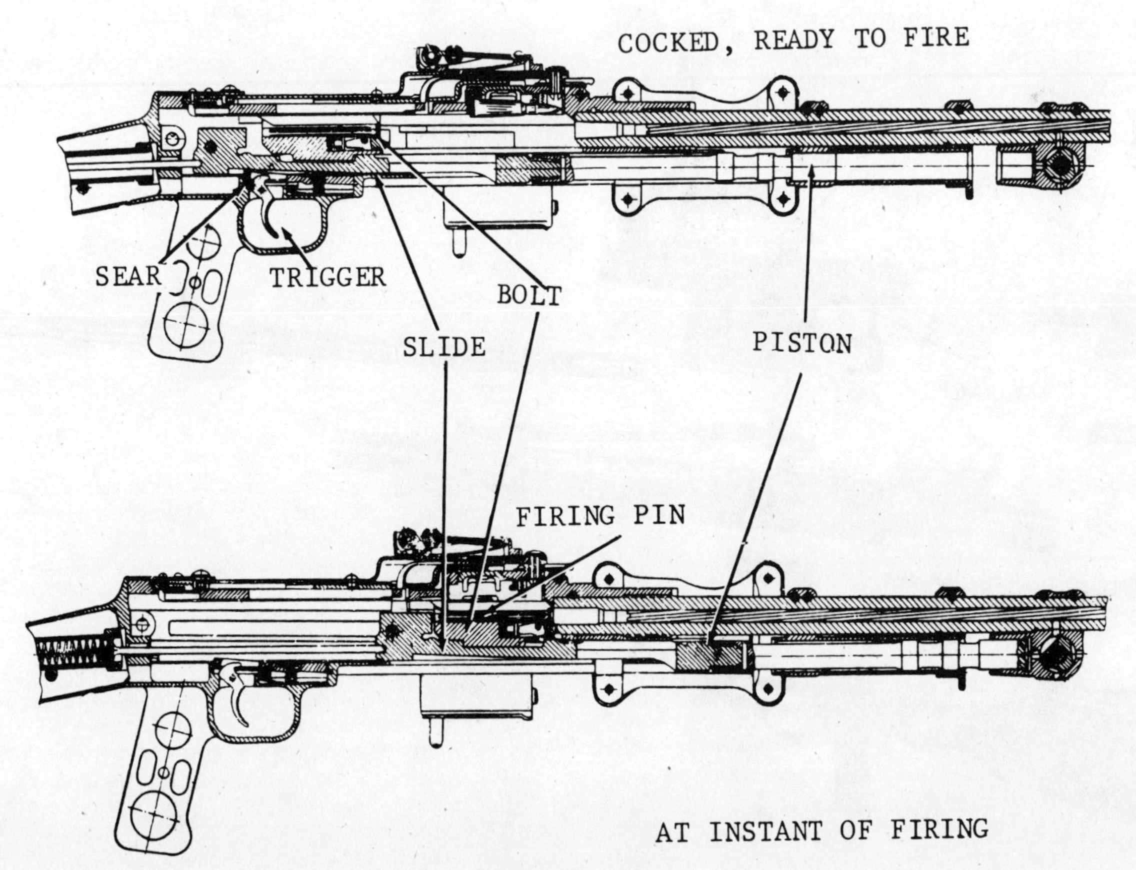

RPD section view.

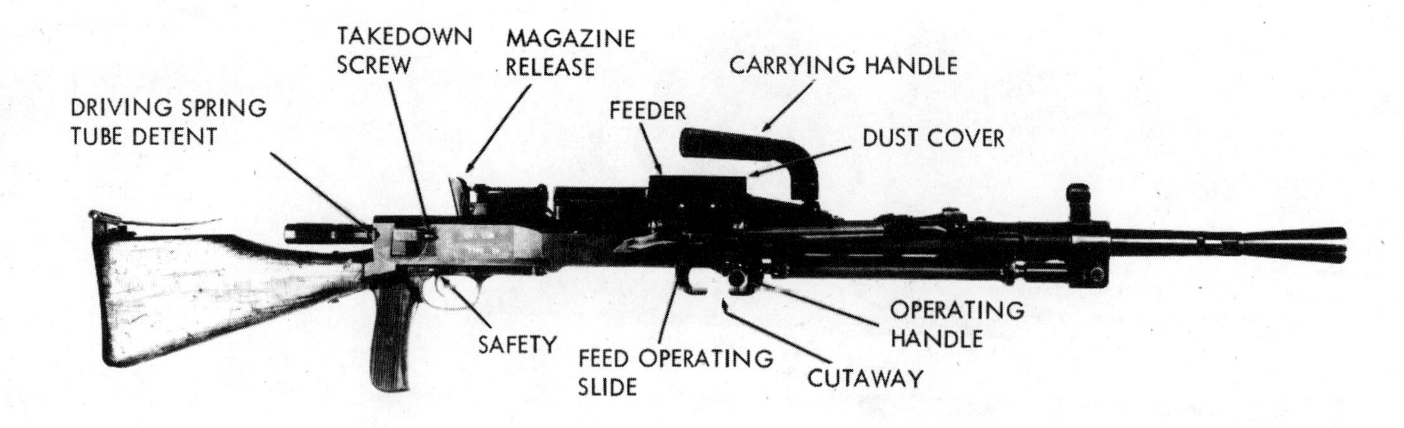

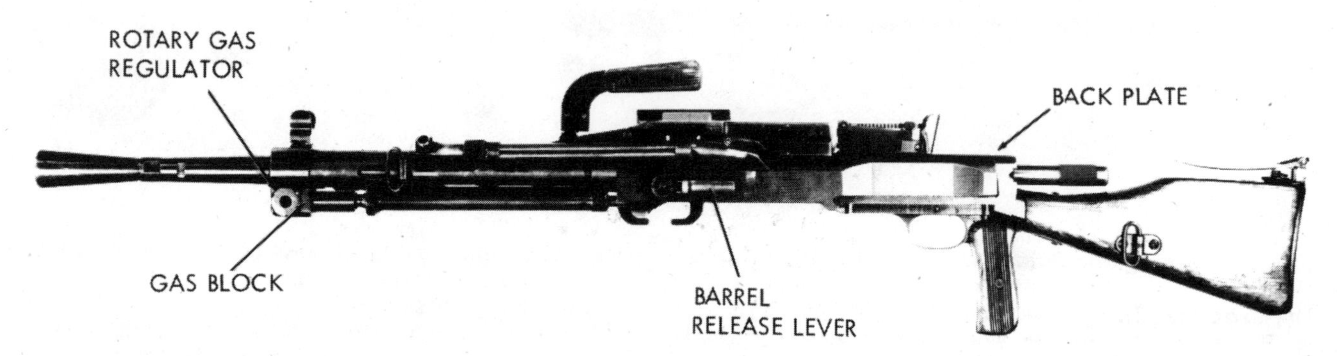

RP46 showing key features.

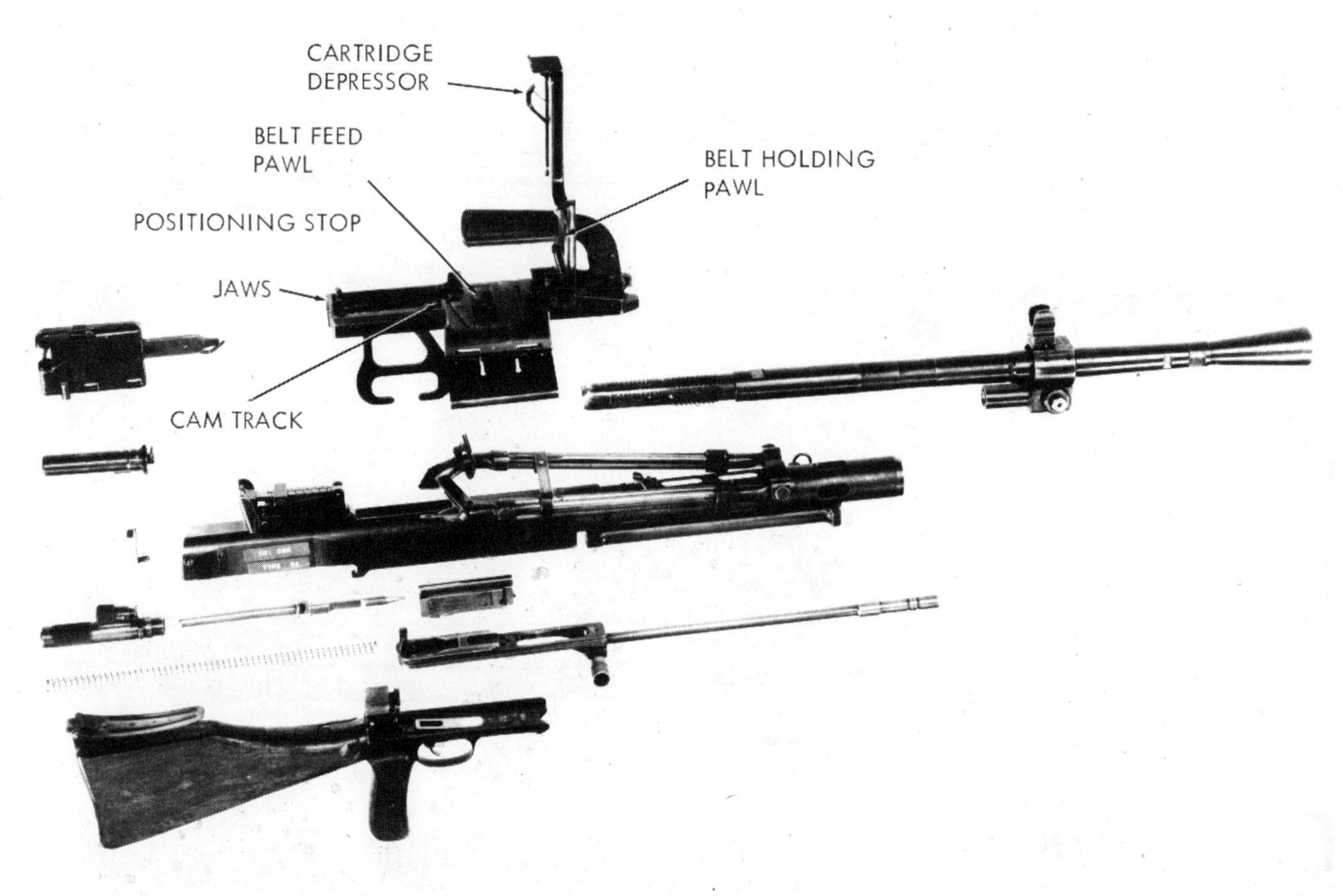

RP46 field stripped.

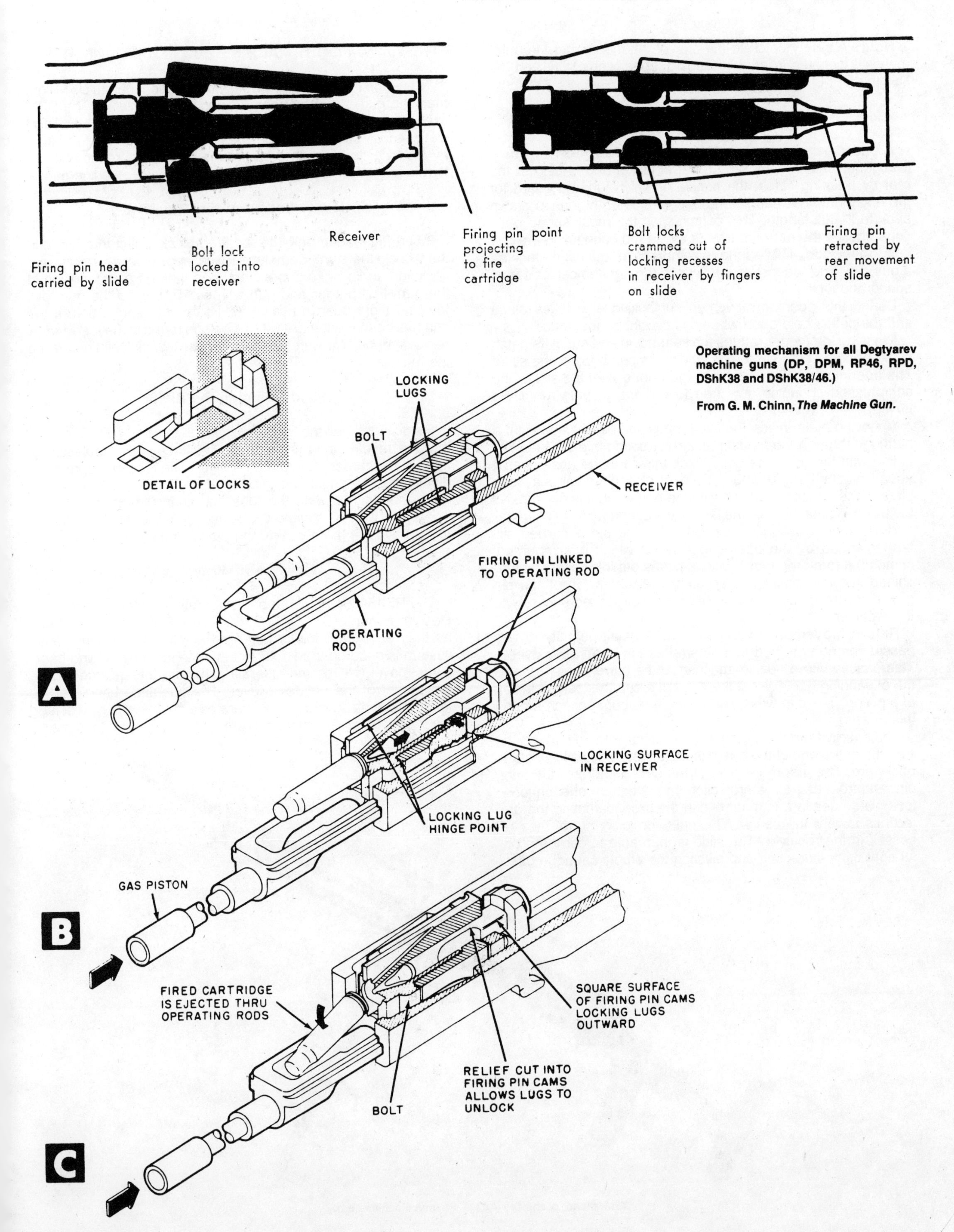

Operating mechanism for all Degtyarev machine guns (DP, DPM, RP46, RPD, DShK38 and DShK38/46.)

From G. M. Chinn, *The Machine Gun*.

Note: A safety catch to the rear of the triggerguard is automatically pressed in as the rifle is gripped ready to fire on the DP; the DPM has a manual safety.

How the DP and DPM Work

A loaded pan magazine is placed on top of the receiver where it engages with a hook on the barrel jacket and is held at the rear by a spring catch, the handle of which forms a guard for the rear sight. The handle on the right side of the gun is drawn back to its full length. This compresses the recoil spring positioned below the barrel; it travels with a rod connecting the gas cup to the slide. This spring is compressed against its lock by a gas cup. The rod moves backwards through the center of this spring and lock.

During this opening movement, unlocking also takes place and the bolt is held open when it is caught by the sear.

Pressing the trigger rotates the sear down and out of its notch in the bottom of the slide. The bolt is carried on top of the slide. The operating rod is attached to the front end of the slide. The compressed spring is now free to pull the moving members forward.

A moving plate inside the fixed magazine pan has brought a cartridge through the medium of a rotor spring into the feed lips in line with the bolt. The bolt strikes this cartridge and drives it ahead into the firing chamber. On each side of the bolt is a loose plate. These plates are flush with the bolt when it is cocked. As the bolt brings up against the face of the cartridge in the chamber and its forward action is stopped, the firing pin is carried still farther ahead by the operating slide on which it is mounted; cams on it force the loose locking plates out into recesses machined into the receiver to accommodate them. The firing pin now discharges the cartridge, with the bolt locked securely to the receiver.

Return Movement of the Action. A small quantity of gas passes through the port in the barrel as the bullet goes over it. This passes into a nozzle attached to the barrel, which directs the expanding gas against the head of a cup that serves in lieu of a piston. The cup walls extend forward about an inch around this nozzle.

The energy from the expanding gas is transmitted through the gas cup at the end of the rod to the attached slide which it starts backward. The first movement of the slide withdraws the firing pin, and from thereon a projection on the bottom of each loose lock plate rides in a cam groove in the slide, camming the projections toward the center. This pulls the locks out of their recesses in the receiver. The slide is then able to carry the bolt straight back extracting and ejecting the empty cartridge case.

THE 7.62mm LIGHT MACHINE GUN MODEL 1946 (RP46)

The RP46 is basically a modification of the 7.62mm DPM light machine gun, which appeared during World War II. The DPM itself was a modification of the DP. While both the DP and DPM were fed by pan type magazines, the RP46 can be fed either by a pan or by the same 250-round nondisintegrating link belt that is used with the 7.62mm Goryunov heavy machine gun.

How to Load and Fire the RP46

Insert the loaded belt into the feed-block, pull it through, and then cock the gun. Sometimes it is necessary to retract the cocking handle twice depending upon the age and condition of the gun. Pull trigger and gun will fire. To unload the gun, pull the rear sight guard to the rear, lift cover, remove belt. If the gun has been fired, there will still be one round in the feed way—remove round. The safety is on when the lever is pointing toward the muzzle.

Field Stripping the RP46

Field stripping of the RP46 is similar to that of the RPD, except that the recoil spring is mounted in a tube which projects out from the right rear side of the receiver. The tube and spring are dismounted by pushing in on the tube lock and turning it to the right, then withdrawing the tube and spring from the receiver. The barrel can be removed by pushing in the latch on the left forward side of the receiver and pulling the barrel out.

How the RP46 Works

The RP46 has the same basic operating system as the other Degtyarev guns (DP, DPM and RPD) and operates the same as they do except for its belt feed mechanism. An ingenious system is used on the RP46 to translate the forward and backward movement of the operating handle into a side-to-side movement of the belt-feed lever. The operating handle on its forward and backward travel operates a double-arm type of feed lever, which transmits this movement to parts in the feed mech-

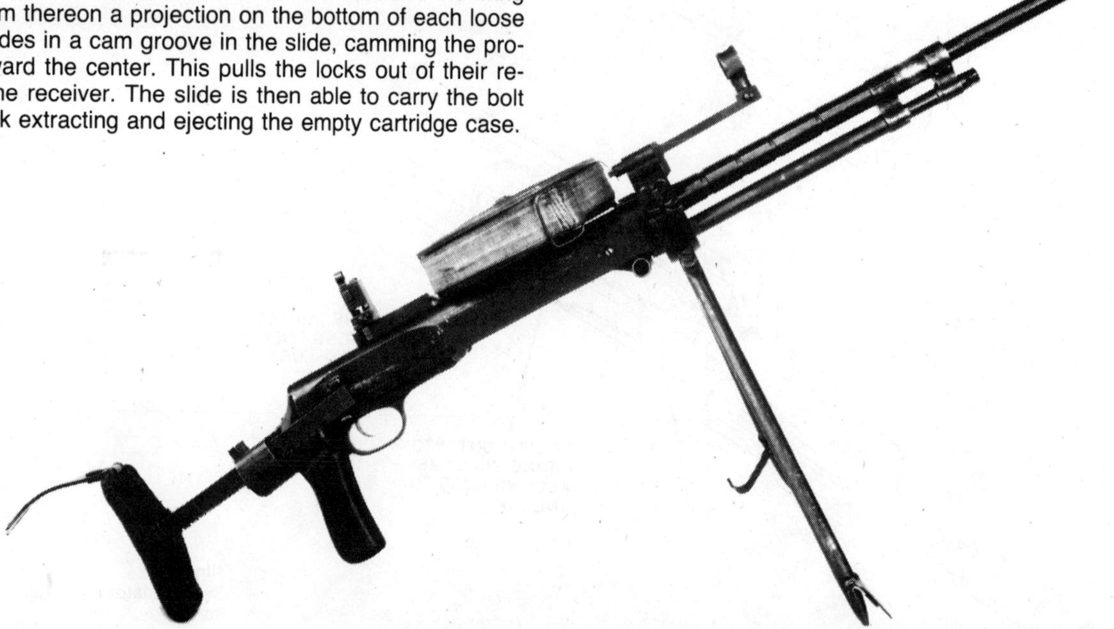

Tank model of the DP, 7.62 × 54Rmm Machine Gun.

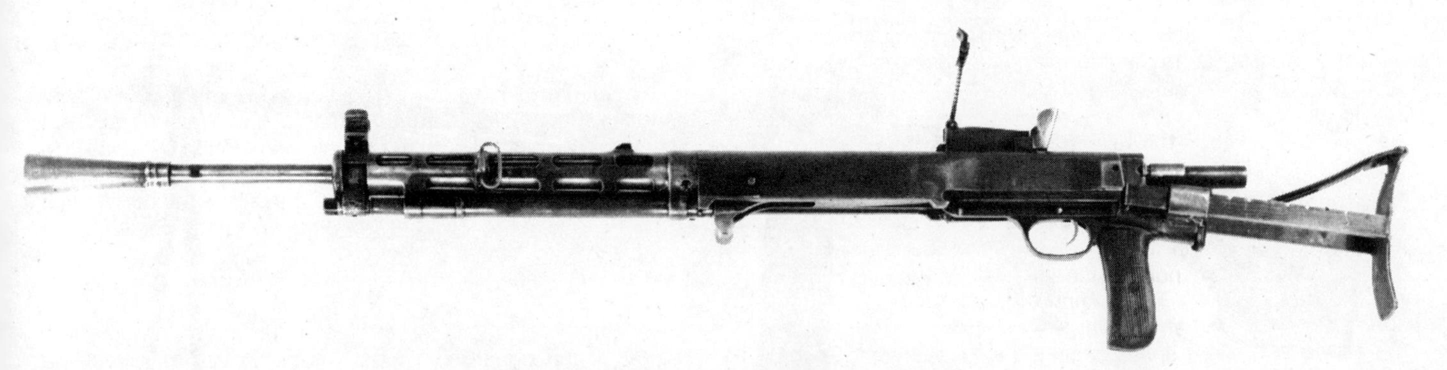

Tank model of the DPM, 7.62 × 54Rmm Machine Gun.

anism; these parts in turn transmit the movement to the belt-feed slide. This system is also used on the Soviet 12.7mm DShK M1938/46 machine gun. Insofar as the bolt parts of the Degtyarev weapons are concerned, there is only one major difference among the lot. In DP, DPM and RP46, the firing pin is mounted in the bolt, and the slide post itself serves as a hammer.

THE 7.62mm RPD LIGHT MACHINE GUN

Loading and Firing the RPD

The link belt of the RPD is loaded by pushing the individual rounds into the belt so that the bottom hook-like section of the link snaps into place in the cartridge extractor groove. The two fifty-round sections are joined together by slipping the tongue of the end link on one belt section through the slots of the starting link on the other belt section. A cartridge is then inserted, locking the two belt sections together.

The belt is then rolled into a tight circle and fitted into the stamped metal drum. The drum is fitted on the weapon by sliding its top dovetail on the mating surfaces that protrude under the forward part of the receiver. The belt loading tab should be protruding from the spring-loaded trap door of the drum, so that it can be inserted in the receiver. The drum is locked in place by pulling down the lock on the underside of the receiver; this lock keeps the drum from moving backwards off the gun during fire.

Cock the gun by pulling operating handle to the rear. On older guns, which have the non-folding operating handle, the handle will remain to the rear. On the newer type guns, which have the folding type operating handle, the handle should be pushed forward after pulling to the rear.

Open the cover by pushing forward on cover latch and lifting the cover. Lay belt on feedway so that the leading cartridge lies beside the cartridge stop. Close cover.

If the trigger is squeezed, the weapon will now fire. The safety catch is located on the right side of the pistol grip butt group, immediately above the trigger. When the catch is forward of the trigger the gun is on safe; to put the gun on fire, rotate the catch to the rear position.

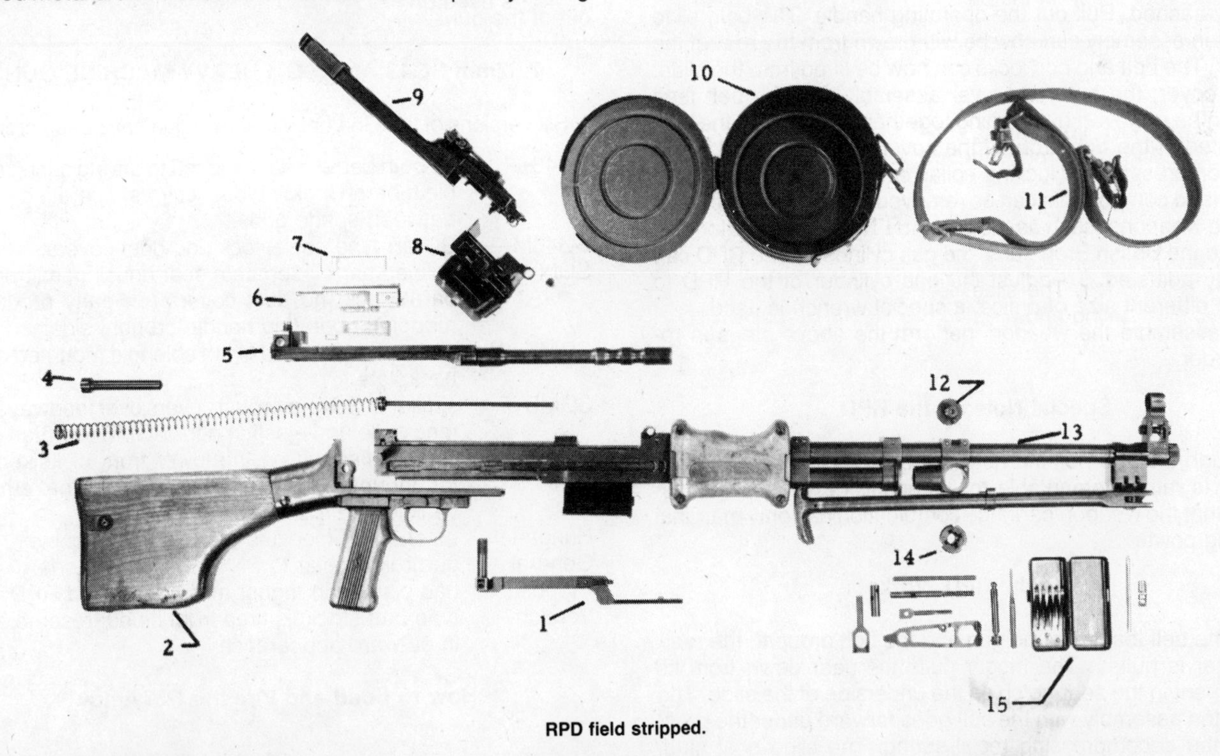

RPD field stripped.

1. **Operating handle**
2. **Butt stock**
3. **Driving spring and rod**
4. **Driving spring guide**
5. **Operating slide and piston**
6. **Bolt**
7. **Bolt locking flap**
8. **Feed tray**
9. **Feed cover**
10. **Drum (open)**
11. **Sling**
12. **Gas regulator screw**
13. **Barrel**
14. **Gas regulator**
15. **Combination tool kit**

RPD Light Machine Gun, right side.

To unload the gun, lift the cover by pushing forward on the cover catch located at the rear of the cover. Lift the link belt out and snap the cover shut.

Field Stripping the RPD

Turn the butt trap cover so that it is at right angles to the butt-stock. Place a screwdriver in the top hole in the stock, and turn the recoil spring plug one quarter turn; this will release the plug and the recoil spring and recoil spring guide. Withdraw the recoil spring and its guide from the gun. The entire butt and pistol grip group can then be removed by forcing out the butt retaining pin, which is located in the lower rear section of the receiver. When the retaining pin is pulled out as far as it will go, slide the butt group rearward till it separates from the receiver. Pull the operating handle to the rear until the cutout point on the handle track is reached. Pull out the operating handle. The bolt, slide and piston assembly can now be withdrawn from the rear of the receiver. The bolt and bolt locks can now be lifted from the slide. Lift the cover; the belt feed lever assembly and the belt feed slide can be removed by pinching together with a pliers the split pin located at the right front of the cover. When the end of the pin is compressed, its locking collar can be removed, and all the belt feed components can be removed. Like many other gas-operated weapons, such as the US and FN Browning automatic rifles and the British Bren guns, the gas cylinder of the RPD can be easily adjusted. To adjust the gas cylinder of the RPD to obtain a different size of orifice, a special wrench is used.

To reassemble the weapon, perform the above steps in reverse order.

Special Note on the RPD

Although there is nothing new in the design of the RPD, the weapon is rather remarkable for its simplicity. There are indications that the weapon/cartridge combination has only marginal operating power.

How the RPD Works

With the belt loaded in the gun and the bolt group to the rear, the trigger is pulled. The trigger pulls the sear down from its engagement in the sear notch on the underside of the slide. The slide piston assembly with the bolt goes forward under the pressure of the decompressing recoil spring. The slide post stud, which operates in the track of the belt feed lever, moves the belt feed lever, which in turn moves the belt feed slide over, indexing the cartridge so that the top of the bolt can engage the cartridge and strip it forward out of the link and downward into the chamber. The locking flaps mounted on the sides of the bolt are cammed out into their locking surfaces in the receiver by the rearward movement of the bolt on the slide, when the bolt abuts the barrel. The outward movement of the locking flaps allows the slide post to strike the rear end of the firing pin, which in turn strikes the primer of the cartridge. Gas from the cartridge enters the gas port in the barrel and forces the piston and slide assembly to the rear. The bolt is also drawn to the rear, and its locking flaps are withdrawn into their unlocked position when the slide post has moved a slight distance to the rear. As the bolt goes to the rear, it carries the empty cartridge case in its face until the ejector strikes the upper part of the rim and knocks the case downward and out of the gun. If the trigger is held to the rear, this process will repeat itself until the 100-round belt is expended. The links are forced out the right side of the receiver by the left-to-right movement of the belt feed slide; when the first fifty rounds are fired, the first fifty-round link belt section falls out of the gun.

7.62mm SG43 AND SGM HEAVY MACHINE GUNS

Six versions of the Goryunov machine gun have been produced:

SG43—	smooth barrel; sear attached to driving spring guide; plain barrel lock; no dust covers; operating handle between spade grips.
SG43M—	SGMB type barrel lock and dust covers.
SGM—	splined barrel; separate sear housing; micrometer barrel lock; no dust covers (on early production models); operating handle on right side.
SGMT—	tank version of SGM with solenoid mounted on the backplate.
SGMB—	similar to SGM with dust covers over feedway ports, feed slide and ejection port; SGMB also has semi-circular flanges on the lower front of its receiver for mounting on the cradle inside the armored vehicle.
Hungarian General Purpose Version—	extensively modified SGM, to fill machine gun role similar to PK GPMG; this model has pistol-grip trigger mechanism and RPD type butt stock; is fired from bipod; resembles PK in outward appearance.

How to Load and Fire the Goryunov

To load the Goryunov, open the top cover by pushing forward the cover latch. Insert the link belt on the feedway, placing the rim of the cartridge in the jaws of the feed carrier. Close the cover and pull the operating handle to the rear as far as it will go. The operating handle of the SG43 is at the rear underside

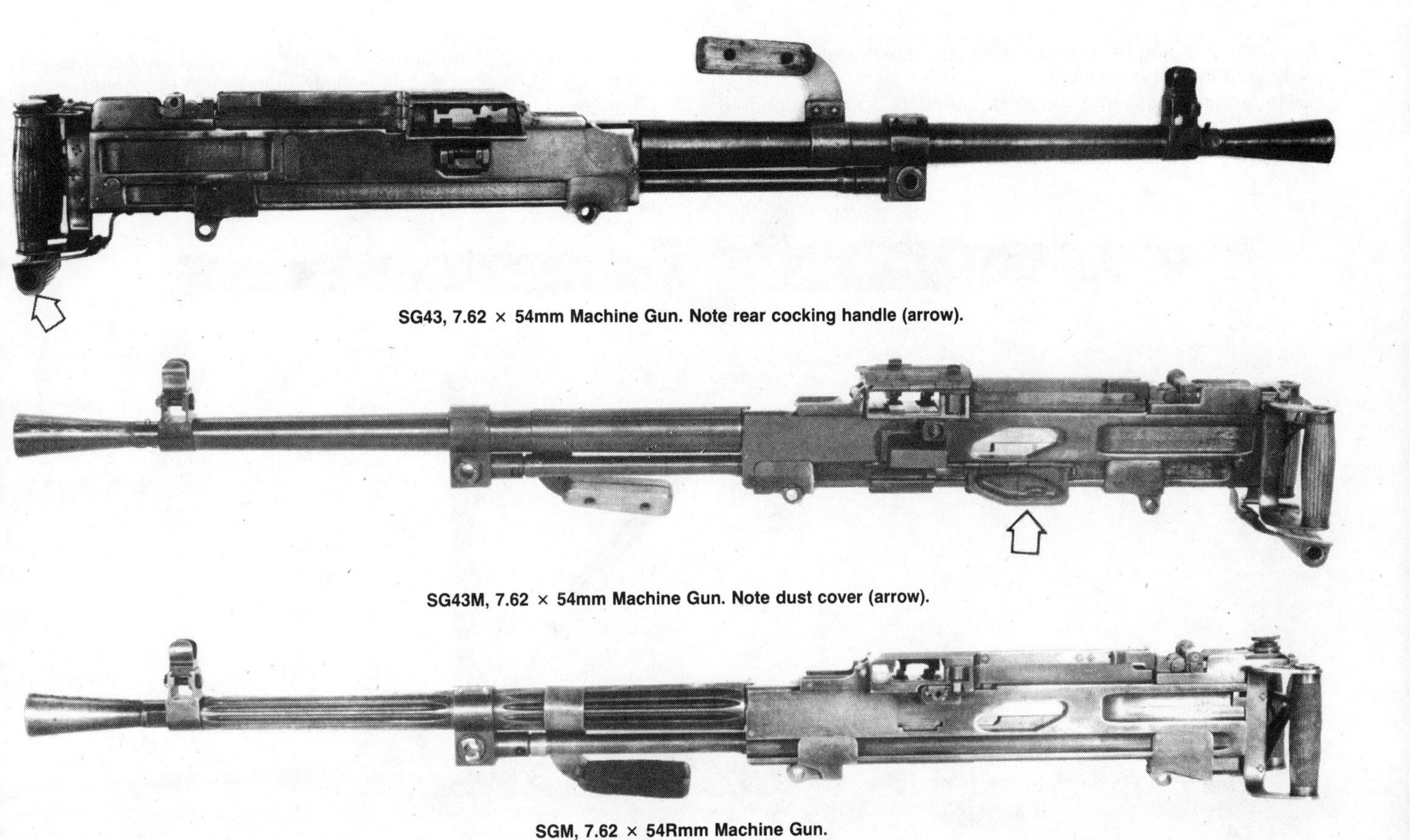

SG43, 7.62 × 54mm Machine Gun. Note rear cocking handle (arrow).

SG43M, 7.62 × 54mm Machine Gun. Note dust cover (arrow).

SGM, 7.62 × 54Rmm Machine Gun.

SGMB, 7.62 × 54Rmm Machine Gun on wheeled mount.

of the gun, under the spade grips. After the operating handle has been retracted, it should be pushed forward again; the bolt remains to the rear since the Goryunov fires from an open bolt. Raise the safety lock with the left thumb and press the upper end of the trigger with the right thumb. The weapon will now fire and will continue to fire until the trigger is released. The weapon can be unloaded by raising the top cover and lifting out the link belt. NOTE: On the SGM, the operating handle projects from the right underside of the gun.

Field Stripping the Goryunov

Open the top cover by pushing the cover catch forward; then lift the feed cover slightly and remove the feed carrier. Move the cover and feed cover to the vertical position. On the SGM, lift the rear sight leaf and with a punch, push out the backplate catch pin and move the catch backward; turn the backplate one quarter-turn to the right, and remove it from the receiver; remove the recoil spring and recoil spring guide. Remove the trigger mechanism from the receiver. Move the slide rearward with the operating handle until the bolt emerges from the receiver; grasp the bolt and slide and remove them from the receiver. Lift the bolt up and out of the slide. Move the operating handle rearward until it rests against the rear clamp of the receiver, and, turning it upward, separate it from the receiver. Move the plate covering the lower opening of the receiver to the rear and remove it from the receiver. Move the belt feed slide to the right and withdraw it from the receiver. To remove the barrel, press on the barrel lock with the thumb, moving the lock as far to the left as it will go; then, grasping the barrel handle, pull the barrel forward and remove it from the receiver.

No further disassembly is recommended. To reassemble the weapon, reverse the above procedure.

The field stripping of the SG43 differs in that the backplate is removed by pulling out its retaining pin, which is located at the bottom rear of the receiver.

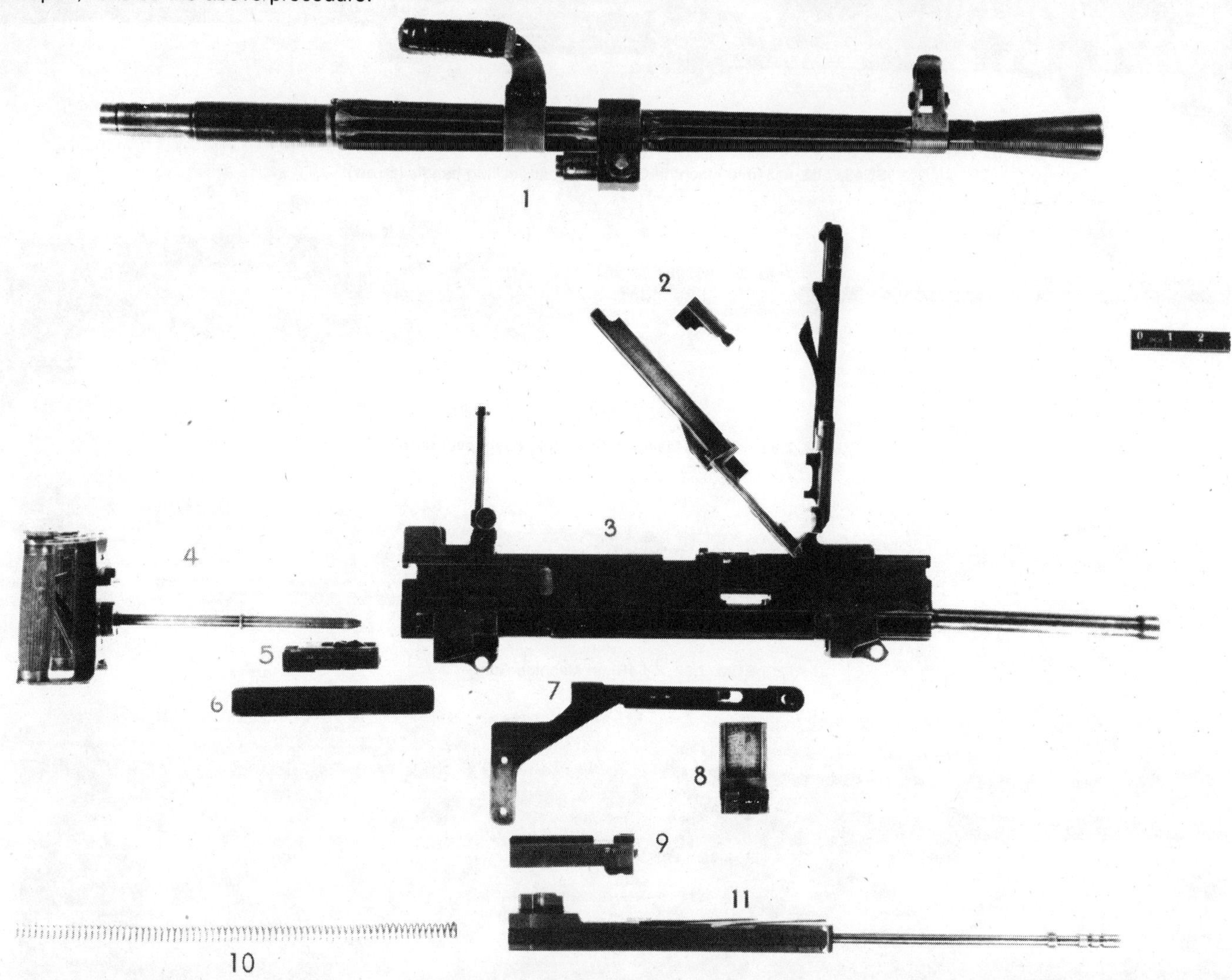

SGM, 7.62 × 54Rmm, Machine Gun, field stripped.

1. Barrel
2. Cartridge gripper
3. Receiver
4. Back plate
5. Sear unit
6. Dust cover
7. Operating handle
8. Feed slide
9. Bolt
10. Driving spring
11. Slide and piston

How the Goryunov Works

In loading, the belt is positioned in the feedway with the cover raised, so that the spring-loaded holders of the feed carrier engage the rim of the first round. The cover is closed and the operating parts retracted by means of the handle provided. The gun is then ready for firing. The feed carrier, which reciprocates in the feed tray (feed cover), operates in a notch in the top of the bolt. As the bolt is retracted, the feed carrier withdraws the cartridge from the belt. A belt holding pawl is located in the cover. The belt feed slide, reciprocating in a cut in the receiver, operates on a cam of the operating slide.

The gun fires from the open bolt position. The bolt is held to the rear by the sear, which is a part of the backplate assembly. The trigger operates through a connector to disengage the sear from the bolt. When the bolt is released, it is driven forward under the energy of the compressed driving spring, which operates against the operating slide. As the round is pushed forward by the bolt, it is forced downward, out of the carrier, and into the feeding tray. From the feeding tray, the round enters the chamber. As the bolt approaches its forward position, a cam on the operating slide forces the rear of the bolt $\frac{3}{16}$ of an inch to the right and into a recess in the receiver. The head of the bolt is recessed at an angle with the center of the bolt to give normal support to the base of the round in firing. As the round is chambered, the spring-loaded extractor, which is located in the left side of the bolt and is pivoted on a pin on the right side of its center, is forced over the rim of the round. After the bolt has been forced into the locked position, the slide moves forward to contact the firing pin. The firing pin strikes the primer, causing ignition of the round.

As the bullet is forced from the barrel by the expanding powder gas, some gas passes through the gas port in the barrel and impinges upon the piston attached to the slide, forcing it to the rear. The slide assembly moves independently of the bolt for a fraction of an inch. At this point, the bolt is forced out of engagement in the receiver and is drawn to the rear.

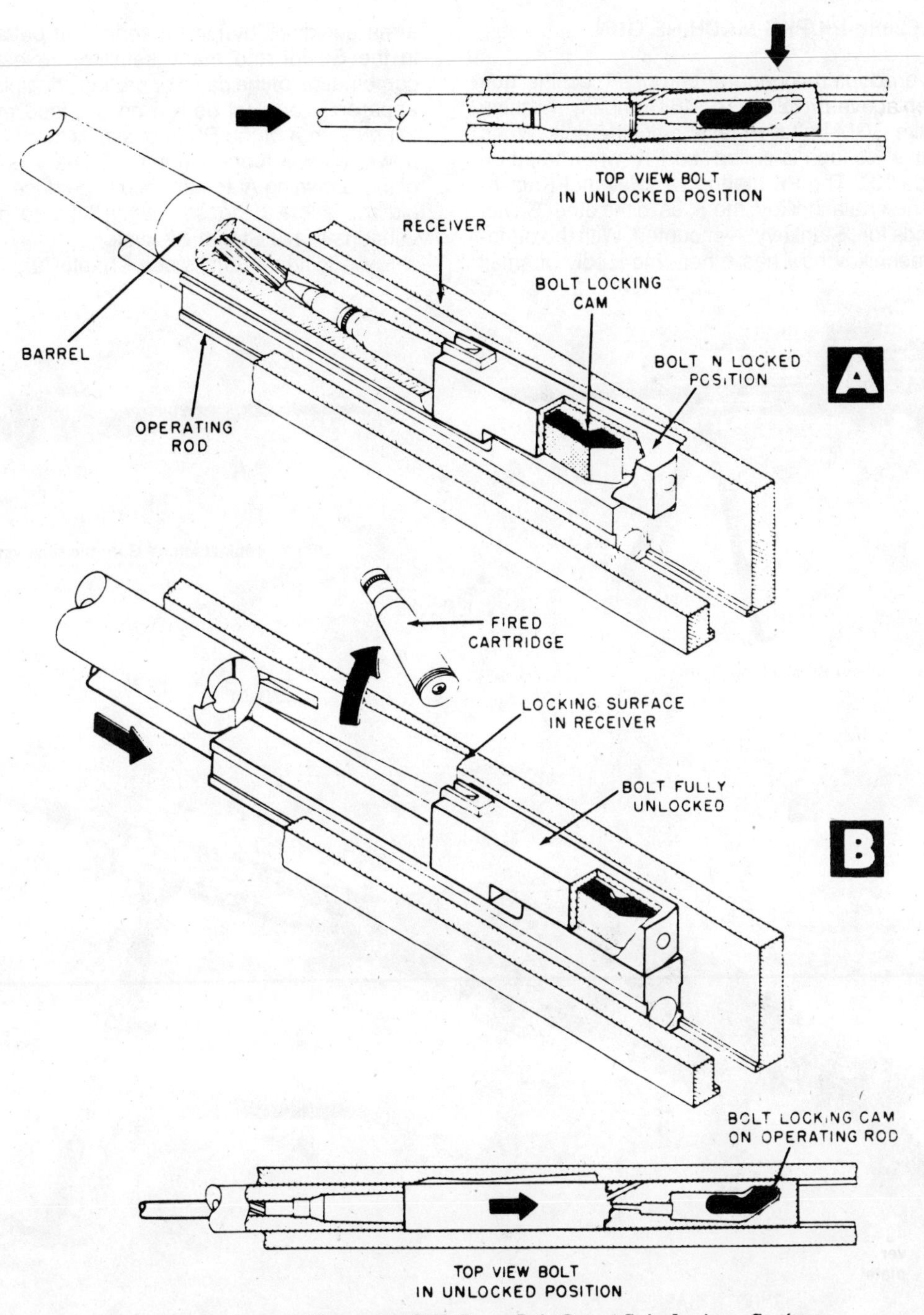

Lug on Operating Rod Cams Bolt Out of Side Locking Surface.

The extractor withdraws the fired case from the chamber. As the operating parts approach the rearward position, the ejector, a pin placed at an angle through the right side of the bolt to contact the base of the cartridge case, contacts the locking recess of the receiver. The ejector and the cartridge case are forced forward. The case, pivoted on the extractor, is ejected through the ejection port in the left side of the receiver.

If the trigger continues to be depressed, the firing cycle will be repeated. However, should the trigger be released, the sear will engage the bolt and hold it to the rear.

Notes on the Goryunov Machine Guns

The idea of a bolt whose rear end locked into the side of the receiver by being turned out of the line of axis of the bore was patented by John Browning many years ago. Browning never did very much with this idea in machine guns, but the Russians did. Goryunov, the Russian developer of this gun, died before his gun was put into service in the Soviet Army.

Since World War II, a modified gun, the SGM, has been adopted by the Soviets. The SGM has a longitudinally fluted barrel, and the barrel lock has been changed to incorporate a provision for headspace adjustment. The barrel lock has been changed by adding a scale, a slide and an Allen-screw type of lock and by having multiple grooves and ridges on the barrel lock and barrel. Thus, as the barrel and/or receiver wears, it is possible to unlock the slide, tighten it in the desired position (which can be determined from the scale) and relock it. This, of course, is not done with every barrel change but only after the components have had enough wear to make a significant difference in headspace.

THE 7.62mm PK/PKS MACHINE GUN

The Soviets have adopted a general purpose machine gun that will probably replace the 7.62mm RP-46 Company machine gun and the 7.62mm SGM battalion-level machine gun. When used on a bipod, this weapon is called the PK; when used on a tripod, it is called PKS. The PK stands for "Pulemet Kalashnikova"—machine gun Kalashnikov; the S, as in the other Soviet machine guns, stands for "Stankovy"—mounted. With the adoption of the PK, Kalashnikov now has a near monopoly of small arms designed by him in service at battalion and lower levels in the Soviet and many satellite armies. The PK is a clever combination of the basic operating principles of the AK with some apparently original design on the feed mechanism. The operating system of the PK is basically that of the AK turned upside down, as was done circa 1955-56 by FN with the mechanism of the Browning Automatic rifle to produce the MAG. Gene Stoner did the same thing somewhat later to produce machine gun versions in his Stoner 63 system.

For additional details see Chapter 2.

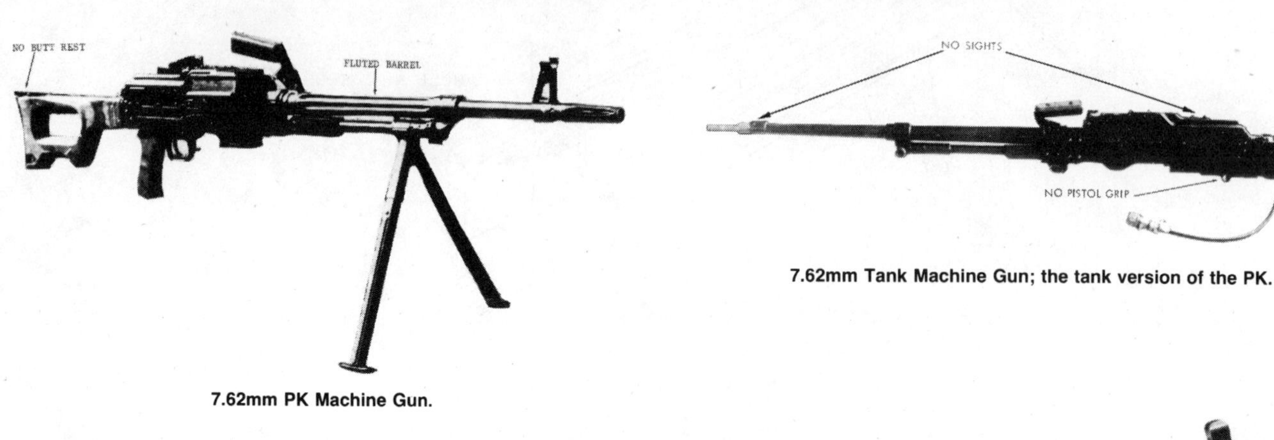

7.62mm PK Machine Gun.

7.62mm Tank Machine Gun; the tank version of the PK.

7.62mm PKM Machine Gun.

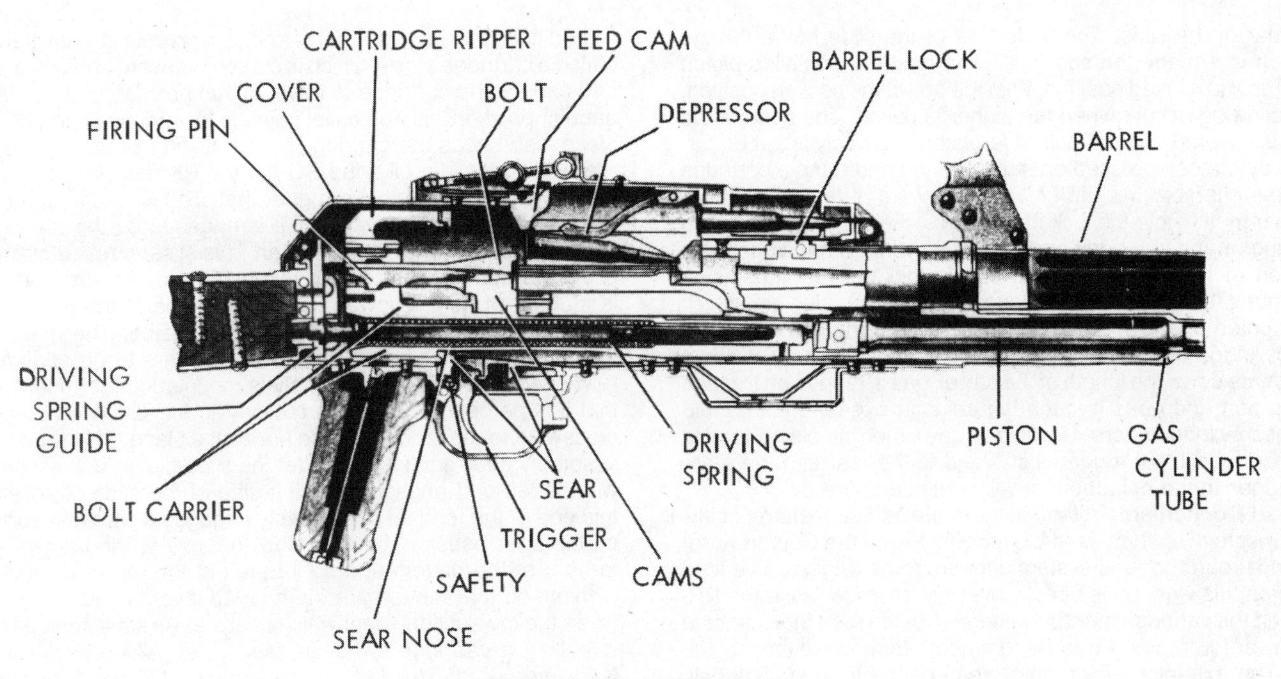

Section view of the PK Machine Gun.

How to Load and Fire the PK/PKS

Open cover by pressing in catch at top rear of cover and lifting cover. Lay cartridge belt in feedway so that first cartridge in belt is flush against the cartridge stop. Close cover and pull operating handle on right side of the receiver all the way to the rear. The weapon is now cocked and pressure on the trigger will produce fire. The safety is located on the receiver at the rear of the trigger. The barrel is removed by sliding out the barrel lock and pulling forward on the carrying handle; it is changed after 500 rounds of sustained fire.

How the PK/PKS Works

The bolt of the PK is similar to that of the AK; it has forward locking lugs and is cammed into and out of its locked position in the barrel by a raised cam lug on the slide. The body of the bolt is mounted in a tunneled-out portion of the slide post at the rear of the slide. The slide is in turn attached to the piston; for ease of disassembly, the slide-piston assembly is articulated. The firing pin is also mounted in the slide post; it rides in a hole in the bolt. The operating spring and operating spring guide, which are seated in the rear of the receiver, go into a hole in

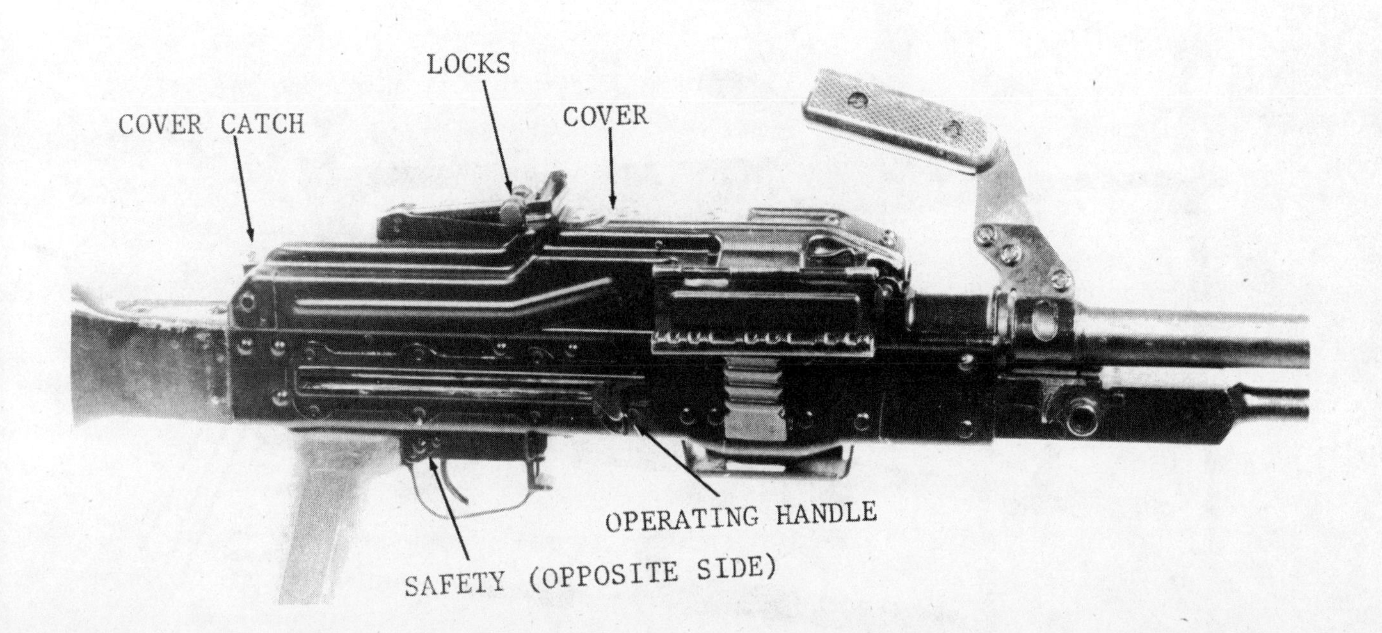

Receiver details of the PKM Machine Gun.

the rear of the slide. The underside of the slide has a cut out section to engage the sear nose and hold the bolt-slide-piston assembly in cocked position. With the bolt in the cocked position, the following occurs when the trigger is pulled. The nose of the sear is lowered and the bolt-slide-piston assembly is forced forward by the compressed operating spring and rams a cartridge into the chamber. As the bolt goes forward, it is cammed in a circular motion into the locked position in the barrel by the action of a lug on the bolt body on the cam-way in the raised cam lug section of the slide. After the bolt is locked, the slide-piston continues forward a slight distance causing the firing pin, which is mounted on the slide, to continue through the bolt and strike the cartridge primer functioning the cartridge. At a point about two-thirds down the length of the barrel, gas is tapped off through a gas port and goes through the adjustable gas regulator into the gas cylinder where it impinges upon the gas piston forcing it to the rear. If the trigger is still held to the rear and there are cartridges in the belt, the gun will continue to fire.

Feed Mechanism. Although there are certain features of this feed mechanism that resemble the RP-46 and the Goryunov, for the most part the feed system appears to be original. The feed plate on this weapon is directly over the chamber as in the RP-46, and the cartridge must be withdrawn to the rear, then directed downward in a position to be rammed into the chamber by the bolt. The extractor, which pulls the cartridge from the link belt, is mounted on the hook-like piece that protrudes forward from the slide post. This piece, the slide post, and the bottom rear section of the piston resemble a square configured letter "C". When the bolt operating handle is pulled to the rear, the extractor pulls a cartridge from the belt and to the rear. In a fashion similar to the RP-46, it is forced downward by a spring-loaded lever mounted on the top cover and presses the cartridge downward. In addition, the base of the cartridge runs into a cam track—called a cartridge stop—which is curved rearward and downward and causes the cartridge to travel in that position and end in an angular position with the bullet pointing toward the chamber; this piece also keeps the cartridge from traveling all the way to the rear with the slide. Like the SG-43, the PK has two covers—a top cover and a feed cover under that. The cartridge is held in the feed lips of this cover until it is engaged by the top of the bolt and rammed into the chamber. This system resembles that of a box magazine if the feed lever pushing down on the cartridge is thought of as a follower and the feed lips of the feed cover are thought of as the feed lips of a magazine. The means of moving the belt across the feedway appears to be original. A bow-shaped feed lever assembly is mounted in the receiver so that it is perpendicular to the receiver in the vertical plane and crosswise to the receiver in the horizontal plane. This feed lever assembly goes completely under the slide-piston-bolt assembly with its left-end engaging the left side of the slide. A roller on this end of the feed lever assembly engages a cammed surface on the slide causing the assembly to move up and down—left to right at the bottom and right to left at the top—in a rocking fashion, so that the left arm of the feed lever is under the slide when the top section—that which operates as a belt feed pawl—is pulling a cartridge on to the feed plate. A short arm of the feed lever—called the feed stud—engages a lip on the right side of the slide and functions as a pivot point for the feed lever. As mentioned, the top portion of the feed lever assembly rises and pushes the cartridge belt from right to left. A spring loaded belt holding pawl, mounted in the cover, then engages the belt and holds it in position. Both the feed and link ejection ports have dust covers. The dust cover on the link ejection port is opened when the bolt is pulled to the rear.

Disassembled view of the PK Machine Gun.

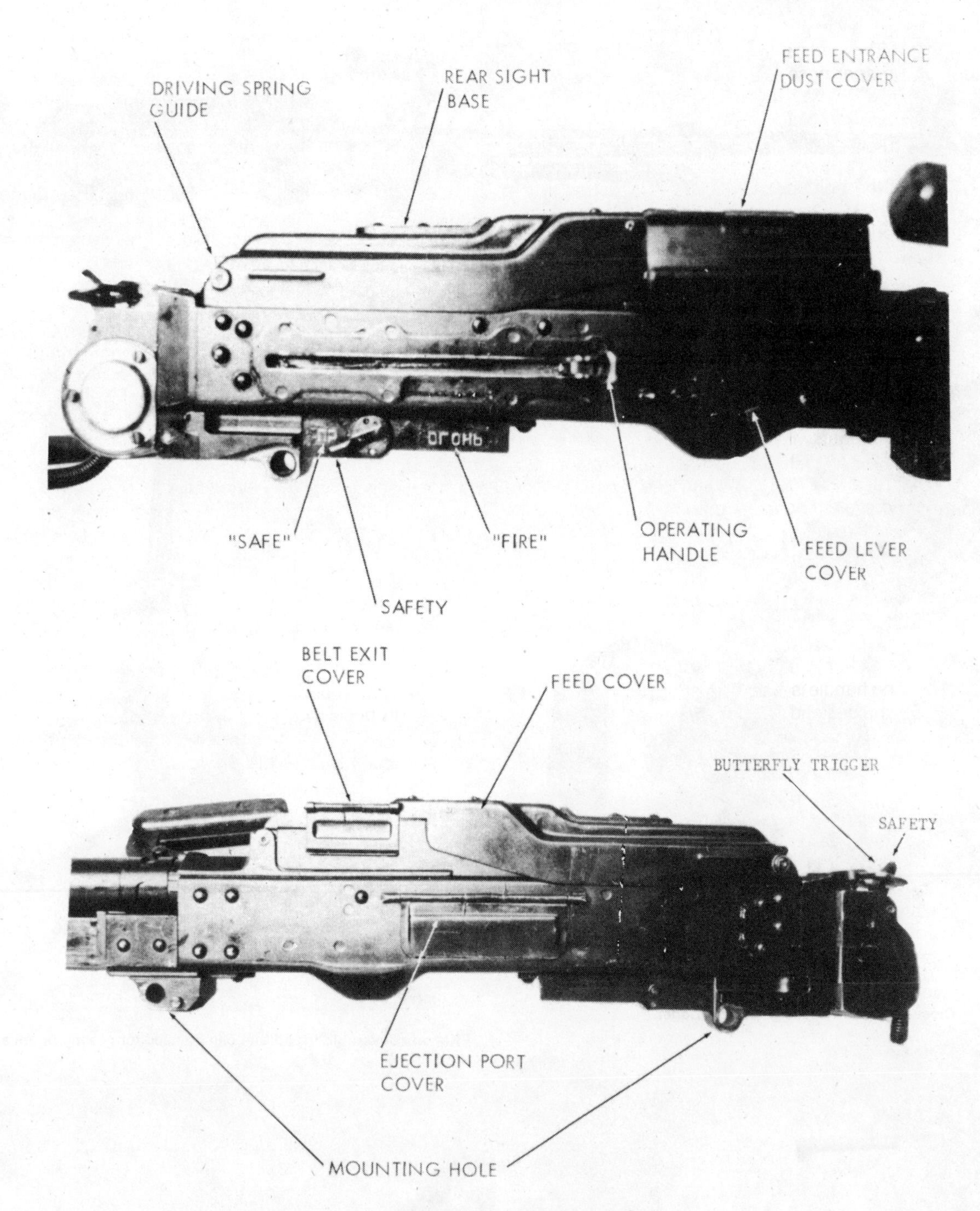

Receiver details of the PKT Tank Machine Gun.

Disassembly and Assembly

To disassemble the weapon:

Clear the gun but do not set at safe or close the cover. Remove the barrel.

Press in the driving spring guide at the rear of the receiver; then ease the guide and spring upward and out of the gun. Grasp the bolt carrier by the cartridge grippers and pull the entire unit rearward, then upward until it comes free of the receiver. Lift the bolt and carrier up and out of the gun.

Pull the bolt forward in the carrier, simultaneously twisting the bolt free of the cam until it comes free of the receiver. Lift the bolt and carrier up and out of the gun.

To reassemble the gun:

First seat the firing pin in its recess in the bolt. Seat the bolt into its hole in the bolt carrier twisting as necessary to engage the firing pin and the bolt with their recess in the carrier.

Start the piston into the gas cylinder tube until the slide can be seated in the receiver. The bolt must be pulled forward in the carrier prior to seating the slide. Pull the trigger and push the bolt carrier fully forward. Insert the driving spring into its tunnel in the slide, and press the guide forward against spring pressure until it can be seated against the rear wall of the receiver. Insert the barrel, slide the barrel lock into position and close the cover.

PKB Machine Gun designed for use as a flexibly mounted weapon on an APC.

Close-up of PKB spade grips and trigger.

PKM on convertible tripod that can be used for ground or antiaircraft fire.

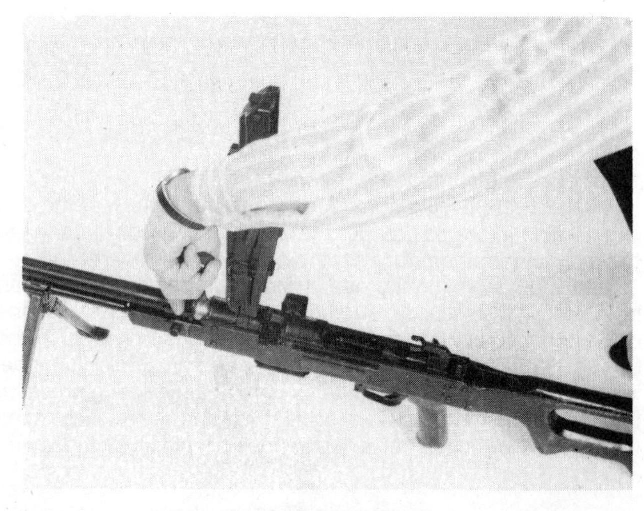

PK quick change barrel.

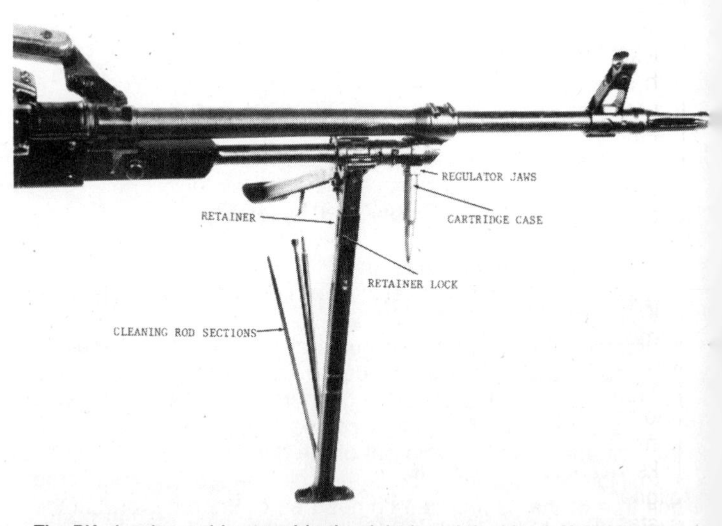

The PK cleaning rod is stored in the right leg of the bipod. Note also that the cartridge is used as the tool for adjusting the gas regulator.

THE 12.7mm DShK M1938 AND M1938/46 HEAVY MACHINE GUNS

The DShK is another of the Degtyarev series of weapons; its basic system of operation is similar to the other Degtyarev guns. The original DShK had a rotating block type of feed; the M1938/46 has a belt-feed lever type similar to the RP46's. These guns are chambered for the Soviet 12.7mm cartridge, which is almost identical in performance to (although not interchangeable with) the US cal. .50 cartridge. The M1938 was the primary Soviet heavy ground gun; in Korea it was frequently used by the North Koreans against aircraft. The weapon has been replaced in the Soviet Army in this role by the 14.5mm ZPU series of weapons. It is still used as antiaircraft armament on tanks and armored personnel carriers and is also used as co-axial armament on tanks. The ground mount for the DShK weapons is a wheeled mount, which can be converted into a tripod mount for AA fire.

Soviet 12.7mm DShK M1938 Heavy Machine Gun.

How to Load and Fire the DShK M1938

Push forward the feed cover latch located at the top rear of the feed cover. Lift the feed cover and place the feed belt on the revolving feed block so that the first round can be put in the upper recess of the feed block. Hold the free end of the ammunition belt with the right hand and press the feed belt against the revolving block. Rapidly rotate the block with belt as far to the right as it can go (the upper recess should rotate 120°). Close the feed cover. Pull reloading handle to the rear until the slide is engaged by the sear. Hold spade grips with both hands and press the trigger slowly with the index fingers of both hands; the weapon will fire. To put the weapon on safe, rotate the safety catch forward.

DShK 12.7mm M1938/46 Heavy Machine Gun.

How to Load and Fire the DShK M1938/46

Loading and firing of the DShK M1938/46 are the same as for the 7.62mm RP46.

Field Stripping the DShK M1938

Set the safety on fire and loosen (one turn) the latch of the machine gun mounting studs on the gun mount. If the bolt is to the rear, move it forward by pulling the trigger. Stand with back to the muzzle in front of the mount axle with the left foot resting on the axle. Grasp the gas piston tube with both hands and pull it as far forward as it will move. Then turn the tube clockwise with both hands until the support of the tube comes out of its grooves in the barrel. Push all the moving parts (bolt and slide group) to the rear until the recoil stop roller emerges from its recess in the receiver.

Unscrew the connecting screw of the rear machine gun locking bracket. Remove the backplate pin and tap the backplate lightly with a wooden mallet or a copper hammer to separate it from the receiver, meanwhile supporting it with one hand. Remove the trigger housing and then withdraw the bolt and slide group from the rear of the receiver. Remove the firing pin and the bolt and bolt locking flaps. To reassemble the weapon, perform the above steps in reverse order.

7mm DShK M1938/46, tripod set for AA fire.

How the DShKs Work

In general, both of the 12.7mm Degtyarev guns operate the same as the 7.62mm guns of this series. The M1938 has a circular type of feed mechanism that operates somewhat like the cylinder of a revolver. The feed lever, which is operated back and forth by the slide, turns the feed drum, which carries the cartridges in its recesses. As the cartridges are turned, their links are stripped, and in the final feed step the cartridge is aligned with the bolt, which rams it into the chamber. The feed system on the M1938/46 is basically the same as that on the RP46.

DShK, field stripped.

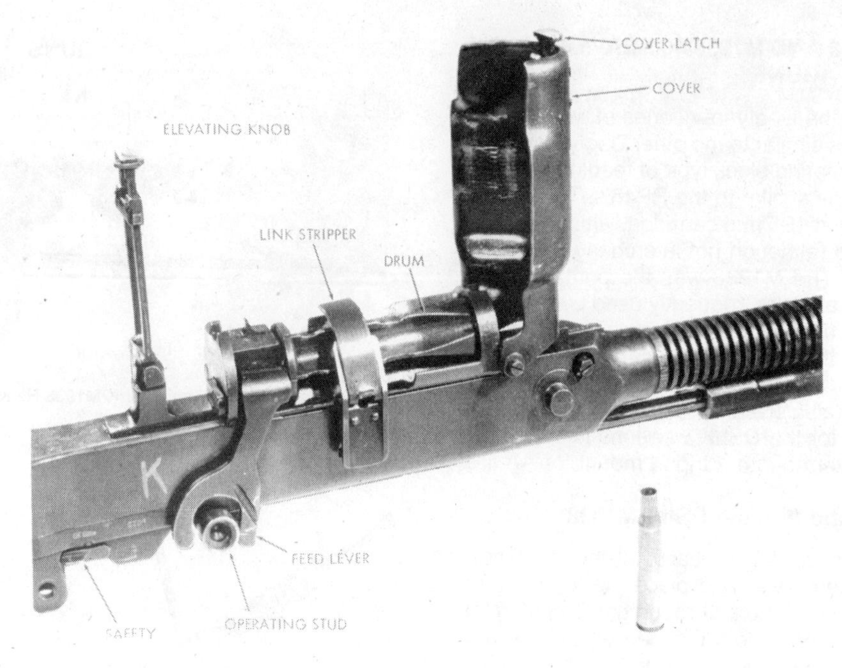

Receiver details of the DShK38 Machine Gun.

CHARACTERISTICS OF PRE-WORLD WAR II AND WORLD WAR II SOVIET 7.62mm GROUND MACHINE GUNS

	Model 1910 Maxim (SPM)	Model 1939 DS	DP	DPM	DT	DTM	SG43
Caliber:	7.62mm.	7.62mm.	7.62mm.	7.62mm.	7.62mm.	7.62mm.	7.62mm.
Type of gun:	Heavy ground gun.	Heavy ground gun.	Squad-automatic	Squad automatic.	Tank gun.	Tank gun.	Heavy ground gun.
System of operation:	Recoil, automatic only.	Gas, automatic only.	Gas, automatic only.	Gas, automatic only.	Gas, automatic only.	Gas, automatic only.	Gas, automatic only.
Weight:	52.47 lb. w/ water.	52.47 lb. w/ water.	26.23 lb. loaded.	26.9 lb.	27.91 lb.	28.46 lb.	30.42 lb.
Mount Weight:	99.71 lb. w/ shield.	99.71 lb. w/ shield.					59.3 lb.
Length overall:	43.6 in.	46 in.	50 in.	50 in.	Stock extended, 46.46 in. Stock retracted, 39.76 in.	Stock extended, 46.44 in. Stock retracted, 39.76 in.	44.09 in.
Barrel length:	28.4 in.	28.4 in.	23.8 in.	23.8 in.	23.5 in.	23.5 in.	28.3 in.
Feed device:	250 rd.	250-rd. web belt or 50-rd. link belt.	47-rd. drum.	47-rd. drum.	60-rd. drum.	60-rd. drum.	250-rd. drum metallic link belt.
Sights:							
Front:	Blade.	Post w/ears.	Post w/ears.	Post w/ears.	None on tanks On bipod: Post w/ears.	None on tanks On bipod: Post w/ears	Blade.
Rear:	Leaf.	Leaf.	Tangent leaf.	Tangent leaf.	Aperture.	Aperture.	Leaf.
Cycle rate of fire:	500-600.	2 Rates: 500-600, 1000-1200.	500-600.	500-600.	600.	600.	600-700.
Muzzle velocity w/light ball:	2822 f.p.s.	2832 f.p.s.	2756 f.p.s.	2756 f.p.s.	2756 f.p.s.	2756 f.p.s.	2832 f.p.s.
Cooling:	Water.	Air.	Air.	Air.	Air.	Air.	Air.

NOTE: An aircraft version of the Degtyarev, the 7.62mm DA, was also used. It has been obsolete for some time. All use the 7.62 × 54mm cartridge.

CHARACTERISTICS OF SOVIET POST-WORLD WAR II MACHINE GUNS

	SGM	RP-46	RPD	RPK	PK/PKS	DShK M1938/46
Caliber:	7.62mm (7.62mm × 54).	7.62mm (7.62 × 54).	7.62mm. (7.62 × 39).	7.62mm. (7.62 × 39).	7.62mm (7.62 × 54).	12.7mm. (12.7 × 108).
System of operation:	Gas, automatic.	Gas, automatic.	Gas, automatic.	Gas, automatic.	Gas, automatic.	Gas, automatic.
Length overall:	44.09 in.	50 in.	40.8 in.	40.9 in.	47.2 in.	62.5 in.
Barrel length:	28.3 in.	23.8 in.	20.5 in.	23.2 in.	25.9 in.	42.1 in.
Feed device:	250-round metallic link belt.	250-round metallic link belt.	100-round metallic link belt in drum.	75-round drum and 40-round box magazine.	100,200 or 250-round metallic link belt	50-round metallic link belt.
Sights: Front:	Blade.	Post w/ears.	Post w/ears.	Post w/ears.	Protected post.	Post w/ears.
Rear:	Leaf.	Tangent.	Tangent.	Tangent.	Tangent w/ notch.	Leaf.
Weight, gun:	29.76 lb.	28.7 lb.	15.6 lb.	w/empty drum: 12.3 lb. w/empty box: 11 lb.	19.8 lb. w/bipod.	78.5 lb.
mount:	50.9 lb.*	—	—	—	16.5 lb.	259 lbs.
Muzzle velocity:	2870 f.p.s.	2750 f.p.s.	2410 f.p.s.	2410 f.p.s.	2700 f.p.s.	2822 f.p.s.
Cyclic rate:	600-700 r.p.m.	600-650 r.p.m.	650-750 r.p.m.	600 r.p.m.	650-700 r.p.m.	540-600 r.p.m.

*A 30.6 lb. tripod mount also exists.

AGS-17 30mm AUTOMATIC GRENADE LAUNCHER

The Soviet 30mm AGS-17 automatic grenade launcher is a belt-fed blowback-operated weapon which, for infantry purposes, is mounted on a tripod. This weapon is the newest weapon to be added to Soviet inventories. The weapon is noteworthy since it is basically an improved version of the US 40mm MK19 Mod 0 machine gun. It represents a departure from the usual Soviet practice of continuous gradual product improvement of infantry weapons and demonstrates their willingness to "borrow" extensively from worldwide technological advances. (See Chapter 5 for more details.)

Field firing of the 30mm AGS-17 Grenade Launcher.

CHARACTERISTICS OF THE AG-17 30mm AUTOMATIC GRENADE LAUNCHER

Limited technical data available at this time is presented below:	
Caliber:	30mm
Weight of launcher:	37.5 lbs.
Combat effective range (est):	1200m
Maximum range:	1730m
Cyclic rate of fire (est):	250–300 rd/min
Practical rate of fire (est):	40–60 rd/min

47 United Arab Republic (Egypt)

SMALL ARMS IN SERVICE

Handguns:	9 × 19mm Parabellum Helwan	Beretta-made; licensed production at Maadi Factory number 54.
Submachine guns:	9 × 19mm Parabellum Port Said	Copy of Carl Gustaf Model 45 made at Maadi Factory no. 54.
Rifles:	7.62 × 39mm Misr (AKM)	USSR-made; production at Maadi Factory no. 54.
	7.62 × 39mm AK47	USSR-made.
	7.62 × 39mm SKS	USSR-made.
	7.92 × 57mm Hakim	Maadi Factory no. 54. Static troops only.
Machine guns:	7.62 × 39mm Suez (RPD)	USSR-made; production at Maadi Factory no. 54.
	7.62 × 54mmR SG-43	USSR-made.
	7.62 × 54mmR Asswan (SGM)	USSR-made; production at Maadi Factory no. 54.
	12.7 × 108mm DShK38/46	USSR-made.

EGYPTIAN SMALL ARMS INDUSTRY

Generally speaking, Egypt obtained all of her weapons from western nations prior to 1954, and most of the weapons from those nations were obtained before that time. In 1954, the UAR made the first of a series of extensive arms purchases from the Soviet Bloc. Egypt has developed a small arms industry of its own and is less dependent on imported small arms. This industry, which was established by Swedish technicians prior to the overthrow of the Egyptian monarchy, originally produced copies of foreign weapons. Cartridges for all UAR service weapons are produced in that country.

EGYPTIAN HANDGUNS

Egypt was originally equipped with British caliber .455 No. 1 Mark VI Webley revolvers and Enfield caliber .38 No. 2 revolvers. After World War II, Egypt adopted the 9mm Parabellum as the standard pistol and submachine gun cartridge. A modification of the Tokarev TT M1933, called the Tokagypt Model 58, chambered for the 9mm NATO cartridge, was developed in Hungary for Egypt, and limited purchases of this weapon were made. The Egyptian authorities were not too pleased with this weapon, and the usage has been confined to police work. The 9mm Beretta Model 1951—covered in detail in the chapter on Italy—is the standard service pistol.

The Egyptian small arms authorities, in conjunction with Beretta, developed a target version of the Model 1951. This weapon has an adjustable target type rear sight and a ramp mounted front sight, a longer barrel than the standard Model 1951 and target type stocks.

9mm Parabellum target type Beretta, Model 1951; developed for Egypt.

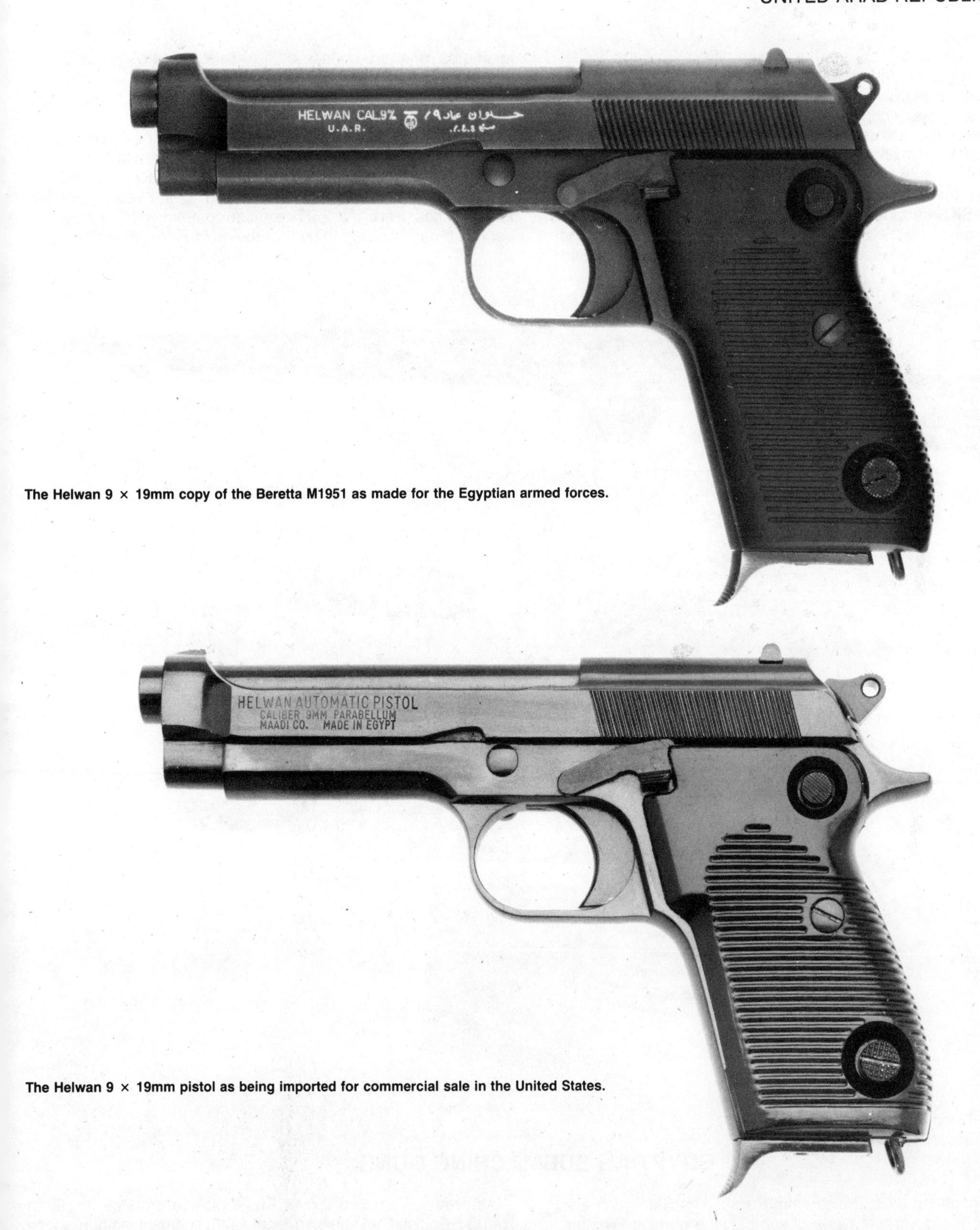

The Helwan 9 × 19mm copy of the Beretta M1951 as made for the Egyptian armed forces.

The Helwan 9 × 19mm pistol as being imported for commercial sale in the United States.

EGYPTIAN RIFLES

The Egyptians used the British caliber .303 rifle No. 1 Mark III and Mark III* rifle from the time of World War I until approximately 1949. At that time, the Royal Egyptian government purchased quantities of the FN self-loading rifle (also known as SAFN) chambered for the 7.92 mm cartridge. These rifles can be identified by the Royal Egyptian crest on the receiver.

In this same time frame, a small arms manufacturing plant was set up in Egypt with the assistance of Swedish technicians. This plant manufactured the 7.92mm Hakim rifle, a modification of the 6.5mm Ljungman Model 42, and the 9mm NATO

Port Said submachine gun. The Hakim differs from the Model 42 in having a full length hand guard, tangent type rear sight, enlarged charger guide, modified magazine catch and the shape of the muzzle brake/compensator. A number of training versions of the Hakim rifle are used by Egypt. Beretta made a caliber .22 version of the Hakim, and Anschutz made a 4.5mm air rifle version of this rifle for training.

In 1954, the UAR procured significant quantities of the Soviet 7.62mm SKS carbine and the Czech 7.62mm Model 52 rifle. The design of the SKS apparently appealed to the Egyptians insofar as its shortness, lightness and permanently attached folding bayonet are concerned. These features of the SKS plus use of the Soviet 7.62mm Model 1943 "intermediate" sized cartridge were incorporated into the design of the Egyptian Rashid rifle.

The Rashid has a modified Ljungman action; the bolt retracting handle on the Rashid is mounted in the right forward section of the action; bolt retraction on the Ljungman is accomplished by pulling the receiver cover forward and to the rear. Very few Rashid rifles were made. The most common rifle in Egyptian service today appears to be the Soviet 7.62mm AK assault rifle.

7.92mm Hakim rifle.

Action of Hakim rifle, field-stripped.

EGYPTIAN SUBMACHINE GUNS

The Egyptians used British 9mm Parabellum Sten guns and also purchased various types of submachine guns in Western Europe, including the 9mm NATO Spanish Star Model Z-45 and the 9mm NATO Beretta Model 38/42 and 38/49.

As previously noted, Egypt was tooled to produce the 9mm NATO Swedish Carl Gustaf (Model 1945) submachine gun, which the Egyptians call the Port Said submachine gun. This weapon is covered in detail under Sweden.

EGYPTIAN MACHINE GUNS

British machine guns formerly predominated in Egypt, and prior to the overthrow of the Egyptian monarchy, some Spanish ALFA 7.92mm M1944 machine guns were procured. Soviet machine guns now predominate in the Egyptian forces.

48 United States

SMALL ARMS IN SERVICE

Handguns:	.45 (11.43 × 23mm) M1911A1	Colt and other contractors.
	.38 revolvers of various types for military police organizations	Colt, Smith & Wesson, Ruger.
	.45 (11.43 × 23 mm) M15 General Officer Pistol	Rock Island Arsenal.
Submachine Guns:	.45 (11.43 × 23mm) M3A1	Government contractors.
	5.56 × 45mm M231 FPW	Bradley Fighting Vehicle firing port weapon.
Rifles:	5.56 × 45mm M16A1	Colt and other contractors.
	5.56 × 45mm NATO M16A2	Colt.
	7.62 × 51mm NATO M14	Springfield Armory and Contractors.
	7.62 × 51mm NATO M21	Sniper rifle.
	7.62 × 51mm NATO M40	USMC sniper rifle.
Machine Guns:	5.56 × 45mm M249	FN Minimi.
	7.62 × 51mm NATO M60 (and variants)	Saco Defense Systems.
	7.62 × 51mm NATO M60E3	Saco Defense Systems, USMC.
	7.62 × 51mm NATO M240 armor machine gun	FN Manufacturing, Inc. Also M240C.
	7.62 × 51mm NATO M73 armor MG	General Electric and Rock Island Arsenal. Also M219.
	12.7 × 99mm (.50) M2 HB	Saco Defense Systems and other contractors during World War II.
	12.7 × 99mm (.50) M85	General Electric and Rock Island Arsenal.
	40 × 53mm, MOD3 Grenade Launcher	Saco Defense Systems.

UNITED STATES HANDGUNS*

The United States adopted the caliber .45 Browning-designed Colt automatic pistol in 1911. All manufacture of this pistol was originally carried on at Colt, but Springfield Armory was tooled to produce the weapon prior to 1914. At the time of United States entry into World War I, 55,553 pistols were on hand. During World War I, Model 1911 pistols were manufactured by Remington Arms and Colt. Springfield did not manufacture the pistol, since top priority was given, at that arsenal, to the manufacture of Model 1903 rifles. Approximately 450,000 Model 1911 pistols were made during World War I by Colt and Remington. Colt was by far the largest producer; Remington produced only 21,265 pistols by the close of December 1918.

Owing to the shortage of Model 1911 pistols, orders were placed with Colt and Smith & Wesson for a heavy frame revolver chambered for the caliber .45 Model 1911 pistol cartridge. The revolvers chosen were the Colt New Service and the Smith & Wesson Hand Ejector models. Both had been in production, chambered for the .455 Webley cartridge, for the United Kingdom. Modifications were made to accommodate the rimless .45 Model 1911 cartridge. Three round, half-moon clips, which fitted in the cartridge case cannelures, were used with these weapons to allow case ejection by use of the ejector rod. They may be loaded and fired without these clips, but the individual cases have to be ejected one by one with a nail or pencil if half-moon clips are not used. These revolvers are conventional swing-out cylinder types; the cylinder latch is pushed forward on the Smith & Wesson to release the cylinder and is pulled back on the Colt. Colt manufactured 151,700 Model 1917 revolvers, and Smith & Wesson manufactured 153,311 of their Model 1917 revolver.

The Colt and Smith & Wesson revolvers were still in use by Military Police and security personnel during World War II but are no longer used in the United States armed forces or as a standard weapon in any army.

THE MODEL 1911A1 PISTOL

After World War I, the Colt automatic was modified. The modifications, adopted on 15 June, 1926, caused a change in nomenclature to: Pistol U.S. Caliber .45 Model 1911A1. The changes were as follows:

(1) The main spring housing of the 1911 is flat and smooth; that of the 1911A1 is arched beyond the line of the grip portion of the receiver and is knurled.

*For additional details see *Handguns of the World*.

(2) The trigger of the 1911 is smooth and is longer than the serrated trigger of the 1911A1.

(3) The tang of the Model 1911A1 is longer than that of the Model 1911.

(4) The 1911A1 front sight is wider than that of the 1911.

(5) Finger clearance cuts were made on the receiver of the 1911A1 immediately behind the trigger; these are not present on the 1911.

(6) Rifling and diameter were reduced, and land height was increased.

During World War II, the Model 1911A1 was manufactured by Colt, Remington Rand, Union Switch and Signal, and the Ithaca Gun Co. Approximately 1,800,000 pistols were made during World War II in addition to the 150,000 or so that were purchased by the United States prior to World War II. Added to the total of caliber .45 Model 1911 pistols made prior to and during World War II, over 2,400,000 .45 Colt Automatics have been made for the United States Government. In addition, hundreds of thousands have been made for commercial sale and export to foreign forces.

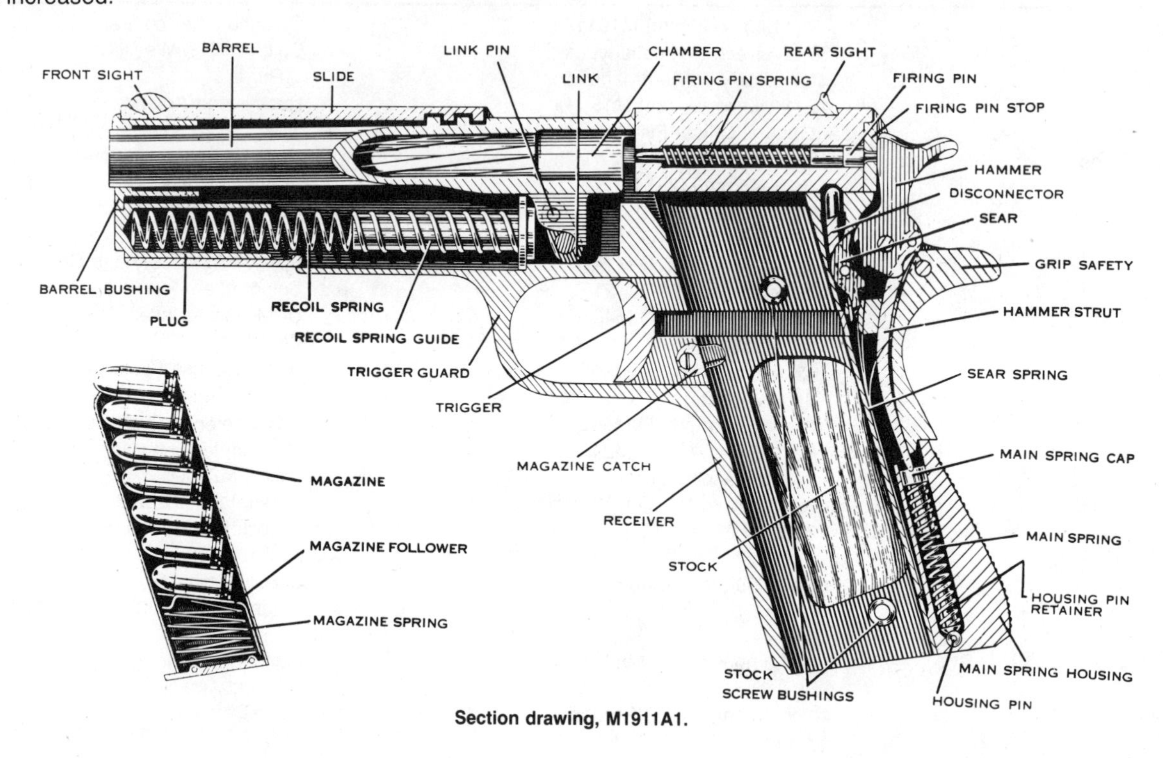

Section drawing, M1911A1.

Field Stripping the M1911A1 Pistol

While no stripping beyond that illustrated is ever necessary to clean and properly care for this pistol, the following instructions will be helpful to those who wish to master every detail.

To Remove Safety Lock. (1) Cock hammer. (2) Grasp thumbpiece of safety lock between thumb and index finger, pull steadily outward and at same time move back and forth.

To Remove Hammer. (1) Lower hammer—do not snap it. (2)Use safety lock to push out hammer pin, removing from left side. (3) Lift out hammer and hammer strut.

To Remove Mainspring Housing. (1) Using hammer strut, push mainspring housing pin out from right side of receiver. (2) Slide housing and its contained spring down out of its guides. (3) Push in on mainspring cap; at the same time push out mainspring cap pin.

To Remove Sear and Disconnector. Using hammer strut, push out sear pin from left side of receiver and remove sear and disconnector.

To Remove Magazine Catch. Press in checkered left end to permit turning catch lock a quarter turn to left out of its seat in receiver, using long leaf of sear spring. Catch, its lock and spring may now be removed. Be careful not to let spring jump away when released.

To Remove Trigger. Pull straight to the rear.

To Remove Slide Stop Plunger, Safety Lock Plunger and Plunger Spring. Draw straight to rear.

Notes on Assembling

Barrel link must be tilted forward and link pin properly in place before it will slide into place in the slide.

Put sear and disconnector together, hold by their lower ends, place them in the receiver, and replace sear pin.

Sear spring should be replaced after sear and disconnector are in place, care being taken that lower end is in its place in the cut in the receiver; upper end of left-hand leaf resting on sear.

Insert mainspring housing until lower end projects about one-eighth inch below frame. Then (1) Replace hammer and pin; (2) Grip safety; (3) Cock hammer and replace safety lock; (4) Lower hammer and push mainspring housing home, and insert pin.

Cock hammer. Insert end of magazine follower to press safety lock plunger home.

When inserting slide stop, make sure that its upper rear end stops on the receiver just below the small slide stop plunger. Then push stop upward and inward with the one motion. This will enable the upper round part of the stop to push the plunger back and let the stop snap into place.

In replacing sear and disconnector, hold the receiver as when firing, then tilt front end down. Insert the sear and the disconnector using the trigger bar as a guide to align the holes in these two parts with the receiver holes. Slight pressure may be applied on the trigger until the holes are properly lined up. When replacing the mainspring housing, it is important that the rear end of the hammer strut be in its place in the mainspring cap.

Instructions for Loading and Firing the M1911A1 Pistol

To remove magazine: Press magazine catch (button). Magazine will normally be ejected and should be caught with left hand. If spring is weak, it may come only part way; withdraw it from handle.

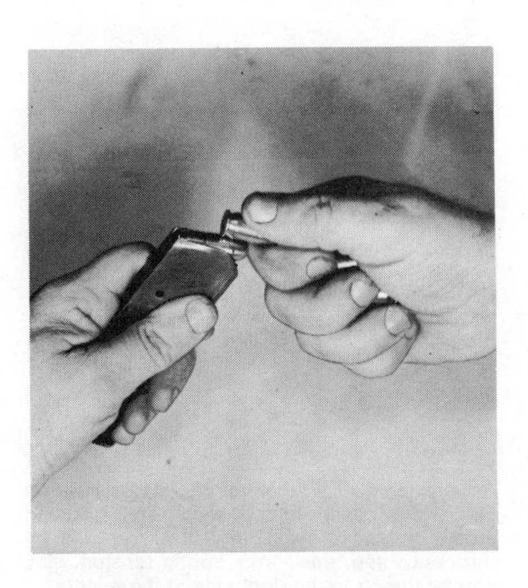

Load magazine: Holding firmly in left hand, press cartridge down in forward end of magazine follower (platform) and slide in under the curved lips of the magazine. Press following cartridges down as illustrated. Any number from 1 to 7 may be inserted.

To load chamber: (1) Holding pistol at height of right shoulder and pistol 6 inches from shoulder, insert loaded magazine and press home until it locks with a click. (2) Grasp slide with thumb and fingers of the left hand, thumb on right side of slide pointing upwards and pull back slide as far as it will go. This compresses the recoil spring, cocks the hammer and permits the magazine spring to push the top cartridge into line with the breech block. (3) Release slide. The recoil spring will drive it forward and feed a cartridge into the chamber; barrel will be forced up on its link and will lock into slide; firing mechanism will engage ready for first shot.

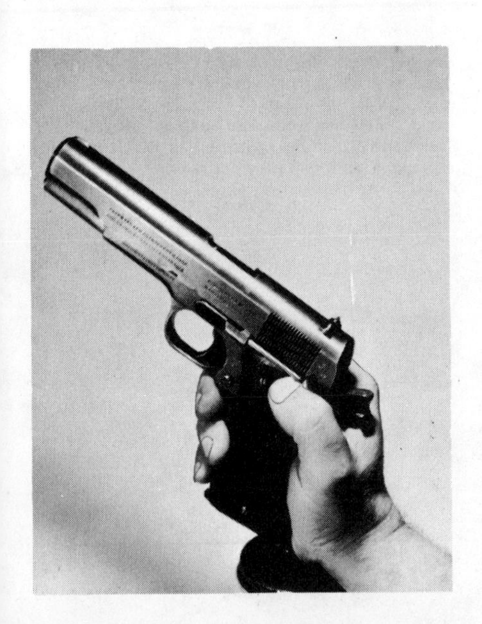

To engage thumb safety: Unless pistol is to fired at once, always push safety lock up into place as soon as chamber is loaded. A stud on the inner face of the thumb safety locks the hammer and sear when the safety is pushed up into the slide. It can be released by simply pushing down on the thumbpiece.

Slide stop: When the last shot has been fired, a section of the front end of the magazine follower, pushed up by the magazine spring, presses against the underside of the slide stop. This forces the stop up into a niche cut in the slide and holds the slide open as an indication that the pistol is empty.

Reloading from open slide: (1) Press magazine catch and extract empty magazine. (2) Insert loaded magazine. (3) Push down on slide stop with right thumb. This will release the slide to drive forward and load the chamber. Note: Slide stop cannot be released while an empty magazine is in the pistol. Slide will go forward only on a loaded magazine or when the magazine has been pulled part way out.

Field Stripping the M1911A1 Pistol

Remove magazine and examine chamber: (1) Press magazine catch and withdraw magazine. (2) Draw back slide and look into chamber through the ejection port to be sure the pistol is empty. Remember that even when the magazine is out, the pistol is still dangerous: there may be a cartridge in the chamber.

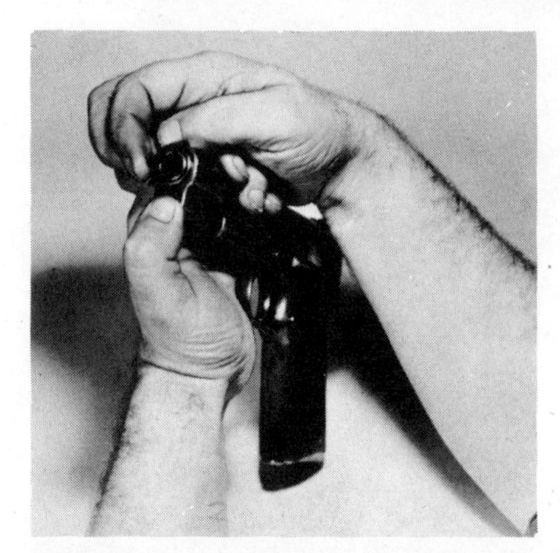

Release tension of recoil spring: (1) Press in on plug which covers end of recoil spring, using thumb or butt of magazine if it is too stiff. (2) Barrel bushing, freed from spring tension, may now be turned to the right side of the pistol.

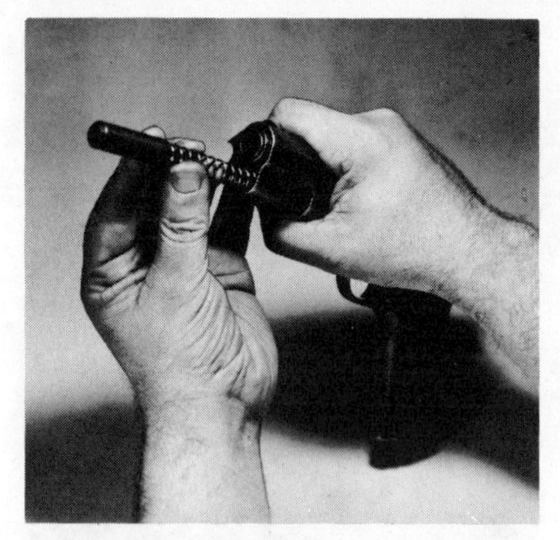

Ease out plug and recoil spring: The spring is very powerful. Take care not to let it fly out of the pistol. Do not withdraw these parts from the pistol yet, as they serve to keep the recoil spring guide in place and make the next step easy.

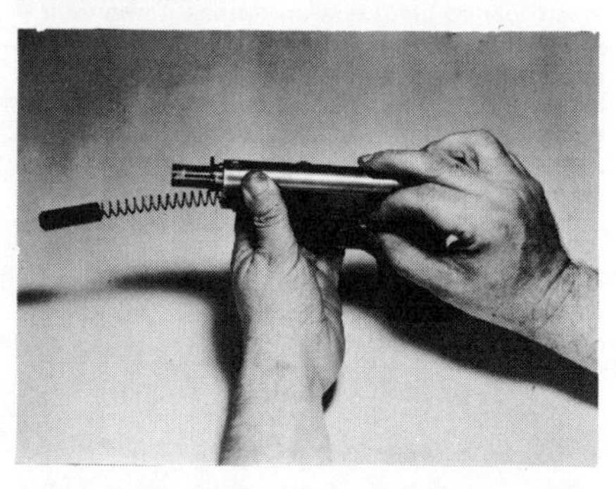

Remove slide stop. (1) Push slide back until the rear edge of the smaller recess in the lower edge of the slide is even with the rear end of the slide stop. (2) Now press from the right side against the protruding pin which is part of the slide stop. This pin passes through the right side of the receiver, then through the barrel link which holds the barrel, then through the left side of the receiver. (3) Now pull slide stop out from left side of pistol.

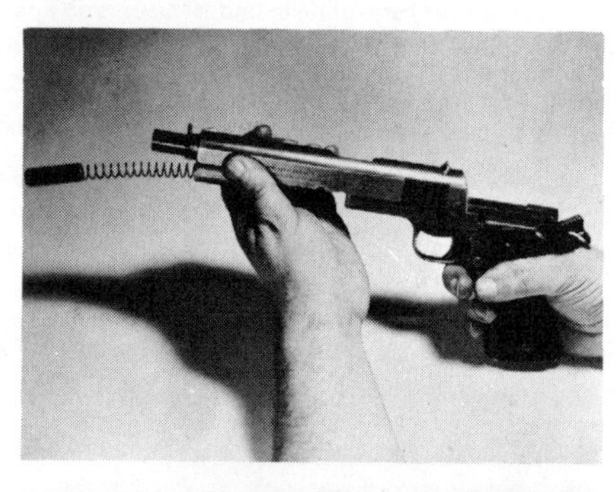

Remove slide and components: Pull slide forward on its guides in the receiver and remove. With the slide will come the barrel, barrel link, barrel bushing, recoil spring and recoil spring guide.

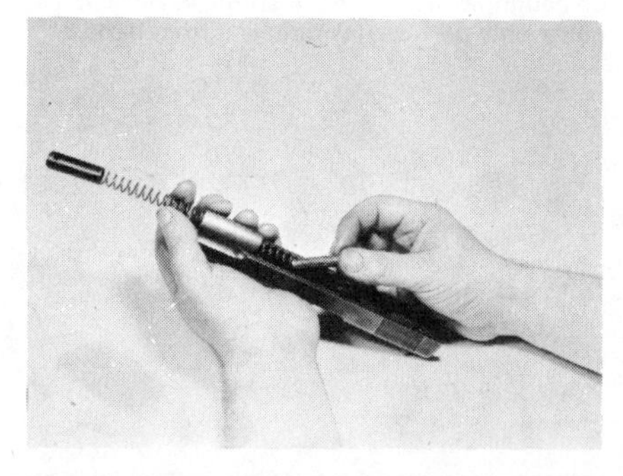

Remove recoil spring guide: (1) The recoil spring guide (on which the recoil spring compresses) may now be lifted out to the rear. (2) The recoil spring and plug are pulled out from the front. (3) The barrel bushing is turned to the left which unlocks it so it can be withdrawn.

Remove barrel: Turn barrel link forward on its pin and withdraw barrel assembly from the front of the slide. Note: Normally no further stripping of this pistol is required.

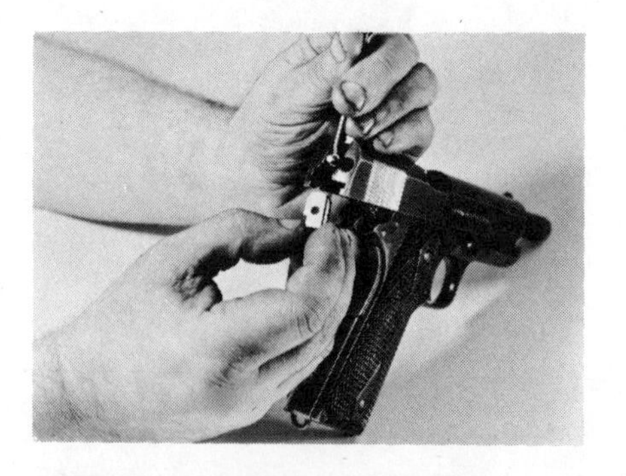

To remove firing pin: (1) Should it be necessary, the firing pin may be easily removed by pressing the pin in against the tension of its spring, at the same time pushing down on the firing pin stop which holds the firing pin in place. This may be done with a nail, match or similar object. (2) Slide the stop down out of its grooves and ease out the firing pin and spring.

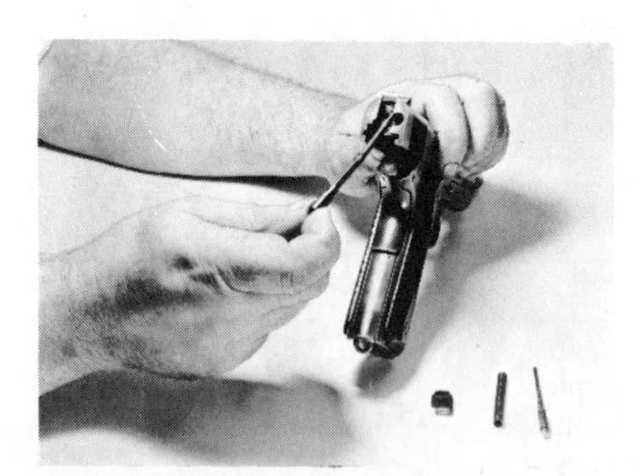

To remove extractor: When the firing pin has been removed the extractor, which is a long piece of spring steel inserted in a hole to the left of the firing pin, may be pried up and pulled out to the rear as illustrated.

SPECIAL PURPOSE HANDGUNS

The United States forces have and do use various commercial pistols for specialized purposes, such as air-crew armament, issue to general officers and issue to security personnel. Among the weapons that have been or are issued for such purposes are the Colt caliber .32 and .380 automatic pistol, the caliber .38 Colt Detective Special Revolver, the caliber .38 Colt Police Positive Revolver, the caliber .38 Colt Special Official Police and the caliber .38 Smith & Wesson Military and Police Revolver. The above revolvers are all chambered for the .38 special cartridge; Smith & Wesson Military and Police revolvers chambered for the .38 S&W cartridge have also been used. Various commercial target type caliber .22 pistols and revolvers are also used for training purposes.

Rock Island Arsenal Caliber .45 General Officer Model Pistol, M15, (top) and older Colt .32 ACP Pistol formerly issued to Generals. More details in Chapter 4.

CHARACTERISTICS OF UNITED STATES SERVICE PISTOLS

	Colt Model 1917	Smith & Wesson Model 1917	Model 1911A1
Caliber:	.45.	.45.	.45.
System of operation:	Double-action revolver.	Double-action revolver.	Recoil, semiautomatic.
Length overall:	10.8 in.	10.8 in.	8.62 in.
Barrel length:	5.5 in.	5.5 in.	5 in.
Feed device:	6-round, revolving cylinder.	6-round, revolving cylinder.	7-round, in-line, detachable box magazine.
Sights: Front:	Blade.	Blade.	Blade.
Rear:	Square notch.	"V" notch.	Square notch.
Weight:	2.5 lb.	2.25 lb.	2.43 lb.
Muzzle velocity:	830 f.p.s.	830 f.p.s.	830 f.p.s.

CURRENT STANDARD US RIFLES

THE 5.56mm M16 and M16A1 RIFLES

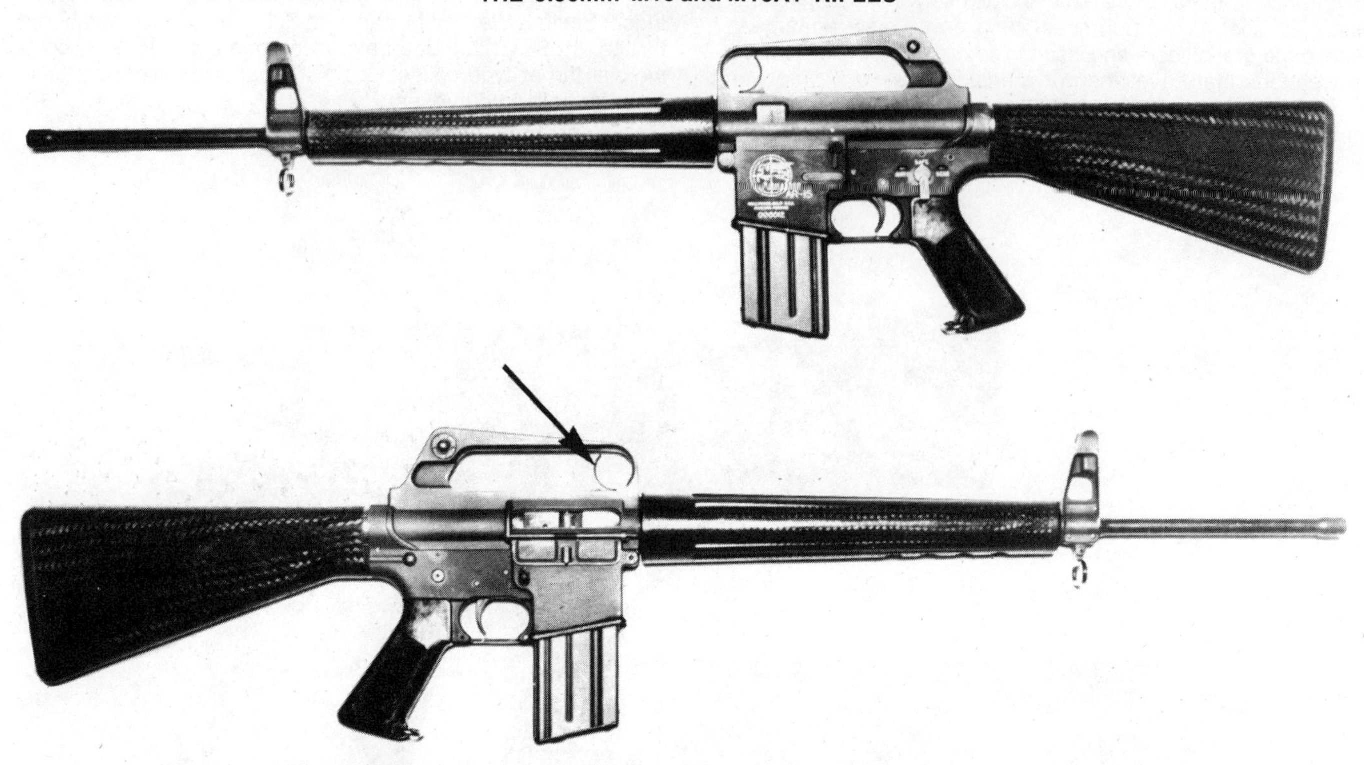

Early Armalite prototype of the AR-15 Rifle. Note the AR-10 type top mounted charging handle, the round handguard and the absence of a flash suppressor.

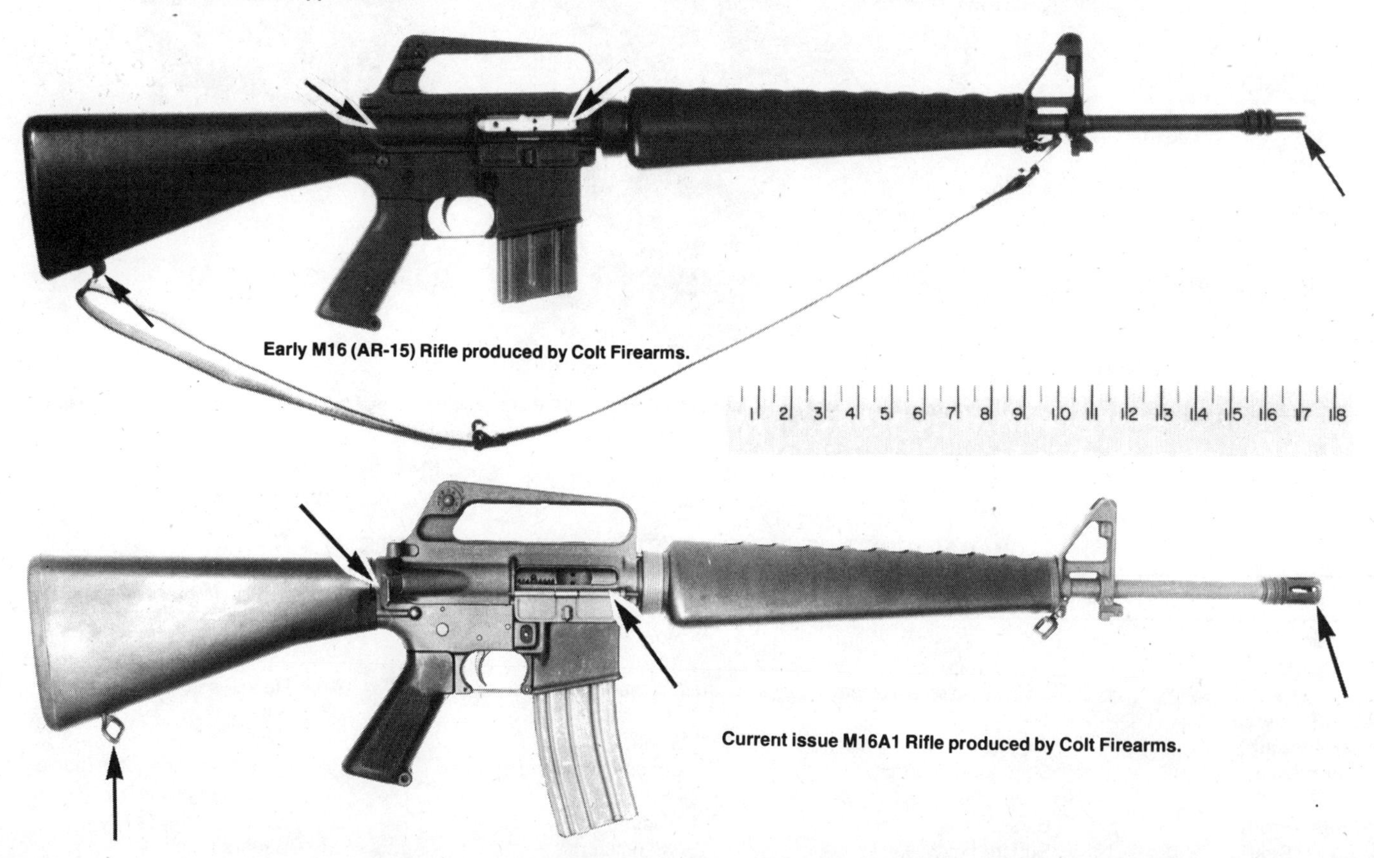

Early M16 (AR-15) Rifle produced by Colt Firearms.

Current issue M16A1 Rifle produced by Colt Firearms.

Note the differences in the above weapons: Left to right, the older models did not have a storage trap in the buttstock and they have a moveable rear sling swivel. Current models have a trap and non-moveable swivel. All M16 and AR-15 Rifles have no forward bolt assist. M16A1s have the bolt assist. Early production weapons have bright chrome plated bolts. Current production M16 and M16A1 Rifles have a dull parkerized bolt. Older weapons have the open flash suppressor, while the new ones have a closed suppressor.

Two views of the product-improved M16A1 rifle. Designated M16A1E1 during the trial period and became the M16A2 in fall of 1982. Note the heavier muzzle section on the barrel, the new handguard, and the adjustable rear sight. See Chapter 1 for more details.

Comparative length of the AK47 and Colt's M16 Carbine (Model 655) with a 14.5-inch barrel. The M16 Carbine is not a standard US issue weapon.

Colt XM177 Submachine Gun.

Colt XM177E2.

There were three variants in the XM177 collapsible stock series. The XM177 had a 10-inch barrel and the basic M16A1 receiver with forward bolt assist. The XM177E2 had an 11.5-inch barrel to permit grenade launching. Absence of the forward bolt assist was the only difference between the E1 and E2, which had that feature. Note: the flash suppressors have been determined to be noise suppressors by the US Bureau of Alcohol, Tobacco and Firearms, and as such, they are registerable and taxable items.

The XM177 series is now obsolete.

XM177E2 showing extended and retracted positions for the sliding buttstock.

CURRENT COLT VARIATIONS OF THE M16A1 THAT ARE NOT US MILITARY STANDARD

The M16A1 Carbine with 14.5 inch barrel. Note that the stock is collapsed.

The M16A1 Submachine Gun with 11-inch barrel with stock extended.

The M16A1 Heavy Barrel Assault Rifle squad automatic weapon.

Field Stripping the AR-15, M16, M16A1 Rifles

The M16 series of weapons all have the same basic components. The AR-15 Sporter will not fire automatically because it does not have the automatic fire components; the M16 and M16A1 differ mainly in that the latter has the forward bolt assist plunger on the right side of the receiver. Briefly described, the major components of the weapon are as follows.

Barrel Group. This group consists of the barrel and barrel extension, front sight assembly, flash suppressor, barrel nut and slip ring assembly and the left and right handguards. The front sight group includes the forward sling assembly, front sight and gas tube assembly and front sight post, which is adjustable for elevation. Inside the handguards are heat resisting shields.

Upper Receiver Group. This group contains the upper receiver, bolt carrier assembly, forward assist mechanism (M16A1 only), charging handle, ejection port cover assembly and mounting provisions for the barrel assembly. The top of the upper receiver takes the form of a carrying handle, which contains the rear sight and provision for mounting a telescope.

Lower Receiver and Buttstock Group. This group includes the lower receiver, pistol grip, lower receiver extension and buttstock. The lower receiver contains the trigger, fire control selector, bolt catch, disconnect, automatic sear and magazine catch. The receiver extension to which the buttstock is fastened contains the buffer assembly and return spring. Both the upper and lower receivers are machined from aluminum forgings. The buttstock and pistol grip are made of a high impact plastic material.

Bolt Carrier-Assembly. This assembly is made up of the bolt carrier, bolt, firing pin, firing pin retaining pin, cam pin, extractor and ejector. The rotary bolt locking system is one of the key mechanical features of the M16 series. Locking lugs on the bolt match up with locking recesses in the barrel extension to lock the weapon closed during the firing part of the operating cycle. The initial force of the cartridge explosion is absorbed by the barrel, barrel extension and bolt.

No special tools are required to field strip and assemble the M16. After clearing the weapon, return the bolt to the forward position, press out the rear takedown pin from the left to the right. Pull the pin until stopped by the detent. Hinge the upper receiver away from the lower receiver. Pull the charging handle to the rear (about 3 inches) and remove the bolt and bolt carrier from the upper receiver. Remove the charging handle by pulling down and to the rear.

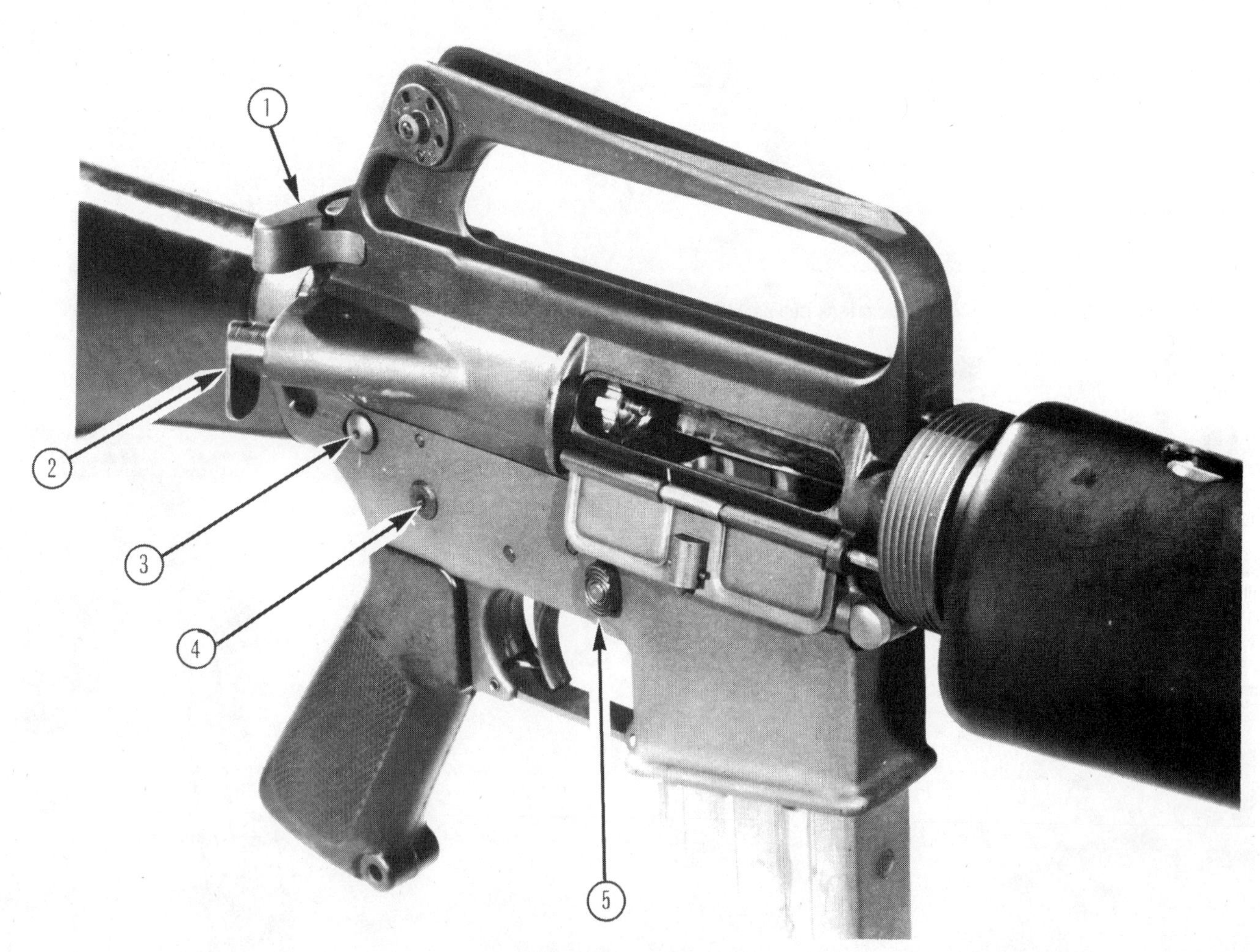

Main Features of the M16A1 Rifle

1. Charging handle
2. Forward bolt assist
3. Takedown pin
4. Selector switch pin
5. Magazine release
6. Magazine catch release
7. Bolt catch
8. Selector switch
9. Charging handle catch.

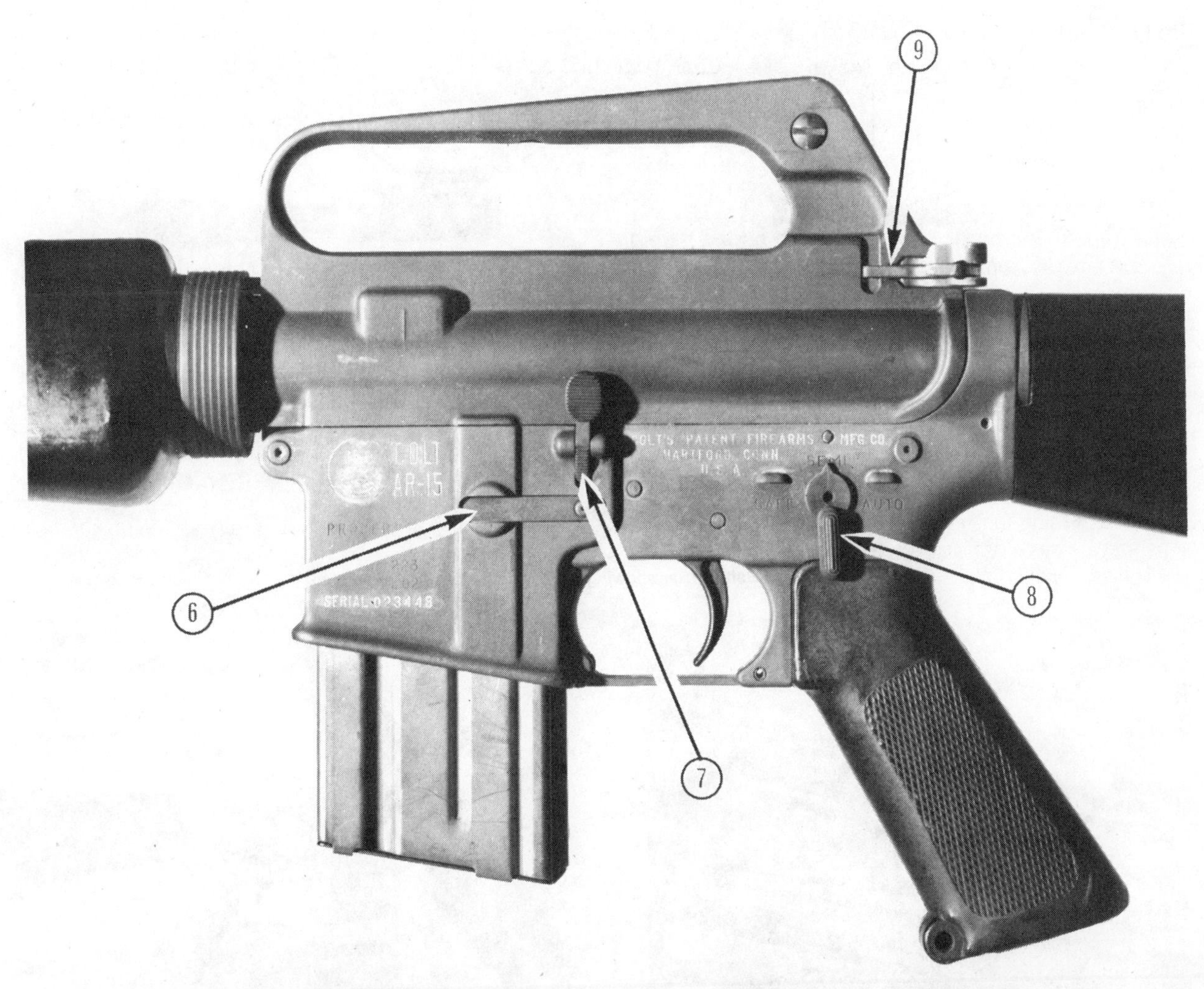

Main features of the M16A1 Rifle continued. Rifle shown is an early Model 602 made for military sales.

For more detailed stripping of the bolt, push out the firing pin retaining pin with the tip of a cartridge projectile. Drop out the firing pin. Rotate the bolt to the right until the cam pin is clear of the bolt carrier key. Rotate the cam pin one quarter turn and remove. Using the firing pin, carefully push out the extractor pin and remove the extractor. (Caution: The extractor is under spring tension.) *Do not* remove the extractor spring from the extractor.

To completely separate the upper and lower receivers, push the pivot pin to the right using the nose of a cartridge projectile. Pull out pin to detent. Separate the receivers.

To remove the buffer, push the buffer to the rear, depress the buffer retainer and slowly ease the buffer forward. (Caution: The buffer is under tension from the operating spring.) As the buffer is moved forward, depress the hammer to permit removal of the buffer and operating spring.

M16 hinged open for removal of operating parts.

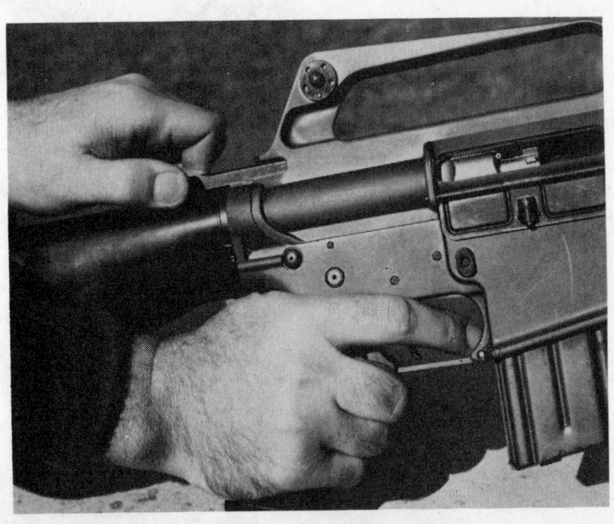

M16 Rifle with charging handle pulled half way to the rear. Note bolt showing through the ejection port.

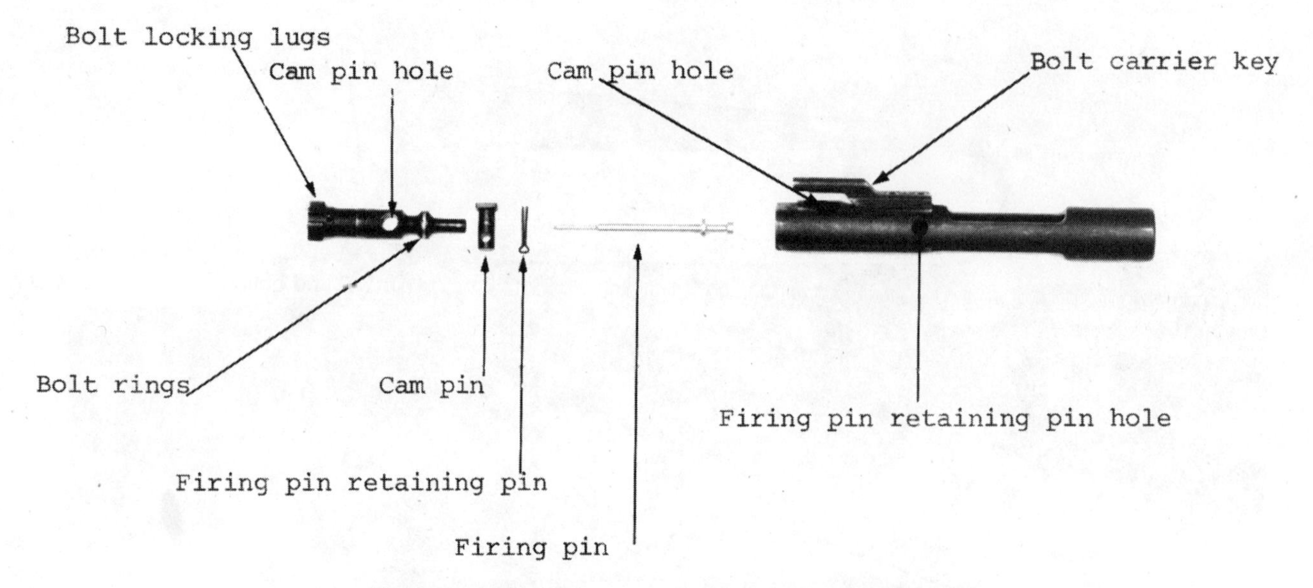

M16A1 Bolt and Bolt Carrier showing major components diassembled.

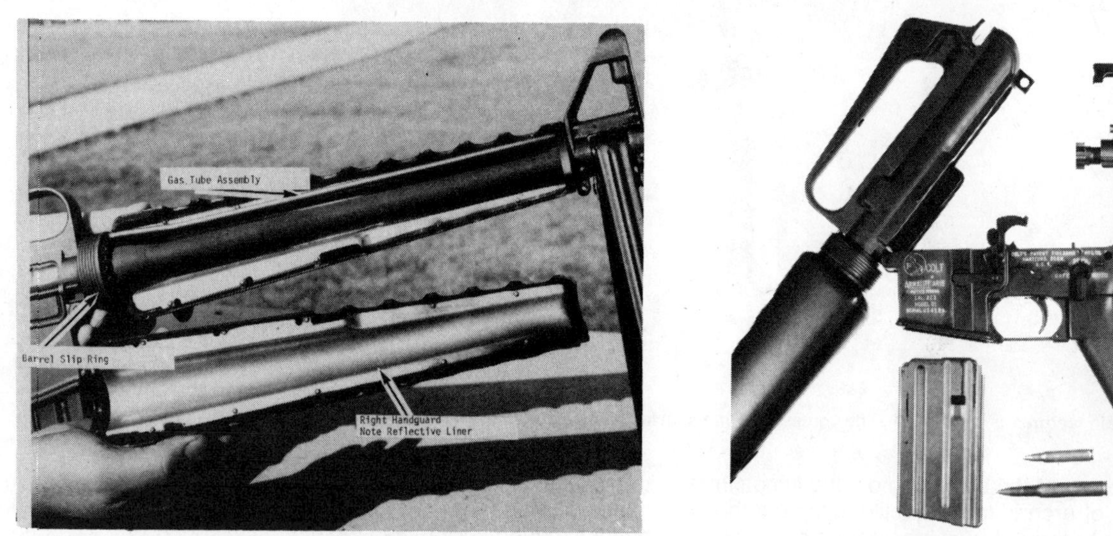

Removal of M16 Rifle Handguards.

Model 601 AR-15 field stripped.

M16A1 disassembled view.

To remove the handguards, pull the handguard slip-ring to the rear until it clears the handguards. Remove the handguards by pulling them out and down.

How the M16 and M16A1 Rifles Work

The cycle of operation of the M16 and M16A1 Rifle is described as follows.

Cocking. The rifle is cocked before firing by pulling the charger handle rearward, which pulls the bolt carrier group to the rear. As the carrier moves rearward, it cocks the hammer. If an empty magazine is installed at the time of cocking, the magazine follower will actuate the bolt catch to hold the carrier to the rear. If a loaded magazine is installed in the gun or the magazine is removed, the bolt catch must be manually operated to hold the bolt to the rear.

Feeding and Chambering. To feed a cartridge into the chamber, the bolt carrier group must be pulled to the rear by the charging handle or held there by the bolt catch. With a loaded magazine installed, the charging handle or the bolt catch is released, and the action spring drives the carrier forward. As the carrier moves forward, the lugs of the bolt pick up a cartridge from the magazine and feed it into the chamber. As the bolt locking lugs enter the barrel extension, the ejector is compressed against the left side of the cartridge head, and the extractor snaps into the extractor groove on the right side of the cartridge.

Locking. When the forward motion of the bolt and cartridge are stopped by the chamber, the bolt carrier continues forward until it is stopped by contact with the rear face of the barrel extension. This last portion of the forward travel of the carrier rotates the bolt through the action of the cam slot in the carrier on the cam pin in the bolt. This engages the bolt lugs with the barrel extension lugs to lock the bolt in battery. The bolt, when so locked, is said to be "closed."

Firing. With the fire control selector, located on the left side of the lower receiver set to either "Auto" or "Semi," the rifle may be fired. When the trigger is pulled, it causes the sear to release the hammer. The hammer spring then drives the hammer against the firing pin, which then strikes the cartridge primer to discharge the chambered round.

Unlocking. As the pressure of the gas generated by the burning propellant drives the projectile down the barrel and past the gas port, a small quantity of the gas is bled off through the gas port, gas tube and bolt carrier key into a cylindrical section in the bolt carrier where it expands and drives the bolt carrier rearward. During the first rearward travel of the carrier, the bolt is rotated by the cam pin acted on by the bolt carrier cam slot. This rotation disengages the bolt lugs from the barrel extension lugs so the bolt is unlocked. The carrier then continues rearward with the unlocked bolt.

Extraction. As the bolt is moved rearward by the carrier, the extractor, which is engaged in the extractor groove of the fired cartridge case and is pinned to the bolt, withdraws the spent case from the chamber.

Ejection. As soon as the extractor has drawn the spent case out of the chamber, the spring loaded ejector, acting against the left side of the case head, pushes the spent case out of the ejection port located on the right side of the upper receiver.

Cocking (after firing). As the carrier group continues rearward to recoil, it compresses the action spring and cocks the hammer. Two different actions now take place dependent upon whether the fire control selector is set on "Semi" (semiautomatic) or "Auto" (automatic). These actions are as follows.

"Semi" (semiautomatic). When the trigger is pulled, the firing action of the rifle is so much faster than human reaction that it would be impossible to release the trigger quickly enough to prevent several shots being fired unless there were a device provided that would limit the shots fired to one. For this reason, a disconnect is used to catch and hold the hammer until the trigger is released and pulled a second time when the fire control selector is in the semiautomatic position. When the trigger is pulled, the disconnect is rotated forward by the action of the disconnect spring. As the hammer is cocked by the recoil action of the carrier group, the hook of the disconnect engages the upper inside notch of the hammer, holding it to the rear.

When the trigger is released, the trigger spring returns the trigger to its normal position rotating the disconnect back with it. The hammer is thus released from the hook on the disconnect. However, before the disconnect hook actually releases the hammer, the trigger sear surface has moved in front of its hammer notch so that the hammer drops from the disconnect sear to the trigger sear. The rifle is then ready for a second shot.

"Auto" (automatic). When the fire control selector is set on "Auto" and the trigger pulled, the trigger sear releases the hammer. The disconnect is prevented from moving forward to engage the hammer by a cam on the fire control selector. After the first shot, as the hammer is being cocked by the recoil action of the carrier group, the notch on the top outside edge of the hammer is engaged by the automatic sear. The hammer is then held in the cocked position by the automatic sear until the bolt carrier strikes the upper edge of the automatic sear in counter-recoil, causing it to release the hammer near the end of the forward travel of the carrier. The hammer then falls to fire the next round. This cycle repeats until the magazine is emptied or the trigger is released. When the trigger is released, the hammer falls from the automatic sear but is held by the trigger sear, thus ending the cycle of automatic fire.

Buffering. The rearward or recoil movement of the carrier group is arrested by the buffer assembly acting against the bottom of the receiver extension.

Counter-recoil. After buffering, the action spring forces the carrier forward toward the chamber.

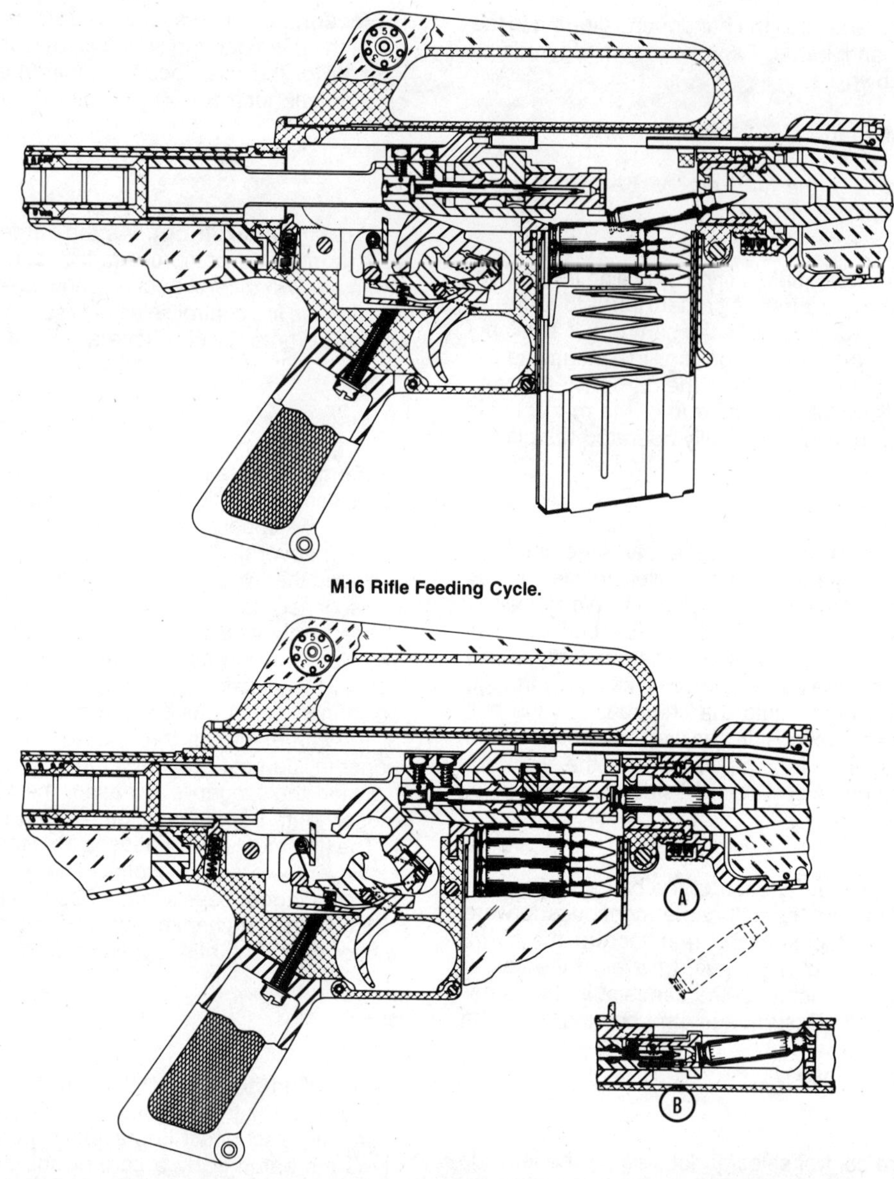

M16 Rifle Feeding Cycle.

M16 Rifle Extraction and Ejection Cycle. In Figure A, extraction begins as the bolt and bolt carrier begin their movement to the rear. Figure B illustrates the ejection process as viewed from the underside of the magazine well. The ejected cartridge moves to the right.

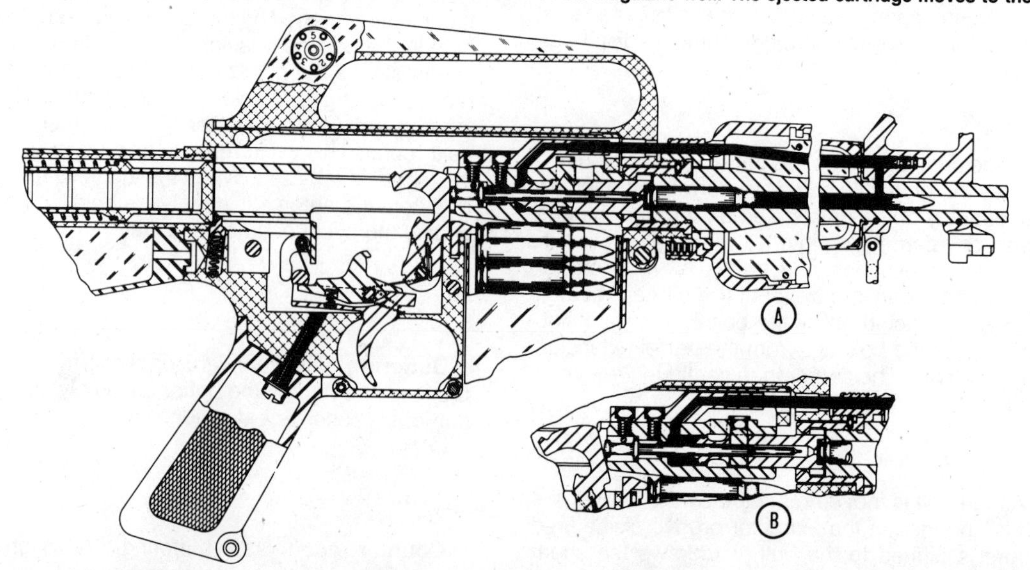

M16 Rifle Firing Cycle. In Figure A, note that the gas (indicated in black) has just begun to fill the cavity inside the bolt carrier as the projectile passes down the barrel. In Figure B, note that the expanding gas inside the bolt carrier is beginning to force the carrier to the rear.

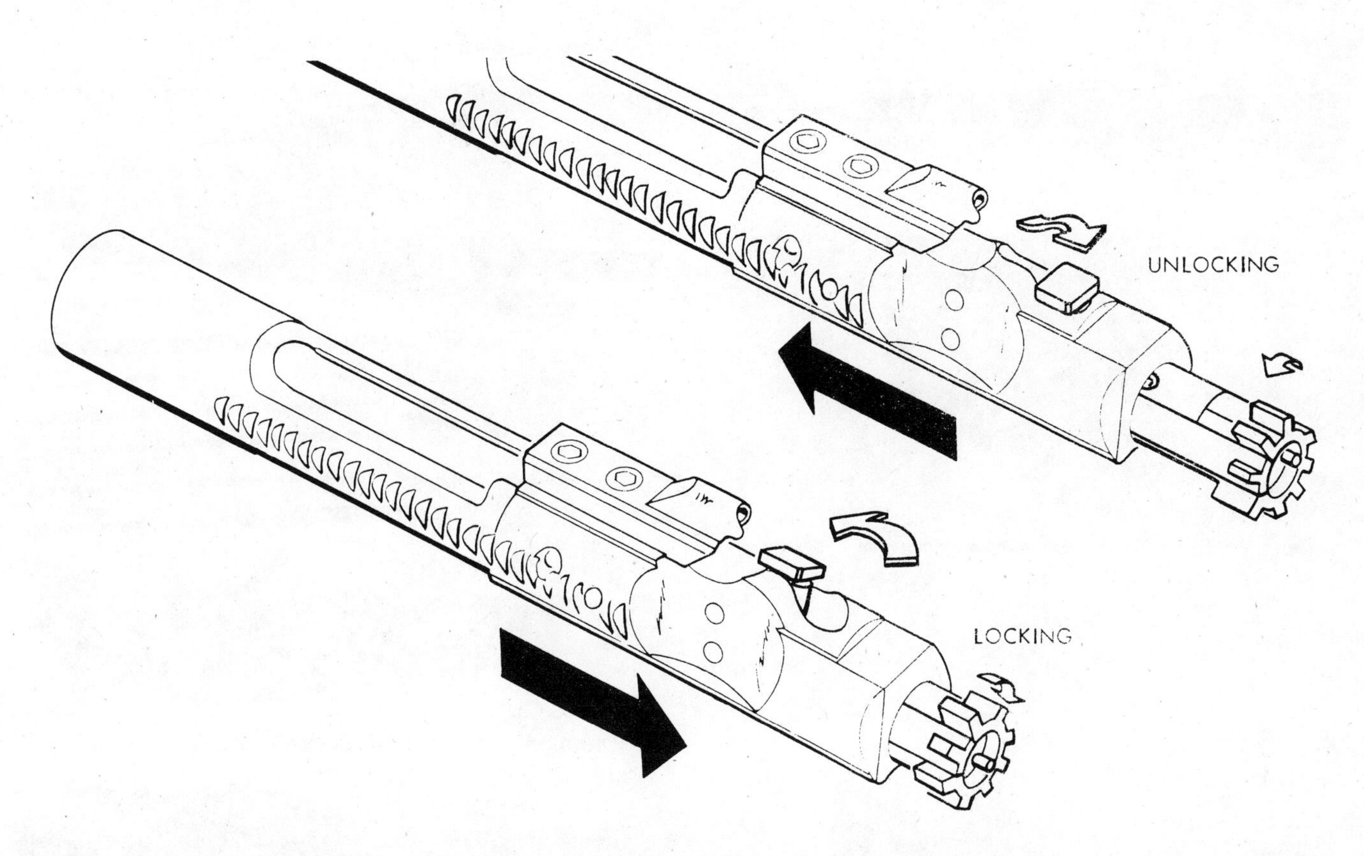

M16 Rifle bolt and bolt carrier showing the locking and unlocking movements of the bolt.

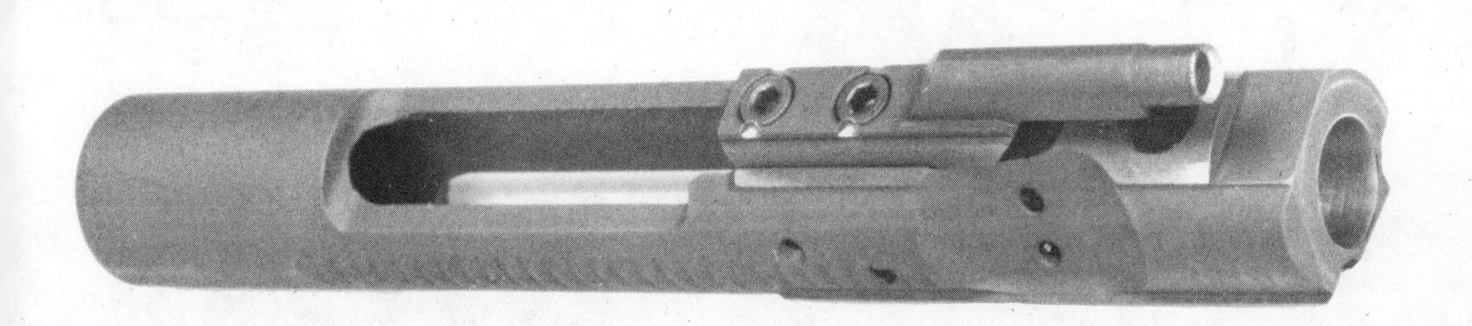

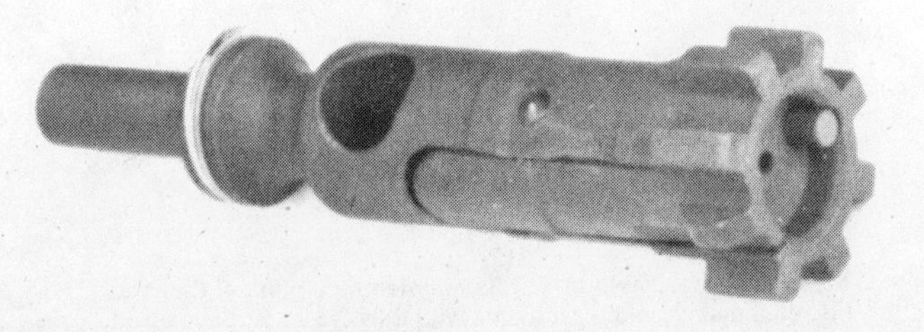

M16A1 bolt and bolt carrier assembly.

Early M16 Rifle Buttplate without storage trap.

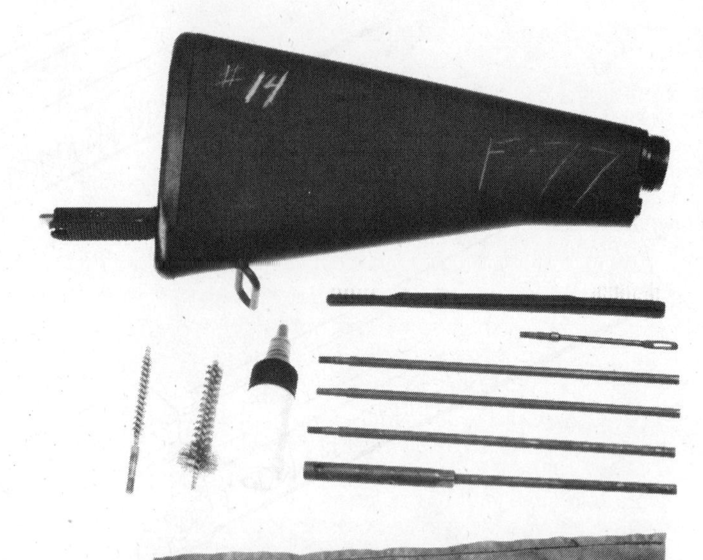

Current issue M16 Buttstock with storage for cleaning equipment.

Very early AR-15 flash suppressor. To the right the current closed suppressor and obsolete open suppressor are illustrated.

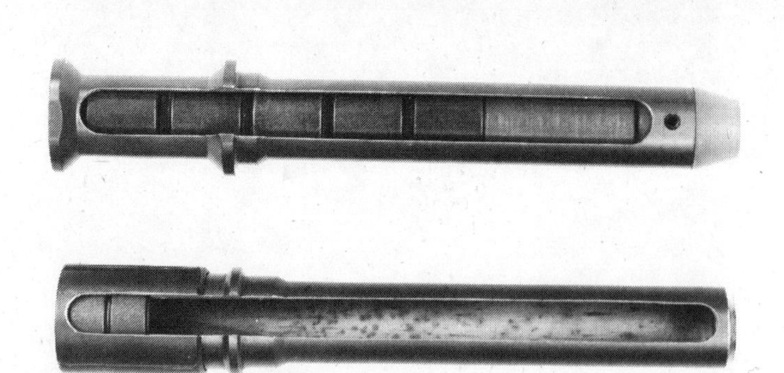

Current buffer with additional mass to slow the cyclic rate is shown on top. The obsolete original issue buffer is shown below.

To the right, the comic book style M16 Rifle maintenance manual that was issued after the initial maintenance problems with the M16 were discovered.

Maintenance

When the M16 was first introduced into service, the proper cleaning tools were not issued with it. This oversight was in part the result of publicity that indicated that the M16 series did not need to be cleaned as often as older weapons. But it was also the consequence of the rifle having been introduced without its having passed through the standard US Army adoption process. The M16s however, did get dirty, and this condition was aggravated by the use of ball type propellant issued by the Army, which left considerable residue in the gas tube, bolt carrier key and bolt carrier. The high humidity of the Southeast Asian theater of operation also led to considerable corrosion. Several major modifications were made to solve the maintenance problems.

1. A new buffer, designed by F. E. Sturtevant, to prevent bolt rebound also helped slow the rate of fire when ball type propellants were used. All M16 and M16A1 Rifles were retrofitted with this buffer.
2. A chrome plated chamber, and later chrome plated bores, reduced corrosion and hence extraction difficulties caused by rusty chambers.
3. A closed prong type flash suppressor reduced the possibility of water being attracted into the barrel. A plastic protective muzzle cap was introduced to keep dirt out of the barrel.
4. A cleaning and lubrication kit was introduced to facilitate maintenance. In addition, a new buttstock with a storage trap was fitted to the M16A1. More explicit cleaning and lubrication instructions were issued.
5. A new 30-shot magazine was introduced to replace the 20-shot feed device.

The past three decades have seen considerable improvement in small arms magazines. Whereas nearly all World War II feed devices were made from steel, Armalite and Colt have pioneered the use of aluminim magazines to reduce total weapon weight. Considerable experimentation has been aimed at producing inexpensive magazines made of synthetics. Such magazines were developed for the M14 Rifle years ago but were not reliable enough for the Army to adopt. The Soviets too have experimented with synthetic magazines for the Kalashnikov Avtomats. Recently, GAPCO of Wilson, North Carolina, introduced a much improved 30-shot plastic magazine fabricated from DuPont Zytel, a fiberglass-reinforced nylon resin. The GAPCO magazine has the advantage of being self-lubricating, and it contains a dye that will not fade, as the finish on aluminum magazines often does. This nylon magazine also demonstrates improved resistance to denting and bending. As Zytel claims higher resistance to temperature and humidity variations, the new GAPCO magazine may herald a new era in small arms magazine design.

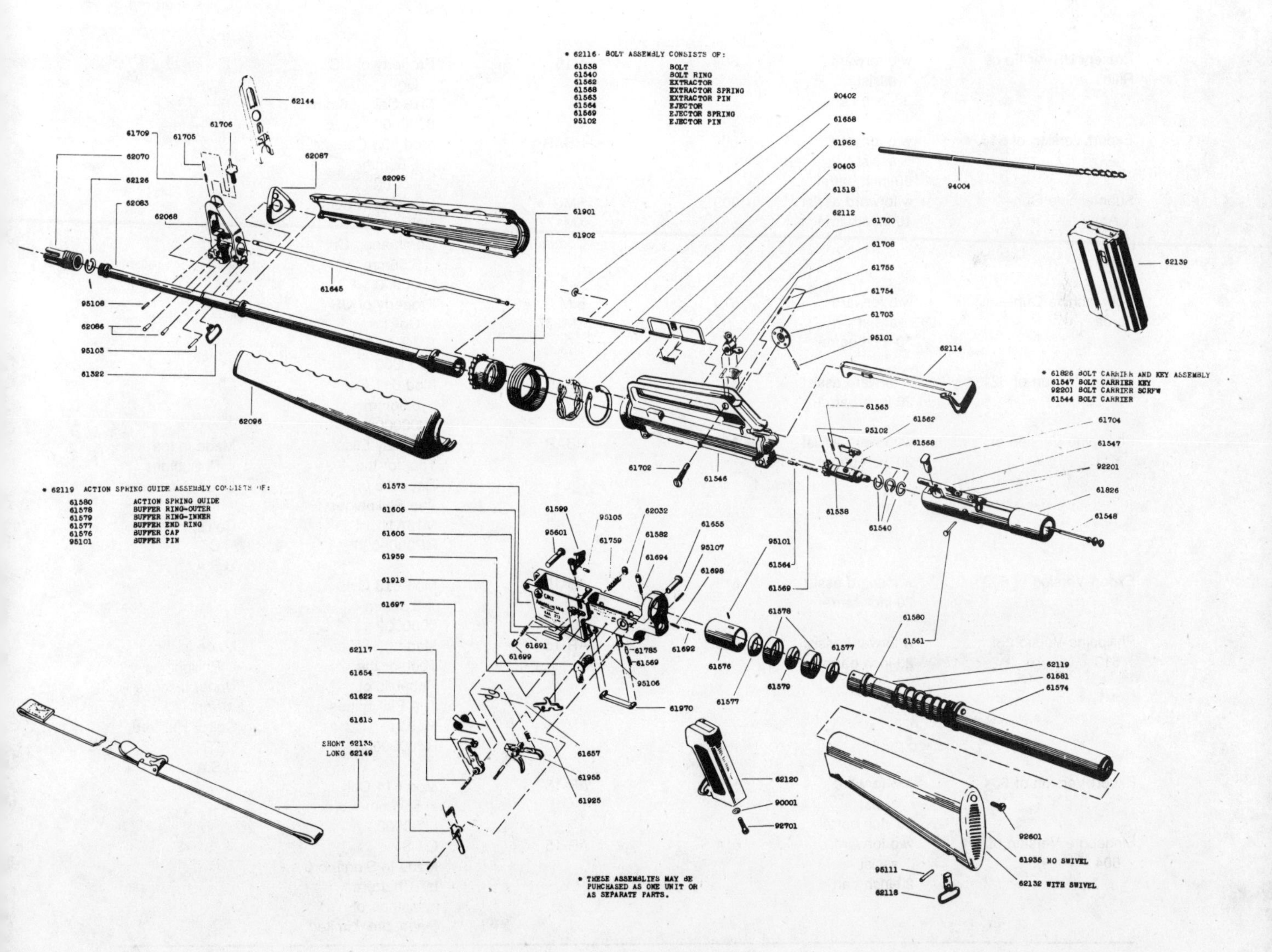

Exploded view of 5.56mm M16 Rifle.

COLT AR-15, M16, M16A1 MODEL NUMBERS

Weapon Description	Feature	Model #	Model Name	Roll Marked (left)	Roll Marked (right)
Early Armalite AR-15	w/o forward assist	NA	AR-15	ArmaLite AR-15 Costa Mesa, Calif. USA Patents Pending 0000000	
Early Military Colt Version	w/o forward assist	601	AR-15	Colt Armalite AR15 Patents Pending Cal. .223 Model 01 Serial 0000000	
Early US Government Purchase	w/o forward assist	602	AR-15	Colt AR-15 Property of US Gov't Cal. .223 Model 02 Serial 0000000	
Present US Army Rifle	w/forward assist 20-inch barrel	603	M16A1	Property of US Gov't M16A1 Cal 5.56mm 0000000	
Korean Version of 603	w/forward assist 20-inch barrel	603K	M16A1	M16A1 K0000000	Made in Korea Under License from Colt's, Hartford, CT. U.S.A.
Present US Air Force Rifle	w/o forward assist 20-inch barrel	604	M16	Property of US Gov't M16 Cal 5.56mm 0000000	
Export Version of 616	w/o forward assist 20-inch barrel	606	HBAR	Mod 606 Cal 5.56mm 0000000	
Submachine Gun—Army	w/forward assist 10-inch barrel	609	SMG #1 XM177E1 Commando	Property of US Gov't Commando Cal 5.56mm 0000000	
Submachine Gun—Air Force	w/o forward assist 10-inch barrel	610	S.M.G. #2 XM177	Property of US Gov't SMG 5.56mm 0000000	
Export Version of 621	w/forward assist 20-inch barrel	611	HBAR	Mod 611 Cal 5.56mm 0000000	
Philippine Version of 611	w/forward assist 20-inch barrel	611P	HBAR	Made by Elisco Tool for the Republic of The Philippines M16A1 RP0000000	Made in the Philippines Under License from Colt's, Hartford, CT U.S.A.
Export Version of 603	w/forward assist 20-inch barrel	613	AR-15	Mod 613 Cal 5.56mm 0000000	
Philippine Version of 613	w/forward assist 20-inch barrel	613P	AR-15	Made by Elisco Tool for the Republic of The Philippines M16A1 RP0000000	Made in the Philippines Under License from Colt's, Hartford, CT. U.S.A.
Export Version of 604	w/o forward assist 20-inch barrel	614	AR-15	Mod 614 Cal 5.56mm 0000000	
Singapore Version of 604	w/o forward assist 20-inch barrel	614-S	AR-15	C I S Made in Singapore by Chartered Industries of Singapore Limited	

COLT AR-15, M16, M16A1 MODEL NUMBERS

Weapon Description	Feature	Model #	Model Name	Roll Marked (left)	Roll Marked (right)
				Under License from Colt Industries Hartford, Conn, USA Patented	
Heavy Barrel Auto Rifle US Government	w/o forward assist 20-inch barrel	616	HBAR	Property of US Gov't Mod 616 Cal 5.56mm 0000000	
Export Version of 609	w/forward assist 10-inch barrel	619	S.M.G.	Mod 619 Cal 5.56mm 0000000	
Export Version of 610	w/o forward assist 10-inch barrel	620	S.M.G.	Mod 620 Cal 5.56mm 0000000	
Heavy Barrel Auto Rifle US Government	w/forward assist 20-inch barrel	621	HBAR	Property of US Gov't Mod 621 Cal 5.56mm 0000000	
Submachine Gun—Army	w/forward assist 11.5-inch barrel	629	XM177E2	Property of US Gov't XM177E2 Cal 5.56mm 0000000	
Submachine Gun—Air Force	w/o forward assist ?-inch barrel	630	Not available		
Export Version of 629	w/forward assist 11.5-inch barrel	639	S.M.G.	Mod 639 Cal 5.56mm 0000000	
Export Version of 630	w/o forward assist	640	S.M.G.	Mod 640 Cal 5.56mm 0000000	
Submachine Gun—Air Force	w/o forward assist ?-inch barrel	649	S.M.G.	Property of US Gov't GAU-5/A/A Cal 5.56mm 0000000	
Colt Carbine	w/forward assist 14.5-inch barrel	651			
Colt Carbine	w/o forward assist 14.5-inch barrel	652			
Colt Carbine	w/forward assist w/sliding buttstock 14.5-inch barrel	653			
Philippine Version of 653	w/forward assist w/sliding buttstock 14.5-inch barrel	653P		Made by Elisco Tool for the Republic of The Philippines M16A1 RP0000000	Made in the Philippines Under License from Colt's, Hartford, CT. U.S.A.
Colt Carbine	w/o forward assist w/sliding buttstock 14.5-inch barrel	654			
Colt Carbine	w/forward assist standard stock 14.5-inch barrel	655			
Colt Automatic Rifle	Closed-Gas System M16 Variant 20-inch barrel	CAR-703 M16A2 (unofficial)			
Semiautomatic Commercial Sporter	w/o forward assist 20-inch barrel	R6000	AR-15	Colt AR-15 Cal. .223 Mode SP1 Ser. SP00000	

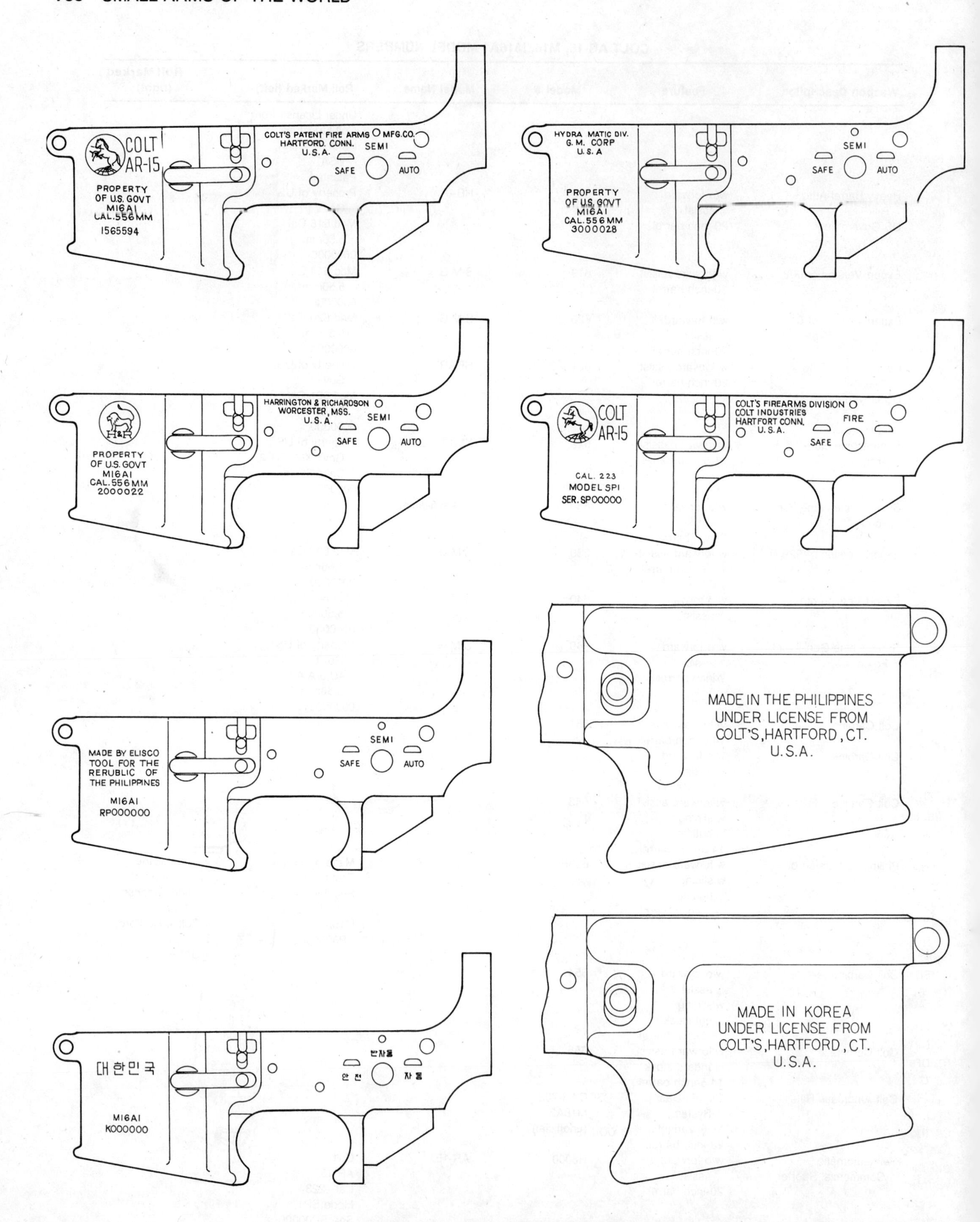

Representative Receiver Markings of the M16 Rifle.

Close-up view of the breech area of an early experimental M16A1E1 with burst control and rear sight adjustable for windage.

Conversion Kit, M261 Caliber .22 Rimfire Adapter for the M16 Series

The appearance of 5.56mm (.223) caliber rifles has made it possible to use inexpensive .22 rimfire ammunition for training. This in turn reduces costs, increases the amount of training and familiarization and permits the firing of the M16 Rifle on indoor ranges. Several organizations have developed .22 caliber conversion units for the M16—Colt Firearms, Military Armaments, US Armament Corporation and Rock Island Small Arms System Laboratory.

All the conversion units embody the same basic principle of replacing the standard bolt with a blow-back .22 caliber mechanism that has a forward extension that looks like the case of the standard 5.56 × 45mm cartridge.

To install the conversion unit, remove the standard bolt as outlined under the M16 above. Then insert the rimfire adapter unit in its place. The rifle is cocked and loaded in the standard manner. The 10-shot magazine adapter fits inside of the standard 30-shot magazine as shown in the illustration.

Caution: When the bolt adapter is not being used with the rifle, NEVER pull the bolt of the unit to the rear and load a cartridge into the chamber of the adapter. If the bolt is allowed to fly forward, the chambered cartridge could explode and thus either damage the adapter or injure personnel in the immediate vicinity.

Two variations of the M261 conversion kit are being manufactured—one for the US Army and one for the US Air Force—by SACO Defense Systems Division of Maremont Corporation.

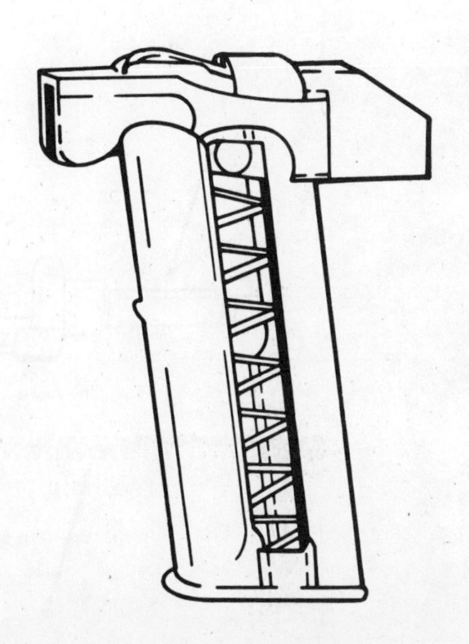

Ten-shot adapter which is inserted into the standard 30-shot magazine.

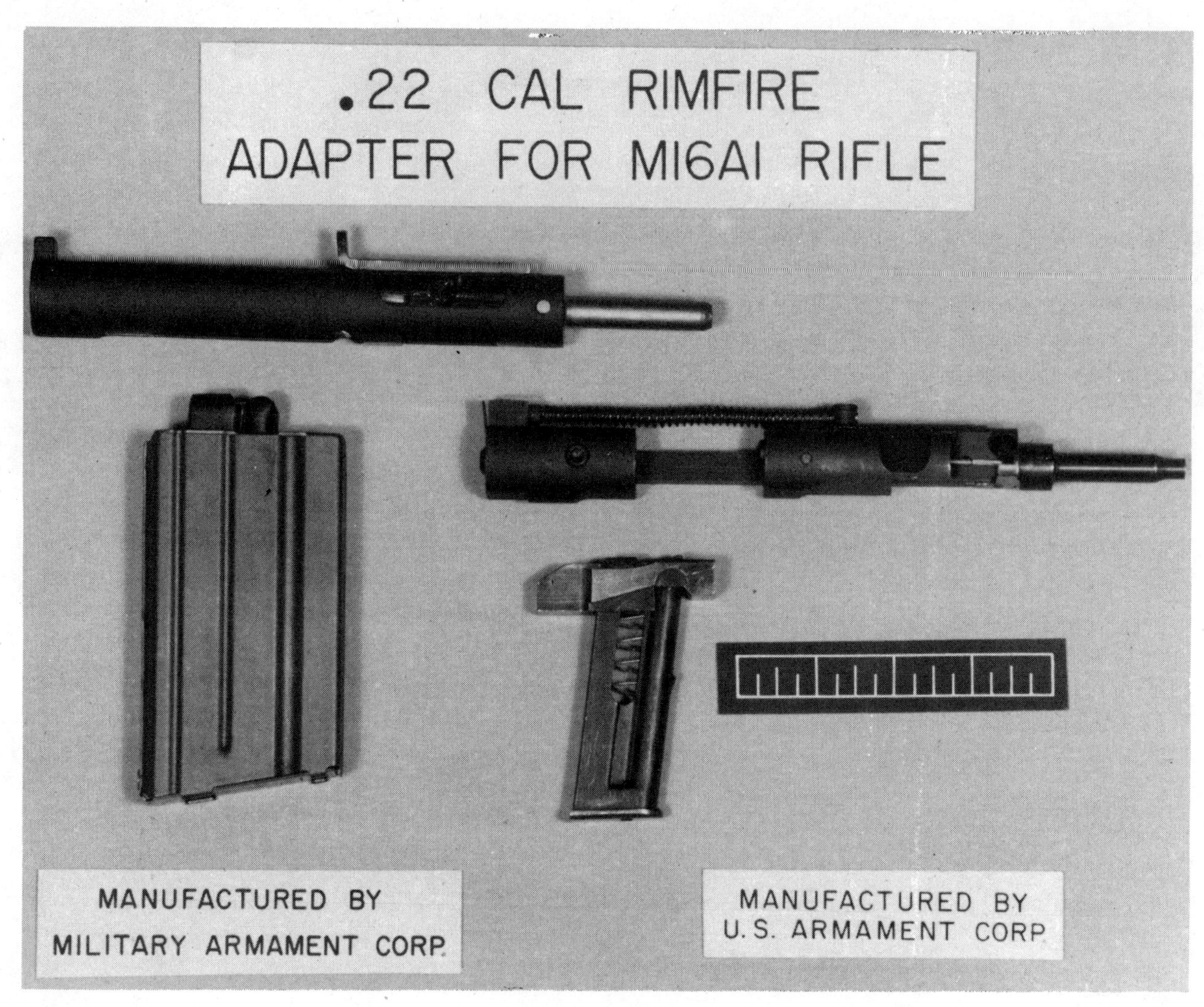

.22 Caliber Rimfire Adapter Units for the M16A1. The M261 is on the right.

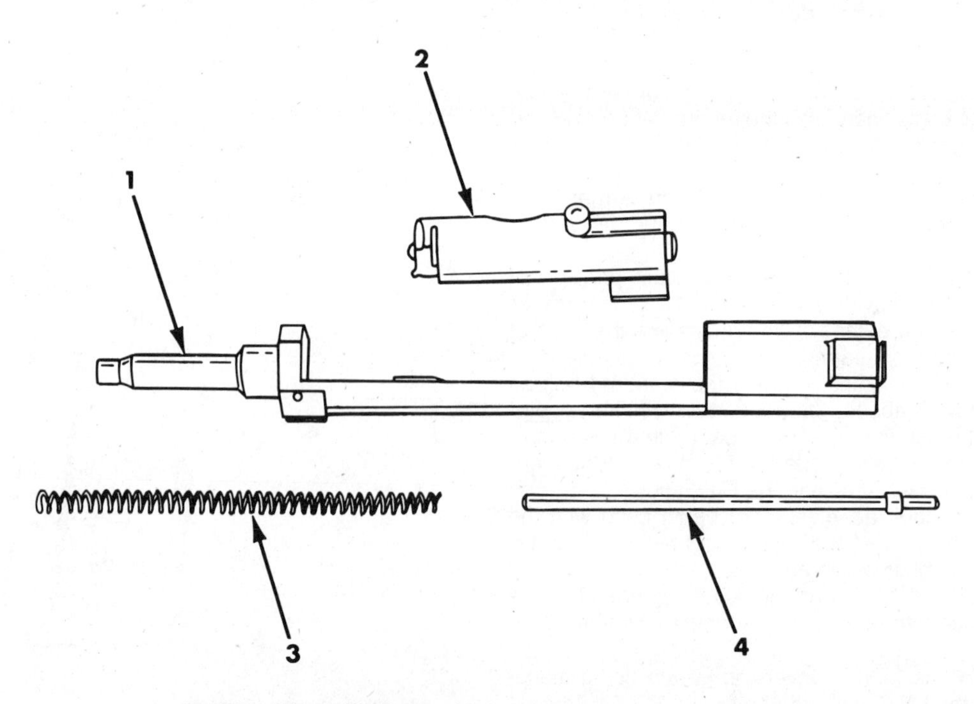

The M261 Adapter Kit disassembled.
1. Barrel and receiver group. 2. Bolt. 3. Guide rod spring. 4. Guide rod.

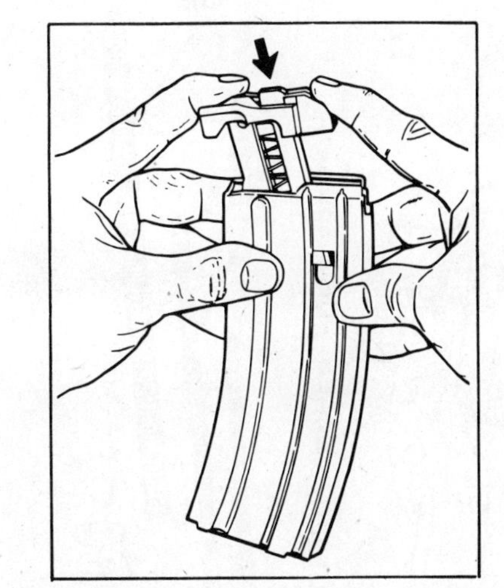

1. POSITION AND DEPRESS MAGAZINE ADAPTER INTO MAGAZINE.

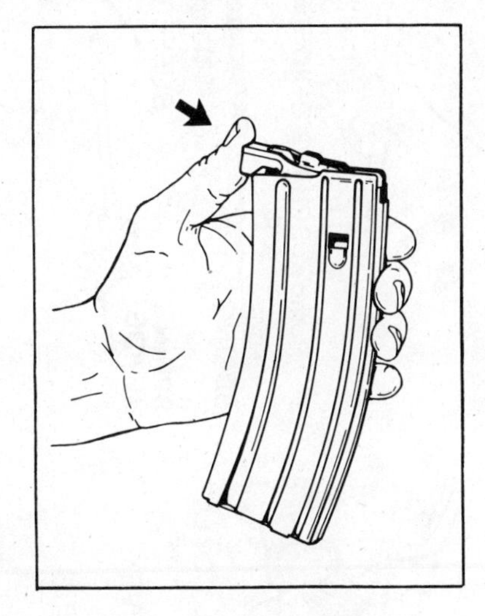

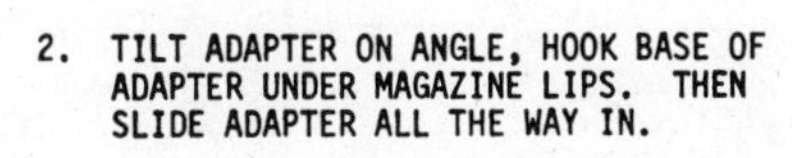

2. TILT ADAPTER ON ANGLE, HOOK BASE OF ADAPTER UNDER MAGAZINE LIPS. THEN SLIDE ADAPTER ALL THE WAY IN.

Installation of the 10-shot magazine adapter into the standard 30-shot magazine.

M231 FIRING PORT WEAPON FOR THE M2 INFANTRY FIGHTING VEHICLE

This new weapon is described in detail in Chapter 5. The following illustrations identify the unique features of the M231.

INSTALLING AND REMOVING THE WEAPON

WARNING
ALWAYS REMOVE MAGAZINE AND BE SURE WEAPON IS CLEARED BEFORE INSTALLING OR REMOVING FROM VEHICLE.

INSTALLING

1. With weapon upright, insert barrel thru firing port hole until BARREL COLLAR engages threads of firing port and PIN GROOVE aligns with QUICK RELEASE PIN.
2. 2. Rotate weapon one full turn (360°) clockwise until QUICK RELEASE PIN locks into place.

REMOVING

1. Pull out on mount QUICK RELEASE PIN and rotate weapon one full turn (360 degrees) counterclockwise so QUICK RELEASE PIN alines with pin groove.
2. Remove weapon.

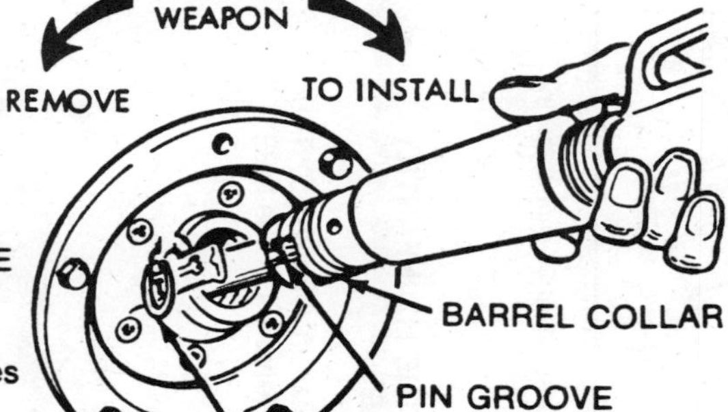

FIELD STRIPPING

DRIVE SPRING AND GUIDE ASSEMBLY

WARNING
Be sure WEAPON is clear and BOLT forward.

1. With BOLT forward, unscrew the SPRING GUIDE counterclockwise from the RECEIVER EXTENSION and remove.

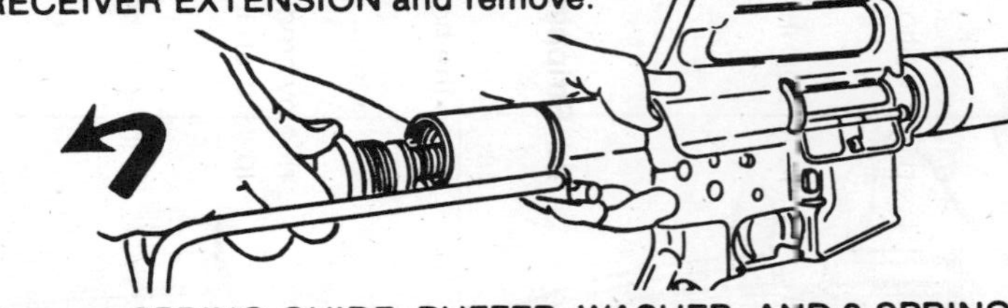

2. Remove SPRING GUIDE, BUFFER, WASHER, AND 3 SPRINGS.

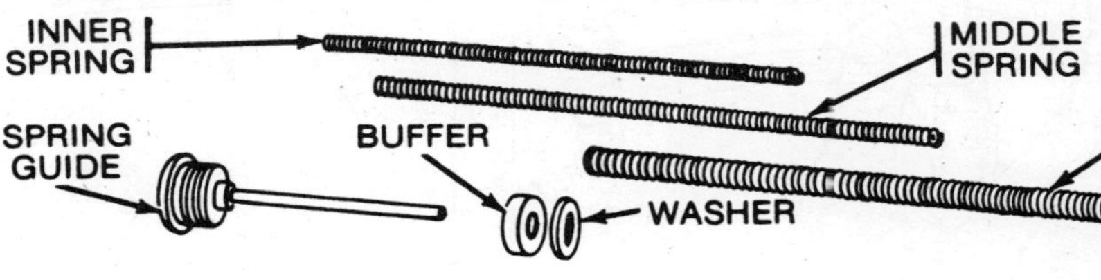

MORE FIELD STRIPPING

BUTTSTOCK LATCH

1. Depress BUTTSTOCK LATCH and pull BUTTSTOCK out to stop.

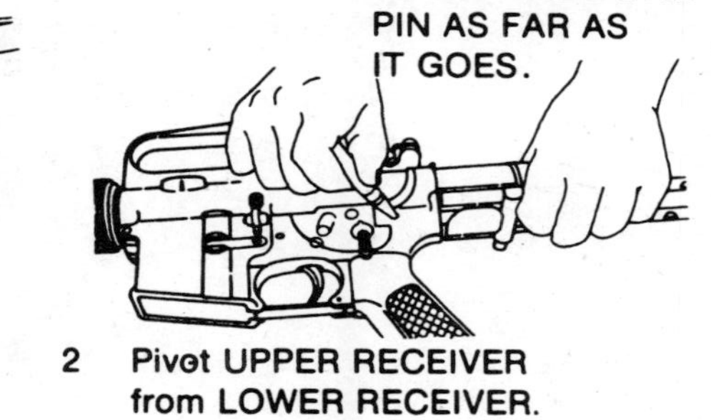

2. Pivot UPPER RECEIVER from LOWER RECEIVER.

3. Push PIVOT PIN as far as it goes.

4. Separate UPPER and LOWER RECEIVERS.

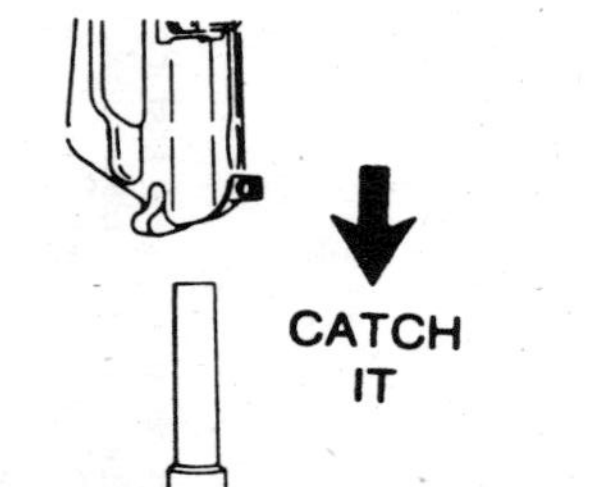

5. Point BARREL up and remove STRIKER.

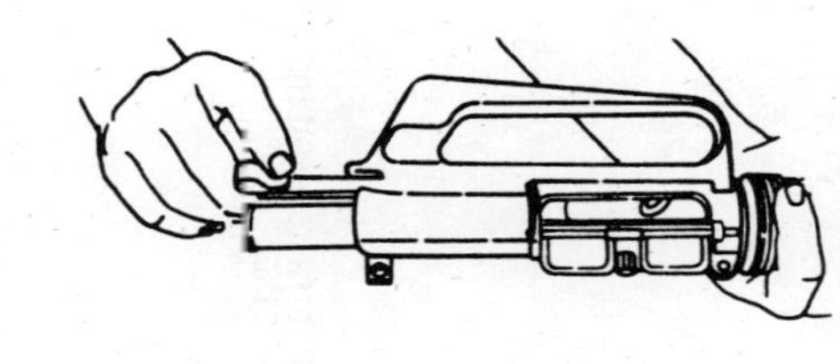

6. Pull back CHARGING HANDLE and BOLT CARRIER.

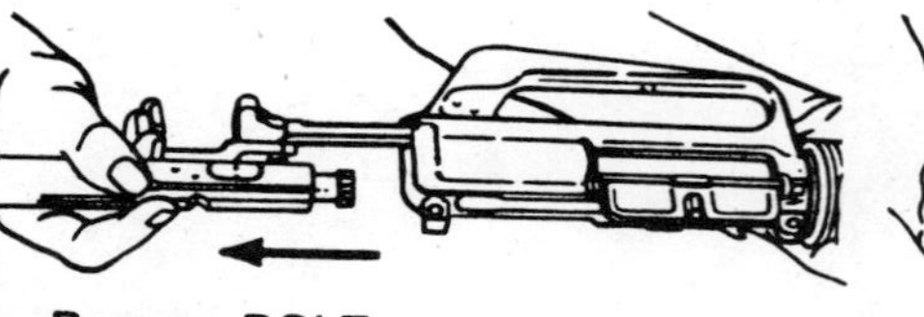
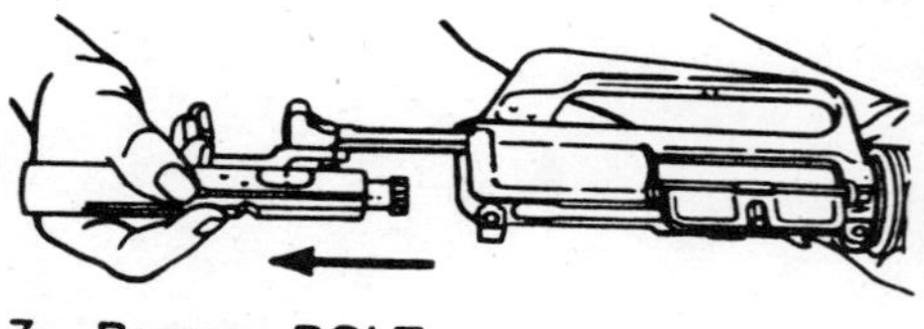

7. Remove BOLT CARRIER.

PULL BACK AND DOWN

8. Remove CHARGING HANDLE.

THE 7.62mm NATO M14 RIFLE

The M14 is a standard rifle of the United States Army. It was produced at Harrington and Richardson Arms Co., Thompson Products (TRW), at the New Haven (Winchester) plant of the Winchester-Western Arms Division of Olin Mathieson Corp. and at Springfield Armory. The M14 is capable of automatic as well as semiautomatic fire, and a certain proportion fitted with bipods serve as squad automatic weapons. The M14 rifle is no longer in production.

The M14 is an evolution of the M1 rifle; in the design of the M14 many of the shortcomings of the M1 have been eradicated. The basic action of the M1 remains, but the troublesome eight-round en bloc clip has gone. The hanging of the gas cylinder on the end of the M1 rifle's barrel gave some accuracy difficulties; these have been overcome in the M14 by moving the gas port and gas cylinder back about eight inches from the muzzle. The gas cutoff and expansion system used on the M14 lends itself to better accuracy because its action is not as abrupt as that of the M1. Various other changes were made to give the Army a basically better weapon than the M1.

How to Load and Fire the M14

Application of Safety. Place the safety in the safe position by cocking the hammer and snapping the safety rearward.

Loading of Rifle. Place the safety in the safe position.

Insert a loaded magazine into the magazine well, front end leading, until the front catch snaps into engagement; then pull backward and upward until the magazine latch snaps into position.

Pull the operating rod handle to its rearmost position and release; this allows the top round to rise and the bolt to move forward, thus stripping and chambering a round from the magazine.

Semiautomatic Fire with Selector Lock. With the selector lock in the rifle, it cannot be fired automatically. Load the rifle and release the safety. The rifle will now fire one round upon each pull of the trigger.

Semiautomatic Fire with Selector. Press in and turn the selector until it snaps into position with its blank face to the rear and its projection downward. The connector assembly is inoperative in this position since the connector is held forward and out of engagement with the operating rod.

Load the rifle and release the safety. The rifle will now fire one round upon each pull of the trigger.

Full Automatic Fire with Selector. Press in and turn the selector until it snaps into position with the face marked "A" to the rear and the projection upward. This rotation of the eccentric selector shaft moves the sear release to the rear into contact with the sear and moves the connector assembly rearward into contact with the operating rod.

Load the rifle and release the safety.

Pull and hold the trigger. The rifle will fire automatically as long as the trigger is squeezed and there is ammunition in the magazine. To cease firing, release the trigger.

Bolt Lock. When the last round of ammunition is fired, the magazine follower engages the bolt lock and raises it into the path of the retracted bolt; this holds the bolt in the open position.

Unloading the Rifle. Place the safety in the safe position.

Grasp the magazine, placing the thumb on the magazine latch, and squeeze the latch. Push the magazine forward and downward to disengage it from the front catch and remove the magazine from the magazine well.

Pull the operating handle rearward to extract and eject a chambered round, and to inspect the chamber. The rifle is now clear.

Gas Shutoff Valve. For semiautomatic and full automatic firing, turn the valve to the open position by pressing in and rotating. The valve is open when the slot in the head of the valve spindle is perpendicular to the barrel. The shutoff valve in the gas cylinder opens and closes the port in the cylinder between the barrel and the gas piston.

Stripping the M14

General Disassembly (Field Stripping). Unload the rifle, remove the magazine, and place the safety in the safe position.

Turn the rifle upside down with the muzzle pointing to the left.

Insert the nose of a cartridge into the hole in the trigger guard and pry upward to unlatch the trigger guard.

Swing the trigger guard upward and lift the trigger group from the stock.

Separate the stock from the rifle by cradling the receiver firmly in one hand and by striking upward sharply on the stock butt with the palm of the other hand.

Turn the barrel and receiver group on its side with the connector assembly upward. Press in and turn the selector until the face marked "A" is toward the rear sight knob and the projection forward (this step applies to rifles modified for selective firing.) Press forward on the connector with the right thumb until the forward end can be lifted off the connector lock. Rotate the connector clockwise, until the slot at the rear end is aligned with the elongated stud on the rear release. Lower slightly the front end of the connector and lift it from the sear release. **Note:** The connector assembly is a semipermanent assembly; it should not be disassembled.

With the barrel and receiver group upside down, pull forward on the operating rod spring, relieving pressure on the connector lock. Pull the lock outward, disconnect the operating rod spring guide, and remove the spring guide and spring. Turn the barrel and receiver group right side up.

Retract the operating rod until the key on its lower surface coincides with the dismount notch in the receiver. Lift the operating rod free and pull to the rear, disengaging it from the operating rod guide.

Grasp the bolt group by the roller and, while sliding it forward, lift it upward and outward to the right front with a slight rotating motion. This rifle is now field stripped, and basic assemblies such as the bolt and the trigger groups may be disassembled, if required.

Bolt Group. With the bolt in the left hand and the thumb over the ejector, insert the blade of a screwdriver between the extractor and the lower cartridge seat flange. Pry the extractor upward to unseat it. The ejector will snap out against the thumb. Lift out the ejector assembly, extractor plunger and spring. Remove the firing pin from the rear of the bolt. **Note:** No attempt should be made to disassemble the roller from the bolt stud.

Barrel and Receiver Group. Disassemble the rear sight as follows. Run the aperture all the way down and record the reading for use in reassembling the sight. Hold the elevating knob and unscrew the nut in the center of the windage knob. Withdraw the elevating knob. Unscrew and remove the windage knob. Pull the aperture up about one-half inch. Place the thumb under the aperture and push upward and forward to remove the aperture, cover and base. Separate the rear sight cover from the rear sight base.

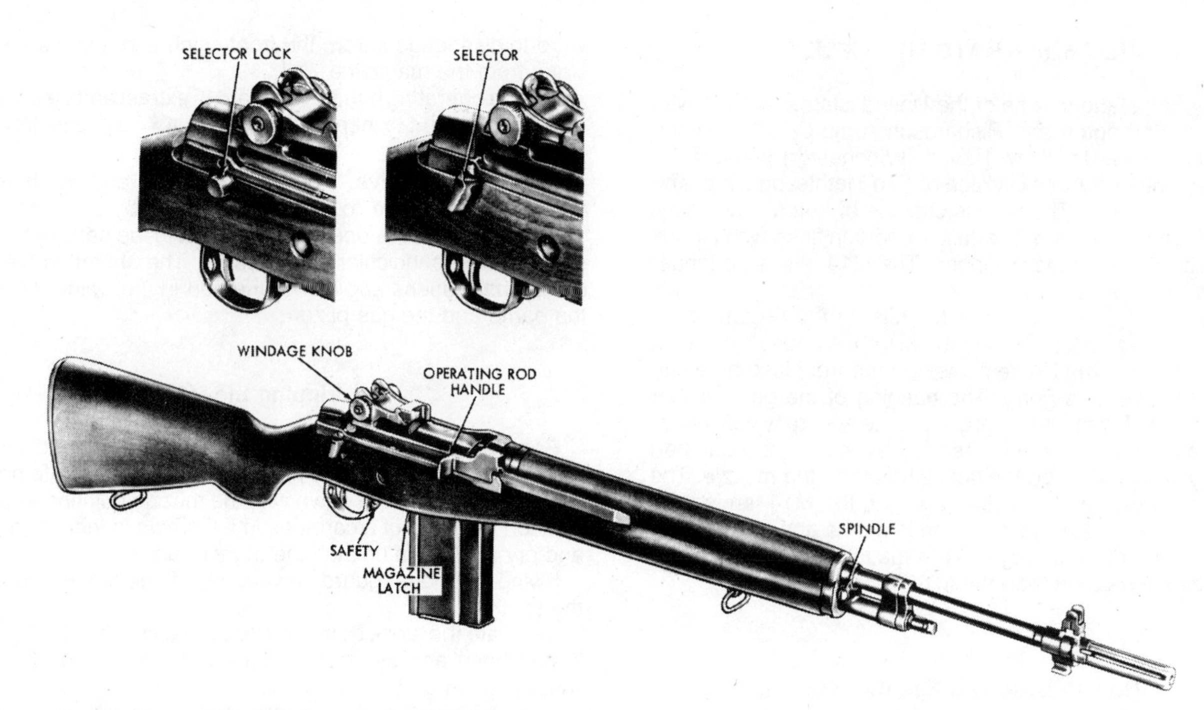

The M14 Rifle as initially produced at Springfield Armory with wood stock and wood handguard. Top left drawing shows blank selector lock, which prevents the rifle from being fired automatically. Top right drawing shows the selector switch that may be substituted to permit selection of fire mode.

Intermediate production version of the M14 Rifle equipped with wood stock, shoulder rest buttplate and ventilated fiberglass reinforced handguard.

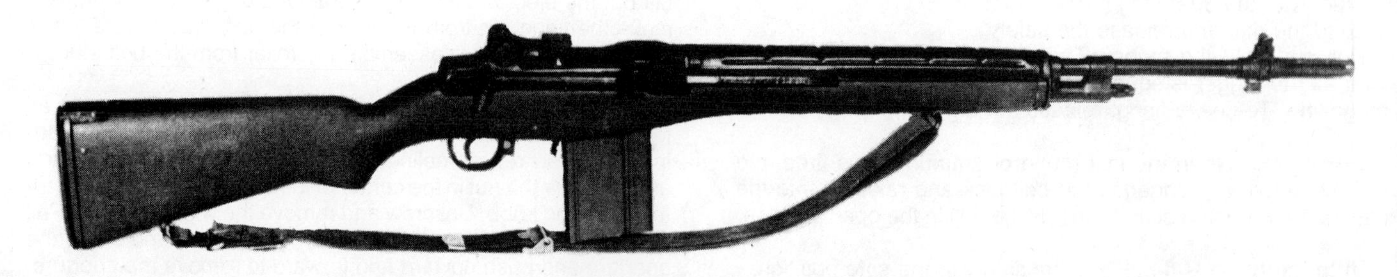

Late production version of the M14 Rifle with fiberglass reinforced synthetic stock and unventilated handguard. M14 Rifles in the field are encountered with both wooden and synthetic stocks.

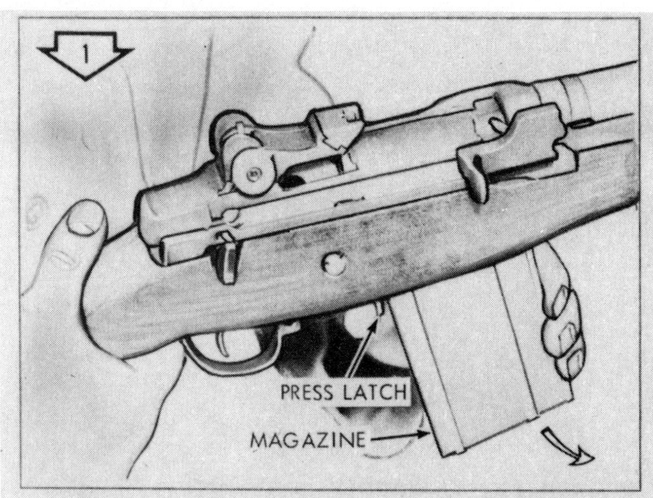

REMOVE MAGAZINE.

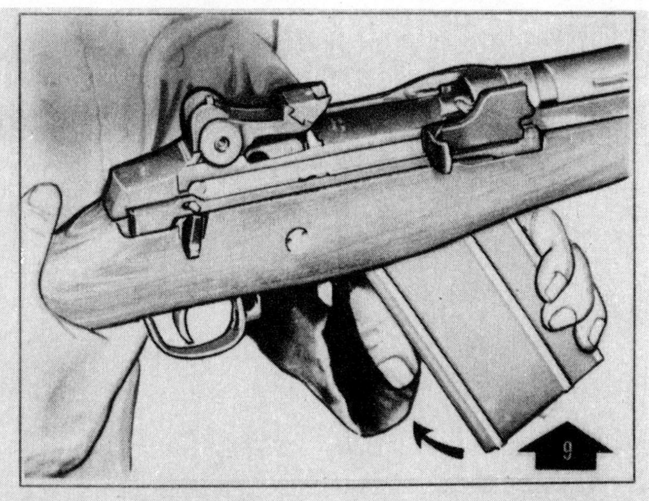

INSTALL MAGAZINE

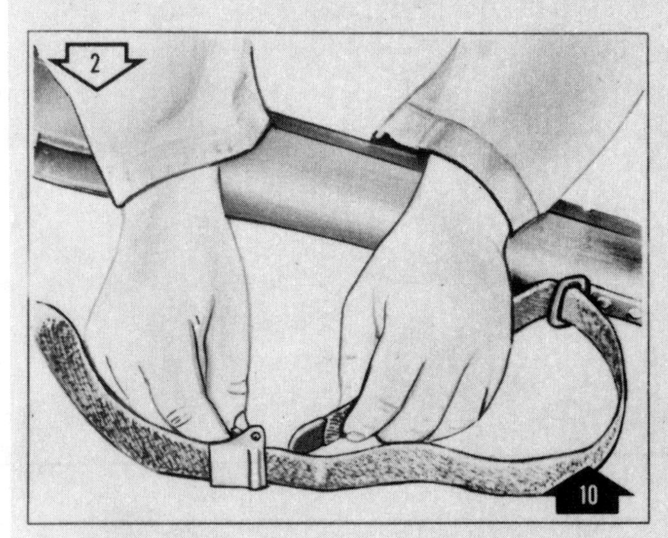

REMOVE/INSTALL SLING.

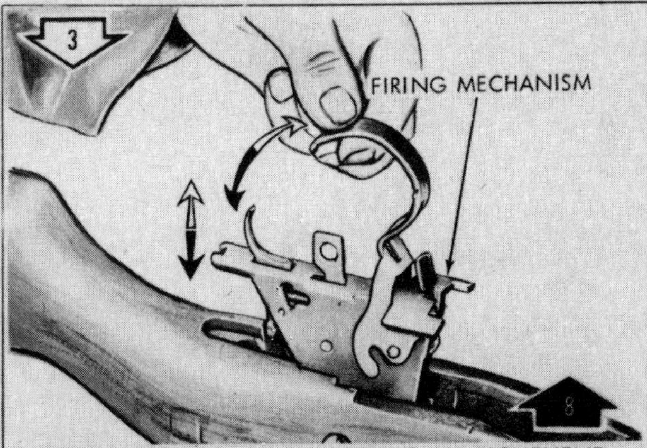

CAUTION: OVER 90 DEGREES ROTATIONAL MOVEMENT, TOWARDS THE MUZZLE, CAN BE FELT WHEN THE COCKING STUD OF THE TRIGGER GUARD ENGAGES POINT AT BASE OF HAMMER, WORKING AGAINST HAMMER SPRING TENSION. THE FIRING MECHANISM SHOULD BE REMOVED BEFORE THIS POSITION IS REACHED. PARTIAL WITHDRAWAL OF FIRING MECHANISM COMBINED WITH THIS ADDED MOVEMENT WILL CAUSE DAMAGE TO THE RIB OR KEEPWAYS ON SIDE OF FIRING MECHANISM HOUSING. THIS WILL RESULT IN DIFFICULT INSTALLATION AND REMOVAL OF FIRING MECHANISM.

REMOVE/INSTALL FIRING MECHANISM.

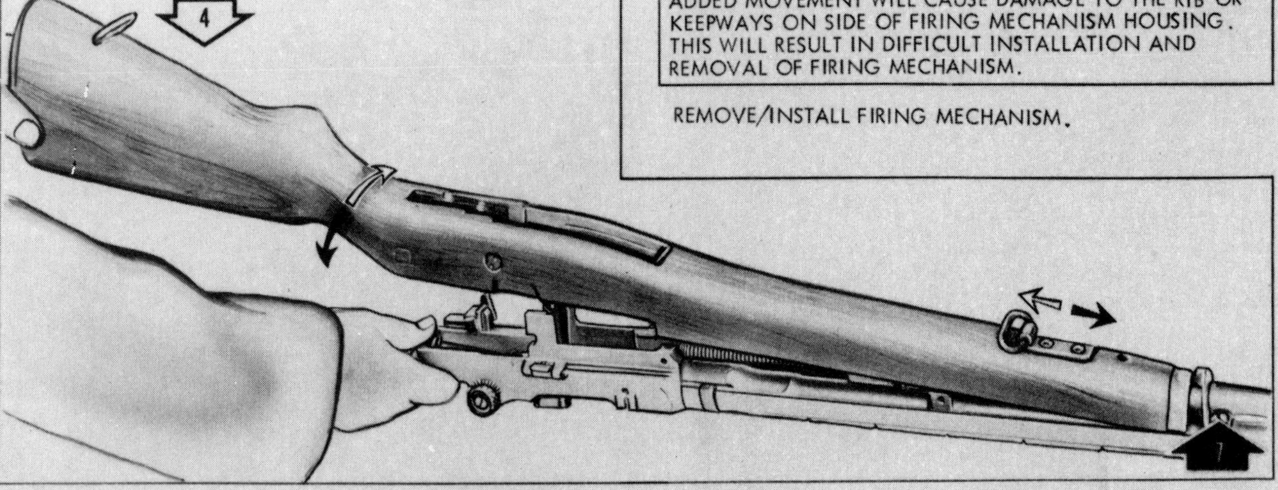

REMOVE/INSTALL STOCK WITH BUTT PLATE ASSEMBLY.

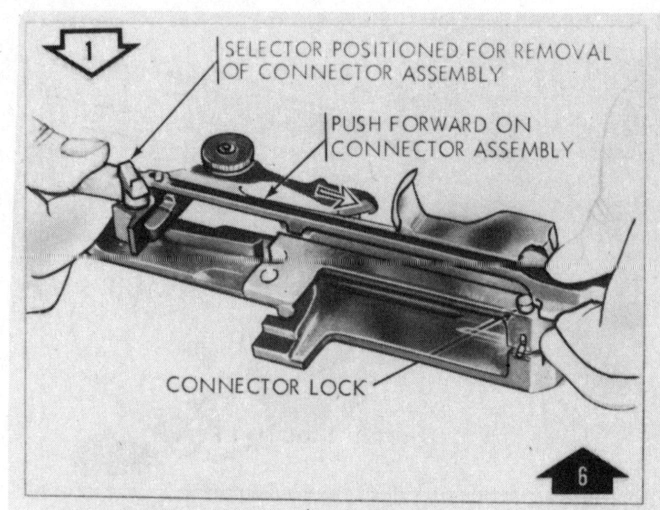

DISENGAGING/ENGAGING CONNECTOR ASSEMBLY.

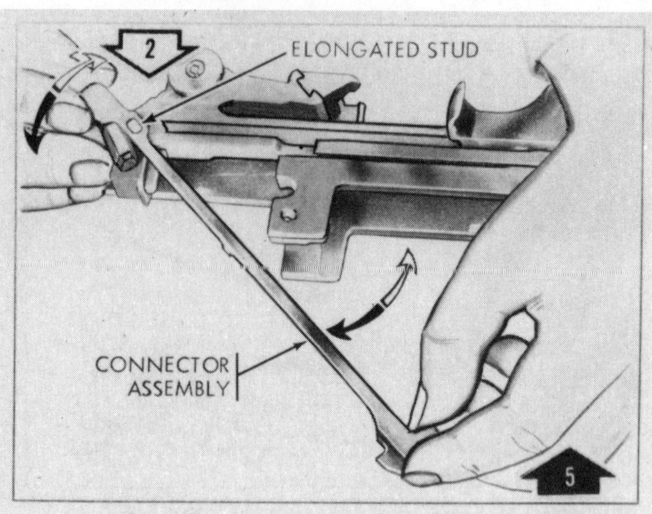

REMOVE/INSTALL CONNECTOR ASSEMBLY.

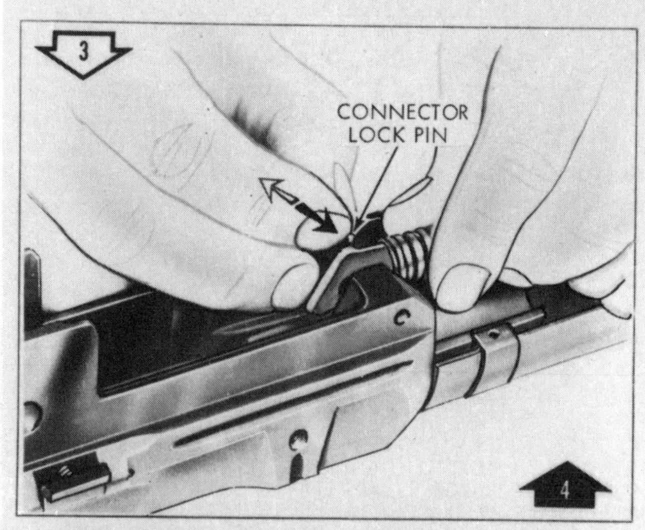

DISENGAGE/ENGAGE CONNECTOR LOCK.

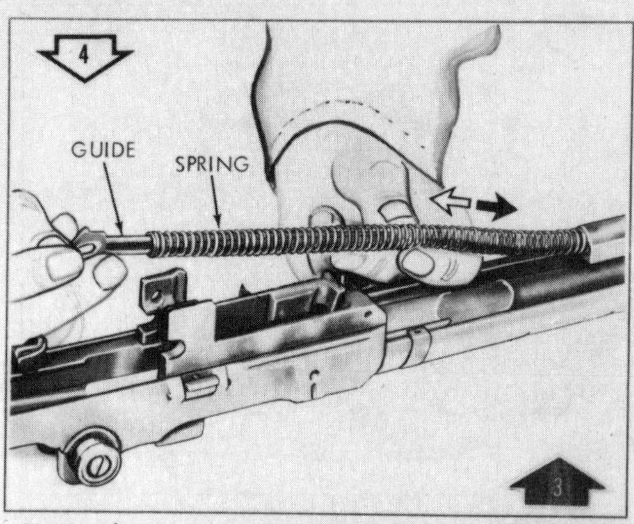

REMOVE/INSTALL OPERATING ROD SPRING GUIDE AND OPERATING ROD SPRING.

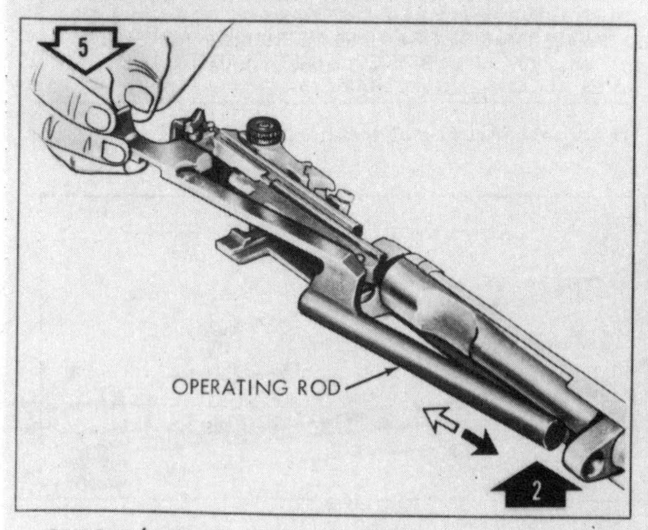

REMOVE/INSTALL OPERATING ROD.

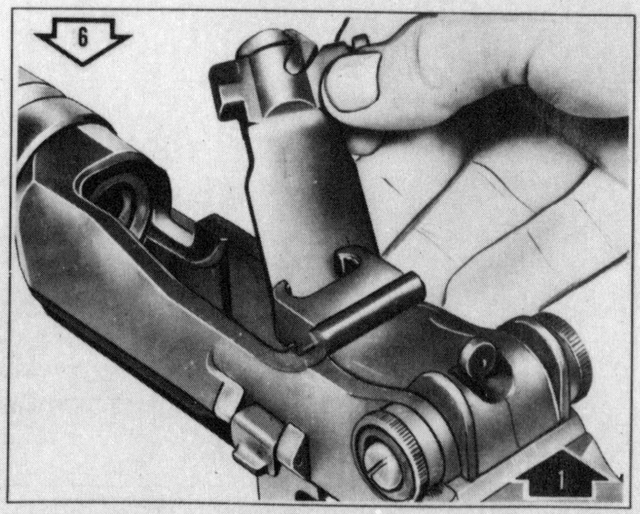

REMOVE/INSTALL BOLT ASSEMBLY.

Loosen the setscrew in the base of the front sight lug on the flash suppressor. Unscrew the flash suppressor nut and slide the flash suppressor forward off the barrel.

Loosen and remove the gas plug, using the gas cylinder plug wrench. Tilt the muzzle down and remove the gas piston from the gas cylinder. Unscrew the gas cylinder lock and slide the lock and the gas cylinder off the barrel.

Slip the front band off the barrel. Push the handguard forward and lift it from the barrel.

Trigger Group. To disassemble the trigger group, close and latch the trigger guard. Squeeze the trigger, allowing the hammer to go forward. Hold the trigger housing group with the first finger of the right hand on the trigger and the thumb against the sear. Place the front of the trigger housing against a firm surface. Squeeze the trigger with the finger and push forward on the sear with the thumb. At the same time, using the tip of a cartridge, push out the trigger pin from left to right. Slowly release the pressure with the finger and thumb; this allows the hammer spring to expand.

Lift out the trigger assembly. Remove and separate the hammer spring plunger, hammer spring, and the hammer spring housing.

Push out the hammer pin from left to right, using the tip of a cartridge. Move the hammer slightly to the rear and lift out.

Unlatch the trigger guard. Push out the stud of the safety from its hole. Remove the safety and safety spring. Slide the trigger guard to the rear until the wings of the trigger guard are aligned with the safety stud hole. Rotate the trigger guard to the right and upward until the hammer stop clears the base of the housing. Remove the trigger guard.

Drive out the magazine latch spring pin with a suitable drift to remove the semipermanently assembled magazine latch and spring.

Assembly of M14. To assemble the rifle, reverse the disassembly procedure. However, the following instructions are provided to facilitate and to insure satisfactory assembly:

To assemble the hand guard, when the gas cylinder and related components are in place, position the front end of the guard in the front band and snap the rear band of the handguard assembly into the barrel grooves. **Note:** The handguard need not be reassembled prior to assembly of the gas cylinder and related components.

To assemble the trigger group to the stock and receiver, cock the hammer and swing the trigger guard to the open position. Insert the assembly into the receiver, and close the trigger guard.

To assemble the gas system, replace front band, gas cylinder and gas cylinder lock. Tighten the lock by hand to its full assembled position and then "back off" until the loop is aligned with the gas cylinder. Assemble the piston and the gas cylinder plug.

Special Note on the M14 Rifle

Although the M14 is a selective-fire weapon, most weapons in the hands of troops will have their selectors locked in the semiautomatic position. When desired, these weapons can be made to deliver selective fire by the removal of the selector lock. This feature has been added to the weapon since combat experience with the M2 carbine and troop tests with earlier prototypes of the M14 indicate that troops keep selective-fire weapons set on full automatic as a matter of course. This limits the effectiveness of the weapon at long ranges, since it is effective in off-hand automatic fire only at ranges up to about 100 yards. It also results in a great expenditure of ammunition with little in the way of results to show for this expenditure. Those weapons equipped with bipods for use as squad automatic weapons will not have their selectors locked and will be capable of selective fire at all times. Production M14s have aluminum butt plates with shoulder support and plastic handguards. Production of the M14 rifle ceased in 1964.

Variations of the M14 Rifle

There have been a number of variations of the M14 rifle produced. Two of these variations have steel folding stocks, one of which folds to the side similar to the M1A1 carbine stock—the Type V—and the other folds under the weapon in a manner similar to the stock of the German MP40 submachine gun and the Soviet AK assault rifle—the Type III.

The M14A1. The M14A1 is a variation of the M14 produced for use as a squad automatic weapon. It was originally developed by the United States Army Infantry Board, Fort Benning, Georgia. Springfield Armory made various changes in the design to ease manufacture and maintenance. The M14A1 has a straight-line stock design with full pistol grip and folding forward handgrip. A compensator, which helps to keep the barrel down in automatic fire, is fitted over the flash suppressor. The stock has a rubber recoil pad and folding shoulder rest, and the M2 bipod has been modified by the addition of a sling swivel and a longer pivot pin. The Browning Automatic Rifle sling is used on this rifle. The selector lever is found on all M14A1 rifles so that they may be used for automatic or semiautomatic fire.

M14 National Match Rifle. A match version of the M14 rifle for use at the National Matches was developed as the result of a requirement set down in 1959. The M14 National Match Rifle cannot be fired full automatic; it has a hooded aperture rear sight, special sight parts, selected barrel and glass bedded action similar to the National Match Rifle.

The M14M Rifle. The M14M rifle was intended for issue to NRA affiliated rifle clubs, for sale through the Director of Civilian Marksmanship. This rifle was modified by welding the selector shaft and lock to eliminate automatic fire capability. The "M" in this rifle's designation stands for "Modified Service." Only a very few M14Ms were fabricated, and their distribution was equally limited.

The M21 Rifle. The M21 is the sniper version of the M14 rifle. It uses a Leatherwood type variable power scope. See Chapter 5 for further details.

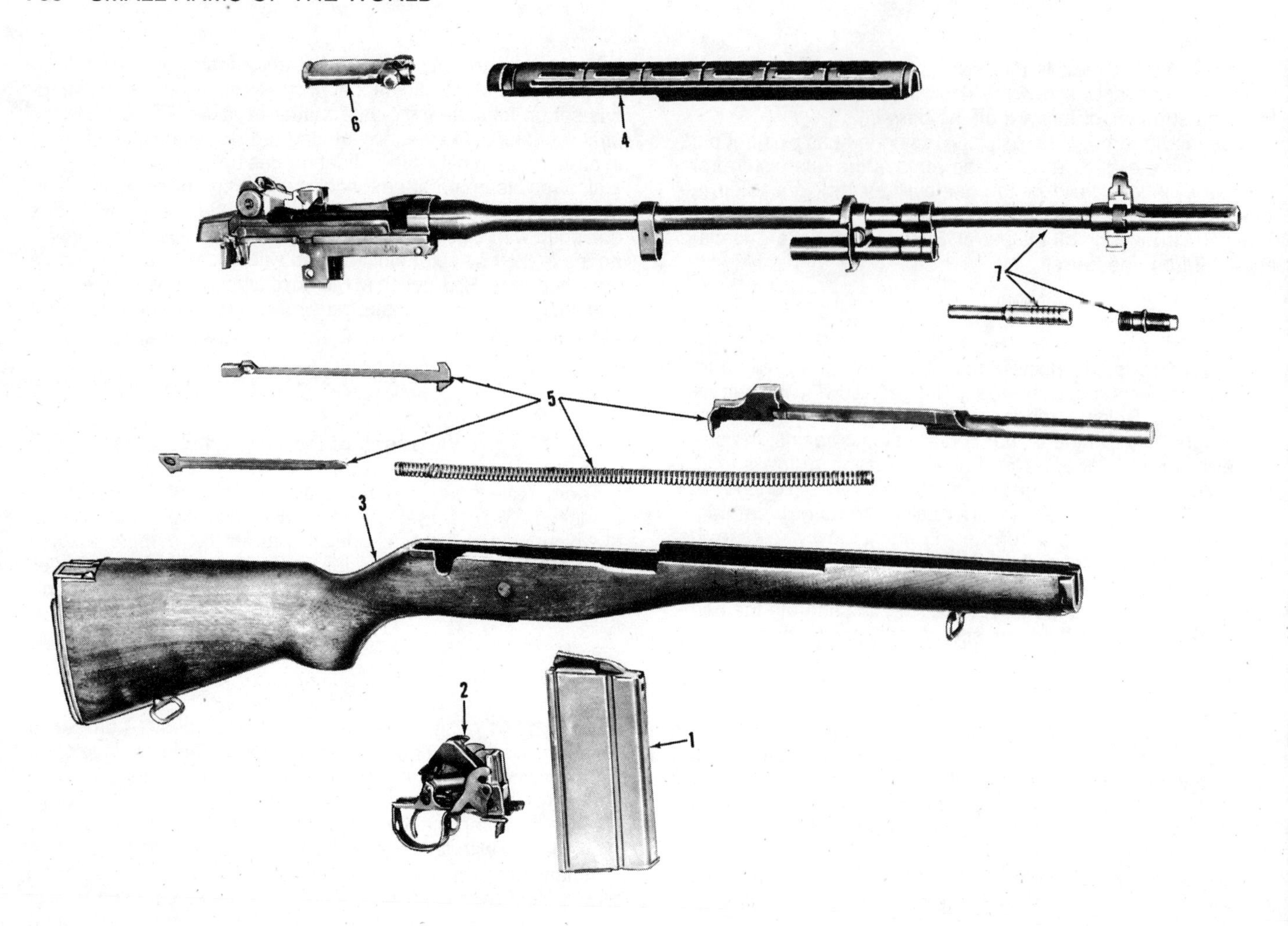

M14 Rifle, field stripped.
1. Magazine 2. Trigger assembly 3. Stock 4. Hand guard 5. Operating rod group 6. Bolt assembly 7. Gas piston, gas plug.

HOW THE M14 WORKS

Semiautomatic operation

The cycle of operation is broken down into eight steps.

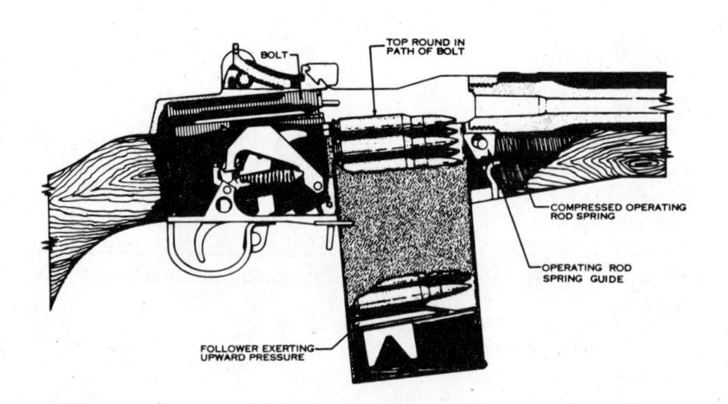

1. Feeding. Feeding takes place when a cartridge is forced into the path of the bolt by the magazine follower, which is under pressure of the magazine spring.

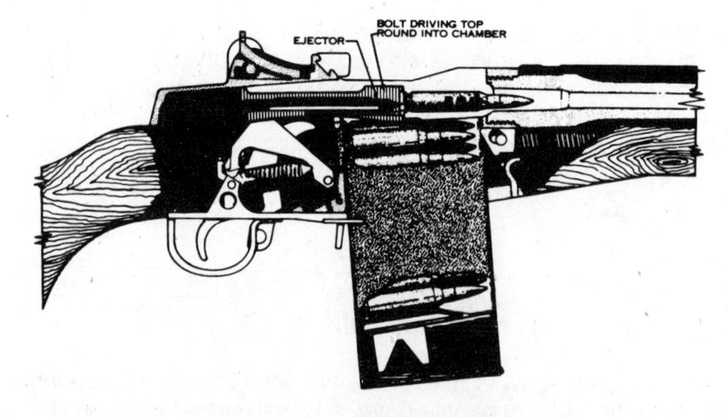

2. Chambering. Chambering occurs when a cartridge is moved from the magazine into the chamber by the bolt, which is propelled forward by the expanding operating spring. Chambering is complete when the extractor snaps into the extracting groove on the cartridge and the ejector is forced into the face of the bolt.

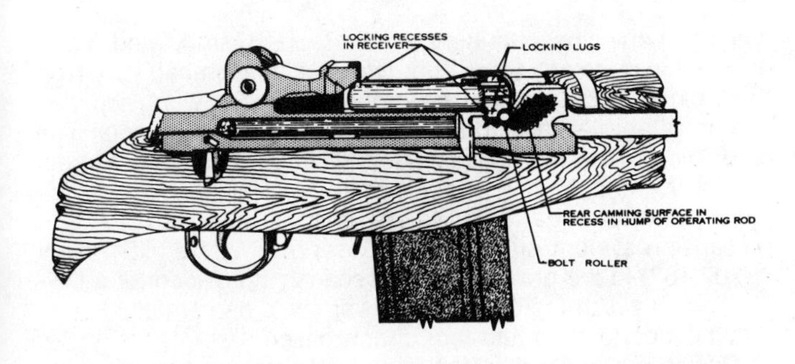

3. Locking. The bolt is locked by the rear camming surface in the hump of the operating rod forcing the bolt roller down. This turning action engages the locking lugs on the bolt with the matching recesses in the receiver.

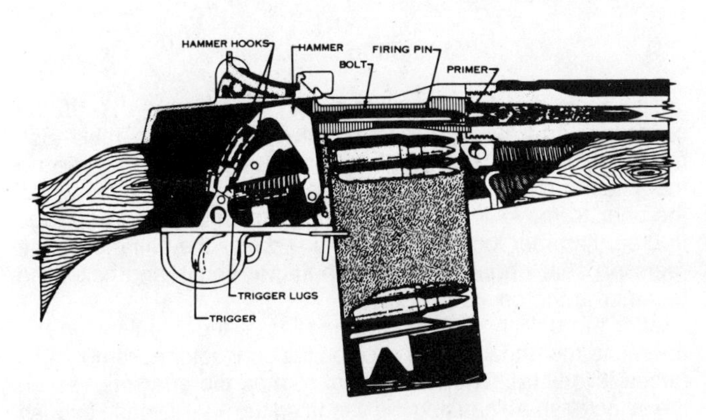

4. Firing. When the trigger is pressed, the trigger lugs are disengaged from the hammer hooks. The hammer is released, moving forward under pressure of the hammer spring and striking the firing pin. As the firing pin moves forward, it in turn strikes the primer, which ignites the propellant.

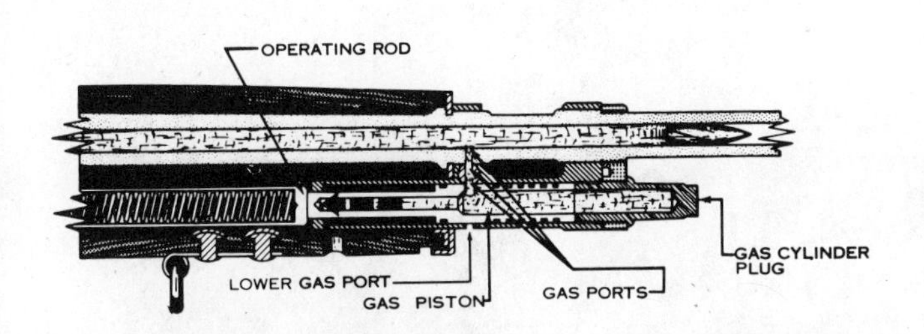

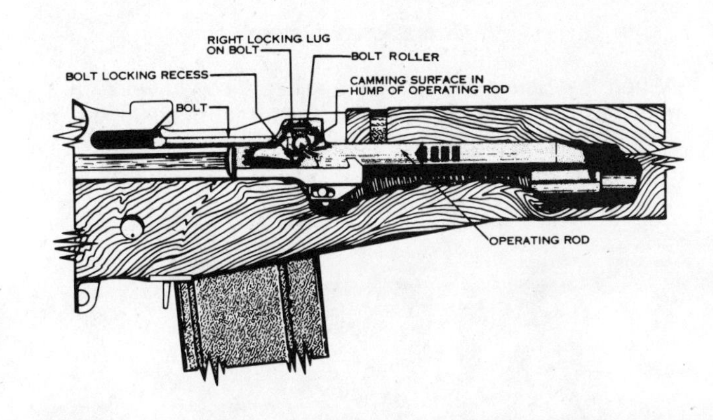

5. Unlocking. After firing, the unlocking cycle begins. As the bullet passes the gas port, a small amount of gas enters the gas cylinder/piston assembly (See fuller description of gas system below.). The gas inside the cylinder expands, and after enough pressure has built up to overcome the tension of the operating rod spring the piston starts its rearward movement. The operating rod travels rearward about 9.5mm (3/8 in.) before unlocking begins. The delay allows the projectile to exit and the residual gas pressure to drop. After the operating rod has moved that short distance, the camming surface inside its hump forces the bolt roller upward, disengaging the locking lugs on the bolt from the locking recesses in the receiver.

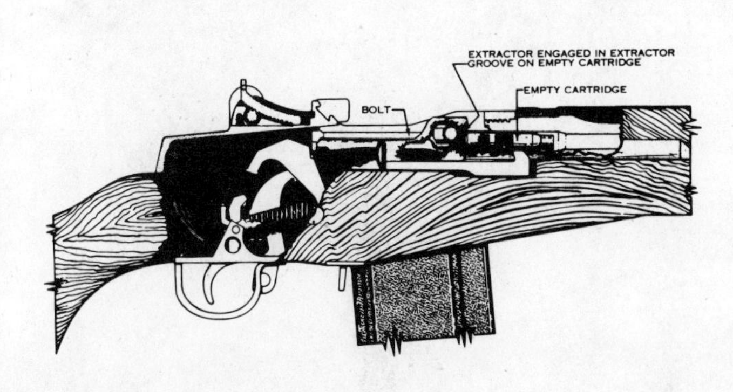

6. Extracting. As the bolt begins its rearward stroke, the extractor pulls the cartridge case from the chamber. The bolt and cartridge continue to the rear together.

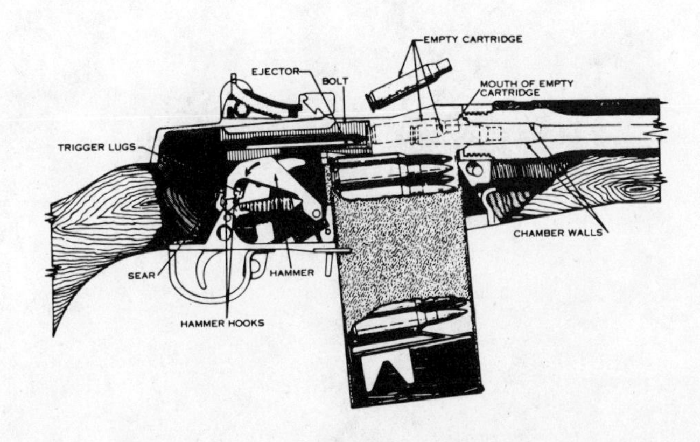

7. Ejecting. As soon as the bolt has completely withdrawn the empty case from the chamber, the compressed ejector plunger pushes the bottom edge of the cartridge base away from the bolt face. As a result, the front (neck) of the cartridge case moves upward and to the right. In rapid succession, the case strikes the lower right corner of the charger guide and the operating rod hump, which aids the right and forward motion of the empty case.

8. Cocking. Cocking occurs as the bolt continues to the rear. The back end of the bolt forces the hammer back and rides over it. The hammer is caught by the sear if the trigger is still held to the rear and by the trigger lugs if the pressure on the trigger has been released.

Automatic operation

The basic cycle of operation is the same except that as the operating rod travels to the rear the connector assembly also moves rearward. That movement rotates the sear release on the selector shaft so that the flange on the sear release allows the sear to move forward into a position where it can engage the rear hammer locks. When the bolt drives the hammer to the rear, the sear engages the rear hammer hooks and holds the hammer in the cocked position.

After the bolt moves forward and locks, the shoulder on the operating rod engages the hook on the connector assembly and forces it forward. That movement rotates the sear release on the selector shaft, causing the flange on the sear release to push the sear to the rear, disengaging it from the hammer hooks. The hammer will then go forward to fire another round. This cycle will be repeated until the trigger is released or the magazine is emptied.

Operational system

When the Springfield Armory engineers began work on the prototypes of the M14 Rifle, they sought a substitute for the gas impingement type operating system John Garand had incorporated into the M1 Rifle. That type of actuation had very high operating stresses since the propellant gases were admitted suddenly at relatively high pressures into the gas cylinder. The short time operating impulse of high intensity placed a great strain on the operating components, especially under adverse firing conditions. Earle M. Harvey adapted the gas cutoff and expansion system originated by Joseph C. White (US Patent 1,907,163) to the prototype M14 because it provided for a flexibility with which the operating power could be controlled as to magnitude, duration and rate of application.

Power to operate the M14 is derived from gas bled from the barrel into a hollow gas piston. The gas flows through ports in the barrel, gas cylinder and piston. Once the piston is filled with the expanding gas and begins its rearward stroke, the gas ports move out of alignment. The flow of additional gas is cut off. The piston, in contact with the operating rod, is driven fully to the rear (a 38mm [1.5 in.] stroke). As the piston nears the end of its rearward travel, the lower (exhaust) port in the bottom of the gas cylinder is uncovered, and the gases trapped in the closed system are vented. The gas system will continue to function in this manner unless the shutoff valve is closed. When that valve is closed, the gas system is rendered inoperative, thus requiring manual operation of the operating rod. The shutoff valve is normally only used when launching grenades from the rifle.

Typical receiver markings on the 7.62mm NATO M14 Rifle. Not shown are the Harrington & Richardson and TRW receivers.

OBSOLETE UNITED STATES RIFLES AND CARBINES

THE CALIBER .30 MODEL 1903A1, 1903A3 and 1903A4 RIFLES

The history of this rifle is given in the introduction to United States Rifles earlier in this chapter.

Differences in 1903 Models

The differences between the various 1903 individual models follow. The basic 1903 with straight stock has a leaf type rear sight and no finger grasping grooves in the fore-end of the stock. The bolt was originally bent straight down but after 1918 was given a slight bend to the rear. These bolts are stamped "NS" on the handle. These rifles were made by Springfield and Rock Island. Receivers made by Springfield with serial numbers below 800,000 are made of Springfield Class "C" steel and should not be used with stepped-up loads. Later receivers made at Springfield are suitable for use with any factory-loaded .30-06 ammunition, except proof loads. Receivers of Rock Island manufacture up to number 285,506 were also manufactured of Class "C" steel and require similar precautions.

M1903 Mark I. This rifle was altered by the fitting of a new sear mechanism and changes in the cutoff; a slot was drilled in the left side of the receiver to accommodate the caliber .30 Pedersen device, which replaced the regular bolt and converted the weapon into a semiautomatic, firing the caliber .30 pistol cartridge. The Pedersen device, or "Pistol caliber .30 Model 1918" as it was known, was never issued, and the Mark I rifles were remodified to be used as standard M1903s.

M1903A1. A pistol grip "type C" stock with grasping grooves was fitted, the butt plate was checkered, and the trigger was serrated. This model was adopted in December 1929.

M1903A2. This is not a shoulder rifle; it is a barreled receiver used as a subcaliber rifle in various artillery pieces.

M1903A3. Adopted on 21 May 1942, this rifle was modified to simplify production. Stamped bands, swivels, butt plate and magazine trigger guard assemblies are used. A one-piece hand guard, simplified front sight and ramp type aperture rear sight, mounted on the receiver bridge, are also found on these rifles. Most of these rifles are fitted with straight stocks; however, a semi-pistol grip stock was also issued. Four, two and occasionally six land-and-groove barrels may be found. These rifles were manufactured by Remington Arms and the L. C. Smith Corona Typewriter Co. A total of 945,846 rifles were made. Remington also produced approximately 345,000 M1903 and M1903 (modified) rifles before production of the M1903A3 began.

M1903A4. This is the sniper version of the M1903A3; a full pistol grip stock is fitted. The bolt handle is cut away to clear the M73B1 (Weaver 330C) 2.5 power scope. No iron sights are fitted. This rifle was adopted in December 1942. Approximately 26,650 were manufactured.

Model 1942. This is a Marine Corps modification of the M1903A1. The rifle is fitted with a 10x Unertl scope.

Total production of the Model 1903 and 1903A1 rifles was approximately 1,295,000 rifles at Springfield, Rock Island and Remington Arms.

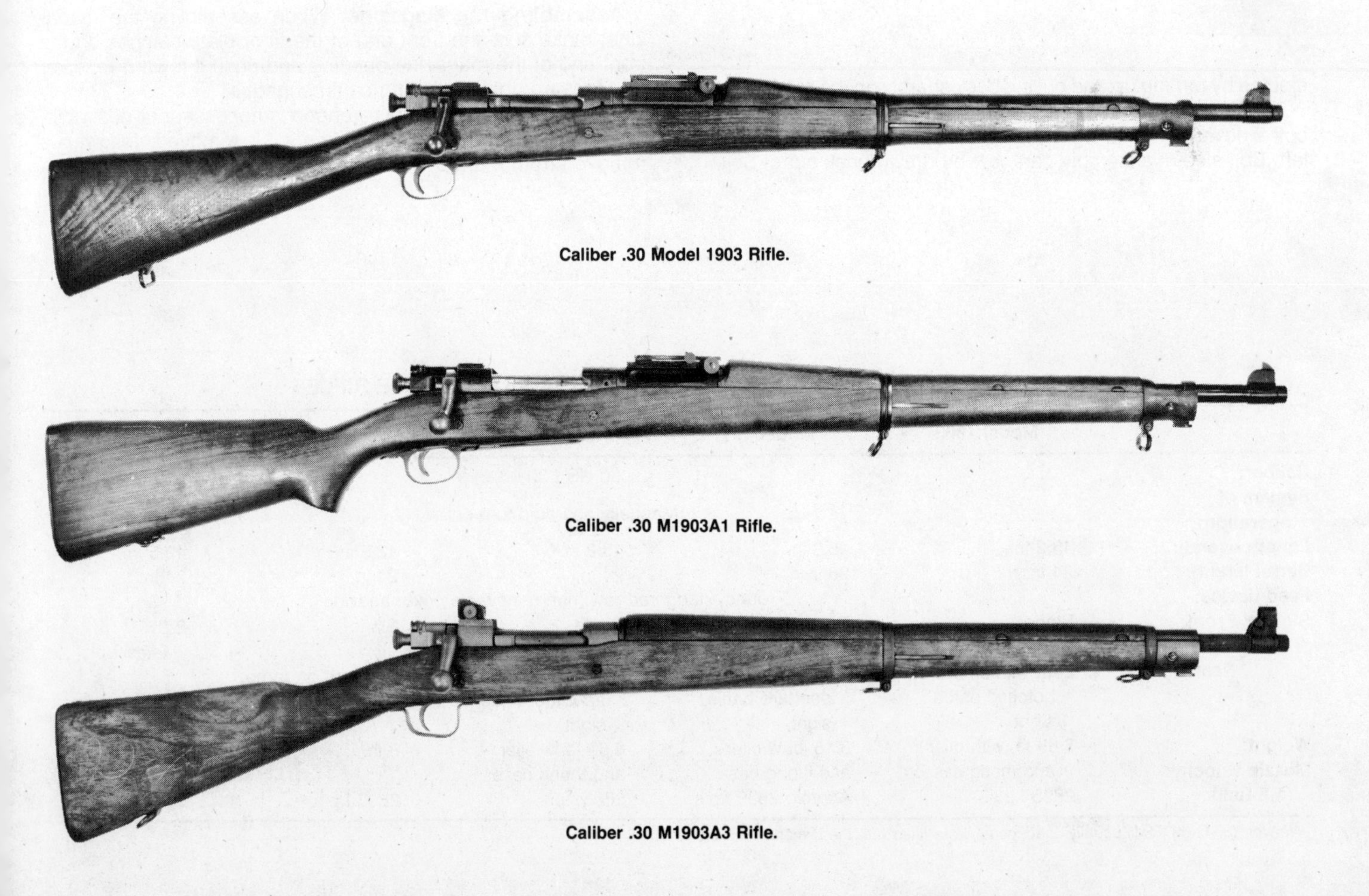

Caliber .30 Model 1903 Rifle.

Caliber .30 M1903A1 Rifle.

Caliber .30 M1903A3 Rifle.

Loading and Firing the Model 1903

As a Single-Shot Rifle. Check the magazine cutoff on the left side of the receiver, making sure that it is down and the word "Off" can be seen. Turn the bolt handle up as far as it will go. This will release the locking lugs from their recesses in the receiver, and permit the bolt to be drawn straight back. Now place a cartridge in the firing chamber, thrust home the bolt to seat the cartridge properly and permit the extractor to snap over the cannelure of the cartridge case, and turn the bolt handle down as far as it will go to lock the piece. Note: This rifle is cocked as the bolt is rotated clockwise. The knob on the cocking piece will project out of its casing when the weapon is at full cock. If desired, the safety lock may now be applied, by turning the safety lock thumb piece at the rear of the bolt over to the right as far as it will go, when the word "Safe" will be seen on its face.

If safety is not set, pressing the trigger will explode the cartridge in the firing chamber, which then may be extracted and

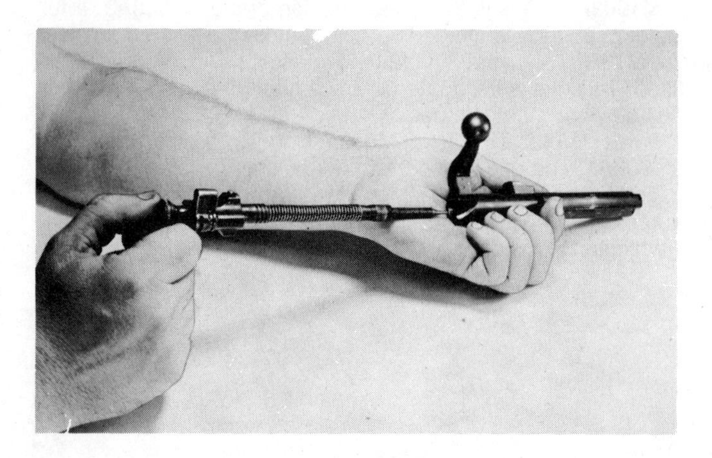

ejected by turning up and pulling back sharply on the bolt handle.

To Dismount Bolt. Holding bolt firmly in left hand, press in bolt sleeve lock with right thumb and unscrew, turning to the left. Bolt sleeve assembly can now be drawn back out of bolt.

Holding firing pin sleeve with left forefinger and thumb, pull back on the cocking piece with the right middle finger and right thumb, and turn the safety lock to the left with the right forefinger to release it. This will relieve part of the tension of the mainspring.

Resting the head of the cocking piece on a firm surface, pull back the firing pin sleeve and remove the striker. Firing pin sleeve, mainspring and firing pin rod may then be withdrawn.

Extractor is removed by turning it to the right and pushing forward.

Assembling Bolt. Holding bolt handle up in left hand, make sure that the extractor collar lug is in line with the safety lock on the bolt and insert the extractor collar lug in its undercut in the extractor, then push extractor until tongue comes in contact with bolt face. Now press extractor hook against a rigid surface to spring it into its groove in the bolt.

See that the safety is down and to the left, and assemble firing pin rod and bolt sleeve. Place the cocking piece against a solid surface, draw back the firing pin sleeve and attach the striker. The firing pin must be cocked before the bolt can be screwed on. This is done by pressing the striker point against a wooden surface (which must not be hard enough to injure it). Force the cocking piece back and engage the safety lock.

Assembled bolt sleeve is now replaced in the bolt and screwed until the bolt sleeve lock engages.

With cutoff still turned to center notch, insert bolt in its guide in the receiver, push down the magazine follower and push the bolt home. Now turn safety lock and cutoff down to the left, and press the trigger.

Stripping the Magazine. Turn rifle upside down, insert nose of bullet in the hole in the rear of the floor plate to depress the spring catch.

Retaining pressure, pull back toward the trigger guard. This will release the spring and the magazine follower, and permit them to be removed from the weapon.

Assembling the Magazine. When assembling the magazine, make sure the front end of the floor plate catches on the front end of the magazine opening and push it toward receiver and forward until the spring catch engages.

No further stripping of this weapon is necessary or desirable.

M1903A3, A4. Floor plate cannot be removed, follower is removed from the top of the magazine.

CHARACTERISTICS OF UNITED STATES BOLT-ACTION SERVICE RIFLES

	Model 1903	Model 1917	Model 1903A1	Model 1903A3	Model 1903A4
Caliber:	.30 US (.30-06)				
System of operation:	Manually operated turn bolt				
Length overall:	43.2 in.	46.3 in.	43.2 in.	43.5 in.	43.5 in.
Barrel length:	24 in.	26 in.	24 in.	24 in.	24 in.
Feed device:	5-round, staggered row, non removable box magazine				
Sights: Front:	Blade	Blade with protecting ears.	Blade.	Blade.	2.2× telescopic*
Rear:	Leaf with aperture, notched battle sight.	Leaf with aperture, aperture battle sight.	Leaf with aperture, aperture battle sight.	Aperture on ramp.	
Weight:	8.69 lb. with oiler and thong case.	8.18 lb. w/oiler and thong case.	8.69 lb. w/oiler and thong case.	8 lb.	9.38 lb.
Muzzle velocity: (M2 Ball):	2805 f.p.s.	Approx. 2830 f.p.s.	2805 f.p.s.	2805 f.p.s.	2805 f.p.s.

*Weaver 330 C (M73B1) usually used, some were fitted with the Lymann Alaskan (M73).

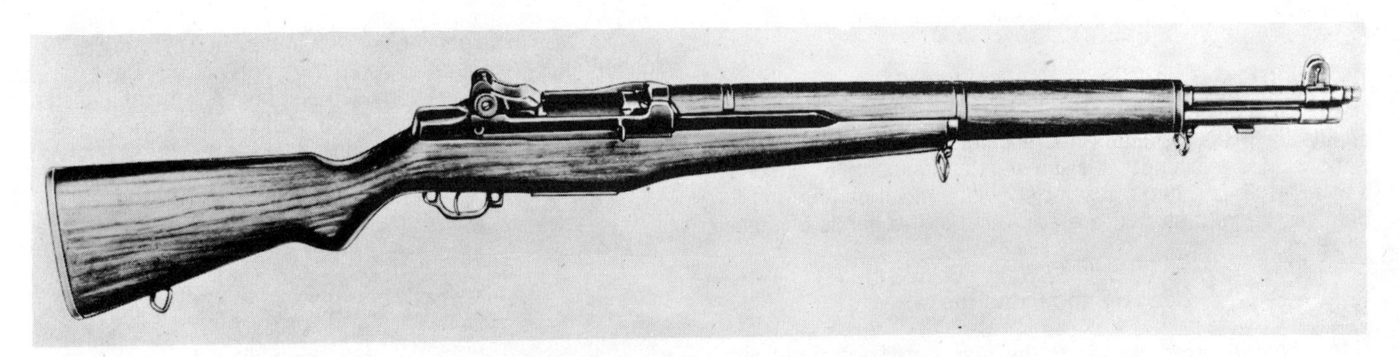

Caliber .30 M1 Rifle (Garand semiautomatic rifle).

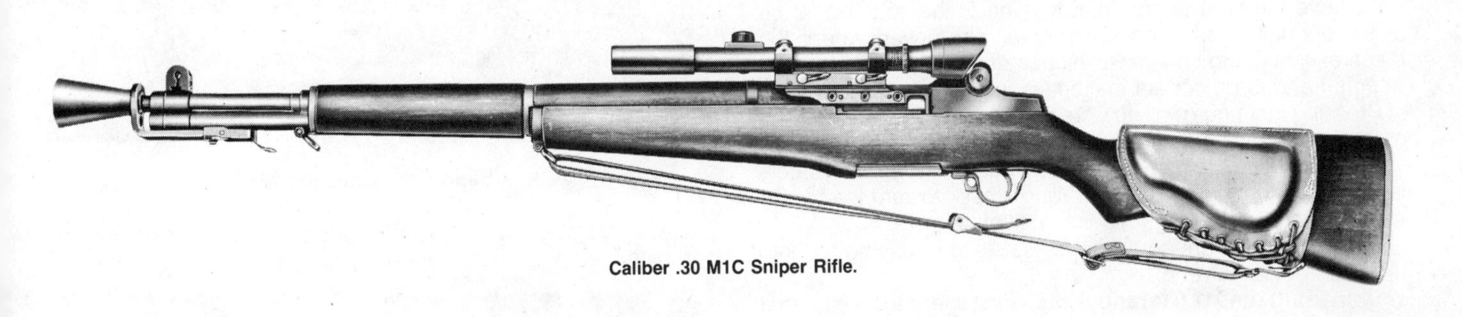

Caliber .30 M1C Sniper Rifle.

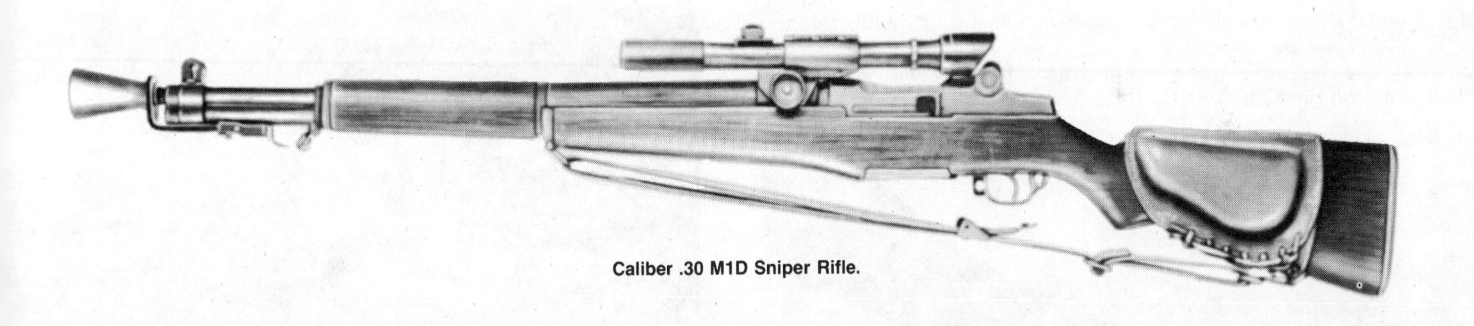

Caliber .30 M1D Sniper Rifle.

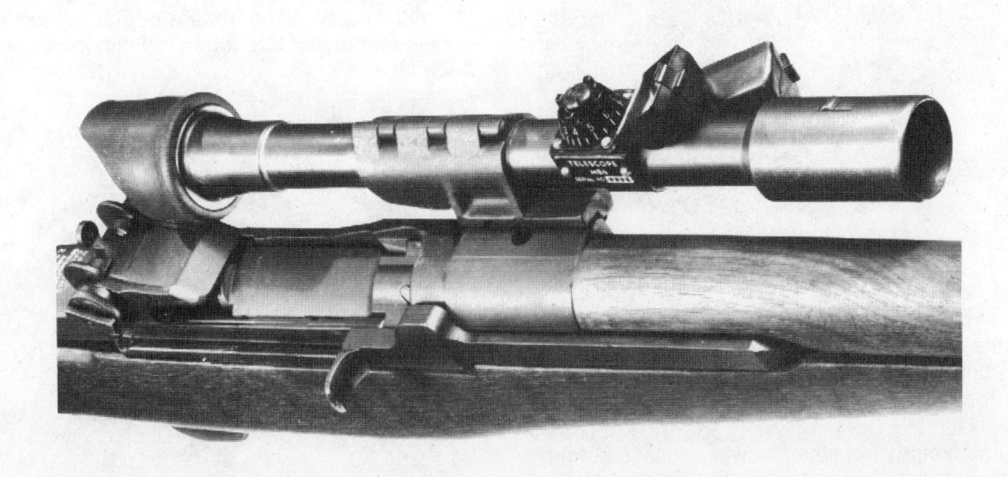

Caliber .30 M1D Sniper Rifle.

RIFLE CALIBER .30 M1

As noted in Chapter 1, the M1 Rifle is now obsolete in the US, but it is still used by a number of other nations and is very popular with US collectors and shooters. The M1D, equipped with the M84 telescope, is still a standard sniper rifle. National Match M1 Rifles, hand assembled with special barrels and fiberglass bedded stocks, are still found in the hands of some shooters.

Loading and Firing the M1

Pull operating rod straight to the rear. It will be caught and held open by the operating rod catch.

Place the loaded clip on top of the magazine follower; with right side of right hand against the operating rod handle press down with right thumb on the clip until it is caught in the receiver by the clip latch.

Remove the right thumb from the line of the bolt and let go of the operating rod handle, which will run forward under the compression of the spring. Push operating rod handle with heel of right hand to be certain that bolt is fully home and locked.

Pressing the trigger will now fire one cartridge. Weapon will be ready for the next pull of the trigger.

If weapon is not to be used at once, set the safety. The safety is in front of the trigger guard. Pulling it back toward the trigger sets it on safe; forward is the fire position.

Note that the cartridge clip is reversible and may be fed into the rifle from either end.

Unloading the M1 (Garand) Rifle. First check to be sure that the safety is off.

Pull the operating rod back sharply and hold it in rear position. This will eject the cartridge that was in the firing chamber. With the left hand, grasp the rifle in front of trigger guard. Hold the butt against the right hip to support it. With the left thumb, release the clip latch.

The clip and whatever cartridges remain in it will now pop up into the right hand.

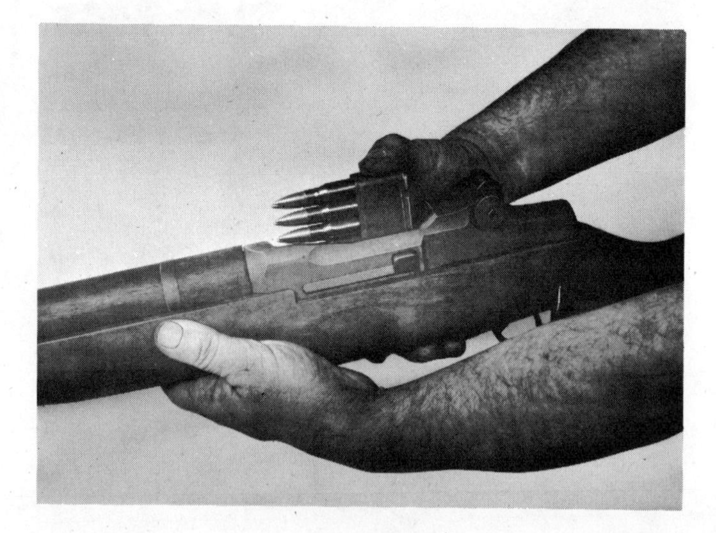

With the right side of the right hand held against the operating rod handle, force the operating rod slightly to the rear. With the right thumb, now push down the magazine follower and permit the bolt to move forward about an inch over the end of the follower.

Remove the thumb smartly from the follower and let go of the operating rod. The action will close under the tension of the spring. Now press the trigger.

If you wish to unload the firing chamber but leave the magazine loaded, pull the operating rod back as described above to eject the cartridge from the firing chamber; then pull the operating rod handle back past its normal rear position, force the clip down, ease the rod far enough forward to let the bolt handle ride over the top of the clip, then let operating rod go forward.

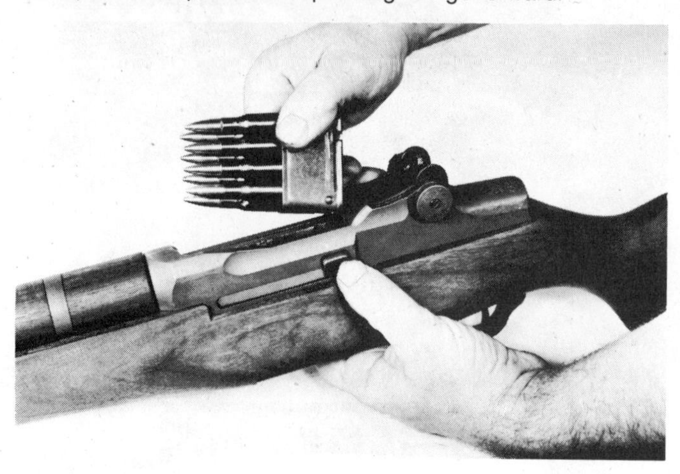

Field Stripping the M1

A thorough knowledge of field stripping is necessary in order to give the rifle the care essential to its correct operation.

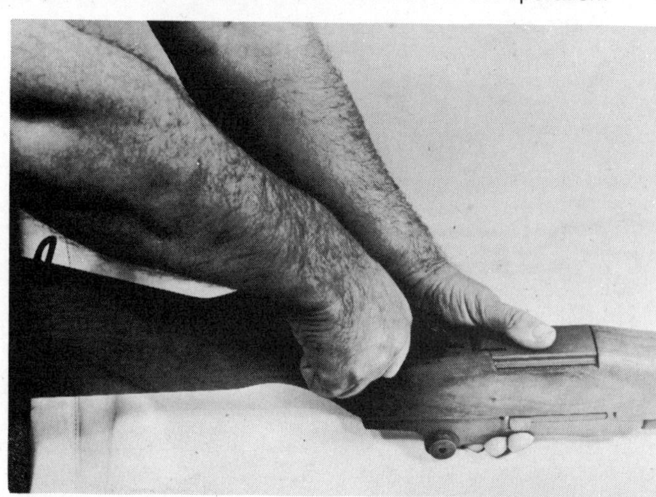

Start by placing the rifle upside down on a firm surface. Holding the rifle with left hand, reset the butt against the left thigh.

With thumb and forefinger, unlatch the trigger guard by pulling back on it.

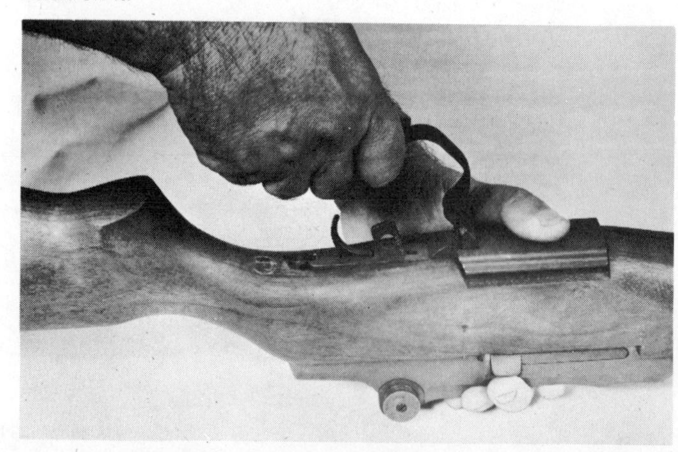

Continue the pressure and pull out the trigger housing group.

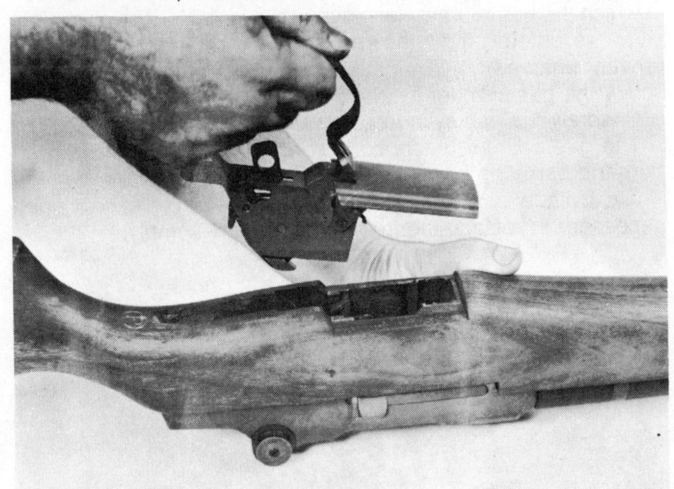

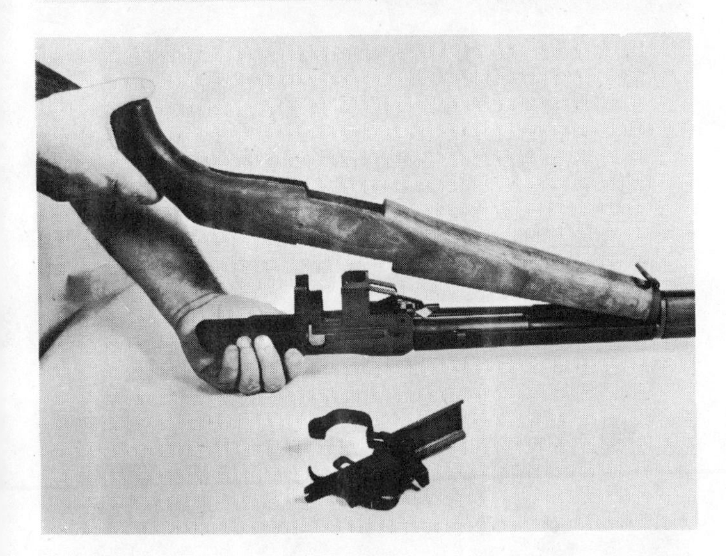

With left hand, grasp rifle over rear sight. With right hand, strike up against the small of the stock, firmly grasping it at the same time. This will separate the barrel and receiver group from the stock group.

To Dismount Barrel and Receiver Group, M1. With the barrel down, grasp the follower rod at the knurled portion with thumb and forefinger and press it toward the muzzle to free it from the follower arm.

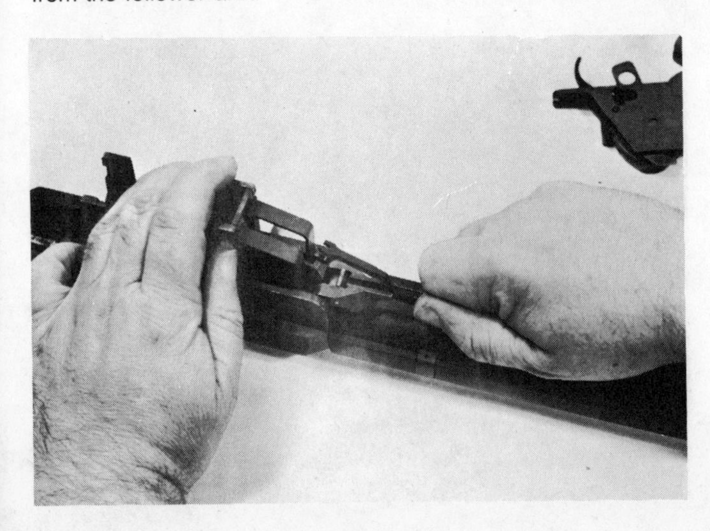

The follower rod and its compensating spring, which is attached, may now be withdrawn to the right. The compensating spring is removed from the follower rod by holding spring with left hand and twisting rod toward the body with the right hand and pulling slightly to the right.

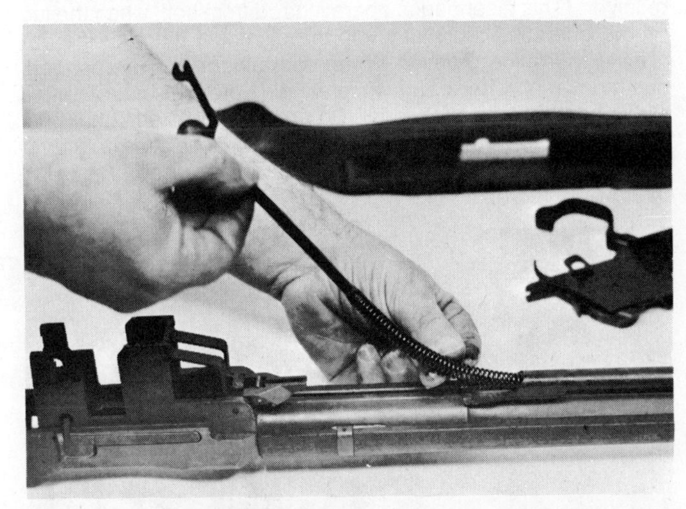

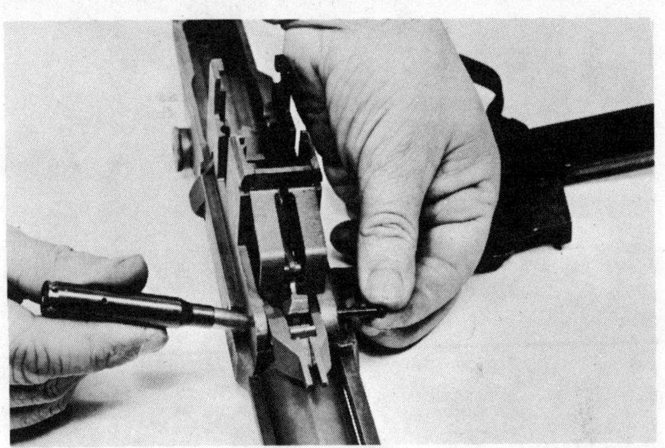

With the point of a bullet, push the follower arm pin from its seat and pull it out with the left hand.

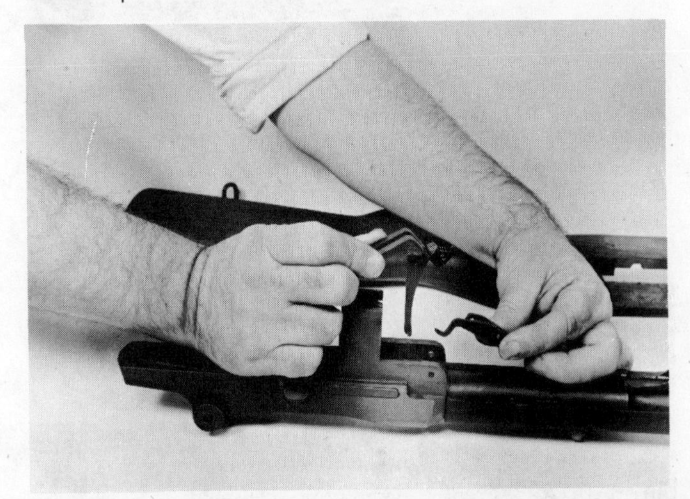

Seize the bullet guide, follower arm and operating rod catch assembly; draw these to the left until they disengage. The three separate parts may now be lifted out. Accelerator pin is riveted in its seat, so do not attempt to remove accelerator from operating catch assembly.

Lift out follower with its slide attached (do not separate follower from slide.)

Holding barrel and receiver assembly with left hand, grasp the operating rod handle with right hand and move it slowly to the rear, meanwhile pulling the rod handle up and away from the receiver. (This disengages operating rod from bolt, when the lug on the operating rod slides into the dismount notch of the operating rod guide groove.) When operating rod is disengaged, pull it down and back and withdraw it. (Note that the operating rod is bent. This is intentional. Do not attempt to straighten it.)

Slide the bolt from the rear to the front by pushing the operating lug on it, and lift it out to the right front with a slight twisting motion.

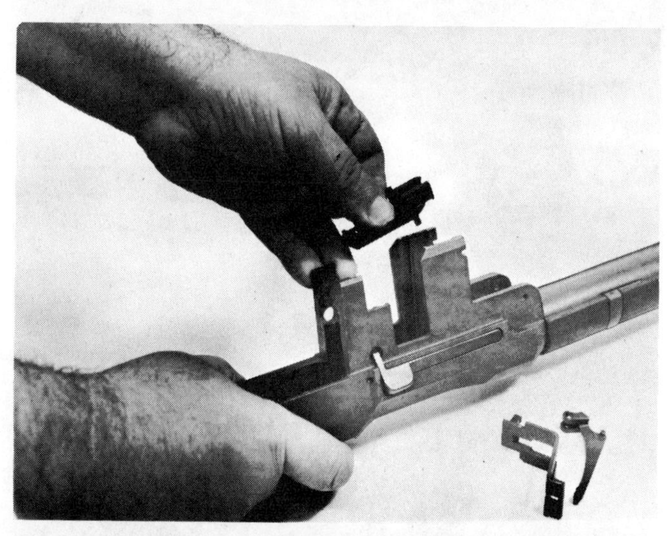

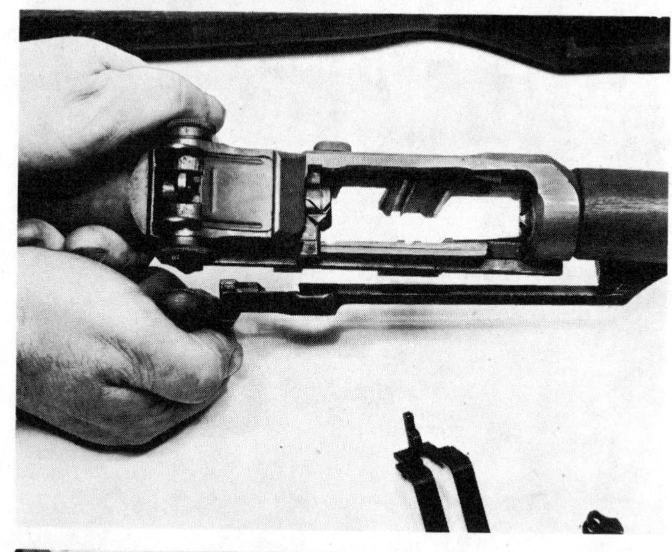

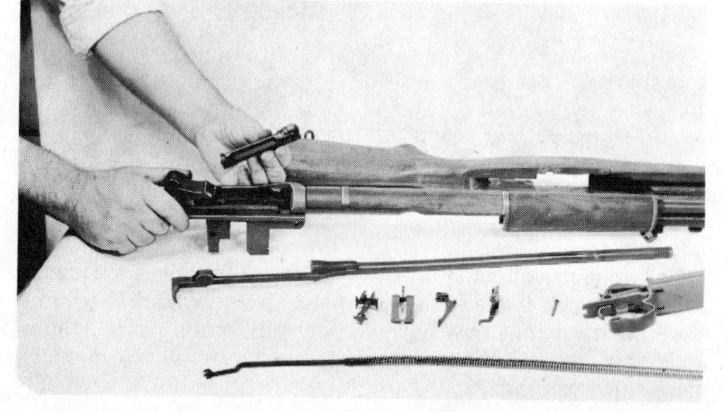

Note on M1 Gas Cylinder. A spline type gas cylinder is used in which the barrel protrudes beyond the cylinder. The front sight screw is entered from the rear of the sight and is sealed to prevent unscrewing. The combination tool must be used to unscrew the gas cylinder lock screw.

Unscrew the gas cylinder lock.

The gas cylinder is tapped toward the muzzle and removed from the barrel.

Gas cylinder assembly should never be removed except when necessary to replace the front handguard assembly.

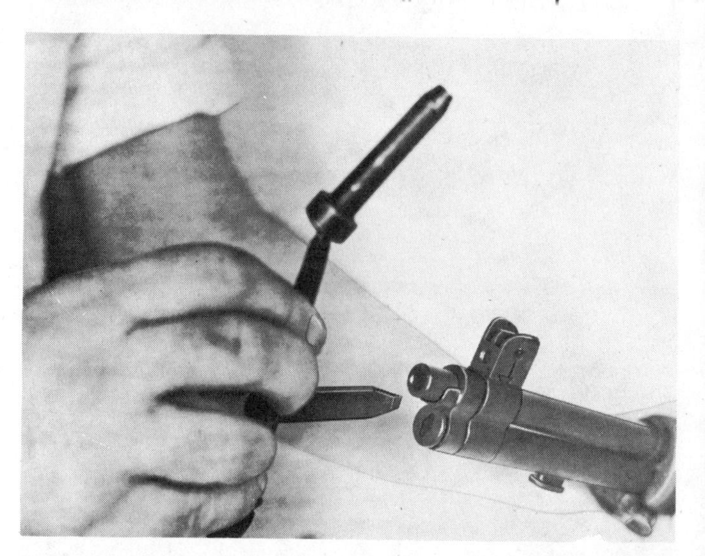

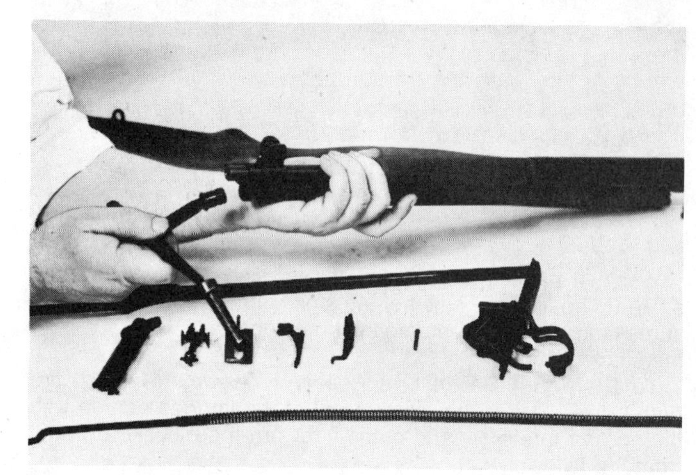

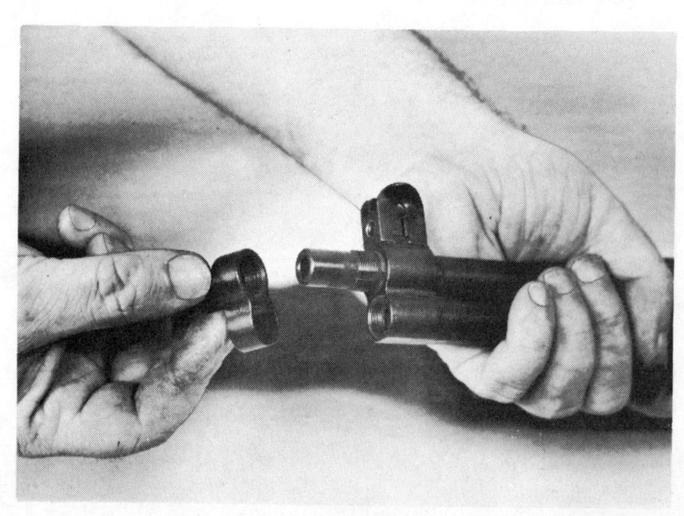

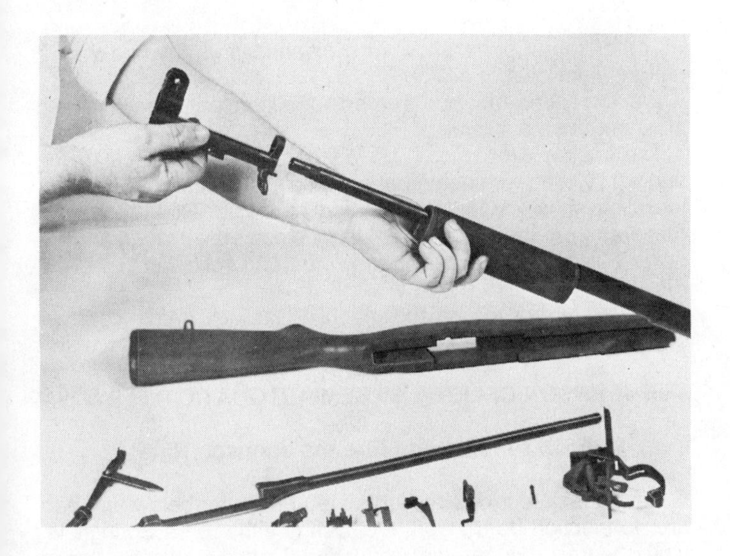

Assembling the M1 Rifle. Replacing gas cylinder, if it has been dismounted, is done by merely reversing the dismounting procedure.

To assemble barrel and receiver group, tilt the barrel and receiver assembly, sight up and muzzle to the front to an angle of about 45°.

Holding the bolt by the right locking lug so the front end of the bolt is somewhat above and to the right of its extreme forward position in the receiver, insert the rear end in its bearing on the bridge of the receiver. Switch it from right to left far enough to let tang of the firing pin clear top of the bridge. Next guide the left locking lug of the bolt into its groove just to the rear of the lug on the left side of the receiver, and start right locking lug into its bearing in the receiver. Now slide bolt back to its extreme rear position.

Turn barrel and receiver assembly in left hand until barrel is down.

Grasp operating rod at the handle and holding it handle up, insert piston head into gas cylinder about ⅜″. Be sure that operating rod handle is to the left of the receiver.

Hold barrel and receiver assembly in left hand and twist to the right until barrel is uppermost.

Adjust operating rod with right hand so that camming recess on its rear end fits over operating lug on bolt. Now press operating rod forward and downward until bolt is seated in its forward position.

With barrel and receiver assembly held barrel down and muzzle to your left, replace the follower with its attached slide so that its guide ribs fit into their grooves in the receiver. (The square hole in the follower must be to the right.) The follower slide rests on the bottom surface of the bolt when the follower is in the correct position.

With left hand replace bullet guide, fitting the shoulders of the guide into their slots in the receiver and the hole in the projecting lug in line with the hole in the receiver.

With left hand replace follower arm passing stud end through bullet guide slot and inserting stud in proper grooves in front end of follower.

Place the forked end of the follower arm in position across the projecting lug on the bullet guide, with pin holes properly aligned.

Insert rear arm of operating rod catch into clearance cut in the bullet guide (be sure its rear end is below the forward stud of the clip latch which projects into the receiver mouth.) Line up the holes in the operating rod catch, the follower arm, and the bullet guide with those in the receiver; and insert the follower arm pin in the side of the receiver toward your body and press the pin home.

Insert operating rod spring into operating rod; assemble follower rod by grasping the spring in left hand and inserting follower rod with right hand, twisting the two together until the spring is fully seated.

Seize the knurled portion of follower rod with thumb and forefinger of left hand with hump down and forked end to the right.

Place front end of follower rod into operating rod spring and push to the left, seating the forked end against the follower arm.

Insert U-shaped flange of stock ferrule in its seat in the lower band.

Pivoting about this group, guide chamber and receiver group and press into position in the stock.

Replace trigger housing group with trigger guard in open position into the stock opening.

Press into position, close and latch trigger guard. This completes reassembly.

How the M1 Works

Starting with the rifle loaded and cocked, the action is as follows. The trigger being pressed, the hammer strikes the firing pin, exploding the cartridge in the chamber. As the bullet passes over the gas port drilled in the under side of the barrel, some of the gas escapes into the cylinder and blasts back against the piston and operating rod with force enough to drive the rod to rear and compress the spring.

During the first $\frac{5}{16}$″ inch of rearward travel, the operating lug slides in a straight section of the recess on the operating rod; after which the cam surface of this recess is brought in contact with the operating lug, which it cams up, thereby rotating the bolt from right to left to unlock its two lugs from their recesses in the receiver.

During the moment of delayed action, the bullet leaves the barrel, and the breech pressure drops to a safe point. The further rotation of the bolt then cams the hammer away from the firing pin and pulls the firing pin back from the bolt. The operating rod continues its backward movement carrying the bolt with it as the lug on the bolt has reached the end of its recess.

During this rearward motion of the bolt, the empty case is withdrawn from the chamber by the extractor positioned in the bolt until it is clear of the breech; at which point the ejector, exerting a steady pressure on the base of the cartridge case, throws it to the right front by the action of its compressed spring.

The rear end of the bolt at this point forces the hammer back, rides over it, compresses the hammer spring and finally stops in the rear end of the receiver.

As the bolt has now cleared the clip, the follower spring forces the cartridges up until the topmost one is in line with the bolt.

The operating rod spring comes into play at this point to pull the action forward.

Forward Movement of the Action. As the bolt moves forward, its lower front base strikes the base of the cartridge case and pushes it into the firing chamber. The hammer, pressed by its spring, rides on the bottom of the bolt. While it tends to rise, it is caught and held by the trigger lugs engaging the hammer hook, if trigger pressure has not been released. Otherwise, the trigger engages the rear hammer hook until letting go the trigger disengages the sear from the hammer. The hammer then slides into engagement with the trigger lugs.

When the bolt nears its forward position, the extractor engages near the rim of the cartridge, and the base of the cartridge forces the ejector into the bolt, compressing the ejector spring.

The rear surface of the cam recess in the operating rod now cams the operating lug down and thereby twists the bolt from left to right until the two lugs lock into their places in the receiver.

The operating rod drives ahead for another $\frac{5}{16}$ inch. The rear end of the straight section of the operating rod recess reaches the operating lug on the bolt, which completes the forward movement and leaves the rifle ready to fire when the trigger is pressed.

This cycle continues as long as there are cartridges in the magazine and the trigger is squeezed.

Care of the M1 Rifle

The rifle must be kept clean and properly lubricated. Failure to do so may result in stoppages at a critical moment. The rifle should be inspected daily.

To Clean the Bore. A clean patch saturated with bore cleaner should be run through the bore a number of times. Plain water, hot or cold, may be used if bore cleaner is lacking. While the bore is still wet, a metal brush should be run through several times to loosen up any material that has not been dissolved by the water. Dry patches should then be pushed through the bore until thoroughly dry. The bore should then be coated with light issue gun oil. Also use the chamber cleaning tool to give the chamber the same attention. Remember that primer fouling in the bore contains a salt that rusts the steel.

To Clean Gas Cylinder. The carbon forming in the gas cylinder varies in amount in different weapons. When the deposit is heavy, the rifle is sluggish in action and may fail to feed. The carbon must be scraped from the exposed surface of the front of the cylinder and the gas cylinder plug and piston head after extensive firing. A knife or similar sharp bladed instrument should be used for this scraping process.

Gas cylinder plugs and grooves in the gas cylinder should be cleaned so they will feed correctly in the plug.

The gas cylinder lock should be removed, and the lock screw inserted in the cylinder far enough to break loose any carbon. Inside the cylinder must be thoroughly wiped clean and oiled at the conclusion of any extensive firing.

When firing is expected to be resumed the next day, tilt the muzzle down and place a few drops of oil into the cylinder between the piston and the walls of the cylinder. Then operate the rod by hand a few times to distribute the oil thoroughly.

Wipe the outside of the gas cylinder and the operating rod and then oil lightly. Should no firing be expected for a week or two, remove the rod and the gas cylinder lock screw (or plug) so that the cylinder is open at both ends. Then clean cylinder with rod and patches exactly as the bore of the rifle is cleaned.

Hold the weapon so that no water gets into the gas port. Do not remove the gas cylinder for cleaning.

Piston head and rod should be cleaned with cleaner or with water and dried thoroughly, while the rod and cylinder should be oiled before assembling. Any carbon present should be removed. Do not use abrasive cloth if it is possible to avoid doing so; should it be used, take proper care that the corners of the plug or lock screw and piston head are not rounded.

Attention to Other Parts of Rifle. Graphite cup grease is used for lubricating bolt lugs, bolt guides, bolt cocking cams, compensating spring, contact surfaces of barrel and operating rod, operating rod cams and springs, and operating rod groove in the receiver.

All other metal parts should be cleaned and covered with a uniform light coat of oil.

Wooden parts must be treated with light coat of raw linseed oil about once a month.

The leather sling should be washed, dried with a clean rag and lightly oiled with neatsfoot oil while it is still damp, whenever the sling shows signs of stiffening or drying. Rust should be removed from the metal parts with a piece of soft wood and oil, never with abrasive. Screw heads must be kept clean to prevent rusting.

Be careful not to use too much oil, as any heavy coat will collect dirt and interfere with operation.

THE JOHNSON CALIBER .30 SEMIAUTOMATIC RIFLE M1941

How to Load and Fire the Johnson Rifle

Lift magazine cover; on right side of receiver below and parallel to ejection port, insert five rounds either on a Springfield-type charger (stripper clip) or singly. If using a charger, insert horizontally into charger guides in charging port. When last round is stripped or fed singly into magazine, magazine cover will close automatically. Raise the operating handle 20° and pull completely to rear; bolt will run forward, chambering a round. The rifle is now loaded and will fire if the trigger is pulled. When the last round is fired, the bolt will remain to the rear; load as directed above and pull bolt handle slightly to the rear and release it—the bolt will run home chambering a cartridge and the weapon is loaded. The Johnson can be loaded with single rounds or chargers with the bolt home on a loaded chamber. It is therefore easy to replenish the magazine at any time. The magazine can be emptied by depressing the magazine cover with the thumb of the right hand. The safety lock lever is located immediately in front of the trigger guard. When the free end of the lever is at an angle to the right of the axis of the barrel, the weapon is on safe. When the lever is to the left of the axis of the barrel, the weapon is on fire.

How to Field Strip the Johnson Rifle

Check rifle to insure that it is empty. With the point of a bulleted round (or a drift), push on the latch plunger of the hinged barrel latch found in the hole in the forward right side of the fore-end and push the barrel rearward. Raise the operating handle with the thumb of the left hand to the unlocked position and withdraw the barrel from the receiver. Disengage the bolt stop plate plunger with the point of a bulleted round and lift out bolt stop plate. Remove bolt stop and disengage the link from the main spring plunger. Raise the operating handle and retract the bolt about two inches. Grasp the knob of the operating handle spindle and pull it outward. Slide the operating handle forward until it is clear of the shoulders in the extractor recess and remove it. Lift out the extractor. Grasp the projecting end of the link and pull it to the rear, withdrawing the bolt through the rear end of the receiver. Rotate the locking cam counterclockwise and remove it from the bolt, remove the firing pin, and push out the link pin and remove the link. Disengage the hammer block pin and push

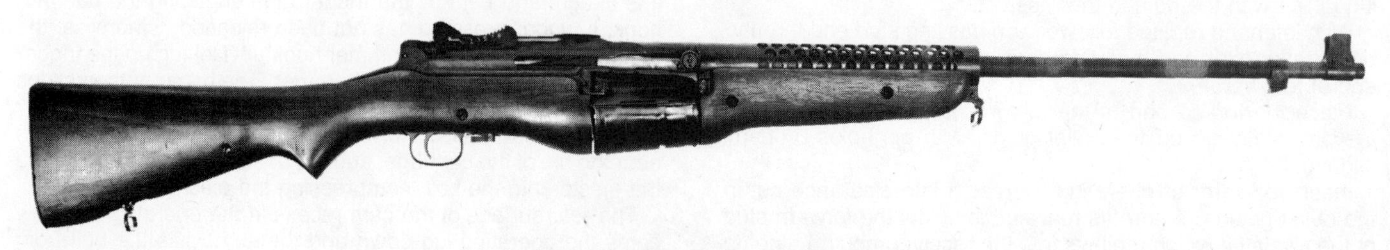

Johnson caliber .30 Semiautomatic Rifle M1941.

it out with point of bullet; pull off the butt stock. Remove the ejector pin and ejector. Hammer should be cocked before removing the butt stock group. Unscrew the front guard screw and the hammer block screw and lift out the hammer group from the stock. Unscrew the rear trigger guard screw and remove the trigger guard and safety assembly. No further disassembly is recommended. To reassemble, follow the above directions in reverse order.

How the Johnson Rifle Works

When the trigger is pressed, the sear disengages from the hammer. The hammer is driven against the firing pin, which protrudes from the rear of the locking cam, and the weapon fires. The barrel recoils against the tension of the barrel recoil spring and the main spring (transmitted through the bolt.) When the bullet is at the muzzle, the barrel has moved rearward about 1/64 inch; when the bullet is about 2 feet from the muzzle the barrel has recoiled about 1/8 inch. The camming arm on the bolt engages the camming face in the receiver and unlocking begins. The Johnson has an 8-lug bolt. When the bullet is about 5 feet from the muzzle, the barrel has recoiled its full 5/8 inch and the bolt has rotated 20° and is unlocked. The rearward motion of the barrel is stopped by a shoulder in the receiver. The bolt moves to the rear independently of the barrel due to inertia and residual pressure in the chamber. The extractor gives the empty case a sharp pull and the bolt receives a sharp blow from the locking cam, which taps the bolt rearward. The rearward movement of the bolt cocks the hammer, and the extracted case is brought into contact with the ejector, which throws the case clear of the rifle. The bolt is halted in its rearward travel by the forward end of the link bringing it up against the bolt stop, when the head of the bolt has passed behind the base of the top cartridge in the magazine.

As the bolt moves forward under the pressure of the main spring, the bolt face picks up a cartridge from the magazine and rams it into the chamber. The locking lugs enter the barrel locking bushing, and the locking cam rotates the bolt 20° to the locked position. Pressure on the trigger must be relaxed between shots. When the last round has been fired, the bolt remains open.

The Johnson is the only recoil-operated military shoulder rifle that has been manufactured in quantity. It appeared soon after the Army had adopted the M1, and at the time its backers claimed that it was far superior to the M1. A series of tests and demonstrations during the period 1939–40, however, indicated that the Johnson was not superior to the M1; Springfield Armory was already tooled up to produce the M1.

Quantities of the 1941 Johnson were used by the US Marines for a limited period of time, and significant quantities were made for the Dutch East Indies. The rotary magazine is the common version of the Johnson Rifle; however, a vertical feed version was made as well.

CARBINE CALIBER .30 M1, M1A1, M2, AND M3

The carbine was developed to replace the pistols in use by noncommissioned officers, special troops and company-grade officers.

Manufacturers of the carbine were:

Winchester: 809,451 M1 carbines, 17,500 M2 carbines, and 1,108 M3 carbines.
Inland Manufacturing Div. of General Motors: a total of 2,625,000 carbines including M1s, M1A1s, M2s, and a few M3s.
Underwood Elliot Fisher: 545,616 carbines.
National Postal Meter: 413,017 carbines.
Rock-ola Manufacturing Corp: 228,500 carbines.
Quality Hardware: 359,662 carbines.
Standard Products: 247,155 carbines.
Saginaw: 739,136 carbines.
IBM: 346,500 carbines.
There were more carbines produced than of any other United States weapon.

The variations of the carbines are as follows:

Carbine M1—semiautomatic, originally made with L type flip over sight, which was replaced with a ramp-mounted aperture adjustable for windage, sporter type stock.
Carbine M1A1—same as M1, but has folding-type metal butt stock.
Carbine M2—selective fire, usually found with fixed wooden stock.
Carbine M3—receiver grooved for Infrared "Snooper Scope," otherwise identical to the M2.

Loading and Firing the M1 Carbine

Load magazine exactly as for automatic pistol with 15 cartridges.

Thrust up into position in the trigger housing until it locks.

Pull back handle of operating slide on right side of gun as far as it will go, opening the action and allowing a cartridge to rise in the magazine in the path of the bolt, and cocking the weapon and compressing the return spring.

Remove hand and permit operating slide to go forward, loading the firing chamber. With heel of hand, push operating slide handle forward to be sure it is fully locked. The weapon is now ready to fire.

Push the button safety in the front end of the trigger guard all the way through to the right. This is the safe position. Pushing the button through to the left side as far as it will go releases the safety.

How the M1 Carbine Works

Starting with the gun loaded and cocked, the action is as follows: The trigger being pressed, the hammer is released to strike the firing pin and discharge the cartridge. As the bullet

The Winchester caliber .30 Light Rifle—prototype of the US M1 Carbine.

passes down the barrel, a minute quantity of gas behind it flows down through a very fine hole bored in the under side of the barrel and escapes into a sealed cylinder where it expands against the head of the piston-like operating slide. This operating slide moves back a short distance until a cam recess engages an operating lug on the bolt. During this time, the bullet has had sufficient time to leave the barrel, and it is safe for the action to open. The extractor fastened in the bolt draws the empty cartridge back to strike the spring-actuated ejector, which hurls it out to the right front of the weapon. The bolt is rotated out of its locking recess, simultaneously turning the hammer away from the rear of the firing pin and forcing the firing pin to draw back inside the bolt. This compresses the hammer spring. The rearward motion of operating slide is completed when rear end of its inertia block strikes against forward end of receiver. Bolt stops when it reaches the end of bolt hole in rear receiver. Boltway is now clear permitting the next round to rise in the magazine in line with the bolt. During this motion, the powerful operating slide spring has been compressed. It now drives the bolt forward loading the chamber. The cam recess in the operating slide again comes into play. Pressing against the bolt operating lug, it rotates it from left to right into its locking recess.

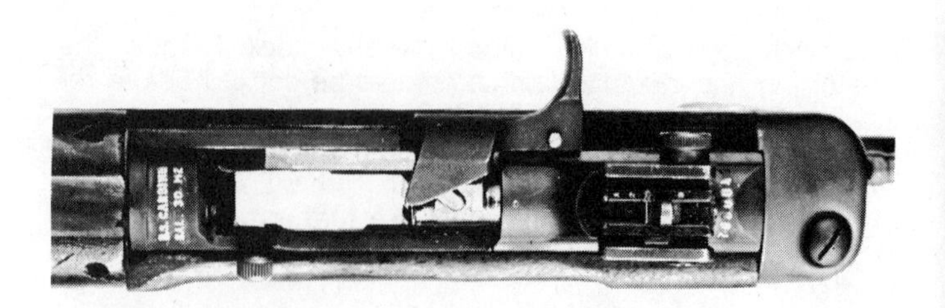

Top view of the M2, shown without magazine. Full auto switch is seen at the receiver above the stock. Externally, the M2 resembles the Carbine M1.

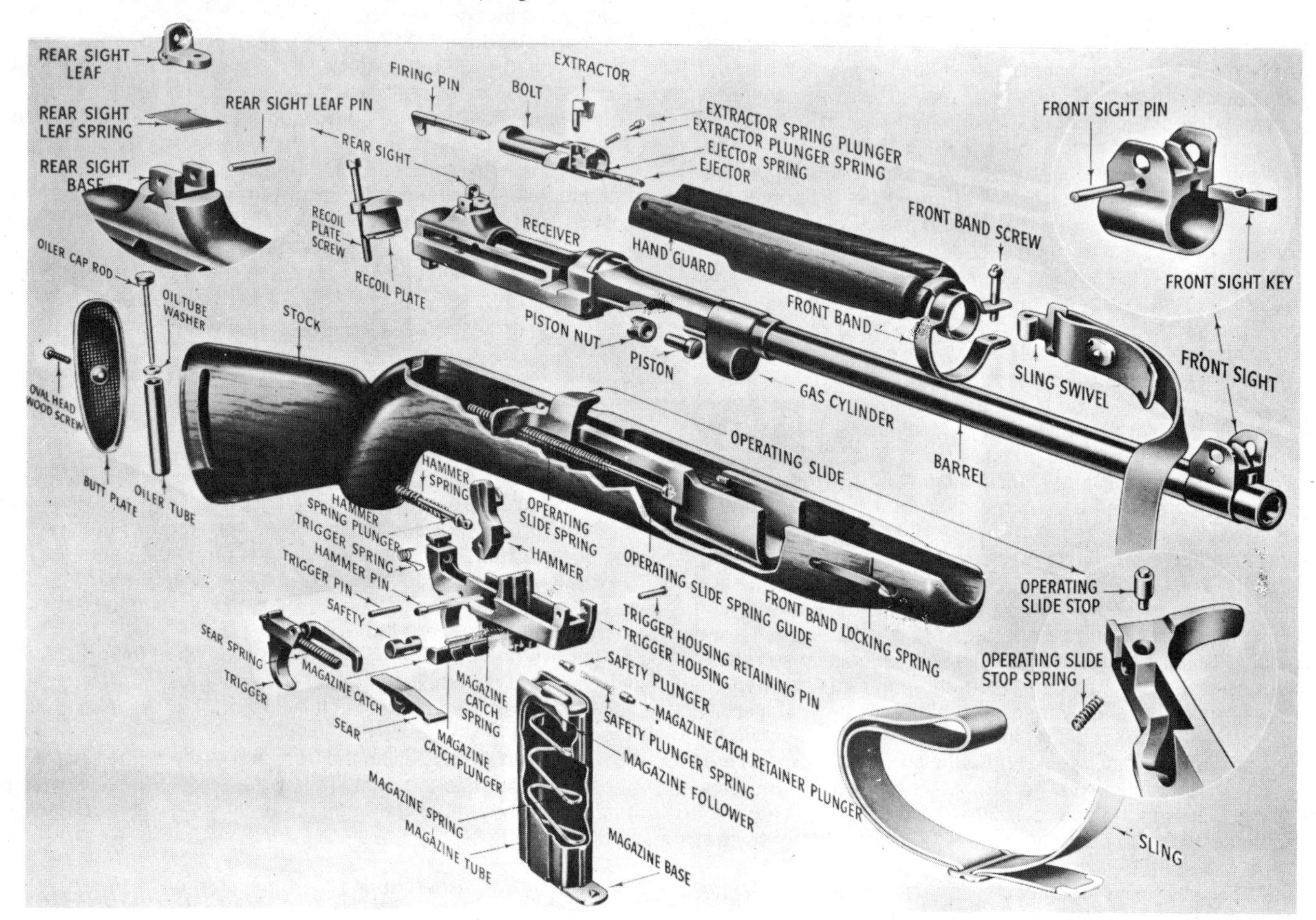

Parts of the M1 Carbine.

Forward movement of operating slide continues until the rear of its inertia block lodges against the piston in the cylinder. This action continues each time the trigger is squeezed until the last cartridge has been fired.

Field Stripping the M1 Carbine

Push the magazine catch to the left (it is positioned just in

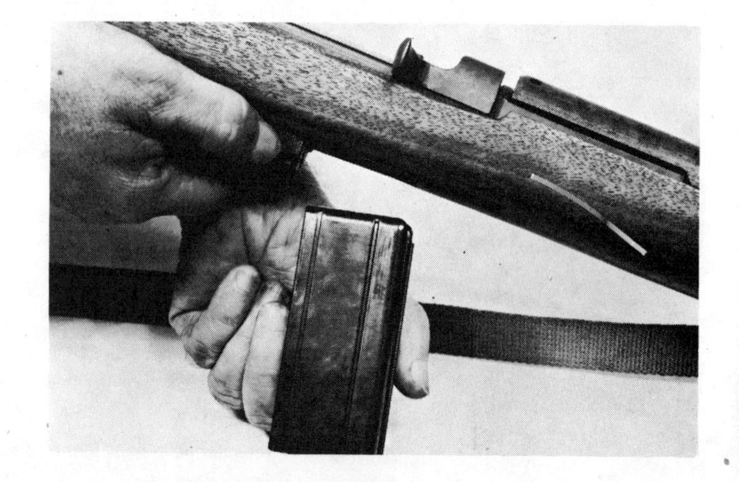

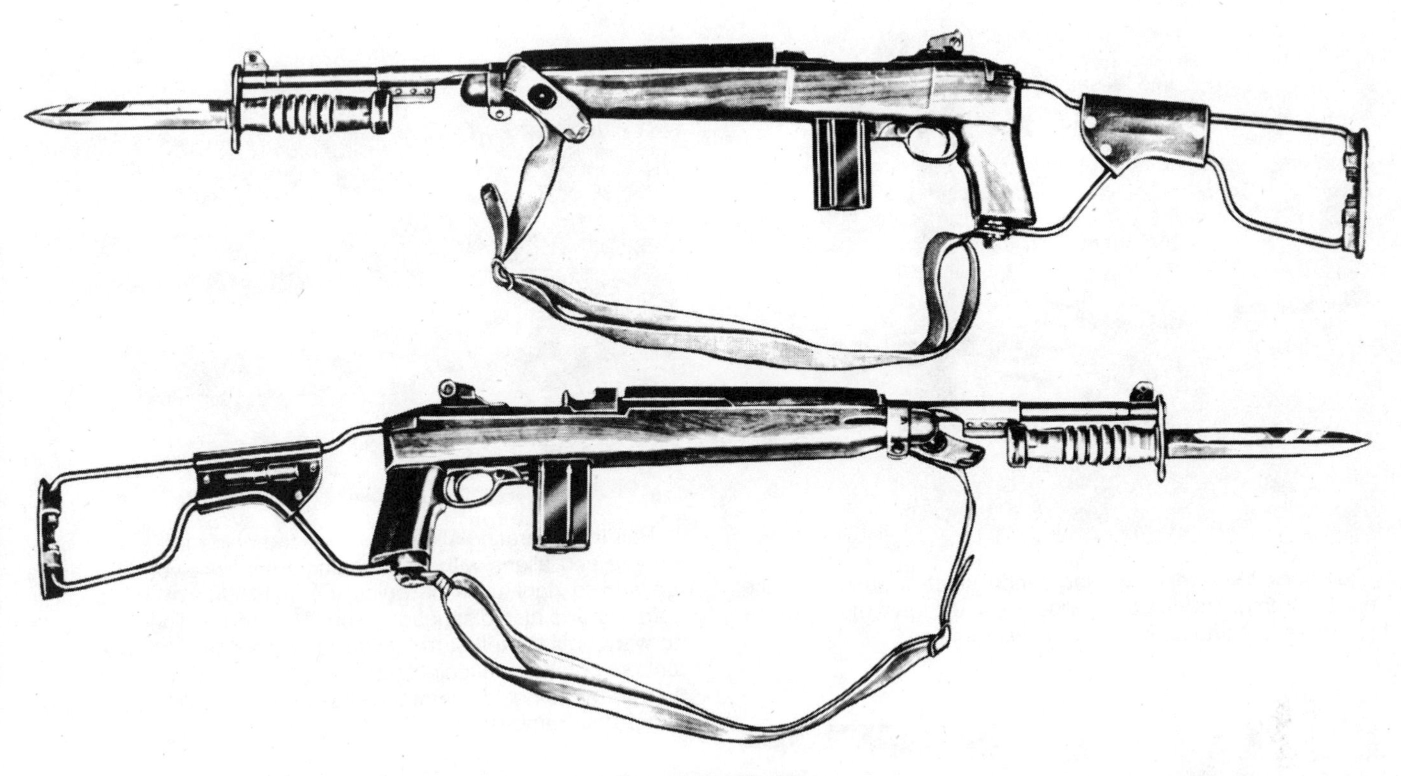

Caliber .30 M1A1 Carbine.

front of the trigger guard on the right side), and withdraw magazine from below.

Draw back bolt to examine chamber and make sure that the weapon is unloaded.

At the end of the wooden fore-end is a sling swivel. Push this back against the fore-end and loosen it by unscrewing the front band screw. A cartridge may be used as a screwdriver.

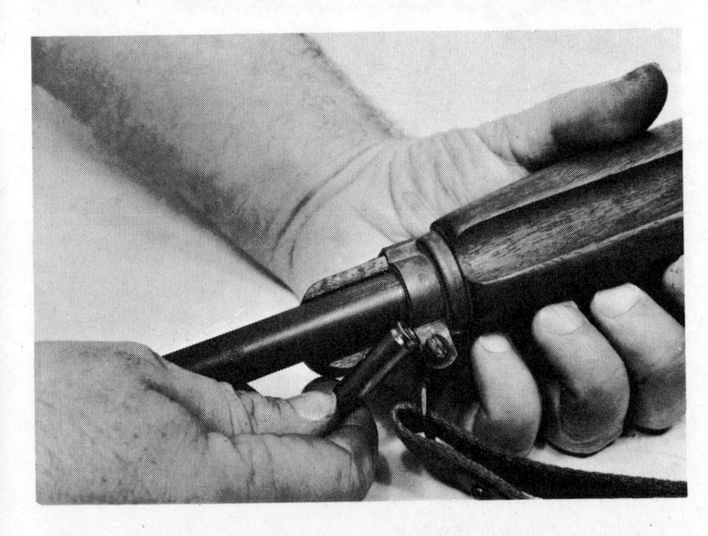

Press the front end of the lock spring toward the rear and slip the front band forward over its locking spring; it will not slip off the barrel unless the front sight is removed.

Now slide the wooden handguard on top of the barrel forward until its liner disengages from the undercut in the forward face of the receiver; it can then be lifted from the barrel.

Holding the stock firmly with the right hand, grasp the barrel near the front end with the left hand and raise it until the lug at the rear of the receiver clears the retaining notch on the face of the recoil plate (the plate just above the pistol grip). The barrel

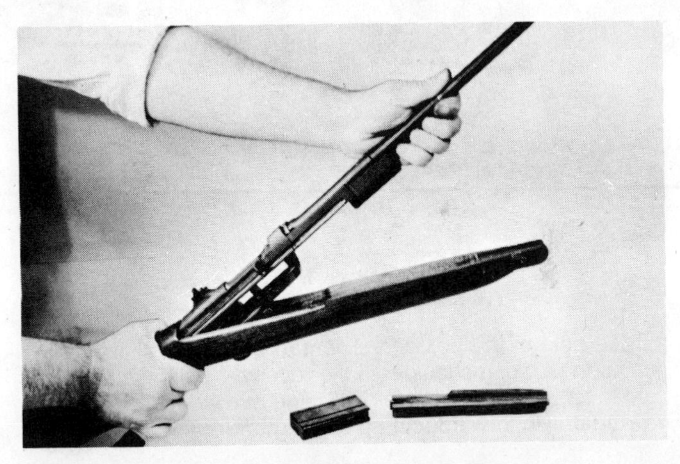

and receiver can now be pulled forward and lifted out of the stock; carrying the trigger housing group with them.

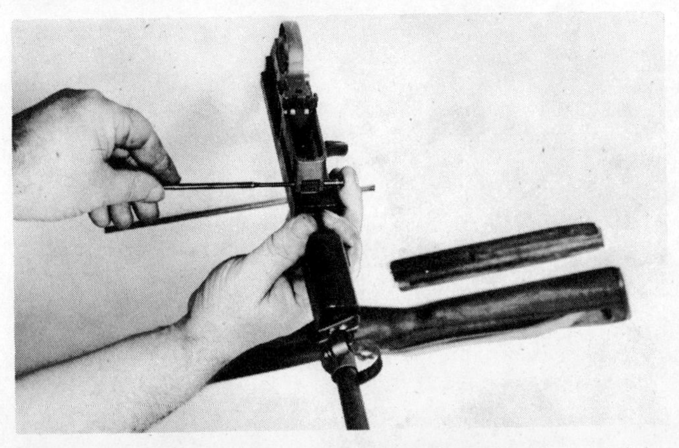

At forward end of trigger guard is the trigger housing retaining pin. Push it out until it clears the lug in the receiver.

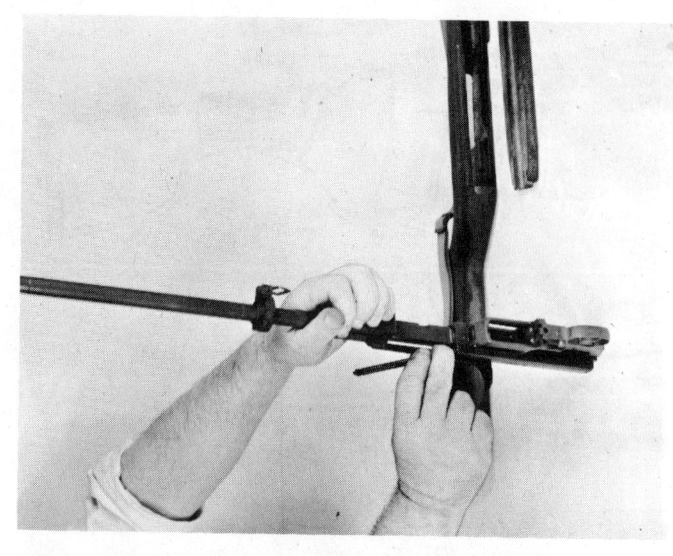

Pull back the operating slide spring guide a short distance until it is free of the operating slide. Pulling it forward and to the right permits it and its spring to be withdrawn.

Take hold of the bolt and slide it to the rear until its face is behind the locking shoulder in the receiver. Twist the bolt from right to left, lift it to an angle of 45° and turn it bottom up. It may now be drawn forward and up out of the receiver.

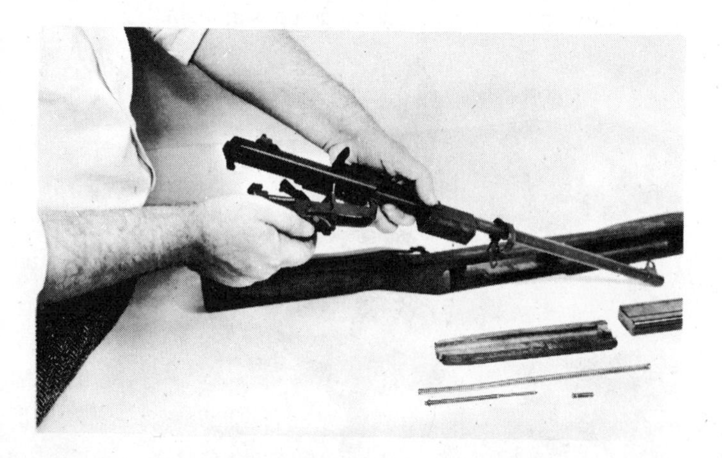

Pull the housing forward until it clears the grooves in the receiver, when it may be lifted out.

Pull the operating slide back until the guide lug at bottom of handle end aligns with dismounting cut in receiver. Lift handle up and to right until the guide lug clears the retaining cut in receiver and also disengages from the bolt lug. Then push slide forward until the left barrel guide lip aligns with clearance cut on bottom of left barrel guide groove. Rotating the slide body so as to free the left guide lip of the slide from its barrel guide groove will permit removing the slide from barrel.

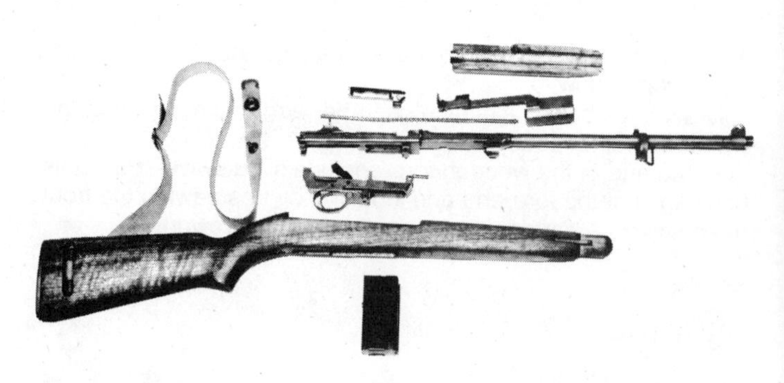

This completes field stripping. To assemble, reverse this procedure.

Caliber .30 M2 and M3 Carbines

The M2 carbine is the selective-fire version of the M1. Parts that differ from those of the M1 carbine are as follows: hammer, sear, trigger housing, operating slide, magazine catch and stock. Added parts are as follows: disconnector group, disconnector lever assembly and selector group. All M2 and M3 carbines and many M1 and M1A1 carbines are fitted with a front band assembly that incorporates a bayonet lug. These carbines use the bayonet knife M4.

The M3 carbine is an M2 with a receiver designed to accommodate an infrared sniperscope. It does not have a conventional rear sight.

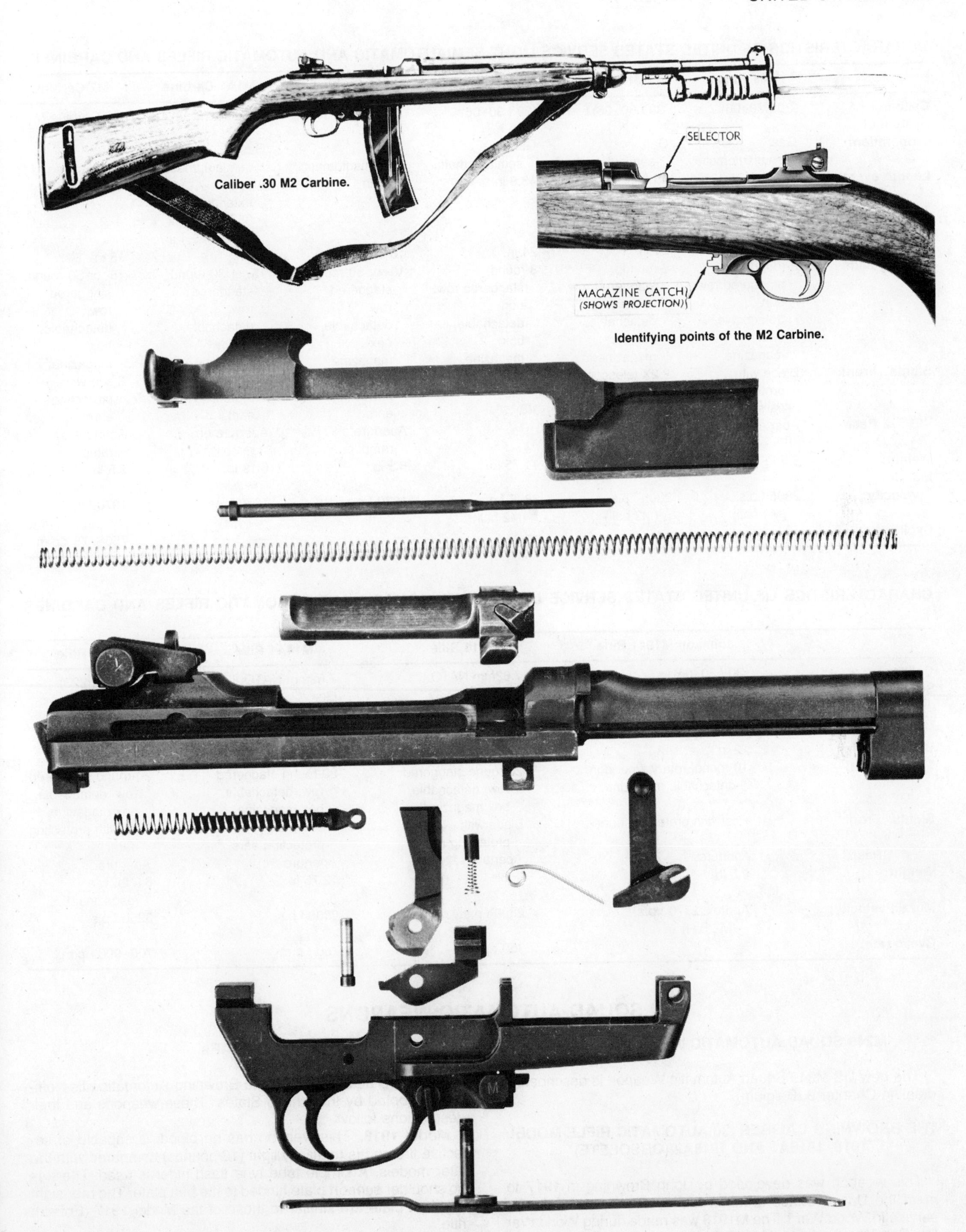

Caliber .30 M2 Carbine.

Identifying points of the M2 Carbine.

Stripped action of the M2 Carbine.

CHARACTERISTICS OF UNITED STATES SERVICE LIGHT SEMIAUTOMATIC AND AUTOMATIC RIFLES AND CARBINES

	M1 Rifle	M1C Rifle	M1D Rifle	M1 Carbine	M1A1 Carbine	M2 Carbine
Caliber:	.30 (.30–06).	.30 (.30–06).	.30 (.30–06)	.30 Carbine M1.	.30 Carbine M1.	.30 Carbine M1.
System of operation:	Gas, semiautomatic	Gas, semiautomatic.	Gas, semiautomatic.	Gas, semiautomatic.	Gas, semiautomatic.	Gas, selective fire
Length overall:	43.6 in.	43.6 in.	43.6 in.	35.6 in.	Stock extended: 35.5 in. Stock folded: 25.4 in.	35.6 in.
Barrel length:	24 in.	24 in.	24 in.	18 in.	18 in.	18 in.
Feed system:	8-round, staggered row non-detachable, box magazine.	8-round, staggered row, non-detachable, box magazine.	8-round, staggered row, non-detachable, box magazine.	15- or 30-round, staggered row, detachable, box magazine.	15- or 30-round, staggered row, detachable, box magazine.	15- or 30-round, staggered row, detachable, box magazine.
Sights: Front:	Blade with protecting ears.	2.2X telescopic.	2.2X telescopic.	Blade with protecting ears.	Blade with protecting ears.	Blade with protecting ears.
Rear:	Aperture.			Aperture on ramp.	Aperture on ramp.	Aperture on ramp.
Weight:	9.5 lb	11.75 lb.	11.75 lb.	5.5 lb.	6.19 lb.	5.5 lb.
Muzzle velocity:	2805 f.p.s. (M2 ball)	2805 f.p.s. (M2 ball)	2805 f.p.s. (M2 ball)	1970 f.p.s.	1970 f.p.s.	1970 f.p.s.
Cyclic rate:	—	—	—	—	—	750–775 r.p.m.

CHARACTERISTICS OF UNITED STATES SERVICE LIGHT SEMIAUTOMATIC AND AUTOMATIC RIFLES AND CARBINES (Cont'd)

	Johnson M1941 Rifle	M14 Rifle	M14A1 Rifle	M16 Rifle
Caliber:	.30 (.30–06).	7.62mm NATO.	7.62mm NATO.	5.56mm (.223).
System of operation:	Recoil, semiautomatic.	Gas, selective fire.	Gas, selective fire.	Gas, selective fire.
Length overall:	45.87 in.	44.14 in.	44.3 in	39 in.
Barrel length:	22 in.	22 in.	22 in.	20 in.
Feed system:	10-round, rotary type, non-detachable magazine.	20-round staggered row, detachable, box magazine.	20-round staggered row, detachable, box magazine.	20-round staggered row, detachable, box magazine.
Sights: Front:	Post with protecting ears.	Blade with protecting ears.	Blade with protecting ears.	Post with protecting ears.
Rear:	Aperture.	Aperture.	Aperture.	Aperture.
Weight:	9.5 lb.	8.7 lb.	12.75 lb.	6.3 lb. w/o magazine.
Muzzle velocity:	Approx. 2770 f.p.s. (M2 Ball)	2800 f.p.s.	2800 f.p.s.	3250 f.p.s.
Cyclic rate:	—	750 r.p.m.	750 r.p.m.	700–900 r.p.m.

SQUAD AUTOMATIC WEAPONS

M249 SQUAD AUTOMATIC WEAPON (MINIMI)

The new US M249 Squad Automatic Weapon is described in detail in Chapter 9 (Belgium).

THE BROWNING CALIBER .30 AUTOMATIC RIFLE MODEL 1918, 1918A1 AND 1918A2 (OBSOLETE)

This weapon was developed by John Browning in 1917 to meet the United States requirement for an automatic rifle for service in World War I. The M1918 was made during World War I by Colt, Winchester and Marlin Rockwell. These concerns made 85,000 weapons before the armistice concluded the war.

Basic BARs

There are actually four basic Browning Automatic rifles officially adopted by the United States. These weapons and their descriptions follow.

Model 1918. This weapon has no bipod, is capable of selective fire and is relatively light (16 pounds) compared with the later models. A simple tube type flash hider is used. There is no shoulder support plate hinged to the butt plate. The rear sight and butt plate are similar to those of the Model 1917 (Enfield) rifle.

Model 1918A1. This model has a shoulder support plate hinged to the butt plate and a bipod attached to the gas cylinder just

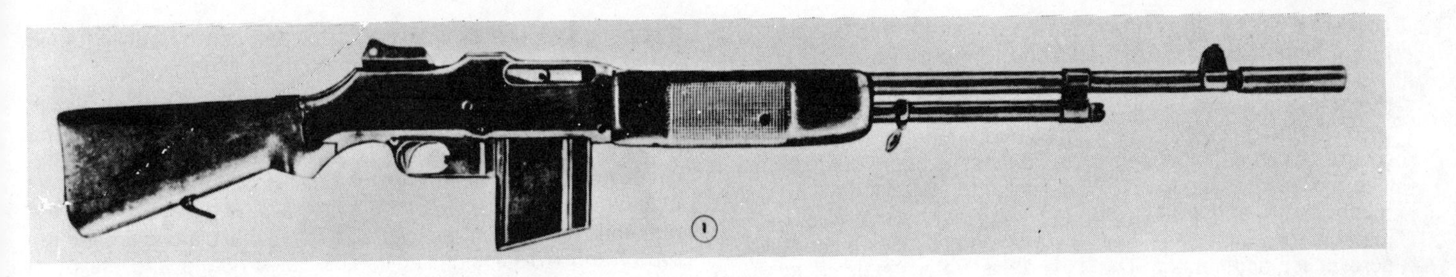

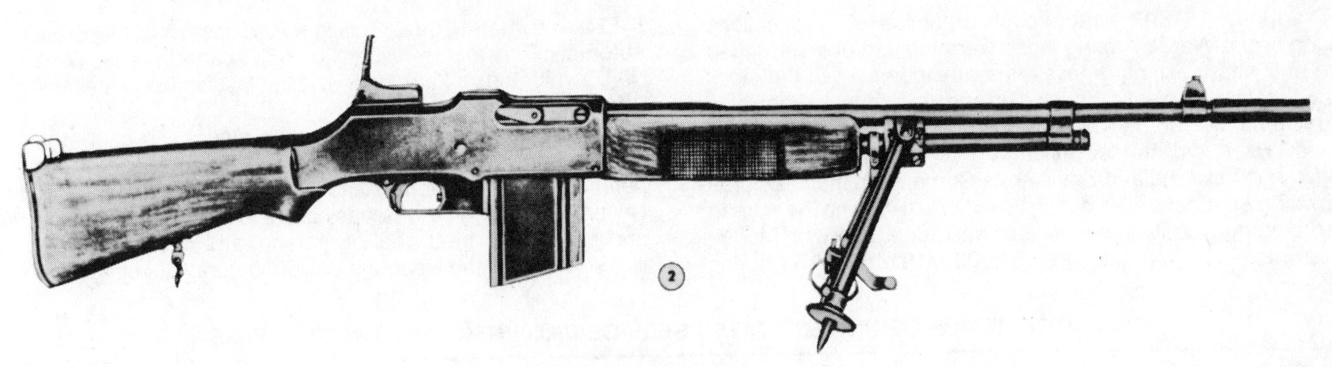

Caliber .30 Browning Automatic Rifles.
(1) Model 1918 (2) Model 1918A1.

forward of the forearm. It fires selective fire, has a tube-type flash hider and uses the same sights as the Model 1918. This modification was adopted in 1937.

Model 1918A2. Adopted shortly before World War II, this weapon, as the Model 1918A1, was orginally made up from Model 1918s and 1918A1s of World War I manufacture. The bipod, which has skid type feet as opposed to the spike type feet of the Model 1918A1, is fitted to the tube type flash hider. The forearm has been cut down in height around the barrel and shortened. As originally made, a removable stock rest, which

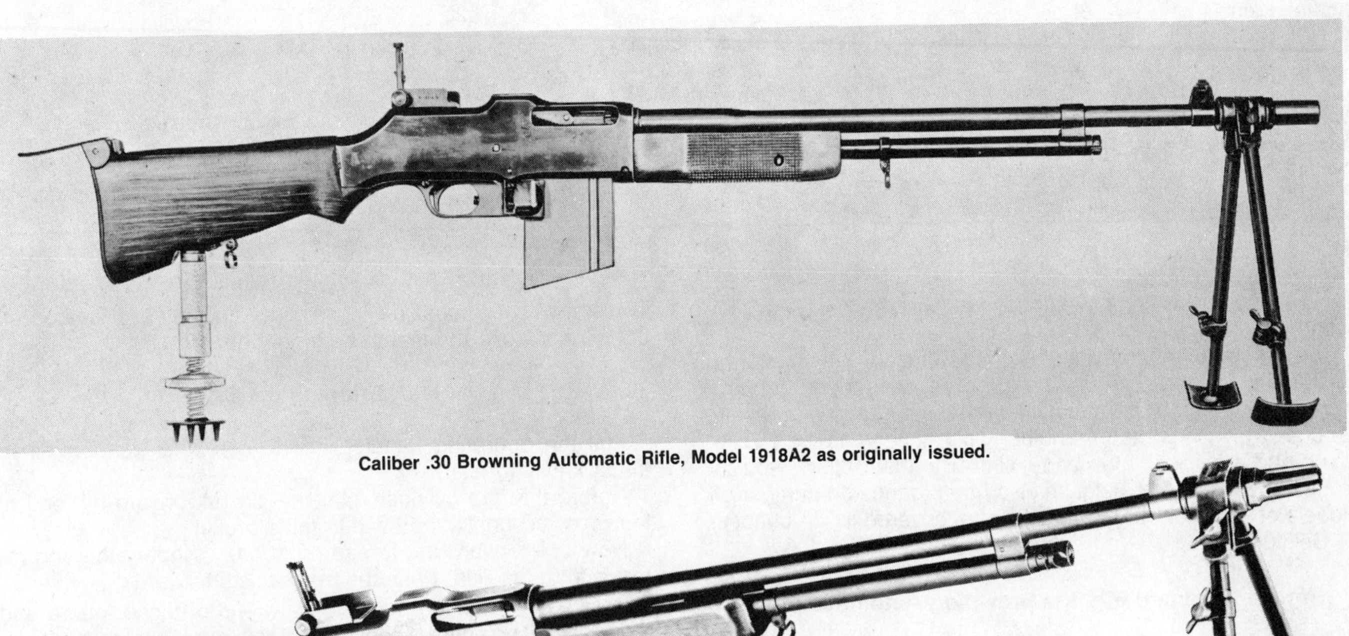

Caliber .30 Browning Automatic Rifle, Model 1918A2 as originally issued.

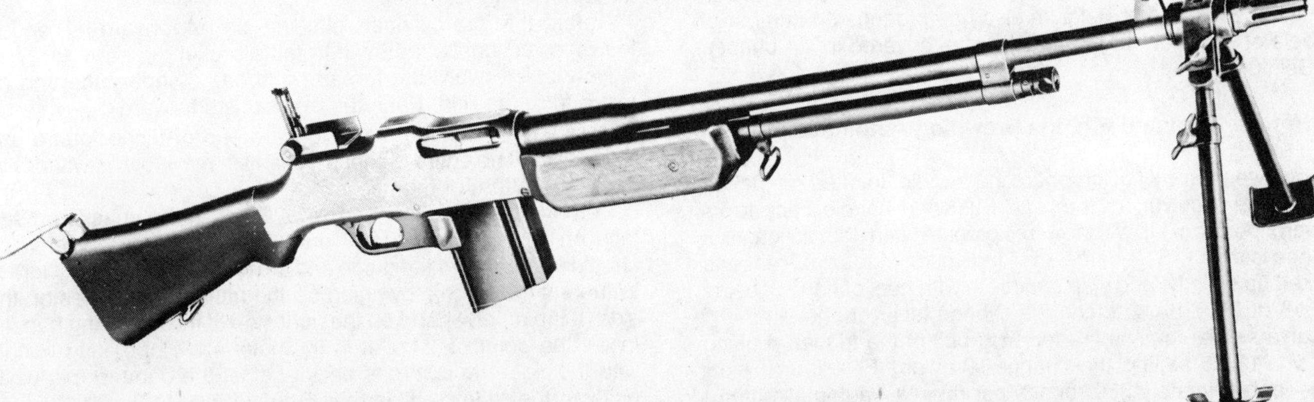

Caliber .30 Model 1918A2 Browning Automatic Rifle of late type. This specimen does not have carrying handle.

fitted in a hole in the buttstock, was used with this weapon. It has a hinged shoulder rest attached to the butt plate as does the M1918A1, but the shoulder rest plate is shorter. There is a metal shield inserted horizontally in the forearm to protect the recoil spring guide from barrel heat, and right and left magazine guards have been attached to the front of the trigger guard body. The rear sight of the M1918A2 is similar to that of the M1918A4 machine gun and the M1922 Browning Automatic rifle. It is adjustable for windage as well as elevation and uses micrometer screws for adjustment. The 1918A2 is not capable of semiautomatic fire. It has two rates of automatic fire and a rate-reducing mechanism.

The Model 1918A2 went through a number of modifications during World War II. Among these were the use of a shortened fore-end with grasping grooves, the abandonment of the stock rest, the use of plastic buttstocks and the development of a carrying handle for the weapon, which, due to the conclusion of World War II, did not see much service until the Korean war. I.B.M. and New England Small Arms Corporation manufactured Browning Automatic rifles during World War II. During the Korean war, a prong type flash suppressor was adopted. Royal McBee Typewriter Corp. manufactured 61,000 M1918A2 BARs during this period. A gas cylinder regulator that can be easily turned by hand was also introduced during this period.

Browning Machine Rifle Model 1922. This weapon, which appeared in very limited numbers, was developed to give the horse cavalry of the twenties lightweight sustained-fire capability. It has a heavy barrel with radial cooling fins, butt swivel attached to the left side of the stock, a wide groove around the buttstock for the butt rest clamp, a bipod that clamps around the barrel and a rear sight adjustable for windage and elevation similar to that used on the Model 1919 machine guns and the M1918A2 BAR. This weapon was declared obsolete about 1940.

T34 Automatic Rifle. This is a modification of the Browning Automatic Rifle for the caliber .30 T65E3 cartridge case (7.62mm NATO) initiated in June 1949. It is the last United States military Browning Automatic rifle.

The Browning Automatic rifle was made with cast steel receivers during World War II and was the subject of much experimentation in materials and methods of manufacture by the United States, since it is basically a difficult weapon to make and requires a great deal of material and machine time. It is, like so many of the weapons designed during its period, built to last a lifetime with commensurate disabilities in cost and tool expenditure.

Colt's Patent Firearms Manufacturing Company manufactured versions of the weapon for sale to police and foreign governments. The Colt Monitor had a shortened barrel and was widely used as a police weapon; another Colt model was the R75A, which had a quick change barrel similar to that of the Swedish Model 37 BAR. FN of Belgium also has produced a number of models of the Browning Automatic rifle, covered in the chapter on Belgium.

CHARACTERISTICS OF UNITED STATES SERVICE BROWNING AUTOMATIC RIFLES

	Model 1918	Model 1918A1	Model 1918A2	Model 1922
Caliber:	.30(.30–06).	.30(.30–06).	.30(.30–06).	.30(.30–06).
System of operation:	Gas, selective fire.	Gas, selective fire.	Gas, automatic only.	Gas, selective fire.
Length overall:	47 in.	47 in.	47.8 in.	Approx. 41 in.
Barrel length:	24 in.	24 in.	24 in.	18 in.
Feed device:	20-round, staggered row, detachable, box magazine.	20-round, staggered row, detachable, box magazine.	20-round, staggered row, detachable, box magazine.	20-round, staggered row, detachable, box magazine.
Sights: Front:	Blade.	Blade.	Blade.	Hooded blade.
Rear:	Leaf w/aperture battle sight w/ aperture.	Leaf w/aperture battle sight w/ aperture.	Leaf w/aperture adjustable for windage.	Leaf w/aperture adjustable for windage.
Weight:	16 lb.	18.5 lb.	19.4 lb.	19.2 lb.
Muzzle velocity:	2805 f.p.s.	2805 f.p.s.	2805 f.p.s.	Approx. 2700 f.p.s.
Cyclic rate:	550 r.p.m.	550 r.p.m.	Slow: 300–450 r.p.m. Fast: 500–650 r.p.m.	550 r.p.m.

How to Load and Fire the Browning Automatic Rifle

Cock weapon by pulling operating handle, located on the left side of the receiver, to the rear. Push the handle back to its forward position—it will now reciprocate with the action. Put change lever on "S"—safe—marking. Insert magazine, rear end cocked up slightly, and slap smartly in with heel of hand. If using M1918 or M1918A1, setting the change lever on the letter "F" will give semiautomatic fire for each pull of the trigger. If using the M1918A2, setting the change lever on "F" will give slow rate—approximately 350 rounds per minute—automatic fire. If the change lever is set on "A", all models will fire automatic fire. To remove the magazine, push magazine release located on the interior front surface of the trigger guard.

Field Stripping the BAR

Pull the operating handle back to cock the weapon. Then thrust it fully forward.

Rotate the gas cylinder retaining pin (at forward left end of the receiver) and withdraw it from its socket.

Now pull forward the forearm and gas cylinder tube and remove from the rifle. Ease the mechanism forward.

Rotate the retaining pin at forward end of trigger guard and withdraw it. The entire trigger mechanism may now be withdrawn from the bottom of the rifle.

Remove the recoil spring guide. Press in the checkered surface on its head and turn it until the ends clear the retaining shoulders; ease out the guide and the recoil spring and withdraw. Withdraw the handle by lining up the hammer pin holes on the side of the receiver and on the right side of the operating handle. Insert the point of a bullet in the hole in the operating handle with the right hand. Press back against the hammer pin while pushing the slide backward with the left hand.

As the two holes register, the pressure of the bullet will force the hammer pin out of the large hole on the left side of the receiver and it may be withdrawn. This will permit the operating handle to be pulled straight to the rear and out of its guide.

Push the hammer forward out of its seat in the slide and lift it out of the weapon.

Pull the slide directly forward out of the receiver, taking care that the link is pushed well down so that slide can clear it. Remove the slide carefully to avoid striking the gas piston or its rings against the gas cylinder tube bracket female.

With the point of a bullet, force out the spring bolt guide from inside the receiver, then lift the bolt, bolt lock and link by pulling slowly to the rear end of the receiver and then lifting them out. The firing pin may now be lifted out of the bolt and the extractor removed by pressing the small end of the cartridge against the claw and exerting upward and frontal pressure. No further stripping is normally necessary or recommended.

How the Basic Browning Automatic Rifle Action Works

Starting with the gun loaded and cocked, the action is as follows. Pressing the trigger pulls down the nose of the sear, disengaging it and permitting the slide to move forward under the action of the recoil spring. The rear end of the slide contains the hammer, which is connected by a link to the bolt. The slide is pulled forward by the compressed recoil spring. During the first quarter-inch of travel of the slide, the front end of its feed rib strikes the base of the top cartridge in the magazine, driving it ahead toward the firing chamber.

When the cartridge has traveled about a quarter of an inch, the bullet strikes the bullet guide on the breech and is deflected upward toward the chamber. This action also guides the front end of the cartridge from under the magazine lip. When the head of the cartridge reaches the part of the magazine where the locking lips are cut away and the opening enlarged, the magazine spring forces it out of the magazine. The base of the cartridge now slides across the face of the bolt and under the extractor; if it fails to position correctly the extractor will snap over its head as the bolt reaches its forward position. At the time the cartridge leaves the magazine, the bullet nose is so far in the chamber that it is guided from that point on.

When the slide is within two inches of its complete forward position, a circular cam surface on the bottom of the bolt lock starts to ride over the rear shoulders of the bolt support, camming up the rear end of the bolt lock. The link pin rises above the line joining the bolt pin and the hammer pin, so that its joint has a tendency to buckle upwards. As the attached bolt is now opposite its locking recess in the receiver, it pivots upward about the bolt lock pin. The link, whose lower edge is attached to the hammer pin, revolves upward and forces the bolt lock up; the rounded surface on the bolt lock, just above this locking face, slips over the locking shoulder in the hump of the receiver and provides a lever thrust, forcing the attached bolt home into final position.

The bolt lock is now above the position of the bolt, and locks firmly in the hump in the receiver as the hammer pin passes beneath the link pin. The firing pin is in the bolt, with a lug on its rear end buried in the slot at the other side of the bolt lock, making it impossible for the firing pin to be struck by the hammer at any time except when the bolt lock is in its recesses in the hump of the receiver. Thus when the hammer pin passes under

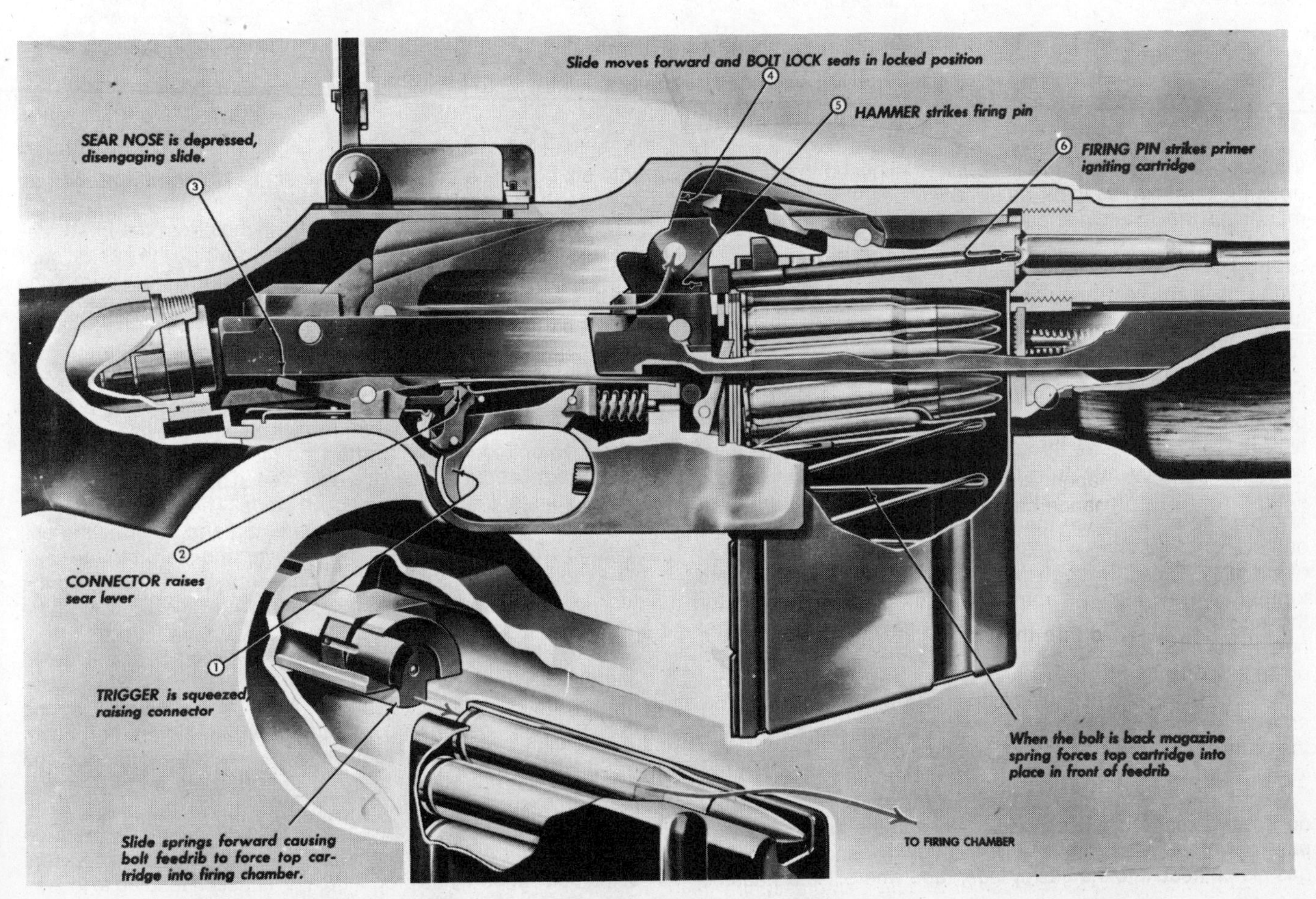

Details of firing action of the Browning Automatic Rifle as trigger is pressed. Follow numbers to study action sequence.

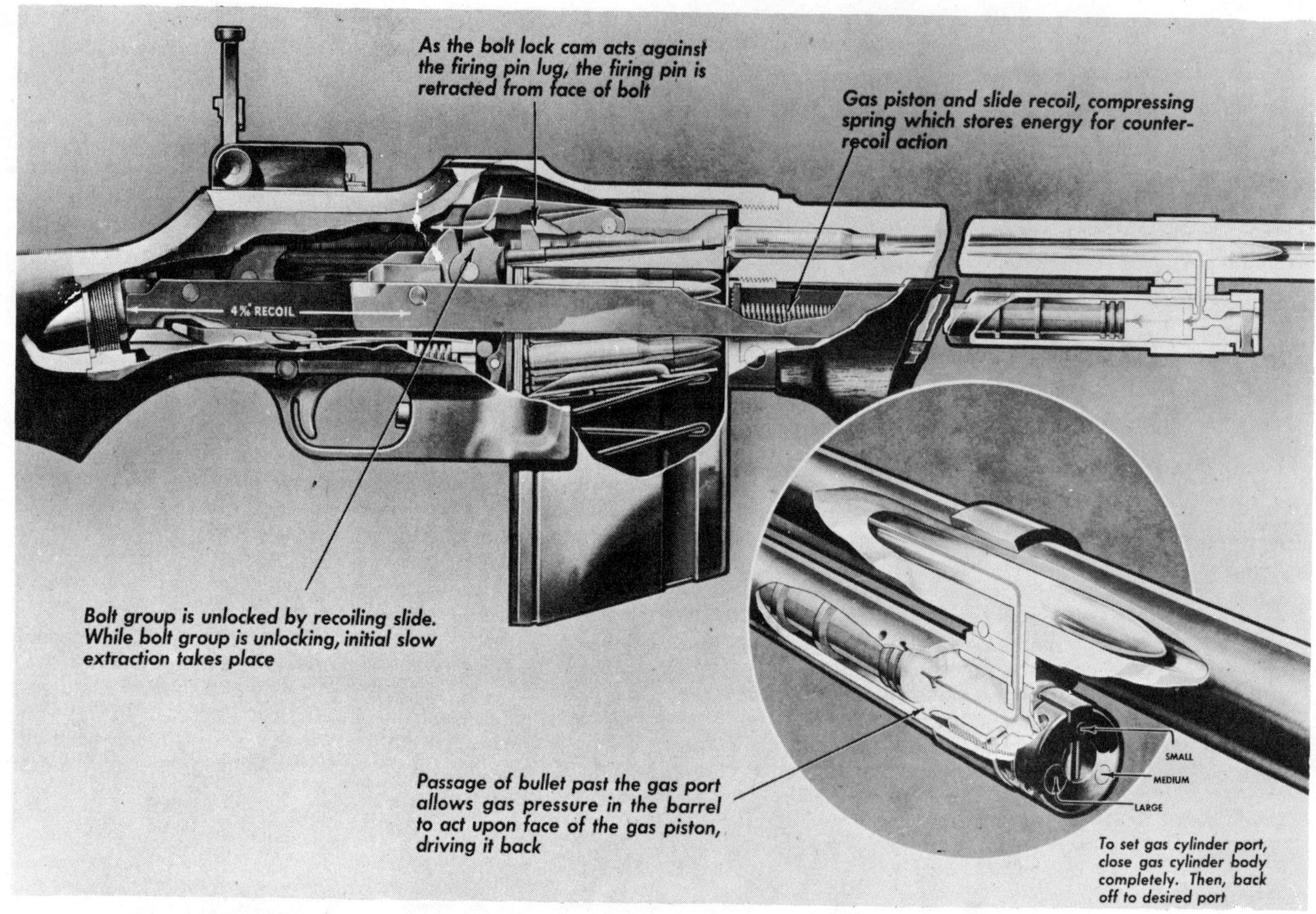

Recoil and unlocking action of the Browning Automatic Rifle M1918A2 showing action beginning as bullet passes over gas port and gas escaping into cylinder which starts the rearward action.

the link pin, the head of the firing pin is exposed to the center rib of the hammer; as the slide still continues forward, the hammer drives the firing pin ahead and explodes the cartridge in the chamber.

The forward motion is now halted when the front end of the slide strikes against a shoulder at the rear end of the gas cylinder tube.

Return Movement of the Action. About 6 inches from the muzzle, a small port is bored in the bottom of the barrel. As the bullet passes over this port, a small amount of gas, still under high pressure, escapes through it and passes through similar ports in the gas cylinder tube bracket, the gas cylinder tube and the gas cylinder. The gas cylinder port is the smallest of these and acts as a throttle on the barrel pressure. The ports in the gas cylinder lead radially into a small well situated in the head of the gas cylinder. Through this well, the pressure is conducted to the gas system plug, through which it acts on the piston for the length of time the bullet is traveling the 6 inches from barrel port to muzzle. This results in a sudden, hard blow, backward against the piston plug.

The gas piston is assembled to the slide, and the sudden blow as it drives the gas piston back also forces back the slide and the parts attached to it and compresses the recoil spring seated in the slide.

After the piston has traveled back a little over ½ inch, bearing rings on its rear and corresponding ones in the gas piston plug pass out of the gas cylinder. The gas now expands around the gas piston head into the gas cylinder tube where it is exhausted into six portholes in the tube placed just at the rear of the gas cylinder tube brackets.

Two rings on the piston about 1-¼-inches from the head prevent most of the gas from traveling back through the gas cylinder tube and also act as bearings to maintain the front end of the piston in the center of the gas cylinder tube after the piston has passed out of the cylinder.

Unlocking Action. As the hammer pin is slightly in advance of the connecting link pin, the initial backward movement of the slide carries the hammer back without moving either the attached bolt lock or bolt; when the movement has progressed far enough (about ⅕") and the high breech pressure has dropped to safe limits, the unlocking action starts. The link is compelled to revolve forward about the hammer pin and so draw the bolt lock down out of a hump in the receiver and start it to the rear. The motion of the bolt and bolt lock is now accelerated as the lock is drawn completely out of its locking recess, locking the shoulders in the receiver.

As the bolt lock is prevented from revolving from below the line of backward travel of the bolt, further rearward travel of all moving parts is in a straight line. Meanwhile, however, during the unlocking motion, a cam surface on the slot in the bottom side of the bolt lock has come in contact with a cam surface on the firing pin lug and has drawn the firing pin away from the base of the bolt.

Also during the backward action, the circular cam surface on the lower part of the bolt lock, operating on the rear shoulders of the bolt support, has produced a lever action tending to loosen the cartridge case in the firing chamber. From that point, the slide and all its moving parts are traveling to the rear at the same speed, carrying along the empty cartridge case held in its seat in the face of the bolt by the extractor (the extractor is positioned in the upper right side of the bolt near the ejection port). Thus as the slide nears the end of its travel and the base of the empty cartridge case strikes the ejector on the left side of the bolt feed rib, the empty case is pivoted about the extractor and through the ejection port. As the front end of the cartridge case passes out of the receiver, it is so pivoted that it strikes the outside of the receiver about an inch to the rear of the ejection port and hence rebounds toward the right front.

The rearward motion is now completed as the end of the slide strikes against the end of the buffer and the sear nose catches in the notch at the underside of the slide and holds the weapon open and ready for the next pull of the trigger. (If the weapon is set for full automatic fire, the sear nose is held depressed, so that it does not stop the slide, which continues forward, firing the weapon in the full automatic cycle.)

The buffer is a tube in the butt of the rifle in which are placed a buffer head against which the slide stops, a friction cup slit to allow for expansion, a steel cone to fit into the cup, and four more cups and cones in series. Behind these is the coil buffer spring and the buffer nut, which is screwed into the end of the tube to form a seat for the spring.

As the rear end of the slide strikes the buffer head, it moves it to the rear, forcing the cups over the cones causing them to expand tightly against the tube, thus producing friction as the cups move back and the buffer spring is compressed. The rearward motion of the slide is therefore checked gradually, and practically no unpleasant rebound occurs. The friction mechanism is returned to its original place by the compression of its spring.

Notes on the BAR

A very important feature of this rifle is that the bolt, bolt lock and link mechanism start back comparatively slowly and do not attain the speed of the slide itself until after the period of high breech pressure passes. This feature is also important in that it does not subject the mechanism to undue strain as the gas pushes the piston back.

There are three different gas ports. The weapon will normally be set to operate on the smallest port. It is properly aligned by screwing in the gas cylinder with combination tool until the shoulder of the cylinder is one turn from the corresponding shoulder of the gas cylinder tube and the smallest circle on the cylinder head is toward the barrel. (To permit setting the regulator, the split pin must be pushed out sufficiently to permit the regulator to be turned on older weapons.)

If the rifle is sluggish from insufficient gas, the cylinder should be set one complete turn on each side of the original setting. However, it is to be noted that the larger ports are provided only for emergency use. They should be utilized only when through lack of oil or accumulation of dirt or carbon, the rifle is sluggish and conditions make it impossible to properly correct these troubles. It is therefore essential that the threads be kept cleaned and oiled and cylinder free to turn at all times.

In field service, at the first sign of insufficient gas, unscrew the cylinder a third of a turn, and line up the medium circle and port with the gas opening.

When gas is insufficient, the weapon may fail to recoil because the port is not properly aligned or is unusually dirty. A very dirty mechanism may also cause such a stoppage. Or the weapon may not recoil far enough to permit complete ejection, or the ejection may be weak. Under some conditions, although this is unusual, it may result in uncontrolled automatic fire.

On the other hand if the gas pressure is too high, the rifle will be speeded up too much causing a pounding, which will interfere with accuracy.. This may also generate excessive heat in the gas operating mechanism.

How the Browning Automatic Rifle Model 1918A2 Works

The Model 1918A2 is a modification of the M1918A1 Browning Automatic rifle. The change lever spring and the carrier have been modified. Several new components have been added. These include a sear release stop lever, a sear, key and head buffers, sear release actuator and actuator spring, actuator stop and buffer head. The bipod is attached to the flash hider. A stock rest has been added and the forearm made lighter. A trap plate has been added between the barrel and the gas cylinder tube.

This gun has been modified to replace the single-shot mechanism. Single shots can be fired in this modified version only by pressing and releasing the trigger rapidly.

However, as a compensating factor, the gun has been designed to fire at a low and a high rate of speed somewhat in the manner of the British Besa Machine Guns.

When the change lever is at "F," the rifle is cocked in normal fashion and the sear engages in a notch in the slide. Pressing the trigger in the usual manner results in controlled automatic fire at a reduced rate, which will be delivered as long as the trigger is held back. There is a distinct difference noticeable when handling the gun when the mechanism is in operation and when it is firing full automatic without it. There is no provision made for semiautomatic fire in this model.

On pressing the trigger, the slide goes forward in normal BAR fashion firing the cartridge; then the slide starts on its rearward movement in the usual manner, driven back by gas expanding into the gas cylinder, but it picks up the sear release and strikes it on the front end. This forces the sear release to the rear until the slide reaches the face of the buffer head, while during this movement it also meets the front end of the actuator. The actuator tube is forced to the rear, meeting the actuator stop. At this point, the actuator reverses its direction of travel moving forward under the tension of the actuator spring, and the slide engages on the sear. The slide remains in engagement with the sear until the actuator reaches its extreme forward position. At that point, the actuator forces the sear release forward forcing it to move through the buffer head, while the foot of the sear release is in contact with the angle surface of the rear of the sear. This cams the sear out of engagement with the slide forcing the slide to go forward at this point to fire the cartridge.

The M1918A2 was the standard squad automatic weapon of World War II and Korea.

UNITED STATES SUBMACHINE GUNS

Although the United States was the third country in the world to develop a submachine gun, this type weapon was not adopted by the United States until about 1928 when it was first used by the Marines in Nicaragua and by the Coast Guard in their war with the rum runners of the prohibition period. The weapon used was the caliber .45 Thompson Model 1928. Developed by General John T. Thompson and the Auto Ordnance Corp., it had made its first public appearance in earlier form in 1919. The first production Thompson was the Model 1921, and the "Tommy Gun" earned a reputation, probably unfairly and mainly due to movies, as a gangster weapon during the age of the "big gang wars" in the United States. The weapon was widely used by police forces, and the attitude seems to have been adopted both in the United States and the United Kingdom that the submachine gun was basically a police weapon. Be that as it may, the first submachine gun purchased by the United States Army was the caliber .45 Thompson Model 1928A1, purchased in limited quantities, principally for use by armored and reconnaissance units. The Thompson was being produced by Colt at this time; the patent owner—Auto Ordnance Corp.—did not have any manufacturing capability at the time and developed only a limited manfacturing capability during World War II. Colt produced approximately 15,000 Thompsons. In 1940, the British government gave large contracts to Auto Ordnance Corp., for the Model 1928A1 Thompson. Auto Ordnance subcontracted most of this order to the Savage Arms Corp., then at Utica, New York. When Lend-Lease came into effect—1941—the United States government took over these contracts, and the Thompson remained in production until 1943. During the course of this contract, modifications were made in the Model 1928A1, and M1 and M1A1 models were produced. Savage made 1,501,000 Thompson M1928A1, M1 and M1A1 submachine guns. Many of these weapons made on the British contract did not make it across the Atlantic because of the German U-boat campaign, at its height when shipments of the Thompson to the United Kingdom were at their greatest. The Thompson, while a reliable weapon, has several outstanding shortcomings. It is overly heavy in relation to its muzzle energy, and more importantly it is expensive in the use of materials, machine time and machine tools, all items that are in short supply during a large war. The Army therefore decided to find or develop a weapon to replace the Thompson. In 1941, a requirement was generated for a new weapon, and a number of guns were submitted in competitive tests to meet this requirement. Among these were the Hyde 109, first tested in 1939 by the Army and Hyde Inland, the ATMED—also designed by George Hyde—the Star, the Atlantic (also a Star design), the United Defense, the Reising 1 and 2, the Olsen, the Owen, the Sten Mark III, the Austen, the Woodhull, the Suomi, the Turner, the Smith & Wesson semiautomatic carbine, the standard Thompsons, the Thompson T2, the German MP38

and Bergmann. The MP38 and the Bergmann were not official competitors but were under study.

The Hyde Inland was adopted as the "substitute standard" caliber .45 submachine gun M2 in April 1942. The M2 was never put in mass production, and the first production models reached Aberdeen Proving Ground for test in May 1943, five months after the M3 had been standardized.

The Reising Gun, in a fixed-stock version, the Model 50, and a folding stock version, the Model 55, was manufactured by Harrington and Richardson for the Marine Corps, the Home Guard and the British Purchasing Commission. Approximately 100,000 weapons were made. Many wound up in the armories of local police departments throughout the United States. It is no longer in military service.

The United Defense gun, known as the U.D. Model 42, was designed at High Standard by Carl Swebilius, and approximately 15,000 were made by Marlin for the United Defense Supply Corporation. This weapon, which was chambered for the 9mm Parabellum cartridge, was used by the OSS and air dropped to various underground organizations. The Dutch government also purchased a few. It was known in the UK as the Marlin.

The weapons that eventually developed into the M3 submachine gun were the T15, a selective-fire weapon, and the T20, an automatic only weapon, which could be converted from caliber .45 to 9mm Parabellum. This weapon was made of stampings with a minimum of machined steel parts and was at least partially the result of studies made of the British Sten. Mr. George Hyde did the basic design work, and the industrial engineering was handled by Mr. Frederick Sampson of the Inland Manufacturing Div. of General Motors Corp. The M3 was adopted in December 1942. In December 1944, the M3A1, a modification of the M3, was adopted as standard.

THE CALIBER .45 M3 AND M3A1 SUBMACHINE GUNS

The M3 submachine gun was adopted in December 1942. It had a number of deficiencies that showed up in field service; these were corrected in the M3A1 submachine gun, standardized in December 1944.

The M3 submachine gun was designed so that by changing the barrel and bolt and adding an adaptor to the magazine, it could be used with 9mm Parabellum cartridges. There is a version of the M3 with silencer built into the barrel. Approximately 1,000 of these were made for the OSS during World War II. Guide Lamp Division of General Motors Corp. produced approximately 646,000 M3 and M3A1 submachine guns during World War II.

A curved barrel was made for use with the M3A1 submachine gun after World War II, and a flash hider was developed for use with both the M3 and M3A1 submachine guns. Approximately 33,200 M3A1s were made by the Ithaca Gun Co. of Ithaca, New York, during the Korean War.

Differences Between M3 and M3A1

The principal differences between these weapons are as follows.

M3. The bolt is pulled to the rear by means of a spring loaded retracting of lever assembly.

M3A1. The bolt has a finger hole in its right front side for cocking the gun. The magazine catch has a guard to prevent it being accidentally depressed. The ejection port and its cover are longer, and the safety lock on the ejection port cover is placed further to the rear. Disassembly grooves were added so that the bolt can be removed without removing the housing assembly. A stock plate and magazine filler were added to the stock, and a larger oil can is fitted inside the grip. The retracting pawl notch is eliminated, and a clearance slot for the cover hinge rivets was added. The barrel ratchet was redesigned to provide a longer contact surface for easier disengagement from the barrel collar.

How the M3 Submachine Gun Works

When a loaded magazine is inserted into the magazine housing and pushed upward until it locks, the cocking handle on the right side of the gun is drawn back to its full extent. This movement raises the bolt cover (which must be opened full to permit proper ejection) and also cocks the bolt against the tension of two recoil springs mounted in the receiver and extending one on each side into the bolt itself.

When the trigger is pressed, the sear releases the bolt. The driving springs force the bolt forward, and the feed guides machined into the bolt strike the rear of the topmost cartridge in the magazine and drive it ahead into the firing chamber.

As the cartridge enters the chamber, the bolt continues forward, and its extractor snaps over the head of the cartridge to fasten in the extracting groove.

The firing pin can now strike the primer and discharge the cartridge.

As the bullet moves out the barrel, the back pressure pushes the empty cartridge case out to the chamber and transmits its force to the bolt face pushing it to the rear.

The bolt assembly as it goes back takes the empty cartridge case with it. The case strikes against the ejector and is thrown out the ejection port.

The driving springs are compressed around their guide rods, and the bolt travels back on tracks in the receiver until the energy is absorbed and the bolt is caught by the sear.

The gun will fire as long as the trigger is held back and there are any cartridges in the magazine. If the bolt cover is pushed down in place when the bolt is back, it acts as a safety by holding the bolt off the sear and also interfering with forward travel. It is also a dust cover.

Special Note on the M3A1 Submachine Gun

The M3A1 is loaded and fired in the same manner as the M3 with the exception that the bolt is retracted by placing the right forefinger in the bolt finger hole and drawing the bolt back until it is engaged by the sear. Disassembly differs from that of the M3 in that the bolt and operating springs can be removed after the barrel is removed without removing the housing.

THE THOMPSON CALIBER .45 MODEL 1928A1, M1, AND M1A1 SUBMACHINE GUNS

The Model 1928A1 Thompson as originally manufactured has a Cutts compensator mounted to the muzzle and a leaf type rear sight adjustable for windage and a barrel with radial cooling fins. Before the end of production of the Model 1928A1, specimens had been produced without compensators, with a simple nonmovable "L" type rear sight and a smooth barrel. All Model 1928A1 Thompsons, however, have a top-mounted actuator, the brass "H" type lock, which operates on the theory of adhesion of different types of metal under pressure (the Blish principle), and breech oiler pads mounted on the receiver. The Model 1928A1 has a removable buttstock and, although usually found with a horizontal fore grip, may be found with a vertical fore grip.

Loading, Firing, and Field Stripping the M3 and M3A1

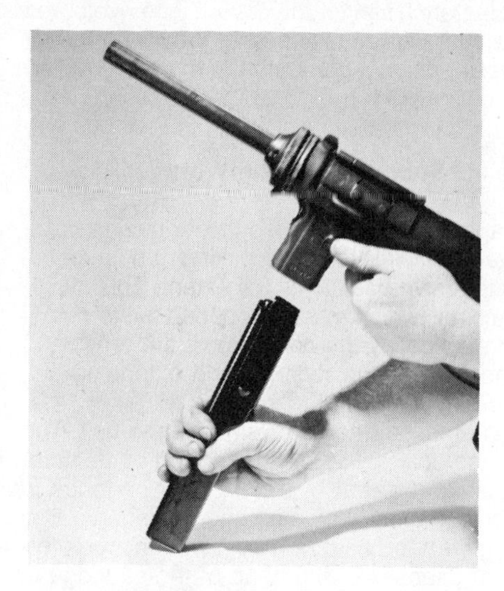

(1) A loaded magazine is inserted in the magazine housing from below until it locks. Pull the cocking handle back as far as it will go to open the ejection port cover to its full extent and draw the bolt back until it is caught and held by the sear.

(2) Release the handle and let it fly forward. If it is desired to carry the gun ready for use but on safety, push the cover down into place. Otherwise the gun is ready to fire on a pull of the trigger.

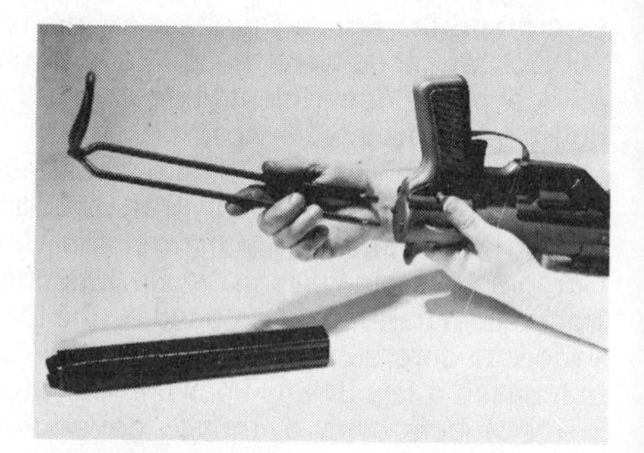

(3) With the magazine out of the weapon, push in the stock catch on the left side of the receiver above the pistol grip with the left thumb and pull the wire stock out of its grooves.

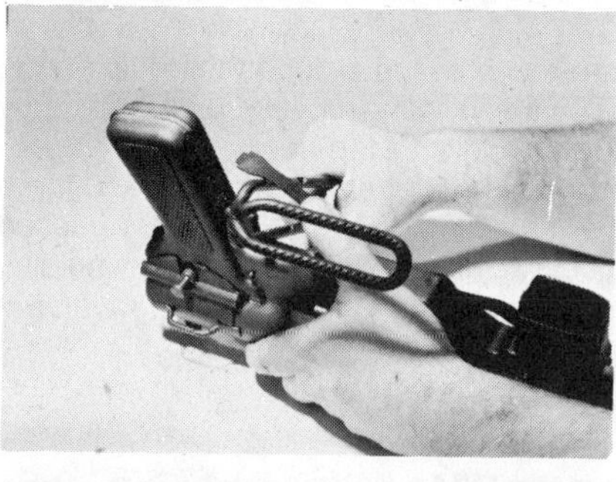

(4) Insert the shoulder end of the wire stock inside the trigger guard to provide a pressure point. Then press down on the lower end of the trigger guard until it springs out of its slot in the pistol grip. Handle this trigger guard carefully as it is of light gauge spring steel. Rotate it toward the muzzle of the gun. This will unhook it so it can be lifted out.

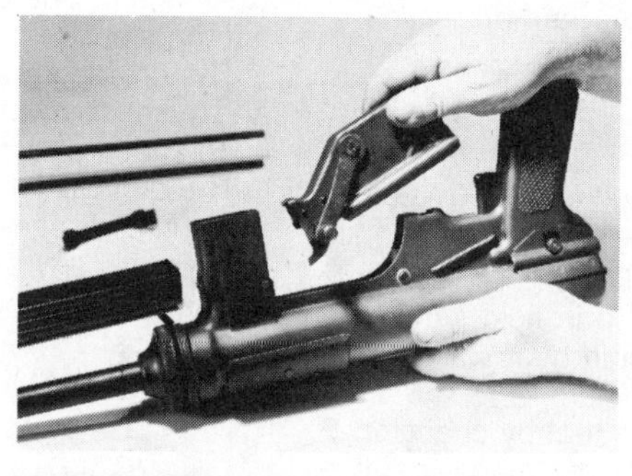

(5) Push down on the housing assembly unit a short distance, and then lift it to the rear until it can be lifted off. This should be done with care to prevent injury to the metal.

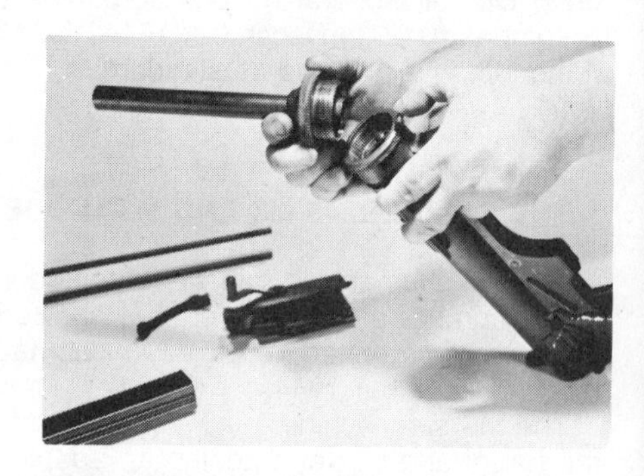

(6) Pull the ratchet catch back with the left thumb and unscrew the barrel assembly from left to right. The barrel and its collar will come out.

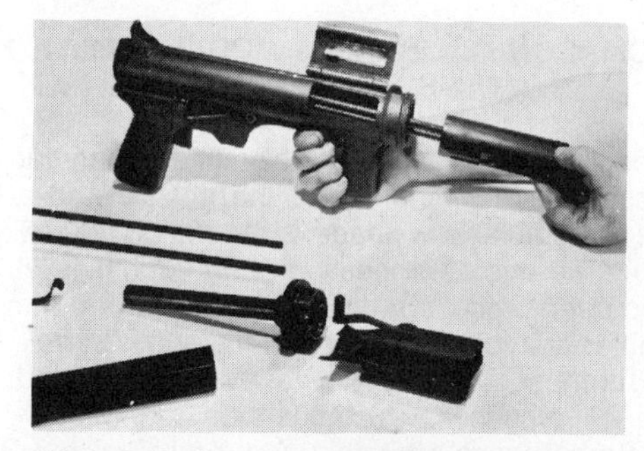

(7) Open the bolt cover and tilt the gun forward. The bolt and the two guide rods and their springs will come forward out the front of the receiver.

(8) No further stripping is normally necessary.

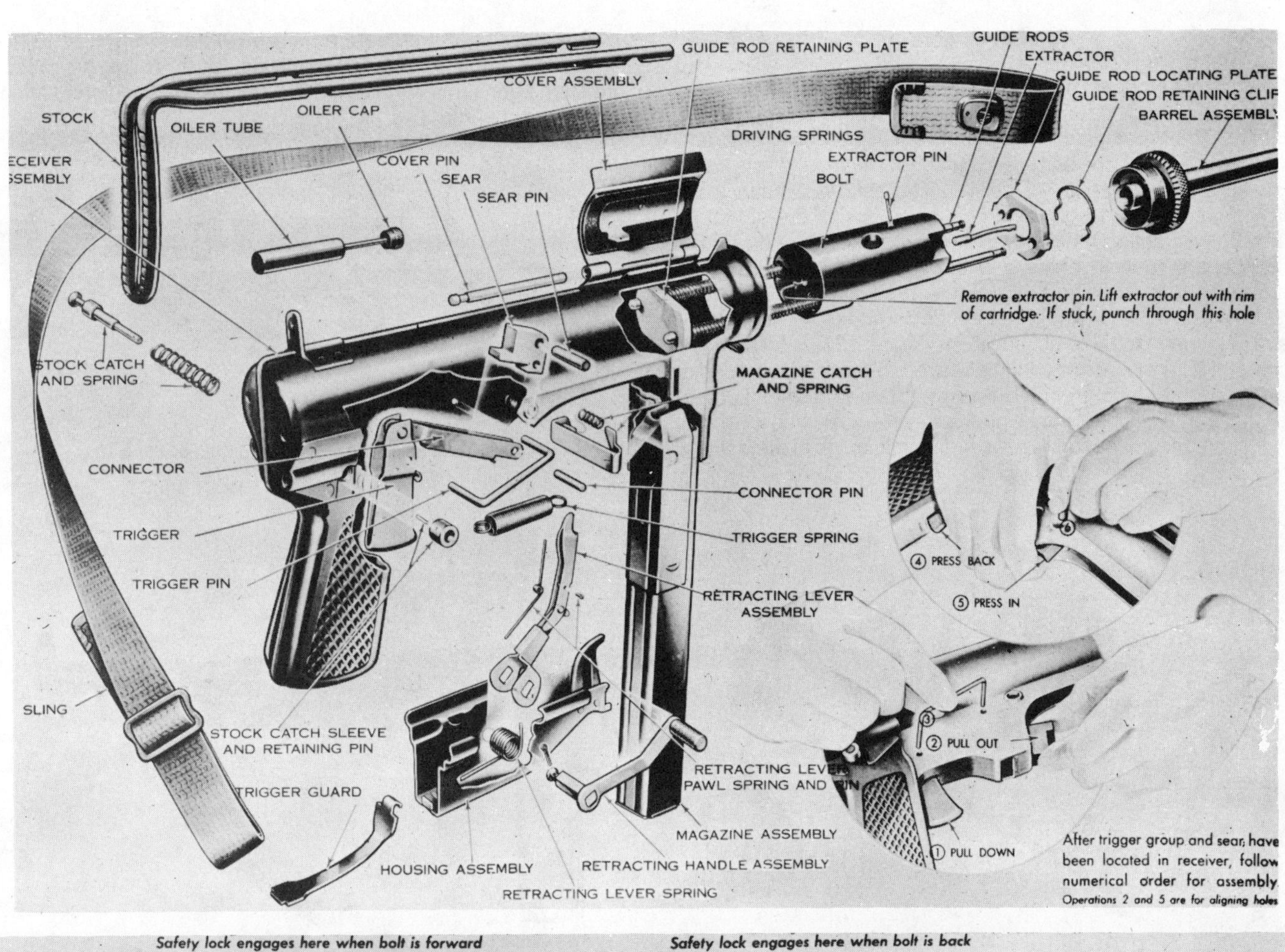
GUIDE ROD RETAINING PLATE
GUIDE RODS
EXTRACTOR
GUIDE ROD LOCATING PLATE
GUIDE ROD RETAINING CLIP
BARREL ASSEMBLY
COVER ASSEMBLY
STOCK
OILER CAP
OILER TUBE
RECEIVER ASSEMBLY
COVER PIN
SEAR
SEAR PIN
DRIVING SPRINGS
EXTRACTOR PIN
BOLT
Remove extractor pin. Lift extractor out with rim of cartridge. If stuck, punch through this hole
STOCK CATCH AND SPRING
MAGAZINE CATCH AND SPRING
CONNECTOR
CONNECTOR PIN
TRIGGER
TRIGGER SPRING
TRIGGER PIN
4 PRESS BACK
5 PRESS IN
RETRACTING LEVER ASSEMBLY
SLING
STOCK CATCH SLEEVE AND RETAINING PIN
2 PULL OUT
3
RETRACTING LEVER PAWL SPRING AND PIN
TRIGGER GUARD
MAGAZINE ASSEMBLY
1 PULL DOWN
HOUSING ASSEMBLY
RETRACTING HANDLE ASSEMBLY
RETRACTING LEVER SPRING
After trigger group and sear have been located in receiver, follow numerical order for assembly. Operations 2 and 5 are for aligning holes

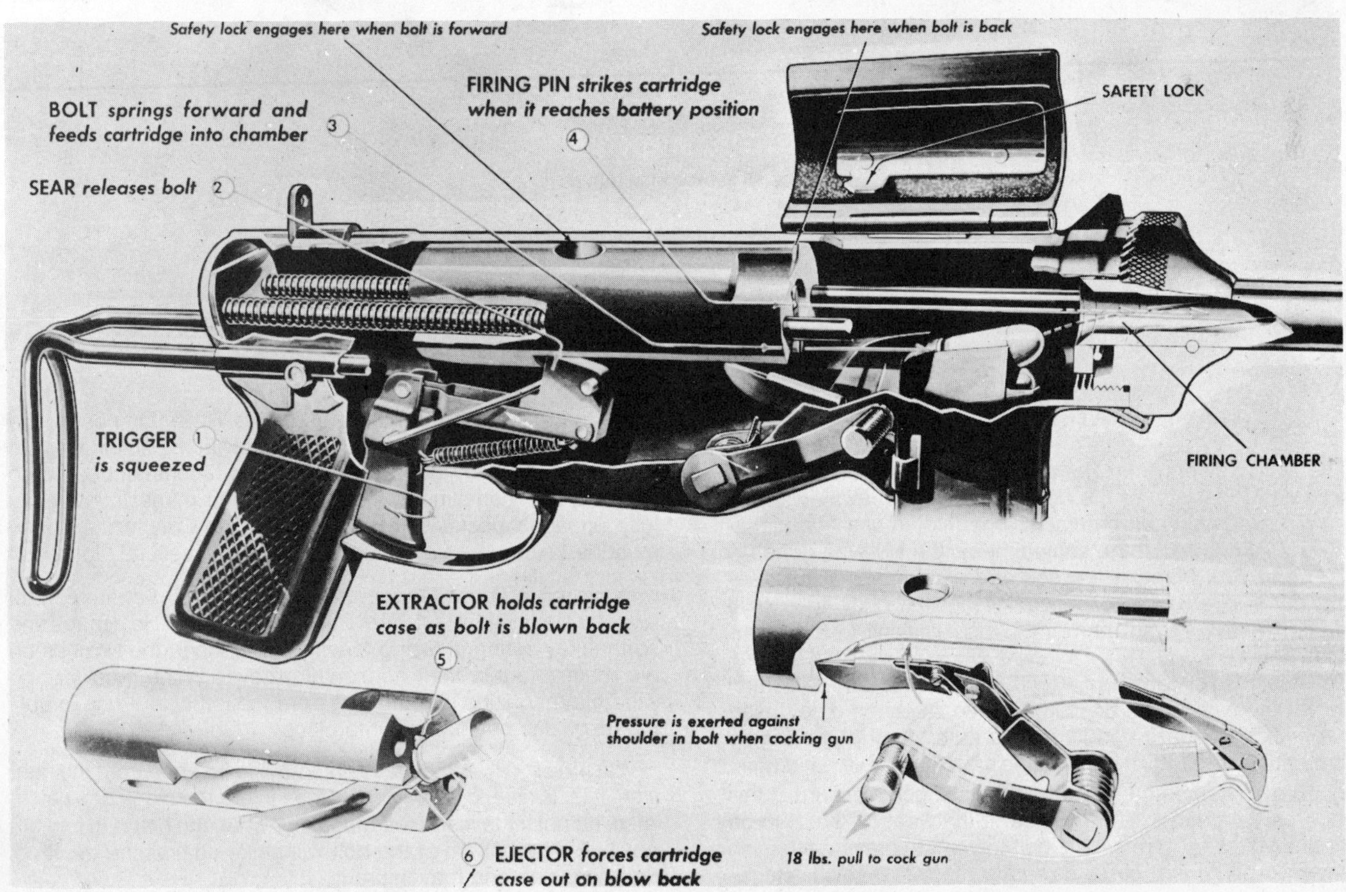
Safety lock engages here when bolt is forward
Safety lock engages here when bolt is back
FIRING PIN strikes cartridge when it reaches battery position 4
SAFETY LOCK
BOLT springs forward and feeds cartridge into chamber 3
SEAR releases bolt 2
TRIGGER 1 is squeezed
FIRING CHAMBER
EXTRACTOR holds cartridge case as bolt is blown back 5
Pressure is exerted against shoulder in bolt when cocking gun
18 lbs. pull to cock gun
6 EJECTOR forces cartridge case out on blow back

M1 Thompson

The M1 Thompson designs were prepared for the simplification of the M1928A1, and in April 1942 the M1 Thompson was adopted. The breech lock, actuator, breech oiler, compensator, radial cooling fins on the barrel and buttstock catch of the M1928A1 were all dropped in the M1 design. The bolt was made a bit heavier, as a result of not being hollowed out for the actuator, and a bolt-retracting handle was fitted into the right side of the bolt where it rode in a track in the receiver. A simple fixed aperture rear sight with sight guards was initially issued; at a later date, the sight guards were dropped. The M1 will not accept the drum type magazine used with the Model 1928A1. The buttstock of the M1 is permanently attached to the trigger housing (frame), and the design of receiver and frame are modified so that the frame slides over a protruding track located on both sides of the receiver and makes a noticeable bulge at the bottom of the receiver, not present on the M1928A1.

The M1, with the exception of the locking and oiling processes, functions the same as the M1928A1. It has a springloaded firing pin and a hammer like the M1928A1.

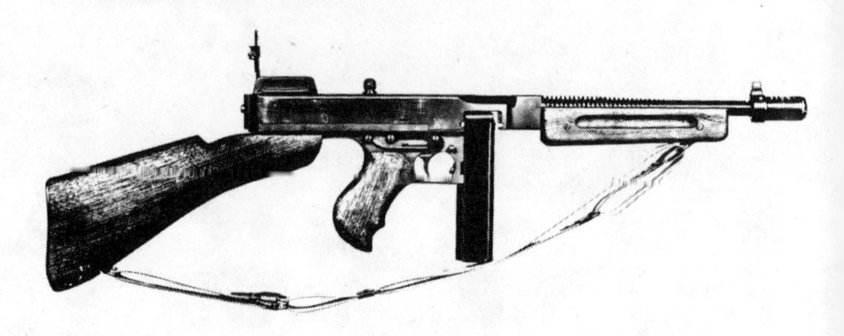

Caliber .45 Model 1928A1 Thompson Submachine Gun.

Caliber .45 Submachine Gun M1.

M1A1 Thompson

The M1A1 differs from the M1 only in having the firing pin machined in the face of the bolt, thereby doing away with the firing pin assembly and hammers. The thirty-round box magazine was introduced at the same time as the M1 type guns.

To Load Box Magazine. Load as for automatic pistol magazine but support base of magazine against body or a solid surface if heavy spring tension makes it difficult to force cartridges down.

To Insert Box Magazine. Cock the gun, set the fire control lever for the type of fire desired. Put the safety on safe. Insert rib at back of magazine in its recess at the front of the trigger guard and push in until the magazine catch engages with a click.

Warning: Remember that when the bolt goes forward in this weapon a cartridge is fired. Hence, if the weapon is not to be fired and you wish to move the bolt forward to prevent straining the recoil spring, first press down the magazine catch and remove the magazine from the gun. **Note:** While it is possible to insert the box type magazine in this gun with the action forward (that is, uncocked), this procedure is not recommended. In so inserting the magazine, make sure that the magazine catch is fully engaged because the overhang of the magazine spring must be taken up before the engagement is securely locked.

Inserting a Drum Magazine. Cock the gun. Set fire control lever for single or full auto fire. Put safety on Safe. Hold magazine so that key spring is facing forward. Now insert the two ribs on the magazine into their horizontal grooves in the receiver and slide the magazine into the gun from the left side. Push in until the magazine catch clicks into place.

Warning: While this magazine may be inserted from the right side, it is unwise to do so as this may injure the magazine catch. Also, do not try to insert the magazine when the bolt is in forward position. The bottom of the bolt will strike against the mouth of the magazine and may injure it.

Loading and Firing the Thompson

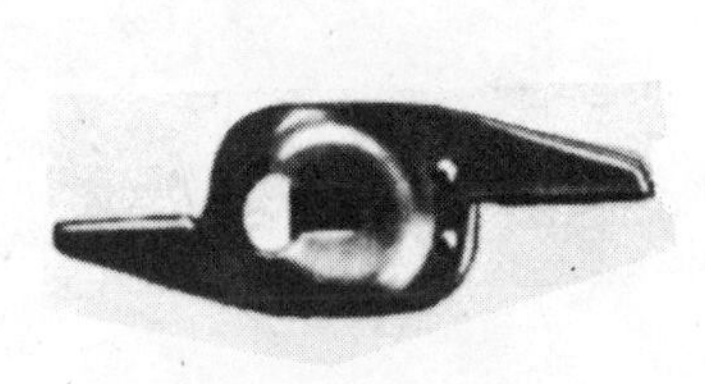

To load the drum magazine raise the flat magazine key spring to disengage its stud and slide the key off via its slot.

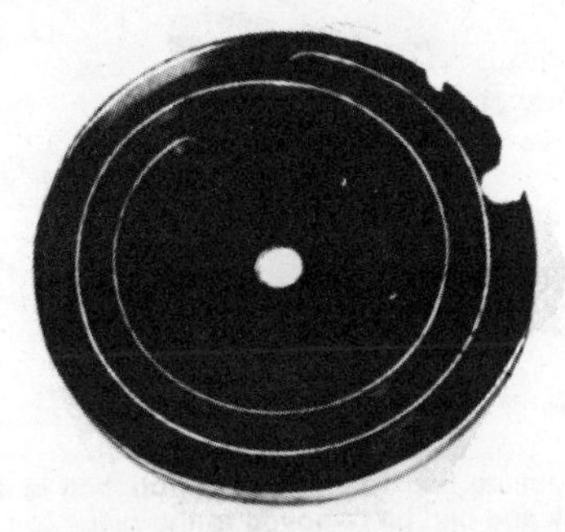

Lift off the magazine cover. Insert 5 cartridges base down in that section of the rotor in which the magazine feed opening is cut.

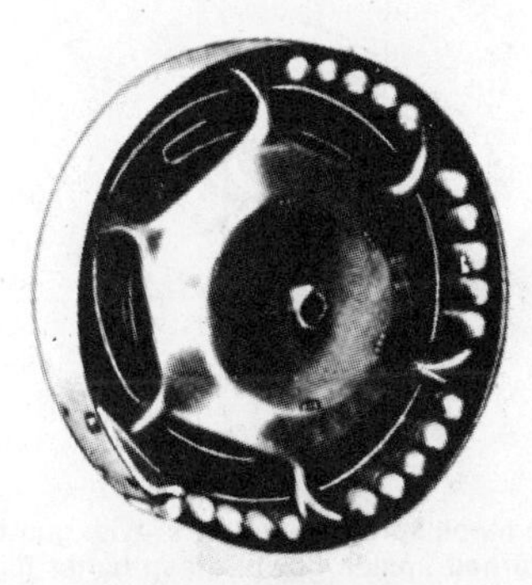

Loading from right to left, place 5 cartridges in each section of the spiral track, taking care to load all outer spirals first.

When correctly loaded, the first four sectors starting left from the magazine opening will contain 10 cartridges each, while the last two will have 5 each.

Warning: Be careful not to insert any cartridges near the loops opposite the two sectors which hold five cartridges each; any cartridges so placed will jam the magazine when the rotor revolves. Now replace the magazine cover. Make sure that the large slot cut in it engages properly with the cover positioning stud. Slide the magazine key into place. Check to be sure that the stud on the spring correctly engages the center piece. Now wind the key from left to right. As it turns you will hear a distinct click. Count the number of clicks. Stamped on the magazine cover you will find the correct number of clicks necessary to indicate sufficient spring tension to work the magazine properly. (The normal number is 9 or 10 for a 50-shot magazine.) Note: If magazine is not to be used at once, wind up only two clicks to assure proper locking of the magazine and prevent straining spring.

Caution: Never rewind a partially empty magazine. This is unnecessary and may break the magazine mainspring.

Field Stripping the Model 1928A1 Thompson

Remove magazine by pressing magazine catch up with the thumb and pulling 20-shot magazine straight down; or sliding 50-shot magazine out to the left.

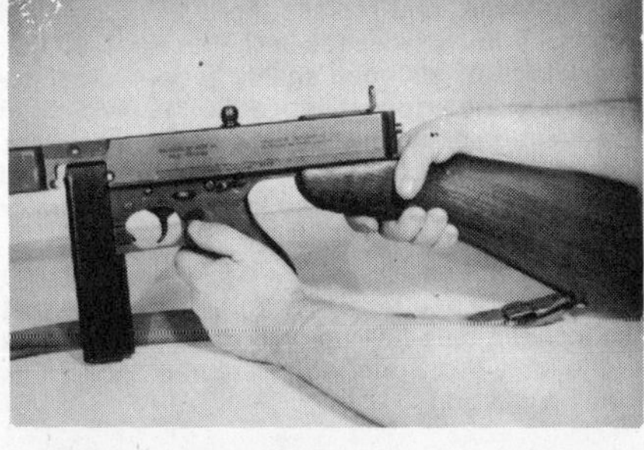

Set safety on "Fire" and set fire control lever on "Full Auto." Remember this can only be done when the weapon is cocked.

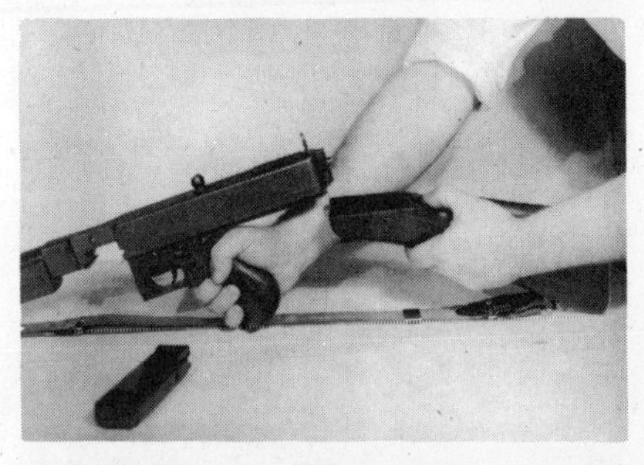

Remove buttstock by pressing its slide catch down and pulling stock straight to the rear out of its guide.

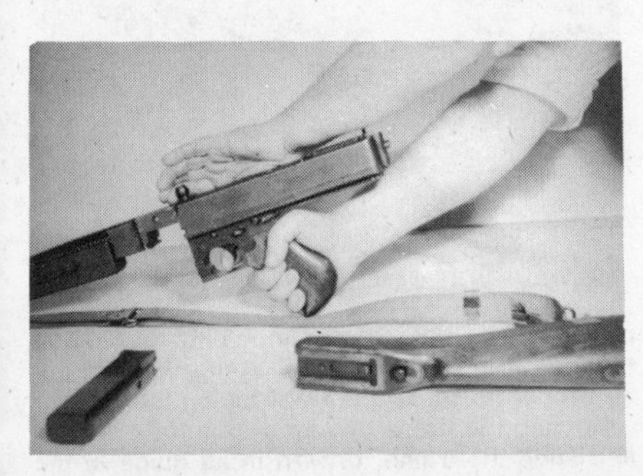

Hold firmly to actuator knob with left hand, pull trigger with right forefinger and ease bolt forward.

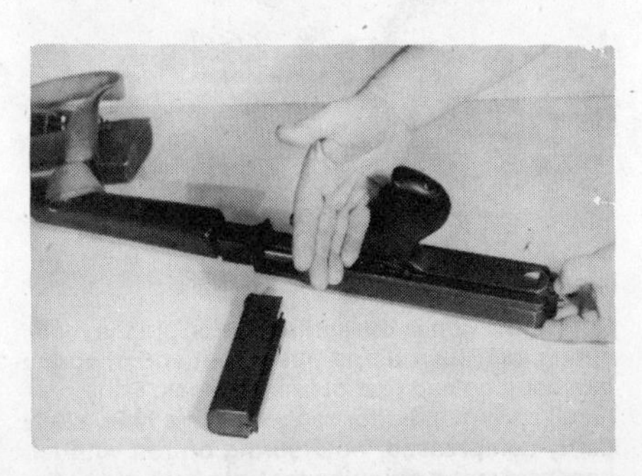

Turn gun upside down on table or knee. Push in the frame latch (the spring plunger on under side of frame behind pistol grip), and tap frame with right hand until it slides back a short distance.

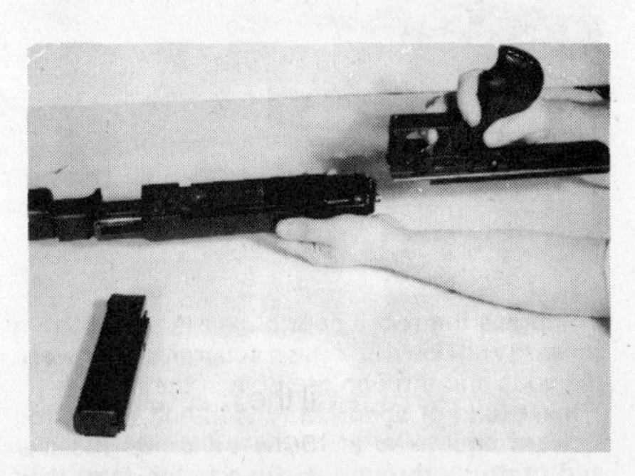

Grasp the rear grip with the right hand and pull the trigger holding the receiver firmly in the left hand, and slide the pistol grip group off out of its grooves.

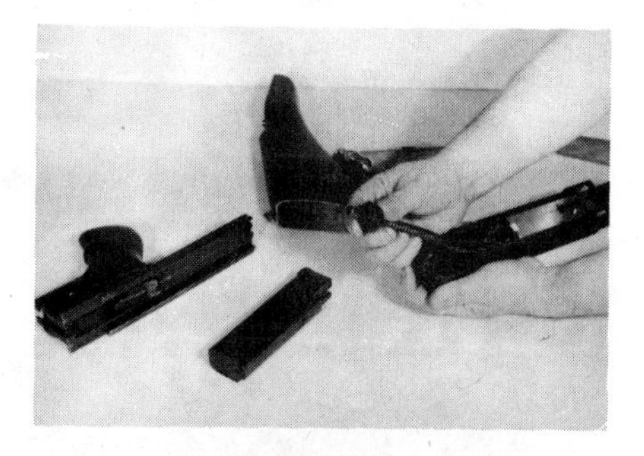

Remove recoil spring, as follows: With gun held firmly, turned upside down, grasp buffer flange with first and second fingers of right hand and pull out with upward and forward motion.

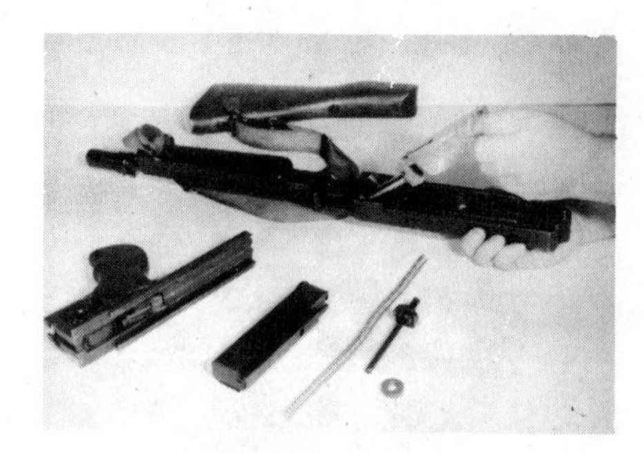

By pulling back on actuator knob, bolt is drawn back and can be removed to the rear.

Actuator is then slipped forward with lock, and lock removed through its grooves in the receiver.

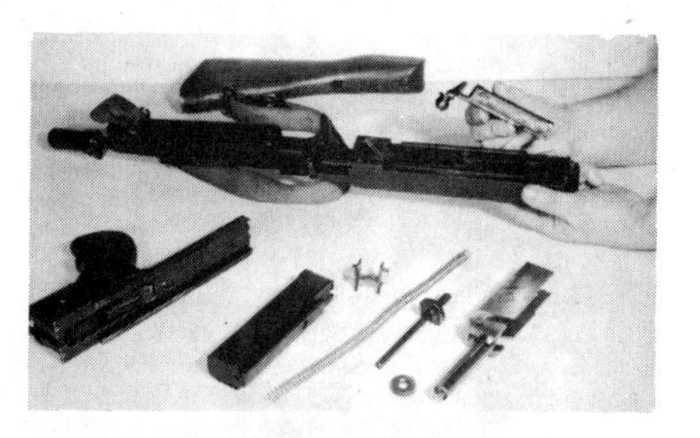

Then actuator itself is removed by sliding it to the rear. This completes field stripping, no further dismounting is necessary.

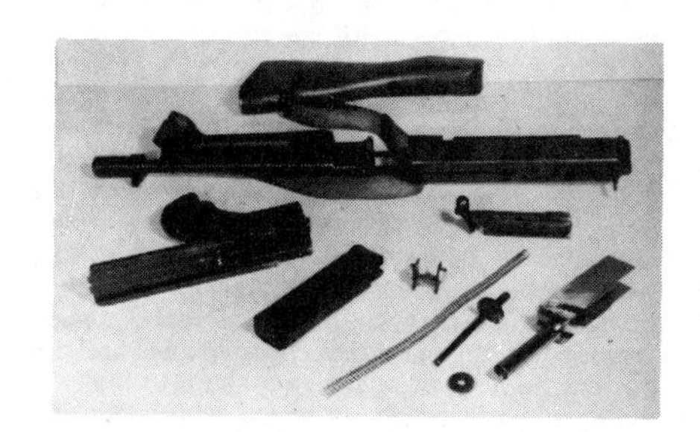

Note: On earlier models of this gun a special tool is required to remove the recoil spring. In this type, the stripping tool is inserted into its hole in the front end of the buffer rod. Then it is pushed in as far as it will go in the direction of the bolt. The rear end of the buffer rod is thus withdrawn from its hole at the rear end of the receiver. By tilting this stripping tool, the buffer may be grasped by the hand, and the recoil spring, fiber buffer disc, and rod will come out with the stripping tool. Buffer rod and spring are to be securely held so they do not fly apart.

Assembling

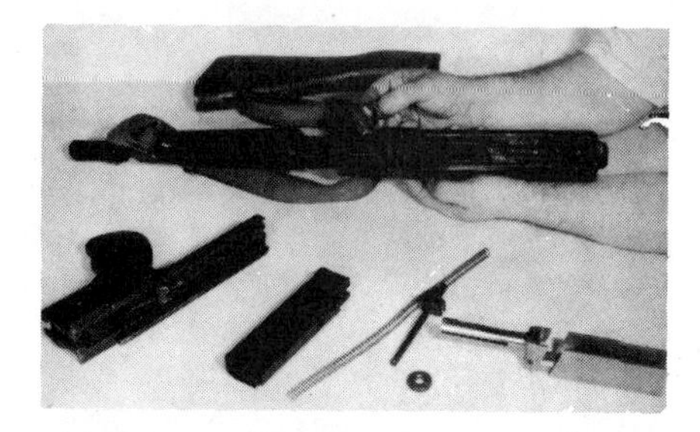

Inserting the actuator knob at its rear position, pull it forward and replace the lock which must be placed in its recesses in the receiver, so that the word "up" stamped on it is in uppermost position and the arrow stamped on it is pointing in the direction of the muzzle. The crosspiece of the lock (the lock is called the "H-Piece" because of its shape) must fit into the jaws of the actuator knob.

Pull the actuator and lock back and insert the bolt. Be sure and insert its bolt-end first so that the inclined cuts line up with the side members of the lock. (Now push the assembly forward as far as it will go.)

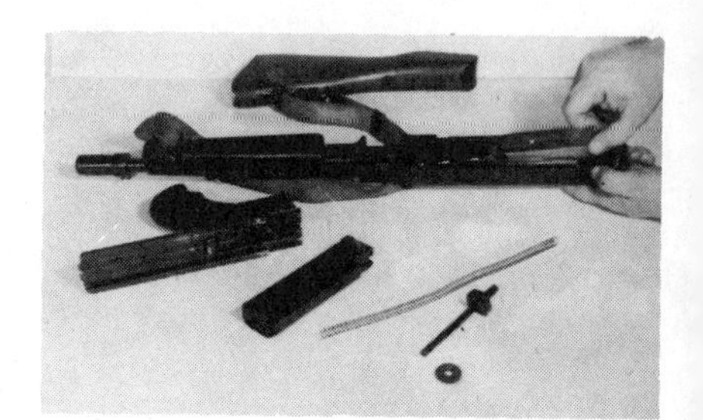

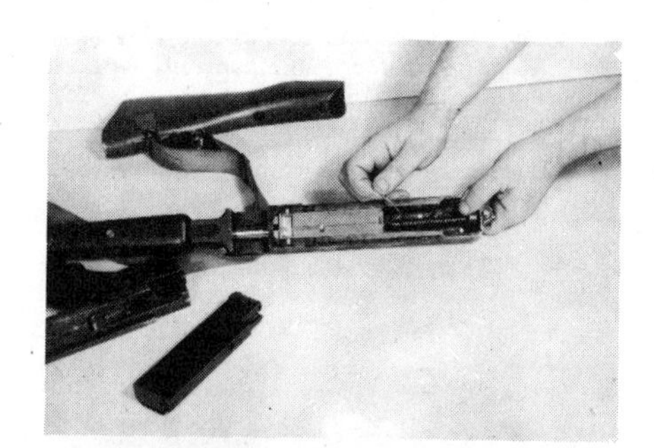

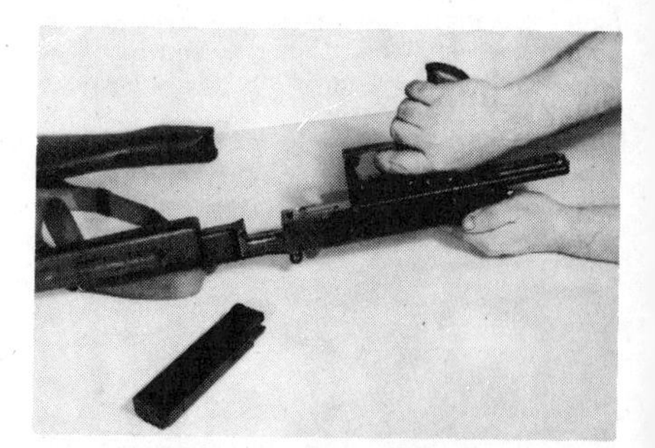

Compress the recoil spring over its rod, push it forward and down and push a nail or clip between the coils and through the hole in the rod.

Insert end of spring in hole in bolt; press rod forward until head of rod will slip into receiver and protrude through its hole to the rear; then withdraw nail.

Note: If the gun is the earlier model, put the recoil spring over its rod and push front end of spring into housing and rear of breech block. Compress recoil spring on buffer rod a little at a time. While partly compressed, hold spring on rod with left hand and insert stripping tool into hole and buffer rod to retain spring in position. Replace fiber buffer disc and insert loose end of recoil spring in its hole in the actuator knob. Now place rear end of buffer rod in its hole at the rear end under receiver. Draw back actuator knob until rear of bolt touches stripping tool. Recoil spring will now enter proper holes. Withdraw stripping tool.

Holding the frame by rear grip pull the trigger and slide the frame forward in its guide in the receiver. Remember that safety must be at "Fire" position and fire control lever at "Full Auto."

Insert undercut of the frame in the buttstock and slide the butt forward until it locks in place.

How the Thompson Gun Works

Starting with the gun loaded and cocked, the action is as follows. When the trigger is pressed, it moves the disconnector up to lift the sear lever. The sear lever raises the forward portion of sear, thus depressing the rear section and disengaging it from the notch on the bolt. The bolt is now free to be driven forward by the coiled recoil spring. If the fire control lever is set for semiautomatic fire, the rocker will act on the disconnector and sear lever to leave the sear free to lodge in the bolt on backward motion.

In its forward motion, the bolt strips the top cartridge from the magazine, forces it into the chamber and drives the lock downward into locked position. The forward end of the bolt is round to fit in the bolt wall of the receiver, and the rear portion is rectangular to fit into the receiver cavity.

The forward motion of the bolt is halted by the rectangular end abutting against the receiver. The lock is an H-shaped piece of steel with lugs on each side whose center is engaged by the actuator.

The hammer is pivoted in the bolt between the H-piece and the receiver bottom, and as the action closes the lower end of this hammer strikes the abutment somewhat in advance of the bolt so that as the cartridge is seated, the upper end of the hammer strikes the firing pin. (The hammer is made so it can strike the firing pin only when the bolt is completely closed.) The extractor snaps over the cannelure of the cartridge case, and the firing pin strikes the primer.

Return Movement of the Action. The residual breech pressure forces the empty cartridge case back against the bolt, which

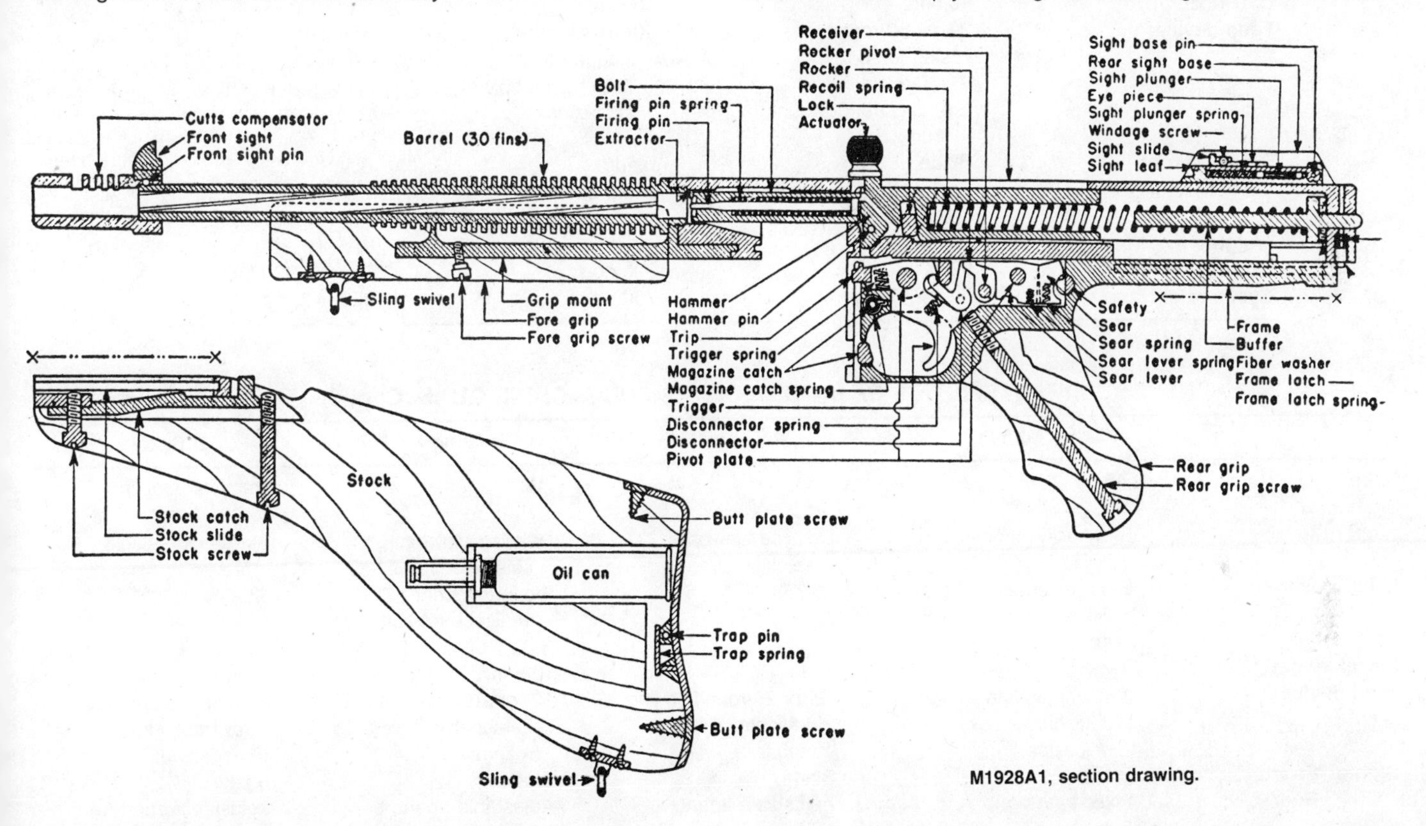

M1928A1, section drawing.

in turn transmits the pressure to the H-piece locking device. This in turn transmits it to the locking surface of the receiver.

How the Lock Works. The lock, or H-piece, is situated in a 70° inclined slot in the bolt, with its lugs engaged in short 45° grooves recessed in the receiver. When engaged in the short 45° inclined slot with the H-piece offering resistance to the backward motion of the bolt, resistance to this lifting action is offered by the forward inclined base of the H-piece meeting the rear face of the 70° inclined slot in the bolt itself. The rising of this H-shaped lock is further resisted because its bridge is thrust up into the slot in the actuator knob, which is set at an angle of 10° from the vertical, pointed to the rear.

The general direction of movement of the locking piece, as a result of the movement of these several components, is upward and backward. Thus the bolt is prevented from moving to the rear while the chamber pressure is dangerous.

Because of the rapidity with which the pressure in the bore rises to its maximum on firing, the bolt is said to be supported by adhesion of the inclined surfaces until the pressure has again dropped materially, which acts as a breech locking factor.

For this adhesion lock to work, it is essential that the engaging surfaces remain constantly lubricated by the oil pads in the receiver.

Note: The actual value of this locking system is dubious. The necessity for constant oiling, incidentally, is a source of jams. Regardless of the real or theoretical value of the locking device, the fact remains that the weight of the parts themselves and the inertia of the recoil spring are sufficient to work the weapon safely and satisfactorily when the locking device is removed from the gun.

The forward end of the recoil spring is housed in a cavity in the actuator. The buffer forms a guide for the rear end of the recoil spring, permitting it to compress in a straight line as the action goes backward. A fiber washer is provided to absorb the shock of recoil, and oil pads in the receiver lubricate the locking lugs and bolt sides during the passage of those pieces.

The extractor, which is positioned on the right forward end of the bolt, draws the empty case out of the firing chamber until it strikes the ejector, which moves into a clearance cut in the bolt path and hurls it out of the ejection port.

If a box type magazine is in the weapon, the magazine spring forces cartridges up in line bringing the next cartridge into position for the forward movement of the bolt. If the drum magazine is being used, springs inside the drum twist the spiral and feed a cartridge into line.

CHARACTERISTICS OF UNITED STATES SUBMACHINE GUNS

	Thompson M1928A1	Thompson M1 and M1A1	M3
Caliber:	.45.	.45.	.45.
System of operation:	Delayed blowback, selective fire.	Blowback, selective fire.	Blowback, automatic.
Length overall:	w/buttstock: 33.75 in. w/o buttstock: 25 in.	32 in.	Stock extended: 29.8 in. Stock retracted: 22.8 in.
Barrel length:	10.5 in.	10.5 in.	8 in.
Feed device:	20- or 30-round, staggered row, detachable, box magazine; 50-round drum.	20- or 30-round staggered row, detachable, box magazine.	30-round, in-line detachable, box magazine.
Sights: Front:	Blade.	Blade.	Blade.
Rear:	Leaf w/aperture notched battle sight.	Fixed aperture.	Fixed aperture.
Weight:	10.75 lb.	10.45 lb.	8.15 lb.
Muzzle velocity:	920 f.p.s.	920 f.p.s.	Approx. 920 f.p.s.
Cyclic rate:	600–725 r.p.m.	700 r.p.m.	350–450 r.p.m.

CHARACTERISTICS OF UNITED STATES SUBMACHINE GUNS (Cont'd)

	M3A1	Reising M50	Reising M55	U.D. M42
Caliber:	.45.	.45.	.45.	9mm Parabellum.
System of operation:	Blowback, automatic.	Delayed blowback, selective fire.	Delayed blowback, selective fire.	Blowback, selective fire.
Length overall:	Stock extended: 29.8 in. Stock retracted: 22.8 in.	35.75 in.	Stock extended: 31. 25 in. Stock retracted: 22.5 in.	32.2 in.
Barrel length:	8 in.	11 in.	10.5 in.	11 in.
Feed device:	30-round, in-line detachable, box magazine.	12- or 20-round, in-line detachable, box magazine.	12- or 20-round, in-line detachable, box magazine.	20-round, staggered row, detachable, box magazine.
Sights: Front:	Blade.	Blade.	Blade.	Blade.
Rear:	Fixed aperture.	Adjustable aperture.	Adjustable aperture.	Adjustable aperture.
Weight:	8 lb.	6.75 lb.	6.25 lb.	9.12 lb.
Muzzle velocity:	Approx. 920 f.p.s.	Approx. 920 f.p.s.	Approx. 920 f.p.s.	1312 f.p.s.
Cyclic rate:	350–450 r.p.m.	550 r.p.m.	450–550 r.p.m.	700 r.p.m.

Field Stripping the M1 and M1A1 Thompson Submachine Gun

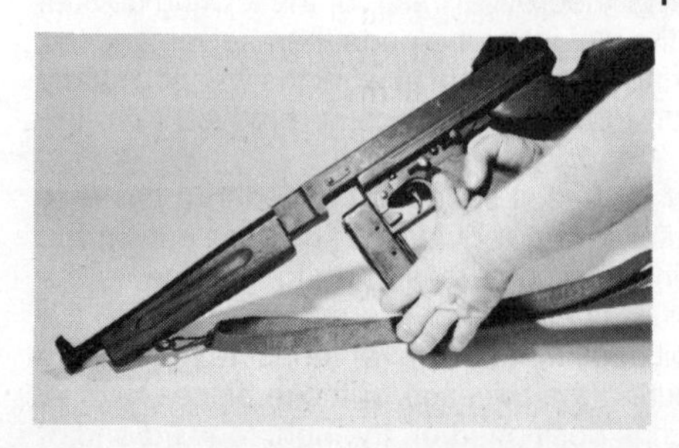

(1) Remove magazine by pressing magazine catch up with the thumb and pulling the box magazine straight down and out of its guide.

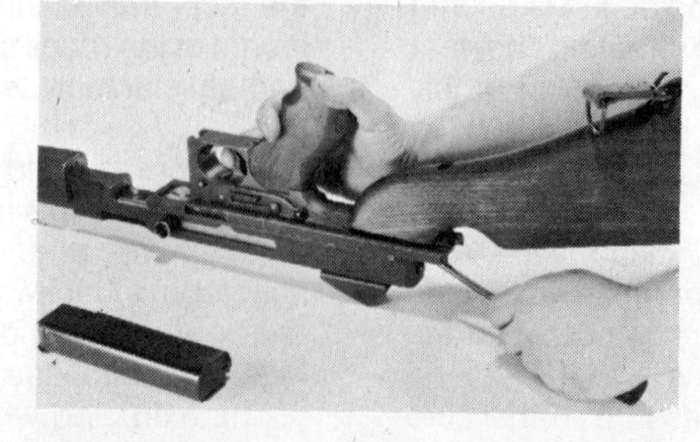

(2) With bolt forward, push in on receiver locking catch.

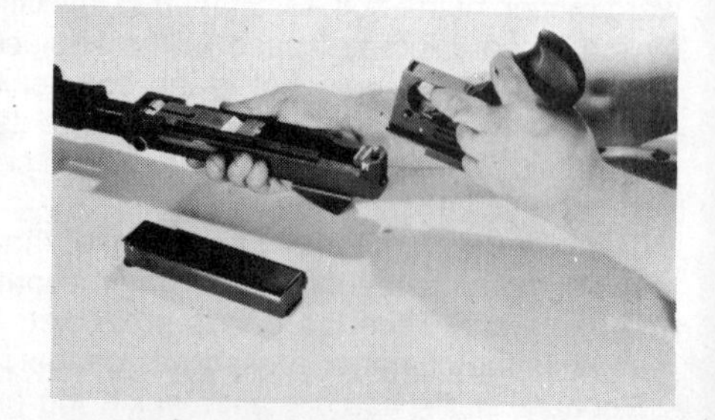

(3) Pressing trigger, draw frame straight back out of its guides in the receiver.

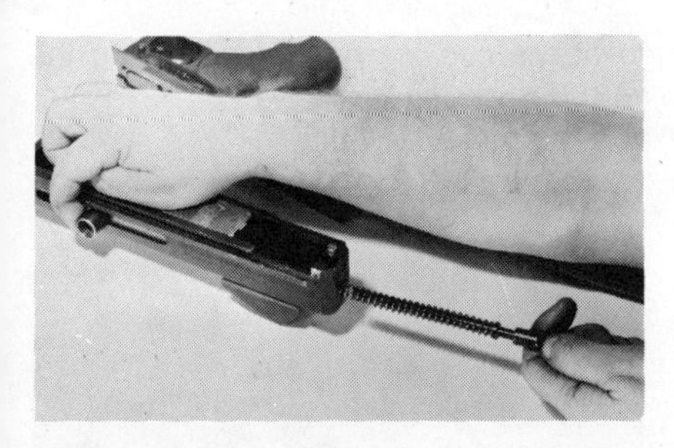

(4) Push the plug protruding from the rear of the receiver (the buffer pilot) in and draw the buffer pad up out of its seat.

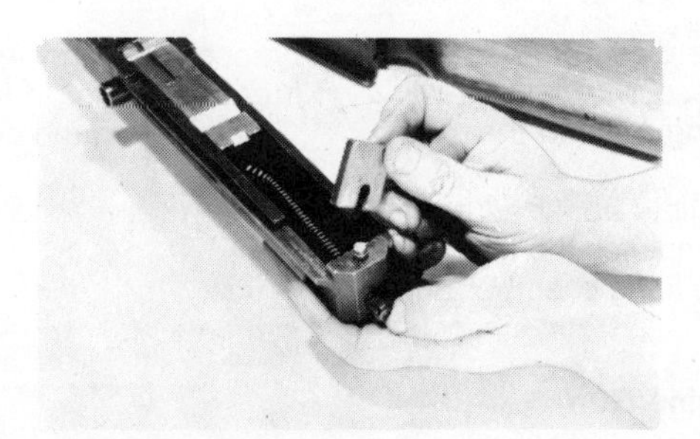

(5) Holding the buffer pilot, pull the bolt back about half way. This will permit removing the recoil spring and the pilot from the rear of the receiver.

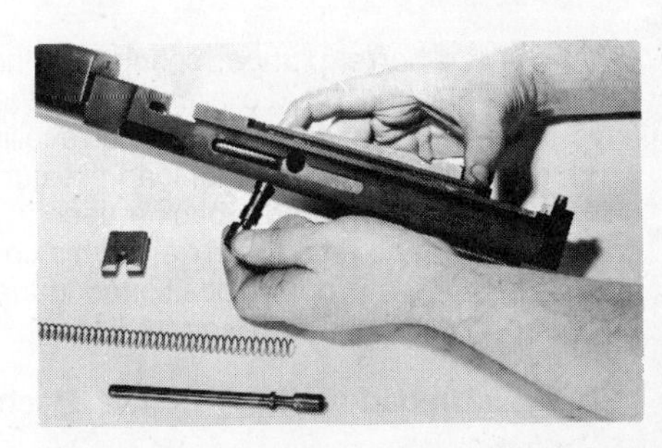

(6) Still holding the weapon upside down, lift the rear end of the bolt when the bolt handle is opposite the low cut in the center of its slot in the receiver.

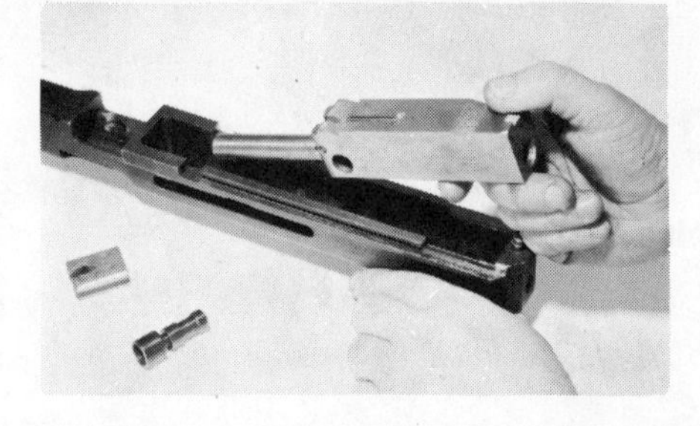

(7) Lift the bolt back and up out of the receiver.

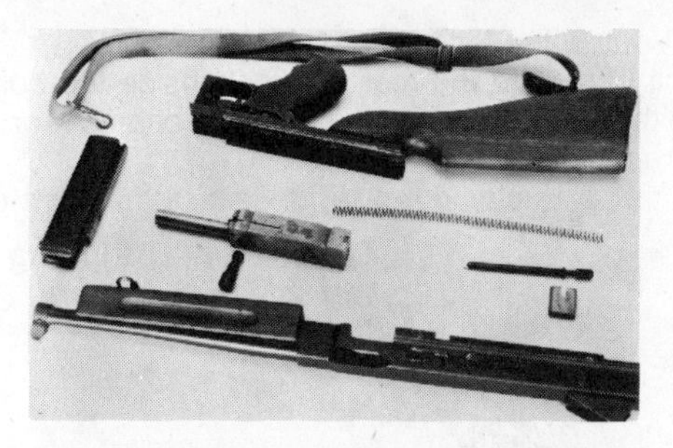

(8) This completes field stripping. No further dismounting is normally necessary.

UNITED STATES MACHINE GUNS

THE CALIBER .30 BROWNING 1917 AND 1917A1 MACHINE GUNS (Obsolete)

All weapons of this type used during World War I were of the Model 1917 type. Wartime service indicated that the bottom plate was weak. This was modified by adding a reinforcing stirrup; 25,000 of these stirrups were mounted on weapons being placed in storage in 1920–21. In 1936, it was determined that further changes were necessary in the weapon, and a remanufacturing program was set up at Rock Island Arsenal. The modified weapon was designated the Model 1917A1.

The principal changes made at that time were:

(1) Fitting of a new bottom plate.
(2) Fitting of a new belt-feed lever.
(3) Fitting of an improved cover-latch assembly.
(4) Fitting of a sight leaf graduated for the caliber .30 M1 and M1906 ball ammunition.
(5) Cover catch assembly that would hold the cover in a fixed position when opened.
(6) The tripod was modified to produce the M1917A1 tripod.

Weapons of new manufacture made during World War II had additional modifications as follows:

(1) A steel end cap was fitted in place of the bronze end caps used with the earlier guns.
(2) A steel trunnion block was fitted in place of the bronze trunnion block used on earlier guns.
(3) An improved steam tube assembly was used.
(4) A new type bunter plate, similar to that of the Model 1919A4, was used.
(5) An improved barrel gland assembly made of non-corrosive material was used.
(6) A recoil plate was fitted in the face of the bolt, i.e., separate ring of steel around the firing pin, which can be replaced as firing pin hole wears, thus obviating the necessity of replacing the complete bolt body.
(7) The sight leaf was graduated for caliber .30 M2 ball ammunition.

The Browning M1917A1 is not used in United States service, but can be found in the armies of other countries.

Loading the M1917A1 Machine Gun

Check the tripod to see that there is no unusual play, that it is firmly seated, that all its jamming handles are tight, and that the splayed feet of the legs are pushed securely into the ground. Check the water jacket and condenser. Make sure that both are full. Then check to see there is no water leakage at the muzzle glands. See that rear barrel packing is water tight and oil or grease it heavily if it is not. Check headspace adjustment. Remember that this is a most important adjustment on this gun. If it is too tight, the gun will refuse to fire; too loose headspace will result in bulged or ruptured cartridge cases, which will jam the gun badly. Check that ammunition belts are loaded uniformly and correctly and that they are clean and dry. Check rear sight for vision and working order. See that all moving parts are lightly oiled and work smoothly by drawing the bolt handle back and permitting it to go forward. All mechanisms should work smoothly and no unusual effort be required to withdraw barrel and recoiling mechanism to the rear. When bolt handle is released, it should position in its fully forward place and bolt should lock home properly. Bore should be inspected to be sure that it is clean.

With ammunition box securely locked in place on the left side of the gun and cover closed, insert the tag of the belt through the feed block as far as it will go to the right and pull the belt sharply to the right.

Pull the bolt handle to the rear as far as it will go. This will compress the barrel plunger spring and the driving rod spring, and when the bolt handle is released, these two springs will drive the action forward, moving the recoiling parts to their forward locked position, and half load the gun.

Now pull the bolt handle back a second time as far as it will go. Release it and let it fly forward. This completes the loading, leaving a catridge properly positioned in the firing chamber and the weapon cocked and ready for firing.

Unloading the M1917A1 Machine Gun

Pull back the cover latch (this is the milled knob on top of the receiver to the rear of the rear sight). The spring controlled section will move backwards and permit you to raise the feed cover.

Lift the belt out to the left and replace it in its box.

Pull the bolt handle to the rear and look inside the bolt way to make sure there is no cartridge in the firing chamber or in the face of the breechblock.

Now push the extractor down to its seat in the front of the breechblock and let go of the cocking handle, permitting the breechblock to go forward.

Snap down the cover and press the trigger.

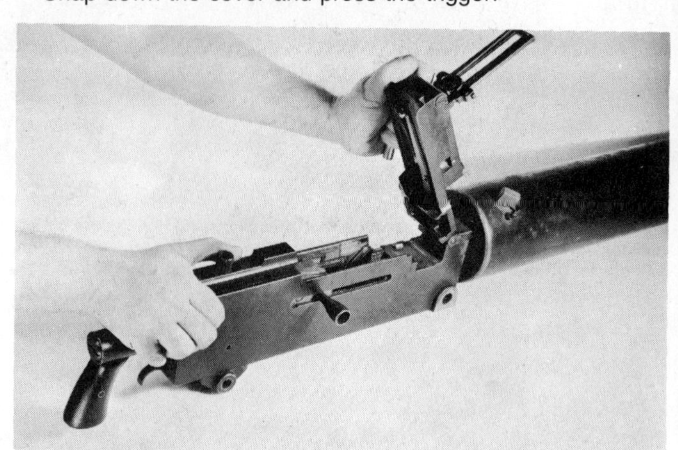

Stripping the 1917A1 Machine Gun

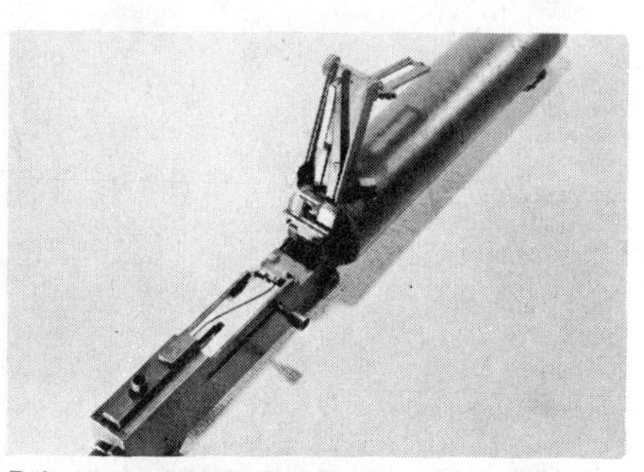

Raise the rear sight. Pull back the cover latch (the knob on top of the receiver behind the rear sight base) and raise the cover.

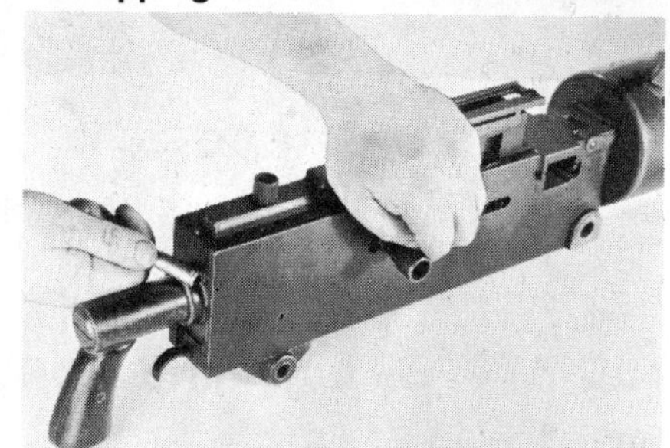

Draw bolt handle back as far as it will go and hold it firmly with the left hand. The driving spring rod protrudes through the back plate of the gun. Insert the base of a cartridge in the slot in the head of the rod. Push the rod in to compress the spring, and turn the rod to the right. This will lock the driving rod and its spring under compression inside the bolt.

Push the bolt handle forward a few inches to draw the driving rod out of the locking hole in the back plate. Then pushing the cover latch forward with the left thumb, raise the pistol grip up and out of its retaining slots in the rear of the receiver.

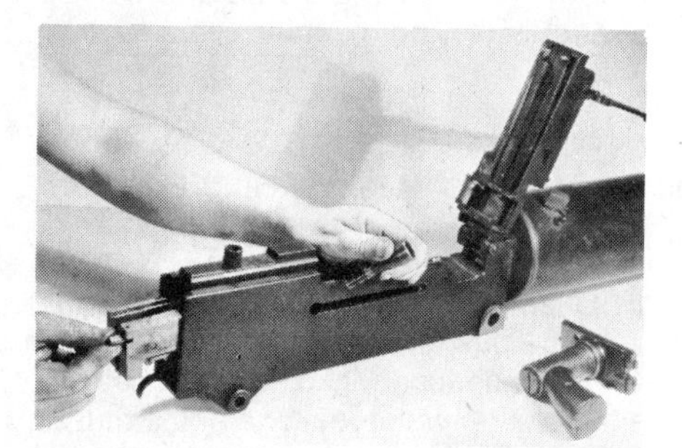

Pull the bolt handle as far as it will go to the rear, at which point it may be pulled to the right out of the bolt and receiver. Reach inside the receiver and grasp the driving rod; then pull the bolt directly to the rear and out of the receiver.

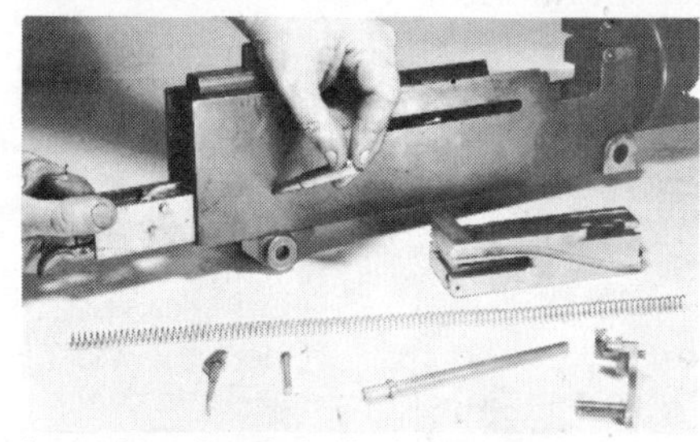

Insert the point of a bullet in the trigger pin locking hole in the lower right side of the receiver and push in the trigger pin against the tension of its spring. This frees the lock frame spacer and other recoiling parts and permits pulling them directly to the rear.

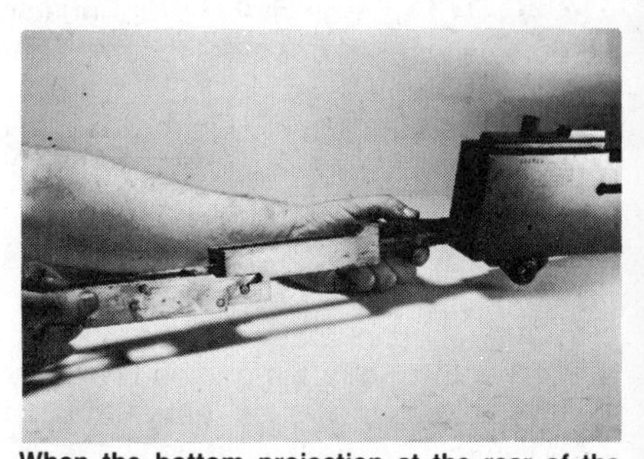

When the bottom projection at the rear of the barrel extension (the part screwed onto the barrel) drops below the bottom of the receiver, pull the combined lock frame spacer, barrel extension, and barrel directly to the rear out of the gun.

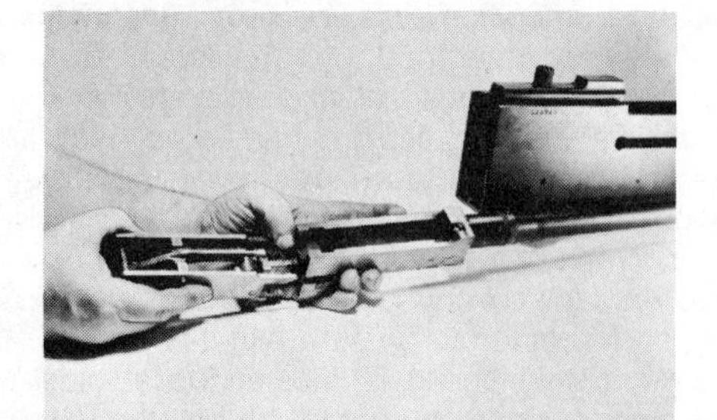

Grasp the lock frame spacer with the right hand, and with the left thumb, push forward on the turned up tips of the accelerator.

This will spring down and forward, separating the lock frame spacer (which holds the trigger and accelerator mechanism) from the barrel extension.

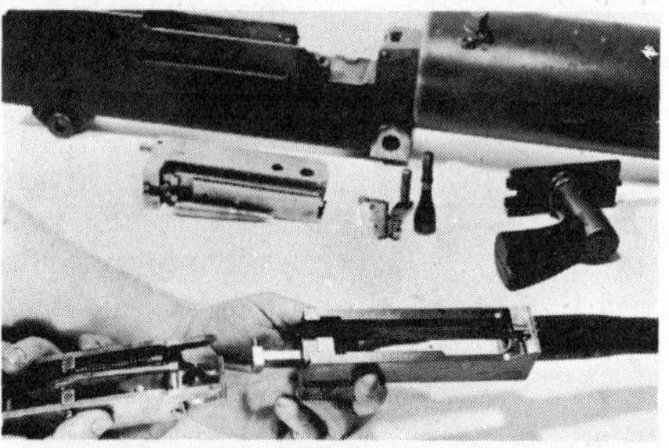

Push the accelerator pin out of the lock frame spacer and remove the accelerator (this is the curved piece of metal with two claws).

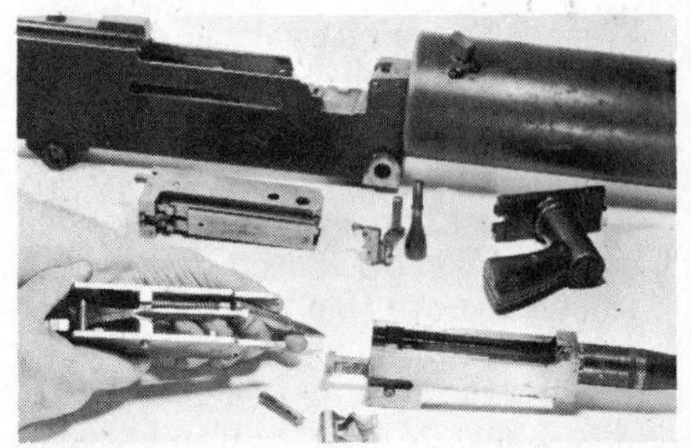

Insert head of cartridge in slit in head of barrel plunger at the left side of the lock frame. Twist it and ease it out. Remember that it is under strong spring tension (if necessary trigger pin may now be pushed out and spring and trigger removed).

Holding the barrel extension with the left hand, use the point of a bullet to start the breechblock pin from left side and remove it from the right, permitting the breechblock (the heavy wedge whose lower front end is beveled) to drop down out of the barrel extension. (Barrel extension may now be unscrewed from the barrel if necessary.) Turn the extractor (the swinging hook-shaped piece on the left forward end of the bolt) up as far as it will go and pull it out of its hole in the bolt.

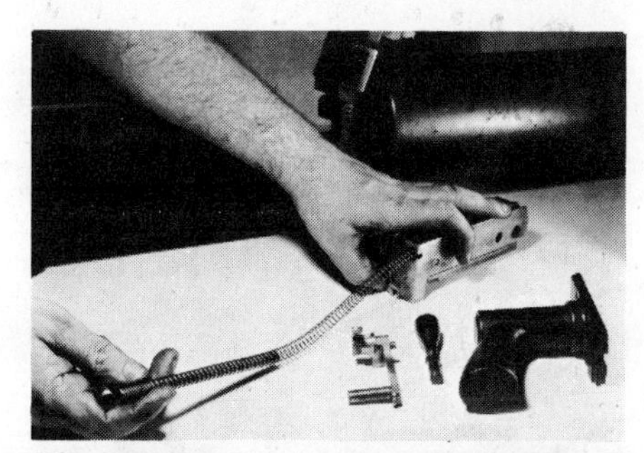

With the base of a cartridge, push in the head of the driving spring rod in the bolt and twist it to the left. Be careful of the very powerful driving spring which will now be free to fly out the back carrying the driving spring rod with it. Grip this firmly and ease it out gradually. Push out the cocking lever pin from the upper front left side of the breechblock and lift out the cocking lever.

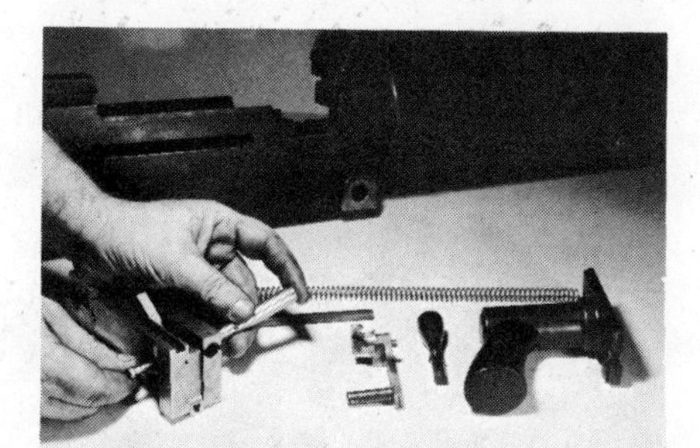

Turning the breechblock upside down, push the sear up with the bullet to release the firing pin spring; turn the breechblock the right way up again and insert the bullet into the slot of the sear spring. Push over to the left, pry the sear spring into a locking recess and the sear drops out. When sear spring is pushed back into normal position its pin may be pushed up and it and the spring removed. The firing pin and its spring may now be dropped out of the back of the breechblock.

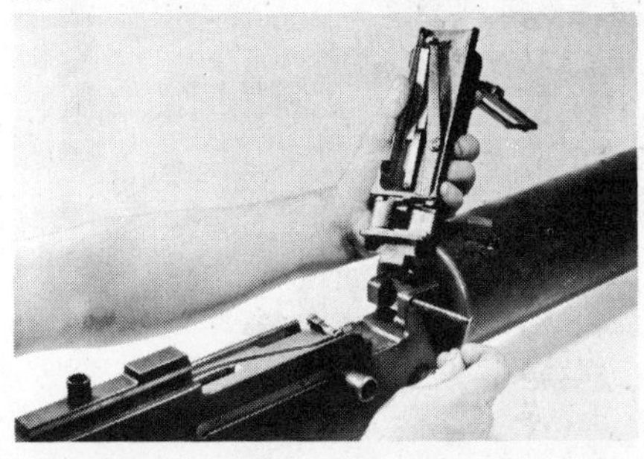

Use the point of a bullet to turn up the cover pin spring at the right side of the receiver just behind the water jacket. Pull the pin out to the right and lift the cover up out of the receiver. Pushing back on the nose of the split pin on the left side of the feed block, control the feed pawl with the left thumb

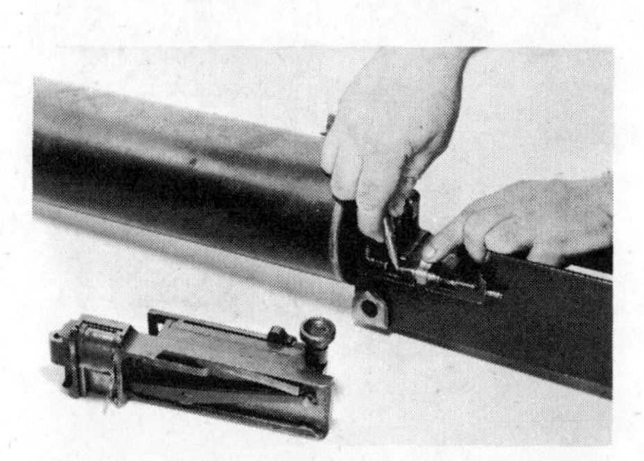

and ease the pawl and its spring up out of the receiver as the locking pin is pulled out. Insert point of bullet between the extractor cam and the long flat piece of metal on the side of the cover opposite the belt feed lever (this is the extractor spring) and pry the extractor spring out of its seat

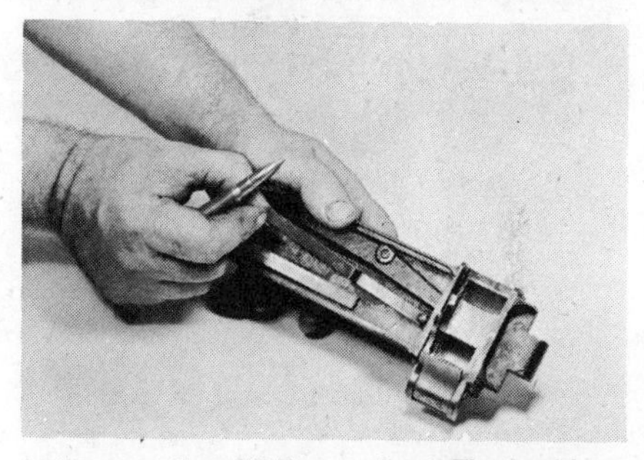

in the cam, then lift it up and out. The feed lever and slide may be removed if necessary by turning feed lever pivot pin spring outwards. This completes stripping the Browning machine gun.

Assembling the M1917A1 Machine Gun

Start by screwing the barrel into the barrel extension, then insert the barrel and barrel extension into receiver. Slide slowly forward until the projection of the barrel extension (which holds the bolt back) is against the bottom plate of the receiver. Holding the lock frame in the right hand, place the accelerator claws between the rear face of the barrel extension and the forward faces of the T-lug extension and at the same time insert the forward projections of the lock frame into their grooves in the barrel extension. Give a quick thrust forward to the lock frame to tip back the accelerator claws and compress the barrel plunger spring; this locks the lock frame to the barrel extension.

Now push the lock frame attached to the barrel extension and barrel farther ahead in the receiver until the trigger pin on the lower right side comes against the side of the receiver. Push in the trigger pin against its spring tension, and the whole assembly can now be pushed fully home while the trigger pin will be forced by its spring into its slot in the right side of the receiver.

Now replace the bolt in the receiver being sure the extractor is in position and that the cocking lever projecting through the top is fully forward.

Insert the bolt handle in the end of the slot in the receiver and into its hole in the bolt; then push it forward far enough so that when the cover latch is pushed forward, the pistol grip and back plate may be slid down in their grooves in tbe receiver. Pull the cover latch back to lock the back plate in position; holding bolt handle back as far as it will go, insert the base of the cartridge into the slit in the driving rod, push in and turn to the left to release the driving rod spring. Now permit the bolt to run forward under the thrust of the driving rod spring.

How the M1917A1 Machine Gun Works

Starting with the weapon loaded and cocked, the action is as follows. The trigger being pressed, the trigger bar disengages from the sear block and allows the striker to be driven forward by the spring in the bolt to fire the cartridge.

As the bullet goes down the barrel, the recoil drives back against the base of the catridge base, which transmits the blow to the bolt face thereby starting the locked recoil action and barrel to the rear.

Locked together, the barrel, barrel extension and bolt recoil about ⅝ of an inch. For the first half of this travel, during the period of high breech pressure, they are securely locked together, and then the front projections of the lock frame (which are set against the sides of the pin passing through the barrel extension and the breechblock) force the breechblock pin down, drawing the breechblock down out of its locking slot on the underside of the breechblock. The bolt is thus released from the barrel extension and so can continue straight to the rear. As the barrel extension itself travels to the rear, the barrel plunger spring is compressed, and the rear of the barrel extension drives the claws of the accelerator back sharply, flipping the accelerator up and backwards on its pin.

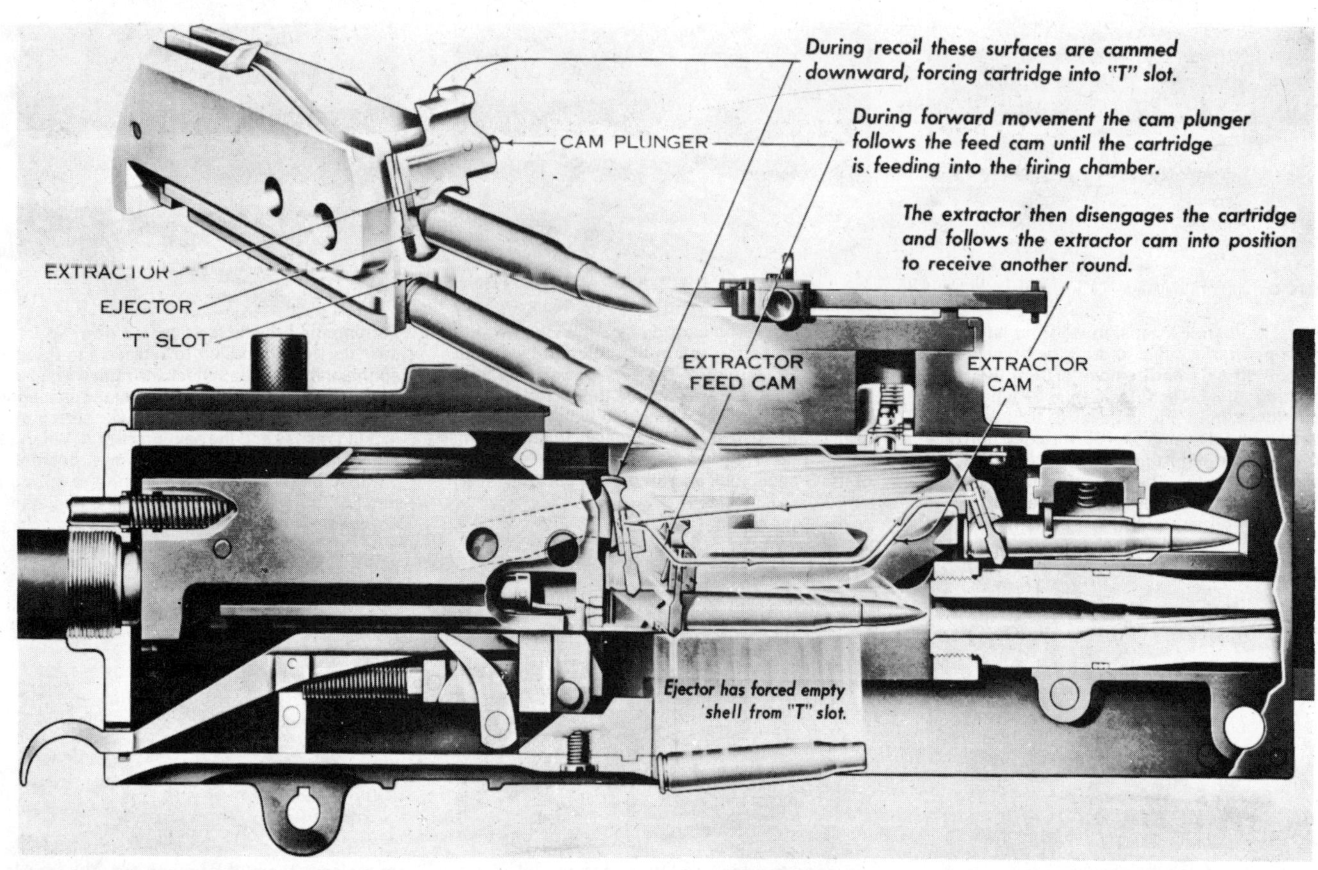

Cross section of parts in firing position.

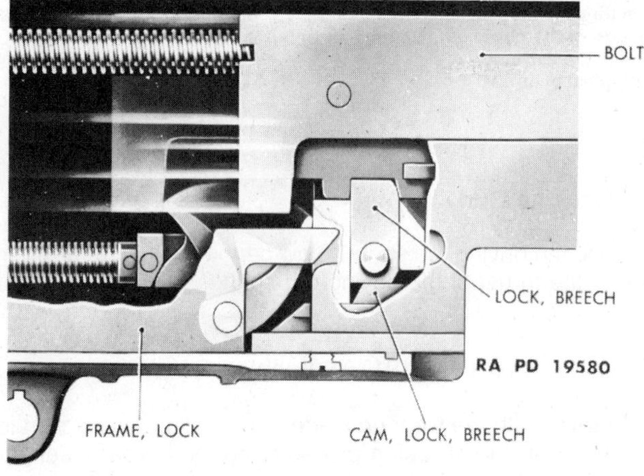

Breech locked.

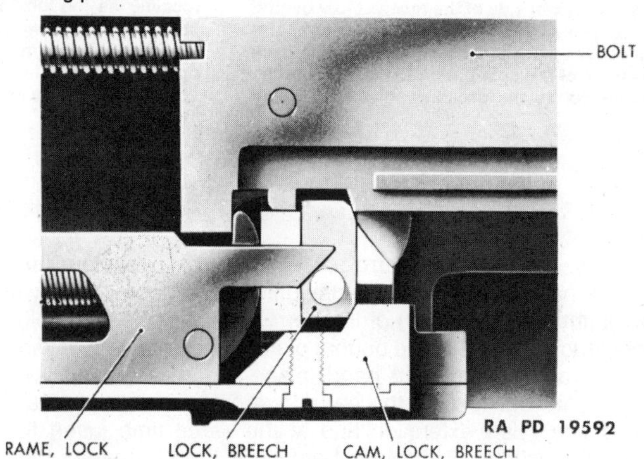

Breech lock beginning to open.

As the accelerator turns, the tips of its claws strike bottom projections on the bolt and thus accelerate the rearward motion of the bolt by transmitting to it the thrust absorbed from the barrel extension, which is now held in rearward position locked to the frame spacer slots. Speeding the rearward motion of the bolt, at the same time that the barrel is slowing up, permits the empty case to be extracted from the chamber without the sudden tug that would normally occur. (This makes special lubrication unnecessary.) The accelerator claws while engaging the shoulder of the T-lug firmly lock the barrel extension in the rearward position to the lock frame. A stop prevents the accelerator from going backwards too far and the barrel plunger spring is held compressed.

During backward motion of the bolt, the driving spring is compressed over the driving spring rod whose head is held securely in the back plate of the receiver. The extractor fitting over the top front of the bolt draws a loaded cartridge from the belt at the same time that the T-slot machined into the face of the bolt draws the empty cartridge case from the firing chamber. The extractor cam plunger (which rides along the top of the extractor cam and the extractor feed cam) is finally forced in by the beveled section of the extractor feed cam. The cover extractor cam thus forces the extractor down and the plunger spring out behind the extractor feed cam.

During the backward movement of the bolt, a stud on the belt feed lever (which is mounted in the front of the cover on a pivot) moves to the right in the cam groove cut in the top of the bolt. Thus the belt feed slide, which is attached to the lever, is moved to the left. The belt feed pawl (located on the underside of the cover above the belt) springs over the left of the first cartridge

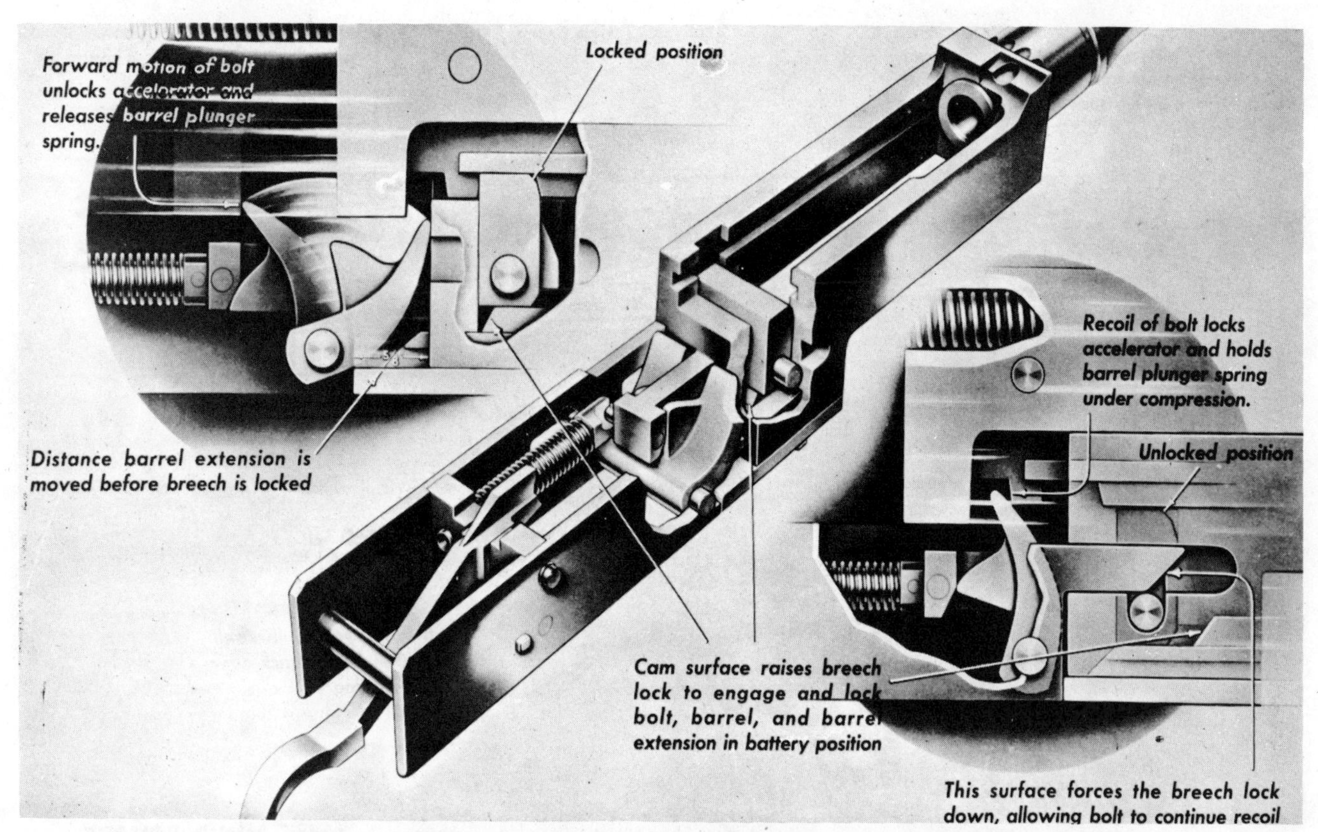

Operation of accelerator mechanism.

(which is being held in position by the belt holding pawl below the belt on the left side of the feed block) and supports the cartridge and the belt to prevent feeding trouble.

The cocking lever is fastened by a pivot pin inside the bolt, and its top protrudes through the top of the bolt and rests inside the top cover of the receiver. Thus as the bolt starts backwards, pressure on the lower part of this lever drives it back revolving it on its axis, and the firing pin spring is compressed drawing back the firing pin. The firing pin engages a notch in the sear (which has been pulled upward by the action of the sear spring).

The rear of the bolt strikes against the buffer plate mounted in the back plate at the upper part of the pistol grip, and the remainder of its energy is absorbed in friction and by the buffer disc. (A brass buffer ring is forced over a plug and expands against the inner wall of the grip.)

The driving spring in the bolt, now fully compressed, reacts to drive the bolt forward. During this forward motion, the upper end of the cocking lever is forced to the rear, thus pulling the lower end away from the rear of the firing pin. The extractor feed cam acts on the extractor cam plunger, forcing the extractor down so that the cartridge it is holding drops down the T-slot in the face of the bolt until it reaches a direct line with the firing chamber. The ejector strikes the empty case in the T-slot expelling it through the bottom of the gun and stopping the live cartridge when properly positioned.

Also during the forward motion of the bolt, the bottom projection strikes the top of the accelerator to swing it forward on its axis pin. This unlocks the barrel extension from the lock frame, permitting those two units to move forward as the bolt acts through the accelerator against the rear end of the barrel extension. The forward motion of the barrel extension and barrel is further assisted by a thrust from the barrel plunger spring as it uncoils. The force passed on by the accelerator from the bolt to the barrel extension is sufficient to guarantee proper timing of the locking action.

The actuating stud on the feed lever fits down in a cam groove in the bolt. Thus as the bolt goes forward, the stud rides in the cam and forces the lever on its pivot to the left; the forward end of the lever, which carried the feed slide, is thus pivoted to the right, bringing with it the belt feed pawl, the belt and the next cartridge. As the motion ends, the cartridge to be fed is held between the cartridge stops and the feedway. The next round to load is pulled over the belt holding pawl, which rises behind it. It is in position to be engaged by the belt feed pawl on the next movement.

The final action of the extractor during forward motion of the bolt is to rise under the influence of its plunger riding along the top of the extractor cam. As the ejector pivots forward, the extractor releases its hold on the cartridge, which is now well into the chamber. The extractor continues to ride upwards and over the base of the next cartridge in the belt in the feedway. Then the flat extractor spring in the top cover forces the extractor down and into the cannelure of this cartridge gripping it ready to pull it back on the next movement to the rear of the bolt.

The breechblock, mounted in the rear of the barrel extension, strikes a cam as the recoiling parts near the firing position and is forced up this cam and into a recess cut in the bottom of the bolt. Thus as the action comes to a complete close, the breechblock firmly locks the barrel extension (into which is screwed the barrel) to the breechblock.

Special Note on M1917A1 Adjustments

It is extremely important that the headspace on this gun be correctly adjusted before firing. To test the adjustment, pull the bolt handle back and let it run forward several times.

If the bolt does not go home fully and smoothly, it indicates too tight space between the face of the bolt and the face of the firing chamber. If the gun is put into use in this condition, it will fire sluggishly, or it may refuse to fire at all.

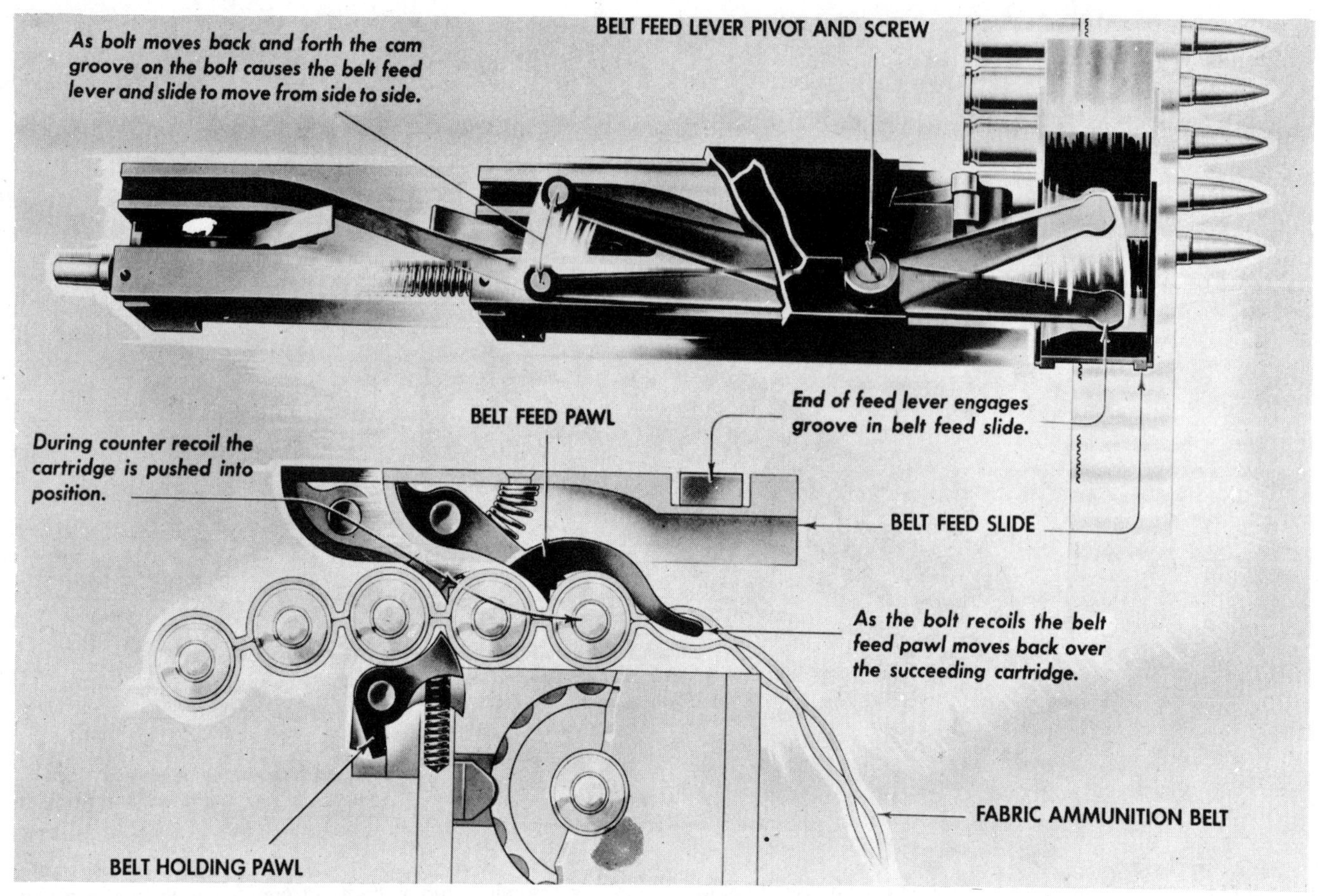

Belt-feed mechanism.

To correct this condition, it will be necessary to strip the weapon and unscrew the barrel one notch, then assemble and test again.

To Test for Loose Headspace. Lift the cover and raise the extractor; then pull the bolt slightly to the rear. If the bolt moves back at all without carrying the barrel and barrel extension with it, then gun is too loose. Fired in this condition, the pressure of the gas in the firing chamber will bulge the head of the empty cartridge case (since it is not fully supported by the bolt), or may rip it off entirely, causing a serious jam.

To correct this condition, screw up the barrel one notch, then retest.

Adjusting Headspace. Screw the barrel into its extension and stop when the first clicking sound is heard (this click is caused by the barrel-locking spring).

Push the breechblock (minus extractor) fully forward on barrel extension.

Push the lock piece up from below to lock the breechblock to the barrel extension; and while holding it firmly, screw up the barrel until resistance is encountered.

Now check to see that barrel locking spring is in a notch and that the lock piece is solidly seated.

Now let go of the lock. If it drops freely, the adjustment is correct. This adjustment should be punched on the barrel, to save time when assembling in future.

Water Leakage. If water leaks from muzzle end, remove the muzzle gland. Wind packing around the barrel. Press it together with combination tool. Then push back on barrel and guide the packing into its seat. Screw the muzzle gland back on. Test by working bolt handle. If there is friction, packing is too tight. This will make a sluggish gun.

If water leaks from breech end, remove barrel and work oiled packing down into barrel cannelure with combination tool. Test as before for undue friction and headspace.

CALIBER .30 BROWNING MODEL 1919A4 AND 1919A6 MACHINE GUNS (Obsolete)

The M1919A4

Two types of the M1919A4 have been issued—the fixed gun and the flexible gun. The fixed gun was widely used on World War II armored vehicles; the flexible gun was mainly used as the infantry company level machine gun. The M1919A4 is used on the caliber .30 M2 mount as a ground gun. The mechanism of the M1919A4 is identical with that of the caliber .30 M1917A1 Browning gun. The principal differences between the M1919A4 and M1917A1 are as follows:

(1) The M1919A4 has a ventilated barrel jacket rather than a water jacket.

(2) The M1919A4 is normally used on a light mount with limited terrain command, classified as a light machine gun by the United States. The M1917A1 is normally used on a heavier mount with much greater terrain command, classed as a heavy machine gun; it was the battalion level gun.

(3) The M1919A4 has a much heavier barrel than the M1917A1 but has a lower sustained rate of fire than the water-cooled gun.

(4) The sights of the M1919A4 and the M1917A1 are different. There are other various minor differences between these guns. The M1919A5 is the same as the M1919A4, except that it has a bolt retracting slide and a different cover detent. These changes on the weapon were necessary to mount it in the World War II M3 light tank. The US Navy has a number of Browning Model 1919A4 machine guns converted to 7.62mm NATO. These weap-

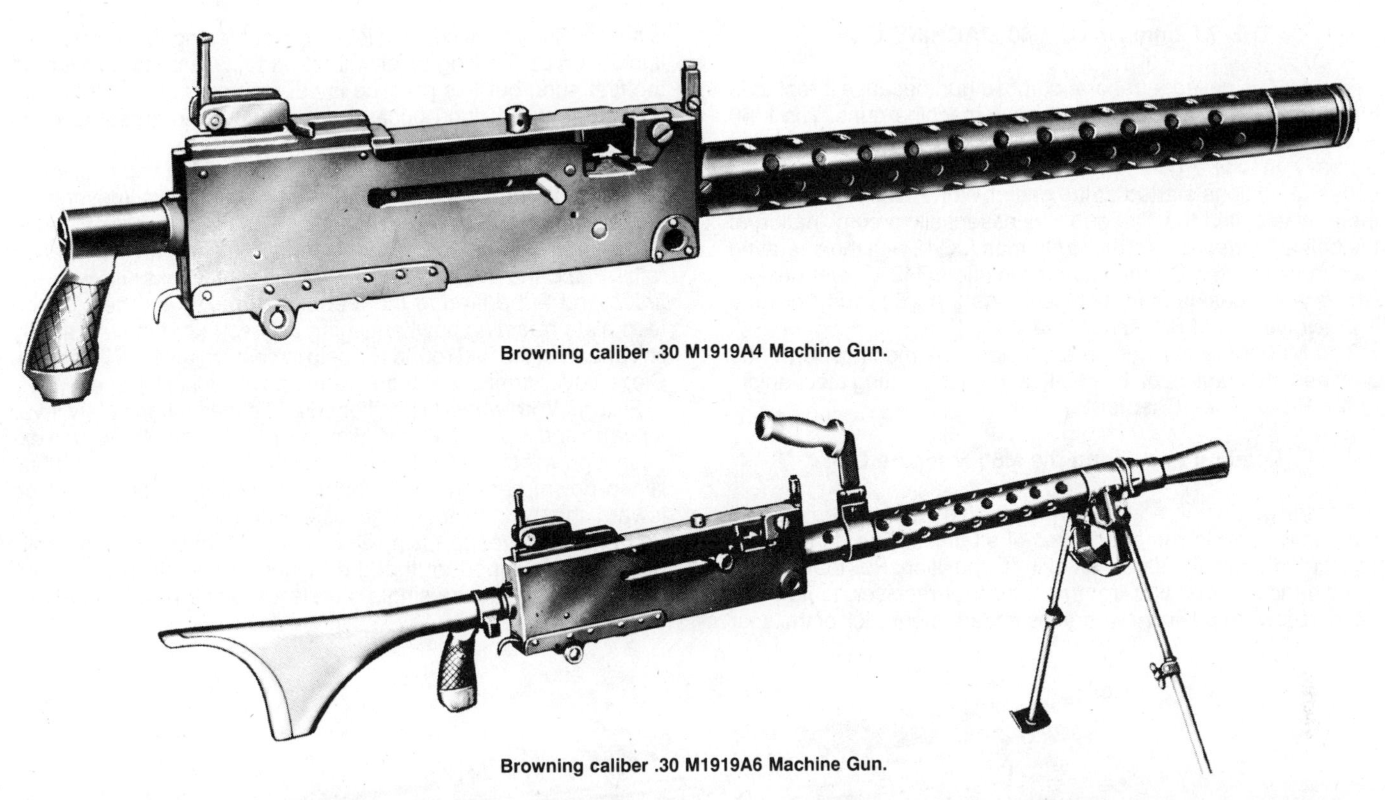
Browning caliber .30 M1919A4 Machine Gun.

Browning caliber .30 M1919A6 Machine Gun.

ons, fitted with a closed-prong type flash suppressor, are called Machine Gun 7.62mm NATO Mark 21 Mod. O by the Navy. A similar weapon is standard in the Canadian Army as the C1 7.62mm Machine Gun.

The M1919A6

The M1919A6 was a wartime modification of the M1919A4 to give the weapon more tactical flexibility. A bipod, shoulder stock and carrying handle were added to the basic Browning machine gun action to create a weapon that was easier to move and to get into action than the tripod-mounted M1919A4. The M1919A6 also has a lighter barrel and a different front barrel bearing than the M1919A4. The lighter barrel of the M1919A6 gives it a higher cyclic rate of fire than the M1919A4.

A post-World War II version of the M1919 Brownings (in addition to the M37) also exists; this is the M1919A4E1. This weapon also has a retracting slide similar to the M1919A5.

Loading and Firing; Field Stripping the M1919 Series

The method of loading and firing and the method of field stripping the M1919 series of guns is, in all essentials, the same as that of the caliber .30 M1917A1 water-cooled Browning gun.

The M1919A4 and the M1919A6 may be used with M1917A1, M2, or M74 tripods but are normally used with the M2 tripod. The M1919A6 is also used on the bipod shown in the photo.

CALIBER .30 TANK MACHINE GUN M37 (Obsolete)

The M37 is basically a modification of the Browning M1919 series of weapons. It was designed to produce a weapon that would have more usefulness in tank mountings than the Browning M1919A4 and A5.

Notes on the M37 Tank Machine Gun

The principal differences between the M37 and the Browing M1919 series of guns are as follows.

The M37 can be fed from either the right or left side. The bolt has a dual track for the belt feed lever stud. By positioning two switches, the ejector and various feed components, the weapon can be easily changed to feed from either side. This arrangement is generally similar to that found on the Browning cal. .30 M2 aircraft gun and the cal. 50 M2 and M3 aircraft guns.

The cover catch has been changed so that it can be opened easily from either side.

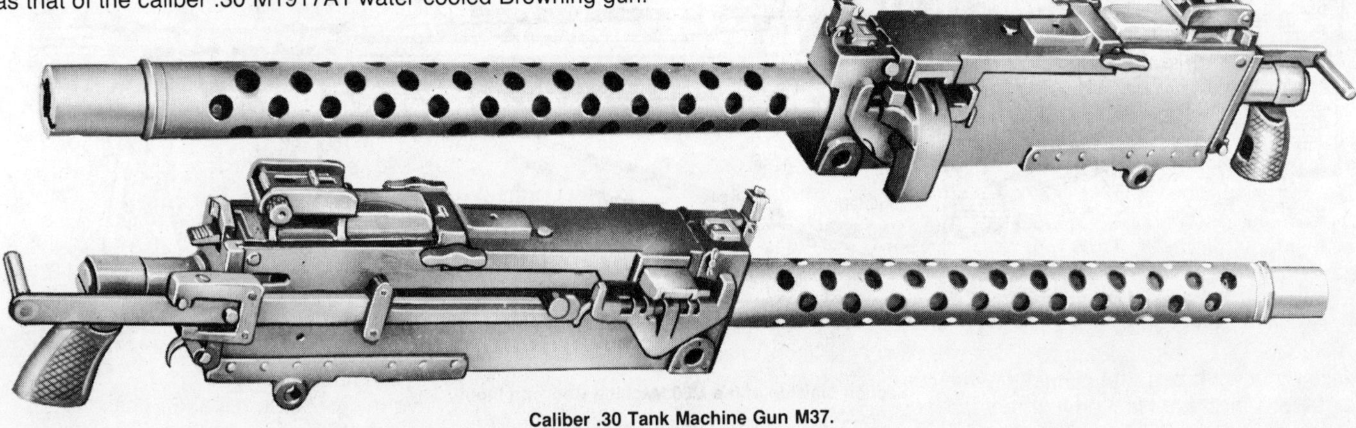
Caliber .30 Tank Machine Gun M37.

THE 7.62mm NATO M60 MACHINE GUN

The M60 is called a general purpose gun because it replaced the cal. .30 Browning light and heavy machine guns. The M60 is used on a bipod as a light machine gun, and on a tripod as a heavy machine gun. The M60 (T-161E3) is the result of a series of designs started at the end of World War II. The first of these was called the T44 and was essentially a combination of the belt feed mechanism of the German MG42 with the operating mechanism of the German automatic rifle FG42. A later design, which was considerably modified, was the T52; and from the T52 evolved the T161 series of guns.

The M60 therefore has, in a considerably modified form, the belt feed mechanism of the MG42 and the operating mechanism of the FG42. (See Chapter 2.)

Loading and Firing the M60 Machine Gun

Cocking. For all normal purposes this weapon, being fully automatic, should remain cocked at all times. The gun can be cocked only with the safety off or in "F" position. Pull the cocking lever handle, extending from right front of receiver, to its maximum rearward position, where the engagement click of the sear is heard. Return the cocking lever forward to engage its retaining latch. NOTE: Cocking lever will return to its latched position on the first shot, but this practice is not recommended, although it is not harmful if done occasionally. Return the safety lever to the "S" or safe position.

Loading. Rounds should be firmly assembled and positioned in their push-through links. Raise feed cover by lifting cover latch at right rear. Feed cover is retained vertically by a torsion spring and detent. The feed plate should remain in place on the receiver rails. Place the linked belt on the feed plate, links up, with the first round to be fired in the feed plate groove and held by the feed plate retaining pawl engaging the second round. An empty link ahead of the first round will help to position the belt, if desired. Close cover firmly, making sure the cover is latched securely.

Firing. With weapon positioned and aimed, push safety lever forward and up, out of the way, with right thumb. Visual examination will show the lever is directed toward the "F" symbol. When down, near the right thumb, the safety lever is directed toward the "S" symbol. Because of the low cyclic rate (550 r.p.m.) of this weapon, single rounds or short bursts can easily be fired. It is important that the trigger be completely released after each shot, to fire single rounds, or to interrupt firing at any time.

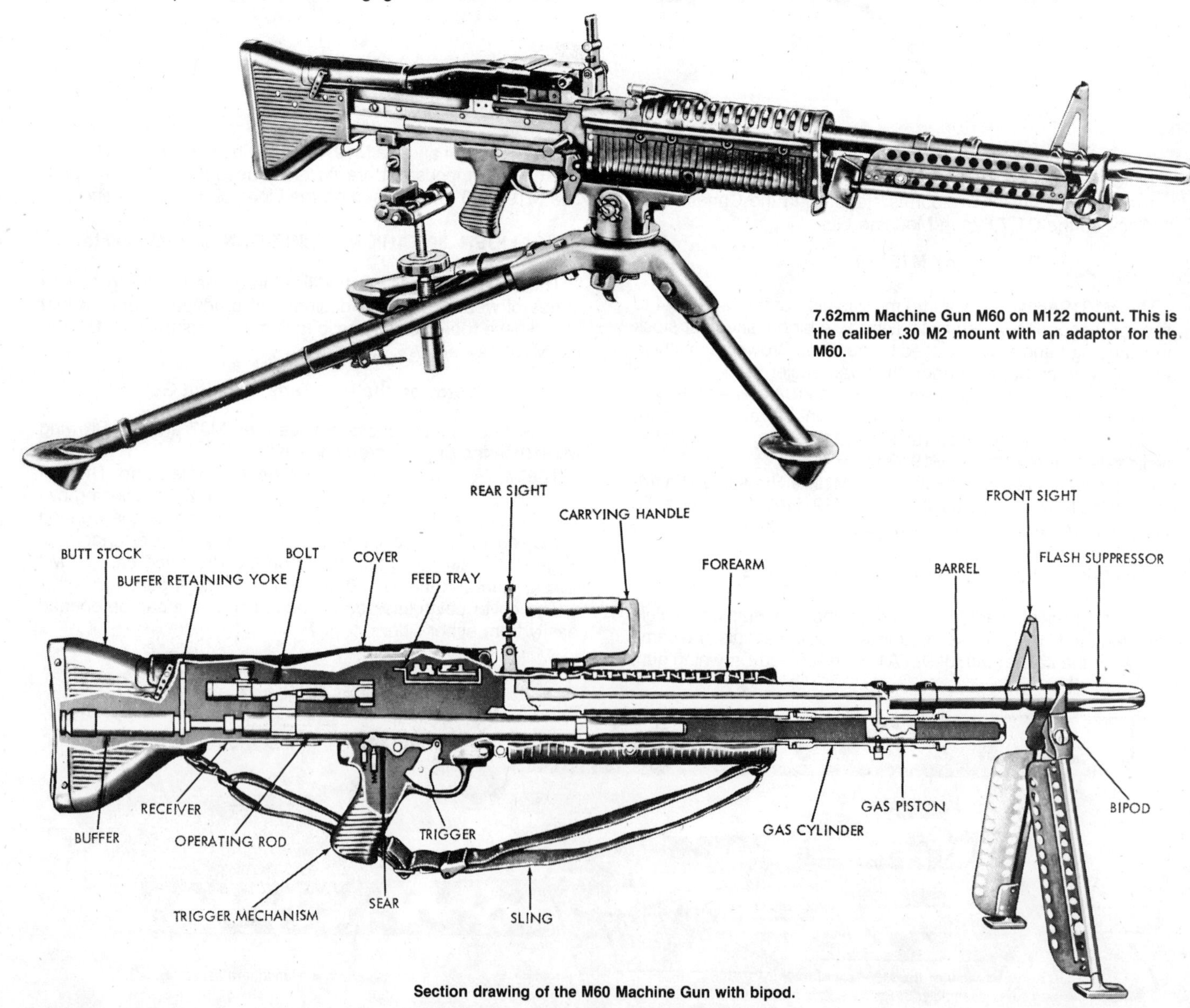

7.62mm Machine Gun M60 on M122 mount. This is the caliber .30 M2 mount with an adaptor for the M60.

Section drawing of the M60 Machine Gun with bipod.

Reloading. After firing, if the belt has been exhausted, the gun will be closed and must be cocked by hand to reload. The last link of the belt will remain in the feed plate. This link can be removed by the alternate method of loading described above; or, if a leading-link tab is provided, the link can be pushed through by utilizing the tab; or, finally, it can be removed when the cover is open for reloading.

Sights. The weapon as furnished is sighted in at 100 yards. A "quick adjustment" type of rear sight is provided. Barrels are zeroed-in by utilizing the lateral and elevating adjustments on the rear sight. For lateral adjustment, turn the knob at the lower left side of the rear sight. If additional lateral adjustment is required, the spring dove-tail (base) can be moved in its dove-tail on the receiver. The receiver should be restaked to retain the sight in the position permanently. For normal use, the spring action of the sight base is sufficient to hold the sight in its place.

For elevating adjustment, a scale slide is provided on the rear sight. Zero the barrel to the aperture; then, after this zeroing is accomplished, the scale can be lined up with the aperture by loosening the lower screw, sliding the scale to its desired location, and tightening the screw securely.

Barrel Changes. Barrels are changed by raising the lever found at the extreme right front of the receiver to the vertical position and withdrawing the barrel. Use the bipod as a handle. Assemble the barrel by inserting from the front, aligning the gas cylinder pilot nut with the receiver extension tube. Press the lock lever down and rearward to the full limit of travel.

Stripping the M60

For ease in cleaning, oiling and inspection, this weapon can be taken down as follows, using the cartridge as a tool. With gun in closed or fired position:

Unlatch and raise feed cover.

Remove butt stock by lifting shoulder rest and insert cartridge in exposed latch hole. Withdraw butt rearward.

Lift lock plate vertically for removal, holding the buffer to prevent drive spring from ejecting it. Withdraw buffer.

Pull operating rod drive spring guide and spring out of the rear opening. Withdraw operating rod and bolt assembly rearward with the cocking lever, pulling the bolt out of the receiver by hand the remaining distance. As the bolt rotating cam is exposed, insert lock plate in front of cam opening to hold firing pin back. The operating rod can then be withdrawn from the bolt.

CAUTION: Roller bearing and pin retainer are staked in place and should not be removed. These parts are normally also held in place by the firing pin; therefore, care should be exercised to prevent their loss.

Remove trigger housing assembly by pressing spring lock flat against the trigger housing at extreme front pin to unlock it from the pin. Rotate the spring lock down and away from the receiver to disengage it from the pin, and withdraw it. Remove the front pin from the left. Remove trigger housing by sliding it forward to disengage it from the receiver.

Feed cover removal is not necessary for field stripping but can be done if desired. With feed cover raised vertically, and feed plate in place on the receiver (at approximately 90° with cover), use bullet nose to remove cover pin spring, and withdraw cover pin spring and cover pin by hand, rocking cover slightly to aid disengagement. Cover and plate are free for removal. A torsional counterweight spring is also retained by the pin and is removable when pin is removed.

Remove barrel by raising lock lever to its vertical position and pulling the barrel out in a forward direction.

Handguard removal is not necessary for field stripping but can be done if desired. Grasp the handguard firmly. Insert a bullet nose in the hole provided at the bottom rear of the handguard.

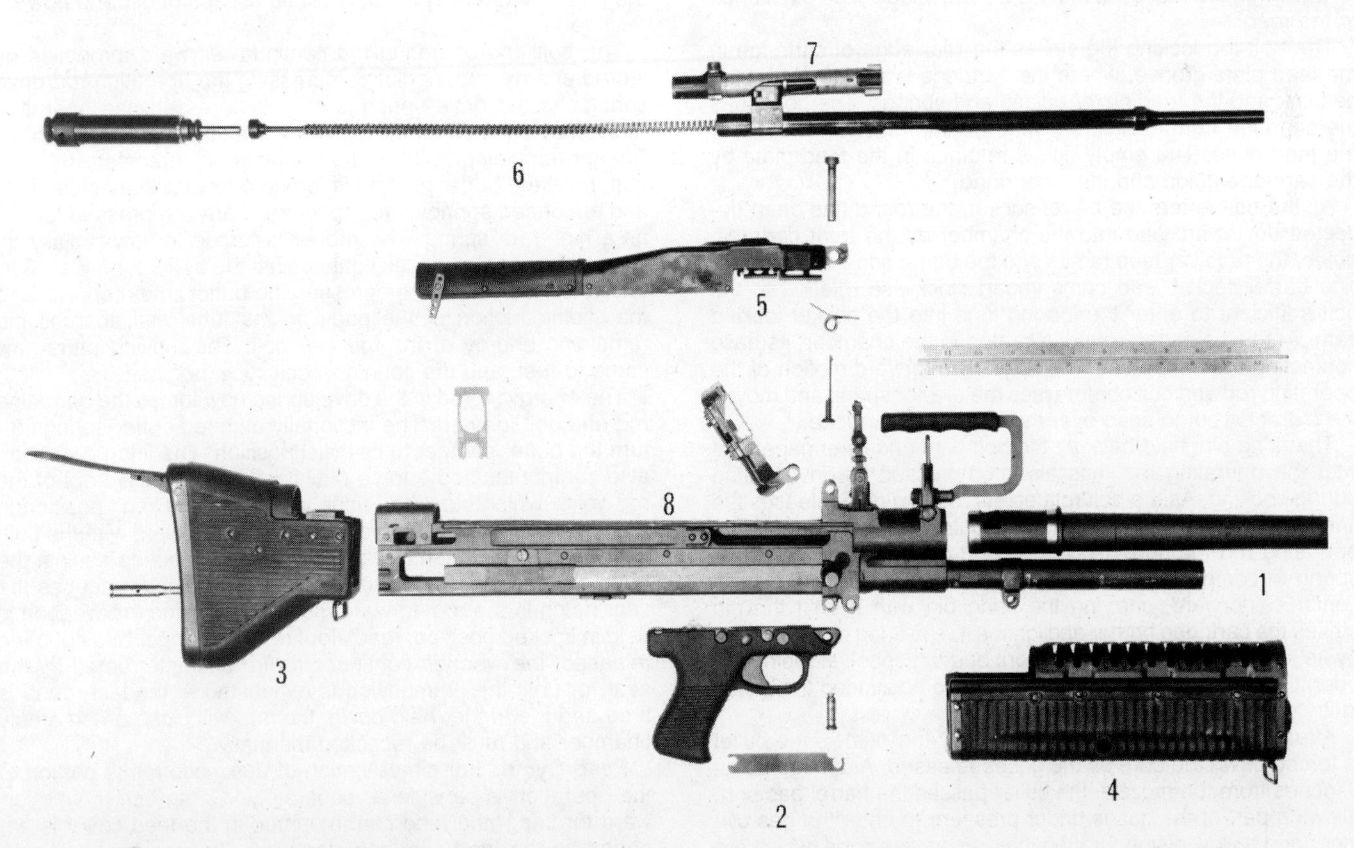

Field stripped view of the M60 Machine Gun. 1. Barrel assembly. 2. Trigger assembly. 3. Butt stock assembly. 4. Handguard assembly. 5. Feed cover assembly. 6. Piston and buffer assembly. 7. Bolt assembly. 8. Receiver assembly.

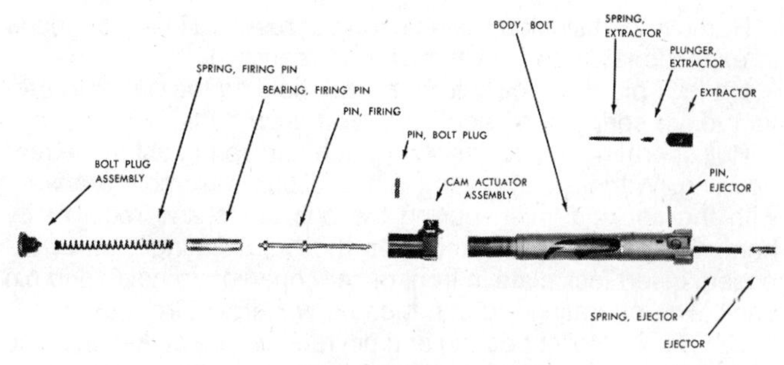

Bolt assembly of the M60.

Push cartridge to disengage spring lock. Slide handguard off to front. It is helpful to position the receiver vertically, with the front extension tube resting on a flat hard surface when disengaging the lock. After unlocking and sliding the handguard forward to disengage it from the receiver, tip the handguard to clear the tripod clevis and remove.

The weapon can be reassembled by reversing the order of the above steps.

How the M60 Works

With the gun loaded, in the cocked position, the following is the sequence of operation.

Release the safety and pull the trigger. The trigger raises the front of the sear, dropping the rear of the sear out of its engagement notch in the operating rod.

The operating rod, released by the sear, is propelled forward by the energy stored in the drive spring by the cocking operation. The bolt, engaged by the operating rod cam yoke, is carried forward with the rod, as is the gas piston, should it be positioned to the rear.

The bolt top locking lug strikes the rear edge of cartridge in the feed plate groove, where the cartridge is positioned by the belt link and the feed cover guides and springs. The bolt strips the cartridge from its link, carrying the cartridge forward out of the feed plate. The empty link is retained in the feed plate by the cartridge guide and the next round.

As the bolt enters the barrel socket, the round has been deflected downward and into the chamber by the front cartridge guide, the receiver feed ramps and the barrel socket feed ramp. The barrel socket lead cams impart clockwise rotation to the bolt sufficient to enter the locking lugs into the socket locking cam. At this point, the round is seated in the chamber, its base contacting the extractor and ejector. The forward motion of the operating rod and bolt compresses the ejector spring and moves the extractor out to snap over the rim of the cartridge.

The firing pin, held back by the bolt cam and its engagement with the operating rod, has also compressed its drive spring during cocking. As the bolt rotates, its rotating cam releases the firing pin. The firing pin spring now contributes its energy to the operating rod through the bolt rotating cam and aids the drive spring to complete the bolt locking rotation. The operating rod continues forward, carrying the firing pin with it until the pin strikes the cartridge primer and ignites it. The firing pin is stopped by its seat in the bolt at the pin front bearing spool, stopping the operating rod. The operating rod has also positioned the piston in its forward position.

After ignition of the powder charge by the primer, the bullet is forced down the bore by the gases released. At a point about 8 inches from the muzzle, the bullet passes the barrel gas port, allowing part of the gases under pressure to enter the gas cylinder and hollow piston through their respective ports, which are aligned with the barrel port.

Gas under pressure, bled from the bore by the gas port, fills the hollow piston until enough pressure is built up to overcome the mass of the piston and operating rod and the load of the operating rod spring. The piston begins its travel rearward, propelled by gas pressure against the forward end of the cylinder. After a very short travel rearward, the piston ports and collector ring are out of line with the barrel gas port, sealing off a measured charge of gas under pressure in the cylinder and hollow piston. By this time, the bullet is well out of the bore, and gas pressure in the bore rapidly falls to help provide easy extraction of the spent round.

The piston, propelled rearward by the trapped expanding hot gases, forces the operating rod rearward to contact the firing pin rear spool. This compresses the firing pin spring withdrawing the firing pin from the cartridge primer as well as compressing the operating rod spring. Rearward motion of the operating rod continues in the "dwell" slot of the bolt rotating cam, until the rod yoke rollers contact the spiral or counterrotating part of the bolt cam.

The bolt now begins its counterclockwise rotation and unlocking, urged by the energy of the operating rod. At full unlock the firing pin spring is fully cocked, the roller of the rod cam yoke contacting the rear of the bolt rotating cam.

Bolt and rod together continue to the rear, propelled by the energy of the expanding gas, withdrawing the spent cartridge from the barrel chamber. As the spent cartridge leaves the chamber, it is moved sideward in the direction of the ejection port; the ejector spins the case sideward to pivot about the extractor lip and disengage, allowing the case to be ejected with force out of the port, and against the ejection deflector which propels it downward.

In this interval, the piston has reached the limit of its travel. Spent gases are exhausted through ports provided in the gas cylinder, uncovered by the piston in its rearward travel.

Air behind the piston is exhausted through a set of ports near the rear of the gas cylinder, allowing escape of dirt and powder residue.

The bolt and operating rod continue rearward, propelled by stored energy and inertia, compressing the operating rod drive spring. As the drive spring is fully compressed, the operating rod contacts the buffer through the drive spring guide rod collar. Energy remaining in the rod and bolt is now transferred to the buffer. As the buffer plunger is forced rearward against its pads and preloaded springs, the rubber pads are compressed, acting as a high-rate spring. The rubber is forced to flow radially, to expand frictional surfaces tight against the buffer tube wall. Further rearward motion compresses the buffer return spring, and the sliding friction of the pads on the tube wall absorbs the remaining energy of the rod and bolt. The moving parts now come to rest, and the counter-recoil cycle begins.

The energy stored in the drive spring now forces the operating rod and bolt forward. The frictionally damped buffer springs return the buffer plunger to its initial position. The firing pin spring also contributes some force until it is halted by the front of the rod yoke, which stops the firing pin in the "cocked" position in the bolt rotating cam. As the rod continues forward with the bolt, if the trigger has been released, the sear is held up against the bottom of the rod by the sear spring. As the sear engages the sear notch in the rod, forward motion stops, and the weapon is held in cocked position, ready for fire. If the trigger has not been released, the weapon continues to fire until interrupted by the sear, or until the ammunition is exhausted. If the last round is fired and the trigger held down, the gun will close on an empty chamber and must be recocked manually.

Feed Cycle. For simplification of description, this portion of the operation is considered separately. As the bolt travels forward for firing, the feed cam mounted in the feed cover is engaged by the feed cam actuator roller attached to the rear of the bolt. Forward motion of the bolt and actuator causes the feed cam to swing to the right, forcing the front of the feed cam

Pin on the bolt actuates the belt feed lever. (Diagram from G. M. Chinn, *The Machine Gun*.)

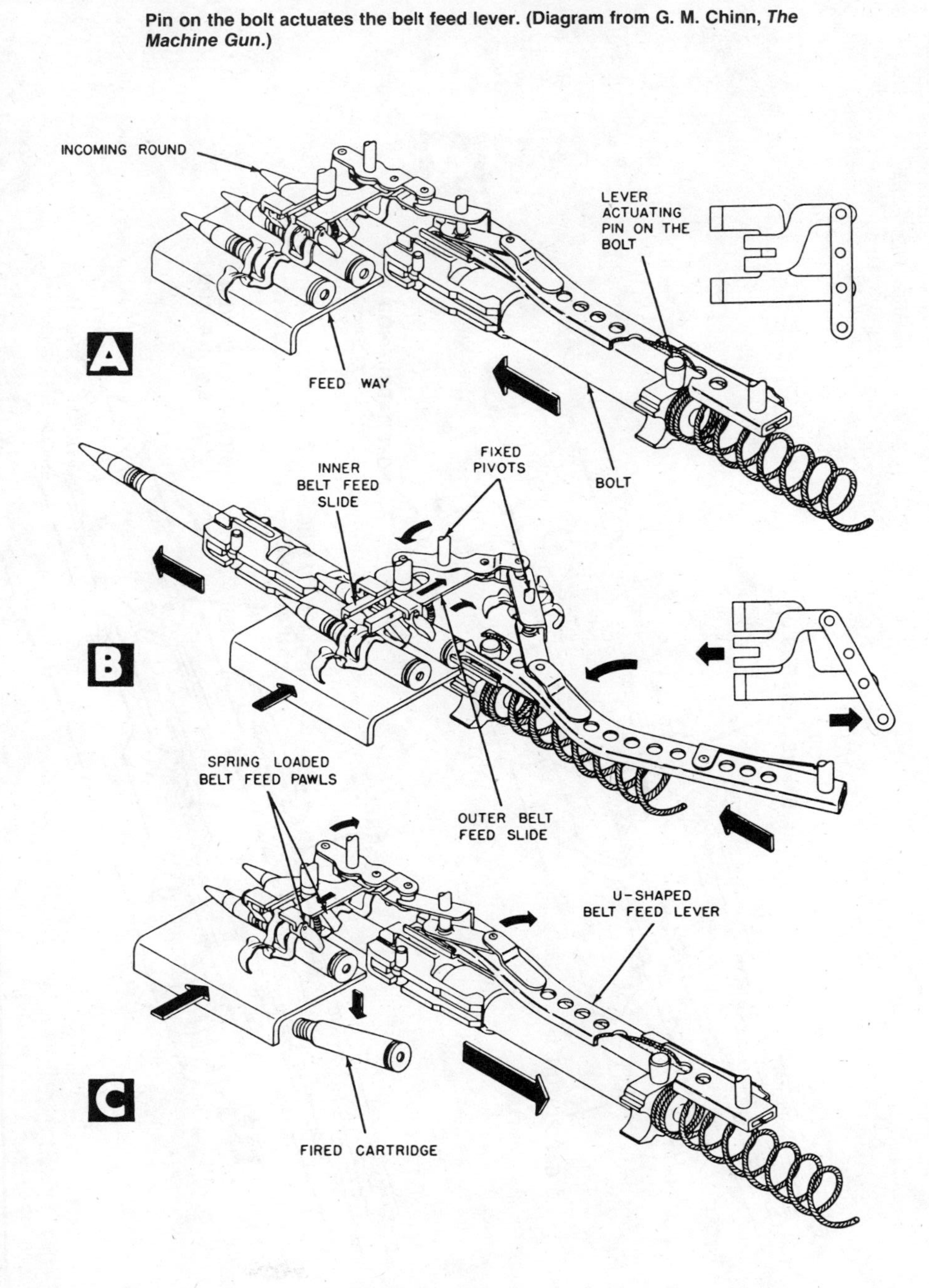

MG42 Feed system adapted to M60 Machine Gun.

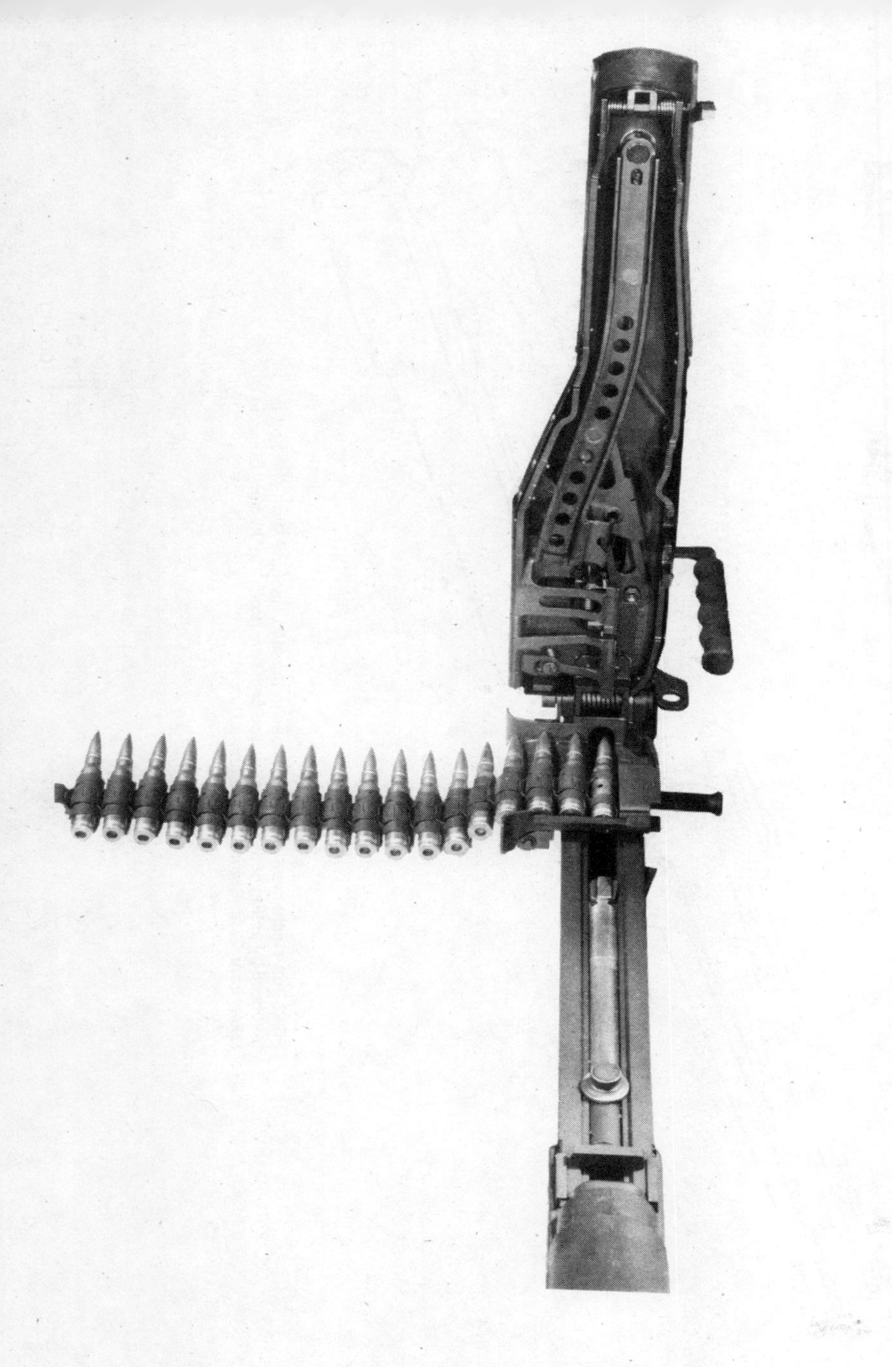

M60 Machine Gun feed system.

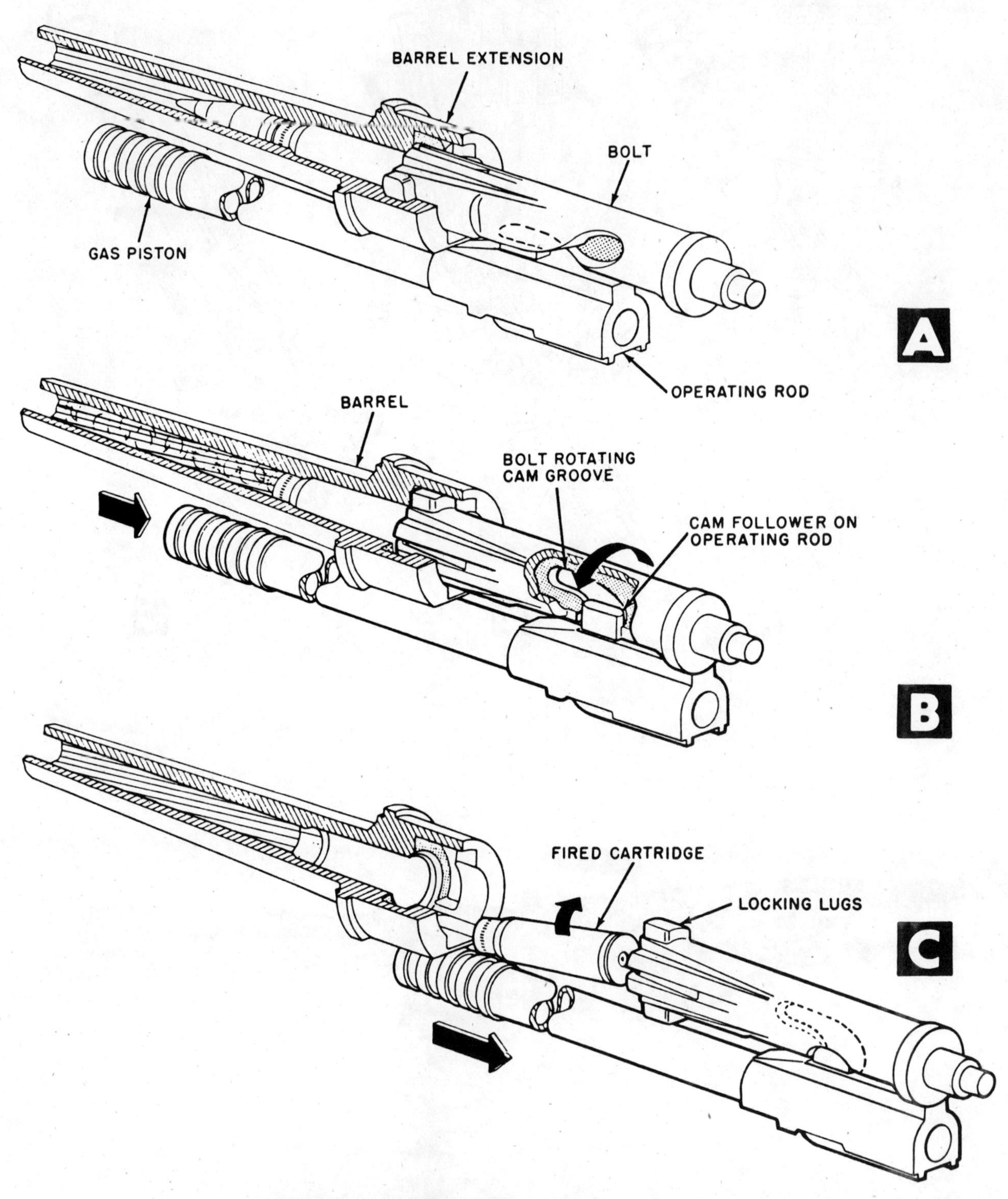

FG42 Bolt Mechanism adapted for use in the M60 Machine Gun. The bolt is unlocked by action of the operating rod on a cam groove in the bolt. Diagram taken from G. M. Chinn, *The Machine Gun.*

Typical receiver markings on the US M60 machine gun.

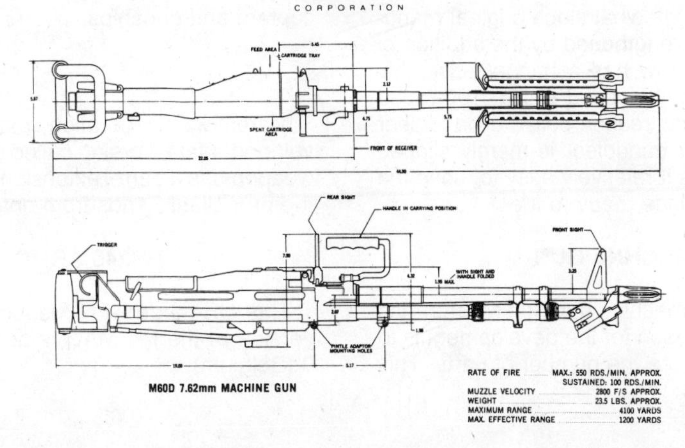

M60D 7.62mm MACHINE GUN

new england group
MAREMONT
CORPORATION

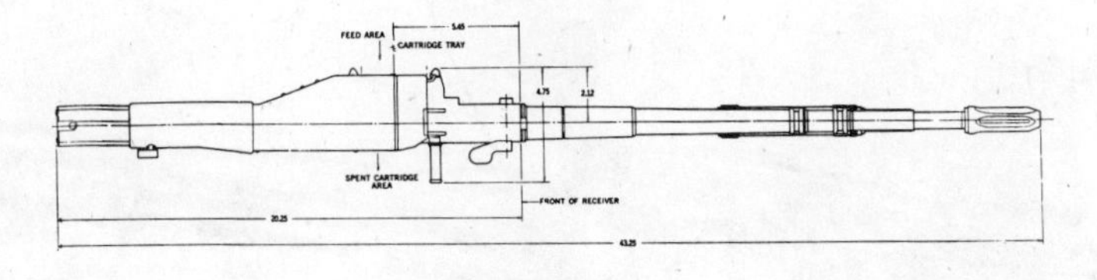

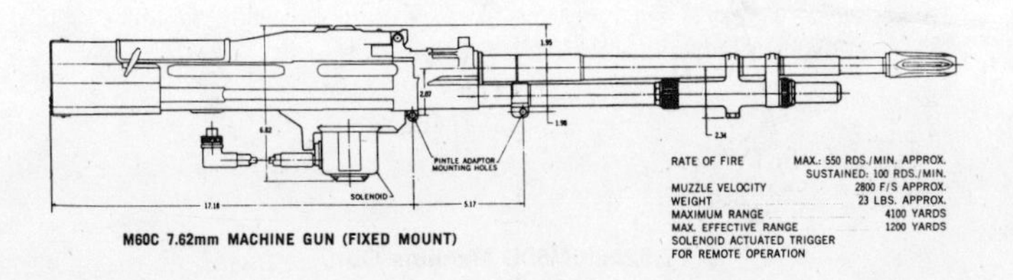

PINTLE ADAPTOR MOUNTING HOLES
SOLENOID
RATE OF FIRE MAX.: 550 RDS./MIN. APPROX.
SUSTAINED: 100 RDS./MIN.
MUZZLE VELOCITY 2800 F/S APPROX.
WEIGHT 23 LBS. APPROX.
MAXIMUM RANGE 4100 YARDS
MAX. EFFECTIVE RANGE 1200 YARDS
SOLENOID ACTUATED TRIGGER
FOR REMOTE OPERATION

M60C 7.62mm MACHINE GUN (FIXED MOUNT)

lever to the left. This carries the cartridge feed pawl plate assembly to the left where the pawls drop over and engage the next or second round for transport and remain there until the round in the plate is stripped and fired.

As the bolt recoils, the actuator forces the feed cam to the left, and the feed pawl assembly transports the round to the right into the feed plate groove. The round is forced down into the groove by the cartridge guides and their springs. The empty link left on the plate is pushed out the port in the feed plate by the new round as it is fed. The feed plate contains a spring loaded retaining pawl that retains the belt, holding the second round when the first round is in the plate groove. Two anti-friction rollers help guide the belt in place and support the hanging belt. The weapon is provided with sufficient reserve power to lift a 100-round belt vertically under all normal operating conditions.

Special Note on the M60 Machine Gun

The M60 is the first United States machine gun to have a true quick change barrel. It was specifically designed for light weight, and components design was simplified for manufacture. Stampings or fabrications were used wherever possible. The quick change barrel, light weight and adaptability for use as either a heavy or light machine gun, as well as relative ease of manufacture, make the M60 superior to the Browning guns it replaces. The performance of the lined and plated barrel of the M60 in sustained fire is exceptional. An added factor of advantage in the case of the M60 versus the Browings is that the M60, like most other weapons with a quick change barrel, has no headspace adjustment problem.

The M60 has been modified somewhat since original manufacture. The receiver has been strengthened by the addition of several pins, and the feed tray now has a hanger assembly pinned to it. The hanger assembly is used with a 100-round ammunition box, for some peculiar reason called a bandoleer in some publications. The box or bandoleer is merely slipped down over the hanger and makes it relatively easy to move the gun around with ammunition in place ready to fire.

7.62mm M60E1 MACHNE GUN

A modified version of the M60 machine gun, the M60E1, has been developed. The principal reason for the development is to simplify barrel change and decrease the number of parts. This weapon differs from the M60 as follows: (1) The barrel does not have the bipod or gas cylinder attached to it. They are attached to the gas cylinder tube; (2) The bipod is attached semi-permanently to the rear of the gas cylinder; (3) The gas cylinder has been simplified and has no threads. It has a "U" shaped key to retain the gas cylinder extension; (4) The operating rod guide tube has a lug that retains the gas cylinder and bipod on the weapon and eliminates the gas cylinder nut; (5) The modified spool type gas piston has no holes; (6) The modified rear sight has the lateral adjustment increased by 20 mils; (7) The modified die-cast feed cover eliminates parts and allows the cover to be closed whether the bolt is in the forward or cocked position; (8) The modified feed tray eliminates parts; (9) The magazine hanger fitted to the left side of the weapon eliminates parts and can be used either with a modified magazine or a modified bandoleer; (10) The new die-cast forearm eliminates parts and eases changing the barrel because the absence of the forearm cover allows the carrying handle to be fitted to tbe barrel; (11) The sling swivels have been relocated to the left side of the forearm and the top rear of the buttstock easing the carrying of the weapon; (12) The carrying handle has been increased in diameter.

M60C AND M60D

The M60C is a modification of the M60 machine gun for use on helicopter armament kits. The weapon has the stock removed and is remotely charged and fired. The M60D also has the stock removed and the pistol grip as well. The trigger has been relocated to the rear, and spade type grips have been fitted. This gun is used on flexible pedestal type and other mounts on helicopters and gunships.

M60E2

Maremont Corporation, manufacturer of the M60, also developed a tank version called the M60E2 (See Chapter 2.). This weapon has a barrel extension and gas evacuator tube that keep the propellant fumes from entering the interior of the tank.

M240 ARMOR MACHINE GUN

This US-made—FN Manufacturing, Inc. of Columbia, SC—version of the FN MAG is described in more detail in Chapter 9 (Belgium).

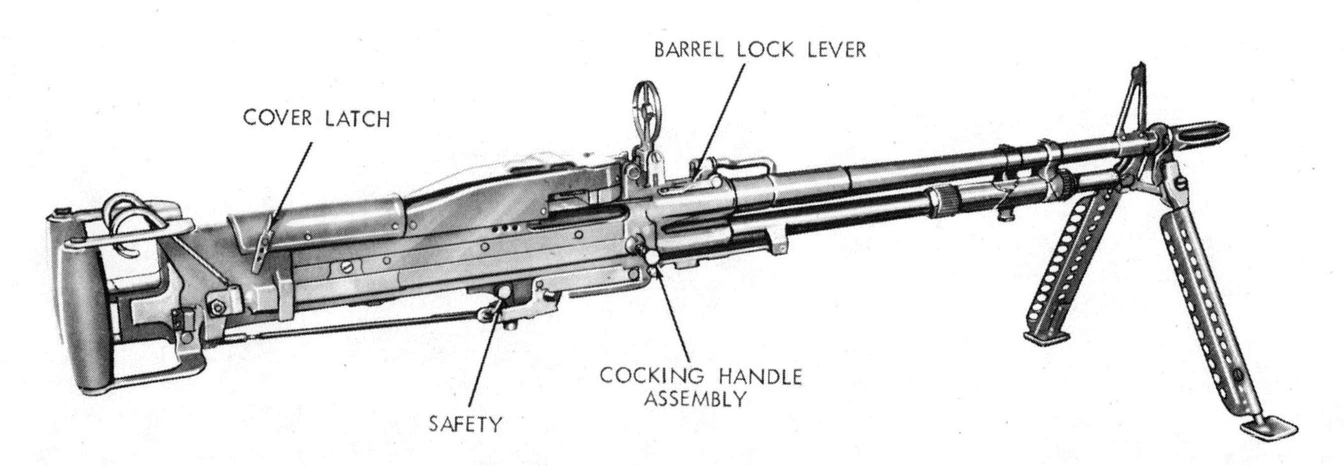

7.62mm M60D Machine Gun.

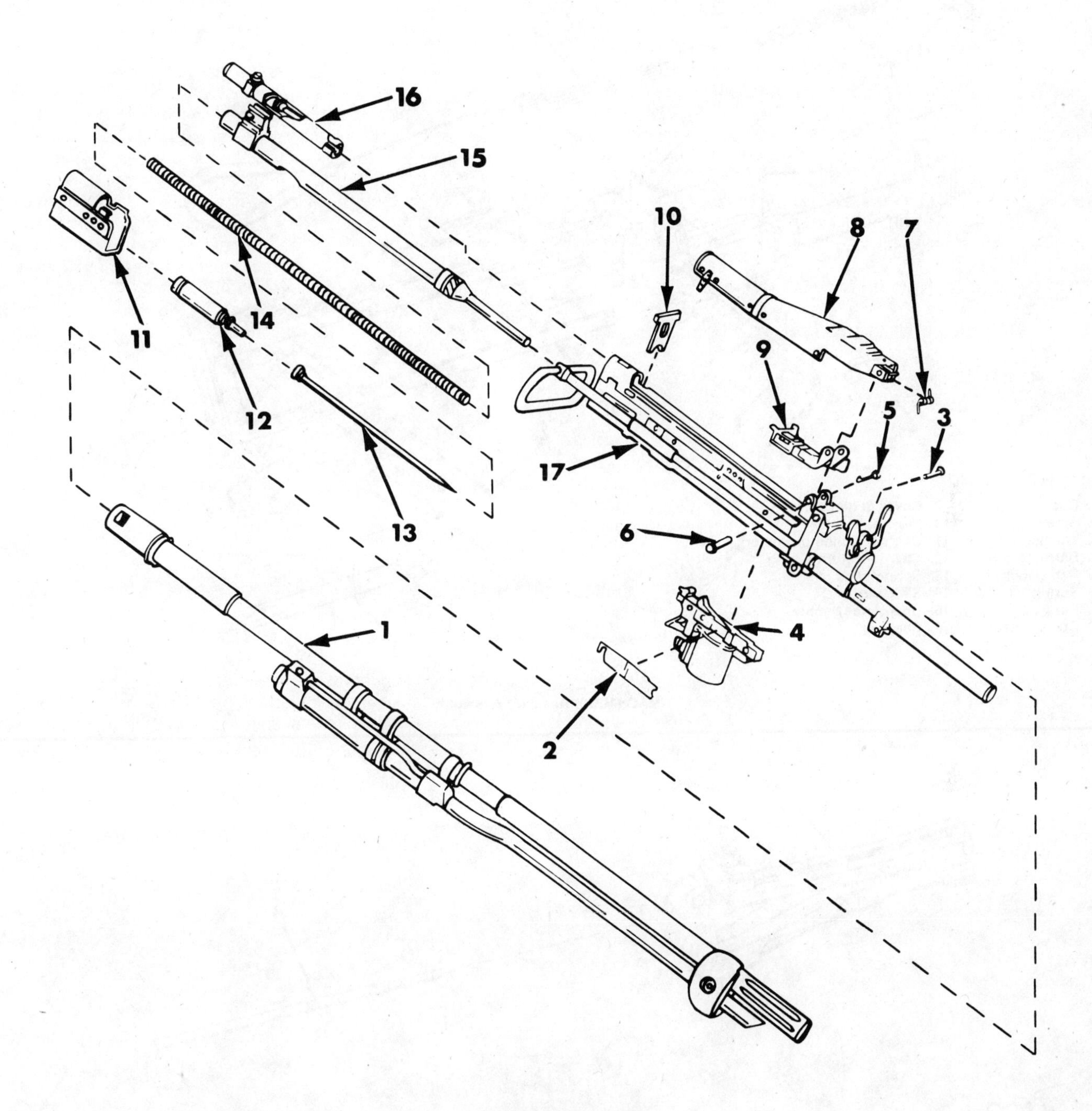

M60E2 Tank Machine Gun.

1. Barrel and evacuator assembly.
2. Leaf spring.
3. Pin.
4. Electric actuator group.
5. Latch.
6. Pin.
7. Spring.
8. Feed cover.
9. Feed tray assembly.
10. Yoke.
11. Backplate assembly.
12. Buffer assembly.
13. Buffer guide.
14. Drive spring.
15. Operating rod.
16. Bolt.
17. Receiver.

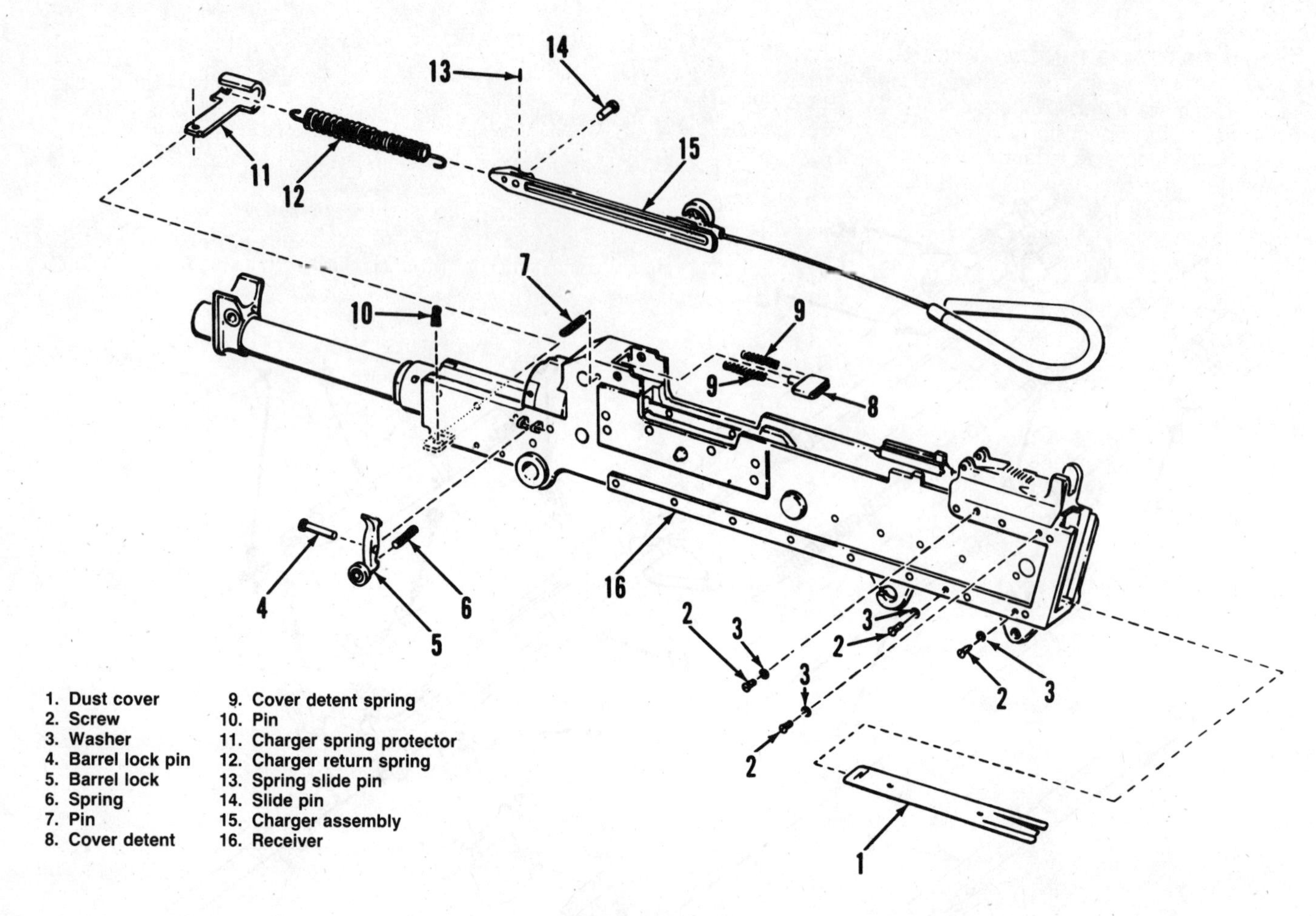

MAG (M240) Receiver Assembly.

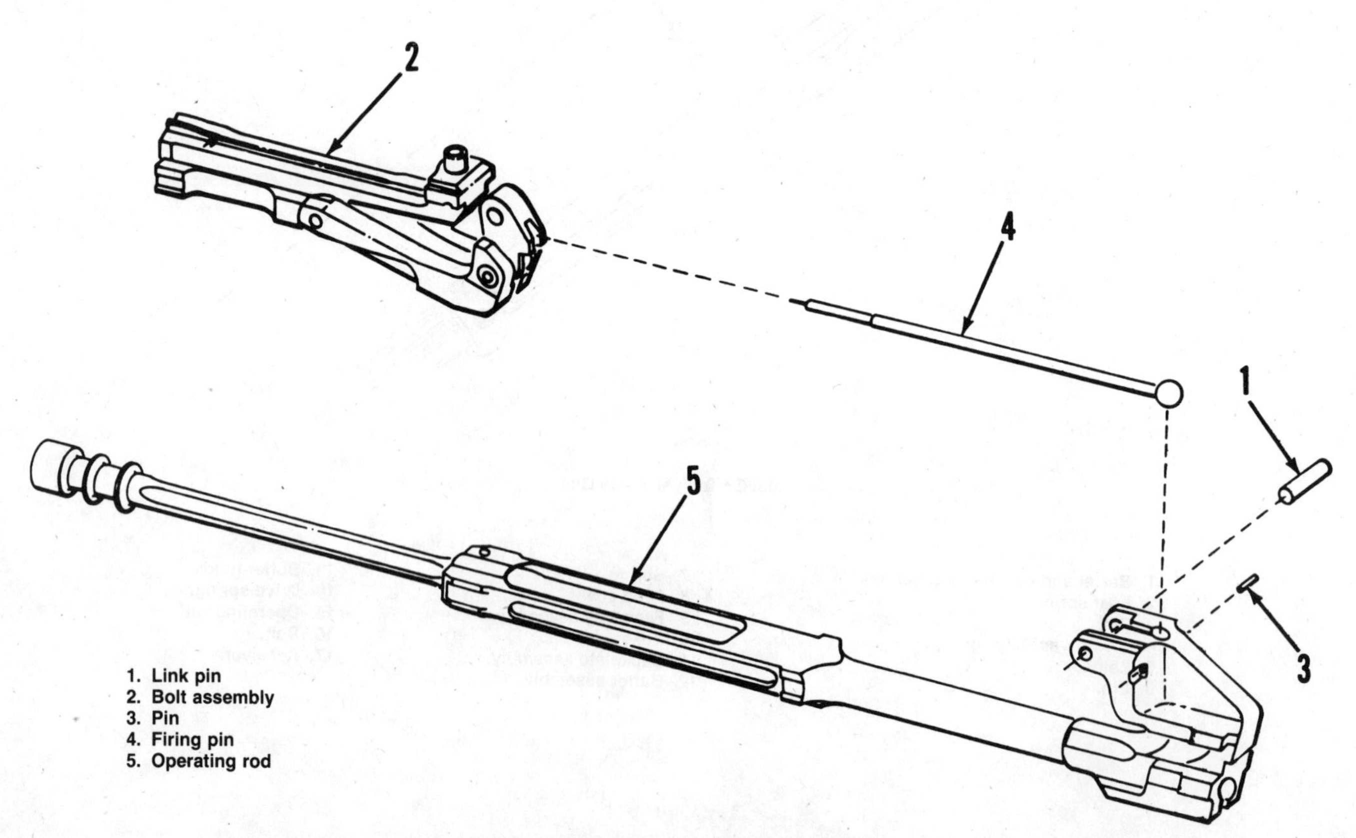

MAG (M240) bolt and operating rod assembly.

7.62mm NATO M73 TANK MACHINE GUN (Obsolete)

The development of the M73 (formerly known as the T197E2) was the result of the need for a rifle-caliber machine gun designed specifically for use in tanks. (See Chapter 2.) The M73 has a quick change barrel and can be fed from either side. The M73 uses the same link, the M13, that is used with the M60 general purpose machine gun. It can be charged and fired by either manual operation of solenoid. The barrel jacket of the M73 is to be attached to the tank in a semipermanent manner, and all the working components of the weapon, plus the barrel, can be removed from inside the tank for cleaning and repair.

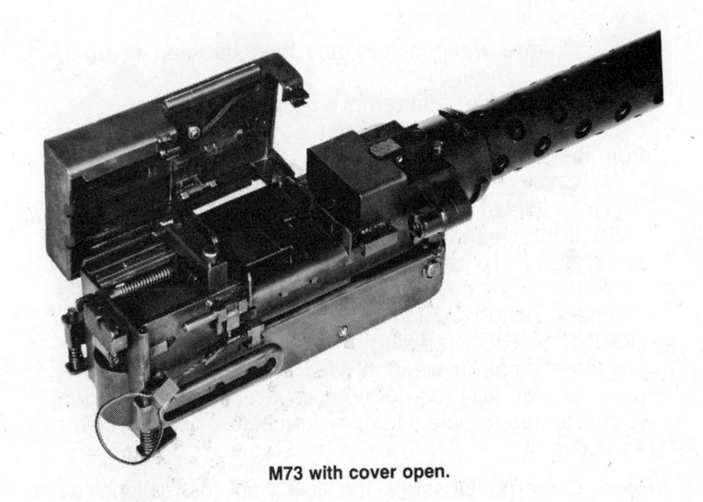

M73 with cover open.

How to Load and Fire the M73

Preparation for Firing. Open cover by pressing the cover latching rod on the side of the receiver from which the belt will be fed. Pivot the cover to an open position.

Retract the action to seared position, using the charging mechanism; make sure that the safety is at the "F" position. NOTE: Mount the charger on either left or right side of the receiver as determined by the mounting conditions of the weapon. Slide the charger connector to the proper side by pressing the retainer and sliding the connector manually.

Quick change barrel of M73.

Slide the safety to the safe position, exposing the letter "S" on the back plate.

Visually inspect the barrel chamber and inside of receiver for assurance against possible obstructions.

To load the weapon, insert the cartridge belt with the first round in the slot of the feed tray (open side of link loops facing downward).

Press the cover latch rod, close the cover, and release the cover latch rod to the lock position.

Firing. Slide the safety in the back plate to the fire position, exposing the letter "F" on the back plate ("S" for safe).

Actuate the solenoid for remote firing.

Press the trigger on the back plate for manual firing.

Immediate Action Procedure (to resume fire after a stoppage). When a stoppage occurs BEFORE COMPLETING A 200-ROUND SERIES (starting from a cool gun), perform the operations listed below in the given order. (If weapon starts after the first operation, do not perform the next, etc.)

(1) Charge weapon fully to sear position and fire.
(2) Repeat above twice if weapon does not respond.

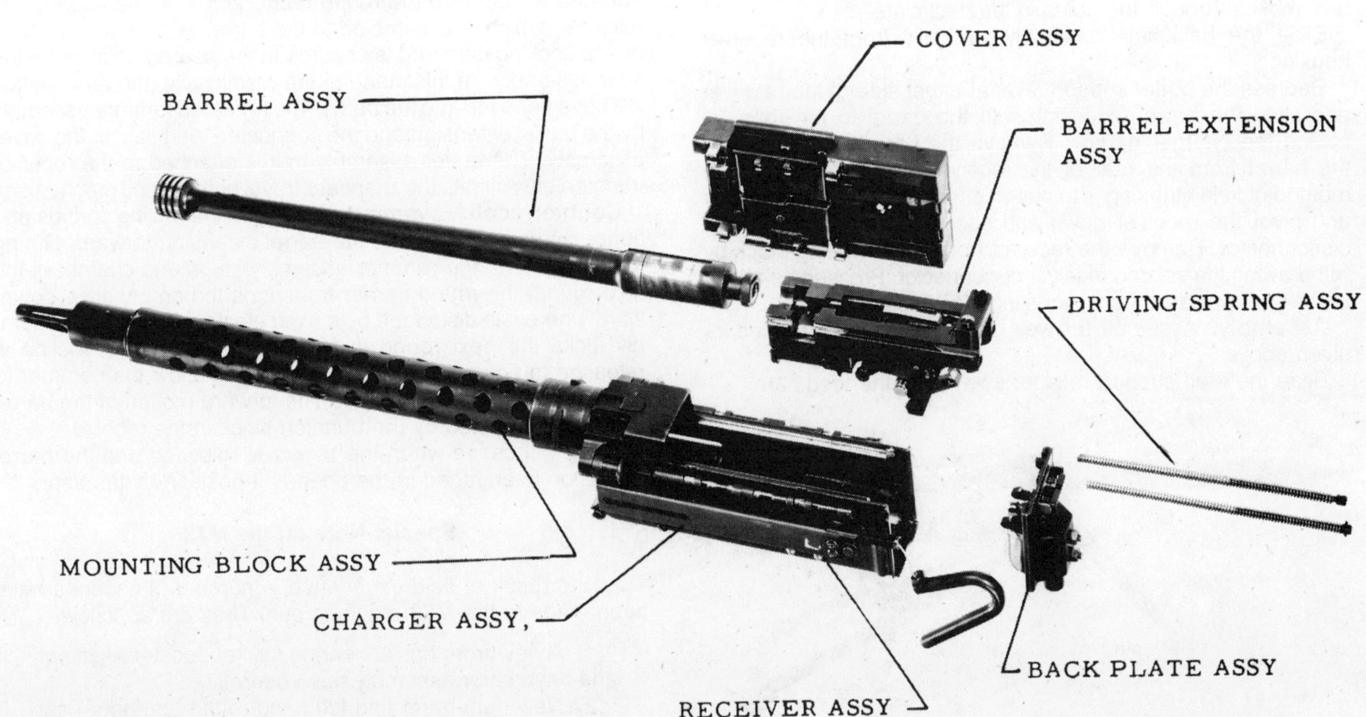

M73, Field stripped.

(3) Charge weapon fully and hold back on charger. Then—

(4) Open cover and remove belted ammunition.

(5) Open feed tray and remove live or spent cartridges from the action.

(6) Charge and hand-function weapon.

(7) Load and fire.

(8) If the weapon still does not fire, inspect for broken parts. NOTE: To prevent the possibility of an open-breech cookoff with the cover open, proceed as follows: When a stoppage occurs AFTER FIRING MORE THAN A 200-ROUND SERIES (starting from a cool gun), charge and fire three times. If weapon does not resume fire at this point, charge fully (do not open cover) and allow the gun to cool to near-ambient temperature before opening cover as above.

Feed Change. Observe the following instructions when changing from left-hand to right-hand feed:

Remove the cover assembly from the receiver.

Swing the feed tray to a vertical position relative to the receiver by disengaging it from one of the cover latching rods. Press the plunger of the round stop, and slide the round stop to engage the locating hole on the left side of the feed tray. In this position, the "R" on the round stop will be adjacent to the "R" on the feed tray ("L" and "L" for left-hand feed).

Follow operations listed below to convert the cover assembly from left to right.

Stripping the M73

Actuate the trigger to insure that the weapon is in the forward position.

Press the cover latch rod at either side of the receiver and raise the cover and feed tray assemblies to clear the latch rod. Release the cover latch rod and allow the cover and feed tray to rest on the rod. Press the second rod, and remove the cover and feed tray from the receiver.

With action forward, press and rotate counterclockwise the driving spring rods. Withdraw the driving spring and guide rod assemblies through the holes in the backplate.

Slide the backplate assembly vertically from the receiver housing.

Depress the buffer support lever at either side of the receiver and slide the barrel extension with the barrel to the rear, by pulling the charger handle. Remove the barrel extension and the barrel from the rear of the receiver. NOTE: An alternate method of field stripping is to pull on either receiver disconnector and pivot the receiver down and about the opposite receiver disconnector. Remove the receiver from the mounting block by withdrawing the second receiver disconnector. Proceed as above and withdraw barrel from mounting block assembly.

Assembly. Assemble the weapon by reversing the procedure given above.

Slide the feed support retainer away from the feed cam.

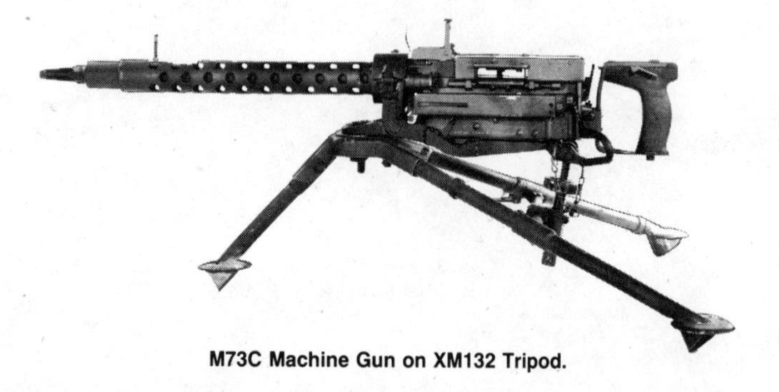

M73C Machine Gun on XM132 Tripod.

Slide the feed cam fully forward and lift it out, retaining it in hand.

Lift out the feed support and retain this in the same hand.

Remove the feed track and slide as a unit. Reassemble this unit to the cover after turning it end for end (180°). This will align the "R" of the feed track with an "R" in the cover ("L" and "L" for left hand feed).

Slide the feed support retainer fully to the opposite side of the cover.

Replace the feed support.

With the "R" side of the feed cam facing up, assemble to the cover in the fully forward position ("R" on cam adjacent to "R" on cover), picking up the feed slide roller during the operation. (Assembly notches prevent improper assembly.) Slide feed cam to rear of cover.

Slide the feed support retainer to the central lock position where it will secure the assembly. NOTE: In changing from right-hand to left-hand feed, reverse the foregoing procedure.

Reassemble the cover to the receiver by placing it in proper position and pressing the two cover latching rods.

How the M73 Works

Operational Power. The energy of recoil is supplied by the momentum of the recoiling parts and a muzzle booster driving the action rearward to buffer contact. Counterrecoil energy is supplied by the driving springs and the buffer spring that were compressed during the recoil cycle. The open-bolt action employs the driving springs alone and counterrecoil for the first shot from the sear position.

Recoil Movement. During the rearward movement of the barrel extension, the attached lever linkage actuates the rammer assembly by means of opening closing cams located in the sides of the receiver. The rearward movement of the barrel extension enables the sliding breechblock to move transversely to the right and away from the base of the chambered cartridge case. The extractor with the rammer grips the rim of the case and removes the spent round from the chamber, carrying it rearward to engage the round carrier grips. At this point, the spent case is transferred from the extractor to the round carrier grips. In the interim, the hammer, which is assembled to the barrel extension, is cocked by the cocking cam and is secured in the seared position by the sear assembly. At this stage of the recoil cycle, the ammunition will have been fed into the path of the retracted rammer assembly by the barrel extension and the connected feed cam in the cover assembly. The buffer assembly that is attached to the receiver trunnion block limits the rearward travel of the recoil mechanism.

Counterrecoil Movement. Energy of the driving springs and buffer return spring forces the barrel extension forward. During this movement, the rammer assembly strips and chambers the next round; the round carrier transports the empty case downward where it is dislodged by a fixed ejector; the breech assembly locks the next round in position; the rate control slide is released to actuate the hammer sear and allow the hammer to fall on the firing pin extension. The forward motion of the barrel extension is limited by the trunnion block in the receiver.

Firing will cease when the trigger is released and the barrel extension is engaged in the open-bolt position by the sear.

Special Note on the M73

As the result of field experience, a number of changes have been made to the M73 machine gun. They are as follows:

(1) A new front barrel bearing, jacket booster assembly and flash hider assembly have been fitted.

(2) New right-hand and left-hand round (cartridge) carriers have been fitted.

(3) A new case carrier link assembly has been fitted.
(4) A new retainer lock has been fitted.
(5) A new sear hammer has been fitted.
(6) Modified springs have been fitted to the barrel extension, the case carrier assembly and the trigger.
(7) The barrel has been modified.
(8) The firing solenoid lever has been modified.
(9) The trigger sear has been modified.
(10) The feed pawl has been modified.
(11) Rings and a washer have been added to the barrel disconnector to ease barrel removal.

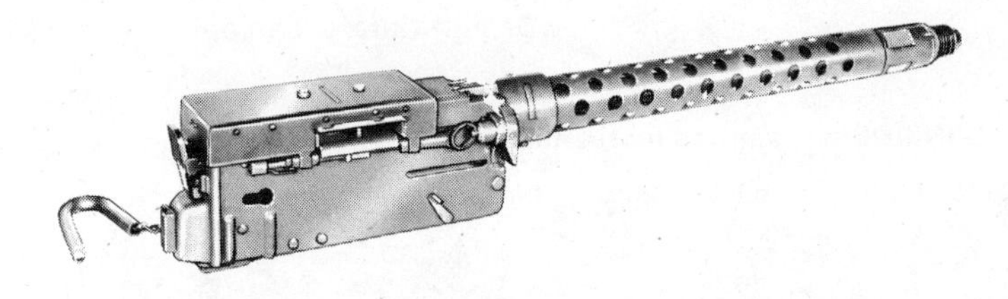

7.62mm M219 Machine Gun.

7.62mm Machine Gun M73C on XM132 Tripod

This is the flexible version of the fixed M73 tank machine gun. It is basically the same gun as the M73, with sights and a pistol grip trigger added. The solenoid, which is normally used to fire the gun in a tank, is integral with the back plate and remains on the M73C. The weapon is loaded as the M73 and is fired as follows.

Slide the safety in the back plate to the fire position—letter "F" exposed. Push down on trigger.

The XM132 tripod mount is the caliber .30 M2 mount with a special adaptor for the M73C. This weapon exists only in prototype form.

7.62mm MACHINE GUN M219 (M73E1)

The M219 is a simplified, product-improved M73. The basic difference in design of the two weapons is in the ejection system. The ejection system has been simplified considerably in that the cartridge case carrier mechanism used in the case ejection cycle of the M73 has been replaced by fixed ejectors located on the underside of the feed tray. The rammer assembly, buffer rod, buffer support tension spring and receiver are modified to be compatible with the fixed ejectors. As indicated in Chapter 2, both the M73 and the M219 Machine Guns will be replaced by the M240 (the FN MAG tank version). The Marine Corps will use the M60E2 as their coaxial gun in the M60 series of Tanks. A complete discussion of the M240 Tank Machine Gun is presented in Chapter 9 (Belgium).

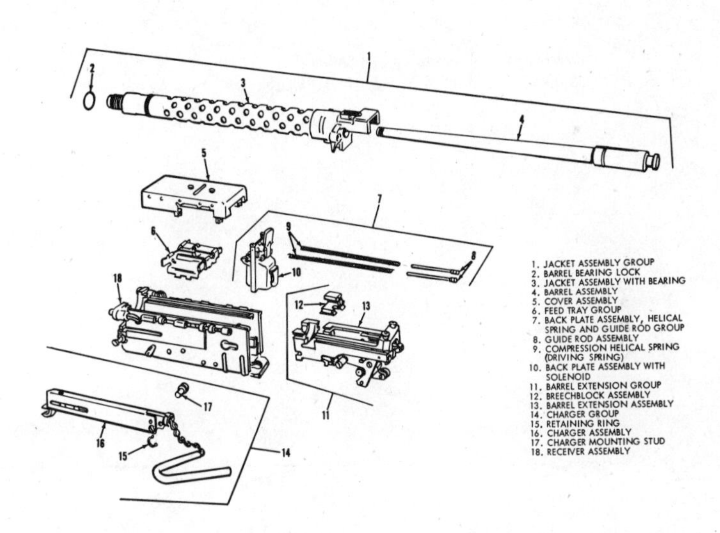

Exploded view 7.62mm M219 Machine Gun.

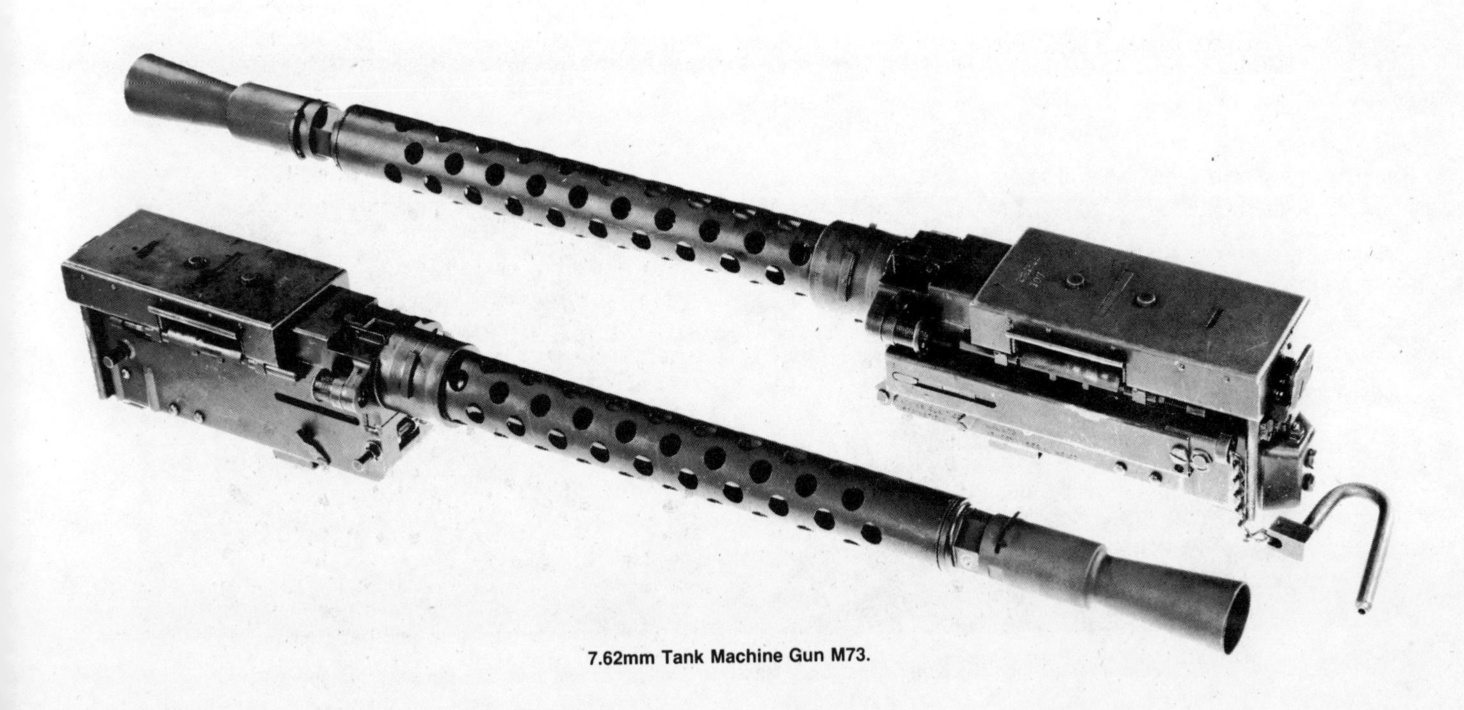

7.62mm Tank Machine Gun M73.

Design differences between the M73 and M219 Armor Machine Guns

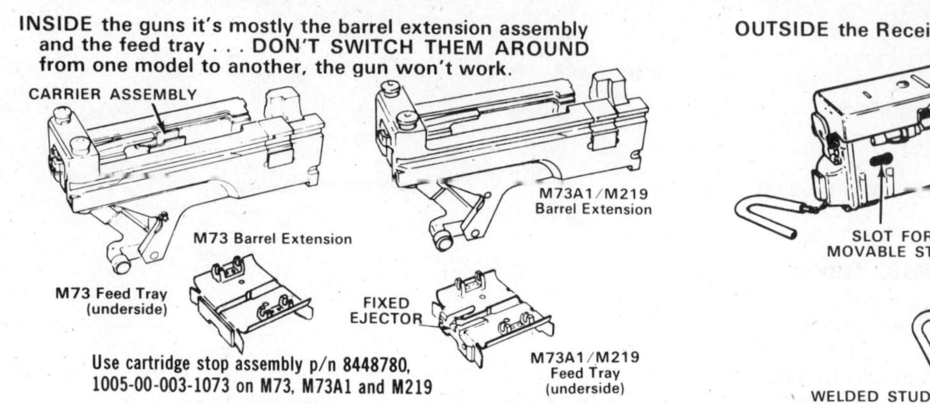

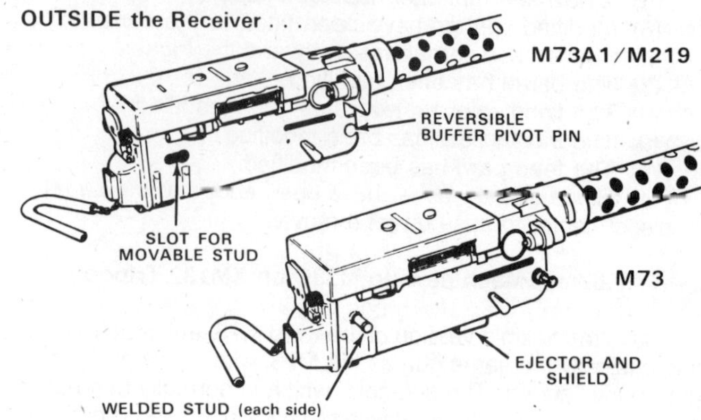

.50 Caliber (12.7 × 99mm) Browning Machine Gun

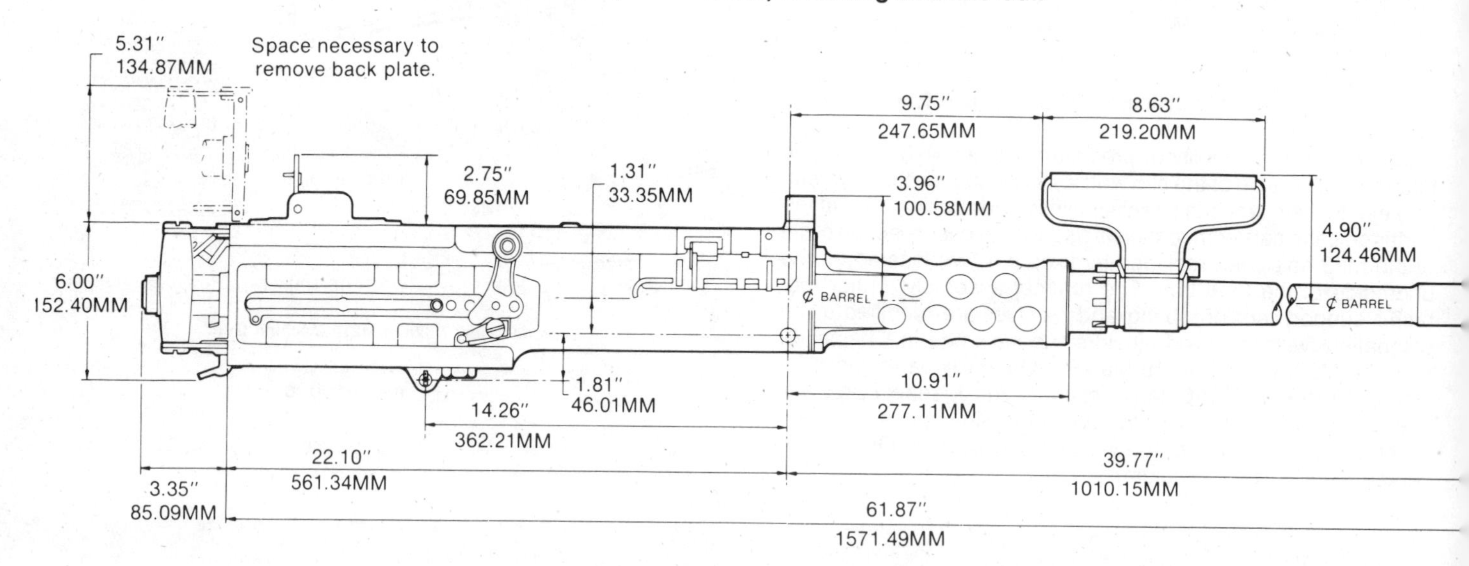

Dimensioned view of the .50 Caliber M2 HB Browning machine gun.

Typical receiver marking on current manufacture M2 HB Browning machine guns. See Chapter 2 for more details.

BROWNING CAL. .50 HEAVY BARREL M2 MACHINE GUN

The caliber .50 M2 Heavy Barrel is the ground gun of the M2 Browning series, which also include the aircraft and watercooled antiaircraft guns. A turret type version of the M2 heavy barrel also exists; it was mainly used on multiple A.A. mounts. The aircraft and water-cooled antiaircraft guns are not too frequently encountered at present, but the heavy barrel is still in wide use throughout the world.

Tripod Mount

A tripod mount is provided for the Browning machine gun caliber .50 HB M2. The tripod assembly weighs about 40 ½ pounds, while the pintle and elevating mechanism assemblies weigh another 4 pounds.

Loading and Firing the M2

To provide for mounting in aircraft or in vehicles where position or space available require a right-hand feed, this gun is fitted with a bolt and feed mechanism that is interchangeable for right- or left-hand feed.

Ammunition box is mounted on the side of the gun and belt fed through the feed block from the side set for feed. Pull belt through as far as it will go, and while retaining grip pull back retracting slide handle as far as it will go and permit it to run forward. This half-loads the gun. Pull the retracting handle back again as far as it will go and release to complete loading.

Note that in this gun the bolt latch release must be locked down before bolt is retracted for loading the gun.

Unlock the bolt latch release by pressing down on it. Pressure on the thumb trigger will now fire the gun. It should be fired only in short bursts.

To Unload the M2. Lift the cover and remove the belt. Pull back the retracting slide handle and look and feel in the feedway, the slot and the chamber to be sure the gun is unloaded.

Release the bolt and let it go forward and then lower the cover.

Press the trigger. If the bolt latch release is unlocked, alternately pressing the trigger, then the bolt latch release will fire the single shots.

If the bolt latch is locked down and the trigger pressed and held, the gun fires until the trigger pressure is released.

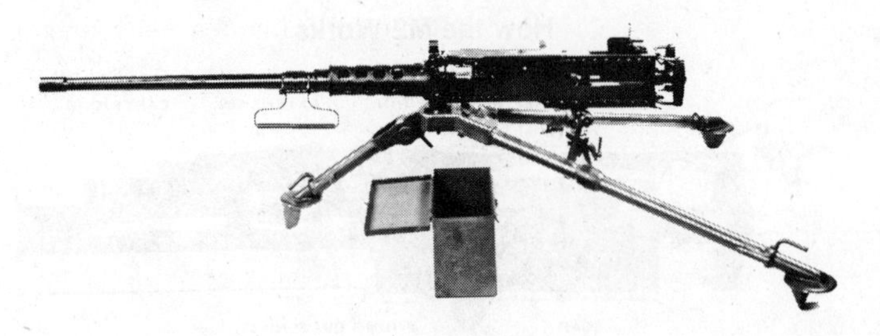

Caliber .50 Browning M2 Heavy Barrel Machine Gun.

Field Stripping the M2

Grasp the barrel handle firmly and unscrew until the barrel is free from the barrel extension, then withdraw it to the front.

Release cover latch and raise cover as far as it will go.

Release back plate latch lock and also the back plate latch; this will permit the back plate to be lifted up out of the top of the receiver.

Push the protruding end of the driving spring rod forward and away from the slide plate and ease out the spring and the rod.

Pull the bolt back until the bolt stop lines up with the hole in the center of the slot in the side plate and then pull the bolt stud out to the right.

The complete bolt may now be removed from the rear of the casing. Driving spring unit need not be removed.

Insert the point of a bullet in the small hole at the rear of the right side plate to compress the oil buffer body spring lock. Oil buffer, barrel extension and the barrel assembly may now be taken back and out of the gun.

Pressing the accelerator forward permits the oil buffer assembly to be detached from the barrel extension.

This completes the field stripping. Cover should not be removed or dismounted except for repairs, as considerable force is required to compress the pawl spring for reassembly, making this a difficult operation.

Assembly of M2. Reverse the dismounting procedure.

Insert oil buffer into the oil buffer body from the rear, making sure that the cross groove in the piston is on the upper side where it can engage the shank of the barrel extension.

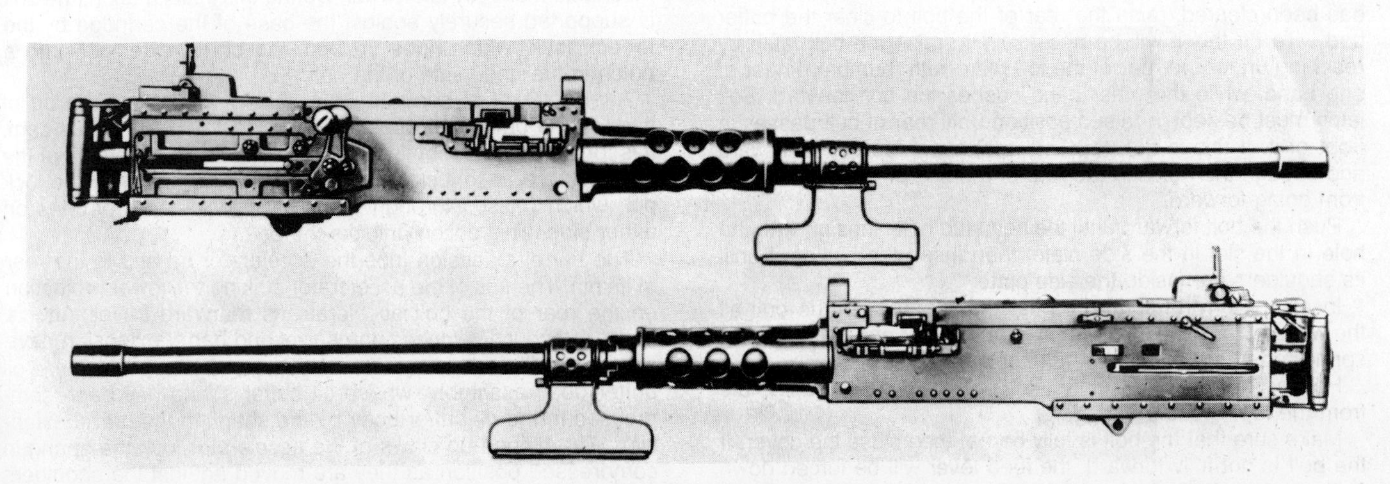

The M2, Heavy Barrel, Flexible.

How the M2 Works

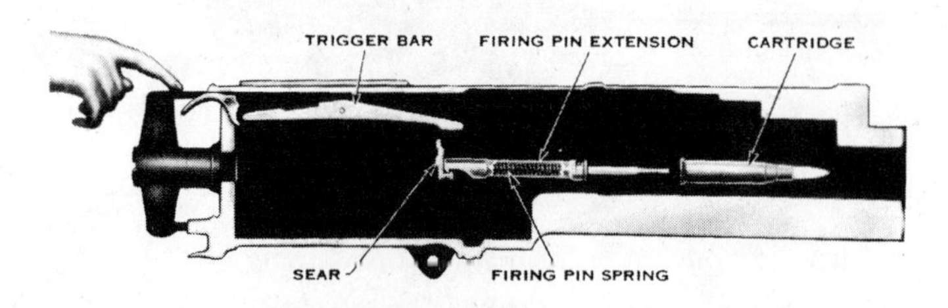

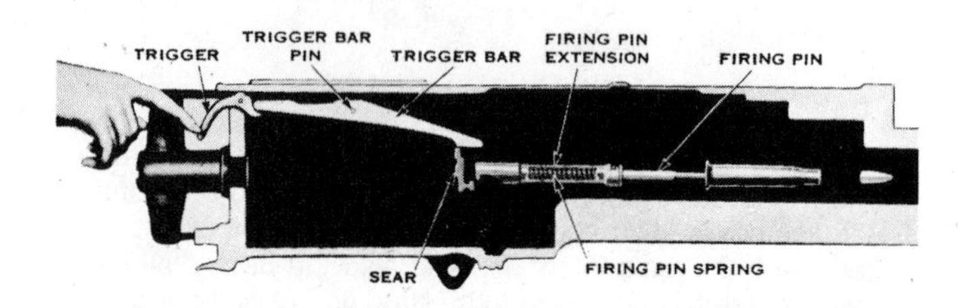

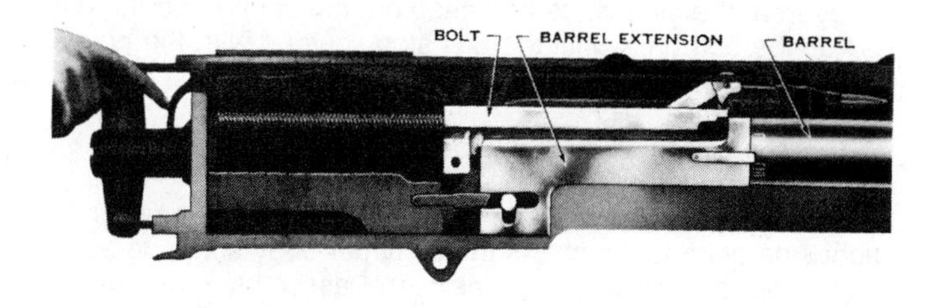

Assemble the buffer and buffer body to the extension. Holding the accelerator up under the barrel extension shank, start the breech lock depressors into their guideway in the barrel extension, and press forward permitting the shank of the barrel extension to engage the cross groove in the piston rod. Thrust sharply forward as far as oil buffer will go. The parts will now lock together and may be assembled into the receiver as a single unit. Press forward until the oil buffer spring locks in its recess in the right side plate.

Insert extractor in bolt and check that cocking lever is fully forward. Then insert bolt into rear of receiver. Press the rear end of the bolt down to elevate the front end just enough to clear the accelerator, otherwise the accelerator will be tripped and will not permit the bolt to be moved forward. When the accelerator has been cleared, raise the rear of the bolt to clear the buffer body. To do this it will be necessary to raise the bolt latch by reaching under the rear of the top plate with thumb or finger of one hand, while the other hand pushes the bolt forward. Bolt latch must be kept in raised position until rear of bolt passes in front of it. If this is not done, its spring will force it downward and engage the notch in the rear of the bolt preventing the bolt from going forward.

Push the bolt forward until the bolt stud hole lines up with the hole in the slot in the side plate; then insert the bolt stud until its shoulders are inside the side plate.

Insert the driving spring rod assembly and push the bolt all the way forward and keep the stud at the rear end of the driving spring rod at the recess in the right side plate.

Holding out the back plate latch lock, insert the back plate from the top. Press the trigger.

Make sure that the bolt is fully home, then close the cover. If the bolt is not fully forward, the feed lever will be forced down in front of the bolt which may result in malfunctioning.

Holding barrel by barrel handle, use both hands to insert it carefully in the front end of the barrel support. Guide the rear end over the breech bearing until it contacts the threads of the barrel extension, then screw it in until definite resistance is met. Now back off two notches to make headspace adjustment.

Headspace Adjustment of M2

As in the case of the Browning cal. .30 machine gun, the headspace adjustment is the most important adjustment on this gun. It is not necessary to remove the barrel to make this adjustment on the caliber .50.

Remember that headspace means that space between the rear end of the barrel and the front face of the bolt; if this space is too wide the gun will function sluggishly or not at all and may pull the head away from the cartridge case causing serious jams; if it is too tight the recoiling parts will not go fully home and the gun may refuse to fire.

Screw the barrel up tight into the barrel extension, then pull back the slide and let bolt go forward to test the action.

If the action does not close fully, unscrew the barrel one notch. Then the test should be made again by pulling the bolt back and then letting it go forward.

The barrel may be unscrewed a notch at a time by pushing with the point of a bullet to rotate the barrel when the cover is raised and the bolt is in rearward position.

Work the bolt by hand several times, and if the breech does not close without effort, unscrew the barrel one notch.

Raise the cover and lift the extractor, then pull the bolt slightly to the rear. If it moves independently of the barrel extension, the adjustment is too loose. Screw the bolt up one notch and then repeat the test. When a dummy cartridge is in the chamber, there must be no rearward motion of the bolt independent of that of the barrel extension before the unlocking action takes place.

Headspace test should be made whenever gun is prepared for firing.

Starting with the gun loaded and cocked, the action is as follows. Pressing the trigger raises the back end of the trigger bar, which pivots on the trigger bar pin and presses its front end down on the top of the sear. The sear is forced down until its notch disengages from the shoulder of the firing pin extension. This permits the firing pin extension and the firing pin to be driven forward by the coiled firing pin spring. The firing pin strikes the primer of the cartridge and explodes the powder.

Recoil Action. As the bullet starts down the barrel, the rearward force of the recoil drives the securely locked recoiling mechanism directly to the rear. During this initial motion, the bolt is supported securely against the base of the cartridge by the breech lock, which rides up from the barrel extension into a notch in the underside of the bolt.

After a travel of about ¾ inch, during which time the bullet has left the barrel, the breech lock is pushed back off its cam. It is forced down out of its locking notch on the underside of the bolt by the breech lock depressors riding up and over the lock pin, which passes through the breech lock and protrudes on either side. This action unlocks the bolt.

The barrel extension trips the accelerator up and to the rear of its pin. The tips of the accelerator striking the lower projection on the rear of the bolt accelerate its rearward travel. After a travel of about 1⅛ inches, the barrel and barrel extension have completed their rearward travel. They are stopped by the oil buffer body assembly, whose oil buffer spring has been compressed in the oil buffer body by the shank in the barrel extension. The flipped-up claws of the accelerator lock the spring in compressed position as they are moved against the shoulders of the barrel extension shank.

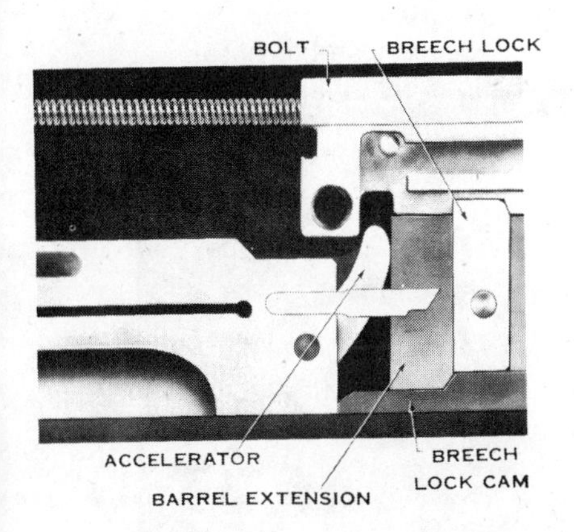

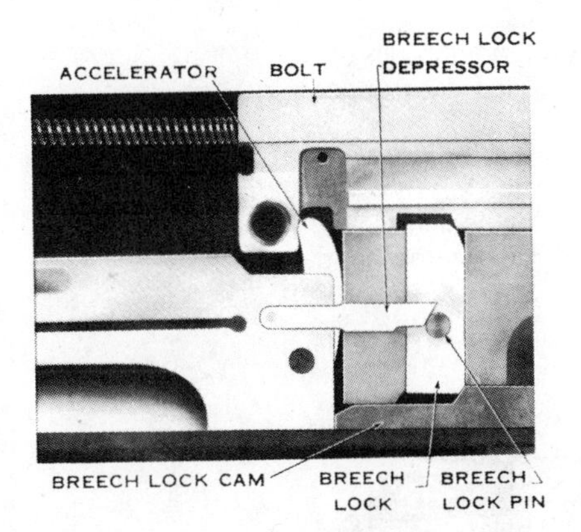

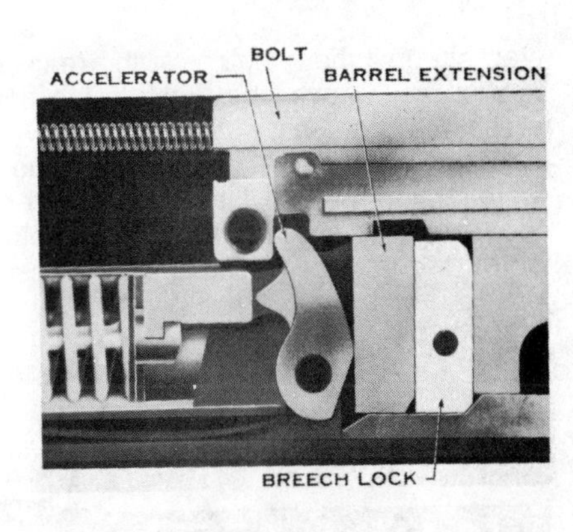

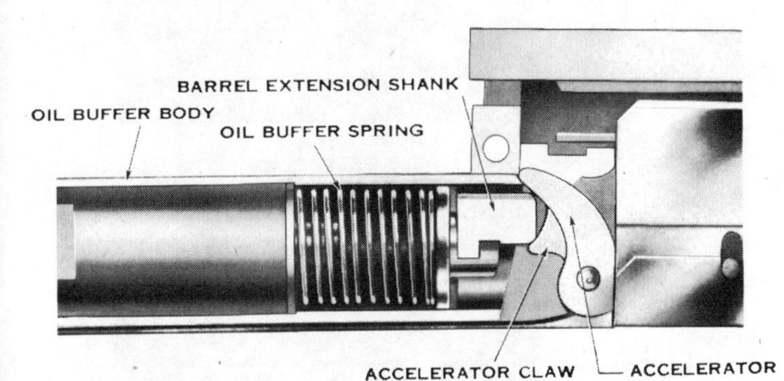

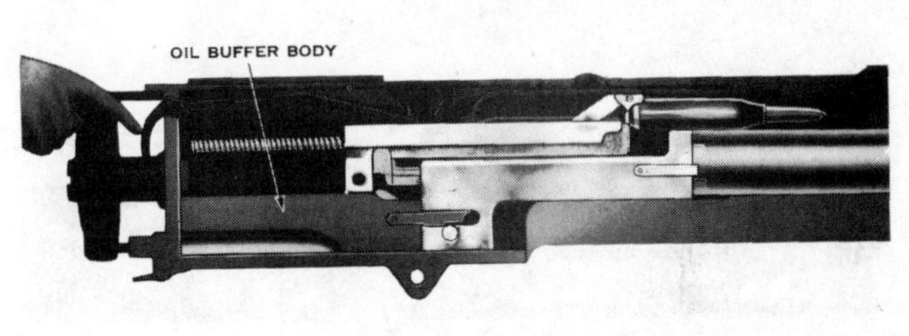

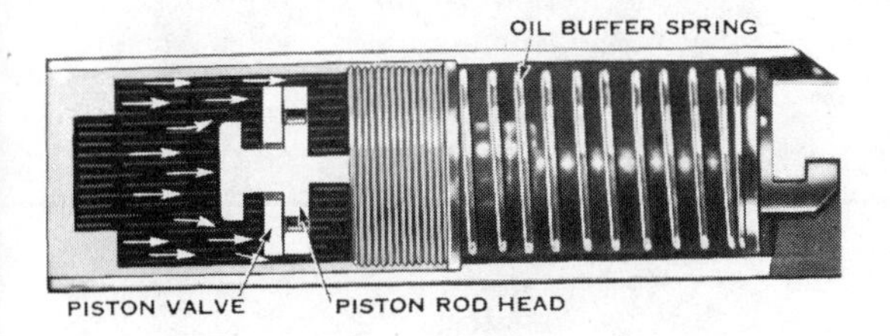

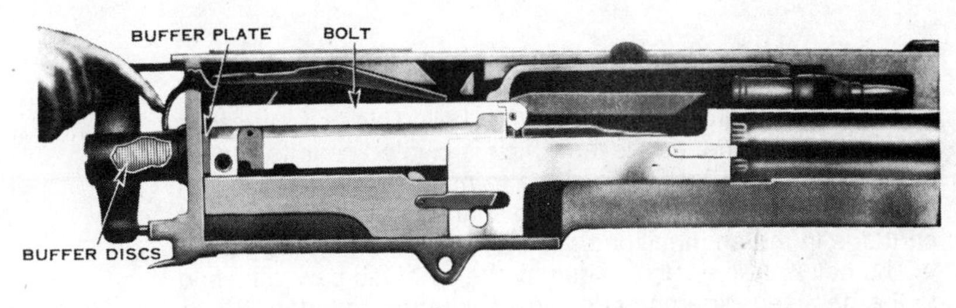

Meanwhile a piston rod head in the oil buffer assembly is forced from front to rear end of the oil buffer tube, and presses against oil in the tube to absorb the rearward shock of recoil until the oil escapes through the front side of the piston. This oil flow is through notches between the edge of the piston rod head and the oil buffer tube. This cushions the recoil and brings the rearward motion to a complete stop when the recoiling functions have been completed.

As the bolt travels to the rear, the driving springs inside it are compressed, and its rearward motion is stopped when the rear of the bolt strikes the buffer plate.

Cocking Action. The tip of the cocking lever protrudes through the top of the bolt where it lies in a V-slot in the top plate bracket. As the bolt starts to recoil, the tip of this cocking lever is pushed forward and its lower end is pivoted to force the firing extension rearward. This compresses the firing pin spring against the sear

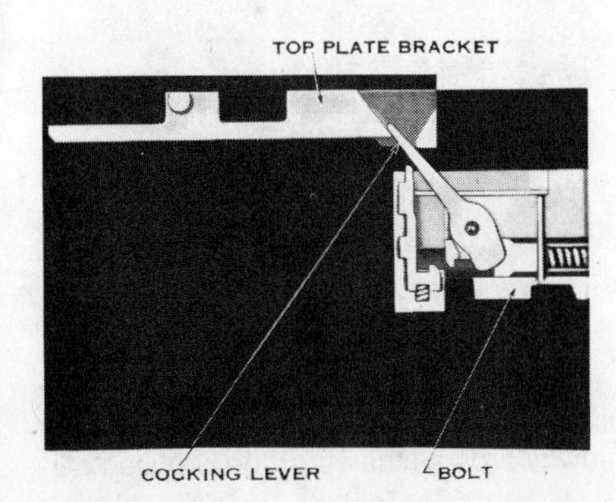

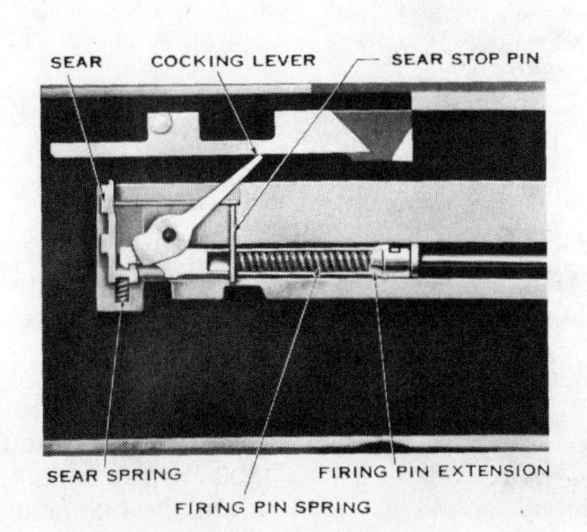

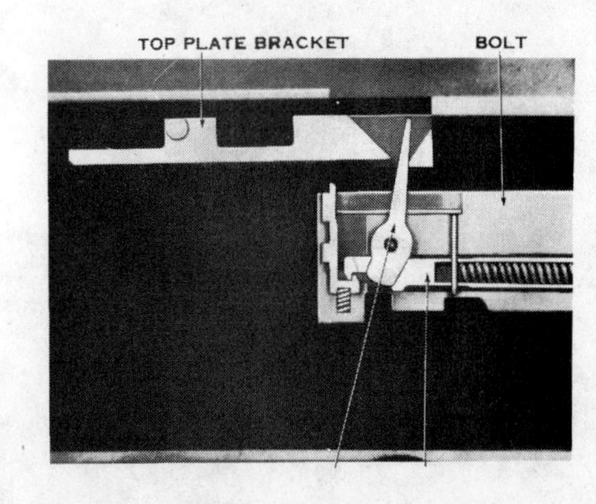

stop pin until the shoulder at the rear end on the firing pin extension hooks over the notch in the bottom of the sear under pressure of the sear spring.

When the bolt goes forward after the completion of the rearward motion, the tip of the cocking lever enters the V-slot in the top plate bracket, thus pivoting the bottom of the cocking lever out of the path of the firing pin extension to release the firing pin.

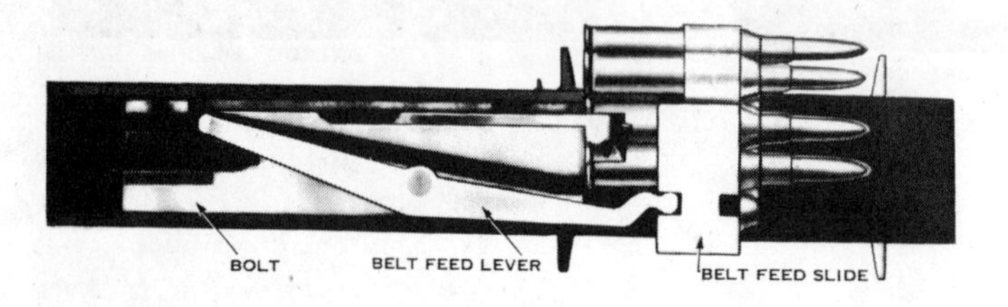

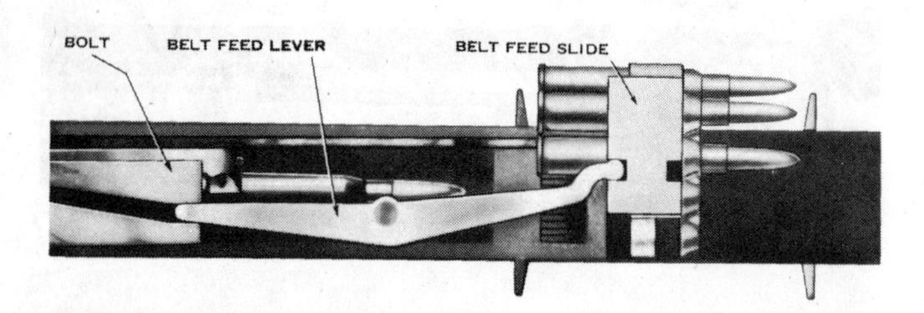

Feeding During the Rearward Motion. As the bolt moves to the rear, the stud at the rear of the belt feed lever is engaged in the diagonal groove on top of the bolt. This bolt stud thus serves to move the feed lever, which is pivoted near its center, and carry the belt feed slide at the front end of the lever out of the side of the gun where its spring snaps it down over the next cartridge in the ammunition belt.

The belt is pulled into the gun by the belt feed pawl attached to the belt-feed slide and ridges over the next cartridge.

As the recoiling motion is completed, the belt-feed slide has traveled far enough to permit the belt-feed pawl to be snapped down by its spring behind the next cartridge, ready to pull the belt forward into the gun on the next motion.

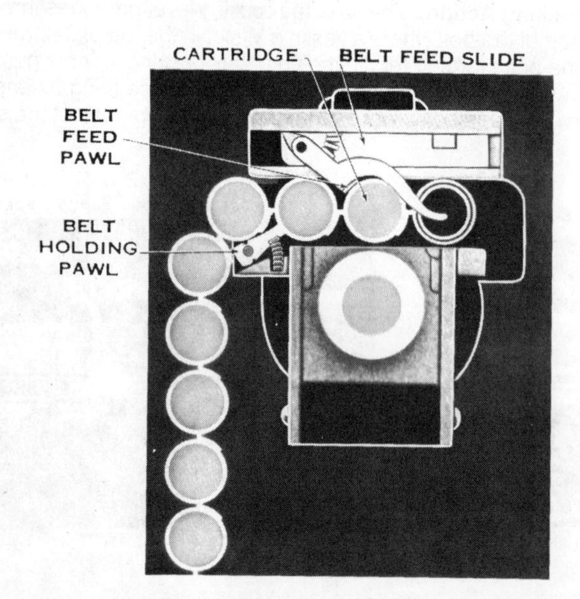

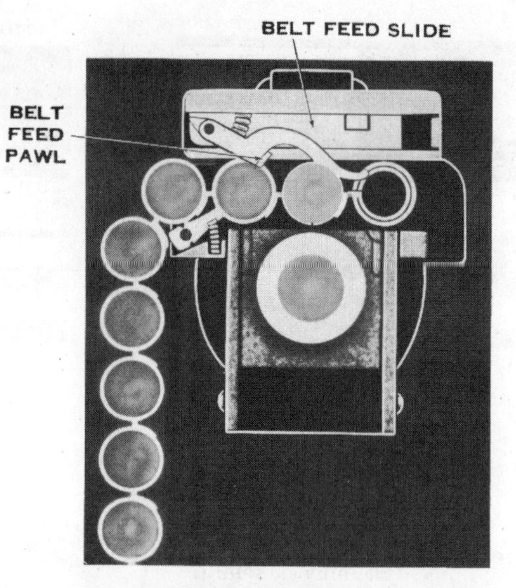

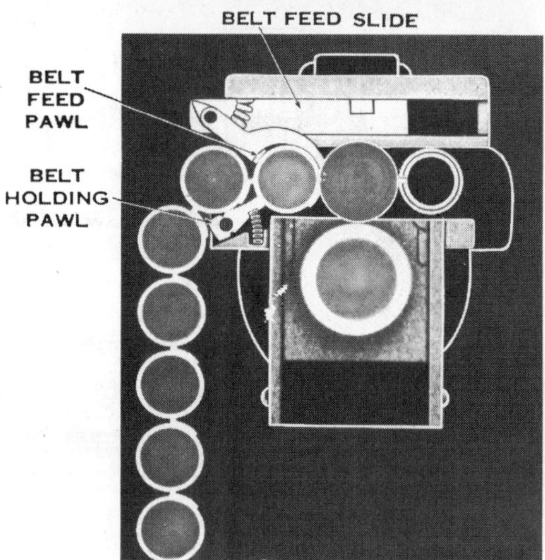

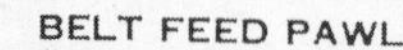

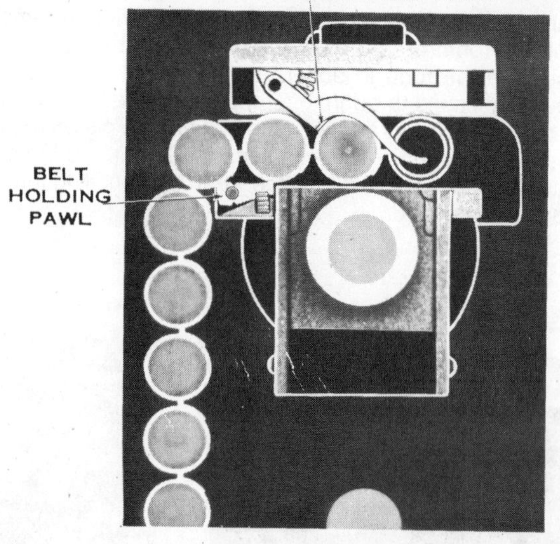

Forward Feeding Motion. As the bolt moves forward, the stud riding in its top pulls on the pivotal belt feed lever. The belt holding pawl is forced downwards as the cartridge is pulled over it and the belt holding pawl snaps up behind the next cartridge.

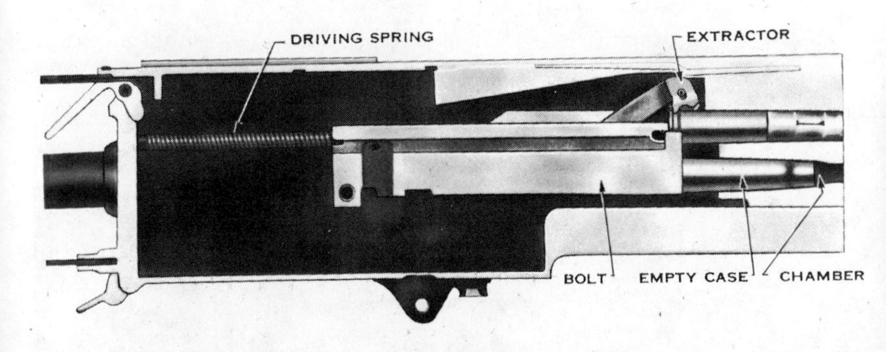

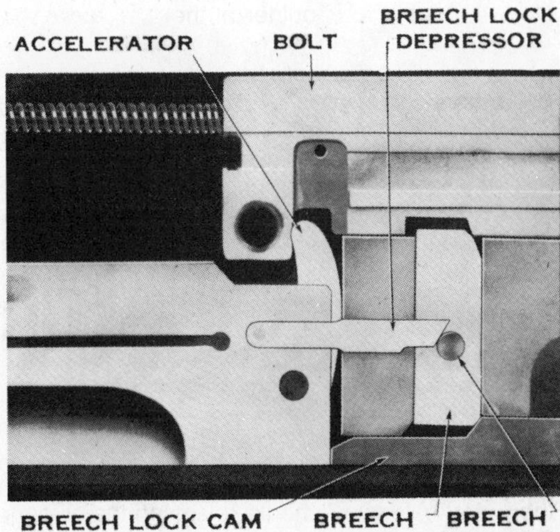

Extraction and Ejection. When the rearward motion starts, the extractor mounted in the side of the bolt and with its head above the bolt level snaps down into the cannelure of the cartridge in the belt, then draws the cartridge back out of the ammunition belt. The empty cartridge case is held in the T-slot in the front face of the bolt and the bolt withdraws it from the chamber.

The top front edge of the breech lock and the front side of the notch in the bolt are beveled to start withdrawals of the empty cartridge case slowly to prevent the case from being torn apart by the sudden jerking motion. As the breech lock is unlatched, the bolt pulls away from the barrel and barrel extension easily enough to prevent rupturing the cartridge case.

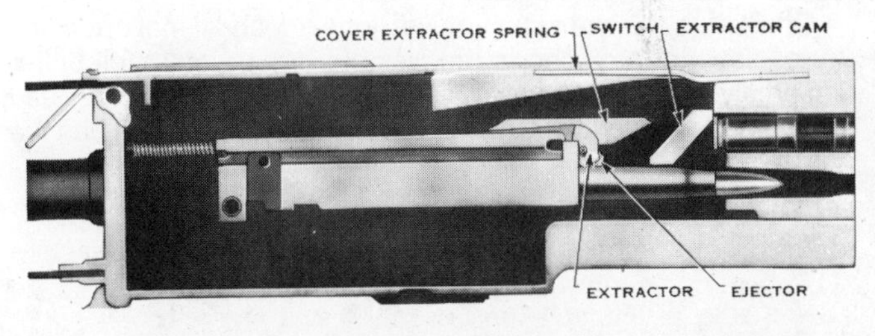

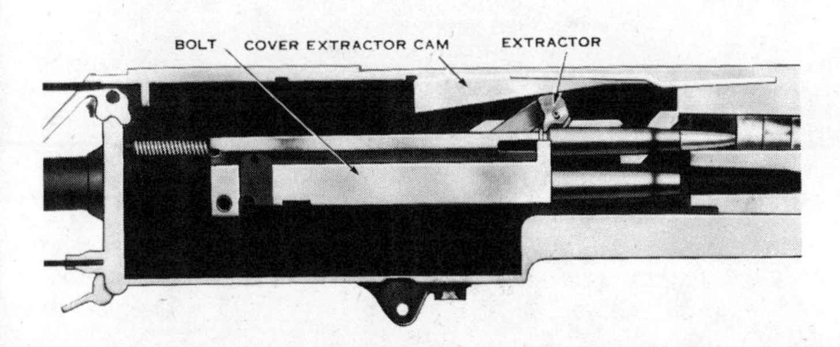

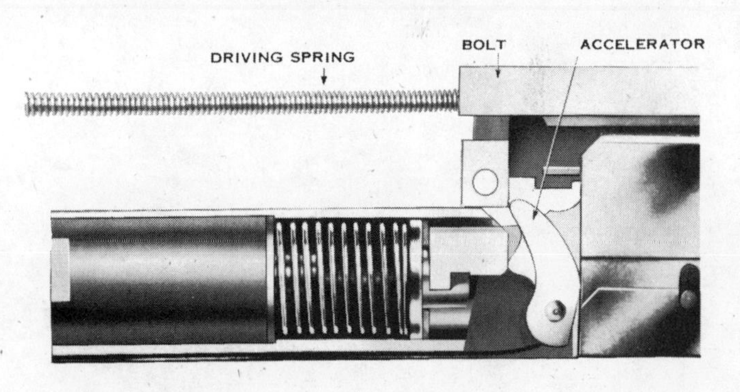

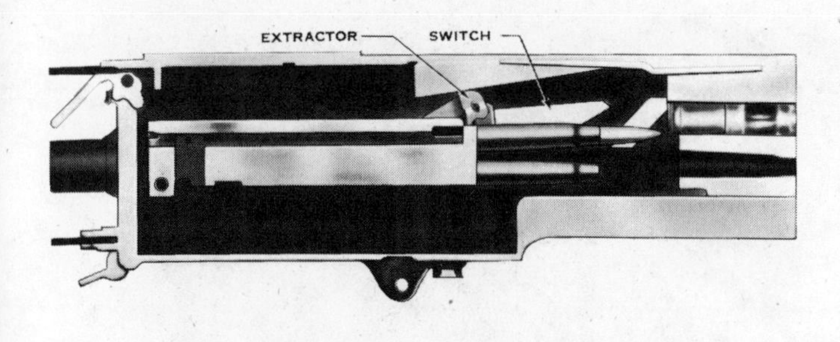

The cam on the inside of the cover forces the head of the extractor down, pushing the loaded cartridge into the mouth of the T-slot in the bolt. A lug on the side of the extractor rides against the top of the switch causing it to pivot downward at the rear; as the recoiling motion comes to an end, the lug on the extractor overrides the end of this switch, permitting it to snap up into normal position.

During this movement, the empty cartridge case drops down out of the T-slot and is expelled through the bottom of the gun.

Forward Motion of the Extractor. As the bolt goes forward, the extractor lug riding under the switch forces the extractor farther down, thus forcing out the empty cartridge case if it has not already dropped out of the gun. A pin in the bolt limits the travel of the extractor and the cartridge, assisted by the ejector, is fed directly into the firing chamber.

When the cartridge is nearly chambered, the extractor rides up its cam compressing the cover extractor spring and is snapped into the cannelure of the next cartridge.

Further Action During Forward Movement. After the recoiling motion has been completed, the compressed driving spring and the compressed buffer disc force the bolt forward. The bolt travels about 5″, when the projection on its bottom strikes the tips of the accelerator, rolling the accelerator forward on its pin.

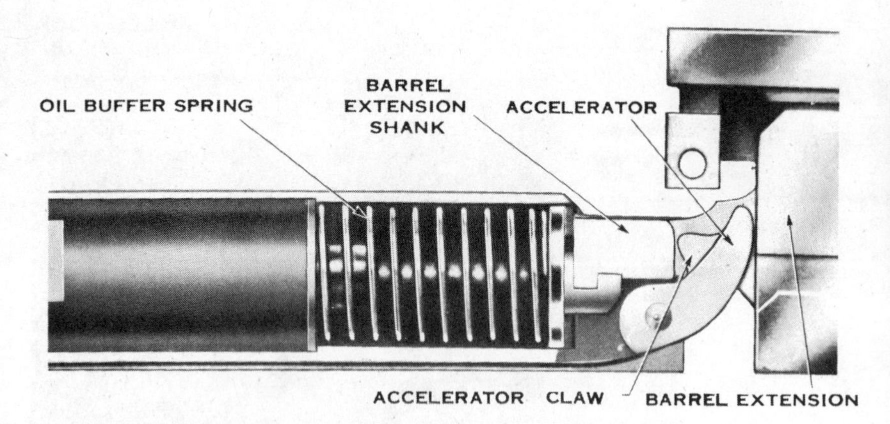

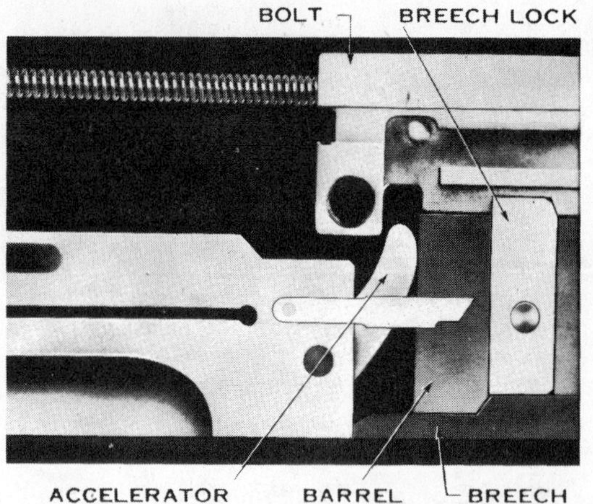

The accelerator claws are pulled away from the shoulder of the barrel extension shank, releasing the oil buffer spring. This spring now shoves the barrel extension on the barrel forward.

As the barrel extension goes forward, the breech lock strikes its cam and is forced upward on its pin. At that moment, the bolt has reached the position where the notch on its underside is directly above the breech lock; the breech lock rides up its cam and engages in this slot in the underside of the bolt. The bolt is locked to the breech end of the barrel just before the recoiling section reaches firing position.

Oil Buffer. As the action moves forward, additional openings for oil flow are provided in the piston rod head of the oil buffer assembly. The piston valve is forced away from the rod head as the parts move forward to uncover these openings. Thus the

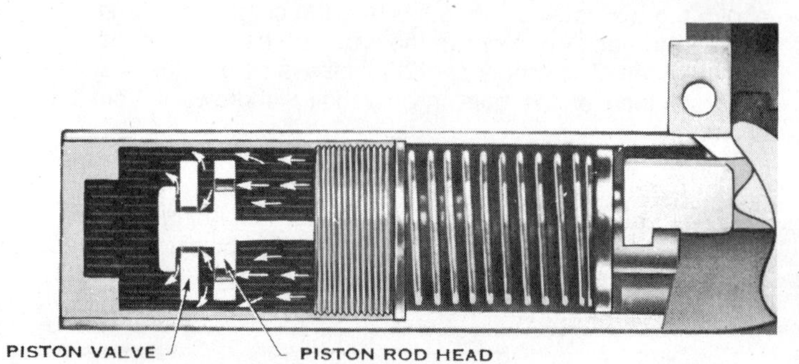

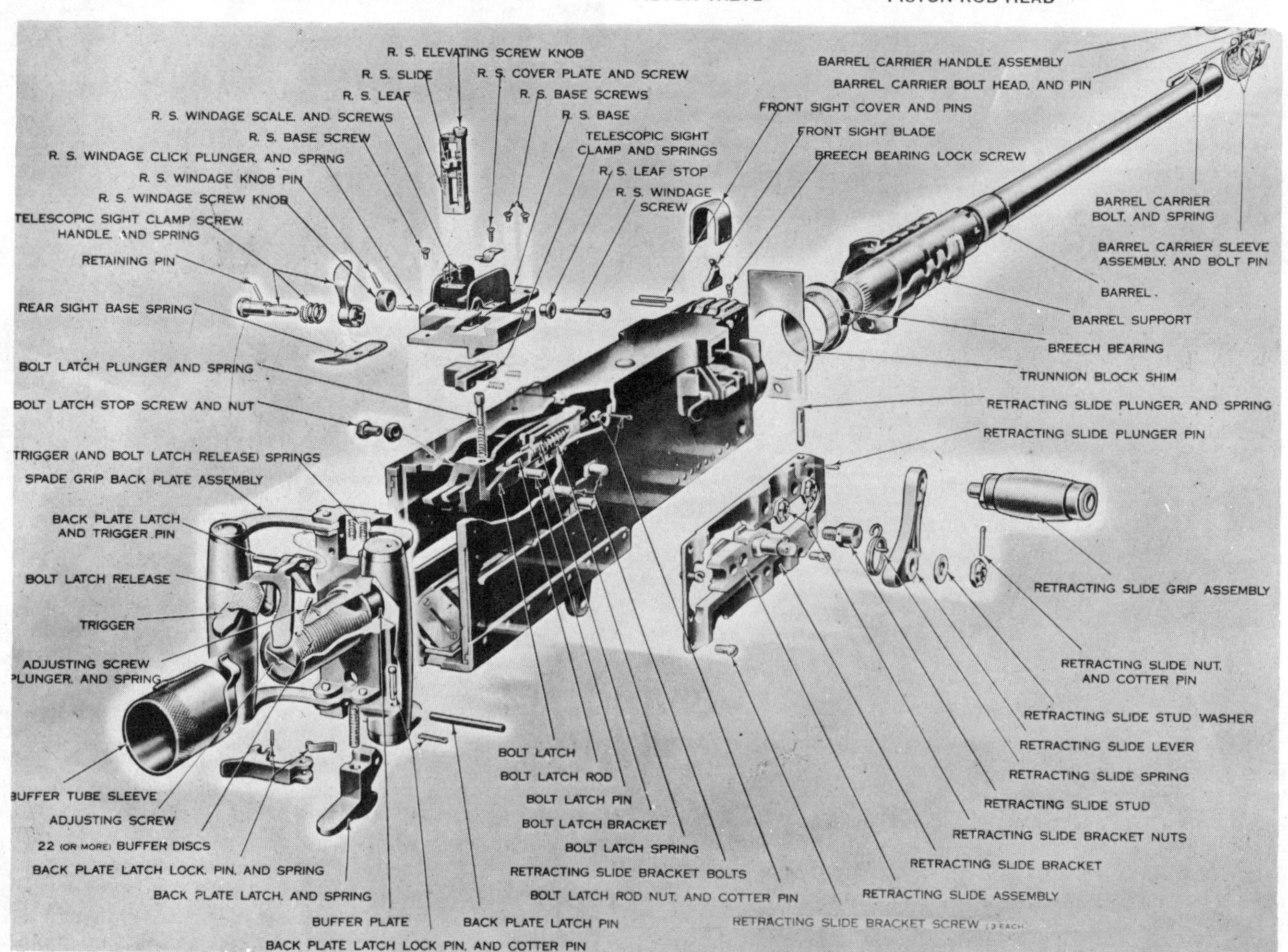

M2 Browning Heavy Barrel Machine Gun.

oil is permitted to escape freely from the opening in the center of the piston valve as well as at the edge of the valve near the tube wall, and so prepare it for the rearward motion.

Oil Buffer Adjustment. The oil buffer provides a method of regulating the speed of fire of this gun. Fire rate may be regulated by turning the oil buffer tube the required number of clicks. Turning the buffer tube to the left opens the oil buffer and permits oil to pass through the large ports, increasing the rate of fire. Turning the buffer tube to the right tightens up the oil buffer allowing it to absorb more recoil and reduce the rate of fire. This tube may be turned by inserting a screwdriver in the slot in the rear of the buffer tube.

Automatic Fire. If the trigger is pressed and held down, the sear is depressed as its tip is pressed against the cam surface of the trigger bar by the forward motion of the bolt just before it completes its forward motion. The notch in the bottom of the sear releases the firing pin extension and firing pin, automatically firing the cartridge as the forward motion is completed and continuing the action as long as the trigger is held and cartridges are fed into the gun.

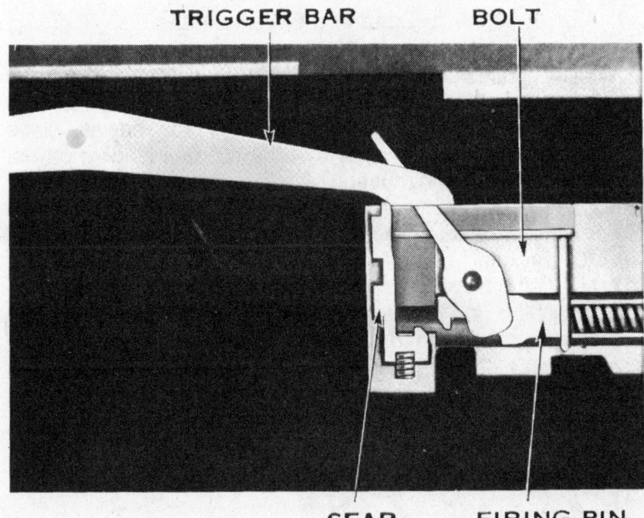

CALIBER .50 TANK MACHINE GUN M85

The M85 was developed by the Aircraft Armaments Corp. of Cockeysville, Md., as the T175E2 machine gun. It fills a requirement of the Armored Forces for a cal. .50 weapon suitable for co-axial or cupola mounting, having a dual rate of fire and quick change barrel and being shorter and lighter than the cal. .50 Browning.

The requirement for a dual rate of fire originated from the desire to have a high rate of fire for use against low-flying aircraft, and a low rate of fire for use against ground targets. While the cal. .50 Browning M2, HB, has an almost ideal rate for use against ground targets, its rate of fire for use against modern attack aircraft leaves something to be desired.

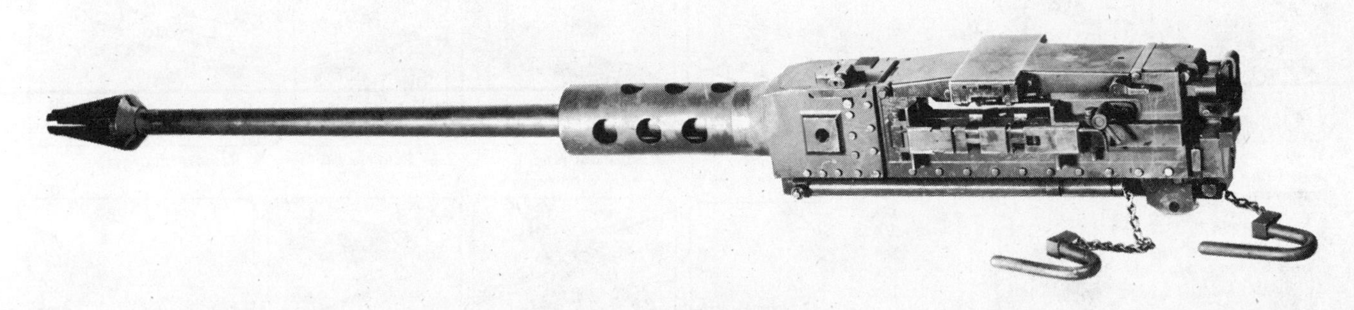

Caliber .50 Tank Machine Gun M85.

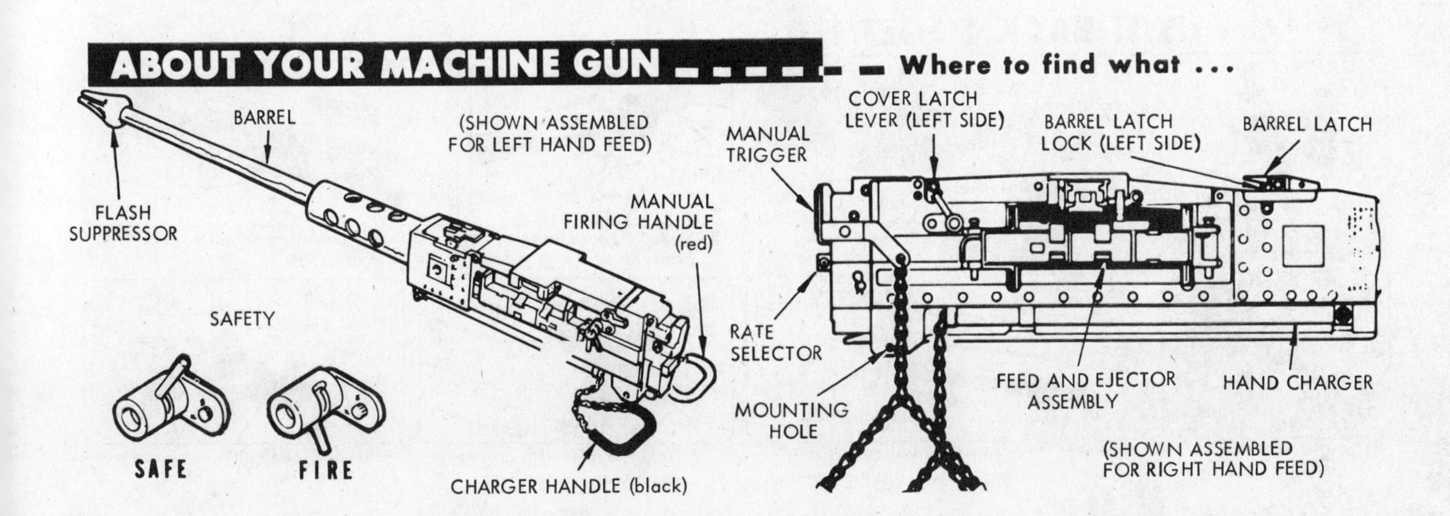

How to Load and Fire the M85

Open the cover and visually inspect the chamber to assure that chamber is clear. The bolt must be in the battery (closed) position before the ammunition belt is set in the feedway. Place the first round in the belt inboard of the two belt holding pawls. The belted ammunition must be placed in the gun with the open side of the links downward. Close the cover; the gun can be loaded without opening the cover if necessary. Pull bolt to the rear with the hand charger. Place safety on safe position. Set the rate selector lever to the proper position for the desired rate of fire. For high rate of fire, turn the lever completely to the left. For low rate of fire, turn the lever completely to the right. Do not change rate selector while firing a burst. Release the safety and fire the gun electrically by depressing the electrical trigger switch or manually by pushing forward on the manual trigger.

How to Field Strip the M85

Check to see that gun is not loaded. With hand charger control, allow the driving spring to slowly return the bolt to battery as follows:

(1) Pull back and hold hand charger.
(2) Depress trigger.
(3) Allow the driving spring to slowly return the charging handle to the original position.

Remove the barrel. Transversely depress the lock on the barrel latch, full depress the barrel latch, rotate the barrel 90° until the head of the "unlock" arrow is in line with the head of the arrow on the barrel support, and pull the barrel forward out of the barrel support. Remove the cover and tray assemblies. Open the cover, withdraw the quick release pins by inserting the rim of a cartridge case in the annular groove at the top of the pin; using the cartridge case as a lever, pry out the pin; withdraw the tray. Depress the latch lock on the lower left side of the back plate, then depress the back plate latch and lift the back plate straight up off the receiver. Remove back plate slowly for the first inch as the preload on the driving spring may cause the buffer assembly to jump out. CAUTION: Do NOT attempt to disassemble gun with bolt in the REAR position. The driving spring is then heavily loaded and may cause injuries. Remove the bolt buffer assembly from the rear of the receiver. Depress the sear block detent, visible through the right side of the receiver, with the nose of a cartridge and withdraw the sear assembly out of the rear of the receiver. Remove the feed and ejector assembly; withdraw the front and rear quick release pins, which fasten the feed and ejector assembly to the receiver. Remove the feed assembly by first pulling its front end out from the side of the receiver, then pull the assembly forward out of the receiver. Pull the barrel extension assembly, with the bolt in it, out through the rear of the receiver. Withdraw the quick release pin from the side of the receiver and lift the accelerator assembly straight up out of the receiver. Disengage the detent from the receiver by pulling on

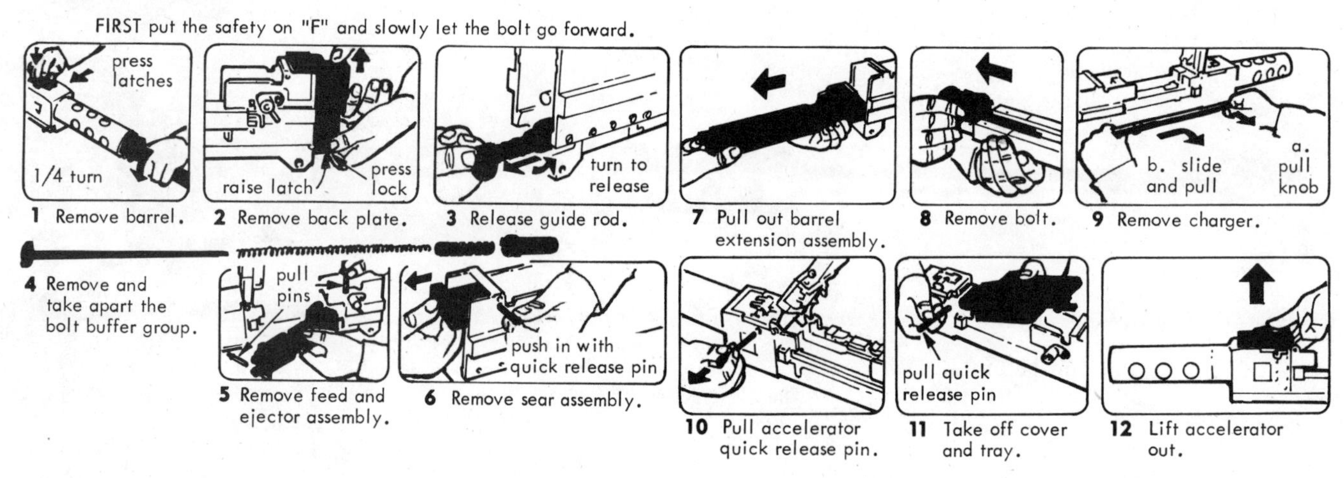

1 Remove barrel. 2 Remove back plate. 3 Release guide rod. 7 Pull out barrel extension assembly. 8 Remove bolt. 9 Remove charger.
4 Remove and take apart the bolt buffer group. 5 Remove feed and ejector assembly. 6 Remove sear assembly. 10 Pull accelerator quick release pin. 11 Take off cover and tray. 12 Lift accelerator out.

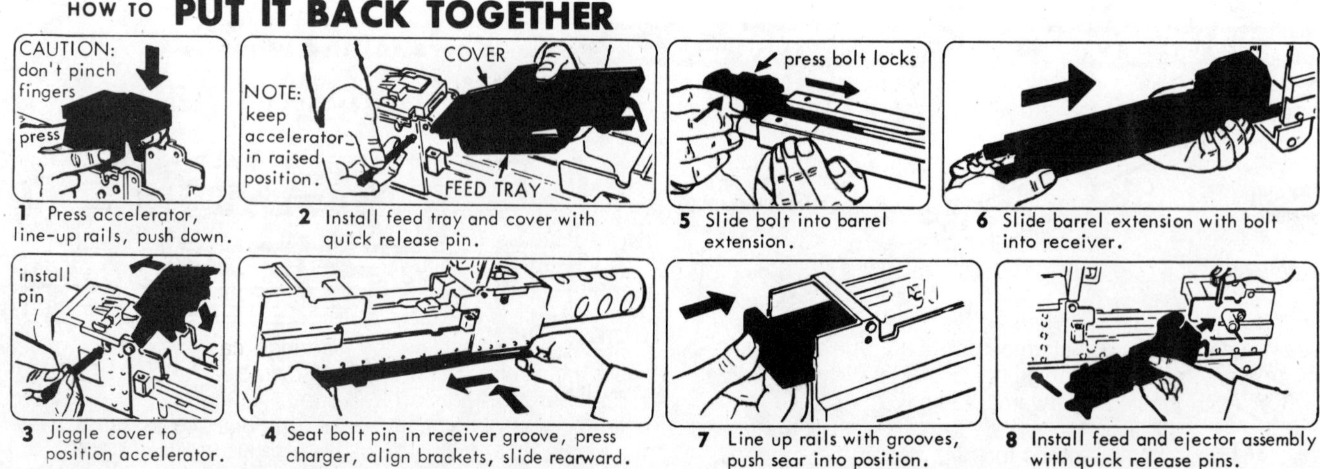

1 Press accelerator, line-up rails, push down. 2 Install feed tray and cover with quick release pin. 5 Slide bolt into barrel extension. 6 Slide barrel extension with bolt into receiver.
3 Jiggle cover to position accelerator. 4 Seat bolt pin in receiver groove, press charger, align brackets, slide rearward. 7 Line up rails with grooves, push sear into position. 8 Install feed and ejector assembly with quick release pins.

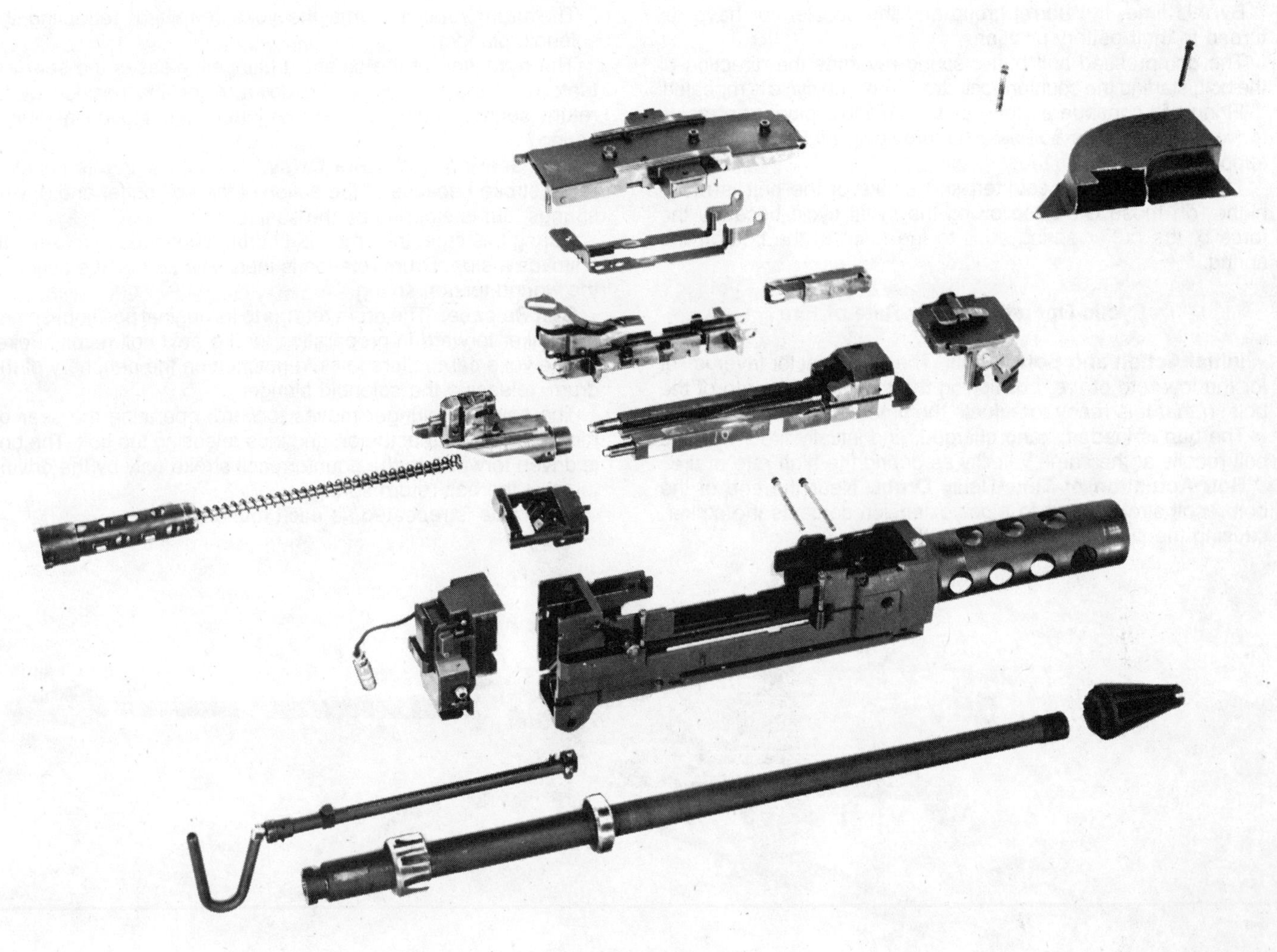

M85, field stripped.

the knurled knob at the front end, and slide the hand charger forward. To reassemble, reverse the procedure given above.

Cyclic Operation—High Rate of Fire

Charging the Gun. To begin the cycle, the loaded gun must be hand charged. During the charging stroke, the feed mechanism is actuated by the bolt, positioning a round on the center line of the gun, in the forward path of the bolt.

The bolt is held in the open-bolt position by the sear, with the bolt-driving spring compressed.

Stripping and Chambering of Round. When the firing switch is depressed, or the hand trigger is pressed, the solenoid plunger pushes the sear actuator forward about its pivot, camming the sear up and out of the bolt notch.

The bolt is driven toward battery position by the compressed bolt-driving spring.

The round in the feedway is picked up by the stripper on the bolt, is stripped through the belt link by the action of the bolt, and is deflected downward into the chamber by the chambering ramp.

Locking and Firing. The forward motion of the bolt block is arrested by the base of the chambered round, and the bottom extractor engages the extraction groove at the rim of the cartridge.

The bolt slide continues forward, wedging the bolt locks in place.

As the bolt slide completes its forward motion, the firing pin protrudes through the bolt block and strikes the cartridge primer, firing the round and initiating the recoil stroke.

Recoil Stroke. The feed mechanism, operated by the recoiling barrel extension through a feed spring and a feed cam, is actuated at the beginning of the recoil stroke.

As the barrel and the barrel extension begin to recoil, the barrel actuates the accelerator, thereby driving the bolt slide rearward at an accelerated speed.

As the bolt slide starts to move rearward, it cams the bolt locks inward, releasing the bolt block, which moves toward the rear, pulling the spent cartridge case out of the chamber and compressing the bolt-driving spring.

Ejection. The ejector mechanism is actuated by the action of the recoiling bolt on the ejector lever.

Just before cartridge ejection, the bottom extractor is disengaged from the extraction groove of the cartridge by the cam grooves in the barrel extension.

As the bolt continues rearward motion, the springloaded stripper is depressed by the income round.

The barrel group and the accelerator now start their return to battery position, driven by the barrel return springs and accelerator return springs.

As the ejector strikes the cartridge case, the case rotates about the side extractor and is thrown clear of the gun.

Bolt Buffing and Counterrecoil. As the bolt nears the end of its recoil stroke, it compresses the bolt buffer spring, which stops the rearward motion of the bolt.

By this time, the barrel group and the accelerator have returned to their battery position.

The compressed bolt buffer spring reverses the direction of the bolt, starting the counterrecoil stroke, and the cycle is repeated.

Firing will continue as long as the solenoid plunger remains forward, holding the bolt sear up, providing that ammunition is supplied.

The velocity of the counterrecoil stroke of the bolt is much higher on those cycles following the initial cycle because the force of the buffer spring adds to the force of the bolt-driving spring.

Cyclic Operation—Low Rate of Fire

Initial Action and Bolt Recoil. The rate selector lever is set for the low rate of fire, positioning the striker in the path of the bolt so that it is ready to actuate the time-delay drum.

The gun is loaded, hand charged, and initially fired, and the bolt recoils at the same velocity as during the high rate of fire.

Bolt Actuation of Time-Delay Drum. Near the end of the bolt recoil stroke, the bolt block extension contacts the striker, causing the time-delay drum to start rotating.

The drum rotation cams the yoke rearward, retracting the solenoid plunger.

The retraction of the solenoid plunger releases the sear actuator and the sear is forced down to locking position by its return spring. Drum rotation continues, winding up a torsion spring.

Bolt Searing and Time Delay. The bolt starts its counterrecoil stroke because of the action of the bolt buffer and driving springs, but is stopped by the sear.

During this time, the time-delay drum continues to rotate until it strikes a stop. Drum rotation is then reversed by the action of the wound torsion spring.

Bolt Release. The drum returns to its original position, driving the striker forward in preparation for the next bolt recoil stroke.

The yoke cam rollers fall into notches on the periphery of the drum, releasing the solenoid plunger.

The solenoid plunger moves forward, operating the sear by means of the sear actuator, and thus releasing the bolt. The bolt is driven forward on the counterrecoil stroke only by the driving force of the bolt-return spring.

This cycle is repeated as each round is fired.

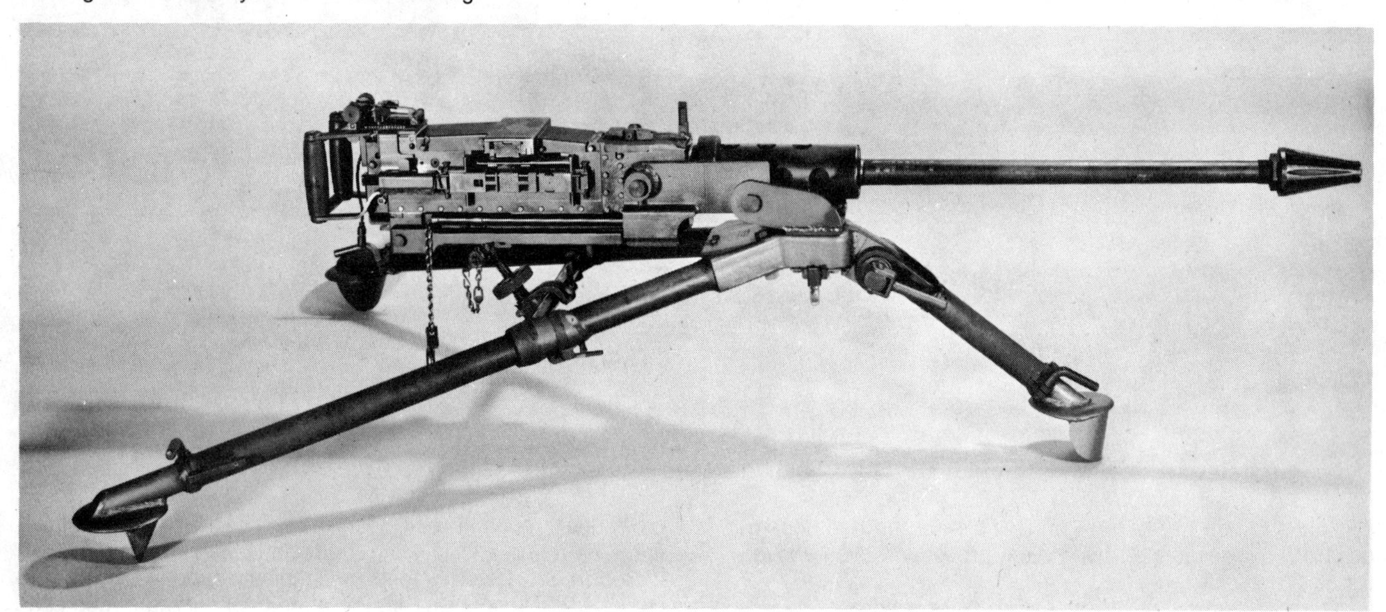

Caliber .50 Machine Gun M85C.

CHARACTERISTICS OF UNITED STATES SERVICE MACHINE GUNS

	Colt M1917	Browning M1917A1	Vickers M1918 Aircraft	M1919A4
Caliber:	.30.	.30.	.30.	.30.
System of operation:	Gas, automatic.	Recoil, automatic.	Recoil with gas assist, automatic.	Recoil automatic.
Length overall:	40.8 in.	38.5 in.	44.19 in.	41 in.
Barrel length:	28 in.	24 in	28.4 in.	24 in.
Feed device:	250-round, fabric belt.	250-round, fabric belt or disintegrating link belt.	250-round, fabric belt or disintegrating link belt.	250-round, fabric belt or disintegrating link belt.
Sights: Front:	Blade.	Blade.	none—aircraft type fitted to the weapons as required.	Blade.
Rear:	Leaf.	Leaf.		Leaf.
Weight: Gun:	35 lb.	41 lb. w/water.	25 lb.	31 lb.
Mount:	61.25 lb.	53.15 lb. (M1917A1).		14 lb. (M2).
Muzzle velocity:	Approx. 2800 f.p.s.	2800 f.p.s.	Approx. 2800 f.p.s.	2800 f.p.s.
Cyclic rate:	480 r.p.m.	450–600 r.p.m.	800–900 r.p.m.	400–550 r.p.m.

CHARACTERISTICS OF UNITED STATES SERVICE MACHINE GUNS (Cont'd)

	M1919A6	M2 Aircraft	M37 Tank	M60	M73 Tank
Caliber:	.30.	.30.	.30.	7.62mm NATO.	7.62mm NATO
System of operation:	Recoil, automatic.	Recoil, automatic.	Recoil, automatic.	Gas, automatic.	Recoil, w/gas assist, automatic.
Length overall:	53 in.	39.9 in.	41.75 in.	43.75 in.	34.75 in.
Barrel length:	24 in.	23.9 in.	24 in.	25.6 in.	22 in.
Feed device:	250-round, fabric belt or disintegrating link belt.	Disintegrating link belt.	Disintegrating link belt.	Disintegrating link belt.	Disintegrating link belt.
Sights: Front:	Blade.	None permanently attached to gun.	Blade.	Blade.	None permanently attached to gun.
Rear:	Leaf.		Leaf.	Leaf.	
Weight: Gun:	32.5 lb.	21.5 lb. (fixed gun). 23 lb. (flexible gun).	31 lb.	23.05 lb.	28 lb.
Mount:	14 lb. (M2)			15 lb. (M122)	
Muzzle velocity:	2800 f.p.s.	2800 f.p.s.	2800 f.p.s.	2800 f.p.s.	2800 f.p.s.
Cyclic rate:	400–500 r.p.m.	1000–1350 f.p.s.	400–550 r.p.m.	600 r.p.m.	450–500 r.p.m.

CHARACTERISTICS OF UNITED STATES SERVICE MACHINE GUNS (Cont'd)

	Johnson M1941	M1921A1 Antiaircraft	M2 Aircraft, Basic	M2 Heavy Barrel
Caliber:	.30.	.50.	.50.	.50.
System of operation:	Recoil, selective fire.	Recoil, automatic.	Recoil, automatic.	Recoil, selective fire.
Length overall:	42 in.	56 in.	56.25 in.	65.1 in.
Barrel length:	22 in.	36 in.	36 in.	45 in.
Feed device:	20-round, in line, detachable box magazine.	250-round, fabric belt or disintegrating link belt	Disintegrating link belt.	Disintegrating link belt.
Sights: Front:	Blade.	Hooded blade.	None permanently attached to gun.	Hooded blade.
Rear:	Adjustable folding aperture.	Leaf.		Leaf.
Weight: Gun:	Approx. 13 lb.	79 lb. w/o water.	61 lb.	84 lb.
Mount:				44 lb. (M3).
Muzzle velocity:	2800 f.p.s.	2840 f.p.s.	2840 f.p.s. (M2 ball).	2930 f.p.s. (M2 ball).
Cyclic rate:	400–450 r.p.m.	Approx. 500 r.p.m.	750–850 r.p.m.	450–550 r.p.m.

CHARACTERISTICS OF UNITED STATES SERVICE MACHINE GUNS (Cont'd)

	M2 Antiaircraft	M3 Aircraft	M85
Caliber:	.50.	.50.	.50.
System of operation:	Recoil, automatic.	Recoil, automatic.	Recoil, automatic.
Length overall:	66 in.	57.25 in.	54.5 in.
Barrel length:	45 in.	36 in.	36 in.
Feed device:	Disintegrating link belt.	Disintegrating link belt.	Disintegrating link belt.
Sights: Front:	Hooded blade.	None permanently attached to gun.	None permanently attached to gun.
Rear:	Leaf (may be found without rear sight).		
Weight: Gun:	121 lb. w/water.	68.75 lb. w/recoil adaptor.	61.5 lb.
Mount:	401 lb. (AA. M3).		
Muzzle velocity:	2930 f.p.s. (M2 ball).	2840 f.p.s. (M2 ball).	2840 f.p.s. (M2 ball).
Cyclic rate:	500–650 r.p.m.	1150–1250 r.p.m.	Low rate: 450 ± 50 r.p.m. High rate: 1050 ± 50 r.p.m.

GRENADE LAUNCHERS

40mm GRENADE LAUNCHER M79

The M79 grenade launcher is a shotgun type weapon designed to fire a high explosive grenade considerably more accurately than a grenade can be fired from a rifle grenade launcher.

CHARACTERISTICS OF M79 GRENADE LAUNCHER

Caliber: 40 × 46mm SR.
System of operation: Single-shot, break-open type.
Weight of launcher (loaded): 6.45 lbs.
Length of launcher: 28.78 in.
Length of barrel: 14 in.
Muzzle velocity: 250 f.p.s.
Sights: Front: Protected blade.
Rear: Leaf, adjustable for windage.

How to Load and Fire the M79

Move barrel locking latch FULLY to the right and break open breech. Moving latch fully to right automatically puts the weapon on "safe." Insert cartridge in chamber until the extractor contacts the rim of the cartridge case. Close the breech, push safety to forward position exposing the letter "F." Pressure on the trigger will now fire the weapon.

How to Field Strip the M79

Under normal conditions, it should not be necessary (for maintenance purposes) to do more than break the weapon using the barrel locking latch. The firing pin retainer in the face of the standing breech can be tightened up from time to time by use of the lugs on the combination wrench supplied with the weapon. If the weapon has been immersed in water or snow, the following procedure should be followed. Remove the fore-end assembly by taking out screw that passes through the rear mounting hole of the front sling swivel. Pull front end of fore-end away from barrel until lug on the rear sight base is clear of the hole in upper surface of fore-end bracket. Keeping lug clear of hole, pull forward on fore-end assembly until it is free of receiver assembly. Operate barrel locking latch and open breech, holding the stock and receiver stationary, move the barrel rearward in the receiver until it is disengaged from the fulcrum pin. Separate barrel from receiver group. From bottom of stock, near front end, remove machine screw, lock washer and flat washer, which secure stock to receiver. Separate stock from receiver. To reassemble, perform the steps listed above in reverse.

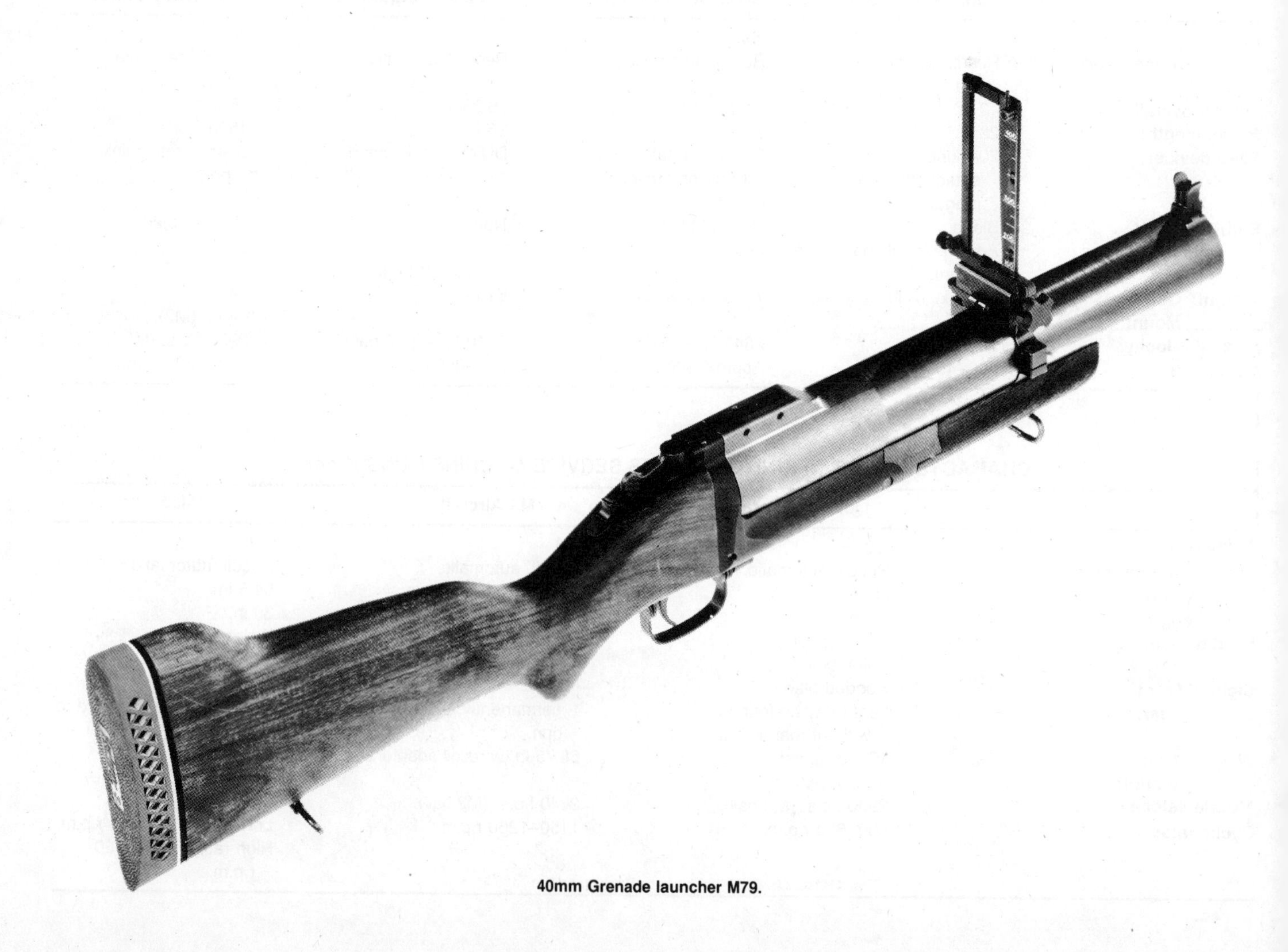

40mm Grenade launcher M79.

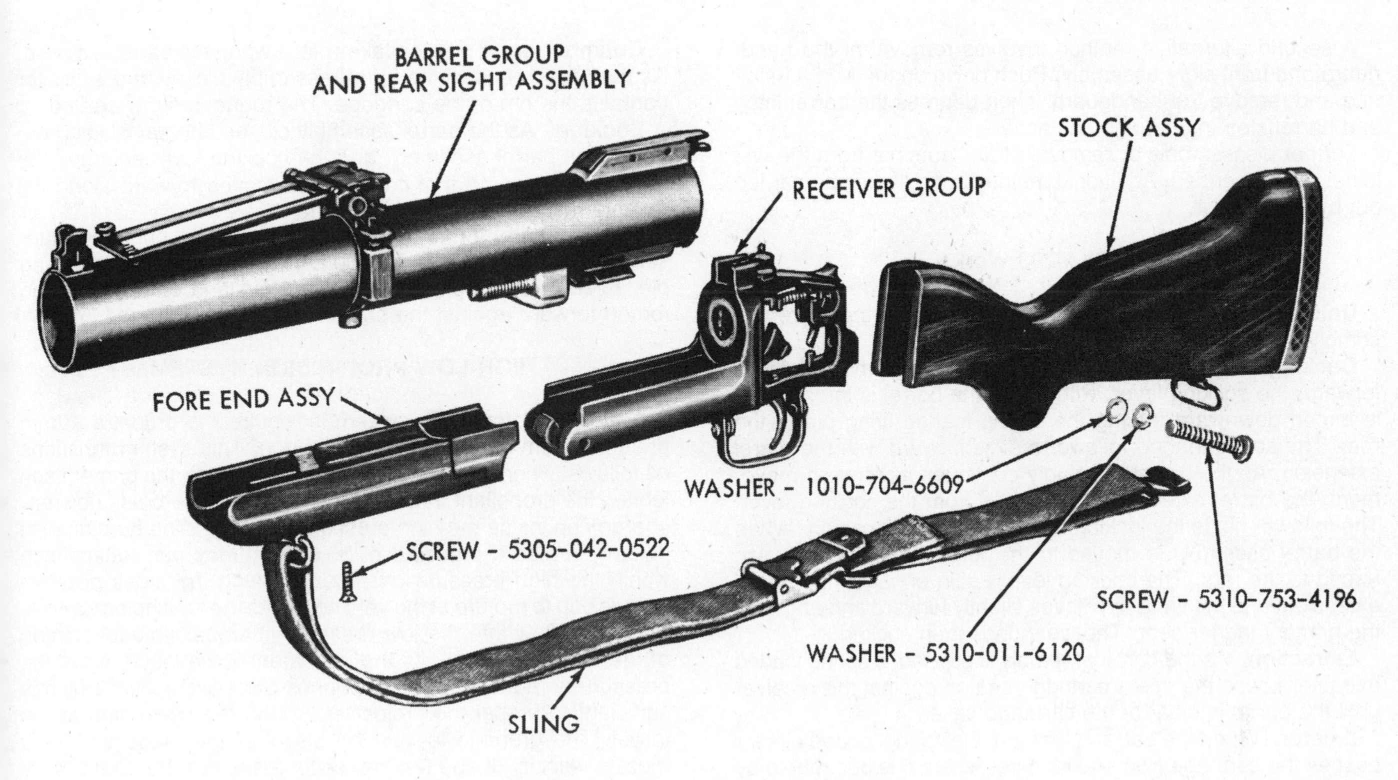

40mm Grenade launcher M79, field-stripped.

40mm GRENADE LAUNCHER XM148

Now obsolete, the XM148 was developed by Colt Firearms to be attached to the M16 and other rifles. This launcher was loaded by sliding the barrel forward, inserting the grenade cartridge and retracting the barrel. The launcher was cocked by pulling the cocking handle to the rear. While the XM148 was well received due to the extra fire-power it gave to the rifle squad, the launcher itself was not considered reliable, simple nor safe enough for type standardization.

CHARACTERISTICS OF XM148

Caliber: 40mm.
System of operation: Single-shot; slide-open type.
Weight of launcher (loaded): 3.5 lbs.
Length of launcher (front of barrel to rear of extended trigger): 16.5 in.
Length of barrel: 10 in.
Muzzle velocity: 250 f.p.s.
Sights: Quadrant sight mounted on adjustable sight slide.

40mm GRENADE LAUNCHER M203

In the spring of 1967, development of a new rifle-attached grenade launcher was begun by the Department of the Army. Called Grenade Launcher Attachment Development (GLAD), this project led to the testing of designs produced by AAI Corporation, Ford Aerospace and Communication Corporation and Aero-Jet General. The AAI launcher, XM203, was standardized as the M203 in August 1969. An initial production contract was carried out by AAI. Beginning in January 1971, Colt has been the sole production source for the M203.

Field Stripping the M203

There are two methods for removing the forward moving barrel assembly.

First, depress the barrel latch and slide the barrel assembly forward. From the muzzle of the M16A1, count back to the fourth hole on the left side of the handguard. Insert the end of a section from the cleaning rod into the fourth hole, depress the barrel stop and slide the barrel assembly off the receiver track.

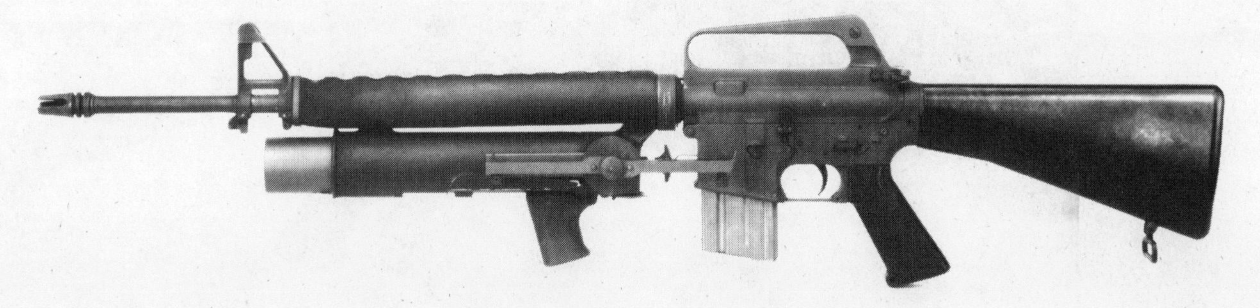

The XM148 Grenade Launcher attached to the M16A1 Rifle.

A second alternative method involves removal of the handguard and front sight assembly. Push down on the M16A1 slip-ring and remove the handguard. Then depress the barrel latch and barrel stop in sequence as above.

Further disassembly or removal of the launcher from the rifle is not recommended. Additional maintenance should be carried out by an armorer.

How the M203 Works

Unlocking. Depress the barrel latch and slide the barrel assembly forward.

Cocking. As the barrel and barrel extension assembly is moved forward, the cocking lever, with which the barrel is interlocked, is forced downward forcing the spring loaded firing pin to the rear. The spring loaded follower moves forward with the barrel extension. As the barrel assembly continues its forward movement, the barrel extension disengages from the cocking lever. The follower holds the locking lever in the down position. When the barrel assembly is moved to the rear, the follower is also forced to the rear. The cocking lever again engages the barrel extension, and the firing pin moves slightly forward and engages the primary trigger sear. The weapon is then cocked.

Extraction. As the barrel assembly is opened, a spring loaded extractor keeps the spent cartridge seated against the receiver until the barrel is clear of the cartridge case.

Ejector. When the barrel is forward, the spring loaded ejector pushes the cartridge from the receiver where it is being held by the extractor.

Loading. While the barrel is open, a cartridge is manually inserted into the breech end of the barrel.

Chambering. This step takes place when the barrel is closed. As the breech end of the barrel assembly closes, the extractor contacts the rim of the cartridge. The round is firmly seated.

Locking. As the barrel assembly closes, the barrel latch engages the barrel assembly, and the cocking lever engages the barrel extension so that it cannot be moved forward along the receiver assembly.

Firing. As the trigger is pulled rearward, the primary trigger sear is disengaged from the bottom sear surface of the firing pin, releasing the spring driven firing pin and causing it to be forced forward against the primer of the cartridge.

HIGH-LOW PROPULSION SYSTEM

A high-low propulsion system is required to propel a 40mm projectile from a shoulder fired weapon. This system functions as follows: when the firing pin strikes the primer, the primer flash ignites the propellant that is contained within the brass powder-charge cup inside the high pressure chamber. The burning propellant creates a pressure of 35,000 pounds per square inch within the high pressure chamber, causing the brass powder-charge cup to rupture at the vent holes. As the vent holes rupture, the gases flow into the low pressure chamber (interior portion of the cartridge case). As the gases enter the larger area, the pressure is reduced to 3,000 pounds per square inch, which is sufficient to propel the projectile through the barrel and to the target. The grenade leaves the barrel of the launcher with a muzzle velocity of 250 feet per second and a right-hand spin of 37,000 revolutions per minute. The spin stabilizes the grenade during flight and provides rotational forces necessary to arm the fuze.

M203, 40mm Grenade Launcher attached to the M16A1 Rifle.

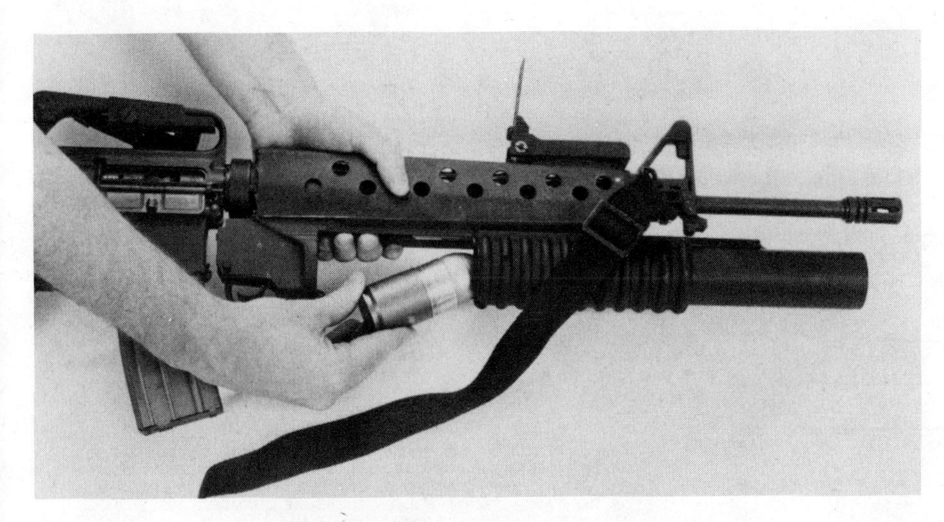

Loading the 40mm M203 Grenade Launcher.

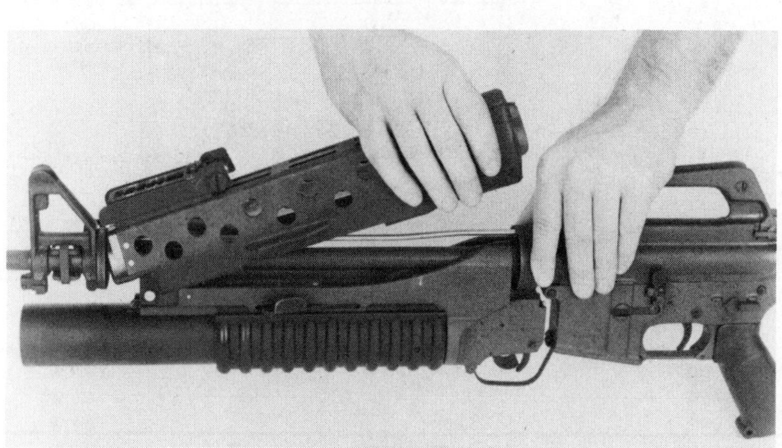

Removing the handguard is the first step in disassembly of the M203 Grenade Launcher.

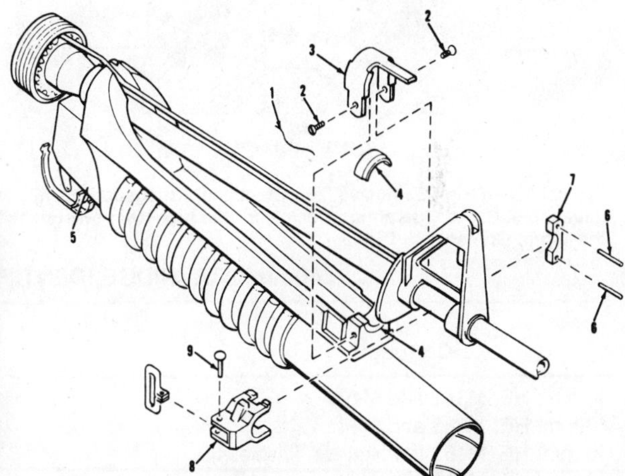

Method of attaching the M203 to the M16A1.

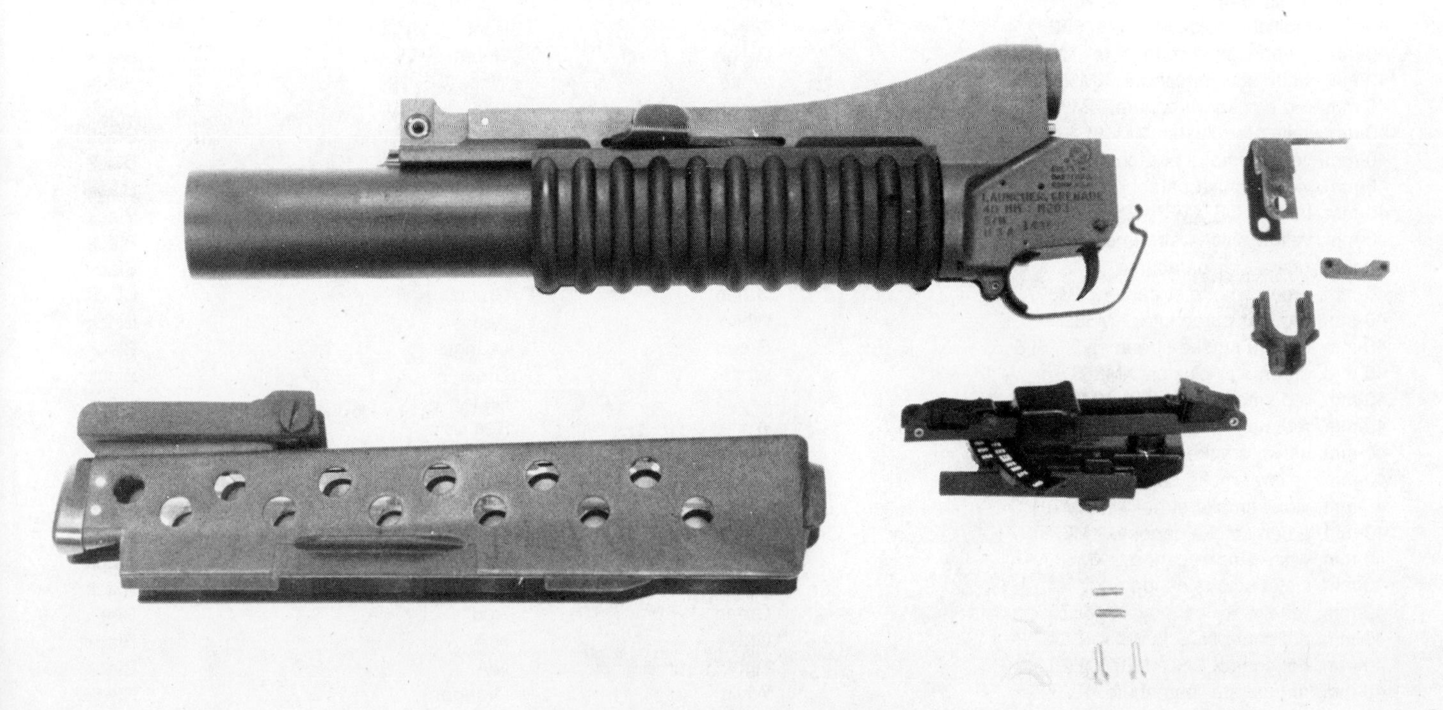

Disassembled M203 Grenade Launcher.

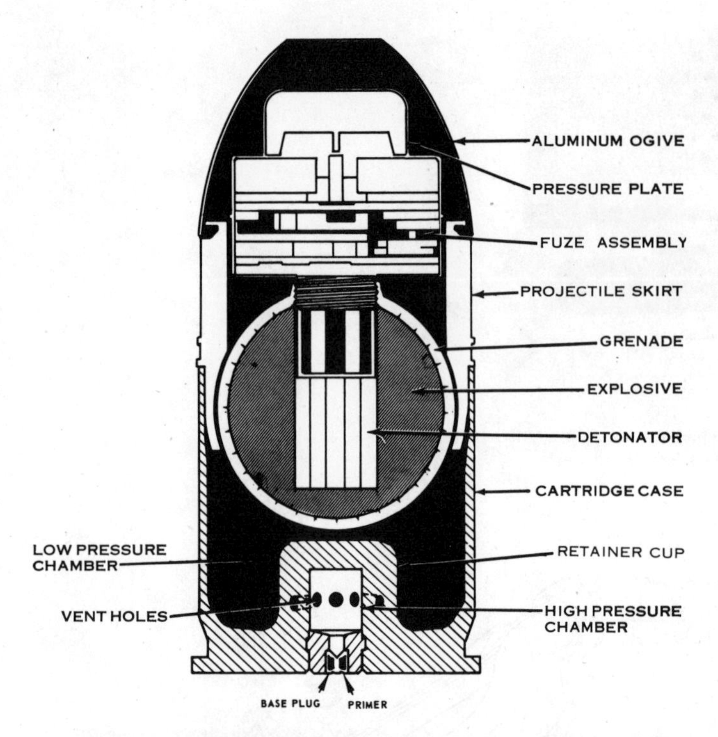

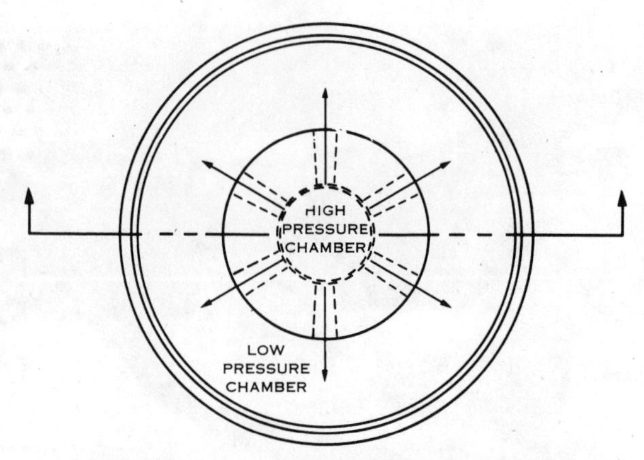

WHEN IGNITED BY THE PRIMER, THE PROPELLANT ENCLOSED IN THE BRASS POWDER CHARGE CUP (HIGH PRESSURE CHAMBER) RUPTURES THE CUP AT THE VENT HOLES AND IS VENTED INTO THE REMAINDER OF THE CARTRIDGE CASE (LOW PRESSURE CHAMBER)

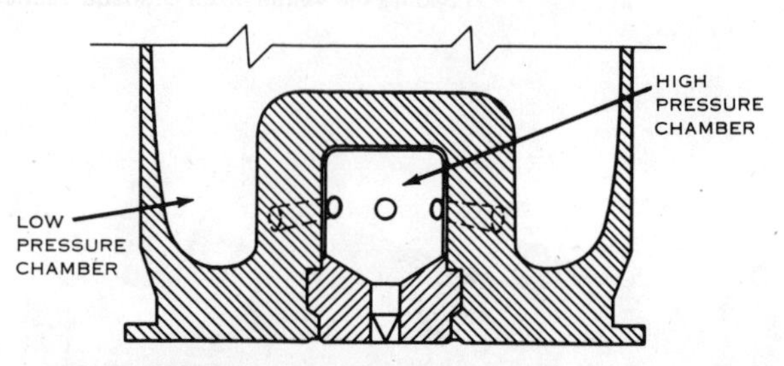

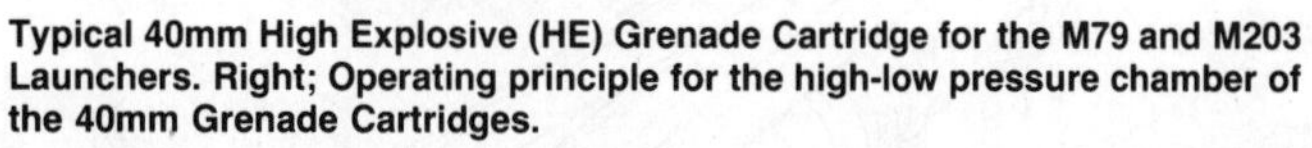

Typical 40mm High Explosive (HE) Grenade Cartridge for the M79 and M203 Launchers. Right; Operating principle for the high-low pressure chamber of the 40mm Grenade Cartridges.

40mm CARTRIDGE IDENTIFICATION (M79 and M203 Grenade Launchers)

Cartridge	Body	Color identification Ogive*	Lettering
40-mm, HE, M381 and M406	Green	Gold	Yellow
40-mm, HE, M386 and M441	Green	Gold	Yellow
40-mm, HE, M463 (smokeless, flashless)	Green	Black	Yellow
40-mm, HE, M397 (airburst)	Green	Gold	Yellow
40-mm, HE, M433 (dual purpose)	Green	Gold	Yellow
40-mm, practice, M382 and M407A1	Green	Silver	Yellow
40-mm, multiple projectile, XM576E1	Green	Black (SABOT)	White
40-mm, multiple projectile, XM576E2	Green	None	White
40-mm, green smoke parachute, XM658	Green	Green	Black
40-mm, white star parachute, XM583	White	White	Black
40-mm, red smoke parachute, XM659	Green	Red	Black
40-mm white star cluster, XM585	White	White	Black
40-mm, yellow smoke parachute, XM660	Green	Yellow	Black
40-mm, violet smoke parachute, XM669	Green	Violet	Black
40-mm, tactical CS,XM651E1	Green	Gray	Black
40-mm, yellow smoke streamer, XM696	Green	Yellow	Black
40-mm, green star parachute, XM661	White	Green	Black
40-mm, green smoke streamer, XM697	Green	Green	Black
40-mm, red star parachute, XM662	White	Red	Black
40-mm, orange smoke streamer, XM698	Green	Orange	Black
40-mm, green star cluster, XM663	White	Green	Black
40-mm, red smoke streamer, XM699	Green	Red	Black
40-mm, red star cluster, XM664	White	Red	Black
40-mm, brown smoke streamer, XM700	Green	Brown	Black
40-mm, yellow smoke canopy, XM676	Green	Yellow	Black
40-mm, violet smoke streamer, XM701	Green	Violet	Black
40-mm, green smoke canopy, XM679	Green	Green	Black
40-mm, white smoke canopy, XM680	Green	White	Black
40-mm, violet smoke canopy, XM681	Green	Violet	Black
40-mm, red smoke canopy, XM682	Green	Red	Black
40-mm, riot control CS, XM674 (E24)**	Gray	N/A	Black
40-mm, riot control CS, XM675 (E25 RS)**	Light Green	N/A	Black
40-mm, orange star parachute, XM695	White	Orange	Black

*The ogive is the nose end of the cartridge.
**Not authorized for use with M203 Grenade launcher.

Two views of the 40 × 53mm High Velocity Mark 19, Mod 3 self-loading Grenade Launcher. This weapon fires the M383 HE, M384 HE, M385 Practice, and M430 Dual Purpose (HEDP) cartridges.

CHARACTERISTICS OF THE MK19 MOD 3 40 × 53mm GRENADE LAUNCHER

Caliber: 40 × 53mm
System of operation: Blowback, belt-fed automatic fire
Length overall: 40.45 in.
Weight: 75.60 lbs.
Feed device: Metallic link belt
Muzzle velocity (f.p.s.): **M383 family:** 790
M430 family: 1200
Maximum range (meters): **M383 family:** 1600
M430 family: 2200
Cyclic rate: 375 r.p.m.

Three parts being produced by RAMO, Inc. for Navy. *Left*, MK19 MOD 3 receiver; *top right*, bolt; *bottom right*, sear housing.

49 Vietnam (Socialist Republic of Vietnam)

The army of North Vietnam was equipped with the basic inventory of Soviet small arms of both Soviet and Chinese manufacture. With the fall of the Saigon government of South Vietnam in 1975, the Socialist Republic acquired a large inventory of US material, which had been shipped to Vietnam during the Americans' two-decade involvement there.

VIETNAMESE SUBMACHINE GUNS

7.62mm Modified MAT 49 Submachine Gun.

7.62mm Modified Type 50 Submachine Gun.

7.62mm MODIFICATION OF THE FRENCH MODEL 49 SUBMACHINE GUN

The French apparently lost considerable quantities of 9mm NATO Model 49 (MAT49) submachine guns. The Vietnamese forces used this weapon as issued until their stocks of captured ammunition ran low. They then rebarreled the weapon for the Soviet 7.62mm Type P pistol cartridge (Chinese Type 50 pistol cartridge). The rebarreled MAT49 has a noticeably longer barrel than the original weapon but in other respects appears the same as the MAT49.

This weapon is loaded, fired and field stripped in the same manner as the MAT49, which is covered in detail in the chapter on France.

MODIFICATION OF THE 7.62mm TYPE 50 SUBMACHINE GUN

This weapon, which appeared in the hands of the Viet Cong as well as the North Vietnamese, is a modification of the 7.62mm Chinese Type 50, a copy of the Soviet Model PPSh M1941. The weapon is made up of components of the Type 50, some of the Type 56 (Chinese copy of the AK) and the wire stock of the French MAT49.

The pistol grip and front sight are the same as those of the 7.62mm Type 56 assault rifle; the receiver and magazine are the same as those of the Type 50 submachine gun. The barrel jacket of the Type 50, reduced in length, is also used.

CHARACTERISTICS OF THE MODIFIED TYPE 50 SUBMACHINE GUN

Caliber: 7.62mm (Type 50 cartridge).
System of operation: Blowback, selective fire.
Length overall: Stock extended: 29.75 in.
Stock retracted: 22.56 in.
Barrel length: Approx. 10.5 in.
Feed device: 35-round, box magazine.
Sights: **Front:** Hooded post.
Rear: "L" type aperture.
Weight: 9 lbs.
Muzzle velocity: 1640 f.p.s.
Cyclic rate: Approx. 900 r.p.m.

SOUTH VIETNAMESE PISTOLS

Prior to the establishment of the Diem government in South Vietnam, there were a number of dissident groups that had, in effect, private armies. The Cao Dai were one of these groups and manufactured arms in rather primitive workshops. Two of these weapons, a copy of the FN Browning Hi-Power and the US Colt .45 Model 1911A1—both chambered for the 9mm NATO cartridge—are shown here. The finish on these pistols is surprisingly good considering the circumstances under which they were made. The quality of the metallurgy is questionable.

Copies of the US caliber .30 M1 rifle were also made by dissident groups. It is not certain that all the parts for these weapons were made in Vietnam, some parts may have been of US manufacture and obtained from stocks of parts originally held by the French.

Although Cambodia is not part of Vietnam, it is geographically contiguous. Therefore, this chapter will include a weapon of Cambodian manufacture. The Cambodians have manufactured a 9mm weapon that combines features of the French M1950 and the USM1911A1 pistols. The Cambodian pistol, although shaped like the French M1950 and having its safety mounted on the slide, has a barrel bushing, recoil-spring-plug-arrangement similar to that of the US M1911A1.

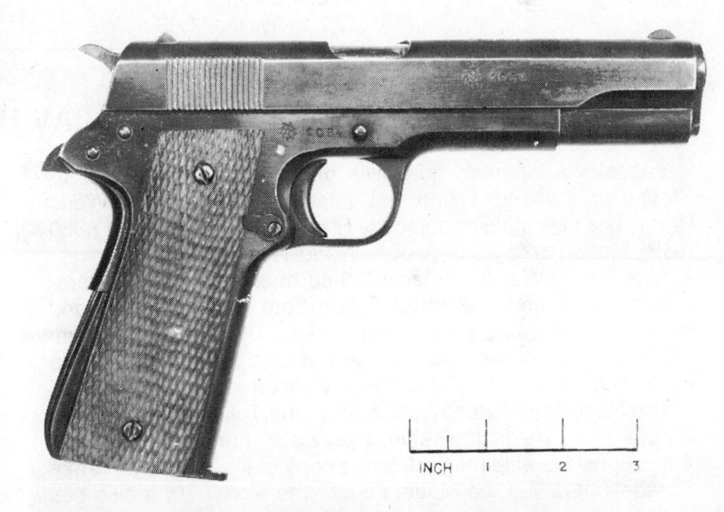

Cao Dai 9mm NATO copy of US M1911A1.

Cao Dai 9mm NATO copy of FN Browning Hi-Power.

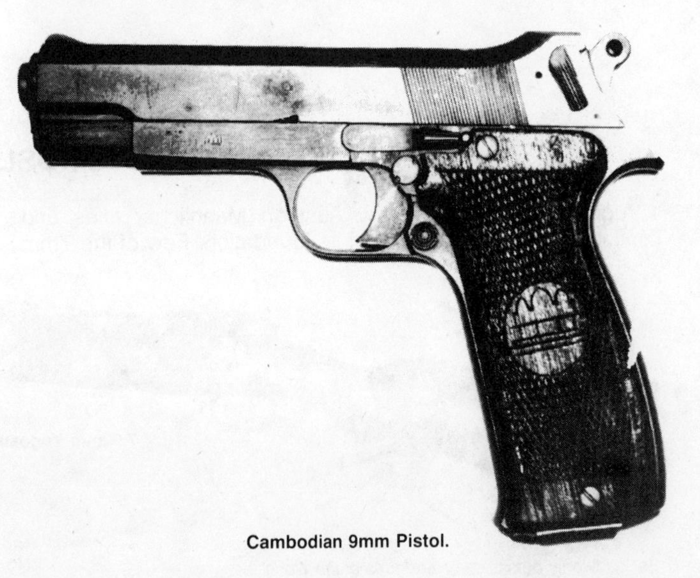
Cambodian 9mm Pistol.

50 Yugoslavia

SMALL ARMS IN SERVICE

Handguns:	7.62 × 25mm M57	Yugoslav copy of TT33. Also manufactured in 9 × 19mm as the M70. Zavodi Crvena Zastava, Kragujevac.
Submachine guns:	7.62 × 25mm M49/57	Yugoslav design made at ZCZ. Obsolete.
	7.62 × 25mm M56	Yugoslav variant of German MP40. Support troops only.
Rifles:	7.62 × 39mm M59/66	Yugoslav version of SKS made at ZCZ.
	7.62 × 39mm M70	Yugoslav version of AK47 made at ZCZ.
	7.62 × 39mm M70A	Folding stock version.
Machine guns:	7.92 × 57mm M53	Yugoslav version of MG42 made at ZCZ.
	7.62 × 39mm M65A	Squad automatic weapon version of Model 70 with fixed barrel. Made at ZCZ.
	7.62 × 39mm M65B	SAW version of Model 70 with quick change barrel. Made at ZCZ.
	7.62 × 39mm M72	Yugoslav version of RPK; improved model of M65. Made at ZCZ.
	12.7 × 99mm M2 HB	US-made.

YUGOSLAV HANDGUNS*

Yugoslavia inherited quantities of Austrian 9mm Model 12 Steyr pistols when the State was established at the end of World War I. The new state adopted the FN Browning 9mm short (.380 ACP) Model 1922 pistol prior to World War II.

After World War II, Yugoslavia had quantities of 9mm Parabellum Luger and P38 pistols taken from German forces and Beretta pistols captured from the Italians. The Tokarev 7.62mm TT M1933 pistol was adopted in the late forties. Yugoslavia manufactures a copy of the Tokarev called the Model 57.

The Yugoslav Model 57 differs from the Tokarev in that it has a nine-round magazine rather than an eight-round magazine. The Yugoslavs also manufacture a copy of the Tokarev in 9mm Parabellum called the Model 65 and the Model 70. It also has a nine-round magazine.

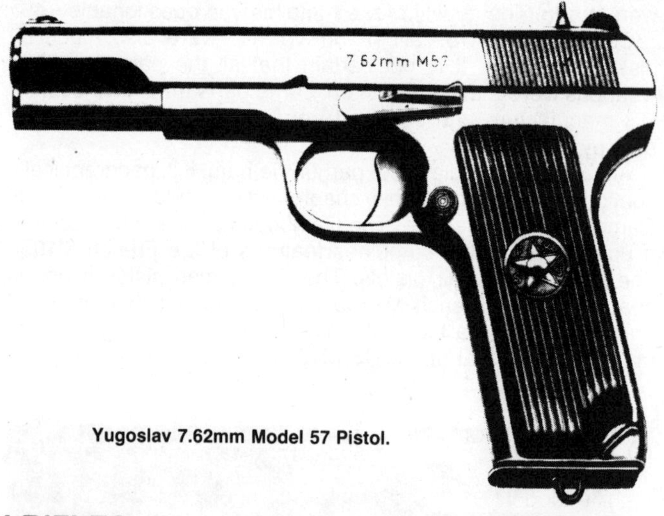

Yugoslav 7.62mm Model 57 Pistol.

YUGOSLAV RIFLES

Yugoslavia obtained many Austrian Mannlicher rifles and Turkish Mausers at the time of its foundation. Few of the 7mm Serbian Mausers (Serbia became part of Yugoslavia), the Model 1889, 99/07, 99/08 or 1910 survived World War II, and they were

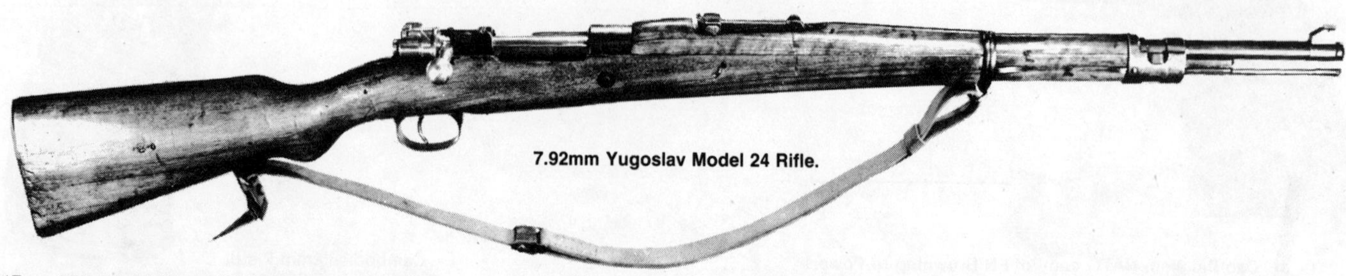
7.92mm Yugoslav Model 24 Rifle.

*For additional details see *Handguns of the World*.

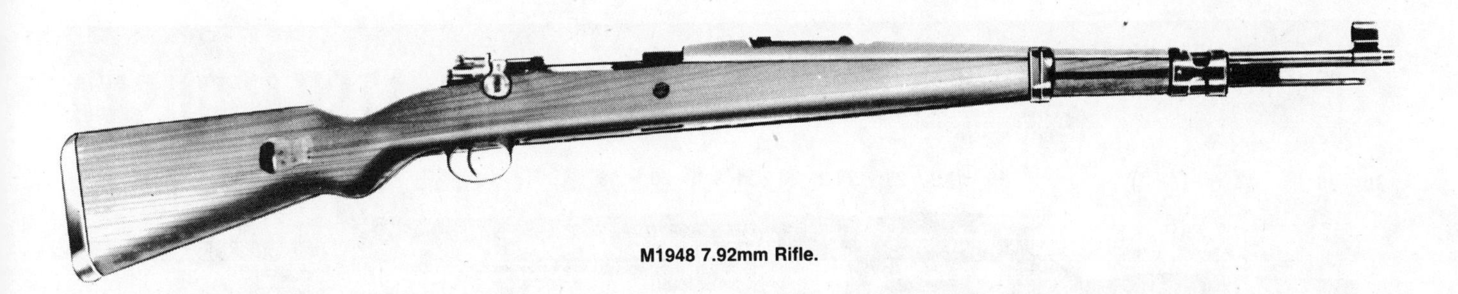

M1948 7.92mm Rifle.

never a major factor in Yugoslav armament. Austrian Model 1895 Mannlichers were converted to 7.92mm and called Model 95M; Turkish Model 1890 and 1893 were also converted to 7.92mm, the conversion being called the Model 90T. French 8mm M1886M93, M1907/15 and Model 1916 rifles and carbines were also procured.

The Czech ZB 7.92mm Model 24 rifle was adopted as standard, and weapons were manufactured at Kragujevac in Yugoslavia, in addition to purchases made from Czechoslovakia. This rifle is the same as the Czech Model 24 rifle, which is covered in detail in the chapter on Czechoslovakia.

After World War II, the Yugoslavs had quantities of 7.92mm Kar 98ks and some Italian rifles. They also were furnished with 7.92mm Model 1944 carbines and Model 1891/30 rifles by the Soviets.

THE YUGOSLAV 7.92mm RIFLE MODEL 1948

The Yugoslavs adopted a slightly modified copy of the Kar 98K as the Model 1948.

The Model is loaded, fired and field stripped in the same manner as the Kar 98K.

CHARACTERISTICS OF THE MODEL 48 RIFLE

Caliber: 7.92mm.
System of operation: Manually operated bolt.
Weight: 8.62 lbs.
Length, overall: 42.9 in.
Barrel length: 23.3 in.
Feed device: 5-round, integral, staggered-row box magazine.
Sights: Front: Hooded blade.
Rear: Tangent w/ramp.
Muzzle velocity: 2600 f.p.s. (approx.).

YUGOSLAV 7.62mm MODEL 59/66 RIFLE

The Model 59/66 is a modified copy of the Soviet SKS. Like the SKS, it is chambered for the 7.62 × 39mm intermediate sized cartridge. The major differences from the SKS are the presence of a gas shut-off valve on the gas cylinder and integral grenade launcher fitted to the barrel. A grenade launcher sight is mounted on the front top of the barrel. The grenade launcher sight pivots upward from the front sight base. As a matter of interest, the diameter of the grenade launcher is 22mm, the same as that of the grenade launchers in service in the Western World but is not the same diameter as the grenade launchers in service in Poland and some of the other Communist countries. The Model 59/66 has a rubber recoil pad fitted to the butt, and its blade type bayonet is slightly longer than that of the SKS. It also has night sighting aids fitted to the front and rear sights. Yugoslavia advertises this weapon for export. With the exception of the gas shut-off valve, which in the on position—turned up—is used for launching grenades, loading, firing and functioning of the Model 59/66 is the same as that of the SKS.

YUGOSLAV 7.62mm MODEL 64, MODEL 70 (64A) AND MODEL 70A (64B) ASSAULT RIFLES

These weapons are all essentially modifications of the Soviet AK assault rifle. They are in the Yugoslav FAZ family of weapons, which includes two light machine guns, the Models 65A and 65B. The Model 64 is essentially the same as the AK but has a longer barrel. As with all the rifles of this series, it has a built in grenade launcher sight that pivots on the gas cylinder. The Model 64 uses a 20-round magazine as opposed to the 30-round magazine normally utilized in AK type weapons. Since the FAZ family of weapons has a bolt stop—bolt holding open device—which works on a notch cut in the magazine, the standard AK magazines cannot be used with these guns.

The Model 70, which was previously called the Model 64A, is essentially the same as the standard wooden stocked AK except for grenade launcher sight and magazine. The Model 70A, which was previously called the Model 64B, is the same as the Model 70 but has a folding steel stock. All of these weapons can be fitted with a 22mm grenade launcher or a short compensator/muzzle brake.

CHARACTERISTICS OF THE MODEL 59/66

Caliber: 7.62mm (7.62 × 39mm)
System of operation: gas, semi-automatic only.
Weight: 9.36 lbs.
Length overall: 43.9 in.
Barrel length: 19.7 in.
Feed device: 10 round, staggered row, integral box magazine.
Sights: Front: hooded post.
Rear: notched tangent.
Muzzle velocity: 2411 f.p.s.

The 7.62 × 39mm Model 59/66 rifle, a Yugoslav version of the SKS.

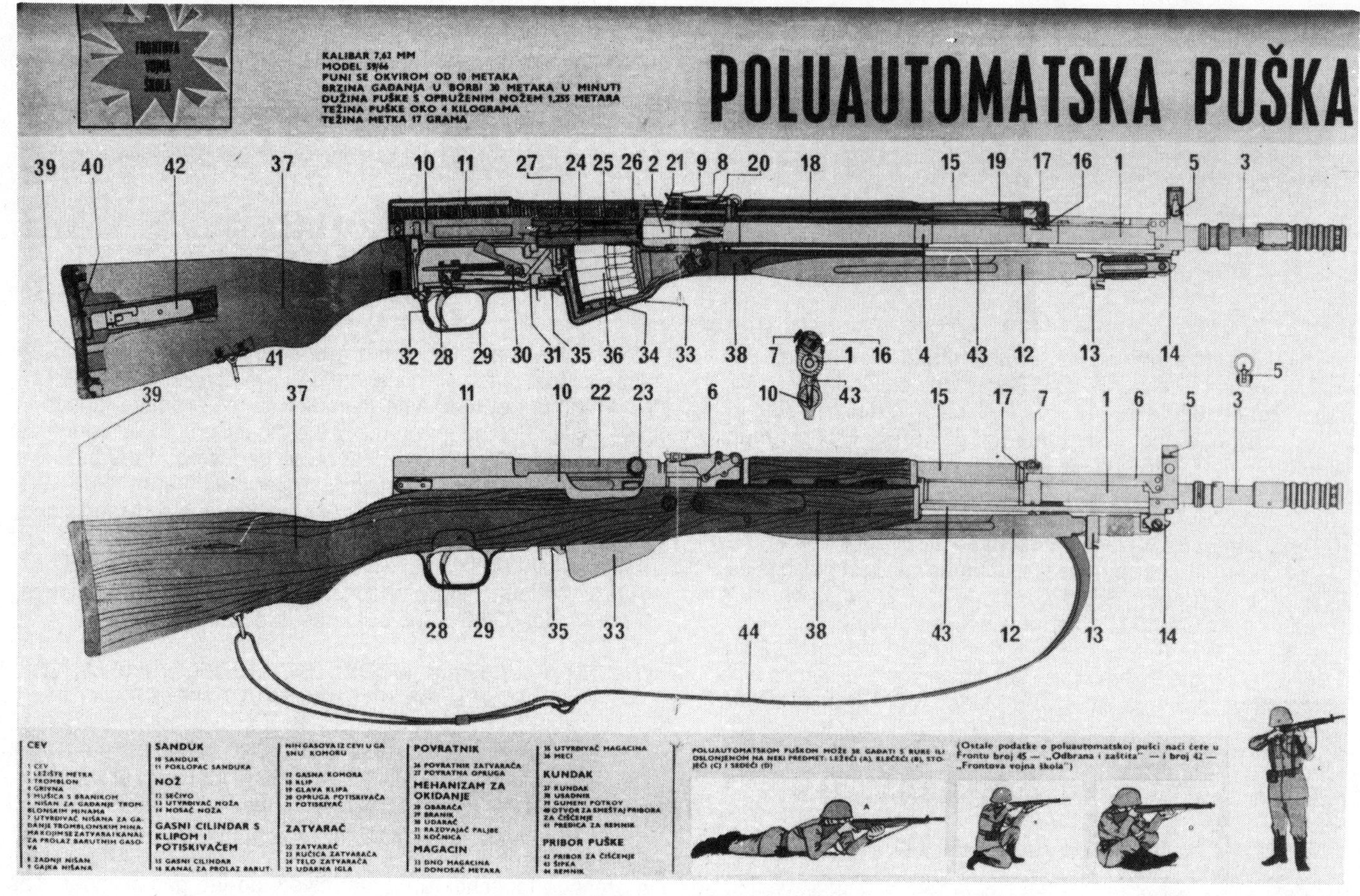

Yugoslav 7.62mm Model 59/69 Rifle.

7.62 × 39mm Model 70A, the Yugoslav copy of the AK47.

7.62 × 39mm Model 70 with AKM-type stamped steel receiver.

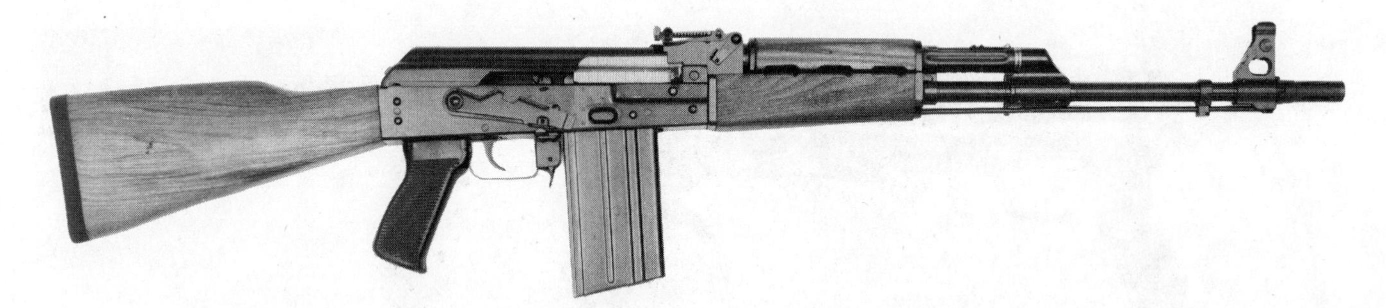

Zastava 7.62 × 51mm NATO caliber M77/82 semiautomatic rifle made for export to the United States.

CHARACTERISTICS OF THE MODEL 64, MODEL 70 AND M70A RIFLES

	Model 64	Model 70	Model 70A
Caliber:		7.62mm (7.62 × 39)	
System of operation:		gas operated selective fire	
Weight:	8.6 lbs.	8.3 lbs.	8.2 lbs.
Length overall: stock fixed:	40.9 in.	37.7 in.	37.7 in.
stock folded:			27.2 in.
Barrel length:	19.7 in.	16.3 in.	16.3 in.
Feed device:	20 round detachable box	30 round detachable box	30 round detachable box
Sights: Front:		hooded post	
Rear:		tangent with notch	
Muzzle velocity:	2395 f.p.s.	2297 f.p.s.	2297 f.p.s.
Cyclic rates:		600–650 r.p.m.	

YUGOSLAV SUBMACHINE GUNS

7.62mm Model 49 Submachine Gun.

Yugoslavia adopted the 9mm Parabellum Vollmer Erma in the mid-thirties. After World War II, quantities of the German 9mm Parabellum MP38 and MP40 were available, as were Italian Beretta submachine guns and British Sten guns. The Soviet Union supplied their 7.62mm PPD and PPSh41 submachine guns to Yugoslavia.

YUGOSLAV 7.62mm SUBMACHINE GUN MODEL 49

Yugoslavia developed a submachine gun of native design chambered for the 7.62 × 25mm Soviet cartridge. The Model 49 is similar in appearance to the PPSh M1941 but differs internally. The bolt is similar to the Beretta Model 38A, and the

7.62mm Model 56 Submachine Gun.

buffer is considerably more complicated than the plastic or rubber piece found on the PPSh M1941. The buffer of the Model 49 has a spring, separate from the operating spring, and split ring assembly which is retained by a collar on the end of the operating spring tube guide. The Model 49 is field stripped by twisting the receiver cap a quarter turn and removing the buffer, operating spring, and bolt assembly.

YUGOSLAV 7.62mm SUBMACHINE GUN MODEL 56

The Yugoslavs have developed a new and somewhat less complicated gun called the 7.62mm Model 56. The Model 56 has a folding stock similar to the German MP40 and uses a knife-type bayonet. The Model 56 submachine gun is now being made in 9mm Parabellum as the Model 65. It is essentially, except for caliber, identical to the Model 56.

YUGOSLAV MACHINE GUNS

7.92mm ZB30J Machine Gun.

Yugoslavia has used the 8mm 07/12 Schwarzlose, the 7.92mm Maxim Model 8M—a conversion of Serbian 7mm and Bulgarian 8 × 50mm Maxim guns to 7.92mm—the 8mm St. Etienne, and various models of the Madsen gun. The principal light machine gun, prior to World War II, was the 7.92mm ZB30J, a slightly modified version of the ZB30. The principal noticeable difference between the ZB30 and ZB30J is the presence of a knurled ring on the barrel just ahead of the receiver on the ZB30J. This shows up clearly in the photograph above. This weapon was made in Yugoslavia as well as in Czechoslovakia.

After World War II, Yugoslavia had 7.92mm MG34 and MG42 machine guns and various Italian machine guns. They were supplied with some 7.62mm DPs and Model 1910 Maxims by the Soviets. They tooled up for the manufacture of the 7.92mm MG42, which is called the Model 53 or "Sarac" in Yugoslavia. This is the current standard machine gun.

THE YUGOSLAV 7.62mm MODEL 65A AND 65B LIGHT MACHINE GUNS

These guns are part of the Yugoslav "FAZ" weapons family. They, as the rifles covered earlier, are based on the Soviet AK in so far as basic design is concerned. The Model 65A and 65B are essentially the same in all basic respects except one: the 65A has a finned, quick change barrel. Both guns have bipods and flash hiders.

CHARACTERISTICS OF YUGOSLAV SUBMACHINE GUNS

	Model 49	Model 56
Caliber:	7.62mm.	7.62mm.
System of operation:	Blowback, selective fire.	Blowback, selective fire.
Length overall:	34.4 in.	Stock extended: 34.25 in. Stock folded: 23.25 in.
Barrel length:	10.5 in.	9.84 in.
Feed device:	32-round, detachable staggered row magazine	
Sights: Front:	Hooded blade.	Hooded blade.
Rear:	L-type with "U" notch.	L-type with "U" notch.
Weight:	9.44 lb.	6.61 lb.
Muzzle velocity:	Approx. 1700 f.p.s.	Approx. 1700 f.p.s.
Cyclic rate:	Approx. 700 r.p.m.	570–620 r.p.m.

A 7.62 × 51mm NATO-caliber M77 squad automatic weapon. This is a variant of the 7.62 × 39mm M72 (RPK).

CHARACTERISTICS OF THE 7.62mm MODEL 65A AND 65B LIGHT MACHINE GUNS

	Model 65A		Model 65B
Caliber:		7.62mm (7.62 × 39)	
System of operation:		gas, selective fire	
Weight:	12.3 lb.		11.4 lb.
Length overall:		41.5 in.	
Barrel length:		approx. 20 in.	
Feed device:		30 round, staggered row detachable box	
Sights: Front:		hooded post	
Rear:		notched tangent	
Muzzle velocity:		2411 f.p.s.	
Cyclic rate:		600–650 r.p.m.	

51 Small Arms for Outdoor Sports

Although *Small Arms of the World* is basically devoted to military firearms, this chapter describes the civilian versions of some of the most popular military weapons, particularly the semiautomatic counterparts of full automatic military weapons. Until the end of World War II, virtually all forms of cartridge era rifles, pistols and revolvers were originally developed for military use and subsequently adapted to sporting purposes—target shooting, hunting and plinking. In the world of rifles, the introduction of military self-loaders in the post-1945 period has altered patterns of rifle production. And in more recent years, the Gun Control Act of 1968 (GCA 68) and the growing numbers of shooters and collectors have had an impact on the sporting rifle market. A closer look at these contemporary developments will benefit both the student of military arms and the users of sporting rifles.

RIFLES

The single most important rifle mechanism, in terms of popularity and quantities produced, has been Peter Paul Mauser's Model 98 bolt action rifle. In addition to the tens of thousands of military Mausers, Springfields and M1917s that have been converted to sporting rifles, a substantial number of Mauser type actions have been and continue to be manufactured. When the 1968 gun control law forbade the further importation of surplus military rifles into the United States, companies such as Interarms arranged for production of modernized versions of the Mauser action. The Mark X rifles and actions, which incorporate a sliding safety mounted on the right rear of the receiver, are marketed by Interarms and manufactured in Yugoslavia. Other companies such as Champlin, Colt-Sauer, Steyr, Mauser and Sturm-Ruger have introduced new designs and manufacturing technologies to the rifle scene during the last decade. One of Ruger's important technological contributions has been the use of investment castings for rifle receivers and other important components. In an era when machining finished parts from forgings has become increasingly costly, investment casting has provided a quality alternative. Thompson-Center Arms and Plainfield Arms have also used this technique in the manufacture of their sporting arms.

In addition to the traditional bolt, lever and pump action rifles, two distinctive classes of sporting rifles have emerged since 1945—military self-loaders and new versions of older patterns. As standard semiautomatic military rifles have increased in price due to their collector appeal and as the newer selective fire weapons have qualified as registerable "firearms" (i.e., automatic weapons) under GCA 68, manufacturers have sought to market "look alike" weapons and semiautomatic versions of military weapons. Chief among the "look alikes" is a semiautomatic-only copy of the M14 Rifle produced by Springfield Armory Inc. (a private firm, not to be confused with the national armory, which was closed in 1968). When it became clear that only limited numbers of the National Match M14s would become available to a few DCM qualified shooting organizations, Springfield Armory Inc. began production of their MIA, and A. R. Sales introduced their Mark IV Sporter. Both Plainfield and Universal have manufactured copies of the US M1 Carbine for commercial and police markets.

Among the commercial sporting versions of military rifles, the leader by far is the Colt AR-15. More than 300,000 of these semiautomatic M16 Rifles have been sold since Colt began to market the sporter model in the early 1960s. Subsequently, Armalite produced the AR-180 version of the AR-18, while Heckler and Koch has sold the HK91 and HK93 models of their G-3 and HK33 rifles. CETME has made the CETME Sport variant of the Spanish military rifle. SIG has marketed a sporterized Model 57 assault rifle, and Beretta has produced limited quantities of semiautomatic BM59 Rifles for sale in the US. The Valmet M62S sold by Interarms was a sporterized military rifle produced to fire a non-standard American cartridge. Chambered for the Soviet 7.62 × 39 M43 cartridge, the M62S is basically a semiautomatic AK47 of extremely high quality manufacture. A newer version with a stamped steel receiver and chambered to fire the US 5.56 × 45mm cartridge was sold by Interarms as the M71S.

Included in the second class of rifles, the newly manufactured versions of old models, is a group of replicas of 19th century guns. The originals are such valued collectors' items that they are no longer used as shooters. Leading the list of these replicas are the Harrington & Richardson reproduction of the Springfield 1873 Trapdoor Rifle and the Navy Arms Company Remington type dropping block and Martini-Henry type dropping block rifles. In addition, several small companies have introduced copies of Sharps, Sharps-Borchardt and Winchester M1885 High Wall rifle actions, while Sturm-Ruger has produced their classic style No. 1 and No. 3 single-shot rifles as the demand has increased from shooters for such high quality specialty firearms.

In rifles, only one design form has a purely sporting heritage—the slide or pump action, which is manually actuated by a sliding fore-end. Slide action rifles offered today generally have the same actions as existing self-loading models. Only the method of applying the actuating power is different.

A few words need to be said about rifle ammunition. Traditionally, military cartridges have been altered by the adoption of special projectiles for both match target and hunting purposes. Whereas the US .30-06 and German 8mm (7.92 × 57mm) cartridges were extremely popular in the early decades of this century, the .308 Winchester (civilian version of the 7.62 × 51mm NATO) and .223 Remington (5.56 × 45mm) currently enjoy wide usage on the sporting scene. As noted in Chapter 1, Gene Stoner started with the .222 Remington cartridge when he began his search for a small caliber military round. At about the same time, Earle M. Harvey at Springfield Armory began work on a modified version of the .222 Remington. That cartridge, the Springfield .224, was subsequently marketed as the .222 Remington Magnum. As with weapons designs, the sporting world has been and will continue to be influenced by military ammunition development.

Following is a representative sample of those rifles popular with shooters today. Unless otherwise noted the rifles illustrated are of the same dimensions and weight of their military counterparts. Chapter references are to the military versions.

SPRINGFIELD ARMORY INCORPORATED

This Geneseo, Illinois–based organization has recently expanded its line of military look-alike rifles to include the M1 and the BM59 rifles. Both of these products are being currently built with surplus Beretta components with the exception of the receiver, which is investment cast in the United States. Receivers of military surplus rifles are no longer importable due to the provisions of the GCA 68. In addition to these rifles, Springfield Armory, Inc. has manufactured in excess of 22,000 M1As, their semiautomatic version of the M14. The M1 and M14 are described in Chapter 48, and the BM59 is described in Chapter 29.

The Springfield Armory, Inc. M1A semiautomatic copy of the M14 Rifle.

Springfield Armory, Inc. M1A-A1 with Beretta-type folding stock.

Springfield Armory, Inc. M1A-E2.

Three versions of the Springfield Armory, Inc. M1 Rifle. *Top*, standard model; *middle*, standard model with commercial scope and mount; *bottom*, M1-D with M84 mount and scope.

7.62 × 51mm NATO BM-59 Ital made by Springfield Armory, Inc.

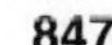

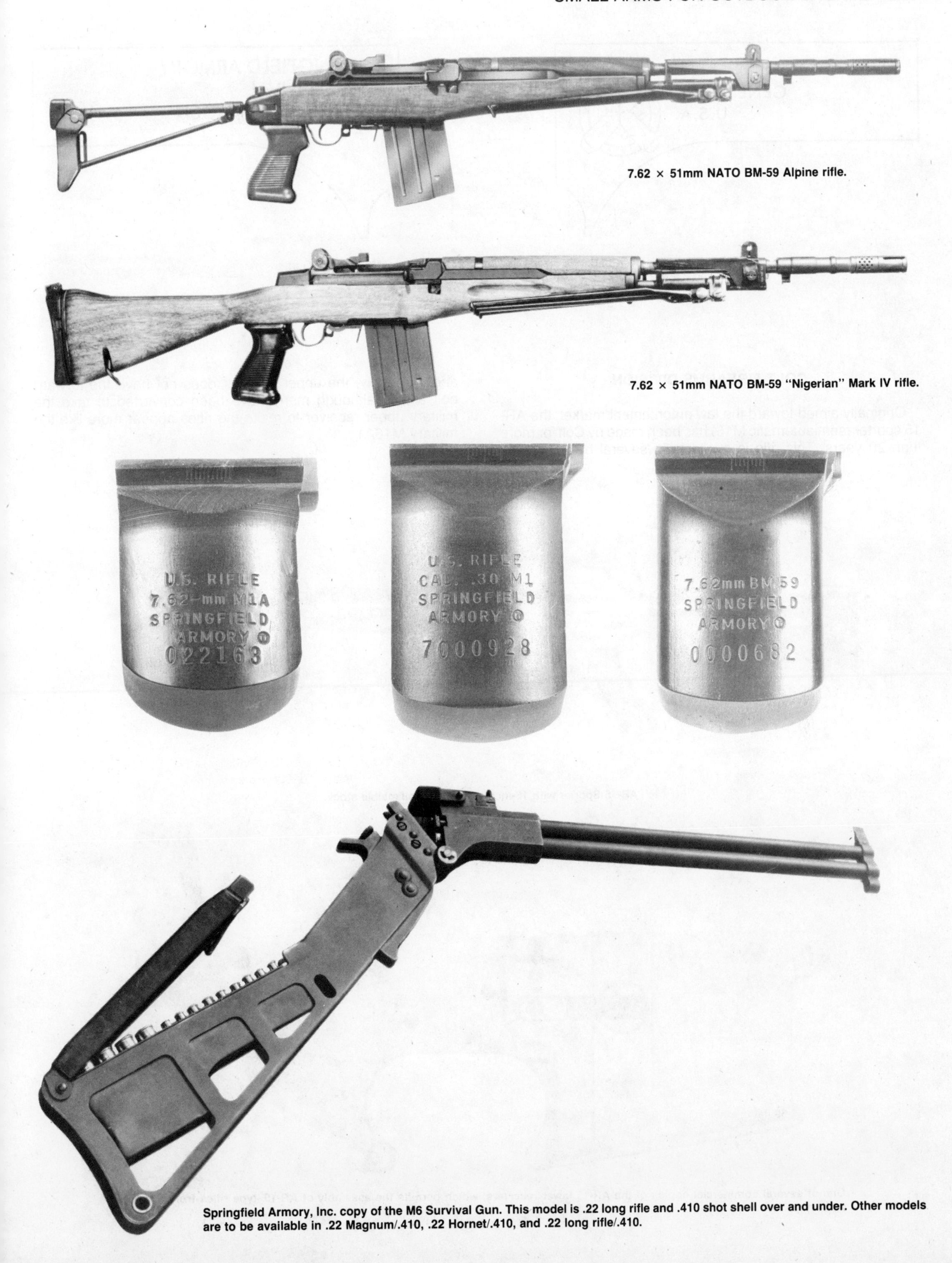

7.62 × 51mm NATO BM-59 Alpine rifle.

7.62 × 51mm NATO BM-59 "Nigerian" Mark IV rifle.

Springfield Armory, Inc. copy of the M6 Survival Gun. This model is .22 long rifle and .410 shot shell over and under. Other models are to be available in .22 Magnum/.410, .22 Hornet/.410, and .22 long rifle/.410.

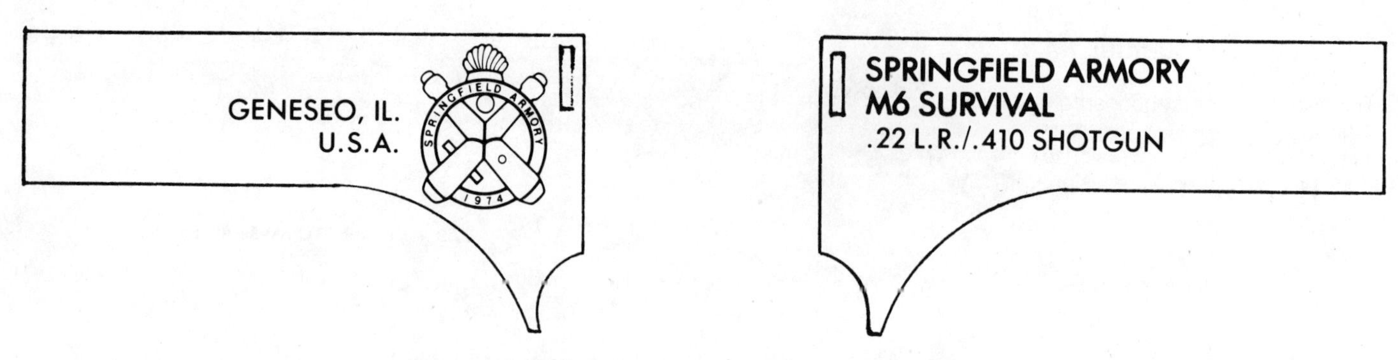

Receiver markings on the Springfield Armory, Inc. M6 Survival Gun.

COLT FIREARMS DIVISION

Originally aimed toward the law enforcement market, the AR-15 sporter (semiautomatic M16) has been made by Colt for more than 20 years. This rifle is available in several barrel lengths and, generally, the upper receiver does not have the forward bolt assist, although many have been converted to take the military upper receiver to make the rifles appear more like the military M16A1.

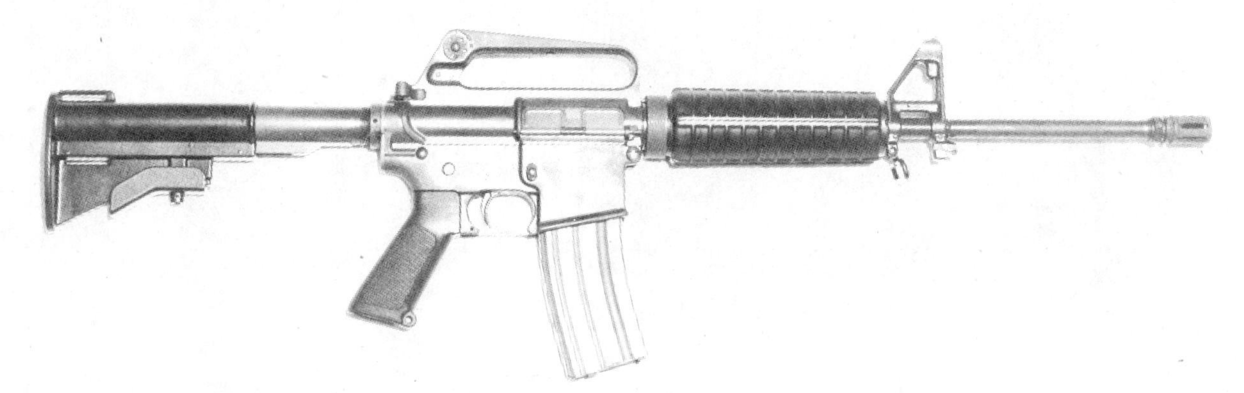

AR-15 Sporter with 16-inch barrel and collapsible stock.

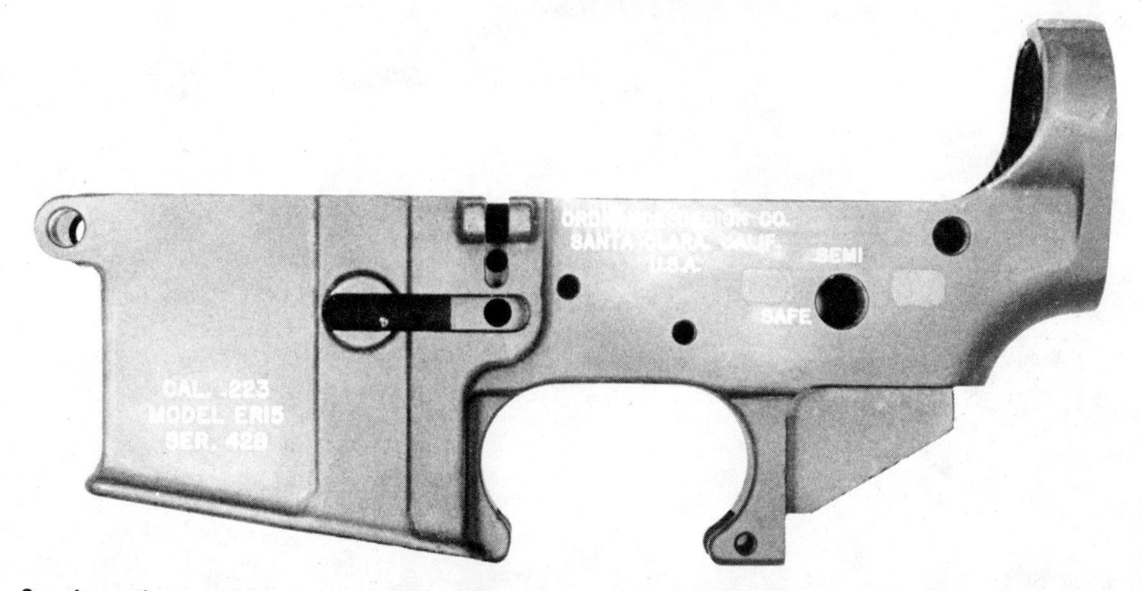

One of several commercial copies of the AR-15 lower receivers, which permits the assembly of AR-15–type rifles from surplus military parts.

FABRIQUE NATIONALE

This Belgian firm introduced a Competition model of the FN FAL in the early 1960s. These first rifles did not meet the US requirements against conversion to full automatic fire. In the mid-1970s, FN introduced a redesigned model of the Competition model. Today these are marketed in the US by Steyr-Daimler-Puch. The Competition models are variants of the standard FAL, FAL PARA, and the FALO heavy barrel model. A semiautomatic version of the 5.56 × 45mm FNC is also available in the United States. See Chapter 9 for details on these rifles.

The early FN Competition 7.62 × 51mm NATO FAL made for the US market.

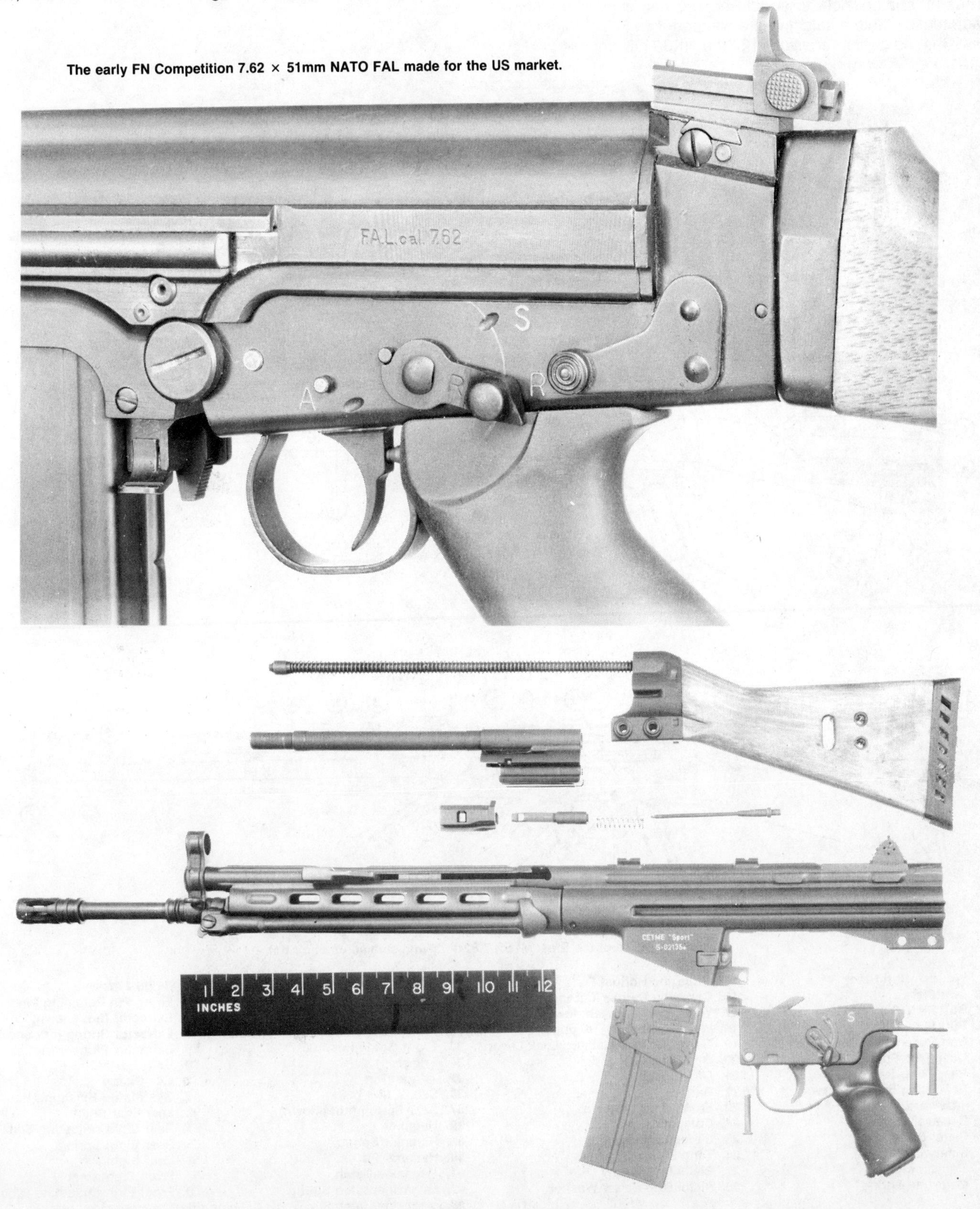

A disassembled view of the 7.62 × 51mm NATO CETME "Sport" rifle.

HECKLER & KOCH

The HK91 is the civilian counterpart of the G3 in 7.62 × 51mm NATO. To make this weapon acceptable under American laws, Heckler & Koch engineers redesigned the trigger assembly housing so that the HK91 and G3 housings will not interchange. Thus, it is impossible to switch the two and convert the semiautomatic rifle to a selective fire weapon. In 5.56 × 45mm, the HK93 is the civilian equivalent of the HK33 military rifle.

CETME

In years past, Mars Equipment Corporation, of Chicago, Illinois, and a few other importers have marketed the "Sport" model of the CETME rifle in 7.62 × 51mm.

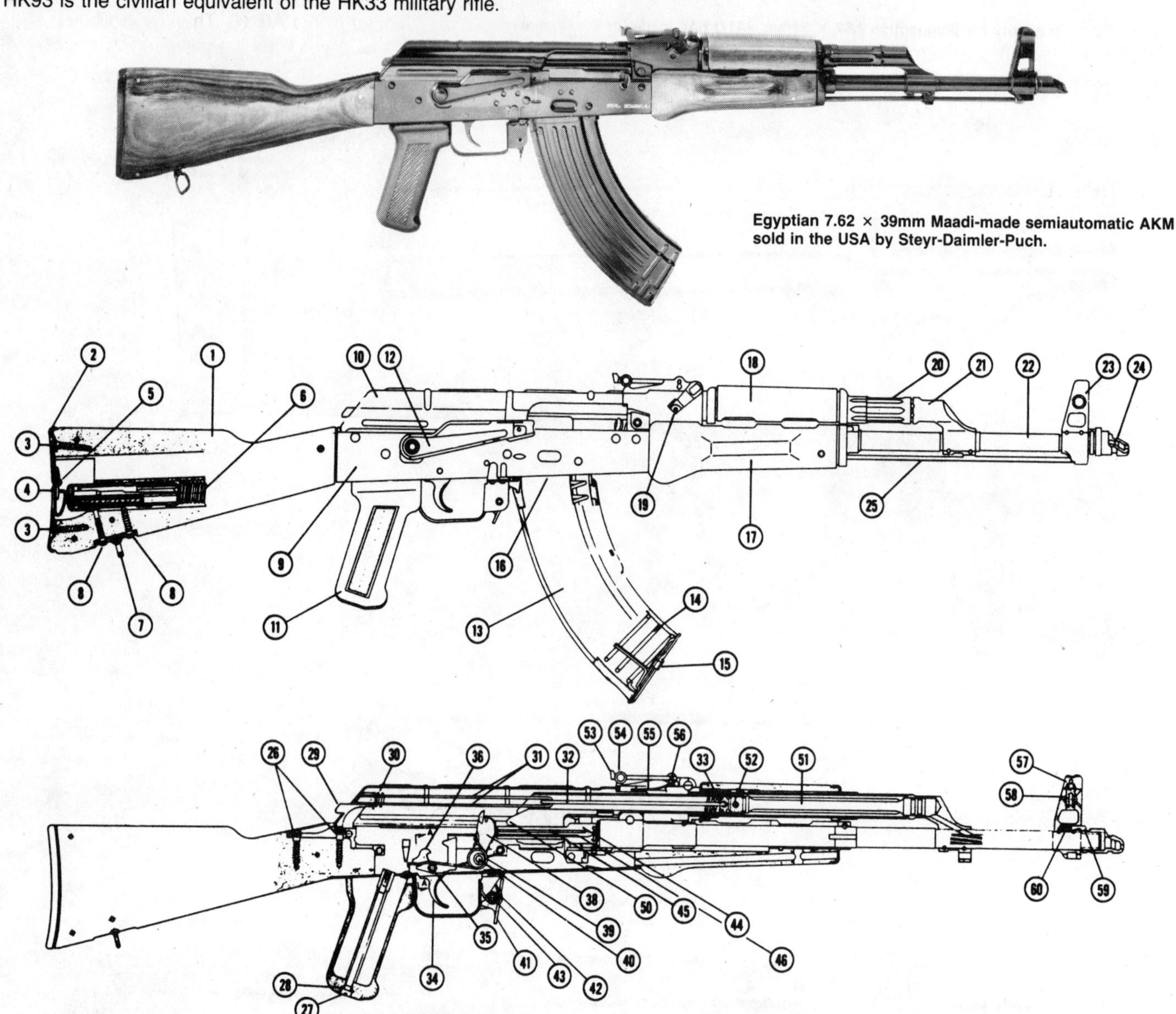

Egyptian 7.62 × 39mm Maadi-made semiautomatic AKM sold in the USA by Steyr-Daimler-Puch.

Section views of the Steyr/Maadi 7.62 × 39mm semiautomatic AKM.

1. Wooden Butt Stock
2. Butt Plate
3. Butt Plate Screws
4. Butt Plate Cover
5. Butt Plate Cover Spring
6. Oiler Trap Spring
7. Rear Sling Swivel
8. Rear Sling Swivel Screws
*9. Receiver Group
10. Receiver Cover
11. Pistol Grip
12. Safety Lever
13. Magazine
14. Magazine Spring
15. Magazine Bottom Plate
16. Cartridge Carrier (not shown)
17. Wooden Handguard, lower
18. Wooden Handguard, upper
19. Gas Cylinder Retaining Lock Lever
20. Gas Cylinder
21. Gas Block
22. Barrel
23. Front Sight Protector
24. Compensator
25. Cleaning Rod
26. Tang Screw (2)
27. Pistol Grip Screw
28. Pistol Grip Screw Washer
29. Rear Guide Retaining Block
30. Recoil Return Spring
31. Rear Recoil Return Spring Guide
32. Front Recoil Return Spring Guide
33. Front Guide Retainer
34. Trigger
35. Trigger Pin
36. Sear
37. Sear Spring (not shown)
38. Hammer
39. Hammer Spring
40. Hammer Pin
41. Magazine Catch
42. Magazine Catch Spring
43. Magazine Catch Pin
44. Bolt
45. Firing Pin
46. Firing Pin Retaining Pin
47. Extractor (not shown)
48. Extractor Spring (not shown)
49. Extractor Plunger (not shown)
50. Bolt Carrier
51. Gas Piston
52. Gas Piston Retaining Pin
53. Leaf Rear Sight
54. Rear Sight Adjusting Slide
55. Rear Sight Spring
56. Rear Sight Pin
57. Front Sight Post
58. Front Sight Adjusting Block
59. Compensator Lock Pin
60. Compensator Lock Pin Spring

*Receiver Group consists of receiver and trigger guard.

STEYR-DAIMLER-PUCH

In addition to selling their handguns and their SSG sniper and target rifles, Steyr-Daimler-Puch, of Secaucus, New Jersey, is importing the Egyptian AKM in semiautomatic only as manufactured by the Maadi Company.

VALMET

This Finnish company has made both 7.62 × 39mm and 5.56 × 45mm caliber rifles patterned after their military AK47s and AKMs. Until recently, these were imported into the USA by Interarms. Currently, Valmet military look-alike rifles are being imported by Odin International, of Alexandria, Virginia. These rifles follow the basic pattern described in Chapter 18.

LEADER DYNAMICS

This Australian firm, see Chapter 1, has been marketing a semiautomatic version of their rifle in the USA through World Public Safety located in Nevada. This rifle has a bolt carrier and bolt patterned after that of the AR-18. This rifle, however, has a three-lug bolt instead of the eight-lug bolt used on the AR-18 and AR-180.

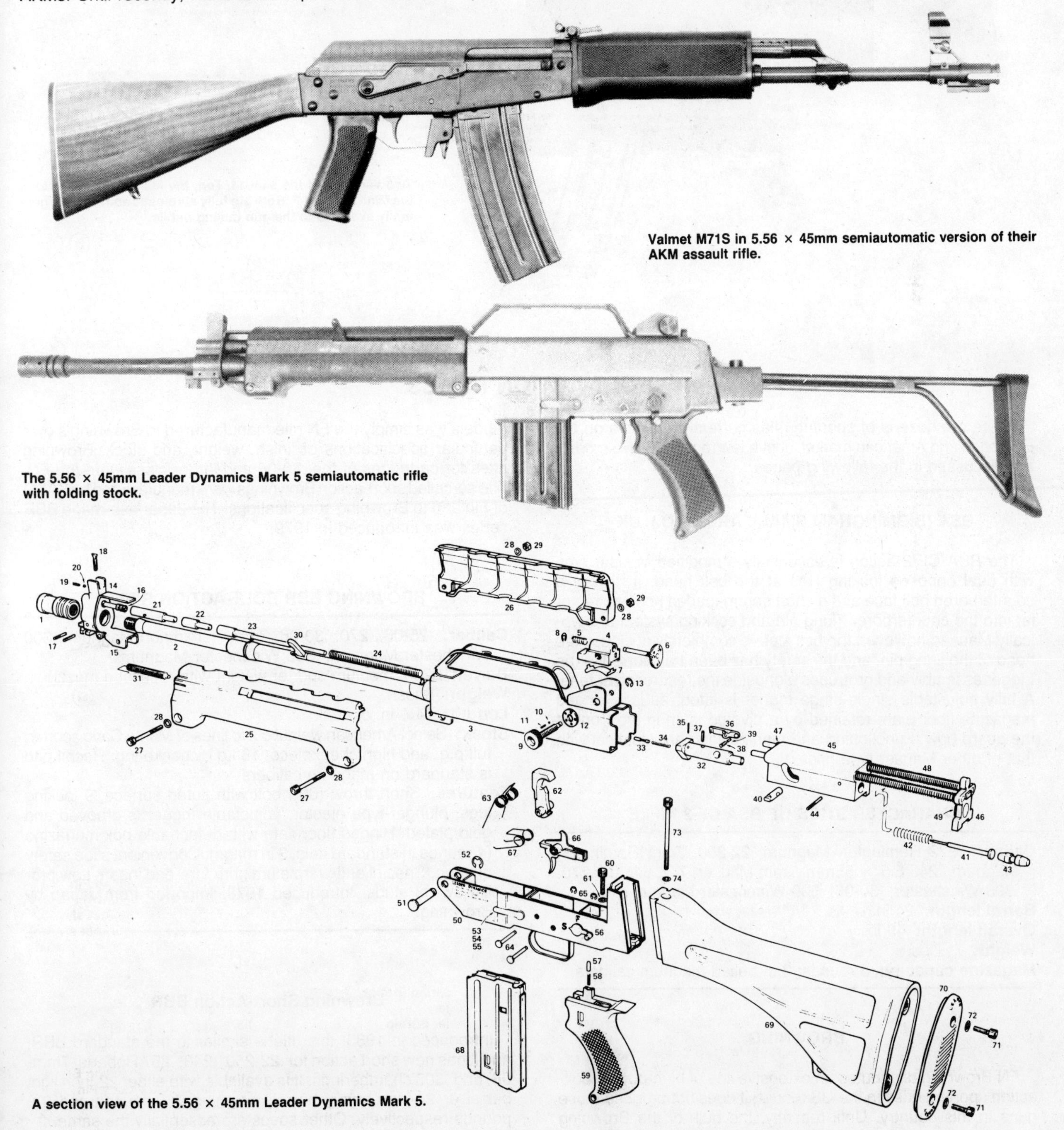

Valmet M71S in 5.56 × 45mm semiautomatic version of their AKM assault rifle.

The 5.56 × 45mm Leader Dynamics Mark 5 semiautomatic rifle with folding stock.

A section view of the 5.56 × 45mm Leader Dynamics Mark 5.

STURM, RUGER & COMPANY

As noted in Chapter 1, Sturm, Ruger and Company introduced a scaled-down version of the 7.62 × 51mm NATO caliber rifle in 5.56 × 45mm, which they call the Mini-14. Large quantities of the semiautomatic version have been marketed in the USA to police and civilian buyers. The accompanying photograph illustrates the standard stock Mini-14 and the folding stock versions. Both of the rifles illustrated are fully automatic.

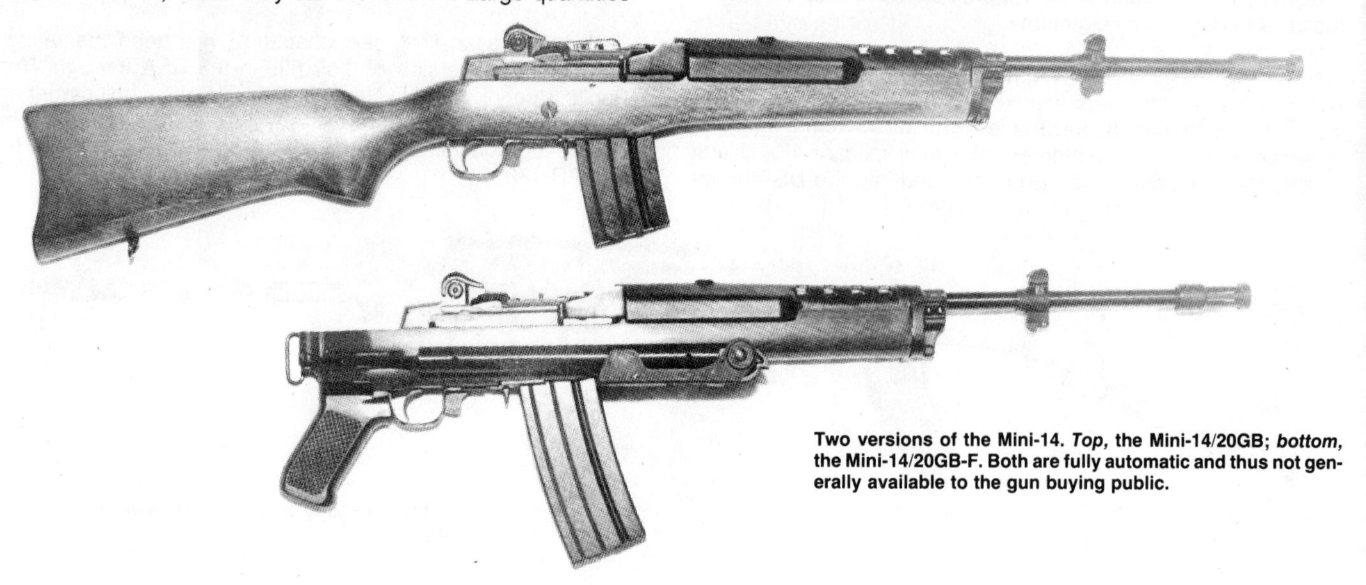

Two versions of the Mini-14. *Top,* the Mini-14/20GB; *bottom,* the Mini-14/20GB-F. Both are fully automatic and thus not generally available to the gun buying public.

BOLT-ACTION RIFLES

There are dozens of sporting rifles currently available on the European and American market. Just a few representative models are discussed in the following pages.

BSA (BIRMINGHAM SMALL ARMS CO.), UK

The BSA CF-2 action is essentially a modified Mauser type with dual opposed locking lugs at the bolt head. It utilizes a counterbored bolt face and a short spring-loaded hook extractor let into the counterbore. Firing pin and cocking system are typically Mauser; however, the bolt sleeve completely encloses the head of the firing pin, and the safety has been transferred to the trigger assembly and protrudes alongside the rear receiver tang. A fully adjustable single-stage trigger is fitted, as is a hinged magazine floor plate retained by a pivoted latch in the front of the guard bow. Functioning and manipulation are identical with that of other Mauser type rifles.

CHARACTERISTICS OF BSA CF-2 RIFLE

Caliber: .222 Remington Magnum .22-250, 7mm Remington Magnum, .243 6.5 × 55mm, 7mm Mauser, 7 × 64mm, .270, .308 Winchester; .30-06, .300 Winchester Magnum.
Barrel length: 24 in.
Overall length: 45 in.
Weight: 7½ lbs.
Magazine capacity: 5 rounds; 3 in belted magnum calibers.

BROWNING

FN Browning distributes an extensive line of high-power, bolt-action sporting rifles in the US though it does not produce those guns in this country. Until recently, the bulk of the Browning models was simply the FN rifle manufactured to Browning's own particular specifications of finish, weight, and stock. Browning rifles for cartridges of the .30-06 and larger class were the FN. The so-called short-action Brownings were manufactured by SAKO of Finland to Browning specifications. The Japanese-made BBR series was introduced in 1979.

BROWNING BBR BOLT-ACTION RIFLE

Caliber: .25-06, .270, .30-06, 7mm Remington Magnum, .300 Winchester Magnum, .338 Winchester Magnum.
Barrel: 22 in. medium sporter weight with recessed muzzle.
Weight: 8 lbs.
Length: 44½ in. overall.
Stock: Select American walnut cut to lines of Monte Carlo sporter; full p.g. and high cheekpiece; 18 l.p.i. checkering. Recoil pad is standard on magnum calibers.
Features: Short throw (60°) bolt with fluted surface, 9 locking lugs, plunger-type ejector, adjustable trigger is grooved and gold plated. Hinged floorplate with detachable box magazine (4 rounds in standard cals, 3 in mags). Convenient slide safety on tang. Special anti-warp aluminum fore-end insert. Low profile swivel studs. Introduced 1978. Imported from Japan by Browning.

Browning Short-Action BBR

Introduced in 1983, this rifle is similar to the standard BBR, and it has new short action for .22-250, .243, .257 Roberts, 7mm-08, and .308 chamberings. It is available with either 22-inch light barrel or 24-inch heavy barrel, weighing 7½ pounds and 9½ pounds respectively. Other specs are essentially the same.

COLT-SAUER

In collaboration with J.P. Sauer & Sohn, W. Germany, Colt introduced the rear-locking Colt-Sauer rifle in late 1972. It is of unusual design, having a non-rotating bolt locked at the rear of the receiver by three retractable locking lugs actuated by a rotating cam on the bolt handle. This new mechanism is very smooth and fast operating. Another key feature of the Colt-Sauer is its patented split receiver, which contributes to the rifle's strength and accuracy. After the barrel is threaded into the split receiver ring, two transverse machine screws are torqued into position, locking the receiver around the barrel. This exclusive attachment method makes the barrel and receiver act as one solid piece of steel. Some additional features include a tang safety that mechanically locks the sear and trigger mechanism; a fully adjustable trigger; an exclusive loaded chamber indicator that shows the hunter when a round has been chambered.

The Colt-Sauer has not been mass produced. Only about 7,000 have been produced annually in this joint venture. Currently there are four models available.

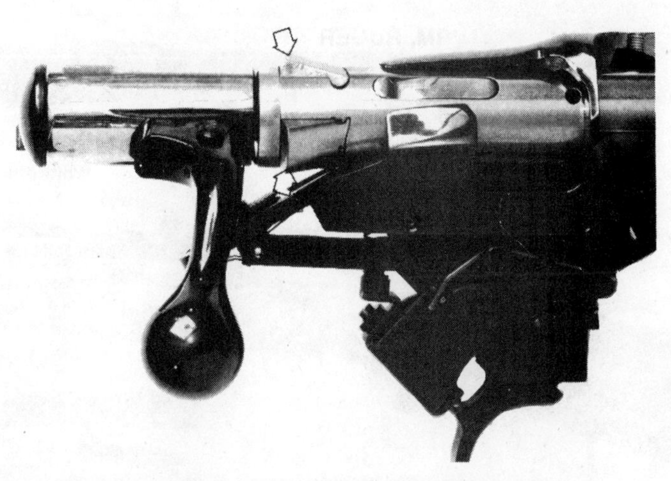

Colt-Sauer rifle with bolt open. Arrows indicate locking lugs.

CHARACTERISTICS OF COLT-SAUER RIFLE

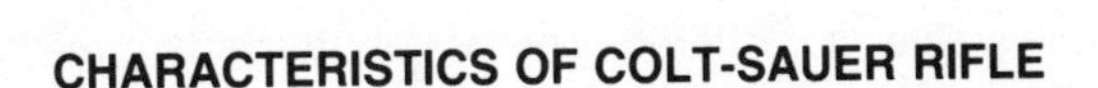

Caliber: .25-06; .270; .30-06; 7mm Remington Magnum; .300 Winchester Magnum; .300 Weatherby Magnum.
Operation by: Manual; rotating bolt handle.
Barrel length: 24 in.
Overall length: 43¾ in.
Weight: 7¾–8½ lbs.
Sights: None; drilled and tapped for scope mounts.
Magazine capacity: 3-round detachable magazine.

CHARACTERISTICS OF COLT-SAUER SHORT ACTION RIFLE

Caliber: .22-250; .243; .308 Winchester (7.62mm NATO).
Operation by: Manual; rotating bolt handle.
Barrel length: 24 in.
Overall length: 43 in.
Weight: 7½–8¼ lbs.
Sights: None; drilled and tapped for scope mounts.
Magazine capacity: 3-round detachable magazine.

Also available are the Grand Alaskan in .375 H&H Magnum and the Grand African in .458 Winchester Magnum. The latter has iron sights; wind and elevation adjustable rear; hooded ramp-style front.

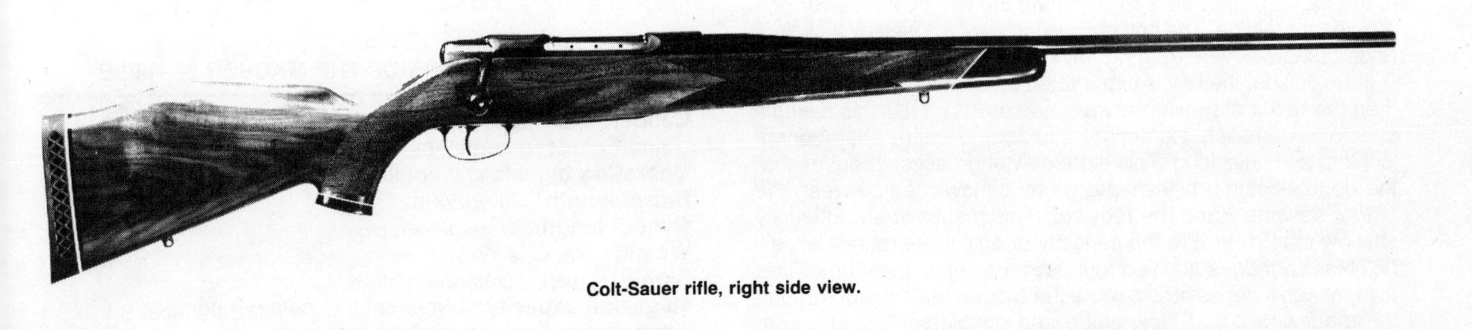

Colt-Sauer rifle, right side view.

HARRINGTON & RICHARDSON

During the early 1960s, Harrington & Richardson added a high-power, bolt-action sporting rifle to its extensive line. This rifle is designated as Model 300, 301, or 330, depending upon finish, barrel and stock. All are built around the SAKO Mauser-type action by Harrington & Richardson. Particulars given under "FN" are applicable. In addition, Harrington & Richardson builds a .17 caliber wildcat rifle on the SAKO L-461 action, for which particulars will be found under "SAKO" elsewhere in this volume. A model 340 was introduced in 1982.

INTERARMS

Interarms currently imports a number of centerfire sporting rifles. The Churchill "One of One Thousand," which was the top of their line for several years, was limited to 1000 rifles. Replacing the Churchill is the Whitworth Express Rifle-African Series. Assembled in the UK using Mark X Mauser actions, the Whitworths are barrelled to handle the heavy bullets of the 7mm Remington Magnum, .300 Winchester Magnum, .375 H & H Magnum, and .458 Winchester Magnum. Interarms also markets five versions of the Mark X Rifle. The latter series will be terminated in 1984, and a yet broader range of Whitworths will be marketed.

Whitworth Express rifle.

Mark X Mauser rifle with Model 4 × 40 telescope.
Mark X Mauser rifle with Model 3-9 × 40 variable telescope.

MAUSER, WEST GERMANY

In 1966, Mauser Werke introduced a radically new, rotating bolt, manually operated, bolt-action magazine rifle. The entire action design is a complete departure from previous Mauser practice. The design is not as new as the date of introduction might indicate; a quite similar prototype was produced by the Mauser plant in WW II. The action is unusually short, measuring only 4½ inches in length. The receiver, as we know it, does not exist. It is replaced by a simple frame containing guide tracks for the bolt and a mortised seat for the quick-detachable barrel. The short bolt body is fitted with a conventional handle near its front end and carries dual, opposed locking lugs. The lugs engage abutments inside a barrel extension, thus locking the bolt to the barrel proper instead of joining the two by a receiver ring as in past practice. The bolt body is largely enclosed by a sliding, nonrotating sleeve engaging tracks on the receiver.

In unlocking, the bolt is first rotated independent of the sleeve; then the two are drawn rearward and thrust forward as a single unit to accomplish extraction, ejection, feeding, chambering, locking, and unlocking. This extremely short action requires that the double-column box magazine by sandwiched between the trigger assembly and the receiver. This results in an unusually shallow magazine with the capacity of only three rounds of .30-06 class cartridges and two rounds in the belted magnums. This also makes it necessary to route the trigger-sear linkage around the magazine box. Single-stage and double-set triggers, both fully adjustable, are offered. The barrel attachment method is unique.

The barrel proper is fitted with a barrel extension that contains locking abutments to engage the locking lugs of the bolt. The lower surface of this extension carries ribs fitting very closely in corresponding grooves in the receiver. And the barrel assembly is secured by a clamp screw. Since the bolt locks into the extension, which is rigidly attached to the barrel, there is no longitudinal stress upon the receiver, so little strength in that direction is necessary. However, accurate alignment is quite important. This method of barrel attachment makes possible quick interchange of barrels in different weights and/or calibers. In fact, the Diplomat Model is supplied cased with barrels of three calibers. Interchangeability of calibers is limited only by magazine shape and size and bolt face dimensions.

CHARACTERISTICS OF THE MAUSER 66 RIFLE

Caliber: Most popular US, from .243 Winchester through .458, and European cartridges.
Operation by: Manual; rotating telescoping bolt.
Barrel length: 20–25½ in.
Overall length: 38–43½ in.
Weight: 6⅝–9⅞ lbs.
Sights: Open, adjustable.
Magazine capacity: 3 rounds; 2 in belted magnums.

Mauser Model 660 rifle.

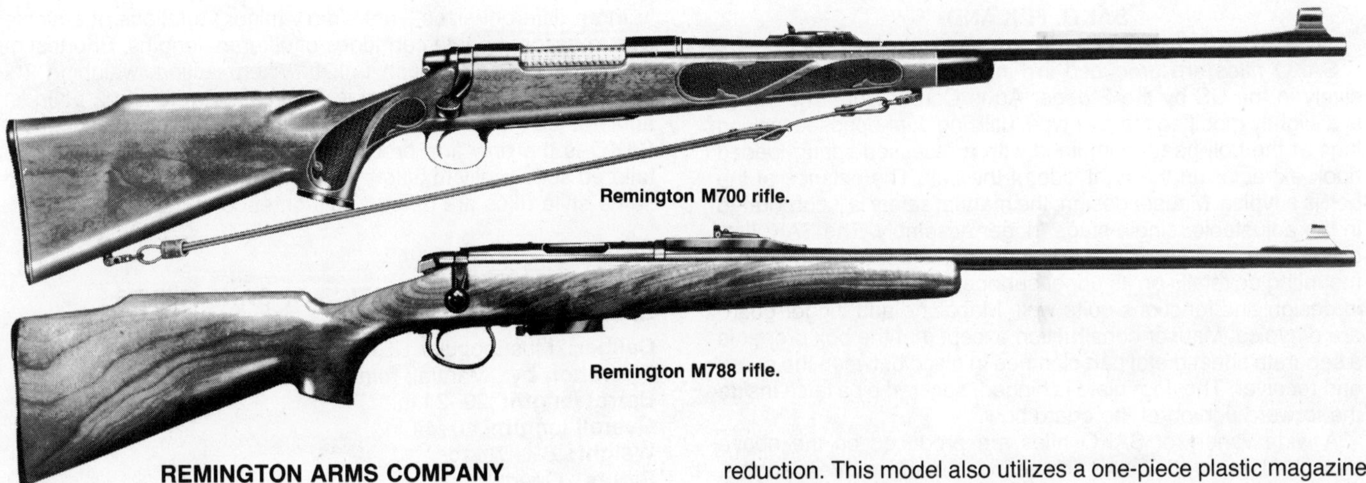

Remington M700 rifle.

Remington M788 rifle.

REMINGTON ARMS COMPANY

The Remington M700 rifle is simply a logical improvement of the M721-722 series introduced in 1948. The M700 makes full use of design features and production techniques that were considered quite advanced, if not almost revolutionary, in sporting arms in 1948. Other manufacturers who first decried such things have since adopted many of the features and techniques introduced at that time by Remington. The bolt is essentially Mauser in character with dual-opposed locking lugs at the head and helical cocking cam at the rear. However, it utilizes a spring-clip type extractor, fully enclosed in a recess inside the bolt face. The extractor design permits complete enclosure of the head of the cartridge case within a counterbore within the face of the bolt. The ejector is a Garand type, spring-loaded plunger in the bolt face. The receiver is round in section, the traditional integral recoil lug being replaced by a barrel bracket protruding downward, sandwiched between barrel and front face of receiver ring.

The magazine is of Mauser type, though simplified by utilizing a separate sheet metal box, sandwiched between floor plate and receiver. The fully adjustable single-stage trigger assembly of the M700 is considered by many to be the best offered on any commercial sporting rifle, and the assembly incorporates a manual safety that lies closely alongside the rear receiver tang and does not interfere with scope mounting.

CHARACTERISTICS OF THE REMINGTON M700 RIFLE

Caliber All popular US.
Operation by: Manual; rotating bolt.
Barrel length: 22–26 in.
Overall length: 42½–46½ in.
Weight: 7–7½ lbs.
Sights: Open, adjustable.
Magazine capacity: 3–5 rounds.

The M40X series of bench-rest and target rifles is built upon the basic M700 action in both single-shot and magazine form.

There is also the M660 Carbine, which possesses all of the basic design features of the M700 but is made shorter to accommodate smaller cartridges and to achieve maximum weight reduction. This model also utilizes a one-piece plastic magazine box/trigger guard.

In 1967, Remington introduced the M788 rifle, built upon an entirely new action designed for maximum rigidity and accuracy and for the most economical production of advanced techniques. The bolt is designed for screw machine production and carries nine small locking lugs in three rows of three just forward of the bolt handle. The balance of the bolt design is mechanically identical to that of the M700. The receiver is simply a thick-wall, steel tube containing ejection and feeding cutouts and abutments against which the locking lugs bear. The trigger and safety mechanism are virtually identical to that of the M700. However, a detachable sheet metal box magazine and stamped trigger guard are utilized. Aside from the differences in location of locking lugs, functioning and manipulation are the same as in the M700. Both the M700 and M788 rifles are available in left-hand versions, and the Model 700 Safari is available in .375 H&H and .458 Winchester Magnum.

CHARACTERISTICS OF THE REMINGTON M788 RIFLE

Caliber: .222, .22-250, 6mm Remington; .243; .308
Operation by: Manual; rear-locking bolt.
Barrel length: 22–24 in.
Overall length: 41½–43½ in.
Weight: 7–7½ lbs.
Sights: Open, adjustable.
Magazine capacity: 4 rounds; 5 in .222 caliber.

REMINGTON MODEL SEVEN BOLT-ACTION RIFLE

Caliber: .222 Remington (5-shot), .243, 7mm-08, 6mm, .308 (4-shot).
Barrel: 18½ in.
Weight: 6¼ lbs.
Length: 37½ in. overall.
Stock: Walnut, with modified Schnabel fore-end. Cut checkering.
Sights: Ramp front, adjustable open rear.
Features: New short-action design; silent side safety; free-floated barrel except for single pressure point at fore-end tip. Introduced 1983.

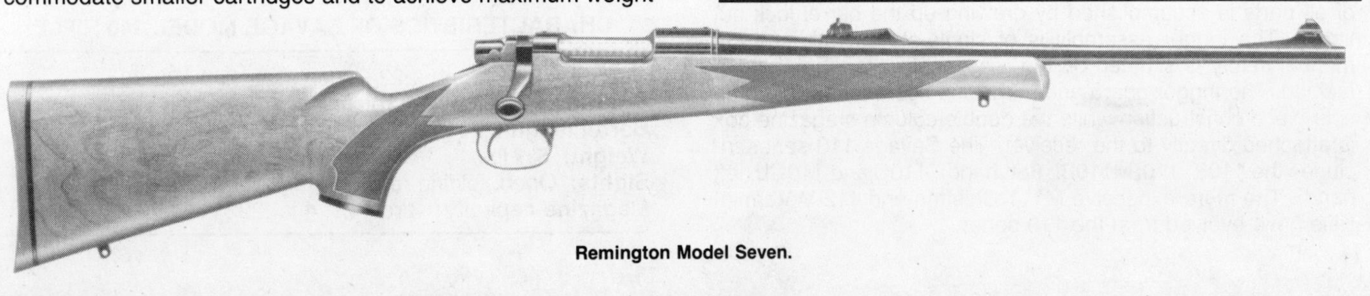

Remington Model Seven.

SAKO, FINLAND

SAKO rifles are produced in Finland and distributed exclusively in the US by the Stoeger Arms Corporation. The action is a slightly modified Mauser type, utilizing dual-opposed locking lugs at the bolt head, combined with a recessed spring-loaded hook extractor on the right side of the bolt. The balance of the bolt is a typical Mauser design; the manual safety is incorporated in the adjustable, single-stage trigger assembly. The SAKO receiver is tubular in section and carries integral tapered scope mounting dovetails on its upper surface. The bolt stop is original in design and functions quite well. Magazine and trigger guard are of typical Mauser construction except that the box proper is a separate sheet metal part clamped in place between the guard and receiver. The floor plate is hinged, secured by a latch inside the forward portion of the guard bow.

A wide variety of SAKO rifles are produced on the above described action; however, the action is made in three lengths. All major parts of the three different actions are dimensionally different so that there is actually a single-action design produced in three different sizes—not simply minor variations of a single size to accommodate cartridges of different lengths. Shortest of the series is the 6½-inch L-461 (Vixen) action, weighing 2⅛ pounds; the L579 (Forester) is 7⅜ inches, weighing 2 pounds; and the L61 (Finnbear) is 8⅜ inches, weighing 2¾ pounds. SAKO is the only firm producing so complete a line of actions tailored specifically to different cartridge lengths. Essentially the same style rifles are offered built on each of the actions.

CHARACTERISTICS OF SAKO RIFLES

Caliber: Most popular US from .222 upward.
Operation by: Manual; rotating bolt.
Barrel length: 20–24 in.
Overall length: 40–44 in.
Weight: 6½–7½ lbs.
Sights: Open.
Magazine capacity: 4 rounds; 3 in belted magnums

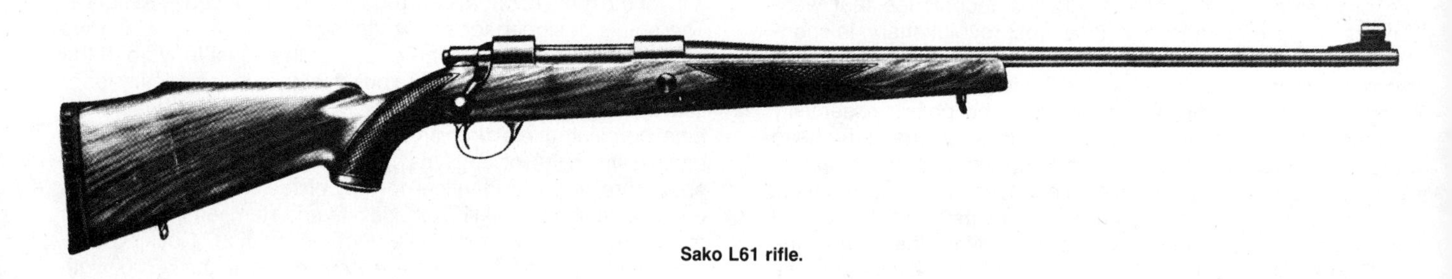

Sako L61 rifle.

SAVAGE ARMS COMPANY

In 1958, Savage introduced the Savage Model 110, a highly modified, Mauser-type bolt-action rifle. The bolt contains an unusual number of parts; the body alone consists of the head, front and rear baffles, friction washer, retaining pin and handle—all separate pieces. The firing mechanism consists of 10 parts, exclusive of the sear. All parts are, however, designed for maximum production economy and result in low overall cost. The design is noteworthy in that there are two baffles to obstruct the rearward flow of gas should a case or primer rupture. The Savage extractor is similar in principle to the Remington enclosed type but is fitted in the outside of the bolt head, with the claw entering the bolt face counterbore through a slot. The cocking piece is housed completely within the bolt, and a pin protrudes from it to engage a Mauser-type cocking cam in the side of the bolt body ahead of the handle. The unusual sear protrudes upward on the right side of the bolt to engage the cocking-piece pin and has an external projection, which serves as a cocking indicator and also as a bolt release.

The receiver is tubular, essentially Mauser form, fitted with a separate recoil lug in Remington fashion. The barrel screws into the receiver, sandwiching the recoil lug between the two, but it is also fitted with a threaded lock nut. The barrel proper does not screw up tightly in assembly as in other makes; final clamping of all parts is accomplished by drawing up the barrel lock nut tightly. The trigger assembly is of single-stage type, and the manual safety is situated on the rear receiver tang in shotgun fashion. The trigger guard and magazine floor plate are of multiple-piece construction while the double-column magazine box is attached directly to the receiver. The Savage 110 series includes the 110E, 110B, 110BL (left hand), 110C and 110CL (left hand). The more expensive M111 Chieftan and 112-V Varmint Rifle have evolved from the 110 series.

CHARACTERISTICS OF SAVAGE MODEL 110 RIFLE

Caliber: .243, .270, .308 Winchester; .30-06; 7mm Remington Magnum.
Operation by: Manual; rotating reciprocating bolt.
Barrel length: 20–22 in.
Overall length: 40½–45 in.
Weight: 6¾–8 lbs.
Sights: Open; folding rear.
Magazine capacity: 4 rounds; 3 in belted magnums.

Savage also produces the Model 340, an economy-priced centerfire rifle of distinctive design. The bolt utilizes a single large locking lug at its head, and the bolt handle functions as a safety lug by virtue of being seated deep into a recess in the receiver. Cocking and extraction are accomplished in Mauser fashion, and the receiver is tubular, requiring a minimum of machining. A pivoted manual safety is fitted to the right rear of the receiver. The trigger is a simple, single-stage design. A detachable, single-column box magazine is installed ahead of the trigger guard.

CHARACTERISTICS OF SAVAGE MODEL 340 RIFLE

Caliber: .22 Hornet; .222 Remington; .223; .30-30.
Operation by: Manual; rotating single-lug bolt.
Barrel length: 20–22 in.
Weight: 6½ lbs.
Sights: Open; folding rear.
Magazine capacity: 3 rounds; 4 in .222 caliber.

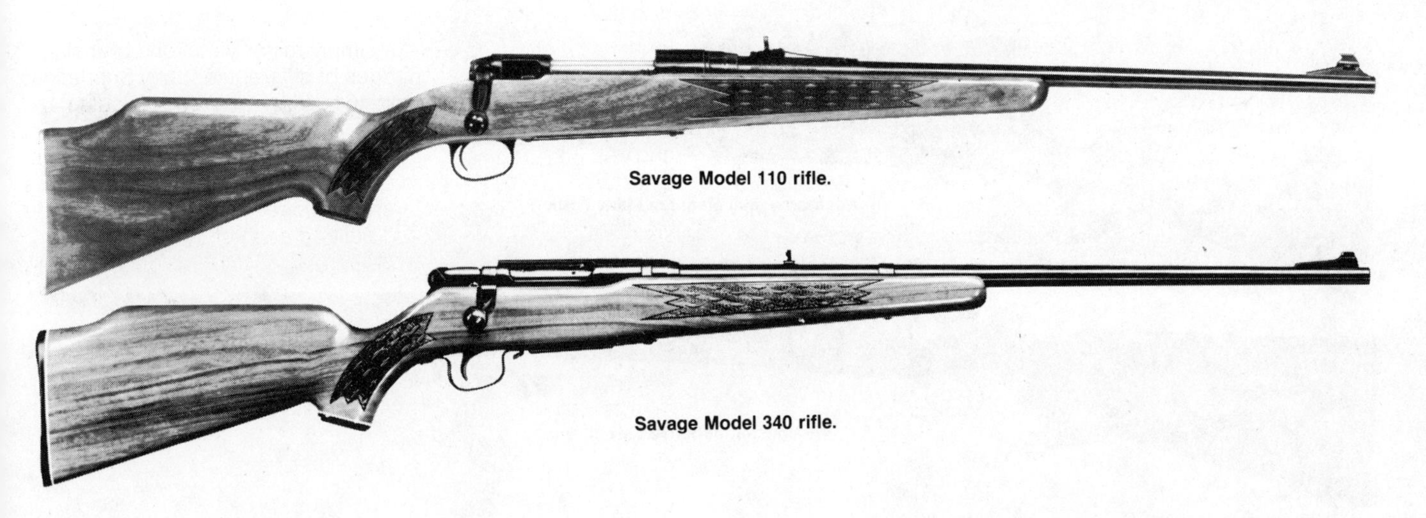

Savage Model 110 rifle.

Savage Model 340 rifle.

STEYR-MANNLICHER, AUSTRIA

Mannlicher sporting rifles have been produced at Steyr-Daimler-Puch in Austria since the early 1900s. The basic sporting rifle is based upon the military M1903 model described in Chapter 22. Numerous improvements have been made on the original action, and the current production model is the Steyr-Mannlicher introduced in 1968. This is a lightweight design intended to follow in the tradition of the Mannlicher-Schoenauer carbine, both in styling and in the more desirable mechanical features. The locking system consists of six rearward lugs locking into the receiver bridge. Cocking is accomplished in the Mauser manner. Only minimum cutouts are made in the receiver to permit ejection and feeding, and the rear of the bolt is closed by a streamlined sleeve which blends into the lines of the receiver. A detachable rotary-type magazine is employed. A fully adjustable single-stage trigger is offered with a European-styled double-set trigger optional. Particularly unusual is the appearance of the barrel, which is hammer-forged around a rifling mandrel. The forging marks on the exterior are retained, resulting in a many-faceted surface, rather than the usual polished barrel finish. The models SL and L are based on a short action suitable for cartridges from the .222 Remington to the .308 Winchester. Two other lengths are being produced—the Model M for .30-06 class cartridges and the Model S for the larger cartridges up through the .375 and .458 Magnums. The Steyr SSG sniper rifle is described in Chapter 8.

CHARACTERISTICS OF THE STEYR-MANNLICHER RIFLES

Caliber: Most popular US and European rifle cartridges.
Operation by: Manual; Mannlicher-style bolt.
Barrel length: 20–25½ in.
Overall length: 40–46 in.
Weight: 7¼–8½ lbs.
Sights: Open; hooded ramp front.
Magazine capacity: 5 rounds.

STURM, RUGER AND COMPANY, INC.

Introduced in 1968, the Ruger M77 was the company's first entry in the bolt-action field. On the surface, it looks more like a cleaned-up M98 Mauser than anything else, but several advance design and production features are incorporated. The bolt is typically Mauser, including the long nonrotating extractor and bolt-guide rib. The firing pin and cocking system is also typically Mauser. The safety, however, is eliminated from the bolt sleeve and placed on the rear receiver tang in shotgun fashion. The receiver is somewhat slab-sided and intricate in contour and carries a Mauser-type bolt stop at the left rear.

Scope-mounting dovetails are formed integrally with the top of the receiver, and Ruger supplies rings to match. The single-stage trigger is quite crisp and fully adjustable. The trigger guard/magazine assembly resembles the Winchester M70 and is held in place by three guard screws. An unusual feature is that the head of the forward guard screw is angled rearward so that tightening it not only pulls the receiver down into the stock, but rearward, seating the recoil lug solidly in the wood. The magazine floor plate is hinged and held in place by a quick-release latch in the front of the trigger guard bow. Five different variants of the Model 77 are being made.

CHARACTERISTICS OF RUGER M77 RIFLE

Caliber: .22-250; .220 Swift 6mm Remington; .243, .308, .284 Winchester; 6.5mm, .350 Remington Magnum; .458 Winchester Magnum.
Operation by: Manual; rotating reciprocating bolt.
Barrel length: 22 in.
Overall length: 42 in.
Weight: 6½ lbs.
Sights: None; furnished with scope bases.
Magazine capacity: 5 rounds; 3 in magnum calibers.

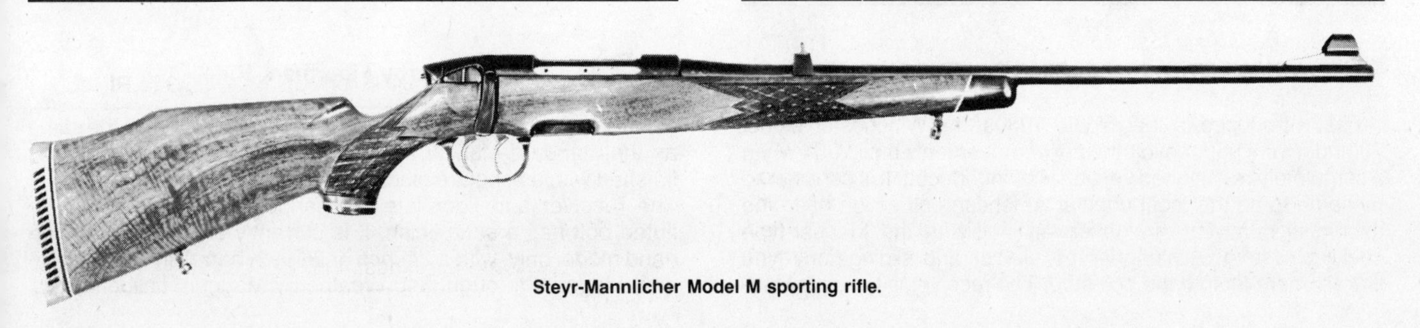

Steyr-Mannlicher Model M sporting rifle.

Weatherby .300 Magnum Mark V rifle.

Winchester Model 70 Target.

WEATHERBY, INC.

The first proprietary Weatherby rifle design, the Mark V, was introduced in 1958. At that time, it represented a radical departure from current action design. The bolt body is a massive cylinder, carrying three small locking lugs, each on a reduced diameter portion at the head. Lugs and bolt body are both of the same major diameter. The body contains longitudinal flutes, which reduce weight and friction and provide escape areas for dust and dirt. The bolt face is deeply counterbored to surround the cartridge-case head completely. The only opening into the counterbore is a slot for the recessed hook extractor. The rear of the bolt is closed by a streamlined sleeve, which completely encloses the cocking piece, except for an indicator tongue that protrudes from underneath the sleeve when the gun is cocked.

A manual safety is pivoted to the right side of the bolt sleeve. A sheet metal box magazine is held between the receiver and Mauser-style trigger guard, to which is fitted a hinged floor plate. The trigger is of single-stage design and is fully adjustable. Weatherby rifles are characterized by high-gloss finishes and unusual stock wood. Also, they are generally chambered only for the Weatherby proprietary cartridges. The Mark V action is made in two sizes, the smaller being adapted to cartridges of the .224 Weatherby and .22-250 class; the larger action handles all other calibers.

CHARACTERISTICS OF WEATHERBY MARK V RIFLE

Caliber: The full line of special Weatherby cartridges; any other on special order.
Operation by: Manual; Weatherby rotating bolt.
Barrel length: 24–26 in.
Overall length: 42½–46½ in.
Weight: 6½–10½ lbs.
Sights: None.
Magazine capacity: 4 rounds; 3 in .378 and .460 calibers.

WINCHESTER REPEATING ARMS COMPANY

First introduced in the middle 1930s, the Winchester Model 70 underwent minor evolutionary improvements until 1964 when a completely redesigned version was introduced. It is considered by some to be the most popular American bolt-action rifle. The design is relatively straightforward, utilizing the Mauser-type cocking system. A plunger-type ejector and sliding/hook-type extractor are let into the bolt face. The receiver is typically Mauser, containing locking abutments at its front and threaded for barrel attachment. The magazine is of basic Mauser type and is considerably modified for economy of fabrication. The trigger mechanism is adjustable. Varients of the Winchester Model 70 include the Model 70A, Model 70 Target rifle, Model 70 African and Model 70 International Army Match Target rifle. Winchester rifles are currently made under license by the United States Repeating Rifle Company.

CHARACTERISTICS OF WINCHESTER M70 RIFLE

Caliber: All popular US.
Operation by: Manual; rotating bolt.
Barrel length: 19–24 in.
Overall length: 39½–44½ in.
Weight: 7½–8½ lbs.
Sights: Open, adjustable.
Magazine capacity: 5 rounds; 3 in belted magnums.

GENERAL

In addition to the individual makes and models of bolt-action rifles already listed and described, a number of firms import rifles marked with their own name but manufactured abroad as minor variations of existing models. These rifles are normally based on one of the basic Mauser M98–type actions. In some instances, actual assembly takes place in the US. The actions may be of new European manufacture, refurbished military actions, or commercial actions purchased in partly finished condition. Generally speaking, these rifles are identical in characteristics, operation and loading and firing procedures to the M98 or FN Mauser or their variations. Some names under which these rifles may be encountered are Colt, Reinhart Fajen, Herter's, L. A. Distributors, Smith & Wesson and Dan Wesson Arms.

Weatherby Fibermark Rifle

Introduced in 1983, the Weatherby Fibermark Rifle is the same as the standard Mark V except the stock is of fiberglass. It is finished with a nonglare black wrinkle finish and black recoil pad. The receiver and floorplate are sandblasted and blued. The fluted bolt has a satin finish. It is currently available in a right-hand model only, with a 24-inch or 26-inch barrel, in .240 Weatherby Magnum through .340 Weatherby Magnum calibers.

LEVER-ACTION RIFLES

BROWNING

One of the newer, and hence more modern, lever-action rifles is Browning's BRL, which features a rotating bolt head with multiple locking lugs, recessed bolt face, and side ejection. For safety, the rifle has a three-position hammer, trigger disconnect system, and inertia firing pin.

CHARACTERISTICS OF BROWNING BLR

Caliber: 22-250; .243; .258 Roberts; 7mm-08; .308; .354 Winchester.
Operation by: Swinging finger lever; reciprocating bolt.
Barrel length: 19⁷⁄₁₆ in.
Overall length: 38⅝ in.
Weight: 7 lbs.
Sights: Open, adjustable; tapped for scope mounts.
Magazine capacity: 4 rounds, detachable.

MARLIN FIREARMS

The design for the Marlin M336 originated in 1889, making it the oldest still produced in the US. It has undergone considerable improvement but has changed little since 1948 when the breech bolt was changed from square to round section. The M36 then became the M336. It utilizes a solid receiver, vertical sliding lock and reciprocating bolt, side ejection, tubular under-barrel magazine, and a two-piece stock. Operation is by a swinging finger lever forming the trigger guard. An unusual feature is the two-piece firing pin, which is aligned by the locking block to prevent firing when the breech is not fully locked. Since 1972, this rifle has also been made in .45-70 as the M1895.

CHARACTERISTICS OF MARLIN M336 RIFLE

Caliber: .30-30 Winchester; .35 Remington; .444 Marlin; .45-70 (M1895). Marlin Model 336 Extra Range Carbine is available in .307 and .356 Winchester.
Operation by: Swinging finger lever; reciprocating, rear-locked bolt.
Barrel length: 20–24 in.
Overall length: 38½–42¼ in.
Weight: 7–7½ lbs.
Sights: Open, adjustable.
Magazine capacity: 5–7 rounds.

The Marlin 1894 is actually a new (1969 introduction), modernized copy of the original M1894 once made for the short .44-40 cartridge. Mechanically, physically and functionally, the M1894 differs from the M336 only in bolt section (square) and its full-length bolt slot in the receiver wall. Operation and handling are as described for the M336. Also available are the Marlin 1895 in .45-70 caliber and the M444 in caliber .444.

CHARACTERISTICS OF MARLIN M1894 RIFLE

Caliber: .44 Magnum.
Operation by: Swinging finger lever; reciprocating bolt.
Barrel length: 20 in.
Overall length: 37½ in.
Weight: 6 lbs.
Sights: Open, adjustable.
Magazine capacity: 10 rounds.

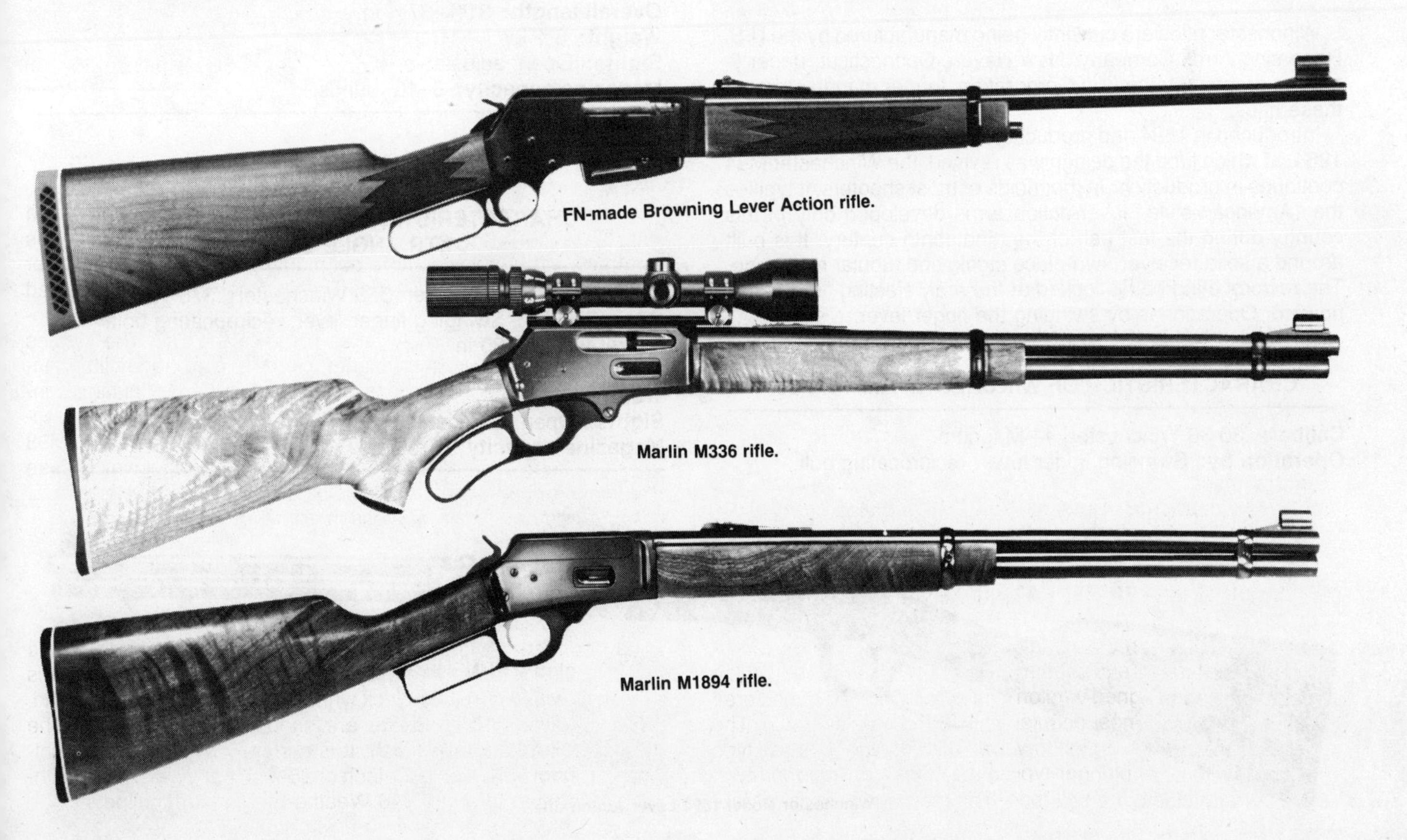

FN-made Browning Lever Action rifle.

Marlin M336 rifle.

Marlin M1894 rifle.

O. F. MOSSBERG & SONS

In 1973, Mossberg introduced a traditional Western-style lever-action carbine design, the M472. In both appearance and design, it greatly resembles the Marlin M336. It is the only exposed-hammer lever-action rifle that utilizes a separate manual safety instead of a safety position on the hammer. It also uses a vertical rising block-locking system similar to the Marlin & Winchester and functions generally the same. The M472 was reintroduced as the Model 479 in 1983.

CHARACTERISTICS OF MOSSBERG M472

Caliber: .30-30 Winchester.
Operation by: Manual; finger lever
Barrel length: 20 in.
Overall length: 38½ in.
Weight: 7¼ lbs.
Sights: Open.
Magazine capacity: 6 rounds.

SAVAGE ARMS

The Savage M99 has been in continuous production since its introduction in 1899, and many minor improvements have been made during that time. Actuated by a swinging finger lever, the M99 has a solid receiver, tipping, rear-locked reciprocating bolt, and rotary magazine. Many minor variations have been produced, and the current M99C utilizes a detachable double-column box magazine. Also currently marketed are the Savage 99-A and 99-E.

CHARACTERISTICS OF SAVAGE M99 RIFLE

Caliber: .243; .284; .308; .358 Winchester; .300 Savage.
Operation by: Swinging finger lever; tipping reciprocating bolt.
Barrel length: 20–22 in.
Overall length: 39¾–41¾ in.
Weight: 6¼–6¾ lbs.
Magazine capacity: Rotary, 5; box, 4; .284 caliber, 3.

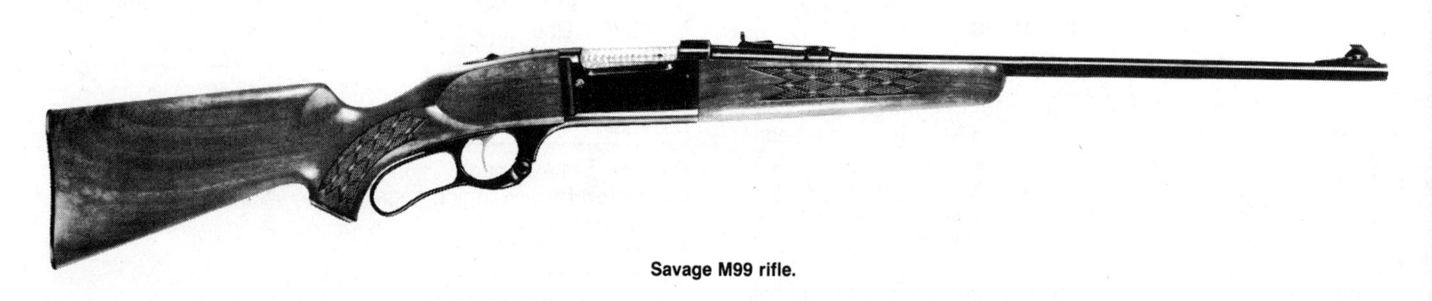

Savage M99 rifle.

U.S. REPEATING ARMS COMPANY

Winchester rifles are currently being manufactured by the U.S. Repeating Arms Company, New Haven, Connecticut, under licenses granted by the Olin Corporation, former manufacturer of these rifles.

Introduced in 1894 and produced virtually without change until 1964, at which time the design was revised, the Winchester M94 continues in production. In the minds of most shooters, it typifies the "American-style" lever-action arms developed only in this country during the last half of the nineteenth century. It is built around a solid receiver, two-piece stock, and tubular magazine. The reciprocating bolt is locked at the rear, ejecting fired cases upward. Operation is by swinging the finger lever.

CHARACTERISTICS OF WINCHESTER M94 RIFLE

Caliber: .30-30 Winchester; 44 Magnum.
Operation by: Swinging finger lever; reciprocating bolt.
Barrel length: 16 and 20 in.
Overall length: 31¾–37¾ in.
Weight: 6½ lbs.
Sights: Open, adjustable.
Magazine capacity: 6–10 rounds.

CHARACTERISTICS OF WINCHESTER M94 XTR ANGLE EJECT

Caliber: .307 Winchester; 356 Winchester; .375 Winchester.
Operation by: Swinging finger lever; reciprocating bolt.
Barrel length: 20 in.
Overall length: 38⅝ in.
Weight: 7 lbs.
Sights: Open, adjustable.
Magazine capacity: 6 rounds.

Winchester Model 1894 Lever Action rifle.

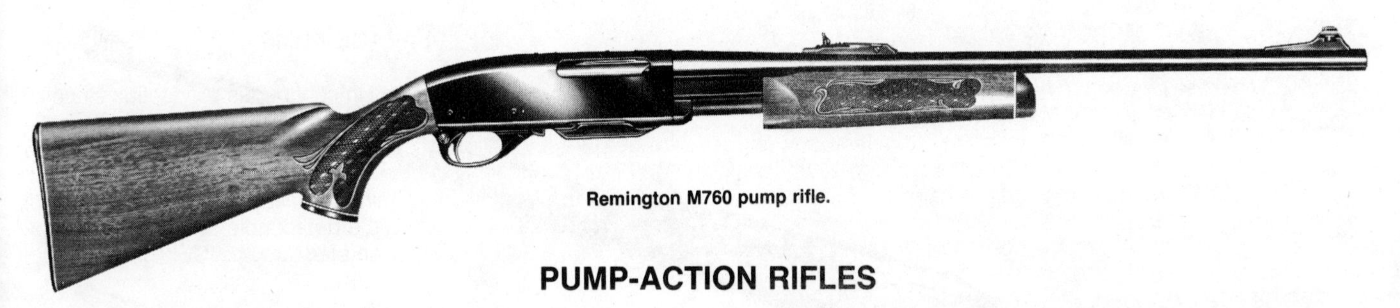

Remington M760 pump rifle.

PUMP-ACTION RIFLES

REMINGTON ARMS COMPANY

Introduced in 1952 to replace the earlier M14 and M141 pump rifles, the Remington M760 was the first such action capable of handling full power cartridges in the .30-06 class. It remains so. Except that power to operate the action is supplied by manual reciprocation of the forestock, it is identical mechanically, physically, and functionally to the Remington M742. Most internal parts are interchangeable between the two.

CHARACTERISTICS OF REMINGTON M760 RIFLE

Caliber: 6mm Remington; .243, .270, .308 Winchester; .30-06.
Operation by: Reciprocating forestock; rotating bolt.
Barrel length: 22 in.
Overall length: 42 in.
Weight: 7¼ lbs.
Sights: Open, adjustable.
Magazine capacity: 4 rounds.

SEMIAUTOMATIC RIFLES

BROWNING

In 1967, the US Browning firm introduced its first centerfire self-loading rifle in this country, the Browning BAR. It is produced by Fabrique Nationale. The design utilizes a short-stroke gas system with a two-piece reciprocating bolt and a characteristically Browning double-hook firing mechanism. It uses a solid receiver, two-piece stock and an unusual hinged box magazine. Present models have the distinction of being the only self-loading design chambered for belted magnum cartridges.

CHARACTERISTICS OF BROWNING BAR

Caliber: .243, .270, .308 Winchester; .30-06; 7mm Remington Magnum; .300 Winchester Magnum; .388 Winchester Magnum.
Operation by: Gas; reciprocating rotating bolt.
Barrel length: 22 in.; 24 in. for magnum calibers.
Overall length: 43½ in; 45½ in. for magnum calibers.
Weight: 7⅜ lbs; 8½ lbs. for magnum calibers.
Sights: Open, adjustable.
Magazine capacity: 4 rounds; 3 in belted magnums.

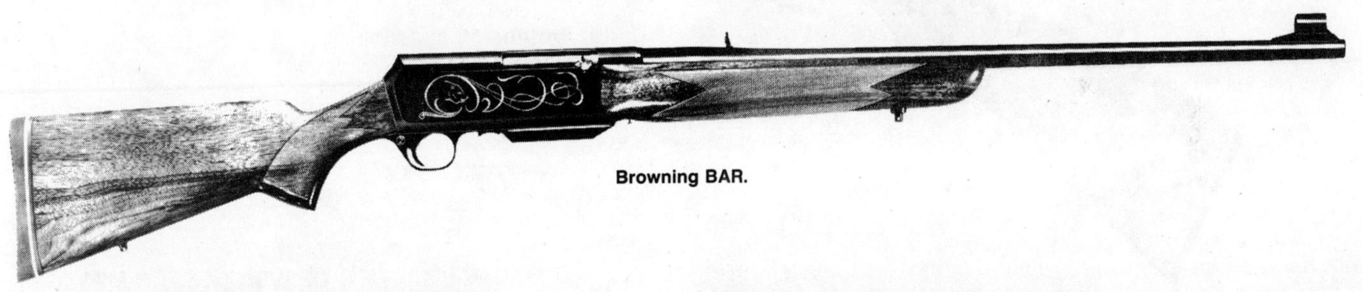

Browning BAR.

HECKLER & KOCH, GERMANY

Building upon their growing family of roller-locked military rifles and pistols, Heckler and Koch has added several semiautomatic, Monte Carlo walnut stocked rifles to their sporting line—the 630, 770, 940 and SL7-SL6.

CHARACTERISTICS OF HK MODEL 630

Caliber: .223.
Operation by: Delayed recoil; roller lock.
Overall length: 39⁷⁄₁₀ in.
Weight: 7³⁄₁₀ lbs.
Sights: Open, adjustable; telescopic.
Magazine capacity: 4 and 10 rounds

CHARACTERISTICS OF HK MODEL 770

Caliber: .308.
Operation by: Delayed recoil; roller lock.
Overall length: 42½ in.
Weight: 7⁷⁄₁₀ lbs.
Sights: Open, adjustable; telescopic.
Magazine capacity: 4 and 10 rounds.

CHARACTERISTICS OF HK MODEL 940

Caliber: .30-06.
Operation by: Delayed recoil; roller lock.
Overall length: 47 in.
Weight: 8.8 lbs.
Sights: Open, adjustable; telescope.
Magazine capacity: 3 and 10 rounds.

CHARACTERISTICS OF HK SL7 AND SL6

Caliber: SL7 in 7.62mm NATO; SL6 in 5.56mm.
Operation by: Delayed recoil; roller lock.
Overall length: 39¾ in.
Weight: 8 lbs.
Sights: Hooded post front, adjustable diopter rear.
Magazine capacity: 10 rounds.

From top to bottom: .223 HK 630; .308 HK 770 and .30-06 940 self-loading rifles.

7.62mm NATO HK SL7 self-loading carbine.

REMINGTON ARMS COMPANY

The Remington M742 was the first successful full power and the first gas-operated semiautomatic to be offered on the American market. It was introduced in 1955 as the M740 and subsequently redesignated the M742 in 1960 as improvements were added. It utilizes a typical under-barrel gas cylinder and piston; the latter connects to twin operating rods that carry a two-piece, reciprocating, multiple-lug bolt at their rear. It utilizes a solid receiver with side ejection and two-piece stock.

An improved version of the 742 was introducd in 1981; this latter rifle is designated the Model Four Auto Rifle.

CHARACTERISTICS OF REMINGTON MODEL FOUR

Caliber: .243 Winchester; 6mm Remington; .270; 7mm Express Remington; .308; and .30-06.
Operation by: Gas; reciprocating rotating bolt.
Barrel length: 22 in.
Overall length: 42 in.
Weight: 7½ lbs.
Magazine capacity: 4 round.

STURM, RUGER AND COMPANY

The Ruger .44 Carbine is the only semiautomatic rifle chambered for the .44 magnum revolver cartridge. Introduced in 1961 as a short-range deer rifle, it is gas operated, utilizing a short-stroke floating piston, heavy operating slide, and short one-piece rotating reciprocating bolt. Uniquely, it is the only centerfire self-loader using a tubular, under-barrel magazine. It is styled after the US M1 Carbine, capitalizing on that gun's popularity.

CHARACTERISTICS OF RUGER .44 AUTOLOADING CARBINE

Caliber: .44 Remington Magnum.
Operation by: Short-stroke gas; rotating bolt.
Barrel length: 18½ in.
Overall length: 37 in.
Weight: 5⅔ lbs.
Sights: Open, adjustable.
Magazine capacity: 4 rounds.

The Ruger 5.56mm Mini-14 semiautomatic rifle is described in Chapter 1.

CHARACTERISTICS OF RUGER MINI-14 CARBINE

Caliber: .223 (5.56mm).
Barrel length: 18½ in.
Overall length: 37¼ in.
Weight: 6.4 lbs.
Magazine capacity: 5 round; but larger capacity military magazines will fit.

"MEDVED" [THE BEAR], USSR

A product of M. T. Kalashnikov's Design Bureau, the "Medved" is a well-made, high-quality, gas-operated sporting rifle. Unlike its military counterpart, the Dragunov (Chapter 43), the Bear, chambered for a larger caliber, has a conventional sporting-type wood stock and a smaller capacity magazine—five shots instead of 10. The soft-nosed, jacketed 9mm bullet (.354 in.) is fired from the 7.62 × 54R case, which has been enlarged to take the larger diameter projectile. Traveling at an instrumented velocity of 2120 f.p.s., the 231.5-grain bullet is quite accurate at normal hunting ranges.

CHARACTERISTICS OF "MEDVED"

Caliber: 9 × 54Rmm
Operation by: Gas; rotating bolt (Kalashnikov type).
Barrel length: 20⅘ in.
Overall length: 43⅜ in.
Weight: 7¼ lbs.
Sights: Open, adjustable; telescopic, 3½x.
Magazine capacity: 5 rounds.

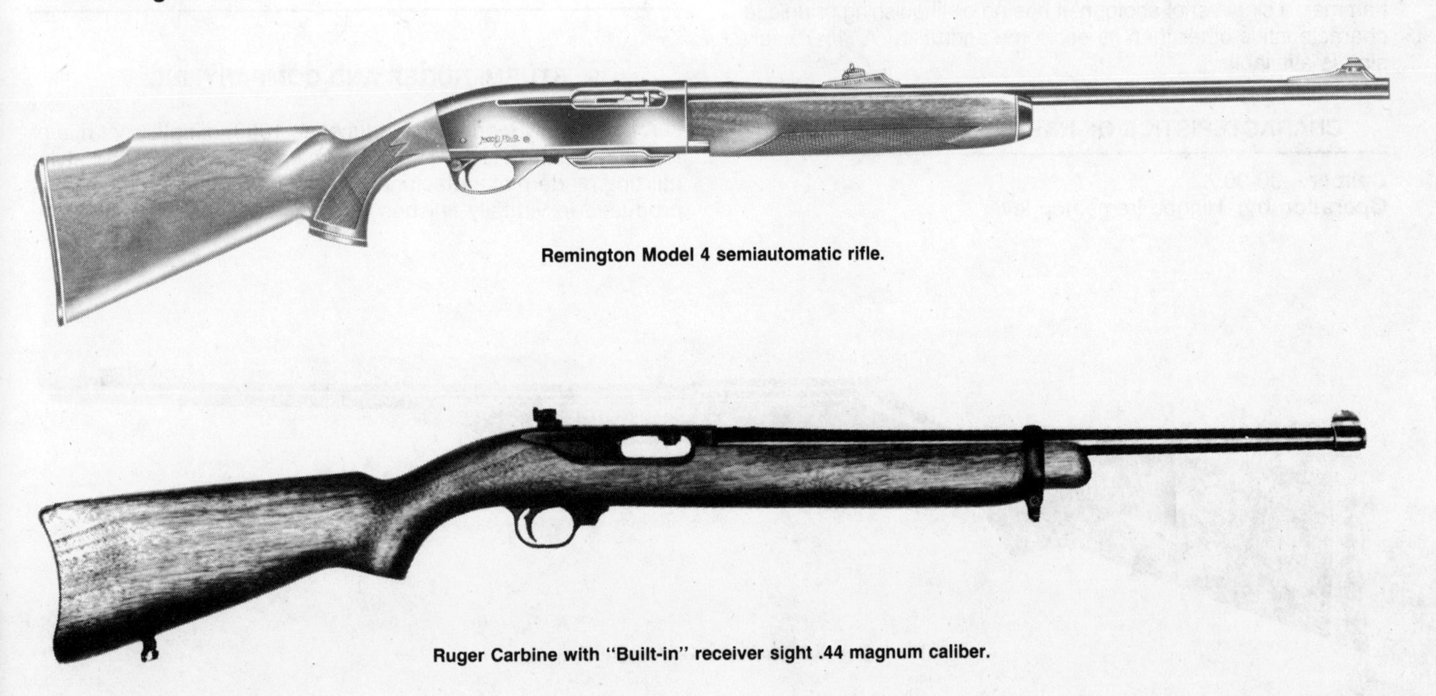

Remington Model 4 semiautomatic rifle.

Ruger Carbine with "Built-in" receiver sight .44 magnum caliber.

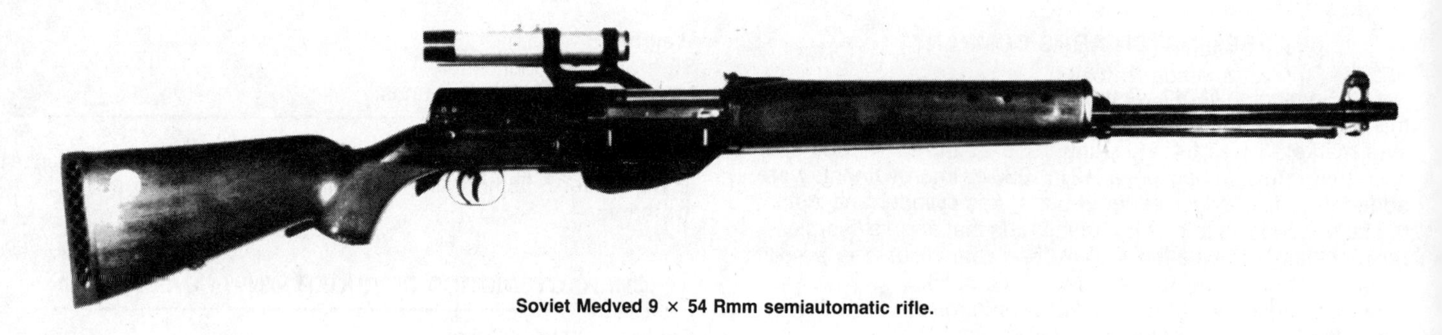

Soviet Medved 9 × 54 Rmm semiautomatic rifle.

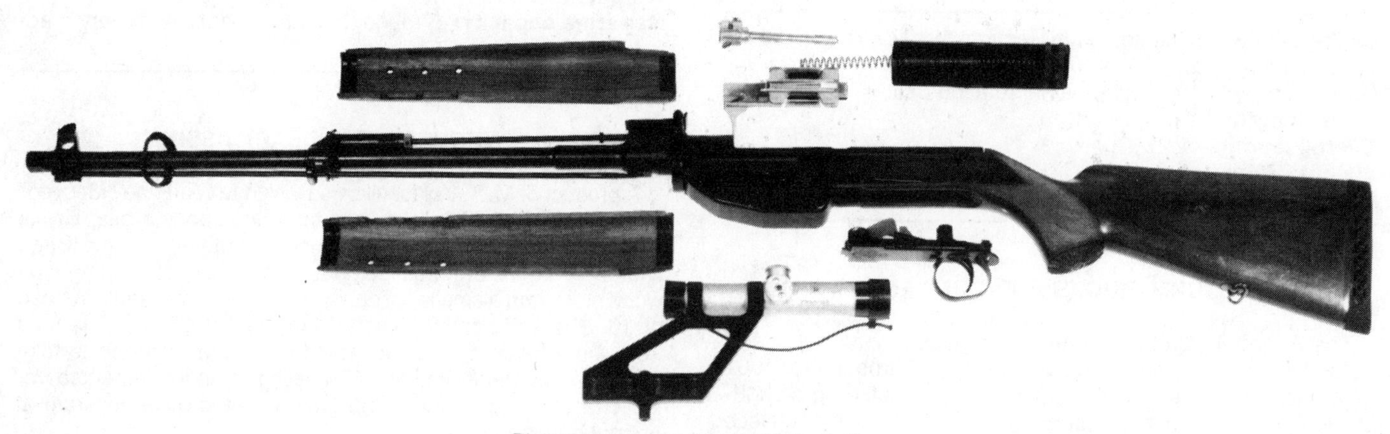

Disassembled view of Soviet Medved rifle.

SINGLE-SHOT RIFLES

HARRINGTON & RICHARDSON

A very basic single-shot caliber .30-30 rifle, the Topper Model 158 is based upon the standard H&R break-open, exposed-hammer, a single-shot shotgun. It has no distinguishing or unique characteristics other than its economy and utility. A .45-70 version is available.

CHARACTERISTICS OF H&R TOPPER MODEL 158

Caliber: .30-30.
Operation by: Hinged frame; top lever.
Barrel length: 24 in,; optional 20-gauge barrel.
Overall length: 37½ in.
Weight: 5½ lbs.
Sights: Open.

STURM, RUGER AND COMPANY, INC.

A completely modern (introduced in 1968), single-shot rifle of top quality, the Ruger No. 1 is designed in the classic style, utilizing modern manufacturing techniques. Even the receiver is produced in virtually finished form by investment casting. The

Harrington & Richardson M158 single-shot rifle.

action combines features of many others but most resembles the British Farquharson. It features a separate bar to support the forestock, a two-piece stock, a through-bolted buttstock, a $\frac{1}{4}$-length barrel rib containing telescope bases, an internal hammer firing mechanism, a cocking indicator, a selective ejector-extractor, a powerful extraction and adaptability to any centerfire sporting cartridge. A .45-70 version is available as the No. 3 rifle.

CHARACTERISTICS OF RUGER NO. 1 RIFLE

Caliber: Any standard US centerfire on special order.
Operation by: Falling-block; under-lever actuated.
Barrel length: 22-26 in.
Overall length: 42 in. (26 in. bbl.).
Weight: $6\frac{3}{4}$–$8\frac{1}{2}$ lbs.
Sights: None; fitted for telescope bases.

SHOTGUNS

As with rifles, there are literally dozens of shotguns marketed in the United States and Europe. Most of the military shotguns described in Chapter 5 are reworked commercial guns. In recent years, the commercial shotguns are basically of the single barrel, double barrel (side-by-side or over and under), pump, and self loading.

SELF-LOADING CARBINES

Many shooters and collectors have long had a fascination with submachine guns, but these have generally been unavailable because they are fully automatic weapons and thus controlled by the revenue taxes associated with the GCA 68. The first of the semiautomatic carbines derived from a submachine gun were the MAC 10 and MAC 11. These guns are marketed as pistols because they have 5.75 and 5.06 inch barrels and because they do not have stocks. These MAC submachine guns were made until mid-1982 when the Bureau of Alcohol, Tobacco and Firearms ordered the manufacturer, RPB Industries, to cease production. The BATF's concern was the ease with which this slam fire open-bolt weapon could be converted to full automatic fire. There have been similar concerns expressed about the Interdynamics KG-9, which is marketed by the F.I.E. Corporation. This latter "pistol" has the receiver molded from high-impact black plastic.

The second type of semiautomatic carbines is exemplified by the long barrel UZI (16.1 inch). These stocked weapons have been converted to closed-bolt firing and equipped with a legal rifle length barrel. A semiautomatic version of the Sterling submachine gun is also available in the United States.

The semiautomatic MAC 10 (*left*) and the selective fire MAC 10 (*right*) both in .45 ACP (11.43 × 23mm).

The 9 × 19mm UZI semiautomatic carbine. Note the 16.1 inch barrel.

HANDGUNS

As in the case of rifle development, many handguns originally appeared as military sidearms. Basic patterns such as the Colt M1911, Colt M1917, Walther P38/P1, FN Browning High Power, SIG 210 and SIG 220 are described under the country of their origin in Part II. More recent military developments, some of which have target applications, are discussed in Chapter 4. The text below presents representative handguns currently available to the sporting public. While some shooters and small arms companies continue to experiment with new calibers, the standard revolver and pistol cartridges of the past two decades still hold sway.

REVOLVERS

CHARTER ARMS

Charter Arms introduced its Undercover Model in the mid-1960s in an effort to capture part of the revolver market held by Colt and Smith & Wesson. Their revolver is a solid frame, swingout cylinder, double/single action revolver of minimum weight and size. The design dispenses with a frame side plate, access being provided by removing the trigger guard. It is unique in that the cylinder may be unlatched by either pressing a thumb piece forward or pulling forward on the extractor rod. In 1973, a .44 Special version was introduced. A .357 magnum target revolver and a .32 Undercoverette were added to the line subsequently.

CHARACTERISTICS OF CHARTER ARMS UNDERCOVER

Caliber: .38 Special; .32 S&W Long; .44 Special.
Operation by: Manual; double or single action.
Barrel length: 2–3 in.
Overall length: 6¼–7¼ in.
Weight: 16–17 oz.
Sights: Open, fixed.
Cylinder capacity: 5 rounds; 6 in .32.

COLT INDUSTRIES

The Firearms Division of Colt Industries is one of the oldest revolver-makers in the world, and their handguns reflect that heritage. Their 11 1977 center-fire revolvers are built on four frame types—P, D, I and J. For convenience, the Colt revolvers will be described by variations within frame type.

Charter Arms revolver.

Model P Revolvers

Known variously as the Colt Single-Action Army, the Peacemaker and the Frontier Six-Shooter, this revolver has been in production in its present form since 1873. The basic action design dates before 1850, and the Model P in .45 Long Colt was the standard US Army sidearm for many years. Produced in 36 calibers during its long career, the Peacemaker is currently manufactured in only two calibers—.357 Magnum and .45 Colt. Colt offers two variants of the Model P—the Single-Action Army, which has the classic lines of the 1873 model, and the New Frontier Single-Action, which also has adjustable target sights. The latter revolver is available only in 5½- and 7½-inch barrels.

CHARACTERISTICS OF THE COLT SINGLE-ACTION ARMY REVOLVER

Caliber: .357 Magnum; .45 Colt.
Operation by: Manual, single-action only.
Barrel length: 4¾ in. to 5½ in.; 7½ in.
Overall length: 10¼ in. to 13 in.
Weight: 36 to 43 oz.
Sights: Open, fixed.
Cylinder capacity: 6 rounds.

The basic popularity of the Model P is reflected in the many copies that have been produced in the past three decades. Those still in production include:

Cattleman—Made for LA Distributors in West Germany, this is a single-action revolver of solid-frame, rod ejection "Frontier" style identical to Colt and Ruger models. Operation and loading-firing instructions for those guns apply.

Hawes Western Marshal—.357 Magnum/9mm; .44 Magnum/.44–40; .45 Long Colt/.45 ACP.

Sturm, Ruger Blackhawk: See below.

Virginian Dragoon—Now made by Interarms in Midland, Virginia, this revolver was formerly manufactured in Switzerland. It is available in .357 Magnum, .44 Magnum, and .45 Colt. Barrel lengths available vary with caliber but include 5-, 6-, 7½-and 8-inch.

CHARACTERISTICS OF THE DEPUTY REVOLVER

Caliber: .22 RF; .22 MRF; .357 Magnum; .44 Magnum.
Operation by: Manual; single-action only.
Barrel length: 4¾in. to 7½ in.
Overall length: 10½ in. to 13½ in.
Weight: 30 to 42 oz.
Sights: Open, fixed. Target type on some versions.
Cylinder capacity: 6 rounds.

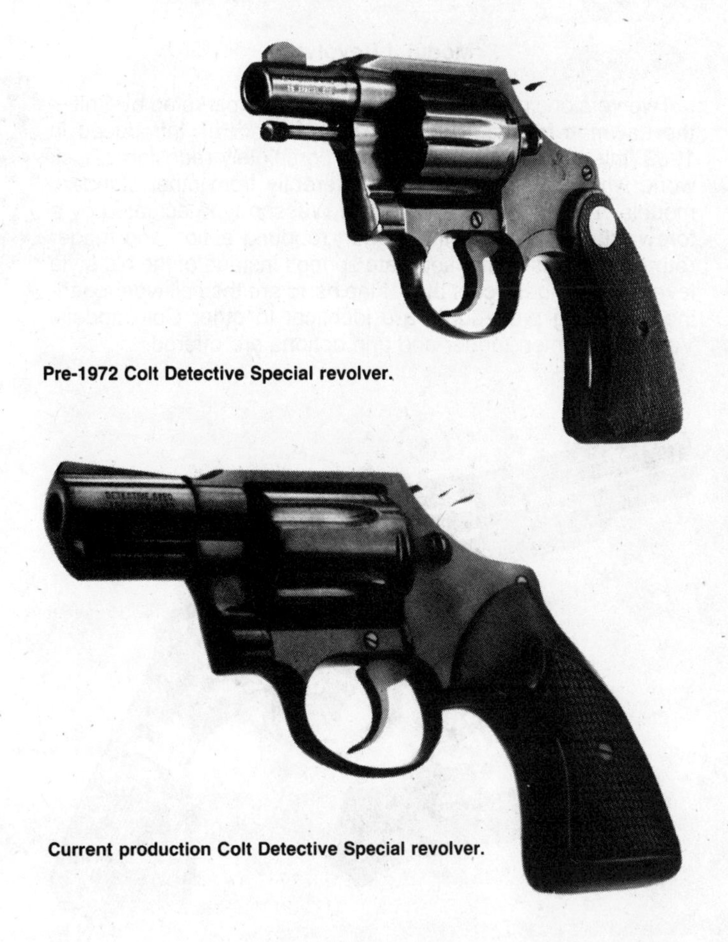

Pre-1972 Colt Detective Special revolver.

Current production Colt Detective Special revolver.

Model D Revolvers

There are six variations of this .38 Special caliber solid-frame, swing-out cylinder, double/single-action revolver—Detective Special, Cobra, Agent, Diamondback, Viper (new 1977) and Police Positive (new 1977 version). This design dates before 1900, and although it has been product improved from time to time the basic frame and lock work remain the same. The most significant mechanical change was the addition of a safety hammer block during the 1920s. (A bar is interposed between hammer face and frame except when the trigger is held fully rearward. This prevents the firing pin from reaching a cartridge or in rebounding-pin models prevents the hammer from reaching the firing pin.)

In 1972, Colt altered the barrel profile of its D Model revolvers, adopting a heavy Python style barrel, which has an ejector rod enclosure and ventilated rib.

CHARACTERISTICS

	Barrel length	Overall length	Weight	Frame
Detective Special:	2 in.	7 in.	22½ oz.	Steel
Cobra:	2 in.	7 in.	16⅝ oz.	Aluminum alloy
Agent:	2 in.	6⅞ in.	16⅜ oz.	Aluminum alloy
Diamondback:	2½–4 in.	7½–9 in.	24–27½ oz.	Steel*
Viper:	4 in.	9 in.	20 oz.	Aluminum alloy
Police Positive:	4 in.	9 in.	26½ oz.	Steel

*Looks like a scaled down Python.

Model I Revolver

Built upon the .41 caliber big brother of the Model D frame, the .357 Magnum caliber Python is one of the leading police handguns in the United States. Available in 2½-, 4- and 6-inch barrel lengths, this steel frame pistol ranges in weight from 33 oz. for the shortest barrel to 43½ oz. for the longest barrel.

Model J Revolvers

Two versions of the Model J are currently marketed by Colt—the Lawman Mark III and the Trooper Mark III. Introduced in 1969, this revolver action contains completely redesigned lock work, which in detail functions differently from other standard models. The bolt is of the Smith & Wesson type actuated by a foreward projection on the trigger; rebound action and trigger return are furnished by separate springs instead of the old style lever. All springs except that of the hand are the coil type. Loading and firing procedures are identical to other Colt models. Various hammer, trigger and grip options are offered.

CHARACTERISTICS OF THE COLT TROOPER MARK III REVOLVER

Caliber: .357 Magnum; .38 Special
Operation by: Manual, double- or single-action.
Barrel length: 4 in. to 6 in.
Overall length: 9½ in. to 11½ in.
Height: 5⅝ in.
Weight: 38⅓ oz. to 42 oz.
Sights: Open, micrometer-adjustable target type.
Cylinder capacity: 6 rounds.

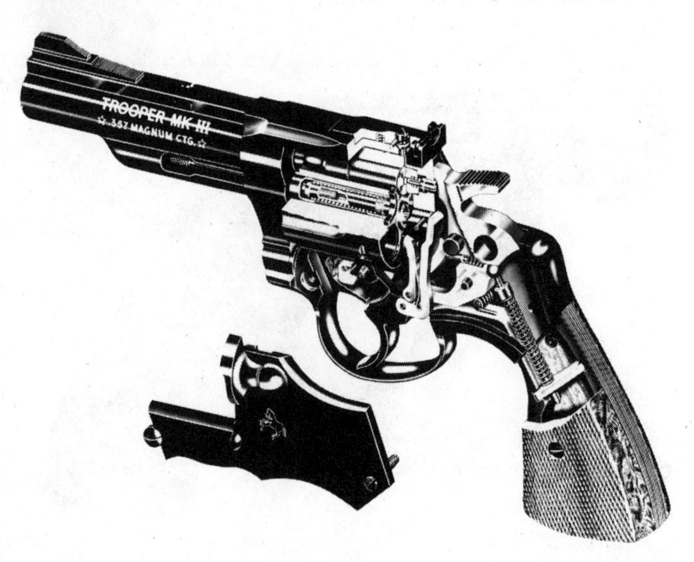

Colt Trooper Mark III revolver.

I Frame Colt .357 Magnum Caliber Python revolver.

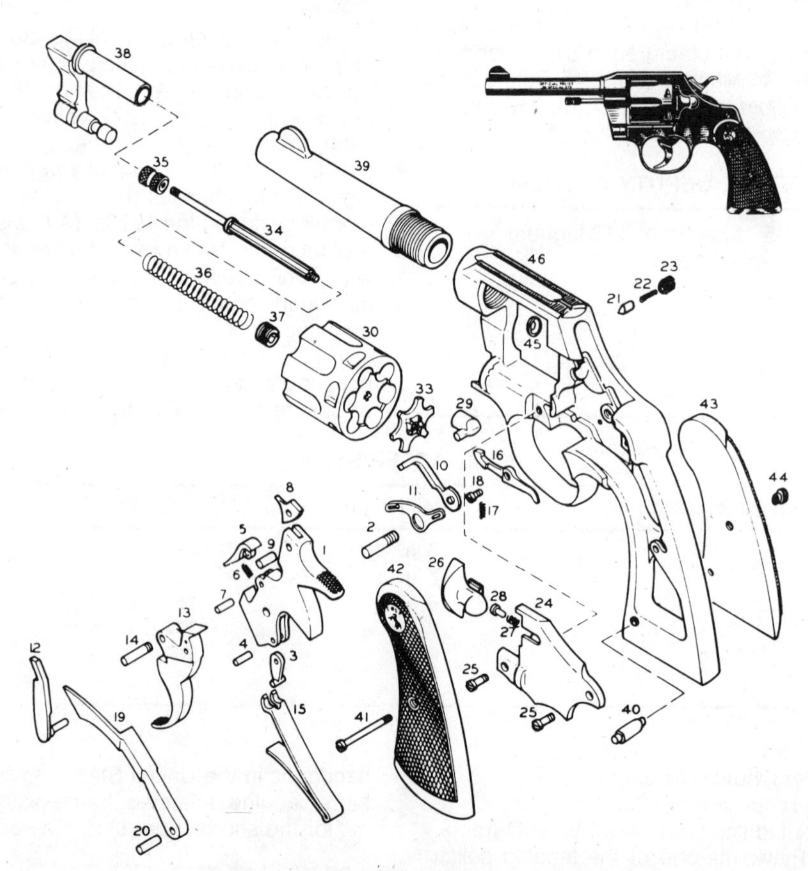

Exploded view of typical Colt Model I revolver.

SMITH & WESSON

Smith & Wesson solid-frame, side-swing, exposed-hammer, double/single-action revolvers are known and respected throughout the world. The basic design is identical in virtually all models, except the Bodyguard and Centennial. The design dates to the turn of the century in the 1902 Military and Police Model with relatively minor improvements in the late 1920s and early 1950s. The various models are produced on three different size frames—the .32 size of the 5-shot guns; the .38 size of the 6-shot guns through .38-.357 caliber; and the .44 size Magnum frame of the 6-shot .41, .44, .45 and heavy .357 guns.

The basic design as represented by the US M1917 .45 revolver and the .38/200 K200 "Victory Model" is covered in Chapters 45 and 10. Present commercial models differ only in minor areas, such as the shortened hammer fall introduced in 1950, micrometer-adjustable sights on target models and frame mounted rebounding firing pins in the rimfire models. Most of the .32 and .38 frame models are available as "Airweight" with aluminum alloy frames.

Smith & Wesson is a pioneer in the development of stainless steel handguns. S&W introduced the M60 "Stainless Chief" in 1965 and in 1972 began delivery of the M64, M66 and M67 revolvers, stainless steel versions of the M10, M15 and M19 respectively. Specifications are identical with their plain steel counterparts.

Characteristics of the Smith & Wesson Revolvers

	.32 Frame (1)	.38 Frame (K) (2)	.44 Frame (M) (3)
Caliber:	.22; .32; .38.	.22; .32; .38; .357.	.357; .44; .45; .41.
Barrel length:	2 in. to 6 in.	2 in. to 8⅜ in.	3½ in. to 8⅜ in.
Overall length:	6¼ in. to 10½ in.	6⅛ in. to 12⅞ in.	9⅜ in. to 13¾ in.
Weight:	*17 to 20 oz.	*28 to 42½ oz.	42 to 51½ oz.
Cylinder capacity:	5 rounds.	6 rounds.	6 rounds.

*Airweight models weigh approximately ⅓ less in 2-in. barrel length.
(1) Terrier, Chief Special, Kit Gun.
(2) M&P, K22-32-38, Combat Masterpiece, Combat Magnum.
(3) .357, .41, and .44 Magnums; .44 Mil; .45 Target, .38/44 Outdoorsman.

S&W Bodyguard, Model 38

Built on the .32 frame and differs from the basic model only in that frame side-walls are extended to surround the hammer. The hammer is altered in profile and the checked spur protrudes through a slot just enough to permit single-action cocking, if desired. It is intended primarily for double-action use.

CHARACTERISTICS OF THE S&W MODEL 38 PISTOL

Caliber: .38 Special.
Operation by: Manual; double- or single-action.
Barrel length: 2 in.
Overall length: 6⅜ in.
Sights: Open, fixed.
Weight: 20½ oz.; Airweight 14½ oz.
Cylinder capacity: 5 rounds.

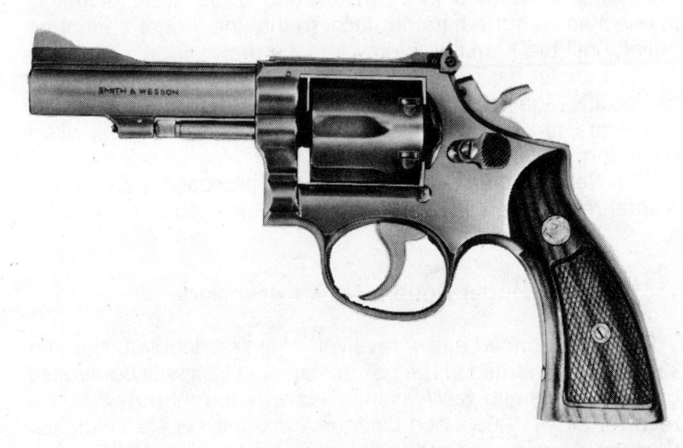

Smith & Wesson Stainless Steel M67.

Smith & Wesson Bodyguard revolver.

STURM, RUGER AND COMPANY INC.

Ruger Blackhawk

This is a modern center-fire revolver of traditional American single-action "Frontier" styling. In appearance and manipulation, it is identical to the Colt SAA. Mechanically, however, it differs considerably. The grip frame is a one-piece casting rather than a two-screw assembly of parts. The lock work, while quite similar and functioning the same as in the Colt, is redesigned to use coil and torsion springs throughout. The entire gun in all versions is designed for modern production techniques, especially precision casting and screw-machine fabrication.

In addition to the Blackhawk, there is a smaller-frame version made for rimfire cartridges and designated "Single Six."

CHARACTERISTICS OF THE RUGER REVOLVERS

	Single Six	Blackhawk
Caliber:	.22 RF; .22 MRF.	.30 Carbine; .357; .41; .44 Magnum.
Operation by:	Manual, single-action only.	—
Barrel length:	5½ in. to 9½ in.	4⅝ in. to 7½ in.
Overall length:	10⅞ in. to 14⅞ in.	10¼ in. to 13⅜ in.
Weight:	36 to 38 oz.	38 to 48 oz.
Sights:	Open, fixed*	Open, adjustable target type
Cylinder capacity:	6 rounds.	6 rounds.

*Super version has target-type adjustable sights.

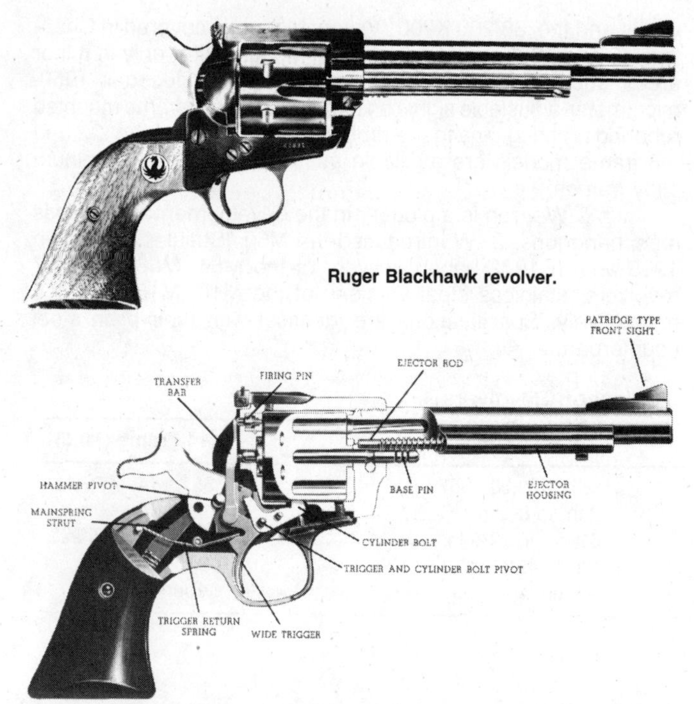

Ruger Blackhawk revolver.

Ruger New Model Super Single-Six. Action cutaway—mechanism shown with trigger pulled and hammer beginning to fall. Transfer bar is held in firing position, between hammer and firing pin. Cylinder bolt is engaged in cylinder locking notch.

Ruger New Model Super Single-Six. Action cutaway—mechanism shown at rest. Transfer bar is not in line between hammer and firing pin. This is the normal carrying position. All six chambers may be loaded.

Ruger New Model Super Blackhawk. Caliber .44 Magnum, 7½-inch barrel.

Ruger produced the Blackhawk (.357 Magnum) and Super Blackhawk (.44 Magnum) revolvers from 1955 and 1959 respectively until 1972. The New Model Blackhawk and Super Blackhawk revolvers, introduced in 1973, have the same outward appearance as the earlier pistols. The internal mechanism, however, has been redesigned for greater safety. Although only the .22 rimfire Single Six is illustrated below, the figures and descriptions are valid for the Blackhawk and Super Blackhawk revolvers as well.

In the Ruger New Model single-actions, the hammer has only two positions—all the way forward, or fully cocked. In the first illustration, the hammer is shown fully forward; it rests directly on the frame. There is no "safety" notch, nor is there a "loading" notch. In this figure, the trigger has been released, and the transfer bar, which is pivoted at its lower end to an arm of the trigger, has been lowered, removing any possible connection between the hammer and the firing pin. In this position, the hammer surrounds but cannot touch the firing pin, and the revolver can be carried with all six chambers loaded.

The second illustration shows the revolver with the trigger pulled and the hammer beginning to fall. The trigger is in its rearmost position, and the transfer bar is raised so that its upper tip is immediately behind the firing pin. When the hammer falls, it will drive both the upper end of the transfer bar and the firing pin forward to ignite the cartridge. As the hammer is cocked, it cams the trigger almost to its extreme rearward position. It is this motion of the trigger that raises the transfer bar. Unless the trigger is held to the rear as the hammer falls, the transfer bar will automatically lower, preventing the revolver from firing.

The firing mechanism of the New Model includes a component usually found only in the double-action revolver—the transfer bar. It is the medium through which the energy of the hammer blow is transmitted to the firing pin to strike the cartridge primer. The hammer itself never touches the firing pin, which, as in all Ruger revolvers, is mounted in the cylinder frame.

The transfer bar is pivoted at its lower end to the trigger. When the trigger is held to its rearmost position, the top of the transfer bar is in line between the hammer and the firing pin. Therefore, unless the trigger is held to the rear, the hammer blow (or in the case of a fall, a blow to the hammer) is not transmitted to the firing pin.

When the hammer is down, the top of the transfer bar is interlocked with the hammer face so that the trigger cannot be pulled until the hammer is drawn to the rear.

The transfer bar system provides an unprecedented measure of security against accidental discharge, and the New Model Revolvers can under all normal conditions be carried with all six chambers loaded.

The New Model revolvers meet the proposed US governmental requirements for hammer safety drop tests.

Ruger Double-Action Revolvers

Ruger introduced a new revolver, their first double-action design, in 1972, aimed at the police market, which was dominated by Colt and Smith & Wesson. (This was the third new police revolver since 1965 when Charter Arms entered the field; Dan Wesson Arms began marketing their contender in 1970.) The Ruger Security-Six has a standard square butt, while the Speed-Six has a round butt. Both are available in blued steel and stainless steel. The Speed-Six is available with only 2¾- and 4-inch barrels, while the Security-Six has a 6-inch barrel option. Wide use of investment castings, a Ruger trademark, and the absence of screws (a coin is all that is required for disassembly) contribute to an excellent design. As with the Ruger single-action revolvers, the transfer bar concept in the trigger mechanism makes the weapon very safe to handle.

CHARACTERISTICS OF THE RUGER SECURITY-SIX

Caliber: .357 Magnum
Operation by: Manual, single/double-action.
Barrel length: 2¾, 4 and 6 in.
Overall length: 8, 9¼ and 11¼ in.

Weight: 33 oz. (4-in. barrel).
Sights: Open, adjustable.
Cylinder capacity: 6 rounds.

There is a third variant in this family of revolvers—the Police Service-Six. Like the Speed-Six, it has fixed sights and is available in .38 Special and 9mm Parabellum calibers.

STURM, RUGER & CO., INC.
Southport, Connecticut, U.S.A.

Ruger Speed-Six, full right view.

DAN WESSON ARMS

The double-action revolver introduced in 1969 by Dan Wesson Arms, Inc. of Manson, Massachusetts, is the third new revolver to recently appear on the US market. Dan Wesson, a direct descendant of Daniel B. Wesson of the original Smith & Wesson firm, has no tie with the Springfield, Massachusetts, company. Conventional in appearance, the Dan Wesson revolvers differ considerably from other pistols available today.

Most unusual is its rapid interchange of barrels. Whereas changing barrels on revolvers has heretofore been a task for a gunsmith, the Dan Wesson people have made it a relatively simple procedure. The barrel proper is a simple, thin-walled, rifled tube threaded on both ends. It is screwed into the receiver against a feeler gage laid across the cylinder face. The gage insures proper cylinder/barrel gap. A separate housing is slipped over the barrel to butt against the frame, and a muzzle nut is tightly turned on the barrel to lock both barrel and housing securely in place. Barrels of 2½-, 4-, 6- and 8-inch lengths, each with its own housing, may be freely exchanged on the same gun in about one minute. As in the current line of Colt revolvers, the ejector rod is protected by the barrel housing.

Dan Wesson revolvers also differ in that the cylinder latches at the front of the frame. The latch moves vertically to act directly on the swing-out crane. Having been designed primarily for double-action work (though capable of single-action), the distance the hammer travels is 25% less than in other contemporary designs.

There are two basic series of Dan Wesson revolvers—.38 Special and .357 Magnum calibers. The standard service revolver model in .38 Special is Model 8 and in .357 Magnum Model 14. The target model in .38 Special is Model 9 and in .357 Magnum Model 15. Model 14–22 designates the .357 service revolver with a 2½-inch barrel. Model 15–28 equals the .357 target model with an 8-inch barrel.

CHARACTERISTICS OF THE DAN WESSON MODEL 14

Caliber: .357 Magnum.
Type: Solid frame, swing-out cylinder, double- and single-action, manually operated.
Barrel length: 2½, 4, 6, 8 in.
Overall length: 9¼ in. with 4-in. barrel.
Height: 5¾ in.
Weight: 34 oz. with 4-in. barrel.
Sights: Open.
Cylinder capacity: 6 rounds.

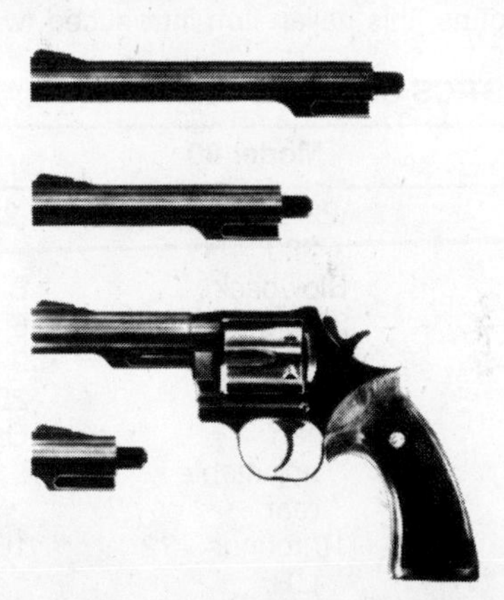

Model 14 Dan Wesson revolver with four interchangeable barrels.

Foreign Copies of Domestic Models

There are numerous foreign manufactured revolvers that are either outright copies or only slightly modified versions of basic modified models. They include copies of both Colt and Ruger single-actions as well as Smith & Wesson and Colt double-actions. Some are encountered regularly in the United States, others not often. Generally speaking, there is sufficient resemblance to domestic models that they are readily recognized. Some are of good quality, others are not, and each should be examined and judged on its own merits.

SEMIAUTOMATIC PISTOLS

BERETTA

With the exception of the Brigadier and M90, all Beretta pistols are based on the M1934 design, described and pictured in Chapter 29. Manipulation and functioning of the Jaguar, Puma and Sable are essentially the same as for the Beretta 9mm Model 1934.

CHARACTERISTICS OF THE BERETTA PISTOL

	Jaguar	Puma	Model 76 Sable	Model 70S*
Caliber:	.22 L.R.	.32 ACP.	.22 L.R.	.380.
Operation by:	Blowback.	Blowback.	Blowback.	Blowback.
Barrel length:	6 in.	6 in.	6 in.	3⅝ in.
Overall length:	8¾ in.	9½ in.	8¾ in.	6½ in.
Sights:	Open, adjustable rear.	Open, adjustable rear.	Open, adjustable rear.	Open, adjustable rear.
Weight:	19 oz.	19 oz.	26 oz.	n/a
Magazine capacity:	10 rounds.	10 rounds.	10 rounds.	7 rounds.

*A similar model called the Cougar cannot under existing laws be imported into the US.

Beretta Brigadier

This is simply a commercial version of the 9mm M1951 pistol adopted by the Italian Army. It is chambered for the 9mm Parabellum (Luger) cartridge and is described and pictured in detail in Chapter 29.

Model 90

In 1971, Beretta introduced a very modern double-action .32 caliber pocket-size autoloader designated M90. Of blowback design, it features a stainless steel removable barrel, concentric recoil spring and exposed hammer. With a 3⅝-inch barrel, it is 6¾ inches overall. The magazine holds 8 rounds.

BERNADELLI

When the 1968 Gun Control Act imposed new standards on imported handguns, this Italian firm introduced two versions of a new centerfire pistol for the US market—the Models 80 and 90. Externally similar to the Walther PPK, these guns have the following characteristics.

CHARACTERISTICS OF BERNADELLI AUTOMATIC PISTOLS

	Model 80	Model 90
Caliber:	.22 L.R.; 32; 380.	.22 L.R.; 32; 380.
Operation by:	Blowback.	Blowback.
Barrel length:	3½ in.	6 in.
Overall length:	6½ in.	9 in.
Weight:	26.8 oz.	25.6 oz.
Sights:	Open, adjustable rear.	Open, adjustable rear.
Magazine capacity:	10 rounds, .22 L.R. 8 rounds, .32 ACP. 7 rounds, .380 ACP.	10 rounds, .22 L.R. 8 rounds, .32 ACP. 7 rounds, .380 ACP.

Model 90 Bernadelli .380 caliber pistol.

COLT INDUSTRIES

In their 1977 line, Colt offered four variants of the M1911A1 automatic pistol—two Commanders and two Mark IV/Series '70 handguns. See US M1911A1 Pistol for mechanical and functional details.

Colt Commander

Introduced in 1951, this is essentially a shortened version of the .45 Government Model. Initially available with an aluminum alloy frame as a spinoff from experimental models developed

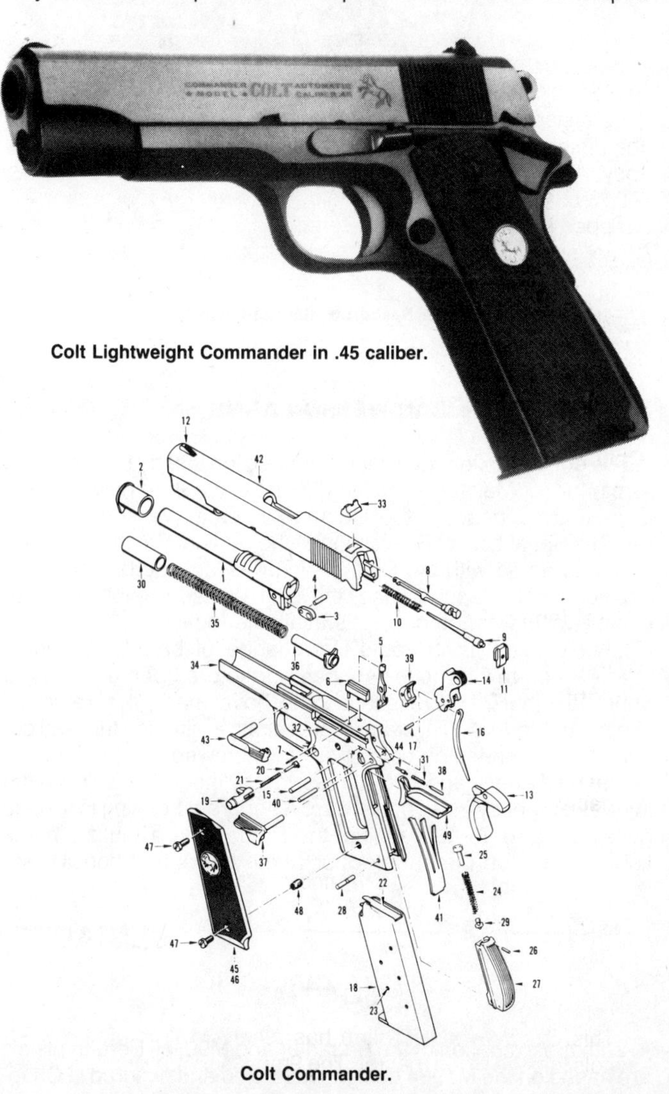

Colt Lightweight Commander in .45 caliber.

Colt Commander.

for the US Army (See Chapter 4.), a steel frame Commander was subsequently introduced. Except for the slide, barrel and recoil spring, parts interchange with those of older models.

CHARACTERISTICS OF COLT COMMANDER PISTOLS

	Lightweight Commander	Combat Commander
Caliber:	.45 ACP.	9mm Parabellum; .38 Super; .45 ACP.
Operation:	Recoil, Browning system.	Recoil, Browning system.
Barrel length:	4¼ in.	4¼ in.
Overall length:	7⅞ in.	7⅞ in.
Height:	5½ in.	5½ in.
Weight:	27½ oz.	36 and 37 oz.
Sights:	Open, fixed.	Open, fixed.
Magazine capacity:	7 rounds.	9 rounds, 9mm Parabellum. 9 rounds, .38 Super. 7 rounds, .45 ACP.

In 1970, Colt introduced their Mark IV/Series '70 versions of the standard government model. Basically identical to the M1911A1, it has an improved collet-type barrel bushing and reverse taper at the end of the barrel to provide a tighter fit when the pistol is ready to fire, thus providing greater inherent accuracy. A match model, the Gold Cup National Match MKIV/Series '70, replaces the older Colt Gold Cup match model, which was dropped in 1970. The MKIV/Series '70 Gold Cup has adjustable target sights, wide, adjustable trigger and target barrel.

CHARACTERISTICS OF COLT MKIV/SERIES '70 PISTOLS

	Government Model	Gold Cup National Match
Caliber:	9mm; .38 Super; .45 ACP.	.45 ACP.
Operation:	Recoil, Browning system.	Recoil, Browning system.
Barrel length:	5 in.	5 in.
Overall length:	8½ in.	8½ in.
Height:	5½ in.	5½ in.
Weight:	39 and 40 oz.	39 oz.
Sights:	Open, fixed.	Open, adjustable.
Magazine capacity:	9 rounds, 9mm. 9 rounds, .38 Super. 7 rounds, .45 ACP.	7 rounds, .45 ACP.

ČESKA ZBROJOVKÁ

This Czechoslovakian firm has recently introduced two new pistols into the world market. The 9mm M1975 military type pistol is described in Chapter 4. Their second handgun is the Model 70 7.65mm Automatic Pistol, similar in operation to Walther's PPK pistol.

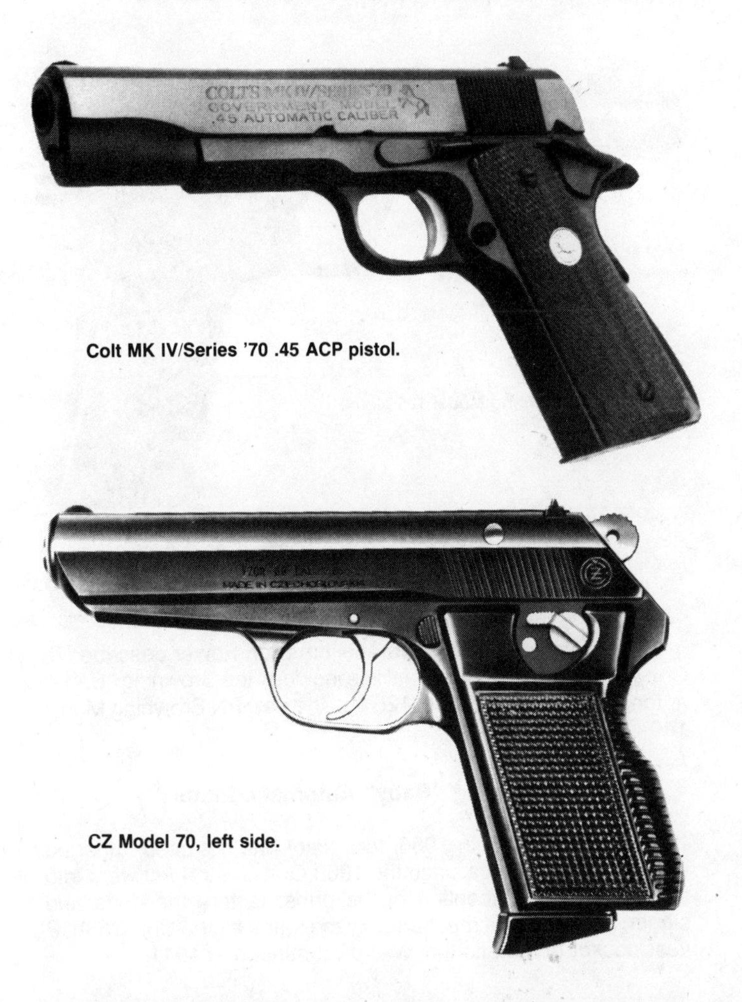

Colt MK IV/Series '70 .45 ACP pistol.

CZ Model 70, left side.

CHARACTERISTICS OF CZ MODEL 70 AUTOMATIC PISTOL

Caliber: .32 ACP.
Operation by: Blowback.
Barrel length: $3\frac{25}{32}$ in.
Overall length: $6\frac{11}{16}$ in.
Weight: 24¾ oz.
Sights: Open, fixed.
Magazine capacity: 8 rounds.

FN Browning "Baby" automatic pistol.

FN Browning Model 125 pistol.

FN Browning 140 D.A. automatic pistol.

FABRIQUE NATIONALE

In addition to their 9mm Parabellum High Power described in Chapter 9, the FN commercial line includes the Browning "Baby" automatic pistol, the Model 125 and the new FN Browning Model 140 D.A.

Browning "Baby" Automatic Pistol

Introduced by FN in 1906, this pistol has not been imported into the United States since the 1968 Gun Control Act went into effect. With the exception of the grips, factory markings and sights, the "Baby" is mechanically the same as the Colt .25 ACP vest pocket automatic that was discontinued in 1941.

CHARACTERISTICS OF FN "BABY" BROWNING AUTO PISTOL

Caliber: .25 ACP.
Operation by: Blowback.
Barrel length: 2⅛ in.
Overall length: 4 in.
Weight: 7 oz.
Sights: Open, fixed.
Magazine capacity: 6 rounds.

Browning Model 125 Automatic Pistol

Basically the M1922 Browning Pistol equipped with adjustable target sights, the Model 125 is still very popular in the world handgun market.

CHARACTERISTICS OF FN BROWNING MODEL 125 AUTOMATIC PISTOL

Caliber: .32 ACP.
Operation by: Blowback.
Barrel length: 4⅜ in.
Overall length: 7 in.
Weight: 24 oz.
Sights: Open, adjustable rear.
Magazine capacity: 9 rounds.

Browning 140 D.A. Automatic Pistol

A very recent addition to the FN product line, the Model 140 D.A. is FN's attempt to capture part of the world military, police and sporting handgun market held by the Colt Commander and the Smith & Wesson Model 39 alloy frame automatic pistols. Like the Model 39, it combines the aluminum alloy frame with a steel slide. This double-action newcomer will undoubtedly make its mark on the handgun world as have FN's other pistols.

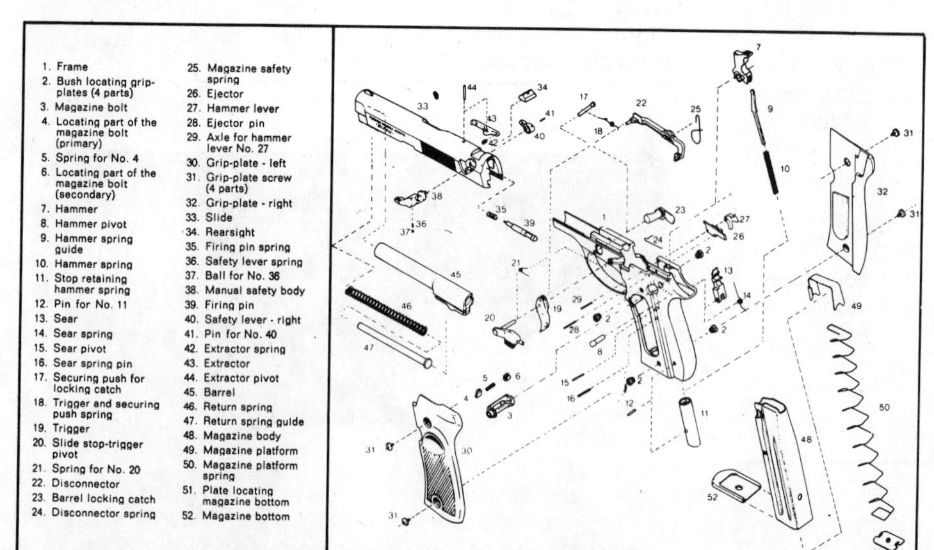

1. Frame
2. Bush locating grip-plates (4 parts)
3. Magazine bolt
4. Locating part of the magazine bolt (primary)
5. Spring for No. 4
6. Locating part of the magazine bolt (secondary)
7. Hammer
8. Hammer pivot
9. Hammer spring guide
10. Hammer spring
11. Stop retaining hammer spring
12. Pin for No. 11
13. Sear
14. Sear spring
15. Sear pivot
16. Sear spring pin
17. Securing push for locking catch
18. Trigger and securing push spring
19. Trigger
20. Slide stop-trigger pivot
21. Spring for No. 20
22. Disconnector
23. Barrel locking catch
24. Disconnector spring
25. Magazine safety spring
26. Ejector
27. Hammer lever
28. Ejector pin
29. Axle for hammer lever No. 27
30. Grip-plate - left
31. Grip-plate screw (4 parts)
32. Grip-plate - right
33. Slide
34. Rearsight
35. Firing pin spring
36. Safety lever spring
37. Ball for No. 36
38. Manual safety body
39. Firing pin
40. Safety lever - right
41. Pin for No. 40
42. Extractor spring
43. Extractor
44. Extractor pivot
45. Barrel
46. Return spring
47. Return spring guide
48. Magazine body
49. Magazine platform
50. Magazine platform spring
51. Plate locating magazine bottom
52. Magazine bottom

Exploded view of FN Browning 140 D.A. automatic pistol.

CHARACTERISTICS OF FN BROWNING 140 D.A. AUTOMATIC PISTOL

Caliber: .32 ACP (7.65mm Browning); 9mm Parabellum.
Operation by: Recoil.
Barrel length: Approx. 4 in.
Overall length: $6\frac{7}{10}$ in.
Weight: $22\frac{1}{4}$ oz.
Sights: Open, fixed.
Magazine capacity: 12 rounds, 7.65mm.
13 rounds, 9mm Parabellum.

HECKLER & KOCH

HK-4

An unusual design, the HK-4's most distinguishing characteristic is that it is sold as a complete 4-caliber set with barrels, springs and magazines for firing .22 L.R., .25 ACP, .32 ACP or .380 ACP. The need for separate slides for each caliber is avoided by a dual purpose firing pin and extractor; the former must be switched from centerfire to rimfire and back again. The extractor automatically adjusts to the different diameter cartridge cases.

Essentially, the HK-4 is a highly refined version of the pre-WW II Mauser HSc 7.65mm pistol. It is constructed by modern methods, the slide being a welded assembly of castings and stampings. The frame is of a light alloy precision casting. The first round may be fired double-action by first lowering the hammer. Disassembly is also unique; depressing the latch in the trigger guard allows the slide and barrel to be lifted off after moving it $\frac{1}{8}$ inch forward.

CHARACTERISTICS OF HK-4 AUTOMATIC PISTOL

Caliber: .22 L.R.; .25 ACP; .32 ACP; .380 ACP.
Operation by: Retarded blowback.
Barrel length: $3\frac{11}{32}$ in.
Overall length: $6\frac{3}{16}$ in.
Weight: $16\frac{9}{10}$ oz.
Sights: Open, fixed.
Magazine capacity: 8 rounds, .22 L.R.
8 rounds, .25 ACP.
8 rounds, .32 ACP.
7 rounds, .380 ACP.

Heckler & Koch also produces a large locked-breech 9mm Parabellum military and police type autoloading pistol designated P9 (single-action) and P9S (double-action) similar in appearance and construction to the HK-4. The frame and slide are of stampings and weldments, and part of the frame is overlaid with plastic. Locking is by the cam and roller system of the G-3 Rifle. See Chapter 21.

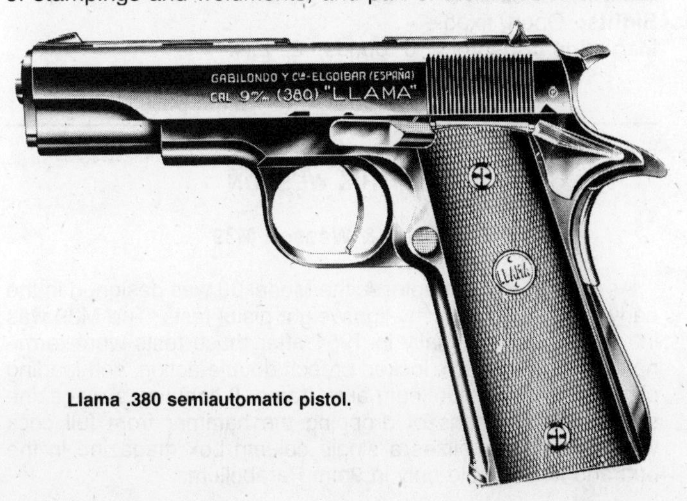

Llama .380 semiautomatic pistol.

LLAMA (GABILONDO & CIA)

These Spanish Llama large-frame pistols in .45 ACP and .38 Colt Super Automatic calibers are extremely close copies of the Colt Government Model described in Chapter 48. The more recent models have ventilated sighting ribs and other minor variations, but the resemblance to the Colt is so great that many parts will interchange. A reduced size version weighing only 20 ounces is produced in .380 ACP but is mechanically and functionally identical to the big guns. An apparently identical Llama in .32 ACP has the locking system deleted and functions as a blowback. It is otherwise identical to the .380, and most parts will interchange.

CHARACTERISTICS OF LLAMA .380 AND .32 ACP PISTOLS

Caliber: .32 ACP; .380 ACP.
Operation by: Blowback, recoil; Browning system.
Barrel length: $3\frac{11}{16}$ in.
Overall length: $6\frac{1}{4}$ in.
Height: $4\frac{3}{8}$ in.
Sights: Open, fixed.
Weight: 21 oz; 20 oz.
Magazine capacity: 8 rounds; 7 rounds.

In addition, premium priced, adjustable sight versions of the .45 and .38 models are offered.

Mauser HSc automatic pistol.

M.A.B. (MANUFACTURE D'ARMES AUTOMATIQUES BAYONNE)

M.A.B.P.15

This is France's most modern pistol design and matches the largest magazine capacity, 15 rounds, of any currently produced handgun. It is of conventional style and construction but uses an unusual rotating barrel locking system. In appearance, it is quite similar to the Browning M1935 and does, in fact, use a magazine identical—except for length—to the Browning. See Chapter 19.

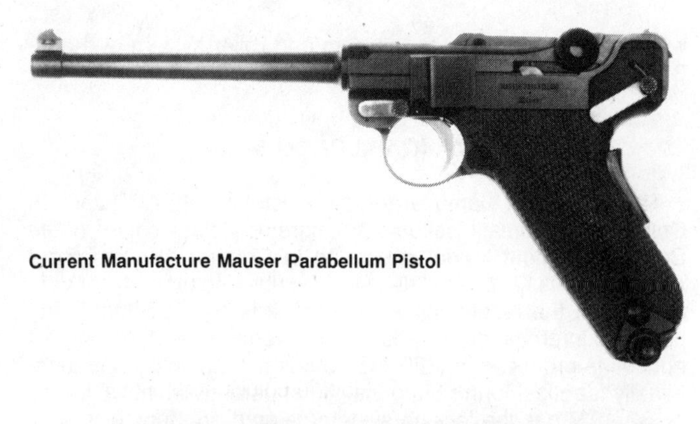

Current Manufacture Mauser Parabellum Pistol

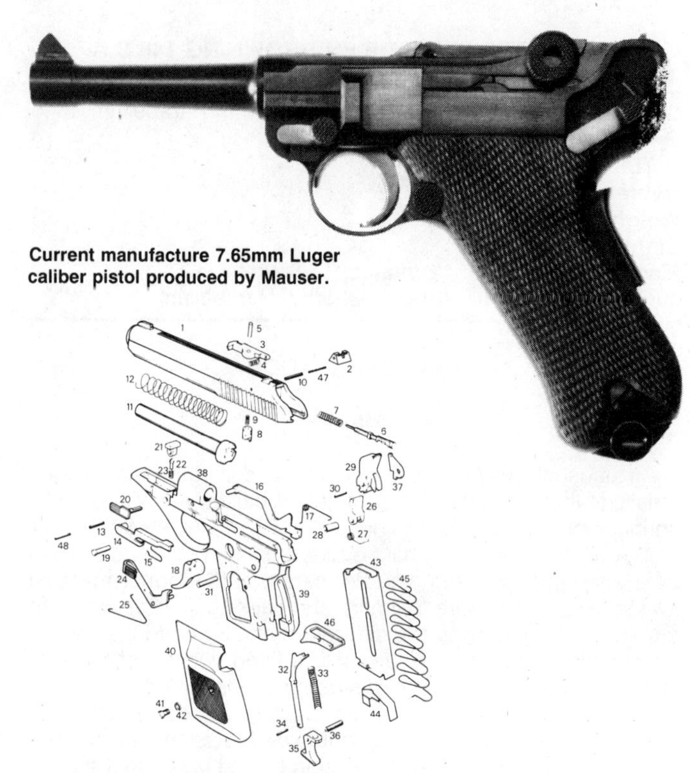

Current manufacture 7.65mm Luger caliber pistol produced by Mauser.

Exploded view of the SIG-Sauer P230 pistol.

SIG-Sauer P230 automatic pistol.

MAUSER

HSc Automatic Pistol

In 1968, the famous West German Mauserwerke again entered the commercial handgun market with a newly produced version of the original HSc pistol. The highly advanced double-action, pocket-size autoloader made its debut during the late 1930s, and hundreds of thousands of 7.65mm specimens saw service in WW II. The new version is offered in .22 L.R., 7.65mm (.32 ACP) and 9mmK (.380 ACP) calibers. Except for the slight differences in the added calibers, the new HSc is as described in Chapter 20.

PARABELLUM

Also in 1968, Mauser introduced newly manufactured Parabellum pistols in 7.65mm (.30 Luger) caliber. This model is identical to pre-World War II "Luger" pistols, but that name cannot be applied to Mauser's weapon since Stoeger Arms owns the name copyright. The new gun is actually made on original Swiss arsenal tooling utilized to produce the M1929 pistol for the Swiss before WW II. The Parabellums are available in 7.65mm Luger in a 6-inch barrel and 9mm Parabellum in a 4-inch barrel. The original design is described in Chapter 21. All Mauser pistols are distributed in the US by Interarms.

SIG (SWISS INDUSTRIAL CO.)

SIG P210

This series of commercial pistols is based on the Swiss SP47/8 (M49) service pistol. A blowback .22 L.R. conversion unit and glossy finished target models are offered. See Chapter 44.

SIG-Sauer P220

See Chapter 4 for a description of this new pistol.

SIG-Sauer P230

A companion to the military P220, this pocket automatic is available in three calibers—.22 L.R., .32 ACP and 9 × 18mm police. The design is notable for its lightweight alloy frame, decocking lever and double-action trigger mechanism. Both the P220 and P230 are manufactured in West Germany.

CHARACTERISTICS OF SIG-SAUER P230 AUTOMATIC PISTOL

Caliber: .22 L.R.; .320 ACP; 9 × 18mm police
Operation by: Blowback.
Barrel length: 3⅗ in.
Overall length: 6⅗ in.
Weight: 18 oz. in 9 × 18mm police
Sights: Open, fixed.
Magazine capacity: 10 rounds, .22 L.R.
8 rounds, .32 ACP.
7 rounds, 9 × 18mm police

SMITH & WESSON

Smith & Wesson M39

As described in Chapter 4, the Model 39 was designed in the early 1950s for US Army lightweight pistol tests. The M39 was introduced commercially in 1954 after those tests were terminated. It is a modern, locked breech double-action, self-loading pistol built on an aluminum alloy frame. It contains a magazine safety and a means of dropping the hammer from full cock without firing. It utilizes a single column box magazine in the butt and is available only in 9mm Parabellum.

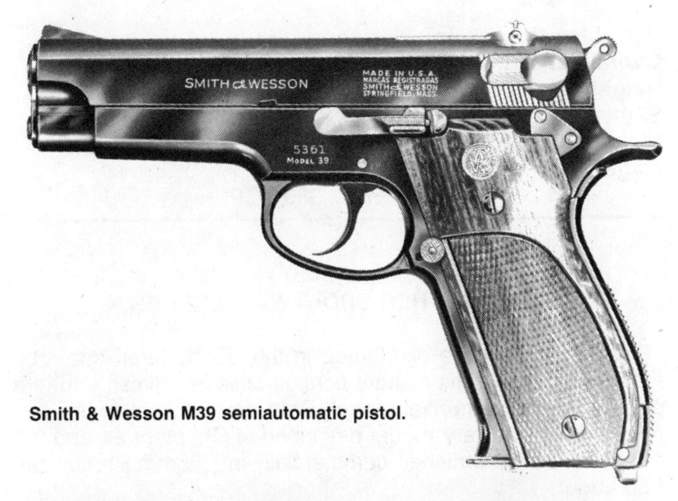

Smith & Wesson M39 semiautomatic pistol.

CHARACTERISTICS OF S&W M39 PISTOL

Caliber: 9mm Parabellum.
Operation by: Recoil; modified Browning system.
Barrel length: 4 in.
Overall length: 7⁷⁄₁₆ in.
Weight: 26½ oz.
Sights: Open, rear adjustable for windage only.
Magazine capacity: 8 rounds.

In 1973, Smith & Wesson introduced an improved version of the M39 pistol, designated the M59. Otherwise the same, it is modified to accept a 15-round, double column magazine in the butt, making its fully loaded capacity 16 rounds. Empty it weighs 30 ounces, loaded 37.

Smith & Wesson M52

This is a longer, heavier version of the M39 intended purely for target shooting. Mechanically and functionally, it is identical to the M39, though much more closely fitted and adjusted. It carries micrometer-adjustable target sights and is chambered only for the .38 Special Wadcutter cartridge.

Smith & Wesson M59 9mm Parabellum automatic pistol.

CHARACTERISTICS OF S&W M52 PISTOL

Caliber: .38 Special Wadcutter.
Operation by: Recoil; modified Browning system.
Barrel length: 5 in.
Overall length: 8⅝ in.
Weight: 41 oz.
Sights: Micrometer-adjustable, target type.
Magazine capacity: 5 rounds.

Smith & Wesson M59

Described in Chapter 5 under the Hush-Puppy silenced pistol project, the M59 was introduced commercially in the early 1970s. The M59 is similar to the M39 in that it has a steel slide and alloy frame, but it has a double row magazine, which holds 14 rounds instead of 8 as in the M39.

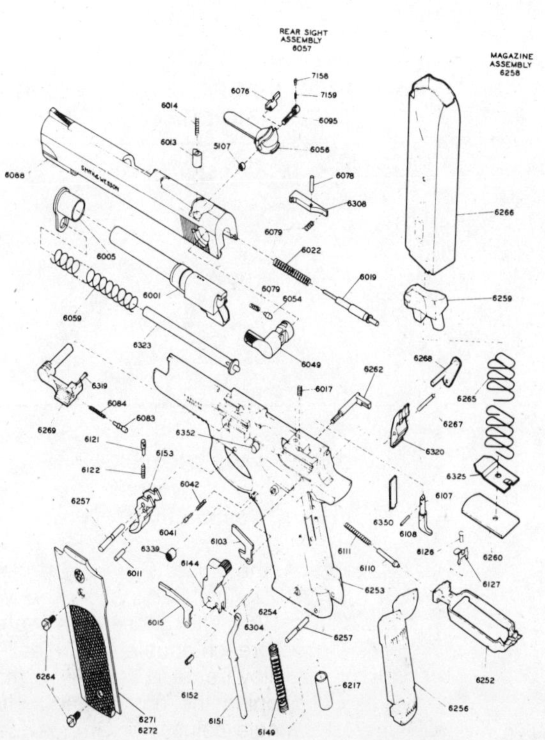

Exploded view of the Smith & Wesson M59 pistol.

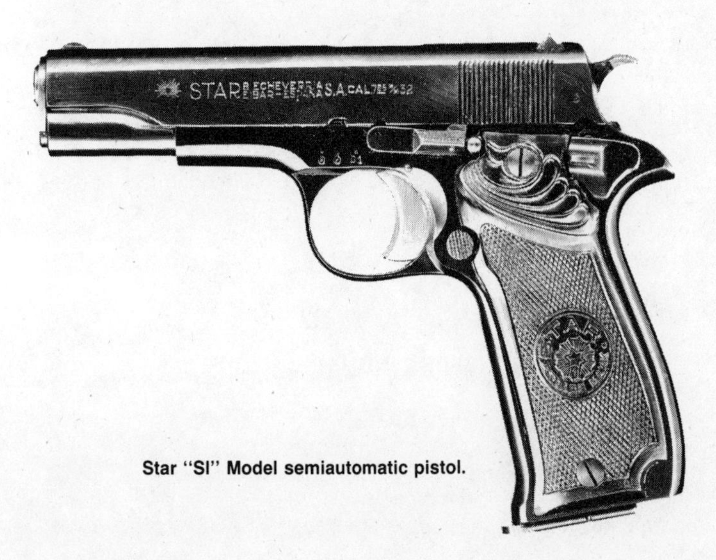

Star "SI" Model semiautomatic pistol.

CHARACTERISTICS OF S&W M59 PISTOL

Caliber: 9mm Parabellum.
Operation by: Recoil; modified Browning system.
Barrel length: 4 in.
Overall length: 7 7/16 in.
Weight: 27 oz.
Sights: Open, fixed front, adjustable rear.
Magazine capacity: 14 rounds.

STAR (BONIFACIO ECHEVERRIA & CIA)

Large caliber Star pistols are described in Chapter 39. Commercial models of the A, B and P are sold widely, especially in the Americas. In addition, a reduced scale version in .380 ACP and .32 ACP is available commercially. Called the S and SI, this model duplicates all features, including the locking system of the larger guns. Smaller Star blowback type pistols are offered in .22, .25, .32 and 380 caliber but are no longer available in the US due to government restrictions.

CHARACTERISTICS OF STAR PISTOLS

Caliber: .32 ACP; .380 ACP.
Operation by: Recoil; Browning system.
Barrel length: 3¾ in.
Overall length: 6½ in.
Height: 4¾ in.
Sights: Open, fixed.
Weight: 22 oz.
Magazine capacity: 8 rounds, .32 ACP.
7 rounds, .380 ACP.

CARL WALTHER SPORTWAFFENFABRIK

Walther pistols are distributed in the US by Interarms. The P38 is available in its military configuration and finish and in a glossy finish commercial model. Both are identical to the aluminum frame military model described in Chapters 21 and 22. The P38 is also furnished commercially in 7.65mm Parabellum (.30 Luger).

More recent is the .22 rimfire P38. It is identical to the others except for a light alloy slide and elimination of the locking system, which converts it to blowback operation.

Walther also now offers the PP Sport Model in .22 L.R. It is simply the basic PP fitted with a 6-inch barrel protruding from the slide and carrying its own removable front sight. An adjustable target type rear sight is fitted, and the hammer has a spur for easier thumb-cocking. See Walther PP in Chapter 22.

Modern military small arms are generally capable of firing several different projectiles. To cover in detail all the variants of military cartridges would require several hundred more pages. This four-page chapter is designed to provide the reader with only a quick guide to contemporary military cartridges.

52 Small Arms Ammunition

TABULAR AMMUNITION DATA

Shown below are charts of pistol cartridges, rifle cartridges, and heavy machine-gun cartridges. In each of the three charts there is shown, opposite each cartridge (and case type) listed, the following data: complete round weight, bullet weight, and propellant weight; and complete round length, case length, bullet diameter, and case diameter. Also shown for each cartridge are some common synonyms and the best known weapons in which each cartridge is used.

The dimensions and weights given are approximate. There are many variations insofar as weights are concerned due to the different bullets used with type variations within a given caliber and the differing propellant weights among different types. Weights given are for the basic ball round in each type. Diameters given are for the point of maximum diameter of bullet and cartridge case.

The data given in the charts should NOT be used as a guide in the reloading of cartridges. There is considerable variation in the performance of the various propellants in use throughout the world. Therefore no presentation such as that given here can serve as a guide to handloading unless the specific military or commercial model designation for the propellant is given.

PISTOL CARTRIDGES

Cartridge and Case Type	Complete Round Weight	Bullet Weight	Propellant Weight	Complete Round Length	Case Length	Bullet Diameter	Case Diameter	Some Common Synonyms	Best Known Weapons in Which Used
7.62mm Pistol Rimless Bottlenecked	167 gr.	87 gr.	7.71 gr.	1.36 in.	.97 in.	.307 in.	.39 in.	7.62mm Type P (Soviet) 7.62mm M48 (Czech)	TT M1933 Pistol, M34/38 SMG, PPD 40 SMG, PPSh 41 SMG, PPS 43 SMG (All Soviet) SMG and M52 Pistol (Czech) and various satellite copies of Soviet pistol and SMGs, can also be used in weapons chambered for 7.63mm Mauser cartridge.—Note that Czech cartridge has a heavier charge.
7.65mm Browning Straight Semi-rim	121.0 gr.	73 gr.	2.47 gr.	.98 in.	.68 in.	.308-.311 in.	.354 in.	.32 ACP, 7.65mm Browning short	Colt Automatic, Walther PP & PPK, (German & various copies), Browning M1910 & M1922, large numbers of pocket automatic pistols throughout the world.
7.65mm Long Rimless Straight	132 gr.	85–88 gr.	5.0 gr.	1.19 in.	.78 in.	.309 in.	.335 in.	7.65 Longue Pour Pistole et Pistolet Mitrailleur .30 cal. Pistol Cartridge M1918 (Pedersen Device Cartridge)	French M1935A and M1935S Pistols and M1938 SMG.
7.65 Luger Rimless Bottlenecked	160.0 gr.	93 gr.	5.2 gr.	1.17 in.	.85 in.	.308 in.	.392 in.	Cal. 30 Luger	Swiss M 06/29 Pistol, some Pre-W.W. II Bergmann and SIG SMGs.
8mm Jap Rimless Bottlenecked	177.5 gr.	102 gr.	5.0 gr.	1.23 in.	.83 in.	.320 in.	.411 in.	8mm Type 14	Japanese Type 14 and Type 94 pistols and Type 100 SMG
8mm Lebel Rimmed Straight	194.5 gr.	120 gr.	14.0 gr. black powder 11.5 gr. smokeless	1.44 in.	1.07 in.	.323 in.	.408 in. (Rim)	8mm M 1892	French M 1892 Revolver.

PISTOL CARTRIDGES

Cartridge and Case Type	Complete Round Weight	Bullet Weight	Propellant Weight	Complete Round Length	Case Length	Bullet Diameter	Case Diameter	Some Common Synonyms	Best Known Weapons in Which Used
9mm Short Rimless Straight	148.8 gr.	95 gr.	3.5 gr.	.98 in.	.68 in.	.356 in.	.372 in.	9mm Browning Short, 9mm Kurz, 9mm Corto, .380 A.C.P., 9mm Court	Colt automatic, Walther PP and PPK and various copies, Hungarian M37, Italian M34 Beretta.
9mm Parabellum Rimless Straight	164. gr.	115 gr.	5.6 gr.	1.17 in.	.76 in.	.356 in.	.457 in.	Pistolen Patrone 08, 9mm M1, 9mm Mk1, 9mm M38, 9mm M39, 9mm Luger	German P08, P38, MP38, MP40, British Browning No. 2 Pistol, Sten SMGs and L2A3 SMG, Fench M50 Pistol and M49 SMG. Belgian Browning HP Pistol and Vigneron SMG, Swedish and Finnish Lahti Pistol and Suomi SMGs. Swedish M45 SMG, Israeli Uzi SMG, Madsen SMGs. Beretta M38, 38/42, 38/49, 5 & 12 SMG, Beretta M951 Pistol.
9mm Largo Rimless Straight	192 gr.	124 gr.	6.6 gr.	1.32 in.	.91 in.	.354 in.	.385 in.	9mm Bergmann Bayard	Spanish Super Star Pistol and Star Model Z45 SMG.
9mm Makarov Rimless Straight	156 gr.	94 gr.	3.7 gr.	.97 in.	.71 in.	.363 in.	.389 in.		Soviet Makarov and Stechkin pistols.
.380 MK II Rimmed Straight	246.5 gr.	181 gr.	5.0 gr.	1.22 in.	.76 in.	.357 in.	.435 in.	.38 S&W and .38 Webley Scott	British Pistol No. 2 (Revolvers) in all Marks, Webley Mark IV Pistol, S & W .38/200.
.45 M1911 Rimless Straight	327. gr.	230 gr.	5 gr.	1.275 in.	.91 in.	.452 in.	.480 in.	.45 ACP, 11.43mm, 11.25mm, 11mm M40	U.S. Pistol M1911 and M1911A1, M1928 A1, M1, M1A1, M3, and M3A1 SMGs. Norway—pistol M1914. Argentina-Ballester.
.455 Webley Rimless Straight	324.0 gr.	224 gr.	7 gr.	1.22 in.	.91 in.	.455 in.	.502 in.	.455 Webley auto.	Webley Automatic Pistol.
.455 Revolver Rimmed Straight	350 gr.	265 gr.	7.5 gr.	1.22 in. to 1.26 in.	.76 in.	.455 in.	.532 in.	.455 Webley revolver .455 in. Mark II	Pistol No. 1 Marks 4, 5, & 6.

RIFLE CARTRIDGES

Cartridge and Case Type	Complete Round Weight	Bullet Weight	Propellant Weight	Complete Round Length	Case Length	Bullet Diameter	Case Diameter	Some Common Synonyms	Best Known Weapons in Which Used
5.56mm Rimless Bottlenecked	182 gr.	55 gr.	25 gr.	2.26 in.	1.76 in.	.224 in.	.378 in.	.223 5.56mm M193	U.S. M16, M16A1, and AR-18 rifles, Stoner 63 System.
6.5mm Dutch Rimmed Bottlenecked	356 gr.	156 gr.	37 gr.	3.03 in.	2.11 in.	.263 in.	.526 in.	6.5mm Rumanian 6.5mm M93 6.5 × 53mm R	
6.5mm Italian Rimless Bottlenecked	350 gr.	162 gr.	35 gr.	3.01 in.	2.06 in.	.2655 in.	.447 in.	6.5mm Mannlicher Carcano, 6.5mm M91-95	Italian M91 rifles and carbines Breda M30 MG.
6.5mm Jap Semi-rim Bottlenecked	326 gr.	139 gr.	33 gr.	2.98 in.	2.00 in.	.262 in.	.476 in.	6.5mm Type 38	Japanese Type 38 rifle, Type 38 and Type 44 carbine, Type 11 and Type 96 MG.
6.5mm Mauser Rimless Bottlenecked	363 gr.	139 gr.	36 gr.	3.15 in.	2.17 in.	.264 in.	.478 in.	6.5mm M94, 6.5mm Norwegian, 6.5mm Swedish, 6.5 × 55mm	Norwegian M94 rifle and M12 carbine. Swedish 94 carbine. M96, M38 and M42 rifles, M21 and M37 automatic rifles (used in some Swedish MGs as well).
7mm Mauser Rimless Bottlenecked	377 gr.	172 gr.	38 gr.	3.00 in.	2.24 in.	.284 in.	.474 in.	7 × 57mm	Spanish M93 rifle and M95 carbine.

RIFLE CARTRIDGES

Cartridge and Case Type	Complete Round Weight	Bullet Weight	Propellant Weight	Complete Round Length	Case Length	Bullet Diameter	Case Diameter	Some Common Synonyms	Best Known Weapons in Which Used
7.5mm French Rimless Bottlenecked	363 gr.	139 gr.	44 gr.	2.99 in.	2.13 in.	.307 in.	.484 in.	7.5mm M1929	French M 07/15, M34, M36, M49, M49/56 rifles and M1924M29 automatic rifle. M31 and M52 MG.
7.5mm Swiss Rimless Bottlenecked	404 gr.	174 gr.	49.35 gr.	3.05 in.	2.18 in.	.308 in.	.496 in.	7.45mm Swiss 755mm M11 7.5mm Schmidt Rubin	Swiss M11 rifle and M11, M31, M31/42 and M31/43 carbines, M57 assault rifle. M11, M25, and M51 machine guns.
7.62mm NATO Rimless Bottlenecked	375 gr.	150 gr.	48 gr.	2.80 in.	2.01 in.	.308 in.	.496 in.	7.62mm Ball M59 7.62 × 51mm 7.62mm M1954 7.62mm OTAN .308 Winchester 7.62mm Ball M85	U.S. M14 rifle, M60 and M73 machine guns. British L2A1 rifle, Belgian FN FAL rifle and FN MAG. M.G. Canadian C1 rifle and C2 LMG. W. German FN and G3 rifles.
7.62mm Russian Rimmed Bottlenecked	348 gr.	148 gr.	50 gr.	3.03 in.	2.11 in.	.311 in.	.564 in.	7.62mm M1908 type L 7.62 × 54mm	Soviet Mosin-Nagant rifles and carbines. Simonov M36 and Tokarev M32, M38, and M40 rifles and carbines. Maxim M1910, DP, DPM, DT, DTM, DA, ShKAS, SG-43, SGM, and RP-46 MGs. Soviet Satellite copies of these weapons.
7.62mm M43 Rimless Bottlenecked	253 gr.	122 gr.	25 gr.	2.20 in.	1.52 in.	.311 in.	.45 in.	7.62mm Russian Short 7.62mm M1934 Type PS 7.62 × 39mm	Soviet AK assault rifle, SKS carbine, RPD LMG.
7.62mm M52 Rimless Bottlenecked	293.7 gr.	132 gr.	27 gr.	2.35 in.	1.77 in.	.310 in.	.442 in.	7.62mm Czech Short	Czech M52 Rifle and LMG.
7.65mm Mauser Rimless Bottlenecked	390 gr.	174 gr.	38.6 gr.	2.95 in.	2.11 in.	.310 in.	.472 in.	7.65mm M30 7.65 × 54mm	Belgian M89 rifles and carbines, M35 and 36 rifles. Argentine M91 and M09 rifles. Turkish M91 and M03 rifles.
7.7mm Rimless Jap Rimless Bottlenecked	415 gr.	182 gr.	43.13 gr.	3.14 in.	2.25 in.	.310 in.	.47 in.	7.7mm Type 99	Japanese Type 99 rifle and LMG.
7.7mm Semi-rim Jap. Semi-rim Bottlenecked	429 gr.	200 gr.	44.18 gr.	3.14 in.	2.25 in.	.310 in.	.49 in.	7.7mm Type 92	Japanese Type 92 MG.
7.92mm Mauser Rimless Bottlenecked	408. gr.	198 gr.	47 gr.	3.15 in.	2.24 in.	.323 in.	.468 in.	7.9mm SS 7.9 × 57mm 7.9 IS 7.9 JS 8mm Mauser .315 Mauser rimless 8 × 57mm M98.	German M98 rifles and carbines. M08, 08/15, 08/18, 15, 17, 34 and 42 MGs, FG-42. Czech M24, M33 rifles and carbines, M26, M30, M30J and M37 MGs. British BESA MGs, Chinese-Brens MK 2.
8mm Austrian Rimmed Bottlenecked	437 gr.	244 gr.	42 gr.	3.00 in.	1.99 in.	.323 in.	.554 in.	8mm M1893 8 × 50mm R	Austrian M90 rifles and carbines. M07/12 MG, M95 rifles and carbines.
8mm M31 Rimmed Bottlenecked	441 gr.	208 gr.	55 gr.	3.02 in.	2.21 in.	.330 in.	.55 in.	8mm M30S 8 × 56mm R	Hungarian M35 rifle and M31 MG. Austrian M30 MG, rebarrelled M95 Mannlichers.
8mm Lebel Rimmed Bottlenecked	429 gr.	198 gr.	43 gr.	2.95 in.	1.98 in.	.328 in.	.629 in.	8mm M 1886 D(AM)	French M86M93, M1890, M92, M07, MO7/15, M16, M17, M18 rifles and carbines. M14 and M15 MGs.
Cal. .30 Carbine Rimless Straight	195 gr.	110 gr.	14.5 gr.	1.67 in.	1.28 in.	.308 in.	.356 in.	Cal. 30 M1 Carbine	U.S. M1 and M2 carbines.

RIFLE CARTRIDGES (continued)

Cartridge and Case Type	Complete Round Weight	Bullet Weight	Propellant Weight	Complete Round Length	Case Length	Bullet Diameter	Case Diameter	Some Common Synonyms	Best Known Weapons in Which Used
Cal. .30-06 Rimless Bottlenecked	396 gr.	150 gr.	50 gr.	3.33 in.	2.49 in.	.308 in.	.469 in.	Cal. .30 M2 Ball Cal. .300 Browning 7.62 × 63mm	U.S. M03, 03A1, 03A2, 03A3, 03A4 Springfield rifles, M17, M1, M1C, M1D rifles. Browning automatic rifle M18, M18A1, 18A2, Browning M17, 17A1, 19A4, 19A5, 19A6, M37 MGs.
Cal. .303 Rimmed Bottlenecked	384 gr.	174 gr.	37.5 gr.	3.04 in.	2.21 in.	.311 in.	.53 in.	.303 in. Mark 7 .303 British	British No. 1, MKs 1, 2, 3, 3*, 4, 5, 6, No. 3, MK 1, 1*, No. 4 MK 1, 1*, 2, No. 5 MK 1, 1* rifles. Lewis MG, Hotchkiss MG, Bren MKs 1, 2, 3, and 4 MGs. Vickers MK 1 MG. Canadian Ross M05 and M10 rifles.

HEAVY MACHINE GUN CARTRIDGES

Cartridge and Case Type	Complete Round Weight	Bullet Weight	Propellant Weight	Complete Round Length	Case Length	Bullet Diameter	Case Diameter	Some Common Synonyms	Best Known Weapons in Which Used
Cal. .50 Rimless Bottlenecked	1800 gr.	709 gr.	240 gr.	5.45 in.	3.90 in.	.511 in.	.804 in.	Cal. .50 Ball M2 Cal. .50 Browning	U.S. M2, M2HB, M3 MGs.
12.7mm Soviet Rimless Bottlenecked	2160 gr.	788 gr.	271 gr.	5.76 in.	4.25 in.	.511 in.	.85 in.	12.7 × 99mm 12.7 × 108mm	Soviet DShK M38, DShk M38/46 and UB MGs.

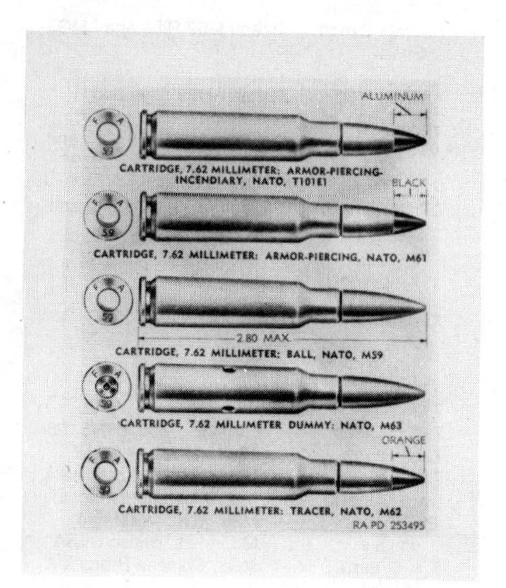

The US family of 7.62 × 51mm NATO cartridges standardized in the late 1950s.

Soviet-type 7.62 × 39mm M43 cartridges. ***Left to right:*** **Ball, Tracer, API, Grenade Blank (Yugoslav), Grenade Blank (Polish), Grenade Blank (PRC).**

53 Selected Bibliography

As interest in the study and collecting of small arms has grown during the past three decades, so has the number of books on specific weapons and weapon groups. Since it has become impossible to describe all of the small arms of the world in a single volume, the editor and publisher decided it would be useful to add a selected bibliography to this edition. For the most part, only newer books have been cited. Readers who wish to locate older volumes are referred to Ray Riling's excellent book, *Guns and Shooting: A Selected Chronological Bibliography,* New York: Greenberg, 1951. Reader recommendations for additional entries to the bibliography that follows, especially European titles, will be appreciated.

JOURNALS FOR STUDENTS OF SMALL ARMS

International Defense Review (Interavia SA, P. O. Box 162, 1212 Cointrin, Geneva, Switzerland). Published in English, French, German and Spanish editions.

National Defense (American Defense Preparedness Association, Union First Bank Building, Washington, D.C. 20005).

BOOKS

Archer, Denis H. R., ed. *Jane's Infantry Weapons, 1976.* London and New York: Macdonald and Jane's Publishers, Ltd., and Franklin Watts, 1976. See also the subsequent yearly volumes edited by Colonel John Weeks.

Belford, James N., and Dunlap, Jack. *The Mauser Self-Loading Pistol.* Alhambra, CA: Borden Publishing Co., 1969.

Boothroyd, Geoffrey. *The Handgun.* New York: Bonanza Books, 1970.

Breathed, John W., Jr., and Schroeder, Joseph J., Jr. *System Mauser: A Pictorial History of the Model 1896 Self-Loading Pistol.* Chicago: Handgun Press, 1967.

Campbell, Clark S. *The '03 Springfields.* Philadelphia: Ray Riling Arms Book Co., 1971.

This revised and updated version of Campbell's 1957 volume is a very helpful guide to the 1903 Springfield family of weapons.

Chinn, George M. *The Machine Gun.* 4 vols. Washington: Government Printing Office, 1951–1954.

These volumes are essential reading for any individual who wishes to understand the history and design characteristics of the world's machine guns.

Datig, Fred A. *The Luger Pistol (Pistole Parabellum); Its History and Development from 1893–1945.* Revised and enlarged edition. NP: Borden Publishing Co., 1962.

Ezell, Edward Clinton. *Handguns of the World: Military revolvers and self-loaders from 1870 to 1945.* Harrisburg, PA: Stackpole Books, 1981.

Federov, V. G. *Evolyutsiya strelkovogo oruzhiya* [Evolution of infantry weapons]. Moscow: unknown, 1938–1939.

This volume is in two parts. Part I is titled "Razvitie ruchnogo ognestrel'nogo oruzhiya ot zaryazhaniya s dula i kremnevogo zamka do magazinnikh vintovok" [The development of hand-operated weapons from muzzle loaders and flintlocks to magazine rifles]; Part II is "Razvitie avtomaticheskogo oruzhiya" [The development of automatic weapons]. With a German introduction, this book was reprinted in 1970 by the Biblio Verlag of Osnabruck, Germany. A classic study, it deserves to be translated into English.

Götz, Hans-Dieter. *Die deutshen Militärgewehre und Maschinenpistolen: 1871–1945.* Stuttgart: Motorbuch-Verlag, 1974. German text.

Greener, W. W. *The Gun and Its Development.* 9th ed. New York: Bonanza Books, 1910.

Hackley, F. W.; Woodin, W. H.; and Scranton, E. L. *History of Modern U.S. Military Ammunition, Vol. I, 1880–1939.* New York and London: Macmillan, 1967.

Hackley, F. W.; Woodin, W. H.; and Scranton, E. L. *History of Modern U.S. Military Ammunition, Vol. II, 1940–1945.* Highland Park, NJ: The Gun Room Press, 1976.

Hatcher, Julian S. *The Book of the Garand.* Washington: Infantry Journal Press, 1948.

An overview of the development of the M1 Rifle.

Hatcher, Julian S. *Hatcher's Notebook.* 2d ed. Harrisburg: The Stackpole Co., 1957 and 1962.

Hicks, James E. *Notes on French Ordnance, 1717 to 1936.* Mt. Vernon, NY: author, 1938.

Hicks, James E. *Notes on German Ordnance for the Collector, 1841 to 1918.* 2d ed. New York: author, 1941.

Hicks, James E. *U.S. Firearms, 1776–1956. Vol. I, Notes on U.S. Ordnance.* La Cañada, CA: James E. Hicks & Sons, 1957.

Hobart, F. W. A., ed. *Jane's Infantry Weapons, 1975.* London and New York: Macdonald and Jane's Publishers, Ltd., and Franklin Watts Inc., 1974.

Hobart, F. W. A. *Pictorial History of the Machine Gun.* London: Ian Allan, 1971.

Hobart, F. W. A. *Pictorial History of the Sub-Machine Gun.* London: Ian Allan, Ltd., 1973.

Hogg, Ian V. *German Pistols and Revolvers, 1871–1945.* New York: Galahad Books, 1971.

Honeycutt, Fred L., Jr., and Anthony, F. Patt. *Military Rifles of Japan, 1897–1945.* Lake Park, FL: F. L. Honeycutt, 1977.

A very useful and well illustrated guide to this subject.

Hughes, James B., Jr. *Mexican Military Arms: The Cartridge Period, 1866–1967.* Houston: Deep River Armory, 1968.

A helpful guide to the small arms used by the Mexican

armed forces.

Johnson, George B., and Lockhoven, Hans Bert. *International Armament: With History, Data, Technical Information and Photographs of over 400 Weapons.* 2 vols. Cologne: International Small Arms Publishers, 1965.

Johnson, Harold E. *Small Arms Identification and Operation Guide—Eurasian Communist Countries.* DST—1110H—394—76. Charlottesville, VA: US Army Foreign Science and Technology Center, 1976.

The latest edition of this series is a very handy guide to foreign small arms.

Johnson, Harold E. *Small Arms Identification and Operation Guide*—Free World. DST-110H-163-76. Charlottesville, VA: US Army Foreign Science and Technology Center, 1976.

Jones, Harry E. *Luger Variations.* Vol. I. Los Angeles: author, 1959.

Catalogue of the many variations of Luger pistols.

Krcma, Vaclav. *The Identification and Registration of Firearms.* Springfield, IL: Charles C. Thomas, 1974.

Leithe, Frederick E. *Japanese Hand Guns.* Alhambra, CA: Borden Publishing Co., 1968.

Identification guide to Japanese handguns through WW II.

Lugs, Jaroslav. *Firearms Past and Present: A Complete Review of Firearms Systems and Their Histories.* 2 vols. London: Grenville, 1975.

This previously little known volume is an invaluable addition to the English language literature on small arms. Of particular significance are the biographies and bibliography contained in Vol. I.

Lugs, Jaroslav. *Handfeuerwaffen: Systematischer Uberblick über die Handfeuerwaffen und ihre Geschicete.* 2 vols. East Berlin: Deutscher Militarverlag, 1962.

German text.

Lugs, Jaroslav. *Rŭcni palne zbraně.* 2 vols. Prague: unknown, 1956.

Czech text.

Mathews, Joseph Howard. *Firearms Identification.* 3 vols. Madison: University of Wisconsin, 1962–1973.

Musgrave, Daniel D., and Nelson, Thomas B. *The World's Assault Rifles & Automatic Carbines.* Vol. II of the World's Weapons Series. Alexandria, VA: T. B. N. Enterprises, 1967.

The first in-depth look at this subject.

Neal, Robert J., and Jinks, Roy G. *Smith & Wesson, 1857–1945.* Cranbury, NJ: 1975.

Nelson, Thomas B., with the assistance of Hans B. Lockhoven. *The World's Submachine Guns* [machine pistols]. Vol. I. Cologne: International Small Arms Publishers, 1963.

The single most comprehensive book on the subject through the date of publication.

Nelson, Thomas B., and Musgrave, Daniel D. *The World's Machine Pistols and Submachine Guns (1964–1980), Vol. IIA.* Alexandria, VA: T.B.N. Enterprises, 1980.

Office of the Chief of Ordnance, Research and Development. *Record of Army Ordnance Research and Development.* 3 vols. Washington: OCO, 1946. Part of Vol. II, Book 1, *Small Arms and Small Arms Ammunition,* has been reprinted as *American Small Arms Research in World War Two. Vol. I: Hand and Shoulder Weapons, Helmets and Body Armor.* Wickenburg, AZ: Normount Technical Publications, 1975.

Olson Ludwig. *Mauser Bolt Rifles.* 3d ed. Montezuma, IA: F. Brownell & Son, 1977.

Owen, J. I., ed. *Brassey's Infantry Weapons of the World, 1974–75.* New York: British Book Center, 1974.

Owen, J. I., ed. *Brassey's Infantry Weapons of the World. 1975.* New York: Hippocrene Books, 1975.

Owen, J. I., ed. Brassey's Infantry Weapons of the World. 2d. ed. New York: Crane Russak and Company, 1979.

Pender, Roy G. III. *Mauser Pocket Pistols, 1910–1946.* Houston: Collectors Press, 1971.

Rankin, James L., with the assistance of Gary Green. *Walther Models PP and PPK, 1929–1945.* Coral Gables, FL: author, 1974.

Rankin, James L., with the assistance of Gary Green. *Walther Vol. II. Engraved, Presentation and Standard Models.* Coral Gables, FL: author, n.d.

Reynolds, E. G. B. *The Lee-Enfield Rifle.* London: Herbert Jenkins, Ltd., 1960.

Schneider, Hugo *et al. Handfeuerwaffen System Vetterli.* Zurich: Stocker-Schmid, AG Dietikon, 1972.

Descriptions and illustrations of the many Vetterli variations. German text.

Sharpe, Philip B. *The Rifle in America.* New York: Funk & Wagnalls Co., 1947.

Useful reference source on American rifles through WW II.

Simone, Gianfranco; Belogi, Ruggero; and Grimaldi, Alessio. *IL91.* Milano: Ravizza, 1970.

A basic introduction to the Model 1891 Carcano Rifle; well illustrated, with a good bibliography. Italian text.

Smith, W. H. B. *Mannlicher Rifles and Pistols: Famous Sporting and Military Weapons.* Harrisburg, PA: The Stackpole Co., 1947.

Smith, W. H. B. *Mauser Rifles and Pistols.* Harrisburg, PA: The Stackpole Co., 1972.

Smith, W. H. B., and Smith, Joseph E. *The Book of Rifles.* 4th ed. Harrisburg, PA: The Stackpole Co., 1972.

First published in 1948, this volume provides an encyclopedic reference on the subject of rifles.

Smith, W. H. B., and Smith, Joseph E. *The W. H. B. Smith Classic Book of Pistols and Revolvers.* rev. ed. Harrisburg, PA: The Stackpole Co., 1968.

This companion to *The Book of Rifles* was first published in 1946. It is a basic reference work for individuals interested in the world's handguns.

Stevens, R. Blake. *North American FALs.* Vol. 1, *The FAL Series.* Toronto: Collector Grade Publications, 1979.

Stevens, R. Blake. *UK and Commonwealth FALs.* Vol. 2, *The FAL Series.* Toronto: Collector Grade Publications, 1980.

Stevens, R. Blake, and Van Rutten, Jean E. *The Metric FAL.* Vol. 3, *The FAL Series.* Toronto: Collector Grade Publications, 1981.

Swearengen, Thomas F. *The World's Fighting Shotguns.* Alexandria, VA: T.B.N. Enterprises, 1979.

Textbook of Small Arms, 1929. London: Holland Press, Ltd., 1961.

US Army Foreign Science and Technology Center. *Small Arms Ammunition Identification Guide.* FSTC–CW–07–7–68. Charlottesville, VA: US Army Foreign Science and Technology Center, 1969.

Wahl, Paul. *Carbine Handbook.* New York: ARCO, 1964.

Walter, John. *Luger: An illustrated history of the handguns of Hugo Borchardt and Georg Luger, 1875 to the present.* London: Arms and Armour Press, 1977.

White, Henry P., and Munhall, Burton D. *Cartridge Headstamp Guide.* Bel Air, MD: H. P. White Laboratory, 1963.

White, Henry P., and Munhall, Burton D. *Center Fire Metric Pistol and Revolver Cartridges.* Vol. I, Cartridge Identification. Washington: Infantry Journal Press, 1948.

Whittington, Robert D. III. *German Pistols and Holsters, 1934–1945: Military—Police—NSDAP.* Dallas: Brownlee Books, 1969.

Williamson, Harold F. *Winchester: The Gun That Won the West.* South Brunswick, New York and London: A. S. Barnes and Co., and Thomas Yoseloff, Ltd., 1952.

Wilson, R. K., with Hogg, I. V. *Textbook of Automatic Pistols.* rev. ed. London, Harrisburg, PA, and Schwäbisch Hall: Arms & Armor Press, Stackpole Books, and Deutsche Waffen Journal, 1975.

First published as *Textbook of Automatic Pistols: Being a Treatise on the History, Development and Functioning of the Modern Military Self-Loading Pistol—Its Special Ammunition—and Their Evolvement into the Sub-Machine Gun Together with a Supplementary Chapter on the Light Machine Gun, 1884–1935.* Plantersville, SC: Small-Arms Technical Publishing Co., 1943.

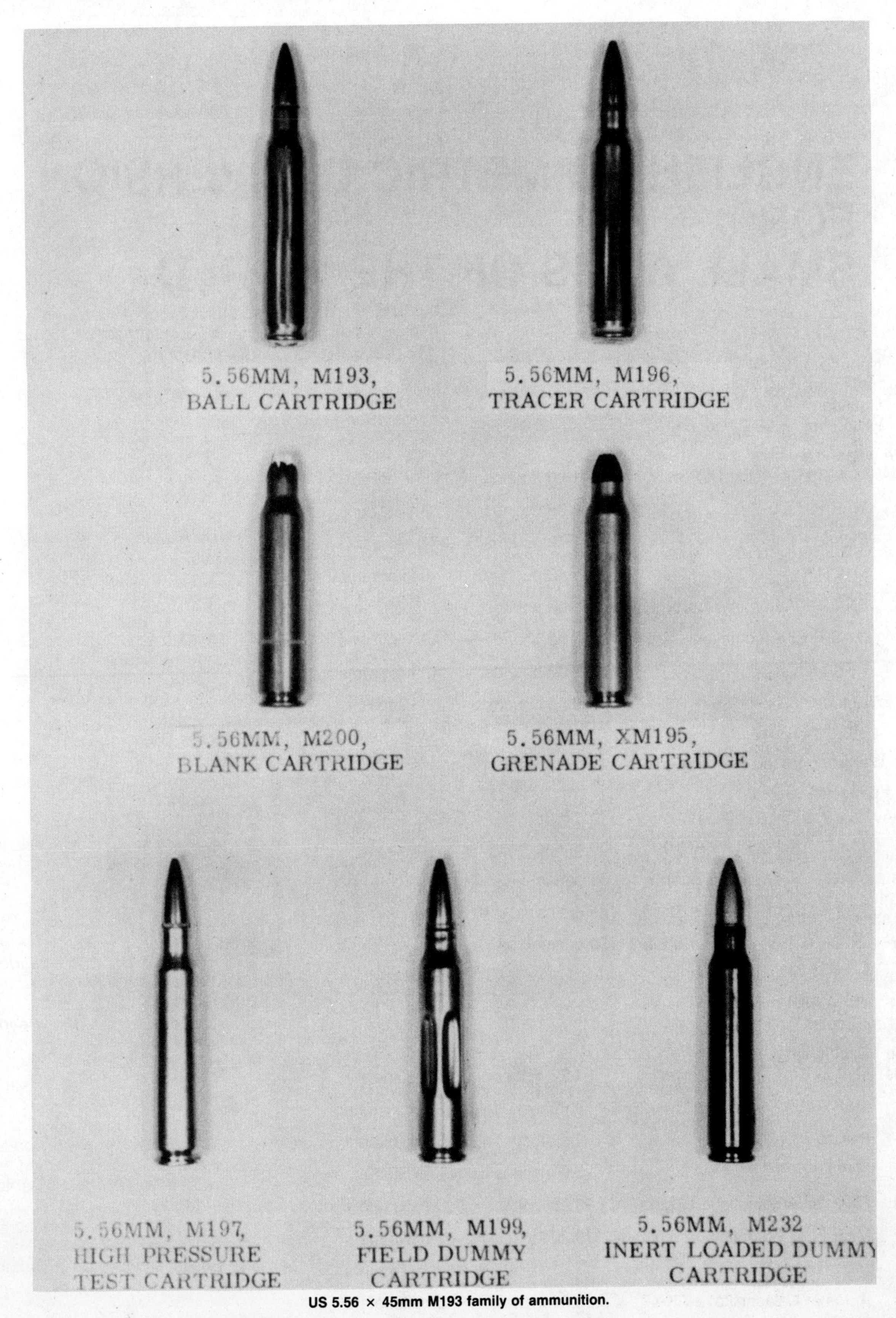

US 5.56 × 45mm M193 family of ammunition.

Appendix

ENGLISH TO METRIC CONVERSION FOR SMALL ARMS OF THE WORLD

Weights: 1 pound = 0.454 kilograms
1 kilogram = 2.2046 pounds
1 ounce = 0.028 kilogram
1 kilogram = 35.27 ounces

Example weapon weights:

Pistols:	M 1935 Browning Hi-Power	1.90 pounds = 0.86 kilogram
	MAB P15S	2.38 pounds = 1.08 kilograms
	Pistole P1 (P38)	1.70 pounds = 0.77 kilogram
	M1911A1	2.43 pounds = 1.10 kilograms
Rifles:	FN Fusil Automatique Leger (FAL)	9.06 pounds = 4.11 kilograms
	MAS Fusil Automatique 4.46mm	7.50 pounds = 3.40 kilograms
	H&K Gewer 3	9.90 pounds = 4.49 kilograms
	Automat Kalashnikova AK-47	10.58 pounds = 4.80 kilograms

Projectile Weights: Bullet weight in grains can be converted to grams by multiplying the grains times 0.0648.

Example projectile weights:

Pistols:	9 × 19mm Parabellum	115 grains = 7.45 grams
	11.43 × 23mm (.45 ACP)	230 grains = 14.90 grams
Rifles:	5.56 × 45mm M193 Ball	55 grains = 3.56 grams
	7.92 × 57mm Mauser Ball	125 grains = 8.10 grams
	.30-06 (7.62 × 63mm) M2 Ball	150 grains = 9.72 grams

Length: 1 inch = 25.4 millimeters
1 millimeter = 0.03937 inch

Example barrel lengths:

Pistols:	M1935 Browning Hi-Power	4.72 inches = 120 millimeters
	MAB P15S	4.60 inches = 116.8 millimeters
	Pistole P1 (P38)	4.92 inches = 125 millimeters
	M1911A1	5.00 inches = 127 millimeters
Rifles:	FN Fusil Automatique Leger (FAL)	20.98 inches = 533 millimeters
	MAS Fusil Automatique 5.56mm	19.20 inches = 488 millimeters
	H&K Gewer 3	17.72 inches = 450 millimeters
	Automat Kalashnikova AK-47	16.34 inches = 415 millimeters

Velocities: 1,000 feet per second (f.p.s.) = 304.8 meters per second (M/S)
2,000 feet per second = 609.6 meters per second
3,000 feet per second = 914.4 meters per second

Example projectile velocities:

Pistols:	9 × 19mm Parabellum	1250 f.p.s.	= 381 M/S
	11.43 × 23mm (.45 ACP)	830 f.p.s.	= 253 M/S
Rifles:	5.56 × 45mm M193 Ball	3250 f.p.s.	= 991 M/S
	7.92 × 57mm Mauser Ball	2750 f.p.s.	= 838 M/S
	.30-06 (7.62 × 63) M2 Ball	2750 f.p.s.	= 838 M/S

The metric unit for foot-pounds of kinetic energy is the Joule (j). To convert foot-pounds to kinetic energy to Joules, multiply by 1.3558179. Therefore, 375 foot-pounds of energy equals 508.4 Joules. Or 1224 foot-pounds equals 1660 Joules.

Caliber: This is one of the precise areas for conversion. Many calibers expressed in English inch dimensions are only approximate; e.g., .38 caliber is actually .357 inch. Therefore, a selected list of standard cartridges with common designation and more precise metric designation is presented below.

Metric		Common
6.35 × 15 SR	=	.25 ACP
7.65 × 17 SR	=	.32 ACP or 7.65 Browning
9 × 17	=	.380 ACP or 9mm Browning
9 × 18	=	9mm Makarov or 9mm Ultra
9 × 19	=	9mm Parabellum or 9mm Luger
7.65 × 20	=	7.65mm French Long
9 × 20 R	=	.38 Smith & Wesson
7.65 × 22	=	7.65 Parabellum or .30 Luger
11.43 × 23	=	.45 ACP
7.62 × 25	=	7.62mm Mauser Pistol, 7.62mm Tokarev, etc.
9 × 29 R	=	.38 Special
7.62 × 33	=	.30 Carbine
7.92 × 33	=	7.92mm Kurz
7.62 × 39	=	7.62mm Soviet M43
5.56 × 45	=	.223 Remington
7.62 × 45	=	7.62mm Czechoslovak M52
4.85 × 49	=	4.85mm UK Assault Rifle
6.5 × 50.5 SR	=	6.5mm Japanese Arisaka
8 × 50.5 R (L)	=	8mm Lebel
8 × 50.5 R (S)	=	8mm Steyr
7.62 × 51	=	7.62mm NATO
7.35 × 52	=	7.35mm Italian Carcano
6.5 × 52.5	=	6.5mm Italian Mannlicher-Carcano
6.5 × 53.5	=	6.5mm Greek Mannlicher
7.65 × 53.5	=	7.65mm Mauser
6.5 × 54 R	=	6.5mm Dutch and Romanian Mannlicher
7.5 × 54	=	7.5mm French M1929
7.62 × 54 R	=	7.62mm Russian
6.5 × 55	=	6.5mm Norwegian Krag and Swedish Mauser
7.5 × 55.5	=	7.5mm Swiss (Schmidt-Rubin)
7.62 × 56 R	=	.303 British Enfield
8 × 56 R	=	8mm Hungarian-Mannlicher
7 × 57	=	7mm Mauser
7.92 × 57	=	7.92mm Mauser
7.92 × 57 R	=	7.92mm Dutch
6.5 × 58	=	6.5mm Mauser-Vergueiro (Portuguese)
7.7 × 58	=	7.7mm Japanese Arisaka
7.7 × 58 SR	=	7.7mm Japanese Type 92
8 × 58 R	=	8mm Danish Krag
8 × 59 RB	=	8mm Breda Machine Gun
7.92 × 61 RB	=	7.92mm Norwegian Machine Gun
7.62 × 63	=	.30-06 or .30 US
8 × 63	=	8mm Swedish Machine Gun
12.7 × 99	=	.50 Browning Machine Gun
12.7 × 108	=	.50 Soviet Machine Gun

Note: All conversions in this table have been made with a Rockwell International Microelectronic Division 51R Converter programmed pocket calculator.

Abbreviations: SR = semi-rimmed
R = rimmed
RB = rebated rim

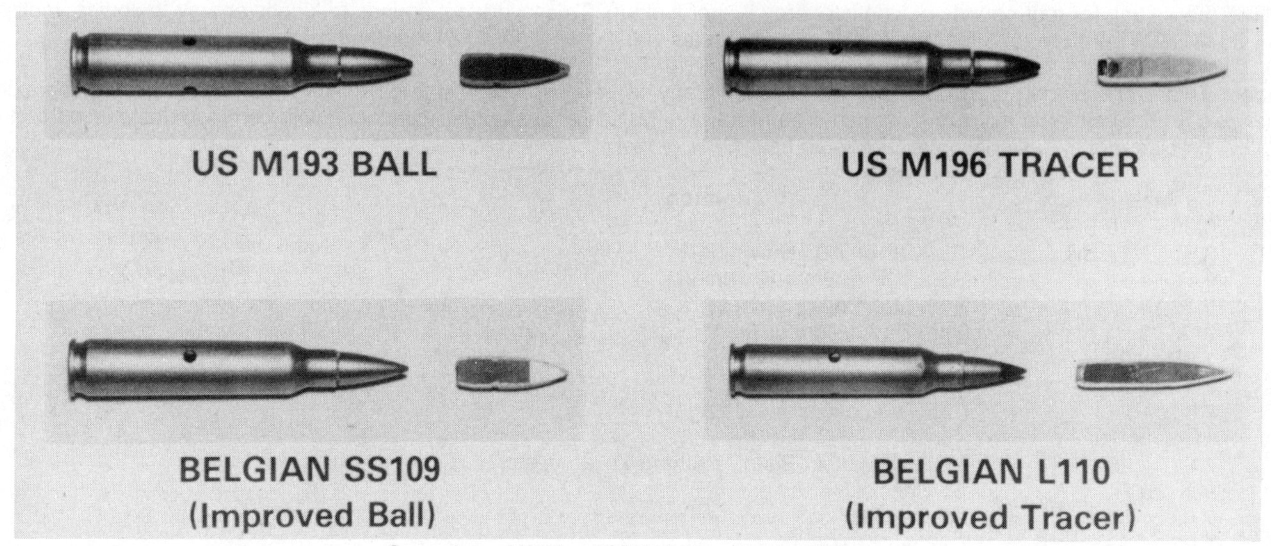

Comparison of M193 and SS109 5.56 × 45mm cartridges.

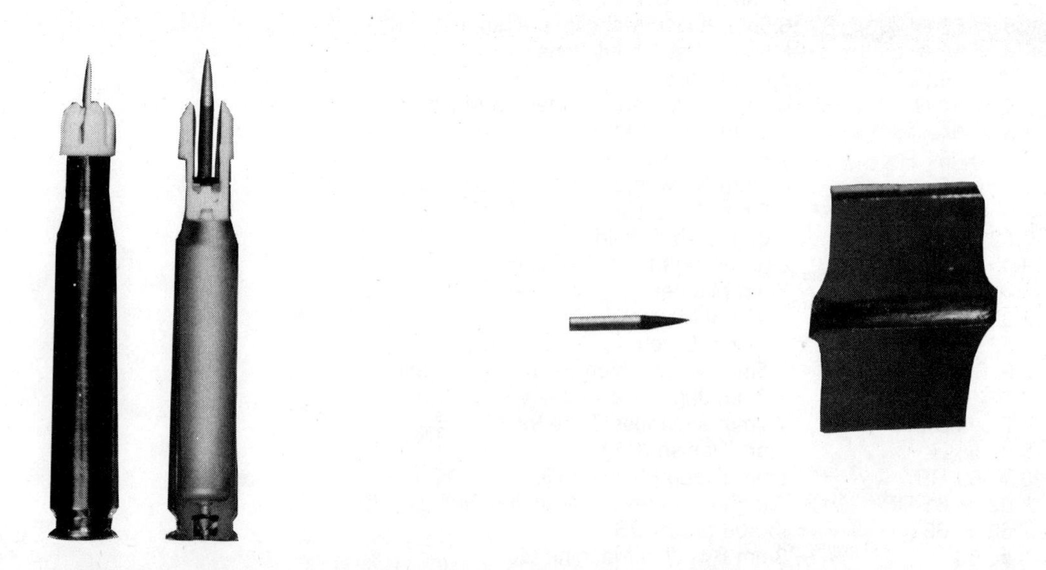

7.62 × 51mm SLAP cartridge developed by Olin Corporation for US ARRADCOM. This first generation Saboted Light Armor Penetrator of Tungsten is capable of penetrating .75 inch (19mm) of rolled homogeneous steel armor plate.

Index